Body systems maintain homeostasis

ENDOCRINE SYSTEM
Acts by means of hormones secreted into the blood to regulate processes that require duration rather than speed—e.g., metabolic activities and water and electrolyte balance.
See Chapters 18 and 19.

INTEGUMENTARY SYSTEM
Serves as a protective barrier between the external environment and the remainder of the body; the sweat glands and adjustments in skin blood flow are important in temperature regulation.
See Chapters 12 and 17.

Keeps internal fluids in

Keeps foreign material out

IMMUNE SYSTEM
Defends against foreign invaders and cancer cells; paves way for tissue repair.
See Chapter 12.

Protects against foreign invaders

MUSCULAR AND SKELETAL SYSTEMS
Support and protect body parts and allow body movement; heat-generating muscle contractions are important in temperature regulation; calcium is stored in the bone.
See Chapters 8, 17, and 19.

Enables the body to interact with the external environment

Exchanges with all other systems.

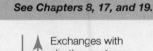

HOMEOSTASIS
A dynamic steady state of the constituents in the internal fluid environment that surrounds and exchanges materials with the cells.
See Chapter 1.
Factors homeostatically maintained are:
- Concentration of nutrient molecules
  See Chapters 16, 17, 18, and 19.
- Concentration of O_2 and CO_2
  See Chapter 13.
- Concentration of waste products
  See Chapter 14.
- pH   *See Chapter 15.*
- Concentration of water, salts, and other electrolytes
  See Chapters 14, 15, 18, and 19.
- Temperature   *See Chapter 17.*
- Volume and pressure
  See Chapters 10, 14, and 15.

Homeostasis is essential for survival of cells

CELLS
Need homeostasis for their own survival and for performing specialized functions essential for survival of the whole body.
See Chapters 1, 2, and 3.
Need a continual supply of nutrients and O_2 and ongoing elimination of acid-forming CO_2 to generate the energy needed to power life-sustaining cellular activities as follows:
Food + O_2 → CO_2 + H_2O + energy
See Chapter 17.

Cells make up body systems

Human Physiology

From Cells to Systems

Fifth Edition

Lauralee Sherwood

*Department of Physiology and
Pharmacology, School of Medicine
West Virginia University*

THOMSON

BROOKS/COLE

Australia • Canada • Mexico • Singapore • Spain
United Kingdom • United States

THOMSON

BROOKS/COLE

Editor in Chief: Michelle Julet
Biology Executive Editor: Nedah Rose
Development Editor: Mary Arbogast
Assistant Editor: Christopher Delgado
Editorial Assistant: Jennifer Keever
Technology Project Manager: Travis Metz
Marketing Manager: Ann Caven
Marketing Assistant: Sandra Perin
Advertising Project Manager: Linda Yip
Project Manager, Editorial Production: Teri Hyde
Print/Media Buyer: Jessica Reed
Permissions Editor: Joohee Lee
Production Service: Thomas E. Dorsaneo, Publishing Consultants

Text Designer: Jeanne Calabrese
Executive Art Director: Rob Hugel
Art Editor: Thomas E. Dorsaneo
Photo Researcher: Myrna Engler
Copy Editor: Linda Purrington
Illustrator: Kelly Collins, Richard Coombs, Cyndie C. H. Wooley, Thompson Type
Cover Designer: Jeanne Calabrese
Cover Image: John Bildbyra/Photonica
Cover Printer: The Lehigh Press, Inc.
Compositor: Thompson Type
Printer: Von Hoffman Corporation

For more information about our products, contact us at:
Thomson Learning Academic Resource Center
1-800-423-0563
For permission to use material from this text, contact us by:
Phone: 1-800-730-2214 **Fax:** 1-800-730-2215
Web: http://www.thomsonrights.com

Library of Congress Control Number: 2002112593

Student Edition: ISBN 0-534-39501-5
International Student Edition: ISBN 0-534-39536-8
Instructor's Edition: ISBN 0-534-39503-1

Brooks/Cole—Thomson Learning
10 Davis Drive
Belmont, CA 94002
USA

Asia
Thomson Learning
5 Shenton Way #01-01
UIC Building
Singapore 068808

Australia
Nelson Thomson Learning
102 Dodds Street
South Melbourne, Victoria 3205
Australia

Canada
Nelson Thomson Learning
1120 Birchmount Road
Toronto, Ontario M1K 5G4
Canada

Europe/Middle East/Africa
Thomson Learning
High Holborn House
50/51 Bedford Row
London WC1R 4LR
United Kingdom

Latin America
Thomson Learning
Seneca, 53
Colonia Polanco
11560 Mexico D.F.
Mexico

Spain/Portugal
Paraninfo
Calle/Magallanes, 25
28015 Madrid, Spain

To my family, for all they have done for me
in the past, all they mean to me in the present,
and all I hope will yet be in the future:

My parents,
Larry (in memoriam) and Lee Sherwood

My husband,
Peter Marshall

My daughters,
Melinda Marple
Allison Hansen

My granddaughters,
Lindsay Marple
Emily Marple

Brief Contents

Chapter 1   Homeostasis: The Foundation of Physiology   2

Chapter 2   Cellular Physiology   22

Chapter 3   The Plasma Membrane and Membrane Potential   56

Chapter 4   Neuronal Physiology   98

Chapter 5   The Central Nervous System   132

Chapter 6   The Peripheral Nervous System:
Afferent Division; Special Senses   184

Chapter 7   The Peripheral Nervous System: Efferent Division   236

Chapter 8   Muscle Physiology   256

Chapter 9   Cardiac Physiology   302

Chapter 10   The Blood Vessels and Blood Pressure   342

Chapter 11   The Blood   390

Chapter 12   The Body Defenses   412

Chapter 13   The Respiratory System   458

Chapter 14   The Urinary System   510

Chapter 15   Fluid and Acid-Base Balance   558

Chapter 16 *The Digestive System* 590

Chapter 17 *Energy Balance and Temperature Regulation* 646

Chapter 18 *Principles of Endocrinology;*
The Central Endocrine Glands 666

Chapter 19 *The Peripheral Endocrine Glands* 700

Chapter 20 *The Reproductive System* 748

Appendix A *The Metric System* A-1

Appendix B *A Review of Chemical Principles* A-3

Appendix C *Storage, Replication, and Expression*
of Genetic Information A-19

Appendix D *Principles of Quantitative Reasoning* A-31

Appendix E *Answers to End-of-Chapter Objective Questions,*
Quantitative Exercises, and Points to Ponder A-35

Appendix F *Text References to Exercise Physiology* A-53

Glossary G-1

Index I-1

Contents

Preface for the Students   xv

Chapter 1

Homeostasis: The Foundation of Physiology   2

Introduction to Physiology   3
  Physiology focuses on mechanisms of action.
  Structure and function are inseparable.
Levels of Organization in the Body   4
  The chemical level: Various atoms and molecules make up the body.
  The cellular level: Cells are the basic units of life.
  The tissue level: Tissues are groups of cells of similar specialization.
  The organ level: An organ is a unit made up of several tissue types.
  The body system level: A body system is a collection of related organs.

CONCEPTS, CHALLENGES, AND CONTROVERSIES
Stem Cell Science and Tissue Engineering:
The Quest to Make Defective Body Parts Like New Again   8

  The organism level: The body systems are packaged together into a functional whole body.
Concept of Homeostasis   10
  Body cells are in contact with a privately maintained internal environment.
  Body systems maintain homeostasis, a dynamic steady state in the internal environment.

A CLOSER LOOK AT EXERCISE PHYSIOLOGY
What Is Exercise Physiology?   13

Homeostatic Control Systems   16
  Homeostatic control systems may operate locally or bodywide.
  Negative feedback opposes an initial change and is widely used to maintain homeostasis.
  Positive feedback amplifies an initial change.
  Feedforward mechanisms initiate responses in anticipation of a change.
  Disruptions in homeostasis can lead to illness and death.

 CHAPTER IN PERSPECTIVE: FOCUS ON HOMEOSTASIS   19

Chapter Summary   19
Review Exercises   20
Points to Ponder   20
PhysioEdge Resources   21

Chapter 2

Cellular Physiology   22

Observations of Cells   23

CONCEPTS, CHALLENGES, AND CONTROVERSIES
HeLa Cells: Problems in a "Growing" Industry   24

An Overview of Cell Structure   24
  The plasma membrane bounds the cell.
  The nucleus contains the DNA.
  The cytoplasm consists of various organelles and the cytosol.
Endoplasmic Reticulum   25
  The rough endoplasmic reticulum synthesizes proteins for secretion and membrane construction.
  The smooth endoplasmic reticulum packages new proteins in transport vesicles.
Golgi Complex   28
  Transport vesicles carry their cargo to the Golgi complex for further processing.
  The Golgi complex packages secretory vesicles for release by exocytosis.
Lysosomes   30
  Lysosomes serve as the intracellular digestive system.
  Extracellular material is brought into the cell by endocytosis for attack by lysosomal enzymes.
  Lysosomes remove useless but not useful parts of the cell.
Peroxisomes   34
  Peroxisomes house oxidative enzymes that detoxify various wastes.
Mitochondria   34
  Mitochondria, the energy organelles, are enclosed by a double membrane.

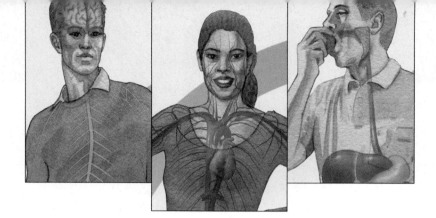

Mitochondria play a major role in generating ATP.
The cell generates more energy in aerobic than in anaerobic conditions.

A CLOSER LOOK AT EXERCISE PHYSIOLOGY
Aerobic Exercise: What For and How Much?   42

The energy stored in ATP is used for synthesis, transport, and mechanical work.
Vaults   42
Vaults may serve as cellular transport vehicles.
Cytosol   43
The cytosol is important in intermediary metabolism, ribosomal protein synthesis, and nutrient storage.
Cytoskeleton   44
Microtubules help maintain asymmetrical cell shapes and play a role in complex cell movements.
Microfilaments are important to cellular contractile systems and as mechanical stiffeners.
Intermediate filaments are important in cell regions subject to mechanical stress.
The cytoskeleton functions as an integrated whole and links other parts of the cell together.

   CHAPTER IN PERSPECTIVE:
FOCUS ON HOMEOSTASIS   51

Chapter Summary   52
Review Exercises   53
Points to Ponder   54
PhysioEdge Resources   55

Chapter 3

*The Plasma Membrane
and Membrane Potential*   56

Membrane Structure and Composition   58
The plasma membrane is a fluid lipid bilayer embedded with proteins.
The lipid bilayer forms the basic structural barrier that encloses the cell.
The membrane proteins perform a variety of specific membrane functions.

CONCEPTS, CHALLENGES, AND CONTROVERSIES
Cystic Fibrosis: A Fatal Defect in Membrane Transport   61

The membrane carbohydrates serve as self-identity markers.
Cell-to-Cell Adhesions   62
The extracellular matrix serves as the biological "glue."
Some cells are directly linked together by specialized cell junctions.
Intercellular Communication
and Signal Transduction   65
Communication between cells is largely orchestrated by extracellular chemical messengers.
Extracellular chemical messengers bring about cell responses primarily by signal transduction.
Some extracellular chemical messengers open chemically gated channels.
Many extracellular chemical messengers activate second messenger pathways.

CONCEPTS, CHALLENGES, AND CONTROVERSIES
Programmed Cell Suicide: A Surprising Example of a
Signal Transduction Pathway   70

Overview of Membrane Transport   71
Unassisted Membrane Transport   72
Particles that can permeate the membrane passively diffuse down their concentration gradient.
Ions that can permeate the membrane also passively move along their electrical gradient.
Osmosis is the net diffusion of water down its own concentration gradient.
Assisted Membrane Transport   78
Carrier-mediated transport is accomplished by a membrane carrier flipping its shape.
Carrier-mediated transport may be passive or active.

A CLOSER LOOK AT EXERCISE PHYSIOLOGY
Exercising Muscles Have a "Sweet Tooth"   81

With vesicular transport, material is moved into or out of the cell wrapped in membrane.
Caveolae may play roles in membrane transport and signal transduction.
Membrane Potential   87
Membrane potential refers to a separation of opposite charges across the membrane.

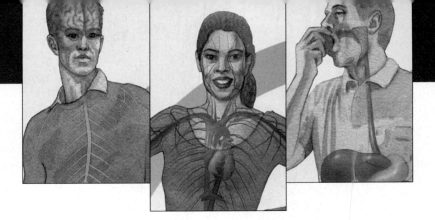

Membrane potential is due to differences in the concentration and permeability of key ions.

 CHAPTER IN PERSPECTIVE:
FOCUS ON HOMEOSTASIS   92

Chapter Summary   93
Review Exercises   94
Points to Ponder   96
PhysioEdge Resources   97

Chapter 4

Neuronal Physiology   98

Introduction   99
  Nerve and muscle are excitable tissues.
  Membrane potential decreases during depolarization and increases during hyperpolarization.
  Electrical signals are produced by changes in ion movement across the plasma membrane.
Graded Potentials   100
  The stronger the triggering event, the larger the resultant graded potential.
  Graded potentials spread by passive current flow.
  Graded potentials die out over short distances.
Action Potentials   103
  During an action potential, the membrane potential rapidly, transiently reverses.
  Marked changes in membrane permeability and ion movement lead to an action potential.
  The $Na^+–K^+$ pump gradually restores the concentration gradients disrupted by action potentials.
  Action potentials are propagated from the axon hillock to the axon terminals.
  Once initiated, action potentials are conducted throughout a nerve fiber.
  The refractory period ensures one-way propagation of the action potential.
  The refractory period also limits the frequency of action potentials.
  Action potentials occur in all-or-none fashion.

The strength of a stimulus is coded by the frequency of action potentials.
Myelination increases the speed of conduction of action potentials.

CONCEPTS, CHALLENGES, AND CONTROVERSIES
Multiple Sclerosis: Myelin—Going, Going, Gone   116

Fiber diameter also influences the velocity of action potential propagation.
Regeneration of Nerve Fibers   116
  Schwann cells guide the regeneration of cut peripheral axons.
  Peripheral but not central nerve fibers can regenerate.
Synapses and Neuronal Integration   117
  Synapses are junctions between two neurons.
  A neurotransmitter carries the signal across a synapse.
  Some synapses excite whereas others inhibit the postsynaptic neuron.
  Each synapse is either always excitatory or always inhibitory.
  Neurotransmitters are quickly removed from the synaptic cleft.
  Some neurotransmitters function through intracellular second messenger systems.
  The grand postsynaptic potential depends on the sum of the activities of all presynaptic inputs.
  Action potentials are initiated at the axon hillock because it has the lowest threshold.
  Neuropeptides act primarily as neuromodulators.
  Presynaptic inhibition or facilitation can selectively alter the effectiveness of a presynaptic input.
  Drugs and diseases can modify synaptic transmission.

CONCEPTS, CHALLENGES, AND CONTROVERSIES
Parkinson's Disease, Pollution, Ethical Problems, and Politics   126

Neurons are linked through complex converging and diverging pathways.

 CHAPTER IN PERSPECTIVE:
FOCUS ON HOMEOSTASIS   128

Chapter Summary   129
Review Exercises   130
Points to Ponder   131
PhysioEdge Resources   131

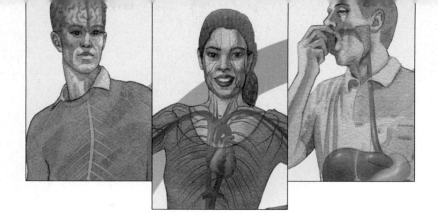

Chapter 5

The Central Nervous System   132

Comparison of the Nervous and Endocrine Systems   133
  The nervous system is a "wired" system, and the endocrine system is a "wireless" system.
  Neural specificity is due to anatomical proximity and endocrine specificity to receptor specialization.
  The nervous and endocrine systems have their own realms of authority but interact functionally.

Organization of the Nervous System   135
  The nervous system is organized into the central nervous system and the peripheral nervous system.
  The three classes of neurons are afferent neurons, efferent neurons, and interneurons.

Protection and Nourishment of the Brain   137
  Glial cells support the interneurons physically, metabolically, and functionally.
  The delicate central nervous system is well protected.
  Three meningeal membranes wrap, protect, and nourish the central nervous system.
  The brain floats in its own special cerebrospinal fluid.
  A highly selective blood-brain barrier carefully regulates exchanges between the blood and brain.

CONCEPTS, CHALLENGES, AND CONTROVERSIES
Strokes: A Deadly Domino Effect   143

  The brain depends on constant delivery of oxygen and glucose by the blood.

Overview of the Central Nervous System   143

Cerebral Cortex   145
  The cerebral cortex is an outer shell of gray matter covering an inner core of white matter.
  The cerebral cortex is organized into layers and functional columns.
  The four pairs of lobes in the cerebral cortex are specialized for different activities.
  The parietal lobes are responsible for somatosensory processing.
  The primary motor cortex is located in the frontal lobes.
  Other brain regions besides the primary motor cortex are important in motor control.
  Somatotopic maps vary slightly between individuals and are dynamic, not static.
  Because of its plasticity, the brain can be remodeled in response to varying demands.

  Different aspects of language are controlled by different regions of the cortex.
  The association areas of the cortex are involved in many higher functions.
  The cerebral hemispheres have some degree of specialization.
  An electroencephalogram is a record of postsynaptic activity in cortical neurons.
  Neurons in different regions of the cerebral cortex may fire in rhythmic synchrony.

Basal Nuclei, Thalamus, and Hypothalamus   154
  The basal nuclei play an important inhibitory role in motor control.
  The thalamus is a sensory relay station and is important in motor control.
  The hypothalamus regulates many homeostatic functions.

The Limbic System and Its Functional Relations with the Higher Cortex   157
  The limbic system plays a key role in emotion.
  The limbic system and higher cortex participate in the control of basic behavior patterns.
  Motivated behaviors are goal directed.
  Norepinephrine, dopamine, and serotonin are neurotransmitters in pathways for emotions and behavior.
  Learning is the acquisition of knowledge as a result of experiences.
  Memory is laid down in stages.
  Memory traces are present in multiple regions of the brain.

CONCEPTS, CHALLENGES, AND CONTROVERSIES
Alzheimer's Disease: A Tale of Beta Amyloid Plaques, Tau Tangles, and Dementia   162

  Short-term memory involves transient changes in synaptic activity.
  Long-term memory involves formation of new, permanent synaptic connections.

Cerebellum   166
  The cerebellum is important in balance and in planning and execution of voluntary movement.

Brain Stem   169
  The brain stem is a vital link between the spinal cord and higher brain regions.
  Sleep is an active process consisting of alternating periods of slow-wave and paradoxical sleep.

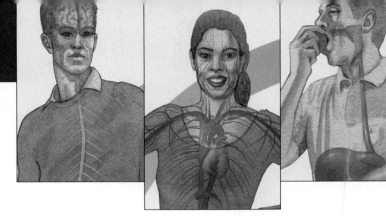

The sleep-wake cycle is controlled by interactions among three neural systems.
The function of sleep is unclear.
Spinal Cord   172
The spinal cord extends through the vertebral canal and is connected to the spinal nerves.
The white matter of the spinal cord is organized into tracts.
Each horn of the spinal cord gray matter houses a different type of neuronal cell body.
Spinal nerves carry both afferent and efferent fibers.

A CLOSER LOOK AT EXERCISE PHYSIOLOGY
Swan Dive or Belly Flop: It's a Matter of CNS Control   177

The spinal cord is responsible for the integration of many basic reflexes.

   CHAPTER IN PERSPECTIVE:
FOCUS ON HOMEOSTASIS   180

Chapter Summary   180
Review Exercises   182
Points to Ponder   183
PhysioEdge Resources   183

Chapter 6

The Peripheral Nervous System: Afferent Division; Special Senses   184

Introduction   185
Visceral afferents carry subconscious input while sensory afferents carry conscious input.
Perception is the conscious awareness of surroundings derived from interpretation of sensory input.

A CLOSER LOOK AT EXERCISE PHYSIOLOGY
Back Swings and Prejump Crouches:
What Do They Share in Common?   186

Receptor Physiology   185
Receptors have differential sensitivities to various stimuli.
A stimulus alters the receptor's permeability, leading to a graded receptor potential.

Receptor potentials may initiate action potentials in the afferent neuron.
Receptors may adapt slowly or rapidly to sustained stimulation.
Each somatosensory pathway is "labeled" according to modality and location.
Acuity is influenced by receptive field size and lateral inhibition.
Pain   192
Stimulation of nociceptors elicits the perception of pain plus motivational and emotional responses.
The brain has a built-in analgesic system.
Eye: Vision   184
Protective mechanisms help prevent eye injuries.

CONCEPTS, CHALLENGES, AND CONTROVERSIES
Acupuncture: Is It for Real?   195

The eye is a fluid-filled sphere enclosed by three specialized tissue layers.
The amount of light entering the eye is controlled by the iris.
The eye refracts the entering light to focus the image on the retina.
Accommodation increases the strength of the lens for near vision.
Light must pass through several retinal layers before reaching the photoreceptors.
Phototransduction by retinal cells converts light stimuli into neural signals.
Rods provide indistinct gray vision at night whereas cones provide sharp color vision during the day.
The sensitivity of the eyes can vary markedly through dark and light adaptation.
Color vision depends on the ratios of stimulation of the three cone types.
Visual information is modified and separated before reaching the visual cortex.
The thalamus and visual cortexes elaborate the visual message.
Visual input goes to other areas of the brain not involved in vision perception.
Some sensory input may be detected by multiple sensory processing areas in the brain.

CONCEPTS, CHALLENGES, AND CONTROVERSIES
"Seeing" with the Tongue   213

Ear: Hearing and Equilibrium 213
  Sound waves consist of alternate regions of compression and rarefaction of air molecules.
  The external ear plays a role in sound localization.
  The tympanic membrane vibrates in unison with sound waves in the external ear.
  The middle ear bones convert tympanic-membrane vibrations into fluid movements in the inner ear.
  The cochlea contains the organ of Corti, the sense organ for hearing.
  Hair cells in the organ of Corti transduce fluid movements into neural signals.
  Pitch discrimination depends on the region of the basilar membrane that vibrates.
  Loudness discrimination depends on the amplitude of the vibration.
  The auditory cortex is mapped according to tone.
  Deafness is caused by defects either in conduction or neural processing of sound waves.
  The vestibular apparatus is important for equilibrium by detecting position and motion of the head.
Chemical Senses: Taste and Smell 224
  Taste receptor cells are located primarily within tongue taste buds.
  Taste discrimination is coded by patterns of activity in various taste bud receptors.
  The olfactory receptors in the nose are specialized endings of renewable afferent neurons.
  Various parts of an odor are detected by different olfactory receptors and sorted into "smell files."
  Odor discrimination is coded by patterns of activity in the olfactory bulb glomeruli.
  The olfactory system adapts quickly, and odorants are rapidly cleared.
  The vomeronasal organ detects pheromones.

 CHAPTER IN PERSPECTIVE:
  FOCUS ON HOMEOSTASIS 232

Chapter Summary 232
Review Exercises 234
Points to Ponder 235
PhysioEdge Resources 235

Chapter 7

The Peripheral Nervous System: Efferent Division 236

Introduction 237
Autonomic Nervous System 237
  An autonomic nerve pathway consists of a two-neuron chain.
  Parasympathetic postganglionic fibers release acetylcholine; sympathetic ones release norepinephrine.
  The autonomic nervous system controls involuntary visceral organ activities.
  The sympathetic and parasympathetic systems dually innervate most visceral organs.
  The adrenal medulla is a modified part of the sympathetic nervous system.
  Several different receptor types are available for each autonomic neurotransmitter.

CONCEPTS, CHALLENGES, AND CONTROVERSIES
The Autonomic Nervous System and Aging: A Fortuitous Find 224

  Many regions of the central nervous system are involved in the control of autonomic activities.
Somatic Nervous System 245
  Motor neurons supply skeletal muscle.
  Motor neurons are the final common pathway.
Neuromuscular Junction 246
  Acetylcholine is the neuromuscular junction neurotransmitter.

A CLOSER LOOK AT EXERCISE PHYSIOLOGY
Loss of Muscle Mass: A Plight of Spaceflight 250

  Acetylcholinesterase terminates acetylcholine activity at the neuromuscular junction.
  The neuromuscular junction is vulnerable to several chemical agents and diseases.

CONCEPTS, CHALLENGES, AND CONTROVERSIES
Botulinum Toxin's Reputation Gets a Facelift 252

 CHAPTER IN PERSPECTIVE:
  FOCUS ON HOMEOSTASIS 253

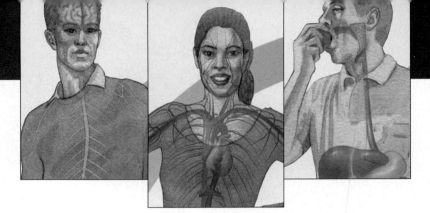

Chapter Summary   253
Review Exercises   254
Points to Ponder   255
PhysioEdge Resources   255

Chapter 8

Muscle Physiology   256

Introduction   257
Structure of Skeletal Muscle   258
  Skeletal muscle fibers are striated due to a highly organized
    internal arrangement.
  Myosin forms the thick filaments.
  Actin is the main structural component of the thin filaments.
Molecular Basis of Skeletal Muscle Contraction   261
  During contraction, cycles of cross-bridge binding and
    bending pull the thin filaments inward.
  Calcium is the link between excitation and contraction.
  Contractile activity far outlasts the electrical activity that
    initiated it.
Skeletal Muscle Mechanics   269
  Whole muscles are groups of muscle fibers bundled together
    and attached to bones.
  Contractions of a whole muscle can be of varying strength.
  The number of fibers contracting within a muscle depends
    on the extent of motor unit recruitment.
  The frequency of stimulation can influence the tension
    developed by each muscle fiber.
  Twitch summation results from a sustained elevation in
    cytosolic calcium.
  There is an optimal muscle length at which maximal
    tension can be developed.
  Muscle tension is transmitted to bone as the contractile
    component tightens the series-elastic component.
  The two primary types of contraction are isotonic and
    isometric.
  The velocity of shortening is related to the load.
  Although muscles can accomplish work, much of the energy
    is converted to heat.
  Interactive units of skeletal muscles, bones, and joints form
    lever systems.

Skeletal Muscle Metabolism and Fiber Types   277
  Muscle fibers have alternate pathways for forming ATP.
  Fatigue may be of muscle or central origin.
  Increased oxygen consumption is necessary to recover from
    exercise.
  There are three types of skeletal muscle fibers based on
    differences in ATP hydrolysis and synthesis.
  Muscle fibers adapt considerably in response to the
    demands placed on them.

A CLOSER LOOK AT EXERCISE PHYSIOLOGY
Are Athletes Who Use Steroids to Gain
Competitive Advantage Really Winners or Losers?   282

Control of Motor Movement   283
  Multiple neural inputs influence motor neuron output.

CONCEPTS, CHALLENGES, AND CONTROVERSIES
Muscular Dystrophy: When One Small Step is a Big Deal   284

  Muscle receptors provide afferent information needed to
    control skeletal muscle activity.
Smooth and Cardiac Muscle   288
  Smooth muscle cells are small and unstriated.
  Smooth muscle cells are turned on by Ca^{2+}-dependent
    phosphorylation of myosin.
  Multiunit smooth muscle is neurogenic.
  Single-unit smooth muscle cells form functional syncytia.
  Single-unit smooth muscle is myogenic.
  Gradation of single-unit smooth muscle contraction differs
    from that of skeletal muscle.
  Smooth muscle can still develop tension yet inherently
    relaxes when stretched.
  Smooth muscle is slow and economical.
  Cardiac muscle blends features of both skeletal and smooth
    muscle.

CHAPTER IN PERSPECTIVE:
FOCUS ON HOMEOSTASIS   297

Chapter Summary   297
Review Exercises   299
Points to Ponder   301
PhysioEdge Resources   301

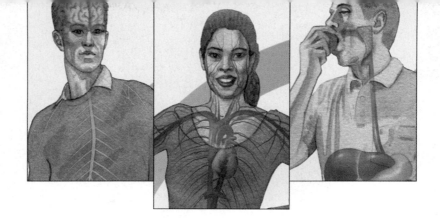

Chapter 9

Cardiac Physiology   302

Introduction   303
Anatomy of the Heart   304
  The heart is a dual pump.
  Heart valves ensure that the blood flows in the proper direction through the heart.
  The heart walls are composed primarily of spirally arranged cardiac muscle fibers.
  Cardiac muscle fibers are interconnected by intercalated discs and form functional syncytia.
  The heart is enclosed by the pericardial sac.
Electrical Activity of the Heart   310
  Cardiac autorhythmic cells display pacemaker activity.
  The sinoatrial node is the normal pacemaker of the heart.
  The spread of cardiac excitation is coordinated to ensure efficient pumping.
  The action potential of cardiac contractile cells shows a characteristic plateau.
  Ca^{2+} entry from the ECF induces a much larger Ca^{2+} release from the sarcoplasmic reticulum.
  Tetanus of cardiac muscle is prevented by a long refractory period.
  The ECG is a record of the overall spread of electrical activity through the heart.
  Various components of the ECG record can be correlated to specific cardiac events.
  The ECG can be used to diagnose abnormal heart rates, arrhythmias, and damage of heart muscle.
Mechanical Events of the Cardiac Cycle   320
  The heart alternately contracts to empty and relaxes to fill.

A CLOSER LOOK AT EXERCISE PHYSIOLOGY
The What, Who, and When of Stress Testing   321

  The two heart sounds are associated with valve closures.
  Turbulent blood flow produces heart murmurs.
Cardiac Output and Its Control   325
  Cardiac output depends on the heart rate and the stroke volume.
  Heart rate is determined primarily by autonomic influences on the SA node.
  Stroke volume is determined by the extent of venous return and by sympathetic activity.
  Increased end-diastolic volume results in increased stroke volume.

  The contractility of the heart is increased by sympathetic stimulation.
  High blood pressure increases the workload of the heart.
  The contractility of the heart is decreased in heart failure.
Nourishing the Heart Muscle   332
  The heart receives most of its own blood supply through the coronary circulation during diastole.
  Atherosclerotic coronary artery disease can deprive the heart of essential oxygen.

CONCEPTS, CHALLENGES, AND CONTROVERSIES
Atherosclerosis: Cholesterol and Beyond   336

 CHAPTER IN PERSPECTIVE:
FOCUS ON HOMEOSTASIS   338

Chapter Summary   338
Review Exercises   339
Points to Ponder   341
PhysioEdge Resources   341

Chapter 10

The Blood Vessels and Blood Pressure   342

Introduction   343
  To maintain homeostasis, reconditioning organs receive blood flow in excess of their own needs.
  Blood flow through vessels depends on the pressure gradient and vascular resistance.
  The vascular tree consists of arteries, arterioles, capillaries, venules, and veins.

CONCEPTS, CHALLENGES, AND CONTROVERSIES
From Humors to Harvey: Historical Highlights in Circulation   347

Arteries   347
  Arteries serve as rapid-transit passageways to the tissues and as a pressure reservoir.
  Arterial pressure fluctuates in relation to ventricular systole and diastole.
  Blood pressure can be measured indirectly by using a sphygmomanometer.

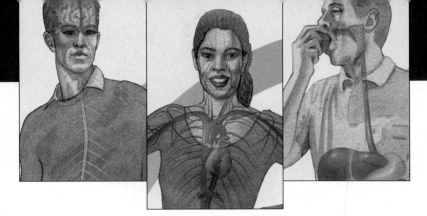

Mean arterial pressure is the main driving force for blood flow.

Arterioles   351

Arterioles are the major resistance vessels.
Local control of arteriolar radius is important in determining the distribution of cardiac output.
Local metabolic influences on arteriolar radius help match blood flow with the tissues' needs.
Local histamine release pathologically dilates arterioles.
Local physical influences on arteriolar radius include temperature changes and stretch.
Extrinsic sympathetic control of arteriolar radius is important in the regulation of blood pressure.
The medullary cardiovascular control center and several hormones regulate blood pressure.

Capillaries   360

Capillaries are ideally suited to serve as sites of exchange.
Water-filled capillary pores permit passage of small, water-soluble substances.
Many capillaries are not open under resting conditions.
Interstitial fluid is a passive intermediary between the blood and cells.
Diffusion across the capillary walls is important in solute exchange.
Bulk flow across the capillary walls is important in extracellular fluid distribution.
The lymphatic system is an accessory route by which interstitial fluid can be returned to the blood.
Edema occurs when too much interstitial fluid accumulates.

Veins   370

Veins serve as a blood reservoir as well as passageways back to the heart.
Venous return is enhanced by a number of extrinsic factors.

Blood Pressure   375

Blood pressure is regulated by controlling cardiac output, total peripheral resistance, and blood volume.
The baroreceptor reflex is an important short-term mechanism for regulating blood pressure.
Other reflexes and responses influence blood pressure.
Hypertension is a serious national public health problem, but its causes are largely unknown.

A CLOSER LOOK AT EXERCISE PHYSIOLOGY
The Ups and Downs of Hypertension and Exercise   382

Orthostatic hypotension results from transient inadequate sympathetic activity.
Circulatory shock can become irreversible.

 CHAPTER IN PERSPECTIVE:
FOCUS ON HOMEOSTASIS   385

Chapter Summary   386
Review Exercises   387
Points to Ponder   388
PhysioEdge Resources   389

Chapter 11

The Blood   390

Introduction   391

Plasma   391

Plasma water is a transport medium for many inorganic and organic substances.
Many of the functions of plasma are carried out by plasma proteins.

Erythrocytes   000

The structure of erythrocytes is well suited to their main function of O_2 transport in the blood.
The bone marrow continuously replaces worn-out erythrocytes.
Erythropoiesis is controlled by erythropoietin from the kidneys.

A CLOSER LOOK AT EXERCISE PHYSIOLOGY
Blood Doping: Is more of a Good Thing Better?   396

Anemia can be caused by a variety of disorders.
Polycythemia is an excess of circulating erythrocytes.

CONCEPTS, CHALLENGES, AND CONTROVERSIES
In Search of a Blood Substitute   398

Leukocytes   398

Leukocytes primarily function as defense agents outside the blood.
There are five types of leukocytes.
Leukocytes are produced at varying rates depending on the changing defense needs of the body.

Platelets and Hemostasis   402

Platelets are cell fragments derived from megakaryocytes.
Hemostasis prevents blood loss from damaged small vessels.

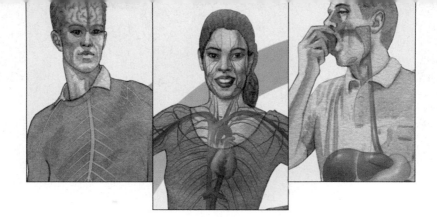

Vascular spasm reduces blood flow through an injured vessel.
Platelets aggregate to form a plug at a vessel defect.
Clot formation results from a triggered chain reaction involving plasma clotting factors.
Fibrinolytic plasmin dissolves clots.
Inappropriate clotting is responsible for thromboembolism.
Hemophilia is the primary condition responsible for excessive bleeding.

 CHAPTER IN PERSPECTIVE: FOCUS ON HOMEOSTASIS   408

Chapter Summary   409
Review Exercises   409
Points to Ponder   410
PhysioEdge Resources   411

Chapter 12
The Body Defenses   412
Introduction   413
Pathogenic bacteria and viruses are the major targets of the immune system.
Leukocytes are the effector cells of the immune system.
Immune responses may be either innate and nonspecific or adaptive and specific.
Innate Immunity   416
Inflammation is a nonspecific response to foreign invasion or tissue damage.
Salicylates and glucocorticoid drugs suppress the inflammatory response.
Interferon transiently inhibits multiplication of viruses in most cells.
Natural killer cells destroy virus-infected cells and cancer cells upon first exposure to them.
The complement system punches holes in micro-organisms.
Adaptive Immunity: General Concepts   424
Adaptive immune responses include antibody-mediated immunity and cell-mediated immunity.
An antigen induces an immune response against itself.
B Lymphocytes: Antibody-Mediated Immunity   425
Antigens stimulate B cells to convert into plasma cells that produce antibodies.

Antibodies are Y-shaped and classified according to properties of their tail portion.
Antibodies largely amplify innate immune responses to promote antigen destruction.
Clonal selection accounts for the specificity of antibody production.
Selected clones differentiate into active plasma cells and dormant memory cells.

CONCEPTS, CHALLENGES, AND CONTROVERSIES
Vaccination: A Victory Over Many Dreaded Diseases   430

The huge repertoire of B cells is built by reshuffling a small set of gene fragments.
Active immunity is self-generated; passive immunity is "borrowed."
Blood types are a form of natural immunity.
Lymphocytes respond only to antigen presented to them by antigen-presenting cells.
T Lymphocytes: Cell-Mediated Immunity   435
T cells bind directly with their targets.
The two types of T cells are cytotoxic T cells and helper T cells.
Cytotoxic T cells secrete chemicals that destroy target cells.
Helper T cells secrete chemicals that amplify the activity of other immune cells.
The immune system is normally tolerant of self-antigens.
Autoimmune diseases arise from loss of tolerance to self-antigens.
The major histocompatibility complex is the code for self-antigens.
Immune surveillance against cancer cells involves an interplay among immune cells and interferon.
A regulatory loop links the immune system with the nervous and endocrine systems.

A CLOSER LOOK AT EXERCISE PHYSIOLOGY
Exercise: A Help or Hindrance to Immune Defense   446

Immune Diseases   446
Immune deficiency diseases result from insufficient immune responses.
Allergies are inappropriate immune attacks against harmless environmental substances.
External Defenses   449
The skin consists of an outer protective epidermis and an inner, connective tissue dermis.

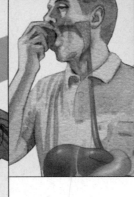

Specialized cells in the epidermis produce keratin and
melanin and participate in immune defense.
Protective measures within body cavities discourage
pathogen invasion into the body.

 CHAPTER IN PERSPECTIVE:
FOCUS ON HOMEOSTASIS 452

Chapter Summary 453
Review Exercises 455
Points to Ponder 456
PhysioEdge Resources 457

Chapter 13

The Respiratory System 458

Introduction 459
The respiratory airways conduct air between the atmosphere
and alveoli.
The gas-exchanging alveoli are thin-walled, inflatable sacs
encircled by pulmonary capillaries.
The lungs occupy much of the thoracic cavity.
A pleural sac separates each lung from the thoracic wall.
Respiratory Mechanics 463
Interrelationships among pressures inside and outside the
lungs are important in ventilation.
The lungs are normally stretched to fill the larger thorax.
Flow of air into and out of the lungs occurs because of
cyclical changes in intra-alveolar pressure.
Airway resistance influences airflow rates.
Airway resistance is abnormally increased with chronic
obstructive pulmonary disease.
Elastic behavior of the lungs is due to elastic connective
tissue and alveolar surface tension.
Pulmonary surfactant decreases surface tension and
contributes to lung stability.
The work of breathing normally requires only about 3% of
total energy expenditure.
The lungs normally operate at about "half full."
Alveolar ventilation is less than pulmonary ventilation
because of the presence of dead space.
Local controls act on the smooth muscle of the airways
and arterioles to match airflow to blood flow.
Gas Exchange 483
Gases move down partial pressure gradients.

Oxygen enters and CO_2 leaves the blood in the lungs
passively down partial pressure gradients.
Factors other than the partial pressure gradient influence
the rate of gas transfer.
Gas exchange across the systemic capillaries also occurs
down partial pressure gradients.
Gas Transport 488
Most O_2 in the blood is transported bound to hemoglobin.
The P_{O_2} is the primary factor determining the percent
hemoglobin saturation.
Hemoglobin promotes the net transfer of O_2 at both the
alveolar and tissue levels.
Factors at the tissue level promote the unloading of O_2
from hemoglobin.
Hemoglobin has a much higher affinity for carbon
monoxide than for O_2.
Most CO_2 is transported in the blood as bicarbonate.
Various respiratory states are characterized by abnormal
blood gas levels.
Control of Respiration 495
Respiratory centers in the brain stem establish a rhythmic
breathing pattern.

CONCEPTS, CHALLENGES, AND CONTROVERSIES
Effects of Heights and Depths on the Body 496

The magnitude of ventilation is adjusted in response to
three chemical factors: P_{O_2}, P_{CO_2}, and H^+.
Decreased arterial P_{O_2} increases ventilation only as an
emergency mechanism.
Carbon dioxide-generated H^+ in the brain is normally the
primary regulator of ventilation.
Adjustments in ventilation in response to changes in
arterial H^+ are important in acid–base balance.
Exercise profoundly increases ventilation, but the
mechanisms involved are unclear.
Ventilation can be influenced by factors unrelated to the
need for gas exchange.

A CLOSER LOOK AT EXERCISE PHYSIOLOGY
How to Find Out How Much Work You're Capable of Doing 504

During apnea, a person "forgets to breathe"; during
dyspnea, a person feels "short of breath."

 CHAPTER IN PERSPECTIVE:
FOCUS ON HOMEOSTASIS 505

Chapter Summary   505
Review Exercises   507
Points to Ponder   508
PhysioEdge Resources   509

Chapter 14

The Urinary System   510

Introduction   511
  The kidneys perform a variety of functions aimed at maintaining homeostasis.
  The kidneys form the urine; the remainder of the urinary system carries the urine to the outside.
  The nephron is the functional unit of the kidney.
  The three basic renal processes are glomerular filtration, tubular reabsorption, and tubular secretion.
Glomerular Filtration   517
  The glomerular membrane is considerably more permeable than capillaries elsewhere.
  The glomerular capillary blood pressure is the major force that induces glomerular filtration.
  Changes in the GFR occur primarily as a result of changes in glomerular capillary blood pressure.
  The GFR can be influenced by changes in the filtration coefficient.
  The kidneys normally receive 20% to 25% of the cardiac output.
Tubular Reabsorption   525
  Tubular reabsorption is tremendous, highly selective, and variable.
  Tubular reabsorption involves transepithelial transport.
  An active Na^+–K^+ ATPase pump in the basolateral membrane is essential for Na^+ reabsorption.
  Aldosterone stimulates Na^+ reabsorption in the distal and collecting tubules.
  Atrial natriuretic peptide inhibits Na^+ reabsorption.
  Glucose and amino acids are reabsorbed by Na^+-dependent secondary active transport.
  In general, actively reabsorbed substances exhibit a tubular maximum.
  Glucose is an example of an actively reabsorbed substance that is not regulated by the kidneys.
  Phosphate is an example of an actively reabsorbed substance that is regulated by the kidneys.
  Active Na^+ reabsorption is responsible for the passive reabsorption of Cl^-, H_2O, and urea.
  In general, unwanted waste products are not reabsorbed.
Tubular Secretion   534
  Hydrogen ion secretion is important in acid–base balance.
  Potassium secretion is controlled by aldosterone.
  Organic anion and cation secretion helps efficiently eliminate foreign compounds from the body.
Urine Excretion and Plasma Clearance   537
  Plasma clearance refers to the volume of plasma cleared of a particular substance per minute.
  The kidneys can excrete urine of varying concentrations depending on the body's state of hydration.
  The medullary vertical osmotic gradient is established by means of countercurrent multiplication.
  Vasopressin-controlled, variable H_2O reabsorption occurs in the final tubular segments.
  Countercurrent exchange within the vasa recta conserves the medullary vertical osmotic gradient.
  Water and solute reabsorption versus excretion are only partially coupled.
  Renal failure has wide-ranging consequences.

A CLOSER LOOK AT EXERCISE PHYSIOLOGY
When Protein in the Urine Does Not Mean Kidney Disease   550

Urine is temporarily stored in the bladder, from which it is emptied by micturition.

CONCEPTS, CHALLENGES, AND CONTROVERSIES
Dialysis: Cellophane Tubing or Abdominal Lining as an Artificial Kidney   551

CHAPTER IN PERSPECTIVE:
FOCUS ON HOMEOSTASIS   552

Chapter Summary   553
Review Exercises   555
Points to Ponder   556
PhysioEdge Resources   557

Chapter 15

Fluid and Acid–Base Balance   558

Balance Concept   559
  The internal pool of a substance is the amount of that substance in the ECF.
  To maintain stable balance of an ECF constituent, its input must equal its output.

Fluid Balance   560
  Body water is distributed between the ICF and ECF compartments.
  The plasma and interstitial fluid are similar in composition but the ECF and ICF differ markedly.
  Fluid balance is maintained by regulating ECF volume and osmolarity.
  Control of ECF volume is important in the long-term regulation of blood pressure.
  Control of salt balance is primarily important in regulating ECF volume.
  The amount of Na^+ filtered is controlled through regulation of the GFR.
  The amount of Na^+ reabsorbed is controlled through the renin-angiotensin-aldosterone system.
  Control of ECF osmolarity prevents changes in ICF volume.

A CLOSER LOOK AT EXERCISE PHYSIOLOGY
A Potentially Fatal Clash: When Exercising Muscles and Cooling Mechanisms Compete for an Inadequate Plasma Volume   566

  During ECF hypertonicity, the cells shrink as H_2O leaves them.
  During ECF hypotonicity, the cells swell as H_2O enters them.
  No water moves into or out of the cells during an ECF isotonic fluid gain or loss.
  Control of water balance by means of vasopressin is important in regulating ECF osmolarity.
  Vasopressin secretion and thirst are largely triggered simultaneously.

Acid–Base Balance   571
  Acids liberate free hydrogen ions, whereas bases accept them.
  The pH designation is used to express $[H^+]$.
  Fluctuations in $[H^+]$ alter nerve, enzyme, and K^+ activity.
  Hydrogen ions are continually added to the body fluids as a result of metabolic activities.
  Chemical buffer systems minimize changes in pH by binding with or yielding free H^+.

  The H_2CO_3:HCO_3^- buffer pair is the primary ECF buffer for non-carbonic acids.
  The protein buffer system is primarily important intracellularly.
  The hemoglobin buffer system buffers H^+ generated from carbonic acid.
  The phosphate buffer system is an important urinary buffer.
  Chemical buffer systems act as the first line of defense against changes in $[H^+]$.
  The respiratory system regulates $[H^+]$ by controlling the rate of CO_2 removal.
  The respiratory system serves as the second line of defense against changes in $[H^+]$.
  The kidneys are a powerful third line of defense against changes in $[H^+]$.
  The kidneys adjust their rate of H^+ excretion depending on the plasma $[H^+]$ or $[CO_2]$.
  The kidneys conserve or excrete HCO_3^- depending on the plasma $[H^+]$.
  The kidneys secrete ammonia during acidosis to buffer secreted H^+.
  Acid–base imbalances can arise from either respiratory dysfunction or metabolic disturbances.
  Respiratory acidosis arises from an increase in $[CO_2]$.
  Respiratory alkalosis arises from a decrease in $[CO_2]$.
  Metabolic acidosis is associated with a fall in $[HCO_3^-]$.
  Metabolic alkalosis is associated with an elevation in $[HCO_3^-]$.

 CHAPTER IN PERSPECTIVE:
FOCUS ON HOMEOSTASIS   586

Chapter Summary   587
Review Exercises   588
Points to Ponder   589
PhysioEdge Resources   589

Chapter 16

The Digestive System   590

Introduction   561
  The digestive system performs four basic digestive processes.
  The digestive tract and accessory digestive organs make up the digestive system.
  The digestive tract wall has four layers.

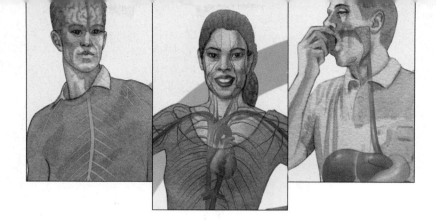

Regulation of digestive function is complex and synergistic.
Receptor activation alters digestive activity through neural reflexes and hormonal pathways.

Mouth   599

The oral cavity is the entrance to the digestive tract.
The teeth are responsible for chewing.
Saliva begins carbohydrate digestion, is important in oral hygiene, and facilitates speech.
Salivary secretion is continuous and can be reflexly increased.
Digestion in the mouth is minimal; no absorption of nutrients occurs.

Pharynx and Esophagus   601

Swallowing is a sequentially programmed all-or-none reflex.
During the oropharyngeal stage of swallowing, food is prevented from entering the wrong passageways.
The pharyngoesophageal sphincter prevents air from entering the digestive tract during breathing.
Peristaltic waves push food through the esophagus.
The gastroesophageal sphincter prevents reflux of gastric contents.
Esophageal secretion is entirely protective.

Stomach   604

The stomach stores food and begins protein digestion.
Gastric filling involves receptive relaxation.
Gastric storage takes place in the body of the stomach.
Gastric mixing takes place in the antrum of the stomach.
Gastric emptying is largely controlled by factors in the duodenum.

A CLOSER LOOK AT EXERCISE PHYSIOLOGY
Pregame Meal: What's In and What's Out?   607

Emotions can influence gastric motility.
The stomach does not actively participate in vomiting.
Gastric digestive juice is secreted by glands located at the base of gastric pits.
Hydrochloric acid activates pepsinogen.
Pepsinogen, once activated, begins protein digestion.
Mucus is protective.
Intrinsic factor is essential for absorption of vitamin B_{12}.
Multiple regulatory pathways influence the parietal and chief cells.
Control of gastric secretion involves three phases.
Gastric secretion gradually decreases as food empties from the stomach into the intestine.
The stomach lining is protected from gastric secretions by the gastric mucosal barrier.

Carbohydrate digestion continues in the body of the stomach; protein digestion begins in the antrum.
The stomach absorbs alcohol and aspirin but no food.

CONCEPTS, CHALLENGES, AND CONTROVERSIES
Ulcers: When Bugs Break the Barrier   615

Pancreatic and Biliary Secretions   615

The pancreas is a mixture of exocrine and endocrine tissue.
The exocrine pancreas secretes digestive enzymes and an aqueous alkaline fluid.
Pancreatic exocrine secretion is regulated by secretin and CCK.
The liver performs various important functions including bile production.
The liver lobules are delineated by vascular and bile channels.
Bile is continuously secreted by the liver and is diverted to the gallbladder between meals.
Bile salts are recycled through the enterohepatic circulation.
Bile salts aid fat digestion and absorption.
Bilirubin is a waste product excreted in the bile.
Bile salts are the most potent stimulus for increased bile secretion.
The gallbladder stores and concentrates bile between meals and empties during meals.
Hepatitis and cirrhosis are the most common liver disorders.

Small Intestine   623

Segmentation contractions mix and slowly propel the chyme.
The migrating motility complex sweeps the intestine clean between meals.
The ileocecal juncture prevents contamination of the small intestine by colonic bacteria.
Small intestine secretions do not contain any digestive enzymes.
The small intestine enzymes complete digestion intracellularly.
The small intestine is remarkably well adapted for its primary role in absorption.
The mucosal lining experiences rapid turnover.
Energy-dependent Na^+ absorption drives passive H_2O absorption.
Carbohydrate and protein are both absorbed by secondary active transport and enter the blood.
Digested fat is absorbed passively and enters the lymph.

Vitamin absorption is largely passive.
Iron and calcium absorption is regulated.
Most absorbed nutrients immediately pass through the liver for processing.
Extensive absorption by the small intestine keeps pace with secretion.
Biochemical balance among the stomach, pancreas, and small intestine is normally maintained.
Diarrhea results in loss of fluid and electrolytes.

CONCEPTS, CHALLENGES, AND CONTROVERSIES
Oral Rehydration Therapy: Sipping a
Simple Solution Saves Lives   636

Large Intestine   637
The large intestine is primarily a drying and storage organ.
Haustral contractions slowly shuffle the colonic contents back and forth.
Mass movements propel colonic contents long distances.
Feces are eliminated by the defecation reflex.
Constipation occurs when the feces become too dry.
Large intestine secretion is entirely protective.
The colon contains myriad beneficial bacteria.
The large intestine absorbs salt and water, converting the luminal contents into feces.
Intestinal gases are absorbed or expelled.
Overview of the Gastrointestinal Hormones   639

CHAPTER IN PERSPECTIVE:
FOCUS ON HOMEOSTASIS   641

Chapter Summary   641
Review Exercises   643
Points to Ponder   644
PhysioEdge Resources   645

Chapter 17

*Energy Balance and
Temperature Regulation*   646

Energy Balance   647
Most food energy is ultimately converted into heat in the body.

The metabolic rate is the rate of energy use.
Energy input must equal energy output to maintain a neutral energy balance.
Food intake is controlled primarily by the hypothalamus.
Obesity occurs when more kilocalories are consumed than are burned up.
Persons suffering from anorexia nervosa have a pathologic fear of gaining weight.

A CLOSER LOOK AT EXERCISE PHYSIOLOGY
What the Scales Don't Tell You   654

Temperature Regulation   654
Internal core temperature is homeostatically maintained at 100°F.
Heat input must balance heat output to maintain a stable core temperature.
Heat exchange takes place by radiation, conduction, convection, and evaporation.
The hypothalamus integrates a multitude of thermosensory inputs.
Shivering is the primary involuntary means of increasing heat production.
The magnitude of heat loss can be adjusted by varying the flow of blood through the skin.
The hypothalamus simultaneously coordinates heat-production and heat-loss mechanisms.
During a fever, the hypothalamic thermostat is "reset" at an elevated temperature.
Hyperthermia can occur unrelated to infection.

CONCEPTS, CHALLENGES, AND CONTROVERSIES
The Extremes of Heat and Cold Can Be Fatal   662

CHAPTER IN PERSPECTIVE:
FOCUS ON HOMEOSTASIS   663

Chapter Summary   663
Review Exercises   664
Points to Ponder   665
PhysioEdge Resources   665

Chapter 18

Principles of Endocrinology; The Central Endocrine Glands 666

General Principles of Endocrinology 667
  Hormones exert a variety of regulatory effects throughout the body.
  Hormones are chemically classified into three categories: peptides, amines, and steroids.
  The mechanisms of hormone synthesis, storage, and secretion vary according to the class of hormone.
  Water-soluble hormones dissolve in the plasma; lipid-soluble hormones are transported by plasma proteins.
  Hormones generally produce their effect by altering intracellular proteins.
  Hydrophilic hormones alter pre-existing proteins via second-messenger systems.
  By stimulating genes, lipophilic hormones promote synthesis of new proteins.
  Hormone actions are greatly amplified at the target cell.
  The effective plasma concentration of a hormone is normally regulated by changes in its rate of secretion.
  The effective plasma concentration of a hormone is influenced by its transport, metabolism, and excretion.
  Endocrine disorders result from hormone excess or deficiency or decreased target-cell responsiveness.
  The responsiveness of a target cell can be varied by regulating the number of hormone-specific receptors.
Pineal Gland and Circadian Rhythms 681
  The suprachiasmatic nucleus is the master biological clock.
  Melatonin helps keep the body's circadian rhythms in time with the light–dark cycle.

CONCEPTS, CHALLENGES, AND CONTROVERSIES
Tinkering with Our Biological Clocks 682

Hypothalamus and Pituitary 683
  The pituitary gland consists of anterior and posterior lobes.
  The hypothalamus and posterior pituitary act as a unit to secrete vasopressin and oxytocin.

A CLOSER LOOK AT EXERCISE PHYSIOLOGY
The Endocrine Response to the Challenge of Combined Heat and Marching Feet 685

  The anterior pituitary secretes six established hormones, many of which are tropic.
  Hypothalamic releasing and inhibiting hormones help regulate anterior-pituitary hormone secretion.
  Target-gland hormones inhibit hypothalamic and anterior pituitary hormone secretion via negative feedback.
Endocrine Control of Growth 689
  Growth depends on growth hormone but is influenced by other factors as well.
  Growth hormone is essential for growth, but it also exerts metabolic effects not related to growth.
  Bone grows in thickness and in length by different mechanisms, both stimulated by growth hormone.
  Growth hormone exerts its growth-promoting effects indirectly by stimulating somatomedins.
  Growth hormone secretion is regulated by two hypophysiotropic hormones.
  Abnormal growth hormone secretion results in aberrant growth patterns.
  Other hormones besides growth hormone are essential for normal growth.

CONCEPTS, CHALLENGES, AND CONTROVERSIES
Growth and Youth in a Bottle? 696

 CHAPTER IN PERSPECTIVE: FOCUS ON HOMEOSTASIS 697

Chapter Summary 697
Review Exercises 698
Points to Ponder 699
PhysioEdge Resources 699

Chapter 19

The Peripheral Endocrine Glands 700

Thyroid Gland 701
  The major cells that secrete thyroid hormone are organized into colloid-filled follicles.
  Thyroid hormone synthesis and storage occur on the thyroglobulin molecule.

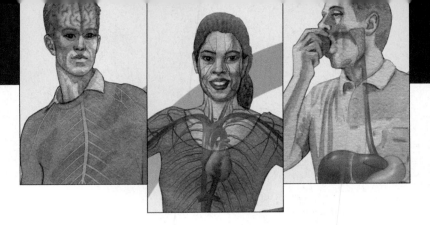

To secrete thyroid hormone, the follicular cells phagocytize thyroglobulin-laden colloid.

Most of the secreted T_4 is converted into T_3 outside the thyroid.

Thyroid hormone is the main determinant of the basal metabolic rate and exerts other effects as well.

Thyroid hormone is regulated by the hypothalamus-pituitary-thyroid axis.

Abnormalities of thyroid function include both hypothyroidism and hyperthyroidism.

A goiter develops when the thyroid gland is overstimulated.

Adrenal Glands   707

Each adrenal gland consists of a steroid-secreting cortex and a catecholamine-secreting medulla.

The adrenal cortex secretes mineralocorticoids, glucocorticoids, and sex hormones.

Mineralocorticoids' major effects are on Na^+ and K^+ balance and blood pressure homeostasis.

Glucocorticoids exert metabolic effects and play a key role in adaptation to stress.

Cortisol secretion is regulated by the hypothalamus-pituitary-adrenal cortex axis.

The adrenal cortex secretes both male and female sex hormones in both sexes.

The adrenal cortex may secrete too much or too little of any of its hormones.

The adrenal medulla is a modified sympathetic postganglionic neuron.

Epinephrine and norepinephrine vary in their affinities for the different adrenergic receptor types.

Epinephrine reinforces the sympathetic nervous system and exerts additional metabolic effects.

Sympathetic stimulation of the adrenal medulla is solely responsible for epinephrine release.

The stress response is a generalized pattern of reactions to any situation that threatens homeostasis.

The multifaceted stress response is coordinated by the hypothalamus.

Activation of the stress response by chronic psychosocial stressors may be harmful.

Endocrine Control of Fuel Metabolism   719

Fuel metabolism includes anabolism, catabolism, and interconversions among energy-rich organic molecules.

Because food intake is intermittent, nutrients must be stored for use between meals.

The brain must be continuously supplied with glucose.

Metabolic fuels are stored during the absorptive state and mobilized during the postabsorptive state.

Lesser energy sources are tapped as needed.

The pancreatic hormones, insulin and glucagon, are most important in regulating fuel metabolism.

Insulin lowers blood glucose, amino acid, and fatty acid levels and promotes their storage.

The primary stimulus for increased insulin secretion is an increase in blood glucose concentration.

The symptoms of diabetes mellitus are characteristic of an exaggerated postabsorptive state.

CONCEPTS, CHALLENGES, AND CONTROVERSIES
Diabetics and Insulin: Some Have It and Some Don't   728

Insulin excess causes brain-starving hypoglycemia.

Glucagon in general opposes the actions of insulin.

Glucagon secretion is increased during the postabsorptive state.

Insulin and glucagon work as a team to maintain blood glucose and fatty acid levels.

Glucagon excess can aggravate the hyperglycemia of diabetes mellitus.

Epinephrine, cortisol, and growth hormone also exert direct metabolic effects.

Endocrine Control of Calcium Metabolism   733

Plasma Ca^{2+} must be closely regulated to prevent changes in neuromuscular excitability.

Control of Ca^{2+} metabolism includes regulation of both Ca^{2+} homeostasis and Ca^{2+} balance.

Parathyroid hormone raises free plasma Ca^{2+} levels by its effects on bone, kidneys, and intestine.

Bone continuously undergoes remodeling.

A CLOSER LOOK AT EXERCISE PHYSIOLOGY
Osteoporosis: The Bane of Brittle Bones   736

PTH's immediate effect is to promote the transfer of Ca^{2+} from bone fluid into plasma.

PTH's chronic effect is to promote localized dissolution of bone to release Ca^{2+} into plasma.

PTH acts on the kidneys to conserve Ca^{2+} and eliminate PO_4^{3-}.

PTH indirectly promotes absorption of Ca^{2+} and PO_4^{3-} by the intestine.

The primary regulator of PTH secretion is the plasma concentration of free Ca^{2+}.

Calcitonin lowers the plasma Ca^{2+} concentration but is not important in the normal control of Ca^{2+} metabolism.

Vitamin D is actually a hormone that increases calcium absorption in the intestine.

Phosphate metabolism is controlled by the same mechanisms that regulate Ca^{2+} metabolism.

Disorders in Ca^{2+} metabolism may arise from abnormal levels of PTH or vitamin D.

 CHAPTER IN PERSPECTIVE: FOCUS ON HOMEOSTASIS   742

Chapter Summary   744
Review Exercises   745
Points to Ponder   746
PhysioEdge Resources   747

Chapter 20

The Reproductive System   748

Introduction   749
The reproductive system includes the gonads, reproductive tract, and accessory sex glands.

Reproductive cells each contain a half set of chromosomes.

Gametogenesis is accomplished by meiosis.

The sex of an individual is determined by the combination of sex chromosomes.

Sexual differentiation along male or female lines depends on the presence or absence of masculinizing determinants.

Male Reproductive Physiology   756
The scrotal location of the testes provides a cooler environment essential for spermatogenesis.

The testicular Leydig cells secrete masculinizing testosterone.

Spermatogenesis yields an abundance of highly specialized, mobile sperm.

Throughout their development, sperm remain intimately associated with Sertoli cells.

LH and FSH from the anterior pituitary control testosterone secretion and spermatogenesis.

Gonadotropin-releasing hormone activity increases at puberty.

The reproductive tract stores and concentrates sperm and increases their fertility.

The accessory sex glands contribute the bulk of the semen.

Prostaglandins are ubiquitous, locally acting chemical messengers.

Sexual Intercourse between Males and Females   766
The male sex act is characterized by erection and ejaculation.

Ejaculation includes emission and expulsion.

The female sexual cycle is very similar to the male cycle.

CONCEPTS, CHALLENGES, AND CONTROVERSIES
Environmental "Estrogens":
Bad News for the Reproductive System   769

Female Reproductive Physiology   770
Complex cycling characterizes female reproductive physiology.

The steps of gametogenesis are the same in both sexes but the timing and outcome sharply differ.

The ovarian cycle consists of alternating follicular and luteal phases.

The follicular phase is characterized by the development of maturing follicles.

The luteal phase is characterized by the presence of a corpus luteum.

The ovarian cycle is regulated by complex hormonal interactions.

Cyclical uterine changes are caused by hormonal changes during the ovarian cycle.

A CLOSER LOOK AT EXERCISE PHYSIOLOGY
Menstrual Irregularities: When Cyclists and
Other Female Athletes Do Not Cycle   780

Fluctuating estrogen and progesterone levels produce cyclical changes in cervical mucus.

Pubertal changes in females are similar to those in males.

Menopause is unique to females.

The oviduct is the site of fertilization.

The blastocyst implants in the endometrium through the action of its trophoblastic enzymes.

The placenta is the organ of exchange between maternal and fetal blood.

CONCEPTS, CHALLENGES, AND CONTROVERSIES
The Ways and Means of Contraception   788

Hormones secreted by the placenta play a critical role in the maintenance of pregnancy.
Maternal body systems respond to the increased demands of gestation.
Changes during late gestation prepare for parturition.
Scientists are closing in on the factors that trigger the onset of parturition.
Parturition is accomplished by a positive feedback cycle.
Lactation requires multiple hormonal inputs.
Breast-feeding is advantageous to both the infant and the mother.
The end is a new beginning.

 CHAPTER IN PERSPECTIVE:
FOCUS ON HOMEOSTASIS 799

Chapter Summary   799
Review Exercises   801
Points to Ponder   802
PhysioEdge Resources   802

Appendix A
The Metric System   A-1

Appendix B
A Review of Chemical Principles   A-3

Appendix C
Storage, Replication, and Expression of Genetic Information   A-19

Appendix D
Principles of Quantitative Reasoning   A-31

Appendix E
Answers to End-of-Chapter Objective Questions, Quantitative Exercises, and Points to Ponder   A-35

Appendix F
Text References to Exercise Physiology   A-53

Glossary   G-1

Index   I-1

Credits   C-1

Boxed Special Features

▍ Concepts, Challenges, and Controversies

Stem Cell Science and Tissue Engineering: The Quest to Make Defective Body Parts Like New Again   8

HeLa Cells: Problems in a "Growing" Industry   24

Cystic Fibrosis: A Fatal Defect in Membrane Transport   61

Programmed Cell Suicide: A Surprising Example of a Signal Transduction Pathway   70

Multiple Sclerosis: Myelin—Going, Going, Gone   116

Parkinson's Disease, Pollution, Ethical Problems, and Politics   126

Strokes: A Deadly Domino Effect   143

Alzheimer's Disease: A Tale of Beta Amyloid Plaques, Tau Tangles, and Dementia   162

Acupuncture: Is It for Real?   195

"Seeing" with the Tongue   213

The Autonomic Nervous System and Aging: A Fortuitous Find   224

Botulinum Toxin's Reputation Gets a Facelift   252

Muscular Dystrophy: When One Small Step is a Big Deal   284

Atherosclerosis: Cholesterol and Beyond   336

From Humors to Harvey: Historical Highlights in Circulation   347

In Search of a Blood Substitute   398

Vaccination: A Victory Over Many Dreaded Diseases   430

Effects of Heights and Depths on the Body   496

Dialysis: Cellophane Tubing or Abdominal Lining as an Artificial Kidney   551

Ulcers: When Bugs Break the Barrier   615

Oral Rehydration Therapy: Sipping a Simple Solution Saves Lives   636

The Extremes of Heat and Cold Can Be Fatal   662

Tinkering with Our Biological Clocks   682

Growth and Youth in a Bottle?   696

Diabetics and Insulin: Some Have It and Some Don't   728

Environmental "Estrogens": Bad News for the Reproductive System   769

The Ways and Means of Contraception   788

▍ A Closer Look at Exercise Physiology

What Is Exercise Physiology?   13

Aerobic Exercise: What For and How Much?   42

Exercising Muscles Have a "Sweet Tooth"   81

Swan Dive or Belly Flop: It's a Matter of CNS Control   177

Back Swings and Prejump Crouches: What Do They Share in Common?   186

Loss of Muscle Mass: A Plight of Spaceflight   250

Are Athletes Who Use Steroids to Gain Competitive Advantage Really Winners or Losers?   282

The What, Who, and When of Stress Testing   321

The Ups and Downs of Hypertension and Exercise   382

Blood Doping: Is more of a Good Thing Better?   396

Exercise: A Help or Hindrance to Immune Defense   446

How to Find Out How Much Work You're Capable of Doing   504

When Protein in the Urine Does Not Mean Kidney Disease   550

A Potentially Fatal Clash: When Exercising Muscles and Cooling Mechanisms Compete for an Inadequate Plasma Volume   566

Pregame Meal: What's In and What's Out?   607

What the Scales Don't Tell You   654

The Endocrine Response to the Challenge of Combined Heat and Marching Feet   685

Osteoporosis: The Bane of Brittle Bones   736

Menstrual Irregularities: When Cyclists and Other Female Athletes Do Not Cycle   780

Preface for Students

GOALS, PHILOSOPHY, AND THEME

My goal is to help students not only learn about how the body works but also to share my enthusiasm for the subject matter. I have been teaching physiology since the mid-1960s and remain awe-struck at the intricacies and efficiency of body function. When a baby first discovers it can control its own hands, it will be fascinated and spend many hours manipulating them in front of its face. Most of us, even infants, have a natural curiosity about how our bodies work. Our bodies are quite miraculous. No machine has been constructed that can take over even a portion of a natural body function as effectively. By capitalizing on students' natural curiosity about themselves, I have sought to make physiology a subject they can enjoy learning.

Even the most tantalizing subject matter, however, can be drudgery to study and difficult to comprehend if not effectively presented. Therefore, this book has a logical, understandable format that is unencumbered by unnecessary details and that emphasizes how each concept is an integral part of the whole subject matter. Too often, students view isolated sections of a physiology course as separate entities; by understanding how each component depends on other components, a student can appreciate the integrated functioning of the human body. The text focuses on the mechanisms of body function from cells to systems and is organized around the central theme of homeostasis—how the body meets changing demands while maintaining the internal constancy necessary for all cells and organs to function.

The text is written with undergraduate students preparing for health-related careers in mind. Its approach and depth are appropriate, however, for other undergraduate student populations. Because it is intended to serve as an introductory text and, for most students, may be their only exposure to a formal physiology text, all aspects of physiology receive broad coverage, yet depth, where needed, is not sacrificed. The scope of this text has been limited by judicious selection of pertinent content that a student can reasonably be expected to assimilate in a one-semester physiology course. Materials were selected for inclusion on a "need to know" basis, not just because a given fact happens to be known. In other words, content is restricted to relevant information needed to understand basic physiologic concepts and to serve as a foundation for future careers in the health professions. "Encyclopedic" peripheral facts have been excluded.

To keep pace with today's rapid advances in the health sciences, students in the health professions must be able to draw on their conceptual understanding of physiology instead of merely recalling isolated facts that soon may be outdated. Therefore, this text is designed to promote understanding of the basic principles and concepts of physiology rather than memorization of details. The text is written in simple, straightforward language, and every effort has been made to assure smooth reading through good transitions, logical reasoning, and integration of ideas throughout the text.

In consideration of the clinical orientation of most students, research methodologies and data are not emphasized, although the material is based on up-to-date evidence. New information based on recent discoveries has been included in all chapters. Students can be assured of the timeliness and accuracy of the material presented. Some controversial ideas and hypotheses are presented to illustrate that physiology is a dynamic, changing discipline.

Because the function of an organ depends on the organ's construction, enough relevant anatomy is integrated within the text to make the inseparable relation between structure and function meaningful.

FEATURES AND LEARNING AIDS

▌ Homeostatic model and chapter opening

A unique, easy-to-follow, pictorial homeostatic model depicting the relationship among cells, systems, and homeostasis is developed in the introductory chapter and presented on the inside front cover as a quick reference. Each chapter begins with a specialized, tailor-made version of this model, accompanied by a brief written introduction, emphasizing how the body system considered in the chapter functionally fits in with the body as a whole. This opening feature is designed to orient the student and help put in perspective the material that follows.

■ Chapter closing focusing on homeostasis

Each chapter concludes with a narrative, **Chapter in Perspective: Focus on Homeostasis,** which helps the students put into perspective how the part of the body just discussed contributes to homeostasis. This capstone feature, the opening homeostatic model, and the introductory comments are designed to work together to facilitate the students' comprehension of the interactions and interdependency of body systems, even though each system is discussed separately.

■ Bulleted chapter summaries

The major points of each chapter are presented in concise, section-by-section bulleted lists in the **Chapter Summary** at each chapter's end, enabling students to focus on the main concepts before moving on.

■ End-of-chapter learning activities

The **Review Exercises** at the end of each chapter include a variety of question formats for students to self-test their knowledge and application of the facts and concepts presented in the chapter. Also available are **Quantitative Exercises** that provide the students with an opportunity to practice calculations that will enhance their understanding of complex relationships. A **Points to Ponder** section features thought-provoking problems that encourage students to analyze and apply what they have learned. The Points to Ponder section is capped off with a **Clinical Consideration,** a mini case study that challenges students to apply their knowledge to a patient's specific symptoms. Finally, the **PhysioEdge Resources** section directs students to the wealth of study aids and ideas for further reading through the human physiology Web site and PhysioEdge CD-ROM. These resources include **Chapter Outlines, Chapter Summaries, Glossary, Flash Cards, Visual Learning Resource, Quizzing, Final Exam, Internet Exercises, InfoTrac College Edition® Exercises, Web References,** and valuable **Case History Exercises.** The CD-ROM has **Concept Tutorials** and **Media Quizzes.**

■ Boxed features

Two types of boxed features highlight the chapters. The **Concepts, Challenges, and Controversies** boxes expose students to high-interest, tangentially relevant information on such diverse topics as environmental impact on the body, aging, ethical issues, new discoveries regarding common diseases, historical perspectives, and body responses to new environments such as those encountered in space flight and deep-sea diving.

The **Closer Look at Exercise Physiology** boxes are included for three reasons: increasing national awareness of the importance of physical fitness, increasing recognition of the value of prescribed therapeutic exercise programs for a variety of conditions, and growing career opportunities related to fitness and exercise.

■ Analogies

Many analogies and frequent references to everyday experiences are included to help students relate to the physiology concepts presented. These useful tools have been drawn in large part from my nearly four decades of teaching experience. Knowing which areas are likely to give students the most difficulty, I have tried to develop links that help these learners relate the new material to something with which they are already familiar.

■ Pathophysiology

Another effective way to keep students' interest is to help them realize they are learning worthwhile and applicable material. Because most students using this text will have health-related careers, frequent references to pathophysiology and clinical physiology demonstrate the content's relevance to their professional goals.

■ Full-color illustrations

The anatomic illustrations, schematic representations, photographs, tables, and graphs are designed to complement and reinforce the written material. Now more three-dimensional and realistic, most of the cellular art and much of the anatomic art are new to this edition. Also, new, numerous process-oriented figures incorporating step-by-step descriptions allow visually oriented students to review processes through figures. Furthermore, flow diagrams are used extensively to help students integrate the written information presented. In flow diagrams, lighter and darker shades of the same color are used to denote a decrease or an increase in a controlled variable, such as blood pressure or the concentration of blood glucose. Also in the flow diagrams, the corners of all physical entities, such as body structures or chemicals, have been rounded to distinguish them from the square corners of all actions. Thorough figure captions are provided to improve understanding of the figures.

Integrated color-coded figure/table combinations enable the students to better visualize what part of the body is responsible for what activities. For example, an anatomic depiction of the brain is integrated with a table of the functions of the major brain components, with each component shown in the same color in the figure and the table.

■ Diversity of human models

A unique feature of this book is that people depicted in the various illustrations are realistic representatives of a cross section of humanity (they were drawn from photographs of real people). Sensitivity to various races, sexes, and ages should enable all students to identify with the material being presented.

■ Feedforward statements as subsection titles

Instead of traditional short, topic titles for each major subsection (for example, "Heart valves"), feedforward statements alert the student to the main point of the subsection to come (for example, "Heart valves ensure that the blood flows in the proper direction through the heart"). New to this edition, more headings have been added to break up the material into more manageable pieces for the student. Large concepts are now presented in smaller chunks.

▌ Cross-references

Cross-references to related material in other chapters enable students to quickly refresh their memories of material already learned in earlier chapters or to proceed if desired to a more in-depth coverage of a particular topic in a later chapter.

▌ Key terms and word derivations

Key terms are defined as they appear in the text. Because physiology is laden with a myriad of new vocabulary words, many of which are rather intimidating at first glance, word derivations are provided as necessary to enhance understanding of new words.

▌ Glossary with phonetic pronunciations

The glossary, which enables students to quickly review key terms when they occur later in the book, includes phonetic pronunciations of the entries.

▌ Appendixes

The appendixes are designed for the most part to help students who need to brush up on some foundation materials that they are assumed to already have had in prerequisite courses.

• *Appendix A*, **The Metric System,** is a conversion table between metric measures and their English equivalents.

• Most undergraduate physiology texts have a chapter on chemistry, yet physiology instructors rarely teach basic chemistry concepts. The decision was made, therefore, to reserve valuable text space for physiological concepts and to provide instead *Appendix B*, **A Review of Chemical Principles,** as a handy reference for students who need a review of basic chemistry concepts that are essential to understanding physiology.

• Likewise, *Appendix C*, entitled **Storage, Replication, and Expression of Genetic Information,** serves as a reference for students or as assigned material if the instructor deems appropriate. It includes a discussion of DNA and chromosomes, protein synthesis, cell division, and mutations.

• *Appendix D*, **Principles of Quantitative Reasoning,** is designed to help students become more comfortable working with equations and translating back and forth between words, concepts, and equations. This appendix supports the Quantitative Exercises at each chapter's end.

• *Appendix E*, **Answers to End-of-Chapter Objective Questions, Quantitative Exercises, and Points to Ponder,** provides answers to all objective learning activities, solutions

to the Quantitative Exercises, and explanations for the Points to Ponder and Clinical Consideration.

• *Appendix F*, **Text References to Exercise Physiology,** pulls together the page numbers for all exercise-related content in the text. Specifically, it includes a listing of "A Closer Look at Exercise Physiology" boxed features by chapter and a listing of exercise references in the text by topic.

ORGANIZATION

There is no ideal organization of physiologic processes into a logical sequence. With the sequence chosen in this book, most chapters build on material presented in immediately preceding chapters, yet each chapter is designed to stand on its own, allowing the instructor flexibility in curriculum design. The general flow is from introductory background information to cells to excitable tissue to organ systems. Every attempt has been made to provide logical transitions from one chapter to the next. For example, Chapter 8, "Muscle Physiology," ends with a discussion of cardiac muscle, which is carried forward into Chapter 9, "Cardiac Physiology." Even topics that seem unrelated in sequence, such as Chapter 12, "Defense Mechanisms of the Body," and Chapter 13, "The Respiratory System," are linked together, in this case, by ending Chapter 12 with a discussion of respiratory defense mechanisms.

Several organizational features warrant specific mention. The most difficult decision in organizing this text was placement of the chapters on the endocrine system. Intermediary metabolism of absorbed nutrient molecules is largely under endocrine control, providing a link from digestion (Chapter 16) and energy balance (Chapter 17) to the endocrine chapters (Chapters 18 and 19). There is merit in placing the chapters on the nervous and endocrine systems in close proximity because of these systems' roles as the body's major regulatory systems. Placing the endocrine system chapters earlier, immediately after the discussion of the nervous system (Chapters 4 through 7), however, would have created two problems. First, it would have disrupted the logical flow of material related to excitable tissue. Second, the endocrine system could not have been covered at the level of depth its importance warrants if it had been discussed before the students were provided the background essential to understanding this system's roles in maintaining homeostasis. Placing the endocrine system chapters late in the book does not mean, however, that students are not exposed to endocrine functions or hormones until near the book's completion. Endocrine control and hormones are defined in Chapter 1, are revisited again in Chapter 3 in the discussion of intercellular communication, and are compared with nervous control in Chapter 5. Specific hormones are introduced in appropriate chapters, such as vasopressin and aldosterone in the chapters on kidney and fluid balance. Chapters 18 and 19 explore the basic characteristics of endocrine glands and hormones as well as the control and functions of specific endocrine secretions.

Uniquely, in this book the skin is covered in the chapter on defense mechanisms of the body, in consideration of the skin's newly recognized immune functions. Bone is also covered more extensively, in the endocrine chapters, than in most undergraduate physiology texts, especially with regard to hormonal control of bone growth and bone's dynamic role in calcium metabolism.

Departure from traditional groupings of material in several important instances has permitted more independent and more extensive coverage of topics that are frequently omitted or buried within chapters concerned with other subject matter. For example, a separate chapter is devoted to fluid balance and acid–base regulation, topics often tucked within the kidney chapter. Another example is the grouping of the autonomic nervous system, motor neurons, and the neuromuscular junction in an independent chapter on the efferent division of the peripheral nervous system, which serves as a link between the nervous system chapters and the muscle chapters.

Although there is a rationale for covering the various aspects of physiology in the order given here, it is by no means the only logical way of presenting the topics. Each chapter is able to stand on its own, especially with the cross-references provided, so that the sequence of presentation can be varied at the instructor's discretion. Some chapters may even be omitted, depending on the students' needs and interests and the time constraints of the course. For example, a cursory explanation of the defense role of the leukocytes is covered in the chapter on blood, so an instructor could choose to omit the more detailed explanations of immune defense in Chapter 12. Similarly, the in-depth coverage of topics in Chapters 2, 6, 15, 17, and 19 could selectively be omitted without sacrificing a student's general appreciation of systems-approach physiology.

ANCILLARIES FOR STUDENTS

▌ Infotrac College Edition®

Available exclusively from Brooks/Cole, this online library offers students unlimited access, at any time of the day, to full-length research articles—updated daily and spanning four years. With InfoTrac, students can search for complete articles from thousands of scholarly and popular periodicals, such as *Physiological Reviews*, *Science*, *American Journal of Sports Medicine*, *Science News*, and *Discover*. A four-month subscription to this password-protected site is offered free to students with each new text purchase. An online student guide correlates with each chapter in this text to InfoTrac articles. This student guide can be accessed free at the following Web site:

http://infotrac.thomsonlearning.com/

▌ Human Physiology Book Companion Web Site

Through this content-rich Web site, students have access to helpful study aids such as Flash Cards, a Glossary, Case Studies, Hypercontents, Quizzes, Internet Activities, InfoTrac College Edition Exercises, Chapter Objectives, Chapter Summaries, and Chapter Lectures. This Web site also features a Visual Learning Resource, where students can find helpful enlargements and informative animations of selected figures from the text. These animations help bring to life some of the physio-

logic processes most difficult to visualize, enhancing student understanding of complex sequences of events. Moreover, with this edition students can use the Web site to review valuable Case Histories that introduce the clinical aspects facilitating the instruction and learning of human physiology.

▌ PhysioEdge CD-ROM

An interactive CD-ROM is packaged with every text. An ideal learning companion, this tool focuses on a limited number of concepts—those that are often the most difficult for students to learn. This testing and diagnostic tool illustrates concepts that are harder to conceptually understand from the text alone. The ability of PhysioEdge to dynamically represent these concepts through animations and robust graphic depictions facilitates learning of the concepts. PhysioEdge is completely integrated with the text. Icons in the text highlight figures associated with activities on the CD-ROM. Students can then explore concepts in greater detail. Quiz questions on the CD-ROM allow students to gauge their understanding, and the program's strong diagnostic component gives immediate feedback on their answers. When a student does well on the quizzes, the PhysioEdge quizzing program is prompted to introduce the student to the next level of difficulty. It's like having a personal tutor available on demand.

▌ WebTutor on WebCT and Blackboard

This online tool completes the outstanding technology package accompanying the text. It offers students additional learning aids to reinforce and clarify complex concepts. Some features include Internet activities, quiz questions with feedback, flashcards, links, and much more. New to WebTutor is access to NewsEdge, an online news service that posts the latest news on the WebTutor site daily. Instructors may request that this option be included in the textbook package at an additional cost. In this case, information regarding access to this Web site is bundled with the text. If an instructor does not request this option, a student may purchase this tool independently by ordering via the following toll-free number or Web site:

Phone: (800) 964-5815 (in the United States) or (813) 282-8807 (outside the United States)

Web site: http://www.thomsonlearning.com

ISBN: 0-534-39506-6
Blackboard ISBN: 0-534-39505-8

▌ Study Guide

Each chapter of this student-oriented supplementary manual, which is correlated with the corresponding chapter in *Human Physiology: From Cells to Systems*, Fifth Edition, contains a chapter overview, a detailed chapter outline, a list of Key Terms, and Review Exercises (multiple choice, true/false, fill-in-the-blank, and matching). This learning resource also offers Points to Ponder, questions that stimulate use of material in the chapter as a starting point to critical thinking and further learning.

Clinical Perspectives, common applications of the physiology under consideration; Experiments of the Day; and simple hands-on activities enhance the learning process. Answers to the Review Questions are provided at the back of the Study Guide.

ISBN: 0-534-39504-X

■ Case Histories

This booklet presents a variety of case histories relevant to human physiology. Questions for students to answer are included after each case history, with the answers being provided at the back of the booklet. These case histories and their related questions may also be found on the companion Web site for students with an option to submit answers electronically.

ISBN:0-534-38110-3

■ Photo Atlas for Anatomy and Physiology

This full-color atlas (with more than 600 photographs) depicts structures in the same colors as they would appear in real life or in a slide. Labels as well as color differentiations within each structure are employed to facilitate identification of the structure's various components. The atlas includes photographs of tissue and organ slides, the human skeleton, commonly used models, cat dissections, cadavers, some fetal pig dissections, and some physiology materials.

ISBN: 0-534-51716-1

■ Art Note Book

This resource includes unlabeled art matching the transparency set for student labeling and notetaking.

ISBN:0-534-39517-1

■ Human Physiology Lab Manual

This manual, which may be required by the instructor in courses that have a laboratory component, contains a variety of exercises that reinforce concepts covered in *Human Physiology: From Cells to Systems*, fifth edition. These laboratory experiences increase student understanding of the subject matter in a straightforward manner, with thorough directions to guide students through the process and relevant questions for students to review, explain, and apply their results.

ISBN:0-534-39507-4

ACKNOWLEDGMENTS

I gratefully acknowledge the many people who helped with the first four editions or this edition of the textbook. A special thank-you goes to four individuals who contributed substantially to the content of the book: Rachel Yeater (Chairwoman, Sports Exercise Program, and Director, Human Performance Laboratory, School of Medicine, West Virginia University), who contributed the material for the boxed features titled "A Closer Look at Exercise Physiology"; Spencer Seager (Chairman, Chemistry Department, Weber State College) who prepared Appendix B, "A Review of Chemical Principles"; and Kim Cooper, Associate Professor, Midwestern University and John Nagy, Professor, Scottsdale Community College, who provided the Quantitative Exercises for the end-of-chapter pedagogy and prepared Appendix D, "Principles of Quantitative Reasoning." One person was especially helpful in this edition's revised art program: anatomist Mark Nielsen, Professor, Department of Biology, University of Utah, who helped oversee the development of the newly rerendered cellular and anatomic art program.

During the book's creation and revision, many colleagues provided assistance. George Hedge, retired Vice Provost for Research, Washington State University, and Robert Goodman, Chairman, Department of Physiology and Pharmacology, West Virginia University, deserve a special note of gratitude for their willingness to share materials used from their publication, *Clinical Endocrine Physiology* (Saunders, 1987), and for their thoughtful reviews of this book's chapters on endocrinology and reproduction. Appreciation is also extended to Elizabeth Walker and Dennis Overman (in memoriam), Department of Anatomy, West Virginia University, and Mark Nielsen, Department of Biology, University of Utah, who provided many custom-made light and electron micrographs for the book.

Others at West Virginia University deserving of recognition for providing resource materials or countless clarifications include James Culberson, William Beresford, Rumy Hilloowala, and Adrienne Salm, Department of Anatomy; Marta Henderson, Division of Medical Technology; Ronald Gaskins, Department of Medicine; Sidney Schochet, Jr., Department of Pathology; Val Vallyathan, Department of Pathology and National Institute of Occupational Safety and Health; Robert Stitzel, Department of Biochemistry and Molecular Pharmacology; Stephen Alway, Department of Human Performance and Applied Exercise Science; and Christine Baylis, Paul Brown, John Connors, Gunter Franz, Wilbert Gladfelter, Linda Huffman, Michael Johnson, Ping Lee, Phillip Miles, Ronald Millechia, William Stauber, and Stanley Yokota, Department of Physiology and Pharmacology.

In addition to the 77 reviewers who carefully evaluated the preceding four editions for accuracy, clarity, and relevance, I express sincere appreciation to the following individuals who served as reviewers for this book. Their comments and suggestions were very helpful while I was considering ways to improve this edition. Delon W. Barfuss, Georgia State University; Parveen Bawa, Simon Fraser University; Corey L. Cleland, James Madison University; Michael B. Ferrari, University of Arkansas; James D. Herman, Texas A&M University; David A. Hood, York University; William F. Jackson, Western Michigan University; Kelly Johnson, University of Kansas; Lisa Parks, North Carolina State University; Karen Swearingen, Mills College; Jeffery W. Walker, University of Wisconsin; Cheryl Watson, Central Connecticut State University; and Edward J. Zambraski, Rutgers University.

I have been very fortunate to work with a highly competent, dedicated team from Brooks/Cole, along with other very capable external suppliers selected by this publishing company. I would like to acknowledge all of their contributions, which collectively made this book possible. It has been a source of comfort and inspiration to know that so many people have been working so diligently in so many ways to make this book become a reality.

From Brooks/Cole, Nedah Rose, Editor, launched this edition and helped shape the new directions it would take. She was especially instrumental in the decision to extensively (and expensively) improve the art program, making it more dynamic and anatomically realistic. Furthermore, Nedah deserves a warm thanks for her willingness to listen and come up with satisfactory solutions whenever a problem arose.

Christopher Delgado, Assistant Editor, oversaw the development of multiple components of the ancillary package that accompanies *Human Physiology: From Cells to Systems*, Fifth Edition, making sure the package was a cohesive whole. I appreciate the efforts of Mary Arbogast, Senior Developmental Editor, for trafficking and evaluating all the paperwork during the development process. Mary offered valuable suggestions and insights that helped the project proceed successfully. I can always count on Mary to lead me in the right direction when I don't know whom to contact about a given issue. And thanks also to Editorial Assistant Jennifer Keever, who coordinated many tasks.

On the production side, Teri Hyde, Production Editor, is a highly capable, responsive individual who takes time to listen and helps smooth over the inevitable rough spots. Despite the fact that Teri simultaneously oversees the complex production process of multiple books, she managed to closely monitor this book's progress and helped make it the best it could possibly be. I felt confident knowing she was in the background, making sure that everything was going according to plan. Teri always responded promptly when I had any questions and was even willing to take a step backward and redo part of the project to "make things right" when something turned out wrong. I am grateful for the creative insight that Rob Hugel, Executive Art Director, provided for revising the chapter openers and for overseeing the development of an aesthetically appealing, meaningful cover. He came up with wonderful images from which we could select to convey the concept of a dynamic, functioning human body. I also thank Permissions Editor, Joohee Lee for tracking down permissions for a myriad of art and other copyrighted materials incorporated in the text—a tedious but absolutely essential task.

With everything finally coming together, Jessica Reed, Print Buyer, oversaw the manufacturing process, coordinating the actual printing of the book. No matter how well a book is conceived, produced, and printed, it would not reach its full potential as an educational tool without being efficiently and effectively marketed. Marketing Manager, Ann Caven, Marketing Communications Marketing Assistant, Sandra Perin, and Linda Yip, Advertising Project Manager, played the lead roles in marketing this text, for which I am most appreciative.

Rounding out the Brooks/Cole team, a hearty note of gratitude is extended to Travis Metz, Technology Project Manager, for his efforts in providing technologically enhanced learning solutions. The new interactive CD-ROM and expanded Web-based features are welcome additions to the multimedia package that accompanies this edition of the text.

I am grateful to Laura Malloy, Hartwick College, Kelly Johnson, University of Kansas, and Ronald Markle, Northern Arizona University, for authoring the highly useful interactive CD-ROM that accompanies the text. Thanks should also be given to Bill Vining, Bill Rohan, and CowTown Productions for their technical development of this most effective learning tool.

Brooks/Cole also did an outstanding job in selecting highly skilled vendors to carry out particular production tasks. First and foremost, it has been my personal and professional pleasure to work with a very capable Production Manager, Thomas E. Dorsaneo, who coordinated the day-to-day management of production. In his competent hands laid the responsibility of seeing that all art, typesetting, page layout, and other related details got done right and in a timely fashion. Tom is a great facilitator. He quickly pinpointed potential problems and figured out ways to solve them. Furthermore, being people-oriented, Tom was always realistic and took into consideration what else was happening in a person's life when developing timelines for completion of particular assignments. Furthermore, Tom offered regular doses of understanding and encouragement. I always felt better after talking with him.

In addition to the role he played directly, Tom also brought on board other highly competent providers who helped bring this text to fruition. Specifically, my appreciation is extended to Myrna Engler, Photo Researcher, who kept searching until we found photos that were "just right." Jeanne Calabrese deserves thanks for the fresh and attractive, yet space-conscious appearance of the text's interior, and for envisioning the book's visually appealing exterior. I further wish to thank Thompson Type for their accurate typesetting, attractive, logical layout design, and execution of the nonanatomic art program.

I also owe a debt of gratitude to the talented artists who rerendered the chapter opener art and many other cellular and anatomic figures, making them more realistic, detailed, and three-dimensional, within especially tight time constraints. They are Richard Coombs, Cyndie C. H. Wooley, and Kelly Collins.

Finally, my love and gratitude go to my family for another year and a half of sacrifice in my family life as this fifth edition was being developed and produced. I want to thank them for their patience and understanding during the times I was working on the book instead of being there with them or for them. My husband, Peter Marshall, deserves special appreciation and recognition for taking over my share of household responsibilities, freeing up time for me to work on this text. I could not have done this, or any of the preceding books, without his help, support, and encouragement.

Thanks to all!

Human Physiology

From Cells to Systems

Fifth Edition

During the minute that it will take you to read this page:

Your eyes will convert the image from this page into electrical signals (nerve impulses) that will transmit the information to your brain for processing.

Besides receiving and processing information such as visual input, your brain will also provide output to your muscles to help maintain your posture, move your eyes across the page as you read, and turn the page as needed. Chemical messengers will carry signals between your nerves and muscles to trigger appropriate muscle contraction.

Your heart will beat 70 times, pumping 5 liters (about 5 quarts) of blood to your lungs and another 5 liters to the rest of your body.

You will breathe in and out about 12 times, exchanging 6 liters of air between the atmosphere and your lungs.

More than 1 liter of blood will flow through your kidneys, which will act on the blood to conserve the "wanted" materials and eliminate the "unwanted" materials in the urine. Your kidneys will produce 1 ml (about a thimbleful) of urine.

Your cells will consume 250 ml (about a cup) of oxygen and produce 200 ml of carbon dioxide.

Your digestive system will be processing your last meal for transfer into your bloodstream for delivery to your cells.

You will use about 2 calories of energy derived from food to support your body's "cost of living," and your contracting muscles will burn additional calories.

Chapter 1

Homeostasis:
The Foundation of Physiology

CONTENTS AT A GLANCE

INTRODUCTION TO PHYSIOLOGY

▋ Definition of physiology

▋ Relationship between structure and function

LEVELS OF ORGANIZATION IN THE BODY

▋ Cells as the basic units of life

▋ Organizational levels of tissues, organs, systems, and organism

CONCEPT OF HOMEOSTASIS

▋ Significance of the internal environment

▋ Necessity of homeostasis

▋ Factors that are homeostatically maintained

▋ Contributions of various systems to homeostasis

HOMEOSTATIC CONTROL SYSTEMS

▋ Components of a homeostatic control system

▋ Negative and positive feedback; feedforward mechanisms

INTRODUCTION TO PHYSIOLOGY

The activities described on the preceding page are a sampling of the processes that occur in our bodies all the time just to keep us alive. We usually take these life-sustaining activities for granted and don't really think about "what makes us tick," but that's what physiology is about. **Physiology** is the study of the functions of the body, or how the body works.

▋ Physiology focuses on mechanisms of action.

There are two approaches to explaining the events that occur in the body, one emphasizing the purpose of a body process and the other the underlying mechanism by which this process occurs. In response to the question "Why do I shiver when I am cold?" one answer would be "To help warm up, because shivering generates heat." This approach, known as a **teleological approach,** explains body functions in terms of meeting a bodily *need,* without considering how this outcome is accomplished. That is, a teleological approach emphasizes the "why" or purpose of body processes. Physiologists focus primarily on a **mechanistic approach** to body function. They view the body as a machine whose *mechanisms* of action can be explained in terms of cause-and-effect sequences of physical and chemical processes—the same types of processes that occur in other components of the universe. That is, physiologists explain the "how" of events that occur in the body. A physiologist's mechanistic explanation of shivering is that when temperature-sensitive nerve cells detect a fall in body temperature, they signal the area in the brain that is responsible for temperature regulation. In response, this brain area activates nerve pathways that ultimately bring about involuntary, oscillating muscle contractions (that is, shivering).

Because those mechanisms of action that are most beneficial to survival have prevailed throughout evolutionary history, it is helpful when studying physiology to predict what mechanistic process would be useful to the body under a particular circumstance. That most bodily mechanisms do serve a useful purpose (having been naturally selected throughout evolutionary time) lets you apply a certain amount of logical reasoning to each new situation you encounter in your study of physiology. If you always try to find the thread of logic in what you are studying, you can avoid a good deal of pure memorization, and, more importantly, you will better understand and assimilate the concepts being presented.

Structure and function are inseparable.

Physiology is closely interrelated with **anatomy,** the study of the structure of the body. Physiological mechanisms are made possible by the structural design and relationships of the various body parts that carry out each of these functions. Just as the functioning of an automobile depends on the shapes, organization, and interactions of its various parts, the structure and function of the human body are inseparable. Therefore, as we tell the story of how the body works we will provide sufficient anatomic background for you to understand the function of the body part being discussed.

Some of the structure–function relationships are obvious. For example, the heart is well designed to receive and pump blood, the teeth to tear and grind food, and the hingelike elbow joint to permit bending of the elbow. Other situations in which form and function are interdependent are more subtle but equally important. Consider the interface between the air and the blood in the lungs as an example: the respiratory airways that carry air from the outside into the lungs branch extensively when they reach the lungs. Tiny air sacs cluster at the ends of the huge number of airway branches. The branching is so extensive that the lungs contain about 300 million air sacs. Similarly, the blood vessels carrying blood into the lungs branch extensively and form dense networks of small vessels that encircle each of the air sacs (see Figure 13-2, p. 461). Because of this structural relationship, the total surface area exposed between the air in the air sacs and the blood in the small vessels is about the size of a tennis court. This tremendous interface is crucial for the lungs' ability to efficiently carry out their function: the transfer of needed oxygen from the air into the blood and the unloading of the waste product carbon dioxide from the blood into the air. The greater the surface area available for these exchanges, the faster the rate of movement of oxygen and carbon dioxide between the air and the blood. This large functional interface packaged within the confines of your lungs is possible only because both the air-containing and blood-containing components of the lungs branch extensively.

LEVELS OF ORGANIZATION IN THE BODY

We now turn our attention to how the body is structurally organized into a whole functional unit, from the chemical level to the total body (● Figure 1-1). These levels of organization make possible life as we know it.

The chemical level: Various atoms and molecules make up the body.

Like all matter on this planet, the human body is a combination of various chemicals. *Atoms* are the smallest building blocks of all nonliving and living matter. The most common atoms in the body—oxygen, carbon, hydrogen, and nitrogen—make up approximately 96% of the total body chemistry. These common atoms and a few others combine to form the *molecules* of life, such as proteins, carbohydrates, fats, and nucleic acids (genetic material, such as deoxyribonucleic acid, or DNA). These important atoms and molecules are the inanimate raw ingredients from which all living things arise. (See Appendix B for a review of this chemical level.)

The cellular level: Cells are the basic units of life.

The mere presence of a particular collection of atoms and molecules does not confer the unique characteristics of life. Instead, these nonliving chemical components must be arranged and packaged in very precise ways to form a living entity. The **cell,** the basic unit of both structure and function in a living being, is the smallest unit capable of carrying out the processes associated with life. Cellular physiology is the focus of Chapter 2.

An extremely thin oily barrier, the *plasma membrane,* encloses the contents of each cell, separating these chemicals from those outside the cell. Because the plasma membrane can control movement of materials into and out of the cell, the cell's interior contains a different combination of atoms and molecules from the mixture of chemicals in the environment that surrounds the cell. Because the plasma membrane and its associated functions are so crucial for carrying out life processes, Chapter 3 is devoted entirely to this structure.

Organisms are independent living entities. The simplest forms of independent life are single-celled organisms such as bacteria and amoebae. Complex multicellular organisms, such as trees and humans, are structural and functional aggregates of trillions of cells (*multi* means "many"). In multicellular organisms, cells are the living building blocks. In the simpler multicellular forms of life—for example, a hydra or a sponge—the cells of the organism may all be similar. However, more complex organisms, such as humans, have many different kinds of cells, such as muscle cells, nerve cells, and gland cells.

Each human organism begins when an egg and sperm unite to form a single cell that starts to multiply and form a growing mass through myriad cell divisions. If cell multiplication were the only process involved in development, all the body cells would be essentially identical, as is the case of the simplest multicellular life forms. During development of complex multicellular organisms such as humans, however, each cell *differentiates*, or becomes specialized to carry out a particular function. As a result of **cell differentiation**, your body is made up of many different specialized types of cells.

Basic cell functions

All cells, whether they exist as solitary cells or as part of a multicellular organism, perform certain basic functions essential for survival of the cell. These basic cell functions include the following:

1. Obtaining food (nutrients) and oxygen (O_2) from the environment surrounding the cell.

2. Performing various chemical reactions that use nutrients and O_2 to provide energy for the cells, as follows:

$$\text{Food} + O_2 \rightarrow CO_2 + H_2O + \text{energy}$$

3. Eliminating to the cell's surrounding environment carbon dioxide (CO_2) and other by-products, or wastes, produced during these chemical reactions.

4. Synthesizing proteins and other components needed for cellular structure, for growth, and for carrying out particular cell functions.

5. Controlling to a large extent the exchange of materials between the cell and its surrounding environment.

6. Moving materials from one part of the cell to another in carrying out cellular activities, with some cells even being able to move in entirety through their surrounding environment.

7. Being sensitive and responsive to changes in the surrounding environment.

8. In the case of most cells, reproducing. Some body cells, most notably nerve cells and muscle cells, lose the ability to reproduce after they are formed during early development. This is why strokes, which result in lost nerve cells in the brain, and heart attacks, which bring about death of heart muscle cells, can be so devastating.

Cells are remarkable in the similarity with which they carry out these basic functions. Thus all cells share many common characteristics.

● **FIGURE 1-1**
Levels of organization in the body

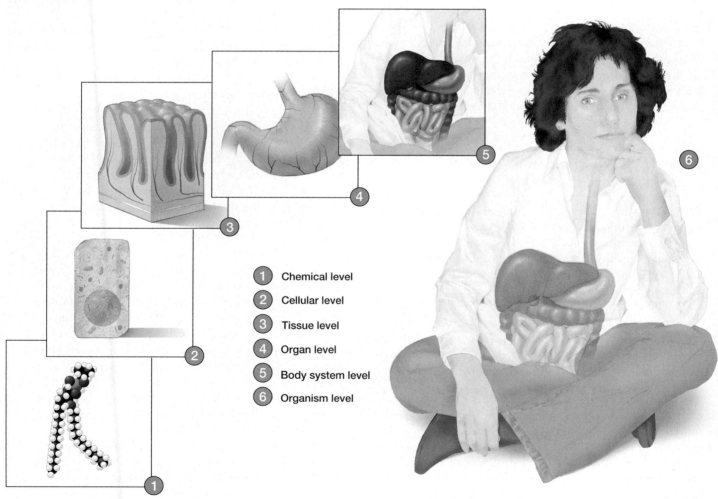

1 Chemical level
2 Cellular level
3 Tissue level
4 Organ level
5 Body system level
6 Organism level

Specialized cell functions

In multicellular organisms, each cell also performs a specialized function, which is usually a modification or elaboration of a basic cell function. Here are a few examples:

• By taking special advantage of their protein-synthesizing ability, the gland cells of the digestive system secrete digestive enzymes that break down ingested food; these enzymes are all proteins.

• The kidney cells are able to selectively retain the substances needed by the body while eliminating unwanted substances in the urine, because of their highly specialized ability to control exchange of materials between the cell and its environment.

• Muscle contraction, which involves selective movement of internal structures to bring about shortening of muscle cells, is an elaboration of the inherent capability of these cells to produce intracellular ("within the cell") movement.

• Capitalizing on the basic ability of cells to respond to changes in their surrounding environment, nerve cells generate and transmit to other regions of the body electrical impulses that relay information about changes to which the nerve cells are responsive. For example, nerve cells in the ear can relay information to the brain about sound in the external environment.

Each cell performs these specialized activities in addition to carrying on the unceasing, fundamental activities required of all cells. The basic cell functions are essential for the survival of each individual cell, whereas the specialized contributions and interactions among the cells of a multicellular organism are essential for the survival of the whole body.

▍The tissue level: Tissues are groups of cells of similar specialization.

Just as a machine does not function unless all its parts are properly assembled, the cells of the body must be specifically organized to carry out the life-sustaining processes of the body as a whole, such as digestion, respiration, and circulation. Cells are progressively organized into tissues, organs, body systems, and finally the whole body.

Cells of similar structure and specialized function combine to form **tissues**, of which there are four *primary types*: muscle, nervous, epithelial, and connective tissue. Each tissue consists of cells of a single specialized type, along with varying amounts of extracellular ("outside the cell") material.

• **Muscle tissue** is composed of cells specialized for contraction and force generation. There are three types of muscle tissue: *skeletal muscle*, which accomplishes movement of the skeleton; *cardiac muscle*, which is responsible for pumping blood out of the heart; and *smooth muscle*, which encloses and controls movement of contents through hollow tubes and organs, such as movement of food through the digestive tract.

• **Nervous tissue** consists of cells specialized for initiating and transmitting electrical impulses, sometimes over long distances. These electrical impulses act as signals that relay information from one part of the body to another. Nervous tissue is found in (1) the brain; (2) the spinal cord; (3) the nerves that signal information about the external environment and about the status of various internal factors in the body that are subject to regulation, such as blood pressure; and (4) the nerves that influence muscle contraction or gland secretion.

• **Epithelial tissue** is made up of cells specialized for the exchange of materials between the cell and its environment. Any substance that enters or leaves the body proper must cross an epithelial barrier. Epithelial tissue is organized into two general types of structures: epithelial sheets and secretory glands. Epithelial cells are joined together very tightly to form sheets of tissue that cover and line various parts of the body. For example, the outer layer of the skin is epithelial tissue, as is the lining of the digestive tract. In general, these epithelial sheets serve as boundaries that separate the body from the external environment and from the contents of cavities that communicate with the external environment, such as the digestive tract lumen. (A **lumen** is the cavity within a hollow organ or tube.) Only selected transfer of materials is permitted between the regions separated by an epithelial barrier. The type and extent of controlled exchange vary, depending on the location and function of the epithelial tissue. For example, very little can be exchanged between the body and external environment across the skin, whereas the epithelial cells lining the digestive tract are specialized for absorption of nutrients.

• **Glands** are epithelial tissue derivatives that are specialized for secretion. **Secretion** is the release from a cell, in response to appropriate stimulation, of specific products that have in large part been synthesized by the cell. Glands are formed during embryonic development by pockets of epithelial tissue that dip inward from the surface and develop secretory capabilities. There are two categories of glands: *exocrine* and *endocrine* (● Figure 1-2). If, during development, the connecting cells between the epithelial surface cells and the secretory gland cells within the depths of the invagination remain intact as a duct between the gland and the surface, an exocrine gland is formed. **Exocrine glands** secrete through ducts to the outside of the body (or into a cavity that communicates with the outside) (*exo* means "external"; crine means "secretion"). Examples are sweat glands and glands that secrete digestive juices. If, in contrast, the connecting cells disappear during development and the secretory gland cells are isolated from the surface, an endocrine gland is formed. **Endocrine glands** lack ducts and release their secretory products, known as *hormones*, internally into the blood within the body (*endo* means "internal"). For example, the parathyroid gland secretes parathyroid hormone into the blood, which transports this hormone to its sites of action at the bones and kidneys.

• **Connective tissue** is distinguished by having relatively few cells dispersed within an abundance of extracellular material. As its name implies, connective tissue connects, supports, and anchors various body parts. It includes such diverse structures as the loose connective tissue that attaches epithelial tis-

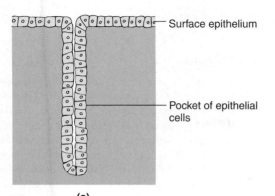

— Surface epithelium

— Pocket of epithelial cells

(a)

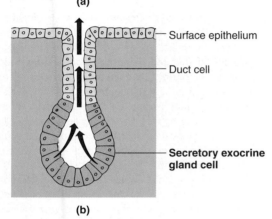

— Surface epithelium

— Duct cell

Secretory exocrine gland cell

(b)

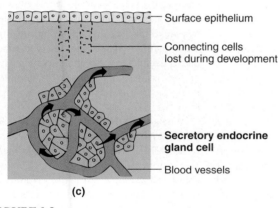

— Surface epithelium

— Connecting cells lost during development

Secretory endocrine gland cell

— Blood vessels

(c)

● **FIGURE 1-2**

Exocrine and endocrine gland formation
(a) Glands arise during development from the formation of pocket-like invaginations of surface epithelial cells. (b) If the cells at the deepest part of the invagination become secretory and release their product through the connecting duct to the surface, an exocrine gland is formed. (c) If the connecting cells are lost and the deepest secretory cells release their product into the blood, an endocrine gland is formed.

sue to underlying structures; tendons, which attach skeletal muscles to bones; bone, which gives the body shape, support, and protection; and blood, which transports materials from one part of the body to another. Except for blood, the cells within connective tissue produce specific molecules that they release into the extracellular spaces between the cells. One such molecule is the rubberband–like protein fiber *elastin*, whose presence facilitates the stretching and recoiling of struc-

tures such as the lungs, which alternately inflate and deflate during breathing.

Muscle, nervous, epithelial, and connective tissue are the primary tissues in a classical sense; that is, each is an integrated collection of cells of the same specialized structure and function. The term *tissue* is also frequently used to refer to the aggregate of various cellular and extracellular components that make up a particular organ (for example, lung tissue or liver tissue).

▌The organ level: An organ is a unit made up of several tissue types.

Organs are composed of two or more types of primary tissue organized together to perform a particular function or functions. The stomach is an example of an organ made up of all four primary tissue types. The tissues that make up the stomach function collectively to store ingested food, move it forward into the rest of the digestive tract, and begin the digestion of protein. The stomach is lined with epithelial tissue that restricts the transfer of harsh digestive chemicals and undigested food from the stomach lumen into the blood. Epithelially derived gland cells in the stomach include exocrine cells, which secrete protein-digesting juices into the lumen, and endocrine cells, which secrete a hormone that helps regulate the stomach's exocrine secretion and muscle contraction. The wall of the stomach contains smooth muscle tissue whose contractions mix ingested food with the digestive juices and push the mixture out of the stomach and into the intestine. The stomach wall also contains nervous tissue, which, along with hormones, controls muscle contraction and gland secretion. These various tissues are all bound together by connective tissue.

Currently several lines of investigation are being hotly pursued in a quest to repair or replace tissues or organs that are no longer able to sufficiently perform their vital functions because of disease, trauma, or age-related changes. (See the accompanying boxed feature ▶ Concepts, Challenges, and Controversies. Each chapter has one or more similar boxed features that explores in greater depth high-interest, tangentially relevant information on such diverse topics as environmental impact on the body, aging, ethical issues, new discoveries regarding common diseases, historical perspectives, and so on.)

▌The body system level: A body system is a collection of related organs.

Groups of organs are further organized into **body systems.** Each system is a collection of organs that perform related functions and interact to accomplish a common activity that is essential for survival of the whole body. For example, the digestive system consists of the mouth, salivary glands, pharynx (throat), esophagus, stomach, pancreas, liver, gallbladder, small intestine, and large intestine. These digestive organs cooperate to ac-

Stem Cell Science and Tissue Engineering: The Quest to Make Defective Body Parts Like New Again

Liver failure, stroke, paralyzing spinal cord injury, diabetes mellitus, damaged heart muscle, arthritis, extensive burns, surgical removal of a cancerous breast, an arm mangled in an accident. Although our bodies are remarkable and normally serve us well, sometimes a body part is defective, injured beyond repair, or lost in situations such as these. In addition to the loss of quality of life for the affected individuals, billions of dollars are spent to treat patients with lost, permanently damaged, or failing organs, accounting for about half of the total health-care costs in the United States. Ideally, when the body suffers an irreparable loss, new, permanent replacement parts could be substituted to restore normal function and appearance. Fortunately, this possibility is moving rapidly from the realm of science fiction to the reality of scientific progress.

The Medical Promise of Stem Cells

The recent isolation of *stem cells* offers incredibly exciting medical promise for repairing or replacing tissues or organs that are diseased, damaged, or worn out. Two categories of stem cells are being explored for their potential in alleviating many conditions associated with tissue or organ failure: embryonic stem cells and tissue-specific stem cells harvested from adults. **Embryonic stem cells** are the "mother cells" resulting from the early divisions of a fertilized egg. These stem cells are undifferentiated cells that ultimately give rise to all the mature, specialized cells of the body, while at the same time renewing themselves. Embryonic stem cells are totipotent (meaning "having total potential"), because they have the potential of generating any cell type in the body if given the appropriate cues. As the embryonic stem cells divide over the course of development, they branch off into various specialty tracks under the guidance of particular genetically controlled chemical signals.

More specifically, the undifferentiated embryonic stem cells give rise to many partially differentiated **tissue-specific stem cells,** each of which becomes committed to generating the highly differentiated, specialized cell types that comprise a particular tissue. For example, tissue-specific muscle stem cells are forced along a lifelong career path of becoming specialized muscle cells. Some tissue-specific stem cells remain in adult tissues and serve as a continual source of new specialized cells in that particular tissue or organ. For example, various partially differentiated stem cells found in the bone marrow generate the different types of blood cells, which are released into the blood as needed after being fully differentiated. Much to investigators' surprise, tissue-specific stem cells have even been found in adult brain and muscle tissue. Even though mature nerve and muscle cells cannot reproduce themselves, researchers recently discovered that, to a limited extent, adult brains and muscles can grow new cells throughout life by means of these persistent stem cells. However, this process is too slow in nerve and muscle tissue to keep pace with major losses, as in a stroke or heart attack. Some investigators are searching for drugs that might spur a person's own tissue-specific stem cells into increased action to make up for damage or loss of that tissue—a feat that is not currently feasible.

More hope is pinned on nurturing stem cells outside of the body for possible transplant into the body. In 1998, for the first time, researchers succeeded in isolating embryonic stem cells and maintaining them indefinitely in an undifferentiated state in culture. With **cell culture,** cells isolated from a living organism continue to thrive and reproduce in laboratory dishes when supplied with appropriate nutrients and supportive materials. Many scientists believe that research involving cultured embryonic stem cells will lead to groundbreaking treatments for a wide variety of disorders.

The medical promise of embryonic stem cells lies in their potential to serve as an all-purpose material that can be coaxed into whatever cell types are needed to patch up the body. Early experiments suggest that these cells have the ability to differentiate into particular cells when exposed to the appropriate chemical signals. Further stem cell research has far-reaching implications that could revolutionize the practice of medicine in the 21st century. As scientists gradually learn to prepare the right cocktail of chemical signals to direct the undifferentiated cells into the desired cell types, they will have the potential to fill in deficits in damaged or dead tissue with healthy cells. Scientists even foresee the ability to grow customized tissues and eventually whole, made-to-order replacement organs, such as livers, hearts, and kidneys.

The Medical Promise of Tissue Engineering

Tissue engineering, another exciting frontier in clinical research, is focused on growing new tissues and even whole, complex organs in the laboratory that can be implanted to serve as permanent replacements for body parts that cannot be repaired. The era of tissue engineering is being ushered in by advances in cell biology, plastic manufacturing, and computer graphics. Using computer-aided designs, very pure, dissolvable plastics are shaped into three-dimensional molds or scaffoldings that mimic the structure of a particular tissue or organ. The plastic mold is then "seeded" with the desired cell types, which are coaxed, through the application of appropriate nourishing and stimulatory chemicals, into multiplying and assembling into the desired tissue. After the biodegradable plastic scaffolding dissolves, only the newly generated tissue remains, ready to be implanted into a patient as a permanent, living replacement part.

What about the source of cells to seed the plastic mold? The immune system is programmed to attack foreign cells, such as bacterial invaders. This system also launches an attack on foreign cells transplanted into the body from another individual. Such an attack brings about rejection of transplanted organs, tissues, or cells unless the transplant recipient is treated with *immunosuppressive drugs* (drugs that suppress the immune system's attack on the transplanted material). An unfortunate side effect of immunosuppressive drugs is the reduced ability of the patient's immune system to defend against potential disease-causing bacteria and other micro-organisms, such as viruses. To prevent rejection by the immune system, without the necessity of lifelong immunosuppressive drugs,

---

complish the breakdown of food into small nutrient molecules that can be absorbed into the blood for distribution to all cells.

The human body has eleven systems: the circulatory, digestive, respiratory, urinary, skeletal, muscular, integumentary, immune, nervous, endocrine, and reproduction systems (● Figure 1-3). Chapters 4 through 20 cover the details of these systems.

▮ **The organism level: The body systems are packaged together into a functional whole body.**

The whole body of a multicellular organism—a single, independently living individual—is composed of the various body

the plastic mold used for tissue engineering could be seeded, if possible, with appropriate cells harvested from the recipient. Often, however, because of the very need for a replacement part, the patient does not have any of the appropriate cells to seed the synthetic scaffold. This is what makes the recent isolation of embryonic stem cells so exciting. Through genetic engineering, these stem cells could be converted into "universal" seed cells that would be immunologically acceptable to any recipient; that is, they could be genetically programmed to not be rejected by any body. Thus the vision of tissue engineers is to create universal replacement parts that could be put into any patient who needs them, without fear of transplant rejection or use of troublesome immunosuppressive drugs.

Following are some of the tissue engineers' early accomplishments and future predictions:

- Engineered skin patches have already been used to treat victims of severe burns.
- Laboratory-grown cartilage has already been successfully implanted in experimental animals. Examples include artificial heart valves, ears, and noses.
- Considerable progress has been made on building artificial bone.
- Tissue-engineered scaffolding to promote nerve regeneration is currently undergoing testing in animals.
- Progress has been made on growing two complicated organs, the liver and the pancreas.
- Another tissue-engineering challenge under development is the growth of completely natural replacement breasts for implantation following surgical removal of cancerous breasts.
- Engineered joints will be used as a living, more satisfactory alternative to the plastic and metal devices used as replacements today.
- Ultimately, complex body parts such as arms and hands will be produced in the laboratory for attachment as needed.

Tissue engineering thus holds the promise that damaged body parts can be replaced with the best alternative, a laboratory-grown version of "the real thing."

Ethical Concerns and Political Issues

Despite this potential, embryonic stem-cell research is fraught with ethical controversy because of the source of these cells. These embryonic stem cells were isolated from embryos from an abortion clinic and from unused embryos from an in-vitro fertility clinic. Opponents of using embryonic stem cells are morally and ethically concerned because embryos are destroyed in the process of harvesting these cells. Proponents claim that these embryos were destined to be destroyed anyway—a decision already made by the parents of the embryos—and that these stem cells have great potential for alleviating much human suffering. Thus embryonic stem-cell science has become inextricably linked with stem cell politics.

Because federal policy currently prohibits use of public funding to support research involving human embryos, the scientists who isolated these embryonic stem cells relied on private money. Public policy makers, scientists, and bioethicists are now faced with balancing a host of ethical issues against the tremendous potential clinical application of embryonic stem-cell research. Such research will proceed at a much faster pace if federal money is available to support it. In a controversial decision, President George W. Bush issued guidelines in August 2001 permitting the use of government funds to support studies using established lines of human embryonic stem cells but not research aimed at deriving new cell lines. This decision was based on the premise that investigations using already established embryonic stem-cell lines would not involve destroying more embryos for scientific purposes. Bush will continue to seek input from his newly formed President's Council on Bioethics, whose only two agenda items for the time being are embryonic stem-cell research and human cloning.

The Search for Noncontroversial Stem Cells

As a possible alternative to using the controversial totipotent embryonic stem cells, other researchers are exploring the possibility of exploiting tissue-specific stem cells harvested from various adult tissues. Until recently, most investigators believed that these adult stem cells could only give rise to the specialized cells of a particular tissue. Although these partially differentiated adult stem cells do not have the complete developmental potential of embryonic stem cells, the adult cells have been coaxed into producing a wider variety of cells than originally thought possible. To name a few examples, provided the right supportive environment, brain stem cells have given rise to blood cells, bone-marrow stem cells to liver and nerve cells, and fat-tissue stem cells to bone, cartilage, and muscle cells. Thus researchers may be able to tap into the more limited but still versatile developmental potential of specialized stem cells in the adult human body. Although embryonic stem cells hold greater potential for developing treatments for a broader range of diseases, adult stem cells are more accessible than embryonic stem cells and their use is not controversial. It might even be possible to harvest stem cells from a patient's own body and manipulate them for use in treating the individual, thus avoiding the issue of transplant rejection. For example, researchers dream of being able to take fat stem cells from a person and transform them into a needed replacement knee joint.

The scientific, political, and ethical dispute over the relative merits of embryonic versus tissue-specific stem cells may soon become a moot point if a recent remarkable discovery is confirmed by further investigation. That is, an adult stem cell has been discovered that is believed to have the same broad potential as embryonic stem cells in its ability to be transformed into every kind of cell in the body. Time will tell whether this is the ultimate stem cell that could provide a noncontroversial, endless source of cells, organs, and tissues for transplants.

Whatever the source, stem cell research promises to revolutionize medicine. According to the Center for Disease Control's National Center for Health Statistics, an estimated 3,000 Americans die every day from conditions that may in the future be treatable with stem cell derivatives.

systems structurally and functionally linked together as an entity that is separate from the external (outside the body) environment. Thus the body is made up of living cells organized into life-sustaining systems.

The different body systems do not act in isolation from each other. Many complex body processes depend on the interplay among multiple systems. For example, regulation of blood pressure depends on coordinated responses among the circulatory, urinary, nervous, and endocrine systems, as you will learn later. Even though physiologists may examine body functions at any level from cells to systems (as indicated in the title of this book), their ultimate goal is to integrate these mechanisms into the big picture of how the entire organism works as a cohesive whole.

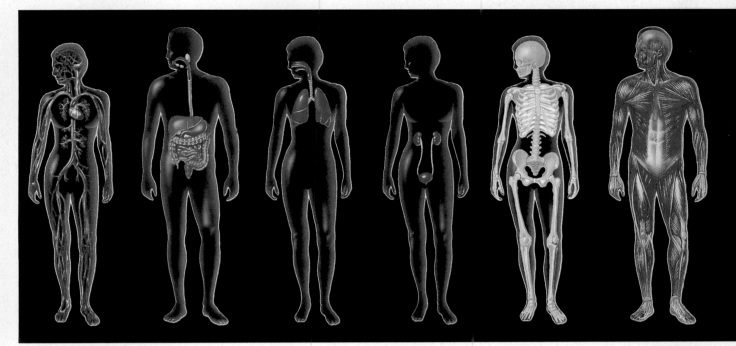

Circulatory system
heart, blood,
blood vessels

Digestive system
mouth, pharynx,
esophagus, stomach,
small intestine, large
intestine, salivary
glands, exocrine
pancreas, liver,
gallbladder

Respiratory system
Nose, pharynx, larynx,
trachea, bronchi, lungs

Urinary system
kidneys, ureters,
urinary bladder,
urethra

Skeletal system
bones, cartilage,
joints

Muscular system
skeletal muscles

● FIGURE 1-3
Components of the body systems

(SOURCE: Adapted from Cecie Starr and Ralph Taggart, *Biology: The Unity and Diversity of Life*, Eighth Edition, Fig. 33.11, p. 552–553. Copyright © 1998 Wadsworth Publishing Company.)

To this point, you have learned what physiology is, its relationship with anatomy, and the structural and functional levels of organization that exist in the human body. We will now focus on how these different levels work together to maintain life.

CONCEPT OF HOMEOSTASIS

If each cell possesses basic survival skills, why can't the body cells live without performing specialized tasks and being organized according to specialization into systems that accomplish functions essential for the whole body's survival? The cell in a multicellular organism must contribute to the survival of the organism as a whole and cannot live and function without contributions from the other body cells, because the vast majority of cells are not in direct contact with the external environment in which the organism lives. A single-celled organism such as an amoeba can directly obtain nutrients and O_2 from its immediate external surroundings and eliminate wastes back into those surroundings. A muscle cell or any other cell in a multicellular organism has the same need for life-supporting nutrient and O_2 uptake and waste elimination, yet the muscle cell cannot directly make these exchanges with the environment surrounding the body because the cell is isolated from this external environment. How is it possible for a muscle cell to make vital exchanges with the external environment with which it has no contact? The key is the presence of a watery **internal environment** with which the body cells are in direct contact and make life-sustaining exchanges. We now turn our attention to the composition, importance, and maintenance of this internal environment.

■ **Body cells are in contact with a privately maintained internal environment.**

The fluid collectively contained within all of the cells of the body is known as **intracellular fluid (ICF)** and the fluid outside the cells is referred to as **extracellular fluid (ECF)** (*intra* means "within"; *extra* means "outside of"). The extracellular fluid is the internal environment of the body. It is the fluid environment in which the cells live. Note that the internal environment is outside the cells but inside the body. By contrast, the intracellular fluid is inside the cells and the external environment is outside the body.

The extracellular fluid is made up of two components: the **plasma,** the fluid portion of the blood; and the **interstitial fluid,** which surrounds and bathes the cells (*inter* means "between"; *stitial* means "that which stands") (● Figure 1-4).

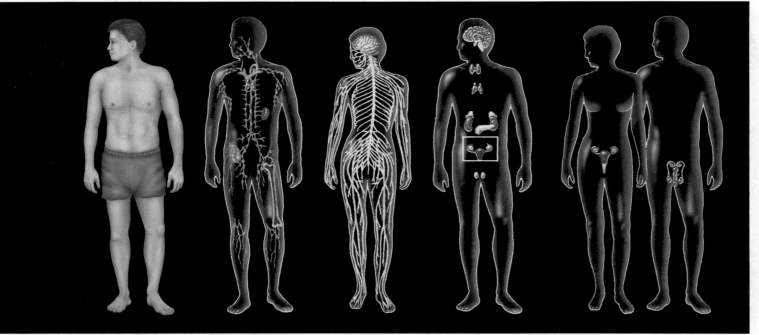

Integumentary system
skin, hair, nails

Immune system
lymph nodes, thymus, bone marrow, tonsils, adenoids, spleen, appendix, and, not shown, white blood cells, gut-associated lymphoid tissue, and skin-associated lymphoid tissue

Nervous system
brain, spinal cord, peripheral nerves, and, not shown, special sense organs

Endocrine system
all hormone-secreting tissues, including hypothalamus, pituitary, thyroid, adrenals, endocrine pancreas, gonads, kidneys, pineal, thymus, and, not shown, parathyroids, intestine, heart, and skin

Reproductive system
Male: testes, penis, prostate gland, seminal vesicles, bulbourethral glands, and associated ducts

Female: ovaries, oviducts, uterus, vagina, breasts

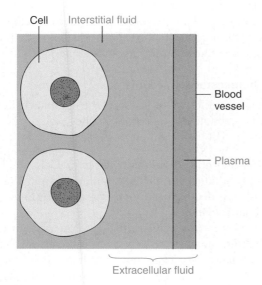

● FIGURE 1-4

Components of the extracellular fluid (internal environment)

No matter how remote a cell is from the external environment, it can make life-sustaining exchanges with its own surrounding internal environment. In turn, various body systems accomplish the transfer of materials between the external environment and the internal environment so that the composition of the internal environment is appropriately maintained to support the life and functioning of the cells. For example, the digestive system transfers the nutrients required by all body cells from the external environment into the plasma. Likewise, the respiratory system transfers O_2 from the external environment into the plasma. The circulatory system distributes these nutrients and O_2 throughout the body. Materials are thoroughly mixed and exchanged between the plasma and the interstitial fluid across the capillaries, the smallest and thinnest of the blood vessels. As a result, the nutrients and O_2 originally obtained from the external environment are delivered to the interstitial fluid that surrounds the cells. The body cells, in turn, pick up these needed supplies from the interstitial fluid. Similarly, wastes produced by the cells are extruded into the interstitial fluid, picked up by the plasma, and transported to the organs that specialize in eliminating these wastes from the internal environment to the external environment. The lungs remove CO_2 from the plasma, and the kidneys remove other wastes for elimination in the urine.

Thus a body cell takes in essential nutrients from and eliminates wastes into its watery surroundings, just as an amoeba does. The major difference is that each body cell must help maintain the composition of the internal environment so that this fluid continuously remains suitable to support the existence of all the body cells. In contrast, an amoeba does nothing to regulate its surroundings.

▌ Body systems maintain homeostasis, a dynamic steady state in the internal environment.

The body cells can live and function only when the extracellular fluid is compatible with their survival; thus the chemical composition and physical state of this internal environment must be maintained within narrow limits. As cells take up nutrients and O_2 from the internal environment, these essential materials must constantly be replenished. Likewise, wastes must constantly be removed from the internal environment so that they do not reach toxic levels. Other aspects of the internal environment that are important for maintaining life, such as temperature, also must be kept relatively constant. Maintenance of a relatively stable internal environment is termed **homeostasis** (*homeo* means "the same"; *stasis* means "to stand or stay").

The functions performed by each body system contribute to homeostasis, thereby maintaining within the body the environment required for the survival and function of all the cells. Cells, in turn, make up body systems. This is the central theme of physiology and of this book: *Homeostasis is essential for the survival of each cell, and each cell, through its specialized activities, contributes as part of a body system to the maintenance of the internal environment shared by all cells* (● Figure 1-5).

The fact that the internal environment must be kept relatively stable does not mean that its composition, temperature, and other characteristics are absolutely unchanging. Both external and internal factors continuously threaten to disrupt homeostasis. When any factor starts to move the internal environment away from optimal conditions, the body systems initiate appropriate counterreactions to minimize the change. For example, exposure to a cold environmental temperature (an external factor) tends to reduce the body's internal temperature. In response, the temperature control center in the brain initiates compensatory measures, such as shivering, to raise body temperature to normal. By contrast, production of extra heat by working muscles during exercise (an internal factor) tends to raise the body's internal temperature. In response, the temperature control center brings about sweating and other compensatory measures to reduce body temperature to normal.

Thus homeostasis should be viewed not as a rigid, fixed state but as a dynamic steady state in which the changes that do occur are minimized by compensatory physiological responses. The term "dynamic" refers to the fact that each homeostatically regulated factor is marked by continuous change, whereas "steady state" implies that these changes do not deviate far from a constant, or steady, level. This situation is comparable to the minor steering adjustments you make as you guide an automobile along a straight course down the highway. Small fluctuations around the optimal level for each factor in the internal environment are normally kept by carefully regulated mechanisms, within the narrow limits compatible with life.

Some compensatory mechanisms are immediate, transient responses to a situation that moves a regulated factor in the internal environment away from the desired level, whereas others are more long-term adaptations that take place in response to prolonged or repeated exposure to a situation that disrupts homeostasis. Long-term adaptations make the body more efficient in responding to an ongoing or repetitive challenge. The body's reaction to exercise includes examples of both short-term compensatory responses and long-term adaptations among the different body systems. (See the accompanying boxed feature ▶ A Closer Look at Exercise Physiology. Most chapters have a boxed feature focusing on exercise physiology. Also, we will be mentioning issues related to exercise physiology throughout the text. Appendix F will help you locate all the references to this important topic.)

Factors homeostatically regulated

As mentioned earlier, many factors of the internal environment must be homeostatically maintained. They include the following. (● Figure 1-6):

1. *Concentration of nutrient molecules.* Cells need a constant supply of nutrient molecules for energy production. Energy, in turn, is needed to support life-sustaining and specialized cellular activities.

2. *Concentration of O_2 and CO_2* to carry out energy-yielding chemical reactions. The CO_2 produced during these chemical reactions must be removed so that acid-forming CO_2 does not increase the acidity of the internal environment.

3. *Concentration of waste products.* Various chemical reactions produce end products that exert a toxic effect on the body's cells if these wastes are allowed to accumulate.

4. *pH.* Changes in the pH (relative amount of acid) adversely affect nerve cell function and wreak havoc with the enzyme activity of all cells.

5. *Concentration of water, salt, and other electrolytes.* Because the relative concentrations of salt (NaCl) and water in the extracellular fluid (internal environment) influence how

● **FIGURE 1-5**

Interdependent relationship of cells, body systems, and homeostasis
The depicted interdependent relationship serves as the foundation for modern-day physiology: *homeostasis is esssential for the survival of cells, body systems maintain homeostasis, and cells make up body systems.*

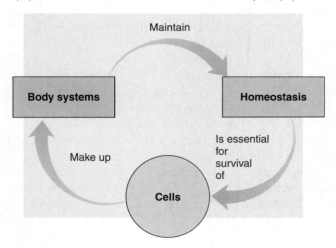

What Is Exercise Physiology?

Exercise physiology is the study of both the functional changes that occur in response to a single session of exercise and the adaptations that occur as a result of regular, repeated exercise sessions. Exercise initially disrupts homeostasis. The changes that occur in response to exercise are the body's attempt to meet the challenge of maintaining homeostasis when increased demands are placed on the body. Exercise often requires prolonged coordination among most body systems, including the muscular, skeletal, nervous, circulatory, respiratory, urinary, integumentary (skin), and endocrine (hormone-producing) systems.

Heart rate is one of the easiest factors to monitor that shows both an immediate response to exercise and long-term adaptation to a regular exercise program. When a person begins to exercise, the active muscle cells use more O_2 to support their increased energy demands. Heart rate increases to deliver more oxygenated blood to the exercising muscles. The heart adapts to regular exercise of sufficient intensity and duration by increasing its strength and efficiency so that it pumps more blood per beat. Because of increased pumping ability, the heart does not have to beat as rapidly to pump a given quantity of blood as it did before physical training.

Exercise physiologists study the mechanisms responsible for the changes that occur as a result of exercise. Much of the knowledge gained from the study of exercise is used to develop appropriate exercise programs to increase the functional capacities of people ranging from athletes to the infirm. The importance of proper and sufficient exercise in disease prevention and rehabilitation is becoming increasingly evident.

---

much water enters or leaves the cells, these concentrations are carefully regulated to maintain the proper volume of the cells. Cells do not function normally when they are swollen or shrunken. Other electrolytes perform a variety of vital functions. For example, the rhythmic beating of the heart depends on a relatively constant concentration of potassium (K^+) in the extracellular fluid.

6. *Temperature.* Body cells function best within a narrow temperature range. If cells are too cold, their functions slow down too much, and worse yet, if they get too hot their structural and enzymatic proteins are impaired or destroyed.

7. *Volume and pressure.* The circulating component of the internal environment, the plasma, must be maintained at adequate volume and blood pressure to ensure bodywide distribution of this important link between the external environment and the cells.

Contributions of the body systems to homeostasis

The 11 body systems contribute to homeostasis in the following important ways (● Figure 1-6):

1. The *circulatory system* is the transport system that carries materials such as nutrients, O_2, CO_2, wastes, electrolytes, and hormones from one part of the body to another.

2. The *digestive system* breaks down dietary food into small nutrient molecules that can be absorbed into the plasma for distribution to the body cells. It also transfers water and electrolytes from the external environment into the internal environment. It eliminates undigested food residues to the external environment in the feces.

3. The *respiratory system* obtains O_2 from and eliminates CO_2 to the external environment. By adjusting the rate of removal of acid-forming CO_2, the respiratory system is also important in maintaining the proper pH of the internal environment.

4. The *urinary system* removes excess water, salt, acid, and other electrolytes from the plasma and eliminates them in the urine, along with waste products other than CO_2.

5. The *skeletal system* provides support and protection for the soft tissues and organs. It also serves as a storage reservoir for calcium (Ca^{2+}), an electrolyte whose plasma concentration must be maintained within very narrow limits. Together with the muscular system, the skeletal system also enables movement of the body and its parts. Furthermore, the bone marrow—the soft interior portion of some types of bone—is the ultimate source of all blood cells.

6. The *muscular system* moves the bones to which the skeletal muscles are attached. From a purely homeostatic view, this system enables an individual to move toward food or away from harm. Furthermore, the heat generated by muscle contraction is important in temperature regulation. In addition, because skeletal muscles are under voluntary control, a person is able to use them to accomplish myriad other movements of his or her own choice. These movements, which range from the fine motor skills required for delicate needlework to the powerful movements involved in weight lifting, are not necessarily directed toward maintaining homeostasis.

7. The *integumentary system* serves as an outer protective barrier that prevents internal fluid from being lost from the body and foreign micro-organisms from entering. This system is also important in the regulation of body temperature. The amount of heat lost from the body surface to the external environment can be adjusted by controlling sweat production and by regulating the flow of warm blood through the skin.

8. The *immune system* defends against foreign invaders and against body cells that have become cancerous. It also paves the way for repair or replacement of injured or worn-out cells.

9. The *nervous system* is one of the two major regulatory systems of the body. In general, it controls and coordinates bodily activities that require swift responses. It is especially important in detecting and initiating reactions to changes in the external environment. Furthermore, it is responsible for higher functions that are not entirely directed toward maintaining homeostasis, such as consciousness, memory, and creativity.

10. The *endocrine system* is the other major regulatory system. In contrast to the nervous system, in general the hormone-secreting glands of the endocrine system regulate activities that require duration rather than speed. This system is especially important in controlling the concentration of nutrients and,

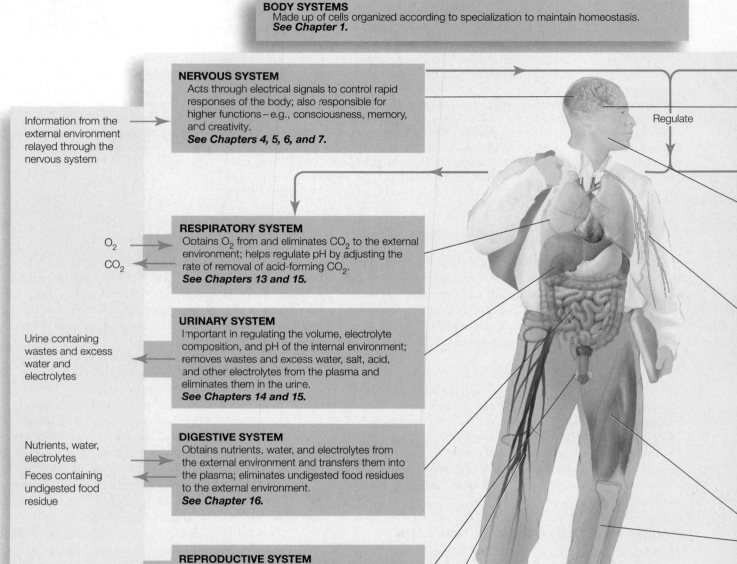

BODY SYSTEMS
Made up of cells organized according to specialization to maintain homeostasis.
See Chapter 1.

NERVOUS SYSTEM
Acts through electrical signals to control rapid responses of the body; also responsible for higher functions – e.g., consciousness, memory, and creativity.
See Chapters 4, 5, 6, and 7.

Information from the external environment relayed through the nervous system

Regulate

RESPIRATORY SYSTEM
Obtains O_2 from and eliminates CO_2 to the external environment; helps regulate pH by adjusting the rate of removal of acid-forming CO_2.
See Chapters 13 and 15.

O_2

CO_2

URINARY SYSTEM
Important in regulating the volume, electrolyte composition, and pH of the internal environment; removes wastes and excess water, salt, acid, and other electrolytes from the plasma and eliminates them in the urine.
See Chapters 14 and 15.

Urine containing wastes and excess water and electrolytes

DIGESTIVE SYSTEM
Obtains nutrients, water, and electrolytes from the external environment and transfers them into the plasma; eliminates undigested food residues to the external environment.
See Chapter 16.

Nutrients, water, electrolytes

Feces containing undigested food residue

REPRODUCTIVE SYSTEM
Not essential for homeostasis, but essential for perpetuation of the species.
See Chapter 20.

Sperm leave male
Sperm enter female

Exchanges with all other systems.

EXTERNAL ENVIRONMENT

CIRCULATORY SYSTEM
Transports nutrients, O_2, CO_2, wastes, electrolytes, and hormones throughout the body.
See Chapters 9, 10, and 11.

● FIGURE 1-6
Role of the body systems in maintaining homeostasis

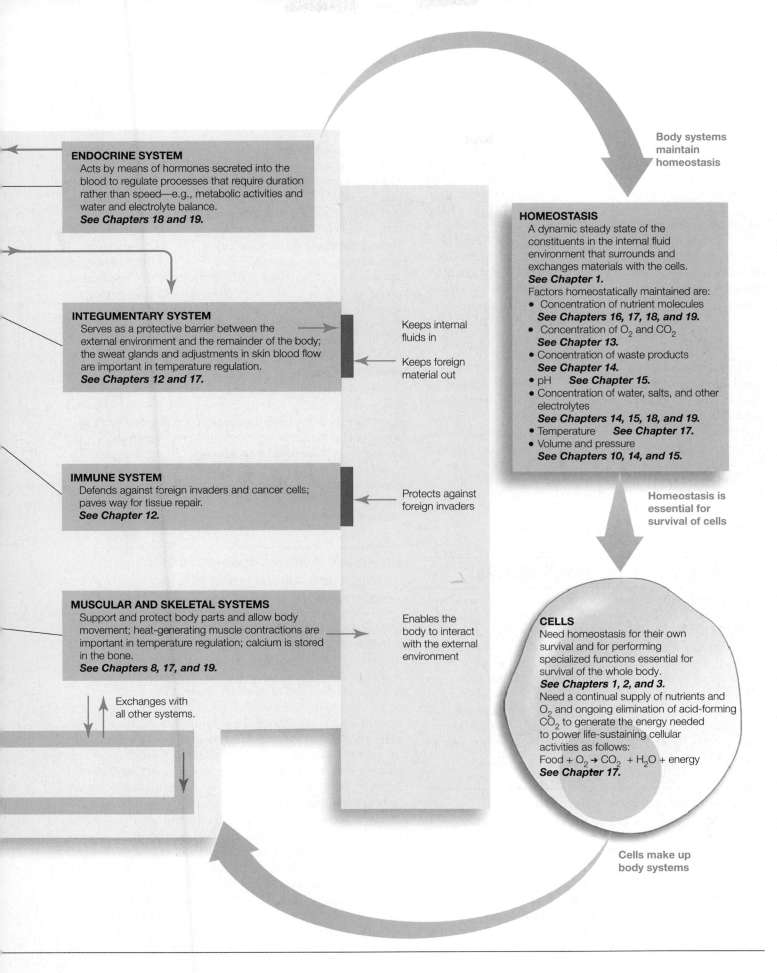

ENDOCRINE SYSTEM
Acts by means of hormones secreted into the blood to regulate processes that require duration rather than speed—e.g., metabolic activities and water and electrolyte balance.
See Chapters 18 and 19.

INTEGUMENTARY SYSTEM
Serves as a protective barrier between the external environment and the remainder of the body; the sweat glands and adjustments in skin blood flow are important in temperature regulation.
See Chapters 12 and 17.

Keeps internal fluids in

Keeps foreign material out

IMMUNE SYSTEM
Defends against foreign invaders and cancer cells; paves way for tissue repair.
See Chapter 12.

Protects against foreign invaders

MUSCULAR AND SKELETAL SYSTEMS
Support and protect body parts and allow body movement; heat-generating muscle contractions are important in temperature regulation; calcium is stored in the bone.
See Chapters 8, 17, and 19.

Enables the body to interact with the external environment

Exchanges with all other systems.

Body systems maintain homeostasis

HOMEOSTASIS
A dynamic steady state of the constituents in the internal fluid environment that surrounds and exchanges materials with the cells.
See Chapter 1.
Factors homeostatically maintained are:
• Concentration of nutrient molecules
  See Chapters 16, 17, 18, and 19.
• Concentration of O_2 and CO_2
  See Chapter 13.
• Concentration of waste products
  See Chapter 14.
• pH   *See Chapter 15.*
• Concentration of water, salts, and other electrolytes
  See Chapters 14, 15, 18, and 19.
• Temperature   *See Chapter 17.*
• Volume and pressure
  See Chapters 10, 14, and 15.

Homeostasis is essential for survival of cells

CELLS
Need homeostasis for their own survival and for performing specialized functions essential for survival of the whole body.
See Chapters 1, 2, and 3.
Need a continual supply of nutrients and O_2 and ongoing elimination of acid-forming CO_2 to generate the energy needed to power life-sustaining cellular activities as follows:
Food + O_2 → CO_2 + H_2O + energy
See Chapter 17.

Cells make up body systems

by adjusting kidney function, controlling the internal environment's volume and electrolyte composition.

11. The *reproductive system* is not essential for homeostasis and therefore is not essential for survival of the individual. It is essential, however, for perpetuation of the species.

As we examine each of these systems in greater detail, always keep in mind that the body is a coordinated whole even though each system provides its own special contributions. It is easy to forget that all the body parts actually fit together into a functioning, interdependent whole body. Accordingly, each chapter begins with a figure and discussion to help you focus on how the body system to be described fits into the body as a whole. In addition, each chapter concludes with a brief overview of the homeostatic contributions of the body system. As a further tool to help you keep track of how all the pieces fit together, ● Figure 1-6 is duplicated on the inside cover as a handy reference.

Also be aware that the functioning whole is greater than the sum of its separate parts. Through specialization, cooperation, and interdependence, cells combine to form an integrated, unique, single living organism with more diverse and complex capabilities than are possessed by any of the cells that make it up. For humans, these capabilities go far beyond the processes needed to maintain life. A cell, or even a random combination of cells, obviously cannot create an artistic masterpiece or design a spacecraft, but body cells working together permit those capabilities in an individual.

You have now learned what homeostasis is and how the functions of different body systems maintain it. Next let's look at the regulatory mechanisms by which the body reacts to changes and controls the internal environment.

HOMEOSTATIC CONTROL SYSTEMS

A **homeostatic control system** is a functionally interconnected network of body components that operate to maintain a given chemical or physical factor in the internal environment relatively constant around an optimal level. To maintain homeostasis, the control system must be able to (1) detect deviations from normal in the internal environmental factor that needs to be held within narrow limits; (2) integrate this information with any other relevant information; and (3) make appropriate adjustments in the activity of the body parts responsible for restoring this factor to its desired value. We now turn our attention to the types of control systems in the body.

▌ Homeostatic control systems may operate locally or bodywide.

Homeostatic control systems can be grouped into two classes—intrinsic and extrinsic controls. **Intrinsic (local) controls** are built into or are inherent in an organ (*intrinsic* means "within"). For example, as an exercising skeletal muscle rapidly uses up O_2 to generate energy to support its contractile activity, the O_2 concentration within the muscle falls. This local chemical change acts directly on the smooth muscle in the walls of the

blood vessels that supply the exercising muscle, causing the smooth muscle to relax so that the vessels dilate, or open widely. As a result, increased blood flows through the dilated vessels into the exercising muscle, bringing in more O_2. This local mechanism contributes to the maintenance of an optimal level of O_2 in the internal fluid environment immediately surrounding the exercising muscle's cells.

Most factors in the internal environment are maintained, however, by **extrinsic controls,** which are regulatory mechanisms initiated outside an organ to alter the activity of the organ (*extrinsic* means "outside of"). Extrinsic control of the organs and body systems is accomplished by the nervous and endocrine systems, the two major regulatory systems of the body. Extrinsic control permits coordinated regulation of several organs toward a common goal; in contrast, intrinsic controls are self-serving for the organ in which they occur. Coordinated, overall regulatory mechanisms are crucial for maintaining the dynamic steady state in the internal environment as a whole. To restore blood pressure to the proper level when it falls too low, for example, the nervous system simultaneously acts on the heart and the blood vessels throughout the body to increase the blood pressure to normal.

To stabilize the physiological factor being regulated, homeostatic control systems must be able to respond to and resist change. The term **feedback** refers to responses made after a change has been detected; the term **feedforward** is used for responses made in anticipation of a change. Intrinsic controls operate on the principle of feedback. Extrinsic controls typically operate on the basis of feedback but also use feedforward mechanisms. We will now take a look at these mechanisms in terms of their type of response.

▌ Negative feedback opposes an initial change and is widely used to maintain homeostasis.

Homeostatic control mechanisms primarily operate on the principle of negative feedback. In **negative feedback,** change in a homeostatically controlled factor triggers a response that seeks to restore the factor to normal by moving the factor in the opposite direction of its initial change.

A common example of negative feedback is control of room temperature. Room temperature is a **controlled variable,** a factor that can vary but is held within a narrow range by a control system. In our example, the control system includes a thermostatic device, a furnace, and all their electrical connections. The room temperature is determined by the activity of the furnace, a heat source that can be turned on or off. To switch on or off appropriately, the control system as a whole must "know" what the *actual* room temperature is, "compare" it with the *desired* room temperature, and "adjust" the output of the heat source to bring the actual temperature to the desired level. Information about the actual room temperature is provided by a thermometer in the thermostat. The thermometer is the **sensor,** which monitors the magnitude of the controlled variable. The desired temperature level, or **set point,** is provided by the thermostat setting. The thermostat acts as an

integrator, or **control center:** It compares the sensor's input with the set point and adjusts the heat output of the furnace to bring about the appropriate effect, or response, to oppose the deviation from the set point. The furnace is the **effector,** the component of the control system commanded to bring about the desired effect. These general components of a negative-feedback control system are summarized in ● Figure 1-7a.

Carefully examine this figure and the guidelines in the footnote, because the symbols and conventions introduced here are used in comparable flow diagrams throughout the text.

Let us look at a typical negative-feedback loop. For example, if in cold weather the room temperature falls below the set point level, the thermostat, through connecting circuitry, activates the furnace, which produces heat to increase the

---

● **FIGURE 1-7**

Negative feedback

(a) Components of a negative-feedback control system. (b) Negative-feedback control of room temperature. (c) Negative-feedback control of body temperature.

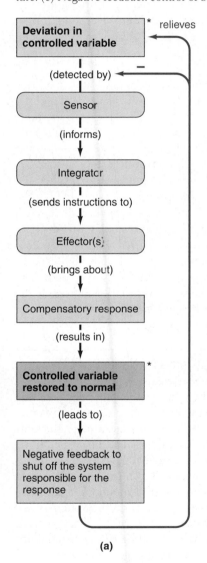

(a)

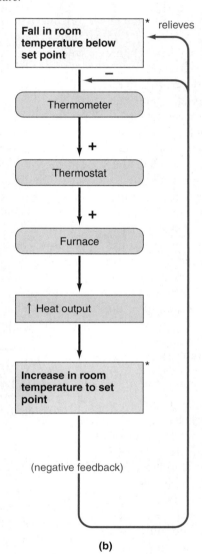

(b)

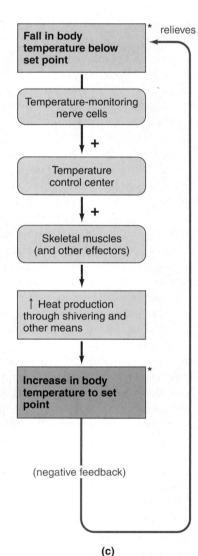

(c)

For flow diagrams throughout the text:
+ = Stimulates or activates
— = Inhibits or shuts off
⬭ = Physical entity, such as body structure or a chemical
▭ = Actions
❙ = Compensatory pathway
❙ = Turning off of compensatory pathway (negative feedback)
* Note that lighter and darker shades of the same color are used to denote respectively a decrease or an increase in a controlled variable.

room temperature (● Figure 1-7b). Once the room temperature reaches the set point, the thermometer no longer detects a deviation from the set point. As a result, the activating mechanism in the thermostat and consequently the furnace are switched off. Thus the heat produced by the furnace counteracts or is "negative" to the original fall in the temperature. If the heat-generating pathway were not shut off once the target temperature is reached, heat production would continue and the room would get hotter and hotter. Overshooting beyond the set point does not occur, because the heat "feeds back" to shut off the thermostat that triggered its output. Thus a negative-feedback control system detects a change in a controlled variable away from the ideal value, initiates mechanisms to correct the situation, then shuts itself off. In this way, the controlled variable does not drift too far below or too far above the set point.

What if the original deviation is a rise in room temperature above the set point because it is hot outside? A heat-producing furnace is of no use in returning the room temperature to the desired level. In this case, the thermostat, through connecting circuitry, can activate the air conditioner, which cools the room air, the opposite effect of the furnace. In negative-feedback fashion, once the set point is reached, the air conditioner is turned off to prevent the room from becoming too cold. Note that if the controlled variable can be deliberately adjusted to oppose a change in one direction only, the variable can move in uncontrolled fashion in the opposite direction. For example, if the house is equipped only with a furnace that produces heat to oppose a fall in room temperature, no mechanism is available to prevent the house from getting too hot in warm weather. However, the room temperature can be kept relatively constant through two opposing mechanisms, one that heats and one that cools the room, despite wide variations in the temperature of the external environment.

Homeostatic negative-feedback systems operate in the same way. For example, when temperature-monitoring nerve cells detect a decrease in body temperature below the desired level, these sensors bring about a sequence of events that culminates in shivering, among other responses, to generate heat and increase the temperature to the proper level (● Figure 1-7c). When the body temperature increases to the set point, the stimulatory signal to the skeletal muscles arising from the temperature-monitoring nerve cells is turned off. As a result, the body temperature does not continue to increase above the set point. In the converse situation, when the temperature-monitoring nerve cells detect a rise in body temperature above normal, cooling mechanisms such as sweating are called into play to reduce the temperature to normal. When the temperature reaches the set point, the cooling mechanisms are shut off. As is the case with body temperature, most homeostatically controlled variables can be moved in either direction by opposing mechanisms as needed.

▌ Positive feedback amplifies an initial change.

In negative feedback, a control system's output is regulated to resist change, so that the controlled variable is maintained at a relatively steady set point. With **positive feedback,** by contrast, the output is continually enhanced or amplified so that the controlled variable continues to be moved in the direction of the initial change. Such action would be comparable to the heat generated by a furnace triggering the thermostat to call for even *more* heat output from the furnace so that the room temperature would continuously rise.

Because the major goal in the body is to maintain stable, homeostatic conditions, positive feedback does not occur nearly as often as negative feedback. Positive feedback does play an important role in certain instances, however, as in the birth of a baby. The hormone oxytocin causes powerful contractions of the uterus. As uterine contractions push the baby against the cervix (the exit from the uterus), the resultant stretching of the cervix triggers a sequence of events that brings about the release of even more oxytocin, which causes even stronger uterine contractions, triggering the release of more oxytocin, and so on. This positive-feedback cycle does not cease until the baby is finally born.

Likewise, all other normal instances of positive feedback in the body include some mechanism for termination of the cycle. However, some abnormal circumstances in the body are characterized by runaway positive-feedback loops that continue to move the body farther and farther from homeostatic balance until death or medical intervention stops this vicious cycle. Such an example occurs during heat stroke. When the temperature-regulating mechanisms are not able to cool the body sufficiently in the face of pronounced environmental heat exposure, the body temperature may rise so high that the temperature control center becomes impaired. Because this control center is no longer functioning normally, its ability to call forth the cooling mechanisms is diminished, so the body temperature soars even higher, causing even further damage to the control center. As a result of this positive-feedback mechanism, body temperature spirals out of control.

▌ Feedforward mechanisms initiate responses in anticipation of a change.

In addition to feedback mechanisms, which bring about a response in *reaction* to a change in a regulated variable, the body less frequently employs feedforward mechanisms, which bring about a response in *anticipation* of a change in a regulated variable. For example, when a meal is still in the digestive tract, a feedforward mechanism increases the secretion of a hormone that will promote the cellular uptake and storage of ingested nutrients after they have been absorbed from the digestive tract. This anticipatory response helps limit the rise in blood nutrient concentration that occurs following nutrient absorption.

▌ Disruptions in homeostasis can lead to illness and death.

Despite control mechanisms, when one or more of the body's systems fail to function properly, homeostasis is disrupted, and all the cells suffer, because they no longer have an optimal environment in which to live and function. Various pathophysiological states ensue, depending on the type and extent of homeostatic disruption. The term **Pathophysiology** refers to the abnormal functioning of the body (altered physiology) associated with disease. When a homeostatic disruption becomes so severe that it is no longer compatible with survival, death results.

vival and normal functioning of cells, and each cell, through its specialized activities, contributes as part of a body system to the maintenance of homeostasis.

This relationship is the foundation of physiology and the central theme of this book. We have already described how cells are organized according to specialization into body systems. How homeostasis is essential for cell survival and how body systems maintain this internal constancy are the topics covered in the remainder of this book. Each chapter will conclude with this capstone feature to facilitate your understanding of the contributions to homeostasis of the system under discussion as well as the interactions and interdependency of the body systems.

CHAPTER IN PERSPECTIVE: FOCUS ON HOMEOSTASIS

In this chapter, you have learned what homeostasis is: a dynamic steady state of the constituents in the internal fluid environment (the extracellular fluid) that surrounds and exchanges materials with the cells. Maintenance of homeostasis is essential for the sur-

CHAPTER SUMMARY

Introduction to Physiology
- Physiology is the study of body functions.
- Physiologists explain body function in terms of the mechanisms of action involving cause-and-effect sequences of physical and chemical processes.
- Physiology and anatomy are closely interrelated because body functions are highly dependent on the structure of the body parts that carry them out.

Levels of Organization in the Body
- The human body is a complex combination of various atoms and molecules.
- These nonliving chemical components are organized in a precise way to form cells, the smallest entities capable of carrying out the processes associated with life.
- Cells are the basic structural and functional living building blocks of the body.
- The basic functions performed by each cell for its own survival include (1) obtaining O_2 and nutrients, (2) performing energy-generating chemical reactions, (3) eliminating wastes, (4) synthesizing proteins and other cellular components, (5) controlling movement of materials between the cell and its environment, (6) moving materials throughout the cell, (7) responding to the environment, and (8) reproducing.
- In addition to its basic cell functions, each cell in a multicellular organism performs a specialized function.
- Combinations of cells of similar structure and specialized function form the four primary tissues of the body: muscle, nervous, epithelial, and connective tissue.
- Glands are derived from epithelial tissue and specialized for secretion. Exocrine glands secrete through ducts to the body surface or a cavity that communicates with the outside; endocrine glands secrete hormones into the blood.
- Organs are combinations of several types of tissues that act together to perform one or more functions. An example is the stomach.
- Body systems are collections of organs that perform related functions and interact to accomplish a common activity essential for survival of the whole body. An example is the digestive system.
- Organ systems combine to form the organism, or whole body.

Concept of Homeostasis
- The fluid inside the cells of the body is called intracellular fluid and the fluid outside the cells is known as extracellular fluid.

- Because most body cells are not in direct contact with the external environment, cell survival depends on maintenance of a relatively stable internal fluid environment with which the cells directly make life-sustaining exchanges.
- The extracellular fluid serves as the internal environment. It consists of the plasma and interstitial fluid.
- Homeostasis is the maintenance of a dynamic steady state in the internal environment.
- The factors of the internal environment that must be homeostatically maintained are its (1) concentration of nutrient molecules; (2) concentration of O_2 and CO_2; (3) concentration of waste products; (4) pH; (5) concentration of water, salt, and other electrolytes; (6) temperature; and (7) volume and pressure.
- The functions performed by the body systems are directed toward maintaining homeostasis.
- The body systems' functions ultimately depend on the specialized activities of the cells composing the system. Thus, homeostasis is essential for each cell's survival, and each cell contributes to homeostasis.

Homeostatic Control Systems
- A homeostatic control system is a network of body components working together to maintain a given factor in the internal environment relatively constant around an optimal set level.
- Homeostatic control systems can be classified as (1) intrinsic (local) controls, which are inherent compensatory responses of an organ to a change; and (2) extrinsic controls, which are responses of an organ that are triggered by factors external to the organ, namely, by the nervous and endocrine systems.
- Both intrinsic and extrinsic control systems generally operate on the principle of negative feedback: A change in a controlled variable triggers a response that drives the variable in the opposite direction of the initial change, thus opposing the change.
- In positive feedback, a change in a controlled variable triggers a response that drives the variable in the same direction as the initial change, thus amplifying the change.
- Feedforward mechanisms are compensatory responses that occur in anticipation of a change.
- Pathophysiological states ensue when one or more of the body systems fail to function properly so that an optimal internal environment can no longer be maintained. Serious homeostatic disruption leads to death.

Homeostasis: The Foundation of Physiology

REVIEW EXERCISES

Objective Questions (Answers on p. A-35)

1. Which of the following activities is *not* carried out by every cell in the body?
   a. obtaining O_2 and nutrients
   b. performing chemical reactions to acquire energy for the cell's use
   c. eliminating wastes
   d. controlling to a large extent exchange of materials between the cell and its external environment
   e. reproducing
2. Which of the following is the proper progression of the levels of organization in the body?
   a. chemicals, cells, organs, tissues, body systems, whole body
   b. chemicals, cells, tissues, organs, body systems, whole body
   c. cells, chemicals, tissues, organs, whole body, body systems
   d. cells, chemicals, organs, tissues, whole body, body systems
   e. cells, tissues, body systems, organs, whole body
3. Which of the following is *not* a type of connective tissue?
   a. bone
   b. blood
   c. the spinal cord
   d. tendons
   e. the tissue that attaches epithelial tissue to underlying structures
4. The term *tissue* can apply either to one of the four primary tissue types or to a particular organ's aggregate of cellular and extracellular components. *(True or false?)*
5. Cells in a multicellular organism have specialized to such an extent that they have little in common with single-celled organisms. *(True or false?)*
6. Cellular specializations are usually a modification or elaboration of one of the basic cell functions *(True or false?)*
7. The four primary types of tissue are _____, _____, _____, and _____.
8. _____ refers to the release from a cell, in response to appropriate stimulation, of specific products that have in large part been synthesized by the cell.
9. _____ glands secrete through ducts to the outside of the body, whereas _____ glands release their secretory products, known as _____, internally into the blood.
10. _____ controls are inherent to an organ, whereas _____ controls are regulatory mechanisms initiated outside of an organ that alter the activity of the organ.

11. Match the following:
   ___ 1. circulatory system
   ___ 2. digestive system
   ___ 3. respiratory system
   ___ 4. urinary system
   ___ 5. muscular and skeletal systems
   ___ 6. integumentary system
   ___ 7. immune system
   ___ 8. nervous system
   ___ 9. endocrine system
   ___ 10. reproductive system

   (a) obtains O_2 and eliminates CO_2
   (b) support and protect body parts and allow movement
   (c) controls, via hormones it secretes, processes that require duration
   (d) transport system
   (e) removes wastes and excess water, salt, and other electrolytes
   (f) essential for perpetuation of species
   (g) obtains nutrients, water, and electrolytes
   (h) defends against foreign invaders and cancer
   (i) acts through electrical signals to control body's rapid responses
   (j) serves as protective barrier between body and external environment

Essay Questions

1. Define *physiology*.
2. What are the basic cell functions?
3. Distinguish between the external environment and the internal environment. What constitutes the internal environment?
4. Of what fluid compartments is the internal environment composed?
5. Define *homeostasis*.
6. Describe the interrelationships among cells, body systems, and homeostasis.
7. What factors of the internal environment must be homeostatically maintained?
8. What is a homeostatic control system? Describe the components of a homeostatic control system.
9. Compare negative and positive feedback.

POINTS TO PONDER

(Explanations on p. A 35)

1. Considering the nature of negative-feedback control and the function of the respiratory system, what effect do you predict that a decrease in CO_2 in the internal environment would have on how rapidly and deeply a person breathes?
2. Would the O_2 levels in the blood be (a) normal, (b) below normal, or (c) elevated in a patient with severe pneumonia in whom exchange of O_2 and CO_2 between the air and blood in the lungs is impaired? Would the CO_2 levels in the same patient's blood be (a) normal, (b) below normal, or (c) elevated? Because CO_2 reacts with H_2O to form carbonic acid (H_2CO_3), would the pa-

tient's blood (a) have a normal pH, (b) be too acidic, or (c) not be acidic enough (that is, be too alkaline), assuming that other compensatory measures have not yet had time to act?
3. The hormone insulin enhances the transport of glucose (sugar) from the blood into most of the body's cells. Its secretion is controlled by a negative-feedback system between the concentration of glucose in the blood and the insulin-secreting cells. Therefore, which of the following statements is correct?
   a. A decrease in blood glucose concentration stimulates insulin secretion, which in turn further lowers the blood glucose concentration.

b. An increase in blood glucose concentration stimulates insulin secretion, which in turn lowers the blood glucose concentration.

c. A decrease in blood glucose concentration stimulates insulin secretion, which in turn increases the blood glucose concentration.

d. An increase in blood glucose concentration stimulates insulin secretion, which in turn further increases the blood glucose concentration.

e. None of the above are correct.

4. Given that most AIDS victims die from overwhelming infections or rare types of cancer, what body system do you think is impaired by HIV (the AIDS virus)?

5. Body temperature is homeostatically regulated around a set point. Based on your knowledge of negative feedback and homeostatic control systems, predict whether narrowing or widening of the blood vessels of the skin will occur when a person is engaged in strenuous exercise. (*Hints:* Muscle contraction generates heat. Narrowing of the vessels supplying an organ decreases blood flow through the organ, whereas widening of the vessels increases blood flow through the organ. The more warm blood flowing through the skin, the greater the loss of heat from the skin to the surrounding environment.)

6. *Clinical Consideration.* Jennifer R. has the "stomach flu" that is going around campus and has been vomiting profusely for the past 24 hours. Not only has she been unable to keep down fluids or food that she has consumed, but she has also lost the acidic digestive juices secreted by the stomach that are normally reabsorbed back into the blood farther down the digestive tract. In what ways might this condition threaten to disrupt the homeostatic maintenance of Jennifer's internal environment? That is, what homeostatically maintained factors are moved away from normal as a result of her profuse vomiting? What organ systems respond to resist these changes?

PHYSIOEDGE RESOURCES

PHYSIOEDGE CD-ROM

PhysioEdge, the CD-ROM packaged with your text, focuses on the concepts students find most difficult to learn. Figures marked with this icon have associated activities on the CD. A diagnostic quiz component allows you to receive immediate feedback on your understanding of the concept and to advance through various levels of difficulty.

PHYSIOEDGE WEB SITE

The Web site for this book contains a wealth of helpful study aids, as well as many ideas for further reading and research. Log on to:

http://www.brookscole.com

Go to the Biology page and select Sherwood's *Human Physiology,* 5th Edition. Select a chapter from the drop-down menu or click on one of these resource areas:

- **Case Histories** provide an introduction to the clinical aspects of human physiology.

- For 2-D and 3-D graphical illustrations and animations of physiological concepts, visit our **Visual Learning Resource.**

- For study and review, **Chapter Outline** gives you an outline of the chapter, **Chapter Summary** allows you to review the chapter's main ideas, and **Glossary** lists concepts and terms for the chapter along with their definitions.

- To test your mastery of important terminology for this chapter, you can use the electronic **Flash Cards,** which may be sorted by definition or by term.

- For testing your knowledge and preparing for in-class examinations, our **Quizzes** pose multiple choice and/or true-false questions based on each chapter.

- **Hypercontents** takes you to an extensive list of current links to Internet sites with news, research, and images related to individual subjects in the chapter.

- **Internet Exercises** are critical thinking questions that involve research on the Internet with starter URLs provided.

- **InfoTrac Exercises** leads you to Critical Thinking Projects that use InfoTrac College Edition® as a research tool.

For more readings, go to InfoTrac College Edition, your online research library, at:

http://infotrac.thomsonlearning.com

Body Systems

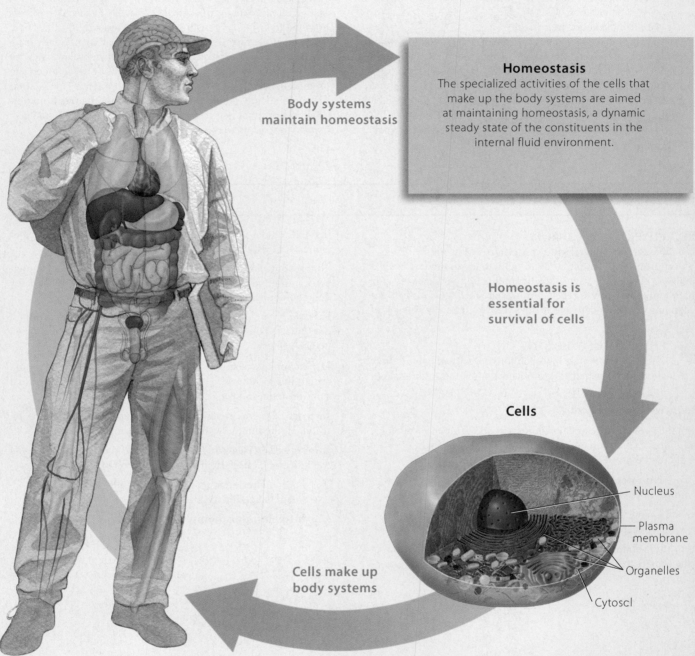

Body systems maintain homeostasis

Homeostasis
The specialized activities of the cells that make up the body systems are aimed at maintaining homeostasis, a dynamic steady state of the constituents in the internal fluid environment.

Homeostasis is essential for survival of cells

Cells

Nucleus

Plasma membrane

Organelles

Cytoscl

Cells make up body systems

Cells are the body's living building blocks. Just as the body as a whole is highly organized, so too is a cell's interior. A cell is made up of three major parts: a **plasma membrane** that encloses the cell; the **nucleus,** which houses the cell's genetic material; and the **cytoplasm,** which is organized into discrete highly specialized *organelles* dispersed throughout a gel-like liquid, the *cytosol*. The cytosol is pervaded by a protein scaffolding, the *cytoskeleton,* that serves as the "bone and muscle" of the cell.

Through the coordinated action of each of these cellular components, every cell is capable of performing certain basic functions essential to its own survival and a specialized task that contributes to the maintenance of homeostasis. Cells are organized according to their specialization into body systems that maintain the stable internal environment essential for the whole body's survival. All body functions ultimately depend on the activities of the individual cells that compose the body.

Chapter 2

Cellular Physiology

CONTENTS AT A GLANCE

OBSERVATIONS OF CELLS

AN OVERVIEW OF CELL STRUCTURE

ENDOPLASMIC RETICULUM

▌ Rough endoplasmic reticulum

▌ Smooth endoplasmic reticulum

GOLGI COMPLEX

▌ Role of the Golgi complex

▌ Secretion by exocytosis

LYSOSOMES

▌ Role of lysosomes

▌ Endocytosis

MITOCHONDRIA

▌ Role of mitochondria

▌ Generation of ATP in aerobic and anaerobic conditions

▌ Uses of ATP

VAULTS

CYTOSOL

CYTOSKELETON

▌ Role of microtubules, microfilaments, and intermediate filaments

Cells are made up of the same chemicals found in nonliving objects on our planet. Even though researchers have analyzed the chemicals of which cells are made, they are unable to organize these molecules into a living cell in a laboratory. Life stems from the complex organization and interaction of these molecules within the cell. Groups of inanimate chemical molecules are structurally organized and function together in unique ways to form a cell, the smallest living entity. Cells, in turn, serve as the living building blocks for the immensely complicated whole body. Thus cells are the bridge between molecules and humans (and all other living organisms). Furthermore, all new cells and all new life arise from division of pre-existing cells, and not from nonliving sources. Because of this continuity of life, the cells of all organisms are fundamentally similar in structure and function. ▲ Table 2-1 summarizes these principles, which are known collectively as the **cell theory.** Modern physiologists are unraveling many of the broader mysteries of how the body works by probing deeper into the molecular structure and organization of the cells that make up the body.

OBSERVATIONS OF CELLS

The cells that compose the human body are so small that they cannot be seen by the unaided eye. The smallest visible particle is about five to ten times larger than a typical human cell, which averages about 10 to 20 micrometers (μm) in diameter. About 100 average-sized cells lined up side by side would stretch a distance of only 1 mm. (one μm = 1 millionth of a meter; 1 mm = 1 thousandth of a meter; 1 m = 39.37 inches. See Appendix A for a comparison of metric units and their English equivalents. This appendix also provides a visual comparison of the size of cells in relation to other selected structures.)

Scientists did not even know about the existence of cells until the microscope was invented in the middle of the 17th century. In the early part of the 19th century, with the development of better light microscopes, researchers learned that all plant and animal tissues are composed of individual cells. The cells of a hummingbird, a human, and a whale are all about the same size. Larger species have a greater number of cells, not larger cells. These early investigators also discovered that cells are filled with a fluid, which, with the microscopic capa-

HeLa Cells: Problems in a "Growing" Industry

Many basic advances in cell physiology, genetics, and cancer research have come about through the use of cells grown, or *cultured,* outside the body. In the middle of the last century, many attempts were made to culture human cells using tissues obtained from biopsies or surgical procedures. These early attempts usually met with failure; the cells died after a few days or weeks in culture, mostly without undergoing cell replication. These difficulties continued until February 1951, when a researcher at Johns Hopkins University received a sample of a cervical cancer from a patient named Henrietta Lacks. Following convention, the culture was named HeLa by combining the first two letters of the donor's first and last names. This cell line not only grew but prospered under culture conditions and represented one of the earliest cell lines successfully grown outside the body.

Researchers were eager to have human cells available on demand to study the effects of drugs, toxic chemicals, radiation, and viruses on human tissue. For example, poliomyelitis virus reproduced well in HeLa cells, providing a breakthrough in the development of a polio vaccine. As cell culture techniques improved, human cell lines were started from other cancers and normal tissues, including heart, kidney, and liver tissues. By the early 1960s, a central collection of cell lines had been established in Washington,

D.C., and cultured human cells were an important tool in many areas of biological research.

But in 1966, geneticist Stanley Gartler made a devastating discovery. He analyzed 18 different human cell lines and found that all had been contaminated and taken over by HeLa cells. Over the next two years, scientists confirmed that 24 of the 32 cell lines in the central repository were actually HeLa cells. Researchers who had spent years studying what they thought were heart or kidney cells had in reality been working with a cervical cancer cell instead. Gartler's discovery meant that hundreds of thousands of experiments performed in laboratories around the world were invalid.

As painful as this lesson was, scientists started over, preparing new cell lines and using new, stricter rules of technique to prevent contamination with HeLa cells. Unfortunately, the problem did not end. In 1974, Walter Nelson-Rees published a paper demonstrating that five cell lines extensively used in cancer research were in fact all HeLa cells. In 1976, 11 additional cell lines, each widely used in research, were also found to be HeLa cells; and in 1981, Nelson-Rees listed 22 more cell lines that were contaminated with HeLa. In all, one-third of all cell lines used in cancer research were apparently really HeLa cells. The result was an enormous waste of dollars and resources.

The invasion of other cultures by HeLa cells is a testament to the ferocious and aggressive

nature of some cancer cells. In cell culture laboratories, rules of sterile technique are supposed to ensure that cross-contamination of one culture with another does not occur. But researchers are only human, and they sometimes make mistakes. For example, perhaps a bottle of culture medium was contaminated through improper handling. Whatever the case, clearly at some point one or more HeLa cells were introduced into cultures where they did not belong.

HeLa cells divide more rapidly than most other human cells, whether normal or cancerous. Because of their rapid growth and division, they use up nutrients more quickly than other cell types. What's more, cells grown in culture are not readily distinguished from one another just by looking; usually biochemical tests are needed for identification. The result is that within a few cycles of transfer, a culture that started out as kidney cells or some other human cell could be taken over completely by the rapidly growing HeLa cells. These cells then crowd out the original cell line, just as cancer cells do in the body.

Henrietta Lacks died long ago from the cervical cancer that started the HeLa line, yet these potent cells continue to live on. Their spread throughout human cell cultures underscores the unrelenting nature of cancer, a disease of uncontrolled growth and resource consumption.

---

bilities of the time, appeared to be a rather uniform, soupy mixture believed to be the elusive "stuff of life." When the technique of electron microscopy was first employed in the 1940s

to observe living matter, an understanding of the great diversity and complexity of the internal structure of cells began to emerge. (Electron microscopes are about 100 times more powerful than light microscopes.) Now, with the availability of even more sophisticated microscopes, biochemical techniques, cell culture technology, and genetic engineering, the concept of the cell as a microscopic bag of formless fluid has given way to our present-day knowledge of the cell as a complex, highly organized, compartmentalized structure. (See the accompanying boxed feature ❯ Concepts, Challenges, and Controversies for a glimpse into an interesting chapter in the history of cell culture.)

AN OVERVIEW OF CELL STRUCTURE

The trillions of cells in a human body are classified into about 200 different cell types based on specific variations in structure and function. Even though there is no such thing as a "typical" cell, because of these diverse structural and functional specializations, different cells share many common features. Most cells have three major subdivisions: the *plasma membrane,* which encloses the cells; the *nucleus,* which con-

---

▲ **TABLE 2-1**
Principles of the Cell Theory

- The cell is the smallest structural and functional unit capable of carrying out life processes.

- The functional activities of each cell depend on the specific structural properties of the cell.

- Cells are the living building blocks of all plant and animal organisms.

- An organism's structure and function ultimately depend on the individual and collective structural characteristics and functional capabilities of its cells.

- All new cells and new life arise only from pre-existing cells.

- Because of this continuity of life, the cells of all organisms are fundamentally similar in structure and function.

---

tains the cell's genetic material; and the *cytoplasm*, the portion of the cell's interior not occupied by the nucleus. These cell subdivisions are described in detail later in this and the following chapter. For now, we will provide an overview of each subdivision.

The plasma membrane bounds the cell.

The **plasma membrane**, or **cell membrane**, is a very thin membranous structure that encloses each cell. This oily barrier separates the cell's contents from its surroundings; it keeps the *intracellular fluid (ICF)* within the cells from mingling with the *extracellular fluid (ECF)* outside of the cells. The plasma membrane is not simply a mechanical barrier to hold in the contents of the cells; it has the ability to selectively control movement of molecules between the ICF and ECF. The plasma membrane can be likened to the gated walls that enclosed ancient cities. Through this structure, the cell can control the entry of food and other needed supplies and the export of products manufactured within, while at the same time guarding against unwanted traffic into or out of the cell. The plasma membrane is discussed thoroughly in Chapter 3.

The nucleus contains the DNA.

The two major parts of the cell's interior are the nucleus and the cytoplasm. The **nucleus,** which is typically the largest single organized cellular component, can be seen as a distinct spherical or oval structure, usually located near the center of the cell. It is surrounded by a double-layered membrane, the **nuclear envelope,** which separates the nucleus from the remainder of the cell. The nuclear envelope is pierced by numerous **nuclear pores** that allow necessary traffic to move between the nucleus and the cytoplasm.

The nucleus houses the cell's genetic material, **deoxyribonucleic acid (DNA),** which has two important functions: (1) directing protein synthesis and (2) serving as a genetic blueprint during cell replication. DNA provides codes, or "instructions," for directing synthesis of specific structural and enzymatic proteins within the cell. By directing the kinds and amounts of various enzymes and other proteins that are produced, the nucleus indirectly governs most cellular activities and serves as the cell's control center.

Three types of **ribonucleic acid (RNA)** play a role in this protein synthesis. First, DNA's genetic code for a particular protein is transcribed into a **messenger RNA** molecule, which exits the nucleus through the nuclear pores. Within the cytoplasm, messenger RNA delivers the coded message to **ribosomal RNA,** which "reads" the code and translates it into the appropriate amino acid sequence for the designated protein being synthesized. Finally, **transfer RNA** transfers the appropriate amino acids within the cytoplasm to their designated site in the protein under construction.

In addition to providing codes for protein synthesis, DNA also serves as a genetic blueprint during cell replication to ensure that the cell produces additional cells just like itself, thus continuing the identical type of cell line within the body. Furthermore, in the reproductive cells, the DNA blueprint serves to pass on genetic characteristics to future generations. (See Appendix C for further details of DNA and RNA function and protein synthesis.)

The cytoplasm consists of various organelles and the cytosol.

The **cytoplasm** is that portion of the cell interior not occupied by the nucleus. It contains a number of distinct, highly organized, membrane-enclosed structures—the *organelles* ("little organs")—dispersed within the *cytosol*, which is a complex, gel-like liquid.

On average, nearly half of the total cell volume is occupied by organelles. Each **organelle** is a separate compartment within the cell that is enclosed by a membrane similar to the plasma membrane. Thus the contents of an organelle are separated from the surrounding cytosol and from the contents of the other organelles. Nearly all cells contain six main types of organelles—the *endoplasmic reticulum, Golgi complex, lysosomes, peroxisomes, mitochondria,* and *vaults* (● Figure 2-1). These organelles are similar in all cells, although there are some variations depending on the specialized capabilities of each cell type. Organelles are like intracellular "specialty shops." Each is a separate internal compartment that contains a specific set of chemicals for carrying out a particular cellular function. This compartmentalization is advantageous because it permits chemical activities that would not be compatible with each other to occur simultaneously within the cell. For example, the enzymes that destroy unwanted proteins in the cell do so within the protective confines of the lysosomes without the risk of destroying essential cellular proteins. Just as organs each play a role that is essential for survival of the whole body, organelles each perform a specialized activity necessary for survival of the whole cell.

The remainder of the cytoplasm not occupied by organelles consists of the **cytosol** ("cell liquid"). The cytosol is made up of a semiliquid, gel-like mass laced with an elaborate protein network known as the *cytoskeleton.* Many of the chemical reactions that are compatible with each other are carried on in the cytosol. The cytoskeletal network gives the cell its shape, provides for its internal organization, and regulates its various movements. (For clarification, the intracellular fluid includes all the fluid inside the cell, including that within cytosol, the organelles, and the nucleus.)

In this chapter, we will examine each of the cytoplasmic components in more detail, concentrating first on the six types of the organelles.

ENDOPLASMIC RETICULUM

The **endoplasmic reticulum (ER)** is an elaborate fluid-filled membranous system distributed extensively throughout the cytosol. It is primarily a protein- and lipid-manufacturing factory. Two distinct types of endoplasmic reticulum—the smooth ER and the rough ER—can be distinguished. The **smooth ER** is a meshwork of tiny interconnected tubules, whereas the **rough ER** projects outward from the smooth ER as stacks of relatively

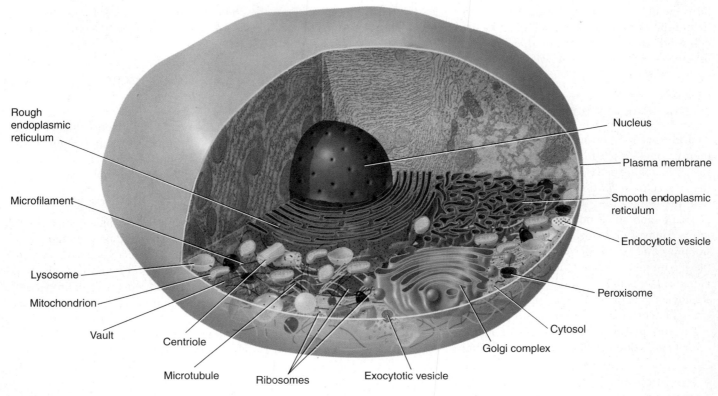

Rough endoplasmic reticulum

Microfilament

Lysosome

Mitochondrion

Vault

Centriole

Microtubule

Ribosomes

Exocytotic vesicle

Golgi complex

Cytosol

Peroxisome

Endocytotic vesicle

Smooth endoplasmic reticulum

Plasma membrane

Nucleus

● **FIGURE 2-1**

Schematic three-dimensional diagram of cell structures visible under an electron microscope

flattened sacs (● Figure 2-2). Even though these two regions differ considerably in appearance and function, they are continuous with each other. In other words, the ER is one continuous organelle with many interconnected channels. The relative amount of smooth and rough ER varies between cells, depending on the activity of the cell.

■ The rough endoplasmic reticulum synthesizes proteins for secretion and membrane construction.

The outer surface of the rough ER membrane is studded with small, dark-staining particles that give it a "rough" or granular appearance under a light microscope. These particles are **ribosomes,** which are ribosomal RNA-protein complexes that synthesize proteins under the direction of nuclear DNA. Messenger RNA carries the genetic message from the nucleus to the ribosome "workbench," where protein synthesis takes place (see p. A-25). Not all ribosomes in the cell are attached to the rough ER. Unattached or "free" ribosomes are dispersed throughout the cytosol.

The rough ER, in association with its ribosomes, synthesizes and releases a variety of new proteins into the ER lumen, the fluid-filled space enclosed by the ER membrane. These proteins serve one of two purposes: (1) Some proteins are destined for export to the cell's exterior as secretory products, such as protein hormones or enzymes. (All enzymes are proteins.)

(2) Other proteins are transported to sites within the cell for use in constructing new cellular membrane (either new plasma membrane or new organelle membrane) or other protein components of organelles. Cellular membranes consist predominantly of lipids (fats) and proteins. The membranous wall of the ER also contains enzymes essential for the synthesis of nearly all the lipids needed to produce new membranes. These newly synthesized lipids enter the ER lumen along with the proteins. Predictably, the rough ER is most abundant in cells specialized for protein secretion (for example, cells that secrete digestive enzymes), or in cells that require extensive membrane synthesis (for example, rapidly growing cells such as immature egg cells).

Within the ER lumen, a newly synthesized protein is folded into its final conformation and may also be modified in other ways, such as being pruned or having sugar molecules attached to it (● Figure 2-3). After this processing within the ER lumen, a new protein is unable to pass out through the ER membrane and therefore becomes permanently separated from the cytosol as soon as it has been synthesized. In contrast to the rough ER ribosomes, the free ribosomes synthesize proteins that are used intracellularly within the cytosol. In this way, newly produced molecules that are destined for export out of the cell or for synthesis of new cellular components (those synthesized by the ER) are physically separated from those that belong in the cytosol (those produced by the free ribosomes).

How do the newly synthesized molecules within the ER lumen get to their destinations at other sites inside the cell or to the outside of the cell, if they cannot pass out through the ER membrane? They do so through the action of the smooth endoplasmic reticulum.

▌ The smooth endoplasmic reticulum packages new proteins in transport vesicles.

The smooth ER does not contain ribosomes, so it is "smooth." Lacking ribosomes, it is not involved in protein synthesis. Instead, it serves a variety of other purposes that vary in different cell types.

In the majority of cells, the smooth ER is rather sparse and serves primarily as a central packaging and discharge site for molecules that are to be transported from the ER. Newly synthesized proteins and lipids pass from the rough ER to gather in the smooth ER. Portions of the smooth ER then "bud off" (that is, bulge on the surface, then are pinched off), forming **transport vesicles** that contain the new molecules enclosed in a spherical capsule of membrane derived from the smooth ER membrane (● Figure 2-4). (A **vesicle** is a fluid-filled, membrane-enclosed intracellular cargo that has been segregated from surrounding contents by means of the membrane forming a sphere around the captured cargo.) Newly synthesized membrane components are rapidly incorporated into the ER membrane itself to replace the membrane that was used to "wrap up" the molecules in the transport vesicle. Transport vesicles move to the Golgi complex, described in the next section, for further processing of their cargo.

In contrast to the sparseness of the smooth ER in most cells, some specialized types of cells have an extensive smooth ER, which has additional responsibilities as follows:

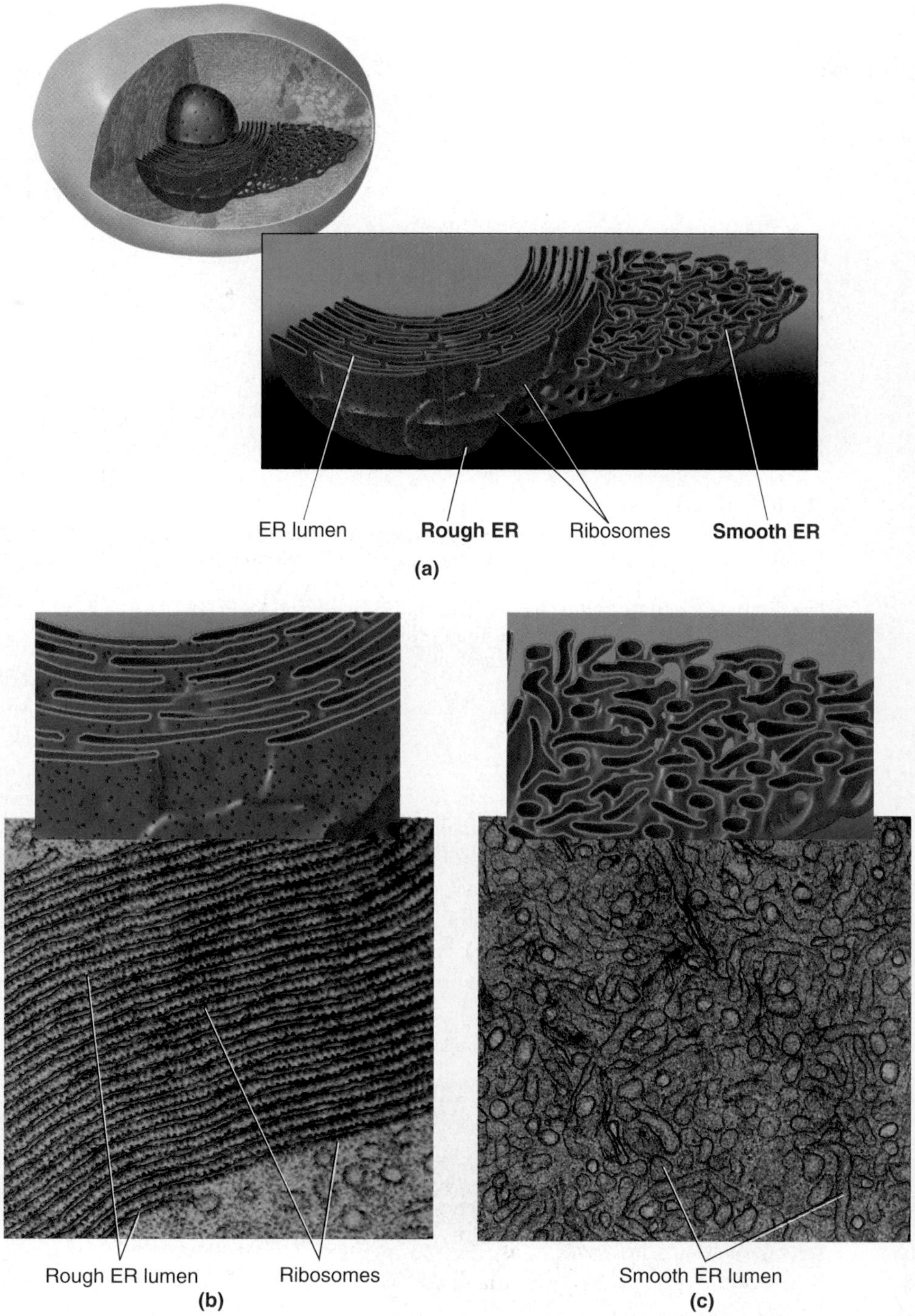

- The smooth ER is abundant in cells that specialize in lipid metabolism—for example, cells that secrete lipid-derived steroid hormones. The membranous wall of the smooth ER, like that of the rough ER, contains enzymes for synthesis of lipids. The lipid-producing enzymes in the membranous wall of the rough ER alone are insufficient to carry out the extensive lipid synthesis necessary to maintain adequate steroid-hormone secretion levels. These cells have an expanded smooth-ER compartment to house the additional enzymes necessary to keep pace with demands for hormone secretion.

- In liver cells, the smooth ER has a special capability. It contains enzymes that are involved in detoxifying harmful sub-

ER lumen   **Rough ER**   Ribosomes   **Smooth ER**
(a)

Rough ER lumen   Ribosomes
(b)

Smooth ER lumen
(c)

● **FIGURE 2-2**

Endoplasmic reticulum (ER)

(a) Schematic three-dimensional representation of the relationship between the smooth ER, which is a meshwork of tiny interconnected tubules, and the rough ER, which is studded with ribosomes and projects outward from the smooth ER as stacks of relatively flattened sacs. (b) Electron micrograph of the rough ER. Note the lay-ers of flattened sacs studded with small dark-staining ribosomes. (c) Electron micrograph of the smooth ER.

stances produced within the body by metabolism or substances that enter the body from the outside in the form of drugs or other foreign compounds. These detoxification enzymes alter toxic substances so that the latter can be eliminated more readily in the urine. The amount of smooth ER available in liver

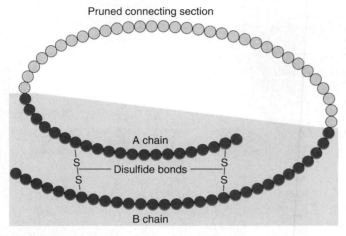

Pruned connecting section

A chain

Disulfide bonds

B chain

Each circle represents a specific amino acid.

● FIGURE 2-3

Formation of an insulin molecule within the endoplasmic reticulum
Insulin is originally formed as a single, long polypeptide chain that is folded back on itself, with the two overlapping ends being joined by two disulfide (sulfur-to-sulfur) bonds. The connecting piece (in gray) between the two overlapping ends is then pruned away, leaving the insulin molecule (in blue) consisting of an A chain and a B chain connected by the disulfide bonds.

cells for the task of detoxification can vary dramatically, depending on the need. For example, if phenobarbital, a sedative drug, is taken in large quantities, the amount of smooth ER in liver cells doubles within a few days, only to return to normal within five days after drug administration ceases.

● Muscle cells have an elaborate, modified smooth ER known as the *sarcoplasmic reticulum*, which stores calcium that plays an important role in the process of muscle contraction (see p. 263).

GOLGI COMPLEX

Closely associated with the endoplasmic reticulum is the **Golgi complex.** Each Golgi complex consists of a stack of flattened, slightly curved, membrane-enclosed sacs, or cisternae (● Figure 2-5). The sacs within each Golgi stack are not physically connected to each other. Note that the flattened sacs are thin in the middle but have dilated, or bulging, edges. The number of Golgi complexes varies, depending on the cell type. Some cells have only one Golgi stack, whereas cells highly specialized for protein secretion may have hundreds of stacks.

▌ Transport vesicles carry their cargo to the Golgi complex for further processing.

The majority of the newly synthesized molecules that have just budded off from the smooth ER enter a Golgi stack. When a transport vesicle carrying its newly synthesized cargo reaches a Golgi stack, the vesicle membrane fuses with the membrane of the sac closest to the center of the cell. The vesicle membrane opens up and becomes integrated into the Golgi membrane, and the contents of the vesicle are released to the interior of the sac (● Figure 2-4).

These newly synthesized raw materials from the ER travel by means of vesicle formation through the layers of the Golgi

● FIGURE 2-4

Overview of the secretion process for proteins synthesized by the endoplasmic reticulum

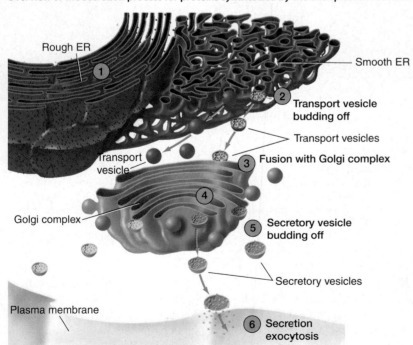

Rough ER

Smooth ER

Transport vesicle budding off

Transport vesicles

Transport vesicle

Fusion with Golgi complex

Golgi complex

Secretory vesicle budding off

Secretory vesicles

Plasma membrane

Secretion exocytosis

1. The rough ER synthesizes proteins to be secreted to the exterior or to be incorporated into the cellular membrane.

2. The smooth ER packages the secretory product into transport vesicles, which bud off and move to the Golgi complex.

3. The transport vesicle fuses with the Golgi complex, opens up, and empties its contents into the closest Golgi sac.

4. As the newly synthesized proteins from the ER travel by vesicular transport through the layers of the Golgi complex, this complex modifies the raw proteins into final form and sorts and directs the finished products to their final destination by varying their wrappers.

5. Secretory vesicles containing the finished protein product bud off the Golgi complex and remain in the cytosol, storing the product until signaled to empty.

6. On appropriate stimulation, the secretory vesicles fuse with the plasma membrane, open, and empty their contents to the cell's exterior. Secretion has occurred by exocytosis, with the secretory product never having come into contact with the cytosol.

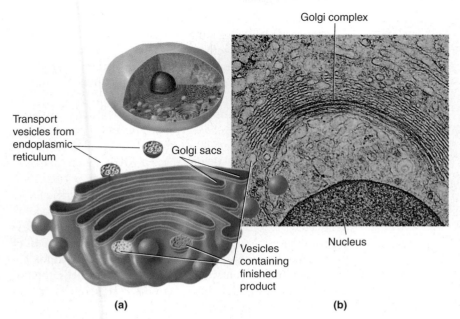

Transport
vesicles from
endoplasmic
reticulum

Golgi sacs

Golgi complex

Nucleus

Vesicles
containing
finished
product

(a)

(b)

● FIGURE 2-5

Golgi complex

(a) Schematic three-dimensional diagram of a Golgi complex. (b) Electron micrograph of a Golgi complex. The vesicles at the dilated edges of the sacs contain finished protein products packaged for distribution to their final destination.

stack, from the innermost sac closest to the ER to the outermost sac near the plasma membrane. During this transit, two important, interrelated functions take place:

1. *Processing the raw materials into finished products.* Within the Golgi complex, the "raw" proteins from the ER are modified into their final form, largely by adjustments made in the sugars attached to the protein. The biochemical pathways that the proteins undergo during their passage through the Golgi complex are elaborate, precisely programmed, and specific for each final product.

2. *Sorting and directing the finished products to their final destinations.* The Golgi complex is responsible for sorting and segregating different types of products according to their function and destination, namely products (1) to be secreted to the cell's exterior, (2) to be used for construction of new plasma membrane, or (3) to be incorporated into other organelles, especially lysosomes.

∎ The Golgi complex packages secretory vesicles for release by exocytosis.

How does the Golgi complex sort and direct finished proteins to the proper destinations inside and outside the cell? Finished products are collected within the dilated edges of the Golgi complex's sacs. The dilated edge of the outermost sac then pinches off to form a membrane-enclosed vesicle that contains the finished product. For each type of product to reach its appropriate site of function, each distinct type of vesicle takes up a specific product before budding off. This vesicle

with its selected cargo then fuses only with the membrane of a particular destination site. Vesicles destined for different sites are wrapped in membranes containing different surface protein molecules that serve as specific "docking markers," which ensure that the vesicles "dock" and "unload" their cargo only at the appropriate "address," or destination within the cell.

Let's examine how the Golgi complex of specialized secretory cells sorts and packages proteins destined for secretion from the cell as an example of how this organelle routes its molecular traffic. Specialized secretory cells include endocrine cells that secrete protein hormones and digestive gland cells that secrete digestive enzymes. Secretory cells package their products destined for export out of the cell in **secretory vesicles.** Secretory vesicles are about 200 times larger than transport vesicles. When pinched off from the Golgi complex, the secretory proteins trapped within the vesicle are in a dilute solution, but the contents become concentrated over time as the secretory vesicle gradually loses water. As a result, the secretory product is concentrated about 200-fold, enabling the cell to secrete large amounts of this product (for example, a specific hormone) on demand as needed. The concentrated secretory proteins remain stored within the secretory vesicles until the cell is stimulated by a specific signal that indicates a need for release of that particular secretory product. On appropriate stimulation, the vesicles move to the cell's periphery. Vesicular contents are quickly released to the cell's exterior as the vesicle fuses with the plasma membrane, opens, and empties its contents to the outside (● Figures 2-4 and 2-6). This mechanism—extrusion to the exterior of substances originating within the cell—is referred to as **exocytosis** (*exo* means "out of"; *cyto* means "cell"). Release of the contents of a secretory vesicle by means of exocytosis constitutes the process of secretion. Secretory vesicles fuse only with the plasma membrane and not with any of the internal membranes that bound organelles, thereby preventing fruitless or even dangerous discharge of secretory products into the organelles.

We will now look in more detail at how secretory vesicles take up specific products to be released into the ECF and then dock only at the plasma membrane. Before budding off from the outermost Golgi sac, the portion of the Golgi membrane that will be used to enclose the secretory vesicle becomes "coated" with a layer of specific proteins from the cytosol (● Figure 2-7). These and associated membrane proteins serve three important functions:

• First, specific proteins on the interior surface of the membrane facing the Golgi lumen act as *recognition markers* for the recognition and attraction of specific molecules that

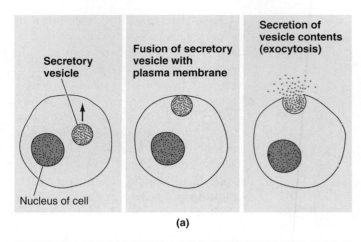

Secretory vesicle

Nucleus of cell

Fusion of secretory vesicle with plasma membrane

Secretion of vesicle contents (exocytosis)

(a)

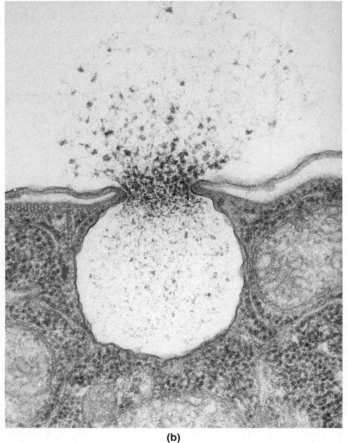

(b)

● FIGURE 2-6

Exocytosis of secretory product
(a) Schematic overview of the process of exocytosis. (b) Transmission electron micrograph of exocytosis.

have been processed in the Golgi lumen. The newly finished proteins destined for secretion contain a unique sequence of amino acids known as a *sorting signal.* Recognition of the right protein's sorting signal by the complementary membrane marker ensures that the proper cargo is captured and packaged as a secretory vesicle is forming and budding off of the outermost Golgi sac.

- Second, *coat proteins* from the cytosol bind with another specific protein facing the outer surface of the membrane. The linking together of these coat proteins causes the surface membrane of the Golgi sac to curve and form a dome-shaped bud around the captured cargo. Eventually the surface membrane closes and pinches off the vesicle.
- After budding off, the vesicle sheds its coat proteins. This uncoating exposes *docking markers,* which are other specific proteins facing the outer surface of the vesicle membrane. These docking markers, known as *v-SNAREs,* can link lock-and-key fashion with another protein marker, a *t-SNARE,* found only on the targeted membrane. In the case of secretory vesicles, the targeted membrane is the plasma membrane, the designated site for secretion to take place. Thus the v-SNAREs of secretory vesicles are able to fuse only with the t-SNAREs of the plasma membrane. Once a vesicle has docked at the appropriate membrane by means of matching SNAREs, the two membranes completely fuse, then the vesicle opens up and empties its contents at the targeted site.

Note that the contents of secretory vesicles never come into contact with the cytosol. From the time these products are first synthesized in the ER until they are released from the cell by exocytosis, they are always wrapped in membrane and thus isolated from the remainder of the cell. By manufacturing its particular secretory protein ahead of time and storing this product in secretory vesicles, a secretory cell has a readily available reserve from which to secrete large amounts of this product on demand. If a secretory cell had to synthesize all its product on the spot as needed for export, the cell would be more limited in its ability to meet varying levels of demand.

Secretory vesicles are formed only by secretory cells. In a similar way, however, the Golgi complex of these and other cell types sorts and packages newly synthesized products for different destinations within the cell. In each case, a particular vesicle captures a specific kind of cargo from among the many proteins in the Golgi lumen, then addresses each shipping container for a distinct destination.

LYSOSOMES

On the average, a cell contains about 300 lysosomes.

▌ Lysosomes serve as the intracellular digestive system.

Lysosomes are membrane-enclosed sacs containing powerful **hydrolytic enzymes,** which catalyze *hydrolysis* reactions (Appendix B, p. A-17). These reactions break down the organic molecules that make up cellular debris and foreign material, such as bacteria that have been brought into the cell. (*Lys* means "breakdown"; *some* means "body." Lysosomes are small bodies within the cell that break down organic molecules with which they come into contact.) The lysosomal enzymes are similar to the hydrolytic enzymes that the digestive system secretes to digest

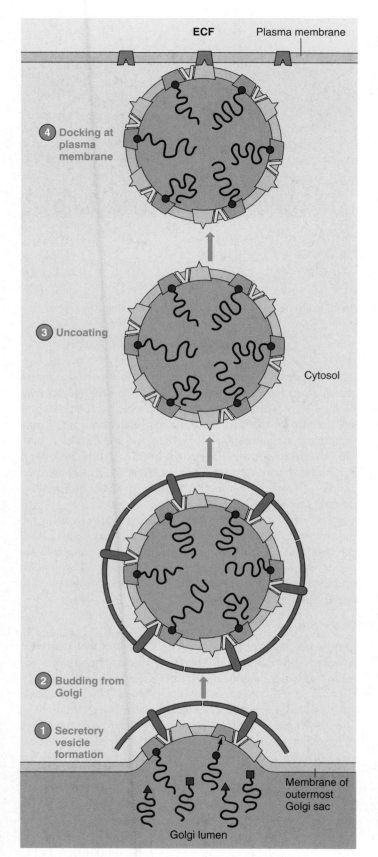

ECF   Plasma membrane

4 Docking at plasma membrane

Cytosol

3 Uncoating

2 Budding from Golgi

1 Secretory vesicle formation

Membrane of outermost Golgi sac

Golgi lumen

1 During secretory vesicle formation, recognition markers in the membrane of the outermost Golgi sac capture the appropriate cargo from the Golgi lumen by binding lock-and-key fashion with the sorting signals of the designated protein molecules to be secreted. The membrane that will wrap the vesicle is coated with a molecule that causes the membrane to curve, forming a dome-shaped bud.

2 The membrane closes beneath the bud, pinching off the secretory vesicle.

3 The vesicle loses its coating, exposing v-SNARE docking markers on the vesicle surface of the membrane.

4 The v-SNAREs bind lock-and-key fashion only with the t-SNARE docking-marker acceptors of the targeted plasma membrane. This specificity ensures that secretory vesicles fuse only with the surface membrane of the cell and empty their contents to the cell's exterior.

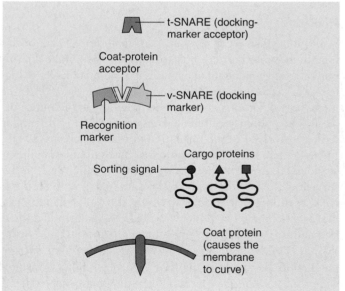

t-SNARE (docking-marker acceptor)

Coat-protein acceptor

v-SNARE (docking marker)

Recognition marker

Sorting signal   Cargo proteins

Coat protein (causes the membrane to curve)

● **FIGURE 2-7**

Secretory vesicle formation and fusion with the plasma membrane

food. Thus lysosomes serve as the intracellular "digestive system."

Instead of having a uniform structure as is characteristic of all other organelles, lysosomes vary in size and shape, depending on the contents they are digesting. Most commonly, lysosomes are small (0.2–0.5 μm in diameter) oval or spherical bodies that appear granular during inactivity (● Figure 2-8). These granules are protein aggregates of the powerful digestive enzymes contained within. The surrounding membrane that confines these enzymes normally prevents them from destroying the cell that houses them. Both the membrane and the enzymes are derived from the Golgi complex. A specific collection of newly synthesized hydrolytic enzymes is captured in a coated vesicle that is pinched off from the Golgi complex to become a new lysosome.

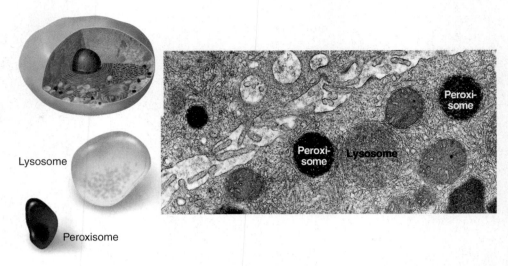

● FIGURE 2-8

Lysosomes and peroxisomes
Electron micrograph showing both lysosomes, which contain hydrolytic enzymes, and peroxisomes, which contain oxidative enzymes.

▌ Extracellular material is brought into the cell by endocytosis for attack by lysosomal enzymes.

Extracellular material to be attacked by lysosomal enzymes is brought inside the cell through the process of **endocytosis** (*endo* means "within"). Endocytosis can be accomplished in three ways—*pinocytosis, receptor-mediated endocytosis,* and *phagocytosis*—depending on the contents of the internalized material.

Pinocytosis

With **pinocytosis** ("cell drinking"), a small droplet of extracellular fluid is internalized. First, the plasma membrane dips inward, forming a pouch that contains a small bit of ECF (● Figure 2-9a). The endocytotic pouch is formed as a result of membrane-deforming coat proteins attaching to the inner surface of the plasma membrane. These coat proteins are similar to those involved in the formation of secretory vesicles. The linking together of the coat proteins causes the plasma membrane to dip inward. The plasma membrane then seals at the surface of the pouch, trapping the contents in a small, intracellular **endocytotic vesicle.** Recently, the molecule responsible for pinching off an endocytotic vesicle has been identified as **dynamin,** which forms rings that wrap around and "wring the neck" of the pouch, severing the vesicle from the surface membrane. Besides bringing ECF into a cell, pinocytosis provides a means to retrieve extra plasma membrane that has been added to the cell surface during exocytosis.

Receptor-mediated endocytosis

Unlike pinocytosis, which involves the nonselective uptake of the surrounding fluid, **receptor-mediated endocytosis** is a highly selective process that enables cells to import specific large molecules that it needs from its environment. Receptor-mediated endocytosis is triggered by the binding of a specific molecule such as a protein to a surface membrane receptor site specific for that protein. This binding causes the plasma membrane at that site to sink in, then seal at the surface, trapping the protein inside the cell (● Figure 2-9b). Cholesterol complexes, vitamin B_{12}, the hormone insulin, and iron are examples of substances selectively taken into cells by receptor-mediated endocytosis.

Unfortunately, some viruses can sneak into cells by exploiting this mechanism. For instance, flu viruses and HIV, the virus that causes AIDS (see p. 436), gain entry to cells via receptor-mediated endocytosis. They do so by binding with membrane receptor sites normally designed to trigger the internalization of a needed molecule.

Phagocytosis

During **phagocytosis** ("cell eating"), large multimolecular particles are internalized. Most body cells perform pinocytosis, many carry out receptor-mediated endocytosis, but only a few specialized cells are capable of phagocytosis. The latter are the "professional" phagocytes, the most notable being certain types of white blood cells that play an important role in the body's defense mechanisms. When a white blood cell encounters a large multimolecular particle, such as a bacterium or tissue debris, it extends surface projections known as **pseudopods** ("false feet") that completely surround or engulf the particle and trap it within an internalized vesicle (● Figure 2-9c and d). A lysosome fuses with the membrane of the internalized vesicle and releases its hydrolytic enzymes into the vesicle, where they safely attack the bacterium or other trapped material without damaging the remainder of the cell. The enzymes largely break down the engulfed material into reusable raw ingredients, such as amino acids, glucose, and fatty acids, that the cell can use. These small products can readily pass through the lysosomal membrane to enter the cytosol.

▌ Lysosomes remove useless but not useful parts of the cell.

Lysosomes can also fuse with aged or damaged organelles to remove these useless parts of the cell. This selective self-digestion, known as **autophagy** (*auto* means "self;" *phag* means "eating")

makes way for new replacement parts. All organelles are renewable. An inherent danger is that lysosomal membranes may inadvertently rupture. Because lysosomes have the potential ability to self-destruct the cell, their discoverer named them "suicide bags." However, two factors make this threat less severe than imagined. First, the lysosomal hydrolytic enzymes function best in an acid environment. The lysosomal mem-

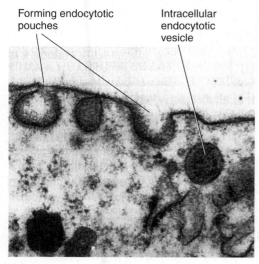

Forming endocytotic pouches

Intracellular endocytotic vesicle

(a) Pinocytosis

● **FIGURE 2-9**

Forms of endocytosis

(a) Electron micrograph of pinocytosis. The surface membrane dips inward to form a pouch, then seals the surface, forming an intracellular vesicle that nonselectively internalizes a bit of ECF. (b) Receptor-mediated endocytosis. When a large molecule such as a protein attaches to a specific surface receptor site, the membrane dips inward and pinches off to selectively internalize the molecule in an intracellular vesicle. (c) Phagocytosis. White blood cells internalize multimolecular particles such as bacteria by extending surface projections known as *pseudopods* that wrap around and seal in the targeted material. A lysosome fuses with the internalized vesicle, releasing enzymes that attack the engulfed material within the confines of the vesicle. (d) Scanning-electron-micrograph series of a white blood cell phagocytizing an old, worn-out red blood cell.

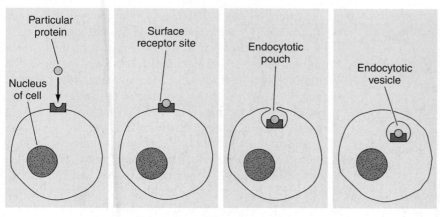

(b) Receptor-mediated endocytosis

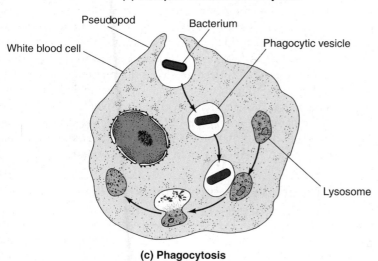

(c) Phagocytosis

Worn-out red blood cell   White blood cell

(d)

brane transports hydrogen ions (acid formers) into the lysosome, making it considerably more acidic than the remainder of the cell. Should the enzymes accidentally leak into the cytosol, they would be less potent than they are in their acidic home. Second, in most instances cells could tolerate the limited damage that would occur if only one or two lysosomes inadvertently ruptured, because most parts of the cell are renewable. The biggest danger is accidental digestion of part of the irreplaceable DNA molecule within the nucleus. Such nuclear damage would alter the cell's genetic properties, and the defect would be perpetuated to all of the cell's progeny.

Some individuals lack the ability to synthesize one or more of the lysosomal enzymes. The result is massive accumulation within the lysosomes of the compound that is normally digested by the missing enzyme. Clinical manifestations often accompany such disorders because the engorged lysosomes interfere with normal cell activity. The nature and severity of the symptoms depend on the type of substance that is accumulating, which in turn depends on what lysosomal enzyme is missing. Among these so-called *storage diseases* is **Tay-Sachs disease.** It is characterized by abnormal accumulation of gangliosides, which are complex molecules found in nerve cells. Profound symptoms of progressive nervous system degeneration result as the accumulation continues.

PEROXISOMES

Typically, several hundred small peroxisomes about one-third to one-half the average size of lysosomes are present in a cell (● Figure 2-8).

▌ Peroxisomes house oxidative enzymes that detoxify various wastes.

Peroxisomes are similar to lysosomes in that they are membrane-enclosed sacs containing enzymes, but unlike the lysosomes, which contain hydrolytic enzymes, peroxisomes house several powerful *oxidative enzymes* and contain most of the cell's *catalase*. (*Peroxi* refers to "hydrogen peroxide;" peroxisomes are intracellular bodies that produce and decompose hydrogen peroxide, as you will learn shortly.)

Oxidative enzymes, as the name implies, use oxygen (O_2), in this case to strip hydrogen from certain organic molecules. This oxidation helps detoxify various wastes produced within the cell or foreign toxic compounds that have entered the cell, such as alcohol consumed in beverages. The major product generated in the peroxisome is *hydrogen peroxide* (H_2O_2), which is formed by molecular oxygen and the hydrogen atoms stripped from the toxic molecule.

Hydrogen peroxide, itself a powerful oxidant, is potentially destructive if it is allowed to accumulate or escape from the confines of the peroxisome. However, peroxisomes also contain an abundance of **catalase,** an antioxidant enzyme that decomposes potent H_2O_2 into harmless H_2O and O_2. This latter reaction is an important safety mechanism that destroys the potentially deadly peroxide at the site of its production, thereby preventing its possible devastating escape into the cytosol.

MITOCHONDRIA

A single cell may contain as few as a hundred or as many as several thousand mitochondria.

▌ Mitochondria, the energy organelles, are enclosed by a double membrane.

Mitochondria are the energy organelles, or "power plants," of the cell; they extract energy from the nutrients in food and transform it into a usable form for cellular activities. Mitochondria generate about 90% of the energy that cells—and, accordingly, the whole body—need to survive and function. The number of mitochondria per cell varies greatly, depending on the energy needs of each particular cell type. In some cell types, the mitochondria are densely compacted in cellular regions that use most of the cell's energy. For example, mitochondria are packed between the contractile units in the muscle cells of the heart.

Mitochondria are rod-shaped or oval structures about the size of bacteria. In fact, considerable evidence indicates that mitochondria are descendants of bacteria that invaded or were engulfed by primitive cells early in evolutionary history and that subsequently became permanent organelles. As part of their separate heritage, mitochondria possess their own DNA, distinct from the DNA housed in the cell's nucleus. Mitochondrial DNA contains the genetic codes for producing many of the molecules the mitochondria need to generate energy. Current research has shown that flaws gradually accumulate in mitochondrial DNA over a person's lifetime; these flaws have been implicated in aging as well as in an array of disorders. Prominent among the mitochondrial diseases are those that become debilitating in later life, such as some forms of chronic degenerative nervous-system and muscle diseases.

Each mitochondrion is enclosed by a double membrane—a smooth outer membrane that surrounds the mitochondrion itself, and an inner membrane that forms a series of infoldings or shelves called **cristae,** which project into an inner cavity filled with a gel-like solution known as the **matrix** (● Figure 2-10). These cristae contain crucial proteins (the electron transport proteins, to be described shortly) that ultimately are responsible for converting much of the energy in food into a usable form. The generous folds of the inner membrane greatly increase the surface area available for housing these important proteins. (Remember that this strategy is a common structure–function relationship: Increasing the surface area of various body structures by means of surface projections or infoldings makes more surface area available to participate in the structure's functions.) The matrix consists of a concentrated mixture of hundreds of different dissolved enzymes (the citric acid cycle enzymes, soon to be described) that prepare nutrient molecules for the final extraction of usable energy by the cristae proteins.

▌ Mitochondria play a major role in generating ATP.

The source of energy for the body is the chemical energy stored in the carbon bonds of ingested food. Body cells are not equipped

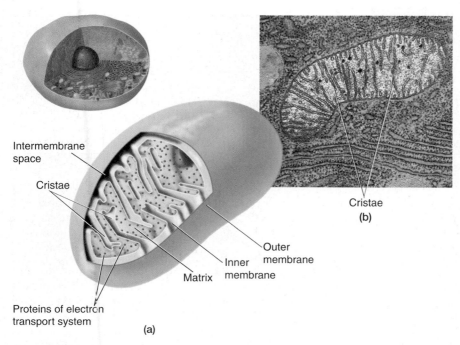

Intermembrane space

Cristae

Proteins of electron transport system

Outer membrane

Inner membrane

Matrix

(a)

Cristae

(b)

● FIGURE 2-10

Mitochondrion

(a) Schematic representation of a mitochondrion. The electron transport proteins embedded in the cristae folds of the mitochondrial inner membrane are ultimately responsible for converting much of the energy of food into a usable form. (b) Electron micrograph of a mitochondrion.

to use this energy directly, however. Instead, they must extract energy from food nutrients and convert it into an energy form that they can use—namely, the high-energy phosphate bonds of **adenosine triphosphate (ATP)**, which consists of adenosine with three phosphate groups attached (*tri* means "three") (see p. A-17). When a high-energy bond such as that binding the terminal phosphate to adenosine is split, a substantial amount of energy is released. Adenosine triphosphate is the universal energy carrier—the common energy "currency" of the body. Cells can "cash in" ATP to pay the energy "price" for running the cellular machinery. To obtain immediate usable energy, cells split the terminal phosphate bond of ATP, which yields **adenosine diphosphate (ADP)**—adenosine with two phosphate groups attached (*di* means "two")—plus inorganic phosphate (P_i) plus energy:

$$ATP \quad \xrightarrow{\text{splitting}} \quad ADP + P_i + \text{energy for use by the cell}$$

In this energy scheme, food can be thought of as the "crude fuel," whereas ATP is the "refined fuel" for operating the body's machinery. Let us elaborate on this fuel conversion process. Dietary food is digested, or broken down, by the digestive system into smaller absorbable units that can be transferred from the digestive tract lumen into the blood (Chapter 16). For example, dietary carbohydrates are broken down primarily into glucose, which is absorbed into the blood. No usable energy is released during the digestion of food. When delivered to the cells by the blood,

the nutrient molecules are transported across the plasma membrane into the cytosol. (Details of how materials cross the membrane are covered in the next chapter.)

We are now going to turn our attention to the steps involved in ATP production within the cell and the role of the mitochondria in these steps. ATP is generated from the sequential dismantling of absorbed nutrient molecules in three different steps: *glycolysis*, the *citric acid cycle*, and the *electron transport chain*. We will use glucose as an example to describe these steps (▲ Table 2-2).

Glycolysis

Among the thousands of enzymes within the cytosol are those responsible for **glycolysis**, a chemical process involving 10 separate sequential reactions that break down the simple six-carbon sugar molecule, glucose, into two pyruvic acid molecules, each of which contains three carbons (*glyc-* means "sweet;" *lysis* means "breakdown"). During glycolysis, some of the energy from the broken chemical bonds of glucose is used to convert ADP into ATP (● Figure 2-11). However, glycolysis is not very efficient in terms of energy extraction; the net yield is only two molecules of ATP per glucose molecule processed. Much of the energy originally contained in the glucose molecule is still locked in the chemical bonds of the pyruvic acid molecules. The low-energy yield of glycolysis is insufficient to support the body's demand for ATP. This is where the mitochondria come into play.

Citric acid cycle

The pyruvic acid produced by glycolysis in the cytosol can be selectively transported into the mitochondrial matrix. Here it is further broken down into a two-carbon molecule, acetic acid, by enzymatic removal of one of the carbons in the form of carbon dioxide (CO_2), which eventually is eliminated from the body as an end product, or waste (● Figure 2-12). During this breakdown process, a carbon–hydrogen bond is disrupted,

● FIGURE 2-11

A simplified summary of glycolysis

Glycolysis involves the breakdown of glucose into two pyruvic acid molecules, with a net yield of two molecules of ATP for every glucose molecule processed.

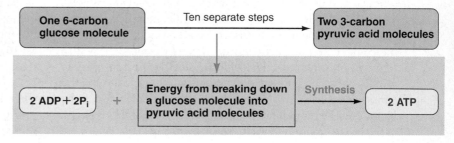

One 6-carbon glucose molecule → Ten separate steps → Two 3-carbon pyruvic acid molecules

2 ADP + 2P_i + Energy from breaking down a glucose molecule into pyruvic acid molecules → Synthesis → 2 ATP

Overview of Cellular Energy Production from Glucose

Reaction	Substance Processed	Location	Energy Yield (per glucose molecule processed)	End Products Available for Further Energy Extraction	Need for Oxygen
Glycolysis	Glucose	Cytosol	2 molecules of ATP	2 pyruvic acid molecules	No; anaerobic
Citric acid cycle	Acetyl CoA, which is derived from pyruvic acid, the end product of glycolysis; 2 acetyl CoA molecules result from the processing of 1 glucose molecule	Mitochondrial matrix	2 molecules of ATP	8 NADH and 2 FADH$_2$ hydrogen carrier molecules	Yes, derived from molecules involved in citric acid cycle reactions
Electron transport chain	High-energy electrons stored in hydrogen atoms in the hydrogen carrier molecules NADH and FADH$_2$ derived from citric acid cycle reactions	Mitochondrial inner-membrane cristae	32 molecules of ATP	None	Yes, derived from molecular oxygen acquired from breathing

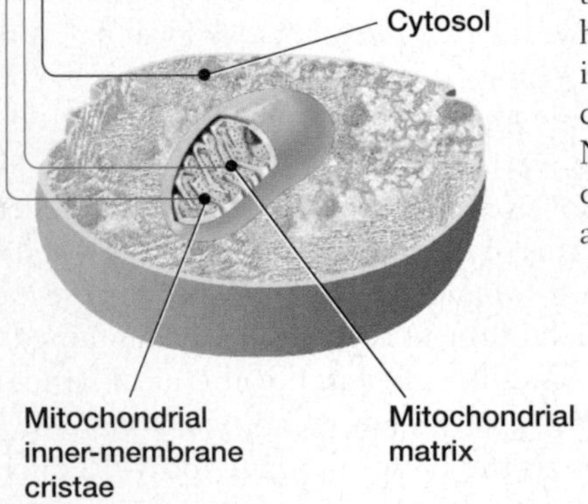

Cytosol

Mitochondrial inner-membrane cristae

Mitochondrial matrix

The mitochondrion is exaggerated and other intracellular components are omitted for better visualization.

so a hydrogen atom is also released. This hydrogen atom is held by a hydrogen carrier molecule, the function of which will be discussed shortly. The acetic acid thus formed combines with coenzyme A, a derivative of pantothenic acid (a B vitamin), producing the compound acetyl coenzyme A (acetyl CoA).

Acetyl CoA then enters the **citric acid cycle,** which consists of a cyclical series of eight separate biochemical reactions that are directed by the enzymes of the mitochondrial matrix. This cycle of reactions can be compared to one revolution around a Ferris wheel. (Keep in mind that ● Figure 2-12 is highly schematic. It depicts a cyclical series of biochemical reactions. The molecules themselves are not physically moved around in a cycle.) On the top of the Ferris wheel, acetyl CoA, a two-carbon molecule, enters a seat already occupied by oxaloacetic acid, a four-carbon molecule. These two molecules link together to form a six-carbon citric acid molecule, and the trip around the citric acid cycle begins. (This cycle is alternatively known as the **Krebs cycle** in honor of its principal discoverer, Sir Hans Krebs, or the **tricarboxylic acid cycle,** because citric acid contains three

carboxylic acid groups.) As the seat moves around the cycle, at each new position, matrix enzymes modify the passenger molecule to form a slightly different molecule. These molecular alterations have the following important consequences:

1. Two carbons are sequentially "kicked off the ride" as they are removed from the six-carbon citric acid molecule, converting it back into the four-carbon oxaloacetic acid, which is now available at the top of the cycle to pick up another acetyl CoA for another revolution through the cycle.

2. The released carbon atoms, which were originally present in the acetyl CoA that entered the cycle, are converted into two molecules of CO_2. This CO_2, as well as the CO_2 produced during the formation of acetic acid from pyruvic acid, passes out of the mitochondrial matrix and subsequently out of the cell to enter the blood. In turn, the blood carries it to the lungs, where it is finally eliminated into the atmosphere through the process of breathing. The oxygen used to make CO_2 from these released carbon atoms is derived from the molecules that were

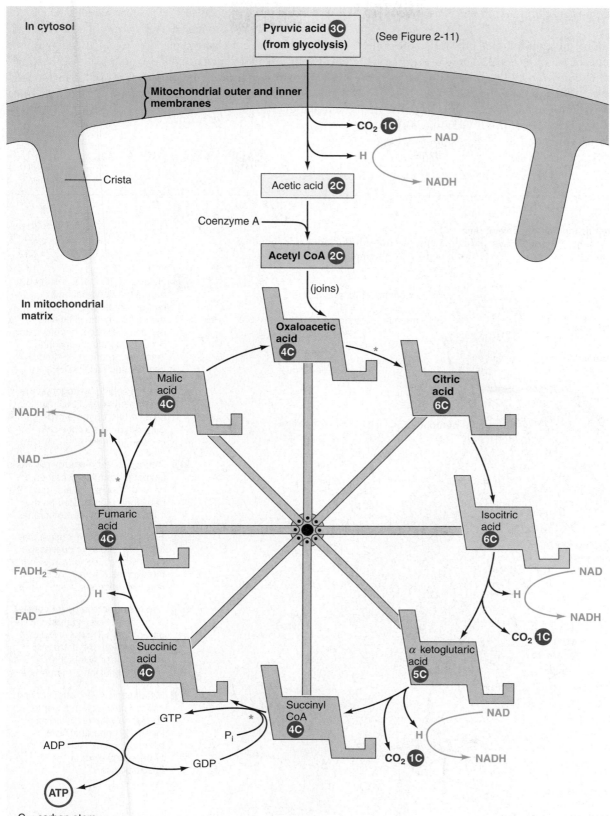

In cytosol

Pyruvic acid ③C
(from glycolysis)

(See Figure 2-11)

Mitochondrial outer and inner membranes

CO_2 ①C

NAD

H

NADH

Crista

Acetic acid ②C

Coenzyme A

Acetyl CoA ②C

(joins)

In mitochondrial matrix

Oxaloacetic acid ④C

Malic acid ④C

Citric acid ⑥C

NADH

H

NAD

Fumaric acid ④C

Isocitric acid ⑥C

NAD

H

NADH

FADH₂

FAD

H

CO_2 ①C

Succinic acid ④C

α ketoglutaric acid ⑤C

GTP

NAD

Succinyl CoA ④C

Pᵢ

H

ADP

GDP

CO_2 ①C

NADH

ATP

C = carbon atom.
* H₂O enters the cycle at the steps marked with an asterisk.

● FIGURE 2-12

Citric acid cycle
A simplified version of the citric acid cycle, showing how the two carbons entering the cycle by means of acetyl CoA are eventually converted to CO_2, with oxaloacetic acid, which accepts acetyl CoA, being regenerated at the end of the cyclical pathway. Also denoted is the release of hydrogen atoms at specific points along the pathway. These hydrogens bind to the hydrogen carrier molecules NAD and FAD for further processing. One molecule of ATP is generated for each molecule of acetyl CoA that enters the citric acid cycle, for a total of two molecules of ATP for each molecule of processed glucose.

involved in the reactions, not from free molecular oxygen supplied by breathing.

3. Hydrogen atoms are also "bumped off" during the cycle at four of the chemical conversion steps. These hydrogens are "caught" by two other compounds that act as hydrogen carrier molecules—**nicotinamide adenine dinucleotide (NAD)**, a derivative of the B vitamin niacin, and **flavine adenine dinucleotide (FAD)**, a derivative of the B vitamin riboflavin. These compounds are converted by the transfer of hydrogen to NADH and FADH$_2$, respectively.

4. One more molecule of ATP is produced for each molecule of acetyl CoA processed. Actually, ATP is not directly produced by the citric acid cycle. The released energy is used to directly link inorganic phosphate to **guanosine diphosphate (GDP)** to form **guanosine triphosphate (GTP)**, a high-energy molecule similar to ATP. The energy from GTP is then transferred to ATP as follows:

$$ADP + GTP = ATP + GDP$$

● **FIGURE 2-13**

ATP synthesis by the mitochondrial inner membrane
(a) ATP synthesis resulting from the passage of high-energy electrons through the mitochondrial electron transport chain. (b) Activation of ATP synthase by movement of H$^+$.

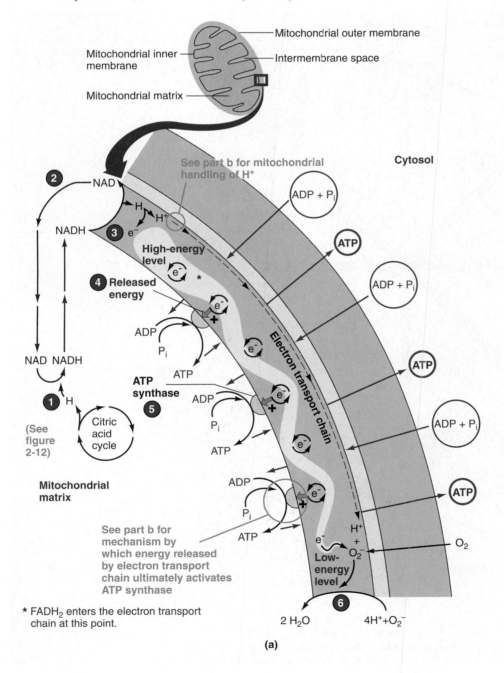

1. Hydrogen (H) that is released during the degradation of carbon-containing nutrient molecules by the citric acid cycle in the mitochondrial matrix is carried to the mitochondrial inner membrane by hydrogen carriers such as NADH.

2. After releasing hydrogen at the inner membrane, NAD shuttles back to pick up more hydrogen generated by the citric acid cycle in the matrix.

3. Meanwhile, high-energy electrons extracted from the hydrogen are passed through the electron transport chain located on the mitochondrial inner membrane.

4. Energy is gradually released as the electrons fall to successively lower energy levels by moving through the electron transport chain of reactions.

5. The released energy triggers a sequence of steps (shown in part b) that ultimately results in the activation of the enzyme ATP synthase within the mitochondrial inner membrane.

6. Molecular oxygen, after serving as the final electron acceptor, combines with the hydrogen ions (H$^+$) generated from hydrogen on extraction of high-energy electrons to produce water.

(a)

Because each glucose molecule is converted into two acetic acid molecules, thus permitting two turns of the citric acid cycle, two more ATP molecules are produced from each glucose molecule.

So far, the cell still does not have much of an energy profit. However, the citric acid cycle is important in preparing the hydrogen carrier molecules for their entry into the next step, the electron transport chain, which produces far more energy than the sparse amount of ATP produced by the cycle itself.

Electron transport chain

Considerable untapped energy is still stored in the released hydrogen atoms, which contain electrons at high energy levels. The "big payoff" comes when NADH and $FADH_2$ enter the **electron transport chain,** which consists of electron carrier molecules located in the inner mitochondrial membrane lining the cristae (● Figure 2-13a). The high-energy electrons are extracted from the hydrogens held in NADH and $FADH_2$ and are transferred through a series of steps from one electron-carrier molecule to another, within the cristae membrane,

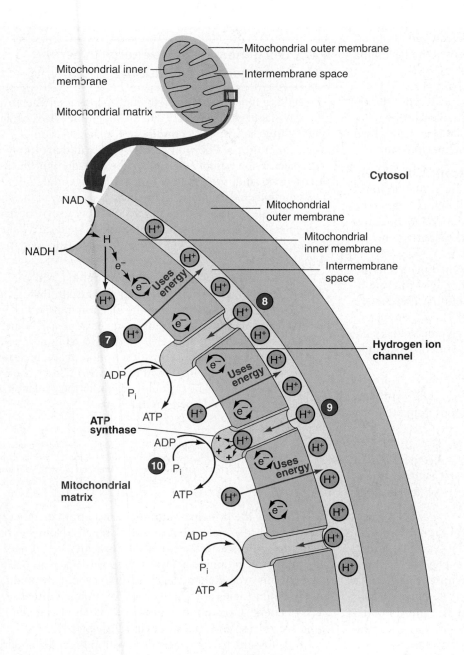

7 Energy released during the transfer of electrons by the electron transport chain is used to transport hydrogen ions from the matrix to the intermembrane space.

8 The resultant buildup of hydrogen ions in the intermembrane space brings about the flow of hydrogen ions from the intermembrane space to the matrix through special channels in the mitochondrial inner membrane.

9 The flow of hydrogen ions through the channel activates ATP synthase, which is located at the matrix end of the channel.

10 Activation of ATP synthase brings about the formation of ATP from ADP and P_i.

(b)

in a kind of assembly line. As a result of giving up hydrogen and electrons within the electron transport chain, NADH and FADH$_2$ are converted back to NAD and FAD. These molecules are now free to pick up more hydrogen atoms released during glycolysis and the citric acid cycle. Thus NAD and FAD serve as the link between the citric acid cycle and the electron transport chain.

The electron carriers are arranged in a specifically ordered fashion on the inner membrane so that the high-energy electrons are progressively transferred through a chain of reactions, with the electrons falling to successively lower energy levels with each step. Ultimately, the electrons are passed to molecular oxygen (O$_2$) derived from the air we breathe. Electrons bound to O$_2$ are in their lowest energy state. Oxygen breathed in from the atmosphere enters the mitochondria to serve as the final electron acceptor of the electron transport chain. This negatively charged oxygen (negative because it has acquired additional electrons) then combines with the positively charged hydrogen ions (positive because they have donated the electrons at the beginning of the electron transport chain) to form water (H$_2$O).

As the electrons move through this chain of reactions to ever lower energy levels, they release energy. Part of the released energy is lost as heat, but some is harnessed by the mitochondrion to synthesize ATP through the following steps, which are collectively known as the **chemiosmotic mechanism**:

1. At three sites in the electron transport chain, the energy released during the transfer of electrons is used to transport hydrogen ions across the inner mitochondrial membrane from the matrix to the space between the inner and outer mitochondrial membranes, the *intermembrane space* (*inter* means "between") (● Figure 2-13b).

2. As a result of this transport process, hydrogen ions are more heavily concentrated in the mitochondrial intermembrane space than in the matrix. Because of this difference in concentration, the transported hydrogen ions have a strong tendency to flow back into the matrix through channels or passageways formed by special proteins within the inner mitochondrial membrane.

3. The channels through which the transported hydrogen ions return to the matrix bear the enzyme **ATP synthase,** which is activated by the flow of hydrogen ions from the intermembrane space to the matrix.

4. On activation, ATP synthase converts ADP + P$_i$ to ATP, providing a rich yield of 32 more ATP molecules for each glucose molecule thus processed. ATP is subsequently transported out of the mitochondrion into the cytosol for use as the cell's energy source.

The harnessing of energy into a useful form as the electrons tumble from a high-energy state to a low-energy state can be likened to a power plant converting the energy of water tumbling down a waterfall into electricity. Because O$_2$ is used in these final steps of energy conversion when a phosphate is added to form ATP, this process is known as **oxidative phosphorylation.** The electron transport chain is also called the **respiratory chain** because it is crucial to **cellular respiration,** a term that refers to the intracellular oxidation of nutrient derivatives.

The series of steps that lead to oxidative phosphorylation might at first seem an unnecessary complication. Why not just directly oxidize, or "burn," food molecules to release their energy? When this process is carried out outside the body, all the energy stored in the food molecule is released explosively in the form of heat (● Figure 2-14). In the body, oxidation of food molecules occurs in many small, controlled steps so that the food molecule's chemical energy is gradually made available for convenient packaging in a storage form that is useful to the cell. The cell, by means of its mitochondria, can more efficiently capture the energy from the food molecules within ATP bonds when it is released in small quantities. In this way, much less of the energy is converted to heat. The heat that is produced is not completely wasted energy; it is used to help maintain body temperature, with any excess heat being eliminated to the environment.

▌ The cell generates more energy in aerobic than in anaerobic conditions.

The cell is a much more efficient energy converter when oxygen is available (● Figure 2-15). In an **anaerobic** ("lack of air," specifically "lack of O$_2$") condition, the degradation of glucose cannot proceed beyond glycolysis. Recall that glycolysis takes place in the cytosol and involves the breakdown of glucose into pyruvic acid, producing a low yield of two molecules of ATP per molecule of glucose. The untapped energy of the glucose molecule remains locked in the bonds of the pyruvic acid molecules, which are eventually converted to lactic acid if they do not enter the pathway that ultimately leads to oxidative phosphorylation.

When sufficient O$_2$ is present—an **aerobic** ("with air" or "with O$_2$") condition—mitochondrial processing (that is, the citric acid cycle in the matrix and the electron transport chain on the cristae) harnesses sufficient energy to generate 34 more molecules of ATP, for a total net yield of 36 ATPs per molecule of glucose processed. (For a description of aerobic exercise, see the accompanying boxed feature, ▶ A Closer Look at Exercise Physiology.) The overall reaction for the oxidation of food molecules to yield energy is as follows:

Food	+	O$_2$	→	CO$_2$	+	H$_2$O	+	ATP
		(necessary for oxidative phosphorylation)		(produced primarily by the citric acid cycle)		(produced by the electron transport chain)		(produced primarily by the electron transport chain)

Glucose, the principal nutrient derived from dietary carbohydrates, is the fuel preference of most cells. However, nutrient molecules derived from fats (fatty acids) and, if necessary, from protein (amino acids) can also participate at specific points in this overall chemical reaction to eventually produce energy. Amino acids are usually used for protein synthesis instead of energy production, but they can be used as fuel if insufficient glucose and fat are available (Chapter 17).

Note that the oxidative reactions within the mitochondria generate energy, unlike the oxidative reactions controlled by

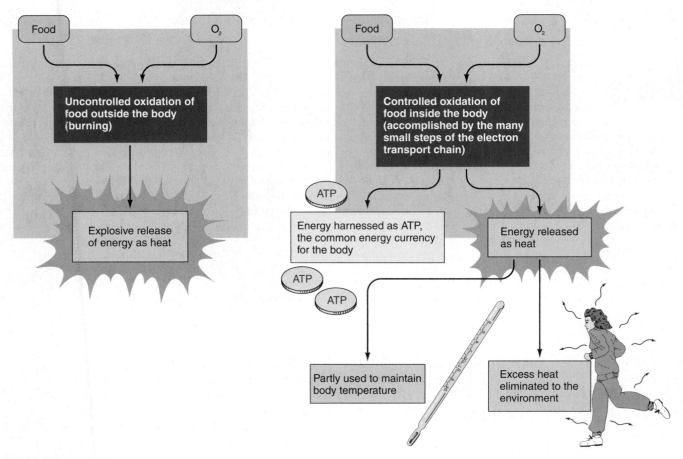

● **FIGURE 2-14**

Uncontrolled versus controlled oxidation of food
Part of the energy that is released as heat when food undergoes uncontrolled oxidation (burning) outside the body is instead harnessed and stored in useful form when controlled oxidation of food occurs inside the body.

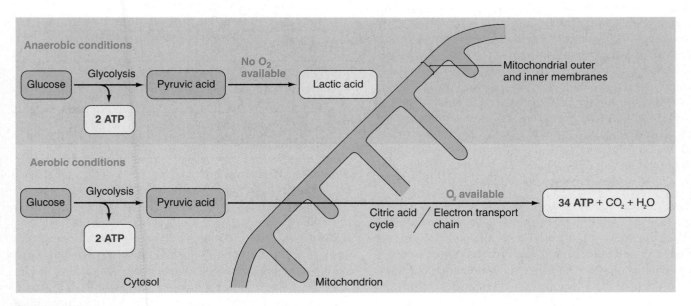

● **FIGURE 2-15**

Comparison of energy yield and products under anaerobic and aerobic conditions
In anaerobic conditions, only 2 ATPs are produced for every glucose molecule processed, but in aerobic conditions a total of 36 ATPs are produced per glucose molecule.

Aerobic Exercise: What For and How Much?

Aerobic ("with O$_2$") **exercise** involves large muscle groups and is performed at a low-enough intensity and for a long-enough period of time that fuel sources can be converted to ATP by using the citric acid cycle and electron transport chain as the predominant metabolic pathway. Aerobic exercise can be sustained for from 15 to 20 minutes to several hours at a time. Short-duration, high-intensity activities such as weight training and the 100-meter dash, which last for a matter of seconds and rely solely on energy stored in the muscles and on glycolysis, are forms of **anaerobic** ("without O$_2$") **exercise.**

Inactivity is associated with increased risk of developing both hypertension (high blood pressure) and coronary artery disease (blockage of the arteries that supply the heart). To reduce the risk of hypertension and coronary artery disease and to improve physical work capacity, the American College of Sports Medicine recommends that an individual participate in aerobic exercise a minimum of three times per week for 20 to 60 minutes. Recent studies have shown the same health benefits are derived whether the exercise is accomplished in one long stretch or is broken down into multiple shorter stints. This is good news, because many individuals find it easier to stick with brief bouts of exercise sprinkled throughout the day.

The intensity of the exercise should be based on a percentage of the individual's maximal capacity to work. The easiest way to establish the proper intensity of exercise and to monitor intensity levels is by checking the heart rate. The estimated maximal heart rate is determined by subtracting the person's age from 220. Significant benefits can be derived from aerobic exercise performed between 70% and 80% of maximal heart rate. For example, the estimated maximal heart rate for a 20-year-old is 200 beats per minute. If this person exercised three times per week for 20 to 60 minutes at an intensity that increased the heart rate to 140 to 160 beats per minute, the participant should significantly improve his or her aerobic work capacity and reduce the risk of cardiovascular disease.

---

the peroxisome enzymes. Both organelles use O$_2$, but for different purposes.

▮ The energy stored within ATP is used for synthesis, transport, and mechanical work.

Once formed, ATP is transported out of the mitochondria and is then available as an energy source as needed within the cell. Cellular activities that require energy expenditure fall into three main categories:

1. *Synthesis of new chemical compounds*, such as protein synthesis by the endoplasmic reticulum. Some cells, especially cells with a high rate of secretion and cells in the growth phase, use up to 75% of the ATP they generate just to synthesize new chemical compounds.

2. *Membrane transport*, such as the selective transport of molecules across the kidney tubules during the process of urine formation. Kidney cells can expend as much as 80% of their ATP currency to operate their selective membrane-transport mechanisms.

3. *Mechanical work*, such as contraction of the heart muscle to pump blood or contraction of skeletal muscles to lift an object. These activities require tremendous quantities of ATP.

As a result of cellular energy expenditure to support these various activities, large quantities of ADP are produced. These energy-depleted ADP molecules enter the mitochondria for "recharging" and then cycle back into the cytosol as energy-rich ATP molecules after participating in oxidative phosphorylation. A single ADP/ATP molecule may shuttle back and forth between the mitochondria and cytosol for this recharging–expenditure cycle thousands of times per day.

The high demands for ATP render glycolysis alone an insufficient as well as inefficient supplier of power for most cells. Were it not for the mitochondria, which house the metabolic machinery for oxidative phosphorylation, our energy capability would be very limited. However, glycolysis does provide cells with a sustenance mechanism that can produce at least some ATP under anaerobic conditions. Skeletal muscle cells in particular take advantage of this ability during short bursts of strenuous exercise, when energy demands for contractile activity outstrip the body's ability to bring adequate O$_2$ to the exercising muscles to support oxidative phosphorylation. Also, red blood cells, which are the only cells that do not contain any mitochondria, rely solely on glycolysis for their limited energy production. The energy needs of red blood cells are low, however, because they also lack a nucleus and therefore are not capable of synthesizing new substances, the biggest energy expenditure for most noncontractile cells.

VAULTS

In addition to the five well-documented organelles, in the early 1990s researchers identified a sixth type of organelle—vaults.

▮ Vaults may serve as cellular transport vehicles.

Vaults, which are three times as large as ribosomes, are shaped like octagonal barrels (● Figure 2-16a and c). Their name comes from their multiple arches, which reminded their discoverers of vaulted or cathedral ceilings. Just like barrels, vaults have a hollow interior. Sometimes vaults are seen in an open state, appearing like pairs of unfolded flowers with each half of the vault bearing eight "petals" attached to a central ring (● Figure 2-16b). A cell may contain thousands of vaults. Why would the presence of these numerous, relatively large organelles have been elusive until recently? The reason is that they do not show up with ordinary staining techniques.

Two clues to the function of vaults may be their octagon shape and their hollow interior. Intriguingly, the nuclear pores are also octagonal and the same size as vaults, leading to spec-

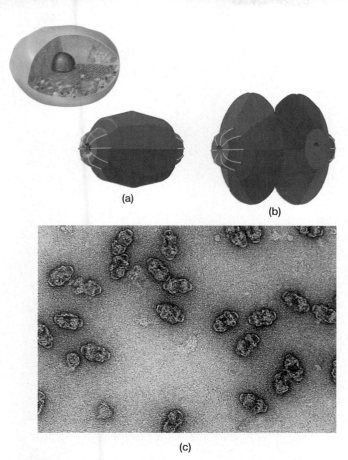

(a)

(b)

(c)

● **FIGURE 2-16**

Vaults

(a) Schematic three-dimensional representation of a vault, an octagonal barrel–shaped organelle believed to transport either messenger RNA or the ribosomal subunits from the nucleus to the cytoplasmic ribosomes. (b) Schematic representation of an opened vault, showing its octagonal structure. (c) Electron micrograph of vaults.

ulation that vaults may be cellular "trucks." According to this proposal, vaults would dock at or enter nuclear pores, pick up molecules synthesized in the nucleus, and deliver their cargo elsewhere in the cell. Ongoing research supports vaults' role in nucleus-to-cytoplasm transport, but what cargo they are carrying has not been determined. One possibility is that vaults may be carrying messenger RNA from the nucleus to the ribosomal sites of protein synthesis within the cytoplasm. Another possibility is that vaults' unknown cargo may be the two subunits that make up ribosomes (see ● Figure C-6). These two subunits are produced in the nucleus, then exit through the nuclear pores by unknown means to reach their sites of action— either attached to the rough ER or in the cytosol. Of interest, the interior of a vault is the right size to accommodate these ribosomal subunits for transport out of the nucleus.

Furthermore, vaults may play an undesirable role in bringing about the multidrug resistance sometimes displayed by cancer cells. Chemotherapy drugs designed to kill cancer cells tend to accumulate in the nuclei of these cells, but some cancer cells develop broad resistance to a wide variety of these drugs. This resistance is a major cause of cancer treatment failure. Researchers have shown that some cancer cells resistant to

chemotherapy produce up to 16 times more than normal quantities of the major vault protein. If further investigation confirms that vaults play a role in drug resistance—perhaps by transporting the drugs from the nucleus to sites for exocytosis from the cancer cells—the exciting possibility exists that interference with this vault activity could improve the sensitivity of cancer cells to chemotherapeutic drugs.

CYTOSOL

Occupying about 55% of the total cell volume, the **cytosol** is the semiliquid portion of the cytoplasm that surrounds the organelles. Its nondescript appearance under an electron microscope gives the mistaken impression that the cytosol is a liquid mixture of uniform consistency, but it is actually a highly organized, gel-like mass with differences in composition and consistency from one part of the cell to another. Furthermore, dispersed throughout the cytosol is a *cytoskeleton*, a protein scaffolding that gives shape to the cell, provides an intracellular organizational framework, and is responsible for various cell movements. For now we will concentrate on the gelatinous portion of the cytosol, then turn our attention to its cytoskeletal component in the next section.

▌ The cytosol is important in intermediary metabolism, ribosomal protein synthesis, and nutrient storage.

Three general categories of activities are associated with the gelatinous portion of the cytosol: (1) enzymatic regulation of intermediary metabolism; (2) ribosomal protein synthesis; and (3) storage of fat, carbohydrate, and secretory vesicles.

Enzymatic regulation of intermediary metabolism

The term **intermediary metabolism** refers collectively to the large set of chemical reactions inside the cell that involve the degradation, synthesis, and transformation of small organic molecules such as simple sugars, amino acids, and fatty acids. These reactions are critical for ultimately capturing energy to be used for cellular activities and for providing the raw materials needed for maintenance of the cell's structure and function and for the cell's growth. All intermediary metabolism occurs in the cytoplasm, with most of it being accomplished in the cytosol. The cytosol contains thousands of enzymes involved in glycolysis and other intermediary biochemical reactions.

Ribosomal protein synthesis

Also dispersed throughout the cytosol are the free ribosomes, which synthesize proteins for use in the cytosol itself. In contrast, recall that the rough ER ribosomes synthesize proteins for secretion and for construction of new cellular components.

Storage of fat and glycogen

Excess nutrients not immediately used for ATP production are converted in the cytosol into storage forms that are readily visible under a light microscope. Such nonpermanent masses of stored material are known as **inclusions.** Inclusions are not surrounded by membrane, and they may or may not be present, depending on the type of cell and the circumstances. The largest and most important storage product is fat. Small fat droplets are present within the cytosol in various cells. In **adipose tissue,** the tissue specialized for fat storage, the stored fat molecules can occupy almost the entire cytosol, where they merge to form one large fat droplet (● Figure 2-17a). The other visible storage product is **glycogen,** the storage form of glucose, which appears as aggregates or clusters dispersed throughout the cell (● Figure 2-17b). Cells vary in their ability to store glycogen, with liver and muscle cells having the greatest stores. When food is not available to provide fuel for the citric acid cycle and electron transport chain, stored glycogen and fat are broken down to release glucose and fatty acids, respectively, which can feed the mitochondrial energy-producing machinery. An average adult has enough glycogen stored to provide sufficient energy for about a day of normal activities, and typically enough fat is stored to provide energy for two months.

Secretory vesicles that have been processed and packaged by the endoplasmic reticulum and Golgi complex also remain in the cytosol, where they are stored until signaled to empty their contents to the outside. In addition, transport and endocytotic vesicles move through the cytosol.

CYTOSKELETON

Different cells in the body have distinct shapes, structural complexities, and functional specializations. Maintenance of the unique characteristics of each cell type necessitates intracellular scaffolding to support and organize the cellular components into an appropriate arrangement and to control their movements. These functions are performed by the **cytoskeleton,** the complex protein network portion of the cytosol that acts as the "bone and muscle" of the cell.

This elaborate network has three distinct elements: (1) *microtubules*, (2) *microfilaments*, and (3) *intermediate filaments*. The different parts of the cytoskeleton are structurally linked and functionally coordinated to provide certain integrated functions for the cell. Because of the complexity of this network and the variety of functions it serves, we address its elements separately. These functions, along with the functions of all other cell structures, are summarized in ▲ Table 2-3 with an emphasis on the components of the cytoplasm.

▌ Microtubules help maintain asymmetric cell shapes and play a role in complex cell movements.

The **microtubules** are the largest of the cytoskeletal elements. They are very slender (22 nm in diameter), long, hollow, unbranched tubes composed primarily of **tubulin,** a small, globular protein molecule (● Figure 2-18a) (1 nanometer (nm) = 1 billionth of a meter). Microtubules are essential for maintaining an asymmetric cell shape, such as that of a nerve cell, whose elongated axon may extend up to a meter in length from

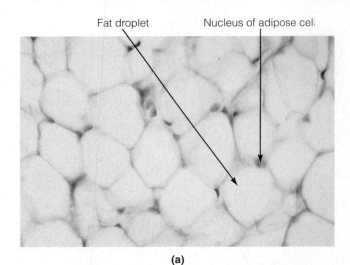

Fat droplet   Nucleus of adipose cell

(a)

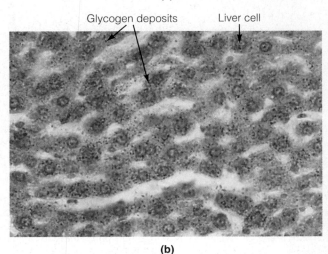

Glycogen deposits   Liver cell

(b)

● **FIGURE 2-17**

Inclusions
(a) Light micrograph depicting fat storage in an adipose cell. Note that the fat droplet occupies almost the entire cytosol. (b) Light micrograph depicting glycogen storage in a liver cell. The red-staining granules throughout the liver cell's cytosol are glycogen deposits.

the origin of the cell body in the spinal cord to the termination of the axon at a muscle (● Figure 2-19). Microtubules, along with specialized intermediate filaments, stabilize this asymmetric axonal extension.

Microtubules also play an important role in coordinating numerous complex cell movements, including (1) transport of secretory vesicles or other materials from one region of the cell to another, (2) movement of specialized cell projections such as cilia and flagella, and (3) distribution of chromosomes during cell division through formation of a mitotic spindle. Let us examine each of these roles.

Transport of secretory vesicles

Axonal transport provides a good example of the importance of an organized system for moving secretory vesicles. In a nerve cell, specific chemicals are released from the terminal end of the elongated axon to influence a muscle or another struc-

Cell Part	Number Per Cell	Structure	Function
Plasma membrane	1	Lipid bilayer studded with proteins and small amounts of carbohydrate	Acts as selective barrier between cellular contents and extracellular fluid; controls traffic in and out of the cell
Nucleus	1	DNA and specialized proteins enclosed by a double-layered membrane	Acts as control center of the cell, providing storage of genetic information; nuclear DNA provides codes for the synthesis of structural and enzymatic proteins that determine the cell's specific nature and serves as blueprint for cell replication
Cytoplasm			
Organelles			
Endoplasmic reticulum	1	Extensive, continuous membranous network of fluid-filled tubules and flattened sacs, partially studded with ribosomes	Forms new cell membrane and other cell components and manufactures products for secretion
Golgi complex	1 to several hundred	Sets of stacked, flattened membranous sacs	Modifies, packages, and distributes newly synthesized proteins
Lysosomes	300 average	Membranous sacs containing hydrolytic enzymes	Serve as digestive system of the cell, destroying foreign substances and cellular debris
Peroxisomes	200 average	Membranous sacs containing oxidative enzymes	Perform detoxification activities
Mitochondria	100–2,000	Rod- or oval-shaped bodies enclosed by two membranes, with the inner membrane folded into cristae that project into an interior matrix	Act as energy organelles; major site of ATP production; contain enzymes for citric acid cycle and electron transport chain
Vaults	Thousands	Shaped like hollow octagonal barrels	Unclear; may transport messenger RNA or ribosomal subunits from nucleus to cytoplasm
Cytosol-gel-like portion			
Intermediary metabolism enzymes	Many	Dispersed within the cytosol	Facilitate intracellular reactions involving the degradation, synthesis, and transformation of small organic molecules
Ribosomes	Many	Granules of RNA and proteins—some attached to rough endoplasmic reticulum, some free in the cytoplasm	Serve as workbenches for protein synthesis
Transport vesicles	Varies	Membrane-enclosed packages of products for construction of new cellular membrane or organelle components	Transport selected proteins and lipids from one organelle to another or to the plasma membrane
Secretory vesicles	Varies	Membrane-enclosed packages of secretory products	Store secretory products until signaled to empty to the outside
Endocytotic vesicles	Varies	Membrane-enclosed packages of material engulfed by the cell	Transfer selected, large, extracellular materials into the cell
Inclusions	Varies	Glycogen granules, fat droplets	Store excess nutrients
Cytosol-cytoskeleton portion			As an integrated whole: is responsible for the shape, rigidity, and spatial geometry of each type of cell (the "bone" of the cell); directs intracellular transport and regulates cellular movements (the "muscle" of the cell)

(continued)

▲ **TABLE 2-3**

Summary of Cell Structures and Functions (*continued*)

Cell Part	Number Per Cell	Structure	Function
Cytosol-cytoskeleton portion (continued)			
Microtubules	Many	Long, slender, hollow tubes composed of tubulin molecules	Maintain asymmetric cell shapes and coordinate complex cell movements, specifically facilitating transport of secretory vesicles within cell, serving as main structural and functional component of cilia and flagella, and forming mitotic spindle during cell division
Microfilaments	Many	Intertwined helical chains of actin molecules; microfilaments composed of myosin molecules also present in muscle cells	Play a vital role in various cellular contractile systems, including muscle contraction and amoeboid movement; serve as a mechanical stiffener for microvilli
Intermediate filaments	Many	Irregular, threadlike proteins	Play a structural role in parts of the cell subject to mechanical stress

● **FIGURE 2-18**

Components of the cytoskeleton

(a) Microtubules, the largest of the cytoskeletal elements, are long, hollow tubes formed by two slightly different variants of globular-shaped tubulin molecules. (b) Most microfilaments, the smallest of the cytoskeletal elements, consist of two chains of actin molecules wrapped around each other. (c) The intermediate filament keratin found in skin is made up of three polypeptide strands wound around each other. The composition of intermediate filaments, which are intermediate in size between the microtubules and microfilaments, varies among different cell types.

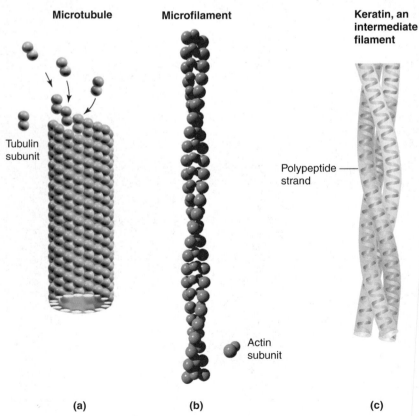

Microtubule   **Microfilament**   **Keratin, an intermediate filament**

Tubulin subunit

Polypeptide strand

Actin subunit

(a)   (b)   (c)

ture that the nerve cell controls. These chemicals are largely produced within the cell body (the part of the nerve cell next to the beginning of the axon), where the nuclear DNA blueprint, endoplasmic reticular factory, and Golgi packaging and distribution outlet are located. Yet these chemicals ultimately function at the end of the axon, which may be a meter away. If these chemicals had to diffuse on their own from the cell body to a distant axon terminal, it would take them about 50 years to get there—obviously an impractical solution. Instead, the microtubules that extend from the beginning to the end of the axon provide a "highway" for vesicular traffic along the axon (● Figure 2-19). A specific **motor protein** attaches to the particle to be transported, then "walks" along the microtubule with the particle riding in "piggyback" fashion (*motor* means "movement"). This process is powered by ATP. **Kinesin,** one such motor protein, consists of two globular heads, a stalk, and a fanlike tail. Kinesin's tail binds to the secretory vesicle to be moved, and its globular heads act like little feet that move one at a time, like the way you walk. The feet alternately attach to one tubulin molecule on the microtubule, bend and push forward, then let go. During this process, the back foot is yanked forward so that it swings ahead of what was the front foot and then attaches to the next tubulin molecule farther down the microtubule. The process is repeated over and over as kinesin moves its cargo to the end of the axon by using each of the tubulin molecules as a stepping stone.

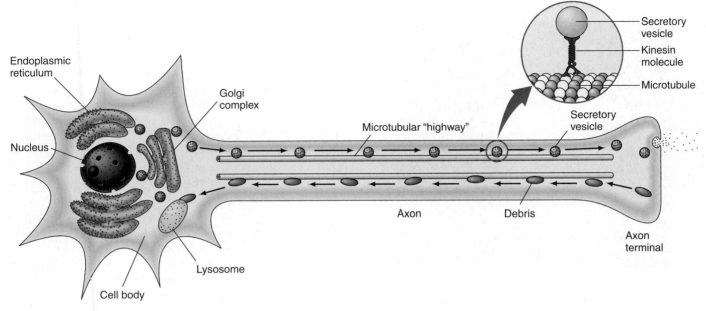

Labels in figure:
Endoplasmic reticulum
Golgi complex
Secretory vesicle
Kinesin molecule
Microtubule
Microtubular "highway"
Nucleus
Secretory vesicle
Lysosome
Axon
Debris
Axon terminal
Cell body

● **FIGURE 2-19**

Two-way vesicular axonal transport facilitated by the microtubular "highway" in a nerve cell
Schematic illustration of a neuron depicting secretory vesicles being transported from the site of production in the cell body along a microtubule "highway" to the terminal end for secretion. Vesicles containing debris are being transported in the opposite direction for degradation in the cell body. The enlargement depicts kinesin, a motor protein, carrying a secretory vesicle down the microtubule by using its "feet" to "step" on one tubulin molecule after another.

Reverse vesicular traffic also occurs along these microtubular highways. Vesicles that contain debris are transported by different motor proteins, most commonly by **dynein**, from the axon terminal to the cell body for degradation by lysosomes, which are confined within the cell body. Again, the driving force depends on ATP. Coincidentally, this reverse axonal transport may also serve as a pathway for the movement of some infectious agents such as herpes virus, poliomyelitis virus, and rabies virus. These viruses travel backward along nerves from their surface site of contamination, such as a break in the skin or an animal bite, to the central nervous system (brain and spinal cord).

Movement of cilia and flagella

Microtubules are also the dominant structural and functional components of cilia and flagella. These specialized protrusions from the cell surface allow a cell to move materials across its surface (in the case of a stationary cell) or to propel itself through its environment (in the case of a motile cell). **Cilia** (meaning "eyelashes") are numerous tiny, hairlike protrusions, whereas a **flagellum** (meaning "whip") is a single, long, whiplike appendage. Even though they project from the surface of the cell, cilia and a flagellum are both intracellular structures—they are covered by the plasma membrane.

Cilia beat or stroke in unison, much like the coordinated efforts of a rowing team. Each cilium exerts a rapid active stroke, which moves material on the cell surface forward. This stroke is followed by a recovery phase in which the cilium more slowly returns to its original position—an action that does not

exert much force, ensuring that the material just pushed forward is not futilely pushed backward.

In humans, ciliated cells are found in the stationary cells that line the respiratory tract, the oviduct of the female reproductive tract, and the fluid-filled ventricles (chambers) of the brain. The coordinated stroking of the thousands of respiratory cilia help keep foreign particles out of the lungs by sweeping outward dust and other inspired particles (● Figure 2-20). In the female reproductive tract, the sweeping action of the cilia that line the oviduct draws the egg (ovum) released from the ovary during ovulation into the oviduct and then guides it toward the uterus (womb). In the brain, the ciliated cells that line the ventricles produce cerebrospinal fluid, which flows through the ventricles and around the brain and spinal cord, cushioning and bathing these fragile neural structures. Beating of the cilia helps promote flow of this supportive fluid.

The only human cells that have a flagellum are sperm. The whiplike motion of the flagellum or "tail" enables a sperm to move through its environment, which is crucial for maneuvering into position to fertilize the female ovum.

Cilia and flagella have the same basic internal structure. Both consist of nine fused pairs of microtubules (doublets) arranged in an outer ring around two single unfused microtubules in the center (● Figure 2-21). This characteristic "nine plus two" array of microtubules extends throughout the length of the motile appendage. A cilium or flagellum originates from a specialized structure inside the main part of the cell, the **basal body.** Each basal body is a short cylinder composed of

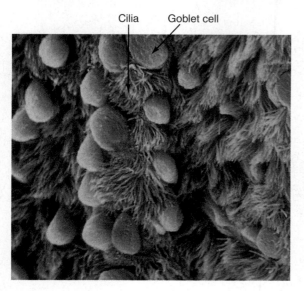

● FIGURE 2-20

Scanning electron micrograph of cilia on cells lining the respiratory tract in humans
The respiratory airways are lined by goblet cells, which secrete a sticky mucus that traps inspired particles, and epithelial cells that bear numerous hairlike cilia. The cilia beat in unison to sweep inspired particles up and out of the airways.

parallel microtubules arranged similarly to those of the cilium or flagellum.

Accessory proteins associated with these microtubules maintain the microtubule's organization and play an essential part in the microtubular movement that causes the entire structure to bend. The motor protein dynein is the most important of these accessory proteins. Dynein molecules form a set of armlike projections from each doublet of microtubules (● Figure 2-21a). The bending movements of cilia and flagella are produced by the sliding of adjacent microtubule doublets past each other. Sliding is accomplished by the dynein arms, which can split ATP and then use the released energy to "crawl" along the neighboring microtubule doublet (similar to kinesin's movement) to cause the bending and stroking. Groups of cilia working together are oriented to beat in the same direction and contract in a synchronized manner, through controlling mechanisms that are poorly understood. These mechanisms appear to involve the single microtubules at the cilium's center and their surrounding accessory proteins.

Formation of the mitotic spindle

Cell division involves two discrete but related activities: *mitosis* (nuclear division) and *cytokinesis* (cytoplasmic division). During **mitosis,** the DNA-containing chromosomes of the nucleus

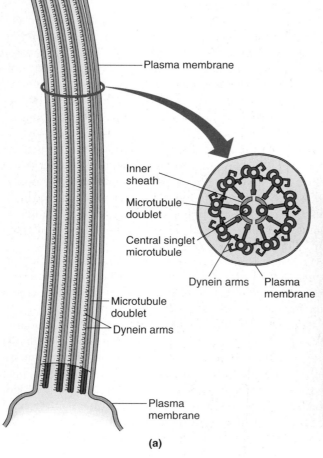

(a)

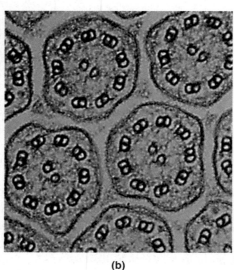

(b)

● FIGURE 2-21

Internal structure of cilia and flagella
(a) Schematic diagram of a cilium in cross section showing the characteristic "nine plus two" arrangement of microtubules along with the dynein arms and other accessory proteins. (b) Electron micrograph of numerous cilia in cross section.

(Source: Figure 2.21(a) adapted from *Molecular Biology of the Cell*, Fig. 10-27, p. 565 by Bruce Alberts, Dennis Bray, Julian Lewis, Martin Raff, Keith Roberts, and James D. Watson. Reproduced by permission of Routledge, Inc., part of The Taylor & Francis Group.)

are replicated, resulting in two identical sets of chromosomes. These duplicate sets of chromosomes are separated and drawn apart to opposite sides of the cell so that the genetic material is evenly distributed in the two halves of the cell (see p. A-27). During **cytokinesis,** the cell constricts in the middle, and the two halves separate into two new daughter cells, each with a full complement of chromosomes (● Figure 2-22).

The replicated chromosomes are pulled apart by a cellular apparatus called the **mitotic spindle,** which is transiently assembled from microtubules only during cell division (see ● Figure C-10, p. A-28). The microtubules of the mitotic spindle are formed by the *centrioles,* a pair of short cylindrical structures that lie at right angles to each other near the nucleus (● Figure 2-1, p. 26). The centrioles also duplicate during cell division. After self-replication, the centriole pairs move toward opposite ends of the cell and form the spindle apparatus between them through a precisely organized assemblage of microtubules. Importantly, some anticancer drugs prevent cancer cells from reproducing by interfering with the microtubules that ordinarily pull the chromosomes to opposite poles during cell division.

Besides their role in mitotic spindle formation, the centrioles also assemble the many microtubules that normally radiate throughout the cytoskeleton. The centrioles are identical in structure to basal bodies. In fact, under some circumstances the centrioles and basal bodies are interconvertible. During development of ciliated human cells, the centriole pair migrates to the region of the cell where the cilia will be formed and duplicates itself to produce the many basal bodies that will form the cilia.

▋ Microfilaments are important to cellular contractile systems and as mechanical stiffeners.

The microfilaments are the smallest (6 nm in diameter) elements of the cytoskeleton visible with a conventional electron microscope. The most obvious microfilaments in most cells are those composed of **actin,** a protein molecule that has a globular shape similar to tubulin. Unlike tubulin, which forms a hollow tube, actin is assembled into two strands twisted around each other to form a microfilament (see ● Figure 2-18b). In muscle cells, another protein called **myosin** forms a different kind of microfilament. In most cells, myosin is not as abundant and does not form such distinct filaments.

Microfilaments serve two functions: (1) They play a vital role in various cellular contractile systems, and (2) they act as mechanical stiffeners for several specific cellular projections.

Microfilaments in cellular contractile systems

Actin-based assemblies are involved in muscle contraction, cell division, and cell locomotion. The most obvious, best organized, and most clearly understood cellular contractile system is that found in muscle. Muscle contains an abundance of actin and myosin microfilaments, which accomplish muscle contraction by sliding past each other, using ATP as an energy source. This ATP-powered microfilament slid-

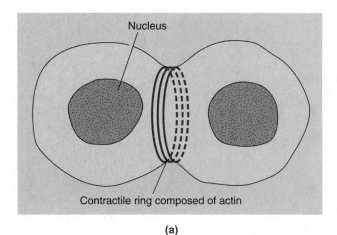

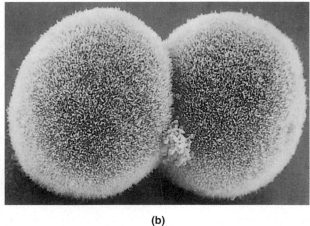

(b)

● **FIGURE 2-22**

Cytokinesis
(a) Schematic illustration of the actin contractile ring squeezing apart the two duplicate cell halves during cytokinesis. (b) Photograph of a cell undergoing cytokinesis.

ing and force development is triggered by a complex sequence of electrical, biochemical, and mechanical events initiated when the muscle cell is stimulated to contract (see Chapter 8 for details).

Nonmuscle cells may also contain "musclelike" assemblies. Some of these microfilament contractile systems are transiently assembled to perform a specific function when needed. A good example is the contractile ring that forms during cytokinesis to split apart the duplicate cell halves. The ring consists of a belt-like bundle of actin filaments located just beneath the plasma membrane in the middle of the cell. When this ring of fibers contracts, it pinches the cell in two (● Figure 2-22a).

Complex actin-based assemblies are also responsible for most cell locomotion. Four types of human cells are capable of moving on their own—sperm, white blood cells, fibroblasts, and skin cells. Sperm move by the flagellar mechanism already described. Motility for the other cells is accomplished by **amoeboid movement,** a cell-crawling process that depends on the activity of their actin filaments, in a mechanism similar to that used by amoebae to maneuver through their environment. When crawling, the motile cell forms the fingerlike extensions known as *pseudopods* at the "front" or leading edge of

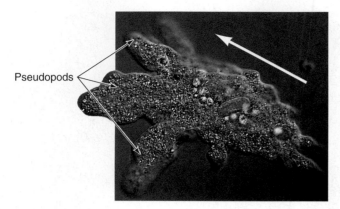

Pseudopods

● FIGURE 2-23
An amoeba undergoing amoeboid movement

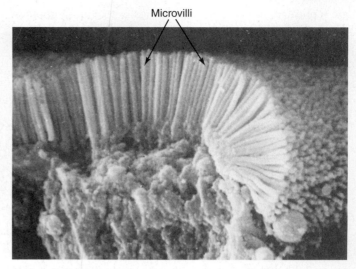

Microvilli

● FIGURE 2-24
Scanning electron micrograph of intestinal microvilli
[*Source:* From *Tissues and Organs: A Text-Atlas of Scanning Electron Microscopy* by Richard G. Kessel and Randy H. Kardon (New York: W. H. Freeman and Company, 1979). Copyright by and reprinted with permission from the authors.]

the cell in the direction of the target (● Figure 2-23). For example, the target that triggers amoeboid movement might be the proximity of food in the case of an amoeba or a bacterium in the case of a white blood cell. Pseudopods are formed as a result of the organized assembly and disassembly of branching actin networks. During amoeboid movement, actin filaments continuously grow at the cell's leading edge through the addition of actin molecules at the front of the actin chain. This filament growth pushes that portion of the cell forward as a pseudopod extension. Simultaneously, actin molecules at the rear of the filament are being disassembled and transferred to the front of the line. Thus the filament does not get any longer; it stays the same length but moves forward through the continuous transfer of actin molecules from the rear to the front of the filament in what is termed *treadmilling* fashion. The cell progressively moves forward through repeated cycles of pseudopod formation at the leading edge.

White blood cells are the most active crawlers in the body. These cells leave the circulatory system and travel by amoeboid movement to areas of infection or inflammation, where they engulf and destroy micro-organisms and cellular debris. Unbelievably, it is estimated that the total distance traveled collectively per day by all your white blood cells while they roam the tissues in their search-and-destroy tactic would circle Earth twice!

Fibroblasts ("fiber formers"), another type of motile cell, move amoeboid fashion into a wound from adjacent connective tissue to help repair the damage and are responsible for scar formation. Skin cells, which are ordinarily stationary, can become modestly mobile and move by amoeboid motion toward a cut to restore the skin surface.

Microfilaments as mechanical stiffeners

Besides their role in cellular contractile systems, the actin filaments' second major function is to serve as mechanical supports or stiffeners for several cellular extensions, of which the most common are microvilli. **Microvilli** are microscopic, nonmotile, hairlike projections from the surface of epithelial cells lining the small intestine and kidney tubules (● Figure 2-24). Their presence greatly increases the surface area available for transferring material across the plasma membrane. In the case

of the small intestine, the microvilli increase the area available for absorbing digested nutrients. In the kidney tubules, microvilli enlarge the absorptive surface that salvages useful substances passing through the kidney so that these materials are saved for the body instead of being eliminated in the urine. Within each microvillus, a core consisting of parallel actin filaments linked together forms a rigid mechanical stiffener that keeps these valuable surface projections intact.

Both microtubules and microfilaments form *stable* structures, such as cilia, flagella, muscle contractile units, and microvilli, and also form *transient* structures, such as mitotic spindles and contractile rings, as the need arises. Pools of unassembled tubulin and actin subunits in the cytosol can be rapidly assembled into organized structures to perform specific activities and then can be disassembled when they are no longer needed.

❚ Intermediate filaments are important in cell regions subject to mechanical stress.

The **intermediate filaments** are intermediate in size between the microtubules and the microfilaments (7–11 nm in diameter)—hence their name. The proteins that compose the intermediate filaments vary between cell types, but in general they appear as irregular, threadlike molecules. These proteins form tough, durable fibers that play a central role in maintaining the structural integrity of a cell and in resisting mechanical stresses externally applied to a cell. In contrast to the other cytoskeletal elements, the intermediate filaments are highly stable structures. There is no evidence for a reversible pool between unassembled and assembled intermediate-filament proteins.

Different types of intermediate filaments are tailored to suit their structural or tension-bearing role in specific cell types.

In general, only one class of intermediate filament is found in a particular cell type. Two important examples follow:

- Neurofilaments are intermediate filaments found in nerve cell axons. Together with microtubules, neurofilaments strengthen and stabilize these elongated cellular extensions.
- Skin cells contain irregular networks of intermediate filaments made of the protein **keratin** (see ● Figure 2-18c). These intracellular filaments interconnect with extracellular filaments that tie adjacent cells together, thereby creating a continuous filamentous network that extends throughout the skin and gives it strength. When the surface skin cells die, their tough keratin skeletons persist to form a protective, waterproof outer layer. Hair and nails are also keratin structures.

Emphasizing the importance of intermediate filaments in some specialized cell types is the fact that intermediate filaments account for up to 85% of the total protein in nerve cells and keratin-producing skin cells, whereas these filaments constitute only about 1% of other cells' total protein on average.

Neurofilament abnormalities are the basis of some neurologic disorders. An important example is **amyotrophic lateral sclerosis (ALS)**, more familiarly known as **Lou Gehrig's disease.** ALS is characterized by progressive degeneration and death of motor neurons, the type of nerve cells that control skeletal muscles. This adult-onset condition leads to gradual loss of control of skeletal muscles, including the muscles of breathing, and ultimately to death, as it did for baseball legend Lou Gehrig. Recent evidence suggests that the underlying problem is an abnormal accumulation and disorganization of neurofilaments. Motor neurons, which have the most neurofilaments, are the most affected. The disorganized neurofilaments are believed to block the axonal transport of crucial materials along the microtubular highways, thus choking off vital supplies from the cell body to the axon terminal.

▌ The cytoskeleton functions as an integrated whole and links other parts of the cell together.

With high-voltage electron microscopy, which provides a three-dimensional view of the internal organization of the cell, a meshwork of exceedingly fine, interlinked filaments can be seen extending throughout the cytoplasm and connecting to the inner layer of the plasma membrane. Some cell biologists believe this meshwork is an artifact that arises during preparation of the specimen, but others think this lattice constitutes intricate interconnections between the cytoskeletal structures as well as various organelles (● Figure 2-25). Collectively, the cytoskeletal elements and their interconnections support the plasma membrane and are responsible for the particular shape, rigidity, and spatial geometry of each different cell type. This internal framework thus acts as the cell's "skeleton."

New studies hint that the cytoskeleton as a whole is not merely a supporting structure that maintains the tensional integrity of the cell but may serve as a mechanical communications system as well. Various components of the cytoskeleton behave as if they are structurally connected or "hard-wired" to

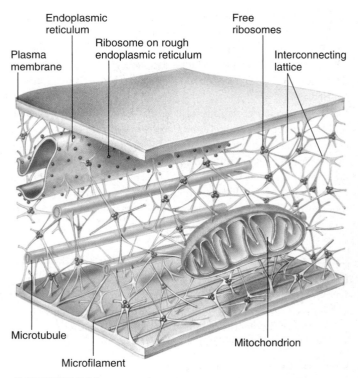

● FIGURE 2-25
Interconnections between cytoskeletal structures and organelles

each other as well as to the surface plasma membrane and to the nucleus. This force-carrying network is speculated to serve as a mechanism for mechanical forces acting on the cell surface to reach all the way from the plasma membrane through the cytoskeleton to ultimately influence gene regulation in the nucleus.

Furthermore, as you have learned, the coordinated action of the cytoskeletal elements is responsible for directing intracellular transport and for regulating numerous cellular movements and thereby also serves as the cell's "muscle."

CHAPTER IN PERSPECTIVE: FOCUS ON HOMEOSTASIS

The ability of cells to perform functions essential for their own survival as well as specialized tasks that contribute to the maintenance of homeostasis within the body ultimately depends on the successful, cooperative operation of the intracellular components. For example, to support life-sustaining activities, all cells must generate energy, in a usable form, from nutrient molecules. Energy is generated intracellularly by chemical reactions that take place within the cytosol and mitochondria.

In addition to being essential for basic cell survival, the organelles and cytoskeleton also participate in many cells' specialized tasks that contribute to homeostasis. Here are several examples:

- Nerve and endocrine cells both release protein chemical messengers that are important in regulatory activities aimed at maintaining homeostasis—for example, chemical messengers released from nerve cells stimulate the respiratory muscles, which accomplish life-sustaining exchanges of O_2 and CO_2 between the body and atmosphere through breathing. These protein chemical messengers (neurotransmitters in nerve cells and hormones in endocrine cells) are all produced by the endoplasmic reticulum and Golgi complex and released by exocytosis from the cell when needed.

● The ability of muscle cells to contract depends on their highly developed cytoskeletal microfilaments sliding past each other. Muscle contraction is responsible for many homeostatic activities, including (1) contraction of the heart muscle, which pumps life-supporting blood throughout the body; (2) contraction of the muscles attached to bones, which enables the body to procure food; and (3) contraction of the muscle in the walls of the stomach and intestine, which moves the food along the digestive tract so that ingested nutrients can be progressively broken down into a form that can be absorbed into the blood for delivery to the cells.

● White blood cells help the body resist infection by making extensive use of lysosomal destruction of engulfed particles as they police the body for microbial invaders. These white blood cells are able to roam the body by means of amoeboid movement, a cell-crawling process accomplished by coordinated assembly and disassembly of actin, one of their cytoskeletal components.

As we begin to examine the various organs and systems, keep in mind that proper cellular functioning is the foundation of all organ activities.

CHAPTER SUMMARY

▮ The complex organization and interaction of the chemicals within a cell confer the unique characteristics of life.
▮ Cells are the living building blocks of the body.

Observations of Cells
▮ Cells are too small for the unaided eye to see.
▮ Using early microscopes, investigators learned that all plant and animal tissues consist of individual cells.
▮ Through more sophisticated techniques, scientists now know that a cell is a complex, highly organized, compartmentalized structure.

An Overview of Cell Structure
▮ Most cells have three major subdivisions: the plasma membrane, the nucleus, and the cytoplasm.
▮ The plasma membrane encloses the cell and separates the intracellular and extracellular fluid.
▮ The nucleus contains the deoxyribonucleic acid (DNA), the cell's genetic material.
▮ Three types of RNA play a role in the protein synthesis coded by DNA—messenger RNA, ribosomal RNA, and transfer RNA.
▮ The cytoplasm consists of cytosol, a complex gel-like mass laced with a cytoskeleton, and organelles, which are highly organized, membrane-enclosed structures dispersed within the cytosol.
▮ The six types of organelles are the endoplasmic reticulum, Golgi complex, lysosomes, peroxisomes, mitochondria, and vaults.

Endoplasmic Reticulum
▮ The endoplasmic reticulum (ER) is a single, complex membranous network that encloses a fluid-filled lumen.
▮ The primary function of the ER is to serve as a factory for synthesizing proteins and lipids to be used for (1) the secretion of special products such as enzymes and hormones to the exterior of the cell and (2) the production of new cellular components, particularly cell membranes.
▮ The two types of endoplasmic reticulum are rough endoplasmic reticulum, which is studded with ribosomes, and smooth endoplasmic reticulum, which lacks ribosomes.
▮ The rough endoplasmic reticular ribosomes synthesize proteins, which are released into the ER lumen so that they are separated from the cytosol. Also entering the lumen are lipids produced within the membranous walls of the ER.
▮ Synthesized products move from the rough ER to the smooth ER, where they are packaged and discharged as transport vesicles.

▮ Transport vesicles are formed as a portion of the smooth ER "buds off," containing a collection of newly synthesized proteins and lipids wrapped in smooth ER membrane.

Golgi Complex
▮ Transport vesicles move to and fuse with the Golgi complex, which consists of stacks of flattened, membrane-enclosed sacs.
▮ The Golgi complex serves a twofold function: (1) to act as a refining plant for modifying into a finished product the newly synthesized molecules delivered to it in crude form from the endoplasmic reticular factory; and (2) to sort, package, and direct molecular traffic to appropriate intracellular and extracellular destinations.
▮ Before budding off of the Golgi complex, a vesicle takes up a specific product that has been processed within the Golgi sacs. The membrane that wraps the vesicle contains docking markers, which ensure that the vesicle docks and unloads its captured cargo only at the appropriate destination within the cell.
▮ The Golgi complex of secretory cells packages proteins to be exported out of the cell in secretory vesicles that are released by exocytosis on appropriate stimulation.

Lysosomes
▮ Lysosomes are membrane-enclosed sacs that contain powerful hydrolytic (digestive) enzymes.
▮ Serving as the intracellular digestive system, lysosomes destroy foreign material such as bacteria that has been internalized by the cell and demolish worn-out cell parts to make way for new replacement parts.
▮ Extracellular material is brought into the cell by endocytosis for attack by lysosomal enzymes.
▮ The three forms of endocytosis are pinocytosis, receptor-mediated endocytosis, and phagocytosis.

Peroxisomes
▮ Peroxisomes are small membrane-enclosed sacs containing powerful oxidative enzymes.
▮ They are specialized for carrying out particular oxidative reactions, including detoxification of various wastes and toxic foreign compounds that have entered the cell.
▮ During these detoxification reactions, peroxisomes generate potent hydrogen peroxide, which they decompose into harmless water and oxygen by means of the catalase they contain.

Mitochondria

▌ The rod-shaped mitochondria are the energy organelles of the cell.

▌ They house the enzymes of the citric acid cycle (in the mitochondrial matrix) and electron transport chain (on the cristae of the mitochondrial inner membrane). Together, these two biochemical steps efficiently convert the energy in food molecules to the usable energy stored in ATP molecules.

▌ During this process, which is known as *oxidative phosphorylation,* the mitochondria use molecular O_2 and produce CO_2 and H_2O as by-products.

▌ A cell is more efficient at converting food energy into ATP when O_2 is available. In an anaerobic (without O_2) condition, a cell can produce only 2 molecules of ATP for every glucose molecule processed by means of glycolysis, which takes place in the cytosol. In an aerobic (with O_2) condition, the mitochondria further degrade the products of glycolysis to yield another 34 molecules of ATP for every glucose molecule processed.

▌ Cells use ATP as an energy source for synthesis of new chemical compounds, for membrane transport, and for mechanical work.

Vaults

▌ Vaults are recently discovered structures shaped like hollow octagonal barrels.

▌ They are the same shape and size as the nuclear pores. Researchers speculate that vaults are cellular trucks that dock at the nuclear pores and pick up cargo for transport from the nucleus.

▌ The leading proposals are that vaults may transport messenger RNA or the ribosomal subunits from the nucleus to the cytoplasmic sites of protein synthesis.

Cytosol

▌ The cytosol contains the enzymes involved in intermediary metabolism and the ribosomal machinery essential for synthesis of these enzymes as well as other cytosolic proteins.

▌ Many cells store unused nutrients within the cytosol in the form of glycogen granules or fat droplets.

▌ Also present in the cytosol are various secretory, transport, and endocytotic vesicles.

Cytoskeleton

▌ Extending throughout the cytosol is the cytoskeleton, which serves as the "bone and muscle" of the cell.

▌ The three types of cytoskeletal elements—microtubules, microfilaments, and intermediate filaments—are each composed of different proteins and perform various roles.

▌ Collectively, the cytoskeletal elements give the cell shape and support, enable it to organize and move its internal structures as needed, and, in some cells, allow movement between the cell and its environment.

REVIEW EXERCISES

Objective Questions (Answers on p. A-35)

1. The barrier that separates and controls movement between the cellular contents and the extracellular fluid is the _____.

2. The chemical that directs protein synthesis and serves as a genetic blueprint is _____, which is found in the _____ of the cell.

3. The cytoplasm consists of _____, which are specialized, membrane-enclosed intracellular compartments, and a gel-like mass known as _____, which contains an elaborate protein network called the _____.

4. Transport vesicles from the _____ fuse with and enter the _____ for modification and sorting.

5. The (*what kind of*) _____ enzymes within the peroxisomes primarily detoxify various wastes produced within the cell or foreign compounds that have entered the cell, generating _____ in the process.

6. The universal energy carrier of the body is _____.

7. The largest cells in the human body can be seen by the unaided eye. (*True or false?*)

8. Amoeboid movement is accomplished by the coordinated assembly and disassembly of microtubules. (*True or false?*)

9. Using the following answer code, indicate which type of ribosome is being described:
   (a) free ribosome
   (b) rough ER-bound ribosome
   ____ 1. synthesizes proteins used to construct new cell membrane
   ____ 2. synthesizes proteins used intracellularly within the cytosol
   ____ 3. synthesizes secretory proteins such as enzymes or hormones

10. Using the following answer code, indicate which form of energy production is being described:
    (a) glycolysis
    (b) citric acid cycle
    (c) electron transport chain
    ____ 1. takes place in the mitochondrial matrix
    ____ 2. produces H_2O as a by-product
    ____ 3. rich yield of ATP
    ____ 4. takes place in the cytosol
    ____ 5. processes acetyl CoA
    ____ 6. located in the mitochondrial inner-membrane cristae
    ____ 7. converts glucose into two pyruvic acid molecules
    ____ 8. uses molecular oxygen

Essay Questions

1. What are a cell's three major subdivisions?

2. State an advantage of organelle compartmentalization.

3. List the six types of organelles.

4. Describe the structure of the endoplasmic reticulum, distinguishing between the rough and the smooth ER. What is the function of each?

5. Compare exocytosis and endocytosis. Define *secretion, pinocytosis, receptor-mediated endocytosis,* and *phagocytosis.*

7. Which organelles serve as the intracellular digestive system? What type of enzymes do they contain? What functions do these organelles serve?

8. Compare lysosomes with peroxisomes.

9. Describe the structure of mitochondria, and explain their role in oxidative phosphorylation.

10. Distinguish between the oxidative enzymes found in peroxisomes and those found in mitochondria.
11. What three categories of cellular activities require energy expenditure?
12. List and describe the functions of each of the components of the cytoskeleton.

Quantitative Exercises (Solutions on p. A-35)
(See Appendix D, Principles of Quantitative Reasoning)
1. Each "turn" of the Krebs cycle
   a. generates 3 NAD^+, 1 FADH, and 2 CO_2
   b. generates 1 GTP, 2 CO_2, and 1 $FADH_2$
   c. consumes 1 pyruvate and 1 oxaloacetate
   d. consumes an amino acid
2. Let's consider how much ATP you synthesize in a day. Assume that you consume 1 mole of O_2 per hour or 24 moles/day (a mole is the number of grams of a chemical equal to its molecular weight). About 6 moles of ATP are produced per mole of O_2 consumed. The molecular weight of ATP is 507. How many grams of ATP do you produce per day at this rate? Given that 1,000 g equal 2.2 pounds, how many pounds of ATP do you produce per day at this rate? (This is under relatively inactive conditions!)
3. Under resting circumstances a person produces about 144 moles ATP per day (73,000 g ATP/day). The amount of *free energy* represented by this amount of ATP can be calculated as follows. Cleavage of the terminal phosphate bond from ATP results in a decrease of free energy of approximately 7,300 cal/mole. This is a crude measure of the energy available to do work that is contained in the terminal phosphate bond of the ATP molecule. How many calories, in the form of ATP, are produced per day by a resting individual, crudely speaking?
4. Calculate the number of cells in the body of an average 68-kg (150-pound) adult. (This will only be accurate to about 1 part in 10 but should give you an idea how this commonly quoted number is arrived at). Assume all cells are spheres 20 μm in diameter. The volume of a sphere can be determined by the equation $v = \frac{4}{3} \pi r^3$. *Hint*: We know that about two-thirds of the water in the body is intracellular, and the density of cells is nearly 1 g/ml. The proportion of the mass made up of water is about 60%.
5. If sucrose is injected into the bloodstream, it tends to stay out of the cells (cells do not use sucrose directly). If it doesn't go into cells, where does it go? In other words, how much "space" is in the body that is not inside some cell? Sucrose can be used to determine this space. Suppose 150 mg of sucrose is injected into a 55-kg woman. If the concentration of sucrose in her blood is 0.015 mg/ml, what is the volume of her extracellular space, assuming that no metabolism is occurring and that the blood sucrose concentration is equal to the sucrose concentration throughout the extracellular space?

POINTS TO PONDER

(Explanations on p. A-36)
1. The stomach has two types of exocrine secretory cells: *chief cells* that secrete an inactive form of a protein-digesting enzyme, *pepsinogen*, and *parietal cells* that secrete *hydrochloric acid* (HCl), which activates pepsinogen. Both these cells have an abundance of mitochondria for ATP production—the chief cells for the energy needed to synthesize pepsinogen, the parietal cells for the energy needed to transport H^+ and Cl^- from the blood into the stomach lumen. Only one of these cell types also has an extensive rough endoplasmic reticulum and abundant Golgi stacks. Would this type be the chief cells or the parietal cells? Why?
2. The poison *cyanide* acts by binding irreversibly to one of the components of the electron transport chain, blocking its action. As a result, the entire electron-transport process comes to a screeching halt, and the cells lose over 94% of their ATP-producing capacity. Considering the types of cellular activities that depend on energy expenditure, what would be the consequences of cyanide poisoning?
3. Hydrogen peroxide, which belongs to a class of very unstable compounds known as *free radicals*, can bring about drastic, detrimental changes in a cell's structure and function by reacting with almost any molecule with which it comes in contact, including DNA. The resultant cellular changes can lead to genetic mutations, cancer, or other serious consequences. Furthermore, some researchers speculate that cumulative effects of more subtle cellular damage resulting from free radical reactions over a period of time might contribute to the gradual deterioration associated with aging. Related to this speculation, studies have shown that longevity decreases in fruit flies in direct proportion to a decrease in a specific chemical found in one of the cellular organelles. Based on your knowledge of how the body rids itself of dangerous hydrogen peroxide, what do you think this chemical in the organelle is?
4. Why do you think a person is able to perform anaerobic exercise (such as lifting and holding a heavy weight) only briefly but can sustain aerobic exercise (such as walking or swimming) for long periods? (*Hint*: Muscles have limited energy stores.)
5. One type of the affliction *epidermolysis bullosa* is caused by a genetic defect that results in production of abnormally weak keratin. Based on your knowledge of the role of keratin, what part of the body do you think would be affected by this condition?
6. *Clinical Consideration.* Kevin S. and his wife have been trying to have a baby for the past three years. On seeking the help of a fertility specialist, Kevin learned that he has a hereditary form of male sterility involving nonmotile sperm. His condition can be traced to defects in the cytoskeletal components of the sperm's flagella. Based on this finding, the physician suspected that Kevin also has a long history of recurrent respiratory tract disease. Kevin confirmed that indeed he has had colds, bronchitis, and influenza more frequently than his friends. Why would the physician suspect that Kevin probably had a history of frequent respiratory disease based on his diagnosis of sterility due to nonmotile sperm?

PHYSIOEDGE RESOURCES

PHYSIOEDGE CD-ROM
Figures marked with the PhysioEdge icon have associated activities on the CD. For this chapter, check out:

Transport Across Membranes
Membrane Potential

The diagnostic quiz allows you to receive immediate feedback on your understanding of the concept and to advance through various levels of difficulty.

CHECK OUT THESE MEDIA QUIZZES:
2.1 Anatomy of a Generic Cell
2.2 Basic Functions of Organelles

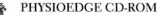
PHYSIOEDGE WEB SITE
The Web site for this book contains a wealth of helpful study aids, as well as many ideas for further reading and research. Log on to:

http://www.brookscole.com

Go to the Biology page and select Sherwood's *Human Physiology*, 5th Edition. Select a chapter from the drop-down menu or click on one of these resource areas:

- **Case Histories** provide an introduction to the clinical aspects of human physiology.

- For 2-D and 3-D graphical illustrations and animations of physiological concepts, visit our **Visual Learning Resource.**

- Resources for study and review include the **Chapter Outline, Chapter Summary, Glossary,** and **Flash Cards.** Use our **Quizzes** to prepare for in-class examinations.

- On-line research resources to consult are **Hypercontents,** with current links to relevant Internet sites; **Internet Exercises** with starter URLs provided; and **InfoTrac Exercises.**

 For more readings, go to InfoTrac College Edition, your online research library, at:

http://infotrac.thomsonlearning.com

Body Systems

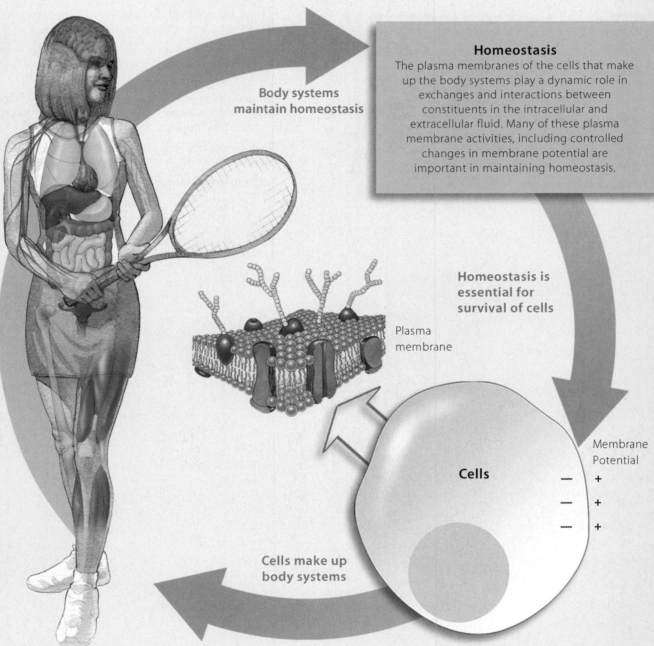

Body systems
maintain homeostasis

Homeostasis
The plasma membranes of the cells that make up the body systems play a dynamic role in exchanges and interactions between constituents in the intracellular and extracellular fluid. Many of these plasma membrane activities, including controlled changes in membrane potential are important in maintaining homeostasis.

Homeostasis is
essential for
survival of cells

Plasma
membrane

Membrane
Potential

Cells

Cells make up
body systems

All cells are enveloped by a **plasma membrane,** a thin, flexible, lipid barrier that separates the contents of the cell from its surroundings. To carry on life-sustaining and specialized activities, each cell must exchange materials across this membrane with the homeostatically maintained internal fluid environment that surrounds it. This discriminating barrier contains specific proteins, some of which enable selective passage of materials. Other membrane proteins serve as receptor sites for interaction with specific chemical messengers in the cell's environment. These messengers control many cellular activities crucial to homeostasis.

Cells have a membrane potential, a slight excess of negative charges lined up along the inside of the membrane and a slight excess of positive charges on the outside. The specialization of nerve and muscle cells depends on the ability of these cells to alter their potential on appropriate stimulation.

Chapter 3

The Plasma Membrane and Membrane Potential

CONTENTS AT A GLANCE

MEMBRANE STRUCTURE AND COMPOSITION

▌ Trilaminar appearance

▌ Membrane composition; fluid mosaic model

▌ Functions of membrane components

CELL-TO-CELL ADHESIONS

▌ Extracellular matrix

▌ Specialized cell junctions: desmosomes, tight junctions, gap junctions

INTERCELLULAR COMMUNICATION AND SIGNAL TRANSDUCTION

▌ Types of cell-to-cell communication

▌ Signal transduction

▌ Channel gating mechanisms

▌ Second-messenger systems

OVERVIEW OF MEMBRANE TRANSPORT

▌ Influence of lipid solubility and size of particle

▌ Active versus passive transport

UNASSISTED MEMBRANE TRANSPORT

▌ Diffusion down concentration gradients

▌ Movement along electrical gradients

▌ Osmosis

ASSISTED MEMBRANE TRANSPORT

▌ Carrier-mediated transport

▌ Vesicular transport

▌ Caveolae

MEMBRANE POTENTIAL

▌ Definition of potential

▌ Ionic basis of resting membrane potential

The survival of every cell depends on the maintenance of intracellular contents unique for that cell type despite the remarkably different composition of the extracellular fluid surrounding it. This difference in fluid composition inside and outside a cell is maintained by the **plasma membrane,** an extremely thin layer of lipids and proteins that forms the outer boundary of every cell and encloses the intracellular contents. In addition to serving as a mechanical barrier that traps needed molecules within the cell, the plasma membrane plays an active role in determining the composition of the cell by selectively permitting specific substances to pass between the cell and its environment. Besides controlling the entry of nutrient molecules and the exit of secretory and waste products, the plasma membrane maintains differences in ion concentrations between the cell's interior and exterior. These ionic differences, as you will learn, are important in the electrical activity of the plasma membrane. Also, the plasma membrane participates in the joining of cells together to form tissues and organs. Furthermore, it plays a key role in the ability of a cell to respond to changes, or signals, in the cell's environment. This ability is important in communication between cells. No matter what the cell type, these common membrane functions are crucial to the cell's survival, to its ability to perform specialized homeostatic activities, and to its ability to function cooperatively and in a coordinated fashion with other cells.

In this chapter, we examine the common structural and functional patterns shared by plasma membranes of all cells. We also explore how many of the functional differences between cell types are due to subtle variations in the composition of their plasma membranes. For example, slight modifications in specific protein components of the plasma membranes of various cell types enable different cells to interact in different ways with essentially the same extracellular fluid environment. To illustrate, thyroid gland cells are the only cells in the body to use iodine. Appropriately, a membrane protein unique to the plasma membranes of thyroid gland cells permits these cells (and no others) to take up iodine from the blood.

MEMBRANE STRUCTURE AND COMPOSITION

The plasma membrane is too thin to be seen under an ordinary light microscope, but with an electron microscope it appears as a **trilaminar structure** consisting of two dark layers separated by a light middle layer (● Figure 3-1) (*tri* means "three;" *lamina* means "layer"). The specific arrangement of the molecules that make up the plasma membrane is responsible for this three-layered "sandwich" appearance.

▌ The plasma membrane is a fluid lipid bilayer embedded with proteins.

The plasma membrane of every cell consists mostly of lipids (fats) and proteins plus small amounts of carbohydrate. The most abundant membrane lipids are phospholipids, with lesser amounts of cholesterol. An estimated billion phospholipid molecules are present in the plasma membrane of a typical human cell. **Phospholipids** have a polar (electrically charged; see p. A-13) head containing a negatively charged phosphate group and two nonpolar (electrically neutral) fatty acid tails

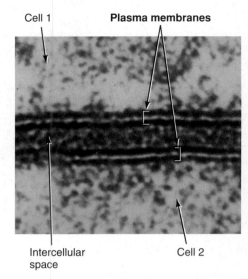

Cell 1   Plasma membranes

Intercellular space   Cell 2

● **FIGURE 3-1**

Trilaminar appearance of a plasma membrane in an electron micrograph
Depicted are the plasma membranes of two adjacent cells. Note the trilaminar structure (that is, two dark layers separated by a light middle layer) of each membrane.

● **FIGURE 3-2**

Structure and organization of phospholipid molecules in a lipid bilayer
(a) Phospholipid molecule. (b) When in contact with water, phospholipid molecules organize themselves into a lipid bilayer with the polar heads interacting with the polar water molecules at each surface and the nonpolar tails all facing the interior of the bilayer. (c) An exaggerated view of the plasma membrane enclosing a cell, separating the ICF from the ECF.

(SOURCE: Adapted from Cecie Starr and Ralph Taggart, *Biology: The Unity and Diversity of Life*, Eighth Edition, Fig. 4-26, p. 56. © 1998 Wadsworth Publishing Company.)

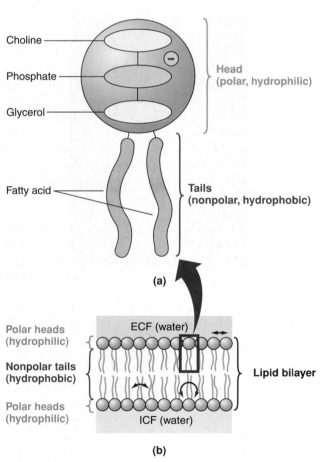

Choline

Phosphate

Glycerol

Head (polar, hydrophilic)

Fatty acid

Tails (nonpolar, hydrophobic)

(a)

Polar heads (hydrophilic)

Nonpolar tails (hydrophobic)

Polar heads (hydrophilic)

ECF (water)

ICF (water)

Lipid bilayer

(b)

⊖ = Negative charge on phosphate group

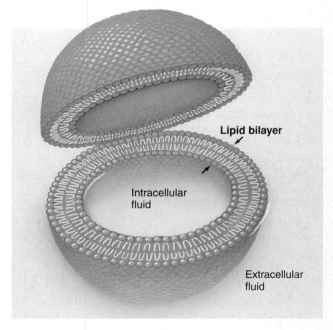

Lipid bilayer

Intracellular fluid

Extracellular fluid

(c)

(● Figure 3-2a). The polar end is hydrophilic ("water loving") because it can interact with water molecules, which are also polar; the nonpolar end is hydrophobic ("water fearing") and will not mix with water. Such two-sided molecules self-assemble into a **lipid bilayer,** a double layer of lipid molecules, when in contact with water (● Figure 3-2b) (*bi* means "two"). The hydrophobic tails bury themselves in the center away from the water, while the hydrophilic heads line up on both sides in contact with the water. The outer surface of the layer is exposed to extracellular fluid (ECF), whereas the inner surface is in contact with the intracellular fluid (ICF) (● Figure 3-2c).

The lipid bilayer is not a rigid structure but instead is fluid in nature, with a consistency more like liquid cooking oil than like solid shortening. The phospholipids, which are not held together by strong chemical bonds, are able to twirl around rapidly as well as move about within their own half of the layer, much like skaters on a crowded skating rink.

Cholesterol also contributes to the fluidity as well as the stability of the membrane. The cholesterol molecules are tucked in between the phospholipid molecules, where they prevent the fatty acid chains from packing together and crystallizing, a process that would drastically reduce membrane fluidity.

Because of its fluid nature, the plasma membrane has structural integrity but at the same time is flexible, enabling the cell to change its shape. For example, muscle cells change shape as they contract, and red blood cells must change shape considerably as they squeeze their way single file through the capillaries, the tiniest of blood vessels.

The **membrane proteins** are attached to or inserted within the lipid bilayer (● Figure 3-3). Some of these proteins extend through the entire thickness of the membrane; they have polar regions at both ends joined by a nonpolar central portion. Other proteins stud only the outer or inner surface; they are anchored by interactions with a protein that spans the membrane or by attachment to the lipid bilayer. The plasma membrane has about 50 times more lipid molecules than protein molecules. However, the proteins account for nearly half of the membrane's mass because the proteins are much larger than the lipids. The fluidity of the lipid bilayer enables many membrane proteins to float freely like "icebergs" in a moving "sea" of lipid, although the cytoskeleton restricts the mobility of proteins that perform a specialized function in a specific area of the cell. This view of membrane structure is known as the **fluid mosaic model,** in reference to the membrane fluidity and the ever-changing

● **FIGURE 3-3**

Fluid mosaic model of plasma membrane structure
The plasma membrane is composed of a lipid bilayer embedded with proteins. Some of these proteins extend through the thickness of the membrane, some are partially submerged in the membrane, and others are loosely attached to the surface of the membrane. Short carbohydrate chains are attached to proteins or lipids on the outer surface only.

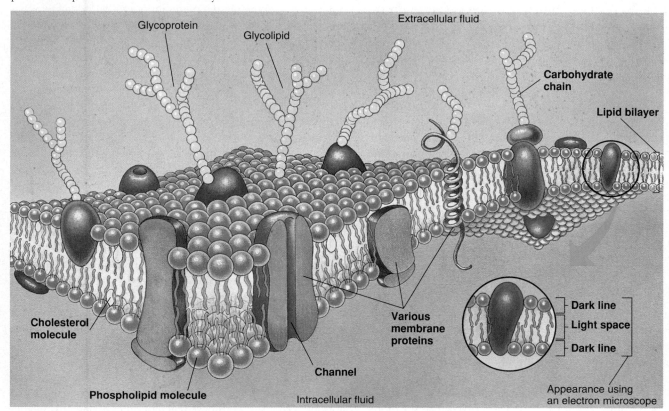

mosaic pattern of the proteins embedded within the lipid bilayer. (A mosaic is a surface decoration made by inlaying small pieces of variously colored tiles to form patterns or pictures.)

The small amount of **membrane carbohydrate** is located only at the outer surface. Thus, your cells are "sugar coated." Short-chain carbohydrates protrude like tiny antennas from the outer surface, bound primarily to membrane proteins and to a lesser extent to lipids. These sugary combinations are known as *glycoproteins* and *glycolipids*, respectively (● Figure 3-3).

This proposed structure can account for the trilaminar appearance of the plasma membrane. When stains are used to help visualize the plasma membrane under an electron microscope, the two dark lines represent the hydrophilic polar regions of the lipid and protein molecules that have taken up the stain. The light space between corresponds to the poorly stained hydrophobic core formed by the nonpolar regions of these molecules.

The different components of the plasma membrane carry out a variety of functions. As a quick preview, the lipid bilayer forms the primary barrier to diffusion, the proteins perform most of the specific membrane functions, and the carbohydrates play an important role in "self"-recognition processes. We will now examine the functions of each of these membrane components in more detail.

■ The lipid bilayer forms the basic structural barrier that encloses the cell.

The lipid bilayer serves three important functions:

1. It forms the basic structure of the membrane. The phospholipids can be visualized as the "pickets" that form the "fence" around the cell.

2. Its hydrophobic interior serves as a barrier to passage of water-soluble substances between the ICF and ECF. Water-soluble substances cannot dissolve in and pass through the lipid bilayer. (However, water molecules themselves are small enough to pass between the molecules that form this barrier.) By means of this barrier, the cell is able to maintain different mixtures and concentrations of solutes inside and outside the cell.

3. It is responsible for the fluidity of the membrane.

■ The membrane proteins perform a variety of specific membrane functions.

Different types of membrane proteins serve the following specialized functions:

1. Some proteins span the membrane to form water-filled pathways, or **channels,** across the lipid bilayer. Their presence enables water-soluble substances that are small enough to enter a channel to pass through the membrane without coming into direct contact with the hydrophobic lipid interior (● Figure 3-3). Channels are highly selective. Their small diameter precludes passage of particles greater than 0.8 nm (40 billionths of an inch) in diameter. Only small ions can fit through channels.

Furthermore, a given channel can selectively attract or repel particular ions. For example, only sodium (Na^+) can pass through Na^+ channels, whereas only potassium (K^+) can pass through K^+ channels. This selectivity is due to specific arrangements of chemical groups on the interior surfaces of the proteins that form the channel walls. Cells vary in the number, kind, and activity of channels they possess. It is even possible for a given channel to be *open* or *closed* to its specific ion as a result of changes in channel shape in response to a controlling mechanism. This is a good example of function depending on structural details. (To learn how a specific channel defect can lead to a devastating disease, see the accompanying boxed feature, ▶ Concepts, Challenges, and Controversies.)

2. Another group of proteins that span the membrane serve as **carrier molecules,** which transfer specific substances across the membrane that are unable to cross on their own. (Thus, channels and carrier molecules are both important in the transport of substances between the ECF and ICF.) Each carrier can transport only a particular molecule or closely related molecules. Cells of different types have different kinds of carriers. As a result, they vary as to which substances they are able to selectively transport across their membranes. For example, only the plasma membranes of thyroid gland cells have carriers for iodine, so these cells alone can transport this element from the blood into the cell interior.

3. Other proteins located on the inner membrane surface are the t-SNAREs, which serve as **docking-marker acceptors** that bind lock-and-key fashion with the v-SNAREs of secretory vesicles (see p. 30). Secretion is initiated as stimulatory signals trigger the fusion of the secretory vesicle membrane with the inner surface of the plasma membrane through interactions between their SNARE proteins. The secretory vesicle subsequently opens up and empties its contents to the outside by exocytosis.

4. Another group of surface-located proteins function as **membrane-bound enzymes** that control specific chemical reactions at either the inner or the outer cell surface. Cells display specialization in the types of enzymes embedded within their plasma membranes. For example, the outer layer of the plasma membrane of skeletal muscle cells contains an enzyme that destroys the chemical messenger that triggers muscle contraction, thus enabling the muscle to relax.

5. Many of the proteins on the outer surface serve as **receptor sites** (or, simply, **receptors**) that "recognize" and bind with specific molecules in the environment of the cell. This binding initiates a series of membrane and intracellular events (to be described later) that alter the activity of the particular cell. In this way, chemical messengers in the blood, such as water-soluble hormones, are able to influence only the specific cells that possess receptors for a given messenger. Even though every cell is exposed to the same messenger via its widespread distribution by the blood, a given messenger has no effect on other cells that do not have receptors for this specific messenger. To illustrate, the anterior pituitary gland secretes into the blood thyroid-stimulating hormone (TSH), which can attach only to the surface of thyroid gland cells to stimulate secretion of thyroid hormone. No other cells have receptors for TSH, so only thyroid cells are influenced by TSH despite its widespread distribution.

Cystic Fibrosis: A Fatal Defect in Membrane Transport

Cystic fibrosis (CF), the most common fatal genetic disease in the United States, strikes 1 in every 2,000 Caucasian children. It is characterized by the production of an abnormally thick, sticky mucus. Most dramatically affected are the respiratory airways and the pancreas.

Respiratory Problems
The presence of the thick, sticky mucus in the respiratory airways makes it difficult to get adequate air in and out of the lungs. Also, because bacteria thrive in the accumulated mucus, CF patients suffer from repeated respiratory infections. They are especially susceptible to *Pseudomonas aeruginosa,* an "opportunistic" bacterium that is frequently present in the environment but usually causes infection only when some underlying problem handicaps the body's defenses. Gradually, the involved lung tissue becomes scarred (fibrotic), making the lungs harder to inflate. This complication increases the work of breathing beyond the extra work required to move air through the clogged airways.

Underlying Cause
During the last decade, researchers found that cystic fibrosis is caused by any one of several different genetic defects that lead to production of a flawed version of a protein known as *cystic fibrosis transmembrane conductance regulator (CFTR)*. CFTR normally helps form and regulate the chloride (Cl^-) channels in the plasma membrane. With CF, the defective CFTR "gets stuck" in the endoplasmic reticulum/Golgi system, which normally manufactures and processes this product and ships it to the plasma membrane (see p. 29). That is, in CF patients the mutated version of CFTR is only partially processed and never makes it to the cell surface. The resultant absence of CFTR protein in the plasma membrane's Cl^- channels leads to membrane impermeability to Cl^-. Because Cl^- transport across the membrane is closely linked to Na^+ transport, cells lining the respiratory airways are unable to absorb salt (NaCl) properly. As a result, salt accumulates in the fluid lining the airways.

What has puzzled researchers is how this Cl^- channel defect and resultant salt accumulation lead to the excess mucus problem. Two recent discoveries have perhaps provided an answer, although these proposals remain to be proven and research into other possible mechanisms continue to be pursued. One group of investigators found that the airway cells produce a natural antibiotic, *defensin,* which normally kills most of the inhaled airborne bacteria. It turns out that defensin cannot function properly in a salty environment. Bathed in the excess salt associated with CF, the disabled antibiotic cannot rid the lungs of inhaled bacteria. This leads to repeated infections. One of the outcomes of the body's response to these infections is excess mucus production. In turn, this mucus serves as a breeding ground for more bacterial growth. The cycle continues as the lung-clogging mucus accumulation and frequency of lung infections grows ever worse. To make matters worse, the excess mucus is especially thick and sticky, making it difficult for the normal ciliary defense mechanisms of the lung to sweep up the bacteria-laden mucus from the lungs (see p. 47 and p. 452). The mucus is thick and sticky because it is underhydrated (has too little water), a problem believed to be linked to the defective salt transport.

The second new study found an additional complicating factor in the CF story. These researchers demonstrated that CFTR appears to serve a dual role as a Cl^- channel and as a membrane receptor that binds with *Pseudomonas aeruginosa* (and perhaps other bacteria). CFTR subsequently destroys the captured bacteria. In the absence of CFTR in the airway cell membranes of CF patients, *P. aeruginosa* is not cleared from the airways as usual. In a double onslaught, these bacteria were shown to trigger the airway cells to produce unusually large amounts of an abnormal, thick, sticky mucus. This mucus promotes more bacterial growth, as the vicious cycle continues.

Pancreatic Problems
Furthermore, in CF patients the pancreatic duct, which carries secretions from the pancreas to the small intestine, becomes plugged with thick mucus. Because the pancreas produces enzymes important in the digestion of food, malnourishment eventually results. In addition, as the pancreatic digestive secretions accumulate behind the blocked pancreatic duct, fluid-filled cysts form in the pancreas, with the affected pancreatic tissue gradually degenerating and becoming fibrotic. The name "cystic fibrosis" aptly describes long-term changes that occur in the pancreas and lungs as the result of a single genetic flaw in CFTR.

Treatment
Treatment consists of physical therapy to help clear the airways of the excess mucus and antibiotic therapy to combat respiratory infections, plus special diets and administration of supplemental pancreatic enzymes to maintain adequate nutrition. Despite this supportive treatment, most CF victims do not survive beyond their early 20s, with most dying from lung complications.

With the recent discovery of the genetic defect responsible for the majority of CF cases, investigators are hopeful of developing a means to correct or compensate for the defective gene. For example, researchers recently reported success in inserting a healthy human CFTR gene into a disabled cold virus that was unable to cause disease but could still penetrate the cells lining the respiratory airways. When this gene-carrying virus was deposited in the lungs of CF patients in a limited clinical study, the stowaway gene produced human CFTR, although in insufficient quantity to remedy the problem. Although of limited success, these preliminary results lead scientists to believe that gene therapy could one day be a potential cure for CF. Scientists are currently searching for a better means of shuttling the healthy gene into the patient's respiratory airways. The treatment would have to be repeated several times per year as old respiratory airway cells die and are replaced by new cells.

Another potential cure being studied is development of drugs that induce the mutated CFTR to be "finished off" and inserted in the plasma membrane. Furthermore, several promising new drug therapy approaches, such as a mucus-thinning aerosol drug that can be inhaled, offer hope of reducing the number of lung infections and extending the life span of CF victims until a cure can be found.

---

6. Still other proteins serve as **cell adhesion molecules (CAMs).** Many CAMs protrude from the outer membrane surface and form loops or hooks that the cells use to grip each other and to grasp the connective tissue fibers that interlace between cells. For example, *cadherins,* one type of CAM, on the surface of adjacent cells interlock in zipper fashion to help hold the cells within tissues and organs together. Some CAMs, such as the *integrins,* span the plasma membrane. Integrins not only serve as a structural link between the outer membrane surface and its extracellular surroundings but also connect the

inner membrane surface to the intracellular cytoskeletal scaffolding. These CAMs mechanically link the cell's external environment and intracellular components. Furthermore, integrins can also relay regulatory signals through the plasma membrane in either direction. Although CAMs originally were believed to serve only as adhesive molecules, investigators have now learned that some of them also act as "signaling molecules." These CAMs participate in signaling cells to grow and in signaling immune system cells to interact with the right kind of other cells in inflammatory responses and wound healing, among other things.

7. Finally, still other proteins on the outer membrane surface, especially in conjunction with carbohydrates (as glycoproteins), are important in the cells' ability to recognize "self" (that is, cells of the same type) and in cell-to-cell interactions.

▌ The membrane carbohydrates serve as self-identity markers.

The short sugar chains on the outer membrane surface serve as self-identity markers that enable cells to identify and interact with each other in the following ways:

1. Different cell types have different markers. The unique combination of sugar chains projecting from the surface membrane proteins serves as the "trademark" of a particular cell type, enabling a cell to recognize others of its own kind. Thus these carbohydrate chains play an important role in recognition of "self" and in cell-to-cell interactions. Cells can recognize other cells of the same type and join together to form tissues. This is especially important during embryonic development. If cultures of embryonic cells of two different types, such as nerve cells and muscle cells, are mixed together, the cells will sort themselves into separate aggregates of nerve cells and muscle cells.

2. Carbohydrate-containing surface markers also appear to be involved in tissue growth, which is normally held within certain limits of cell density. Cells do not "trespass" across the boundaries of neighboring tissues; that is, they do not overgrow their own territory.

CELL-TO-CELL ADHESIONS

In multicellular organisms such as humans, plasma membranes not only serve as the outer boundaries of all cells but also participate in cell-to-cell adhesions. These adhesions bind groups of cells together into tissues and package them further into organs. The life-sustaining activities of the body systems depend not only on the functions of the individual cells of which they are made but also on how these cells live and work together in tissue and organ communities.

Organization of cells into appropriate groupings is at least partially attributable to the carbohydrate chains on the membrane surface. Once arranged, cells are held together by three different means: (1) the extracellular matrix, (2) cell adhesion molecules in the cells' plasma membranes, and (3) specialized cell junctions.

▌ The extracellular matrix serves as the biological "glue."

Tissues are not made up solely of cells, and many cells within a tissue are not in direct physical contact with neighboring cells. Instead, they are held together by the **extracellular matrix** (**ECM**), an intricate meshwork of fibrous proteins embedded in a watery, gel-like substance composed of complex carbohydrates. The ECM serves as the biological "glue." The watery gel provides a pathway for diffusion of nutrients, wastes, and other water-soluble traffic between the blood and tissue cells. It is usually called the *interstitial fluid* (see p. 10). Interwoven within this gel are three major types of protein fibers: collagen, elastin, and fibronectin.

1. **Collagen** forms cablelike fibers or sheets that provide tensile strength (resistance to longitudinal stress). In *scurvy*, a condition caused by vitamin C deficiency, these fibers are not properly formed. As a result, the tissues, especially those of the skin and blood vessels, become very fragile. This leads to bleeding in the skin and mucous membranes, which is especially noticeable in the gums.

2. **Elastin** is a rubberlike protein fiber most abundant in tissues that must be capable of easily stretching and then recoiling after the stretching force is removed. It is found, for example, in the lungs, which stretch and recoil as air moves in and out.

3. **Fibronectin** promotes cell adhesion and holds cells in position. Reduced amounts of this protein have been found within certain types of cancerous tissue, possibly accounting for the fact that cancer cells do not adhere well to each other but tend to break loose and metastasize (spread elsewhere in the body).

The extracellular matrix is secreted by local cells present in the matrix. The relative amount of extracellular matrix compared to cells varies greatly among tissues. For example, the extracellular matrix is scant in epithelial tissue but is the predominant component of connective tissue. Most of this abundant matrix in connective tissue is secreted by **fibroblasts** ("fiber formers"). The exact composition of ECM components varies for different tissues, thus providing distinct local environments for the various cell types in the body. In some tissues, the matrix becomes highly specialized to form such structures as cartilage or tendons or, on appropriate calcification, the hardened structures of bones and teeth.

New information suggests that the ECM is not just a passive scaffolding for cellular attachment but also helps regulate the behavior and functions of the cells with which it interacts. Investigators have shown that cells are able to function normally and indeed even to survive only when in association with their normal matrix components. The matrix is especially influential in cell growth and differentiation. In the body, only circulating blood cells are designed to survive and function without attaching to the ECM.

▌ Some cells are directly linked together by specialized cell junctions.

In tissues where the cells lie in close proximity to each other, some tissue cohesion is provided by the cell adhesion molecules, or CAMs. As you just learned, these special loop- and hook-shaped surface membrane proteins "Velcro" adjacent cells to each other. In addition, some cells within given types of tissues are directly linked together by one of three types of specialized cell junctions: (1) desmosomes (adhering junctions), (2) tight junctions (impermeable junctions), or (3) gap junctions (communicating junctions). Let's examine these junctions in more detail.

Desmosomes

Desmosomes act like "spot rivets" that anchor together two closely adjacent but nontouching cells. A desmosome consists of two components: (1) a pair of dense, buttonlike cytoplasmic thickenings known as *plaque* located on the inner surface of each of the two adjacent cells; and (2) strong glycoprotein filaments that extend across the space between the two cells and attach to the plaque on both sides (● Figure 3-4). These intercellular filaments bind adjacent plasma membranes together so that they resist being pulled apart. Thus desmosomes are adhering junctions.

Desmosomes are distributed widely throughout the body. They are most abundant in tissues that are subject to considerable stretching, such as the epithelial tissue of the skin, the cardiac muscle tissue of the heart, and the smooth muscle tissue of the uterus. In these tissues, functional groups of cells are riveted together by desmosomes that extend from one cell to the next, then from that cell to the next, and so on. Furthermore, intermediate cytoskeletal filaments, such as tough keratin filaments in the skin (see p. 51), stretch across the interior of these cells and attach to the desmosome plaques located on opposite sides of the cell's inner surface. This arrangement forms a continuous network of strong fibers extending throughout the tissue, both through the cells and between the cells, much like a continuous line of people firmly holding hands. This interlinking fibrous network provides tensile strength, reducing the chances of the tissue being torn when stretched.

Tight junctions

At **tight junctions,** adjacent cells firmly adhere to each other at points of direct contact to seal off passageway between the two cells. They are found primarily in sheets of epithelial tissue. Epithelial tissue covers the surface of the body and lines all its internal cavities. All epithelial sheets serve as highly selective barriers between two compartments that have considerably different chemical compositions. For example, the epithelial sheet lining the digestive tract separates the food and potent

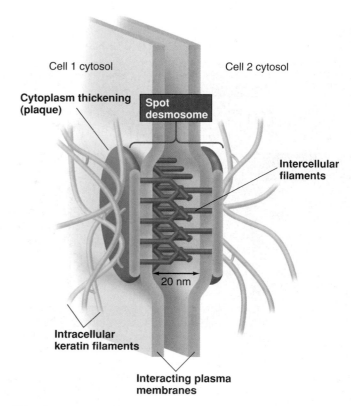

● **FIGURE 3-4**
Spot desmosome
Spot desmosomes are adhering junctions that anchor cells together in tissues subject to considerable stretching.

digestive juices within the inner cavity (lumen) from the blood vessels that lie on the other side. It is important that only completely digested food particles and not undigested food particles or digestive juices move across the epithelial sheet from the lumen to the blood. Accordingly, the lateral (side) edges of the adjacent cells in the epithelial sheet are joined together in a tight seal near their luminal border by "kiss sites," sites of direct fusion of *junctional proteins* on the outer surfaces of the two interacting plasma membranes (● Figure 3-5). These tight junctions are impermeable and thus prevent materials from passing between the cells. Passage across the epithelial barrier, therefore, must take place *through* the cells, not *between* them. This traffic across the cell is regulated by means of the channels and carriers present. If the cells were not joined by tight junctions, uncontrolled exchange of molecules could take place between the compartments by unpoliced traffic through the spaces between adjacent cells. Tight junctions thus prevent undesirable leaks within epithelial sheets. (See Chapters 14 and 16 for specific details on epithelial transport in the kidneys and digestive tract, respectively.)

Gap junctions

At a **gap junction,** as the name implies, a gap exists between adjacent cells, which are linked by small connecting tunnels known as **connexons.** Connexons are formed by the joining of

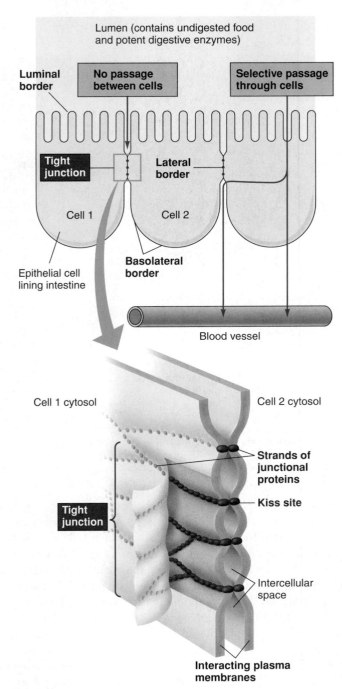

● FIGURE 3-5

Tight junction
Tight junctions are impermeable junctions that join the lateral edges of epithelial cells near their luminal borders, thus preventing materials from passing *between* the cells. Only regulated passage of materials can occur *through* these cells, which form highly selective barriers that separate two compartments of highly different chemical composition.

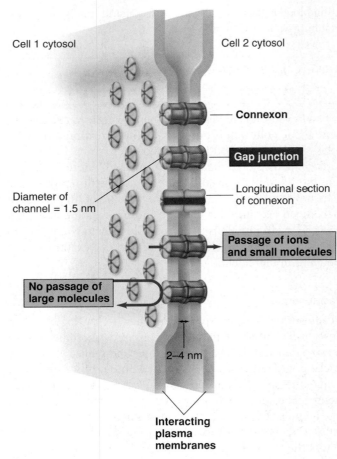

● FIGURE 3-6

Gap junction
Gap junctions are communicating junctions consisting of connexons, small connecting tunnels that permit movement of charge-carrying ions and small molecules between two adjacent cells.

proteins. Through these specialized anatomic arrangements, ions (electrically charged particles) and small molecules can be directly exchanged between interacting cells without ever entering the ECF.

Gap junctions are especially abundant in cardiac muscle and smooth muscle. In these tissues, movement of ions between cells through gap junctions plays an important role in transmitting electrical activity throughout an entire muscle mass. Because this electrical activity brings about contraction, the presence of gap junctions enables synchronized contraction of a whole muscle mass, such as the pumping chamber of the heart.

Many tissues that are not electrically active also are connected by gap junctions. The gap junctions in these nonmuscle tissues link the involved cells together metabolically by permitting unrestricted passage of small metabolically active molecules between the cells. Furthermore, these gap junctions also serve as avenues for the direct transfer of small signaling molecules from one cell to the next. Such transfer permits the cells connected by gap junctions to directly communicate with each other. This communication provides one

proteins that extend outward from each of the adjacent plasma membranes (● Figure 3-6). Gap junctions are communicating junctions. The small diameter of the tunnels permits small water-soluble particles to pass between the connected cells but precludes passage of large molecules such as vital intracellular

possible mechanism by which cooperative cell activity may be coordinated.

We are now going to turn our attention to other means by which cells "talk to each other."

INTERCELLULAR COMMUNICATION AND SIGNAL TRANSDUCTION

The ability of cells to communicate with each other is essential for coordination of their diverse activities to maintain homeostasis as well as to control growth and development of the body as a whole.

▌ Communication between cells is largely orchestrated by extracellular chemical messengers.

There are three types of intercellular ("between cell") communication (● Figure 3-7):

1. The most intimate means of intercellular communication is through gap junctions.

2. The presence of signaling molecules on the surface membrane of some cells permits them to directly link up transiently and interact with certain other cells in a specialized way. This is the means by which the phagocytes of the body's defense system specifically recognize and selectively destroy only undesirable cells, such as bacterial invaders, while leaving the body's own cells alone. One of the primary ways in which antibodies help defend us against invading bacteria is by attaching to the bacterial surface to form a coat that can be recognized by specific receptor sites in the plasma membrane of the phagocytic white blood cells. Such "marked" bacteria are quickly engulfed and destroyed (see p. 418).

3. The most common means by which cells communicate with each other is through **extracellular chemical messengers,** of which there are four types: *paracrines, neurotransmitters, hormones,* and *neurohormones.* In each case, a specific chemical messenger is synthesized by specialized cells to serve a designated purpose. On being released into the ECF by appropriate stimulation, these signaling agents act on other particular

● **FIGURE 3-7**

Types of intercellular communication
Gap junctions and transient direct linkup of cells are both means of direct communication between cells. Paracrines, neurotransmitters, hormones, and neurohormones are all extracellular chemical messengers that accomplish indirect communication between cells. These chemical messengers differ in their source and the distance they travel to reach their target cells.

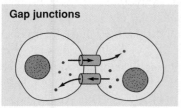

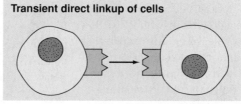

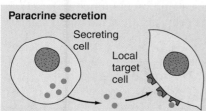

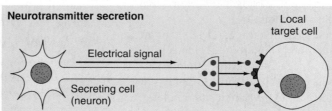

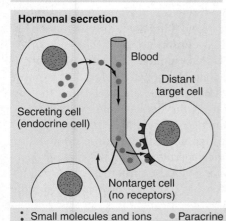

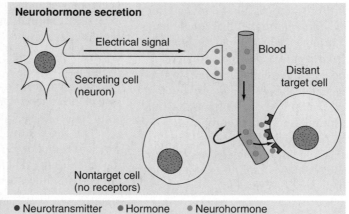

cells, the messenger's **target cells,** in a prescribed manner. To exert its effect, an extracellular chemical messenger, also known as a **ligand,** must bind with target cell receptors specific for it.

The four types of chemical messengers differ in their source and the distance and means by which they get to their site of action as follows:

- **Paracrines** are local chemical messengers whose effect is exerted only on neighboring cells in the immediate environment of their site of secretion. Because paracrines are distributed by simple diffusion, their action is restricted to short distances. They do not gain entry to the blood in any significant quantity because they are rapidly inactivated by locally existing enzymes. One example of a paracrine is *histamine*, which is released from a specific type of connective tissue cell during an inflammatory response within an invaded or injured tissue (see p. 416). Among other things, histamine dilates (opens more widely) the blood vessels in the vicinity to increase blood flow to the tissue. This action brings additional blood-borne combat supplies into the affected area.

Paracrines must be distinguished from chemicals that influence neighboring cells after being nonspecifically released during the course of cellular activity. For example, an increased local concentration of CO_2 in an exercising muscle is among the factors that promote local dilation of the blood vessels supplying the muscle. The resultant increased blood flow helps to meet the more active tissue's increased metabolic demands. However, CO_2 is produced by all cells and is not specifically released to accomplish this particular response, so it and similar nonspecifically released chemicals are not considered paracrines.

- Nerve cells (*neurons*) communicate directly with the cells they innervate (their target cells) by releasing **neurotransmitters,** which are very short-range chemical messengers, in response to electrical signals. Like paracrines, neurotransmitters diffuse from their site of release across a narrow extracellular space to act locally on only an adjoining target cell, which may be another neuron, a muscle, or a gland.

- **Hormones** are long-range chemical messengers that are specifically secreted into the blood by endocrine glands in response to an appropriate signal. The blood carries the messengers to other sites in the body, where they exert their effects on their target cells some distance away from their site of release. Only the target cells of a particular hormone have membrane receptors for binding with this hormone. Nontarget cells are not influenced by any blood-borne hormones that reach them.

- **Neurohormones** are hormones released into the blood by *neurosecretory neurons*. Like ordinary neurons, neurosecretory neurons can respond to and conduct electrical signals. Instead of directly innervating target cells, however, a neurosecretory neuron releases its chemical messenger, a neurohormone, into the blood on appropriate stimulation. The neurohormone is then distributed through the blood to the target cells. Thus, like endocrine cells, neurosecretory neurons release blood-borne chemical messengers, whereas ordinary neurons secrete short-range neurotransmitters into a confined space.

In every case, extracellular chemical messengers are released from one cell type and interact with other target cells to bring about a desired effect in the target cells. How do these chemical messengers bring about the desired cell response? Binding of extracellular chemical messengers to target membrane receptors brings about a wide range of responses in different cells through only a few remarkably similar pathways, to which we now turn our attention.

∎ Extracellular chemical messengers bring about cell responses primarily by signal transduction.

The term **signal transduction** refers to the process by which incoming signals (instructions from extracellular chemical messengers) are conveyed to the target cell's interior for execution. (A *transducer* is a device that receives energy from one system and transmits it in a different form to another system. For example, your radio receives radio waves sent out from the broadcast station and transmits these signals in the form of sound waves that can be detected by your ears.) Extracellular chemical messengers that cannot gain entry to the target cell, such as protein hormones delivered by the blood or neurotransmitters released from nerve endings, signal the cell to perform a given response by first binding with specialized membrane receptors specific for that given messenger. This combination of extracellular messenger with membrane receptor triggers a sequence of intracellular events that ultimately controls a particular cellular activity important in the maintenance of homeostasis, such as membrane transport, secretion, metabolism, or contraction.

Despite the wide range of possible responses, binding of the receptor with the extracellular chemical messenger (the **first messenger**) brings about the desired intracellular response by only two general means: (1) by opening or closing specific channels in the membrane to regulate the movement of particular ions into or out of the cell or (2) by transferring the signal to an intracellular chemical messenger (the **second messenger**), which in turn triggers a preprogrammed series of biochemical events within the cell. Because of the universal nature of these events, let us examine each more closely.

∎ Some extracellular chemical messengers open chemically gated channels.

Membrane channels may be either *leak channels* or *gated channels*. **Leak channels** are open all the time, thus permitting unregulated leakage of their chosen ion across the membrane through the channels. **Gated channels,** in contrast, behave as if they have "gates" that can be opened or closed in response to various controlling stimuli. Gated channels allow the passage of their chosen ion when they have been triggered to open. Thus ionic passage through a gated channel is regulated.

The three different types of gated channels depend on which of the following mechanisms brings about channel opening and closing: (1) binding of a chemical messenger to a specific membrane receptor that is in close association with the channel, (2) changes in the electrical status of the plasma membrane, and (3) stretching or other mechanical deformation of the channel. For now we will concentrate on channel regulation by means of chemical messengers (*chemically gated channels*) and defer discussion of the other two methods of controlling channels until later.

An extracellular chemical messenger can alter a channel through one of two mechanisms. For some channels, the receptor binding site on the plasma membrane is part of the channel itself. When an extracellular messenger binds with its receptor, the channel opens. For other channels, the receptor is a separate protein located near the channel. In this case, binding of the extracellular messenger with its receptor activates membrane-bound intermediaries known as *G proteins*, which in turn open (or in some instances close) the appropriate adjacent channel. (The mechanism of action of G proteins will be discussed shortly.)

By opening or closing specific channels, extracellular messengers can regulate the flow of particular ions across the membrane. This ionic movement can be responsible for two different cellular events:

1. In nerve and muscle cells, electrical signals can be produced in response to the opening of chemically gated channels for Na^+, K^+, or both. When these channels open, the resultant small, short-lived movement of these charge-carrying ions across the membrane through these open channels generates electrical signals. This important regulatory mechanism serves as a foundation for nerve and muscle physiology.

2. In some cells, a transient flow of calcium (Ca^{2+}) into the cell through opened Ca^{2+} channels triggers a change in the shape and function of specific intracellular proteins, which leads to the cell's response. For example, the activity of a specific enzymatic protein might be increased.

On completion of the response, the extracellular messenger is removed from the receptor site, and the chemically gated channels close once again. The ions that moved across the membrane through opened channels to trigger the response are returned to their original location by special membrane carriers.

Cytosolic Ca^{2+} may be increased in another way besides entry from the ECF through opened membrane channels; intracellular Ca^{2+} stores can be released in a second-messenger pathway, as described in the next section.

▍ Many extracellular chemical messengers activate second-messenger pathways.

Many extracellular chemical messengers that cannot actually enter their target cells bring about the desired intracellular response by another means than opening chemically gated channels. These first messengers issue their orders by triggering a "Psst, pass it on" process. Binding of the first messenger to a membrane receptor serves as a signal for activating an intracellular second messenger. The second messenger ultimately relays the orders through a series of biochemical intermediaries to particular intracellular proteins that carry out the dictated response, such as changes in cellular metabolism or secretory activity. The intracellular pathways activated by a second messenger in response to binding of the first messenger to a surface receptor are remarkably similar among different cells despite the diversity of ultimate responses to that signal. The

variability in response depends on the specialization of the cell, not on the mechanism used.

There are two major second-messenger pathways: One uses **cyclic adenosine monophosphate (cyclic AMP, or cAMP)** as a second messenger, and the other employs Ca^{2+} in this role. Let's examine each of these second-messenger pathways in more detail.

Cyclic AMP second-messenger pathway

In the cyclic AMP pathway, binding of an appropriate extracellular messenger to its surface receptor eventually activates the enzyme **adenylyl cyclase** (step 1 in ● Figure 3-8) on the inner surface of the membrane. A membrane-bound "middleman," a **G protein,** acts as an intermediary between the receptor and adenylyl cyclase. G proteins are found on the inner surface of the plasma membrane. An unactivated G protein consists of a complex of alpha (α), beta (β), and gamma (γ) subunits. A number of different G proteins with varying alpha subunits have been identified. The different G proteins are activated in response to binding of various first messengers to surface receptors. When a first messenger binds with its receptor, the receptor attaches to the appropriate G protein, resulting in activation of the alpha subunit. Once activated, the alpha subunit breaks away from the G protein complex and moves along the inner surface of the plasma membrane until it reaches an **effector protein.** An effector protein is either an ion channel or an enzyme. The alpha subunit links up with the effector protein and alters its activity. Depending on the outcome signaled by the first messenger, the G protein either opens or closes a particular channel or activates or inhibits a particular enzyme. Researchers have identified more than 300 different receptors that convey instructions of extracellular messengers through the membrane to effector proteins by means of G proteins.

In the cyclic AMP pathway, the effector protein is the enzyme adenylyl cyclase, which is located on the cytoplasmic side of the plasma membrane. Adenylyl cyclase induces the conversion of intracellular ATP to cAMP by cleaving off two of the phosphates (step 2). (This is the same ATP used as the common energy currency in the body.) Cyclic AMP activates a specific intracellular enzyme, **protein kinase A (PKA; also known as cyclic AMP-dependent protein kinase)** (step 3). Protein kinase A in turn **phosphorylates** (attaches a phosphate group from ATP to) a specific intracellular protein (step 4), such as an enzyme important in a particular metabolic pathway. Phosphorylation causes the protein to change its shape and function (either activating or inhibiting it) (step 5). This altered protein brings about the desired response (step 6). For example, a given ion channel may be opened or closed or the activity of a particular enzymatic protein that regulates a specific metabolic event may be increased or decreased. After the response is accomplished, the alpha subunit rejoins the beta and gamma subunits to restore the inactive G-protein complex. Intracellular enzymes inactivate the other participating chemicals so that the response can be terminated. Otherwise, once triggered the response would go on indefinitely until the cell ran out of necessary supplies.

Note that in this signal transduction pathway, the steps involving the extracellular first messenger, the receptor, the G

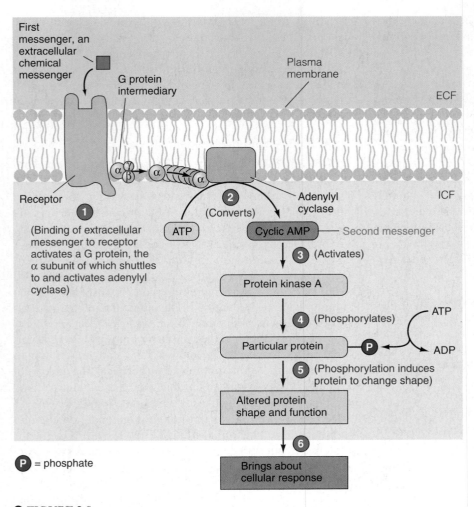

First messenger, an extracellular chemical messenger

G protein intermediary

Plasma membrane

ECF

Receptor

1

(Binding of extracellular messenger to receptor activates a G protein, the α subunit of which shuttles to and activates adenylyl cyclase)

ATP

2 (Converts)

Adenylyl cyclase

ICF

Cyclic AMP — Second messenger

3 (Activates)

Protein kinase A

4 (Phosphorylates)

ATP

Particular protein — P

ADP

5 (Phosphorylation induces protein to change shape)

Altered protein shape and function

6

Brings about cellular response

P = phosphate

1. Binding of an extracellular messenger, the *first messenger*, to a surface membrane receptor activates by means of a G protein intermediary the membrane-bound enzyme adenylyl cyclase.

2. Adenylyl cyclase converts intracellular ATP into cyclic AMP.

3. Cyclic AMP acts as an intracellular *second messenger*, triggering the desired cellular response by activating protein kinase A.

4. Protein kinase A in turn phosphorylates a particular intracellular protein.

5. Phosphorylation induces a change in the shape and function of the protein.

6. The altered protein then accomplishes the cellular response dictated by the extracellular messenger.

● FIGURE 3-8
Activation of the cyclic AMP second-messenger system by an extracellular messenger

protein complex, and the effector protein occur *in the plasma membrane* and lead to activation of the second messenger. The extracellular messenger cannot gain entry into the cell to "personally" deliver its message to the proteins that carry out the desired response. Instead, it initiates membrane events that activate an intracellular messenger, cAMP. The second messenger then triggers a chain reaction of biochemical events *inside the cell* that leads to the cellular response.

Different types of cells have different proteins available for phosphorylation and modification by protein kinase A. Therefore, a *common second messenger, cAMP, can induce widely differing responses in different cells*, depending on what proteins are modified. Cyclic AMP can be thought of as a commonly used molecular "switch" that can "turn on" (or "turn off") different cellular events, depending on the unique specialization of a particular cell type. This can be likened to being able to either illuminate or cool off a room depending on whether the wall switch you flip on is wired to a device specialized to light up (a chandelier) or one specialized to create air movement (a ceiling fan). In the body, the variable responsiveness once the switch is turned on is due to the genetically programmed differences in the sets of proteins within different cells. For exam-

ple, activation of the cAMP system brings about modification of heart rate in the heart, stimulation of the formation of female sex hormones in the ovaries, breakdown of stored glucose in the liver, control of water conservation during urine formation in the kidneys, creation of some simple memory traces in the brain, and perception of a sweet taste by a taste bud.

Calcium second-messenger pathway

Instead of cAMP, some cells use Ca^{2+} as a second messenger. In such cases, binding of the first messenger to the surface receptor eventually leads by means of G proteins to activation of the enzyme **phospholipase C,** a protein effector that is bound to the inner side of the membrane (step 1 in ● Figure 3-9). This enzyme breaks down **phosphatidylinositol bisphosphate** (abbreviated PIP_2), a component of the tails of the phospholipid molecules within the membrane itself. The products of PIP_2 breakdown are **diacylglycerol (DAG)** and **inositol trisphosphate (IP_3)** (step 2). IP_3 is the fragment responsible for mobilizing intracellular Ca^{2+} stores to increase cytosolic Ca^{2+} (step 3). Calcium then takes over the role of second messenger, ultimately bringing about the response dictated by the first messenger. Many of the Ca^{2+}-dependent cellular events are

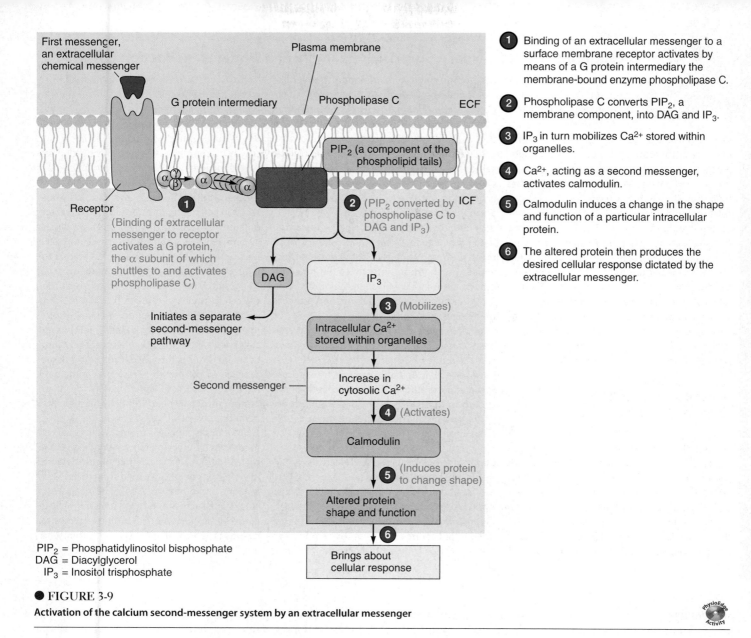

First messenger, an extracellular chemical messenger

Plasma membrane

G protein intermediary

Phospholipase C

ECF

PIP$_2$ (a component of the phospholipid tails)

Receptor

(Binding of extracellular messenger to receptor activates a G protein, the α subunit of which shuttles to and activates phospholipase C)

2 (PIP$_2$ converted by phospholipase C to DAG and IP$_3$)

ICF

DAG

IP$_3$

3 (Mobilizes)

Initiates a separate second-messenger pathway

Intracellular Ca^{2+} stored within organelles

Second messenger

Increase in cytosolic Ca^{2+}

4 (Activates)

Calmodulin

5 (Induces protein to change shape)

Altered protein shape and function

6

Brings about cellular response

PIP$_2$ = Phosphatidylinositol bisphosphate
DAG = Diacylglycerol
IP$_3$ = Inositol trisphosphate

1 Binding of an extracellular messenger to a surface membrane receptor activates by means of a G protein intermediary the membrane-bound enzyme phospholipase C.

2 Phospholipase C converts PIP$_2$, a membrane component, into DAG and IP$_3$.

3 IP$_3$ in turn mobilizes Ca^{2+} stored within organelles.

4 Ca^{2+}, acting as a second messenger, activates calmodulin.

5 Calmodulin induces a change in the shape and function of a particular intracellular protein.

6 The altered protein then produces the desired cellular response dictated by the extracellular messenger.

● **FIGURE 3-9**

Activation of the calcium second-messenger system by an extracellular messenger

triggered by activation of **calmodulin**, an intracellular Ca^{2+}-binding protein (step 4). Activation of calmodulin by Ca^{2+} is similar to activation of protein kinase A by cAMP. From here the patterns of the two pathways are similar. The activated calmodulin alters the shape and function of another cellular protein (step 5), either activating or inhibiting it. This altered protein brings about the ultimate desired cellular response (step 6). (Simultaneously, the other PIP$_2$ breakdown product, DAG, sets off another second-messenger pathway. DAG activates *protein kinase C (PKC)*, which in turn brings about a given cellular response by phosphorylating particular cellular proteins.)

The cAMP and Ca^{2+} pathways frequently overlap in bringing about a particular cellular activity. For example, cAMP and Ca^{2+} can influence each other. Calcium-activated calmodulin can regulate adenylyl cyclase and thus influence cAMP, whereas protein kinase A may phosphorylate and thereby change the activity of Ca^{2+} channels or carriers.

Even though the Ca^{2+} and cAMP effects can become complexly intertwined, it is still notable that a great many diverse cellular events can be traced to a surprisingly small number of pathways: channel effects, one of the two major second-messenger pathways, or some combination of these. It would not be correct, however, to leave the impression that these are the only possible pathways. For example, in a few cells **cyclic guanosine monophosphate (cyclic GMP)** serves as a second messenger in a system analogous to the cAMP system. An example is the signal transduction pathway involved in vision.

(See the accompanying boxed feature, ▶ Concepts, Challenges, and Controversies, for a description of a surprising signal-transduction pathway—one that causes a cell to kill itself.)

Amplification by a second-messenger pathway

Several remaining points about receptor activation and the ensuing events merit attention. First, considering the number of steps involved in a second-messenger relay chain, you might

Programmed Cell Suicide: A Surprising Example of a Signal Transduction Pathway

In the vast majority of cases, the signal transduction pathways triggered by the binding of an extracellular chemical messenger to a cell's surface membrane receptor are aimed at promoting proper functioning, growth, survival, or reproduction of the cell. By contrast, every cell has an unusual built-in pathway that, if triggered, causes the cell to commit suicide by activating intracellular protein-snipping enzymes, which slice the cell into small, disposable pieces. Such intentional programmed cell death is termed **apoptosis.** (This term means "dropping off," in reference to the dropping off of cells that are no longer useful, much as autumn leaves drop off trees.) Apoptosis is a normal part of life: Individual cells that have become superfluous or disordered are triggered to self-destruct for the greater good of maintaining the whole body's health.

Roles of Apoptosis

Following are examples of the vital roles played by this intrinsic sacrificial program:

- *Predictable self-elimination of selected cells is a normal part of development.* Certain unwanted cells produced during development are programmed to kill themselves as the body is sculpted into its final form. For example, apoptosis deliberately prunes the embryonic ducts capable of forming a male reproductive tract during the development of a female. Likewise, apoptosis carves fingers from a mitten-shaped developing hand by eliminating the weblike membranes between them.

- *Apoptosis is important in tissue turnover in the adult body.* Optimal functioning of most tissues depends on a balance between controlled production of new cells and regulated cellular self-destruction. This balance maintains the proper number of cells in a given tissue while ensuring a controlled supply of fresh cells that are at their peak of performance.

- *Programmed cell death plays an important role in the immune system.* Apoptosis provides a means to remove cells infected with harmful viruses. Furthermore, infection-fighting white blood cells that have finished their prescribed function and are no longer needed, execute themselves.

- *Undesirable cells that threaten homeostasis are typically culled from the body by apoptosis.* Included in this hit list are aged cells, cells that have suffered irreparable damage by exposure to radiation or other poisons, and cells that have somehow gone awry. Many mutated cells are eliminated by this means before they become fully cancerous.

Comparison of Apoptosis and Necrosis

Apoptosis is not the only means by which a cell can meet its demise, but it is the neatest way. Apoptosis is a controlled, intentional, tidy way of removing individual cells that are no longer needed or that pose a threat to the body. The other form of cell death, **necrosis** (meaning "make dead"), is uncontrolled, accidental, messy murder of useful cells that have been severely injured by an agent external to the cell, as by a physical blow, O_2 deprivation, or disease. For example, heart muscle cells deprived of their O_2 supply by complete blockage of the blood vessels supplying them during a heart attack die as a result of necrosis (see p. 320).

Even though necrosis and apoptosis both result in cell death, the steps involved are very different. In necrosis, the dying cells are passive victims, whereas in apoptosis the cells actively participate in their own deaths. In necrosis, the injured cell is unable to pump out Na^+ as usual. As a result, water streams in by osmosis, causing the cell to swell and rupture. Typically in necrosis, the insult that prompted cell death injures many cells in the vicinity, so many neighboring cells swell and rupture together. Release of intracellular contents into the surrounding tissues initiates an inflammatory response at the damaged site (see p. 416). Unfortunately, the inflammatory response can potentially harm healthy neighboring cells.

By contrast, apoptosis targets individual cells for destruction, leaving the surrounding cells intact. A cell signaled to commit suicide detaches itself from its neighbors, then shrinks instead of swelling and bursting. As its lethal weapon, the suicidal cell activates a cascade of normally inactive intracellular protein-cutting enzymes, the **caspases,** which kill the cell from within. When a cell has been signaled to undergo apoptosis,

---

wonder why so many cell types use the same complex system to accomplish such a wide range of functions. The multiple steps of a second-messenger pathway are actually advantageous, because a **cascading** (multiplying) effect of these pathways greatly amplifies the initial signal (● Figure 3-10). Amplification means that the magnitude of the output of a system is much greater than the input. Binding of one extracellular chemical-messenger molecule to a receptor activates a number of adenylyl cyclase molecules (let us arbitrarily say 10), each of which activates many (in our example, 100) cAMP molecules. Each cAMP molecule then acts on a single protein kinase A, which phosphorylates and thereby influences many (again, let us say 100) specific proteins, such as enzymes. Each enzyme, in turn, is responsible for producing many (perhaps 100) molecules of a particular product, such as a secretory product. The result of this cascade of events, with one event triggering the next event in sequence, is a tremendous amplification of the initial signal. In the hypothetical example in ● Figure 3-10, one chemical-messenger molecule has been responsible for inducing a yield of 10 million molecules of a secretory product. In this way, very low concentrations of hormones and other chemical messengers can trigger pronounced cellular responses. Therefore, multistepped second-messenger systems are very efficient. Such amplification through a cascading effect is another common functional pattern seen throughout the body.

Modifications of second-messenger pathways

Although membrane receptors serve as links between extracellular first messengers and intracellular second messengers in the regulation of specific cellular activities, the receptors themselves are also frequently subject to regulation. In many instances, the number and affinity (attraction of a receptor for its chemical messenger) can be altered, depending on the circumstances. For example, the number of receptors for the hormone insulin can be deliberately decreased in response to a chronic elevation of insulin in the blood. Later, when we cover the endocrine system, you will learn more about this mechanism to regulate the responsiveness of a target cell to its hormone.

Many disease processes can be linked to malfunctioning receptors or defects in one of the components of the ensuing

the mitochondria become leaky, permitting *cytochrome c* to leak out into the cytosol. Cytochrome c, a component of the electron transport chain, usually participates in oxidative phosphorylation to produce ATP. Outside its typical mitochondrial environment, cytochrome c activates the caspase cascade. Once unleashed, the caspases act like molecular scissors to systematically dismantle the cell. Snipping protein after protein, they chop up the nucleus, disassembling its life-essential DNA, then break down the internal shape-holding cytoskeleton, and finally fragment the cell itself into disposable membrane-enclosed packets. Importantly, the contents of the dying cell remain wrapped by plasma membrane throughout the entire self-execution process, thus avoiding the spewing of potentially harmful intracellular contents characteristic of necrosis. No inflammatory response is triggered, so no neighboring cells are harmed. Instead, cells in the vicinity swiftly engulf and destroy the apoptotic cell fragments by phagocytosis. The breakdown products are recycled for other purposes as needed. Meanwhile, the tissue as a whole has continued to function normally, while the targeted cell has unobtrusively killed itself.

Control of Apoptosis

If every cell contains the potent self-destructive caspases, what normally keeps these powerful enzymes under control (that is, in inactive form) in cells that are useful to the body and deserve to live? Likewise, what activates the death-wielding

caspase cascade in unwanted cells destined to eliminate themselves? Given the importance of these life-or-death decisions, it is not surprising that multiple pathways tightly control whether a cell is "to be or not to be." A cell normally receives a constant stream of "survival signals," which reassure the cell that it is useful to the body, that all is right in the internal environment surrounding the cell, and that everything is in good working order within the cell. These signals include tissue-specific growth factors, certain hormones, and appropriate contact with neighboring cells and the extracellular matrix. These extracellular survival signals trigger intracellular pathways that block activation of the caspase cascade, thus restraining the cell's death machinery. Most cells are programmed to commit suicide if they do not receive their normal reassuring survival signals. The usual safeguards are removed, and the lethal protein-snipping enzymes are unleashed. For example, withdrawal of growth factors or detachment from the extracellular matrix causes a cell to promptly execute itself.

Furthermore, cells display "death receptors" in their plasma membrane that receive specific extracellular "death signals," such as a particular hormone or a specific chemical messenger from white blood cells. Activation of death pathways by these signals can override the life-saving pathways triggered by the survival signals. The death-signal transduction pathway swiftly ignites the internal apoptotic machinery, driving the cell to

its own demise. Likewise, the self-execution machinery is set in motion when a cell suffers irreparable intracellular damage. Thus some signals block apoptosis whereas others promote it. Whether a cell lives or dies depends on which of these competing signals dominates at any given time. Although all cells possess the same death machinery, they vary in the specific signals that induce them to commit suicide.

Considering that every cell's life hangs in delicate balance at all times, it is not surprising that faulty control of apoptosis—resulting in either too much or too little cell suicide—appears to participate in many major diseases. Excessive apoptotic activity is believed to contribute to the brain cell death associated with Alzheimer's disease, Parkinson's disease, and stroke as well as to the premature demise of important infection-fighting cells in AIDS. Conversely, too little apoptosis most likely plays a role in cancer. Evidence suggests that cancer cells fail to respond to the normal extracellular signals that promote cell death. Because these cells neglect to die on cue, they grow in unchecked fashion, forming a chaotic, out-of-control mass.

Apoptosis is currently one of the hottest topics of investigation. Researchers are scrambling to sort out the multiple factors involved in the signal transduction pathways controlling this process. Their hope is to find ways to tinker with the apoptotic machinery to find badly needed new therapies to treat a variety of big killers.

signal-transduction pathways. For example, defective receptors are responsible for the extreme muscular weakness that characterizes *myasthenia gravis*. With this disease, affected skeletal muscle receptors are unable to respond to the chemical messenger released by nerves that normally triggers muscle contraction. In contrast, in the intestinal disease *cholera* the receptors are all functional, but the toxin produced by the cholera pathogen prevents the normal inactivation of cAMP in intestinal cells. In these cells, cAMP induces fluid secretion into the lumen. The severe diarrhea characteristic of cholera is caused by the continued secretion of this fluid into the gut, triggered by the continual presence of cAMP. (Interestingly, some cholera bacteria cause disease, whereas others do not. The difference is a virus that infects the cholera bacterium. This virus causes the bacterium to produce the cholera toxin. Without the viral stowaway, the cholera bacterium is harmless!)

We have now examined the structure and composition of the plasma membrane, the ways in which plasma membranes of adjacent cells link together to form different tissues, and how cells communicate with each other. We are now going to turn our attention to the topic of membrane transport, focusing on

how the plasma membrane can selectively control what enters and exits the cell.

OVERVIEW OF MEMBRANE TRANSPORT

Anything that passes between a cell and the surrounding extracellular fluid must be able to penetrate the plasma membrane. If a substance can cross the membrane, the membrane is said to be **permeable** to that substance; if a substance is unable to pass, the membrane is **impermeable** to it. The plasma membrane is **selectively permeable** in that it permits some particles to pass through while excluding others.

Two properties of particles influence whether they are able to permeate the plasma membrane without any assistance: (1) the relative solubility of the particle in lipid and (2) the size of the particle. Highly lipid-soluble particles are able to dissolve in the lipid bilayer and pass through the membrane. Uncharged or nonpolar molecules (such as O_2, CO_2, and fatty acids) are highly lipid soluble and readily permeate the membrane. Charged particles (ions such as Na^+ and K^+) and polar

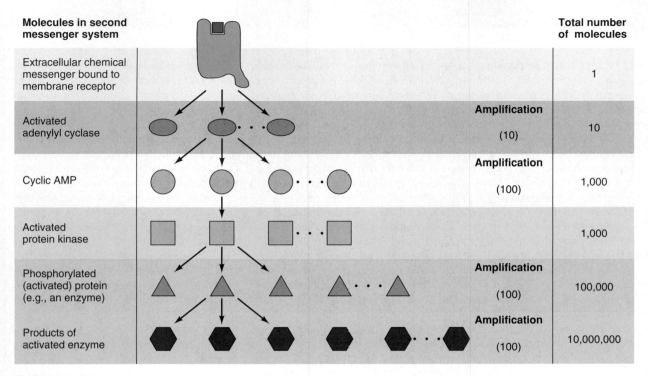

Molecules in second messenger system		Total number of molecules
Extracellular chemical messenger bound to membrane receptor		1
Activated adenylyl cyclase	Amplification (10)	10
Cyclic AMP	Amplification (100)	1,000
Activated protein kinase		1,000
Phosphorylated (activated) protein (e.g., an enzyme)	Amplification (100)	100,000
Products of activated enzyme	Amplification (100)	10,000,000

● FIGURE 3-10

Amplification of the initial signal by a second-messenger system
Through amplification, very low concentrations of extracellular chemical messengers, such as hormones, can trigger pronounced cellular responses.

molecules (such as glucose and proteins) have low lipid solubility but are very soluble in water. The lipid bilayer serves as an impermeable barrier to particles poorly soluble in lipid. For water-soluble (and thus lipid-insoluble) ions less than 0.8 nm in diameter, the protein channels serve as an alternate route for passage across the membrane. Only ions for which specific channels are available and open can permeate the membrane.

Particles that have low lipid solubility and are too large for channels cannot permeate the membrane on their own. Yet some of these particles must cross the membrane for the cell to survive and function. Glucose is an example of a large, poorly lipid soluble particle that must gain entry to the cell but cannot permeate by dissolving in the lipid bilayer or passing through a channel. Cells have several means of assisted transport to move particles across the membrane that must enter or leave the cell but cannot do so unaided, as you will learn shortly.

Even if a particle is capable of permeating the membrane by virtue of its lipid solubility or its ability to fit through a channel, some force is needed to produce its movement across the membrane (a particle does not just decide it wants to be on the other side). Two general types of forces are involved in accomplishing transport across the membrane: (1) forces that do not require the cell to expend energy to produce movement (**passive forces**) and (2) forces requiring cellular energy (ATP) expenditure to transport a substance across the membrane (**active forces**).

We will now examine the various methods of membrane transport, indicating whether each is an unassisted or assisted means of transport and whether each is a passive or active transport mechanism.

UNASSISTED MEMBRANE TRANSPORT

Molecules (or ions) that can penetrate the plasma membrane on their own are passively driven across the membrane by two forces: diffusion down a concentration gradient and/or movement along an electrical gradient. We will first examine diffusion down a concentration gradient.

▌ Particles that can permeate the membrane passively diffuse down their concentration gradient.

All molecules (or ions) are in continuous random motion at temperatures above absolute zero as a result of heat energy. This motion is most evident in liquids and gases, where the individual molecules have more room to move before colliding with another molecule. Each molecule moves separately and randomly in any direction. As a consequence of this haphazard movement, the molecules frequently collide, bouncing off each other in different directions like billiard balls striking each other. The greater the molecular concentration of a substance in a solution, the greater the likelihood of collisions. Consequently, molecules within a particular space tend to become evenly distributed over time. Such uniform spreading

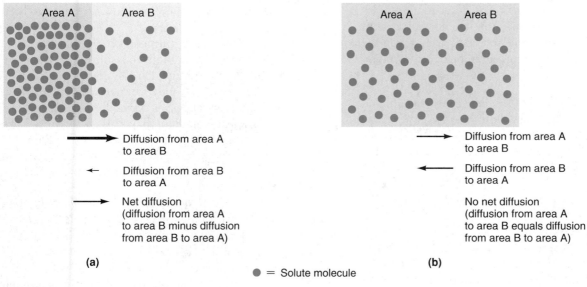

Diffusion from area A
to area B

Diffusion from area B
to area A

Net diffusion
(diffusion from area A
to area B minus diffusion
from area B to area A)

(a)

Diffusion from area A
to area B

Diffusion from area B
to area A

No net diffusion
(diffusion from area A
to area B equals diffusion
from area B to area A)

(b)

● = Solute molecule

● **FIGURE 3-11**

Diffusion
(a) Diffusion down a concentration gradient. (b) Steady state, with no net diffusion occurring.

out of molecules due to their random intermingling is known as **diffusion** (*diffusus* means "to spread out"). To illustrate, in ● Figure 3-11a, the concentration differs between area A and area B in a solution. Such a difference in concentration between two adjacent areas is referred to as a **concentration gradient** (or **chemical gradient**). Random molecular collisions will occur more frequently in area A because of its greater concentration of molecules. For this reason, more molecules will bounce from area A into area B than in the opposite direction. In both areas, the individual molecules will move randomly and in all directions, but the net movement of molecules by diffusion will be from the area of higher concentration to the area of lower concentration.

The term **net diffusion** refers to the difference between two opposing movements. If 10 molecules move from A to B while 2 molecules simultaneously move from B to A, the net diffusion is 8 molecules moving from A to B. Molecules will spread in this way until the substance is uniformly distributed between the two areas and a concentration gradient no longer exists (● Figure 3-11b). At this point, even though movement is still taking place, no *net* diffusion is occurring because the opposing movements exactly counterbalance each other; that is, are in equilibrium. Movement of molecules from A to B will be exactly matched by movement of molecules from B to A. This situation is known as a **steady state**.

What happens if different concentrations of a substance are separated by a plasma membrane? If the substance can permeate the membrane, net diffusion of the substance will occur through the membrane down its con-

centration gradient from the area of high concentration to the area of low concentration (● Figure 3-12a). No energy is required for this movement, so it is a *passive* mechanism of membrane transport. O_2 is transferred across the lung membrane by this means. The blood carried to the lungs is low in O_2, having given up O_2 to the body tissues for cellular metabolism. The air in the lungs, in contrast, is high in O_2 because it is continuously exchanged with fresh air by the process of breathing. Because of this concentration gradient, net diffusion of O_2 occurs from the lungs into the blood as blood flows through the lungs. Thus as blood leaves the lungs for delivery to the tissues, it is high in O_2.

The process of diffusion is crucial to the survival of every cell and plays an important role in many specialized homeo-

● **FIGURE 3-12**

Diffusion through a membrane
(a) Net diffusion across the membrane down a concentration gradient. (b) No diffusion through the membrane despite the presence of a concentration gradient.

If a substance can
permeate the membrane:

If the membrane is
impermeable to a substance:

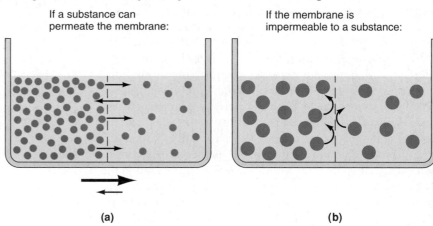

(a)   **(b)**

static functions carried out by cells. The following examples illustrate the widespread use of diffusion. Do not worry about the details of this list now; it is only intended to show you the range of activities that depend on this common pattern of membrane transport.

- The exchange of O_2 and CO_2 between the blood and air in the lungs is accomplished by diffusion.
- Water and lipid-soluble nutrients, such as fatty acids and fat-soluble vitamins, are absorbed from the digestive tract into the blood by diffusion.
- Water and several other plasma components, including some wastes, diffuse across the kidney tubules during the process of urine formation.
- All cells rely on diffusion for taking up O_2 and lipid-soluble nutrients from the interstitial fluid. (Because these supplies reach the interstitial fluid by diffusion from the blood across the capillary wall, then promptly diffuse into the cell from the interstitial fluid across the plasma membrane, we often lump these steps together and simply speak of diffusion of these supplies from the blood into the cell.) Likewise, cells get rid of CO_2 by this same process in reverse.
- Many regulatory hormones are lipid soluble. Instead of binding with surface membrane receptors to exert their effects as water-soluble protein hormones do, the lipid-soluble hormones diffuse into the cells they are controlling.
- Diffusion of Na^+ and K^+ through open channels across the plasma membrane of nerve and muscle cells plays a key role in generating electrical signals in these cells. The specialized functions of both nerve and muscle cells depend on these electrical signals.

Fick's law of diffusion

Several factors in addition to the concentration gradient influence the rate of net diffusion across a membrane. Here are the most important of these factors, whose effects collectively make up **Fick's law of diffusion** (▲ Table 3-1):

1. *The magnitude* (or *steepness*) *of the concentration gradient.* The greater the difference in concentration, the faster the rate of net diffusion. For example, during exercise the working muscles produce CO_2 more rapidly as a result of burning additional fuel to produce the extra ATP they need to power the stepped-up, energy-demanding contractile activity. The resultant increase in CO_2 level in the muscles creates a greater-than-normal CO_2 difference between the muscles and the blood supplying the muscles. Because of this larger gradient, more CO_2 than usual enters the blood passing through the exercising muscles before equilibrium is achieved. As this blood with its increased CO_2 load reaches the lungs, a greater-than-normal CO_2 difference exists between the blood and the air sacs in the lungs. Accordingly, more CO_2 than normal diffuses from the blood into the air sacs before equilibrium is achieved. This extra CO_2 is subsequently breathed out to the environment. Thus any additional CO_2 produced by exercising muscles is eliminated from the body as a result of the increased transfer of CO_2 from the muscles to the blood and from the blood to the air sacs (and subsequently to the out-side) simply as a result of the increase in CO_2 concentration gradient.

2. *The permeability of the membrane to the substance.* The more permeable the membrane is to a substance, the more rapidly the substance can diffuse down its concentration gradient. Of course, if the membrane is impermeable to the substance, no diffusion can take place across the membrane, even though a concentration gradient may exist (● Figure 3-12b). For example, because the plasma membrane is impermeable to the vital intracellular proteins, they are unable to escape from the cell, even though they are in much greater concentration in the ICF than in the ECF.

3. *The surface area of the membrane across which diffusion is taking place.* The larger the surface area available, the greater the rate of diffusion it can accommodate. Various strategies are used throughout the body for increasing the membrane surface area across which diffusion and other types of transport take place. For example, absorption of nutrients in the small intestine is enhanced by the presence of microvilli, which greatly increase the available absorptive surface in contact with the nutrient-rich contents of the small intestine lumen (see p. 50). Conversely, abnormal loss of membrane surface area decreases the rate of net diffusion. For example, in *emphysema* O_2 and CO_2 exchange between the air and blood in the lungs is reduced. In this condition, the walls of the air sacs break down, resulting in less surface area available for diffusion of these gases.

4. *The molecular weight of the substance.* Lighter molecules such as O_2 and CO_2 bounce farther on collision than heavier molecules do. Consequently, O_2 and CO_2 diffuse rapidly, permitting rapid exchanges of these gases across the lung membranes.

▲ TABLE 3-1

Factors Influencing the Rate of Net Diffusion of a Substance across a Membrane (Fick's Law of Diffusion)

Factor	Effect on Rate of Net Diffusion
↑ Concentration gradient of substance (ΔC)	↑
↑ Permeability of membrane to substance (P)	↑
↑ Surface area of membrane (A)	↑
↑ Molecular weight of substance (MW)	↓
↑ Distance (thickness) (ΔX)	↓

Modified Fick's equation:

$$\text{Net rate of diffusion (Q)} = \frac{\Delta C \cdot P \cdot A}{MW \cdot \Delta X}$$

$$\left[\frac{P}{\sqrt{MW}} = \text{diffusion coefficient (D)} \right]$$

$$\text{Restated } Q \propto \frac{\Delta C \cdot A \cdot D}{\Delta X}$$

5. *The distance through which diffusion must take place.* The greater the distance, the slower the rate of diffusion. Accordingly, membranes across which diffusing particles must travel are normally relatively thin, such as the membranes separating the air and blood in the lungs. Thickening of this air–blood interface (as in *pneumonia*, for example) slows down exchange of O_2 and CO_2. Furthermore, diffusion is efficient only for short distances between cells and their surroundings. It becomes an inappropriately slow process for distances of more than a few centimeters. To illustrate, it would take months or even years for O_2 to diffuse from the surface of the body to the cells in the interior. Instead, the circulatory system provides a network of tiny vessels that deliver and pick up materials at every "block" of a few cells, with diffusion accomplishing short local exchanges between the blood and surrounding cells.

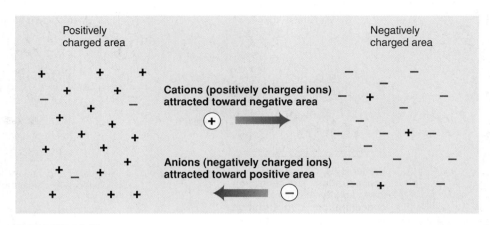

Positively charged area

Negatively charged area

Cations (positively charged ions) attracted toward negative area

Anions (negatively charged ions) attracted toward positive area

● **FIGURE 3-13**
Movement along an electrical gradient

Ions that can permeate the membrane also passively move along their electrical gradient.

Movement of ions (electrically charged particles that have either lost or gained an electron) is also affected by their electrical charge. Like charges (those with the same kind of charge) repel each other, and opposite charges attract each other. If a relative difference in charge exists between two adjacent areas, the positively charged ions (*cations*) tend to move toward the more negatively charged area, whereas the negatively charged ions (*anions*) tend to move toward the more positively charged area (● Figure 3-13). A difference in charge between two adjacent areas thus produces an **electrical gradient** that promotes the movement of ions toward the area of opposite charge. Because the cell does not have to expend energy for ions to move into or out of the cell along an electrical gradient, this method of membrane transport is *passive*. When an electrical gradient exists between the ICF and ECF, only ions that can permeate the plasma membrane are able to move along this gradient.

Both an electrical and a concentration (chemical) gradient may be acting on a particular ion at the same time. The net effect of simultaneous electrical and concentration gradients on this ion is called an **electrochemical gradient.** Later in this chapter you will learn how electrochemical gradients contribute to the electrical properties of the plasma membrane.

Osmosis is the net diffusion of water down its own concentration gradient.

Water can readily permeate the plasma membrane. It can pass between the phospholipid molecules and through **aquaporins,** which are channels for the passage of water, formed by membrane proteins in some cell types. The driving force for diffusion of water across the membrane is the same as for any

other diffusing molecule, namely its concentration gradient. Usually the term *concentration* refers to the density of the solute (dissolved substance) in a given volume of water. It is important to recognize, however, that adding a solute to pure water in essence decreases the water concentration. In general, one molecule of a solute displaces one molecule of water.

Compare the water and solute concentrations in the two containers in ● Figure 3-14. The container in part a of the figure is full of pure water, so the water concentration is 100% and the solute concentration is 0%. In part b, solute has replaced 10% of the water molecules. The water concentration is now 90%, and the solute concentration is 10%—a lower water concentration and a higher solute concentration than in part a. Note that as the solute concentration increases, the water concentration decreases correspondingly.

If solutions of unequal solute concentration (and hence unequal water concentration) are separated by a membrane that permits passage of water, such as the plasma membrane, water will diffuse passively down its own concentration gradient

● **FIGURE 3-14**

Relationship between solute and water concentration in a solution (a) Pure water. (b) Solution.

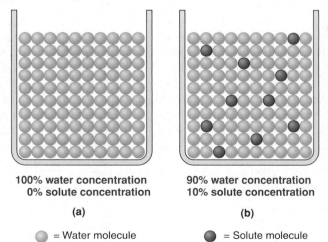

100% water concentration
0% solute concentration

(a)

90% water concentration
10% solute concentration

(b)

○ = Water molecule   ● = Solute molecule

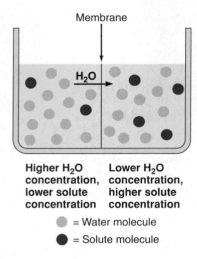

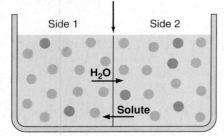

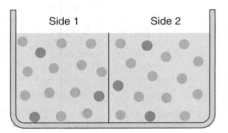

● FIGURE 3-15

Osmosis

Osmosis is the net diffusion of water down its own concentration gradient (to the area of higher solute concentration).

from the area of higher water concentration (lower solute concentration) to the area of lower water concentration (higher solute concentration) (● Figure 3-15). This net diffusion of water is known as **osmosis**. Because solutions are always referred to in terms of concentration of solute, *water moves by osmosis to the area of higher solute concentration*. Despite the impression that the solutes are "drawing," or attracting, water, osmosis is nothing more than the diffusion of water down its own concentration gradient across the membrane.

Thus far in our discussion of osmosis, we have ignored any solute movement. Let us compare the results of osmosis when the solute can and cannot permeate the membrane:

Osmosis when a membrane separates unequal solutions of a penetrating solute

Assume that solutions of unequal solute concentration are separated by a membrane that permits passage of both water and the solute. Because the membrane is permeable to the solute as well as to water, the solute is able to move down its own concentration gradient in the opposite direction of the net water movement (● Figure 3-16). This movement continues until both the solute and water are evenly distributed across the membrane. With all concentration gradients gone, osmosis ceases. The final volume of the compartments when the steady state is achieved is the same as at the onset. Water and solute molecules have merely exchanged places between the two compartments until their distributions have equalized; that is, an equal number of water molecules have moved from side 1 to side 2 as solute molecules have moved from side 2 to side 1.

Osmosis when a membrane separates unequal solutions of a nonpenetrating solute

Now assume that solutions of unequal solute concentration are separated by a membrane that is permeable to water but

● FIGURE 3-16

Movement of water and a penetrating solute unequally distributed across a membrane

excludes passage of the solute. Because the membrane is impermeable to the solute, the solute is not able to cross the membrane down its concentration gradient (● Figure 3-17). At first the concentration gradients are identical to those in the previous example. However, even though net diffusion of water takes place from side 1 to side 2, the solute cannot move. As a result of water movement alone, the volume of side 2 increases while the volume of side 1 correspondingly decreases. Loss of water from side 1 increases the solute concentration on side 1, whereas addition of water to side 2 reduces the solute concentration on that side. Eventually the concentrations of water and solute on the two sides of the membrane become equal, and net diffusion of water ceases. Unlike the situation in which the solute can also permeate, diffusion of water alone has resulted in a change in the final volumes of the two compartments. The side originally containing the greater solute concentration has a larger volume, having gained water.

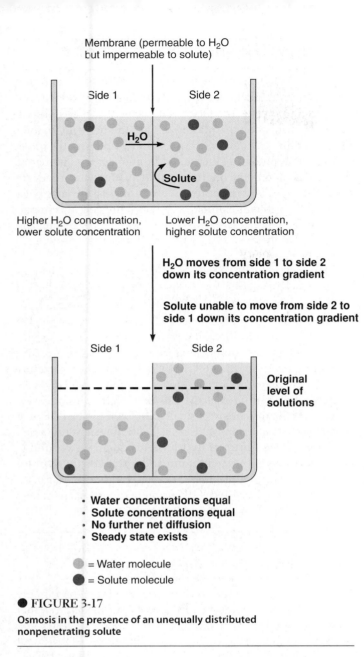

● FIGURE 3-17
Osmosis in the presence of an unequally distributed nonpenetrating solute

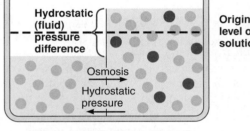

● FIGURE 3-18
Osmosis when pure water is separated from a solution containing a nonpenetrating solute

Osmosis when a membrane separates pure water from a solution of a nonpenetrating solute

What will happen if a nonpenetrating solute is present on side 2 and pure water is present on side 1 (● Figure 3-18)? Osmosis occurs from side 1 to side 2, but the concentrations between the two compartments can never become equal. No matter how dilute side 2 becomes because of water diffusing into it, it can never become pure water, nor can side 1 ever acquire any solute. Because equilibrium is impossible to achieve, does net diffusion of water (osmosis) continue until all the water has left side 1? No. As the volume expands in compartment 2, a difference in hydrostatic pressure between the two compartments is created, and it opposes osmosis. **Hydrostatic (fluid) pressure** is the pressure exerted by a standing, or stationary, fluid on an object—in this case the plasma membrane (*hydro* means "fluid;" *static* means "standing"). The hydrostatic pres-

sure exerted by the larger volume of fluid on side 2 is greater than the hydrostatic pressure exerted on side 1. This hydrostatic pressure difference tends to push fluid from side 2 to side 1.

The **osmotic pressure** of a solution is a measure of the tendency for water to move into that solution because of its relative concentration of nonpenetrating solutes and water. Net movement of water continues until the opposing hydrostatic pressure exactly counterbalances the osmotic pressure. The magnitude of the osmotic pressure is equal to the magnitude of the opposing hydrostatic pressure necessary to completely stop osmosis. The greater the concentration of nonpenetrating solute → the lower the concentration of water → the greater the drive for water to move by osmosis from pure water into the solution → the greater the opposing pressure required to stop the osmotic flow → the greater the osmotic pressure of the solution. Therefore, a solution with a high concentration of nonpenetrating solute exerts greater osmotic pressure than a solution with a lower concentration of nonpenetrating solute does.

Osmosis is another important functional pattern that is widely used throughout the body. Osmosis is the major force responsible for the net movement of water into and out of cells. Approximately 100 times the volume of water in a cell crosses the plasma membrane every second. Yet body cells normally do not experience any net gain (swelling) or loss (shrinking) of volume, because the concentration of nonpenetrating solutes in the ECF is normally carefully regulated (primarily by the kidneys) so that the osmotic pressure in the ECF is the same as the osmotic pressure within the cells.

Tonicity

The **tonicity** of a solution is the effect the solution has on cell volume—whether the cell remains the same size, swells, or shrinks—when the solution surrounds the cell. The tonicity of a solution is determined by its concentration of nonpenetrating solutes. Solutes that can penetrate the plasma membrane quickly become equally distributed between the ECF and ICF, so they do not contribute to osmotic differences. An **isotonic solution** (*iso* means "same") has the same concentration of nonpenetrating solutes as normal body cells. When a cell is bathed in an isotonic solution, no water enters or leaves the cell by osmosis, so cell volume remains constant. For this reason, the ECF is normally kept isotonic so that no net diffusion of water occurs across the plasma membranes of body cells. This is important because cells, especially brain cells, do not function properly if they are swollen or shrunken.

Any change in the concentration of nonpenetrating solutes in the ECF produces a corresponding change in the water concentration difference across the plasma membrane. The resultant osmotic movement of water brings about changes in cell volume. The easiest way to demonstrate this phenomenon is to place red blood cells in solutions with varying concentrations of nonpenetrating solutes. Normally the plasma in which red blood cells are suspended has the same osmotic activity as the fluid inside these cells, so that the cells maintain a constant volume. If red blood cells are placed in a dilute or **hypotonic solution** (*hypo* means "below"), a solution with a lower (below normal) concentration of nonpenetrating solutes (and therefore a higher concentration of water), water enters the cells by osmosis. Net gain of water by the cells causes them to swell, perhaps to the point of rupturing, or *lysing*. If, in contrast, red blood cells are placed in a concentrated or **hypertonic solution** (*hyper* means "above"), a solution with a greater (above normal) concentration of nonpenetrating solutes (and therefore a lower concentration of water), the cells shrink as they lose water by osmosis. Thus it is crucial that the concentration of nonpenetrating solutes in the ECF quickly be restored to normal should the ECF become hypotonic (as with ingesting too much water) or hypertonic (as with losing too much water through severe diarrhea). (See pp. 565–571 for further details about the important homeostatic mechanisms that maintain the normal concentration of nonpenetrating solutes in the ECF.)

ASSISTED MEMBRANE TRANSPORT

All the kinds of transport we have discussed thus far—diffusion down concentration gradients, movement along electrical gradients, and osmosis—produce net movement of molecules capable of permeating the plasma membrane by virtue of their lipid solubility or small size. Large, poorly lipid-soluble molecules such as proteins, glucose, and amino acids cannot cross the plasma membrane on their own no matter what forces are acting on them. This impermeability ensures that the large polar intracellular proteins cannot escape from the cell. It is important that these proteins stay in the cell where they belong and can carry out their life-sustaining functions, such as serving as metabolic enzymes.

However, because large, poorly lipid-soluble molecules cannot cross the plasma membrane on their own, the cell must provide mechanisms for deliberately transporting these types of molecules into or out of the cell as needed. For example, the cell must usher into the cell essential nutrients, such as glucose for energy and amino acids for the synthesis of proteins, and transport out of the cell metabolic wastes and secretory products, such as water-soluble protein hormones and digestive enzymes. Furthermore, passive diffusion alone cannot always account for the movement of small ions. Cells use two different mechanisms to accomplish these selective transport processes: *carrier-mediated transport* for transfer of small water-soluble substances across the membrane and *vesicular transport* for movement of large molecules and even multimolecular particles between the ECF and ICF. We will examine each of these methods of membrane transport in turn.

▌ Carrier-mediated transport is accomplished by a membrane carrier flipping its shape.

Carrier proteins span the thickness of the plasma membrane and are able to undergo reversible changes in shape so that specific binding sites can alternately be exposed at either side of the membrane. That is, the carrier "flip-flops" so that binding sites located in the interior of the carrier are alternately exposed to the ECF and the ICF. ● Figure 3-19 is a schematic representation of how this **carrier-mediated transport** takes place. As the molecule to be transported attaches to a binding site within the interior of the carrier on one side of the membrane (step 1), this binding causes the carrier to flip its shape so that the same site is now exposed to the other side of the membrane (step 2). Then, having been moved in this way from one side of the membrane to the other, the bound molecule detaches from the carrier (step 3). After the passenger detaches, the carrier reverts back to its original shape (step 4).

Carrier-mediated transport systems display three important characteristics that determine the kind and amount of material that can be transferred across the membrane: *specificity*, *saturation*, and *competition*.

1. **Specificity.** Each carrier protein is specialized to transport a specific substance, or at most a few closely related chemical compounds. For example, amino acids cannot bind to glucose

carriers, although several similar amino acids may be able to use the same carrier. Cells vary in the types of carriers they possess, thus permitting transport selectivity among cells. A number of inherited diseases involve defects in transport systems for a particular substance. *Cysteinuria* (cysteine in the urine) is such a disease involving defective cysteine carriers in the kidney membranes. This transport system normally removes cysteine from the fluid destined to become urine and returns this essential amino acid to the blood. When this carrier is malfunctional, large quantities of cysteine remain in the urine, where it is relatively insoluble and tends to precipitate. This is one cause of urinary stones.

2. **Saturation.** A limited number of carrier binding sites are available within a particular plasma membrane for a specific substance. Therefore, there is a limit to the amount of a substance that a carrier can transport across the membrane in a given time. This limit is known as the **transport maximum** **(T_m)**. Until the T_m is reached, the number of carrier binding sites occupied by a substance and, accordingly, the substance's rate of transport across the membrane are directly related to its concentration. The more of a substance available to be transported, the more will be transported. When the T_m is reached, the carrier is saturated (all binding sites are occupied), and the rate of the substance's transport across the membrane is maximal. Further increases in the substance's concentration are not accompanied by corresponding increases in the rate of transport (● Figure 3-20).

As an analogy, consider a ferryboat that can maximally carry 100 people across a river during one trip in an hour. If 25 people are on hand to board the ferry, 25 will be transported that hour. Doubling the number of people on hand to 50 will double the rate of transport to 50 people that hour. Such a direct relationship will exist between the number of people waiting to board (the concentration) and the rate of transport until the ferry is fully occupied (its T_m is reached). The ferry can maximally transport 100 people per hour. Even if

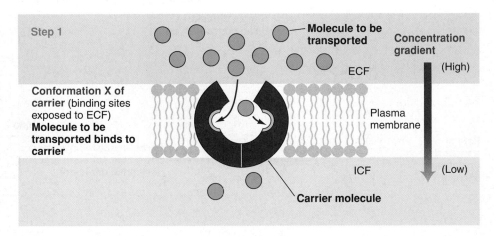

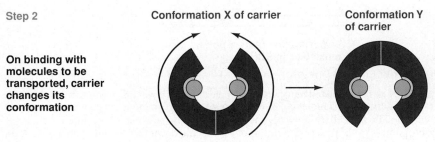

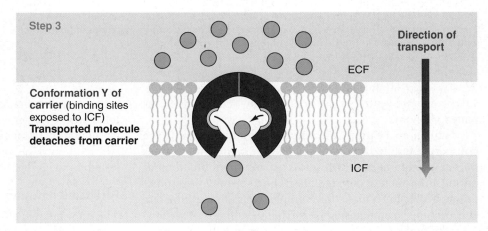

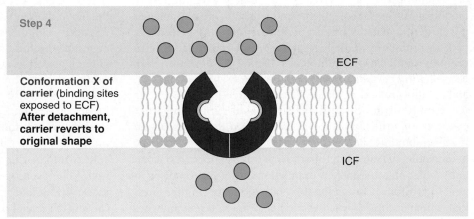

● **FIGURE 3-19**

Schematic representation of carrier-mediated transport: facilitated diffusion

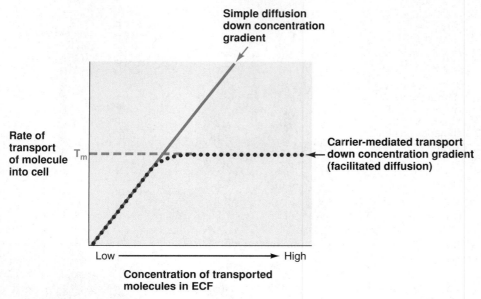

● FIGURE 3-20

Comparison of carrier-mediated transport and simple diffusion down a concentration gradient
With simple diffusion of a molecule down its concentration gradient, the rate of transport of the molecule into the cell is directly proportional to the extracellular concentration of the molecule. With carrier-mediated transport of a molecule down its concentration gradient, the rate of transport of the molecule into the cell is directly proportional to the extracellular concentration of the molecule until the carrier is saturated, at which time the rate of transport reaches a maximal value (transport maximum, or T_m). The rate of transport does not increase with further increases in the ECF concentration of the molecule.

150 people are waiting to board, still only 100 will be transported per hour.

Saturation of carriers is a critical rate-limiting factor in the transport of selected substances across the kidney membranes during urine formation and across the intestinal membranes during absorption of digested foods. Furthermore, it is sometimes possible to regulate (for example, by hormones) the rate of carrier-mediated transport by varying the affinity (attraction) of the binding site for its passenger or by varying the number of binding sites. For example, the hormone insulin greatly increases the carrier-mediated transport of glucose into most cells of the body by promoting an increase in the number of glucose carriers in the cell's plasma membranes. Deficiency of insulin (*diabetes mellitus*) drastically impairs the body's ability to take up and use glucose as the primary energy source.

3. **Competition.** Several closely related compounds may compete for a ride across the membrane on the same carrier. If a given binding site can be occupied by more than one type of molecule, the rate of transport of each substance is less when both molecules are present than when either is present by itself. To illustrate, assume the ferry has 100 seats (binding sites) that can be occupied by either men or women. If only men are waiting to board, up to 100 men can be transported during each trip; the same holds true if only women are waiting to board. If, however, both men and women are waiting to board, they will compete for the available seats so that fewer men and fewer women will be transported than when either group is present alone. Fifty of each might make the trip, although the total number of people transported will still be the same, 100 people. In

other words, when a carrier is able to transport two closely related substances, such as the amino acids glycine and alanine, the presence of both diminishes the rate of transfer of either.

▌ Carrier-mediated transport may be passive or active.

Carrier-mediated transport takes two forms, depending on whether energy must be supplied to complete the process: *facilitated diffusion* (not requiring energy) and *active transport* (requiring energy). **Facilitated diffusion** uses a carrier to facilitate (assist) the transfer of a particular substance across the membrane "downhill" from high to low concentration. This process is passive and does not require energy because movement occurs naturally down a concentration gradient. The unassisted diffusion that we described earlier is sometimes called *simple diffusion*, to distinguish it from facilitated diffusion. **Active transport,** in contrast, requires the carrier to expend energy to transfer its passenger "uphill" against a concentration gradient, from an area of lower concentration to an area of higher concentration. An analogous situation is a car on a hill. To move the car downhill requires no energy; it will coast from the top down. Driving the car uphill, however, requires the use of energy (gasoline).

Facilitated diffusion

The most notable example of facilitated diffusion is the transport of glucose into cells. Glucose is in higher concentration in the blood than in the tissues. Fresh supplies of this nutrient are regularly added to the blood by eating and from reserve energy stores in the body. Simultaneously, the cells metabolize glucose almost as rapidly as it enters the cells from the blood. As a result, a continuous gradient exists for net diffusion of glucose into the cells. However, glucose cannot cross cell membranes on its own. Being polar, it is not lipid soluble, and it is too large to fit through a channel. Without the glucose carrier molecules to facilitate membrane transport of glucose, the cells would be deprived of glucose, their preferred source of fuel. (The accompanying boxed feature, ▶ A Closer Look at Exercise Physiology, describes the effect of exercise on glucose carriers in skeletal muscle cells.)

The carrier binding sites involved in facilitated diffusion can bind with their passenger molecules when exposed to either side of the membrane (● Figure 3-19). Passenger binding triggers the carrier to flip its conformation and drop off the passenger on the opposite side of the membrane. Because passengers are more likely to bind with the carrier on the high-concentration side than on the low-concentration side, the net

Exercising Muscles Have a "Sweet Tooth"

During exercise, muscle cells use more glucose and other nutrient fuels than usual to power their increased contractile activity. The rate of glucose transport into exercising muscle may increase more than 10-fold during moderate or intense physical activity. Glucose uptake by cells is accomplished by glucose carriers in the plasma membrane. Cells maintain an intracellular pool of additional carriers that can be inserted into the plasma membrane as the need for glucose uptake increases. In many cells, including resting muscle cells, facilitated diffusion of glucose into the cells depends on the hormone insulin. Insulin promotes the insertion of glucose carriers in the plasma membranes of insulin-dependent cells. Because plasma insulin levels fall during exercise, however, insulin is not responsible for the increased transport of glucose into exercising muscles. Instead, researchers have shown that muscle cells insert more glucose carriers in

their plasma membranes in response to exercise. This has been demonstrated in rats that have undergone physical training.

Exercise influences glucose transport into cells in yet another way. Regular aerobic exercise (see p. 277) has been shown to increase both the affinity (degree of attraction) and number of plasma membrane receptor sites that bind specifically with insulin. This adaptation results in an increase in insulin sensitivity; that is, the cells are more responsive than normal to a given level of circulating insulin.

Because insulin enhances the facilitated diffusion of glucose into most cells, an exercise-induced increase in insulin sensitivity is one of the factors that makes exercise a beneficial therapy for controlling diabetes mellitus. In this disorder, glucose entry into most cells is impaired as a result of inadequate insulin action (see Chapter 19). Plasma levels of glucose become

elevated because glucose remains in the plasma instead of being transported into the cells. In the Type I form of the disease, not enough insulin is produced to meet the body's need for glucose uptake. Regular aerobic exercise reduces the amount of insulin that must be injected to promote glucose uptake and reduce the blood glucose level toward normal. In the Type II form of the disease, insulin is produced, but insulin's target cells have decreased sensitivity to its presence. By increasing the cells' responsiveness to the insulin available, regular aerobic exercise helps drive glucose into the cells, where it can be used for energy production, instead of remaining in the plasma, where it leads to detrimental consequences for the body.

---

movement always proceeds down the concentration gradient from higher to lower concentration. As is characteristic of mediated transport, the rate of facilitated diffusion is limited by saturation of the carrier binding sites, unlike the rate of simple diffusion, which is always directly proportional to the concentration gradient (● Figure 3-20).

Active transport

Active transport also involves the use of a protein carrier to transfer a specific substance across the membrane, but in this case the carrier transports the substance against its concentration gradient. For example, the uptake of iodine by thyroid gland cells necessitates active transport because 99% of the iodine in the body is concentrated in the thyroid. To move iodine from the blood, where its concentration is low, into the thyroid, where its concentration is high, requires expenditure of energy to drive the carrier. Specifically, energy in the form of ATP is required in active transport to vary the affinity of the binding site when exposed on opposite sides of the plasma membrane. In contrast, the affinity of the binding site in facilitated diffusion is the same when exposed to either the outside or the inside of the cell.

With active transport, the binding site has a greater affinity for its passenger on the low-concentration side as a result of *phosphorylation* of the carrier on this side (● Figure 3-21, step 1). The carrier exhibits ATPase activity in that it splits the terminal phosphate from an ATP molecule to yield ADP plus a free inorganic phosphate (see p. 35). The phosphate group is then attached to the carrier. This phosphorylation and the binding of the passenger on the low-concentration side cause the carrier protein to flip its conformation so that the passenger is now exposed to the high-concentration side of the membrane (● Figure 3-21, step 2). The change in carrier shape is accompanied

by *dephosphorylation*; that is, the phosphate group detaches from the carrier. Removal of phosphate reduces the affinity of the binding site for the passenger, so the passenger is released on the high-concentration side. The carrier then returns to its original conformation. Thus ATP energy is used in the phosphorylation–dephosphorylation cycle of the carrier. It alters the affinity of the carrier's binding sites on opposite sides of the membrane so that transported particles are moved uphill from an area of low concentration to an area of higher concentration. These active transport mechanisms are frequently called **pumps,** analogous to water pumps that require energy to lift water against the downward pull of gravity.

The simplest active-transport systems pump a single type of passenger. An example is the **hydrogen ion (H^+) pump** used by specialized stomach cells to transport H^+ into the stomach lumen in association with the secretion of hydrochloric acid during digestion of a meal. This pump moves H^+ against a tremendous gradient: The concentration of H^+ in the stomach lumen is 3 to 4 million times greater than in the blood.

Na^+–K^+ pump

Other more complicated active-transport mechanisms involve the transfer of two different passengers, either simultaneously in the same direction or sequentially in opposite directions. For example, the plasma membrane of all cells contains a sequentially active **Na^+–K^+ ATPase pump (Na^+–K^+ pump** for short). This carrier transports Na^+ out of the cell, concentrating it in the ECF, and picks up K^+ from the outside, concentrating it in the ICF (● Figure 3-22). Splitting of ATP through ATPase activity and the subsequent phosphorylation of the carrier on the intracellular side increases the carrier's affinity for Na^+ and induces a change in carrier shape, leading to the drop off of Na^+ on the exterior. The subsequent de-

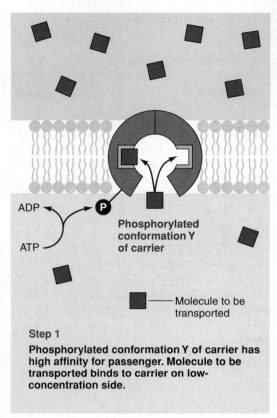

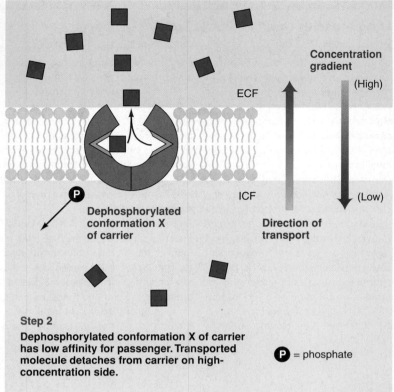

ADP ⟵ P
ATP

Phosphorylated conformation Y of carrier

Dephosphorylated conformation X of carrier

Concentration gradient
(High)

ECF

ICF

(Low)

Direction of transport

◼ —— Molecule to be transported

Step 1
Phosphorylated conformation Y of carrier has high affinity for passenger. Molecule to be transported binds to carrier on low-concentration side.

Step 2
Dephosphorylated conformation X of carrier has low affinity for passenger. Transported molecule detaches from carrier on high-concentration side.

P = phosphate

● FIGURE 3-21

Active transport
The energy of ATP is required in the phosphorylation–dephosphorylation cycle of the carrier to transport the molecule uphill from a region of low concentration to a region of high concentration.

● FIGURE 3-22

Na⁺–K⁺ ATPase pump
The plasma membrane of all cells contains an active transport carrier, the Na⁺–K⁺ ATPase pump, which uses energy in the carrier's phosphyorylation–dephosphyorylation cycle to sequentially transport Na⁺ out of the cell and K⁺ into the cell against these ions' concentration gradients.

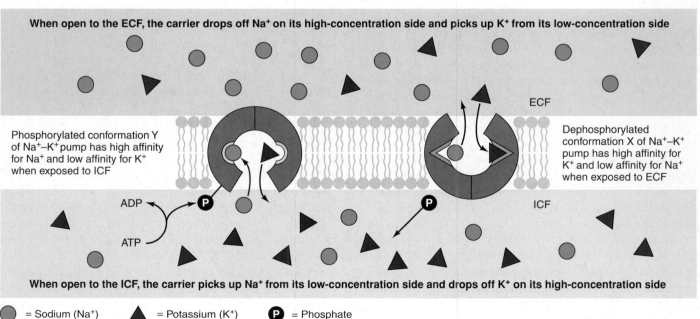

When open to the ECF, the carrier drops off Na⁺ on its high-concentration side and picks up K⁺ from its low-concentration side

ECF

Phosphorylated conformation Y of Na⁺–K⁺ pump has high affinity for Na⁺ and low affinity for K⁺ when exposed to ICF

Dephosphorylated conformation X of Na⁺–K⁺ pump has high affinity for K⁺ and low affinity for Na⁺ when exposed to ECF

ADP ⟵ P
ATP

ICF

When open to the ICF, the carrier picks up Na⁺ from its low-concentration side and drops off K⁺ on its high-concentration side

⬤ = Sodium (Na⁺)   ▲ = Potassium (K⁺)   P = Phosphate

phosphorylation of the carrier increases its affinity for K^+ on the extracellular side and restores the original carrier conformation, thereby transferring K^+ into the cytoplasm. There is not a direct exchange of Na^+ for K^+, however. The Na^+–K^+ pump moves three Na^+ out of the cell for every two K^+ it pumps in. (To appreciate the magnitude of active Na^+–K^+ pumping that takes place, consider that a single nerve cell membrane contains perhaps one million Na^+–K^+ pumps capable of transporting about 200 million ions per second.)

The Na^+–K^+ pump plays three important roles:

1. It establishes Na^+ and K^+ concentration gradients across the plasma membrane of all cells; these gradients are critically important in the ability of nerve and muscle cells to generate electrical signals essential to their functioning (a topic discussed more thoroughly in the next chapter).

2. It helps regulate cell volume by controlling the concentrations of solutes inside the cell and thus minimizing osmotic effects that would induce swelling or shrinking of the cell.

3. The energy used to run the Na^+–K^+ pump also indirectly serves as the energy source for the cotransport of glucose and amino acids across intestinal and kidney cells. This process is known as *secondary active transport.*

Secondary active transport

Unlike most cells of the body, the intestinal and kidney cells actively transport glucose and amino acids by moving them uphill from low to high concentration. The intestinal cells transport these nutrients from inside the intestinal lumen into the blood, concentrating them in the blood until none of these molecules are left in the lumen to be lost in the feces. The kidney cells save these nutrient molecules for the body by transporting them out of the fluid that is to become urine, moving them against a concentration gradient into the blood. However, energy is not directly supplied to the carrier in these instances. The carriers that transport glucose against its concentration gradient from the lumen in the intestine and kidneys are distinct from the glucose facilitated-diffusion carriers. The luminal carriers in intestinal and kidney cells are **cotransport carriers** in that they have two binding sites, one for Na^+ and one for the nutrient molecule. The Na^+–K^+ pumps in these cells are located in the basolateral membrane (the membrane at the base of the cell opposite the lumen and along the lateral edge of the cell below the tight junction; see ● Figure 3-5). More Na^+ is present in the lumen than inside the cells because the energy-requiring Na^+–K^+ pump transports Na^+ out of the cell at the basolateral membrane, keeping the intracellular Na^+ concentration low. Because of this Na^+ concentration difference, more Na^+ binds to the luminal cotransport carrier when it is exposed to the outside (● Figure 3-23). Binding of Na^+ to the cotransport carrier increases the carrier's affinity for its other passenger (for example, glucose), so the carrier has a high affinity for glucose when exposed to the outside. When both Na^+ and glucose are bound to the carrier, it undergoes a change in shape and opens to the inside of the cell. Both Na^+ and glucose are released to the interior, Na^+ because of the lower intracellular Na^+ concentration, and glucose because of the reduced affinity of the binding site on release of Na^+.

The movement of Na^+ into the cell by this cotransport carrier is downhill because the intracellular Na^+ concentration is low, but the movement of glucose is uphill because glucose becomes concentrated in the cell. The released Na^+ is quickly pumped out by the active Na^+–K^+ transport mechanism, keeping the level of intracellular Na^+ low. The energy expended in this process is not directly used to run the cotransport carrier, because phosphorylation is not required to alter the affinity of the binding site to glucose. Instead, the establishment of a Na^+ concentration gradient by a primary active transport mechanism (the Na^+–K^+ pump) drives this secondary active transport mechanism (Na^+-glucose cotransport carrier) to move glucose against its concentration gradient. With **primary active transport,** energy is *directly* required to move a substance uphill. The term "active transport" without a qualifier typically means primary active transport. With **secondary active transport,** energy is required in the entire process, but it is *not directly* required to run the pump. Rather, it uses "secondhand" energy stored in the form of an **ion concentration gradient** (for example, a Na^+ gradient) to move the cotransported molecule uphill. This is a very efficient interaction. The cotransported molecule is essentially getting a free ride, because Na^+ must be pumped out anyway to maintain the electrical and osmotic integrity of the cell.

The glucose that is carried into the cell across the luminal border by secondary active transport then passively moves out of the cell across the basolateral border down its concentration gradient and enters the blood. This movement is accomplished by facilitated diffusion, mediated by another carrier in the plasma membrane. This passive carrier is identical to the one that transports glucose into other cells, but in intestinal and kidney cells it transports glucose out of the cell. The difference depends on the direction of the glucose concentration gradient. In the case of intestinal and kidney cells, glucose is in higher concentration inside the cells.

Before leaving the topic of carrier-mediated transport, think about all the activities that rely on carrier assistance. All cells depend on carriers for the uptake of glucose and amino acids, which serve as the major energy source and structural building blocks, respectively. Na^+–K^+ pumps are essential for generating cellular electrical activity and for ensuring that cells have an appropriate intracellular concentration of osmotically active solutes. Active transport, either primary or secondary or both, is used extensively to accomplish the specialized functions of the nervous and digestive systems as well as the kidneys and all types of muscle.

▌ With vesicular transport, material is moved into or out of the cell wrapped in membrane.

The special carrier-mediated transport systems embedded in the plasma membrane can selectively transport ions and small polar molecules. But what about large polar molecules or even multimolecular materials that must leave or enter the cell, such as during secretion of protein hormones (large polar molecules) by endocrine cells or during ingestion of invading

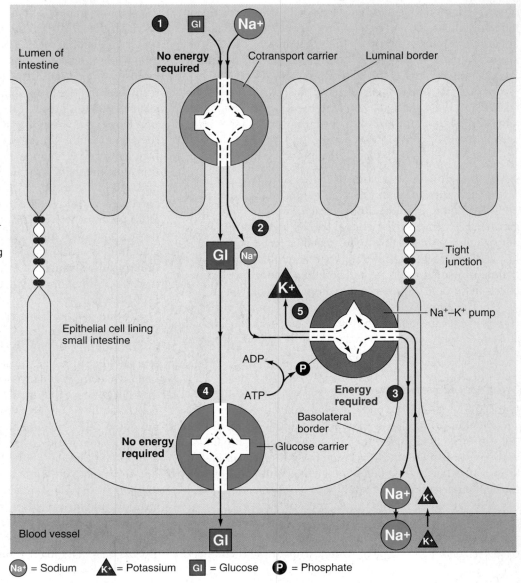

1. A cotransport carrier at the luminal border simultaneously transfers glucose against a concentration gradient and Na^+ down a concentration gradient from the lumen into the cell.

2. No energy is directly used by the cotransport carrier to move glucose uphill. Instead, operation of the cotransport carrier is driven by the Na^+ concentration gradient (low Na^+ in ICF compared to lumen) established by the energy-using Na^+–K^+ pump.

3. The Na^+–K^+ pump actively transports Na^+ out of the cell at the basolateral border, keeping the ICF Na^+ concentration lower than the luminal concentration.

4. After entering the cell by secondary active transport, glucose is transported down its concentration gradient from the cell into the blood by facilitated diffusion, mediated by a passive glucose carrier at the basal border.

5. The Na^+–K^+ pump also actively transports K^+ into the cell, maintaining a high intracellular K^+ concentration, but this action has no influence on secondary active transport.

Labels in figure:
Lumen of intestine
No energy required
Cotransport carrier
Luminal border
Tight junction
Na^+–K^+ pump
Epithelial cell lining small intestine
ADP
ATP
Energy required
Basolateral border
Glucose carrier
No energy required
Blood vessel

Na^+ = Sodium K^+ = Potassium   Gl = Glucose   P = Phosphate

● **FIGURE 3-23**

Secondary active transport
Glucose (as well as amino acids) is transported across intestinal and kidney cells against its concentration gradient by means of secondary active transport.

bacteria (multimolecular particles) by white blood cells? These materials are unable to cross the plasma membrane, even with assistance: They are too large for channels, and no carriers exist for them (they would not even fit into a carrier molecule). These large particles are transferred between the ICF and ECF not by crossing the membrane but by being wrapped in a membrane-enclosed vesicle, a process known as **vesicular transport.** Vesicular transport requires energy expenditure by the cell, so this is an *active* method of membrane transport. Energy is needed to accomplish vesicle formation and vesicle movement within the cell. Transport into the cell in this manner is termed *endocytosis*, whereas transport out of the cell is called *exocytosis*. An important feature of vesicular transport is that the materials sequestered within the vesicles

never mix with the cytosol. The vesicles are designed to recognize and fuse only with a targeted membrane, ensuring a directed transfer of designated large polar molecules or multimolecular materials between the cell's interior and exterior.

Endocytosis

To review, in **endocytosis** the plasma membrane surrounds the substance to be ingested, then fuses over the surface, pinching off a membrane-enclosed vesicle so that the engulfed material is trapped within the cell (see p. 33). Recall that there are three forms of endocytosis, depending on the nature of the material internalized: pinocytosis (nonselective uptake of ECF), receptor-mediated endocytosis (selective uptake of a large molecule), and phagocytosis (selective uptake of a multimolecular particle).

Once inside the cell, an engulfed vesicle has two possible destinies:

1. In most instances, lysosomes fuse with the vesicle to degrade and release its contents into the intracellular fluid.

2. In some cells, the endocytotic vesicle bypasses the lysosomes and travels to the opposite side of the cell, where it releases its contents by exocytosis. This provides a pathway to shuttle intact particles through the cell. Such vesicular traffic is one means by which materials are transferred through the thin cells lining the capillaries, across which exchanges are made between the blood and surrounding tissues.

Exocytosis

In **exocytosis,** almost the reverse of endocytosis occurs. A membrane-enclosed vesicle formed within the cell fuses with the plasma membrane, then opens up and releases its contents to the exterior (see p. 30). Materials packaged for export by the endoplasmic reticulum and Golgi complex are externalized by exocytosis.

Exocytosis serves two different purposes:

1. It provides a mechanism for secreting large polar molecules, such as protein hormones and enzymes, that are unable to cross the plasma membrane. In this case, the vesicular contents are highly specific and are released only on receipt of appropriate signals.

2. It enables the cell to add specific components to the membrane, such as selected carriers, channels, or receptors, depending on the cell's needs. In such cases, the composition of the membrane surrounding the vesicle is important, and the contents may be merely a sampling of ICF.

The rate of endocytosis and exocytosis must be kept in balance to maintain a constant membrane surface area and cell volume. More than 100% of the plasma membrane may be used in an hour to wrap internalized vesicles in a cell actively involved in endocytosis, necessitating rapid replacement of surface membrane by exocytosis. In contrast, when a secretory cell is stimulated to secrete it may temporarily insert up to 30 times its surface membrane through exocytosis. This added membrane must be specifically retrieved by an equivalent level of endocytotic activity. Thus, through exocytosis and endocytosis, portions of the membrane are constantly being restored, retrieved, and generally recycled.

There may be one more way in which cells can exchange materials with their environment. Caveolae, recently investigated membranous structures, are believed to accomplish this method of membrane transport. Let's see how.

▮ Caveolae may play roles in membrane transport and signal transduction.

The outer surface of the plasma membrane is not smooth but instead is dimpled with tiny cavelike indentations known as **caveolae** ("tiny caves"). These small flask-shaped pits have been observed for more than 40 years through electron microscopy (● Figure 3-24). Caveolae were not considered to

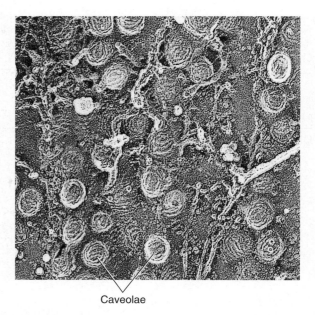

Caveolae

● **FIGURE 3-24**

Caveolae
Using a new technology involving making microscopic replicas of molds of rapidly frozen membranes, the caveolae, which are small cavelike indentations in the surface membrane, appear as small bumps on the surface of the mold. This is analogous to the "valleys" in a Jell-O mold becoming the "ridges" of the molded gelatin.

---

have functional significance, however, until further investigation in the mid-1990s suggested that they (1) provide a new route for transport into the cell and (2) serve as a "switchboard" for relaying signals from many extracellular chemical messengers into the cell's interior.

An abundance of membrane proteins, including a variety of receptors, cluster in these tiny chambers. Some of these receptors appear to play a role in a new form of cellular uptake of small molecules and ions. The best-studied example is the transport of the B vitamin folic acid into the cell. When folic acid binds with its receptors, which are concentrated in the portions of the plasma membrane that form the caveolae, the extracellular openings of these tiny caves close off. The high concentration of folic acid within a closed caveolar compartment encourages the movement of this vitamin across the caveolar membrane into the cytoplasm. Cellular uptake through the cyclic opening and closing of caveolae has been termed **potocytosis.** Potocytosis is believed to be an uptake mechanism for selected small molecules and ions, in contrast to receptor-mediated endocytosis, which transports selected large molecules into the cell (see p. 32). Furthermore, unlike endocytosis in which the internalized vesicle fuses with a lysosome, caveolar vesicles formed in potocytosis do not fuse with these organelles. Thus the contents of endocytotic vesicles are degraded within the cell, but the contents of caveolar vesicles remain intact within the cell.

Besides serving as uptake vehicles, caveolae also appear to be important sites for signal transduction. Many membrane receptors important in signal transduction are concentrated in the caveolae. These tiny membrane caves are thought to be important in cell-to-cell communication because they gather and

The Plasma Membrane and Membrane Potential

▲ **TABLE 3-2**
Characteristics of the Methods of Membrane Transport

Methods of Transport	Substances Involved	Energy Requirements and Force-Producing Movement	Limit to Transport
Diffusion			
Through lipid bilayer	Nonpolar molecules of any size (e.g., O_2, CO_2, fatty acids)	Passive; molecules move down concentration gradient (from high to low concentration)	Continues until the gradient is abolished (steady state with no net diffusion)
Through protein channel	Specific small ions (e.g., Na^+, K^+, Ca^{2+}, Cl^-)	Passive; ions move down electro-chemical gradient (from high to low concentration and attraction of ion to area of opposite charge)	Continues until there is no net movement and a steady state is established
Special case of osmosis	Water only	Passive; water moves down its own concentration gradient (water moves to area of lower water concentration, i.e., higher solute concentration)	Continues until concentration difference is abolished or until stopped by an opposing hydrostatic pressure or until cell is destroyed
Carrier-Mediated Transport			
Facilitated diffusion	Specific polar molecules for which a carrier is available (e.g., glucose)	Passive; molecules move down concentration gradient (from high to low concentration)	Displays a transport maximum (T_m); carrier can become saturated
Primary active transport	Specific ions or polar molecules for which carriers are available (e.g., Na^+, K^+, amino acids)	Active; ions move against concentration gradient (from low to high concentration); requires ATP	Displays a transport maximum; carrier can become saturated
Secondary active transport	Specific polar molecules and ions for which cotransport carriers are available (e.g., glucose, amino acids, some ions)	Active; molecules move against concentration gradient (from low to high concentration); driven directly by ion gradient (usually Na^+) established by ATP-requiring primary pump	Displays a transport maximum; cotransport carrier can become saturated
Vesicular Transport			
Endocytosis			
Pinocytosis	Small volume of ECF fluid; also important in membrane recycling	Active; plasma membrane dips inward and pinches off at surface, forming an internalized vesicle	Control poorly understood
Receptor-mediated endocytosis	Specific large polar molecule (e.g., protein)	Active; plasma membrane dips inward and pinches off at surface, forming an internalized vesicle	Necessitates binding to specific receptor site on membrane surface
Phagocytosis	Multimolecular particles (e.g., bacteria and cellular debris)	Active; cell extends pseudopods that surround particle, forming an internalized vesicle	Necessitates binding to specific receptor site on membrane surface
Exocytosis	Secretory products (e.g., hormones and enzymes) as well as large molecules passing through cell intact; also important in membrane recycling	Active; increase in cytosolic Ca^{2+} induces fusion of vesicle with plasma membrane; vesicle opens up and releases contents to outside	Secretion triggered by specific neural or hormonal stimuli; other controls involved in transcellular traffic and membrane recycling not known
Potocytosis	Small molecules and ions (e.g., folic acid)	Active; involves cyclic opening and closing of caveolae	Control poorly understood

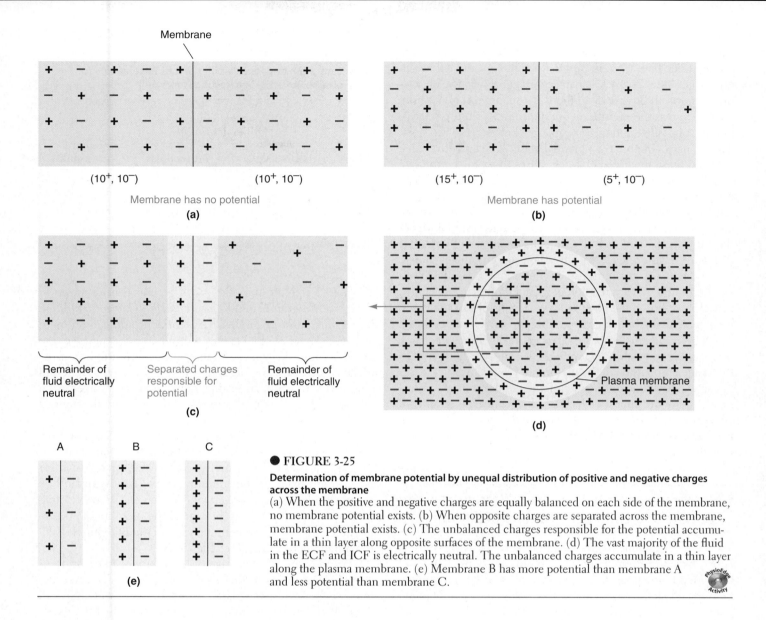

Membrane

(10⁺, 10⁻)   (10⁺, 10⁻)

Membrane has no potential

(a)

(15⁺, 10⁻)   (5⁺, 10⁻)

Membrane has potential

(b)

Remainder of fluid electrically neutral

Separated charges responsible for potential

Remainder of fluid electrically neutral

(c)

Plasma membrane

(d)

A   B   C

(e)

● **FIGURE 3-25**

Determination of membrane potential by unequal distribution of positive and negative charges across the membrane

(a) When the positive and negative charges are equally balanced on each side of the membrane, no membrane potential exists. (b) When opposite charges are separated across the membrane, membrane potential exists. (c) The unbalanced charges responsible for the potential accumulate in a thin layer along opposite surfaces of the membrane. (d) The vast majority of the fluid in the ECF and ICF is electrically neutral. The unbalanced charges accumulate in a thin layer along the plasma membrane. (e) Membrane B has more potential than membrane A and less potential than membrane C.

transmit into the cell signals carried by chemical messengers released by other cells.

Our discussion of membrane transport is now complete; ▲ Table 3-2 summarizes the pathways by which materials can pass between the ECF and ICF.

The selective transport of K⁺ and Na⁺ is responsible for the electrical properties of cells. We turn our attention to this topic next.

MEMBRANE POTENTIAL

The plasma membranes of all living cells have a membrane potential, or are polarized electrically.

▌ Membrane potential is a separation of opposite charges across the plasma membrane.

The term *membrane potential* refers to a separation of charges across the membrane or to a difference in the relative number

of cations and anions in the ICF and ECF. Recall that opposite charges tend to attract each other and like charges tend to repel each other. Work must be performed (energy expended) to separate opposite charges after they have come together. Conversely, when oppositely charged particles have been separated, the electrical force of attraction between them can be harnessed to perform work when the charges are permitted to come together again. This is the basic principle underlying electrically powered devices. Because separated charges have the "potential" to do work, a separation of charges across the membrane is referred to as a *membrane potential*. Potential is measured in units of volts (the same unit used for the voltage in electrical devices), but because the membrane potential is relatively low, the unit used is the **millivolt (mV)** (1 mV = 1/1,000 volt).

Because the concept of potential is fundamental to understanding much of physiology, especially nerve and muscle physiology, it is important to understand clearly what this term means. The membrane in ● Figure 3-25a is electrically neutral. An equal number of positive (+) and negative (−) charges are on each side of the membrane, so no membrane potential exists. In ● Figure 3-25b, some of the positive charges from the right side have been moved to the left. Now the left has an excess of positive charges, leaving an excess of negative charges on the right. In other words, there is a separation of opposite charges across the membrane, or a difference in the relative number of positive and negative charges between the two sides. That is, now a membrane potential exists. The attractive force between these separated charges causes them to accumulate in a thin layer along the outer and inner surfaces of the plasma membrane (● Figure 3-25c). These separated charges represent only a small fraction of the total number of charged particles (ions) present in the ICF and ECF. The vast majority of the fluid inside and outside the cells is electrically neutral (● Figure 3-25d). The electrically balanced ions can be ignored, because they do not contribute to membrane potential. Thus an almost insignificant fraction of the total number of charged particles present in the body fluids is responsible for the membrane potential. Note that the membrane *itself* is not charged. The term *membrane potential* refers to the difference in charge between the wafer-thin regions of ICF and ECF lying next to the inside and outside of the membrane, respectively.

The magnitude of the potential depends on the degree of separation of the opposite charges: The greater the number of charges separated, the larger the potential. Therefore, in ● Figure 3-25e membrane B has more potential than A and less potential than C.

▮ Membrane potential is due to differences in the concentration and permeability of key ions.

The cells of *excitable tissues*—namely nerve cells and muscle cells—have the ability to produce rapid, transient changes in their membrane potential when excited. These brief fluctuations in potential serve as electrical signals. The constant membrane potential present in the cells of nonexcitable tissues and those of excitable tissues when they are at rest—that is, when they are not producing electrical signals—is known as the **resting membrane potential.** We will concentrate now on the generation and maintenance of the resting membrane potential and will examine the changes that take place in excitable tissues during electrical signaling in later chapters.

The unequal distribution of a few key ions between the ICF and ECF and their selective movement through the plasma membrane are responsible for the electrical properties of the membrane. In the body, electrical charges are carried by ions. The ions primarily responsible for the generation of the resting membrane potential are Na^+, K^+, and A^-. The last refers to the large, negatively charged (anionic) intracellular proteins. Other ions (calcium, magnesium, chloride, bicarbonate, and phosphate, to name a few) do not make a direct contribution to the resting electrical properties of the plasma membrane in most cells, even though they play other important roles in the body.

The concentrations and relative permeabilities of the ions critical to membrane electrical activity are compared in

▲ **TABLE 3-3**

Concentration and Permeability of Ions Responsible for Membrane Potential in a Resting Nerve Cell

Ion	Concentration (Millimoles/Liter)		Relative permeability
	Extracellular	Intracellular	
Na^+	150	15	1
K^+	5	150	50–75
A^-	0	65	0

▲ Table 3-3. Note that Na^+ *is in greater concentration in the extracellular fluid and* K^+ *is in much higher concentration in the intracellular fluid.* These concentration differences are maintained by the $Na^+–K^+$ pump at the expense of energy. Because the plasma membrane is virtually impermeable to A^-, these large, negatively charged proteins are found *only inside* the cell. After they have been synthesized from amino acids transported into the cell, they remain trapped within the cell.

In addition to the active carrier mechanism, Na^+ and K^+ can passively cross the membrane through protein channels specific for them. It is usually much easier for K^+ than for Na^+ to get through the membrane, because typically the membrane has many more channels open for passive K^+ traffic than for passive Na^+ traffic across the membrane. At resting potential in a nerve cell, the membrane is about 50 to 75 times more permeable to K^+ than to Na^+.

Armed with a knowledge of the relative concentrations and permeabilities of these ions, we can now analyze the forces acting across the plasma membrane. This analysis will be broken down as follows: We will consider first the direct contributions of the $Na^+–K^+$ pump to membrane potential; second, the effect that the movement of K^+ alone would have on membrane potential; third, the effect of Na^+ alone; and finally, the situation that exists in the cells when both K^+ and Na^+ effects are taking place concurrently. Remember throughout this discussion that the *concentration gradient for* K^+ *will always be outward* and the *concentration gradient for* Na^+ *will always be inward*, because the $Na^+–K^+$ pump maintains a higher concentration of K^+ inside the cell and a higher concentration of Na^+ outside the cell. Also, note that because K^+ and Na^+ are both cations (positively charged), the *electrical gradient for both of these ions will always be toward the negatively charged side of the membrane.*

Effect of the sodium–potassium pump on membrane potential

About 20% of the membrane potential is directly generated by the $Na^+–K^+$ pump. This active transport mechanism pumps three Na^+ out for every two K^+ it transports in. Because Na^+ and K^+ are both positive ions, this unequal transport generates a membrane potential, with the outside becoming relatively more positive than the inside as more positive ions are transported out than in. However, most of the membrane potential—the remaining 80%—is caused by the passive diffusion of K^+ and Na^+ down concentration gradients. Thus most of the Na^+–K^+ pump's role in producing membrane potential is indirect, through its critical contribution to maintaining the concentration gradients directly responsible for the ion movements that generate most of the potential.

Effect of the movement of potassium alone on membrane potential: K^+ equilibrium potential

Let's consider a hypothetical situation characterized by (1) the concentrations that exist for K^+ and A^- across the plasma membrane, (2) free permeability of the membrane to K^+ but not to A^-, and (3) no potential as yet present. The concentration gradient for K^+ would tend to move this ion out of the cell (● Figure 3-26). Because the membrane is permeable to K^+, this ion would readily pass through. As potassium ions moved to the outside, they would carry their positive charge with them, so more positive charges would be on the outside. At the same time, negative charges in the form of A^- would be left behind on the inside, similar to the situation shown in ● Figure 3-25b. (Remember that the large protein anions cannot diffuse out, despite a tremendous concentration gradient.) A membrane potential would now exist. Because an electrical gradient would also be present, K^+, being a positively charged ion, would be attracted toward the negatively charged interior and repelled by the positively charged exterior. Thus two opposing forces would now be acting on K^+: the concentration gradient tending to move K^+ out of the cell and the electrical gradient tending to move these same ions into the cell.

Initially the concentration gradient would be stronger than the electrical gradient, so net diffusion of K^+ out of the cell would continue, and the membrane potential would increase. As more and more K^+ moved down its concentration gradient and out of the cell, however, the opposing electrical gradient would also become greater as the outside became increasingly more positive and the inside more negative. One might think that the outward concentration gradient for K^+ would gradually decrease as K^+ leaves the cell down this gradient. Surprisingly, however, the K^+ concentration gradient would remain essentially constant despite the outward movement of K^+. The reason is that only infinitesimal movement of K^+ out of the cell would bring about rather large changes in membrane potential. Accordingly, such an extremely few K^+ ions present in the cell would have to leave to establish an opposing electrical gradient that the K^+ concentration inside and outside the cell would remain essentially unaltered. As K^+ would continue to move out down its unchanging concentration gradient, the inward electrical gradient would continue to increase in strength. Net outward diffusion would gradually be reduced as the strength of the electrical gradient approached that of the concentration gradient. Finally, when these two forces exactly balanced each other (that is, when they were in equilibrium), no further net movement of K^+ would occur. The potential that would exist at this equilibrium is known as the **K^+ equilibrium potential** (E_{K^+}). At this point, a large concentration gradient for K^+ would still exist, but no more net movement of K^+ would occur out of the cell down this concentration gradient because of the exactly equal opposing electrical gradient (● Figure 3-26).

The membrane potential at E_{K^+} is -90 mV. It is not really a negative potential. By convention, *the sign always designates the polarity of the excess charge on the inside of the membrane.* A membrane potential of -90 mV means that the potential is of a magnitude of 90 mV, with the inside being negative relative to the outside. A potential of $+90$ mV would

---

● **FIGURE 3-26**
Equilibrium potential for K^+

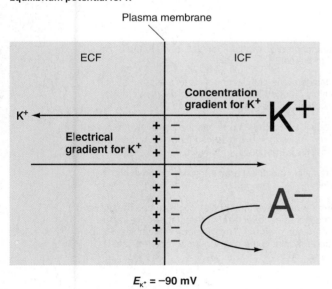

$E_{K^+} = -90$ mV

① The concentration gradient for K^+ tends to push this ion out of the cell.

② The outside of the cell becomes more + as the positively charged K^+ ions move to the outside down their concentration gradient.

③ The membrane is impermeable to the large intracellular protein anion (A^-). The inside of the cell becomes more − as the positively charged K^+ ions move out, leaving behind the negatively charged A^-.

④ The resulting electrical gradient tends to move K^+ into the cell.

⑤ No further net movement of K^+ occurs when the inward electrical gradient exactly counterbalances the outward concentration gradient. The membrane potential at this equilibrium point is the equilibrium potential for K^+ (E_{K^+}) at -90mV.

have the same strength, but in this case the inside would be more positive than the outside.

The equilibrium potential for a given ion of differing concentrations across a membrane can be calculated by means of the **Nernst equation** as follows:

$$E = 61 \log \frac{C_o}{C_i}$$

where

E = equilibrium potential for ion in mV

61 = a constant that incorporates the universal gas constant (R), absolute temperature (T), the ion's valence (z), and an electrical constant known as Faraday (F); $61 = RT/zF$

C_o = concentration of the ion outside the cell in millimoles/liter (millimolars; mM)

C_i = concentration of the ion inside the cell in mM

Given that the ECF concentration of K^+ is 5 mM and the ICF concentration is 150 mM,

$$E_{K^+} = 61 \log \frac{5 \text{ mM}}{150 \text{ mM}}$$

$$= 61 \log \frac{1}{30}$$

Because the log of $\frac{1}{30} = -1.477$,

$$E_{K^+} = 61(-1.477) = -90 \text{ mV}$$

Because 61 is a constant, the equilibrium potential is essentially a measure of the membrane potential (that is, the magnitude of the electrical gradient) that exactly counterbalances the concentration gradient that exists for the ion (that is,

the ratio between the ion's concentration outside and inside the cell). Note that the larger the concentration gradient for an ion, the greater the ion's equilibrium potential. A comparably greater opposing electrical gradient would be required to counterbalance the larger concentration gradient.

Effect of movement of sodium alone on membrane potential: Na⁺ equilibrium potential

A similar hypothetical situation could be developed for Na^+ alone (● Figure 3-27). The concentration gradient for Na^+ would move this ion into the cell, producing a buildup of positive charges on the interior of the membrane and leaving negative charges unbalanced outside (primarily in the form of chloride, Cl^-; Na^+ and Cl^-—that is, salt—are the predominant ECF ions). Net diffusion inward would continue until equilibrium was established by the development of an opposing electrical gradient that exactly counterbalanced the concentration gradient. At this point, given the concentrations for Na^+, the **Na⁺ equilibrium potential (E_{Na^+})** would be +60 mV. In this case the inside of the cell would be positive, in contrast to the equilibrium potential for K^+. The magnitude of E_{Na^+} is somewhat less than for E_{K^+} (60 mV compared to 90 mV) because the concentration gradient for Na^+ is not as large (▲ Table 3-3); thus, the opposing electrical gradient (membrane potential) is not as great at equilibrium.

Concurrent potassium and sodium effects on membrane potential

Neither K^+ nor Na^+ exists alone in the body fluids, so equilibrium potentials are not present in the body cells. They exist only in hypothetical or experimental conditions. In a living cell, the effects of both K^+ and Na^+ must be taken into account. *The greater the permeability of the plasma membrane*

● **FIGURE 3-27**

Equilibrium potential for Na⁺

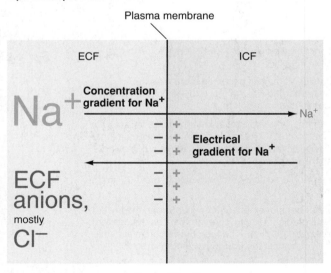

Plasma membrane

① The concentration gradient for Na^+ tends to push this ion into the cell.

② The inside of the cell becomes more + as the positively charged Na^+ ions move to the inside down their concentration gradient.

③ The outside becomes more − as the positively charged Na^+ ions move in, leaving behind in the ECF unbalanced negatively charged ions, mostly Cl^-.

④ The resulting electrical gradient tends to move Na^+ out of the cell.

⑤ No further net movement of Na^+ occurs when the outward electrical gradient exactly counterbalances the inward concentration gradient. The membrane potential at this equilibrium point is the equilibrium potential for Na^+ (E_{Na^+}) at +60 mV.

$E_{Na^+} = +60 \text{ mV}$

for a given ion, the greater the tendency for that ion to drive the membrane potential toward the ion's own equilibrium potential. Because the membrane at rest is 50 to 75 times more permeable to K^+ than to Na^+, K^+ passes through more readily than Na^+; thus K^+ influences the resting membrane potential to a much greater extent than Na^+ does. Recall that K^+ acting alone would establish an equilibrium potential of -90 mV. The membrane is somewhat permeable to Na^+, however, so some Na^+ enters the cell in a limited attempt to reach its equilibrium potential. This Na^+ influx neutralizes, or cancels, some of the potential produced by K^+ alone.

To better understand this concept, let's assume that each separated pair of charges in ● Figure 3-28 represents 10 mV of potential. (This is not technically correct because, in reality, many separated charges must be present to account for a potential of 10 mV.) In this simplified example, nine separated pluses and minuses, with the minuses on the inside, would represent the E_{K^+} of -90 mV. Superimposing the slight influence of Na^+ on this K^+-dominated membrane, assume that two sodium ions enter the cell down the Na^+ concentration and electrical gradients. (Note that the electrical gradient for Na^+ is now inward in contrast to the outward electrical gradient for Na^+ at E_{Na^+}. At E_{Na^+}, the inside of the cell is positive as a result of the inward movement of Na^+ down its concentration gradient. In a resting nerve cell, however, the inside is negative because of the dominant influence of K^+ on membrane potential. Thus both the concentration and electrical gradients now favor the inward movement of Na^+.) The inward movement of these two positively charged sodium ions neutralizes some of the potential established by K^+, so now

only seven pairs of charges are separated, and the potential is -70 mV. This is the resting membrane potential of a typical nerve cell. The resting potential is much closer to E_{K^+} than to E_{Na^+} because of the greater permeability of the membrane to K^+, but it is slightly less than E_{K^+} (-70 mV is a lower potential than -90 mV) because of the weak influence of Na^+.

Balance of passive leaks and active pumping at resting membrane potential

At resting potential, neither K^+ nor Na^+ is at equilibrium. A potential of -70 mV does not exactly counterbalance the concentration gradient for K^+; it takes a potential of -90 mV to do that. Thus there is a continual tendency for K^+ to passively exit through its leak channels. In the case of Na^+, the concentration and electrical gradients do not even oppose each other; they both favor the inward movement of Na^+. Therefore, Na^+ continually leaks inward down its electrochemical gradient, but only slowly, because of its low permeability; that is, because of the sparsity of Na^+ leak channels.

Because such leaking goes on all the time, why doesn't the intracellular concentration of K^+ continue to fall and the concentration of Na^+ inside the cell progressively increase? This does not happen because of the Na^+–K^+ pump. This active transport mechanism counterbalances the rate of leakage (● Figure 3-29). At resting potential, the pump transports back into the cell essentially the same number of potassium ions that have leaked out and simultaneously transports to the outside the sodium ions that have leaked in. Because the pump offsets the leaks, the concentration gradients for K^+ and Na^+

● **FIGURE 3-28**

Effect of concurrent K^+ and Na^+ movement on establishing the resting membrane potential

(1) The Na^+–K^+ pump actively transports Na^+ out of and K^+ into the cell, keeping the concentration of Na^+ high in the ECF and the concentration of K^+ high in the ICF.

(2) Given the concentration gradients that exist across the plasma membrane, K^+ tends to drive the membrane potential to K^+'s equilibrium potential (-90 mV), whereas Na^+ tends to drive the membrane potential to Na^+'s equilibrium potential ($+60$ mV).

(3) However, K^+ exerts the dominant effect on the resting membrane potential because the membrane is more permeable to K^+. As a result, the resting potential (-70 mV) is much closer to E_{K^+} than to E_{Na^+}.

(4) During the establishment of resting potential, the relatively large net diffusion of K^+ outward does not produce a potential of -90 mV because the resting membrane is slightly permeable to Na^+ and the relatively small net diffusion of Na^+ inward neutralizes (in gray shading) some of the potential that would be created by K^+ alone, bringing the resting potential to -70 mV, slightly less than E_{K^+}.

(5) The negatively charged intracellular proteins (A^-) that cannot permeate the membrane remain unbalanced inside the cell during the net outward movement of the positively charged ions, so the inside of the cell is more negative than the outside.

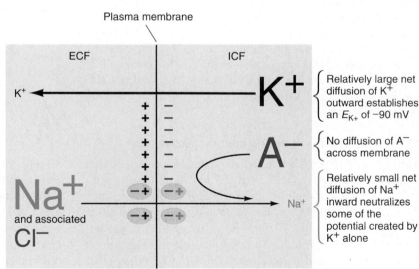

Resting membrane potential = -70 mV

(A^- = Large intracellular anionic proteins)

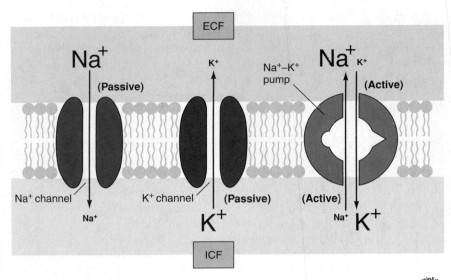

● **FIGURE 3-29**

Counterbalance between passive Na⁺ and K⁺ leaks and the active Na⁺–K⁺ pump
At resting membrane potential, the passive leaks of Na⁺ and K⁺ down their electrochemical gradients are counterbalanced by the active Na⁺–K⁺ pump, so that there is no net movement of Na⁺ and K⁺, and the membrane potential remains constant.

remain constant across the membrane. Thus not only is the Na⁺–K⁺ pump initially responsible for the Na⁺ and K⁺ concentration differences across the membrane, but it also maintains these differences.

As just discussed, it is the presence of these concentration gradients, together with the difference in permeability of the membrane to these ions, that accounts for the resting membrane potential. In this resting state, the potential remains constant. There is no net movement of any ions. All passive forces are exactly balanced by active forces. A steady state exists, even though there is still a strong concentration gradient for both K⁺ and Na⁺ in opposite directions, as well as a slight excess of positive charges in the ECF accompanied by a corresponding slight excess of negative charges in the ICF (enough to account for a potential of the magnitude of 70 mV). At this point, although movement across the membrane is taking place by means of passive leaks and active pumping, the exchange of charges between the ICF and ECF is exactly balanced, with the potential that has been established by these forces remaining constant.

Chloride movement at resting membrane potential

Thus far, we have largely ignored one other ion present in high concentration in the ECF, namely Cl⁻. Chloride is the principal ECF anion. Its equilibrium potential is −70 mV, exactly the same as the resting membrane potential. Movement alone of negatively charged Cl⁻ into the cell down its concentration gradient would produce an opposing electrical gradient, with the inside negative compared to the outside. When physiologists were first examining the ionic effects that could account for the membrane potential, they were tempted to think that Cl⁻ movements and establishment of the Cl⁻ equilib-

rium potential could be solely responsible for producing the identical resting membrane potential. Actually, the reverse is the case. The membrane potential is responsible for driving the distribution of Cl⁻ across the membrane.

Most cells are highly permeable to Cl⁻ but have no active transport mechanisms for this ion. With no active forces acting on it, Cl⁻ passively distributes itself to achieve an individual state of equilibrium. In this case, Cl⁻ is driven out of the cell, establishing an inward concentration gradient that exactly counterbalances the outward electrical gradient (that is, the resting membrane potential) produced by K⁺ and Na⁺ movement. Thus the concentration difference for Cl⁻ between the ECF and ICF is brought about passively by the presence of the membrane potential, rather than being maintained by an active pump, as is the case for K⁺ and Na⁺. Therefore, in most cells Cl⁻ does not influence membrane potential; instead, membrane potential passively influences the Cl⁻ distribution. (Some specialized cells have an active Cl⁻ pump, with subsequent movement of Cl⁻ accounting for part of the potential.)

Specialized use of membrane potential in nerve and muscle cells

Nerve and muscle cells have developed a specialized use for membrane potential. They are able to rapidly and transiently alter their membrane permeabilities to the involved ions in response to appropriate stimulation, thereby bringing about fluctuations in membrane potential. The rapid fluctuations in potential are responsible for producing nerve impulses in nerve cells and for triggering contraction in muscle cells. These activities are the focus of the next five chapters. Even though all cells display a membrane potential, its significance in other cells is uncertain, although the potential of some secretory cells is believed to be linked to their level of secretory activity.

CHAPTER IN PERSPECTIVE: FOCUS ON HOMEOSTASIS

All cells of the body must obtain vital materials such as nutrients and O₂ from the surrounding ECF and must transfer to the ECF wastes to be eliminated as well as secretory products such as chemical messengers and digestive enzymes. Thus transport of materials across the plasma membrane between the ECF and ICF is essential for cell survival, and the constituents in the ECF must be homeostatically maintained in order to support these life-sustaining exchanges.

Many cell types use membrane transport to carry out their specialized activities geared toward maintaining homeostasis. Here are several examples:

1. Absorption of nutrients from the digestive tract lumen involves the transport of these energy-giving molecules across the membranes of the cells lining the tract.

2. Exchange of O_2 and CO_2 between the air and blood in the lungs involves the transport of these gases across the membranes of the cells lining the lungs' air sacs and blood vessels.

3. Urine formation is accomplished by the selective transfer of materials between the blood and the fluid within the kidney tubules across the membranes of the cells lining the tubules.

4. The beating of the heart is triggered by cyclical changes in the transport of Na^+, K^+, and Ca^{2+} across the heart cells' membranes.

5. Secretion of chemical messengers such as neurotransmitters from nerve cells and hormones from endocrine cells involves the transport of these regulatory products to the ECF on appropriate stimulation.

In addition to providing selective transport of materials between the ECF and ICF, the plasma membrane contains receptors for binding with specific chemical messengers that regulate various cell activities, many of which are specialized activities aimed toward maintaining homeostasis. For example, the hormone vasopressin, which is secreted in response to a water deficit in the body, binds with receptors in the plasma membrane of a specific type of kidney cell. This binding triggers these cells to conserve water during urine formation, thus helping alleviate the water deficit that initiated the response.

All living cells have a membrane potential, with the cell's interior being slightly more negative than the fluid surrounding the cell when the cell is electrically at rest. The specialized activities of nerve and muscle cells depend on these cells' ability to change their membrane potential rapidly on appropriate stimulation. These transient, rapid changes in potential in nerve cells serve as electrical signals or nerve impulses, which provide a means to transmit information along nerve pathways. This information is used to accomplish homeostatic adjustments, such as restoring blood pressure to normal when signaled that it has fallen too low.

Rapid changes in membrane potential in muscle cells trigger muscle contraction, the specialized activity of muscle. Muscle contraction contributes to homeostasis in many ways, including the pumping of blood by the heart and moving food through the digestive tract.

CHAPTER SUMMARY

Membrane Structure and Composition

■ All cells are bounded by a plasma membrane, a thin lipid bilayer in which proteins are interspersed and to which carbohydrates are attached on the outer surface.

■ The electron microscopic appearance of the plasma membrane as a trilaminar structure (two dark lines separated by a light interspace) is believed to be caused by the arrangement of the molecules composing it. The phospholipids orient themselves to form a bilayer with a hydrophobic interior (light interspace) sandwiched between the hydrophilic outer and inner surfaces (dark lines).

■ This lipid bilayer forms the structural boundary of the cell, serving as a barrier for water-soluble substances and being responsible for the fluid nature of the membrane.

■ Cholesterol molecules tucked between the phospholipids contribute to the fluidity and stability of the membrane.

■ Membrane proteins, which vary in type and distribution among cells, serve as (1) channels for passage of small ions across the membrane; (2) carriers for transport of specific substances in or out of the cell; (3) docking-marker acceptors for fusion with and subsequent exocytosis of secretory vesicles; (4) membrane-bound enzymes that govern specific chemical reactions; (5) receptors for detecting and responding to chemical messengers that alter cell function; and (6) cell adhesion molecules that help hold cells together, and serve as a structural link between the extracellular surroundings and intracellular cytoskeleton.

■ The membrane carbohydrates, short sugar chains that project from the outer surface only, serve as self-identity markers. They are important in recognition of "self" in cell-to-cell interactions such as tissue formation and tissue growth.

Cell-to-Cell Adhesions

■ Special cells locally secrete a complex extracellular matrix, which serves as a biological "glue" between the cells of a tissue.

■ The extracellular matrix consists of a watery, gel-like substance interspersed with three major types of protein fibers: collagen, elastin, and fibronectin.

■ Many cells are further joined by specialized cell junctions, of which there are three types: desmosomes, tight junctions, and gap junctions.

■ Desmosomes serve as adhering junctions to hold cells together mechanically and are especially important in tissues subject to a great deal of stretching.

■ Tight junctions actually fuse cells together to seal off passage between cells, thereby permitting only regulated passage of materials through the cells. These impermeable junctions are found in the epithelial sheets that separate compartments with very different chemical compositions.

■ Gap junctions are communicating junctions between two adjacent but not touching cells. Cells joined by gap junctions are connected by small tunnels that permit exchange of ions and small molecules between the cells. Such movement of ions plays a key role in the spread of electrical activity to synchronize contraction in heart and smooth muscle.

Intercellular Communication and Signal Transduction

■ Intercellular communication is accomplished by (1) gap junctions, (2) transient direct link up and interaction between cells, and (3) extracellular chemical messengers.

■ Cells communicate with each other to carry out various coordinated activities largely by dispatching extracellular chemical messengers, which act on particular target cells to bring about the desired response.

■ There are four types of extracellular chemical messengers, depending on their source and the distance and means by which they get to their site of action: (1) paracrines (local chemical messengers); (2) neurotransmitters (very short-range chemical messengers released by neurons); (3) hormones (long-range chemical messengers secreted into the blood by endocrine glands; and (4) neurohormones (long-range chemical messengers secreted into the blood by neurosecretory neurons).

■ Transfer of the signal carried by the extracellular messenger into the cell for execution is known as *signal transduction.*

■ Attachment of an extracellular chemical messenger that cannot gain entry to the cell such as a protein hormone (the first messenger) to a membrane receptor initiates one of several related intracellular pathways to bring about the desired response. Chemical messengers trigger cellular responses by two major methods: (1) opening or closing specific channels or (2) activating an intracellular messenger (the second messenger).

■ Two commonly employed second messengers are cyclic AMP and Ca^{2+}. Once activated, these second messengers initiate a similar cascade of intracellular events that ultimately lead to a change in the shape and function of particular proteins to cause the appropriate cellular response.

■ The multiple steps of a second-messenger pathway greatly amplify the initial signal.

Membrane Transport

■ Materials can pass between the ECF and ICF by unassisted and assisted means.

■ Transport mechanisms may also be passive (the particle moves across the membrane without the cell expending energy) or active (the cell expends energy to bring about movement of the particle across the membrane).

■ Lipid-soluble particles and ions can cross the membrane unassisted. Nonpolar (lipid-soluble) molecules of any size can dissolve in and passively pass through the lipid bilayer down concentration gradients. Small ions traverse the membrane passively down electrochemical gradients through protein channels specific for the ion.

■ Osmosis is a special case of water passively moving down its own concentration gradient to an area of higher solute concentration.

■ Carrier mechanisms are important for the assisted transfer of small polar molecules and for selected movement of ions across the membrane.

■ In carrier-mediated transport, the particle is transported across by specific membrane carrier proteins. Carrier-mediated transport may be passive and move the particle down its concentration gradient (facilitated diffusion), or active and move the particle against its concentration gradient (active transport).

■ A given carrier can move a single specific substance in one direction, two substances in opposite directions, or two substances in the same direction. Primary active transport requires the direct use of ATP to drive the pump, whereas secondary active transport is driven by an ion concentration gradient established by a primary active transport system.

■ Large polar molecules and multimolecular particles can leave or enter the cell by being wrapped in a piece of membrane to form vesicles that can be internalized (endocytosis) or externalized (exocytosis).

■ Cells are differentially selective in what enters or leaves because they possess varying numbers and kinds of channels, carriers, and mechanisms for vesicular transport.

■ Large polar molecules (too large for channels and not lipid soluble) for which there are no special transport mechanisms, are unable to permeate.

Membrane Potential

■ All cells have a membrane potential, which is a separation of opposite charges across the plasma membrane.

■ The Na^+–K^+ pump makes a small direct contribution to membrane potential through its unequal transport of positive ions; it transports more Na^+ ions out than K^+ ions in.

■ The primary role of the Na^+–K^+ pump, however, is to actively maintain a greater concentration of Na^+ outside the cell and a greater concentration of K^+ inside the cell. These concentration gradients tend to passively move K^+ out of the cell and Na^+ into the cell.

■ Because the resting membrane is much more permeable to K^+ than to Na^+, substantially more K^+ leaves the cell than Na^+ enters. This results in an excess of positive charges outside the cell and leaves an unbalanced excess of negative charges inside in the form of large protein anions (A^-) that are trapped within the cell. When the resting membrane potential of -70 mV is achieved, no further net movement of K^+ and Na^+ takes place, because any further leaking of these ions down their concentration gradients is quickly reversed by the Na^+–K^+ pump.

■ The distribution of Cl^- across the membrane is passively driven by the established membrane potential so that Cl^- is concentrated in the ECF.

REVIEW EXERCISES

Objective Questions (Answers on p. A-36)

1. The nonpolar tails of the phospholipid molecules bury themselves in the interior of the plasma membrane. (*True or false?*)

2. The hydrophobic regions of the molecules composing the plasma membrane correspond to the two dark layers of this structure visible under an electron microscope. (*True or false?*)

3. Second-messenger systems ultimately bring about the desired cellular response by inducing a change in the shape and function of particular intracellular proteins. (*True or false?*)

4. Through its unequal pumping, the Na^+–K^+ pump is directly responsible for separating sufficient charges to establish a resting membrane potential of -70 mV. (*True or false?*)

5. The two general ways in which interaction of a chemical messenger with a membrane receptor can bring about the desired intracellular responses are _____ and _____.

6. A common membrane-bound intermediary between the receptor and the effector protein within the plasma membrane is the _____.

7. At resting membrane potential, there is a slight excess of _____ (*positive/negative*) charges on the inside of the membrane, with a corresponding slight excess of _____ charges on the outside.

8. Using the following answer code, indicate which membrane component is responsible for the function in question:

(a) lipid bilayer
(b) proteins
(c) carbohydrates
1. channel formation
2. barrier to passage of water-soluble substances
3. receptor sites
4. membrane fluidity
5. recognition of "self"
6. membrane-bound enzymes
7. structural boundary
8. carriers

9. Using the following answer code, indicate the direction of net movement in each case:
(a) movement from high to low concentration
(b) movement from low to high concentration
1. simple passive diffusion
2. facilitated diffusion
3. primary active transport
4. Na^+ during secondary active transport
5. cotransported molecule during secondary active transport
6. water with regard to the water concentration gradient during osmosis
7. water with regard to the solute concentration gradient during osmosis

10. Using the following answer code, indicate the type of cell junction described:
(a) gap junction
(b) tight junction
(c) desmosome
1. adhering junction
2. impermeable junction
3. communicating junction
4. consists of connexons, which permit passage of ions and small molecules between cells
5. consists of interconnecting fibers, which spot-rivet adjacent cells
6. consists of an actual fusion of proteins on the outer surfaces of two interacting cells
7. important in tissues subject to mechanical stretching
8. important in synchronizing contractions within heart and smooth muscle by allowing spread of electrical activity between the cells composing the muscle mass
9. important in preventing passage between cells in epithelial sheets that separate compartments of two different chemical compositions

Essay Questions

1. Describe the fluid mosaic model of membrane structure.
2. What are the functions of the three major types of protein fibers in the extracellular matrix?
3. List and describe the types of intercellular communication.
4. Compare the cAMP and Ca^{2+} second-messenger pathways.
5. What two properties of a particle influence whether it can permeate the plasma membrane?
6. List and describe the methods of membrane transport. Indicate what types of substances are transported by each method, and state whether each is a passive or active means of transport.
7. As stated by Fick's law of diffusion, what factors influence the rate of net diffusion across a membrane?
8. State three important roles of the Na^+–K^+ pump.

9. Describe the contribution of each of the following to the establishment and maintenance of membrane potential: (a) the Na^+–K^+ pump; (b) passive movement of K^+ across the membrane; (c) passive movement of Na^+ across the membrane; and (d) the large intracellular anions.

Quantitative Exercises (Solutions on p. A-36)
(See Appendix D, "Principles of Quantitative Reasoning.")

1. When using the Nernst equation for an ion that has a valence other than 1, you must divide the potential by the valence. Thus for Ca^{2+} the Nernst equation becomes

$$E = \frac{61 \text{ mV}}{z} \log \frac{C_o}{C_i}$$

Use this equation to calculate the Nernst (equilibrium) potentials from the following sets of data:
a. Given $[Ca^{2+}]_o = 1$ mM, $[Ca^{2+}]_i = 100$ nM, find $E_{Ca^{2+}}$
b. Given $[Cl^-]_o = 110$ mM, $[Cl^-]_i = 10$ mM, find E_{Cl^-}

2. One of the important uses of the Nernst equation is in describing the flow of ions across cell membranes. Ions move under the influence of two forces, the concentration gradient (given in electrical units by the Nernst equation) and the electrical gradient (given by the membrane voltage). This is summarized by *Ohm's law*, as follows:

$$I_x = G_x(V_m - E_x)$$

This equation describes the movement of ion x across the membrane. I is the current in amperes (A); G is the conductance, a measure of the permeability of x, in Siemens (S), which is $\Delta I/\Delta V$; V_m is the membrane voltage; and E_x is the equilibrium potential of ion x. Not only does this equation tell how large the current is, it also tells what direction the current is flowing. By convention, a negative value of the current represents either a positive ion entering the cell or a negative ion leaving the cell. The opposite is true of a positive value of the current.
a. Using the following information, calculate the magnitude of I_{Na^+}.

$$[Na^+]_o = 145 \text{ mM}, [Na^+]_i = 15 \text{ mM}$$
$$G_{Na^+} = 1 \text{ nS}, V_m = -70 \text{ mV}$$

b. Is Na^+ entering or leaving the cell?
c. Is Na^+ moving with or against the concentration gradient? Is it moving with or against the electrical gradient?

3. Another important use of the Nernst equation is in determining the resting membrane potential of a cell. The cell resting-membrane potential is a weighted average of the equilibrium potentials of all permeant ions. The weighting factor is the relative permeability (conductance) to that ion. For a cell permeable only to Na^+ and K^+, this equation is

$$V_m = \{G_{Na^+}/G_T\}E_{Na^+} + \{G_{K^+}/G_T\}E_{K^+}$$

In this equation, G_T is the total conductance (in this case, $G_T = G_{Na^+} + G_{K^+}$).
a. Given the following information, calculate V_m:

$$G_{Na^+} = 1 \text{ nS}, G_{K^+} = 5.3 \text{ nS}, E_{Na^+} = 59.1 \text{ mV},$$
$$E_{K^+} = -94.4 \text{ mV}$$

b. What would happen to V_m if $[K^+]_o$ were increased to 150 mM?

POINTS TO PONDER

(Explanations on p. A-36)

1. Assume that a membrane permeable to Na$^+$ but not to Cl$^-$ separates two solutions. The concentration of sodium chloride on side 1 is much higher than on side 2. Which of the following ionic movements would occur?

    a. Na$^+$ would move until its concentration gradient is dissipated (until the concentration of Na$^+$ on side 2 is the same as the concentration of Na$^+$ on side 1).

    b. Cl$^-$ would move down its concentration gradient from side 1 to side 2.

    c. A membrane potential, negative on side 1, would develop.

    d. A membrane potential, positive on side 1, would develop.

    e. None of the above are correct.

2. Compared to resting potential, would the membrane potential become more negative or more positive if the membrane were more permeable to Na$^+$ than to K$^+$?

3. Which of the following methods of transport is being used to transfer the substance into the cell in the accompanying graph?

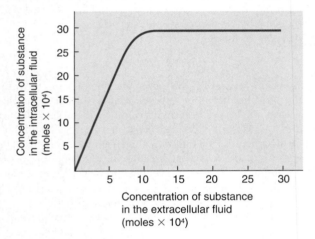

    a. diffusion down a concentration gradient

    b. osmosis

    c. facilitated diffusion

    d. active transport

    e. vesicular transport

    f. It is impossible to tell with the information provided.

4. Colostrum, the first milk that a mother produces, contains an abundance of antibodies, large protein molecules. These maternal antibodies help protect breast-fed infants from infections until the babies are capable of producing their own antibodies. By what means would you suspect these maternal antibodies are transported across the cells lining a newborn's digestive tract into the bloodstream?

5. The rate at which the Na$^+$–K$^+$ pump operates is not constant but is controlled by a combined effect of changes in ICF Na$^+$ concentration and ECF K$^+$ concentration. Do you think an increase in both ICF Na$^+$ and ECF K$^+$ concentrations would accelerate or slow down the Na$^+$–K$^+$ pump? What would be the benefit of this response? Before you reply, consider the following additional information about Na$^+$ and K$^+$ movement across the membrane. Not only do Na$^+$ and K$^+$ slowly and passively leak through their channels in a resting cell, but during an electrical impulse, known as an *action potential*, Na$^+$ rapidly and passively enters the cell; this movement is followed by a rapid, passive outflow of K$^+$. (These ion movements, which result from rapid changes in membrane permeability, bring about rapid, pronounced changes in membrane potential. This sequence of rapid potential changes—an action potential—serves as an electrical signal for conveying information along a nerve pathway.)

6. ***Clinical Consideration.*** When William H. was helping victims following a devastating earthquake in a region that was not prepared to swiftly set up adequate temporary shelter, he developed severe diarrhea. He was diagnosed as having cholera, a disease transmitted through unsanitary water supplies that have been contaminated by fecal material from infected individuals. In this condition, an increase in cAMP in the intestinal cells opens the Cl$^-$ channels in the luminal membranes of these cells, thereby increasing the secretion of Cl$^-$ from the cells into the intestinal tract lumen. By what mechanisms would Na$^+$ and water be secreted into the lumen in accompaniment with Cl$^-$ secretion? How does this secretory response account for the severe diarrhea that is characteristic of cholera? (See p. 71 for the underlying defect in cholera.)

PHYSIOEDGE RESOURCES

PHYSIOEDGE CD-ROM
Figures marked with the PhysioEdge icon have associated activities on the CD. For this chapter, check out:

Transport Across Membranes
Membrane Potential

The diagnostic quiz allows you to receive immediate feedback on your understanding of the concept and to advance through various levels of difficulty.

You will also find Media Quizzes related to Figures 3-5 and 3-6.

CHECK OUT THESE MEDIA QUIZZES:
3.1 The Plasma Membrane and Cell–Cell Connections
3.2 Means of Transmembrane Exchange
3.3 Signaling at Cell Membranes and Membrane Potential

PHYSIOEDGE WEB SITE
The Web site for this book contains a wealth of helpful study aids, as well as many ideas for further reading and research. Log on to:

http://www.brookscole.com

Go to the Biology page and select Sherwood's *Human Physiology*, 5th Edition. Select a chapter from the drop-down menu or click on one of these resource areas:

- **Case Histories** provide an introduction to the clinical aspects of human physiology.

- For 2-D and 3-D graphical illustrations and animations of physiological concepts, visit our **Visual Learning Resource.**

- Resources for study and review include the **Chapter Outline, Chapter Summary, Glossary,** and **Flash Cards.** Use our **Quizzes** to prepare for in-class examinations.

- On-line research resources to consult are **Hypercontents,** with current links to relevant Internet sites; **Internet Exercises** with starter URLs provided; and **InfoTrac Exercises.**

 For more readings, go to InfoTrac College Edition, your on-line research library, at:

http://infotrac.thomsonlearning.com

Nervous System

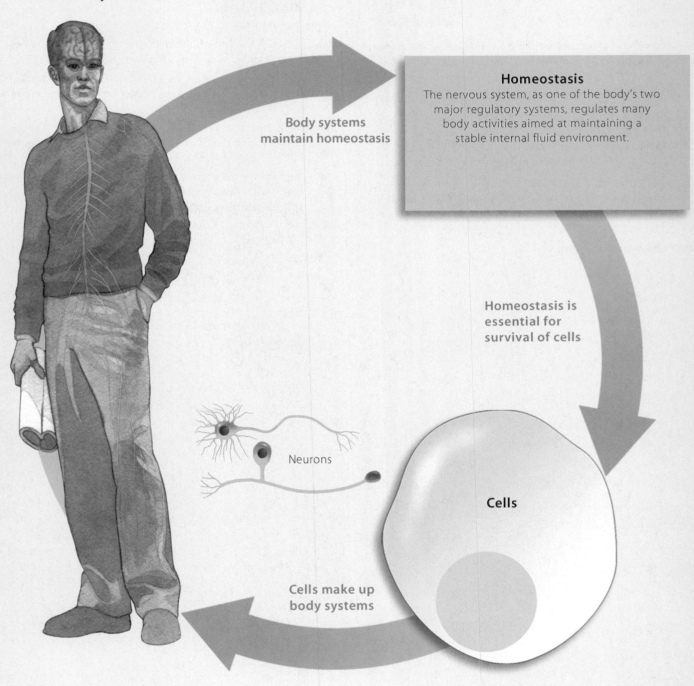

Homeostasis
The nervous system, as one of the body's two major regulatory systems, regulates many body activities aimed at maintaining a stable internal fluid environment.

Body systems maintain homeostasis

Homeostasis is essential for survival of cells

Neurons

Cells

Cells make up body systems

Nerve cells, or neurons, make up the nervous system, one of the two major regulatory systems of the body. The nervous system exerts control over much of the body's muscular and glandular activities, most of which are directed toward maintaining homeostasis. Neurons are specialized for rapid electrical and chemical signaling. They are able to process, initiate, code, and conduct changes in their membrane potential as a means of rapidly transmitting a message throughout their length. Moreover, neurons have developed chemical means of passing this information through intricate nerve pathways from neuron to neuron as well as to muscles and glands.

Chapter 4

Neuronal Physiology

CONTENTS AT A GLANCE

INTRODUCTION

GRADED POTENTIALS

▌ Grading of graded potentials

▌ Spread of graded potentials

ACTION POTENTIALS

▌ Changes in membrane potential during an action potential

▌ Changes in membrane permeability and ion movement during an action potential

▌ Propagation of action potentials; contiguous conduction

▌ Refractory period

▌ All or none law

▌ Role of myelin; saltatory conduction

REGENERATION OF NERVE FIBERS

SYNAPSES AND NEURONAL INTEGRATION

▌ Events at a synapse; role of neurotransmitters

▌ Excitatory and inhibitory synapses

▌ Grand postsynaptic potential; summation

▌ Action potential initiation at the axon hillock

▌ Neuropeptides as neuromodulators

▌ Presynaptic inhibition or facilitation

▌ Convergence and divergence

INTRODUCTION

All body cells display a membrane potential, which is a separation of positive and negative charges across the membrane, as discussed in the preceding chapter. This potential is related to the uneven distribution of Na^+, K^+, and large intracellular protein anions between the intracellular fluid (ICF) and extracellular fluid (ECF), and to the differential permeability of the plasma membrane to these ions (see pp. 87–92).

▌ Nerve and muscle are excitable tissues.

Two types of cells, *nerve cells* and *muscle cells*, have developed a specialized use for this membrane potential. They are able to undergo transient, rapid changes in their membrane potentials. These fluctuations in potential serve as electrical signals. The constant membrane potential that exists when a nerve or muscle cell is not displaying rapid changes in potential is referred to as the *resting potential*. In Chapter 3 you learned that the resting potential of a typical nerve cell is -70 mV.

Nerve and muscle are considered **excitable tissues** because when excited they change their resting potential to produce electrical signals. Nerve cells use these electrical signals to receive, process, initiate, and transmit messages. In muscle cells, these electrical signals initiate contraction. Thus electrical signals are critical to the function of the nervous system as well as all muscles. In this chapter, we will consider how neurons undergo changes in potential to accomplish their function. Muscle cells are discussed in later chapters.

▌ Membrane potential decreases during depolarization and increases during hyperpolarization.

Before understanding what electrical signals are and how they are created, you must become familiar with the following terms, used to describe changes in potential, as graphically represented on ● Figure 4-1:

1. **Polarization:** Charges are separated across the plasma membrane, so that the membrane has potential. Any time the value of the membrane potential is other than 0 mV, in either

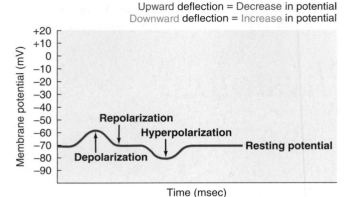

Upward **deflection** = Decrease in potential
Downward **deflection** = Increase in potential

● FIGURE 4-1

Types of changes in membrane potential

the positive or negative direction, the membrane is in a state of polarization. Recall that the magnitude of the potential is directly proportional to the number of positive and negative charges separated by the membrane and that the sign of the potential (+ or −) always designates whether excess positive or excess negative charges are present, respectively, on the inside of the membrane.

2. **Depolarization:** A change in potential that makes the membrane less polarized (less negative) than at resting potential. Depolarization decreases membrane potential, moving it closer to 0 mV (for example, a change from −70 mV to −60 mV); fewer charges are separated than at resting potential.

3. **Repolarization:** The membrane returns to resting potential after having been depolarized.

4. **Hyperpolarization:** A change in potential that makes the membrane more polarized (more negative) than at resting potential. Hyperpolarization increases membrane potential, moving it even farther from 0 mV (for instance, a change from −70 mV to −80 mV); more charges are separated than at resting potential.

One possibly confusing point should be clarified. On the device used for recording rapid changes in potential, a *decrease* in potential (that is, the inside being less negative than at resting) is represented as an *upward* deflection, whereas an *increase* in potential (that is, the inside being more negative than at resting) is represented by a *downward* deflection.

▌Electrical signals are produced by changes in ion movement across the plasma membrane.

Changes in membrane potential are brought about by changes in ion movement across the membrane. For example, if the net inward flow of positively charged ions increases compared to the resting state, the membrane becomes depolarized (less negative inside). By contrast, if the net outward flow of positively charged ions increases compared to the resting state, the membrane becomes hyperpolarized (more negative inside).

Changes in ion movement in turn are brought about by changes in membrane permeability in response to *triggering events*. Depending on the type of electrical signal, a triggering

event might be (1) a stimulus, such as sound waves stimulating specialized nerve cells in your ear; (2) a change in the electrical field in the vicinity of an excitable membrane; (3) an interaction of a chemical messenger with a surface receptor on a nerve or muscle cell membrane; or (4) a spontaneous change of potential caused by inherent imbalances in the leak–pump cycle. (You will learn more about the nature of these various triggering events as our discussion of electrical signals continues.)

Because the water-soluble ions responsible for carrying charge cannot penetrate the plasma membrane's lipid bilayer, these charges can only cross the membrane through channels specific for them. There are two types of channels: *leak channels*, which are open all the time; and *gated channels*, which can be opened or closed in response to specific triggering events. Thus, triggering events alter membrane permeability and consequently also alter ion flow across the membrane by opening or closing the gates guarding particular ion channels. These ion movements redistribute charge across the membrane, causing membrane potential to fluctuate.

There are two basic forms of electrical signals: (1) *graded potentials*, which serve as short-distance signals; and (2) *action potentials*, which signal over long distances. We are now going to examine these types of signals in more detail and then will explore how nerve cells use these signals to convey messages.

GRADED POTENTIALS

Graded potentials are local changes in membrane potential that occur in varying grades or degrees of magnitude or strength. For example, membrane potential could change from −70 mV to −60 mV (a 10-mV graded potential) or from −70 mV to −50 mV (a 20-mV graded potential).

▌The stronger a triggering event, the larger the resultant graded potential.

Graded potentials are usually produced by a specific triggering event that causes gated ion channels to open in a specialized region of the excitable cell membrane. Most commonly, gated Na^+ channels open, leading to the inward movement of Na^+ down its concentration and electrical gradients. The resultant depolarization—the graded potential—is confined to this small, specialized region of the total plasma membrane.

The magnitude of this initial graded potential (that is, the difference between the new potential and the resting potential) is related to the magnitude of the triggering event: *The stronger the triggering event, the more gated channels that open, the greater the positive charge entering the cell, and the larger the depolarizing graded potential at the point of origin. Also, the longer the duration of the triggering event, the longer the duration of the graded potential* (● Figure 4-2).

▌Graded potentials spread by passive current flow.

When a graded potential occurs locally in a nerve or muscle cell membrane, the remainder of the membrane is still at rest-

The magnitude and duration of a graded potential
The magnitude and duration of a graded potential
depend directly on the strength and duration
of the triggering event, such as a stimulus.

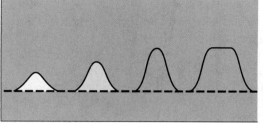

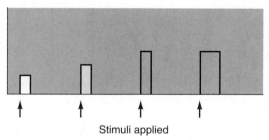

ing potential. The temporarily depolarized region
is called an *active area*. Note from ● Figure 4-3
that inside the cell, the active area is relatively
more positive than the neighboring *inactive areas*
that are still at resting potential. Outside the cell,
the active area is relatively less positive than these
adjacent areas. Because of this difference in potential, electrical charges, in this case carried by
ions, passively flow between the active and adjacent resting regions on both the inside and out-

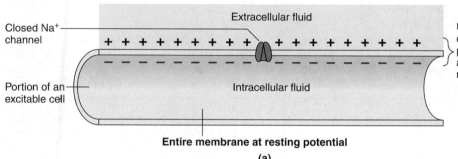

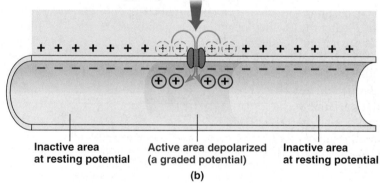

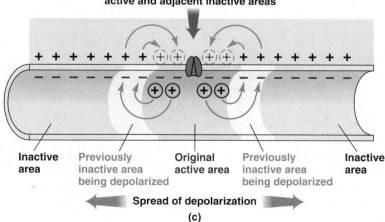

● FIGURE 4-3

**Current flow
during a graded
potential**
(a) The membrane of
an excitable cell at resting potential. (b) A triggering event opens Na$^+$
channels, leading to
the Na$^+$ entry that
brings about depolar-
ization. The adjacent
inactive areas are still
at resting potential.
(c) Local current flow
occurs between the
active and adjacent
inactive areas. This
local current flow
results in depolariza-
tion of the previously
inactive areas. In this
way, the depolarization
spreads away from its
point of origin.

side of the membrane. Any flow of electrical charges is called a **current.** By convention, the direction of current flow is always designated by the direction in which the positive charges are moving (● Figure 4-3c). On the inside, positive charges flow through the ICF away from the relatively more positive depolarized active region toward the more negative adjacent resting regions. Similarly, outside the cell positive charges flow through the ECF from the more positive adjacent inactive regions toward the relatively more negative active region. Ion movement (that is, current) is occurring *along* the membrane between regions next to each other on the same side of the membrane. This flow is in contrast to ion movement *across* the membrane through ion channels, with which you are more familiar.

As a result of local current flow between an active depolarized area and an adjacent inactive area, potential in the previously inactive area alters. Positive charges have flowed into this adjacent area on the inside, while simultaneously positive charges have flowed out of this area on the outside. Thus at this adjacent site the inside is more positive (or less negative) and the outside is less positive (or more negative) than before (● Figure 4-3c). Stated differently, the previously inactive adjacent region has been depolarized; thus the graded potential has spread. This area's potential now differs from that of the inactive region immediately next to it on the other side, inducing further current flow at this new site, and so on. In this manner, current spreads in both directions away from the initial site of the potential change.

The amount of current that flows between two areas depends on the difference in potential between the areas and on the resistance of the material through which the charges are moving. **Resistance** is the hindrance to electrical charge movement. The greater the difference in potential, the greater the current flow. The lower the resistance, the greater the current flow. *Conductors* have low resistance, providing little hindrance to current flow. Electrical wires and the ICF and ECF are all good conductors, so current readily flows through them. *Insulators* have high resistance and greatly hinder movement of charge. The plastic surrounding electrical wires has high resistance, as do body lipids. Thus current does not flow across the plasma membrane's lipid bilayer. Current, carried by ions, can move across the membrane only through ion channels.

▌ Graded potentials die out over short distances.

The passive current flow between active and adjacent inactive areas is similar to the means by which current is carried through electrical wires. We know from experience that current leaks out of an electrical wire with dangerous results unless the wire is covered with an insulating material such as plastic. (People can get an electric shock if they touch a bare wire.) Likewise, current is lost across the cell membrane as charge-carrying ions leak through the "uninsulated" parts of the membrane, that is, through open channels. Because of this current loss, the magnitude of the local current progressively diminishes with

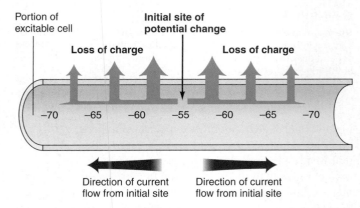

* Numbers refer to the local potential in mV at various points along the membrane.

● **FIGURE 4-4**

Current loss across the plasma membrane
Leakage of charge-carrying ions across the plasma membrane results in progressive loss of current with increasing distance from the initial site of potential change.

increasing distance from the initial site of origin (● Figure 4-4). Thus the magnitude of the graded potential continues to decrease the farther it moves away from the initial active area. Another way of saying this is that the spread of a graded potential is *decremental* (gradually decreases). Note that in ● Figures 4-4 and 4-5, the magnitude of the initial potential change is 15 mV (a change from −70 mV to −55 mV), then decreases as it moves along the membrane to a potential change of 10 mV (from −70 mV to −60 mV), and continues to diminish the farther it moves away from the initial active area, until there is no longer a potential change. In this way, these local currents die out within a few millimeters from the initial site of potential change and consequently can function as signals for only very short distances.

● **FIGURE 4-5**

Decremental spread of graded potentials
Because of leaks in current, the magnitude of a graded potential continues to decrease as it passively spreads from the initial active area. The potential dies out altogether within a few millimeters of its site of initiation.

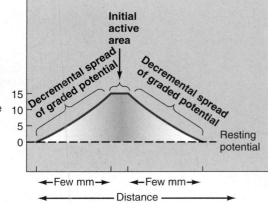

Although graded potentials have limited signaling distance, they are critically important to the body's function, as explained in later chapters. The following are all graded potentials: *postsynaptic potentials, receptor potentials, end-plate potentials, pacemaker potentials,* and *slow-wave potentials.* These terms are unfamiliar to you now, but you will become well acquainted with them as we continue discussing nerve and muscle physiology. We are including this list here because it is the only place all these graded potentials will be grouped together. For now it's enough to say that for the most part, excitable cells produce one of these types of graded potentials in response to a triggering event. In turn, graded potentials can initiate *action potentials,* the long-distance signals, in an excitable cell.

ACTION POTENTIALS

Action potentials are brief, rapid, large (100 mV) changes in membrane potential during which the potential actually reverses, so that the inside of the excitable cell transiently becomes more positive than the outside. As with a graded potential, a single action potential involves only a small portion of the total excitable cell membrane. Unlike graded potentials, however, action potentials are conducted, or propagated, throughout the entire membrane in *nondecremental* fashion; that is, they do not diminish in strength as they travel from their site of initiation throughout the remainder of the cell membrane. Thus action potentials can serve as faithful long-distance signals. Think about the nerve cell that brings about contraction of muscle cells in your big toe (see ● Figure 2-19, p. 47). If you want to wiggle your big toe, commands are sent from your brain down your spinal cord to initiate an action potential at the beginning of this nerve cell, which is located in the spinal cord. This action potential travels in undiminishing fashion all the way down the nerve cell's long axon, which runs through your leg to terminate on your big-toe muscle cells. The signal has not weakened or died off, instead being preserved at full strength from beginning to end.

Let's now consider the changes in potential during an action potential and the permeability and ion movements responsible for generating this potential change before turning our attention to the means by which action potentials spread throughout the cell membrane in undiminishing fashion.

▌ During an action potential, the membrane potential rapidly, transiently reverses.

If of sufficient magnitude, a graded potential can initiate an action potential before the graded potential dies off. (Later you will discover the means by which this initiation is accomplished for the various types of graded potentials.) Typically, the portion of the excitable membrane where graded potentials are produced in response to a triggering event does not undergo action potentials. Instead, the graded potential, by electrical or chemical means, brings about depolarization of adjacent portions of the membrane where action potentials can take place.

For convenience in this discussion, we will now jump from the triggering event to the depolarization of the membrane portion that is to undergo an action potential, without considering the involvement of the intervening graded potential.

To initiate an action potential, a triggering event causes the membrane to depolarize from the resting potential of −70 mV (● Figure 4-6). Depolarization proceeds slowly at first until it reaches a critical level known as **threshold potential,** typically between −50 and −55 mV. At threshold potential, an explosive depolarization takes place. A recording of the potential at this time shows a sharp upward deflection to +30 mV as the potential rapidly moves toward 0 mV, then reverses itself so that the inside of the cell becomes positive compared to the outside. Just as rapidly, the membrane repolarizes, dropping back to resting potential. Often the forces that repolarize the membrane push the potential too far, causing a brief **after hyperpolarization,** during which the inside of the membrane briefly becomes even more negative than normal (for example, −80 mV) before the resting potential is restored.

The entire rapid change in potential from threshold to peak reversal and then back to resting is called the *action potential.* Unlike the variable duration of a graded potential, the duration of an action potential is always the same in a given excitable cell. In a nerve cell, an action potential lasts for only 1 msec (0.001 sec). It lasts longer in muscle, with the duration depending on the muscle type. The portion of the action potential during which the potential is reversed (between 0 mV and +30 mV) is called the **overshoot.** Often an action potential is referred to as a **spike,** because of its spikelike recorded appearance. Alternatively, when an excitable membrane is triggered to undergo an action potential, it is said to **fire.** Thus, the terms *action potential, spike,* and *firing* all refer to the same phenomenon of rapid reversal of membrane potential.

● **FIGURE 4-6**

Changes in membrane potential during an action potential

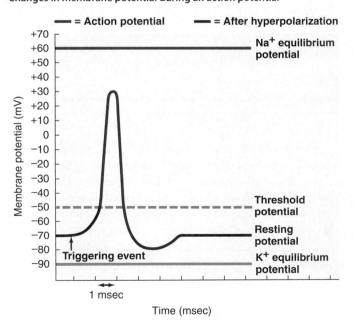

Voltage-Gated Sodium Channel

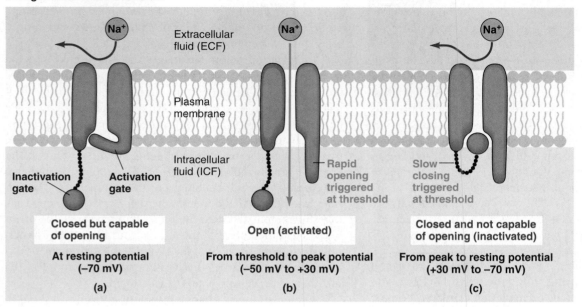

Extracellular fluid (ECF)

Plasma membrane

Intracellular fluid (ICF)

Inactivation gate   **Activation gate**

Rapid opening triggered at threshold

Slow closing triggered at threshold

Closed but capable of opening

Open (activated)

Closed and not capable of opening (inactivated)

At resting potential (−70 mV)

From threshold to peak potential (−50 mV to +30 mV)

From peak to resting potential (+30 mV to −70 mV)

(a)   (b)   (c)

Voltage-Gated Potassium Channel

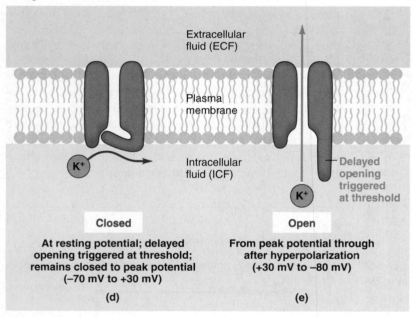

Extracellular fluid (ECF)

Plasma membrane

Intracellular fluid (ICF)

Delayed opening triggered at threshold

Closed

Open

At resting potential; delayed opening triggered at threshold; remains closed to peak potential (−70 mV to +30 mV)

From peak potential through after hyperpolarization (+30 mV to −80 mV)

(d)   (e)

● **FIGURE 4-7**
Conformations of voltage-gated sodium and potassium channels

If threshold potential is not reached by the initial triggered depolarization, no action potential takes place. Thus, threshold is a critical all-or-none point. Either the membrane is depolarized to threshold and an action potential takes place, or threshold is not reached in response to the depolarizing event and no action potential occurs.

▮ Marked changes in membrane permeability and ion movement lead to an action potential.

How is the membrane potential, which is usually maintained at a constant resting level, thrown out of balance to such an extent as to produce an action potential? Recall that K$^+$ makes the greatest contribution to the establishment of the resting potential, because the membrane at rest is considerably more permeable to K$^+$ than to Na$^+$ (see p. 88). During an action potential, marked changes in membrane permeability to Na$^+$ and K$^+$ take place, permitting rapid fluxes of these ions down their electrochemical gradients. These ion movements carry the current responsible for the potential changes that occur during an action potential.

Two specific types of channels are of major importance in the development of an action potential: voltage-gated Na$^+$ channels and voltage-gated K$^+$ channels. Gated channels have gates that can alternately be open, permitting ion passage through the channel, or closed, preventing ion passage through the channel. Gate opening and closing results from a change in the three-dimensional conformation (shape) of the protein that forms the gated channel. There are three kinds of gated channels, depending on the factor that induces the change in channel conformation: (1) **voltage-gated channels,** which open or close in response to changes in membrane potential; (2) **chemically gated channels,** which change conformation in response to the binding of a specific chemical messenger with a membrane receptor in close association with the channel; and (3) **mechanically gated channels,** which respond to stretching or other mechanical deformation.

Voltage-gated Na$^+$ and K$^+$ channels

Voltage-gated Na$^+$ and K$^+$ channels are the ones involved in action potentials. Voltage-gated channels are composed of proteins that have a number of charged groups. The electric field (potential) surrounding the channels can exert a distorting force on the channel structure as charged portions of the channel proteins are electrically attracted or repelled by charges in the fluids surrounding the membrane. Unlike the majority of membrane proteins, which remain stable despite fluctuations in membrane potential, the voltage-gated channel proteins are especially sensitive to voltage changes. Small distortions in channel shape induced by potential changes can cause them to flip to another conformation. Here again is an example of how subtle changes in structure can profoundly influence function.

The voltage-gated Na^+ channel has two gates: an *activation gate* and an *inactivation gate* (● Figure 4-7). The activation gate guards the channel by opening and closing like a hinged door. The inactivation gate consists of a ball-and-chain–like sequence of amino acids. This gate is open when the ball is dangling free on its chain and closed when the ball binds to its receptor located at the channel opening, thus blocking the opening. Both gates must be open to permit passage of Na^+ through the channel, and closure of either gate prevents passage. This voltage-gated Na^+ channel can exist in three different conformations: (1) *closed but capable of opening* (activation gate closed, inactivation gate open, ● Figure 4-7a); (2) *open*, or *activated* (both gates open, ● Figure 4-7b); and (3) *closed and not capable of opening* (activation gate open, inactivation gate closed, ● Figure 4-7c).

The voltage-gated K^+ channel is simpler. It has only one gate, which can be either open or closed (● Figure 4-7d and e). These voltage-gated Na^+ and K^+ channels exist in addition to the $Na^+–K^+$ pump and the leak channels for these ions (described in Chapter 3).

Changes in permeability and ion movement during an action potential

At resting potential (-70 mV), all the voltage-gated Na^+ and K^+ channels are closed, with the Na^+ channels' activation gates being closed and their inactivation gates being open; that is, the voltage-gated Na^+ channels are in their "closed but capable of opening" conformation. Therefore, passage of Na^+ and K^+ does not occur through these voltage-gated channels at resting potential. However, because of the presence of many K^+ leak channels and very few Na^+ leak channels, the resting membrane is 50 to 75 times more permeable to K^+ than to Na^+.

When a membrane starts to depolarize toward threshold as a result of a triggering event, the activation gates of some of its voltage-gated Na^+ channels open. Now both gates of these activated channels are open. Because both the concentration and electrical gradients for Na^+ favor its movement into the cell, Na^+ starts to move in. The inward movement of positively charged Na^+ depolarizes the membrane further, thereby opening even more voltage-gated Na^+ channels and allowing more Na^+ to enter, and so on, in a positive-feedback cycle (● Figure 4-8).

At threshold potential, there is an explosive increase in Na^+ permeability, which is symbolized as P_{Na^+}, as the membrane swiftly becomes 600 times more permeable to Na^+ than to K^+. Each individual channel is either closed or open and cannot be partially open. However, the delicately poised gating mechanisms of the various voltage-gated Na^+ channels are jolted open by slightly different voltage changes. During the early depolarizing phase, more and more of the Na^+ chan-

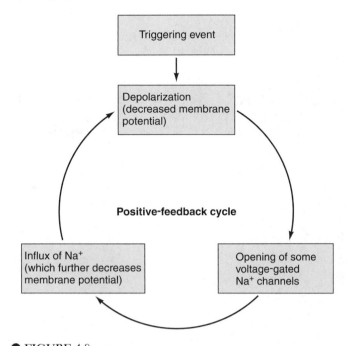

● FIGURE 4-8
Positive-feedback cycle responsible for opening Na^+ channels at threshold

nels open as the potential progressively decreases. At threshold, enough Na^+ gates have opened to set off the positive feedback cycle that rapidly causes the remaining Na^+ gates to swing open. Now Na^+ permeability dominates the membrane, in contrast to the K^+ domination at resting potential. Thus at threshold Na^+ rushes into the cell, rapidly eliminating the internal negativity and even making the inside of the cell more positive than the outside in an attempt to drive the membrane potential to the Na^+ equilibrium potential (which is $+60$ mV; see p. 90) (● Figure 4-9). The potential reaches $+30$ mV, close to the Na^+ equilibrium potential. The potential does not become any more positive, because, at the peak of the action potential, the Na^+ channels start to close to the inactivated state, and P_{Na^+} starts to fall to its low resting value.

What causes the Na^+ channels to close? When the membrane potential reaches threshold, two closely related events take place in the gates of each Na^+ channel. First the activation gates are triggered to *open rapidly*, in response to the depolarization, converting the channel to its open (activated) conformation (● Figure 4-7b). Surprisingly, this channel opening initiates the process of channel closing. The conformational change that opens the channel also allows the inactivation gate's ball to bind to its receptor at the channel opening, thereby physically blocking the mouth of the channel. However, this closure process takes time, so the inactivation gate *closes slowly* compared to the rapidity of channel opening. Meanwhile, during the 0.5-msec delay after the activation gate opens and before the inactivation gate closes, both gates are open, and Na^+ rushes into the cell through these open channels, bringing the action potential to its peak. Then the inactivation gate closes, membrane permeability to Na^+ plummets to its low resting value, and further Na^+ entry is prevented.

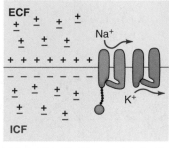

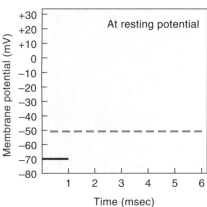

At resting potential

At resting potential, all voltage-gated Na⁺ and K⁺ channels are closed. Only the unbalanced charges separated across the membrane contribute to potential.

(a)

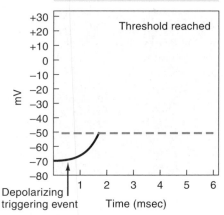

Threshold reached

Depolarizing triggering event

After a depolarizing triggering event brings the membrane to the threshold potential of -50 mV, the Na⁺ activation gates open.

(b)

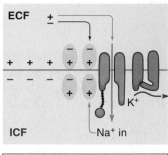

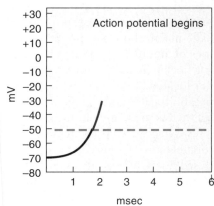

Action potential begins

The resultant movement of Na⁺ inward neutralizes negative charges inside the cell. Inward movement of Na⁺ leaves behind on the outside the negative charges (primarily Cl⁻) with which Na⁺ had been paired. These negative charges neutralize positive charges that had been contributing to membrane potential, making the outside progressively less positive.

(c)

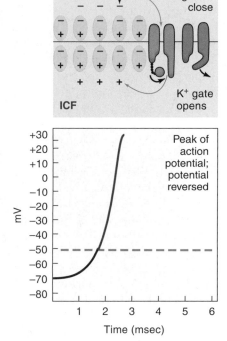

Peak of action potential; potential reversed

Further inward movement of Na⁺ reverses the potential, with the inside becoming positive and the outside becoming negative as the action potential peaks. At the peak of the action potential, the Na⁺ inactivation gates begin to close and the K⁺ gates open. Entry of Na⁺ ceases, and K⁺ starts to leave the cell.

(e)

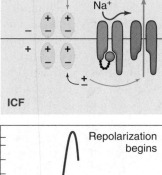

Repolarization begins

Outward movement of K⁺ leaves behind negative charges (A⁻) inside the cell and neutralizes negative charges outside; as a consequence, the inside becomes progressively less positive and the outside less negative until 0 mV is reached.

(f)

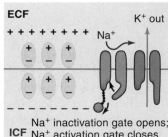

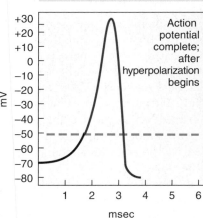

Action potential complete; after hyperpolarization begins

Continued outward movement of K⁺ restores the resting membrane potential, with the potential reversing back, so that the inside is once again negative and the outside positive. At resting potential, the Na⁺ inactivation gates open, and the activation gates close, reset to respond to another triggering event. Further outward movement of K⁺ through the still-open K⁺ gates briefly hyperpolarizes the membrane.

(g)

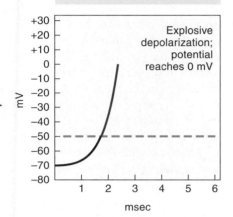

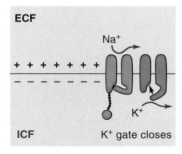

+30
+20 — Explosive
+10 — depolarization;
0 — potential
-10 — reaches 0 mV

Continued inward movement of Na⁺ progressively reduces the potential (the inside becoming less negative and the outside less positive) until 0 mV is reached.

(d)

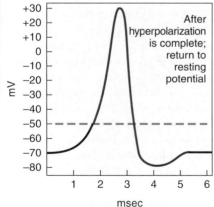

After hyperpolarization is complete; return to resting potential

Then the K⁺ gates close, and the membrane returns to resting potential.

(h)

● FIGURE 4-9

Ion movements responsible for changes in membrane potential during an action potential

The channel remains in this inactivated conformation until the membrane potential has been restored to its resting value.

Simultaneous with inactivation of Na⁺ channels at the peak of the action potential, the voltage-gated K⁺ channels open. Opening of the K⁺ channel gate is a delayed voltage-gated response triggered by the initial depolarization to threshold. Thus three action-potential-related events occur at threshold: (1) the rapid opening of the Na⁺ activation gates, which permits Na⁺ to enter, moving the potential from threshold to its positive peak; (2) the slow closing of the Na⁺ inactivation gates, which halts further Na⁺ entry after a brief time delay, thus keeping the potential from rising any further; and (3) the slow opening of the K⁺ gates, which, as you will see, is in large part responsible for returning the potential from its peak back to resting.

The membrane potential would gradually return to resting after closure of the Na⁺ channels as K⁺ continued to leak out but no further Na⁺ entered. However, the return to resting is hastened by the opening of K⁺ gates at the peak of the action potential. Opening of the voltage-gated K⁺ channels greatly increases the K⁺ permeability (designated P_{K^+}) to about 300 times the resting P_{Na^+}. This marked increase in P_{K^+} causes K⁺ to rush out of the cell down its concentration and electrical gradients, carrying positive charges back to the outside. Note that at the peak of the action potential, the positive potential inside the cell tends to repel the positive K⁺ ions, so the electrical gradient for K⁺ is outward, unlike at resting potential. The outward movement of K⁺ rapidly restores the negative resting potential (● Figure 4-9g).

To review (● Figure 4-10), *the rising phase of the action potential* (from threshold to +30 mV) *is due to Na⁺ influx* (Na⁺ entering the cell) induced by an explosive increase in

● FIGURE 4-10

Permeability changes and ion fluxes during an action potential

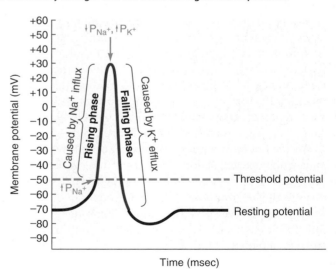

P_{Na^+} at threshold. *The falling phase* (from +30 mV to resting potential) *is brought about largely by* K^+ *efflux* (K^+ leaving the cell) caused by the marked increase in P_{K^+} occurring simultaneously with the inactivation of the Na^+ channels at the peak of the action potential.

As the potential returns to resting, the changing voltage shifts the Na^+ channels to their "closed but capable of opening" conformation, with the activation gate closed and the inactivation gate open. Now the channel is reset, ready to respond to another triggering event. The newly opened voltage-gated K^+ channels also close, so the membrane returns to the resting number of open K^+ leak channels. Typically, the voltage-gated K^+ channels are slow to close. As a result of this persistent increased permeability to K^+, more K^+ may leave than is necessary to bring the potential to resting. This slight excessive K^+ efflux makes the interior of the cell transiently even more negative than resting potential, causing the after hyperpolarization.

▌ The Na^+–K^+ pump gradually restores the concentration gradients disrupted by action potentials.

At the completion of an action potential, the membrane potential has been restored to its resting condition, but the ion distribution has been altered slightly. Sodium has entered the cell during the rising phase, and a comparable amount of K^+ has left during the falling phase. The Na^+–K^+ pump restores these ions to their original locations in the long run, but not after each action potential.

The active pumping process takes much longer to restore Na^+ and K^+ to their original locations than it takes for the passive fluxes of these ions during an action potential. However, the membrane does not need to wait until the Na^+–K^+ pump slowly restores the concentration gradients before it can undergo another action potential. Actually, the movement of only relatively few of the total number of Na^+ and K^+ ions present causes the large swings in potential that occur during an action potential. Only about 1 out of 100,000 K^+ ions present in the cell leaves during an action potential, while a comparable number of Na^+ ions enters from the ECF. The movement of this extremely small proportion of the total Na^+ and K^+ during a single action potential produces dramatic 100-mV changes in potential (between −70 mV and +30 mV), but only infinitesimal changes in the ICF and ECF concentrations of these ions. Much more K^+ is still inside the cell than outside, and Na^+ is still predominantly an extracellular cation. Consequently, the Na^+ and K^+ concentration gradients still exist, so repeated action potentials can occur without the pump having to keep pace to restore the gradients.

Were it not for the pump, of course, even tiny fluxes accompanying repeated action potentials would eventually "run down" the concentration gradients so that further action potentials would be impossible. If the concentrations of Na^+ and K^+ were equal between the ECF and ICF, changes in permeability to these ions would not bring about ion fluxes, so no change in potential would occur. Thus the Na^+–K^+ pump is critical to maintaining the concentration gradients in the long

run. However, it does not have to perform its role between action potentials, nor is it directly involved in the ion fluxes or potential changes that occur during an action potential.

▌ Action potentials are propagated from the axon hillock to the axon terminals.

A single action potential involves only a small patch of the total surface membrane of an excitable cell. But if action potentials are to serve as long-distance signals, they cannot be merely isolated events occurring in a limited area of a nerve or muscle cell membrane. Mechanisms must exist to conduct or spread the action potential throughout the entire cell membrane. Furthermore, the signal must be transmitted from one cell to the next cell (for example, along specific nerve pathways). To explain how these mechanisms are accomplished, we will first begin with a brief look at neuronal structure. Then we will examine how an action potential (nerve impulse) is conducted throughout a nerve cell, before turning our attention to how the signal is passed to another cell.

Neuronal structure

A single nerve cell, or **neuron,** typically consists of three basic parts: the *cell body*, the *dendrites*, and the *axon*, although there are variations in structure, depending on the location and function of the neuron. The nucleus and organelles are housed in the **cell body,** from which numerous extensions known as **dendrites** typically project like antennae to increase the surface area available for receiving signals from other nerve cells (● Figure 4-11). Some neurons have up to 400,000 of these elongated surface extensions. Dendrites carry signals *toward* the cell body. In most neurons the plasma membrane of the dendrites and cell body contains protein receptors for binding chemical messengers from other neurons. Therefore, the dendrites and cell body are the neuron's *input zone*, because these components receive and integrate incoming signals. This is the region where graded potentials are produced in response to triggering events, in this case, incoming chemical messengers.

The **axon,** or **nerve fiber,** is a single, elongated, tubular extension that conducts action potentials *away from* the cell body and eventually terminates at other cells. The axon frequently gives off side branches, or **collaterals,** along its course. The first portion of the axon plus the region of the cell body from which the axon leaves is known as the **axon hillock.** The axon hillock is the neuron's *trigger zone*, because it is the site where action potentials are triggered, or initiated, by the graded potential if it is of sufficient magnitude. The action potentials are then conducted along the axon from the axon hillock to the typically highly branched ending at the **axon terminals.** These terminals release chemical messengers that simultaneously influence numerous other cells with which they come into close association. Functionally, therefore, the axon is the *conducting zone* of the neuron, and the axon terminals constitute its *output zone*. (The major exception to this typical neuronal structure and functional organization is neurons specialized to carry sensory information, a topic described in a later chapter.)

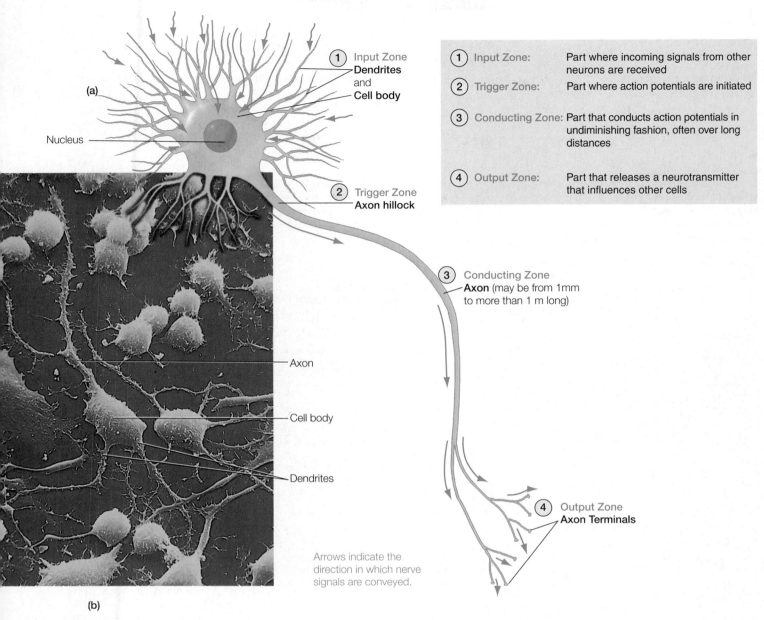

(a)

Nucleus

① Input Zone
Dendrites
and
Cell body

② Trigger Zone
Axon hillock

③ Conducting Zone
Axon (may be from 1mm
to more than 1 m long)

Axon

Cell body

Dendrites

④ Output Zone
Axon Terminals

Arrows indicate the
direction in which nerve
signals are conveyed.

(b)

①	Input Zone:	Part where incoming signals from other neurons are received
②	Trigger Zone:	Part where action potentials are initiated
③	Conducting Zone:	Part that conducts action potentials in undiminishing fashion, often over long distances
④	Output Zone:	Part that releases a neurotransmitter that influences other cells

● **FIGURE 4-11**

Anatomy of a neuron (nerve cell)
(a) Most but not all neurons consist of the basic parts schematically represented in the figure. (b) An electron micrograph highlighting the cell body, dendrites, and part of the axon of a neuron within the central nervous system.

Axons vary in length from less than a millimeter in neurons that communicate only with neighboring cells to longer than a meter in neurons that communicate with distant parts of the nervous system or with peripheral organs. For example, the axon of the nerve cell innervating your big toe must traverse the distance from the origin of its cell body within the spinal cord in the lower region of your back all the way down your leg to your toe.

Action potentials can be initiated only in portions of the membrane that have an abundance of voltage-gated Na^+ channels that can be triggered to open by a depolarizing event. Typically, regions of excitable cells where graded potentials take place do not undergo action potentials, because voltage-gated Na^+ channels are sparse there. Therefore, sites specialized for graded potentials do not undergo action potentials, even though they might be depolarized considerably. However, graded potentials can, before dying out, trigger action potentials in adjacent portions of the membrane by bringing these more sensitive regions to threshold through local current flow spreading from the site of the graded potential. In a typical neuron, for example, graded potentials are generated in the dendrites and cell body in response to incoming signals. If these graded potentials have sufficient magnitude by the time they have spread to the axon hillock, they initiate an action potential at this triggering zone.

■ Once initiated, action potentials are conducted throughout a nerve fiber.

Once an action potential is initiated at the axon hillock, no further triggering event is necessary to activate the remainder of the nerve fiber. The impulse is automatically conducted throughout the neuron without further stimulation by one of two methods of propagation: *contiguous conduction* or *saltatory conduction*.

Contiguous conduction involves the spread of the action potential along every patch of membrane down the length of

---

● **FIGURE 4-12**

Contiguous conduction

Local current flow between the active area at the peak of an action potential and the adjacent inactive area still at resting potential reduces the potential in this contiguous inactive area to threshold, which triggers an action potential in the previously inactive area. The original active area returns to resting potential, and the new active area induces an action potential in the next adjacent inactive area by local current flow as the cycle repeats itself down the length of the axon.

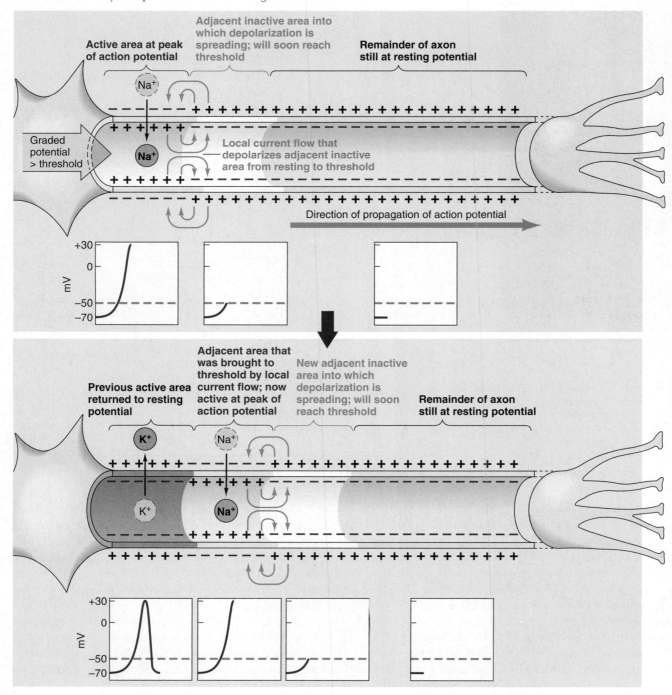

the axon (*contiguous* means "touching" or "next to in sequence.") This process is illustrated in ● Figure 4-12. You are viewing a schematic representation of a longitudinal section of the axon hillock and the portion of the axon immediately beyond it. The membrane at the axon hillock is at the peak of an action potential. The inside of the cell is positive in this active area, because Na^+ has already rushed into the nerve cell at this point. The remainder of the axon, still at resting potential and negative inside, is considered inactive. For the action potential to spread from the active to the inactive areas, the inactive areas must somehow be depolarized to threshold before they can undergo an action potential. This depolarization is accomplished by local current flow between the area already undergoing an action potential and the adjacent inactive area, similar to the current flow responsible for the spread of graded potentials. Because opposite charges attract, current can flow locally between the active area and the neighboring inactive area on both the inside and the outside of the membrane. This local current flow in effect neutralizes or eliminates some of the unbalanced charges in the inactive area; that is, it reduces the number of opposite charges separated across the membrane, reducing the potential in this area. This depolarizing effect quickly brings the involved inactive area to threshold, at which time the voltage-gated Na^+ channels in this region of the membrane are all thrown open, leading to an action potential in this previously inactive area. Meanwhile, the original active area returns to resting potential as a result of K^+ efflux.

In turn, beyond the new active area is another inactive area, so the same thing happens again. This cycle repeats itself in a chain reaction until the action potential has spread to the end of the axon. *Once an action potential is initiated in one part of a nerve cell membrane, a self-perpetuating cycle is initiated so that the action potential is propagated along the rest of the fiber automatically.* In this way, the axon is similar to a fire-cracker fuse that needs to be lit at only one end. Once ignited, the fire spreads down the fuse; it is not necessary to hold a match to every separate section of the fuse.

Note that the original action potential does not travel along the membrane. Instead, it triggers an identical new action potential in the adjacent area of the membrane, with this process being repeated along the axon's length. An analogy is the "wave" at a stadium. Each section of spectators stands up (the rising phase of an action potential), then sits down (the falling phase) in sequence one after another as the wave moves around the stadium. The wave, not individual spectators, travels around the stadium. Similarly, new action potentials arise sequentially down the axon. Each new action potential in the conduction process is a fresh local event that depends on the induced permeability changes and electrochemical gradients, which are virtually identical down the length of the axon. Therefore, the last action potential at the end of the axon is identical to the original one, no matter how long the axon. Thus an action potential is spread along the axon in undiminished fashion. In this way, action potentials can serve as faithful long-distance signals without attenuation or distortion.

This nondecremental propagation of an action potential contrasts with the decremental spread of a graded potential, which dies out over a very short distance because it cannot regenerate itself. ▲ Table 4-1 summarizes the differences between graded potentials and action potentials, some of which are yet to be discussed.

■ **The refractory period ensures one-way propagation of the action potential.**

What ensures the one-way propagation of an action potential away from the initial site of activation? Note from ● Figure 4-13

▲ TABLE 4-1
Comparison of Graded Potentials and Action Potentials

Graded Potentials	Action Potentials
Graded potential change; magnitude varies with magnitude of triggering event	All-or-none membrane response; magnitude of triggering event coded in frequency rather than amplitude of action potentials
Duration varies with duration of triggering event	Constant duration
Decremental conduction; magnitude diminishes with distance from initial site	Propagated throughout membrane in undiminishing fashion
Passive spread to neighboring inactive areas of membrane	Self-regeneration in neighboring inactive areas of membrane
No refractory period	Refractory period
Can be summed	Summation impossible
Can be depolarization or hyperpolarization	Always depolarization and reversal of charges
Triggered by stimulus, by combination of neurotransmitter with receptor, or by spontaneous shifts in leak–pump cycle	Triggered by depolarization to threshold, usually through spread of graded potential
Occurs in specialized regions of membrane designed to respond to triggering event	Occurs in regions of membrane with abundance of voltage-gated Na^+ channels

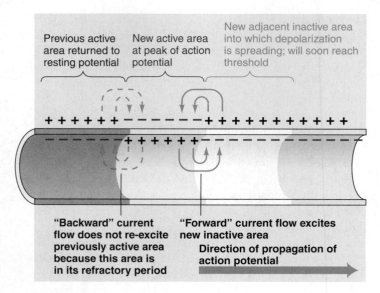

● FIGURE 4-13

Value of the refractory period
"Backward" current flow is prevented by the refractory period. During an action potential and slightly beyond, an area cannot be re-stimulated by normal events to undergo another action potential. Thus the refractory period ensures that an action potential can be propagated only in the forward direction along the axon.

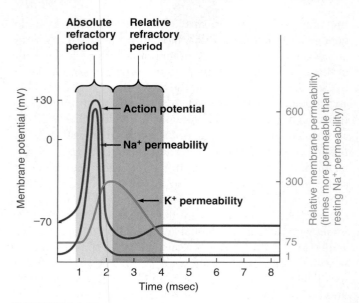

● FIGURE 4-14

Absolute and relative refractory periods
During the absolute refractory period, the portion of the membrane that has just undergone an action potential cannot be restimulated. This period corresponds to the time during which the Na$^+$ gates are not in their resting conformation. During the relative refractory period, the membrane can be restimulated only by a stronger stimulus than is usually necessary. This period corresponds to the time during which the K$^+$ gates opened during the action potential have not yet closed.

that once the action potential has been regenerated at a new neighboring site (now positive inside) and the original active area has returned to resting (once again negative inside), the close proximity of opposite charges between these two areas is conducive to local current flow taking place in the backward direction, as well as in the forward direction into as yet unexcited portions of the membrane. If such backward current flow were able to bring the just inactivated area to threshold, another action potential would be initiated here, which would spread both forward and backward, initiating still other action potentials, and so on. But if action potentials were to move in both directions, the situation would be chaotic, with numerous action potentials bouncing back and forth along the axon until the nerve cell eventually fatigued. Fortunately, neurons are saved from this fate of oscillating action potentials by the **refractory period,** during which a new action potential cannot be initiated by normal events in a region that has just undergone an action potential.

The refractory period has two components: the *absolute refractory period* and the *relative refractory period.* During the time that a particular patch of axonal membrane is undergoing an action potential, it is incapable of initiating another action potential, no matter how strongly it is stimulated by a triggering event. This time period when a recently activated patch of membrane is completely refractory (meaning "stubborn," or unresponsive) to further stimulation is known as the **absolute refractory period** (● Figure 4-14). Once the voltage-gated Na$^+$ channels have flipped to their open, or activated, state, they cannot be triggered to open again in response to another depolarizing triggering event, no matter how strong, until resting potential is restored and the channels are reset to their orig-

inal positions. Accordingly, the absolute refractory period lasts the entire time from opening of the voltage-gated Na$^+$ channels' activation gates at threshold, through closure of their inactivation gates at the peak of the action potential, until the return to resting potential when the channels' activation gates close and inactivation gates open once again; that is, until the channels are in their "closed but capable of opening" conformation. Only then can they respond to another depolarization with an explosive increase in P$_{Na^+}$ to initiate another action potential. Because of this absolute refractory period, one action potential must be over before another can be initiated at the same site. Action potentials cannot overlap or be added one on top of another "piggyback fashion."

Following the absolute refractory period is a **relative refractory period,** during which a second action potential can be produced only by a triggering event considerably stronger than is usually necessary. The relative refractory period occurs after the action potential is completed because of a twofold effect: lingering inactivation of the voltage-gated Na$^+$ channels, and slowness of the voltage-gated K$^+$ channels that opened at the peak of the action potential to close. During this time, fewer than normal voltage-gated Na$^+$ channels are in a position to be jolted open by a depolarizing triggering event. Simultaneously, K$^+$ is still leaving through its slow-to-close channels during the after hyperpolarization. The less-than-normal Na$^+$ entry in response to another triggering event is opposed by a persistent hyperpolarizing outward leak of K$^+$ through its still-open channels, and thus a greater-than-normal depolarizing triggering event is needed to bring the membrane to threshold during the relative refractory period.

By the time the original site has recovered from its refractory period and is capable of being restimulated by normal current flow, the action potential has been rapidly propagated in the forward direction only and is so far away that it can no longer influence the original site. Thus, *the refractory period ensures the one-way propagation of the action potential down the axon away from the initial site of activation.*

▌ The refractory period also limits the frequency of action potentials.

The refractory period is also responsible for setting an upper limit on the frequency of action potentials; that is, it determines the maximum number of new action potentials that can be initiated and propagated along a fiber in a given period of time. The original site must recover from its refractory period before a new action potential can be triggered to follow the preceding action potential. The length of the refractory period varies for different types of neurons. The longer the refractory period, the greater the delay before a new action potential can be initiated and the lower the frequency with which a nerve cell can respond to repeated or ongoing stimulation.

▌ Action potentials occur in all-or-none fashion.

If any portion of the neuronal membrane is depolarized to threshold, an action potential is initiated and relayed along the membrane in undiminished fashion. Furthermore, once threshold has been reached the resultant action potential always goes to maximal height. The reason for this effect is that the changes in voltage during an action potential result from ion movements down concentration and electrical gradients, and these gradients are not affected by the strength of the depolarizing triggering event. A triggering event stronger than one necessary to bring the membrane to threshold does not produce a larger action potential. However, a triggering event that fails to depolarize the membrane to threshold does not trigger an action potential at all. Thus *an excitable membrane either responds to a triggering event with a maximal action potential that spreads nondecrementally throughout the membrane, or it does not respond with an action potential at all.* This property is called the **all-or-none law.**

This all-or-none concept is analogous to firing a gun. Either the trigger is not pulled sufficiently to fire the bullet (threshold is not reached), or it is pulled hard enough to elicit the full firing response of the gun (threshold is reached). Squeezing the trigger harder does not produce a greater explosion. Just as it is not possible to fire a gun halfway, it is not possible to cause a halfway action potential.

The threshold phenomenon allows some discrimination between important and unimportant stimuli or other triggering events. Stimuli too weak to bring the membrane to threshold do not initiate action potentials and therefore do not clutter up the nervous system by transmitting insignificant signals.

▌ The strength of a stimulus is coded by the frequency of action potentials.

How is it possible to differentiate between two stimuli of varying strengths when both stimuli bring the membrane to threshold and generate action potentials of the same magnitude? For example, how can one distinguish between touching a warm object or touching a very hot object if both trigger identical action potentials in a nerve fiber relaying information about skin temperature to the central nervous system? The answer lies in the *frequency* with which the action potentials are generated. A stronger stimulus does not produce a larger action potential, but it does trigger a greater *number* of action potentials per second. In addition, a stronger stimulus in a region causes more neurons to reach threshold, increasing the total information sent to the central nervous system.

Once initiated, the velocity, or speed, with which an action potential travels down the axon depends on two factors: (1) whether the fiber is myelinated and (2) the diameter of the fiber. Contiguous conduction occurs in unmyelinated fibers. In this case, as you just learned, each individual action potential initiates an identical new action potential in the next contiguous (bordering) segment of the axon membrane so that every portion of the membrane undergoes an action potential as this electrical signal is conducted from the beginning to the end of the axon. A faster method of propagation, *saltatory conduction*, takes place in myelinated fibers. We are next going to see how a myelinated fiber compares with an unmyelinated fiber, then see how saltatory conduction compares with contiguous conduction.

▌ Myelination increases the speed of conduction of action potentials.

Myelinated fibers, as the name implies, are covered with myelin at regular intervals along the length of the axon (● Figure 4-15a). **Myelin** is composed primarily of lipids. Because the water-soluble ions responsible for carrying current across the membrane cannot permeate this thick lipid barrier, the myelin coating acts as an insulator, just like plastic around an electrical wire, to prevent current leakage across the myelinated portion of the membrane. Myelin is not actually a part of the nerve cell but consists of separate myelin-forming cells that wrap themselves around the axon in jelly-roll fashion (● Figure 4-15b and c). These myelin-forming cells are **oligodendrocytes** in the central nervous system (the brain and spinal cord) and **Schwann cells** in the peripheral nervous system (the nerves running between the central nervous system and the various regions of the body). The lipid composition of myelin is due to the presence of layer on layer of the lipid bilayer that composes the plasma membrane of these myelin-forming cells. Between the myelinated regions, at the **nodes of Ranvier,** the axonal membrane is bare and exposed to the ECF. It is only at these bare spaces that current can flow across the membrane to produce action potentials (● Figure 4-15d). Voltage-gated Na^+ channels are concentrated at the nodes, whereas the

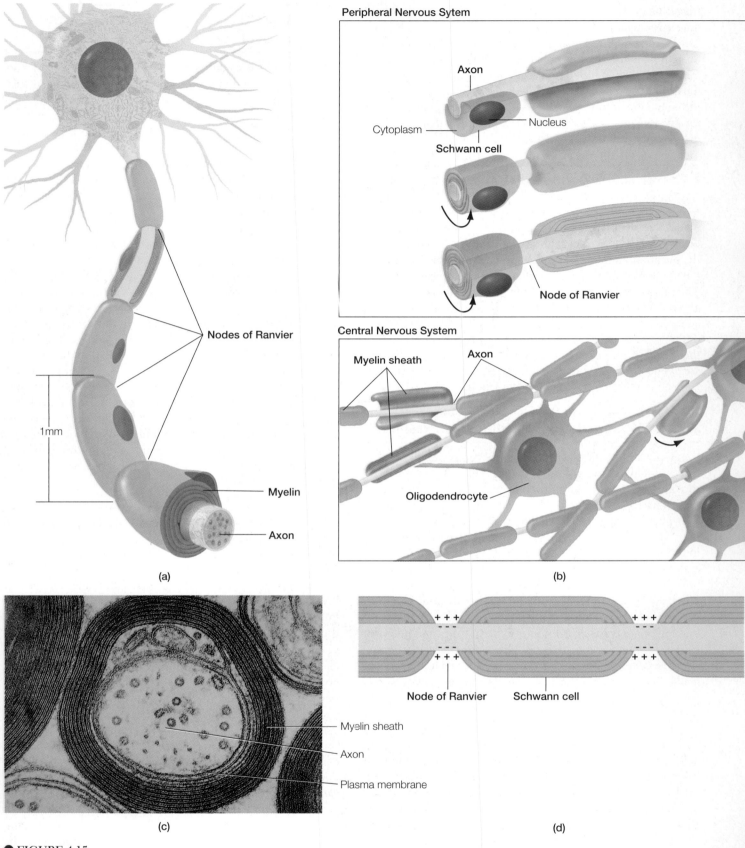

Peripheral Nervous Sytem

Axon

Cytoplasm

Nucleus

Schwann cell

Node of Ranvier

Nodes of Ranvier

1mm

Myelin

Axon

(a)

Central Nervous System

Myelin sheath

Axon

Oligodendrocyte

(b)

Myelin sheath

Axon

Plasma membrane

(c)

+ + +   + + +
– – –   – – –

– – –   – – –
+ + +   + + +

Node of Ranvier   Schwann cell

(d)

● FIGURE 4-15

Myelinated fibers
(a) A myelinated fiber is surrounded by myelin at regular intervals. The intervening unmyelinated regions are known as nodes of Ranvier. (b) In the peripheral nervous system, each patch of myelin is formed by a separate Schwann cell that wraps itself jelly-roll fashion around the nerve fiber. In the central nervous system, each of the several processes of a myelin-forming oligodendrocyte forms a patch of myelin around a separate nerve fiber. (c) An electron micrograph of a myelinated fiber in cross section. (d) Membrane potential exists only at the nodes of Ranvier, where the bare axon is exposed to the ECF. No charges exist across the insulated myelinated regions.

myelin-covered regions are almost devoid of these special passageways. By contrast, an unmyelinated fiber has a high density of voltage-gated Na$^+$ channels throughout its entire length. As you now know, action potentials can be generated only at portions of the membrane furnished with an abundance of these channels. Spread of action potentials in myelinated fibers is called *saltatory conduction*, which we will now examine.

The nodes are usually about 1 mm apart, short enough that local current from an active node can reach an adjacent node before dying off. When an action potential occurs at one node, opposite charges attract from the adjacent inactive node,

reducing its potential to threshold so that it undergoes an action potential, and so on. Consequently, in a myelinated fiber, the impulse "jumps" from node to node, skipping over the myelinated sections of the axon (● Figure 4-16); this process is called **saltatory conduction** (*saltere* means "to jump or leap"). Saltatory conduction propagates action potentials more rapidly than does contiguous conduction, because the action potential does not have to be regenerated at myelinated sections but must be regenerated within every section of an unmyelinated axonal membrane from beginning to end. Myelinated fibers conduct impulses about 50 times faster than unmyelinated

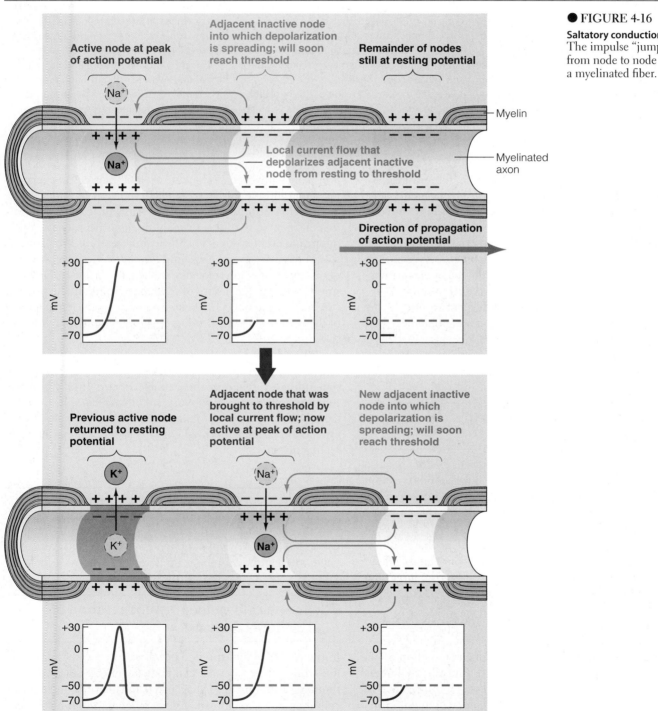

● **FIGURE 4-16**

Saltatory conduction
The impulse "jumps"
from node to node in
a myelinated fiber.

Multiple Sclerosis: Myelin—Going, Going, Gone

Multiple sclerosis (MS) is a pathophysiological condition in which nerve fibers in various locations throughout the nervous system become demyelinated (lose their myelin). MS is an autoimmune disease, in which the body's defense system erroneously attacks the myelin sheath surrounding myelinated nerve fibers (auto means "self"; immune means "defense against"). The condition afflicts about 1 in 1,000 people in the United States.

Many investigators believe that MS arises from a combination of genetic and environmental factors. Some people are genetically predisposed to developing the condition. They have increased susceptibility to other factors that may trigger the disease. The leading theory is that MS may be a result of an earlier infection with a particular form of herpes virus, *HHV-6*. This virus causes *roseola,* a relatively mild babyhood condition characterized by a fever and rash. Over 90% of infants get this disease, after which HHV-6 may remain dormant in nerve fibers. In a recent study, researchers found that over 70% of the MS patients studied showed evidence of active infection with HHV-6, which apparently became reactivated years later in these individuals. MS typically begins between the ages of 20 and 40. Interestingly, this virus may act as a "molecular mimic" of myelin in those genetically predisposed; that is, the virus shares some common structural features with the person's myelin. This means that antibodies produced against this virus may mistakenly attack and inflame myelin instead.

Loss of myelin as a result of this misguided immune attack slows transmission of impulses in the affected neurons. A hardened scar known as a *sclerosis* (meaning "hard") forms at the multiple sites of myelin damage. These scars further interfere with and can eventually block action potential propagation in the underlying axons. The symptoms of MS vary considerably, depending on the extent and location of the myelin damage. The most common symptoms include visual problems, tingling and numbness, muscle weakness, impaired coordination, and gradual paralysis. The early stage of the disease is often characterized by cycles of relapse and recovery, while the later chronic stage is marked by slow, progressive worsening of symptoms. MS is debilitating but not fatal.

Currently there is no cure and no really effective treatment for MS, although researchers are scrambling to find means to prevent, halt, or even reverse its debilitating symptoms.

---

fibers of comparable size. As a general rule, the most urgent types of information are transmitted via myelinated fibers, whereas nervous pathways carrying less urgent information are unmyelinated.

In addition to permitting action potentials to travel faster, a second advantage of myelination is that it conserves energy. Because the ion fluxes associated with action potentials are confined to the nodal regions, the energy-consuming $Na^+–K^+$ pump must restore fewer ions to their respective sides of the membrane following propagation of an action potential.

The accompanying boxed feature, ▶ Concepts, Challenges and Controversies, examines the myelin-destroying disease multiple sclerosis and the resultant debilitation.

▎ Fiber diameter also influences the velocity of action potential propagation.

Besides the effect of myelination, the diameter of the fiber also influences the speed with which an axon can conduct action potentials. The magnitude of current flow (that is, the amount of charge that moves) depends not only on the difference in potential between two adjacent electrically charged regions but also on the resistance or hindrance to electrical charge movement between the two regions. When fiber diameter increases, the resistance to local current decreases. Thus the larger the diameter of the nerve fiber, the faster it can propagate action potentials.

Large myelinated fibers, such as those supplying skeletal muscles, can conduct action potentials at a speed of up to 120 meters/sec (360 miles/hr), compared with a conduction velocity of 0.7 meters/sec (2 miles/hr) in small unmyelinated fibers such as those supplying the digestive tract. This difference in speed of propagation is related to the urgency of the information being conveyed. A signal to skeletal muscles to execute a particular movement (for example, to prevent you from falling as you trip on something) must be transmitted more rapidly than a signal to modify a slow-acting digestive process. Without myelination, axon diameters within urgent nerve pathways would have to be very large and cumbersome to achieve the necessary conduction velocities. Indeed, this is the case in many invertebrates. In the course of vertebrate evolution, the need for very large nerve fibers has been overcome by the development of the myelin sheath, allowing economic, rapid long-distance signaling.

The presence of myelinating cells can be either of tremendous benefit or tremendous detriment when an axon is cut, depending on whether the damage occurs in a peripheral nerve or in the central nervous system (CNS). Next we will discuss the regeneration of damaged nerve fibers, a matter of crucial importance in spinal cord injuries or other trauma affecting nerves.

REGENERATION OF NERVE FIBERS

Whether or not a severed axon regenerates depends on its location. Cut axons in the peripheral nervous system can regenerate, whereas those in the central nervous system cannot.

▎ Schwann cells guide the regeneration of cut peripheral axons.

In the case of a cut axon in a peripheral nerve, the portion of the axon farthest from the cell body degenerates, and the surrounding Schwann cells phagocytize the debris. The Schwann cells themselves remain and form a **regeneration tube** to guide the regenerating nerve fiber to its proper destination. The remaining portion of the axon connected to the cell body starts to grow and move forward within the Schwann cell column by amoeboid movement (see p. 49). Physiologists believe that the growing axon tip "sniffs" its way forward in the proper direction, guided by a chemical secreted into the regeneration tube by the Schwann cells. Successful fiber regeneration is responsible for the return of sensation and movement after a period of time following traumatic peripheral nerve injuries, although regeneration is not always successful.

▮ Oligodendrocytes inhibit regeneration of cut central axons.

Fibers in the CNS, which are myelinated by oligodendrocytes rather than Schwann cells, do not have this regenerative ability. Actually, the axons themselves have the ability to regenerate, but the oligodendrocytes surrounding them synthesize certain proteins that inhibit axonal growth, in sharp contrast to the nerve growth–promoting action of the Schwann cells that myelinate peripheral axons. Nerve growth in the brain and spinal cord appears to be controlled by a delicate balance between *nerve growth–enhancing* and *nerve growth–inhibiting proteins.* During fetal development, nerve growth in the CNS is possible as the brain and spinal cord are being formed. Researchers speculate that the nerve growth inhibitors, which are produced late in fetal development in the myelin sheaths surrounding central nerve fibers, may normally serve as "guardrails" to keep new nerve endings from straying outside their proper paths. The growth-inhibiting action of oligodendrocytes may thus serve to stabilize the enormously complex structure of the CNS.

Growth inhibition is a disadvantage, however, when central axons need to be mended, as when the spinal cord has been severed accidentally. Damaged central fibers show immediate signs of repairing themselves after an injury, but within several weeks they start to degenerate, and scar tissue forms at the site of injury. Any recovery is consequently halted. Therefore, damaged neuronal fibers in the brain and spinal cord never regenerate.

Research aimed toward promoting regeneration of central axons

In the future, however, with the help of exciting new findings, it may be possible to induce significant regeneration of damaged fibers in the CNS. Here are some current lines of research:

• Scientists have been able to induce significant nerve regeneration in rats with severed spinal cords by *chemically blocking the nerve growth inhibitors,* thereby allowing the nerve growth enhancers to promote abundant sprouting of new nerve fibers at the site of injury. One of the nerve growth inhibitors, dubbed *Nogo,* was recently identified. Now investigators are trying to encourage axon regrowth in experimental animals with spinal cord injuries by using an antibody to Nogo.

• Other researchers are experimentally *using peripheral nerve grafts* to bridge the defect at an injured spinal-cord site.

These grafts contain the nurturing Schwann cells, which release nerve growth–enhancing proteins.

• Another promising avenue under study involves *transplanting olfactory ensheathing glia* into the damaged site. Olfactory neurons, the cells that carry information about odors to the brain, are replaced on a regular basis, unlike most neurons. The growing axons of these newly generated neurons enter the brain to form functional connections with appropriate neurons in the brain. This capability is promoted by the special olfactory ensheathing glia, which wrap around and myelinate the olfactory axons. Early experimental evidence suggests that transplants of these special myelin-forming cells might help induce axonal regeneration in the CNS.

• Another avenue of hope is the recent discovery of *neural stem cells* (see pp. 8–9 and 139). These cells might someday be implanted into a damaged spinal cord and coaxed into multiplying and differentiating into mature, functional neurons that will replace those lost.

• Still other investigators are exploring other promising avenues of spurring repair of central axonal pathways, with the goal of enabling victims of spinal cord injuries to walk again.

You have now examined the changes in potential during an action potential and the underlying changes in permeability and ion movements responsible for the generation of action potentials. You have further seen how an action potential is propagated along the axon and learned about the factors that influence the speed of this propagation. But what happens when an action potential reaches the end of the axon? We are going to now turn our attention to this topic.

SYNAPSES AND NEURONAL INTEGRATION

When the action potential reaches the axon terminals, they release a chemical messenger that alters the activity of the cells at which the neuron terminates. A neuron may terminate at one of three structures: a muscle, a gland, or another neuron. Therefore, depending on where a neuron terminates, it can cause a muscle cell to contract, a gland cell to secrete, another neuron to convey an electrical message along a nerve pathway, or some other function. When a neuron terminates on a muscle or a gland, the neuron is said to **innervate,** or supply, the structure. The junctions between nerves and the muscles and glands that they innervate will be described later. For now we will concentrate on the junction between two neurons—a **synapse.**

▮ Synapses are junctions between two neurons.

Typically, a synapse involves a junction between an axon terminal of one neuron, known as the *presynaptic neuron,* and the dendrites or cell body of a second neuron, known as the *postsynaptic neuron.* Less frequently, axon-to-axon or dendrite-to-dendrite connections occur. (*Pre* means "before" and *post* means "after;" the presynaptic neuron lies before the synapse and the postsynaptic neuron lies after the synapse.) The dendrites and to a lesser extent the cell body of most neurons receive thousands of synaptic inputs, which are axon terminals from many other neurons. It has been estimated that some neurons within the central nervous system receive as many as 100,000 synaptic inputs (● Figure 4-17).

The anatomy of one of these thousands of synapses is shown in ● Figure 4-18a. The axon terminal of the **presynaptic neuron,** which conducts its action potentials *toward* the synapse, ends in a slight swelling, the **synaptic knob.** The synaptic knob contains **synaptic vesicles,** which store a specific chemical messenger, a **neurotransmitter** that has been synthesized and packaged by the presynaptic neuron. The synaptic knob comes into close proximity to, but does not actually directly contact, the **postsynaptic neuron,** the neuron whose action potentials are propagated *away* from the synapse. The space between the presynaptic and postsynaptic neurons, the **synaptic cleft,** is too wide for the direct spread of current from one cell to the other and therefore prevents action potentials from electrically passing between the neurons. The portion of the postsynaptic membrane immediately underlying the synaptic knob is referred to as the **subsynaptic membrane** (*sub* means "under").

Synapses operate in one direction only; that is, the presynaptic neuron brings about changes in membrane potential of the postsynaptic neuron, but the postsynaptic neuron does not directly influence the potential of the presynaptic neuron. The reason for this becomes readily apparent when one examines the events that occur at a synapse.

▌A neurotransmitter carries the signal across a synapse.

The events that occur at a synapse are summarized next and illustrated in ● Figure 4-18:

1. When an action potential in a presynaptic neuron has been propagated to the axon terminal (step 1 in ● Figure 4-18), this change in potential triggers the opening of voltage-gated Ca^{2+} channels in the synaptic knob.

2. Because Ca^{2+} is much more highly concentrated in the ECF, this ion flows into the synaptic knob through the opened channels (step 2).

3. Ca^{2+} induces the release of a neurotransmitter from some of the synaptic vesicles into the synaptic cleft (step 3). The release is accomplished by exocytosis (see p. 30).

4. The released neurotransmitter diffuses across the cleft and binds with specific protein receptor sites on the subsynaptic membrane (step 4).

5. This binding triggers the opening of specific ion channels in the subsynaptic membrane, changing the permeability of the postsynaptic neuron (step 5). These are chemically gated channels, in contrast to the voltage-gated channels responsible for the action potential and for the Ca^{2+} influx into the synaptic knob.

Because only the presynaptic terminal can release a neurotransmitter and only the subsynaptic membrane of the postsynaptic neuron has receptor sites for the neurotransmitter, the synapse can operate only in the direction from presynaptic to postsynaptic neuron.

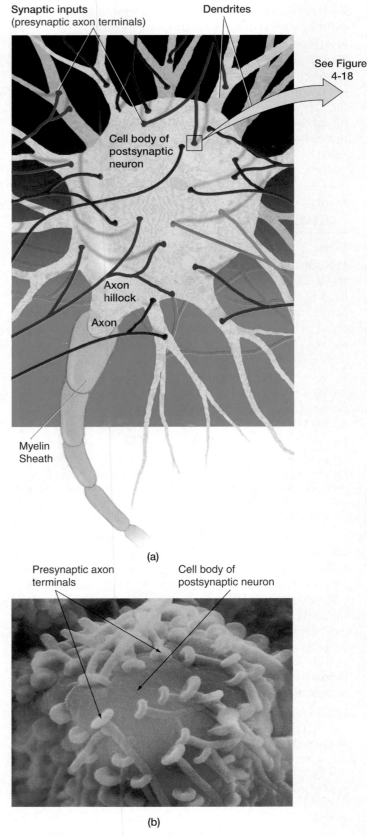

(a)

(b)

● **FIGURE 4-17**

Synaptic inputs to a postsynaptic neuron
(a) Schematic representation of synaptic inputs (presynaptic axon terminals) to the dendrites and cell body of a single postsynaptic neuron. (b) Electron micrograph showing multiple presynaptic axon terminals to a single postsynaptic cell body.

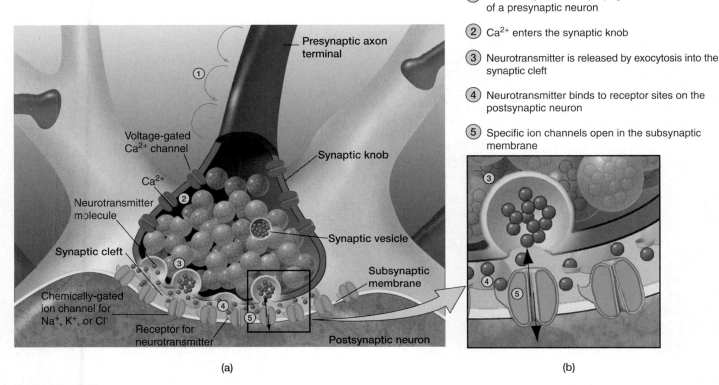

1. An action potential is propagated to the terminal of a presynaptic neuron

2. Ca²⁺ enters the synaptic knob

3. Neurotransmitter is released by exocytosis into the synaptic cleft

4. Neurotransmitter binds to receptor sites on the postsynaptic neuron

5. Specific ion channels open in the subsynaptic membrane

Presynaptic axon terminal

Voltage-gated Ca²⁺ channel

Ca²⁺

Synaptic knob

Neurotransmitter molecule

Synaptic vesicle

Synaptic cleft

Subsynaptic membrane

Chemically-gated ion channel for Na⁺, K⁺, or Cl⁻

Receptor for neurotransmitter

Postsynaptic neuron

(a)

(b)

● FIGURE 4-18

Synaptic structure and function

(a) Schematic representation of the structure of a single synapse. The circled numbers designate the sequence of events that take place at a synapse. (b) A blow-up depicting the release by exocytosis of neurotransmitter from the presynaptic axon terminal and its subsequent binding with receptor sites specific for it on the subsynaptic membrane of the postsynaptic neuron.

▌ Some synapses excite whereas others inhibit the postsynaptic neuron.

Each presynaptic neuron can release only one neurotransmitter; however, different neurons vary in the neurotransmitter they release. On binding with their subsynaptic receptor sites, different neurotransmitters cause different permeability changes. There are two types of synapses, depending on the permeability changes induced in the postsynaptic neuron by the combination of a specific neurotransmitter with its receptor sites: *excitatory synapses* and *inhibitory synapses*.

Excitatory synapses

At an **excitatory synapse,** the response to the binding of a neurotransmitter to the receptor is the opening of nonspecific cation channels within the subsynaptic membrane that permit simultaneous passage of Na⁺ and K⁺ through them. (These are a different type of channel from those you have encountered before.) Thus permeability to both these ions is increased at the same time. How much of each ion diffuses through an open cation channel depends on their electrochemical gradients. At resting potential, both the concentration and electrical gradients for Na⁺ favor its movement into the postsynaptic neuron, whereas only the concentration gradient for K⁺ favors its movement outward. Therefore, the permeability change induced at an excitatory synapse results in the movement of a few K⁺ ions out of the postsynaptic neuron, while a relatively larger number of Na⁺ ions simultaneously enter this neuron. The result is a net movement of positive ions into the cell. This makes the inside of the membrane slightly less negative than at resting potential, thus producing a *small depolarization* of the postsynaptic neuron.

Activation of one excitatory synapse can rarely depolarize the postsynaptic neuron sufficiently to bring it to threshold. Too few channels are involved at a single subsynaptic membrane to permit adequate ion flow to reduce the potential to threshold. This small depolarization, however, does bring the membrane of the postsynaptic neuron closer to threshold, increasing the likelihood that threshold will be reached (in response to further excitatory input) and an action potential will occur. That is, the membrane is now more excitable (easier to bring to threshold) than when at rest. Accordingly, this postsynaptic potential change occurring at an excitatory synapse is called an **excitatory postsynaptic potential,** or **EPSP** (● Figure 4-19a).

Inhibitory synapses

At an **inhibitory synapse,** the binding of a different released chemical messenger with its receptor sites increases the permeability of the subsynaptic membrane to either K⁺ or Cl⁻. In either case, the resulting ion movements bring about a *small*

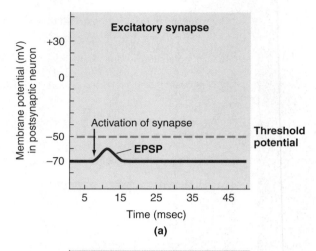

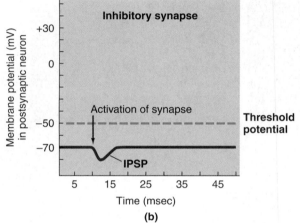

● FIGURE 4-19

Postsynaptic potentials
(a) Excitatory synapse. An excitatory postsynaptic potential (EPSP) brought about by activation of an excitatory presynaptic input brings the postsynaptic neuron closer to threshold potential. (b) Inhibitory synapse. An inhibitory postsynaptic potential (IPSP) brought about by activation of an inhibitory presynaptic input moves the postsynaptic neuron farther from threshold potential.

---

hyperpolarization of the postsynaptic neuron—that is, greater internal negativity. In the case of increased P_{K^+}, more positive charges leave the cell via K^+ efflux, leaving more negative charges behind on the inside; in the case of increased P_{Cl^-}, negative charges enter the cell in the form of Cl^- ions, because Cl^- concentration is higher outside the cell. This small hyperpolarization moves the membrane potential even farther away from threshold (● Figure 4-19b), lessening the likelihood that the postsynaptic neuron will reach threshold and undergo an action potential. That is, the membrane is now less excitable (harder to bring to threshold by excitatory input) than when it is at resting potential. The membrane is said to be *inhibited* under these circumstances, and the small hyperpolarization of the postsynaptic cell is called an **inhibitory postsynaptic potential,** or **IPSP.**

In cells where the equilibrium potential for Cl^- exactly equals the resting potential, an increased P_{Cl^-} does not result in a hyperpolarization because there is no driving force to produce Cl^- movement. Opening of Cl^- channels in these cells

tends to hold the membrane at resting potential, reducing the likelihood that threshold will be reached.

Synaptic delay

This conversion of the electrical signal in the presynaptic neuron (an action potential) to an electrical signal in the postsynaptic neuron (either an EPSP or IPSP) by chemical means (via the neurotransmitter-receptor combination) takes time. This **synaptic delay** is usually about 0.5 to 1 msec. In a neural pathway, chains of neurons often must be traversed. The more complex the pathway, the more synaptic delays, and the longer the *total reaction time* (the time required to respond to a particular event).

❚ Each synapse is either always excitatory or always inhibitory.

Many different chemicals are known or suspected to serve as neurotransmitters (▲ Table 4-2). Even though neurotransmitters vary from synapse to synapse, the same neurotransmitter is always released at a particular synapse. Furthermore, binding of a neurotransmitter with its appropriate subsynaptic receptors always leads to the same change in permeability and resultant change in potential of the postsynaptic membrane. That is, the response to a given neurotransmitter-receptor combination is always constant. *Each synapse is either always excitatory or always inhibitory.* It does not give rise to an EPSP under one circumstance and produce an IPSP at another time. Some neurotransmitters (for example, *glutamate*) always induce EPSPs, whereas others (for example, *glycine*) always induce IPSPs. Still other neurotransmitters (for example, *norepinephrine*) may even produce an EPSP at one synapse and an IPSP at a different synapse; that is, different permeability changes in the postsynaptic neuron can occur in response to the binding of the same neurotransmitter to the subsynaptic receptor sites of different postsynaptic neurons.

Most of the time, each axon terminal releases only one neurotransmitter. Recent evidence suggests, however, that in some cases two different neurotransmitters can be released simultaneously from a single axon terminal. For example, glycine and *gamma-aminobutyric acid (GABA)*, both of which produce inhibitory responses, can be packaged and released from the same synaptic vesicles. Scientists speculate that the fast-acting glycine and more slowly acting GABA may complement each other in the control of activities that depend on precise timing—for example, coordination of complex movements.

❚ Neurotransmitters are quickly removed from the synaptic cleft.

As long as the neurotransmitter remains bound to the receptor sites, the alteration in membrane permeability responsible for the EPSP or IPSP continues. The neurotransmitter must be inactivated or removed after it has produced the appropriate response in the postsynaptic neuron, however, so that the postsynaptic "slate" is "wiped clean," leaving it ready to receive ad-

▲ TABLE 4-2
Some Known or Suspected Neurotransmitters and Neuropeptides

Classical Neurotransmitters (small, rapid-acting molecules)	
Acetylcholine	Histamine
Dopamine	Glycine
Norepinephrine	Glutamate
Epinephrine	Aspartate
Serotonin	Gamma-aminobutyric acid (GABA)

Neuropeptides (large, slow-acting molecules)	
β-endorpin	Motilin
Adrenocorticotropic hormone (ACTH)	Insulin
α-melancoyte-stimulating hormone (MSH)	Glucagon
Thyrotropin-releasing hormone (TRH)	Angiotensin II
Gonadotropin-releasing hormone (GnRH)	Bradykinin
Somatostatin	Vasopressin
Vasoactive intestinal polypeptide (VIP)	Oxytocin
Cholecystokinin (CCK)	Carnosine
Gastrin	Bombesin
Substance P	Neurotensin

ditional messages from the same or other presynaptic inputs. Thus, after combining with the postsynaptic receptor, chemical transmitters are removed and the response is terminated. Several mechanisms can remove the neurotransmitter: It may diffuse away from the synaptic cleft, be inactivated by specific enzymes within the subsynaptic membrane, or be actively taken back up into the axon terminal by transport mechanisms in the presynaptic membrane. Once the neurotransmitter is taken back up, it can be stored and released another time (recycled) in response to a subsequent action potential, or destroyed by enzymes within the synaptic knob. The method employed depends on the particular synapse.

▌ Some neurotransmitters function through intracellular second-messenger systems.

Most, but not all, neurotransmitters function by changing the conformation of chemically gated channels, thereby altering membrane permeability and ion fluxes across the postsynaptic membrane. Synapses involving these rapid responses are considered "fast" synapses. Another mode of synaptic transmission used by some neurotransmitters, such as *serotonin*, involves the activation of intracellular second messengers, such as cyclic AMP (cAMP), within the postsynaptic neuron (see p. 67). Synapses that lead to responses mediated by second messengers are known as "slow" synapses, because these responses take

longer and often last longer than those accomplished by fast synapses. Activation of cAMP can induce both short- and long-term effects. In the short term, cAMP can lead to opening of specific ionic gates, a task that the other neurotransmitter-receptor combinations do directly and more rapidly. The gating effects can be either excitatory or inhibitory. In addition, cAMP may trigger more long-term changes in the postsynaptic cell, even to the extent of altering the cell's genetic expression. Such long-term cellular changes are believed to be linked to neuronal growth and development, and they may play a role in learning and memory.

▌ The grand postsynaptic potential depends on the sum of the activities of all presynaptic inputs.

The events that occur at a single synapse result in either an EPSP or an IPSP at the postsynaptic neuron. But if a single EPSP is inadequate to bring the postsynaptic neuron to threshold and an IPSP moves it even farther from threshold, how can an action potential be initiated in the postsynaptic neuron? The answer lies in the thousands of presynaptic inputs that a typical neuronal cell body receives from many other neurons. Some of these presynaptic inputs may be carrying sensory information from the environment; some may be signaling internal changes in homeostatic balance; others may be transmitting signals from control centers in the brain; and still others may arrive carrying other bits of information. At any given time, any number of these presynaptic neurons (probably hundreds) may be firing and thus influencing the postsynaptic neuron's level of activity. The total potential in the postsynaptic neuron, the **grand postsynaptic potential (GPSP)**, is a composite of all EPSPs and IPSPs occurring at approximately the same time.

The postsynaptic neuron can be brought to threshold in two ways: (1) *temporal summation* and (2) *spatial summation*. To illustrate these methods of summation, we will examine the possible interactions of three presynaptic inputs—two excitatory inputs (Ex1 and Ex2) and one inhibitory input (In1)—on a hypothetical postsynaptic neuron (● Figure 4-20). The recording shown in the figure represents the potential in the postsynaptic cell. Bear in mind during our discussion of this simplified version that many thousands of synapses are actually interacting in the same way on a single cell body and its dendrites.

Temporal summation

Suppose that Ex1 has an action potential that causes an EPSP in the postsynaptic neuron. Because EPSPs (as well as IPSPs) are graded potentials, the EPSP spreads only a short distance before dying off. If another action potential occurs later in Ex1, an EPSP of the same magnitude takes place (panel A in ● Figure 4-20). Next suppose that Ex1 has two action potentials in close succession (panel B). The first action potential in Ex1 produces an EPSP in the postsynaptic membrane. While the postsynaptic membrane is still partially depolarized from this first EPSP, the second action potential in Ex1 produces a second EPSP. The second EPSP adds on to the first EPSP, bringing

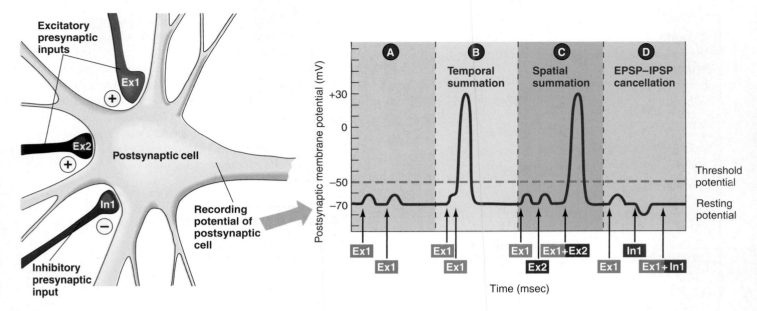

Panel **A** If an excitatory presynaptic input (Ex1) is stimulated a second time after the first EPSP in the postsynaptic cell has died off, a second EPSP of the same magnitude will occur.

Panel **B** If, however, Ex1 is stimulated a second time before the first EPSP has died off, the second EPSP will add onto, or sum with, the first EPSP, resulting in *temporal summation*, which may bring the postsynaptic cell to threshold.

Panel **C** The postsynaptic cell may also be brought to threshold by *spatial summation* of EPSPs that are initiated by simultaneous activation of two (Ex1 and Ex2) or more excitatory presynaptic inputs.

Panel **D** Simultaneous activation of an excitatory (Ex1) and inhibitory (In1) presynaptic input does not change the postsynaptic potential, because the resultant EPSP and IPSP cancel each other out.

● **FIGURE 4-20**

Determination of the grand postsynaptic potential by the sum of activity in the presynaptic inputs
Two excitatory (Ex1 and Ex2) and one inhibitory (In1) presynaptic inputs terminate on this hypothetical postsynaptic neuron. The potential of the postsynaptic neuron is being recorded.

the membrane to threshold, so an action potential occurs in the postsynaptic neuron. Graded potentials do not have a refractory period, so this additive effect is possible.

The summing of several EPSPs occurring very close together in time because of successive firing of a single presynaptic neuron is known as **temporal summation** (*tempus* means "time"). In reality, the situation is much more complex than just described. The sum of up to 50 EPSPs might be needed to bring the postsynaptic membrane to threshold. Each action potential in a single presynaptic neuron triggers the emptying of a certain number of synaptic vesicles. The amount of neurotransmitter released and the resultant magnitude of the change in postsynaptic potential are thus directly related to the frequency of presynaptic action potentials. One way, then, in which the postsynaptic membrane can be brought to threshold is through rapid, repetitive excitation from a single persistent input.

Spatial summation

Let us now see what happens in the postsynaptic neuron if both excitatory inputs are stimulated simultaneously (panel C). An action potential in either Ex1 or Ex2 will produce an EPSP in

the postsynaptic neuron; however, neither of these alone brings the membrane to threshold to elicit a postsynaptic action potential. But simultaneous action potentials in Ex1 and Ex2 produce EPSPs that add to each other, bringing the postsynaptic membrane to threshold, so an action potential does occur. The summation of EPSPs originating simultaneously from several different presynaptic inputs (that is, from different points in "space") is known as **spatial summation**. A second way, therefore, to elicit an action potential in a postsynaptic cell is through concurrent activation of several excitatory inputs. Again, in reality up to 50 simultaneous EPSPs are required to bring the postsynaptic membrane to threshold.

Similarly, IPSPs can undergo temporal and spatial summation. As IPSPs add together, however, they progressively move the potential further from threshold.

Cancellation of concurrent EPSPs and IPSPs

If an excitatory and an inhibitory input are simultaneously activated, the concurrent EPSP and IPSP more or less cancel each other out. The extent of cancellation depends on their respective magnitudes. In most cases, the postsynaptic membrane potential remains close to resting (panel D).

Importance of postsynaptic neuronal integration

The magnitude of the GPSP depends on the sum of activity in all the presynaptic inputs and in turn determines whether or not the neuron will undergo an action potential to pass information on to the cells at which the neuron terminates. The following oversimplified real-life example demonstrates the benefits of this neuronal integration. The explanation is not completely accurate technically, but the principles of summation are accurate.

Assume for simplicity's sake that urination is controlled by a postsynaptic neuron that innervates the urinary bladder. When this neuron fires, the bladder contracts. (Actually, voluntary control of urination is accomplished by postsynaptic integration at the neuron controlling the external urethral sphincter rather than the bladder itself.) As the bladder starts to fill with urine and becomes stretched, a reflex is initiated that ultimately produces EPSPs in the postsynaptic neuron responsible for causing bladder contraction. Partial filling of the bladder does not cause sufficient excitation to bring the neuron to threshold, so urination does not take place (panel A of ● Figure 4-20). As the bladder becomes progressively filled, the frequency of action potentials progressively increases in the presynaptic neuron that signals the postsynaptic neuron of the extent of bladder filling (Ex1 in panel B of ● Figure 4-20). When the frequency becomes great enough that the EPSPs are temporally summed to threshold, the postsynaptic neuron undergoes an action potential that stimulates bladder contraction.

What if the time is inopportune for urination to take place? IPSPs can be produced at the bladder postsynaptic neuron by presynaptic inputs originating in higher levels of the brain responsible for voluntary control (In1 panel D of ● Figure 4-20). These "voluntary" IPSPs in effect cancel out the "reflex" EPSPs triggered by stretching of the bladder. Thus the postsynaptic neuron remains at resting potential and does not have an action potential, so the bladder is prevented from contracting and emptying even though it is full.

What if a person's bladder is only partially filled, so that the presynaptic input originating from this source is insufficient to bring the postsynaptic neuron to threshold to cause bladder contraction, and yet he or she needs to supply a urine specimen for laboratory analysis? The person can voluntarily activate an excitatory presynaptic neuron (Ex2 in panel C of ● Figure 4-20). The EPSPs originating from this neuron and the EPSPs of the reflex-activated presynaptic neuron (Ex1) are spatially summed to bring the postsynaptic neuron to threshold. This achieves the action potential necessary to stimulate bladder contraction, even though the bladder is not full.

This example illustrates the importance of postsynaptic neuronal integration. Each postsynaptic neuron in a sense "computes" all the input it receives and makes a "decision" about whether to pass the information on (that is, whether threshold is reached and an action potential is transmitted down the axon). In this way, neurons serve as complex computational devices, or integrators. The dendrites function as the primary processors of incoming information. They receive and tally the signals coming in from all the presynaptic neurons. Each neuron's output in the form of frequency of action potentials to other cells (muscle cells, gland cells, or other neurons) reflects the balance of activity in the inputs it receives via EPSPs or IPSPs from the thousands of other neurons that terminate on it. Each postsynaptic neuron filters out and does not pass on information it receives that is not significant enough to bring it to threshold. If every action potential in every presynaptic neuron that impinges on a particular postsynaptic neuron were to cause an action potential in the postsynaptic neuron, the neuronal pathways would be overwhelmed with trivia. Only if an excitatory presynaptic signal is reinforced by other supporting signals through summation will the information be passed on. Furthermore, interaction of postsynaptic potentials provides a way for one set of signals to offset another set (IPSPs negating EPSPs). This allows a fine degree of discrimination and control in determining what information will be passed on.

Let us now see why action potentials are initiated at the axon hillock.

▌ Action potentials are initiated at the axon hillock because it has the lowest threshold.

Threshold potential is not uniform throughout the postsynaptic neuron. The lowest threshold is present at the axon hillock, because this region has a much greater density of voltage-gated Na^+ channels than anywhere else in the neuron. This greater density of these voltage-sensitive channels makes the axon hillock considerably more responsive to changes in potential than the dendrites or remainder of the cell body. The latter regions have a significantly higher threshold than the axon hillock. Because of local current flow, changes in membrane potential (EPSPs or IPSPs) occurring anywhere on the dendrites or cell body spread throughout the dendrites, cell body, and axon hillock. When summation of EPSPs takes place, the lower threshold of the axon hillock is reached first, whereas the dendrites and cell body at the same potential are still considerably below their own, much higher thresholds. Therefore, an action potential originates in the axon hillock and is propagated from there to the end of the axon.

▌ Neuropeptides act primarily as neuromodulators.

Researchers recently discovered that in addition to the classical neurotransmitters just described, some neurons also release *neuropeptides* (see ▲ Table 4-2). Neuropeptides differ from classical neurotransmitters in several important ways (▲ Table 4-3). **Classical neurotransmitters** are small, rapid-acting molecules that typically trigger the opening of specific ion channels to bring about a change in potential in the postsynaptic neuron (an EPSP or IPSP) within a few milliseconds or less. Most classical neurotransmitters are synthesized and packaged locally in synaptic vesicles in the cytosol of the axon terminal. These chemical messengers are primarily amino acids or closely related compounds.

▲ **TABLE 4-3**
Comparison of Classical Neurotransmitters and Neuropeptides

Characteristic	Classical Neurotransmitters	Neuropeptides
Size	Small; one amino acid or similar chemical	Large; 2 to 40 amino acids in length
Site of Synthesis	Cytosol of synaptic knob	Endoplasmic reticulum and Golgi complex in cell body; travel to synaptic knob by axonal transport
Site of Storage	In small synaptic vesicles in axon terminal	In large dense-core vesicles in axon terminal
Site of Release	Axon terminal	Axon terminal; may be cosecreted with neurotransmitter
Speed and Duration of Action	Rapid, brief response	Slow, prolonged response
Site of Action	Subsynaptic membrane of postsynaptic cell	Nonsynaptic sites on either presynaptic or postsynaptic cell at much lower concentrations than classical neurotransmitters
Effect	Usually alter potential of postsynaptic cell by opening specific ion channels	Usually enhance or suppress synaptic effectiveness by long-term changes in neurotransmitter synthesis or postsynaptic receptor sites (act as neuromodulators)

Neuropeptides are larger molecules made up of anywhere from two to about 40 amino acids. They are synthesized in the neuronal cell body in the endoplasmic reticulum and Golgi complex (see p. 28), and are subsequently moved by axonal transport along the microtubular highways to the axon terminal (see p. 47). Neuropeptides are not stored within the small synaptic vesicles with the classical neurotransmitter, but instead are packaged in large **dense-core vesicles,** which are also present in the axon terminal. The dense core vesicles undergo Ca^{2+}-induced exocytosis and release neuropeptides at the same time that the neurotransmitter is released from the synaptic vesicles. An axon terminal typically releases only a single classical neurotransmitter, but the same terminal may also contain one or more neuropeptides that are cosecreted along with the neurotransmitter.

Even though neuropeptides are currently the subject of intense investigation, our knowledge about their functions and control is still sketchy. They are known to diffuse locally and act on other adjacent neurons at much lower concentrations than do classical neurotransmitters, and they bring about slower, more prolonged responses. Some neuropeptides released at synapses may function as true neurotransmitters, but most are believed to function as neuromodulators.

Neuromodulators are chemical messengers that do not cause the formation of EPSPs or IPSPs, but rather bring about long-term changes that subtly *modulate*—depress or enhance—the action of the synapse. They bind to neuronal receptors at nonsynaptic sites—that is, not at the subsynaptic membrane—and they often activate second-messenger systems. Neuromodulators may act at either presynaptic or postsynaptic sites. For example, a neuromodulator may influence the level of an enzyme involved in the synthesis of a neurotransmitter by a presynaptic neuron, or it may alter the sensitivity of the postsynaptic neuron to a particular neurotransmitter by causing long-term changes in the number of subsynaptic receptor sites for the neurotransmit-

ter. Thus neuromodulators delicately fine-tune the synaptic response. The effect may last for days or even months or years. Whereas neurotransmitters are involved in rapid communication between neurons, neuromodulators are involved with more long-lasting events, such as learning and motivation.

Interestingly, the synaptically released neuromodulators include many substances that also have distinctly different roles as hormones released into the blood from endocrine tissues. For example, *cholecystokinin (CCK)* is a well-known hormone released from the small intestine. Among other digestive activities, CCK causes the gallbladder to contract and release bile into the intestine, as will be described more fully in Chapter 16. CCK has also been found in axon terminal vesicles in the brain, where it is believed to cause the feeling of no longer being hungry. In many instances, neuropeptides are named for their first-discovered role as hormones, as is the case with cholecystokinin (*chole* means "bile"; *cysto* means "bladder"; *kinin* means "contraction"). It appears that a number of chemical messengers are quite versatile and can assume different roles, depending on their source, their distribution, and their interaction with different cell types.

▌ Presynaptic inhibition or facilitation can selectively alter the effectiveness of a presynaptic input.

Besides neuromodulation, presynaptic inhibition or facilitation is another means of depressing or enhancing synaptic effectiveness. Sometimes a third neuron can influence activity between a presynaptic ending and a postsynaptic neuron. A presynaptic axon terminal (labeled A in ● Figure 4-21) may itself be innervated by another axon terminal (labeled B). The neurotransmitter released from modulatory terminal B binds with receptor sites on terminal A. This binding alters the amount of neurotransmitter released from terminal A in response to action potentials. If the amount of neurotransmitter released from A is reduced, the phenomenon is known as **presynaptic inhibition.** If the release of neurotransmitter is enhanced, the effect is called **presynaptic facilitation.**

Let's look more closely at how this process works. You know that Ca^{2+} entry into an axon terminal causes the release of neurotransmitter by exocytosis of synaptic vesicles. The amount of neurotransmitter released from terminal A depends on how much Ca^{2+} enters this terminal in response to an action potential. Ca^{2+} entry into terminal A, in turn, can be influenced by activity in modulatory terminal B. Let's use presynaptic inhibition to illustrate (● Figure 4-21). The amount of neurotransmitter released from presynaptic terminal A, an excitatory input in our example, influences the potential in the postsynaptic neuron at which it terminates (labeled C in the figure). Firing of A, alone, brings about an EPSP in postsynaptic neuron C. Now consider that B is stimulated simultaneously with A. When neurotransmitter from terminal B binds on terminal A, Ca^{2+} entry into terminal A is reduced. Less Ca^{2+} entry means less neurotransmitter release from A. Note that modulatory neuron B can suppress neurotransmitter release from A only when A is firing. If this presynaptic inhibition by B prevents A from releasing its neurotransmitter, the formation of EPSPs on postsynaptic membrane C from input A is specifically prevented. As a result, no change in the potential of the postsynaptic neuron occurs despite action potentials in A.

Could the same thing be accomplished by the simultaneous production of an IPSP through activation of an inhibitory input to negate an EPSP produced by activation of A? Not quite. Activation of an inhibitory input to cell C in ● Figure 4-21 would produce an IPSP in cell C, but this IPSP could cancel out not only an EPSP from excitatory input A, but also any EPSPs produced by other excitatory terminals, such as terminal D in the figure. The entire postsynaptic membrane is hyperpolarized by IPSPs, thereby negating (canceling) excitatory information fed into any part of the cell from any presynaptic input. By contrast, presynaptic inhibition (or presynaptic facilitation) works in a much more specific way than does the action of inhibitory inputs to the postsynaptic cell. Presynaptic inhibition provides a means by which certain inputs to the postsynaptic neuron can be *selectively* inhibited without affecting the contributions of any other inputs. For example, firing of B specifically prevents the formation of an EPSP in the postsynaptic neuron from excitatory presynaptic neuron A but does not have any influence on other excitatory presynaptic inputs. Excitatory input D can still produce an EPSP in the postsynaptic neuron even when B is firing. This type of neuronal integration is another means by which electrical signaling between nerve cells can be carefully fine-tuned.

▮ Drugs and diseases can modify synaptic transmission.

The vast majority of drugs that influence the nervous system perform their function by altering synaptic mechanisms. Synaptic drugs may block an undesirable effect or enhance a desir-

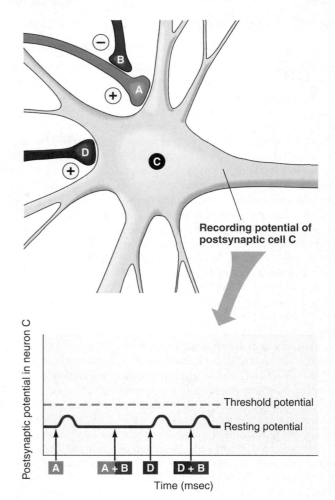

Recording potential of postsynaptic cell C

● FIGURE 4-21
Presynaptic inhibition
A, an excitatory terminal ending on postsynaptic cell C, is itself innervated by inhibitory terminal B. Stimulation of terminal A alone produces an EPSP in cell C, but simultaneous stimulation of terminal B prevents the release of excitatory neurotransmitter from terminal A. Consequently, no EPSP is produced in cell C despite the fact that terminal A has been stimulated. Such presynaptic inhibition selectively depresses activity from terminal A without suppressing any other excitatory input to cell C. Stimulation of excitatory terminal D produces an EPSP in cell C even though inhibitory terminal B is simultaneously stimulated because terminal B only inhibits terminal A.

able effect. Possible drug actions include (1) altering the synthesis, axonal transport, storage, or release of a neurotransmitter; (2) modifying neurotransmitter interaction with the postsynaptic receptor; (3) influencing neurotransmitter reuptake or destruction; and (4) replacing a deficient neurotransmitter with a substitute transmitter.

For example, the socially abused drug **cocaine** blocks the reuptake of the neurotransmitter *dopamine* at presynaptic terminals. It does so by binding competitively with the dopamine reuptake transporter, which is a protein molecule that picks up released dopamine from the synaptic cleft and shuttles it back to the axon terminal. With cocaine occupying the dopamine transporter, dopamine remains in the synaptic cleft longer than usual

Parkinson's Disease, Pollution, Ethical Problems, and Politics

In 1982, Dr. J. W. Langston, a California neurologist, was confronted with a medical puzzle unlike any he had ever seen. Several young men and women were admitted to the hospital in a rigid, stuporous state as if they had been frozen in their tracks. Confined to their beds, the patients lay immobile day after day. They could neither talk nor feed themselves and could not move their limbs.

Langston began an intensive study of the victims and found that they all were drug addicts. Each had injected meperidine (MPPP), a synthetic form of heroin. The MPPP was contaminated with a slightly different drug, MPTP, which a naturally occurring enzyme in the body converts into a paralyzing chemical, MPP$^+$. Made in a basement by one of California's small-time drug pushers, the synthetic heroin is one of the "designer drugs" available on the black market today. Sloppy chemical technique resulted in the contamination.

Researchers have shown that MPP$^+$ destroys dopamine-secreting cells in a part of the brain called the *substantia nigra.* Axons from these cells terminate in the *basal nuclei,* another region of the brain involved in the coordination of slow, sustained movements, inhibition of muscle tone, and suppression of useless patterns of movement. Reduced dopamine activity in the basal nuclei was responsible for the puzzling symptoms. Most prominently, loss of inhibition of muscle tone led to the pronounced muscular rigidity experienced by these patients.

Symptoms of Parkinson's Disease

A gradual destruction of these same dopamine-secreting cells is responsible for **Parkinson's disease (PD),** a movement disorder. The symptoms, which develop slowly and progressively as dopamine activity in the basal nuclei gradually diminishes, usually begin with involuntary tremors at rest, such as involuntary rhythmic shaking of the hands or head. As the disease worsens, patients speak in slow monotone, become increasingly stiff, and walk with a shuffling, stooped gait. In later stages, memory and thinking become severely impaired. Ultimately, debilitating rigidity ensues, similar to that seen in the MPTP victims.

Underlying Cause?

In most instances, the underlying cause of the neuronal death responsible for the symptoms is unknown. Researchers have identified several defective genes in hereditary forms of the condition, yet 80% of PD patients lack a family history of the disease. Scientists suspect that environmental factors are responsible for instigating the vast majority of these cases. Mounting evidence suggests that the neuronal death is caused by excessive accumulation of highly reactive, unstable compounds called *free radicals.* Free radicals are destructive because they are electron deficient. They wreak their havoc by snatching electrons from other molecules, causing cellular damage in the process. Dopamine-secreting neurons may be particularly susceptible to free-radical destruction because dopamine has the ability to promote the production of free radicals. What investigators do not know is what environmental or perhaps internal trigger causes the buildup of the free radicals.

Intriguingly, MPP$^+$, which destroyed the dopamine-secreting cells of the drug addicts, can cause an increase in free-radical production. MPP$^+$ is very similar to paraquat, a commonly used agricultural pesticide. This observation led to the hypothesis that PD may be linked to pesticide use. In fact, epidemiological studies have shown a remarkable correlation between the use of pesticides and the incidence of PD. Other chemical pollutants are also suspected of contributing to the destruction of the dopamine-secreting brain cells. PD was unheard of before the industrial revolution. As the industrial revolution and associated environmental pollution spread, the incidence of the disease rose sharply, underscoring the fact that to protect our health we must also protect the quality of our environment. Currently, an estimated 1 million people in the United States have PD.

Instead of an MPP$^+$-like chemical pollutant as the underlying culprit, other scientists propose that overactive *microglia,* the brain's immune cells, might be central players in PD. Unusually active microglia have been discovered in the substantia nigra of patients with this condition. These microglia can initiate a chain of events resulting in an increase in free radicals. We have few clues at present as to what causes the microglia to become overactive—whether that be an environmental factor or something gone awry internally.

Treatment

Current therapies for PD focus mainly on symptomatic relief. A standard treatment is the administration of *levodopa (*L-dopa), a precursor of dopamine. Dopamine itself cannot be administered because it is unable to cross the blood–brain barrier (discussed in the following chap-

---

and continues to interact with its postsynaptic receptor sites. The result is prolonged activation of neural pathways that use this chemical as a neurotransmitter. Among these pathways are those that play a role in emotional responses, especially feelings of pleasure. In essence, when cocaine is present the neural switches in the pleasure pathway are locked in the "on" position.

Cocaine is addictive because it causes permanent molecular adaptations of the involved neurons such that they are unable to transmit normally across synapses without increasingly higher doses of the drug. Because the postsynaptic cells are incessantly stimulated for an extended time, they permanently adapt to "expecting" this high level of stimulation; that is, they are "hooked" on the drug. When the cocaine molecules diffuse away, the sense of pleasure evaporates, because the normal level of dopamine activity does not sufficiently "satisfy" the overly needy demands of the postsynaptic cells for stimula-

tion. Cocaine users reaching this low become frantic and profoundly depressed. Only more cocaine makes them feel good again. But repeated use of cocaine modifies responsiveness to the drug. After prolonged heavy use of cocaine, the pleasurable effects of the drug diminish and become increasingly overshadowed by unpleasant side effects. Furthermore, the amount of cocaine needed to overcome the devastating crashes progressively increases. Cocaine is abused by millions who have become addicted to its mind-altering properties, with devastating social and economic effects.

Whereas cocaine abuse leads to excessive dopamine activity, **Parkinson's disease** is attributable to a deficiency of dopamine in a region of the brain involved in controlling complex movements. (For further details on Parkinson's disease, see the accompanying boxed feature, ❯ Concepts, Challenges, and Controversies.)

ter), but L-dopa can enter the brain from the blood. Once inside the brain, L-dopa is converted into dopamine, thus substituting for the deficient neurotransmitter. This therapy greatly alleviates the symptoms associated with the deficit in most patients.

Unfortunately, however, with prolonged use the beneficial effects of L-dopa tend to diminish and troublesome side effects develop. Therefore, researchers have been seeking other means to manage this disabling condition. In recent years, other new drugs that enhance or mimic dopamine activity have been discovered. All the drugs used to treat PD relieve symptoms in the early stages of the disease but do nothing to stop the relentless destruction of neurons.

In people with advanced PD whose symptoms cannot be controlled by medication, surgeons sometimes remove the *globus pallidus*, a region of the brain that becomes overactive in the condition. Dopamine normally inhibits the globus pallidus in a check-and-balance system involved in muscle control. With the loss of the dopamine-releasing neurons, the globus pallidus becomes overactive, because it is less inhibited than normal. This results in many symptoms of exaggerated muscle activity, such as muscular rigidity. Removing the globus pallidus relieves these symptoms but carries the risks of major brain surgery.

In a newer technology for treating the condition, scientists have taken a different tack for interfering with the overactive globus pallidus: an electrical stimulator implanted in this region. Delivery of rapidly oscillating electrical impulses to the globus pallidus disrupts ongoing electrical activity in this brain region. This disruption disables the globus pallidus. The result is more normal muscle control.

Researchers are currently seeking ways not only to better compensate for neuronal loss but also to halt or reverse the progress of the disease itself. Two different approaches toward this goal are under intense investigation—transplant of fetal cells in the brain to replace the lost neurons and administration of neurotrophic ("nerve nourishing") factors to halt or reverse the loss of neurons.

The first alternative approach—harvesting dopamine-secreting brain cells from aborted fetuses and transplanting them into the brains of patients with PD—has stirred considerable controversy. Opponents of the procedure are concerned that, among other issues, the medical use of aborted fetuses will encourage and justify the practice of abortion. Proponents argue that fetal tissue obtained from a legitimate medical procedure should not be wasted when the tissue could be used to potentially treat or cure a number of diseases, including PD. During the Reagan and George Bush (senior) presidencies, this politically heated debate led to a ban on federally funded research using cells from aborted fetuses, a ban that was lifted by President Clinton during his first week in office.

With the ban lifted, fetal cell transplants proceeded forward. The mixed results of these early attempts, however, have not only fueled the old debates about whether or not these transplants should continue but have further led to new debates among scientists over the most appropriate methodologies for administering the fetal cells. Although mounting evidence indicates that fetal cells can set up shop and replace damaged dopamine-secreting tissue in the adult brain, outcomes have been inconsistent. A third of patients have experienced remarkable improvement following surgery, another third have shown only moderate gains, and the remaining third have had no long-lasting benefit. Furthermore, about 15% of the younger transplant recipients who had shown some early signs of improvement began about a year after surgery to exhibit disturbing, irreversible side effects, such as uncontrollable jerking and writhing. Researchers speculate that these side effects may be attributed to possible overgrowth of the transplanted cells, leading to questions about how much tissue to transplant and how best to transplant it.

Also attracting considerable interest is a newly discovered brain protein *GDNF*, a neurotropic factor that maintains and nourishes neurons. In experimental animals, GDNF stops the ongoing loss of dopamine-secreting cells and even stimulates the sprouting of additional fibers in neurons that have survived, the end result being an improvement in symptoms. Although yet unproven in treating humans, researchers are hopeful that delivery of GDNF into the brain via gene therapy or other means may halt or even reverse the progressive neural degeneration characteristic of the disease.

Thus although considerable progress has been made in recent years in understanding and treating PD, no ideal method for tackling this disorder has yet been identified. Yet researchers remain optimistic about finding ways to reduce the suffering of PD victims.

Synaptic transmission is also vulnerable to neural toxins, which may cause nervous system disorders by acting at either presynaptic or postsynaptic sites. For example, two different neural poisons, strychnine and tetanus toxin, act at different synaptic sites to block inhibitory impulses while leaving excitatory inputs unchecked. **Strychnine** competes with one of the inhibitory neurotransmitters, glycine, at the postsynaptic receptor site. This poison combines with the receptor but does not directly alter the potential of the postsynaptic cell in any way. Instead, strychnine blocks the receptor so that it is not available for interaction with glycine when the latter is released from the inhibitory presynaptic ending. Thus postsynaptic inhibition (formation of IPSPs) is abolished in nerve pathways that use glycine as an inhibitory neurotransmitter. Unchecked excitatory pathways lead to convulsions, muscle spasticity, and death.

Tetanus toxin, in contrast, prevents the release of another inhibitory neurotransmitter, gamma-aminobutyric acid (GABA), from inhibitory presynaptic inputs terminating at neurons that supply skeletal muscles. Unchecked excitatory inputs to these neurons result in uncontrolled muscle spasms. These spasms occur especially in the jaw muscles early in the disease, giving rise to the common name of *lockjaw* for this condition. Later they progress to the muscles responsible for breathing, at which point death occurs.

Strychnine and tetanus toxin poisoning have similar outcomes, but one poison (strychnine) blocks specific postsynaptic inhibitory receptors, whereas the other (tetanus toxin) prevents the presynaptic release of a specific inhibitory neurotransmitter.

Other drugs and diseases that influence synaptic transmission are too numerous to mention, but as these examples illustrate, any site along the synaptic pathway is vulnerable to in-

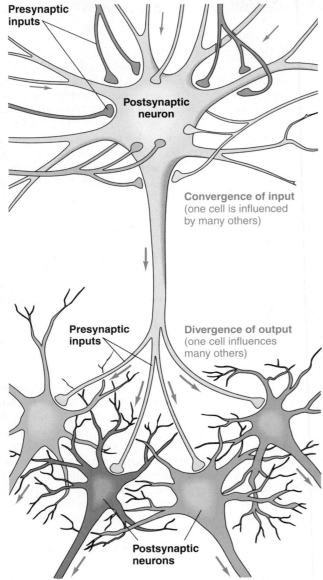

Presynaptic inputs

Postsynaptic neuron

Convergence of input
(one cell is influenced by many others)

Presynaptic inputs

Divergence of output
(one cell influences many others)

Postsynaptic neurons

Arrows indicate direction in which information is being conveyed.

● **FIGURE 4-22**
Convergence and divergence

terference, either pharmacologic (drug induced) or pathologic (disease induced).

▌ Neurons are linked through complex converging and diverging pathways.

Two important relationships exist between neurons: convergence and divergence. A given neuron may have many other neurons synapsing on it. Such a relationship is known as **convergence** (● Figure 4-22). Through this converging input, a single cell is influenced by thousands of other cells. This single cell, in turn, influences the level of activity in many other cells by divergence of output. The term **divergence** refers to the branching of axon terminals so that a single cell synapses with and influences many other cells.

Note that a particular neuron is postsynaptic to the neurons converging on it but presynaptic to the other cells at which it terminates. Thus the terms *presynaptic* and *postsynaptic* refer only to a single synapse. Most neurons are presynaptic to one group of neurons and postsynaptic to another group.

There are an estimated 100 billion neurons and 10^{14} (100 quadrillion) synapses in the brain alone! When you consider the vast and intricate interconnections possible between these neurons through converging and diverging pathways, you can begin to imagine how complex the wiring mechanism of our nervous system really is. Even the most sophisticated computers are far less complex than the human brain. The "language" of the nervous system—that is, all communication between neurons—is in the form of graded potentials, action potentials, neurotransmitter signaling across synapses, and other nonsynaptic forms of chemical chatter. All activities for which the nervous system is responsible—every sensation you feel, every command to move a muscle, every thought, every emotion, every memory, every spark of creativity—all depend on the patterns of electrical and chemical signaling between neurons along these complexly wired neural pathways.

CHAPTER IN PERSPECTIVE: FOCUS ON HOMEOSTASIS

Nerve cells are specialized to receive, process, encode, and rapidly transmit information from one part of the body to another. The information is transmitted over intricate nerve pathways by propagation of action potentials along the nerve cell's length as well as by chemical transmission of the signal from neuron to neuron at synapses and from neuron to muscles and glands through other neurotransmitter–receptor interactions at these junctions.

Collectively, the nerve cells make up the nervous system, one of the two major regulatory systems of the body. Many of the activities controlled by the nervous system are geared toward maintaining homeostasis. Some neuronal electrical signals convey information about changes to which the body must respond in order to maintain homeostasis—for example, information about a fall in blood pressure. Other neuronal electrical signals convey messages to muscles and glands to stimulate appropriate responses to counteract these changes—for example, adjustments in heart and blood vessel activity to restore blood pressure to normal when it starts to fall.

The specialization of muscle cells, contraction, also depends on their ability to undergo action potentials. Action potentials trigger muscle contractions, many of which are important in maintaining homeostasis. For example, beating of the heart, mixing of ingested food with digestive enzymes, and shivering to generate heat when the body is cold are all accomplished by muscle contractions.

CHAPTER SUMMARY

Introduction

▪ Nerve and muscle cells are known as *excitable* tissues because they can rapidly alter their membrane permeabilities and thus undergo transient membrane potential changes when excited. These rapid changes in potential serve as electrical signals.

▪ Compared to resting potential, a membrane becomes depolarized when its potential is reduced and hyperpolarized when its potential is increased.

▪ Changes in potential are brought about by triggering events that alter membrane permeability, in turn leading to changes in ion movement across the membrane.

▪ There are two kinds of potential change: (1) graded potentials, which serve as short-distance signals, and (2) action potentials, the long-distance signals.

Graded Potentials

▪ Graded potentials occur in a small specialized region of an excitable cell membrane.

▪ The magnitude of a graded potential varies directly with the magnitude of the triggering event.

▪ Graded potentials passively spread decrementally by local current flow and die out over a short distance.

Action Potentials

▪ During an action potential, depolarization of the membrane to threshold potential triggers sequential changes in permeability caused by conformational changes in voltage-gated Na^+ and K^+ channels.

▪ These permeability changes bring about a brief reversal of membrane potential, with Na^+ influx being responsible for the rising phase (from -70 mV to $+30$ mV), followed by K^+ efflux during the falling phase (from peak back to resting potential).

▪ Before an action potential returns to resting, it regenerates an identical new action potential in the area next to it by means of current flow that brings the previously inactive area to threshold. This self-perpetuating cycle continues until the action potential has spread throughout the cell membrane in undiminished fashion.

▪ There are two types of action potential propagation: (1) contiguous conduction in unmyelinated fibers, in which the action potential spreads along every portion of the membrane; and (2) the more rapid saltatory conduction in myelinated fibers, where the impulse jumps over the sections of the fiber covered with insulating myelin.

▪ The Na^+–K^+ pump gradually restores the ions that moved during propagation of the action potential to their original location to maintain the concentration gradients.

▪ It is impossible to restimulate the portion of the membrane where the impulse has just passed until it has recovered from its refractory period. The refractory period ensures the one-way propagation of action potentials away from the original site of activation.

▪ Action potentials occur either maximally in response to stimulation or not at all.

▪ Variable strengths of stimuli are coded by varying the frequency of action potentials, not their magnitude.

Regeneration of Nerve Fibers

▪ Schwann cells guide the regeneration of severed peripheral axons.

▪ Oligodendrocytes inhibit the regeneration of severed central axons.

Synapses and Neuronal Integration

▪ The primary means by which one neuron directly interacts with another neuron is through a synapse.

▪ Most neurons have four different functional parts:

1. The dendrite/cell body region is specialized to serve as the postsynaptic component that binds with and responds to neurotransmitters released from other neurons. The plasma membrane in this region has an abundance of chemically gated channels for binding with specific neurotransmitters.

2. The axon hillock is specialized for initiation of action potentials in response to graded potential changes induced by binding of a neurotransmitter with receptors on the dendrite/cell body region. The axon hillock has the lowest threshold and thus reaches threshold first in response to an excitatory graded potential change, because it has the highest density of voltage-gated Na^+ channels.

3. The axon, or nerve fiber, is specialized to conduct action potentials in undiminished fashion from the axon hillock to the axon terminals. The axon can conduct action potentials because it has voltage-gated Na^+ and K^+ channels throughout its length.

4. The axon terminal is specialized to serve as the presynaptic component, which releases a neurotransmitter that influences other postsynaptic cells in response to action potential propagation down the axon. Neurotransmitter is released because the axon terminals have voltage-gated Ca^{2+} channels that open in response to an action potential. The resultant Ca^{2+} entry triggers release of the neurotransmitter by exocytosis of synaptic vesicles.

▪ Released neurotransmitter combines with receptor sites on the postsynaptic neuron with which the presynaptic axon terminal interacts. This combination alters the postsynaptic cell in one of two ways:

1. The most typical response is the opening of chemically gated channels. This mechanism occurs at fast synapses. (a) If nonspecific cation channels that permit passage of both Na^+ and K^+ are opened, the resultant ionic fluxes cause an EPSP, a small depolarization that brings the postsynaptic cell closer to threshold. (b) However, the likelihood that the postsynaptic neuron will reach threshold is diminished when an IPSP, a small hyperpolarization, is produced as a result of the opening of either K^+ or Cl^- channels, or both.

2. In an alternate synaptic mechanism, an intracellular second-messenger system, such as cyclic AMP, is activated by the neurotransmitter-receptor combination. This mechanism occurs at slow synapses. Cyclic AMP can bring about channel opening or can have more prolonged effects within the cell, including alterations of the cell's genetic expression.

▪ Even though there are a number of different neurotransmitters, each synapse always releases the same neurotransmitter to produce a given response when combined with a particular receptor.

▪ The response is terminated when the neurotransmitter is removed from the synaptic cleft by methods specific for the synapse.

▪ Many neurons cosecrete larger, more slowly acting neuropeptides along with the classical neurotransmitter. The neuropeptides largely function as neuromodulators at nonsynaptic sites on either the presynaptic or the postsynaptic neuron to enhance or suppress synaptic effectiveness.

▪ The interconnecting synaptic pathways between various neurons are incredibly complex, due to convergence of neuronal input and divergence of its output. Usually, many presynaptic inputs converge on a single neuron and jointly control its level of excitability. This same neuron, in turn, diverges to synapse with and influence the excitability of many other cells.

■ Each neuron thus has the task of computing an output to numerous other cells from a complex set of inputs to itself. Depending on the combination of signals it is receiving from its various presynaptic inputs, at any given time a neuron may react by (1) firing action potentials along its axon, (2) remaining at rest and not passing any signals along, or (3) having its level of excitability increased or reduced.

■ If the dominant activity is in its excitatory inputs, the postsynaptic cell is likely to be brought to threshold and have an action potential. This can be accomplished by either (1) temporal summation (EPSPs from single, repetitively firing, presynaptic inputs occurring so close together in time that they add together) or (2) spatial summation (adding of EPSPs occurring simultaneously from several different presynaptic inputs).

■ If inhibitory inputs dominate, the postsynaptic potential is brought farther than usual away from threshold.

■ If excitatory and inhibitory activity to the postsynaptic neuron is balanced, the membrane remains close to resting.

■ Numerous factors may alter synaptic effectiveness: Some are built-in mechanisms to fine-tune neural responsiveness, some are deliberate pharmacologic manipulations to achieve a desired result, and some are accidents caused by poisons or disease processes.

REVIEW EXERCISES

Objective Questions (Answers on p. A-37)

1. Conformational changes in channel proteins brought about by voltage changes are responsible for opening and closing Na^+ and K^+ gates during the generation of an action potential. (*True or false?*)

2. The Na^+–K^+ pump restores the membrane to resting potential after it reaches the peak of an action potential. (*True or false?*)

3. Following an action potential, there is more K^+ outside the cell than inside because of the efflux of K^+ during the falling phase. (*True or false?*)

4. Postsynaptic neurons can either excite or inhibit presynaptic neurons. (*True or false?*)

5. The one-way propagation of action potentials away from the original site of activation is ensured by the _____.

6. The _____ is the site of action potential initiation in most neurons because it has the lowest threshold.

7. A junction in which electrical activity in one neuron influences the electrical activity in another neuron by means of a neurotransmitter is called a _____.

8. Summing of EPSPs occurring very close together in time as a result of repetitive firing of a single presynaptic input is known as _____.

9. Summing of EPSPs occurring simultaneously from several different presynaptic inputs is known as _____.

10. The neuronal relationship where synapses from many presynaptic inputs act on a single postsynaptic cell is called _____, whereas the relationship in which a single presynaptic neuron synapses with and thereby influences the activity of many postsynaptic cells is known as _____.

11. Using the following answer code, indicate which potential is being described:
    (a) graded potential
    (b) action potential
    ___ 1. behaves in all-or-none fashion
    ___ 2. magnitude of the potential change varies with the magnitude of the triggering response
    ___ 3. decremental spread away from the original site
    ___ 4. spreads throughout the membrane in nondiminishing fashion
    ___ 5. serves as a long-distance signal
    ___ 6. serves as a short-distance signal

Essay Questions

1. What are the two types of excitable tissue?

2. Define the following terms: *polarization, depolarization, hyperpolarization, repolarization, resting membrane potential,* *threshold potential, action potential, refractory period, all-or-none law.*

3. Compare the three kinds of channels in terms of the factor that induces the change in channel conformation.

4. Describe the permeability changes and ion fluxes that occur during an action potential.

5. Compare contiguous conduction and saltatory conduction.

6. Compare the events that occur at excitatory and inhibitory synapses.

7. Distinguish between classical neurotransmitters and neuropeptides. Explain what a neuromodulator is.

8. Discuss the possible outcomes of the grand postsynaptic potential brought about by interactions between EPSPs and IPSPs.

9. Distinguish between presynaptic inhibition and an inhibitory postsynaptic potential.

Quantitative Exercises (Solutions on p. A-37)
(See Appendix D, "Principles of Quantitative Reasoning")

1. The following calculations give some insight into action potential conduction.
    a. How long would it take for an action potential to travel 0.6 m along the axon of an unmyelinated neuron of the digestive tract?
    b. How long would it take for an action potential to travel the same distance along the axon of a large myelinated neuron innervating a skeletal muscle?
    c. Suppose there were two synapses in a 0.6-m nerve tract and the delay at each synapse is 1 msec. How long would it take an action potential/chemical signal to travel the 0.6 m now, for both the myelinated and unmyelinated neurons?
    d. What if there were five synapses?

2. Suppose point A is 1 m from point B. Compare the following situations:
    a. A single axon spans the distance from A to B, and its conduction velocity is 60 m/sec;
    b. Three neurons span the distance from A to B, all three neurons have the same conduction velocity, and the synaptic delay at both synapses (draw a picture) is 1 msec. What are the conduction velocities of the three neurons in this second situation if the total conduction time in both cases is the same?

3. One can predict what the Na^+ current produced by the Na^+–K^+ pump is with the following equation[1]:

[1]F. C. Hoppensteadt, and C. S. Peskin, *Mathematics in Medicine and the Life Sciences* (New York: Springer 1992, equation 7.4.35, p. 178).

$$p = \frac{kT}{q} \left(\frac{G_{Na^+} G_{K^+}}{G_{Na^+} + G_{K^+}} \right) \log \frac{G_{K^+}[Na^+]_o}{G_{Na^+}[K^+]_i}$$

where p is the sodium pump current, G is the conductance of the membrane to the indicated ion, $[x]_o$ and $[x]_i$ are the concentrations of ion x outside and inside the cell respectively, k is Boltzmann's constant, T is the temperature in Kelvins, and q is the elementary charge constant. Suppose $kT/q = 25$ mV, $G_{Na^+} = 3.3$ μS/cm², $G_{K^+} = 240$ μS/cm², $[Na^+]_o = 145$ mM, and $[K^+]_i = 4$ mM. What is the pump current for sodium, in μA/cm²?

POINTS TO PONDER

(Explanations on p. A-37)

1. Which of the following would occur if a neuron were experimentally stimulated simultaneously at both ends?
   a. The action potentials would pass in the middle and travel to the opposite ends.
   b. The action potentials would meet in the middle and then be propagated back to their starting positions.
   c. The action potentials would stop as they met in the middle.
   d. The stronger action potential would override the weaker action potential.
   e. Summation would occur when the action potentials met in the middle, resulting in a larger action potential.
2. Compare the expected changes in membrane potential of a neuron stimulated with a *subthreshold stimulus* (a stimulus not sufficient to bring a membrane to threshold), a *threshold stimulus* (a stimulus just sufficient to bring the membrane to threshold), and a *suprathreshold stimulus* (a stimulus larger than that necessary to bring the membrane to threshold).
3. Assume you touched a hot stove with your finger. Contraction of the biceps muscle causes flexion (bending) of the elbow, whereas contraction of the triceps muscle causes extension (straightening) of the elbow. What pattern of postsynaptic potentials (EPSPs and IPSPs) would you expect to be initiated as a reflex in the cell bodies of the neurons controlling these muscles to pull your hand away from the painful stimulus?

   Now assume your finger is being pricked to obtain a blood sample. The same *withdrawal reflex* would be initiated. What pattern of postsynaptic potentials would you voluntarily produce in the neurons controlling the biceps and triceps to keep your arm extended in spite of the painful stimulus?

4. *Schizophrenia* is believed to be caused by excessive dopamine activity in a particular region of the brain. Explain why symptoms of schizophrenia sometimes occur as a side effect in patients being treated for Parkinson's disease.
5. Assume presynaptic excitatory neuron A terminates on a postsynaptic cell near the axon hillock and presynaptic excitatory neuron B terminates on the same postsynaptic cell on a dendrite located on the side of the cell body opposite the axon hillock. Explain why rapid firing of presynaptic neuron A could bring the postsynaptic neuron to threshold through temporal summation, thus initiating an action potential, whereas firing of presynaptic neuron B at the same frequency and the same magnitude of EPSPs may not bring the postsynaptic neuron to threshold.
6. **Clinical Consideration.** Becky N. was apprehensive as she sat in the dentist's chair awaiting the placement of her first silver amalgam (the "filling" in a cavity in a tooth). Before preparing the tooth for the amalgam by drilling away the decayed portion of the tooth, the dentist injected a local anesthetic in the nerve pathway supplying the region. As a result, Becky, much to her relief, did not feel any pain during the drilling and filling procedure. Local anesthetics block Na^+ channels. Explain how this action prevents the transmission of pain impulses to the brain.

PHYSIOEDGE RESOURCES

PHYSIOEDGE CD-ROM
Figures marked with the PhysioEdge icon have associated activities on the CD. For this chapter, check out:

Membrane Potential
Neuronal Physiology

The diagnostic quiz allows you to receive immediate feedback on your understanding of the concept and to advance through various levels of difficulty.

CHECK OUT THESE MEDIA QUIZZES:
4.1 Basics of a Neuron
4.2 Graded Potentials and Actions Potentials
4.3 Synapses and Neuronal Integration

PHYSIOEDGE WEB SITE
The Web site for this book contains a wealth of helpful study aids, as well as many ideas for further reading and research. Log on to:

http://www.brookscole.com

Go to the Biology page and select Sherwood's *Human Physiology*, 5th Edition. Select a chapter from the drop-down menu or click on one of these resource areas:

- **Case Histories** provide an introduction to the clinical aspects of human physiology. Check out:

  #15: A Stiff Baby
  #16: And a Limp Baby

- For 2-D and 3-D graphical illustrations and animations of physiological concepts, visit our **Visual Learning Resource.**

- Resources for study and review include the **Chapter Outline, Chapter Summary, Glossary,** and **Flash Cards.** Use our **Quizzes** to prepare for in-class examinations.

- On-line research resources to consult are **Hypercontents,** with current links to relevant Internet sites; **Internet Exercises** with starter URLs provided; and **InfoTrac Exercises.**

   For more readings, go to InfoTrac College Edition, your on-line research library, at:

  http://infotrac.thomsonlearning.com

**Nervous System
(Central Nervous System)**

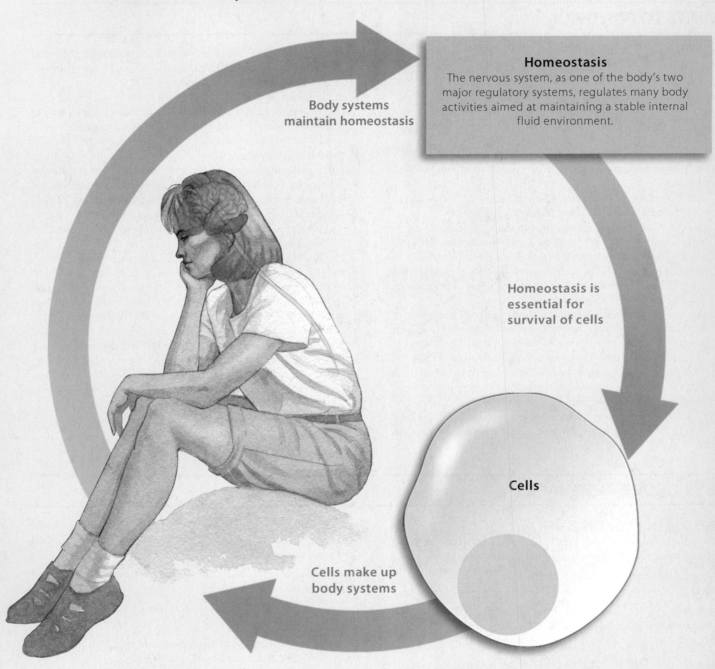

**Body systems
maintain homeostasis**

Homeostasis
The nervous system, as one of the body's two major regulatory systems, regulates many body activities aimed at maintaining a stable internal fluid environment.

Homeostasis is essential for survival of cells

Cells

Cells make up body systems

The **nervous system** is one of the two major regulatory systems of the body, the other being the endocrine system. A complex interactive network of three basic types of nerve cells—afferent neurons, efferent neurons, and interneurons—constitutes the excitable cells of the nervous system. The **central nervous system (CNS)** is composed of the brain and spinal cord, which receive input about the external and internal environment from the afferent neurons. The CNS sorts and processes this input, then initiates appropriate directions in the efferent neurons, which carry the instructions to glands or muscles to bring about the desired response—some type of secretion or movement. Many of these neurally controlled activities are directed toward maintaining homeostasis. In general, the nervous system acts by means of its electrical signals (action potentials) to control the rapid responses of the body.

Chapter 5

The Central Nervous System

CONTENTS AT A GLANCE

COMPARISON OF THE NERVOUS AND ENDOCRINE SYSTEM

ORGANIZATION OF THE NERVOUS SYSTEM

▌ Central nervous system; peripheral nervous system

▌ The three classes of neurons

PROTECTION AND NOURISHMENT OF THE BRAIN

▌ Neuroglia; meninges; cerebrospinal fluid

▌ Blood-brain barrier

OVERVIEW OF THE CENTRAL NERVOUS SYSTEM

CEREBRAL CORTEX

▌ Cortical structure

▌ Sensory perception

▌ Motor control

▌ Language ability

▌ Association areas

▌ Cerebral specialization

BASAL NUCLEI, THALAMUS, AND HYPOTHALAMUS

THE LIMBIC SYSTEM AND ITS FUNCTIONAL RELATIONS WITH THE HIGHER CORTEX

▌ Limbic system

▌ Emotion; behavioral patterns

▌ Memory

CEREBELLUM

BRAIN STEM

▌ Components and functions of the brain stem

▌ Consciousness; sleep-wake cycle

SPINAL CORD

▌ Anatomy of the spinal cord

▌ Spinal reflexes

The way humans act and react depends on complex, organized, discrete neuronal processing. Many of the basic life-supporting neuronal patterns, such as those controlling respiration and circulation, are similar in all individuals. However, there must be subtle differences in neuronal integration between someone who is a talented composer and someone who cannot carry a tune, or between someone who is a math wizard and someone who struggles with long division. Some differences in the nervous systems of individuals are genetically endowed. The rest, however, are due to environmental encounters and experiences. When the immature nervous system develops according to its genetic plan, an overabundance of neurons and synapses is formed. Depending on external stimuli and the extent these pathways are used, some are retained, firmly established, and even enhanced, whereas others are eliminated. A case in point is **amblyopia** (lazy eye), in which the weaker of the two eyes is not used for vision. A lazy eye that does not get appropriate visual stimulation during a critical developmental period will almost completely and permanently lose the power of vision. The functionally blind eye itself is completely normal; the defect lies in the lost neuronal connections in the brain's visual pathways. If, however, the weak eye is forced to work, by covering the stronger eye with a patch during the sensitive developmental period, the weaker eye will retain full vision. The maturation of the nervous system truly does involve instances of "use it or lose it." Once the nervous system has matured, ongoing modifications still occur as we continue to learn from our unique set of experiences. For example, the act of reading this page is somehow altering the neuronal activity of your brain as you (it is hoped) tuck the information away in your memory.

COMPARISON OF THE NERVOUS AND ENDOCRINE SYSTEMS

The nervous and endocrine systems are the two main regulatory systems of the body. The **nervous system** swiftly transmits electrical impulses to the skeletal muscles and the exocrine glands that it innervates. The **endocrine system** secretes hormones into the blood for delivery to distant sites of action. Although these two systems differ in many respects, they have

Comparison of the Nervous System and the Endocrine System

Property	Nervous System	Endocrine System
Anatomic Arrangement	A "wired" system; specific structural arrangement between neurons and their target cells; structural continuity in the system	A "wireless" system; endocrine glands widely dispersed and not structurally related to one another or to their target cells
Type of Chemical Messenger	Neurotransmitters released into synaptic cleft	Hormones released into blood
Distance of Action of Chemical Messenger	Very short distance (diffuses across synaptic cleft)	Long distance (carried by blood)
Means of Specificity of Action on Target Cell	Dependent on close anatomic relationship between nerve cells and their target cells	Dependent on specificity of target cell binding and responsiveness to a particular hormone
Speed of Response	Rapid (milliseconds)	Slow (minutes to hours)
Duration of Action	Brief (milliseconds)	Long (minutes to days or longer)
Major Functions	Coordinates rapid, precise responses	Controls activities that require long duration rather than speed
Influence on Other Major Control System?	Yes	Yes

much in common (▲ Table 5-1). They both ultimately alter their **target cells** (their sites of action) by releasing chemical messengers (neurotransmitters in the case of nerve cells, hormones in the case of endocrine cells), which interact in particular ways with specific receptors (particular plasma membrane proteins) of the target cells. Let's examine the anatomic distinctions between these two systems and the different ways in which they accomplish specificity of action.

▌ The nervous system is a "wired" system, and the endocrine system is a "wireless" system.

Anatomically, the nervous and endocrine systems are quite different. In the nervous system, each nerve cell terminates directly on its specific target cells; that is, the nervous system is "wired" in a very specific way into highly organized, distinct anatomic pathways for transmission of signals from one part of the body to another. Information is carried along chains of neurons to the desired destination through action potential propagation coupled with synaptic transmission (Chapter 4). In contrast, the endocrine system is a "wireless" system in that the endocrine glands are not anatomically linked with their target cells. Instead, the endocrine chemical messengers are secreted into the blood and delivered to distant target sites. In fact, the components of the endocrine system itself are not anatomically interconnected; the endocrine glands are scattered throughout the body (see ● Figure 18-1, p. 668). These glands constitute a system in a functional sense, however, because they all secrete hormones and many interactions take place between various endocrine glands.

▌ Neural specificity is due to anatomic proximity and endocrine specificity to receptor specialization.

As a result of their anatomic differences, the nervous and endocrine systems accomplish specificity of action by distinctly different means. Specificity of neural communication depends on nerve cells having a close anatomic relationship with their target cells, so that each neuron has a very narrow range of influence. A neurotransmitter is released for restricted distribution only to specific adjacent target cells, then is swiftly inactivated by enzymes at the nerve–target cell juncture or is taken back up by the nerve terminal before it is able to gain access to the blood. The target cells for a particular neuron have receptors for the neurotransmitter, but so do many other cells in other locations, and they could respond to this same mediator if it were delivered to them. For example, the entire system of nerve cells supplying all of your body's skeletal muscles (motor neurons) use the same neurotransmitter, *acetylcholine (Ach)*, and all of your skeletal muscles bear complementary ACh receptors (Chapter 7). Yet you are able to specifically wiggle your big toe without influencing any of your other muscles because ACh can be discretely released from the motor neurons that are specifically wired to the muscles controlling your toe. If ACh were indiscriminately released into the blood, as are the hormones of the wireless endocrine system, all the skeletal muscles would simultaneously respond by contracting, because they all have identical receptors for ACh. This does not happen, of course, because of the precise wiring patterns that provide direct lines of communication between motor neurons and their target cells.

This specificity is in sharp contrast to the way specificity of communication is built into the endocrine system. Because hormones travel in the blood, they are able to reach virtually all tissues. Yet despite this ubiquitous distribution, only specific target cells are able to respond to each hormone. Specificity of hormonal action depends on specialization of target cell receptors. For a hormone to exert its effect, the hormone must first bind with receptors specific for it that are located only on or in the hormone's target cells. Target cell receptors are highly discerning in their binding function. They will recognize and bind only a certain hormone, even though they are exposed simultaneously to many other blood-borne hormones, some of which are structurally very similar to the one that they discriminately bind. A receptor recognizes a specific hormone because the conformation of a portion of the receptor molecule matches a unique portion of its binding hormone in "lock-and-key" fashion. Binding of a hormone with target cell receptors initiates a reaction (or series of reactions) that culminates in the hormone's final effect. The hormone cannot influence any other cells, because they lack the right binding receptors.

■ The nervous and endocrine systems have their own realms of authority but interact functionally.

The nervous and endocrine systems are specialized for controlling different types of activities. In general, the nervous system is responsible for coordinating rapid, precise responses. It is especially important in the body's interactions with the external environment. Neural signals in the form of action potentials are rapidly propagated along nerve cell fibers, resulting in the release at the nerve terminal of a neurotransmitter that has to diffuse only a microscopic distance to its target cell before a response is effected. A neurally mediated response is not only rapid but brief; the action is quickly brought to a halt as the neurotransmitter is swiftly removed from the target site. This permits either termination of the response, almost immediate repetition of the response, or rapid initiation of an alternate response, depending on the circumstances (for example, the swift changes in commands to muscle groups needed to coordinate walking). This mode of action makes neural communication extremely rapid and precise. The target tissues of the nervous system are the muscles and glands, especially exocrine glands, of the body.

The endocrine system, in contrast, is specialized to control activities that require duration rather than speed, such as regulating organic metabolism and H_2O and electrolyte balance; promoting smooth, sequential growth and development; controlling reproduction; and regulating red blood cell production. The endocrine system responds more slowly to its triggering stimuli than does the nervous system for several reasons. First, the endocrine system must depend on blood flow to convey its hormonal messengers over long distances. Second, hormones' mechanism of action at their target cells is more complex than that of neurotransmitters and thus requires more time before a response occurs. The ultimate effect of some hormones cannot

be detected until a few hours after they bind with target cell receptors. Also, because of the receptors' high affinity for their respective hormone, hormones often remain bound to receptors for some time, thus prolonging their biological effectiveness. Furthermore, unlike the brief, neurally induced responses that come to a halt almost immediately after the neurotransmitter is removed, endocrine effects usually last for some time after the hormone's withdrawal. Neural responses to a single burst of neurotransmitter release usually last only milliseconds to seconds, whereas the alterations in target cells induced by hormones range from minutes to days or, in the case of growth-promoting effects, even a lifetime. Thus, hormonal action is relatively slow and prolonged, making endocrine control particularly suitable for the regulation of metabolic activities that require long-term stability.

Although the endocrine and nervous systems have their own areas of specialization, they are intimately interconnected functionally. Some nerve cells do not release neurotransmitters at synapses but instead terminate at blood vessels and release their chemical messengers (neurohormones) into the blood, where these chemicals act as hormones (see p. 66). A given messenger may even be a neurotransmitter when released from a nerve ending and a hormone when secreted by an endocrine cell (see p. 669). The nervous system directly or indirectly controls the secretion of many hormones (see Chapters 18 and 19). At the same time, many hormones act as neuromodulators, altering synaptic effectiveness and thereby influencing the excitability of the nervous system (see p. 124). The presence of certain key hormones is even essential for the proper development and maturation of the brain during fetal life. Furthermore, in many instances the nervous and endocrine systems both influence the same target cells in supplementary fashion. For example, these two major regulatory systems both contribute to the regulation of the circulatory and digestive systems. Thus, many important regulatory interfaces exist between the nervous and endocrine systems; the study of these relationships is known as **neuroendocrinology.**

For now we will concentrate on the nervous system and will examine the endocrine system in more detail in later chapters. Throughout the text we will continue to point out the numerous ways these two regulatory systems interact, so that the body is a coordinated whole, even though each system has its own realm of authority.

ORGANIZATION OF THE NERVOUS SYSTEM

■ The nervous system is organized into the central nervous system and the peripheral nervous system.

The nervous system is organized into the **central nervous system (CNS),** consisting of the brain and spinal cord, and the **peripheral nervous system (PNS),** consisting of nerve fibers that carry information between the CNS and other parts of the body (the periphery) (● Figure 5-1). The PNS is further subdi-

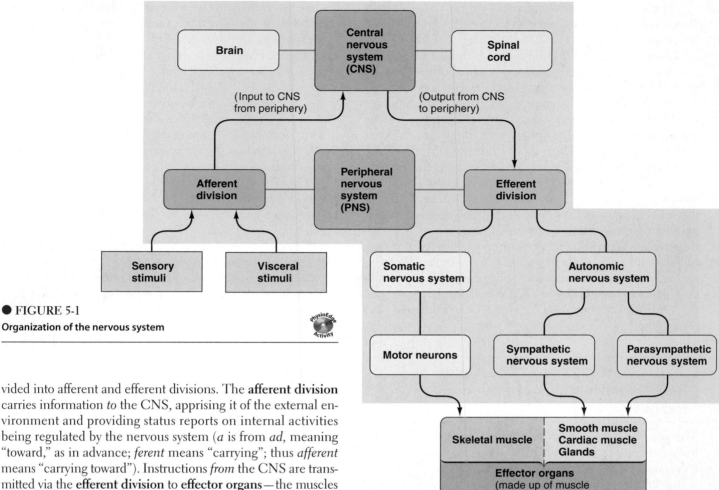

● **FIGURE 5-1**

Organization of the nervous system

vided into afferent and efferent divisions. The **afferent division** carries information *to* the CNS, apprising it of the external environment and providing status reports on internal activities being regulated by the nervous system (*a* is from *ad*, meaning "toward," as in advance; *ferent* means "carrying"; thus *afferent* means "carrying toward"). Instructions *from* the CNS are transmitted via the **efferent division** to **effector organs**—the muscles or glands that carry out the orders to bring about the desired effect (*e* is from *ex*, meaning "from," as in exit; thus *efferent* means "carrying from"). The efferent nervous system is divided into the **somatic nervous system,** which consists of the fibers of the motor neurons that supply the skeletal muscles, and the **autonomic nervous system** fibers, which innervate smooth muscle, cardiac muscle, and glands. The latter system is further subdivided into the **sympathetic nervous system** and the **parasympathetic nervous system,** both of which innervate most of the organs supplied by the autonomic system.

It is important to recognize that all these "nervous systems" are really subdivisions of a single, integrated nervous system. They are arbitrary divisions based on differences in the structure, location, and functions of the various diverse parts of the whole nervous system.

▌ **The three classes of neurons are afferent neurons, efferent neurons, and interneurons.**

Three classes of neurons make up the nervous system: *afferent neurons, efferent neurons,* and *interneurons.* The afferent division of the peripheral nervous system is composed of **afferent neurons,** which are shaped differently from efferent neurons and interneurons (● Figure 5-2). At its peripheral ending, an

afferent neuron has a **sensory receptor** that generates action potentials in response to a particular type of stimulus. (This stimulus-sensitive afferent neuronal receptor should not be confused with the special protein receptors that bind chemical messengers and are found in the plasma membrane of all cells.) The afferent neuron cell body, which is devoid of dendrites and presynaptic inputs, is located adjacent to the spinal cord. A long *peripheral axon,* commonly called the *afferent fiber,* extends from the receptor to the cell body, and a short *central axon* passes from the cell body into the spinal cord. Action potentials are initiated at the receptor end of the peripheral axon in response to a stimulus and are propagated along the peripheral axon and central axon toward the spinal cord. The terminals of the central axon diverge and synapse with other neurons within the spinal cord, thus disseminating information about the stimulus. Afferent neurons lie primarily within the peripheral nervous system. Only a small portion of their central axon endings project into the spinal cord to relay peripheral signals.

Efferent neurons also lie primarily in the peripheral nervous system (● Figure 5-2). The cell bodies of efferent neurons originate in the CNS, where many centrally located presynaptic inputs converge on them to influence their outputs to the effector organs. Efferent axons (*efferent fibers*) leave the CNS to course their way to the muscles or glands they innervate, conveying their integrated output for the effector organs to put into effect. (An autonomic nerve pathway actually consists of a two-neuron chain between the CNS and the effector organ.)

Interneurons lie entirely within the CNS. About 99% of all neurons belong to this category. The human CNS is estimated to have over 100 billion interneurons! These neurons serve two main roles. First, as their name implies, they lie between the afferent and efferent neurons and are important in the integration of peripheral responses to peripheral information (*inter* means "between"). For example, on receiving information through afferent neurons that you are touching a hot object, appropriate interneurons signal efferent neurons that transmit to your hand and arm muscles the message, "Pull the hand away from the hot object!" The more complex the required action, the greater the number of interneurons interposed between the afferent message and efferent response. Second, interconnections between interneurons themselves are responsible for the abstract phenomena associated with the "mind," such as thoughts, emotions, memory, creativity, intellect, and motivation. These activities are the least understood functions of the nervous system.

With this brief introduction to the types of neurons and their location in the various divisions of the nervous system, we will now turn our attention to the central nervous system, followed in the next two chapters by a discussion of the two divisions of the peripheral nervous system.

PROTECTION AND NOURISHMENT OF THE BRAIN

About 90% of the cells within the CNS are not neurons but **glial cells** or **neuroglia.** Despite their large numbers, the glial cells occupy only about half the volume of the brain, because they do not branch as extensively as the neurons do.

▌ Glial cells support the interneurons physically, metabolically, and functionally.

Unlike neurons, glial cells do not initiate or conduct nerve impulses. They are important in the viability of the CNS, however. For much of the time since their discovery in the 19th century, the glial cells were thought to be passive "mortar" that physically supported the functionally important neurons. In the last decade, however, the varied and important roles of these dynamic cells have become apparent. Without the glial cells, the neurons could not grow, nourish themselves, nor establish and use synapses effectively. The glial cells serve as the connective tissue of the CNS and as such help support the neurons both physically and metabolically. They homeostatically maintain the composition of the specialized extracellular environ-

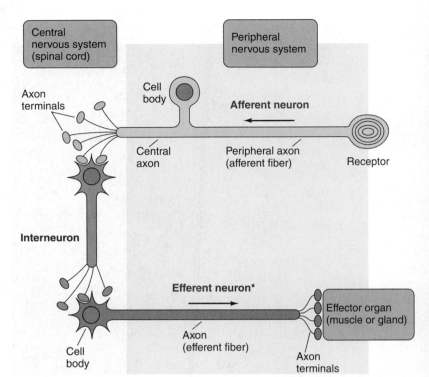

* Efferent autonomic nerve pathways consist of a two-neuron chain between the CNS and the effector organ.

● **FIGURE 5-2**

Structure and location of the three classes of neurons

ment surrounding the neurons within the narrow limits optimal for normal neuronal function. Furthermore, they actively modulate synaptic function.

The four major types of glial cells in the CNS are *astrocytes, oligodendrocytes, microglia,* and *ependymal cells* (● Figure 5-3 and ▲ Table 5-2 on page 139).

Astrocytes

Named for their starlike shape (*astro* means "star," *cyte* means "cell") (● Figure 5-4), **astrocytes** are the most abundant glial cells. They fill a number of critical functions:

1. As the main "glue" (*glia* means "glue") of the CNS, astrocytes hold the neurons together in proper spatial relationships.

2. Astrocytes serve as a scaffold to guide neurons to their proper final destination during fetal brain development.

3. These glial cells induce the small blood vessels of the brain to undergo the anatomic and functional changes that are responsible for establishing the blood–brain barrier, a highly selective barricade between the blood and brain that we will soon describe in greater detail.

4. Astrocytes are important in the repair of brain injuries and in neural scar formation.

5. They play a role in neurotransmitter activity. Astrocytes take up glutamate and gamma-aminobutyric acid (GABA), excitatory and inhibitory neurotransmitters, respectively, thus bringing the actions of these chemical messengers to a halt.

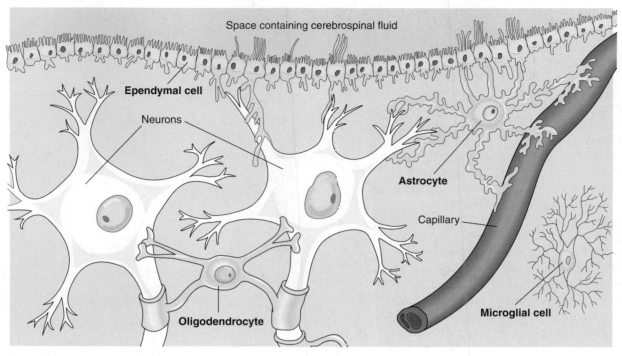

● FIGURE 5-3

Glial cells of the central nervous system
The glial cells include the astrocytes, oligodendrocytes, microglia, and ependymal cells.

Furthermore, they degrade the sopped-up chemical messengers into raw materials for the neurons to use in making more of these neurotransmitters.

● FIGURE 5-4

Astrocytes
Note the starlike shape of these astrocytes, which have been grown in tissue culture.

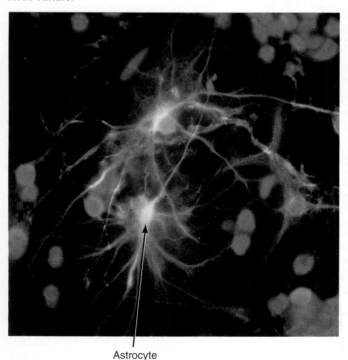

Astrocyte

6. Astrocytes take up excess K^+ from the brain ECF when high action-potential activity outpaces the ability of the Na^+–K^+ pump to return the effluxed K^+ to the neurons. (Recall that K^+ leaves a neuron during the falling phase of an action potential; see p. 108.) By taking up excess K^+, the astrocytes help maintain the proper brain ECF ion concentration to sustain normal neural excitability. If K^+ levels in brain ECF were allowed to rise, the resultant lower K^+-concentration gradient between the neuronal ICF and surrounding ECF would reduce the neuronal membrane closer to threshold, even at rest. This would increase the excitability of the brain. In fact, an elevation in brain-ECF K^+ concentration may be one of the factors responsible for the brain cells' explosive convulsive discharge that occurs during epileptic seizures.

7. In recent discoveries, astrocytes appear to enhance synapse formation and to strengthen synaptic transmission. Astrocytes are now believed to communicate chemically with each other and with neurons in two ways. First, gap junctions (see p. 63) have been identified between astrocytes themselves and between astrocytes and neurons. Chemical signals could pass directly between these cells without entering the surrounding ECF by means of these small connecting tunnels. Second, astrocytes possess receptors for the same neurotransmitters that neurons do. Researchers have shown that binding of the common neuronally released neurotransmitter glutamate to astrocyte receptors causes these cells to release stored calcium ions. This calcium unleashing in turn appears to strengthen the synaptic activity of the neurons, such as by increasing the release of neurotransmitter. Evidence suggests that this two-directional extracellular signaling plays an important role in synaptic transmission and the processing of information in the brain. In fact,

▲ TABLE 5-2
Functions of Glial Cells

Type of Glial Cell	Functions
Astrocytes	Physically support neurons in proper spatial relationships
	Serve as a scaffold during fetal brain development
	Induce formation of blood–brain barrier
	Form neural scar tissue
	Take up and degrade released neurotransmitters into raw materials for synthesis of more neurotransmitters by neurons
	Take up excess K^+ to help maintain proper brain ECF ion concentration and normal neural excitability
	Enhance synapse formation and strengthen synaptic transmission via chemical signaling with neurons
Oligodendrocytes	Form myelin sheaths in CNS
Microglia	Play a role in defense of brain as phagocytic scavengers
Ependymal Cells	Line internal cavities of brain and spinal cord
	Contribute to formation of cerebrospinal fluid
	Serve as neural stem cells with the potential to form new neurons and glial cells

some neuroscientists have ventured to say that synapses should be considered "three-party" junctures involving the glial cells as well as the traditional synapse members, the presynaptic and postsynaptic neurons. This point of view is indicative of the increasingly important role being placed on the astrocytes in synapse function. Some researchers even speculate that glial modulation of synaptic activity may be important in memory and learning.

Oligodendrocytes

Oligodendrocytes form the insulative myelin sheaths around axons in the CNS. An oligodendrocyte has several elongated projections, each of which is wrapped jelly-roll fashion around a section of an interneuronal axon to form a patch of myelin (see ● Figure 4-15b, p. 114; and ● Figure 5-3).

Microglia

Microglia are the immune defense cells of the CNS. These scavengers are "cousins" of monocytes, a type of white blood cell that leaves the blood and sets up residence as a front-line defense agent in various tissues throughout the body. Microglia are derived from the same tissue that gives rise to mono-

cytes, and during embryonic development, they migrate to the CNS. There they remain stationary until activated by an infection or injury. In the resting state, microglia are wispy cells with many long branches that radiate outward. Recent evidence suggests that resting microglia are not just waiting watchfully. They are thought to release low levels of growth factors, such as *nerve growth factor*, that help neurons and other glial cells survive and thrive. When trouble occurs in the CNS, microglia retract their branches, round up, and become highly mobile, moving toward the affected area to remove any foreign invaders or tissue debris. Activated microglia release destructive chemicals for assault against their target. Researchers increasingly suspect that excessive release of these chemicals from overzealous microglia may damage the neurons they are meant to protect, thus contributing to the insidious neuronal damage seen in stroke, Alzheimer's disease, multiple sclerosis, the dementia (mental failing) of AIDS, and other *neurodegenerative diseases*.

Ependymal cells

Ependymal cells line the internal cavities of the CNS. As the nervous system develops embryonically from a hollow neural tube, the original central cavity of this tube is maintained and modified to form the **ventricles** of the brain (● Figure 5-5) and the **central canal** of the spinal cord. The ependymal cells lining the ventricles contribute to the formation of cerebrospinal fluid, a topic to be discussed shortly. Ependymal cells are one of the few cell types to bear cilia (see p. 47). Beating of ependymal cilia contributes to the flow of cerebrospinal fluid throughout the ventricles.

Importantly, exciting new research has identified a totally different role for the ependymal cells: They serve as neural stem cells with the potential of forming not only other glial cells but new neurons as well (see p. 8). Traditional view has long held that new neurons are not produced anywhere in the mature brain. Then, in the late 1990s, scientists discovered that new neurons are produced in one restricted site, namely, in a specific part of the hippocampus, a structure important for learning and memory (see p. 161). Neurons in the rest of the brain are considered irreplaceable. But the discovery that

● **FIGURE 5-5**
The ventricles of the brain

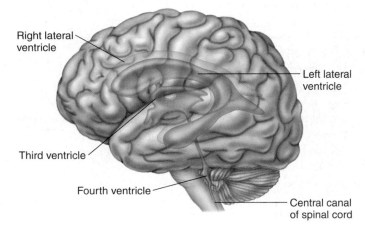

Right lateral ventricle

Left lateral ventricle

Third ventricle

Fourth ventricle

Central canal of spinal cord

ependymal cells are precursors for new neurons suggests that the adult brain has more potential for repairing damaged regions than previously assumed. Currently there is no evidence that the brain spontaneously repairs itself following neuron-losing insults such as head trauma, strokes, and neurodegenerative disorders. Apparently most brain regions are not able to activate this mechanism for replenishing neurons, probably because the appropriate "cocktail" of supportive chemicals is not present. Researchers are hopeful that by probing into why these ependymal cells are dormant and how they might be activated will lead to the possibility of unlocking the brain's latent capacity for self-repair.

Unlike neurons, glial cells do not lose the ability to undergo cell division, so most brain tumors of neural origin consist of glial cells (**gliomas**). Neurons themselves do not form tumors because they are unable to divide and multiply. Brain tumors of nonneural origin are of two types: (1) those that metastasize (spread) to the brain from other sites and (2) **meningiomas,** which originate from the meninges, the protective membranes covering the central nervous system. We are next going to examine the meninges and other means by which the central nervous system is protected.

▊ The delicate central nervous tissue is well protected.

Central nervous tissue is very delicate. This characteristic, coupled with the fact that damaged nerve cells cannot be replaced, makes it imperative that this fragile, irreplaceable tissue be well protected. Four major features help protect the CNS from injury:

1. It is enclosed by hard, bony structures. The *cranium (skull)* encases the brain, and the *vertebral column* surrounds the spinal cord.
2. Three protective and nourishing membranes, the *meninges,* lie between the bony covering and the nervous tissue.
3. The brain "floats" in a special cushioning fluid, the *cerebrospinal fluid.*
4. A highly selective *blood–brain barrier* limits access of blood-borne materials into the vulnerable brain tissue.

The role of the first of these protective devices, the bony covering, is self-evident. The latter three protective mechanisms warrant further discussion.

▊ Three meningeal membranes wrap, protect, and nourish the central nervous system.

The **meninges,** the three membranes that wrap the central nervous system are, from the outermost to the innermost layer, the *dura mater,* the *arachnoid mater,* and the *pia mater* (● Figure 5-6). (*Mater* means "mother," indicative of the protective and supportive role played by these membranes.)

The **dura mater** is a tough, inelastic covering consisting of two layers (*dura* means "tough"). Usually, these layers adhere closely, but in some regions they are separated to form blood-filled cavities, **dural sinuses,** or in the case of the larger cavities, **venous sinuses.** Venous blood draining from the brain empties into these sinuses to be returned to the heart. Cerebrospinal fluid also re-enters the blood at one of these sinus sites.

The **arachnoid mater** is a delicate, richly vascularized layer with a "cobwebby" appearance (*arachnoid* means "spider-like"). The space between the arachnoid layer and the underlying pia mater, the **subarachnoid space,** is filled with CSF. Protrusions of arachnoid tissue, the **arachnoid villi,** penetrate through gaps in the overlying dura and project into the dural sinuses. CSF is reabsorbed across the surfaces of these villi into the blood circulating within the sinuses.

The innermost meningeal layer, the **pia mater,** is the most fragile (*pia* means "gentle"). It is highly vascular and closely adheres to the surfaces of the brain and spinal cord, following every ridge and valley. In certain areas it dips deeply into the brain to bring a rich blood supply into close contact with the ependymal cells lining the ventricles. This relationship is important in the formation of CSF, a topic to which we now turn our attention.

▊ The brain floats in its own special cerebrospinal fluid.

Cerebrospinal fluid (CSF) surrounds and cushions the brain and spinal cord. The CSF has about the same density as the brain itself, so the brain essentially floats or is suspended in its special fluid environment. The major function of CSF is to serve as a shock-absorbing fluid to prevent the brain from bumping against the interior of the hard skull when the head is subjected to sudden, jarring movements.

Cerebrospinal fluid is formed primarily by the **choroid plexuses** found in particular regions of the ventricle cavities of the brain. Choroid plexuses consist of richly vascularized, cauliflower-like masses of pia mater tissue that dip into pockets formed by ependymal cells. Cerebrospinal fluid is formed as a result of selective transport mechanisms across the membranes of the choroid plexuses. The composition of CSF differs from that of plasma. For example, CSF is lower in K^+ and higher in Na^+, making it an ideal environment for the movement of these ions down concentration gradients, a process essential for conduction of nerve impulses (see pp.110–111).

Once CSF is formed, it flows through the four interconnected ventricles within the interior of the brain and through the spinal cord's narrow central canal, which is continuous with the last ventricle. Cerebrospinal fluid escapes through small openings from the fourth ventricle at the base of the brain to enter the subarachnoid space and subsequently flows between the meningeal layers over the entire surface of the brain and spinal cord (● Figure 5-6). When the CSF reaches the upper regions of the brain, it is reabsorbed from the subarachnoid space into the venous blood through the arachnoid villi.

Flow of CSF through this system is facilitated by ciliary beating along with circulatory and postural factors that result in a CSF pressure of about 10 mm Hg. Reduction of this pressure by removal of even a few milliliters (ml) of CSF during a spinal tap for laboratory analysis may produce severe headaches.

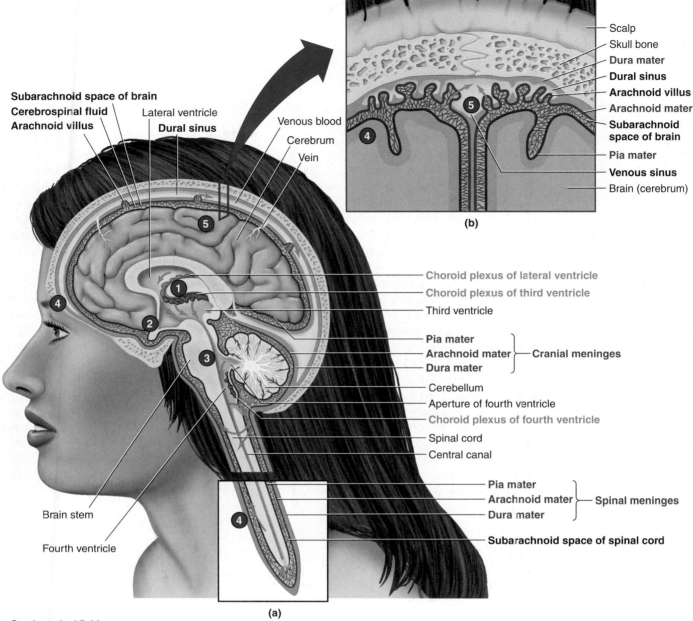

Subarachnoid space of brain
Cerebrospinal fluid
Arachnoid villus
Lateral ventricle
Dural sinus
Venous blood
Cerebrum
Vein

Scalp
Skull bone
Dura mater
Dural sinus
Arachnoid villus
Arachnoid mater
Subarachnoid space of brain
Pia mater
Venous sinus
Brain (cerebrum)

(b)

Choroid plexus of lateral ventricle
Choroid plexus of third ventricle
Third ventricle
Pia mater
Arachnoid mater — **Cranial meninges**
Dura mater
Cerebellum
Aperture of fourth ventricle
Choroid plexus of fourth ventricle
Spinal cord
Central canal

Pia mater
Arachnoid mater — **Spinal meninges**
Dura mater
Subarachnoid space of spinal cord

Brain stem
Fourth ventricle

(a)

Cerebrospinal fluid

1. is produced by the choroid plexuses,

2. circulates throughout the ventricles,

3. exits the fourth ventricle at the base of the brain,

4. flows in the subarachnoid space between the meningeal layers, and

5. is finally reabsorbed from the subarachnoid space into the venous blood across the arachnoid villi.

● **FIGURE 5-6**

Relationship of the meninges and cerebrospinal fluid to the brain and spinal cord
(a) Brain, spinal cord, and meninges in sagittal section. The arrows and circled numbers with accompanying explanations indicate the direction of flow of cerebrospinal fluid (in yellow). (b) Frontal section in the region between the two cerebral hemispheres of the brain, depicting the meninges in greater detail.

Through the ongoing processes of formation, circulation, and reabsorption, the entire volume of CSF of about 125 to 150 ml is replaced more than three times a day. If any one of these processes is defective so that excess CSF accumulates, **hydrocephalus** ("water on the brain") occurs. The resulting increase in CSF pressure can lead to brain damage and mental retardation if untreated. Treatment consists of surgically shunting the excess CSF to veins elsewhere in the body.

❚ A highly selective blood–brain barrier carefully regulates exchanges between the blood and brain.

The brain is carefully shielded from harmful changes in the blood by a highly selective blood–brain barrier. Throughout the body, exchange of materials between the blood and the surrounding interstitial fluid can take place only across the walls of capillaries, the smallest of blood vessels. Unlike the rather free exchange across capillaries elsewhere, permissible exchanges across brain capillaries are strictly limited. Changes in most plasma constituents do not easily influence the composition of brain interstitial fluid, because only selected, carefully regulated exchanges can be made across this barrier. For example, even if the K^+ level in the blood is doubled little change occurs in the K^+ concentration of the fluid bathing the central neurons. This is beneficial because alterations in interstitial fluid K^+ would be detrimental to neuronal function.

The blood–brain barrier consists of both anatomic and physiological factors. Capillary walls throughout the body are formed by a single layer of cells. Usually, all plasma components (except the large plasma proteins) can be freely exchanged between the blood and the surrounding interstitial fluid through holes or pores between the cells making up the capillary wall. In the brain capillaries, however, the cells are joined by *tight junctions* (see p. 63), which completely seal the capillary wall so that nothing can be exchanged across the wall by passing between the cells (● Figure 5-7). The only possible exchanges are through the capillary cells themselves. Lipid-soluble substances such as O_2, CO_2, alcohol, and steroid hormones penetrate these cells easily by dissolving in their lipid plasma membrane. Small water molecules also diffuse through readily, apparently by passing between the phospholipid molecules that compose the plasma membrane. All other substances exchanged between the blood and brain interstitial fluid, including such essential materials as glucose, amino acids, and ions, are transported by highly selective membrane-bound carriers. Accordingly, transport across

the capillary walls between the cells is *anatomically prevented,* and transport through the cells is *physiologically restricted.* Together, these mechanisms constitute the blood–brain barrier.

The brain capillaries are surrounded by astrocyte processes, which at one time were thought to be physically responsible for the blood–brain barrier. Evidence now indicates that the astrocytes have two roles regarding the blood–brain barrier: (1) They appear to signal the cells forming the brain capillaries to "get tight." Capillary cells do not have an inherent ability to form tight junctions; they do so only at the command of a signal within their neural environment. (2) Astrocytes are believed to participate in the cross-cellular transport of some substances, such as K^+.

The blood–brain barrier protects the delicate brain and spinal cord from chemical fluctuations in the blood and minimizes the possibility that potentially harmful blood-borne substances might reach the central neural tissue. It further prevents certain circulating hormones that could also act as neurotransmitters from reaching the brain, where they could produce uncontrolled nervous activity. On the negative side, the blood–brain barrier limits the use of drugs for the treatment of brain and spinal cord disorders, because many drugs are unable to penetrate this barrier.

Certain areas of the brain are not subject to the blood–brain barrier, most notably a portion of the hypothalamus. Functioning of the hypothalamus depends on its "sampling" the blood and adjusting its controlling output accordingly to maintain homeostasis. Part of this output is in the form of hormones that must enter hypothalamic capillaries to be transported to their sites of action. Appropriately, these hypothalamic capillaries are not sealed by tight junctions.

● **FIGURE 5-7**

Blood–brain barrier
Unlike most capillaries in the body, the cells forming the walls of brain capillaries are joined by tight junctions that prevent materials from passing between the cells. The only passage across brain capillaries is through the cells that form the capillary walls. With the exception of lipid-soluble substances and water, passage of all other materials through these cells is physiologically regulated by carrier-mediated systems, which are not present in capillaries elsewhere.

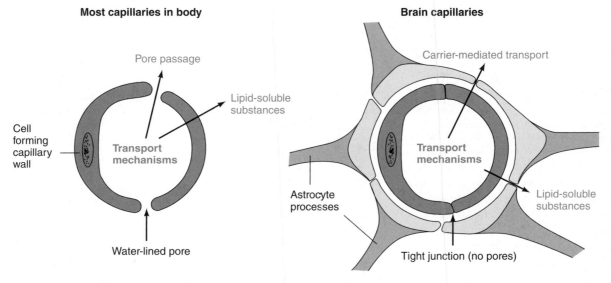

Most capillaries in body

Brain capillaries

Pore passage

Carrier-mediated transport

Lipid-soluble substances

Cell forming capillary wall

Transport mechanisms

Transport mechanisms

Lipid-soluble substances

Astrocyte processes

Water-lined pore

Tight junction (no pores)

Capillaries in cross section

Strokes: A Deadly Domino Effect

The most common cause of brain damage is **cerebrovascular accidents (strokes).** When a brain (cerebral) blood vessel is blocked by a clot or rupture, the brain tissue being supplied by that vessel is deprived of its vital O_2 and glucose supply. The result is damage and usually death of the deprived tissue. Recently, researchers have learned that neural damage (and the subsequent loss of neural function) extends well beyond the blood-deprived area as a result of a neurotoxic effect that leads to the death of additional nearby cells. Whereas the initial blood-deprived cells die by necrosis (unintentional cell death), the doomed neighbors undergo apoptosis (deliberate cell suicide; see p. 70). The initial O_2-starved cells release excessive amounts of glutamate, a common excitatory neurotransmitter. Glutamate or other neurotransmitters are normally released in small amounts from neurons as a means of chemical communication between brain cells. The excitatory overdose of glutamate from the damaged brain cells binds with and overexcites surrounding neurons. Specifically, glutamate binds with excitatory receptors known as *NMDA receptors,* which function as Ca^{2+} channels. As a result of toxic activation of these receptor-channels, they remain open for too long, permitting too much Ca^{2+} to rush into the affected neighboring neurons. This elevated intracellular Ca^{2+} triggers these cells to self-destruct. Cell-damaging free radicals (see p. 126) are produced during this process. Adding to the injury, researchers speculate that the Ca^{2+} apoptotic signal may spread from these dying cells to abutting healthy cells through gap junctions, cell-to-cell conduits that allow Ca^{2+} and other small ions to diffuse freely between cells. This action kills even more neuronal victims. Thus the majority of neurons that die following a stroke are originally unharmed cells that commit suicide in response to the chain of reactions unleashed by the toxic release of glutamate from the initial site of O_2 deprivation.

Until the last decade, physicians could do nothing to halt the inevitable neuronal loss following a stroke, leaving patients with an unpredictable mix of neural deficits. Treatment was limited to rehabilitative therapy after the damage was already complete. In recent years, armed with the new knowledge about the underlying factors in stroke-related neuronal death, the medical community has been seeking ways to halt the cell-killing domino effect. The goal, of course, is to limit the extent of neuronal damage and thus minimize or even prevent clinical symptoms such as paralysis. In the early 1990s, doctors started administering clot-dissolving drugs within the first three hours after the onset of a stroke to restore blood flow through blocked cerebral vessels. Clot busters were the first drugs used to treat strokes, but they are only the beginning of new stroke therapies. Other methods are currently under investigation to prevent adjacent nerve cells from succumbing to the neurotoxic release of glutamate. These include blocking the NMDA receptors that initiate the death-wielding chain of events in response to glutamate, halting the apoptosis pathway that results in self-execution, and blocking the gap junctions that permit the Ca^{2+} death messenger to spread to adjacent cells. These tactics hold much promise for treating strokes, which are the most prevalent cause of adult disability and the third leading cause of death in the United States. However, to date no new neuroprotective drugs have been found that do not cause serious side effects.

■ The brain depends on constant delivery of oxygen and glucose by the blood.

Even though many substances in the blood never actually come in contact with the brain tissue, the brain, more than any other tissue, is highly dependent on a constant blood supply. Unlike most tissues, which can resort to anaerobic metabolism to produce ATP in the absence of O_2 for at least short periods (see p. 40), the brain cannot produce ATP in the absence of O_2. Furthermore, in contrast to most tissues, which can use other sources of fuel for energy production in lieu of glucose, the brain normally uses only glucose but does not store any of this nutrient. Therefore, the brain is absolutely dependent on a continuous, adequate blood supply of O_2 and glucose. Brain damage results if this organ is deprived of its critical O_2 supply for more than four to five minutes or if its glucose supply is cut off for more than ten to fifteen minutes. The most common cause of inadequate blood supply to the brain is a stroke. (See the accompanying boxed feature, ❱ Concepts, Challenges, and Controversies, for details.)

OVERVIEW OF THE CENTRAL NERVOUS SYSTEM

The central nervous system consists of the brain and spinal cord. The brain has an estimated 100 billion neurons, which are assembled into complex networks that enable us to (1) subconsciously regulate our internal environment by neural means; (2) experience emotions; (3) voluntarily control our movements; (4) perceive (be consciously aware of) our own bodies and our surroundings; and (5) engage in other higher cognitive processes such as thought and memory. The term **cognition** refers to the act or process of "knowing," including both awareness and judgment.

No part of the brain acts in isolation of other brain regions because networks of neurons are anatomically linked by synapses, and neurons throughout the brain communicate extensively with each other by electrical and chemical means. However, neurons that work together to ultimately accomplish a given function tend to be organized within a discrete location. Therefore, even though the brain is a functional whole, it is organized into different regions. The parts of the brain can be arbitrarily grouped in various ways based on anatomic distinctions, functional specialization, and evolutionary development. We will use the following grouping (▲ Table 5-3):

1. Brain stem
2. Cerebellum
3. Forebrain
   a. Diencephalon
      (1) Hypothalamus
      (2) Thalamus
   b. Cerebrum
      (1) Basal nuclei
      (2) Cerebral cortex

Overview of Structures and Functions of the Major Components of the Brain

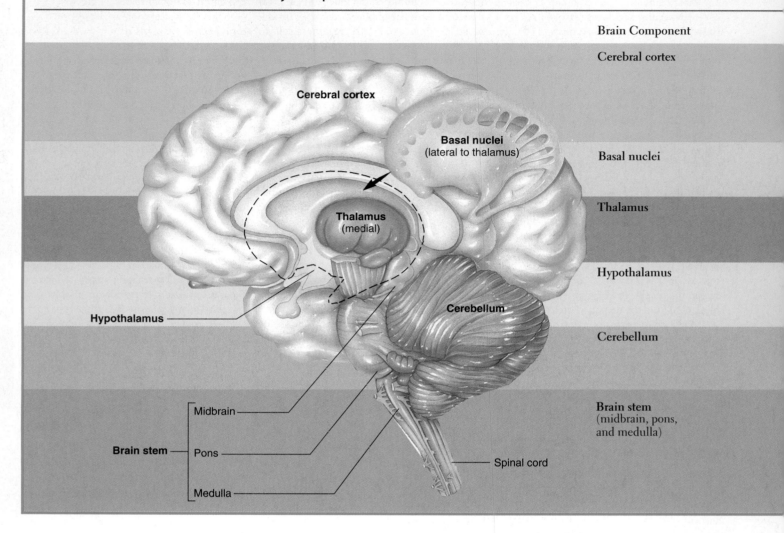

The order in which these components are listed generally represents both their anatomic location (from bottom to top) and their complexity and sophistication of function (from the least specialized, oldest level to the newest, most specialized level).

A primitive nervous system consists of comparatively few interneurons interspersed between afferent and efferent neurons. During evolutionary development, the interneuronal component progressively expanded, formed more complex interconnections, and became localized at the head end of the nervous system, forming the brain. Newer, more sophisticated layers of the brain were added on to the older, more primitive layers. The human brain represents the present peak of development.

The *brain stem,* the oldest region of the brain, is continuous with the spinal cord (▲ Table 5-3 and ● Figure 5-8b). It consists of the midbrain, pons, and medulla. The brain stem controls many of the life-sustaining processes, such as respiration, circulation, and digestion, which are common to many of the lower vertebrate forms. These processes are often referred to as "vegetative" functions because, with the loss of higher brain functions, these lower brain levels, in accompaniment

with appropriate supportive therapy such as provision of adequate nourishment, can still sustain the functions essential for survival. Because the person has no awareness or control of that life, however, the condition is sometimes referred to as "being a vegetable."

Attached at the top rear portion of the brain stem is the *cerebellum,* which is concerned with maintaining proper position of the body in space and subconscious coordination of motor activity (movement). The cerebellum also plays a key role in learning skilled motor tasks, such as a dance routine. On top of the brain stem, tucked within the interior of the cerebrum, is the *diencephalon.* It houses two brain components: the *hypothalamus,* which controls many homeostatic functions important in maintaining stability of the internal environment, and the *thalamus,* which performs some primitive sensory processing.

On top of this "cone" of lower brain regions is the *cerebrum,* whose "scoop" gets progressively larger and more highly convoluted (that is, has tortuous ridges delineated by deep grooves or folds) the more advanced the vertebrate species is. The cere-

Major Functions

1. Sensory perception
2. Voluntary control of movement
3. Language
4. Personality traits
5. Sophisticated mental events, such as thinking, memory, decision making, creativity, and self-consciousness

1. Inhibition of muscle tone
2. Coordination of slow, sustained movements
3. Suppression of useless patterns of movement

1. Relay station for all synaptic input
2. Crude awareness of sensation
3. Some degree of consciousness
4. Role in motor control

1. Regulation of many homeostatic functions, such as temperature control, thirst, urine output, and food intake
2. Important link between nervous and endocrine systems
3. Extensive involvement with emotion and basic behavioral patterns

1. Maintenance of balance
2. Enhancement of muscle tone
3. Coordination and planning of skilled voluntary muscle activity

1. Origin of majority of peripheral cranial nerves
2. Cardiovascular, respiratory, and digestive control centers
3. Regulation of muscle reflexes involved with equilibrium and posture
4. Reception and integration of all synaptic input from spinal cord; arousal and activation of cerebral cortex
5. Role in sleep–wake cycle

brum is most highly developed in humans, where it constitutes about 80% of the total brain weight. The outer layer of the cerebrum is the highly convoluted *cerebral cortex*, which caps an inner core that houses the *basal nuclei*. The myriad convolutions of the human cerebral cortex give it the appearance of a much-folded walnut (● Figure 5-8a). The cortex is perfectly smooth in many lower mammals. Without these surface wrinkles, the human cortex would take up to three times the area it does, and, accordingly, would not fit like a cover over the underlying structures. The increased neural circuitry housed in the extra cerebral cortical area not found in lower species is responsible for much of our unique human abilities. The cerebral cortex plays a key role in the most sophisticated neural functions, such as voluntary initiation of movement, final sensory perception, conscious thought, language, personality traits, and other factors we associate with the mind or intellect. It is the highest, most complex integrating area of the brain.

Each of these regions of the central nervous system will be discussed in turn, starting with the highest level, the cerebral cortex, and moving down to the lowest level, the spinal cord.

CEREBRAL CORTEX

The **cerebrum,** by far the largest portion of the human brain, is divided into two halves, the right and left **cerebral hemispheres** (● Figure 5-8a). They are connected to each other by the **corpus callosum,** a thick band consisting of an estimated 300 million neuronal axons traversing between the two hemispheres (● Figure 5-8b; also see ● Figure 5-16, p. 155). The corpus callosum is the body's "information superhighway." The two hemispheres communicate and cooperate with each other by means of constant information exchange through this neural connection.

▌ The cerebral cortex is an outer shell of gray matter covering an inner core of white matter.

Each hemisphere is composed of a thin outer shell of *gray matter*, the **cerebral cortex,** covering a thick central core of *white matter* (see ● Figure 5-16, p. 155). Another region of gray matter, the *basal nuclei*, is located deep within the white matter. Throughout the entire CNS, **gray matter** consists predominantly of densely packaged neuronal cell bodies and their dendrites as well as most glial cells. Bundles or tracts of myelinated nerve fibers (axons) constitute the white matter; its white appearance is due to the lipid composition of the myelin. The gray matter can be viewed as the "computers" of the CNS and the white matter as the "wires" that connect the computers to each other. Integration of neural input and initiation of neural output take place at synapses within the gray matter. The fiber tracts in the white matter transmit signals from one part of the cerebral cortex to another or between the cortex and other regions of the CNS. Such communication between different areas of the cortex and elsewhere facilitates integration of their activity. This integration is essential for even a relatively simple task such as picking a flower. Vision of the flower is received by one area of the cortex, reception of its fragrance takes place in another area, and movement is initiated by still another area. More subtle neuronal responses, such as appreciation of the flower's beauty and the urge to pick it, are poorly understood but undoubtedly extensively involve interconnecting fibers between different cortical regions.

▌ The cerebral cortex is organized into layers and functional columns.

The cerebral cortex is organized into six well-defined layers based on varying distributions of the cell bodies and locally associated fibers of several distinctive cell types. These layers are organized into functional vertical columns that extend perpendicularly about 2 mm from the cortical surface down through the thickness of the cortex to the underlying white matter. The neurons within a given column are believed to function as a "team," with each cell being involved in different aspects of the same specific activity—for example, perceptual processing of the same stimulus from the same location.

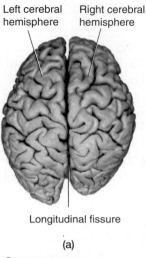

Left cerebral hemisphere   Right cerebral hemisphere

Longitudinal fissure

(a)

● **FIGURE 5-8**

Brain of human cadaver
(a) Dorsal view looking down on the top of the brain. Note that the deep longitudinal fissure divides the cerebrum into the right and left cerebral hemispheres. (b) Sagittal view of the right half of the brain. All major brain regions are visible from this midline interior view. The corpus callosum serves as a neural bridge between the two cerebral hemispheres.

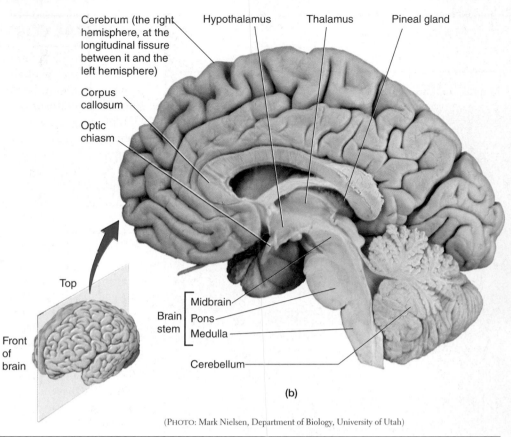

Cerebrum (the right hemisphere, at the longitudinal fissure between it and the left hemisphere)

Corpus callosum

Optic chiasm

Hypothalamus   Thalamus   Pineal gland

Top

Front of brain

Brain stem { Midbrain / Pons / Medulla

Cerebellum

(b)

(PHOTO: Mark Nielsen, Department of Biology, University of Utah)

The functional differences between various areas of the cortex result from different layering patterns within the columns and from different input–output connections, not from the presence of unique cell types or different neuronal mechanisms. For example, those regions of the cortex responsible for perception of senses have an expanded layer 4, a layer rich in **stellate cells,** which are responsible for initial processing of sensory input to the cortex. In contrast, the cortical areas that control output to skeletal muscles have a thickened layer 5, which contains an abundance of large **pyramidal cells.** These cells send fibers down the spinal cord from the cortex to terminate on the efferent motor neurons that innervate the skeletal muscles.

▍ The four pairs of lobes in the cerebral cortex are specialized for different activities.

We are now going to consider the locations of the major functional areas of the cerebral cortex. Throughout this discussion, keep in mind that even though a discrete activity is ultimately attributed to a particular region of the brain, no part of the brain functions in isolation. Each part depends on complex interplay among numerous other regions for both incoming and outgoing messages.

The anatomical landmarks used in cortical mapping are certain deep folds that divide each half of the cortex into four major lobes: the *occipital, temporal, parietal,* and *frontal lobes* (● Figure 5-9). Refer to the basic functional map of the cortex in ● Figure 5-10a during the following discussion of the major activities attributed to various regions of these lobes.

The **occipital lobes,** which are located posteriorly (at the back of the head), are responsible for initially processing visual input. Sound sensation is initially received by the **temporal lobes,** located laterally (on the sides of the head) (● Figure 5-10a

● **FIGURE 5-9**

Cortical lobes
Each half of the cerebral cortex is divided into the occipital, temporal, parietal, and frontal lobes, as depicted in this schematic lateral view of the brain.

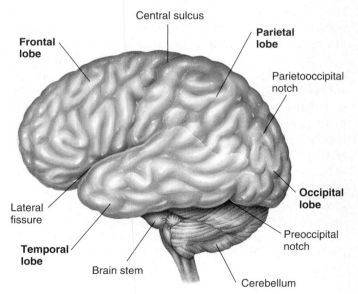

Central sulcus

Frontal lobe

Parietal lobe

Parietooccipital notch

Occipital lobe

Lateral fissure

Preoccipital notch

Temporal lobe

Brain stem

Cerebellum

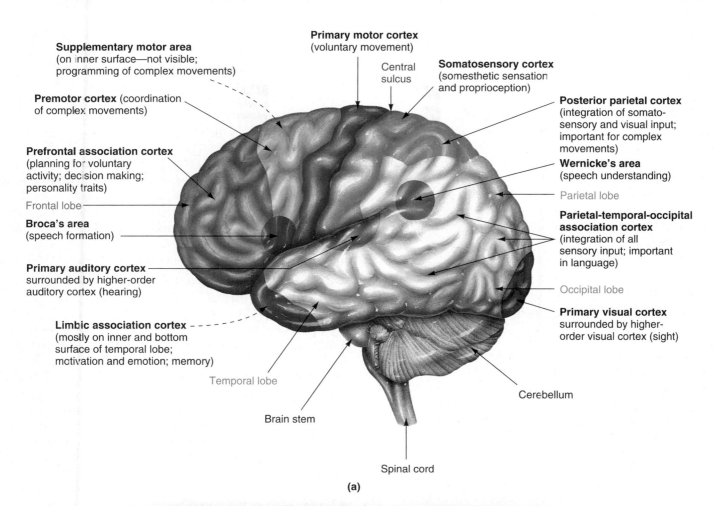

Supplementary motor area
(on inner surface—not visible;
programming of complex movements)

Primary motor cortex
(voluntary movement)

Central
sulcus

Somatosensory cortex
(somesthetic sensation
and proprioception)

Premotor cortex (coordination
of complex movements)

Posterior parietal cortex
(integration of somato-
sensory and visual input;
important for complex
movements)

Prefrontal association cortex
(planning for voluntary
activity; decision making;
personality traits)

Wernicke's area
(speech understanding)

Frontal lobe

Parietal lobe

Broca's area
(speech formation)

**Parietal-temporal-occipital
association cortex**
(integration of all
sensory input; important
in language)

Primary auditory cortex
surrounded by higher-order
auditory cortex (hearing)

Occipital lobe

Limbic association cortex
(mostly on inner and bottom
surface of temporal lobe;
motivation and emotion; memory)

Primary visual cortex
surrounded by higher-
order visual cortex (sight)

Temporal lobe

Cerebellum

Brain stem

Spinal cord

(a)

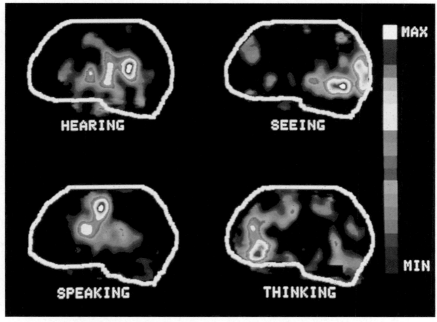

(b)

● FIGURE 5-10

Functional areas of the cerebral cortex

(a) Various regions of the cerebral cortex are primarily responsible for various aspects of neural processing, as indicated in this schematic lateral view of the brain. (b) Different areas of the brain "light up" on positron emission tomography (PET) scans as a person performs different tasks. PET scans detect the magnitude of blood flow in various regions of the brain. Because more blood flows into a particular region of the brain when it is more active, neuroscientists can use PET scans to "take pictures" of the brain at work on various tasks.

and b). You will learn more about the functions of these regions in Chapter 6 when we discuss vision and hearing.

The parietal lobes and frontal lobes, located on the top of the head, are separated by a deep infolding, the **central sulcus,** which runs roughly down the middle of the lateral surface of each hemisphere. The **parietal lobes** lie to the rear of the central sulcus on each side, and the **frontal lobes** lie in front of it. The parietal lobes are primarily responsible for receiving and processing sensory input. The frontal lobes are responsible for three main functions: (1) voluntary motor activity, (2) speaking ability, and (3) elaboration of thought. We are next going to examine the role of the parietal lobes in sensory perception, then turn our attention to the functions of the frontal lobes.

▌ The parietal lobes are responsible for somatosensory processing.

Sensations from the surface of the body, such as touch, pressure, heat, cold, and pain are collectively known as **somesthetic sensations** (*somesthetic* means "body feelings"). The means by which afferent neurons detect and relay information to the CNS regarding these sensations will be covered in Chapter 6 when we explore the afferent division of the peripheral nervous system in detail. Within the CNS, this information is "projected" (transmitted along specific neural pathways to higher brain levels) to the **somatosensory cortex.** The somatosensory cortex is located in the front portion of each parietal lobe immediately behind the central sulcus (● Figures 5-10a and 5-11a). It is the site for initial cortical processing and perception of somesthetic input as well as proprioceptive input. **Proprioception** is the awareness of body position.

Each region within the somatosensory cortex receives somesthetic and proprioceptive input from a specific area of the body. This distribution of cortical sensory processing is depicted in ● Figure 5-11b. Note that on this so-called **sensory homunculus** (*homunculus* means "little man"), the body is represented upside down on the somatosensory cortex and, more importantly, *different parts of the body are not equally represented.* The size of each body part in this homunculus is indicative of the relative proportion of the somatosensory cortex devoted to that area. The exaggerated size of the face, tongue, hands, and genitalia is indicative of the high degree of sensory perception associated with these body parts.

The somatosensory cortex on each side of the brain for the most part receives sensory input from the opposite side of the body, because most of the ascending pathways carrying sensory information up the spinal cord cross over to the opposite side before eventually terminating in the cortex (see ● Figure 5-30b, p. 175). Thus damage to the somatosensory cortex in the left hemisphere produces sensory deficits on the right side of the body, whereas sensory losses on the left side are associated with damage to the right half of the cortex.

Simple awareness of touch, pressure, temperature, or pain is detected by the thalamus, a lower level of the brain, but the somatosensory cortex goes beyond pure recognition of sensations to fuller sensory perception. The thalamus makes you aware that something hot versus something cold is touching your body, but it does not tell you where or of what intensity. The somatosensory cortex localizes the source of sensory input and perceives the level of intensity of the stimulus. It also is capable of spatial discrimination, so it can discern shapes of objects being held and can distinguish subtle differences in similar objects that come into contact with the skin.

The somatosensory cortex, in turn, projects this sensory input via white matter fibers to adjacent higher sensory areas for even further elaboration, analysis, and integration of sensory information. These higher areas are important in the perception of complex patterns of somatosensory stimulation—for example, simultaneous appreciation of the texture, firmness, temperature, shape, position, and location of an object you are holding.

▌ The primary motor cortex is located in the frontal lobes.

The area in the rear portion of the frontal lobe immediately in front of the central sulcus and adjacent to the somatosensory cortex is the **primary motor cortex** (● Figures 5-10a and 5-12a). It confers voluntary control over movement produced by skeletal muscles. As in sensory processing, the motor cortex on each side of the brain primarily controls muscles on the opposite side of the body. Neuronal tracts originating in the motor cortex of the left hemisphere cross over before passing down the spinal cord to terminate on efferent motor neurons that trigger skeletal muscle contraction on the right side of the body (see ● Figure 5-30c, p. 175). Accordingly, damage to the motor cortex on the left side of the brain produces paralysis on the right side of the body, and the converse is also true.

Stimulation of different areas of the primary motor cortex brings about movement in different regions of the body. Like the sensory homunculus for the somatosensory cortex, the **motor homunculus,** which depicts the location and relative amount of motor cortex devoted to output to the muscles of each body part, is upside down and distorted (● Figure 5-12b). The fingers, thumbs, and muscles important in speech, especially those of the lips and tongue, are grossly exaggerated, indicative of the fine degree of motor control with which these body parts are endowed. Compare this to how little brain tissue is devoted to the trunk, arms, and lower extremities, which are not capable of such complex movements. Thus the extent of representation in the motor cortex is proportional to the precision and complexity of motor skills required of the respective part.

▌ Other brain regions besides the primary motor cortex are important in motor control.

Even though signals from the primary motor cortex terminate on the efferent neurons that trigger voluntary skeletal muscle contraction, the motor cortex is not the only region of the brain involved with motor control. First, lower brain regions and the spinal cord control involuntary skeletal muscle activity, such

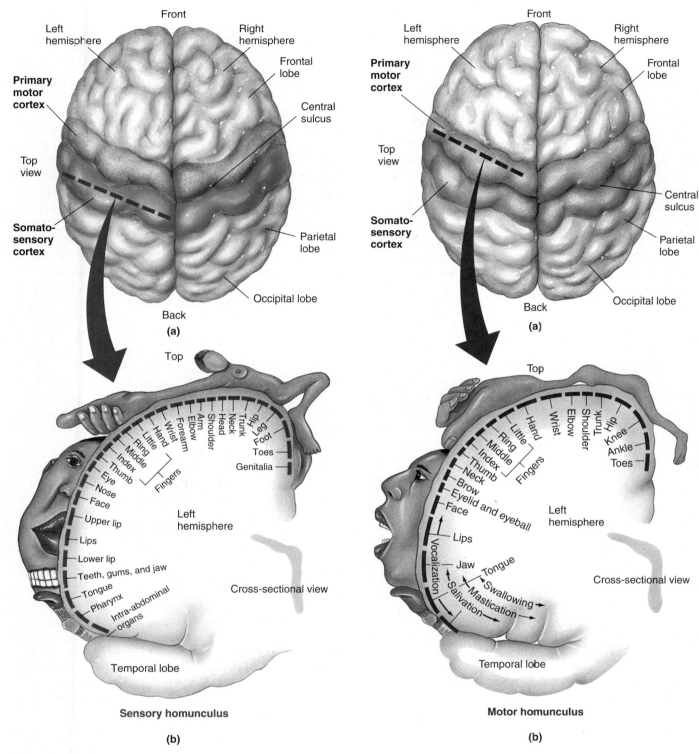

● **FIGURE 5-11**

Somatotopic map of the somatosensory cortex
(a) Top view of cerebral hemispheres. (b) Sensory homunculus showing the distribution of sensory input to the somatosensory cortex from different parts of the body. The distorted graphic representation of the body parts indicate the relative proportion of the somatosensory cortex devoted to reception of sensory input from each area.

● **FIGURE 5-12**

Somatotopic map of the primary motor cortex
(a) Top view of cerebral hemispheres. (b) Motor homunculus showing the distribution of motor output from the primary motor cortex to different parts of the body. The distorted graphic representation of the body parts is indicative of the relative proportion of the primary motor cortex devoted to controlling skeletal muscles in each area.

as the maintenance of posture. Some of these same regions also play an important role in monitoring and coordinating voluntary motor activity that has been set in motion by the pri-

mary motor cortex. Second, although fibers originating from the motor cortex can activate motor neurons to bring about muscle contraction, the motor cortex itself does not *initiate* voluntary movement. The motor cortex is activated by a widespread pattern of neuronal discharge, the **readiness potential,** which occurs about 750 msec before specific electrical activity is detectable in the motor cortex. The higher motor areas of the brain believed to be involved in this voluntary decision-making period include the *supplementary motor area*, the *premotor cortex*, and the *posterior parietal cortex* (● Figure 5-10a). These higher areas all command the primary motor cortex. Furthermore, a subcortical region of the brain, the *cerebellum*, plays an important role in the planning, initiation, and timing of certain kinds of movement by sending input to the motor areas of the cortex.

These other four regions of the brain are important in programming and coordinating complex movements involving simultaneous contraction of many muscles. Even though electrical stimulation of the primary motor cortex brings about contraction of particular muscles, no purposeful coordinated movement can be elicited, just as pulling on isolated strings of a puppet does not produce any meaningful movement. A puppet displays purposeful movements only when a skilled puppeteer manipulates the strings in a coordinated manner. In the same way, these four regions (and perhaps other areas as yet undetermined) develop a **motor program** for the specific voluntary task and then "pull" the appropriate pattern of "strings" in the primary motor cortex to bring about the sequenced contraction of appropriate muscles to accomplish the desired complex movement.

The **supplementary motor area** lies on the medial (inner) surface of each hemisphere anterior to (in front of) the primary motor cortex. It appears to play a preparatory role in programming complex sequences of movement. Stimulation of various regions of this motor area brings about complex patterns of movement, such as opening or closing the hand. Lesions here do not result in paralysis, but they do interfere with performance of more complex, useful integrated movements.

The **premotor cortex,** located on the lateral surface of each hemisphere in front of the primary motor cortex, is important in orienting the body and arms toward a specific target. In order to command the primary motor cortex to bring about the appropriate skeletal muscle contraction to accomplish the desired movement, the premotor cortex must be informed of the body's momentary position in relation to the target. The premotor cortex is guided by sensory input processed by the **posterior parietal cortex,** a region that lies posterior to (in back of) the primary somatosensory cortex. These two higher motor areas have many anatomic interconnections and appear to be closely interrelated functionally. When either of these areas is damaged, the individual cannot process complex sensory information to accomplish purposeful movement in a spatial context. Such patients, for example, cannot successfully manipulate silverware when eating.

Even though these higher motor areas command the primary motor cortex and are important in preparing for the execution of deliberate, meaningful movement, we cannot say that voluntary movement is actually initiated by these areas. This pushes the question of how and where voluntary activity is initiated one step further. Probably no single area is responsible; undoubtedly, numerous pathways can ultimately bring about deliberate movement.

For example, think about the neural systems called into play during the simple act of picking up an apple to eat. Because of your memory, you know the fruit is located in a bowl on the kitchen counter. Sensory systems, coupled with your knowledge based on past experience, enable you to distinguish the apple from the other varieties of fruit in the bowl. On receiving this integrated sensory information, motor systems issue commands to the exact muscles of the body in the proper sequence to enable you to move to the fruit bowl and pick up the targeted apple. During execution of this act, minor adjustments in the motor command are made as needed, based on continual updating provided by sensory input about the position of the body relative to the goal. Then there is the issue of motivation and behavior. Why are you reaching for the apple in the first place? Is it because you are hungry (detected by a neural system in the hypothalamus) or because of a more complex behavioral scenario unrelated to a basic hunger drive, such as the fact that you started to think about food because you just saw someone eating on television? Why did you select an apple rather than a banana when both are in the fruit bowl and you like the taste of both, and so on? Thus initiation and execution of purposeful voluntary movement actually include a complex neuronal interplay involving output from the motor regions guided by integrated sensory information and ultimately dependent on motivational systems and elaboration of thought. All this plays against a background of memory stores from which meaningful decisions about desirable movements can be made.

▌Somatotopic maps vary slightly between individuals and are dynamic, not static.

Although the general organizational pattern of sensory and motor somatotopic ("body representation") maps of the cortex is similar in all people, the precise distribution is unique for each individual. Just as each of us has two eyes, a nose, and a mouth and yet no two faces have these features arranged in exactly the same way, so it is with brains. Furthermore, an individual's somatotopic mapping is not "carved in stone" but is subject to constant subtle modifications based on use. The general pattern is governed by genetic and developmental processes, but the individual cortical architecture appears capable of being influenced by **use-dependent competition** for cortical space. For example, when monkeys were encouraged to use their middle fingers instead of their other fingers to press a bar for food, after only several thousand bar presses the "middle finger area" of the motor cortex was greatly expanded and encroached upon territory previously devoted to the other fingers. Similarly, modern neuroimaging techniques reveal that the left hand of a right-handed string musician is represented by a larger area of the somatosensory cortex, the touch-sensing region of the cortex, than is the left hand of a person who does not play a string instrument. In this way, the musician's left-hand fingers develop a greater "feel" for the instrument as they skillfully manipulate the strings.

Other regions of the brain besides the somatosensory cortex and motor cortex can also be modified by experience. We are now going to turn our attention to this plasticity of the brain.

■ Because of its plasticity, the brain can be remodeled in response to varying demands.

The brain displays a degree of **plasticity,** that is, an ability to change or be functionally remodeled in response to the demands placed on it. The term *plasticity* is used to describe this ability because plastics can be manipulated into any desired shape to serve a particular purpose. The ability of the brain to be modified as needed is more pronounced in the early developmental years, but even adults retain some plasticity. When an area of the brain associated with a particular activity is destroyed, other areas of the brain may gradually assume some or all of the responsibilities of the damaged region. The underlying molecular mechanisms responsible for the brain's plasticity are only beginning to be unraveled. Current evidence suggests that the formation of new neural pathways (not new neurons, but new connections between existing neurons) in response to changes in experience are mediated in part by alterations in dendritic shape resulting from modifications in certain cytoskeletal elements (see p. 44). As its dendrites become more branched and elongated, a neuron is able to receive and integrate more signals from other neurons. Thus the precise synaptic connections between neurons are not fixed but can be modified by experience.

The gradual modification of each person's brain by a unique set of experiences provides a biological basis for individuality. Even though the particular architecture of your own rather plastic brain has been and continues to be influenced by your unique experiences, it is important to realize that what you do and do not do cannot totally shape the organization of your cortex and other parts of the brain. Some limits are genetically established. Also, there are developmental limits on the extent to which modeling can be influenced by patterns of use. For example, some cortical regions maintain their plasticity throughout life, especially the ability to add new memory stores and learn, but other cortical regions can be modified by use for only a specified time after birth before becoming permanently fixed. The length of this critical developmental period varies for different cortical regions.

■ Different aspects of language are controlled by different regions of the cortex.

Language ability is an excellent example of early cortical plasticity coupled with later permanence. Unlike the sensory and motor regions of the cortex, which are present in both hemispheres, the areas of the brain responsible for language ability are found in only one hemisphere—the left hemisphere, in the vast majority of the population. However, if a child under the age of 2 accidentally suffers damage to the left hemisphere, language functions are transferred to the right hemisphere with no delay in language development but at the expense of less obvious nonverbal abilities for which the right hemisphere is normally responsible. Up to about the age of 10, after damage to the left hemisphere, language ability can usually be reestablished in the right hemisphere following a temporary period of loss. If damage occurs beyond the early teens, however, language ability is permanently impaired, even though some limited restoration may be possible. The regions of the brain involved in comprehending and expressing language apparently are permanently assigned before adolescence.

Even in normal individuals, there is evidence for early plasticity and later permanence in language development. Infants can distinguish between and articulate the entire range of speech sounds, but each language uses only a portion of these sounds. As children mature, they often lose the ability to distinguish between or express speech sounds that are not important in their native language. For example, Japanese children can distinguish between the sounds of "r" and "l," but many Japanese adults cannot perceive the difference between them.

Roles of Broca's area and Wernicke's area

Language is a complex form of communication in which written or spoken words symbolize objects and convey ideas. It involves the integration of two distinct capabilities—namely, *expression* (speaking ability) and *comprehension*—each of which is related to a specific area of the cortex. The primary areas of cortical specialization for language are Broca's area and Wernicke's area. **Broca's area,** which is responsible for speaking ability, is located in the left frontal lobe in close association with the motor areas of the cortex that control the muscles necessary for articulation (● Figures 5-10a and b and 5-13). **Wernicke's area,** located in the left cortex at the juncture of the parietal, temporal, and occipital lobes, is concerned with language comprehension. It plays a critical role in understanding both spoken and written messages. Furthermore, it is responsible for formulating coherent patterns of speech that are transferred via a bundle of fibers to Broca's area, which in turn controls articulation of this speech. Wernicke's area receives input from the visual cortex in the occipital lobe, a pathway important in reading comprehension and in describing objects seen, as well as from the auditory cortex in the temporal lobe, a pathway essential for understanding spoken words. According to the leading model of language, precise interconnecting pathways between these localized cortical areas are involved in the various aspects of speech (● Figure 5-13). Wernicke's area also receives input from the somatosensory cortex, a pathway important in the ability to read Braille.

Because various aspects of language are localized in different regions of the cortex, damage to specific regions of the brain can result in selective disturbances of language. Damage to Broca's area results in a failure of word formation, although the patient can still understand the spoken and written word. Such individuals know what they want to say but are unable to express themselves. Even though they can move their lips and tongue, they cannot establish the proper motor command to articulate the desired words. In contrast, patients with a lesion in Wernicke's area cannot understand words they see or hear. They are able to speak fluently even though their perfectly articulated words make no sense. They cannot attach meaning to words or choose appropriate words to convey their thoughts. Such language disorders caused by damage to specific cortical areas are known as **aphasias,** most of which result from strokes. Aphasias should not be confused with **speech impediments,**

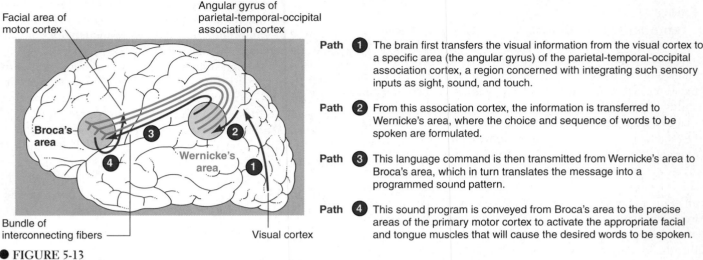

Path **1** The brain first transfers the visual information from the visual cortex to a specific area (the angular gyrus) of the parietal-temporal-occipital association cortex, a region concerned with integrating such sensory inputs as sight, sound, and touch.

Path **2** From this association cortex, the information is transferred to Wernicke's area, where the choice and sequence of words to be spoken are formulated.

Path **3** This language command is then transmitted from Wernicke's area to Broca's area, which in turn translates the message into a programmed sound pattern.

Path **4** This sound program is conveyed from Broca's area to the precise areas of the primary motor cortex to activate the appropriate facial and tongue muscles that will cause the desired words to be spoken.

● **FIGURE 5-13**

Cortical pathway for speaking a written word or naming a visual object
The red arrows and circled numbers with accompanying explanation indicate the pathway used to speak about something seen. Similarly, appropriate muscles of the hand can be commanded to write the desired words.

---

which are caused by a defect in the mechanical aspect of speech, such as weakness or incoordination of the muscles controlling the vocal apparatus.

Dyslexia, another language disorder, is a difficulty in learning to read because of inappropriate interpretation of words. The problem arises from developmental abnormalities in connections between the visual and language areas of the cortex or within the language areas themselves; that is, the person is born with "faulty wiring" within the language-processing system. Emerging evidence suggests that dyslexia stems from a deficit in phonological processing, meaning an impaired ability to break down written words into their underlying phonetic components. Dyslexics have difficulty decoding and thus identifying and assigning meaning to words. The condition is in no way related to intellectual ability.

▌ The association areas of the cortex are involved in many higher functions.

The motor, sensory, and language areas account for only about half of the total cerebral cortex. The remaining areas, called **association areas,** are involved in higher functions. There are three association areas: (1) the *prefrontal association cortex*, (2) the *parietal-temporal-occipital association cortex*, and (3) the *limbic association cortex* (● Figure 5-10). At one time the association areas were called "silent" areas, because stimulation does not produce any observable motor response or sensory perception. (During brain surgery, typically the patient remains awake and only local anesthetic is used along the cut scalp. This is possible because the brain itself is insensitive to pain. Before cutting into this precious, nonregenerative tissue, the neurosurgeon explores the exposed region with a tiny stimulating electrode. The patient is asked to describe what happens with each stimulation—the flick of a finger, a prickly feeling on the bottom

of the foot, nothing? In this way, the surgeon can ascertain the appropriate landmarks on the neural map before making an incision.)

The **prefrontal association cortex** is the front portion of the frontal lobe just anterior to the premotor cortex. The roles attributed to this region are (1) planning for voluntary activity, (2) decision making (that is, weighing consequences of future actions and choosing between different options for various social or physical situations) (● Figure 5-10b), (3) creativity, and (4) personality traits. To carry out these highest of neural functions, the prefrontal association cortex is the site of operation of *working memory* where the brain temporarily stores and actively manipulates information used in reasoning and planning. You will learn more about working memory later.

Stimulation of the prefrontal association cortex does not produce any observable effects, but deficits in this area result in changes in personality and social behavior. Because damage to the prefrontal lobe was known to produce these changes, about 60 years ago prefrontal lobotomy (surgical removal) was used for treating violent individuals or others with "bad" personality traits or social behavior in the hopes that their personality change would be for the better. Of course, the other functions of the prefrontal association cortex were lost in the process (fortunately, the technique was used only a short time).

The **parietal-temporal-occipital association cortex** is found at the interface of the three lobes for which it is named. In this strategic location, it pools and integrates somatic, auditory, and visual sensations projected from these three lobes for complex perceptual processing. It enables us to "get the complete picture" of the relationship of various parts of our bodies with the external world. For example, it integrates visual information with proprioceptive input to enable you to place what you are seeing in proper perspective, such as realizing that a bottle is in an upright position despite the angle from which you view it (that is, whether you are standing up, lying down, or hanging upside down from a tree branch). This region is also involved in the language pathway connecting Wernicke's area to the visual and auditory cortices.

The **limbic association cortex** is located mostly on the bottom and adjoining inner portion of each temporal lobe. This area is concerned primarily with motivation and emotion and is extensively involved in memory.

The cortical association areas are all interconnected by bundles of fibers within the cerebral white matter. Collectively, the association areas integrate diverse information for purposeful action. An oversimplified basic sequence of linkage between the various functional areas of the cortex is schematically represented in ● Figure 5-14.

▌ The cerebral hemispheres have some degree of specialization.

The cortical areas described thus far appear to be equally distributed in both the right and left hemispheres, except for the language areas, which are found only on one side, usually the left. The left side is also most commonly the dominant hemisphere for fine motor control. Thus most people are right-handed, because the left side of the brain controls the right side of the body. Furthermore, each hemisphere is somewhat specialized in the types of mental activities it carries out best. The **left cerebral hemisphere** excels in the performance of logical, analytic, sequential, and verbal tasks, such as math, language forms, and philosophy. In contrast, the **right cere-**

bral hemisphere excels in nonlanguage skills, especially spatial perception and artistic and musical talents. Whereas the left hemisphere tends to process information in a fine-detail, fragmentary way, the right hemisphere views the world in a big-picture, holistic way. Normally, much sharing of information occurs between the two hemispheres so that they complement each other, but in many individuals the skills associated with one hemisphere appear to be more strongly developed. Left cerebral hemisphere dominance tends to be associated with "thinkers," whereas the right hemispheric skills dominate in "creators."

▌ An electroencephalogram is a record of postsynaptic activity in cortical neurons.

Extracellular current flow arising from electrical activity within the cerebral cortex can be detected by placing recording electrodes on the scalp to produce a graphic record known as an **electroencephalogram** or **EEG.** These "brain waves" for the most part are not due to action potentials but instead represent the momentary collective postsynaptic potential activity (that is, EPSPs and IPSPs; see p. 119) in the cell bodies and dendrites located in the cortical layers under the recording electrode.

Electrical activity can always be recorded from the living brain, even during sleep and unconscious states, but the waveforms vary, depending on the degree of activity of the cerebral cortex. Often the waveforms appear irregular, but sometimes distinct patterns in the wave's amplitude and frequency can be observed. A dramatic example of this is illustrated in ● Figure 5-15, in which the EEG waveform recorded over the occipital (visual) cortex changes markedly in response to simply opening and closing the eyes.

The EEG has three major uses:

1. It is often used as a *clinical tool in the diagnosis of cerebral dysfunction.* Diseased or damaged cortical tissue often gives rise to altered EEG patterns. One of the most common neurologic diseases accompanied by a distinctively abnormal EEG is **epilepsy.** Epileptic seizures occur when a large collection of neurons abnormally undergo synchronous action potentials that produce stereotypical, involuntary spasms and alterations in behavior. Different underlying problems, including genetic defects and traumatic brain injuries, can lead to the neuronal

● **FIGURE 5-14**

Schematic linking of various regions of the cortex

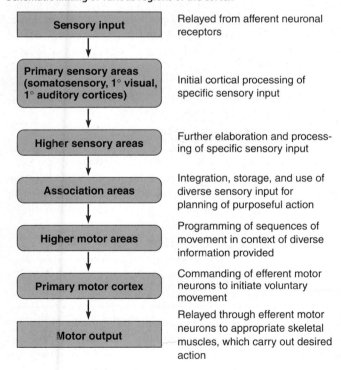

Sensory input	Relayed from afferent neuronal receptors
Primary sensory areas (somatosensory, 1° visual, 1° auditory cortices)	Initial cortical processing of specific sensory input
Higher sensory areas	Further elaboration and processing of specific sensory input
Association areas	Integration, storage, and use of diverse sensory input for planning of purposeful action
Higher motor areas	Programming of sequences of movement in context of diverse information provided
Primary motor cortex	Commanding of efferent motor neurons to initiate voluntary movement
Motor output	Relayed through efferent motor neurons to appropriate skeletal muscles, which carry out desired action

For simplicity, a number of interconnections have been omitted.

● **FIGURE 5-15**

Replacement of an alpha rhythm on an EEG with a beta rhythm when the eyes are opened

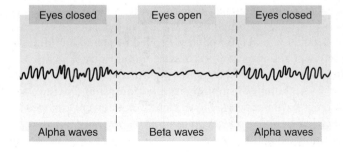

hyperexcitability that characterizes epilepsy. Typically there is too little inhibitory compared to excitatory activity, as with compromised functioning of the inhibitory neurotransmitter GABA or prolonged action of the excitatory neurotransmitter glutamate. The seizures may be partial or generalized, depending on the location and extent of the abnormal neuronal discharge. Each type of seizure displays different EEG features.

2. The EEG is also used to *distinguish various stages of sleep*, as described later in this chapter.

3. The EEG finds further use in the *legal determination of brain death*. Even though a person may have stopped breathing and the heart may have stopped pumping blood, it is often possible to restore and maintain respiratory and circulatory activity if resuscitative measures are instituted soon enough. Yet because of the susceptibility of the brain to O_2 deprivation, irreversible brain damage may have already occurred before lung and heart function have been re-established, resulting in the paradoxical situation of a dead brain in a living body. The determination of whether a comatose patient being maintained by artificial respiration and other supportive measures is alive or dead has important medical, legal, and social implications. The need for viable organs for modern transplant surgery has made the timeliness of such life/death determinations of utmost importance. Physicians, lawyers, and the American public in general have accepted the notion of brain death—that is, a brain that is not functioning, with no possibility of recovery—as the determinant of death under such circumstances. Brain-dead people make good organ donors because the organs are still being supplied by circulating blood and thus are in better shape than those obtained from a person whose heart has stopped beating. The most widely accepted indication of brain death is *electrocerebral silence*—an essentially flat EEG. This must be coupled with other stringent criteria, such as absence of eye reflexes, to guard against a false terminal diagnosis in individuals with a flat EEG that is due to causes that can be reversed, as in certain kinds of drug intoxication.

▌ Neurons in different regions of the cerebral cortex may fire in rhythmic synchrony.

Most information about the electrical activity of the brain has been gleaned not from EEG studies but from direct recordings of individual neurons in experimental animals engaged in various activities. Following surgical implantation of an extremely thin recording microelectrode into a single neuron within a specific region of the cerebral cortex, scientists have been able to observe changes in the electrical activity of the neuron as the animal engages in particular motor tasks or encounters various sensations. Through these studies, investigators have concluded that neural information is coded by changes in the frequency of action potentials in specific neurons: The greater the triggering event, the greater the firing rate of the neuron.

Even though this finding is important, single-neuron recordings have not been able to identify concurrent changes in electrical activity in a group of neurons working together to accomplish a particular activity. As an analogy, consider if you

tried to record a concert by using a single microphone that could pick up only the sounds produced by one musician. You would get a very limited impression of the performance by hearing only the changes in notes and tempo as played by this one individual. You would miss the richness of the melody and rhythm being performed in synchrony by the entire orchestra. Similarly, by recording from single neurons and detecting their changes in firing rates, scientists have been overlooking a parallel information mechanism involving changes in the relative timing of action potential discharges among a functional group, or *assembly*, of neurons. Exciting new studies involving simultaneous recordings from multiple neurons suggest that interacting neurons may transiently fire together for fractions of a second. Although the theory is still controversial, many neuroscientists believe the brain encodes information not just by changing the firing rates of individual neurons but also by changing the patterns of these brief neural synchronizations. That is, groups of neurons may be communicating, or sending messages about what's happening, by changing their pattern of synchronous firing.

Neurons within an assembly that fire together may be widely scattered. For example, when you view a bouncing ball, different visual units initially process different aspects of this object—its shape, its color, its movement, and so on. Somehow all these separate processing pathways must be integrated, or "bound together" for you to "see" the bouncing ball as a whole unit without stopping to contemplate its many separate features. The solution to the longtime mystery of how the brain accomplishes this integration might lie in the synchronous firing of neurons in separate regions of the brain that are functionally linked together by virtue of being responsive to different aspects of the same objects, such as the bouncing ball.

We are now going to shift our attention to the **subcortical regions** of the brain, which interact extensively with the cortex in the performance of their functions (*subcortical* means "under the cortex"). These regions include the *basal nuclei*, located in the cerebrum, and the *thalamus* and *hypothalamus*, located in the diencephalon.

BASAL NUCLEI, THALAMUS, AND HYPOTHALAMUS

The **basal nuclei** (also known as **basal ganglia**) consist of several masses of gray matter located deep within the cerebral white matter (see ▲ Table 5-3 and ● Figure 5-16). In the central nervous system, a **nucleus** (plural, **nuclei**) is a functional aggregation of neuronal cell bodies.

▌ The basal nuclei play an important inhibitory role in motor control.

The basal nuclei play a complex role in the control of movement in addition to having nonmotor functions that are less understood. In particular, the basal nuclei are important in (1) inhibiting muscle tone throughout the body (proper muscle tone is normally maintained by a balance of excitatory and

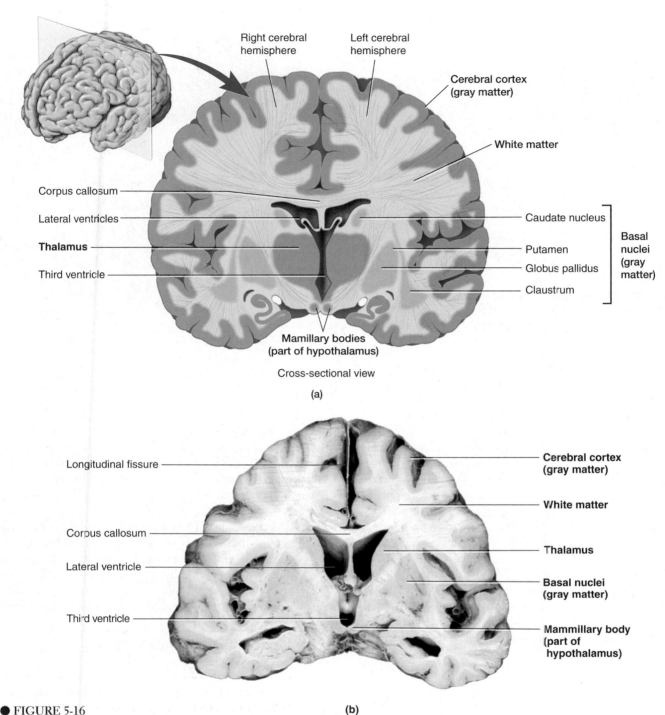

Right cerebral hemisphere

Left cerebral hemisphere

Cerebral cortex (gray matter)

White matter

Corpus callosum

Lateral ventricles

Thalamus

Third ventricle

Caudate nucleus

Putamen

Globus pallidus

Claustrum

Basal nuclei (gray matter)

Mamillary bodies (part of hypothalamus)

Cross-sectional view

(a)

Longitudinal fissure

Corpus callosum

Lateral ventricle

Third ventricle

Cerebral cortex (gray matter)

White matter

Thalamus

Basal nuclei (gray matter)

Mammillary body (part of hypothalamus)

(b)

● **FIGURE 5-16**

Frontal section of the brain

(a) Schematic frontal section of the brain. The cerebral cortex, an outer shell of gray matter, surrounds an inner core of white matter. Deep within the cerebral white matter are several masses of gray matter, the basal nuclei. The ventricles are cavities in the brain through which the cerebrospinal fluid flows. The thalamus forms the walls of the third ventricle. (b) Photograph of a frontal section of the brain of a cadaver.

(PHOTO: Mark Nielsen, Department of Biology, University of Utah)

inhibitory inputs to the neurons that innervate skeletal muscles), (2) selecting and maintaining purposeful motor activity while suppressing useless or unwanted patterns of movement, and (3) helping monitor and coordinate slow, sustained contractions, especially those related to posture and support. The basal nuclei do not directly influence the efferent motor neu-

rons that bring about muscle contraction but act instead by modifying ongoing activity in motor pathways.

To accomplish these complex integrative roles, the basal nuclei receive and send out much information, as is indicated by the tremendous number of fibers linking them to other regions of the brain. One important pathway consists of strategic

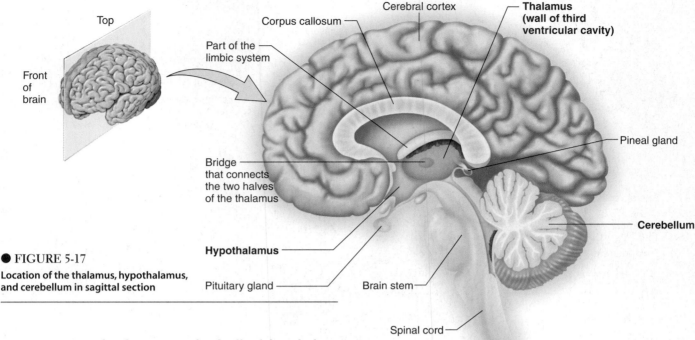

Top

Front
of
brain

Cerebral cortex

Corpus callosum

**Thalamus
(wall of third
ventricular cavity)**

Part of the
limbic system

Pineal gland

Bridge
that connects
the two halves
of the thalamus

Cerebellum

Hypothalamus

Pituitary gland

Brain stem

Spinal cord

● FIGURE 5-17
Location of the thalamus, hypothalamus,
and cerebellum in sagittal section

interconnections that form a complex feedback loop linking the cerebral cortex (especially its motor regions), the basal nuclei, and the thalamus. It is speculated that the thalamus positively reinforces voluntary motor behavior initiated by the cortex, whereas the basal nuclei modulate this activity by exerting an inhibitory effect on the thalamus to eliminate antagonistic or unnecessary movements. The basal nuclei also exert an inhibitory effect on motor activity by acting through neurons in the brain stem.

The importance of the basal nuclei in motor control is evident in diseases involving this region, the most common of which is **Parkinson's disease (PD).** This condition is associated with a deficiency of dopamine, an important neurotransmitter in the basal nuclei (see p. 126). As a result of the basal nuclei lacking sufficient dopamine to exert their normal roles, three types of motor disturbances characterize PD: (1) increased muscle tone, or rigidity; (2) involuntary, useless, or unwanted movements, such as *resting tremors* (for example, hands rhythmically shaking, making it difficult or impossible to hold a cup of coffee); and (3) slowness in initiating and carrying out different motor behaviors. For those who suffer from PD, it is difficult to stop ongoing activities. If sitting down, they tend to remain seated, and if they get up, they do so very slowly.

▮ The thalamus is a sensory relay station and is important in motor control.

Deep within the brain near the basal nuclei is the **diencephalon,** a midline structure that forms the walls of the third ventricular cavity, one of the spaces through which cerebrospinal fluid flows. The diencephalon consists of two main parts, the *thalamus* and the *hypothalamus* (see ▲ Table 5-3 and ● Figures 5-8b, 5-16, and 5-17).

The **thalamus** serves as a "relay station" and synaptic integrating center for preliminary processing of all sensory input on its way to the cortex. It screens out insignificant signals and

routes the important sensory impulses to appropriate areas of the somatosensory cortex, as well as to other regions of the brain. Along with the brain stem and cortical association areas, the thalamus is important in our ability to direct attention to stimuli of interest. For example, parents can sleep soundly through the noise of outdoor traffic but become instantly aware of their baby's slightest whimper. The thalamus is also capable of crude awareness of various types of sensation but cannot distinguish their location or intensity. Some degree of consciousness resides here as well. As described in the preceding section, the thalamus also plays an important role in motor control by positively reinforcing voluntary motor behavior initiated by the cortex.

▮ The hypothalamus regulates many homeostatic functions.

The **hypothalamus** is a collection of specific nuclei and associated fibers that lie beneath the thalamus. It is an integrating center for many important homeostatic functions and serves as an important link between the autonomic nervous system and the endocrine system. Specifically, the hypothalamus (1) controls body temperature; (2) controls thirst and urine output; (3) controls food intake; (4) controls anterior pituitary hormone secretion; (5) produces posterior pituitary hormones; (6) controls uterine contractions and milk ejection; (7) serves as a major autonomic nervous system coordinating center, which in turn affects all smooth muscle, cardiac muscle, and exocrine glands; and (8) plays a role in emotional and behavioral patterns.

The hypothalamus is the area of the brain most notably involved in the direct regulation of the internal environment. For example, when the body is cold, the hypothalamus initiates internal responses to increase heat production (such as shivering) and to decrease heat loss (such as constricting the skin blood vessels to reduce the flow of warm blood to the body surface, where heat could be lost to the external environment). Other areas of the brain, such as the cerebral cortex, act more indirectly to regulate the internal environment. For example, a person who feels cold is motivated to voluntarily put on warmer clothing, close the window, turn up the thermostat, and so on. Even these voluntary behavioral activities are strongly influenced by the hypothalamus, which, as a part of the limbic system, functions in conjunction with the cortex in controlling emotions and motivated behavior. We are now going to turn our attention to the limbic system and its functional relations with the higher cortex.

THE LIMBIC SYSTEM AND ITS FUNCTIONAL RELATIONS WITH THE HIGHER CORTEX

The **limbic system** is not a separate structure but refers to a ring of forebrain structures that surround the brain stem and are interconnected by intricate neuronal pathways (● Figure 5-18). It includes portions of each of the following: the lobes of the cerebral cortex (especially the limbic association cortex), the basal nuclei, the thalamus, and the hypothalamus. This complex interacting network is associated with emotions, basic survival and sociosexual behavioral patterns, motivation, and learning. Let's examine each of these brain functions in further detail.

▐ The limbic system plays a key role in emotion.

The concept of **emotion** encompasses subjective emotional feelings and moods (such as anger, fear, and happiness) plus the overt physical responses that occur in association with these feelings. These responses include specific behavioral patterns (for example, preparation for attack or defense when angered by an adversary) and observable emotional expressions (for example, laughing, crying, or blushing). Evidence points to a central role for the limbic system in all aspects of emotion. Stimulation of specific regions within the limbic system of humans during brain surgery produces various vague subjective sensations described by the patient as joy, satisfaction, or pleasure in one region and discouragement, fear, or anxiety in another. For example, the **amygdala** on the interior underside of the temporal lobe (● Figure 5-18) is an especially important region for processing inputs that give rise to the sensation of fear. In humans and to an undetermined extent in other species, higher levels of the cortex are additionally crucial for conscious awareness of emotional feelings.

▐ The limbic system and higher cortex participate in the control of basic behavioral patterns.

Basic behavioral patterns controlled at least in part by the limbic system include those aimed at survival of the individual

(attack, searching for food) and those directed toward perpetuation of the species (sociosexual behaviors conducive to mating). In experimental animals, stimulation of the limbic system brings about complex and even bizarre behaviors. For example, stimulation in one area can elicit responses of anger and rage in a normally docile animal, whereas stimulation in another area results in placidity and tameness, even in an otherwise vicious animal. Stimulation in yet another limbic area can induce sexual behaviors such as copulatory movements.

Role of the hypothalamus in basic behavioral patterns

The relationships among the hypothalamus, limbic system, and higher cortical regions regarding emotions and behavior are still not well understood. It appears that the extensive involvement of the hypothalamus in the limbic system is responsible for the involuntary internal responses of various body systems in preparation for appropriate action to accompany a particular emotional state. For example, the hypothalamus controls the increased heart rate and respiratory rate, elevation of blood pressure, and diversion of blood to skeletal muscles that occur in anticipation of attack or when angered. These preparatory changes in internal state require no conscious control.

Role of the higher cortex in basic behavioral patterns

In executing complex behavioral activities such as attack, flight, or mating, the individual (animal or human) must interact with the external environment. Higher cortical mechanisms are called into play to connect the limbic system and hypothalamus with the outer world so that appropriate overt behaviors are manifested. At the simplest level, the cortex provides the neural mechanisms necessary for implementing the appropriate skeletal muscle activity required to approach or avoid an

---

● FIGURE 5-18

Limbic system
This partially transparent view of the brain reveals the structures composing the limbic system.

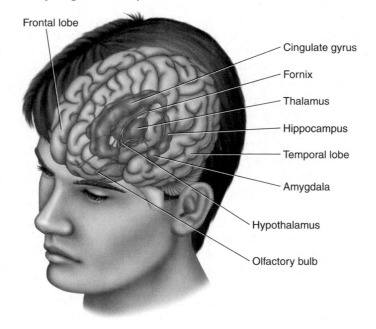

Frontal lobe

Cingulate gyrus

Fornix

Thalamus

Hippocampus

Temporal lobe

Amygdala

Hypothalamus

Olfactory bulb

adversary, participate in sexual activity, or display emotional expression. For example, the stereotypical sequence of movement for the universal human emotional expression of smiling is apparently preprogrammed in the cortex and can be called forth by the limbic system. One can also voluntarily call forth the smile program, as when posing for a picture. Even individuals blind from birth have normal facial expressions; that is, they do not learn to smile by observation. Smiling means the same thing in every culture, despite widely differing environmental experiences. Such behavior patterns shared by all members of a species are believed to be more abundant in lower animals.

Higher cortical levels are also capable of reinforcing, modifying, or suppressing basic behavioral responses so that actions can be guided by planning, strategy, and judgment based on an understanding of the situation. Even if you were angry at someone and your body was internally preparing for attack, you probably would judge that an attack would be inappropriate and could consciously suppress the external manifestation of this basic emotional behavior. Thus the higher levels of the cortex, particularly the prefrontal and limbic association areas, are important in conscious learned control of innate behavioral patterns. Using fear as an example, exposure to an aversive experience appears to call two parallel tracks into play for processing this emotional stimulus: a fast track in which the lower-level amygdala plays a key role and a slower track mediated primarily by the higher-level prefrontal cortex. The fast track permits a rapid, rather crude, instinctive response ("gut reaction") and is essential for the "feeling" of being afraid. The slower track involving the prefrontal cortex permits a more refined response to the aversive stimulus based on a rational analysis of the current situation compared to stored past experiences. The prefrontal cortex formulates plans and guides behavior, suppressing amygdala-induced responses that may be inappropriate for the situation at hand.

Reward and punishment centers

An individual tends to reinforce behaviors that have proved to be gratifying and suppress behaviors that have been associated with unpleasant experiences. Certain regions of the limbic system have been designated as "**reward**" and "**punishment**" **centers,** because stimulation in these respective areas gives rise to pleasant or unpleasant sensations. When a self-stimulating device is implanted in a reward center, an experimental animal will self-deliver up to 5,000 stimulations per hour and will even shun food when starving, in preference for the pleasure derived from self-stimulation. In contrast, when the device is implanted in a punishment center animals will avoid stimulation at all costs. Reward centers are found most abundantly in regions involved in mediating the highly motivated behavioral activities of eating, drinking, and sexual activity.

▌ Motivated behaviors are goal directed.

Motivation is the ability to direct behavior toward specific goals. Some goal-directed behaviors are aimed at satisfying specific identifiable physical needs related to homeostasis. **Homeostatic drives** represent the subjective urges associated with specific bodily needs that motivate appropriate behavior to satisfy those needs. As an example, the sensation of thirst accompanying a water deficit in the body drives an individual to drink to satisfy the homeostatic need for water. However, whether water, a soft drink, or another beverage is chosen as the thirst quencher is unrelated to homeostasis. Much human behavior is not dependent on purely homeostatic drives related to simple tissue deficits such as thirst. Human behavior is influenced by experience, learning, and habit, shaped in a complex framework of unique personal gratifications blended with cultural expectations. To what extent, if any, motivational drives unrelated to homeostasis, such as the drive to pursue a particular career or win a certain race, are involved with the reinforcing effects of the reward and punishment centers is unknown. Indeed, some individuals motivated toward a particular goal may even deliberately "punish" themselves in the short term to achieve their long-range gratification (for example, the temporary pain of training in preparation for winning a competitive athletic event).

▌ Norepinephrine, dopamine, and serotonin are neurotransmitters in pathways for emotions and behavior.

The underlying neurophysiological mechanisms responsible for the psychological observations of emotions and motivated behavior largely remain a mystery, although the neurotransmitters *norepinephrine, dopamine,* and *serotonin* all have been implicated. Norepinephrine and dopamine, both chemically classified as *catecholamines,* are known to be transmitters in the regions that elicit the highest rates of self-stimulation in animals equipped with do-it-yourself devices. A number of **psychoactive drugs** affect moods in humans, and some of these drugs have also been shown to influence self-stimulation in experimental animals. For example, increased self-stimulation is observed after the administration of drugs that increase catecholamine synaptic activity, such as *amphetamine,* an "upper" drug.

Psychoactive drug abuse and tolerance

Although most psychoactive drugs are used therapeutically to treat various mental disorders, others, unfortunately, are abused. Many of the abused drugs act by enhancing the effectiveness of dopamine in the "pleasure" pathways, thus initially giving rise to an intense sensation of pleasure. An example is *cocaine,* which blocks the reuptake of dopamine at synapses (see p. 126). Users of cocaine and other addictive drugs typically develop tolerance to the drug. The term **tolerance** refers to a *desensitization* to the drug so that the user needs greater quantities of the drug to achieve the same effect. With prolonged use of cocaine, the number of dopamine receptors in the brain is reduced in response to the glut of the abused substance. As a result of this desensitization, the user must steadily increase the dosage of the drug to get the same "high," or sensation of pleasure. Over the course of abuse, the user often no longer can derive pleasure from the drug but suffers unpleasant *withdrawal symptoms* once the effect of the drug has worn off. The user typically becomes **addicted** to the drug, compulsively seeking out and taking the drug at all costs, first to experience the pleasurable sensations and later to avoid the negative withdrawal symptoms, even when the drug no longer provides pleasure.

Examples of mental disorders associated with defects in limbic-system neurotransmitters

Schizophrenia—a mental disorder characterized by delusions, hallucinations, blunted or inappropriate emotions, pronounced apathy, and social withdrawal—results from excess dopamine transmission. An unanswered question has been the anatomic site of the abnormal dopamine transmission. Studies identifying the regions of the brain that extensively use dopamine as a neurotransmitter have narrowed down the possible sites. Among these are neural networks linked to the limbic system. Because the limbic system is extensively involved in emotions and in triggering behavior appropriate to environmental circumstances—functions that are abnormal in schizophrenic patients—it is believed that this is the defective site in schizophrenia.

By contrast, a functional deficiency of serotonin or norepinephrine or both is implicated in **depression,** a disorder characterized by a pervasive negative mood accompanied by a generalized loss of interests, an inability to experience pleasure, and suicidal tendencies. All effective antidepressant drugs increase the available concentration of these neurotransmitters in the CNS. *Prozac,* the most widely prescribed drug in American psychiatry, is illustrative. It blocks the reuptake of released serotonin, thus prolonging serotonin activity at synapses. The fact that serotonin and norepinephrine are synaptic messengers in the limbic regions of the brain involved in pleasure and motivation suggests that the pervasive sadness and lack of interest (no motivation) in depressed patients are related at least in part to disruption of these regions by deficiencies or decreased effectiveness of these neurotransmitters.

Researchers are optimistic that as our understanding of the molecular mechanisms of mental disorders is expanded in the future, many psychiatric problems can be corrected or better managed through drug intervention, a hope of great medical significance.

the same way again to the same stimulus as a consequence of this experience. Conversely, if a particular response is accompanied by punishment, the animal is less likely to repeat the same response to the same stimulus. When behavioral responses that give rise to pleasure are reinforced or those accompanied by punishment are avoided, learning has taken place. Housebreaking a puppy is an example. If the puppy is praised when it urinates outdoors but scolded when it wets the carpet, it will soon learn the acceptable place to empty its bladder. Thus learning is a change in behavior that occurs as a result of experiences. It is highly dependent on the organism's interaction with its environment. The only limits to the effects that environmental influences can have on learning are the biological constraints imposed by species-specific and individual genetic endowments.

▌ Memory is laid down in stages.

Memory is the storage of acquired knowledge for later recall. Learning and memory form the basis by which individuals adapt their behavior to their particular external circumstances. Without these mechanisms, planning for successful interactions and intentional avoidance of predictably disagreeable circumstances would be impossible.

The neural change responsible for retention or storage of knowledge is known as the **memory trace.** Generally, concepts, not verbatim information, are stored. As you read this page, you are storing the concept discussed, not the specific words. Later, when you retrieve the concept from memory, you will convert it into your own words. It is possible, however, to memorize bits of information word by word.

Storage of acquired information is accomplished in at least two stages: short-term memory and long-term memory (● Figure 5-19 and ▲ Table 5-4). **Short-term memory** lasts for seconds to hours, whereas **long-term memory** is retained for days to years. The process of transferring and fixing short-term memory traces into long-term memory stores is known as **consolidation.** Stored knowledge is of no use unless it can be retrieved and used to influence current or future behavior.

▌ Learning is the acquisition of knowledge as a result of experiences.

Learning is the acquisition of knowledge or skills as a consequence of experience, instruction, or both. It is widely believed that rewards and punishments are integral parts of many types of learning. If an animal is rewarded on responding in a particular way to a stimulus, the likelihood increases that the animal will respond in

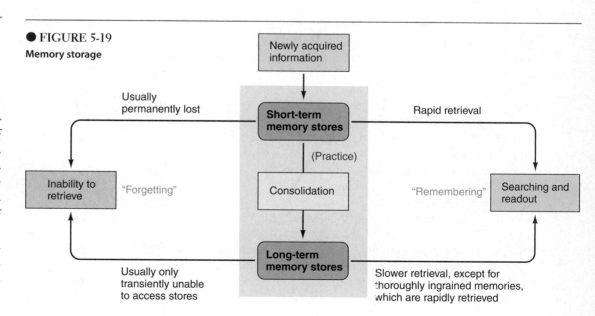

● **FIGURE 5-19**
Memory storage

▲ TABLE 5-4
Comparison of Short-Term and Long-Term Memory

Characteristic	Short-Term Memory	Long-Term Memory
Time of Storage after Acquisition of New Information	Immediate	Later; must be transferred from short-term to long-term memory through consolidation; enhanced by practice or recycling of information through short-term mode
Capacity of Storage	Limited	Very large
Retrieval Time (remembering)	Rapid retrieval	Slower retrieval, except for thoroughly ingrained memories, which are rapidly retrieved
Inability to Retrieve (forgetting)	Permanently forgotten; memory fades quickly unless consolidated into long-term memory	Usually only transiently unable to access; relatively stable memory trace
Mechanism of Storage	Involves transient modifications in functions of pre-existing synapses, such as altering amount of neurotransmitter released	Involves relatively permanent functional or structural changes between existing neurons, such as formation of new synapses; synthesis of new proteins plays a key role

A recently developed concept is that of a **working memory,** or what has been called "the erasable blackboard of the mind." Working memory temporarily holds and interrelates various pieces of information that are relevant to a current mental task. Through your working memory, you briefly hold and process data for immediate use—both newly acquired information and related, previously stored knowledge that is transiently brought forth into working memory—so that you can evaluate the incoming data in context. This integrative function is crucial to your ability to reason, plan, and make judgments. By comparing and manipulating new and old information within your working memory, you are able to comprehend what you are reading, to carry on a conversation, to calculate a restaurant tip in your head, to find your way home, and to know that you should put on warm clothing if you see snow outside. In short, working memory enables people to string thoughts together in a logical sequence and plan for future action.

Comparison of short-term and long-term memory

Newly acquired information is initially deposited in short-term memory, which has a limited capacity for storage. Information in short-term memory has one of two eventual fates. Either it is soon forgotten (for example, forgetting a telephone number after you have looked it up and finished dialing), or it is transferred into the more permanent long-term memory mode through *active practice* or *rehearsal*. The recycling of newly acquired information through short-term memory increases the likelihood that the information will be consolidated into long-term memory. (Therefore, when you cram for an exam, your long-term retention of the information is poor!) This relationship can be likened to developing photographic film. The originally developed image (short-term memory) will rapidly fade unless it is chemically fixed (consolidated) to provide a more enduring image (long-term memory). Sometimes only parts of memories are fixed, while others fade away. Information of in-

terest or importance to the individual is more likely to be recycled and fixed in long-term stores, whereas less important information is quickly erased.

The storage capacity of the long-term memory bank is much larger than the capacity of short-term memory. Different informational aspects of long-term memory traces seem to be processed and codified, then stored in conjunction with other memories of the same type; for example, visual memories are stored separately from auditory memories. This organization facilitates future searching of memory stores to retrieve desired information. For example, in remembering a woman you once met you may use various recall cues from different storage pools, such as her name, her appearance, the fragrance she wore, an incisive point she made, or the song playing in the background.

Because long-term memory stores are larger, it often takes longer to retrieve information from long-term memory than from short-term memory. *Remembering* is the process of retrieving specific information from memory stores; *forgetting* is the inability to retrieve stored information. Information lost from short-term memory is permanently forgotten, but information in long-term storage is frequently forgotten only transiently. Often you are only temporarily unable to access the information—for example, being unable to remember an acquaintance's name, then having it suddenly "come to you" later.

Some forms of long-term memory involving information or skills used on a daily basis are essentially never forgotten and are rapidly accessible, such as knowing your own name or being able to write. Even though long-term memories are relatively stable, stored information may be gradually lost or modified over time unless it is thoroughly ingrained as a result of years of practice.

Amnesia

Occasionally, individuals suffer from a lack of memory that involves whole portions of time rather than isolated bits of information. This condition, known as **amnesia,** occurs in two forms. The most common form, *retrograde* ("going backward") *amnesia,* is the inability to recall recent past events. It usually follows a traumatic event that interferes with electrical activity of the brain, such as a concussion or stroke. If a person is knocked unconscious, the content of short-term memory is essentially erased, resulting in loss of memory about activities that occurred within about the last half hour before the event. Severe trauma may interfere with access to recently acquired information in long-term stores as well.

Anterograde ("going forward") *amnesia,* conversely, is the inability to store memory in long-term storage for later retrieval. It is usually associated with lesions of the medial portions of the temporal lobes, which are generally considered critical regions for memory consolidation. Individuals suffering from this condition may be able to recall things they learned before the onset of their problem, but they are unable to establish new permanent memories. New information is lost as quickly as it fades from short-term memory. In one case study, the individual could not remember where the bathroom was in his new home but still had total recall of his old home.

▮ Memory traces are present in multiple regions of the brain.

What parts of the brain are responsible for memory? There is no single "memory center" in the brain. Instead, the neurons involved in memory traces are widely distributed throughout the subcortical and cortical regions of the brain. The regions of the brain most extensively implicated in memory include the hippocampus and associated structures of the medial (inner) temporal lobes, the limbic system, the cerebellum, the prefrontal cortex, and other regions of the cerebral cortex.

The hippocampus and declarative memories

The **hippocampus,** the elongated, seahorse–shaped medial portion of the temporal lobe that is part of the limbic system (see ● Figure 5-18), plays a vital role in short-term memory involving the integration of various related stimuli and is also crucial for consolidation into long-term memory (*hippocampus* means "horse-headed sea monster"). The hippocampus is believed to store new long-term memories only temporarily and then transfer them to other cortical sites for more permanent storage. The sites for long-term storage of various types of memories are only beginning to be identified by neuroscientists.

The hippocampus and surrounding regions play an especially important role in **declarative memories**—the "what" memories of specific people, places, objects, facts, and events that often result after only one experience and that can be declared in a statement such as "I saw the Statue of Liberty last summer" or conjured up in a mental image. Declarative memories require conscious recall. The hippocampus and associated structures are especially important in maintaining a durable record of the everyday episodic events in our lives. Individuals with hippocampal damage are profoundly forgetful of facts critical to daily functioning. Interestingly, extensive damage in the hippocampus region is evident in Alzheimer's patients during autopsy. (For an expanded discussion of Alzheimer's disease, see the accompanying boxed feature, ❱ Concepts, Challenges, and Controversies.)

The cerebellum and procedural memories

In contrast to the role of the hippocampus and surrounding regions of the temporal lobes and limbic system in declarative memories, the cerebellum and relevant cortical regions such as the primary motor cortex, somatosensory cortex, and visual processing areas play an essential role in the "how to" **procedural memories** involving motor skills gained through repetitive training, such as memorizing a particular dance routine. The cortical areas important for a given procedural memory are the specific motor or sensory systems engaged in performing the routine. In contrast to declarative memories, which are consciously recollected from previous experiences, procedural memories can be brought forth without conscious effort. For example, an ice skater during a competition typically performs best by "letting the body take over" the routine instead of thinking about exactly what needs to be done next. The distinct localization in different parts of the brain of these two types of memory is apparent in individuals who have temporal/limbic lesions. They are able to perform a skill, such as playing a piano, but the next day they have no recollection that they played it.

The prefrontal cortex and working memory

The major orchestrator of *working memory* is the prefrontal association cortex. The prefrontal cortex not only serves as a temporary storage site for holding relevant data online but also is largely responsible for the so-called executive functions involving manipulation and integration of the information for planning, problem-solving, and organizing activities. The prefrontal cortex carries out these complex reasoning functions in cooperation with all the brain's sensory regions, which are linked to the prefrontal cortex through neural connections. Researchers have identified different storage bins in the prefrontal cortex, depending on the nature of the current relevant data. For example, working memory involving spatial cues is in a prefrontal location distinct from working memory involving verbal cues or cues about an object's appearance. One recent fascinating proposal suggests that how intelligent a person is may be determined by the capacity of his or her working memory to temporarily hold and relate a variety of relevant data.

Another question besides the "where" of memory is the "how" of memory. Despite a vast amount of psychological data, only a few tantalizing scraps of physiologic evidence concerning the cellular basis of memory traces are available. Obviously, some change must take place within the neural circuitry of the brain to account for the altered behavior that follows learning. A single memory does not reside in a single neuron but rather in changes in the pattern of signals transmitted across synapses within a vast neuronal network. Different mechanisms are responsible for short-term and long-term memory. Short-term memory involves transient modifications in the function of pre-existing synapses, such as a temporary alter-

Alzheimer's Disease: A Tale of Beta Amyloid Plaques, Tau Tangles, and Dementia

"I can't remember where I put my keys. I must be getting Alzheimer's." The incidence and awareness of **Alzheimer's disease (AD),** which is characterized in its early stages by loss of recent memories, have become so commonplace that people sometimes jest about having it when they can't remember something. AD is no joking matter, however.

Incidence

About 4 million Americans currently have AD, but because it is an age-related condition and the population is aging, the incidence is expected to climb. About 0.1% of those between 60 and 65 years of age are afflicted with the disease, but the incidence rises to 30% to 47% among those over age 85. According to the *National Institute of Aging Report to Congress,* the most rapidly growing segment of our population by percentage is the over-85 age group.

Symptoms

AD accounts for about two-thirds of the cases of *senile dementia,* which is a generalized age-related diminution of mental abilities. In the earliest stages, only short-term memory is impaired, but as the disease progresses, even firmly entrenched long-term memories, such as recognition of family members, are lost. Confusion, disorientation, and personality changes characterized by irritability and emotional outbursts are common. Higher mental abilities gradually deteriorate as the patient inexorably loses the ability to read, write, and calculate. Language ability and speech are often impaired. In later stages, AD victims become childlike and are unable to feed, dress, and groom themselves. Patients usually die in a severely debilitated state 4 to 12 years after the onset of the disease.

First described nearly a century ago by Alois Alzheimer, a German neurologist, the condition can only be confirmed at autopsy on finding the characteristic brain lesions associated with the disease. Currently, AD is diagnosed prior to death by a process of elimination; that is, all other disorders that could produce dementia, such as a stroke or brain tumor, must be ruled out. A battery of cognitive tests is sometimes used to support a probable diagnosis of AD. Although scientists have been searching various avenues for a reliable test for diagnosing AD in patients, to date they have not been successful.

Characteristic Brain Lesions

The characteristic brain lesions, extracellular *neuritic (senile) plaques* and intracellular *neurofibrillary tangles,* are dispersed throughout the cerebral cortex and are especially abundant in the hippocampus. A **neuritic plaque** consists of a central core of extracellular, waxy, fibrous protein known as **beta amyloid (Aβ)** surrounded by degenerating dendritic and axonal nerve endings. **Neurofibrillary tangles** are dense bundles of abnormal, paired helical filaments that accumulate in the cell bodies of affected neurons. AD is also characterized by degeneration of cell bodies of certain neurons in the basal forebrain. The acetylcholine-releasing axons of these neurons normally terminate in the cerebral cortex and hippocampus, so the loss of these neurons results in a deficiency of acetylcholine in these areas. Neuron death and loss of synaptic communication are responsible for the ensuing dementia.

Underlying Pathology

Much progress has been made in understanding the pathology underlying the condition in recent years. **Amyloid precursor protein (APP)** is a structural component of neuronal plasma membranes. It is especially abundant in presynaptic terminal endings. APP can be cleaved at different locations to produce different products. Cleavage of APP at one site yields a secretory product (*sAPPα*) that is released from the presynaptic terminal. This secretory product is believed to exert a normal physiological role at the postsynaptic neuron, possibly playing a role in learning and memory processes. Cleavage of APP at an alternative site yields beta amyloid (*Aβ*). Depending on the exact site of cleavage, two different variants of Aβ are produced and released from the neuron. Normally about 90% of the Aβ is the 40-amino-acid-long version (*Aβ40*), a soluble and harmless form of this product. The other 10% is the dangerous, plaque-forming version that contains 42 amino acids (*Aβ42*). Aβ42 forms thin, insoluble filaments that readily aggregate into Aβ plaques. Furthermore, Aβ42 appears to be neurotoxic. The balance between these APP products can be shifted by mutations in APP, other genetic defects, age-related or pathologic changes in the brain, or perhaps environmental factors. The end result is reduced production of sAPPα and Aβ40 and increased production of plaque-producing Aβ42.

Aβ formation is seen early in the course of the disease, with neurofibrillary tangles developing somewhat later. Most evidence points to these abnormal Aβ deposits as being responsible for the many subsequent events that lead to neuronal death and ensuing dementia. AD does not "just happen" in old age. Instead, it probably results from a host of gradual, insidious processes that occur over the course of years or decades. Although not all the pieces of the puzzle have been identified, the following is a possible scenario based on the findings to date. The deposited Aβ has been shown to be directly toxic to neurons. Furthermore, the gradual buildup of Aβ plaques attracts microglia to the plaque sites. These immune cells of the brain launch an inflammatory attack on the plaque, releasing toxic chemicals that can damage surrounding "innocent bystander" neurons in the process.

These inflammatory assaults, along with the direct toxicity of the deposited Aβ, also appear to induce changes in the neuronal cytoskeleton that lead to the formation of the neurofibrillary tangles. Some researchers speculate that death of neurons could result from the development of nerve cell–clogging tangles. The protein **tau** normally associates with tubulin molecules in the formation of microtubules, which serve as axonal "highways" for transport of materials back and forth between the cell body and the axon terminal (see p. 46). Tau molecules act like "railroad ties" anchoring the "rails" of tubulin molecules within the microtubule. If tau molecules become hyperphosphorylated (have too many phosphate groups attached), they cannot interact with tubulin. Research suggests that Aβ binds to Ach receptors on the surface of nerve cells, triggering a chain of intracellular events that leads to tau hyperphosphorylation. When not bound to tubulin, the incapacitated tau molecules intertwine, forming paired helical filaments that aggregate to form neurofibrillary tangles. More importantly, just as train tracks start to fall apart if too many ties are missing, the microtubules start to break down as increasing numbers of tau molecules can no longer do their job. The resultant loss of the neuron's transport system can lead to the death of the cell. Tangle

---

ation in the amount of neurotransmitter released in response to stimulation within affected nerve pathways. Long-term memory, in contrast, involves relatively permanent functional or structural changes between existing neurons in the brain. Let's look at each of these types of memory in more detail.

▊ Short-term memory involves transient changes in synaptic activity.

Ingenious experiments in the sea snail, *Aplysia,* have shown that two forms of short-term memory—habituation and sensi-

formation correlates more closely than amyloid plaque formation does with symptoms of dementia, probably because the tangles are a later event in the development of the disease.

Other factors also play a role in the complex story of AD, but exactly where they fit in is unclear. According to a leading proposal, Aβ causes excessive influx of Ca^{2+}, which triggers a chain of biochemical events that kills the cells. Brain cells that have an abundance of glutamate NMDA receptors, most notably the hippocampal cells involved in long-term potentiation (see p. 165) are especially vulnerable to glutamate toxicity. Loss of hippocampal memory-forming capacity is a hallmark feature of AD. Some evidence suggests that the injured neurons commit apoptosis (cell suicide; see p. 70), but the self-induced death seems to be a much slower process than typical apoptosis. Findings indicate that Aβ can activate the self-execution biochemical pathway. Other studies suggest that cell-damaging free radicals (see p. 126) are produced during the course of the disease. All these neuron-destroying pathways eventually lead to neuronal loss and the gradual development of symptoms.

Possible Causes

The underlying cause that triggers abnormal Aβ formation in AD is unknown in the majority of cases. Many investigators believe the condition has many underlying causes. Both genetic and environmental factors have been implicated in an increased risk of acquiring AD. About 15% of the cases are linked to specific, known genetic defects that run in families and cause early onset of the disease. Individuals with this *familial Alzheimer's disease* typically develop clinical symptoms in their 40s or 50s. Two different types of genetic defects have been implicated in familial AD: mutations in the gene that codes for APP and those that code for presenilins. Interestingly, **presenilins** have recently been shown to be γ secretase, one of the APP-cleaving enzymes. These hereditary mutations all influence APP processing, with the result being increased production of Aβ42. Elevated levels of this dangerous form of Aβ lead to accelerated plaque development.

The other 85% of AD patients do not begin to manifest symptoms until later in life, somewhere between 65 to 85 years of age. Two specific gene traits have been identified that increase

an individual's vulnerability of acquiring *late-onset Alzheimer's disease*: the *apoE gene* and the *A2M gene*. Scientists have discovered a strong link between developing AD and having a particular version of a gene that codes for *apolipoprotein E*. This protein, which is normally found in the blood, has several forms. The version of the gene associated with increased AD risk codes for the **apolipoprotein E-4 (apoE-4)** form of the protein. An individual may have no, one, or two copies of the gene that codes for the apoE-4 form of the protein. About 15% of the U.S. population has one apoE-4 gene and 1% has two. Those with one of these genes have a five times greater risk of developing AD, and those who inherit two of these genes are 17 times more likely to develop the condition as people with two genes for apoE-3, the most common form of the protein. Researchers suspect that apoE-4 may predispose an individual to Aβ plaque deposition. Interestingly, apoE-4 binds to Aβ deposited in the neuritic plaques. For reasons that are still unclear, people with apoE-4 tend to have larger plaques than those with other forms of apoE. Furthermore, studies suggest that apoE-3 helps prevent phosphates from attaching to tau, whereas apoE-4 fails to keep phosphates away. Thus the rate of both microtubule failure and the formation of neurofibrillary tangles could increase as the number of copies of apoE-4 increases. Some think that apoE-4 does not cause AD but instead hastens its onset, which is brought on by other factors.

Researchers recently identified another gene that increases an individual's risk of developing late-onset AD—a variant of the gene encoding **A2M protein.** The more common form of A2M protein binds to Aβ and helps clear it out of the space between neurons. The variant form of A2M is unable to remove Aβ as efficiently, leading to Aβ accumulation and deposition. Increasingly, investigators suspect that improper clearance of Aβ may play just an important role as Aβ production in plaque development. Even a slight imbalance between production and removal of Aβ due to a decrease in the rate of degradation of this product could lead to Aβ buildup. An estimated 30% of the population carries the A2M variant gene.

Not everyone with genetic tendencies for AD develops the disease. Furthermore, many develop the illness with no apparent genetic

predisposition. Obviously, some other factors in addition to genetic vulnerability must be at play in producing the condition. Hormonal imbalances may play a role. In particular, research findings suggest that *cortisol*, the stress hormone, increases the propensity to developing the condition, whereas *estrogen*, the female sex hormone, seems to protect against the onslaught of the disease.

In addition, investigators have been searching for possible environmental triggers, but none have been found to date. Although aluminum from pots and pans and aerosol antiperspirants was implicated for a while, it is no longer believed to be a factor in the development of the disease.

Treatment

The only drugs specifically approved for treatment of AD raise the levels of acetylcholine (the deficient neurotransmitter) in the brain. For example, *tacrine* (cognex) inhibits the enzyme that normally clears released acetylcholine from the synapse. Even though these drugs transiently improve the symptoms in some patients, they do nothing to halt or slow down the relentless destruction of neurons and further deterioration of the patient's condition. Several other agents are also used to treat AD. Antioxidants hold some promise of thwarting free-radical damage. Aspirin and other anti-inflammatory drugs may slow the course of AD by blocking the inflammatory components of the condition.

As researchers continue to unravel the underlying factors, the likelihood of finding a means to block the gradual, relentless progression of the disease increases. For example, investigators are cautiously optimistic about a newly developed AD vaccine targeted against Aβ plaque. Also, the search is on for new drugs that might block the cleavage of Aβ42 from APP or might inhibit the aggregation of this amyloid into dangerous plaques, thus halting AD in its tracks in the earliest stages.

Prevention or treatment of AD cannot come too soon in view of the tragic toll the condition is taking on its victims, their families, and society. The cost of custodial care for AD patients is currently estimated at $100 billion annually and will continue to rise as a greater percentage of our population ages and becomes afflicted with the condition.

tization—are due to modification of different channel proteins in presynaptic terminals of specific afferent neurons involved in the pathway that mediates the behavior being modified. This modification in turn brings about changes in transmitter release. **Habituation** is a decreased responsiveness to repetitive

presentations of an indifferent stimulus—that is, one that is neither rewarding nor punishing. **Sensitization** is increased responsiveness to mild stimuli following a strong or noxious stimulus. *Aplysia* reflexly withdraws its gill when its siphon, a breathing organ at the top of its gill, is touched. Afferent neurons responding to touch of the siphon (presynaptic neurons) directly synapse on efferent motor neurons (postsynaptic neurons) controlling gill withdrawal. The snail becomes habituated when its siphon is repeatedly touched; that is, it learns to ignore the stimulus and no longer withdraws its gill in response. Sensitization, a more complex form of learning, takes place in *Aplysia* when it is given a hard bang on the siphon. Subsequently, the snail withdraws its gill more vigorously in response to even mild touch. Interestingly, these different forms of learning affect the same site—the synapse between a siphon afferent and a gill efferent—in opposite ways. Habituation depresses this synaptic activity, whereas sensitization enhances it (● Figure 5-20). These transient modifications persist for as long as the time course of the memory.

Mechanism of habituation

In habituation, closing of Ca^{2+} channels reduces Ca^{2+} entry into the presynaptic terminal, which leads to a decrease in neurotransmitter release. As a consequence, the postsynaptic potential is reduced compared to normal, resulting in a decrease or absence of the behavioral response controlled by the postsynaptic efferent neuron (gill withdrawal). Thus the memory for habituation in *Aplysia* is stored in the form of modification of specific Ca^{2+} channels. With no further training, this reduced responsiveness lasts for several hours. A similar process is responsible for short-term habituation in other invertebrates studied. This suggests that Ca^{2+} channel modification might be a general mechanism of habituation, although in higher species the involvement of intervening interneurons makes the process somewhat more complicated. Habituation is probably the most common form of learning and is believed to be the first learning process to take place in human infants. By learning to ignore indifferent stimuli, the animal or person is free to attend to other more important stimuli.

Mechanism of sensitization

Sensitization in *Aplysia* has also been shown to involve channel modification, but a different channel and mechanism are involved. In contrast to habituation, Ca^{2+} entry into the presynaptic terminal is enhanced in sensitization. The subsequent increase in neurotransmitter release produces a larger postsynaptic potential, resulting in a more vigorous gill withdrawal response. Sensitization does not have a direct effect on the presynaptic Ca^{2+} channels. Instead, it indirectly enhances Ca^{2+} entry via presynaptic facilitation (see p. 124). The neurotransmitter serotonin is released from a facilitating interneuron that synapses on the presynaptic terminal to bring about increased release of presynaptic neurotransmitter in response to an action potential. It does so by triggering activation of a cyclic AMP second messenger (see p. 67) within the presynaptic terminal, which ultimately brings about blocking of K^+ channels. This blockage prolongs the action potential in the presynaptic terminal. Remember that K^+ efflux through opened K^+ channels hastens the return to resting potential (repolarization) during the falling phase of the action potential. Because the presence of a local action potential is responsible for opening of Ca^{2+} channels in the terminal, a prolonged ac-

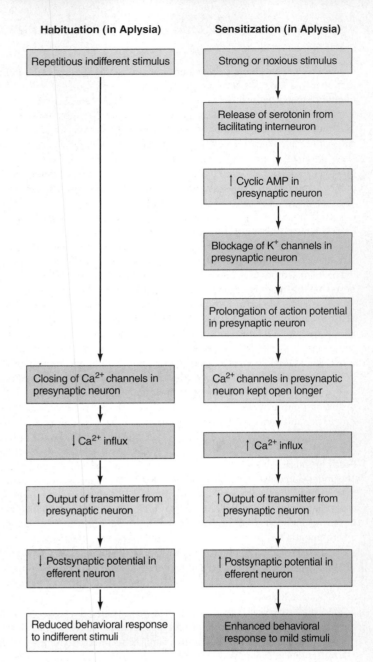

● **FIGURE 5-20**

Habituation and sensitization in *Aplysia*
Researchers have shown that in the sea snail *Aplysia* two forms of short-term memory—habituation and sensitization—result from opposite changes in neurotransmitter release from the same presynaptic neuron, caused by different transient channel modifications.

tion potential permits the greater Ca^{2+} influx associated with sensitization.

Thus, existing synaptic pathways may be functionally interrupted (habituated) or enhanced (sensitized) during simple learning. It is speculated that much of short-term memory is similarly a temporary modification of already existing processes. Several lines of research suggest that the cyclic AMP cascade, especially activation of protein kinase, plays an important role, at least in elementary forms of learning and memory.

Further studies have revealed that declarative memories, which involve conscious awareness and are more complex than habituation and sensitization, are initially stored by means of more persistent changes in activity of existing synapses. Specifically, initial storage of declarative information appears to be accomplished by means of long-term potentiation, to which we now turn our attention.

Mechanism of long-term potentiation.

The term **long-term potentiation (LTP)** refers to prolonged increase in the strength of existing synaptic connections in activated pathways following brief periods of repetitive stimulation. LTP has been shown to last for days or even weeks—long enough for this short-term memory to be consolidated into more permanent long-term memory. LTP is especially prevalent in the hippocampus, a site critical for converting short-term memories into long-term memories. When LTP occurs, simultaneous activation of both the presynaptic and postsynaptic neurons across a given excitatory synapse results in long-lasting modifications that enhance the ability of the presynaptic neuron to excite the postsynaptic neuron. Because both the presynaptic and postsynaptic neurons must be active at the same time, the development of LTP is restricted to the pathway that is stimulated. Pathways between other inactive presynaptic inputs and the same postsynaptic cell are not affected. With LTP, signals from a given presynaptic neuron to a postsynaptic neuron become stronger with repeated use. Keep in mind that strengthening of synaptic activity results in the formation of more EPSPs in the postsynaptic neuron in response to chemical signals from this particular excitatory presynaptic input. This increased excitatory responsiveness is ultimately translated into more action potentials being sent along this postsynaptic cell to other neurons.

Enhanced synaptic transmission could theoretically result from either changes in the postsynaptic neuron (such as increased responsiveness to the neurotransmitter) or in the presynaptic neuron (such as increased release of neurotransmitter). The underlying mechanisms for LTP are still the subject of much research and debate. Most likely, multiple mechanisms are involved in this complex phenomenon. Based on current scientific evidence, here are two plausible proposals, one involving a postsynaptic change and the other a presynaptic modification (● Figure 5-21). The first proposed mechanism involves increased responsiveness of the postsynaptic cell. In this proposal, the presynaptic neuron releases glutamate, an excitatory neurotransmitter, that binds with **NMDA receptors,** one of several types of glutamate receptors in the postsynaptic neuron's plasma membrane. Because these receptors act as Ca^{2+} channels, their activation and opening leads to Ca^{2+} entry into the postsynaptic cell. Activation of NMDA receptors does not lead to EPSP formation.

Calcium activates a Ca^{2+}-dependent second-messenger pathway in the postsynaptic neuron resulting in phosphorylation and thus increased availability of yet another type of glutamate receptor, the **AMPA receptor.** The AMPA receptor is the one that results in the formation of EPSPs in response to

● FIGURE 5-21

Possible pathways for long-term potentiation

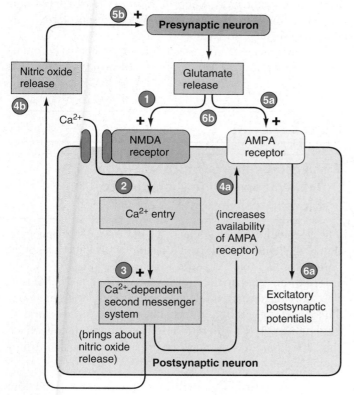

1. Glutamate released from an activated presynaptic neuron binds with NMDA receptors on the postsynaptic neuron, resulting in opening of receptor-related Ca^{2+} channels.

2. Calcium enters through these opened channels.

3. Calcium entry activates a second-messenger system that can initiate two separate intracellular pathways that enhance the effectiveness of this synapse.

4a. In the first pathway, the Ca^{2+}-dependent second-messenger system increases the availability of AMPA receptors.

4b. In an alternate pathway, the Ca^{2+}-dependent second-messenger system causes the postsynaptic neuron to release nitric oxide.

5a. These receptors also bind with glutamate.

5b. This retrograde messenger enhances the release of more glutamate from the presynaptic neuron.

6a. This binding results in the formation of EPSPs. Thus this pathway strengthens the synapse by increasing postsynaptic excitatory responsiveness.

6b. This positive feedback strengthens the signaling process at this synapse.

glutamate activation. Because of this array of new AMPA receptors, the postsynaptic cell exhibits a greater response to subsequent release of glutamate from the presynaptic cell. This heightened sensitivity of the postsynaptic neuron to glutamate from the presynaptic cell helps maintain LTP.

Another proposal involves increased neurotransmitter release from the presynaptic cell as a major contributor to LTP. In this proposed mechanism, activation of the Ca^{2+}-dependent second-messenger pathway in the postsynaptic neuron causes this postsynaptic cell to release a retrograde ("going backward") factor that diffuses to the presynaptic neuron. Here, the retrograde factor activates a second-messenger system in the presynaptic neuron, ultimately enhancing the release of more neurotransmitter from the presynaptic neuron. This positive feedback strengthens the signaling process at this synapse, helping sustain LTP. Note that in this mechanism for the development of LTP, a chemical factor from the postsynaptic neuron influences the presynaptic neuron, just the opposite direction of neurotransmitter activity at a synapse. Synapses traditionally operate unidirectionally, with the presynaptic neuron releasing a neurotransmitter that influences the postsynaptic neuron. The cell body and dendrites of the postsynaptic neuron do not contain any transmitter vesicles. Thus the retrograde factor is distinct from classical neurotransmitters or neuropeptides. Most investigators believe that the retrograde messenger is **nitric oxide,** a chemical that recently has been found to exert a bewildering array of other functions in the body. These other functions range from dilation of blood vessels in the penis during erection to destruction of foreign invaders by the immune system (see p. 356).

In addition to the Ca^{2+}-dependent second-messenger pathway, studies suggest a regulatory role for the cAMP second-messenger pathway in the development and maintenance of LTP. Participation of cAMP may hold a key to linking short-term memory to long-term memory consolidation.

▌ Long-term memory involves formation of new, permanent synaptic connections.

Whereas short-term memory involves transient strengthening of pre-existing synapses, long-term memory storage requires the activation of specific genes that control protein synthesis. These proteins, in turn, are needed for the formation of new synaptic connections. Thus, long-term memory storage involves rather permanent physical changes in the brain.

Studies comparing the brains of experimental animals reared in a sensory-deprived environment with those raised in a sensory-rich environment demonstrate readily observable microscopic differences. The animals afforded more environmental interactions—and supposedly, therefore, more opportunity to learn—displayed greater branching and elongation of dendrites in nerve cells in regions of the brain suspected to be involved with memory storage. Greater dendritic surface area presumably provides more sites for synapses. Thus, long-term memory may be stored at least in part by a particular pattern of dendritic branching and synaptic contacts.

A positive regulatory protein, **CREB,** is the molecular switch that activates (turns on) the genes important in long-term memory storage. Another related molecule, **CREB2,** has recently been identified. CREB2 is a repressor of CREB-facilitated protein synthesis. Formation of enduring memories involves not only the activation of positive regulatory factors (CREB) that favor memory storage but also the turning off of inhibitory constraining factors (CREB2) that prevent memory storage. The shifting balance between positive and repressive factors is believed to ensure that only information relevant to the individual, not everything encountered, is put into long-term storage.

How is CREB activated (and CREB2 inhibited)? No one knows for sure, but some researchers suggest the interesting possibility that the conversion from short-term to long-term memory may involve the turning on of CREB by cAMP. This second messenger plays a regulatory role in LTP as well as in simpler forms of short-term memory such as sensitization. Furthermore, cAMP can activate CREB, which ultimately leads to new protein synthesis and the consolidation of long-term memory. Consolidation of both declarative and procedural memories depends on CREB (and presumably CREB2). CREB serves as a common molecular switch for converting both these categories of memory from the transient short-term to the permanent long-term mode.

The CREB regulatory proteins regulate a group of genes, the **immediate early genes (IEGs),** which play a critical role in memory consolidation. These genes govern the synthesis of the proteins that encode long-term memory. The exact role that these critical new long-term memory proteins might play remains speculative. They may be needed for structural changes in dendrites or used for synthesis of more neurotransmitters or additional receptor sites. Alternatively, they may accomplish long-term modification of neurotransmitter release by sustaining biochemical events initially activated by short-term memory processes.

To complicate the issue even further, numerous hormones and neuropeptides are known to affect learning and memory processes.

CEREBELLUM

▌ The cerebellum is important in balance and in planning and executing of voluntary movement.

The cerebellum, which is attached to the back of the upper portion of the brain stem, lies underneath the occipital lobe of the cortex (see ▲ Table 5-3 and ● Figures 5-8b and 5-17). It consists of three functionally distinct parts, which are believed to have developed successively during evolution (● Figure 5-22). These parts have different sets of inputs and outputs, and accordingly, each has different functions, although collectively they are concerned primarily with subconscious control of motor activity.

1. The **vestibulocerebellum** is important for the maintenance of balance and controls eye movement.

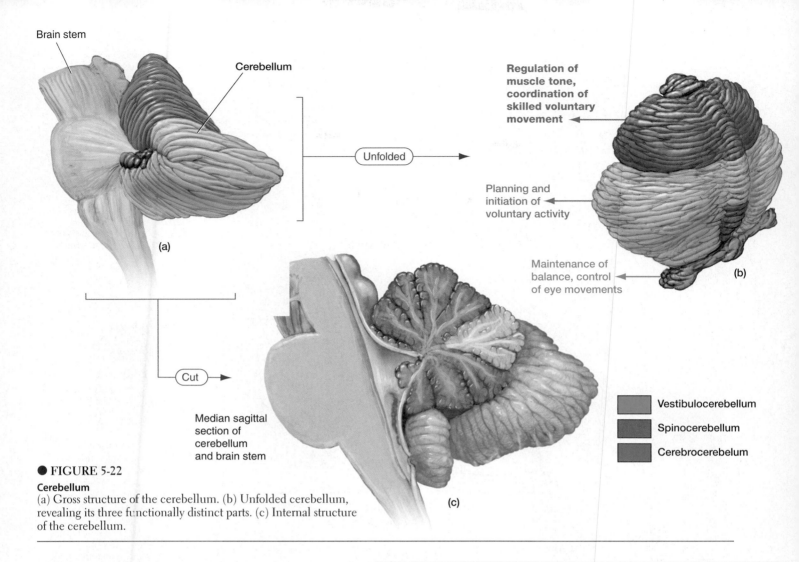

● FIGURE 5-22

Cerebellum
(a) Gross structure of the cerebellum. (b) Unfolded cerebellum, revealing its three functionally distinct parts. (c) Internal structure of the cerebellum.

Labels in figure (a): Brain stem; Cerebellum; (a); Unfolded; Cut; Median sagittal section of cerebellum and brain stem; (c)

Labels in figure (b): Regulation of muscle tone, coordination of skilled voluntary movement; Planning and initiation of voluntary activity; Maintenance of balance, control of eye movements; (b)

Legend: Vestibulocerebellum; Spinocerebellum; Cerebrocerebellum

2. The **spinocerebellum** enhances muscle tone and coordinates skilled, voluntary movements. This brain region is especially important in ensuring the accurate timing of various muscle contractions to coordinate movements involving multiple joints. For example, the movements of your shoulder, elbow, and wrist joints must be synchronized even during the simple act of reaching for a pencil. When cortical motor areas send messages to muscles for the execution of a particular movement, the spinocerebellum is informed of the intended motor command. This region also receives input from peripheral receptors that apprise it of the body movements and positions that are actually taking place. The spinocerebellum essentially acts as "middle management," comparing the "intentions" or "orders" of the higher centers with the "performance" of the muscles and then correcting any "errors" or deviations from the intended movement (● Figure 5-23). The spinocerebellum even appears to be able to predict the position of a body part in the next fraction of a second during a complex movement and to make adjustments accordingly. If you are reaching for a pencil, for example, this region "puts on the brakes" soon enough to stop the forward movement of your hand at the intended location rather than allowing you to overshoot your target. These ongoing adjustments,

● FIGURE 5-23

Role of the spinocerebellum in subconscious control of voluntary motor activity
In coordinating rapid, phasic motor activity, the spinocerebellum compares the "intentions" of higher motor centers with the "performance" of the muscles and corrects any "errors" by making the necessary adjustments to accomplish the intended movement.

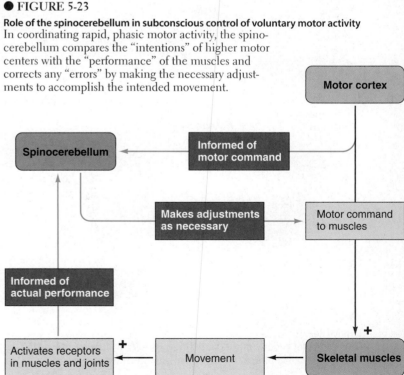

Flowchart labels: Motor cortex; Spinocerebellum; Informed of motor command; Makes adjustments as necessary; Motor command to muscles; Informed of actual performance; Activates receptors in muscles and joints; Movement; Skeletal muscles

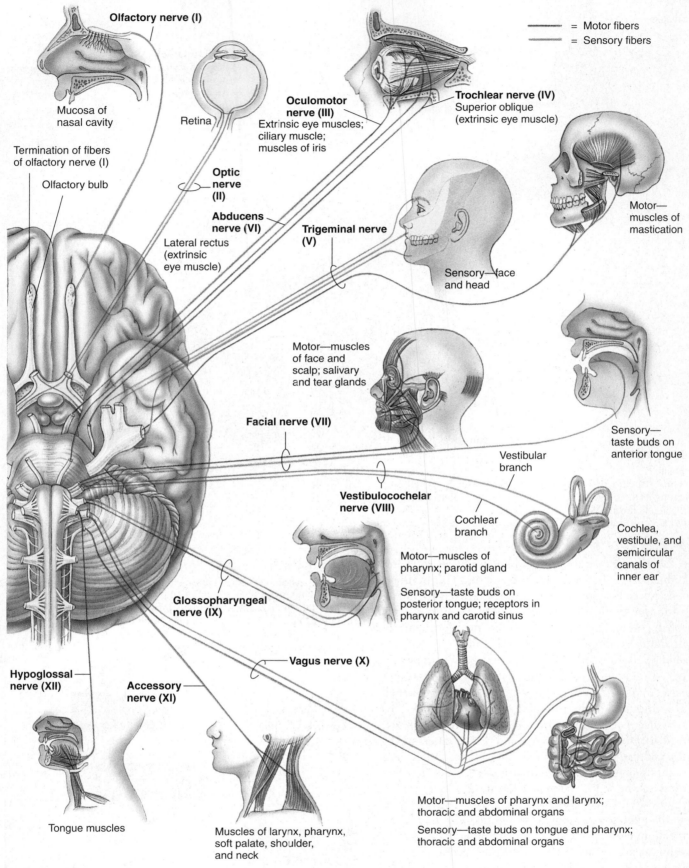

= Motor fibers
= Sensory fibers

Olfactory nerve (I)

Mucosa of nasal cavity

Retina

Oculomotor nerve (III)
Extrinsic eye muscles; ciliary muscle; muscles of iris

Trochlear nerve (IV)
Superior oblique (extrinsic eye muscle)

Termination of fibers of olfactory nerve (I)

Olfactory bulb

Optic nerve (II)

Abducens nerve (VI)

Lateral rectus (extrinsic eye muscle)

Trigeminal nerve (V)

Motor— muscles of mastication

Sensory—face and head

Motor—muscles of face and scalp; salivary and tear glands

Facial nerve (VII)

Sensory— taste buds on anterior tongue

Vestibular branch

Vestibulocochelar nerve (VIII)

Cochlear branch

Cochlea, vestibule, and semicircular canals of inner ear

Glossopharyngeal nerve (IX)

Motor—muscles of pharynx; parotid gland

Sensory—taste buds on posterior tongue; receptors in pharynx and carotid sinus

Hypoglossal nerve (XII)

Accessory nerve (XI)

Vagus nerve (X)

Tongue muscles

Muscles of larynx, pharynx, soft palate, shoulder, and neck

Motor—muscles of pharynx and larynx; thoracic and abdominal organs

Sensory—taste buds on tongue and pharynx; thoracic and abdominal organs

● **FIGURE 5-24**

Cranial nerves
Inferior (underside) view of the brain, showing the attachments of the 12 pairs of cranial nerves to the brain and many of the structures innervated by those nerves.

which ensure smooth, precise, directed movement, are especially important for rapidly changing (phasic) activities like typing, playing the piano, or running.

3. The **cerebrocerebellum** plays a role in the planning and initiation of voluntary activity by providing input to the cortical motor areas. This is also the cerebellar region involved in procedural memories.

All the following range of symptoms that characterize cerebellar disease are referable to a loss of these functions: poor balance, nystagmus (rhythmic, oscillating eye movements), reduced muscle tone but no paralysis, inability to perform rapid movements smoothly, and inability to stop and start skeletal muscle action quickly. The latter gives rise to an *intention tremor* characterized by oscillating to-and-fro movements of a limb as it approaches its intended destination. As a person with cerebellar damage attempts to pick up a pencil, he or she may overshoot the pencil and then rebound excessively, repeating this to-and-fro process until success is finally achieved. No tremor is observed except in the performance of intentional activity, in contrast to the resting tremor associated with disease of the basal nuclei.

The cerebellum and basal nuclei both monitor and adjust motor activity commanded from the motor cortex, and like the basal nuclei, the cerebellum does not have any direct influence on the efferent motor neurons. Both function indirectly by modifying the output of major motor systems of the brain. Even though the cerebellum and basal nuclei both help to subconsciously coordinate voluntary motor activity, they play different roles. The cerebellum helps maintain balance; helps coordinate fast, phasic motor activity; and enhances muscle tone. The basal nuclei are important in coordinating slow, sustained movement related to posture and support, and they also function in inhibiting muscle tone.

The motor command for a particular voluntary activity arises from the motor cortex, but coordination of the actual execution of that activity is accomplished subconsciously by these subcortical regions. To illustrate, you can voluntarily decide that you want to walk, but you do not have to consciously think about the specific sequence of movements that will have to be performed to accomplish this intentional act. Accordingly, much of voluntary activity is actually involuntarily regulated.

You will learn more about motor control when we cover skeletal muscle physiology in Chapter 8. For now, we are going to move on to the remaining part of the brain, the brain stem.

BRAIN STEM

The **brain stem** consists of the **medulla, pons,** and **midbrain** (see ▲ Table 5-3 and ● Figure 5-8b).

▌ The brain stem is a vital link between the spinal cord and higher brain regions.

All incoming and outgoing fibers traversing between the periphery and higher brain centers must pass through the brain stem, with incoming fibers relaying sensory information to the brain and outgoing fibers carrying command signals from the

brain for efferent output. A few fibers merely pass through, but most synapse within the brain stem for important processing. Thus the brain stem is a critical connecting link between the remainder of the brain and the spinal cord.

The functions of the brain stem include the following:

1. The majority of the twelve pairs of **cranial nerves** arise from the brain stem (● Figure 5-24). With one major exception, these nerves supply structures in the head and neck with both sensory and motor fibers. They are important in sight, hearing, taste, smell, sensation of the face and scalp, eye movement, chewing, swallowing, facial expressions, and salivation. The major exception is cranial nerve X, the **vagus nerve.** Instead of innervating regions in the head, most of the branches of the vagus nerve supply organs in the thoracic and abdominal cavities. The vagus is the major nerve of the parasympathetic nervous system.

2. Collected within the brain stem are neuronal clusters, or "centers," that control heart and blood vessel function, respiration, and many digestive activities.

3. This region also plays a role in modulating the sense of pain.

4. The brain stem plays a role in the regulation of muscle reflexes involved in equilibrium and posture.

5. A widespread network of interconnected neurons called the **reticular formation** runs throughout the entire brain stem and into the thalamus. This network receives and integrates all incoming sensory synaptic input. Ascending fibers originating in the reticular formation carry signals upward to arouse and activate the cerebral cortex (● Figure 5-25). These fibers com-

---

● **FIGURE 5-25**

The reticular activating system
The reticular formation, a widespread network of neurons within the brain stem (in red), receives and integrates all synaptic input. The reticular activating system, which promotes cortical alertness and helps direct attention toward specific events, consists of ascending fibers (in blue) that originate in the reticular formation and carry signals upward to arouse and activate the cerebral cortex.

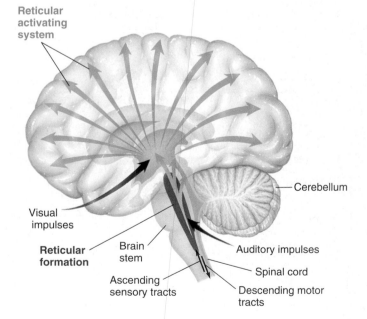

pose the **reticular activating system (RAS)**, which controls the overall degree of cortical alertness and is important in the ability to direct attention. In turn, fibers descending from the cortex, especially its motor areas, can activate the RAS.

6. The centers responsible for sleep traditionally have been considered to be housed within the brain stem, although recent evidence suggests that the sleep-promoting centers may be located in the hypothalamus.

We are now going to examine sleep and the other states of consciousness in further detail.

■ Sleep is an active process consisting of alternating periods of slow-wave and paradoxical sleep.

The term **consciousness** refers to subjective awareness of the external world and self, including awareness of the private inner world of one's own mind—that is, awareness of thoughts, perceptions, dreams, and so on. Even though the final level of awareness resides in the cerebral cortex and a crude sense of awareness is detected by the thalamus, conscious experience depends on the integrated functioning of many parts of the nervous system.

The following states of consciousness are listed in decreasing order of level of arousal, based on the extent of interaction between peripheral stimuli and the brain:

- maximum alertness
- wakefulness
- sleep (several different types)
- coma

Maximum alertness depends on attention-getting sensory input that "energizes" the RAS and subsequently the level of activity of the CNS as a whole. At the other extreme, coma refers to the total unresponsiveness of a living person to external stimuli, caused either by brain stem damage that interferes with the RAS or by widespread depression of the cerebral cortex, such as that accompanying O_2 deprivation.

The **sleep–wake** cycle is a normal cyclical variation in awareness of surroundings. In contrast to being awake, sleeping individuals are not consciously aware of the external world, but they do have inward conscious experiences such as dreams. Furthermore, they can be aroused by external stimuli, such as an alarm going off.

Sleep is an *active* process, not just the absence of wakefulness. The brain's overall level of activity is not reduced during sleep. During certain stages of sleep, O_2 uptake by the brain is even increased above normal waking levels.

There are two types of sleep, characterized by different EEG patterns and different behaviors: *slow-wave sleep* and *paradoxical*, or *REM*, sleep (▲ Table 5-5).

EEG patterns during sleep

Slow-wave sleep occurs in four stages, each displaying progressively slower EEG waves of higher amplitude (hence, "slow-wave" sleep) (● Figure 5-26). At the onset of sleep, an individual moves from the light sleep of stage 1 to the deep sleep of stage 4 of slow-wave sleep during a period of 30 to 45 minutes, then reverses through the same stages in the same amount of time. A 10- to 15-minute episode of **paradoxical sleep** punctuates the end of each slow-wave sleep cycle. Paradoxically, the EEG pattern during this time abruptly becomes similar to that of a wide-awake, alert individual, even though the person is still asleep (hence, "paradoxical" sleep) (● Figure 5-26). Following the paradoxical episode, the stages of slow-wave sleep are repeated once again.

A person cyclically alternates between the two types of sleep throughout the night. In a normal sleep cycle, a person always passes through slow-wave sleep before entering para-

---

▲ **TABLE 5-5**
Comparison of Slow-Wave and Paradoxical Sleep

Characteristic	Type of Sleep	
	Slow-wave sleep	Paradoxical sleep
EEG	Displays slow waves	Similar to EEG of alert, awake person
Motor Activity	Considerable muscle tone; frequent shifting	Abrupt inhibition of muscle tone; no movement
Heart Rate, Respiratory Rate, Blood Pressure	Minor reductions	Irregular
Dreaming	Rare (mental activity is extension of waking-time thoughts)	Common
Arousal	Sleeper easily awakened	Sleeper hard to arouse but apt to wake up spontaneously
Percentage of Sleeping Time	80%	20%
Other Important Characteristics	Has four stages; sleeper must pass through this type of sleep first	Rapid eye movements

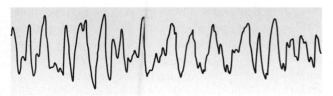

Slow-wave sleep, stage 4

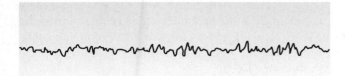

Paradoxical sleep

Awake, eyes open

● FIGURE 5-26

EEG patterns during different types of sleep
Note that the EEG pattern during paradoxical sleep is similar to that of an alert, awake person, whereas the pattern during slow-wave sleep displays distinctly different waves.

doxical sleep. On the average, paradoxical sleep occupies 20% of total sleeping time throughout adolescence and most of adulthood. Infants spend considerably more time in paradoxical sleep. In contrast, paradoxical as well as deep stage 4 slow-wave sleep declines in the elderly. Individuals who require less total sleeping time than normal spend proportionately more time in paradoxical and deep stage 4 slow-wave sleep and less time in the lighter stages of slow-wave sleep.

Behavioral patterns during sleep

In addition to distinctive EEG patterns, the two types of sleep are distinguished by behavioral differences. It is difficult to pinpoint exactly when an individual drifts from drowsiness into slow-wave sleep. In this type of sleep, the person still has considerable muscle tone and frequently shifts body position. Only minor reductions in respiratory rate, heart rate, and blood pressure occur. During this time the sleeper can be easily awakened and rarely dreams. The mental activity associated with slow-wave sleep is less visual than dreaming. It is more conceptual and plausible—like an extension of waking-time thoughts concerned with everyday events—and it is less likely to be recalled. The major exception is nightmares, which occur during stages 3 and 4. People who walk and talk in their sleep do so during slow-wave sleep.

The behavioral pattern accompanying paradoxical sleep is marked by abrupt inhibition of muscle tone throughout the body. The muscles are completely relaxed with no movement taking place. Paradoxical sleep is further characterized by *rapid*

eye movements, which give it the alternative name REM sleep. Heart rate and respiratory rate become irregular and blood pressure may fluctuate. Another characteristic of REM sleep is *dreaming*. The rapid eye movements are not related to "watching" the dream imagery. The eye movements are driven in a locked oscillating pattern uninfluenced by dream content.

Brain imaging of volunteers during REM sleep shows heightened activity in the higher-level visual processing areas and limbic system (the seat of emotions), coupled with reduced activity in the prefrontal cortex (the seat of reasoning). This pattern of activity lays the groundwork for the characteristics of dreaming: internally generated visual imagery reflecting activation of the person's "emotional memory bank" with little guidance or interpretation from the complex thinking areas. As a result, dreams are often charged with intense emotions, a distorted sense of time, and bizarre content that is uncritically accepted as real, with little reflection about all the strange happenings.

▌ The sleep–wake cycle is controlled by interactions among three neural systems.

The sleep–wake cycle as well as the various stages of sleep are due to the cyclical interplay of three different neural systems: (1) an *arousal system*, which is part of the reticular activating system, (2) a *slow-wave sleep center*; and (3) a *paradoxical sleep center*. The pattern of interaction among these three neural regions, which brings about the fairly predictable cyclical sequence between being awake and passing alternately between the two types of sleep, is the subject of intense investigation. Nevertheless, the molecular mechanisms controlling the sleep–wake cycle remain poorly understood.

The normal cycle can easily be interrupted, with the arousal system more readily overriding the sleep systems than vice versa; that is, it is easier to stay awake when you are sleepy than to fall asleep when you are wide awake. The arousal system can be activated by afferent sensory input (for example, a person has difficulty falling asleep when it is noisy) or by input descending to the brain stem from higher brain regions. Intense concentration or strong emotional states, such as anxiety or excitement, can keep a person from falling asleep, just as motor activity, such as getting up and walking around, can arouse a drowsy person.

▌ The function of sleep is unclear.

Even though humans spend about a third of their lives sleeping, the reason sleep is needed largely remains a mystery. Sleep is not accompanied by a *reduction* in neural activity (that is, the brain cells are not "resting"), as once was suspected, but rather by a profound *change* in activity. One widely accepted proposal holds that sleep provides the brain "catch-up" time to restore biochemical or physiological processes that have progressively degraded during wakefulness. The most direct evidence that supports this proposal is the potential role of *adenosine* as a neural sleep factor. Adenosine, the backbone of adenosine triphosphate (ATP), the body's energy currency, is generated during the awake state by metabolically active neurons and glial cells. Thus the brain's extracellular concentration of adenosine continues to rise the longer a person has been awake. Adenosine, which acts as a neuromodulator, has been shown experimentally to inactivate the arousal center. This action can bring on slow-wave sleep. Injections of adenosine induce apparently normal sleep, whereas *caffeine*, which blocks adenosine receptors in the brain, revives drowsy people by removing adenosine's inhibitory influence on the arousal center. Adenosine levels diminish during sleep, presumably because the brain uses this adenosine as a raw ingredient for replenishing its limited energy stores. Thus the body's need for sleep may stem from the brain's periodic need to replenish diminishing energy stores. Because adenosine reflects the level of brain cell activity, the concentration of this chemical in the brain may serve as a gauge of how much energy has been depleted.

Some scientists alternatively speculate that sleep, especially paradoxical sleep, is necessary to allow the brain to "shift gears," to accomplish the long-term structural and chemical adjustments necessary for learning and memory, especially consolidation of procedural memories. This theory might explain why infants require so much sleep. Their highly plastic brains are rapidly undergoing profound synaptic modifications in response to environmental stimulation. In contrast, mature individuals, in whom neural changes are less dramatic, sleep less.

Not much is known about the brain's need for the two types of sleep, although a specified amount of paradoxical sleep appears to be required. Individuals experimentally deprived of paradoxical sleep for a night or two by being aroused every time the paradoxical EEG pattern appeared suffered hallucinations and spent proportionally more time in paradoxical sleep during subsequent undisturbed nights, as if to make up for lost time.

An unusual sleep disturbance is **narcolepsy.** It is characterized by brief (5- to 30-minute) irresistible sleep attacks during the day. A person suffering from this condition suddenly falls asleep during any ongoing activity, often without warning. Narcolepsy is suspected of being linked to abnormal activation of neurons in the paradoxical sleep center during the waking state. Narcoleptic patients, in fact, can enter into paradoxical sleep directly without the normal prerequisite passage through slow-wave sleep.

We have now completed our discussion of the brain and are going to shift our attention to the other component of the central nervous system, the spinal cord.

SPINAL CORD

The **spinal cord** is a long, slender cylinder of nerve tissue that extends from the brain stem. It is about 45 cm long (18 inches) and 2 cm in diameter (about the size of your thumb).

▌ The spinal cord extends through the vertebral canal and is connected to the spinal nerves.

Exiting through a large hole in the base of the skull, the spinal cord is enclosed by the protective vertebral column as it descends through the vertebral canal (● Figure 5-27). Paired *spinal nerves* emerge from the spinal cord through spaces formed between the bony, winglike arches of adjacent vertebrae. The spinal nerves are named according to the region of the vertebral column from which they emerge (● Figure 5-28): there are eight pairs of *cervical* (neck) *nerves* (namely C1–C8), twelve *thoracic* (chest) *nerves*, five *lumbar* (abdominal) *nerves*, five *sacral* (pelvic) *nerves*, and one *coccygeal* (tailbone) *nerve*.

During development, the vertebral column grows about 25 cm longer than the spinal cord. Because of this differential growth, segments of the spinal cord that give rise to various spinal nerves are not aligned with the corresponding intervertebral spaces. Most spinal nerve roots must descend along the cord before emerging from the vertebral column at the corresponding space. The spinal cord itself extends only to the level of the first or second lumbar vertebra (about waist level), so the nerve roots of the remaining nerves are greatly elongated, to exit the vertebral column at their appropriate space. The thick bundle of elongated nerve roots within the lower vertebral canal is known as the **cauda equina** ("horse's tail") because of its appearance (● Figure 5-28). This region is the site where "spinal taps" are performed to obtain a sample of CSF. Inserting of a needle into the canal below the level of the second lumbar vertebra does not run the risk of penetrating the spinal cord. The needle pushes aside the nerve roots of the cauda equina so that the sample of the surrounding fluid can be withdrawn safely.

● FIGURE 5-27

Location of the spinal cord relative to the vertebral column

(SOURCE: Adapted from Cecie Starr and Ralph Taggart, *Biology: The Unity and Diversity of Life*, Eighth Edition, Fig. 35.9a, p. 577. Copyright 1998 Wadsworth Publishing Company.)

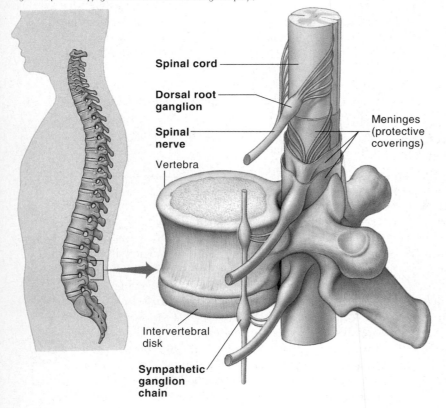

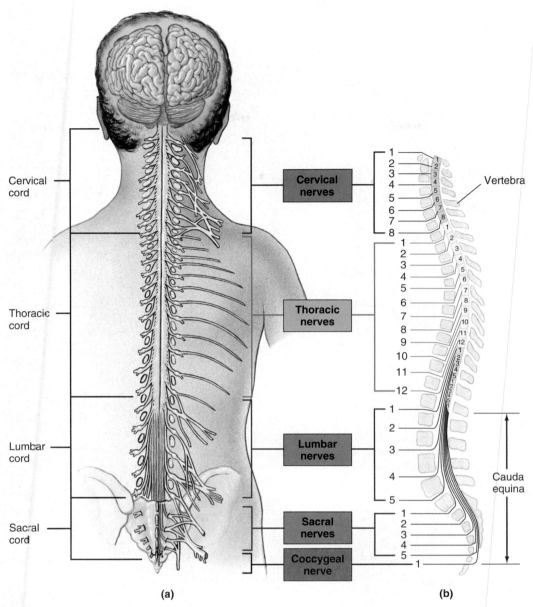

Cervical cord

Thoracic cord

Lumbar cord

Sacral cord

Cervical nerves

Thoracic nerves

Lumbar nerves

Sacral nerves

Coccygeal nerve

Vertebra

Cauda equina

(a)   (b)

● **FIGURE 5-28**

Spinal nerves
There are 31 pairs of spinal nerves named according to the region of the vertebral column from which they emerge. Because the spinal cord is shorter than the vertebral column, spinal nerve roots must descend along the cord before emerging from the vertebral column at the corresponding intervertebral space, especially those beyond the level of the first lumbar vertebra (L1). Collectively these rootlets are called the cauda equina, literally "horse's tail." (a) Posterior view of the brain, spinal cord, and spinal nerves (on the right side only). (b) Lateral view of the spinal cord and spinal nerves emerging from the vertebral column.

▌ The white matter of the spinal cord is organized into tracts.

Although there are some slight regional variations, the cross-sectional anatomy of the spinal cord is generally the same throughout its length (● Figure 5-29). In contrast to the gray matter in the brain, the gray matter in the spinal cord forms a butterfly-shaped region on the inside and is surrounded by the outer white matter. As in the brain, the cord gray matter consists primarily of neuronal cell bodies and their dendrites, short interneurons, and glial cells. The white matter is organized into tracts, which are bundles of nerve fibers (axons of long interneurons) with a similar function. The bundles are grouped into columns that extend the length of the cord. Each of these tracts begins or ends within a particular area of the brain, and each is specific in the type of information that it transmits. Some are **ascending** (cord to brain) **tracts** that transmit to the brain

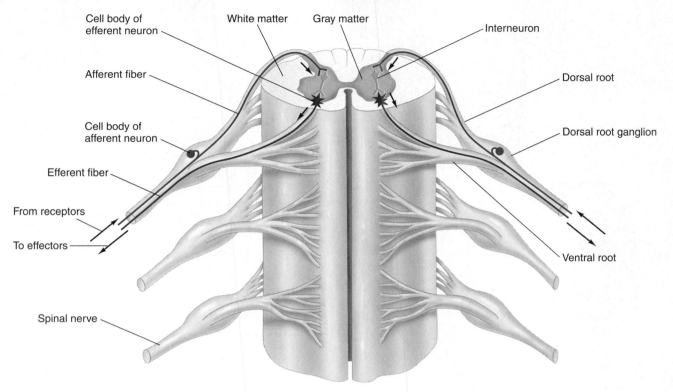

● FIGURE 5-29

Spinal cord in cross section

Schematic representation of the spinal cord in cross section showing the relationship between the spinal cord and spinal nerves. The afferent fibers enter through the dorsal root, and the efferent fibers exit through the ventral root. Afferent and efferent fibers are enclosed together within a spinal nerve.

● FIGURE 5-30

Ascending and descending tracts in the white matter of the spinal cord

(a) Location and functions of the major ascending and descending tracts in the spinal cord in cross section. *Continued*

(SOURCE: Adapted from Cecie Starr and Ralph Taggart, *Biology: The Unity and Diversity of Life*, Eighth Edition, Fig. 34.10, p. 566. Copyright 1998 Wadsworth Publishing Company.)

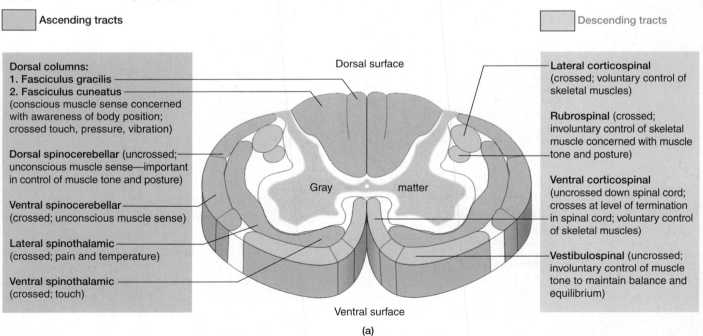

(a)

signals derived from afferent input. Others are **descending** (brain to cord) **tracts** that relay messages from the brain to efferent neurons (● Figure 5-30).

The tracts are generally named for their origin and termination. For example, the **ventral corticospinal tract** is a descending pathway; its cell bodies originate primarily in the motor region of the cerebral cortex, and its axons travel down the ventral (toward the front) portion of the spinal cord and terminate within the spinal cord on the cell bodies of efferent motor neurons supplying skeletal muscles. In contrast, the **lateral spinothalamic tract** is an ascending pathway that originates in the spinal cord and runs at the lateral (toward the side)

● **FIGURE 5-30**

Ascending and descending tracts in the white matter of the spinal cord *(continued)*
(b) Cord-to-brain pathways of several ascending tracts (fasciculus cuneatus and ventral spinocerebellar tract).
(c) Brain-to-cord pathways of several descending tracts (lateral corticospinal and ventral corticospinal tracts).

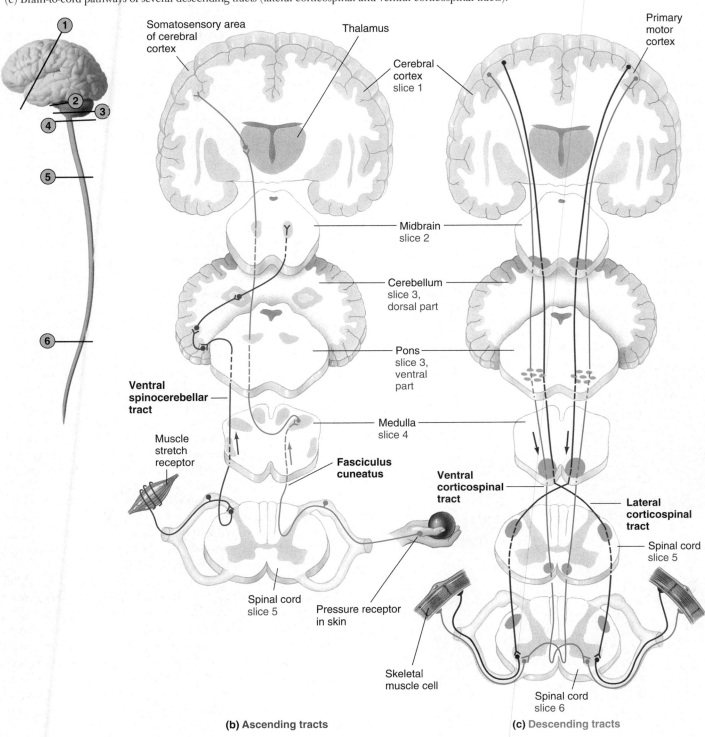

(b) Ascending tracts

(c) Descending tracts

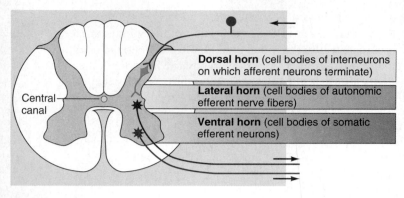

● FIGURE 5-31
Regions of the gray matter

Within the figure:
Central canal

Dorsal horn (cell bodies of interneurons on which afferent neurons terminate)

Lateral horn (cell bodies of autonomic efferent nerve fibers)

Ventral horn (cell bodies of somatic efferent neurons)

margin of the cord until it synapses in the thalamus. Sensory information about pain and temperature that has been delivered to the spinal cord by afferent input from various regions of the body is carried within the spinothalamic tract up the spinal cord to the thalamus, which in turn sorts and relays the information to the somatosensory cortex. Because various types of signals are carried in different tracts within the spinal cord, damage to particular areas of the cord can interfere with some functions while other functions remain intact.

● FIGURE 5-32
Structure of a nerve
Diagrammatic view of a nerve, showing neuronal axons (both afferent and efferent fibers) bundled together into connective-tissue–wrapped fascicles. A nerve consists of a group of fascicles enclosed by a connective tissue covering and following the same pathway. The photograph is a scanning electron micrograph of several nerve fascicles in cross section.

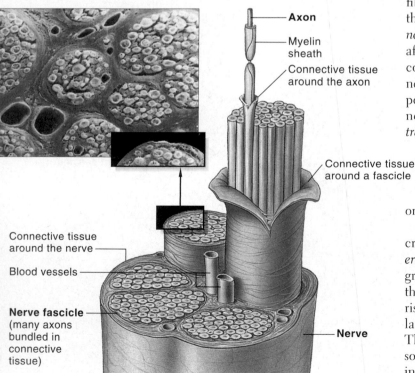

Labels within the figure:
Axon
Myelin sheath
Connective tissue around the axon
Connective tissue around a fascicle
Connective tissue around the nerve
Blood vessels
Nerve fascicle (many axons bundled in connective tissue)
Nerve

▌Each horn of the spinal cord gray matter houses a different type of neuronal cell body.

The centrally located gray matter is also functionally organized (● Figure 5-31). The central canal, which is filled with CSF, lies in the center of the gray matter. Each half of the gray matter is arbitrarily divided into a **dorsal (posterior)** (toward the back) **horn,** a **ventral (anterior) horn,** and a **lateral horn.** The dorsal horn contains cell bodies of interneurons on which afferent neurons terminate. The ventral horn contains cell bodies of the efferent motor neurons supplying skeletal muscles. Autonomic nerve fibers supplying cardiac and smooth muscle and exocrine glands originate at cell bodies found in the lateral horn.

▌Spinal nerves carry both afferent and efferent fibers.

Spinal nerves connect with each side of the spinal cord by a **dorsal root** and a **ventral root** (● Figure 5-29). Afferent fibers carrying incoming signals enter the spinal cord through the dorsal root; efferent fibers carrying outgoing signals leave through the ventral root. The cell bodies for the afferent neurons at each level are clustered together in a **dorsal root ganglion.** (A collection of neuronal cell bodies located outside of the CNS is called a *ganglion,* whereas a functional collection of cell bodies within the CNS is referred to as a *center* or a *nucleus.*) The cell bodies for the efferent neurons originate in the gray matter and send axons out through the ventral root.

The dorsal and ventral roots at each level join to form a **spinal nerve** that emerges from the vertebral column (● Figure 5-29). A spinal nerve contains both afferent and efferent fibers traversing between a particular region of the body and the spinal cord. Note the relationship between a *nerve* and a *neuron.* A **nerve** is a bundle of peripheral neuronal axons, some afferent and some efferent, enclosed by a connective tissue covering and following the same pathway (● Figure 5-32). A nerve does not contain a complete nerve cell, only the axonal portions of many neurons. (By this definition, there are no nerves in the CNS! Bundles of axons in the CNS are called *tracts.*) The individual fibers within a nerve generally do not have any direct influence on each other. They travel together for convenience, just as many individual telephone lines are carried within a telephone cable, yet any particular connection can be private without interference or influence from other lines in the cable.

The 31 pairs of spinal nerves, along with the 12 pairs of cranial nerves that arise from the brain, constitute the *peripheral nervous system.* After they emerge, the spinal nerves progressively branch to form a vast network of peripheral nerves that supply the tissues. Each segment of the spinal cord gives rise to a pair of spinal nerves that ultimately supply a particular region of the body with both afferent and efferent fibers. Thus the location and extent of sensory and motor deficits associated with spinal cord injuries can be clinically important in determining the level and extent of the cord injury.

Swan Dive or Belly Flop: It's a Matter of CNS Control

Sport skills must be learned. Much of the time, strong basic reflexes must be overridden in order to perform the skill. Learning to dive into water, for example, is very difficult initially. Strong head-righting reflexes controlled by sensory organs in the neck and ears initiate a straightening of the neck and head before the beginning diver enters the water, causing what is commonly known as a "belly flop." In a backward dive, the head-righting reflex causes the beginner to land on his or her back or even in a sitting position. To perform any motor skill that involves body inversions, somersaults, back flips, or other abnormal postural movements, the person must learn to consciously inhibit basic postural reflexes. This is accomplished by having the person concentrate on specific body positions during the movement. For example, to perform a somersault, the person must concentrate on keeping the chin tucked and grabbing the knees. After the skill is performed repeatedly, new synaptic patterns are formed in the CNS, and the new or conditioned response substitutes for the natural reflex responses. Sport skills must be practiced until the movement becomes automatic; then the athlete is free during competition to think about strategy or the next move to be performed in a routine.

With reference to sensory input, each specific region of the body surface supplied by a particular spinal nerve is called a **dermatome**. These same spinal nerves also carry fibers that branch off to supply internal organs, and sometimes pain originating from one of these organs is "referred" to the corresponding dermatome supplied by the same spinal nerve. **Referred pain** originating in the heart, for example, may appear to come from the left shoulder and arm. The mechanism responsible for referred pain is not completely understood. Inputs arising from the heart presumably share a pathway to the brain in common with inputs from the left upper extremity. The higher perception levels, being more accustomed to receiving sensory input from the left arm than from the heart, may interpret the input from the heart as having arisen from the left arm.

▮ The spinal cord is responsible for the integration of many basic reflexes.

The spinal cord is strategically located between the brain and afferent and efferent fibers of the peripheral nervous system; this location enables the spinal cord to fulfill its two primary functions: (1) serving as a link for transmission of information between the brain and the remainder of the body and (2) integrating reflex activity between afferent input and efferent output without involving the brain. This type of reflex activity is known as a *spinal reflex*.

A **reflex** is any response that occurs automatically without conscious effort. There are two types of reflexes: (1) **simple,** or basic, **reflexes,** which are built-in, unlearned responses, such as pulling the hand away from a burning hot object; and (2) **acquired,** or **conditioned, reflexes,** which are a result of practice and learning, such as a pianist striking a particular key on seeing a given note on the music staff. The musician does this automatically, but only after considerable conscious training effort. (For a discussion of the role of acquired reflexes in many sport skills see the accompanying boxed feature, ▶ A Closer Look at Exercise Physiology.)

Reflex arc

The neural pathway involved in accomplishing reflex activity is known as a **reflex arc,** which typically includes five basic components:

1. receptor
2. afferent pathway
3. integrating center
4. efferent pathway
5. effector

The **receptor** responds to a **stimulus,** which is a detectable physical or chemical change in the environment of the receptor. In response to the stimulus, the receptor produces an action potential that is relayed by the **afferent pathway** to the **integrating center** for processing. Usually the integrating center is the CNS. The spinal cord and brain stem are responsible for integrating basic reflexes, whereas higher brain levels usually process acquired reflexes. The integrating center processes all information available to it from this receptor as well as from all other inputs, then "makes a decision" about the appropriate response. The instructions from the integrating center are transmitted via the **efferent pathway** to the **effector**—a muscle or gland—which carries out the desired response. Unlike conscious behavior, in which any one of a number of responses is possible, a reflex response is predictable, because the pathway between the receptor and effector is always the same.

Withdrawal reflex

A basic **spinal reflex** is one integrated by the spinal cord; that is, all components necessary for linking afferent input to efferent response are present within the spinal cord. The **withdrawal reflex** can serve to illustrate a basic spinal reflex (● Figure 5-33). When a person touches a hot stove (or receives another painful stimulus), a withdrawal reflex is initiated to pull the hand away from the stove (to withdraw from the painful stimulus). The skin has different receptors for warmth, cold, light touch, pressure, and pain. Even though all information is sent to the CNS by way of action potentials, the CNS can distinguish between various stimuli because different receptors and consequently different afferent pathways are activated by different stimuli. When a receptor is stimulated sufficiently to reach threshold, an action potential is generated in the afferent neuron. The stronger the stimulus, the greater the frequency of action potentials generated and propagated to the CNS. Once the afferent neuron enters the spinal cord, it diverges to synapse with the following different interneurons (the numbers correspond to those on ● Figure 5-33).

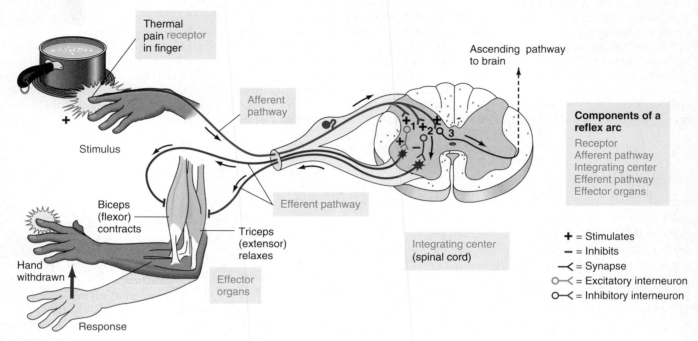

Thermal pain receptor in finger

Afferent pathway

Ascending pathway to brain

Stimulus

Efferent pathway

Components of a reflex arc
Receptor
Afferent pathway
Integrating center
Efferent pathway
Effector organs

Biceps (flexor) contracts

Triceps (extensor) relaxes

Integrating center (spinal cord)

Hand withdrawn

Effector organs

+ = Stimulates
– = Inhibits
≺ = Synapse
○─≺ = Excitatory interneuron
●─≺ = Inhibitory interneuron

Response

● **FIGURE** 5-33

The withdrawal reflex
When a painful stimulus activates a receptor in the finger, action potentials are generated in the corresponding afferent pathway, which propagates the electrical signals to the CNS. Once the afferent neuron enters the spinal cord, it diverges and terminates on three different types of interneurons (only one of each type is depicted): (1) excitatory interneurons, which in turn stimulate the efferent motor neurons to the biceps, causing the arm to flex and pull the hand away from the painful stimulus; (2) inhibitory interneurons, which inhibit the efferent motor neurons to the triceps, thus preventing counterproductive contraction of this antagonistic muscle; and (3) interneurons that carry the signal up the spinal cord via an ascending pathway to the brain for awareness of pain, memory storage, and so on.

1. An excited afferent neuron stimulates excitatory interneurons that in turn stimulate the efferent motor neurons supplying the biceps, the muscle in the arm that flexes (bends) the elbow joint. Contraction of the biceps pulls the hand away from the hot stove.

2. The afferent neuron also stimulates inhibitory interneurons that in turn inhibit the efferent neurons supplying the triceps to prevent it from contracting. The triceps is the muscle in the arm that extends (straightens out) the elbow joint. When the biceps is contracting to flex the elbow, it would be counterproductive for the triceps to be contracting. Therefore, built into the withdrawal reflex is inhibition of the muscle that antagonizes (opposes) the desired response. This type of neuronal connection involving stimulation of the nerve supply to one muscle and simultaneous inhibition of the nerves to its antagonistic muscle is known as **reciprocal innervation.**

3. The afferent neuron stimulates still other interneurons that carry the signal up the spinal cord to the brain via an ascending pathway. Only when the impulse reaches the sensory area of the cortex is the person aware of the pain, its location,

and the type of stimulus. Also, when the impulse reaches the brain, the information can be stored as memory, and the person can start thinking about the situation—how it happened, what to do about it, and so on. All this activity at the conscious level is above and beyond the basic reflex.

As is characteristic of all spinal reflexes, the brain can modify the withdrawal reflex. Impulses may be sent down descending pathways to the efferent neurons supplying the involved muscles to override the input from the receptors, actually preventing the biceps from contracting in spite of the painful stimulus. When your finger is being pricked to obtain a blood sample, pain receptors are stimulated, initiating the withdrawal reflex. However, knowing that you must be brave and not pull your hand away, you can consciously override the reflex by sending IPSPs via descending pathways to the motor neurons supplying the biceps and EPSPs to those supplying the triceps. The activity in these efferent neurons depends on the sum of activity of all their synaptic inputs. Because the neurons supplying the biceps are now receiving more IPSPs from the brain (voluntary) than EPSPs from the afferent pain pathway (reflex), these neurons are inhibited and do not reach threshold. Therefore, the biceps is not stimulated to contract and withdraw the hand. Simultaneously, the neurons to the triceps are receiving more EPSPs from the brain than IPSPs via the reflex arc, so they reach threshold, fire, and consequently stimulate the triceps to contract. Thus the arm is kept extended in spite of the painful stimulus. In this way, the withdrawal reflex has been voluntarily overridden.

Stretch reflex

Only one reflex is simpler than the withdrawal reflex: the **stretch reflex,** in which an afferent neuron originating at a stretch-detecting receptor in a skeletal muscle terminates directly on the efferent neuron supplying the same skeletal muscle to cause it to contract and counteract the stretch. (You will learn more about the role of this reflex in Chapter 8). The stretch reflex is a **monosynaptic** ("one synapse") reflex, because the only synapse in the reflex arc is the one between the afferent neuron and the efferent neuron. The withdrawal reflex and all other reflexes are **polysynaptic** ("many synapses"), because interneurons are interposed in the reflex pathway and, therefore, a number of synapses are involved.

Other reflex activity

Spinal reflex action is not necessarily limited to motor responses on the side of the body to which the stimulus is applied. Assume that a person steps on a tack instead of burning a finger. A reflex arc is initiated to withdraw the injured foot from the painful stimulus, while the opposite leg simultaneously prepares to suddenly bear all the weight so that the person does not lose balance or fall (● Figure 5-34). Unimpeded bending of the injured extremity's knee is accomplished by concurrent reflex stimulation of the muscles that flex the knee and inhibition of the muscles that extend the knee. At the same time, unimpeded extension of the opposite limb's knee is accomplished by activation of pathways that cross over to the opposite side of the spinal cord to reflexly stimulate this knee's extensors and inhibit its flexors. This **crossed extensor reflex** ensures that the opposite limb will be in a position to bear the weight of the body as the injured limb is withdrawn from the stimulus.

Besides protective reflexes such as the withdrawal reflex and simple postural reflexes such as the crossed extensor reflex, basic spinal reflexes also mediate emptying of pelvic organs (for example, urination, defecation, and expulsion of semen). All spinal reflexes can be voluntarily overridden at least temporarily by higher brain centers.

● **FIGURE 5-34**

The crossed extensor reflex coupled with the withdrawal reflex
(a) The withdrawal reflex, which causes flexion of the injured extremity to withdraw from a painful stimulus. (b) The crossed extensor reflex, which extends the opposite limb to support the full weight of the body.

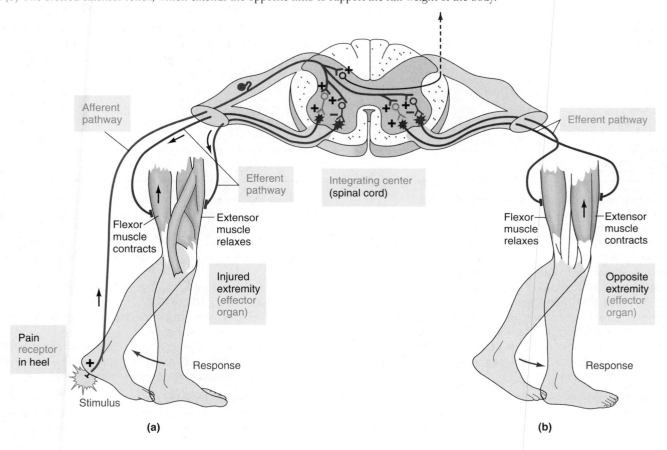

Not all reflex activity involves a clear-cut reflex arc, although the basic principles of a reflex (that is, an automatic response to a detectable change) are still present. Pathways for unconscious responsiveness digress from the typical reflex arc in two general ways:

1. *Responses mediated at least in part by hormones.* A particular reflex may be mediated solely by either neurons or hormones or may involve a pathway using both.

2. *Local responses that do not involve either nerves or hormones.* For example, the blood vessels in an exercising muscle dilate because of local metabolic changes, thereby increasing blood flow to match the active muscle's metabolic needs.

CHAPTER IN PERSPECTIVE: FOCUS ON HOMEOSTASIS

To interact in appropriate ways with the external environment to sustain the body's viability, such as in acquiring food, and to make the internal adjustments necessary to maintain homeostasis, the body must be informed about any changes taking place in the external and internal environment and must be able to process this information and send messages to various muscles and glands to accomplish the desired results. The nervous system, one of the body's two major regulatory systems, plays a central role in this life-sustaining communication. The central nervous system (CNS), which consists of the brain and spinal cord, receives information about the external and internal environment by means of the afferent peripheral nerves. After sorting, processing, and integrating this input, the CNS sends directions, by means of efferent peripheral nerves, to bring about appropriate muscular contractions and glandular secretions.

With its swift electrical signaling system, the nervous system is especially important in controlling the rapid responses of the body. Many neurally controlled muscular and glandular activities are aimed toward maintaining homeostasis. The CNS is the main site of integration between the afferent input and efferent output. It is responsible for linking the appropriate response to a particular input so that conditions compatible with life are maintained in the body. For example, when informed by the afferent nervous system that blood pressure has fallen, the CNS sends appropriate commands to the heart and blood vessels to increase the blood pressure to normal. Likewise, when informed that the body is overheated the CNS promotes secretion of sweat by the sweat glands. Evaporation of sweat helps cool the body to normal temperature. Were it not for this processing and integrative ability of the CNS, maintaining homeostasis in an organism as complex as a human would be impossible.

At the simplest level, the spinal cord integrates many basic protective and evacuative reflexes that do not require conscious participation, such as withdrawal from a painful stimulus and emptying of the urinary bladder. In addition to serving as a more complex integrating link between afferent input and efferent output, the brain is also responsible for the initiation of all voluntary movement, for complex perceptual awareness of the external environment and self, for language, and for abstract neural phenomena such as thinking, learning, remembering, consciousness, emotions, and personality traits. All neural activity—from the most private thoughts to commands for motor activity, from enjoying a concert to retrieving memories from the distant past—is ultimately attributable to propagation of action potentials along individual nerve cells and chemical transmission between cells.

The nervous system has become progressively more complex during evolutionary development. Newer, more complicated, and more sophisticated layers of the brain have been piled on top of older, more primitive regions. Many of the basic activities necessary for survival are built into the older parts of the brain. The newer, higher levels progressively modify, enhance, or nullify actions coordinated by lower levels in a hierarchy of command, and they also add new capabilities. Many of these higher neural activities are not aimed toward maintaining life, but they add immeasurably to the quality of being alive.

CHAPTER SUMMARY

Comparison of the Nervous and Endocrine Systems

▮ The nervous and endocrine systems are the two main regulatory systems of the body.

▮ The nervous system is anatomically "wired" to its target organs, whereas the "wireless" endocrine system secretes blood-borne hormones that reach distant target organs.

▮ Specificity of neural action depends on the anatomical proximity of the neurotransmitter-releasing neuronal terminal to its target organ. Specificity of endocrine action depends on specialization of target-cell receptors for a specific circulating hormone.

▮ In general, the nervous system coordinates rapid responses, whereas the endocrine system regulates activities that require duration rather than speed.

Organization of the Nervous System

▮ The nervous system consists of the central nervous system (CNS), which includes the brain and spinal cord, and the peripheral nervous system, which includes the nerve fibers carrying information to (afferent division) and from (efferent division) the CNS.

▮ Three classes of neurons—afferent neurons, efferent neurons, and interneurons—compose the excitable cells of the nervous system.

▮ Afferent neurons apprise the CNS of conditions in both the external and internal environment.

▮ Efferent neurons carry instructions from the CNS to effector organs, namely muscles and glands.

▮ Interneurons are responsible for integrating afferent information and formulating an efferent response, as well as for all higher mental functions associated with the "mind."

Protection and Nourishment of the Brain

▮ Glial cells form the connective tissue within the CNS and physically, metabolically, and functionally support the neurons.

- The four types of glial cells are the astrocytes, oligodendrocytes, microglia, and ependymal cells.
- The brain is provided with several protective devices, which is important because neurons cannot divide to replace damaged cells.
- The brain is wrapped in three layers of protective membranes—the meninges—and is further surrounded by a hard bony covering.
- Cerebrospinal fluid flows within and around the brain to cushion it against physical jarring.
- Protection against chemical injury is conferred by a blood–brain barrier that limits access of blood-borne substances to the brain.
- The brain depends on a constant blood supply for delivery of O_2 and glucose because it is unable to generate ATP in the absence of either of these substances.

Overview of the Central Nervous System

- Even though no part of the brain acts in isolation of other brain regions, the brain is organized into networks of neurons within discrete locations that are ultimately responsible for carrying out particular tasks.
- The parts of the brain from the lowest, most primitive level to the highest, most sophisticated level are the brain stem, cerebellum, hypothalamus, thalamus, basal nuclei, and cerebral cortex.

Cerebral Cortex

- The cerebral cortex is the outer shell of gray matter that caps an underlying core of white matter; the white matter consists of bundles of nerve fibers that interconnect various cortical regions with other areas. The cortex itself consists primarily of neuronal cell bodies, dendrites, and glial cells.
- Ultimate responsibility for many discrete functions is known to be localized in particular regions of the cortex as follows: (1) the occipital lobes house the visual cortex, (2) the auditory cortex is found in the temporal lobes, (3) the parietal lobes are responsible for reception and perceptual processing of somatosensory (somesthetic and proprioceptive) input, and (4) voluntary motor movement is set into motion by the frontal lobe, where the primary motor cortex and higher motor areas are located.
- Language ability depends on the integrated activity of two primary language areas—Broca's area and Wernicke's area—located in the left cerebral hemisphere only.
- The association areas are regions of the cortex not specifically assigned to processing sensory input or commanding motor output or language ability. These areas provide an integrative link between diverse sensory information and purposeful action; they also play a key role in higher brain functions such as memory and decision making. The association areas include the prefrontal association cortex, the parietal-temporal-occipital association cortex, and the limbic association cortex.

Basal Nuclei, Thalamus, and Hypothalamus

- The subcortical brain structures, which include the basal nuclei, thalamus, and hypothalamus, interact extensively with the cortex in performing their functions.
- The basal nuclei inhibit muscle tone; coordinate slow, sustained postural contractions; and suppress useless patterns of movement.
- The thalamus serves as a relay station for preliminary processing of sensory input on its way to the cortex. It also accomplishes a crude awareness of sensation and some degree of consciousness.
- The hypothalamus regulates many homeostatic functions, in part through its extensive control of the autonomic nervous system and endocrine system.

The Limbic System and Its Functional Relations with the Higher Cortex

- The limbic system, which includes portions of the hypothalamus and other forebrain structures that encircle the brain stem, is responsible for emotion as well as basic, inborn behavioral patterns related to survival and perpetuation of the species. It also plays an important role in motivation and learning.
- There are two types of memory: (1) a short-term memory with limited capacity and brief retention, coded by enhancement of activity at pre-existing synapses, such as increased neurotransmitter release and increased responsiveness of the postsynaptic cell to the neurotransmitter; and (2) a long-term memory with large storage capacity and enduring memory traces, involving relatively permanent structural or functional changes, such as the formation of new synapses, between existing neurons. Enhanced protein synthesis underlies these long-term changes.
- The hippocampus and associated structures are especially important in declarative or "what" memories of specific objects, facts, and events.
- The cerebellum and associated structures are especially important in procedural or "how to" memories of motor skills gained through repetitive training.
- The prefrontal association cortex is the site of working memory, which temporarily holds currently relevant data—both new information and knowledge retrieved from memory stores—and manipulates and relates them to accomplish the higher-reasoning processes of the brain.

Cerebellum

- The vestibulocerebellum helps to maintain balance.
- The spinocerebellum enhances muscle tone and helps coordinate voluntary movement, especially fast, phasic motor activities.
- The cerebrocerebellum plays a role in initiating voluntary movement and in storing procedural memories.

Brain Stem

- The brain stem is an important link between the spinal cord and higher brain levels.
- It is the origin of the cranial nerves; contains centers that control cardiovascular, respiratory, and digestive function; regulates postural muscle reflexes; controls the overall degree of cortical alertness; and plays a key role in the sleep–wake cycle.
- The prevailing state of consciousness depends on the cyclical interplay between an arousal system (the reticular activating system) located in the brain stem along with a slow-wave sleep center and a paradoxical sleep center both in the hypothalamus.
- Sleep is an active process, not just the absence of wakefulness. While sleeping, a person cyclically alternates between slow-wave sleep and paradoxical sleep.
- Slow-wave sleep is characterized by slow waves on the EEG and little change in behavior pattern from the waking state except for not being consciously aware of the external world.
- Paradoxical, or REM, sleep is characterized by an EEG pattern similar to that of an alert, awake individual. Rapid eye movements, dreaming, and abrupt changes in behavior pattern occur.

Spinal Cord

- The spinal cord has two vital functions. First, it serves as the neuronal link between the brain and the peripheral nervous system. All communication up and down the spinal cord is located in ascending and descending tracts in the cord's outer white matter. Second, it is the integrating center for spinal reflexes, including some of the basic protective and postural reflexes and those involved with the emptying of the pelvic organs.

- The basic reflex arc includes a receptor, an afferent pathway, an integrating center, an efferent pathway, and an effector.
- The centrally located gray matter of the spinal cord contains the interneurons interposed between the afferent input and efferent output as well as the cell bodies of efferent neurons.

- Afferent and efferent fibers, which carry signals to and from the spinal cord, respectively, are bundled together into spinal nerves. These nerves supply specific body regions and are attached to the spinal cord in paired fashion throughout its length.

REVIEW EXERCISES

Objective Questions (Answers on p. A-38)

1. The major function of the CSF is to nourish the brain. *(True or false?)*
2. The brain can perform anaerobic metabolism in emergencies when O_2 supplies are low. *(True or false?)*
3. Damage to the left cerebral hemisphere brings about paralysis and loss of sensation on the left side of the body. *(True or false?)*
4. The hands and structures associated with the mouth have a disproportionately large share of representation in both the sensory and motor cortexes. *(True or false?)*
5. The left cerebral hemisphere specializes in artistic and musical ability, whereas the right side excels in verbal and analytical skills. *(True or false?)*
6. The specific function a particular cortical region will carry out is permanently determined during embryonic development. *(True or false?)*
7. _____ is a decreased responsiveness to an indifferent stimulus that is repeatedly presented.
8. The process of transferring and fixing short-term memory traces into long-term memory stores is known as _____.
9. Afferent fibers enter through the _____ root of the spinal cord, and efferent fibers leave through the _____ root.
10. List the five components of a basic reflex arc:
    1. _____ 2. _____ 3. _____ 4. _____
    5. _____.
11. Using the following answer code, indicate which neurons are being described (a characteristic may apply to more than one class of neurons):
    (a) afferent neurons
    (b) efferent neurons
    (c) interneurons
    1. have receptor at peripheral endings
    2. lie entirely within the CNS
    3. lie primarily within the peripheral nervous system
    4. innervate muscles and glands
    5. cell body is devoid of presynaptic inputs
    6. predominant type of neuron
    7. responsible for thoughts, emotions, memory, etc.

12. Match the following:
    ___ 1. consists of nerves carrying information between the periphery and the CNS
    ___ 2. consists of the brain and spinal cord
    ___ 3. division of the peripheral nervous system that transmits signals to the CNS
    ___ 4. division of the peripheral nervous system that transmits signals from the CNS
    ___ 5. supplies skeletal muscles
    ___ 6. supplies smooth muscle, cardiac muscles, and glands.

    (a) somatic nervous system
    (b) autonomic nervous system
    (c) central nervous system
    (d) peripheral nervous system
    (e) efferent division
    (f) afferent division

Essay Questions

1. Compare the nervous and endocrine systems.
2. Discuss the function of each of the following: astrocytes, oligodendrocytes, ependymal cells, microglia, cranium, vertebral column, meninges, cerebrospinal fluid, and blood–brain barrier.
3. Compare the composition of white and gray matter.
4. Draw and label the major functional areas of the cerebral cortex, indicating the functions attributable to each area.
5. Discuss the function of each of the following parts of the brain: thalamus, hypothalamus, basal nuclei, limbic system, cerebellum, and brain stem.
6. Define *somesthetic sensations* and *proprioception*.
7. What is an electroencephalogram?
8. Discuss the roles of Broca's area and Wernicke's area in language.
9. Compare short-term and long-term memory.
10. What is the reticular activating system?
11. Compare slow-wave and paradoxical sleep.
12. Draw and label a cross section of the spinal cord.
13. Distinguish between a monosynaptic and a polysynaptic reflex.

POINTS TO PONDER

(Explanations on p. A-38)

1. Special studies designed to assess the specialized capacities of each cerebral hemisphere have been performed on "split-brain" patients. These are individuals in whom the corpus callosum—the bundle of fibers that links the two halves of the brain together—has been surgically cut to prevent the spread of epileptic seizures from one hemisphere to the other. Even though no overt changes in behavior, intellect, or personality occur in these patients because both hemispheres individually receive the same information, deficits are observable with tests designed to restrict information to one brain hemisphere at a time. One such test involves limiting a visual stimulus to only half of the brain. Because of a crossover in the nerve pathways from the eyes to the occipital cortex, the visual information to the right of a midline point is transmitted to only the left half of the brain, whereas visual information to the left of this point is received by only the right half of the brain. A split-brain patient presented with a visual stimulus that reaches only the left hemisphere accurately describes the object seen, but when a visual stimulus is presented to only the right hemisphere, the patient denies having seen anything. The right hemisphere does receive the visual input, however, as demonstrated by nonverbal tests. Even though a split-brain patient denies having seen anything after an object is presented to the right hemisphere, he or she can correctly match the object by picking it out from among a number of objects, usually to the patient's surprise. What is your explanation of this finding?

2. The hormone insulin enhances the carrier-mediated transport of glucose into most of the body's cells but not into brain cells. The uptake of glucose from the blood by neurons is not dependent on insulin. Knowing the brain's need for a continuous supply of blood-borne glucose, what effect do you predict that insulin excess would have on the brain?

3. Which of the following symptoms are most likely to occur as the result of a severe blow to the back of the head?
   a. paralysis
   b. hearing impairment
   c. visual disturbances
   d. burning sensations
   e. personality disorders

4. Give examples of conditioned reflexes you have acquired.

5. Under what circumstances might it be inadvisable to administer a clot-dissolving drug to a stroke victim?

6. *Clinical Consideration.* Julio D., who had recently retired, was enjoying an afternoon of playing golf when suddenly he experienced a severe headache and dizziness. These symptoms were quickly followed by numbness and partial paralysis on the upper right side of his body, accompanied by an inability to speak. After being rushed to the emergency room, Julio was diagnosed as having suffered a stroke. Based on the observed neurologic impairment, what areas of his brain were affected?

PHYSIOEDGE RESOURCES

 PHYSIOEDGE CD-ROM
Figures marked with the PhysioEdge icon have associated activities on the CD. For this chapter, check out the Media Quizzes related to Figures 5-1, 5-2, 5-9, 5-10, and 5-16.

CHECK OUT THESE MEDIA QUIZZES:
5.1 The Nervous and Endocrine Systems
5.2 The Cerebral Cortex
5.3 Subcortical Structures

 PHYSIOEDGE WEB SITE
The Web site for this book contains a wealth of helpful study aids, as well as many ideas for further reading and research. Log on to:

http://www.brookscole.com

Go to the Biology page and select Sherwood's *Human Physiology*, 5th Edition. Select a chapter from the drop-down menu or click on one of these resource areas:

- **Case Histories** provide an introduction to the clinical aspects of human physiology. Check out:

   #13: A Critical Twenty-four Hours
   #14: One Disease, Two Outcomes

- For 2-D and 3-D graphical illustrations and animations of physiological concepts, visit our **Visual Learning Resource.**

- Resources for study and review include the **Chapter Outline, Chapter Summary, Glossary,** and **Flash Cards.** Use our **Quizzes** to prepare for in-class examinations.

- On-line research resources to consult are **Hypercontents,** with current links to relevant Internet sites; **Internet Exercises** with starter URLs provided; and **InfoTrac Exercises.**

 For more readings, go to InfoTrac College Edition, your on-line research library, at:

   http://infotrac.thomsonlearning.com

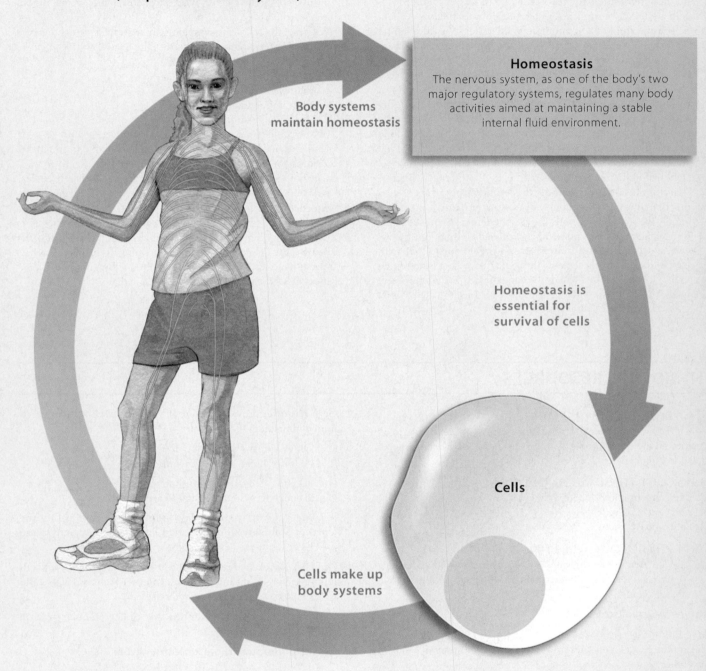

Nervous System
(Peripheral Nervous System)

Body systems maintain homeostasis

Homeostasis
The nervous system, as one of the body's two major regulatory systems, regulates many body activities aimed at maintaining a stable internal fluid environment.

Homeostasis is essential for survival of cells

Cells

Cells make up body systems

The nervous system, one of the two major regulatory systems of the body, consists of the central nervous system (CNS), composed of the brain and spinal cord, and the **peripheral nervous system (PNS)**, composed of the afferent and efferent fibers that relay signals between the CNS and periphery (other parts of the body).

The **afferent division** of the peripheral nervous system detects, encodes, and transmits peripheral signals to the CNS for processing. It is the communication link by which the CNS is informed about the internal and external environment. This input to the controlling centers of the CNS is essential in maintaining homeostasis. To make appropriate adjustments, the CNS has to "know" what is going on. Afferent input is also used to plan for voluntary actions unrelated to homeostasis.

Chapter 6

The Peripheral Nervous System: Afferent Division; Special Senses

CONTENTS AT A GLANCE

INTRODUCTION

RECEPTOR PHYSIOLOGY

▌ Receptor types

▌ Receptor potential; receptor adaptation

▌ Labeled lines for afferent input

▌ Acuity; receptive field; lateral inhibition

PAIN

▌ Receptors and mechanisms of pain

▌ Built-in analgesic system

EYE: VISION

▌ Light

▌ Refractive structures; accommodation

▌ Phototransduction

▌ Comparison of rod and cone vision

▌ Visual pathways; visual processing

EAR: HEARING AND EQUILIBRIUM

▌ Sound waves

▌ Roles of the external ear and middle ear

▌ Sound transduction by the organ of Corti

▌ Auditory pathway

▌ Vestibular apparatus

CHEMICAL SENSES: TASTE AND SMELL

INTRODUCTION

The peripheral nervous system consists of nerve fibers that carry information between the CNS and other parts of the body. The afferent division of the peripheral nervous system sends information about the internal and external environment to the CNS.

▌ Visceral afferents carry subconscious input whereas sensory afferents carry conscious input.

Afferent information about the internal environment, such as the blood pressure and the concentration of CO_2 in the body fluids, never reaches the level of conscious awareness, but this input is essential for determining the appropriate efferent output to maintain homeostasis. The incoming pathway for subconscious information derived from the internal viscera (organs) is called a **visceral afferent.** Afferent input that does reach the level of conscious awareness is known as *sensory information*, and the incoming pathway is considered a **sensory afferent.** Sensory information is categorized as either (1) **somatic** (body sense) **sensation** arising from the body surface, including *somesthetic sensation* from the skin and *proprioception* from the muscles, joints, skin, and inner ear (see p. 148) or (2) **special senses**, including *vision, hearing, taste,* and *smell.* (See the accompanying boxed feature, ▶ A Closer Look at Exercise Physiology, for a description of the usefulness of proprioception in athletic performance.) Final processing of sensory input by the CNS not only is essential for interaction with the environment for basic survival (for example, food procurement and defense from danger) but also adds immeasurably to the richness of life.

▌ Perception is the conscious awareness of surroundings derived from interpretation of sensory input.

Perception is our conscious interpretation of the external world as created by the brain from a pattern of nerve impulses delivered to it from sensory receptors. Is the world as we perceive it,

Back Swings and Prejump Crouches: What Do They Share in Common?

Proprioception, the sense of the body's position in space, is critical to any movement and is especially important in athletic performance, whether it be a figure skater performing triple jumps on ice, a gymnast performing a difficult floor routine, or a football quarterback throwing perfectly to a spot 60 yards down field. To control skeletal muscle contraction to achieve the desired movement, the CNS must be continuously apprised of the results of its action, through sensory feedback.

A number of receptors provide proprioceptive input. Muscle proprioceptors provide feedback information on muscle tension and length. Joint proprioceptors provide feedback on joint acceleration, angle, and direction of movement. Skin proprioceptors inform the CNS of weight-bearing pressure on the skin. Proprioceptors in the inner ear, along with those in neck muscles, provide information about head and neck position so that the CNS can orient the head correctly.

For example, neck reflexes facilitate essential trunk and limb movements during somersaults, and divers and tumblers use strong movements of the head to maintain spins.

The most complex and probably one of the most important proprioceptors is the muscle spindle (see p. 286). Muscle spindles are found throughout a muscle but tend to be concentrated in its center. Each spindle lies parallel to the muscle fibers within the muscle. The spindle is sensitive to both the muscle's rate of change in length and the final length achieved. If a muscle is stretched, each muscle spindle within the muscle is also stretched, and the afferent neuron whose peripheral axon terminates on the muscle spindle is stimulated. The afferent fiber passes into the spinal cord and synapses directly on the motor neurons that supply the same muscle. Stimulation of the stretched muscle as a result of this stretch reflex causes the muscle to contract sufficiently to relieve the stretch.

Older persons or those with weak quadriceps (thigh) muscles unknowingly take advantage of the muscle spindle by pushing on the center of the thighs when they get up from a sitting position. Contraction of the quadriceps muscle extends the knee joint, thus straightening the leg. The act of pushing on the center of the thighs when getting up slightly stretches the quadriceps muscle in both limbs, stimulating the muscle spindles. The resultant stretch reflex aids in contraction of the quadriceps muscles and helps the person assume a standing position.

In sports, people use the muscle spindle to advantage all the time. To jump high, as in basketball jumpball, an athlete starts by crouching down. This action stretches the quadriceps muscles and increases the firing rate of their spindles, thus triggering the stretch reflex that reinforces the quadriceps muscles' contractile response so that these extensor muscles of the legs gain additional power. The same is true for crouch starts in running events. The backswing in tennis, golf, and baseball similarly provides increased muscular excitation through reflex activity initiated by stretched muscle spindles.

---

reality? The answer is a resounding no. Our perception is different from what is really "out there," for several reasons. First, humans have receptors that detect only a limited number of existing energy forms. We perceive sounds, colors, shapes, textures, smells, tastes, and temperature but are not informed of magnetic forces, polarized light waves, radio waves, or x rays because we do not have receptors to respond to the latter energy forms. What is not detected by receptors, the brain will never know. Our response range is limited even for the energy forms for which we do have receptors. For example, dogs can hear a whistle whose pitch is above our level of detection. Second, the information channels to our brains are not high-fidelity recorders. During precortical processing of sensory input, some features of stimuli are accentuated and others are suppressed or ignored. Third, the cerebral cortex further manipulates the data, comparing the sensory input with other incoming information as well as with memories of past experiences to extract the significant features—for example, sifting out a friend's words from the hubbub of sound in a school cafeteria. In the process, the cortex often fills in or distorts the information to abstract a logical perception; that is, it "completes the picture." As a simple example, you "see" a white square in ● Figure 6-1 even though there is no white square but merely right-angle wedges taken out of four red circles. Optical illusions illustrate how the brain interprets reality according to its own rules. Do you see two faces in profile or a wineglass in ● Figure 6-2? You can alternately see one or the other out of identical visual input. Thus our perceptions do not replicate reality. Other species, equipped with different types of receptors and

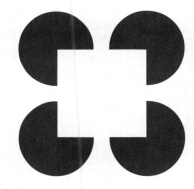

● FIGURE 6-1

Do you "see" a white square that is not really there?

● FIGURE 6-2

Variable perceptions from the same visual input
Do you see two faces in profile, or a wineglass?

sensitivities and with different neural processing, perceive a markedly different world from what we perceive.

RECEPTOR PHYSIOLOGY

A **stimulus** is a change detectable by the body. Stimuli exist in a variety of energy forms, or **modalities,** such as heat, light, sound, pressure, and chemical changes. Afferent neurons have **receptors** at their peripheral endings that respond to stimuli in both the external world and internal environment. Because the only way afferent neurons can transmit information to the CNS about these stimuli is via action potential propagation, receptors must convert these other forms of energy into electrical signals (action potentials). This energy conversion process is known as **transduction.**

▌ Receptors have differential sensitivities to various stimuli.

Each type of receptor is specialized to respond more readily to one type of stimulus, its **adequate stimulus,** than to other stimuli. For example, receptors in the eye are most sensitive to light, receptors in the ear to sound waves, and warmth receptors in the skin to heat energy. We cannot "see" with our ears or "hear" with our eyes because of this differential sensitivity of receptors. Some receptors can respond weakly to stimuli other than their adequate stimulus, but even when activated by a different stimulus, a receptor still gives rise to the sensation usually detected by that receptor type. As an example, the adequate stimulus for eye receptors (photoreceptors) is light, to which they are exquisitely sensitive, but these receptors can also be activated to a lesser degree by mechanical stimulation. When hit in the eye, a person often "sees stars," because the mechanical pressure stimulates the photoreceptors. Thus the sensation perceived depends on the type of receptor stimulated rather than on the type of stimulus. However, because receptors typically are activated by their adequate stimulus, the sensation usually corresponds to the stimulus modality.

Types of receptors according to their adequate stimulus

Depending on the type of energy to which they ordinarily respond, receptors are categorized as follows:

- **Photoreceptors** are responsive to visible wavelengths of light.
- **Mechanoreceptors** are sensitive to mechanical energy. Examples include skeletal muscle receptors sensitive to stretch, the receptors in the ear containing fine hair cells that are bent as a result of sound waves, and blood pressure–monitoring baroreceptors.
- **Thermoreceptors** are sensitive to heat and cold.
- **Osmoreceptors** detect changes in the concentration of solutes in the body fluids and the resultant changes in osmotic activity (see p. 75).
- **Chemoreceptors** are sensitive to specific chemicals. Chemoreceptors include the receptors for smell and taste, as well as those located deeper within the body that detect O_2 and CO_2 concentrations in the blood or the chemical content of the digestive tract.
- **Nociceptors,** or **pain receptors,** are sensitive to tissue damage such as pinching or burning or to distortion of tissue. Intense stimulation of any receptor is also perceived as painful.

Some sensations are compound sensations in that their perception arises from central integration of several simultaneously activated primary sensory inputs. For example, the perception of wetness comes from touch, pressure, and thermal receptor input; there is no such thing as a "wet receptor."

Uses for information detected by receptors

The information detected by receptors is conveyed via afferent neurons to the CNS, where it is used for a variety of purposes:

- First, afferent input is essential for the control of efferent output, both for regulating motor behavior in accordance with external circumstances and for coordinating internal activities directed toward maintenance of homeostasis. At the most basic level, afferent input provides information (of which the individual may or may not be consciously aware) for the CNS to use in directing activities necessary for survival. On a broader level, we could not interact successfully with our environment or with each other without sensory input.
- Second, processing of sensory input by the reticular activating system in the brain stem is critical for cortical arousal and consciousness (see p. 169).
- Third, central processing of sensory information gives rise to our perceptions of the world around us.
- Fourth, selected information delivered to the CNS may be stored for future reference.
- Finally, sensory stimuli can have a profound impact on our emotions. The smell of just-baked apple pie, the sensuous feel of silk, the sight of a loved one, hearing bad news—sensory input can gladden, sadden, arouse, calm, anger, frighten or evoke any other range of emotions.

We will next examine how adequate stimuli initiate action potentials that ultimately are used for these purposes.

▌ A stimulus alters the receptor's permeability, leading to a graded receptor potential.

A receptor may be either (1) a specialized ending of the afferent neuron or (2) a separate cell closely associated with the peripheral ending of the neuron. Stimulation of a receptor alters its membrane permeability, usually by causing a nonselective opening of all small ion channels. The means by which this permeability change takes place is individualized for each receptor type. Because the electrochemical driving force is greater for Na^+ than for other small ions at resting potential, the predominant effect is an inward flux of Na^+, which depolarizes the receptor membrane (see p. 100). (There are exceptions; for example, photoreceptors are hyperpolarized upon stimulation.)

This local depolarizing change in potential is known as a **receptor potential** in the case of a separate receptor or as a **generator potential** if the receptor is a specialized ending of an afferent neuron. The receptor (or generator) potential is a graded potential whose amplitude and duration can vary, depending on the strength and the rate of application or removal of the stimulus (see p. 100). The stronger the stimulus, the greater the permeability change and the larger the receptor potential. As is true of all graded potentials, receptor potentials have no refractory period, so summation in response to rapidly successive stimuli is possible. Because the receptor region has a very high threshold, action potentials do not take place at the receptor itself. For long-distance transmission, the receptor po-

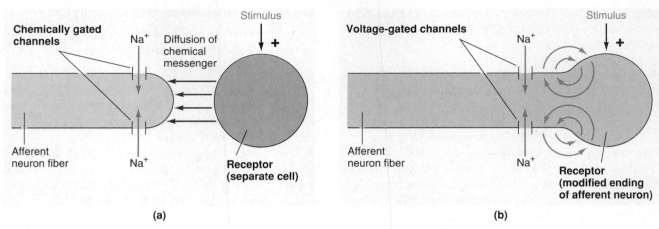

(a)   **(b)**

● **FIGURE 6-3**

Conversion of receptor and generator potentials into action potentials
(a) Receptor potential. The chemical messenger released from a separate receptor initiates an action potential in the fiber by opening chemically gated Na^+ channels. (b) Generator potential. The local current flow between the depolarized receptor ending and the afferent fiber initiates an action potential in the fiber by opening voltage-gated Na^+ channels.

tential must be converted into action potentials that can be propagated along the afferent fiber.

▌ Receptor potentials may initiate action potentials in the afferent neuron.

If of sufficient magnitude, a receptor (or generator) potential initiates an action potential in the afferent neuron membrane adjacent to the receptor by triggering the opening of Na^+ channels in this region. The means by which the Na^+ channels are opened differ depending on whether the receptor is a separate cell or a specialized afferent ending.

- In the case of a separate receptor, a receptor potential triggers the release of a chemical messenger that diffuses across the small space separating the receptor from the ending of the afferent neuron, similar to a synapse (● Figure 6-3a). Binding of the chemical messenger with specific protein receptor sites on the afferent neuron membrane opens chemically gated Na^+ channels (see p. 104).
- In the case of a specialized afferent ending, local current flow between the activated receptor ending undergoing a generator potential and the cell membrane adjacent to the receptor brings about opening of voltage-gated Na^+ channels in this adjacent region (● Figure 6-3b).

In either case, if the magnitude of the resulting ionic flux is sufficient to bring the adjacent membrane to threshold, an action potential is initiated that is self-propagated along the afferent fiber to the CNS. (For convenience, from here on we will refer to both receptor potentials and generator potentials as receptor potentials.)

Note that the site of initiation of action potentials differs in an afferent neuron from that in an efferent neuron or interneuron. In the latter two types of neurons, action potentials are initiated at the axon hillock located at the beginning of the

axon adjacent to the cell body (see p. 108). By contrast, action potentials are initiated at the peripheral end of an afferent nerve fiber adjacent to the receptor, a long distance from the cell body (● Figure 6-4).

The intensity of the stimulus is reflected by the magnitude of the receptor potential. In turn, the larger the receptor potential, the greater the frequency of action potentials generated in the afferent neuron. A larger receptor potential cannot bring about a larger action potential (all-or-none law), but it can induce more rapid firing of action potentials (see p. 113).

● **FIGURE 6-4**

Comparison of the site of initiation of an action potential in the three types of neurons

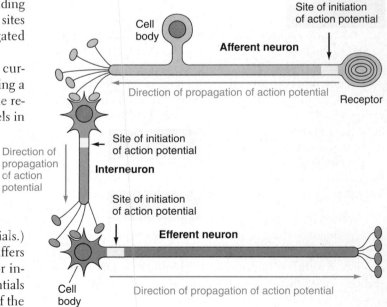

Stimulus strength is also reflected by the size of the area stimulated. Stronger stimuli usually affect larger areas, so a correspondingly greater population of receptors respond. For example, a light touch does not activate as many pressure receptors in the skin as a more forceful touch applied to the same area. Stimulus intensity is therefore distinguished both by the frequency of action potentials generated in the afferent neuron and by the number of receptors activated within the area.

▌ Receptors may adapt slowly or rapidly to sustained stimulation.

Stimuli of the same intensity do not always bring about receptor potentials of the same magnitude from the same receptor. Some receptors have the ability to diminish the extent of their depolarization despite sustained stimulus strength, a phenomenon known as **adaptation.** Subsequently, the frequency of action potentials generated in the afferent neuron decreases. That is, the receptor "adapts" to the stimulus by no longer responding to it to the same degree.

Types of receptors according to their speed of adaptation

There are two types of receptors—*tonic receptors* and *phasic receptors*—based on their speed of adaptation. **Tonic receptors** do not adapt at all or adapt slowly (● Figure 6-5a). These receptors are important in situations where maintained information about a stimulus is valuable. Examples of tonic receptors are muscle stretch receptors, which monitor muscle length, and joint proprioceptors, which measure the degree of joint flexion. To maintain posture and balance, the CNS must be continually apprised of the degree of muscle length and joint position. It is important, therefore, that these receptors do not adapt to a stimulus but continue to generate action potentials to relay this information to the CNS.

Phasic receptors, in contrast, are rapidly adapting receptors. The receptor rapidly adapts by no longer responding to a maintained stimulus, but when the stimulus is removed, the receptor typically responds with a slight depolarization known as the **off response** (● Figure 6-5b). Phasic receptors are useful in situations where it is important to signal a change in stimulus intensity rather than to relay status quo information. Rapidly adapting receptors include *tactile (touch) receptors* in the skin that signal changes in pressure on the skin surface. Because these receptors adapt rapidly, you are not continually conscious of wearing your watch, rings, and clothing. When you put something on, you soon become accustomed to it because of these receptors' rapid adaptation. When you take the item off, you are aware of its removal because of the off response.

Mechanism of adaptation in the Pacinian corpuscle

The mechanism by which adaptation is accomplished varies for different receptors and has not been fully elucidated for all receptor types. One receptor type that has been extensively studied is the **Pacinian corpuscle,** a rapidly adapting skin receptor that detects pressure and vibration. Adaptation in a

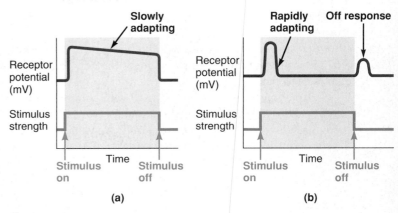

(a)   **(b)**

● **FIGURE 6-5**

Tonic and phasic receptors
(a) Tonic receptor. This receptor type does not adapt at all or adapts slowly to a sustained stimulus and thus provides continuous information about the stimulus. (b) Phasic receptor. This receptor type adapts rapidly to a sustained stimulus and frequently exhibits an off response when the stimulus is removed. Thus the receptor signals changes in stimulus intensity rather than relaying status quo information.

Pacinian corpuscle is believed to involve both mechanical and electrochemical components. The mechanical component depends on the physical properties of this receptor. A Pacinian corpuscle is a specialized receptor ending that consists of concentric layers of connective tissue resembling layers of an onion wrapped around the peripheral terminal of an afferent neuron (● Figure 6-6). When pressure is first applied to the Pacinian corpuscle, the underlying terminal responds with a receptor potential of a magnitude that reflects the intensity of the stimulus. As the stimulus continues, the pressure energy is dissipated because it causes the receptor layers to slip (just as steady pressure on a peeled onion causes its layers to slip). Because this physical effect filters out the steady component of the applied pressure, the underlying neuronal ending no longer responds with a receptor potential; that is, adaptation has occurred. Also contributing to adaptation is the electrochemical component, which involves changes in ionic movement across

● **FIGURE 6-6**

Pacinian corpuscle
The Pacinian corpuscle consists of concentric layers of connective tissue wrapped around the peripheral terminal of an afferent neuron.

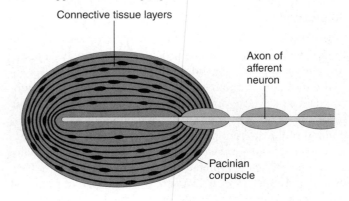

the receptor membrane. For reasons unknown, in a Pacinian corpuscle the Na^+ channels that opened in response to the stimulus are slowly inactivated, reducing the inward flow of Na^+ ions that was largely responsible for the depolarizing receptor potential.

Adaptation should not be confused with habituation (see p. 164). Although both these phenomena involve decreased neural responsiveness to repetitive stimuli, they operate at different points in the neural pathway. Adaptation is a receptor adjustment in the PNS, whereas habituation involves a modification in synaptic effectiveness in the CNS.

▌ Each somatosensory pathway is "labeled" according to modality and location.

On reaching the spinal cord, afferent information has two possible destinies: (1) it may become part of a reflex arc, bringing about an appropriate effector response, or (2) it may be relayed upward to the brain via ascending pathways for further processing and possible conscious awareness. Pathways conveying conscious somatic sensation, the **somatosensory pathways,** consist of discrete chains of neurons synaptically interconnected in a particular sequence to accomplish progressively more sophisticated processing of the sensory information. The afferent neuron with its peripheral receptor that first detects the stimulus is known as a **first-order sensory neuron.** It synapses on a **second-order sensory neuron,** either in the spinal cord or the medulla, depending on which sensory pathway is involved. This neuron then synapses on a **third-order sensory neuron** in the thalamus, and so on. With each step, the input is processed further. A particular sensory modality detected by a specialized receptor type is sent over a specific afferent and ascending pathway (a neural pathway committed to that modality) to excite a defined area in the somatosensory cortex. That is, a particular sensory input is **projected** to a specific region of the cortex (see ● Figure 5-30b, p. 175, for an example). Thus different types of incoming information are kept separated within specific **labeled lines** between the periphery and the cortex. In this way, even though all information is propagated to the CNS via the same type of signal (action potentials), the brain can decode the type and location of the stimulus. ▲ Table 6-1 summarizes how the CNS is informed of the type (what?), location (where?), and intensity (how much?) of a stimulus.

Phantom pain

Activation of a sensory pathway at any point gives rise to the same sensation that would be produced by stimulation of the receptors in the body part itself. This phenomenon has served as the traditional explanation for **phantom pain**—for example, pain perceived as originating in the foot by a person whose leg has been amputated at the knee. Irritation of the severed endings of the afferent pathways in the stump can trigger action potentials that, on reaching the foot region of the somatosensory cortex, are interpreted as pain in the missing foot. New evidence suggests that in addition, the sensation of phantom pain may arise from the recently documented extensive remod-

▲ **TABLE 6-1**
Coding of Sensory Information

Stimulus Property	Mechanism of Coding
Type of Stimulus (stimulus modality)	Distinguished by the type of receptor activated and the specific pathway over which this information is transmitted to a particular area of the cerebral cortex
Location of Stimulus	Distinguished by the location of the activated receptor field and the pathway that is subsequently activated to transmit this information to the area of the somatosensory cortex representing that particular location
Intensity of Stimulus (stimulus strength)	Distinguished by the frequency of action potentials initiated in an activated afferent neuron and the number of receptors (and afferent neurons) activated

eling of the brain region that originally handled sensations from the severed limb. This "remapping" of the "vacated" area of the brain is speculated to somehow lead to signals from elsewhere being misinterpreted as pain arising from the missing extremity.

▌ Acuity is influenced by receptive field size and lateral inhibition.

Each somatosensory neuron responds to stimulus information only within a circumscribed region of the skin surface surrounding it; this region is known as its **receptive field.** The size of a receptive field varies inversely with the density of receptors in the region; the more closely receptors of a particular type are spaced, the smaller the area of skin each monitors. The smaller the receptive field in a region, the greater its **acuity** or **discriminative ability.** Compare the tactile discrimination in your fingertips with that in your elbow by "feeling" the same object with both. You are able to discern more precise information about the object with your richly innervated fingertips because the receptive fields there are small; as a result, each neuron signals information about small, discrete portions of the object's surface. An estimated 17,000 tactile mechanoreceptors are present in the fingertips and palm of each hand. In contrast, the skin over the elbow is served by relatively few sensory endings with larger receptive fields. Subtle differences within each large receptive field cannot be detected (● Figure 6-7). The distorted cortical representation of various body parts in the sensory homunculus (see p. 148) corresponds precisely with the innervation density; more cortical space is allotted for sensory reception from areas with smaller receptive fields and, accordingly, greater tactile discriminative ability.

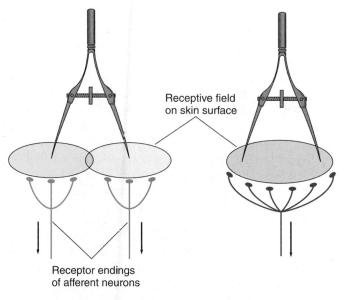

Receptive field
on skin surface

Receptor endings
of afferent neurons

Two receptive fields stimulated by
the two points of stimulation:
Two points felt

Only one receptive field stimulated
by the two points of stimulation
the same distance apart as in (a):
One point felt

(a)

(b)

● **FIGURE 6-7**

Comparison of discriminative ability of regions with small versus large receptive fields
The relative tactile acuity of a given region can be determined by the *two-point threshold-of-discrimination test*. If the two points of a pair of calipers applied to the surface of the skin stimulate two different receptive fields, two separate points will be felt. If the two points touch the same receptive field, they will be perceived as only one point. By adjusting the distance between the caliper points, one can determine the minimal distance at which the two points can be recognized as two rather than one, which is a reflection of the size of the receptive fields in the region. With this technique, it is possible to plot the discriminative ability of the body surface. The two-point threshold ranges from 2 mm in the fingertip (enabling one to read Braille, where the raised dots are spaced 2.5 mm apart) to 48 mm in the poorly discriminative skin of the calf of the leg. (a) Region with small receptive fields. (b) Region with large receptive fields.

Besides receptor density, a second factor influencing acuity is **lateral inhibition.** You can appreciate the importance of this phenomenon by slightly indenting the surface of your skin with the point of a pencil (● Figure 6-8a). The receptive field is excited immediately under the center of the pencil point where the stimulus is most intense, but the surrounding receptive fields are also stimulated, only to a lesser extent because they are less distorted. If information from these marginally excited afferent fibers in the fringe of the stimulus area were to reach the cortex, localization of the pencil point would be blurred. To facilitate localization and sharpen contrast, lateral inhibition occurs within the CNS (● Figure 6-8b). The most strongly activated signal pathway originating from the center of the stimulus area inhibits the less excited pathways from the fringe areas. This occurs via inhibitory interneurons that pass laterally between ascending fibers serving neighboring receptive fields. Blockage of further transmission in the weaker inputs increases the contrast between wanted and unwanted informa-

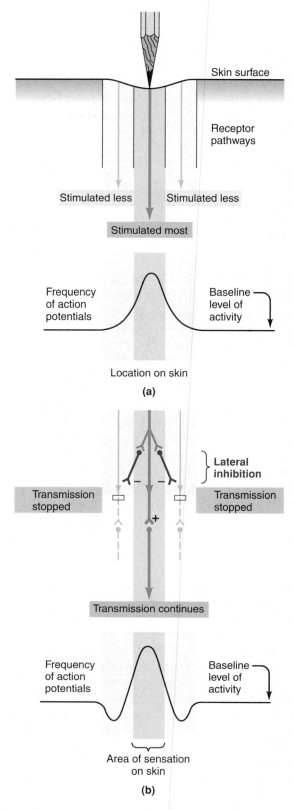

● **FIGURE 6-8**

Lateral inhibition
(a) The receptor at the site of most intense stimulation is activated to the greatest extent. Surrounding receptors are also stimulated but to a lesser degree. (b) The most intensely activated receptor pathway halts transmission of impulses in the less intensely stimulated pathways through lateral inhibition. This process facilitates localization of the site of stimulation.

The Peripheral Nervous System: Afferent Division; Special Senses

tion so that the pencil point can be precisely localized. The extent of lateral inhibitory connections within sensory pathways varies for different modalities. Those with the most lateral inhibition—touch and vision—bring about the most accurate localization.

Having completed our general discussion of receptor physiology, we are now going to examine one important somatic sensation in greater detail—pain.

PAIN

Pain is primarily a protective mechanism meant to bring to conscious awareness the fact that tissue damage is occurring or is about to occur.

▌ Stimulation of nociceptors elicits the perception of pain plus motivational and emotional responses.

Unlike other somatosensory modalities, the sensation of pain is accompanied by motivated behavioral responses (such as withdrawal or defense) as well as emotional reactions (such as crying or fear). Also, unlike other sensations, the subjective perception of pain can be influenced by other past or present experiences (for example, heightened pain perception accompanying fear of the dentist or lowered pain perception in an injured athlete during a competitive event). Thus pain is a complex perception whose processing is not fully understood.

Categories of pain receptors

There are three categories of pain receptors, or nociceptors: **Mechanical nociceptors** respond to mechanical damage such as cutting, crushing, or pinching; **thermal nociceptors** respond to temperature extremes, especially heat; and **polymodal nociceptors** respond equally to all kinds of damaging stimuli, including irritating chemicals released from injured tissues. None of the nociceptors have specialized receptor structures; they are all naked nerve endings. Because of their value to survival, nociceptors do not adapt to sustained or repetitive stimulation. However, all nociceptors can be sensitized by the presence of *prostaglandins*, which greatly enhance the receptor response to noxious stimuli (that is, it hurts more when prostaglandins are present). Prostaglandins are a special group of fatty acid derivatives that are cleaved from the lipid bilayer of the plasma membrane and act locally on being released (see p. 765). Tissue injury, among other things, can lead to the local release of prostaglandins. These chemicals act on the nociceptors' peripheral endings to lower their threshold for activation. Aspirin-like drugs inhibit the synthesis of prostaglandins, accounting at least in part for the **analgesic** (pain-relieving) properties of these drugs.

Fast and slow afferent pain fibers

Pain impulses originating at nociceptors are transmitted to the CNS via one of two types of afferent fibers (▲ Table 6-2). Signals arising from mechanical and thermal nociceptors are transmitted over small, myelinated **A-delta fibers** at rates of up to 30 meters/sec (the **fast pain pathway**). Impulses from polymodal nociceptors are carried by small, unmyelinated **C fibers** at a much slower rate of 12 meters/sec (the **slow pain pathway**). Think about the last time you cut or burned your finger. You undoubtedly felt a sharp twinge of pain at first, with a more diffuse, disagreeable pain commencing shortly thereafter. Pain typically is perceived initially as a brief, sharp, prickling sensation that is easily localized; this is the fast pain pathway originating from specific mechanical or heat nociceptors. This feeling is followed by a dull, aching, poorly localized sensation that persists for a longer time and is more unpleasant; this is the slow pain pathway, which is activated by chemicals, especially **bradykinin**, a normally inactive substance that is activated by enzymes released into the ECF from damaged tissue. Bradykinin and related compounds not only provoke pain, presumably by stimulating the polymodal nociceptors, but they also contribute to the inflammatory response to tissue injury (Chapter 12). The persistence of these chemicals might explain the long-lasting, aching pain that continues after removal of the mechanical or thermal stimulus that caused the tissue damage.

Interestingly, the peripheral receptors of afferent C fibers are activated by **capsaicin**, the ingredient in hot chili peppers that gives them their fiery zing. (In addition to binding with pain receptors, capsaicin also binds with the thermal receptors that are normally activated by heat—hence the burning sensation when eating hot peppers.) Ironically, local application of capsaicin can actually reduce clinical pain, most likely by overstimulating and damaging the nociceptors with which it binds.

Higher-level processing of pain input

The primary afferent pain fibers synapse with specific second-order interneurons in the dorsal horn of the spinal cord. In response to stimulus-induced action potentials, afferent pain fibers release neurotransmitters that influence these next neurons in line. The two best known of these pain neurotransmitters are *substance P* and *glutamate*. **Substance P** activates ascending pathways that transmit nociceptive signals to higher

▲ TABLE 6-2 Characteristics of Pain	
Fast Pain	**Slow Pain**
Occurs on stimulation of mechanical and thermal nociceptors	Occurs on stimulation of polymodal nociceptors
Carried by small, myelinated A-delta fibers	Carried by small, unmyelinated C fibers
Produces sharp, prickling sensation	Produces dull, aching, burning sensation
Easily localized	Poorly localized
Occurs first	Occurs second; persists for longer time; more unpleasant

levels for further processing (● Figure 6-9a). Ascending pain pathways have poorly understood destinations in the *somatosensory cortex*, the *thalamus*, and the *reticular formation*. The role of the cortex in pain perception is not clear, although it is probably important at least in localizing the pain. Pain can still be perceived in the absence of the cortex, presumably at the level of the thalamus. The reticular formation increases the level of alertness associated with the noxious encounter. Interconnections from the thalamus and reticular formation to the *hypothalamus* and *limbic system* elicit the behavioral

and emotional responses accompanying the painful experience. The limbic system appears to be especially important in perceiving the unpleasant aspects of pain.

Glutamate, the other neurotransmitter released from primary afferent pain terminals, is a major excitatory neurotransmitter (see p. 120). Glutamate acts on two different plasma-membrane receptors on the dorsal horn neurons, with two different outcomes. First, binding of glutamate with its *AMPA receptors* leads to permeability changes that ultimately result in the generation of action potentials in the dorsal horn cell.

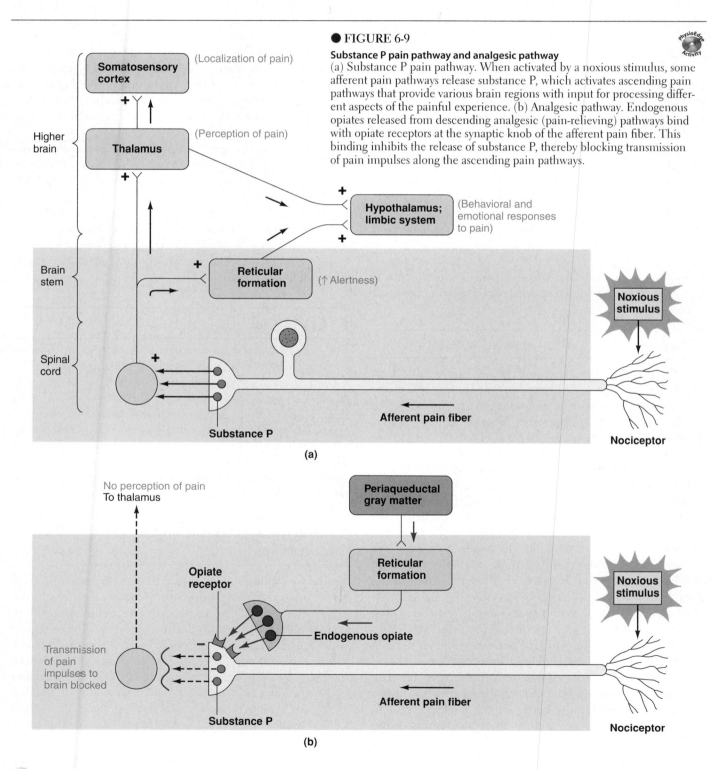

● FIGURE 6-9

Substance P pain pathway and analgesic pathway
(a) Substance P pain pathway. When activated by a noxious stimulus, some afferent pain pathways release substance P, which activates ascending pain pathways that provide various brain regions with input for processing different aspects of the painful experience. (b) Analgesic pathway. Endogenous opiates released from descending analgesic (pain-relieving) pathways bind with opiate receptors at the synaptic knob of the afferent pain fiber. This binding inhibits the release of substance P, thereby blocking transmission of pain impulses along the ascending pain pathways.

These action potentials transmit the pain message to higher centers. Second, binding of glutamate with its *NMDA receptors* leads to Ca^{2+} entry into the dorsal horn cell. This pathway is not involved in the transmission of pain messages. Instead, Ca^{2+} initiates second-messenger systems that make the dorsal horn neuron more excitable than usual (see p. 165). This hyperexcitability contributes in part to the exaggerated sensitivity of an injured area to subsequent exposure to painful or even normally nonpainful stimuli, such as a light touch. Think about how exquisitely sensitive your sunburned skin is, even to clothing. Other mechanisms in addition to glutamate-induced hyperexcitability of the dorsal horn neurons also contribute to supersensitivity of an injured area. For example, responsiveness of the pain-sensing peripheral receptors can be boosted so that they react more vigorously to subsequent stimuli. This exaggerated sensitivity presumably serves a useful purpose by discouraging activities that could cause further damage or interfere with healing of the injured area. Usually this hypersensitivity resolves as the injury heals.

Chronic pain, sometimes excruciating, occasionally occurs in the absence of tissue injury. In contrast to the pain accompanying peripheral injury, which serves as a normal protective mechanism to warn of impending or actual damage to the body, abnormal chronic pain states result from damage within the pain pathways in the peripheral nerves or the CNS. That is, pain is perceived because of abnormal signaling within the pain pathways in the absence of peripheral injury or typical painful stimuli. For example, strokes that damage ascending pathways can lead to an abnormal, persistent sensation of pain. Abnormal chronic pain is sometimes categorized as *neuropathic pain*.

▋ The brain has a built-in analgesic system.

In addition to the chain of neurons connecting peripheral nociceptors with higher CNS structures for pain perception, the CNS also contains a built-in pain-suppressing or **analgesic system** that suppresses transmission in the pain pathways as they enter the spinal cord. Two regions are known to be a part of this descending analgesic pathway. Electrical stimulation of the *periaqueductal gray matter* (gray matter surrounding the cerebral aqueduct, a narrow canal that connects the third and fourth ventricular cavities) results in profound analgesia, as does stimulation of the *reticular formation* within the brain stem. This analgesic system suppresses pain by blocking the release of substance P from afferent pain fiber terminals (● Figure 6-9b).

Specifically, the analgesic system depends on the presence of **opiate receptors.** It has long been known that **morphine,** a derivative of the opium poppy, is a powerful analgesic. It seemed highly unlikely that the body would be endowed with opiate receptors only to interact with chemicals derived from a flower! A search was therefore undertaken to discover the substances that normally bind with these opiate receptors. The result was the discovery of **endogenous opiates** (morphinelike substances)—the **endorphins, enkephalins,** and **dynorphin**—which are important in the body's natural analgesic system. These endogenous opiates serve as analgesic neurotransmitters; they are released from the descending analgesic pathway and bind with opiate receptors on the afferent pain fiber terminal. This binding suppresses the release of substance P via presynaptic inhibition, thereby blocking further transmission of the pain signal (see p. 124). Morphine binds to these same opiate receptors, which accounts for its analgesic properties.

The endogenous opiate system is normally dormant. It is unclear how this natural pain-suppressing mechanism is activated. Factors known to modulate pain include exercise, stress, and acupuncture. Endorphins are believed to be released during prolonged exercise and presumably are responsible for the "runner's high." Some types of stress also induce analgesia via the opiate pathway and other, less understood nonopiate mechanisms. It is sometimes disadvantageous for a stressed organism to display the normal reaction to pain. For example, when two male lions are fighting for dominance of the group, withdrawing, escaping, or resting when injured would mean certain defeat. (See the accompanying boxed feature, ◗ Concepts, Challenges, and Controversies, for an examination of how acupuncture relieves pain.)

We have now completed our coverage of somatic sensation. As you are now aware, somatic sensation is detected by widely distributed receptors that provide information about the body's interactions with the environment in general. In contrast, each of the special senses has highly localized, extensively specialized receptors that respond to unique environmental stimuli. The special senses include **vision, hearing, taste,** and **smell,** to which we now turn our attention, starting with vision.

EYE: VISION

The eyes capture the patterns of illumination in the environment as an "optical picture" on a layer of light-sensitive cells, the *retina*, much as a camera captures an image on film. Just as film can be developed into a visual likeness of the original image, the coded image on the retina is transmitted through a series of progressively more complex steps of visual processing until it is finally consciously perceived as a visual likeness of the original image. Before considering the steps involved in the process of vision, we are first going to examine how the eyes are protected from injury.

▋ Protective mechanisms help prevent eye injuries.

Several mechanisms help protect the eyes from injury. Except for its anterior (front) portion, the eyeball is sheltered by the bony socket in which it is positioned. The **eyelids** act like shutters to protect the anterior portion of the eye from environmental insults. They close reflexly to cover the eye under threatening circumstances, such as rapidly approaching objects, dazzling light, and instances when the exposed surface of the eye or eyelashes are touched. Frequent spontaneous blinking of the eyelids helps disperse the lubricating, cleansing, bactericidal ("germ-killing") **tears.** Tears are produced continuously

Acupuncture: Is It for Real?

It sounds like science fiction. How can a needle inserted in the hand relieve a toothache? **Acupuncture analgesia (AA)** the technique of relieving pain by the insertion and manipulation of threadlike needles at key points, has been practiced in China for over 2,000 years but is relatively new to Western medicine and still remains controversial in our country.

A history of mistrust and misunderstanding

Traditional Chinese teaching holds that disease can occur when the normal patterns of flow of healthful energy (called Qi; pronounced "chee") become disrupted, with acupuncture correcting this imbalance and restoring health. Many Western scientists have been skeptical because, until recently, the phenomenon could not be explained on the basis of any known, logical, physiological principles, although a tremendous body of anecdotal evidence in support of the effectiveness of AA existed in China.

According to a leading expert in acupuncture, the technique was not embraced in Western culture because of a clash in philosophies between West and East:

> Western medical science is quick to reject a phenomenon if it does not fit the current scientific theories. Chinese Taoism had a distaste for explanatory theories and chose instead merely to observe phenomena in order to be in harmony with mother nature. If a needle in the hand cured a toothache, that was sufficient for Chinese Taoism. For Western medicine acupuncture was impossible and hence was relegated to the wastebasket of placebo effects.*

The term *placebo effect* refers to a chemical or technique that brings about a desired response through the power of suggestion or distraction rather than through any direct action. The placebo effect was first documented in 1945 when a physician injected patients with what the patients thought was morphine for pain relief, but

*Gabriel Stux and Bruce Pomeranz, *Basics of Acupuncture*, 2nd ed. (New York: Springer, 1994), p. 1.

some received sugar (a placebo) instead. Pain was relieved in 70% of those who actually received morphine; surprisingly, 35% of those who received sugar, but believed they were receiving morphine, also reported pain relief.

Because the Chinese were content with anecdotal evidence for the success of AA, this phenomenon did not come under close scientific scrutiny until the last several decades when European and American scientists started studying it. As a result of these efforts, an impressive body of rigorous scientific investigation supports the contention that AA really works (that is, by a physiological rather than a placebo/psychological effect). Furthermore, its mechanisms of action have become apparent. Indeed, more is known about the underlying physiological mechanisms of AA than about those of many conventional medical techniques, such as gas anesthesia.

Effectiveness of acupuncture

AA has been proven effective in treating chronic pain and to exert a real physical effect in that it is more effective than placebo controls. In fact, AA compares favorably with morphine for treating chronic pain. In controlled clinical studies, 55% to 85% of patients were helped by AA. (By comparison, 70% of patients benefit from morphine therapy.) Pain relief was reported by only 30% to 35% of placebo controls (individuals who thought they were receiving proper AA treatment, but in whom needles were inserted in the wrong places or not deep enough).

Mechanism of action

The overwhelming body of evidence supports the *acupuncture endorphin hypothesis* as the primary mechanism of AA's action. According to this hypothesis, acupuncture needles activate specific afferent nerve fibers, which send impulses to the central nervous system. Here the incoming impulses activate three centers (a spinal cord center, a midbrain center, and a hormonal center, the hypothalamus-anterior pituitary unit) to cause analgesia. All three centers have been shown to block pain transmission

through use of endorphins and closely related compounds. Several other neurotransmitters, such as serotonin and norepinephrine, as well as cortisol, the major hormone released during stress, are implicated as well. (Pain relief in placebo controls is believed to occur as a result of such persons subconsciously activating their own built-in analgesic system.)

Acupuncture in the United States

In the United States, AA has not been used in mainstream medicine, even by physicians who have been convinced by scientific evidence that the technique is valid. AA methodology has traditionally not been taught in our medical colleges, and the techniques take time to learn. Also, AA is much more time-consuming than using drugs. Western physicians who have been trained to use drugs to solve most pain problems are generally reluctant to scrap their known methods for an unfamiliar, time-consuming technique. However, acupuncture is gaining favor as an alternative treatment for relief of chronic pain, especially because analgesic drugs can have troublesome side effects. After decades of being spurned by the vast majority of the U.S. medical community, acupuncture started gaining respectability following a 1997 report issued by an expert panel convened by the National Institutes of Health (NIH). This report, based on an evaluation of published scientific studies, concluded that acupuncture is effective as an alternative or adjunct to conventional treatment for many kinds of pain and nausea. Now that acupuncture has been sanctioned by NIH, some medical insurers have taken the lead in paying for this now scientifically legitimate treatment, and some of the nation's medical schools are beginning to incorporate the technique into their curricula. There are also 40 nationally accredited acupuncture schools for nonphysicians. Of the 13,000 licensed acupuncture practitioners in the United States, only 3,000 of them are physicians.

---

by the **lacrimal gland** in the upper lateral corner under the eyelid. This eye-washing fluid flows across the anterior surface of the eye and drains into tiny canals in the corner of each eye (● Figure 6-10a), eventually emptying into the back of the nasal passageway. This drainage system cannot handle the profuse tear production during crying, so the tears overflow from the eyes. The eyes are also equipped with protective **eyelashes,** which trap fine airborne debris such as dust before it can fall into the eye.

■ The eye is a fluid-filled sphere enclosed by three specialized tissue layers.

Each **eye** is a spherical, fluid-filled structure enclosed by three layers. From outermost to innermost, these are (1) the *sclera/cornea*; (2) the *choroid/ciliary body/iris*; and (3) the *retina* (● Figure 6-10b). Most of the eyeball is covered by a tough outer layer of connective tissue, the **sclera,** which forms the visible

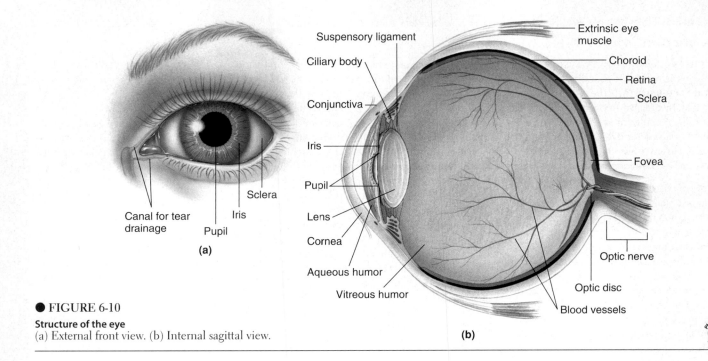

Suspensory ligament
Ciliary body
Conjunctiva
Iris
Pupil
Lens
Cornea
Aqueous humor
Vitreous humor

Extrinsic eye muscle
Choroid
Retina
Sclera
Fovea
Optic nerve
Optic disc
Blood vessels

Canal for tear drainage
Pupil
Iris
Sclera

(a)

(b)

● **FIGURE 6-10**
Structure of the eye
(a) External front view. (b) Internal sagittal view.

white part of the eye (● Figure 6-10a). Anteriorly, the outer layer consists of the transparent **cornea** through which light rays pass into the interior of the eye. The middle layer underneath the sclera is the highly pigmented **choroid,** which contains many blood vessels that nourish the retina. The choroid layer becomes specialized anteriorly to form the *ciliary body*

and *iris*, which will be described shortly. The innermost coat under the choroid is the **retina,** which consists of an outer pigmented layer and an inner nervous tissue layer. The latter contains the **rods** and **cones,** the photoreceptors that convert light energy into nerve impulses. Like the black walls of a photographic studio, the pigment in the choroid and retina absorbs light after it strikes the retina to prevent reflection or scattering of light within the eye.

The interior of the eye consists of two fluid-filled cavities, separated by an elliptical shaped **lens,** all of which are transparent to permit light to pass through the eye from the cornea to the retina. The larger posterior (rear) cavity between the lens and retina contains a semifluid, jellylike substance, the **vitreous humor.** The vitreous humor is important in maintaining the spherical shape of the eyeball. The anterior cavity between the cornea and lens contains a clear watery fluid, the **aqueous humor.** The aqueous humor carries nutrients for the cornea and lens, both of which lack a blood supply. Blood vessels in these structures would impede the passage of light to the photoreceptors.

Aqueous humor is produced at a rate of about 5 ml/day by a capillary network within the **ciliary body,** a specialized anterior derivative of the choroid layer. This fluid drains into a canal at the edge of the cornea and eventually enters the blood (● Figure 6-11). If aqueous humor is not drained as rapidly as it forms (for example, because of a blockage in the drainage canal), the excess will accumulate in the anterior cavity, causing the pressure to rise within the eye. This condition is known as **glaucoma.** The excess aqueous humor pushes the lens backward into the vitreous humor, which in turn is pushed against the inner

● **FIGURE 6-11**
Formation and drainage of aqueous humor
Aqueous humor is formed by a capillary network in the ciliary body, then drains into the canal of Schlemm, and eventually enters the blood.

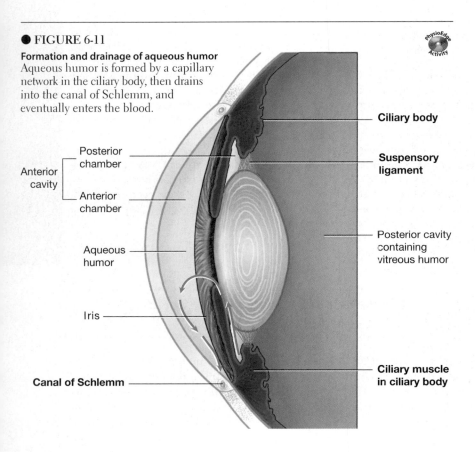

Anterior cavity
Posterior chamber
Anterior chamber
Aqueous humor
Iris
Canal of Schlemm

Ciliary body
Suspensory ligament
Posterior cavity containing vitreous humor
Ciliary muscle in ciliary body

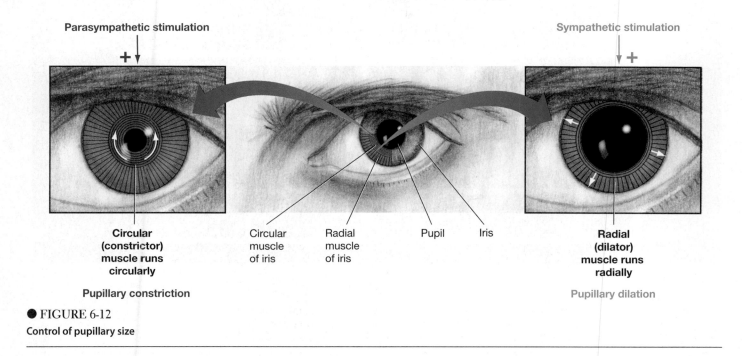

Parasympathetic stimulation

Sympathetic stimulation

| Circular (constrictor) muscle runs circularly | Circular muscle of iris | Radial muscle of iris | Pupil | Iris | Radial (dilator) muscle runs radially |

Pupillary constriction

Pupillary dilation

● FIGURE 6-12
Control of pupillary size

---

neural layer of the retina. This compression causes retinal and optic nerve damage that can lead to blindness if the condition is not treated.

▌ The amount of light entering the eye is controlled by the iris.

Not all the light passing through the cornea reaches the light-sensitive photoreceptors because of the presence of the iris, a thin, pigmented smooth muscle that forms a visible ringlike structure within the aqueous humor (● Figure 6-10a and b). The pigment in the iris is responsible for eye color. The varied flecks, lines, pits, and other nuances of the iris are unique for each individual, making the iris the basis of the latest identification technology. Recognition of iris patterns by a video camera that captures iris images and translates the landmarks into a computerized code is more foolproof than fingerprinting or even DNA testing.

The round opening in the center of the iris through which light enters the interior portions of the eye is the **pupil.** The size of this opening can be adjusted by variable contraction of the iris muscles to admit more or less light as needed, much as the shutter controls the amount of light entering a camera. The iris contains two sets of smooth muscle networks, one *circular* (the muscle fibers run in a ringlike fashion within the iris) and the other *radial* (the fibers project outward from the pupillary margin like bicycle spokes) (● Figure 6-12). Because muscle fibers shorten when they contract, the pupil gets smaller when the **circular** (or **constrictor**) **muscle** contracts and forms a smaller ring. This reflex pupillary constriction occurs in bright light to decrease the amount of light entering the eye. When the **radial** (or **dilator**) **muscle** shortens, the size of the pupil increases. Such pupillary dilation occurs in dim light to allow the entrance of more light. Iris muscles are controlled by the

autonomic nervous system. Parasympathetic nerve fibers innervate the circular muscle, and sympathetic fibers supply the radial muscle.

▌ The eye refracts the entering light to focus the image on the retina.

Light is a form of electromagnetic radiation composed of particle-like individual packets of energy called **photons** that travel in wavelike fashion. The distance between two wave peaks is known as the **wavelength** (● Figure 6-13). The photoreceptors in the eye are sensitive only to wavelengths between 400 and 700 nanometers (nm; billionths of a meter between peaks). This **visible light** is only a small portion of the total electromagnetic spectrum (● Figure 6-14). Light of different

---

● FIGURE 6-13

Properties of an electromagnetic wave
A wavelength is the distance between two wave peaks. The term *intensity* refers to the amplitude of the wave.

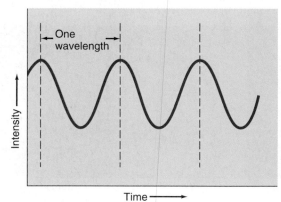

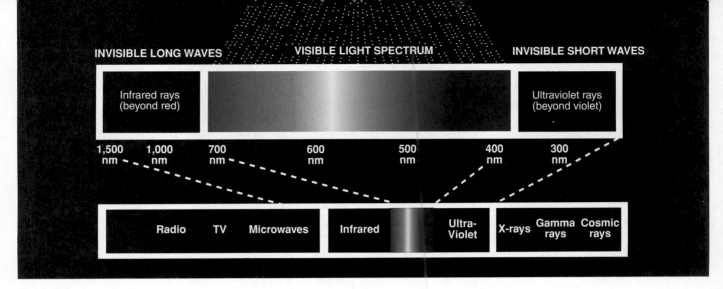

● FIGURE 6-14

Electromagnetic spectrum
The wavelengths in the electromagnetic spectrum range from 10^4m (10 km—for example, long radio waves) to less than 10^{-14}m (quadrillionths of a meter—for example, gamma and cosmic rays). The visible spectrum includes wavelengths ranging from 400 to 700 nanometers (nm; billionths of a meter).

wavelengths in this visible band is perceived as different color sensations. Short wavelengths are sensed as violet and blue; long wavelengths are interpreted as orange and red.

In addition to having variable wavelengths, light energy also varies in **intensity;** that is, the amplitude, or height, of the wave (● Figure 6-13). Dimming a bright red light does not change its color; it just becomes less intense or less bright.

Light waves *diverge* (radiate outward) in all directions from every point of a light source. The forward movement of a light wave in a particular direction is known as a **light ray.** Divergent light rays reaching the eye must be bent inward to be focused back into a point (the **focal point**) on the light-sensitive retina to provide an accurate image of the light source (● Figure 6-15).

Process of refraction

The bending of a light ray (**refraction**) occurs when the ray passes from a medium of one density into a medium of a dif-

ferent density (● Figure 6-16). Light travels faster through air than through other transparent media such as water and glass. When a light ray enters a medium of greater density, it is slowed down (the converse is also true). The course of direction of the ray changes if it strikes the surface of the new medium at any angle other than perpendicular. Thus two factors contribute to the degree of refraction: the comparative densities of the two media (the greater the difference in density, the greater the degree of bending) and the angle at which the light strikes the second medium (the greater the angle, the greater the refraction).

With a curved surface such as a lens, the greater the curvature, the greater the degree of bending and the stronger the lens. When a light ray strikes the curved surface of any object of greater density, the direction of refraction depends on the angle of the curvature (● Figure 6-17). A **convex** surface curves outward (like the outer surface of a ball) while a **concave** surface curves inward (like a cave). Convex surfaces converge light rays, bringing them closer together. Because convergence is essential for bringing an image to a focal point, refractive surfaces of the eye are convex. Concave surfaces diverge light rays (spread them farther apart). A concave lens is useful for correcting certain refractive errors of the eye, such as nearsightedness.

Eye's refractive structures

The two structures most important in the eye's refractive ability are the *cornea* and the *lens*. The curved corneal surface, the first structure light passes through as it enters the eye, contributes most extensively to the eye's total refractive ability because the difference in density at the air/corneal interface is much greater than the differences in density between the lens and the fluids surrounding it.

● FIGURE 6-15

Focusing of diverging light rays
Diverging light rays must be bent inward to be focused.

Point source of light   Light rays   Eye structures that bend light rays   Light rays focused on retina

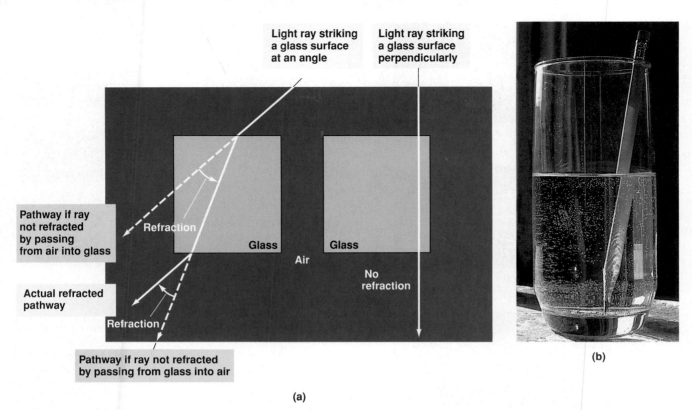

Light ray striking a glass surface at an angle

Light ray striking a glass surface perpendicularly

Pathway if ray not refracted by passing from air into glass

Refraction

Glass

Glass

Air

No refraction

Actual refracted pathway

Refraction

Pathway if ray not refracted by passing from glass into air

(a)

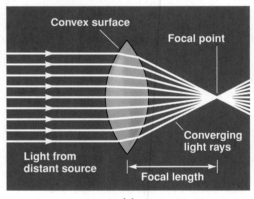

(b)

● **FIGURE 6-16**

Refraction

A light ray is bent (refracted) when it strikes the surface of a medium of different density from the one in which it had been traveling (for example, moving from air into glass) at any angle other than perpendicular to the new medium's surface. (b) The pencil in the glass of water *appears* to bend. What is happening, though, is that the light rays coming to the camera (or your eyes) are bent as they pass through the water, then the glass, and then the air. Consequently the pencil appears distorted.

● **FIGURE 6-17**

Refraction by convex and concave lenses

(a) A lens with a convex surface, which converges the rays (brings them closer together). (b) A lens with a concave surface, which diverges the rays (spreads them farther apart).

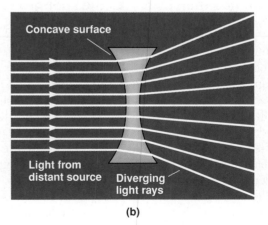

Convex surface

Focal point

Light from distant source

Converging light rays

Focal length

(a)

Concave surface

Light from distant source

Diverging light rays

(b)

In **astigmatism**, the curvature of the cornea is uneven, so light rays are unequally refracted. The refractive ability of a person's cornea remains constant because the curvature of the cornea never changes. In contrast, the refractive ability of the lens can be adjusted by changing its curvature as needed for near or far vision.

Rays from light sources more than 20 feet away are considered to be parallel by the time they reach the eye. By contrast, light rays originating from near objects are still diverging when they reach the eye. For a given refractive ability of the eye, it takes a greater distance behind the lens to bring the divergent rays of a near source to a focal point than to bring the parallel rays of a far source to a focal point (● Figure 6-18 a and b). However, in a particular eye the distance between the lens and the retina always remains the same. Therefore, a greater distance beyond the lens is not available for bringing near objects into focus. Yet the refractive structures of the eye must bring both near and far light sources into focus on the retina for clear vision. If an image is focused before it reaches the retina or is not yet focused when it reaches the retina, it will be blurred (● Figure 6-19). To bring both near and far light sources into focus on the retina (that is, in the same dis-

The Peripheral Nervous System: Afferent Division; Special Senses

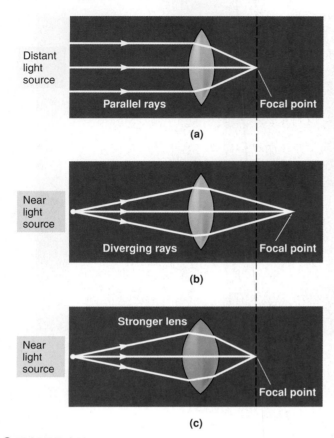

● FIGURE 6-18

Focusing of distant and near sources of light
The rays from a distant (far) light source (more than 20 feet from the eye) are parallel by the time the rays reach the eye. (b) The rays from a near light source (less than 20 feet from the eye) are still diverging when they reach the eye. A longer distance is required for a lens of a given strength to bend the diverging rays from a near light source into focus compared to the parallel rays from a distant light source. (c) To focus both a distant and a near light source in the same distance (the distance between the lens and retina), a stronger lens must be used for the near source.

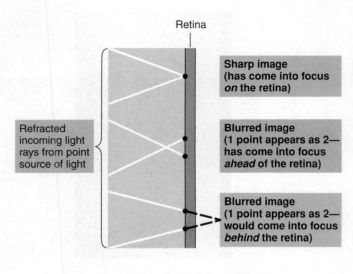

● = Points of stimulation of the retina

● FIGURE 6-19

Comparison of images that do and do not come into focus on the retina

tance), a stronger lens must be used for the near source (● Figure 6-18c). Let's see how the strength of the lens can be adjusted as needed.

▌ Accommodation increases the strength of the lens for near vision.

The ability to adjust the strength of the lens is known as **accommodation.** The strength of the lens depends on its shape, which in turn is regulated by the ciliary muscle.

The **ciliary muscle** is part of the ciliary body, an anterior specialization of the choroid layer. The ciliary body has two major components: the ciliary muscle and the capillary network that produces the aqueous humor (see ● Figure 6-11). The ciliary muscle is a circular ring of smooth muscle attached to the lens by **suspensory ligaments** (● Figure 6-20).

When the ciliary muscle is relaxed, the suspensory ligaments are taut, and they pull the lens into a flattened, weakly

refractive shape (● Figure 6-20c). As the muscle contracts, its circumference decreases, slackening the tension in the suspensory ligaments (● Figure 6-20d). When the lens is subjected to less tension by the suspensory ligaments, it assumes a more spherical shape because of its inherent elasticity. The greater curvature of the more rounded lens increases its strength, causing greater bending of light rays. In the normal eye, the ciliary muscle is relaxed and the lens is flat for far vision, but the muscle contracts to allow the lens to become more convex and stronger for near vision. The ciliary muscle is controlled by the autonomic nervous system. Sympathetic stimulation causes the ciliary muscle to relax for far vision, whereas parasympathetic stimulation causes the muscle to contract for near vision.

The lens is an elastic structure consisting of transparent fibers. These fibers occasionally become opaque so that light rays cannot pass through, a condition known as a **cataract.** The defective lens can usually be surgically removed and vision restored by an implanted artificial lens or compensating eyeglasses.

Throughout life, only cells at the outer edges of the lens are replaced. Cells in the center of the lens are in double jeopardy. Not only are they oldest, but they also are the farthest away from the aqueous humor, the lens's nutrient source. With advancing age, these nonrenewable central cells die and become stiff. With loss of elasticity, the lens can no longer assume the spherical shape required to accommodate for near vision. This age-related reduction in accommodative ability, **presbyopia,** affects most people by middle age (45 to 50), requiring them to resort to corrective lenses for near vision (reading).

Other common vision disorders are *nearsightedness (myopia)* and *farsightedness (hyperopia).* In a normal eye (**emmetropia**) (● Figure 6-21a), a far light source is focused on the retina without accommodation, whereas the strength of the lens is increased by accommodation to bring a near source into focus. In **myopia** (● Figure 6-21b1), because the eyeball is too long

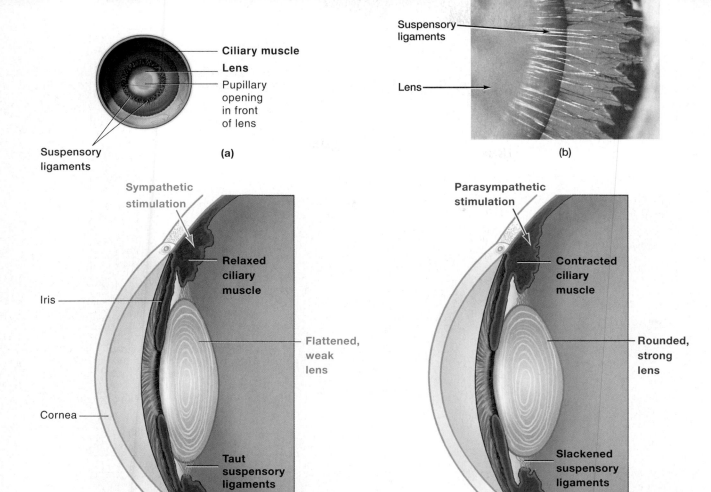

Ciliary muscle

Lens

Pupillary opening in front of lens

Suspensory ligaments

(a)

Suspensory ligaments

Lens

(b)

Sympathetic stimulation

Iris

Cornea

Relaxed ciliary muscle

Flattened, weak lens

Taut suspensory ligaments

(c)

Parasympathetic stimulation

Contracted ciliary muscle

Rounded, strong lens

Slackened suspensory ligaments

(d)

● **FIGURE 6-20**

Mechanism of accommodation
(a) Schematic representation of suspensory ligaments extending from the ciliary muscle to the outer edge of the lens. (b) Scanning electron micrograph showing the suspensory ligaments attached to the lens. (c) When the ciliary muscle is relaxed, the suspensory ligaments are taut, putting tension on the lens so that it is flat and weak. (d) When the ciliary muscle is contracted, the suspensory ligaments become slack, reducing the tension on the lens. The lens can then assume a stronger, rounder shape because of its elasticity.

or the lens is too strong, a near light source is brought into focus on the retina without accommodation (even though accommodation is normally used for near vision), whereas a far light source is focused in front of the retina and is blurry. Thus a myopic individual has better near vision than far vision, a condition that can be corrected by a concave lens (● Figure 6-21b2). With **hyperopia** (● Figure 6-21c1), either the eyeball is too short or the lens is too weak. Far objects are focused on the retina only with accommodation, whereas near objects are focused behind the retina even with accommodation and, accordingly, are blurry. Thus a hyperopic individual has better far vision than near vision, a condition that can be corrected by a convex lens (● Figure 6-21c2). Such vision tends to get worse as the person gets older because of loss of accommodative ability with the onset of presbyopia.

▌ Light must pass through several retinal layers before reaching the photoreceptors.

The major function of the eye is to focus light rays from the environment on the *rods* and *cones*, the photoreceptor cells of the retina. The photoreceptors then transform the light energy into electrical signals for transmission to the CNS.

The receptor-containing portion of the retina is actually an extension of the CNS and not a separate peripheral organ.

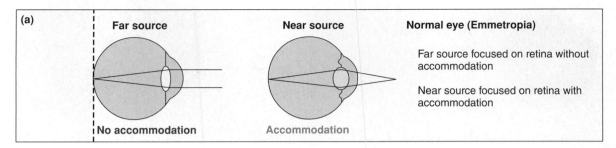

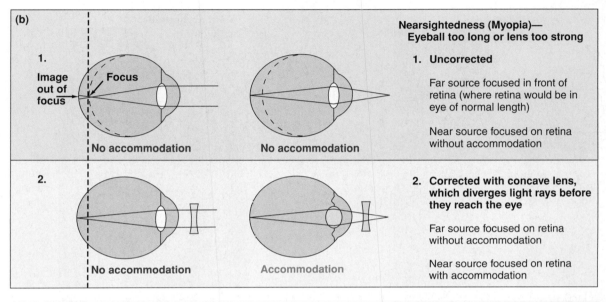

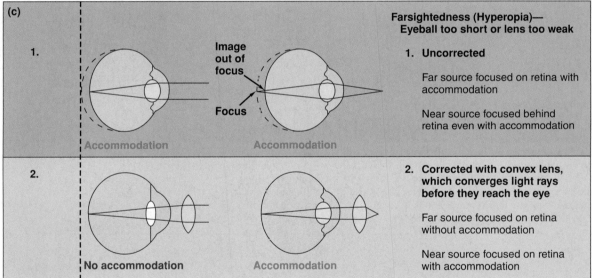

● FIGURE 6-21

Emmetropia, myopia, and hyperopia
This figure compares far vision and near vision in the (a) normal eye with (b) nearsightedness and (c) farsightedness in both their (1) uncorrected and (2) corrected states. The vertical dashed line represents the normal distance of the retina from the cornea; that is, the site at which an image is brought into focus by the refractive structures in a normal eye.

During embryonic development, the retinal cells "back out" of the nervous system, so the retinal layers, surprisingly, are facing backward! The neural portion of the retina consists of three layers of excitable cells (● Figure 6-22): (1) the outermost layer (closest to the choroid) containing the **rods** and **cones,** whose light-sensitive ends face the choroid (away from the incoming light); (2) a middle layer of **bipolar cells;** and (3) an inner layer of **ganglion cells.** Axons of the ganglion cells join together to

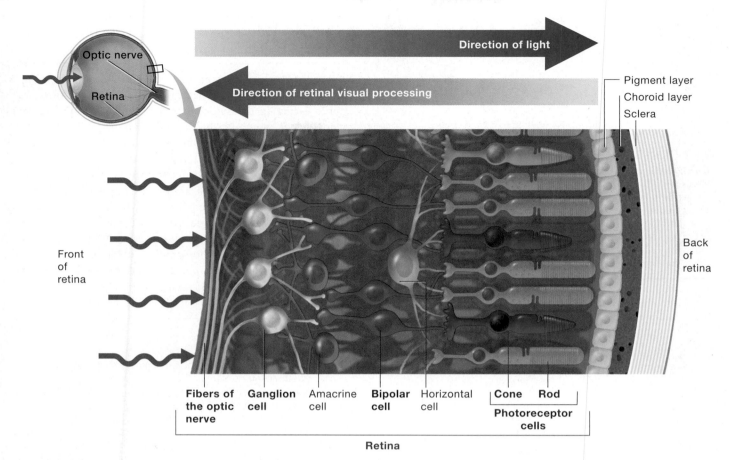

Direction of light

Direction of retinal visual processing

Optic nerve

Retina

Pigment layer
Choroid layer
Sclera

Front
of
retina

Back
of
retina

Fibers of the optic nerve   **Ganglion cell**   Amacrine cell   **Bipolar cell**   Horizontal cell   **Cone**   **Rod**

Photoreceptor cells

Retina

● FIGURE 6-22

Retinal layers
The retinal visual pathway extends from the photoreceptor cells (rods and cones, whose light-sensitive ends face the choroid *away from* the incoming light) to the bipolar cells to the ganglion cells. The horizontal and amacrine cells act locally for retinal processing of visual input.

form the **optic nerve,** which leaves the retina slightly off center. The point on the retina at which the optic nerve leaves and through which blood vessels pass is the **optic disc** (● Figures 6-10b and 6-23). This region is often referred to as the **blind spot;** no image can be detected in this area because it is devoid of rods and cones. We are normally not aware of the blind spot, because central processing somehow "fills in" the missing spot. You can discover the existence of your own blind spot by a simple demonstration (● Figure 6-24).

Light must pass through the ganglion and bipolar layers before reaching the photoreceptors in all areas of the retina except the fovea. In the **fovea,** which is a pinhead-sized depression located in the exact center of the retina, the bipolar and ganglion cell layers are pulled aside so that light strikes the photoreceptors directly (● Figure 6-10b). This feature, coupled with the fact that only cones (which have greater acuity or discriminative ability than the rods) are found here, makes the fovea the point of most distinct vision. Thus we turn our eyes so that the object at which we are looking is focused on the fovea. The area immediately surrounding the fovea, the **macula lutea,** also has a high concentration of cones and fairly high acuity (● Figure 6-23). Macular acuity, however, is less than that of the fovea, because of the overlying ganglion and bipolar cells in the macula.

● FIGURE 6-23

View of the retina seen through an ophthalmoscope
With an ophthalmoscope, a lighted viewing instrument, it is possible to view the optic disc (blind spot) and macula lutea within the retina of the rear of the eye.

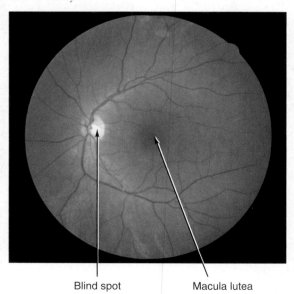

Blind spot   Macula lutea

● FIGURE 6-24

Demonstration of the blind spot
Discover the blind spot in your left eye by closing your right eye and holding the book about 4 inches from your face. While focusing on the cross, gradually move the book away from you until the circle vanishes from view. At this time, the image of the circle is striking the blind spot of your left eye. You can similarly discover the blind spot in your right eye by closing your left eye and focusing on the circle. The cross will disappear when its image strikes the blind spot of your right eye.

▌ Phototransduction by retinal cells converts light stimuli into neural signals.

Photoreceptors (rod and cone cells) consist of three parts (● Figure 6-25a): (1) an *outer segment*, which lies closest to the eye's exterior, facing the choroid, and detects the light stimulus; (2) an *inner segment*, which lies in the middle of the photoreceptor's length and contains the metabolic machinery of the cell; and (3) a *synaptic terminal*, which lies closest to the eye's interior, facing the bipolar cells, and transmits the signal generated in the photoreceptor upon light stimulation to these next cells in the visual pathway. The outer segment, which is rod shaped in rods and cone shaped in cones (● Figure 6-25a), is composed of stacked, flattened, membranous discs containing an abundance of light-sensitive *photopigment* molecules. Each retina has about 150 million photoreceptors, and over a billion photo-

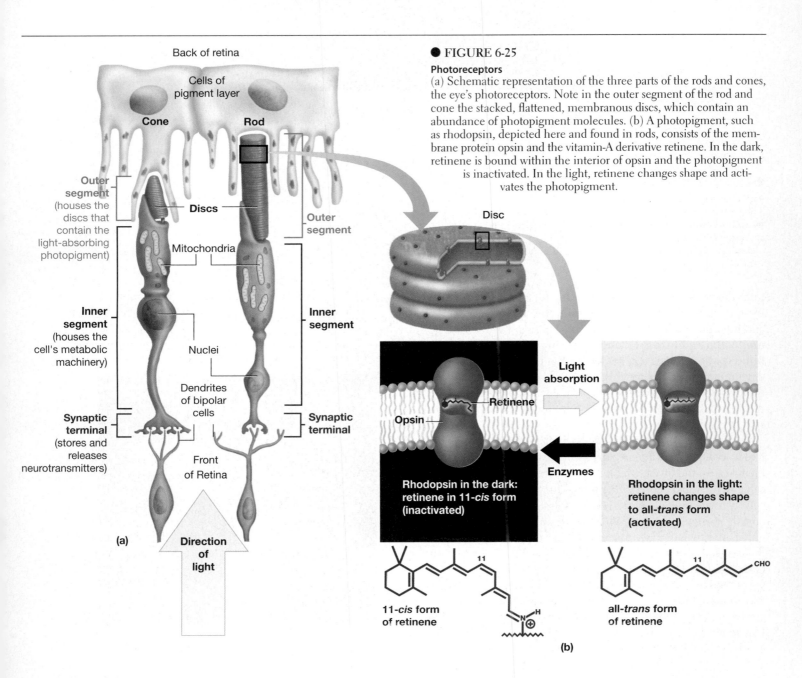

● FIGURE 6-25

Photoreceptors
(a) Schematic representation of the three parts of the rods and cones, the eye's photoreceptors. Note in the outer segment of the rod and cone the stacked, flattened, membranous discs, which contain an abundance of photopigment molecules. (b) A photopigment, such as rhodopsin, depicted here and found in rods, consists of the membrane protein opsin and the vitamin-A derivative retinene. In the dark, retinene is bound within the interior of opsin and the photopigment is inactivated. In the light, retinene changes shape and activates the photopigment.

pigment molecules may be packed into the outer segment of each photoreceptor.

Photopigments undergo chemical alterations when activated by light. Through a series of steps, this light-induced change and subsequent activation of the photopigment bring about a receptor potential that ultimately leads to the generation of action potentials, which transmit this information to the brain for visual processing. A photopigment consists of two components: **opsin,** a protein that is an integral part of the disc membrane; and **retinene,** a derivative of vitamin A that is bound within the interior of the opsin molecule (● Figure 6-25b). Retinene is the light-absorbing part of the photopigment. There are four different photopigments, one in the rods and one in each of three types of cones. These four photopigments differentially absorb various wavelengths of light. **Rhodopsin,** the rod photopigment, cannot discriminate between various wavelengths in the visible spectrum; it absorbs all visible wavelengths. Therefore, rods provide vision only in shades of gray by detecting different intensities, not different colors. The photopigments in the three types of cones—**red, green,** and **blue cones**—respond selectively to various wavelengths of light, making color vision possible.

Phototransduction, the process of converting light stimuli into electrical signals, is basically the same for all photoreceptors, but the mechanism is contrary to the usual means by which receptors respond to their adequate stimulus. Receptors typically *depolarize* when stimulated, but photoreceptors *hyperpolarize* on light absorption. Let's first examine the status of the photoreceptors in the dark, then consider what happens when they are exposed to light.

Photoreceptor activity in the dark

The plasma membrane of a photoreceptor's outer segment contains chemically gated Na^+ channels. Unlike other chemically gated channels that respond to extracellular chemical messengers, these channels respond to an internal second messenger, **cyclic GMP** or **cGMP** (cyclic guanosine monophosphate). Binding of cGMP to these Na^+ channels keeps them open. In the absence of light, the concentration of cGMP is high (● Figure 6-26a). Unlike most receptors, therefore, the Na^+ channels of a photoreceptor are open in the absence of stimulation, that is, in the dark. The resultant passive inward Na^+ leak depolarizes the photoreceptor. The passive spread of this depolarization from the outer segment (where the Na^+ channels are located) to the synaptic terminal (where the photoreceptor's neurotransmitter is stored) keeps the synaptic terminal's voltage-gated Ca^{2+} channels open. Calcium entry triggers the release of neurotransmitter from the synaptic terminal while in the dark.

Photoreceptor activity in the light

On exposure to light, the concentration of cGMP is decreased through a series of biochemical steps triggered by photopigment activation (● Figure 6-26b). When retinene absorbs light, it changes shape but still remains bound to opsin (● Figure 6-25b). This change in conformation activates the photopigment. Rod and cone cells contain a G protein called **transducin** (see p. 67). The activated photopigment activates transducin,

which in turn activates the intracellular enzyme phosphodiesterase. This enzyme degrades cGMP, thus decreasing the concentration of this second messenger in the photoreceptor. During the light excitation process, the reduction in cGMP permits the chemically gated Na^+ channels to close. This channel closure stops the depolarizing Na^+ leak and causes membrane hyperpolarization. This hyperpolarization, which is the receptor potential, passively spreads from the outer segment to the synaptic terminal of the photoreceptor. Here the potential changes lead to closure of the voltage-gated Ca^{2+} channels and a subsequent reduction in neurotransmitter release from the synaptic terminal. Thus photoreceptors are *inhibited by their adequate stimulus* (hyperpolarized by light) and *excited in the absence of stimulation* (depolarized by darkness). The hyperpolarizing potential and subsequent decrease in neurotransmitter release are graded according to the intensity of light. The brighter the light, the greater the hyperpolarizing response and the greater the reduction in neurotransmitter release.

Further retinal processing of light input

How does the retina signal the brain about light stimulation through such an inhibitory response? The photoreceptors synapse with bipolar cells. These cells in turn terminate on the ganglion cells, whose axons form the optic nerve for transmission of signals to the brain. The neurotransmitter released from the photoreceptors' synaptic terminal has an *inhibitory* action on the bipolar cells. The reduction in neurotransmitter release that accompanies light-induced receptor hyperpolarization decreases this inhibitory action on the bipolar cells. Removal of inhibition has the same effect as direct excitation of the bipolar cells. The greater the illumination on the receptor cells, the greater the removal of inhibition from the bipolar cells and the greater in effect the excitation of these next cells in the visual pathway to the brain.

Bipolar cells display graded potentials similar to the photoreceptors. Action potentials do not originate until the ganglion cells, the first neurons in the chain that must propagate the visual message over long distances to the brain.

The altered photopigments are restored to their original conformation in the dark by enzyme-mediated mechanisms (● Figure 6-25b). Subsequently, the membrane potential and rate of neurotransmitter release of the photoreceptor are returned to their unexcited state, and no further action potentials are transmitted to the visual cortex (● Figure 6-26a).

Researchers are currently working on an ambitious and still highly speculative microelectronic chip that would serve as a partial substitute retina. Their hope is that the device will be able to restore at least some sight in people who are blinded by loss of photoreceptor cells but whose ganglion cells and optic pathways remain healthy. For example, if successful the chip could benefit people with **macular degeneration,** the leading cause of blindness in the western hemisphere. This condition is characterized by loss of photoreceptors in the macula in association with advancing age. The envisioned "vision chip" would bypass the photoreceptor step altogether: Images received by means of a camera mounted on eyeglasses would be translated by the chip into electrical signals detectable by the ganglion cells and transmitted on for further optical processing.

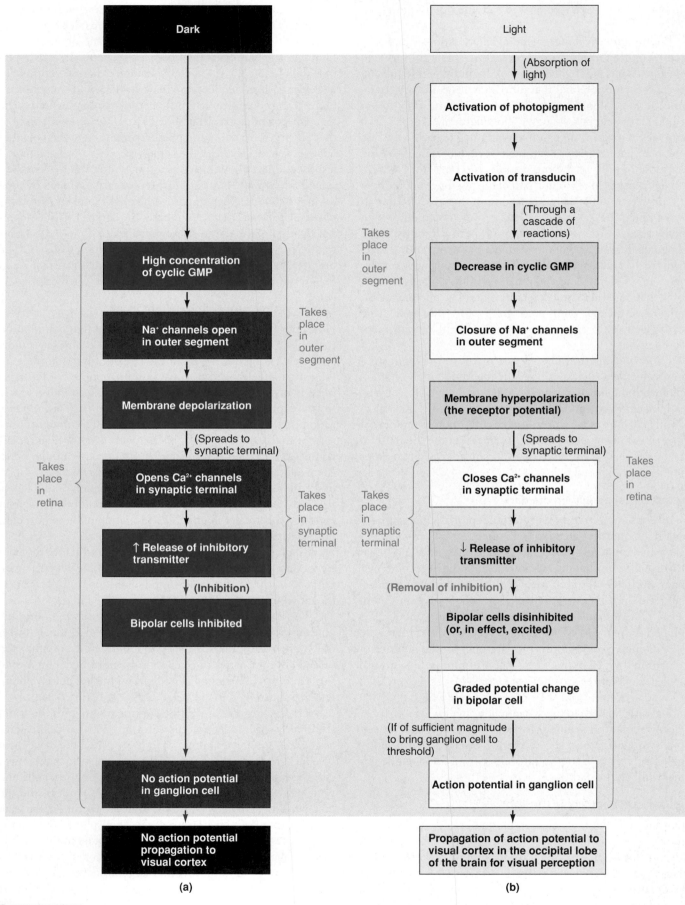

● FIGURE 6-26

Phototransduction and initiation of an action potential in the visual pathway
(a) Events occurring in a photoreceptor in response to the dark that prevent action potentials from being initiated in the visual pathway. (b) Events occurring in a photoreceptor in response to a light stimulus that initiate an action potential in the visual pathway (phototransduction).

Rods provide indistinct gray vision at night, whereas cones provide sharp color vision during the day.

The retina contains more than 30 times more rods than cones (100 million rods compared to 3 million cones per eye). Cones are most abundant in the center of the retina, in the macula. From this point outward, the concentration of cones decreases and the concentration of rods increases. Rods are most abundant in the periphery. Because of differential absorption of various wavelengths of light, cones provide color vision whereas rods provide vision only in shades of gray. The capabilities of the rods and cones also differ in other respects because of a difference in the "wiring patterns" between these photoreceptor types and other retinal neuronal layers (▲ Table 6-3). Cones have low sensitivity to light, being "turned on" only by bright daylight, but they have high acuity (sharpness; ability to distinguish between two nearby points). Thus cones provide sharp vision with high resolution for fine detail. Humans use cones for day vision, which is in color and distinct. Rods, in contrast, have low acuity but high sensitivity, so they respond to the dim light of night. We are able to see at night with our rods but at the expense of color and distinctness. Let us see how wiring patterns influence sensitivity and acuity.

There is little convergence of neurons in the retinal pathways for cone output (see p. 128). Each cone generally has a private line connecting it to a particular ganglion cell. In contrast, there is much convergence in rod pathways. Output from more than 100 rods may converge via bipolar cells on a single ganglion cell.

Before a ganglion cell can have an action potential, the cell must be brought to threshold through influence of the graded potentials in the receptors to which it is wired. Because a single-cone ganglion cell is influenced by only one cone, only bright daylight is intense enough to induce a sufficient receptor potential in the cone to ultimately bring the ganglion cell to threshold. The abundant convergence in the rod visual pathways, in contrast, offers good opportunities for summation of subthreshold events in a rod ganglion cell (see p. 121).

Whereas a small receptor potential induced by dim light in a single cone would not be sufficient to bring its ganglion cell to threshold, similar small receptor potentials induced by the same dim light in multiple rods converging on a single ganglion cell would have an additive effect to bring the rod ganglion cell to threshold. Because rods can bring about action potentials in response to small amounts of light, they are much more sensitive than cones. However, because cones have private lines into the optic nerve, each cone transmits information about an extremely small receptive field on the retinal surface. Cones are thus able to provide highly detailed vision at the expense of sensitivity. With rod vision, acuity is sacrificed for sensitivity. Because many rods share a single ganglion cell, once an action potential is initiated it is impossible to discern which of the multiple rod inputs were activated to bring the ganglion cell to threshold. Objects appear fuzzy when rod vision is used, because of this poor ability to distinguish between two nearby points.

The sensitivity of the eyes can vary markedly through dark and light adaptation.

The eyes' sensitivity to light depends on the amount of photopigment present in the rods and cones. When you go from bright sunlight into darkened surroundings, you cannot see anything at first, but gradually you begin to distinguish objects as a result of the process of **dark adaptation.** Breakdown of photopigments during exposure to sunlight tremendously decreases photoreceptor sensitivity. For example, a reduction in rhodopsin content of only 0.6% from its maximum value decreases rod sensitivity approximately 3,000 times. In the dark, the photopigments broken down during light exposure are gradually regenerated. As a result, the sensitivity of your eyes gradually increases so that you can begin to see in the darkened surroundings. However, only the highly sensitive, rejuvenated rods are "turned on" by the dim light.

Conversely, when you move from the dark to the light (for example, leaving a movie theater and entering the bright sunlight), your eyes are very sensitive to the dazzling light at first. With little contrast between lighter and darker parts, the entire image appears bleached. As some of the photopigments are rapidly broken down by the intense light, the sensitivity of the eyes decreases and normal contrasts can once again be detected, a process known as **light adaptation.** The rods are so sensitive to light that sufficient rhodopsin is broken down in bright light to essentially "burn out" the rods; that is, after the rod photopigments have already been broken down by the bright light, they are no longer able to respond to the light. Furthermore, a central neural adaptive mechanism switches the eye from the rod system to the cone system on exposure to bright light. Therefore, only the less sensitive cones are used for day vision.

It is estimated that our eyes' sensitivity can change as much as 1 million times as they adjust to various levels of illumination through dark and light adaptation. These adaptive measures are also enhanced by pupillary reflexes that adjust the amount of available light permitted to enter the eye.

▲ TABLE 6-3
Properties of Rod Vision and Cone Vision

Rods	Cones
100 million per retina	3 million per retina
Vision in shades of gray	Color vision
High sensitivity	Low sensitivity
Low acuity	High acuity
Night vision	Day vision
Much convergence in retinal pathways	Little convergence in retinal pathways
More numerous in periphery	Concentrated in fovea

Because retinene, one of the photopigment components, is a derivative of vitamin A, adequate amounts of this nutrient must be available for the ongoing resynthesis of photopigments. **Night blindness** occurs as a result of dietary deficiencies of vitamin A. Although photopigment concentrations in both rods and cones are reduced in this condition, there is still sufficient cone photopigment to respond to the intense stimulation of bright light, except in the most severe cases. However, even modest reductions in rhodopsin content can decrease the sensitivity of rods so much that they cannot respond to dim light. The person can see in the day using cones but cannot see at night because the rods are no longer functional. Thus carrots are "good for your eyes" because they are rich in vitamin A.

▌ Color vision depends on the ratios of stimulation of the three cone types.

Vision depends on stimulation of retinal photoreceptors by light. Certain objects in the environment such as the sun, fire, and lightbulbs, emit light. But how do we see objects such as chairs, trees, and people, which do not emit light? The pigments in various objects selectively absorb particular wavelengths of light transmitted to them from light-emitting sources, and the unabsorbed wavelengths are reflected from the objects' surfaces. These reflected light rays enable us to see the objects. An object perceived as blue absorbs the longer red and green wavelengths of light and reflects the shorter blue wavelengths, which can be absorbed by the photopigment in the eyes' blue cones, thereby activating them.

Each cone type is most effectively activated by a particular wavelength of light in the range of color indicated by its name—blue, green, or red. However, cones also respond in varying degrees to other wavelengths (● Figure 6-27). **Color vision,** our perception of the many colors of the world, depends on the three cone types' various *ratios of stimulation* in response to different wavelengths. A wavelength perceived as blue does not stimulate red or green cones at all but excites blue cones maximally (the percentage of maximal stimulation for red, green, and blue cones, respectively, is 0:0:100). The sensation of yellow, in comparison, arises from a stimulation ratio of 83:83:0, red and green cones each being stimulated 83% of maximum while blue cones are not excited at all. The ratio for green is 31:67:36, and so on, with various combinations giving rise to the sensation of all the different colors. White is a mixture of all wavelengths of light, whereas black is the absence of light.

The extent to which each of the cone types is excited is coded and transmitted in separate parallel pathways to the brain. A distinct color vision center in the primary visual cortex combines and processes these inputs to generate the perception of color, taking into consideration the object in comparison with its background. The concept of color is therefore in the mind of the beholder. Most of us agree on what color we see because we have the same types of cones and use similar neural pathways for comparing their output. Occasionally, however, individuals lack a particular cone type, so their color vision is a product of the differential sensitivity of only two types

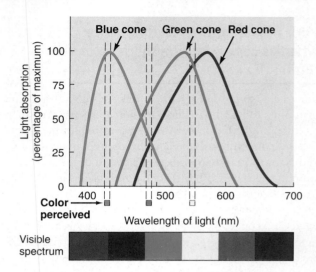

Color perceived	Percent of maximum stimulation		
	Red cones	Green cones	Blue cones
◼	0	0	100
◼	31	67	36
◻	83	83	0

● **FIGURE 6-27**

Sensitivity of the three types of cones to different wavelengths
The ratios of stimulation of the three cone types are shown for three sample colors.

of cones, a condition known as **color blindness.** Not only do color-defective individuals perceive certain colors differently, but they are also unable to distinguish as many varieties of colors (● Figure 6-28). For example, people with certain color defects are unable to distinguish between red and green. At a

● **FIGURE 6-28**

Color blindness chart
People with red-green color blindness cannot detect the number 29 in this chart.

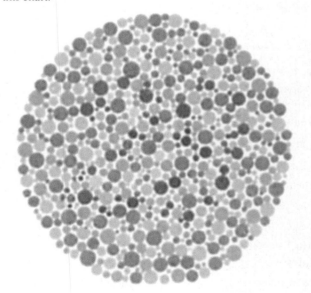

traffic light they can tell which light is "on" by its intensity, but they must rely on the position of the bright light to know whether to stop or go.

Although the three-cone system has been accepted as the standard model of color vision for more than two centuries, new evidence suggests that perception of color may be more complex. DNA studies have shown that men with normal color vision have a variable number of genes coding for cone pigments. For example, many had multiple genes (from two up to four) for red light detection and could distinguish more subtle differences in colors in this long-wavelength range than those with single copies of red cone genes. This finding will undoubtedly lead to a reevaluation of how the various photopigments contribute to color vision.

● **FIGURE 6-29**
Inversion of the image on the retina

Labels: Sclera, Choroid, Iris, Pupil, Lens, Fovea, Optic disc, Retina

Visual information is modified and separated before reaching the visual cortex.

The field of view that can be seen without moving the head is known as the **visual field.** The information that reaches the visual cortex in the occipital lobe is not a replica of the visual field for several reasons:

1. The image detected on the retina at the onset of visual processing is upside down and backward because of bending of the light rays (● Figure 6-29). Once it is projected to the brain, the inverted image is interpreted as being in its correct orientation.

2. The information transmitted from the retina to the brain is not merely a point-to-point record of photoreceptor activation. Before the information reaches the brain, the retinal neuronal layers beyond the rods and cones reinforce selected information and suppress other information to enhance contrast. One mechanism of retinal processing is lateral inhibition, by which strongly excited cone pathways suppress activity in surrounding pathways of weakly stimulated cones. This increases the dark–bright contrast to enhance the sharpness of boundaries.

Another mechanism of retinal processing involves differential activation of two types of ganglion cells, **on-center** and **off-center ganglion cells.** The receptive field of a cone ganglion cell is determined by the field of light detection by the cone with which it is linked. On-center and off-center ganglion cells respond in opposite ways, depending on the relative comparison of illumination between the center and periphery of their receptive fields. Think of the receptive field as a doughnut. An on-center ganglion cell increases its rate of firing when light is most intense at the center of its receptive field (that is, when the doughnut hole is lit up). In contrast, an off-center cell increases its firing rate when the periphery of its receptive field is most intensely illuminated (that is, when the donut itself is lit up). This is useful for enhancing the difference in light level between one small area at the center of a receptive field and the illumination immediately around it. By emphasizing differences in relative brightness, this mechanism helps define contours of images, but in so doing, information about absolute brightness is sacrificed (● Figure 6-30).

3. Various aspects of visual information such as form, color, depth, and movement are separated and projected in parallel pathways to different regions of the cortex. Only when these separate bits of processed information are integrated by higher visual regions is a reassembled picture of the visual scene perceived. This is similar to the blobs of paint on an artist's palette versus the finished portrait; the separate pigments do not represent a portrait of a face until they are appropriately integrated on a canvas.

Patients with lesions in specific visual processing regions of the brain may be unable to completely combine components of a visual impression. For example, a person may be unable to discern movement of an object but have reasonably good vision for shape, pattern, and color. Sometimes the defect can be remarkably specific, like being unable to recognize familiar faces while retaining the ability to recognize inanimate objects.

4. Because of the pattern of wiring between the eyes and the visual cortex, the left half of the cortex receives information only from the right half of the visual field as detected by both eyes, and the right half receives input only from the left half of the visual field of both eyes.

● **FIGURE 6-30**

Example of the outcome of retinal processing by on-center and off-center ganglion cells
Note that the gray circle surrounded by black appears brighter than the one surrounded by white, even though the two circles are identical (same shade and size). Retinal processing by on-center and off-center ganglion cells is largely responsible for enhancing differences in relative (rather than absolute) brightness, which helps define contours.

As a result of refraction, light rays from the left half of the visual field fall on the right half of the retina of both eyes (the medial or inner half of the left retina and the lateral or outer half of the right retina) (● Figure 6-31a). Similarly, rays from the right half of the visual field reach the left half of each retina (the lateral half of the left retina and the medial half of the right retina). Each optic nerve exiting the retina carries information from both halves of the retina it serves. This information is separated as the optic nerves meet at the **optic chiasm** located underneath the hypothalamus (*chiasm* means "cross") (see ● Figure 5-8b, p. 146). Within the optic chiasm, the fibers from the medial half of each retina cross to the opposite side, but those from the lateral half remain on the original side. The reorganized bundles of fibers leaving the optic chiasm are known as **optic tracts.** Each optic tract carries information from the lateral half of one retina and the medial half of the other retina. Therefore, this partial crossover brings together from the two eyes fibers that carry information from the same half of the visual field. Each optic tract, in turn, delivers to the half of the brain on its same side information about the opposite half of the visual field. A knowledge of these pathways can facilitate diagnosis of visual defects arising from interruption of the visual pathway at various points (● Figure 6-31b).

Before we move on to how the brain processes visual information, take a look at ▲ Table 6-4, which summarizes the functions of the various components of the eyes.

▌ The thalamus and visual cortexes elaborate the visual message.

The first stop in the brain for information in the visual pathway is the *lateral geniculate nucleus* in the thalamus (● Figure 6-31a). It separates information received from the eyes and relays it via fiber bundles known as **optic radiations** to different zones in the cortex, each of which processes different aspects of the visual stimulus (for example, color, form, depth, movement). This sorting process is no small task, because each optic nerve contains more than a million fibers carrying information from the photoreceptors in one retina. This is more than all the afferent fibers carrying somatosensory input from all the other regions of the body! Researchers estimate that hundreds of millions of neurons occupying about 30% of the cortex participate in visual processing, compared to 8% devoted to touch perception and 3% to hearing. Yet the connections in the visual pathways are precise. The lateral geniculate nucleus and each of the zones in the cortex that processes visual information have a topographical map representing the retina point for point. As with the somatosensory cortex, the neural maps of the retina are distorted. The fovea, the retinal region capable of greatest acuity, has much greater representation in the neural map than do the more peripheral regions of the retina.

Depth perception

Although each half of the visual cortex receives information simultaneously from the same part of the visual field as received by both eyes, the messages from the two eyes are not identical. Each eye views an object from a slightly different vantage point,

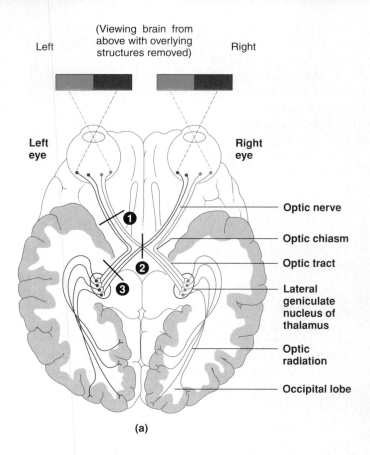

(a)

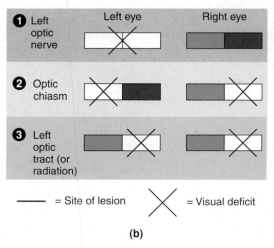

(b)

● FIGURE 6-31

The visual pathway and visual deficits associated with lesions in the pathway
(a) Visual pathway. Note that the left half of the visual cortex in the occipital lobe receives information from the right half of the visual field of both eyes (in blue), and the right half of the cortex receives information from the left half of the visual field of both eyes (in red). (b) Visual deficits with specific lesions in the visual pathway. Each visual deficit illustrated is associated with a lesion at the corresponding numbered point in the visual pathway in part (a).

---

even though the overlap is tremendous (● Figure 6-32). The overlapping area seen by both eyes at the same time is known as the **binocular** ("two-eyed") field of vision, which is important

Functions of the Major Components of the Eye

Structure	Location	Function
(in alphabetical order)		
Aqueous Humor	Anterior cavity between cornea and lens	Clear watery fluid that is continually formed and carries nutrients to the cornea and lens
Bipolar Cells	Middle layer of nerve cells in retina	Important in retinal processing of light stimulus
Blind Spot	Point slightly off-center on retina where optic nerve exits; is devoid of photoreceptors (also known as *optic disc*)	Route for passage of optic nerve and blood vessels
Choroid	Middle layer of eye	Pigmented to prevent scattering of light rays in eye; contains blood vessels that nourish retina; anteriorly specialized to form ciliary body and iris
Ciliary Body	Specialized anterior derivative of the choroid layer; forms a ring around the outer edge of the lens	Produces aqueous humor and contains ciliary muscle
Ciliary Muscle	Circular muscular component of ciliary body; attached to lens by means of suspensory ligaments	Important in accommodation
Cones	Photoreceptors in outermost layer of retina	Responsible for high acuity, color, and day vision
Cornea	Anterior clear outermost layer of eye	Contributes most extensively to eye's refractive ability
Fovea	Exact center of retina	Region with greatest acuity
Ganglion Cells	Inner layer of nerve cells in retina	Important in retinal processing of light stimulus; form optic nerve
Iris	Visible pigmented ring of muscle within aqueous humor	Varies size of pupil by variable contraction; responsible for eye color
Lens	Between aqueous humor and vitreous humor; attaches to ciliary muscle by suspensory ligaments	Provides variable refractive ability during accommodation
Macula Lutea	Area immediately surrounding the fovea	Has high acuity because of abundance of cones
Optic Disc	(*see entry for blind spot*)	
Optic Nerve	Leaves each eye at optic disc (blind spot)	First part of visual pathway to the brain
Pupil	Anterior round opening in middle of iris	Permits variable amounts of light to enter eye
Retina	Innermost layer of eye	Contains the photoreceptors (rods and cones)
Rods	Photoreceptors in outermost layer of retina	Responsible for high-sensitivity, black-and-white, and night vision
Sclera	Tough outer layer of eye	Protective connective tissue coat; forms visible white part of eye; anteriorly specialized to form cornea
Suspensory Ligaments	Suspended between ciliary muscle and lens	Important in accommodation
Vitreous Humor	Between lens and retina	Semifluid, jellylike substance that helps maintain spherical shape of eye

for **depth perception.** Like other areas of the cortex, the primary visual cortex is organized into functional columns, each processing information from a small region of the retina. Independent alternating columns are devoted to information about the same point in the visual field from the right and left eyes. The brain uses the slight disparity in the information received from the two eyes to estimate distance, allowing us to perceive three-dimensional objects in spatial depth. Some depth perception is possible using only one eye, based on experience and compar-

ison with other cues. For example, if your one-eyed view includes a car and a building and the car is much larger, you correctly interpret that the car must be closer to you than is the building.

Sometimes the two views are not successfully merged. This condition may occur for two reasons: (1) The eyes are not both focused on the same object simultaneously, because of defects of the external eye muscles that make fusion of the two eyes' visual fields impossible; or (2) the binocular information is improperly integrated during visual processing. The result

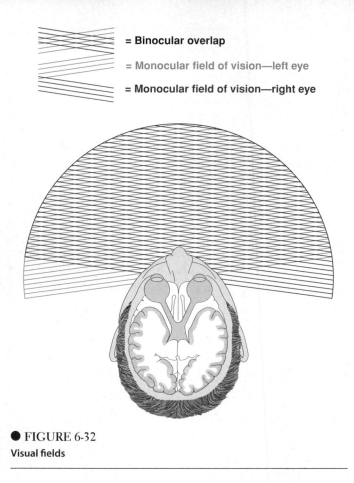

= Binocular overlap

= Monocular field of vision—left eye

= Monocular field of vision—right eye

● **FIGURE 6-32**
Visual fields

is double vision or **diplopia,** a condition in which the disparate views from both eyes are seen simultaneously.

Hierarchy of visual cortical processing

Within the cortex, visual information is first processed in the primary visual cortex, then is projected to higher-order visual areas for even more complex processing and abstraction. The cortex contains a hierarchy of visual cells that respond to increasingly complex stimuli. Three types of visual cortical neurons have been identified based on the complexity of stimulus requirements needed for the cell to respond; these are called **simple, complex,** and **hypercomplex cells.** Simple and complex cells are stacked on top of each other within the cortical columns of the primary visual cortex, whereas hypercomplex cells are found in the higher visual processing areas. Unlike a retinal cell, which responds to the amount of light, a cortical cell fires only when it receives a particular pattern of illumination for which it is programmed. These patterns are built up by converging connections that originate from closely aligned photoreceptor cells in the retina. For example, some simple cells fire only when a bar is viewed vertically in a specific location, others when a bar is horizontal, and others at various oblique orientations. Movement of a critical axis of orientation becomes important for response by some of the complex cells. Hypercomplex cells add a new dimension to visual processing by responding only to particular edges, corners, and curves. Each level of cortical visual neurons has increasingly greater capac-

ity for abstraction of information built up from the increasing convergence of input from lower-order neurons. In this way, the dotlike pattern of photoreceptors stimulated to varying degrees by varying light intensities in the retinal image is transformed in the cortex into information about depth, position, orientation, movement, contour, and length. Other aspects of this information, such as color perception, are processed simultaneously through a similar hierarchical organization. How and where the entire image is finally put together is still unresolved.

▌ Visual input goes to other areas of the brain not involved in vision perception.

Not all fibers in the visual pathway terminate in the visual cortexes. Some are projected to other regions of the brain for purposes other than direct vision perception. Following are examples of nonsight activities dependent on input from the retina:

1. Control of pupil size.
2. Synchronization of biological clocks to cyclical variations in light intensity (for example, the sleep–wake cycle synchronized to the night–day cycle).
3. Contribution to cortical alertness and attention.
4. Control of eye movements.

In the latter regard, each eye is equipped with a set of six **external eye muscles** that position and move the eye so that it can better locate, see, and track objects. Eye movements are among the fastest, most discretely controlled movements of the body.

▌ Some sensory input may be detected by multiple sensory processing areas in the brain.

Before shifting gears to another sense—hearing—we should mention a controversial new theory regarding the senses that challenges the prevailing view that the separate senses feed into distinct brain regions that handle only one sense. A growing body of evidence suggests that the brain regions devoted almost exclusively to a certain sense, such as the visual cortex for visual input and the somatosensory cortex for touch input, actually receive a variety of sensory signals. Therefore, tactile and auditory signals also arrive in the visual cortex. For example, one study using new brain imaging techniques showed that people blind from birth use the visual cortex when they read Braille, even though they are not "seeing" anything. The tactile input from their fingers reaches the visual area of the brain as well as the somatosensory cortex. This input helps them "visualize" the patterns of the Braille bumps.

Also reinforcing the notion that central processing of different types of sensory input overlaps to some extent, scientists recently discovered *multisensory neurons*—brain cells that react to multiple sensory inputs instead of just to one. No one knows whether these cells are rare or commonplace in the brain. (See the accompanying boxed feature, ▶ Concepts, Challenges, and Controversies, for one way in which researchers

"Seeing" with the Tongue

Although each type of sensory input is received primarily by a distinct brain region responsible for perception of that modality, the regions of the brain involved with perceptual processing receive sensory signals from a variety of sources. Thus the visual cortex receives sensory input not only from the eyes but from the body surface and ears as well. One group of scientists is exploiting in an unusual but exciting way this sharing of sensory input by multiple regions of the brain. In this research, blind or sighted-but-blindfolded volunteers are able to crudely perceive shapes and features in space by means of a tongue display unit. When this device, which consists of a grid of electrodes, is positioned on the tongue, it translates images detected by a camera into a pattern of electrical signals that activate touch receptors on the tongue (see the accompanying figure). The pattern of "tingling" on the tongue as a result of the light-induced electrical signals corresponds with the image recorded by the camera. With practice, the visual cortex interprets this alternate sensory input as a visual image. As one of the investigators who developed this technique claims, a person sees with the visual cortex, not with the eyes. Any means of sending signals to the visual cortex can be perceived as a visual image. For example, one blind participant in the study saw the flickering of a candle flame for the first time by means of this tongue device.

The tongue is a better choice than the skin for receiving this light-turned-tactile input because the saliva is an electrically conducive fluid that readily conducts the current generated in the device by the visual input. Furthermore, the tongue is densely populated with tactile receptors, opening up the possibility that the tongue can provide higher acuity of visual input than

the skin could. This feature will be important if such a device is ever used to help the visually impaired. The researchers' goal is to improve the resolution of the device by increasing the number of in-the-mouth electrodes. Even so, the perceived image will still be crude because the acuity afforded by this device can never come close to matching that provided by the eyes' small receptive fields.

Although using the tongue as a surrogate eye could never provide anywhere near the same vision as a normal eye, the hope is that this technique will afford the blind a means to make out doorways, to see objects as vague shapes, and to track motion. Even this limited visual input would enable a sightless person to get around

easier and improve the quality of his or her life. The device's developers plan to shrink the size of the unit so that it will fit inconspicuously in the user's mouth and be connected by a wireless link to a miniature camera mounted on eyeglasses. Such a trimmed-down unit would be practical to use and cosmetically acceptable.

Alternative methods to provide at least limited sight to the blind are under exploration, including implanting "vision" microchips in the eyes or brain, as you already learned (see p. 205), or converting visual images into "soundscapes" that are transmitted to a blind person's ears for visual interpretation.

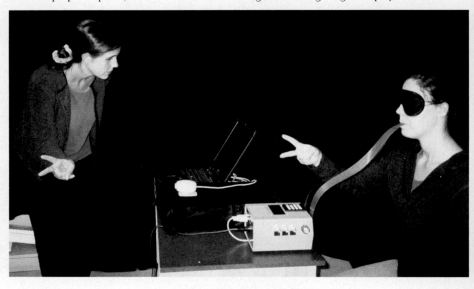

A blindfolded subject mimics hand gestures being recorded by a video camera (white box beside laptop computer) and transmitted to an image-translating tongue display unit.

are exploiting this sharing of sensory input by multiple regions of the brain.)

For the remainder of the chapter, we will concentrate on the mainstream function of the other special senses.

EAR: HEARING AND EQUILIBRIUM

Each **ear** consists of three parts: the *external*, the *middle*, and the *inner ear* (● Figure 6-33). The external and middle portions of the ear transmit airborne sound waves to the fluid-filled inner ear, amplifying the sound energy in the process. The inner ear houses two different sensory systems: the *cochlea*, which contains the receptors for conversion of sound waves into nerve impulses, making hearing possible; and the *vestibular apparatus*, which is necessary for the sense of equilibrium.

▋ Sound waves consist of alternate regions of compression and rarefaction of air molecules.

Hearing is the neural perception of sound energy. Hearing involves two aspects: identification of the sounds ("what") and their localization ("where"). We will first examine the characteristics of sound waves, then how the ears and brain process sound input to accomplish hearing.

Sound waves are traveling vibrations of air that consist of regions of high pressure caused by compression of air molecules alternating with regions of low pressure caused by rarefaction of the molecules (● Figure 6-34a). Any device capable of producing such a disturbance pattern in air molecules is a source of sound. A simple example is a tuning fork. When a tuning fork is struck, its prongs vibrate. As a prong of the fork

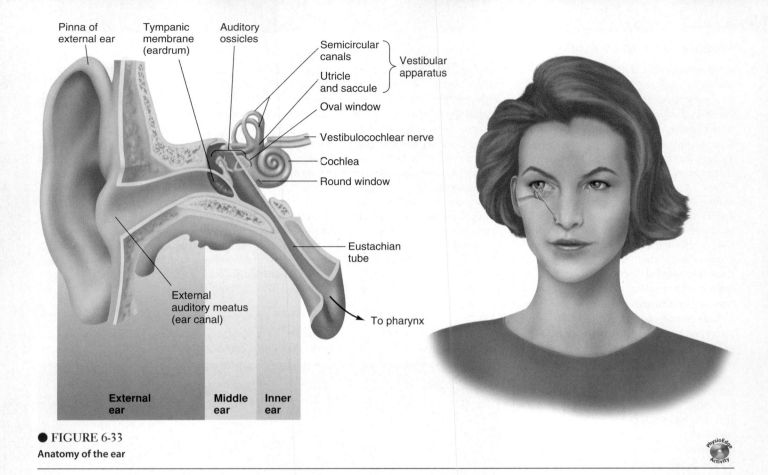

● FIGURE 6-33
Anatomy of the ear

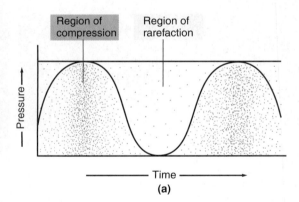

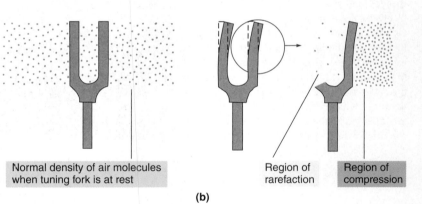

● FIGURE 6-34
Formation of sound waves
(a) Sound waves are alternating regions of compression and rarefaction of air molecules. (b) A vibrating tuning fork sets up sound waves as the air molecules ahead of the advancing arm of the tuning fork are compressed while the molecules behind the arm are rarefied. (c) Disturbed air molecules bump into molecules beyond them, setting up new regions of air disturbance more distant from the original source of sound. In this way, sound waves travel progressively farther from the source, even though each individual air molecule travels only a short distance when it is disturbed. The sound wave dies out when the last region of air disturbance is too weak to disturb the region beyond it.

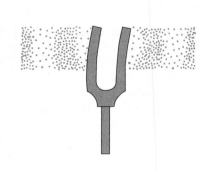

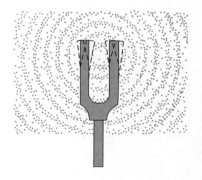

moves in one direction (● Figure 6-34b), air molecules ahead of it are pushed closer together, or compressed, increasing the pressure in this area. Simultaneously, as the prong moves forward the air molecules behind the prong spread out or are rarefied, lowering the pressure in that region. As the prong moves in the opposite direction, an opposite wave of compression and rarefaction is created. Even though individual molecules are moved only short distances as the tuning fork vibrates, alternating waves of compression and rarefaction spread out considerable distances in a rippling fashion. Disturbed air molecules disturb other molecules in adjacent regions, setting up new regions of compression and rarefaction, and so on (● Figure 6-34c). Sound energy is gradually dissipated as sound waves travel farther from the original sound source. The intensity of the sound decreases, until it finally dies out when the last sound wave is too weak to disturb the air molecules around it.

Sound waves can also travel through media other than air, such as water. They do so less efficiently, however; greater pressures are required to cause movements of fluid than movements of air because of the fluid's greater inertia (resistance to change).

Sound is characterized by its pitch (tone), intensity (loudness), and timbre (quality) (● Figure 6-35):

- The **pitch** or **tone** of a sound (for example, whether it is a C or a G note) is determined by the *frequency* of vibrations. The greater the frequency of vibration, the higher the pitch. Human ears can detect sound waves with frequencies from 20 to 20,000 cycles per second but are most sensitive to frequencies between 1,000 and 4,000 cycles per second.
- The **intensity,** or **loudness,** of a sound depends on the *amplitude* of the sound waves, or the pressure differences between a high-pressure region of compression and a low-pressure region of rarefaction. Within the hearing range, the greater the amplitude, the louder the sound. Human ears can detect a wide range of sound intensities, from the slightest whisper to the painfully loud takeoff of a jet. Loudness is expressed in **decibels (dB)**, which are a logarithmic measure of intensity compared with the faintest sound that can be heard—the **hearing threshold.** Because of the logarithmic relationship, every 10 decibels indicates a tenfold increase in loudness. A few examples of common sounds illustrate the magnitude of these increases (▲ Table 6-5). Note that the rustle of leaves at 10 dB is 10 times louder than hearing threshold, but the sound of a jet taking off at 150 dB is a quadrillion (a million billion) times, not 150 times, louder than the faintest audible sound. Sounds greater than 100 dB can permanently damage the sensitive sensory apparatus in the cochlea.
- The **timbre,** or **quality,** of a sound depends on its *overtones,* which are additional frequencies superimposed on top of the fundamental pitch or tone. A tuning fork has a pure tone, but most sounds lack purity. For example, complex mixtures of overtones impart different sounds to different instruments playing the same note (a C note on a trumpet sounds different from C on a piano). Overtones are likewise responsible for the characteristic differences in voices. Timbre enables the listener to distinguish the source of sound waves, because each source produces a different pattern of overtones. Thanks to timbre, you can tell whether it is your mother or girlfriend calling on the telephone before you say the wrong thing.

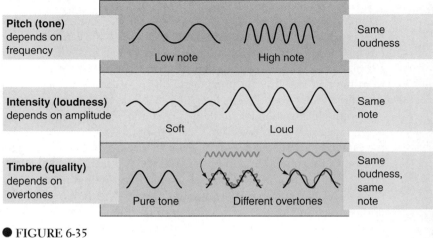

● **FIGURE 6-35**
Properties of sound waves

■ The external ear plays a role in sound localization.

The specialized receptors for sound are located in the fluid-filled inner ear. Airborne sound waves must therefore be channeled toward and transferred into the inner ear, compensating in the process for the loss in sound energy that naturally occurs as sound waves pass from air into water. This function is performed by the external ear and the middle ear.

The **external ear** (● Figure 6-33) consists of the *pinna (ear), external auditory meatus (ear canal),* and *tympanic membrane (eardrum).* The **pinna,** a prominent skin-covered flap of

▲ TABLE 6-5
Relative Magnitude of Common Sounds

Sound	Loudness in Decibels (dB)	Comparison to Faintest Audible Sound (Hearing Threshold)
Rustle of Leaves	10 dB	10 times louder
Ticking of Watch	20 dB	100 times louder
Hush of Library	30 dB	1 thousand times louder
Normal Conversation	60 dB	1 million times louder
Food Blender	90 dB	1 billion times louder
Loud Rock Concert	120 dB	1 trillion times louder
Takeoff of Jet Plane	150 dB	1 quadrillion times louder

cartilage, collects sound waves and channels them down the external ear canal. Many species (dogs, for example) can cock their ears in the direction of sound to collect more sound waves, but human ears are relatively immobile. Because of its shape, the pinna partially shields sound waves that approach the ear from the rear and thus helps a person distinguish whether a sound is coming from directly in front or behind.

Sound localization for sounds approaching from the right or left is determined by two cues. First, the sound wave reaches the ear closer to the sound source slightly before it arrives at the farther ear. Second, the sound is less intense as it reaches the farther ear, because the head acts as a sound barrier that partially disrupts the propagation of sound waves. The auditory cortex integrates all these cues to determine the location of the sound source. It is difficult to localize sound with only one ear. Recent evidence suggests that the auditory cortex pinpoints the location of a sound by differences in the timing of neuronal firing patterns, not by any spatially organized map such as the one projected on the visual cortex point-for-point from the retina that enables the location of a visual object to be identified.

The entrance to the **ear canal** is guarded by fine hairs. The skin lining the canal contains modified sweat glands that produce **cerumen** (earwax), a sticky secretion that traps fine foreign particles. Together the hairs and earwax help prevent airborne particles from reaching the inner portions of the ear canal, where they could accumulate or injure the tympanic membrane and interfere with hearing.

■ The tympanic membrane vibrates in unison with sound waves in the external ear.

The **tympanic membrane,** which is stretched across the entrance to the middle ear, vibrates when struck by sound waves. The alternating higher- and lower-pressure regions of a sound wave cause the exquisitely sensitive eardrum to bow inward and outward in unison with the wave's frequency.

For the membrane to be free to move as sound waves strike it, the resting air pressure on both sides of the tympanic membrane must be equal. The outside of the eardrum is exposed to atmospheric pressure that reaches it through the ear canal. The inside of the eardrum facing the middle ear cavity is also exposed to atmospheric pressure via the **eustachian (auditory) tube,** which connects the middle ear to the **pharynx** (back of the throat) (● Figure 6-33). The eustachian tube is normally closed, but it can be pulled open by yawning, chewing, and swallowing. Such opening permits air pressure within the middle ear to equilibrate with atmospheric pressure so that pressures on both sides of the tympanic membrane are equal. During rapid external pressure changes (for example, during air flight), the eardrum bulges painfully as the pressure outside the ear changes, while the pressure in the middle ear remains unchanged. Opening the eustachian tube by yawning allows the pressure on both sides of the tympanic membrane to equalize, relieving the pressure distortion as the eardrum "pops" back into place. Infections originating in the throat sometimes spread through the eustachian tube to the middle ear. The resultant

fluid accumulation in the middle ear not only is painful but also interferes with conduction of sound across the middle ear.

■ The middle ear bones convert tympanic membrane vibrations into fluid movements in the inner ear.

The **middle ear** transfers the vibratory movements of the tympanic membrane to the fluid of the inner ear. This transfer is facilitated by a moveable chain of three small bones or **ossicles** (the **malleus, incus,** and **stapes**) that extend across the middle ear (● Figure 6-36a). The first bone, the malleus, is attached to the tympanic membrane, and the last bone, the stapes, is attached to the **oval window,** the entrance into the fluid-filled cochlea. As the tympanic membrane vibrates in response to sound waves, the chain of bones is set into motion at the same frequency, transmitting this frequency of movement from the tympanic membrane to the oval window. The resultant pressure on the oval window with each vibration produces wavelike movements in the inner ear fluid at the same frequency as the original sound waves. However, as noted earlier, greater pressure is required to set fluid in motion. Two mechanisms related to the ossicular system amplify the pressure of the airborne sound waves to set up fluid vibrations in the cochlea. First, because the surface area of the tympanic membrane is much larger than that of the oval window, pressure is increased as force exerted on the tympanic membrane is conveyed to the oval window (pressure = force/unit area). Second, the lever action of the ossicles provides an additional mechanical advantage. Together, these mechanisms increase the force exerted on the oval window by 20 times what it would be if the sound wave struck the oval window directly. This additional pressure is sufficient to set the cochlear fluid in motion.

Several tiny muscles in the middle ear contract reflexly in response to loud sounds (over 70 dB), causing the tympanic membrane to tighten and limiting movement of the ossicular chain. This reduced movement of middle ear structures diminishes the transmission of loud sound waves to the inner ear to protect the delicate sensory apparatus from damage. This reflex response is relatively slow, however, happening at least 40 msec after exposure to a loud sound. It thus provides protection only from prolonged loud sounds, not from sudden sounds like an explosion. Taking advantage of this reflex, World War II antiaircraft guns were designed to make a loud prefiring sound to protect the gunner's ears from the much louder boom of the actual firing.

■ The cochlea contains the organ of Corti, the sense organ for hearing.

The pea-sized, snail-shaped **cochlea,** the "hearing" portion of the inner ear, is a coiled tubular system lying deep within the temporal bone (● Figure 6-33). It is easier to understand the functional components of the cochlea by "unrolling" it, as shown in ● Figure 6-36a. The cochlea is divided throughout most of its length into three fluid-filled longitudinal compart-

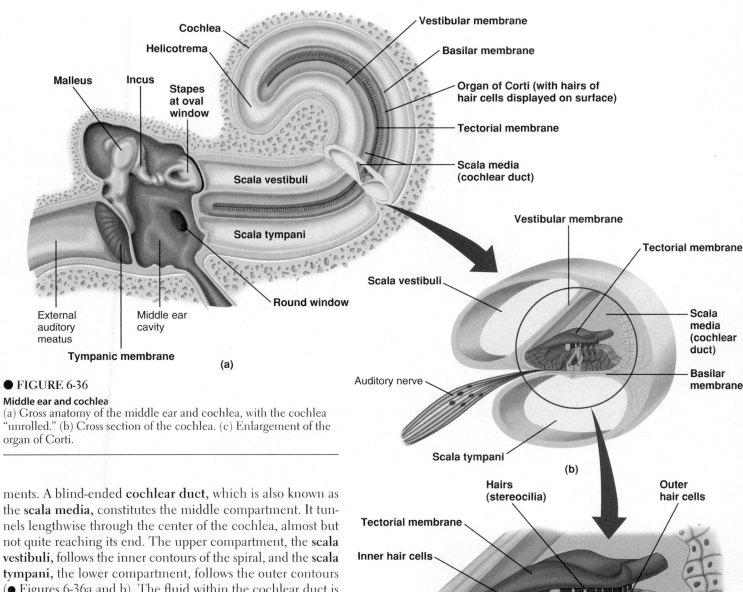

● **FIGURE 6-36**

Middle ear and cochlea
(a) Gross anatomy of the middle ear and cochlea, with the cochlea "unrolled." (b) Cross section of the cochlea. (c) Enlargement of the organ of Corti.

ments. A blind-ended **cochlear duct,** which is also known as the **scala media,** constitutes the middle compartment. It tunnels lengthwise through the center of the cochlea, almost but not quite reaching its end. The upper compartment, the **scala vestibuli,** follows the inner contours of the spiral, and the **scala tympani,** the lower compartment, follows the outer contours (● Figures 6-36a and b). The fluid within the cochlear duct is called **endolymph** (● Figure 6-37a). The scala vestibuli and scala tympani both contain a slightly different fluid, the **perilymph.** The region beyond the tip of the cochlear duct where the fluid in the upper and lower compartments is continuous is called the **helicotrema.** The scala vestibuli is sealed from the middle ear cavity by the oval window, to which the stapes is attached. Another small membrane-covered opening, the **round window,** seals the scala tympani from the middle ear. The thin **vestibular membrane** forms the ceiling of the cochlear duct and separates it from the scala vestibuli. The **basilar membrane** forms the floor of the cochlear duct, separating it from the scala tympani. The basilar membrane is especially important because it bears the **organ of Corti,** the sense organ for hearing.

▮ Hair cells in the organ of Corti transduce fluid movements into neural signals.

The organ of Corti, which rests on top of the basilar membrane throughout its full length, contains **hair cells** that are the re-

ceptors for sound. The 16,000 hair cells within each cochlea are arranged in four parallel rows along the length of the basilar membrane: one row of **inner hair cells** and three rows of **outer hair cells** (● Figure 6-36c). Protruding from the surface of each hair cell are about 100 hairs known as **stereocilia,** which are actin-stiffened microvilli (see p. 50). Hair cells generate neural signals when their surface hairs are mechanically deformed in association with fluid movements in the inner ear. These stereocilia are mechanically embedded in the **tectorial membrane,** an awninglike projection overhanging the organ of Corti throughout its length (● Figure 6-36b and c).

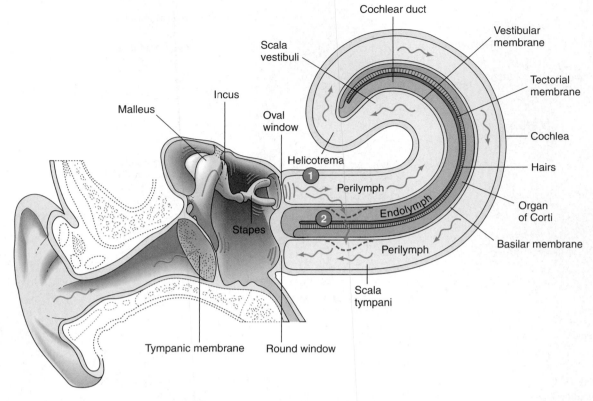

Fluid movement within the perilymph set up by vibration of the oval window follows two pathways:

Pathway **1** : Through the scala vestibuli, around the heliocotrema, and through the scala tympani, causing the round window to vibrate. This pathway just dissipates sound energy.

Pathway **2** : A "shortcut" from the scala vestibuli through the basilar membrane to the scala tympani. This pathway triggers activation of the receptors for sound by bending the hairs of hair cells as the organ of Corti on top of the vibrating basilar membrane is displaced in relation to the overlying tectorial membrane.

(a)

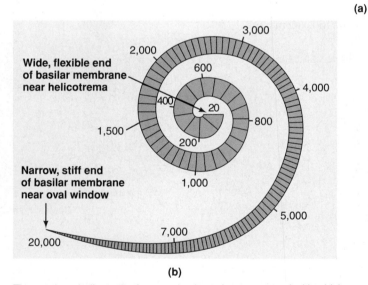

(b)

The numbers indicate the frequencies in cycles per second with which different regions of the basilar membrane maximally vibrate.

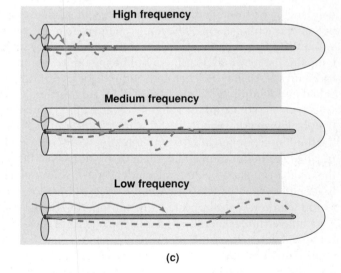

(c)

● FIGURE 6-37

Transmission of sound waves
(a) Fluid movement within the perilymph set up by vibration of the oval window follows two pathways, one dissipating sound energy and the other initiating the receptor potential. (b) Different regions of the basilar membrane vibrate maximally at different frequencies. (c) The narrow, stiff end of the basilar membrane nearest the oval window vibrates best with high-frequency pitches. The wide, flexible end of the basilar membrane near the helicotrema vibrates best with low-frequency pitches.

The pistonlike action of the stapes against the oval window sets up pressure waves in the upper compartment. Because fluid is incompressible, pressure is dissipated in two ways as the stapes causes the oval window to bulge inward: (1) displacement of the round window and (2) deflection of the basilar membrane (● Figure 6-37a). In the first of these pathways, the pressure wave pushes the perilymph forward in the upper compartment, then around the helicotrema, and into the lower compartment, where it causes the round window to bulge outward into the middle ear cavity to compensate for the pressure increase. As the stapes rocks backward and pulls the oval window outward toward the middle ear, the perilymph shifts in the opposite direction, displacing the round window inward. This pathway does not result in sound reception; it just dissipates pressure.

Pressure waves of frequencies associated with sound reception take a "shortcut." Pressure waves in the upper compartment are transferred through the thin vestibular membrane, into the cochlear duct, and then through the basilar membrane into the lower compartment, where they cause the round window to alternately bulge outward and inward. The main difference in this pathway is that transmission of pressure waves through the basilar membrane causes this membrane to move up and down, or vibrate, in synchrony with the pressure wave. Because the organ of Corti rides on the basilar membrane, the hair cells also move up and down as the basilar membrane oscillates.

Role of the inner hair cells

The inner and outer hair cells differ in function. The inner hair cells are the ones that transform the mechanical forces of sound (cochlear fluid vibration) into the electrical impulses of hearing (action potentials propagating auditory messages to the cerebral cortex). Because the stereocilia of these receptor cells are embedded in the stiff, stationary tectorial membrane, they are bent back and forth when the oscillating basilar membrane shifts their position in relationship to the tectorial membrane (● Figure 6-38). This back-and-forth mechanical defor-

mation of the hairs alternately opens and closes mechanically gated ion channels (see p. 104) in the hair cell, resulting in alternating depolarizing and hyperpolarizing potential changes—the receptor potential—at the same frequency as the original sound stimulus.

The inner hair cells communicate via a chemical synapse with the terminals of afferent nerve fibers making up the **auditory (cochlear) nerve.** Depolarization of these hair cells (when the basilar membrane is deflected upward) increases their rate of neurotransmitter release, which steps up the rate of firing in the afferent fibers. Conversely, the firing rate decreases as these hair cells release less neurotransmitter when they are hyperpolarized upon displacement in the opposite direction.

Thus the ear converts sound waves in the air into oscillating movements of the basilar membrane that bends the hairs of the receptor cells back and forth. This shifting mechanical deformation of the hairs alternately opens and closes the receptor cells' channels, bringing about graded potential changes in the receptor that lead to changes in the rate of action potentials propagated to the brain. In this way, sound waves are translated into neural signals that can be perceived by the brain as sound sensations (● Figure 6-39).

Role of the outer hair cells

Whereas the inner hair cells send auditory signals to the brain over afferent fibers, the outer hair cells do not signal the brain about incoming sounds. Instead, the outer hair cells actively and rapidly elongate in response to changes in membrane potential, a behavior known as *electromotility*. These changes in length are believed to amplify or accentuate the motion of the basilar membrane. An analogy would be a person deliberately pushing the pendulum of a grandfather clock in time with its swing to accentuate its motion. Such modification of basilar membrane movement is speculated to improve and tune the stimulation of the inner hair cells. Thus the outer hair cells enhance the response of the inner hair cells, the real auditory sensory receptors, making them exquisitely sensitive to sound intensity and highly discriminatory between various pitches of sound.

● FIGURE 6-38

Bending of hairs on deflection of the basilar membrane

The stereocilia (hairs) from the hair cells of the basilar membrane are embedded in the overlying tectorial membrane. These hairs are bent when the basilar membrane is deflected in relation to the stationary tectorial membrane. This bending opens channels, leading to ion movements that result in a receptor potential.

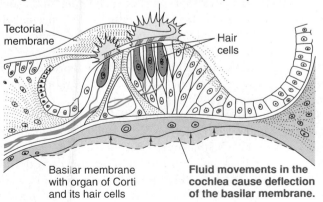

Tectorial membrane

Hair cells

Basilar membrane with organ of Corti and its hair cells

Fluid movements in the cochlea cause deflection of the basilar membrane.

▌ Pitch discrimination depends on the region of the basilar membrane that vibrates.

Pitch discrimination (that is, the ability to distinguish between various frequencies of incoming sound waves) depends on the shape and properties of the basilar membrane, which is narrow and stiff at its oval window end and wide and flexible at its helicotrema end (● Figure 6-37b). Different regions of the basilar membrane naturally vibrate maximally at different frequencies; that is, each frequency displays peak vibration at a different position along the membrane. The narrow end nearest the oval window vibrates best with high-frequency pitches, whereas the wide end nearest the helicotrema vibrates maximally with low-frequency tones (● Figure 6-37c). The pitches in between are sorted out precisely along the length of the membrane from higher to lower frequency. As a sound wave of a particular frequency is set up in the cochlea by oscillation of the stapes, the wave travels to the region of the basilar mem-

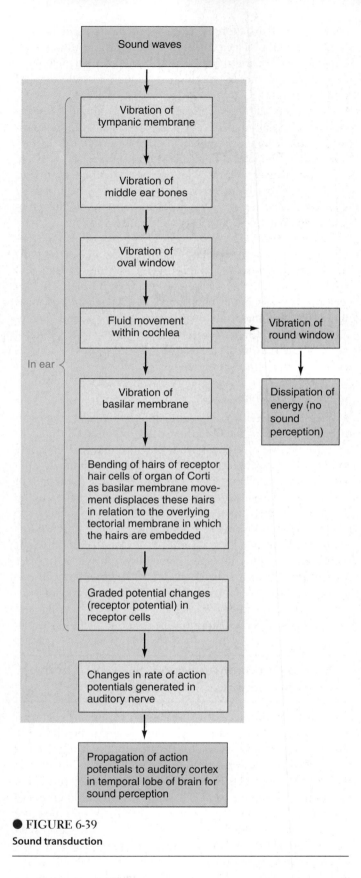

● FIGURE 6-39
Sound transduction

The hair cells in the region of peak vibration of the basilar membrane undergo the most mechanical deformation and accordingly are the most excited. This information is propagated to the CNS, which interprets the pattern of hair cell stimulation as a sound of a particular frequency. Modern techniques have determined that the basilar membrane is so fine-tuned that the peak membrane response to a single pitch probably extends no more than the width of a few hair cells.

Overtones of varying frequencies cause many points along the basilar membrane to vibrate simultaneously but less intensely than the fundamental tone, enabling the CNS to distinguish the timbre of the sound (**timbre discrimination**).

▌ Loudness discrimination depends on the amplitude of vibration.

Intensity (loudness) discrimination depends on the amplitude of vibration. As sound waves originating from louder sound sources strike the eardrum, they cause it to vibrate more vigorously (that is, bulge in and out to a greater extent) but at the same frequency as a softer sound of the same pitch. The greater tympanic membrane deflection is converted into a greater amplitude of basilar membrane movement in the region of peak responsiveness. The CNS interprets greater basilar membrane oscillation as a louder sound.

The auditory system is so sensitive and can detect sounds so faint that the distance of basilar membrane deflection is comparable to only a fraction of the diameter of a hydrogen atom, the smallest of atoms. No wonder very loud sounds, which cannot be sufficiently attenuated by protective middle ear reflexes (for example, the sounds of a typical rock concert), can set up such violent vibrations of the basilar membrane that irreplaceable hair cells are actually sheared off or permanently distorted, leading to partial hearing loss (● Figure 6-40).

▌ The auditory cortex is mapped according to tone.

Just as various regions of the basilar membrane are associated with particular tones, the **primary auditory cortex** in the temporal lobe is also *tonotopically* organized. Each region of the basilar membrane is linked to a specific region of the primary auditory cortex. Accordingly, specific cortical neurons are activated only by particular tones; that is, each region of the auditory cortex becomes excited only in response to a specific tone detected by a selected portion of the basilar membrane.

The afferent neurons that pick up the auditory signals from the inner hair cells exit the cochlea via the auditory nerve. The neural pathway between the organ of Corti and the auditory cortex involves several synapses en route, the most notable of which are in the brain stem and *medial geniculate nucleus* of the thalamus. The brain stem uses the auditory input for alertness and arousal. The thalamus sorts and relays the signals upward. Unlike the visual pathways, auditory signals from each ear are transmitted to both temporal lobes because the fibers partially cross over in the brain stem. For this reason, a disrup-

brane that naturally responds maximally to that frequency. The energy of the pressure wave is dissipated with this vigorous membrane oscillation, so the wave dies out at the region of maximal displacement.

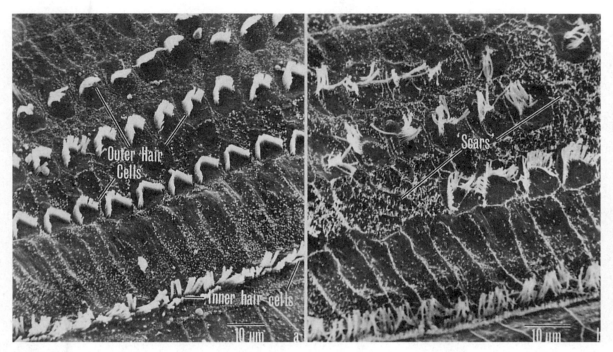

● **FIGURE 6-40**

Loss of hair cells caused by loud noises
Injury and loss of hair cells caused by intense noise. Portions of the organ of Corti, with its three rows of outer hair cells and one row of inner hair cells, from the inner ear of (a) a normal guinea pig and (b) a guinea pig after a 24-hour exposure to noise at 120 decibels SPL, a level approached by loud rock music.
(Scanning electron micrographs by R. S. Preston and J. E. Hawkins, Kresge Hearing Research Institute, University of Michigan.)

tion of the auditory pathways on one side beyond the brain stem does not affect hearing in either ear to any extent.

The primary auditory cortex appears to perceive discrete sounds, whereas the surrounding higher-order auditory cortex integrates the separate sounds into a coherent, meaningful pattern. Think about the complexity of the task accomplished by your auditory system. When you are at a concert, your organ of Corti responds to the simultaneous mixture of the instruments, the applause and hushed talking of the audience, and the background noises in the theater. You can distinguish these separate parts of the many sound waves reaching your ears and can pay attention to those of importance to you.

▌ Deafness is caused by defects either in conduction or neural processing of sound waves.

Loss of hearing, or **deafness,** may be temporary or permanent, partial or complete. Deafness is classified into two types—*conductive deafness* and *sensorineural deafness*—depending on the part of the hearing mechanism that fails to function adequately. **Conductive deafness** occurs when sound waves are not adequately conducted through the external and middle portions of the ear to set the fluids in the inner ear in motion. Possible causes include physical blockage of the ear canal with earwax, rupture of the eardrum, middle ear infections with accompanying fluid accumulation, or restriction of the ossicu-

lar movement because of bony adhesions between the stapes and the oval window. In **sensorineural deafness,** the sound waves are transmitted to the inner ear, but they are not translated into nerve signals that are interpreted by the brain as sound sensations. The defect can lie in the organ of Corti or the auditory nerves or, rarely, in the ascending auditory pathways or auditory cortex.

One of the most common causes of partial hearing loss, **neural presbycusis,** is a degenerative, age-related process that occurs as hair cells "wear out" with use. Over time, exposure to even ordinary modern-day sounds eventually damages hair cells, so that on average adults have lost more than 40% of their cochlear hair cells by age 65. The hair cells that process high-frequency sounds are the most vulnerable to destruction.

Hearing aids are helpful in conductive deafness but are less beneficial for sensorineural deafness. These devices increase the intensity of airborne sounds and may modify the sound spectrum and tailor it to the patient's particular pattern of hearing loss at higher or lower frequencies. For the sound to be perceived, however, the receptor cell–neural pathway system must still be intact.

In recent years, **cochlear implants** have become available. These electronic devices, which are surgically implanted, transduce sound signals into electrical signals that can directly stimulate the auditory nerve, thus bypassing a defective cochlear system. Cochlear implants cannot restore normal hearing, but they do permit recipients to recognize sounds. Success ranges from an ability to "hear" a phone ringing to being able

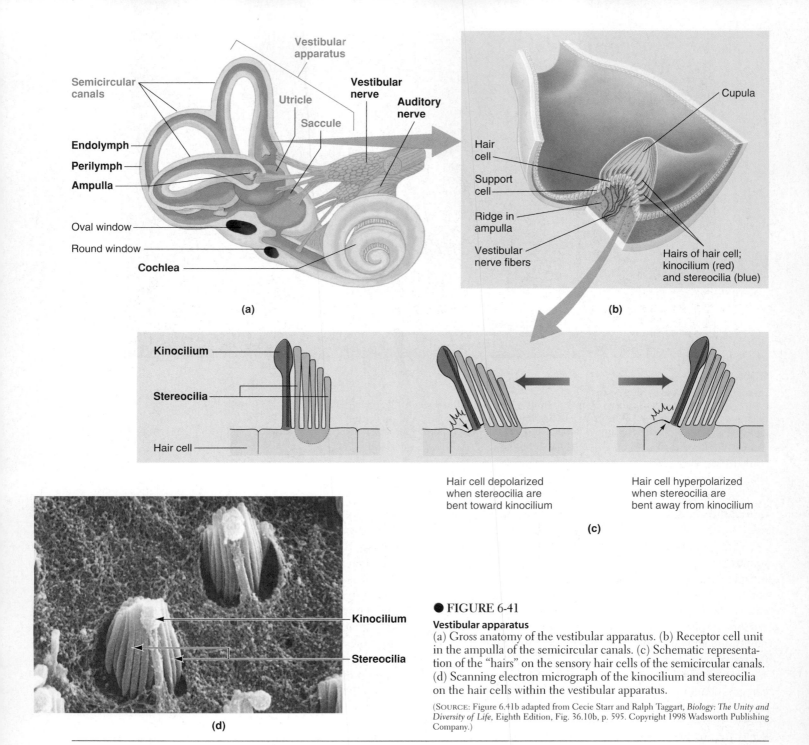

● FIGURE 6-41

Vestibular apparatus
(a) Gross anatomy of the vestibular apparatus. (b) Receptor cell unit in the ampulla of the semicircular canals. (c) Schematic representation of the "hairs" on the sensory hair cells of the semicircular canals. (d) Scanning electron micrograph of the kinocilium and stereocilia on the hair cells within the vestibular apparatus.

(SOURCE: Figure 6.41b adapted from Cecie Starr and Ralph Taggart, *Biology: The Unity and Diversity of Life*, Eighth Edition, Fig. 36.10b, p. 595. Copyright 1998 Wadsworth Publishing Company.)

to carry on a conversation over the telephone (without visual cues, such as reading lips).

Exciting new findings suggest that in the future it may be possible to restore hearing by stimulating an injured inner ear to repair itself. Scientists have long considered the hair cells of the inner ear irreplaceable. Thus hearing loss resulting from hair cell damage due to the aging process or exposure to loud noises is considered permanent. Encouraging new studies suggest, to the contrary, that hair cells in the inner ear have the latent ability to regenerate in response to an appropriate chemical signal. Researchers are currently trying to develop a drug

that will spur regrowth of hair cells, thus repairing inner ear damage and hopefully restoring hearing.

The vestibular apparatus is important for equilibrium by detecting position and motion of the head.

In addition to its cochlear-dependent role in hearing, the inner ear has another specialized component, the **vestibular apparatus,** which provides information essential for the sense of equilibrium and for coordinating head movements with eye and postural movements. The vestibular apparatus consists of two sets of structures lying within a tunneled-out region of the temporal bone near the cochlea—the *semicircular canals* and the *otolith organs*, namely the *utricle* and *saccule* (● Figure 6-41a).

The vestibular apparatus detects changes in position and motion of the head. As in the cochlea, all components of the vestibular apparatus contain endolymph and are surrounded by perilymph. Also, similar to the organ of Corti, the vestibular components each contain hair cells that respond to mechanical deformation triggered by specific movements of the endolymph. Also, like the auditory hair cells, the vestibular receptors may be either depolarized or hyperpolarized, depending on the direction of the fluid movement. Unlike the auditory system, however, much of the information provided by the vestibular apparatus does not reach the level of conscious awareness.

Role of the semicircular canals

The **semicircular canals** detect rotational or angular acceleration or deceleration of the head, such as when starting or stopping spinning, somersaulting, or turning the head. Each ear contains three semicircular canals arranged three-dimensionally in planes that lie at right angles to each other. The receptive hair cells of each semicircular canal are situated on top of a ridge located in the **ampulla,** a swelling at the base of the canal (● Figures 6-41a and b). The hairs are embedded in an overlying, caplike, gelatinous layer, the **cupula,** which protrudes into the endolymph within the ampulla. The cupula sways in the

direction of fluid movement, much like seaweed leaning in the direction of the prevailing tide.

Acceleration or deceleration during rotation of the head in any direction causes endolymph movement in at least one of the semicircular canals, because of their three-dimensional arrangement. As the head starts to move, the bony canal and the ridge of hair cells embedded in the cupula move with the head. Initially, however, the fluid within the canal, not being attached to the skull, does not move in the direction of the rotation but lags behind because of its inertia. (Because of inertia, a resting object remains at rest, and a moving object continues to move in the same direction unless the object is acted on by some external force that induces change.) When the endolymph is left behind as the head starts to rotate, the endolymph that is in the same plane as the head movement is in effect shifted in the opposite direction from the movement (similar to your body tilting to the right as the car in which you are riding suddenly turns to the left) (● Figure 6-42). This fluid movement causes the cupula to lean in the opposite direction from the head movement, bending the sensory hairs embedded in it. If the head movement continues at the same rate in the same direction, the endolymph catches up and moves in unison with the head so that the hairs return to their unbent position. When the head slows down and stops, the reverse situation occurs. The endolymph briefly continues to move in the direction of the rotation while the head decelerates to a stop. As a result, the cupula and its hairs are transiently bent in the direction of the preceding spin, which is opposite to the way they were bent during acceleration. When the endolymph gradually comes to a halt, the hairs straighten

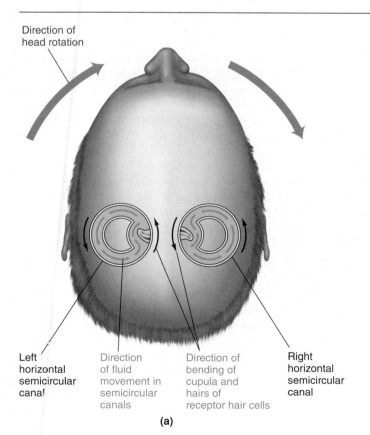

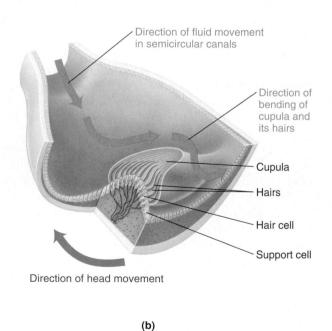

● **FIGURE 6-42**
Activation of the hair cells in the semicircular canals

(a)

(b)

again. Thus the semicircular canals detect changes in the rate of rotational movement of the head. They do not respond when the head is motionless or during circular motion at a constant speed.

The hairs of a vestibular hair cell consist of 20 to 50 microvilli—the stereocilia—and one cilium, the **kinocilium** (● Figure 6-41c and d) (see p. 47). Each hair cell is oriented so that it depolarizes when its stereocilia are bent toward the kinocilium; bending in the opposite direction hyperpolarizes the cell. The hair cells form a chemically mediated synapse with terminal endings of afferent neurons whose axons join with those of the other vestibular structures to form the **vestibular nerve.** This nerve unites with the auditory nerve from the cochlea to form the **vestibulocochlear nerve.** Depolarization of the hair cells increases the rate of firing in the afferent fibers; conversely, when the hair cells are hyperpolarized, the frequency of action potentials in the afferent fibers is reduced.

Role of the otolith organs

The **otolith organs** provide information about the position of the head relative to gravity and also detect changes in the rate of linear motion (moving in a straight line regardless of direction). The otolith organs, the **utricle** and the **saccule,** are sac-like structures housed within a bony chamber situated between the semicircular canals and the cochlea (● Figure 6-41a). The hairs of the receptive hair cells in these sense organs also protrude into an overlying gelatinous sheet, whose movement displaces the hairs and results in changes in hair cell potential. Many tiny crystals of calcium carbonate—the **otoliths** ("ear stones")—are suspended within the gelatinous layer, making it heavier and giving it more inertia than the surrounding fluid (● Figure 6-43). When a person is in an upright position, the hairs within the utricles are oriented vertically and the saccule hairs are lined up horizontally.

Let's look at the *utricle* as an example. Its otolith-embedded, gelatinous mass shifts positions and bends the hairs in two ways:

1. When the head is tilted in any direction other than vertical (that is, other than straight up and down), the hairs are bent in the direction of the tilt because of the gravitational force exerted on the top-heavy gelatinous layer (● Figure 6-43b). Within the utricles on each side of the head, some of the hair cell bundles are oriented to depolarize and others to hyperpolarize when the head is in any position other than upright. The CNS thus receives different patterns of neural activity depending on head position with respect to gravity.

2. The utricle hairs are also displaced by any change in horizontal linear motion (such as moving straight forward, backward, or to the side). As a person starts to walk forward (● Figure 6-43c), the top-heavy otolith membrane at first lags behind the endolymph and hair cells because of its greater inertia. The hairs are thus bent to the rear, in the opposite direction of the forward movement of the head. If the walking pace is maintained, the gelatinous layer soon catches up and moves at the same rate as the head so that the hairs are no longer bent. When the person stops walking, the otolith sheet continues to move forward briefly as the head slows and stops, bending the hairs toward the front. Thus the hair cells of the utricle detect horizontally directed linear acceleration and deceleration, but they do not provide information about movement in a straight line at constant speed.

The *saccule* functions similarly to the utricle, except that it responds selectively to the tilting of the head away from a horizontal position (such as getting up from bed) and to vertically directed linear acceleration and deceleration (such as jumping up and down or riding in an elevator).

Signals arising from the various components of the vestibular apparatus are carried through the vestibulocochlear nerve to the **vestibular nuclei,** a cluster of neuronal cell bodies in the brain stem, and to the cerebellum. Here the vestibular information is integrated with input from the skin surface, eyes, joints, and muscles for (1) maintaining balance and desired posture; (2) controlling the external eye muscles so that the eyes remain fixed on the same point, despite movement of the head; and (3) perceiving motion and orientation (● Figure 6-44).

Some individuals, for poorly understood reasons, are especially sensitive to particular motions that activate the vestibular apparatus and cause symptoms of dizziness and nausea; this sensitivity is called **motion sickness.** Occasionally, fluid imbalances within the inner ear lead to **Ménière's disease.** Not surprisingly, because both the vestibular apparatus and cochlea contain the same inner ear fluids, both vestibular and auditory symptoms occur with this condition. An afflicted individual suffers transient attacks of severe vertigo (dizziness) accompanied by pronounced ringing in the ears and some loss of hearing. During these episodes, the person cannot stand upright and reports feeling as though self or surrounding objects in the room are spinning around.

▲ Table 6-6, p. 227, summarizes the functions of the major components of the ear.

CHEMICAL SENSES: TASTE AND SMELL

Unlike the photoreceptors of the eye and the mechanoreceptors of the ear, the receptors for taste and smell are chemoreceptors, which generate neural signals on binding with particular chemicals in their environment. The sensations of taste and smell in association with food intake influence the flow of digestive juices and affect appetite. Furthermore, stimulation of taste or smell receptors induces pleasurable or objectionable sensations and signals the presence of something to seek (a nutritionally useful, good-tasting food) or to avoid (a potentially toxic, bad-tasting substance). Thus, the chemical senses provide a "quality control" checkpoint for substances available for ingestion. In lower animals, smell also plays a major role in finding direction, in seeking prey or avoiding predators, and in sexual attraction to a mate. The sense of smell is less sensitive in humans and much less important in influencing our behavior (although millions of dollars are spent annually on perfumes and deodorants to make us smell better and thereby be more socially attractive). We will first examine the mechanism of taste (**gustation**) and then turn our attention to smell (**olfaction**).

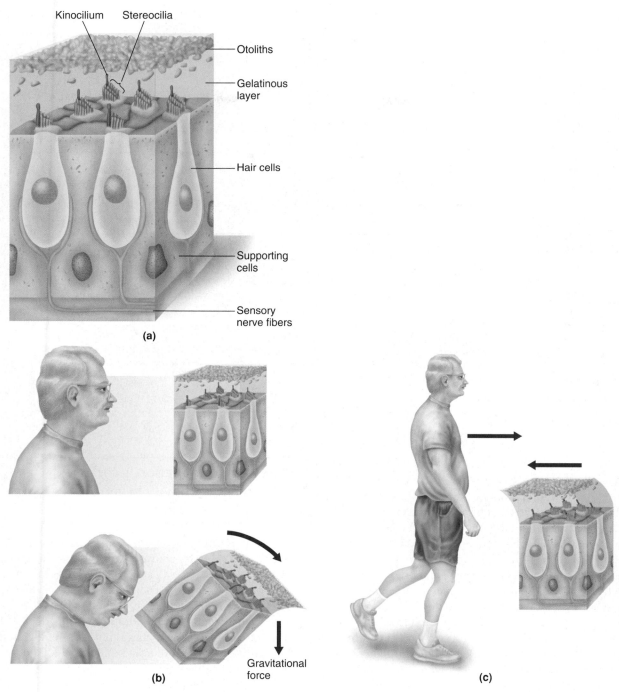

Kinocilium   Stereocilia

— Otoliths

— Gelatinous layer

— Hair cells

— Supporting cells

— Sensory nerve fibers

(a)

(b)

Gravitational force

(c)

● **FIGURE 6-43**

Utricle

(a) Receptor unit in utricle. (b) Activation of the utricle by a change in head position. (c) Activation of the utricle by horizontal linear accleration.

▌ Taste receptor cells are located primarily within tongue taste buds.

The chemoreceptors for taste sensation are packaged in taste buds, about 10,000 of which are present in the oral cavity and throat, with the greatest percentage on the upper surface of the tongue (● Figure 6-45, p. 228). A **taste bud** consists of about 50 long, spindle-shaped *taste receptor cells* packaged with *supporting cells* in an arrangement like slices of an orange. Each taste bud has a small opening, the **taste pore**, through which fluids in the mouth come into contact with the surface of its receptor cells.

Taste receptor cells are modified epithelial cells with many surface folds, or microvilli, that protrude slightly through the taste pore, greatly increasing the surface area exposed to

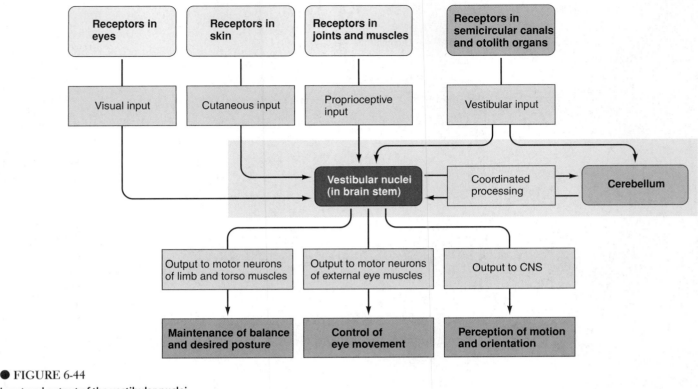

● **FIGURE 6-44**
Input and output of the vestibular nuclei

the oral contents. The plasma membrane of the microvilli contains receptor sites that bind selectively with chemical molecules in the environment. Only chemicals in solution—either ingested liquids or solids that have been dissolved in saliva—can attach to receptor cells and evoke the sensation of taste. Binding of a taste-provoking chemical, a **tastant,** with a receptor cell alters the cell's ionic channels to produce a depolarizing receptor potential. This receptor potential, in turn, initiates action potentials within terminal endings of afferent nerve fibers with which the receptor cell synapses.

Most receptors are carefully sheltered from direct exposure to the environment, but the taste receptor cells, by virtue of their task, frequently come into contact with potent chemicals. Unlike the eye or ear receptors, which are irreplaceable, taste receptors have a life span of about 10 days. Epithelial cells surrounding the taste bud differentiate first into supporting cells and then into receptor cells to constantly renew the taste bud components.

Terminal afferent endings of several cranial nerves synapse with taste buds in various regions of the mouth. Signals in these sensory inputs are conveyed via synaptic stops in the brain stem and thalamus to the **cortical gustatory area,** a region in the parietal lobe adjacent to the "tongue" area of the somatosensory cortex. Unlike most sensory input, the gustatory pathways are primarily uncrossed. The brain stem also projects fibers to the hypothalamus and limbic system, presumably to add affective dimensions, such as whether the taste is pleasant or unpleasant, and to process behavioral aspects associated with taste and smell.

■ **Taste discrimination is coded by patterns of activity in various taste bud receptors.**

We can discriminate among thousands of different taste sensations, yet all tastes are varying combinations of four **primary tastes:** *salty, sour, sweet,* and *bitter.* A growing number of sensory physiologists suggest a fifth primary taste—*umami,* a meaty or savory taste—should be added to the list.

Each receptor cell responds in varying degrees to all four (or perhaps five) primary tastes but is generally preferentially responsive to one of the taste modalities (● Figure 6-46). The richness of fine taste discrimination beyond the primary tastes depends on subtle differences in the stimulation patterns of all the taste buds in response to various substances, similar to the variable stimulation of the three cone types that gives rise to the range of color sensations.

Receptor cells use different pathways to bring about a depolarizing receptor potential in response to each of the five tastant categories:

• **Salt taste** is stimulated by chemical salts, especially NaCl (table salt). Direct entry of positively charged Na^+ ions through specialized Na^+ channels in the receptor cell membrane, a movement that reduces the cell's internal negativity, is responsible for receptor depolarization in response to salt.
• **Sour taste** is caused by acids, which contain a free hydrogen ion, H^+. The citric acid content of lemons, for example,

▲ TABLE 6-6

Functions of the Major Components of the Ear

Structure	Location	Function
External Ear		**Collects and transfers sound waves to the middle ear**
Pinna (ear)	Skin-covered flap of cartilage located on each side of the head	Collects sound waves and channels them down the ear canal; contributes to sound localization
External auditory meatus (ear canal)	Tunnels from the exterior through the temporal bone to the tympanic membrane	Directs sound waves to tympanic membrane; contains filtering hairs and secretes earwax to trap foreign particles
Tympanic membrane (eardrum)	Thin membrane that separates the external ear and the middle ear	Vibrates in synchrony with sound waves that strike it, setting middle ear bones in motion
Middle Ear		**Transfers vibrations of the tympanic membrane to the fluid in the cochlea**
Malleus, incus, stapes	Movable chain of bones that extends across the middle ear cavity; malleus attaches to the tympanic membrane, and stapes attaches to the oval window	Oscillate in synchrony with tympanic membrane vibrations and set up wavelike movements in the cochlear perilymph at the same frequency
Inner Ear: Cochlea		**Houses sensory system for hearing**
Oval window	Thin membrane at the entrance to the cochlea; separates the middle ear from the scala vestibuli	Vibrates in unison with movement of stapes, to which it is attached; oval window movement sets cochlear perilymph in motion
Scala vestibuli	Upper compartment of the cochlea, a snail-shaped tubular system that lies deep within the temporal bone	Contains perilymph that is set in motion by oval window movement driven by oscillation of middle ear bones
Scala tympani	Lower compartment of the cochlea	Contains perilymph that is continuous with the scala vestibuli
Cochlear duct (scala media)	Middle compartment of the cochlea; a blind-ended tubular compartment that tunnels through the center of the cochlea	Contains endolymph; houses the basilar membrane
Basilar membrane	Forms the floor of the cochlear duct	Vibrates in unison with perilymph movements; bears the organ of Corti, the sense organ for hearing
Organ of Corti	Rests on top of the basilar membrane throughout its length	Contains hair cells, the receptors for sound, which undergo receptor potentials when bent as a result of fluid movement in cochlea
Tectorial membrane	Stationary membrane that overhangs the organ of Corti and within which the surface hairs of the receptor hair cells are embedded	Site at which the embedded hairs of the receptor cells are bent and undergo receptor potentials as the vibrating basilar membrane moves in relation to the stationary tectorial membrane
Round window	Thin membrane that separates the scala tympani from the middle ear	Vibrates in unison with fluid movements in perilymph to dissipate pressure in cochlea; does not contribute to sound reception
Inner Ear: Vestibular Apparatus		**Houses sensory systems for equilibrium, and provides input essential for maintenance of posture and balance**
Semicircular canals	Three semicircular canals arranged three-dimensionally in planes at right angles to each other near the cochlea	Detect rotational or angular acceleration or deceleration
Utricle	Saclike structure in a bony chamber between the cochlea and semicircular canals	Detects (1) changes in head position away from vertical and (2) horizontally directed linear acceleration and deceleration
Saccule	Lies next to the utricle	Detects (1) changes in head position away from horizontal and (2) vertically directed linear acceleration and deceleration

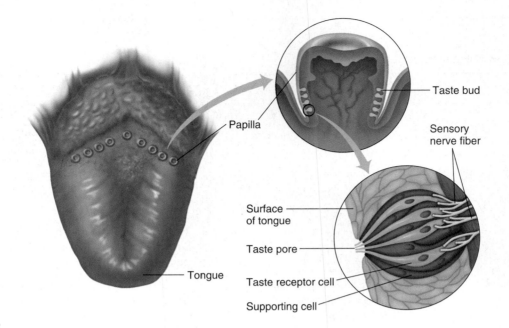

● **FIGURE 6-45**

Location and structure of the taste buds
Taste buds are located primarily along the edges of moundlike papillae on the upper surface of the tongue. The receptor cells and supporting cells of a taste bud are arranged like slices of an orange.

accounts for their distinctly sour taste. Depolarization of the receptor cell by sour tastants occurs because H^+ blocks K^+ channels in the receptor cell membrane. The resultant decrease in the passive movement of positively charged K^+ ions out of the cell reduces the internal negativity, producing a depolarizing receptor potential.

• **Sweet taste** is evoked by the particular configuration of glucose. From an evolutionary perspective, we crave sweet foods because they supply necessary calories in a readily usable form. However, other organic molecules with similar structures but no calories, such as saccharin, aspartame, and other artificial sweeteners, can also interact with sweet receptor binding sites. Binding of glucose or another chemical with the taste cell receptor activates a G protein, which turns on the cAMP second-messenger pathway in the taste cell (see p. 67). The second-messenger pathway ultimately results in phosphorylation and blockage of K^+ channels in the receptor cell membrane, leading to a depolarizing receptor potential.

• **Bitter taste** is elicited by a more chemically diverse group of tastants than the other taste sensations. For example, alkaloids (such as caffeine, nicotine, strychnine, morphine, and other toxic plant derivatives), as well as poisonous substances, all taste bitter, presumably as a protective mechanism to discourage ingestion of these potentially dangerous compounds. Taste cells that detect bitter flavors possess fifty to a hundred bitter receptors, each of which responds to a different bitter flavor. Because each receptor cell has this diverse family of bitter receptors, a wide variety of unrelated chemicals all taste bitter despite their diverse structures. This mechanism expands the ability of the taste receptor to detect a wide range of potentially harmful chemicals. The first G protein in taste—**gustducin**—was identified in one of the bitter signaling pathways. This G protein, which sets off a second-messenger pathway in the taste cell, is very similar to the visual G protein, transducin.

• **Umami taste,** which was first identified and named by a Japanese researcher, is triggered by amino acids such as glutamate. The presence of amino acids serves as a marker for a desirable, nutritionally protein-rich food. Glutamate binds to a G protein–coupled receptor and activates a second-messenger system, but the details of this pathway are unknown. This pathway is responsible for the distinctive taste of the flavor additive

● **FIGURE 6-46**

Relative responsiveness of different taste buds to different stimuli
Each taste bud responds in varying degrees to the four primary tastes. The pattern of stimulation for four different taste buds is depicted.

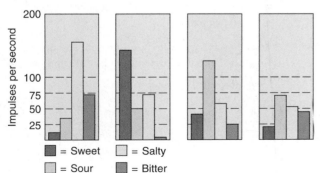

monosodium glutamate (MSG), which is especially popular in Asian dishes.

Taste perception is also influenced by information derived from other receptors, especially odor. When you temporarily lose your sense of smell because of swollen nasal passageways during a cold, your sense of taste is also markedly reduced, even though your taste receptors are unaffected by the cold. Other factors affecting taste include temperature and texture of the food as well as psychological factors associated with past experiences with the food. How the cortex accomplishes the complex perceptual processing of taste sensation is currently not known.

▌ The olfactory receptors in the nose are specialized endings of renewable afferent neurons.

The **olfactory** (smell) **mucosa**, a 3-centimeter-square patch of mucosa located in the ceiling of the nasal cavity, contains three cell types: *olfactory receptors, supporting cells*, and *basal cells* (● Figure 6-47). The supporting cells secrete mucus, which coats the nasal passages. The basal cells are precursors for new olfactory receptor cells, which are replaced about every two months. This is remarkable because, unlike the other special sense receptors, **olfactory receptors** are specialized endings of

afferent neurons, not separate cells. The entire neuron, including its afferent axon projecting into the brain, is replaced. These are the only neurons generally considered to undergo cell division (although recent evidence suggests that new neurons can be produced in the hippocampus, a part of the brain important in the formation of long-term memories; see p. 139). The axons of the olfactory receptor cells collectively form the **olfactory nerve.**

The receptor portion of an olfactory receptor cell consists of an enlarged knob bearing several long cilia that extend like a tassle to the surface of the mucosa. These cilia contain the binding sites for attachment of **odorants,** molecules that can be smelled. During quiet breathing, odorants typically reach the sensitive receptors only by diffusion because the olfactory mucosa is above the normal path of air flow. The act of sniffing enhances this process by drawing the air currents upward within the nasal cavity so that a greater percentage of the odoriferous molecules in the air come into contact with the olfactory mucosa. Odorants also reach the olfactory mucosa during eating by wafting up to the nose from the mouth through the pharynx (back of the throat).

To be smelled, a substance must be (1) sufficiently volatile (easily vaporized) that some of its molecules can enter the nose in the inspired air and (2) sufficiently water soluble that it can dissolve in the mucous layer coating the olfactory mucosa. As with taste receptors, to be detected by olfactory receptors molecules must be dissolved.

● FIGURE 6-47
Location and structure of the olfactory receptors

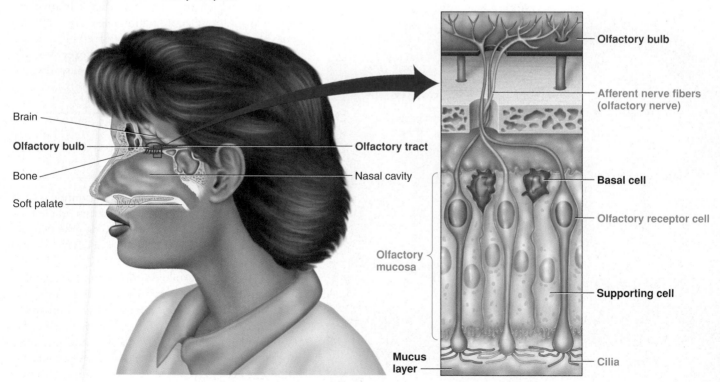

▌Various parts of an odor are detected by different olfactory receptors and sorted into "smell files."

The human nose contains 5 million olfactory receptors, of which there are 1,000 different types. During smell detection, an odor is "dissected" into various components. Each receptor responds to only one discrete component of an odor rather than to the whole odorant molecule. Accordingly, each of the various parts of an odor are detected by one of the thousand different receptors, and a given receptor can respond to a particular odor component shared in common by different scents. Compare this to the three cone types for coding color vision and the taste buds that respond differentially to only four (or five) primary tastes to accomplish coding for taste discrimination.

Binding of an appropriate scent signal to an olfactory receptor activates a G protein, triggering a cascade of cAMP-dependent intracellular reactions that leads to opening of Na^+ channels. The resultant ion movement brings about a depo-

larizing receptor potential that generates action potentials in the afferent fiber. The frequency of the action potentials depends on the concentration of the stimulating chemical molecules.

The afferent fibers arising from the receptor endings in the nose pass through tiny holes in the flat bone plate separating the olfactory mucosa from the overlying brain tissue (● Figure 6-47). They immediately synapse in the **olfactory bulb,** a complex neural structure containing several different layers of cells that are functionally similar to the retinal layers of the eye. The twin olfactory bulbs, one on each side, are about the size of small grapes. Each olfactory bulb is lined by small ball-like neural junctions known as **glomeruli** ("little balls") (● Figure 6-48). Within each glomerulus, the terminals of receptor cells carrying information about a particular scent component synapse with the next cells in the olfactory pathway, the **mitral cells.** Because each glomerulus receives signals only from receptors that detect a particular odor component, the glomeruli serve as "smell files." The separate components of an odor are sorted into different glomeruli, one component per file. Thus the glomeruli, which serve as the first relay station in the brain for processing olfactory information, play a key role in organizing scent perception.

The mitral cells on which the olfactory receptors terminate in the glomeruli refine the smell signals and relay them to the brain for further processing. Fibers leaving the olfactory bulb travel in two different routes (● Figure 6-49): (1) a *subcortical route* going primarily to regions of the limbic system, especially the lower medial sides of the temporal lobes (considered the **primary olfactory cortex**), and (2) a *thalamic-cortical route.* Until recently, the subcortical route was thought to be the only olfactory pathway. This route, which includes hypothalamic involvement, permits close coordination between smell and behavioral reactions associated with feeding, mating, and direction orienting. As with the other senses, the thalamic-cortical route is important for conscious perception and fine discrimination of smell.

▌Odor discrimination is coded by patterns of activity in the olfactory bulb glomeruli.

Because each given odorant activates multiple receptors and glomeruli in response to its various odor components, odor discrimination is based on different patterns of glomeruli activated by various scents. In this way, the cortex can distinguish over 10,000 different scents. This mechanism for sorting out and distinguishing different odors is very effective. A noteworthy example is our ability to detect methyl mercaptan (garlic odor) at a concentration of 1 molecule per 50 billion molecules in the air! This substance is added to odorless natural gas to enable us to detect potentially lethal gas leaks. Despite this impressive sensitivity, humans have a poor sense of smell compared to other

● **FIGURE 6-48**

Processing of scents in the olfactory bulb
Each of the glomeruli lining the olfactory bulb receives synaptic input from only one type of olfactory receptor, which, in turn, responds to only one discrete component of an odorant. Thus the glomeruli sort and file the various components of an odoriferous molecule before relaying the smell signal to the mitral cells and higher brain levels for further processing.

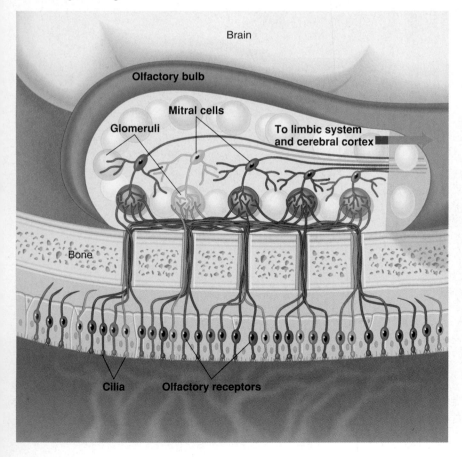

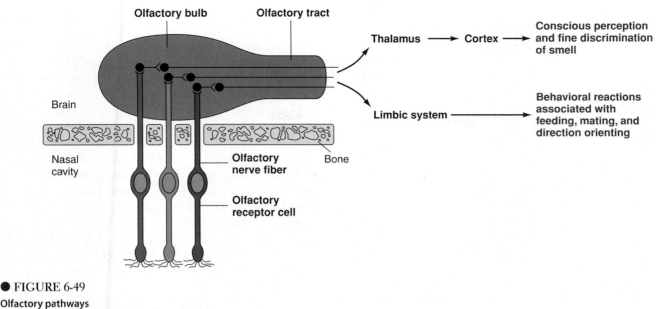

● FIGURE 6-49

Olfactory pathways
See Figures 6-47 and 6-48 for a location reference point.

species. By comparison, dogs' sense of smell is hundreds of times more sensitive than that of humans. Bloodhounds, for example, have about 4 billion olfactory receptor cells compared to our 5 million such cells, accounting for bloodhounds' superior scent-sniffing ability.

▌ The olfactory system adapts quickly, and odorants are rapidly cleared.

Although the olfactory system is sensitive and highly discriminating, it is also quickly adaptive. Our sensitivity to a new odor diminishes rapidly after a short period of exposure to it, even though the odor source continues to be present. This reduced sensitivity does not involve receptor adaptation, as has been thought for years; actually, the olfactory receptors themselves adapt slowly. It apparently involves some sort of adaptation process in the CNS. Adaptation is specific for a particular odor, and responsiveness to other odors remains unchanged.

What clears the odorants away from their binding sites on the olfactory receptors so that the sensation of smell doesn't "linger" after the source of the odor is removed? Several "odor-eating" enzymes have recently been discovered in the olfactory mucosa that could serve as molecular janitors, clearing away the odoriferous molecules so that they do not continue to stimulate the olfactory receptors. Interestingly, these odorant-clearing enzymes are very similar chemically to detoxification enzymes found in the liver. (These liver enzymes inactivate potential toxins absorbed from the digestive tract; see p. 27). This resemblance may not be coincidental. Researchers speculate that the nasal enzymes may serve the dual purpose of clearing the olfactory mucosa of old odorants and transforming potentially harmful chemicals into harmless molecules. Such detoxification would serve a very useful purpose, considering the open passageway between the olfactory mucosa and the brain.

▌ The vomeronasal organ detects pheromones.

In addition to the olfactory mucosa, the nose contains another sense organ, the **vomeronasal organ (VNO),** which is common in animals but until recently was thought to be nonexistent in humans. The VNO is located about half an inch inside the human nose next to the vomer bone, hence its name. It detects **pheromones,** chemical signals passed subconsciously from one individual to another. Binding of a pheromone to its receptor on the surface of a neuron in the VNO triggers an action potential that travels through nonolfactory pathways to the limbic system, the brain region that governs emotional responses and sociosexual behaviors. These signals never reach the higher levels of conscious awareness. In animals, the VNO is known as the "sexual nose" for its role in governing reproductive and social behaviors, such as identifying and attracting a mate.

Scientists have now proved the existence of pheromones in humans. Although the role of the VNO in human behavior has not been validated, researchers suspect that it is responsible for spontaneous "feelings" between people, either "good chemistry," such as "love at first sight," or "bad chemistry," such as "getting bad vibes" from someone you just met. They speculate that pheromones in humans subtly influence sexual activity, compatibility with others, or group behavior, similar to the role they play in animals. Because messages conveyed by the VNO are believed to bypass cortical consciousness, the response to the largely odorless pheromones is not a distinct, discrete perception, such as smelling a favorite fragrance, but more like an inexplicable impression.

CHAPTER IN PERSPECTIVE: FOCUS ON HOMEOSTASIS

To maintain a life-sustaining stable internal environment, adjustments must constantly be made to compensate for myriad external and internal factors that continuously threaten to disrupt homeostasis, such as external exposure to cold or internal acid production. Many of these adjustments are directed by the nervous system, one of the body's two major regulatory systems. The central nervous system (CNS), the integrating and decision-making component of the nervous system, must continuously be informed of "what's happening" in both the internal and the external environment so that it can command appropriate responses in the organ systems to maintain the body's viability. In other words, the CNS must know what changes are taking place before it can respond to these changes.

The afferent division of the peripheral nervous system is the communication link by which the CNS is informed about the internal and external environment. The afferent division detects, encodes, and transmits peripheral signals to the CNS for processing. Afferent input is necessary for arousal, perception, and determination of efferent output.

Afferent information about the internal environment, such as the CO_2 level in the blood, never reaches the level of conscious awareness, but this input to the controlling centers of the CNS is essential for the maintenance of homeostasis. Afferent input reaching the level of conscious awareness is known as *sensory information*

and includes somesthetic and proprioceptive sensation (body sense) and special senses (vision, hearing, taste, and smell).

The body sense receptors are distributed over the entire body surface as well as throughout the joints and muscles. Afferent signals from these receptors provide information about what's happening directly to each specific body part in relation to the external environment (that is, the "what," "where," and "how much" of stimulatory inputs to the body's surface and the momentary position of the body in space). In contrast, each special sense organ is restricted to a single site in the body. Rather than providing information about a specific body part, a special sense organ provides a specific type of information about the external environment that is useful to the body as a whole. For example, through their ability to detect, extensively analyze, and integrate patterns of illumination in the external environment, the eyes and visual processing system enable us to "see" our surroundings. The same integrative effect could not be achieved if photoreceptors were scattered over the entire body surface as are the touch receptors.

Sensory input (both body sense and special senses) enables a complex multicellular organism such as a human to interact in meaningful ways with the external environment in procuring food, defending against danger, and engaging in other behavioral actions geared toward maintaining homeostasis. In addition to providing information essential for interactions with the external environment for basic survival, the perceptual processing of sensory input adds immeasurably to the richness of life, such as enjoyment of a good book, concert, or meal.

CHAPTER SUMMARY

Receptor Physiology

▮ Receptors are specialized peripheral endings of afferent neurons. Each type of receptor responds to its adequate stimulus (a change in the energy form or modality to which it is responsive), translating the energy form of the stimulus into electrical signals, the language of the nervous system. This conversion process is known as *transduction*.

▮ A stimulus brings about a graded, depolarizing receptor potential by altering the receptor's membrane permeability. Receptor potentials, if of sufficient magnitude, generate action potentials in the afferent neuronal membrane adjacent to the receptor by opening Na^+ channels in this region. These action potentials are self-propagated along the afferent neuron to the CNS.

▮ The strength and rate of change of the stimulus are reflected in the magnitude of the receptor potential, which in turn determines the frequency of action potentials generated in the afferent neuron.

▮ The magnitude of the receptor potential is also influenced by the extent of receptor adaptation, which refers to a reduction in receptor potential in spite of sustained stimulation. (1) Tonic receptors adapt slowly or not at all and thus provide continuous information about the stimuli they monitor. (2) Phasic receptors adapt rapidly and frequently exhibit off responses, thereby providing information about changes in the energy form they monitor.

▮ Discrete labeled-line pathways lead from the receptors to the CNS so that information about the type and location of the stimuli can be deciphered by the CNS, even though all the information arrives in the form of action potentials.

▮ What the brain perceives from its input is an abstraction and not reality. The only stimuli that can be detected are those for which receptors are present. Furthermore, as sensory signals ascend through progressively more complex processing, some of the information may be suppressed, whereas other parts of it may be enhanced.

▮ The term *receptive field* refers to the area surrounding a receptor within which the receptor can detect stimuli. The acuity, or discriminative ability, of a body region varies inversely with the size of its receptive fields and also depends on the extent of lateral inhibition in the afferent pathways arising from receptors in the region.

Pain

▮ Painful experiences are elicited by noxious mechanical, thermal, or chemical stimuli and consist of two components: the perception of pain coupled with emotional and behavioral responses to it.

▮ Three categories of nociceptors, or pain receptors, respond to these stimuli: mechanical nociceptors, thermal nociceptors, and polymodal nociceptors. The latter respond to all kinds of damaging stimuli, including chemicals such as bradykinin released from injured tissues.

- Pain signals are transmitted over two afferent pathways: a fast pathway that carries sharp, prickling pain signals; and a slow pathway that carries dull, aching, persistent pain signals.
- Afferent pain fibers terminate in the spinal cord on ascending pathways that transmit the signal to the brain for processing.
- Descending pathways from the brain use endogenous opiates to suppress the release of substance P, a pain-signaling neurotransmitter from the afferent pain-fiber terminal. Thus these descending pathways block further transmission of the pain signal and serve as a built-in analgesic system.

Eye: Vision

- The eye is a specialized structure housing the light-sensitive receptors essential for vision perception—namely the rods and cones found in its retinal layer.
- The iris controls the size of the pupil, thereby adjusting the amount of light permitted to enter the eye.
- The cornea and lens are the primary refractive structures that bend the incoming light rays to focus the image on the retina. The cornea contributes most to the total refractive ability of the eye. The strength of the lens can be adjusted through action of the ciliary muscle to accommodate for differences in near and far vision.
- The rod and the cone photoreceptors are activated when the photopigments they contain differentially absorb various wavelengths of light. Light absorption causes a biochemical change in the photopigment that is ultimately converted into a change in the rate of action potential propagation in the visual pathway leaving the retina. The conversion of light stimuli into electrical signals is known as *phototransduction*.
- The visual message is transmitted via a complex crossed and uncrossed pathway to the visual cortex in the occipital lobe of the brain for perceptual processing.
- Cones display high acuity but can be used only for day vision because of their low sensitivity to light. Different ratios of stimulation of three cone types by varying wavelengths of light lead to color vision.
- Rods provide only indistinct vision in shades of gray, but because they are very sensitive to light, they can be used for night vision.
- The eyes' sensitivity is increased during dark adaptation by the regeneration of rod photopigments that had been broken down during preceding light exposure and is decreased during light adaptation by the rapid breakdown of cone photopigments.

Ear: Hearing and Equilibrium

- The ear performs two unrelated functions: (1) hearing, which involves the external ear, middle ear, and cochlea of the inner ear; and (2) sense of equilibrium, which involves the vestibular apparatus of the inner ear. In contrast to the photoreceptors of the eye, the ear receptors located in the inner ear—the hair cells in the cochlea and vestibular apparatus—are mechanoreceptors.
- Hearing depends on the ear's ability to convert airborne sound waves into mechanical deformations of receptive hair cells, thereby initiating neural signals.
- Sound waves consist of high-pressure regions of compression alternating with low-pressure regions of rarefaction of air molecules. The pitch (tone) of a sound is determined by the frequency of its waves, the loudness (intensity) by the ampli-

tude of the waves, and the timbre (quality) by its characteristic overtones.

- Sound waves are funneled through the external ear canal to the tympanic membrane, which vibrates in synchrony with the waves.
- Middle ear bones bridging the gap between the tympanic membrane and the inner ear amplify the tympanic movements and transmit them to the oval window, whose movement sets up traveling waves in the cochlear fluid.
- These waves, which are at the same frequency as the original sound waves, set the basilar membrane in motion. Various regions of this membrane selectively vibrate more vigorously in response to different frequencies of sound.
- On top of the basilar membrane are the receptive hair cells of the organ of Corti, whose hairs are bent as the basilar membrane is deflected up and down in relation to the overhanging stationary tectorial membrane in which the hairs are embedded.
- This mechanical deformation of specific hair cells in the region of maximal basilar membrane vibration is transduced into neural signals that are transmitted to the auditory cortex in the temporal lobe of the brain for sound perception.
- The vestibular apparatus in the inner ear consists of (1) the semicircular canals, which detect rotational acceleration or deceleration in any direction, and (2) the utricle and saccule, which detect changes in the rate of linear movement in any direction and provide information important for determining head position in relation to gravity.
- Neural signals are generated in response to mechanical deformation of hair cells caused by specific movement of fluid and related structures within these vestibular sense organs. This information is important for the sense of equilibrium and for maintaining posture.

Chemical Senses: Taste and Smell

- Taste and smell are chemical senses. In both cases, attachment of specific dissolved molecules to binding sites on the receptor membrane causes receptor potentials that, in turn, set up neural impulses that signal the presence of the chemical.
- Taste receptors are housed in taste buds on the tongue; olfactory receptors are located in the olfactory mucosa in the upper part of the nasal cavity. Both sensory pathways include two routes: one to the limbic system for emotional and behavioral processing and one through the thalamus to the cortex for conscious perception and fine discrimination.
- Taste and olfactory receptors are continuously renewed, unlike visual and hearing receptors, which are irreplaceable.
- The four classic primary tastes are salt, sour, sweet, and bitter, along with a likely fifth primary taste, umami, a meaty or savory taste. Taste discrimination beyond the primary tastes depends on the patterns of stimulation of the taste buds, each of which responds in varying degrees to the different primary tastes. Salty and sour tastants bring about receptor potentials in taste buds that respond to them by directly affecting membrane channels, whereas the other three categories of tastants act through second-messenger systems to bring about receptor potentials.
- There are 1,000 different types of olfactory receptors, each of which responds to only one discrete component of an odor, an odorant. Odorants act through second-messenger systems to trigger receptor potentials. The afferent signals arising from the olfactory receptors are sorted according to scent component by the glomeruli within the olfactory bulb. Odor discrimination depends on the patterns of activation of the glomeruli.

REVIEW EXERCISES

Objective Questions (Answers on p. A-38)

1. Conversion of the energy forms of stimuli into electrical energy by the receptors is known as _____.
2. The type of stimulus to which a particular receptor is most responsive is called its _____.
3. The sensation perceived depends on the type of receptor stimulated rather than on the type of stimulus. (*True or false?*)
4. All afferent information is sensory information. (*True or false?*)
5. Off-center ganglion cells increase their rate of firing when a beam of light strikes the periphery of their receptive field. (*True or false?*)
6. During dark adaptation, rhodopsin is gradually regenerated to increase the sensitivity of the eyes. (*True or false?*)
7. An optic nerve carries information from the lateral and medial halves of the same eye, whereas an optic tract carries information from the lateral half of one eye and the medial half of the other. (*True or false?*)
8. Displacement of the round window generates neural impulses that are perceived as sound sensations. (*True or false?*)
9. Hair cells in different regions of the organ of Corti and neurons in different regions of the auditory cortex are activated by different tones. (*True or false?*)
10. Each taste receptor responds to just one of the four primary tastes. (*True or false?*)
11. Rapid adaptation to odors results from adaptation of the olfactory receptors. (*True or false?*)
12. Match the following:

   _____ 1. layer that contains the photoreceptors
   _____ 2. point from which the optic nerve leaves the retina
   _____ 3. forms the white part of the eye
   _____ 4. thalamic structure that processes visual input
   _____ 5. colored diaphragm of muscle that controls the amount of light entering the eye
   _____ 6. contributes most to the eye's refractive ability
   _____ 7. supplies nutrients to the lens and cornea
   _____ 8. produces aqueous humor
   _____ 9. contains the vascular supply for the retina and a pigment that minimizes scattering of light within the eye
   _____ 10. has adjustable refractive ability
   _____ 11. portion of the retina with greatest acuity
   _____ 12. point at which fibers from the medial half of each retina cross to the opposite side

   (a) choroid
   (b) aqueous humor
   (c) fovea
   (d) lateral geniculate nucleus
   (e) cornea
   (f) retina
   (g) lens
   (h) optic disc; blind spot
   (i) iris
   (j) ciliary body
   (k) optic chiasm
   (l) sclera

13. Using the following answer code, indicate which properties apply to taste and/or smell:
   (a) applies to taste
   (b) applies to smell
   (c) applies to both taste and smell
   _____ 1. Receptors are separate cells that synapse with terminal endings of afferent neurons.
   _____ 2. Receptors are specialized endings of afferent neurons.
   _____ 3. Receptors are regularly replaced.
   _____ 4. Specific chemicals in the environment attach to special binding sites on the receptor surface, leading to a depolarizing receptor potential.
   _____ 5. There are two processing pathways: a limbic system route and a thalamic-cortical route.
   _____ 6. Four (or possibly five) different receptor types are used.
   _____ 7. A thousand different receptor types are used.
   _____ 8. Information from receptor cells is filed and sorted by neural junctions called *glomeruli*.

Essay Questions

1. List and describe the receptor types according to their adequate stimulus.
2. Compare tonic and phasic receptors.
3. Explain how acuity is influenced by receptive field size and by lateral inhibition.
4. Compare the fast and slow pain pathways.
5. Describe the built-in analgesic system of the brain.
6. Describe the process of phototransduction.
7. Compare the functional characteristics of rods and cones.
8. What are sound waves? What is responsible for the pitch, intensity, and timbre of a sound?
9. Describe the function of each of the following parts of the ear: pinna, ear canal, tympanic membrane, ossicles, oval window, and the various parts of the cochlea. Include a discussion of how sound waves are transduced into action potentials.
10. Discuss the functions of the semicircular canals, the utricle, and the saccule.
11. Describe the location, structure, and activation of the receptors for taste and smell.
12. Compare the processes of color vision, hearing, taste, and smell discrimination.

Quantitative Exercises (Solutions on p. A-38)

1. Calculate the difference in the time it takes for an action potential to travel 1.3 m between the slow (12 m/sec) and fast (30 m/sec) pain pathways.
2. Have you ever noticed that humans have circular pupils, whereas cats' pupils are more rectangular? The following calculations will help you understand the implication of this difference. For simplicity, assume a constant intensity of light.
   a. If the diameter of a human's circular pupil were decreased by half on contraction of the constrictor muscle of the iris, by what percentage would the amount of light allowed into the eye be decreased?
   b. If a cat's rectangular pupil were decreased by half along one axis only, by what percentage would the amount of light allowed into the eye be decreased?
   c. Comparing these calculations, do humans or cats have more precise control over the amount of light falling on the retina?

3. A decibel is the unit of *sound level*, β, defined as follows:
$$\beta = (10 \text{ dB}) \log_{10}(I/I_o),$$
where I is *sound intensity*, or the rate at which sound waves transmit energy per unit area. The units of I are watts per square meter (W/m^2). I_o is a constant intensity close to the human hearing threshold, namely 10^{-12} W/m^2.
   a. For the following sound levels, calculate the corresponding sound intensities:
      (1) 20 dB (a whisper)
      (2) 70 dB (a car horn)
      (3) 120 dB (a low-flying jet)
      (4) 170 dB (a space shuttle launch)
   b. Explain why the sound levels of these sounds increase by the same increment (that is, each sound is 50 dB higher than the one preceding it), yet the incremental increases in sound intensities that you calculated are so different. What implications does this have for the performance of the human ear?

POINTS TO PONDER

(Explanations on p. A-39)

1. Patients with certain nerve disorders are unable to feel pain. Why is this disadvantageous?
2. Ophthalmologists frequently instill eye drops in their patients' eyes to bring about pupillary dilation, which makes it easier for the physician to view the eye's interior. In what way would the drug in the eye drops affect autonomic nervous system activity in the eye in order to cause the pupils to dilate?
3. A patient complains of not being able to see the right half of the visual field with either eye. At what point in the patient's visual pathway is the defect?
4. Explain how middle ear infections interfere with hearing. Of what value are the "tubes" that are sometimes surgically placed in the eardrums of patients with a history of repeated middle ear infections accompanied by chronic fluid accumulation?
5. Explain why your sense of smell is reduced when you have a cold, even though the cold virus does not directly adversely affect the olfactory receptor cells.
6. **Clinical Consideration.** Suzanne J. complained to her physician of bouts of dizziness. The physician asked her whether by "dizziness" she meant a feeling of lightheadedness, as if she felt she were going to faint (a condition known as *syncope*), or a feeling that she or surrounding objects in the room were spinning around (a condition known as *vertigo*). Why is this distinction important in the differential diagnosis of her condition? What are some possible causes of each of these symptoms?

PHYSIOEDGE RESOURCES

 PHYSIOEDGE CD-ROM
Figures marked with the PhysioEdge icon have associated activities on the CD. For this chapter, check out the Media Quizzes related to Figures 6-9, 6-10, 6-11, 6-20, and 6-33.

CHECK OUT THESE MEDIA QUIZZES:
6.1 Pain
6.2 The Eye
6.3 The Ear
6.4 Taste

 PHYSIOEDGE WEB SITE
The Web site for this book contains a wealth of helpful study aids, as well as many ideas for further reading and research. Log on to:

http://www.brookscole.com

Go to the Biology page and select Sherwood's *Human Physiology*, 5th Edition. Select a chapter from the drop-down menu or click on one of these resource areas:

- **Case Histories** provide an introduction to the clinical aspects of human physiology.
- For 2-D and 3-D graphical illustrations and animations of physiological concepts, visit our **Visual Learning Resource**.
- Resources for study and review include the **Chapter Outline, Chapter Summary, Glossary,** and **Flash Cards**. Use our **Quizzes** to prepare for in-class examinations.
- On-line research resources to consult are **Hypercontents**, with current links to relevant Internet sites; **Internet Exercises** with starter URLs provided; and **InfoTrac Exercises**.

For more readings, go to InfoTrac College Edition, your on-line research library, at:

http://infotrac.thomsonlearning.com

Nervous System
(Peripheral Nervous Systems)

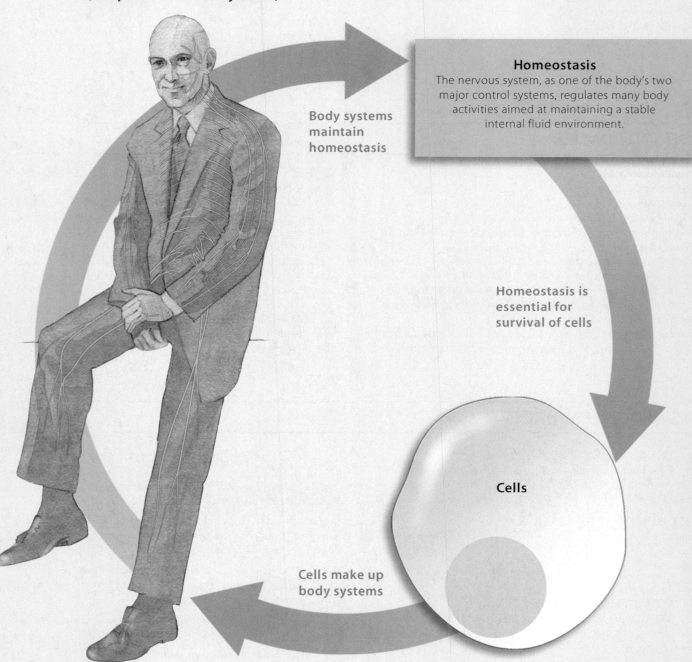

Body systems maintain homeostasis

Homeostasis
The nervous system, as one of the body's two major control systems, regulates many body activities aimed at maintaining a stable internal fluid environment.

Homeostasis is essential for survival of cells

Cells

Cells make up body systems

The nervous system, one of the two major regulatory systems of the body, consists of the central nervous system (CNS), composed of the brain and spinal cord, and the **peripheral nervous system,** composed of the afferent and efferent fibers that relay signals between the CNS and periphery (other parts of the body).

Once informed by the afferent division of the peripheral nervous system that a change in the internal or external environment is threatening homeostasis, the CNS makes appropriate adjustments to maintain homeostasis. The CNS makes these adjustments by controlling the activities of effector organs (muscles and glands) by transmitting signals *from* the CNS to these organs through the **efferent division** of the peripheral nervous system.

Chapter 7

The Peripheral Nervous System: Efferent Division

CONTENTS AT A GLANCE

INTRODUCTION

AUTONOMIC NERVOUS SYSTEM

▌ Anatomy and neurotransmitters of autonomic fibers

▌ Dominance patterns of sympathetic and parasympathetic systems

▌ Autonomic receptor types

▌ CNS control of autonomic activity

SOMATIC NERVOUS SYSTEM

▌ Motor neurons

▌ Final common pathway

NEUROMUSCULAR JUNCTION

▌ Events at the neuromuscular junction; role of acetylcholine

▌ Role of acetylcholinesterase

▌ Influence of specific chemical agents and diseases

INTRODUCTION

The efferent division of the peripheral nervous system is the communication link by which the central nervous system controls the activities of muscles and glands, the effector organs that carry out the intended effects, or actions. The CNS regulates these effector organs by initiating action potentials in the cell bodies of efferent neurons whose axons terminate on these organs. Cardiac muscle, smooth muscle, most exocrine glands, and some endocrine glands are innervated by the **autonomic nervous system,** the involuntary branch of the peripheral efferent division. Skeletal muscle is innervated by the **somatic nervous system,** the voluntary branch of the efferent division. Efferent output typically influences either movement or secretion, as illustrated in ▲ Table 7-1, which provides examples of the effects of neural control on various effector organs composed of different types of muscle and gland tissue. Much of this efferent output is directed toward maintaining homeostasis. The efferent output to skeletal muscles is also directed toward voluntarily controlled nonhomeostatic activities, such as riding a bicycle. (It is important to realize that many effector organs are also subject to hormonal control and/or intrinsic control mechanisms; see p. 16.)

How many different neurotransmitters would you guess are released from the various efferent neuronal terminals to elicit essentially all the neurally controlled effector organ responses? Only two—acetylcholine and norepinephrine! Acting independently, these neurotransmitters bring about such diverse effects as salivary secretion, bladder contraction, and voluntary motor movements. These effects are a prime example of how the same chemical messenger may elicit a multiplicity of responses from various tissues, depending on specialization of the effector organs.

AUTONOMIC NERVOUS SYSTEM

▌ An autonomic nerve pathway consists of a two-neuron chain.

Each autonomic nerve pathway extending from the CNS to an innervated organ consists of a two-neuron chain (● Figure 7-1).

Examples of the Influence of Efferent Output on Movement and Secretion by Effector Organs

Category of Influence	Examples of Effector Organs with Different Types of Tissues	Sample Outcome in Response to Efferent Output
Influence on Movement	Heart (cardiac muscle)	Increased pumping of blood when the blood pressure falls too low
	Stomach (smooth muscle)	Delayed emptying of the stomach until the intestine is ready to process the food
	Diaphragm—a respiratory muscle (skeletal muscle)	Augmented breathing in response to exercise
Influence on Secretion	Sweat glands (exocrine glands)	Initiation of sweating on exposure to a hot environment
	Endocrine pancreas (endocrine gland)	Increased secretion of insulin, a hormone that puts excess nutrients in storage following a meal

The cell body of the first neuron in the series is located in the CNS. Its axon, the **preganglionic fiber,** synapses with the cell body of the second neuron, which lies within a ganglion. (Recall that a ganglion is a cluster of neuronal cell bodies located outside the CNS.) The axon of the second neuron, the **postganglionic fiber,** innervates the effector organ.

The autonomic nervous system consists of two subdivisions—the **sympathetic** and the **parasympathetic nervous systems** (● Figure 7-2). Sympathetic nerve fibers originate in the thoracic and lumbar regions of the spinal cord (see p. 173). Most sympathetic preganglionic fibers are very short, synapsing with cell bodies of postganglionic neurons within ganglia that lie in a **sympathetic ganglion chain** (also called the **sympathetic trunk**) located along either side of the spinal cord (see ● Figure 5-27, p. 172). Long postganglionic fibers originating in the ganglion chain terminate on the effector organs. Some preganglionic fibers pass through the ganglion chain without synapsing. Instead, they terminate later in sympathetic **collateral ganglia** located about halfway between the CNS and the innervated organs, with postganglionic fibers traveling the remainder of the distance.

Parasympathetic preganglionic fibers arise from the cranial (brain) and sacral (lower spinal cord) areas of the CNS. These fibers are long in comparison to sympathetic preganglionic fibers because they do not end until they reach **terminal ganglia** that lie in or near the effector organs. Very short postganglionic fibers terminate on the cells of an organ itself.

▌Parasympathetic postganglionic fibers release acetylcholine; sympathetic ones release norepinephrine.

Sympathetic and parasympathetic preganglionic fibers release the same neurotransmitter, **acetylcholine (ACh),** but the postganglionic endings of these two systems release different neurotransmitters (the neurotransmitters that influence the effector organs). Parasympathetic postganglionic fibers release acetylcholine. Accordingly, they, along with all autonomic preganglionic fibers, are called **cholinergic fibers.** Most sympathetic postganglionic fibers, in contrast, are called **adrenergic fibers** because they release **noradrenaline,** commonly known as **nor-**

● FIGURE 7-1

Autonomic nerve pathway

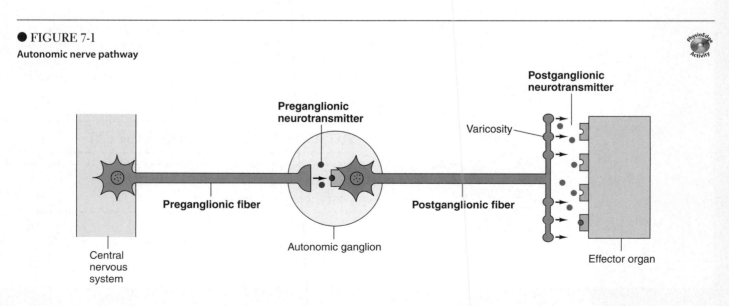

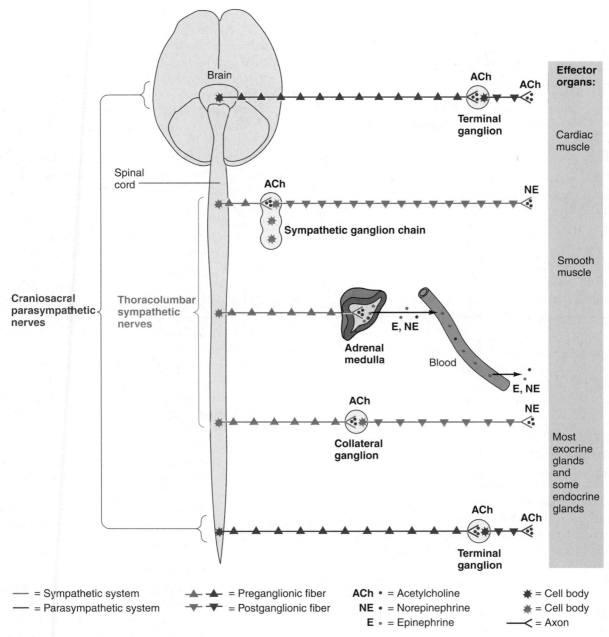

Brain

Spinal cord

Craniosacral parasympathetic nerves

Thoracolumbar sympathetic nerves

ACh **ACh**
Terminal ganglion

ACh **NE**
Sympathetic ganglion chain

E, NE
Adrenal medulla
Blood
E, NE

ACh **NE**
Collateral ganglion

ACh **ACh**
Terminal ganglion

Effector organs:

Cardiac muscle

Smooth muscle

Most exocrine glands and some endocrine glands

— = Sympathetic system
— = Parasympathetic system
▲—▲ = Preganglionic fiber
▼—▼ = Postganglionic fiber
ACh • = Acetylcholine
NE • = Norepinephrine
E • = Epinephrine
✳ = Cell body
✳ = Cell body
—< = Axon

● **FIGURE 7-2**

Autonomic nervous system
The sympathetic nervous system, which originates in the thoracolumbar regions of the spinal cord, has short cholinergic (acetylcholine-releasing) preganglionic fibers and long adrenergic (norepinephrine-releasing) postganglionic fibers. The parasympathetic nervous system, which originates in the brain and sacral region of the spinal cord, has long cholinergic preganglionic fibers and short cholinergic postganglionic fibers. In most instances, sympathetic and parasympathetic postganglionic fibers both innervate the same effector organs. The adrenal medulla is a modified sympathetic ganglion, which releases epinephrine and norepinephrine into the blood.

---

epinephrine.[1] Both acetylcholine and norepinephrine also serve as chemical messengers elsewhere in the body (▲ Table 7-2).

Postganglionic autonomic fibers do not end in a single terminal swelling like a synaptic knob. Instead, the terminal branches of autonomic fibers contain numerous swellings, or **varicosities,** that simultaneously release neurotransmitter over a large area of the innervated organ rather than on single cells (● Figure 7-1 and ● Figure 8-34, p. 295). This diffuse release

[1]*Noradrenaline (norepinephrine)* is chemically very similar to *adrenaline (epinephrine),* the primary hormone product secreted by the adrenal medulla gland. Because a U.S. pharmaceutical company marketed this product for use as a drug under the trade name Adrenalin, the scientific community in this country prefers the alternate name "epinephrine" as a generic term for this chemical messenger, and accordingly, "noradrenaline" is known as "norepinephrine." In most other English-speaking countries, however, "adrenaline" and "noradrenaline" are the terms of choice.

▲ TABLE 7-2
Sites of Release for Acetylcholine and Norepinephrine

Acetylcholine	Norepinephrine
All preganglionic terminals of the autonomic nervous system	Most sympathetic postganglionic terminals
All parasympathetic postganglionic terminals	Adrenal medulla
Sympathetic postganglionic terminals at sweat glands and some blood vessels in skeletal muscle	Central nervous system
Terminals of efferent neurons supplying skeletal muscle (motor neurons)	
Central nervous system	

of neurotransmitter, coupled with the fact that any resulting change in electrical activity is spread throughout a smooth or cardiac muscle mass via gap junctions (see p. 63), means that whole organs instead of discrete cells are typically influenced by autonomic activity.

The autonomic nervous system controls involuntary visceral organ activities.

The autonomic nervous system regulates visceral activities normally outside the realm of consciousness and voluntary control, such as circulation, digestion, sweating, and pupillary size. It is not entirely true, however, that an individual has no control over activities governed by the autonomic system. Visceral afferent information usually does not reach the conscious level, so individuals have no way of consciously controlling the resultant efferent output. With the technique of **biofeedback,** however, people are provided with a conscious signal regarding visceral afferent information. This signal, which may be a sound, a light, or a graphic display on a computer screen, enables them to exert some voluntary control over events that are normally considered subsconscious activities. For example, individuals have learned to consciously lower their blood pressure when they "hear" that it is elevated via special devices that convert blood pressure levels into sound signals. Such biofeedback techniques are gaining wider acceptance and usage.

The sympathetic and parasympathetic nervous systems dually innervate most visceral organs.

Most visceral organs are innervated by both sympathetic and parasympathetic nerve fibers (● Figure 7-3). ▲ Table 7-3 summarizes the major effects of these autonomic branches. Although the details of this wide array of autonomic responses are described more fully in later chapters that discuss the indi-

vidual organs involved, several general concepts can be derived now. As you can see from the table, the sympathetic and parasympathetic nervous systems generally exert opposite effects in a particular organ. Sympathetic stimulation increases the heart rate, whereas parasympathetic stimulation decreases it; sympathetic stimulation slows down movement within the digestive tract, whereas parasympathetic stimulation enhances digestive motility. Note that both systems increase the activity of some organs and reduce the activity of others.

Rather than memorizing a list such as that presented in ▲ Table 7-3, it is better to logically deduce the actions of the two systems based on an understanding of the circumstances under which each system dominates. Usually, both systems are partially active; that is, normally some level of action potential activity exists in both the sympathetic and the parasympathetic fibers supplying a particular organ. This ongoing activity is called **sympathetic** or **parasympathetic tone** or **tonic activity.** Under given circumstances, activity of one division can dominate the other. *Sympathetic dominance* to a particular organ exists when the sympathetic fibers' rate of firing to that organ increases above tonic level, coupled with a simultaneous decrease below tonic level in the parasympathetic fibers' frequency of action potentials to the same organ. The reverse situation is true for *parasympathetic dominance.* Shifts in balance between sympathetic and parasympathetic activity can be accomplished discretely for individual organs to meet specific demands (for example, sympathetically induced dilation of the pupil in dim light; see p. 197), or a more generalized, widespread discharge of one autonomic system in favor of the other can be elicited to control bodywide functions. Massive widespread discharges take place more frequently in the sympathetic system. The value of this potential for massive sympathetic discharge is evident considering the circumstances during which this system usually dominates.

Times of sympathetic dominance

The sympathetic system promotes responses that prepare the body for strenuous physical activity in the face of emergency or stressful situations, such as a physical threat from the outside environment. This response is typically referred to as a **fight-or-flight response,** because the sympathetic system readies the body to fight against or flee from the threat. Think about the body resources needed in such circumstances. The heart beats more rapidly and more forcefully, blood pressure is elevated because of generalized constriction of the blood vessels, the respiratory airways open wide to permit maximal air flow, glycogen (stored sugar) and fat stores are broken down to release extra fuel into the blood, and blood vessels supplying skeletal muscles dilate (open more widely). All these responses are aimed at providing increased flow of oxygenated, nutrient-rich blood to the skeletal muscles in anticipation of strenuous physical activity. Furthermore, the pupils dilate and the eyes adjust for far vision, enabling the person to visually assess the entire threatening scene. Sweating is promoted in anticipation of excess heat production by the physical exertion. Because digestive and urinary activities are not essential in meeting the threat, the sympathetic system inhibits these activities.

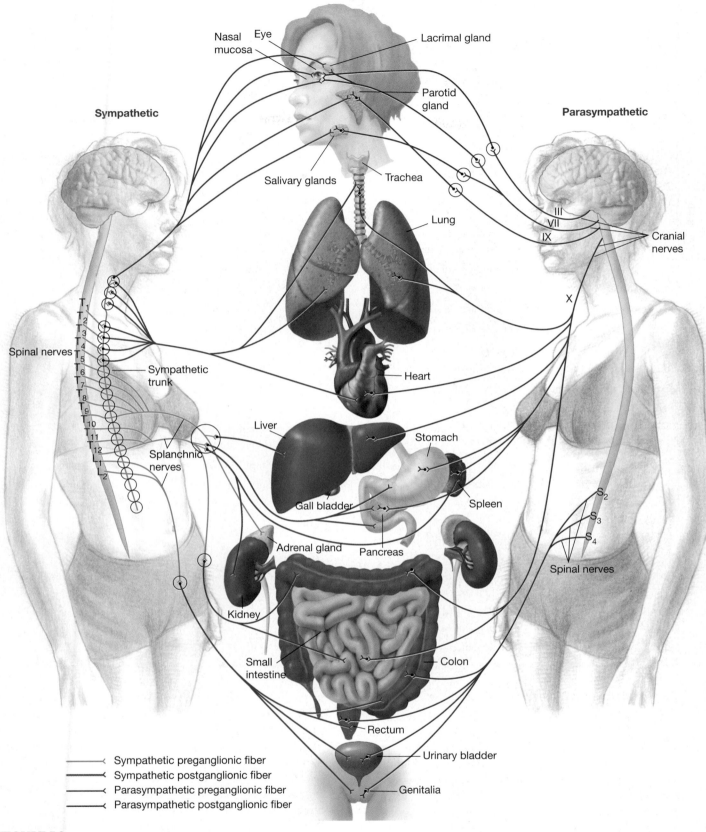

Sympathetic

Parasympathetic

Nasal mucosa
Eye
Lacrimal gland
Parotid gland
Salivary glands
Trachea
Lung
Heart
Liver
Stomach
Gall bladder
Spleen
Adrenal gland
Pancreas
Kidney
Small intestine
Colon
Rectum
Urinary bladder
Genitalia

III
VII
IX
Cranial nerves
X

Spinal nerves

T₁ T₂ T₃ T₄ T₅ T₆ T₇ T₈ T₉ T₁₀ T₁₁ T₁₂ L₁ L₂

Sympathetic trunk

Splanchnic nerves

S₂
S₃
S₄
Spinal nerves

Sympathetic preganglionic fiber
Sympathetic postganglionic fiber
Parasympathetic preganglionic fiber
Parasympathetic postganglionic fiber

● FIGURE 7-3

Schematic representation of the structures innervated by the sympathetic and parasympathetic nervous systems

Effects of the Autonomic Nervous System on Various Organs

Organ	Effect of Sympathetic Stimulation	Effect of Parasympathetic Stimulation
Heart	Increased rate, increased force of contraction (of whole heart)	Decreased rate, decreased force of contraction (of atria only)
Blood Vessels	Constriction	Dilation of vessels supplying the penis and clitoris only
Lungs	Dilation of bronchioles (airways)	Constriction of bronchioles
	Inhibition (?) of mucus secretion	Stimulation of mucus secretion
Digestive Tract	Decreased motility (movement)	Increased motility
	Contraction of sphincters (to prevent forward movement of contents)	Relaxation of sphincters (to permit forward movement of contents)
	Inhibition (?) of digestive secretions	Stimulation of digestive secretions
Gallbladder	Relaxation	Contraction (emptying)
Urinary Bladder	Relaxation	Contraction (emptying)
Eye	Dilation of pupil	Constriction of pupil
	Adjustment of eye for far vision	Adjustment of eye for near vision
Liver (Glycogen Stores)	Glycogenolysis (glucose released)	None
Adipose Cells (Fat Stores)	Lipolysis (fatty acids released)	None
Exocrine Glands		
Exocrine pancreas	Inhibition of pancreatic exocrine secretion	Stimulation of pancreatic exocrine secretion (important for digestion)
Sweat glands	Stimulation of secretion by most sweat glands	Stimulation of secretion by some sweat glands
Salivary glands	Stimulation of small volume of thick saliva rich in mucus	Stimulation of large volume of watery saliva rich in enzymes
Endocrine Glands		
Adrenal medulla	Stimulation of epinephrine and norepinephrine secretion	None
Endocrine pancreas	Inhibition of insulin secretion; stimulation of glucagon secretion	Stimulation of insulin and glucagon secretion
Genitals	Ejaculation and orgasmic contractions (males); orgasmic contractions (females)	Erection (caused by dilation of blood vessels in penis [male] and clitoris [female])
Brain Activity	Increased alertness	None

Times of parasympathetic dominance

The parasympathetic system dominates in quiet, relaxed situations. Under such nonthreatening circumstances, the body can be concerned with its own "general housekeeping" activities, such as digestion and emptying of the urinary bladder. The parasympathetic system promotes these types of bodily functions while slowing down those activities that are enhanced by the sympathetic system. There is no need, for example, to have the heart beating rapidly and forcefully when the person is in a tranquil setting.

Advantage of dual autonomic innervation

What is the advantage of dual innervation of organs with nerve fibers whose actions oppose each other? It enables precise control over an organ's activity, similar to having both an accelerator and a brake to control the speed of a car. If an animal suddenly darts across the road as you are driving, you could eventually stop if you simply took your foot off the accelerator, but you might stop too slowly to avoid hitting the animal. If you simultaneously apply the brake as you lift up on the accelerator, however, you can come to a more rapid, controlled stop. In a similar manner, a sympathetically accelerated heart rate could gradually be reduced to normal following a stressful situation by decreasing the rate of firing in the cardiac sympathetic nerve (letting up on the accelerator), but the heart rate can be reduced more rapidly by simultaneously increasing activity in the parasympathetic supply to the heart (applying the brake). Indeed, the two divisions of the autonomic nervous system are usually reciprocally controlled; increased activity in one division is accompanied by a corresponding decrease in the other.

There are several exceptions to the general rule of dual reciprocal innervation by the two branches of the autonomic nervous system; the most notable are the following:

• *Innervated blood vessels* (most arterioles and veins are innervated, arteries and capillaries are not) receive only sympathetic nerve fibers. Regulation is accomplished by increasing or decreasing the firing rate above or below the tonic level in these sympathetic fibers. The only blood vessels to receive both sympathetic and parasympathetic fibers are those supplying the penis and clitoris. The precise vascular control this dual innervation affords these organs is important in accomplishing erection.

• Most *sweat glands* are innervated only by sympathetic nerves. The postganglionic fibers of these nerves are unusual because they secrete acetylcholine rather than norepinephrine.

• *Salivary glands* are innervated by both autonomic divisions, but unlike elsewhere, sympathetic and parasympathetic activity is not antagonistic. Both stimulate salivary secretion, but the saliva's volume and composition differ, depending on which autonomic branch is dominant.

You will learn more about these exceptions in later chapters.

A wide variety of autonomic malfunctions accompany aging. Preliminary studies are providing clues to the age-related decline in autonomic control. (To learn more about the autonomic nervous system and aging, see the accompanying boxed feature, ▶ Concepts, Challenges, and Controversies.)

We are now going to turn our attention to the adrenal medulla, a unique endocrine component of the sympathetic nervous system.

▌ The adrenal medulla is a modified part of the sympathetic nervous system.

There are two *adrenal glands*, one lying above the kidney on each side (*ad* means "next to"; *renal* means "kidney"). The adrenal glands are endocrine glands, each consisting of an outer portion, the *adrenal cortex*, and an inner portion, the *adrenal medulla*. The **adrenal medulla** is considered a modified sympathetic ganglion that does not give rise to postganglionic fibers. Instead, on stimulation by the preganglionic fiber that originates in the CNS it secretes hormones into the blood (● Figures 7-2 and 7-4). Not surprisingly, the hormones are identical or similar to postganglionic sympathetic neurotransmitters. About 20% of the adrenal medullary hormone output is norepinephrine, and the remaining 80% is the closely related substance **epinephrine (adrenaline)** (see footnote 1, p. 239). These hormones, in general, reinforce activity of the sympathetic nervous system.

▌ Several different receptor types are available for each autonomic neurotransmitter.

Because each autonomic neurotransmitter and medullary hormone stimulate activity in some tissues but inhibit activity in others, the particular responses must depend on specialization of the tissue cells rather than on properties of the chemicals

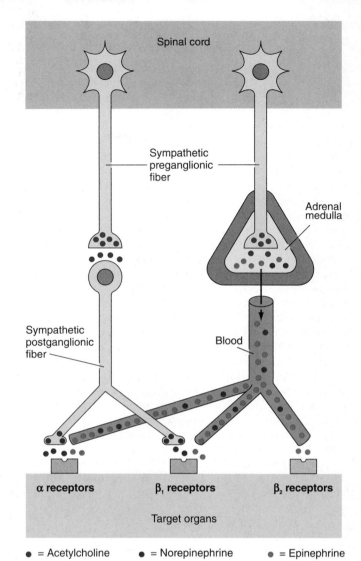

● = Acetylcholine   ● = Norepinephrine   ● = Epinephrine

● FIGURE 7-4

Comparison of the release and binding to receptors of epinephrine and norepinephrine
Norepinephrine is released both as a neurotransmitter from sympathetic postganglionic fibers and as a hormone from the adrenal medulla. Beta$_1$ (β_1) receptors bind equally with both norepinephrine and epinephrine, whereas beta$_2$ (β_2) receptors bind primarily with epinephrine and alpha (α) receptors of both subtypes have a greater affinity for norepinephrine than for epinephrine.

themselves. Responsive tissue cells possess one or more of several different types of plasma membrane receptor proteins for these chemical messengers. Binding of a neurotransmitter to a receptor induces the tissue-specific response.

Cholinergic receptors

Two types of acetylcholine (cholinergic) receptors—*nicotinic* and *muscarinic* receptors—have been identified on the basis of their response to particular drugs (▲ Table 7-4). **Nicotinic receptors** are found on the postganglionic cell bodies in all autonomic ganglia. (These receptors are so-named because they are activated by the tobacco plant derivative nicotine.) Nicotinic receptors respond to acetylcholine released from both

The Autonomic Nervous System and Aging: A Fortuitious Find

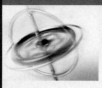

Often in scientific research, an investigator unexpectedly comes across an important finding while studying a totally different phenomenon. This happened with Robert Schmidt, a neuropathologist, who accidentally stumbled across the first evidence of the effects of aging on the autonomic nervous system while investigating a different problem—what might be responsible for the intestinal disorders that often accompany diabetes mellitus. Because nerve damage is known to be one of the long-term complications of diabetes, Schmidt suspected that damage to the autonomic nerves that help regulate the intestine might be the cause of the intestinal symptoms. On microscopic examination of the sympathetic ganglia supplying the intestine in diabetic rats, Schmidt was astonished to see that some of the preganglionic fibers (axons) near their terminal ending were swollen to more than 30 times their normal size. Thinking this dramatic pathology must somehow be linked to diabetes, he examined the same ganglia in nondiabetic control rats for comparison, fully expecting to find normal preterminal axons. To his surprise, some of the axon endings in these normal rats were as swollen as those in the diabetic rats.

Puzzling over his results, Schmidt finally thought of one thing that the two groups had in common. They were both old. Most studies on rats involved young specimens, but Schmidt's animals were old because he was investigating a long-term effect of diabetes.

Schmidt has demonstrated comparable evidence of this autonomic aging in humans. In autopsy examinations of sympathetic ganglia in 56 people who had died between the ages of 15 and 93 from various causes, Schmidt found very few enlarged preterminal axons in persons younger than 60, whereas the number of swollen axons increased dramatically in those over 60. In fact, abnormal preterminal axons were significantly more frequent in older humans than in aging rats.

Further study revealed that the abnormal sympathetic axons contain an overabundance of disoriented neurofilaments, cytoskeletal components that normally serve as internal scaffolding to help maintain a nerve cell's shape (see p. 51). Neurofilaments are normally degraded in the axon terminals for recycling, but this process is apparently impaired in connection with the age-related autonomic damage.

Not all of the preganglionic fibers in a sympathetic ganglion are swollen and packed with neurofilaments. Only terminals that contain a recently discovered neuromodulator, **neuropeptide Y (NPY),** are affected (see p. 124). The role that NPY plays in sympathetic ganglia is unknown, but other investigators have found evidence that NPY is a cotransmitter in sympathetic control of vascular (blood vessel) smooth muscle, among other actions.

The autonomic nervous system helps control many body functions such as regulation of blood pressure and movement of food through the digestive tract via its influence on smooth muscle. Aging is often accompanied by autonomic malfunctions, such as becoming lightheaded when standing up too quickly because of inadequate vascular control or constipation due to sluggish bowel motility. These age-related autonomic dysfunctions are assuming greater significance as life expectancy and consequently the number of elderly are increasing.

Normal function of the smooth muscle of the intestine and blood vessels depends on complex, incompletely understood control mechanisms involving the interaction of multiple neural elements and hormones. One can speculate that an abnormality in NPY neuromodulation of these activities might contribute to age-related autonomic malfunctions, although this remains to be confirmed. According to Schmidt, "We predict that when the function of NPY-containing processes in human prevertebral (sympathetic) ganglia is understood, that function will be compromised in the elderly."

▲ TABLE 7-4

Location of Nicotinic and Muscarinic Cholinergic Receptors

Type of Receptor	Site of Receptor	Respond to Acetylcholine Released from:
Nicotinic Receptors	All autonomic ganglia	Sympathetic and parasympathetic preganglionic fibers
	Motor end plates of skeletal muscle fibers	Motor neurons
	Some CNS cell bodies and dendrites	Some CNS presynaptic terminals
Muscarinic Receptors	Effector cells (cardiac muscle, smooth muscle, glands)	Parasympathetic postganglionic fibers
	Some CNS cell bodies and dendrites	Some CNS presynaptic terminals

sympathetic and parasympathetic preganglionic fibers. Binding of acetylcholine to these receptors brings about opening of cation channels in the postganglionic cell that permit passage of both Na^+ and K^+. Because of the greater electrochemical gradient for Na^+ than for K^+, more Na^+ enters the cell than K^+ leaves, bringing about a depolarization that leads to initiation of an action potential in the postganglionic cell.

Muscarinic receptors (activated by the mushroom poison muscarine) are found on effector cell membranes (smooth muscle, cardiac muscle, and glands). They bind with acetylcholine released from parasympathetic postganglionic fibers. There are five subtypes of muscarinic receptors, all of which are linked to G proteins that activate second-messenger systems that lead to the target cell response (see p. 67).

Adrenergic receptors

There are two major classes of adrenergic receptors for norepinephrine and epinephrine: **alpha (α) and beta (β) receptors,** which are further subclassified into α_1 and α_2 and β_1 and β_2 **receptors.** These various receptor types are distinctly distributed among the effector organs. Receptors of the β_2 type bind primarily with epinephrine, whereas β_1 receptors have about equal affinities for norepinephrine and epinephrine, and α receptors of both subtypes have a greater sensitivity to norepinephrine than to epinephrine (● Figure 7-4).

All adrenergic receptors are coupled to G proteins, but the ensuing pathway differs for the various receptor types. Activation of both β_1 and β_2 receptors brings about the target cell response by means of the cyclic AMP second-messenger system (see p. 67). Stimulation of α_1 receptors elicits the desired response via the Ca^{2+} second-messenger system (see p. 68). By contrast, binding of a neurotransmitter to an α_2 receptor blocks cyclic AMP production in the target cell.

Accordingly, activation of α_2 receptors brings about an inhibitory response in the effector organ, such as decreased smooth muscle contraction in the digestive tract. Activation of α_1 receptors, in contrast, usually brings about an excitatory response in the effector organ—for example, arteriolar constriction caused by increased contraction of the smooth muscle in the walls of these blood vessels. Alpha$_1$ receptors are present in most sympathetic target tissues. Stimulation of β_1 receptors, which are found primarily in the heart, also causes an excitatory response, namely increased rate and force of cardiac contraction. The response to β_2 receptor activation is generally inhibitory, such as arteriolar or bronchiolar (respiratory airway) dilation caused by relaxation of the smooth muscle in the walls of these tubular structures.

Autonomic agonists and antagonists

Drugs are available that selectively enhance or mimic (**agonists**) or block (**antagonists**) autonomic responses at each of the receptor types. Some are only of experimental interest, but others are very important therapeutically. For example, *atropine* blocks the effect of acetylcholine at muscarinic receptors but does not affect nicotinic receptors. Because the acetylcholine released at both parasympathetic and sympathetic preganglionic fibers combines with nicotinic receptors, blockage at nicotinic synapses would knock out both of these autonomic branches. By acting selectively to interfere with acetylcholine action only at muscarinic junctions, which are the sites of parasympathetic postganglionic action, atropine effectively blocks parasympathetic effects but does not influence sympathetic activity at all. This principle is used to suppress salivary and bronchial secretions before surgery to reduce the risk of a patient inhaling these secretions into the lungs.

Likewise, drugs that act selectively at α and β adrenergic receptor sites to either activate or block specific sympathetic effects are widely used. *Salbutamol* is an excellent example. It selectively activates β_2 adrenergic receptors at low doses, making it possible to dilate the bronchioles in the treatment of asthma without undesirably stimulating the heart (the heart has mostly β_1 receptors). Other drugs that act selectively at α and β receptors are beneficial in manipulating blood pressure and heart rate in the treatment of hypertension and cardiac arrhythmias.

∎ Many regions of the central nervous system are involved in the control of autonomic activities.

Messages from the CNS are delivered to cardiac muscle, smooth muscle, and glands via the autonomic nerves, but what regions of the CNS regulate autonomic output?

- Some autonomic reflexes, such as urination, defecation, and erection, are integrated at the spinal cord level, but all these spinal reflexes are subject to control by higher levels of consciousness.

- The medulla within the brain stem is the region most directly responsible for autonomic output. Centers for controlling cardiovascular, respiratory, and digestive activity via the autonomic system are located there.

- The hypothalamus plays an important role in integrating the autonomic, somatic, and endocrine responses that automatically accompany various emotional and behavioral states. For example, the increased heart rate, blood pressure, and respiratory activity associated with anger or fear are brought about by the hypothalamus acting through the medulla.

- Autonomic activity can also be influenced by the prefrontal association cortex through its involvement with emotional expression characteristic of the individual's personality. An example is blushing when embarrassed, which is caused by dilation of blood vessels supplying the skin of the cheeks. Such responses are mediated through hypothalamic-medullary pathways.

▲ Table 7-5 summarizes the main distinguishing features of the sympathetic and parasympathetic nervous systems.

SOMATIC NERVOUS SYSTEM

∎ Motor neurons supply skeletal muscle.

Skeletal muscle is innervated by **motor neurons,** the axons of which constitute the somatic nervous system. The cell bodies of motor neurons are located within the ventral horn of the spinal cord. Unlike the two-neuron chain of autonomic nerve fibers, the axon of a motor neuron is continuous from its origin in the spinal cord to its termination on skeletal muscle. Motor neuron axon terminals release acetylcholine, which brings about excitation and contraction of the innervated muscle cells. Motor neurons can only stimulate skeletal muscles, in contrast to autonomic fibers, which can either stimulate or inhibit their effector organs. Inhibition of skeletal muscle activity can be accomplished only within the CNS through inhibitory synaptic input to the dendrites and cell bodies of the motor neurons supplying that particular muscle.

∎ Motor neurons are the final common pathway.

Motor neuron dendrites and cell bodies are influenced by many converging presynaptic inputs, both excitatory and inhibitory. Some of these inputs are part of spinal reflex pathways originating with peripheral sensory receptors. Others are part of descending pathways originating within the brain. Areas of the brain that exert control over skeletal muscle movements include the motor regions of the cortex, the basal nuclei, the cerebellum, and the brain stem (see p. 148 and p. 285; also see ● Figure 5-30, p. 174, for specific examples of these descending motor pathways).

Distinguishing Features of the Sympathetic and Parasympathetic Nervous System

Feature	Sympathetic System	Parasympathetic System
Origin of Preganglionic Fiber	Thoracic and lumbar regions of spinal cord	Brain and sacral region of spinal cord
Origin of Postganglionic Fiber (Location of Ganglion)	Sympathetic ganglion chain (near spinal cord) or collateral ganglia (about halfway between spinal cord and effector organs)	Terminal ganglia (in or near effector organs)
Length and Type of Fiber	Short cholinergic preganglionic fibers	Long cholinergic preganglionic fibers
	Long adrenergic postganglionic fibers (most)	Short cholinergic postganglionic fibers
	Long cholinergic postganglionic fibers (few)	
Effector Organs Innervated	Cardiac muscle, almost all smooth muscle, most exocrine glands, and some endocrine glands	Cardiac muscle, most smooth muscle, most exocrine glands, and some endrocrine glands
Types of Receptors for Neurotransmitters	α_1, α_2, β_1, β_2	Nicotinic, muscarinic
Dominance	Dominates in emergency "fight-or-flight" situations; prepares body for strenuous physical activity	Dominates in quiet, relaxed situations; promotes "general housekeeping" activities such as digestion

Motor neurons are considered the **final common pathway,** because the only way any other parts of the nervous system can influence skeletal muscle activity is by acting on these motor neurons. The level of activity in a motor neuron and its subsequent output to the skeletal muscle fibers it innervates depend on the relative balance of EPSPs and IPSPs (see p. 121) brought about by its presynaptic inputs originating from these diverse sites in the brain.

The somatic system is considered to be under voluntary control, but much of the skeletal muscle activity involving posture, balance, and stereotypical movements is subconsciously controlled. You may decide you want to start walking, but you do not have to consciously bring about the alternate contraction and relaxation of the involved muscles because these movements are involuntarily coordinated by lower brain centers.

The cell bodies of the crucial motor neurons may be selectively destroyed by **poliovirus.** The result is paralysis of the muscles innervated by the affected neurons. **Amyotrophic lateral sclerosis (ALS),** also known as **Lou Gehrig's disease,** is the most common motor neuron disease. This incurable condition is characterized by degeneration and eventually death of motor neurons. The result is gradual loss of motor control, progressive paralysis, and finally, death within one to five years of onset. The underlying problem appears to be pathologic changes in neurofilaments that block axonal transport of crucial materials (see p. 51).

Before turning our attention to the junction between a motor neuron and the muscle cells it innervates, we are going to pull together in tabular form two groups of information we have been examining. ▲ Table 7-6 summarizes the features of the two divisions of the efferent nervous system discussed in

this chapter. ▲ Table 7-7 compares the three types of neurons that have been examined in the last three chapters.

NEUROMUSCULAR JUNCTION

An action potential in a motor neuron is rapidly propagated from the cell body within the CNS to the skeletal muscle along the large myelinated axon (efferent fiber) of the neuron. As the axon approaches a muscle, it divides into many terminal branches and loses its myelin sheath. Each of these axon terminals forms a special junction, a **neuromuscular junction,** with one of the many muscle cells that compose the whole muscle (● Figure 7-5). A single muscle cell, referred to as a **muscle fiber,** is long and cylindrical in shape. The axon terminal is enlarged into a knoblike structure, the **terminal button,** which fits into a shallow depression, or groove, in the underlying muscle fiber (● Figure 7-6). Some scientists alternatively call the neuromuscular junction a motor end plate. However, we will reserve the term **motor end plate** for the specialized portion of the muscle cell membrane immediately under the terminal button.

▌ Acetylcholine is the neuromuscular junction neurotransmitter.

Nerve and muscle cells do not actually come into direct contact at a neuromuscular junction. The space, or cleft, between these two structures is too large to permit electrical transmission of an impulse between them (that is, the action potential cannot "jump" that far). Therefore, just as at a neuronal synapse (see p. 118), a chemical messenger is used to carry the signal

▲ TABLE 7-6

Comparison of the Autonomic Nervous System and the Somatic Nervous System

Feature	Autonomic Nervous System	Somatic Nervous System
Site of Origin	Brain or lateral horn of spinal cord	Ventral horn of spinal cord for most; those supplying muscles in head originate in brain
Number of Neurons from Origin in CNS to Effector Organ	Two-neuron chain (preganglionic and postganglionic)	Single neuron (motor neuron)
Organs Innervated	Cardiac muscle, smooth muscle, exocrine and some endocrine glands	Skeletal muscle
Type of Innervation	Most effector organs dually innervated by the two antagonistic branches of this system (sympathetic and parasympathetic)	Effector organs innervated only by motor neurons
Neurotransmitter at Effector Organs	May be acetylcholine (parasympathetic terminals) or norepinephrine (sympathetic terminals)	Only acetylcholine
Effects on Effector Organs	Either stimulation or inhibition (antagonistic actions of two branches)	Stimulation only (inhibition possible only centrally through IPSPs on cell body of motor neuron)
Types of Control	Under involuntary control; may be voluntarily controlled with biofeedback techniques and training	Subject to voluntary control; much activity subconsciously coordinated
Higher Centers Involved in Control	Spinal cord, medulla, hypothalamus, prefrontal association cortex	Spinal cord, motor cortex, basal nuclei, cerebellum, brain stem

between the neuron terminal and the muscle fiber. This neurotransmitter is acetylcholine (ACh.)

Release of ACh at the neuromuscular junction

Each terminal button contains thousands of vesicles that store ACh. Propagation of an action potential to the axon terminal (step 1, ● Figure 7-5) triggers the opening of voltage-gated Ca^{2+} channels in the terminal button (see p. 104). Opening of Ca^{2+} channels permits Ca^{2+} to diffuse into the terminal button from its higher extracellular concentration (step 2), which in turn causes the release of ACh by exocytosis from several hundred of the vesicles into the cleft (step 3).

Formation of an end-plate potential

The released ACh diffuses across the cleft and binds with specific receptor sites, which are specialized membrane proteins unique to the motor end-plate portion of the muscle fiber membrane (step 4). (These cholinergic receptors are of the nicotinic type.) Binding of ACh with these receptor sites induces the opening of chemically gated channels in the motor end plate. These channels permit a small amount of cation traffic through them (both Na^+ and K^+) but no anions (step 5). Because the permeability of the end-plate membrane to Na^+ and K^+ on opening of these channels is essentially equal, the relative movement of these ions through the channels depends on their electrochemical driving forces. Recall that at resting potential, the net driving force for Na^+ is much greater than that for K^+, because the resting potential is much closer to the K^+ equilibrium potential. Both the concentration and electrical

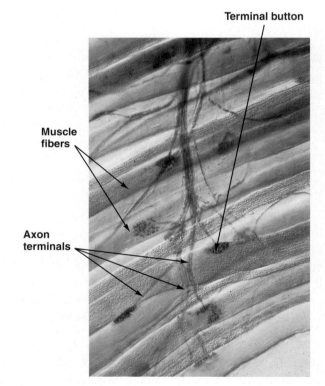

Terminal button

Muscle fibers

Axon terminals

● FIGURE 7-5

Motor neuron innervating skeletal muscle cells
When a motor neuron reaches a skeletal muscle, it divides into many terminal branches, each of which forms a neuromuscular junction with a single muscle cell (muscle fiber).

Comparison of Types of Neurons

Feature	Afferent neuron	Efferent Neuron		Interneuron
		Autonomic nervous system	**Somatic nervous system**	
Origin, Structure, Location	Receptor at peripheral ending; elongated peripheral axon, which travels in peripheral nerve; cell body located in dorsal root ganglion; short central axon entering spinal cord	Two-neuron chain; first neuron (preganglionic fiber) originating in CNS and terminating on a ganglion; second neuron (postganglionic fiber) originating in the ganglion and terminating on the effector organ	Cell body of motor neuron lying in spinal cord; long axon traveling in peripheral nerve and terminating on the effector organ	Various shapes; lying entirely within CNS; some cell bodies originating in brain, with long axons traveling down the spinal cord in descending pathways; some originating in spinal cord, with long axons traveling up the cord to the brain in ascending pathways; others forming short local connections
Termination	Interneurons*	Effector organs (cardiac muscle, smooth muscle, exocrine and some endocrine glands)	Effector organs (skeletal muscle)	Other interneurons and efferent neurons
Function	Carries information about the external and internal environment to CNS	Carries instructions from CNS to effector organs	Carries instructions from CNS to effector organs	Processes and integrates afferent input; initiates and coordinates efferent output; responsible for thought and other higher mental functions
Convergence of Input on Cell Body	No (only input is through receptor)	Yes	Yes	Yes
Effect of Input to Neuron	Can only be excited (through receptor potential induced by stimulus; must reach threshold for action potential)	Can be excited or inhibited (through EPSPs and IPSPs at first neuron; must reach threshold for action potential)	Can be excited or inhibited (through EPSPs and IPSPs; must reach threshold for action potential)	Can be excited or inhibited (through EPSPs and IPSPs; must reach threshold for action potential)
Site of Action Potential Initiation	First excitable portion of membrane adjacent to receptor	Axon hillock	Axon hillock	Axon hillock
Divergence of Output	Yes	Yes	Yes	Yes
Effect of Output on Structure on Which It Terminates	Only excites	Postganglionic fiber either excites or inhibits	Only excites	Either excites or inhibits

*The afferent neuron terminates directly on the alpha motor neuron in the case of the monosynaptic stretch reflex (see p. 179).

gradients for Na^+ are inward, whereas the outward concentration gradient for K^+ is almost, but not quite, balanced by the opposing inward electrical gradient. As a result, when ACh triggers the opening of these channels, considerably more Na^+ moves inward than K^+ outward, bringing about a depolarization of the motor end plate. This potential change is known as the **end-plate potential (EPP)**. It is similar to an EPSP (excitatory postsynaptic potential; see p. 119), except that the magnitude of an EPP is much larger for the following reasons: (1) more neurotransmitter is released from a terminal button than from a presynaptic knob in response to an action potential; (2) the motor end plate has a larger surface area bearing a

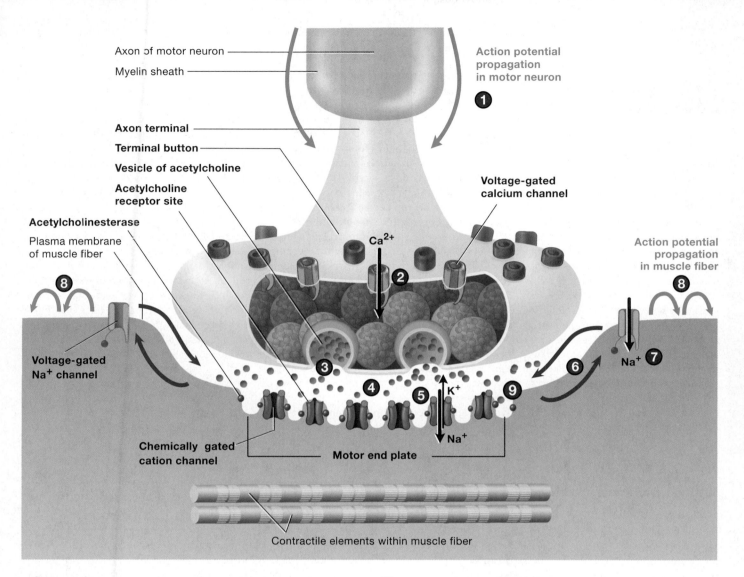

Axon of motor neuron

Myelin sheath

Action potential propagation in motor neuron ❶

Axon terminal

Terminal button

Vesicle of acetylcholine

Acetylcholine receptor site

Acetylcholinesterase

Plasma membrane of muscle fiber

Voltage-gated calcium channel

Ca^{2+} ❷

Action potential propagation in muscle fiber ❽

❽

❸

❹   ❺ K^+ ❾

Na^+

Na^+ ❼

❻

Voltage-gated Na^+ channel

Chemically gated cation channel

Motor end plate

Contractile elements within muscle fiber

❶ An action potential in a motor neuron is propagated to the terminal button.

❷ The presence of an action potential in the terminal button triggers the opening of voltage-gated Ca^{2+} channels and the subsequent entry of Ca^{2+} into the terminal button.

❸ Ca^{2+} triggers the release of acetylcholine by exocytosis from a portion of the vesicles.

❹ Acetylcholine diffuses across the space separating the nerve and muscle cells and binds with receptor sites specific for it on the motor end plate of the muscle cell membrane.

❺ This binding brings about the opening of cation channels, leading to a relatively large movement of Na^+ into the muscle cell compared to a smaller movement of K^+ outward.

❻ The result is an end-plate potential. Local current flow occurs between the depolarized end plate and adjacent membrane.

❼ This local current flow opens voltage-gated Na^+ channels in the adjacent membrane.

❽ The resultant Na^+ entry reduces the potential to threshold, initiating an action potential, which is propagated throughout the muscle fiber.

❾ Acetylcholine is subsequently destroyed by acetylcholinesterase, an enzyme located on the motor end-plate membrane, terminating the muscle cell's response.

● FIGURE 7-6
Events at a neuromuscular junction

higher density of neurotransmitter receptor sites, and accordingly has more sites for binding with neurotransmitter than a subsynaptic membrane has; and (3) many more ion channels are opened in response to the neurotransmitter-receptor com-

plex in the motor end plate. This permits a greater net influx of positive ions and a larger depolarization. As with an EPSP, an EPP is a graded potential, whose magnitude depends on the amount and duration of ACh at the end plate.

Loss of Muscle Mass: A Plight of Space Flight

Skeletal muscles are a case of "use it or lose it." Stimulation of skeletal muscles by motor neurons is essential not only to induce the muscles to contract but also to maintain their size and strength. Muscles that are not routinely stimulated gradually atrophy, or diminish in size and strength.

Our skeletal muscles are important in supporting our upright posture in the face of gravitational forces in addition to moving body parts. When humans entered the weightlessness of space, it became apparent that the muscular system required the stress of work or gravity to maintain its size and strength. In 1991, the space shuttle *Columbia* was launched for a nine-day mission dedicated among other things to comprehensive research on physiological changes brought on by weightlessness. The three female and four male astronauts aboard suffered a dramatic and significant 25% reduction of mass in their weight-bearing muscles. The effort required to move the body is remarkably less in space than on Earth, and there is no need for active muscular opposition to gravity. Furthermore, the muscles used to move around the confines of a space capsule differ from those used for walking down the street. As a result, some muscles rapidly undergo what is known as *functional atrophy*.

The muscles most affected are those in the lower extremities, the gluteal (buttocks) muscles, the extensor muscles of the neck and back, and the muscles of the trunk—that is, the muscles used for antigravity support on the ground. Changes include a decrease in muscle volume and mass, decrease in strength and endurance, increased breakdown of muscle protein, and loss of muscle nitrogen (an important component of muscle protein). The exact biological mechanisms that induce muscle atrophy are unknown, but a majority of scientists believe that the lack of customary forcefulness of contraction plays a major role. This atrophy presents no problem while within the space capsule, but such loss of muscle mass must be restricted if astronauts are to perform heavy work during space walks and are to resume normal activities on their return to Earth.

Space programs in the United States and the former Soviet Union have employed intervention techniques that emphasize both diet and exercise in an attempt to prevent muscle atrophy. Faithful performance of vigorous, carefully designed physical exercise several hours daily has helped reduce the severity of functional atrophy. Studies of nitrogen and mineral balances, however, suggest that muscle atrophy continues to progress during exposure to weightlessness despite efforts to prevent it. Furthermore, only half of the muscle mass was restored in the *Columbia* crew after the astronauts had been back on the ground a length of time equal to their flight. These and other findings suggest that further muscle-preserving interventions will be necessary for extended stays in space. An upcoming International Space Station slated for completion in 2004 will have nearly five times more cubic feet of work space than the Mir or Skylab stations. This additional space will include sophisticated laboratory equipment for further studies on the effect of weightlessness on the body, not only on muscle but on other systems as well.

Initiation of an action potential

The motor end-plate region itself does not have a threshold potential, so an action potential cannot be initiated at this site. However, an EPP brings about an action potential in the rest of the muscle fiber, as follows. The neuromuscular junction is usually located in the middle of the long, cylindrical muscle fiber. When an EPP takes place, local current flow occurs between the depolarized end plate and the adjacent, resting cell membrane in both directions (step 6), opening voltage-gated Na^+ channels and thus reducing the potential to threshold in the adjacent areas (step 7). The subsequent action potential initiated at these sites is propagated throughout the muscle fiber membrane by contiguous conduction (step 8) (see p. 110). The spread occurs in both directions, away from the motor end plate toward both ends of the fiber. This electrical activity triggers contraction of the muscle fiber. Thus, by means of ACh, an action potential in a motor neuron brings about an action potential and subsequent contraction in the muscle fiber. (See the accompanying boxed feature, ▶ A Closer Look at Exercise Physiology, to examine the importance of motor neuron stimulation in maintaining the integrity of skeletal muscles.)

Unlike synaptic transmission, the magnitude of an EPP is normally enough to cause an action potential in the muscle cell. Therefore, one-to-one transmission of an action potential typically occurs at a neuromuscular junction; one action potential in a nerve cell triggers one action potential in a muscle cell that it innervates. Other comparisons of neuromuscular junctions with synapses can be found in ▲ Table 7-8.

▌ Acetylcholinesterase terminates acetylcholine activity at the neuromuscular junction.

To ensure purposeful movement, electrical activity and the resultant contraction of the muscle cell turned on by motor neuron action potentials must be switched off promptly when there is no longer a signal from the motor neuron. The muscle cell's electrical response is turned off by an enzyme present in the motor end-plate membrane, **acetylcholinesterase (AChE),** which inactivates ACh.

As a result of diffusion, many of the released ACh molecules come into contact with and bind to the receptor sites, which are located on the surface of the motor end-plate membrane. However, some of the ACh molecules bind with AChE, which is also located at the end-plate surface. Being quickly inactivated, this ACh never contributes to the end-plate potential. The acetylcholine that does bind with receptor sites does so very briefly (for about 1 millionth of a second), then detaches. Some of the detached ACh molecules quickly rebind with receptor sites, keeping the end-plate channels open, but some randomly contact AChE instead this time around and are inactivated (step 9). As this process is repeated, more and more ACh is inactivated until it has been virtually removed from the cleft within a few milliseconds after its release. Removal of ACh terminates the EPP, so the remainder of the muscle cell membrane returns to resting potential. Now the muscle cell can relax. Or, if sustained contraction is essential for the desired movement, another motor neuron action potential leads to the

Comparison of a Synapse and a Neuromuscular Junction

Similarities	Differences
Both consist of two excitable cells separated by a narrow cleft that prevents direct transmission of electrical activity between them.	A synapse is a junction between two neurons. A neuromuscular junction exists between a motor neuron and a skeletal muscle fiber.
The axon terminals of both store chemical messengers (neurotransmitters) that are released by the Ca^{2+}-induced exocytosis of storage vesicles when an action potential reaches the terminal.	There is a one-to-one transmission of action potentials at a neuromuscular junction, whereas one action potential in a presynaptic neuron cannot by itself bring about an action potential in a postsynaptic neuron. An action potential in a postsynaptic neuron occurs only when the summation of EPSPs brings the membrane to threshold.
In both, binding of the neurotransmitter with receptor sites in the membrane of the cell underlying the axon terminal opens specific channels in the membrane, permitting ionic movements that alter the membrane potential of the cell.	A neuromuscular junction is always excitatory (an EPP); a synapse may be either excitatory (an EPSP) or inhibitory (an IPSP). The inhibition of skeletal muscles cannot be accomplished at the neuromuscular junction; it can take place only in the CNS through IPSPs at the cell body of the motor neuron.
The resultant change in membrane potential in both cases is a graded potential.	

release of more ACh, which keeps the contractile process going. By removing contraction-inducing ACh from the motor end plate, AChE permits the choice of allowing relaxation to take place (no more ACh released) or keeping the contraction going (more ACh released), depending on the body's momentary needs.

▍ The neuromuscular junction is vulnerable to several chemical agents and diseases.

Several chemical agents and diseases are known to affect the neuromuscular junction by acting at different sites in the transmission process (▲ Table 7-9). Two well-known toxins—*black widow spider venom* and *botulinum toxin*—alter the release of ACh, but in opposite directions.

Black widow spider venom

The venom of black widow spiders exerts its deadly effect by causing an explosive release of ACh from the storage vesicles, not only at neuromuscular junctions but at all cholinergic sites. All cholinergic sites undergo prolonged depolarization, the most detrimental consequence of which is respiratory failure. Breathing is accomplished by alternate contraction and relaxation of skeletal muscles, particularly the diaphragm. Respiratory paralysis occurs as a result of prolonged depolarization of the diaphragm. During this so-called *depolarization block*, the voltage-gated Na^+ channels are trapped in their inactivated state (see p. 105), thus prohibiting the initiation of new action potentials and resultant contraction of the diaphragm. Thus the victim cannot breathe.

Botulinum toxin

Botulinum toxin, in contrast, exerts its lethal blow by blocking the release of ACh from the terminal button in response to a motor neuron action potential. *Clostridium botulinum* toxin is

responsible for **botulism,** a form of food poisoning. It most frequently results from improperly canned foods contaminated with clostridial bacteria that survive and multiply, producing their toxin in the process. When this toxin is consumed, it prevents muscles from responding to nerve impulses. Death is due

▲ TABLE 7-9

Examples of Chemical Agents and Diseases That Affect the Neuromuscular Junction

Mechanism	Chemical Agent or Disease
Alters Release of Acetylcholine	
Causes explosive release of acetylcholine	Black widow spider venom
Blocks release of acetylcholine	*Clostridium botulinum* toxin
Blocks Acetylcholine Receptor Sites	
Reversibly binds with acetylcholine receptor sites	Curare
Autoimmune (self-produced) antibodies inactivate acetylcholine receptor sites	Myasthenia gravis
Prevents Inactivation of Acetylcholine	
Irreversibly inhibits acetylcholinesterase	Organophosphates (certain pesticides and military nerve gases)
Temporarily inhibits acetylcholinesterase	Neostigmine

Botulinum Toxin's Reputation Gets a Facelift

The powerful toxin produced by *Clostridium botulinum* is responsible for the deadly food poisoning, *botulism*. Yet this dreaded, highly lethal poison has been put to use as a treatment for alleviating specific movement disorders and, more recently, has been added to the list of tools used by cosmetic surgeons to fight wrinkles.

During the last decade, botulinum toxin, marketed in therapeutic doses as *Botox,* has offered welcome relief to sufferers of a number of painful, disruptive neuromuscular diseases known categorically as **dystonias.** These conditions are characterized by spasms (excessive, sustained, involuntarily produced muscle contractions) that result in involuntary twisting or abnormal postures, depending on the body part affected. For example, painful neck spasms that twist the head to one side occur as a result of *spasmodic torticollis* (*tortus* means "twisted"; *collum* means "neck"), the most common dystonia. The problem is believed to arise from too little inhibitory compared to excitatory input to the motor neurons that supply the affected muscle. The reasons for this imbalance in motor neuron input

are unknown. The end result of excessive motor neuron activation is sustained, disabling contraction of the muscle supplied by the overactive motor neurons. Fortunately, injection of miniscule amounts of botulinum toxin into the affected muscle causes a reversible, partial paralysis of the muscle. Botulinum toxin interferes with the release of muscle-contraction-causing acetylcholine from the overactive motor neurons at the neuromuscular junctions in the treated muscle. The goal is to inject just enough botulinum toxin to alleviate the troublesome spasmodic contractions but not enough to eliminate the normal contractions needed for ordinary movements. The therapeutic dose is considerably less than the amount of toxin needed to induce even mild symptoms of botulinum poisoning. Botulinum toxin is eventually cleared away, so its muscle-relaxing effects wear off after three to six months, at which time the treatment must be repeated.

The first dystonia for which Botox was approved as a treatment by the Food and Drug Administration (FDA) was *blepharospasm* (*blepharo* means "eyelid"). In this condition, sustained and involuntary contractions of the mus-

cles around the eye nearly permanently close the eyelids.

Botulinum toxin's potential as a treatment option for cosmetic surgeons was accidentally discovered when physicians noted that injections used to counter abnormal eye muscle contractions also decreased the appearance of wrinkles in the treated areas. It turns out that frown lines, crow's feet, and furrowed brows are caused by facial muscles that have become overactivated, or permanently contracted, as a result of years of performing certain repetitive facial expressions. By relaxing these muscles, botulinum toxin temporarily smoothes out these age-related wrinkles. Botox has recently received FDA approval as an antiwrinkle treatment. The agent is considered an excellent alternative to facelift surgery for combating lines and creases. However, as with its therapeutic use to treat dystonias, the costly injections of botulinum toxin must be repeated every three to six months to maintain the desired effect in appearance. Furthermore, Botox does not work against the fine, crinkly wrinkles associated with years of excessive sun exposure, because these wrinkles are caused by skin damage, not by contracted muscles.

---

to respiratory failure caused by the inability to contract the diaphragm. Botulinum toxin is one of the most lethal poisons known; ingesting less than 0.0001 mg can kill an adult. (See the accompanying boxed feature, ▶ Concepts, Challenges, and Controversies to learn about a surprising new wrinkle in the botulinum toxin story.)

Curare

Other chemicals interfere with neuromuscular junction activity by blocking the effect of released ACh. The best known example is **curare,** which reversibly binds to the ACh receptor sites on the motor end plate. Unlike ACh, however, curare does not alter membrane permeability, nor is it inactivated by AChE. When ACh receptor sites are occupied by curare, ACh cannot combine with these sites to open the channels that would permit the ionic movement responsible for an EPP. Consequently, because muscle action potentials cannot occur in response to nerve impulses to these muscles, paralysis ensues. When sufficient curare is present to effectively block a significant number of ACh receptor sites, the person dies from respiratory paralysis caused by an inability to contract the diaphragm. Curare was used in the past as a deadly arrowhead poison. Curare and related drugs have also been used medically during surgery to help achieve more complete skeletal muscle relaxation with less anesthetic. Under these circumstances, the amount of the agent is carefully administered, and facilities are available to maintain respiration artificially, if necessary, until the effects of the drug wear off.

Organophosphates

Organophosphates are a group of chemicals that modify neuromuscular junction activity in yet another way—namely, by irreversibly inhibiting AChE. Inhibition of AChE prevents the inactivation of released ACh. Death from organophosphates is also due to respiratory failure, because the diaphragm is unable to repolarize and return to resting conditions, then contract again to bring in a fresh breath of air. These toxic agents are found in some pesticides and military nerve gases.

Myasthenia gravis

Myasthenia gravis, a disease involving the neuromuscular junction, is characterized by extreme muscular weakness (*myasthenia* means "muscular weakness"; *gravis* means "severe"). It is an autoimmune condition (*autoimmune* means "immunity against self") in which the body erroneously produces antibodies against its own motor end-plate ACh receptors. Consequently, not all the released ACh molecules can find a functioning receptor site with which to bind. As a result, much of the ACh is destroyed by AChE without ever having an opportunity to interact with a receptor site and contribute to the EPP. Treatment consists of administering a drug such as **neostigmine** that inhibits AChE temporarily (in contrast to the toxic organophosphates, which irreversibly block this enzyme). This drug prolongs the action of ACh at the neuromuscular junction by permitting it to build up for the short term. The resultant EPP is of sufficient magnitude to initiate an action potential and subsequent contraction in the muscle fiber, as it normally would.

CHAPTER IN PERSPECTIVE: FOCUS ON HOMEOSTASIS

The nervous system, along with the other major regulatory system, the endocrine system, is responsible for controlling most of the body's muscular and glandular activities. Whereas the afferent division of the peripheral nervous system detects and carries information to the central nervous system (CNS) for processing and decision making, the efferent division of the peripheral nervous system carries directives from the CNS to the effector organs (muscles and glands), which carry out the intended response. Much of this efferent output is directed toward maintaining homeostasis.

The autonomic nervous system, which is the efferent branch that innervates smooth muscle, cardiac muscle, and glands, plays a major role in the following range of homeostatic activities:

- regulation of blood pressure
- control of digestive juice secretion and of digestive tract contractions that mix ingested food with the digestive juices
- control of sweating to help maintain body temperature

The somatic nervous system, the efferent branch that innervates skeletal muscle, contributes to homeostasis by stimulating the following activities:

- Skeletal muscle contractions that enable the body to move in relation to the external environment contribute to homeostasis by moving the body toward food or away from harm.
- Skeletal muscle contraction also accomplishes breathing to maintain appropriate levels of O_2 and CO_2 in the body.
- Shivering is a skeletal muscle activity important in the maintenance of body temperature.

In addition, efferent output to skeletal muscles accomplishes many movements that are not aimed at maintaining a stable internal environment but nevertheless enrich our lives and enable us to engage in activities that contribute to society, such as dancing, building bridges, or performing surgery.

CHAPTER SUMMARY

Introduction
- The CNS controls muscles and glands by transmitting signals to these effector organs through the efferent division of the peripheral nervous system.
- There are two types of efferent output: the autonomic nervous system, which is under involuntary control and supplies cardiac and smooth muscle as well as most exocrine and some endocrine glands; and the somatic nervous system, which is subject to voluntary control and supplies skeletal muscle.

Autonomic Nervous System
- The autonomic nervous system consists of two subdivisions—the sympathetic and parasympathetic nervous systems.
- An autonomic nerve pathway consists of a two-neuron chain. The preganglionic fiber originates in the CNS and synapses with the cell body of the postganglionic fiber in a ganglion outside the CNS. The postganglionic fiber terminates on the effector organ.
- All preganglionic fibers and parasympathetic postganglionic fibers release acetylcholine. Sympathetic postganglionic fibers release norepinephrine.
- The same neurotransmitter elicits different responses in different tissues. Thus the response depends on specialization of the tissue cells, not on the properties of the messenger.
- Tissues innervated by the autonomic nervous system possess one or more of several different receptor types for the postganglionic chemical messengers. Cholinergic receptors include nicotinic and muscarinic receptors; adrenergic receptors include alpha and beta receptors.
- A given autonomic fiber either excites or inhibits activity in the organ it innervates.
- Most visceral organs are innervated by both sympathetic and parasympathetic nerve fibers, which in general produce opposite effects in a particular organ. Dual innervation of visceral organs by both branches of the autonomic nervous system permits precise control over an organ's activity.

- The sympathetic system dominates in emergency or stressful situations and promotes responses that prepare the body for strenuous physical activity (for "fight" or "flight"). The parasympathetic system dominates in quiet, relaxed situations and promotes body maintenance activities such as digestion.
- Autonomic activities are controlled by multiple areas of the CNS, including the spinal cord, medulla, hypothalamus, and prefrontal association cortex.

Somatic Nervous System
- The somatic nervous system consists of the axons of motor neurons, which originate in the spinal cord and terminate on skeletal muscle.
- Acetylcholine, the neurotransmitter released from a motor neuron, stimulates muscle contraction.
- Motor neurons are the final common pathway by which various regions of the CNS exert control over skeletal muscle activity. The areas of the CNS that influence skeletal muscle activity by acting through the motor neurons are the spinal cord, motor regions of the cortex, basal nuclei, cerebellum, and brain stem.

Neuromuscular Junction
- Each axon terminal of a motor neuron forms a neuromuscular junction with a single muscle cell (fiber).
- Because these structures do not make direct contact, signals are passed between the nerve terminal and muscle fiber by means of the chemical messenger acetylcholine (ACh).
- An action potential in the axon terminal causes the release of ACh from its storage vesicles. The released ACh diffuses across the space separating the nerve and muscle cell and binds to special receptor sites on the underlying motor end plate of the muscle cell membrane. This combination of ACh with the receptor sites triggers the opening of specific channels in the motor end plate. The subsequent ion movements depolarize the motor end plate, producing the end-plate potential (EPP).

- Local current flow between the depolarized end plate and adjacent muscle-cell membrane brings these adjacent areas to threshold, initiating an action potential that is propagated throughout the muscle fiber.

- This muscle action potential triggers muscle contraction.
- Acetylcholinesterase inactivates ACh, terminating the EPP and, subsequently, the action potential and resultant contraction.

REVIEW EXERCISES

Objective Questions (Answers on p. A-39)

1. Sympathetic preganglionic fibers originate in the thoracic and lumbar segments of the spinal cord. *(True or false?)*
2. Action potentials are transmitted on a one-to-one basis at both a neuromuscular junction and a synapse. *(True or false?)*
3. The sympathetic nervous system
   a. is always excitatory.
   b. innervates only tissues concerned with protecting the body against challenges from the outside environment.
   c. has short preganglionic and long postganglionic fibers.
   d. is part of the afferent division of the peripheral nervous system.
   e. is part of the somatic nervous system.
4. Acetylcholinesterase
   a. is stored in vesicles in the terminal button.
   b. combines with receptor sites on the motor end plate to bring about an end-plate potential.
   c. is inhibited by organophosphates.
   d. is the chemical transmitter at the neuromuscular junction.
   e. paralyzes skeletal muscle by strongly binding with acetylcholine receptor sites.
5. The two divisions of the autonomic nervous system are the _____ nervous system, which dominates in fight-or-flight situations, and the _____ nervous system, which dominates in quiet, relaxed situations.
6. The _____ is a modified sympathetic ganglion that does not give rise to postganglionic fibers but instead secretes hormones similar or identical to sympathetic postganglionic neurotransmitters into the blood.
7. Using the following answer code, identify the autonomic transmitter being described:
   (a) acetylcholine
   (b) norepinephrine
   ___ 1. secreted by all preganglionic fibers
   ___ 2. secreted by sympathetic postganglionic fibers
   ___ 3. secreted by parasympathetic postganglionic fibers
   ___ 4. secreted by the adrenal medulla
   ___ 5. secreted by motor neurons
   ___ 6. binds to muscarinic or nicotinic receptors
   ___ 7. binds to α or β receptors
8. Using the following answer code, indicate which type of efferent output is being described:
   (a) characteristic of the somatic nervous system
   (b) characteristic of the autonomic nervous system
   ___ 1. composed of two-neuron chains
   ___ 2. innervates cardiac muscle, smooth muscle, and glands
   ___ 3. innervates skeletal muscle
   ___ 4. consists of the axons of motor neurons
   ___ 5. exerts either an excitatory or an inhibitory effect on its effector organs
   ___ 6. dually innervates its effector organs
   ___ 7. exerts only an excitatory effect on its effector organs

Essay Questions

1. Distinguish between preganglionic and postganglionic fibers.
2. Compare the origin, preganglionic and postganglionic fiber length, and neurotransmitters of the sympathetic and parasympathetic nervous systems.
3. What is the advantage of dual innervation of many organs by both branches of the autonomic nervous system?
4. Distinguish among the following types of receptors: nicotinic receptors, muscarinic receptors, α_1 receptors, α_2 receptors, β_1 receptors, and β_2 receptors.
5. What regions of the CNS regulate autonomic output?
6. Why are motor neurons called the "final common pathway"?
7. Describe the sequence of events that occurs at a neuromuscular junction.
8. Discuss the effect each of the following has at the neuromuscular junction: black widow spider venom, botulinum toxin, curare, organophosphates, myasthenia gravis, and neostigmine.

Quantitative Exercises (Solutions on p. A-39)

1. When a muscle fiber is activated at the neuromuscular junction, tension does not begin to rise until about 1 msec after initiation of the action potential in the muscle fiber. Many things are occurring during this delay, one time-consuming event being the diffusion of acetylcholine across the neuromuscular junction. The following equation can be used to calculate how long this diffusion takes:

$$t = \frac{x^2}{2D}$$

In this equation, x is the distance covered, D is the diffusion coefficient, and t is the time it takes for diffusion of the substance across the distance x. In this example, x is the width of the cleft between the neuronal axon terminal and the muscle fiber at the neuromuscular junction (assume 200 nm) and D is the diffusion coefficient of acetylcholine (assume 1×10^{-5} cm^2/sec). How long does it take the acetylcholine to diffuse across the neuromuscular junction?

POINTS TO PONDER

(Explanations on p. A-39)

1. Explain why epinephrine, which causes arteriolar constriction (narrowing) in most tissues, is frequently administered in conjunction with local anesthetics.

2. Would skeletal muscle activity be affected by atropine? Why or why not?

3. Considering that you can voluntarily control the emptying of your urinary bladder by contracting (preventing emptying) or relaxing (permitting emptying) your external urethral sphincter, a ring of muscle that guards the exit from the bladder, of what type of muscle is this sphincter composed and what branch of the nervous system supplies it?

4. The venom of certain poisonous snakes contains α bungarotoxin, which binds tenaciously to acetylcholine receptor sites on the motor end-plate membrane. What would the resultant symptoms be?

5. Explain how destruction of motor neurons by poliovirus or amyotrophic lateral sclerosis can be fatal.

6. *Clinical Consideration.* Christopher K. experienced chest pains when he climbed the stairs to his fourth-floor office or played tennis but had no symptoms when not physically exerting himself. His condition was diagnosed as *angina pectoris* (*angina* means "pain"; *pectoris* means "chest"), heart pain that occurs whenever the blood supply to the heart muscle cannot meet the muscle's need for oxygen delivery. This condition usually is caused by narrowing of the blood vessels supplying the heart by cholesterol-containing deposits. Most people with this condition do not have any pain at rest but experience bouts of pain whenever the heart's need for oxygen increases, such as during exercise or emotionally stressful situations that increase sympathetic nervous activity. Christopher obtains immediate relief of angina attacks by promptly taking a vasodilator drug such as *nitroglycerin*, which relaxes the smooth muscle in the walls of his narrowed heart vessels. Consequently, the vessels open more widely and more blood can flow through them. For prolonged treatment, Christopher's doctor has indicated that he will experience fewer and less severe angina attacks if he takes a β_1-blocker drug, such as *metoprolol*, on a regular basis. Explain why.

PHYSIOEDGE RESOURCES

PHYSIOEDGE CD-ROM

Figures marked with the PhysioEdge icon have associated activities on the CD. For this chapter, check out the Media Quizzes related to Figures 7-1 and 7-6.

CHECK OUT THESE MEDIA QUIZZES:

7.1 Autonomic Nervous System
7.2 Neuromuscular Junction

PHYSIOEDGE WEB SITE

The Web site for this book contains a wealth of helpful study aids, as well as many ideas for further reading and research. Log on to:

http://www.brookscole.com

Go to the Biology page and select Sherwood's *Human Physiology*, 5th Edition. Select a chapter from the drop-down menu or click on one of these resource areas:

- **Case Histories** provide an introduction to the clinical aspects of human physiology.

- For 2-D and 3-D graphical illustrations and animations of physiological concepts, visit our **Visual Learning Resource**.

- Resources for study and review include the **Chapter Outline**, **Chapter Summary**, **Glossary**, and **Flash Cards**. Use our **Quizzes** to prepare for in-class examinations.

- On-line research resources to consult are **Hypercontents**, with current links to relevant Internet sites; **Internet Exercises** with starter URLs provided; and **InfoTrac Exercises**.

 For more readings, go to InfoTrac College Edition, your on-line research library, at:

http://infotrac.thomsonlearning.com

Muscular System

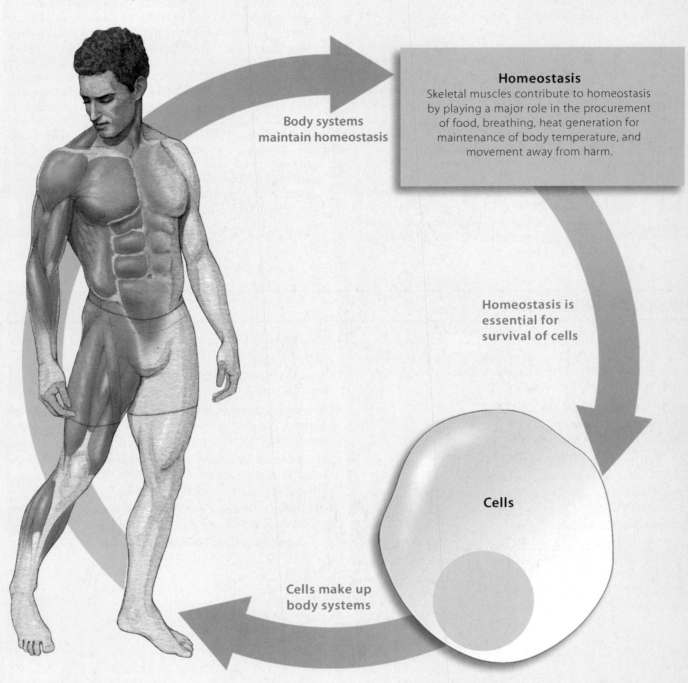

Homeostasis
Skeletal muscles contribute to homeostasis by playing a major role in the procurement of food, breathing, heat generation for maintenance of body temperature, and movement away from harm.

Body systems
maintain homeostasis

Homeostasis is
essential for
survival of cells

Cells

Cells make up
body systems

Muscles are the contraction specialists of the body. **Skeletal muscle** attaches to the skeleton. Contraction of skeletal muscles moves bones to which they are attached, allowing the body to perform a variety of motor activities. Skeletal muscles that support homeostasis include those important in acquiring, chewing, and swallowing food and those essential for breathing. Skeletal muscle contraction is also used to move the body away from harm. Heat-generating muscle contractions are important in temperature regulation. Skeletal muscles are also used for nonhomeostatic activities, such as dancing or operating a computer. **Smooth muscle** is found in the walls of hollow organs and tubes. Controlled contraction of smooth muscle regulates movement of blood through blood vessels, food through the digestive tract, air through respiratory airways, and urine to the exterior. **Cardiac muscle** is found only in the walls of the heart, whose contraction pumps life-sustaining blood throughout the body.

Chapter 8

Muscle Physiology

CONTENTS AT A GLANCE

INTRODUCTION

STRUCTURE OF SKELETAL MUSCLE

▮ Levels of organization in muscle

▮ Thick- and thin-filament composition

MOLECULAR BASIS OF SKELETAL MUSCLE CONTRACTION

▮ Sliding-filament mechanism

▮ Excitation–contraction coupling

SKELETAL MUSCLE MECHANICS

▮ Gradation of contraction—recruitment, summation, length-tension relationship

▮ Isotonic and isometric contractions

▮ Load-velocity relationship

▮ Muscle efficiency; heat production

▮ Lever systems of muscles and bones

SKELETAL MUSCLE METABOLISM AND FIBER TYPES

▮ Pathways supplying ATP

▮ Muscle fatigue

▮ Oxygen debt

▮ Muscle fiber types

CONTROL OF MOTOR MOVEMENT

▮ Inputs influencing motor neuron output

▮ Local reflexes; muscle spindles; Golgi tendon organs

SMOOTH AND CARDIAC MUSCLE

▮ Multiunit and single-unit smooth muscle

▮ Myogenic activity

▮ Factors modifying smooth muscle activity; gradation of smooth muscle contraction

▮ Cardiac muscle

INTRODUCTION

Almost all living cells possess rudimentary intracellular machinery for moving and redistributing various cellular components, such as during cell division. White blood cells use intracellular contractile proteins to propel themselves through their environment. The contraction specialists of the body, however, are the muscle cells. By moving specialized intracellular components, muscle cells are able to develop tension and shorten, that is, to contract. Recall that there are three types of muscle: *skeletal muscle*, *cardiac muscle*, and *smooth muscle* (see p. 6). Through their highly developed ability to contract, groups of muscle cells working together within a muscle are able to produce movement and to do work. Controlled contraction of muscles allows (1) purposeful movement of the whole body or parts of the body in relation to the environment (such as walking or waving your hand), (2) manipulation of external objects (such as driving a car or moving a piece of furniture), (3) propulsion of contents through various hollow internal organs (such as circulation of blood or movement of materials through the digestive tract), and (4) emptying the contents of certain organs to the external environment (such as urination or giving birth).

Muscle comprises the largest group of tissues in the body, accounting for approximately half of the body's weight. Skeletal muscle alone makes up about 40% of body weight in men and 32% in women, with smooth and cardiac muscle making up another 10% of the total weight. Although the three muscle types are structurally and functionally distinct, they can be classified in two different ways according to their common characteristics (● Figure 8-1). First, muscles are categorized as *striated* (skeletal and cardiac muscle) or *unstriated* (smooth muscle), depending on whether alternating dark and light bands, or striations, can be seen when the muscle is viewed under a light microscope. Second, muscles are categorized as *voluntary* (skeletal muscle) or *involuntary* (cardiac and smooth muscle), depending respectively on whether they are innervated by the somatic nervous system and are subject to voluntary control or are innervated by the autonomic nervous system and are not subject to voluntary control (see p. 237).

Most of this chapter is devoted to a detailed examination of the most abundant and best understood muscle, skeletal muscle. Skeletal muscles make up the muscular system. We

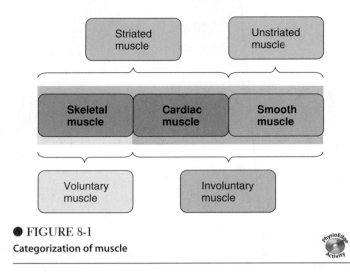

● **FIGURE 8-1**
Categorization of muscle

will begin with a discussion of skeletal muscle structure, then examine how it works from the molecular through the cellular level and finally to the whole muscle. The chapter concludes with a discussion of the unique properties of smooth and cardiac muscle in comparison to skeletal muscle. Smooth and cardiac muscle are found throughout the body systems as components of the hollow organs and tubes.

STRUCTURE OF SKELETAL MUSCLE

A single skeletal muscle cell, known as a **muscle fiber,** is relatively large, elongated, and cylinder-shaped, measuring from 10 to 100 micrometers (μm) in diameter and up to 750,000 μm or 2.5 feet, in length (1 μm = 1 millionth of a meter). A skeletal muscle consists of a number of muscle fibers lying parallel to each other and bundled together by connective tissue (● Figure 8-2a). The fibers usually extend the entire length of the muscle. During embryonic development, the huge skeletal-muscle fibers are formed by the fusion of many smaller cells; thus one striking feature is the presence of multiple nuclei in a single muscle cell. Another feature is the abundance of mitochondria, the energy-generating organelles, as would be expected with the high energy demands of a tissue as active as skeletal muscle.

▌ Skeletal muscle fibers are striated by a highly organized internal arrangement.

The predominant structural feature of a skeletal muscle fiber is the presence of numerous **myofibrils.** These specialized contractile elements, which constitute 80% of the volume of the muscle fiber, are cylindrical intracellular structures 1μm in diameter that extend the entire length of the muscle fiber (● Figure 8-2b). Each myofibril consists of a regular arrangement of highly organized cytoskeletal elements—the thick and the thin filaments (● Figure 8-2c). The **thick filaments,** which are 12 to 18 nm in diameter and 1.6 μm in length, are special assemblies of the protein *myosin*, whereas the **thin filaments,** which are 5 to 8 nm in diameter and 1.0 μm long, are made up primarily of the protein *actin* (● Figure 8-2d). These same

proteins are found in all other cells of the body but in smaller quantities and in a less-organized form. The levels of organization in a skeletal muscle can be summarized as follows:

$$\underset{\text{(an organ)}}{\overset{\text{Whole}}{\text{muscle}}} \rightarrow \underset{\text{(a cell)}}{\overset{\text{muscle}}{\text{fiber}}} \rightarrow \underset{\substack{\text{(a specialized} \\ \text{intracellular} \\ \text{structure)}}}{\text{myofibril}} \rightarrow \underset{\substack{\text{(cytoskeletal} \\ \text{elements)}}}{\overset{\text{thick}}{\underset{\text{filaments}}{\text{and thin}}}} \rightarrow \underset{\text{(proteins)}}{\overset{\text{myosin}}{\underset{\text{actin}}{\text{and}}}}$$

A and I bands

Viewed with a light microscope, a myofibril displays alternating dark bands (the A bands) and light bands (the I bands) (● Figure 8-3a). The bands of all the myofibrils lined up parallel to each other collectively lead to the striated or striped appearance of a skeletal muscle fiber (● Figure 8-3b). Alternate stacked sets of thick and thin filaments that slightly overlap each other are responsible for the A and I bands (● Figure 8-2c).

An **A band** consists of a stacked set of thick filaments along with the portions of the thin filaments that overlap on both ends of the thick filaments. The thick filaments are found only within the A band and extend its entire width; that is, the two ends of the thick filaments within a stack define the outer limits of a given A band. The lighter area within the middle of the A band, where the thin filaments do not reach, is known as the **H zone.** Only the central portions of the thick filaments are found in this region. A system of supporting proteins holds the thick filaments together vertically within each stack. These proteins can be seen as the **M line,** which extends vertically down the middle of the A band within the center of the H zone.

The **I band** consists of the remaining portion of the thin filaments that do not project into the A band. Thus the I band contains only thin filaments but not their entire length.

Visible in the middle of each I band is a dense, vertical **Z line.** The area between two Z lines is called a **sarcomere,** which is the functional unit of skeletal muscle. A **functional unit** of any organ is the smallest component that can perform all the functions of that organ. Accordingly, a sarcomere is the smallest component of a muscle fiber that is capable of contraction. The Z line is a flattened, disclike cytoskeletal structure that connects the thin filaments of two adjoining sarcomeres. Each relaxed sarcomere is about 2.5 μm in width and consists of one whole A band and half of each of the two I bands located on either side. During growth, a muscle increases in length by adding new sarcomeres, not by increasing the size of each sarcomere.

Cross bridges

With an electron microscope, fine **cross bridges** can be seen extending from each thick filament toward the surrounding thin filaments in the regions where the thick and thin filaments overlap (● Figures 8-2c and 8-4a). Three-dimensionally, the thin filaments are arranged hexagonally around the thick filaments. Cross bridges project from each thick filament in all six directions toward the surrounding thin filaments. Each thin filament, in turn, is surrounded by three thick filaments (● Figure 8-4b). To give you an idea of the magnitude of these filaments, a single muscle fiber may contain an estimated 16 bil-

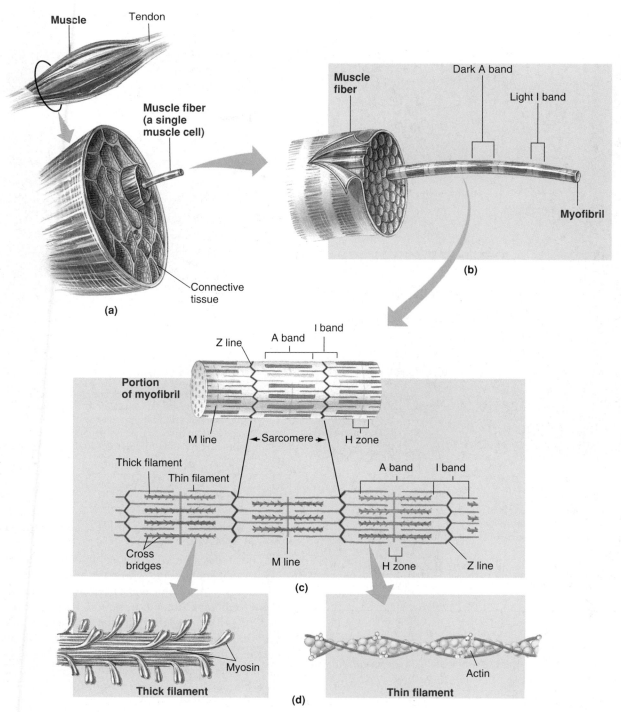

● FIGURE 8-2

Levels of organization in a skeletal muscle
(a) Enlargement of a cross section of a whole muscle. (b) Enlargement of a myofibril within a muscle fiber. (c) Cytoskeletal components of a myofibril. (d) Protein components of thick and thin filaments.

---

lion thick and 32 billion thin filaments, all arranged in this very precise pattern within the myofibrils.

▌ Myosin forms the thick filaments.

Each thick filament is composed of several hundred myosin molecules packed together in a specific arrangement. A **myosin** molecule is a protein consisting of two identical subunits, each

shaped somewhat like a golf club (● Figure 8-5a). The protein's tail ends are intertwined around each other, with the two globular heads projecting out at one end. The two halves of each thick filament are mirror images made up of myosin molecules lying lengthwise in a regular staggered array, with their tails oriented toward the center of the filament and their globular heads protruding outward at regularly spaced intervals (● Figure 8-5b). These heads form the cross bridges between the

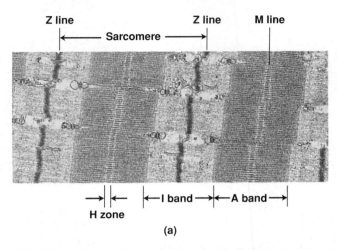

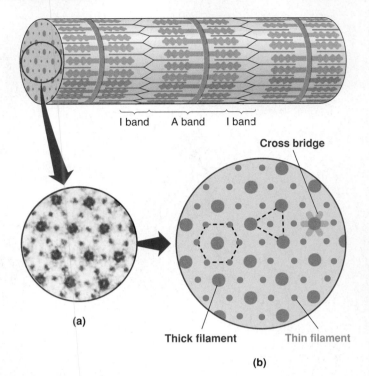

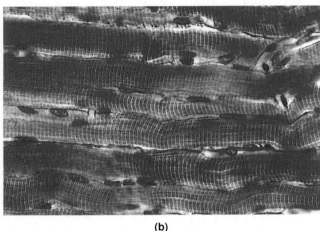

(b)

● FIGURE 8-3

Light-microscope view of skeletal muscle components
(a) High-power light-microscope view of a myofibril. (b) Low-power light-microscope view of skeletal muscle fibers. Note striated appearance.

[SOURCE: Reprinted with permission from Sydney Schochet Jr., M.D., Professor, Department of Pathology, School of Medicine, West Virginia University: *Diagnostic Pathology of Skeletal Muscle and Nerve* (Stamford, Connecticut: Appleton & Lange, 1986), ● Figure 1-13.]

thick and thin filaments. Each cross bridge has two important sites crucial to the contractile process: (1) an actin-binding site and (2) a myosin ATPase (ATP-splitting) site.

▌ Actin is the main structural component of the thin filaments.

Thin filaments are composed of three proteins: *actin, tropomyosin,* and *troponin* (● Figure 8-6). **Actin** molecules, the primary structural proteins of the thin filament, are spherical in shape. The backbone of a thin filament is formed by actin molecules joined into two strands and twisted together, like two strings of pearls wrapped around each other. Each actin molecule has a special binding site for attachment with a myosin cross bridge. By a mechanism to be described shortly, binding of actin and myosin molecules at the cross bridges results in energy-consuming contraction of the muscle fiber. Accordingly, actin and myosin are often referred to as **contractile pro-**

● FIGURE 8-4

Cross-sectional arrangement of thick and thin filaments
(a) Electron micrograph cross section through the A band in the region of thick and thin filament overlap. Note the fine cross bridges extending from the thick filaments. (b) Schematic representation of the geometric relation among thick and thin filaments and cross bridges.

teins, even though, as you will see, neither myosin nor actin actually contracts.

In a relaxed muscle fiber, contraction does not take place; actin is not able to bind with cross bridges because of the position of the two other types of protein within the thin filament—tropomyosin and troponin. **Tropomyosin** molecules are threadlike proteins that lie end-to-end alongside the groove of the actin spiral. In this position, tropomyosin covers the actin sites that bind with the cross bridges, thus blocking the interaction that leads to muscle contraction. The other thin filament component, **troponin,** is a protein complex consisting of three polypeptide units: one that binds to tropomyosin, one that binds to actin, and a third that can bind with Ca^{2+}.

When troponin is not bound to Ca^{2+}, this protein stabilizes tropomyosin in its blocking position over actin's cross-bridge binding sites (● Figure 8-7a). When Ca^{2+} binds to troponin, the shape of this protein is changed in such a way that tropomyosin slips away from its blocking position toward the center of the groove (● Figure 8-7b). With tropomyosin out of the way, actin and myosin can bind and interact at the cross bridges, resulting in muscle contraction. Tropomyosin and troponin are often called **regulatory proteins** because of their role in covering (preventing contraction) or exposing (permitting contraction) the binding sites for cross-bridge interaction between actin and myosin.

MOLECULAR BASIS OF SKELETAL MUSCLE CONTRACTION

Several important links in the contractile process remain to be discussed. How does cross-bridge interaction between actin and myosin bring about muscle contraction? How does a muscle action potential trigger this contractile process? What is the source of the Ca^{2+} that physically repositions troponin and tropomyosin to permit cross-bridge binding? We will turn our attention to these topics in this section.

▌ During contraction, cycles of cross-bridge binding and bending pull the thin filaments inward.

The thin filaments on each side of a sarcomere slide inward over the stationary thick filaments toward the A band's center during contraction (● Figure 8-8). As they slide inward, the thin filaments pull the Z lines to which they are attached closer together, so the sarcomere shortens. As all the sarcomeres throughout the muscle fiber's length shorten simultaneously, the entire fiber becomes shorter. This is known as the **sliding filament mechanism** of muscle contraction. The H zone, the region in the center of the A band where the thin filaments do not reach, becomes smaller as the thin filaments approach each other when they slide more deeply inward. The I band, which consists of the portions of the thin filaments that do not overlap with the thick filaments, decreases in width as the thin filaments further overlap the thick filaments during their inward slide. The thin filaments themselves do not change length during muscle fiber shortening. The width of the A band remains unchanged during contraction, because its width is determined by the length of the thick filaments, and the thick filaments do not change length during the shortening process. Note that neither the thick nor thin filaments decrease in length to shorten the sarcomere. Instead, contraction is accomplished by the thin filaments within each sarcomere sliding closer together between the thick filaments.

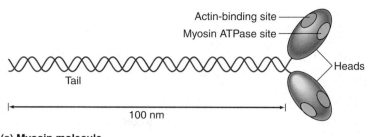

(a) Myosin molecule

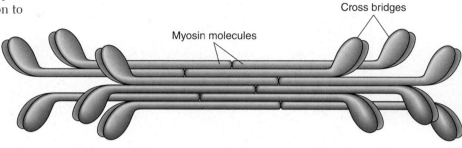

(b) Thick filament

● FIGURE 8-5

Structure of myosin molecules and their organization within a thick filament
(a) Myosin molecule. Each myosin molecule consists of two identical golf club–shaped subunits with their tails intertwined and their globular heads, each of which contains an actin-binding site and a myosin ATPase site, projecting out at one end. (b) Thick filament. A thick filament is made up of myosin molecules lying lengthwise parallel to each other. Half are oriented in one direction and half in the opposite direction. The globular heads, which protrude at regular intervals along the thick filament, form the cross bridges.

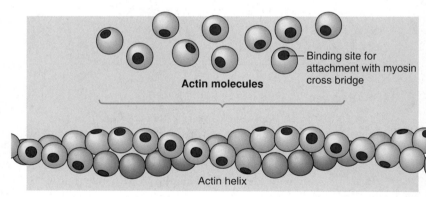

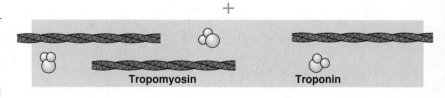

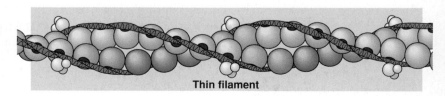

● FIGURE 8-6

Composition of a thin filament
The main structural component of a thin filament is two chains of spherical actin molecules that are twisted together. Troponin molecules (which consist of three small spherical subunits) and threadlike tropomyosin molecules are arranged to form a ribbon that lies alongside the groove of the actin helix and physically covers the binding sites on actin molecules for attachment with myosin cross bridges. (The thin filaments shown here are not drawn in proportion to the thick filaments in ● Figure 8-5. Thick filaments are two to three times larger in diameter than thin filaments.)

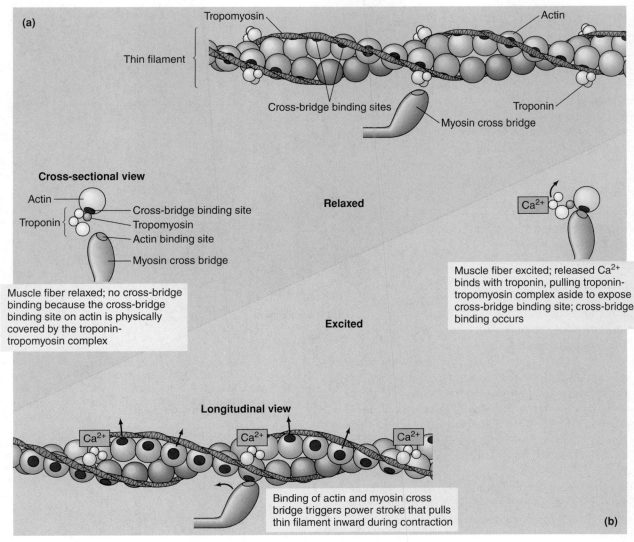

● FIGURE 8-7
Role of calcium in turning on cross bridges

Power stroke

The thin filaments are pulled inward relative to the stationary thick filaments by cross-bridge activity. During contraction, with the tropomyosin and troponin "chaperones" pulled out of the way by Ca^{2+}, the myosin cross bridges from a thick filament are able to bind with the actin molecules in the surrounding thin filaments. Let's concentrate on a single cross-bridge interaction (● Figure 8-9a). The two myosin heads of each myosin molecule act independently, with only one head attaching to actin at a given time. When myosin and actin make contact at a cross bridge, the conformation of the bridge is altered so that it bends inward as if it were on a hinge, "stroking" toward the center of the sarcomere, similar to the stroking of a boat oar. This so-called **power stroke** of a cross bridge pulls the thin filament to which it is attached inward. A single power stroke pulls the thin filament inward only a small percentage of the total shortening distance. Complete shortening is accomplished by repeated cycles of cross-bridge binding and bending.

At the end of one cross-bridge cycle, the link between the myosin cross bridge and actin molecule is broken. The cross bridge returns to its original conformation and binds to the next actin molecule positioned behind its previous actin partner. The cross bridge bends once again to pull the thin filament in further, then detaches and repeats the cycle. Repeated cycles of cross-bridge power strokes successively pull in the thin filaments, much like pulling in a rope hand over hand.

Because of the orientation of the myosin molecules within a thick filament (● Figure 8-9b), all the cross bridges' power strokes are directed toward the center of the sarcomere, so that all six of the surrounding thin filaments on each end of the sarcomere are pulled inward simultaneously (● Figure 8-9c). The cross bridges aligned with given thin filaments do not all stroke in unison, however. At any time during contraction, part of the cross bridges are attached to the thin filaments and are stroking, while others are returning to their original conformation in preparation for binding with another actin molecule. Thus, some cross bridges are "holding on" to the thin filaments, whereas others "let go" to bind with new actin. Were it not for this asynchronous cycling of the cross bridges, the thin filaments would be able to slip back toward their resting position between strokes.

How is this cross-bridge cycling switched on by muscle excitation? The term **excitation–contraction coupling** refers to the series of events linking muscle excitation (the presence of an action potential in a muscle fiber) to muscle contraction (cross-bridge activity that causes the thin filaments to slide closer together to produce sarcomere shortening). We will now turn our attention to excitation–contraction coupling.

Calcium is the link between excitation and contraction.

Skeletal muscles are stimulated to contract by release of acetylcholine (ACh) at neuromuscular junctions between motor neuron terminals and muscle fibers. Recall that the binding of ACh with the motor end plate of a muscle fiber brings about permeability changes in the muscle fiber that result in an action potential that is conducted over the entire surface of the muscle cell membrane (see p. 249). Two membranous structures within the muscle fiber play an important role in linking this excitation to contraction–transverse tubules and the sarcoplasmic reticulum. Let's examine the structure and function of each.

Spread of the action potential down the T tubules

At each junction of an A band and I band, the surface membrane dips into the muscle fiber to form a **transverse tubule (T tubule),** which runs perpendicularly from the surface of the muscle cell membrane into the central portions of the muscle fiber (● Figure 8-10). Because the T tubule membrane is continuous with the surface membrane, an action potential on the surface membrane also spreads down into the T tubule, providing a means of rapidly transmitting the surface electric activity into the central portions of the fiber. The presence of a local action potential in the T tubules induces permeability changes in a separate membranous network within the muscle fiber, the sarcoplasmic reticulum.

Release of calcium from the sarcoplasmic reticulum

The **sarcoplasmic reticulum** is a modified endoplasmic reticulum (see p. 28) that consists of a fine network of interconnected compartments surrounding each myofibril like a mesh sleeve (● Figure 8-10). This membranous network encircles the myofibril throughout its length, but is not continuous. Separate segments of sarcoplasmic reticulum are wrapped around each A band and each I band. The ends of each segment expand to form saclike regions, the **lateral sacs** (alternatively known as **terminal cisternae),** which are separated from the adjacent T tubules by a slight gap (● Figures 8-10 and 8-11). The sarcoplasmic reticulum's lateral sacs store Ca^{2+}. Spread of an action potential down a T tubule triggers release of Ca^{2+} from the sarcoplasmic reticulum into the cytosol.

How is a change in T tubule potential linked with the release of Ca^{2+} from the lateral sacs? An orderly arrangement of **foot proteins** extends from the sarcoplasmic reticulum and

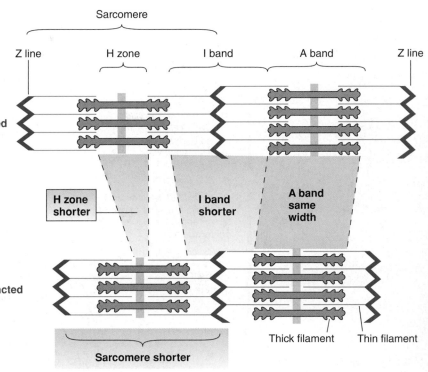

● **FIGURE 8-8**

Changes in banding pattern during shortening
During muscle contraction, each sarcomere shortens as the thin filaments slide closer together between the thick filaments so that the Z lines are pulled closer together. The width of the A bands does not change as a muscle fiber shortens, but the I bands and H zones become shorter.

spans the gap between the lateral sac and T tubule. Each foot protein contains four subunits arranged in a specific pattern (● Figure 8-11a). These foot proteins not only bridge the gap but also serve as Ca^{2+}-release channels. These foot-protein Ca^{2+} channels are also known as **ryanodine receptors** because they are locked in the open position by the plant chemical ryanodine.

Half of the sarcoplasmic reticulum's foot proteins are "zipped together" with complementary receptors on the T tubule side of the junction. These T tubule receptors, which are made up of four subunits in exactly the same pattern as the foot proteins, are located like mirror images in contact with every other foot protein protruding from the sarcoplasmic reticulum (● Figure 8-11b and c). These T tubule receptors are known as **dihydropyridine receptors** because they are blocked by the drug dihydropyridine. These dihydropyridine receptors are voltage-gated sensors. When an action potential is propagated down the T tubule, the local depolarization activates the voltage-gated dihydropyridine receptors. These activated T tubule receptors in turn trigger the opening of the directly abutting Ca^{2+}-release channels (alias ryanodine receptors alias foot proteins) in the adjacent lateral sacs of the sarcoplasmic reticulum. Opening of the half of the Ca^{2+}-release channels that are in direct contact with the dihydropyridine receptors triggers the opening of the other half of the Ca^{2+}-release channels that are not directly associated with the T tubule receptors.

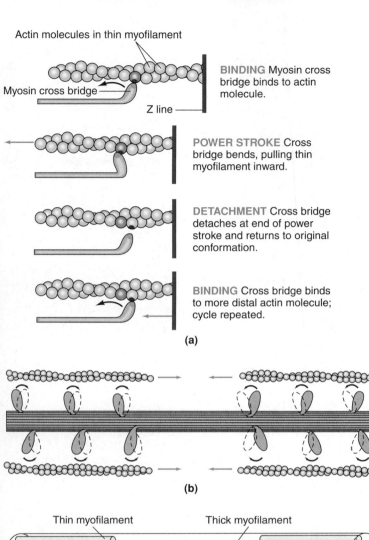

Actin molecules in thin myofilament

Myosin cross bridge

Z line

BINDING Myosin cross bridge binds to actin molecule.

POWER STROKE Cross bridge bends, pulling thin myofilament inward.

DETACHMENT Cross bridge detaches at end of power stroke and returns to original conformation.

BINDING Cross bridge binds to more distal actin molecule; cycle repeated.

(a)

(b)

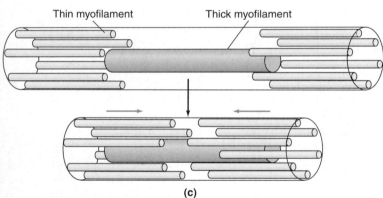

Thin myofilament Thick myofilament

(c)

● **FIGURE 8-9**

Cross-bridge activity
(a) During each cross-bridge cycle, the cross bridge binds with an actin molecule, bends to pull the thin filament inward during the power stroke, then detaches and returns to its resting conformation, ready to repeat the cycle. (b) The power strokes of all cross bridges extending from a thick filament are directed toward the center of the thick filament. (c) Each thick filament is surrounded on each end by six thin filaments, all of which are pulled inward simultaneously through cross-bridge cycling during muscle contraction.

Calcium is released into the surrounding cytosol from the lateral sacs through all these open Ca^{2+}-release channels. By slightly repositioning the troponin and tropomyosin mole-

cules, this released Ca^{2+} exposes the binding sites on the actin molecules so that they can link with the myosin cross bridges at their complementary binding sites (● Figure 8-12).

ATP-powered cross-bridge cycling

Recall that a myosin cross bridge has two special sites, an actin-binding site and an ATPase site. The latter is an enzymatic site that can bind the energy carrier *adenosine triphosphate (ATP)* and split it into *adenosine diphosphate (ADP)* and *inorganic phosphate (P_i)*, yielding energy in the process. Magnesium (Mg^{2+}) must be attached to ATP before myosin ATPase can split the ATP. The breakdown of ATP occurs on the myosin cross bridge before the bridge ever links with an actin molecule (step 1 in ● Figure 8-13). The ADP and P_i remain tightly bound to the myosin, and the generated energy is stored within the cross bridge to produce a high-energy form of myosin. To use an analogy, the cross bridge is "cocked" like a gun, ready to be fired when the trigger is pulled. When the muscle fiber is excited, Ca^{2+} pulls the troponin-tropomyosin complex out of its blocking position so that the energized (cocked) myosin cross bridge can bind with an actin molecule (step 2a). This contact between myosin and actin "pulls the trigger," causing the cross-bridge bending responsible for the power stroke (step 3). The mechanism by which the chemical energy released from ATP is stored within the myosin cross bridge and then translated into the mechanical energy of the power stroke is not known. Inorganic phosphate is released from the cross bridge during the power stroke. After the power stroke is complete, ADP is released .

When the muscle is not excited and Ca^{2+} is not released, troponin and tropomyosin remain in their blocking position, so that actin and the myosin cross bridges do not bind and no power stroking takes place (step 2b).

When inorganic phosphate and ADP are released from myosin following contact with actin and the subsequent power stroke, the myosin ATPase site is free for attachment of another ATP molecule. The actin and myosin remain linked together at the cross bridge until a fresh molecule of ATP attaches to myosin at the end of the power stroke. Attachment of the new ATP molecule permits detachment of the cross bridge, which returns to its unbent conformation, ready to start another cycle (step 4a). The newly attached ATP is then split by myosin ATPase, energizing the myosin cross bridge once again (step 1). On binding with another actin molecule, the energized cross bridge again bends, and so on, successively pulling the thin filament inward to accomplish contraction.

Rigor mortis

Note that fresh ATP must attach to myosin to permit the cross-bridge link between myosin and actin to be broken at the end of a cycle, even though the ATP is not split during this dissociation process. The necessity for ATP in the separation of myosin and actin is amply demonstrated by the phenomenon of **rigor mortis**. This "stiffness of death" is a generalized locking in place of the skeletal muscles that begins 3 to 4 hours after death and becomes complete in about 12 hours. Following

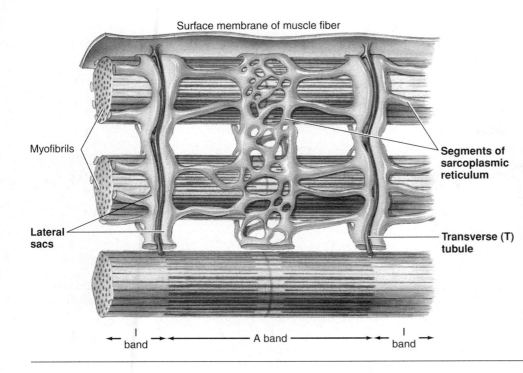

Surface membrane of muscle fiber

Myofibrils

Lateral sacs

Segments of sarcoplasmic reticulum

Transverse (T) tubule

I band — A band — I band

● **FIGURE 8-10**

The T tubules and sarcoplasmic reticulum in relationship to the myofibrils

The transverse (T) tubules are membranous perpendicular extensions of the surface membrane that dip deep into the muscle fiber at the junctions between the A and I bands of the myofibrils. The sarcoplasmic reticulum is a fine, membranous network that runs longitudinally and surrounds each myofibril, with separate segments encircling each A band and I band. The ends of each segment are expanded to form lateral sacs that lie next to the adjacent T tubules.

● **FIGURE 8-11**

Relationship between a T tubule and the adjacent lateral sacs of the sarcoplasmic reticulum

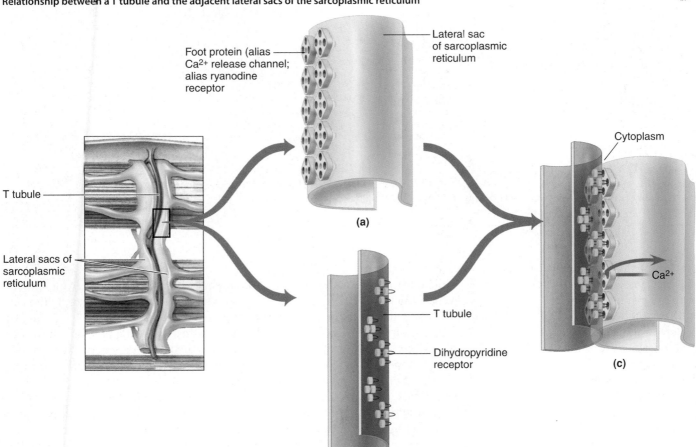

Foot protein (alias Ca²⁺ release channel; alias ryanodine receptor

Lateral sac of sarcoplasmic reticulum

T tubule

Lateral sacs of sarcoplasmic reticulum

Cytoplasm

Ca²⁺

T tubule

Dihydropyridine receptor

(a)

(b)

(c)

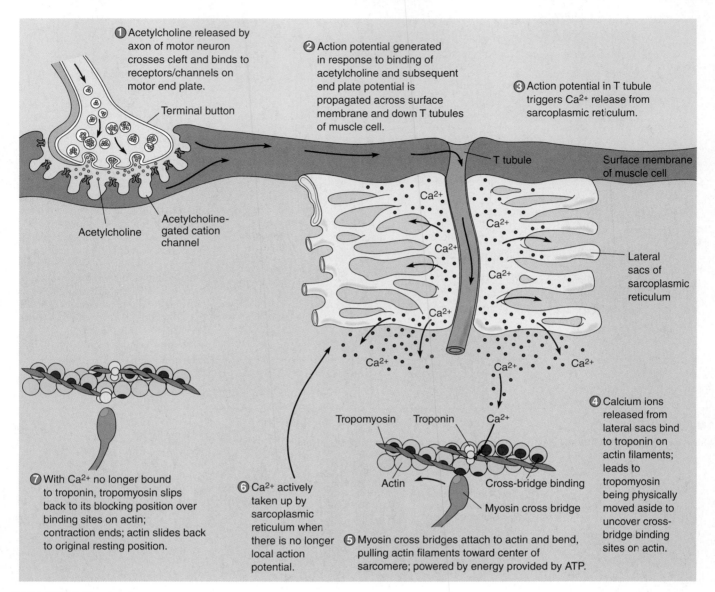

● FIGURE 8-12

Calcium release in excitation–contraction coupling
Steps ❶ through ❺ depict the events that couple neurotransmitter release and subsequent electrical excitation of the muscle cell with muscle contraction. Steps ❻ and ❼ depict events associated with muscle relaxation.

death, the cytosolic concentration of Ca^{2+} begins to rise, most likely because the inactive muscle-cell membrane is unable to keep out extracellular Ca^{2+} and perhaps also because Ca^{2+} leaks out of the lateral sacs. This Ca^{2+} moves the regulatory proteins aside, permitting actin to bind with the myosin cross bridges, which were already charged with ATP before death. Because dead cells cannot produce any more ATP, actin and myosin, once bound, are unable to detach because of the absence of fresh ATP. The thick and thin filaments thus remain linked together by the immobilized cross bridges, resulting in the stiffened condition of dead muscles (step 4b). During the next several days, rigor mortis gradually subsides as the proteins involved in the rigor complex begin to degrade.

Relaxation

How is **relaxation** normally accomplished in a living muscle? Just as an action potential in a muscle fiber turns on the contractile process by triggering the release of Ca^{2+} from the lateral sacs into the cytosol, the contractile process is turned off when Ca^{2+} is returned to the lateral sacs on cessation of local electrical activity. The sarcoplasmic reticulum possesses an energy-consuming carrier, a Ca^{2+}–ATPase pump, which actively transports Ca^{2+} from the cytosol and concentrates it in the lateral sacs. When acetylcholinesterase removes ACh from the neuromuscular junction, the muscle-fiber action potential ceases. When there is no longer a local action potential in the T tubules to trigger the release of Ca^{2+}, the ongoing activity

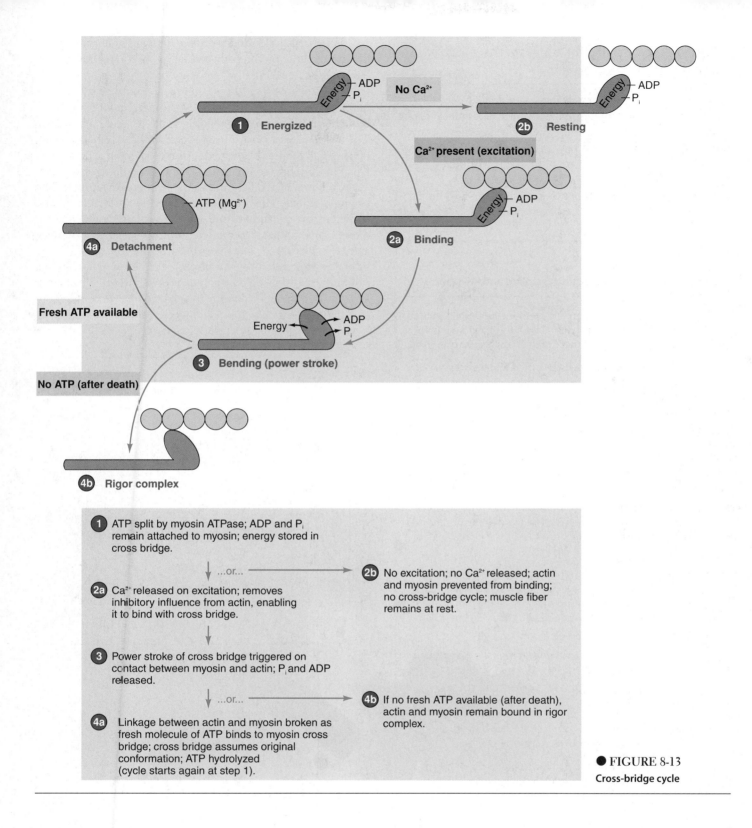

1. ATP split by myosin ATPase; ADP and P_i remain attached to myosin; energy stored in cross bridge.

...or...

2a. Ca^{2+} released on excitation; removes inhibitory influence from actin, enabling it to bind with cross bridge.

2b. No excitation; no Ca^{2+} released; actin and myosin prevented from binding; no cross-bridge cycle; muscle fiber remains at rest.

3. Power stroke of cross bridge triggered on contact between myosin and actin; P_i and ADP released.

...or...

4b. If no fresh ATP available (after death), actin and myosin remain bound in rigor complex.

4a. Linkage between actin and myosin broken as fresh molecule of ATP binds to myosin cross bridge; cross bridge assumes original conformation; ATP hydrolyzed (cycle starts again at step 1).

● FIGURE 8-13
Cross-bridge cycle

of the sarcoplasmic reticulum's Ca^{2+} pump returns the released Ca^{2+} back into its lateral sacs. Removal of cytosolic Ca^{2+} allows the troponin-tropomyosin complex to slip back into its blocking position, so that actin and myosin are no longer able to bind at the cross bridges. The thin filaments, freed from cycles of cross-bridge attachment and pulling, are able to return passively to their resting position. Relaxation has occurred.

▲ Table 8-1 summarizes the steps of excitation–contraction coupling and relaxation. We are now going to compare the duration of contractile activity to the duration of excitation before we shift gears to discuss skeletal muscle mechanics.

Contractile activity far outlasts the electrical activity that initiated it.

A single action potential in a skeletal muscle fiber lasts only 1 to 2 msec. The onset of the resultant contractile response lags behind the action potential because the entire excitation–contraction coupling process must take place before cross-bridge activity begins. In fact, the action potential is completed before the contractile apparatus even becomes operational. This time delay of a few milliseconds between stimulation and the onset of contraction is known as the **latent period** (● Figure 8-14).

▲ TABLE 8-1
Steps of Excitation–Contraction Coupling and Relaxation

1. Acetylcholine released from the terminal of a motor neuron initiates an action potential in the muscle cell that is propagated over the entire surface of the muscle cell membrane.

2. The surface electric activity is carried into the central portions of the muscle fiber by the T tubules.

3. Spread of the action potential down the T tubules triggers the release of stored Ca^{2+} from the adjacent lateral sacs of the sarcoplasmic reticulum.

4. Released Ca^{2+} binds with troponin and changes its shape so that the troponin-tropomysin complex is physically pulled aside, uncovering actin's cross-bridge binding sites.

5. Exposed actin sites bind with myosin cross bridges, which have previously been energized by the splitting of ATP into $ADP + P_i +$ energy by the myosin ATPase site on the cross bridges.

6. Binding of actin and myosin at a cross bridge causes the cross bridge to bend, producing a power stroke that pulls the thin filament inward. Inward sliding of all the thin filaments surrounding a thick filament shortens the sarcomere (causes muscle contraction).

7. P_i is released from the cross bridge during the power stroke; ADP is released after the power stroke is complete.

8. Attachment of a new molecule of ATP permits detachment of the cross bridge, which returns to its original conformation.

9. Splitting of the fresh ATP molecule by myosin ATPase energizes the cross bridge once again.

10. If Ca^{2+} is still present so that the troponin-tropomyosin complex remains pulled aside, the cross bridges go through another cycle of binding and bending, pulling the thin filament in even further.

11. When there is no longer a local action potential and Ca^{2+} has been actively returned to its storage site in the sarcoplasmic reticulum's lateral sacs, the troponin-tropomyosin complex slips back into its blocking position, actin and myosin no longer bind at the cross bridges, and the thin filaments passively slide back to their resting position as relaxation takes place.

Time is also required for the generation of tension within the muscle fiber produced by means of the sliding interactions between the thick and thin filaments through cross-bridge activity. The time from the onset of contraction until peak tension is developed—the **contraction time**—averages about 50 msec, although this time varies, depending on the type of muscle fiber. The contractile response does not cease until the lateral sacs have taken up all the Ca^{2+} released in response to the action potential. This reuptake of Ca^{2+} is also time consuming. Even after Ca^{2+} is removed, it takes time for the filaments to return to their resting positions. The time from peak tension until relaxation is complete, the **relaxation time,** usually lasts another 50 msec or more. Consequently, the entire contractile response to a single action potential may last up to 100 msec or more; this is considerably longer than the duration of the action potential that initiated it (100 msec compared to 1 to 2 msec). This fact is important in the body's ability to produce muscle contractions of variable strength, as you will discover in the next section.

● FIGURE 8-14
Relationship of an action potential to the resultant muscle twitch

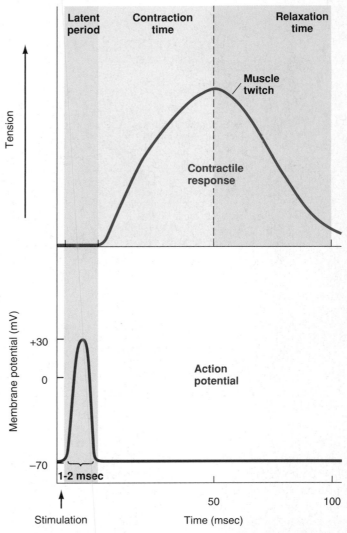

The duration of the action potential is not drawn to scale but is exaggerated.

SKELETAL MUSCLE MECHANICS

Thus far we have described the contractile response in a single muscle fiber. In the body, groups of muscle fibers are organized into whole muscles. We will now turn our attention to contraction of whole muscles.

▌ Whole muscles are groups of muscle fibers bundled together and attached to bones.

Each person has about 600 skeletal muscles, which range in size from the delicate external eye muscles that control eye movements and contain only a few hundred fibers, to the large, powerful leg muscles that contain several hundred thousand fibers.

Each muscle is covered by a sheath of connective tissue that penetrates from the surface into the muscle to envelop each individual fiber and divide the muscle into columns or bundles. The connective tissue extends beyond the ends of the muscle to form tough, collagenous **tendons** that attach the muscle to bones. A tendon may be quite long, attaching to a bone some distance from the fleshy portion of the muscle. For example, some of the muscles involved in finger movement are found in the forearm, with long tendons extending down to attach to the bones of the fingers. (You can readily observe the movement of these tendons on the top of your hand when you wiggle your fingers.) This arrangement permits greater dexterity; the fingers would be much thicker and more awkward if all the muscles involved in finger movement were actually located in the fingers.

▌ Contractions of a whole muscle can be of varying strength.

A single action potential in a muscle fiber produces a brief, weak contraction known as a **twitch**, which is too short and too weak to be useful and normally does not take place in the body. Muscle fibers are arranged into whole muscles, where they can function cooperatively to produce contractions of variable grades of strength stronger than a twitch. In other words, the force exerted by the same muscle can be made to vary, depending on whether the person is picking up a piece of paper, a book, or a 50-pound weight. Two primary factors can be adjusted to accomplish gradation of whole-muscle tension: (1) *the number of muscle fibers contracting within a muscle* and (2) *the tension developed by each contracting fiber.* We will discuss each of these factors in turn.

▌ The number of fibers contracting within a muscle depends on the extent of motor unit recruitment.

Because the greater the number of fibers contracting, the greater the total muscle tension, larger muscles consisting of more muscle fibers are obviously capable of generating more tension than are smaller muscles with fewer fibers.

Each whole muscle is innervated by a number of different motor neurons. When a motor neuron enters a muscle, it branches, with each axon terminal supplying a single muscle fiber (● Figure 8-15). One motor neuron innervates a number of muscle fibers, but each muscle fiber is supplied by only one motor neuron. When a motor neuron is activated, all the muscle fibers it supplies are stimulated to contract simultaneously. This team of concurrently activated components—one motor neuron plus all the muscle fibers it innervates—is called a **motor unit.** The muscle fibers that compose a motor unit are dispersed throughout the whole muscle; thus their simultaneous contraction results in an evenly distributed, although weak, contraction of the whole muscle. Each muscle consists of a number of intermingled motor units. For a weak contraction of the whole muscle, only one or a few of its motor units are activated. For stronger and stronger contractions, more and more motor units are recruited, or stimulated to contract, a phenomenon known as **motor unit recruitment.**

How much stronger the contraction will be with the recruitment of each additional motor unit depends on the size of the motor units (that is, the number of muscle fibers controlled by a single motor neuron) (● Figure 8-16). The number of muscle fibers per motor unit and the number of motor units per muscle vary widely, depending on the specific function of the muscle. For muscles that produce precise, delicate movements, such as the external eye muscles and the hand muscles, a single motor unit may contain as few as a dozen muscle fibers. Because so few muscle fibers are involved with each motor unit, recruitment of each additional motor unit results in only a small additional increment in the whole muscle's strength of contraction. These small motor units allow a very fine degree of control over muscle tension. In contrast, in mus-

● FIGURE 8-15

Schematic representation of motor units in a skeletal muscle

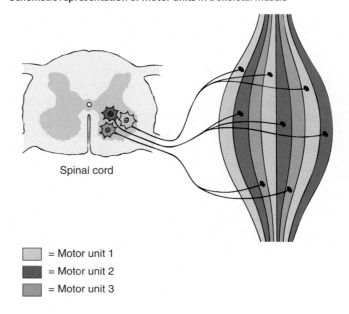

Spinal cord

☐ = Motor unit 1
■ = Motor unit 2
▨ = Motor unit 3

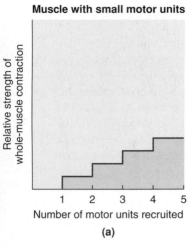

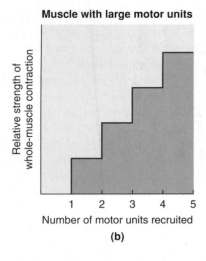

Muscle with small motor units

Relative strength of whole-muscle contraction

Number of motor units recruited
1 2 3 4 5

(a)

Muscle with large motor units

Relative strength of whole-muscle contraction

Number of motor units recruited
1 2 3 4 5

(b)

● **FIGURE 8-16**

Comparison of motor unit recruitment in muscles with small motor units and muscles with large motor units
(a) Small incremental increases in strength of contraction occur during motor unit recruitment in muscles with small motor units because only a few additional fibers are called into play as each motor unit is recruited. (b) Large incremental increases in strength of contraction occur during motor unit recruitment in muscles with large motor units, because so many additional fibers are stimulated with the recruitment of each additional motor unit.

cles designed for powerful, coarsely controlled movement, such as those of the legs, a single motor unit may contain 1,500 to 2,000 muscle fibers. Recruitment of motor units in these muscles results in large incremental increases in whole-muscle tension. More powerful contractions occur at the expense of less precisely controlled gradations. Thus the number of muscle fibers participating in the whole muscle's total contractile effort depends on the number of motor units recruited and the number of muscle fibers per motor unit in that muscle.

To delay or prevent **fatigue** (inability to maintain muscle tension at a given level) during a sustained contraction involving only a portion of a muscle's motor units, as is necessary in muscles supporting the weight of the body against the force of gravity, **asynchronous recruitment of motor units** takes place. The body alternates motor unit activity, like shifts at a factory, to give motor units that have been active an opportunity to rest while others take over. Changing of the shifts is carefully coordinated, so that the sustained contraction is smooth rather than jerky. Asynchronous motor unit recruitment is possible only for submaximal contractions, during which only some of the motor units are required to maintain the desired level of tension. During maximal contractions, when participation of all the muscle fibers is essential, it is impossible to alternate motor unit activity to prevent fatigue. This is one reason why you cannot support a heavy object as long as one that is light.

Furthermore, the type of muscle fiber that is activated varies with the extent of gradation. Most muscles consist of a mixture of fiber types that differ metabolically, some being more resistant to fatigue than others. During weak or moderate endurance-type activities (aerobic exercise), the motor units most resistant to fatigue are recruited first. The last fibers to be called into play in the face of demands for further increases in

tension are those that fatigue rapidly. An individual can therefore engage in endurance activities for prolonged periods of time but can only briefly maintain bursts of all-out, powerful effort. Of course, even the muscle fibers most resistant to fatigue will eventually fatigue if required to maintain a certain level of sustained tension.

■ **The frequency of stimulation can influence the tension developed by each muscle fiber.**

Whole-muscle tension depends not only on the number of muscle fibers contracting but also on the tension developed by each contracting fiber. Various factors influence the extent to which tension can be developed. These factors include the following:

1. The frequency of stimulation
2. The length of the fiber at the onset of contraction
3. The extent of fatigue
4. The thickness of the fiber

We will now examine the effect of frequency of stimulation. (The other factors are discussed in later sections.)

Twitch summation and tetanus

Even though a single action potential in a muscle fiber produces only a twitch, contractions with longer duration and greater tension can be achieved by repetitive stimulation of the fiber. Let us see what happens when a second action potential occurs in a muscle fiber. If the muscle fiber has completely relaxed before the next action potential takes place, a second twitch of the same magnitude as the first occurs (● Figure 8-17a). The same excitation–contraction events take place each time, resulting in identical twitch responses. If, however, the muscle fiber is stimulated a second time before it has completely relaxed from the first twitch, a second action potential occurs that causes a second contractile response, which is added "piggyback" on top of the first twitch (● Figure 8-17b). The two twitches resulting from the two action potentials add together, or sum, to produce greater tension in the fiber than that produced by a single action potential. This **twitch summation** is similar to temporal summation of EPSPs at the postsynaptic neuron (see p. 121).

Twitch summation is possible only because the duration of the action potential (1 to 2 msec) is much shorter than the duration of the resultant twitch (100 msec). Once an action potential has been initiated, a brief refractory period occurs during which another action potential cannot be initiated (see p. 112). It is therefore impossible to achieve summation of action potentials. The membrane must return to resting potential and recover from its refractory period before another action potential can occur. However, because the action potential and refractory period are over long before the resultant muscle twitch is completed, the muscle fiber may be restimulated while some contractile activity still exists to produce summation of the mechanical response.

If the muscle fiber is stimulated so rapidly that it does not have a chance to relax at all between stimuli, a smooth, sustained contraction of maximal strength known as **tetanus** occurs (● Figure 8-17c). A tetanic contraction is usually three to four times stronger than a single twitch. (This normal physiologic tetanus should not be confused with the disease tetanus; see p. 127.)

▮ Twitch summation results from a sustained elevation in cytosolic calcium.

What is the mechanism of twitch summation and tetanus at the cellular level? The tension produced by a contracting muscle fiber increases as a result of greater cross-bridge cycling. As the frequency of action potentials increases, the resultant tension development increases until a maximum tetanic contraction is achieved. Sufficient Ca^{2+} is released in response to a single action potential to interact with all the troponin within the cell. As a result, all the cross bridges are free to participate in the contractile response. How, then, can repetitive action potentials bring about a greater contractile response? The difference depends on how long enough Ca^{2+} is available.

The cross bridges will remain active and continue to cycle as long as sufficient Ca^{2+} is present to keep the troponin-tropomyosin complex away from the cross-bridge binding sites on actin. Each troponin-tropomyosin complex spans a distance of seven actin molecules. Thus, binding of Ca^{2+} to one troponin molecule leads to the uncovering of only seven cross-bridge binding sites on the thin filament.

As soon as Ca^{2+} is released in response to an action potential, the sarcoplasmic reticulum starts pumping Ca^{2+} back into the lateral sacs. As the cytosolic Ca^{2+} concentration declines with the reuptake of Ca^{2+} by the lateral sacs, less Ca^{2+} is present to bind with troponin, so some of the troponin-tropomyosin complexes slip back into their blocking positions. Consequently, not all the cross-bridge binding sites remain available to participate in the cycling process during a single twitch induced by a single action potential. Because not all the cross bridges find a binding site, the resultant contraction during a single twitch is not of maximal strength.

If action potentials and twitches occur far enough apart in time for all the released Ca^{2+} from the first contractile response to be pumped back into the lateral sacs between the action potentials, an identical twitch response will occur as a result of the second action potential. If, however, a second action potential occurs and more Ca^{2+} is released while the Ca^{2+} that was released in response to the first action potential is being taken back up, the cytosolic Ca^{2+} concentration remains elevated. This prolonged availability of Ca^{2+} in the cytosol permits more of the cross bridges to continue participating in the cycling process for a longer time. As a result, tension development increases correspondingly. As the frequency of action potentials increases, the duration of elevated cytosolic Ca^{2+} concentration increases, and contractile activity likewise increases until a maximum tetanic contraction is reached. With tetanus, the maximum number of cross-bridge binding sites remain uncovered so that cross-bridge cycling, and consequently tension development, are at their peak.

Because skeletal muscle must be stimulated by motor neurons to contract, the nervous system plays a key role in regulating the strength of contraction. The two main factors subject to control to accomplish gradation of contraction are the *number of motor units stimulated* and the *frequency of their stimula-*

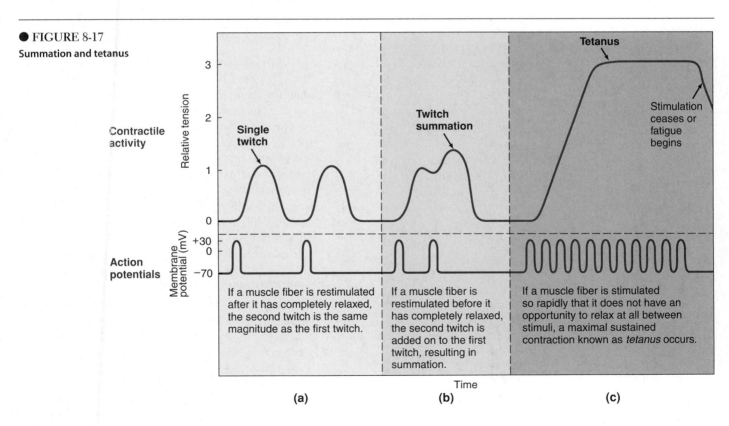

● **FIGURE 8-17**
Summation and tetanus

(a) If a muscle fiber is restimulated after it has completely relaxed, the second twitch is the same magnitude as the first twitch.

(b) If a muscle fiber is restimulated before it has completely relaxed, the second twitch is added on to the first twitch, resulting in summation.

(c) If a muscle fiber is stimulated so rapidly that it does not have an opportunity to relax at all between stimuli, a maximal sustained contraction known as *tetanus* occurs.

tion. The areas of the brain responsible for directing motor activity use a combination of tetanic contractions and precisely timed shifts of asynchronous motor unit recruitment to execute smooth rather than jerky contractions.

Additional factors not directly under nervous control also influence the tension developed during contraction. Among these is the length of the fiber at the onset of contraction, to which we now turn our attention.

There is an optimal muscle length at which maximal tension can be developed.

A relationship exists between the length of the muscle before the onset of contraction and the tetanic tension that each contracting fiber can subsequently develop at that length. For every muscle there is an **optimal length** (l_o) at which maximal force can be achieved on a subsequent tetanic contraction. The tension that can be achieved during tetanus at the optimal muscle length is greater than the tetanic tension that can be achieved when the contraction begins with the muscle less than or greater than its optimal length. This **length–tension relationship** can be explained by the sliding filament mechanism of muscle contraction.

At optimal length (l_o)

At l_o, when maximum tension can be developed (point A in ● Figure 8-18), the thin filaments optimally overlap the regions of the thick filaments from which the cross bridges project. At this length, a maximal number of cross-bridge sites are accessible to the actin molecules for binding and bending. The central region of thick filaments, where the thin filaments do not overlap at l_o, is devoid of cross bridges; only myosin tails are found here.

At lengths greater than l_o

At greater lengths, as when a muscle is passively stretched (point B), the thin filaments are pulled out from between the thick filaments, decreasing the number of actin sites available for cross-bridge binding; that is, some of the actin sites and cross bridges no longer "match up," so they "go unused." When less cross-bridge activity can occur, less tension can be developed. In fact, when the muscle is stretched to about 70% longer than its l_o (point C) the thin filaments are completely pulled out from between the thick filaments so that no cross-bridge activity and consequently no contraction can occur.

At lengths less than l_o

If a muscle is shorter than l_o before contraction (point D), less tension can be developed for three reasons:

1. The thin filaments from the opposite sides of the sarcomere become overlapped, which limits the opportunity for the cross bridges to interact with actin.

2. The ends of the thick filaments become forced against the Z lines, so further shortening is impeded.

3. Besides these two mechanical factors, at muscle lengths less than 80% of l_o not as much Ca^{2+} is released during excitation–contraction coupling for reasons unknown. Furthermore, by an unknown mechanism the ability of Ca^{2+} to bind to troponin and pull the troponin-tropomyosin complex aside is reduced at shorter muscle lengths. Consequently, fewer actin sites are uncovered for participation in cross-bridge activity.

Limitations on muscle length

The extremes in muscle length that prevent the development of tension occur only under experimental conditions, when a muscle is removed and stimulated at various lengths. In the body the muscles are so positioned that their relaxed length is approximately their optimal length; thus they are capable of achieving near-maximal tetanic contraction most of the time. Because

● FIGURE 8-18

Length–tension relationship

Maximal tetanic contraction can be achieved when a muscle fiber is at its optimal length (l_o) before the onset of contraction, because this is the point of optimal overlap of thick-filament cross bridges and thin-filament cross-bridge binding sites (point A). The percentage of maximal tetanic contraction that can be achieved decreases when the muscle fiber is longer or shorter than l_o before contraction. When it is longer, fewer thin-filament binding sites are accessible for binding with thick-filament cross bridges, because the thin filaments are pulled out from between the thick filaments (points B and C). When the fiber is shorter, fewer thin-filament binding sites are exposed to thick-filament cross bridges because the thin filaments overlap (point D). Also, further shortening and tension development are impeded as the thick filaments become forced against the Z lines (point D). In the body, the resting muscle length is at l_o. Furthermore, because of restrictions imposed by skeletal attachments, muscles cannot vary beyond 30% of their l_o in either direction (the range screened in light green). At the outer limits of this range, muscles are still able to achieve about 50% of their maximal tetanic contraction.

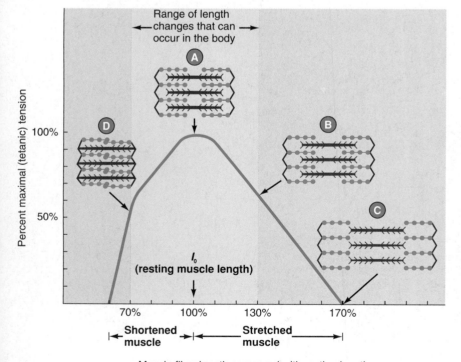

Muscle fiber length compared with resting length

▲ TABLE 8-2
Determinants of Whole-Muscle Tension in Skeletal Muscle

Number of Fibers Contracting	Tension Developed by Each Contracting Fiber
Number of motor units recruited*	Frequency of stimulation (twitch summation and tetanus)*
Number of muscle fibers per motor unit	Length of fiber at onset of contraction (length–tension relationship)

Number of muscle fibers available to contract
    Size of muscle (number of muscle fibers in muscle)
    Presence of disease (e.g., muscular dystrophy)
    Extent of recovery from traumatic losses
	Extent of fatigue
	    Duration of activity
	    Amount of asynchronous recruitment of motor units
	    Type of fiber (fatigue-resistant oxidative or fatigue-prone glycolytic)

	Thickness of fiber
	    Pattern of neural activity (hypertrophy, atrophy)
	    Amount of testosterone (larger fibers in males than females)

*Factors controlled to accomplish gradation of contraction.

of limitations imposed by attachment to the skeleton, a muscle cannot be stretched or shortened more than 30% of its resting optimal length, and usually it deviates much less than 30% from normal length. Even at the outer limits (130% and 70% of l_o), the muscles are still able to generate half their maximum tension.

The factors we have discussed thus far that influence how much tension can be developed by a contracting muscle fiber—the frequency of stimulation and the muscle length at the onset of contraction—can vary from contraction to contraction. Other determinants of muscle fiber tension—the metabolic capability of the fiber relative to resistance to fatigue and the thickness of the fiber—do not vary from contraction to contraction but depend on the fiber type and can be modified over a period of time. After we complete our discussion of skeletal muscle mechanics, we will consider these other factors in the next section, on skeletal muscle metabolism and fiber types. ▲ Table 8-2 summarizes all the determinants of whole-muscle tension in a skeletal muscle.

▌ Muscle tension is transmitted to bone as the contractile component tightens the series-elastic component.

Tension is produced internally within the sarcomeres, considered the **contractile component** of the muscle, as a result of cross-bridge activity and the resultant sliding of filaments. However, the sarcomeres are not attached directly to the bones. Instead, the tension generated by these contractile elements must be transmitted to the bone via the connective tissue and tendons before the bone can be moved. Connective tissue, as well as other components of the muscle, such as intracellular elastic proteins, exhibits a certain degree of passive elasticity. These noncontractile tissues are referred to as the **series-elastic component** of the muscle; they behave like a stretchy spring placed between the internal tension-generating elements and the bone that is to be moved against an external load (● Figure 8-19). Shortening of the sarcomeres stretches the series-

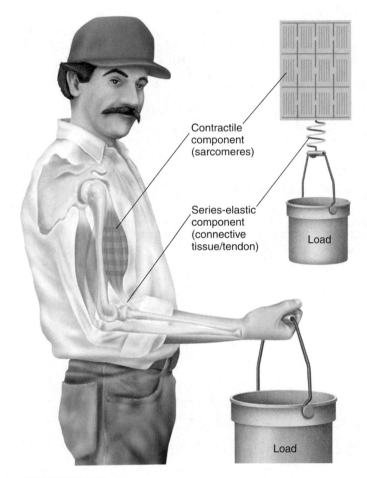

Contractile component (sarcomeres)

Series-elastic component (connective tissue/tendon)

Load

Load

● **FIGURE 8-19**

Relationship between the contractile component and the series-elastic component in transmitting muscle tension to bone
Muscle tension is transmitted to the bone by means of the stretching and tightening of the muscle's elastic connective tissue and tendon as a result of sarcomere shortening brought about by cross-bridge cycling.

elastic component. Muscle tension is transmitted to the bone by means of this tightening of the series-elastic component. This force applied to the bone is responsible for moving the bone against a load.

A muscle is typically attached to at least two different bones across a joint by means of tendons that extend from each end of the muscle (● Figure 8-20). When the muscle shortens during contraction, the position of the joint is changed as one bone is moved in relation to the other—for example, *flexion* of the elbow joint by contraction of the biceps muscle and *extension* of the elbow by contraction of the triceps. The end of the muscle attached to the more stationary part of the skeleton is called the **origin**, and the end attached to the skeletal part that moves is referred to as the **insertion.**

Not all muscle contractions result in muscle shortening and movement of bones, however. For a muscle to shorten during contraction, the tension developed in the muscle must exceed the forces that oppose movement of the bone to which the muscle's insertion is attached. In the case of elbow flexion, the opposing force, or **load,** is the weight of an object being lifted. When you flex your elbow without lifting any external object, there is still a load, albeit a minimal one—the weight of your forearm being moved against the force of gravity.

▮ The two primary types of contraction are isotonic and isometric.

There are two primary types of contraction, depending on whether the muscle changes length during contraction. In an **isotonic contraction,** muscle tension remains constant as the muscle changes length. In an **isometric contraction,** the muscle is prevented from shortening, so tension develops at constant muscle length. The same internal events occur in both isotonic and isometric contractions: the tension-generating contractile process is turned on by muscle excitation; the cross bridges start cycling; and filament sliding shortens the sarcomeres, which stretches the series-elastic component to exert force on the bone at the site of the muscle's insertion.

Considering your biceps as an example, assume that you are going to lift an object. When the tension developing in your biceps becomes great enough to overcome the weight of the object in your hand, you can lift the object, with the whole muscle shortening in the process. Because the weight of the object does not change as it is lifted, the muscle tension remains constant throughout the period of shortening. This is an *isotonic* (literally, "constant tension") contraction. Isotonic contractions are used for body movements and for moving external objects.

What happens if you try to lift an object too heavy for you (that is, if the tension you are capable of developing in your arm muscles is less than that required to lift the load)? In this case, the muscle cannot shorten and lift the object but remains at constant length despite the development of tension, so an *isometric* ("constant length") contraction occurs. In addition to occurring when the load is too great, isometric contractions also take place when the tension developed in the muscle is deliberately less than that needed to move the load. In this case, the goal is to keep the muscle at fixed length although it is capable of developing more tension. These submaximal isometric contractions are important for maintaining posture (such as keeping the legs stiff while standing) and for support-

---

● **FIGURE 8-20**
Extension and flexion of the elbow joint

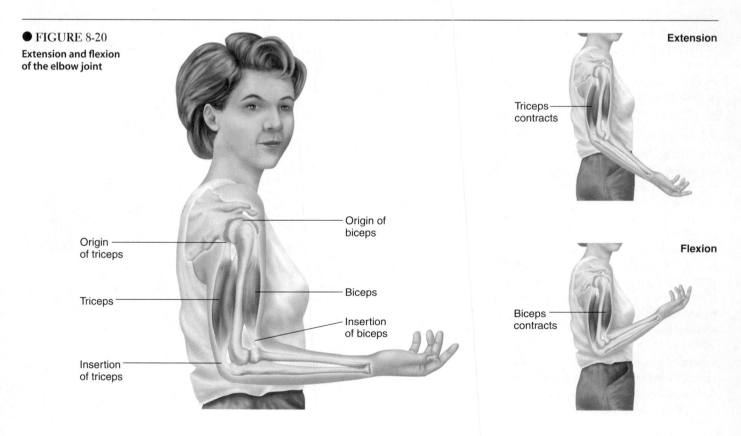

Origin of triceps

Triceps

Insertion of triceps

Origin of biceps

Biceps

Insertion of biceps

Extension

Triceps contracts

Flexion

Biceps contracts

ing objects in a fixed position. During a given movement, a muscle may shift between isotonic and isometric contractions. For example, when you pick up a book to read, your biceps undergoes an isotonic contraction while you are lifting the book, but the contraction becomes isometric as you stop to hold the book in front of you.

Concentric and eccentric isotonic contractions

There are actually two types of isotonic contraction—*concentric* and *eccentric*. In both cases the muscle changes length at constant tension. With **concentric contractions**, however, the muscle shortens, whereas with **eccentric contractions** the muscle lengthens because it is being stretched by an external force while contracting. With an eccentric contraction, the contractile activity is resisting the stretch. An example is lowering a load to the ground. During this action, the muscle fibers in the biceps are lengthening but are still contracting in opposition to being stretched. This tension supports the weight of the object.

Other contractions

The body is not limited to pure isotonic and isometric contractions. Muscle length and tension frequently vary throughout a range of motion. Think about drawing a bow and arrow. The tension of the biceps muscle continuously increases to overcome the progressively increasing resistance as the bow is stretched further. At the same time, the muscle progressively shortens as the bow is drawn farther back. Such a contraction occurs at neither constant tension nor constant length.

Some skeletal muscles do not attach to bones at both ends but still produce movement. For example, the muscles of the tongue are not attached at the free end. Isotonic contractions of the tongue muscles maneuver the free unattached portion of the tongue to facilitate speech and eating. The external eye muscles attach to the skull at their origin but to the eye at their insertion. Isotonic contractions of these muscles produce the eye movements that enable us to track moving objects, read, and so on. A few skeletal muscles that are completely unattached to bone actually prevent movement. These are the voluntarily controlled rings of skeletal muscles, known as **sphincters,** that guard the exit of urine and feces from the body by isotonically contracting.

▌ The velocity of shortening is related to the load.

The load is also an important determinant of the **velocity,** or speed, of shortening (● Figure 8-21). The greater the load, the lower the velocity at which a single muscle fiber (or a constant number of contracting fibers within a muscle) shortens during an isotonic tetanic contraction. The velocity of shortening is maximal when there is no external load, progressively decreases with an increasing load, and falls to zero (no shortening—isometric contraction) when the load cannot be overcome by maximal tetanic tension. You have frequently experienced this load–velocity relationship. You can lift light objects requiring little muscle tension quickly, whereas you can lift very heavy

objects only slowly, if at all. This relationship between load and shortening velocity is a fundamental property of muscle, presumably because it takes the cross bridges longer to stroke against a greater load.

▌ Although muscles can accomplish work, much of the energy is converted to heat.

Muscle accomplishes work in a physical sense only when an object is moved. **Work** is defined as force multiplied by distance. **Force** can be equated to the muscle tension required to overcome the load (the weight of the object). The amount of work accomplished by a contracting muscle therefore depends on how much an object weighs and how far it is moved. In an isometric contraction when no object is moved, the muscle contraction's efficiency as a producer of external work is zero. All energy consumed by the muscle during the contraction is converted to heat. In an isotonic contraction, the muscle's efficiency is about 25%. Of the energy consumed by the muscle during the contraction, 25% is realized as external work whereas the remaining 75% is converted to heat.

Much of this heat is not really wasted in a physiological sense because it is used in maintaining the body temperature. In fact, shivering—a form of involuntarily induced skeletal muscle contraction—is a well-known mechanism for increasing heat production on a cold day. Heavy exercise on a hot day, in contrast, may overheat the body, because the normal heat-loss mechanisms may be unable to compensate for this increase in heat production (Chapter 17).

▌ Interactive units of skeletal muscles, bones, and joints form lever systems.

Most skeletal muscles are attached to bones across joints, forming lever systems. A **lever** is a rigid structure capable of moving around a pivot point known as a **fulcrum.** In the body, the

● **FIGURE 8-21**

Load–velocity relationship
The velocity of shortening decreases as the load increases.

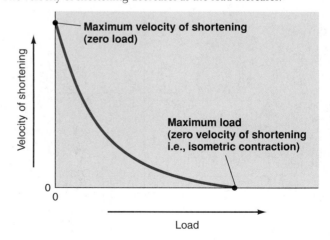

bones function as levers, the joints serve as fulcrums, and the skeletal muscles provide the force to move the bones. The portion of a lever between the fulcrum and the point where an upward force is applied is called the **power arm;** the portion between the fulcrum and the downward force exerted by a load is known as the **load arm** (● Figure 8-22a).

The most common type of lever system in the body is exemplified by flexion of the elbow joint. Skeletal muscles, such as the biceps whose contraction flexes the elbow joint, consist of many parallel (side-by-side) tension-generating fibers that can exert a large force at their insertion but shorten only a small distance and at relatively slow velocity. The lever system of the elbow joint amplifies the slow, short movements of the biceps to produce more rapid movements of the hand that cover a greater distance.

Consider how an object weighing 5 kg is lifted by the hand (● Figure 8-22b). When the biceps contracts, it exerts an upward force at the point where it inserts on the forearm bone about 5 cm away from the elbow joint, the fulcrum. Thus the power arm of this lever system is 5 cm long. The length of the load arm, the distance from the elbow joint to the hand, averages 35 cm. In this case, the load arm is seven times longer than the power arm, which enables the load to be moved a distance

seven times greater than the shortening distance of the muscle (while the biceps shortens a distance of 1 cm, the hand moves the load a distance of 7 cm) and at a velocity seven times greater (the hand moves 7 cm during the same length of time the biceps shortens 1 cm).

The disadvantage of this lever system is that at its insertion the muscle must exert a force seven times greater than the load. The product of the length of the power arm times the upward force applied must equal the product of the length of the load arm times the downward force exerted by the load. Because the load arm times the downward force is 35 cm × 5 kg, the power arm times the upward force must be 5 cm × 35 kg (the force that must be exerted by the muscle to be in mechanical equilibrium). Thus skeletal muscles typically work at a mechanical disadvantage in that they must exert a considerably greater force than the actual load to be moved. Nevertheless, the amplification of velocity and distance afforded by the lever arrangement enables muscles to move loads faster over greater distances than would otherwise be possible. This amplification provides valuable maneuverability and speed.

We are now going to shift our attention from muscle mechanics to the metabolic means by which muscles power these movements.

● **FIGURE 8-22**

Lever systems of muscles, bones, and joints
(a) Schematic representation of the most common type of lever system in the body, showing the location of the fulcrum, the upward force, the downward force, the power arm, and the load arm. (b) Flexion of the elbow joint as an example of lever action in the body. Note that the lever ratio (length of the power arm to length of the load arm) is 1:7 (5 cm:35 cm), which amplifies the distance and velocity of movement seven times [distance moved by the muscle (extent of shortening) = 1 cm, distance moved by the hand = 7 cm; velocity of muscle shortening = 1 cm/unit of time, hand velocity = 7 cm/unit of time], but at the expense of the muscle having to exert seven times the force of the load (muscle force = 35 kg, load = 5 kg).

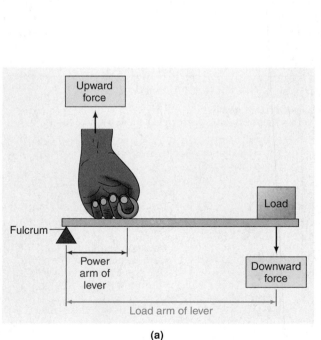

(a)

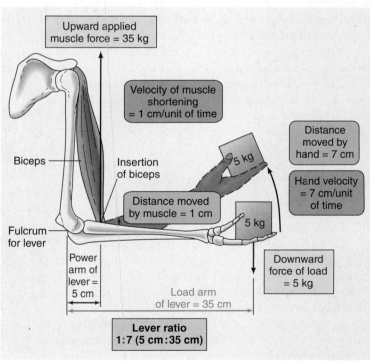

(b)

SKELETAL MUSCLE METABOLISM AND FIBER TYPES

Three different steps in the contraction–relaxation process require ATP:

1. Splitting of ATP by myosin ATPase provides the energy for the power stroke of the cross bridge.
2. Binding (but not splitting) of a fresh molecule of ATP to myosin permits detachment of the bridge from the actin filament at the end of a power stroke so that the cycle can be repeated. This ATP is subsequently split to provide energy for the next stroke of the cross bridge.
3. The active transport of Ca^{2+} back into the sarcoplasmic reticulum during relaxation depends on energy derived from the breakdown of ATP.

▋ Muscle fibers have alternate pathways for forming ATP.

Because ATP is the only energy source that can be directly used for these activities, ATP must constantly be supplied for contractile activity to continue. Only limited stores of ATP are immediately available in muscle tissue, but three pathways supply additional ATP as needed during muscle contraction: (1) transfer of a high-energy phosphate from creatine phosphate to ADP, (2) oxidative phosphorylation (the citric acid cycle and electron transport system), and (3) glycolysis.

Creatine phosphate

Creatine phosphate is the first energy storehouse tapped at the onset of contractile activity (● Figure 8-23 ⓐ). Like ATP, creatine phosphate contains a high-energy phosphate group, which can be donated directly to ADP to form ATP. Just as energy is released when the terminal phosphate bond in ATP is split, similarly energy is released when the bond between phosphate and creatine is broken. The energy released from the hydrolysis of creatine phosphate, along with the phosphate, can be donated directly to ADP to form ATP. This reaction, which is catalyzed by the muscle-cell enzyme **creatine kinase,** is reversible; energy and phosphate from ATP can be transferred to creatine to form creatine phosphate:

$$\text{Creatine phosphate} + \text{ADP} \underset{}{\overset{\text{creatine kinase}}{\rightleftharpoons}} \text{creatine} + \text{ATP}$$

As the energy reserves are built up in a resting muscle, the increased concentration of ATP favors the transfer of the high-energy phosphate group to creatine phosphate, in accordance with the law of mass action (see p. 488). A rested muscle contains about five times as much creatine phosphate as ATP. Thus, most energy is stored in muscle in creatine phosphate pools. At the onset of contraction, when the meager reserves of ATP are rapidly used, the reaction is reversed. Additional ATP is quickly formed by the transfer of energy and phosphate from creatine phosphate to ADP. Because only one enzymatic reaction is involved in this energy transfer, ATP can be formed rapidly (within a fraction of a second) by using creatine phosphate.

Thus creatine phosphate is the first source for supplying additional ATP when exercise begins. Muscle ATP levels actually remain fairly constant early in contraction, but creatine phosphate stores become depleted. In fact, short bursts of high-intensity contractile effort, such as high jumps, sprints, or weight lifting, are supported primarily by ATP derived at the expense of creatine phosphate. Other energy systems do not have a chance to become operable before the activity is over. Creatine phosphate stores typically power the first minute or less of exercise.

Some athletes hoping to gain a competitive edge take oral creatine supplements to boost their performance in short-term, high-intensity activities lasting less than a minute. (We naturally get creatine in our diets, especially in meat.) Loading the muscles with extra creatine means larger creatine phosphate stores—that is, larger energy stores that can translate into a small edge in performance of activities requiring short, explosive bursts of energy. Yet creatine supplements should be used with caution, because the long-term health effects are unknown. Also, extra creatine stores are of no use in activities of longer duration that rely on more long-term energy-supplying mechanisms.

Oxidative phosphorylation

If the energy-dependent contractile activity is to be continued, the muscle shifts to the alternate pathways of oxidative phosphorylation and glycolysis to form ATP. These multistepped pathways require time to pick up their rates of ATP formation to match the increased demands for energy, time that has been provided by the immediate supply of energy from the one-step creatine phosphate system.

Oxidative phosphorylation takes place within the muscle mitochondria if sufficient O_2 is present. Oxygen is required to support the mitochondrial electron-transport chain, which efficiently harnesses energy captured from the breakdown of nutrient molecules and uses it to generate ATP (see p. 39). This pathway is fueled by glucose or fatty acids, depending on the intensity and duration of the activity (● Figure 8-23 ⓑ). Although it provides a rich yield of 36 ATP molecules for each glucose molecule processed, oxidative phosphorylation is relatively slow because of the number of steps involved.

During light exercise (such as walking) to moderate exercise (such as jogging or swimming), muscle cells are able to form enough ATP through oxidative phosphorylation to keep pace with the modest energy demands of the contractile machinery for prolonged periods of time. To sustain ongoing oxidative phosphorylation, the exercising muscles depend on delivery of adequate O_2 and nutrients to maintain their activity. Activity that can be supported in this way is known as **endurance-type exercise,** or **aerobic** ("with O_2") **exercise.**

The O_2 required for oxidative phosphorylation is primarily delivered by the blood. Increased O_2 is made available to muscles during exercise by several mechanisms: Deeper, more rapid breathing brings in more O_2; the heart contracts more rapidly and forcefully to pump more oxygenated blood to the tissues; more blood is diverted to the exercising muscles by dilation of the blood vessels supplying them; and the hemoglobin molecules that carry the O_2 in the blood release more O_2 in exercising muscles. (These mechanisms are discussed further in later chapters.) Furthermore, some types of muscle fibers have an abundance of **myoglobin,** which is similar to

hemoglobin. Myoglobin can store small amounts of O_2, but more importantly, it increases the rate of O_2 transfer from the blood into muscle fibers.

Glucose and fatty acids, ultimately derived from ingested food, are also delivered to the muscle cells by the blood. In ad-

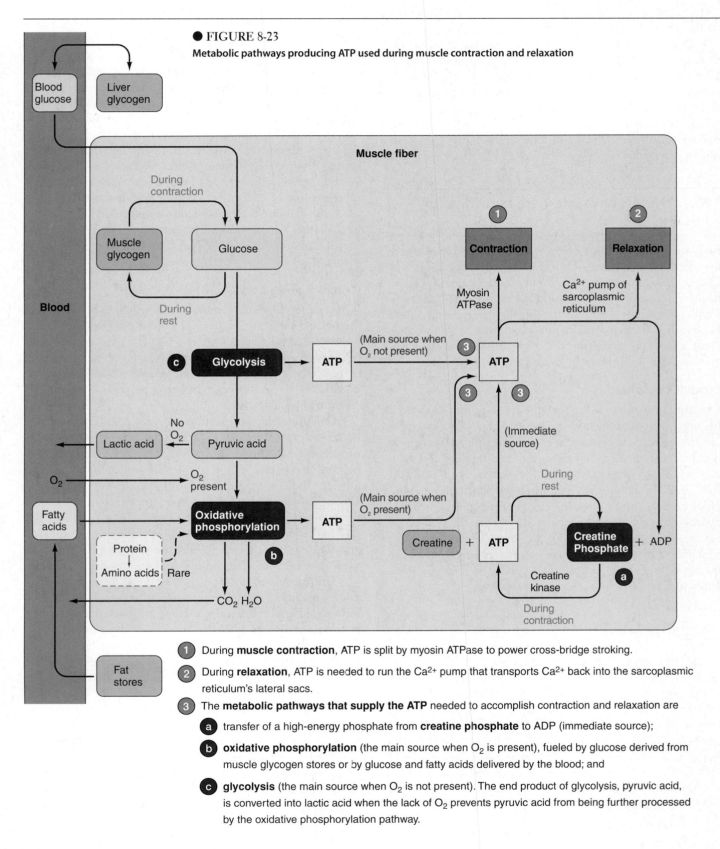

● **FIGURE 8-23**

Metabolic pathways producing ATP used during muscle contraction and relaxation

① During **muscle contraction**, ATP is split by myosin ATPase to power cross-bridge stroking.

② During **relaxation**, ATP is needed to run the Ca^{2+} pump that transports Ca^{2+} back into the sarcoplasmic reticulum's lateral sacs.

③ The **metabolic pathways that supply the ATP** needed to accomplish contraction and relaxation are

  ⓐ transfer of a high-energy phosphate from **creatine phosphate** to ADP (immediate source);

  ⓑ **oxidative phosphorylation** (the main source when O_2 is present), fueled by glucose derived from muscle glycogen stores or by glucose and fatty acids delivered by the blood; and

  ⓒ **glycolysis** (the main source when O_2 is not present). The end product of glycolysis, pyruvic acid, is converted into lactic acid when the lack of O_2 prevents pyruvic acid from being further processed by the oxidative phosphorylation pathway.

dition, muscle cells are able to store limited quantities of glucose in the form of glycogen (chains of glucose). Furthermore, up to a point the liver can store excess ingested carbohydrates as glycogen, which can be broken down to release glucose into the blood for use between meals. *Carbohydrate loading* — increasing carbohydrate intake prior to a competition — is a tactic used by some athletes in the hopes of boosting performance in endurance events such as marathons. However, once the muscle and liver glycogen stores are filled, excess ingested carbohydrates (or any other energy-rich nutrient) are converted to body fat.

Glycolysis

There are cardiovascular limits to the amount of O_2 that can be delivered to a muscle. In near-maximal contractions, the blood vessels that course through the muscle are compressed almost closed by the powerful contraction, severely limiting the O_2 available to the muscle fibers. Furthermore, even when O_2 is available, the relatively slow oxidative-phosphorylation system may not be able to produce ATP rapidly enough to meet the muscle's needs during intense activity. A skeletal muscle's energy consumption may increase up to 100-fold when going from rest to high-intensity exercise. When O_2 delivery or oxidative phosphorylation cannot keep pace with the demand for ATP formation as the intensity of exercise increases, the muscle fibers rely increasingly on glycolysis to generate ATP (● Figure 8-23 c) (see p. 35). The chemical reactions of **glycolysis** yield products for ultimate entry into the oxidative phosphorylation pathway, but glycolysis can also proceed alone in the absence of further processing of its products by oxidative phosphorylation. During glycolysis, a glucose molecule is broken down into two **pyruvic acid** molecules, yielding two ATP molecules in the process. Pyruvic acid can be further degraded by oxidative phosphorylation to extract more energy. However, glycolysis alone has two advantages over the oxidative phosphorylation pathway: (1) glycolysis can form ATP in the absence of O_2 (operating *anaerobically*, that is, "without O_2"), and (2) it can proceed more rapidly than oxidative phosphorylation. Although glycolysis extracts considerably fewer ATP molecules from each nutrient molecule processed, it can proceed so much more rapidly that it can outproduce oxidative phosphorylation over a given period of time if sufficient glucose is present.

Lactic acid production

Even though anaerobic glycolysis provides a means of performing intense exercise when the O_2 delivery/oxidative phosphorylation capacity is exceeded, using this pathway has two consequences. First, large amounts of nutrient fuel must be processed, because glycolysis is much less efficient than oxidative phosphorylation in converting nutrient energy into the energy of ATP. (Glycolysis yields a net of 2 ATP molecules for each glucose molecule degraded, whereas the oxidative phosphorylation pathway can extract 36 molecules of ATP from each glucose molecule.) Muscle cells can store limited quantities of glucose in the form of glycogen, but anaerobic glycolysis rapidly depletes the muscle's glycogen supplies. Second, the end product of anaerobic glycolysis, pyruvic acid, is converted to **lactic acid** when it cannot be further processed by the oxidative phosphorylation pathway. Lactic acid accumulation has been implicated in the muscle soreness that occurs during the time that intense exercise is actually taking place. (The delayed-onset pain and stiffness that begin the day after unaccustomed muscular exertion, however, are probably caused by reversible structural damage.) Furthermore, lactic acid picked up by the blood is responsible for the metabolic acidosis accompanying intense exercise. Both depletion of energy reserves and the fall in muscle pH caused by lactic acid accumulation are believed to play a role in the onset of muscle fatigue, a topic to which we turn our attention in the next section. Therefore, **high-intensity exercise,** or **anaerobic exercise,** can be sustained for only a short duration, in contrast to the body's prolonged ability to sustain aerobic activities.

▮ Fatigue may be of muscle or central origin.

Contractile activity in a particular skeletal muscle cannot be maintained at a given level indefinitely. Eventually, the tension in the muscle declines as fatigue sets in. There are two different types of fatigue: muscle fatigue and central fatigue.

Muscle fatigue occurs when an exercising muscle can no longer respond to stimulation with the same degree of contractile activity. The underlying causes of muscle fatigue are unclear. The primary implicated factors include (1) a local increase in inorganic phosphate resulting from the breakdown of creatine phosphate; (2) accumulation of lactic acid, which may inhibit key enzymes in the energy-producing pathways or excitation–contraction coupling process; and (3) depletion of energy reserves. The time of onset of fatigue varies with the type of muscle fiber, some fibers being more resistant to fatigue than others, and with the intensity of the exercise, more rapid onset of fatigue being associated with high-intensity activities.

Central fatigue occurs when the CNS no longer adequately activates the motor neurons supplying the working muscles. The person slows down or stops exercising even though the muscles are still able to perform. Central fatigue often is psychologically based. During strenuous exercise, central fatigue may stem from discomfort associated with the activity; it takes strong motivation (a will to win) to deliberately persevere when in pain. In less strenuous activities, central fatigue may reduce physical performance in association with boredom and monotony (such as assembly line work) or tiredness (lack of sleep). The mechanisms involved in central fatigue are poorly understood. In some cases, central fatigue may stem from biochemical insufficiencies within the brain.

Neuromuscular fatigue — the inability of active motor neurons to synthesize acetylcholine rapidly enough to sustain chemical transmission of action potentials from the motor neurons to the muscles — can be produced experimentally but does not occur under normal physiological conditions. Even though ACh secretion may not be able to keep pace with the rate of stimulation during fast, powerful activities, the magnitude of the somewhat reduced end-plate potential is still sufficient to generate action potentials in the muscle cell membrane (see

p. 247). Only individuals with marginal neuromuscular activity, such as victims of myasthenia gravis, can experience neuromuscular fatigue (see p. 252).

▌Increased oxygen consumption is necessary to recover from exercise.

A person continues to breathe deeply and rapidly for a period of time after exercising. The necessity for the elevated O_2 uptake during recovery from exercise is due to a variety of factors. The best known is repayment of an **oxygen debt** that was incurred during exercise, when contractile activity was being supported by ATP derived from nonoxidative sources such as creatine phosphate and anaerobic glycolysis. Oxygen is needed for recovery of the energy systems. During exercise, the creatine phosphate stores of active muscles are reduced, lactic acid may accumulate, and glycogen stores may be tapped; the extent of these effects depends on the intensity and duration of the activity. During the recovery period, fresh supplies of ATP are formed by oxidative phosphorylation using the newly acquired O_2, which is provided by the sustained increase in respiratory activity after exercise has stopped. Most of this ATP is used to resynthesize creatine phosphate to restore its reserves. This can be accomplished in a matter of a few minutes. Any accumulated lactic acid is converted back into pyruvic acid, part of which is used by the oxidative phosphorylation system for ATP production. The remainder of the pyruvic acid is converted back into glucose by the liver. Most of this glucose in turn is used to replenish the glycogen stores drained from the muscles and liver during exercise. These biochemical transformations involving pyruvic acid require O_2 and take several hours for completion. Thus, as the O_2 debt is repaid, the creatine phosphate system is restored, lactic acid is removed, and glycogen stores are at least partially replenished.

Unrelated to increased O_2 uptake is the need to restore nutrients after grueling exercise, such as marathon races, in which glycogen stores are severely depleted. In such cases, long-term recovery can take a day or more, because the exhausted energy stores require nutrient intake for full replenishment. Therefore, depending on the type and duration of activity, recovery can be complete within a few minutes or can require more than a day.

Part of the extra O_2 uptake during recovery is not directly related to the repayment of energy stores but instead is the result of a general metabolic disturbance following exercise. For example, the local increase in muscle temperature arising from heat-generating contractile activity speeds up the rate of all chemical reactions in the muscle tissue, including those dependent on O_2. Likewise, the secretion of epinephrine, a hormone that increases O_2 consumption by the body, is elevated during exercise. Until the circulating level of epinephrine returns to its pre-exercise state, O_2 uptake is increased above normal.

We have been looking at the contractile and metabolic activities of skeletal muscle fibers in general. Yet not all skeletal muscle fibers use these mechanisms to the same extent. We are next going to examine the different types of muscle fibers

based on their speed of contraction and how they are metabolically equipped to generate ATP.

▌There are three types of skeletal muscle fibers, based on differences in ATP hydrolysis and synthesis.

Classified by their biochemical capacities, there are three major types of muscle fibers (▲ Table 8-3):

1. **Slow-oxidative (type I) fibers**
2. **Fast-oxidative (type IIa) fibers**
3. **Fast-glycolytic (type IIb**, also known as **Type IIx) fibers**

As their names imply, the two main differences among these fiber types are their speed of contraction (slow or fast) and the type of enzymatic machinery they primarily use for ATP formation (oxidative or glycolytic).

Fast versus slow fibers

Fast fibers have higher myosin ATPase (ATP-splitting) activity than slow fibers. The higher the ATPase activity, the more rapidly ATP is split and the faster the rate at which energy is made available for cross-bridge cycling. The result is a fast twitch, compared to the slower twitches of those fibers that split ATP more slowly. Thus two factors determine the speed

▲ **TABLE 8-3**
Characteristics of Skeletal Muscle Fibers

Characteristic	Type of Fiber		
	Slow-oxidative (Type I)	Fast-oxidative (Type IIa)	Fast-glycolytic (Type IIb)
Myosin-ATPase Activity	Low	High	High
Speed of Contraction	Slow	Fast	Fast
Resistance to Fatigue	High	Intermediate	Low
Oxidative Phosphorylation Capacity	High	High	Low
Enzymes for Anaerobic Glycolysis	Low	Intermediate	High
Mitochondria	Many	Many	Few
Capillaries	Many	Many	Few
Myoglobin Content	High	High	Low
Color of Fiber	Red	Red	White
Glycogen Content	Low	Intermediate	High

with which a muscle contracts: the load (load–velocity relationship) and the myosin ATPase activity of the contracting fibers (fast or slow twitch).

Oxidative versus glycolytic fibers

Fiber types also differ in ATP-synthesizing ability. Those with a greater capacity to form ATP are more resistant to fatigue. Some fibers are better equipped for oxidative phosphorylation, whereas others rely primarily on anaerobic glycolysis for synthesizing ATP. Because oxidative phosphorylation yields considerably more ATP from each nutrient molecule processed, it does not readily deplete energy stores. Furthermore, it does not result in lactic acid accumulation. Oxidative types of muscle fibers are therefore more resistant to fatigue than are glycolytic fibers.

Other related characteristics distinguishing these three fiber types are summarized in ▲ Table 8-3. As would be expected, the oxidative fibers, both slow and fast, contain an abundance of mitochondria, the organelles that house the enzymes involved in the oxidative phosphorylation pathway. Because adequate oxygenation is essential to support this pathway, these fibers are richly supplied with capillaries. Oxidative fibers also have a high myoglobin content. Myoglobin not only helps support oxidative fibers' O_2 dependency, but it also imparts a red color to them, just as oxygenated hemoglobin is responsible for the red color of arterial blood. Accordingly, these muscle fibers are referred to as **red fibers.**

In contrast, the fast fibers specialized for glycolysis contain few mitochondria but have a high content of glycolytic enzymes instead. Also, to supply the large amounts of glucose necessary for glycolysis, they contain an abundance of stored glycogen. Because the glycolytic fibers need relatively less O_2 to function, they receive only a meager capillary supply compared with the oxidative fibers. The glycolytic fibers contain very little myoglobin and therefore are pale in color, so they are sometimes called **white fibers.** (The most readily observable comparison between red and white fibers is the dark and white meat in poultry.)

Genetic endowment of muscle fiber types

In humans, most of the muscles contain a mixture of all three fiber types; the percentage of each type is largely determined by the type of activity for which the muscle is specialized. Accordingly, a high proportion of slow-oxidative fibers are found in muscles specialized for maintaining low-intensity contractions for long periods of time without fatigue, such as the muscles of the back and legs that support the body's weight against the force of gravity. A preponderance of fast-glycolytic fibers are found in the arm muscles, which are adapted for performing rapid, forceful movements such as lifting heavy objects.

The percentage of these various fibers not only differs between muscles within an individual but also varies considerably among individuals. Those genetically endowed with a higher percentage of the fast-glycolytic fibers are good candidates for power and sprint events, whereas those with a greater proportion of slow-oxidative fibers are more likely to be successful in endurance activities such as marathon races.

Of course, success in any event depends on many factors other than genetic endowment, such as the extent and type of training and the level of dedication. Indeed, the mechanical and metabolic capabilities of muscle fibers can be modified considerably in response to the patterns of demands placed on them. Let us see how.

▌ Muscle fibers adapt considerably in response to the demands placed on them.

Not only is the nerve supply to a skeletal muscle essential for initiating contraction, but the motor neurons supplying a skeletal muscle are also important in maintaining the muscle's integrity and chemical composition. Different types of exercise produce different patterns of neuronal discharge to the muscle involved. Depending on the pattern of neural activity, long-term adaptive changes occur in the muscle fibers, enabling them to respond most efficiently to the types of demands placed on the muscle. Two types of changes can be induced in muscle fibers: changes in their ATP-synthesizing capacity and changes in their diameter.

Improvement in oxidative capacity

Regular endurance (aerobic) exercise, such as long-distance jogging or swimming, induces metabolic changes within the oxidative fibers, which are the ones primarily recruited during aerobic exercise. These changes enable the muscles to use O_2 more efficiently. For example, mitochondria, the organelles that house the enzymes involved in the oxidative phosphorylation pathway, increase in number in the oxidative fibers. In addition, the number of capillaries supplying blood to these fibers increases. Muscles so adapted are better able to endure prolonged activity without fatiguing, but they do not change in size.

Muscle hypertrophy

The actual size of the muscles can be increased by regular bouts of anaerobic, short-duration, high-intensity resistance training, such as weight lifting. The resulting muscle enlargement comes primarily from an increase in diameter (**hypertrophy**) of the fast-glycolytic fibers that are called into play during such powerful contractions. Most of the fiber thickening results from increased synthesis of myosin and actin filaments, which permits a greater opportunity for cross-bridge interaction and subsequently increases the muscle's contractile strength. The resultant bulging muscles are better adapted to activities that require intense strength for brief periods, but endurance has not been improved.

Influence of testosterone

Men's muscle fibers are thicker, and accordingly, their muscles are larger and stronger than those of women, even without weight training, because of the actions of testosterone, a steroid hormone secreted primarily in males. Testosterone promotes the synthesis and assembly of myosin and actin. This

Are Athletes Who Use Steroids to Gain Competitive Advantage Really Winners or Losers?

The testing of athletes for drugs, and the much publicized exclusion from competition of those found to be using substances outlawed by sports federations, have stirred considerable controversy. One such group of drugs are **anabolic androgenic steroids** (*anabolic* means "buildup of tissues," *androgenic* means "male producing," and *steroids* are a class of hormone). These agents are closely related to testosterone, the natural male sex hormone, which is responsible for promoting the increased muscle mass characteristic of males.

Although their use is outlawed (possession of anabolic steroids without a prescription became a federal offense in 1991), these agents are taken by many athletes who specialize in power events such as weight lifting and sprinting in the hopes of increasing muscle mass and, accordingly, muscle strength. Both male and female athletes have resorted to using these substances in an attempt to gain a competitive edge. Anabolic steroids are also taken by bodybuilders. There are an estimated 1 million anabolic steroid abusers in the United States. Unfortunately, use of these agents has spread into our nation's high schools and even younger age groups. In a survey of 17,000 high school youths by the National Institute of Drug Abuse, 5% of the respondents reported current or past steroid use to "build muscles" or to "improve appearance." The manager of the steroid abuse hotline of the National Steroid Research Center reports having calls for help from abusers as young as 12 years old.

Studies have confirmed that steroids can increase muscle mass when used in large amounts and coupled with heavy exercise. One reputable study demonstrated an average 8.9-pound gain of lean muscle in bodybuilders who used steroids during a 10-week period. Anecdotal evidence suggests that some steroid users have added as much as 40 pounds of muscle in a year.

The adverse effects of these drugs, however, outweigh any benefits derived. In females, who normally lack potent androgenic hormones, anabolic steroid drugs not only promote "male-type" muscle mass and strength but also "masculinize" the users in other ways, such as by inducing growth of facial hair and by lowering the voice. More importantly, in both males and females, these agents adversely affect the reproductive and cardiovascular systems and the liver, may have an impact on behavior, and may be addictive.

Adverse Effects on the Reproductive System

In males, testosterone secretion and sperm production by the testes are normally controlled by hormones from the anterior pituitary gland. In negative-feedback fashion, testosterone inhibits secretion of these controlling hormones, so that a constant circulating level of testosterone is maintained. The anterior pituitary is similarly inhibited by androgenic steroids taken as a drug. As a result, because the testes do not receive their normal stimulatory input from the anterior pituitary, testosterone secretion and sperm production decrease and the testes shrink. This hormone abuse also may set the stage for cancer of the testes and prostate gland.

In females, inhibition of the anterior pituitary by the androgenic drugs represses the hormonal output that controls ovarian function. The result is failure to ovulate, menstrual irregularities, and decreased secretion of "feminizing" female sex hormones. Their decline results in diminution in breast size and other female characteristics.

Adverse Effects on the Cardiovascular System

Use of anabolic steroids induces several cardiovascular changes that increase the risk of developing atherosclerosis, which in turn is associated with an increased incidence of heart attacks and strokes (see p. 333). Among these adverse cardiovascular effects are (1) a reduction in high-density lipoproteins (HDL), the "good" cholesterol carriers that help remove cholesterol from the body, and (2) an elevation in blood pressure. Animal studies have also demonstrated damage to the heart muscle itself.

Adverse Effects on the Liver

Liver dysfunction is common with high steroid intake, because the liver, which normally inactivates steroid hormones and prepares them for urinary excretion, is overloaded by the excess steroid intake. The incidence of liver cancer is also increased.

Adverse Effects on Behavior

Although the evidence is still controversial, anabolic steroid use appears to promote aggressive, even hostile behavior—the so-called *roid rages*.

Addictive Effects

A troubling new concern is the addiction to anabolic steroids of some who abuse these drugs. In one study involving face-to-face interviews, 14% of steroid users were judged on the basis of their responses to be addicted. In another survey, using anonymous, self-administered questionnaires, 57% of steroid users qualified as being addicted. This apparent tendency to become chemically dependent on steroids is alarming because the potential for adverse effects on health increases with long-term, heavy use, the kind of use that would be expected in someone hooked on the drug.

Thus, for health reasons, without even taking into account the legal and ethical issues, people should not use anabolic steroids. However, the problem appears to be worsening. Currently, the international black market for anabolic steroids is estimated at $1 billion per year.

---

fact has led some athletes, both males and females, to the dangerous practice of taking this or closely related steroids to increase their athletic performance. (To explore this topic further, see the accompanying boxed feature, ❯ A Closer Look at Exercise Physiology.)

Interconversion between fast muscle fiber types

Apparently, all the muscle fibers within a single motor unit are of the same fiber type. This pattern usually is established early in life, but the two types of fast-twitch fibers are interconvertible, depending on training efforts. Regular endurance activities can convert fast-glycolytic fibers into fast-oxidative fibers, whereas fast-oxidative fibers can be shifted to fast-glycolytic fibers in response to power events such as weight training. Adaptive changes that take place in skeletal muscle gradually reverse to their original state over a period of months if the regular exercise program that induced these changes is discontinued.

Slow and fast fibers are not interconvertible, however. Although training can induce changes in muscle fibers' metabolic support systems, whether a fiber is fast or slow twitch depends on the fiber's nerve supply. Slow-twitch fibers are supplied by motor neurons that exhibit a low-frequency pattern of electrical activity, whereas fast-twitch fibers are innervated by motor neurons that display intermittent rapid bursts of electrical activity. Experimental switching of motor neurons supplying slow muscle fibers with those supplying fast fibers gradually reverses the speed at which these fibers contract.

Muscle atrophy

At the other extreme, if a muscle is not used, its actin and myosin content decreases, its fibers become smaller, and the muscle accordingly **atrophies** (decreases in mass) and becomes weaker. Muscle atrophy can result in two ways. **Disuse atrophy** occurs when a muscle is not used for a long period of time even though the nerve supply is intact, as when a cast or brace must be worn or during prolonged bed confinement. **Denervation atrophy** occurs after the nerve supply to a muscle is lost. If the muscle is stimulated electrically until innervation can be re-established, such as during regeneration of a severed peripheral nerve, atrophy can be diminished but not entirely prevented. Contractile activity itself obviously plays an important role in preventing atrophy; however, poorly understood factors released from active nerve endings, perhaps packaged with the ACh vesicles, apparently contribute to the integrity and growth of muscle tissue.

Limited repair of muscle

When a muscle is damaged, limited repair is possible, even though muscle cells cannot divide mitotically to replace damaged cells. A small population of **myoblasts**, the same undifferentiated stem cells that formed the muscle during embryonic development, remains close to the muscle surface (*myo* means "muscle"; *biast* means "former"). When a muscle fiber is damaged, it can be replaced by fusion of a group of these myoblasts to form a large, multinucleated cell, which immediately begins to synthesize and assemble the intracellular machinery characteristic of the muscle. With extensive injury, this limited mechanism is not adequate to completely replace all the lost fibers. In that case, the remaining fibers often hypertrophy to compensate.

Transplantation of myoblasts provides one of several glimmers of hope for victims of **muscular dystrophy,** a hereditary pathologic condition characterized by progressive degeneration of contractile elements, which are ultimately replaced by fibrous tissue. (See the accompanying boxed feature, ▶ Concepts, Challenges, and Controversies, for further information on this devastating condition.)

For the remaining section on skeletal muscle, we are going to examine the central and local mechanisms involved in regulating the motor activity performed by these muscles.

CONTROL OF MOTOR MOVEMENT

Particular patterns of motor unit output are responsible for motor activity, ranging from maintenance of posture and balance to stereotypical locomotor movements, such as walking, to individual, highly skilled motor activity, such as gymnastics. Control of any motor movement, no matter what its level of complexity, depends on converging input to the motor neurons of specific motor units. The motor neurons in turn trigger contraction of the muscle fibers within their respective motor

units by means of the events that occur at the neuromuscular junction.

▌ Multiple neural inputs influence motor unit output.

Three levels of input control motor neuron output (● Figure 8-24):

1. *Input from afferent neurons,* usually through intervening interneurons, at the level of the spinal cord—that is, spinal reflexes (see p. 177).

2. *Input from the primary motor cortex.* Fibers originating from cell bodies of pyramidal cells within the primary motor cortex (see p. 148) descend directly without synaptic interruption to terminate on motor neurons (or on local interneurons that terminate on motor neurons). These fibers make up the **corticospinal** (or **pyramidal**) **motor system.**

3. *Input from the brain stem* as part of the multineuronal motor system. The pathways composing the **multineuronal** (or **extrapyramidal**) **motor system** include a number of synapses that involve many regions of the brain (*extra* means "outside of;" *pyramidal* refers to the pyramidal system). The final link in multineuronal pathways is the brain stem, especially the reticular formation, which in turn is influenced by motor regions of the cortex, the cerebellum, and the basal nuclei. In addition, the motor cortex itself is interconnected with the thalamus as well as with premotor and supplementary motor areas, all part of the multineuronal system.

The only brain regions that directly influence motor neurons are the primary motor cortex and brain stem; the other involved brain regions indirectly regulate motor activity by adjusting motor output from the motor cortex and brain stem. A number of complex interactions take place between these various brain regions; the most important are represented in ● Figure 8-24. (See Chapter 5 for further discussion of the specific roles and interactions of these brain regions.)

Spinal reflexes involving afferent neurons are important in maintaining posture and in the execution of basic protective movements, such as the withdrawal reflex. The corticospinal system primarily mediates performance of fine, discrete, voluntary movements of the hands and fingers, such as those required for doing intricate needlework. Premotor and supplementary motor areas, with input from the cerebrocerebellum, plan the voluntary motor command that is issued to the appropriate motor neurons by the primary motor cortex through this descending system. The multineuronal system, in contrast, is primarily concerned with regulation of overall body posture involving involuntary movements of large muscle groups of the trunk and limbs. Considerable complex interaction and overlapping of function exist between the corticospinal and multineuronal systems. To voluntarily manipulate your fingers to do needlework, for example, you subconsciously assume a particular posture of your arms that enables you to hold your work.

Some of the inputs converging on motor neurons are excitatory, whereas others are inhibitory. Coordinated movement

Muscular Dystrophy: When One Small Step Is a Big Deal

Hope of treatment is on the horizon for **muscular dystrophy (MD),** a fatal muscle-wasting disease that primarily strikes boys and relentlessly leads to their death before age 20.

Symptoms

Muscular dystrophy encompasses a variety of hereditary pathologic conditions, which have in common a progressive degeneration of contractile elements and their replacement by fibrous tissue. The gradual muscle wasting is characterized by progressive weakness over a period of years. Typically, an MD patient begins to show symptoms of muscle weakness at about 2 to 3 years of age, becomes wheelchair bound when he is 10 to 12 years old, and dies within the next 10 years, either from respiratory failure when his respiratory muscles become too weak or from heart failure when his heart becomes too weak.

Cause

The disease is caused by a recessive genetic defect on the X sex chromosome, of which males have only one copy. (Males have XY sex chromosomes; females have XX sex chromosomes). If a male inherits from his mother an X chromosome bearing the defective dystrophic gene, he is destined to develop the disease, which affects one out of every 3,500 boys worldwide. To acquire the condition, females must inherit a dystrophic-carrying X gene from both parents, a much rarer occurrence.

The defective gene responsible for *Duchenne muscular dystrophy (DMD)*, the most common and most devastating form of the disease, was pinpointed in 1986. The gene normally produces **dystrophin,** a large protein that provides structural stability to the muscle cell's plasma membrane. Dystrophin is part of a complex of membrane-associated proteins that form a mechanical link between actin, a major component of the muscle cell's internal cytoskeleton, and the extracellular matrix, an external support network (see p. 62). This mechanical reinforcement of the plasma membrane enables the muscle cell to withstand the stresses and strains encountered during contraction and stretching.

Dystrophic muscles are characterized by a lack of dystrophin. Even though this protein represents only 0.002% of the total amount of skeletal muscle protein, its presence is crucial in maintaining the integrity of the muscle cell

membrane. The absence of dystrophin permits a constant leakage of Ca^{2+} into the muscle cells. This Ca^{2+} activates proteases, protein-snipping enzymes that harm the muscle fibers. The resultant damage leads to the muscle wasting and ultimate fibrosis that characterize the disorder.

With the discovery of the dystrophin gene and its deficiency in DMD came the hope that scientists could somehow replenish this missing protein in the muscles of the disease's young victims. Although the disease is still considered untreatable and fatal, several lines of research are being pursued vigorously to intervene in the relentless muscle loss.

Gene-therapy approach

One approach is a possible "gene fix." With gene therapy, healthy genes are usually delivered to the defective cells by means of viruses. Viruses operate by invading a body cell and micromanaging the cell's genetic machinery. In this way, the virus directs the host cell to synthesize the proteins needed for viral replication. With gene therapy, the desired gene is inserted into an incapacitated virus that cannot cause disease but can still enter the target cell and take over genetic commands.

One of the big challenges for gene therapy for DMD is the enormity of the dystrophin gene. This gene, being over 3 million base pairs long, is the largest gene ever found. It will not fit inside the viruses usually used to deliver genes to cells—they only have enough space for a gene one-thousandth the size of the dystrophin gene. Therefore, researchers have created a minigene that is a thousand times smaller than the dystrophin gene but still contains the essential components for directing the synthesis of dystrophin. This stripped-down minigene can fit inside the viral carrier. Injection of these agents has stopped and even reversed the progression of MD in experimental animals. Gene-therapy clinical trials in humans have not been completed.

Cell transplant approach

Another approach involves injecting myoblasts into the dystrophic muscles. *Myoblasts* are the undifferentiated cells that fuse to form the large, multinucleated skeletal-muscle cells during embryonic development. After development, a small group of these immature myoblasts remains close to the muscle surface. Myoblasts can fuse together to form a new skeletal-muscle cell to replace a damaged cell. When the loss of mus-

cle cells is extensive, however, as in MD, this limited mechanism is not adequate to replace all the lost fibers.

One therapeutic approach for MD under study involves the transplantation of dystrophin-producing myoblasts harvested from muscle biopsies of healthy donors into the patient's dwindling muscles. The hope is that the healthy myoblasts will fuse with the patient's dystrophin-deficient muscle fibers and produce the missing dystrophin. Although proponents of this technique report positive outcomes, most researchers in the field remain skeptical of what they believe are exaggerated and misleading claims of its success.

Other investigators are hopeful that some day it will be possible to program recently discovered stem cells into muscle cells for transplant into patients who have lost muscle (see p. 8).

Utrophin approach

An alternative strategy that holds considerable promise for treating MD is upregulation of **utrophin,** a naturally occurring protein in muscle that is closely related to dystrophin. Eighty percent of the amino acid sequence for dystrophin and utrophin is identical, but these two proteins normally have different functions. Whereas dystrophin is dispersed throughout the muscle cell's surface membrane, where it contributes to the membrane's structural stability, utrophin is concentrated at the motor end plate. Here, utrophin plays a role in anchoring the acetylcholine receptors.

When researchers genetically engineered dystrophin-deficient mice that produced extra amounts of utrophin, this utrophin upregulation compensated in large part for the absent dystrophin; that is, the additional utrophin dispersed throughout the muscle-cell membrane, where it assumed dystrophin's responsibilities. The result was improved intracellular Ca^{2+} homeostasis, enhanced muscle strength, and a marked reduction in the microscopic signs of muscle degeneration. Researchers are now scrambling to find a drug that will entice muscle cells to overproduce utrophin in humans, in the hopes of preventing or even repairing the muscle wasting that characterizes this devastating condition.

These steps toward an eventual treatment mean that hopefully one day the afflicted boys will be able to take steps on their own instead of being destined to wheelchairs and early death.

---

depends on an appropriate balance of activity in these inputs. The following types of motor abnormalities result from defective motor control:

• If an inhibitory system originating in the brain stem is disrupted, muscles become hyperactive because of the unopposed activity in excitatory inputs to motor neurons. This con-

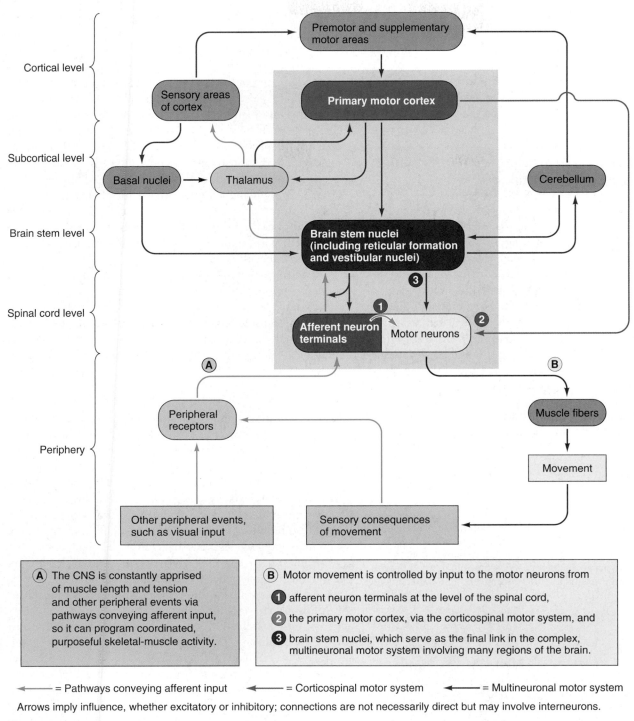

The figure contains the following labels:

Cortical level

Premotor and supplementary motor areas

Sensory areas of cortex

Primary motor cortex

Subcortical level

Basal nuclei → Thalamus

Cerebellum

Brain stem level

Brain stem nuclei (including reticular formation and vestibular nuclei)

❸

Spinal cord level

Afferent neuron terminals ❶ Motor neurons ❷

Ⓐ

Ⓑ

Peripheral receptors

Muscle fibers

Periphery

Movement

Other peripheral events, such as visual input

Sensory consequences of movement

Ⓐ The CNS is constantly apprised of muscle length and tension and other peripheral events via pathways conveying afferent input, so it can program coordinated, purposeful skeletal-muscle activity.

Ⓑ Motor movement is controlled by input to the motor neurons from

❶ afferent neuron terminals at the level of the spinal cord,

❷ the primary motor cortex, via the corticospinal motor system, and

❸ brain stem nuclei, which serve as the final link in the complex, multineuronal motor system involving many regions of the brain.

◄── = Pathways conveying afferent input   ◄── = Corticospinal motor system   ◄── = Multineuronal motor system

Arrows imply influence, whether excitatory or inhibitory; connections are not necessarily direct but may involve interneurons.

● **FIGURE 8-24**
Motor control

dition, characterized by increased muscle tone and augmented limb reflexes, is known as **spastic paralysis.**

• In contrast, loss of excitatory input, such as that accompanying destruction of descending excitatory pathways exiting the primary motor cortex, brings about **flaccid paralysis.** In this condition, the muscles are relaxed and the person cannot voluntarily contract muscles, although spinal reflex activity is still present. Damage to the primary motor cortex on one side of the brain, as with a stroke, leads to flaccid paralysis on the opposite half of the body (**hemiplegia,** or paralysis of one side of the body). Disruption of all descending pathways, as in traumatic severance of the spinal cord, is accompanied by flaccid paralysis below the level of the damaged region—**quadriplegia** (paralysis of all four limbs) in upper spinal cord damage and **paraplegia** (paralysis of the legs) in lower spinal cord injury.

• Destruction of motor neurons—either their cell bodies or efferent fibers—causes flaccid paralysis and lack of reflex responsiveness in the affected muscles.

- Damage to the cerebellum or basal nuclei does not result in paralysis but instead in uncoordinated, clumsy activity and inappropriate patterns of movement. These regions normally smooth out activity initiated voluntarily.

- Damage to higher cortical regions involved in planning motor activity results in the inability to establish appropriate motor commands to accomplish desired goals.

■ Muscle receptors provide afferent information needed to control skeletal muscle activity.

Coordinated, purposeful skeletal muscle activity depends on afferent input from a variety of sources. At a simple level, afferent signals indicating that your finger is touching a hot stove trigger reflex contractile activity in appropriate arm muscles to withdraw the hand from the injurious stimulus. At a more complex level, if you are going to catch a ball, the motor systems of your brain must program sequential motor commands that will move and position your body correctly for the catch, using predictions of the ball's direction and rate of movement provided by visual input. Many muscles acting simultaneously or alternately at different joints are called into play to shift your body's location and position rapidly, while maintaining your balance in the process. It is critical to have ongoing input about your body position with respect to the surrounding environment, as well as the position of your various body parts in relationship

to each other. This information is necessary for establishing a neuronal pattern of activity to perform the desired movement. To appropriately program muscle activity, your CNS must know the starting position of your body. Further, it must be constantly apprised of the progression of movement it has initiated so that it can make adjustments as needed. Your brain receives this information, which is known as *proprioceptive input* (see p. 148), from receptors in your eyes, joints, vestibular apparatus, and skin, as well as from the muscles themselves.

You can demonstrate your joint and muscle proprioceptive receptors in action by closing your eyes and bringing the tips of your right and left index fingers together at any point in space. You can do so without seeing where your hands are because your brain is informed of the position of your hands and other body parts at all times by afferent input from the joint and muscle receptors.

Two types of muscle receptors—*muscle spindles* and *Golgi tendon organs*—monitor changes in muscle length and tension. Muscle length is monitored by muscle spindles, whereas changes in muscle tension are detected by Golgi tendon organs. Both these receptor types are activated by muscle stretch, but they are designed to convey different types of information. Let us see how.

Muscle spindle structure

Muscle spindles, which are distributed throughout the fleshy part of a skeletal muscle, consist of collections of specialized muscle fibers known as **intrafusal fibers,** which lie within spindle-shaped connective tissue capsules parallel to the "ordinary" **extrafusal fibers** (*fusus* means "spindle") (● Figure 8-25). Unlike an ordinary extrafusal skeletal muscle fiber, which contains contractile elements (myofibrils) throughout its entire length, an intrafusal fiber has a noncontractile central portion, with the contractile elements being limited to both ends.

Each muscle spindle has its own private efferent and afferent nerve supply. The efferent neuron that innervates a muscle spindle's intrafusal fibers is known as a **gamma motor neuron,** whereas the motor neurons that supply the ordinary extrafusal fibers are designated as **alpha motor neurons.** Two types of afferent sensory endings terminate on the intrafusal fibers and serve as muscle spindle receptors, both of which are activated by stretch. The **primary (annulospiral) endings** are wrapped around the central portion of the intrafusal fibers; they detect changes in the length of the fibers during stretching as well as the speed with which it occurs. The **secondary (flower-spray) endings,** which are clustered at the end segments of many of the intrafusal fibers, are sensitive only to changes in length. Muscle spindles play a key role in the stretch reflex.

Stretch reflex

Whenever the whole muscle is passively stretched, the intrafusal fibers within its

● **FIGURE 8-25**

Muscle spindle
A muscle spindle consists of a collection of specialized intrafusal fibers that lie within a connective tissue capsule parallel to the ordinary extrafusal skeletal-muscle fibers. The muscle spindle is innervated by its own gamma motor neuron and is supplied by two types of afferent sensory terminals, the primary (annulospiral) endings and the secondary (flower-spray) endings, both of which are activated by stretch.

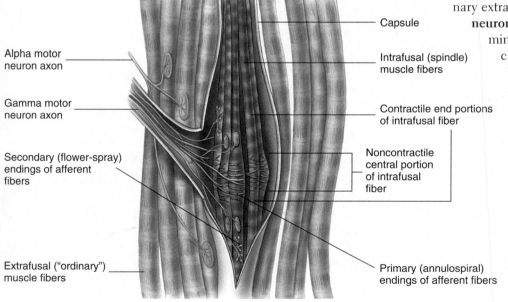

Alpha motor neuron axon

Gamma motor neuron axon

Secondary (flower-spray) endings of afferent fibers

Extrafusal ("ordinary") muscle fibers

Capsule

Intrafusal (spindle) muscle fibers

Contractile end portions of intrafusal fiber

Noncontractile central portion of intrafusal fiber

Primary (annulospiral) endings of afferent fibers

muscle spindles are likewise stretched, increasing the rate of firing in the afferent nerve fibers whose sensory endings terminate on the stretched spindle fibers. The afferent neuron directly synapses on the alpha motor neuron that innervates the extrafusal fibers of the same muscle, resulting in contraction of that muscle (● Figure 8-26a, pathway ① → ②). This **stretch reflex** serves as a local negative-feedback mechanism to resist any passive changes in muscle length so that optimal resting length can be maintained.

The classic example of the stretch reflex is the **patellar tendon,** or **knee-jerk, reflex** (● Figure 8-27). The extensor muscle of the knee is the *quadriceps femoris,* which forms the anterior (front) portion of the thigh and is attached just below the knee to the tibia (shinbone) by the *patellar tendon.* Tapping this tendon with a rubber mallet passively stretches the quadriceps muscle, activating its spindle receptors. The resultant stretch reflex brings about contraction of this extensor muscle, causing the knee to extend and raise the foreleg in the well-known knee-jerk fashion. This test is routinely performed as a preliminary assessment of nervous system function. A normal knee jerk indicates to a physician that a number of neural and muscular components—muscle spindle, afferent input, motor neurons, efferent output, neuromuscular junctions, and the muscles themselves—are functioning normally. It also indicates the presence of an appropriate balance of excitatory and inhibitory input to the motor neurons from higher brain levels. Muscle jerks may be absent or depressed with loss of higher-level excitatory inputs or may be greatly exaggerated with loss of inhibitory input to the motor neurons from higher brain levels.

The primary purpose of the stretch reflex is to resist the tendency for the passive stretch of extensor muscles caused by gravitational forces when a person is standing upright. Whenever the knee joint tends to buckle because of gravity, the quadriceps muscle is stretched. The resultant enhanced contraction of this extensor muscle brought about by the stretch reflex quickly straightens out the knee, holding the limb extended so that the person remains standing.

Coactivation of gamma and alpha motor neurons

Gamma motor neurons initiate contraction of the muscular end regions of intrafusal fibers (● Figure 8-26a, pathway ③). This contractile response is too weak to have any influence on whole-muscle tension, but it does have an important localized effect on the muscle spindle itself. If there were no compensating mechanisms, shortening of the whole muscle by alpha motor-neuron stimulation of extrafusal fibers would cause

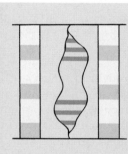
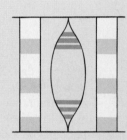

① Afferent input from sensory endings of muscle spindle fiber

② Alpha motor neuron output to regular skeletal-muscle fiber

① → ② Stretch reflex pathway

③ Gamma motor-neuron output to contractile end portions of spindle fiber

④ Descending pathways coactivating alpha and gamma motor neurons

(a)

(b) Relaxed muscle; spindle fiber sensitive to stretch of muscle

(c) Contracted muscle in hypothetical situation of no spindle coactivation; slackened spindle fiber not sensitive to stretch of muscle

(d) Contracted muscle in normal situation of spindle coactivation; contracted spindle fiber sensitive to stretch of muscle

● **FIGURE 8-26**

Muscle spindle function

(a) Pathways involved in the monosynaptic stretch reflex and coactivation of alpha and gamma motor neurons. (b) Status of a muscle spindle when the muscle is relaxed. (c) Status of a muscle spindle in the hypothetical situation of a muscle being contracted on alpha motor-neuron stimulation in the absence of spindle coactivation. (d) Status of a muscle spindle in the actual physiological situation when both the muscle and muscle spindle are contracted on alpha and gamma motor-neuron coactivation.

slack in the spindle fibers so that they would be less sensitive to stretch and therefore not as effective as muscle length detectors (● Figures 8-26b and c). **Coactivation** of the gamma motor-neuron system along with the alpha motor-neuron system during reflex and voluntary contractions (● Figure 8-26a, pathway ④) takes the slack out of the spindle fibers as the whole muscle shortens, permitting these receptor structures to

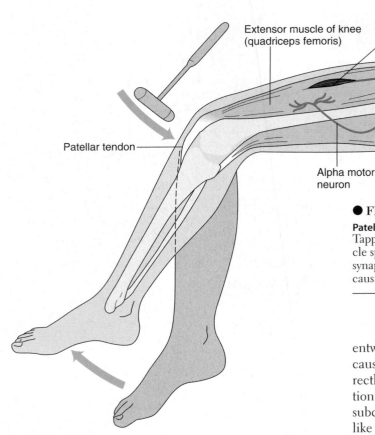

Extensor muscle of knee
(quadriceps femoris)

Muscle
spindle

Patellar tendon

Alpha motor
neuron

● **FIGURE 8-27**

Patellar tendon reflex (a stretch reflex)
Tapping the patellar tendon with a rubber mallet stretches the muscle spindles in the quadriceps femoris muscle. The resultant monosynaptic stretch reflex results in contraction of this extensor muscle, causing the characteristic knee-jerk response.

maintain their high sensitivity to stretch over a wide range of muscle lengths. When gamma motor-neuron stimulation triggers simultaneous contraction of both end muscular portions of an intrafusal fiber, the noncontractile central portion is pulled in opposite directions, tightening this region and taking out the slack (● Figure 8-26d). Whereas the extent of alpha motor-neuron activation depends on the intended strength of the motor response, the extent of simultaneous gamma motor-neuron activity to the same muscle depends on the anticipated distance of shortening.

Golgi tendon organs

In contrast to muscle spindles, which lie within the belly of the muscle, **Golgi tendon organs** are located in the tendons of the muscle, where they are able to respond to changes in the muscle's tension rather than to changes in its length. Because a number of factors determine the tension developed in the whole muscle during contraction (for example, frequency of stimulation or length of the muscle at the onset of contraction), it is essential that motor control systems be apprised of the tension actually achieved so that adjustments can be made if necessary.

The Golgi tendon organs consist of endings of afferent fibers entwined within bundles of connective tissue fibers that make up the tendon. When the extrafusal muscle fibers contract, the resultant pull on the tendon tightens the connective tissue bundles, which in turn increase the tension exerted on the bone to which the tendon is attached. In the process, the

entwined Golgi-organ afferent receptor endings are stretched, causing the afferent fibers to fire; the frequency of firing is directly related to the tension developed. This afferent information is sent to the brain for processing, much of which is used subconsciously for the smooth execution of motor activity. Unlike afferent information from the muscle spindles, afferent information from the Golgi tendon organ reaches the level of conscious awareness. You are aware of the tension within a muscle but you are not aware of its length.

Traditionally, the Golgi tendon organ was thought to trigger a protective spinal reflex that halted further contraction and brought about sudden reflex contraction when the muscle tension became great enough, thus helping prevent damage to the muscle or tendon from excessive, tension-developing muscle contractions. Scientists now believe, however, that this receptor is a pure sensor and does not initiate any reflexes. Other unknown mechanisms are apparently involved in inhibiting further contraction to prevent tension-induced damage.

Having completed our discussion of skeletal muscle, we are now going to examine smooth and cardiac muscle.

SMOOTH AND CARDIAC MUSCLE

The two other types of muscle—smooth muscle and cardiac muscle—share some basic properties with skeletal muscle, but each also displays unique characteristics (▲ Table 8-4). The three muscle types have several features in common. First, they all have a specialized contractile apparatus made up of thin actin filaments that slide relative to stationary thick myosin filaments in response to a rise in cytosolic Ca^{2+} to accomplish contraction. Second, they all directly use ATP as the energy source for cross-bridge cycling. However, the structure and organization of fibers within these different muscle types vary, as do their mechanisms of excitation and the means by which excitation and contraction are coupled. Furthermore, there are important distinctions in the contractile response itself. We will spend the remainder of this chapter highlighting unique features of smooth and cardiac muscle as compared with skeletal muscle, reserving a more detailed discussion of their function for chapters devoted to organs containing these muscle types.

Smooth muscle cells are small and unstriated.

The majority of smooth muscle cells are found in the walls of hollow organs and tubes. Their contraction exerts pressure on and regulates the forward movement of the contents of these structures.

Both smooth and skeletal muscle cells are elongated, but in contrast to their large, cylindrical, multinucleated skeletal-muscle counterparts, smooth muscle cells are spindle shaped, have a single nucleus, and are considerably smaller (2 to 10 μm in diameter and 50 to 400 μm long). Also unlike skeletal muscle cells, a single smooth-muscle cell does not extend the full length of a muscle. Instead, groups of smooth-muscle cells are typically arranged in sheets (● Figure 8-28a).

Three types of filaments are found in a smooth muscle cell: (1) thick myosin filaments, which are longer than those found in skeletal muscle; (2) thin actin filaments, which contain tropomyosin but lack the regulatory protein troponin; and (3) filaments of intermediate size, which do not directly participate in the contractile process but serve as part of the cytoskeletal framework that supports the shape of the cell. Smooth muscle filaments do not form myofibrils and are not arranged in the sarcomere pattern found in skeletal muscle. Thus smooth muscle cells do not display the banding or striation found in skeletal muscle, giving rise to the term *smooth* for this muscle type.

Lacking sarcomeres, smooth muscle does not have Z lines as such, but has **dense bodies** containing the same protein constituent found in Z lines (● Figure 8-28b). Dense bodies are positioned throughout the smooth muscle cell as well as being attached to the internal surface of the plasma membrane. The dense bodies are held in place by a scaffold of intermediate filaments. The actin filaments are anchored to the dense bodies. Considerably more actin is present in smooth muscle cells than in skeletal muscle cells, with 10 to 15 thin filaments for each thick myosin filament in smooth muscle compared to 2 thin filaments for each thick filament in skeletal muscle.

The thick- and thin-filament contractile units are oriented slightly diagonally from side to side within the smooth muscle cell in an elongated, diamond-shaped lattice, rather than running parallel with the long axis as myofibrils do in skeletal muscle (● Figure 8-29a, p. 292). Relative sliding of the thin filaments past the thick filaments during contraction causes the filament lattice to reduce in length and expand from side to side. As a result, the whole cell shortens and bulges out between the points where the thin filaments are attached to the inner surface of the plasma membrane (● Figure 8-29b).

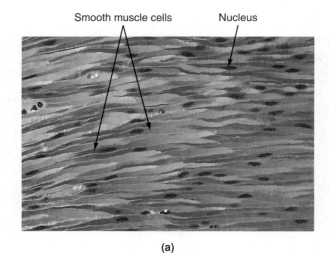

(a)

(b)

● **FIGURE 8-28**

Microscopic view of smooth muscle cells
(a) Low-power light micrograph of smooth muscle cells. Note the spindle shape and single, centrally located nucleus. (b) Electron micrograph of smooth muscle cells at 14,000× magnification. Note the presence of dense bodies and lack of banding.

Smooth muscle cells are turned on by Ca²⁺-dependent phosphorylation of myosin.

If the thin filaments of smooth muscle cells do not contain troponin, and tropomyosin does not block actin's cross-bridge binding sites, what prevents actin and myosin from binding at the cross bridges in the resting state, and how is cross-bridge activity switched on in the excited state? Smooth muscle myosin can interact with actin only when the myosin is *phosphorylated* (that is, has a phosphate group attached to it). During excitation, the increased cytosolic Ca^{2+} acts as an intracellular messenger, initiating a chain of biochemical events that re-

Comparison of Muscle Types

Characteristic	Type of Muscle			
	Skeletal	Multiunit smooth	Single-unit smooth	Cardiac
Location	Attached to skeleton	Large blood vessels, eye, and hair follicles	Walls of hollow organs in digestive, reproductive, and urinary tracts and in small blood vessels	Heart only
Function	Movement of body in relation to external environment	Varies with structure involved	Movement of contents within hollow organs	Pumps blood out of heart
Mechanism of Contraction	Sliding filament mechanism	Sliding filament mechanism	Sliding filament mechanism	Sliding filament mechanism
Innervation	Somatic nervous system (alpha motor neurons)	Autonomic nervous system	Autonomic nervous system	Autonomic nervous system
Level of Control	Under voluntary control; also subject to subconscious regulation	Under involuntary control	Under involuntary control	Under involuntary control
Initiation of Contraction	Neurogenic	Neurogenic	Myogenic (pacemaker potentials and slow-wave potentials)	Myogenic (pacemaker potentials)
Role of Nervous Stimulation	Initiates contraction; accomplishes gradation	Initiates contraction; contributes to gradation	Modifies contraction; can excite or inhibit; contributes to gradation	Modifies contraction; can excite or inhibit; contributes to gradation
Modifying Effect of Hormones	No	Yes	Yes	Yes
Presence of Thick Myosin and Thin Actin Filaments	Yes	Yes	Yes	Yes
Striated by Orderly Arrangement of Filaments	Yes	No	No	Yes
Presence of Troponin and Tropomyosin	Yes	Tropomyosin only	Tropomyosin only	Yes
Presence of T Tubules	Yes	No	No	Yes

(continued)

sults in phosphorylation of myosin (● Figure 8-30). Smooth muscle Ca^{2+} binds with **calmodulin,** an intracellular protein found in most cells that is structurally similar to troponin (see p. 69). This Ca^{2+}-calmodulin complex binds to and activates another protein, **myosin kinase,** which in turn phosphorylates myosin. Phosphorylated myosin then binds with actin so that cross-bridge cycling can begin. Thus smooth muscle is triggered to contract by a rise in cytosolic Ca^{2+}, similar to what happens in skeletal muscle. In smooth muscle, however, Ca^{2+} ultimately turns on the cross bridges by inducing a *chemical* change in myosin in the *thick* filaments, whereas in

skeletal muscle it exerts its effects by invoking a *physical* change at the *thin* filaments (● Figure 8-31). Recall that in skeletal muscle Ca^{2+} moves troponin and tropomyosin from their blocking position, so that actin and myosin are free to bind with each other.

The means by which excitation brings about an increase in cytosolic Ca^{2+} concentration in smooth muscle cells also differs from that for skeletal muscle. A smooth muscle cell has no T tubules and a poorly developed sarcoplasmic reticulum. The increased cytosolic Ca^{2+} that triggers the contractile response comes from two sources: Most Ca^{2+} enters from the

Comparison of Muscle Types *(continued)*

Characteristic	Type of Muscle			
	Skeletal	Multiunit smooth	Single-unit smooth	Cardiac
Level of Development of Sarcoplasmic Reticulum	Well developed	Poorly developed	Poorly developed	Moderately developed
Cross Bridges Turned on by Ca^{2+}	Yes	Yes	Yes	Yes
Source of Increased Cytosolic Ca^{2+}	Sarcoplasmic reticulum	Extracellular fluid and sarcoplasmic reticulum	Extracellular fluid and sarcoplasmic reticulum	Extracellular fluid and sarcoplasmic reticulum
Site of Ca^{2+} Regulation	Troponin in thin filaments	Myosin in thick filaments	Myosin in thick filaments	Troponin in thin filaments
Mechanism of Ca^{2+} Action	Physically repositions troponin-tropomyosin complex to uncover actin cross-bridge binding sites	Chemically brings about phosphorylation of myosin cross bridges so they can bind with actin	Chemically brings about phosphorylation of myosin cross bridges so they can bind with actin	Physically repositions troponin-tropomyosin complex
Presence of Gap Junctions	No	Yes (very few)	Yes	Yes
ATP Used Directly by Contractile Apparatus	Yes	Yes	Yes	Yes
Myosin ATPase Activity; Speed of Contraction	Fast or slow, depending on type of fiber	Very slow	Very slow	Slow
Means by Which Gradation Accomplished	Varying number of motor units contracting (motor unit recruitment) and frequency at which they're stimulated (twitch summation)	Varying number of muscle fibers contracting and varying cytosolic Ca^{2+} concentration in each fiber by autonomic and hormonal influences	Varying cytosolic Ca^{2+} concentration through myogenic activity and influences of the autonomic nervous system, hormones, mechanical stretch, and local metabolites	Varying length of fiber (depending on extent of filling of the heart chambers) and varying cytosolic Ca^{2+} concentration through autonomic, hormonal, and local metabolite influence
Presence of Tone in Absence of External Stimulation	No	No	Yes	No
Clear-cut Length–Tension Relationship	Yes	No	No	Yes

ECF, but some is released intracellularly from the sparse sarcoplasmic-reticulum stores. Unlike their role in skeletal muscle cells, voltage-gated dihydropyridine receptors in the plasma membrane of smooth muscle cells function as Ca^{2+} channels. When these surface-membrane channels are opened, Ca^{2+} enters down its concentration gradient from the ECF. The entering Ca^{2+} triggers the opening of Ca^{2+} channels in the sarcoplasmic reticulum, so that small additional amounts of Ca^{2+} are released from this meager source. Because smooth muscle cells are so much smaller in diameter than skeletal muscle fibers, the preponderance of Ca^{2+} entering from the ECF is able to influence cross-bridge activity, even in the central portions of the cell, without the necessity of an elaborate T tubule–sarcoplasmic reticulum mechanism.

Relaxation is accomplished by removal of Ca^{2+} as it is actively transported out across the plasma membrane and back into the sarcoplasmic reticulum. When Ca^{2+} is removed, myosin is dephosphorylated (the phosphate is removed) and can no longer interact with actin, so the muscle relaxes.

We still have not addressed the question of how smooth muscle becomes excited to contract; that is, what opens the Ca^{2+} channels in the plasma membrane? Smooth muscle is

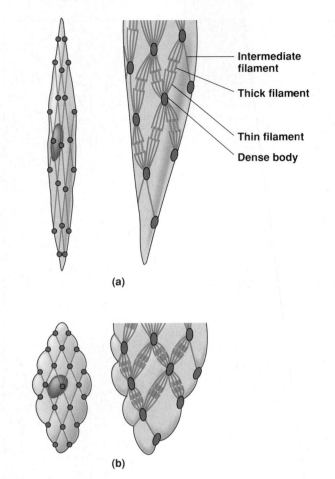

(a)

(b)

● FIGURE 8-29

Schematic representation of the arrangement of thick and thin filaments in a smooth muscle cell in contracted and relaxed states (a) Relaxed smooth muscle cell. (b) Contracted smooth muscle cell.

Intermediate filament
Thick filament
Thin filament
Dense body

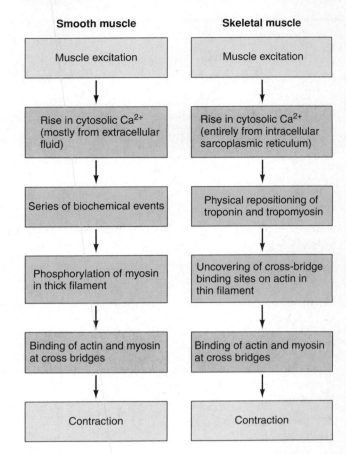

● FIGURE 8-31

Comparison of the role of calcium in bringing about contraction in smooth muscle and skeletal muscle

grouped into two categories—*multiunit* and *single-unit smooth muscle*—based on differences in how the muscle fibers become excited. Let's compare these two types of smooth muscle.

● FIGURE 8-30

Calcium activation of myosin in smooth muscle

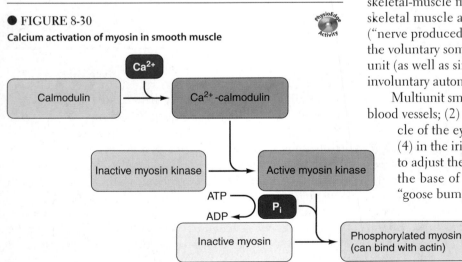

▌ Multiunit smooth muscle is neurogenic.

Multiunit smooth muscle exhibits properties partway between skeletal muscle and single-unit smooth muscle. As the name implies, a multiunit smooth muscle consists of multiple discrete units that function independently of each other and must be separately stimulated by nerves to contract, similar to skeletal-muscle motor units. Thus contractile activity in both skeletal muscle and multiunit smooth muscle is **neurogenic** ("nerve produced"). Whereas skeletal muscle is innervated by the voluntary somatic nervous system (motor neurons), multiunit (as well as single-unit) smooth muscle is supplied by the involuntary autonomic nervous system.

Multiunit smooth muscle is found (1) in the walls of large blood vessels; (2) in large airways to the lungs; (3) in the muscle of the eye that adjusts the lens for near or far vision; (4) in the iris of the eye, which alters the size of the pupil to adjust the amount of light entering the eye; and (5) at the base of hair follicles, contraction of which causes "goose bumps."

Single-unit smooth muscle cells form functional syncytia.

Most smooth muscle is of the **single-unit** variety. It is alternatively called **visceral smooth muscle,** because it is found in the walls of the hollow organs or viscera (for example, the digestive, reproductive, and urinary tracts and small blood vessels). The term "single-unit smooth muscle" derives from the fact that the muscle fibers that make up this type of muscle become excited and contract as a single unit. The muscle fibers in single-unit smooth muscle are electrically linked by gap junctions (see p. 63). When an action potential occurs anywhere within a sheet of single-unit smooth muscle, it is quickly propagated via these special points of electrical contact throughout the entire group of interconnected cells, which then contract as a single, coordinated unit. Such a group of interconnected muscle cells that function electrically and mechanically as a unit is known as a **functional syncytium** (plural, *syncytia; syn* means "together"; *cyt* means "cell").

Thinking about the role of the uterus during labor will help you appreciate the significance of this arrangement. Muscle cells composing the uterine wall act as a functional syncytium. They repetitively become excited and contract as a unit during labor, exerting a series of coordinated "pushes" that are eventually responsible for delivering the baby. Independent, uncoordinated contractions of individual muscle cells in the uterine wall would not exert the uniformly applied pressure needed to expel the baby. A similar situation applies for single-unit smooth muscle elsewhere in the body.

Single-unit smooth muscle is myogenic.

Single-unit smooth muscle is **self-excitable** rather than requiring nervous stimulation for contraction. Clusters of specialized smooth muscle cells within a functional syncytium display spontaneous electrical activity; that is, they are able to undergo action potentials without any external stimulation. In contrast to the other excitable cells we have been discussing (such as neurons, skeletal muscle fibers, and multiunit smooth muscle), the self-excitable cells of single-unit smooth muscle do not maintain a constant resting potential. Instead, their membrane potential inherently fluctuates without any influence by factors external to the cell. Two major types of spontaneous depolarizations displayed by self-excitable cells are pacemaker potentials and slow-wave potentials.

Pacemaker potentials

With **pacemaker potentials,** the membrane potential gradually depolarizes on its own because of shifts in passive ionic fluxes accompanying automatic changes in channel permeability (● Figure 8-32a). When the membrane has depolarized to threshold, an action potential is initiated. After repolarizing, the membrane potential once again depolarizes to threshold, cyclically continuing in this manner to self-generate action potentials.

Slow-wave potentials

Slow-wave potentials are gradually alternating hyperpolarizing and depolarizing swings in potential caused by automatic cyclical changes in the rate at which sodium ions are actively transported across the membrane (● Figure 8-32b). The potential is moved farther from threshold during each hyperpolarizing swing and closer to threshold during each depolarizing swing. If threshold is reached, a burst of action potentials occurs at the peak of a depolarizing swing. Threshold is not always reached, however, so the oscillating slow-wave potentials can continue without generating action potentials. Whether threshold is reached depends on the starting point of the membrane potential at the onset of its depolarizing swing. The starting point, in turn, is influenced by neural and local factors. (We have now discussed all the means by which excitable tissues can be brought to threshold. ▲ Table 8-5 summarizes the different triggering events that can initiate action potentials in various excitable tissues.)

● **FIGURE 8-32**

Self-generated electrical activity in smooth muscle
(a) With pacemaker potentials, the membrane gradually depolarizes to threshold on a regular periodic basis without any nervous stimulation. These regular depolarizations cyclically trigger self-induced action potentials. (b) In slow-wave potentials, the membrane gradually undergoes self-induced hyperpolarizing and depolarizing swings in potential. A burst of action potentials occurs if a depolarizing swing brings the membrane to threshold.

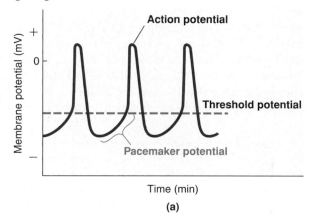

(a)

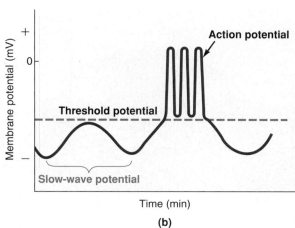

(b)

Various Means of Initiating Action Potentials in Excitable Tissues

Method of Depolarizing the Membrane to Threshold Potential	Type of Excitable Tissue Involved	Description of This Triggering Event
Summation of excitatory postsynaptic potentials (EPSPs) (see p. 119)	Efferent neurons, interneurons	Temporal or spatial summation (see p. 121) of slight depolarizations (EPSPs) of the dendrite/cell body end of the neuron brought about by changes in channel permeability in response to binding of excitatory neurotransmitter with surface membrane receptors
Receptor potential (see p. 187)	Afferent neurons	Typically a depolarization of the afferent neuron's receptor initiated by changes in channel permeability in response to the neuron's adequate stimulus (see p. 187)
End-plate potential (see p. 247)	Skeletal muscle	Depolarization of the motor end plate (see p. 248) brought about by changes in channel permeability in response to binding of the neurotransmitter acetylcholine with receptors on the end-plate membrane
Pacemaker potential (see p. 293)	Smooth muscle, cardiac muscle	Gradual depolarization of the membrane on its own because of shifts in passive ionic fluxes accompanying automatic changes in channel permeability
Slow-wave potential (see p. 293)	Smooth muscle (in digestive tract only)	Gradual alternating hyperpolarizing and depolarizing swings in potential caused by automatic cyclical changes in active ionic transport across the membrane

Myogenic activity

Self-excitable smooth muscle cells are specialized to initiate action potentials, but they are not equipped to contract. Only a very small portion of the total number of cells in a functional syncytium are noncontractile, pacemaker cells. These cells are typically clustered together in a specific location. The vast majority of smooth muscle cells in a functional syncytium are specialized to contract but cannot self-initiate action potentials. However, once an action potential is initiated by a self-excitable smooth muscle cell, it is conducted to the remaining contractile, nonpacemaker cells of the functional syncytium via gap junctions, so that the entire group of connected cells contracts as a unit without any nervous input (● Figure 8-33). Such nerve-independent contractile activity initiated by the muscle itself is called **myogenic** ("muscle-produced") **activity,** in contrast to the neurogenic activity of skeletal muscle and multiunit smooth muscle.

● **FIGURE 8-33**

Functional syncytium
An action potential spontaneously initiated in a pacemaker smooth muscle cell spreads to surrounding nonpacemaker cells through gap junctions, exciting the entire sheet of connected smooth muscle cells to contract as single, coordinated unit.

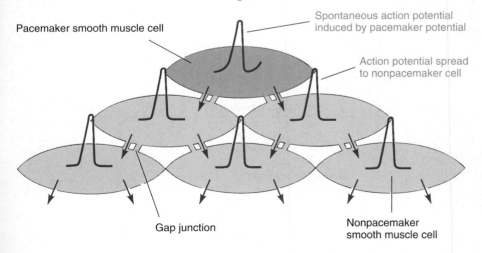

Pacemaker smooth muscle cell

Spontaneous action potential induced by pacemaker potential

Action potential spread to nonpacemaker cell

Gap junction

Nonpacemaker smooth muscle cell

▌ Gradation of single-unit smooth muscle contraction differs from that of skeletal muscle.

Single-unit smooth muscle differs from skeletal muscle in the way contraction is graded. Gradation of skeletal muscle contraction is entirely under neural control, primarily involving motor unit recruitment and twitch summation. In smooth muscle, the gap junctions ensure that an entire smooth-muscle mass contracts as a single unit, making it impossible to vary the number of muscle fibers contracting. Only the tension of the fibers can be modified to achieve varying strengths of contraction of the whole organ. The portion of cross bridges activated and the tension subsequently developed in single-unit smooth muscle can be graded by varying the cytosolic Ca^{2+} concentration. A single excitation in smooth muscle does not cause all the cross bridges to switch on, in contrast to skeletal muscles, where a single action potential triggers the release of sufficient Ca^{2+} to permit all cross bridges to cycle. As Ca^{2+} concentration increases in smooth muscle, more cross bridges are brought into play, and greater tension develops.

Smooth muscle tone

Many single-unit smooth muscle cells have sufficient levels of cytosolic Ca^{2+} to maintain a low level of tension, or **tone**, even in the absence of action potentials. A sudden drastic change in Ca^{2+}, such as accompanies a myogenically induced action potential, brings about a contractile response superimposed on the ongoing tonic tension. Besides self-induced action potentials, a number of other factors, including autonomic neurotransmitters, can influence contractile activity and the development of tension in smooth muscle cells by altering their cytosolic Ca^{2+} concentration.

Modification of smooth muscle activity by the autonomic nervous system

Smooth muscle is typically innervated by both branches of the autonomic nervous system. In single-unit smooth muscle, this nerve supply does not *initiate* contraction, but it can *modify* the rate and strength of contraction, either enhancing or retarding the inherent contractile activity of a given organ. Recall that the isolated motor end-plate region of a skeletal muscle fiber interacts with ACh released from a single axon terminal of a motor neuron. In contrast, the receptor proteins that bind with autonomic neurotransmitters are dispersed throughout the entire surface membrane of a smooth muscle cell. Smooth muscle cells are sensitive to varying degrees and in varying ways to autonomic neurotransmitters, depending on their distribution of cholinergic and adrenergic receptors (see p. 243).

Each terminal branch of a postganglionic autonomic fiber travels across the surface of one or more smooth-muscle cells, releasing neurotransmitter from the vesicles within its multiple **varicosities** (bulges) as an action potential passes along the terminal (● Figure 8-34). The neurotransmitter diffuses to the many receptor sites specific for it on the cells underlying the terminal. Thus, in contrast to the discrete one-to-one relationship at motor end plates, a given smooth-muscle cell can be influenced by more than one type of neurotransmitter, and each autonomic terminal can influence more than one smooth-muscle cell.

Other factors influencing smooth muscle activity

Other factors (besides autonomic neurotransmitters) can influence the rate and strength of both multiunit and single-unit smooth muscle contraction, including mechanical stretch, certain hormones, local metabolites, and specific drugs. Recent evidence suggests that the abundant caveolae found in smooth muscle may serve as sites of integration of the extracellular signals that modify smooth muscle contractility (see p. 85). All these factors ultimately act by modifying the permeability of Ca^{2+} channels in the plasma membrane, the sarcoplasmic reticulum, or both, through a variety of mechanisms. Thus, smooth muscle is subject to more external influences than is skeletal muscle, even though smooth muscle is capable of contracting on its own, whereas skeletal muscle is not.

For now, we are going to consider only the effect of mechanical stretch on smooth muscle contractility in more detail as we look at the length–tension relationship in smooth muscle. We will defer an examination of the extracellular chemical influences on smooth muscle contractility to other chap-

● **FIGURE 8-34**

Schematic representation of innervation of smooth muscle by autonomic postganglionic nerve terminals

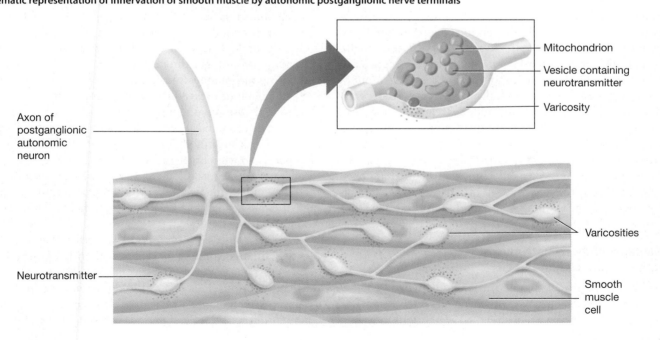

- Mitochondrion
- Vesicle containing neurotransmitter
- Varicosity
- Axon of postganglionic autonomic neuron
- Varicosities
- Neurotransmitter
- Smooth muscle cell

ters when we discuss regulation of the various organs that contain smooth muscle.

▍ Smooth muscle can still develop tension yet inherently relaxes when stretched.

The relationship between the length of the muscle fibers before contraction and the tension that can be developed on a subsequent contraction is less closely linked in smooth muscle than in skeletal muscle. The range of lengths over which a smooth muscle fiber is able to develop near-maximal tension is much greater than for skeletal muscle. Smooth muscle can still develop considerable tension even when stretched up to 2.5 times its resting length, for two probable reasons. First, in contrast to skeletal muscle in which the resting length is at l_o, in smooth muscle the resting (nonstretched) length is much shorter than the l_o. Therefore, smooth muscle can be stretched considerably before even reaching its optimal length. Second, the thin filaments still overlap the much longer thick filaments even in the stretched-out position, so that cross-bridge interaction and tension development can still take place. In contrast, the thick and thin filaments of skeletal muscle are completely pulled apart and no longer able to interact when the muscle is stretched only three-fourths longer than its resting length.

The ability of a considerably stretched smooth-muscle fiber to still develop tension is important, because the smooth muscle fibers within the wall of a hollow organ are progressively stretched as the volume of the organ's contents increases. Consider the urinary bladder as an example. Even though the muscle fibers in the urinary bladder are stretched as the bladder gradually fills with urine, they still maintain their tone and are even capable of developing further tension in response to inputs regulating bladder emptying. If considerable stretching prevented development of tension, as it does in skeletal muscle, a filled bladder would not be capable of contracting to empty.

Stress relaxation response

When a smooth muscle is suddenly stretched, it initially increases its tension, much like the tension created in a stretched rubber band. The muscle quickly adjusts to this new length, however, and inherently relaxes to the tension level prior to the stretch, probably as a consequence of rearrangement of cross-bridge attachments. Smooth-muscle cross bridges detach comparatively slowly. On sudden stretching, it is speculated that any attached cross bridges would strain against the stretch, contributing to a passive (not actively generated) increase in tension. As these cross bridges detach, the filaments would be permitted to slide into an unstrained stretched position, restoring the tension to its original level. This inherent property of smooth muscle is called the **stress relaxation response.**

Advantages of the smooth muscle length–tension relationship

These two responses of smooth muscle to being stretched — being able to develop tension even when considerably stretched and inherently relaxing when stretched — are highly advanta-geous. They enable smooth muscle to exist at a variety of lengths with little change in tension. As a result, a hollow organ enclosed by smooth muscle can accommodate variable volumes of contents with little change in the pressure exerted on the contents except when the contents are to be pushed out of the organ. At that time, the tension is deliberately increased by fiber shortening. It is possible for smooth muscle fibers to contract to half their normal length, enabling hollow organs to dramatically empty their contents on increased contractile activity; thus smooth-muscled viscera can easily accommodate large volumes but can empty to practically zero volume. This length range in which smooth muscle normally functions (anywhere from 0.5 to 2.5 times the normal length) is considerably greater than the limited length range within which skeletal muscle remains functional.

Smooth muscle contains an abundance of connective tissue, which resists being stretched. Unlike skeletal muscle in which the skeletal attachments restrict how far the muscle can be stretched, this connective tissue prevents smooth muscle from being overstretched and thus places an upper limit on the volume capacity of a smooth-muscled hollow organ.

▍ Smooth muscle is slow and economical.

A smooth muscle contractile response proceeds at a more leisurely pace than does a skeletal muscle twitch. A single smooth muscle contraction may last as long as 3 seconds (3,000 msec), compared to the maximum of 100 msec required for a single contractile response in skeletal muscle. The rate of ATP splitting by myosin ATPase is much slower in smooth muscle, so cross-bridge activity and filament sliding occur more slowly. Smooth muscle also relaxes more slowly because of a slower rate of Ca^{2+} removal. Slowness should not be equated with weakness, however. Smooth muscle is able to generate the same contractile tension per unit of cross-sectional area as skeletal muscle, but it does so more slowly and at considerably less energy expense. Because of the low rate of cross-bridge cycling during much of smooth muscle contraction, cross bridges are maintained in the attached state for a longer period of time during each cycle compared with skeletal muscle; that is, the cross bridges "latch onto" the thin filaments for a longer time each cycle. This so-called **latch phenomenon** enables smooth muscle to maintain tension with comparatively less ATP consumption, because each cross-bridge cycle uses up one molecule of ATP. Smooth muscle is therefore an economical contractile tissue, making it well suited for long-term sustained contractions with little energy consumption and without fatigue. Unlike the rapidly changing demands placed on our skeletal muscles as we maneuver through and manipulate our external environment, our smooth muscle activities are geared for long-term duration and slower adjustments to change.

Because of its slowness and the less ordered arrangement of its filaments, smooth muscle has often been mistakenly viewed as a poorly developed version of skeletal muscle. Actually, smooth muscle is just as highly specialized for the demands placed on it — that is, being able to economically maintain tension for prolonged periods without fatigue and being able to accommodate considerable variations in the volume of contents it encloses with little change in tension. It is an extremely adaptive, efficient tissue.

Nutrient and O_2 delivery are generally adequate to support the smooth-muscle contractile process. Smooth muscle can use a wide variety of nutrient molecules for ATP production. There are no energy storage pools comparable to creatine phosphate in smooth muscle; they are not necessary. Oxygen delivery is usually adequate to keep pace with the low rate of oxidative phosphorylation needed to provide ATP for the energy-efficient smooth muscle. If necessary, anaerobic glycolysis can sustain adequate ATP production if O_2 supplies are diminished.

▌ Cardiac muscle blends features of both skeletal and smooth muscle.

Cardiac muscle, found only in the heart, shares structural and functional characteristics with both skeletal and single-unit smooth muscle. Like skeletal muscle, cardiac muscle is striated, with its thick and thin filaments highly organized into a regular banding pattern. Cardiac thin filaments contain troponin and tropomyosin, which constitute the site of Ca^{2+} action in turning on cross-bridge activity, as in skeletal muscle. Also similar to skeletal muscle, cardiac muscle has a clear length–tension relationship. Like the oxidative skeletal-muscle fibers, cardiac muscle cells have an abundance of mitochondria and myoglobin. They also possess T tubules and a moderately well-developed sarcoplasmic reticulum.

As in smooth muscle, Ca^{2+} enters the cytosol from both the ECF and the sarcoplasmic reticulum during cardiac excitation. Ca^{2+} entry from the ECF occurs through voltage-gated dihydropyridine receptors, which also act as Ca^{2+} channels in the T tubule membrane. This Ca^{2+} entry from the ECF triggers the release of Ca^{2+} intracellularly from the sarcoplasmic reticulum. Like single-unit smooth muscle, the heart displays pacemaker (but not slow-wave) activity, initiating its own action potentials without any external influence. Cardiac cells are interconnected by gap junctions that enhance the spread of action potentials throughout the heart, just as in single-unit smooth muscle. Also similarly, the heart is innervated by the autonomic nervous system, which, along with certain hormones and local factors, can modify the rate and strength of contraction.

Unique to cardiac muscle, the cardiac fibers are joined together in a branching network, and the action potentials of cardiac muscle have a much longer duration before repolarizing. Further details and the importance of cardiac muscle's features are addressed in the next chapter.

CHAPTER IN PERSPECTIVE: FOCUS ON HOMEOSTASIS

The skeletal muscles comprise the muscular system itself. Cardiac and smooth muscle are part of the organs that make up other body systems. Cardiac muscle is found only in the heart, which is part of the circulatory system. Smooth muscle is found in the walls of hollow organs and tubes, including the blood vessels in the circulatory system, airways in the respiratory system, bladder in the urinary system, stomach and intestines in the digestive system, and uterus and ductus deferens (the duct that provides a route of exit for sperm from the testes) in the reproductive system.

Contraction of skeletal muscles accomplishes movement of the body parts in relation to each other and movement of the whole body in relation to the external environment. Thus these muscles permit us to move through and manipulate our external environment. At a very general level, some of these movements are aimed at maintaining homeostasis, such as moving the body toward food or away from harm. Examples of more specific homeostatic functions accomplished by skeletal muscles include the chewing and swallowing of food for further breakdown in the digestive system into usable energy-producing nutrient molecules (the mouth and throat muscles are all skeletal muscles) and the process of breathing to obtain O_2 and eliminate CO_2 (the respiratory muscles are all skeletal muscles). Generation of heat by contracting skeletal muscles also serves as the major source of heat production in maintaining body temperature. The skeletal muscles further accomplish many nonhomeostatic activities that enable us to work and play— for example, operating a computer or riding a bicycle— so that we may contribute to society and enjoy ourselves.

All the other systems of the body, except the immune (defense) system, depend on their nonskeletal muscle components to enable them to accomplish their homeostatic functions. For example, contraction of cardiac muscle in the heart pushes life-sustaining blood forward into the blood vessels, and contraction of smooth muscle in the stomach and intestines pushes the ingested food through the digestive tract at a rate appropriate for the digestive juices secreted along the route to break down the food into usable units.

CHAPTER SUMMARY

Introduction
▌ Muscles are the contraction specialists of the body. Through their highly developed ability to move specialized cytoskeletal components, they are able to develop tension, shorten, produce movement, and accomplish work.

▌ The three types of muscle—skeletal, cardiac, and smooth—are categorized in two different ways according to common characteristics. (1) Skeletal and cardiac muscle are *striated*, whereas

smooth muscle is *unstriated*. (2) Skeletal muscle is *voluntary*, whereas cardiac and smooth muscle are *involuntary*.

Structure of Skeletal Muscle
▌ Skeletal muscles are made up of bundles of long, cylindrical muscle cells known as *muscle fibers*, wrapped in connective tissue.

▌ Muscle fibers are packed with myofibrils, each myofibril consisting of alternating, slightly overlapping stacked sets of thick and thin filaments. This arrangement leads to a skeletal muscle fiber's striated microscopic appearance, which consists of alternating dark A bands and light I bands.

- Thick filaments are composed of the protein myosin. Cross bridges made up of the myosin molecules' globular heads project from each thick filament toward the surrounding thin filaments.

- Thin filaments are composed primarily of the protein actin, which has the ability to bind and interact with the myosin cross bridges to bring about contraction. Accordingly, myosin and actin are known as *contractile proteins*. However, in the resting state two other regulatory proteins, tropomyosin and troponin, lie across the surface of the thin filament to prevent this cross-bridge interaction.

Molecular Basis of Skeletal Muscle Contraction

- Excitation of a skeletal muscle fiber by its motor neuron brings about contraction through a series of events that results in the thin filaments sliding closer together between the thick filaments.

- This sliding filament mechanism of muscle contraction is switched on by the release of Ca^{2+} from the lateral sacs of the sarcoplasmic reticulum.

- Calcium release occurs in response to the spread of a muscle-fiber action potential into the central portions of the fiber by means of the T tubules.

- Released Ca^{2+} binds to the troponin-tropomyosin complex of the thin filament, causing a slight repositioning of the complex to uncover actin's cross-bridge binding sites.

- After the exposed actin attaches to a myosin cross bridge, the molecular interaction between actin and myosin releases the energy within the myosin head that was stored from the prior splitting of ATP by the myosin ATPase site. This released energy powers cross-bridge stroking.

- During a power stroke, an activated cross bridge bends toward the center of the thick filament, "rowing" in the thin filament to which it is attached.

- With the addition of a fresh ATP molecule to the myosin cross bridge, myosin and actin detach, the cross bridge returns to its original shape, and the cycle is repeated.

- Repeated cycles of cross-bridge activity slide the thin filaments inward step by step.

- When there is no longer a local action potential, the lateral sacs actively take up the Ca^{2+}, troponin and tropomyosin slip back into their blocking position, and relaxation occurs.

- The entire contractile response lasts about 100 times longer than the action potential.

Skeletal Muscle Mechanics

- Gradation of whole-muscle contraction can be accomplished by (1) varying the number of muscle fibers contracting within the muscle and (2) varying the tension developed by each contracting fiber.

- The greater the number of active muscle fibers, the greater the whole-muscle tension. The number of fibers contracting depends on (1) the size of the muscle (the number of muscle fibers present), (2) the extent of motor unit recruitment (how many motor neurons supplying the muscle are active), and (3) the size of each motor unit (how many muscle fibers are activated simultaneously by a single motor neuron).

- Also, the greater the tension developed by each contracting fiber, the stronger the contraction of the whole muscle. Two readily variable factors having an effect on the fiber tension are (1) the frequency of stimulation, which determines the extent of twitch summation, and (2) the length of the fiber before the onset of contraction.

- The term *twitch summation* refers to the increase in tension accompanying repetitive stimulation of the muscle fiber. After undergoing an action potential, the muscle cell membrane recovers from its refractory period and can be restimulated again while some contractile activity triggered by the first action potential still remains. As a result, the contractile responses (twitches) induced by the two rapidly successive action potentials are able to sum, increasing the tension developed by the fiber. If the muscle fiber is stimulated so rapidly that it does not have a chance to start relaxing between stimuli, a smooth, sustained maximal (maximal for the fiber at that length) contraction known as *tetanus* takes place.

- The tension developed on a tetanic contraction also depends on the length of the fiber at the onset of contraction. At the optimal length (l_o), which is the resting muscle length, there is maximal opportunity for cross-bridge interaction, because of the optimal overlap of thick and thin filaments; thus the greatest tension can be developed. At lengths shorter or longer than l_o, less tension can be developed on contraction, primarily because a portion of the cross bridges are unable to participate.

- The two primary types of muscle contraction—isometric (constant length) and isotonic (constant tension)—depend on the relationship between muscle tension and the load. The load is the force opposing the contraction, that is, the weight of an object being lifted. (1) If tension is less than the load, the muscle cannot shorten and lift the object but remains at constant length, producing an isometric contraction. (2) In an isotonic contraction, the tension exceeds the load so the muscle can shorten and lift the object, maintaining constant tension throughout the period of shortening.

- The velocity, or speed, of shortening is inversely proportional to the load.

- The amount of work accomplished by a contracting muscle equals the magnitude of the load times the distance the load is moved. The amount of energy consumed by a contracting muscle that is realized as external work varies from 0% to 25%, with the remaining energy being converted to heat.

- The bones, muscles, and joints form lever systems. The most common type amplifies the velocity and distance of muscle shortening to increase the speed and range of motion of the body part moved by the muscle. This increased maneuverability is accomplished at the expense of the muscle having to exert considerably more force than the load.

Skeletal Muscle Metabolism and Fiber Types

- Three biochemical pathways furnish the ATP needed for muscle contraction: (1) the transfer of high-energy phosphates from stored creatine phosphate to ADP, providing the first source of ATP at the onset of exercise; (2) oxidative phosphorylation, which efficiently extracts large amounts of ATP from nutrient molecules if sufficient O_2 is available to support this system; and (3) glycolysis, which can synthesize ATP in the absence of O_2 but uses large amounts of stored glycogen and produces lactic acid in the process.

- Fatigue is of two types: (1) muscle fatigue, which occurs when an exercising muscle can no longer respond to neural stimulation with the same degree of contractile activity, and (2) central fatigue, which occurs when the CNS no longer adequately activates the motor neurons.

- The three types of muscle fibers are classified by the pathways they use for ATP synthesis (oxidative or glycolytic) and the rapidity with which they split ATP and subsequently contract (slow twitch or fast twitch): (1) slow-oxidative fibers, (2) fast-oxidative fibers, and (3) fast-glycolytic fibers.

■ Muscle fibers adapt in response to different demands placed on them. Regular endurance exercise promotes improved oxidative capacity in oxidative fibers, whereas high-intensity resistance training promotes hypertrophy of fast glycolytic fibers. The two types of fast-twitch fibers are interconvertible, depending on the type and extent of training. Fast- and slow-twitch fibers are not interconvertible. Muscles atrophy when not used.

Control of Motor Movement

■ Control of any motor movement depends on the level of activity in the presynaptic inputs that converge on the motor neurons supplying various muscles. These inputs come from three sources: (1) spinal reflex pathways, which originate with afferent neurons; (2) the corticospinal (pyramidal) motor system, which originates at the pyramidal cells in the primary motor cortex and is concerned primarily with discrete, intricate movements of the hands; and (3) the multineuronal (extrapyramidal) motor system, which originates in the brain stem and is mostly involved with postural adjustments and involuntary movements of the trunk and limbs. The final motor output from the brain stem is influenced by the cerebellum, basal nuclei, and cerebral cortex.

■ Establishment and adjustment of motor commands depend on continuous afferent input, especially feedback about changes in muscle length (monitored by muscle spindles) and muscle tension (monitored by Golgi tendon organs).

■ When a whole muscle is passively stretched, the accompanying stretch of its muscle spindles triggers the stretch reflex, which results in reflex contraction of that muscle. This reflex resists any passive changes in muscle length.

Smooth and Cardiac Muscle

■ The thick and thin filaments of smooth muscle are not arranged in an orderly pattern, so the fibers are not striated.

■ In smooth muscle, cytosolic Ca^{2+}, which enters from the extracellular fluid as well as being released from sparse intracellular stores, activates cross-bridge cycling by initiating a series of biochemical reactions that result in phosphorylation of the myosin cross bridges to enable them to bind with actin.

■ Multiunit smooth muscle is neurogenic, requiring stimulation of individual muscle fibers by its autonomic nerve supply to trigger contraction.

■ Single-unit smooth muscle is myogenic; it can initiate its own contraction without any external influence, as a result of spontaneous depolarizations to threshold potential brought about by automatic shifts in ionic fluxes. Only a few of the smooth muscle cells in a functional syncytium are self-excitable. The two major types of spontaneous depolarizations displayed by self-excitable smooth muscle cells are pacemaker potentials and slow-wave potentials.

■ Once an action potential is initiated within a self-excitable smooth muscle cell, this electrical activity spreads by means of gap junctions to the surrounding cells within the functional syncytium, so that the entire sheet becomes excited and contracts as a unit.

■ The level of tension in single-unit smooth muscle depends on the level of cytosolic Ca^{2+}. Many single-unit smooth muscle cells have sufficient cytosolic Ca^{2+} to maintain a low level of tension known as *tone*, even in the absence of action potentials.

■ The autonomic nervous system as well as hormones and local metabolites can modify the rate and strength of the self-induced smooth-muscle contractions. All these factors influence smooth muscle activity by altering the cytosolic Ca^{2+} concentration.

■ Smooth muscle does not have a clear-cut length–tension relationship. It is able to develop tension when considerably stretched and inherently relaxes when stretched.

■ Smooth muscle contractions are energy efficient, enabling this type of muscle to economically sustain long-term contractions without fatigue. This economy, coupled with the fact that single-unit smooth muscle is able to exist at a variety of lengths with little change in tension, makes single-unit smooth muscle ideally suited for its task of forming the walls of distensible hollow organs.

■ Cardiac muscle is found only in the heart. It has highly organized striated fibers, like skeletal muscle. Like single-unit smooth muscle, some cardiac muscle fibers are capable of generating action potentials, which are spread throughout the heart with the aid of gap junctions.

REVIEW EXERCISES

Objective Questions (Answers on p. A-40)

1. On completion of an action potential in a muscle fiber, the contractile activity initiated by the action potential ceases. (*True or false?*)

2. The velocity at which a muscle shortens depends entirely on the ATPase activity of its fibers. (*True or false?*)

3. When a skeletal muscle is maximally stretched, it can develop maximal tension on contraction, because the actin filaments can slide in a maximal distance. (*True or false?*)

4. A pacemaker potential always initiates an action potential. (*True or false?*)

5. A slow-wave potential always initiates an action potential. (*True or false?*)

6. Smooth muscle can develop tension even when considerably stretched, because the thin filaments still overlap with the long thick filaments. (*True or false?*)

7. A(n) _____ contraction is an isotonic contraction in which the muscle shortens, whereas the muscle lengthens in a(n) _____ isotonic contraction.

8. _____ motor neurons supply extrafusal muscle fibers, whereas intrafusal fibers are innervated by _____ motor neurons.

9. The two types of atrophy are _____ and _____.

10. Which of the following provide(s) direct input to alpha motor neurons? (*Indicate all correct answers.*)
    a.   primary motor cortex   d.   basal nuclei
    b.   brain stem   e.   spinal reflex pathways
    c.   cerebellum

11. Which of the following is *not* involved in bringing about muscle relaxation?
    a.   reuptake of Ca^{2+} by the sarcoplasmic reticulum
    b.   no more ATP

c. no more action potential
d. removal of ACh at the end plate by acetylcholinesterase
e. filaments sliding back to their resting position

12. Match the following (with reference to skeletal muscle):

1. Ca^{2+}
2. T tubule
3. ATP
4. lateral sac of the sarcoplasmic reticulum
5. myosin
6. troponin-tropomyosin complex
7. actin

(a) cyclically binds with the myosin cross bridges during contraction
(b) has ATPase activity
(c) supplies energy for the power stroke of a cross bridge
(d) rapidly transmits the action potential to the central portion of the muscle fiber
(e) stores Ca^{2+}
(f) pulls the troponin-tropomyosin complex out of its blocking position
(g) prevents actin from interacting with myosin when the muscle fiber is not excited

13. Using the answer code at the right, indicate what happens in the banding pattern during contraction:

1. thick myofilament
2. thin myofilament
3. A band
4. I band
5. H zone
6. sarcomere

(a) remains the same size during contraction
(b) decreases in length (shortens) during contraction

Essay Questions

1. Describe the levels of organization in a skeletal muscle.
2. What is responsible for the striated appearance of skeletal muscles? Describe the arrangement of thick and thin filaments that gives rise to the banding pattern.
3. What is the functional unit of skeletal muscle?
4. Describe the composition of thick and thin filaments.
5. Describe the sliding filament mechanism of muscle contraction. How do cross-bridge power strokes bring about shortening of the muscle fiber?
6. Compare the excitation–contraction coupling process in skeletal muscle with that in smooth muscle.
7. By what means can gradation of skeletal muscle contraction be accomplished?
8. What is a motor unit? Compare the size of motor units in finely controlled muscles with those specialized for coarse, powerful contractions. Describe motor unit recruitment.
9. Explain the phenomenon of twitch summation and tetanus.
10. What effect does a skeletal muscle fiber's length at the onset of contraction have on the strength of the subsequent contraction?
11. Compare isotonic and isometric contractions.
12. Describe the role of each of the following in powering skeletal muscle contraction: ATP, creatine phosphate, oxidative phosphorylation, and glycolysis. Distinguish between aerobically and anaerobically supported exercise.

13. Compare the three types of skeletal muscle fibers.
14. What are the roles of the corticospinal system and multineuronal system in the control of motor movement?
15. Describe the structure and function of muscle spindles and Golgi tendon organs.
16. Distinguish between multiunit and single-unit smooth muscle.
17. Differentiate between neurogenic and myogenic muscle activity.
18. How can smooth muscle contraction be graded?
19. Compare the contractile speed and relative energy expenditure of skeletal muscle with that of smooth muscle.
20. In what ways is cardiac muscle functionally similar to skeletal muscle and to single-unit smooth muscle?

Quantitative Exercises (Solutions on p. A-40)

1. Consider two individuals each throwing a baseball, one a weekend athlete and the other a professional pitcher.
   a. Given the following information, calculate the velocity of the ball as it leaves the amateur's hand:
      • The distance from his shoulder socket (humeral head) to the ball is 70 cm.
      • The distance from his humeral head to the points of insertion of the muscles moving his arm forward (we must simplify here, because the shoulder is such a complex joint) is 9 cm.
      • The velocity of muscle shortening is 2.6 m/sec.
   b. The professional pitcher throws the ball 85 miles per hour. If his points of insertion are also 9 cm from the humeral head and the distance from his humeral head to the ball is 90 cm, how much faster did the professional pitcher's muscles shorten compared to the amateur's?

2. The velocity at which a muscle shortens is related to the force that it can generate in the following way:

$$v = b(F_0 - F)/(F + a)^1$$

where v is the velocity of shortening, and F_0 can be thought of as an "upper load limit," or the maximum force a muscle can generate against a resistance. The parameter a is inversely proportional to the cross-bridge cycling rate, and b is proportional to the number of sarcomeres in line in a muscle. Draw the resistance (force)–velocity curve predicted by this equation by plotting the points $F = 0$ and $F = F_0$. Values of v are on the vertical axis; values of F are on the horizontal axis; a, b and F_0 are constants.
   a. Notice that the curve generated from this equation is the same as that in ● Figure 8-21, p. 275. Why does the curve have this shape? That is, what does the shape of the curve tell you about muscle performance in general?
   b. What happens to the resistance (force)-velocity curve when F_0 is increased? When the cross-bridge cycling rate is increased? When the size of the muscle is increased? How will each of these changes affect the performance of the muscle?

[1]F. C. Hoppensteadt and C. S. Peskin, *Mathematics in Medicine and the Life Sciences* (New York: Springer, 1992), equation 9.1.1, p. 199.

POINTS TO PONDER

(Explanations on p. A 40)

1. Why does regular aerobic exercise provide more cardiovascular benefit than weight training does? (*Hint:* The heart responds to the demands placed on it in a way similar to skeletal muscle.)

2. If the biceps muscle of a child inserts 4 cm from the elbow, and the length of the arm from the elbow to the hand is 28 cm, how much force must the biceps generate in order for the child to lift an 8-kg stack of books with one hand?

3. Put yourself in the position of the scientists who discovered the sliding filament mechanism of muscle contraction by considering what molecular changes must be involved to account for the observed alterations in the banding pattern during contraction. If you were comparing a relaxed and contracted muscle fiber under a high-power light microscope (see ● Figure 8-3a, p. 260), how could you determine that the thin filaments do not change in length during muscle contraction? You cannot see or measure a single thin filament at this magnification. (*Hint:* What landmark in the banding pattern represents each end of the thin filament? If these landmarks are the same distance apart in a relaxed and contracted fiber, then the thin filaments must not change in length.)

4. What type of off-the-snow training would you recommend for a competitive downhill skier versus a competitive cross-country skier? What adaptive skeletal muscle changes would you hope to accomplish in the athletes in each case?

5. When the bladder is filled and the micturition (urination) reflex is initiated, the nervous supply to the bladder promotes contraction of the bladder and relaxation of the external urethral sphincter, a ring of muscle that guards the exit from the bladder. If the time is inopportune for bladder emptying when the micturition reflex is initiated, the external urethral sphincter can be voluntarily tightened to prevent urination despite the fact that the bladder is contracting. Using your knowledge of the muscle types and their innervation, of what types of muscle are the bladder and the external urethral sphincter composed, and what branch of the efferent division of the peripheral nervous system supplies each of these muscles?

6. *Clinical Consideration.* Jason W. is waiting impatiently for the doctor to finish removing the cast from his leg, which he broke the last day of school six weeks ago. Summer vacation is half over, and he hasn't been able to swim, play baseball, or participate in any of his favorite sports. When the cast is finally off, Jason's excitement is replaced with concern when he sees that the injured limb is noticeably smaller in diameter than his normal leg. What is the explanation for this reduction in size? How can the leg be restored to its normal size and functional ability?

PHYSIOEDGE RESOURCES

 PHYSIOEDGE CD-ROM
Figures marked with the PhysioEdge icon have associated activities on the CD. For this chapter, check out:

Skeletal Muscle Contraction

The diagnostic quiz allows you to receive immediate feedback on your understanding of the concept and to advance through various levels of difficulty.

CHECK OUT THESE MEDIA QUIZZES:
8.1 Muscle Types
8.2 Skeletal Muscle Structure
8.3 Skeletal Muscle Mechanics
8.4 Smooth Muscle

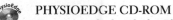 **PHYSIOEDGE WEB SITE**
The Web site for this book contains a wealth of helpful study aids, as well as many ideas for further reading and research. Log on to:

http://www.brookscole.com

Go to the Biology page and select Sherwood's *Human Physiology,* 5th Edition. Select a chapter from the drop-down menu or click on one of these resource areas:

- **Case Histories** provide an introduction to the clinical aspects of human physiology.

- For 2-D and 3-D graphical illustrations and animations of physiological concepts, visit our **Visual Learning Resource.**

- Resources for study and review include the **Chapter Outline, Chapter Summary, Glossary,** and **Flash Cards.** Use our **Quizzes** to prepare for in-class examinations.

- On-line research resources to consult are **Hypercontents,** with current links to relevant Internet sites; **Internet Exercises** with starter URLs provided; and **InfoTrac Exercises.**

 For more readings, go to InfoTrac College Edition, your on-line research library, at:

http://infotrac.thomsonlearning.com

Circulatory System (Heart)

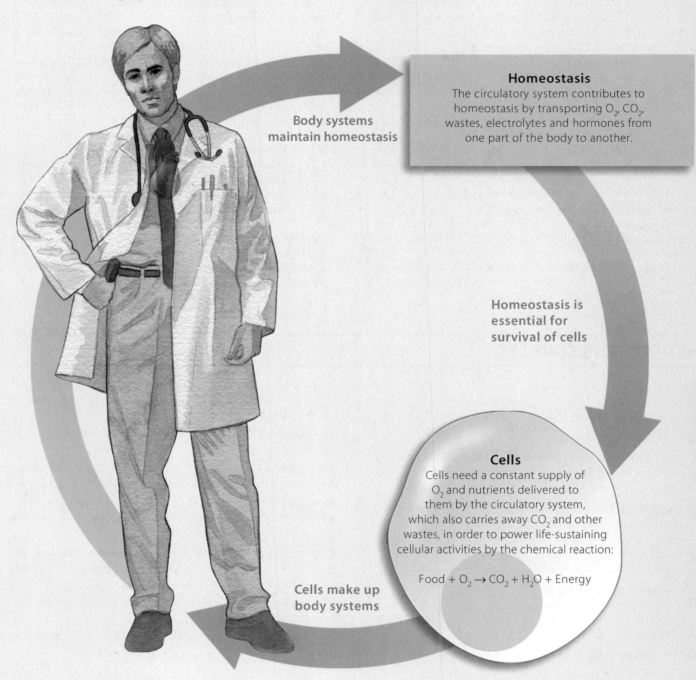

Body systems maintain homeostasis

Homeostasis
The circulatory system contributes to homeostasis by transporting O_2, CO_2, wastes, electrolytes and hormones from one part of the body to another.

Homeostasis is essential for survival of cells

Cells
Cells need a constant supply of O_2 and nutrients delivered to them by the circulatory system, which also carries away CO_2 and other wastes, in order to power life-sustaining cellular activities by the chemical reaction:

$$Food + O_2 \rightarrow CO_2 + H_2O + Energy$$

Cells make up body systems

The maintenance of homeostasis depends on essential materials such as O_2 and nutrients being continually picked up from the external environment and delivered to the cells and on waste products being continually removed. Homeostasis also depends on the transfer of hormones, which are important regulatory chemical messengers, from their site of production to their site of action. The circulatory system, which contributes to homeostasis by serving as the body's transport system, consists of the heart, blood vessels, and blood.

All body tissues constantly depend on the life-supporting blood flow provided to them by the contraction or beating of the heart. The heart drives the blood through the blood vessels for delivery to the tissues in sufficient amounts, whether the body is at rest or engaging in vigorous exercise.

Chapter 9

Cardiac Physiology

CONTENTS AT A GLANCE

INTRODUCTION

ANATOMY OF THE HEART

▌ Location of the heart

▌ The heart as a dual pump

▌ Heart valves

▌ Heart walls; cardiac muscle

▌ Pericardial sac

ELECTRICAL ACTIVITY OF THE HEART

▌ Pacemaker activity

▌ Spread of cardiac excitation

▌ Action potential of cardiac contractile cells

▌ Cardiac refractory period

▌ Electrocardiography

MECHANICAL EVENTS OF THE CARDIAC CYCLE

▌ Electrical, pressure, and volume relationships during diastole and systole

▌ Heart sounds

CARDIAC OUTPUT AND ITS CONTROL

▌ Determinants of cardiac output

▌ Control of heart rate

▌ Control of stroke volume

NOURISHING THE HEART MUSCLE

▌ Coronary circulation

▌ Coronary artery disease

INTRODUCTION

From just a matter of days following conception until death, the beat goes on. In fact, throughout an average human life span, the heart contracts about 3 billion times, never stopping to rest except for a fraction of a second between beats. Within about three weeks after conception, the heart of the developing embryo starts to function. It is believed to be the first organ to become functional. At this time the human embryo is only a few millimeters long, about the size of a capital letter on this page.

Why does the heart develop so early, and why is it so crucial throughout life? It is because the circulatory system is the transport system of the body. A human embryo, having very little yolk available as food, depends on the prompt establishment of a circulatory system that can interact with the maternal circulation to pick up and distribute to the developing tissues the supplies so critical for survival and growth. Thus begins the story of the circulatory system, which continues throughout life to be a vital pipeline for transporting materials on which the cells of the body are absolutely dependent.

The **circulatory system** consists of three basic components:

1. The **heart** serves as the pump that imparts pressure to the blood to establish the pressure gradient needed for blood to flow to the tissues. Like all liquids, blood flows down a pressure gradient from an area of higher pressure to an area of lower pressure. This chapter focuses on cardiac physiology (*cardia* means "heart").

2. The **blood vessels** serve as the passageways through which blood is directed and distributed from the heart to all parts of the body and subsequently returned to the heart (Chapter 10).

3. The **blood** serves as the transport medium within which materials being transported are dissolved or suspended (Chapter 11).

The blood travels continuously through the circulatory system to and from the heart through two separate vascular (blood vessel) loops, both originating and terminating at the heart (● Figure 9-1). The **pulmonary circulation** consists of a closed loop of vessels carrying blood between the heart and lungs (*pulmo* means "lung). The **systemic circulation** consists of a circuit of vessels carrying blood between the heart and other body systems.

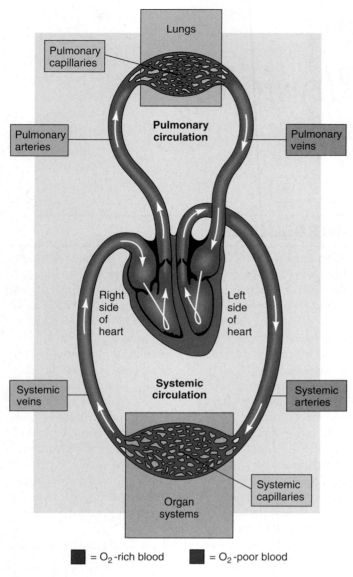

● FIGURE 9-1

Pulmonary and systemic circulation in relation to the heart
The circulatory system consists of two separate vascular loops:
the pulmonary circulation, which carries blood between the heart
and lungs; and the systemic circulation, which carries blood
between the heart and organ systems.

ANATOMY OF THE HEART

The heart is a hollow, muscular organ about the size of a
clenched fist. It is located in the **thoracic** (chest) **cavity** approximately midline between the **sternum** (breastbone) anteriorly
and the **vertebrae** (backbone) posteriorly (● Figure 9-2a). The
midline location of the heart brings up a potentially confusing
point. Place your hand over your heart as if to recite the Pledge
of Allegiance. Where did you place your hand? People usually
place their hand on the left side of the chest, even though the
heart is actually in the middle of the chest. The heart has a
broad base at the top and tapers to a pointed tip known as the

apex at the bottom. It is situated at an angle under the sternum
so that its base lies predominantly to the right and the apex to
the left of the sternum. When the heart beats, especially when
it contracts forcefully, the apex actually thumps against the inside of the chest wall on the left side. Because we become aware
of the beating heart through the apex beat on the left side of the
chest, we tend to think—erroneously—that the entire heart is
on the left.

The fact that the heart is positioned between two bony structures, the sternum and vertebrae, makes it possible to manu-
ally drive blood out of the heart when it is not pumping effectively, by rhythmically depressing the sternum (● Figure 9-2b).
This maneuver compresses the heart between the sternum and
vertebrae so that blood is squeezed out into the blood vessels
to maintain blood flow to the tissues. In many instances, this
external cardiac compression, which is part of **cardiopulmonary
resuscitation (CPR),** serves as a lifesaving measure until appropriate therapy can be instituted to restore the heart to normal function.

▮ The heart is a dual pump.

Even though anatomically the heart is a single organ, the right
and left sides of the heart function as two separate pumps. The
heart is divided into right and left halves and has four chambers, an upper and a lower chamber within each half (● Figure 9-3a). The upper chambers, the **atria** (**atrium,** singular),
receive blood returning to the heart and transfer it to the lower
chambers, the **ventricles,** which pump the blood from the heart.
The vessels that return blood from the tissues to the atria are
veins, and those that carry blood away from the ventricles to
the tissues are **arteries.** The two halves of the heart are separated by the **septum,** a continuous muscular partition that prevents mixture of blood from the two sides of the heart. This
separation is extremely important, because the right half of
the heart is receiving and pumping O_2-poor blood while the
left side of the heart receives and pumps O_2-rich blood.

The complete circuit of blood flow

Let us examine how the heart functions as a dual pump, by tracing a drop of blood through one complete circuit (● Figure 9-3a
and b). Blood returning from the systemic circulation enters
the right atrium via large veins known as the **venae cavae.** The
drop of blood entering the right atrium has returned from the
body tissues, where O_2 has been extracted from it and CO_2 has
been added to it. This partially deoxygenated blood flows from
the right atrium into the right ventricle, which pumps it out
through the **pulmonary artery** to the lungs. Thus the *right side
of the heart receives blood from the systemic circulation and
pumps it into the pulmonary circulation.*

Within the lungs, the drop of blood loses its extra CO_2
and picks up a fresh supply of O_2 before being returned to the
left atrium via the **pulmonary veins.** This O_2-rich blood returning to the left atrium subsequently flows into the left ventricle, the pumping chamber that propels the blood to all body
systems except the lungs; that is, *the left side of the heart re-*

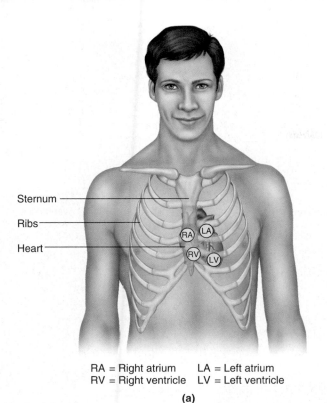

Sternum
Ribs
Heart

RA = Right atrium   LA = Left atrium
RV = Right ventricle LV = Left ventricle

(a)

● **FIGURE 9-2**

Location and external compression of the heart within the thoracic cavity
(a) Location of the heart within the thoracic cavity. (b) External
cardiac compression during cardiopulmonary resuscitation. Manual
compression of the heart between the sternum anteriorly and the
vertebrae posteriorly forces blood out of a nonfunctioning heart.

*ceives blood from the pulmonary circulation and pumps it into
the systemic circulation.* The large artery carrying blood away
from the left ventricle is the **aorta.** Major arteries branch from
the aorta to supply the various tissues of the body.

In contrast to the pulmonary circulation, in which all the
blood flows through the lungs, the systemic circulation may
be viewed as a series of parallel pathways. Part of the blood
pumped out by the left ventricle goes to the muscles, part to
the kidneys, part to the brain, and so on. Thus the output of
the left ventricle is distributed so that each part of the body receives a fresh blood supply; the same arterial blood does not
pass from tissue to tissue. Accordingly, the drop of blood we
are tracing goes to only one of the systemic tissues. Tissue cells
take O_2 from the blood and use it to oxidize nutrients for energy
production; in the process, the tissue cells form CO_2 as a waste
product that is added to the blood (see p. 5 and p. 40). The drop
of blood, now partially depleted of O_2 content and increased
in CO_2 content, returns to the right side of the heart, which
once again will pump it to the lungs. One circuit is complete.

Comparison of the right and left pumps

Both sides of the heart simultaneously pump equal amounts
of blood. The volume of O_2-poor blood being pumped to the
lungs by the right side of the heart soon becomes the same vol-

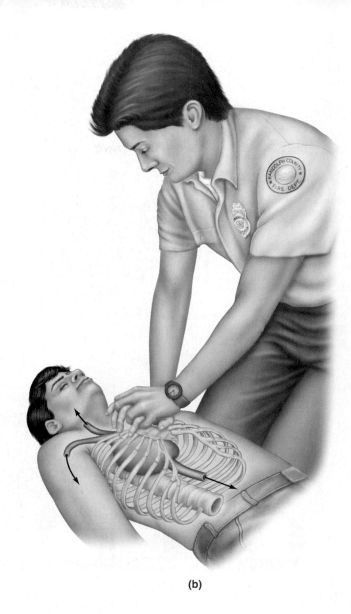

(b)

ume of O_2-rich blood being delivered to the tissues by the left
side of the heart. The pulmonary circulation is a low-pressure,
low-resistance system, whereas the systemic circulation is a
high-pressure, high-resistance system. Pressure is the force exerted on the vessel walls by the blood pumped into the vessels
by the heart. Resistance is the opposition to blood flow, largely
caused by friction between the flowing blood and the vessel wall.
Even though the right and left sides of the heart pump the same
amount of blood, the left side performs more work, because it
pumps an equal volume of blood at a higher pressure into a
higher-resistance system. Accordingly, the heart muscle on the
left side is much thicker than the muscle on the right side, making the left side a stronger pump (● Figure 9-3c).

▌ Heart valves ensure that the blood flows in the proper direction through the heart.

Blood flows through the heart in one fixed direction from veins
to atria to ventricles to arteries. The presence of 4 one-way
heart valves ensures this unidirectional flow of blood. The

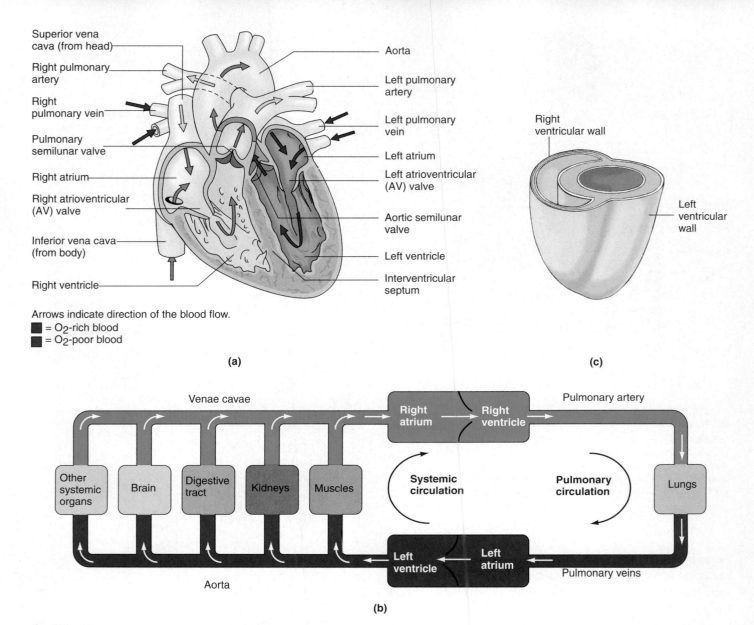

Superior vena cava (from head)

Right pulmonary artery

Right pulmonary vein

Pulmonary semilunar valve

Right atrium

Right atrioventricular (AV) valve

Inferior vena cava (from body)

Right ventricle

Aorta

Left pulmonary artery

Left pulmonary vein

Left atrium

Left atrioventricular (AV) valve

Aortic semilunar valve

Left ventricle

Interventricular septum

Arrows indicate direction of the blood flow.
■ = O₂-rich blood
■ = O₂-poor blood

(a)

Right ventricular wall

Left ventricular wall

(c)

Venae cavae

Right atrium → Right ventricle

Pulmonary artery

Other systemic organs | Brain | Digestive tract | Kidneys | Muscles

Systemic circulation

Pulmonary circulation

Lungs

Left ventricle ← Left atrium

Pulmonary veins

Aorta

(b)

● FIGURE 9-3

Blood flow through and pump action of the heart
(a) Blood flow through the heart. (b) Dual pump action of the heart. The right side of the heart receives O₂-poor blood from the systemic circulation and pumps it into the pulmonary circulation. The left side of the heart receives O₂-rich blood from the pulmonary circulation and pumps it into the systemic circulation. (The relative volume of blood flowing through each organ is not drawn to scale.) (c) Comparison of the thickness of the right and left ventricular walls. Note that the left ventricular wall is much thicker than the right wall.

valves are positioned so that they open and close passively because of pressure differences, similar to a one-way door (● Figure 9-4). A forward pressure gradient (that is, a greater pressure behind the valve) forces the valve open, much as you open a door by pushing on one side of it, whereas a backward pressure gradient (that is, a greater pressure in front of the valve) forces the valve closed, just as you apply pressure to the opposite side of the door to close it. Note that a backward gradient can force the valve closed but cannot force it to swing open in the opposite direction; that is, heart valves are not like swinging, saloon-type doors.

AV valves between the atria and ventricles

Two of the heart valves, the **right** and **left atrioventricular (AV) valves,** are positioned between the atrium and the ventricle on the right and left sides, respectively (● Figure 9-5a). These valves allow blood to flow from the atria into the ventricles during ventricular filling (when atrial pressure exceeds ventricular pressure) but prevent the backflow of blood from the ventricles into the atria during ventricular emptying (when ventricular pressure greatly exceeds atrial pressure). If the rising ventricular pressure did not force the AV valves to close as the ventricles contracted to empty, much of the blood would inefficiently be forced back into the atria and veins instead of being pumped into the arteries. The right AV valve is also called the **tricuspid valve** (*tri* means "three"), because it consists of three cusps or leaflets (● Figure 9-5b). Likewise, the left AV valve, which consists of two cusps, is often called the **bicuspid valve** (*bi* means "two") or, alternatively, the **mitral valve** (because of its physical resemblance to a mitre, or bishop's headgear).

The edges of the AV valve leaflets are fastened by tough, thin, fibrous cords of tendinous-type tissue, the **chordae tendineae,** which prevent the valves from being everted. That is, the chordae tendineae prevent the AV valve from being forced by the high ventricular pressure to open in the opposite direction into the atria. These cords extend from the edges of each cusp and attach to small, nipple-shaped **papillary muscles,** which protrude from the inner surface of the ventricular walls (*papilla* means "nipple"). When the ventricles contract, the papillary muscles also contract, pulling downward on the chordae tendineae. This pulling exerts tension on the closed AV valve cusps to hold them in position, thus helping them remain tightly sealed in the face of a strong backward pressure gradient (● Figure 9-5c).

Semilunar valves between the ventricles and major arteries

The two remaining heart valves, the **aortic** and **pulmonary valves,** are located at the juncture where the major arteries leave the ventricles (● Figure 9-5a). They are known as **semilunar valves** because they are composed of three cusps, each resembling a shallow half-moon-shaped pocket (*semi* means "half;" *lunar* means "moon") (● Figure 9-5b). These valves are forced open when the left and right ventricular pressures exceed the pressure in the

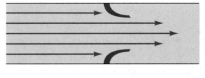

When pressure is greater behind the valve, it opens.

Valve opened

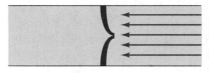

When pressure is greater in front of the valve, it closes. Note that when pressure is greater in front of the valve, it does not open in the opposite direction; that is, it is a one-way valve.

Valve closed; does not open in opposite direction

● **FIGURE 9-4**

Mechanism of valve action

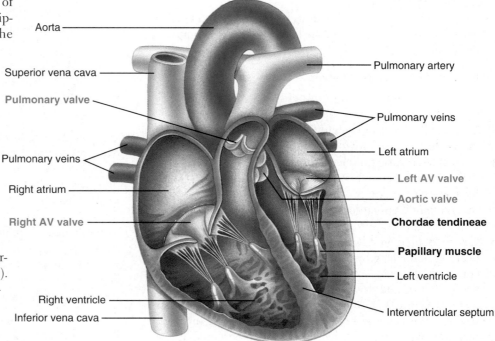

(a)

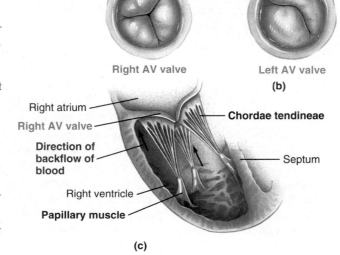

Right AV valve **Left AV valve** **Aortic or pulmonary valve**

(b)

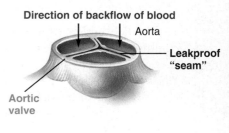

(c) (d)

● **FIGURE 9-5**

Heart valves
(a) Longitudinal section of the heart, depicting the location of the four heart valves. (b) The heart valves in closed position, viewed from above. (c) Prevention of eversion of the AV valves. Eversion of the AV valves is prevented by tension on the valve leaflets exerted by the chordae tendineae when the papillary muscles contract. (d) Prevention of eversion of the semilunar valves. When the semilunar valves are swept closed, their upturned edges fit together in a deep, leakproof seam that prevents valve eversion.

Cardiac Physiology

aorta and pulmonary arteries, respectively, during ventricular contraction and emptying. Closure results when the ventricles relax and ventricular pressures fall below the aortic and pulmonary artery pressures. The closed valves prevent blood from flowing from the arteries back into the ventricles from which it has just been pumped.

The semilunar valves are prevented from everting by the anatomical structure and positioning of the cusps. When a backward pressure gradient is created on ventricular relaxation, the back surge of blood fills the pocketlike cusps and sweeps them into a closed position, with their unattached upturned edges fitting together in a deep, leakproof seam (● Figure 9-5d).

No valves between the atria and veins

Even though there are no valves between the atria and veins, backflow of blood from the atria into the veins usually is not a significant problem for two reasons: (1) atrial pressures usually are not much higher than venous pressures, and (2) the sites where the venae cavae enter the atria are partially compressed during atrial contraction.

Fibrous skeleton of the valves

Four interconnecting rings of dense connective tissue provide a firm base for attachment of the four heart valves (● Figure 9-6). This **fibrous skeleton**, which separates the atria from the ventricles, also provides a fairly rigid structure for attachment of the cardiac muscle. The atrial muscle mass is anchored above the rings, and the ventricular muscle mass is attached to the bottom of the rings.

It might seem rather surprising that the inlet valves to the ventricles (the AV valves) and the outlet valves from the ventricles (the semilunar valves) all lie on the same plane through the heart, as delineated by the fibrous skeleton. This relationship comes about because the heart forms from a single tube that bends on itself and twists on its axis during embryonic development (● Figure 9-7). Although this turning and twisting makes studying the structural relationships of the heart more difficult, the twisted structure has functional importance in that it helps the heart pump more efficiently. We will see how by turning our attention to the portion of the heart that actually generates the forces responsible for blood flow, the cardiac muscle within the heart walls.

▌ The heart walls are composed primarily of spirally arranged cardiac muscle fibers.

The heart wall consists of three distinct layers:

- The **endocardium** is a thin inner layer of **endothelium,** a unique type of epithelial tissue that lines the entire circulatory system (*endo* means "within"; *cardia* means "heart").
- The **myocardium,** the middle layer, composed of cardiac muscle, constitutes the bulk of the heart wall (*myo* means "muscle").
- The **epicardium** is a thin external membrane covering the heart (*epi* means "on").

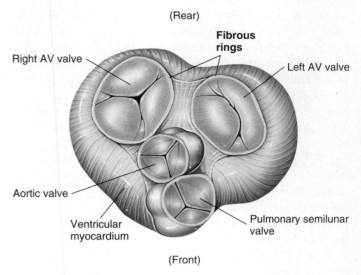

● **FIGURE 9-6**

Fibrous skeleton of the heart
A view of the heart from above, with the atria and major vessels removed to show the heart valves and fibrous rings. Note that the inlet and outlet valves to the ventricle all lie on the same plane through the heart.

The myocardium consists of interlacing bundles of cardiac muscle fibers arranged spirally around the circumference of the heart. The spiral arrangement is due to the heart's complex twisting during development. As a result of this arrangement, when the ventricular muscle contracts and shortens, the diameter of the ventricular chambers is reduced while the apex is simultaneously pulled upward toward the top of the heart in a rotating manner. This exerts a "wringing" effect, efficiently exerting pressure on the blood within the enclosed chambers and directing it upward toward the openings of the major arteries that exit at the base of the ventricles.

▌ Cardiac muscle fibers are interconnected by intercalated discs and form functional syncytia.

The individual cardiac muscle cells are interconnected to form branching fibers, with adjacent cells joined end to end at specialized structures known as **intercalated discs.** Within an intercalated disc, there are two types of membrane junctions: desmosomes and gap junctions (● Figure 9-8). A *desmosome*, a type of adhering junction that mechanically holds cells together, is particularly abundant in tissues, such as the heart, that are subject to considerable mechanical stress (see p. 63). At intervals along the intercalated disc, the opposing membranes approach each other very closely to form *gap junctions*, which are areas of low electrical resistance that allow action potentials to spread from one cardiac cell to adjacent cells (see p. 64). Some cardiac-muscle cells are capable of generating action potentials without any nervous stimulation. When one of the cardiac cells spontaneously undergoes an action potential, the electrical impulse spreads to all the other cells that are joined by gap junctions in the surrounding muscle mass, so

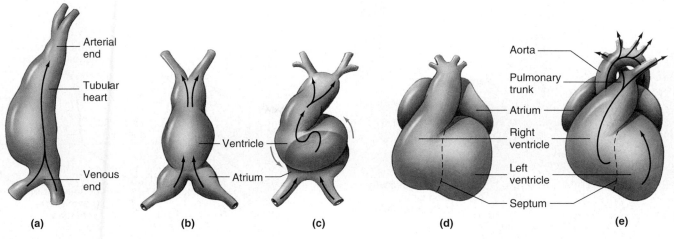

● FIGURE 9-7

Twisting of the embryonic heart on its axis during development
The black arrows within the heart indicate the direction of blood flow. The blue arrows outside the heart depict the direction the embryonic heart twists during development.

that they become excited and contract as a single, functional syncytium (see p. 293). The atria and the ventricles each form a functional syncytium and contract as separate units. The synchronous contraction of the muscle cells composing the walls of each of these chambers produces the force necessary to eject the enclosed blood.

There are no gap junctions between the atrial and ventricular contractile cells, and furthermore, these muscle masses are separated by the electrically nonconductive fibrous skeleton that surrounds the valves. However, an important, specialized conduction system is present to facilitate and coordinate the transmission of electrical excitation from the atria to the ventricles to ensure synchronization between atrial and ventricular pumping.

Because of both the syncytial nature of cardiac muscle and the conduction system between the atria and ventricles, an impulse spontaneously generated in one part of the heart spreads throughout the entire heart. Therefore, unlike skeletal muscle, where graded contractions can be produced by varying the number of muscle cells that are contracting within the muscle (recruitment of motor units), either all the cardiac muscle fibers contract or none of them do. A "half-hearted" contraction is not possible. Gradation of cardiac contraction is accomplished by varying the strength of contraction of all the cardiac muscle cells. You will learn more about this process in a later section.

▌The heart is enclosed by the pericardial sac.

The heart is enclosed in the double-walled, membranous **pericardial sac** (*peri* means "around") (● Figure 9-9). The outer layer of the sac is a tough, fibrous membrane that is attached to the connective tissue partition that separates the lungs. This attachment anchors the heart, so that it remains properly positioned within the chest. The portion of the outer fibrous covering of the pericardial sac that adheres to the surface of the heart forms the epicardium. The sac is lined by a membrane

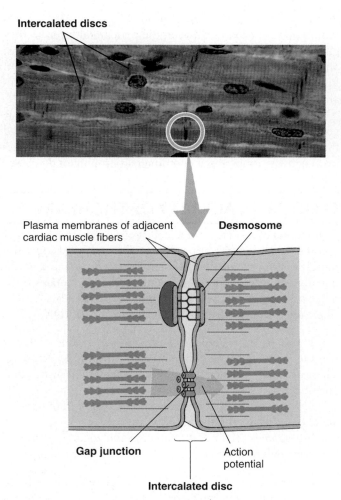

● FIGURE 9-8

Organization of cardiac muscle fibers
Adjacent cardiac muscle cells are joined end to end by intercalated discs, which contain two types of specialized junctions: desmosomes, which act as spot rivets mechanically holding the cells together; and gap junctions, which permit action potentials to spread from one cell to adjacent cells.

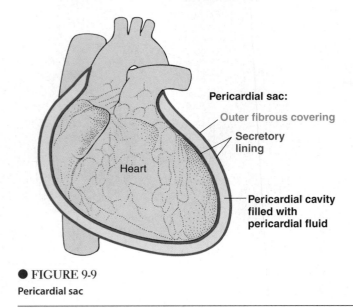

● FIGURE 9-9
Pericardial sac

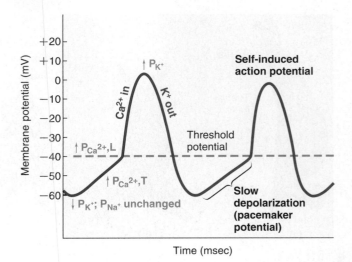

● FIGURE 9-10
Pacemaker activity of cardiac autorhythmic cells

that secretes a thin **pericardial fluid,** which provides lubrication to prevent friction between the pericardial layers as they glide over each other with every beat of the heart. **Pericarditis,** an inflammation of the pericardial sac that results in a painful friction rub between the two pericardial layers, occurs occasionally because of viral or bacterial infection.

Building on this foundation of heart structure, we are now going to explain how action potentials are initiated and spread throughout the heart, followed by a discussion of how this electrical activity brings about coordinated pumping of the heart.

ELECTRICAL ACTIVITY OF THE HEART

Contraction of cardiac muscle cells to bring about ejection of blood is triggered by action potentials sweeping across the muscle cell membranes. The heart contracts, or beats, rhythmically as a result of action potentials that it generates by itself, a property known as **autorhythmicity** (*auto* means "self"). There are two specialized types of cardiac muscle cells:

1. Ninety-nine percent of the cardiac muscle cells are **contractile cells,** which do the mechanical work of pumping. These working cells normally do not initiate their own action potentials.

2. In contrast, the small but extremely important remainder of the cardiac cells, the **autorhythmic cells,** do not contract but instead are specialized for initiating and conducting the action potentials responsible for contraction of the working cells.

Let us examine the role of the specialized autorhythmic cells in the origin and spread of the heartbeat.

▮ Cardiac autorhythmic cells display pacemaker activity.

In contrast to nerve and skeletal muscle cells, in which the membrane remains at constant resting potential unless the cell is stimulated, the cardiac autorhythmic cells do not have a resting potential. Instead, they display **pacemaker activity;** that is, their membrane potential slowly depolarizes, or drifts, between action potentials until threshold is reached, at which time the membrane fires or has an action potential (● Figure 9-10; see also p. 293). Through repeated cycles of drift and fire, these autorhythmic cells cyclically initiate action potentials, which then spread throughout the heart to trigger rhythmic beating without any nervous stimulation.

Pacemaker potential and action potential in autorhythmic cells

Complex interactions of several different ionic mechanisms are responsible for the **pacemaker potential,** which is an autorhythmic cell membrane's slow drift to threshold. The most important changes in ion movement that give rise to the pacemaker potential are (1) a decreased outward K^+ current coupled with a constant inward Na^+ current and (2) an increased inward Ca^{2+} current.

To elaborate, the initial phase of the slow depolarization to threshold is caused by a cyclical decrease in the passive outward flux of K^+ superimposed on a slow, unchanging inward leak of Na^+. In cardiac autorhythmic cells, permeability to K^+ does not remain constant between action potentials as it does in nerve and skeletal muscle cells. Instead, membrane permeability to K^+ decreases between action potentials, because K^+ channels slowly close at negative potentials. This slow closure gradually diminishes the outflow of positive potassium ions down their concentration gradient. Also, unlike nerve and skeletal muscle cells, cardiac autorhythmic cells do not have voltage-gated Na^+ channels. Instead, they have channels that are always open and thus permeable to Na^+ at negative potentials. As a result, a small passive influx of Na^+ continues unchanged at the same time the rate of K^+ efflux slowly declines. Thus the inside gradually becomes less negative; that is, the membrane gradually depolarizes and drifts toward threshold.

In the second half of the pacemaker potential, a transient Ca^{2+} channel (Ca^{2+}, T), one of two types of voltage-gated

Ca^{2+} channels, opens. As the slow depolarization proceeds, this channel is opened before the membrane reaches threshold. The resultant brief influx of Ca^{2+} further depolarizes the membrane, bringing it to threshold.

Once threshold is reached, the rising phase of the action potential occurs in response to activation of a longer-lasting, voltage-gated Ca^{2+} channel (Ca^{2+}, L) and a subsequently large influx of Ca^{2+}. The Ca^{2+}-induced rising phase of a cardiac pacemaker cell differs from nerve and skeletal muscle cells, where Na^+ influx rather than Ca^{2+} influx swings the potential in the positive direction.

The falling phase is due, as usual, to the K^+ efflux that occurs when K^+ permeability increases as a result of activation of voltage-gated K^+ channels. After the action potential is over, slow closure of these K^+ channels initiates the next slow depolarization to threshold.

■ The sinoatrial node is the normal pacemaker of the heart.

The specialized noncontractile cardiac cells capable of autorhythmicity are found in the following specific locations (● Figure 9-11):

1. The **sinoatrial node (SA node)**, a small specialized region in the right atrial wall near the opening of the superior vena cava.

2. The **atrioventricular node (AV node)**, a small bundle of specialized cardiac muscle cells located at the base of the right atrium near the septum, just above the junction of the atria and ventricles.

3. The **bundle of His (atrioventricular bundle)**, a tract of specialized cells that originates at the AV node and enters the interventricular septum. Here, it divides to form the right and left bundle branches that travel down the septum, curve around the tip of the ventricular chambers, and travel back toward the atria along the outer walls.

4. **Purkinje fibers,** small terminal fibers that extend from the bundle of His and spread throughout the ventricular myocardium much like small twigs of a tree branch.

Normal pacemaker activity

Because these various autorhythmic cells have different rates of slow depolarization to threshold, the rates at which they are normally capable of generating action potentials also differ (▲ Table 9-1). Comparing two autorhythmic cells (● Figure 9-12), cell A has a faster rate of depolarization and thus reaches threshold more quickly and generates action potentials more frequently than cell B.

The heart cells with the fastest rate of action potential initiation are localized in the SA node. Once an action potential occurs in any cardiac muscle cell, it is propagated throughout the rest of the myocardium via gap junctions and the specialized conduction system. Therefore, the SA node, which normally exhibits the fastest rate of autorhythmicity, at 70 to 80 action potentials per minute, drives the rest of the heart at this rate and is known as the **pacemaker** of the heart. That is, the entire heart becomes excited, triggering the contractile cells to contract and the heart to beat at the pace or rate set by SA node autorhythmicity, normally at 70 to 80 beats per minute. The other autorhythmic tissues are unable to assume their own naturally slower rates, because they are activated by action potentials originating in the SA node before they are able to reach threshold at their own, slower rhythm.

The following analogy demonstrates how the SA node drives the remainder of the heart at its own pace. Suppose that a train consists of 100 cars, 3 of which are engines capable of moving on their own; the other 97 cars must be pulled in order to move (● Figure 9-13a). One engine (the SA node) can travel at 70 miles/hour (mph) on its own, another engine (the AV node) at 50 mph, and the last engine (the Purkinje fibers) at 30 mph. If all these cars are joined together, the engine capable of traveling at 70 mph will pull the remainder of the cars at that speed. The engines that can travel at lower speeds on their own will be pulled at a faster speed by the fastest engine and will therefore be unable to assume their own slower rate as long as they are being driven by a faster engine. The other 97 cars (nonautorhythmic, contractile cells), being unable to move on their

● **FIGURE 9-11**

Specialized conduction system of the heart

PhysioEdge Activity

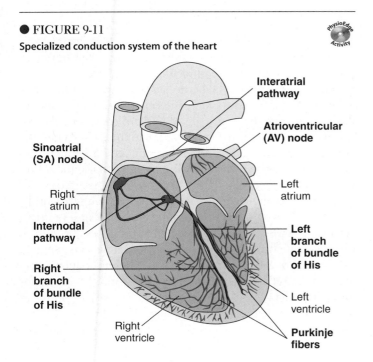

Sinoatrial (SA) node

Right atrium

Internodal pathway

Right branch of bundle of His

Right ventricle

Interatrial pathway

Atrioventricular (AV) node

Left atrium

Left branch of bundle of His

Left ventricle

Purkinje fibers

▲ TABLE 9-1

Normal Rate of Action Potential Discharge in Autorhythmic Tissues of the Heart

Tissue	Action Potentials Per Minute*
SA node (normal pacemaker)	70–80
AV node	40–60
Bundle of His and Purkinje fibers	20–40

*In the presence of parasympathetic tone; see p. 240 and p. 327.

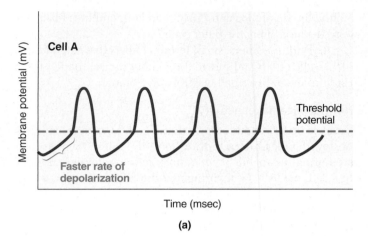

(a)

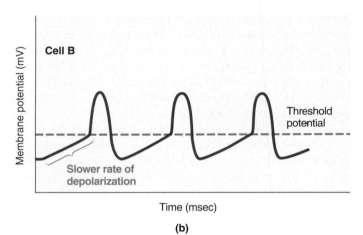

(b)

● **FIGURE 9-12**

Different autorhythmic rates
(a) Recording from autorhythmic cell A. (b) Recording from autorhythmic cell B. Because cell A has a faster rate of depolarization, it reaches threshold more quickly than cell B and therefore generates action potentials more rapidly.

own, will likewise travel at whatever speed the fastest engine pulls them.

Abnormal pacemaker activity

If for some reason the fastest engine breaks down (SA node damage), the next fastest engine (AV node) takes over and the entire train travels at 50 mph; that is, if the SA node becomes nonfunctional, the AV node assumes pacemaker activity (● Figure 9-13b). The non-SA nodal autorhythmic tissues are **latent pacemakers** that can take over, although at a lower rate, should the normal pacemaker fail.

If conduction of the impulse becomes blocked between the atria and the ventricles, the atria continue at the typical rate of 70 beats per minute, and the ventricular tissue, not being driven by the faster SA nodal rate, assumes its own much slower autorhythmic rate of about 30 beats per minute, initiated by the ventricular autorhythmic cells (Purkinje fibers). This situation is comparable to a breakdown of the second engine (AV node) so that the lead engine (SA node) becomes disconnected from the slow third engine (Purkinje fibers) and the remainder

of the cars (● Figure 9-13c). The lead engine continues at 70 mph while the remainder of the train proceeds at 30 mph. Such a phenomenon, known as **complete heart block,** occurs when the conducting tissue between the atria and ventricles is damaged and becomes nonfunctional. A ventricular rate of 30 beats per minute will support only a very sedentary existence; in fact, the patient usually becomes comatose.

In circumstances of abnormally low heart rate, as in SA node failure or heart block, an **artificial pacemaker** can be used. Such an implanted device rhythmically generates impulses that spread throughout the heart to drive both the atria and ventricles at the typical rate of 70 beats per minute.

Occasionally an area of the heart, such as a Purkinje fiber, becomes overly excitable and depolarizes more rapidly than the SA node. (The slow engine suddenly goes faster than the lead engine; see ● Figure 9-13d). This abnormally excitable area, an **ectopic focus,** initiates a premature action potential that spreads throughout the rest of the heart before a normal action potential can be initiated by the SA node. An occasional abnormal impulse from an ectopic focus produces a **premature beat,** or **extrasystole.** If the ectopic focus continues to discharge at its more rapid rate, pacemaker activity is shifted from the SA node to the ectopic focus. The heart rate abruptly becomes greatly accelerated and continues this rapid rate for a variable time period until the ectopic focus returns to normal. Such overly irritable areas may be associated with organic heart disease, but more frequently they occur in response to anxiety, lack of sleep, or excess caffeine, nicotine, or alcohol consumption.

We will now turn our attention to how an action potential, once initiated, is conducted throughout the heart.

■ The spread of cardiac excitation is coordinated to ensure efficient pumping.

Once initiated in the SA node, an action potential spreads throughout the rest of the heart. For efficient cardiac function, the spread of excitation should satisfy three criteria:

1. *Atrial excitation and contraction should be complete before the onset of ventricular contraction.* Complete ventricular filling requires that atrial contraction precede ventricular contraction. During the period of cardiac relaxation, the AV valves are open, so that venous blood entering the atria continues to flow directly into the ventricles. Almost 80% of ventricular filling occurs by this means before atrial contraction. When the atria do contract, additional blood is squeezed into the ventricles to complete ventricular filling. Ventricular contraction then occurs to eject blood from the heart into the arteries.

If the atria and ventricles were to contract simultaneously, the AV valves would be closed immediately, because ventricular pressures would greatly exceed atrial pressures. The ventricles have much thicker walls and, accordingly, can generate more pressure. Atrial contraction would be unproductive, because the atria could not squeeze blood into the ventricles through closed valves. Therefore, to ensure complete filling of the ventricles—to obtain the remaining 20% of ventricular filling that occurs during atrial contraction—the atria must be-

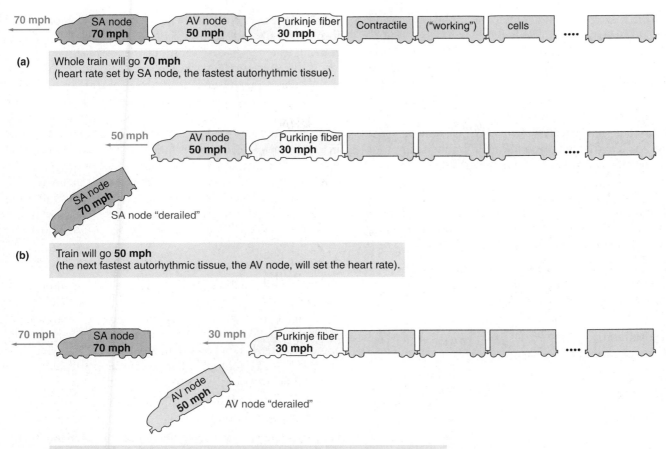

(a) Whole train will go **70 mph** (heart rate set by SA node, the fastest autorhythmic tissue).

(b) Train will go **50 mph** (the next fastest autorhythmic tissue, the AV node, will set the heart rate).

SA node "derailed"

(c) First part of train will go **70 mph**; last part will go **30 mph** (atria will be driven by SA node; ventricles will assume own, much slower rhythm).

AV node "derailed"

Ectopic focus

(d) Train will be driven by ectopic focus, which is now going faster than the SA node (the whole heart will be driven more rapidly by an abnormal pacemaker).

● **FIGURE 9-13**

Analogy of pacemaker activity
(a) Normal pacemaker activity by the SA node. (b) Takeover of pacemaker activity by the AV node when the SA node is nonfunctional. (c) Takeover of ventricular rate by the slower ventricular autorhythmic tissue in the condition of heart block even though the SA node is still functioning. (d) Takeover of pacemaker activity by an ectopic focus.

come excited and contract before ventricular excitation and contraction. During a normal heartbeat, atrial contraction occurs about 160 msec before ventricular contraction.

2. *Excitation of cardiac muscle fibers should be coordinated to ensure that each heart chamber contracts as a unit to accomplish efficient pumping.* If the muscle fibers in a heart chamber were to become excited and contract randomly rather than contracting simultaneously in a coordinated fashion, they would

be unable to eject blood. A smooth, uniform ventricular contraction is essential to squeeze out the blood. As an analogy, assume you have a basting syringe full of water. If you merely poke a finger here or there into the rubber bulb of the syringe, you will not eject much water. However, if you compress the bulb in a smooth, coordinated fashion, you can squeeze out the water.

In a similar manner, contraction of isolated cardiac muscle fibers is not successful in pumping blood. Such random, uncoordinated excitation and contraction of the cardiac cells is known as **fibrillation.** Ventricular fibrillation rapidly causes death, because the heart is not able to pump blood into the arteries. This condition can often be corrected by **electrical defibrillation,** in which a very strong electrical current is applied on the chest wall. When this current reaches the heart, it stimulates (depolarizes) all parts of the heart simultaneously. Usually the first part of the heart to recover is the SA node, which takes over pacemaker activity, once again initiating impulses that trigger the synchronized contraction of the remainder of the heart.

3. *The pair of atria and pair of ventricles should be functionally coordinated so that both members of the pair contract simultaneously.* This coordination permits synchronized pumping of blood into the pulmonary and systemic circulation.

The normal spread of cardiac excitation is carefully orchestrated to ensure that these criteria are met and the heart functions efficiently as follows (● Figure 9-14):

Atrial excitation

An action potential originating in the SA node first spreads throughout both atria, primarily from cell to cell via gap junctions. In addition, several poorly delineated, specialized conduction pathways hasten conduction of the impulse through the atria (● Figure 9-11):

• The **interatrial pathway** extends from the SA node within the right atrium to the left atrium. Because this pathway rapidly transmits the action potential from the SA node to the pathway's termination in the left atrium, a wave of excitation can spread across the gap junctions throughout the left

atrium at the same time a similar spread is being accomplished throughout the right atrium. This ensures that both atria become depolarized to contract simultaneously.

• The **internodal pathway** extends from the SA node to the AV node. The AV node is the only point of electrical contact between the atria and ventricles; in other words, because the atria and ventricles are structurally connected by electrically nonconductive fibrous tissue, the only way an action potential in the atria can spread to the ventricles is by passing through the AV node. The internodal conduction pathway directs the spread of an action potential originating at the SA node to the AV node to ensure sequential contraction of the ventricles following atrial contraction. Hastened by this pathway, the action potential arrives at the AV node within 30 msec of SA node firing.

Transmission between the atria and the ventricles

The action potential is conducted relatively slowly through the AV node. This slowness is advantageous because it allows time for complete ventricular filling. The impulse is delayed about 100 msec (the **AV nodal delay**), which enables the atria to become completely depolarized and to contract, emptying their contents into the ventricles, before ventricular depolarization and contraction occur.

Ventricular excitation

Following the AV nodal delay, the impulse travels rapidly down the bundle of His and throughout the ventricular myocardium via the Purkinje fibers. The network of fibers in this ventricular conduction system is specialized for rapid propagation of action potentials. Its presence hastens and coordinates the spread of ventricular excitation to ensure that the ventricles contract as a unit. The action potential is transmitted through the entire Purkinje fiber system within 30 msec.

Although this system carries the action potential rapidly to a large number of cardiac muscle cells, it does not terminate on every cell. The impulse quickly spreads from the excited cells to the remainder of the ventricular muscle cells by means of gap junctions.

The ventricular conduction system is more highly organized and more important than the interatrial and internodal conduction pathways. Because the ventricular mass is so much larger than the atrial mass, it is crucial that a rapid conduction system be present to hasten the spread of excitation in the ventricles. Purkinje fibers are able to transmit an action potential six times faster than the ventricular syncytium of contractile cells could. If the entire ventricular depolarization process depended on the cell-to-cell spread of the impulse via gap junctions, the ventricular tissue immediately adjacent to the AV node would become excited and contract before the impulse had even passed to the apex of the heart. This, of course, would not allow efficient pumping. The rapid conduction of the action potential down the bundle of His and its swift, diffuse distribution throughout the Purkinje network lead to almost simultaneous activation of the ventricular myocardial cells in both ventricular chambers, which ensures a single, smooth, coordinated contraction that can efficiently eject blood into both the systemic and pulmonary circulations at the same time.

● **FIGURE 9-14**

Spread of cardiac excitation
An action potential initiated at the SA node first spreads throughout both atria. Its spread is facilitated by two specialized atrial conduction pathways, the interatrial and internodal pathways. The AV node is the only point where an action potential can spread from the atria to the ventricles. From the AV node, the action potential spreads rapidly throughout the ventricles, hastened by a specialized ventricular conduction system consisting of the bundle of His and Purkinje fibers.

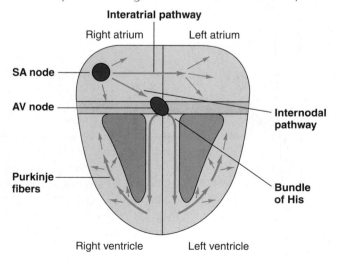

Interatrial pathway

Right atrium — Left atrium

SA node

AV node

Internodal pathway

Purkinje fibers

Bundle of His

Right ventricle — Left ventricle

The action potential of cardiac contractile cells shows a characteristic plateau.

The action potential in cardiac contractile cells, although initiated by the nodal pacemaker cells, varies considerably in ionic mechanisms and shape from the SA node potential (compare ● Figures 9-10 and 9-15). Unlike autorhythmic cells, the membrane of contractile cells remains essentially at rest at about −90 mV until excited by electrical activity propagated from the pacemaker. Once the membrane of a ventricular myocardial contractile cell is excited, an action potential is generated by a complicated interplay of permeability changes and membrane potential changes as follows (● Figure 9-15):

1. During the rising phase of the action potential, the membrane potential rapidly becomes reversed to a positive value of +30 mV as a result of an explosive increase in membrane permeability to Na^+ and a subsequent massive Na^+ influx. Thus far, the process is the same as in neurons and skeletal muscle cells (see p. 107). The Na^+ permeability then rapidly plummets to its low resting value, but, unique to these cardiac muscle cells, the membrane potential is maintained at this positive level for several hundred milliseconds, producing a *plateau phase* of the action potential. In contrast, the short action potential of neurons and skeletal muscle cells lasts 1 to 2 msec.

2. The sudden change in voltage occurring during the rising phase of the action potential brings about two voltage-dependent permeability changes that are responsible for maintaining this plateau: activation of slow L-type Ca^{2+} channels

and a marked decrease in K^+ permeability. Opening of the Ca^{2+} channels results in a slow, inward diffusion of Ca^{2+}, because Ca^{2+} is in greater concentration in the ECF. This continued influx of positively charged Ca^{2+} prolongs the positivity inside the cell and is primarily responsible for the plateau portion of the action potential. This effect is enhanced by the concomitant decrease in K^+ permeability. The resultant reduction in outflux of positively charged K^+ prevents rapid repolarization of the membrane and thus contributes to prolongation of the plateau phase.

3. The rapid falling phase of the action potential results from inactivation of the Ca^{2+} channels and activation of K^+ channels. The decrease in Ca^{2+} permeability diminishes the slow, inward movement of positive Ca^{2+}, whereas the sudden increase in K^+ permeability simultaneously promotes rapid outward diffusion of positive K^+. Thus rapid repolarization at the end of the plateau is accomplished primarily by K^+ efflux, which once again makes the inside of the cell more negative than the outside and restores membrane potential to resting.

Let's now see how this action potential brings about contraction.

Ca²⁺ entry from the ECF induces a much larger Ca²⁺ release from the sarcoplasmic reticulum.

In cardiac contractile cells, the L-type Ca^{2+} channels are located primarily in the T tubules. As you just learned, these voltage-gated channels are opened in response to a local action potential. Thus, unlike in skeletal muscle, Ca^{2+} diffuses into the cytosol from the ECF across the T tubule membrane during a cardiac action potential. This small amount of entering Ca^{2+} triggers the opening of nearby Ca^{2+}-release channels in the adjacent lateral sacs of the sarcoplasmic reticulum. In this way, Ca^{2+} entering the cytosol from the ECF induces a much larger release of Ca^{2+} into the cytosol from the intracellular stores (● Figure 9-16). The resultant local bursts of Ca^{2+} release, known as Ca^{2+} *sparks*, from the sarcoplasmic reticulum collectively increase the cytosolic Ca^{2+} pool sufficiently to turn on the contractile machinery. This extra supply of Ca^{2+} from the sarcoplasmic reticulum is responsible for the long period of cardiac contraction, which lasts about three times longer than a single skeletal muscle fiber contraction (300 msec compared to 100 msec). This increased contractile time ensures adequate time to eject the blood.

Role of cytosolic Ca²⁺ in cardiac excitation–contraction coupling

As in skeletal muscle, the role of Ca^{2+} within the cytosol is to bind with the troponin-tropomyosin complex and physically pull it aside so that cross-bridge cycling and contraction can take place (● Figure 9-16) (see p. 260). However, unlike skeletal muscle, in which sufficient Ca^{2+} is always released to turn on all of the cross bridges, in cardiac muscle the extent of cross-bridge activity varies with the amount of cytosolic Ca^{2+}. As you will learn, various regulatory factors can alter the amount of cytosolic Ca^{2+}.

● **FIGURE 9-15**

Action potential in contractile cardiac muscle cells
The action potential in cardiac contractile cells differs considerably from the action potential in cardiac autorhythmic cells (compare with Figure 9-10).

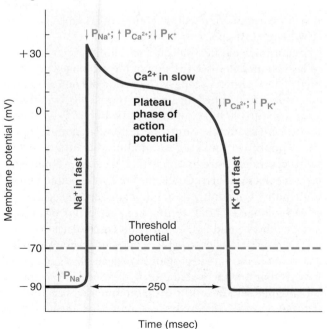

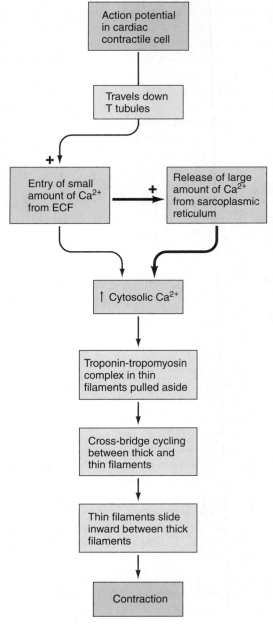

● **FIGURE 9-16**
Excitation–contraction coupling in cardiac contractile cells

Removal of Ca^{2+} from the cytosol by energy-dependent mechanisms in both the plasma membrane and sarcoplasmic reticulum restores the blocking action of troponin and tropomyosin, so that contraction ceases and the heart muscle relaxes.

Influence of altered ECF K^+ and Ca^{2+} concentrations

It is not surprising that changes in the ECF concentration of K^+ and Ca^{2+} can have profound effects on the heart. Abnormal levels of K^+ are most important clinically, followed to a lesser extent by Ca^{2+} imbalances. Changes in K^+ concentration in the ECF alter the K^+ concentration gradient between the ICF and ECF. Normally, there is substantially more K^+ inside the cells than in the ECF, but with elevated ECF K^+

levels this gradient is reduced. Associated with this change is a reduction in "resting" potential (that is, the membrane is less negative on the inside than normal because less K^+ leaves). Among the consequences is a tendency to develop ectopic foci as well as cardiac arrhythmias. Also, because the magnitude of voltage change from the reduced "resting" state to the peak of the action potential is less than from the normal "resting" state to the peak, the resultant diminution of the action potentials' intensity causes the heart to become weak, flaccid, and dilated. At the extreme, with K^+ levels elevated two to three times the normal value, the weakened heart may actually stop pumping.

A rise in ECF Ca^{2+} concentration, in contrast, augments the strength of cardiac contraction by prolonging the plateau phase of the action potential and by increasing the cytosolic concentration of Ca^{2+}. The heart tends to contract spastically, with little time to rest between contractions. Some drugs that alter cardiac function do so by influencing Ca^{2+} movement across the myocardial cell membranes. For example, Ca^{2+} blocking agents, such as *verapamil*, block Ca^{2+} influx during an action potential, thereby reducing the force of cardiac contraction. Other drugs, such as *digitalis*, increase cardiac contractility by inducing an accumulation of cytosolic Ca^{2+}.

▌ Tetanus of cardiac muscle is prevented by a long refractory period.

Like other excitable tissues, cardiac muscle has a refractory period. During the refractory period, which occurs immediately after the initiation of an action potential, an excitable membrane's responsiveness is totally abolished, making it impossible for another action potential to be generated. In skeletal muscle, the refractory period is very short compared with the duration of the resultant contraction, so the fiber can be restimulated again before the first contraction is complete to produce summation of contractions. Rapidly repetitive stimulation that does not allow the muscle fiber to relax between stimulations results in a sustained, maximal contraction known as *tetanus* (see ● Figure 8-17, p. 271).

In contrast, cardiac muscle has a long refractory period that lasts about 250 msec because of the prolonged action potential. This is almost as long as the period of contraction initiated by the action potential; a cardiac muscle fiber contraction averages about 300 msec in duration (● Figure 9-17). Consequently, cardiac muscle cannot be restimulated until contraction is almost over, making summation of contractions and tetanus of cardiac muscle impossible. This is a valuable protective mechanism, because the pumping of blood requires alternate periods of contraction (emptying) and relaxation (filling). A prolonged tetanic contraction would prove fatal. The heart chambers could not be filled and emptied again.

The chief factor responsible for the long refractory period is inactivation, during the prolonged plateau phase, of the Na^+ channels that were activated during the initial Na^+ influx of the rising phase. Not until the membrane recovers from this inactivation process (when the membrane has already repolar-

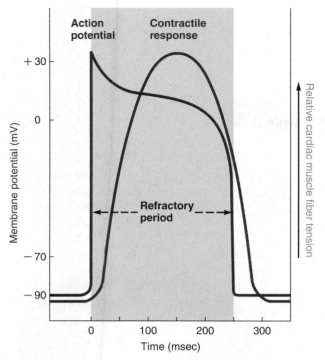

Action potential

Contractile response

Membrane potential (mV)

+30

0

−70

−90

Refractory period

Relative cardiac muscle fiber tension

0   100   200   300

Time (msec)

● **FIGURE 9-17**

Relationship of an action potential and the refractory period to the duration of the contractile response in cardiac muscle

ized to resting), can the Na^+ channels be activated once again to begin another action potential.

The ECG is a record of the overall spread of electrical activity through the heart.

The electrical currents generated by cardiac muscle during depolarization and repolarization (see p. 100) spread into the tissues surrounding the heart and are conducted through the body fluids. A small portion of this electrical activity reaches the body surface, where it can be detected using recording electrodes. The record produced is an **electrocardiogram**, or **ECG**. (Alternatively, the term **EKG** is used, based on the ancient Greek word "kardia" instead of "cardia" for *heart.*)

Three important points should be remembered when considering what an ECG represents:

1. An ECG is a recording of that portion of the electrical activity induced in the body fluids by the cardiac impulse that reaches the surface of the body, not a direct recording of the actual electrical activity of the heart.

2. The ECG is a complex recording representing the overall spread of activity throughout the heart during depolarization and repolarization. It is not a recording of a single action potential in a single cell at a single point in time. The record at any given time represents the sum of electrical activity in all the cardiac muscle cells, some of which may be undergoing action potentials while others may not yet be activated. For example, immediately after firing of the SA node, the atrial

cells are undergoing action potentials while the ventricular cells are still at rest. At a later point, the electrical activity will have spread to the ventricular cells while the atrial cells will be repolarizing. Therefore, the overall pattern of cardiac electrical activity varies with time as the impulse passes throughout the heart.

3. The recording represents comparisons in voltage detected by electrodes at two different points on the body surface, not the actual potential. For example, the ECG does not record a potential at all when the ventricular muscle is either completely depolarized or completely repolarized; both electrodes are "viewing" the same potential, so no difference in potential between the two electrodes is recorded.

The exact pattern of electrical activity recorded from the body surface depends on the orientation of the recording electrodes. Electrodes may be loosely thought of as "eyes" that "see" electrical activity and translate it into a visible recording, the ECG record. Whether an upward deflection or downward deflection is recorded is determined by the orientation of electrodes with respect to the current flow in the heart. For example, the spread of excitation across the heart is seen differently from the right arm and from the left foot, and both of these are seen differently from a recording directly over the heart. Even though the same electrical events are occurring in the heart, different waveforms representing the same electrical activity result when this activity is recorded by electrodes at different points on the body.

To provide standard comparisons, ECG records routinely consist of 12 conventional electrode systems, or leads. When an electrocardiograph machine is connected between recording electrodes at two points on the body, the specific arrangement of each pair of connections is called a **lead**. The 12 different leads each record electrical activity in the heart from different locations—six different electrical arrangements from the limbs and six chest leads at various sites around the heart. The same 12 leads are routinely used in all ECG recordings to provide a common basis for comparison and for recognizing deviations from normal (● Figure 9-18).

Various components of the ECG record can be correlated to specific cardiac events.

Interpretation of the wave configurations recorded from each lead depends on a thorough knowledge of the sequence of the spread of cardiac excitation and the position of the heart relative to the placement of the electrodes. A normal ECG exhibits three distinct waveforms: the P wave, the QRS complex, and the T wave (● Figure 9-19). (The letters do not signify anything other than the orderly sequence of the waves. The founder of the technique simply started in the middle of the alphabet when naming the waves.)

- The **P wave** represents atrial depolarization.
- The **QRS complex** represents ventricular depolarization.
- The **T wave** represents ventricular repolarization.

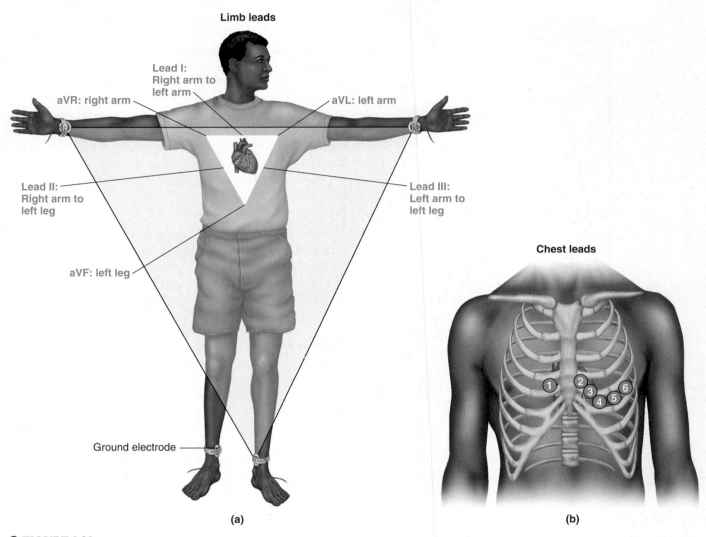

● FIGURE 9-18

Electrocardiogram leads
(a) Limb leads. The six limb leads include leads I, II, III, aVR, aVL, and aVF. Leads I, II, and III are bipolar leads because two recording electrodes are used. The tracing records the *difference* in potential between the two electrodes. For example, lead I records the difference in potential detected at the right arm and left arm. The electrode placed on the right leg serves as a ground and is not a recording electrode. The aVR, aVL, and aVF leads are unipolar leads. Even though two electrodes are used, only the actual potential under one electrode, the exploring electrode, is recorded. The other electrode is set at zero potential and serves as a neutral reference point. For example, the aVR lead records the potential reaching the right arm in comparison to the rest of the body. (b) Chest leads. The six chest leads, V_1 through V_6, are also unipolar leads. The exploring electrode mainly records the electrical potential of the cardiac musculature immediately beneath the electrode in six different locations surrounding the heart.

The following important points about the ECG record should also be noted:

1. Firing of the SA node does not generate sufficient electrical activity to reach the surface of the body, so no wave is recorded for SA nodal depolarization. Therefore, the first recorded wave, the P wave, occurs when the impulse spreads across the atria.

2. In a normal ECG, there is no separate wave for atrial repolarization. The electrical activity associated with atrial repolarization normally occurs simultaneously with ventricular depolarization and is masked by the QRS complex.

3. The P wave is much smaller than the QRS complex, because the atria have a much smaller muscle mass than the ventricles and consequently generate less electrical activity.

4. At three times no current is flowing in the heart musculature and the ECG remains at baseline:

    a. During the AV nodal delay. This delay is represented by the interval of time between the end of the P wave and the onset of the QRS wave; this interval is known as the **PR segment.** (It is called the PR segment rather than the PQ segment because the Q deflection is small and sometimes absent, whereas the R deflection is the dominant wave of the complex.) Current

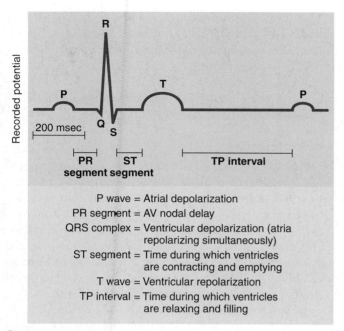

P wave = Atrial depolarization
PR segment = AV nodal delay
QRS complex = Ventricular depolarization (atria repolarizing simultaneously)
ST segment = Time during which ventricles are contracting and emptying
T wave = Ventricular repolarization
TP interval = Time during which ventricles are relaxing and filling

● **FIGURE 9-19**
Electrocardiogram waveforms in lead II

is flowing through the AV node, but the magnitude is too small to be detected by the ECG electrodes.

b. When the ventricles are completely depolarized and the cardiac contractile cells are undergoing the plateau phase of their action potential before they repolarize again, represented by the **ST segment.** This segment is the interval between QRS and T; it coincides with the time during which ventricular activation is complete and the ventricles are contracting and emptying.

c. When the heart muscle is completely at rest and ventricular filling is taking place, after the T wave and before the next P wave; this time segment is called the **TP interval.**

▌ The ECG can be used to diagnose abnormal heart rates, arrhythmias, and damage of heart muscle.

Because electrical activity triggers mechanical activity, abnormal electrical patterns are usually accompanied by abnormal contractile activity of the heart. Thus evaluation of ECG patterns can provide useful information about the status of the heart. The principal deviations from normal that can be ascertained through electrocardiography are (1) abnormalities in rate, (2) abnormalities in rhythm, and (3) cardiac myopathies (● Figure 9-20).

Abnormalities in rate

The distance between two consecutive QRS complexes on an ECG record is calibrated to the beat-to-beat heart rate. A rapid heart rate of more than 100 beats per minute is known as **tachycardia** (*tachy* means "fast"), whereas a slow heart rate of fewer

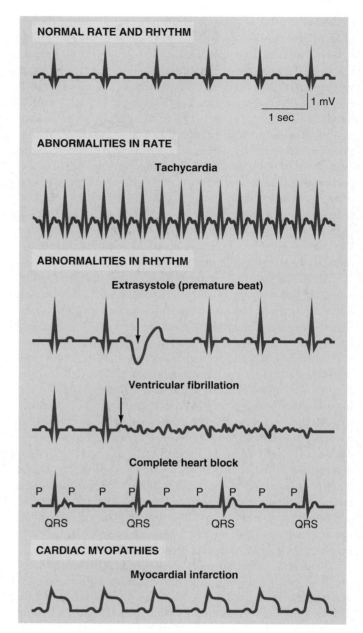

● **FIGURE 9-20**
Representative heart conditions detectable through electrocardiography

than 60 beats per minutes is referred to as **bradycardia** (*brady* means "slow").

Abnormalities in rhythm

The term *rhythm* refers to the regularity of the ECG waves. Any variation from the normal rhythm and sequence of excitation of the heart is termed an **arrhythmia.** It may result from the presence of ectopic foci, alterations in SA node pacemaker activity, or interference with conduction. Heart rate is also often altered. *Extrasystoles* or *premature beats* originating from an ectopic focus are common deviations from normal rhythm. Other abnormalities in rhythm easily detected on an ECG include atrial flutter, atrial fibrillation, ventricular fibrillation, and heart block.

Atrial flutter is characterized by a rapid but regular sequence of atrial depolarizations at rates between 200 to 380 beats per minute. The ventricles rarely keep pace with the racing atria. Because the conducting tissue's refractory period is longer than that of the atrial muscle, the AV node is unable to respond to every impulse that converges on it from the atria. Maybe only one out of every two or three atrial impulses successfully passes through the AV node to the ventricles. Such a situation is referred to as a *2:1 or 3:1 rhythm.* The fact that not every atrial impulse reaches the ventricle in atrial flutter is important, because it precludes a rapid ventricular rate of more than 200 beats per minute. Such a high rate would not allow adequate time for ventricular filling between beats. In such a case, the output of the heart would be reduced to the extent that loss of consciousness or even death could result because of decreased blood flow to the brain.

Atrial fibrillation is characterized by rapid, irregular, uncoordinated depolarizations of the atria with no definite P waves. Accordingly, atrial contractions are chaotic and asynchronized. Because impulses reach the AV node erratically, the ventricular rhythm is also very irregular. The QRS complexes are normal in shape but occur sporadically. Variable lengths of time between ventricular beats are available for ventricular filling. Some ventricular beats come so close together that little filling can occur between beats. When less filling occurs, the subsequent contraction is weaker. In fact, some of the ventricular contractions may be too weak to eject enough blood to produce a palpable wrist pulse. In this situation, if the heart rate is determined directly, either by the apex beat or via the ECG, and the pulse rate is taken concurrently at the wrist, the heart rate will exceed the pulse rate. Such a difference in heart rate and pulse rate is known as a **pulse deficit.** Normally, the heart rate coincides with the pulse rate, because each cardiac contraction initiates a pulse wave as it ejects blood into the arteries.

Ventricular fibrillation is a very serious rhythmic abnormality in which the ventricular musculature exhibits uncoordinated, chaotic contractions. Multiple impulses travel erratically in all directions around the ventricles. The ECG tracing in ventricular fibrillations is very irregular with no detectable pattern or rhythm. When contractions are so disorganized, the ventricles are ineffectual as pumps. If circulation is not restored in less than four minutes through external cardiac compression or electrical defibrillation, irreversible brain damage occurs, and death is imminent.

Another type of arrhythmia, **heart block,** arises from defects in the cardiac conducting system. The atria still beat regularly, but the ventricles occasionally fail to be stimulated and thus do not contract following atrial contraction. Impulses between the atria and ventricles can be blocked to varying degrees. In some forms of heart block, only every second or third atrial impulse is passed to the ventricles. This is known as *2:1 or 3:1 block,* which can be distinguished from the 2:1 or 3:1 rhythm associated with atrial flutter by the rates involved. In heart block, the atrial rate is normal but the ventricular rate is considerably below normal, whereas in atrial flutter the atrial rate is very high, in accompaniment with a normal or above-normal ventricular rate. *Complete heart block* is characterized by complete dissociation between atrial and ventricular activity, with impulses from the atria not being conducted to the ventricles at all. The atrial beat continues to be governed by the SA node, but the ventricles generate their own impulses at a rate much slower than the atria. On the ECG, the P waves exhibit a normal rhythm. The QRS and T waves also occur regularly but at a much slower rate than the P waves and are completely independent of P wave rhythm.

Cardiac myopathies

Abnormal ECG waves are also important in recognizing and assessing **cardiac myopathies** (damage of the heart muscle). **Myocardial ischemia** is inadequate delivery of oxygenated blood to the heart tissue. Actual death, or **necrosis,** of heart muscle cells, occurs when a blood vessel supplying that area of the heart becomes blocked or ruptured. This condition is termed **acute myocardial infarction,** commonly known as a **heart attack.** Abnormal QRS waveforms can be seen when a portion of the heart muscle becomes necrotic. In addition to ECG changes, because damaged heart muscle cells release characteristic enzymes into the blood, the level of these enzymes in the blood provides a further index of the extent of myocardial damage.

Interpretation of an ECG is a complex task requiring extensive knowledge and training. The foregoing discussion is not by any means intended to make you an ECG expert but seeks to give you an appreciation of the ways in which the ECG can be used as a diagnostic tool, as well as to present an overview of some of the more common abnormalities of heart function. (For a further use of the ECG, see the accompanying boxed feature, ▶ A Closer Look at Exercise Physiology.)

MECHANICAL EVENTS OF THE CARDIAC CYCLE

The mechanical events of the cardiac cycle—contraction, relaxation, and the resultant changes in blood flow through the heart—are brought about by the rhythmic changes in cardiac electrical activity.

■ The heart alternately contracts to empty and relaxes to fill.

The cardiac cycle consists of alternate periods of **systole** (contraction and emptying) and **diastole** (relaxation and filling). The atria and ventricles go through separate cycles of systole and diastole. Contraction occurs as a result of the spread of excitation across the heart, whereas relaxation follows the subsequent repolarization of the cardiac musculature.

The following discussion correlates various events that occur concurrently during the cardiac cycle, including ECG features, pressure changes, volume changes, valve activity, and heart sounds. Reference to ● Figure 9-21 will facilitate this discussion. Only the events on the left side of the heart are described, but keep in mind that identical events are occurring on the right side of the heart, except that the pressures are lower. To complete one full cardiac cycle, our discussion will begin and end with ventricular diastole.

The What, Who, and When of Stress Testing

Stress tests, or **graded exercise tests,** are conducted primarily to aid in diagnosing or quantifying heart or lung disease and to evaluate the functional capacity of asymptomatic individuals. The tests are usually given on motorized treadmills or bicycle ergometers (stationary, variable-resistance bicycles). Workload intensity (how hard the subject is working) is adjusted by progressively increasing the speed and incline of the treadmill or by progressively increasing the pedaling frequency and resistance on the bicycle. The test starts at a low intensity and continues until a prespecified workload is achieved, physiologic symptoms occur, or the subject is too fatigued to continue.

During diagnostic testing, the patient is monitored with an ECG, and blood pressure is taken each minute. A test is considered positive if ECG abnormalities occur (such as ST segment depression, inverted T waves, or dangerous arrhythmias) or if physical symptoms such as chest pain develop. A test that is interpreted as positive in a person who does not have heart disease is called a *false positive test.* In men, false positives occur only about 10% to 20% of the time, so the diagnostic stress test for men has a *specificity* of 80% to 90%. Women have a greater frequency of false positive tests, with a corresponding lower specificity of about 70%.

The *sensitivity* of a test means that people with disease are correctly identified and there are few false negatives. The sensitivity of the stress test is reported to be 60% to 80%; that is, if 100 people with heart disease were tested, 60 to 80 would be correctly identified, but 20 to 40 would have a false negative test. Although stress testing is now an important diagnostic tool, it is just one of several tests used to determine the presence of coronary artery disease.

Stress tests are also conducted on people not suspected of having heart or lung disease to determine their present functional capacity. These functional tests are administered in the same way as diagnostic tests, but they are conducted by exercise physiologists and a physician need not be present. These tests are used to establish safe exercise prescriptions, to aid athletes in establishing optimal training programs, and as research tools to evalute the effectiveness of a particular training regimen. Functional stress testing is becoming more prevalent as more people are joining hospital- or community-based wellness programs for disease prevention.

Early ventricular diastole

During early ventricular diastole, the atrium is still also in diastole. This stage corresponds to the TP interval on the ECG—the interval after ventricular repolarization and before another atrial depolarization. Because of the continuous inflow of blood from the venous system into the atrium, atrial pressure slightly exceeds ventricular pressure even though both chambers are relaxed (point 1 in ● Figure 9-21). Because of this pressure differential, the AV valve is open, and blood flows directly from the atrium into the ventricle throughout ventricular diastole (heart A in ● Figure 9-21). As a result, the ventricular volume slowly continues to rise even before atrial contraction takes place (point 2).

Late ventricular diastole

Late in ventricular diastole, the SA node reaches threshold and fires. The impulse spreads throughout the atria, which is recorded on the ECG as the P wave (point 3). Atrial depolarization brings about atrial contraction, which squeezes more blood into the ventricle, causing a rise in the atrial pressure curve (point 4). The excitation–contraction coupling process is taking place during the short delay between the P wave and the rise in atrial pressure. The corresponding rise in ventricular pressure (point 5) that occurs simultaneous to the rise in atrial pressure is due to the additional volume of blood added to the ventricle by atrial contraction (point 6 and heart B). Throughout atrial contraction, atrial pressure still slightly exceeds ventricular pressure, so the AV valve remains open.

End of ventricular diastole

Ventricular diastole ends at the onset of ventricular contraction. By this time, atrial contraction and ventricular filling are completed. The volume of blood in the ventricle at the end of diastole (point 7) is known as the **end-diastolic volume (EDV),** which averages about 135 ml. No more blood will be added to the ventricle during this cycle. Therefore, the end-diastolic volume is the maximum amount of blood that the ventricle will contain during this cycle.

Ventricular excitation and onset of ventricular systole

Following atrial excitation, the impulse passes through the AV node and specialized conduction system to excite the ventricle. Simultaneously, atrial contraction is occurring. By the time ventricular activation is complete, atrial contraction is already accomplished. The QRS complex represents this ventricular excitation (point 8), which induces ventricular contraction. The ventricular pressure curve sharply increases shortly after the QRS complex, signaling the onset of ventricular systole (point 9). The slight delay between the QRS complex and the actual onset of ventricular systole is the time required for the excitation-contraction coupling process to occur. As ventricular contraction begins, ventricular pressure immediately exceeds atrial pressure. This backward pressure differential forces the AV valve closed (point 9).

Isovolumetric ventricular contraction

After ventricular pressure exceeds atrial pressure and the AV valve has closed, the ventricular pressure must continue to increase before it exceeds aortic pressure to open the aortic valve. Therefore, between closure of the AV valve and opening of the aortic valve, there is a brief period of time when the ventricle remains a closed chamber (point 10). Because all valves are closed, no blood can enter or leave the ventricle during this time. This interval is termed the period of **isovolumetric ventricular contraction** (*isovolumetric* means "constant volume and length") (heart C). Because no blood enters or leaves the

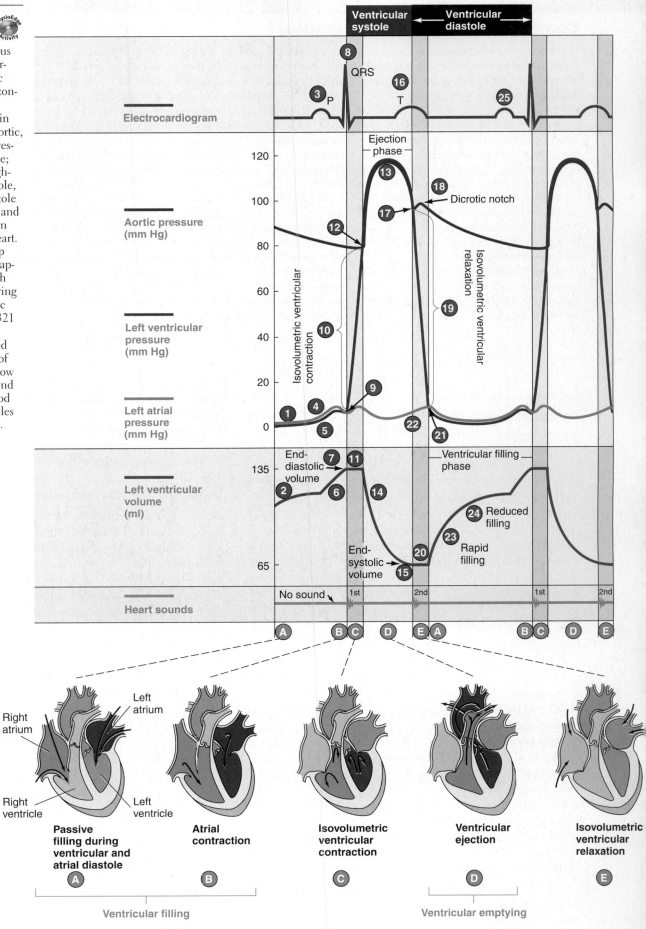

● **FIGURE 9-21**

Cardiac cycle
This graph depicts various events that occur concurrently during the cardiac cycle. Follow each horizontal strip across to see the changes that take place in the electrocardiogram; aortic, ventricular, and atrial pressures; ventricular volume; and heart sounds throughout the cycle. Late diastole, one full systole and diastole (one full cardiac cycle), and another systole are shown for the left side of the heart. Follow each vertical strip downward to see what happens simultaneously with each of these factors during each phase of the cardiac cycle. See the text (pp. 321 and 323) for a detailed explanation of the circled numbers. The sketches of the heart illustrate the flow of O₂-poor (dark blue) and O₂-rich (bright red) blood in and out of the ventricles during the cardiac cycle.

ventricle, the ventricular chamber remains at constant volume, and the muscle fibers remain at constant length. This isovolumetric condition is similar to an isometric contraction in skeletal muscle. During the period of isovolumetric ventricular contraction, ventricular pressure continues to increase as the volume remains constant (point 11).

Ventricular ejection

When ventricular pressure exceeds aortic pressure (point 12), the aortic valve is forced open and ejection of blood begins (heart D). The aortic pressure curve rises as blood is forced into the aorta from the ventricle faster than blood is draining off into the smaller vessels at the other end (point 13). The ventricular volume decreases substantially as blood is rapidly pumped out (point 14). Ventricular systole includes both the period of isovolumetric contraction and the ventricular ejection phase.

End of ventricular systole

The ventricle does not empty completely during ejection. Normally, only about half the blood contained within the ventricle at the end of diastole is pumped out during the subsequent systole. The amount of blood remaining in the ventricle at the end of systole when ejection is complete is known as the **end-systolic volume (ESV)**, which averages about 65 ml (point 15). This is the least amount of blood that the ventricle will contain during this cycle.

The amount of blood pumped out of each ventricle with each contraction is known as the **stroke volume (SV)**; it is equal to the end-diastolic volume minus the end-systolic volume; in other words, the difference between the volume of blood in the ventricle before contraction and the volume after contraction is the amount of blood ejected during the contraction. In our example, the end-diastolic volume is 135 ml, the end-systolic volume is 65 ml, and the stroke volume is 70 ml.

Ventricular repolarization and onset of ventricular diastole

The T wave signifies ventricular repolarization occurring at the end of ventricular systole (point 16). As the ventricle starts to relax on repolarization, ventricular pressure falls below aortic pressure and the aortic valve closes (point 17). Closure of the aortic valve produces a disturbance or notch on the aortic pressure curve known as the **dicrotic notch** (point 18). No more blood leaves the ventricle during this cycle, because the aortic valve has closed.

Isovolumetric ventricular relaxation

When the aortic valve closes, the AV valve is not yet open, because ventricular pressure still exceeds atrial pressure, so no blood can enter the ventricle from the atrium. Therefore, all valves are once again closed for a brief period of time known as **isovolumetric ventricular relaxation** (point 19 and heart E). The muscle fiber length and chamber volume (point 20) remain constant. No blood leaves or enters as the ventricle continues to relax and the pressure steadily falls.

Ventricular filling

When the ventricular pressure falls below the atrial pressure, the AV valve opens (point 21), and ventricular filling occurs once again. Ventricular diastole includes both the period of isovolumetric ventricular relaxation and the ventricular filling phase.

Atrial repolarization and ventricular depolarization occur simultaneously, so the atria are in diastole throughout ventricular systole. Blood continues to flow from the pulmonary veins into the left atrium. As this incoming blood pools in the atrium, atrial pressure rises continuously (point 22). When the AV valve opens at the end of ventricular systole, the blood that accumulated in the atrium during ventricular systole pours rapidly into the ventricle (heart A again). Ventricular filling thus occurs rapidly at first (point 23) because of the increased atrial pressure resulting from the accumulation of blood in the atria. Then ventricular filling slows down (point 24) as the accumulated blood has already been delivered to the ventricle, and atrial pressure starts to fall. During this period of reduced filling, blood continues to flow from the pulmonary veins into the left atrium and through the open AV valve into the left ventricle. During late ventricular diastole, when ventricular filling is proceeding slowly, the SA node fires again (point 25), and the cardiac cycle starts over.

It is significant that much of ventricular filling occurs early in diastole during the rapid-filling phase. During times of rapid heart rate, the length of diastole is reduced to a much greater extent than is the length of systole. For example, if the heart rate increases from 75 to 180 beats per minute, the duration of diastole decreases about 75%, from 500 msec to 125 msec. This greatly reduces the time available for ventricular relaxation and filling. However, because much ventricular filling is accomplished during early diastole, filling is not seriously impaired during periods of increased heart rate, such as during exercise (● Figure 9-22). There is a limit, however, to how rapidly the heart can beat without decreasing the period of diastole to the point that ventricular filling is severely impaired. At heart rates greater than 200 beats per minute, diastolic time is too short to allow adequate ventricular filling. With inadequate filling, the resultant cardiac output is deficient. Normally, ventricular rates do not exceed 200 beats per minute, because the relatively long refractory period of the AV node will not allow impulses to be conducted to the ventricles more frequently than this.

▮ The two heart sounds are associated with valve closures.

Two major heart sounds normally can be heard with a stethoscope during the cardiac cycle. The **first heart sound** is low-pitched, soft, and relatively long—often said to sound like "lub." The **second heart sound** has a higher pitch and is shorter and sharper—often said to sound like "dup." Thus, one normally hears "lub-dup-lub-dup-lub-dup . . . ." The first heart sound is associated with closure of the AV valves, whereas the second sound is associated with closure of the semilunar valves (● Figure 9-21). Opening of valves does not produce any sound.

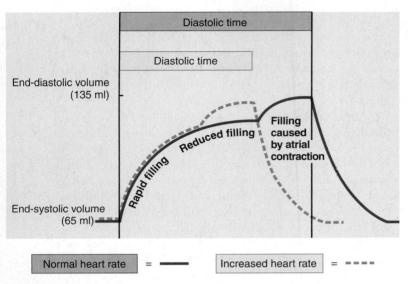

Diastolic time

Diastolic time

End-diastolic volume
(135 ml)

Rapid filling

Reduced filling

Filling caused by atrial contraction

End-systolic volume
(65 ml)

| Normal heart rate | = —— | Increased heart rate | = ---- |

● **FIGURE 9-22**

Ventricular filling profiles during normal and rapid heart rates
Because much of ventricular filling occurs early in diastole during the rapid-filling phase, filling is not seriously impaired when diastolic time is reduced as a result of an increase in heart rate.

The sounds are caused by vibrations set up within the walls of the ventricles and major arteries during valve closure, not by the valves snapping shut. Because closure of the AV valves occurs at the onset of ventricular contraction, when ventricular pressure first exceeds atrial pressure, the first heart sound signals the onset of ventricular systole. Closure of the semilunar valves occurs at the onset of ventricular relaxation, as the left and right ventricular pressures fall below the aortic and pulmonary artery pressures, respectively. The second heart sound, therefore, signals the onset of ventricular diastole.

▌ Turbulent blood flow produces heart murmurs.

Abnormal heart sounds, or **murmurs,** are usually (but not always) associated with cardiac disease. Murmurs not involving heart pathology, so-called **functional murmurs,** are more common in young people.

Blood normally flows in a *laminar* fashion; that is, layers of the fluid slide smoothly over each other (*lamina* means "layer"). Laminar flow does not produce any sound. When blood flow becomes turbulent, however, a sound can be heard (● Figure 9-23). Such an abnormal sound is due to vibrations created in the surrounding structures by the turbulent flow.

Stenotic and insufficient valves

The most common cause of turbulence is valve malfunction, either a stenotic or an insufficient valve. A **stenotic valve** is a stiff, narrowed valve that does not open completely. Blood must be forced through the constricted opening at tremendous velocity, resulting in turbulence that produces an abnormal whistling sound similar to the sound produced when you force air rapidly through narrowed lips to whistle.

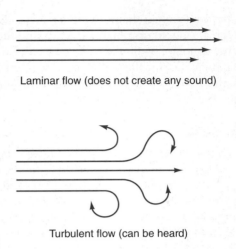

Laminar flow (does not create any sound)

Turbulent flow (can be heard)

● **FIGURE 9-23**
Comparison of laminar and turbulent flow

An **insufficient valve** is one that cannot close completely, usually because the valve edges are scarred and do not fit together properly. Turbulence is produced when blood flows backward through the insufficient valve and collides with blood moving in the opposite direction, creating a swishing or gurgling murmur. Such backflow of blood is known as **regurgitation.** An insufficient heart valve is often called a **leaky valve,** because it allows blood to leak back through at a time when the valve should be closed.

Most often, both valvular stenosis and insufficiency are caused by **rheumatic fever,** an autoimmune ("immunity against self") disease triggered by a streptococcus bacterial infection. Antibodies formed against toxins produced by these bacteria interact with many of the body's own tissues, resulting in immunological damage. The heart valves are among the most susceptible tissues in this regard. Large, hemorrhagic, fibrous lesions form along the inflamed edges of an affected heart valve, causing the valve to become thickened, stiff, and scarred. Sometimes the leaflet edges permanently adhere to each other. Depending on the extent and specific nature of the lesions, the valve may become either stenotic or insufficient or some degree of both.

Timing of murmurs

The valve involved and the type of defect can usually be detected by the *location* and *timing* of the murmur. Each heart valve may be heard best at a specific location on the chest. Noting the location at which a murmur is loudest helps the diagnostician determine which valve is involved.

The timing of the murmur refers to the part of the cardiac cycle during which the murmur is heard. Recall that the first heart sound signals the onset of ventricular systole, and the second heart sound signals the onset of ventricular diastole. Thus a murmur occurring between the first and second heart sounds (lub-murmur-dup, lub-murmur-dup) signifies a **systolic murmur.** A **diastolic murmur,** in contrast, occurs between the second and first heart sound (lub-dup-murmur, lub-dup-murmur). The sound of the murmur characterizes it as either

Timing and Type of Murmur Associated with Various Heart Valve Disorders

Pattern Heard on Auscultation	Type of Valve Defect	Timing of Murmur	Valve Disorder	Comment
Lub-Whistle-Dup	Stenotic	Systolic	Stenotic semilunar valve	A whistling systolic murmur signifies that a valve that should be open during systole (a semilunar valve) does not open completely.
Lub-Dup-Whistle	Stenotic	Diastolic	Stenotic AV valve	A whistling diastolic murmur signifies that a valve that should be open during diastole (an AV valve) does not open completely.
Lub-Swish-Dup	Insufficient	Systolic	Insufficient AV valve	A swishy systolic murmur signifies that a valve that should be closed during systole (an AV valve) does not close completely.
Lub-Dup-Swish	Insufficient	Diastolic	Insufficient semilunar valve	A swishy diastolic murmur signifies that a valve that should be closed during diastole (a semilunar valve) does not close completely.

a stenotic (whistling) murmur or an insufficient (swishy) murmur. Armed with these facts, one can determine the cause of a valvular murmur (▲ Table 9-2). As an example, a whistling murmur (denoting a stenotic valve) occurring between the first and second heart sounds (denoting a systolic murmur) signifies the presence of stenosis in a valve that should be open during systole. It could be either the aortic or the pulmonary semilunar valve through which blood is being ejected. Identifying which of these valves is stenotic is accomplished by determining the location over which the murmur is best heard.

The main concern with heart murmurs, of course, is not the murmur itself but the accompanying detrimental circulatory consequences caused by the defect.

CARDIAC OUTPUT AND ITS CONTROL

▌ Cardiac output depends on the heart rate and the stroke volume.

Cardiac output (CO) is the volume of blood pumped by *each ventricle* per minute (not the total amount of blood pumped by the heart). During any period of time, the volume of blood flowing through the pulmonary circulation is equivalent to the volume flowing through the systemic circulation. Therefore, the cardiac output from each ventricle normally is identical, although on a beat-to-beat basis, minor variations may occur. The two determinants of cardiac output are *heart rate* (beats per minute) and *stroke volume* (volume of blood pumped per beat or stroke).

The average resting heart rate is 70 beats per minute, established by SA node rhythmicity, and the average resting stroke volume is 70 ml per beat, producing an average cardiac output of 4,900 ml/min, or close to 5 liters/min:

$$\text{Cardiac output (CO)} = \text{heart rate} \times \text{stroke volume}$$
$$= 70 \text{ beats/min} \times 70 \text{ ml/beat}$$
$$= 4,900 \text{ ml/min} \approx 5 \text{ liters/min}$$

Because the body's total blood volume averages 5 to 5.5 liters, each half of the heart pumps the equivalent of the entire blood volume each minute. In other words, each minute the right ventricle normally pumps 5 liters of blood through the lungs, and the left ventricle pumps 5 liters of blood through the systemic circulation. At this rate, each half of the heart would pump about 2.5 million liters of blood in just one year. Yet this is only the resting cardiac output! During exercise the cardiac output can increase to 20 to 25 liters per minute, and outputs as high as 40 liters per minute have been recorded in trained athletes during heavy exercise. The difference between the cardiac output at rest and the maximum volume of blood the heart is capable of pumping per minute is known as the **cardiac reserve.**

How can cardiac output vary so tremendously, depending on the demands of the body? You can readily answer this question by thinking about how your own heart pounds rapidly (increased heart rate) and forcefully (increased stroke volume) when you engage in strenuous physical activities (need for increased cardiac output). Thus the regulation of cardiac output depends on the control of both heart rate and stroke volume, topics that are discussed next.

▌ Heart rate is determined primarily by autonomic influences on the SA node.

The SA node is normally the pacemaker of the heart, because it has the fastest spontaneous rate of depolarization to threshold. Recall that this automatic gradual reduction of membrane potential between beats is due to a complex interplay of ion movements involving a reduction in K^+ permeability, a constant Na^+ permeability, and an increased Ca^{2+} permeability. When the SA node reaches threshold, an action potential is initiated that spreads throughout the heart, inducing the heart to contract or have a "heartbeat." This happens about 70 times per minute, setting the average heart rate at 70 beats per minute.

The heart is innervated by both divisions of the autonomic nervous system, which can modify the rate (as well as the strength) of contraction, even though nervous stimulation is not required to initiate contraction. The parasympathetic nerve to the heart, the **vagus nerve,** primarily supplies the atrium, especially the SA and AV nodes. Parasympathetic innervation of the ventricles is sparse. The cardiac sympathetic nerves also supply the atria, including the SA and AV nodes, and richly innervate the ventricles as well.

Both the parasympathetic and sympathetic nervous system bring about their effects on the heart by altering the activity of the cyclic AMP second-messenger system in the innervated cardiac cells. Acetylcholine released from the vagus nerve binds to a muscarinic receptor and is coupled to an inhibitory G protein that reduces activity of the cyclic AMP pathway (see p. 244 and p. 67). By contrast, the sympathetic neurotransmitter norepinephrine binds with a β_1 adrenergic receptor and is coupled to a stimulatory G protein that accelerates the cyclic AMP pathway in the target cells (see p. 245).

Let's examine the specific effects that parasympathetic and sympathetic stimulation have on the heart (▲ Table 9-3).

Effect of parasympathetic stimulation on the heart

- The parasympathetic nervous system's influence on the SA node is to decrease the heart rate (● Figure 9-24). Acetylcholine released on increased parasympathetic activity increases the permeability of the SA node to K^+ by slowing the closure of K^+ channels. As a result, the rate at which spontaneous action potentials are initiated is reduced through a twofold effect:

1. Enhanced K^+ permeability hyperpolarizes the SA node membrane because more positive potassium ions leave than normal, making the inside even more negative. Because the "resting" potential starts even farther away from threshold, it takes longer to reach threshold.

2. The enhanced K^+ permeability induced by vagal stimulation also opposes the automatic reduction in K^+ permeability that is responsible for initiating the gradual depolarization of the membrane to threshold. This countering effect decreases the rate of spontaneous depolarization, prolonging the time required to drift to threshold. Therefore, the SA node reaches threshold and fires less frequently, decreasing the heart rate.

- Parasympathetic influence on the AV node decreases the node's excitability, prolonging transmission of impulses to the ventricles even longer than the usual AV nodal delay. This effect is brought about by increasing K^+ permeability, which hyperpolarizes the membrane, thereby retarding the initiation of excitation in the AV node.

- Parasympathetic stimulation of the atrial contractile cells shortens the action potential, an effect believed to be caused by a reduction in the slow inward current carried by Ca^{2+}; that is, the plateau phase is reduced. As a result, atrial contraction is weakened.

- The parasympathetic system has little effect on ventricular contraction, because of the sparsity of parasympathetic innervation to the ventricles.

▲ TABLE 9-3

Effects of the Autonomic Nervous System on the Heart and Structures That Influence the Heart

Area Affected	Effect of Parasympathetic Stimulation	Effect of Sympathetic Stimulation
SA Node	Decreases the rate of depolarization to threshold; decreases the heart rate	Increases the rate of depolarization to threshold; increases the heart rate
AV Node	Decreases excitability; increases the AV nodal delay	Increases excitability; decreases the AV nodal delay
Ventricular Conduction Pathway	No effect	Increases excitability; hastens conduction through the bundle of His and Purkinje cells
Atrial Muscle	Decreases contractility; weakens contraction	Increases contractility; strengthens contraction
Ventricular Muscle	No effect	Increases contractility; strengthens contraction
Adrenal Medulla (an Endocrine Gland)	No effect	Promotes adrenomedullary secretion of epinephrine, a hormone that augments the sympathetic nervous system's actions on the heart
Veins	No effect	Increases venous return, which increases the strength of cardiac contraction through the Frank-Starling mechanism

Thus the heart is more "leisurely" under parasympathetic influence—it beats less rapidly, the time between atrial and ventricular contraction is stretched out, and atrial contraction is weaker. These actions are appropriate considering that the parasympathetic system controls heart action in quiet, relaxed situations when the body is not demanding an enhanced cardiac output.

Effect of sympathetic stimulation on the heart

• In contrast, the sympathetic nervous system, which controls heart action in emergency or exercise situations, when there is a need for greater blood flow, speeds up the heart rate through its effect on the pacemaker tissue. The main effect of sympathetic stimulation on the SA node is to increase its rate of depolarization so that threshold is reached more rapidly (● Figure 9-24 and ▲ Table 9-3). Norepinephrine released from the sympathetic nerve endings decreases K^+ permeability by accelerating inactivation of the K^+ channels. With fewer positive potassium ions leaving, the inside of the cell becomes less negative, creating a depolarizing effect. This swifter drift to threshold under sympathetic influence permits a greater frequency of action potentials and a correspondingly more rapid heart rate.

• Sympathetic stimulation of the AV node reduces the AV nodal delay by increasing conduction velocity, presumably by enhancing the slow, inward Ca^{2+} current.

• Similarly, sympathetic stimulation speeds up the spread of the action potential throughout the specialized conduction pathway.

• In the atrial and ventricular contractile cells, both of which have an abundance of sympathetic nerve endings, sympathetic stimulation increases contractile strength so that the heart beats more forcefully and squeezes out more blood. This effect is brought about by increasing Ca^{2+} permeability, which enhances the slow Ca^{2+} influx and intensifies Ca^{2+} participation in the excitation–contraction coupling process.

The overall effect of sympathetic stimulation on the heart, therefore, is to improve its effectiveness as a pump by increasing the heart rate, decreasing the delay between atrial and ventricular contraction, decreasing conduction time throughout the heart, and increasing the force of contraction; that is, sympathetic stimulation "revs up" the heart.

Control of heart rate

Thus, as is typical of the autonomic nervous system, parasympathetic and sympathetic effects on heart rate are antagonistic (oppose each other). At any given moment, the heart rate is determined largely by the existing balance between the inhibitory effects of the vagus nerve and the stimulatory effects of the cardiac sympathetic nerves. Under resting conditions, parasympathetic discharge is dominant. In fact, if all autonomic nerves to the heart were blocked, the resting heart rate would increase from its average value of 70 beats per minute to about 100 beats per minute, which is the inherent rate of the SA node's spontaneous discharge when not subjected to any nervous influence. (We use 70 beats per minute as the normal rate of SA node discharge because this is the average rate under normal

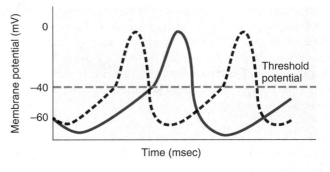

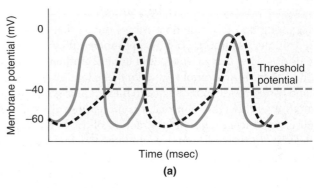

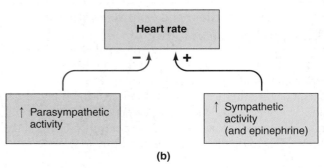

● **FIGURE 9-24**

Autonomic control of SA node activity and heart rate
(a) Autonomic influence on SA node potential. Parasympathetic stimulation decreases the rate of SA nodal depolarization so that the membrane reaches threshold more slowly and has fewer action potentials, whereas sympathetic stimulation increases the rate of depolarization of the SA node so that the membrane reaches threshold more rapidly and has more frequent action potentials. (b) Control of heart rate by the autonomic nervous system. Because each SA node action potential ultimately leads to a heartbeat, increased parasympathetic activity decreases the heart rate, whereas increased sympathetic activity increases the heart rate.

conditions in the body.) Alterations in the heart rate beyond this resting level in either direction can be accomplished by shifting the balance of autonomic nervous stimulation. Heart rate is increased by simultaneously increasing sympathetic and decreasing parasympathetic activity; a reduction in heart rate is brought about by a concurrent rise in parasympathetic activity and decline in sympathetic activity. The relative level of activity in these two autonomic branches to the heart in turn is primarily coordinated by the *cardiovascular control center* located in the brain stem.

Although autonomic innervation is the primary means by which heart rate is regulated, other factors affect it as well. The most important of these is epinephrine, a hormone that is secreted into the blood from the adrenal medulla on sympathetic stimulation and that acts on the heart in a manner similar to norepinephrine to increase the heart rate. Epinephrine therefore reinforces the direct effect that the sympathetic nervous system has on the heart.

▌ Stroke volume is determined by the extent of venous return and by sympathetic activity.

The other component that determines the cardiac output is stroke volume, the amount of blood pumped out by each ventricle during each beat. Two types of controls influence stroke volume: (1) *intrinsic control* related to the extent of venous return and (2) *extrinsic control* related to the extent of sympathetic stimulation of the heart. Both factors increase stroke volume by increasing the strength of contraction of the heart (● Figure 9-25). Let us examine each of these factors in more detail to see how they influence the stroke volume.

▌ Increased end-diastolic volume results in increased stroke volume.

As more blood is returned to the heart, the heart pumps out more blood, but the relationship is not quite as simple as it appears, because the heart does not eject all the blood it contains. The direct correlation between end-diastolic volume and stroke volume constitutes the **intrinsic control** of stroke volume, which refers to the heart's inherent ability to vary the stroke

---

● **FIGURE 9-25**
Intrinsic and extrinsic control of stroke volume

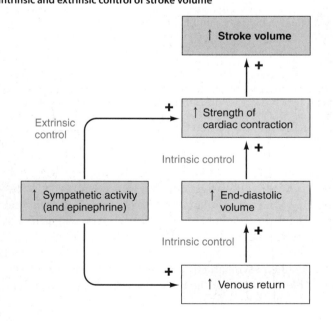

---

volume. This intrinsic control depends on the length–tension relationship of cardiac muscle, which is similar to that of skeletal muscle. For skeletal muscle, the resting muscle length is approximately the optimal length at which maximal tension can be developed during a subsequent contraction. When the skeletal muscle is longer or shorter than this optimal length, the subsequent contraction is weaker (see ● Figure 8-18, p. 272). For cardiac muscle, the resting cardiac muscle fiber length is less than optimal length. Therefore, the length of cardiac muscle fibers normally varies along the ascending limb of the length–tension curve. An increase in cardiac muscle fiber length, by moving closer to the optimal length, increases the contractile tension of the heart on the following systole (● Figure 9-26).

Unlike skeletal muscle, the length–tension curve of cardiac muscle normally does not operate at lengths that fall within the region of the descending limb. That is, within physiological limits cardiac muscle does not get stretched beyond its optimal length to the point that contractile strength diminishes with further stretching.

Frank-Starling law of the heart

What causes cardiac muscle fibers to vary in length before contraction? Skeletal muscle length can vary before contraction because of the positioning of the skeletal parts to which the muscle is attached, but cardiac muscle is not attached to any bones. The main determinant of cardiac muscle fiber length is the degree of diastolic filling. An analogy is a balloon filled with water—the more water you put in, the larger the balloon becomes, and the more it is stretched. Likewise, the greater the extent of diastolic filling, the larger the end-diastolic volume, and the more the heart is stretched. The more the heart is stretched, the longer the initial cardiac-fiber length before contraction. The increased length results in a greater force on the subsequent cardiac contraction and consequently, in a greater stroke volume. This intrinsic relationship between end-diastolic volume and stroke volume is known as the **Frank-Starling law of the heart**. Stated simply, the law says that the heart normally pumps all the blood returned to it; increased venous return results in increased stroke volume. In ● Figure 9-26, assume that the end-diastolic volume increases from point A to point B. You can see that this increase in end-diastolic volume is accompanied by a corresponding increase in stroke volume from point A^1 to point B^1. The extent of filling is referred to as the **preload**, because it is the workload imposed on the heart before contraction begins.

Advantages of the cardiac length–tension relationship

The built-in relationship matching stroke volume with venous return has two important advantages. First, one of the most important functions served by this intrinsic mechanism is equalization of output between the right and left sides of the heart, so that the blood pumped out by the heart is equally distributed between the pulmonary and systemic circulation. If, for example, the right side of the heart ejects a larger stroke volume, more blood enters the pulmonary circulation, so venous return to the left side of the heart is increased accord-

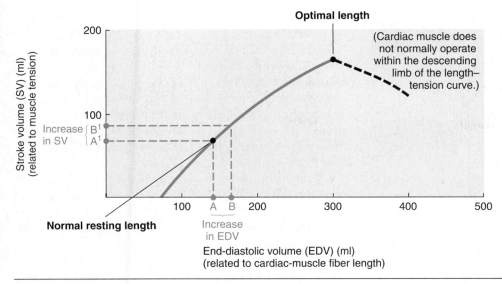

Optimal length

(Cardiac muscle does not normally operate within the descending limb of the length–tension curve.)

Normal resting length

Increase in EDV

End-diastolic volume (EDV) (ml)
(related to cardiac-muscle fiber length)

● **FIGURE 9-26**

Intrinsic control of stroke volume (Frank-Starling curve) The cardiac muscle fiber's length, which is determined by the extent of venous filling, is normally less than the optimal length for developing maximal tension. Therefore, an increase in end-diastolic volume (that is, an increase in venous return), by moving the cardiac muscle fiber length closer to optimal length, increases the contractile tension of the fibers on the next systole. A stronger contraction squeezes out more blood. Thus, as more blood is returned to the heart and the end-diastolic volume increases, the heart automatically pumps out a correspondingly larger stroke volume.

ingly. The increased end-diastolic volume of the left side of the heart causes it to contract more forcefully, so it too pumps out a larger stroke volume. In this way, equality of output of the two ventricular chambers is maintained. If such equalization did not happen, excessive damming of blood would occur in the venous system preceding the ventricle with the lower output.

Second, when a larger cardiac output is needed, such as during exercise, venous return is increased through action of the sympathetic nervous system and other mechanisms to be described in the next chapter. The resultant increase in end-diastolic volume automatically increases stroke volume correspondingly. Because exercise also increases heart rate, these two factors act together to increase the cardiac output so that more blood can be delivered to the exercising muscles.

Mechanism of the cardiac length–tension relationship

Although the length–tension relationship in cardiac muscle fibers depends to a degree on the extent of overlap of thick and thin filaments, similar to the length–tension relationship in skeletal muscle, the key factor relating cardiac muscle fiber length to tension development is a length dependence of myofilament Ca^{2+} sensitivity. Specifically, as a cardiac muscle fiber's length increases along the ascending limb of the length–tension curve, the lateral spacing between the thick and thin filaments is reduced. Stated differently, as a cardiac muscle fiber is stretched as a result of greater ventricular filling, its myofilaments are pulled closer together. As a result of this reduction in distance between the thick and thin filaments, more cross-bridge interactions between myosin and actin can take place when Ca^{2+} pulls the troponin-tropomyosin complex away from actin's cross-bridge binding sites—that is, myofilament Ca^{2+} sensitivity increases. Thus the length–tension relationship in cardiac muscle depends not on muscle fiber length per se, but on the resultant variations in the lateral spacing between the myosin and actin filaments.

We are now going to shift our attention from intrinsic control to extrinsic control of stroke volume.

❚ The contractility of the heart is increased by sympathetic stimulation.

In addition to intrinsic control, stroke volume is also subject to **extrinsic control** by factors originating outside the heart, the most important of which are actions of the cardiac sympathetic nerves and epinephrine (▲ Table 9-3). Sympathetic stimulation and epinephrine enhance the heart's **contractility**, which is the strength of contraction at any given end-diastolic volume. In other words, on sympathetic stimulation, the heart contracts more forcefully and squeezes out a greater percentage of the blood it contains, leading to more complete ejection. This increased contractility is due to the increased Ca^{2+} influx triggered by norepinephrine and epinephrine. The extra cytosolic Ca^{2+} allows the myocardial fibers to generate more force through greater cross-bridge cycling than they would without sympathetic influence. Normally, the end-diastolic volume is 135 ml and the end-systolic volume is 65 ml for a stroke volume of 70 ml (● Figure 9-27a). Under sympathetic influence, for the same end-diastolic volume of 135 ml, the end-systolic volume might be 35 ml and the stroke volume 100 ml (● Figure 9-27b). In effect, sympathetic stimulation shifts the Frank-Starling curve to the left (● Figure 9-28). Depending on the extent of sympathetic stimulation, the curve can be shifted to varying degrees, up to a maximal increase in contractile strength of about 100% greater than normal.

Sympathetic stimulation increases stroke volume not only by strengthening cardiac contractility but also by enhancing venous return (● Figure 9-27c). Sympathetic stimulation constricts the veins, which squeezes more blood forward from the veins to the heart, increasing the end-diastolic volume and subsequently increasing the stroke volume even further.

Summary of factors affecting stroke volume and cardiac output

The strength of cardiac muscle contraction and accordingly the stroke volume can thus be graded by (1) varying the initial length of the muscle fibers, which in turn depends on the de-

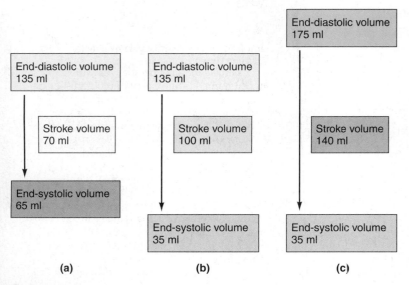

(a)   (b)   (c)

● **FIGURE 9-27**

Effect of sympathetic stimulation on stroke volume
(a) Normal stroke volume. (b) Stroke volume during sympathetic stimulation.
(c) Stroke volume with combination of sympathetic stimulation and increased
end-diastolic volume.

gree of ventricular filling before contraction (intrinsic control); and (2) varying the extent of sympathetic stimulation (extrinsic control) (● Figure 9-25). This is in contrast to gradation of skeletal muscle. In skeletal muscle, twitch summation and recruitment of motor units are employed to produce variable strength of muscle contraction, but these mechanisms are not applicable to cardiac muscle. Twitch summation is impossible because of the long refractory period. Recruitment of motor units is not possible because the heart muscle cells are arranged into functional syncytia that become excited and contract with

● FIGURE 9-28

Shift of the Frank-Starling curve to the left by sympathetic stimulation
For the same end-diastolic volume (point A), there is a larger stroke volume (from point B to point C) on sympathetic stimulation as a result of increased contractility of the heart. The Frank-Starling curve is shifted to the left by variable degrees, depending on the extent of sympathetic stimulation.

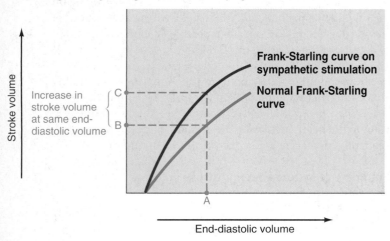

every beat, instead of into distinct motor units that can be discretely activated.

All the factors that determine the cardiac output by influencing the heart rate or stroke volume are summarized in ● Figure 9-29. Note that sympathetic stimulation increases the cardiac output by increasing both heart rate and stroke volume. Sympathetic activity to the heart increases, for example, during exercise when the working skeletal muscles need increased delivery of O_2-laden blood to support their high rate of ATP consumption.

We are next going to examine how the afterload influences the ability of the heart to pump out blood, then how a failing heart is unable to pump out sufficient blood, before turning to the final section of this chapter, on how the heart muscle is nourished.

▌ High blood pressure increases the workload of the heart.

When the ventricles contract, to force open the semilunar valves they must generate sufficient pressure to exceed the blood pressure in the major arteries. The arterial blood pressure is called the **afterload**, because it is the workload imposed on the heart after the contraction has begun. If the arterial blood pressure is chronically elevated (high blood pressure) or if the exit valve is stenotic, the ventricle must generate more pressure to eject

● **FIGURE** 9-29

Control of cardiac output
Because cardiac output equals heart rate times stroke volume, this figure is a composite of ● Figure 9-24 (control of heart rate) and ● Figure 9-25 (control of stroke volume).

blood. For example, instead of generating the normal pressure of 120 mm Hg, the ventricular pressure may need to rise as high as 400 mm Hg to force blood through a narrowed aortic valve.

The heart may be able to compensate for a sustained increase in afterload by enlarging (through hypertrophy or enlargement of the cardiac muscle fibers; see p. 281). This enables it to contract more forcefully and maintain a normal stroke volume despite an abnormal impediment to ejection. A diseased heart or a heart weakened with age may not be able to compensate completely, however; in that case, heart failure ensues. Even if the heart is initially able to compensate for a chronic increase in afterload, the sustained extra workload placed on the heart can eventually cause pathologic changes in the heart that lead to heart failure. In fact, a chronically elevated afterload is one of the two major factors that cause heart failure, a topic to which we now turn our attention.

The contractility of the heart is decreased in heart failure.

The term **heart failure** refers to the inability of the cardiac output to keep pace with the body's demands for supplies and removal of wastes. Either one or both ventricles may progressively weaken and fail. When a failing ventricle is unable to pump out all the blood returned to it, the veins behind the failing ventricle become congested with blood. Heart failure may occur for a variety of reasons, but the two most common are (1) damage to the heart muscle as a result of a heart attack or impaired circulation to the cardiac muscle and (2) prolonged pumping against a chronically increased afterload, as with a stenotic semilunar valve or a sustained elevation in blood pressure. Heart failure presently affects about 5 million Americans, nearly 50% of whom will die within four years of diagnosis.

Prime defect in heart failure

The prime defect in heart failure is a decrease in cardiac contractility; that is, the cardiac muscle cells contract less effectively. The intrinsic ability of the heart to develop pressure and eject a stroke volume is reduced so that the heart operates on a lower length–tension curve (● Figure 9-30a). The Frank-Starling curve is shifted downward and to the right such that for a given end-diastolic volume, a failing heart will pump out a smaller stroke volume than a normal healthy heart.

Compensatory measures for heart failure

Two major compensatory measures help restore the stroke volume to normal in the early stages of heart failure. First, sympathetic activity to the heart is reflexly increased, which increases the contractility of the heart toward normal (● Figure 9-30b). Sympathetic stimulation can help compensate only for a limited period of time, however. The heart becomes

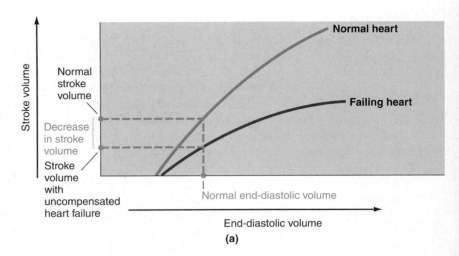

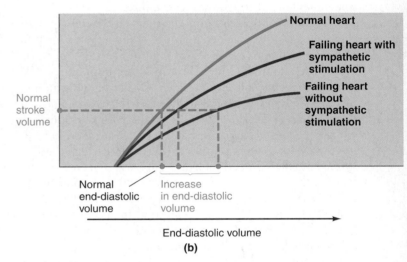

● **FIGURE 9-30**

Compensated heart failure
(a) Shift of the Frank-Starling curve downward and to the right in a failing heart. Because its contractility is decreased, the failing heart pumps out a smaller stroke volume at the same end-diastolic volume than a normal heart does. (b) Compensations for heart failure. Reflex sympathetic stimulation shifts the Frank-Starling curve of a failing heart to the left, increasing the contractility of the heart toward normal. A compensatory increase in end-diastolic volume as a result of blood volume expansion further increases the strength of contraction of the failing heart. Operating at a longer cardiac muscle fiber length, a compensated failing heart is able to eject a normal stroke volume.

less responsive to norepinephrine after prolonged exposure, and furthermore, the norepinephrine stores in the sympathetic nerve terminals in the heart become depleted. Second, when cardiac output is reduced, the kidneys, in a compensatory attempt to improve their reduced blood flow, retain extra salt and water in the body during urine formation to expand the blood volume. The increase in circulating blood volume increases the end-diastolic volume. The resultant stretching of the cardiac muscle fibers enables the weakened heart to pump out a normal stroke volume (● Figure 9-30b). The heart is now pumping out the blood returned to it but is operating at a longer cardiac muscle fiber length.

Decompensated heart failure

As the disease progresses and the contractility of the heart deteriorates further, the heart reaches a point at which it is no longer able to pump out a normal stroke volume (that is, cannot pump out all the blood returned to it) despite compensatory measures. At this point, the heart slips from compensated heart failure to a state of decompensated heart failure. During this state, the cardiac muscle fibers are stretched to the point that they are operating in the descending limb of the length-tension curve. *Backward failure* occurs as blood that cannot enter and be pumped out by the heart continues to dam up in the venous system. *Forward failure* occurs simultaneously as the heart fails to pump an adequate amount of blood forward to the tissues because the stroke volume becomes progressively smaller. The congestion in the venous system is the reason this condition is sometimes termed **congestive heart failure.**

Left-sided failure has more serious consequences than right-sided failure. Backward failure of the left side leads to pulmonary edema (excess tissue fluid in the lungs) because blood dams up in the lungs. This fluid accumulation in the lungs reduces exchange of O_2 and CO_2 between the air and blood in the lungs, leading to reduced arterial oxygenation and an elevation of acid-forming CO_2 in the blood. In addition, one of the more serious consequences of left-sided forward failure is an inadequate blood flow to the kidneys, which causes a twofold problem. First, vital kidney function is depressed; and second, the kidneys retain even more salt and water in the body during urine formation as they attempt to expand the plasma volume even further to improve their reduced blood flow. Excessive fluid retention further exacerbates the already existing problems of venous congestion.

Treatment of congestive heart failure therefore includes measures that reduce salt and water retention and increase urinary output as well as drugs that enhance the contractile ability of the weakened heart—digitalis, for example.

NOURISHING THE HEART MUSCLE

Cardiac muscle cells contain an abundance of mitochondria, the O_2-dependent energy organelles. In fact, up to 40% of the cell volume of cardiac muscle cells is occupied by mitochondria, indicative of how much the heart depends on O_2 delivery and aerobic metabolism to generate the energy necessary for contraction (see p. 40). Cardiac muscle also has an abundance of myoglobin, which stores limited amounts of O_2 within the heart for immediate use.

▌The heart receives most of its own blood supply through the coronary circulation during diastole.

Although all the blood passes through the heart, the heart muscle is unable to extract O_2 or nutrients from the blood within its chambers for two reasons. First, the watertight endocardial lining does not permit blood to pass from the chamber into the myocardium. Second, the heart walls are too thick to permit diffusion of O_2 and other supplies from the blood in the chamber to the individual cardiac cells. Therefore, like other tissues of the body, heart muscle must receive blood through blood vessels, specifically by means of the **coronary circulation.** The coronary arteries branch from the aorta just beyond the aortic valve (see ● Figure 9-35, p. 336), and the coronary veins empty into the right atrium.

The heart muscle receives most of its blood supply during diastole. Blood flow to the heart muscle cells is substantially reduced during systole for two reasons. First, the major branches of the coronary arteries are compressed by the contracting myocardium, and, second, the entrance to the coronary vessels is partially blocked by the open aortic valve. Thus most coronary arterial flow (about 70%) occurs during diastole, driven by the aortic blood pressure, with flow declining as aortic pressure drops. Only about 30% of coronary arterial flow occurs during systole (● Figure 9-31).

The limited time for coronary blood flow becomes especially important during rapid heart rates, when diastolic time is substantially reduced. Just when increased demands are placed on the heart to pump more rapidly, it has less time to provide O_2 and nourishment to its own musculature to accomplish the increased workload.

Matching of coronary blood flow to heart muscle's O_2 needs

Nevertheless, under normal circumstances the heart muscle does receive adequate blood flow to support its activities, even during exercise, when the rate of coronary blood flow increases up to five times its resting rate. Let's see how.

Increased delivery of blood to the cardiac cells is accomplished primarily by vasodilation, or enlargement, of the coronary vessels, which allows more blood to flow through them,

---

● **FIGURE 9-31**

Coronary blood flow
Most coronary blood flow occurs during diastole because the coronary vessels are compressed almost completely closed during systole.

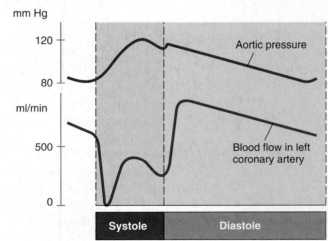

especially during diastole. The increased coronary blood flow is necessary to meet the heart's increased O_2 requirements because the heart, unlike most other tissues, is unable to remove much additional O_2 from the blood passing through its vessels to support increased metabolic activities. Most other tissues under resting conditions extract only about 25% of the O_2 available from the blood flowing through them, leaving a considerable O_2 reserve that can be drawn on when a tissue has increased O_2 needs; that is, the tissue can immediately increase the O_2 available to it by removing a greater percentage of O_2 from the blood passing through it. In contrast, the heart, even under resting conditions, removes up to 65% of the O_2 available in the coronary vessels, far more than is withdrawn by other tissues. This leaves little O_2 in reserve in the coronary blood should cardiac O_2 demands increase. Therefore, the primary means by which more O_2 can be made available to the heart muscle is by increasing coronary blood flow.

Coronary blood flow is adjusted primarily in response to changes in the heart's O_2 requirements. The major link that coordinates coronary blood flow with myocardial O_2 needs is *adenosine*, which is formed from adenosine triphosphate (ATP) during cardiac metabolic activity. Increased formation and release of adenosine from the cardiac cells occur (1) when there is a cardiac O_2 deficit or (2) when cardiac activity is increased and the heart accordingly requires more O_2 and is using more ATP as an energy source. The released adenosine induces dilation of the coronary blood vessels, thereby allowing more O_2-rich blood to flow to the more active cardiac cells to meet their increased O_2 demand (● Figure 9-32). This matching of O_2 delivery with O_2 needs is critical because the heart muscle depends on oxidative processes to generate energy. The heart cannot obtain sufficient ATP through anaerobic metabolism.

● **FIGURE 9-32**

Matching of coronary blood flow to the oxygen need of cardiac muscle cells

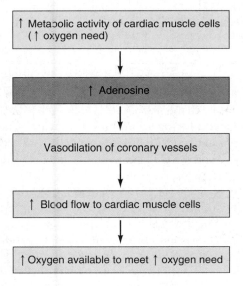

Nutrient supply to the heart

Although the heart has little ability to support its energy needs by means of anaerobic metabolism and must rely heavily on its O_2 supply, it can tolerate wide variations in its nutrient supply. As fuel sources, the heart primarily uses free fatty acids and, to a lesser extent, glucose and lactate, depending on their availability. Because the cardiac muscle is remarkably adaptable and can shift metabolic pathways to use whatever nutrient is available, the primary danger of insufficient coronary blood flow is not fuel shortage but O_2 deficiency.

▌ Atherosclerotic coronary artery disease can deprive the heart of essential oxygen.

Adequacy of coronary blood flow is relative to the heart's O_2 demands at any given moment. In the normal heart, coronary blood flow increases correspondingly as O_2 demands rise. With **coronary artery disease,** however, it may not be possible for coronary blood flow to keep pace with rising O_2 needs. The term *coronary artery disease* refers to pathologic changes within the walls of the coronary arteries that diminish blood flow through these vessels. A given rate of coronary blood flow may be adequate at rest but insufficient on physical exertion or other stressful situations.

Complications of coronary artery disease, including heart attacks, make it the single leading cause of death in the United States. Coronary artery disease is the underlying cause of about 50% of all deaths in this country.

Coronary artery disease can cause myocardial ischemia and possibly lead to acute myocardial infarction by three mechanisms: (1) profound vascular spasm of the coronary arteries, (2) the formation of atherosclerotic plaques, and (3) thromboembolism. We will discuss each in turn.

Vascular spasm

Vascular spasm is an abnormal spastic constriction that transiently narrows the coronary vessels; it is most often triggered by exposure to cold, physical exertion, or anxiety. Vascular spasms are associated with the early stages of coronary artery disease. The condition is reversible and usually of insufficient duration to produce damage to the cardiac muscle.

Recent evidence suggests that reduced O_2 availability in the coronary vessels causes the release of **platelet-activating factor (PAF)** from the endothelium (blood vessel lining). PAF, which exerts a variety of biological actions, was named for its first discovered effect, activating platelets. Among its other effects, PAF, once released from the endothelium, diffuses to the underlying vascular smooth muscle and causes it to contract, bringing about vascular spasm.

Development of atherosclerosis

Atherosclerosis is a progressive, degenerative arterial disease that leads to occlusion (gradual blockage) of affected vessels, reducing blood flow through them. Atherosclerosis is characterized by the formation of plaques within arterial walls. An **atherosclerotic plaque** consists of a lipid-rich core covered by an abnormal overgrowth of smooth muscle cells, topped off by a collagen-rich connective tissue cap. As this plaque forms beneath the vessel lining, it forms a bulge that protrudes into the vessel lumen (● Figure 9-33).

Although not all the contributing factors have been identified, in recent years investigators have sorted out the following complex sequence of events involved in the gradual development of atherosclerosis:

1. Atherosclerosis is believed to start with injury to the blood vessel wall, which triggers an *inflammatory response* that sets the stage for the buildup of plaque. Normally inflammation is a protective response that fights infection and promotes repair of damaged tissue (see p. 416). However, when the cause of the injury persists within the vessel wall, the sustained, low-grade inflammatory response over a course of decades can insidiously lead to arterial plaque formation and heart disease. Suspected artery-abusing agents that may set off the vascular inflammatory response included oxidized cholesterol, free radicals, high blood pressure, homocysteine, or even bacteria and viruses that damage blood vessel walls. The most common triggering agent appears to be oxidized cholesterol. (For a further discussion of the role of cholesterol and other factors in the development of atherosclerosis, see the accompanying boxed feature, ❱ Concepts, Challenges, and Controversies.)

2. Typically, the initial stage of atherosclerosis is characterized by the accumulation beneath the endothelium of excessive amounts of **low-density lipoprotein (LDL)**, the so-called "bad" cholesterol, in combination with a protein carrier. As LDL accumulates within the vessel wall, this cholesterol product becomes oxidized, primarily by oxidative wastes produced by the blood vessel cells. These wastes are *free radicals*, very unstable electron-deficient particles that are highly reactive. Antioxidant vitamins that prevent LDL oxidation, such as *vitamin E*, *vitamin C*, and *beta-carotene*, have been shown to slow plaque deposition.

3. In response to the presence of oxidized LDL and/or other irritants, the endothelial cells produce chemicals that attract *monocytes*, a type of white blood cell, to the site. These immune cells trigger a local inflammatory response.

4. Once they leave the blood and enter the vessel wall, monocytes settle down permanently, enlarge, and become large phagocytic cells called *macrophages*. Macrophages voraciously phagocytize (see p. 32) the oxidized LDL until these cells become so packed with fatty droplets that they appear foamy under a microscope. Now called **foam cells**, these greatly engorged macrophages accumulate beneath the vessel lining and form a visible **fatty streak**, the earliest form of an atherosclerotic plaque.

5. Thus the earliest stage of plaque formation is characterized by the accumulation beneath the endothelium of a cholesterol-rich deposit. The disease progresses as smooth muscle cells within the blood vessel wall migrate from the muscular layer of the blood vessel to a position on top of the lipid accumulation, just beneath the endothelium. This migration is triggered by chemicals released at the inflammatory site. At

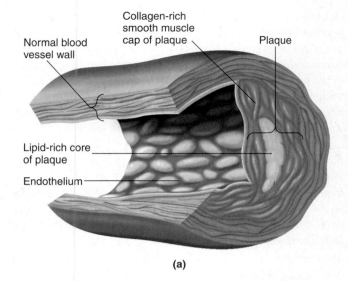

(a)

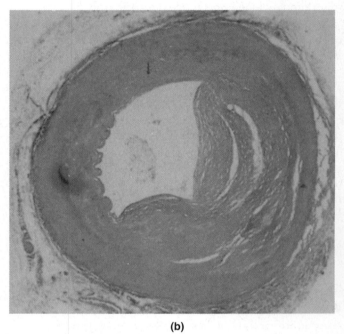

(b)

● **FIGURE 9-33**
Atherosclerotic plaque
(a) Schematic representation of the components of a plaque. (b) Photomicrograph of a severe atherosclerotic plaque in a coronary vessel.

their new location, the smooth muscle cells continue to divide and enlarge, producing **atheromas**, which are benign (noncancerous) tumors of smooth muscle cells within the blood vessel walls. Together the lipid-rich core and overlying smooth muscle form a maturing plaque.

6. The plaque progressively bulges into the lumen of the vessel as it continues to develop. The protruding plaque narrows the opening through which blood can flow.

7. Further contributing to vessel narrowing, oxidized LDL inhibits the release of *nitric oxide* from the endothelial cells. Nitric oxide is a local chemical messenger that causes relaxation of the underlying layer of normal smooth-muscle cells

within the vessel wall. Relaxation of these smooth muscle cells causes the vessel to dilate. Because of reduced nitric oxide release, vessels damaged by developing plaques cannot dilate as readily as normal.

8. A thickening plaque also interferes with nutrient exchange for the cells located within the involved arterial wall, leading to degeneration of the wall in the vicinity of the plaque. The damaged area is invaded by *fibroblasts* (scar-forming cells), which form a connective tissue cap over the plaque. (The term *sclerosis* means excessive growth of fibrous connective tissue, hence the term *atherosclerosis* for this condition characterized by atheromas and sclerosis, along with abnormal lipid accumulation.)

9. In the later stages of the disease, Ca^{2+} often precipitates in the plaque. A vessel so afflicted becomes hard and poorly distensible.

Thromboembolism and other complications of atherosclerosis

Atherosclerosis attacks arteries throughout the body, but the most serious consequences involve damage to the vessels of the brain and heart. In the brain, atherosclerosis is the prime cause of strokes, whereas in the heart it brings about myocardial ischemia and its complications. The following are potential complications of coronary atherosclerosis:

- *Angina pectoris.* Gradual enlargement of the protruding plaque continues to narrow the vessel lumen and progressively diminishes coronary blood flow, triggering increasingly frequent bouts of transient myocardial ischemia as the ability to match blood flow with cardiac O_2 needs becomes more limited. Although the heart cannot normally be "felt," pain is associated with myocardial ischemia. Such cardiac pain, known as **angina pectoris** ("pain of the chest"), can be felt beneath the sternum and is often referred to the left shoulder and down the left arm (see p. 177). The symptoms of angina pectoris recur whenever cardiac O_2 demands become too great in relation to the coronary blood flow—for example, during exertion or emotional stress. The pain is thought to result from stimulation of cardiac nerve endings by the accumulation of lactic acid when the heart shifts to its limited ability to perform anaerobic me-

tabolism (see p. 40). The ischemia associated with the characteristically brief anginal attacks is usually temporary and reversible and can be relieved by rest, administration of vasodilator drugs such as *nitroglycerin*, or both. Nitroglycerin brings about coronary vasodilation by being metabolically converted to nitric oxide, which in turn relaxes the vascular smooth muscle.

- *Thromboembolism.* The enlarging atherosclerotic plaque can break through the weakened endothelial lining that covers it, exposing blood to the underlying collagen in the collagen-rich connective tissue cap of the plaque. Foam cells release chemicals that can weaken the fibrous cap of a plaque by breaking down the connective tissue fibers. Plaques with thick fibrous caps are considered stable because they are not likely to rupture. However, plaques with thinner fibrous caps are considered unstable because they are likely to rupture and trigger clot formation.

Blood platelets (formed elements of the blood involved in plugging vessel defects and in clot formation) normally do not adhere to smooth, healthy vessel linings. However, when platelets come into contact with collagen at the site of vessel damage, they stick to the site and contribute to the formation of a blood clot. Furthermore, foam cells produce a potent clot promoter. Such an abnormal clot attached to a vessel wall is known as a **thrombus.** The thrombus may enlarge gradually until it completely blocks the vessel at that site, or the continued flow of blood past the thrombus may break it loose from its attachment. Such a freely floating clot, or **embolus,** may completely plug a smaller vessel as it flows downstream ($\bullet$ Figure 9-34). Thus, through **thromboembolism** atherosclerosis can result in a gradual or sudden occlusion of a coronary vessel (or any other vessel).

- *Heart attack.* When a coronary vessel is completely plugged, the cardiac tissue served by the vessel soon dies from O_2 deprivation, and a heart attack occurs, unless the area can be supplied with blood from nearby vessels.

Sometimes a deprived area is fortunate enough to receive blood from more than one pathway. **Collateral circulation** exists when small terminal branches from adjacent blood vessels nourish the same area. These accessory vessels cannot develop suddenly following an abrupt blockage but may be lifesaving if already developed. Such alternate vascular pathways often de-

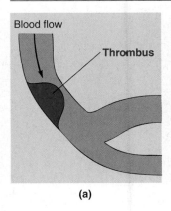

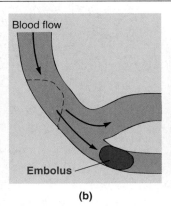

(a)   (b)   (c)

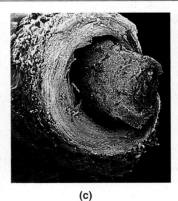

$\bullet$ **FIGURE 9-34**

Consequences of thromboembolism (a) A thrombus may enlarge gradually until it completely occludes the vessel at that site. (b) A thrombus may break loose from its attachment, forming an embolus that may completely occlude a smaller vessel downstream. (c) Scanning electron micrograph of a vessel completely occluded by a thromboembolic lesion.

Atherosclerosis: Cholesterol and Beyond

The cause of atherosclerosis is still not entirely clear. Certain high-risk factors have been associated with an increased incidence of atherosclerosis and coronary heart disease. Included among them are genetic predisposition, obesity, advanced age, smoking, hypertension, lack of exercise, high blood concentrations of C-reactive protein, elevated levels of homocysteine, infectious agents, and, most notoriously, elevated cholesterol levels in the blood.

Sources of Cholesterol

There are two sources of cholesterol for the body: (1) dietary intake of cholesterol, with animal products such as egg yolk, red meats, and butter being especially rich in this lipid (animal fats contain cholesterol, whereas plant fats typically do not); and (2) manufacture of cholesterol by cells, especially liver cells.

"Good" versus "Bad" Cholesterol

Actually, it is not the total blood-cholesterol level but the amount of cholesterol bound to various plasma protein carriers that is most important with regard to the risk of developing atherosclerotic heart disease. Because cholesterol is a lipid, it is not very soluble in blood. Most cholesterol in the blood is attached to specific plasma-protein carriers in the form of lipoprotein complexes, which are soluble in blood. There are three major lipoproteins, named for their density of protein as compared to lipid: (1) **high-density lipoproteins (HDL),** which contain the most protein and least cholesterol; (2) **low-density lipopro-**teins (LDL), which contain less protein and more cholesterol; and (3) **very-low-density lipoproteins (VLDL),** which contain the least protein and most lipid, but the lipid they carry is neutral fat, not cholesterol.

Cholesterol carried in LDL complexes has been termed "bad" cholesterol, because cholesterol is transported *to* the cells, including those lining the blood vessel walls, by means of LDL. The propensity toward developing atherosclerosis substantially increases with elevated levels of LDL. The presence of oxidized LDL within an arterial wall is a major trigger for the inflammatory process that leads to the development of atherosclerotic plaques (see p. 334).

In contrast, cholesterol carried in HDL complexes has been dubbed "good" cholesterol, because HDL removes cholesterol *from* the cells and transports it to the liver for partial elimination from the body. Not only does HDL help remove excess cholesterol from the tissues, but in addition, it protects by inhibiting oxidation of LDL. The risk of atherosclerosis is inversely related to the concentration of HDL in the blood; that is, elevated levels of HDL are associated with a low incidence of atherosclerotic heart disease.

Some other factors known to influence atherosclerotic risk can be related to HDL levels; for example, cigarette smoking lowers HDL, whereas regular exercise raises HDL.

Cholesterol Uptake by Cells

Unlike most lipids, cholesterol is not used as metabolic fuel by cells. Instead, it serves as an essential component of plasma membranes. In addition, a few special cell types use cholesterol as a precursor for the synthesis of secretory products, such as steroid hormones and bile salts. Although most cells can synthesize some of the cholesterol needed for their own plasma membranes, they cannot manufacture sufficient amounts and therefore must rely on supplemental cholesterol being delivered by the blood.

Cells accomplish cholesterol uptake from the blood by synthesizing receptor proteins specifically capable of binding LDL and inserting these receptors into the plasma membrane. When an LDL particle binds to one of the membrane receptors, the cell engulfs the particle by endocytosis, receptor and all (see p. 32). Within the cell, lysosomal enzymes break down the LDL to free the cholesterol, making it available to the cell for synthesis of new cellular membrane. The LDL receptor, which is also freed within the cell, is recycled back to the surface membrane.

If too much free cholesterol accumulates in the cell, there is a shutdown of both the synthesis of LDL receptor proteins (so that less cholesterol is taken up) and the cell's own cholesterol synthesis (so that less new cholesterol is made). Faced with a cholesterol shortage, in contrast, the cell makes more LDL receptors so that it can engulf more cholesterol from the blood.

Maintenance of Blood Cholesterol Level and Cholesterol Metabolism

The maintenance of a blood-borne cholesterol supply to the cells involves an interaction between dietary cholesterol and the synthesis of cholesterol by the liver. When the amount of dietary cholesterol is increased, hepatic (liver) synthesis of cholesterol is turned off because cho-

---

velop over a period of time when an atherosclerotic constriction progresses slowly, or they may be induced by sustained demands on the heart through a regular aerobic exercise program.

In the absence of collateral circulation, the extent of the damaged area during a heart attack depends on the size of the blocked vessel. The larger the vessel occluded, the greater the area deprived of its blood supply. As ● Figure 9-35 illustrates, a blockage at point A in the coronary circulation would cause more extensive damage than would a blockage at point B. Because there are only two major coronary arteries, complete blockage of either one of these main branches results in extensive myocardial damage. Left coronary-artery blockage is most devastating because this vessel is responsible for supplying 85% of the cardiac tissue.

A heart attack has four possible outcomes: immediate death, delayed death from complications, full functional recovery, or recovery with impaired function (▲ Table 9-4).

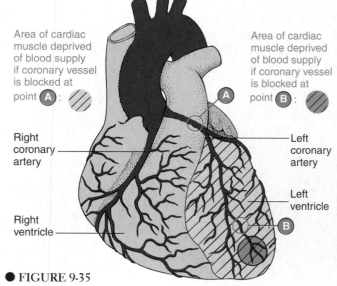

● **FIGURE 9-35**

Extent of myocardial damage as a function of the size of the occluded vessel

lesterol in the blood directly inhibits a hepatic enzyme essential for cholesterol synthesis. Thus, as more cholesterol is ingested, less is produced by the liver. Conversely, when cholesterol intake from food is reduced, the liver synthesizes more of this lipid because the inhibitory effect of cholesterol on the crucial hepatic enzyme is removed. In this way, the blood concentration of cholesterol is maintained at a fairly constant level despite changes in cholesterol intake; thus it is difficult to significantly reduce cholesterol levels in the blood by decreasing cholesterol intake.

HDL transports cholesterol to the liver. The liver secretes cholesterol as well as cholesterol-derived bile salts into the bile. Bile enters the intestinal tract, where bile salts participate in the digestive process. Most of the secreted cholesterol and bile salts are subsequently reabsorbed from the intestine into the blood to be recycled to the liver. However, the cholesterol and bile salts not reclaimed by absorption are eliminated in the feces and lost from the body.

Thus the liver has a primary role in determining total blood-cholesterol levels, and the interplay between LDL and HDL determines the traffic flow of cholesterol between the liver and the other cells. Whenever these mechanisms are altered, blood cholesterol levels may be affected in such a way as to influence the individual's predisposition to atherosclerosis.

Varying the intake of dietary fatty acids may alter total blood-cholesterol levels by influencing one or more of the mechanisms involving cholesterol balance. The blood cholesterol level tends to be raised by ingesting saturated fatty acids found predominantly in animal fats, be-cause these fatty acids stimulate the synthesis of cholesterol and inhibit its conversion to bile salts. In contrast, ingesting polyunsaturated fatty acids, the predominant fatty acids of most plants, tend to reduce blood-cholesterol levels by enhancing the elimination of both cholesterol and cholesterol-derived bile salts in the feces.

Other Risk Factors Besides Cholesterol
Despite the strong links between cholesterol and heart disease, over half of all patients with heart attacks have a normal cholesterol profile and no other well-established risk factors. Clearly, other factors are involved in the development of coronary artery disease in these individuals. These same factors may also contribute to the development of atherosclerosis in those with unfavorable cholesterol levels. Following are other possible risk factors being investigated:

- Elevated blood levels of the amino acid **homocysteine** have recently been implicated as a strong predictor for heart disease, independent of the person's cholesterol/lipid profile. Homocysteine is formed as an intermediate product during the metabolism of the essential dietary amino acid *methionine*. Investigators believe that homocysteine contributes to atherosclerosis by promoting the proliferation of vascular smooth muscle cells, an early step in the development of this artery-clogging condition. Furthermore, homocysteine appears to damage the endothelial cells and may cause the oxidation of LDL, both of which can contribute to plaque formation. Three B vitamins—*folic acid, vitamin B₁₂, and vitamin B₆*—all play key roles in pathways that clear homocysteine from the

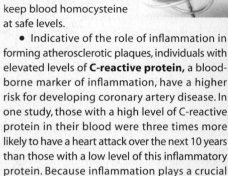

blood. Therefore, these B vitamins are all needed to keep blood homocysteine at safe levels.

- Indicative of the role of inflammation in forming atherosclerotic plaques, individuals with elevated levels of **C-reactive protein,** a blood-borne marker of inflammation, have a higher risk for developing coronary artery disease. In one study, those with a high level of C-reactive protein in their blood were three times more likely to have a heart attack over the next 10 years than those with a low level of this inflammatory protein. Because inflammation plays a crucial role in the development of atherosclerosis, anti-inflammatory drugs, such as aspirin, help prevent heart attacks. Furthermore, aspirin protects against heart attacks through its role in inhibiting clot formation.

- Accumulating data suggest that an infectious agent may be the underlying culprit in a significant number of cases of atherosclerotic disease. Among the leading suspects are respiratory infection–causing *Chlamydia pneumoniae,* cold sore–causing *herpes virus,* and gum disease–causing bacteria. Importantly, if a link between infections and coronary artery disease can be confirmed, antibiotics may be added to the regimen of heart disease prevention strategies.

As you can see, the relationship between atherosclerosis, cholesterol, and other factors is far from clear. Much research concerning this complex disease is currently in progress, because the incidence of atherosclerosis is so high and its consequences are potentially fatal.

▲ TABLE 9-4
Possible Outcomes of Acute Myocardial Infarction (Heart Attack)

Immediate Death	Delayed Death from Complications	Full Functional Recovery	Recovery with Impaired Function
Acute cardiac failure occurring because the heart is too weakened to pump effectively to support the body tissues Fatal ventricular fibrillation brought about by damage to the specialized conducting tissue or induced by O₂ deprivation	Fatal rupture of the dead, degenerating area of the heart wall Slowly progressing congestive heart failure occurring because the weakened heart is unable to pump out all the blood returned to it	Replacement of the damaged area with a strong scar, accompanied by enlargement of the remaining normal contractile tissue to compensate for the lost cardiac musculature	Persistence of permanent functional defects, such as bradycardia or conduction blocks, caused by destruction of irreplaceable autorhythmic or conductive tissues

CHAPTER IN PERSPECTIVE: FOCUS ON HOMEOSTASIS

Survival depends on continual delivery of needed supplies to all the cells throughout the body and on ongoing removal of wastes generated by the cells. Furthermore, regulatory chemical messengers, such as hormones, must be transported from their site of production to their site of action, where they control a variety of activities, most of which are directed toward maintaining a stable internal environment.

The circulatory system contributes to homeostasis by serving as the body's transport system. It provides a means of rapidly moving materials from one part of the body to another. Without the circulatory system, materials would not get where they need to go to support life-sustaining activities nearly rapidly enough. For example, O_2 would take months to years to diffuse from the surface of the body to internal organs, yet O_2 and other substances can be picked up by the blood and delivered to all the cells in a few seconds through the heart's swift pumping action.

The heart serves as a dual pump to continuously circulate blood between the lungs, where O_2 is picked up, and the other body tissues, which use O_2 to support their energy-generating chemical re-

actions. As blood is pumped through the various tissues, other substances besides O_2 are also exchanged between the blood and tissues. For example, the blood picks up nutrients as it flows through the digestive organs, and other tissues remove nutrients from the blood as it flows through them.

Although all the body tissues constantly depend on the life-supporting blood flow provided to them by the heart, the heart itself is quite an independent organ. It is able to take care of many of its own needs without any outside influence. Contraction of this magnificent muscle is self-generated through a carefully orchestrated interplay of changing ionic permeabilities. Local mechanisms within the heart ensure that blood flow to the cardiac muscle normally meets the heart's need for O_2. In addition, the heart has built-in capabilities to vary its strength of contraction, depending on the amount of blood returned to it. The heart does not act entirely autonomously, however. It is innervated by the autonomic nervous system and is influenced by the hormone epinephrine, both of which can vary the rate and contractility of the heart, depending on the body's needs for blood delivery. Furthermore, as with all tissues, the cells that compose the heart depend on the other body systems for maintenance of a stable internal environment in which they can survive and function.

CHAPTER SUMMARY

Introduction
- The circulatory system is the transport system of the body.
- The three basic components of the circulatory system are the heart (the pump), the blood vessels (the passageways), and the blood (the transport medium).

Anatomy of the Heart
- The heart is basically a dual pump that provides the driving pressure for blood flow through the pulmonary and systemic circulations.
- The heart has four chambers: Each half of the heart consists of an atrium, or venous input chamber, and a ventricle, or arterial output chamber.
- Four heart valves direct the blood in the proper direction and prevent it from flowing in the reverse direction.
- Contraction of the spirally arranged cardiac muscle fibers produces a wringing effect important for efficient pumping. Also important for efficient pumping is the fact that the muscle fibers in each chamber act as a functional syncytium, contracting as a coordinated unit.

Electrical Activity of the Heart
- The heart is self-excitable, initiating its own rhythmic contractions.
- One percent of the cardiac muscle cells are autorhythmic cells, which do not contract but are specialized to initiate and conduct action potentials. These cells display a pacemaker potential, a slow drift to threshold potential, as a result of a complex interplay of inherent changes in ion movement across the membrane. The other 99% of the cardiac cells are contractile cells that contract in response to the spread of an action potential initiated by autorhythmic cells.
- The cardiac impulse originates at the SA node, the pacemaker of the heart, which has the fastest rate of spontaneous depolarization to threshold.

- Once initiated, the action potential spreads throughout the right and left atria, partially facilitated by specialized conduction pathways but mostly by cell-to-cell spread of the impulse through gap junctions.
- The impulse passes from the atria into the ventricles through the AV node, the only point of electrical contact between these chambers. The action potential is delayed briefly at the AV node, ensuring that atrial contraction precedes ventricular contraction to allow complete ventricular filling.
- The impulse then travels rapidly down the interventricular septum via the bundle of His and is rapidly dispersed throughout the myocardium by means of the Purkinje fibers. The remainder of the ventricular cells are activated by cell-to-cell spread of the impulse through gap junctions.
- Thus the atria contract as a single unit, followed after a brief delay by a synchronized ventricular contraction.
- The action potentials of cardiac contractile cells exhibit a prolonged positive phase, or plateau, accompanied by a prolonged period of contraction, which ensures adequate ejection time. This plateau is primarily due to activation of slow Ca^{2+} channels.
- Because a long refractory period occurs in conjunction with this prolonged plateau phase, summation and tetanus of cardiac muscle are impossible, thereby ensuring the alternate periods of contraction and relaxation essential for pumping of blood.
- The spread of electrical activity throughout the heart can be recorded from the surface of the body. This record, the ECG, can provide useful information about the status of the heart.

Mechanical Events of the Cardiac Cycle
- The cardiac cycle consists of three important events:
  1. The generation of electrical activity as the heart autorhythmically depolarizes and repolarizes.

2. Mechanical activity consisting of alternate periods of systole (contraction and emptying) and diastole (relaxation and filling), which are initiated by the rhythmic electrical cycle.

3. Directional flow of blood through the heart chambers, guided by valvular opening and closing induced by pressure changes that are generated by mechanical activity.

▌ Valve closing gives rise to two normal heart sounds. The first heart sound is caused by closure of the atrioventricular (AV) valves and signals the onset of ventricular systole. The second heart sound is due to closure of the aortic and pulmonary valves at the onset of diastole.

▌ As seen in ● Figure 9-21 on p. 322, the atrial pressure curve remains low throughout the entire cardiac cycle, with only minor fluctuations (normally varying between 0 and 8 mm Hg). The aortic pressure curve remains high the entire time, with moderate fluctuations (normally varying between a systolic pressure of 120 mm Hg and a diastolic pressure of 80 mm Hg). The ventricular pressure curve fluctuates dramatically because ventricular pressure must be below the low atrial pressure during diastole to allow the AV valve to open so filling can take place, and to force the aortic valve open to allow emptying to occur, it must be above the high aortic pressure during systole. Therefore, ventricular pressure normally varies from 0 mm Hg during diastole to slightly more than 120 mm Hg during systole.

▌ The end-diastolic volume is the volume of blood in the ventricle when filling is complete at the end of diastole. The end-systolic volume is the volume of blood remaining in the ventricle when ejection is complete at the end of systole. The stroke volume is the volume of blood pumped out by each ventricle each beat.

▌ Defective valve function produces turbulent blood flow, which is audible as a heart murmur. Abnormal valves may be either stenotic and not open completely or insufficient and not close completely.

Cardiac Output and Its Control

▌ Cardiac output, the volume of blood ejected by each ventricle each minute, is determined by the heart rate times the stroke volume.

▌ Heart rate is varied by altering the balance of parasympathetic and sympathetic influence on the SA node. Parasympathetic stimulation slows the heart rate, and sympathetic stimulation speeds it up.

▌ Stroke volume depends on (1) the extent of ventricular filling, with an increased end-diastolic volume resulting in a larger stroke volume by means of the length–tension relationship (Frank-Starling law of the heart, a form of intrinsic control), and (2) the extent of sympathetic stimulation, with increased sympathetic stimulation resulting in increased contractility of the heart, that is, increased strength of contraction and increased stroke volume at a given end-diastolic volume (extrinsic control).

Nourishing the Heart Muscle

▌ Cardiac muscle is supplied with oxygen and nutrients by blood delivered to it by the coronary circulation, not by blood within the heart chambers.

▌ Most coronary blood flow occurs during diastole, because the coronary vessels are compressed by the contracting heart muscle during systole.

▌ Coronary blood flow is normally varied to keep pace with cardiac oxygen needs.

▌ Coronary blood flow may be compromised by the development of atherosclerotic plaques, which can lead to ischemic heart disease ranging in severity from mild chest pain on exertion to fatal heart attacks.

REVIEW EXERCISES

Objective Questions (Answers on p. A-41)

1. Adjacent cardiac muscle cells are joined end to end at specialized structures known as _____, which contain two types of membrane junctions: _____ and _____ .

2. _____ is an abnormally slow heart rate, whereas _____ is a rapid heart rate.

3. The link that coordinates coronary blood flow with myocardial oxygen needs is _____.

4. The left ventricle is a stronger pump than the right ventricle because more blood is needed to supply the body tissues than to supply the lungs. (*True or false?*)

5. The heart lies in the left half of the thoracic cavity. (*True or false?*)

6. The only point of electrical contact between the atria and ventricles is the fibrous skeletal rings. (*True or false?*)

7. The atria and ventricles each act as a functional syncytium. (*True or false?*)

8. Which of the following is the proper sequence of cardiac excitation?
   a. SA node → AV node → atrial myocardium → bundle of His → Purkinje fibers → ventricular myocardium.
   b. SA node → atrial myocardium → AV node → bundle of His → ventricular myocardium → Purkinje fibers.
   c. SA node → atrial myocardium → ventricular myocardium → AV node → bundle of His → Purkinje fibers.
   d. SA node → atrial myocardium → AV node → bundle of His → Purkinje fibers → ventricular myocardium.

9. What percentage of ventricular filling is normally accomplished before atrial contraction begins?
   a. 0%
   b. 20%
   c. 50%
   d. 80%
   e. 100%

10. Sympathetic stimulation of the heart
    a. increases the heart rate.
    b. increases the contractility of the heart muscle.
    c. shifts the Frank-Starling curve to the left.
    d. Both (a) and (b) above are correct.
    e. All of the above are correct.

11. Circle the correct choice in each instance to complete the statements: During ventricular filling, ventricular pressure must be (*greater than/less than*) atrial pressure, whereas during ventricular ejection ventricular pressure must be (*greater than/less than*) aortic pressure. Atrial pressure is always (*greater than/less than*)

aortic pressure. During isovolumetric ventricular contraction and relaxation, ventricular pressure is (greater than/less than) atrial pressure and (greater than/less than) aortic pressure.

12. Circle the correct choice in each instance to complete the statement: The first heart sound is associated with closure of the (AV/semilunar) valves and signals the onset of (systole/diastole), whereas the second heart sound is associated with closure of the (AV/semilunar) valves and signals the onset of (systole/diastole).

13. Match the following:
    1. receives O_2-poor blood from the venae cavae
    2. prevent backflow of blood from the ventricles to the atria
    3. pumps O_2-rich blood into the aorta
    4. prevent backflow of blood from the arteries into the ventricles
    5. pumps O_2-poor blood into the pulmonary artery
    6. receives O_2-rich blood from the pulmonary veins
    7. permit AV valves to function as one-way valves

    (a) AV valves
    (b) semilunar valves
    (c) left atrium
    (d) left ventricle
    (e) right atrium
    (f) right ventricle
    (g) chordae tendineae and papillary muscles

14. Match the following:
    1. characterized by uncoordinated excitation and contraction of the cardiac cells
    2. caused by AV nodal damage
    3. an overly irritable area that takes over pacemaker activity

    (a) heart block
    (b) ectopic focus
    (c) ventricular fibrillation

Essay Questions

1. What are the three basic components of the circulatory system?
2. Trace a drop of blood through one complete circuit of the circulatory system.
3. Describe the location and function of the four heart valves. What prevents eversion of each of these valves?

4. What are the three layers of the heart wall? Describe the distinguishing features of the structure and arrangement of cardiac muscle cells. What are the two specialized types of cardiac muscle cells?
5. Why is the SA node the pacemaker of the heart?
6. Describe the normal spread of cardiac excitation. What is the significance of the AV nodal delay? Why is the ventricular conduction system important?
7. Compare the changes in membrane potential associated with an action potential in a nodal pacemaker cell with those in a myocardial contractile cell. What is responsible for the plateau phase?
8. Why is tetanus of cardiac muscle impossible? Why is this advantageous?
9. Draw and label the waveforms of a normal ECG. What electrical event does each component of the ECG represent?
10. Describe the mechanical events (that is, pressure changes, volume changes, valve activity, and heart sounds) that occur during the cardiac cycle. Correlate these mechanical events with the changes that take place in electrical activity.
11. Distinguish between a stenotic and an insufficient valve.
12. Define the following: end-diastolic volume, end-systolic volume, stroke volume, heart rate, cardiac output, and cardiac reserve.
13. Discuss autonomic nervous system control of heart rate.
14. Describe the intrinsic and extrinsic control of stroke volume.
15. By what means is the heart muscle provided with blood? Why does the heart receive most of its own blood supply during diastole?
16. What are the pathologic changes and consequences of coronary artery disease?
17. Discuss the sources, transport, and elimination of cholesterol in the body. Distinguish between "good" cholesterol and "bad" cholesterol.

Quantitative Exercises (Solutions on p. A-41)

1. During heavy exercise, the cardiac output can increase to as much as 35 liters/min. If stroke volume could not increase above the normal value of 70 ml, what heart rate would be necessary to achieve this cardiac output? Is such a heart rate physiologically possible?
2. How much blood remains in the heart after systole if the stroke volume is 85 ml and the end-diastolic volume is 125 ml?

POINTS TO PONDER

(Explanations on p. A-41)

1. The stroke volume ejected on the next heartbeat after a premature beat is usually larger than normal. Can you explain why? (*Hint:* At a given heart rate, the interval between a premature beat and the next normal beat is longer than the interval between two normal beats.)
2. Trained athletes usually have lower resting heart rates than normal (for example, 50 beats/min in an athlete compared to 70 beats/min in a sedentary individual). Considering that the resting cardiac output is 5,000 ml/min in both trained athletes and sedentary individuals, what is responsible for the bradycardia of trained athletes?
3. During fetal life, because of the tremendous resistance offered by the collapsed, nonfunctioning lungs, the pressures in the right half of the heart and pulmonary circulation are higher than in the left half of the heart and systemic circulation, the opposite of the situation after birth. Also in the fetus, a vessel known as the **ductus arteriosus** connects the pulmonary artery and aorta as

these major vessels both leave the heart. When blood is pumped out by the heart, blood is shunted from the pulmonary artery into the aorta through the ductus arteriosus, thus bypassing the nonfunctional lungs. What force is driving blood to flow in this direction through the ductus arteriosus?

At birth, the ductus arteriosus normally collapses and eventually degenerates into a thin, ligamentous strand. On occasion, this fetal bypass fails to close properly at birth, leading to a *patent ductus arteriosus*. In what direction would blood flow through a patent ductus arteriosus? What possible outcomes would you predict might occur as a result of this blood flow?

4. Through what regulatory mechanisms is a transplanted heart, which does not have any innervation, able to adjust the cardiac output to meet the body's changing needs?
5. There are two branches of the bundle of His, the right and left bundle branches, each of which travels down its respective side of the ventricular septum (see ● Figure 9-11, p. 311). Occa-

sionally conduction through one of these branches becomes blocked (so-called *bundle-branch block*). In this case, the wave of excitation spreads out from the terminals of the intact branch and eventually depolarizes the whole ventricle, but the normally stimulated ventricle becomes completely depolarized a considerable time before the ventricle on the side of the defective bundle branch. For example, if the left bundle branch is blocked, the right ventricle will be completely depolarized two to three times more rapidly than the left ventricle. What effect would this defect have on the heart sounds?

6. *Clinical Consideration.* In a physical exam, Rachel B.'s heart rate was rapid and very irregular. Furthermore, her heart rate determined directly by listening to her heart with a stethoscope exceeded the pulse rate taken concurrently at her wrist. No definite P waves could be detected on Rachel's ECG. The QRS complexes were normal in shape but occurred sporadically. Given these findings, what is the most likely diagnosis of Rachel's condition? Explain why the condition is characterized by a rapid, irregular heartbeat. Would cardiac output be seriously impaired by this condition? Why or why not? What accounts for the pulse deficit?

PHYSIOEDGE RESOURCES

PHYSIOEDGE CD-ROM
Figures marked with the PhysioEdge icon have associated activities on the CD. For this chapter, check out:

Cardiovascular Physiology

The diagnostic quiz allows you to receive immediate feedback on your understanding of the concept and to advance through various levels of difficulty.

CHECK OUT THESE MEDIA QUIZZES:
9.1 The Electrocardiogram
9.2 The Heart: A Dual Pump
9.3 The Heart: Cardiac Output

PHYSIOEDGE WEB SITE
The Web site for this book contains a wealth of helpful study aids, as well as many ideas for further reading and research. Log on to:

http://www.brookscole.com

Go to the Biology page and select Sherwood's *Human Physiology*, 5th Edition. Select a chapter from the drop-down menu or click on one of these resource areas:

- **Case Histories** provide an introduction to the clinical aspects of human physiology. Check out:

  #7: Why Am I So Tired?
  #9: Endocarditis
  #10: Blue Baby
  #11: Congestive Heart Failure

- For 2-D and 3-D graphical illustrations and animations of physiological concepts, visit our **Visual Learning Resource.**

- Resources for study and review include the **Chapter Outline, Chapter Summary, Glossary,** and **Flash Cards.** Use our **Quizzes** to prepare for in-class examinations.

- On-line research resources to consult are **Hypercontents,** with current links to relevant Internet sites; **Internet Exercises** with starter URLs provided; and **InfoTrac Exercises.**

For more readings, go to InfoTrac College Edition, your online research library, at:

http://infotrac.thomsonlearning.com

Cardiovascular System
(Blood Vessels)

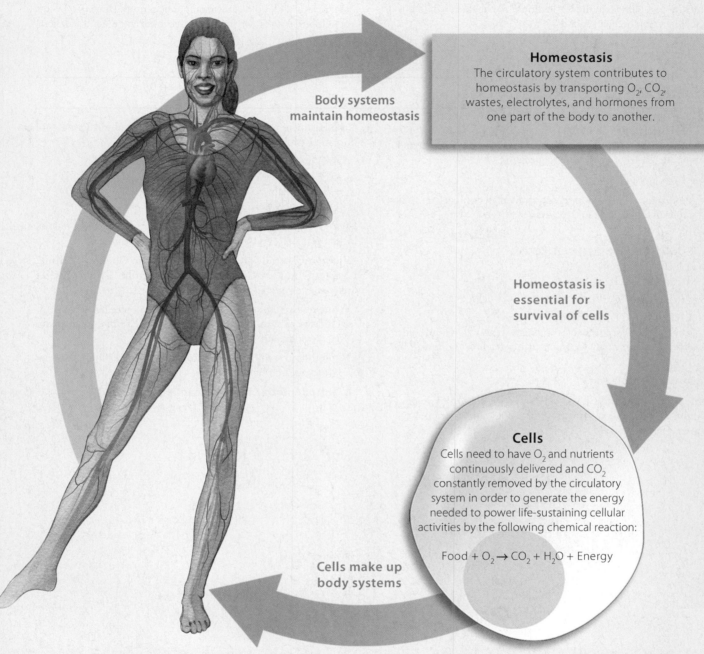

Body systems maintain homeostasis

Homeostasis
The circulatory system contributes to homeostasis by transporting O_2, CO_2, wastes, electrolytes, and hormones from one part of the body to another.

Homeostasis is essential for survival of cells

Cells
Cells need to have O_2 and nutrients continuously delivered and CO_2 constantly removed by the circulatory system in order to generate the energy needed to power life-sustaining cellular activities by the following chemical reaction:

$$Food + O_2 \rightarrow CO_2 + H_2O + Energy$$

Cells make up body systems

The **circulatory system** contributes to homeostasis by serving as the body's transport system. The blood vessels transport and distribute blood pumped through them by the heart to meet the body's needs for O_2 and nutrient delivery, waste removal, and hormonal signaling. The highly elastic **arteries** transport blood from the heart to the tissues and serve as a pressure reservoir to continue driving blood forward when the heart is relaxing and filling. The **mean arterial blood pressure** is closely regulated to ensure adequate blood delivery to the tissues. The amount of blood that flows through a given tissue depends on the caliber (internal diameter) of the highly muscular **arterioles** that supply the tissue. Arteriolar caliber is subject to control so that the distribution of the cardiac output can be constantly readjusted to best serve the body's needs at the moment. The thin-walled, pore-lined **capillaries** are the actual site of exchange between the blood and the surrounding tissues. The highly distensible **veins** return the blood from the tissues to the heart and also serve as a blood reservoir.

Chapter 10

The Blood Vessels and Blood Pressure

CONTENTS AT A GLANCE

INTRODUCTION

❙ Blood-vessel types

❙ Flow, pressure, resistance relationships

ARTERIES

❙ Passageways to the tissues

❙ Role as a pressure reservoir

❙ Arterial pressure

ARTERIOLES

❙ Major resistance vessels

❙ Control of arteriolar radius

❙ Role in distribution of cardiac output

❙ Role in maintenance of arterial blood pressure

CAPILLARIES

❙ Sites of exchange

❙ Diffusion across the capillary wall

❙ Bulk flow across the capillary wall

❙ Formation and function of lymph

❙ Edema

VEINS

❙ Passageways to the heart

❙ Role as a blood reservoir

❙ Venous return

BLOOD PRESSURE

❙ Factors influencing mean arterial pressure

❙ Baroreceptor reflex

❙ Hypertension

❙ Hypotension; circulatory shock

INTRODUCTION

The majority of body cells are not in direct contact with the external environment, yet these cells must make exchanges with this environment, such as picking up O_2 and nutrients and eliminating wastes. Furthermore, chemical messengers must be transported between cells to accomplish integrated activity. To achieve these long-distance exchanges, the cells are linked with each other and with the external environment by vascular (blood vessel) highways. Blood is transported to all parts of the body through a system of vessels that brings fresh supplies to the vicinity of all cells while removing their wastes.

To review, all blood pumped by the right side of the heart passes through the pulmonary circulation to the lungs for O_2 pickup and CO_2 removal. The blood pumped by the left side of the heart into the systemic circulation is parceled out in various proportions to the systemic organs through a parallel arrangement of vessels that branch from the aorta (● Figure 10-1) (see p. 305). This arrangement ensures that all organs receive blood of the same composition; that is, one organ does not receive "leftover" blood that has passed through another organ. Because of this parallel arrangement, the flow of blood through each systemic organ can be independently adjusted as needed.

In this chapter, we will first examine some general principles regarding blood flow patterns and the physics of blood flow. Then we will turn our attention to the roles of the various types of blood vessels through which blood flows. We will conclude with a discussion of how blood pressure is regulated to ensure adequate delivery of blood to the tissues.

❙ To maintain homeostasis, reconditioning organs receive blood flow in excess of their own needs.

Blood is constantly "reconditioned" so that its composition remains relatively constant despite an ongoing drain of supplies to support metabolic activities and the continual addition of wastes from the tissues. The organs that recondition the blood normally receive substantially more blood than is necessary to meet their basic metabolic needs, so that they can perform homeostatic adjustments to the blood. Large percentages of the cardiac output are distributed to the digestive tract (to pick

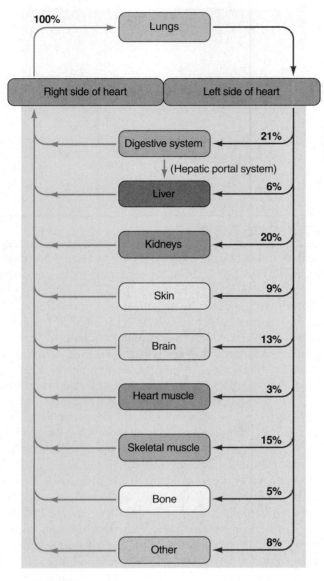

100%

Lungs

Right side of heart | Left side of heart

Digestive system   21%

↓ (Hepatic portal system)

Liver   6%

Kidneys   20%

Skin   9%

Brain   13%

Heart muscle   3%

Skeletal muscle   15%

Bone   5%

Other   8%

● FIGURE 10-1

Distribution of cardiac output at rest
The lungs receive all the blood pumped out by the right side of the heart, whereas the systemic organs each receive a portion of the blood pumped out by the left side of the heart. The percentage of pumped blood received by the various organs under resting conditions is indicated. This distribution of cardiac output can be adjusted as needed.

up nutrient supplies), to the kidneys (to eliminate metabolic wastes and adjust water and electrolyte composition), and to the skin (to eliminate heat). Blood flow to the other organs—heart, skeletal muscles, and so on—is solely for the purpose of supplying these tissues' metabolic needs and can be adjusted according to their level of activity. For example, during exercise additional blood is delivered to the active muscles to meet their increased metabolic needs.

Because reconditioning organs—that is, digestive organs, kidneys, and skin—receive blood flow in excess of their own needs, they can withstand temporary reductions in blood flow

much better than can the other organs that do not have this extra margin of blood supply. The brain in particular suffers irreparable damage when transiently deprived of blood supply. Without O_2, permanent brain damage occurs after only four minutes. Therefore, a high priority in the overall operation of the circulatory system is the constant delivery of adequate blood to the brain, which can least tolerate a disruption in its blood supply.

In contrast, the reconditioning organs can tolerate significant reductions in blood flow for a considerable length of time. In fact, they frequently do so. For example, during exercise some of the blood that normally flows through the digestive organs and kidneys is diverted instead to the skeletal muscles. Likewise, blood flow through the skin is markedly restricted during exposure to cold to conserve body heat.

Later in the chapter, you will see how the distribution of cardiac output is adjusted according to the body's momentary needs . For now, we are going to concentrate on the factors that influence blood flow through a given blood vessel.

▌ Blood flow through vessels depends on the pressure gradient and vascular resistance.

The **flow rate** of blood through a vessel (that is, the volume of blood passing through per unit of time) is directly proportional to the pressure gradient and inversely proportional to vascular resistance:

$$F = \frac{\Delta P}{R}$$

where

F = flow rate of blood through a vessel

ΔP = pressure gradient

R = resistance of blood vessels

Pressure gradient

The **pressure gradient**—the difference in pressure between the beginning and end of a vessel—is the main driving force for flow through the vessel; that is, blood flows from an area of higher pressure to an area of lower pressure down a pressure gradient. Contraction of the heart imparts pressure to the blood, but because of frictional losses (resistance), the pressure decreases as blood flows through a vessel. Because pressure drops throughout the vessel's length, it is higher at the beginning than at the end. This establishes a pressure gradient for forward flow of blood through the vessel. The greater the pressure gradient forcing blood through a vessel, the greater the rate of flow through that vessel (● Figure 10-2a). Think of a garden hose attached to a faucet. If you turn on the faucet slightly, a small stream of water will flow out of the end of the hose, because the pressure is slightly greater at the beginning than at the end of the hose. If you open the faucet all the way, the pressure gradient increases tremendously, so that water flows through the hose much faster, and spurts from the end of the hose. Note that the *difference* in pressure between the

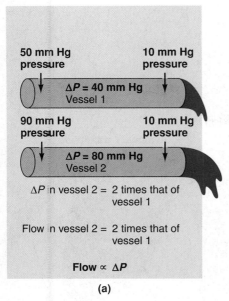

50 mm Hg pressure   10 mm Hg pressure

ΔP = 40 mm Hg
Vessel 1

90 mm Hg pressure   10 mm Hg pressure

ΔP = 80 mm Hg
Vessel 2

ΔP in vessel 2 = 2 times that of vessel 1

Flow in vessel 2 = 2 times that of vessel 1

Flow $\propto \Delta P$

(a)

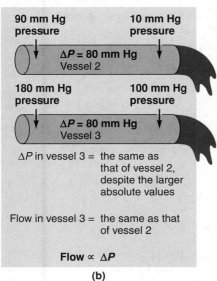

90 mm Hg pressure   10 mm Hg pressure

ΔP = 80 mm Hg
Vessel 2

180 mm Hg pressure   100 mm Hg pressure

ΔP = 80 mm Hg
Vessel 3

ΔP in vessel 3 = the same as that of vessel 2, despite the larger absolute values

Flow in vessel 3 = the same as that of vessel 2

Flow $\propto \Delta P$

(b)

● **FIGURE 10-2**

Relationship of flow to the pressure gradient in a vessel
(a) As the difference in pressure (ΔP) between the two ends of a vessel increases, the flow rate increases proportionately. (b) Flow rate is determined by the difference in pressure between the two ends of a vessel, not the absolute pressures.

two ends of a vessel, not the absolute pressures within the vessel, determines flow rate (● Figure 10-2b).

Resistance

The other factor influencing flow rate through a vessel is the **resistance,** which is a measure of the hindrance or opposition to blood flow through a vessel caused by friction between the moving fluid and the stationary vascular walls. As resistance to flow increases, it is more difficult for blood to pass through the vessel, so flow decreases (as long as the pressure gradient remains unchanged). When resistance increases, the pressure gradient must increase correspondingly to maintain the same flow

rate. Accordingly, when the vessels offer more resistance to flow, the heart must work harder to maintain adequate circulation.

Resistance to blood flow depends on three factors: (1) viscosity of the blood, (2) vessel length, and (3) vessel radius, which is by far the most important. The term **viscosity** (designated as η) refers to the friction developed between the molecules of a fluid as they slide over each other during flow of the fluid. The greater the viscosity, the greater the resistance to flow. In general, the thicker a liquid, the more viscous it is. For example, molasses flows more slowly than water because molasses has greater viscosity. The most important factor that determines the blood viscosity is the number of circulating red blood cells. Normally, this factor is relatively constant and thus not important in the control of resistance. Occasionally, however, blood viscosity and, accordingly, resistance to flow are altered because of an abnormal number of red blood cells. Blood flow is more sluggish than normal when excessive red blood cells are present.

Because blood "rubs" against the lining of the vessels as it flows past, the greater the vessel surface area in contact with the blood, the greater the resistance to flow. Surface area is determined by both the length (L) and radius (r) of the vessel. At a constant radius, the longer the vessel, the greater the surface area and the greater the resistance to flow. Because vessel length remains constant in the body, it is not a variable factor in the control of vascular resistance.

Therefore, the major determinant of resistance to flow is the vessel's radius. Fluid passes more readily through a large vessel than through a smaller vessel. The reason is that a given volume of blood comes into contact with much more of the surface area of a small-radius vessel than of a larger-radius vessel, resulting in greater resistance (● Figure 10-3a).

Furthermore, a slight change in the radius of a vessel brings about a notable change in flow, because the resistance is inversely proportional to the fourth power of the radius (multiplying the radius by itself 4 times):

$$R \propto \frac{1}{r^4}$$

Thus doubling the radius decreases the resistance 16 times ($r^4 = 2 \times 2 \times 2 \times 2 = 16$; $R \propto 1/16$) and therefore increases flow through the vessel 16-fold (at the same pressure gradient) (● Figure 10-3b). The converse is also true. Only 1/16th as much blood flows through a vessel at the same driving pressure when its radius is halved. Importantly, the radius of arterioles is subject to regulation and is the most important factor in the control of resistance to blood flow throughout the vascular tree.

Poiseuille's law

The factors that affect the flow rate through a vessel are integrated in an idealized way in **Poiseuille's law** as follows:

$$\text{flow rate} = \frac{\pi \Delta P r^4}{8 \eta L}$$

The significance of the relationships among flow, pressure, and resistance, as largely determined by vessel radius, will be-

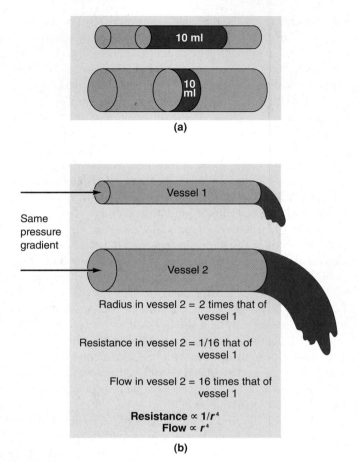

(a)

(b)

Radius in vessel 2 = 2 times that of vessel 1

Resistance in vessel 2 = 1/16 that of vessel 1

Flow in vessel 2 = 16 times that of vessel 1

$$\text{Resistance} \propto 1/r^4$$
$$\text{Flow} \propto r^4$$

● **FIGURE 10-3**

Relationship of resistance and flow to the vessel radius
(a) The same volume of blood comes into contact with a greater surface area of a small-radius vessel compared to a larger-radius vessel. Accordingly, the smaller-radius vessel offers more resistance to blood flow, because the blood "rubs" against a larger surface area. (b) Doubling the radius decreases the resistance to 1/16 and increases the flow 16 times, because the resistance is inversely proportional to the fourth power of the radius.

come even more important as we embark on a voyage through the vessels in the next section.

▌ The vascular tree consists of arteries, arterioles, capillaries, venules, and veins.

The systemic and pulmonary circulations each consist of a closed system of vessels (● Figure 10-4). (For the history leading up to the conclusion that blood vessels form a closed system, see the accompanying boxed feature, ▶ Concepts, Challenges, and Controversies.) These vascular loops each consist of a continuum of different blood vessel types that begins and ends with the heart as follows. Looking specifically at the systemic circulation, **arteries,** which carry blood from the heart to the tissues, branch into a "tree" of progressively smaller vessels, with the various branches delivering blood to different regions of the body. When a small artery reaches the organ it is supplying, it branches into numerous **arterioles.** The volume of blood flowing through an organ

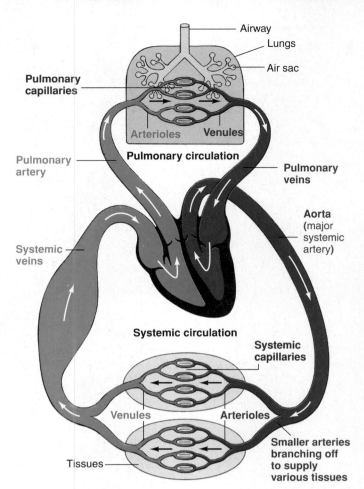

For simplicity, only two capillary beds within two organs are illustrated.

● **FIGURE 10-4**

Basic organization of the cardiovascular system
Arteries progressively branch as they carry blood from the heart to the tissues. A separate small arterial branch delivers blood to each of the various organs. As a small artery enters the organ it is supplying, it branches into arterioles, which further branch into an extensive network of capillaries. The capillaries rejoin to form venules, which further unite to form small veins that leave the organ. The small veins progressively merge as they carry blood back to the heart.

can be adjusted by regulating the caliber (internal diameter) of the organ's arterioles. Arterioles branch further within the organs into **capillaries,** the smallest of vessels, across which all exchanges are made with surrounding cells. Capillary exchange is the entire purpose of the circulatory system; all other activities of the system are directed toward ensuring an adequate distribution of replenished blood to capillaries for exchange with all cells. Capillaries rejoin to form small **venules,** which further merge to form small **veins** that leave the organs. The small veins progressively unite to form larger veins that eventually empty into the heart. The arterioles, capillaries, and venules are collectively referred to as the **microcirculation,** because they are only visible through a microscope. (Venules and small veins serve no function other than carrying blood from the capillaries to the large

From Humors to Harvey: Historical Highlights in Circulation

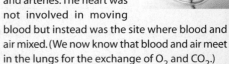

Today even grade school children know that blood is pumped by the heart and continually circulates throughout the body in a system of blood vessels. Furthermore, it is accepted without question that the blood picks up O_2 in the lungs from the air we breathe and delivers it to the body tissues. This common knowledge was unknown for most of human history, however. Even though the function of blood was described as early as the fifth century B.C., our modern concept of circulation did not develop until 1628, more than 2,000 years later, when William Harvey published his now classical study, *De Motu Cordis et Sanquinis in Animalibus* ("On the Motion of the Heart and of Blood in Animals").

Ancient Greeks believed everything material in the universe consisted of just four elements: earth, air, fire, and water. Extending this view to the human body, they thought these four elements took the form of four "humors": *black bile* (representing earth), *blood* (representing air), *yellow bile* (representing fire), and *phlegm* (representing water). According to the Greeks, disease resulted when one humor was out of normal balance with the rest. The "cure" was logical: Drain off whichever humor was in excess, to restore normal balance. Because the easiest humor to drain off was the blood, bloodletting became standard procedure for treating many illnesses—a practice that persisted well into the Renaissance (which began in the 1300s and extended into the 1500s).

Although the ancient Greek notion of the four humors was erroneous, their concept of the necessity of balance within the body was remarkably accurate. As we now know, life depends on homeostasis, maintenance of the proper balance among all elements of the internal environment.

Aristotle (384–322 B.C.), a biologist as well as a philosopher, was among the first to rightly describe the heart at the center of a system of blood vessels. However, he thought the heart was both the seat of intellect (the brain was not identified as the seat of intellect until over a century later) and a furnace that heated the blood. He considered this warmth the vital force of life, because the body cools quickly at death. Aristotle also erroneously theorized that the "furnace" was ventilated by breathing, with air serving as a cooling agent. Aristotle could observe with his eyes the arteries and veins in cadavers but did not have a microscope with which to observe capillaries. (The microscope was not invented until the 17th century.) Thus he did not think arteries and veins were directly connected.

In the third century B.C., Erasistratus, a Greek many considered the first "physiologist," proposed that the liver used food to make blood, which the veins delivered to the other organs. He believed the arteries contained air, not blood. According to his view, *pneuma* ("air"), a living force, was taken in by the lungs, which transferred it to the heart. The heart transformed the air into a "vital spirit" that the arteries carried to the other organs.

Galen (A.D. 130–206), a prolific, outspoken, dogmatic Roman physician, philosopher, and scholar, expanded on the work of Erasistratus and others who had preceded him. Galen further elaborated on the pneumatic theory. He proposed three fundamental members in the body, from lowest to highest: liver, heart, and brain. Each was dominated by a special *pneuma, or* "spirit." (In Greek, *pneuma* encompassed the related ideas of "air," "breath," and "spirit.") Like Erasistratus, Galen believed that the liver made blood from food, taking on a "natural" or "physical" spirit *(pneuma physicon)* in the process. The newly formed blood then proceeded through veins to organs. The natural spirit, which Galen considered a vapor rising from the blood, controlled the functions of nutrition, growth, and reproduction. Once its spirit supply was depleted, the blood moved in the opposite direction through the same venous pathways, returning to the liver to be replenished. When the natural spirit was carried in the venous blood to the heart, it mixed with air that was breathed in and transferred from the lungs to the heart. Contact with air in the heart transformed the natural spirit into a higher-level spirit, the "vital" spirit *(pneuma zotikon)*. The vital spirit, which was carried by the arteries, conveyed heat and life throughout the body. The vital spirit was transformed further into a yet higher "animal" or "psychical" spirit *(pneuma psychicon)* in the brain. This ultimate spirit regulated the brain, nerves, feelings, and so on. Thus, according to Galenic theory, the veins and arteries were conduits for carrying different levels of pneuma, and there was no direct connection between the veins and arteries. The heart was not involved in moving blood but instead was the site where blood and air mixed. (We now know that blood and air meet in the lungs for the exchange of O_2 and CO_2.)

Galen was one of the first to understand the need for experimentation, but unfortunately, his impatience and his craving for philosophical and literary fame led him to expound comprehensive theories that were not always based on the time-consuming collection of evidence. Even though his assumptions about bodily structure and functions often were incorrect, his theories were convincing because they seemed a logical way of pulling together what was known at the time. Furthermore, the sheer quantity of his writings helped establish him as an authority. In fact, his writings remained the anatomic and physiologic "truth" for nearly 15 centuries, throughout the Middle Ages and well into the Renaissance. So firmly entrenched was Galenic doctrine that people who challenged its accuracy risked their lives by being declared secular heretics.

Not until the Renaissance and the revival of classical learning did independent-minded European investigators begin to challenge Galen's theories. Most notably, the English physician William Harvey (1578–1657) revolutionized the view of the roles played by the heart, blood vessels, and blood. Through careful observations, experimentation, and deductive reasoning, Harvey was the first to correctly identify the heart as a pump that repeatedly moves a small volume of blood forward in one fixed direction in a circular path through a closed system of blood vessels (the *circulatory system*). He also correctly proposed that blood travels to the lungs to mix with air (instead of air traveling to the heart to mix with blood). Even though he could not *see* physical connections between arteries and veins, he speculated on their existence. Not until the discovery of the microscope later in the century was the existence of these connections, the capillaries, confirmed, by Marcello Malpighi (1628–1694).

veins and will not be discussed further.) The pulmonary circulation consists of the same vessel types, except that all the blood in this loop goes between the heart and lungs. In discussing the vessel types in this chapter, we will refer to their roles in the systemic circulation, starting with systemic arteries.

ARTERIES

The consecutive segments of the vascular tree are specialized to perform specific tasks (▲ Table 10-1).

▲ TABLE 10-1
Features of Blood Vessels

Feature	Vessel Type					
	Aorta	Large arterial branches	Arterioles	Capillaries	Large veins	Venae cavae
Number	One	Several hundred	Half a million	Ten billion	Several hundred	Two
Wall Thickness	2 mm (2,000 μm)	1 mm (1,000 μm)	20 μm	1 μm	0.5 mm (500 μm)	1.5 mm (1,500 μm)
Internal Radius	1.25 cm (12,500 μm)	0.2 cm (2,000 μm)	30 μm	3.5 μm	0.5 cm (5,000 μm)	3 cm (30,000 μm)
Total Cross-Sectional Area	4.5 cm^2	20 cm^2	400 cm^2	6,000 cm^2	40 cm^2	18 cm^2
Special Features	Thick, highly elastic, walls; large radii		Highly muscular, well-innervated walls; small radii	Thin walled; large total cross-sectional area	Thin walled; highly distensible; large radii	
Functions	Passageway from the heart to the tissues; serve as a pressure reservoir		Primary resistance vessels; determine the distribution of cardiac output	Site of exchange; determine the distribution of extracellular fluid between the plasma and interstitial fluid	Passageway to the heart from the tissues; serve as a blood reservoir	

▮ Arteries serve as rapid-transit passageways to the tissues and as a pressure reservoir.

Arteries are specialized (1) to serve as rapid-transit passageways for blood from the heart to the tissues (because of their large radii, arteries offer little resistance to blood flow) and (2) to act as a **pressure reservoir** to provide the driving force for blood when the heart is relaxing.

Let us expand on the role of the arteries as a pressure reservoir. The heart alternately contracts to pump blood into the arteries and then relaxes to refill from the veins. No blood is pumped out when the heart is relaxing and refilling. However, capillary flow does not fluctuate between cardiac systole and diastole; that is, blood flow is continuous through the capillaries supplying the tissues. The driving force for the continued flow of blood to the tissues during cardiac relaxation is provided by the elastic properties of the arterial walls.

All vessels are lined with a thin layer of smooth, flattened endothelial cells that are continuous with the endocardial lining of the heart. A thick wall made up of smooth muscle and connective tissue surrounds the arteries' endothelial lining. Arterial connective tissue contains an abundance of two types of connective tissue fibers; *collagen fibers*, which provide tensile strength against the high driving pressure of blood ejected from the heart, and *elastin fibers*, which give the arterial walls elasticity so that they behave much like a balloon (● Figure 10-5).

As the heart pumps blood into the arteries during ventricular systole, a greater volume of blood enters the arteries from the heart then leaves them to flow into smaller vessels downstream because the smaller vessels have a greater resistance to flow. The arteries' elasticity enables them to expand to temporarily hold this excess volume of ejected blood, storing some of the pressure energy imparted by cardiac contraction in their stretched walls—just as a balloon expands to accommodate the extra volume of air that you blow into it (● Figure 10-6a). When the heart relaxes and ceases pumping blood into the arteries, the stretched arterial walls passively recoil, like an inflated balloon that is released. This recoil pushes the excess blood contained in the arteries into the vessels downstream, ensuring continued blood flow to the tissues when the heart is relaxing and not pumping blood into the system (● Figure 10-6b).

▮ Arterial pressure fluctuates in relation to ventricular systole and diastole.

Blood pressure, the force exerted by the blood against a vessel wall, depends on the volume of blood contained within the vessel and the **compliance,** or **distensibility,** of the vessel walls (how easily they can be stretched). If the volume of blood entering the arteries were equal to the volume of blood leaving the arteries during the same period, arterial blood pressure would remain constant. This is not the case, however. During ventricular systole, a stroke volume of blood enters the arteries

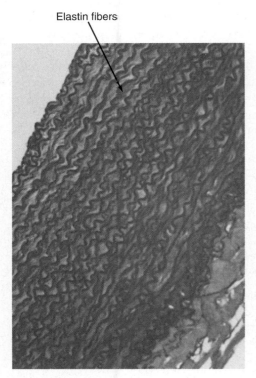

Elastin fibers

● FIGURE 10-5

Elastin fibers in an artery
Light micrograph of a portion of the aorta wall in cross section, showing numerous wavy elastin fibers, common to all arteries.

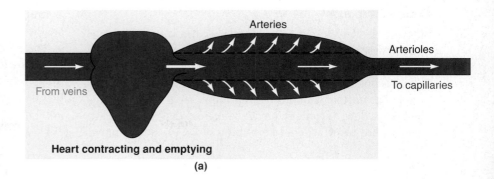

Arteries

Arterioles

From veins

To capillaries

Heart contracting and emptying

(a)

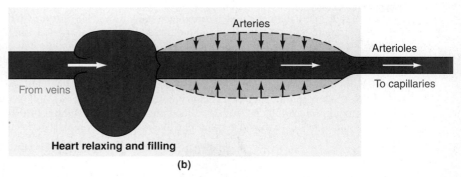

Arteries

Arterioles

From veins

To capillaries

Heart relaxing and filling

(b)

● FIGURE 10-6

Arteries as a pressure reservoir
Because of their elasticity, arteries act as a pressure reservoir. (a) The elastic arteries distend during cardiac systole as more blood is ejected into them than drains off into the narrow, high-resistance arterioles downstream. (b) The elastic recoil of arteries during cardiac diastole continues driving the blood forward when the heart is not pumping.

from the ventricle while only about one-third as much blood leaves the arteries to enter the arterioles. During diastole, no blood enters the arteries, while blood continues to leave, driven by elastic recoil. The maximum pressure exerted in the arteries when blood is ejected into them during systole, the **systolic pressure,** averages 120 mm Hg. The minimum pressure within the arteries when blood is draining off into the remainder of the vessels during diastole, the **diastolic pressure,** averages 80 mm Hg. The arterial pressure does not fall to 0 mm Hg, because the next cardiac contraction occurs and refills the arteries before all the blood drains off (● Figure 10-7; also see ● Figure 9-21, p. 322).

▌ Blood pressure can be measured indirectly by using a sphygmomanometer.

The changes in arterial pressure throughout the cardiac cycle can be measured directly by connecting a pressure-measuring device to a needle inserted in an artery. However, it is more convenient and reasonably accurate to measure the pressure indirectly through use of a **sphygmomanometer,** an externally applied inflatable cuff attached to a pressure gauge. When the cuff is wrapped around the upper arm and then inflated with air, the pressure of the cuff is transmitted through the tissues to the underlying brachial artery, the main vessel carrying blood to the forearm (● Figure 10-8). The technique involves balancing

the pressure in the cuff against the pressure in the artery. When cuff pressure is greater than the pressure in the vessel, the vessel is pinched closed so that no blood flows through it. When blood pressure is greater than cuff pressure, the vessel is open and blood flows through.

● FIGURE 10-7

Arterial blood pressure
The systolic pressure is the peak pressure exerted in the arteries when blood is pumped into them during ventricular systole. The diastolic pressure is the lowest pressure exerted in the arteries when blood is draining off into the vessels downstream during ventricular diastole. The pulse pressure is the difference between systolic and diastolic pressure. The mean pressure is the average pressure throughout the cardiac cycle.

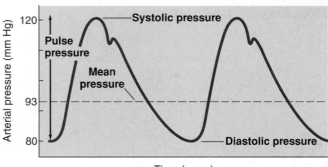

Korotkoff sounds

During the determination of blood pressure, a stethoscope is placed over the brachial artery at the inside bend of the elbow just below the cuff. No sound can be detected either when blood is not flowing through the vessel or when blood is flowing in the normal, smooth laminar flow (see p. 324). Turbulent blood flow, in contrast, creates vibrations that can be heard. The sounds heard when determining blood pressure, known as **Korotkoff sounds**, are distinct from the heart sounds associated with valve closure heard when listening to the heart with a stethoscope.

At the onset of a blood pressure determination, the cuff is inflated to a pressure greater than systolic blood pressure so that the brachial artery collapses. Because the externally applied pressure is greater than the peak internal pressure, the artery remains completely pinched closed throughout the en-

● **FIGURE 10-8**

Sphygmomanometry
(a) Use of a sphygmomanometer in determining blood pressure. The pressure in the inflatable cuff can be varied to prevent or permit blood flow in the underlying brachial artery. Turbulent blood flow can be detected through use of a stethoscope, whereas smooth laminar flow and no flow are inaudible. (b) Pattern of sounds in relation to cuff pressure compared with blood pressure. The numbers on the illustration refer to the following key points during a blood pressure determination: 1 Cuff pressure exceeds blood pressure throughout the cardiac cycle. No sound is heard. 2 The first sound is heard at peak systolic pressure. 3 Intermittent sounds are heard as blood pressure cyclically exceeds cuff pressure. 4 The last sound is heard at minimum diastolic pressure. 5 Blood pressure exceeds cuff pressure throughout the cardiac cycle. No sound is heard. (c) Blood flow through the brachial artery in relation to cuff pressure and sounds.

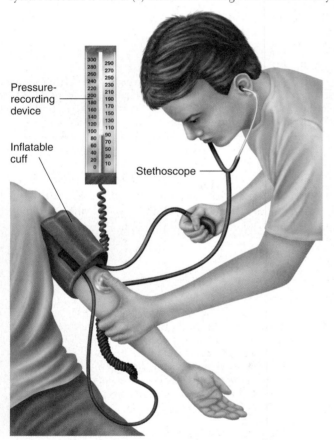

Pressure-recording device

Inflatable cuff

Stethoscope

(a)

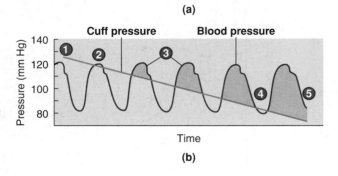

Cuff pressure   Blood pressure

(b)

When blood pressure is 120/80:

When cuff pressure is greater than 120 mm Hg:

No blood flows through the vessel.

No sound is heard.

When cuff pressure is between 120 and 80 mm Hg:

Blood flow through the vessel is turbulent whenever blood pressure exceeds cuff pressure.

Intermittent sounds are heard as blood pressure fluctuates throughout the cardiac cycle.

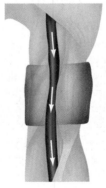

When cuff pressure is less than 80 mm Hg:

Blood flows through the vessel in smooth, laminar fashion.

No sound is heard.

(c)

tire cardiac cycle; no sound can be heard, because no blood is passing through (point 1 in ● Figure 10-8b).

As air in the cuff is slowly released, the pressure in the cuff is gradually reduced. When the cuff pressure falls to just below the peak systolic pressure, the artery transiently opens a bit when the blood pressure reaches this peak. Blood escapes through the partially occluded artery for a brief interval before the arterial pressure falls below the cuff pressure and the artery collapses once again. This spurt of blood is turbulent, so it can be heard. Thus the highest cuff pressure at which the *first sound* can be heard is indicative of the *systolic blood pressure* (point 2). As the cuff pressure continues to fall, blood intermittently spurts through the artery and produces a sound with each subsequent cardiac cycle whenever the arterial pressure exceeds the cuff pressure (point 3).

When the cuff pressure finally falls below diastolic pressure, the brachial artery is no longer pinched closed during any part of the cardiac cycle, and blood can flow uninterrupted through the vessel (point 5). With the return of nonturbulent blood flow, no further sounds can be heard. Therefore, the highest cuff pressure at which the *last sound* can be detected indicates the *diastolic pressure* (point 4).

In clinical practice, arterial blood pressure is expressed as systolic pressure over diastolic pressure, with the average blood pressure being 120/80 (120 over 80) mm Hg.

Pulse pressure

The pulse that can be felt in an artery lying close to the surface of the skin is due to the difference between systolic and diastolic pressures. This pressure difference is known as the **pulse pressure.** When the blood pressure is 120/80, pulse pressure is 40 mm Hg (120 mm Hg − 80 mm Hg).

▌ Mean arterial pressure is the main driving force for blood flow.

The **mean arterial pressure** is the *average pressure* responsible for driving blood forward into the tissues throughout the cardiac cycle. Contrary to what you might expect, mean arterial pressure is not the halfway value between systolic and diastolic pressure (for example, with a blood pressure of 120/80, mean pressure is not 100 mm Hg). The reason is that arterial pressure remains closer to diastolic than to systolic pressure for a longer portion of each cardiac cycle. At resting heart rate, about two-thirds of the cardiac cycle is spent in diastole and only one-third in systole. As an analogy, if a race car traveled 80 miles per hour (mph) for 40 minutes and 120 mph for 20 minutes, its average speed would be 93 mph, not the halfway value of 100 mph.

Similarly, a good approximation of the mean arterial pressure can be determined using the following formula:

Mean arterial pressure =
diastolic pressure + 1/3 pulse pressure

At 120/80, mean arterial pressure =
80 mm Hg + (1/3) 40 mm Hg = 93 mm Hg

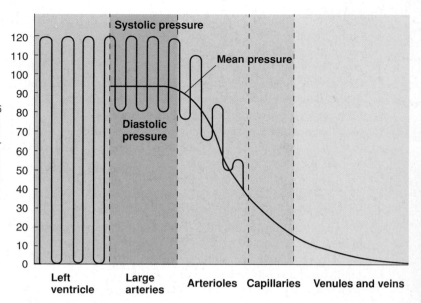

● FIGURE 10-9

Pressures throughout the systemic circulation
Left ventricular pressure swings between a low pressure of 0 mm Hg during diastole to a high pressure of 120 mm Hg during systole. Arterial blood pressure, which fluctuates between a peak systolic pressure of 120 mm Hg and a low diastolic pressure of 80 mm Hg each cardiac cycle, is of the same magnitude throughout the large arteries. Because of the arterioles' high resistance, the pressure drops precipitously and the systolic-to-diastolic swings in pressure are converted to a nonpulsatile pressure when blood flows through the arterioles. The pressure continues to decline but at a slower rate as blood flows through the capillaries and venous system.

The mean arterial pressure, not the systolic or diastolic pressures, is monitored and regulated by blood pressure reflexes described later in the chapter.

Because arteries offer little resistance to flow, only a negligible amount of pressure energy is lost in them because of friction. Therefore, arterial pressure—systolic, diastolic, pulse, or mean—is essentially the same throughout the arterial tree (● Figure 10-9).

Blood pressure exists throughout the entire vascular tree, but when discussing a person's "blood pressure" without qualifying which blood vessel type is being referred to, the term is tacitly understood to mean the pressure in the arteries.

ARTERIOLES

▌ Arterioles are the major resistance vessels.

When an artery reaches the organ it is supplying, it branches into numerous arterioles, whose radii are small enough to offer considerable resistance to flow. In fact, the arterioles are the major resistance vessels in the vascular tree. (Even though the capillaries have smaller radii than the arterioles, you will see later how collectively the capillaries do not offer as much resistance to flow as the arteriolar level of the vascular tree does.) In contrast to the low resistance of the arteries, the high degree of arteriolar resistance causes a marked drop in mean pressure as the blood flows through these vessels. On average, the pressure falls from 93 mm Hg, the mean arterial pressure (the pressure of the blood entering the arterioles), to 37 mm Hg, the pressure of the blood leaving the arterioles and entering the capillaries (● Figure 10-9). This decline in pressure helps establish the pressure differential that encourages the flow of blood from the heart to the various organs downstream. If no pressure drop occurred in the arterioles, the pressure at the end of the arterioles (that is, at the beginning of the capillaries) would be equal to the mean arterial pressure. No pressure gradient would exist to drive the blood from the heart to the tissue capillary beds.

In addition to causing a marked drop in mean blood pressure at the arteriolar level, arteriolar resistance is also responsible for converting the pulsatile systolic-to-diastolic pressure swings in the arteries into the nonfluctuating pressure present in the capillaries.

The radii (and, accordingly, the resistances) of arterioles supplying individual organs can be adjusted independently to accomplish two functions: (1) to variably distribute the cardiac output among the systemic organs, depending on the body's momentary needs, and (2) to help regulate arterial blood pressure. Before considering how such adjustments are important in accomplishing these two functions, we will discuss the mechanisms involved in adjusting arteriolar resistance.

Vasoconstriction and vasodilation

Unlike arteries, arteriolar walls contain very little elastic connective tissue. However, they do have a thick layer of smooth muscle that is richly innervated by sympathetic nerve fibers. The smooth muscle is also sensitive to many local chemical changes and to a few circulating hormones. The smooth muscle layer runs circularly around the arteriole (● Figure 10-10a) so when it contracts, the vessel's circumference (and its radius) becomes smaller, increasing resistance and decreasing the flow through that vessel. **Vasoconstriction** is the term applied to such narrowing of a vessel (● Figure 10-10c). The term **vasodilation** refers to enlargement in the circumference and radius of a vessel as a result of relaxation of its smooth muscle layer (● Figure 10-10d). Vasodilation leads to decreased resistance and increased flow through that vessel.

Vascular tone

Arteriolar smooth muscle normally displays a state of partial constriction known as **vascular tone**, which establishes a baseline of arteriolar resistance (● Figure 10-10b). Two factors are responsible for vascular tone. First, arteriolar smooth muscle has considerable myogenic activity; that is, its membrane potential fluctuates without any neural or hormonal influences, leading to self-induced contractile activity (see p. 294). Second, the sympathetic fibers supplying most arterioles continually release norepinephrine, which further enhances the vascular tone.

This ongoing tonic activity makes it possible to either increase or decrease the level of contractile activity to accomplish vasoconstriction or vasodilation, respectively. Were it not for tone, it would be impossible to reduce the tension in an arteriolar wall to accomplish vasodilation; only varying degrees of vasoconstriction would be possible.

A variety of factors can influence the level of contractile activity in arteriolar smooth muscle, thereby substantially changing resistance to flow in these vessels. These factors fall into two categories: *local (intrinsic) controls*, which are important in determining the distribution of cardiac output; and *extrinsic controls*, which are important in blood pressure regulation. We will look at each of these controls in turn.

▌ Local control of arteriolar radius is important in determining the distribution of cardiac output.

The fraction of the total cardiac output delivered to each organ is not always constant; it varies, depending on the demands for blood at the time. The amount of the cardiac output received by each organ is determined by the number and caliber of the arterioles supplying that area. Recall that $F = \Delta P/R$. Because blood is delivered to all tissues at the same mean arterial pressure, the driving force for flow is identical for each organ. Therefore, differences in flow to various organs are completely determined by differences in the extent of vascularization and by differences in resistance offered by the arterioles supplying each organ. On a moment-to-moment basis, the distribution of cardiac output can be varied by differentially adjusting arteriolar resistance in the various vascular beds.

As an analogy, consider a pipe carrying water, with a number of adjustable valves located throughout its length (● Figure 10-11). Assuming that water pressure in the pipe is constant, differences in the amount of water flowing into a beaker under each valve depend entirely on which valves are open and to what extent. No water enters beakers under closed valves (high resistance), and more water flows into beakers under valves that are opened completely (low resistance) than into beakers under valves that are only partially opened (moderate resistance).

Similarly, more blood flows to areas whose arterioles offer the least resistance to its passage. During exercise, for example, not only is cardiac output increased, but also, because of vasodilation in skeletal muscle and in the heart, a greater percentage of the pumped blood is diverted to these organs to support their increased metabolic activity. Simultaneously, blood flow to the digestive tract and kidneys is reduced, as a result of arteriolar vasoconstriction in these organs (● Figure 10-12). Only the blood supply to the brain remains remarkably constant no matter what activity the person is engaged in, be it vigorous physical activity, intense mental concentration, or sleep. Although the *total* blood flow to the brain remains constant, new imaging techniques demonstrate that differences in regional blood flow occur within the brain in close correlation with local neural activity patterns (see p. 147).

Local (intrinsic) controls are changes within a tissue that alter the radii of the vessels and hence adjust blood flow through the tissue by directly affecting the smooth muscle of the tissue's arterioles. Local influences may be either chemical or physical in nature. Local chemical influences on arteriolar radius include (1) local metabolic changes and (2) histamine release. Local physical influences include (1) local application of heat or cold and (2) myogenic response to stretch. We are going to examine the role and mechanism of each of these local influences.

Normal arteriolar tone

Cross section
of arteriole

(b)

Vasoconstriction
(increased contraction
of circular smooth
muscle in the arteriolar
wall, which leads to
increased resistance
and decreased
flow through the vessel)

(c)

Caused by:
↑ Myogenic activity
↑ Oxygen (O₂)
↓ Carbon dioxide (CO₂)
and other metabolites
↑ Endothelin
↑ Sympathetic stimulation
Vasopressin; angiotensin II
Cold

Vasodilation
(decreased contraction
of circular smooth
muscle in the arteriolar
wall, which leads to
decreased resistance
and increased flow
through the vessel)

(d)

Caused by:
↓ Myogenic activity
↓ O₂
↑ CO₂ and other metabolites
↑ Nitric oxide
↓ Sympathetic stimulation
Histamine release
Heat

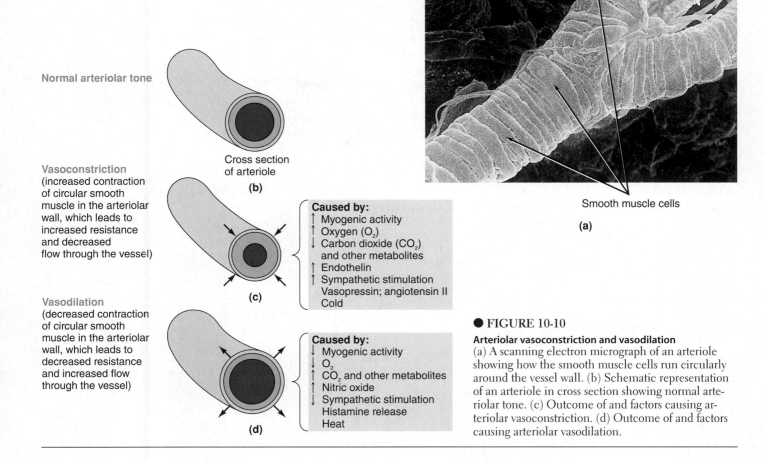

Smooth muscle cells

(a)

● **FIGURE 10-10**

Arteriolar vasoconstriction and vasodilation
(a) A scanning electron micrograph of an arteriole
showing how the smooth muscle cells run circularly
around the vessel wall. (b) Schematic representation
of an arteriole in cross section showing normal arteriolar tone. (c) Outcome of and factors causing arteriolar vasoconstriction. (d) Outcome of and factors
causing arteriolar vasodilation.

▌ Local metabolic influences on arteriolar radius help match blood flow with the tissues' needs.

The most important local chemical influences
on arteriolar smooth muscle are related to metabolic changes within a given tissue. The influence of these local changes on arteriolar radius
is important in matching the blood flow through
a tissue with the tissue's metabolic needs. Local metabolic
controls are especially important in skeletal muscle and in
the heart, the tissues whose metabolic activity and need for
blood supply normally vary most extensively, and in the brain,
whose overall metabolic activity and need for blood supply
remain constant. Local controls help maintain the constancy
of blood flow to the brain.

Active hyperemia

Arterioles lie within the tissue they are supplying and can be
acted on by local factors within the tissue. During increased
metabolic activity, such as when a skeletal muscle is contracting during exercise, the local concentrations of a number of
the tissue's chemicals change. For example, the local O₂ concentration decreases as the actively metabolizing cells use up
more O₂ to support oxidative phosphorylation for ATP pro-

From pump
(heart)

● **FIGURE 10-11**

Flow rate as a function of resistance

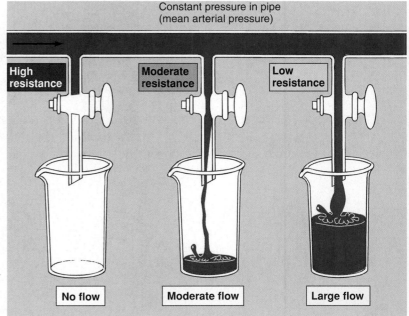

Constant pressure in pipe
(mean arterial pressure)

High
resistance

Moderate
resistance

Low
resistance

No flow

Moderate flow

Large flow

Control valves = Arterioles

duction (see p. 40). This and other local chemical changes produce local arteriolar dilation by triggering relaxation of the arteriolar smooth muscle in the vicinity. Local arteriolar vasodilation subsequently increases blood flow to that particular area, a response called **active hyperemia** (*hyper* means "above normal"; *emia* means "blood"). When cells are more active metabolically, they need more blood to bring in O_2 and nutrients and to remove metabolic wastes. The increased blood flow meets these increased local needs.

Conversely, when a tissue, such as a relaxed muscle, is less active metabolically and thus has reduced needs for blood de-

livery, the resultant local chemical changes (for example, increased local O_2 concentration) bring about local arteriolar vasoconstriction and a subsequent reduction in blood flow to the area. Local metabolic changes can thus adjust blood flow as needed without involving nerves or hormones.

Local metabolic changes that influence arteriolar radius

A variety of local chemical changes act together in a cooperative, redundant manner to bring about these "selfish" local adjustments in arteriolar caliber that match a tissue's blood flow

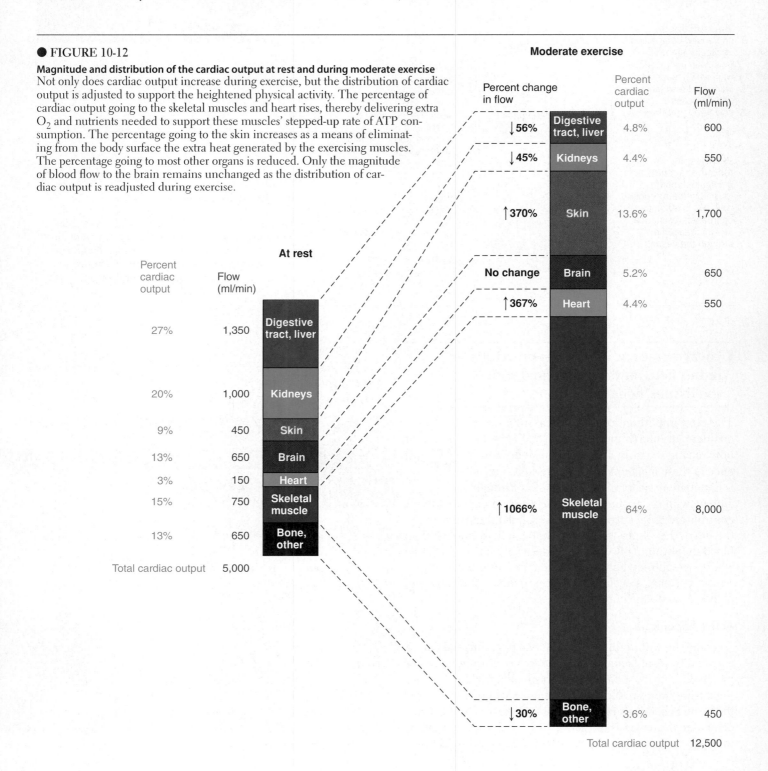

● FIGURE 10-12

Magnitude and distribution of the cardiac output at rest and during moderate exercise
Not only does cardiac output increase during exercise, but the distribution of cardiac output is adjusted to support the heightened physical activity. The percentage of cardiac output going to the skeletal muscles and heart rises, thereby delivering extra O_2 and nutrients needed to support these muscles' stepped-up rate of ATP consumption. The percentage going to the skin increases as a means of eliminating from the body surface the extra heat generated by the exercising muscles. The percentage going to most other organs is reduced. Only the magnitude of blood flow to the brain remains unchanged as the distribution of cardiac output is readjusted during exercise.

with its needs. Specifically, the following local chemical factors produce relaxation of arteriolar smooth muscles:

- *Decreased O_2*.
- *Increased CO_2*. More CO_2 is generated as a by-product during the stepped-up pace of oxidative phosphorylation that accompanies increased activity.
- *Increased acid*. More carbonic acid is generated from the increased CO_2 produced as the metabolic activity of a cell increases. Also, lactic acid accumulates if the glycolytic pathway is used for ATP production (see p. 40).
- *Increased K^+*. Repeated action potentials that outpace the ability of the Na^+–K^+ pump to restore the resting concentration gradients (see p. 108) result in an increase in K^+ in the tissue fluid of an actively contracting muscle or a more active region of the brain.
- *Increased osmolarity*. Osmolarity (the concentration of osmotically active solutes) increases during elevated cellular metabolism because of increased formation of osmotically active particles.
- *Adenosine release*. Especially in cardiac muscle, adenosine is released in response to increased metabolic activity or O_2 deprivation (see p. 333).
- *Prostaglandin release*. Prostaglandins are local chemical messengers derived from fatty acid chains within the plasma membrane (see p. 765).

The relative contributions of these various chemical changes (and possibly others) in the local metabolic control of blood flow in systemic arterioles are still being investigated.

Local vasoactive mediators

These local factors do not act directly on vascular smooth muscle to change its contractile state. Instead, the **endothelial cells,** the single layer of specialized epithelial cells that line the lumen of all blood vessels, release chemical mediators that play a key role in the local regulation of arteriolar caliber. Until recently, endothelial cells were regarded as little more than a passive barrier between the blood and the remainder of the vessel wall. It is now known that endothelial cells are active participants in a variety of vessel-related activities, some of which will be described elsewhere (▲ Table 10-2). Among these functions, endothelial cells release locally acting chemical messengers in response to chemical changes in their environment (such as a reduction in O_2) or physical changes (such as stretching of the vessel wall). These local vasoactive ("acting on vessels") mediators act on the underlying smooth muscle to alter its state of contraction.

Among the best studied of these local vasoactive mediators is **endothelial-derived relaxing factor (EDRF),** which causes local arteriolar vasodilation by inducing relaxation of arteriolar smooth muscle in the vicinity. It does so by inhibiting the entry of contraction-inducing Ca^{2+} into these smooth muscle cells (see p. 291). EDRF has been identified surprisingly as **nitric oxide (NO),** a small, highly reactive, short-lived gas molecule that once was known primarily as a toxic air pollutant. Yet studies have revealed an astonishing number of biological roles for NO, which is produced in numerous other tissues besides en-

---

▲ **TABLE 10-2**
Functions of Endothelial Cells

- Line the blood vessels and heart chambers; serve as a physical barrier between the blood and the remainder of the vessel wall

- Secrete vasoactive substances in response to local chemical and physical changes; these substances cause relaxation (vasodilation) or contraction (vasoconstriction) of the underlying smooth muscle

- Secrete substances that stimulate new vessel growth and proliferation of smooth muscle cells in vessel walls

- Participate in the exchange of materials between the blood and surrounding tissues across capillaries through vesicular transport (see p. 85)

- Influence formation of platelet plugs, clotting, and clot dissolution (see Chapter 11)

- Participate in the determination of capillary permeability by contracting to vary the size of the pores between adjacent endothelial cells

---

dothelial cells. In fact, it appears that NO serves as one of the body's most important messenger molecules, as exemplified by the range of functions that have been identified for this chemical and are listed in ▲ Table 10-3. As you can see, very few areas of the body are *not* influenced by this versatile intercellular messenger molecule.

The endothelial cells release other important chemicals besides EDRF/NO. **Endothelin,** another endothelial vasoactive substance, causes arteriolar smooth-muscle contraction and is one of the most potent vasoconstrictors yet identified. Still other chemicals, released from the endothelium in response to chronic changes in blood flow to an organ, trigger long-term vascular changes that permanently influence blood flow to a region. Some chemicals, for example, stimulate angiogenesis (new vessel growth).

▮ Local histamine release pathologically dilates arterioles.

Histamine is another local chemical mediator that influences arteriolar smooth muscle, but it is not released in response to local metabolic changes and is not derived from the endothelial cells. Although histamine normally does not participate in controlling blood flow, it is important in certain pathological conditions. Histamine is synthesized and stored within special connective tissue cells in many tissues and in certain types of circulating white blood cells. When tissues are injured or during allergic reactions, histamine is released in the damaged region. By promoting relaxation of arteriolar smooth muscle, histamine is the major cause of vasodilation in an injured area. The resultant increase in blood flow into the area is responsible for the redness and contributes to the swelling associated with inflammatory responses (see Chapter 12 for further details).

▲ TABLE 10-3
Functions of Nitric Oxide (NO)

- Causes relaxation of arteriolar smooth muscle. By means of this action, NO plays an important role in controlling blood flow through the tissues and in maintaining mean arterial blood pressure.

- Dilates the arterioles of the penis and clitoris, thus serving as the direct mediator of erection of these reproductive organs. Erection is accomplished by rapid engorgement of these organs with blood.

- Used as chemical warfare against bacteria and cancer cells by macrophages, large phagocytic cells of the immune system.

- Interferes with platelet function and blood clotting at sites of vessel damage.

- Serves as a novel type of neurotransmitter in the brain and elsewhere. Unlike classical neurotransmitters, NO is not stored in vesicles and released by exocytosis, and it does not bind with receptors on its target cell. NO is synthesized on demand in the neuron terminal, then diffuses out of the terminal and into the adjacent target cell, where it brings about its effect.

- Plays a role in the changes underlying memory.

- By promoting relaxation of digestive-tract smooth muscle, helps regulate peristalsis, a type of contraction that pushes digestive tract contents forward.

- Relaxes the smooth muscle cells in the airways of the lungs, helping keep these passages open to facilitate movement of air in and out of the lungs.

- May play a role in relaxation of skeletal muscle.

▌ Local physical influences on arteriolar radius include temperature changes and stretch.

Among the physical influences on arteriolar smooth muscle, the effect of temperature changes is exploited clinically but the myogenic response to stretch is most important physiologically. Let's examine each of these effects.

Local heat or cold application

Heat application, by causing localized arteriolar vasodilation, is a useful therapeutic agent for promoting increased blood flow to an area. Conversely, applying ice packs to an inflamed area produces vasoconstriction, which reduces swelling by counteracting histamine-induced vasodilation.

Myogenic responses to stretch

Arteriolar smooth muscle responds to being passively stretched by myogenically increasing its tone, thereby acting to resist the initial passive stretch. Conversely, a decrease in arteriolar stretching is accompanied by a reduction in myogenic vessel tone. Endothelial-derived vasoactive substances may also contribute

to these mechanically induced responses. The extent of passive stretch varies with the volume of blood delivered to the arterioles from the arteries. An increase in mean arterial pressure drives more blood forward into the arterioles and stretches them further, whereas arterial occlusion blocks blood flow into the arterioles and reduces arteriolar stretch.

Myogenic responses, coupled with metabolically induced responses, are important in reactive hyperemia and pressure autoregulation, topics to which we now turn our attention.

Reactive hyperemia

When the blood supply to a region is completely occluded, arterioles in the region dilate because of (1) myogenic relaxation, which occurs in response to the diminished stretch accompanying no blood flow, and (2) changes in local chemical composition. Many of the same chemical changes occur in a blood-deprived tissue that occur during metabolically induced active hyperemia. When a tissue's blood supply is blocked, O_2 levels decrease in the deprived tissue; the tissue continues to consume O_2, but no fresh supplies are being delivered. Meanwhile, the concentrations of CO_2, acid, and other metabolites rise. Even though their production does not increase as it does when a tissue is more active metabolically, these substances accumulate in the tissue when the normal amounts produced are not "washed away" by blood.

After the occlusion is removed, blood flow to the previously deprived tissue is transiently much higher than normal because the arterioles are widely dilated. This postocclusion increase in blood flow, called **reactive hyperemia**, can take place in any tissue. Such a response is beneficial for rapidly restoring the local chemical composition to normal. Of course, prolonged blockage of blood flow leads to irreversible tissue damage.

Pressure autoregulation

When mean arterial pressure falls (for example, because of hemorrhage or a weakened heart), the driving force is reduced, so blood flow to tissues decreases. This is a milder version of what happens during vessel occlusion. The resultant changes in local metabolites and the reduced stretch in the arterioles collectively bring about arteriolar dilation to restore tissue blood flow to normal despite the reduced driving pressure. On the negative side, widespread arteriolar dilation reduces the mean arterial pressure still further, which aggravates the problem. Conversely, in the presence of sustained elevations in mean arterial pressure (hypertension), local chemical and myogenic influences triggered by the initial increased flow of blood to tissues bring about an increase in arteriolar tone and resistance. This greater degree of vasoconstriction subsequently reduces tissue blood flow toward normal despite the elevated blood pressure (● Figure 10-13). **Pressure autoregulation** is the term for these local arteriolar mechanisms that keep tissue blood flow fairly constant despite rather wide deviations in mean arterial driving pressure.

Active hyperemia, reactive hyperemia, and histamine release all deliberately increase blood flow to a particular tissue for a specific purpose by inducing local arteriolar vasodilation. In contrast, pressure autoregulation is a means by which each

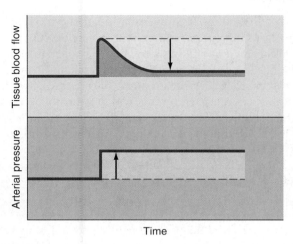

● FIGURE 10-13

Pressure autoregulation of tissue blood flow
Even though blood flow through a tissue immediately increases in response to a rise in arterial pressure, the tissue blood flow is reduced gradually as a result of pressure autoregulation within the tissue, despite a sustained increase in arterial pressure.

tissue resists alterations in its own blood flow secondary to changes in mean arterial pressure by making appropriate adjustments in arteriolar radius.

This completes our discussion of the local control of arteriolar radius. We are now going to shift our attention to extrinsic control of arteriolar radius.

▌ Extrinsic sympathetic control of arteriolar radius is important in the regulation of blood pressure.

Extrinsic control of arteriolar radius includes both neural and hormonal influences, the effects of the sympathetic nervous system being the most important. Sympathetic nerve fibers supply arteriolar smooth muscle everywhere in the systemic circulation except in the brain. Recall that a certain level of ongoing sympathetic activity contributes to vascular tone. Increased sympathetic activity produces generalized arteriolar vasoconstriction, whereas decreased sympathetic activity leads to generalized arteriolar vasodilation. These widespread changes in arteriolar resistance bring about changes in mean arterial blood pressure because of their influence on total peripheral resistance as follows.

Influence of total peripheral resistance on mean arterial pressure

To find the effect of changes in arteriolar resistance on mean arterial pressure, the formula $F = \Delta P/R$ applies to the entire circulation as well as to a single vessel:

- F: Looking at the circulatory system as a whole, flow (F) through all the vessels in either the systemic or pulmonary circulation is equal to the cardiac output.
- ΔP: The pressure gradient (ΔP) for the entire systemic circulation is the mean arterial pressure. (ΔP equals the differ-

ence in pressure between the beginning and the end of the systemic circulatory system. The beginning pressure is the mean arterial pressure as the blood leaves the left ventricle at an average of 93 mm Hg. The end pressure in the right atrium is 0 mm Hg. Therefore, ΔP = 93 mm Hg minus 0 mm Hg = 93 mm Hg, which is equivalent to the mean systemic pressure.)

- R: The total resistance (R) offered by all the systemic peripheral vessels together is the **total peripheral resistance**. By far the greatest percentage of the total peripheral resistance is due to arteriolar resistance, because arterioles are the primary resistance vessels.

Therefore, for the entire systemic circulation, rearranging

$$F = \Delta P/R$$

to

$$\Delta P = F \times R$$

gives us the equation

$$\text{Mean arterial pressure} = \text{cardiac output} \times \text{total peripheral resistance}$$

Thus the extent of total peripheral resistance offered collectively by all the systemic arterioles influences the mean arterial blood pressure immensely. A dam provides an analogy to this relationship. At the same time a dam restricts the flow of water downstream, it increases the pressure upstream by elevating the water level in the reservoir behind the dam. Similarly, generalized, sympathetically induced vasoconstriction reflexly reduces blood flow downstream to the tissue cells while elevating the upstream mean arterial pressure, thereby increasing the main driving force for blood flow to all the organs.

These effects seem counterproductive. Why increase the driving force for flow to the organs by increasing arterial blood pressure while reducing flow to the organs by narrowing the vessels supplying them? In effect, the sympathetically induced arteriolar responses help maintain the appropriate driving pressure head (that is, the mean arterial pressure) to all organs. The extent to which each organ actually receives blood flow is determined by local arteriolar adjustments that override the sympathetic constrictor effect. If all arterioles were dilated, blood pressure would fall substantially, so there would not be an adequate driving force for blood flow. An analogy is the pressure head for water in the pipes in your home. If the water pressure is adequate, you can selectively obtain satisfactory water flow at any of the faucets by turning the appropriate handle to the open position. If the water pressure in the pipes is too low, however, you cannot obtain satisfactory flow at any faucet, even if you turn the handle to the maximally open position. Tonic sympathetic activity thus constricts most vessels (with the exception of those in the brain) to help maintain a pressure head on which organs can draw as needed through local mechanisms that control arteriolar radius.

Norepinephrine's influence on arteriolar smooth muscle

The norepinephrine released from sympathetic nerve endings combines with α_1-adrenergic receptors on arteriolar smooth muscle to bring about vasoconstriction (see p.245). Cerebral

The Blood Vessels and Blood Pressure

(brain) arterioles are the only ones that do not have α_1 receptors, so no vasoconstriction occurs in the brain. It is important that cerebral arterioles are not reflexly constricted by neural influences, because brain blood flow must remain constant to meet the brain's continual need for O_2, no matter what is going on elsewhere in the body. Cerebral vessels are almost entirely controlled by local mechanisms that maintain a constant blood flow to support a constant level of brain metabolic activity. In fact, reflex vasconstrictor activity in the remainder of the cardiovascular system is aimed at maintaining an adequate pressure head for blood flow to the vital brain.

Thus sympathetic activity contributes in an important way to the maintenance of mean arterial pressure, assuring an adequate driving force for blood flow to the brain at the expense of organs and tissues that can better withstand reduced blood flow. Other tissues that really need additional blood, such as active muscles (including active heart muscle), obtain it through local controls that override the sympathetic effect.

Local controls overriding sympathetic vasoconstriction

Skeletal and cardiac muscles have the most powerful local control mechanisms with which to override generalized sympathetic vasoconstriction. For example, if you are pedaling a bicycle the increased activity in the skeletal muscles of your legs brings about an overriding local, metabolically induced vasodilation in those particular muscles, despite the generalized sympathetic vasoconstriction that accompanies exercise. As a result, more blood flows through your leg muscles but not through your inactive arm muscles.

No parasympathetic innervation to arterioles

There is no significant parasympathetic innervation to arterioles, with the exception of the abundant parasympathetic vasodilator supply to the arterioles of the penis and clitoris. The rapid, pro-

fuse vasodilation induced by parasympathetic stimulation in these organs (by means of promoting release of NO) is largely responsible for accomplishing erection. Vasodilation elsewhere is produced by decreasing sympathetic vasoconstrictor activity below its tonic level.[1] When mean arterial pressure becomes elevated above normal, reflex reduction in sympathetic vasoconstrictor activity accomplishes generalized arteriolar vasodilation to help restore the driving pressure down toward normal.

▌The medullary cardiovascular control center and several hormones regulate blood pressure.

The main region of the brain responsible for adjusting sympathetic output to the arterioles is the **cardiovascular control center** in the medulla of the brain stem. This is the integrating center for blood pressure regulation. Several other brain regions also influence blood distribution, the most notable being the hypothalamus, which, as part of its temperature-regulating function, controls blood flow to the skin to adjust heat loss to the environment.

In addition to neural reflex activity, several hormones also extrinsically influence arteriolar radius. These hormones include the adrenal medullary hormones *epinephrine* and *norepinephrine*, which generally reinforce the sympathetic nervous system in most tissues, as well as *vasopressin* and *angiotensin II*, which are important in controlling fluid balance.

Influence of epinephrine and norepinephrine

Sympathetic stimulation of the adrenal medulla causes this endocrine gland to release epinephrine and norepinephrine. Adrenal medullary norepinephrine combines with the same α_1 receptors as sympathetically released norepinephrine to produce generalized vasoconstriction. However, epinephrine, the more abundant of the adrenal medullary hormones, combines with both β_2 and α_1 receptors but has a much greater affinity for the β_2 receptors (▲ Table 10-4). Activation of β_2 receptors produces vasodilation, but not all tissues have β_2 receptors; they are most abundant in the arterioles of the heart and skeletal muscles. During sympathetic discharge, the released epinephrine combines with the β_2 receptors in the heart and skeletal muscle to reinforce local vasodilatory mechanisms in these tissues. Arterioles in digestive organs and kidneys, in contrast, are equipped only with α receptors. Therefore, the arterioles of these organs undergo more profound vasoconstriction during generalized sympathetic discharge than those in heart and skeletal muscle do. Lacking β_2 receptors, the digestive organs and kidneys do not experience an overriding vasodilatory response on top of the α_1 receptor–induced vasoconstriction.

Influence of vasopressin and angiotensin II

The two other hormones that extrinsically influence arteriolar tone are vasopressin and angiotensin II. Vasopressin is primar-

---

▲ **TABLE 10-4**
Arteriolar Smooth-Muscle Adrenergic Receptors

Characteristic	Receptor Type	
	α_1	β_2
Location of the Receptor	All arteriolar smooth muscle except in the brain	Arteriolar smooth muscle in the heart and skeletal muscles
Chemical Mediator	Norepineprine from sympathetic fibers and the adrenal medulla	Epinephrine from the adrenal medulla (greater affinity for this receptor
	Epinephrine from the adrenal medulla (less affinity for this receptor)	
Arteriolar Smooth-Muscle Response	Vasoconstriction	Vasodilation

---

[1]A portion of the skeletal muscle fibers in some species is supplied by sympathetic cholinergic (ACh-releasing) fibers that bring about vasodilation in anticipation of exercise. However, the existence of such sympathetic vasodilator fibers in humans remains questionable.

ily involved in maintaining water balance by regulating the amount of water the kidneys retain for the body during urine formation. Angiotensin II is part of a hormonal pathway, the *renin-angiotensin-aldosterone pathway*, which is important in regulating the body's salt balance. This pathway promotes salt conservation during urine formation and also leads to water retention, because salt exerts a water-holding osmotic effect in the ECF. Thus both these hormones play important roles in maintaining fluid balance in the body, which in turn is an important determinant of plasma volume and blood pressure.

In addition, both vasopressin and angiotensin II are potent vasoconstrictors. Their role in this regard is especially important during hemorrhage. A sudden loss of blood reduces the plasma volume, which triggers increased secretion of both of these hormones to help restore plasma volume. Their vasoconstrictor effect also helps maintain blood pressure despite abrupt loss of plasma volume. (The functions and control of these hormones are discussed more thoroughly in later chapters.)

This completes our discussion of the various factors that affect total peripheral resistance, the most important of which are controlled adjustments in arteriolar radius. These factors are summarized in ● Figure 10-14.

We are now going to turn our attention to the next vessels in the vascular tree, the capillaries.

● **FIGURE 10-14**

Factors affecting total peripheral resistance
The primary determinant of total peripheral resistance is the adjustable arteriolar radius. Two major categories of factors influence arteriolar radius: (1) local (intrinsic) control, which is primarily important in matching blood flow through a tissue with the tissue's metabolic needs and is mediated by local factors acting on the arteriolar smooth muscle, and (2) extrinsic control, which is important in the regulation of blood pressure and is mediated primarily by sympathetic influence on arteriolar smooth muscle.

Major factors affecting arteriolar radius

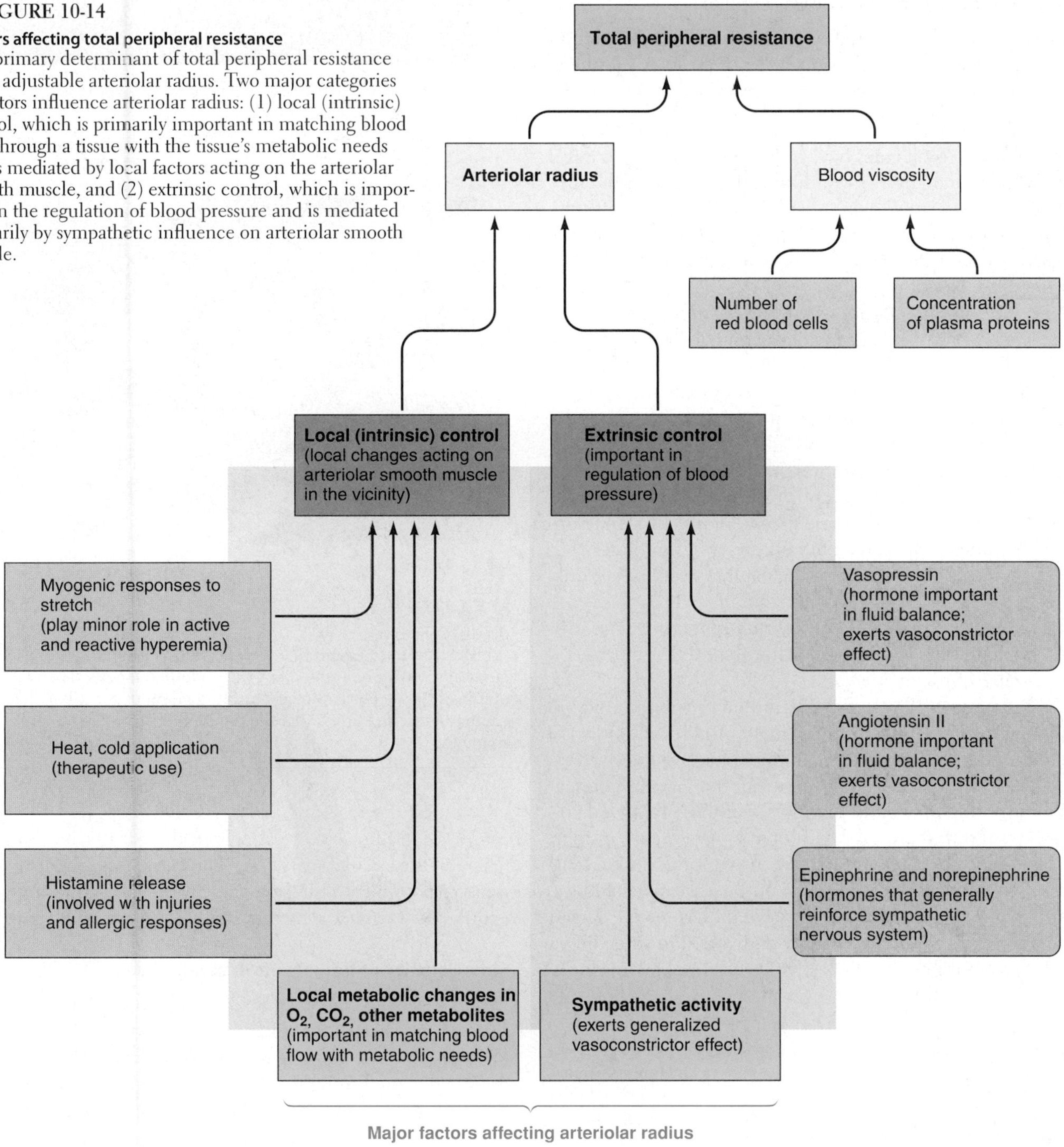

CAPILLARIES

Capillaries are ideally suited to serve as sites of exchange.

Capillaries, the sites for exchange of materials between the blood and tissues,[2] branch extensively to bring blood within the reach of every cell. There are no carrier-mediated transport systems across capillaries, with the exception of those in the brain that play a role in the blood–brain barrier (see p. 142). Exchange of materials across capillary walls is accomplished primarily by the process of diffusion.

Factors enhancing diffusion across capillaries

Capillaries are ideally suited to enhance diffusion in accordance with Fick's law of diffusion (see p. 74). They minimize diffusion distances while maximizing surface area and time available for exchange, as follows:

1. Diffusing molecules have only a short distance to travel between the blood and surrounding cells because of the thin capillary wall and small capillary diameter, coupled with the close proximity of every cell to a capillary. This short distance is important because the rate of diffusion slows down as the diffusion distance increases.

   a. Capillary walls are very thin (1 μm in thickness; in comparison, the diameter of a human hair is 100 μm). Capillaries are composed of only a single layer of flattened endothelial cells—essentially the lining of the other vessel types. No smooth muscle or connective tissue is present (● Figure 10-15a).

   b. Each capillary is so narrow (7 mm average diameter) that red blood cells (8 mm diameter) have to squeeze through single file (● Figure 10-15b). Consequently, plasma contents are either in direct contact with the inside of the capillary wall or are only a short diffusing distance from it.

   c. Because of extensive capillary branching, it is estimated that no cell is farther than 0.01 cm (4/1000 inch) from a capillary.

2. Because capillaries are distributed in such incredible numbers (estimates range from 10 to 40 billion capillaries), a tremendous total surface area is available for exchange (an estimated 600 m²). In spite of this large number of capillaries, at any point in time they contain only 5% of the total blood volume (250 ml out of a total of 5,000 ml). As a result, a small volume of blood is exposed to an extensive surface area. If all the capillary surfaces were stretched out in a flat sheet and the volume of blood contained within the capillaries were spread over the top, this would be roughly equivalent to spreading a half pint of paint over the floor of a high school gymnasium. Imagine how thin the paint layer would be!

---

[2]Actually, some exchange takes place across the other microcirculatory vessels, especially the postcapillary venules. The entire vasculature is a continuum and does not abruptly change from one vascular type to another. When the term *capillary exchange* is used, it tacitly refers to all exchange at the microcirculatory level, the majority of which occurs across the capillaries.

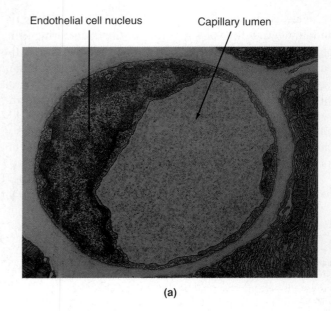

Endothelial cell nucleus   Capillary lumen

(a)

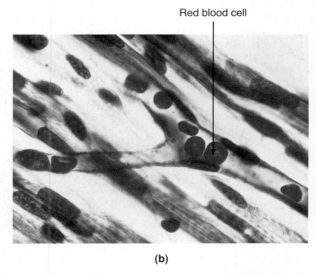

Red blood cell

(b)

● FIGURE 10-15

Capillary anatomy
(a) Electron micrograph of a cross section of a capillary. The capillary wall consists of a single layer of endothelial cells. The nucleus of one of these cells is shown. (b) Photograph of a capillary bed. The capillaries are so narrow that the red blood cells must pass through single file.

---

3. Blood flows more slowly in the capillaries than elsewhere in the circulatory system. The extensive capillary branching is responsible for this slow velocity of blood flow through the capillaries. Let's see why blood slows down in the capillaries.

Slow velocity of flow through capillaries

First, we need to clarify a potentially confusing point. The term *flow* can be used in two different contexts—flow rate and velocity of flow. The *flow rate* refers to the *volume* of blood per unit of time flowing through a given segment of the circulatory system (this is the flow we have been talking about in relation to the pressure gradient and resistance). The *velocity of flow* refers to the linear *speed*, or distance per unit of time, with which blood flows forward through a given segment of the circulatory system. Because the circulatory system is a closed system, the volume of blood flowing through any level of the system must equal the cardiac output. For example, if the heart pumps out 5 liters of blood per minute, and 5 l/min return to the heart, then 5 l/min must flow through the arteries, arterioles, capillaries, and veins. Therefore, the flow rate is the same at all levels of the circulatory system.

However, the velocity with which blood flows through the different segments of the vascular tree varies because velocity of flow is inversely proportional to the total cross-sectional area of all the vessels at any given level of the circulatory system as follows:

$$\begin{array}{l}\text{velocity of flow} \\ \text{at a given level}\end{array} = \dfrac{\text{flow rate}}{\begin{array}{c}\text{total cross-sectional} \\ \text{area of vessels} \\ \text{at that level}\end{array}}$$

Even though the cross-sectional area of each capillary is extremely small compared to that of the large aorta, the total cross-sectional area of all the capillaries added together is about 1,300 times greater than the cross-sectional area of the aorta because there are so many capillaries (▲ Table 10-1). Accordingly, blood slows considerably as it passes through the capillaries (● Figure 10-16). This slow velocity allows adequate time for exchange of nutrients and metabolic end products between blood and tissues, which is the sole purpose of the entire circulatory system. As the capillaries rejoin to form veins, the total cross-sectional area is once again reduced, and the velocity of blood flow increases as blood returns to the heart.

As an analogy, consider a river (the arterial system) that widens into a lake (the capillaries), then narrows into a river again (the venous system) (● Figure 10-17). The flow rate is the same throughout the length of this body of water, that is, identical volumes of water are flowing past all the points along the bank of the river and lake. However, the velocity of flow is slower in the wide lake than in the narrow river because the identical volume of water, now spread out over a larger cross-sectional area, moves forward a much shorter distance in the

● FIGURE 10-16

Comparison of blood flow rate and velocity of flow in relation to total cross-sectional area
The blood flow rate (red curve) is identical through all levels of the circulatory system and is equal to the cardiac output (5 liters/min at rest). The velocity of flow (purple curve) varies throughout the vascular tree and is inversely proportional to the total cross-sectional area (green curve) of all the vessels at a given level. Note that the velocity of flow is slowest in the capillaries, which have the largest total cross-sectional area.

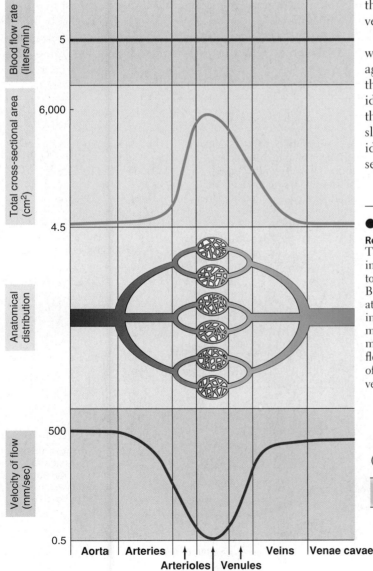

● FIGURE 10-17

Relationship between total cross-sectional area and velocity of flow
The three dark blue areas represent equal volumes of water. During one minute, this volume of water moves forward from points A to points C. Therefore, an identical volume of water flows past points B1, B2, and B3 during this minute; that is, the flow rate is the same at all points along the length of this body of water. However, during that minute the identical volume of water moves forward a much shorter distance in the wide lake (A2 to C2) than in the much narrower river (A1 to C1 and A3 to C3). Thus velocity of flow is much slower in the lake than in the river. Similarly, velocity of flow is much slower in the capillaries than in the arterial and venous systems.

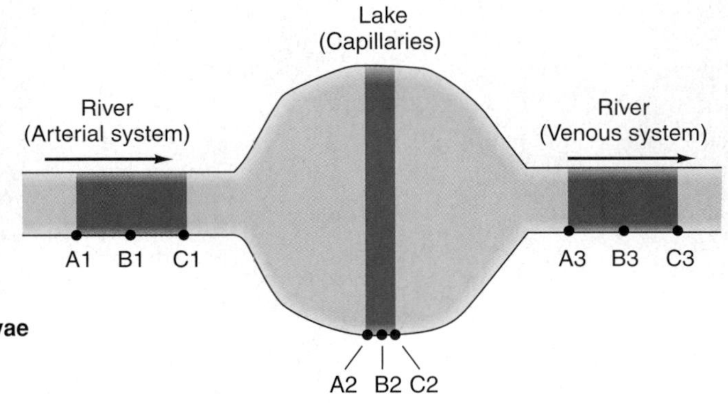

wide lake than in the narrow river during a given period of time. You could readily observe the forward movement of water in the swift-flowing river, but the forward motion of water in the lake would be unnoticeable.

Also, because of the capillaries' tremendous total cross-sectional area the resistance offered by all the capillaries is much lower than that offered by all the arterioles, even though each capillary has a smaller radius than each arteriole. For this reason, the arterioles contribute more to total peripheral resistance. Furthermore, arteriolar caliber (and, accordingly resistance) is subject to control, whereas capillary caliber cannot be adjusted.

▌ Water-filled capillary pores permit passage of small, water-soluble substances.

Diffusion across capillary walls also depends on the walls' permeability to the materials being exchanged. The endothelial cells forming the capillary walls fit together in jigsaw-puzzle fashion, but the closeness of the fit varies considerably between organs. In most capillaries, narrow, water-filled clefts, or **pores,** are present at the junctions between the cells (● Figure 10-18). These pores permit passage of water-soluble substances. Lipid-soluble substances, such as O_2 and CO_2, can readily pass

through the endothelial cells themselves by dissolving in the lipid bilayer barrier.

The size of the capillary pores varies from organ to organ. At one extreme, the endothelial cells in brain capillaries are joined by tight junctions so that pores are nonexistent. These junctions prevent transcapillary passage of materials between the cells and thus constitute part of the protective blood–brain barrier. In most tissues, small, water-soluble substances such as ions, glucose, and amino acids can readily pass through the water-filled clefts, but large, non–lipid-soluble materials such as plasma proteins are excluded from passage. At the other extreme, liver capillaries have such large pores that even proteins pass through readily. This is adaptive, because the liver's functions include synthesis of plasma proteins and the metabolism of protein-bound substances such as cholesterol. These proteins must all pass through the liver's capillary walls. The leakiness of various capillary beds is therefore a function of how tightly the endothelial cells are joined, which varies according to the different organs' needs.

The capillary wall has traditionally been considered a passive sieve, like a brick wall with permanent gaps in the mortar acting as pores. Recent studies, however, suggest that even under normal conditions, changes in endothelial cells are involved in actively regulating capillary membrane permeability; that is, in response to appropriate signals, the "bricks" can

---

● **FIGURE 10-18**

Exchanges across the capillary wall
(a) Slitlike gaps between adjacent endothelial cells form pores within the capillary wall. (b) As depicted in this schematic representation of a cross section of a capillary wall, small water-soluble substances are exchanged between the plasma and the interstitial fluid by passing through the water-filled pores, whereas lipid-soluble substances are exchanged across the capillary wall by passing through the endothelial cells. Proteins to be moved across are exchanged by vesicular transport. The plasma proteins generally cannot escape from the plasma across the capillary wall.

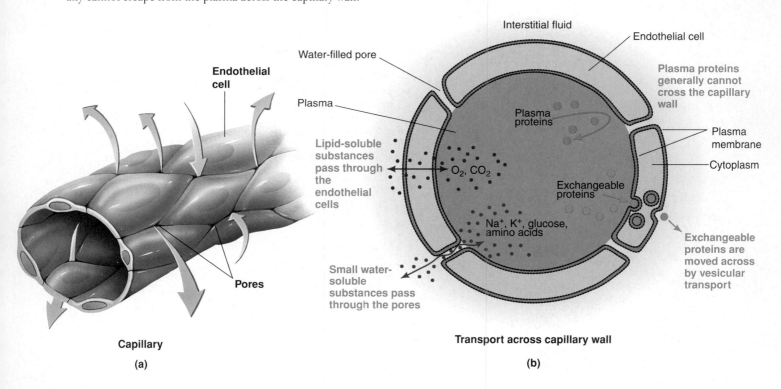

readjust themselves to vary the size of the holes. Thus, the degree of leakiness does not necessarily remain constant for a given capillary bed. For example, researchers speculate that histamine increases capillary permeability by triggering contractile responses in endothelial cells to widen the intercellular gaps. This is not a muscular contraction, because no smooth muscle cells are present in capillaries. It is due to an actin-myosin contractile apparatus in the nonmuscular capillary endothelial cells. Because of these enlarged gaps, the affected capillary wall is leakier. As a result, normally retained plasma proteins escape into the surrounding tissue, where they exert an osmotic effect. Along with histamine-induced vasodilation, the resultant additional local fluid retention contributes to inflammatory swelling.

Vesicular transport also plays a limited role in the passage of materials across the capillary wall. Large non–lipid-soluble molecules such as protein hormones that must be exchanged between the blood and surrounding tissues are transported from one side of the capillary wall to the other in endocytotic-exocytotic vesicles (see p. 85).

▮ Many capillaries are not open under resting conditions.

The branching and reconverging arrangement within capillary beds varies somewhat, depending on the tissue. Capillaries typically branch either directly from an arteriole or from a thoroughfare channel known as a **metarteriole,** which runs between an arteriole and a venule. Likewise, capillaries may rejoin at either a venule or a metarteriole (● Figure 10-19).

Unlike the true capillaries within a capillary bed, metarterioles are sparsely surrounded by wisps of spiraling smooth-muscle cells. These cells also form **precapillary sphincters,** each of which consists of a ring of smooth muscle around the entrance to a capillary as it arises from a metarteriole.[3]

Role of precapillary sphincters

Precapillary sphincters are not innervated, but they have a high degree of myogenic tone and are sensitive to local metabolic changes. They act as stopcocks to control blood flow through the particular capillary that each one guards. Arterioles perform a similar function for a small group of capillaries. Capillaries themselves have no smooth muscle, so they cannot actively participate in regulating their own blood flow.

Generally, tissues that are more metabolically active have a greater density of capillaries. Muscles, for example, have relatively more capillaries than their tendinous attachments. Only about 10% of the precapillary sphincters in a resting muscle are open at any moment, however, so blood is flowing through only about 10% of the muscle's capillaries. As chemical concentrations start to change in a region of the muscle tissue supplied by closed-down capillaries, the precapillary sphincters and arterioles in the region relax. Restoration of the chemical concentrations to normal as a result of increased

[3]Although generally accepted, the existence of precapillary sphincters in humans has not been conclusively established.

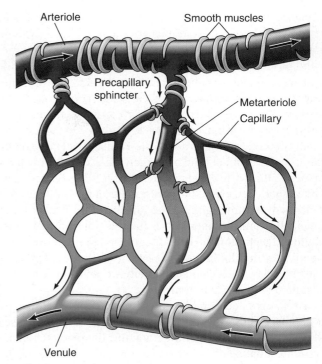

● **FIGURE 10-19**

Capillary bed
Capillaries branch either directly from an arteriole or from a metarteriole, a thoroughfare channel between an arteriole and venule. Capillaries rejoin at either a venule or a metarteriole. Metarterioles are surrounded by smooth muscle cells, which also form precapillary sphincters that encircle capillaries as they arise from a metarteriole.

blood flow to that region removes the impetus for vasodilation, so the precapillary sphincters close once again and the arterioles return to normal tone. In this way, blood flow through any given capillary is often intermittent, as a result of arteriolar and precapillary sphincter action working in concert.

When the muscle as a whole becomes more active, a greater percentage of the precapillary sphincters relax, simultaneously opening up more capillary beds, while concurrent arteriolar vasodilation increases total flow to the organ. As a result of more blood flowing through more open capillaries, the total volume and surface area available for exchange increase, and the diffusion distance between the cells and an open capillary decreases (● Figure 10-20). Thus, blood flow through a particular tissue (assuming a constant blood pressure) is regulated by (1) the degree of resistance offered by the arterioles in the organ, controlled by sympathetic activity and local factors; and (2) the number of open capillaries, controlled by action of the same local metabolic factors on precapillary sphincters.

▮ Interstitial fluid is a passive intermediary between the blood and cells.

Exchanges between blood and the tissue cells are not made directly. Interstitial fluid, the true internal environment in immediate contact with the cells, acts as the go-between (● Figure 10-21). Only 20% of the ECF circulates as plasma. The remaining 80% consists of interstitial fluid, which bathes all the cells in the body. Cells exchange materials directly with the interstitial fluid. The type and extent of exchange is governed by the properties of the cellular plasma membranes. Movement across the plasma membrane may be either passive (that is, by diffusion down electrochemical gradients or by facilitated diffusion) or active (that is, by active carrier-mediated transport or by vesicular transport) (see ▲ Table 3-2, p. 86).

In contrast, exchanges across the capillary wall between the plasma and interstitial fluid are largely passive. The only transport across this barrier that requires energy is the limited vesicular transport. Because of the permeability of the capillary walls, exchange is so thorough that the interstitial fluid takes on essentially the same composition as the incoming arterial blood, with the exception of the large plasma proteins that usually do not escape from the blood. Therefore, when we speak of exchanges between blood and tissue cells, we tacitly include interstitial fluid as a passive intermediary.

Exchanges between blood and surrounding tissues across the capillary walls are accomplished by two means: (1) passive diffusion down concentration gradients, the primary mechanism for exchange of individual solutes; and (2) bulk flow, a

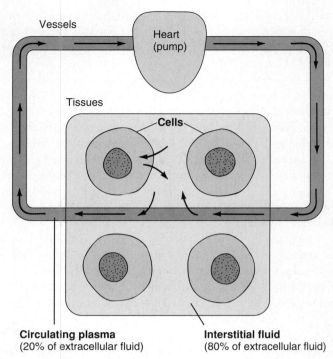

Circulating plasma (20% of extracellular fluid) **Interstitial fluid** (80% of extracellular fluid)

● **FIGURE 10-21**

Interstitial fluid acting as an intermediary between blood and cells

process that accomplishes the totally different function of determining the distribution of the ECF volume between the vascular and interstitial fluid compartments. We will examine each of these mechanisms in more detail, starting with diffusion.

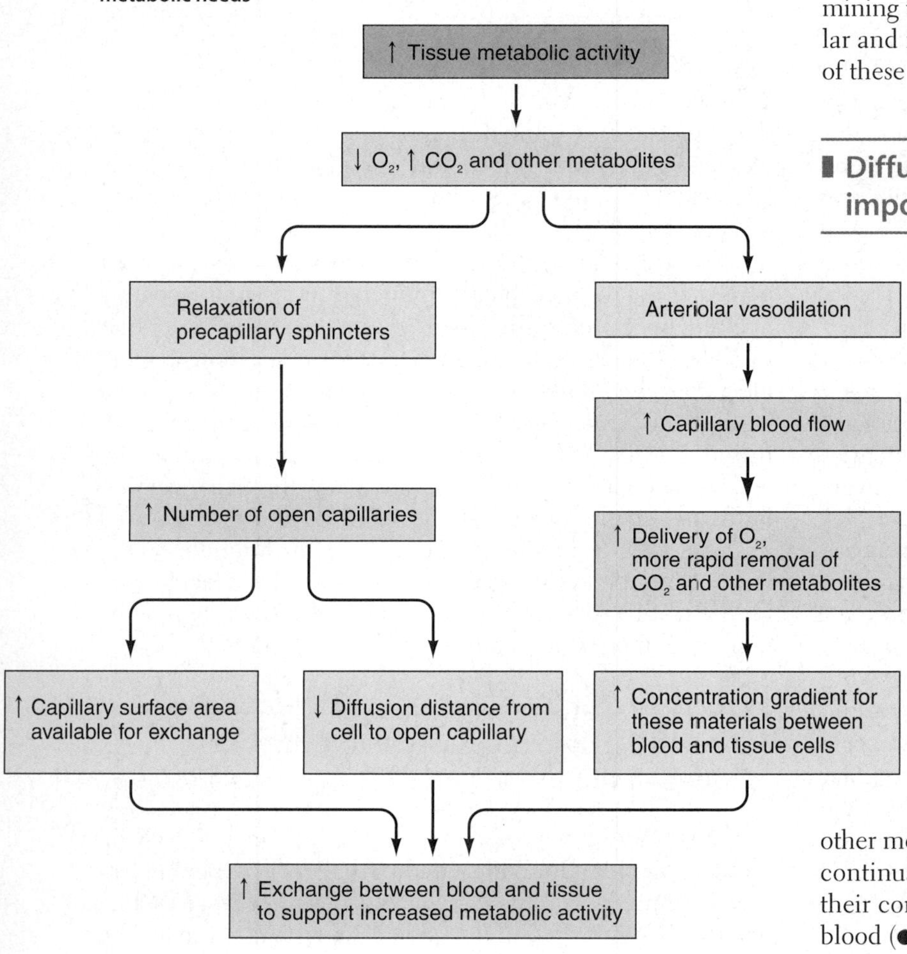

▌ Diffusion across the capillary walls is important in solute exchange.

Because there are no carrier-mediated transport systems in most capillary walls, solutes cross primarily by diffusion down concentration gradients. The chemical composition of arterial blood is carefully regulated to maintain the concentrations of individual solutes at levels that will promote each solute's movement in the appropriate direction across the capillary walls. The reconditioning organs primarily contribute to this homeostatic process, continuously adding nutrients and O_2 and removing CO_2 and other wastes as blood passes through them. Meanwhile, cells are constantly using up supplies and generating metabolic wastes. As cells use up O_2 and glucose, the blood constantly brings in fresh supplies of these vital materials, maintaining concentration gradients that favor the net diffusion of these substances from blood to cells. Simultaneously, ongoing net diffusion of CO_2 and other metabolic wastes from cells to blood is maintained by the continual production of these wastes at the cellular level and their constant removal from the tissue level by the circulating blood (● Figure10-22).

● **FIGURE 10-20**

Complementary action of precapillary sphincters and arterioles in adjusting blood flow through a tissue in response to changing metabolic needs

↑ Tissue metabolic activity

↓ O_2, ↑ CO_2 and other metabolites

Relaxation of precapillary sphincters

↑ Number of open capillaries

↑ Capillary surface area available for exchange

↓ Diffusion distance from cell to open capillary

Arteriolar vasodilation

↑ Capillary blood flow

↑ Delivery of O_2, more rapid removal of CO_2 and other metabolites

↑ Concentration gradient for these materials between blood and tissue cells

↑ Exchange between blood and tissue to support increased metabolic activity

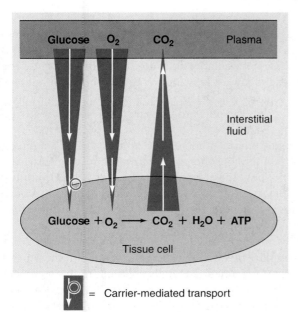

● FIGURE 10-22

Independent exchange of individual solutes down their own concentration gradients across the capillary wall

○ = Carrier-mediated transport

Because the capillary wall does not limit the passage of any constituent except plasma proteins, the extent of exchanges for each solute is independently determined by the magnitude of its concentration gradient between the blood and surrounding cells. As cells increase their level of activity, they use up more O_2 and produce more CO_2, among other things. This creates larger concentration gradients for O_2 and CO_2 between cells and blood, so more O_2 diffuses out of the blood into the cells and more CO_2 proceeds in the opposite direction to help support the increased metabolic activity.

▌Bulk flow across the capillary walls is important in extracellular fluid distribution.

The second means by which exchange is accomplished across capillary walls is bulk flow. A volume of protein-free plasma actually filters out of the capillary, mixes with the surrounding interstitial fluid, and is subsequently reabsorbed. This process is called **bulk flow** because the various constituents of the fluid are moving together in bulk, or as a unit, in contrast to the discrete diffusion of individual solutes down concentration gradients.

The capillary wall acts like a sieve, with the fluid moving through its water-filled pores. When pressure inside the capillary exceeds pressure on the outside, fluid is pushed out through the pores in a process known as **ultrafiltration.** The majority of the plasma proteins are retained on the inside during this process because of the pores' filtering effect, although a few do escape. Because all other constituents in the plasma are dragged along as a unit with the volume of fluid leaving the capillary, the filtrate is essentially a protein-free plasma. When inward-driving pressures exceed outward pressures across the capillary wall, net inward movement of fluid from the interstitial fluid compartment into the capillaries takes place through the pores, a process known as **reabsorption.**

Forces influencing bulk flow

Bulk flow occurs because of differences in the hydrostatic and colloid osmotic pressures between the plasma and interstitial fluid. Even though pressure differences exist between plasma and surrounding fluid elsewhere in the circulatory system, only the capillaries have pores that allow fluids to pass through. Four forces influence fluid movement across the capillary wall (● Figure 10-23):

1. **Capillary blood pressure** (P_C) is the fluid or hydrostatic pressure exerted on the inside of the capillary walls by the

● FIGURE 10-23

Bulk flow across the capillary wall
Schematic representation of ultrafiltration and reabsorption as a result of imbalances in the forces acting across the capillary wall.

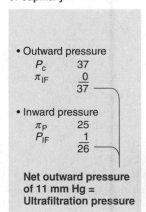

Forces at arteriolar end of capillary

- Outward pressure

P_C	37
π_{IF}	0
	37

- Inward pressure

π_P	25
P_{IF}	1
	26

Net outward pressure of 11 mm Hg = Ultrafiltration pressure

11 mm Hg (ultrafiltration)

Interstitial fluid

Initial lymphatic vessel

9 mm Hg (reabsorption)

P_{IF} (1) π_{IF} (0)

P_C (37) π_p (25) π_p (25) P_C (17)

From arteriole   **To venule**

Blood capillary

Forces at venular end of capillary

- Outward pressure

P_C	17
π_{IF}	0
	17

- Inward pressure

π_P	25
P_{IF}	1
	26

Net inward pressure of 9 mm Hg = Reabsorption pressure

All values are given in mm Hg.

blood. This pressure tends to force fluid *out of* the capillaries into the interstitial fluid. By the level of the capillaries, mean blood pressure has dropped substantially because of frictional losses in pressure in the high-resistance arterioles upstream. On the average, the hydrostatic pressure is 37 mm Hg at the arteriolar end of a tissue capillary and has declined even further to 17 mm Hg at the venular end (see ● Figure 10-9, p. 351).

2. **Plasma-colloid osmotic pressure (π_P)**, also known as *oncotic pressure*, is a force caused by the colloidal dispersion of plasma proteins (see p. A-10); it encourages fluid movement *into* the capillaries. Because the plasma proteins remain in the plasma rather than entering the interstitial fluid, a protein concentration difference exists between the plasma and the interstitial fluid. Accordingly, there is also a water concentration difference between these two regions. The plasma has a higher protein concentration and a lower water concentration than the interstitial fluid does. This difference exerts an osmotic effect that tends to move water from the area of higher water concentration in the interstitial fluid to the area of lower water concentration (or higher protein concentration) in the plasma (see p. 75). The other plasma constituents do not exert an osmotic effect because they readily pass through the capillary wall, so their concentrations are equal in the plasma and interstitial fluid. The plasma-colloid osmotic pressure averages 25 mm Hg.

3. **Interstitial fluid hydrostatic pressure (P_{IF})** is the fluid pressure exerted on the outside of the capillary wall by the interstitial fluid. This pressure tends to force fluid *into* the capillaries. Because of the difficulties encountered in measuring interstitial fluid hydrostatic pressure, the actual value of the pressure is a controversial issue. It is either at, slightly above, or slightly below atmospheric pressure. For purposes of illustration, we will say it is 1 mm Hg above atmospheric pressure.

4. **Interstitial fluid–colloid osmotic pressure (π_{IF})** is another force that does not normally contribute significantly to bulk flow. The small fraction of plasma proteins that leak across the capillary walls into the interstitial spaces are normally returned to the blood by means of the lymphatic system. Therefore, the protein concentration in the interstitial fluid is extremely low, and the interstitial fluid–colloid osmotic pressure is very close to zero. If plasma proteins pathologically leak into the interstitial fluid, however, as they do when histamine widens the intercellular clefts during tissue injury, the leaked proteins exert an osmotic effect that tends to promote movement of fluid *out of* the capillaries into the interstitial fluid.

Therefore, the two pressures that tend to force fluid out of the capillary are capillary blood pressure and interstitial fluid–colloid osmotic pressure. The two opposing pressures that tend to force fluid into the capillary are plasma-colloid osmotic pressure and interstitial fluid hydrostatic pressure. We are now prepared to analyze the fluid movement that occurs across a capillary wall because of imbalances in these opposing physical forces (● Figure 10-23).

Net exchange of fluid across the capillary wall

Net exchange at a given point across the capillary wall can be calculated using the following equation:

$$\text{Net exchange pressure} = \underbrace{(P_C + \pi_{IF})}_{\text{(outward pressure)}} - \underbrace{(\pi_P + P_{IF})}_{\text{(inward pressure)}}$$

A positive net exchange pressure (when the outward pressure exceeds the inward pressure) represents an ultrafiltration pressure. A negative net exchange pressure (when the inward pressure exceeds the outward pressure) represents a reabsorption pressure.

At the arteriolar end of the capillary, the outward pressure totals 37 mm Hg, whereas the inward pressure totals 26 mm Hg, for a net outward pressure of 11 mm Hg. Ultrafiltration takes place at the beginning of the capillary as this outward pressure gradient forces a protein-free filtrate through the capillary pores.

By the time the venular end of the capillary is reached, the capillary blood pressure has dropped, but the other pressures have remained essentially constant. At this point the outward pressure has fallen to a total of 17 mm Hg, whereas the total inward pressure is still 26 mm Hg, for a net inward pressure of 9 mm Hg. Reabsorption of fluid takes place as this inward pressure gradient forces fluid back into the capillary at its venular end.

Ultrafiltration and reabsorption, collectively known as *bulk flow*, are thus due to a shift in the balance between the passive physical forces acting across the capillary wall. No active forces or local energy expenditures are involved in the bulk exchange of fluid between the plasma and surrounding interstitial fluid. With only minor contributions from the interstitial fluid forces, ultrafiltration occurs at the beginning of the capillary because capillary blood pressure exceeds plasma-colloid osmotic pressure, whereas by the end of the capillary, reabsorption takes place because blood pressure has fallen below osmotic pressure.

It is important to realize that we have taken "snapshots" at two points—at the beginning and at the end—in a hypothetical capillary. Actually, blood pressure gradually diminishes along the length of the capillary, so that progressively diminishing quantities of fluid are filtered out in the first half of the vessel and progressively increasing quantities of fluid are reabsorbed in the last half (● Figure 10-24). Even this situation is idealized. The pressures used in this figure are average values and controversial at that. Some capillaries have such high blood pressure that filtration actually occurs throughout their entire length, whereas others have such low hydrostatic pressure that reabsorption takes place throughout their length.

In fact, a recent theory that has received considerable attention is that net filtration occurs throughout the length of all *open* capillaries, whereas net reabsorption occurs throughout the length of all *closed* capillaries. According to this theory, when the precapillary sphincter is relaxed, capillary blood pressure exceeds the plasma osmotic pressure even at the venular end of the capillary, promoting filtration throughout the length. When the precapillary sphincter is closed, the reduction in blood flow through the capillary diminishes capillary blood pressure below the plasma osmotic pressure even at the beginning of the capillary, so reabsorption takes place all along the capillary. Whichever mechanism is involved, the net effect is the same. A protein-free filtrate exits the capillaries and is ultimately reabsorbed.

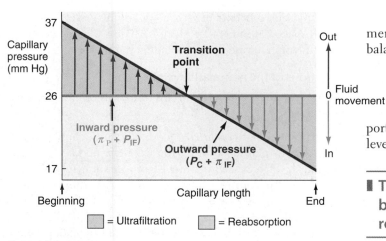

● FIGURE 10-24

Net filtration and net reabsorption along the vessel length
The inward pressure ($\pi_P + P_{IF}$) remains constant throughout the length of the capillary whereas the outward pressure ($P_C + \pi_{IF}$) progressively declines throughout the capillary's length. In the first half of the vessel, where the declining outward pressure still exceeds the constant inward pressure, progressively diminishing quantities of fluid are filtered out (depicted by the upward red arrows). In the last half of the vessel, progressively increasing quantities of fluid are reabsorbed (depicted by the downward blue arrows) as the declining outward pressure falls farther below the constant inward pressure.

Role of bulk flow

Bulk flow does not play an important role in the exchange of individual solutes between blood and tissues, because the quantity of solutes moved across the capillary wall by bulk flow is extremely small compared to the much larger transfer of solutes by diffusion. Thus ultrafiltration and reabsorption are not important in the exchange of nutrients and wastes. Bulk flow is extremely important, however, in regulating the distribution of ECF between the plasma and interstitial fluid. Maintenance of proper arterial blood pressure depends in part on an appropriate volume of circulating blood. If plasma volume is reduced (for example, by hemorrhage), blood pressure falls. The resultant lowering of capillary blood pressure alters the balance of forces across the capillary walls. Because the net outward pressure is decreased while the net inward pressure remains unchanged, extra fluid is shifted from the interstitial compartment into the plasma as a result of reduced filtration and increased reabsorption. The extra fluid soaked up from the interstitial fluid provides additional fluid for the plasma, temporarily compensating for the loss of blood. Meanwhile, reflex mechanisms acting on the heart and blood vessels (to be described later) also come into play to help maintain blood pressure until long-term mechanisms, such as thirst (and its satisfaction) and reduction of urinary output, can restore the fluid volume to completely compensate for the loss.

Conversely, if the plasma volume becomes overexpanded, as with excessive fluid intake, the resultant elevation in capillary blood pressure forces extra fluid from the capillaries into the interstitial fluid, temporarily relieving the expanded plasma volume until the excess fluid can be eliminated from the body by long-term measures, such as increased urinary output.

These internal fluid shifts between the two ECF compartments occur automatically and immediately whenever the balance of forces acting across the capillary walls is changed; they provide a temporary mechanism to help keep the plasma volume fairly constant. In the process of restoring the plasma volume to an appropriate level, the interstitial fluid volume fluctuates, but it is much more important that the plasma volume be maintained at a constant level to ensure that the circulatory system functions effectively.

▮ The lymphatic system is an accessory route by which interstitial fluid can be returned to the blood.

Even under normal circumstances, slightly more fluid is filtered out of the capillaries into the interstitial fluid than is reabsorbed from the interstitial fluid back into the plasma. On average, the net ultrafiltration pressure starts at 11 mm Hg at the beginning of the capillary, whereas the net reabsorption pressure only reaches 9 mm Hg by the vessel's end (● Figure 10-23). Because of this pressure differential, on average more fluid is filtered out of the first half of the capillary than is reabsorbed in its last half. The extra fluid filtered out as a result of this filtration–reabsorption imbalance is picked up by the **lymphatic system.** This system consists of an extensive network of one-way vessels that provide an accessory route by which fluid can be returned from the interstitial fluid to the blood. The lymphatic system functions much like a storm sewer that picks up and carries away excess rainwater so that it does not accumulate and flood an area.

Pickup and flow of lymph

Small, blind-ended terminal lymph vessels known as **initial lymphatics** permeate almost every tissue of the body (● Figure 10-25a). The endothelial cells forming the walls of initial lymphatics slightly overlap like shingles on a roof, with their overlapping edges being free instead of attached to the surrounding cells. This arrangement creates one-way, valvelike openings in the vessel wall. Fluid pressure on the outside of the vessel pushes the innermost edge of a pair of overlapping edges inward, creating a gap between the edges (that is, opening the valve). This opening permits interstitial fluid to enter (● Figure 10-25b). Once interstitial fluid enters a lymphatic vessel, it is called **lymph.** Fluid pressure on the inside forces the overlapping edges together, closing the valves so that lymph does not escape. These lymphatic valvelike openings are much larger than the pores in blood capillaries. Consequently, large particulates in the interstitial fluid, such as escaped plasma proteins and bacteria, can gain access to initial lymphatics but are excluded from blood capillaries.

Initial lymphatics converge to form larger and larger **lymph vessels,** which eventually empty into the venous system near the point where the blood enters the right atrium (● Figure 10-26a). Because there is no "lymphatic heart" to provide driving pressure, you may wonder how lymph is directed from the tissues toward the venous system in the thoracic cavity. Lymph flow is accomplished by two mechanisms. First, lymph

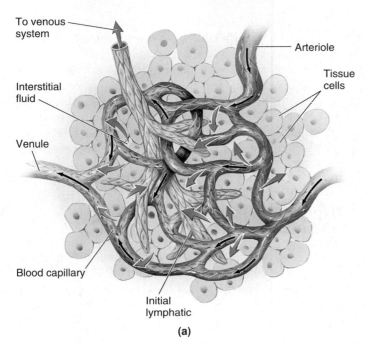

(a)

Labels on figure (a):
- To venous system
- Arteriole
- Interstitial fluid
- Tissue cells
- Venule
- Blood capillary
- Initial lymphatic

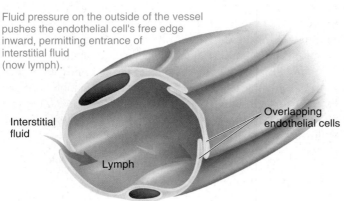

Fluid pressure on the outside of the vessel pushes the endothelial cell's free edge inward, permitting entrance of interstitial fluid (now lymph).

- Interstitial fluid
- Overlapping endothelial cells
- Lymph

Fluid pressure on the inside of the vessel forces the overlapping edges together so that lymph cannot escape.

(b)

● **FIGURE 10-25**

Initial lymphatics

(a) Relationship between initial lymphatics and blood capillaries. Blind-ended initial lymphatics pick up excess fluid filtered by blood capillaries and return it to the venous system in the chest. (b) Arrangement of endothelial cells in an initial lymphatic. Note that the overlapping edges of the endothelial cells create valvelike openings in the vessel wall.

vessels beyond the initial lymphatics are surrounded by smooth muscle, which contracts rhythmically as a result of myogenic initiation of action potentials. When this muscle is stretched because the vessel is distended with lymph, the muscle inherently contracts more forcefully, thereby pushing the lymph through the vessel. This intrinsic "lymph pump" is the major force for propelling lymph. Stimulation of lymphatic smooth muscle by the sympathetic nervous system further increases the pumping activity of the lymph vessels. Second, because lymph vessels lie between skeletal muscles, contraction of

these muscles squeezes the lymph out of the vessels. One-way valves spaced at intervals within the lymph vessels direct the flow of lymph toward its venous outlet in the chest.

Functions of the lymphatic system

Following are the most important functions of the lymphatic system:

- *Return of excess filtered fluid.* Normally, capillary filtration exceeds reabsorption by about 3 liters per day (20 liters filtered, 17 liters reabsorbed) (● Figure 10-26b). Yet the entire blood volume is only 5 liters, and only 2.75 liters of that is plasma. (Blood cells make up the remainder of the blood volume.) With an average cardiac output, 7,200 liters of blood pass through the capillaries daily under resting conditions (more when cardiac output increases). Even though only a small fraction of the filtered fluid is not reabsorbed by the blood capillaries, the cumulative effect of this process being repeated with every heartbeat results in the equivalent of more than the entire plasma volume being left behind in the interstitial fluid each day. Obviously, this fluid must be returned to the circulating plasma, and this task is accomplished by the lymph vessels. The average rate of flow through the lymph vessels is 3 liters per day, compared with 7,200 liters per day through the circulatory system.
- *Defense against disease.* The lymph percolates through **lymph nodes** located en route within the lymphatic system. Passage of this fluid through the lymph nodes is an important aspect of the body's defense mechanism against disease. For example, bacteria picked up from the interstitial fluid are destroyed by special phagocytic cells located within the lymph nodes (see Chapter 12).
- *Transport of absorbed fat.* The lymphatic system is important in the absorption of fat from the digestive tract. The end products of the digestion of dietary fats are packaged by cells lining the digestive tract into fatty particles that are too large to gain access to the blood capillaries but can easily enter the initial lymphatics (see Chapter 16).
- *Return of filtered protein.* Most capillaries permit leakage of some plasma proteins during filtration. These proteins cannot readily be reabsorbed back into the blood capillaries but can easily gain access to the initial lymphatics. If the proteins were allowed to accumulate in the interstitial fluid rather than being returned to the circulation via the lymphatics, the interstitial fluid–colloid osmotic pressure (an outward pressure) would progressively increase while the plasma–colloid osmotic pressure (an inward pressure) would progressively fall. As a result, filtration forces would gradually increase and reabsorption forces would gradually decrease, resulting in progressive accumulation of fluid in the interstitial spaces at the expense of loss of plasma volume.

▋ Edema occurs when too much interstitial fluid accumulates.

Occasionally, excessive interstitial fluid does accumulate when one of the physical forces acting across the capillary walls becomes abnormal for some reason. Swelling of the tissues because of excess interstitial fluid is known as **edema.** The causes of edema can be grouped into four general categories:

1. A *reduced concentration of plasma proteins* causes a decrease in plasma-colloid osmotic pressure. Such a drop in the major inward pressure allows excess fluid to be filtered out whereas less-than-normal amounts of fluid are reabsorbed; hence extra fluid remains in the interstitial spaces. Edema caused by a decreased concentration of plasma proteins can arise in several different ways: excessive loss of plasma proteins in the urine caused by kidney disease, reduced synthesis of plasma proteins as a result of liver disease (the liver synthesizes almost all plasma proteins), a diet deficient in protein, or significant loss of plasma proteins from large burned surfaces.

2. *Increased permeability of the capillary walls* allows more plasma proteins than usual to pass from the plasma into the surrounding interstitial fluid—for example, via histamine-induced widening of the capillary pores during tissue injury or allergic reactions. The resultant fall in plasma-colloid osmotic pressure decreases the effective inward pressure, whereas the resultant rise in interstitial fluid-colloid osmotic pressure caused by the excess protein in the interstitial fluid increases the effective outward force. This imbalance contributes in part to the localized edema associated with injuries (for example, blisters) and allergic responses (for example, hives).

3. *Increased venous pressure,* as when blood dams up in the veins, is accompanied by an increased capillary blood pressure, because the capillaries drain into the veins. This elevation in outward pressure across the capillary walls is largely responsible for the edema seen with congestive heart failure (see p. 332). Regional edema can also occur because of localized restriction of venous return. An example is the swelling often occurring in the legs and feet during pregnancy. The enlarged uterus compresses the major veins that drain the lower extremities as these vessels enter the abdominal cavity. The resultant damming of blood in these veins causes a rise in blood pressure in the capillaries of the legs and feet, which promotes regional edema of the lower extremities.

4. *Blockage of lymph vessels* produces edema because the excess filtered fluid is retained in the interstitial fluid rather than being returned to the blood through the lymphatics. The protein accumulation in the interstitial fluid compounds the problem through its osmotic effect. Local lymph blockage can occur, for example, in the arms of women whose major lymphatic drainage channels from the arm have been blocked as a result of lymph node removal during surgery for breast cancer. More

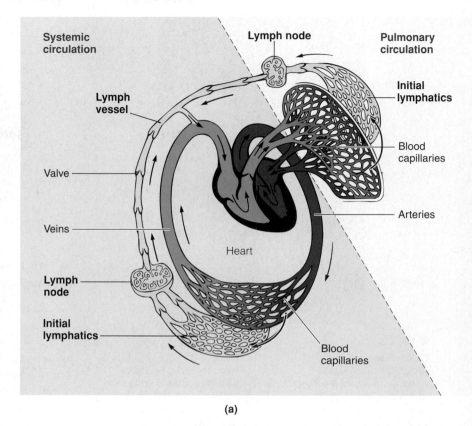

(a)

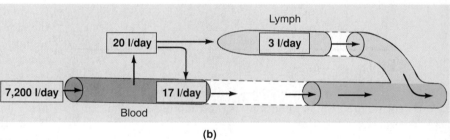

(b)

● **FIGURE 10-26**

Lymphatic sysem
(a) Lymph empties into the venous system near its entrance to the right atrium. (b) Lymph flow averages 3 liters per day, whereas blood flow averages 7,200 liters per day.

widespread lymph blockage occurs with *filariasis,* a mosquito-borne parasitic disease that is found predominantly in tropical coastal regions. In this condition, small, threadlike filaria worms infect the lymph vessels, where their presence prevents proper lymph drainage. The affected body parts, particularly the scrotum and extremities, become grossly edematous. The condition is often called *elephantiasis* because of the elephant-like appearance of the swollen extremities (● Figure 10-27).

Whatever the cause of edema, an important consequence is a reduction in exchange of materials between the blood and cells. As excess interstitial fluid accumulates, the distance between the blood and cells across which nutrients, O_2, and wastes must diffuse is increased, so the rate of diffusion decreases. Therefore, cells within edematous tissues may not be adequately supplied.

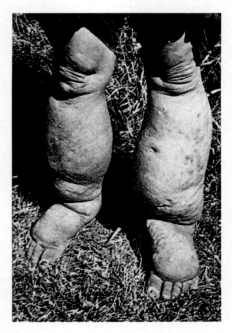

● **FIGURE 10-27**

Elephantiasis
This tropical condition is caused by a mosquito-borne parasitic worm that invades the lymph vessels. As a result of the interference with lymph drainage, the affected body parts, usually the extremities, become grossly edematous, appearing elephant-like.

VEINS

The venous system completes the circulatory circuit. Blood leaving the capillary beds enters the venous system for transport back to the heart.

▌ Veins serve as a blood reservoir as well as passageways back to the heart.

Veins have large radii, so they offer little resistance to flow. Furthermore, because the total cross-sectional area of the venous system gradually decreases as smaller veins converge into progressively fewer but larger vessels, the velocity of blood flow increases as the blood approaches the heart.

In addition to serving as low-resistance passageways to return blood from the tissues to the heart, systemic veins also serve as a *blood reservoir*. Because of their storage capacity, veins are often referred to as **capacitance vessels.** Veins have much thinner walls with less smooth muscle than arteries do. Also, in contrast to arteries, veins have very little elasticity, because venous connective tissue contains considerably more collagen fibers than elastin fibers. Unlike arteriolar smooth muscle, venous smooth muscle has little inherent myogenic tone. Because of these features, veins are highly distensible, or stretchable, and have little elastic recoil. They easily distend to accommodate additional volumes of blood with only a small increase in ve-

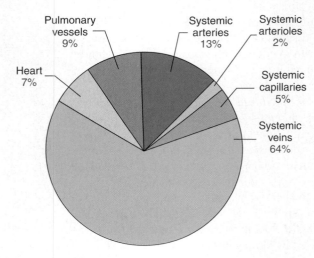

● **FIGURE 10-28**
Percentage of total blood volume in different parts of the circulatory system

nous pressure. Arteries stretched by an excess volume of blood recoil because of the elastic fibers in their walls, driving the blood forward. Veins containing an extra volume of blood simply stretch to accommodate the additional blood without tending to recoil. In this way veins serve as a **blood reservoir;** that is, when demands for blood are low, the veins can store extra blood in reserve because of their passive distensibility. Under resting conditions, the veins contain more than 60% of the total blood volume (● Figure 10-28).

Let's clarify a possible point of confusion. Contrary to a common misconception, the blood stored in the veins is not being held in a stagnant holding tank. Normally all the blood is circulating all the time. When the body is at rest and many of the capillary beds are closed, the capacity of the venous reservoir is increased as extra blood bypasses the capillaries and enters the veins. When this extra volume of blood stretches the veins, the blood moves forward through the veins more slowly because the total cross-sectional area of the veins has been increased as a result of the stretching. Therefore, the blood spends more time in the veins. As a result of this slower transit time through the veins, the veins are essentially storing the extra volume of blood because it is not moving forward as quickly to the heart to be pumped out again.

When the stored blood is needed, such as during exercise, extrinsic factors (soon to be described) reduce the capacity of the venous reservoir and drive the extra blood from the veins to the heart so that it can be pumped to the tissues. Increased venous return leads to an increased cardiac stroke volume in accordance with the Frank-Starling law of the heart (see p. 328). In contrast, if too much blood pools in the veins instead of being returned to the heart, cardiac output is abnormally diminished. Thus a delicate balance exists between the capacity of the veins, the extent of venous return, and the cardiac output. We will now turn our attention to the factors that affect venous capacity and contribute to venous return.

Venous return is enhanced by a number of extrinsic factors.

Venous capacity (the volume of blood that the veins can accommodate) depends on the distensibility of the vein walls (how much they can stretch to hold blood) and the influence of any externally applied pressure squeezing inwardly on the veins. At a constant blood volume, as venous capacity increases, more blood spends a longer time in the veins instead of being returned to the heart. Such venous storage decreases the effective circulating volume, the volume of blood being returned to and pumped out of the heart. Conversely, when venous capacity decreases, more blood is returned to the heart and is subsequently pumped out. Thus changes in venous capacity directly influence the magnitude of venous return, which in turn is an important (although not the only) determinant of effective circulating blood volume. The magnitude of the total blood volume is also influenced on a short-term basis by passive shifts in bulk flow between the vascular and interstitial fluid compartments and on a long-term basis by factors that control total ECF volume, such as salt and water balance.

The term **venous return** refers to the volume of blood entering each atrium per minute from the veins. Recall that the magnitude of flow through a vessel is directly proportional to the pressure gradient. Much of the driving pressure imparted to

the blood by cardiac contraction has been lost by the time the blood reaches the venous system, because of frictional losses along the way, especially during passage through the high-resistance arterioles. By the time the blood enters the venous system, mean pressure averages only 17 mm Hg (● Figure 10-9, p. 351). However, because atrial pressure is near 0 mm Hg, a small but adequate driving pressure still exists to promote the flow of blood through the large-radius, low-resistance veins. If atrial pressure becomes pathologically elevated, as in the presence of a leaky AV valve, the venous-to-atrial pressure gradient is decreased, reducing venous return and causing blood to dam up in the venous system. Elevated atrial pressure is thus one cause of congestive heart failure (see p. 331).

In addition to the driving pressure imparted by cardiac contraction, five other factors enhance venous return: sympathetically induced venous vasoconstriction, skeletal muscle activity, the effect of venous valves, respiratory activity, and the effect of cardiac suction (● Figure 10-29). Most of these secondary factors affect venous return by influencing the pressure gradient between the veins and the heart. We will examine each in turn.

Effect of sympathetic activity on venous return

Veins are not very muscular and have little inherent tone, but venous smooth muscle is abundantly supplied with sympa-

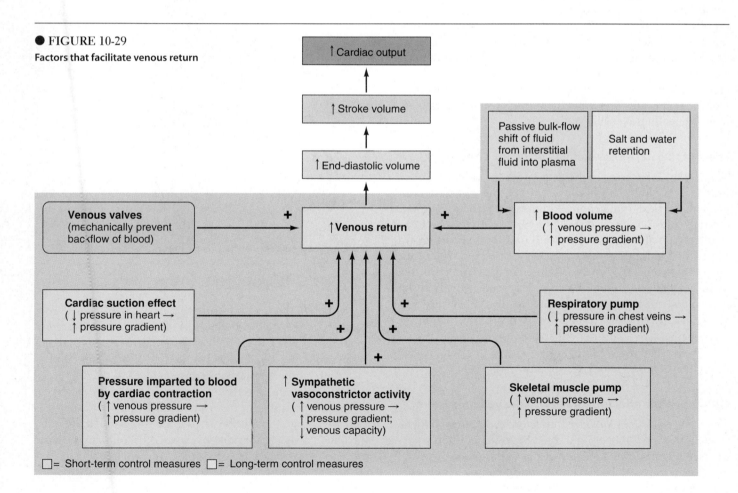

● **FIGURE 10-29**
Factors that facilitate venous return

□ = Short-term control measures   □ = Long-term control measures

thetic nerve fibers. Sympathetic stimulation produces venous vasoconstriction, which modestly elevates venous pressure; this, in turn, increases the pressure gradient to drive more of the stored blood from the veins into the right atrium. The veins normally have such a large diameter that the moderate vasoconstriction accompanying sympathetic stimulation has little effect on resistance to flow. Even when constricted, the veins still have a relatively large diameter and are still low-resistance vessels.

In addition to mobilizing the stored blood, venous vasoconstriction enhances venous return by decreasing venous capacity. With the filling capacity of the veins reduced, less blood draining from the capillaries remains in the veins but continues to flow instead toward the heart. The increased venous return initiated by sympathetic stimulation leads to increased cardiac output because of the increase in end-diastolic volume. Sympathetic stimulation of the heart also increases cardiac output by increasing the heart rate and increasing the heart's contractility (see p. 327 and p. 329). As long as sympathetic activity remains elevated, as during exercise, the increased cardiac output, in turn, helps sustain the increased venous return initiated in the first place by sympathetically induced venous vasoconstriction. More blood being pumped out by the heart means greater return of blood to the heart, because the reduced-capacity veins do not stretch to store any of the extra blood being pumped into the vascular system.

It is important to recognize the different outcomes of vasoconstriction in arterioles and veins. Arteriolar vasoconstriction immediately *reduces* flow through these vessels because of their increased resistance (less blood can enter and flow through a narrowed arteriole), whereas venous vasoconstriction immediately *increases* flow through these vessels because of their decreased capacity (narrowing of veins squeezes out more of the blood that is already present in the veins, thus increasing blood flow through these vessels).

Effect of skeletal muscle activity on venous return

Many of the large veins in the extremities lie between skeletal muscles, so when the muscles contract, the veins are compressed. This external venous compression decreases venous capacity and increases venous pressure, in effect squeezing fluid contained in the veins forward toward the heart (● Figure 10-30). This pumping action, known as the **skeletal muscle pump,** is one way extra blood stored in the veins is returned to the heart during exercise. Increased muscular activity pushes more blood out of the veins and into the heart. Increased sympathetic activity and the resultant venous vasoconstriction also accompany exercise, further enhancing venous return.

The skeletal muscle pump also counters the effect of gravity on the venous system. Let's see how.

Countering the effects of gravity on the venous system

The average pressures provided thus far for various regions of the vascular tree are for a person in the horizontal position. When a person is lying down, the force of gravity is uniformly applied, so it does not have to be taken into consideration. When a person stands up, however, gravitational effects are not uniform.

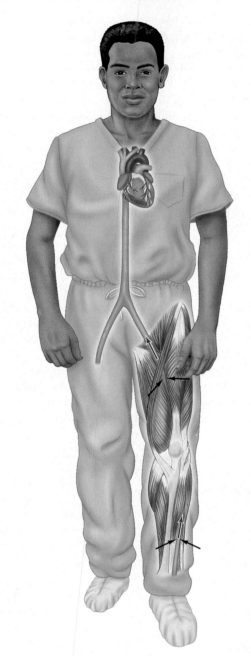

● **FIGURE 10-30**
Skeletal muscle pump enhancing venous return

In addition to the usual pressure that results from cardiac contraction, vessels below the level of the heart are subjected to pressure caused by the weight of the column of blood extending from the heart to the level of the vessel (● Figure 10-31a).

There are two important consequences of this increased pressure. First, the distensible veins yield under the increased hydrostatic pressure, further expanding so that their capacity is increased. Even though the arteries are subjected to the same gravitational effects, they do not expand like the veins, because arteries are not nearly as distensible. Much of the blood entering from the capillaries tends to pool in the expanded lower-leg veins instead of returning to the heart. Because venous return

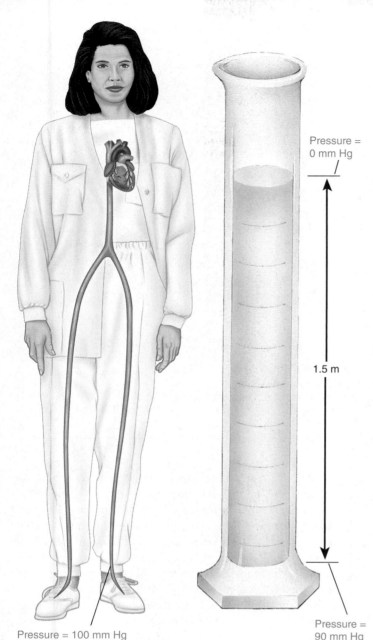

Pressure =
0 mm Hg

1.5 m

Pressure =
90 mm Hg

Pressure = 100 mm Hg

90 mm Hg caused by gravitational effect
10 mm Hg caused by pressure imparted
by cardiac contraction

(a)

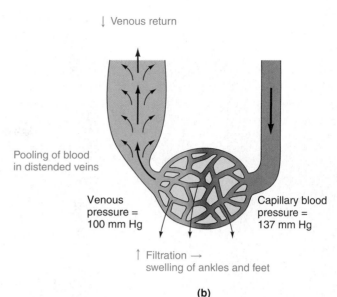

↓ Venous return

Pooling of blood
in distended veins

Venous
pressure =
100 mm Hg

Capillary blood
pressure =
137 mm Hg

↑ Filtration →
swelling of ankles and feet

(b)

● **FIGURE 10-31**

Effect of gravity on venous pressure
(a) In an upright adult, the blood in the vessels extending between
the heart and foot is equivalent to a 1.5-m column of blood. The
pressure exerted by this column of blood as a result of the effect of
gravity is 90 mm Hg. The pressure imparted to the blood by the heart
has declined to about 10 mm Hg in the lower-leg veins because of
frictional losses in preceding vessels. Together these pressures produce a venous pressure of 100 mm Hg in the ankle and foot veins.
Similarly, the capillaries in the region are subjected to these same
gravitational effects. (b) Because of the increased pressure the gravitational effect causes, blood pools in the distended veins, resulting
in decreased venous return. Filtration also increases across the capillary walls, resulting in swollen ankles and feet, unless compensatory
measures can counteract the effect of gravity.

is reduced, cardiac output is decreased and the effective circulating volume is reduced. Second, the marked increase in capillary blood pressure resulting from the effect of gravity causes
excessive fluid to filter out of capillary beds in the lower extremities, producing localized edema (that is, swollen feet and ankles) (● Figure 10-31b).

Two compensatory measures normally counteract these
gravitational effects. First, the resultant fall in mean arterial
pressure that occurs when a person moves from a lying-down
to an upright position triggers sympathetically induced venous
vasoconstriction, which drives some of the pooled blood forward.
Second, the skeletal muscle pump "interrupts" the column of

blood by completely emptying given vein segments intermittently so that a particular portion of a vein is not subjected to
the weight of the entire venous column from the heart to its
level (● Figures 10-30 and 10-32). Reflex venous vasoconstriction cannot completely compensate for gravitational
effects without the assistance of skeletal muscle activity.
When a person stands still for a long time, therefore, blood flow
to the brain is reduced because of the decline in effective circulating volume, despite reflexes aimed at maintaining mean
arterial pressure. Reduced flow of blood to the brain, in turn,
leads to fainting, which returns the person to a horizontal position, thereby eliminating the gravitational effects on the vascular system and enabling effective circulation to be restored.
For this reason, it is counterproductive to try to hold upright
someone who has fainted. Fainting is the remedy to the problem, not the problem itself.

Because the skeletal muscle pump facilitates venous return,
it is advisable when you are working at a desk to get up periodically and to move around when you are on your feet. The mild
muscular activity "gets the blood moving." It is further recommended that people who must be on their feet for long periods
of time use elastic stockings that apply a continuous gentle external compression, similar to the effect of skeletal muscle con-

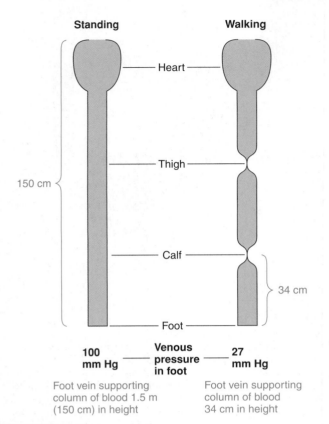

● FIGURE 10-32

Effect of contraction of the skeletal muscles of the legs in counteracting the effects of gravity
Contraction of skeletal muscles (as in walking) completely empties given vein segments, interrupting the column of blood that the lower veins must support.

(SOURCE: Adapted from *Physiology of the Heart and Circulation*, Fourth Edition, by R. C. Little and W. C. Little. Copyright © 1989 Year Book Medical Publishers, Inc. Reprinted by permission of the author and Mosby-Year Book, Inc.)

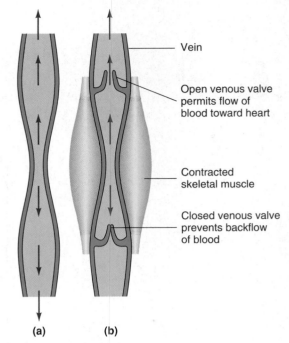

● FIGURE 10-33

Function of venous valves
(a) When a tube is squeezed in the middle, fluid is pushed in both directions. (b) Venous valves permit the flow of blood only toward the heart.

traction, to further counter the effect of gravitational pooling of blood in the leg veins.

Effect of venous valves on venous return

Venous vasoconstriction and external venous compression both drive blood in the direction of the heart. Yet if you squeeze a fluid-filled tube in the middle, fluid is pushed in both directions from the point of constriction (● Figure 10-33a). Then why isn't blood driven backward as well as forward by venous vasoconstriction and the skeletal muscle pump? Blood can only be driven forward because the large veins are equipped with one-way valves spaced at 2- to 4-cm intervals; these valves permit blood to move forward toward the heart but prevent it from moving back toward the tissues (● Figure 10-33b). These venous valves also play a role in counteracting the gravitational effects of upright posture by helping minimize the backflow of blood that tends to occur when a person stands up and by temporarily supporting portions of the column of blood when the skeletal muscles are relaxed.

Varicose veins occur when the venous valves become incompetent and can no longer support the column of blood above them. People predisposed to this condition usually have an inherited overdistensibility and weakness of their vein walls.

Aggravated by frequent, prolonged standing, the veins become so distended as blood pools in them that the edges of the valves can no longer meet to form a seal. Varicosed superficial leg veins become visibly overdistended and tortuous. Contrary to what might be expected, the chronic pooling of blood in the pathologically distended veins does not reduce cardiac output, because there is a compensatory increase in total circulating blood volume. Instead, the most serious consequence of varicosed veins is the possibility of abnormal clot formation in the sluggish, pooled blood. Particularly dangerous is the risk that these clots may break loose and block small vessels elsewhere, especially the pulmonary capillaries.

Effect of respiratory activity on venous return

As a result of respiratory activity, the pressure within the chest cavity averages 5 mm Hg less than atmospheric pressure. As the venous system returns blood to the heart from the lower regions of the body, it travels through the chest cavity, where it is exposed to this subatmospheric pressure. Because the venous system in the limbs and abdomen is subjected to normal atmospheric pressure, an externally applied pressure gradient exists between the lower veins (at atmospheric pressure) and the chest veins (at 5 mm Hg less than atmospheric pressure). This pressure difference squeezes blood from the lower veins to the chest veins, promoting increased venous return (● Figure 10-34). This mechanism of facilitating venous return is known as the **respiratory pump** because it results from respiratory activity. Increased respiratory activity as well as the effects of the skeletal muscle pump and venous vasoconstriction all enhance venous return during exercise.

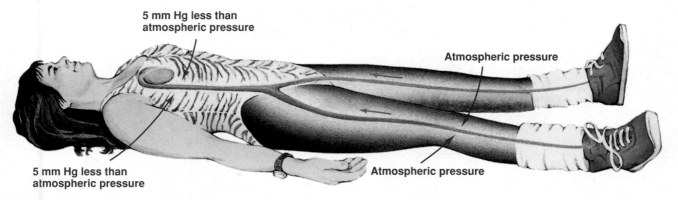

5 mm Hg less than atmospheric pressure

Atmospheric pressure

5 mm Hg less than atmospheric pressure

Atmospheric pressure

● **FIGURE 10-34**

Respiratory pump enhancing venous return
As a result of respiratory activity, the pressure surrounding the chest veins is lower than the pressure surrounding the veins in the extremities and abdomen. This establishes an externally applied pressure gradient on the veins that drives blood toward the heart.

Effect of cardiac suction on venous return

The extent of cardiac filling does not depend entirely on factors affecting the veins. The heart plays a role in its own filling. During ventricular contraction, the AV valves are drawn downward, enlarging the atrial cavities. As a result, the atrial pressure transiently drops below 0 mm Hg, thus increasing the vein-to-atria pressure gradient so that venous return is enhanced. In addition, the rapid expansion of the ventricular chambers during ventricular relaxation appears to create a transient negative pressure in the ventricles so that blood is "sucked in" from the atria and veins; that is, the negative ventricular pressure increases the vein-to-atria-to-ventricle pressure gradient, further enhancing venous return. Thus the heart functions as a "suction pump" to facilitate cardiac filling.

BLOOD PRESSURE

▮ **Blood pressure is regulated by controlling cardiac output, total peripheral resistance, and blood volume.**

Mean arterial blood pressure is the main driving force for propelling blood to the tissues. This pressure must be closely regulated for two reasons. First, it must be high enough to ensure sufficient driving pressure; without this pressure, the brain and other tissues will not receive adequate flow, no matter what local adjustments are made in the resistance of the arterioles supplying them. Second, the pressure must not be so high that it creates extra work for the heart and increases the risk of vascular damage and possible rupture of small blood vessels.

Determinants of mean arterial pressure

Elaborate mechanisms involving the integrated action of the various components of the circulatory system and other body systems are vital in the regulation of this all-important mean arterial pressure (● Figure 10-35). Remember from an earlier

discussion that the two determinants of mean arterial pressure are cardiac output and total peripheral resistance:

Mean arterial pressure =
cardiac output × total peripheral resistance

(Do not confuse this equation, which indicates the *determinants* of mean arterial pressure, namely the magnitude of both the cardiac output and total peripheral resistance, with the equation used to *calculate* mean arterial pressure, namely mean arterial pressure = diastolic pressure + 1/3 pulse pressure.)

Recall that a number of factors, in turn, determine cardiac output (● Figure 9-29, p. 330) and total peripheral resistance (● Figure 10-14, p. 359). Thus one can quickly appreciate the complexity of blood pressure regulation. Let's work our way through ● Figure 10-35, reviewing all the factors that affect mean arterial blood pressure. Even though we've covered all these factors before, it is useful to pull them all together. The circled numbers in the text correspond to the numbers in the figure and indicate the portion of the figure being discussed.

- Mean arterial pressure depends on cardiac output and total peripheral resistance (1 on ● Figure 10-35).
- Cardiac output depends on heart rate and stroke volume 2 .
- Heart rate depends on the relative balance of parasympathetic activity 3 , which decreases heart rate, and sympathetic activity (tacitly including epinephrine throughout this discussion) 4 , which increases heart rate.
- Stroke volume increases in response to sympathetic activity 5 (extrinsic control of stroke volume).
- Stroke volume also increases as venous return increases 6 (intrinsic control of stroke volume by means of the Frank-Starling law of the heart).
- Venous return is enhanced by sympathetically induced venous vasoconstriction 7 , the skeletal muscle pump 8 , the respiratory pump 9 , and cardiac suction 10 .
- The effective circulating blood volume also influences how much blood is returned to the heart 11 . The blood vol-

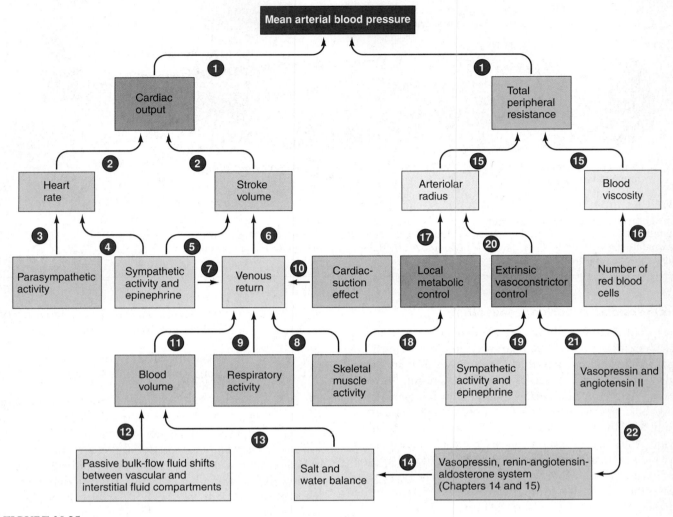

● FIGURE 10-35

Determinants of mean arterial blood pressure
Note that this figure is basically a composite of ● Figure 9-29, p. 330, "Control of cardiac output"; ● Figure 10-14, p. 359, "Factors affecting total peripheral resistance"; and ● Figure 10-29, p. 371, "Factors that facilitate venous return." See the text for a discussion of the circled numbers.

ume depends in the short term on the magnitude of passive bulk-flow fluid shifts between the plasma and interstitial fluid across the capillary walls 12. In the long term, the blood volume depends on salt and water balance 13, which are hormonally controlled by the renin-angiotensin-aldosterone system and vasopressin, respectively 14.

• The other major determinant of mean arterial blood pressure, total peripheral resistance, depends on the radius of all arterioles as well as blood viscosity 15. The major factor determining blood viscosity is the number of red blood cells 16. However, arteriolar radius is the more important factor determining total peripheral resistance.

• Arteriolar radius is influenced by local (intrinsic) metabolic controls that match blood flow with metabolic needs 17. For example, local changes that take place in active skeletal muscles cause local arteriolar vasodilation and increased blood flow to these muscles 18.

• Arteriolar radius is also influenced by sympathetic activity 19, an extrinsic control mechanism that causes arterio-

lar vasoconstriction 20 to increase total peripheral resistance and mean arterial blood pressure.

• Arteriolar radius is also extrinsically controlled by the hormones vasopressin and angiotensin II, which are potent vasoconstrictors 21 as well as being important in salt and water balance 22.

Altering any of the pertinent factors that influence blood pressure will produce a change in blood pressure, unless a compensatory change in another variable keeps the blood pressure constant. Blood flow to any given tissue depends on the driving force of the mean arterial pressure and on the degree of vasoconstriction of the tissue's arterioles. Because mean arterial pressure depends on the cardiac output and the degree of arteriolar vasoconstriction, if the arterioles in one tissue dilate, the arterioles in other tissues will have to constrict to maintain an adequate arterial blood pressure. An adequate pressure is needed to provide a driving force to push blood not only to the vasodilated tissue but also to the brain, which depends on a constant blood supply. Thus the cardiovascular variables must be continuously juggled to maintain a constant blood pressure in spite of tissues' varying needs for blood.

Short-term and long-term control measures

Mean arterial pressure is constantly monitored by **baroreceptors** (pressure sensors) within the circulatory system. When deviations from normal are detected, multiple reflex responses are initiated to return the arterial pressure to its normal value. *Short-term* (within seconds) adjustments are accomplished by alterations in cardiac output and total peripheral resistance, mediated by means of autonomic nervous system influences on the heart, veins, and arterioles. *Long-term* (requiring minutes to days) control involves adjusting total blood volume by restoring normal salt and water balance through mechanisms that regulate urine output and thirst (Chapters 14 and 15). The magnitude of the total blood volume, in turn, has a profound effect on cardiac output and mean arterial pressure. Let us now turn our attention to the short-term mechanisms involved in the ongoing regulation of this pressure.

■ The baroreceptor reflex is an important short-term mechanism for regulating blood pressure.

Any change in mean blood pressure triggers an autonomically mediated **baroreceptor reflex** that influences the heart and blood vessels to adjust cardiac output and total peripheral resistance in an attempt to restore blood pressure to normal. Like any reflex, the baroreceptor reflex includes a receptor, an afferent pathway, an integrating center, an efferent pathway, and effector organs.

The most important receptors involved in moment-to-moment regulation of blood pressure, the **carotid sinus** and **aortic arch baroreceptors,** are mechanoreceptors sensitive to changes in both mean arterial pressure and pulse pressure. Their responsiveness to fluctuations in pulse pressure enhances their sensitivity as pressure sensors, because small changes in systolic or diastolic pressure may alter the pulse pressure without changing the mean pressure. These baroreceptors are strategically located (● Figure 10-36) to provide critical information about arterial blood pressure in the vessels leading to the brain (the carotid sinus baroreceptor) and in the major arterial trunk before it gives off branches that supply the rest of the body (the aortic arch baroreceptor).

The baroreceptors constantly provide information about blood pressure; in other words, they continuously generate action potentials in response to the ongoing pressure within the arteries. When arterial pressure (either mean or pulse pressure) increases, the receptor potential of these baroreceptors increases, thus increasing the rate of firing in the corresponding afferent neurons. Conversely, when blood pressure decreases the rate of firing generated in the afferent neurons by the baroreceptors decreases (● Figure 10-37).

The integrating center that receives the afferent impulses about the state of arterial pressure is the **cardiovascular control center,**[4] located in the medulla within the brain stem. The efferent pathway is the autonomic nervous system. The cardiovascular control center alters the ratio between sympathetic and parasympathetic activity to the effector organs (the heart and blood vessels). To show how autonomic changes alter arterial blood pressure, ● Figure 10-38 provides a review of the major effects of parasympathetic and sympathetic stimulation on the heart and blood vessels.

Let us fit all the pieces of the baroreceptor reflex together now by tracing the reflex activity that occurs to compensate for an elevation or fall in blood pressure. If for any reason arterial pressure becomes elevated above normal (● Figure 10-39a), the carotid sinus and aortic arch baroreceptors increase the rate of firing in their respective afferent neurons. On being informed by increased afferent firing that arterial pressure has become too high, the cardiovascular control center responds by

---

[4]The cardiovascular control center is sometimes divided into cardiac and vasomotor centers, which are occasionally further classified into smaller subdivisions, such as cardioacceleratory and cardioinhibitory centers and vasoconstrictor and vasodilator areas. Because these regions are highly interconnected and functionally interrelated, we will refer to them collectively as the *cardiovascular control center.*

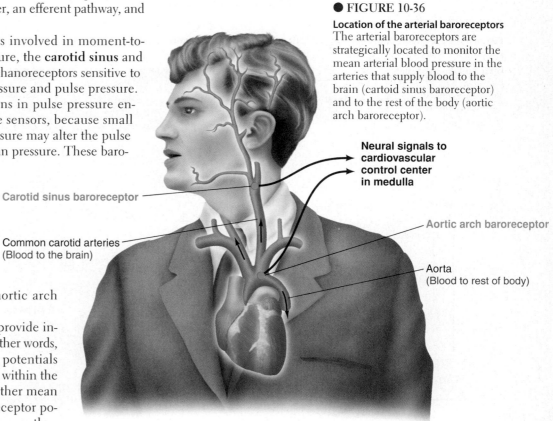

● **FIGURE 10-36**

Location of the arterial baroreceptors The arterial baroreceptors are strategically located to monitor the mean arterial blood pressure in the arteries that supply blood to the brain (cartoid sinus baroreceptor) and to the rest of the body (aortic arch baroreceptor).

Neural signals to cardiovascular control center in medulla

Carotid sinus baroreceptor

Common carotid arteries (Blood to the brain)

Aortic arch baroreceptor

Aorta (Blood to rest of body)

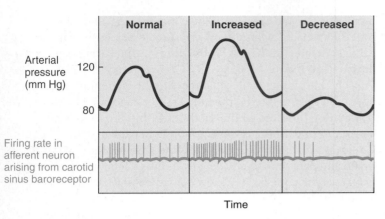

● FIGURE 10-37

Firing rate in the afferent neuron from the carotid sinus baroreceptor in relation to the magnitude of mean arterial pressure

decreasing sympathetic and increasing parasympathetic activity to the cardiovascular system. These efferent signals decrease heart rate, decrease stroke volume, and produce arteriolar and venous vasodilation, which in turn lead to a decrease in cardiac output and a decrease in total peripheral resistance, with a subsequent decrease in blood pressure back toward normal.

Conversely, when blood pressure falls below normal (● Figure 10-39b) baroreceptor activity decreases, inducing the cardiovascular center to increase sympathetic cardiac and vasoconstrictor nerve activity while decreasing its parasympathetic output. This efferent pattern of activity leads to an increase in

heart rate and stroke volume, coupled with arteriolar and venous vasoconstriction. These changes result in an increase in both cardiac output and total peripheral resistance, producing an elevation in blood pressure back toward normal.

▌ Other reflexes and responses influence blood pressure.

Besides the baroreceptor reflex, whose sole function is blood pressure regulation, several other reflexes and responses influence the cardiovascular system even though they primarily are concerned with the regulation of other body functions. Some of these other influences deliberately move arterial pressure away from its normal value temporarily, overriding the baroreceptor reflex to accomplish a particular goal. These factors include the following:

1. Left atrial volume receptors and hypothalamic osmoreceptors are primarily important in water and salt balance in the body; thus they affect the long-term regulation of blood pressure by controlling the plasma volume.

2. Chemoreceptors located in the carotid and aortic arteries, in close association with but distinct from the baroreceptors, are sensitive to low O_2 or high acid levels in the blood. These chemoreceptors' main function is to reflexly increase respiratory activity to bring in more O_2 or to blow off more acid-forming CO_2, but they also reflexly increase blood pressure by sending excitatory impulses to the cardiovascular center.

● FIGURE 10-38

Summary of the effects of the parasympathetic and sympathetic nervous systems on factors that influence mean arterial blood pressure

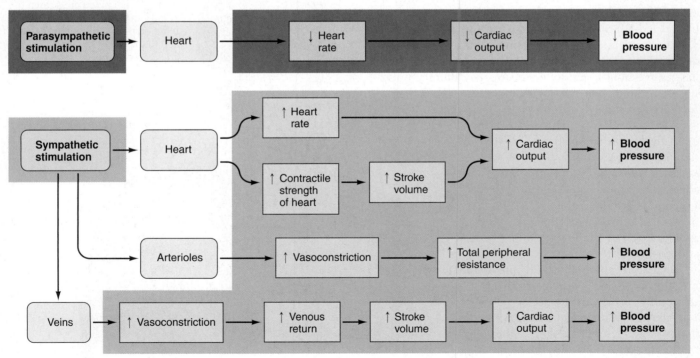

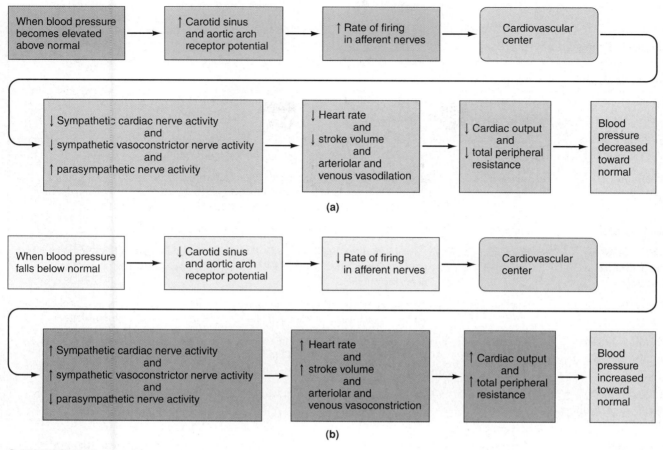

● FIGURE 10-39

Baroreceptor reflexes to restore the blood pressure to normal
(a) Baroreceptor reflex in response to an elevation in blood pressure. (b) Baroreceptor reflex in response
to a fall in blood pressure.

3. Cardiovascular responses associated with certain behaviors and emotions are mediated through the cerebral cortex–hypothalamic pathway and appear preprogrammed. These responses include the widespread changes in cardiovascular activity accompanying the generalized sympathetic fight-or-flight response, the characteristic marked increase in heart rate and blood pressure associated with sexual orgasm, and the localized cutaneous vasodilation characteristic of blushing.

4. Pronounced cardiovascular changes accompany exercise, including a substantial increase in skeletal muscle blood flow (see ● Figure 10-12, p. 354), a significant increase in cardiac output, a fall in total peripheral resistance (because of widespread vasodilation in skeletal muscles despite generalized arteriolar vasoconstriction in most organs), and a modest increase in mean arterial pressure (▲ Table 10-5). Evidence suggests that discrete exercise centers yet to be identified induce the appropriate cardiac and vascular changes at the onset of exercise or even in anticipation of exercise. These effects are then reinforced by afferent inputs to the medullary cardiovascular center from chemoreceptors in exercising muscles as well as by local mechanisms important in maintaining vasodilation in active muscles. The baroreceptor reflex further modulates these cardiovascular responses.

5. Hypothalamic control over cutaneous (skin) arterioles for the purpose of temperature regulation takes precedence over control that the cardiovascular center has over these same vessels for the purpose of blood pressure regulation. As a result, blood pressure can fall when the skin vessels are widely dilated to eliminate excess heat from the body, even though the baroreceptor responses are calling for cutaneous vasoconstriction to help maintain adequate total peripheral resistance.

6. Vasoactive substances released from the endothelial cells play a role in regulating blood pressure. For example, EDRF/NO normally exerts an ongoing vasodilatory effect.

Despite all these control measures, sometimes blood pressure is not maintained at the appropriate level. We will next examine blood pressure abnormalities.

■ Hypertension is a serious national public health problem, but its causes are largely unknown.

Sometimes blood-pressure control mechanisms do not function properly or are unable to completely compensate for changes that have taken place. Blood pressure may be above the normal

▲ **TABLE 10-5**
Cardiovascular Changes during Exercise

Cardiovascular Variable	Change	Comment
Heart Rate	Increases	Occurs as a result of increased sympathetic and decreased parasympathetic activity to the SA node
Venous Return	Increases	Occurs as a result of sympathetically induced venous vasoconstriction and increased activity of the skeletal muscle pump and respiratory pump
Stroke Volume	Increases	Occurs both as a result of increased venous return by means of the Frank-Starling mechanism (unless diastolic filling time is significantly reduced by a high heart rate) and as a result of a sympathetically induced increase in myocardial contractility
Cardiac Output	Increases	Occurs as a result of increases in both heart rate and stroke volume
Blood Flow to Active Skeletal Muscles and Heart Muscle	Increases	Occurs as a result of locally controlled arteriolar vasodilation, which is reinforced by the vasodilatory effects of epinephrine and overpowers the weaker sympathetic vasoconstrictor effect
Blood Flow to the Brain	Unchanged	Occurs because sympathetic stimulation has no effect on brain arterioles; local control mechanisms maintain constant cerebral blood flow whatever the circumstances
Blood Flow to the Skin	Increases	Occurs because the hypothalamic temperature control center induces vasodilation of skin arterioles; increased skin blood flow brings heat produced by exercising muscles to the body surface where the heat can be lost to the external environment
Blood Flow to the Digestive System, Kidneys, and Other Organs	Decreases	Occurs as a result of generalized sympathetically induced arteriolar vasoconstriction
Total Peripheral Resistance	Decreases	Occurs because resistance in the skeletal muscles, heart, and skin decreases to a greater extent than resistance in the other organs increases
Mean Arterial Blood Pressure	Increases (modest)	Occurs because cardiac output increases to a greater extent than total peripheral resistance decreases

range (**hypertension** if above 140/90 mm Hg) or below normal (**hypotension** if less than 100/60 mm Hg). Hypotension in its extreme form is *circulatory shock*. We will first examine hypertension and then conclude this chapter with a discussion of hypotension and shock.

There are two broad classes of hypertension, secondary hypertension and primary hypertension, depending on the cause.

Secondary hypertension

A definite cause for hypertension can be established in only 10% of the cases. Hypertension that occurs secondary to another primary problem is called **secondary hypertension.** Here are examples of each of the four categories of secondary hypertension:

1. *Renal hypertension.* For example, atherosclerotic lesions protruding into the lumen of a renal artery or external compression of the vessel by a tumor may reduce blood flow through the kidney. The kidney responds by initiating the hormonal pathway involving angiotensin II. This pathway promotes salt and water retention during urine formation, thus in-

creasing the blood volume to compensate for the reduced renal blood flow. Recall that angiotensin II is also a powerful vasoconstrictor. Although these two effects (increased blood volume and angiotensin-induced vasoconstriction) are compensatory mechanisms to improve blood flow through the narrowed renal artery, they also are responsible for elevating the arterial pressure as a whole.

2. *Cardiovascular hypertension.* For example, when the elasticity of the arteries is reduced because of the loss of elastin fibers or the presence of calcified atherosclerotic plaques, arterial pressure is chronically increased as a result of this "hardening of the arteries."

3. *Endocrine hypertension.* For example, a *pheochromocytoma* is an adrenal medullary tumor that secretes excessive epinephrine and norepinephrine. Abnormally elevated levels of these hormones bring about a high cardiac output and generalized peripheral vasoconstriction, both of which contribute to the hypertension characteristic of this disorder.

4. *Neurogenic hypertension.* An example is the hypertension caused by erroneous blood pressure control arising from a defect in the cardiovascular control center.

Primary hypertension

The underlying cause is unknown in the remaining 90% of hypertension cases. Such hypertension is known as **primary (essential** or **idiopathic) hypertension.** Primary hypertension is a catchall category for elevated blood pressure caused by a variety of unknown causes rather than by a single disease entity. There is a strong genetic tendency to develop primary hypertension, which can be hastened or worsened by contributing factors such as obesity, stress, smoking, or dietary habits. Consider the following range of potential causes for primary hypertension that are currently being investigated.

- *Defects in salt management by the kidneys.* Disturbances in kidney function too minor to produce outward signs of renal disease could nevertheless insidiously lead to gradual accumulation of salt and water in the body, resulting in progressive elevation of arterial pressure.
- *Excessive salt intake.* Because salt osmotically retains water, thus expanding the plasma volume and contributing to the long-term control of blood pressure, excessive ingestion of salt could theoretically contribute to hypertension. Yet controversy continues over whether or not restricting salt intake should be recommended as a means of preventing and treating high blood pressure. The research data to date have been inconclusive and subject to conflicting interpretations.
- *Diets low in fruits, vegetables, and dairy products (that is, low in K^+ and Ca^{2+}).* Dietary factors other than salt have been shown to markedly affect blood pressure. The DASH (Dietary Approaches to Stop Hypertension) studies found that a low-fat diet rich in fruits, vegetables, and dairy products could lower blood pressure in people with mild hypertension as much as any single drug treatment. Research indicates that the high K^+ intake associated with eating abundant fruits and vegetables may lower blood pressure by relaxing arteries. Furthermore, inadequate Ca^{2+} intake from dairy products has been identified as the most prevalent dietary pattern among individuals with untreated hypertension, although Ca^{2+}'s role in regulating blood pressure is unclear.
- *Plasma membrane abnormalities such as defective Na^+–K^+ pumps.* Such defects, by altering the electrochemical gradient across plasma membranes, could change the excitability and contractility of the heart and the smooth muscle in blood vessel walls in such a way as to lead to high blood pressure. In addition, the Na^+–K^+ pump is crucial to salt management by the kidneys. A genetic defect in the Na^+–K^+ pump of hypertensive-prone laboratory rats was the first gene–hypertension link to be discovered.
- *Variation in the gene that encodes for angiotensinogen.* Angiotensinogen is part of the hormonal pathway that produces the potent vasoconstrictor angiotensin II and promotes salt and water retention. One variant of the gene in humans appears to be associated with a higher incidence of hypertension. Researchers speculate that the suspect version of the gene leads to a slight excess production of angiotensinogen, thus increasing activity of this blood pressure–raising pathway. This is the first gene–hypertension link discovered in humans.
- *Endogenous digitalis-like substances.* Such substances act similarly to the drug digitalis (see p. 316) to increase cardiac contractility as well as to constrict blood vessels and reduce salt elimination in the urine, all of which could cause chronic hypertension.
- *Abnormalities in EDRF/NO, endothelin, or other locally acting vasoactive chemicals.* For example, a shortage of NO has been discovered in the blood vessel walls of some hypertensive patients, leading to an impaired ability to accomplish blood pressure–lowering vasodilation. Furthermore, an underlying abnormality in the gene that codes for endothelin, a locally acting vasoconstrictor, has been strongly implicated as a possible cause of hypertension, especially among African Americans.
- *Excess vasopressin.* Recent experimental evidence suggests that hypertension may result from a malfunction of the vasopressin-secreting cells of the hypothalamus. Vasopressin is a potent vasoconstrictor and also promotes water retention.

Whatever the underlying defect, once initiated, hypertension appears to be self-perpetuating. Constant exposure to elevated blood pressure predisposes vessel walls to the development of atherosclerosis, which further elevates blood pressure.

Adaptation of baroreceptors during hypertension

The baroreceptors do not respond to bring the blood pressure back to normal during hypertension because they adapt, or are "reset," to operate at a higher level. In the presence of chronically elevated blood pressure, the baroreceptors still function to regulate blood pressure, but they maintain it at a higher mean pressure.

Complications of hypertension

Hypertension imposes stresses on both the heart and the blood vessels. The heart has an increased workload because it is pumping against an increased total peripheral resistance, whereas blood vessels may be damaged by the high internal pressure, particularly when the vessel wall is weakened by the degenerative process of atherosclerosis. Complications of hypertension include congestive heart failure caused by the heart's inability to pump continuously against a sustained elevation in arterial pressure, strokes caused by rupture of brain vessels, and heart attacks caused by rupture of coronary vessels. Spontaneous hemorrhage caused by bursting of small vessels elsewhere in the body may also occur but with less serious consequences; an example is the rupture of blood vessels in the nose, resulting in nosebleeds. Another serious complication of hypertension is renal failure caused by progressive impairment of blood flow through damaged renal blood vessels. Furthermore, retinal damage caused by changes in the blood vessels supplying the eyes may result in progressive loss of vision.

Until complications occur, hypertension is symptomless, because the tissues are adequately supplied with blood. Therefore, unless blood pressure measurements are made on a routine basis the condition can go undetected until a precipitous complicating event. When you become aware of these potential complications of hypertension and consider that 25% of

The Ups and Downs of Hypertension and Exercise

When blood pressure is up, one way to bring it down is to increase the level of physical activity. Research studies have demonstrated that participation in aerobic activities protects against the development of hypertension. Furthermore, exercise can be used as a therapy to reduce hypertension once it has already developed.

Antihypertensive medication is available to lower blood pressure in severely hypertensive patients, but sometimes undesirable side effects occur. The side effects of diuretics include electrolyte imbalances, inability to handle glucose normally, and raised blood-cholesterol levels. The side effects of drugs that manipulate total peripheral resistance include increased blood-triglyceride levels, lower HDL-cholesterol

levels (the "good" form of cholesterol), weight gain, sexual dysfunction, and depression.

Patients with mild hypertension, arbitrarily defined as a diastolic blood pressure between 90 and 100 mm Hg and a systolic pressure of 160 mm Hg, pose a dilemma for physicians. The risks of taking the drugs may outweigh the benefits gained from lowering the blood pressure. Because of the drug therapy's possible side effects, nondrug treatment of mild hypertension may be most beneficial. The most common nondrug therapies are weight reduction, salt restriction, and exercise.

Although losing weight will almost always reduce blood pressure, research has shown that weight reduction programs usually result in the loss of only 12 pounds, and the overall long-term success in keeping the weight off is only about

20%. Salt restriction is beneficial for many hypertensives, but adherence to a low-salt diet is difficult for many people because fast foods and foods prepared in restaurants usually contain high amounts of salt.

The preponderance of evidence in the literature suggests that moderate aerobic exercise performed for 15 to 60 minutes three times per week is a beneficial therapy in most cases of mild to moderate hypertension. It is wise, therefore, to include a regular aerobic-exercise program along with other therapeutic measures to optimally reduce high blood pressure. If more convenient, the total exercise time on a given day may even be split up into smaller sessions and still provide the same benefits.

---

all adults in America are estimated to have chronic elevated blood pressure, you can appreciate the magnitude of this national health problem.

Treatment of hypertension

Once hypertension is detected, therapeutic intervention can reduce the course and severity of the problem. Dietary management, including weight loss, along with a variety of drugs that manipulate salt and water management or autonomic activity on the cardiovascular system can be used to treat hypertension. No matter what the original cause, agents that reduce the plasma volume or total peripheral resistance (or both) will decrease the blood pressure toward normal. Furthermore, a regular aerobic exercise program can be employed to help reduce high blood pressure. (For details, see the accompanying boxed feature, ▶ A Closer Look at Exercise Physiology.)

We are now going to examine the other extreme, hypotension, looking first at transient orthostatic hypotension, then at the (more serious) circulatory shock.

▮ Orthostatic hypotension results from transient inadequate sympathetic activity.

Hypotension, or low blood pressure, occurs either when there is a disproportion between vascular capacity and blood volume (in essence, too little blood to fill the vessels) or when the heart is too weak to impart sufficient driving pressure to the blood.

The most common situation in which hypotension occurs transiently is orthostatic hypotension. **Orthostatic (postural) hypotension** is a transient hypotensive condition resulting from insufficient compensatory responses to the gravitational shifts in blood that occur when a person moves from a horizontal to a vertical position, especially following prolonged

bed rest. When a person moves from lying down to standing up, pooling of blood in the leg veins from gravity reduces venous return, subsequently decreasing stroke volume, lowering cardiac output and blood pressure. This fall in blood pressure is normally detected by the baroreceptors, which initiate immediate compensatory responses to restore blood pressure to its proper level. When a long-bedridden patient first starts to rise, however, these reflex compensatory adjustments are temporarily lost or reduced because of disuse. Sympathetic control of the leg veins is inadequate, so when the patient first stands up blood pools in the lower extremities. The resultant orthostatic hypotension and decrease in blood flow to the brain are responsible for the dizziness or actual fainting that occurs. Because postural compensatory mechanisms are depressed during prolonged bed confinement, patients sometimes are put on a tilt table so that they can be moved gradually from a horizontal to an upright position. This allows the body to adjust slowly to the gravitational shifts in blood.

▮ Circulatory shock can become irreversible.

When blood pressure falls so low that adequate blood flow to the tissues can no longer be maintained, the condition known as **circulatory shock** occurs. Circulatory shock is categorized into four main types (● Figure 10-40):

1. *Hypovolemic shock* is caused by a fall in blood volume, which occurs either directly through severe hemorrhage or indirectly through loss of fluids derived from the plasma (for example, severe diarrhea, excessive urinary losses, or extensive sweating).

2. *Cardiogenic shock* is due to a weakened heart's failure to pump blood adequately.

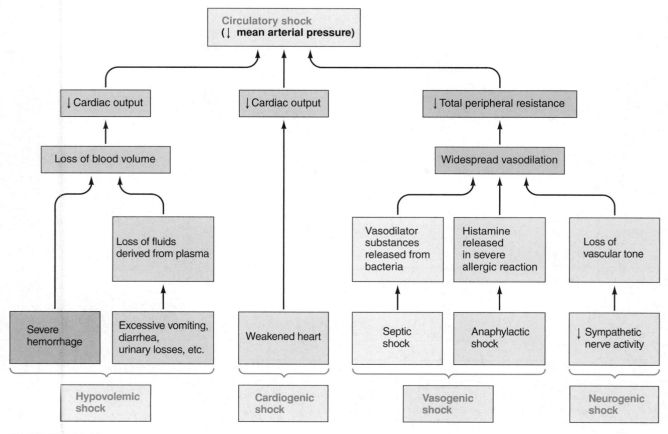

● **FIGURE 10-40**

Causes of circulatory shock
Circulatory shock, which occurs when mean arterial blood pressure falls so low that adequate blood flow to the tissues can no longer be maintained, may result from (1) extensive loss of blood volume (hypovolemic shock), (2) failure of the heart to pump blood adequately (cardiogenic shock), (3) widespread arteriolar vasodilation (vasogenic shock), or (4) neurally defective vasoconstrictor tone (neurogenic shock).

3. *Vasogenic shock* is caused by widespread vasodilation triggered by the presence of vasodilator substances. There are two types of vasogenic shock: septic and anaphylactic. *Septic shock*, which may accompany massive infections, is due to vasodilator substances released from the infective agents. Similarly, extensive histamine release accompanying severe allergic reactions can cause widespread vasodilation in *anaphylactic shock*.

4. *Neurogenic shock* also involves generalized vasodilation but not by means of the release of vasodilator substances. In this case, loss of sympathetic vascular tone leads to generalized vasodilation. This undoubtedly is responsible for the shock accompanying crushing injuries when blood loss has not been sufficient to cause hypovolemic shock. Deep, excruciating pain apparently inhibits sympathetic vasoconstrictor activity.

We will now examine the consequences of and compensations for shock, using hemorrhage as an example (● Figure 10-41). This figure may look intimidating, but we will work through it step by step. It is an important example that pulls together many of the principles discussed in this chapter. As before, the cir-

cled numbers in the text correspond to the numbers in the figure and indicate the portion of the figure being discussed.

Consequences and compensations of shock

• Following severe loss of blood, the resultant reduction in circulating blood volume leads to a decrease in venous return ① and a subsequent fall in cardiac output and arterial blood pressure. (Note the blue boxes, which indicate consequences of hemorrhage.)

• Compensatory measures immediately attempt to maintain adequate blood flow to the brain. (Note the pink boxes, which indicate compensations for hemorrhage.)

• The baroreceptor reflex response to the fall in blood pressure brings about increased sympathetic and decreased parasympathetic activity to the heart ②. The result is an increase in heart rate ③ to offset the reduced stroke volume ④ brought about by the loss of blood volume. With severe fluid loss, the pulse is weak because of the reduced stroke volume but rapid because of the increased heart rate.

• As a result of increased sympathetic activity to the veins, generalized venous vasoconstriction occurs ⑤, increasing venous return by means of the Frank-Starling mechanism ⑥.

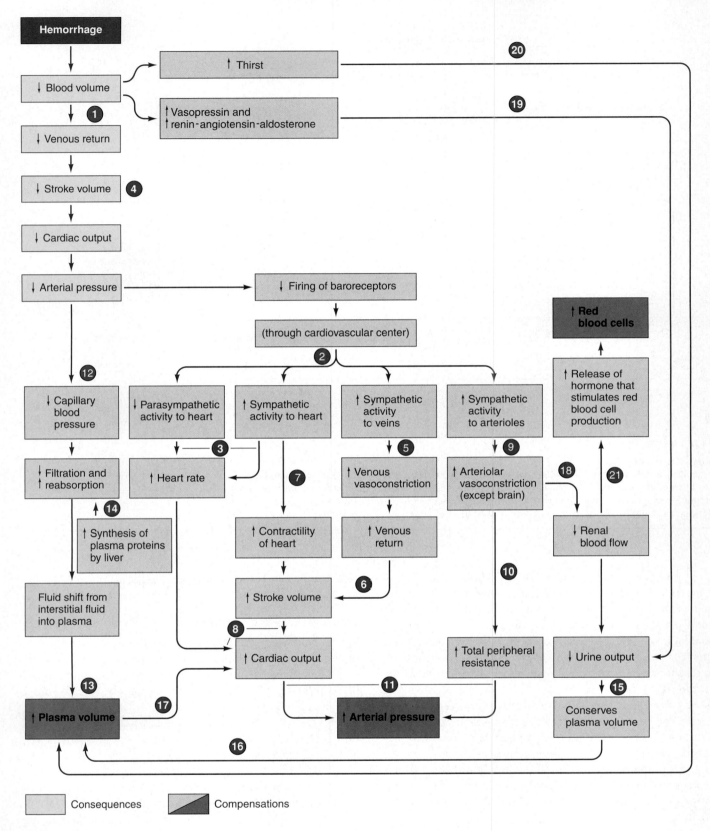

● **FIGURE 10-41**

Consequences and compensations of hemorrhage
The reduction in blood volume resulting from hemorrhage leads to a fall in arterial pressure. (Note the blue boxes, representing consequences of hemorrhage). A series of compensations ensue (light pink boxes) that ultimately restore plasma volume, arterial pressure, and the number of red blood cells toward normal (dark pink boxes). Refer to the text (pp. 383–385) for an explanation of the circled numbers and a detailed discussion of the compensations.

- Simultaneously, sympathetic stimulation of the heart increases the heart's contractility 7 so that it beats more forcefully and ejects a greater volume of blood, likewise increasing the stroke volume.
- The increase in heart rate and increase in stroke volume collectively lead to an increase in cardiac output 8.
- Sympathetically-induced generalized arteriolar vasoconstriction 9 leads to an increase in total peripheral resistance 10.
- Together, the increase in cardiac output and total peripheral resistance bring about a compensatory increase in arterial pressure 11.
- The original fall in arterial pressure is also accompanied by a fall in capillary blood pressure 12, which results in fluid shifts from the interstitial fluid into the capillaries to expand the plasma volume 13. This response is sometimes termed **autotransfusion,** because it restores the plasma volume as a transfusion does.
- This ECF fluid shift is enhanced by plasma protein synthesis by the liver during the next few days following hemorrhage 14. The plasma proteins exert a colloid osmotic pressure that helps retain extra fluid in the plasma.
- Urinary output is reduced, thereby conserving water that normally would have been lost from the body 15. This additional fluid retention helps expand the reduced plasma volume 16. Expansion of plasma volume further augments the increase in cardiac output brought about by the baroreceptor reflex 17. Reduction in urinary output results from decreased renal blood flow caused by compensatory renal arteriolar vasoconstriction 18. The reduced plasma volume also triggers increased secretion of the hormone vasopressin and activation of the salt- and water-conserving renin-angiotensin-aldosterone hormonal pathway, which brings about a further reduction in urinary output 19.
- Increased thirst is also stimulated by a fall in plasma volume 20. The resultant increased fluid intake helps restore plasma volume.
- Over a longer course of time (a week or more), lost red blood cells are replaced through increased red blood cell production triggered by a reduction in O_2 delivery to the kidneys 21.

Irreversible shock

These compensatory mechanisms are often insufficient in the face of substantial fluid loss. Even if they can maintain an adequate blood pressure level, the short-term measures cannot continue indefinitely. Ultimately, fluid volume must be replaced from the outside through drinking, transfusion, or a combination of both. Blood supply to kidneys, digestive tract, skin, and other organs can be compromised to maintain blood flow to the brain only so long before organ damage begins to occur. A point may be reached at which blood pressure continues to drop rapidly because of tissue damage, despite vigorous therapy. This condition is frequently termed **irreversible shock,** in contrast to **reversible shock,** which can be corrected by compensatory mechanisms and effective therapy.

Although the exact mechanism underlying irreversibility is not currently known, many logical possibilities could contribute to the unrelenting, progressive circulatory deterioration that characterizes irreversible shock. Metabolic acidosis arises when lactic acid production increases as blood-deprived tissues resort to anaerobic metabolism. Acidosis deranges the enzymatic systems responsible for energy production, limiting the capability of the heart and other tissues to produce ATP. Prolonged depression of kidney function results in electrolyte imbalances that may lead to cardiac arrhythmias. The blood-deprived pancreas releases a chemical that is toxic to the heart (**myocardial toxic factor**), further weakening the heart. Vasodilator substances build up within ischemic organs, inducing local vasodilation that overrides the generalized reflex vasoconstriction. As cardiac output progressively declines because of the heart's diminishing effectiveness as a pump and total peripheral resistance continues to fall, hypotension becomes increasingly more severe. This causes further cardiovascular failure, which leads to a further decline in blood pressure. Thus when shock progresses to the point that the cardiovascular system itself starts to fail, a vicious positive-feedback cycle ensues that ultimately results in death.

CHAPTER IN PERSPECTIVE: FOCUS ON HOMEOSTASIS

Homeostatically, the blood vessels serve as passageways to transport blood to and from the cells for the purpose of O_2 and nutrient delivery, waste removal, distribution of fluid and electrolytes, and hormonal signaling, among other things. Cells soon die if deprived of their blood supply, with brain cells succumbing within four minutes. Blood is constantly recycled and reconditioned as it travels through the various organs by means of the vascular highways. Hence the body needs only a very small volume of blood to maintain the appropriate chemical composition of the entire internal fluid environment on which the cells depend for their survival. For example, O_2 is continually picked up by the blood in the lungs and constantly delivered to all the body cells.

The smallest of blood vessels, the capillaries, are the actual site of exchange between the blood and surrounding cells. Capillaries bring homeostatically maintained blood within 0.01 cm of every cell in the body; this proximity is critical because beyond a few centimeters, materials cannot diffuse rapidly enough to support life-sustaining activities. Oxygen that would take months to years to diffuse from the lungs to all the cells of the body is continuously delivered at the "doorstep" of every cell, where diffusion can efficiently accomplish short local exchanges between the capillaries and surrounding cells. Likewise, hormones must be rapidly transported through the circulatory system from their sites of production in endocrine glands to their sites of action in other parts of the body. These chemical messengers could not diffuse nearly rapidly enough to their target organs to effectively exert their controlling effects, many of which are aimed toward maintaining homeostasis.

The remainder of the circulatory system is designed to transport blood to and from the capillaries. The arteries and arterioles distribute blood pumped by the heart to the capillaries for life-sustaining exchanges to take place, and the venules and veins collect blood from the capillaries and return it to the heart, where the process is repeated.

CHAPTER SUMMARY

Introduction

▌ Materials can be exchanged between various parts of the body and with the external environment by means of the blood vessel network that transports blood to and from all tissues.

▌ Organs that replenish nutrient supplies and remove metabolic wastes from the blood receive a greater percentage of the cardiac output than is warranted by their metabolic needs. These "reconditioning" organs can better tolerate reductions in blood supply than can organs that receive blood solely for the purpose of meeting their own metabolic needs.

▌ The brain is especially vulnerable to reductions in its blood supply. Therefore, the maintenance of adequate flow to this vulnerable organ is a high priority in circulatory function.

▌ The flow rate of blood through a vessel is directly proportional to the pressure gradient and inversely proportional to the resistance. The higher pressure at the beginning of a vessel is established by the pressure imparted to the blood by cardiac contraction. The lower pressure at the end is due to frictional losses as flowing blood rubs against the vessel wall.

▌ Resistance, the hindrance to blood flow through a vessel, is influenced most by the vessel's radius. Resistance is inversely proportional to the fourth power of the radius, so small changes in radius profoundly influence flow. As the radius increases, resistance decreases and flow increases.

▌ Blood flows in a closed loop between the heart and the tissues. The arteries transport blood from the heart throughout the body. The arterioles regulate the amount of blood that flows through each organ. The capillaries are the actual site where materials are exchanged between the blood and surrounding tissue. The veins return the blood from the tissues to the heart.

Arteries

▌ Arteries are large-radius, low-resistance passageways from the heart to the tissues.

▌ They also serve as a pressure reservoir. Because of their elasticity, arteries expand to accommodate the extra volume of blood pumped into them by cardiac contraction and then recoil to continue driving the blood forward when the heart is relaxing.

▌ Systolic pressure is the peak pressure exerted by the ejected blood against the vessel walls during cardiac systole. Diastolic pressure is the minimum pressure in the arteries when blood is draining off into the vessels downstream during cardiac diastole.

▌ The average driving pressure throughout the cardiac cycle is the mean arterial pressure, which can be estimated using the following formula: Mean arterial pressure = diastolic pressure + 1/3 pulse pressure.

Arterioles

▌ Arterioles are the major resistance vessels. Their high resistance produces a large drop in mean pressure between the arteries and capillaries. This decline enhances blood flow by contributing to the pressure differential between the heart and the tissues.

▌ Tone, a baseline of contractile activity, is maintained in arterioles at all times.

▌ Arteriolar vasodilation, an expansion of arteriolar caliber above tonic level, decreases resistance and increases blood flow through the vessel, whereas vasoconstriction, a narrowing of the vessel, increases resistance and decreases flow.

▌ Arteriolar caliber is subject to two types of control mechanisms: local (intrinsic) controls and extrinsic controls.

▌ Local controls primarily involve local chemical changes associated with changes in the level of metabolic activity in a tissue. These changes in local metabolic factors cause the release of vasoactive mediators from the endothelium in the vicinity. These vasoactive mediators act on the underlying arteriolar smooth muscle to bring about an appropriate change in the caliber of the arterioles supplying the tissue. By adjusting the resistance to blood flow in this manner, the local control mechanism adjusts blood flow to the tissue to match the momentary metabolic needs of the tissue.

▌ Adjustments in arteriolar caliber can be accomplished independently in different tissues by local control factors. Such adjustments are important in determining the distribution of cardiac output.

▌ Extrinsic control is accomplished primarily by sympathetic nerve influence and to a lesser extent by hormonal influence over arteriolar smooth muscle. Extrinsic controls are important in maintaining mean arterial blood pressure. Arterioles are richly supplied with sympathetic nerve fibers, whose increased activity produces generalized vasoconstriction and a subsequent increase in mean arterial pressure. Decreased sympathetic activity produces generalized arteriolar vasodilation, which lowers mean arterial pressure. These extrinsically controlled adjustments of arteriolar caliber help maintain the appropriate pressure head for driving blood forward to the tissues.

Capillaries

▌ The thin-walled, small-radius, extensively branched capillaries are ideally suited to serve as sites of exchange between the blood and surrounding tissues. Anatomically, the surface area for exchange is maximized and diffusion distance is minimized in the capillaries. Furthermore, because of their large total cross-sectional area the velocity of blood flow through capillaries is relatively slow, providing adequate time for exchanges to take place.

▌ Two types of passive exchanges—diffusion and bulk flow—take place across capillary walls.

▌ Individual solutes are exchanged primarily by diffusion down concentration gradients. Lipid-soluble substances pass directly through the single layer of endothelial cells lining a capillary, whereas water-soluble substances pass through water-filled pores between the endothelial cells. Plasma proteins generally do not escape.

▌ Imbalances in physical pressures acting across capillary walls are responsible for bulk flow of fluid through the pores back and forth between the plasma and interstitial fluid. (1) Fluid is forced out of the first portion of the capillary (ultrafiltration), where outward pressures (mainly capillary blood pressure) exceed inward pressures (mainly plasma-colloid osmotic pressure). (2) Fluid is returned to the capillary along its last half, when outward pressures fall below inward pressures. The reason for the shift in balance down the length of the capillary is the continuous decline in capillary blood pressure while the plasma-colloid osmotic pressure remains constant.

▌ Bulk flow is responsible for the distribution of extracellular fluid between the plasma and the interstitial fluid.

▌ Normally, slightly more fluid is filtered than is reabsorbed. The extra fluid, any leaked proteins, and tissue contaminants such as bacteria are picked up by the lymphatic system. Bacteria are destroyed as lymph passes through the lymph nodes en route to being returned to the venous system.

Veins

- Veins are large-radius, low-resistance passageways for return of blood from the tissues to the heart.
- In addition, veins can accommodate variable volumes of blood and therefore act as a blood reservoir. The capacity of veins to hold blood can change markedly with little change in venous pressure. Veins are thin-walled, highly distensible vessels that can passively stretch to store a larger volume of blood.
- The primary force responsible for venous flow is the pressure gradient between the veins and atrium (that is, what remains of the driving pressure imparted to the blood by cardiac contraction).
- Venous return is enhanced by sympathetically induced venous vasoconstriction and by external compression of the veins resulting from contraction of surrounding skeletal muscles, both of which drive blood out of the veins.
- One-way venous valves ensure that blood is driven toward the heart and prevented from flowing back toward the tissues.
- Venous return is also enhanced by the respiratory pump and the cardiac suction effect. Respiratory activity produces a less-than-atmospheric pressure in the chest cavity, thus establishing an external pressure gradient that encourages flow from the lower veins that are exposed to atmospheric pressure to the chest veins that empty into the heart. In addition, slightly negative pressures created within the atria during ventricular systole and within the ventricles during ventricular diastole exert a suction-

ing effect that further enhances venous return and facilitates cardiac filling.

Blood Pressure

- Regulation of mean arterial pressure depends on control of its two main determinants, cardiac output and total peripheral resistance.
- Control of cardiac output in turn depends on regulation of heart rate and stroke volume, whereas total peripheral resistance is determined primarily by degree of arteriolar vasoconstriction.
- Short-term regulation of blood pressure is accomplished primarily by the baroreceptor reflex. Carotid sinus and aortic arch baroreceptors continuously monitor mean arterial pressure. When they detect a deviation from normal, they signal the medullary cardiovascular center, which responds by adjusting autonomic output to the heart and blood vessels to restore the blood pressure to normal.
- Long-term control of blood pressure involves maintaining proper plasma volume through the kidneys' control of salt and water balance.
- Blood pressure can be abnormally high (hypertension) or abnormally low (hypotension). Severe sustained hypotension resulting in generalized inadequate blood delivery to the tissues is known as *circulatory shock*.

REVIEW EXERCISES

Objective Questions (Answers on p. A-42)

1. In general, the parallel arrangement of the vascular system enables each organ to receive its own separate arterial blood supply. *(True or false?)*
2. More blood flows through the capillaries during cardiac systole than during diastole. *(True or false?)*
3. The capillaries contain only 5% of the total blood volume at any point in time. *(True or false?)*
4. The same volume of blood passes through the capillaries in a minute as passes through the aorta, even though the velocity of blood flow is much slower in the capillaries. *(True or false?)*
5. Because capillary walls have no carrier transport systems, all capillaries are equally permeable. *(True or false?)*
6. Because of gravitational effects, venous pressure in the lower extremities is greater when a person is standing up than when the person is lying down. *(True or false?)*
7. Which of the following functions is (are) attributable to arterioles? *(Indicate all correct answers.)*
   a. responsible for a significant decline in mean pressure, which helps establish the driving pressure gradient between the heart and tissues
   b. site of exchange of materials between the blood and surrounding tissues
   c. main determinant of total peripheral resistance
   d. determine the pattern of distribution of cardiac output
   e. play a role in regulating mean arterial blood pressure
   f. convert the pulsatile nature of arterial blood pressure into a smooth, nonfluctuating pressure in the vessels further downstream
   g. act as a pressure reservoir

8. Using the following answer code, indicate what kind of compensatory changes occur in the factors in question to restore the blood pressure to normal in response to hypovolemic hypotension resulting from severe hemorrhage:
   (a) increased
   (b) decreased
   (c) no effect
   ___1. rate of afferent firing generated by the carotid sinus and aortic arch baroreceptors
   ___2. sympathetic output by the cardiovascular center
   ___3. parasympathetic output by the cardiovascular center
   ___4. heart rate
   ___5. stroke volume
   ___6. cardiac output
   ___7. arteriolar radius
   ___8. total peripheral resistance
   ___9. venous radius
   ___10. venous return
   ___11. urinary output
   ___12. fluid retention within the body
   ___13. fluid movement from the interstitial fluid into the plasma across the capillaries
9. Using the following answer code, indicate whether these factors increase or decrease venous return:
   (a) increases venous return
   (b) decreases venous return
   (c) has no effect on venous return
   ___1. sympathetically induced venous vasoconstriction
   ___2. skeletal muscle activity
   ___3. gravitational effects on the venous system

____4.   respiratory activity
____5.   increased atrial pressure associated with a leaky AV valve
____6.   ventricular pressure change associated with diastolic recoil

Essay Questions

1. Compare blood flow through reconditioning organs and through organs that do not recondition the blood.
2. Discuss the relationships among flow rate, pressure gradient, and vascular resistance. What is the major determinant of resistance to flow?
3. Describe the structure and major functions of each segment of the vascular tree.
4. How do the arteries serve as a pressure reservoir?
5. Describe the indirect technique of measuring arterial blood pressure by means of a sphygmomanometer.
6. Define *vasoconstriction* and *vasodilation*.
7. Discuss the local and extrinsic controls that regulate arteriolar resistance.
8. What is the primary means by which individual solutes are exchanged across the capillary walls? What forces are responsible for bulk flow across the capillary walls? Of what importance is bulk flow?
9. How is lymph formed? What are the functions of the lymphatic system?
10. Define *edema*, and discuss its possible causes.
11. How do veins serve as a blood reservoir?
12. Compare the effect of vasoconstriction on the rate of blood flow in arterioles and veins.
13. Discuss the factors that determine mean arterial pressure.
14. Review the effects on the cardiovascular system of parasympathetic and sympathetic stimulation.
15. Differentiate between secondary hypertension and primary hypertension. What are the potential consequences of hypertension?
16. Define *circulatory shock*. What are its consequences and compensations? What is irreversible shock?

Quantitative Exercises (Solutions on p. A-42)

1. Recall that the flow rate of blood equals the pressure gradient divided by the total peripheral resistance of the vascular system. The conventional unit of resistance in physiological systems is expressed in PRU (peripheral resistance unit), which is defined as (1 liter/min)/(1 mm Hg). At rest, Tom's total peripheral resistance is about 20 PRU. Last week while playing racquetball, his cardiac output increased to 30 liters/min and his mean arterial pressure increased to 120 mm Hg. What was his total peripheral resistance during the game?
2. Systolic pressure rises as a person ages. By age 85, an average male has a systolic pressure of 180 mm Hg and a diastolic pressure of 90 mm Hg.
   a. What is the mean arterial pressure of this average 85-year-old male?
   b. From your knowledge of capillary dynamics, predict the result at the capillary level of this age-related change in mean arterial pressure if no homeostatic mechanisms were operating. (Recall that mean arterial pressure is about 93 mm Hg at age 20.)
3. Compare the flow rates in the systemic and the pulmonary circulations of an individual with the following measurements:
   systemic mean arterial pressure = 95 mm Hg
   systemic resistance = 19 PRU
   pulmonary mean arterial pressure = 20 mm Hg
   pulmonary resistance = 4 PRU
4. Which of the following changes would increase the resistance in an arteriole? Explain.
   a. a longer length
   b. a smaller caliber
   c. increased sympathetic stimulation
   d. increased blood viscosity
   e. all of the above

POINTS TO PONDER

(Explanations on p. A-42)

1. During coronary bypass surgery, a piece of vein is removed from the patient's leg and surgically attached within the coronary circulatory system so that blood is detoured around an occluded coronary artery segment. For an extended period of time following surgery, why must the patient wear an elastic support stocking on the limb from which the vein was removed?
2. Assume a person has a blood pressure recording of 125/77:
   a. What is the systolic pressure?
   b. What is the diastolic pressure?
   c. What is the pulse pressure?
   d. What is the mean arterial pressure?
   e. Would any sound be heard when the pressure in an external cuff around the arm was 130 mm Hg? *(yes or no)*
   f. Would any sound be heard when cuff pressure was 118 mm Hg?
   g. Would any sound be heard when cuff pressure was 75 mm Hg?

3. A classmate who has been standing still for several hours working on a laboratory experiment suddenly faints. What is the probable explanation? What would you do if the person next to him tried to get him up?
4. A drug applied to a piece of excised arteriole causes the vessel to relax, but an isolated piece of arteriolar muscle stripped from the other layers of the vessel fails to respond to the same drug. What is the probable explanation?
5. Explain how each of the following antihypertensive drugs would lower arterial blood pressure:
   a. drugs that block alpha$_1$-adrenergic receptors (for example, *phentolamine*)
   b. drugs that block beta-adrenergic receptors (for example, *propranolol*) (*Hint*: More important than the action of these drugs on the heart and blood vessels in reducing arterial blood pressure is their inhibitory effect on the hormonal pathway involving angiotensin II that promotes salt and water conservation.)

c. drugs that directly relax arteriolar smooth muscle (for example, *hydralazine*)

d. diuretic drugs that increase urinary output (for example, *furosemide*)

e. drugs that block release of norepinephrine from sympathetic endings (for example, *guanethidine*)

f. drugs that act on the brain to reduce sympathetic output (for example, *clonidine*)

g. drugs that block Ca^{2+} channels (for example, *verapamil*)

h. drugs that interfere with the production of angiotensin II (for example, *captopril*)

6. ***Clinical Consideration.*** Li-Ying C. has just been diagnosed as having hypertension secondary to a *pheochromocytoma*, a tumor of the adrenal medulla that secretes excessive epinephrine. Explain how this condition leads to secondary hypertension by describing the effect that excessive epinephrine would have on various factors that determine arterial blood pressure.

PHYSIOEDGE RESOURCES

PHYSIOEDGE CD-ROM
Figures marked with the PhysioEdge icon have associated activities on the CD. For this chapter, check out:

Cardiovascular Physiology

The diagnostic quiz allows you to receive immediate feedback on your understanding of the concept and to advance through various levels of difficulty.

CHECK OUT THESE MEDIA QUIZZES:
10.1 Blood Flow and Total Cross Sectional Area
10.2 Arteriolar Resistance and Flow in Capillaries

PHYSIOEDGE WEB SITE
The Web site for this book contains a wealth of helpful study aids, as well as many ideas for further reading and research. Log on to:

http://www.brookscole.com

Go to the Biology page and select Sherwood's *Human Physiology*, 5th Edition. Select a chapter from the drop-down menu or click on one of these resource areas:

- **Case Histories** provide an introduction to the clinical aspects of human physiology. Check out:

    #8: Toxic Shock Syndrome

- For 2-D and 3-D graphical illustrations and animations of physiological concepts, visit our **Visual Learning Resource**.

- Resources for study and review include the **Chapter Outline, Chapter Summary, Glossary,** and **Flash Cards.** Use our **Quizzes** to prepare for in-class examinations.

- On-line research resources to consult are **Hypercontents**, with current links to relevant Internet sites; **Internet Exercises** with starter URLs provided; and **InfoTrac Exercises**.

  For more readings, go to InfoTrac College Edition, your on-line research library, at:

    http://infotrac.thomsonlearning.com

Blood

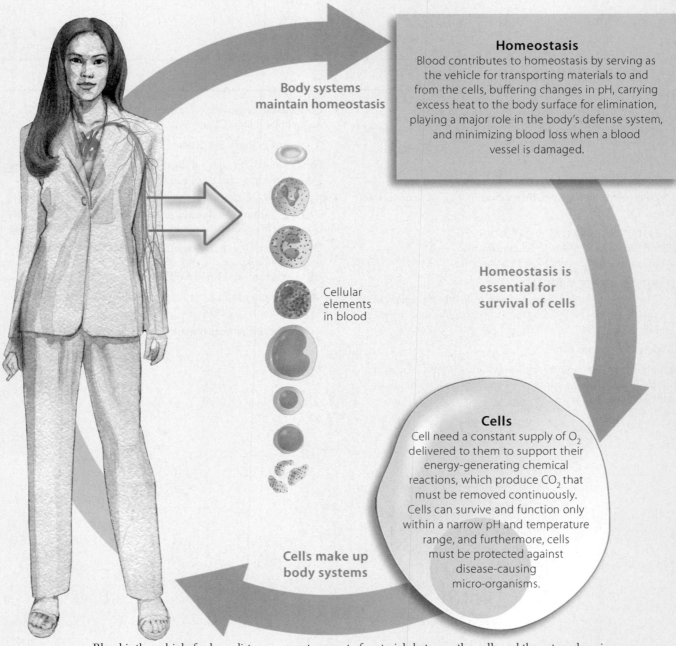

Body systems
maintain homeostasis

Homeostasis
Blood contributes to homeostasis by serving as the vehicle for transporting materials to and from the cells, buffering changes in pH, carrying excess heat to the body surface for elimination, playing a major role in the body's defense system, and minimizing blood loss when a blood vessel is damaged.

Cellular elements in blood

Homeostasis is essential for survival of cells

Cells
Cell need a constant supply of O_2 delivered to them to support their energy-generating chemical reactions, which produce CO_2 that must be removed continuously. Cells can survive and function only within a narrow pH and temperature range, and furthermore, cells must be protected against disease-causing micro-organisms.

Cells make up body systems

Blood is the vehicle for long-distance, mass transport of materials between the cells and the external environment or between the cells themselves. Such transport is essential for maintaining homeostasis. Blood consists of a complex liquid **plasma** in which the cellular elements—*erythrocytes*, *leukocytes*, and *platelets*—are suspended.

Erythrocytes (red blood cells or **RBCs)** are essentially plasma membrane—enclosed bags of **hemoglobin** that transport O_2 in the blood. **Leukocytes (white blood cells** or **WBCs),** the immune system's mobile defense units, are transported in the blood to sites of injury or invasion by disease-causing micro-organisms.

Given the importance of blood, it is imperative that mechanisms exist to minimize blood loss when a vessel is injured. **Platelets (thrombocytes)** are important in **hemostasis,** the stopping of bleeding from an injured vessel.

Chapter 11

The Blood

CONTENTS AT A GLANCE

INTRODUCTION

PLASMA

▌ Composition and functions of plasma

▌ Plasma proteins

ERYTHROCYTES

▌ Structure and function of erythrocytes

▌ Erythropoiesis

LEUKOCYTES

▌ Types and functions of leukocytes

▌ Leukocyte production

PLATELETS AND HEMOSTASIS

▌ Platelet structure and function

▌ Hemostasis

INTRODUCTION

Blood represents about 8% of total body weight and has an average volume of 5 liters in women and 5.5 liters in men. It consists of three types of specialized cellular elements, *erythrocytes*, *leukocytes*, and *platelets*, suspended in the complex liquid *plasma* (▲ Table 11-1).

The constant movement of blood as it flows through the blood vessels keeps its cellular elements rather evenly dispersed within the plasma. However, if a sample of whole blood is placed in a test tube and treated to prevent clotting, the heavier cellular elements slowly settle to the bottom and the lighter plasma rises to the top. This process can be hastened by centrifugation, which rapidly packs the cells in the bottom of the tube (● Figure 11-1). Because over 99% of the cells are erythrocytes, the **hematocrit,** or **packed cell volume,** essentially represents the percentage of total blood volume occupied by erythrocytes. The hematocrit averages 42% for women and slightly higher, 45%, for men. Plasma accounts for the remaining volume. Accordingly, the average blood volume occupied by plasma is 58% for women and 55% for men.

The white blood cells and platelets, which are colorless and less dense than red cells, are packed in a thin, cream-colored layer, the *"buffy coat,"* on top of the packed red cell column. They represent less than 1% of the total blood volume.

We will first consider the properties of the largest portion of the blood, the plasma, before turning our attention to the cellular elements.

PLASMA

Plasma, being a liquid, is composed of 90% water.

▌ Plasma water is a transport medium for many inorganic and organic substances.

Plasma water serves as a medium for materials being carried in the blood. Also, because water has a high capacity to hold heat, plasma can absorb and distribute much of the heat generated metabolically within tissues, whereas the temperature of the blood itself undergoes only small changes. As the blood travels close to the surface of the skin, heat energy not needed to maintain body temperature is eliminated to the environment.

Blood Constituents and Their Functions

Constituent	Functions
Plasma	
Water	Transport medium; carries heat
Electrolytes	Membrane excitability; osmotic distribution of fluid between the extracellular and intracellular fluid; buffering of pH changes
Nutrients, wastes, gases, hormones	Transported in blood; the blood gas CO_2 plays a role in acid–base balance
Plasma proteins	In general, exert an osmotic effect that is important in the distribution of extracellular fluid between the vascular and interstitial compartments; buffering of pH changes
Albumins	Transport many substances; make the greatest contribution to colloid osmotic pressure
Globulins	
Alpha and beta	Transport many substances; clotting factors; inactive precursor molecules
Gamma	Antibodies
Fibrinogen	Inactive precursor for the fibrin meshwork of a clot
Cellular Elements	
Erythrocytes	Transport O_2 and CO_2 (mainly O_2)
Leukocytes	
Neutrophils	Phagocytes that engulf bacteria and debris
Eosinophils	Attack parasitic worms; important in allergic reactions
Basophils	Release histamine, which is important in allergic reactions, and heparin, which helps clear fat from the blood, and may function as an anticoagulant
Monocytes	In transit to become tissue macrophages
Lymphocytes	
B lymphocytes	Production of antibodies
T lymphocytes	Cell-mediated immune responses
Platelets	Hemostasis

A large number of inorganic and organic substances are dissolved in the plasma. Inorganic constituents account for approximately 1% of plasma weight. The most abundant electrolytes (ions) in the plasma are Na^+ and Cl^-, which are the

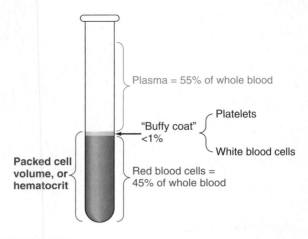

● **FIGURE 11-1**

Hematocrit
The values given are for men. The average hematocrit for women is 42%, with plasma occupying 58% of the blood volume.

components of common salt. There are lesser amounts of HCO_3^-, K^+, Ca^{2+}, and others. The most notable functions of these ions are their roles in membrane excitability, osmotic distribution of fluid between the ECF and cells, and buffering of pH changes; these functions are discussed elsewhere.

The most plentiful organic constituents by weight are the *plasma proteins*, which compose 6% to 8% of plasma's total weight. We examine them more thoroughly in the next section. The remaining small percentage of plasma is occupied by other organic substances, including *nutrients* (for example, glucose, amino acids, lipids, and vitamins), *waste products* (creatinine, bilirubin, and nitrogenous substances such as urea); *dissolved gases* (O_2 and CO_2), and *hormones*. Most of these substances are merely being transported in the plasma. For example, endocrine glands secrete hormones into the plasma, which transports these chemical messengers to their sites of action.

▌ Many of the functions of plasma are carried out by plasma proteins.

The **plasma proteins** are the one group of plasma constituents not present just for the ride. These important components normally remain in the plasma, where they perform many valuable functions. Here are the most important of these functions, which are elaborated on elsewhere in the text:

1. Unlike other plasma constituents that are dissolved in the plasma water, the plasma proteins exist in a colloidal dispersion (see p. A-10). Furthermore, because they are the largest of the plasma constituents plasma proteins usually do not exit through the narrow pores in the capillary walls to enter the interstitial fluid. By virtue of their presence as a colloidal dispersion in the plasma and their absence in the interstitial fluid, plasma proteins establish an osmotic gradient between the blood and interstitial fluid. This colloid osmotic pressure is the primary force responsible for preventing excessive loss of plasma from the capillaries into the interstitial fluid and thus helps maintain plasma volume (see p. 366).

2. Plasma proteins are partially responsible for the plasma's capacity to buffer changes in pH (see p. 576).

3. The three groups of plasma proteins—*albumins, globulins,* and *fibrinogen*—are classified according to their various physical and chemical properties. In addition to the general functions just listed, each type of plasma protein performs specific tasks as follows:

    a. **Albumins,** the most abundant of the plasma proteins, bind many substances (for example, bilirubin, bile salts, and penicillin) for transport through the plasma and contribute most extensively to the colloid osmotic pressure by virtue of their numbers.

    b. There are three subclasses of **globulins: alpha (α), beta (β),** and **gamma (γ).**

        (1) Specific alpha and beta globulins bind and transport a number of substances in the plasma, such as thyroid hormone, cholesterol, and iron.

        (2) Many of the factors involved in the blood-clotting process are alpha or beta globulins.

        (3) Inactive precursor protein molecules, which are activated as needed by specific regulatory inputs, belong to the alpha globulin group (for example, the alpha globulin angiotensinogen is activated to angiotensin, which plays an important role in regulating salt balance in the body; see p. 564).

        (4) The gamma globulins are the immunoglobulins (antibodies), which are crucial to the body's defense mechanism (see p. 425).

    c. **Fibrinogen** is a key factor in blood-clotting.

The plasma proteins are synthesized by the liver, with the exception of the gamma globulins, which are produced by lymphocytes, one of the types of white blood cells.

ERYTHROCYTES

Each milliliter of blood contains about 5 billion **erythrocytes** (**red blood cells** or **RBCs**) on average, commonly reported clinically in a **red blood cell count** as 5 million cells per cubic millimeter (mm^3).

▌ The structure of erythrocytes is well suited to their main function of O_2 transport in the blood.

The shape and content of erythrocytes are ideally suited to carry out the primary function of these blood cellular elements, namely transporting O_2 and to a lesser extent CO_2 and hydrogen ion in the blood. We are now going to turn our attention to these form-and-function relationships.

Erythrocyte shape

Erythrocytes are flat, disc-shaped cells indented in the middle on both sides, like a doughnut with a flattened center instead of a hole (they are biconcave discs 8 mm in diameter, 2 μm thick at the outer edges, and 1 μm thick in the center) (● Figure 11-2). This unique shape contributes in two ways to the ef-

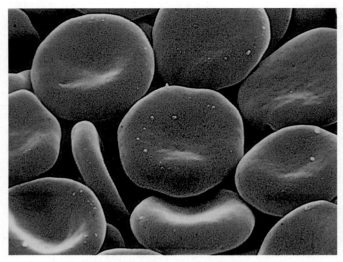

● **FIGURE 11-2**
Anatomical characteristics of erythrocytes
Appearance of erythrocytes under a scanning electron microscope. Note their biconcave shape.

ficiency with which erythrocytes perform their main function of O_2 transport in the blood. (1) The biconcave shape provides a larger surface area for diffusion of O_2 across the membrane than a spherical cell of the same volume would provide. (2) The thinness of the cell enables O_2 to diffuse rapidly between the exterior and innermost regions of the cell.

Another structural feature that facilitates RBCs' transport function is the flexibility of their membrane. Red blood cells, whose diameter is normally 8 mm, can deform amazingly as they squeeze single file through capillaries as narrow as 3 mm in diameter. Because they are extremely pliant, RBCs can travel through the narrow, tortuous capillaries to deliver their O_2 cargo at the tissue level without rupturing in the process.

The most important anatomic feature that enables erythrocytes to transport O_2 is the hemoglobin they contain. Let's look at this unique molecule in more detail.

Presence of hemoglobin

Hemoglobin is found only in red blood cells. A **hemoglobin** molecule consists of two parts: (1) the **globin portion,** a protein made up of four highly folded polypeptide chains, and (2) four iron-containing, nonprotein groups known as **heme** groups, each of which is bound to one of the polypeptides (● Figure 11-3). Each of the four iron atoms can combine reversibly with one molecule of O_2; thus each hemoglobin molecule can pick up four O_2 passengers in the lungs. Because O_2 is poorly soluble in the plasma, 98.5% of the O_2 carried in the blood is bound to hemoglobin (see p. 488).

Hemoglobin is a pigment (that is, it is naturally colored). Because of its iron content, it appears reddish when combined with O_2 and bluish when deoxygenated. Thus fully oxygenated arterial blood is red in color, and venous blood, which has lost some of its O_2 load at the tissue level, has a bluish cast.

In addition to carrying O_2, hemoglobin can also combine with the following:

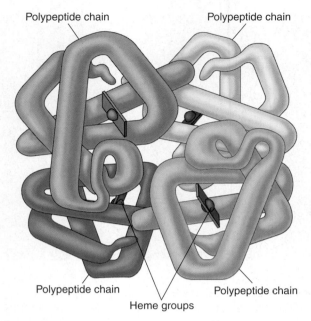

● **FIGURE 11-3**

Hemoglobin molecule
A hemoglobin molecule consists of four highly folded polypeptide chains (the globin portion) and four iron-containing heme groups.

Polypeptide chain

Polypeptide chain

Polypeptide chain

Polypeptide chain

Heme groups

1. *Carbon dioxide.* Hemoglobin contributes to the transport of this gas from the tissues back to the lungs (see p. 493).
2. *The acidic hydrogen-ion portion (H^+) of ionized carbonic acid,* which is generated at the tissue level from CO_2. Hemoglobin buffers this acid so that it minimally alters the pH of the blood (see p. 577).
3. *Carbon monoxide (CO).* This gas is not normally in the blood but if inhaled, preferentially occupies the O_2-binding sites on hemoglobin, causing carbon monoxide poisoning (see p. 492).
4. *Nitric oxide (NO).* In the lungs, the vasodilator nitric oxide binds to a sulfur atom within the hemoglobin molecule to form SNO. This nitric oxide is released at the tissues, where it relaxes and dilates the local arterioles (see p. 355). This vasodilation helps ensure that the O_2-rich blood can make its vital rounds and also helps stabilize blood pressure.

Therefore, hemoglobin plays the key role in O_2 transport while contributing significantly to CO_2 transport and the pH-buffering capacity of blood. Furthermore, by toting along its own vasodilator hemoglobin aids delivery of the O_2 it is carrying.

Lack of nucleus and organelles

To maximize its hemoglobin content, a single erythrocyte is stuffed with more than 250 million hemoglobin molecules, to the exclusion of almost everything else. (That means each RBC has the capacity to carry over a billion O_2 molecules!) Erythrocytes contain no nucleus, organelles, or ribosomes. These structures are extruded during the cell's development to make room for more hemoglobin. Thus an RBC is mainly a plasma membrane–enclosed sac full of hemoglobin.

Key erythrocyte enzymes

Only a few crucial, nonrenewable enzymes remain within a mature erythrocyte: These are glycolytic enzymes and carbonic anhydrase. The **glycolytic enzymes** are necessary for generating the energy needed to fuel the active transport mechanisms involved in maintaining proper ionic concentrations within the cell. Ironically, even though erythrocytes are the vehicles for transporting O_2 to all other tissues of the body, for energy production they themselves cannot use the O_2 they are carrying. Erythrocytes, lacking the mitochondria that house the enzymes for oxidative phosphorylation, must rely entirely on glycolysis for ATP formation (see p. 35).

The other important enzyme within RBCs, **carbonic anhydrase,** is critical in CO_2 transport. This enzyme catalyzes a key reaction that ultimately leads to the conversion of metabolically produced CO_2 into **bicarbonate ion (HCO_3^-)**, which is the primary form in which CO_2 is transported in the blood. Thus erythrocytes contribute to CO_2 transport in two ways—by means of its carriage on hemoglobin and its carbonic anhydrase-induced conversion to HCO_3^-.

▌ The bone marrow continuously replaces worn-out erythrocytes.

Each of us has a total of 25 to 30 trillion RBCs streaming through our blood vessels at any given time (100,000 times more than the entire U.S. population!) Yet these vital gas-transport vehicles are short-lived and must be replaced at the average rate of 2 to 3 million cells per second.

Erythrocytes' short life span

The price RBCs pay for their generous content of hemoglobin to the exclusion of the usual specialized intracellular machinery is a shortened life span. Without DNA and RNA, red blood cells cannot synthesize proteins for cellular repair, growth, and division or for renewal of enzyme supplies. Equipped only with initial supplies synthesized before extrusion of their nucleus, organelles, and ribosomes, RBCs survive an average of only 120 days, in contrast to nerve and muscle cells, which last a person's entire life. During its short life span of four months, each erythrocyte travels about 700 miles as it circulates through the vasculature.

As a red blood cell ages, its nonreparable plasma membrane becomes fragile and prone to rupture as the cell squeezes through tight spots in the vascular system. Most old RBCs meet their final demise in the **spleen,** because this organ's narrow, winding capillary network is a tight fit for these fragile cells. The spleen lies in the upper left part of the abdomen. In addition to removing most of the old erythrocytes from circulation, the spleen has a limited ability to store healthy erythrocytes in its pulpy interior; serves as a reservoir site for platelets; and contains an abundance of lymphocytes, a type of white blood cell.

Erythropoiesis

Because erythrocytes cannot divide to replenish their own numbers, the old ruptured cells must be replaced by new cells produced in an erythrocyte factory—the **bone marrow**—which is the soft, highly cellular tissue that fills the internal cavities of bones. The bone marrow normally generates new red blood cells, a process known as **erythropoiesis,** at a rate to keep pace with the demolition of old cells.

During intrauterine development, erythrocytes are produced first by the yolk sac and then by the developing liver and spleen, until the bone marrow is formed and takes over erythrocyte production exclusively. In children, most bones are filled with **red bone marrow** that is capable of blood cell production. As a person matures, however, fatty **yellow bone marrow** that is incapable of erythropoiesis gradually replaces red marrow, which remains only in a few isolated places, such as the sternum (breastbone), ribs, and upper ends of the long limb bones.

Red marrow not only produces RBCs but also is the ultimate source for leukocytes and platelets as well. Undifferentiated **pluripotent stem cells** reside in the red marrow, where they continuously divide and differentiate to give rise to each of the types of blood cells (see p. 8 and p. 400). These stem cells, the source of all blood cells, have now been isolated. The search has been arduous, because stem cells represent less than 0.1% of all cells in the bone marrow. Although a great deal of research remains to be done, this recent discovery may hold a key to a cure for a host of blood and immune disorders as well as some types of cancer and genetic diseases.

The different types of immature blood cells, along with the stem cells, are intermingled in the red marrow at various stages of development. Once mature, the blood cells are released into the rich supply of capillaries that permeate the red marrow. Regulatory factors act on the *hemopoietic* ("blood-producing") red marrow to govern the type and number of cells generated and discharged into the blood. Of the blood cells, the mechanism for regulating RBC production is the best understood. We will consider it next.

▮ Erythropoiesis is controlled by erythropoietin from the kidneys.

The number of circulating erythrocytes normally remains fairly constant, indicative that erythropoiesis must be closely regulated. Because O_2 transport in the blood is the erythrocytes' primary function, you might logically suspect that the primary stimulus for increased erythrocyte production would be reduced O_2 delivery to the tissues. You would be correct, but low O_2 levels do not stimulate erythropoiesis by acting directly on the red bone marrow. Instead, reduced O_2 delivery to the kidneys stimulates them to secrete the hormone **erythropoietin** into the blood, and this hormone in turn stimulates erythropoiesis by the bone marrow (● Figure 11-4).

Erythropoietin acts on derivatives of undifferentiated stem cells that are already committed to becoming RBCs, stimulating their proliferation and maturation into mature erythrocytes. This increased erythropoietic activity elevates the number of circulating RBCs, thereby increasing the O_2-carrying capacity of the blood and restoring O_2 delivery to the tissues to normal. Once normal O_2 delivery to the kidneys is achieved, erythropoietin secretion is turned down until needed once again. In this way, erythrocyte production is normally balanced against destruction or loss of these cells so that O_2-carrying capacity in the blood remains fairly constant. In response to severe loss of RBCs, as in hemorrhage or abnormal destruction of young circulating erythrocytes, the rate of erythropoiesis can be increased to more than six times the normal level. (For a discussion of erythropoietin abuse by some athletes, see the accompanying boxed feature, ▶ A Closer Look at Exercise Physiology.)

The preparation of an erythrocyte for its departure from the marrow involves several steps, such as synthesis of hemoglobin and extrusion of the nucleus and organelles. Cells clos-

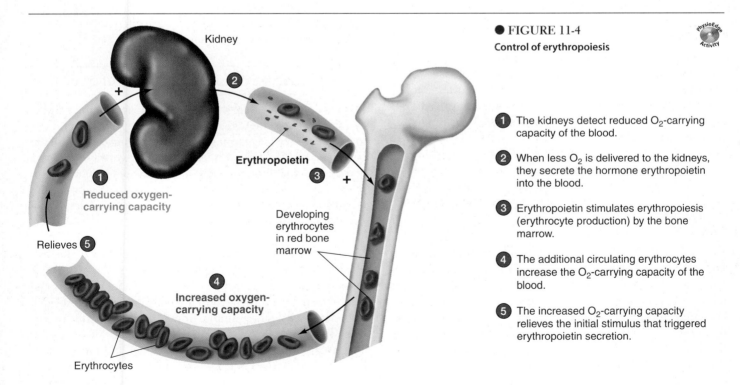

● **FIGURE 11-4**

Control of erythropoiesis

Kidney

Erythropoietin

① Reduced oxygen-carrying capacity

Developing erythrocytes in red bone marrow

Relieves ⑤

Increased oxygen-carrying capacity

Erythrocytes

① The kidneys detect reduced O_2-carrying capacity of the blood.

② When less O_2 is delivered to the kidneys, they secrete the hormone erythropoietin into the blood.

③ Erythropoietin stimulates erythropoiesis (erythrocyte production) by the bone marrow.

④ The additional circulating erythrocytes increase the O_2-carrying capacity of the blood.

⑤ The increased O_2-carrying capacity relieves the initial stimulus that triggered erythropoietin secretion.

Blood Doping: Is More of a Good Thing Better?

Exercising muscles require a continual supply of O_2 for generating energy to sustain endurance activities (see p. 277). **Blood doping** is a technique designed to temporarily increase the O_2-carrying capacity of the blood in an attempt to gain a competitive advantage. Blood doping involves removing blood from an athlete, then promptly reinfusing the plasma but freezing the RBCs for reinfusion one to seven days before a competitive event. One to four units of blood (one unit equals 450 ml) are usually withdrawn at three- to eight-week intervals before the competition. In the periods between blood withdrawals, increased erythropoietic activity restores the RBC count to a normal level.

Reinfusion of the stored RBCs temporarily increases the red blood cell count and hemoglobin level above normal. Theoretically, blood doping would benefit endurance athletes by improving the blood's O_2-carrying capacity. If too many red cells were infused, however, performance could suffer because the increased blood viscosity would decrease blood flow.

Research indicates that, in a standard laboratory exercise test, athletes who have used blood doping may realize a 5% to 13% increase in aerobic capacity; a reduction in heart rate during exercise compared to the rate during the same exercise in the absence of blood doping; improved performance; and reduced lactic acid levels in the blood. (Lactic acid is produced when muscles resort to less efficient anaerobic glycolysis for energy production—see page 279.)

Blood doping, although probably effective, is illegal in both collegiate athletics and Olympic competition for ethical and medical reasons. Of concern, as with the use of any banned performance-enhancing product, is the loss of fair competition. Furthermore, the practice has even been implicated in the deaths of some athletes. The prohibitive regulations are very difficult to enforce, however. Current procedures permit only the testing of urine, not blood, for the use of illegal drugs among athletes. Blood doping cannot be detected by means of a urine sample. The only way of exposing the practice of blood doping is through witnesses or self-admission.

The recent development of synthetic erythropoietin exacerbates the problem of blood doping. Injection of this product stimulates RBC production and thus temporarily increases the O_2-carrying capacity of the blood. Rigorous studies have demonstrated that synthetic erythropoietin may improve an endurance athlete's performance by 7% to 10%. Using erythropoietin is more convenient for the athlete than blood doping and minimizes the number of witnesses involved, thereby reducing the probability of disclosure. Although formally banned, erythropoietin is not detectable using the technology that sports officials currently have available to catch drug cheats.

Erythropoietin is now widely used among competitors in cycling, cross-country skiing, and long-distance running and swimming. This practice is ill advised, however, not only because of legal and ethical implications but because of the dangers of increasing blood viscosity. Synthetic erythropoietin is believed responsible for the deaths of 20 European cyclists since 1987. Unfortunately, too many athletes are willing to take the risks.

---

est to maturity need a few days to be "finished off" and released into the blood in response to erythropoietin, and less developed and newly proliferated cells may take up to several weeks to reach maturity. Therefore, the time required for complete replacement of lost RBCs depends on how many are needed to return the number to normal. (When you donate blood, your circulating erythrocyte supply is replenished in less than a week.)

Reticulocytes

When demands for RBC production are high (for example, following hemorrhage), the bone marrow may release large numbers of immature erythrocytes, known as **reticulocytes,** into the blood to quickly meet the need. These immature cells can be recognized by staining techniques that make visible the residual ribosomes and organelle remnants that have not yet been extruded. Their presence above the normal level of 0.5% to 1.5% of the total number of circulating erythrocytes indicates a high rate of erythropoietic activity. At very rapid rates, more than 30% of the circulating red cells may be at the immature reticulocyte stage.

Synthetic erythropoietin

The gene that directs erythropoietin synthesis has been identified, so this hormone can now be produced in a laboratory. Laboratory-produced erythropoietin has become biotechnology's current single biggest money maker, with sales exceeding $1 billion annually. This hormone is often used to boost RBC production in patients with suppressed erythropoietic activity, such as those undergoing chemotherapy for cancer. (Chemotherapy drugs interfere with the rapid cell division characteristic of both

cancer cells and developing RBCs.) Furthermore, the ready availability of this hormone has profoundly diminished the need for blood transfusions. For example, transfusion of a surgical patient's own predonated blood, coupled with administration of erythropoietin to stimulate further RBC production, has reduced the use of donor blood by as much as 50% in some hospitals. (See the boxed feature on pp. 398–99, ▶ Concepts, Challenges, and Controversies, for an update on other alternatives to whole-blood transfusions under investigation.)

▌ Anemia can be caused by a variety of disorders.

Despite control measures, O_2-carrying capacity cannot always be maintained to meet tissue needs. The term **anemia** refers to a reduction below normal in the O_2-carrying capacity of the blood and is characterized by a low hematocrit (● Figure 11-5a and b). It can be brought about by a decreased rate of erythropoiesis, excessive losses of erythrocytes, or a deficiency in the hemoglobin content of erythrocytes. The various causes of anemia can be grouped into six categories:

1. **Nutritional anemia** is caused by a dietary deficiency of a factor needed for erythropoiesis. The production of RBCs depends on an adequate supply of essential raw ingredients, some of which are not synthesized in the body but must be provided by dietary intake. For example, **iron deficiency anemia** occurs when not enough iron is available for synthesis of hemoglobin. The usual numbers of RBCs are produced, but they contain less hemoglobin than normal and can transport less O_2.

2. **Pernicious anemia** is caused by an inability to absorb adequate amounts of ingested vitamin B_{12} from the digestive tract. Vitamin B_{12} is essential for normal RBC production and maturation. It is found abundantly in a variety of foods. The problem is a deficiency of **intrinsic factor,** a special substance secreted by the lining of the stomach (see p. 611). Vitamin B_{12} can be absorbed from the intestinal tract only when this nutrient is bound to intrinsic factor. When intrinsic factor is deficient, insufficient amounts of ingested vitamin B_{12} are absorbed. The resulting impairment of RBC production and maturation leads to anemia.

3. **Aplastic anemia** is caused by failure of the bone marrow to produce adequate numbers of RBCs, even though all ingredients necessary for erythropoiesis are available. Reduced erythropoietic capability can be caused by destruction of red bone marrow by toxic chemicals (such as benzene or certain drugs, especially chloramphenicol), heavy exposure to radiation (fallout from a nuclear bomb explosion, for example, or excessive exposure to x rays), invasion of the marrow by cancer cells, or chemotherapy for cancer. The destructive process may selectively reduce the marrow's output of erythrocytes, or it may reduce the productive capability for leukocytes and platelets as well. The anemia's severity depends on the extent to which erythropoietic tissue is destroyed, with severe losses being fatal.

4. **Renal anemia** may be a consequence of kidney disease. Because erythropoietin from the kidneys is the primary stimulus for promoting erythropoiesis, inadequate erythropoietin secretion as a result of kidney disease leads to insufficient RBC production.

5. **Hemorrhagic anemia** is caused by the loss of substantial quantities of blood. The loss can be either acute, such as a bleeding wound, or chronic, such as excessive menstrual flow.

6. **Hemolytic anemia** is caused by the rupture of excessive numbers of circulating erythrocytes. **Hemolysis,** the rupture of RBCs, occurs either because the cells are defective, as in sickle cell disease, or because otherwise normal cells are induced to rupture by external factors. **Sickle cell disease** is the best-known example among various hereditary abnormalities of erythrocytes that make these cells very fragile. It affects about 1 in 650 African Americans. In this condition, a defective type of hemoglobin polymerizes or joins together to form rigid chains that cause the RBC to become stiff and assume an unnatural crescent or sickle shape (● Figure 11-6). These deformed RBCs tend to clump together and can block blood flow through small vessels, thus leading to tissue damage. Furthermore, the defective erythrocytes are fragile and prone to rupture, even as young cells, as they travel through the narrow splenic capillaries. Despite an accelerated rate of erythropoiesis triggered by the constant excessive loss of RBCs, production may not be able to keep pace with the rate of destruction, and anemia may result.

Normally formed cells can also hemolyze prematurely if attacked by various external factors. An example is **malaria,** which is caused by protozoan parasites introduced into a victim's blood by the bite of a carrier mosquito. These parasites selectively invade RBCs, where they multiply to the point that

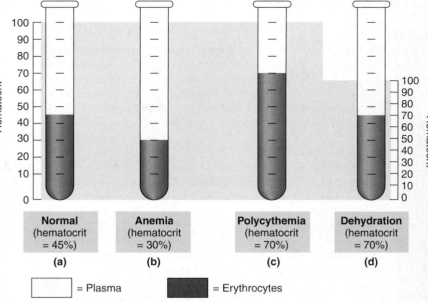

= Plasma   = Erythrocytes

● **FIGURE 11-5**

Hematocrit under various circumstances
(a) Normal hematocrit. (b) The hematocrit is lower than normal in anemia, because of too few circulating erythrocytes, and (c) above normal in polycythemia, because of excess circulating erythrocytes. (d) The hematocrit can also be elevated in dehydration when the normal number of circulating erythrocytes is concentrated in a reduced plasma volume.

the mass of malarial organisms ruptures the cells, releasing hundreds of new active parasites that quickly invade other RBCs. As this cycle continues and more erythrocytes are destroyed, the anemic condition progressively worsens.

▌Polycythemia is an excess of circulating erythrocytes.

Polycythemia, in contrast to anemia, is characterized by an excess of circulating RBCs and an elevated hematocrit (● Figure 11-5c). There are two general types of polycythemia, de-

● **FIGURE 11-6**
Sickle-shaped red blood cell

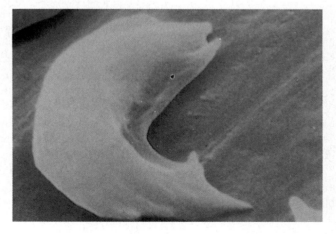

In Search of a Blood Substitute

One of the hottest medical contests of the last two decades has been the race to develop a universal substitute for human blood that is safe, inexpensive, and disease free, and that has a long shelf life.

Need for a blood substitute

A blood transfusion is administered on average every 3 seconds in the United States alone. With only about 5% of the population now donating blood, regional shortages of particular blood types necessitate the shipping and sharing of blood among areas. Medical personnel anticipate serious widespread shortages of blood in the near future, because the number of blood donors continues to decline at the same time that the number of senior citizens, the group of people who most often need transfusions, continues to grow. The benefits for society of a safe blood substitute that could be administered without regard for the recipient's blood type are great, as will be the profits for the manufacturer of the first successful product. Experts estimate that the world market for a good blood substitute may be as much as $10 billion per year.

The search for an alternative to whole-blood transfusions was given new impetus by the rising incidence of AIDS and the concomitant concern over the safety of the nation's blood supply. Infectious diseases such as AIDS and viral hepatitis can be transmitted from infected blood donors to recipients of blood transfusions. Fur-

thermore, restrictions have been placed on potential donors who lived or traveled in Europe during the time mad cow disease hit the beef industry. Although careful screening of our blood supply minimizes the possibility that infectious diseases will be transmitted through transfusion, the public remains wary and would welcome a safe substitute.

Eliminating the risk of disease transmission is only one advantage of finding an alternative to whole-blood transfusion. Whole blood must be kept refrigerated, and even then it has a shelf life of only 42 days. Also, transfusion of whole blood requires blood typing and cross matching, which cannot be done at the scene of an accident or on a battlefield.

Major approaches

The goal is not to find a replacement for whole blood but to duplicate its O_2-carrying capacity. The biggest need for blood transfusions is to replace acute blood loss in accident victims, surgical patients, and wounded soldiers. These individuals require short-term replenishment of blood's O_2-carrying capacity until their own bodies synthesize replacement erythrocytes. The many other important elements in blood are not as immediately critical in sustaining life as is the hemoglobin within the RBCs. Problematically, red blood cells are the whole blood component that requires refrigeration, has a short shelf life, and bears the markers for the various blood types.

Therefore, the search for a blood substitute has focused on two major possibilities: (1) hemoglobin products that exist outside an RBC and can be stored at room temperature for up to six months to a year, and (2) chemically synthesized products that serve as artificial hemoglobin by dissolving large amounts of O_2 when O_2 levels are high (as in the lungs) and releasing it when O_2 levels are low (as in the tissues). A variety of potential blood substitutes are in various stages of development. Some have reached the stage of clinical trials, but no products have yet reached the market, although they are getting close. Let us examine each of these major approaches.

Hemoglobin products

By far the greatest number of research efforts have focused on manipulating the structure of hemoglobin so that it can be effectively and safely administered as a substitute for whole-blood transfusions. If appropriately stabilized and suspended in saline solution, hemoglobin could be injected to bolster O_2-carrying capacity of the recipients' blood no matter what their blood type. The following strategies are among those being pursued to develop a hemoglobin product:

● One problem is that hemoglobin behaves differently when it is outside RBCs. "Naked" hemoglobin splits into halves that do not release O_2 for tissue use as readily as normal hemoglo-

pending on the circumstances triggering the excess RBC production: primary polycythemia and secondary polycythemia.

Primary polycythemia is caused by a tumorlike condition of the bone marrow in which erythropoiesis proceeds at an excessive, uncontrolled rate instead of being subject to the normal erythropoietin regulatory mechanism. The RBC count may reach 11 million cells/mm^3 (normal is 5 million cells/mm^3), and the hematocrit may be as high as 70% to 80% (normal is 42% to 45%). No benefit is derived from the extra O_2-carrying capacity of the blood, because O_2 delivery is more than adequate with normal RBC numbers. Inappropriate polycythemia has adverse effects, however. The excessive number of red cells increases the blood's viscosity up to five to seven times normal, causing the blood to flow very sluggishly, which may actually reduce O_2 delivery to the tissues. The increased viscosity also increases the total peripheral resistance, which may elevate the blood pressure, thus increasing the workload of the heart, unless blood-pressure control mechanisms are able to compensate.

Secondary polycythemia, in contrast, is an appropriate erythropoietin-induced adaptive mechanism to improve the blood's O_2-carrying capacity in response to a prolonged reduction in O_2 delivery to the tissues. It occurs normally in people

living at high altitudes, where less O_2 is available in the atmospheric air, or people in whom O_2 delivery to the tissues is impaired by chronic lung disease or cardiac failure. The red cell count in secondary polycythemia is usually lower than that in primary polycythemia, typically averaging 6 to 8 million cells/mm^3. The price paid for improved O_2 delivery is an increased viscosity of the blood.

An elevated hematocrit can occur when the body loses fluid but not erythrocytes, as in dehydration accompanying heavy sweating or profuse diarrhea (● Figure 11-5d). This is not a true polycythemia, however, because the number of circulating RBCs is not increased. A normal number of erythrocytes is simply concentrated in a smaller plasma volume. This condition is sometimes termed **relative polycythemia**.

LEUKOCYTES

Leukocytes (white blood cells or **WBCs)** are the mobile units of the body's immune defense system. **Immunity** is the body's ability to resist or eliminate potentially harmful foreign materials or abnormal cells. The leukocytes and their derivatives

bin does. Also, these hemoglobin fragments can cause kidney damage. A cross-binding reagent has been developed that keeps hemoglobin molecules intact when they are outside the confines of red blood cells, thus surmounting one major obstacle to administering free hemoglobin.

- Some products under investigation are derived from outdated, donated human blood. Instead of discarding the blood, its hemoglobin is extracted, purified, sterilized, and chemically stabilized. However, this strategy relies on the continued practice of collecting human blood donations.

- Several products use cows' blood as a starting point. Bovine hemoglobin is readily available from slaughterhouses, is cheap, and can be treated for administration to humans. A big concern with these products is the potential of introducing into humans unknown disease-causing microbes that might be lurking in the bovine products.

- A potential candidate as a blood substitute is genetically engineered hemoglobin that bypasses the ongoing need for human donors or the risk of spreading disease from cows to humans. Genetic engineers can insert the gene for human hemoglobin into bacteria, which act as a "factory" to produce the desired hemoglobin product. A drawback for genetically engineered hemoglobin is the high cost involved in operating the facilities.

- One promising strategy encapsulates hemoglobin within liposomes—membrane-wrapped containers—similar to real hemoglobin-stuffed, plasma membrane-enclosed RBCs. These so-called *neo red cells* await further investigation.

Synthetic O_2 carriers

Other researchers are pursuing the development of chemical-based strategies that rely on *perfluorocarbons (PFCs)*, which are synthetic O_2-carrying compounds. PFCs are completely inert, chemically synthesized molecules that can dissolve large quantities of O_2 in direct proportion to the amount of O_2 breathed in. Because they are derived from a nonbiological source, PFCs cannot transmit disease. This, coupled with their low cost, makes them attractive as a blood substitute. Yet use of PFCs is not without risk. Their administration can cause flulike symptoms, and because of poor excretion, they may be retained and accumulate in the body. Ironically, PFC administration poses the danger of causing O_2 toxicity by delivering too much O_2 to the tissues in uncontrolled fashion (see p. 495).

Tactics to reduce need for donated blood

Other tactics besides blood substitutes aimed toward reducing the need for donated blood include the following:

- By changing surgical practices, the medical community has reduced the need for transfusions. These blood-saving methods include recycling a patient's own blood during surgery (collecting lost blood, then reinfusing it); using less invasive and therefore less bloody surgical techniques; and treating the patient with blood-enhancing erythropoietin prior to surgery.

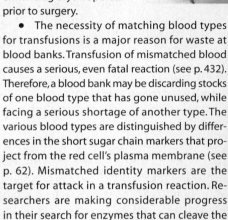

- The necessity of matching blood types for transfusions is a major reason for waste at blood banks. Transfusion of mismatched blood causes a serious, even fatal reaction (see p. 432). Therefore, a blood bank may be discarding stocks of one blood type that has gone unused, while facing a serious shortage of another type. The various blood types are distinguished by differences in the short sugar chain markers that project from the red cell's plasma membrane (see p. 62). Mismatched identity markers are the target for attack in a transfusion reaction. Researchers are making considerable progress in their search for enzymes that can cleave the identity markers away from RBCs, thus converting them all into a type that could be safely transfused into anyone. Such a product would reduce the current wastage.

- Other investigators are seeking ways to prolong the life of RBCs, either in a blood bank or in patients, thus reducing the need for fresh, transfusable blood.

As this list of strategies attests, considerable progress has been made toward developing a safe, effective alternative to whole-blood transfusions. Yet after more than two decades of intense effort, considerable challenges remain, and no ideal solution has been found.

---

(1) defend against invasion by **pathogens** (disease-causing micro-organisms such as bacteria and viruses) by phagocytizing (see p. 32) the foreigners or causing their destruction by more subtle means, (2) identify and destroy cancer cells that arise within the body, and (3) function as a "cleanup crew" that removes the body's "litter" by phagocytizing debris resulting from dead or injured cells. The latter is essential for wound healing and tissue repair.

▊ Leukocytes primarily function as defense agents outside the blood.

To carry out their functions, the leukocytes largely use a "seek out and attack" strategy; that is, they go to sites of invasion or tissue damage. The main reason WBCs are present in the blood is to be rapidly transported from their site of production or storage to wherever they are needed. Therefore, although the specific circulating leukocytes are introduced here to round out our discussion of blood, we will leave a more detailed discussion of their phagocytic and immunologic functions, which take place primarily in the tissues, for the next chapter.

▊ There are five types of leukocytes.

Leukocytes lack hemoglobin (in contrast to erythrocytes) so they are colorless (that is, "white") unless specifically stained for microscopic visibility. Unlike erythrocytes, which are of uniform structure, identical function, and constant number, leukocytes vary in structure, function, and number. There are five different types of circulating leukocytes—neutrophils, eosinophils, basophils, monocytes, and lymphocytes—each with a characteristic structure and function. They are all somewhat larger than erythrocytes.

The five types of leukocytes fall into two main categories, depending on the appearance of their nuclei and the presence or absence of granules in their cytoplasm when viewed microscopically (● Figure 11-7). Neutrophils, eosinophils, and basophils are categorized as **polymorphonuclear** ("many-shaped nucleus") **granulocytes** ("granule-containing cells"). Their nuclei are segmented into several lobes of varying shapes, and their cytoplasm contains an abundance of membrane-enclosed granules. The three types of granulocytes are distinguished on the basis of the varying affinity of their granules for dyes: *eosino-*

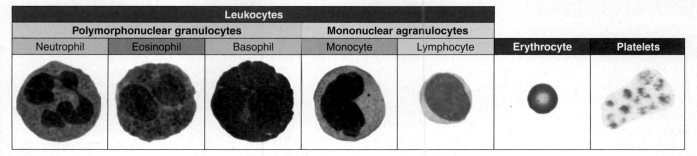

Leukocytes					Erythrocyte	Platelets
Polymorphonuclear granulocytes			Mononuclear agranulocytes			
Neutrophil	Eosinophil	Basophil	Monocyte	Lymphocyte		

● FIGURE 11-7

Normal blood cellular elements

phils have an affinity for the red dye eosin, *basophils* preferentially take up a basic blue dye, and *neutrophils* are neutral, showing no dye preference. Monocytes and lymphocytes are known as **mononuclear** ("single nucleus") **agranulocytes** ("cells lacking granules"). Both have a single, large, nonsegmented nucleus and few granules. *Monocytes* are the larger of the two and have an oval or kidney-shaped nucleus. *Lymphocytes*, the smallest of the leukocytes, characteristically have a large spherical nucleus that occupies most of the cell.

▌ Leukocytes are produced at varying rates depending on the changing defense needs of the body.

All leukocytes ultimately originate from the same undifferentiated pluripotent stem cells in the red bone marrow that also give rise to erythrocytes and platelets (● Figure 11-8). The cells destined to become leukocytes eventually differentiate into various committed cell lines and proliferate under the influence of appropriate stimulating factors. Granulocytes and monocytes are produced only in the bone marrow, which releases these mature leukocytes into the blood. Lymphocytes are originally derived from precursor cells in the bone marrow, but most new lymphocytes are actually produced by lymphocytes already residing in the **lymphoid** (lymphocyte-containing) **tissues,** such as the lymph nodes and tonsils.

The total number of leukocytes normally ranges from 5 to 10 million cells per milliliter of blood, with an average of 7 million cells/ml, expressed as an average **white blood cell count** of 7,000/mm^3. Leukocytes are the least numerous of the cellular elements in the blood (about 1 white blood cell for every 700 red blood cells), not because fewer are produced but because they are merely in transit while in the blood. Normally, approximately two-thirds of the circulating leukocytes are granulocytes, mostly neutrophils, whereas one-third are agranulocytes, predominantly lymphocytes (▲ Table 11-2). However, the total number of white cells and the percentage of each type may vary considerably to meet changing defense needs. Depending on the type and extent of assault the body is combating, different types of leukocytes are selectively produced at varying rates. Chemical messengers arising from invaded or damaged tissues or from activated leukocytes themselves govern the rates of production of the various leukocytes. Specific

● FIGURE 11-8

Blood cell production (hemopoiesis)

All the blood cell types ultimately originate from the same undifferentiated pluripotent stem cells in the bone marrow.

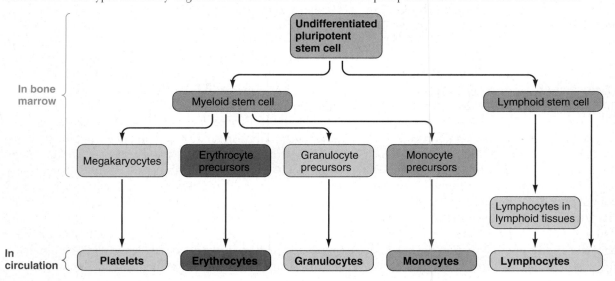

TABLE 11-2
Typical Human Blood Cell Count

Total Erythrocytes = 5,000,000,000 cells/ml blood
Red blood cell (RBC) count = 5,000,000/mm^3

Total Leukocytes = 7,000,000 cells/ml blood
White blood cell (WBC) count = 7,000/mm^3

Differential White Blood Cell Count
(percentage distribution of types of leukocytes)

Polymorphonuclear granulocytes		Mononuclear agranulocytes	
Neutrophils	60–70%	Lymphocytes	25–33%
Eosinophils	1–4%	Monocytes	2–6%
Basophils	0.25–0.5%		

Total Platelets = 250,000,000/ml blood
Platelet count = 250,000/mm^3

---

hormones analogous to erythropoietin are required to direct the differentiation and proliferation of each cell type. Some of these hormones have been identified and can be produced in the laboratory; an example is **granulocyte colony–stimulating factor,** which stimulates increased replication and release of granulocytes, especially neutrophils, from the bone marrow. This achievement has opened up the ability to administer these hormones as a powerful new therapeutic tool to bolster a person's normal defense against infection or cancer.

Functions and life spans of leukocytes

Among the granulocytes, **neutrophils** are phagocytic specialists. They invariably are the first defenders on the scene of bacterial invasion and, accordingly, are very important in inflammatory responses. Furthermore, they scavenge to clean up debris. As might be expected in view of these functions, an increase in circulating neutrophils (**neutrophilia**) typically accompanies acute bacterial infections. In fact, a differential WBC count (a determination of the proportion of each type of leukocyte present) can be useful in making an immediate, reasonably accurate prediction of whether an infection, such as pneumonia or meningitis, is of bacterial or viral origin. Obtaining a definitive answer as to the causative agent by culturing a sample of the infected tissue's fluid takes several days. Because an elevated neutrophil count is highly indicative of bacterial infection, it is appropriate to initiate antibiotic therapy long before the true causative agent is actually known. (Bacteria generally succumb to antibiotics, whereas viruses do not.)

Eosinophils are specialists of another type. An increase in circulating eosinophils (**eosinophilia**) is associated with allergic conditions (such as asthma and hay fever) and with internal parasite infestations (for example, worms). Eosinophils obviously cannot engulf a much larger parasitic worm, but they do attach to the worm and secrete substances that kill it.

Basophils are the least numerous and most poorly understood of the leukocytes. They are quite similar structurally and functionally to **mast cells,** which never circulate in the blood but instead are dispersed in the connective tissue throughout the body. It was once believed that basophils became mast cells by migrating from the circulatory system, but researchers have shown that basophils arise from the bone marrow, whereas mast cells are derived from precursor cells located in the connective tissue. Both basophils and mast cells synthesize and store *histamine* and *heparin*, powerful chemical substances that can be released on appropriate stimulation. Histamine release is important in allergic reactions, whereas heparin hastens the removal of fat particles from the blood after a fatty meal. Heparin can also prevent blood clotting (coagulation), but whether it plays a physiologic role as an anticoagulant is still being debated.

Once released into the blood from the bone marrow, a granulocyte usually remains in transit in the blood for less than a day before leaving the blood vessels to enter the tissues, where it survives another three to four days unless it dies sooner in the line of duty.

Among the agranulocytes, **monocytes,** like neutrophils, are destined to become professional phagocytes. They emerge from the bone marrow while still immature and circulate for only a day or two before settling down in various tissues throughout the body. At their new residences, monocytes continue to mature and greatly enlarge, becoming the large tissue phagocytes known as **macrophages** (*macro* means "large;" *phage* means "eater"). A macrophage's life span may range from months to years unless it is destroyed sooner while performing its phagocytic activity. A phagocytic cell can ingest only a limited amount of foreign material before it succumbs.

Lymphocytes provide immune defense against targets for which they are specifically programmed. There are two types of lymphocytes, B lymphocytes and T lymphocytes. **B lymphocytes** produce **antibodies,** which circulate in the blood. An antibody binds with and marks for destruction (by phagocytosis or other means) the specific kinds of foreign matter, such as bacteria, that induced production of the antibody. **T lymphocytes** do not produce antibodies; instead they directly destroy their specific target cells by releasing chemicals that punch holes in the victim cell. This process is known as a **cell-mediated immune response.** The target cells of T lymphocytes include body cells invaded by viruses and cancer cells. Lymphocytes have life spans estimated at 100 to 300 days. During this period, the majority of them continually recycle among the lymphoid tissues, lymph, and blood, spending only a few hours at a time in the blood. Therefore, only a small proportion of the total lymphocytes are in transit in the blood at any given moment.

Abnormalities in leukocyte production

Even though levels of circulating leukocytes may vary, changes in these levels are normally controlled and adjusted according to the body's needs. However, abnormalities in leukocyte production can occur that are not subject to control; that is, either too few or too many WBCs can be produced. The bone marrow can greatly slow down or even stop its production of white blood cells when it is exposed to certain toxic physical agents (such as radiation) or chemical agents (such as benzene and anticancer drugs). The most serious consequence is the reduction in professional phagocytes (neutrophils and macrophages), which leads to a notable reduction in the body's defense capabilities against invading micro-organisms. The only defense still available when the bone marrow fails is the immune capabilities of the lymphocytes produced by the lymphoid organs.

In **infectious mononucleosis,** not only does the number of lymphocytes (but not other leukocytes) in the blood increase, but many of the lymphocytes are atypical in structure. This condition, which is caused by the *Epstein-Barr virus*, is characterized by pronounced fatigue, a mild sore throat, and low-grade fever. Full recovery usually requires a month or more.

Surprisingly, one of the major consequences of **leukemia,** a cancerous condition that involves uncontrolled proliferation of WBCs, is inadequate defense capabilities against foreign invasion. In leukemia, the WBC count may reach as high as 500,000/mm^3, compared with the normal 7,000/mm^3, but because the majority of these cells are abnormal or immature, they cannot perform their normal defense functions. Another devastating consequence of leukemia is displacement of the other blood cell lines in the bone marrow. This results in anemia because of a reduction in erythropoiesis and in internal bleeding because of a deficit of platelets. Platelets play a critical role in preventing bleeding from the myriad tiny breaks that normally occur in small blood vessel walls. Consequently, overwhelming infections or hemorrhage are the most common causes of death in leukemic patients. The next section will examine the platelets' role in greater detail to show how they normally minimize the threat of hemorrhage.

PLATELETS AND HEMOSTASIS

In addition to erythrocytes and leukocytes, **platelets (thrombocytes)** are a third type of cellular element present in the blood.

▌ Platelets are cell fragments derived from megakaryocytes.

Platelets are not whole cells but small cell fragments (about 2 to 4 μm in diameter) that are shed off the outer edges of extraordinarily large (up to 60 μm in diameter) bone marrow–bound cells known as **megakaryocytes** (● Figure 11-9). A single megakaryocyte typically produces about 1,000 platelets. Megakaryocytes are derived from the same undifferentiated stem cells that give rise to the erythrocytic and leukocytic cell lines. Platelets are essentially detached vesicles containing portions of megakaryocyte cytoplasm wrapped in plasma membrane.

An average of 250,000,000 platelets are normally present in each milliliter of blood (range of 150,000 to 350,000/mm^3). Platelets remain functional for an average of 10 days, at which time they are removed from circulation by the tissue macrophages, especially those in the spleen and liver, and are replaced by newly released platelets from the bone marrow. The hormone **thrombopoietin** produced by the liver increases the number of megakaryocytes in the bone marrow and stimulates

each megakaryocyte to produce more platelets. The factors that control thrombopoietin secretion and regulate the platelet level are currently under investigation.

Platelets do not leave the blood as WBCs do, but at any given time about one-third of them are in storage in blood-filled spaces in the spleen. These stored platelets can be released from the spleen into the circulating blood as needed (for example, during hemorrhage) by sympathetically induced splenic contraction.

Because platelets are cell fragments, they lack nuclei. However, they are equipped with organelles and cytosolic enzyme systems for generating energy and synthesizing secretory products, which they store in numerous granules dispersed throughout the cytosol. Furthermore, platelets contain high concentrations of actin and myosin, which enable them to contract. Their secretory and contractile abilities are important in hemostasis, a topic to which we now turn.

▌ Hemostasis prevents blood loss from damaged small vessels.

Hemostasis is the arrest of bleeding from a broken blood vessel—that is, the stopping of hemorrhage. (Be sure not to confuse this with the term *homeostasis*.) For bleeding to take place from a vessel, there must be a break in the vessel wall, and the pressure inside the vessel must be greater than the pressure outside it to force the blood out through the defect. The body's inherent hemostatic mechanisms normally are adequate to seal defects and stop loss of blood through small damaged capillaries, arterioles, and venules. These small vessels are frequently ruptured by the minor traumas of everyday life; such traumas are the most common source of bleeding, although we often are not even aware that any damage has taken place. The hemostatic mechanisms normally keep blood loss from these minor vascular traumas to a minimum.

● **FIGURE 11-9**
Photomicrograph of a megakaryocyte shedding off platelets

Platelets   Megakaryocyte

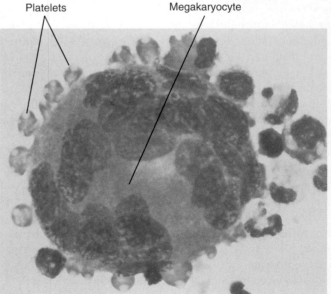

The much rarer occurrence of bleeding from medium- to large-size vessels usually cannot be stopped by the body's hemostatic mechanisms alone. Bleeding from a severed artery is more profuse and therefore more dangerous than venous bleeding, because the outward driving pressure is greater in the arteries (that is, arterial blood pressure is considerably higher than venous pressure). First-aid measures for a severed artery include application over the wound of external pressure of greater magnitude than the arterial blood pressure to temporarily halt the bleeding until the torn vessel can be surgically closed. Hemorrhage from a traumatized vein can often be stopped simply by elevating the bleeding body part to reduce gravity's effects on pressure in the vein (see p. 372). If the accompanying drop in venous pressure is not sufficient to stop the bleeding, mild external compression is usually adequate.

Hemostasis involves three major steps: (1) *vascular spasm*, (2) *formation of a platelet plug*, and (3) *blood coagulation (clotting)*. Platelets play a pivotal role in hemostasis. They obviously play a major part in forming a platelet plug, but they contribute significantly to the other two steps as well.

Vascular spasm reduces blood flow through an injured vessel.

A cut or torn blood vessel immediately constricts as a result of both an inherent vascular response to injury and sympathetically induced vasoconstriction. This constriction, or **vascular spasm,** slows blood flow through the defect and thus minimizes blood loss. Also, as the opposing endothelial (inner) surfaces of the vessel are pressed together by this initial vascular spasm, they become sticky and adhere to each other, further sealing off the damaged vessel. These physical measures alone cannot completely prevent further blood loss, but they minimize blood flow through the break in the vessel until the other hemostatic measures are able to actually plug up the defect.

Platelets aggregate to form a plug at a vessel defect.

Platelets normally do not adhere to the smooth endothelial surface of blood vessels, but when this lining is disrupted because of vessel injury, platelets become activated by the exposed collagen, which is a fibrous protein present in the underlying connective tissue. When activated, platelets quickly adhere to the collagen and form a hemostatic **platelet plug** at the site of the defect. Once platelets start aggregating, they release several important chemicals from their storage granules (● Figure 11-10). Among these chemicals is *adenosine diphosphate (ADP)*, which causes the surface of nearby circulating platelets to become sticky, so that they adhere to the first layer of aggregated platelets. These newly aggregated platelets release more ADP, which causes more platelets to pile on, and so on; thus a plug of platelets is rapidly built up at the defect site, in a positive-feedback fashion.

Given the self-perpetuating nature of platelet aggregation, why doesn't the platelet plug continue to develop and expand over the surface of the adjacent normal vessel lining? A key reason why this does not happen is that ADP and other chemicals

● **FIGURE 11-10**

Formation of a platelet plug
Platelets aggregate at a vessel defect through a positive-feedback mechanism involving the release of adenosine diphosphate (ADP) from platelets, which stick to exposed collagen at the site of the injury. Platelets are prevented from aggregating at the adjacent normal vessel lining by the release of prostacyclin and nitric oxide from the undamaged endothelial cells.

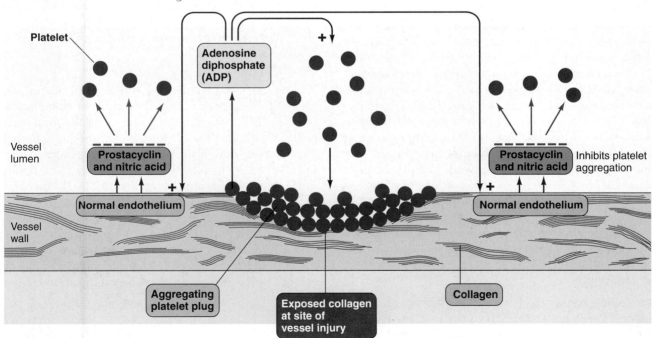

released by the activated platelets stimulate the release of *prostacyclin* and *nitric oxide* from the adjacent normal endothelium. Both these chemicals profoundly inhibit platelet aggregation. Thus the platelet plug is limited to the defect and does not spread to the nearby undamaged vascular tissue (● Figure 11-10).

The aggregated platelet plug not only physically seals the break in the vessel but also performs three other important roles. (1) The actin-myosin complex within the aggregated platelets contracts to compact and strengthen what was originally a fairly loose plug. (2) The chemicals released from the platelet plug include several powerful vasoconstrictors (serotonin, epinephrine, and thromboxane A_2), which induce profound constriction of the affected vessel to reinforce the initial vascular spasm. (3) The platelet plug releases other chemicals that enhance blood coagulation, the next step of hemostasis. Although the platelet-plugging mechanism alone is often sufficient to seal the myriad minute tears in capillaries and other small vessels that occur many times daily, larger holes in vessels require the formation of a blood clot to completely stop the bleeding.

Clot formation results from a triggered chain reaction involving plasma clotting factors.

Blood coagulation, or **clotting,** is the transformation of blood from a liquid into a solid gel. Formation of a clot on top of the platelet plug strengthens and supports the plug, reinforcing the seal over a break in a vessel. Furthermore, as blood in the vicinity of the vessel defect solidifies, it can no longer flow. Clotting is the body's most powerful hemostatic mechanism. It is required to stop bleeding from all but the most minute defects.

Clot formation

The ultimate step in clot formation is the conversion of **fibrinogen,** a large, soluble plasma protein produced by the liver and normally always present in the plasma, into **fibrin,** an insoluble, threadlike molecule. The conversion into fibrin is catalyzed by the enzyme **thrombin** at the site of the vessel injury. Fibrin molecules adhere to the damaged vessel surface, forming a loose, netlike meshwork that traps the cellular elements of the blood, including aggregating platelets. The resultant mass, or **clot,** typically appears red because of the abundance of trapped RBCs, but the foundation of the clot is formed from fibrin derived from the plasma (● Figure 11-11). Except for platelets, which play an important role in ultimately bringing about the conversion of fibrinogen to fibrin, clotting can take place in the absence of all other cellular elements in the blood.

The original fibrin web is rather weak, because the fibrin strands are only loosely interlaced. However, chemical linkages rapidly form between adjacent strands to strengthen and stabilize the clot meshwork. This cross-linkage process is catalyzed by a clotting factor known as **factor XIII (fibrin-stabilizing factor),** which normally is present in the plasma in inactive form.

Roles of thrombin

Thrombin, in addition to (1) converting fibrinogen into fibrin, also (2) activates factor XIII to stabilize the resultant fibrin mesh, (3) acts in a positive-feedback fashion to facili-

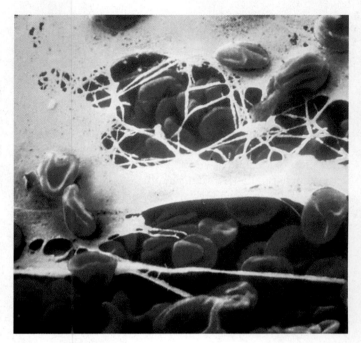

● FIGURE 11-11
Erythrocytes trapped in the fibrin meshwork of a clot

tate its own formation, and (4) enhances platelet aggregation, which in turn is essential to the clotting process (● Figure 11-12).

Because thrombin's action converts the ever-present fibrinogen molecules in the plasma into a blood-stanching clot, thrombin must normally be absent from the plasma except in the vicinity of vessel damage. Otherwise, blood would *always* be coagulated—a situation incompatible with life. How can thrombin normally be absent from the plasma, yet be readily available to trigger fibrin formation when a vessel is injured? The solution lies in thrombin's existence in the plasma in the form of an inactive precursor called **prothrombin.** What converts prothrombin into thrombin when blood clotting is desirable? This conversion involves the clotting cascade.

The clotting cascade

Yet another activated plasma clotting factor, **factor X,** is responsible for converting prothrombin to thrombin; factor X itself is normally present in the blood in inactive form and must be converted into its active form by still another activated factor, and so on. Altogether, 12 plasma clotting factors participate in essential steps that lead to the final conversion of fibrinogen into a stabilized fibrin mesh (● Figure 11-13). These factors are designated by roman numerals in the order in which they were discovered, not the order in which they participate in the clotting process.[1] Most of these clotting factors are plasma proteins synthesized by the liver. One consequence of liver disease is that clotting time is prolonged by reduced production of clotting factors. Normally, these factors are always present in

---

[1]The term *factor VI* is no longer used. What once was considered a separate factor VI has now been determined to be an activated form of factor V.

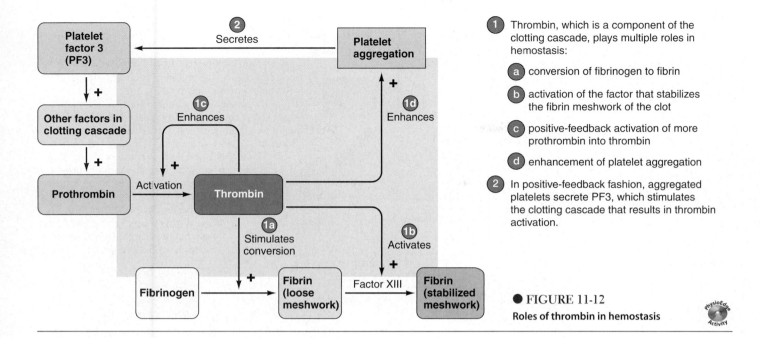

1 Thrombin, which is a component of the clotting cascade, plays multiple roles in hemostasis:

- **a** conversion of fibrinogen to fibrin

- **b** activation of the factor that stabilizes the fibrin meshwork of the clot

- **c** positive-feedback activation of more prothrombin into thrombin

- **d** enhancement of platelet aggregation

2 In positive-feedback fashion, aggregated platelets secrete PF3, which stimulates the clotting cascade that results in thrombin activation.

● **FIGURE 11-12**
Roles of thrombin in hemostasis

---

the plasma in an inactive form, similar to fibrinogen and prothrombin. In contrast to fibrinogen, which is converted into insoluble fibrin strands, prothrombin and the other precursors, when converted to their active form, act as proteolytic (protein-splitting) enzymes. These enzymes activate another specific factor in the clotting sequence. Once the first factor in the sequence is activated, it in turn activates the next factor, and so on, in a series of sequential reactions known as the **clotting cascade,** until thrombin catalyzes the final conversion of fibrinogen into fibrin. Several of these steps require the presence of plasma Ca^{2+} and *platelet factor 3 (PF3)*, a phospholipid secreted by the aggregated platelet plug. Thus, platelets also contribute to clot formation.

Intrinsic and extrinsic pathways

The clotting cascade may be triggered by the *intrinsic pathway* or the *extrinsic pathway*:

- The **intrinsic pathway** precipitates clotting within damaged vessels as well as clotting of blood samples in test tubes. All elements necessary to bring about clotting by means of the intrinsic pathway are present in the blood. This pathway, which involves seven separate steps (shown in blue in ● Figure 11-13), is set off when **factor XII (Hageman factor)** is activated by coming into contact with either exposed collagen in an injured vessel or a foreign surface such as a glass test tube. Remember that exposed collagen also initiates platelet aggregation. Thus formation of a platelet plug and the chain reaction leading to clot formation are simultaneously set in motion when a vessel is damaged. Furthermore, these complementary hemostatic mechanisms reinforce each other. The aggregated platelets secrete PF3, which is essential for the clotting cascade that in turn enhances further platelet aggregation (● Figures 11-12 and 11-14).

- The **extrinsic pathway** takes a shortcut and requires only four steps (shown in gray in ● Figure 11-13). This pathway, which requires contact with tissue factors external to the blood, initiates clotting of blood that has escaped into the tissues. When a tissue is traumatized, it releases a protein complex known as **tissue thromboplastin.** Tissue thromboplastin directly activates factor X, thereby bypassing all preceding steps of the intrinsic pathway. From this point on, the two pathways are identical.

The intrinsic and extrinsic mechanisms usually operate simultaneously. When tissue injury involves rupture of vessels, the intrinsic mechanism stops blood in the injured vessel, whereas the extrinsic mechanism clots the blood that escaped into the tissue before the vessel was sealed off. Typically, clots are fully formed in three to six minutes.

Clot retraction

Once a clot is formed, contraction of the platelets trapped within the clot shrinks the fibrin mesh, pulling the edges of the damaged vessel closer together. During **clot retraction,** fluid is squeezed from the clot. This fluid, which is essentially plasma minus fibrinogen and other clotting precursors that have been removed during the clotting process, is called **serum.**

Amplification during clotting process

Although a clotting process that involves so many steps may seem inefficient, the advantage is the amplification accomplished during many of the steps. One molecule of an activated factor can activate perhaps a hundred molecules of the next factor in the sequence, each of which can activate many more molecules of the next factor, and so on. In this way, large numbers of the final factors involved in clotting are rapidly activated as a result of the initial activation of only a few molecules in the beginning step of the sequence.

How then is the clotting process, once initiated, confined to the site of vessel injury? If the activated clotting factors were allowed to circulate, they would induce inappropriate widespread clotting that would plug up vessels throughout the body.

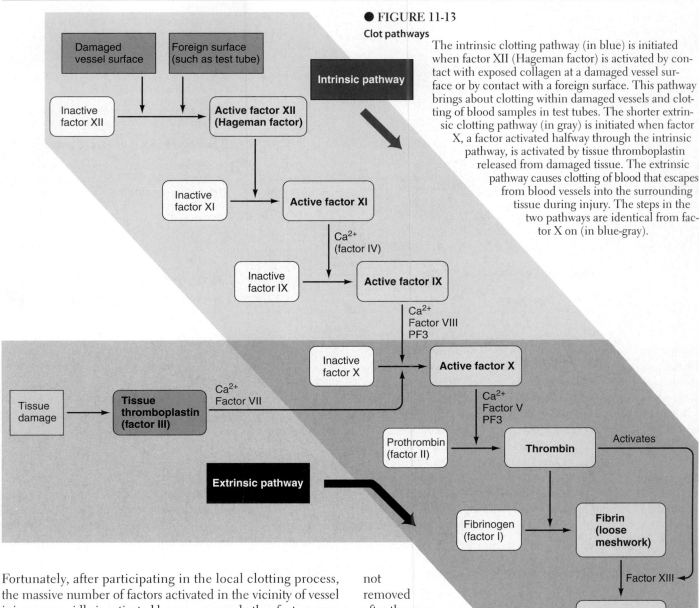

● FIGURE 11-13
Clot pathways

The intrinsic clotting pathway (in blue) is initiated when factor XII (Hageman factor) is activated by contact with exposed collagen at a damaged vessel surface or by contact with a foreign surface. This pathway brings about clotting within damaged vessels and clotting of blood samples in test tubes. The shorter extrinsic clotting pathway (in gray) is initiated when factor X, a factor activated halfway through the intrinsic pathway, is activated by tissue thromboplastin released from damaged tissue. The extrinsic pathway causes clotting of blood that escapes from blood vessels into the surrounding tissue during injury. The steps in the two pathways are identical from factor X on (in blue-gray).

Fortunately, after participating in the local clotting process, the massive number of factors activated in the vicinity of vessel injury are rapidly inactivated by enzymes and other factors present in the plasma or tissue.

▌ Fibrinolytic plasmin dissolves clots.

A clot is not meant to be a permanent solution to vessel injury. It is a transient device to stop bleeding until the vessel can be repaired.

Vessel repair

The aggregated platelets secrete a chemical that is at least partially responsible for the invasion of fibroblasts ("fiber formers") from the surrounding connective tissue into the wounded area of the vessel. Fibroblasts form a scar at the vessel defect.

Clot dissolution

Simultaneous with the healing process, the clot, which is no longer needed to prevent hemorrhage, is slowly dissolved by a fibrinolytic (fibrin-splitting) enzyme called **plasmin.** If clots were

not removed after they performed their hemostatic function, the vessels, especially the small ones that endure tiny ruptures on a regular basis, would eventually become obstructed by clots.

Plasmin, like the clotting factors, is a plasma protein produced by the liver and present in the blood in an inactive precursor form, **plasminogen.** Plasmin is activated in a cascade fashion by many factors, among them factor XII (Hageman factor), which also triggers the chain reaction leading to clot formation (● Figure 11-15). When a clot is being formed, activated plasmin becomes trapped in the clot and subsequently dissolves it by slowly breaking down the fibrin meshwork.

Phagocytic white blood cells gradually remove the products of clot dissolution. You have observed the slow removal of blood that has clotted after escaping into the tissue layers of

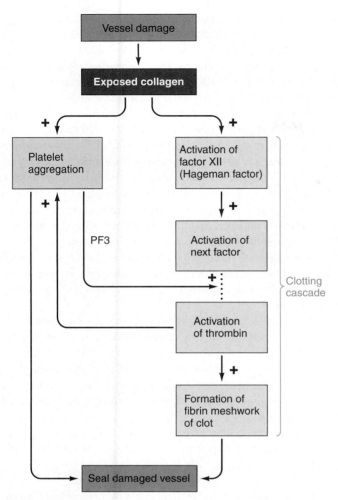

● FIGURE 11-14

Concurrent platelet aggregation and clot formation
Exposed collagen at the site of vessel damage simultaneously initiates platelet aggregation and the clotting cascade. These two hemostatic mechanisms positively reinforce each other as they seal the damaged vessel.

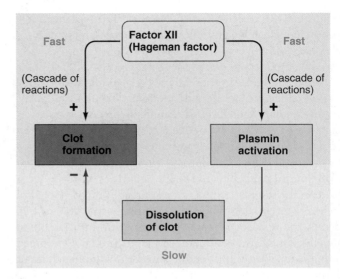

● FIGURE 11-15

Role of factor XII in clot formation and dissolution
Activation of factor XII (Hageman factor) simultaneously initiates a fast cascade of reactions that result in clot formation and a fast cascade of reactions that result in plasmin activation. Plasmin, which is trapped in the clot, subsequently slowly dissolves the clot. This action removes the clot when it is no longer needed after the vessel has been repaired.

your skin following an injury. The black-and-blue marks of such bruised skin result from deoxygenated clotted blood within the skin; this blood is eventually cleared by plasmin action, followed by the phagocytic cleanup crew.

Prevention of inappropriate clot formation

In addition to removing clots that are no longer needed, plasmin functions continually to prevent clots from forming inappropriately. Throughout the vasculature, small amounts of fibrinogen are constantly being converted into fibrin, triggered by unknown mechanisms. Clots do not develop, however, because the fibrin is quickly disposed of by plasmin that is activated by **tissue plasminogen activator (tPA)** derived from the tissues, especially the lungs. Normally, the low level of fibrin formation is counterbalanced by a low level of fibrinolytic activity, so inappropriate clotting does not occur. Only when a vessel is damaged, additional factors precipitate the explosive chain reaction that leads to more extensive fibrin formation and results in local clotting at the site of injury.

Naturally occurring anticoagulants that interfere with one or more steps in the clotting process exist in the body. Whether they play a physiologic role in clot prevention is still being debated. The best known of these is **heparin,** which is present in basophils and mast cells. It is used extensively as an anticoagulant drug as well as to prevent clotting of blood samples drawn for clinical analysis.

Genetically engineered tPA and other similar chemicals that trigger clot dissolution are frequently used to limit damage to cardiac muscle during heart attacks. Administration of a clot-busting drug within the first hours after a clot has blocked a coronary (heart) vessel often dissolves the clot in time to restore blood flow to the cardiac muscle supplied by the blocked vessel before the muscle dies of O_2 deprivation. In recent years, tPA and related drugs have also been used successfully to dissolve a stroke-causing clot within a cerebral (brain) blood vessel, thus minimizing the loss of irreplaceable brain tissue following a stroke.

▌ Inappropriate clotting is responsible for thromboembolism.

Despite protective measures, clots occasionally form in intact vessels. Abnormal or excessive clot formation within blood vessels—what has been dubbed "hemostasis in the wrong place"— can compromise blood flow to vital organs. The body's clotting and anticlotting systems normally function in a check-and-balance manner. Acting in concert, they permit prompt formation of "good" blood clots, thus minimizing blood loss from damaged vessels, while preventing "bad" clots from forming and blocking blood flow in intact vessels. An abnormal intravascular clot attached to a vessel wall is known as a **thrombus**, and freely floating clots are called **emboli.** An enlarging thrombus narrows and can eventually completely occlude the vessel in which it forms. By entering and completely plugging a smaller vessel, a circulating embolus can suddenly block blood flow.

Several factors, acting independently or simultaneously, can cause *thromboembolism:* (1) Roughened vessel surfaces associated with atherosclerosis can lead to thrombus formation (see p. 335). (2) Imbalances in the clotting–anticlotting systems can trigger clot formation. (3) Slow-moving blood is more apt to clot, probably because small quantities of fibrin accumulate in the stagnant blood, for example, in blood pooled in varicosed leg veins. (4) Widespread clotting is occasionally triggered by the release of tissue thromboplastin into the blood from large amounts of traumatized tissue. Similar widespread clotting can occur in **septicemic shock,** in which bacteria or their toxins initiate the clotting cascade.

▮ Hemophilia is the primary condition responsible for excessive bleeding.

In contrast to inappropriate clot formation in intact vessels, the opposite hemostatic disorder is failure of clots to form promptly in injured vessels, resulting in life-threatening hemorrhage from even relatively mild traumas. The most common cause of excessive bleeding is **hemophilia,** which is caused by a deficiency of one of the factors in the clotting cascade. Although a deficiency of any of the clotting factors could block the clotting process, 80% of all hemophiliacs lack the genetic ability to synthesize factor VIII.

In contrast to the more profuse bleeding that accompanies defects in the clotting mechanism, people with a deficiency of platelets continuously develop hundreds of small, confined hemorrhagic areas throughout the body tissues as blood is permitted to leak from tiny breaks in the small blood vessels before coagulation takes place. Platelets normally are the primary sealers of these ever-occurring minute ruptures. In the skin of a platelet-deficient person, the diffuse capillary hemorrhages are visible as small, purplish blotches, giving rise to the term **thrombocytopenia purpura** ("the purple of thrombocyte deficiency") for this condition. (Recall that the term *thrombocytes* is an alternate name for platelets.)

Vitamin K deficiency can also cause a bleeding tendency. Vitamin K, commonly known as the blood-clotting vitamin, is essential for normal clot formation. Researchers recently figured out vitamin K's role in the clotting process. In a complex sequence of biochemical events, vitamin K combines with O_2, releasing free energy that is ultimately used in the activation processes of the clotting cascade.

CHAPTER IN PERSPECTIVE: FOCUS ON HOMEOSTASIS

Blood contributes to homeostasis in a variety of ways. First, the composition of the interstitial fluid, the true internal environment that surrounds and directly exchanges materials with the cells, depends on the composition of the blood plasma. Because of the thorough

exchange that occurs between the interstitial and vascular compartments, the interstitial fluid has the same composition as the plasma with the exception of plasma proteins, which cannot escape through the capillary walls. Thus the blood serves as the vehicle for rapid, long-distance, mass transport of materials to and from the cells, and the interstitial fluid serves as the go-between.

Homeostasis depends on the blood carrying materials such as O_2 and nutrients to the cells as rapidly as the cells consume these supplies and carrying materials such as metabolic wastes away from the cells as rapidly as the cells produce these products. It also depends on the blood carrying hormonal messengers from their site of production to their distant site of action. Once a substance enters the blood, it can be transported throughout the body within seconds, whereas diffusion of the substance over long distances in a large multicellular organism such as a human would take months to years—a situation incompatible with life. Diffusion can, however, effectively accomplish short local exchanges of materials between the blood and surrounding cells through the intervening interstitial fluid.

The blood has special transport capabilities that enable it to move its cargo efficiently throughout the body. For example, life-sustaining O_2 is poorly soluble in water, but the blood is equipped with O_2-carrying specialists, the erythrocytes (red blood cells), which are stuffed full of hemoglobin, a complex molecule that transports O_2. Likewise, homeostatically important water-insoluble hormonal messengers are shuttled in the blood by plasma protein carriers.

Specific components of the blood perform the following additional homeostatic activities that are unrelated to blood's transport function:

- Blood helps maintain the proper pH in the internal environment by buffering changes in the body's acid–base load.
- Blood helps maintain body temperature by absorbing heat produced by heat-generating tissues such as contracting skeletal muscles and distributing it throughout the body. Excess heat is carried by the blood to the body surface for elimination to the external environment.
- Electrolytes in the plasma are important in membrane excitability, which is critical for nerve and muscle function.
- Electrolytes in the plasma are important in the osmotic distribution of fluid between the extracellular and intracellular fluid. Plasma proteins play a critical role in distributing extracellular fluid between the plasma and interstitial fluid.
- Through their hemostatic functions, the platelets and clotting factors minimize the loss of life-sustaining blood following vessel injury.
- The leukocytes (white blood cells), their secretory products, and certain types of plasma proteins, such as antibodies, constitute the immune defense system. This system defends the body against invading disease-causing agents, destroys cancer cells, and paves the way for wound healing and tissue repair by clearing away debris from dead or injured cells. These actions indirectly contribute to homeostasis by helping keep the organs that directly maintain homeostasis healthy.

CHAPTER SUMMARY

Introduction

- Blood consists of three types of cellular elements—erythrocytes (red blood cells), leukocytes (white blood cells), and platelets (thrombocytes)—suspended in the liquid plasma.

- The 5- to 5.5-liter volume of blood in an adult consists of 42% to 45% erythrocytes, less than 1% leukocytes and platelets, and 55% to 58% plasma. The percentage of whole-blood volume occupied by erythrocytes is known as the *hematocrit*.

Plasma

- Plasma is a complex liquid consisting of 90% water that serves as a transport medium for substances being carried in the blood.

- The most abundant inorganic constituents in plasma are Na^+ and Cl^-. The most plentiful organic constituents in plasma are plasma proteins.

- All plasma constituents are freely diffusible across the capillary walls except the plasma proteins, which remain in the plasma, where they perform a variety of important functions. Plasma proteins include the albumins, globulins, and fibrinogen.

Erythrocytes

- Erythrocytes are specialized for their primary function of O_2 transport in the blood. They do not contain a nucleus, organelles, or ribosomes but instead are packed full of hemoglobin, an iron-containing molecule that can loosely and reversibly bind with O_2. Because O_2 is poorly soluble in blood, hemoglobin is indispensable for O_2 transport.

- Hemoglobin also contributes to CO_2 transport and buffering of blood by reversibly binding with CO_2 and H^+.

- Unable to replace cell components, erythrocytes are destined to a short life span of about 120 days.

- Undifferentiated pluripotent stem cells in the red bone marrow give rise to all cellular elements of the blood. Erythrocyte production (erythropoiesis) by the marrow normally keeps pace with the rate of erythrocyte loss, keeping the red cell count constant. Erythropoiesis is stimulated by erythropoietin, a hormone secreted by the kidneys in response to reduced O_2 delivery.

Leukocytes

- Leukocytes are the defense corps of the body. They attack foreign invaders, destroy abnormal cells that arise in the body, and clean up cellular debris.

- Each of the five types of leukocytes has a different task: (1) Neutrophils, the phagocytic specialists, are important in engulfing bacteria and debris. (2) Eosinophils specialize in attacking parasitic worms and play a key role in allergic responses. (3) Basophils release two chemicals: histamine, which is also important in allergic responses, and heparin, which helps clear fat particles from the blood. (4) Monocytes, on leaving the blood, set up residence in the tissues and greatly enlarge to become the large tissue phagocytes known as *macrophages*. (5) Lymphocytes provide immune defense against bacteria, viruses, and other targets for which they are specifically programmed. Their defense tools include the production of antibodies (for B lymphocytes) and cell-mediated immune responses (for T lymphocytes).

- Leukocytes are present in the blood only while in transit from their site of production and storage in the bone marrow (and also in the lymphoid tissues in the case of the lymphocytes) to their site of action in the tissues. At any given time, the majority of the leukocytes are out in the tissues on surveillance missions or performing actual combative activities.

- All leukocytes have a limited life span and must be replenished by ongoing differentiation and proliferation of precursor cells. The total number and percentage of each of the different types of leukocytes produced vary depending on the momentary defense needs of the body.

Platelets and Hemostasis

- Platelets are cell fragments derived from large megakaryocytes in the bone marrow.

- Platelets play a role in hemostasis, the arrest of bleeding from an injured vessel. The three main steps in hemostasis are (1) vascular spasm, (2) platelet plugging, and (3) clot formation.

- Vascular spasm reduces blood flow through an injured vessel.

- Aggregation of platelets at the site of vessel injury quickly plugs the defect. Platelets start to aggregate on contact with exposed collagen in the damaged vessel wall.

- Clot formation reinforces the platelet plug and converts blood in the vicinity of a vessel injury into a nonflowing gel.

- The majority of factors necessary for clotting are always present in the plasma in inactive precursor form. When a vessel is damaged, exposed collagen initiates a cascade of reactions involving successive activation of these clotting factors, ultimately converting fibrinogen into fibrin.

- Fibrin, an insoluble threadlike molecule, is laid down as the meshwork of the clot; the meshwork in turn entangles blood cellular elements to complete clot formation.

- Blood that has escaped into the tissues clots on exposure to tissue thromboplastin, which sets the clotting process into motion.

- When no longer needed, clots are dissolved by plasmin, a fibrinolytic factor also activated by exposed collagen.

REVIEW EXERCISES

Objective Questions (Answers on p. A-43)

1. Blood can absorb metabolic heat while undergoing only small changes in temperature. *(True or false?)*
2. Hemoglobin can carry only O_2. *(True or false?)*
3. Erythrocytes originate from the same undifferentiated stem cells as leukocytes and platelets. *(True or false?)*
4. Erythrocytes are unable to use the O_2 they contain for their own ATP formation. *(True or false?)*
5. White blood cells spend the majority of their time in the blood. *(True or false?)*
6. The body's defense capabilities are reduced in leukemia even though there are an excessive number of leukocytes. *(True or false?)*
7. Which type of leukocyte is produced primarily in lymphoid tissue? _____
8. The majority of plasma clotting factors are synthesized by the _____.
9. Which of the following is *not* a function served by the plasma proteins?
   a. facilitating retention of fluid in the blood vessels
   b. playing an important role in blood clotting

c. binding and transporting certain hormones in the blood
d. transporting O_2 in the blood
e. serving as antibodies
f. contributing to the buffering capacity of the blood

10. Which of the following is *not* directly triggered by exposed collagen in an injured vessel?
   a. initial vascular spasm
   b. platelet aggregation
   c. activation of the clotting cascade
   d. activation of plasminogen

11. Match the following (*an answer may be used more than once*):
   _____ 1. causes platelets to aggregate in positive-feedback fashion
   _____ 2. activates prothrombin
   _____ 3. fibrinolytic enzyme
   _____ 4. inhibits platelet aggregation
   _____ 5. first factor activated in intrinsic clotting pathway
   _____ 6. forms the meshwork of the clot
   _____ 7. stabilizes the clot
   _____ 8. activates fibrinogen
   _____ 9. activated by tissue thromboplastin

   (a) prostacyclin
   (b) plasmin
   (c) ADP
   (d) fibrin
   (e) thrombin
   (f) factor X
   (g) factor XII
   (h) factor XIII

12. Match the following blood abnormalities with their causes:
   _____ 1. deficiency of intrinsic factor
   _____ 2. insufficient amount of iron to synthesize adequate hemoglobin
   _____ 3. destruction of bone marrow
   _____ 4. abnormal loss of blood
   _____ 5. tumorlike condition of bone marrow
   _____ 6. inadequate erythropoietin secretion
   _____ 7. excessive rupture of circulating erythrocytes
   _____ 8. associated with living at high altitudes

   (a) hemolytic anemia
   (b) aplastic anemia
   (c) nutritional anemia
   (d) hemorrhagic anemia
   (e) pernicious anemia
   (f) renal anemia
   (g) primary polycythemia
   (h) secondary polycythemia

Essay Questions

1. What is the average blood volume in women and in men?
2. What is the normal percentage of blood occupied by erythrocytes and by plasma? What is the hematocrit? What is the buffy coat?

3. What is the composition of plasma?
4. List the three major groups of plasma proteins, and state their functions.
5. Describe the structure and functions of erythrocytes.
6. Why can erythrocytes survive for only about 120 days?
7. Describe the process and control of erythropoiesis.
8. Compare the structure and functions of the five types of leukocytes.
9. Discuss the derivation of platelets.
10. Describe the three steps of hemostasis, including a comparison of the intrinsic and extrinsic pathways by which the clotting cascade is triggered.
11. Compare plasma and serum.
12. What normally prevents inappropriate clotting in vessels?

Quantitative Exercises (Solutions on p. A-43)

1. The normal concentration of hemoglobin in blood (as measured clinically) is 15 g/100 ml of blood.
   a. Given that 1 mole of hemoglobin weighs 66,000 grams, what is the concentration of hemoglobin in millimoles (mM)?
   b. Each hemoglobin molecule can bind four molecules of O_2. What is the concentration of O_2 bound to hemoglobin at maximal saturation (in mM)?
   c. Given that 1 mole of an ideal gas occupies 22.4 liters, what is the maximal carrying capacity of normal blood for O_2 (usually expressed in ml of O_2/liter of blood)?

2. Assume that the blood sample in Figure 11-5b, p. 397, is from a patient with hemorrhagic anemia. Given a normal blood volume of 5 liters, a normal red blood cell concentration of 5 billion/ml, and an RBC production rate of 3 million cells/second, how long will it take the body to return the hematocrit to normal?

3. Note that in the blood sample in Figure 11-5c from a patient with polycythemia, the hematocrit has increased to 70%. An increased hematocrit increases blood viscosity, which in turn increases total peripheral resistance and increases the workload on the heart. The effect of hematocrit (h) on relative blood viscosity (v, viscosity relative to that of water) is given approximately by the following equation:

$$v = 1.5 \times \exp(2h)$$

Note that in this equation h is the hematocrit as a fraction, not a percent. Given a normal hematocrit of 0.40, what percent increase in viscosity would result from the polycythemia in Figure 11-5c? What percent change would this cause in total peripheral resistance?

POINTS TO PONDER

(Explanations on p. A-43)

1. A person has a hematocrit of 62. Can you conclude from this finding that the person has polycythemia? Explain.
2. There are different forms of hemoglobin. *Hemoglobin A* is normal adult hemoglobin. The abnormal form *hemoglobin S* causes RBCs to warp into fragile, sickle-shaped cells. Fetal RBCs contain *hemoglobin F*, the production of which stops soon after birth. Now researchers are trying to goad the genes that direct hemoglobin F synthesis back into action as a means of treating sickle cell anemia. Explain how turning on these fetal genes could be a useful remedy. (Indeed, the first effective drug therapy recently approved for the treatment of sickle cell anemia, *hydroxyurea*, acts on the bone marrow to boost the production of fetal hemoglobin.)

3. Low on the list of popular animals are vampire bats, leeches, and ticks, yet these animals may someday indirectly save your life. Scientists are currently examining the "saliva" of these blood-sucking creatures in search of new chemicals that might limit cardiac muscle damage in heart attack victims. What do you suspect the nature of these sought-after chemicals is?
4. With the screening methods currently employed by blood banks, about one out of 225,000 units of blood used for transfusions may harbor HIV, the virus that causes AIDS (see p. 436). HIV-

contaminated blood that slips through the screening process comes largely from recently infected donors. The screening tests now employed only detect antibodies against HIV, which do not appear in the blood for nearly a month following HIV infection. Therefore, a window of about a month exists during which donated blood may be infectious but still pass the screening process. An estimated 10 people are infected with HIV annually by this means. Tests are available for detecting HIV earlier than the blood banks' antibody-based method, such as by screening for the presence of a specific protein on the surface of HIV. These tests are costly, however, and currently are used only for research. If blood banks employed these more sensitive tests, they could detect about half of the HIV-contaminated blood that currently goes to patients. The estimated cost of implementing these more expensive tests is somewhere between $70 million and $200 million. Do you think the health care system should assume this additional financial burden to prevent four or five cases of transfusion-delivered HIV infection per year?

5. *Porphyria* is a genetic disorder that shows up in about one in every 25,000 individuals. Affected individuals lack certain enzymes that are part of a metabolic pathway leading to formation of heme, which is the iron-containing group of hemoglobin. An accumulation of porphyrins, which are intermediates of the pathway, causes a variety of symptoms, especially after exposure to sunlight. Lesions and scars form on the skin. Hair grows thickly on the face and hands. As gums retreat from teeth, the elongated canine teeth take on a fanglike appearance. Symptoms worsen on exposure to a variety of substances, including garlic and alcohol. Affected individuals avoid sunlight and aggravating substances, and get injections of heme from normal red blood cells.

If you are familiar with vampire stories, which date from the Middle Ages or earlier, speculate on how they may have evolved among superstitious folk who did not have medical knowledge of porphyria.

6. *Clinical Consideration.* Linda P. has just been diagnosed as having pneumonia. Her white blood cell count is 7,200/mm^3, with 67% of the white blood cells being neutrophils. It will take several days to obtain a definitive answer as to the causative agent by culturing a sample of discharges from her respiratory system. Based on the WBC count, do you think that Linda should be given antibiotics immediately, long before the causative agent is actually known? Are antibiotics likely to combat her infection?

PHYSIOEDGE RESOURCES

PHYSIOEDGE CD-ROM

Figures marked with the PhysioEdge icon have associated activities on the CD. For this chapter, check out the Media Quizzes related to Figures 11-1, 11-9, 11-10, 11-11, and 11-12.

CHECK OUT THESE MEDIA QUIZZES:
11.1 Blood
11.2 Abnormal Hemostasis
11.3 Thrombocytes and Hemostasis

PHYSIOEDGE WEB SITE

The Web site for this book contains a wealth of helpful study aids, as well as many ideas for further reading and research. Log on to:

http://www.brookscole.com

Go to the Biology page and select Sherwood's *Human Physiology*, 5th Edition. Select a chapter from the drop-down menu or click on one of these resource areas:

- **Case Histories** provide an introduction to the clinical aspects of human physiology. Check out:

    #17: Pneumococcal Bacteremia

- For 2-D and 3-D graphical illustrations and animations of physiological concepts, visit our **Visual Learning Resource.**

- Resources for study and review include the **Chapter Outline, Chapter Summary, Glossary,** and **Flash Cards.** Use our **Quizzes** to prepare for in-class examinations.

- On-line research resources to consult are **Hypercontents,** with current links to relevant Internet sites; **Internet Exercises** with starter URLs provided; and **InfoTrac Exercises.**

  For more readings, go to InfoTrac College Edition, your on-line research library, at:

http://infotrac.thomsonlearning.com

Immune System
Integumentary System (Skin)

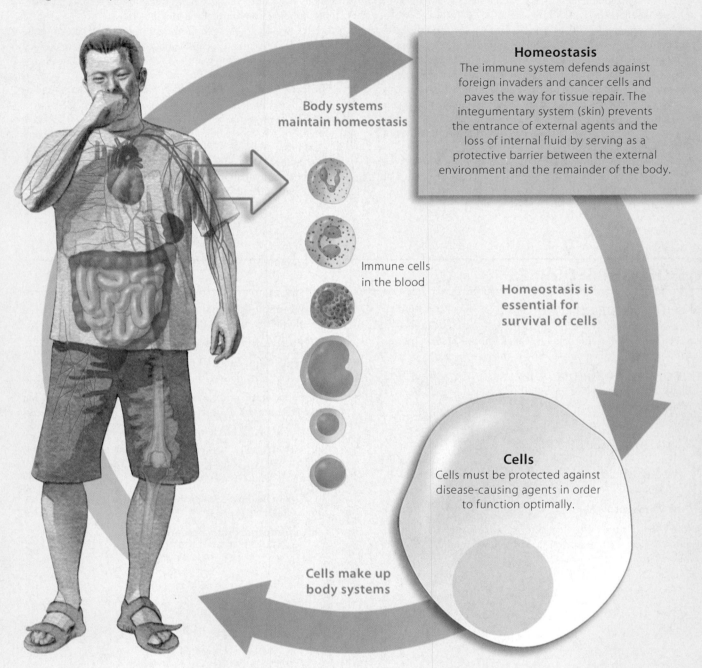

Body systems maintain homeostasis

Immune cells in the blood

Homeostasis
The immune system defends against foreign invaders and cancer cells and paves the way for tissue repair. The integumentary system (skin) prevents the entrance of external agents and the loss of internal fluid by serving as a protective barrier between the external environment and the remainder of the body.

Homeostasis is essential for survival of cells

Cells
Cells must be protected against disease-causing agents in order to function optimally.

Cells make up body systems

Humans are constantly coming into contact with external agents that could be harmful if they entered the body. The most serious are disease-causing micro-organisms. If bacteria or viruses do gain entry, the body is equipped with a complex, multifaceted internal defense system—the **immune system**—that provides continual protection against invasion by foreign agents. The body surfaces that are exposed to the external environment, such as the **skin**, serve as a first line of defense to resist penetration by foreign micro-organisms. The immune system also protects against cancer and paves the way for repair of damaged tissues.

The immune system indirectly contributes to homeostasis by helping maintain the health of organs that directly contribute to homeostasis.

Chapter 12

The Body Defenses

CONTENTS AT A GLANCE

INTRODUCTION

▮ Bacteria and viruses as targets of the immune system

▮ Leukocytes as effector cells of the immune system

▮ Comparison of innate and adaptive immunity

INNATE IMMUNITY

▮ Inflammation

▮ Interferon

▮ Natural killer cells

▮ Complement system

ADAPTIVE IMMUNITY: GENERAL CONCEPTS

▮ Types of adaptive immunity

▮ Antigens

B LYMPHOCYTES: ANTIBODY-MEDIATED IMMUNITY

▮ Role of antibodies

▮ Clonal selection theory

▮ Active, passive, and natural immunity

▮ Antigen-presenting cells

T LYMPHOCYTES: CELL-MEDIATED IMMUNITY

▮ Cytotoxic and helper T cells

▮ Tolerance

▮ Major histocompatibility complex; class I and class II MHC self-antigens

▮ Immune surveillance against cancer

IMMUNE DISEASES

▮ Immune deficiency diseases

▮ Inappropriate immune attacks; allergies

EXTERNAL DEFENSES

▮ Structure and function of skin

▮ Role of the skin in immunity

▮ Protective measures within body cavities that communicate with the external environment

INTRODUCTION

Immunity is the body's ability to resist or eliminate potentially harmful foreign materials or abnormal cells. The following activities are attributable to the **immune system,** an internal defense system which plays a key role in recognizing and either destroying or neutralizing materials within the body that are foreign to the "normal self":

1. Defense against invading **pathogens** (disease-producing micro-organisms such as bacteria and viruses).

2. Removal of "worn-out" cells (such as aged red blood cells) and tissue debris (for example, tissue damaged by trauma or disease). The latter is essential for wound healing and tissue repair.

3. Identification and destruction of abnormal or mutant cells that have originated in the body. This function, termed *immune surveillance*, is the primary internal-defense mechanism against cancer.

4. Inappropriate immune responses that lead either to *allergies*, which occur when the body turns against a normally harmless environmental chemical entity, or to *autoimmune diseases*, which happen when the defense system erroneously produces antibodies against a particular type of the body's own cells.

5. Rejection of tissue cells of foreign origin, which constitutes the major obstacle to successful organ transplantation.

▮ Pathogenic bacteria and viruses are the major targets of the immune system.

The primary foreign enemies against which the immune system defends are bacteria and viruses. **Bacteria** are nonnucleated, single-celled micro-organisms self-equipped with all machinery essential for their own survival and reproduction. Pathogenic bacteria that invade the body cause tissue damage and produce disease largely by releasing enzymes or toxins that physically injure or functionally disrupt affected cells and organs. The disease-producing power of a pathogen is known as its **virulence.**

In contrast to bacteria, **viruses** are not self-sustaining cellular entities. They consist only of nucleic acids (genetic material—DNA or RNA) enclosed by a protein coat. Because they lack cellular machinery for energy production and protein synthesis, viruses cannot carry out metabolism and reproduce unless they invade a **host cell** (a body cell of the infected individual) and take over the cellular biochemical facilities for their own purposes. Not only do viruses sap the host cell's energy resources, but the viral nucleic acids also direct the host cell to synthesize proteins needed for viral replication.

When a virus becomes incorporated into a host cell, the body's own defense mechanisms may destroy the cell, because they no longer recognize it as a "normal self" cell. Other ways in which viruses can lead to cellular damage or death are by depleting essential cellular components, dictating that the cell produce substances that are toxic to the cell, or transforming the cell into a cancer cell.

▌ Leukocytes are the effector cells of the immune system.

The leukocytes (white blood cells, or WBCs) and their derivatives, along with a variety of plasma proteins, are responsible for the different immune defense strategies. As a brief review, the functions of the five types of leukocytes are as follows (see pp. 398–402):

1. **Neutrophils** are highly mobile phagocytic specialists that engulf and destroy unwanted materials.
2. **Eosinophils** secrete chemicals that destroy parasitic worms and are involved in allergic manifestations.
3. **Basophils** release histamine and heparin and also are involved in allergic manifestations.
4. **Monocytes** are transformed into **macrophages**, which are large, tissue-bound phagocytic specialists.
5. **Lymphocytes** are of two types.
   a. **B lymphocytes** are transformed into plasma cells, which secrete antibodies that indirectly lead to the destruction of foreign material.
   b. **T lymphocytes** are responsible for cell-mediated immunity involving direct destruction of virus-invaded cells and mutant cells through nonphagocytic means.

A given leukocyte is present in the blood only transiently. Most leukocytes are out in the tissues, on defense missions. As a result, the immune system's effector cells are widely dispersed throughout the body and can defend in any location.

Lymphoid tissues

Almost all leukocytes originate from common precursor stem cells in the bone marrow and are subsequently released into the blood. The only exception is lymphocytes, which arise in part from lymphocyte colonies in various lymphoid tissues originally populated by cells derived from bone marrow (see p. 400).

Lymphoid tissues, collectively, are the tissues that produce, store, or process lymphocytes. These include the bone marrow, lymph nodes, spleen, thymus, tonsils, adenoids, appendix, and aggregates of lymphoid tissue in the lining of the digestive tract called **Peyer's patches** or **gut-associated lymphoid tissue (GALT)** (● Figure 12-1). Lymphoid tissues are strategically located to intercept invading micro-organisms before they have

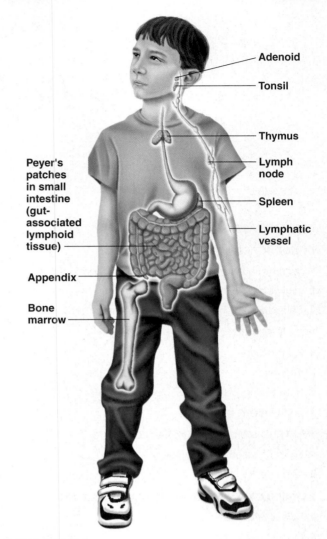

● FIGURE 12-1

Lymphoid tissues
The lymphoid tissues, which are dispersed throughout the body, produce, store, or process lymphocytes.

a chance to spread very far. For example, lymphocytes populating the *tonsils* and *adenoids* are situated advantageously to respond to inhaled microbes, whereas micro-organisms invading through the digestive system immediately encounter lymphocytes in the *appendix* and *GALT*. Potential pathogens that gain access to lymph are filtered through *lymph nodes*, where they are exposed to lymphocytes as well as to macrophages that line lymphatic passageways. The *spleen*, the largest lymphoid tissue, performs immune functions on the blood similar to those lymph nodes perform on lymph. Through actions of its lymphocyte and macrophage population, the spleen clears blood that passes through it of micro-organisms and other foreign matter and also removes worn-out red blood cells (see p. 394). The *thymus* and *bone marrow* play important roles in processing T and B lymphocytes, respectively, to prepare them to carry out their specific immune strategies. ▲ Table 12-1 summarizes the major functions of the various lymphoid tissues, some described in Chapter 11 and others of which are discussed later in this chapter.

▲ **TABLE 12-1**
Functions of Lymphoid Tissues

Lymphoid Tissue	Function
Bone Marrow	Origin of all blood cells
	Site of maturational processing for B lymphocytes
Lymph Nodes, Tonsils, Adenoids, Appendix, Gut-Associated Lymphoid Tissue	Exchange lymphocytes with the lymph (remove, store, produce, and add them)
	Resident lymphocytes produce antibodies and sensitized T cells, which are released into the lymph
	Resident macrophages remove microbes and other particulate debris from the lymph
Spleen	Exchanges lymphocytes with the blood (removes, stores, produces, and adds them)
	Resident lymphocytes produce antibodies and sensitized T cells, which are released into the blood
	Resident macrophages remove microbes and other particulate debris, most notably worn-out red blood cells, from the blood
	Stores a small percentage of red blood cells, which can be added to the blood by splenic contraction as needed
Thymus	Site of maturational processing for T lymphocytes
	Secretes the hormone thymosin

We are now going to turn our attention to the two major components of the immune system's response to foreign invaders and other targets—innate and adaptive immune responses. In the process, we will further examine the roles of each type of leukocyte.

▌ Immune responses may be either innate and nonspecific, or adaptive and specific.

Protective immunity is conferred by the complementary actions of two separate but interdependent components of the immune system: the *innate immune system* and the *adaptive or acquired immune system*. The responses of these two systems differ in their timing and in the selectivity of the defense mechanisms. The **innate immune system** encompasses the body's *nonspecific* immune responses that come into play immediately on exposure to a threatening agent. These nonspecific responses are inherent (innate or built-in) defense mechanisms that nonselectively defend against foreign or abnormal material of any type, even on initial exposure to it. Such responses provide a first line of defense against a wide range of threats, including infectious agents, chemical irritants, and tissue in-jury from mechanical trauma and burns. Everyone is born with essentially the same innate immune-response mechanisms, although there are some subtle genetic differences. The **adaptive** or **acquired immune system,** in contrast, relies on *specific* immune responses selectively targeted against a particular foreign material to which the body has already been exposed and has had an opportunity to prepare for an attack aimed discriminatingly at the enemy. The adaptive immune system thus takes considerably more time to mount and takes on one specific foe at a time. The targets of the adaptive immune system vary among people, depending on the types of immune assaults encountered by each individual. The innate and adaptive immune systems work in harmony to contain, then eliminate, harmful agents.

Innate immune system

The components of the innate system are always on guard, ready to unleash a limited, rather crude, repertoire of defense mechanisms at any and every invader. Of the immune effector cells, the neutrophils and macrophages—both phagocytic specialists—are especially important in innate defense. Several groups of plasma proteins also play key roles, as you will see shortly. The various nonspecific immune responses are set in motion in response to generic molecular patterns associated with threatening agents, such as the carbohydrates typically found in bacterial cell walls but not found in mammalian cells.

The responding phagocytic cells are studded with recently discovered **toll-like receptors** (**TLRs**), a type of plasma membrane protein. One investigator in the field has called TLRs the "eyes of the innate immune system." These immune sensors recognize and bind with the telltale pathogen markers, allowing the effector cells of the innate system to "see" pathogens as distinct from self-cells. A TLR's recognition of a pathogen triggers the phagocyte to engulf and destroy the infectious micro-organism. Moreover, activation of the TLR induces the phagocytic cell to secrete chemicals, some of which contribute to inflammation, an important innate response to microbial invasion.

Still other chemicals secreted by the phagocytes are important in recruiting cells of the adaptive immune system. Thus TLRs link the innate and adaptive branches of the immune system. Furthermore, foreign particles are deliberately marked for phagocytic ingestion by being coated with antibodies produced by the B cells of the adaptive immune system—another link between the innate and adaptive branches. These are but a few examples of how various components of the immune system are highly interactive and interdependent. The most significant cooperative relationships among the immune effectors are pointed out throughout this chapter.

The innate mechanisms provide us with a rapid but limited and nonselective response to unfriendly challenges of all kinds, much like medieval guardsmen lashing out with brute force weapons at any enemy approaching the walls of the castle they are defending. Innate immunity largely contains and limits the spread of infection. These nonspecific responses are important for keeping the foe at bay until the adaptive immune system, with its highly selective weapons, can be prepared to take over and mount strategies to eliminate the villain.

Adaptive immune system

The specific responses of the adaptive immune system are mediated by the B and T lymphocytes. Each B and T lymphocyte can recognize and defend against only one particular type of foreign material, such as one kind of bacterium. Among the millions of B and T lymphocytes in the body, only the ones specifically equipped to recognize the unique molecular features of a particular infectious agent are called into action to discriminatingly defend against this agent, similar to modern, specially trained military personnel called into active duty to accomplish a very specific task. The chosen lymphocytes multiply, expanding the pool of specialists that can launch a highly targeted attack against the invader. The adaptive immune system is our ultimate weapon against most pathogens. The repertoire of activated and expanded B and T cells is constantly changing in response to the various pathogens encountered. Thus the adaptive or acquired immune system adapts to wage battle against the specific pathogens in the person's environment. Furthermore, this system acquires an ability to more efficiently eradicate a particular foe when rechallenged by the same pathogen in the future. It does so by establishing a pool of memory cells as a result of an encounter with a given pathogen, so that on subsequent exposure to the same agent, it can more swiftly defend against the invader.

We will first examine in more detail the innate immune responses before looking more closely at adaptive immunity.

INNATE IMMUNITY

Innate defenses include the following:

1. *Inflammation*, a nonspecific response to tissue injury in which the phagocytic specialists—neutrophils and macrophages—play a major role, along with supportive input from other immune-cell types.

2. *Interferon*, a family of proteins that nonspecifically defend against viral infection.

3. *Natural killer cells*, a special class of lymphocyte-like cells that spontaneously and nonspecifically lyse (rupture) and thereby destroy virus-infected host cells and cancer cells.

4. *The complement system*, a group of inactive plasma proteins that, when sequentially activated, bring about destruction of foreign cells by attacking their plasma membranes.

We will discuss each of these in turn, beginning with inflammation.

▌ Inflammation is a nonspecific response to foreign invasion or tissue damage.

The term **inflammation** refers to an innate, nonspecific series of highly interrelated events that are set into motion in response to foreign invasion, tissue damage, or both. The ultimate goal of inflammation is to bring to the invaded or injured area phagocytes and plasma proteins that can (1) isolate, destroy, or inactivate the invaders; (2) remove debris; and (3) prepare for subsequent healing and repair. The overall inflammatory response is remarkably similar no matter what the triggering event (be it

bacterial invasion, chemical injury, or mechanical trauma), although some subtle differences may be evident, depending on the injurious agent or the site of damage. The following sequence of events typically occurs during the inflammatory response. As an example we will use bacterial entry into a break in the skin.

Defense by resident tissue macrophages

On bacterial invasion through a break in the external barrier of skin, the macrophages already present in the area immediately begin phagocytizing the foreign microbes. Although the resident macrophages are usually not present in sufficient numbers to meet the challenge alone, they defend against infection during the first hour or so, before other mechanisms can be mobilized. Macrophages are usually rather stationary, gobbling debris and contaminants that come their way, but when necessary, they become mobile and migrate to sites of battle against invaders.

Localized vasodilation

Almost immediately on microbial invasion, arterioles within the area dilate, increasing blood flow to the site of injury. This localized vasodilation is primarily induced by histamine released in the area of tissue damage from mast cells, a type of cell found in the connective tissue that is similar to circulating basophils (see p. 335 and p. 401). Increased local delivery of blood brings to the site more phagocytic leukocytes and plasma proteins, both of which are crucial to the defense response.

Increased capillary permeability

Released histamine also increases the capillaries' permeability by enlarging the capillary pores (the spaces between the endothelial cells; see p.362) so that plasma proteins that normally are prevented from leaving the blood can escape into the inflamed tissue (see p. 363).

Localized edema

Accumulation of leaked plasma proteins in the interstitial fluid elevates the local interstitial fluid colloid osmotic pressure. Furthermore, increased capillary blood pressure accompanies the increased local blood flow. Because both these pressures tend to move fluid out of the capillaries, these changes favor enhanced filtration and reduced reabsorption of fluid across the involved capillaries. The end result of this shift in fluid balance is localized edema (see p. 369). Thus the familiar swelling that accompanies inflammation is due to histamine-induced vascular changes. Likewise, the other well-known gross manifestations of inflammation such as redness and heat are largely attributable to the enhanced flow of warm arterial blood to the damaged tissue. Pain is caused both by local distension within the swollen tissue and by the direct effect of locally produced substances on the receptor endings of afferent neurons that supply the area. These observable characteristics of the inflammatory process (swelling, redness, heat, and pain) are coincidental to the primary purpose of the vascular changes in the injured area—to increase the number of leukocytic phagocytes and crucial plasma proteins in the area (● Figure 12-2).

Walling off of the inflamed area

The leaked plasma proteins most critical to the immune response are those involved in the complement system and kinin system (described later) as well as clotting and anticlotting factors (see p. 404). On exposure to tissue thromboplastin in the injured tissue and to specific chemicals secreted by phagocytes on the scene, fibrinogen—the final factor in the clotting system—is converted into fibrin. Fibrin forms interstitial fluid clots in the spaces around the bacterial invaders and damaged cells. This walling off of the injured region from the surrounding tissues prevents or at least delays the spread of bacterial invaders and their toxic products. Later, the more slowly activated anticlotting factors gradually dissolve the clots after they are no longer needed.

Emigration of leukocytes

Within an hour after the injury, the involved area is teeming with leukocytes that have left the vessels. Neutrophils arrive first, followed during the next 8 to 12 hours by the slower-moving monocytes. The latter swell and mature into macrophages during another 8- to 12-hour period. Once neutrophils or monocytes leave the bloodstream, they never recycle back to the blood.

Leukocyte emigration from the blood into the tissues involves the following steps:

• Blood-borne leukocytes, especially neutrophils and monocytes, stick to the inner endothelial lining of capillaries in the affected tissue, a process called **margination.** *Selectins,* a type of cell adhesion molecule (CAM; see p. 61) that protrudes from the inner endothelial lining, cause leukocytes that are flowing by in the bloodstream to slow down and roll along the interior of the vessel, much as the nap of a carpet slows down a child's rolling toy car. This slowing down allows the leukocytes enough time to check for local activating factors—"SOS signals" from nearby injured or infected tissues. When present, these activating factors cause the leukocytes to adhere firmly to the endothelial lining by means of interaction with another type of CAM, the *integrins.*

• Soon the adhered leukocytes start leaving by a mechanism known as **diapedesis.** Assuming *amoeba-like behavior* (see p. 49), an adhered leukocyte pushes a long, narrow projection through a capillary pore; then the remainder of the cell flows forward into the projection (● Figure 12-3). In this way, the leukocyte is able to wriggle its way in about a minute through the capillary pore—even though it is much larger than the pore. Outside the vessel, the leukocyte moves in an amoeboid fashion toward the site of tissue damage and bacterial invasion. Neutrophils arrive on the inflammatory scene earliest because they are more mobile than monocytes.

• Phagocytic cells are guided in their direction of migration by **chemotaxis**; that is, they are attracted to certain chemical mediators, known as *chemotaxins* or *chemokines,* released at the site of damage. Binding of chemotaxins with protein receptors on the plasma membrane of a phagocytic cell increases Ca^{2+} entry into the cell. Calcium, in turn, switches on the cellular contractile apparatus that leads to amoeba-like crawling. Because the concentration of chemotaxins progressively increases toward the site of injury, phagocytic cells move unerringly toward this site along a chemotaxin concentration gradient.

Leukocyte proliferation

Resident tissue macrophages as well as leukocytes that exited from the blood and migrated to the inflammatory site are soon joined by new phagocytic recruits from the bone marrow. Within a few hours after the onset of the inflammatory response, the number of neutrophils in the blood may increase up to four to five times that of normal. This increase is due partly to transfer

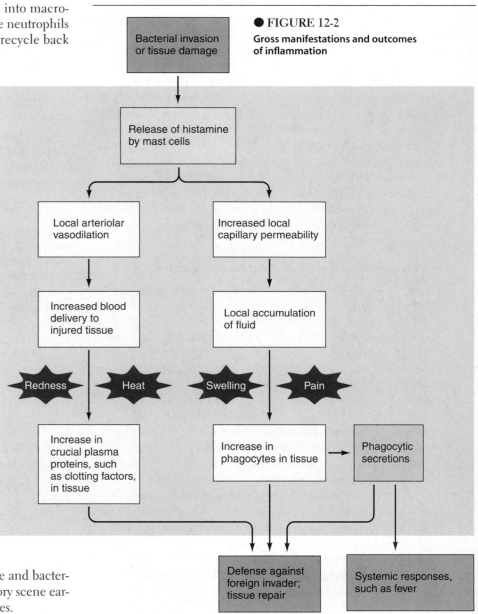

● **FIGURE 12-2**

Gross manifestations and outcomes of inflammation

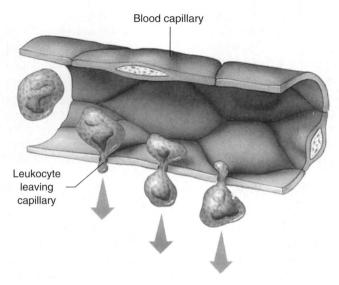

● **FIGURE 12-3**

Leukocyte emigration from the blood
Leukocytes emigrate from the blood into the tissues by assuming amoeba-like behavior and squeezing through the capillary pores, a process known as *diapedesis*.

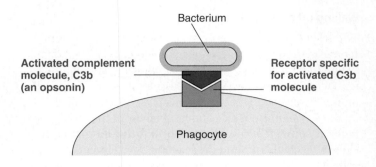

Structures are not drawn to scale.

● **FIGURE 12-4**

Mechanism of opsonin action
One of the activated complement molecules, C3b, links a foreign cell, such as a bacterium, and a phagocytic cell by nonspecifically binding with the foreign cell and specifically binding with a receptor on the phagocyte. This link ensures that the foreign victim does not escape before it can be engulfed by the phagocyte.

into the blood of large numbers of preformed neutrophils stored in the bone marrow and partly to increased production of new neutrophils by the bone marrow. A slower-commencing but longer-lasting increase in monocyte production by the bone marrow also occurs, making larger numbers of these macrophage precursor cells available. In addition, multiplication of resident macrophages adds to the pool of these important immune cells. Proliferation of new neutrophils, monocytes, and macrophages and mobilization of stored neutrophils are stimulated by various chemical mediators released from the inflamed region.

Marking of bacteria for destruction by opsonins

Obviously, phagocytes must be able to distinguish between normal cells and foreign or abnormal cells before accomplishing their destructive mission. Otherwise they could not selectively engulf and destroy only unwanted materials. First, as you already learned, the phagocyte's TLRs trigger phagocytosis of infiltrators that display standard bacterial cell wall components or other targeted molecular patterns. Second, foreign particles are deliberately marked for phagocytic ingestion by being coated with chemical mediators generated by the immune system. Such body-produced chemicals that make bacteria more susceptible to phagocytosis are known as **opsonins**. The most important opsonins are antibodies and one of the activated proteins of the complement system.

An opsonin enhances phagocytosis by linking the foreign cell to a phagocytic cell (● Figure 12-4). One portion of an opsonin molecule binds nonspecifically to the surface of an invading bacterium, whereas another portion of the opsonin molecule binds to receptor sites specific for it on the phagocytic cell's plasma membrane. This link ensures that the bacterial victim does not have a chance to "get away" before the phagocyte can perform its lethal attack.

Leukocytic destruction of bacteria

Neutrophils and macrophages clear the inflamed area of infectious and toxic agents as well as tissue debris; this clearing action is the primary function of the inflammatory response. They accomplish this by both phagocytic and nonphagocytic means.

Phagocytosis involves the engulfment and intracellular degradation (breakdown) of foreign particles and tissue debris. Macrophages can engulf a bacterium in less than 0.01 second. Recall that phagocytic cells contain an abundance of lysosomes, which are organelles filled with hydrolytic enzymes. After a phagocyte has internalized targeted material, a lysosome fuses with the membrane that encloses the engulfed matter and releases its hydrolytic enzymes within the confines of the vesicle, where the enzymes begin breaking down the entrapped material (see p. 32). Phagocytes eventually die from the accumulation of toxic by-products from foreign particle degradation or from inadvertent release of destructive lysosomal chemicals into the cytosol. Neutrophils usually succumb after phagocytizing from 5 to 25 bacteria, whereas macrophages survive much longer and can engulf up to 100 or more bacteria. Indeed, the longer-lived macrophages even clear the area of dead neutrophils in addition to other tissue debris. The **pus** that forms in an infected wound is a collection of these phagocytic cells, both living and dead; necrotic (dead) tissue liquified by lysosomal enzymes released from the phagocytes; and bacteria.

Mediation of the inflammatory response by phagocyte-secreted chemicals

Microbe-stimulated phagocytes release many chemicals that function as mediators of the inflammatory response. These chemical mediators induce a broad range of interrelated immune activities, varying from local responses to the systemic manifestations that accompany microbe invasion. The following are among the most important functions of phagocytic secretions:

1. Some of the chemicals, which are highly destructive, directly kill microbes by nonphagocytic means. For example,

macrophages secrete *nitric oxide*, a multipurpose chemical that is toxic to nearby microbes (see p. 355). As a more subtle means of destruction, neutrophils secrete **lactoferrin**, a protein that tightly binds with iron, making it unavailable for use by invading bacteria. Bacterial multiplication depends on high concentrations of available iron.

2. Phagocytic secretions stimulate the release of histamine from mast cells in the vicinity. Histamine, in turn, induces the local vasodilation and increased vascular permeability that accompany inflammation.

3. Some phagocytic chemical mediators trigger both the clotting and anticlotting systems to first enhance the walling-off process and then facilitate the gradual dissolution of the fibrous clot after it is no longer needed.

4. Other secretions split **kininogens**, which are inactive precursor plasma proteins synthesized by the liver, into active **kinins**. In particular, **kallikrein**, which is released by neutrophils, can accomplish this activation. Once kinins are generated by kallikrein, they augment a variety of inflammatory events. Specifically, kinins

   a. stimulate several key steps in the complement system

   b. reinforce the vascular changes induced by histamine

   c. activate nearby pain receptors and are thus partially responsible for the soreness associated with inflammation

   d. act as powerful chemotaxins to induce phagocyte migration into the affected area

In positive-feedback fashion, the newly arriving neutrophils release kallikrein, which can generate still more kinins that chemotactically entice more neutrophils to join the battle (● Figure 12-5).

5. One chemical in particular released by macrophages, **endogenous pyrogen (EP)**, induces the development of fever (*endogenous* means "from within the body;" *pyro* means "fire" or "heat;" *gen* means "production"). This response occurs especially when the invading organisms have spread into the bloodstream. Endogenous pyrogen causes release within the hypothalamus of *prostaglandins*, locally acting chemical messengers that "turn up" the hypothalamic "thermostat" that regulates body temperature. The function of the resultant elevation in body temperature in fighting infection remains unclear. The fact that fever is such a common systemic manifestation of inflammation suggests that the higher body temperature plays an important beneficial role in the overall inflammatory response. Recent evidence strongly supports this supposition. For example, higher temperatures augment the process of phagocytosis and increase the rate at which the many enzyme-dependent inflammatory activities proceed. One theory proposes that an elevated body temperature increases bacterial requirements for iron while reducing the plasma concentration of iron, thus interfering with bacterial multiplication. Resolving the controversial issue of whether a fever can be beneficial is extremely important in view of the widespread use of drugs that suppress fever.

Although a mild fever may possibly be beneficial, there is no doubt that an extremely high fever can be detrimental, particularly by harming the central nervous system. Young children, whose temperature-regulating mechanisms are not as stable as those of more mature individuals, occasionally have convulsions in association with high fevers.

6. One chemical mediator secreted by macrophages, **leukocyte endogenous mediator (LEM)**, decreases the plasma concentration of iron by altering iron metabolism within the liver, spleen, and other tissues. This action reduces the amount of iron available to support bacterial multiplication. Evidence suggests that LEM and EP are the same substance or at least very closely related.

7. LEM also stimulates **granulopoiesis**, the synthesis and release of neutrophils and other granulocytes by the bone marrow. This effect is especially prominent in response to bacterial infections.

8. Furthermore, LEM stimulates the release of **acute-phase proteins** from the liver. This collection of proteins, which have not yet been sorted out by scientists, exert a multitude of wide-ranging effects associated with the inflammatory process, tissue repair, and immune cell activities.

9. Another effect of phagocytic secretions is to enhance the proliferation and differentiation of both B and T lymphocytes, which, in turn, are responsible for antibody production and cell-mediated immunity, respectively. Specifically, **interleukin 1 (IL-1)**, a secretory product released by macrophages, is responsible for this effect on lymphocytes. Amazingly, IL-1 is identical to (or closely related to) EP and LEM. Apparently, the same chemical substance is responsible for a diverse array of effects throughout the body, all of which are geared toward defending the body against infection or tissue injury. In fact, the release of EP/LEM/IL-1 can be elicited by stressful situations not associated with microbe invasion (for example, during endurance exercise). Accordingly, this mediator may be part of a generalized nonspecific protective response.

This list of events that are augmented by chemicals secreted by phagocytes is not complete, but it serves to illustrate the diversity and complexity of responses these mediators elicit. As noted later, there are other important macrophage—

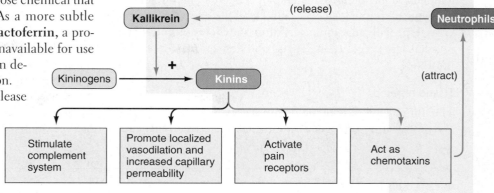

● **FIGURE 12-5**

Relationship between the kinin system and neutrophils in an inflammatory response
The green arrows indicate a positive-feedback loop between the kinin system and neutrophils during an inflammatory response.

lymphocyte interactions that are not dependent on the release of chemicals from phagocytic cells. Thus the effect that phagocytes, especially macrophages, ultimately have on microbial invaders far exceeds their "engulf and destroy" tactics (● Figure 12-6).

Tissue repair

The ultimate purpose of the inflammatory process is to isolate and destroy injurious agents and to clear the area for tissue repair. In some tissues (for example, skin, bone, and liver), the healthy organ-specific cells surrounding the injured area undergo cell division to replace the lost cells, often accomplishing perfect repair. In nonregenerative tissues such as nerve and muscle, however, lost cells are replaced by **scar tissue.** Fibroblasts, a type of connective tissue cell, start to divide rapidly in the vicinity and secrete large quantities of the protein collagen, which fills in the region vacated by the lost cells and results in the formation of scar tissue (see p. 62). Even in a tissue as read-

● **FIGURE 12-6**

Phagocyte responsibilities
Phagocytes not only destroy foreign particles or damaged cells by phagocytosis but also secrete chemical mediators that destroy the targeted material by nonphagocytic means and augment many aspects of inflammation and other immune processes. The pink boxes represent final outcomes of phagocytic actions.

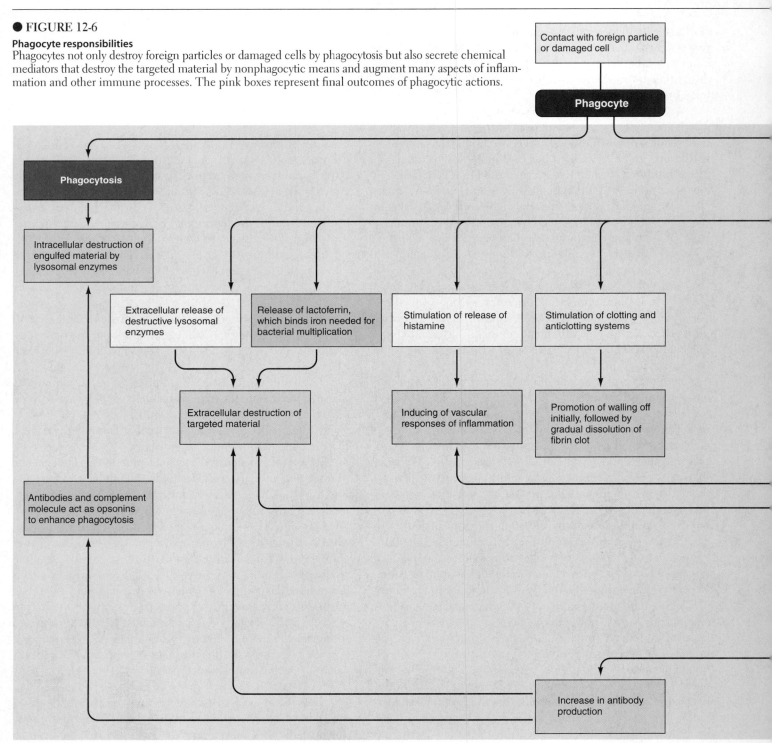

EP/LEM/IL-1 = Endogenous pyrogen/leukocyte endogenous mediator/interleukin 1

ily replaceable as the skin, scar formation sometimes takes place when complex underlying structures, such as hair follicles and sweat glands, are permanently destroyed by deep wounds.

▌Salicylates and glucocorticoid drugs suppress the inflammatory response.

Numerous drugs can suppress the inflammatory process; the most effective are the *salicylates* and related compounds (aspirin-type drugs) and *glucocorticoids* (drugs similar to the steroid hormone cortisol, which is secreted by the adrenal cortex; see p. 708). Salicylates interfere with the inflammatory response by decreasing histamine release, resulting in a reduction in swelling, redness, and pain. Furthermore, salicylates reduce fever by inhibiting the production of prostaglandins, the local mediators of endogenous pyrogen-induced fever.

Glucocorticoids, which are potent anti-inflammatory drugs, suppress almost every aspect of the inflammatory response. In addition, they destroy lymphocytes within lymphoid tissue and reduce antibody production. These therapeutic agents are useful for treating undesirable immune responses, such as allergic reactions (for example, poison ivy rash and asthma) and the inflammation associated with arthritis. Unfortunately, however, by suppressing inflammatory and other immune responses that localize and eliminate bacteria, such therapy also reduces the body's ability to resist infection. For this reason, glucocorticoids should be administered discriminately.

Is naturally secreted cortisol likewise counterproductive to the immune defense system? Traditionally, cortisol has not been considered to display anti-inflammatory activity at normal blood concentrations. Instead, anti-inflammatory action has been attributed only to blood concentrations that are higher than the

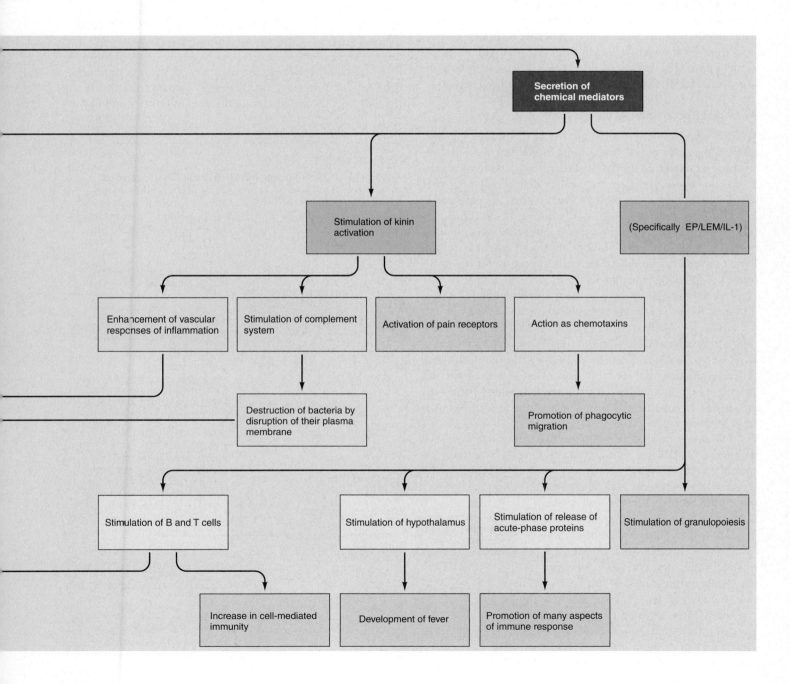

normal physiologic range brought about by the administration of exogenous ("from outside the body") cortisol-like drugs. Recent evidence, however, suggests that cortisol, whose secretion is increased in response to any stressful situation, does exert anti-inflammatory activity even at normal physiological levels. According to this proposal, the anti-inflammatory effect of cortisol modulates stress-activated immune responses, preventing them from overshooting, and thus protecting us against damage by potentially overreactive defense mechanisms.

We are now going to shift our attention from inflammation to interferon, another component of innate immunity.

▋ Interferon transiently inhibits multiplication of viruses in most cells.

Besides the inflammatory response, another innate defense mechanism is the release of **interferon** from virus-infected cells. Interferon briefly provides nonspecific resistance to viral infections by transiently interfering with replication of the same or unrelated viruses in other host cells. In fact, interferon was named for its ability to "interfere" with viral replication.

Antiviral effect of interferon

When a virus invades a cell, the presence of viral nucleic acid induces the cell's genetic machinery to synthesize interferon, which is secreted into the extracellular fluid. Interferon acts as a "whistle-blower," forewarning healthy cells of potential viral attack and helping them prepare to resist such an attack. Specifically, once released from a virus-infected cell, interferon binds with receptors on the plasma membranes of healthy neighboring cells or even distant cells that it reaches through the bloodstream, signaling these cells to prepare for the possibility of impending viral attack.

Interferon does not have a direct antiviral effect; instead, it triggers the production of virus-blocking enzymes by potential host cells. When interferon binds with these other cells, they synthesize enzymes that can break down viral messenger RNA (see p. A-22) and inhibit protein synthesis. Both these processes are essential for viral replication. Although viruses are still able to invade these forewarned cells, these pathogens are unable to govern cellular protein synthesis for their own replication (● Figure 12-7).

The newly synthesized inhibitory enzymes remain inactive within the tipped-off potential host cell unless it is actually invaded by a virus, at which time the enzymes are activated by the presence of viral nucleic acid. This activation requirement protects the cell's own messenger RNA and protein-synthesizing machinery from unnecessary inhibition by these enzymes should viral invasion not occur. Because activation can take place only during a limited time span, this is a short-term defense mechanism.

Interferon is released nonspecifically from any cell infected by any virus and, in turn, can induce temporary self-protective activity against many different viruses in any other cells that it

reaches. Thus, it provides a general, rapidly responding defense strategy against viral invasion until more specific but slower-responding immune mechanisms come into play.

In addition to facilitating inhibition of viral replication, interferon reinforces other immune activities (▲ Table 12-2). For example, it enhances macrophage phagocytic activity and stimulates the production of antibodies. Interferon also exerts anticancer as well as antiviral effects.

Anticancer effects of interferon

Interferon slows cell division and suppresses tumor growth. Furthermore, it markedly enhances the actions of cell-killing cells—the *natural killer cells* and a special type of T lymphocyte, *cytotoxic T cells*—which attack and destroy both virus-infected cells and cancer cells.

▋ Natural killer cells destroy virus-infected cells and cancer cells on first exposure to them.

Natural killer (NK) cells are naturally occurring, lymphocyte-like cells that nonspecifically destroy virus-infected cells and cancer cells by directly lysing their membranes upon first ex-

● **FIGURE 12-7**

Mechanism of action of interferon in preventing viral replication Interferon, which is released from virus-infected cells, binds with other uninvaded host cells and induces these cells to produce inactive enzymes capable of blocking viral replication. These inactive enzymes are activated only if a virus subsequently invades one of these prepared cells.

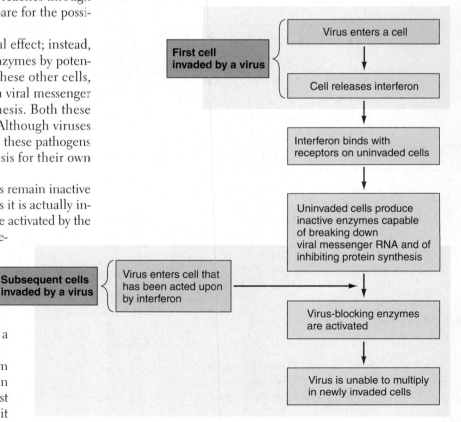

▲ TABLE 12-2
Functions of Interferon

Functions	Antiviral Effect	Anticancer Effect
Released by virus-infected cells and transiently interferes with viral replication in other host cells by inducing them to produce enzymes that destroy viral messenger RNA and inhibit virus-directed protein synthesis	✓	
Enhances macrophage phagocytic activity	✓	✓
Stimulates production of antibodies	✓	
Enhances actions of natural killer cells and cytotoxic T cells	✓	✓
Slows cell division and suppresses tumor growth		✓

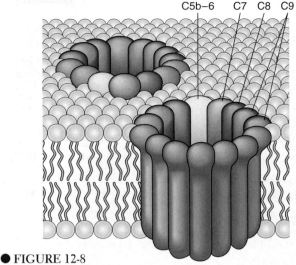

C5b–6 C7 C8 C9

● **FIGURE 12-8**

Membrane attack complex (MAC) of the complement system
Activated complement proteins C5, C6, C7, C8, and a number of C9s aggregate to form a porelike channel in the plasma membrane of the target cell. The resultant leakage leads to destruction of the cell.

posure to them. Their mode of action and major targets are similar to those of cytotoxic T cells, but the latter can fatally attack only the specific types of virus-infected cells and cancer cells to which they have been previously exposed. Furthermore, following exposure cytotoxic T cells require a maturation period before they can launch their lethal assault. The NK cells provide an immediate, nonspecific defense against virus-invaded cells and cancer cells before the more specific and more abundant cytotoxic T cells become functional.

▮ The complement system punches holes in micro-organisms.

The **complement system** is another defense mechanism brought into play nonspecifically in response to invading organisms. This system can be activated in two ways:

1. By exposure to particular carbohydrate chains present on the surfaces of micro-organisms but not found on human cells, a nonspecific innate immune response.

2. By exposure to antibodies produced against a specific foreign invader, an adaptive immune response.

In addition to bringing about direct lysis of the invader, the powerful complement cascade reinforces other general inflammatory tactics. In fact, the system derives its name from the fact that it "complements" the action of antibodies; it is the primary mechanism activated by antibodies to kill foreign cells. It does so by forming membrane attack complexes that punch holes in the victim cells.

Formation of the membrane attack complex

In the same tradition as the clotting and anticlotting systems and the kinin system, the complement system consists of plasma proteins that are produced by the liver and circulate in the blood in inactive form. Once the first component, C1, is activated, it activates the next component, C2, and so on, in a sequential cascade of activation reactions. The five final components, C5 through C9, assemble into a large, doughnut-shaped protein complex, the **membrane attack complex (MAC),** which attacks the surface membrane of nearby micro-organisms by imbedding itself so that a large channel is created through the microbial surface membrane (● Figure 12-8). In other words, the parts make a hole. This hole-punching technique makes the membrane extremely leaky; the resulting osmotic flux of water into the victim cell causes it to swell and burst. This complement-induced lysis is the major means of directly killing microbes without phagocytizing them.

Augmentation of inflammation

Unlike the other cascade systems, in which the sole function of the various components is activation of the next precursor in the sequence, several activated proteins in the complement cascade perform additional important functions on their own. Besides the direct destruction of foreign cells accomplished by the membrane attack complex, various other activated complement components augment the inflammatory process by the following:

• *Serving as chemotaxins,* which attract and guide professional phagocytes to the site of complement activation (that is, the site of microbial invasion)

• *Acting as opsonins* by binding with microbes and thereby enhancing their phagocytosis

• *Promoting vasodilation and increased vascular permeability* to increase blood flow to the invaded areas

• *Stimulating the release of histamine* from mast cells in the vicinity, which in turn enhances the local vascular changes characteristic of inflammation

• *Activating kinins,* which further reinforce inflammatory reactions

Several activated components in the cascade are very unstable. Because these unstable components can perpetuate the sequence only in the immediate vicinity in which they are activated before they decompose, the complement attack is confined to the surface membrane of the microbe whose presence initiated activation of the system. Nearby host cells are thus spared from lytic attack.

We have now completed our discussion of innate immunity and are going to turn our attention to adaptive immunity.

ADAPTIVE IMMUNITY: GENERAL CONCEPTS

A specific adaptive immune response is a selective attack aimed at limiting or neutralizing a particular offending target for which the body has been specially prepared following prior exposure to it.

▌ Adaptive immune responses include antibody-mediated immunity and cell-mediated immunity.

There are two classes of adaptive immune responses: **antibody-mediated**, or **humoral**, **immunity**, involving the production of antibodies by B lymphocyte derivatives known as *plasma cells*; and **cell-mediated immunity**, involving the production of *activated T lymphocytes*, which directly attack unwanted cells.

Lymphocytes can specifically recognize and selectively respond to an almost limitless variety of foreign agents as well as cancer cells. The recognition and response processes are different for B and T lymphocytes (B and T cells). In general, B cells recognize free-existing foreign invaders such as bacteria and their toxins and a few viruses, which they combat by secreting antibodies specific for the invaders. T cells specialize in recognizing and destroying body cells gone awry, including virus-infected cells and cancer cells. We will examine each of these processes in detail in the upcoming sections. For now, we are going to explore the different life histories of B and T lymphocytes.

Origins of B and T cells

Both types of lymphocytes, like all blood cells, are derived from common stem cells in the bone marrow. Whether a lymphocyte and all its progeny are destined to be B or T cells depends on the site of final differentiation and maturation of the original cell in the lineage (● Figure 12-9). B cells differentiate and mature in the bone marrow. As for T cells, during fetal life and early childhood, some of the immature lymphocytes from the bone marrow migrate through the blood to the thymus, where they undergo further processing to become T lymphocytes (named for their site of maturation). The **thymus** is a lymphoid tissue located midline within the chest cavity above the heart in the space between the lungs (see Figure 12-1, p. 414).

On being released into the blood from either the bone marrow or the thymus, mature B and T cells take up residence and establish lymphocyte colonies in the peripheral lymphoid tissues. Here, on appropriate stimulation, they undergo cell di-

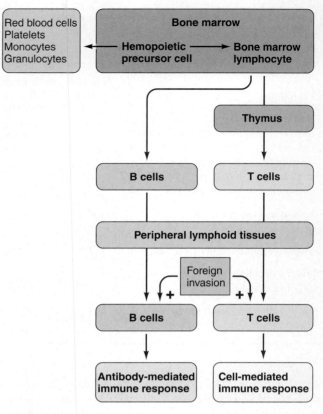

● FIGURE 12-9

Origins of B and T cells
B cells are derived from lymphocytes that matured and differentiated in the bone marrow, whereas T cells are derived from lymphocytes that originated in the bone marrow but matured and differentiated in the thymus. New B and T cells are produced by colonies of B and T cells in peripheral lymphoid tissues.

vision to produce new generations of either B or T cells, depending on their ancestry. After early childhood, most new lymphocytes are derived from these peripheral lymphocyte colonies rather than from the bone marrow.

Each of us has an estimated total of 2 trillion lymphocytes, which, if aggregated together in a mass, would be comparable to the size of the brain or liver. At any one time, the majority of these lymphocytes are concentrated in the various strategically located lymphoid tissues, but both B and T cells continually circulate among the lymph, blood, and body tissues, where they remain on constant surveillance.

Role of thymosin

Because most of the migration and differentiation of T cells occurs early in development, the thymus gradually atrophies and becomes less important as the individual matures. It does, however, continue to produce **thymosin**, a hormone important in maintaining the T cell lineage. Thymosin enhances proliferation of new T cells within the peripheral lymphoid tissues and augments the immune capabilities of existing T cells. Recent evidence indicates that secretion of thymosin decreases after about 30 to 40 years of age. This decline has been implicated as a contributing factor in aging. It is further speculated that diminishing T cell capacity with advancing age may be linked to increased susceptibility to viral infections and cancer, because T cells play an especially important role in defense against viruses and cancer.

Let's now see how lymphocyes detect their selected target.

▮ An antigen induces an immune response against itself.

Both B and T cells must be able to specifically recognize unwanted cells and other material to be destroyed or neutralized, as being distinct from the body's own normal cells. The presence of antigens enables lymphocytes to make this distinction. An **antigen** is a large, complex, unique molecule that triggers a specific immune response against itself when it gains entry into the body. In general, the more complex a molecule, the greater its antigenicity. Foreign proteins are the most common antigens because of their size and structural complexity, although other macromolecules, such as large polysaccharides, can also act as antigens. Antigens may exist as isolated molecules, such as bacterial toxins, or they may be an integral part of a multimolecular structure, such as being present on the surface of an invading foreign microbe.

We are now going to see how B cells respond to their targeted antigen, followed by an examination of T cells' response to their antigen.

B LYMPHOCYTES: ANTIBODY-MEDIATED IMMUNITY

Each B and T cell has receptors on its surface for binding with one particular type of the multitude of possible antigens. These receptors are the "eyes of the adaptive immune system," although a given lymphocyte can "see" only one unique antigen. This is in contrast to the TLRs of the innate effector cells, which recognize generic trademarks characteristic of all microbial invaders. Furthermore, lymphocytes cannot respond directly to antigen. New incoming antigen must first be processed and presented to them by *antigen-presenting cells*, a process we will describe in detail later.

▮ Antigens stimulate B cells to convert into plasma cells that produce antibodies.

In the case of B cells, binding with processed and presented antigen induces the cell to differentiate into a **plasma cell**. A plasma cell produces antibodies that can combine with the specific type of antigen that stimulated activation of the plasma cell. During differentiation into a plasma cell, a B cell swells as the rough endoplasmic reticulum (the site for synthesis of proteins to be exported) greatly expands (● Figure 12-10). Because antibodies are proteins, plasma cells essentially become prolific protein factories, producing up to 2,000 antibody molecules per second. So great is the commitment of a plasma cell's protein-synthesizing machinery to antibody production that it cannot maintain protein synthesis for its own viability and growth. Consequently, it dies after a brief (five- to seven-day), highly productive life span.

Antibodies are secreted into the blood or lymph, depending on the location of the activated plasma cells, but all antibodies eventually gain access to the blood, where they are known as **gamma globulins,** or **immunoglobulins.**

Antibody subclasses

Antibodies are grouped into the following five subclasses based on differences in their biological activity:

- **IgM** immunoglobulin serves as the B cell surface receptor for antigen attachment and is secreted in the early stages of plasma cell response.

- **IgG,** the most abundant immunoglobulin in the blood, is produced copiously when the body is subsequently exposed to the same antigen.

Together, IgM and IgG antibodies are responsible for most specific immune responses against bacterial invaders and a few types of viruses.

- **IgE** helps protect against parasitic worms and is the antibody mediator for common allergic responses, such as hay fever, asthma, and hives.

- **IgA** immunoglobulins are found in secretions of the digestive, respiratory, and genitourinary systems, as well as in milk and tears.

- **IgD** is present on the surface of many B cells, but its function is uncertain.

● **FIGURE 12-10**

Comparison of an unactivated B cell and a plasma cell
Electron micrograph of (a) an unactivated B cell, or small lymphocyte, and (b) a plasma cell. A plasma cell is an activated B cell. It is filled with an abundance of rough endoplasmic reticulum distended with antibody molecules.

(Contributed by Dr. Dorothea Zucker-Franklin, New York University Medical Center.)

Plasma cell

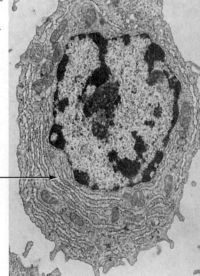

Endoplasmic reticulum

(b)

Unactivated B cell

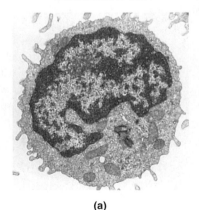

(a)

Note that this classification is based on different ways in which antibodies function. It does not imply that there are only five different antibodies. Within each functional subclass are millions of different antibodies, each able to bind only with a specific antigen.

▌ Antibodies are Y-shaped and classified according to properties of their tail portion.

Antibody proteins of all five subclasses are composed of four interlinked polypeptide chains—two long, heavy chains and two short, light chains—arranged in the shape of a Y (● Figure 12-11). Characteristics of the arm regions of the Y determine the *specificity* of the antibody (that is, with what antigen the antibody can bind). Properties of the tail portion of the antibody determine the *functional properties* of the antibody (what the antibody does once it binds with antigen).

An antibody has two identical antigen-binding sites, one at the tip of each arm. These **antigen-binding fragments (Fab)** are unique for each different antibody, so that each antibody can interact only with an antigen that specifically matches it, much like a lock and key. The tremendous variation in the antigen-binding fragments of different antibodies is responsible for the extremely large number of unique antibodies that are capable of binding specifically with millions of different antigens.

● **FIGURE 12-11**

Antibody structure
An antibody is Y-shaped. It is able to bind only with the specific antigen that "fits" its antigen-binding sites (Fab) on the arm tips. The tail region (Fc) binds with particular mediators of antibody-induced activities.

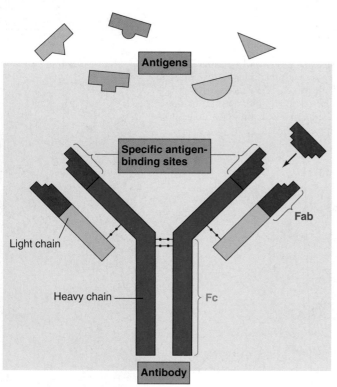

In contrast to these variable Fab regions at the arm tips, the tail portion of every antibody within each immunoglobulin subclass is identical. The tail, the antibody's so-called **constant (Fc) region,** contains binding sites for particular mediators of antibody-induced activities, which vary among the different subclasses. In fact, differences in the constant region are the basis for distinguishing between the different immunoglobulin subclasses. For example, the constant tail region of IgG antibodies, when activated by antigen binding in the Fab region, binds with phagocytic cells and serves as an opsonin to enhance phagocytosis. In comparison, the constant tail region of IgE antibodies attaches to mast cells and basophils, even in the absence of antigen. When the appropriate antigen gains entry to the body and binds with the attached antibodies, this triggers the release of histamine from the affected mast cells and basophils. Histamine, in turn, induces the allergic manifestations that follow.

▌ Antibodies largely amplify innate immune responses to promote antigen destruction.

Immunoglobulins cannot directly destroy foreign organisms or other unwanted materials on binding with antigens on their surfaces. Instead, antibodies exert their protective influence in one of two general ways: (1) physical hindrance of antigens or (2) amplification of innate immune responses (● Figure 12-12).

Neutralization and agglutination

Antibodies can physically hinder some antigens from exerting their detrimental effects. For example, by combining with bacterial toxins antibodies can prevent these harmful chemicals from interacting with susceptible cells. This process is known as **neutralization.** Similarly, antibodies can bind with surface antigens on some types of viruses, preventing these viruses from entering cells, where they could exert their damaging effects. Sometimes multiple antibody molecules can cross-link numerous antigen molecules into chains or lattices of antigen-antibody complexes. The process in which foreign cells, such as bacteria or mismatched transfused red blood cells, bind together in such a clump is known as **agglutination.** When linked antigen-antibody complexes involve soluble antigens, such as tetanus toxin, the lattice can become so large that it precipitates out of solution. (**Precipitation** is the process in which a substance separates from a solution.) Within the body, these physical hindrance mechanisms play only a minor protective role against invading agents. However, the tendency for certain antigens to agglutinate or precipitate on forming large complexes with antibodies specific for them is useful clinically and experimentally for detecting the presence of particular antigens or antibodies. Pregnancy diagnosis tests, for example, employ this principle to detect, in urine, the presence of a hormone secreted soon after conception.

Amplification of innate immune responses

Antibodies' most important function by far is to profoundly augment the innate immune responses already initiated by the

Neutralization

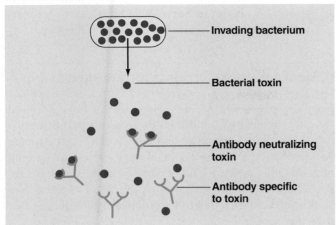

- Invading bacterium
- Bacterial toxin
- Antibody neutralizing toxin
- Antibody specific to toxin

Agglutination (clumping of antigenic cells) and **precipitation** (if soluble antigen-antibody complex is too large to stay in solution)

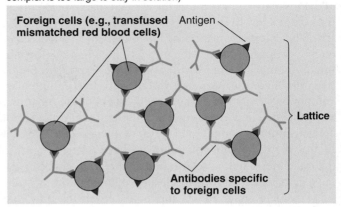

- Foreign cells (e.g., transfused mismatched red blood cells)
- Antigen
- Lattice
- Antibodies specific to foreign cells

Activation of complement system

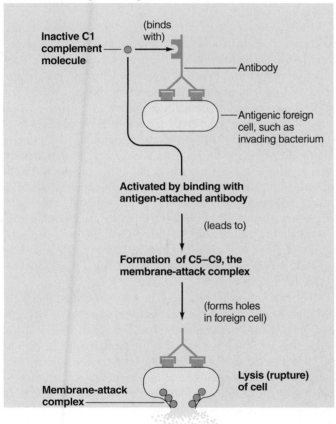

- Inactive C1 complement molecule
- (binds with)
- Antibody
- Antigenic foreign cell, such as invading bacterium
- **Activated by binding with antigen-attached antibody**
- (leads to)
- **Formation of C5–C9, the membrane-attack complex**
- (forms holes in foreign cell)
- **Membrane-attack complex**
- **Lysis (rupture) of cell**

Enhancement of phagocytosis (opsonization)

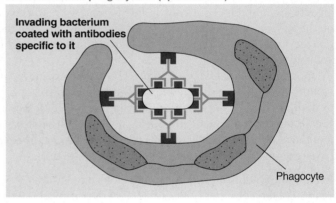

- Invading bacterium coated with antibodies specific to it
- Phagocyte

Stimulation of killer cells

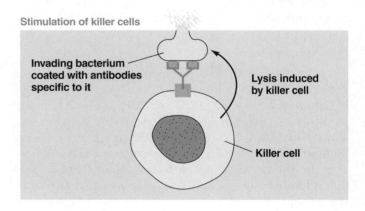

- Invading bacterium coated with antibodies specific to it
- Lysis induced by killer cell
- Killer cell

Structures are not drawn to scale.

● **FIGURE 12-12**

How antibodies help eliminate invading microbes
Antibodies physically hinder antigens through (1) neutralization or (2) agglutination and precipitation. Antibodies amplify innate immune responses by (1) activating the complement system, (2) enhancing phagocytosis by acting as opsonins, and (3) stimulating killer cells.

invaders. Antibodies mark or identify foreign material as targets for actual destruction by the complement system, phagocytes, or killer cells while enhancing the activity of these other defense systems as follows:

1. *Activation of the complement system.* When an appropriate antigen binds with an antibody, receptors on the tail portion of the antibody are able to bind with and activate C1, the first component of the complement system. This sets off

the cascade of events leading to formation of the membrane attack complex, which is specifically directed at the membrane of the invading cell that bears the antigen that initiated the activation process. In fact, antibody is the most powerful activator of the complement system. The biochemical attack subsequently unleashed against the invader's membrane is the most important mechanism by which antibodies exert their protective influence. Furthermore, various activated complement components enhance virtually every aspect of the inflammatory process. The same complement system is activated by an antigen–antibody complex regardless of the type of antigen. Although the binding of antigen to antibody is highly specific, the outcome, which is determined by the antibody's constant tail region, is identical for all activated antibodies within a given subclass; for example, all IgG antibodies activate the same complement system.

2. *Enhancement of phagocytosis.* As mentioned previously, antibodies, especially IgG, act as opsonins. The tail portion of an antigen-bound IgG antibody is able to bind with a receptor on the surface of a phagocyte and subsequently to promote the phagocytosis of the antigen-containing victim attached to the antibody.

3. *Stimulation of killer (K) cells.* The binding of antibody to antigen can also induce attack of the antigen-bearing cell by a **killer (K) cell.** K cells are similar to NK cells except that K cells require the target cell to be coated with antibodies before they can destroy it by lysing its plasma membrane. K cells have receptors for the constant tail portion of antibodies.

In these ways, antibodies, although unable to directly destroy invading bacteria or other undesirable material, bring about destruction of the antigens to which they are specifically attached by amplifying other nonspecific lethal defense mechanisms.

Immune complex disease

Occasionally, an overzealous antigen–antibody response can inadvertently cause damage to normal cells as well as to invading foreign cells. Typically, antigen–antibody complexes, formed in response to foreign invasion, are removed by phagocytic cells after having revved up nonspecific defense strategies. If large numbers of these complexes are continuously produced, however, the phagocytes cannot clear away all the immune complexes formed. Antigen–antibody complexes that are not removed continue to activate the complement system, among other things. Excessive amounts of activated complement and other inflammatory agents may "spill over," damaging the surrounding normal cells as well as the unwanted cells. Furthermore, destruction is not necessarily restricted to the initial site of inflammation. Antigen–antibody complexes may circulate freely and become trapped in the kidneys, joints, brain, small vessels of the skin, and elsewhere, causing widespread inflammation and tissue damage. Such damage produced by immune complexes is referred to as an **immune complex disease,** which can be a complicating outcome of bacterial, viral, or parasitic infection.

More insidiously, immune complex disease can also stem from overzealous inflammatory activity prompted by immune complexes formed by "self-antigens" (proteins synthesized by the person's own body) and antibodies erroneously produced against them. *Rheumatoid arthritis* develops in this way.

■ Clonal selection accounts for the specificity of antibody production.

Consider the diversity of foreign molecules a person can potentially encounter during a lifetime. Yet each B cell is preprogrammed to respond to only one of these millions of different antigens. Other antigens cannot combine with the same B cell and induce it to secrete different antibodies. The astonishing implication is that each of us is equipped with millions of different preformed B lymphocytes, at least one for every possible antigen that we might ever encounter—including those specific for synthetic substances that do not exist in nature. The clonal selection theory proposes how a "matching" B cell responds to its antigen.

Early researchers in immunologic theory believed antibodies to be "made to order" whenever a foreign antigen gained entry to the body. In contrast, the currently accepted **clonal selection theory** proposes that diverse B lymphocytes are produced during fetal development, each capable of synthesizing antibody against a particular antigen before ever being exposed to it. All offspring of a particular ancestral B lymphocyte form a family of identical cells, or a **clone,** that is committed to producing the same specific antibody. B cells remain dormant, not actually secreting their particular antibody product until (or unless) they come into contact with the appropriate antigen. Lymphocytes that have not yet been exposed to their specific antigen are known as **naive lymphocytes.** When an antigen gains entry to the body, it "selects" (that is, activates) the particular clone of B cells that bear receptors on their surface uniquely specific for that antigen, hence the term "clonal selection theory" (● Figure 12-13).

The first antibodies produced by a newly formed B cell are IgM immunoglobulins, which are inserted into the cell's plasma membrane rather than being secreted. Here they serve as receptor sites for binding with a specific kind of antigen, almost like "advertisements" for the kind of antibody the cell can produce. Binding of the appropriate antigen to a B cell amounts to "placing an order" for the manufacture and secretion of large quantities of that particular antibody.

■ Selected clones differentiate into active plasma cells and dormant memory cells.

Antigen binding causes the activated B-cell clone to multiply and differentiate into two cell types—*plasma cells* and *memory cells.* Most progeny are transformed into plasma cells, which are prolific producers of customized antibodies that contain the same antigen-binding sites as the surface receptors. However, plasma cells switch to producing IgG antibodies, which are secreted rather than remaining membrane bound. In the blood, the secreted antibodies combine with invading free (not

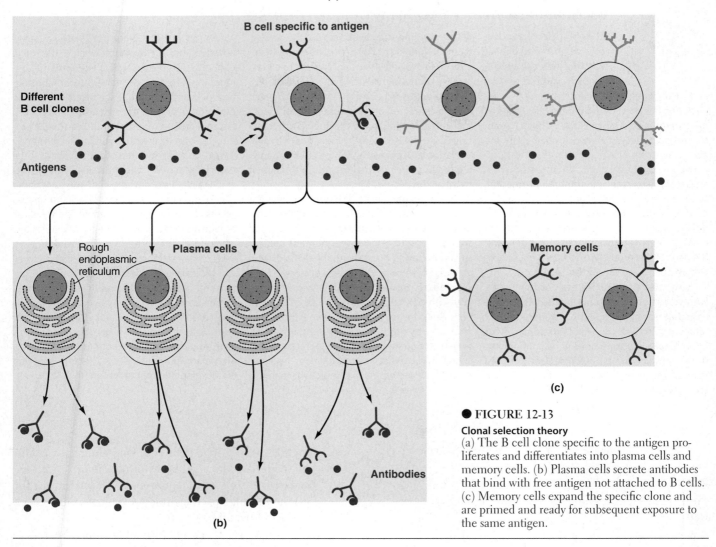

(a)

B cell specific to antigen

Different
B cell clones

Antigens

Rough
endoplasmic
reticulum

Plasma cells

Memory cells

Antibodies

(b)

(c)

● **FIGURE 12-13**

Clonal selection theory
(a) The B cell clone specific to the antigen proliferates and differentiates into plasma cells and
memory cells. (b) Plasma cells secrete antibodies
that bind with free antigen not attached to B cells.
(c) Memory cells expand the specific clone and
are primed and ready for subsequent exposure to
the same antigen.

---

bound to lymphocytes) antigen, marking it for destruction by
the complement system, phagocytic ingestion, or other means.

Not all the new B lymphocytes produced by the specifically activated clone differentiate into antibody-secreting plasma
cells. A small proportion become **memory cells,** which do not
participate in the current immune attack against the antigen
but instead remain dormant and expand the specific clone.
Should the person ever be re-exposed to the same antigen, these
memory cells are primed and ready for even more immediate
action than were the original lymphocytes in the clone.

Even though each of us has essentially the same original
pool of different B-lymphocyte clones, the pool gradually becomes appropriately biased to respond most efficiently to each
person's particular antigenic environment. Those clones specific for antigens to which a person is never exposed remain dormant for life, whereas those specific for antigens in the individual's environment typically become expanded and enhanced
by forming highly responsive memory cells. The different naive
clones provide protection against unknown new pathogens,
and the evolving populations of memory cells protect against
the recurrence of infections encountered in the past.

Primary and secondary responses

During initial contact with a microbial antigen, the antibody
response is delayed for several days until plasma cells are
formed and does not reach its peak for a couple of weeks (● Figure 12-14). This response is known as the **primary response.**
Meanwhile, symptoms characteristic of the particular microbial invasion persist until either the invader succumbs to the
mounting specific immune attack against it or the infected individual dies. After reaching the peak, the antibody levels gradually decline over a period of time, although some circulating
antibody from this primary response may persist for a prolonged
period. Long-term protection against the same antigen, however, is primarily attributable to the memory cells. If the same
antigen ever reappears, the long-lived memory cells launch a
more rapid, more potent, and longer-lasting **secondary response**
than occurred during the primary response. This swifter, more
powerful immune attack is frequently adequate to prevent or
minimize overt infection on subsequent exposures to the same
microbe, forming the basis of long-term immunity against a specific disease.

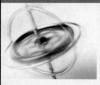

Vaccination: A Victory Over Many Dreaded Diseases

Modern society has come to hope and even expect that vaccines can be developed to protect us from almost any dreaded infectious disease. This expectation has been brought into sharp focus by our current frustration over the inability to date to develop a successful vaccine against HIV, the virus that causes AIDS.

Nearly 2,500 years ago, our ancestors were aware of the existence of immune protection. Writing about a plague that struck Athens in 430 B.C., Thucydides observed that the same person was never attacked twice by this disease. However, the ancients did not understand the basis of this protection, so they were unable to manipulate it to their advantage.

Early attempts at deliberately acquiring lifelong protection against smallpox, a dreaded disease that was highly infectious and frequently fatal (up to 40% of the sick died), consisted of intentionally exposing oneself by coming into direct contact with a person suffering from a milder form of the disease. The hope was to protect against a future fatal bout of smallpox by

deliberately inducing a mild case of the disease. By the beginning of the 17th century, this technique had evolved into using a needle to extract small amounts of pus from active smallpox pustules (the fluid-filled bumps on the skin, which leave a characteristic depressed scar or "pock" mark after healing) and introducing this infectious material into healthy individuals. This inoculation process was accomplished by applying the pus directly to slight cuts in the skin or by inhaling dried pus.

Edward Jenner, an English physician, was the first to demonstrate that immunity against cowpox, a disease similar to but less serious than smallpox, could also protect humans against smallpox. Having observed that milkmaids who acquired cowpox seemed to be protected from smallpox, Jenner in 1796 inoculated a healthy boy with pus he had extracted from cowpox boils. After the boy recovered, Jenner (not being restricted by modern ethical standards of research on human subjects) deliberately inoculated him with what was considered to be a normally fatal dose of smallpox infectious material. The boy survived.

Jenner's results were not taken seriously, however, until a century later when, in the 1880s, Louis Pasteur, the first great experimental immunologist, extended Jenner's technique. Pasteur demonstrated that the disease-inducing capability of organisms could be greatly reduced (attenuated) so that they could no longer produce disease but would still induce antibody formation when introduced into the body—the basic principle of modern vaccines. His first vaccine was against anthrax, a deadly disease of sheep and cows. Pasteur isolated and heated anthrax bacteria, then injected these attenuated organisms into a group of healthy sheep. A few weeks later at a gathering of fellow scientists, Pasteur injected these vaccinated sheep as well as a group of unvaccinated sheep with fully potent anthrax bacteria. The result was dramatic— all the vaccinated sheep survived, but all the unvaccinated sheep died. Pasteur's notorious public demonstrations such as this, coupled with his charismatic personality, caught the attention of physicians and scientists of the time, sparking the development of modern immunology.

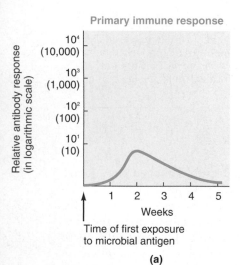

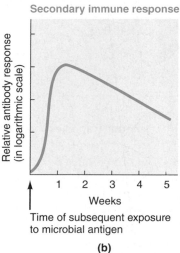

● FIGURE 12-14

Primary and secondary immune responses
(a) Primary response on first exposure to a microbial antigen. (b) Secondary response on subsequent exposure to the same microbial antigen. The primary response does not peak for a couple of weeks, whereas the secondary response peaks in a week. The magnitude of the secondary response is 100 times that of the primary response. (The relative antibody response is in the logarithmic scale.)

The original antigenic exposure that induces the formation of memory cells can occur through either actually having the disease or being vaccinated (● Figure 12-15). During vaccination, the individual is deliberately exposed to a pathogen

that has been stripped of its disease-inducing capability but can still induce antibody formation against it. (For the early history of vaccination development, see the accompanying boxed feature, ❒ Concepts, Challenges, and Controversies.)

Memory cells are not formed for some diseases, so no lasting immunity is conferred by an initial exposure, as in the case of "strep throat." The course and severity of the disease are the same each time a person is reinfected with a microbe that the immune system does not "remember," regardless of the number of prior exposures.

▌ The huge repertoire of B cells is built by reshuffling a small set of gene fragments.

Considering the millions of different antigens against which each of us has the potential to actively produce antibodies, how is it possible for an individual to have such a tremendous diversity of B lymphocytes, each capable of producing a different antibody? Antibodies are proteins that are produced in accordance with a nuclear DNA blueprint. Because all cells of the body, including the antibody-producing cells, contain the same nuclear DNA, it is hard to imagine how enough DNA could be packaged within the nuclei of every cell to code for the millions of different antibodies (a different portion of the genetic code being used by each B cell clone), along with all the other genetic instructions used by other cells. Actually, only a relatively small number of gene fragments code for antibody synthesis, but during B cell development these frag-

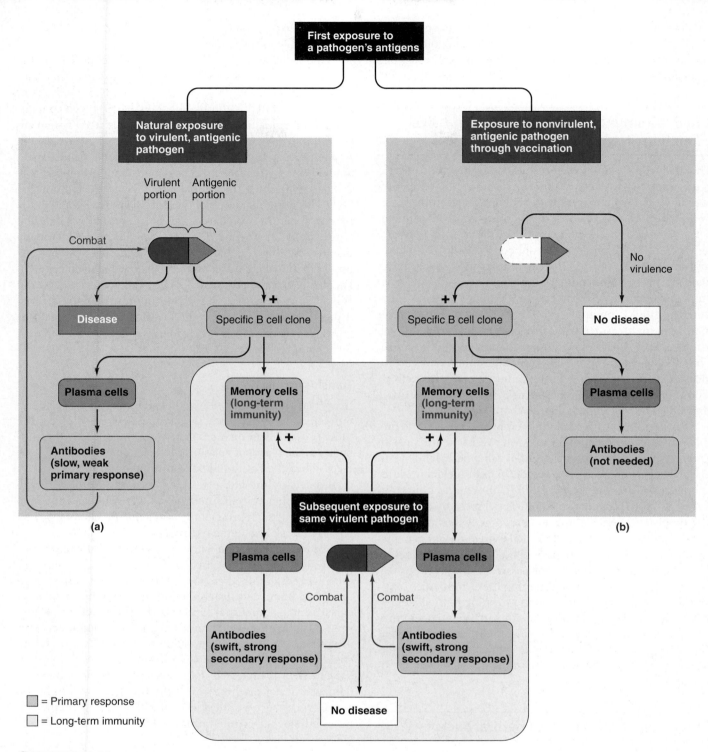

● FIGURE 12-15

Means of acquiring long-term immunity
Long-term immunity against a pathogen can be acquired through having the disease or being vaccinated against it. (a) Exposure to a virulent (disease-producing) pathogen. (b) Vaccination with a modified pathogen that is no longer virulent (that is, can no longer produce disease) but is still antigenic. In both cases, long-term memory cells are produced that mount a swift, secondary response that prevents or minimizes symptoms on a subsequent natural exposure to the same virulent pathogen.

ments are cut, reshuffled, and spliced in a vast number of different combinations. Each different combination gives rise to a unique B cell clone. Antibody genes are later even further diversified by somatic mutation (see p. A-30). The antibody genes of already-formed B cells are highly prone to mutations in the region that codes for the variable antigen-binding sites on the antibodies. Each different mutant cell in turn gives rise to a new clone. Thus the great diversity of antibodies is made possible by the reshuffling of a small set of gene fragments during B cell development, as well as by further somatic mutation in already-formed B cells. In this way, a huge antibody repertoire is possible using only a modest share of the genetic blueprint.

■ Active immunity is self-generated; passive immunity is "borrowed."

The production of antibodies as a result of exposure to an antigen is referred to as **active immunity** against that antigen. A second way in which an individual can acquire antibodies is by the direct transfer of antibodies actively formed by another person (or animal). The immediate "borrowed" immunity conferred on receipt of preformed antibodies is known as **passive immunity.** Such transfer of antibodies of the IgG class normally occurs from the mother to the fetus across the placenta during intrauterine development. In addition, a mother's colostrum (first milk) contains IgA antibodies that provide further protection for breast-fed babies. Passively transferred antibodies are usually broken down in less than a month, but meanwhile the newborn is provided important immune protection (essentially the same as its mother's) until it can begin actively mounting its own immune responses. Antibody-synthesizing ability does not develop for about a month after birth.

Passive immunity is sometimes employed clinically to provide immediate protection or to bolster resistance against an extremely virulent infectious agent or potentially lethal toxin to which a person has been exposed (for example, rabies virus, tetanus toxin in nonimmunized individuals, and poisonous snake venom). Typically, the administered preformed antibodies have been harvested from another source (often nonhuman) that has been exposed to an attenuated form of the antigen. Frequently, horses or sheep are used in the deliberate production of antibodies to be collected for passive immunizations. Although injection of serum containing these antibodies (**antiserum** or **antitoxin**) is beneficial in providing immediate protection against the specific disease or toxin, the recipient may develop an immune response against the injected antibodies themselves, because they are foreign proteins. The result may be a severe allergic reaction to the treatment, a condition known as **serum sickness.**

■ Blood types are a form of natural immunity.

Certain antibodies were once thought to occur naturally in the blood. Antibodies associated with blood types are the classic example of "natural antibodies," although natural immunity is actually a special case of actively acquired immunity. Let's see how.

ABO blood types

The surface membranes of human erythrocytes contain inherited antigens that vary depending on blood type. With the major blood group system, the **ABO system,** the erythrocytes of people with type A blood contain A antigens, those with type B blood contain B antigens, those with type AB blood have both

A and B antigens, and those with type O blood do not have any A or B red blood cell surface antigens.

Antibodies against erythrocyte antigens not present on the body's own erythrocytes begin to appear in human plasma after a baby is about 6 months of age. Accordingly, the plasma of type A blood contains anti-B antibodies, type B blood contains anti-A antibodies, no antibodies related to the ABO system are present in type AB blood, and both anti-A and anti-B antibodies are present in type O blood. Typically, one would expect antibody production against A or B antigen to be induced only if blood containing the alien antigen were injected into the body. However, high levels of these antibodies are found in the plasma of persons who have never been exposed to a different type of blood. Consequently, these were considered naturally occurring antibodies, that is, produced without any known exposure to the antigen. Scientists now know that people are unknowingly exposed at an early age to small amounts of A- and B-like antigens associated with common intestinal bacteria. Antibodies produced against these foreign antigens coincidentally also interact with a nearly identical antigen for a foreign blood group, even on first exposure to it.

Transfusion reaction

If a person is administered blood of an incompatible type, two different antigen–antibody interactions take place. By far the more serious consequences arise from the effect of the antibodies in the recipient's plasma on the incoming donor erythrocytes. The effect of the donor's antibodies on the recipient's erythrocyte-bound antigens is less important unless a large amount of blood is transfused, because the donor's antibodies are so diluted by the recipient's plasma that little red blood cell damage takes place in the recipient.

Antibody interaction with erythrocyte-bound antigen may result in agglutination (clumping) or hemolysis (rupture) of the attacked red blood cells. Agglutination and hemolysis of donor red blood cells by antibodies in the recipient's plasma can lead to a sometimes fatal **transfusion reaction** (● Figure 12-16). Agglutinated clumps of incoming donor cells can plug small blood vessels. In addition, one of the most lethal consequences of mismatched transfusions is acute kidney failure caused by the release of large amounts of hemoglobin from damaged donor erythrocytes. If the free hemoglobin in the plasma rises above a critical level, it will precipitate in the kidneys and block the urine-forming structures, leading to acute kidney shutdown.

Universal blood donors and recipients

Because type O individuals have no A or B antigens, their erythrocytes will not be attacked by either anti-A or anti-B antibodies, so they are considered **universal donors.** Their blood can be transfused into people of any blood type. However, type O individuals can receive only type O blood, because the anti-A and anti-B antibodies in their plasma will attack either A or B antigens in incoming blood. In contrast, type AB individuals are called **universal recipients.** Lacking both anti-A and anti-B antibodies, they can accept donor blood of any type, although

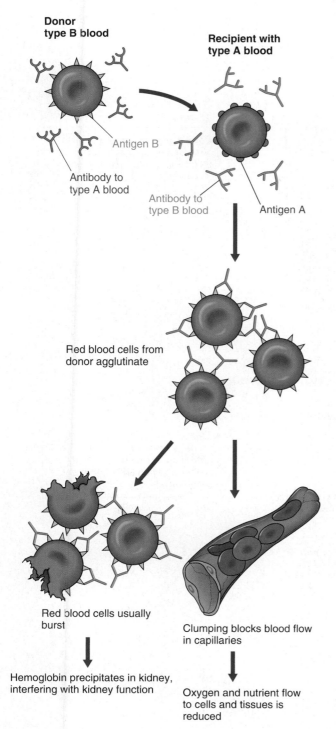

Donor type B blood

Recipient with type A blood

Antigen B

Antibody to type A blood

Antibody to type B blood

Antigen A

Red blood cells from donor agglutinate

Red blood cells usually burst

Clumping blocks blood flow in capillaries

Hemoglobin precipitates in kidney, interfering with kidney function

Oxygen and nutrient flow to cells and tissues is reduced

● **FIGURE 12-16**

Transfusion reaction
A transfusion reaction resulting from type B blood being transfused into a recipient with type A blood.

they can donate blood only to other AB people. Because their erythrocytes have both A and B antigens, their cells would be attacked if transfused into individuals with antibodies against either of these antigens.

The terms *universal donor* and *universal recipient* are somewhat misleading, however. In addition to the ABO system,

many other erythrocyte antigens and plasma antibodies can cause transfusion reactions, the most important of which is the Rh factor.

Other blood-group systems

People who have the **Rh factor** (an erythrocyte antigen first observed in rhesus monkeys, hence the designation *Rh*) are said to have *Rh-positive* blood, whereas those lacking the Rh factor are considered *Rh-negative*. In contrast to the ABO system, no naturally occurring antibodies develop against the Rh factor. Anti-Rh antibodies are produced only by Rh-negative individuals when (and if) they are first exposed to the foreign Rh antigen present in Rh-positive blood. A subsequent transfusion of Rh-positive blood could produce a transfusion reaction in such a sensitized Rh-negative person. Rh-positive individuals, in contrast, never produce antibodies against the Rh factor that they themselves possess. Therefore, Rh-negative people should be given only Rh-negative blood, whereas Rh-positive can safely receive either Rh-negative or Rh-positive blood. The Rh factor is of particular medical importance when an Rh-negative mother develops antibodies against the erythrocytes of an Rh-positive fetus she is carrying, a condition known as **erythroblastosis fetalis,** or **hemolytic disease of the newborn** (see p. 456).

Except in extreme emergencies, it is safest to individually cross-match blood before a transfusion is undertaken even though the ABO and Rh typing is already known, because there are approximately 12 other minor human erythrocyte antigen systems. Compatibility is determined by mixing the red blood cells from the potential donor with plasma from the recipient. If no clumping occurs, the blood is considered an adequate match for transfusion.

In addition to being an important consideration in transfusions, the various blood-group systems are also of legal importance in disputed paternity cases, because the erythrocyte antigens are inherited. In recent years, however, DNA "fingerprinting" has become a more definitive test.

▌ Lymphocytes respond only to antigens presented to them by antigen-presenting cells.

B cells typically cannot perform their task of antibody production without assistance from macrophages or other antigen-presenting cells and, in most cases, from T cells as well (● Figure 12-17). Relevant B-cell clones cannot recognize and produce antibodies in response to "raw" foreign antigens entering the body; before reacting to it, a B cell clone must be formally "introduced" to the antigen.

Antigen presentation

Using macrophages as an example of an **antigen-presenting cell,** invading organisms or other antigens are first engulfed by macrophages. These large phagocytes cluster around the appropriate B-cell clone and handle the formal introduction. During phagocytosis, the macrophage processes the raw antigen intracellularly and then "presents" the processed antigen

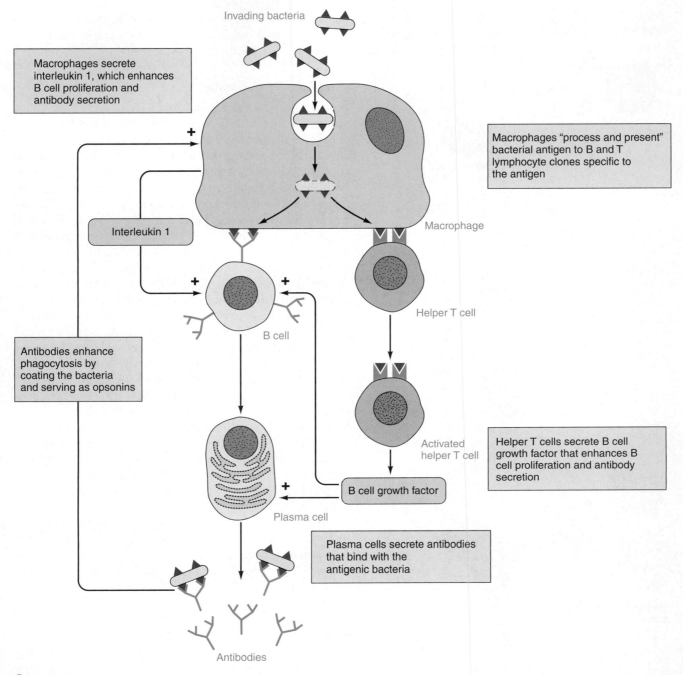

Macrophages secrete interleukin 1, which enhances B cell proliferation and antibody secretion

Invading bacteria

Interleukin 1

Macrophages "process and present" bacterial antigen to B and T lymphocyte clones specific to the antigen

Macrophage

Helper T cell

Antibodies enhance phagocytosis by coating the bacteria and serving as opsonins

B cell

Activated helper T cell

Helper T cells secrete B cell growth factor that enhances B cell proliferation and antibody secretion

B cell growth factor

Plasma cell

Plasma cells secrete antibodies that bind with the antigenic bacteria

Antibodies

● FIGURE 12-17

Synergistic interactions among macrophages, B cells, and helper T cells
B and T cells cannot react to a newly entering foreign antigen until the antigen has been processed and presented to them by macrophages or other antigen-presenting cells. Macrophages also secrete inter-leukin 1, which stimulates proliferation of the activated B cells. These B cells are transformed into plasma cells, which produce antibodies against the antigen. Activated helper T cells secrete B cell growth factor, which further stimulates B cell proliferation and antibody production. The antibodies not only lead to the demise of the foreign antigen but also serve as opsonins to enhance phagocytosis by the macrophages.

by exposing it on the outer surface of the macrophage's plasma membrane in such a way that the adjacent B cells can recognize and be activated by it. Specifically, when a macrophage engulfs a foreign microbe, it digests the microbe into antigenic peptides (small protein fragments). Each antigenic peptide is then bound to an **MHC molecule,** which is synthesized within the endoplasmic reticulum–Golgi complex. An MHC molecule

has a deep groove into which a variety of antigenic peptides can bind, depending on what the macrophage has engulfed. Loading of the antigenic peptide onto an MHC molecule takes place in a newly discovered specialized organelle within antigen-presenting cells, the **compartment for peptide loading.** The MHC molecule then transports the bound antigen to the cell surface where it is presented to passing lymphocytes.

In addition, these antigen-presenting macrophages secrete *interleukin 1*, a multipurpose chemical mediator that enhances the differentiation and proliferation of the now-activated B cell clone. Interleukin 1 (also known as endogenous pyrogen, or leukocyte endogenous mediator) is also largely responsible for the fever and malaise accompanying many infections. In collaborative fashion, activated lymphocytes secrete antibodies that, among other things, enhance further phagocytic activity.

Many antigens are similarly presented to T cells by macrophages and by closely related dendritic cells. **Dendritic cells** are specialized antigen-presenting cells that act as sentinels in almost every tissue. They are especially abundant in the skin and mucosal linings of the lungs and digestive tract—strategic locations where microbes are likely to enter the body. After exposure to the appropriate antigen, dendritic cells leave their tissue home and migrate through the lymphatic system to lymph nodes, where they cluster and activate T cells.

One specialized class of T lymphocytes, called helper T cells, help B cells on being activated by macrophage-presented antigen. The helper T cells secrete a chemical mediator, **B-cell growth factor,** which further contributes to B cell function in concert with the interleukin 1 that macrophages secreted. Therefore, mutually supportive interactions among macrophages, B cells, and helper T cells synergistically reinforce the phagocyte-antibody immune attack against the foreign intruder. ▲ Table 12-3 summarizes the innate and adaptive immune strategies that defend against bacterial invasion.

We are now going to turn our attention to the other roles of T cells besides enhancing B cell activity.

T LYMPHOCYTES: CELL-MEDIATED IMMUNITY

As important as B lymphocytes and their antibody products are in specific defense against invading bacteria and other foreign material, they represent only half of the body's specific immune defense corps. The T lymphocytes are equally important in defense against most viral infections and also play an important regulatory role in immune mechanisms. ▲ Table 12-4 compares the properties of these two adaptive effector cells, summarizing what you have already learned about B cells and previewing features you are about to learn about T cells.

∎ T cells bind directly with their targets.

Whereas B cells and antibodies defend against conspicuous invaders in the extracellular fluid, T cells defend against covert invaders that hide out inside cells where antibodies and the complement system cannot reach them. Unlike B cells, which secrete antibodies that can attack antigen at long distances, T cells do not secrete antibodies. Instead, they must directly contact their targets, a process known as *cell-mediated immunity*. T cells of the killer type release chemicals that destroy targeted cells that they contact, such as virus-infected cells and cancer cells.

Like B cells, T cells are clonal and exquisitely antigen specific. On its plasma membrane, each T cell bears unique receptor proteins, similar although not identical to the surface

---

▲ TABLE 12-3
Innate and Adaptive Immune Responses to Bacterial Invasion

Innate Immune Mechanisms	Adaptive Immune Mechanisms
Inflammation Engulfment of invading bacteria by resident tissue macrophages	Processing and presenting of bacterial antigen by macrophages to B cells specific to the antigen
Histamine-induced vascular responses to enhance delivery of increased blood flow to the area, bringing in additional immune-effector cells and plasma proteins	Proliferation and differentiation of the activated B cell clone into plasma cells and memory cells
Walling off of the invaded area by a fibrin clot	Secretion by plasma cells of customized antibodies, which specifically bind to invading bacteria
Emigration of neutrophils and monocytes/macrophages to the area to engulf and destroy foreign invaders and to remove cellular debris	Enhancement by interleukin 1 secreted by macrophages
	Enhancement by helper T cells, which have been activated by the same bacterial antigen processed and presented to them by macrophages
Secretion by phagocytic cells of chemical mediators, which enhance both innate and adaptive immune responses and induce local and systemic symptoms associated with an infection	Binding of antibodies to invading bacteria and enhancement of innate mechanisms that lead to the bacteria's destruction
	Action as opsonins to enhance phagocytic activity
Nonspecific activation of the complement system Formation of a hole-punching membrane attack complex that lyses bacterial cells	Activation of lethal complement system
	Stimulation of killer cells, which directly lyse bacteria
Enhancement of many steps of inflammation	Persistence of memory cells capable of responding more rapidly and more forcefully should the same bacteria be encountered again

▲ TABLE 12-4
B versus T Lymphocytes

Characteristic	B Lymphocytes	T Lymphocytes
Ancestral Origin	Bone marrow	Bone marrow
Site of Maturational Processing	Bone marrow	Thymus
Receptors for Antigen	Antibodies inserted in the plasma membrane serve as surface receptors; highly specific	Surface receptors present but differing from antibodies; highly specific
Bind with	Extracellular antigens such as bacteria, free viruses, and other circulating foreign material	Foreign antigen in association with self-antigen, such as virus-infected cells
Antigen Must Be Processed and Presented by macrophages	Yes	Yes
Types of Active Cells	Plasma cells	Cytotoxic T cells, helper T cells
Formation of Memory Cells	Yes	Yes
Type of Immunity	Antibody-mediated immunity	Cell-mediated immunity
Secretory Product	Antibodies	Cytokines
Function	Help eliminate free foreign invaders by enhancing nonspecific immune responses against them; provide immunity against most bacteria and a few viruses	Lyse virus-infected cells and cancer cells; provide immunity against most viruses and fungi and a few bacteria; aid B cells in antibody production
Life Span	Short	Long

receptors on B cells. Immature lymphocytes acquire their T cell receptors in the thymus during their differentiation into T cells. Unlike B cells, T cells are activated by foreign antigen only when it is on the surface of a cell that also carries a marker of the individual's own identity; that is, both foreign antigens and **self-antigens** must be on a cell's surface before a T cell can bind with it (with one important exception being whole transplanted foreign cells). During thymic education, T cells learn to recognize foreign antigens only in combination with the person's own tissue antigens—a lesson passed on to all T cells' future progeny. The importance of this dual antigen requirement and the nature of the self-antigens will be described shortly.

A delay of a few days generally follows exposure to the appropriate antigen before **sensitized,** or **activated,** T cells are prepared to launch a cell-mediated immune attack. When exposed to a specific antigen combination, cells of the complementary T cell clone proliferate and differentiate for several days, yielding large numbers of activated effector T cells that carry out various cell-mediated responses.

▌ The two types of T cells are cytotoxic T cells and helper T cells.

There are two main subpopulations of T cells, depending on their roles when activated by antigen:

1. **Cytotoxic T cells (killer T cells** or **CD8 cells),** which destroy host cells harboring anything foreign and thus bearing foreign antigen, such as body cells invaded by viruses, cancer cells that have mutated proteins resulting from malignant transformations, and transplanted cells.

2. **Helper T cells (CD4 cells),** which enhance the development of antigen-stimulated B cells into antibody-secreting plasma cells, enhance activity of the appropriate cytotoxic cells, and activate macrophages. Helper T cells do not directly participate in immune destruction of invading pathogens. Instead, they modulate activities of other immune cells.

Helper T cells are by far the most numerous T cells, making up 60% to 80% of circulating T cells. Because of the important role these cells play in "turning on" the full power of all the other activated lymphocytes and macrophages, helper T cells may constitute the immune system's "master switch." It is for this reason that **acquired immune deficiency syndrome (AIDS),** caused by the **human immunodeficiency virus (HIV),** is so devastating to the immune defense system. The AIDS virus selectively invades helper T cells, destroying or incapacitating the cells that normally orchestrate much of the immune response (● Figure 12-18). The virus also invades macrophages, further crippling the immune system, and sometimes enters brain cells as well, leading to the dementia (severe impairment of intellectual capacity) noted in some AIDS victims.

Memory T cells

Like B cells, T cells form a memory pool and display both primary and secondary responses. Primary responses tend to be initiated in the lymphoid tissues, where naive lymphocytes and antigen-presenting cells interact. During a few-week period

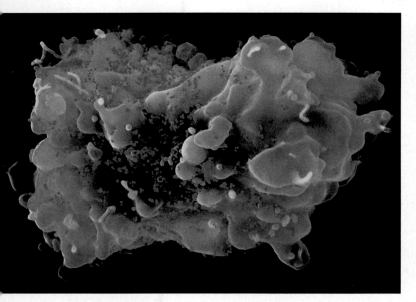

AIDS virus

Human immunodeficiency virus (HIV) (in gray), the AIDS-causing virus, on a helper T lymphocyte, HIV's primary target.

after the infection is cleared, over 90% of the huge number of effector T cells generated during the primary response die by means of *apoptosis* (cell suicide; see p. 70). To stay alive, activated T lymphocytes require the continued presence of their specific antigen and appropriate stimulatory signals. Once the foe succumbs, the vast majority of the now superfluous T lymphocytes commit suicide because their supportive antigen and stimulatory signals are withdrawn. Elimination of most of the effector T cells following a primary response is essential to prevent congestion in the lymphoid tissues. (Such paring down is not needed for B cells—those that become plasma cells and not memory B cells on antigen stimulation rapidly work themselves to death producing antibodies.) The remaining surviving effector T cells become long-lived memory T cells that migrate to all areas of the body, where they are poised for a swift secondary response to the same pathogen in the future.

▌ Cytotoxic T cells secrete chemicals that destroy target cells.

Cytotoxic T cells are microscopic "hit men." The targets of these destructive cells most frequently are host cells infected with viruses. When a virus invades a body cell, as it must to survive, the cell breaks down the envelope of proteins surrounding the virus and loads a fragment of this viral antigen piggyback onto a newly synthesized self-antigen. This self-antigen and viral antigen complex is inserted into the host cell's surface membrane, where it serves as a red flag indicating the cell is harboring the invader (● Figure 12-19, steps ①and ②). To attack the intracellular virus, cytotoxic T cells must destroy the infected host cell in the process. Cytotoxic T cells of the clone specific for this particular virus recognize and bind to the viral antigens and self-antigens on the surface of the infected cell

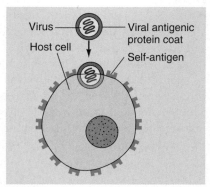

step ① A virus invades a host cell.

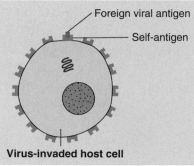

step ② The viral antigen is displayed on the surface of the host cell alongside the cell's self-antigen.

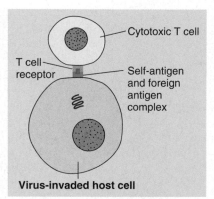

step ③ The cytotoxic T cell recognizes and binds with a specific foreign antigen (viral antigen) in association with the self-antigen.

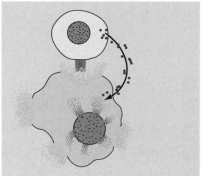

step ④ The cytotoxic T cell releases chemicals that destroy the attacked cell before the virus can enter the nucleus and start to replicate.

(● Figure 12-19, step ③). Thus sensitized by viral antigen, a cytotoxic T cell either directly kills the victim cell by releasing chemicals that lyse the attacked cell before viral replication can begin (● Figure 12-19, step ④) or indirectly destroys the infected cell by signaling it to commit suicide.

The direct means by which cytotoxic T cells as well as NK cells destroy a targeted cell is by releasing **perforin** molecules, which penetrate into the target cell's surface membrane and join together to form porelike channels (● Figure 12-20). This technique of killing a cell by punching holes in its membrane is similar to the method employed by the membrane attack complex of the complement cascade. This contact-dependent mechanism of killing has been termed the "kiss of death." Cytotoxic T cells can also indirectly bring about death of infected host cells by releasing **granzymes**, which are enzymes similar to digestive enzymes. Granzymes enter the target cell through the perforin channels. Once inside, these chemicals trigger the virus-infected cell to self-destruct through apoptosis.

The virus released on destruction of the host cell is directly destroyed in the extracellular fluid by phagocytic cells, neu-

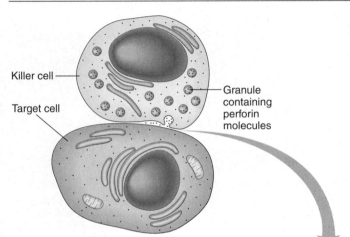

● **FIGURE 12-20**

Mechanism of killing by killer cells
(a) Details of the killing process. (b) Enlargement of perforin-formed pores in a target cell. Note the similarity to the membrane-attack complex formed by complement molecules (see Figure 12-8, p. 423).

(SOURCE: Adapted from the illustration by Dana Burns-Pizer in "How Killer Cells Kill," by John Ding-E Young and Zanvil A. Cohn. Copyright © 1988 *Scientific American*, Scientific American, Inc. All rights reserved.)

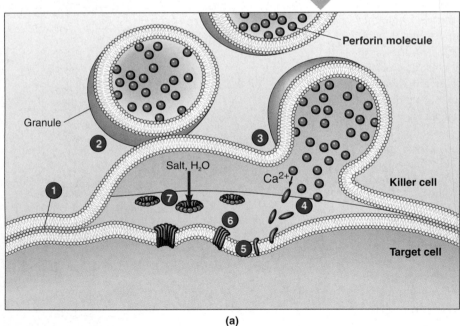

(a)

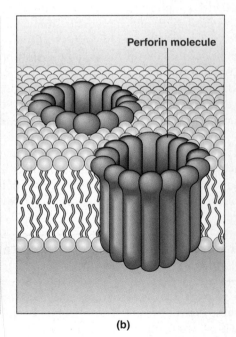

(b)

① The killer cell binds to its target cell.

② As a result of this binding, the killer cell's perforin-containing granules fuse with the plasma membrane.

③ The granules disgorge their perforin by exocytosis into a small pocket of intercellular space between the killer cell and its target.

④ On exposure to Ca^{2+} in this ECF space, the individual perforin molecules change from a spherical to a cylindrical shape.

⑤ The remodeled perforin molecules bind to the target cell membrane and insert into it.

⑥ Individual perforin molecules group together like staves of a barrel to form pores.

⑦ The pores admit salt and H_2O, causing the target cell to swell and burst.

tralizing antibodies, and the complement system. Meanwhile, the cytotoxic T cell, which has not been harmed in the process, can move on to kill other infected host cells.

The surrounding healthy cells replace the lost cells by means of cell division. Usually, to halt a viral infection only some of the host cells must be destroyed. If the virus has had a chance to multiply, however, with replicated virus leaving the original cell and spreading to other host cells, the cytotoxic T-cell defense mechanism may sacrifice so many of the host cells that serious malfunction may ensue.

Recall that other nonspecific defense mechanisms also come into play to combat viral infections, most notably NK cells, interferon, macrophages, and the complement system. As usual, an intricate web of interplay exists among the immune defenses that are launched against viral invaders (▲ Table 12-5).

Antiviral defense in the nervous system

The usual method of destroying virus-infected host cells is not appropriate for the nervous system. If cytotoxic T cells destroyed virus-infected neurons, the lost cells could not be replaced, because neurons cannot reproduce. Fortunately, virus-infected neurons are spared from extermination by the immune system, but how, then, are neurons protected from viruses? Immunologists long thought that the only antiviral defenses for neurons were those aimed at free viruses in the extracellular fluid. Surprising new research has revealed, however, that antibodies not only target viruses for destruction in the extracellular fluid but can also eliminate viruses inside neurons. It is unclear whether antibodies actually enter the neurons and interfere directly with viral replication (neurons have been shown to take up antibodies near their synaptic endings) or bind with the surface of nerve cells and trigger intracellular changes that stop viral replication. The fact that some viruses, such as the herpesvirus, persist for years in nerve cells, occasionally "flaring up" to produce symptoms, demonstrates that the antibodies' intraneuronal mechanism does not provide a foolproof antiviral defense for neurons.

■ Helper T cells secrete chemicals that amplify the activity of other immune cells.

In contrast to cytotoxic T cells, helper T cells are not killer cells. Instead, helper T cells secrete chemicals classified as cytokines that "help," or augment, nearly all aspects of the immune response.

Cytokines

Exposure to antigen frequently activates both the B and T cell mechanisms simultaneously. Just as the helper T cells can modulate the secretion of antibody by B cells, antibodies may either enhance or block the ability of cytotoxic T cells to destroy a victim cell, depending on the circumstances. Most effects that lymphocytes exert on other immune cells are mediated by the secretion of chemical messengers. All chemicals other than antibodies that leukocytes secrete are collectively called **cytokines,** most of which are produced by helper T cells. Unlike antibodies, cytokines do not interact directly with the antigen responsible for inducing their production. Instead,

---

▲ **TABLE 12-5**
Defenses against Viral Invasion

When the virus is free in the extracellular fluid,

Macrophages

Destroy the free virus by phagocytosis.

Process and present the viral antigen to both B and T cells.

Secrete interleukin 1, which activates B and T cell clones specific to the viral antigen.

Plasma Cells Derived from B Cells Specific to the Viral Antigen Secrete Antibodies That

Neutralize the virus to prevent its entry into a host cell.

Activate the complement cascade that directly destroys the free virus and enhances phagocytosis of the virus by acting as an opsonin.

When the virus has entered a host cell (which it must do to survive and multiply, with the replicated viruses leaving the original host cell to enter the extracellular fluid in search of other host cells),

Interferon

Is secreted by virus-infected cells.

Binds with and prevents viral replication in other host cells.

Enhances the killing power of macrophages, natural killer cells, and cytotoxic T cells.

Natural Killer Cells

Nonspecifically lyse virus-infected host cells.

Cytotoxic T Cells

Are specifically sensitized by the viral antigen; lyse the infected host cells before the virus has a chance to replicate.

Helper T Cells

Secrete cytokines, which enhance cytotoxic T-cell activity and B-cell antibody production.

When a virus-infected cell is destroyed, the free virus is released into the extracellular fluid, where it is attacked directly by macrophages, antibodies, and the activated complement components.

---

cytokines spur other immune cells into action to help ward off the invader. The following are among the best known of the helper T-cell cytokines:

1. As noted earlier, helper T cells secrete *B-cell growth factor,* which enhances the antibody-secreting ability of the activated B-cell clone. Antibody secretion is greatly reduced in the absence of helper T cells.

2. Helper T cells similarly secrete **T-cell growth factor,** also known as **interleukin 2 (IL-2),** which augments the activity of cytotoxic T cells and even of other helper T cells responsive to

the invading antigen. In typical interplay fashion, interleukin 1 secreted by macrophages not only enhances the activity of both the appropriate B- and T-cell clones but also stimulates secretion of interleukin 2 by activated helper T cells. (The 16 known interleukins that mediate interactions between various leukocytes—*interleukin* means "between leukocytes"—were numbered in the order of their discovery.)

3. Some chemicals secreted by T cells act as *chemotaxins* to lure more neutrophils and macrophages-to-be to the invaded area.

4. Once macrophages are attracted to the area, **macrophage-migration inhibition factor,** another important cytokine released from helper T cells, keeps these large phagocytic cells in the region by inhibiting their outward migration. As a result, a great number of chemotactically attracted macrophages accumulate in the infected area. This factor also confers greater phagocytic power on the gathered macrophages. These so-called **angry macrophages** have more powerful destructive ability. They are especially important in defending against the bacteria that cause tuberculosis, because such microbes can survive simple phagocytosis by nonactivated macrophages.

5. Some cytokines secreted by helper T cells activate eosinophils and promote the development of IgE antibodies for defense against parasitic worms.

T helper 1 and T helper 2 cells

Recent studies have demonstrated the existence of two subsets of helper T cells—**T helper 1 (T$_H$1)** and **T helper 2 (T$_H$2) cells.** T$_H$1 and T$_H$2 cells augment different patterns of immune responses by secreting different types of cytokines. T$_H$1 cells rally a cell-mediated (cytotoxic T cell) response, which is appropriate for infections with intracellular microbes, such as viruses, whereas T$_H$2 cells promote antibody-mediated immunity by B cells and rev up eosinophil activity for defense against parasitic worms.

Helper T cells produced in the thymus are in a naive state until they encounter the antigen they are primed to recognize. Whether a naive helper T cell becomes a T$_H$1 or T$_H$2 cell depends on which cytokines are secreted by the dendritic cell or macrophage as it presents the antigen to the naive T cell. **Interleukin 12 (IL-12)** drives a naive T cell specific for the antigen to become a T$_H$1 cell, whereas **interleukin 4 (IL-4)** favors the development of a naive cell into a T$_H$2 cell. Thus the antigen-presenting cells of the nonspecific immune system can influence the whole tenor of the specific immune response by determining whether the T$_H$1 or T$_H$2 cellular subset dominates. In the usual case, the secreted cytokines promote the appropriate specific immune response against the particular threat at hand.

What normally prevents the adaptive immune system from unleashing its powerful defense capabilities against the body's own self-antigens? We will examine this issue next.

▮ The immune system is normally tolerant of self-antigens.

The term **tolerance** refers to the phenomenon of preventing the immune system from attacking the person's own tissues.

During the genetic "cut, shuffle, and paste process" that goes on during lymphocyte development, some B and T cells are by chance formed that could react against the body's own tissue antigens. If these lymphocyte clones were allowed to function, they would destroy the individual's own body. Fortunately, the immune system normally does not produce antibodies or activated T cells against the body's own self-antigens but instead directs its destructive tactics only at foreign antigens.

At least five different mechanisms are involved in tolerance:

1. *Clonal deletion.* In response to continuous exposure to body antigens early in development, lymphocyte clones specifically capable of attacking these self-antigens in most cases are permanently destroyed. This **clonal deletion** is accomplished by triggering apoptosis of immature cells that would react with the body's own proteins. This physical elimination is the major mechanism by which tolerance is developed.

2. *Clonal anergy.* The premise of **clonal anergy** is that a lymphocyte must receive two specific simultaneous signals to be activated (turned on), one from its compatible antigen and a stimulatory cosignal molecule known as **B7** found only on the surface of an antigen-presenting cell. Both signals are present for foreign antigens, which are introduced to lymphocytes by antigen-presenting cells. Once a B or T cell is turned on by finding its matching antigen in accompaniment with the cosignal, the cell no longer needs the cosignal to interact with other cells. For example, an activated cytotoxic T cell can destroy any virus-invaded cell that bears the viral antigen even though the infected cell does not possess the cosignal. In contrast, these dual signals—antigen plus cosignal—never are present for self-antigens because these antigens are not handled by cosignal-bearing antigen-presenting cells. The first exposure to a single signal from a self-antigen turns *off* the compatible T cell, rendering the cell unresponsive to further exposure to the antigen instead of spurring the cell to proliferate. This reaction is referred to as *clonal anergy* (*anergy* means "lack of energy"), because T cells are being inactivated (that is, "become lazy") rather than activated by their antigens. Clonal anergy is a backup to clonal deletion. Anergized lymphocyte clones survive but they are unable to function.

3. *Receptor editing.* A newly identified means of ridding the body of self-reactive B cells is **receptor editing.** With this mechanism, once a B cell that bears a receptor for one of the body's own antigens encounters the self-antigen, the B cell escapes death or a lifetime of anergy by swiftly changing its antigen receptor to a nonself version. In this way, an originally self-reactive B cell survives but is "rehabilitated" so that it will never target the body's own tissues again.

4. *Antigen sequestering.* Some self-molecules are normally hidden from the immune system, because they never come into direct contact with the extracellular fluid in which the immune cells and their products circulate. An example of such a segregated antigen is thyroglobulin, a complex protein sequestered within the hormone-secreting structures of the thyroid gland.

5. *Immune privilege.* A few tissues, most notably the testes and the eyes, have **immune privilege,** because they escape immune attack even when they are transplanted in an unrelated

individual. Scientists recently discovered that the cellular plasma membranes in these immune-privileged tissues possess a specific molecule that triggers apoptosis of approaching activated lymphocytes that could attack the tissues.

▎ Autoimmune diseases arise from loss of tolerance to self-antigens.

Occasionally, the immune system fails to distinguish between self-antigens and foreign antigens, and unleashes its deadly powers against one or more of the body's own tissues. A condition in which the immune system fails to recognize and tolerate self-antigens associated with particular tissues is known as an **autoimmune disease.** Autoimmunity underlies more than 80 diseases, many of which are well known. Examples include multiple sclerosis, rheumatoid arthritis, and Type I diabetes mellitus. About 50 million Americans suffer from some type of autoimmune disease, with the incidence being about three times higher in females than in males.

Autoimmune diseases may arise from a number of different causes:

1. Exposure of normally inaccessible self-antigens sometimes induces an immune attack against these antigens. Because the immune system is usually never exposed to hidden self-antigens, it does not "learn" to tolerate them. Inadvertent exposure of these normally inaccessible antigens to the immune system because of tissue disruption caused by injury or disease can lead to a rapid immune attack against the affected tissue, just as if these self-proteins were foreign invaders. *Hashimoto's disease,* which involves the production of antibodies against thyroglobulin and the destruction of the thyroid gland's hormone-secreting capacity, is one such example.

2. Normal self-antigens may be modified by factors such as drugs, environmental chemicals, viruses, or genetic mutations so that they are no longer recognized and tolerated by the immune system.

3. Exposure of the immune system to a foreign antigen structurally almost identical to a self-antigen may induce the production of antibodies or activated T lymphocytes that not only interact with the foreign antigen but also cross-react with the closely similar body antigen. An example of this molecular mimicry is the streptococcal bacteria responsible for "strep throat." The bacteria possess antigens that are structurally very similar to self-antigens in the tissue covering the heart valves of some individuals, in which case the antibodies produced against the streptococcal organisms may also bind with this heart tissue. The resultant inflammatory response is responsible for the heart valve lesions associated with *rheumatic fever.*

4. New studies hint at another possible trigger of autoimmune diseases, one that could explain why a whole host of these disorders are more common in women than in men. Traditionally, scientists have speculated that the sex bias of autoimmune diseases was somehow related to hormonal differences. Recent findings suggest, however, that the higher incidence of these self-destructive conditions in females may be a legacy of pregnancy. Researchers have learned that fetal cells, which often gain access to the mother's bloodstream during the trauma of labor and delivery, sometimes linger in the mother for decades after the pregnancy. The investigators believe that the immune system typically clears these cells from the mother's body following childbirth, but studies involving one particular autoimmune disease demonstrated that those women with the condition were more likely than healthy women to have persistent fetal cells in their blood. The persistence of similar but not identical fetal antigens that were not wiped out early on as being foreigners may somehow trigger a gradual, more subtle immune attack that eventually turns against the mother's own closely related antigens.

What is the nature of the self-antigens that the immune system learns to recognize as markers of a person's own cells? That is the topic of the next section.

▎ The major histocompatibility complex is the code for self-antigens.

Self-antigens are plasma membrane–bound glycoproteins (proteins with sugar attached) known as **MHC molecules** because their synthesis is directed by a group of genes called the **major histocompatibility complex** or **MHC.** These are the same MHC molecules that escort engulfed foreign antigen to the cell surface for presentation by antigen-presenting cells. The MHC genes are the most variable ones in humans. More than 100 different MHC molecules have been identified in human tissue, but each individual has a code for only 3 to 6 of these possible antigens. Because of the tremendous number of different combinations possible, the exact pattern of MHC molecules varies from one individual to another, much like a "biochemical fingerprint" or "molecular identification card," except in identical twins, who have the same MHC self-antigens.

The major histocompatibility (*histo* means "tissue"; *compatibility* means "ability to get along") complex was so named because these genes and the self-antigens they encode were first discerned in relation to tissue typing (similar to blood typing), which is done to obtain the most compatible matches for tissue grafting and transplantation. However, the transfer of tissue from one individual to another does not normally occur in nature. The natural function of MHC antigens lies in their ability to direct the responses of T cells, not in their artificial role in the rejection of transplanted tissue.

MHC molecules alone on a cell surface signal to immune cells, "Leave me alone, I'm one of you." T cells typically bind with MHC self-antigens only when they are in association with a foreign antigen, such as a viral protein, also displayed on the cell surface in a groove on the top of the MHC molecule. Thus T cell receptors bind only with body cells making the statement—by bearing both self- and nonself-antigens on their surface—"I, one of your own kind, have been invaded. Here's a description of the enemy I am housing within." Only T cells that specifically match up with both the self- and foreign antigen can bind with the infected cell.

Loading of foreign peptide on MHC molecule

Unlike B cells, T cells cannot bind with foreign antigen that is not in association with self-antigen. It would be futile for T cells to bind with free, extracellular antigen—they cannot defend against foreign material unless it is intracellular. A foreign protein first must be enzymatically broken down within a body cell into small fragments known as **peptides.** These antigenic peptides are inserted into the binding groove of a newly synthesized MHC molecule before the MHC–foreign antigen complex travels to the surface membrane. Once displayed at the cell surface, the combined presence of these self- and nonself-antigens alerts the immune system to the presence of an undesirable agent within the cell. Highly specific T cell receptors fit a particular MHC–foreign antigen complex in complementary fashion. This binding arrangement can be likened to a hot dog in a bun, with the MHC molecule being the bottom of the bun, the T cell receptor the bun's top, and the foreign antigen the hot dog (● Figure 12-21). In the case of cytotoxic T cells, the outcome of this binding is destruction of the infected body cell. Because cytotoxic T cells do not bind to MHC self-antigens in the absence of foreign antigen, normal body cells are protected from lethal immune attack.

Class I and Class II MHC glycoproteins

T cells become active only when they match a given MHC–foreign peptide combination. In addition to having to fit a specific foreign peptide, the T cell receptor must also match the appropriate MHC protein. Each individual has two main classes of MHC-encoded molecules that are differentially recognized by cytotoxic T and helper T cells—class I and class II MHC glycoproteins, respectively (● Figure 12-22). The class I and II markers serve as signposts to guide cytotoxic and helper T cells to the precise cellular locations where their immune capabilities can be most effective.

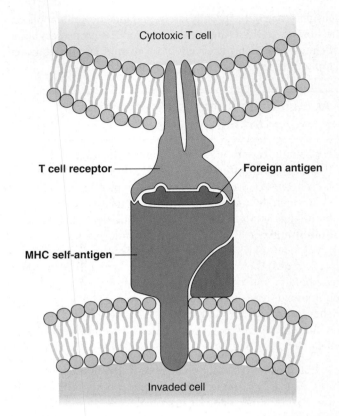

● FIGURE 12-21
Binding of a T cell receptor with an MHC self-antigen and foreign antigen complex

● FIGURE 12-22

Distinctions between class I and class II major histocompatibility complex (MHC) glycoproteins
Specific binding requirements for the two types of T cells ensure that these cells bind only with the targets that they can influence. Cytotoxic T cells can recognize and bind with foreign antigen only when the antigen is in association with class I MHC glycoproteins, which are found on the surface of all body cells. This requirement is met when a virus invades a body cell, whereupon the cell is destroyed by the cytotoxic T cells. Helper T cells, which enhance the activities of other immune cells, can recognize and bind with foreign antigen only when it is in association with class II MHC glycoproteins, which are found only on the surface of these other immune cells.

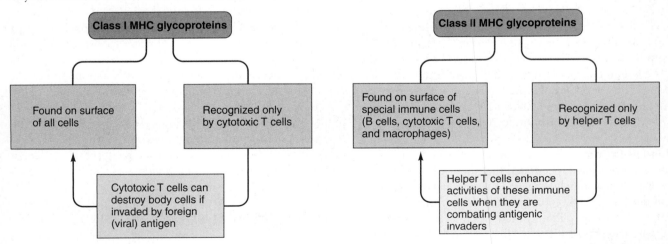

Cytotoxic T cells can respond to foreign antigen only in association with **class I MHC glycoproteins,** which are found on the surface of virtually all nucleated body cells. To carry out their role of dealing with pathogens that have invaded host cells, it is appropriate that cytotoxic T cells bind only with cells of the organism's own body that viruses have infected—that is, with foreign antigen in association with self-antigen. Furthermore, these deadly T cells can also link up with any cancerous body cell, because class I MHC molecules also display mutated cellular proteins characteristic of these abnormal cells. Because any nucleated body cell can be invaded by viruses or become cancerous, essentially all cells display class I MHC glycoproteins, enabling cytotoxic T cells to attack any virus-invaded host cell or any cancer cell.

In contrast, **class II MHC glycoproteins,** which are recognized by helper T cells, are restricted to the surface of a few special types of immune cells. That is, a helper T cell can bind with foreign antigen only when it is found on the surfaces of immune cells with which the helper T cell interacts. These include the macrophages, which present antigen to helper T cells, as well as B cells and cytotoxic T cells, whose activities are enhanced by helper T cells. The capabilities of helper T cells would be squandered if these cells were able to bind with body cells other than these special immune cells. In this way, the specific binding requirements for the two types of T cells help ensure the appropriate T cell responses.

Transplant rejection

T cells do bind with MHC antigens present on the surface of *transplanted cells* in the absence of foreign viral antigen. The ensuing destruction of the transplanted cells is responsible for the rejection of transplanted or grafted tissues. Presumably, some of the recipient's T cells "mistake" the MHC antigens of the donor cells for a closely resembling combination of a conventional viral foreign antigen complexed with the recipient's MHC self-antigens.

To minimize the rejection phenomenon, the tissues of donor and recipient are matched according to MHC antigens as closely as possible. Therapeutic procedures to suppress the immune system then follow. In the past, the primary immunosuppressive tools included radiation therapy and drugs aimed at destroying the actively multiplying lymphocyte populations, plus anti-inflammatory drugs that suppressed growth of all lymphoid tissue. However, these measures not only suppressed the T cells that were primarily responsible for rejecting transplanted tissue but also depleted the antibody-secreting B cells. Unfortunately, the treated individual was left with little specific immune protection against bacterial and viral infections. In recent years, new therapeutic agents have become extremely useful in selectively depressing T cell–mediated immune activity while leaving B-cell humoral immunity essentially intact. For example, *cyclosporin* blocks interleukin 2, the cytokine secreted by helper T cells that is required for expansion of the selected cytotoxic T-cell clone. Furthermore, a new technique under investigation may completely prevent rejection of transplanted tissues even from an unmatched donor. This technique involves the use of tailor-made antibodies that block spe-

cific facets of the rejection process. If proven safe and effective, the technique will have a tremendous impact on tissue transplantation.

CD47: A new self-antigen

Researchers recently identified a new self-antigen in addition to the MHC-encoded markers. Macrophages have been shown to recognize **CD47,** a self-marker on red blood cells. Scientists expect to discover other self-antigens in the future.

Let's now look in more detail at the role of T cells in defending against cancer.

▌ Immune surveillance against cancer cells involves an interplay among immune cells and interferon.

Besides destroying virus-infected host cells, another important function of the T cell system is recognizing and destroying newly arisen, potentially cancerous tumor cells before they have a chance to multiply and spread, a process known as **immune surveillance.** At least once a day on average, your immune system destroys a mutated cell that could potentially become cancerous. Any normal cell may be transformed into a cancer cell if mutations occur within its genes responsible for controlling cell division and growth. Such mutations may occur by chance alone or, more frequently, by exposure to **carcinogenic** (cancer-causing) factors such as ionizing radiation, certain environmental chemicals, certain viruses, or physical irritants.

Benign and malignant tumors

Cellular multiplication and growth are normally under strict control, but the regulatory mechanisms are largely unknown. Cell multiplication in an adult is generally restricted to replacing lost cells. Furthermore, cells normally respect their own place and space in the body's society of cells. If a cell that has been transformed into a tumor cell manages to escape destruction, however, it defies the normal controls on its proliferation and position. Unrestricted multiplication of a single tumor cell results in a **tumor** that consists of a clone of cells identical to the original mutated cell.

If the mass is slow growing, stays put in its original location, and does not infiltrate the surrounding tissue, it is considered a **benign** tumor. In contrast, the transformed cell may multiply rapidly and form an invasive mass that lacks the "altruistic" behavior characteristic of normal cells. Such invasive tumors are known as **malignant tumors,** or **cancer.** Malignant tumor cells usually do not adhere well to the neighboring normal cells, so that often some of the cancer cells break away from the parent tumor. These "emigrant" cancer cells are transported through the blood to new territories, where they continue to proliferate, forming multiple malignant tumors. The term **metastasis** is applied to this spreading of cancer to other parts of the body.

If a malignant tumor is detected early, before it has metastasized, it can be removed surgically. Once cancer cells have dispersed and seeded multiple cancerous sites, surgical elimi-

Normal ciliated respiratory airway cells   Cancer cells

● **FIGURE 12-23**

Comparison of normal and cancerous cells in the large respiratory airways
The normal cells display specialized cilia, which constantly contract in whiplike motion to sweep debris and micro-organisms from the respiratory airways so they do not gain entrance to the deeper portions of the lungs. The cancerous cells are not ciliated, so they are unable to perform this specialized defense task.

nation of the malignancy is impossible. In this case, agents that interfere with rapidly dividing and growing cells, such as certain chemotherapeutic drugs, are used in an attempt to destroy the malignant cells. Unfortunately, these agents also harm normal body cells, especially rapidly proliferating cells such as blood cells and the cells lining the digestive tract.

Untreated cancer is eventually fatal in most cases, for several interrelated reasons. The uncontrollably growing malignant mass crowds out normal cells by vigorously competing with them for space and nutrients, yet the cancer cells cannot take over the functions of the cells they are destroying. Cancer cells typically remain immature and do not become specialized, often resembling embryonic cells instead (● Figure 12-23). Such poorly differentiated malignant cells lack the ability to perform the specialized functions of the normal cell type from which they mutated. Affected organs gradually become disrupted to the point that they are no longer able to perform their life-sustaining functions, and the person dies.

Genetic mutations that do not lead to cancer

Even though many body cells undergo mutations throughout a person's lifetime, most of these mutations do not result in malignancy, for three reasons:

1. Only a fraction of the mutations involve loss of control over the cell's growth and multiplication. More frequently, other facets of cellular function are altered.

2. A cell usually becomes cancerous only after an accumulation of multiple independent mutations. This requirement contributes at least in part to the much higher incidence of cancer in older individuals, in whom mutations have had more time to accumulate in a single cell lineage. Alternatively, a few cancers are caused by tumor viruses, which permanently alter particular DNA sequences of the cells they invade.

3. Potentially cancerous cells that do arise are usually destroyed by the immune system early in their development. Presumably, the immune system recognizes cancer cells because they bear new and different surface antigens alongside the cell's normal self-antigens, because of either genetic mutation or invasion by a tumor virus.

Effectors of immune surveillance

Immune surveillance against cancer depends on an interplay among three types of immune cells—*cytotoxic T cells, NK cells,* and *macrophages*—as well as *interferon.* Not only are all three of these immune cell types able to attack and destroy cancer cells directly, but all of them also secrete interferon. Interferon in turn inhibits multiplication of cancer cells and increases the killing ability of the immune cells (● Figure 12-24).

Because NK cells do not require prior exposure and sensitization to a cancer cell before being able to launch a lethal attack, they are the first line of defense against cancer. In addition, cytotoxic T cells take aim at abnormal cancer cells bearing mutated cellular proteins on their surface in conjunction with the class I MHC molecules. Cytotoxic T cells are believed to be

● **FIGURE 12-24**

Immune surveillance against cancer
Anticancer interactions of cytotoxic T cells, natural killer cells, macrophages, and interferon.

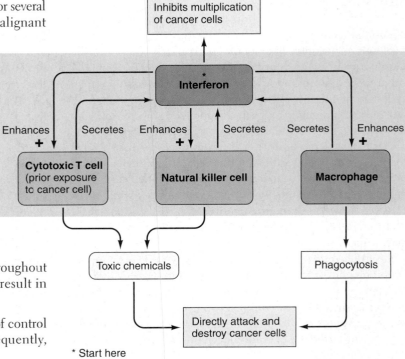

* Start here

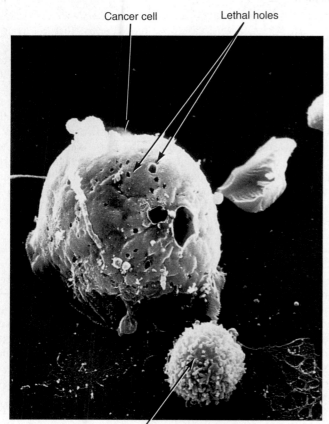

Cancer cell   Lethal holes

Cytotoxic T cell

● FIGURE 12-25

A cytotoxic T cell destroying a cancer cell
On contacting a cancer cell with which it can specifically bind,
a cytotoxic T cell releases toxic chemicals such as perforin, which
destroy the cancer cell.

especially important in defending against the few kinds of
virus-induced cancer. On contacting a cancer cell, both these
killer cells release perforin and other toxic chemicals that destroy the targeted mutant cell (● Figure 12-25). Macrophages,
in addition to clearing away the remains of the dead victim cell,
are able to engulf and destroy cancer cells intracellularly.

The fact that cancer does sometimes occur means that
cancer cells occasionally escape these immune mechanisms.
Some cancer cells are believed to survive by evading immune
detection, for example, by failing to display identifying antigens on their surface or by being surrounded by counterproductive **blocking antibodies** that interfere with T cell function. Although B cells and antibodies are not believed to play
a direct role in cancer defense, B cells, on viewing a mutant
cancer cell as an alien to normal self, may produce antibodies
against it. These antibodies, for unknown reasons, do not activate the complement system, which could destroy the cancer
cells. Instead, the antibodies are able to bind with the antigenic sites on the cancer cell, "hiding" these sites from recognition by cytotoxic T cells. The coating of a tumor cell by
blocking antibodies thus protects the harmful cell from attack
by deadly T cells. A new finding reveals that still other suc-

cessful cancer cells thwart immune attack by turning on their
pursuers. They induce the T cells that bind with them to commit suicide.

A regulatory loop links the immune system with the nervous and endocrine systems.

From the preceding discussion, it is obvious that complex controlling factors operate within the immune system itself. Until
recently, the immune system was believed to function independently of other control systems in the body. Investigations
now indicate, however, that the immune system both influences and is influenced by the two major regulatory systems,
the nervous and endocrine systems. For example, interleukin 1
can turn on the stress response by activating a sequence of nervous and endocrine events that result in the secretion of cortisol, one of the major hormones released during stress. This
linkage between a mediator of the immune response and a mediator of the stress response is appropriate. Cortisol mobilizes
the body's nutrient stores so that metabolic fuel is readily available to keep pace with the body's energy demands at a time
when the person is sick and may not be eating enough (or, in
the case of an animal, may not be able to search for food). Furthermore, cortisol mobilizes amino acids, which serve as building blocks to repair any tissue damage sustained during the
encounter that triggered the immune response.

In the reverse direction, lymphocytes and macrophages are
responsive to blood-borne signals from the nervous system and
from certain endocrine glands. These important immune cells
possess receptors for a wide variety of neurotransmitters, hormones, and other chemical mediators. For example, cortisol
and other chemical mediators of the stress response have a profound immunosuppressive effect, inhibiting many functions of
lymphocytes and macrophages and decreasing the production
of cytokines. Thus a negative-feedback loop appears to exist
between the immune system and the nervous and endocrine
systems. Cytokines released by immune cells enhance the neurally and hormonally controlled stress response, whereas cortisol and related chemical mediators released during the stress response suppress the immune system. In large part because stress
suppresses the immune system, stressful physical, psychological, and social life events are linked with increased susceptibility to infections and cancer. Thus the body's resistance to disease *can* be influenced by the person's mental state—a case of
"mind over matter."

There are other important links between the immune and
nervous systems in addition to the cortisol connection. For example, many immune system organs such as the thymus, spleen,
and lymph nodes are innervated by the sympathetic nervous
system, the branch of the nervous system called into play during stress-related "fight-or-flight" situations (see p. 240). In the
reverse direction, immune system secretions act on the brain
to produce fever and other general symptoms that accompany
infections. (For a discussion of the possible effects of exercise
on immune defense, see the accompanying boxed feature, ▶ A
Closer Look at Exercise Physiology.)

Exercise: A Help or Hindrance to Immune Defense?

For years people who engage in moderate exercise regimens have claimed they have fewer colds when they are in good aerobic condition. In contrast, elite athletes and their coaches have often complained about the number of upper respiratory infections that the athletes seem to contract at the height of their competitive seasons. The results of recent scientific studies lend support to both these claims. The impact of exercise on immune defense depends on the intensity of the exercise.

Animal studies have shown that high-intensity exercise after experimentally induced infection results in more severe infection. Moderate exercise performed prior to infection or to tumor implantation, in contrast, results in less severe infection and slower tumor growth in experimental animals.

Recent studies on humans lend further support to the hypothesis that exhaustive exercise suppresses immune defense whereas moderate exercise stimulates the immune system. A survey of 2,300 runners competing in the 1987 Los Angeles Marathon indicated that those who trained more than 60 miles a week had twice the number of respiratory infections of those who trained less than 20 miles a week in the two months preceding the race. In another study, 10 elite athletes were asked to run on a treadmill for three hours at the same pace they would run in competition. Blood tests after the run indicated that natural killer cell activity had decreased by 25% to 50%, and this decrease lasted for six hours. The runners also showed a 60% increase in the stress hormone cortisol, which is known to suppress immunity. Other studies have shown that athletes have lower resting salivary IgA levels compared with control subjects and that their respiratory mucosal immunoglobulins are decreased after prolonged exhaustive exercise. These results suggest a lower resistance to respiratory infection following high-intensity exercise. Because of these results, researchers in the field recommend that athletes keep exposure to respiratory viruses to a minimum by avoiding crowded places or anyone with a cold or flu for the first six hours after strenuous competition.

However, a study evaluating the effects of a moderate exercise program in which a group of women walked 45 minutes a day, 5 days a week, for 15 weeks found that the walkers' antibody levels and natural killer cell activity increased throughout the exercise program. Other studies using moderate exercise on stationary bicycles in subjects over the age of 65 showed increases in natural killer cell activity as large as those found in young people.

Unfortunately, the few studies conducted on those infected with human immunodeficiency virus (HIV, the AIDS virus) have not found an improvement in immune function with exercise. The studies have shown that HIV-positive patients can gain strength through resistance training and improve psychological well-being through exercise and that they suffer no detrimental effects from moderate exercise.

IMMUNE DISEASES

Abnormal functioning of the immune system can lead to immune diseases in two general ways: *immune deficiency diseases* (too little immune response) and *inappropriate immune attacks* (too much or mistargeted immune response).

Immune deficiency diseases result from insufficient immune responses.

Immune deficiency diseases occur when the immune system fails to respond adequately to foreign invasion. The condition may be congenital (present at birth) or acquired (nonhereditary), and it may specifically involve impairment of either antibody-mediated immunity, cell-mediated immunity, or both. In a rare hereditary condition known as **severe combined immunodeficiency**, both B and T cells are lacking. Its victims have extremely limited defenses against pathogenic organisms and usually die in infancy unless maintained in a germ-free environment (that is, live in a "bubble"). Acquired (nonhereditary) immune deficiency states can arise from inadvertent destruction of lymphoid tissue during prolonged therapy with anti-inflammatory agents, such as cortisol derivatives, or from cancer therapy aimed at destroying rapidly dividing cells (which unfortunately include lymphocytes as well as cancer cells). The most recent and tragically the most common acquired immune deficiency disease is AIDS, which, as described earlier, is caused by HIV, a virus that invades and incapacitates the critical helper T cells.

Let's now look at inappropriate immune attacks.

Allergies are inappropriate immune attacks against harmless environmental substances.

Inappropriate adaptive immune attacks cause reactions harmful to the body. These include (1) *autoimmune responses*, in which the immune system turns against one of the body's own tissues; (2) *immune complex diseases*, which involve overexuberant antibody responses that "spill over" and damage normal tissue; and (3) *allergies*. The first two conditions have been described earlier in this chapter, so we will now concentrate on allergies.

An **allergy** is the acquisition of an inappropriate specific immune reactivity, or **hypersensitivity**, to a normally harmless environmental substance, such as dust or pollen. The offending agent is known as an **allergen**. Subsequent re-exposure of a sensitized individual to the same allergen elicits an immune attack, which may vary from a mild, annoying reaction to a severe, body-damaging reaction that may even be fatal.

Allergic responses are classified into two different categories: immediate hypersensitivity and delayed hypersensitivity (▲ Table 12-6). In **immediate hypersensitivity**, the allergic response appears within about 20 minutes after a sensitized individual is exposed to an allergen. In **delayed hypersensitivity**, the reaction is not generally manifested until a day or so following exposure. The difference in timing is due to the different mediators involved. A particular allergen may activate either a B cell or a T cell response. Immediate allergic reactions involve B cells and are elicited by antibody interactions with an allergen; delayed reactions involve T cells and the more slowly responding process of cell-mediated immunity against the aller-

▲ TABLE 12-6
Immediate versus Delayed Hypersensitivity Reactions

Characteristic	Immediate Hypersensitivity Reaction	Delayed Hypersensitivity Reaction
Time of onset of symptoms after exposure to the allergen	Within 20 minutes	Within one to three days
Type of immune response	Antibody-mediated immunity against the allergen	Cell-mediated immunity against the allergen
Immune effectors involved	B cells, IgE antibodies, mast cells, basophils, histamine, slow-reactive substance of anaphylaxis, eosinophil chemotactic factor	T cells
Allergies commonly involved	Hay fever, asthma, hives, anaphylactic shock in extreme cases	Contact allergies such as allergies to poison ivy, cosmetics, and household cleaning agents
Treatment	Antihistamines (partially effective); adrenergic drugs to counteract the effects of histamine and slow-reactive substance of anaphylaxis; anti-inflammatory agents such as cortisol derivatives	Anti-inflammatory agents such as cortisol derivatives

gen. Let us examine the causes and consequences of each of these reactions in more detail.

Triggers for immediate hypersensitivity

In immediate hypersensitivity, the antibodies involved and the events that ensue on exposure to an allergen differ from the typical antibody-mediated response to bacteria. The most common allergens that provoke immediate hypersensitivities are pollen grains, bee stings, penicillin, certain foods, molds, dust, feathers, and animal fur. (Actually, people allergic to cats are not allergic to the fur itself. The true allergen is in the cat's saliva, which is deposited on the fur during licking. Likewise, people are not allergic to dust or feathers per se, but to tiny mites that inhabit the dust or feathers and eat the scales constantly being shed from the skin.) For unclear reasons, these allergens bind to and elicit the synthesis of IgE antibodies rather than the IgG antibodies associated with bacterial antigens. IgE antibodies are the least plentiful immunoglobulin, but their presence spells trouble. Without IgE antibodies, there would be no immediate hypersensitivity. When an individual with an allergic tendency is first exposed to a particular allergen, compatible helper T cells secrete **interleukin 4 (IL-4)**, which prods compatible B cells to synthesize IgE antibodies specific for the allergen. During this initial **sensitization** period, no symptoms are evoked, but memory cells are formed that are primed for a more powerful response on subsequent re-exposure to the same allergen.

In contrast to the antibody-mediated response elicited by bacterial antigens, IgE antibodies do not freely circulate. Instead, their tail portions attach to **mast cells** and **basophils.** Recall that mast cells are connective tissue–bound "cousins" of circulating basophils, both of which produce and store an arsenal of potent inflammatory chemicals, such as histamine, in preformed granules (see p. 401). Mast cells are most plentiful in regions that come into contact with the external environment, such as the skin, the outer surface of the eyes, and the linings of the respiratory system and digestive tract. Bind-

ing of an appropriate allergen with the outreached arm regions of the IgE antibodies that are lodged tail first in a mast cell or basophil triggers the rupture of the cell's preformed granules. As a result, histamine and other chemical mediators spew forth into the surrounding tissue.

A single mast cell (or basophil) may be coated with a number of different IgE antibodies, each able to bind with a different allergen. Thus the mast cell can be triggered to release its chemical products by any one of a number of different allergens (● Figure 12-26).

Chemical mediators of immediate hypersensitivity

These released chemicals are responsible for the reactions that characterize immediate hypersensitivity. The following are among the most important chemicals released during immediate allergic reactions:

1. **Histamine,** which brings about vasodilation and increased capillary permeability as well as increased mucus production.

2. **Slow-reactive substance of anaphylaxis (SRS-A),** which induces prolonged and profound contraction of smooth muscle, especially of the small respiratory airways. SRS-A is a leukotriene, a locally acting mediator similar to prostaglandins (see p. 765).

3. **Eosinophil chemotactic factor,** which specifically attracts eosinophils to the area. Interestingly, eosinophils release enzymes that inactivate SRS-A and may also inhibit histamine, perhaps serving as an "off switch" to limit the allergic response.

Symptoms of immediate hypersensitivity

Symptoms of immediate hypersensitivity vary depending on the site, allergen, and mediators involved. Most frequently, the reaction is localized to the body site in which the IgE-bearing cells first come into contact with the allergen. If the reaction is limited to the upper respiratory passages after a person inhales an allergen such as ragweed pollen, the released chemi-

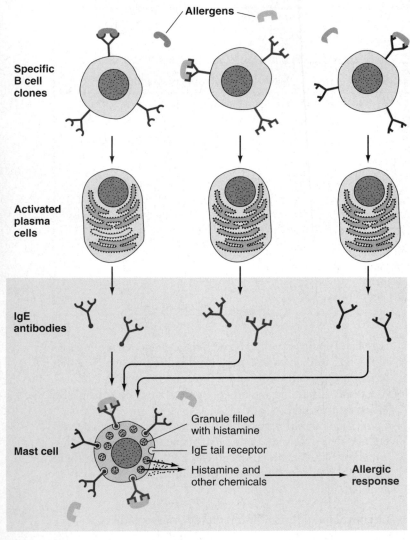

Specific
B cell
clones

Allergens

Activated
plasma
cells

IgE
antibodies

Mast cell

Granule filled
with histamine

IgE tail receptor

Histamine and
other chemicals

Allergic
response

● FIGURE 12-26

Role of IgE antibodies and mast cells in immediate hypersensitivity.
B cell clones are converted into plasma cells, which secrete IgE antibodies on contact with the allergen for which they are specific. The Fc tail portion of all IgE antibodies, regardless of the specificity of their Fab arm regions, binds to receptor proteins specific for IgE tails on mast cells and basophils. Unlike B cells, each mast cell bears a variety of antibody surface receptors for binding different allergens. When an allergen combines with the IgE receptor specific for it on the surface of a mast cell, the mast cell releases histamine and other chemicals by exocytosis. These chemicals elicit the allergic response.

Treatment of immediate hypersensitivity

Treatment of localized immediate allergic reactions with antihistamines often offers only partial relief of the symptoms, because some of the manifestations are invoked by other chemical mediators not blocked by these drugs. For example, antihistamines are not particularly effective in treating asthma, the most serious symptoms of which are invoked by SRS-A. Adrenergic drugs (which mimic the sympathetic nervous system) are helpful through their vasoconstrictor-bronchodilator actions in counteracting the effects of both histamine and SRS-A. Anti-inflammatory drugs such as cortisol derivatives are often used as the primary treatment for ongoing allergen-induced inflammation, such as that associated with asthma.

Anaphylactic shock

A life-threatening systemic reaction can occur if the allergen becomes blood-borne or if very large amounts of chemicals are released from the localized site into the circulation. When large amounts of these chemical mediators gain access to the blood, the extremely serious systemic (involving the entire body) reaction known as **anaphylactic shock** occurs. Severe hypotension (see p. 383) that can lead to circulatory failure results from widespread vasodilation and a massive shift of plasma fluid into the interstitial spaces as a result of a generalized increase in capillary permeability. Concurrently, pronounced bronchiolar constriction occurs and can lead to respiratory failure. The victim may suffocate because of an inability to move air through the narrowed airways. Unless countermeasures, such as injection of a vasoconstrictor-bronchodilator drug, are undertaken immediately, anaphylactic shock is frequently fatal. This reaction is why even a single bee sting or a single dose of penicillin can be so dangerous in individuals sensitized to these allergens.

Immediate hypersensitivity and absence of parasitic worms

Although the immediate hypersensitivity response differs considerably from the typical IgG antibody response to bacterial infections, it is strikingly similar to the immune response elicited by parasitic worms. Shared characteristics of the immune reactions to allergens and parasitic worms include the production of IgE antibodies and increased basophil and eosinophil activity. This finding has led to the proposal that harmless allergens somehow trigger an immune response designed to fight worms. Mast cells are concentrated in areas where parasitic worms (and allergens) could come into contact with the body. Parasitic worms can penetrate the skin or digestive tract or can attach to the digestive tract lining. Some worms migrate through the lungs during a part of their life cycle. Scientists suspect that the IgE response helps ward off these invaders through the following means. The inflammatory response in the skin could wall off parasitic worms attempting to burrow in. Coughing and sneezing could expel worms that migrated to the lungs. Diarrhea could help flush out worms before

cals bring about the symptoms characteristic of **hay fever**—for example, nasal congestion caused by histamine-induced localized edema and sneezing and runny nose caused by increased mucus secretion in response to local irritation. If the reaction is concentrated primarily within the bronchioles (the small respiratory airways that lead to the tiny air sacs within the lungs), **asthma** results. Contraction of the smooth muscle in the walls of the bronchioles narrows or constricts these passageways, making breathing difficult. Localized swelling in the skin because of allergy-induced histamine release causes **hives**. An allergic reaction in the digestive tract in response to an ingested allergen can lead to diarrhea.

they could penetrate or attach to the digestive tract lining. Interestingly, epidemiologic studies suggest that the incidence of allergies in a country rises as the presence of parasites decreases. Thus, superfluous immediate hypersensitivity responses to normally harmless allergens may represent a pointless marshaling of a honed immune-response system "with nothing better to do" in the absence of parasitic worms.

Delayed hypersensitivity

Some allergens invoke delayed hypersensitivity, a T cell–mediated immune response, rather than an immediate, B cell–IgE antibody response. Among these allergens are poison ivy toxin and certain chemicals to which the skin is frequently exposed, such as cosmetics and household cleaning agents. Most commonly, the response is characterized by a delayed skin eruption that reaches its peak intensity one to three days following contact with an allergen to which the T system has previously been sensitized. To illustrate, poison ivy toxin does not harm the skin on contact, but it activates T cells specific for the toxin, including formation of a memory component. On subsequent exposure to the toxin, activated T cells diffuse into the skin within a day or two, combining with the poison ivy toxin that is present. The resultant interaction gives rise to the tissue damage and discomfort typically associated with the condition. The best relief is obtained from application of anti-inflammatory preparations, such as those containing cortisol derivatives.

▲ Table 12-6 summarizes the distinctions between immediate and delayed hypersensitivities. This completes our discussion of the internal immune defense system. We are now going to turn our attention to external defenses that thwart entry of foreign invaders as a first line of defense.

EXTERNAL DEFENSES

The body's defenses against foreign microbes are not limited to the intricate, interrelated immune mechanisms that destroy the micro-organisms that have actually invaded the body. In addition to the internal immune defense system, the body is equipped with external defense mechanisms designed to prevent microbial penetration wherever body tissues are exposed to the external environment. The most obvious external defense is the **skin,** or **integument,** which covers the outside of the body (*integere* means "to cover").

▌ The skin consists of an outer protective epidermis and an inner, connective tissue dermis.

The skin, which is the largest organ of the body, not only serves as a mechanical barrier between the external environment and the underlying tissues but is dynamically involved in defense mechanisms and other important functions as well. The skin consists of two layers, an outer *epidermis* and an inner *dermis* (● Figure 12-27).

Epidermis

The **epidermis** consists of numerous layers of epithelial cells. The inner epidermal layers are composed of cube-shaped cells that are living and rapidly dividing, whereas the cells in the outer layers are dead and flattened. The epidermis has no direct blood supply. Its cells are nourished only by diffusion of nutrients from a rich vascular network in the underlying der-

● **FIGURE 12-27**
Anatomy of the skin
The skin consists of two layers, a keratinized outer epidermis and a richly vascularized inner connective tissue dermis. Special infoldings of the epidermis form the sweat glands, sebaceous glands, and hair follicles. The epidermis contains four types of cells: keratinocytes, melanocytes, Langerhans cells, and Granstein cells. The skin is anchored to underlying muscle or bone by the hypodermis, a loose, fat-containing layer of connective tissue.

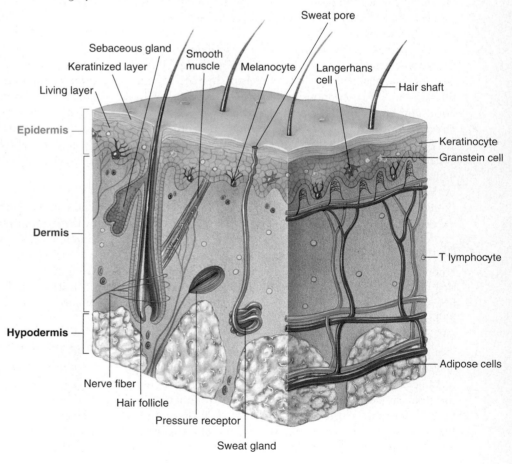

mis. The newly forming cells in the inner layers constantly push the older cells closer to the surface, farther and farther from their nutrient supply. This, coupled with the fact that the outer layers are continuously subjected to pressure and "wear and tear," causes these older cells to die and become flattened. Epidermal cells are tightly bound together by spot desmosomes (see p. 63), which interconnect with intracellular keratin filaments (see p. 51) to form a strong, cohesive covering. During maturation of a keratin-producing cell, keratin filaments progressively accumulate and cross-link with each other within the cytoplasm. As the outer cells die, this fibrous keratin core remains, forming flattened, hardened scales that provide a tough, protective **keratinized layer.** As the scales of the outermost keratinized layer slough or flake off through abrasion, they are continuously replaced by means of cell division in the deeper epidermal layers. The rate of cell division, and consequently the thickness of this keratinized layer, varies in different regions of the body. It is thickest in the areas where the skin is subjected to the most pressure, such as the bottom of the feet.

The keratinized layer is airtight, fairly waterproof, and impervious to most substances. It serves to resist passage in both directions between the body and the external environment. For example, it minimizes loss of water and other vital constituents from the body. This protective layer's value in holding in body fluids becomes obvious in severe burns. Not only can bacterial infections occur in the unprotected underlying tissue, but even more serious are the systemic consequences of loss of body water and plasma proteins, which escape from the exposed, burned surface. The resultant circulatory disturbances can be life threatening.

Likewise, the skin barrier impedes passage into the body of most materials that come into contact with the body surface, including bacteria and toxic chemicals. In many instances, the skin modifies compounds that come into contact with it. For example, epidermal enzymes can convert many potential carcinogens into harmless compounds. Some materials, however, especially lipid-soluble substances, are able to penetrate intact skin through the lipid bilayers of the plasma membranes of the epidermal cells. Drugs that can be absorbed by the skin, such as nicotine or estrogen, are sometimes administered in the form of a cutaneous "patch" impregnated with the drug.

Dermis

Under the epidermis is the **dermis,** a connective tissue layer that contains many elastin fibers (for stretch) and collagen fibers (for strength), as well as an abundance of blood vessels and specialized nerve endings. The dermal blood vessels not only supply both the dermis and epidermis but also play a major role in temperature regulation. The caliber of these vessels, and hence the volume of blood flowing through them, is subject to control to vary the amount of heat exchange between these skin surface vessels and the external environment (Chapter 17). Receptors at the peripheral endings of afferent nerve fibers in the dermis detect pressure, temperature, pain, and other somatosensory input. Efferent nerve endings in the der-

mis control blood vessel caliber, hair erection, and secretion by the skin's exocrine glands.

Skin's exocrine glands and hair follicles

Special infoldings of the epidermis into the underlying dermis form the skin's exocrine glands—the sweat glands and sebaceous glands—as well as the hair follicles. **Sweat glands,** which are located over the majority of the body, release a dilute salt solution through small openings, the sweat pores, onto the surface of the body. Evaporation of this sweat cools the skin and is important in temperature regulation. The amount of sweat produced is subject to regulation and depends on the environmental temperature, the amount of heat-generating muscular activity, and various emotional factors (for example, a person often sweats when nervous). A special type of sweat gland located in the axilla (armpit) and pubic region produces a protein-rich sweat that supports the growth of surface bacteria, which give rise to a characteristic odor. In contrast, most sweat, as well as the secretions from the sebaceous glands, contains chemicals that are generally highly toxic to bacteria.

The cells of the **sebaceous glands** produce an oily secretion known as **sebum** that is released into adjacent hair follicles. From there the oily sebum flows to the surface of the skin, oiling both the hairs and the outer keratinized layers of the skin, helping to waterproof them and prevent them from drying and cracking. Chapped hands or lips indicate insufficient protection by sebum. The sebaceous glands are particularly active during adolescence, causing the oily skin common among teenagers.

Each **hair follicle** is lined by special keratin-producing cells, which secrete keratin and other proteins that form the hair shaft. Hairs increase the sensitivity of the skin's surface to tactile (touch) stimuli. In some other species, this function is more finely tuned. For example, the whiskers on a cat are exquisitely sensitive in this regard. An even more important role of hair in hairier species is heat conservation, but this function is not significant in us relatively hairless humans. Like hair, the nails are another special keratinized product derived from living epidermal structures, the nail beds.

Hypodermis

The skin is anchored to the underlying tissue (muscle or bone) by the **hypodermis** (*hypo* means "below"), also known as **subcutaneous tissue** (*sub* means "under," *cutaneous* means "skin"), a loose layer of connective tissue. Most fat cells in the body are housed within the hypodermis. These subcutaneous fat deposits throughout the body are collectively referred to as **adipose tissue.**

▌ Specialized cells in the epidermis produce keratin and melanin and participate in immune defense.

The epidermis contains four distinct resident cell types—*melanocytes, keratinocytes, Langerhans cells,* and *Granstein cells*—plus transient T lymphocytes that are scattered throughout the epidermis and dermis. Each of these resident cell types performs specialized functions.

Melanocytes

Melanocytes produce the pigment **melanin,** which they disperse to surrounding skin cells. The amount and type of melanin, which can vary among black, brown, yellow, and red pigments, are responsible for the different shades of skin color of the various races. Melanin is produced through complex biochemical pathways in which the melanocyte enzyme *tyrosinase* plays a key role. Recent studies have demonstrated that most people, regardless of skin color, have enough tyrosinase that, if fully functional, could result in enough melanin to make their skin very black. In those with lighter skin, however, two genetic factors prevent this melanocyte enzyme from functioning at full capacity: (1) Much of the tyrosinase produced is in an inactive form, and (2) various inhibitors that block tyrosinase action are produced. As a result, less melanin is produced.

In addition to hereditary determination of melanin content, the amount of this pigment can be increased transiently in response to exposure to ultraviolet (UV) light rays from the sun. This additional melanin, the outward appearance of which constitutes a "tan," performs the protective function of absorbing harmful UV rays.

Keratinocytes

The most abundant epidermal cells are the **keratinocytes,** which, as the name implies, are specialists in keratin production. As they die, they form the outer protective keratinized layer. They are also responsible for generating hair and nails. A surprising, recently discovered function is that keratinocytes are also important immunologically. They secrete interleukin 1 (a product also secreted by macrophages), which influences the maturation of T cells that tend to localize in the skin. Interestingly, the epithelial cells of the thymus have been shown to bear anatomic, molecular, and functional similarities to those of the skin. Apparently, some postthymic steps in T cell maturation take place in the skin under keratinocyte guidance.

Other immune cells of the skin

The two other epidermal cell types also play a role in immunity. **Langerhans cells,** which migrate to the skin from the bone marrow, are dendritic cells that serve as antigen-presenting cells. Thus the skin not only serves as a mechanical barrier but actually alerts regulatory lymphocytes if the barrier is breached by invading micro-organisms. Langerhans cells present antigen to helper T cells, facilitating their responsiveness to skin-associated antigens. In contrast, it appears that **Granstein cells** serve as a "brake" on skin-activated immune responses. These cells are the most recently discovered and least understood of the skin's immune cells. Significantly, Langerhans cells are more susceptible to damage by UV radiation (as from the sun) than Granstein cells are. Losing Langerhans cells as a result of exposure to UV radiation can detrimentally lead to a predominant suppressor signal rather than the normally dominant helper signal, leaving the skin more vulnerable to microbial invasion and cancer cells.

The various epidermal components of the immune system are collectively termed **skin-associated lymphoid tissue,** or **SALT.** Recent research suggests that the skin probably plays an even more elaborate role in adaptive immune defense than described here. This is appropriate, because the skin serves as a major interface with the external environment.

Vitamin D synthesis by the skin

The epidermis also synthesizes vitamin D in the presence of sunlight. The cell type that produces vitamin D is undetermined. Vitamin D, which is derived from a precursor molecule closely related to cholesterol, promotes the absorption of Ca^{2+} from the digestive tract into the blood (Chapter 16). Dietary supplements of vitamin D are usually required because typically the skin is not exposed to sufficient sunlight to produce adequate amounts of this essential chemical.

▌ Protective measures within body cavities discourage pathogen invasion into the body.

The human body's defense system must guard against entry of potential pathogens not only through the outer surface of the body but also through the internal cavities that communicate directly with the external environment—namely, the digestive system, the genitourinary system, and the respiratory system. These systems employ various strategies to destroy micro-organisms entering through these routes.

Defenses of the digestive system

Saliva secreted into the mouth at the entrance of the digestive system contains an enzyme that lyses certain ingested bacteria. "Friendly" bacteria that live on the back of the tongue convert food-derived nitrate into nitrite, which is swallowed. Acidification of nitrite on reaching the highly acidic stomach generates nitric oxide, which is toxic to a variety of micro-organisms. Furthermore, many of the surviving bacteria that are swallowed are killed directly by the strongly acidic gastric juice that they encounter in the stomach. Farther down the tract, the intestinal lining is endowed with gut-associated lymphoid tissue. These defensive mechanisms are not 100% effective, however. Some bacteria do manage to survive and reach the large intestine (the last portion of the digestive tract), where they continue to flourish. Surprisingly, this normal microbial population provides a natural barrier against infection within the lower intestine. These harmless resident flora competitively suppress the growth of potential pathogens that have managed to escape the antimicrobial measures of earlier parts of the digestive tract. On occasion, orally administered antibiotic therapy against one infection within the body may actually induce another infection in the intestinal tract. By knocking out some of the normal intestinal flora, an antibiotic may permit an antibiotic-resistant pathogenic species to overgrow.

Defenses of the genitourinary systems

Within the genitourinary (reproductive and urinary) system, would-be invaders encounter hostile conditions in the acidic urine and acidic vaginal secretions. The genitourinary organs also produce a sticky mucus, which, like flypaper, entraps small invading particles. Subsequently, the particles are either engulfed by phagocytes or are swept out as the organ empties (for example, they are flushed out with urine flow).

Defenses of the respiratory system

The respiratory system is likewise equipped with several important defense mechanisms against inhaled particulate matter. The respiratory system is the largest surface of the body that comes into direct contact with the increasingly polluted external environment. The surface area of the respiratory system that is exposed to the air is 30 times that of the skin. Larger airborne particles are filtered out of the inhaled air by hairs at the entrance of the nasal passages. Lymphoid tissues, the *tonsils* and *adenoids*, provide immunological protection against inhaled pathogens near the beginning of the respiratory system. Farther down in the respiratory airways, millions of tiny hair-like projections known as cilia constantly beat in an outward direction (see p. 47). The respiratory airways are coated with a layer of thick, sticky mucus secreted by epithelial cells within the airway lining. This mucus sheet, laden with any inspired particulate debris (such as dust) that adheres to it, is constantly moved upward to the throat by ciliary action. This moving "staircase" of mucus is known as the **mucus escalator.** The dirty mucus is either expectorated (spit out) or in most cases swallowed without the person even being aware of it; any undigestible foreign particulate matter is subsequently eliminated in the feces. Besides keeping the lungs clean, this mechanism is an important defense against bacterial infection, because many bacteria enter the body on dust particles. Also contributing to defense against respiratory infections are IgA antibodies secreted in the mucus. In addition, an abundance of phagocytic specialists called the **alveolar macrophages** scavenge within the air sacs (alveoli) of the lungs. Further respiratory defenses include coughs and sneezes. These commonly experienced reflex mechanisms involve forceful outward expulsion of material in an attempt to remove irritants from the trachea (*coughs*) or nose (*sneezes*).

Cigarette smoking suppresses these normal respiratory defenses. The smoke from a single cigarette can paralyze the cilia for several hours, with repeated exposure eventually leading to ciliary destruction. Failure of ciliary activity to sweep out a constant stream of particulate-laden mucus enables inhaled carcinogens to remain in contact with the respiratory airways for prolonged periods. Furthermore, cigarette smoke incapacitates alveolar macrophages. Not only do particulates in cigarette smoke overwhelm the macrophages, but certain components of cigarette smoke have a direct toxic effect on the macrophages,

reducing their ability to engulf foreign material. In addition, noxious agents in tobacco smoke irritate the mucous linings of the respiratory tract, resulting in excess mucus production, which may partially obstruct the airways. "Smoker's cough" is an attempt to dislodge this excess stationary mucus. These and other direct toxic effects on lung tissue lead to the increased incidence of lung cancer and chronic respiratory diseases associated with cigarette smoking. Air pollutants include some of the same substances found in cigarette smoke and can similarly affect the respiratory system.

We will examine the respiratory system in greater detail in the next chapter.

CHAPTER IN PERSPECTIVE: FOCUS ON HOMEOSTASIS

We could not survive beyond early infancy were it not for the body's defense mechanisms. These mechanisms resist and eliminate potentially harmful foreign agents with which we continuously come into contact in our hostile external environment and also destroy abnormal cells that often arise within the body. Homeostasis can be optimally maintained, and thus life sustained, only if the body cells are not physically injured or functionally disrupted by pathogenic micro-organisms or are not replaced by abnormally functioning cells, such as traumatized cells or cancer cells. The immune defense system—a complex, multifaceted, interactive network of leukocytes, their secretory products, and plasma proteins—contributes indirectly to homeostasis by keeping other cells alive so that they can perform their specialized activities to maintain a stable internal environment. The immune system protects the other healthy cells from foreign agents that have gained entrance to the body, eliminates newly arisen cancer cells, and clears away dead and injured cells to pave the way for replacement with healthy new cells.

The skin contributes indirectly to homeostasis by serving as a protective barrier between the external environment and the remainder of the body cells. It helps prevent harmful foreign agents such as pathogens and toxic chemicals from entering the body and helps prevent the loss of precious internal fluids from the body. The skin also contributes directly to homeostasis by helping maintain body temperature by means of the sweat glands and adjustments in skin blood flow. The amount of heat carried to the body surface for dissipation to the external environment is determined by the volume of warmed blood flowing through the skin.

Other systems that have internal cavities in contact with the external environment, such as the digestive, genitourinary, and respiratory systems, also have defense capabilities to prevent harmful external agents from entering the body through these avenues.

CHAPTER SUMMARY

Introduction

▪ Foreign invaders and newly arisen mutant cells are immediately confronted with multiple interrelated defense mechanisms aimed at destroying and eliminating anything that is not part of the normal self.

▪ These mechanisms, collectively referred to as *immunity,* include both innate and adaptive immune responses. Innate immune responses are nonspecific responses that nonselectively defend against foreign material even on initial exposure to it. Adaptive immune responses are specific responses selectively targeted against particular invaders for which the body has been specially prepared after a prior exposure.

▪ The most common invaders are bacteria and viruses. Bacteria are self-sustaining, single-celled organisms, which produce disease by virtue of the destructive chemicals they release. Viruses are protein-coated nucleic acid particles, which invade host cells and take over the cellular metabolic machinery for their own survival to the detriment of the host cell.

▪ Leukocytes and their derivatives are the major effector cells of the immune system and are reinforced by a number of different plasma proteins. Leukocytes include neutrophils, eosinophils, basophils, monocytes, and lymphocytes.

▪ Leukocytes are produced in the bone marrow, then circulate transiently in the blood. They spend most of their time on defense missions in the tissues.

▪ Some lymphocytes are also produced and differentiated and perform their defense activities within lymphoid tissues strategically located at likely points of foreign infiltration.

▪ In addition to defending against microbes and mutant cells, the immune cells also clean up cellular debris, preparing the way for tissue repair.

Innate Immunity

▪ Innate immune responses, which form a first line of defense against atypical cells (foreign, mutant, or injured cells) within the body, include inflammation, interferon, natural killer cells, and the complement system.

▪ Inflammation is a nonspecific response to foreign invasion or tissue damage mediated largely by the professional phagocytes (neutrophils and monocytes-turned-macrophages) and their secretions.

▪ The phagocytic cells destroy foreign and damaged cells both by phagocytosis and by the release of lethal chemicals.

▪ Histamine-induced vasodilation and increased permeability of local vessels at the site of invasion or injury permit enhanced delivery of more phagocytic leukocytes and inactive plasma protein precursors crucial to the inflammatory process, such as clotting factors and components of the complement system. These vascular changes are also largely responsible for the observable local manifestations of inflammation—swelling, redness, heat, and pain.

▪ The chemicals released from the phagocytes on the scene augment inflammation, induce systemic manifestations such as fever, and enhance adaptive immune responses.

▪ Interferon is nonspecifically released by virus-infected cells and transiently inhibits viral multiplication in other cells to which it binds. Interferon further exerts anticancer effects by slowing division and growth of tumor cells as well as by enhancing the power of killer cells.

▪ Natural killer (NK) cells nonspecifically lyse and destroy virus-infected cells and cancer cells on first exposure to them.

▪ On being activated by microbes themselves at the site of invasion or by antibodies produced against the microbes, the complement system directly destroys the foreign invaders by lysing their membranes and also augments other aspects of the inflammatory process. The complement system lyses the targeted cells by forming a hole-punching membrane attack complex that inserts into the victim cell's membrane, leading to osmotic rupture of the cell.

Adaptive Immunity: General Concepts

▪ Not only is the adaptive immune system able to recognize foreign molecules as different from self-molecules—so that destructive immune reactions are not unleashed against the body itself—but it can also distinguish between millions of different foreign molecules. Lymphocytes, the effector cells of adaptive immunity, are each uniquely equipped with surface membrane receptors that are able to bind lock-and-key fashion with only one specific complex foreign molecule, which is known as an antigen.

▪ There are two broad classes of adaptive immune responses: antibody-mediated immunity and cell-mediated immunity. In both instances, the ultimate outcome of a particular lymphocyte binding with a specific antigen is destruction of the antigen, but the effector cells, stimuli, and tactics involved are different. Plasma cells derived from B lymphocytes (B cells) are responsible for antibody-mediated immunity, whereas T lymphocytes (T cells) accomplish cell-mediated immunity.

▪ B cells develop from a lineage of lymphocytes that originally matured within the bone marrow. The T cell lineage arises from lymphocytes that migrated from the bone marrow to the thymus to complete their maturation.

B Lymphocytes: Antibody-Mediated Immunity

▪ Each B cell recognizes specific free extracellular antigen that is not associated with cell-bound self-antigens, such as that found on the surface of bacteria.

▪ The activated B cell differentiates into a plasma cell, which is specialized to secrete freely circulating antibodies that besiege the freely existing invading bacteria (or other foreign substance) that induced their production

▪ Antibodies are Y-shaped molecules. The antigen-binding sites on the tips of each arm of the antibody determine with what specific antigen the antibody can bind. Properties of the antibody's tail portion determine what the antibody does once it binds with antigen.

▪ There are five subclasses of antibodies, depending on differences in the biological activity of their tail portion: IgM, IgG, IgE, IgA, and IgD immunoglobulins.

▪ Antibodies do not directly destroy material. Instead, they exert their protective effect by physically hindering antigens through neutralization or agglutination or by intensifying lethal innate immune responses already called into play by the foreign invasion. Antibodies activate the complement system, enhance phagocytosis, and stimulate killer cells.

▪ After being activated by antigen associated with a foreign invader, a B cell rapidly proliferates, producing a clone of its own kind that can specifically wage battle against the invader. Most lymphocytes in the expanded B-cell clone become plasma cells that participate in the primary response against the invader.

∎ Some of the newly developed lymphocytes do not participate in the attack but become memory cells that lie in waiting, ready to launch a swifter and more forceful secondary response should the same foreigner ever invade the body again.

∎ The tremendous variation in antigen-detecting ability between different lymphocytes arises from the shuffling around of a few different gene segments, coupled with a high incidence of somatic mutation, during lymphocyte development.

∎ Both B and T cells can recognize and bind with antigen only when it has been processed and presented to them by antigen-presenting cells, such as macrophages and dendritic cells.

T Lymphocytes: Cell-Mediated Immunity

∎ T cells accomplish cell-mediated immunity by being in direct contact with their targets.

∎ There are two types of T cells: cytotoxic T cells and helper T cells.

∎ The targets of cytotoxic T cells are virally invaded cells and cancer cells, which they destroy by releasing perforin molecules that form a lethal hole-punching complex that inserts into the membrane of the victim cell or by releasing granzymes that trigger the victim cell to undergo apoptosis.

∎ Helper T cells bind with other immune cells and release chemicals that augment the activity of these other cells. Chemicals other than antibodies released by leukocytes are known as *cytokines,* most of which are secreted by helper T cells.

∎ Like B cells, T cells bear receptors that are antigen specific, undergo clonal selection, exert primary and secondary responses, and form memory pools for long-lasting immunity against targets to which they have already been exposed.

∎ Those lymphocytes produced by chance that are able to attack the body's own antigen-bearing cells are eliminated or suppressed so that they are prevented from functioning. In this way, the body is able to "tolerate" (not attack) its own antigens. Tolerance is accomplished by clonal deletion, clonal anergy, receptor editing, antigen sequestering, and immune privilege.

∎ The major self-antigens on the surface of body cells are known as MHC molecules, which are coded for by the major histocompatibility complex (MHC), a group of genes with DNA sequences unique for each individual.

∎ B and T cells have different targets because their requirements for antigen recognition differ. B cells recognize freely circulating antigen, such as bacteria, that can lead to antigen destruction at long distances. T cells, in contrast, have a dual binding requirement of foreign antigen in association with MHC molecules (self-antigens) on the surface of one of the body's own cells.

∎ The presence of class I or class II MHC self-antigens on the surface of these foreign antigen–bearing host cells causes the two different types of T cells to differentially interact with them.

1. Cytotoxic T cells are able to bind only with virus-infected host cells or cancer cells, which always bear the class I MHC self-antigen in association with foreign or abnormal antigen. On binding with the abnormal host cell, these T cells release toxic substances that kill the dangerous body cell.

2. Helper T cells can only bind with other T cells, B cells, and macrophages that have encountered foreign antigen. These immune cells bear the class II MHC self-marker in association with foreign antigen. Subsequently, helper T cells enhance the immune powers of these other effector cells by secreting specific chemical mediators.

∎ Such differential activation of the various types of lymphocytes assures that the appropriate specific immune response ensues to dispose of the particular enemy efficiently.

∎ Moreover, B cells, the various T cells, and macrophages reinforce each other's defense strategies, primarily by releasing a number of important secretory products.

∎ In a process known as immune surveillance, natural killer cells, cytotoxic T cells, macrophages, and the interferon that they collectively secrete normally eradicate newly arisen cancer cells before they have a chance to spread.

Immune Diseases

∎ Immune diseases are of two types: immune deficiency diseases (insufficient immune responses) or inappropriate immune attacks (excessive or mistargeted immune responses).

∎ With immune deficiency diseases, the immune system fails to defend normally against bacterial or viral infections through a deficit of B or T cells, respectively.

∎ With inappropriate immune attacks, the immune system becomes overzealous. There are three categories of inappropriate attacks:

1. In autoimmune disease, the immune system erroneously turns against one of the person's own tissues that it no longer recognizes and tolerates as self.

2. With immune complex diseases, body tissues are inadvertently destroyed as an overabundance of antigen-antibody complexes activates excessive quantities of lethal complement, which destroys surrounding normal cells as well as the antigen.

3. Allergies, or hypersensitivities, occur when the immune system inappropriately launches a symptom-producing, body-damaging attack against an allergen, a normally harmless environmental antigen. (a) Immediate hypersensitivities involve the production of IgE antibodies by B cells that trigger the release of powerful inflammatory chemicals from mast cells and basophils to bring about a swift response to the allergen. (b) Delayed hypersensitivities involve a more slowly responding cell-mediated, symptom-producing response by T cells against the allergen.

External Defenses

∎ The body surfaces exposed to the outside environment—both the outer covering of skin and the linings of internal cavities that communicate with the external environment—serve not only as mechanical barriers to deter would-be pathogenic invaders but also play an active role in thwarting entry of bacteria and other unwanted materials.

∎ The skin consists of two layers: an outer vascular, keratinized epidermis and an inner, connective tissue dermis.

∎ The epidermis contains four cell types:

1. Melanocytes produce a pigment, melanin, the color and amount of which is responsible for the varying shades of skin color. Melanin protects the skin by absorbing harmful UV radiation.

2. The most abundant cells are the keratinocytes, producers of the tough keratin that forms the outer protective layer of the skin. This physical barrier discourages bacteria and other harmful environmental agents from entering the body and prevents water and other valuable body substances from escaping. Keratinocytes further serve immunologically by secreting interleukin 1, which enhances postthymic T cell maturation within the skin.

3. Langerhans cells also function in specific immunity by presenting antigen to helper T cells.

4. Granstein cells suppress skin-activated immune responses.

▪ The dermis contains (1) blood vessels, which nourish the skin and play an important role in regulating body temperature; (2) sensory nerve endings, which provide information about the external environment; and (3) several exocrine glands and hair follicles, which are formed by specialized invaginations of the overlying epithelium.

▪ The skin's exocrine glands include sebaceous glands, which produce sebum, an oily substance that softens and waterproofs the skin, and sweat glands, which produce cooling sweat. Hair follicles produce hairs, the distribution and function of which are minimal in humans.

▪ In addition, the skin synthesizes vitamin D in the presence of sunlight.

▪ Besides the skin, the other main routes by which potential pathogens enter the body are (1) the digestive system, which is defended by an antimicrobial salivary enzyme, destructive acidic gastric secretions, gut-associated lymphoid tissue, and harmless colonic resident flora; (2) the genitourinary system, which is protected by destructive acidic and particle-entrapping mucus secretions; and (3) the respiratory system, whose defense depends on alveolar macrophage activity and on secretion of a sticky mucus that traps debris, which is subsequently swept out by ciliary action. Other respiratory defenses include nasal hairs, which filter out large inspired particles; reflex cough and sneeze mechanisms, which expel irritant materials from the trachea and nose, respectively; and the tonsils and adenoids, which defend immunologically.

REVIEW EXERCISES

Objective Questions (Answers on p. A-44)

1. The complement system can only be activated by antibodies. (*True or false?*)
2. Specific adaptive immune responses are accomplished by neutrophils. (*True or false?*)
3. Damaged tissue is always replaced by scar tissue. (*True or false?*)
4. Active immunity against a particular disease can be acquired only by actually having the disease. (*True or false?*)
5. A secondary response has a more rapid onset, is more potent, and has a longer duration than a primary response. (*True or false?*)
6. _____ are receptors on the plasma membrane of phagocytes that recognize and bind with telltale molecular patterns which are present on the surface of micro-organisms but absent from human cells.
7. The complement system's _____ forms a doughnut-shaped complex that imbeds in a microbial surface membrane, causing osmotic lysis of the victim cell.
8. _____ is a collection of phagocytic cells, necrotic tissue, and bacteria.
9. _____ is the localized response to microbial invasion or tissue injury that is accompanied by swelling, heat, redness, and pain.
10. A chemical that enhances phagocytosis by serving as a link between a microbe and the phagocytic cell is known as a(n) _____.
11. _____, collectively, are all the chemical messengers other than antibodies secreted by lymphocytes.
12. Which of the following statements concerning leukocytes is/are *incorrect*?
    a. Monocytes are transformed into macrophages.
    b. T lymphocytes are transformed into plasma cells that secrete antibodies.
    c. Neutrophils are highly mobile phagocytic specialists.
    d. Basophils release histamine.
    e. Lymphocytes arise in large part from lymphoid tissues.
13. Match the following:
    (a) complement system
    (b) natural killer cells
    (c) interferon
    (d) inflammation
    ____ 1. a family of proteins that nonspecifically defend against viral infection
    ____ 2. a response to tissue injury in which neutrophils and macrophages play a major role

    ____ 3. a group of plasma proteins that, when activated, bring about destruction of foreign cells by attacking their plasma membranes
    ____ 4. lymphocyte-like entities that spontaneously lyse tumor cells and virus-infected host cells
14. Using the following answer code, indicate whether the numbered characteristics of the adaptive immune system apply to antibody-mediated immunity or cell-mediated immunity (or both):
    (a) antibody-mediated immunity
    (b) cell-mediated immunity
    (c) both antibody-mediated and cell-mediated immunity
    ____ 1. involves secretion of antibodies
    ____ 2. mediated by B cells
    ____ 3. mediated by T cells
    ____ 4. accomplished by thymus-educated lymphocytes
    ____ 5. triggered by the binding of specific antigens to complementary lymphocyte receptors
    ____ 6. involves formation of memory cells in response to initial exposure to an antigen
    ____ 7. primarily aimed against virus-infected host cells
    ____ 8. protects primarily against bacterial invaders
    ____ 9. directly destroys targeted cells
    ____ 10. involved in rejection of transplanted tissue
    ____ 11. requires binding of a lymphocyte to a free extracellular antigen
    ____ 12. requires dual binding of a lymphocyte with both foreign antigen and self-antigens present on the surface of a host cell
15. Using the following answer code, indicate whether the numbered characteristics apply to the epidermis or dermis:
    (a) epidermis
    (b) dermis
    ____ 1. is the inner layer of skin
    ____ 2. has layers of epithelial cells that are dead and flattened
    ____ 3. has no direct blood supply
    ____ 4. contains sensory nerve endings
    ____ 5. contains keratinocytes
    ____ 6. contains melanocytes
    ____ 7. contains rapidly dividing cells
    ____ 8. is mostly connective tissue

Essay Questions

1. Distinguish between bacteria and viruses.
2. Summarize the functions of each of the lymphoid tissues.
3. Distinguish between innate and adaptive immune responses.
4. Compare the life history of B cells and of T cells.
5. What is an antigen?
6. Describe the structure of an antibody. List and describe the five subclasses of immunoglobulins.
7. In what ways do antibodies exert their effect?
8. Describe the clonal selection theory.
9. Compare the functions of B cells and T cells. What are the roles of the two types of T cells?
10. Summarize the functions of macrophages in immune defense.
11. What mechanisms are involved in tolerance?
12. What is the importance of class I and class II MHC glycoproteins?
13. Describe the factors that contribute to immune surveillance against cancer cells.
14. Distinguish among immune deficiency disease, autoimmune disease, immune complex disease, immediate hypersensitivity, and delayed hypersensitivity.
15. What are the immune functions of the skin?

Quantitative Exercises (Solutions on p. A-44)

1. As a result of the innate immune response to an infection, for example from a cut on the skin, capillary walls near the site of infection become very permeable to plasma proteins that normally remain in the blood. These proteins diffuse into the interstitial fluid, raising the interstitial fluid–colloid osmotic pressure. This increased colloid osmotic pressure causes fluid to leave the circulation and accumulate in the tissue, forming a welt, or wheal. This process is referred to as the *wheal response*. The wheal response is mediated in part by histamine secreted from mast cells in the area of infection. The histamine binds to receptors, called *H-1 receptors*, on capillary endothelial cells. The histamine signal is transduced via the Ca^{2+} second-messenger pathway involving phospholipase C (see p. 68). In response to this signal, the capillary endothelial cells contract (via internal actin-myosin interaction), which causes a widening of the intercellular gaps (pores) between the capillary endothelial cells (see p. 363). In addition, substance P (see p. 192) also contributes to pore widening. Plasma proteins can pass through these widened pores and leave the capillaries. Looking at Figure 10-23, p. 365, compare the magnitude of the wheal response (that is, the extent of localized edema) if p_{IF} were raised (a) from 0 mm Hg to 5 mm Hg and (b) from 0 mm Hg to 10 mm Hg. In both cases, compare the net exchange pressure (NEP) at the arteriolar end of the capillary, the venular end of the capillary, and the average NEP. (Assume the other forces acting across the capillary wall remain unchanged.)

POINTS TO PONDER

(Explanations on p. A-44)

1. Compare the defense mechanisms that come into play in response to bacterial and viral pneumonia.
2. Why does the frequent mutation of the HIV (the AIDS virus) make it difficult to develop a vaccine against this virus?
3. What impact would failure of the thymus to develop embryonically have on the immune system after birth?
4. Medical researchers are currently working on ways to "teach" the immune system to view foreign tissue as "self." What useful clinical application will the technique have?
5. When someone looks at you, are the cells of your body that person is viewing dead or alive?
6. *Clinical Consideration.* Heather L., who has Rh-negative blood, has just given birth to her first child, who has Rh-positive blood. Both mother and baby are fine, but the doctor administers an Rh immunoglobulin preparation so that any future Rh-positive babies Heather has will not suffer from hemolytic disease of the newborn (see p. 433). During gestation (pregnancy), fetal and maternal blood do not mix. Instead, materials are exchanged between these two circulatory systems across the placenta, a special organ that develops during gestation from both maternal and fetal structures (see p. 786). Red blood cells are unable to cross the placenta, but antibodies can cross. During the birthing process, a small amount of the infant's blood may enter the maternal circulation.

   a. Why did Heather's first-born child not have hemolytic disease of the newborn; that is, why didn't maternal antibodies against the Rh factor attack the fetal Rh-positive red blood cells during gestation?

   b. Why would any subsequent Rh-positive babies Heather might carry be likely to develop hemolytic disease of the newborn if she were not treated with Rh immunoglobulin?

   c. How would administration of Rh immunoglobulin immediately following Heather's first pregnancy with an Rh-positive child prevent hemolytic disease of the newborn in a subsequent pregnancy with another Rh-positive child? Similarly, why must Rh immunoglobulin be administered to Heather following the birth of every Rh-positive child she bears?

   d. Suppose Heather were not treated with Rh immunoglobulin following the birth of her first Rh-positive child and a second Rh-positive child developed hemolytic disease of the newborn. Would administration of Rh immunoglobulin to Heather immediately following the second birth prevent this condition in a third Rh-positive child? Why or why not?

PHYSIOEDGE RESOURCES

 PHYSIOEDGE CD-ROM
PhysioEdge focuses on concepts students find difficult to learn. Figures marked with this icon have associated activities on the CD.

CHECK OUT THESE MEDIA QUIZZES:
12.1 The Body's Defenses
12.2 Inflammation
12.3 Basics of Specific Immunity

 PHYSIOEDGE WEB SITE
The Web site for this book contains a wealth of helpful study aids, as well as many ideas for further reading and research. Log on to:

http://www.brookscole.com

Go to the Biology page and select Sherwood's *Human Physiology*, 5th Edition. Select a chapter from the drop-down menu or click on one of these resource areas:

- **Case Histories** provide an introduction to the clinical aspects of human physiology. Check out:

#18: "It's Nothing—Just a Bee Sting"
#19: A Close Call
#20: David—Life in a Germ-Free World
#21: AIDS
#22: Acne Can Be Controlled

- For 2-D and 3-D graphical illustrations and animations of physiological concepts, visit our **Visual Learning Resource.**

- Resources for study and review include the **Chapter Outline, Chapter Summary, Glossary,** and **Flash Cards.** Use our **Quizzes** to prepare for in-class examinations.

- On-line research resources to consult are **Hypercontents,** with current links to relevant Internet sites; **Internet Exercises** with starter URLs provided; and **InfoTrac Exercises.**

 For more readings, go to InfoTrac College Edition, your on-line research library, at:

http://infotrac.thomsonlearning.com

Respiratory System

Body systems maintain homeostasis

Homeostasis

The respiratory system contributes to homeostasis by obtaining O_2 from and eliminating CO_2 to the external environment. It helps regulate the pH of the internal environment by adjusting the rate of removal of acid-forming CO_2.

Homeostasis is essential for survival of cells

Cells

Cells need a constant supply of O_2 delivered to them to support their energy-generating chemical reactions, which produce CO_2 that must be removed continuously. Furthermore, CO_2 generates carbonic acid with which the body must continuously deal in order to maintain the proper pH in the internal environment. Cells can survive only within a narrow pH range.

Cells make up body systems

Energy is essential for sustaining life-supporting cellular activities, such as protein synthesis and active transport across plasma membranes. The cells of the body need a continual supply of O_2 to support their energy-generating chemical reactions. The CO_2 produced during these reactions must be eliminated from the body at the same rate as produced, to prevent dangerous fluctuations in pH (that is, to maintain the acid–base balance), because CO_2 generates carbonic acid.

Respiration involves the sum of the processes that accomplish ongoing passive movement of O_2 from the atmosphere to the tissues to support cellular metabolism, as well as the continual passive movement of metabolically produced CO_2 from the tissues to the atmosphere. The **respiratory system** contributes to homeostasis by exchanging O_2 and CO_2 between the atmosphere and the blood. The blood transports O_2 and CO_2 between the respiratory system and tissues.

Chapter 13

The Respiratory System

CONTENTS AT A GLANCE

INTRODUCTION

▮ Internal and external respiration

▮ Anatomic considerations

RESPIRATORY MECHANICS

▮ Pressure considerations

▮ Respiratory cycle

▮ Airway resistance

▮ Elastic behavior of the lungs

▮ Work of breathing

▮ Lung volumes and capacities

▮ Pulmonary and alveolar ventilation

▮ Local controls to match airflow to blood flow

GAS EXCHANGE

▮ Concept of partial pressure

▮ Gas exchange across the pulmonary and systemic capillaries

GAS TRANSPORT

▮ Oxygen transport

▮ Carbon dioxide transport

▮ Abnormalities in blood gas content

CONTROL OF RESPIRATION

▮ Respiratory control centers in the brain stem

▮ Generation of respiratory rhythm

▮ Chemical inputs affecting the magnitude of ventilation

INTRODUCTION

The primary function of respiration is to obtain O_2 for use by the body's cells and to eliminate the CO_2 the cells produce.

▮ The respiratory system does not participate in all steps in respiration.

Most people think of respiration as the process of breathing in and breathing out. In physiology, however, respiration has a much broader meaning. Respiration encompasses two separate but related processes: internal respiration and external respiration.

Internal respiration

The term **internal** or **cellular respiration** refers to the intracellular metabolic processes carried out within the mitochondria, which use O_2 and produce CO_2 during the derivation of energy from nutrient molecules (see p. 40). The **respiratory quotient (RQ)**, the ratio of CO_2 produced to O_2 consumed, varies depending on the foodstuff consumed. When carbohydrate is being used, the RQ is 1; that is, for every molecule of O_2 consumed, one molecule of CO_2 is produced. For fat utilization, the RQ is 0.7; for protein, it is 0.8. On a typical American diet consisting of a mixture of these three nutrients, resting O_2 consumption averages about 250 ml/min, and CO_2 production averages about 200 ml/min, for an average RQ of 0.8:

$$RQ = \frac{CO_2 \text{ produced}}{O_2 \text{ consumed}} = \frac{200 \text{ ml/min}}{250 \text{ ml/min}} = 0.8$$

External respiration

The term **external respiration** refers to the entire sequence of events involved in the exchange of O_2 and CO_2 between the external environment and the cells of the body. External respiration, the topic of this chapter, encompasses four steps (● Figure 13-1):

1. Air is alternately moved in and out of the lungs so that exchange of air can occur between the atmosphere (external environment) and the air sacs (*alveoli*) of the lungs. This exchange is accomplished by the mechanical act of **breathing**,

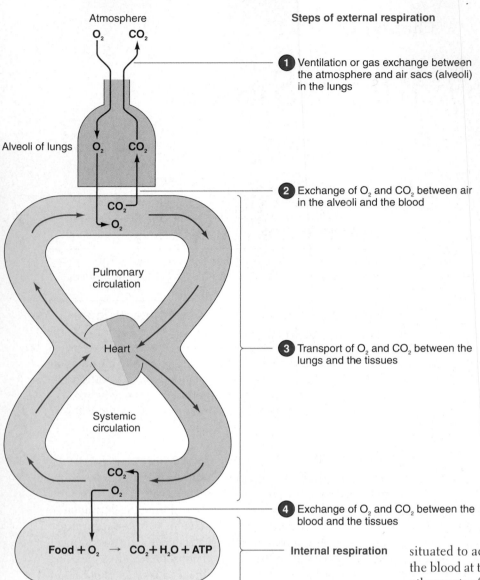

Steps of external respiration

Atmosphere
O_2 CO_2

Alveoli of lungs O_2 CO_2

1. Ventilation or gas exchange between the atmosphere and air sacs (alveoli) in the lungs

CO_2
O_2

Pulmonary circulation

2. Exchange of O_2 and CO_2 between air in the alveoli and the blood

Heart

3. Transport of O_2 and CO_2 between the lungs and the tissues

Systemic circulation

CO_2
O_2

4. Exchange of O_2 and CO_2 between the blood and the tissues

Food + O_2 → CO_2 + H_2O + ATP   **Internal respiration**

Tissue cell

● **FIGURE 13-1**

External and internal respiration
External respiration encompasses the steps involved in the exchange of O_2 and CO_2 between the external environment and the tissue cells (steps 1 through 4). Internal respiration encompasses the intracellular metabolic reactions involving the use of O_2 to derive energy (ATP) from food, producing CO_2 as a by-product.

or **ventilation.** The rate of ventilation is regulated so that the flow of air between the atmosphere and the alveoli is adjusted according to the body's metabolic needs for O_2 uptake and CO_2 removal.

2. Oxygen and CO_2 are exchanged between air in the alveoli and blood within the pulmonary (*pulmonary* means "lung") capillaries by the process of diffusion.

3. Oxygen and CO_2 are transported by the blood between the lungs and the tissues.

4. Exchange of O_2 and CO_2 takes place between the tissues and the blood by the process of diffusion across the systemic (tissue) capillaries.

The respiratory system does not accomplish all the steps of respiration; it is involved only with ventilation and the exchange of O_2 and CO_2 between the lungs and blood (steps 1 and 2). The circulatory system carries out the remaining steps.

Nonrespiratory functions of the respiratory system

The respiratory system additionally performs the following nonrespiratory functions:

- It provides a route for water loss and heat elimination. Inspired (inhaled) atmospheric air is humidified and warmed by the respiratory airways before it is expired. Moistening of inspired air is essential to prevent the alveolar linings from drying out. Oxygen and CO_2 cannot diffuse through dry membranes.
- It enhances venous return (see the "respiratory pump," p. 374).
- It contributes to the maintenance of normal acid–base balance by altering the amount of H^+-generating CO_2 exhaled (see p. 578).
- It enables speech, singing, and other vocalization.
- It defends against inhaled foreign matter (see p. 452).
- It removes, modifies, activates, or inactivates various materials passing through the pulmonary circulation. All blood returning to the heart from the tissues must pass through the lungs before being returned to the systemic circulation. The lungs, therefore, are uniquely situated to act on specific materials that have been added to the blood at the tissue level before they have a chance to reach other parts of the body by means of the arterial system. For example, prostaglandins, a collection of chemical messengers released in numerous tissues to mediate particular local responses (see p. 765), may spill into the blood, but they are *inactivated* during passage through the lungs so that they cannot exert systemic effects. By contrast, the lungs *activate* angiotensin II, a hormone that plays an important role in regulating the concentration of Na^+ in the ECF (see p. 528).
- The nose, a part of the respiratory system, serves as the organ of smell (see p. 229).

The respiratory airways conduct air between the atmosphere and alveoli.

The **respiratory system** includes the respiratory airways leading into the lungs, the lungs themselves, and the structures of the thorax (chest) involved in producing movement of air through the airways into and out of the lungs. The **respiratory airways** are tubes that carry air between the atmosphere and the alveoli, the latter being the only site where exchange of gases can take place between air and blood. The airways begin with the **nasal passages (nose)** (● Figure 13-2a). The nasal passages open

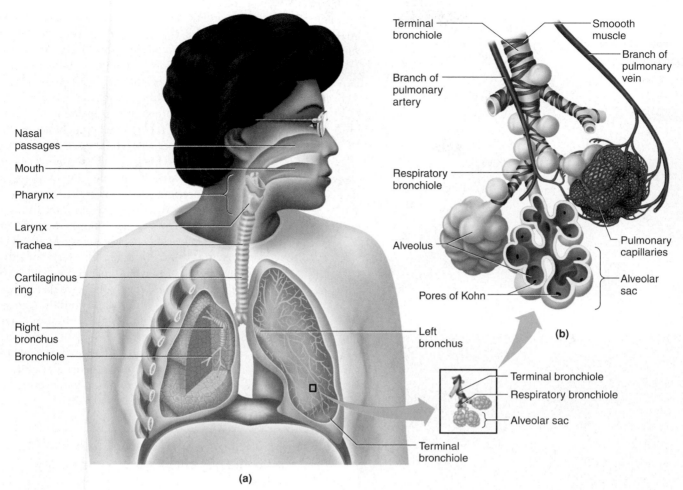

(a)

● **FIGURE 13-2**

Anatomy of the respiratory system
(a) The respiratory airways. (b) Enlargement of the alveoli (air sacs) at the terminal end of the airways. Most alveoli are clustered in grapelike arrangements at the end of the terminal bronchioles.

(SOURCE: Adapted from Cecie Starr and Ralph Taggart, *Biology: The Unity and Diversity of Life*, Eighth Edition, Fig. 41.10a, p. 696. Copyright 1998 Wadsworth.)

into the **pharynx (throat)**, which serves as a common passageway for both the respiratory and the digestive systems. Two tubes lead from the pharynx—the **trachea (windpipe)**, through which air is conducted to the lungs, and the **esophagus,** the tube through which food passes to the stomach. Air normally enters the pharynx through the nose, but it can enter by the mouth as well when the nasal passages are congested; that is, you can breathe through your mouth when you have a cold. Because the pharynx serves as a common passageway for food and air, reflex mechanisms exist to close off the trachea during swallowing so that food enters the esophagus and not the airways. The esophagus remains closed except during swallowing to prevent air from entering the stomach during breathing.

The **larynx,** or **voice box,** is located at the entrance of the trachea. The anterior protrusion of the larynx forms the "Adam's apple." The **vocal folds,**[1] two bands of elastic tissue that lie across the opening of the larynx, can be stretched and positioned

● **FIGURE 13-3**

Vocal folds
Photograph of the vocal folds as viewed from above at the laryngeal opening.

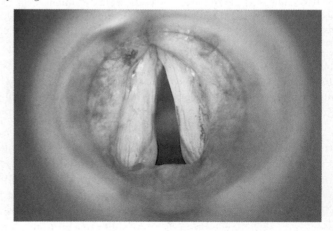

[1]The commonly used term *vocal cords* has been replaced by the more descriptive term *vocal folds* in recent years.

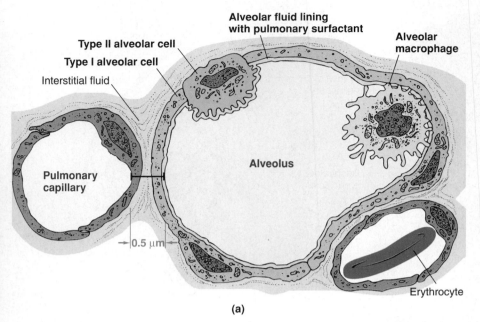

Type II alveolar cell

Type I alveolar cell

Interstitial fluid

Alveolar fluid lining
with pulmonary surfactant

Alveolar
macrophage

Pulmonary
capillary

Alveolus

0.5 μm

Erythrocyte

(a)

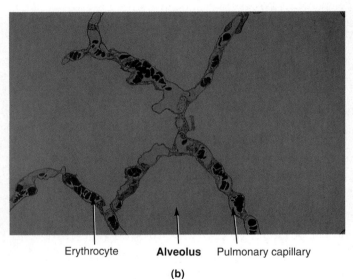

Erythrocyte   **Alveolus**   Pulmonary capillary

(b)

● **FIGURE 13-4**

Alveolus and associated pulmonary capillaries
(a) A schematic representation of a detailed electron microscope view of an alveolus and surrounding capillaries. A single layer of flattened Type I alveolar cells forms the alveolar walls. Type II alveolar cells embedded within the alveolar wall secrete pulmonary surfactant. Wandering alveolar macrophages are found within the alveolar lumen. (The size of the cells and respiratory membrane is exaggerated compared to the size of the alveolar and pulmonary capillary lumens. The diameter of an alveolus is actually about 600 times larger than the intervening space between air and blood.) (b) A transmission electron micrograph showing several alveoli and the close relationship of the capillaries surrounding them.

in different shapes by laryngeal muscles (● Figure 13-3). As air is moved past the taut vocal folds, they vibrate to produce the many different sounds of speech. The lips, tongue, and soft palate modify the sounds into recognizable sound patterns. During swallowing, the vocal folds assume a function not related to speech; they are brought into tight apposition to each other to close off the entrance to the trachea.

Beyond the larynx, the trachea divides into two main branches, the right and left **bronchi,** which enter the right and left lungs, respectively. Within each lung, the bronchus continues to branch into progressively narrower, shorter, and more numerous airways, much like the branching of a tree. The smaller branches are known as **bronchioles.** Clustered at the ends of the terminal bronchioles are the **alveoli,** the tiny air sacs where gas exchange between air and blood takes place (● Figure 13-2b).

To permit airflow in and out of the gas-exchanging portions of the lungs, the continuum of conducting airways from the entrance through the terminal bronchioles to the alveoli must remain open. The trachea and larger bronchi are fairly rigid, nonmuscular tubes encircled by a series of cartilaginous rings that prevent these tubes from compressing. The smaller bronchioles have no cartilage to hold them open. Their walls contain smooth muscle that is innervated by the autonomic nervous system and is sensitive to certain hormones and local chemicals. These factors, by varying the degree of contraction of bronchiolar smooth muscle and hence the caliber of these small terminal airways, regulate the amount of air passing between the atmosphere and each cluster of alveoli.

▌The gas-exchanging alveoli are thin-walled, inflatable air sacs encircled by pulmonary capillaries.

The lungs are ideally structured for their function of gas exchange. Recall that according to Fick's law of diffusion, the shorter the distance through which diffusion must take place, the greater the rate of diffusion. Also, the greater the surface area across which diffusion can take place, the greater the rate of diffusion (see p. 74).

The alveoli are clusters of thin-walled, inflatable, grapelike sacs at the terminal branches of the conducting airways (● Figure 13-2b). The alveolar walls consist of a single layer of flattened **Type I alveolar cells** (● Figure 13-4a). The walls of the dense network of pulmonary capillaries encircling each alveolus are also only one cell thick. The interstitial space between an alveolus and the surrounding capillary network forms an extremely thin barrier, with only 0.5 μm separating the air in the alveoli from the blood in the pulmonary capillaries. (A sheet of tracing paper is about 50 times thicker than this air-to-blood barrier.) The thinness of this barrier facilitates gas exchange.

Furthermore, the alveolar air–blood interface presents a tremendous surface area for exchange. The lungs contain about 300 million alveoli, each about 300 μm in diameter. So dense are the pulmonary capillary networks that each alveolus is encircled by an almost continuous sheet of blood (● Figure 13-4b). The total surface area thus exposed between alveolar air and pulmonary capillary blood is about 75 m^2 (about the size of a tennis court). In contrast, if the lungs consisted of a single hollow chamber of the same dimensions instead of being divided into myriad alveolar units, the total surface area would be only about 0.01 m^2.

In addition to the thin, wall-forming Type I cells, the alveolar epithelium also contains **Type II alveolar cells** (● Figure 13-4a). These cells secrete *pulmonary surfactant,* a phospholipoprotein complex that facilitates lung expansion (to be described later). Furthermore, defensive alveolar macrophages are present within the lumen of the air sacs (see p. 452).

Minute **pores of Kohn** exist in the walls between adjacent alveoli (● Figure 13-2b). Their presence permits airflow between adjoining alveoli, a process known as **collateral ventilation.** These passageways are especially important in allowing fresh air to enter an alveolus whose terminal conducting airway is blocked because of disease.

▌ The lungs occupy much of the thoracic cavity.

There are two **lungs,** each divided into several lobes and each supplied by one of the bronchi. The lung tissue itself consists of the series of highly branched airways, the alveoli, the pulmonary blood vessels, and large quantities of elastic connective tissue. The only muscle within the lungs is the smooth muscle in the walls of the arterioles and the walls of the bronchioles, both of which are subject to control. No muscle is present within the alveolar walls to cause them to inflate and deflate during the breathing process. Instead, changes in lung volume (and accompanying changes in alveolar volume) are brought about through changes in the dimensions of the thoracic cavity. You will learn about this mechanism after we complete our discussion of respiratory anatomy.

The lungs occupy most of the volume of the **thoracic (chest) cavity,** the only other structures in the chest being the heart and associated vessels, the esophagus, the thymus, and some nerves. The outer chest wall (**thorax**) is formed by 12 pairs of curved **ribs,** which join the **sternum** (breastbone) anteriorly and the **thoracic vertebrae** (backbone) posteriorly. The rib cage provides bony protection for the lungs and heart. The **diaphragm,** which forms the floor of the thoracic cavity, is a large, dome-shaped sheet of skeletal muscle that completely separates the thoracic cavity from the abdominal cavity. It is penetrated only by the esophagus and blood vessels traversing between the thoracic and abdominal cavities. The thoracic cavity is enclosed at the neck by muscles and connective tissue. The only communication between the thorax and the atmosphere is through the respiratory airways into the alveoli. Like the lungs, the chest wall contains considerable amounts of elastic connective tissue.

▌ A pleural sac separates each lung from the thoracic wall.

A double-walled, closed sac called the **pleural sac** separates each lung from the thoracic wall and other surrounding structures (● Figure 13-5). The interior of the pleural sac is known as the **pleural cavity.** In the illustration, the dimensions of the pleural cavity are greatly exaggerated to aid visualization; in reality the layers of the pleural sac are in close contact with one another. The surfaces of the pleura secrete a thin **intrapleural fluid** (*intra* means "within"), which lubricates the pleural surfaces as they slide past each other during respiratory

movements. **Pleurisy,** an inflammation of the pleural sac, is accompanied by painful breathing, because each inflation and each deflation of the lungs cause a "friction rub."

RESPIRATORY MECHANICS

Air tends to move from a region of higher pressure to a region of lower pressure, that is, down a pressure gradient.

▌ Interrelationships among pressures inside and outside the lungs are important in ventilation.

Air flows in and out of the lungs during the act of breathing by moving down alternately reversing pressure gradients established between the alveoli and the atmosphere by cyclical respiratory muscle activity. Three different pressure considerations are important in ventilation (● Figure 13-6):

1. **Atmospheric (barometric) pressure** is the pressure exerted by the weight of the air in the atmosphere on objects on Earth's surface. At sea level it equals 760 mm Hg (● Fig-

---

● FIGURE 13-5

Pleural sac
(a) Pushing a lollipop into a water-filled balloon produces a relationship analogous to that between each double-walled, closed pleural sac and the lung that it surrounds and separates from the thoracic wall. (b) Schematic representation of the relationship of the pleural sac to the lungs and thorax. One layer of the pleural sac, the *visceral pleura,* closely adheres to the surface of the lung (*viscus* means "organ") then reflects back on itself to form another layer, the *parietal pleura,* which lines the interior surface of the thoracic wall (*paries* means "wall"). The relative size of the pleural cavity between these two layers is grossly exaggerated for the purpose of visualization.

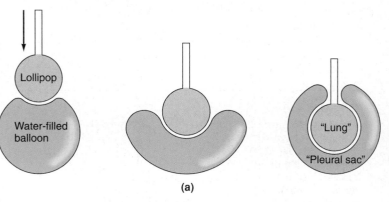

(a)

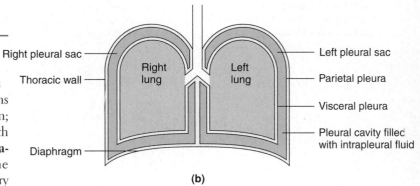

(b)

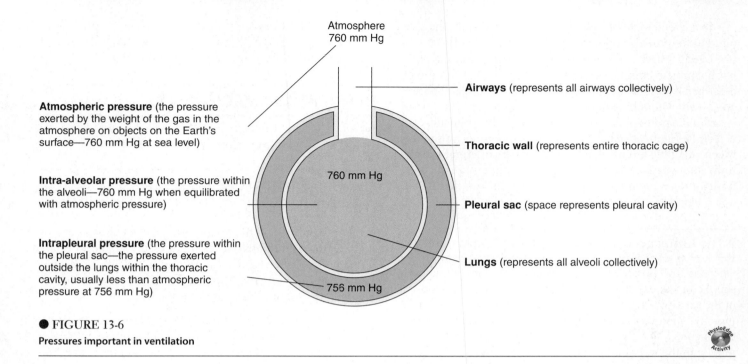

Atmospheric pressure (the pressure exerted by the weight of the gas in the atmosphere on objects on the Earth's surface—760 mm Hg at sea level)

Intra-alveolar pressure (the pressure within the alveoli—760 mm Hg when equilibrated with atmospheric pressure)

Intrapleural pressure (the pressure within the pleural sac—the pressure exerted outside the lungs within the thoracic cavity, usually less than atmospheric pressure at 756 mm Hg)

Atmosphere
760 mm Hg

Airways (represents all airways collectively)

Thoracic wall (represents entire thoracic cage)

760 mm Hg

Pleural sac (space represents pleural cavity)

Lungs (represents all alveoli collectively)

756 mm Hg

● **FIGURE 13-6**
Pressures important in ventilation

ure 13-7). Atmospheric pressure diminishes with increasing altitude above sea level as the column of air above Earth's surface correspondingly decreases. Minor fluctuations in atmospheric pressure occur at any height because of changing weather conditions (that is, when barometric pressure is rising or falling).

2. **Intra-alveolar pressure,** also known as **intrapulmonary pressure,** is the pressure within the alveoli. Because the alveoli communicate with the atmosphere through the conducting airways, air quickly flows down its pressure gradient any time intra-alveolar pressure differs from atmospheric pressure; the airflow continues until the two pressures equilibrate (become equal).

3. **Intrapleural pressure** is the pressure within the pleural sac. Also known as **intrathoracic pressure,** it is the pressure

exerted outside the lungs within the thoracic cavity. The intrapleural pressure is usually less than atmospheric pressure, averaging 756 mm Hg at rest. Just as blood pressure is recorded using atmospheric pressure as a reference point (that is, a systolic blood pressure of 120 mm Hg is 120 mm Hg greater than the atmospheric pressure of 760 mm Hg or in reality 880 mm Hg), 756 mm Hg is sometimes referred to as a pressure of -4 mm Hg. However, there is really no such thing as an absolute negative pressure. A pressure of -4 mm Hg is just negative when compared with the normal atmospheric pressure of 760 mm Hg. To avoid confusion, we will use absolute positive values throughout our discussion of respiration.

Intrapleural pressure does not equilibrate with atmospheric or intra-alveolar pressure, because there is no direct communication between the pleural cavity and either the atmosphere or the lungs. Because the pleural sac is a closed sac with no openings, air cannot enter or leave despite any pressure gradients that might exist between it and surrounding regions.

▌ The lungs are normally stretched to fill the larger thorax.

The thoracic cavity is larger than the unstretched lungs, because the thoracic wall grows more rapidly than the lungs during development. However, two forces—the *intrapleural fluid's cohesiveness* and the *transmural pressure gradient*—hold the thoracic wall and lungs in close apposition, stretching the lungs to fill the larger thoracic cavity.

Intrapleural fluid's cohesiveness

The polar water molecules in the intrapleural fluid resist being pulled apart because of their attraction to each other. The resultant cohesiveness of the intrapleural fluid tends to hold the pleural surfaces together. Thus the intrapleural fluid can be con-

● **FIGURE 13-7**

Atmospheric pressure The pressure exerted on objects by the atmospheric air above Earth's surface at sea level can push a column of mercury to a height of 760 mm. Therefore, atmospheric pressure at sea level is 760 mm Hg.

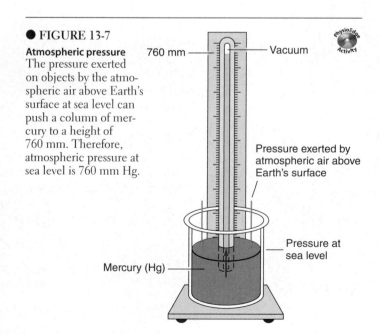

760 mm — Vacuum

Pressure exerted by atmospheric air above Earth's surface

Pressure at sea level

Mercury (Hg)

sidered very loosely as a "stickiness" or "glue" between the lining of the thoracic wall and the lung. Have you ever tried to pull apart two smooth surfaces held together by a thin layer of liquid, such as two wet glass slides? If so, you know that the two surfaces act as if they were stuck together by the thin layer of water. Even though you can easily slip the slides back and forth relative to each other (just as the intrapleural fluid facilitates movement of the lungs against the interior surface of the chest wall), you can pull the slides apart only with great difficulty, because the molecules within the intervening liquid resist being separated. This relationship is partly responsible for the fact that changes in thoracic dimension are always accompanied by corresponding changes in lung dimension; that is, when the thorax expands, the lungs—being stuck to the thoracic wall by virtue of the intrapleural fluid's cohesiveness—do likewise. An even more important reason that the lungs follow the movements of the chest wall is the transmural pressure gradient that exists across the lung wall.

Transmural pressure gradient

The intra-alveolar pressure, equilibrated with atmospheric pressure at 760 mm Hg, is greater than the intrapleural pressure of 756 mm Hg, so a greater pressure is pushing outward than is pushing inward across the lung wall. This net outward pressure differential, the **transmural pressure gradient**, pushes out on the lungs, stretching, or distending them (*trans* means "across"; *mural* means "wall") (● Figure 13-8). Because of this pressure gradient, the lungs are always forced to expand to fill the thoracic cavity.

A similar transmural pressure gradient exists across the thoracic wall. The atmospheric pressure pushing inward on the thoracic wall is greater than the intrapleural pressure pushing outward on this same wall, so the chest wall tends to be "squeezed in" or compressed compared to what it would be in an unrestricted state. The effect of the transmural pressure gradient across the lung wall is much more pronounced, however, because the highly distensible lungs are influenced by this modest pressure differential to a much greater extent than the more rigid thoracic wall is.

Reason why the intrapleural pressure is subatmospheric

Because neither the thoracic wall nor the lungs are in their natural position when they are held in apposition to each other, they constantly try to assume their own inherent dimensions. The stretched lungs have a tendency to pull inward away from the thoracic wall, whereas the compressed thoracic wall tends to move outward away from the lungs. The transmural pressure gradient and intrapleural fluid's cohesiveness, however, prevent these structures from pulling away from each other except to the slightest degree. Even so, the resultant ever-so-slight expansion of the pleural cavity is sufficient to drop the pressure in this cavity by 4 mm Hg, bringing the intrapleural pressure to the subatmospheric level of 756 mm Hg. This pressure drop occurs because the pleural cavity is filled with fluid, which cannot expand to fill the slightly larger volume. Therefore, a vacuum exists in the infinitesimal space in the slightly expanded pleural cavity not occupied by intrapleural fluid, producing a small drop in intrapleural pressure below atmospheric pressure.

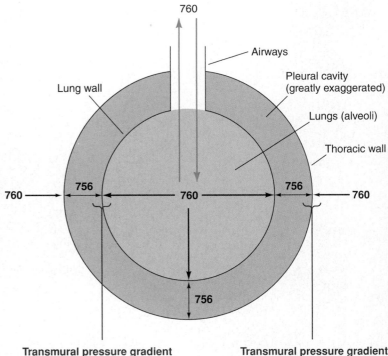

Transmural pressure gradient
across lung wall =
intra-alveolar pressure minus
intrapleural pressure

Transmural pressure gradient
across thoracic wall =
atmospheric pressure minus
intrapleural pressure

Numbers are mm Hg pressure.

● FIGURE 13-8

Transmural pressure gradient
Across the lung wall, the intra-alveolar pressure of 760 mm Hg pushes outward, while the intrapleural pressure of 756 mm Hg pushes inward. This 4-mm Hg difference in pressure constitutes a transmural pressure gradient that pushes out on the lungs, stretching them to fill the larger thoracic cavity. Across the thoracic wall, the atmospheric pressure of 760 mm Hg pushes inward, while the intrapleural pressure of 756 mm Hg pushes outward. This 4-mm Hg difference in pressure constitutes a transmural pressure gradient that pushes inward and compresses the thoracic wall.

Note the interrelationship between the transmural pressure gradient and the subatmospheric intrapleural pressure. The lungs are stretched and the thorax is compressed by a transmural pressure gradient that exists across their walls because of the presence of a subatmospheric intrapleural pressure. The intrapleural pressure in turn is subatmospheric, because the stretched lungs and compressed thorax tend to pull away from each other, slightly expanding the pleural cavity and dropping the intrapleural pressure below atmospheric pressure.

Pneumothorax

Normally, air does not enter the pleural cavity, because there is no communication between the cavity and either the atmosphere or the alveoli. However, if the chest wall is punctured (for example, by a stab wound or a broken rib), air flows down its pressure gradient from the higher atmospheric pressure and rushes into the pleural space (● Figure 13-9a). The abnormal condition of air entering the pleural cavity is known as **pneumothorax** ("air in the chest"). Intrapleural and intra-alveolar pressure are now both equilibrated with atmospheric pressure, so a

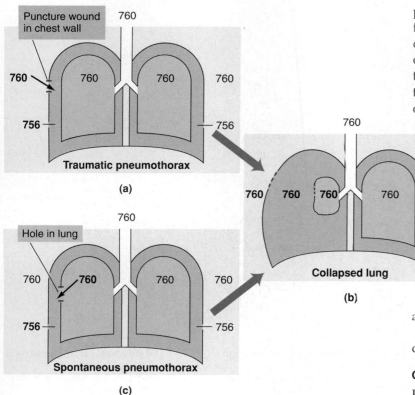

Puncture wound in chest wall

760

760

760 | 760 | 760 | 760

756 | 756

Traumatic pneumothorax

(a)

760

760 | 760 | 760 | 760

Collapsed lung

756

(b)

Hole in lung

760

760 | **760** | 760 | 760

756 | 756

Spontaneous pneumothorax

(c)

Numbers are mm Hg pressure.

● **FIGURE 13-9**

Pneumothorax
(a) Traumatic pneumothorax. A puncture in the chest wall permits air from the atmosphere to flow down its pressure gradient and enter the pleural cavity, abolishing the transmural pressure gradient. (b) When the transmural pressure gradient is abolished, the lung collapses to its unstretched size, and the chest wall springs outward. (c) Spontaneous pneumothorax. A hole in the lung wall permits air to move down its pressure gradient and enter the pleural cavity from the lungs, abolishing the transmural pressure gradient. As with traumatic pneumothorax, the lung collapses to its unstretched size.

transmural pressure gradient no longer exists across either the lung wall or the chest wall. With no force present to stretch the lung, it collapses to its unstretched size (● Figure 13-9b). (The intrapleural fluid's cohesiveness cannot hold the lungs and thoracic wall in apposition in the absence of the transmural pressure gradient.) The thoracic wall likewise springs outward to its unrestricted dimensions, but this has much less serious consequences than collapse of the lung. Similarly, pneumothorax and lung collapse can occur if air enters the pleural cavity through a hole in the lung produced, for example, by a disease process (● Figure 13-9c).

▌ Flow of air into and out of the lungs occurs because of cyclical changes in intra-alveolar pressure.

Because air flows down a pressure gradient, the intra-alveolar pressure must be less than atmospheric pressure for air to flow into the lungs during inspiration. Similarly, the intra-alveolar

pressure must be greater than atmospheric pressure for air to flow out of the lungs during expiration. Intra-alveolar pressure can be changed by altering the volume of the lungs, in accordance with Boyle's law. **Boyle's law** states that at any constant temperature, the pressure exerted by a gas varies inversely with the volume of the gas (● Figure 13-10); that is, as the volume of a gas increases, the pressure exerted by the gas decreases proportionately. Conversely, the pressure increases proportionately as the volume decreases. Changes in lung volume, and accordingly intra-alveolar pressure, are brought about indirectly by respiratory muscle activity.

The respiratory muscles that accomplish breathing do not act directly on the lungs to change their volume. Instead, these muscles change the volume of the thoracic cavity, causing a corresponding change in lung volume because the thoracic wall and lungs are linked together by the intrapleural fluid's cohesiveness and the transmural pressure gradient.

Let us follow the changes that occur during one respiratory cycle—that is, one breath in (**inspiration**) and out (**expiration**).

Onset of inspiration: Contraction of inspiratory muscles

Before the beginning of inspiration, the respiratory muscles are relaxed, no air is flowing, and intra-alveolar pressure is equal to atmospheric pressure. The major **inspiratory muscles**—the muscles that contract to accomplish an inspiration during quiet breathing—include the *diaphragm* and *external intercostal muscles* (● Figure 13-11). At the onset of inspiration, these muscles are stimulated to contract, resulting in enlargement of the thoracic cavity. The major inspiratory muscle is the **diaphragm,** a sheet of skeletal muscle that forms the floor of the thoracic cavity and is innervated by the **phrenic nerve.** The relaxed diaphragm assumes a dome shape that protrudes upward into the thoracic cavity. When the diaphragm contracts (on stimulation by the phrenic nerve), it descends downward, enlarging the volume of the thoracic cavity by increasing its vertical dimension (● Figure 13-12a). The abdominal wall, if relaxed, bulges outward during inspiration as the descending diaphragm pushes the abdominal contents downward and forward. Seventy-five percent of the enlargement of the thoracic cavity during quiet inspiration is attributed to contraction of the diaphragm.

Two sets of **intercostal muscles** lie between the ribs (*inter* means "between"; *costa* means "rib"). These are the *external intercostal muscles* lying on top of the *internal intercostal muscles.* Contraction of the **external intercostal muscles,** whose fibers run downward and forward between adjacent ribs, enlarges the thoracic cavity in both the lateral (side-to-side) and anteroposterior (front-to-back) dimensions. When the external intercostals contract, they elevate the ribs and subsequently the sternum upward and outward (● Figures 13-12a and b). **Intercostal nerves** activate these intercostal muscles.

Before inspiration, at the end of the preceding expiration, intra-alveolar pressure is equal to atmospheric pressure so no air is flowing into or out of the lungs (● Figure 13-13a). As the thoracic cavity enlarges, the lungs are also forced to expand to fill the larger thoracic cavity. As the lungs enlarge, the intra-alveolar pressure drops because the same number of air molecules now occupy a larger lung volume. In a typical inspiratory excursion, the intra-alveolar pressure drops 1 mm Hg to 759 mm Hg (● Figure 13-13b). Because the intra-alveolar pressure is now less than atmospheric pressure, air flows into the lungs down the pressure gradient from higher to lower pressure. Air continues to enter the lungs until no further gradient exists—that is, until intra-alveolar pressure equals atmospheric pressure. Thus lung expansion is not caused by movement of air into the lungs; instead, air flows into the lungs, because of the fall in intra-alveolar pressure brought about by lung expansion.

During inspiration, the intrapleural pressure falls to 754 mm Hg as a result of expansion of the thorax. The resultant increase in the transmural pressure gradient during inspiration ensures that the lungs are stretched to fill the expanded thoracic cavity.

Role of accessory inspiratory muscles

Deeper inspirations (more air breathed in) can be accomplished by contracting the diaphragm and external intercostal muscles more forcefully and by bringing the **accessory inspiratory muscles** into play to further enlarge the thoracic cavity. Contraction of these accessory muscles, which are located in the neck (● Figure 13-11), raises the sternum and elevates the first two ribs, enlarging the upper portion of the thoracic cavity. As the thoracic cavity increases even further in volume than under resting conditions, the lungs likewise expand even more, dropping the intra-alveolar pressure even further. Consequently, a larger inward flow of air occurs before equilibration with atmospheric pressure is achieved; that is, a deeper breath occurs.

Onset of expiration:
Relaxation of inspiratory muscles

At the end of inspiration, the inspiratory muscles relax. The diaphragm assumes its original dome-shaped position when it relaxes. The elevated rib cage falls because of gravity when the external intercostals relax. With no forces causing expansion of the chest wall (and accordingly expansion of the lungs), the chest wall and stretched lungs recoil to their preinspiratory size because of their elastic properties, much as a stretched balloon would on release (● Figure 13-12c). As the lungs recoil and become smaller in volume, the intra-alveolar pressure rises, because the greater number of air molecules contained within the larger lung volume at the end of inspiration are now compressed into a smaller volume. In a resting expiration, the intra-alveolar pressure increases about 1 mm Hg above atmospheric level to 761 mm Hg (● Figure 13-13c). Air now leaves the lungs down its pressure gradient from high intra-alveolar pressure to lower atmospheric pressure. Outward flow of air ceases when

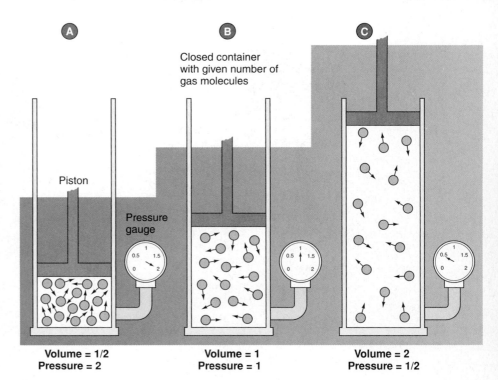

Volume = 1/2
Pressure = 2

Volume = 1
Pressure = 1

Volume = 2
Pressure = 1/2

● **FIGURE 13-10**

Boyle's law
Each container has the same number of gas molecules. Given the random motion of gas molecules, the likelihood of a gas molecule striking the interior wall of the container and exerting pressure varies inversely with the volume of the container at any constant temperature. The gas in container B exerts more pressure than the same gas in larger container C but less pressure than the same gas in smaller container A. This relationship is stated as Boyle's law: $P_1V_1 = P_2V_2$. As the volume of a gas increases, the pressure of the gas decreases proportionately; conversely, the pressure increases proportionately as the volume decreases.

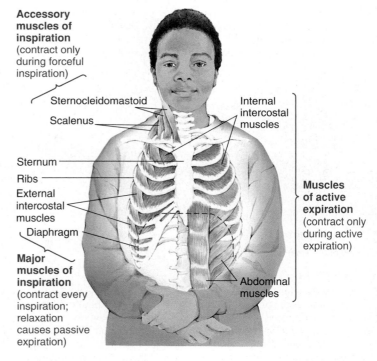

Accessory muscles of inspiration (contract only during forceful inspiration)

Sternocleidomastoid
Scalenus

Sternum
Ribs
External intercostal muscles
Diaphragm

Major muscles of inspiration (contract every inspiration; relaxation causes passive expiration)

Internal intercostal muscles

Muscles of active expiration (contract only during active expiration)

Abdominal muscles

● **FIGURE 13-11**

Anatomy of the respiratory muscles

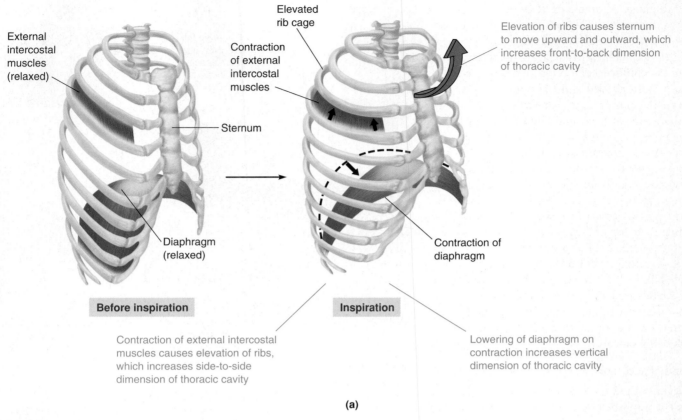

● FIGURE 13-12

Respiratory muscle activity during inspiration and expiration
(a) Inspiration, during which the diaphragm descends on contraction, increasing the vertical dimension of the thoracic cavity. Contraction of the external intercostal muscles elevates the ribs and subsequently the sternum to enlarge the thoracic cavity from front to back and from side to side. *(continued)*

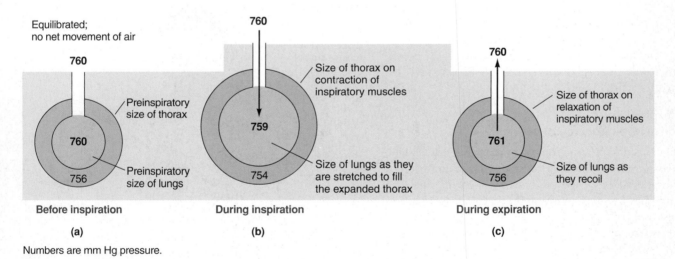

Numbers are mm Hg pressure.

● FIGURE 13-13

Changes in lung volume and intra-alveolar pressure during inspiration and expiration
(a) Before inspiration, at the end of the preceding expiration. Intra-alveolar pressure is equilibrated with atmospheric pressure and no air is flowing. (b) Inspiration. As the lungs increase in volume during inspiration, the intra-alveolar pressure decreases, establishing a pressure gradient that favors the flow of air into the alveoli from the atmosphere; that is, an inspiration occurs. (c) Expiration. As the lungs recoil to their preinspiratory size on relaxation of the inspiratory muscles, the intra-alveolar pressure increases, establishing a pressure gradient that favors the flow of air out of the alveoli into the atmosphere; that is, an expiration occurs.

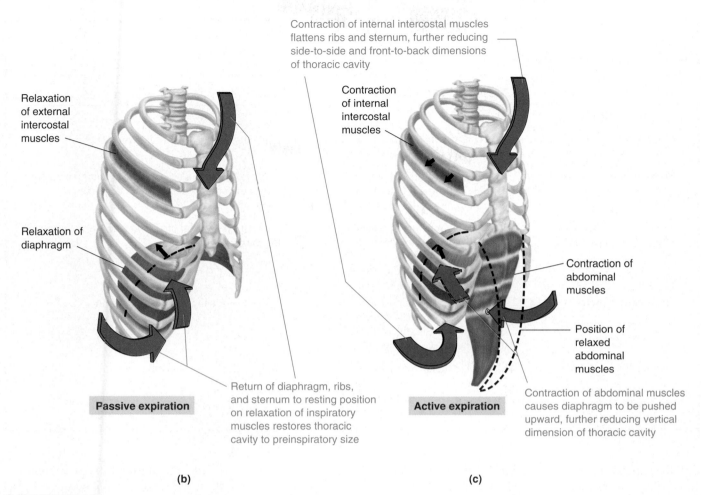

Relaxation
of external
intercostal
muscles

Relaxation of
diaphragm

Passive expiration

Contraction of internal intercostal muscles
flattens ribs and sternum, further reducing
side-to-side and front-to-back dimensions
of thoracic cavity

Contraction
of internal
intercostal
muscles

Contraction of
abdominal
muscles

Position of
relaxed
abdominal
muscles

Active expiration

Return of diaphragm, ribs,
and sternum to resting position
on relaxation of inspiratory
muscles restores thoracic
cavity to preinspiratory size

Contraction of abdominal muscles
causes diaphragm to be pushed
upward, further reducing vertical
dimension of thoracic cavity

(b)

(c)

● **FIGURE 13-12**

Respiratory muscle activity during inspiration and expiration *(continued)*
(b) Quiet passive expiration, during which the diaphragm relaxes, reducing the volume of the thoracic
cavity from its peak inspiratory size. As the external intercostal muscles relax, the elevated rib cage falls
because of the force of gravity. This also reduces the volume of the thoracic cavity. (c) Active expiration,
during which contraction of the abdominal muscles increases the intra-abdominal pressure, exerting an
upward force on the diaphragm. This reduces the vertical dimension of the thoracic cavity further than it
is reduced during quiet passive expiration. Contraction of the internal intercostal muscles decreases the
front-to-back and side-to-side dimensions by flattening the ribs and sternum.

intra-alveolar pressure becomes equal to atmospheric pressure
and a pressure gradient no longer exists. ● Figure 13-14 summarizes the intra-alveolar and intrapleural pressure changes
that take place during one respiratory cycle.

Because the diaphragm is the major inspiratory muscle
and its relaxation also causes expiration, paralysis of the intercostal muscles alone does not seriously influence quiet breathing. Disruption of diaphragm activity caused by nerve or muscle disorders, however, leads to respiratory paralysis. Fortunately,
the phrenic nerve arises from the spinal cord in the neck region
(cervical segments 3, 4, and 5) and then descends to the diaphragm at the floor of the thorax, instead of arising from the
thoracic region of the cord as might be expected. For this reason, individuals completely paralyzed below the neck by traumatic severance of the spinal cord are still able to breathe,
even though they have lost use of all other skeletal muscles in
their trunk and limbs.

Forced expiration: Contraction of expiratory muscles

During quiet breathing, expiration is normally a *passive* process,
because it is accomplished by elastic recoil of the lungs on relaxation of the inspiratory muscles, with no muscular exertion
or energy expenditure required. In contrast, inspiration is *always*
active, because it is brought about only by contraction of inspiratory muscles at the expense of energy use. To empty the lungs
more completely and more rapidly than is accomplished during
quiet breathing, as during the deeper breaths accompanying exercise, expiration does become active. The intra-alveolar pressure must be increased even further above atmospheric pressure
than can be accomplished by simple relaxation of the inspiratory muscles and elastic recoil of the lungs. To produce such a
forced, or **active**, **expiration**, expiratory muscles must contract
to further reduce the volume of the thoracic cavity and lungs
(● Figures 13-11 and 13-12c and ▲ Table 13.1). The most important **expiratory muscles** are (unbelievable as it may seem at

Actions of the Respiratory Muscles

Muscles	Result of Muscle Contraction	Timing of Stimulation to Contract
Inspiratory Muslces		
Diaphragm	Descends downward, increasing the vertical dimension of the thoracic cavity	Every inspiration; the primary muscle of inspiration
External Intercostal Muscles	Elevate ribs upward and outward, enlarging the thorax in both the front-to-back and side-to-side dimensions	Every inspiration; play a secondary complementary role to the primary action of the diaphragm
Neck Muscles (Scalenus, Sternocleiodomastoid)	Raise the sternum and elevate the first two ribs, enlarging the upper portion of the thoracic cavity	Only during forceful inspiration; accessory inspiratory muscles
Expiratory Muscles		
Abdominal Muscles	Increase the intra-abdominal pressure, which exerts an upward force on the diaphragm to decrease the vertical dimension of the thoracic cavity	Only during active (forced) expiration
Internal Intercostal Muscles	Flatten the thorax by pulling the ribs downward and inward, decreasing the front-to-back and side-to-side dimensions of the thoracic cavity	Only during active (forced) expiration

first) the *muscles of the abdominal wall*. As the abdominal muscles contract, the resultant increase in intra-abdominal pressure exerts an upward force on the diaphragm, pushing it further up into the thoracic cavity than its relaxed position, thus decreasing the vertical dimension of the thoracic cavity even more. The other expiratory muscles are the **internal intercostal muscles,** whose contraction pulls the ribs downward and inward, flattening the chest wall and further decreasing the size of the thoracic cavity; this action is just the opposite of that of the external intercostal muscles (● Figure 13-12c).

● **FIGURE 13-14**

Intra-alveolar and intrapleural pressure changes throughout the respiratory cycle

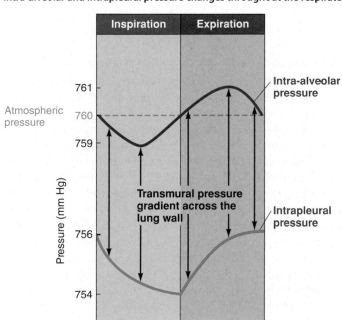

- During inspiration, intra-alveolar pressure is < atmospheric pressure.

- During expiration, intra-alveolar pressure is > atmospheric pressure.

- At the end of both inspiration and expiration, intra-alveolar pressure is equal to atmospheric pressure, because the alveoli are in direct communication with the atmosphere and air continues to flow down its pressure gradient until the two pressures equilibrate.

- Throughout the respiratory cycle, intrapleural pressure is < intra-alveolar pressure. Thus a transmural pressure gradient always exists, and the lung is always stretched to some degree, even during expiration.

As the active contraction of the expiratory muscles further reduces the volume of the thoracic cavity, the lungs also become further reduced in volume because they do not have to be stretched as much to fill the smaller thoracic cavity; that is, they are permitted to recoil to an even smaller volume. The intra-alveolar pressure increases further as the air in the lungs is confined within this smaller volume. The differential between intra-alveolar and atmospheric pressure is even greater now than during passive expiration, so more air leaves down the pressure gradient before equilibration is achieved. In this way, the lungs are emptied more completely during forceful, active expiration than during quiet, passive expiration.

During forceful expiration, the intrapleural pressure exceeds atmospheric pressure, but the lungs do not collapse. Because the intra-alveolar pressure is also increased correspondingly, a transmural pressure gradient still exists across the walls of the lungs, keeping them stretched to fill the thoracic cavity. For example, if the pressure within the thorax increases 10 mm Hg, the intrapleural pressure becomes 766 mm Hg and the intra-alveolar pressure becomes 770 mm Hg—still a 4-mm Hg pressure difference.

■ Airway resistance influences airflow rates.

Thus far we have discussed airflow in and out of the lungs as a function of the magnitude of the pressure gradient between the alveoli and the atmosphere. However, just as flow of blood through the blood vessels depends not only on the pressure gradient but also on the resistance to the flow offered by the vessels, so it is with airflow:

$$F = \frac{\Delta P}{R}$$

where

F = airflow rate

ΔP = difference between atmospheric and intra-alveolar pressure (pressure gradient)

R = resistance of airways, determined by their radii

The primary determinant of resistance to airflow is the radius of the conducting airways. We ignored airway resistance in our preceding discussion of pressure gradient–induced airflow rates because in a healthy respiratory system, the radius of the conducting system is large enough that resistance remains extremely low. Therefore, the pressure gradient between the alveoli and the atmosphere is usually the primary factor determining the airflow rate. Indeed, the airways normally offer such low resistance that only very small pressure gradients of 1 to 2 mm Hg need be created to achieve adequate rates of airflow in and out of the lungs. (By comparison, it would take a pressure gradient 250 times greater to move air through a smoker's pipe than through the respiratory airways at the same flow rate.)

Normally, modest adjustments in airway size can be accomplished by autonomic nervous system regulation to suit the body's needs. Parasympathetic stimulation, which occurs in quiet, relaxed situations when the demand for airflow is low,

promotes bronchiolar smooth muscle contraction, which increases airway resistance by producing **bronchoconstriction** (a reduction in bronchiolar caliber). In contrast, sympathetic stimulation and to a greater extent its associated hormone, epinephrine, bring about **bronchodilation** (an increase in bronchiolar caliber) and decreased airway resistance by promoting bronchiolar smooth muscle relaxation (▲ Table 13-2). Thus during periods of sympathetic domination, when increased demands for O_2 uptake are actually or potentially placed on the body, bronchodilation ensures that the pressure gradients established by respiratory muscle activity can achieve maximum airflow rates with minimum resistance. Because of this bronchodilator action, epinephrine or similar drugs are useful therapeutic tools to counteract airway constriction in patients with bronchial spasms.

Resistance becomes an extremely important impediment to airflow when airway lumens become abnormally narrowed by disease. We have all transiently experienced the effect that increased airway resistance has on breathing when we have a cold. We know how difficult it is to produce an adequate airflow rate through a "stuffy nose" when the nasal passages are narrowed by swelling and mucus accumulation. More serious is chronic obstructive pulmonary disease, to which we now turn our attention.

■ Airway resistance is abnormally increased with chronic obstructive pulmonary disease.

Chronic obstructive pulmonary disease (COPD) is a group of lung diseases characterized by increased airway resistance resulting from the narrowing of the lumen of the lower airways. When airway resistance increases, a larger pressure gradient must be established to maintain even a normal airflow rate. For example, if resistance is doubled by narrowing of airway lumens, ΔP must be doubled through increased respiratory muscle exertion to induce the same flow rate of air in and out of the lungs as a normal individual accomplishes during quiet breathing. Accordingly, patients with COPD must work harder to breathe.

Chronic obstructive pulmonary disease encompasses three chronic (long-term) diseases: *chronic bronchitis*, *asthma*, and *emphysema*.

Chronic bronchitis

Chronic bronchitis is a long-term inflammatory condition of the lower respiratory airways, generally triggered by frequent exposure to irritating cigarette smoke, polluted air, or allergens. In response to the chronic irritation, the airways become narrowed by prolonged edematous thickening of the airway linings, coupled with overproduction of thick mucus. Despite frequent coughing associated with the chronic irritation, the plugged mucus often cannot be satisfactorily removed, especially because the irritants immobilize the ciliary mucus escalator (see p. 452). Pulmonary bacterial infections frequently occur, because the accumulated mucus serves as an excellent medium for bacterial growth.

▲ TABLE 13-2
Factors Affecting Airway Resistance

Status of Airways	Effect on Resistance	Factors Producing the Effect
Bronchoconstriction	↓ radius, ↑ resistance to airflow	*Pathological factors:* Allergy-induced spasm of the airways caused by 　Slow-reactive substance of anaphylaxis 　Histamine Physical blockage of the airways caused by 　Excess mucus 　Edema of the walls 　Airway collapse *Physiological control factors:* Neural control: parasympathetic stimulation Local chemical control: ↓ CO_2 concentration
Bronchodilation	↑ radius, ↓ resistance to airflow	*Pathological factors:* none *Physiological control factors:* Neural control: sympathetic stimulation (minimal effect) Hormonal control: epinephrine Local chemical control: ↑ CO_2 concentration

Asthma

In **asthma**, airway obstruction is due to (1) thickening of airways' walls, brought about by inflammation and histamine-induced edema (see p. 369); (2) plugging of the airways by excessive secretion of very thick mucus; and (3) airway hyperresponsiveness, characterized by profound constriction of the smaller airways caused by trigger-induced spasm of the smooth muscle in the walls of these airways (see p. 447). Triggers that lead to these inflammatory changes and the exaggerated bronchoconstrictor response include repeated exposure to allergens (like dust mites or pollen), irritants (as in cigarette smoke), and infections. A growing number of studies suggest that long-term infections with *Chlamydia pneumoniae*, a common cause of lung infections, may underlie up to half of the adult cases of asthma. In severe asthmatic attacks, pronounced clogging and narrowing of the airways can cut off all airflow, leading to death. An estimated 15 million people in the United States have asthma, with the number steadily climbing. The prevalence of this condition has doubled in the last 20 years. Asthma is the most common chronic childhood disease, affecting 5 million children in our country alone. Scientists are unsure why asthma incidence is rising. However, they have made considerable progress in working out the cascade of events that lead to asthmatic symptoms, with the aim of developing drugs to intervene in the process.

Emphysema

Emphysema is characterized by collapse of the smaller airways and breakdown of alveolar walls. This irreversible condition can arise in two different ways. Most commonly, emphysema results from excessive release of destructive enzymes such as *trypsin* from alveolar macrophages as a defense mechanism in response to chronic exposure to inhaled cigarette smoke. The lungs are normally protected from damage by these enzymes by α_1-*antitrypsin*, a protein that inhibits trypsin. Excessive secretion of these destructive enzymes in response to chronic irritation, however, can overwhelm the protective capability of α_1-antitrypsin so that these enzymes destroy not only foreign materials but lung tissue as well. Loss of lung tissue leads to breakdown of alveolar walls and collapse of small airways, the characteristics of emphysema.

Less frequently, emphysema arises from a genetic inability to produce α_1-antitrypsin so that the lung tissue has no protection from trypsin. The unprotected lung tissue gradually disintegrates under the influence of even small amounts of macrophage-released enzymes, in the absence of chronic exposure to inhaled irritants.

Difficulty in expiring

When airway resistance is increased by COPD of any type, expiration is more difficult than inspiration. The smaller airways, lacking the cartilaginous rings that hold the larger airways open, are held open by the same transmural pressure gradient that distends the alveoli. Expansion of the thoracic cavity during inspiration indirectly dilates the airways even further than their expiratory dimensions, like alveolar expansion, so airway resistance is lower during inspiration than during expiration. In a normal individual, the airway resistance is always so low that the slight variation between inspiration and expiration is not noticeable. When airway resistance has substantially increased, however, as during an asthmatic attack, the difference is quite noticeable. Thus an asthmatic has more difficulty expiring than inspiring, giving rise to the characteristic "wheeze" as air is forced out through the narrowed airways.

In normal individuals, the smaller airways collapse and further outflow of air is halted only at very low lung volumes (● Figure 13-15). Because of this airway collapse, the lungs can never be emptied completely. In COPD, especially emphysema, the smaller airways may routinely collapse during expiration, preventing further outflow of air through these passageways (● Figure 13-15d).

▌ Elastic behavior of the lungs is due to elastic connective tissue and alveolar surface tension.

During the respiratory cycle, the lungs alternately expand during inspiration and recoil during expiration. What properties of the lungs enable them to behave like balloons, being stretchable and then snapping back to their resting position when the stretching forces are removed? Two interrelated concepts are involved in pulmonary elasticity: *elastic recoil* and *compliance*.

The term **elastic recoil** refers to how readily the lungs rebound after having been stretched. It is responsible for the lungs returning to their preinspiratory volume when the inspiratory muscles relax at the end of inspiration.

The term **compliance** refers to how much effort is required to stretch or distend the lungs; it is analogous to how hard you have to work to blow up a balloon. (By comparison, 100 times more distending pressure is required to inflate a child's toy balloon than to inflate the lungs.) Specifically, compliance is a measure of the magnitude of change

in lung volume accomplished by a given change in the transmural pressure gradient, the force that stretches the lungs. A highly compliant lung stretches farther for a given increase in the pressure difference than does a less compliant lung. Stated another way, the lower the compliance of the lungs, the larger the transmural pressure gradient that must be created during inspiration to produce normal lung expansion. In turn, a greater-than-normal transmural pressure gradient during inspiration can be achieved only by making the intrapleural pressure more subatmospheric than usual. This is accomplished by greater expansion of the thorax through more vigorous contraction of

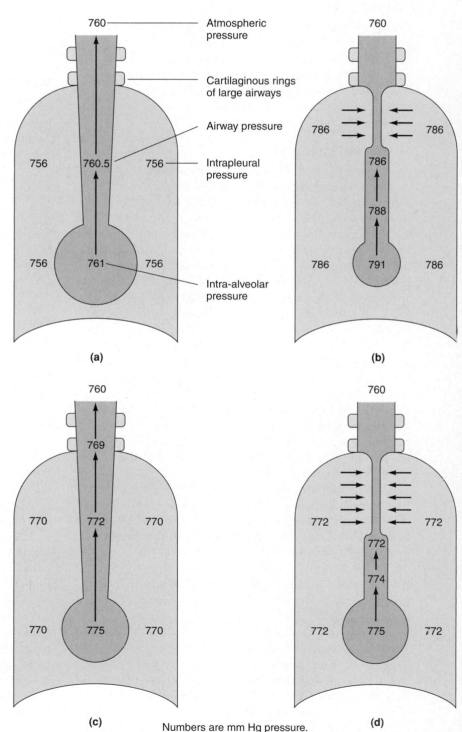

Numbers are mm Hg pressure.

● FIGURE 13-15

Airway collapse during forced expiration
(a) Normal quiet breathing, during which airway resistance is low, so there is little frictional loss of pressure within the airways. Intrapleural pressure remains less than airway pressure throughout the length of the airways, so the airways remain open. (b) Maximal forced expiration, during which both intra-alveolar and intrapleural pressures are markedly increased. When frictional losses cause the airway pressure to fall below the surrounding elevated intrapleural pressure, the small nonrigid airways are compressed closed, blocking further expiration of air through the airway. In normal individuals, this dynamic compression of airways occurs only at very low lung volumes. (c) Routine exercise. Even though intrapleural pressure is elevated during the active expiration accompanying routine vigorous activity, the airway pressure does not drop below the intrapleural pressure until the level at which the airways are held open by cartilaginous rings, so airway collapse does not occur. (d) Obstructive lung disease. Premature airway collapse occurs for two reasons: (1) the pressure drop along the airways is magnified as a result of increased airway resistance, and (2) the intrapleural pressure is higher than normal because of the loss, as in emphysema, of lung tissue that is responsible for the lung's tendency to recoil and pull away from the thoracic wall. Excessive air trapped in the alveoli behind the compressed bronchiolar segments reduces the amount of gas exchanged between the alveoli and the atmosphere. Therefore, less alveolar air is "freshened" with each breath when airways collapse at higher lung volumes in patients with obstructive lung disease.

the inspiratory muscles. Therefore, the less compliant the lungs, the more work required to produce a given degree of inflation. A poorly compliant lung is referred to as a "stiff" lung, because it lacks normal stretchability. Respiratory compliance can be decreased by a number of factors, as in *pulmonary fibrosis*, where normal lung tissue is replaced with scar-forming fibrous connective tissue as a result of chronically breathing in asbestos fibers or similar irritants.

Pulmonary elastic behavior depends mainly on two factors: *highly elastic connective tissue* in the lungs and *alveolar surface tension*.

Pulmonary elastic connective tissue

Pulmonary connective tissue contains large quantities of elastin fibers (see p. 62). Not only do these fibers exhibit elastic properties themselves, but they are arranged into a meshwork that amplifies their elastic behavior, much like threads in stretch-knit fabric. The entire piece of fabric (or lung) is stretchier and tends to bounce back to its original shape more than do the individual threads (elastin fibers).

Alveolar surface tension

An even more important factor influencing elastic behavior of the lungs is the **alveolar surface tension** displayed by the thin liquid film that lines each alveolus. At an air–water interface, the water molecules at the surface are more strongly attracted to other surrounding water molecules than to the air above the surface. This unequal attraction produces a force known as *surface tension* at the surface of the liquid. Surface tension is responsible for a twofold effect. First, the liquid layer resists any force that increases its surface area; that is, it opposes expansion of the alveolus, because the surface water molecules oppose being pulled apart. Accordingly, the greater the surface tension, the less compliant the lungs. Second, the liquid surface area tends to shrink as small as possible, because the surface water molecules, being preferentially attracted to each other, try to get as close together as possible. Thus the surface tension of the liquid lining an alveolus tends to reduce alveolus size, squeezing in on the air inside (● Figure 13-16). This property, along with the rebound of the stretched elastin fibers, is responsible for the lungs' elastic recoil back to their preinspiratory size when inspiration is over.

With emphysema, elastic recoil is decreased by loss of elastin fibers and the reduction in alveolar surface tension resulting from the breakdown of alveolar walls. The decrease in elastic recoil contributes, along with the increased airway resistance, to the patient's difficulty in expiration.

▌ Pulmonary surfactant decreases surface tension and contributes to lung stability.

The cohesive forces between water molecules are so strong that if the alveoli were lined with water alone, the surface tension would be so great that the lungs would collapse. The recoil force attributable to the elastin fibers and high surface tension would exceed the opposing stretching force of the transmural pressure gradient. Furthermore, the lungs would be very poorly

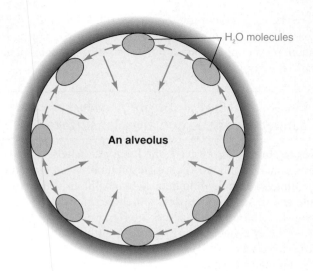

● **FIGURE 13-16**

Alveolar surface tension
The attractive forces between the water (H_2O) molecules in the liquid film that lines the alveolus are responsible for surface tension. Because of its surface tension, an alveolus (1) resists being stretched, (2) tends to be reduced in surface area or size, and (3) tends to recoil after being stretched.

---

compliant, so exhausting muscular efforts would be required to accomplish stretching and inflation of the alveoli. Two factors oppose the tendency for alveoli to collapse, thereby maintaining alveolar stability and reducing the work of breathing. These are *pulmonary surfactant* and *alveolar interdependence*.

Pulmonary surfactant

The tremendous surface tension of pure water is normally counteracted by **pulmonary surfactant,** a complex mixture of lipids and proteins secreted by the Type II alveolar cells (● Figure 13-4a, p. 462). Pulmonary surfactant intersperses between the water molecules in the fluid lining the alveoli and lowers the alveolar surface tension, because the cohesive force between a water molecule and an adjacent pulmonary surfactant molecule is very low. By lowering the alveolar surface tension, pulmonary surfactant provides two important benefits: (1) it increases pulmonary compliance, reducing the work of inflating the lungs; and (2) it reduces the lungs' tendency to recoil, so they do not collapse as readily.

Pulmonary surfactant's role in reducing the tendency of alveoli to recoil, thereby discouraging alveolar collapse, is important in helping maintain lung stability. The division of the lung into myriad tiny air sacs provides the advantage of a tremendously increased surface area for exchange of O_2 and CO_2, but it also presents the problem of maintaining the stability of all these alveoli. Recall that the pressure generated by alveolar surface tension is directed inward, squeezing in on the air in the alveoli. If you visualize the alveoli as spherical bubbles, according to the the **law of LaPlace,** the magnitude of the inward-directed collapsing pressure is directly proportional to the surface tension and inversely proportional to the radius of the bubble:

$$P = \frac{2T}{r}$$

where

P = inward-directed collapsing pressure

T = surface tension

r = radius of bubble (alveolus)

Because the collapsing pressure is inversely proportional to the radius, the smaller the alveolus, the smaller its radius and the greater its tendency to collapse at a given surface tension. Accordingly, if two alveoli of unequal size but the same surface tension are connected by the same terminal airway, the smaller alveolus—because it generates a larger collapsing pressure—has a tendency to collapse and empty its air into the larger alveolus (● Figure 13-17a).

Small alveoli normally do not collapse and blow up larger alveoli, however, because pulmonary surfactant reduces the surface tension of small alveoli more than that of larger alveoli. Pulmonary surfactant decreases surface tension to a greater degree in small alveoli than in larger alveoli because the surfactant molecules are closer together in the smaller alveoli. The larger an alveolus, the more spread out are its surfactant molecules and the less effect they have on reducing surface tension. The surfactant-induced lower surface tension of small alveoli offsets the effect of their smaller radii in determining the inward-directed pressure. Therefore, the presence of surfactant causes the collapsing pressure of small alveoli to become comparable to that of

larger alveoli and minimizes the tendency for small alveoli to collapse and empty their contents into larger alveoli (● Figure 13-17b). Pulmonary surfactant therefore helps stabilize the sizes of the alveoli and helps keep them open and available to participate in gas exchange.

Alveolar interdependence

A second factor that contributes to alveolar stability is the interdependence among neighboring alveoli. Each alveolus is

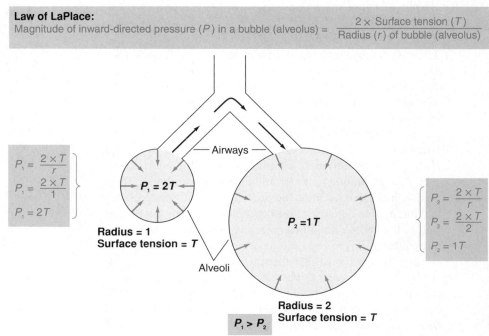

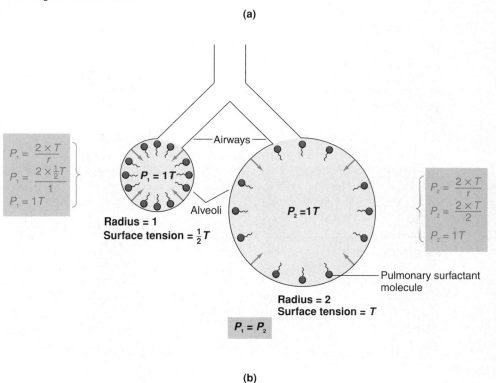

● FIGURE 13-17

Role of pulmonary surfactant in counteracting the tendency for small alveoli to collapse into larger alveoli
(a) According to the law of LaPlace, if two alveoli of unequal size but the same surface tension are connected by the same terminal airway, the smaller alveolus—because it generates a larger inward-directed collapsing pressure—has a tendency (without pulmonary surfactant) to collapse and empty its air into the larger alveolus. (b) Pulmonary surfactant reduces the surface tension of a smaller alveolus more than that of a larger alveolus. This reduction in surface tension offsets the effect of the smaller radius in determining the inward-directed pressure. Consequently, the collapsing pressures of the small and large alveoli are comparable. Therefore, in the presence of pulmonary surfactant a small alveolus does not collapse and empty its air into the larger alveolus.

surrounded by other alveoli and interconnected with them by connective tissue. If an alveolus starts to collapse, the surrounding alveoli are stretched as their walls are pulled in the direction of the caving-in alveolus (● Figure 13-18a). In turn, these neighboring alveoli, by recoiling in resistance to being stretched, exert expanding forces on the collapsing alveolus and thereby help keep it open (● Figure 13-18b). This phenomenon, which can be likened to a stalemated "tug of war" between adjacent alveoli, is termed **alveolar interdependence.**

The opposing forces acting on the lung (that is, the forces keeping the alveoli open and the countering forces that promote alveolar collapse) are summarized in ▲ Table 13-3.

Newborn respiratory distress syndrome

The developing fetal lungs normally do not have the ability to synthesize pulmonary surfactant until late in pregnancy. Especially in an infant born prematurely, pulmonary surfactant may be insufficient to reduce the alveolar surface tension to manageable levels. The resulting collection of symptoms are referred to as **newborn respiratory distress syndrome.** The infant needs very strenuous inspiratory efforts to overcome the high surface tension in an attempt to inflate the poorly compliant lungs. Adding to the problem, the work of breathing is further increased because the alveoli, in the absence of surfactant, tend to collapse almost completely during each expiration. It is more difficult (requires a greater transmural pressure differential) to expand a collapsed alveolus by a given volume than to increase an already partially expanded alveolus by the same volume. The situation is analogous to blowing up a new balloon. It takes more effort to blow in that first breath of air when starting to blow up a new balloon than to blow additional breaths into the already partially expanded balloon. With newborn respiratory distress syndrome, it is as though with every breath the infant must start blowing up a

Forces Keeping the Alveoli Open	Forces Promoting Alveolar Collapse
Transmural pressure gradient	Elasticity of stretched pulmonary connective tissue fibers
Pulmonary surfactant (which opposes alveolar surface tension)	
	Alveolar surface tension
Alveolar interdependence	

▲ TABLE 13-3
Opposing Forces Acting on the Lung

new balloon. Lung expansion may require transmural pressure gradients of 20 to 30 mm Hg (compared to the normal 4 to 6 mm Hg) to overcome the tendency of surfactant-deprived alveoli to collapse.

Worse yet, the newborn's muscles are still weak. The respiratory distress associated with surfactant deficiency may soon lead to death if breathing efforts become exhausting or inadequate to support sufficient gas exchange.

This life-threatening condition affects 30,000 to 50,000 newborns, primarily premature infants, each year in the United States. Until the surfactant-secreting cells mature sufficiently, the condition is treated by surfactant replacement. In addition, drugs can hasten the maturation process.

▌ The work of breathing normally requires only about 3% of total energy expenditure.

During normal quiet breathing, the respiratory muscles must work during inspiration to expand the lungs against their elastic forces and to overcome airway resistance, whereas expira-

● FIGURE 13-18

Alveolar interdependence
(a) When an alveolus (in pink) in a group of interconnected alveoli starts to collapse, the surrounding alveoli are stretched by the collapsing alveolus. (b) As the neighboring alveoli recoil in resistance to being stretched, they pull outward on the collapsing alveolus. This expanding force pulls the collapsing alveolus open.

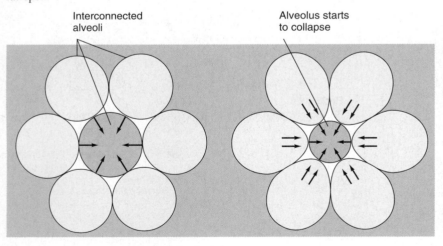

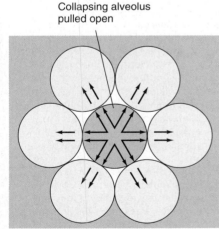

(a)

(b)

tion is passive. Normally, the lungs are highly compliant and airway resistance is low, so only about 3% of the total energy expended by the body is used for quiet breathing. The work of breathing may be increased in four different situations:

1. *When pulmonary compliance is decreased*, such as with pulmonary fibrosis, more work is required to expand the lungs.

2. *When airway resistance is increased*, such as with COPD, more work is required to achieve the greater pressure gradients necessary to overcome the resistance so that adequate airflow can occur.

3. *When elastic recoil is decreased*, as with emphysema, passive expiration may be inadequate to expel the volume of air normally exhaled during quiet breathing. Thus the abdominal muscles must work to aid in emptying the lungs, even when the person is at rest.

4. *When there is a need for increased ventilation*, such as during exercise, more work is required to accomplish both a greater depth of breathing (a larger volume of air moving in and out with each breath) and a faster rate of breathing (more breaths per minute).

During strenuous exercise, the amount of energy required to power pulmonary ventilation may increase up to 25-fold. However, because total energy expenditure by the body increases up to 15- to 20-fold during heavy exercise, the energy used for increased ventilation still represents only about 5% of total energy expended. In contrast, in patients with poorly compliant lungs or obstructive lung disease, the energy required for breathing even at rest may increase to 30% of total energy expenditure. In such cases, the individual's exercise ability is severely limited, as breathing itself becomes exhausting.

▌ The lungs normally operate at about "half full."

On average, in healthy young adults, the maximum air that the lungs can hold is about 5.7 liters in males (4.2 liters in females). Anatomic build, age, the distensibility of the lungs, and the presence or absence of respiratory disease affect this total lung capacity. Normally, during quiet breathing the lungs are nowhere near maximally inflated nor are they deflated to their minimum volume. Thus the lungs normally remain moderately inflated throughout the respiratory cycle. At the end of a normal quiet expiration, the lungs still contain about 2,200 ml of air. During each typical breath under resting conditions, about 500 ml of air are inspired and the same quantity is expired, so during quiet breathing the lung volume varies between 2,200 ml at the end of expiration to 2,700 ml at the end of inspiration (● Figure 13-19a). During maximal expiration, lung volume can decrease to 1,200 ml in males (1,000 ml in females), but the lungs can never be completely deflated, because the small airways collapse during forced expirations at low lung volumes, blocking further outflow (see ● Figure 13-15b, p. 473).

An important outcome of not being able to empty the lungs completely is that even during maximal expiratory efforts, gas exchange can still continue between blood flowing through the lungs and the remaining alveolar air. As a result, the gas content of the blood leaving the lungs for delivery to the tissues normally remains remarkably constant throughout the respiratory cycle. By contrast, if the lungs were to completely fill and empty with each breath, wide fluctuations would occur in O_2 uptake and CO_2 removal by the blood. Another advantage of the lungs not completely emptying with each breath is a reduction in

● **FIGURE 13-19**

Variations in lung volume

(a) Normal range and extremes of lung volume in a healthy young adult male. (b) Normal spirogram of a healthy young adult male. (The residual volume cannot be measured with a spirometer but must be determined by another means.)

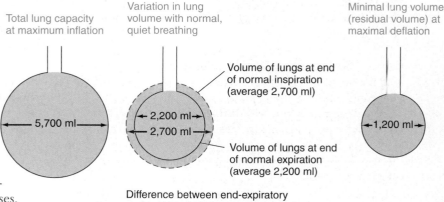

Total lung capacity at maximum inflation — 5,700 ml

Variation in lung volume with normal, quiet breathing — 2,200 ml / 2,700 ml

Volume of lungs at end of normal inspiration (average 2,700 ml)

Volume of lungs at end of normal expiration (average 2,200 ml)

Difference between end-expiratory and end-inspiratory volume equals tidal volume (average 500 ml)

Minimal lung volume (residual volume) at maximal deflation — 1,200 ml

(a)

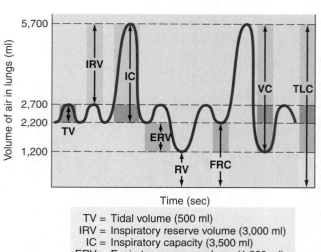

TV = Tidal volume (500 ml)
IRV = Inspiratory reserve volume (3,000 ml)
IC = Inspiratory capacity (3,500 ml)
ERV = Expiratory reserve volume (1,000 ml)
RV = Residual volume (1,200 ml)
FRC = Functional residual capacity (2,200 ml)
VC = Vital capacity (4,500 ml)
TLC = Total lung capacity (5,700 ml)

(b)

Values are average for a healthy young adult male; values for females are somewhat lower.

the work of breathing. Recall that it takes less effort to inflate a partially inflated alveolus than a totally collapsed one.

The changes in lung volume that occur with different respiratory efforts can be measured using a spirometer. Let's see what this device is and the various lung volumes and capacities it measures.

Lung volumes and capacities

Basically, a **spirometer** consists of an air-filled drum floating in a water-filled chamber. As the person breathes air in and out of the drum through a tube connecting the mouth to the air chamber, the drum rises and falls in the water chamber (● Figure 13-20). This rise and fall can be recorded as a **spirogram,** which is calibrated to volume changes. The pen records inspiration as an upward deflection and expiration as a downward deflection.

● Figure 13-19b is a hypothetical example of a spirogram in a healthy young adult male. Generally, the values are lower for females. The following lung volumes and lung capacities (a lung capacity is the sum of two or more lung volumes) can be determined:

- **Tidal volume (TV).** The volume of air entering or leaving the lungs during a single breath. Average value under resting conditions = 500 ml.
- **Inspiratory reserve volume (IRV).** The extra volume of air that can be maximally inspired over and above the typical resting tidal volume. The IRV is accomplished by maximal contraction of the diaphragm, external intercostal muscles, and accessory inspiratory muscles. Average value = 3,000 ml.
- **Inspiratory capacity (IC).** The maximum volume of air that can be inspired at the end of a normal quiet expiration (IC = IRV + TV). Average value = 3,500 ml.
- **Expiratory reserve volume (ERV).** The extra volume of air that can be actively expired by maximal contraction of the expiratory muscles beyond that normally passively expired at the end of a typical resting tidal volume. Average value = 1,000 ml.
- **Residual volume (RV).** The minimum volume of air remaining in the lungs even after a maximal expiration. Average value = 1,200 ml. The residual volume cannot be measured directly with a spirometer, because this volume of air does not move in and out of the lungs. It can be determined indirectly, however, through gas dilution techniques involving inspiration of a known quantity of a harmless tracer gas such as helium.
- **Functional residual capacity (FRC).** The volume of air in the lungs at the end of a normal passive expiration (FRC = ERV + RV). Average value = 2,200 ml.
- **Vital capacity (VC).** The maximum volume of air that can be moved out during a single breath following a maximal inspiration. The subject first inspires maximally, then expires maximally (VC = IRV + TV + ERV). The VC represents the maximum volume change possible within the lungs (● Fig-

---

● FIGURE 13-20

A spirometer

A spirometer is a device that measures the volume of air breathed in and out; it consists of an air-filled drum floating in a water-filled chamber. As a person breathes air in and out of the drum through a connecting tube, the resultant rise and fall of the drum are recorded as a spirogram, which is calibrated to the magnitude of the volume change.

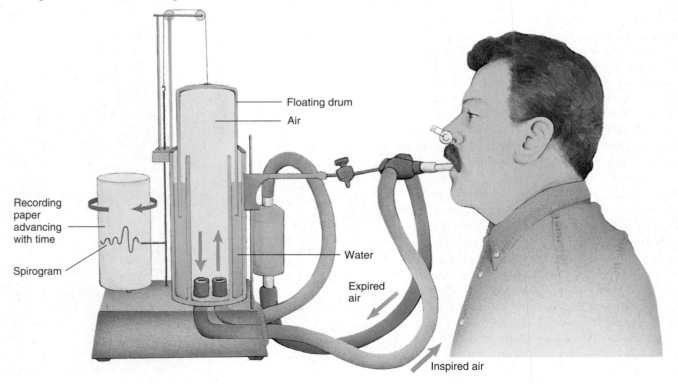

ure 13-21). It is rarely used, because the maximal muscle contractions involved become exhausting, but it is useful in ascertaining the functional capacity of the lungs. Average value = 4,500 ml.

• **Total lung capacity (TLC).** The maximum volume of air that the lungs can hold (TLC = VC + RV). Average value = 5,700 ml.

• **Forced expiratory volume in one second (FEV₁).** The volume of air that can be expired during the first second of expiration in a VC determination. Usually, FEV_1 is about 80% of VC; that is, normally 80% of the air that can be forcibly expired from maximally inflated lungs can be expired within 1 second. This measurement gives an indication of the maximal airflow rate that is possible from the lungs.

Respiratory dysfunction

Measurement of the lungs' various volumes and capacities is of more than pure academic interest, because such determinations are useful to the diagnostician in various respiratory disease states. Two general categories of respiratory dysfunction yield abnormal results during spirometry—*obstructive lung disease* and *restrictive lung disease* (● Figure 13-22). However, these are not the only categories of respiratory dysfunction, nor is spirometry the only pulmonary function test. Other conditions affecting respiratory function include (1) diseases im-

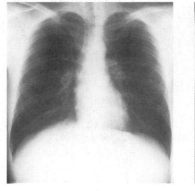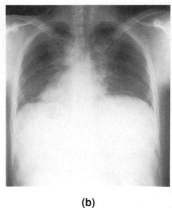

(a)   (b)

● **FIGURE 13-21**

X rays of lungs showing maximum volume change
(a) Maximum volume of the lungs at maximum inspiration. (b) Minimum volume of the lungs at maximum expiration. The difference between these two volumes is the vital capacity, which is the maximum volume of air that can be moved out during a single breath following a maximum inspiration.

pairing diffusion of O_2 and CO_2 across the pulmonary membranes; (2) reduced ventilation because of mechanical failure, as with neuromuscular disorders affecting the respiratory mus-

● **FIGURE 13-22**

Abnormal spirograms associated with obstructive and restrictive lung diseases
(a) Spirogram in obstructive lung disease. Because a patient with obstructive lung disease experiences more difficulty emptying the lungs than filling them, the total lung capacity (TLC) is essentially normal, but the functional residual capacity (FRC) and the residual volume (RV) are elevated as a result of the additional air trapped in the lungs following expiration. Because the RV is increased, the vital capacity (VC) is reduced. With more air remaining in the lungs, less of the TLC is available to be used in exchanging air with the atmosphere. Another common finding is a markedly reduced forced expiratory volume in one second (FEV_1) because the airflow rate is reduced by the airway obstruction. Even though both the VC and the FEV_1 are reduced, the FEV_1 is reduced more markedly than is the VC. As a result, the $FEV_1/VC\%$ is much lower than the normal 80%; that is, much less than 80% of the reduced VC can be blown out during the first second. (b) Spirogram in restrictive lung disease. In this disease the lungs are less compliant than normal. Total lung capacity, inspiratory capacity, and VC are reduced, because the lungs cannot be expanded as normal. The percentage of the VC that can be exhaled within 1 second is the normal 80% or an even higher percentage, because air can flow freely in the airways. Therefore, the $FEV_1/VC\%$ is particularly useful in distinguishing between obstructive and restrictive lung disease. Also, in contrast to obstructive lung disease, the RV is usually normal in restrictive lung disease.

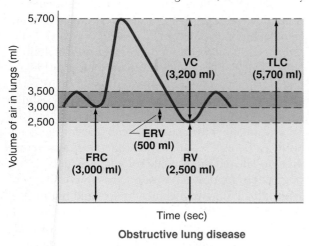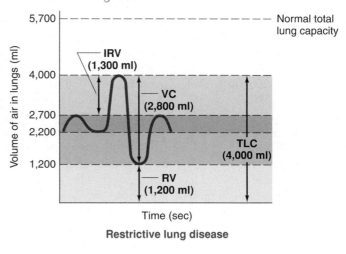

Obstructive lung disease

(a)

Restrictive lung disease

(b)

cles, or because of depression of the respiratory control center by alcohol, drugs, or other chemicals; (3) failure of adequate pulmonary blood flow; or (4) ventilation/perfusion abnormalities involving a poor matching of air and blood so that efficient gas exchange cannot occur. Some lung diseases are actually a complex mixture of different types of functional disturbances. To determine what abnormalities are present, the diagnostician relies on a variety of respiratory function tests in addition to spirometry, including x-ray examination, blood gas determinations, and tests to measure the diffusion capacity of the alveolar capillary membrane.

▌Alveolar ventilation is less than pulmonary ventilation because of the presence of dead space.

Various changes in lung volume represent only one factor in the determining **pulmonary,** or **minute, ventilation,** which is the volume of air breathed in and out in one minute. The other important factor is **respiratory rate,** which averages 12 breaths per minute.

$$\text{Pulmonary ventilation} = \text{tidal volume} \times \text{respiratory rate}$$
$$\text{(ml/min)} \qquad\qquad \text{(ml/breath)} \qquad\qquad \text{(breaths/min)}$$

At an average tidal volume of 500 ml/breath and a respiratory rate of 12 breaths/minute, pulmonary ventilation is 6,000 ml, or 6 liters, of air breathed in and out in one minute under resting conditions. For a brief period of time, a healthy young adult male can voluntarily increase his total pulmonary ventilation 25-fold, to 150 liters/min. To increase pulmonary ventilation, both tidal volume and respiratory rate increase, but depth of breathing increases more than frequency of breathing. It is usually more advantageous to have a greater increase in tidal volume than in respiratory rate, because of anatomic dead space.

Anatomic dead space

Not all the inspired air gets down to the site of gas exchange in the alveoli. Part remains in the conducting airways, where it is not available for gas exchange. The volume of the conducting passages in an adult averages about 150 ml. This volume is considered **anatomic dead space,** because air within these conducting airways is useless for exchange. Anatomic dead space greatly affects efficiency of pulmonary ventilation. In effect, even though 500 ml of air are moved in and out with each breath, only 350 ml are actually exchanged between the atmosphere and the alveoli, because of the 150 ml occupying the anatomic dead space.

Looking at ● Figure 13-23, note that at the end of inspiration, the respiratory airways are filled with 150 ml of fresh atmospheric air from the inspiration. During the subsequent expiration, 500 ml of air are expired to the atmosphere. The first 150 ml expired are the fresh air that was retained in the airways and never used. The remaining 350 ml expired are "old" alveolar air that has participated in gas exchange with the blood. During the same expiration, 500 ml of gas also leave the alveoli. The first 350 ml are expired to the atmosphere; the other 150 ml of old alveolar air never reach the outside but remain in the conducting airways.

On the next inspiration, 500 ml of gas enter the alveoli. The first 150 ml to enter the alveoli are the old alveolar air that remained in the dead space during the preceding expiration. The other 350 ml entering the alveoli are fresh air inspired from the atmosphere. Simultaneously, 500 ml of air enter from the atmosphere. The first 350 ml of atmospheric air reach the alveoli; the other 150 ml remain in the conducting airways to be expired without benefit of being exchanged with the blood, as the cycle repeats itself.

Alveolar ventilation

Because the amount of atmospheric air that reaches the alveoli and is actually available for exchange with the blood is more important than the total amount breathed in and out, **alveolar ventilation**—the volume of air exchanged between the atmosphere and the alveoli per minute—is more important than pulmonary ventilation. In determining alveolar ventilation, the amount of wasted air moved in and out through the anatomic dead space must be taken into account, as follows:

$$\text{Alveolar ventilation} = (\text{tidal volume} - \text{dead space volume}) \times \text{respiratory rate}$$

With average resting values,

$$\text{Alveolar ventilation} = (500 \text{ ml/breath} - 150 \text{ ml dead space volume}) \times 12 \text{ breaths/min}$$

$$= 4{,}200 \text{ ml/min}$$

Thus with quiet breathing, alveolar ventilation is 4,200 ml/min, whereas pulmonary ventilation is 6,000 ml/min.

Effect of breathing patterns on alveolar ventilation

To understand how important dead space volume is in determining the magnitude of alveolar ventilation, examine the effect of various breathing patterns on alveolar ventilation, as shown in ▲ Table 13-4. If a person deliberately breathes deeply (for example, a tidal volume of 1,200 ml) and slowly (for example, a respiratory rate of 5 breaths/min), pulmonary ventilation is 6,000 ml/min, the same as during quiet breathing at rest, but alveolar ventilation increases to 5,250 ml/min compared to the resting rate of 4,200 ml/min. In contrast, if a person deliberately breathes shallowly (for example, a tidal volume of 150 ml) and rapidly (a frequency of 40 breaths/min), pulmonary ventilation would still be 6,000 ml/min; however, alveolar ventilation would be 0 ml/min. In effect, the person would only be drawing air in and out of the anatomic dead space without any atmospheric air being exchanged with the alveoli, where it could be useful. The individual could voluntarily maintain such a breathing pattern for only a few minutes before losing consciousness, at which time normal breathing would resume.

The value of reflexly bringing about a larger increase in depth of breathing than in rate of breathing when pulmonary ventilation increases during exercise, should now be apparent.

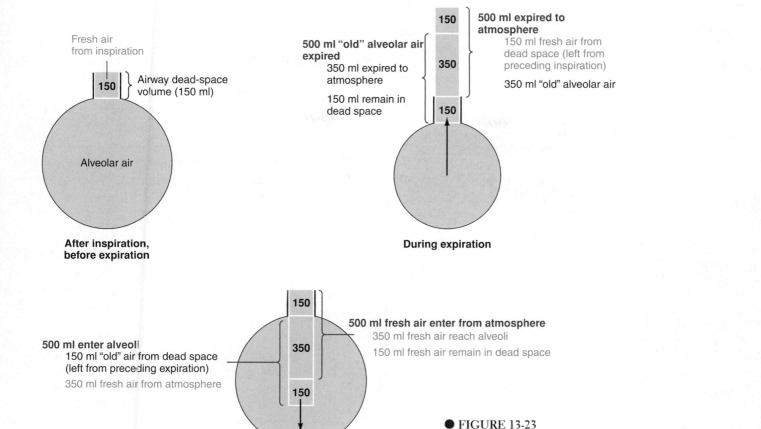

Fresh air from inspiration

150 Airway dead-space volume (150 ml)

Alveolar air

After inspiration, before expiration

500 ml expired to atmosphere

150 ml fresh air from dead space (left from preceding inspiration)

350 ml "old" alveolar air

150

500 ml "old" alveolar air expired
350 ml expired to atmosphere
150 ml remain in dead space

350

150

During expiration

500 ml enter alveoli
150 ml "old" air from dead space (left from preceding expiration)
350 ml fresh air from atmosphere

150

350

150

500 ml fresh air enter from atmosphere
350 ml fresh air reach alveoli
150 ml fresh air remain in dead space

During inspiration

☐ "Old" alveolar air that has exchanged O_2 and CO_2 with the blood

☐ Fresh atmospheric air that has not exchanged O_2 and CO_2 with the blood

The numbers in the figure represent ml of air.

● **FIGURE 13-23**

Effect of dead space volume on exchange of tidal volume between the atmosphere and the alveoli
Even though 500 ml of air move in and out between the atmosphere and the respiratory system and 500 ml move in and out of the alveoli with each breath, only 350 ml are actually exchanged between the atmosphere and the alveoli because of the anatomic dead space (the volume of air in the respiratory airways).

It is the most efficient means of elevating alveolar ventilation. When tidal volume is increased, the entire increase goes toward elevating alveolar ventilation, whereas an increase in respiratory rate does not go entirely toward increasing alveolar ventilation. When respiratory rate is increased, the frequency with which air is wasted in the dead space also increases, because a portion of *each* breath must move in and out of the dead space. As needs vary, ventilation is normally adjusted to a tidal volume and respiratory rate that meet those needs most efficiently in terms of energy cost.

▲ TABLE 13-4

Effect of Different Breathing Patterns on Alveolar Ventilation

Breathing Pattern	Tidal Volume (ml/breath)	Respiratory Rate (breaths/min)	Dead Space Volume (ml)	Pulmonary Ventilation (ml/min)*	Alveolar Ventilation (ml/min)**
Quiet breathing at rest	500	12	150	6,000	4,200
Deep, slow breathing	1,200	5	150	6,000	5,250
Shallow, rapid breathing	150	40	150	6,000	0

*Equals tidal volume × respiratory rate.
**Equals (tidal volume − dead space volume) × respiratory rate.

Alveolar dead space

We have assumed that all the atmospheric air entering the alveoli participates in exchanges of O_2 and CO_2 with pulmonary blood. However, the match between air and blood is not always perfect, because not all alveoli are equally ventilated with air and perfused with blood. Any ventilated alveoli that do not participate in gas exchange with blood because they are inadequately perfused are considered **alveolar dead space.** In normal people, alveolar dead space is quite small and of little importance, but it can increase to even lethal levels in several types of pulmonary disease.

Next you will learn why alveolar dead space is minimal in healthy individuals.

Local controls act on the smooth muscle of the airways and arterioles to match airflow to blood flow.

When discussing the role of airway resistance in determining airflow rate in and out of the lungs, we were referring to the overall resistance of all the airways collectively. However, the resistance of individual airways supplying specific alveoli can be adjusted independently in response to changes in the airway's local environment. This situation is comparable to the control of systemic arterioles. Recall that overall systemic arteriolar resistance (that is, total peripheral resistance) is an important determinant of the blood pressure gradient that drives blood flow throughout the systemic circulatory system (see p. 357). Yet the caliber of individual arterioles supplying various tissues can be adjusted locally to match the tissues' differing metabolic needs (see p. 354).

Effect of CO_2 on bronchiolar smooth muscle

Like arteriolar smooth muscle, bronchiolar smooth muscle is sensitive to local changes within its immediate environment, particularly to local CO_2 levels. If an alveolus is receiving too little airflow (ventilation) in comparison to its blood flow (perfusion), CO_2 levels will increase in the alveolus and surrounding tissue as the blood drops off more CO_2 than is exhaled into the atmosphere. This local increase in CO_2 acts directly on the bronchiolar smooth muscle involved to induce the airway supplying the underaerated alveolus to relax. The resultant decrease in airway resistance leads to an increased airflow (for the same ΔP) to the involved alveolus, so its airflow now matches its blood supply (● Figure 13-24). The converse is also true. A localized decrease in CO_2 associated with an alveolus that is receiving too much air for its blood supply, directly increases contractile activity of the airway smooth muscle involved, constricting the airway supplying this overaerated alveolus. The result is reduced airflow to the overaerated alveolus.

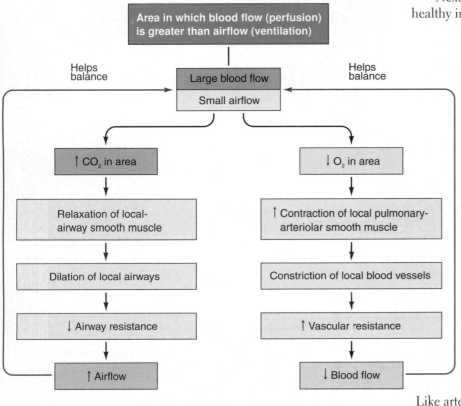

● FIGURE 13-24

Local controls to match airflow and blood flow to an area of the lung

Effect of O_2 on pulmonary arteriolar smooth muscle

Simultaneously, a similar locally induced effect on pulmonary vascular smooth muscle also takes place, to maximally match blood flow to airflow. Just as in the systemic circulation, distribution of the cardiac output to different alveolar capillary networks can be controlled by adjusting the resistance to blood flow through specific pulmonary arterioles. If blood flow is greater than airflow to a given alveolus, the O_2 level in the alveolus and surrounding tissues falls below normal as the overabundance of blood extracts more O_2 than usual from the alveolus. The local decrease in O_2 concentration causes vasoconstriction of the pulmonary arteriole supplying this particular capillary bed, thus reducing blood flow to match the smaller airflow. Conversely, an increase in alveolar O_2 concentration caused by a mismatched large airflow and small blood flow brings about pulmonary vasodilation, which increases blood flow to match the larger airflow.

This local effect of O_2 on pulmonary arteriolar smooth muscle is, appropriately, just the opposite of its effect on systemic arteriolar smooth muscle (▲ Table 13-5). In the systemic circulation, a decrease in O_2 in a tissue causes localized vasodilation to increase blood flow to the deprived area, and vice versa, which is important in matching blood supply to local metabolic needs.

The two mechanisms for matching airflow and blood flow function concurrently, so normally very little air or blood is wasted in the lung. Because of gravitational effects, some regional differences in ventilation and perfusion exist from the top to the bottom of the lung. Nevertheless, airflow and blood flow at a particular alveolar interface are usually matched as much as possible by these local controls to accomplish efficient exchange of O_2 and CO_2.

We have now completed our discussion of respiratory mechanics—all the factors involved in ventilation. We are now going to examine gas exchange between the alveolar air and blood and then between the blood and systemic tissues.

GAS EXCHANGE

The ultimate purpose of breathing is to provide a continual supply of fresh O_2 for pickup by the blood and to constantly remove CO_2 unloaded from the blood. The blood acts as a transport system for O_2 and CO_2 between the lungs and tissues, with the tissue cells extracting O_2 from the blood and eliminating CO_2 into it.

■ Gases move down partial pressure gradients.

Gas exchange at both the pulmonary capillary and the tissue capillary levels involves simple passive diffusion of O_2 and CO_2 down *partial pressure gradients*. No active transport mechanisms exist for these gases. Let us see what partial pressure gradients are and how they are established.

Partial pressures

Atmospheric air is a mixture of gases; typical dry air contains about 79% nitrogen (N_2) and 21% O_2, with almost negligible percentages of CO_2, H_2O vapor, other gases, and pollutants. Altogether, these gases exert a total atmospheric pressure of 760 mm Hg at sea level. This total pressure is equal to the sum of the pressures that each gas in the mixture partially contributes. The pressure exerted by a particular gas is directly proportional to the percentage of that gas in the total air mixture. Every gas molecule, no matter what its size, exerts the same amount of pressure; for example, a N_2 molecule exerts the same pressure as an O_2 molecule. Because 79% of the air consists of N_2 molecules, 79% of the 760 mm Hg atmospheric pressure, or 600 mm Hg, is exerted by the N_2 molecules. Similarly, because O_2 represents 21% of the atmosphere, 21% of the 760 mm Hg atmospheric pressure, or 160 mm Hg, is exerted by O_2 (● Figure 13-25). The individual pressure exerted independently by a particular gas within a mixture of gases is known as its **partial pressure**, designated by P_{gas}. Thus the par-

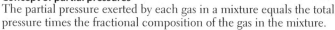

● FIGURE 13-25

Concept of partial pressures
The partial pressure exerted by each gas in a mixture equals the total pressure times the fractional composition of the gas in the mixture.

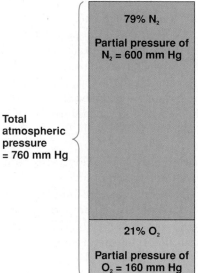

▲ TABLE 13-5

Effects of Local Changes in O_2 on the Pulmonary and Systemic Arterioles

	Effect of a Local Change in O_2	
Vessels	Decreased O_2	Increased O_2
Pulmonary Arterioles	Vasoconstriction	Vasodilation
Systemic Arterioles	Vasodilation	Vasoconstriction

tial pressure of O_2 in atmospheric air, P_{O_2}, is normally 160 mm Hg. The atmospheric partial pressure of CO_2, P_{CO_2}, is negligible at 0.03 mm Hg.

Gases dissolved in a liquid such as blood or another body fluid also exert a partial pressure. The greater the partial pressure of a gas in a liquid, the more of that gas dissolved.

Partial pressure gradients

A difference in partial pressure between capillary blood and surrounding structures is known as a **partial pressure gradient**. Partial pressure gradients exist between the alveolar air and pulmonary capillary blood. Similarly, partial pressure gradients exist between systemic capillary blood and surrounding tissues. A gas always diffuses down its partial pressure gradient from the area of higher partial pressure to the area of lower partial pressure, similar to diffusion down a concentration gradient.

▌ Oxygen enters and CO_2 leaves the blood in the lungs passively down partial pressure gradients.

We are first going to consider the magnitude of alveolar P_{O_2} and P_{CO_2} and then look at the partial pressure gradients that move these two gases between the alveoli and incoming pulmonary capillary blood.

Alveolar P_{O_2} and P_{CO_2}

Alveolar air is not of the same composition as inspired atmospheric air, for two reasons. First, as soon as atmospheric air enters the respiratory passages, exposure to the moist airways saturates it with H_2O. Like any other gas, water vapor exerts a partial pressure. At body temperature, the partial pressure of H_2O vapor is 47 mm Hg. Humidification of inspired air in effect "dilutes" the partial pressure of the inspired gases by 47 mm Hg, because the sum of the partial pressures must total the atmospheric pressure of 760 mm Hg. In moist air, $P_{H_2O} = 47$ mm Hg, $P_{N_2} = 563$ mm Hg, and $P_{O_2} = 150$ mm Hg.

Second, alveolar P_{O_2} is also lower than atmospheric P_{O_2} because fresh inspired air is mixed with the large volume of old air that remained in the lungs and dead space at the end of the preceding expiration (the functional residual capacity). Only about one-seventh of the total alveolar air is replaced by fresh atmospheric air with each normal breath. Thus at the end of inspiration, less than 15% of the air in the alveoli is fresh air. As a result of humidification and the small turnover of alveolar air, the average alveolar P_{O_2} is 100 mm Hg, compared to the atmospheric P_{O_2} of 160 mm Hg.

It is logical to think that alveolar P_{O_2} would increase during inspiration with the arrival of fresh air and would decrease during expiration. Only small fluctuations of a few mm Hg occur, however, for two reasons. First, only a small proportion of the total alveolar air is exchanged with each breath. The relatively small volume of high-P_{O_2} air that is inspired is quickly mixed with the much larger volume of retained alveolar air, which has a lower P_{O_2}. Thus the O_2 in the inspired air can only slightly

elevate the total alveolar P_{O_2}. Even this potentially small elevation of P_{O_2} is diminished for another reason. Oxygen is continually moving by passive diffusion down its partial pressure gradient from the alveoli into the blood. The O_2 arriving in the alveoli in the newly inspired air simply replaces the O_2 diffusing out of the alveoli into the pulmonary capillaries. Therefore, the alveolar P_{O_2} remains relatively constant at about 100 mm Hg throughout the respiratory cycle. Because the pulmonary blood P_{O_2} equilibrates with the alveolar P_{O_2}, the P_{O_2} of the blood leaving the lungs likewise remains fairly constant at this same value. Accordingly, the amount of O_2 in the blood available to the tissues varies only slightly during the respiratory cycle.

A similar situation exists in reverse for CO_2. Carbon dioxide, which is continually produced by the body tissues as a metabolic waste product, is constantly added to the blood at the level of the systemic capillaries. In the pulmonary capillaries, CO_2 diffuses down its partial pressure gradient from the blood into the alveoli and is subsequently removed from the body during expiration. As with O_2, alveolar P_{CO_2} remains fairly constant throughout the respiratory cycle but at a lower value of 40 mm Hg.

P_{O_2} and P_{CO_2} gradients across the pulmonary capillaries

As the blood passes through the lungs, it picks up O_2 and gives up CO_2 simply by diffusion down partial pressure gradients that exist between the blood and the alveoli. Ventilation constantly replenishes alveolar O_2 and removes CO_2, thus maintaining the appropriate partial pressure gradients between the blood and alveoli. The blood entering the pulmonary capillaries is systemic venous blood pumped to the lungs through the pulmonary arteries. This blood, having just returned from the body tissues, is relatively low in O_2, with a P_{O_2} of 40 mm Hg, and is relatively high in CO_2, with a P_{CO_2} of 46 mm Hg. As this blood flows through the pulmonary capillaries, it is exposed to alveolar air (● Figure 13-26). Because the alveolar P_{O_2} at 100 mm Hg is higher than the P_{O_2} of 40 mm Hg in the blood entering the lungs, O_2 diffuses down its partial pressure gradient from the alveoli into the blood until no further gradient exists. As the blood leaves the pulmonary capillaries, it has a P_{O_2} equal to alveolar P_{O_2} at 100 mm Hg.

The partial pressure gradient for CO_2 is in the opposite direction. Blood entering the pulmonary capillaries has a P_{CO_2} of 46 mm Hg, whereas alveolar P_{CO_2} is only 40 mm Hg. Carbon dioxide diffuses from the blood into the alveoli until blood P_{CO_2} equilibrates with alveolar P_{CO_2}. Thus, the blood leaving the pulmonary capillaries has a P_{CO_2} of 40 mm Hg. After leaving the lungs, the blood, which now has a P_{O_2} of 100 mm Hg and a P_{CO_2} of 40 mm Hg, is returned to the heart to be subsequently pumped out to the body tissues as systemic arterial blood.

Note that blood returning to the lungs from the tissues still contains O_2 (P_{O_2} of systemic venous blood = 40 mm Hg) and that blood leaving the lungs still contains CO_2 (P_{CO_2} of systemic arterial blood = 40 mm Hg). The additional O_2 carried in the blood beyond that normally given up to the tissues represents an immediately available O_2 reserve that can be tapped by the tissue cells whenever their O_2 demands increase. The

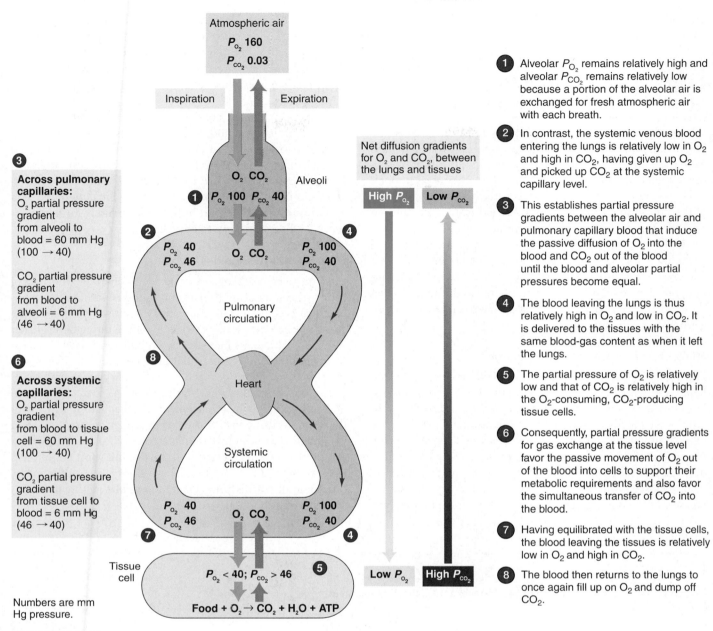

Atmospheric air

P_{O_2} 160

P_{CO_2} 0.03

Inspiration   Expiration

① Alveolar P_{O_2} remains relatively high and alveolar P_{CO_2} remains relatively low because a portion of the alveolar air is exchanged for fresh atmospheric air with each breath.

Net diffusion gradients for O_2 and CO_2, between the lungs and tissues

O_2 CO_2 Alveoli

① P_{O_2} 100 P_{CO_2} 40

② In contrast, the systemic venous blood entering the lungs is relatively low in O_2 and high in CO_2, having given up O_2 and picked up CO_2 at the systemic capillary level.

High P_{O_2} Low P_{CO_2}

③

Across pulmonary capillaries:
O_2 partial pressure gradient from alveoli to blood = 60 mm Hg (100 → 40)

CO_2 partial pressure gradient from blood to alveoli = 6 mm Hg (46 → 40)

② P_{O_2} 40 P_{CO_2} 46

O_2 CO_2

④ P_{O_2} 100 P_{CO_2} 40

Pulmonary circulation

③ This establishes partial pressure gradients between the alveolar air and pulmonary capillary blood that induce the passive diffusion of O_2 into the blood and CO_2 out of the blood until the blood and alveolar partial pressures become equal.

④ The blood leaving the lungs is thus relatively high in O_2 and low in CO_2. It is delivered to the tissues with the same blood-gas content as when it left the lungs.

⑧

Heart

⑥

Across systemic capillaries:
O_2 partial pressure gradient from blood to tissue cell = 60 mm Hg (100 → 40)

CO_2 partial pressure gradient from tissue cell to blood = 6 mm Hg (46 → 40)

Systemic circulation

⑤ The partial pressure of O_2 is relatively low and that of CO_2 is relatively high in the O_2-consuming, CO_2-producing tissue cells.

⑥ Consequently, partial pressure gradients for gas exchange at the tissue level favor the passive movement of O_2 out of the blood into cells to support their metabolic requirements and also favor the simultaneous transfer of CO_2 into the blood.

⑦ P_{O_2} 40 P_{CO_2} 46

O_2 CO_2

④ P_{O_2} 100 P_{CO_2} 40

⑦ Having equilibrated with the tissue cells, the blood leaving the tissues is relatively low in O_2 and high in CO_2.

Tissue cell

$P_{O_2} < 40; P_{CO_2} > 46$ **⑤**

Low P_{O_2} High P_{CO_2}

⑧ The blood then returns to the lungs to once again fill up on O_2 and dump off CO_2.

Numbers are mm Hg pressure.

Food + O_2 → CO_2 + H_2O + ATP

● FIGURE 13-26

Oxygen and CO_2 exchange across pulmonary and systemic capillaries caused by partial pressure gradients

CO_2 remaining in the blood even after passage through the lungs plays an important role in the acid–base balance of the body, because CO_2 generates carbonic acid. Furthermore, arterial P_{CO_2} is important in driving respiration. This mechanism will be described later.

The amount of O_2 picked up in the lungs matches the amount extracted and used by the tissues. When the tissues metabolize more actively (for example, during exercise), they extract more O_2 from the blood, reducing the systemic venous P_{O_2} even lower than 40 mm Hg—for example, to a P_{O_2} of 30 mm Hg. When this blood returns to the lungs, a larger-than-normal P_{O_2} gradient exists between the newly entering blood and alveolar air. The difference in P_{O_2} between the alveoli and blood is now 70 mm Hg (alveolar P_{O_2} of 100 mm Hg and blood P_{O_2}

of 30 mm Hg), compared to the normal P_{O_2} gradient of 60 mm Hg (alveolar P_{O_2} of 100 mm Hg and blood P_{O_2} of 40 mm Hg). Therefore, more O_2 diffuses from the alveoli into the blood down the larger partial pressure gradient before blood P_{O_2} equals alveolar P_{O_2}. This additional transfer of O_2 into the blood replaces the increased amount of O_2 consumed, so O_2 uptake matches O_2 use even when O_2 consumption increases. At the same time that more O_2 is diffusing from the alveoli into the blood because of the increased partial pressure gradient, ventilation is stimulated so that O_2 enters the alveoli more rapidly from the atmosphere to replace the O_2 diffusing into the blood.

Similarly, the amount of CO_2 given up to the alveoli from the blood matches the amount of CO_2 picked up at the tissues. Once again, an increased ventilation associated with increased activity ensures that increased CO_2 delivered to the alveoli is blown off to the atmosphere.

Factors other than the partial pressure gradient influence the rate of gas transfer.

We have been discussing diffusion of O_2 and CO_2 between the blood and the alveoli as if these gases' partial pressure gradients were the sole determinants of their rates of diffusion. According to Fick's law of diffusion, the diffusion rate of a gas through a sheet of tissue also depends on the surface area and thickness of the membrane through which the gas is diffusing and on the diffusion coefficient of the particular gas (▲ Table 13-6). Normally, changes in the rate of gas exchange are determined primarily by changes in partial pressure gradients between blood and alveoli, because these other factors are relatively constant under resting conditions. However, under circumstances when these other factors do change, these changes alter the rate of gas transfer in the lungs.

Effect of surface area on gas exchange

During exercise, the surface area available for exchange can be physiologically increased to enhance the rate of gas transfer. During resting conditions, some of the pulmonary capillaries are typically closed, because the normally low pressure of the pulmonary circulation is inadequate to keep all the capillaries open. During exercise, when the pulmonary blood pressure is raised by increased cardiac output, many of the previously closed pulmonary capillaries are forced open. This increases the surface area of blood available for exchange. Fur-

thermore, the alveolar membranes are stretched further than normal during exercise because of the larger tidal volumes (deeper breathing). Such stretching increases the alveolar surface area and decreases the thickness of the alveolar membrane. Collectively, these changes expedite gas exchange during exercise.

Several pathologic conditions can markedly reduce the pulmonary surface area and, in turn, decrease the rate of gas exchange. Most notably, in emphysema surface area is reduced because many alveolar walls are lost, resulting in larger but fewer chambers (● Figure 13-27). Loss of surface area for exchange is likewise associated with collapsed regions of the lung and also results when part of the lung tissue is surgically removed—for example, in treating lung cancer.

Effect of thickness on gas exchange

Inadequate gas exchange can also occur when the thickness of the barrier separating the air and blood is pathologically increased. As the thickness increases, the rate of gas transfer decreases, because a gas takes longer to diffuse through the greater thickness. Thickness increases in (1) *pulmonary edema*, an excess accumulation of interstitial fluid between the alveoli and pulmonary capillaries caused by pulmonary inflammation or left-sided congestive heart failure (see p. 332); (2) *pulmonary fibrosis* involving replacement of delicate lung tissue with thick fibrous tissue in response to certain chronic irritants; and (3) *pneumonia*, which is characterized by inflammatory fluid accumulation within or around the alveoli. Most commonly, pneumonia is due to bacterial or viral infection of the lungs, but it may also arise from accidental *aspiration* (breathing in) of food, vomitus, or chemical agents.

▲ TABLE 13-6
Factors That Influence the Rate of Gas Transfer across the Alveolar Membrane

Factor	Influence on the Rate of Gas Transfer Across the Alveolar Membrane	Comments
Partial Pressure Gradients of O_2 and CO_2	Rate of transfer ↑ as partial pressure gradient ↑	Major determinant of the rate of transfer
Surface Area of the Alveolar Membrane	Rate of transfer ↑ as surface area ↑	Surface area remains constant under resting conditions
		Surface area ↑ during exercise as more pulmonary capillaries open up when the cardiac output increases and the alveoli expand as breathing becomes deeper
		Surface area ↓ with pathologic conditions such as emphysema and lung collapse
Thickness of the Barrier Separating the Air and Blood Across the Alveolar Membrane	Rate of transfer ↓ as thickness ↑	Thickness normally remains constant
		Thickness ↑ with pathologic conditions such as pulmonary edema, pulmonary fibrosis, and pneumonia
Diffusion Coefficient (Solubility of the Gas in the Membrane)	Rate of transfer ↑ as diffusion coefficient ↑	Diffusion coefficient for CO_2 is 20 times that of O_2, offsetting the smaller partial pressure gradient for CO_2; therefore, approximately equal amounts of CO_2 and O_2 are transferred across the membrane

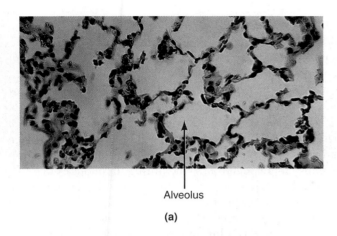

Alveolus

(a)

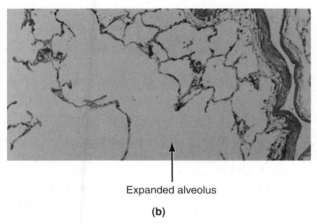

Expanded alveolus

(b)

● **FIGURE 13-27**

Comparison of normal and emphysematous lung tissue
(a) Photomicrograph of lung tissue from a normal individual. Each of the smallest clear spaces is an alveolar lumen. (b) Photomicrograph of lung tissue from a patient with emphysema. Note the loss of alveolar walls in the emphysematous lung tissue, resulting in larger but fewer alveolar chambers.

Effect of the diffusion coefficient on gas exchange

The rate of gas transfer is directly proportional to the diffusion coefficient (D), a constant value related to the solubility of a particular gas in the lung tissues and to its molecular weight ($D \propto \text{sol}\sqrt{\text{mw}}$). The diffusion coefficient for CO_2 is 20 times that of O_2 because CO_2 is much more soluble in body tissues than O_2 is. The rate of CO_2 diffusion across the respiratory membranes is therefore 20 times more rapid than that of O_2 for a given partial pressure gradient. This difference in diffusion coefficients is normally offset by the difference in partial pressure gradients that exist for O_2 and CO_2 across the alveolar capillary membrane. The CO_2 partial pressure gradient is 6 mm Hg (P_{CO_2} of 46 mm Hg in the blood; P_{CO_2} of 40 mm Hg in the alveoli), compared to the O_2 gradient of 60 mm Hg (P_{O_2} of 100 mm Hg in the alveoli; P_{O_2} of 40 mm Hg in the blood).

Normally, approximately equal amounts of O_2 and CO_2 are exchanged—a respiratory quotient's worth. Even though a given volume of blood spends three-fourths of a second passing through the pulmonary capillary bed, P_{O_2} and P_{CO_2} are usually both equilibrated with alveolar partial pressures by time the blood has traversed only one-third the length of the pulmonary capillaries. This means that the lung normally has enormous diffusion reserves, a fact that becomes extremely important during heavy exercise. The time the blood spends in transit in the pulmonary capillaries is decreased as pulmonary blood flow increases with the greater cardiac output that accompanies exercise. Even when less time is available for exchange, blood P_{O_2} and P_{CO_2} are normally able to equilibrate with alveolar levels because of the lungs' diffusion reserves.

In a diseased lung in which diffusion is impeded because the surface area is decreased or the blood–air barrier is thickened, O_2 transfer is usually more seriously impaired than CO_2 transfer, because of the larger CO_2 diffusion coefficient. By the time the blood reaches the end of the pulmonary capillary network, it is more likely to have equilibrated with alveolar P_{CO_2} than with alveolar P_{O_2}, because CO_2 can diffuse more rapidly through the respiratory barrier. In milder conditions, diffusion of both O_2 and CO_2 might remain adequate at rest, but during exercise, when pulmonary transit time is decreased, the blood gases, especially O_2, may not have completely equilibrated with the alveolar gases before the blood leaves the lungs.

▌ Gas exchange across the systemic capillaries also occurs down partial pressure gradients.

Just as they do at the pulmonary capillaries, O_2 and CO_2 move between the systemic capillary blood and the tissue cells by simple passive diffusion down partial pressure gradients. Refer again to ● Figure 13-26. The arterial blood that reaches the systemic capillaries is essentially the same blood that left the lungs by means of the pulmonary veins, because the only two places in the entire circulatory system at which gas exchange can take place are the pulmonary capillaries and the systemic capillaries. The arterial P_{O_2} is 100 mm Hg, and the arterial P_{CO_2} is 40 mm Hg, the same as alveolar P_{O_2} and P_{CO_2}.

P_{O_2} and P_{CO_2} gradients across the systemic capillaries

The cells constantly consume O_2 and produce CO_2 through oxidative metabolism. Cellular P_{O_2} averages about 40 mm Hg and P_{CO_2} about 46 mm Hg, although these values are highly variable, depending on the level of cellular metabolic activity. Oxygen moves by diffusion down its partial pressure gradient from the entering systemic capillary blood (P_{O_2} = 100 mm Hg) into the adjacent cells (P_{O_2} = 40 mm Hg) until equilibrium is reached. Therefore, the P_{O_2} of venous blood leaving the systemic capillaries is equal to the tissue P_{O_2} at an average of 40 mm Hg.

The reverse situation exists for CO_2. Carbon dioxide rapidly diffuses out of the cells (P_{CO_2} = 46 mm Hg) into the entering capillary blood (P_{CO_2} = 40 mm Hg) down the partial pressure gradient created by the ongoing production of CO_2. Transfer of CO_2 continues until blood P_{CO_2} equilibrates with tissue P_{CO_2}.[2] Accordingly, the blood leaving the systemic cap-

---

[2]Actually, the partial pressures of the systemic blood gases never completely equilibrate with tissue P_{O_2} and P_{CO_2}. Because the cells are constantly consuming O_2 and producing CO_2, the tissue P_{O_2} is always slightly less than the P_{O_2} of the blood leaving the systemic capillaries, and the tissue P_{CO_2} always slightly exceeds the systemic venous P_{CO_2}.

illaries has an average P_{CO_2} of 46 mm Hg. This systemic venous blood, which is relatively low in O_2 (P_{O_2} = 40 mm Hg) and relatively high in CO_2 (P_{CO_2} = 46 mm Hg), returns to the heart and is subsequently pumped to the lungs as the cycle repeats itself.

The more actively a tissue is metabolizing, the lower the cellular P_{O_2} falls and the higher the cellular P_{CO_2} rises. As a consequence of the larger blood-to-cell partial pressure gradients, more O_2 diffuses from the blood into the cells, and more CO_2 moves in the opposite direction before blood P_{O_2} and P_{CO_2} achieve equilibrium with the surrounding cells. Thus the amount of O_2 transferred to the cells and the amount of CO_2 carried away from the cells both depend on the rate of cellular metabolism.

Net diffusion of O_2 and CO_2 between the alveoli and tissues

Net diffusion of O_2 occurs first between the alveoli and the blood and then between the blood and the tissues because of the O_2 partial pressure gradients created by continuous use of O_2 in the cells and continuous replenishment of fresh alveolar O_2 provided by alveolar ventilation. Net diffusion of CO_2 occurs in the reverse direction, first between the tissues and the blood and then between the blood and the alveoli, because of the CO_2 partial pressure gradients created by continuous production of CO_2 in the cells and the continuous removal of alveolar CO_2 through the process of alveolar ventilation (● Figure 13-26).

Now let's see how O_2 and CO_2 are transported in the blood between the alveoli and the tissues.

GAS TRANSPORT

Oxygen picked up by the blood at the lungs must be transported to the tissues for cell use. Conversely, CO_2 produced at the cellular level must be transported to the lungs for elimination.

▌ Most O_2 in the blood is transported bound to hemoglobin.

Oxygen is present in the blood in two forms: physically dissolved and chemically bound to hemoglobin (▲ Table 13-7).

▲ TABLE 13-7
Methods of Gas Transport in the Blood

Gas	Method of Transport in Blood	Percentage Carried in This Form
O_2	Physically dissolved	1.5
	Bound to hemoglobin	98.5
CO_2	Physically dissolved	10
	Bound to hemoglobin	30
	As bicarbonate (HCO_3^-)	60

Physically dissolved O_2

Very little O_2 physically dissolves in plasma water, because O_2 is poorly soluble in body fluids. The amount dissolved is directly proportional to the P_{O_2} of the blood; the higher the P_{O_2}, the more O_2 dissolved. At a normal arterial P_{O_2} of 100 mm Hg, only 3 ml of O_2 can dissolve in 1 liter of blood. Thus only 15 ml of O_2/min can dissolve in the normal pulmonary blood flow of 5 liters/min (the resting cardiac output). Even under resting conditions, the cells consume 250 ml of O_2/min, and consumption may increase up to 25-fold during strenuous exercise. To deliver the O_2 required by the tissues even at rest, the cardiac output would have to be 83.3 liters/min if O_2 could only be transported in dissolved form. Obviously, there must be an additional mechanism for transporting O_2 to the tissues. This mechanism is *hemoglobin (Hb)*. Only 1.5% of the O_2 in the blood is dissolved; the remaining 98.5% is transported in combination with Hb. *The O_2 bound to Hb does not contribute to the P_{O_2} of the blood*; thus blood P_{O_2} is not a measure of the total O_2 content of the blood but only of the dissolved portion of O_2.

Oxygen bound to hemoglobin

Hemoglobin, an iron-bearing protein molecule contained within the red blood cells, can form a loose, easily reversible combination with O_2 (see p. 393). When not combined with O_2, Hb is referred to as **reduced hemoglobin**, or **deoxyhemoglobin**; when combined with O_2, it is called **oxyhemoglobin** (**HbO_2**):

$$Hb + O_2 \leftrightharpoons HbO_2$$

reduced hemoglobin   oxyhemoglobin

We need to answer several questions about the role of Hb in O_2 transport. What determines whether O_2 and Hb are combined or dissociated (separated)? Why does Hb combine with O_2 in the lungs and release O_2 at the tissues? How can a variable amount of O_2 be released at the tissues, depending on the level of tissue activity? How can we talk about O_2 transfer between blood and surrounding tissues in terms of O_2 partial pressure gradients when 98.5% of the O_2 is bound to Hb and thus does not contribute to the P_{O_2} of the blood at all?

▌ The P_{O_2} is the primary factor determining the percent hemoglobin saturation.

Each of the four atoms of iron within the heme portions of a hemoglobin molecule can combine with an O_2 molecule, so each Hb molecule can carry up to four molecules of O_2. Hemoglobin is considered *fully saturated* when all the Hb present is carrying its maximum O_2 load. The **percent hemoglobin (% Hb) saturation**, a measure of the extent to which the Hb present is combined with O_2, can vary from 0% to 100%.

The most important factor determining the % Hb saturation is the P_{O_2} of the blood, which in turn is related to the concentration of O_2 physically dissolved in the blood. According to the **law of mass action**, if the concentration of one substance involved in a reversible reaction is increased, the reaction is driven toward the opposite side. Conversely, if the concentration of one substance is decreased, the reaction is driven toward that side. Applying this law to the reversible reaction involving Hb and O_2 ($Hb + O_2 \leftrightarrows HbO_2$), when the blood P_{O_2} increases, as in the pulmonary capillaries, the reaction is driven toward the right side of the equation, increasing formation of HbO_2 (increased % Hb saturation). When the blood P_{O_2} decreases, as in the systemic capillaries, the reaction is driven toward the left side of the equation and oxygen is released from Hb as HbO_2 dissociates (decreased % Hb saturation). Thus because of the difference in P_{O_2} at the lungs and other tissues, Hb automatically "loads up" on O_2 in the lungs, where ventilation is continually providing fresh supplies of O_2 and "unloads" it in the tissues, which are constantly using up O_2.

O_2–Hb dissociation curve

The relationship between blood P_{O_2} and % Hb saturation is not linear, however, a point that is very important physiologically. Doubling the partial pressure does not double the % Hb saturation. Rather, the relationship between these variables follows an S-shaped curve, the O_2–Hb dissociation (or saturation) curve (● Figure 13-28). At the upper end, between a blood P_{O_2} of 60 and 100 mm Hg, the curve flattens off, or plateaus. Within this pressure range, a rise in P_{O_2} produces

only a small increase in the extent to which Hb is bound with O_2. In the P_{O_2} range of 0 to 60 mm Hg, in contrast, a small change in P_{O_2} results in a large change in the extent to which Hb is combined with O_2, as shown by the steep lower part of the curve. Both the upper plateau and lower steep portion of the curve have physiological significance.

Significance of the plateau portion of the O_2–Hb curve

The plateau portion of the curve is in the blood P_{O_2} range that exists at the pulmonary capillaries where O_2 is being loaded onto Hb. The systemic arterial blood leaving the lungs, having equilibrated with alveolar P_{O_2}, normally has a P_{O_2} of 100 mm Hg. Looking at the O_2–Hb curve, note that at a blood P_{O_2} of 100 mm Hg, Hb is 97.5% saturated. Therefore, the Hb in the systemic arterial blood normally is almost fully saturated.

If the alveolar P_{O_2} and consequently the arterial P_{O_2} fall below normal, there is little reduction in the total amount of O_2 transported by the blood until the P_{O_2} falls below 60 mm Hg, because of the plateau region of the curve. If the arterial P_{O_2} falls 40%, from 100 to 60 mm Hg, the concentration of dissolved O_2 as reflected by the P_{O_2} is likewise reduced 40%. At a blood P_{O_2} of 60 mm Hg, however, the % Hb saturation is still remarkably high at 90%. Accordingly, the total O_2 content of the blood is only slightly decreased despite the 40% reduction in P_{O_2}, because Hb is still carrying an almost full load of O_2, and, as mentioned before, the vast majority of O_2 is transported by Hb rather than being dissolved. However, even if the blood P_{O_2} is greatly increased—say, to 600 mm Hg—by breathing pure O_2, very little additional O_2 is added to the blood. A small extra amount of O_2 dissolves, but the % Hb saturation can be maximally increased by only another 2.5%, to 100% saturation. Therefore, in the P_{O_2} range between 60 and 600 mm Hg or even higher, there is only a 10% difference in the amount of O_2 carried by Hb. Thus the plateau portion of the O_2–Hb curve provides a good margin of safety in O_2-carrying capacity of the blood.

Arterial P_{O_2} may be reduced by pulmonary diseases accompanied by inadequate ventilation or defective gas exchange or by circulatory disorders that result in inadequate blood flow to the lungs. It may also fall in healthy individuals under two circumstances: (1) at high altitudes, where the total atmospheric pressure and hence the P_{O_2} of the inspired air are reduced, or (2) in O_2-deprived environments at sea level, such as if someone were accidentally locked in a vault. Unless the arterial P_{O_2} becomes markedly reduced (falls below 60 mm Hg) in either pathologic conditions or abnormal circumstances, near-normal amounts of O_2 can still be carried to the tissues.

Significance of the steep portion of the O_2–Hb curve

The steep portion of the curve between 0 and 60 mm Hg is in the blood P_{O_2} range that exists at the systemic capillaries, where O_2 is unloaded from Hb. In the systemic capillaries, the blood equilibrates with the surrounding tissue cells at an average P_{O_2} of 40 mm Hg. Note on ● Figure 13-28 that at a P_{O_2} of 40 mm Hg, the % Hb saturation is 75%. The blood arrives in the tissue capillaries at a P_{O_2} of 100 mm Hg with 97.5% Hb saturation. Because Hb can only be 75% saturated at the P_{O_2} of 40 mm

● FIGURE 13-28

physioEdge Activity

Oxygen–hemoglobin (O_2–Hb) dissociation (saturation) curve
The percent hemoglobin saturation (the scale on the left side of the graph) depends on the P_{O_2} of the blood. The relationship between these two variables is depicted by an S-shaped curve with a plateau region between a blood P_{O_2} of 60 and 100 mm Hg and a steep portion between 0 and 60 mm Hg. Another way of expressing the effect of blood P_{O_2} on the amount of O_2 bound with hemoglobin is the volume percent of O_2 in the blood (ml of O_2 bound with hemoglobin in each 100 ml of blood). That relationship is represented by the scale on the right side of the graph.

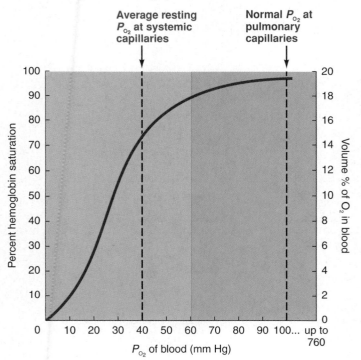

Hg in the systemic capillaries, nearly 25% of the HbO_2 must dissociate, yielding reduced Hb and O_2. This released O_2 is free to diffuse down its partial pressure gradient from the red blood cells through the plasma and interstitial fluid into the tissue cells.

The Hb in the venous blood returning to the lungs is still normally 75% saturated. If the tissue cells are metabolizing more actively, the P_{O_2} of the systemic capillary blood falls (for example, from 40 to 20 mm Hg) because the cells are consuming O_2 more rapidly. Note on the curve that this drop of 20 mm Hg in P_{O_2} decreases the % Hb saturation from 75% to 30%; that is, about 45% more of the total HbO_2 than normal gives up its O_2 for tissue use. The normal 60 mm Hg drop in P_{O_2} from 100 to 40 mm Hg in the systemic capillaries causes about 25% of the total HbO_2 to unload its O_2. In comparison, a further drop in P_{O_2} of only 20 mm Hg results in an additional 45% of the total HbO_2 unloading its O_2, because the O_2 partial pressures in this range are operating in the steep portion of the curve. In this range, only a small drop in systemic capillary P_{O_2} can automatically make large amounts of O_2 immediately available to meet the O_2 needs of more actively metabolizing tissues. As much as 85% of the Hb may give up its O_2 to actively metabolizing cells during strenuous exercise. In addition to this more thorough withdrawal of O_2 from the blood, even more O_2 is made available to actively metabolizing cells, such as exercising muscles, by circulatory and respiratory adjust-ments that increase the flow rate of oxygenated blood through the active tissues.

▮ Hemoglobin promotes the net transfer of O_2 at both the alveolar and tissue levels.

We still have not really clarified the role of Hb in gas exchange. Because blood P_{O_2} depends entirely on the concentration of *dissolved* O_2, we could ignore the O_2 bound to Hb in our earlier discussion of O_2 being driven from the alveoli to the blood by a P_{O_2} gradient. However, Hb does play a crucial role in permitting the transfer of large quantities of O_2 before blood P_{O_2} equilibrates with the surrounding tissues (● Figure 13-29).

Role of hemoglobin at the alveolar level

Hemoglobin acts as a "storage depot" for O_2, removing O_2 from solution as soon as it enters the blood from the alveoli. Because only dissolved O_2 contributes to P_{O_2}, the O_2 stored in Hb cannot contribute to blood P_{O_2}. When systemic venous blood enters the pulmonary capillaries, its P_{O_2} is considerably lower than the alveolar P_{O_2}, so O_2 immediately diffuses into the blood, raising blood P_{O_2}. As soon as the P_{O_2} of the blood increases, the percentage of Hb that can bind with O_2 likewise increases, as indicated by the O_2–Hb curve. Consequently, most of the O_2 that has diffused into the blood combines with

● **FIGURE 13-29**

Hemoglobin facilitating a large net transfer of O_2 by acting as a storage depot to keep P_{O_2} low
(a) In the hypothetical situation in which no hemoglobin is present in the blood, the alveolar P_{O_2} and the pulmonary capillary blood P_{O_2} are at equilibrium. (b) Hemoglobin has been added to the pulmonary capillary blood. As the Hb starts to bind with O_2, it removes O_2 from solution. Because only dissolved O_2 contributes to blood P_{O_2}, the blood P_{O_2} falls below that of the alveoli, even though the same number of O_2 molecules are present in the blood as in part (a). By "soaking up" some of the dissolved O_2, Hb favors the net diffusion of more O_2 down its partial pressure gradient from the alveoli to the blood. (c) Hemoglobin is fully saturated with O_2, and the alveolar and blood P_{O_2} are at equilibrium again. The blood P_{O_2} resulting from dissolved O_2 is equal to the alveolar P_{O_2}, despite the fact that the total O_2 content in the blood is much greater than in part (a) when blood P_{O_2} was equal to alveolar P_{O_2} in the absence of Hb.

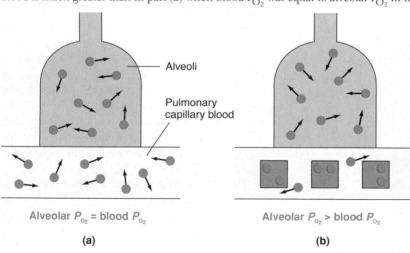
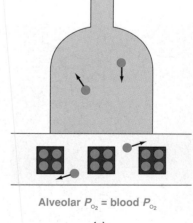

 = O_2 molecule    = Partially saturated hemoglobin molecule   = Fully saturated hemoglobin molecule

Hb and no longer contributes to blood P_{O_2}. As O_2 is removed from solution by combining with Hb, blood P_{O_2} falls to about the same level it was when the blood entered the lungs, although the total quantity of O_2 in the blood actually has increased. Because the blood P_{O_2} is once again considerably below alveolar P_{O_2}, more O_2 diffuses from the alveoli into the blood, only to be soaked up by Hb again.

Even though we have considered this process stepwise, for clarity, net diffusion of O_2 from alveoli to blood occurs continuously until Hb becomes as completely saturated with O_2 as it can be at that particular P_{O_2}. At a normal P_{O_2} of 100 mm Hg, Hb is 97.5% saturated. Thus by soaking up O_2, Hb keeps blood P_{O_2} low and prolongs the existence of a partial pressure gradient so that a large net transfer of O_2 into the blood can take place. Not until Hb can store no more O_2 (that is, Hb is maximally saturated for that P_{O_2}) does all the O_2 transferred into the blood remain dissolved and directly contribute to the P_{O_2}. Only now does the blood P_{O_2} rapidly equilibrate with the alveolar P_{O_2} and bring further O_2 transfer to a halt, but this point is not reached until Hb is already loaded to the maximum extent possible. Once the blood P_{O_2} equilibrates with the alveolar P_{O_2}, no further O_2 transfer can take place, no matter how little or how much total O_2 has already been transferred.

Role of hemoglobin at the tissue level

The reverse situation occurs at the tissue level. Because the P_{O_2} of blood entering the systemic capillaries is considerably higher than the P_{O_2} of the surrounding tissue, O_2 immediately diffuses from the blood into the tissues, lowering blood P_{O_2}. When blood P_{O_2} falls, Hb must unload some stored O_2, because the % Hb saturation is reduced. As the O_2 released from Hb dissolves in the blood, the blood P_{O_2} increases and once again exceeds the P_{O_2} of the surrounding tissues. This favors further movement of O_2 out of the blood, although the total quantity of O_2 in the blood has already fallen. Only when Hb can no longer release any more O_2 into solution (when Hb is unloaded to the greatest extent possible for the P_{O_2} existing at the systemic capillaries) can blood P_{O_2} fall as low as in surrounding tissue. At this time, further transfer of O_2 stops. Hemoglobin, because it stores a large quantity of O_2 that can be freed by a slight reduction in P_{O_2} at the systemic capillary level, permits the transfer of tremendously much more O_2 from the blood into the cells than would be possible in its absence.

Thus Hb plays an important role in the *total quantity* of O_2 that the blood can pick up in the lungs and drop off in the tissues. If Hb levels fall to one-half of normal, as in a severely anemic patient (see p. 396), the O_2-carrying capacity of the blood falls by 50% even though the arterial P_{O_2} is the normal 100 mm Hg with 97.5% Hb saturation. Only half as much Hb is available to be saturated, emphasizing once again how critical Hb is in determining how much O_2 can be picked up at the lungs and made available to tissues.

Factors at the tissue level promote the unloading of O_2 from hemoglobin.

Even though the primary factor determining the % Hb saturation is the P_{O_2} of the blood, other factors can affect the affinity, or bond strength, between Hb and O_2 and, accordingly, can shift the O_2–Hb curve (that is, change the % Hb saturation at a given P_{O_2}). These other factors are CO_2, acidity, temperature, and 2,3-bisphosphoglycerate, which we will examine separately. The O_2–Hb dissociation curve with which you are already familiar (● Figure 13-28) is a typical curve at normal arterial CO_2 and acidity levels, normal body temperature, and normal 2,3-bisphosphoglycerate concentration.

Effect of CO_2 on % Hb saturation

An increase in P_{CO_2} shifts the O_2–Hb curve to the right (● Figure 13-30). The % Hb saturation still depends on the P_{O_2}, but for any given P_{O_2} less O_2 and Hb can be combined. This effect is important, because the P_{CO_2} of the blood increases in the systemic capillaries as CO_2 diffuses down its gradient from the cells into the blood. The presence of this additional CO_2 in the blood in effect decreases the affinity of Hb for O_2, so Hb unloads even more O_2 at the tissue level than it would if the reduction in P_{O_2} in the systemic capillaries were the only factor affecting % Hb saturation.

Effect of acid on % Hb saturation

An increase in acidity also shifts the curve to the right. Because CO_2 generates carbonic acid (H_2CO_3), the blood becomes

● FIGURE 13-30

Effect of increased P_{CO_2}, H^+, temperature, and 2,3-bisphosphoglycerate on the O_2–Hb curve
Increased P_{CO_2}, acid, temperature, and 2,3-bisphosphoglycerate, as found at the tissue level, shift the O_2–Hb curve to the right. As a result, less O_2 and Hb can be combined at a given P_{O_2}, so that more O_2 is unloaded from Hb for use by the tissues.

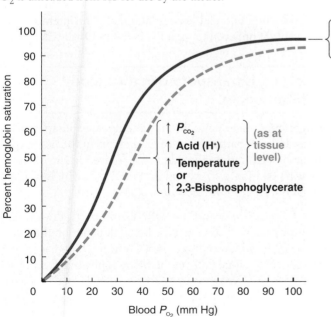

more acidic at the systemic capillary level as it picks up CO_2 from the tissues. The resulting reduction in Hb affinity for O_2 in the presence of increased acidity aids in releasing even more O_2 at the tissue level for a given P_{O_2}. In actively metabolizing cells, such as exercising muscles, not only is more carbonic acid–generating CO_2 produced, but lactic acid also may be produced if the cells resort to anaerobic metabolism (see p. 40 and p. 279). The resultant local elevation of acid in the working muscles facilitates further unloading of O_2 in the very tissues that need the most O_2.

Bohr effect

The influence of CO_2 and acid on the release of O_2 is known as the **Bohr effect.** Both CO_2 and the hydrogen ion (H^+) component of acids can combine reversibly with Hb at sites other than the O_2-binding sites. The result is a change in the molecular structure of Hb that reduces its affinity for O_2. (Note that the % Hb saturation refers only to the extent to which Hb is combined with O_2, not the extent to which it is bound with CO_2, H^+, or other molecules. Indeed, the % Hb saturation decreases when CO_2 and H^+ bind with Hb, because their presence on Hb facilitates increased release of O_2 from Hb.)

Effect of temperature on % Hb saturation

In a similar manner, a rise in temperature shifts the O_2–Hb curve to the right, resulting in more unloading of O_2 at a given P_{O_2}. An exercising muscle or other actively metabolizing cell produces heat. The resulting local rise in temperature enhances O_2 release from Hb for use by more active tissues.

Comparison of these factors at the tissue and pulmonary levels

As you just learned, increases in CO_2, acidity, and temperature at the tissue level, all of which are associated with increased cellular metabolism and increased O_2 consumption, enhance the effect of a drop in P_{O_2} in facilitating the release of O_2 from Hb. These effects are largely reversed at the pulmonary level, where the extra acid-forming CO_2 is blown off and the local environment is cooler. Appropriately, therefore, Hb has a higher affinity for O_2 in the pulmonary capillary environment, enhancing the effect of raised P_{O_2} in loading O_2 onto Hb.

Effect of 2,3-bisphosphoglycerate on % Hb saturation

The preceding changes take place in the *environment* of the red blood cells, but a factor *inside* the red blood cells can also affect the degree of O_2–Hb binding: **2,3-bisphosphoglycerate (BPG).** This erythrocyte constituent, which is produced during red blood cell metabolism, can bind reversibly with Hb and reduce its affinity for O_2, just as CO_2 and H^+ do. Thus an increased level of BPG, like the other factors, shifts the O_2–Hb curve to the right, enhancing O_2 unloading as the blood flows through the tissues.

BPG production by red blood cells gradually increases whenever Hb in the arterial blood is chronically undersaturated—that is, when arterial HbO_2 is below normal. This con-

dition may occur in people living at high altitudes or in those suffering from certain types of circulatory or respiratory diseases or anemia. By helping unload O_2 from Hb at the tissue level, increased BPG helps maintain O_2 availability for tissue use under circumstances associated with decreased arterial O_2 supply.

However, unlike the other factors—which normally are present only at the tissue level and thus shift the O_2–Hb curve to the right only in the systemic capillaries, where the shift is advantageous in unloading O_2—BPG is present in the red blood cells throughout the circulatory system and, accordingly, shifts the curve to the right to the same degree in both the tissues and the lungs. As a result, BPG decreases the ability to load O_2 at the pulmonary level, which is the negative side of increased BPG production.

▌ Hemoglobin has a much higher affinity for carbon monoxide than for O_2.

Carbon monoxide (CO) and O_2 compete for the same binding sites on Hb, but Hb's affinity for CO is 240 times that of its affinity for O_2. The combination of CO and Hb is known as **carboxyhemoglobin (HbCO).** Because Hb preferentially latches onto CO, even small amounts of CO can tie up a disproportionately large share of Hb, making Hb unavailable for O_2 transport. Even though the Hb concentration and P_{O_2} are normal, the O_2 content of the blood is seriously reduced.

Fortunately, CO is not a normal constituent of inspired air. It is a poisonous gas produced during the incomplete combustion (burning) of carbon products such as automobile gasoline, coal, wood, and tobacco. Carbon monoxide is especially dangerous because it is so insidious. If CO is being produced in a closed environment so that its concentration continues to increase (for example, in a parked car with the motor running and windows closed), it can reach lethal levels without the victim ever being aware of the danger. Because it is odorless, colorless, tasteless, and nonirritating, carbon monoxide is not detectable. Furthermore, for reasons described later, the victim has no sensation of breathlessness and makes no attempt to increase ventilation, even though the cells are O_2-starved.

▌ Most CO_2 is transported in the blood as bicarbonate.

When arterial blood flows through the tissue capillaries, CO_2 diffuses down its partial pressure gradient from the tissue cells into the blood. Carbon dioxide is transported in the blood in three ways (● Figure 13-31 and ▲ Table 13-7, which summarizes gas transport in the blood):

1. *Physically dissolved.* As with dissolved O_2, the amount of CO_2 physically dissolved in the blood depends on the P_{CO_2}. Because CO_2 is more soluble than O_2 in the blood, a greater proportion of the total CO_2 in the blood is physically dissolved compared to O_2. Even so, only 10% of the blood's total CO_2 content is carried this way at the normal systemic venous P_{CO_2} level.

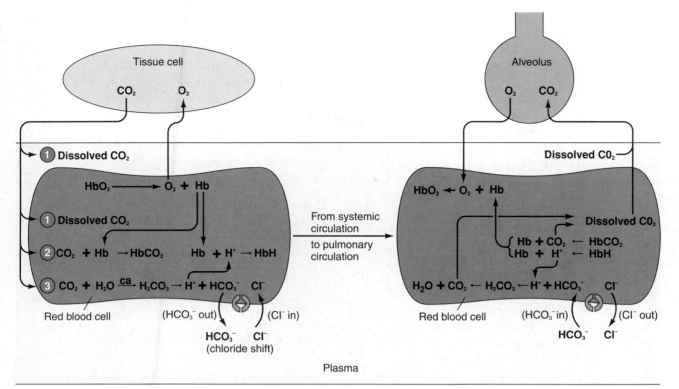

ca = Carbonic anhydrase

● **FIGURE 13-31**

Carbon dioxide transport in the blood

Carbon dioxide (CO_2) picked up at the tissue level is transported in the blood to the lungs in three ways: ① physically dissolved, ② bound to hemoglobin (Hb), and ③ as bicarbonate ion (HCO_3^-). Hemoglobin is present only in the red blood cells, as is carbonic anhydrase, the enzyme that catalyzes the production of HCO_3^-. The H^+ generated during the production of HCO_3^- also binds to Hb. Bicarbonate moves by facilitated diffusion down its concentration gradient out of the red blood cell into the plasma, and chloride (Cl^-) moves by means of the same passive carrier into the red blood cell down the electrical gradient created by the outward diffusion of HCO_3^-.

The reactions that occur at the tissue level are reversed at the pulmonary level, where CO_2 diffuses out of the blood to enter the alveoli.

2. *Bound to hemoglobin.* Another 30% of the CO_2 combines with Hb to form **carbamino hemoglobin ($HbCO_2$).** Carbon dioxide binds with the globin portion of Hb in contrast to O_2, which combines with the heme portions. Reduced Hb has a greater affinity for CO_2 than HbO_2 does. The unloading of O_2 from Hb in the tissue capillaries therefore facilitates the picking up of CO_2 by Hb.

3. *As bicarbonate.* By far the most important means of CO_2 transport is as **bicarbonate (HCO_3^-),** with 60% of the CO_2 being converted into HCO_3^- by the following chemical reaction, which takes place within the red blood cells:

$$CO_2 + H_2O \stackrel{\text{carbonic anhydrase}}{\rightleftharpoons} H_2CO_3 \rightleftharpoons H^+ + HCO_3^-$$

In the first step, CO_2 combines with H_2O to form **carbonic acid (H_2CO_3).** This reaction can occur very slowly in the plasma, but it proceeds swiftly within the red blood cells be-

cause of the presence of the erythrocyte enzyme **carbonic anhydrase,** which catalyzes (speeds up) the reaction. As is characteristic of acids, some of the carbonic acid molecules spontaneously dissociate into hydrogen ions (H^+) and bicarbonate ions (HCO_3^-). The one carbon and two oxygen atoms of the original CO_2 molecule are thus present in the blood as an integral part of HCO_3^-. This is beneficial because HCO_3^- is more soluble in the blood than CO_2 is.

Chloride shift

As this reaction proceeds, HCO_3^- and H^+ start to accumulate within the red blood cells in the systemic capillaries. The red cell membrane has a $HCO_3^- - Cl^-$ carrier that passively facilitates the diffusion of these ions in opposite directions across the membrane. The membrane is relatively impermeable to H^+. Consequently, HCO_3^-, but not H^+, diffuses down its concentration gradient out of the erythrocytes into the plasma. Because HCO_3^- is a negatively charged ion, the efflux of HCO_3^- unaccompanied by a comparable outward diffusion of positively charged ions creates an electrical gradient (see p. 75). Chloride ions (Cl^-), the dominant plasma anions, diffuse into the red blood cells down this electrical gradient to restore electric neutrality. This inward shift of Cl^- in exchange for the efflux of CO_2-generated HCO_3^- is known as the **chloride (Cl^-) shift**.

Haldane effect

After the dissociation of H_2CO_3, most of the accumulated H^+ within the erythrocytes becomes bound to Hb. As with CO_2, reduced Hb has a greater affinity for H^+ than HbO_2 does. Therefore, unloading O_2 facilitates Hb pickup of CO_2-generated H^+. Because only free, dissolved H^+ contributes to the acidity of a solution, the venous blood would be considerably more acidic than the arterial blood if Hb did not mop up most of the H^+ generated at the tissue level.

The fact that removing O_2 from Hb increases the ability of Hb to pick up CO_2 and CO_2-generated H^+ is known as the **Haldane effect**. The Haldane effect and Bohr effect work in synchrony to facilitate O_2 liberation and the uptake of CO_2 and CO_2-generated H^+ at the tissue level. Increased CO_2 and H^+ cause increased O_2 release from Hb by the Bohr effect; increased O_2 release from Hb in turn causes increased CO_2 and H^+ uptake by Hb through the Haldane effect. The entire process is very efficient. Reduced Hb must be carried back to the lungs to refill on O_2 anyway. Meanwhile, after O_2 is released, Hb picks up new passengers—CO_2 and H^+—that are going in the same direction to the lungs.

The reactions at the tissue level as CO_2 enters the blood from the tissues are reversed once the blood reaches the lungs and CO_2 leaves the blood to enter the alveoli (● Figure 13-31).

❚ Various respiratory states are characterized by abnormal blood gas levels.

Abnormalities in arterial P_{O_2}

The term **hypoxia** refers to insufficient O_2 at the cellular level. (▲ Table 13-8 is a glossary of terms used to describe various states associated with respiratory abnormalities.) There are four general categories of hypoxia:

1. *Hypoxic hypoxia* is characterized by a low arterial blood P_{O_2} accompanied by inadequate Hb saturation. It is caused by (a) a respiratory malfunction involving inadequate gas exchange, typified by a normal alveolar P_{O_2} but a reduced arterial P_{O_2}, or (b) exposure to high altitude or to a suffocating environment where atmospheric P_{O_2} is reduced so that alveolar and arterial P_{O_2} are likewise reduced.

2. *Anemic hypoxia* is a reduced O_2-carrying capacity of the blood. It can be brought about by (a) a decrease in circulating red blood cells, (b) an inadequate amount of Hb within the red blood cells, or (c) CO poisoning. In all cases of anemic hypoxia, the arterial P_{O_2} is at a normal level, but the O_2 content of the arterial blood is lower than normal because of the reduction in available Hb.

3. *Circulatory hypoxia* arises when too little oxygenated blood is delivered to the tissues. Circulatory hypoxia can be restricted to a limited area by a local vascular spasm or blockage. Or the body may experience circulatory hypoxia in general, from congestive heart failure or circulatory shock. The arterial P_{O_2} and O_2 content are typically normal, but too little oxygenated blood reaches the cells.

4. In *histotoxic hypoxia*, O_2 delivery to the tissues is normal, but the cells cannot use the O_2 available to them. The classic example is *cyanide poisoning*. Cyanide blocks cellular enzymes essential for internal respiration.

Hyperoxia, an above-normal arterial P_{O_2}, cannot occur when a person is breathing atmospheric air at sea level. However, breathing supplemental O_2 can increase alveolar and consequently arterial P_{O_2}. Because more of the inspired air is

---

▲ **TABLE 13-8**
Miniglossary of Clinically Important Respiratory States

Apnea Transient cessation of breathing

Asphyxia O_2 starvation of tissues, caused by a lack of O_2 in the air, respiratory impairment, or inability of the tissues to use O_2

Cyanosis Blueness of the skin resulting from insufficiently oxygenated blood in the arteries

Dyspnea Difficult or labored breathing

Eupnea Normal breathing

Hypercapnia Excess CO_2 in the arterial blood

Hyperpnea Increased pulmonary ventilation that matches increased metabolic demands, as in exercise

Hyperventilation Increased pulmonary ventilation in excess of metabolic requirements, resulting in decreased P_{CO_2} and respiratory alkalosis

Hypocapnia Below-normal CO_2 in the arterial blood

Hypoventilation Underventilation in relation to metabolic requirements, resulting in increased P_{CO_2} and respiratory acidosis

Hypoxia Insufficent O_2 at the cellular level

  Anemic hypoxia Reduced O_2-carrying capacity of the blood

  Circulatory hypoxia Too little oxygenated blood delivered to the tissues; also known as *stagnant hypoxia*

  Histotoxic hypoxia Inability of the cells to use O_2 available to them

  Hypoxic hypoxia Low arterial blood P_{O_2} accompanied by inadequate Hb saturation

Respiratory arrest Permanent cessation of breathing (unless clinically corrected)

Suffocation O_2 deprivation as a result of an inability to breathe oxygenated air

---

O_2, more of the total pressure of the inspired air is attributable to the O_2 partial pressure, so more O_2 dissolves in the blood before arterial P_{O_2} equilibrates with alveolar P_{O_2}. Even though arterial P_{O_2} increases, the *total* blood O_2 content does not significantly increase, because Hb is nearly fully saturated at the normal arterial P_{O_2}. In certain pulmonary diseases associated with a reduced arterial P_{O_2}, however, breathing supplemental O_2 can help establish a larger alveoli-to-blood driving gradient, improving arterial P_{O_2}. In contrast, far from being advantageous, a markedly elevated arterial P_{O_2} can be dangerous. If the arterial P_{O_2} is too high, **oxygen toxicity** can occur. Even though the total O_2 content of the blood is only slightly increased, exposure to a high P_{O_2} can damage some cells. In particular, brain damage and damage to the retina, causing blindness, are associated with O_2 toxicity. Therefore, O_2 therapy must be administered cautiously.

Abnormalities in arterial P_{CO_2}

The term **hypercapnia** refers to excess CO_2 in arterial blood; it is caused by **hypoventilation** (ventilation inadequate to meet metabolic needs for O_2 delivery and CO_2 removal). With most lung diseases, CO_2 accumulation in arterial blood occurs concurrently with an O_2 deficit, because both O_2 and CO_2 exchange between lungs and atmosphere are equally affected (● Figure 13-32). However, when a decrease in arterial P_{O_2} is due to reduced pulmonary diffusing capacity, as in pulmonary edema or emphysema, O_2 transfer suffers more than CO_2 transfer because the diffusion coefficient for CO_2 is 20 times that of O_2. As a result, in these circumstances hypoxic hypoxia occurs much more readily than hypercapnia.

● **FIGURE 13-32**

Effects of hyperventilation and hypoventilation on arterial P_{O_2} and P_{CO_2}

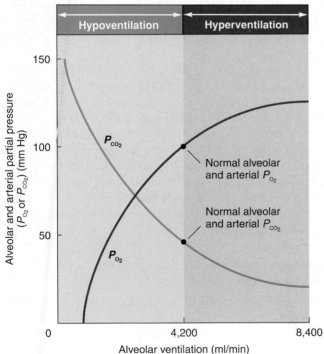

Hypocapnia, below-normal arterial P_{CO_2} levels, is brought about by hyperventilation. **Hyperventilation** occurs when a person "overbreathes," that is, when the rate of ventilation is in excess of the body's metabolic needs for CO_2 removal. As a result, CO_2 is blown off to the atmosphere more rapidly than it is produced in the tissues, and arterial P_{CO_2} falls. Hyperventilation can be triggered by anxiety states, fever, and aspirin poisoning. Alveolar P_{O_2} increases during hyperventilation as more fresh O_2 is delivered to the alveoli from the atmosphere than the blood extracts from the alveoli for tissue consumption, and arterial P_{O_2} increases correspondingly (● Figure 13-32). However, because Hb is almost fully saturated at the normal arterial P_{O_2}, very little additional O_2 is added to the blood. Except for the small extra amount of dissolved O_2, blood O_2 content remains essentially unchanged during hyperventilation.

Increased ventilation is not synonymous with hyperventilation. Increased ventilation that matches an increased metabolic demand, such as the increased need for O_2 delivery and CO_2 elimination during exercise, is termed **hyperpnea**. During exercise, alveolar P_{O_2} and P_{CO_2} remain constant, with the increased atmospheric exchange just keeping pace with the increased O_2 consumption and CO_2 production.

Consequences of abnormalities in arterial blood gases

The consequences of reduced O_2 availability to the tissues during hypoxia are apparent. The cells need adequate O_2 supplies to sustain energy-generating metabolic activities. The consequences of abnormal blood CO_2 levels are less obvious. Changes in blood CO_2 concentration primarily affect acid–base balance. Hypercapnia elevates production of carbonic acid. The subsequent generation of excess H^+ produces an acidic condition termed *respiratory acidosis*. Conversely, less-than-normal amounts of H^+ are generated through carbonic acid formation in conjunction with hypocapnia. The resultant alkalotic (less acidic than normal) condition is called *respiratory alkalosis* (Chapter 15). (To learn about the effects of mountain climbing and deep sea diving on blood gases, see the accompanying boxed feature, ◗ Concepts, Challenges, and Controversies.)

CONTROL OF RESPIRATION

Like the heartbeat, breathing must occur in a continuous, cyclical pattern to sustain life processes. Cardiac muscle must rhythmically contract and relax to alternately empty blood from the heart and fill it again. Similarly, inspiratory muscles must rhythmically contract and relax to alternately fill the lungs with air and empty them. Both these activities are accomplished automatically without conscious effort. However, the underlying mechanisms and control of these two systems are remarkably different.

■ Respiratory centers in the brain stem establish a rhythmic breathing pattern.

Whereas the heart can generate its own rhythm by means of its intrinsic pacemaker activity, the respiratory muscles, being

Effects of Heights and Depths on the Body

Our bodies are optimally equipped for existence at normal atmospheric pressure. Ascent into mountains high above sea level or descent into the depths of the ocean can have adverse effects on the body.

Effects of High Altitude on the Body

The atmospheric pressure progressively declines as altitude increases. At 18,000 ft above sea level, the atmospheric pressure is only 380 mm Hg—half of its normal sea level value. Because the proportion of O_2 and N_2 in the air remains the same, the P_{O_2} of inspired air at this altitude is 21% of 380 mm Hg, or 80 mm Hg, with alveolar P_{O_2} being even lower at 45 mm Hg. At any altitude above 10,000 ft, the arterial P_{O_2} falls into the steep portion of the O_2–Hb curve, below the safety range of the plateau region. As a result, the % Hb saturation in the arterial blood declines precipitously with further increases in altitude.

People who rapidly ascend to altitudes of 10,000 ft or more experience symptoms of **acute mountain sickness** attributable to hypoxic hypoxia and the resultant hypocapnia-induced alkalosis. The increased ventilatory drive to obtain more O_2 causes respiratory alkalosis, because acid-forming CO_2 is blown off more rapidly than it is produced. Symptoms of mountain sickness include fatigue, nausea, loss of appetite, labored breathing, rapid heart rate (triggered by hypoxia as a compensatory measure to increase circulatory delivery of available O_2 to the tissues), and nerve dysfunction characterized by poor judgment, dizziness, and incoordination.

Despite these acute responses to high altitude, millions of people live at elevations above 10,000 ft, with some villagers even residing in the Andes at altitudes higher than 16,000 ft. How do they live and function normally? They do so through the process of **acclimatization.** When a person remains at high altitude, the acute compensatory responses of increased ventilation and increased cardiac output are gradually replaced over a period of days by more slowly developing compensatory measures that permit adequate oxygenation of the tissues and restoration of normal acid–base balance. Red blood cell (RBC) production is increased, stimulated by erythropoietin in response to reduced O_2 delivery to the kidneys (see p. 395). The rise in the number of RBCs increases the O_2-carrying capacity of the blood. Hypoxia also promotes the synthesis of BPG within the RBCs, so that O_2 is unloaded from Hb more easily at the tissues. The number of capillaries within the tissues is increased, reducing the distance that O_2 must diffuse from the blood to reach the cells. Furthermore, acclimatized cells are able to use O_2

---

skeletal muscles, require nervous stimulation to induce contraction. The rhythmic pattern of breathing is established by cyclical neural activity to the respiratory muscles. In other words, the pacemaker activity that establishes the rhythmicity of breathing resides in the respiratory control centers in the brain, not in the lungs or respiratory muscles themselves. The nerve supply to the heart, not being necessary to initiate the heartbeat, serves only to modify the rate and strength of cardiac contraction. In contrast, the nerve supply to the respiratory system is absolutely essential in maintaining breathing and in reflexly adjusting the level of ventilation to match changing needs for O_2 uptake and CO_2 removal. Furthermore, unlike cardiac activity, which is not subject to voluntary control, respiratory activity can be voluntarily modified to accomplish speaking, singing, whistling, playing a wind instrument, or holding one's breath while swimming.

Components of neural control of respiration

Neural control of respiration involves three distinct components: (1) factors responsible for generating the alternating inspiration/expiration rhythm, (2) factors that regulate magnitude of ventilation (that is, rate and depth of breathing) to match body needs, and (3) factors that modify respiratory activity to serve other purposes. The latter modifications may be either voluntary, as in the breath control required for speech, or involuntary, as in the respiratory maneuvers involved in a cough or sneeze.

Respiratory control centers housed in the brain stem are responsible for generating the rhythmic pattern of breathing. The primary respiratory control center, the *medullary respiratory center*, consists of several aggregations of neuronal cell bodies within the medulla that provide output to the respiratory muscles. In addition, two other respiratory centers lie higher in the brain stem in the pons—the *apneustic center* and *pneumotaxic center*. These pontine centers influence output from the medullary respiratory center (● Figure 13-33). Following is a description of how these various regions interact to establish respiratory rhythmicity.

Inspiratory and expiratory neurons in the medullary center

We rhythmically breathe in and out during quiet breathing because of alternate contraction and relaxation of the inspiratory muscles, namely the diaphragm and external intercostal muscles, supplied by the phrenic nerve and intercostal nerves, respectively. The cell bodies for the neuronal fibers composing

more efficiently through an increase in the number of mitochondria, the energy organelles (see p. 34). The kidneys restore the arterial pH to nearly normal by conserving acid that normally would have been lost in the urine.

These compensatory measures are not without undesirable tradeoffs. For example, the greater number of RBCs increases blood viscosity (makes the blood "thicker"), thereby increasing resistance to blood flow. As a result, the heart has to work harder to pump blood through the vessels (see p. 398).

Effects of Deep-Sea Diving on the Body

When a deep-sea diver descends underwater, the body is exposed to greater than atmospheric pressure. Pressure rapidly increases with sea depth as a result of the weight of the water. Pressure is already doubled by about 30 ft below sea level. The air provided by scuba equipment is delivered to the lungs at these high pressures. Recall that (1) the amount of a gas in solution is directly proportional to the partial pressure of the gas and (2) air is composed of 79% N_2. Nitrogen is poorly soluble in body tissues, but the high P_{N_2} that occurs during deep-sea diving causes more of this gas than normal to dissolve in the body

tissues. The small amount of N_2 dissolved in the tissues at sea level has no known effect, but as more N_2 dissolves at greater depths, **nitrogen narcosis,** or **"rapture of the deep,"** ensues. Nitrogen narcosis is believed to result from a reduction in the excitability of neurons due to the

highly lipid-soluble N_2 dissolving in their lipid membranes. At 150 ft under water, divers experience a feeling of euphoria and become drowsy, similar to the effect of having a few cocktails. At lower depths, divers become weak and clumsy, and at 350 to 400 ft, they lose consciousness. *Oxygen toxicity* resulting from the high P_{O_2} is another possible detrimental effect of descending deep under water.

Another problem associated with deep-sea diving occurs during ascent. If a diver who has been submerged long enough for a significant amount of N_2 to dissolve in the tissues suddenly ascends to the surface, the rapid reduction in P_{N_2} causes N_2 to quickly come out of solution and form bubbles of gaseous N_2 in the body. The consequences depend on the amount and location of the bubble formation. This condition is called **decompression sickness** or **"the bends,"** because the victim often bends over in pain. Decompression sickness can be prevented by slowly ascending to the surface or by gradually decompressing in a decompression tank so that the excess N_2 can slowly escape through the lungs without bubble formation.

---

these nerves are located in the spinal cord. Impulses originating in the medullary center terminate on these motor neuron cell bodies (● Figure 13-34). When these motor neurons are activated, they in turn stimulate the inspiratory muscles, leading to inspiration; when these neurons are not firing, the inspiratory muscles relax, and expiration takes place.

The **medullary respiratory center** consists of two neuronal clusters known as the dorsal respiratory group and the ventral respiratory group (● Figure 13-33).

- The **dorsal respiratory group (DRG)** consists mostly of *inspiratory neurons* whose descending fibers terminate on the motor neurons that supply the inspiratory muscles. When the DRG inspiratory neurons fire, inspiration takes place; when they cease firing, expiration occurs. Expiration is brought to an end as the inspiratory neurons once again reach threshold and fire. The DRG has important interconnections with the ventral respiratory group.

- The **ventral respiratory group (VRG)** is composed of *inspiratory neurons* and *expiratory neurons*, both of which remain inactive during normal quiet breathing. This region is called into play by the DRG as an "overdrive" mechanism during periods when demands for ventilation are increased. It is especially important in active expiration. No impulses are gen-

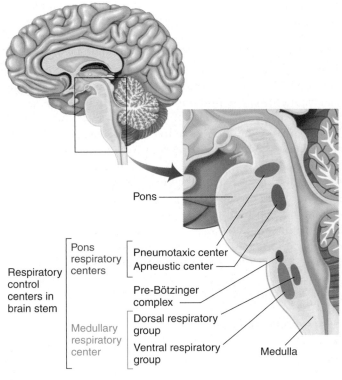

Pons

Respiratory control centers in brain stem

Pons respiratory centers

Medullary respiratory center

Pneumotaxic center
Apneustic center

Pre-Bötzinger complex

Dorsal respiratory group

Ventral respiratory group

Medulla

● **FIGURE 13-33**
Respiratory control centers in the brain stem

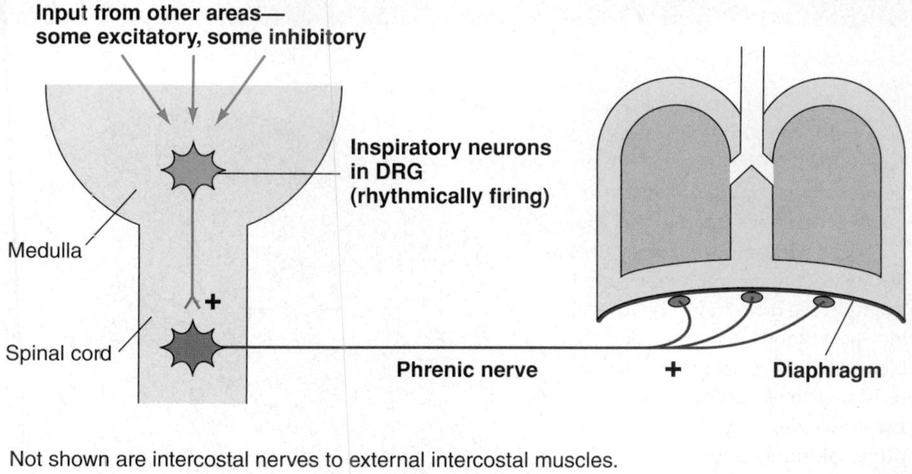

● FIGURE 13-34

Schematic representation of medullary dorsal respiratory group (DRG) control of inspiration Inspiration takes place when the inspiratory neurons are firing and activating the motor neurons that supply the inspiratory muscles. Expiration takes place when the inspiratory neurons cease firing, so that the motor neurons supplying the inspiratory muscles are no longer activated.

Not shown are intercostal nerves to external intercostal muscles.

erated in the descending pathways from the expiratory neurons during quiet breathing. Only during active expiration do the expiratory neurons stimulate the motor neurons supplying the expiratory muscles (the abdominal and internal intercostal muscles). Furthermore, the VRG inspiratory neurons, when stimulated by the DRG, rev up inspiratory activity when demands for ventilation are high.

Generation of respiratory rhythm

Until recently, the DRG was generally thought to be responsible for the basic rhythm of ventilation. However, generation of respiratory rhythm is now widely believed to lie in the **pre-Bötzinger complex,** a region located near the upper (head) end of the medullary respiratory center. A network of neurons in this region display pacemaker activity, undergoing self-induced action potentials similar to the SA node of the heart. The rate at which the DRG inspiratory neurons rhythmically fire is believed to be driven by synaptic input from this complex.

Influences from the pneumotaxic and apneustic centers

The pontine centers exert "fine-tuning" influences over the medullary center to help produce normal, smooth inspirations and expirations. The **pneumotaxic center** sends impulses to the DRG that help "switch off" the inspiratory neurons, limiting the duration of inspiration. In contrast, the **apneustic center** prevents the inspiratory neurons from being switched off, thus providing an extra boost to the inspiratory drive. In this check-and-balance system, the pneumotaxic center dominates over the apneustic center, helping halt inspiration and letting expiration occur normally. Without the pneumotaxic brakes, the breathing pattern consists of prolonged inspiratory gasps abruptly interrupted by very brief expirations. This abnormal pattern of breathing is known as **apneusis;** hence, the center responsible for this type of breathing is the apneustic center. Apneusis occurs in certain types of severe brain damage.

Hering-Breuer reflex

When the tidal volume is large (greater than 1 liter), as during exercise, the **Hering-Breuer reflex** is triggered to prevent over-inflation of the lungs. **Pulmonary stretch receptors** located within the smooth muscle layer of the airways are activated by the stretching of the lungs at large tidal volumes. Action potentials from these stretch receptors travel through afferent nerve fibers to the medullary center and inhibit the inspiratory neurons. This negative feedback from the highly stretched lungs themselves helps cut inspiration short before the lungs become overinflated.

▌ The magnitude of ventilation is adjusted in response to three chemical factors: P_{O_2}, P_{CO_2}, and H^+.

No matter how much O_2 is extracted from the blood or how much CO_2 is added to it at the tissue level, the P_{O_2} and P_{CO_2} of the systemic arterial blood leaving the lungs are held remarkably constant, indicating that arterial blood gas content is precisely regulated. Arterial blood gases are maintained within normal range by varying magnitude of ventilation (rate and depth of breathing) to match the body's needs for O_2 uptake and CO_2 removal. If more O_2 is extracted from the alveoli and more CO_2 is dropped off by the blood because the tissues are metabolizing more actively, ventilation increases correspondingly to bring in more fresh O_2 and blow off more CO_2.

The medullary respiratory center receives inputs that provide information about the body's needs for gas exchange. It responds by sending appropriate signals to the motor neurons supplying the respiratory muscles, to adjust the rate and depth of ventilation to meet those needs. The two most obvious signals to increase ventilation are decreased arterial P_{O_2} or increased arterial P_{CO_2}. Intuitively, you would suspect that if O_2 levels in the arterial blood declined or if CO_2 accumulated, ventilation would be stimulated to obtain more O_2 or to eliminate excess CO_2. These two factors do indeed influence the magnitude of ventilation, but not to the same degree nor through the same pathway. Also, a third chemical factor, H^+, notably influences the level of respiratory activity. We will examine the role of each of these important chemical factors in the control of ventilation (▲ Table 13-9).

Influence of Chemical Factors on Respiration

Chemical Factor	Effect on the Peripheral Chemoreceptors	Effect on the Central Chemoreceptors
↓ P_{O_2} in the Arterial Blood	Stimulates only when the arterial P_{O_2} has fallen to the point of being life-threatening (<60 mm Hg); an emergency mechanism	Directly depresses the central chemoreceptors and the respiratory center itself when <60mm Hg
↑ P_{CO_2} in the Arterial Blood (↑ H^+ in the Brain ECF)	Weakly stimulates	Strongly stimulates; is the dominant control of ventilation (Levels >70–80 mm Hg directly depress the respiratory center and central chemoreceptors)
↑ H^+ in the Arterial Blood	Stimulates; important in acid–base balance	Does not affect; cannot penetrate the blood–brain barrier

▌ Decreased arterial P_{O_2} increases ventilation only as an emergency mechanism.

Arterial P_{O_2} is monitored by **peripheral chemoreceptors** known as the **carotid bodies** and **aortic bodies,** which are located at the bifurcation of the common carotid arteries on both the right and left sides and in the arch of the aorta, respectively (● Figure 13-35). These chemoreceptors respond to specific changes in the chemical content of the arterial blood that bathes them. They are distinctly different from the carotid sinus and aortic arch baroreceptors located in the same vicinity. The lat-

● FIGURE 13-35

Location of the peripheral chemoreceptors
The carotid bodies are located in the carotid sinus, and the aortic bodies are located in the aortic arch.

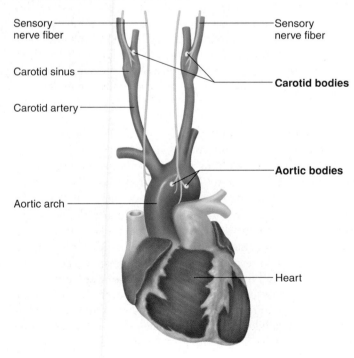

Sensory nerve fiber

Sensory nerve fiber

Carotid sinus

Carotid bodies

Carotid artery

Aortic bodies

Aortic arch

Heart

ter monitor pressure changes rather than chemical changes and are important in regulating systemic arterial blood pressure (see p. 377).

Effect of a large decrease in P_{O_2} on the peripheral chemoreceptors

The peripheral chemoreceptors are not sensitive to modest reductions in arterial P_{O_2}. The arterial P_{O_2} must fall below 60 mm Hg (>40% reduction) before the peripheral chemoreceptors respond by sending afferent impulses to the medullary inspiratory neurons, thereby reflexly increasing ventilation. Because arterial P_{O_2} only falls below 60 mm Hg in the unusual circumstances of severe pulmonary disease or reduced atmospheric P_{O_2}, it does not play a role in the normal ongoing regulation of respiration. This fact might seem surprising at first thought, because a primary function of ventilation is to provide enough O_2 for uptake by the blood. However, there is no need to increase ventilation until the arterial P_{O_2} falls below 60 mm Hg, because of the safety margin in % Hb saturation afforded by the plateau portion of the O_2–Hb curve. Hemoglobin is still 90% saturated at an arterial P_{O_2} of 60 mm Hg, but the % Hb saturation drops precipitously when the P_{O_2} falls below this level. Therefore, reflex stimulation of respiration by the peripheral chemoreceptors serves as an important emergency mechanism in dangerously low arterial P_{O_2} states. Indeed, this reflex mechanism is a lifesaver, because a low arterial P_{O_2} tends to directly depress the respiratory center, as it does all the rest of the brain (● Figure 13-36).

Because the peripheral chemoreceptors respond to the P_{O_2} of the blood, *not* the total O_2 content of the blood, O_2 content in the arterial blood can fall to dangerously low or even fatal levels without the peripheral chemoreceptors ever responding to reflexly stimulate respiration. Remember that only physically dissolved O_2 contributes to blood P_{O_2}. The total O_2 content in the arterial blood can be reduced in anemic states, in which O_2-carrying Hb is reduced, or in CO poisoning, when the Hb is preferentially bound to this molecule rather than to O_2. In both cases, arterial P_{O_2} is normal, so respiration is not stimulated, even though O_2 delivery to the tissues may be so reduced that the person dies from cellular O_2 deprivation.

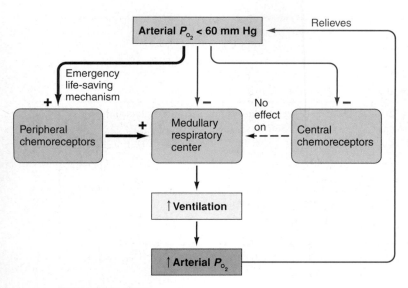

● FIGURE 13-36
Effect of threateningly low arterial P_{O_2} (<60 mm Hg) on ventilation

Direct effect of a large decrease in P_{O_2} on the respiratory center

Except for the peripheral chemoreceptors, the level of activity in all nervous tissue becomes reduced in the face of O_2 deprivation. Were it not for stimulatory intervention of the peripheral chemoreceptors when the arterial P_{O_2} falls threateningly low, a vicious cycle ending in cessation of breathing would ensue. Direct depression of the respiratory center by the markedly low arterial P_{O_2} would further reduce ventilation, leading to an even greater fall in arterial P_{O_2}, which would even further depress the respiratory center until ventilation ceased and death occurred.

▍ Carbon dioxide–generated H^+ in the brain is normally the primary regulator of ventilation.

In contrast to arterial P_{O_2}, which does not contribute to the minute-to-minute regulation of respiration, arterial P_{CO_2} is the most important input regulating the magnitude of ventilation under resting conditions. This role is appropriate, because changes in alveolar ventilation have an immediate and pronounced effect on arterial P_{CO_2}. By contrast, changes in ventilation have little effect on % Hb saturation and O_2 availability to the tissues until the arterial P_{O_2} falls by more than 40%. Even slight alterations from normal in arterial P_{CO_2} induce a significant reflex effect on ventilation. An increase in arterial P_{CO_2} reflexly stimulates the respiratory center, with the resultant increase in ventilation promoting elimination of the excess CO_2 to the atmosphere. Conversely, a fall in arterial P_{CO_2} reflexly reduces the respiratory drive. The subsequent decrease in ventilation allows metabolically produced CO_2 to accumulate so that P_{CO_2} can be returned to normal.

Effect of increased P_{CO_2} on the central chemoreceptors

Surprisingly, given the key role of arterial P_{CO_2} in regulating respiration, no important receptors monitor arterial P_{CO_2}

per se. The carotid and aortic bodies are only weakly responsive to changes in arterial P_{CO_2}, so they play only a minor role in reflexly stimulating ventilation in response to elevation in arterial P_{CO_2}. More important in linking changes in arterial P_{CO_2} to compensatory adjustments in ventilation are the **central chemoreceptors,** located in the medulla near the respiratory center. These central chemoreceptors do not monitor CO_2 itself; however, they are sensitive to changes in CO_2-induced H^+ concentration in the brain extracellular fluid (ECF) that bathes them.

Movement of materials across the brain capillaries is restricted by the blood–brain barrier (see p. 141). Because this barrier is readily permeable to CO_2, any increase in arterial P_{CO_2} causes a similar rise in brain ECF P_{CO_2} as CO_2 diffuses down its pressure gradient from the cerebral blood vessels into the brain ECF. The increased P_{CO_2} within the brain ECF correspondingly raises the concentration of H^+ according to the law of mass action as it applies to this reaction: $CO_2 + H_2O \leftrightharpoons H_2CO_3 \leftrightharpoons H^+ + HCO_3^-$. An elevation in H^+ concentration in the brain ECF directly stimulates the central chemoreceptors, which in turn increase ventilation by stimulating the respiratory center through synaptic connections (● Figure 13-37). As the excess CO_2 is subsequently blown off, the arterial P_{CO_2} and the P_{CO_2} and H^+ concentration of the brain ECF return to normal. Conversely, a decline in arterial P_{CO_2} below normal is paralleled by a fall in P_{CO_2} and H^+ in the brain ECF, the result of which is a central chemoreceptor-mediated decrease in ventilation. As CO_2 produced by cellular metabolism is consequently allowed to accumulate, arterial P_{CO_2} and P_{CO_2} and H^+ of the brain ECF are restored toward normal.

Unlike CO_2, H^+ is not readily able to permeate the blood–brain barrier, so H^+ present in the plasma cannot gain access to the central chemoreceptors. Accordingly, the central chemoreceptors are responsive only to H^+ generated within the brain ECF itself as a result of CO_2 entry. Thus the major mechanism controlling ventilation under resting conditions is specifically aimed at regulating the brain ECF H^+ concentration, which in turn directly reflects the arterial P_{CO_2}. Unless there are extenuating circumstances such as reduced availability of O_2 in the inspired air, arterial P_{O_2} is coincidentally also maintained at its normal value by the brain ECF H^+ ventilatory driving mechanism.

The powerful influence of the central chemoreceptors on the respiratory center is responsible for your inability to deliberately hold your breath for more than about a minute. While you hold your breath, metabolically produced CO_2 continues to accumulate in your blood and subsequently to build up the H^+ concentration in your brain ECF. Finally, the increased P_{CO_2}–H^+ stimulant to respiration becomes so powerful that central-chemoreceptor excitatory input overrides voluntary inhibitory input to respiration, so breathing resumes despite deliberate attempts to prevent it. Breathing resumes long before arterial P_{O_2} falls to the threateningly low levels that trigger the peripheral chemoreceptors. Therefore, you cannot deliberately hold your breath long enough to create a dangerously high level of CO_2 or low level of O_2 in the arterial blood.

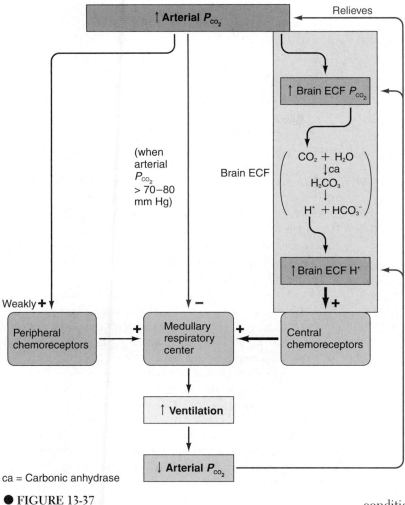

● FIGURE 13-37
Effect of increased arterial P_{CO_2} on ventilation

Direct effect of a large increase in P_{CO_2} on the respiratory center

In contrast to the normal reflex stimulatory effect of the increased P_{CO_2}–H^+ mechanism on respiratory activity, very high levels of CO_2 directly depress the entire brain, including the respiratory center, just as very low levels of O_2 do. Up to a P_{CO_2} of 70 to 80 mm Hg, progressively higher P_{CO_2} levels induce correspondingly more vigorous respiratory efforts in an attempt to blow off the excess CO_2. A further increase in P_{CO_2} beyond 70 to 80 mm Hg, however, does not further increase ventilation but actually depresses the respiratory neurons. For this reason, CO_2 must be removed and O_2 supplied in closed environments such as closed-system anesthesia machines, submarines, or space capsules. Otherwise, CO_2 could reach lethal levels, not only because it depresses respiration but also because it produces severe respiratory acidosis.

Loss of sensitivity to P_{CO_2} with lung disease

During prolonged hypoventilation caused by certain types of chronic lung disease, an elevated P_{CO_2} occurs simultaneously with a markedly reduced P_{O_2}. In most cases, the elevated P_{CO_2}

(acting via the central chemoreceptors) and the reduced P_{O_2} (acting via the peripheral chemoreceptors) are *synergistic*; that is, the combined stimulatory effect on respiration exerted by these two inputs together is greater than the sum of their independent effects.

However, some patients with severe chronic lung disease lose their sensitivity to an elevated arterial P_{CO_2}. In a prolonged increase in H^+ generation in the brain ECF, from long-standing CO_2 retention, enough HCO_3^- may cross the blood–brain barrier to buffer, or "neutralize," the excess H^+. The additional HCO_3^- combines with the excess H^+, removing it from solution so that it no longer contributes to free H^+ concentration. When brain ECF HCO_3^- concentration rises, brain ECF H^+ concentration returns to normal although arterial P_{CO_2} and brain ECF P_{CO_2} remain high. The central chemoreceptors are no longer aware of the elevated P_{CO_2}, because the brain ECF H^+ is normal. Because the central chemoreceptors no longer reflexly stimulate the respiratory center in response to the elevated P_{CO_2}, the drive to eliminate CO_2 is blunted in such patients; that is, their level of ventilation is abnormally low considering their high arterial P_{CO_2}. In these patients, the hypoxic drive to ventilation becomes their primary respiratory stimulus, in contrast to normal individuals, in whom the arterial P_{CO_2} level is the dominant factor governing the magnitude of ventilation. Ironically, administering O_2 to such patients to relieve the hypoxic condition can markedly depress their drive to breathe by elevating the arterial P_{O_2} and removing the primary driving stimulus for respiration. Thus O_2 therapy must be administered cautiously in patients with long-term pulmonary diseases.

■ Adjustments in ventilation in response to changes in arterial H^+ are important in acid–base balance.

Changes in arterial H^+ concentration cannot influence the central chemoreceptors, because H^+ does not readily cross the blood–brain barrier. However, the aortic and carotid body peripheral chemoreceptors are highly responsive to fluctuations in arterial H^+ concentration, in contrast to their weak sensitivity to deviations in arterial P_{CO_2} and their unresponsiveness to arterial P_{O_2} until it falls 40% below normal.

Any change in arterial P_{CO_2} brings about a corresponding change in the H^+ concentration of the blood as well as of the brain ECF. These CO_2-induced H^+ changes in the arterial blood are detected by the peripheral chemoreceptors; the result is reflexly stimulated ventilation in response to increased arterial H^+ concentration and depressed ventilation in association with decreased arterial H^+ concentration. However, these changes in ventilation mediated by the peripheral chemoreceptors are far less important than the powerful central-chemoreceptor mechanism in adjusting ventilation in response to changes in CO_2-generated H^+ concentration.

The peripheral chemoreceptors do play a major role in adjusting ventilation in response to alterations in arterial H^+ concentration unrelated to fluctuations in P_{CO_2}. In many situations, even though P_{CO_2} is normal, arterial H^+ concentration is changed by the addition or loss of noncarbonic acid from the body. For example, arterial H^+ concentration increases during diabetes mellitus because excess H^+-generating keto acids are abnormally produced and added to the blood. A rise in arterial H^+ concentration reflexly stimulates ventilation by means of the peripheral chemoreceptors. Conversely, the peripheral chemoreceptors reflexly suppress respiratory activity in response to a fall in arterial H^+ concentration resulting from nonrespiratory causes. Changes in ventilation by this mechanism are extremely important in regulating the body's acid–base balance. Changing the magnitude of ventilation can vary the amount of H^+-generating CO_2 eliminated. The resultant adjustment in the amount of H^+ added to the blood from CO_2 can compensate for the nonrespiratory-induced abnormality in arterial H^+ concentration that first elicited the respiratory response (● Figure 13-38). (See Chapter 15 for further details.)

▮ Exercise profoundly increases ventilation, but the mechanisms involved are unclear.

Alveolar ventilation may increase up to 20-fold during heavy exercise to keep pace with the increased demand for O_2 up-

---

● **FIGURE 13-38**

Effect of increased arterial non–carbonic-acid-generated hydrogen ion (non–CO_2-H^+) on ventilation

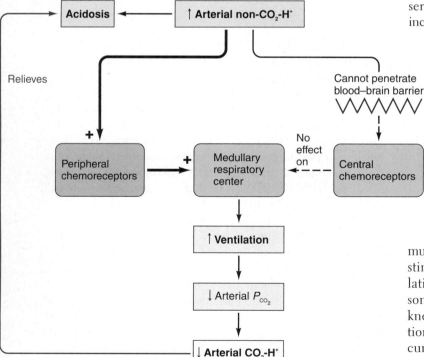

take and CO_2 output. (▲ Table 13-10 highlights changes in O_2- and CO_2-related variables during exercise.) The cause of increased ventilation during exercise is still largely speculative. It would seem logical that changes in the "big three" chemical factors—decreased P_{O_2}, increased P_{CO_2}, and increased H^+—could account for the increase in ventilation. This does not appear to be the case, however.

- Despite the marked increase in O_2 use during exercise, arterial P_{O_2} does not decrease but remains normal or may actually increase slightly. This is because the increase in alveolar ventilation keeps pace with or even slightly exceeds the stepped-up rate of O_2 consumption.
- Likewise, despite the marked increase in CO_2 production during exercise, arterial P_{CO_2} does not increase but remains normal or decreases slightly. This is because the extra CO_2 is removed as rapidly or even more rapidly than it is produced by the increase in ventilation.
- During mild or moderate exercise, H^+ concentration does not increase, because H^+-generating CO_2 is held constant. During heavy exercise, H^+ concentration does increase somewhat from release of H^+-generating lactic acid into the blood by anaerobic metabolism in the exercising muscles. Even so, the elevation in H^+ concentration resulting from lactic acid formation is not enough to account for the large increase in ventilation accompanying exercise.

Some investigators argue that the constancy of the three chemical regulatory factors during exercise shows that ventilatory responses to exercise are actually being controlled by these factors—particularly by P_{CO_2}, because it is normally the dominant control during resting conditions. According to this reasoning, how else could alveolar ventilation be increased in exact proportion to CO_2 production, thereby keeping the P_{CO_2} constant? This proposal, however, cannot account for the observation that during heavy exercise, alveolar ventilation may increase relatively more than CO_2 production increases, thereby actually causing a slight decline in P_{CO_2}. Also, ventilation increases abruptly at the onset of exercise (within seconds), long before changes in arterial blood gases could become important influences on the respiratory center (which requires a matter of minutes).

Factors that may increase ventilation during exercise

Researchers have suggested that a number of other factors, including the following, play a role in the ventilatory response to exercise:

1. *Reflexes originating from body movements.* Joint and muscle receptors excited during muscle contraction reflexly stimulate the respiratory center, abruptly increasing ventilation. Even passive movement of the limbs (for example, someone else alternately flexing and extending a person's knee) may increase ventilation several-fold through activation of these receptors, even though no actual exercise is occurring. Thus the mechanical events of exercise are believed to play an important role in coordinating respiratory activ-

Oxygen and Carbon-Dioxide-Related Variables during Exercise

O_2- or CO_2-Related Variable	Change	Comment
O_2 Use	Marked increase	Active muscles are oxidizing nutrient molecules more rapidly to meet their increased energy needs.
CO_2 Production	Marked increase	More actively metabolizing muscles produce more CO_2.
Alveolar Ventilation	Marked increase	By mechanisms not completely understood, alveolar ventilation keeps pace with or even slightly exceeds the increased metabolic demands during exercise.
Arterial P_{O_2}	Normal or slight ↑	Despite a marked increase in O_2 use and CO_2 production during exercise, alveolar ventilation keeps pace with or even slightly exceeds the stepped-up rate of O_2 consumption and CO_2 production.
Arterial P_{CO_2}	Normal or slight ↓	
O_2 Delivery to Muscles	Marked increase	Although arterial P_{O_2} remains normal, O_2 delivery to muscles is greatly increased by the increased blood flow to exercising muscles accomplished by increased cardiac output coupled with local vasodilation of active muscles.
O_2 Extraction by Muscles	Marked increase	Increased use of O_2 lowers the P_{O_2} at the tissue level, which results in more O_2 unloading from hemoglobin; this is enhanced by ↑ P_{CO_2}, ↑ H^+, and ↑ temperature.
CO_2 Removal from Muscles	Marked increase	The increased blood flow to exercising muscles removes the excess CO_2 produced by these more actively metabolizing tissues.
Arterial H^+ Concentration		
Mild to moderate exercise	Normal	Because carbonic acid–generating CO_2 is held constant in arterial blood, arterial H^+ concentration does not change.
Heavy exercise	Modest increase	In heavy exercise, when muscles resort to anaerobic metabolism, lactic acid is added to the blood.

ity with the increased metabolic requirements of the active muscles.

2. *Increase in body temperature.* Much of the energy generated during muscle contraction is converted to heat rather than to actual mechanical work. Heat loss mechanisms such as sweating frequently cannot keep pace with the increased heat production that accompanies increased physical activity, so body temperature often rises slightly during exercise. Because raised body temperature stimulates ventilation, this exercise-related heat production undoubtedly contributes to the respiratory response to exercise. For the same reason, increased ventilation often accompanies a fever.

3. *Epinephrine release.* The adrenal medullary hormone epinephrine also stimulates ventilation. The level of circulating epinephrine rises during exercise in response to the sympathetic nervous system discharge that accompanies increased physical activity.

4. *Impulses from the cerebral cortex.* Especially at the onset of exercise, the motor areas of the cerebral cortex are believed to simultaneously stimulate the medullary respiratory neurons and activate the motor neurons of the exercising muscles. This is similar to the cardiovascular adjustments initiated by the motor cortex at the onset of exercise. In this way, the motor region of the brain calls forth increased ventilatory and circulatory responses to support the increased physical activity it is about

to orchestrate. These anticipatory adjustments are feedforward regulatory mechanisms; that is, they occur *before* any homeostatic factors actually change (see p. 18). This is in contrast to the more common case in which regulatory adjustments take place *after* a factor has altered, to restore homeostasis.

None of these factors or combinations of factors are fully satisfactory in explaining the abrupt and profound effect exercise has on ventilation, nor can they completely account for the high degree of correlation between respiratory activity and the body's needs for gas exchange during exercise. (For a discussion of how O_2 consumption during exercise can be measured to determine a person's maximum work capacity, see the accompanying boxed feature, ▶ A Closer Look at Exercise Physiology.)

▮ Ventilation can be influenced by factors unrelated to the need for gas exchange.

Respiratory rate and depth can be modified for reasons other than the need to supply O_2 or remove CO_2. Here are some examples of involuntary influences in this category:

- Protective reflexes such as sneezing and coughing temporarily govern respiratory activity in an effort to expel irritant materials from respiratory passages.

How to Find Out How Much Work You're Capable of Doing

The best single predictor of a person's work capacity is the determination of **maximal O₂ consumption**, or **max VO₂**, which is the maximum volume of O_2 the person is capable of using per minute to oxidize nutrient molecules for energy production. Max VO₂ is measured by having the person engage in exercise, usually on a treadmill or bicycle ergometer (a stationary bicycle with variable resistance). The workload is incrementally increased until the person becomes exhausted. Expired air samples collected during the last minutes of exercise, when O_2 consumption is at a maximum because the person is working as hard as possible, are analyzed for the percentage of O_2 and CO_2 they contain. Furthermore, the volume of air expired is measured. Equations are then employed to determine the amount of O_2 consumed, taking into account the percentages of O_2 and CO_2 in the inspired air, the total volume of air expired, and the percentages of O_2 and CO_2 in the exhaled air.

Maximal O_2 consumption depends on three systems. The respiratory system is essential for ventilation and exchange of O_2 and CO_2 between the air and blood in the lungs. The circulatory system is required to deliver O_2 to the working muscles. Finally, the muscles must have the oxidative enzymes available to use the O_2 once it has been delivered.

Regular aerobic exercise can improve max VO₂ by making the heart and respiratory system more efficient, thereby delivering more O_2 to the working muscles. Exercised muscles themselves become better equipped to use O_2 once it is delivered. The number of functional capillaries increases, as do the number and size of mitochondria, which contain the oxidative enzymes.

Maximal O_2 consumption is measured in liters per minute and then converted into milliliters per kilogram of body weight per minute so that large and small people can be compared. As would be expected, athletes have the highest values for maximal O_2 consumption. The max VO₂ for male cross-country skiers has been recorded to be as high as 94 ml O_2/kg/min. Distance runners maximally consume between 65 and 85 ml O_2/kg/min, and football players have max VO₂ values between 45 and 65 ml O_2/kg/min, depending on the position they play. Sedentary young men maximally consume between 25 and 45 ml O_2/kg/min. Female values for max VO₂ are 20% to 25% lower than for males when expressed as ml/kg/min of total body weight. The difference in max VO₂ between females and males is only 8% to 10% when expressed as ml/kg/min of lean body weight, however, because females generally have a higher percentage of body fat (the female sex hormone estrogen promotes fat deposition).

Available norms are used to classify people as being low, fair, average, good, or excellent in aerobic capacity for their age group. Exercise physiologists use max VO₂ measurements to prescribe or adjust training regimens to help people achieve their optimal level of aerobic conditioning.

- Inhalation of particularly noxious agents frequently triggers immediate cessation of ventilation.
- Pain originating anywhere in the body reflexly stimulates the respiratory center (for example, one "gasps" with pain).
- Involuntary modification of breathing also occurs during the expression of various emotional states, such as laughing, crying, sighing, and groaning. The emotionally induced modifications are mediated through connections between the limbic system in the brain (which is responsible for emotions) and the respiratory center.
- In addition, the respiratory center is reflexly inhibited during swallowing, when the airways are closed to prevent food from entering the lungs.

Humans also have considerable voluntary control over ventilation. Voluntary control of breathing is accomplished by the cerebral cortex, which does not act on the respiratory center in the brain stem but instead sends impulses directly to the motor neurons in the spinal cord that supply the respiratory muscles. We can voluntarily hyperventilate ("overbreathe") or, at the other extreme, hold our breath, but only for a brief period of time. The resulting chemical changes in the arterial blood directly and reflexly influence the respiratory center, which in turn overrides the voluntary input to the motor neurons of the respiratory muscles. Other than these extreme forms of deliberately controlling ventilation, we also control our breathing to perform such voluntary acts as speaking, singing, and whistling.

▌ During apnea, a person "forgets to breathe"; during dyspnea, a person feels "short of breath."

Apnea is the transient interruption of ventilation, with the expectation that breathing will resume spontaneously. If breathing does not resume, the condition is called **respiratory arrest.** Because ventilation is normally decreased and the central chemoreceptors are less sensitive to the arterial P_{CO_2} drive during sleep, especially REM sleep (see p. 170), apnea is most likely to occur during this time. Victims of **sleep apnea** may stop breathing for a few seconds or up to one or two minutes as many as 500 times a night. Mild sleep apnea is not dangerous unless the sufferer has pulmonary or circulatory disease, which can be exacerbated by recurrent bouts of apnea.

Sudden infant death syndrome

In exaggerated cases of sleep apnea, the victim may be unable to recover from an apneic period, and death results. This is the case in **sudden infant death syndrome (SIDS)**, or "crib death." With this tragic form of sleep apnea, a previously healthy 2- to 4-month-old infant is found dead in his or her crib for no apparent reason. The underlying cause of SIDS is the subject of intense investigation. Most evidence suggests that the baby "forgets to breathe" because the respiratory control mechanisms are immature, either in the brain stem or in the chemoreceptors that monitor the body's respiratory status. For example, on autopsy more than half the victims have poorly developed carotid bodies, the more important of the peripheral chemoreceptors. Alternatively, some researchers believe the condition may be triggered by an initial cardiovascular failure rather than by an initial cessation of breathing. Still other investigators propose that some cases may be due to aspiration of gastric (stomach) juice containing the bacterium *Helicobacter pylori*. In one study, this micro-organism was present in 88% of infants who died of SIDS. Scientists speculate that *H. pylori* may lead to the production of ammonia, which can be lethal if it gains access to the blood from the lungs.

Whatever the underlying cause, certain risk factors make babies more vulnerable to SIDS. Among them are sleeping position (an almost 40% higher incidence of SIDS is associated with sleeping on the abdomen rather than on the back or side) and exposure to nicotine during fetal life or after birth. Infants whose mothers smoked during pregnancy or who breathe cigarette smoke in the home are three times more likely to die of SIDS than those not exposed to smoke.

Dyspnea

People who have **dyspnea** have the subjective sensation that they are not getting enough air; that is, they feel "short of breath." Dyspnea is the mental anguish associated with the unsatiated desire for more adequate ventilation. It often accompanies the labored breathing characteristic of obstructive lung disease or the pulmonary edema associated with congestive heart failure. In contrast, during exercise a person can breathe very hard without experiencing the sensation of dyspnea, because such exertion is not accompanied by a sense of anxiety over the adequacy of ventilation. Surprisingly, dyspnea is not directly related to chronic elevation of arterial P_{CO_2} or reduction of P_{O_2}. The subjective feeling of air hunger may occur even when alveolar ventilation and the blood gases are normal. Some individuals experience dyspnea when they *perceive* that they are short of air even though this is not actually the case, such as in a crowded elevator.

CHAPTER IN PERSPECTIVE: FOCUS ON HOMEOSTASIS

The respiratory system contributes to homeostasis by obtaining O_2 from and eliminating CO_2 to the external environment. All body cells ultimately need an adequate supply of O_2 to use in oxidizing nutrient molecules to generate ATP. Brain cells, which are especially dependent on a continual supply of O_2, die if deprived of O_2 for more than four minutes. Even cells that can resort to anaerobic ("without O_2") metabolism for energy production, such as strenuously exercising muscles, can do so only transiently by incurring an O_2 debt that ultimately must be repaid (see p. 280).

As a result of these energy-yielding metabolic reactions, the body produces large quantities of CO_2 that must be eliminated. Because CO_2 and H_2O form carbonic acid, adjustments in the rate of CO_2 elimination by the respiratory system are important in regulating acid–base balance in the internal environment. Cells can survive only within a narrow pH range.

CHAPTER SUMMARY

Introduction

▮ Internal respiration encompasses the intracellular metabolic reactions that use O_2 and produce CO_2 during energy-yielding oxidation of nutrient molecules.

▮ External respiration encompasses the various steps in the transfer of O_2 and CO_2 between the external environment and tissue cells. The respiratory and circulatory systems function together to accomplish external respiration.

▮ The respiratory system exchanges air between the atmosphere and the lungs through the process of ventilation.

▮ Respiratory airways conduct air from the atmosphere to the alveoli, the gas-exchanging portion of the lungs.

▮ Exchange of O_2 and CO_2 between the air in the lungs and the blood in the pulmonary capillaries takes place across the extremely thin walls of the air sacs, or alveoli. The alveolar walls are formed by Type I alveolar cells. Type II alveolar cells secrete pulmonary surfactant.

▮ The lungs are housed within the closed compartment of the thorax, the volume of which can be changed by contractile activity of surrounding respiratory muscles.

▮ Each lung is surrounded by a double-walled, closed sac, the pleural sac.

Respiratory Mechanics

▮ Ventilation, or breathing, is the process of cyclically moving air in and out of the lungs, so that old alveolar air that has already participated in exchanging O_2 and CO_2 with the pulmonary capillary blood can be exchanged for fresh atmospheric air.

▮ Ventilation is mechanically accomplished by alternately shifting the direction of the pressure gradient for airflow between the atmosphere and the alveoli through the cyclical expansion and recoil of the lungs. When intra-alveolar pressure decreases as a result of lung expansion during inspiration, air flows into the lungs from the higher atmospheric pressure. When intra-alveolar pressure increases as a result of lung recoil during expiration, air flows out of the lungs toward the lower atmospheric pressure.

▮ Alternate contraction and relaxation of the inspiratory muscles (primarily the diaphragm) indirectly produce periodic inflation and deflation of the lungs by cyclically expanding and compressing the thoracic cavity, with the lungs passively following its movements.

▮ The lungs follow the movements of the thoracic cavity by virtue of the intrapleural fluid's cohesiveness and the transmural pressure gradient across the lung wall. The transmural pressure gradient exists because the intrapleural pressure is subatmospheric and thus less than the intra-alveolar pressure.

▮ Because energy is required for contracting inspiratory muscles, inspiration is an active process, but expiration is passive during quiet breathing because it is accomplished by elastic recoil of the lungs on relaxation of inspiratory muscles, at no energy expense.

▮ For more forceful active expiration, contraction of the expiratory muscles (namely, the abdominal muscles) further decreases the size of the thoracic cavity and lungs, which further increases the intra-alveolar-to-atmospheric pressure gradient.

- The larger the gradient between the alveoli and the atmosphere in either direction, the larger the airflow rate, because air continues to flow until the intra-alveolar pressure equilibrates with atmospheric pressure.
- Besides being directly proportional to the pressure gradient, airflow rate is also inversely proportional to airway resistance. Because airway resistance, which depends on the caliber of the conducting airways, is normally very low, airflow rate usually depends primarily on the pressure gradient between the alveoli and the atmosphere.
- If airway resistance is pathologically increased by chronic obstructive pulmonary disease, the pressure gradient must be correspondingly increased by more vigorous respiratory muscle activity to maintain a normal airflow rate.
- The lungs can be stretched to varying degrees during inspiration and then recoil to their preinspiratory size during expiration because of their elastic behavior.
  1. The term *pulmonary compliance* refers to the distensibility of the lungs—how much they stretch in response to a given change in the transmural pressure gradient, the stretching force exerted across the lung wall.
  2. The term *elastic recoil* refers to the phenomenon of the lungs snapping back to their resting position during expiration.
- Pulmonary elastic behavior depends on the elastic connective tissue meshwork within the lungs and on alveolar surface tension/pulmonary surfactant interaction. Alveolar surface tension, which is due to the attractive forces between the surface water molecules in the liquid film lining each alveolus, tends to resist the alveolus being stretched on inflation (decreases compliance) and tends to return it back to a smaller surface area during deflation (increases lung rebound).
- If the alveoli were lined by water alone, the surface tension would be so great that the lungs would be poorly compliant and would tend to collapse. Pulmonary surfactant intersperses between the water molecules and lowers the alveolar surface tension, thereby increasing the compliance of the lungs and counteracting the tendency for alveoli to collapse.
- The lungs can be filled to over 5.5 liters on maximal inspiratory effort or emptied to about one liter on maximal expiratory effort. Normally, however, the lungs operate at "half full." The lung volume typically varies from about 2 to 2.5 liters as an average tidal volume of 500 ml of air is moved in and out with each breath.
- The amount of air moved in and out of the lungs in one minute, the pulmonary ventilation, is equal to tidal volume times respiratory rate.
- Not all the air moved in and out is available for O_2 and CO_2 exchange with the blood, because part occupies the conducting airways, known as the *anatomic dead space*. Alveolar ventilation, the volume of air exchanged between the atmosphere and the alveoli in one minute, is a measure of the air actually available for gas exchange with the blood. Alveolar ventilation equals (tidal volume minus the dead space volume) times respiratory rate.

Gas Exchange

- Oxygen and CO_2 move across body membranes by passive diffusion down partial pressure gradients.
- The partial pressure of a gas in air is that portion of the total atmospheric pressure contributed by this individual gas, which in turn is directly proportional to the percentage of this gas in the air. The partial pressure of a gas in blood depends on the amount of this particular gas dissolved in the blood.
- Net diffusion of O_2 occurs first between the alveoli and the blood and then between the blood and the tissues as a result of the O_2 partial pressure gradients created by continuous use of O_2 in the cells and continuous replenishment of fresh alveolar O_2 provided by ventilation.
- Net diffusion of CO_2 occurs in the reverse direction, first between the tissues and the blood and then between the blood and the alveoli, as a result of the CO_2 partial pressure gradients created by continuous production of CO_2 in the cells and continuous removal of alveolar CO_2 through the process of ventilation.
- Factors other than the partial pressure gradient that influence the rate of gas exchange are the surface area and thickness of the membrane across which the gas is diffusing and the diffusion coefficient of the gas in the membrane, according to Fick's law of diffusion.

Gas Transport

- Because O_2 and CO_2 are not very soluble in the blood, they must be transported primarily by mechanisms other than simply being physically dissolved.
- Only 1.5% of the O_2 is physically dissolved in the blood, with 98.5% chemically bound to hemoglobin (Hb).
- The primary factor that determines the extent to which Hb and O_2 are combined (the % Hb saturation) is the P_{O_2} of the blood, depicted by an S-shaped curve known as the O_2–Hb dissociation curve.
  1. The relationship between blood P_{O_2} and % Hb saturation is such that in the P_{O_2} range found in the pulmonary capillaries (the plateau portion of the curve), Hb is still almost fully saturated even if the blood P_{O_2} falls as much as 40%. This provides a margin of safety by ensuring near-normal O_2 delivery to the tissues despite a substantial reduction in arterial P_{O_2}.
  2. In the P_{O_2} range in the systemic capillaries (the steep portion of the curve), Hb unloading increases greatly in response to a small local decline in blood P_{O_2} associated with increased cellular metabolism. In this way, more O_2 is provided to match the increased tissue needs.
- Carbon dioxide picked up at the systemic capillaries is transported in the blood by three methods: (1) 10% is physically dissolved, (2) 30% is bound to Hb, and (3) 60% takes the form of bicarbonate (HCO_3^-).
- The erythrocyte enzyme carbonic anhydrase catalyzes conversion of CO_2 to HCO_3^- according to the reaction $CO_2 + H_2O \leftrightharpoons H_2CO_3 \leftrightharpoons H^+ + HCO_3^-$. The carbon and oxygen originally present in CO_2 are now part of the bicarbonate ion. The generated H^+ binds to Hb. These reactions are all reversed in the lungs as CO_2 is eliminated to the alveoli.

Control of Respiration

- Ventilation involves two distinct aspects, both subject to neural control: (1) rhythmic cycling between inspiration and expiration and (2) regulation of ventilation magnitude, which in turn depends on control of respiratory rate and depth of tidal volume.
- Respiratory rhythm is established by a complex neuronal network, the pre-Bötzinger complex, that displays pacemaker activity and drives the inspiratory neurons located in the dorsal respiratory group (DRG) of the respiratory control center in the medulla of the brain stem. When these inspiratory neurons fire, impulses ultimately reach the inspiratory muscles to bring about inspiration.

- When the inspiratory neurons cease firing, the inspiratory muscles relax and expiration takes place. If active expiration is to occur, the expiratory muscles are activated by output at this time from the medullary expiratory neurons located in the ventral respiratory group of the medullary respiratory control center.
- This basic rhythm is smoothed out by a balance of activity in the apneustic and pneumotaxic centers located higher in the brain stem in the pons. The apneustic center prolongs inspiration, whereas the more powerful pneumotaxic center limits inspiration.
- Three chemical factors play a role in determining magnitude of ventilation: P_{CO_2}, P_{O_2}, and H^+ concentration of the arterial blood.
- The dominant factor in the minute-to-minute regulation of ventilation is the arterial P_{CO_2}. An increase in arterial P_{CO_2} is the most potent chemical stimulus for increasing ventilation. Changes in arterial P_{CO_2} alter ventilation primarily by bringing about corresponding changes in the brain ECF H^+ concentration, to which the central chemoreceptors are exquisitely sensitive.
- The peripheral chemoreceptors are responsive to an increase in arterial H^+ concentration, which likewise reflexly brings about increased ventilation. The resultant adjustment in arterial H^+-generating CO_2 is important in maintaining the acid–base balance of the body.
- The peripheral chemoreceptors also reflexly stimulate the respiratory center in response to a marked reduction in arterial P_{O_2} (<60 mm Hg). This response serves as an emergency mechanism to increase respiration when the arterial P_{O_2} levels fall below the safety range provided by the plateau portion of the O_2–Hb curve.

REVIEW EXERCISES

Objective Questions (Answers on p. A-45)

1. Breathing is accomplished by alternate contraction and relaxation of muscles within the lung tissue. *(True or false?)*
2. The alveoli normally empty completely during maximal expiratory efforts. *(True or false?)*
3. Alveolar ventilation does not always increase when pulmonary ventilation increases. *(True or false?)*
4. Oxygen and CO_2 have equal diffusion coefficients. *(True or false?)*
5. Hemoglobin has a higher affinity for O_2 than for any other substance. *(True or false?)*
6. Rhythmicity of breathing is brought about by pacemaker activity displayed by the respiratory muscles. *(True or false?)*
7. The expiratory neurons send impulses to the motor neurons controlling the expiratory muscles during normal quiet breathing. *(True or false?)*
8. The three forces that tend to keep the alveoli open are _____ _____, and _____.
9. The two forces that promote alveolar collapse are _____, and _____.
10. _____ is a measure of the magnitude of change in lung volume accomplished by a given change in the transmural pressure gradient.
11. _____ is the phenomenon of the lungs snapping back to their resting size after having been stretched.
12. _____ is the erythrocytic enzyme that catalyzes the conversion of CO_2 into HCO_3^-.
13. Which of the following reactions take(s) place at the pulmonary capillaries?
    a. $Hb + O_2 \rightarrow HbO_2$
    b. $CO_2 + H_2O \rightarrow H_2CO_3 \rightarrow H^+ + HCO_3^-$
    c. $Hb + CO_2 \rightarrow HbCO_2$
    d. $HbH \rightarrow Hb + H^+$
14. Using the following answer code, indicate which chemoreceptors are being described:
    (a) peripheral chemoreceptors
    (b) central chemoreceptors
    (c) both peripheral and central chemoreceptors
    (d) neither peripheral nor central chemoreceptors
    1. stimulated by an arterial P_{O_2} of 80 mm Hg
    2. stimulated by an arterial P_{O_2} of 55 mm Hg
    3. directly depressed by an arterial P_{O_2} of 55 mm Hg
    4. weakly stimulated by an elevated arterial P_{CO_2}
    5. strongly stimulated by an elevated brain ECF H^+ concentration induced by an elevated arterial P_{CO_2}
    6. stimulated by an elevated arterial H^+ concentration
15. Indicate the O_2 and CO_2 partial-pressure relationships that are important in gas exchange by circling *(greater than)* >, *(less than)* <, or = *(equal to)* as appropriate in each of the following statements;
    a. P_{O_2} in blood entering the pulmonary capillaries is (>, <, or =) P_{O_2} in the alveoli.
    b. P_{CO_2} in blood entering the pulmonary capillaries is (>, <, or =) P_{CO_2} in the alveoli.
    c. P_{O_2} in the alveoli is (>, <, or =) P_{O_2} in blood leaving the pulmonary capillaries.
    d. P_{CO_2} in the alveoli is (>, <, or =) P_{CO_2} in blood leaving the pulmonary capillaries.
    e. P_{O_2} in blood leaving the pulmonary capillaries is (>, <, or =) P_{O_2} in blood entering the systemic capillaries.
    f. P_{CO_2} in blood leaving the pulmonary capillaries is (>, <, or =) P_{CO_2} in blood entering the systemic capillaries.
    g. P_{O_2} in blood entering the systemic capillaries is (>, <, or =) P_{O_2} in the tissue cells.
    h. P_{CO_2} in blood entering the systemic capillaries is (>, <, or =) P_{CO_2} in the tissue cells.
    i. P_{O_2} in the tissue cells is (>, <, or =) or approximately =) P_{O_2} in blood leaving the systemic capillaries.
    j. P_{CO_2} in the tissue cells is (>, <, or approximately =) P_{CO_2} in blood leaving the systemic capillaries.
    k. P_{O_2} in blood leaving the systemic capillaries is (>, <, or =) P_{O_2} in blood entering the pulmonary capillaries.
    l. P_{CO_2} in blood leaving the systemic capillaries is (>, <, or =) P_{CO_2} in blood entering the pulmonary capillaries.

Essay Questions

1. Distinguish between internal and external respiration. List the steps in external respiration.
2. Describe the components of the respiratory system. What is the site of gas exchange?
3. Compare atmospheric, intra-alveolar, and intrapleural pressures.
4. Why are the lungs normally stretched even during expiration?
5. Explain why air enters the lungs during inspiration and leaves during expiration.
6. Why is inspiration normally active and expiration normally passive?

7. Why does airway resistance become an important determinant of airflow rates in chronic obstructive pulmonary disease?
8. Explain pulmonary elasticity in terms of elastic recoil and compliance. What are the source and function of pulmonary surfactant?
9. Define the various lung volumes and capacities.
10. Compare pulmonary ventilation and alveolar ventilation. What is the consequence of anatomic and alveolar dead space?
11. What determines the partial pressures of a gas in air and in blood?
12. List the methods of O_2 and CO_2 transport in the blood.
13. What is the primary factor that determines the percent hemoglobin saturation? What is the significance of the plateau and the steep portions of the O_2–Hb dissociation curve?
14. How does hemoglobin promote the net transfer of O_2 from the alveoli to the blood?
15. Explain the Bohr and Haldane effects.
16. Define the following: *hypoxic hypoxia, anemic hypoxia, circulatory hypoxia, histotoxic hypoxia, hypercapnia, hypocapnia, hyperventilation, hypoventilation, hyperpnea, apnea,* and *dyspnea.*
17. What are the locations and functions of the three respiratory control centers? Distinguish between the DRG and the VRG.
18. What brain region establishes the rhythmicity of breathing?

Quantitative Exercises (Solutions on p. A-45)
1. The two curves in ● Figure 13-32, on p. 495, show partial pressures for O_2 and CO_2 at various alveolar ventilation rates. These curves can be calculated from the following two equations:

$$P_{AO_2} = P_{IO_2} - (V_{O_2}/V_A)\ 863 \text{ mm Hg}$$
$$P_{ACO_2} = (V_{CO_2}/V_A)\ 863 \text{ mm Hg}$$

In these equations, P_{AO_2} = the partial pressure of O_2 in the alveoli, P_{ACO_2} = the partial pressure CO_2 in the alveoli, P_{IO_2} = the partial pressure of O_2 in the inspired air, V_{O_2} = the rate of O_2 consumption by the body, V_{CO_2} = the rate of CO_2 production by the body, V_A = the rate of alveolar ventilation, and 863 mm Hg is a constant that accounts for atmospheric pressure and temperature.

John is in training for a marathon tomorrow and just ate a meal of pasta (assume this is pure carbohydrate, which is metabolized with an RQ of 1). His alveolar ventilation rate is 3.0 liters/min, and he is consuming O_2 at a rate of 300 ml/min. What is the value of John's P_{ACO_2}?
2. Assume you are flying in an airplane that is cruising at 18,000 feet, where the pressure outside the plane is 380 mm Hg.
   a. Calculate the partial pressure of O_2 in the air outside the plane, ignoring water vapor pressure.
   b. If the plane depressurized, what would be the value of your P_{AO_2}? Assume that the ratio of your O_2 consumption to ventilation was not changed (that is, equaled 0.06), and note that under these conditions the constant in the equation that accounts for atmospheric pressure and temperature decreases from 863 mm Hg to 431.5 mm Hg.
   c. Calculate your P_{ACO_2}, assuming that your CO_2 production and ventilation rates remained unchanged at 200 ml/min and 4.2 liters/min, respectively.
3. A student has a tidal volume of 350 ml. While breathing at a rate of 12 breaths/min, her alveolar ventilation is 80% of her pulmonary ventilation. What is her anatomic dead space volume?

POINTS TO PONDER

(Explanations on p. A-45)
1. Why is it important that airplane interiors are pressurized (that is, the pressure is maintained at sea-level atmospheric pressure even though the atmospheric pressure surrounding the plane is substantially lower)? Explain the physiological value of using O_2 masks if the pressure in the airplane interior cannot be maintained.
2. Would hypercapnia accompany the hypoxia produced in each of the following situations? Explain why or why not.
   a. cyanide poisoning
   b. pulmonary edema
   c. restrictive lung disease
   d. high altitude
   e. severe anemia
   f. congestive heart failure
   g. obstructive lung disease
3. If a person lives one mile above sea level at Denver, Colorado, where the atmospheric pressure is 630 mm Hg, what would the P_{O_2} of the inspired air be once it is humidified in the respiratory airways before it reaches the alveoli?
4. Based on what you know about the control of respiration, explain why it is dangerous to voluntarily hyperventilate to lower the arterial P_{CO_2} before going underwater. The purpose of the hyperventilation is to stay under longer before P_{CO_2} rises above normal and drives the swimmer to surface for a breath of air.

5. If a person whose alveolar membranes are thickened by disease has an alveolar P_{O_2} of 100 mm Hg and an alveolar P_{CO_2} of 40 mm Hg, which of the following values of systemic arterial blood gases are most likely to exist?
   a. P_{O_2} = 105 mm Hg P_{CO_2} = 35 mm Hg
   b. P_{O_2} = 100 mm Hg P_{CO_2} = 40 mm Hg
   c. P_{O_2} = 90 mm Hg P_{CO_2} = 45 mm Hg
   If the person is administered 100% O_2, will the arterial P_{O_2} increase, decrease, or remain the same? Will the arterial P_{CO_2} increase, decrease, or remain the same?
6. ***Clinical Consideration.*** Keith M., a former heavy cigarette smoker, has severe emphysema. What effect does this condition have on his airway resistance? How does this change in airway resistance influence Keith's inspiratory and expiratory efforts? Describe how his respiratory muscle activity and intra-alveolar pressure changes compare to normal to accomplish a normal tidal volume. How would his spirogram compare to normal? What influence would Keith's condition have on gas exchange in his lungs? What blood gas abnormalities are likely to be present? Would it be appropriate to administer O_2 to Keith to relieve his hypoxic condition?

PHYSIOEDGE RESOURCES

PHYSIOEDGE CD-ROM

Figures marked with the PhysioEdge icon have associated activities on the CD. For this chapter, check out:

Respiratory Mechanics

The diagnostic quiz allows you to receive immediate feedback on your understanding of the concept and to advance through various levels of difficulty.

CHECK OUT THESE MEDIA QUIZZES:

13.1 Anatomy of the Respiratory System
13.2 Mechanics of Ventilation
13.3 Gas Transport and Exchange
13.4 Control of Ventilation, Lung Volumes and Terms

PHYSIOEDGE WEB SITE

The Web site for this book contains a wealth of helpful study aids, as well as many ideas for further reading and research. Log on to:

http://www.brookscole.com

Go to the Biology page and select Sherwood's *Human Physiology*, 5th Edition. Select a chapter from the drop-down menu or click on one of these resource areas:

- **Case Histories** provide an introduction to the clinical aspects of human physiology. Check out:

  #1: Fighting for Every Breath
  #2: Asthma and Influenza: A Dangerous Combination
  #3: When Help Becomes Harm

- For 2-D and 3-D graphical illustrations and animations of physiological concepts, visit our **Visual Learning Resource.**

- Resources for study and review include the **Chapter Outline, Chapter Summary, Glossary,** and **Flash Cards.** Use our **Quizzes** to prepare for in-class examinations.

- On-line research resources to consult are **Hypercontents,** with current links to relevant Internet sites; **Internet Exercises** with starter URLs provided; and **InfoTrac Exercises.**

 For more readings, go to InfoTrac College Edition, your on-line research library, at:

http://infotrac.thomsonlearning.com

Urinary System

Body systems maintain homeostasis

Homeostasis
The urinary system contributes to homeostasis by helping regulate the volume, electrolyte composition and pH of the internal environment and by eliminating metabolic waste products.

Homeostasis is essential for survival of cells

Cells
The concentrations of salt, acids, and other electrolytes must be closely regulated because even small changes can have a profound impact on cell function. Also, the waste products that are continually generated by cells as they perform life-sustaining chemical reactions must be removed because these wastes are toxic if allowed to accumulate.

Cells make up body systems

The survival and proper functioning of cells depend on the maintenance of stable concentrations of salt, acids, and other electrolytes in the internal fluid environment. Cell survival also depends on continual removal of the toxic metabolic wastes that cells produce as they perform life-sustaining chemical reactions. The kidneys play a major role in maintaining homeostasis by regulating the concentration of many plasma constituents, especially electrolytes and water, and by eliminating all metabolic wastes (except CO_2, which is removed by the lungs). As plasma repeatedly filters through the kidneys, they retain constituents of value for the body and eliminate undesirable or excess materials in urine. Of special importance is the kidneys' ability to regulate the volume and osmolarity (solute concentration) of the internal fluid environment by controlling salt and water balance. Also crucial is their ability to help regulate pH by controlling elimination of acid and base in urine.

Chapter 14

The Urinary System

CONTENTS AT A GLANCE

INTRODUCTION

▊ Functions of the kidneys

▊ Anatomical considerations

▊ Basic renal processes

GLOMERULAR FILTRATION

▊ Properties of the glomerular membrane

▊ Forces responsible for glomerular filtration

▊ Magnitude and regulation of the GFR

▊ Renal blood flow; filtration fraction

TUBULAR REABSORPTION

▊ Transepithelial transport

▊ Active versus passive reabsorption

▊ Process and control of active Na^+ reabsorption

▊ Secondary active reabsorption of glucose and amino acids

▊ Tubular maximum; renal threshold

▊ Regulated reabsorption of PO_4^{3-} and Ca^{2+}

▊ Passive reabsorption of Cl^-, H_2O, and urea

TUBULAR SECRETION

▊ Hydrogen-ion secretion

▊ Potassium-ion secretion

▊ Organic anion and cation secretion

URINE EXCRETION AND PLASMA CLEARANCE

▊ Urine excretion rate

▊ Plasma clearance

▊ Excretion of urine of varying concentrations; medullary countercurrent system

▊ Vasopressin-controlled H_2O reabsorption

▊ Renal failure

▊ Micturition

INTRODUCTION

Exchanges between the cells and the ECF could notably alter the composition of the small, private internal fluid environment were it not for mechanisms that maintain its stability.

▊ The kidneys perform a variety of functions aimed at maintaining homeostasis.

The kidneys, in concert with the hormonal and neural inputs that control their function, are primarily responsible for maintaining the stability of ECF volume, electrolyte composition, and osmolarity (solute concentration). By adjusting the quantity of water and various plasma constituents that are either conserved for the body or eliminated in the urine, the kidneys can maintain water and electrolyte balance within the very narrow range compatible with life, despite a wide range of intake and losses of these constituents through other avenues. The kidneys not only adjust for wide variations in ingestion of H_2O, salt, and other electrolytes, but they also adjust urinary output of these ECF constituents to compensate for abnormal losses through heavy sweating, vomiting, diarrhea, or hemorrhage. Thus urine composition varies widely as the kidneys do what they can to maintain homeostasis.

When the ECF has a surplus of water or a particular electrolyte such as salt (NaCl), the kidneys can eliminate the excess in urine. If there is a deficit, the kidneys cannot actually provide additional quantities of the depleted constituent, but they can limit urinary losses of the material in short supply and thus conserve it until the person can ingest more of the depleted substance. Accordingly, the kidneys can compensate more efficiently for excesses than for deficits. In fact, in some instances the kidneys cannot completely halt the loss of a particular valuable substance in the urine, even though the substance may be in short supply. A prime example is the case of a water (H_2O) deficit. Even if a person is not consuming any H_2O, the kidneys must put out about half a liter of H_2O in the urine each day to accomplish another major role as the body's "cleaners."

In addition to the kidneys' important regulatory role in maintaining fluid and electrolyte balance, they are the primary route for eliminating potentially toxic metabolic wastes and foreign compounds from the body. These wastes cannot be eliminated as solids; they must be excreted in solution, thus obligating the kidneys to produce a minimum volume of around 500 ml of waste-filled urine per day. Because H_2O eliminated in urine is derived from the blood plasma, a person stranded without H_2O eventually urinates him- or herself to death: The plasma volume falls to a fatal level as H_2O is inexorably removed to accompany the wastes.

Overview of kidney functions

The kidneys perform the following specific functions, most of which help preserve the constancy of the internal fluid environment:

1. *Maintaining H_2O balance in the body.*
2. *Maintaining the proper osmolarity* of body fluids, primarily through regulating H_2O balance. This function is important to prevent osmotic fluxes into or out of the cells, which could lead to detrimental swelling or shrinking of the cells, respectively.
3. *Regulating the quantity and concentration of most ECF ions,* including Na^+, Cl^-, K^+, H^+, HCO_3^-, Ca^{2+}, Mg^{2+}, SO_4^{2-}, and PO_4^{3-}. Even minor fluctuations in the ECF concentrations of some of these electrolytes can have profound influences. For example, changes in the ECF concentration of K^+ can potentially lead to fatal cardiac dysfunction.
4. *Maintaining proper plasma volume,* thereby contributing significantly to the long-term regulation of arterial blood pressure. This function is accomplished through the kidneys' regulatory role in salt (Na^+ and Cl^-) and H_2O balance.
5. *Helping maintain the proper acid–base balance* of the body by adjusting urinary output of H^+ and HCO_3^-.
6. *Excreting (eliminating) the end products (wastes) of bodily metabolism* such as urea, uric acid, and creatinine. If allowed to accumulate, these wastes are toxic, especially to the brain.
7. *Excreting many foreign compounds* such as drugs, food additives, pesticides, and other exogenous nonnutritive materials that have entered the body.
8. *Producing erythropoietin,* a hormone that stimulates red blood cell production.
9. *Producing renin,* an enzymatic hormone that triggers a chain reaction important in salt conservation by the kidneys.
10. *Converting vitamin D into its active form.*

▌ The kidneys form the urine; the remainder of the urinary system carries the urine to the outside.

The **urinary system** consists of the urine-forming organs—the **kidneys**—and the structures that carry the urine from the kidneys to the outside for elimination from the body (● Figure 14-1a). The kidneys are a pair of bean-shaped organs that lie in the back of the abdominal cavity, one on each side of the vertebral column, slightly above the waistline. Each kidney is supplied by a **renal artery** and a **renal vein,** which, respectively, enter and leave the kidney at the medial indentation that gives

this organ its beanlike form. The kidney acts on the plasma flowing through it to produce urine, conserving materials to be retained in the body and eliminating unwanted materials into the urine.

After urine is formed, it drains into a central collecting cavity, the **renal pelvis,** located at the medial inner core of each kidney (● Figure 14-1b). From there urine is channeled into the **ureter,** a smooth muscle-walled duct that exits at the medial border in close proximity to the renal artery and vein. There are two ureters, one carrying urine from each kidney to the single urinary bladder.

The **urinary bladder,** which temporarily stores urine, is a hollow, distensible sac whose volume can be adjusted by varying the contractile state of the smooth muscle within its walls. Periodically, urine is emptied from the bladder to the outside through another tube, the **urethra.** The urethra in females is straight and short, passing directly from the neck of the bladder to the outside (● Figure 14-2; see also ● Figure 20-2, p. 752). In males the urethra is much longer and follows a curving course from the bladder to the outside, passing through both the prostate gland and the penis (● Figures 14-1a and 14-2b; see also ● Figure 20-1, p. 751). The male urethra serves the dual function of providing both a route for eliminating urine from the bladder and a passageway for semen from the reproductive organs. The prostate gland lies below the neck of the bladder and completely encircles the urethra. Prostatic enlargement, which often occurs during middle to older age, can partially or completely occlude the urethra, thereby impeding the flow of urine.

The parts of the urinary system beyond the kidneys merely serve as ductwork to transport urine to the outside. Once formed by the kidneys, urine is not altered in composition or volume as it moves downstream through the remainder of the urinary system.

▌ The nephron is the functional unit of the kidney.

Each kidney is composed of about 1 million microscopic functional units known as **nephrons,** which are bound together by connective tissue. Recall that a functional unit is the smallest unit within an organ capable of performing all of that organ's functions. Because the primary function of the kidneys is to produce urine and, in so doing, maintain constancy in the ECF composition, a nephron is the smallest unit capable of urine formation.

The arrangement of nephrons within the kidneys gives rise to two distinct regions—an outer region called the **renal cortex,** which appears granular, and an inner region known as the **renal medulla,** which is made up of striated triangles, the **renal pyramids** (● Figure 14-1b).

Knowledge of the structural arrangement of an individual nephron is essential for understanding the distinction between the cortical and medullary regions of the kidney and, more importantly, for understanding renal function. Each nephron consists of a *vascular component* and a *tubular component,*

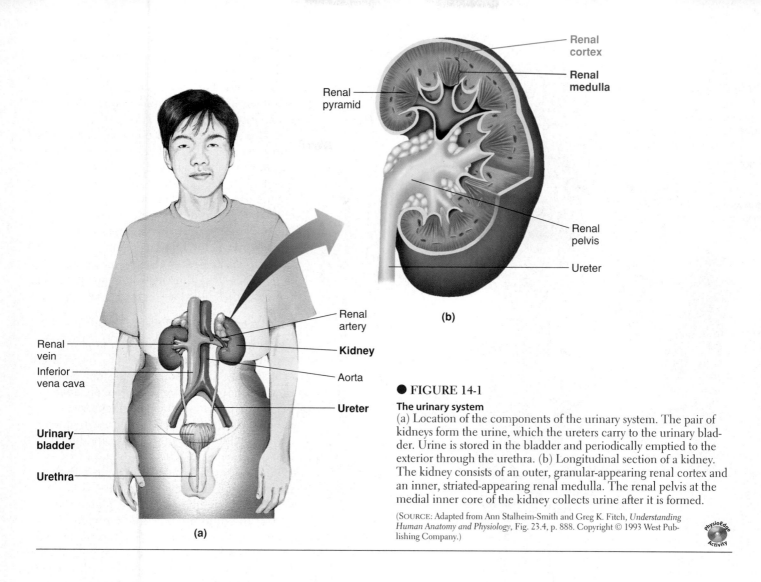

(a)

Renal cortex

Renal medulla

Renal pyramid

Renal pelvis

Ureter

(b)

Renal artery

Kidney

Aorta

Ureter

Renal vein

Inferior vena cava

Urinary bladder

Urethra

● **FIGURE 14-1**

The urinary system
(a) Location of the components of the urinary system. The pair of kidneys form the urine, which the ureters carry to the urinary bladder. Urine is stored in the bladder and periodically emptied to the exterior through the urethra. (b) Longitudinal section of a kidney. The kidney consists of an outer, granular-appearing renal cortex and an inner, striated-appearing renal medulla. The renal pelvis at the medial inner core of the kidney collects urine after it is formed.

(SOURCE: Adapted from Ann Stalheim-Smith and Greg K. Fitch, *Understanding Human Anatomy and Physiology*, Fig. 23.4, p. 888. Copyright © 1993 West Publishing Company.)

both of which are intimately related structurally and functionally (● Figure 14-3).

Vascular component of the nephron

The dominant portion of the vascular component of the nephron is the **glomerulus,** a ball-like tuft of capillaries through which part of the water and solutes is filtered from the blood passing through. This filtered fluid, which is almost identical in composition to the plasma, then passes through the tubular component of the nephron, where it is modified by various transport processes that convert it into urine.

On entering the kidney, the renal artery systematically subdivides to ultimately form many small vessels known as **afferent arterioles,** one of which supplies each nephron. Within the nephron, the afferent arteriole delivers blood to the glomerulus. The glomerular capillaries rejoin to form another arteriole, the **efferent arteriole,** through which blood that was not filtered into the tubular component leaves the glomerulus (● Figure 14-4). The efferent arterioles are the only arterioles in the body that drain from capillaries. Typically, arterioles break up into capillaries that rejoin to form venules. At the glomerular capillaries, no O_2 or nutrients are extracted from the

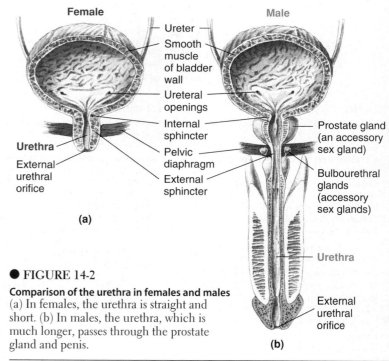

Female   **Male**

Ureter

Smooth muscle of bladder wall

Ureteral openings

Internal sphincter

Prostate gland (an accessory sex gland)

Urethra

External urethral orifice

Pelvic diaphragm

External sphincter

Bulbourethral glands (accessory sex glands)

(a)

Urethra

External urethral orifice

(b)

● **FIGURE 14-2**

Comparison of the urethra in females and males
(a) In females, the urethra is straight and short. (b) In males, the urethra, which is much longer, passes through the prostate gland and penis.

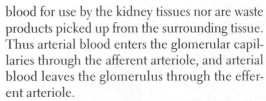

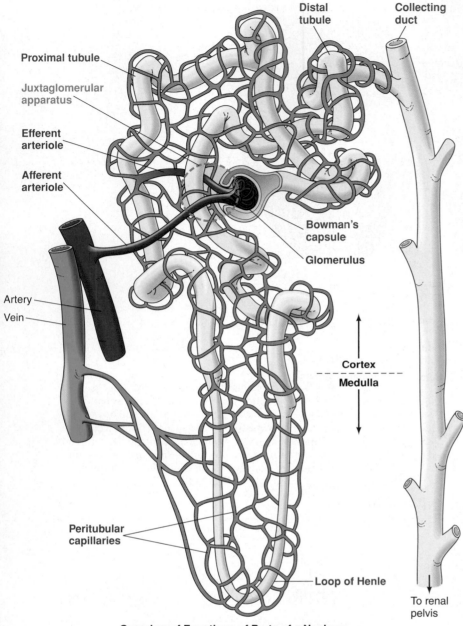

Overview of Functions of Parts of a Nephron

Vascular component
- Afferent arteriole—carries blood to the glomerulus
- Glomerulus—a tuft of capillaries that filters a protein-free plasma into the tubular component
- Efferent arteriole—carries blood from the glomerulus
- Peritubular capillaries—supply the renal tissue; involved in exchanges with the fluid in the tubular lumen

Combined vascular/tubular component
- Juxtaglomerular apparatus—produces substances involved in the control of kidney function

Tubular component
- Bowman's capsule—collects the glomerular filtrate
- Proximal tubule—uncontrolled reabsorption and secretion of selected substances occur here
- Loop of Henle—establishes an osmotic gradient in the renal medulla that is important in the kidney's ability to produce urine of varying concentration
- Distal tubule and collecting duct—variable, controlled reabsorption of Na^+ and H_2O and secretion of K^+ and H^+ occur here; fluid leaving the collecting duct is urine, which enters the renal pelvis

● **FIGURE 14-3**

A nephron
A schematic representation of a cortical nephron, the most abundant type of nephron in humans.

blood for use by the kidney tissues nor are waste products picked up from the surrounding tissue. Thus arterial blood enters the glomerular capillaries through the afferent arteriole, and arterial blood leaves the glomerulus through the efferent arteriole.

The efferent arteriole quickly subdivides into a second set of capillaries, the **peritubular capillaries,** which supply the renal tissue with blood and are important in exchanges between the tubular system and blood during conversion of the filtered fluid into urine. These peritubular capillaries, as their name implies, are intertwined around the tubular system (*peri* means "around"). The peritubular capillaries rejoin to form venules that ultimately drain into the renal vein, by which blood leaves the kidney.

Tubular component of the nephron

The tubular component of each nephron is a hollow, fluid-filled tube formed by a single layer of epithelial cells. Even though the tubule is continuous from its beginning in close proximity to the glomerulus to its ending at the renal pelvis, it is arbitrarily divided into various segments based on differences in structure and function that occur along its length (● Figures 14-3 and 14-5). The tubular component begins with **Bowman's capsule,** an expanded, double-walled invagination that cups around the glomerulus to collect the fluid filtered from the glomerular capillaries. From Bowman's capsule, the filtered fluid passes into the **proximal tubule,** which lies entirely within the cortex and is highly coiled or convoluted throughout much of its course. The next segment, the **loop of Henle,** forms a sharp **U**-shaped or hairpin loop that dips into the renal medulla. The *descending limb* of Henle's loop plunges from the cortex into the medulla; the *ascending limb* traverses back up into the cortex. The ascending limb returns to the glomerular region of its own nephron, where it passes through the fork formed by the afferent and efferent arterioles. Both the tubular and vascular cells at this point are specialized to form the **juxtaglomerular apparatus,** a structure that lies next to the glomerulus (*juxta* means "next to"). This specialized region plays an important role in regulating kidney function. Beyond the juxtaglomerular apparatus, the tubule once again becomes highly coiled to form the **distal tubule,** which also lies entirely within the cortex. The distal tubule empties into a **collecting duct** or **tubule,** with each collecting duct draining fluid from up to eight separate nephrons. Each collecting duct plunges down through the medulla to empty its fluid contents (which have now been converted into urine) into the renal pelvis.

Cortical and juxtamedullary nephrons

The two types of nephrons—*cortical nephrons* and *juxtamedullary nephrons*—are distinguished by the location and length of some of their structures (● Figure 14-5). All nephrons originate in the cortex, but the glomeruli of **cortical nephrons** lie in the outer layer of the cortex, whereas the glomeruli of **juxtamedullary nephrons** lie in the inner layer of the cortex, adjacent to the medulla. (Note the distinction between *juxtamedullary* nephrons and *juxtaglomerular* apparatus.) The presence of all glomeruli and associated Bowman's capsules in the cortex is responsible for this region's granular appearance. These two nephron types differ most markedly in their loops of Henle. The hairpin loop of cortical nephrons dips only slightly into the medulla. In contrast, the loop of juxtamedullary nephrons plunges through the entire depth of the medulla. Furthermore, the peritubular capillaries of juxtamedullary nephrons form hairpin vascular loops known as **vasa recta** ("straight vessels"), which run in close association with the long loops of Henle. In cortical nephrons, the peritubular capillaries do not form vasa recta but instead entwine around these nephrons'

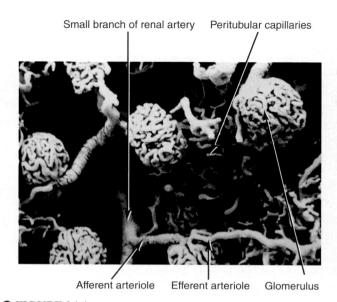

● **FIGURE 14-4**

Scanning electron micrograph of a glomerulus and associated arterioles

● **FIGURE 14-5**

Comparison of juxtamedullary and cortical nephrons

The glomeruli of cortical nephrons lie in the outer cortex, whereas the glomeruli of juxtamedullary nephrons lie in the inner part of the cortex next to the medulla. The loops of Henle of cortical nephrons dip only slightly into the medulla, but the juxtamedullary nephrons have long loops of Henle that plunge deep into the medulla. The juxtamedullary nephrons' peritubular capillaries form hairpin loops known as *vasa recta*.

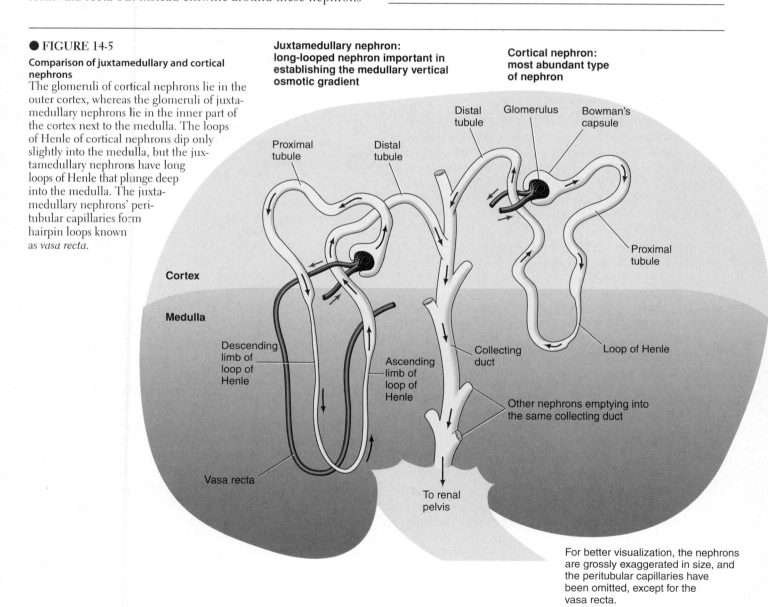

Juxtamedullary nephron:
long-looped nephron important in establishing the medullary vertical osmotic gradient

Cortical nephron:
most abundant type of nephron

For better visualization, the nephrons are grossly exaggerated in size, and the peritubular capillaries have been omitted, except for the vasa recta.

short loops of Henle. As they course through the medulla, the collecting ducts of both cortical and juxtamedullary nephrons run parallel to the ascending and descending limbs of the juxtamedullary nephrons' long loops of Henle and vasa recta. This parallel arrangement of tubules and vessels creates the medullary tissue's striated appearance. More importantly, as you will see, this arrangement—coupled with the permeability and transport characteristics of the long loops of Henle and vasa recta—plays a key role in the kidneys' ability to produce urine of varying concentrations, depending on the needs of the body. About 80% of the nephrons in humans are of the cortical type. Species with greater urine-concentrating abilities than humans, such as the desert rat, have a greater proportion of juxtamedullary nephrons.

▌ The three basic renal processes are glomerular filtration, tubular reabsorption, and tubular secretion.

Three basic processes are involved in the formation of urine: *glomerular filtration*, *tubular reabsorption*, and *tubular secretion*. To aid in visualizing the relationships among these renal processes, it is useful to unwind the nephron schematically, as in ● Figure 14-6.

Process of glomerular filtration

As blood flows through the glomerulus, filtration of protein-free plasma occurs through the glomerular capillaries into Bowman's capsule. Normally, about 20% of the plasma that enters the glomerulus is filtered. This process, known as **glomerular filtration,** is the first step in urine formation. On average, 125 ml of glomerular filtrate (filtered fluid) are formed collectively through all the glomeruli each minute. This amounts to 180 liters (about 47.5 gallons) each day. Considering that the average plasma volume in an adult is 2.75 liters, this means that the entire plasma volume is filtered by the kidneys about 65 times per day. If everything filtered were to pass out in the urine, the total plasma volume would be urinated in less than half an hour! This does not happen, however, because the kidney tubules and peritubular capillaries are intimately related throughout their lengths, so that materials can be transferred between the fluid inside the tubules and the blood within the peritubular capillaries.

Process of tubular reabsorption

As the filtrate flows through the tubules, substances of value to the body are returned to the peritubular capillary plasma. This selective movement of substances from inside the tubule (the tubular lumen) into the blood is referred to as **tubular reabsorption.** Reabsorbed substances are not lost from the body in the urine but are carried instead by the peritubular capillaries to the venous system and then to the heart to be recirculated again. Of the 180 liters of plasma filtered per day, 178.5 liters on the average are reabsorbed. The remaining 1.5 liters left in the tubules passes into the renal pelvis to be eliminated as urine. In general, substances that the body needs to conserve are selec-

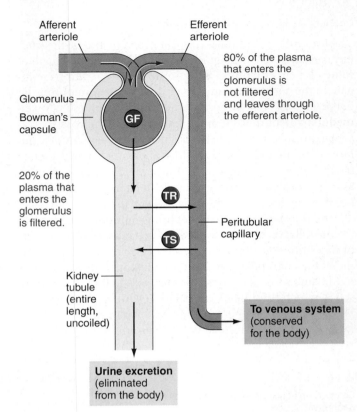

GF = Glomerular filtration—nondiscriminant filtration of a protein-free plasma from the glomerulus into Bowman's capsule

TR = Tubular reabsorption—selective movement of filtered substances from the tubular lumen into the peritubular capillaries

TS = Tubular secretion—selective movement of nonfiltered substances from the peritubular capillaries into the tubular lumen

● FIGURE 14-6

Basic renal processes
Anything filtered or secreted but not reabsorbed is excreted in the urine and lost from the body. Anything filtered and subsequently reabsorbed, or not filtered at all, enters the venous blood and is saved for the body.

tively reabsorbed, whereas unwanted substances that must be eliminated remain in the urine.

Process of tubular secretion

The third renal process, **tubular secretion,** is the selective transfer of substances from the peritubular capillary blood into the tubular lumen. It provides a second route for substances to enter the renal tubules from the blood, the first being by glomerular filtration. Only about 20% of the plasma flowing through the glomerular capillaries is filtered into Bowman's capsule; the remaining 80% flows on through the efferent arteriole into the peritubular capillaries. A few substances may be discriminately transferred by tubular secretion from the plasma in the peritubular capillaries into the tubular lumen. Tubular secretion provides a mechanism for more rapidly eliminating selected substances from the plasma by extracting an additional quantity of a particular substance from the 80% of unfiltered plasma in

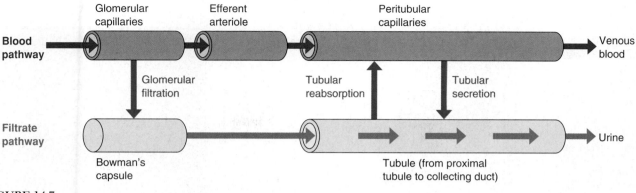

● FIGURE 14-7

Pathways traveled by blood and filtrate as urine is formed in the nephron

the peritubular capillaries and by adding it to the quantity of the substance already present in the tubule as a result of filtration.

Urine excretion

Urine excretion is the elimination of substances from the body in urine. It is not really a separate process but rather is the result of the first three processes. All plasma constituents that are filtered or secreted but are not reabsorbed remain in the tubules and pass into the renal pelvis to be excreted as urine and eliminated from the body (● Figures 14-6 and 14-7). (Do not confuse *excretion* with *secretion*.) Note that anything filtered and subsequently reabsorbed, or not filtered at all, enters the venous blood from the peritubular capillaries and thus is conserved for the body instead of being excreted in urine, despite passing through the kidneys.

The "big picture" of the basic renal processes

Glomerular filtration is largely an indiscriminate process. With the exception of blood cells and plasma proteins, all constituents within the blood—H_2O, nutrients, electrolytes, wastes, and so on—nonselectively enter the tubular lumen as a bulk unit during filtration. That is, of the 20% of the plasma that is filtered at the glomerulus, everything present in that portion of plasma enters Bowman's capsule except for the plasma proteins. The highly discriminating tubular processes then go to work on the filtrate to return to the blood a fluid of the composition and volume necessary to maintain the constancy of the internal fluid environment. The unwanted filtered material is left behind in the tubular fluid to be excreted as urine. Glomerular filtration can be thought of as pushing a portion of plasma, with all its essential components as well as those that need to be eliminated from the body, onto a tubular "conveyor belt" that terminates at the renal pelvis, which is the collecting point for urine within the kidney. All plasma constituents that enter this conveyor belt and are not subsequently returned to the plasma by the end of the line are spilled out of the kidney as urine. It is up to the tubular system to salvage by reabsorption the filtered materials that need to be preserved for the body, while leaving behind substances that must be excreted. In addition, some substances are not only filtered but

are also secreted onto the tubular conveyor belt, so that the amounts of these substances excreted in urine are greater than the amounts that were filtered. For many substances, these renal processes are subject to physiologic control. Thus the kidneys handle each constituent in the plasma in a characteristic manner by a particular combination of filtration, reabsorption, and secretion.

The kidneys act only on the plasma, yet ECF consists of both plasma and interstitial fluid. The interstitial fluid is actually the true internal fluid environment of the body, because it is the only component of the ECF that comes into direct contact with the cells. However, because of the free exchange between plasma and interstitial fluid across the capillary walls (with the exception of plasma proteins), interstitial fluid composition reflects the composition of plasma. Thus, by performing their regulatory and excretory roles on the plasma, the kidneys maintain the proper interstitial fluid environment for optimal cell function. Most of the remainder of this chapter will be devoted to considering how the basic renal processes are accomplished and the mechanisms by which they are carefully regulated to help maintain homeostasis.

GLOMERULAR FILTRATION

Fluid filtered from the glomerulus into Bowman's capsule must pass through the three layers that make up the **glomerular membrane** (● Figure 14-8): (1) the wall of the glomerular capillaries, (2) the basement membrane, and (3) the inner layer of Bowman's capsule. Collectively, these layers function as a fine molecular sieve that retains the blood cells and plasma proteins but permits H_2O and solutes of small molecular dimension to filter through. Let's consider each layer in more detail.

▌ The glomerular membrane is considerably more permeable than capillaries elsewhere.

The *glomerular capillary wall* consists of a single layer of flattened endothelial cells. It is perforated by many large pores that make it over 100 times more permeable to H_2O and solutes than capillaries elsewhere in the body.

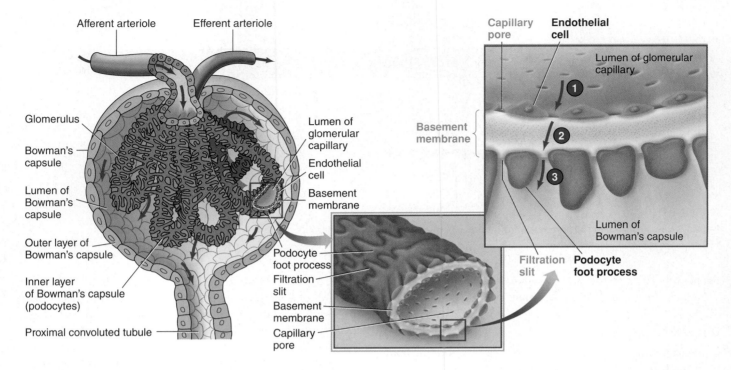

To be filtered, a substance must pass through

1 the pores between the endothelial cells of the glomerular capillary

2 an acellular basement membrane

3 the filtration slits between the foot processes of the podocytes of the inner layer of Bowman's capsule

● **FIGURE 14-8**
Layers of the glomerular membrane

The *basement membrane* is an acellular gelatinous layer composed of collagen and glycoproteins that is sandwiched between the glomerulus and Bowman's capsule. The collagen provides structural strength, and the glycoproteins discourage the filtration of small plasma proteins. The larger plasma proteins cannot be filtered, because they cannot fit through the capillary pores, but the pores are just barely large enough to permit passage of albumin, the smallest of plasma proteins. However, because the glycoproteins are negatively charged, they repel albumin and other plasma proteins, which are also negatively charged. Therefore, plasma proteins are almost completely excluded from the filtrate, with less than 1% of the albumin molecules escaping into Bowman's capsule. Some renal diseases characterized by excessive albumin in the urine (*albuminuria*) are due to disruption of the negative charges within the glomerular membrane, which causes the membrane to be more permeable to albumin even though the size of the pores remains constant.

The final layer of the glomerular membrane is the *inner layer of Bowman's capsule*. It consists of **podocytes,** octopus-like cells that encircle the glomerular tuft. Each podocyte bears many elongated foot processes (*podo* means "foot") that interdigitate with foot processes of adjacent podocytes, much as you interlace your fingers between each other when you cup your hands around a ball (● Figure 14-9). The narrow slits between adjacent foot processes, known as **filtration slits,** provide a pathway through which fluid exiting the glomerular capillaries can enter the lumen of Bowman's capsule.

Thus the route that filtered substances take across the glomerular membrane is completely extracellular—first through capillary pores, then through the acellular basement membrane, and finally through capsular filtration slits (● Figure 14-8).

■ **The glomerular capillary blood pressure is the major force that induces glomerular filtration.**

To accomplish glomerular filtration, a force must drive a portion of the plasma in the glomerulus through the openings in the glomerular membrane. No active transport mechanisms or local energy expenditures are involved in moving fluid from the plasma across the glomerular membrane into Bowman's capsule. Passive physical forces similar to those acting across capillaries elsewhere are responsible for glomerular filtration. Because the glomerulus is a tuft of capillaries, the same principles of fluid dynamics apply here that are responsible for ultrafiltration across other capillaries (see p. 365), except for two important differences: (1) The glomerular capillaries are much more permeable than capillaries elsewhere, so more fluid is filtered for a given filtration pressure; and (2) the balance of

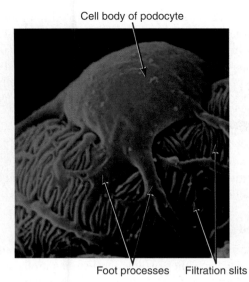

Cell body of podocyte

Foot processes   Filtration slits

● FIGURE 14-9

Bowman's capsule podocytes with foot processes and filtration slits
Note the filtration slits between adjacent foot processes on this scanning electron micrograph. The podocytes and their foot processes encircle the glomerular capillaries.

forces across the glomerular membrane is such that filtration occurs throughout the entire length of the capillaries. In contrast, the balance of forces in other capillaries shifts, so that filtration occurs in the beginning portion of the vessel but reabsorption occurs toward the vessel's end.

Forces involved in glomerular filtration

Three physical forces are involved in glomerular filtration (▲ Table 14-1): glomerular capillary blood pressure, plasma-colloid osmotic pressure, and Bowman's capsule hydrostatic pressure. Let's examine the role of each.

1. The *glomerular capillary blood pressure* is the fluid pressure exerted by the blood within the glomerular capillaries. It ultimately depends on contraction of the heart (the source of energy that produces glomerular filtration) and the resistance to blood flow offered by the afferent and efferent arterioles. The glomerular capillary blood pressure, at an estimated average value of 55 mm Hg, is higher than capillary blood pressure elsewhere. The reason for the higher pressure in the glomerular capillaries is the larger diameter of the afferent arteriole compared to the efferent arteriole. Because blood can more readily enter the glomerulus through the wide afferent arteriole than it can leave through the narrower efferent arteriole, glomerular capillary blood pressure is maintained high as a result of blood damming up in the glomerular capillaries. Furthermore, because of the high resistance offered by the efferent arterioles, blood pressure does not have the same tendency to decrease along the length of the glomerular capillaries as it does along other capillaries. This elevated, nondecremental glomerular blood pressure tends to push fluid out of the glomerulus into Bowman's capsule along the glomerular capillaries' entire length, and it is the major force responsible for producing glomerular filtration.

Force	Effect	Magnitude (mm Hg)
Glomerular Capillary Blood Pressure	Favors filtration	55
Plasma-Colloid Osmotic Pressure	Opposes filtration	30
Bowman's Capsule Hydrostatic Pressure	Opposes filtration	15
Net Filtration Pressure (Difference Between Force Favoring Filtration and Forces Opposing Filtration)	Favors filtration	10

$$55 - (30 + 15) = 10$$

Whereas glomerular capillary blood pressure *favors* filtration, the two other forces acting across the glomerular membrane (plasma-colloid osmotic pressure and Bowman's capsule hydrostatic pressure) *oppose* filtration.

2. *Plasma-colloid osmotic pressure* is caused by the unequal distribution of plasma proteins across the glomerular membrane. Because plasma proteins cannot be filtered, they are present in the glomerular capillaries but are absent in Bowman's capsule. Accordingly, the concentration of H_2O is higher in Bowman's capsule than in the glomerular capillaries. The resultant tendency for H_2O to move by osmosis down its own concentration gradient from Bowman's capsule into the glomerulus opposes glomerular filtration. This opposing osmotic force averages 30 mm Hg, which is slightly higher than across other capillaries. It is higher because considerably more H_2O is filtered out of the glomerular blood, so the concentration of plasma proteins is higher than elsewhere.

3. *Bowman's capsule hydrostatic pressure,* the pressure exerted by the fluid in this initial part of the tubule, is estimated to be about 15 mm Hg. This pressure, which tends to push fluid out of Bowman's capsule, opposes the filtration of fluid from the glomerulus into Bowman's capsule.

Glomerular filtration rate

As can be seen in ▲ Table 14-1, the forces acting across the glomerular membrane are not in balance. The total force favoring filtration is attributable to the glomerular capillary blood pressure at 55 mm Hg. The total of the two forces opposing filtration is 45 mm Hg. The net difference favoring filtration (10 mm Hg of pressure) is referred to as the **net filtration pressure.** This modest pressure is responsible for forcing large volumes of fluid from the blood through the highly permeable glomerular membrane. The actual rate of filtration, the **glomerular filtration rate (GFR)**, depends not only on the net filtration pressure but also on how much glomerular surface area is available for penetration and how permeable the glomerular membrane is (that is, how "holey" it is). These properties of the glomerular membrane are collectively referred to as the **filtration coefficient (K_f)**. Accordingly,

$$GFR = K_f \times \text{net filtration pressure}$$

Normally, about 20% of the plasma that enters the glomerulus is filtered at the net filtration pressure of 10 mm Hg, producing collectively through all glomeruli 180 liters of glomerular filtrate each day for an average GFR of 125 ml/min in males and 160 liters of filtrate per day for an average GFR of 115 ml/min in females.

■ Changes in the GFR occur primarily as a result of changes in glomerular capillary blood pressure.

Because the net filtration pressure responsible for inducing glomerular filtration is simply due to an imbalance of opposing physical forces between the glomerular capillary plasma and Bowman's capsule fluid, alterations in any of these physical forces can affect the GFR. We will examine the effect that changes in each of these physical forces have on the GFR.

Unregulated influences on the GFR

Plasma-colloid osmotic pressure and Bowman's capsule hydrostatic pressure are not subject to regulation and under normal conditions do not vary substantially. However, they can change pathologically and thus inadvertently affect the GFR. Because plasma-colloid osmotic pressure opposes filtration, a decrease in plasma protein concentration, by reducing this pressure, leads to an increase in the GFR. An uncontrollable reduction in plasma protein concentration might occur, for example, in severely burned patients who lose a large quantity of protein-rich, plasma-derived fluid through the exposed burned surface of their skin. Conversely, in situations in which the plasma-colloid osmotic pressure is elevated, such as in cases of dehydrating diarrhea, the GFR is reduced.

Bowman's capsule hydrostatic pressure can become uncontrollably elevated, and filtration subsequently can decrease, in the presence of a urinary tract obstruction, such as a kidney stone or prostatic enlargement. A damming up of fluid behind the obstruction elevates capsular hydrostatic pressure.

Controlled adjustments in the GFR

Unlike plasma-colloid osmotic pressure and Bowman's capsule hydrostatic pressure—which may be uncontrollably altered in various disease states and, thereby inappropriately alter the GFR—glomerular capillary blood pressure can be controlled to adjust the GFR to suit the body's needs. Assuming that all other factors remain constant, as the glomerular capillary blood pressure goes up, the net filtration pressure increases and the GFR increases correspondingly. The magnitude of the glomerular capillary blood pressure depends on the rate of blood flow within each of the glomeruli. The amount of blood flowing into a glomerulus per minute is determined largely by the magnitude of the mean systemic arterial blood pressure and the resistance offered by the nephron's arterioles. If resistance increases in the *afferent* arteriole, less blood flows into the glomerulus, resulting in a *decrease* in the GFR. However, if resistance increases in the *efferent* arteriole, blood dams up even more in the glomerulus behind the narrowed exit route. As a result, GFR *increases*. Two major control mechanisms regulate the GFR, both directed at adjusting glomerular blood flow by regulating caliber and thus resistance of the afferent arteriole. These mechanisms are (1) autoregulation, which is aimed at preventing spontaneous changes in GFR, and (2) extrinsic sympathetic control, which is aimed at long-term regulation of arterial blood pressure. Adjustments in efferent arteriolar resistance contribute to a smaller extent to the regulation of GFR and will not be considered.

Mechanisms responsible for autoregulation of GFR

Because the arterial blood pressure is the force that drives blood into the glomerulus, the glomerular capillary blood pressure and, accordingly, the GFR would increase in direct proportion to an increase in arterial pressure if everything else remained constant (● Figure 14-10). Similarly, a fall in arterial blood pressure would be accompanied by a decline in GFR. Such spontaneous, inadvertent changes in GFR are largely prevented by intrinsic regulatory mechanisms initiated by the kidneys themselves, a process known as **autoregulation** (*auto* means "self"). The kidneys can, within limits, maintain a constant blood flow into the glomerular capillaries (and thus a constant glomerular capillary

● FIGURE 14-10

Direct effect of arterial blood pressure on the glomerular filtration rate (GFR)

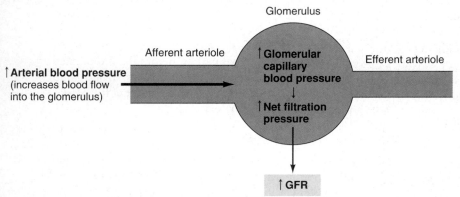

blood pressure and a stable GFR) despite changes in the driving arterial pressure. They do so primarily by altering afferent arteriolar caliber, thereby adjusting resistance to flow through these vessels. For example, if the GFR increases as a direct result of a rise in arterial pressure, the net filtration pressure and GFR can be reduced to normal by constriction of the afferent arteriole, which decreases the flow of blood into the glomerulus (● Figure 14-11a). This local adjustment lowers the glomerular blood pressure and the GFR to normal.

Conversely, when GFR falls in the presence of a decline in arterial pressure, glomerular pressure can be increased to normal by vasodilation of the afferent arteriole, which allows more blood to enter despite the reduction in driving pressure (● Figure 14-11b). The resultant buildup of glomerular blood volume increases glomerular blood pressure, which in turn brings the GFR back up to normal.

The exact mechanisms responsible for accomplishing autoregulatory responses are not completely understood. Currently, two intrarenal mechanisms are thought to contribute to autoregulation: (1) a *myogenic* mechanism, which responds to changes in pressure within the nephron's vascular component; and (2) a *tubuloglomerular feedback mechanism*, which senses changes in flow through the nephron's tubular component.

• The **myogenic mechanism** is a common property of vascular smooth muscle (*myogenic* means "muscle produced"). Arteriolar vascular smooth muscle contracts inherently in response to the stretch accompanying increased pressure within the vessel (see p. 356). Accordingly, the afferent arteriole automatically constricts on its own when it is stretched because of an increased arterial driving pressure. This response helps limit blood flow into the glomerulus to normal despite the elevated arterial pressure. Conversely, inherent relaxation of an unstretched afferent arteriole when pressure within the vessel is reduced increases blood flow into the glomerulus despite the fall in arterial pressure.

• The **tubuloglomerular feedback mechanism** involves the *juxtaglomerular apparatus*, which is the specialized combination of tubular and vascular cells where the tubule, after having bent back on itself, passes through the angle formed by the afferent and efferent arterioles as they join the glomerulus (● Figures 14-3 and 14-12). The smooth muscle cells within the wall of the afferent arteriole in this region are specialized to form **granular cells,** so called because they contain many secretory granules. Specialized tubular cells in this region are collectively known as the **macula densa.** The macula densa cells detect changes in the rate at which fluid is flowing past them through the tubule.

If the GFR is increased secondary to an elevation in arterial pressure, more fluid than normal is filtered and reaches the distal tubule (● Figure 14-13). In response, the macula densa cells trigger the release of locally acting vasoactive chemicals from the juxtaglomerular apparatus, which in turn cause the adjacent afferent arteriole to constrict, reducing glomerular blood flow and returning GFR to normal. The exact nature of these locally released vasoactive chemicals is currently unclear. Several chemicals have been identified, some of which are vasoconstrictors (for example, endothelin) and others va-

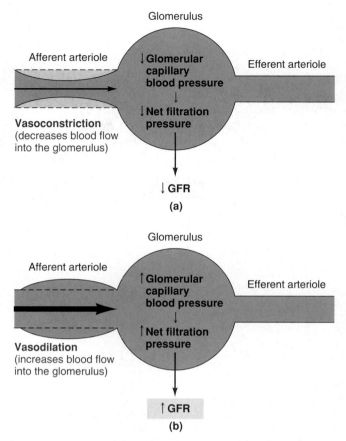

● **FIGURE 14-11**
Adjustments of afferent arteriole caliber to alter the GFR
(a) Arteriolar adjustment to reduce the GFR. (b) Arteriolar adjustment to increase the GFR.

sodilators (for example, bradykinin), but their exact contributions are yet to be determined. In the opposite situation, when the macula densa cells detect that the flow rate of fluid through the tubules is low because of a spontaneous decline in GFR accompanying a fall in arterial pressure, these cells bring about afferent arteriolar vasodilation by altering the rate of secretion of the relevant vasoactive chemicals. The resultant increase in glomerular flow rate restores the GFR to normal. Thus, by means of the juxtaglomerular apparatus, the tubule of a nephron is able to monitor the rate of movement of fluid through it and adjust the GFR accordingly. This tubuloglomerular feedback mechanism is initiated by the tubule to help each nephron regulate the rate of filtration through its own glomerulus.

Importance of autoregulation of the GFR

The myogenic and tubuloglomerular feedback mechanisms work in unison to autoregulate the GFR within the mean arterial blood pressure range of 80 to 180 mm Hg. Within this wide range, intrinsic autoregulatory adjustments of afferent arteriolar resistance can compensate for changes in arterial pressure, thus preventing inappropriate fluctuations in GFR, even though glomerular pressure tends to change in the same direction as arterial pressure. Normal mean arterial pressure is 93 mm Hg, so this range encompasses the transient changes in blood pressure that accompany daily activities unrelated to the need for

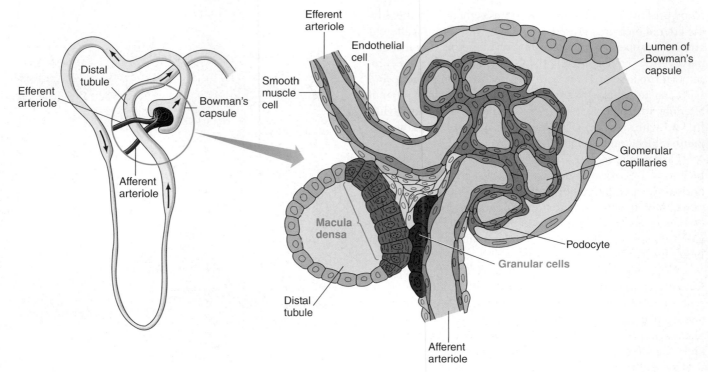

● FIGURE 14-12

The juxtaglomerular apparatus
The juxtaglomerular apparatus consists of specialized vascular cells (the granular cells) and specialized tubular cells (the macula densa) at a point where the distal tubule passes through the fork formed by the afferent and efferent arterioles of the same nephron.

the kidneys to regulate H_2O and salt excretion, such as the normal elevation in blood pressure accompanying exercise. Autoregulation is important because unintentional shifts in GFR could lead to dangerous imbalances of fluid, electrolytes, and wastes. Because at least a certain portion of the filtered fluid is always excreted, the amount of fluid excreted in the urine is automatically increased as the GFR increases. If autoregulation did not occur, the GFR would increase, and H_2O and solutes would be lost needlessly, as a result of the rise in arterial pressure accompanying heavy exercise. If by contrast the GFR were too low the kidneys could not eliminate enough wastes, excess electrolytes, and other materials that should be excreted. Autoregulation thus greatly blunts the direct effect that changes in arterial pressure would otherwise have on GFR and subsequently on H_2O, solute, and waste excretion.

When changes in mean arterial pressure fall outside the autoregulatory range, these mechanisms cannot compensate. Therefore, dramatic changes in mean arterial pressure (<80 mm Hg or >180 mm Hg) directly cause the glomerular capillary pressure and, accordingly the GFR, to decrease or increase in proportion to the change in arterial pressure.

Importance of extrinsic sympathetic control of the GFR

In addition to the intrinsic autoregulatory mechanisms designed to keep the GFR constant in the face of fluctuations in arterial blood pressure, the GFR can be *changed on purpose*—even when the mean arterial blood pressure is within the autoregulatory range—by extrinsic-control mechanisms that override

the autoregulatory responses. Extrinsic control of GFR, which is mediated by sympathetic nervous system input to the afferent arterioles, is aimed at regulating arterial blood pressure. The parasympathetic nervous system does not exert any influence on the kidneys.

If plasma volume is decreased—for example, because of hemorrhage—the resultant fall in arterial blood pressure is detected by the arterial carotid sinus and aortic arch baroreceptors, which initiate neural reflexes to increase blood pressure toward normal (see p. 377). These reflex responses are coordinated by the cardiovascular control center in the brain stem and are mediated primarily through increased sympathetic activity to the heart and blood vessels. Although the resulting increase in both cardiac output and total peripheral resistance helps increase blood pressure toward normal, the plasma volume is still reduced. In the long term, the plasma volume must be restored to normal. One compensation for a depleted plasma volume is reduced urine output, so that more fluid than normal is conserved for the body. This reduced urine output is accomplished in part by reducing the GFR; if less fluid is filtered, less is available to be excreted.

Role of the baroreceptor reflex in extrinsic control of the GFR

No new mechanism is required to decrease the GFR. It is reduced by the baroreceptor reflex response to a fall in blood pressure (● Figure 14-14). During this reflex, sympathetically induced vasoconstriction occurs in most arterioles throughout the body as a compensatory mechanism to increase total peripheral resistance. Among the arterioles that constrict in response to the baroreceptor reflex are the afferent arterioles carrying blood to the glomeruli. The afferent arterioles are innervated with sympathetic vasoconstrictor fibers to a far greater extent than are the efferent arterioles. When the afferent arterioles constrict from increased sympathetic activity, less blood flows into the glomeruli than normal, lowering the glomerular capillary blood pressure (● Figure 14-11a). The resultant decrease in GFR in turn reduces urine volume. In this way, some of the H₂O and salt that would otherwise

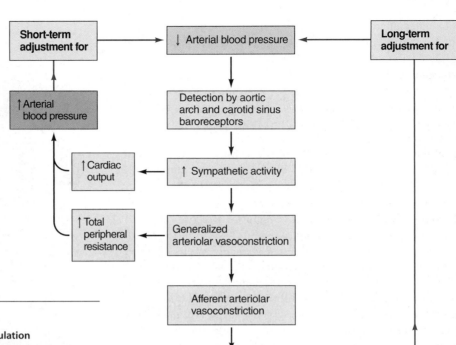

● FIGURE 14-13

Tubuloglomerular feedback mechanism of autoregulation

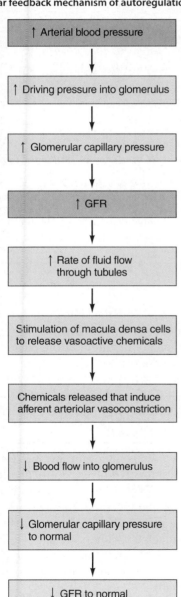

● FIGURE 14-14

Baroreceptor reflex influence on the GFR in long-term regulation of arterial blood pressure

have been lost in urine are saved for the body, helping in the long term to restore the plasma volume to normal so that short-term cardiovascular adjustments that have been made are no longer necessary. Other mechanisms, such as increased tubular reabsorption of H₂O and salt as well as increased thirst (described more thoroughly elsewhere), also contribute to long-term maintenance of blood pressure, despite a loss of plasma volume, by helping to restore plasma volume.

Conversely, if blood pressure is elevated (for example, because of an expansion of plasma volume following ingestion of excessive fluid), the opposite responses occur. When the baroreceptors detect a rise in blood pressure, sympathetic vasoconstrictor activity to the arterioles, including the renal afferent arterioles, is reflexly reduced, allowing afferent arteriolar vasodilation to occur. As more blood enters the glomeruli through the dilated afferent arterioles, glomerular capillary blood pressure rises, increasing the GFR (● Figure 14-11b). As more fluid is filtered, more fluid is available to be eliminated in urine. Contributing to the increase in urine volume is a hormonally adjusted reduction in the tubular reabsorption of H_2O and salt. By these two renal mechanisms—increased glomerular filtration and decreased tubular reabsorption of H_2O and salt—urine volume is increased, and the excess fluid is eliminated from the body. A reduction in thirst and fluid intake also contributes to restoring an elevated blood pressure to normal.

▌ The GFR can be influenced by changes in the filtration coefficient.

Thus far we have discussed changes in the GFR as a result of changes in net filtration pressure. The rate of glomerular filtration, however, depends on the filtration coefficient (K_f) as well as on the net filtration pressure. For years K_f was considered a constant, except in disease situations in which the glomerular membrane becomes leakier than usual. Exciting new research to the contrary indicates that K_f is subject to change under physiologic control. Both factors on which K_f depends—the surface area and the permeability of the glomerular membrane—can be modified by contractile activity within the membrane.

The surface area available for filtration within the glomerulus is represented by the inner surface of the glomerular capillaries that comes into contact with blood. Each tuft of glomerular capillaries is held together by **mesangial cells.** These cells contain contractile elements (that is, actinlike filaments). Contraction of these mesangial cells closes off a portion of the filtering capillaries, reducing the surface area available for filtration within the glomerular tuft. When the net filtration pressure remains unchanged, this reduction in K_f decreases GFR. Sympathetic stimulation causes the mesangial cells to contract, thus providing a second mechanism (besides promoting afferent arteriolar vasoconstriction) by which sympathetic activity can decrease the GFR. In addition, several hormones and local chemical mediators involved in the control of tubuloglomerular feedback or tubular reabsorption also influence mesangial cell contractile activity.

The podocytes also possess actinlike contractile filaments, whose contraction or relaxation can, respectively, decrease or increase the number of filtration slits open in the inner membrane of Bowman's capsule by changing the shapes and proximities of the foot processes (● Figure 14-15). The number of slits is a determinant of permeability; the more open slits, the greater the permeability. Contractile activity of the podocytes, which in turn affects permeability and the K_f, appears to be under physiologic control.

As the full extent of various regulatory mechanisms is becoming understood, it is evident that control of glomerular filtration (both through adjustments in afferent arteriolar resistance and K_f) and control of tubular reabsorption are complexly interrelated.

Before turning our attention to the process of tubular reabsorption, we are first going to examine the percentage of the

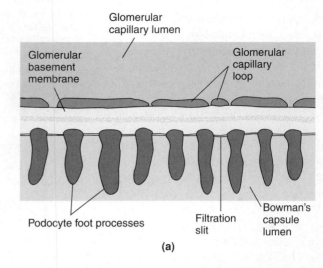

(a)

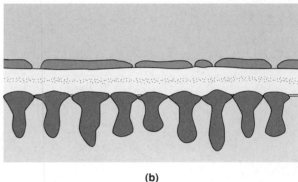

(b)

● **FIGURE 14-15**

Change in the number of open filtration slits caused by podocyte relaxation and contraction
(a) Podocyte relaxation narrows the bases of the foot processes, increasing the number of fully open intervening filtration slits spanning a given area. (b) Podocyte contraction flattens the foot processes and thus decreases the number of intervening filtration slits.

(Source: Adapted from Federation Proceedings, Vol. 42, p. 3046–3052, 1983. Reprinted by permission.)

cardiac output that goes to the kidneys. This will further reinforce the concept of how much blood flows through the kidneys and how much of that fluid is filtered and subsequently acted on by the tubules.

▌ The kidneys normally receive 20% to 25% of the cardiac output.

At the average net filtration pressure and K_f, 20% of the plasma that enters the kidneys is converted into glomerular filtrate. Thus, at an average GFR of 125 ml/min, the total renal plasma flow must average about 625 ml/min. Because 55% of whole blood consists of plasma (that is, hematocrit = 45), the total flow of blood through the kidneys averages 1,140 ml/min. This quantity is about 22% of the total cardiac output of 5 liters (5,000 ml)/min, although the kidneys compose less than 1% of total body weight.

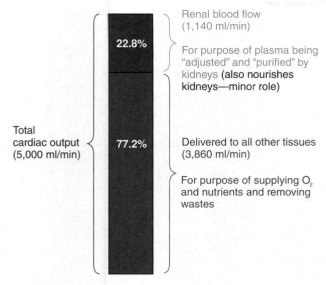

● **FIGURE 14-16**
Percentage of the cardiac output distributed to the kidneys

The kidneys need to receive such a seemingly disproportionate share of the cardiac output because they must continuously perform their regulatory and excretory functions on the huge volumes of plasma delivered to them to maintain stability in the internal fluid environment. Most of the blood goes to the kidneys not to supply the renal tissue but to be adjusted and purified by the kidneys. On the average, 20% to 25% of the blood pumped out by the heart each minute "goes to the cleaners" instead of serving its normal purpose of exchanging materials with the tissues (● Figure 14-16). Only by continuously processing such a large proportion of the blood are the kidneys able to precisely regulate the volume and electrolyte composition of the internal environment and adequately eliminate the large quantities of metabolic waste products that are constantly produced.

TUBULAR REABSORPTION

All plasma constituents except the proteins are indiscriminately filtered together through the glomerular capillaries. In addition to waste products and excess materials that the body must eliminate, the filtered fluid also contains nutrients, electrolytes, and other substances that the body cannot afford to lose in urine. Indeed, through ongoing glomerular filtration, greater quantities of these materials are filtered per day than are even present in the entire body. It is important that the essential materials that are filtered be returned to the blood by the process of *tubular reabsorption*, the discrete transfer of substances from the tubular lumen into the peritubular capillaries.

▌Tubular reabsorption is tremendous, highly selective, and variable.

Tubular reabsorption is a highly selective process. All constituents except plasma proteins are at the same concentration in the glomerular filtrate as in the plasma. In most cases, the quantity of each material that is reabsorbed is the amount required to maintain the proper composition and volume of the internal fluid environment. In general, the tubules have a high reabsorptive capacity for substances needed by the body and little or no reabsorptive capacity for substances of no value (▲ Table 14-2). Accordingly, only a small percentage, if any, of filtered plasma constituents that are useful to the body are present in the urine, most having been reabsorbed and returned to the blood. Only excess amounts of essential materials such as electrolytes are excreted in the urine. For the essential plasma constituents regulated by the kidneys, absorptive capacity may vary depending on the body's needs. In contrast, a large percentage of filtered waste products is present in urine. These wastes, which are useless or even potentially harmful to the body if allowed to accumulate, are not reabsorbed to any extent. Instead, they remain in the tubules to be eliminated in urine. As H_2O and other valuable constituents are reabsorbed, the waste products remaining in the tubular fluid become highly concentrated.

Of the 125 ml/min filtered, typically 124 ml/min are reabsorbed. Considering the magnitude of glomerular filtration, the extent of tubular reabsorption is tremendous: The tubules typically reabsorb 99% of the filtered H_2O (47 gallons/day), 100% of the filtered sugar (2.5 pounds/day), and 99.5% of the filtered salt (0.36 pounds/day).

▌Tubular reabsorption involves transepithelial transport.

Throughout its entire length, the tubule wall is one cell-layer thick and is in close proximity to a surrounding peritubular capillary (● Figure 14-17). Adjacent tubular cells do not come into contact with each other except where they are joined by tight junctions (see p. 63) at their lateral edges near their *luminal membranes*, which face the tubular lumen. Interstitial fluid lies in the gaps between adjacent cells—the **lateral spaces**—as well as between the tubules and capillaries. The *basolateral*

▲ **TABLE 14-2**
Fate of Various Substances Filtered by the Kidneys

Substance	Average Percentage of Filtered Substance Reabsorbed	Average Percentage of Filtered Substance Excreted
Water	99	1
Sodium	99.5	0.5
Glucose	100	0
Urea (a waste product)	50	50
Phenol (a waste product)	0	100

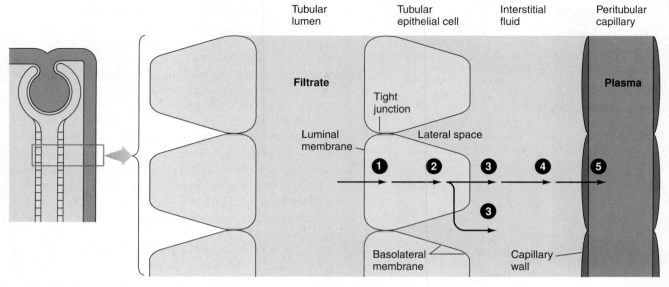

To be reabsorbed (move from the filtrate to the plasma), a substance must traverse five distinct barriers:

1 the luminal cell membrane   **3** the basolateral cell membrane   **5** the capillary wall

2 the cytosol   **4** the interstital fluid

● **FIGURE 14-17**
Steps of transepithelial transport

---

membrane faces the interstitial fluid at the base and lateral edges of the cell.

The tight junctions largely prevent substances from moving *between* the cells, so materials must pass *through* the cells to leave the tubular lumen and gain entry to the blood. To be reabsorbed, a substance must traverse five distinct barriers (● Figure 14-17):

- *Step 1.* It must leave the tubular fluid by crossing the luminal membrane of the tubular cell.
- *Step 2.* It must pass through the cytosol from one side of the tubular cell to the other.
- *Step 3.* It must traverse the basolateral membrane of the tubular cell to enter the interstitial fluid.
- *Step 4.* It must diffuse through the interstitial fluid.
- *Step 5.* It must penetrate the capillary wall to enter the blood plasma.

This entire sequence of steps is known as **transepithelial** ("across the epithelium") **transport.**

Passive versus active reabsorption

The two types of tubular reabsorption—*passive reabsorption* and *active reabsorption*—depend on whether local energy expenditure is required for transfer of a particular substance. In **passive reabsorption,** *all* steps in the transepithelial transport of a substance from the tubular lumen to the plasma are passive; that is, no energy is expended for the substance's net movement, which occurs down electrochemical or osmotic gradients (see p. 75). In contrast, **active reabsorption** takes place

if any one of the steps in the transepithelial transport of a substance requires energy, even if the four other steps are passive. With active reabsorption, net movement of the substance from the tubular lumen to the plasma occurs against an electrochemical gradient. Substances that are actively reabsorbed are of particular importance to the body, such as glucose, amino acids, and other organic nutrients, as well as Na^+ and other electrolytes such as PO_4^{3-}. Rather than specifically describing the reabsorptive process for each of the many filtered substances that are returned to the plasma, we will provide illustrative examples of the general mechanisms involved, after first highlighting the unique and important case of Na^+ reabsorption.

▌ An active Na^+–K^+ ATPase pump in the basolateral membrane is essential for Na^+ reabsorption.

Sodium reabsorption is unique and complex. Of the total energy requirement of the kidneys, 80 percent is used for Na^+ transport, indicating the importance of this process. Unlike most filtered solutes, Na^+ is reabsorbed throughout most of the tubule, but to varying extents in different regions. Of the Na^+ filtered, 99.5% is normally reabsorbed. Of the Na^+ reabsorbed, on average 67% is reabsorbed in the proximal tubule, 25% in the loop of Henle, and 8% in the distal and collecting tubules. Sodium reabsorption plays different important roles in each of these segments, as will become apparent as our discussion continues. Here is a preview of these roles.

- Sodium reabsorption in the *proximal tubule* plays a pivotal role in reabsorption of glucose, amino acids, H_2O, Cl^-, and urea.
- Sodium reabsorption in the ascending limb of the *loop of Henle*, along with Cl^- reabsorption, plays a critical role in the kidneys' ability to produce urine of varying concentrations and volumes, depending on the body's need to conserve or eliminate H_2O.
- Sodium reabsorption in the *distal portion of the nephron* is variable and subject to hormonal control. It is important in regulation of ECF volume and long-term control of arterial blood pressure and is also linked in part to K^+ secretion and H^+ secretion.

Sodium is reabsorbed throughout the tubule with the exception of the descending limb of the loop of Henle. You will learn about the significance of this exception later. Throughout all Na^+-reabsorbing segments of the tubule, the active step in Na^+ reabsorption involves the energy-dependent Na^+–K^+ ATPase carrier located in the tubular cell's basolateral membrane (● Figure 14-18). This carrier is the same one that is present in all cells and actively extrudes Na^+ from the cell (see p. 81). As this basolateral pump transports Na^+ out of the tubular cell into the lateral space, it keeps the intracellular Na^+ concentration low while it simultaneously builds up the concentration of Na^+ in the lateral space; that is, it moves Na^+ *against* a concentration gradient. Because the intracellular Na^+ concentration is kept low by basolateral pump activity, a concentration gradient is established that favors the diffusion of Na^+ from its higher concentration in the tubular lumen across the luminal border into the tubular cell. The nature of the luminal Na^+ channels and/or transport carriers that permit movement of Na^+ from the lumen into the cell varies for different parts of the tubule, but in each case, movement of Na^+ across the luminal membrane is always a passive step. For example, in the proximal tubule, Na^+ crosses the luminal border by means of a cotransport carrier that simultaneously moves Na^+ and an organic nutrient such as glucose from the lumen into the cell. You will learn more about this cotransport process shortly. By contrast, in the collecting tubule, Na^+ crosses the luminal border through a Na^+ channel. Once Na^+ enters the cell across the luminal border by whatever means, it is actively extruded to the lateral space by the basolateral Na^+-K^+ pump. This step is the same throughout the tubule. Sodium continues to diffuse down a concentration gradient from its high concentration in the lateral space into the surrounding interstitial fluid and finally into the peritubular capillary blood. Thus net transport of Na^+ from the tubular lumen into the blood occurs at the expense of energy. We are first going to consider the importance of regulating Na^+ reabsorption in the distal portion of the nephron and examine how this control is accomplished. Later we will explore in further detail the roles of Na^+ reabsorption in the proximal tubule and in the loop of Henle.

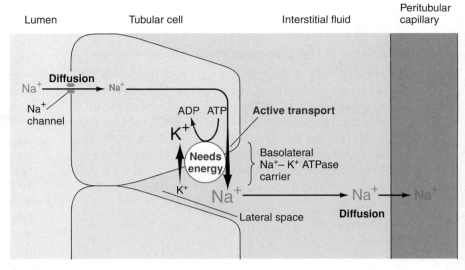

● FIGURE 14-18

Sodium reabsorption
The basolateral Na^+–K^+ ATPase carrier actively transports Na^+ from the tubular cell into the interstitial fluid within the lateral space. This process establishes a concentration gradient for diffusion of Na^+ from the lumen into the tubular cell and from the lateral space into the peritubular capillary, accomplishing net transport of Na^+ from the tubular lumen into the blood at the expense of energy.

▌ Aldosterone stimulates Na^+ reabsorption in the distal and collecting tubules.

In the proximal tubule and loop of Henle, a constant percentage of the filtered Na^+ is reabsorbed regardless of the **Na^+ load** (*total amount* of Na^+ in the body fluids, *not the concentration* of Na^+ in the body fluids). In the distal portion of the tubule, the reabsorption of a small percentage of the filtered Na^+ is subject to hormonal control. The extent of this controlled, discretionary reabsorption is inversely related to the magnitude of the Na^+ load in the body. If there is too much Na^+, little of this controlled Na^+ is reabsorbed; instead, it is lost in urine, thereby removing excess Na^+ from the body. If Na^+ is depleted, however, most or all of this controlled Na^+ is reabsorbed, conserving for the body Na^+ that otherwise would be lost in urine.

The Na^+ load in the body is reflected by the ECF volume. Sodium and its accompanying anion Cl^- account for more than 90% of the ECF's osmotic activity. (NaCl is common table salt.) Recall that osmotic pressure can be thought of loosely as a force that attracts and holds H_2O (see p. 76). When the Na^+ load is above normal and the ECF's osmotic activity is therefore increased, the extra Na^+ "holds" extra H_2O, expanding the ECF volume. Conversely, when the Na^+ load is below normal, thereby decreasing ECF osmotic activity, less H_2O than normal can be held in the ECF, so the ECF volume is reduced. Because plasma is a component of the ECF, the most important consequence of a change in ECF volume is the corresponding change in blood pressure accompanying expansion (↑blood pressure) or reduction (↓blood pressure) of the plasma volume. Thus, long-term control of arterial blood pressure ultimately depends on Na^+-regulating mechanisms. We will now turn our attention to these mechanisms.

Activation of the renin-angiotensin-aldosterone system

The granular cells of the juxtaglomerular apparatus (● Figure 14-12) secrete a hormone, **renin,** into the blood in response to factors that signal a fall in NaCl/ECF volume/blood pressure. This function is in addition to the role the juxtaglomerular apparatus plays in autoregulation, and renin is distinct from the local vasoactive chemicals that influence glomerular blood flow. Specifically, the following three inputs to the granular cells increase renin secretion:

1. The granular cells themselves function as *intrarenal baroreceptors.* They are sensitive to pressure changes within the afferent arteriole. When the granular cells detect a fall in blood pressure, they secrete more renin.

2. The macula densa cells in the tubular portion of the juxtaglomerular apparatus are sensitive to the NaCl moving past them through the tubular lumen. In response to a fall in NaCl, the macula densa cells trigger the granular cells to secrete more renin.

3. The granular cells are innervated by the sympathetic nervous system. When blood pressure falls below normal, the baroreceptor reflex increases sympathetic activity. As part of this reflex response, increased sympathetic activity stimulates the granular cells to secrete more renin.

These interrelated signals for increased renin secretion all indicate the need to expand the plasma volume to increase the arterial pressure to normal on a long-term basis. Through a complex series of events, increased renin secretion brings about increased Na^+ reabsorption by the distal portion of the tubule. Chloride always passively follows Na^+ down the electrical gradient established by sodium's active movement. The ultimate benefit of this salt retention is that it osmotically promotes H_2O retention, which helps restore the plasma volume, thus being important in the long-term control of blood pressure.

Let us examine the mechanism by which renin secretion ultimately leads to increased Na^+ reabsorption—the **renin-angiotensin-aldosterone system** (● Figure 14-19). Once secreted into the blood, renin acts as an enzyme to activate **angiotensinogen** into **angiotensin I.** Angiotensinogen is a plasma protein synthesized by the liver and always present in the plasma in high concentration. On passing through the lungs via the pulmonary circulation, angiotensin I is converted into **angiotensin II** by **angiotensin-converting enzyme (ACE),** which is abundant in the pulmonary capillaries. Angiotensin II is the primary stimulus for the secretion of the hormone **aldosterone** from the adrenal cortex. The adrenal cortex is an endocrine gland that produces several different hormones, each of which is secreted in response to different stimuli.

Functions of the renin-angiotensin-aldosterone system

Among its actions, *aldosterone* increases Na^+ reabsorption by the distal and collecting tubules. It does so by promoting the insertion of additional Na^+ channels into the luminal membranes and additional Na^+–K^+ ATPase carriers into the basolateral membranes of the distal and collecting tubular cells.

The net result is a greater passive inward flux of Na^+ into the tubular cells from the lumen and increased active pumping of Na^+ out of the cells into the plasma—that is, an increase in Na^+ reabsorption, with Cl^- following passively. The renin-angiotensin-aldosterone system thus promotes salt retention and a resultant H_2O retention and elevation of arterial blood pressure. Acting in negative-feedback fashion, this system alleviates the factors that triggered the initial release of renin—namely, salt depletion, plasma volume reduction, and decreased arterial blood pressure.

In addition to stimulating aldosterone secretion, *angiotensin II* is also a potent constrictor of the systemic arterioles, thereby directly increasing blood pressure by increasing total peripheral resistance (see p. 359). Furthermore, it stimulates thirst (increasing fluid intake) and stimulates vasopressin (a hormone that increases H_2O retention by the kidneys), both of which contribute to plasma volume expansion and elevation of arterial pressure. (As you will learn later, other mechanisms related to the long-term regulation of blood pressure and ECF osmolarity are also important in controlling thirst and vasopressin secretion.)

The opposite situation exists when the Na^+ load, ECF and plasma volume, and arterial blood pressure are above normal. Under these circumstances, renin secretion is inhibited. Consequently, because angiotensinogen is not activated to angiotensin I and II, aldosterone secretion is not stimulated. Without aldosterone, the small aldosterone-dependent portion of Na^+ reabsorption in the distal segments of the tubule does not occur. Instead, this nonreabsorbed Na^+ is lost in urine. In the absence of aldosterone, the ongoing loss of this small percentage of filtered Na^+ can rapidly remove excess Na^+ from the body. Even though only about 8% of the filtered Na^+ is dependent on aldosterone for reabsorption, this small loss, multiplied manyfold as the entire plasma volume is filtered through the kidneys many times per day, can lead to a sizable loss of Na^+.

In the complete absence of aldosterone, 20 g of salt may be excreted per day. With maximum aldosterone secretion, all the filtered Na^+ (and, accordingly, all the filtered Cl^-) is reabsorbed, so salt excretion in the urine is zero. The amount of aldosterone secreted, and consequently the relative amount of salt conserved versus salt excreted, usually varies between these extremes, depending on the body's needs. For example, an "average" salt consumer typically excretes about 10 g of salt per day in the urine, a heavy salt consumer excretes more, and someone who has lost considerable salt during heavy sweating excretes less urinary salt. By varying the amount of renin and aldosterone secreted in accordance with the salt-determined fluid load in the body, the kidneys can finely adjust the amount of salt conserved or eliminated. In doing so, they maintain the salt load and ECF volume/arterial blood pressure at a relatively constant level despite wide variations in salt consumption and abnormal losses of salt-laden fluid.

Role of the renin-angiotensin-aldosterone system in various diseases

Some cases of hypertension (high blood pressure) are due to abnormal increases in renin-angiotensin-aldosterone activ-

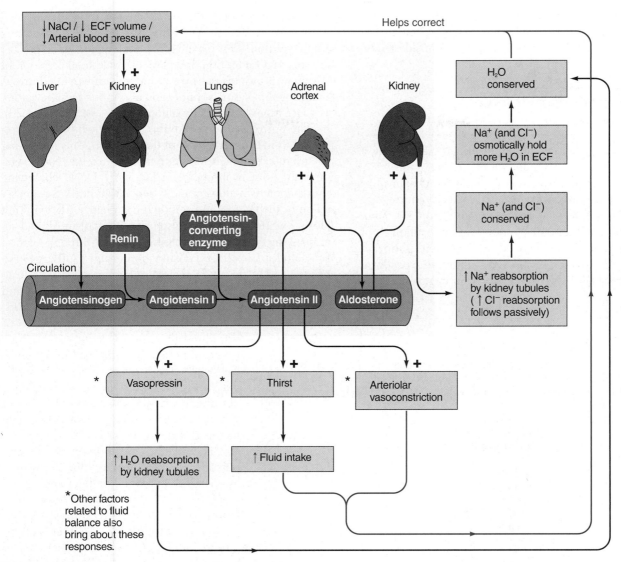

● **FIGURE 14-19**

Renin-angiotensin-aldosterone system
The kidneys secrete the hormone renin in response to reduced NaCl, ECF volume, and arterial blood pressure. Renin activates angiotensinogen, a plasma protein produced by the liver, into angiotensin I. Angiotensin I is converted into angiotensin II by angiotensin-converting enzyme (ACE) produced in the lungs. Angiotensin II stimulates the adrenal cortex to secrete the hormone aldosterone, which stimulates Na^+ reabsorption by the kidneys. The resulting retention of Na^+ exerts an osmotic effect that holds more H_2O in the ECF. Together, the conserved Na^+ and H_2O help correct the original stimuli that activated this hormonal system. Angiotensin II also exerts other effects that help rectify the original stimuli.

ity. This system is also responsible in part for the fluid retention and edema accompanying congestive heart failure. Because of the failing heart, the cardiac output is reduced and arterial blood pressure is low despite a normal or even expanded plasma volume. When a fall in blood pressure is due to a failing heart rather than a reduced salt/fluid load in the body, the salt- and fluid-retaining reflexes triggered by the low blood pressure are inappropriate. Sodium excretion may fall to virtually zero despite continued salt ingestion and accumulation in the body. The resulting expansion of the ECF produces edema and exacerbates the congestive heart failure because the weakened heart cannot pump the additional plasma volume.

Drugs that affect Na^+ reabsorption

Because their salt-retaining mechanisms are being inappropriately triggered, patients with congestive heart failure are placed on low-salt diets. Often they are treated with **diuretics**, which are therapeutic agents that cause *diuresis* (increased urinary output) and thus promote loss of fluid from the body. Many of these drugs function by inhibiting tubular reabsorption of Na^+. As more Na^+ is excreted, more H_2O is also lost from the body, thus helping to remove the excess ECF. Diuretics are also useful for treating hypertension.

ACE inhibitor drugs, which block the action of angiotensin-converting enzyme (ACE), are also beneficial in the treatment of congestive heart failure as well as certain cases of hypertension. By blocking the generation of angiotensin II, ACE inhibitors halt the ultimate salt- and fluid-conserving actions and arteriolar constrictor effects of the renin-angiotensin-aldosterone system.

▌ Atrial natriuretic peptide inhibits Na⁺ reabsorption.

While the renin-angiotensin-aldosterone system exerts the most powerful influence on the renal handling of Na^+, this Na^+-retaining system is opposed by a Na^+-losing system that involves the hormone **atrial natriuretic peptide (ANP)** and several similar, recently identified natriuretic hormones from the brain. (*Natriuretic* means "inducing excretion of large amounts of sodium in the urine.") The heart, in addition to its pump action, produces ANP, which is stored in granules in specialized atrial myocardial cells. ANP is released from the cardiac atria when these specialized muscle cells are mechanically stretched by expansion of the ECF volume, including the circulating plasma volume. This expansion, which occurs as a result of Na^+ and H_2O retention, increases arterial blood pressure. In turn, ANP promotes natriuresis and accompanying diuresis, decreasing the plasma volume, and also influences the cardiovascular system to lower the blood pressure (● Figure 14-20).

The primary action of ANP is to directly inhibit Na^+ reabsorption in the distal parts of the nephron, thus increasing Na^+ excretion in the urine. ANP further increases Na^+ excretion in the urine by inhibiting two steps of the Na^+-conserving renin-angiotensin-aldosterone system. ANP inhibits renin secretion by the kidneys and acts on the adrenal cortex to inhibit aldosterone secretion. ANP also promotes natriuresis and accompanying diuresis by increasing the GFR through dilation of the afferent arterioles, raising glomerular capillary blood pressure, and by relaxing the glomerular mesangial cells, leading to an increase in K_f. As more salt and water are filtered, more salt and water are excreted in the urine. Besides indirectly lowering blood pressure by reducing the Na^+ load and hence the fluid load in the body, ANP also directly lowers blood pressure by decreasing the cardiac output and reducing peripheral vascular resistance by means of inhibiting sympathetic nervous activity to the heart and blood vessels.

The relative contributions of ANP and possibly other salt-losing, blood pressure–lowering factors in the maintenance of salt and H_2O balance and blood pressure regulation are presently being intensely investigated. Importantly, derangements of this system could logically contribute to hypertension. In fact, recent studies suggest that a deficiency of the counterbalancing natriuretic system may underlie some cases of long-term hypertension by leaving the powerful Na^+-conserving system unopposed. The resulting salt retention, especially in association with high salt intake, could expand ECF volume and elevate blood pressure.

We are now going to shift our attention to the reabsorption of other filtered solutes. However, we will still continue to discuss Na^+ reabsorption, because the reabsorption of many other solutes is linked in some way to Na^+ reabsorption.

● **FIGURE 14-20**

Atrial natriuretic peptide
The atria secrete the hormone atrial natriuretic peptide (ANP) in response to being stretched by Na^+ retention, expansion of the ECF volume, and increase in arterial blood pressure. Atrial natriuretic peptide in turn promotes natriuretic, diuretic, and hypotensive effects to help correct the original stimuli that resulted in its release.

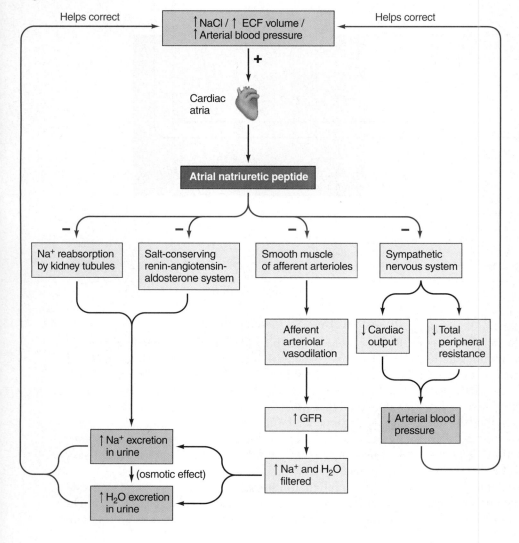

▌ Glucose and amino acids are reabsorbed by Na⁺-dependent secondary active transport.

Large quantities of nutritionally important organic molecules such as glucose and amino acids are filtered each day. Because these substances normally are completely reabsorbed back into the blood by energy- and Na^+-dependent mechanisms located in the proximal tubule, none of these materials are usually excreted in the urine. This rapid and thorough reabsorption early in the tubules protects against the loss of these important organic nutrients.

Even though glucose and amino acids are actively moved uphill against their concentration gradients from the tubular lumen into the blood until their concentration in the tubular fluid is virtually zero, no energy is directly used to operate the glucose or amino acid carriers. Glucose and amino acids are transferred by means of secondary active transport. With this process, specialized cotransport carriers located only in the proximal tubule simultaneously transfer both Na^+ and the specific organic molecule from the lumen into the cell (see p. 83). This luminal cotransport carrier is the means by which Na^+ passively crosses the luminal membrane in the proximal tubule. The lumen-to-cell Na^+ concentration gradient maintained by the energy-consuming basolateral $Na^+–K^+$ pump drives this cotransport system and pulls the organic molecule against its concentration gradient without the direct expenditure of energy. Because the overall process of glucose and amino acid reabsorption depends on the use of energy, these organic molecules are considered to be actively reabsorbed, even though energy is not used directly to transport them across the membrane. In essence, glucose and amino acids get a "free ride" at the expense of energy already used in the reabsorption of Na^+. Secondary active transport requires the presence of Na^+ in the lumen; without Na^+ the cotransport carrier is inoperable. Once transported into the tubular cells, glucose and amino acids passively diffuse down their concentration gradients across the basolateral membrane into the plasma, facilitated by a carrier that is not dependent on energy.

▌ In general, actively reabsorbed substances exhibit a tubular maximum.

All actively reabsorbed substances bind with plasma membrane carriers that transfer them across the membrane against a concentration gradient. Each carrier is specific for the types of substances it can transport; for example, the glucose cotransport carrier cannot transport amino acids, or vice versa. Because a limited number of each carrier type are present in the cells lining the tubules, there is an upper limit on how much of a particular substance that can be actively transported from the tubular fluid in a given period of time. The maximum reabsorption rate is reached when all the carriers specific for a particular substance are fully "occupied" or saturated, so they cannot handle any additional passengers at that time (see p. 79). This transport maximum is designated as the **tubular maximum,** or T_m. Any quantity of a substance filtered beyond its T_m is not reabsorbed, and escapes instead into the urine. With the exception of Na^+, all actively reabsorbed substances have a tubular maximum.(Even though individual Na^+ transport carriers can become saturated, the tubules as a whole do not display a tubular maximum for Na^+, because aldosterone promotes the synthesis of more active $Na^+–K^+$ carriers in the distal and collecting tubular cells as needed.)

The plasma concentrations of some but not all substances that display carrier-limited reabsorption are regulated by the kidneys. How can the kidneys regulate some actively reabsorbed substances but not others, when the renal tubules limit the quantity of each of these substances that can be reabsorbed and returned to the plasma? We will compare glucose, a substance that has a T_m but *is not regulated* by the kidneys, with phosphate, a T_m-limited substance that *is regulated* by the kidneys.

▌ Glucose is an example of an actively reabsorbed substance that is not regulated by the kidneys.

The normal plasma concentration of glucose is 100 mg of glucose/100 ml of plasma. Because glucose is freely filterable at the glomerulus, it passes into Bowman's capsule at the same concentration it has in the plasma. Accordingly, 100 mg of glucose are present in every 100 ml of plasma filtered. With 125 ml of plasma normally being filtered each minute (average GFR = 125 ml/min), 125 mg of glucose pass into Bowman's capsule with this filtrate every minute. The quantity of any substance filtered per minute, known as its **filtered load,** can be calculated as follows:

$$\text{Filtered load of a substance} = \text{plasma concentration} \times \text{GFR of the substance}$$

$$\text{Filtered load of glucose} = 100 \text{ mg }/100 \text{ ml} \times 125 \text{ ml/min}$$

$$= 125 \text{ mg/min}$$

At a constant GFR, the filtered load of glucose is directly proportional to the plasma glucose concentration. Doubling the plasma glucose concentration to 200 mg/100 ml doubles the filtered load of glucose to 250 mg/min, and so on (● Figure 14-21).

Tubular maximum for glucose

The T_m for glucose averages 375 mg/min; that is, the glucose carrier mechanism is capable of actively reabsorbing up to 375 mg of glucose per minute before it reaches its maximum transport capacity. At a normal plasma glucose concentration of 100 mg/100 ml, the 125 mg of glucose filtered per minute can readily be reabsorbed by the glucose carrier mechanism, because the filtered load is well below the T_m for glucose. Ordinarily, therefore, no glucose appears in the urine. Not until the filtered load of glucose exceeds 375 mg/min is the T_m reached. When more glucose is filtered per minute than can be reabsorbed because the T_m is exceeded, the maximum amount is reabsorbed, while the rest remains in the filtrate to be excreted. Accordingly, the plasma glucose concentration must be greater than 300 mg/100 ml—more than three times the normal value—before glucose starts spilling into the urine.

Renal threshold for glucose

The plasma concentration at which the T_m of a particular substance is reached and the substance first starts appearing in urine is known as the **renal threshold.** At the normal T_m of 375 mg/min and GFR of 125 ml/min, the renal threshold for

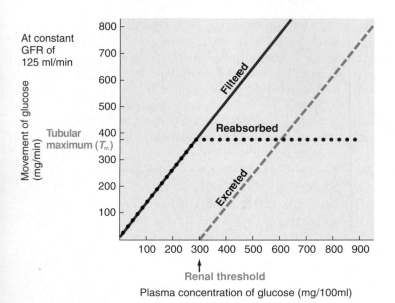

At constant GFR of 125 ml/min

Movement of glucose (mg/min)

Tubular maximum (T_m)

Filtered

Reabsorbed

Excreted

Renal threshold

Plasma concentration of glucose (mg/100ml)

Plasma concentration of substance × GFR = Amount of substance filtered

For example, 100 mg glucose/100 ml plasma × 125 ml plasma filtered/min = 125 mg glucose filtered/min

Plasma Concentration (mg/100ml)	GFR (ml/min)	Amount Filtered (mg/min)	T_m (mg/min)	Amount Reabsorbed (mg/min)	Amount Excreted (mg/min)
a. 80	125	100	375	100	0
b. 100	125	125	375	125	0
c. 200	125	250	375	250	0
d. 300	125	375	375	375	0
e. 400	125	500	375	375	125
f. 500	125	625	375	375	250

● **FIGURE 14-21**

Renal handling of glucose as a function of plasma glucose concentration
At a constant GFR, the quantity of glucose filtered per minute is directly proportional to the plasma concentration of glucose. All the filtered glucose can be reabsorbed up to the tubular maximum (T_m). If the amount of glucose filtered per minute exceeds the T_m, the maximum amount of glucose is reabsorbed (a T_m worth), and the rest remains in the filtrate to be excreted in urine. The renal threshold is the plasma concentration at which the T_m is reached and glucose first starts appearing in the urine.

glucose is 300 mg/100 ml. Beyond the T_m, reabsorption remains constant at its maximum rate, and any further increase in filtered load is accompanied by a directly proportional increase in amount of the substance excreted. For example, at a plasma glucose concentration of 400 mg/100 ml, the filtered load of glucose is 500 mg/min, 375 mg/min of which can be reabsorbed (a T_m worth) and 125 mg/min of which are excreted in urine. At a plasma glucose concentration of 500 mg/100 ml, the filtered load is 625 mg/min, still only 375 mg/min can be reabsorbed, and 250 mg/min spill into urine (● Figure 14-21).

The plasma glucose concentration can become extremely high in diabetes mellitus, an endocrine disorder involving deficiency of insulin, a pancreatic hormone. This hormone is important in facilitating transport of glucose into many body cells.

In insulin deficiency, glucose that cannot gain entry into cells remains in the plasma, elevating the plasma glucose concentration. Consequently, although glucose does not normally appear in urine, it is found in the urine of people with diabetes when the plasma glucose concentration exceeds the renal threshold, even though renal function has not changed.

What happens when plasma glucose concentration falls below normal? The renal tubules, of course, reabsorb all the filtered glucose, because the glucose reabsorptive capacity is far from being exceeded. The kidneys cannot do anything to raise a low plasma glucose level to normal. They simply return all the filtered glucose to the plasma.

Reason why the kidneys do not regulate glucose

The kidneys do not influence plasma glucose concentration over a wide range of values from abnormally low levels up to three times the normal level. Because the T_m for glucose is well above the normal filtered load, the kidneys usually conserve all the glucose, thereby protecting against loss of this important nutrient in urine. The kidneys do not regulate glucose, because they do not maintain glucose at some specific plasma concentration; instead, this concentration is normally regulated by endocrine and liver mechanisms, with the kidneys merely maintaining whatever plasma glucose concentration is set by these other mechanisms (except when excessively high levels overwhelm the kidney's reabsorptive capacity). The same general principle holds true for other organic plasma nutrients, such as amino acids and water-soluble vitamins.

▌ Phosphate is an example of an actively reabsorbed substance that is regulated by the kidneys.

The kidneys do directly contribute to the regulation of many electrolytes, such as phosphate (PO_4^{3-}) and calcium (Ca^{2+}), because renal thresholds of these inorganic ions equal their normal plasma concentrations. The transport carriers for these electrolytes are located in the proximal tubule. We will use PO_4^{3-} as an example. Our diets are generally rich in PO_4^{3-}, but because the tubules can reabsorb up to the normal plasma concentration's worth of PO_4^{3-} and no more, the excess ingested PO_4^{3-} is quickly spilled into urine, restoring the plasma concentration to normal. The greater the amount of PO_4^{3-} ingested beyond the body's needs, the greater the amount excreted. In this way, the kidneys maintain the desired plasma PO_4^{3-} concentration while eliminating any excess PO_4^{3-} ingested.

Unlike the reabsorption of organic nutrients, the reabsorption of PO_4^{3-} and Ca^{2+} is also subject to hormonal control. Parathyroid hormone can alter the renal thresholds for PO_4^{3-} and Ca^{2+}, thus adjusting the quantity of these electrolytes conserved, depending on the body's momentary needs (Chapter 19).

■ Active Na⁺ reabsorption is responsible for the passive reabsorption of Cl⁻, H₂O, and urea.

Not only is secondary active reabsorption of glucose and amino acids linked to the basolateral Na^+–K^+ pump, but passive reabsorption of Cl^-, H_2O, and urea, also depends on this active Na^+ reabsorption mechanism.

Chloride reabsorption

The negatively charged chloride ions are passively reabsorbed down the electrical gradient created by the active reabsorption of the positively charged sodium ions. For the most part, chloride ions pass between, not through, the tubular cells. The amount of Cl^- reabsorbed is determined by the rate of active Na^+ reabsorption, instead of being directly controlled by the kidneys.

Water reabsorption

Water is passively reabsorbed by osmosis throughout the length of the tubule. Of the H_2O filtered, 80% is obligatorily reabsorbed in the proximal tubules and loops of Henle, osmotically following solute reabsorption. This reabsorption occurs regardless of the H_2O load in the body and is not subject to regulation. Variable amounts of the remaining 20% are reabsorbed in the distal portions of the tubule; the extent of reabsorption in the distal and collecting tubules is subject to direct hormonal control, depending on the body's state of hydration.

During reabsorption, H_2O passes through **aquaporins,** or **water channels,** formed by specific plasma membrane proteins in the tubular cells. Different types of water channels are present in various parts of the nephron. The water channels in the proximal tubule are always open, accounting for the high H_2O permeability of this region. The channels in the distal parts of the nephron, in contrast, are regulated by the hormone *vasopressin,* accounting for the variable H_2O reabsorption in this region.

The main driving force for H_2O reabsorption in the proximal tubule is a compartment of hypertonicity in the lateral spaces between the tubular cells that is established by the active extrusion of Na^+ by the basolateral pump (● Figure 14-22). As a result of this pump activity, the concentration of Na^+ rapidly diminishes in the tubular fluid and tubular cells while it simultaneously increases in the localized region within the lateral spaces. This osmotic gradient induces the passive net flow of H_2O from the lumen into the lateral spaces, either through the cells or intercellularly through "leaky" tight junctions. The accumulation of fluid in the lateral spaces results in a buildup of hydrostatic (fluid) pressure, which flushes H_2O out of the lateral spaces into the interstitial fluid and finally into the peritubular capillaries. Water also osmotically follows other preferentially reabsorbed solutes such as glucose (which is also Na^+ dependent), but the direct influence of Na^+ reabsorption on passive H_2O reabsorption is quantitatively more important.

This return of filtered H_2O to the plasma is enhanced by the fact that the plasma-colloid osmotic pressure is greater in the peritubular capillaries than elsewhere. The concentration of plasma proteins, which is responsible for the plasma-colloid osmotic pressure, is elevated in the blood entering the peritubular capillaries because of the extensive filtration of H_2O through the glomerular capillaries upstream. The plasma proteins left behind in the glomerulus are concentrated into a smaller volume of plasma H_2O, increasing the plasma-colloid osmotic pressure of the unfiltered blood that leaves the glomerulus and enters the peritubular capillaries. This force tends to "pull" H_2O into the peritubular capillaries, simultaneous with the "push" of the hydrostatic pressure in the lateral spaces that drives H_2O toward the capillaries. By these means, 65% of the filtered H_2O—117 liters per day—is passively reabsorbed by the end of the proximal tubule. Neither the proximal tubule nor indeed any other portion of the tubule directly requires energy for this tremendous reabsorption of H_2O.

Another 15% of the filtered H_2O is obligatorily reabsorbed from the loop of Henle. Adjustable reabsorption of the remaining 20% of the filtered H_2O is accomplished in the distal and collecting tubules under control of vasopressin. The mechanisms responsible for H_2O reabsorption beyond the proximal tubule will be described later.

Urea reabsorption

In addition to Cl^- and H_2O, the passive reabsorption of urea is also indirectly linked to active Na^+ reabsorption. **Urea** is a waste product resulting from the breakdown of protein. The osmotically induced reabsorption of H_2O in the proximal tubule secondary to active Na^+ reabsorption produces a concentration gradient for urea that favors passive reabsorption of this nitrogenous waste, as follows (● Figure 14-23). Because of extensive reabsorption of H_2O in the proximal tubule, the original 125 ml/min of filtrate are progressively reduced until

● **FIGURE 14-22**

Water reabsorption in the proximal tubule
The force for H_2O reabsorption is the compartment of hypertonicity in the lateral spaces established by active extrusion of Na^+ by the basolateral pump. The dashed arrows show the direction of osmotic movement of H_2O.

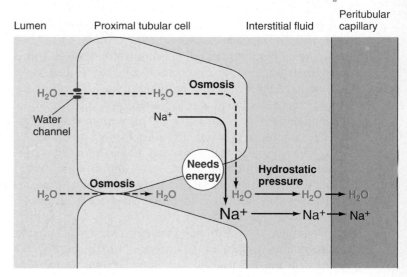

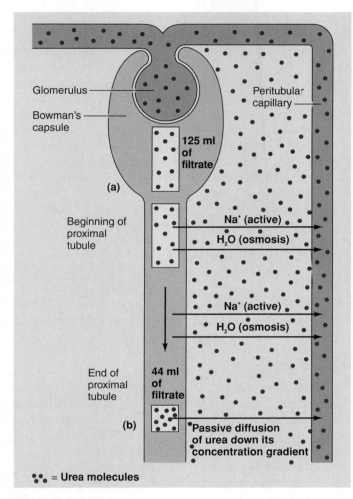

● FIGURE 14-23

Passive reabsorption of urea at the end of the proximal tubule
(a) In Bowman's capsule and at the beginning of the proximal tubule, urea is at the same concentration as in the plasma and surrounding interstitial fluid. (b) By the end of the proximal tubule, 65% of the original filtrate has been reabsorbed, concentrating the filtered urea in the remaining filtrate. This establishes a concentration gradient favoring passive reabsorption of urea.

only 44 ml/min of fluid remain in the lumen by the end of the proximal tubule (with 65% of the H_2O in the original filtrate, or 81 ml/min, having been reabsorbed). Substances that have been filtered but not reabsorbed become progressively more concentrated in the tubular fluid as H_2O is reabsorbed while they are left behind. Urea is one such substance. Urea's concentration as it is filtered at the glomerulus is identical to its concentration in the plasma entering the peritubular capillaries. The quantity of urea present within the 125 ml of filtered fluid at the beginning of the proximal tubule, however, is concentrated almost threefold in the small volume of only 44 ml left at the end of the proximal tubule. As a result, the urea concentration within the tubular fluid becomes considerably greater than the plasma urea concentration in the adjacent capillaries. Therefore, a concentration gradient is created for urea to passively diffuse from the tubular lumen into the peritubular capillary plasma. Because the walls of the proximal tubules are

only moderately permeable to urea, only about 50% of the filtered urea is passively reabsorbed by this means.

Even though only half of the filtered urea is eliminated from the plasma with each pass through the nephrons, this removal rate is adequate. The urea concentration in the plasma becomes elevated only in impaired kidney function, when much less than half of the urea is removed. An elevated urea level was one of the first chemical characteristics to be identified in the plasma of patients with severe renal failure. Accordingly, clinical measurement of **blood urea nitrogen (BUN)** came into use as a crude assessment of kidney function. It is now known that the most serious consequences of renal failure are not attributable to the retention of urea, which itself is not especially toxic, but rather to the accumulation of other substances that are not adequately excreted because of their failure to be properly secreted—most notably H^+ and K^+. Health professionals still often refer to renal failure as **uremia** ("urea in the blood"), indicating excess urea in the blood, even though urea retention is not this condition's major threat.

▌ In general, unwanted waste products are not reabsorbed.

The other filtered waste products besides urea, such as *phenol* and *creatinine*, are likewise concentrated in the tubular fluid as H_2O leaves the filtrate to enter the plasma, but they are not passively reabsorbed, as urea is. Urea molecules, being the smallest of the waste products, are the only wastes passively reabsorbed by this concentrating effect. Even though the other wastes are also concentrated in the tubular fluid, they cannot leave the lumen down their concentration gradients to be passively reabsorbed, because they cannot permeate the tubular wall. Therefore, the waste products, failing to be reabsorbed, generally remain in the tubules and are excreted in urine in highly concentrated form. This excretion of metabolic wastes is not subject to physiologic control. When renal function is normal, however, the excretory processes proceed at a satisfactory rate even though they are not controlled.

We have now completed our discussion of tubular reabsorption and are going to shift our attention to the other basic renal process carried out by the tubules—tubular secretion.

TUBULAR SECRETION

Like tubular reabsorption, tubular secretion involves transepithelial transport, but now the steps are reversed. By providing a second route of entry into the tubules for selected substances, *tubular secretion*, the discrete transfer of substances from the peritubular capillaries into the tubular lumen, is a supplemental mechanism that hastens elimination of these compounds from the body. Anything that gains entry to the tubular fluid, whether by glomerular filtration or tubular secretion, and fails to be reabsorbed is eliminated in the urine.

The most important substances secreted by the tubules are *hydrogen ion (H^+), potassium (K^+), and organic anions and cations*, many of which are compounds foreign to the body.

■ Hydrogen ion secretion is important in acid–base balance.

Renal H^+ secretion is extremely important in regulating acid–base balance in the body. Hydrogen ion can be added to the filtered fluid by being secreted by the proximal, distal, and collecting tubules. The extent of H^+ secretion depends on the acidity of the body fluids. When the body fluids are too acidic, H^+ secretion increases. Conversely, H^+ secretion is reduced when the H^+ concentration in the body fluids is too low. (See Chapter 15 for further detail.)

■ Potassium secretion is controlled by aldosterone.

Potassium ion is selectively moved in opposite directions in different parts of the tubule; it is actively reabsorbed in the proximal tubule and actively secreted in the distal and collecting tubules. Potassium reabsorption early in the tubule occurs in a constant, unregulated fashion, whereas K^+ secretion later in the tubule is variable and subject to regulation. Because the filtered K^+ is almost completely reabsorbed in the proximal tubule, most K^+ in the urine is derived from controlled K^+ secretion in the distal portions of the nephron rather than from filtration.

During K^+ depletion, K^+ secretion in the distal portions of the nephron is reduced to a minimum, so only the small percentage of filtered K^+ that escapes reabsorption in the proximal tubule is excreted in the urine. In this way, K^+ that normally would have been lost in urine is conserved for the body. However, when plasma K^+ levels are elevated, K^+ secretion is adjusted so that just enough K^+ is added to the filtrate for elimination to reduce the plasma K^+ concentration to normal. Thus K^+ secretion, not the filtration or reabsorption of K^+, is varied in a controlled fashion to regulate the rate of K^+ excretion and maintain the desired plasma K^+ concentration.

Mechanism of K^+ secretion

Potassium secretion in the distal and collecting tubules is coupled to Na^+ reabsorption by means of the energy-dependent basolateral Na^+–K^+ pump (● Figure 14-24). This pump not only moves Na^+ out of the cell into the lateral space but also transports K^+ from the lateral space into the tubular cells. The resulting high intracellular K^+ concentration favors net diffusion of K^+ from the cells into the tubular lumen. Movement across the luminal membrane occurs passively through the large number of K^+ channels in this barrier in the distal and collecting tubules. By keeping the interstitial fluid concentration of K^+ low as it transports K^+ into the tubular cells from the surrounding interstitial fluid, the basolateral pump encourages passive diffusion of K^+ out of the peritubular capillary plasma

into the interstitial fluid. Potassium exiting the plasma in this manner is later pumped into the cells, from which it diffuses into the lumen. In this way, the basolateral pump actively induces the net secretion of K^+ from the peritubular capillary plasma into the tubular lumen.

Because K^+ secretion is linked with Na^+ reabsorption by means of the Na^+–K^+ pump, why isn't K^+ secreted throughout the Na^+-reabsorbing segments of the tubule instead of taking place only in the distal parts of the nephron? The answer lies in the location of the passive K^+ channels. In the distal and collecting tubules, the K^+ channels are concentrated in the luminal membrane, providing a route for K^+ pumped into the cell to exit into the lumen, thus being secreted. In the other tubular segments, the K^+ channels are located primarily in the basolateral membrane. As a result, K^+ pumped into the cell from the lateral space by the Na^+–K^+ pump simply diffuses back out into the lateral space through these channels. This K^+ recycling permits the ongoing operation of the Na^+–K^+ pump to accomplish Na^+ reabsorption with no local net effect on K^+.

Control of K^+ secretion

Several factors can alter the rate of K^+ secretion, the most important being aldosterone. This hormone stimulates K^+ secretion by the tubular cells late in the nephron simultaneous to enhancing these cells' reabsorption of Na^+. An elevation in plasma K concentration directly stimulates the adrenal cortex to increase its output of aldosterone, which in turn promotes the secretion and ultimate urinary excretion and elimination of excess K^+. Conversely, a decline in plasma K^+ concentration

---

● FIGURE 14-24

Potassium secretion
The basolateral pump simultaneously transports Na^+ into the lateral space and K^+ into the tubular cell. In the parts of the tubule that secrete K^+, this ion leaves the cell through channels located in the luminal border, thus being secreted. (In the parts of the tubule that do not secrete K^+, the K^+ pumped into the cell during the process of Na^+ reabsorption leaves the cell through channels located in the basolateral border, thus being retained in the body.)

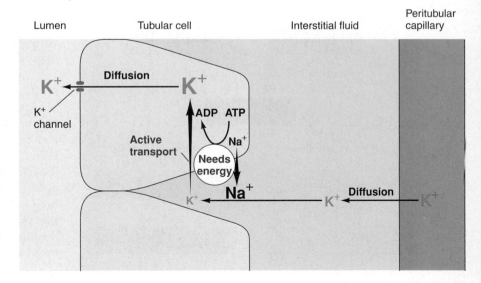

causes a reduction in aldosterone secretion and a corresponding decrease in aldosterone-stimulated renal K^+ secretion.

Note that a rise in plasma K^+ concentration directly stimulates aldosterone secretion by the adrenal cortex, whereas a fall in plasma Na^+ concentration stimulates aldosterone secretion by means of the complex renin-angiotensin pathway. Thus aldosterone secretion can be stimulated by two separate pathways (● Figure 14-25). No matter what the stimulus, however, increased aldosterone secretion always promotes simultaneous Na^+ reabsorption and K^+ secretion. For this reason, K^+ secretion can be inadvertently stimulated as a result of increased aldosterone activity brought about by Na^+ depletion, ECF volume reduction, or a fall in arterial blood pressure totally unrelated to K^+ balance. The resulting inappropriate loss of K^+ can lead to K^+ deficiency.

Effect of H^+ secretion on K^+ secretion

Another factor that can inadvertently alter the magnitude of K^+ secretion is the acid–base status of the body. The basolateral pump in the distal portions of the nephron can secrete either K^+ or H^+ in exchange for reabsorbed Na^+. An increased rate of secretion of either K^+ or H^+ is accompanied by a decreased rate of secretion of the other ion. Normally the kidneys secrete a preponderance of K^+, but when the body fluids are too acidic and H^+ secretion is increased as a compensatory measure, K^+ secretion is correspondingly reduced. This reduced secretion leads to inappropriate K^+ retention in the body fluids.

Importance of regulating plasma K^+ concentration

Except in the overriding circumstances of K^+ imbalances inadvertently induced during renal compensations for Na^+ or ECF volume deficits or acid–base imbalances, the kidneys usually exert a fine degree of control over plasma K^+ concentration. This is extremely important, because even minor fluctuations in plasma K^+ concentration can have detrimental consequences. Potassium plays a key role in the membrane electrical activity of excitable tissues. Both increases or decreases in the plasma (ECF) K^+ concentration can alter the intracellular-to-extracellular K^+ concentration gradient, which in turn can change the resting membrane potential. A rise in ECF K^+ concentration leads to a reduction in resting potential and a subsequent increase in excitability, especially of heart muscle. This cardiac overexcitability can lead to a rapid heart rate and even fatal cardiac arrhythmias. Conversely, a fall in ECF K^+ concentration results in hyperpolarization of nerve and muscle cell membranes, which reduces their excitability. The manifestations of ECF K^+ depletion are skeletal muscle weakness, diarrhea and abdominal distension caused by smooth muscle dysfunction, and abnormalities in cardiac rhythm and impulse conduction.

● **FIGURE 14-25**

Dual control of aldosterone secretion of K^+ and Na^+

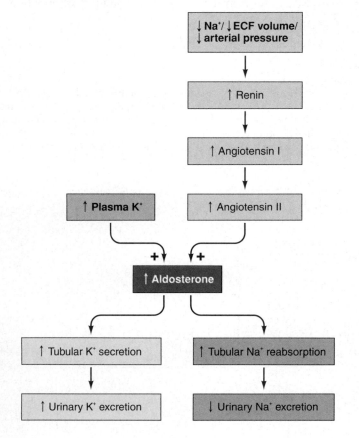

▌ Organic anion and cation secretion helps efficiently eliminate foreign compounds from the body.

The proximal tubule contains two distinct types of secretory carriers, one for the secretion of organic anions and a separate system for secretion of organic cations. These systems serve several important functions.

1. By adding more of a particular type of organic ion to the quantity that has already gained entry to the tubular fluid by means of glomerular filtration, these organic secretory pathways facilitate excretion of these substances. Included among these organic ions are certain blood-borne chemical messengers such as prostaglandins, histamine, and norepinephrine, that, having served their purpose, must be rapidly removed from the blood so that their biological activity is not unduly prolonged.

2. In some important instances, organic ions are poorly soluble in water. To be transported in blood, they are extensively but not irreversibly bound to plasma proteins. Because they are attached to plasma proteins, these substances cannot be filtered through the glomeruli. Tubular secretion facilitates elimination of these nonfilterable organic ions in urine. Even though a given organic ion is largely bound to plasma proteins, a small percentage of the ion always exists in free or unbound form in the plasma. Removal of this free organic ion by secretion permits "unloading" of some of the bound ion, which is then free to be secreted. This, in turn, encourages the unloading of even more organic ion, and so on.

3. Most important is the ability of the organic-ion secretory systems to eliminate many foreign compounds from the body. The organic ion systems can secrete a large number of different organic ions, both those produced endogenously (within the body) and those foreign organic ions that have gained access to the body fluids. This nonselectivity permits these organic-ion secretory systems to hasten removal of many foreign organic chemicals, including food additives, environmental pollutants (for example, pesticides), drugs, and other nonnutritive organic substances that have entered the body. Even though this mechanism helps rid the body of potentially harmful foreign compounds, it is not subject to physiologic adjustments. The carriers cannot pick up their secretory pace when confronting an elevated load of these organic ions.

However, the liver plays an important role in helping rid the body of foreign compounds. Many foreign organic chemicals are not ionic in their original form, so they cannot be secreted by the organic ion systems. The liver converts these foreign substances into an anionic form that facilitates their secretion by the organic anion system and thus accelerates their elimination.

Many drugs, such as penicillin and nonsteroidal anti-inflammatory drugs (NSAIDs), are eliminated from the body by means of the proximal tubule organic-ion secretory systems. To keep the plasma concentration of these drugs at effective levels, the dosage must be repeated on a regular, frequent basis to keep pace with the rapid removal of these compounds in urine.

Summary of reabsorptive and secretory processes

This completes our discussion of the reabsorptive and secretory processes that occur across the proximal and distal portions of the nephron. These processes are summarized in ▲ Table 14-3. To generalize, the proximal tubule does most of the reabsorbing. This "mass reabsorber" transfers much of the filtered water and needed solutes back into the blood in unregulated fashion. Similarly, the proximal tubule is the major site of secretion, with the exception of K^+ secretion. The distal and collecting tubules then determine the final amounts of H_2O, Na^+, K^+, and H^+ that are excreted in urine and thus eliminated from the body. They do so by "fine-tuning" the amount of Na^+ and H_2O reabsorbed and the amount of K^+ and H^+ secreted. These processes in the distal part of the nephron are all subject to control, depending on the body's momentary needs. The unwanted filtered waste products are left behind to be eliminated in urine, along with excess amounts of filtered or secreted nonwaste products that fail to be reabsorbed.

We will next focus on the end result of the basic renal processes—what's left in the tubules to be excreted in urine, and, as a consequence, what has been cleared from plasma.

URINE EXCRETION AND PLASMA CLEARANCE

Typically, of the 125 ml of plasma filtered per minute, 124 ml/min are reabsorbed, so the final quantity of urine formed aver-

▲ TABLE 14-3

Summary of Transport across Proximal and Distal Portions of the Nephron

Proximal Tubule	
Reabsorption	Secretion
67% of filtered Na^+ actively reabsorbed; not subject to control; Cl^- follows passively	Variable H^+ secretion depending on acid–base status of body
All filtered glucose and amino acids reabsorbed by secondary active transport; not subject to control	Organic ion secretion; not subject to control
Variable amounts of filtered PO_4^{3-} and other electrolytes reabsorbed; subject to control	
65% of filtered H_2O osmotically reabsorbed; not subject to control	
50% of filtered urea passively reabsorbed; not subject to control	
Almost all filtered K^+ reabsorbed; not subject to control	

Distal Tubule and Collecting Duct	
Reabsorption	Secretion
Variable Na^+ reabsorption, controlled by aldosterone; Cl^- follows passively	Variable H^+ secretion, depending on acid–base status of body
Variable H_2O reabsorption, controlled by vasopressin	Variable K^+ secretion, controlled by aldosterone

ages 1 ml/min. Thus, of the 180 liters filtered per day, 1.5 liters of urine are excreted.

Urine contains high concentrations of various waste products plus variable amounts of the substances regulated by the kidneys, with any excess quantities having spilled into the urine. Useful substances are conserved by reabsorption, so they do not appear in the urine.

A relatively small change in the quantity of filtrate reabsorbed can bring about a large change in the volume of urine formed. For example, a reduction of less than 1% in the total reabsorption rate, from 124 to 123 ml/min, increases the urinary excretion rate by 100%, from 1 to 2 ml/min.

▌Plasma clearance is the volume of plasma cleared of a particular substance per minute.

Compared to plasma entering the kidneys through the renal arteries, plasma leaving the kidneys through the renal veins lacks the materials that were left behind to be eliminated in urine.

By excreting substances in urine, kidneys clean or "clear" the plasma flowing through them of these substances. The **plasma clearance** of any substance is defined as the volume of plasma completely cleared of that substance by the kidneys per minute.[1] It does not refer to the *amount of the substance* removed but to the *volume of plasma* from which that amount was removed. Plasma clearance is actually a more useful measure than urine excretion; it is more important to know what effect urine excretion has on removing materials from body fluids than to know the volume and composition of discarded urine. Plasma clearance expresses the kidneys' effectiveness in removing various substances from the internal fluid environment.

Plasma clearance can be calculated for any plasma constituent as follows:

$$
\begin{array}{l}
\text{Clearance rate} \\
\text{of a substance} \\
\text{(ml/min)}
\end{array}
=
\frac{
\begin{array}{c}
\text{urine concentration} \\
\text{of the substance} \\
\text{(quantity/ml urine)}
\end{array}
\times
\begin{array}{c}
\text{urine flow rate} \\
\text{(ml/min)}
\end{array}
}{
\begin{array}{c}
\text{plasma concentration of the substance} \\
\text{(quantity/ml plasma)}
\end{array}
}
$$

The plasma clearance rate varies for different substances, depending on how the kidneys handle each substance. Let's compare the clearance rates for substances handled in different ways by the kidneys.

If a substance is filtered but not reabsorbed or secreted, its plasma clearance rate equals the GFR.

Assume that a plasma constituent, substance X, is freely filterable at the glomerulus but is not reabsorbed or secreted. As 125 ml/min of plasma are filtered and subsequently reabsorbed, the quantity of substance X originally contained within the 125 ml is left behind in the tubules to be excreted. Thus 125 ml of plasma are cleared of substance X each minute (● Figure 14-26a). (Of the 125 ml/min of plasma filtered, 124 ml/min of the filtered fluid are returned, through reabsorption, to the plasma minus substance X, thus clearing this 124 ml/min of substance X. In addition, the 1 ml/min of fluid lost in urine is eventually replaced by an equivalent volume of ingested H_2O that is already clear of substance X. Therefore, 125 ml of plasma cleared of substance X are, in effect, returned to the plasma for every 125 ml of plasma filtered per minute.)

There is no endogenous chemical with the characteristics of substance X. All substances naturally present in the plasma, even wastes, are reabsorbed or secreted to some extent. However, **inulin** (do not confuse with insulin), a harmless foreign carbohydrate produced by Jerusalem artichokes, is freely filtered and not reabsorbed or secreted—an ideal substance X.

[1]Actually, plasma clearance is an artificial concept, because when a particular substance is excreted in the urine, that substance's concentration in the plasma as a whole is uniformly decreased as a result of thorough mixing in the circulatory system. However, it is useful for comparative purposes to consider clearance in effect as the volume of plasma that would have contained the total quantity of the substance (at the substance's concentration prior to excretion) that the kidneys excreted in one minute; that is, the hypothetical volume of plasma completely cleared of that substance per minute.

Inulin can be injected and its plasma clearance determined as a clinical means of ascertaining the GFR. Because all glomerular filtrate formed is cleared of inulin, the volume of plasma cleared of inulin per minute equals the volume of plasma filtered per minute—that is, the GFR.

$$
\begin{array}{l}
\text{Clearance rate} \\
\text{for inulin}
\end{array}
=
\frac{30 \text{ mg/ml urine} \times 1.25 \text{ ml urine/min}}{0.30 \text{ mg/ml plasma}}
$$

$$
= 125 \text{ ml plasma/min}
$$

Although determination of inulin plasma clearance is accurate and straightforward, it is not very convenient, because inulin must be infused continuously throughout the determination to maintain a constant plasma concentration. Therefore, the plasma clearance of an endogenous substance, **creatinine**, is often used instead to find a rough estimate of the GFR. Creatinine, an end product of muscle metabolism, is produced at a relatively constant rate. It is freely filtered and not reabsorbed but is slightly secreted. Accordingly, creatinine clearance is not a completely accurate reflection of the GFR, but it does provide a close approximation and can be more readily determined than inulin clearance.

If a substance is filtered and reabsorbed but not secreted, its plasma clearance rate is always less than the GFR.

Some or all of a reabsorbable substance that has been filtered is returned to the plasma. Because less than the filtered volume of plasma will have been cleared of the substance, the plasma clearance rate of a reabsorbable substance is always less than the GFR. For example, the plasma clearance for glucose is normally zero. All the filtered glucose is reabsorbed along with the rest of the returning filtrate, so none of the plasma is cleared of glucose (● Figure 14-26b).

For a substance that is partially reabsorbed, such as urea, only part of the filtered plasma is cleared of that substance. With about 50% of the filtered urea being passively reabsorbed, only half of the filtered plasma, or 62.5 ml, is cleared of urea each minute (● Figure 14-26c).

If a substance is filtered and secreted but not reabsorbed, its plasma clearance rate is always greater than the GFR.

Tubular secretion allows the kidneys to clear certain materials from the plasma more efficiently. Only 20% of the plasma entering the kidneys is filtered. The remaining 80% passes unfiltered into the peritubular capillaries. The only means by which this unfiltered plasma can be cleared of any substance during this trip through the kidneys before being returned to the general circulation is by secretion. An example is H^+. Not only is filtered plasma cleared of nonreabsorbable H^+, but the plasma from which H^+ is secreted is also cleared of H^+. For example, if the quantity of H^+ secreted is equivalent to the quantity of H^+ present in 25 ml of plasma, the clearance rate for H^+ will be 150 ml/min at the normal GFR of 125 ml/min. Every minute 125 ml of plasma will lose its H^+ through filtration and failure of reabsorption, and 25 more ml of plasma will lose its H^+ through secretion. The plasma clearance for a secreted substance is always greater than the GFR (● Figure 14-26d).

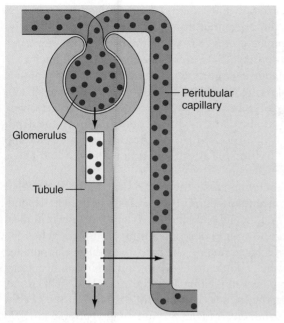

Peritubular
capillary

Glomerulus

Tubule

In
urine

For a substance filtered and
not reabsorbed or secreted,
such as inulin, all of the
filtered plasma is cleared of
the substance.

(a)

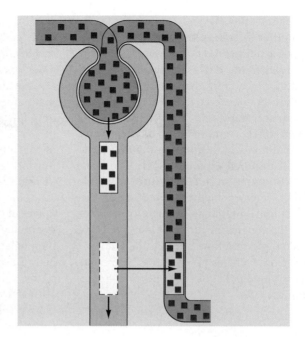

For a substance filtered,
not secreted, and completely
reabsorbed, such as glucose,
none of the filtered plasma
is cleared of the substance.

(b)

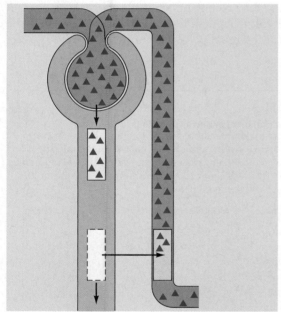

For a substance filtered,
not secreted, and partially
reabsorbed, such as urea, only
a portion of the filtered plasma
is cleared of the substance.

(c)

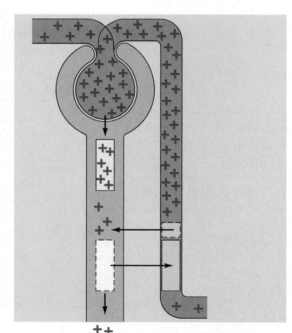

For a substance filtered and secreted
but not reabsorbed, such as hydrogen ion, all of the filtered plasma is
cleared of the substance, and the
peritubular plasma from which the
substance is secreted is also cleared.

(d)

● FIGURE 14-26
Plasma clearance for substances handled in different ways by the kidneys

Just as inulin can be used clinically to determine the GFR, plasma clearance of another foreign compound, the organic anion **para-aminohippuric acid (PAH),** can be used to measure renal plasma flow. Like inulin, PAH is freely filterable and nonreabsorbable. It differs, however, in that all the PAH in the plasma that escapes filtration is secreted from the peritubular capillaries by means of the organic anion secretory pathway in the proximal tubule. Thus PAH is removed from *all* the plasma that flows through the kidneys—both from plasma that is filtered and subsequently reabsorbed without its PAH, and from unfiltered plasma that continues on in the peritubular capillaries and loses its PAH by means of active secretion into the tubules. Because all the plasma that flows through the kidneys is cleared of PAH, the plasma clearance for PAH is a reasonable estimate of the rate of plasma flow through the kidneys. Typically, renal plasma flow averages 625 ml/min, for a renal blood flow (plasma plus blood cells) of 1,140 ml/min—over 20% of the cardiac output.

Filtration fraction

Knowing PAH clearance (renal plasma flow) and inulin clearance (GFR), you can easily determine the **filtration fraction,** the fraction of the plasma flowing through the glomeruli that is filtered into the tubules:

$$\frac{\text{Filtration}}{\text{fraction}} = \frac{\text{GFR (plasma inulin clearance)}}{\text{renal plasma flow (plasma PAH clearance)}}$$

$$= \frac{125 \text{ ml/min}}{625 \text{ ml/min}} = 20\%$$

Thus 20% of the plasma that enters the glomeruli is typically filtered.

▪ The kidneys can excrete urine of varying concentrations depending on the body's state of hydration.

Having considered how the kidneys deal with a variety of solutes in the plasma, we will now concentrate on renal handling of plasma H_2O. The ECF osmolarity (solute concentration) depends on the relative amount of H_2O compared to solute. At normal fluid balance and solute concentration, the body fluids are **isotonic** at an osmolarity of 300 milliosmols/liter (mosm/liter) (see pp. 78 and A-10). If too much H_2O is present relative to the solute load, the body fluids are **hypotonic,** which means they are too dilute at an osmolarity less than 300 mosm/liter. However, if a H_2O deficit exists relative to the solute load, the body fluids are too concentrated or are **hypertonic,** having an osmolarity greater than 300 mosm/liter.

Knowing that the driving force for H_2O reabsorption throughout the entire length of the tubules is an osmotic gradient between the tubular lumen and surrounding interstitial fluid, you would expect, given osmotic considerations, that the kidneys could not excrete urine more or less concentrated than the body fluids. Indeed, this would be the case if the interstitial fluid surrounding the tubules in the kidneys were identical in osmolarity to the remaining body fluids. Water reabsorption

would proceed only until the tubular fluid equilibrated osmotically with the interstitial fluid, and the body would have no way to eliminate excess H_2O when the body fluids were hypotonic, or to conserve H_2O in the presence of hypertonicity.

Fortunately, a large vertical osmotic gradient is uniquely maintained in the interstitial fluid of the medulla of each kidney. The concentration of the interstitial fluid progressively increases from the cortical boundary down through the depth of the renal medulla until it reaches a maximum of 1,200 mosm/liter in humans at the junction with the renal pelvis (● Figure 14-27).

This gradient enables the kidneys to produce urine that ranges in concentration from 100 to 1,200 mosm/liter, depending on the body's state of hydration. When the body is in ideal fluid balance, 1 ml/min of isotonic urine is formed. When the body is overhydrated (too much H_2O), the kidneys can produce a large volume of dilute urine (up to 25 ml/min and hypotonic at 100 mosm/liter), thus eliminating the excess H_2O in the urine. Conversely, the kidneys can put out a small volume of concentrated urine (down to 0.3 ml/min and hypertonic at 1,200 mosm/liter) when the body is dehydrated (too little H_2O), thus conserving H_2O for the body.

Unique anatomic arrangements and complex functional interactions between the various nephron components present in the renal medulla are responsible for establishing and using the vertical osmotic gradient. Recall that the hairpin loop of Henle dips only slightly into the medulla in the cortical nephrons, but in the juxtamedullary nephrons the loop plunges through the entire depth of the medulla so that the tip of the loop lies near the renal pelvis (● Figure 14-5, p. 515). Also,

---

● FIGURE 14-27

Vertical osmotic gradient in the renal medulla
The osmolarity of the interstitial fluid throughout the renal cortex is isotonic at 300 mosm/liter, but the osmolarity of the interstitial fluid in the renal medulla increases progressively from 300 mosm/liter at the boundary with the cortex to a maximum of 1,200 mosm/liter at the junction with the renal pelvis.

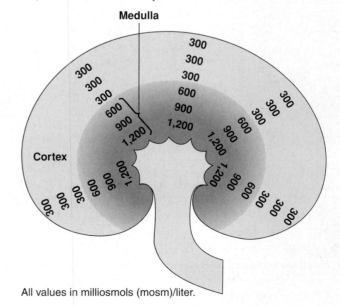

All values in milliosmols (mosm)/liter.

the vasa recta of the juxtamedullary nephrons follow the same deep hairpin loop as the long loop of Henle. Flow in both the long loops of Henle and the vasa recta is considered countercurrent, because the flow in the two closely adjacent limbs of the loop moves in opposite directions. Also running through the medulla in the descending direction only, on their way to the renal pelvis, are the collecting tubules that serve both types of nephrons. This arrangement, coupled with the permeability and transport characteristics of these tubular segments, plays a key role in the kidneys' ability to produce urine of varying concentrations, depending on the body's needs for water conservation or elimination. Briefly, the juxtamedullary nephrons' long loops of Henle *establish the vertical osmotic gradient*, their vasa recta *prevent the dissolution of this gradient* while providing blood to the renal medulla, and the collecting tubules of all nephrons *use the gradient*, in conjunction with the hormone vasopressin, to produce urine of varying concentrations. Collectively, this entire functional organization is known as the **medullary countercurrent system.** We will examine each of its facets in greater detail.

■ The medullary vertical osmotic gradient is established by countercurrent multiplication.

We will follow the filtrate through a long-looped nephron to see how this structure establishes a vertical osmotic gradient in the medulla. Immediately after the filtrate is formed, uncontrolled osmotic reabsorption of filtered H_2O occurs in the proximal tubule secondary to active Na^+ reabsorption. As a result, by the end of the proximal tubule about 65% of the filtrate has been reabsorbed, but the 35% remaining in the tubular lumen still has the same osmolarity as the body fluids. Therefore, the fluid entering the loop of Henle is still isotonic. An additional 15% of the filtered H_2O is obligatorily reabsorbed from the loop of Henle during the establishment and maintenance of the vertical osmotic gradient, with the osmolarity of the tubular fluid being altered in the process.

Properties of the descending and ascending limbs of a long Henle's loop

The following functional distinctions between the descending limb of a long Henle's loop (which carries fluid from the proximal tubule down into the depths of the medulla) and the ascending limb (which carries fluid up and out of the medulla into the distal tubule) are crucial for establishing the incremental osmotic gradient in the medullary interstitial fluid.

The *descending limb*

1. is highly permeable to H_2O.
2. does not actively extrude Na^+. (That is, it does not reabsorb Na^+. It is the only segment of the entire tubule that does not do so.)

The *ascending limb*

1. actively transports NaCl out of the tubular lumen into the surrounding interstitial fluid.
2. is always impermeable to H_2O, so salt leaves the tubular fluid without H_2O osmotically following along.

Mechanism of countercurrent multiplication

The close proximity and countercurrent flow of the two limbs allow important interactions to occur between them. Even though the flow of fluids is continuous through the loop of Henle, we will visualize what happens step by step, much like an animated film run so slowly that each individual frame can be viewed.

- *Initial scene* (● Figure 14-28a). Before the vertical osmotic gradient is established, the medullary interstitial fluid concentration is uniformly 300 mosm/liter, as is the remainder of the body fluids.
- *Step 1* (● Figure 14-28b). The active salt pump in the ascending limb can transport NaCl out of the lumen until the surrounding interstitial fluid is 200 mosm/liter more concentrated than the tubular fluid in this limb. When the ascending limb pump starts actively extruding salt, the medullary interstitial fluid becomes hypertonic. Water cannot follow osmotically from the ascending limb, because this limb is impermeable to H_2O. However, net diffusion of H_2O does occur from the descending limb into the interstitial fluid. The tubular fluid entering the descending limb from the proximal tubule is isotonic. Because the descending limb is highly permeable to H_2O, net diffusion of H_2O occurs by osmosis out of the descending limb into the more concentrated interstitial fluid. The passive movement of H_2O out of the descending limb continues until the osmolarities of the fluid in the descending limb and interstitial fluid become equilibrated. Thus the tubular fluid entering the loop of Henle immediately starts to become more concentrated as it loses H_2O. At equilibrium, the osmolarity of the ascending limb fluid is 200 mosm/liter and the osmolarities of the interstitial fluid and descending limb fluid are equal at 400 mosm/liter.
- *Step 2* (● Figure 14-28c). If we now advance the entire column of fluid in the loop of Henle several "frames," a mass of 200-mosm/liter fluid exits from the top of the ascending limb into the distal tubule, and a new mass of isotonic fluid at 300 mosm/liter enters the top of the descending limb from the proximal tubule. At the bottom of the loop, a comparable mass of 400-mosm/liter fluid from the descending limb moves forward around the tip into the ascending limb, placing it opposite a 400-mosm/liter region in the descending limb. Note that the 200-mosm/liter concentration difference has been lost at both the top and the bottom of the loop.
- *Step 3* (● Figure 14-28d). The ascending limb pump again transports NaCl out while H_2O passively leaves the descending limb until a 200-mosm/liter difference is re-established between the ascending limb and both the interstitial fluid and descending limb at each horizontal level. Note, however, that the concentration of tubular fluid is progressively increasing in the descending limb and progressively decreasing in the ascending limb.
- *Step 4* (● Figure 14-28e). As the tubular fluid is advanced still further, the 200-mosm/liter concentration gradient is disrupted once again at all horizontal levels.
- *Step 5* (● Figure 14-28f). Again, active extrusion of NaCl from the ascending limb, coupled with the net diffusion of H_2O out of the descending limb, re-establishes the 200-mosm/liter gradient at each horizontal level.

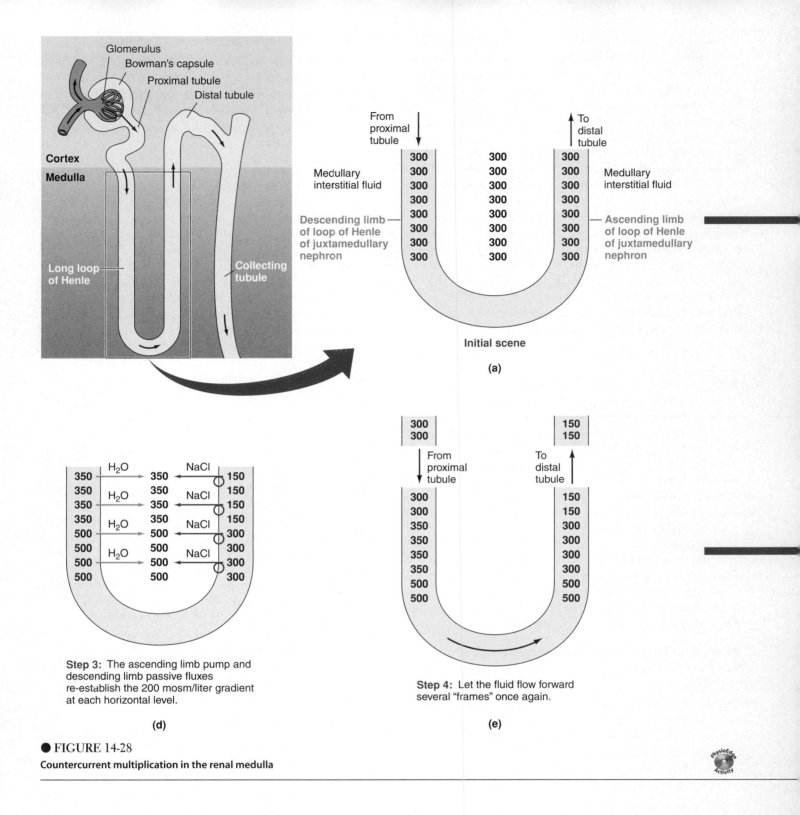

Step 3: The ascending limb pump and descending limb passive fluxes re-establish the 200 mosm/liter gradient at each horizontal level.

(d)

Step 4: Let the fluid flow forward several "frames" once again.

(e)

● FIGURE 14-28

Countercurrent multiplication in the renal medulla

• *Steps 6 and on* (● Figure 14-28g). As the fluid flows slightly forward again and this stepwise process continues, the fluid in the descending limb becomes progressively more hypertonic until it reaches a maximum concentration of 1,200 mosm/liter at the bottom of the loop, four times the normal concentration of body fluids. Because the interstitial fluid always achieves equilibrium with the descending limb, an incremental vertical concentration gradient ranging from 300 to 1,200 mosm/liter is likewise established in the medullary in-

terstitial fluid. In contrast, the concentration of the tubular fluid progressively decreases in the ascending limb as salt is pumped out but H_2O is unable to follow. In fact, the tubular fluid even becomes hypotonic as it leaves the ascending limb to enter the distal tubule at a concentration of 100 mosm/liter, one-third the normal concentration of body fluids.

Note that although a gradient of only 200 mosm/liter exists between the ascending limb and the surrounding fluids at

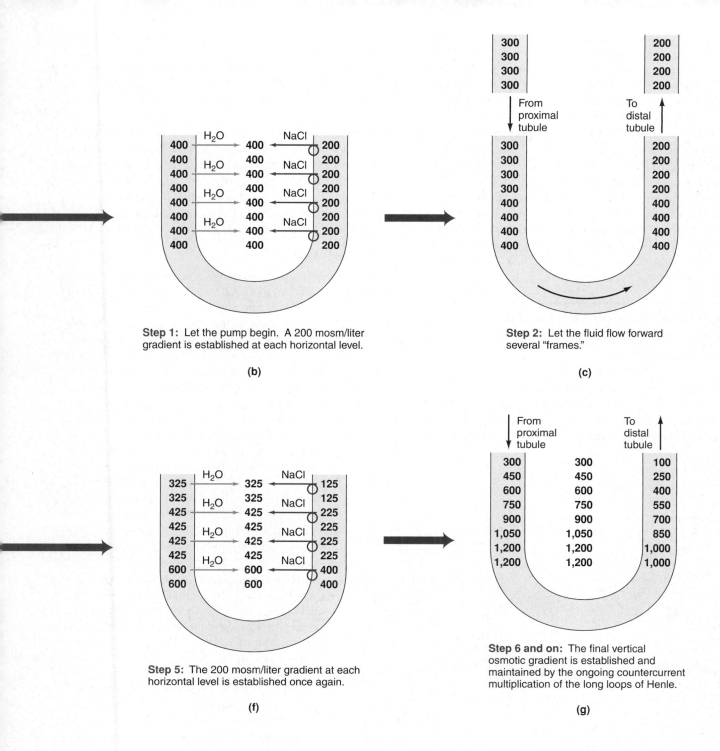

Step 1: Let the pump begin. A 200 mosm/liter gradient is established at each horizontal level.

(b)

Step 2: Let the fluid flow forward several "frames."

(c)

Step 5: The 200 mosm/liter gradient at each horizontal level is established once again.

(f)

Step 6 and on: The final vertical osmotic gradient is established and maintained by the ongoing countercurrent multiplication of the long loops of Henle.

(g)

each medullary horizontal level, a much larger vertical gradient exists from the top to the bottom of the medulla. Even though the ascending limb pump can generate a gradient of only 200 mosm/liter, this effect is multiplied into a large vertical gradient because of the countercurrent flow within the loop. This concentrating mechanism accomplished by the loop of Henle is known as **countercurrent multiplication.**

We have artificially described countercurrent multiplication in a "stop-and-flow," stepwise fashion to facilitate under-

standing. It is important to realize that once the incremental medullary gradient is established, it remains constant, because of the continuous flow of fluid coupled with the ongoing ascending limb active transport activity and accompanying descending limb passive fluxes.

Benefits of countercurrent multiplication

If you consider only what happens to the tubular fluid as it flows through the loop of Henle, the whole process seems an exercise in futility. The isotonic fluid that enters the loop becomes progressively more concentrated as it flows down the descending limb, achieving a maximum concentration of 1,200 mosm/liter, only to become progressively more dilute as it flows up the ascending limb, finally leaving the loop at a minimum concentration of 100 mosm/liter. What is the point of concentrating the fluid fourfold and then turning around and diluting it until it leaves at one-third the concentration at which it entered? Such a mechanism offers two benefits. First, it establishes a vertical osmotic gradient in the medullary interstitial fluid. This gradient, in turn, is used by the collecting ducts to concentrate the tubular fluid so that a urine *more concentrated* than normal body fluids can be excreted. Second, the fact that the fluid is hypotonic as it enters the distal portions of the tubule enables the kidneys to excrete a urine *more dilute* than normal body fluids. Let us see how.

▍Vasopressin-controlled, variable H_2O reabsorption occurs in the final tubular segments.

After obligatory H_2O reabsorption from the proximal tubule (65% of the filtered H_2O) and loop of Henle (15% of the filtered H_2O), 20% of the filtered H_2O remains in the lumen to enter the distal and collecting tubules for variable reabsorption that is under hormonal control. This is still a large volume of filtered H_2O subject to regulated reabsorption; 20% × GFR (180 liters/day) = 36 liters per day to be reabsorbed to varying extents, depending on the body's state of hydration. This is more than 13 times the amount of plasma H_2O in the entire circulatory system.

The fluid leaving the loop of Henle enters the distal tubule at 100 mosm/liter, so it is hypotonic to the surrounding isotonic (300 mosm/liter) interstitial fluid of the renal cortex through which the distal tubule passes. The distal tubule then empties into the collecting tubule, which is bathed by progressively increasing concentrations (300 to 1,200 mosm/liter) of surrounding interstitial fluid as it descends through the medulla.

Role of vasopressin

For H_2O absorption to occur across a segment of the tubule, two criteria must be met: (1) An osmotic gradient must exist across the tubule, and (2) the tubular segment must be permeable to H_2O. The distal and collecting tubules are *impermeable* to H_2O except in the presence of **vasopressin**, also known as **antidiuretic hormone** (*anti* means "against"; *diuretic* means "increased urine output"),[2] which increases their permeability to H_2O. Vasopressin is produced by several specific neuronal cell bodies in the *hypothalamus*, part of the brain, then stored in the *posterior pituitary gland*, which is attached to the hypothalamus by a thin stalk. The hypothalamus controls release of vasopressin from the posterior pituitary into the blood. In negative-feedback fashion, vasopressin secretion is stimulated by a H_2O deficit, when the ECF is too concentrated (that is, hypertonic) and H_2O must be conserved for the body, and inhibited by a H_2O excess, when the ECF is too dilute (that is, hypotonic) and surplus H_2O must be eliminated in urine.

Vasopressin reaches the basolateral membrane of the tubular cells lining the distal and collecting tubules through the circulatory system. Here it binds with receptors specific for it (● Figure 14-29). This binding activates the cyclic AMP (cAMP) second-messenger system within the tubular cells (see p. 67), which ultimately increases permeability of the opposite luminal membrane to H_2O by promoting insertion of aquaporins in this membrane. Without these aquaporins, the luminal membrane is impermeable to H_2O. Once H_2O enters the tubular cells from the filtrate through these vasopressin-regulated luminal water channels, it passively leaves the cells down the osmotic gradient across the cells' basolateral membrane (which is always permeable to H_2O) to enter interstitial fluid. By permitting more H_2O to permeate from the lumen into the tubular cells, these additional luminal channels thus

● FIGURE 14-29

Mechanism of action of vasopressin

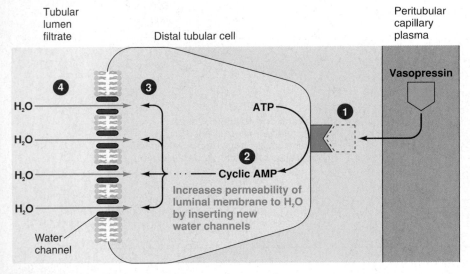

① Blood-borne vasopressin binds with its receptor sites on the basolateral membrane of a distal or collecting tubule cell.

② This binding activates the cyclic AMP second-messenger system within the cell.

③ Cyclic AMP increases the opposite luminal membrane's permeability to H_2O by promoting the insertion of water channels in this membrane. This membrane is impermeable to water in the absence of vasopressin.

④ Water enters the tubular cell from the tubular lumen through the water channels and subsequently enters the blood, in this way being reabsorbed.

[2]Even though textbooks traditionally have tended to use the name *antidiuretic hormone* for this hormone, especially when discussing its actions on the kidney, investigators in the field now prefer *vasopressin*.

increase H_2O reabsorption from the filtrate into interstitial fluid. The tubular response to vasopressin is graded; the more vasopressin present, the more water channels inserted, and the greater the permeability of the distal and collecting tubules to H_2O. The increase in luminal membrane water channels is not permanent, however. The channels are retrieved when vasopressin secretion decreases and cAMP activity is similarly decreased. Accordingly, H_2O permeability is reduced when vasopressin secretion decreases.

Vasopressin influences H_2O permeability only in the distal and collecting tubules. It has no influence over the 80% of the filtered H_2O that is obligatorily reabsorbed without control in the proximal tubule and loop of Henle. The ascending limb of Henle's loop is always impermeable to H_2O, even in the presence of vasopressin.

Regulation of H_2O reabsorption in response to a H_2O deficit

When vasopressin secretion increases in response to a H_2O deficit and the permeability of the distal and collecting tubules to H_2O accordingly increases, the hypotonic tubular fluid entering the distal tubules can lose progressively more H_2O by osmosis into the interstitial fluid as the tubular fluid first flows through the isotonic cortex and then is exposed to the ever-increasing osmolarity of the medullary interstitial fluid as it plunges toward the renal pelvis (● Figure 14-30a). As the 100 mosm/liter tubular fluid enters the distal tubule and is exposed to a surrounding interstitial fluid of 300 mosm/liter, H_2O leaves the tubular fluid by osmosis across the now permeable tubular cells until the tubular fluid reaches a maximum concentration of 300 mosm/liter by the end of the distal tubule. As this 300-mosm/liter tubular fluid progresses farther into the collecting tubule, it is exposed to even higher osmolarity in the surrounding medullary interstitial fluid. Consequently, the tubular fluid loses more H_2O by osmosis and becomes further concentrated, only to move farther forward and be exposed to an even higher interstitial fluid osmolarity and lose even more H_2O, and so on.

Under the influence of maximum levels of vasopressin, it is possible to concentrate the tubular fluid up to 1,200 mosm/liter by the end of the collecting tubules. No further modification of the tubular fluid occurs beyond the collecting tubule, so what remains in the tubules at this point is urine. As a result of this extensive vasopressin-promoted reabsorption of H_2O in the late segments of the tubule, a small volume of urine concentrated up to 1,200 mosm/liter can be excreted. As little as 0.3 ml of urine may be formed each minute, less than one-third the normal urine flow rate of 1 ml/min. The reabsorbed H_2O entering the medullary interstitial fluid is picked up by the peritubular capillaries and returned to the general circulation, thus being conserved for the body.

Although vasopressin promotes H_2O conservation by the body, it cannot completely halt urine production, even when a person is not taking in any H_2O, because a minimum volume of H_2O must be excreted with the solute wastes. Collectively, the waste products and other constituents eliminated in the urine average 600 mosm each day. Because the maximum urine concentration, is 1,200 mosm/liter, the minimum vol-

ume of urine that is required to excrete these wastes is 500 ml/day (600 mosm of wastes/day ÷ 1,200 mosm/liter of urine = 0.5 liter, or 500 ml/day, or 0.3 ml/min). Thus, under maximal vasopressin influence, 99.7% of the 180 liters of plasma H_2O filtered per day is returned to the blood, with an obligatory H_2O loss of half a liter.

The kidneys' ability to tremendously concentrate urine to minimize H_2O loss when necessary is possible only because of the presence of the osmotic gradient in the medulla. If this gradient did not exist, the kidneys could not produce a urine more concentrated than the body fluids no matter how much vasopressin was secreted, because the only driving force for H_2O reabsorption is a concentration differential between the tubular fluid and the interstitial fluid.

Regulation of H_2O reabsorption in response to a H_2O excess

Conversely, when a person consumes large quantities of H_2O, the excess H_2O must be removed from the body without simultaneously losing solutes that are critical for maintaining homeostasis. Under these circumstances, no vasopressin is secreted, so the distal and collecting tubules remain impermeable to H_2O. The tubular fluid entering the distal tubule is hypotonic (100 mosm/liter), having lost salt without an accompanying loss of H_2O in the ascending limb of Henle's loop. As this hypotonic fluid passes through the distal and collecting tubules (● Figure 14-30b), the medullary osmotic gradient cannot exert any influence because of the late tubular segments' impermeability to H_2O. In other words, none of the H_2O remaining in the tubules can leave the lumen to be reabsorbed, even though the tubular fluid is less concentrated than the surrounding interstitial fluid. Thus in the absence of vasopressin, the 20% of the filtered fluid that reaches the distal tubule is not reabsorbed. Meanwhile, excretion of wastes and other urinary solutes remains constant. The net result is a large volume of dilute urine, which helps rid the body of excess H_2O. Urine osmolarity may be as low as 100 mosm/liter, the same as in the fluid entering the distal tubule. Urine flow may be increased up to 25 ml/min in the absence of vasopressin, compared to the normal urine production of 1 ml/min.

The ability to produce urine less concentrated than the body fluids depends on the fact that the tubular fluid is hypotonic as it enters the distal portion of the nephron. This dilution is accomplished in the ascending limb as NaCl is actively extruded but H_2O cannot follow. Therefore, the loop of Henle, by simultaneously establishing the medullary osmotic gradient and diluting the tubular fluid before it enters the distal segments, plays a key role in allowing the kidneys to excrete urine that ranges in concentration from 100 to 1,200 mosm/liter.

▌ Countercurrent exchange within the vasa recta conserves the medullary vertical osmotic gradient.

Obviously, the renal medulla must be supplied with blood to nourish the tissues in this area as well as to transport water that

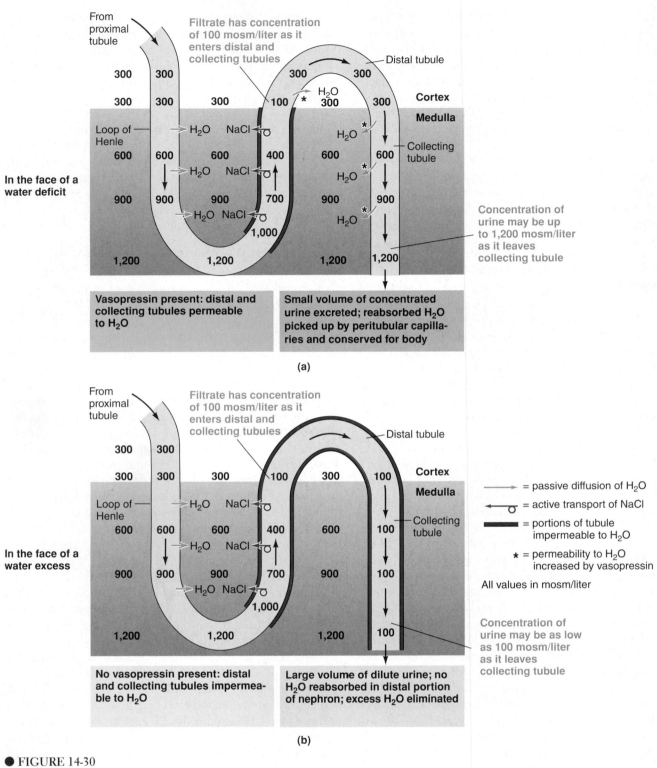

From proximal tubule

Filtrate has concentration of 100 mosm/liter as it enters distal and collecting tubules

Distal tubule

300 300 300 300

300 300 300 100 ∗ H₂O 300 300 **Cortex**

Medulla

Loop of Henle → H₂O NaCl ← H₂O ∗ Collecting tubule

600 600 600 400 600 600

↓ → H₂O NaCl ← H₂O ∗

900 900 900 700 900 900

→ H₂O NaCl ← H₂O ∗

1,000

1,200 1,200 1,200 1,200

In the face of a water deficit

Concentration of urine may be up to 1,200 mosm/liter as it leaves collecting tubule

Vasopressin present: distal and collecting tubules permeable to H₂O

Small volume of concentrated urine excreted; reabsorbed H₂O picked up by peritubular capillaries and conserved for body

(a)

From proximal tubule

Filtrate has concentration of 100 mosm/liter as it enters distal and collecting tubules

Distal tubule

300 300

300 300 300 100 300 100 **Cortex**

Medulla

Loop of Henle → H₂O NaCl ← Collecting tubule

600 600 600 400 600 100

↓ → H₂O NaCl ←

900 900 900 700 900 100

→ H₂O NaCl ←

1,000

1,200 1,200 1,200 100

In the face of a water excess

= passive diffusion of H₂O

= active transport of NaCl

= portions of tubule impermeable to H₂O

∗ = permeability to H₂O increased by vasopressin

All values in mosm/liter

Concentration of urine may be as low as 100 mosm/liter as it leaves collecting tubule

No vasopressin present: distal and collecting tubules impermeable to H₂O

Large volume of dilute urine; no H₂O reabsorbed in distal portion of nephron; excess H₂O eliminated

(b)

● FIGURE 14-30
Excretion of urine of varying concentration depending on the body's needs

is reabsorbed by the loops of Henle and collecting tubules back to the general circulation. In doing so, however, it is important that circulation of blood through the medulla does not disturb the vertical gradient of hypertonicity established by the loops of Henle. Consider the situation if blood were to flow straight through from the cortex to the inner medulla and then

directly into the renal vein (● Figure 14-31a). Because capillaries are freely permeable to NaCl and H₂O, the blood would progressively pick up salt and lose H₂O through passive fluxes down concentration and osmotic gradients as it flowed through the depths of the medulla. Isotonic blood entering the medulla, on equilibrating with each medullary level, would leave the med-

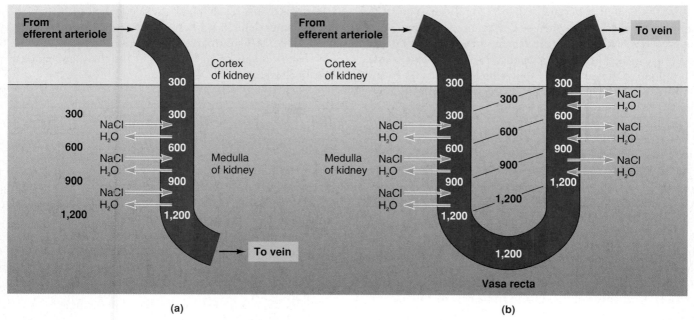

All values in mosm/liter.

● **FIGURE 14-31**

Countercurrent exchange in the renal medulla
(a) Hypothetical pattern of blood flow. If the blood supply to the renal medulla flowed straight through from the cortex to the inner medulla, the blood would be isotonic on entering but very hypertonic on leaving, having picked up salt and lost H_2O as it equilibrated with the surrounding interstitial fluid at each incremental horizontal level. It would be impossible to maintain the vertical osmotic gradient, because the salt pumped out by the ascending limb of Henle's loop would be continuously flushed away by blood flowing through the medulla. (b) Actual pattern of blood flow. Blood equilibrates with the interstitial fluid at each incremental horizontal level in both the descending limb and the ascending limb of the vasa recta, so blood is isotonic as it enters and leaves the medulla. This countercurrent exchange prevents dissolution of the medullary osmotic gradient while providing blood to the renal medulla.

ulla very hypertonic at 1,200 mosm/liter. It would be impossible to establish and maintain the medullary hypertonic gradient, because the NaCl pumped into the medullary interstitial fluid would continuously be carried away by the circulation.

This dilemma is avoided by the hairpin construction of the vasa recta, which, by looping back through the concentration gradient in reverse, allows the blood to leave the medulla and enter the renal vein essentially isotonic to incoming arterial blood (● Figure 14-31b). As blood passes down the descending limb of the vasa recta, equilibrating with the progressively increasing concentration of the surrounding interstitial fluid, it picks up salt and loses H_2O until it is very hypertonic by the bottom of the loop. Then, as blood flows up the ascending limb, salt diffuses back out into the interstitium, and H_2O re-enters the vasa recta as progressively decreasing concentrations are encountered in the surrounding interstitial fluid. This passive exchange of solutes and H_2O between the two limbs of the vasa recta and the interstitial fluid is known as **countercurrent exchange.** Unlike countercurrent multiplication, it does not *establish* the concentration gradient. Rather, it *prevents the dissolution* of the gradient. Because blood enters and leaves the medulla at the same osmolarity as a result of countercurrent exchange, the medullary tissue is nourished with blood, yet the incremental gradient of hypertonicity in the medulla is preserved.

▮ Water and solute reabsorption versus excretion are only partially coupled.

It is important to distinguish between H_2O reabsorption that mandatorily follows solute reabsorption and reabsorption of "free" H_2O not linked to solute reabsorption.

• *In the tubular segments that are permeable to H_2O, solute reabsorption is* **always** *accompanied by comparable H_2O reabsorption because of osmotic considerations.* Therefore, the total volume of H_2O reabsorbed is determined in large part by the total mass of solute reabsorbed; this is especially true of NaCl, because it is the most abundant solute in the ECF.

• *Solute excretion is* **always** *accompanied by comparable H_2O excretion because of osmotic considerations.* This fact is responsible for the obligatory excretion of at least a minimal volume of H_2O, even when a person is severely dehydrated.

For the same reason, when excess unreabsorbed solute is present in the tubular fluid, its presence exerts an osmotic effect to hold excessive H_2O in the lumen. This phenomenon is known as osmotic diuresis. Diuresis is increased urinary excretion, of which there are two types: osmotic diuresis and water diuresis. **Osmotic diuresis** involves increased excretion of both H_2O and solute caused by excess unreabsorbed solute in the tubular fluid, such as occurs in diabetes mellitus. The large quantity of unreabsorbed glucose that remains in the tubular fluid in people with diabetes osmotically drags H_2O with it into the urine. Some diuretic drugs act by blocking specific solute reabsorption so that extra H_2O spills into the urine along with the unreabsorbed solute. **Water diuresis,** in contrast, is increased urinary output of H_2O with little or no increase in excretion of solutes.

• *A loss or gain of pure H_2O that is not accompanied by comparable solute deficit or excess in the body (that is, "free" H_2O) leads to changes in ECF osmolarity.* Such an imbalance between H_2O and solute is corrected by partially dissociating H_2O reabsorption from solute reabsorption in the distal portions of the nephron through the combined effects of vasopressin secretion and the medullary osmotic gradient. Through this mechanism, free H_2O can be reabsorbed without comparable solute reabsorption to ameliorate hypertonicity of the body fluids. Conversely, to rid the body of excess pure H_2O a large quantity of free H_2O can be excreted unaccompanied by comparable solute excretion (that is, water diuresis), thus correcting for hypotonicity of the body fluids. Water diuresis is normally a compensatory response to ingestion of too much H_2O.

Excessive water diuresis accompanies alcohol ingestion. Because alcohol inhibits vasopressin secretion, the kidneys inappropriately lose too much H_2O. Typically, more fluid is lost in the urine than is consumed in the alcoholic beverage, so the body becomes dehydrated despite substantial fluid ingestion.

▲ Table 14-4 summarizes how various tubular segments of the nephron handle Na^+ and H_2O and the significance of these processes.

▌ Renal failure has wide-ranging consequences.

Urine excretion and the resultant clearance of wastes and excess electrolytes from the plasma are crucial for maintaining homeostasis. When the functions of both kidneys are so disrupted that they cannot perform their regulatory and excretory functions sufficiently to maintain homeostasis, **renal failure** is said to exist. Renal failure has a variety of causes, some of which begin elsewhere in the body and affect renal function secondarily. Among the causes are the following:

1. *Infectious organisms,* either blood-borne or gaining entrance to the urinary tract through the urethra.
2. *Toxic agents,* such as lead, arsenic, pesticides, or even long-term exposure to high doses of aspirin.
3. *Inappropriate immune responses,* such as *glomerulonephritis,* which occasionally follows streptococcal throat infections as antigen-antibody complexes leading to localized inflammatory damage are deposited in the glomeruli (see p. 428).

▲ TABLE 14-4
Handling of Sodium and Water by Various Tubular Segments of the Nephron

Tubular Segment	Na$^+$ Reabsorption		H$_2$O Reabsorption	
	Percentage of reabsorption in this segment	Distinguishing features	Percentage of reabsorption in this segment	Distinguishing features
Proximal Tubule	67	Active; uncontrolled; plays a pivotal role in the reabsorption of glucose, amino acids, Cl$^-$, H$_2$O, and urea	65	Passive; obligatory osmotic reabsorption following active Na$^+$ reabsorption
Loop of Henle	25	Active, uncontrolled; Na$^+$ along with Cl$^-$ reabsorption from the ascending limb helps establish the medullary interstitial vertical osmotic gradient, which is important in the kidneys' ability to produce urine of varying concentrations and volumes, depending on the body's needs	15	Passive; obligatory osmotic reabsorption from the descending limb as the ascending limb extrudes NaCl into the interstitial fluid (that is, reabsorbs NaCl)
Distal and Collecting Tubules	8	Active; variable and subject to aldosterone control; important in the regulation of ECF volume and long-term control of blood pressure; linked to K$^+$ secretion and H$^+$ secretion	20	Passive; not linked to solute reabsorption; variable quantities of "free" H$_2$O reabsorption subject to vasopressin control; driving force is the vertical osmotic gradient in the medullary interstitial fluid established by the long loops of Henle; important in regulating ECF osmolarity

4. *Obstruction of urine flow* by kidney stones, tumors, or an enlarged prostate gland, with back pressure reducing glomerular filtration as well as damaging renal tissue.

5. An *insufficient renal blood supply* that leads to inadequate filtration pressure. This can occur secondary to circulatory disorders such as heart failure, hemorrhage, shock, or narrowing and hardening of the renal arteries by atherosclerosis.

Although these conditions may have different origins, almost all can cause some degree of nephron damage. The glomeruli and tubules may be independently affected or both may be dysfunctional. Regardless of cause, renal failure can manifest itself either as *acute renal failure*, characterized by a sudden onset with a rapid reduction in urine formation until less than the essential minimum of around 500 ml of urine is being produced per day, or *chronic renal failure*, characterized by slow, progressive, insidious loss of renal function. A person may die from acute renal failure, or the condition may be reversible

and lead to full recovery. Chronic renal failure, in contrast, is not reversible. Gradual, permanent destruction of renal tissue eventually proves fatal. Chronic renal failure is insidious because up to 75% of the kidney tissue can be destroyed before the loss of kidney function is even noticeable. Because of the abundant reserve of kidney function, only 25% of the kidney tissue is needed to adequately maintain all the essential renal excretory and regulatory functions. With less than 25% of functional kidney tissue remaining, however, renal insufficiency becomes apparent. *End-stage renal failure* ensues when 90% of kidney function has been lost.

We will not sort out the different stages and symptoms associated with various renal disorders, but ▲ Table 14-5, which summarizes the potential consequences of renal failure, will give you an idea of the broad effects that kidney impairment can have. The extent of these effects should not be surprising, considering the central role the kidneys play in maintaining homeostasis. When the kidneys cannot maintain a normal in-

---

▲ **TABLE 14-5**
Potential Ramifications of Renal Failure

Uremic toxicity caused by retention of waste products

    Nausea, vomiting, diarrhea, and ulcers caused by a toxic effect on the digestive system

    Bleeding tendency arising from a toxic effect on platelet function

    Mental changes—such as reduced alertness, insomnia, and shortened attention span, progressing to convulsions and coma—caused by toxic effects on the central nervous system

    Abnormal sensory and motor activity caused by a toxic effect on the peripheral nerves

Metabolic acidosis* caused by the inability of the kidneys to adequately secrete H^+ that is continually being added to the body fluids as a result of metabolic activity

    Altered enzyme activity caused by the action of too much acid on enzymes

    Depression of the central nervous system caused by the action of too much acid interfering with neuronal excitability

Potassium retention* resulting from inadequate tubular secretion of K^+

    Altered cardiac and neural excitability as a result of changing the resting membrane potential of excitable cells

Sodium imbalances caused by the inability of the kidneys to adjust Na^+ excretion to balance the changes in Na^+ consumption

    Elevated blood pressure, generalized edema, and congestive heart failure if too much Na^+ is consumed

    Hypotension and, if severe enough, circulatory shock if too little Na^+ is consumed

Phosphate and calcium imbalances arising from impaired reabsorption of these electrolytes

    Disturbances in skeletal structures caused by abnormalities in deposition of calcium phosphate crystals, which harden bone

Loss of plasma proteins as a result of increased "leakiness" of the glomerular membrane

    Edema caused by a reduction in plasma-colloid osmotic pressure

Inability to vary urine concentration as a result of impairment of the countercurrent system

    Hypotonicity of body fluids if too much H_2O is ingested

    Hypertonicity of body fluids if too little H_2O is ingested

Hypertension arising from the combined effects of salt and fluid retention and vasoconstrictor action of excess angiotensin II

Anemia caused by inadequate erythropoietin production

Depression of the immune system, most likely caused by toxic levels of wastes and acids

    Increased susceptibility to infections

---

*Among the most life-threatening consequences of renal failure.

When Protein in the Urine Does Not Mean Kidney Disease

Urinary loss of proteins usually signifies kidney disease (nephritis). However, a urinary protein loss similar to that of nephritis often occurs following exercise, but the condition is harmless, transient, and reversible. The term *athletic pseudonephritis* is used to describe this postexercise (after exercise) proteinuria (protein in the urine). Studies indicate that 70% to 80% of athletes have proteinuria after very strenuous exercise. This condition occurs in participants in both noncontact and contact sports, so it does not arise from physical trauma to the kidneys. In one study, subjects who engaged in maximal short-term running excreted more protein than when they were bicycling, rowing, or swimming at the same work intensity. The reason for this difference is unknown.

Usually, only a very small fraction of the plasma proteins that enter the glomerulus is filtered; those that are filtered are reabsorbed in the tubules, so normally no plasma proteins appear in the urine. Two basic mechanisms can cause proteinuria: (1) increased glomerular permeability with no change in tubular reabsorption; or (2) impairment of tubular reabsorption. Research has shown the proteinuria that occurs during mild to moderate exercise results from changes in glomerular permeability, whereas the proteinuria during short-term exhaustive exercise seems to be caused by both increased glomerular permeability and tubular dysfunction.

This reversible kidney dysfunction is believed to result from circulatory and hormonal changes that occur with exercise. Several studies have shown that renal blood flow is reduced during exercise as the renal vessels are constricted and blood is diverted to the exercising muscles. This reduction is positively correlated with exercise intensity. With intense exercise, the renal blood flow may be reduced to 20% of normal. As a result, glomerular blood flow is also reduced, but not to the same extent as renal blood flow, presumably because of autoregulatory mechanisms. Some investigators propose that decreased glomerular blood flow enhances diffusion of proteins into the tubular lumen, because as the more slowly flowing blood spends more time in the glomerulus, a greater proportion of the plasma proteins has time to escape through the glomerular membrane. Hormonal changes that occur with exercise may also affect glomerular permeability. For example, renin injection is a well-recognized way to experimentally induce proteinuria. Plasma renin activity increases during strenuous exercise and may contribute to postexercise proteinuria. It is also hypothesized that maximal tubular reabsorption is reached during severe exercise, which could impair protein reabsorption.

---

ternal environment, widespread disruption of cellular activities can bring about abnormal function in other organ systems, as well. By the time end-stage renal failure occurs, literally every body system has become impaired to some extent. (One symptom of renal disease occurs during strenuous exercise, but it is transient and harmless. See the accompanying boxed feature, ▶ A Closer Look at Exercise Physiology.)

Because chronic renal failure is irreversible and eventually fatal, treatment is aimed at maintaining renal function by alternative methods, such as dialysis and kidney transplantation. (For further explanation of these procedures, see the other boxed feature, ▶ Concepts, Challenges, and Controversies.)

This finishes our discussion of kidney function. For the remainder of the chapter, we will focus on the "plumbing" that stores and carries the urine formed by the kidneys to the outside.

▍ Urine is temporarily stored in the bladder, from which it is emptied by micturition.

Once urine has been formed by the kidneys, it is transmitted through the ureters to the urinary bladder. Urine does not flow through the ureters by gravitational pull alone. Peristaltic contractions of the smooth muscle within the ureteral wall propel the urine forward from the kidneys to the bladder. The ureters penetrate the wall of the bladder obliquely, coursing through the wall several centimeters before they open into the bladder cavity. This anatomic arrangement prevents backflow of urine from the bladder to the kidneys when pressure builds up in the bladder. As the bladder fills, the ureteral ends within its wall are compressed closed. Urine can still enter, however, because ureteral contractions generate enough pressure to overcome the resistance and push urine through the occluded ends.

Role of the bladder

The bladder wall consists of smooth muscle lined by a special type of epithelium. It was once assumed that the bladder was an inert sac. However, both the epithelium and the smooth muscle actively participate in the bladder's ability to accommodate large fluctuations in urine volume. The epithelial lining can increase and decrease in surface area by the orderly process of membrane recycling as the bladder alternately fills and empties. Membrane-enclosed cytoplasmic vesicles are inserted by exocytosis into the surface area during bladder filling; then the vesicles are withdrawn by endocytosis to shrink the surface area following emptying (see p. 84). As is characteristic of smooth muscle, bladder muscle can stretch tremendously without building up bladder wall tension. In addition, the highly folded bladder wall flattens out during filling to increase bladder storage capacity. Because urine is continuously being formed by the kidneys, the bladder must have sufficient storage capacity to preclude the necessity for continual evacuation of the urine.

The bladder smooth muscle is richly supplied by parasympathetic fibers, stimulation of which causes bladder contraction. If the passageway through the urethra to the outside is open, bladder contraction brings about emptying of urine from the bladder. The exit from the bladder, however, is guarded by two sphincters, the *internal urethral sphincter* and the *external urethral sphincter*.

Role of the urethral sphincters

A **sphincter** is a ring of muscle that, when contracted, closes off passage through an opening. The **internal urethral sphincter**—which is composed of smooth muscle and, accordingly, is under involuntary control—is not really a separate muscle but instead consists of the last portion of the bladder. Although it is not a true sphincter, it performs the same function as a sphincter. When

Dialysis: Cellophane Tubing or Abdominal Lining as an Artificial Kidney

Because chronic renal failure is irreversible and eventually fatal, treatment is aimed at maintaining renal function by alternative methods, such as dialysis and kidney transplantation. The process of dialysis bypasses the kidneys to maintain normal fluid and electrolyte balance and remove wastes artificially. In the original method of dialysis, **hemodialysis,** a patient's blood is pumped through cellophane tubing that is surrounded by a large volume of fluid similar in composition to normal plasma. Following dialysis, the blood is returned to the patient's circulatory system. Like capillaries, cellophane is highly permeable to most plasma constituents but is impermeable to plasma proteins. As blood flows through the tubing, solutes move across the cellophane down their individual concentration gradients; plasma proteins, however, remain in the blood. Urea and other wastes, which are absent in the dialysis fluid, diffuse out of the plasma into the surrounding fluid, cleansing the blood of these wastes. Plasma constituents that are not regulated by the kidneys and are at normal concentration, such as glucose, do not move across the cellophane into the dialysis fluid, because there is no driving force to produce their movement. (The dialysis fluid's glucose concentration is the same as normal plasma glucose concentration.) Electrolytes, such as K^+ and PO_4^{3-}, which are higher than their normal plasma concentrations because the diseased kidneys cannot eliminate excess quantities of

these substances, move out of the plasma until equilibrium is achieved between the plasma and the dialysis fluid. Because the dialysis fluid's solute concentrations are maintained at normal plasma values, the solute concentration of the blood returned to the patient following dialysis is essentially normal. Hemodialysis is repeated as often as necessary to maintain the plasma composition within an acceptable level. Typically, it is done three times per week for several hours at each session.

In a more recent method of dialysis, **continuous ambulatory peritoneal dialysis (CAPD),** the peritoneal membrane (the lining of the abdominal cavity) is used as the dialysis membrane. With this method, 2 liters of dialysis fluid are inserted into the patient's abdominal cavity through a permanently implanted catheter. Urea, K^+, and other wastes and excess electrolytes diffuse from the plasma across the peritoneal membrane into the dialysis fluid, which is drained off and replaced several times a day. The CAPD method offers several advantages: The patient can self-administer it, the patient's blood is continuously purified and adjusted, and the patient can engage in normal activities while dialysis is being accomplished. One drawback is the increased risk of peritoneal infections.

Although dialysis can remove metabolic wastes and foreign compounds and help maintain fluid and electrolyte balance within acceptable limits, this plasma-cleansing technique can-

not make up for the failing kidneys' reduced ability to produce hormones (erythropoietin and renin) and to activate vitamin D. One new experimental technique incorporates living kidney cells derived from pigs within a dialysis-like machine. Standard ultrafiltration technology like that used in hemodialysis purifies and adjusts the plasma as usual. Importantly, the living cells not only help maintain even better control of plasma constituents, especially K^+, but also add the deficient renal hormones to the plasma passing through the machine and activate vitamin D as well. This promising new technology has not yet been tested in large-scale clinical trials.

For now, transplanting a healthy kidney from a donor is another option for treating chronic renal failure. A kidney is one of the few transplants that can be provided by a living donor. Because 25% of the total kidney tissue can maintain the body, both the donor and the recipient have ample renal function with only one kidney each. The biggest problem with transplantation is the possibility that the patient's immune system will reject the organ. Risk of rejection can be minimized by matching the tissue types of the donor and the recipient as closely as possible (the best donor choice is usually a close relative), coupled with immunosuppressive drugs.

---

the bladder is relaxed, the anatomic arrangement of the internal urethral sphincter region closes the outlet of the bladder.

Farther down the passageway, the urethra is encircled by a layer of skeletal muscle, the **external urethral sphincter.** This sphincter is reinforced by the entire **pelvic diaphragm,** a skeletal muscle sheet that forms the floor of the pelvis and helps support the pelvic organs (see ● Figure 14-2). The motor neurons that supply the external sphincter and pelvic diaphragm fire continuously at a moderate rate unless they are inhibited, keeping these muscles tonically contracted so they prevent urine from escaping through the urethra. Normally, when the bladder is relaxed and filling, closure of both the internal and external urethral sphincters prevents urine from dribbling out. Furthermore, because they are skeletal muscles, the external sphincter and pelvic diaphragm are under voluntary control. They can be deliberately tightened to prevent urination from occurring even when the bladder is contracting and the internal sphincter is open.

Micturition reflex

Micturition, or **urination,** the process of bladder emptying, is governed by two mechanisms: the micturition reflex and voluntary control. The **micturition reflex** is initiated when stretch re-

ceptors within the bladder wall are stimulated (● Figure 14-32). The bladder in an adult can accommodate up to 250 to 400 ml of urine before the tension within its walls begins to rise sufficiently to activate the stretch receptors (● Figure 14-33). The greater the distension beyond this, the greater the extent of receptor activation. Afferent fibers from the stretch receptors carry impulses into the spinal cord and eventually, by interneurons, stimulate the parasympathetic supply to the bladder and inhibit the motor neuron supply to the external sphincter. Parasympathetic stimulation of the bladder causes it to contract. No special mechanism is required to open the internal sphincter; changes in the shape of the bladder during contraction mechanically pull the internal sphincter open. Simultaneously, the external sphincter relaxes as its motor neuron supply is inhibited. Now both sphincters are open, and urine is expelled through the urethra by the force of bladder contraction. This micturition reflex, which is entirely a spinal reflex, governs bladder emptying in infants. As soon as the bladder fills sufficiently to trigger the reflex, the baby automatically wets.

Voluntary control of micturition

In addition to triggering the micturition reflex, bladder filling also gives rise to the conscious urge to urinate. The perception

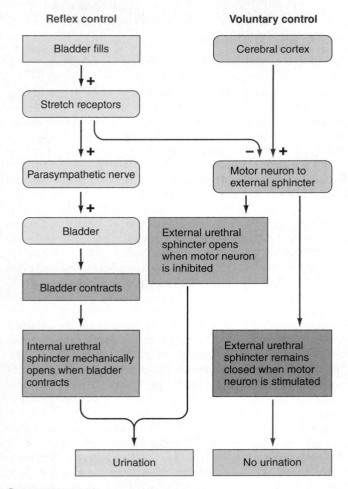

● FIGURE 14-32
Reflex and voluntary control of micturition

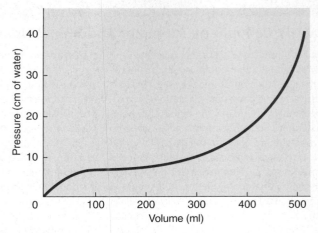

● FIGURE 14-33
Pressure changes within the urinary bladder as the bladder fills with urine

of bladder fullness appears before the external sphincter reflexly relaxes, "warning" that micturition is imminent. As a result, voluntary control of micturition, learned during toilet training in early childhood, can override the micturition reflex so that bladder emptying can take place at the person's convenience rather than when bladder filling first activates the stretch receptors. If the time when the micturition reflex is initiated is inopportune for urination, the person can voluntarily prevent bladder emptying by deliberate tightening the external sphincter and pelvic diaphragm. Voluntary excitatory impulses from the cerebral cortex override the reflex inhibitory input from the stretch receptors to the involved motor neurons (the relative balance of EPSPs and IPSPs), keeping these muscles contracted so that no urine is expelled.

Urination cannot be delayed indefinitely. As the bladder continues to fill, reflex input from the stretch receptors increases with time. Finally, reflex inhibitory input to the external-sphincter motor neuron becomes so powerful that it can no longer be overridden by voluntary excitatory input, so the sphincter relaxes and the bladder uncontrollably empties.

Micturition can also be deliberately initiated, even though the bladder is not distended, by voluntarily relaxing the external sphincter and pelvic diaphragm. Lowering of the pelvic floor allows the bladder to drop downward, which simultaneously pulls open the internal urethral sphincter and stretches the bladder wall. The subsequent activation of the stretch receptors brings about bladder contraction by means of the micturition reflex. Voluntary bladder emptying may be further assisted by contracting the abdominal wall and respiratory diaphragm. The resultant increase in intra-abdominal pressure "squeezes down" on the bladder to facilitate its emptying.

Urinary incontinence

Urinary incontinence, or inability to prevent discharge of urine, occurs when descending pathways in the spinal cord that mediate voluntary control of the external sphincter and pelvic diaphragm are disrupted, as in spinal cord injury. Because the components of the micturition reflex arc are still intact in the lower spinal cord, bladder emptying becomes governed by an uncontrollable spinal reflex, as in infants. A lesser degree of incontinence characterized by urine escaping when the bladder pressure suddenly increases transiently, such as during coughing or sneezing, can result from impaired sphincter function. This is common in women who have borne children or in men whose sphincters have been injured during prostate surgery.

CHAPTER IN PERSPECTIVE: FOCUS ON HOMEOSTASIS

The kidneys contribute to homeostasis more extensively than any other single organ. They regulate the electrolyte composition, volume, osmolarity, and pH of the internal environment and eliminate all the waste products of bodily metabolism, with the exception of respiration-removed CO_2. They accomplish these regulatory functions by eliminating in the urine substances the body doesn't need, such as metabolic wastes and excess quantities of ingested salt or water, while conserving useful substances. The kidneys can maintain the plasma constituents they regulate within the narrow range compatible with life, despite wide variations in intake and losses of these substances through other avenues. Illustrating the magnitude of the kidneys' task, about a quarter of the blood pumped into the systemic circulation goes to the kidneys to be "adjusted" and "purified," with only three-quarters of the blood being used to supply all the other tissues.

The kidneys contribute to homeostasis in the following specific ways:

▌ Regulatory Functions

- The kidneys regulate the quantity and concentration of most ECF electrolytes, including those important in maintaining proper neuromuscular excitability.

- They contribute to maintenance of proper pH by eliminating excess H^+ (acid) or HCO_3^- (base) in the urine.

- They help maintain proper plasma volume, which is important in long-term regulation of arterial blood pressure, by controlling salt balance in the body. The ECF volume, including plasma volume, reflects total salt load in the ECF, because Na^+ and its attendant anion Cl^- are responsible for over 90% of the ECF's osmotic (water-holding) activity.

- The kidneys maintain water balance in the body, which is important in maintaining proper ECF osmolarity (concentration of solutes). This role is important in maintaining stability of cell volume by preventing cells from swelling or shrinking as water osmotically moves in or out of the cells, respectively.

▌ Excretory functions

- The kidneys excrete the end products of metabolism in urine. If allowed to accumulate, these wastes are toxic to cells.

- The kidneys also excrete many foreign compounds that enter the body.

▌ Hormonal functions

- The kidneys produce erythropoietin, the hormone that stimulates bone marrow to produce red blood cells. This action contributes to homeostasis by helping maintain the optimal O_2 content of blood. Over 98% of O_2 in blood is bound to hemoglobin within red blood cells.

- They also produce renin, the hormone that initiates the renin-angiotensin-aldosterone pathway for controlling renal tubular Na^+ reabsorption, which is important in long-term maintenance of plasma volume and arterial blood pressure.

▌ Metabolic functions

- The kidneys help convert vitamin D into its active form. Vitamin D is essential for Ca^{2+} absorption from the digestive tract. Calcium, in turn, exerts a wide variety of homeostatic functions.

CHAPTER SUMMARY

Introduction

▌ The kidneys eliminate unwanted plasma constituents in urine while conserving materials of value to the body.

▌ The urine-forming functional unit of the kidneys is the nephron, which is composed of interrelated vascular and tubular components.

▌ The vascular component consists of two capillary networks in series, the first being the glomerulus, a tuft of capillaries that filters large volumes of protein-free plasma into the tubular component. The second capillary network consists of the peritubular capillaries, which wind around the tubular component. The peritubular capillaries nourish the renal tissue and participate in exchanges between the tubular fluid and plasma.

▌ The tubular component begins with Bowman's capsule, which cups around the glomerulus to catch the filtrate, then continues a specific tortuous course to ultimately empty into the renal pelvis. As the filtrate passes through various regions of the tubule, cells lining the tubules modify it, returning to the plasma only those materials necessary for maintaining proper ECF composition and volume. What is left behind in the tubules is excreted as urine.

▌ The kidneys perform three basic processes in carrying out their regulatory and excretory functions: (1) glomerular filtration, the nondiscriminating movement of protein-free plasma from blood into the tubules; (2) tubular reabsorption, the selective transfer of specific constituents in filtrate back into the blood of the peritubular capillaries; and (3) tubular secretion, the highly specific movement of selected substances from peritubular capillary blood into the tubular fluid. Everything filtered or secreted but not reabsorbed is excreted as urine.

Glomerular Filtration

▌ Glomerular filtrate is produced as a portion of the plasma flowing through each glomerulus is passively forced under pressure through the glomerular membrane into the lumen of the underlying Bowman's capsule.

▌ The net filtration pressure that induces filtration is caused by an imbalance in physical forces acting across the glomerular membrane. A high glomerular capillary blood pressure favoring filtration outweighs the combined opposing forces of plasma-colloid osmotic pressure and Bowman's capsule hydrostatic pressure.

▌ Typically, 20% to 25% of the cardiac output is delivered to the kidneys to be acted on by renal regulatory and excretory processes.

▌ Of the plasma flowing through the kidneys, normally 20% is filtered through the glomeruli, producing an average glomerular filtration rate (GFR) of 125 ml/min. This filtrate is identical in composition to plasma except for plasma proteins held back by the glomerular membrane.

▌ The GFR can be deliberately altered by changing the glomerular capillary blood pressure by sympathetic influence on the afferent arterioles. Afferent arteriolar vasoconstriction decreases flow of blood into the glomerulus, lowering glomerular blood pressure and GFR. Conversely, afferent arteriolar vasodilation leads to increased glomerular blood flow and a rise in GFR.

- Sympathetic control of the GFR is part of the baroreceptor reflex response that compensates for changed arterial blood pressure. As the GFR is altered, the amount of fluid lost in urine is changed correspondingly, adjusting plasma volume as needed to help restore blood pressure to normal on a long-term basis.

Tubular Reabsorption

- After a protein-free plasma is filtered through the glomerulus, the tubules handle each substance discretely, so that even though the concentrations of all constituents in the initial glomerular filtrate are identical to their concentrations in the plasma (with the exception of plasma proteins), the concentrations of different constituents are variously altered as the filtered fluid flows through the tubular system.
- The reabsorptive capacity of the tubular system is tremendous. Over 99% of the filtered plasma is returned to the blood through reabsorption. On average, 124 ml out of the 125 ml filtered per minute are reabsorbed.
- Tubular reabsorption involves transepithelial transport from the tubular lumen into the peritubular capillary plasma. This process may be active (requiring energy) or passive (using no energy).
- The pivotal event to which most reabsorptive processes are linked in some way is the active reabsorption of Na^+. An energy-dependent Na^+–K^+ ATPase carrier located in the basolateral membrane of almost all tubular cells transports Na^+ out of the cells into the lateral spaces between adjacent cells. This transport of Na^+ induces the net reabsorption of Na^+ from the tubular lumen to the peritubular capillary plasma, most of which takes place in the proximal tubules.
- Early in the nephron, Na^+ reabsorption occurs in constant unregulated fashion, but in the distal and collecting tubules, the reabsorption of a small percentage of the filtered Na^+ is variable and subject to control. The extent of this controlled Na^+ reabsorption depends primarily on the complex renin-angiotensin-aldosterone system.
- Because Na^+ and its attendant anion, Cl^-, are the major osmotically active ions in the ECF, the ECF volume is determined by the Na^+ load in the body. In turn, the plasma volume, which reflects the total ECF volume, is important in the long-term determination of arterial blood pressure. Whenever the Na^+ load, ECF volume, plasma volume, and arterial blood pressure are below normal, the kidneys secrete renin, an enzymatic hormone that triggers a series of events ultimately leading to increased secretion of aldosterone from the adrenal cortex. Aldosterone increases Na^+ reabsorption from the distal portions of the tubule, thus correcting for the original reduction in Na^+, ECF volume, and blood pressure.
- By contrast, sodium reabsorption is inhibited by atrial natriuretic peptide, a hormone released from the cardiac atria in response to expansion of the ECF volume and a subsequent increase in blood pressure.
- In addition to driving the reabsorption of Na^+, the energy used to supply the Na^+–K^+ ATPase carrier is also ultimately responsible for the reabsorption of organic nutrient molecules from the proximal tubule by secondary active transport. Specific cotransport carriers located at the luminal border of the proximal tubular cell are driven by the Na^+ concentration gradient to selectively transport glucose or an amino acid from the luminal fluid into the tubular cell, from which the nutrient eventually enters the plasma.
- The other electrolytes besides Na^+ that are actively reabsorbed by the tubules, such as PO_4^{3-} and Ca^{2+}, have their own independently functioning carrier systems within the proximal tubule.

- Because these carriers, like the organic-nutrient cotransport carriers, can become saturated, each exhibits a maximal carrier-limited transport capacity, or T_m. Once the filtered load of an actively reabsorbed substance exceeds the T_m, reabsorption proceeds at a constant maximal rate, with the additional filtered quantity of the substance being excreted in urine.
- Active Na^+ reabsorption also drives the passive reabsorption of Cl^-, H_2O, and urea.
- Chloride is passively reabsorbed down the electrical gradient established by active Na^+ reabsorption.
- Water is passively reabsorbed as a result of the osmotic gradient created by active Na^+ reabsorption. Sixty-five percent of the filtered H_2O is reabsorbed from the proximal tubule in this unregulated fashion.
- This extensive reabsorption of H_2O increases the concentration of other substances remaining in the tubular fluid, most of which are filtered waste products. The small urea molecules are the only waste products that can passively permeate the tubular membranes. Accordingly, urea is the only waste product partially reabsorbed as a result of this concentration effect; about 50% of the filtered urea is reabsorbed.
- The other waste products, which are not reabsorbed, remain in urine in highly concentrated form.

Tubular Secretion

- Tubular secretion also involves transepithelial transport, in this case from the peritubular capillary plasma into the tubular lumen.
- By tubular secretion, the kidney tubules can selectively add some substances to the quantity already filtered. Secretion of substances hastens their excretion in urine.
- The most important secretory systems are for (1) H^+, which is important in regulating acid–base balance; (2) K^+, which keeps the plasma K^+ concentration at an appropriate level to maintain normal membrane excitability in muscles and nerves; and (3) organic ions, which accomplishes more efficient elimination of foreign organic compounds from the body.
- Hydrogen ion is secreted in the proximal, distal, and collecting tubules. K^+ is secreted only in the distal and collecting tubules under control of aldosterone. Organic ions are secreted only in the proximal tubule.

Urine Excretion and Plasma Clearance

- Of the 125 ml/min filtered in the glomeruli, normally only 1 ml/min remains in the tubules to be excreted as urine.
- Only wastes and excess electrolytes not wanted by the body are left behind, dissolved in a given volume of H_2O to be eliminated in urine.
- Because the excreted material is removed or "cleared" from the plasma, the term *plasma clearance* refers to the volume of plasma cleared of a particular substance each minute by renal activity.
- The kidneys can excrete urine of varying volumes and concentrations to either conserve or eliminate H_2O, depending on whether the body has a H_2O deficit or excess, respectively. The kidneys can produce urine ranging from 0.3 ml/min at 1,200 mosm/liter to 25 ml/min at 100 mosm/liter by reabsorbing variable amounts of H_2O from the distal portions of the nephron.
- This variable reabsorption is made possible by a vertical osmotic gradient ranging from 300 to 1,200 mosm/liter in the medullary interstitial fluid, established by the long loops of Henle of the juxtamedullary nephrons by means of the countercurrent system.
- This vertical osmotic gradient, to which the hypotonic (100 mosm/liter) tubular fluid is exposed as it passes through the distal por-

tions of the nephron, establishes a passive driving force for progressive reabsorption of H_2O from the tubular fluid, but the actual extent of H_2O reabsorption depends on the amount of vasopressin (antidiuretic hormone) secreted.

■ Vasopressin increases permeability of the distal and collecting tubules to H_2O; they are impermeable to H_2O in its absence. Vasopressin secretion increases in response to a H_2O deficit, and H_2O reabsorption increases accordingly. Vasopressin secretion is inhibited in response to a H_2O excess, reducing H_2O reabsorption. In this way, adjustments in vasopressin-controlled H_2O reabsorption help correct any fluid imbalances.

■ Once formed, urine is propelled by peristaltic contractions through the ureters from the kidneys to the urinary bladder for temporary storage.

■ The bladder can accommodate up to 250 to 400 ml of urine before stretch receptors within its wall initiate the micturition reflex. This reflex causes involuntary emptying of the bladder by simultaneous bladder contraction and opening of both the internal and external urethral sphincters.

■ Micturition can transiently be voluntarily prevented until a more opportune time for bladder evacuation by deliberate tightening of the external sphincter and surrounding pelvic diaphragm.

REVIEW EXERCISES

Objective Questions (Answers on p. A-46)

1. Part of the kidneys' energy supply is used to accomplish glomerular filtration. *(True or false?)*
2. Sodium reabsorption is under hormonal control throughout the length of the tubule. *(True or false?)*
3. Glucose and amino acids are reabsorbed by secondary active transport. *(True or false?)*
4. Solute excretion is always accompanied by comparable H_2O excretion. *(True or false?)*
5. Water excretion can occur without comparable solute excretion. *(True or false?)*
6. The functional unit of the kidneys is the _____.
7. _____ is the only ion actively reabsorbed in the proximal tubule and actively secreted in the distal and collecting tubules.
8. The daily minimum volume of obligatory H_2O loss that must accompany excretion of wastes is _____ ml.
9. Indicate whether each of the following factors would (a) increase or (b) decrease the GFR, if everything else remained constant.
   ____ 1. a rise in Bowman's capsule pressure resulting from ureteral obstruction by a kidney stone
   ____ 2. a fall in plasma protein concentration resulting from loss of these proteins from a large burned surface of skin
   ____ 3. a dramatic fall in arterial blood pressure following severe hemorrhage (<80 mm Hg)
   ____ 4. afferent arteriolar vasoconstriction
   ____ 5. tubuloglomerular feedback response to a reduction in tubular flow rate
   ____ 6. myogenic response of an afferent arteriole stretched as a result of an increased driving blood pressure
   ____ 7. ↑sympathetic activity to the afferent arterioles
   ____ 8. contraction of mesangial cells
   ____ 9. contraction of podocytes
10. Which of the following filtered substances is normally *not* present in the urine at all?
    a. Na^+
    b. PO_4^{3-}
    c. urea
    d. H^+
    e. glucose
11. Reabsorption of which of the following substances is *not* linked in some way to active Na^+ reabsorption?
    a. glucose
    b. PO_4^{3-}
    c. H_2O

    d. urea
    e. Cl^-

In questions 12–14, indicate the proper sequence through which fluid flows as it traverses the structures in question by writing the identifying letters in the proper order in the blanks.

12. a. ureter   ____ ____ ____ ____ ____
    b. kidney
    c. urethra
    d. bladder
    e. renal pelvis
13. a. efferent arteriole   ____ ____ ____ ____ ____ ____
    b. peritubular capillaries
    c. renal artery
    d. glomerulus
    e. afferent arteriole
    f. renal vein
14. a. loop of Henle   ____ ____ ____ ____ ____ ____ ____
    b. collecting duct
    c. Bowman's capsule
    d. proximal tubule
    e. renal pelvis
    f. distal tubule
    g. glomerulus
15. Using the following answer code, indicate what the osmolarity of the tubular fluid is at each of the designated points in a nephron:
    (a) isotonic (300 mosm/liter)
    (b) hypotonic (100 mosm/liter)
    (c) hypertonic (1,200 mosm/liter)
    (d) ranging from hypotonic to hypertonic (100 mosm/liter to 1,200 mosm/liter)
    1. Bowman's capsule
    2. end of proximal tubule
    3. tip of Henle's loop of juxtamedullary nephron (at the bottom of the U-turn)
    4. end of Henle's loop of juxtamedullary nephron (before entry into distal tubule)
    5. end of collecting duct

Essay Questions

1. List the functions of the kidneys.
2. Describe the anatomy of the urinary system. Describe the components of a nephron.
3. Describe the three basic renal processes; indicate how they relate to urine excretion. Distinguish between *secretion* and *excretion*.

4. Discuss the forces involved in glomerular filtration. What is the average GFR?
5. How is GFR regulated as part of the baroreceptor reflex?
6. Why do the kidneys receive a seemingly disproportionate share of the cardiac output? What percentage of renal blood flow is normally filtered?
7. List the steps in transepithelial transport.
8. Distinguish between active and passive reabsorption.
9. Describe all the tubular transport processes that are linked to the basolateral Na^+-K^+ ATPase carrier.
10. Describe the renin-angiotensin-aldosterone system. What are the source and function of atrial natriuretic peptide?
11. To what do the terms *tubular maximum* (T_m) and *renal threshold* refer? Compare two substances that display a T_m, one substance that *is* and one that *is not* regulated by the kidneys.
12. What is the importance of tubular secretion? What are the most important secretory processes?
13. What is the average rate of urine formation?
14. Define *plasma clearance*.
15. What is responsible for the presence of a vertical osmotic gradient in the medullary interstitial fluid? Of what importance is this gradient?
16. Discuss the function of vasopressin.
17. Describe the transfer of urine to, the storage of urine in, and the emptying of urine from the bladder.

Quantitative Exercises (Solutions on p. A-46)
1. Two patients are voiding protein in their urine. To determine whether or not this proteinuria indicates a serious problem, a physician injects small amounts of inulin and PAH into each patient. Recall that inulin is freely filtered and neither secreted nor reabsorbed in the nephron and that PAH at this concentration is completely removed from the blood by tubular secretion. The data collected are given in the following table, where $[x]_u$ is the concentration of substance X (either inulin or PAH) in the urine (in mM); $[x]_p$ is the concentration of X in the plasma, and v_u is the flow rate of urine (in ml/min).

Patient	$[I]_u$	$[I]_p$	$[PAH]_u$	$[PAH]_p$	v_u
1	25	2	186	3	10
2	31	1.5	300	4.5	6

a. Calculate each patient's GFR and renal plasma flow.
b. Calculate the renal blood flow for each patient, assuming both have a hematocrit of 0.45.
c. Calculate the filtration fraction for each patient.
d. Which of the values calculated for each patient are within the normal range? Which values are abnormal? What could be causing these deviations from normal?
2. What is the filtered load of sodium if inulin clearance is 125 ml/min and the sodium concentration in plasma is 145mM?
3. Calculate a patient's rate of urine production, given that his inulin clearance is 125 ml/min and his urine and plasma concentrations of inulin are 300 mg/liter and 3 mg/liter, respectively.
4. If the urine concentration of a substance is 7.5 mg/ml of urine, its plasma concentration is 0.2 mg/ml of plasma, and the urine flow rate is 2 ml/min, what is the clearance rate of the substance? Is the substance being reabsorbed or secreted by the kidneys?

POINTS TO PONDER

(Explanations on p. A-47)
1. The long-looped nephrons of animals adapted to survive with minimal water consumption, such as desert rats, have relatively much longer loops of Henle than humans have. Of what benefit would these longer loops be?
2. If the plasma concentration of substance X is 200 mg/100 ml and the GFR is 125 ml/min, the filtered load of this substance is _____ .

   If the T_m for substance X is 200 mg/min, how much of the substance will be reabsorbed at a plasma concentration of 200 mg/100 ml and a GFR of 125 ml/min? _____ How much of substance X will be excreted? _____
3. Conn's syndrome is an endocrine disorder brought about by a tumor of the adrenal cortex that secretes excessive aldosterone in uncontrolled fashion. Given what you know about the functions of aldosterone, describe what the most prominent features of this condition would be.
4. Because of a mutation, a child was born with an ascending limb of Henle that was water permeable. What would be the minimum/maximum urine osmolarities (in units of mosm/liter) the child could produce?
   a. 100/300
   b. 300/1,200
   c. 100/100
   d. 1,200/1,200
   e. 300/300
5. An accident victim suffers permanent damage of the lower spinal cord and is paralyzed from the waist down. Describe what governs bladder emptying in this individual.
6. *Clinical Consideration.* Marcus T. has noted a gradual decrease in his urine flow rate and is now experiencing difficulty in initiating micturition. He needs to urinate frequently, and often he feels as if his bladder is not empty even though he has just urinated. Analysis of Marcus's urine reveals no abnormalities. Are his urinary tract symptoms most likely caused by kidney disease, a bladder infection, or prostate enlargement?

PHYSIOEDGE RESOURCES

 PHYSIOEDGE CD-ROM
Figures marked with the PhysioEdge icon have associated activities on the CD. For this chapter, check out:

Renal Physiology

The diagnostic quiz allows you to receive immediate feedback on your understanding of the concept and to advance through various levels of difficulty.

CHECK OUT THESE MEDIA QUIZZES:
14.1 The Urinary System
14.2 The Kidney
14.3 The Nephron

 PHYSIOEDGE WEB SITE
The Web site for this book contains a wealth of helpful study aids, as well as many ideas for further reading and research. Log on to:

http://www.brookscole.com

Go to the Biology page and select Sherwood's *Human Physiology,* 5th Edition. Select a chapter from the drop-down menu or click on one of these resource areas:

- **Case Histories** provide an introduction to the clinical aspects of human physiology. Check out:

    #11: Congestive Heart Failure
    #12: Urinary Tract Infection

- For 2-D and 3-D graphical illustrations and animations of physiological concepts, visit our **Visual Learning Resource.**

- Resources for study and review include the **Chapter Outline, Chapter Summary, Glossary,** and **Flash Cards.** Use our **Quizzes** to prepare for in-class examinations.

- On-line research resources to consult are **Hypercontents,** with current links to relevant Internet sites; **Internet Exercises** with starter URLs provided; and **InfoTrac Exercises.**

 For more readings, go to InfoTrac College Edition, your on-line research library, at:

http://infotrac.thomsonlearning.com

Systems of Major Importance in Maintaining Fluid and Acid-Base Balance

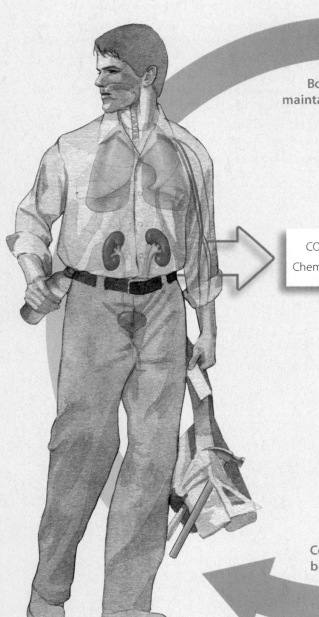

Body systems maintain homeostasis

Homeostasis
The kidneys, in conjunction with hormones involved in salt and water balance, are responsible for maintaining the volume and osmolarity of the extracellular fluid (internal environment). The kidneys, along with the respiratory system and chemical buffer systems in the body fluids, also contribute to homeostasis by maintaining the proper pH in the internal environment.

$$CO_2 + H_2O \overset{ca}{\leftrightarrows} H_2CO_3 \leftrightarrows H^+ + HCO_3^-$$
Chemical buffer systems in body fluids

Homeostasis is essential for survival of cells

Cells
The volume of circulating blood must be maintained to help ensure adequate pressure to drive life-sustaining blood to the cells. The osmolarity of fluid surrounding the cells must be closely regulated to prevent detrimental osmotic movement of water between the cells and ECF. The pH of the internal environment must remain stable because pH changes alter neuromuscular excitability and enzyme activity, among other serious consequences.

Cells make up body systems

Homeostasis depends on maintaining a balance between the input and the output of all constituents present in the internal fluid environment. Regulation of **fluid balance** involves two separate components: *control of ECF volume*, of which circulating plasma volume is a part, and *control of ECF osmolarity* (solute concentration). The kidneys control ECF volume by maintaining **salt balance** and control ECF osmolarity by maintaining **water balance**. The kidneys maintain this balance by adjusting the output of salt and water in the urine as needed to compensate for variable input and abnormal losses of these constituents.

Similarly, the kidneys help maintain **acid–base balance** by adjusting the urinary output of hydrogen ion (acid) and bicarbonate ion (base) as needed. Also contributing to acid–base balance are the lungs, which can adjust the rate at which they excrete hydrogen ion–generating CO_2, and the chemical buffer systems in the body fluids.

Chapter 15

Fluid and Acid–Base Balance

CONTENTS AT A GLANCE

BALANCE CONCEPT

❚ Internal pool of a substance

❚ Balance of input and output

FLUID BALANCE

❚ Body-fluid compartments

❚ Importance of regulating ECF volume

❚ Control of ECF volume by regulating salt balance

❚ Importance of regulating ECF osmolarity

❚ Control of ECF osmolarity by regulating water balance

ACID–BASE BALANCE

❚ Acid–base chemistry; pH

❚ Consequences of changes in [H^+]

❚ Sources of H^+

❚ Lines of defense against changes in [H^+]

❚ Chemical buffer systems

❚ Respiratory control of pH

❚ Renal control of pH

❚ Acid–base imbalances

BALANCE CONCEPT

The cells of complex multicellular organisms are able to survive and function only within a very narrow range of composition of the extracellular fluid (ECF), the internal fluid environment that bathes them.

❚ The internal pool of a substance is the amount of that substance in the ECF.

The quantity of any particular substance in the ECF is considered a readily available internal **pool**. The amount of the substance in the pool may be increased either by transferring more in from the external environment (most commonly by ingestion) or by metabolically producing it within the body (● Figure 15-1). Substances may be removed from the body by being excreted to the outside or by being used up in a metabolic reaction. If the quantity of a substance is to remain stable within the body, its **input** through ingestion or metabolic production must be balanced by an equal **output** through excretion or metabolic consumption. This relationship, known as the **balance concept,** is extremely important in the maintenance of homeostasis. Not all input and output pathways are applicable for every body fluid constituent. For example, salt is not synthesized or consumed by the body, so the stability of salt concentration in the body fluids depends entirely on a balance between salt ingestion and salt excretion.

Exchanges between the pool and other internal sites

The ECF pool can further be altered by transferring a particular ECF constituent into storage within the cells or bones. If the body as a whole has a surplus or deficit of a particular stored substance, the storage site can be expanded or partially depleted to maintain the ECF concentration of the substance within homeostatically prescribed limits. For example, after absorption of a meal, when more glucose is entering the plasma than is being consumed by the cells, the surfeit of glucose can be temporarily stored in muscle and liver cells, in the form of glycogen. This storage depot can then be tapped between meals as necessary to maintain the plasma glucose level when no new nutrients are being added to the blood by eating. This internal storage capacity is limited, however. Although an internal ex-

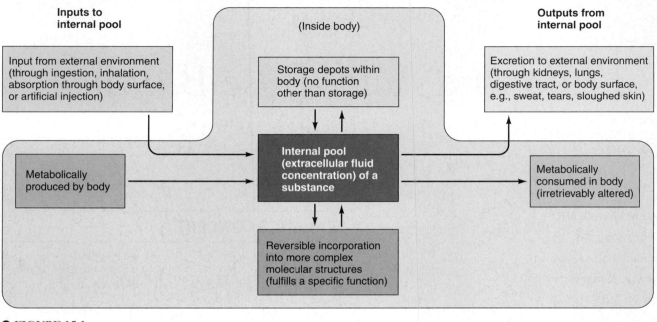

● FIGURE 15-1

Inputs to and outputs from the internal pool of a body constituent

change between the ECF and a storage depot can temporarily restore the plasma concentration of a particular substance to normal, in the long run any excess or deficit of that constituent must be compensated for by appropriate adjustments in total body input or output.

Another possible internal exchange between the pool and the remainder of the body is the reversible incorporation of certain plasma constituents into more complex molecular structures. For example, iron is incorporated into hemoglobin within the red blood cells during their synthesis but is released intact back into the body fluids when the red cells degenerate. This process differs from metabolic consumption of a substance, in which the substance is irretrievably converted into another form—for example, glucose converted into CO_2 plus H_2O plus energy. It also differs from storage in that the latter serves no purpose other than storage, whereas reversible incorporation into a more complex structure serves a specific purpose.

▌ To maintain stable balance of an ECF constituent, its input must equal its output.

When total body input of a particular substance equals its total body output, a **stable balance** exists. When the gains via input for a substance exceed its losses via output, a **positive balance** exists. The result is an increase in the total amount of the substance in the body. In contrast, when losses for a substance exceed its gains, a **negative balance** exists and the total amount of the substance in the body decreases.

Changing the magnitude of any of the input or output pathways for a given substance can alter its plasma concentration. To maintain homeostasis, any change in input must be balanced by a corresponding change in output (for example, increased salt intake must be matched by a corresponding increase in salt output in the urine), and, conversely, increased losses must be compensated for by increased intake. Thus maintaining a stable balance requires control. However, not all input and output pathways are regulated to maintain balance. Generally, input of various plasma constituents is poorly controlled or not controlled at all. We frequently ingest salt and H_2O, for example, not because we *need* them but because we *want* them, so the intake of salt and H_2O is highly variable. Likewise, hydrogen ion (H^+) is uncontrollably generated internally and added to the body fluids. Salt, H_2O, and H^+ can also be lost to the external environment to varying degrees through the digestive tract (vomiting), skin (sweating), and elsewhere without regard for salt, H_2O, or H^+ balance in the body. Compensatory adjustments in the urinary excretion of these substances are responsible for maintaining the body fluid's volume and salt and acid composition within the extremely narrow homeostatic range compatible with life despite the wide variations in input and unregulated losses of these plasma constituents.

This chapter is devoted to discussing the regulation of fluid balance (maintenance of salt and H_2O balance) and acid–base balance (maintenance of H^+ balance).

FLUID BALANCE

Water is by far the most abundant component of the human body, constituting 60% of body weight on average but ranging from 40% to 80%. The H_2O content of an individual remains fairly constant over a period of time, largely because of the kidneys' efficiency in regulating H_2O balance, but the percentage of body H_2O varies from person to person. The reason for the wide range in body H_2O among individuals is the variability in the amount of their adipose tissue (fat). Adipose tissue has a low H_2O percentage compared to other tissues. Plasma, as you might suspect, is more than 90% H_2O. Even the soft tissues such as skin, muscles, and internal organs consist of 70% to 80% H_2O. The relatively drier skeleton is only 22% H_2O. Fat, however, is the driest tissue of all, having only 10% H_2O content. Accordingly, a high body H_2O percentage is associated with leanness and a low body H_2O percentage with obesity, because a larger proportion of the body consists of relatively dry fat in overweight individuals.

The percentage of body H_2O is also influenced by the sex and age of the individual. Women have a lower body H_2O percentage than men, primarily because the female sex hormone, estrogen, promotes fat deposition in the breasts, buttocks, and elsewhere. This not only gives rise to the typical female figure but also endows women with a higher proportion of adipose tissue and, therefore, a lower body H_2O proportion. The percentage of body H_2O also decreases progressively with age.

▍ Body water is distributed between the ICF and ECF compartments.

Body H_2O is distributed between two major fluid compartments: fluid within the cells, **intracellular fluid (ICF)**, and fluid surrounding the cells, **extracellular fluid (ECF)** (▲ Table 15-1). (The terms "H_2O" and "fluid" are commonly used interchangeably. Although this usage is not entirely accurate, because it ignores the solutes in body fluids, it is acceptable when discussing total volume of fluids, because the major proportion of these fluids consists of H_2O.)

Proportion of H_2O in the major fluid compartments

The ICF compartment comprises about two-thirds of the total body H_2O. Even though each cell contains its own unique mixture of constituents, these trillions of minute fluid compartments are similar enough to be considered collectively as one large fluid compartment.

The remaining third of the body H_2O found in the ECF compartment is further subdivided into plasma and interstitial fluid. The **plasma,** which makes up about a fifth of the ECF volume, is the fluid portion of blood. The **interstitial fluid,** which represents the other four-fifths of the ECF compartment, is the fluid that lies in the spaces between cells. It bathes and makes exchanges with tissue cells.

Minor ECF compartments

Quantitatively, two other minor categories are included in the ECF compartment: lymph and transcellular fluid. **Lymph** is fluid being returned from interstitial fluid to plasma by means of the lymphatic system, where it is filtered through lymph nodes for immune defense purposes (see p. 368). **Transcellular fluid** consists of a number of small specialized fluid volumes, all of which are secreted by specific cells into a particular body cavity to perform some specialized function. Transcellular fluid

▲ TABLE 15-1
Classification of Body Fluid

Compartment	Volume of Fluid (in Liters)	Percentage of Body Fluid	Percentage of Body Weight
Total Body Fluid	42	100%	60%
Intracellular Fluid (ICF)	28	67	40
Extracellular Fluid (ECF)	14	33	20
Plasma	2.8	6.6 (20% of ECF)	4
Interstitial fluid	11.2	26.4 (80% of ECF)	16
Lymph	Negligible	Negligible	Negligible
Transcellular fluid	Negligible	Negligible	Negligible

includes *cerebrospinal fluid* (surrounding, cushioning, and nourishing the brain and spinal cord); *intraocular fluid* (maintaining the shape of and nourishing the eye); *synovial fluid* (lubricating and serving as a shock absorber for the joints); *pericardial, intrapleural,* and *peritoneal fluids* (lubricating movements of the heart, lungs, and intestines, respectively); and the *digestive juices* (digesting ingested foods).

Although these fluids are extremely important functionally, they represent an insignificant fraction of total body H_2O. Furthermore, the transcellular compartment as a whole usually does not reflect changes in the body's fluid balance. For example, cerebrospinal fluid does not decrease in volume when the body as a whole is experiencing a negative H_2O balance. This is not to say that these fluid volumes never change. Localized changes in a particular transcellular fluid compartment can occur pathologically (such as too much intraocular fluid accumulating in the eyes of people with glaucoma—see p. 196), but such a localized fluid disturbance does not affect the fluid balance of the body. Therefore, the transcellular compartment can usually be ignored when dealing with problems of fluid balance. The main exception to this generalization is when digestive juices are abnormally lost from the body during heavy vomiting or diarrhea, which can bring about a fluid imbalance.

▍ The plasma and interstitial fluid are similar in composition, but the ECF and ICF differ markedly.

Several barriers separate the body fluid compartments, limiting the movement of H_2O and solutes between the various compartments to differing degrees.

The barrier between plasma and interstitial fluid: blood vessel walls

The two components of the ECF—plasma and interstitial fluid—are separated by the walls of the blood vessels. However, H_2O and all plasma constituents with the exception of plasma proteins are continuously and freely exchanged between plasma and interstitial fluid by passive means across the thin, pore-lined capillary walls. Accordingly, plasma and interstitial fluid are nearly identical in composition, except that interstitial fluid lacks plasma proteins. Any change in one of these ECF compartments is quickly reflected in the other compartment, because they are constantly mixing.

The barrier between the ECF and ICF: cellular plasma membranes

In contrast to the very similar composition of vascular and interstitial fluid compartments, the composition of the ECF differs considerably from that of the ICF (● Figure 15-2). Each cell is surrounded by a highly selective plasma membrane that permits passage of certain materials while excluding others. Movement through the membrane barrier occurs by both passive and active means and may be highly discriminating. Among the major differences between the ECF and ICF are (1) the presence of cellular proteins in the ICF that cannot permeate the enveloping membranes to leave the cells and (2) the unequal distribution of Na^+ and K^+ and their attendant anions as a result of the action of the membrane-bound Na^+–K^+ ATPase pump that is present in all cells. This pump actively transports Na^+ out of and K^+ into cells; for this reason, Na^+ is the primary ECF cation, and K^+ is primarily found in the ICF. This unequal distribution of Na^+ and K^+, coupled with differences in membrane permeability to these ions, is responsible for the electrical properties of cells, including initiation and propagation of action potentials in excitable tissues (see Chapters 3 and 4).

Except for the extremely small, electrically imbalanced portion of the total intracellular and extracellular ions that are involved in membrane potential, the majority of the ECF and ICF ions are electrically balanced. In the ECF, Na^+ is accompanied primarily by the anion Cl^- (chloride) and to a lesser extent by HCO_3^- (bicarbonate). The major intracellular anions are PO_4^{3-} (phosphate) and the negatively charged proteins trapped within the cell.

All cells are freely permeable to H_2O. The movement of H_2O between the plasma and the interstitial fluid across capillary walls is governed by relative imbalances between capillary blood pressure (a fluid, or hydrostatic, pressure) and colloid osmotic pressure (see p. 366). In contrast, the net transfer of H_2O between the interstitial fluid and the ICF across the cellular plasma membranes occurs as a result of osmotic effects

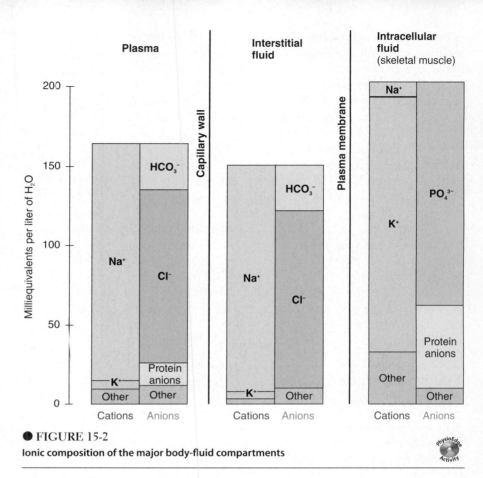

● **FIGURE 15-2**
Ionic composition of the major body-fluid compartments

alone. The hydrostatic pressures of the interstitial fluid and ICF are both extremely low and fairly constant.

▌ Fluid balance is maintained by regulating ECF volume and osmolarity.

Extracellular fluid serves as an intermediary between the cells and the external environment. All exchanges of H_2O and other constituents between the ICF and the external world must occur through the ECF. Water added to the body fluids always enters the ECF compartment first, and fluid always leaves the body by way of the ECF.

Plasma is the only fluid that can be acted on directly to control its volume and composition. It is the fluid that circulates through all the reconditioning organs that perform homeostatic adjustments (see p. 343). However, because of the free exchange across the capillary walls, if volume and composition of the plasma are regulated, volume and composition of the interstitial fluid bathing the cells are likewise regulated. Thus any control mechanism that operates on plasma in effect regulates the entire ECF. The ICF in turn is influenced by changes in the ECF to the extent permitted by the permeability of membrane barriers surrounding cells.

The factors regulated to maintain fluid balance in the body are ECF volume and ECF osmolarity. Although regulation of these two factors is closely interrelated, both being dependent on the relative NaCl and H_2O load in the body, the reasons why they are closely controlled are significantly different:

1. *Extracellular fluid volume* must be closely regulated to help *maintain blood pressure*. Maintenance of *salt balance* is of primary importance in long-term regulation of ECF volume.

2. *Extracellular fluid osmolarity* must be closely regulated to *prevent swelling or shrinking of cells*. Maintenance of *water balance* is of primary importance in regulating ECF osmolarity.

We will examine each of these factors in more detail.

Control of ECF volume is important in the long-term regulation of blood pressure.

A reduction in ECF volume causes a fall in arterial blood pressure by decreasing plasma volume. Conversely, expanding ECF volume raises arterial blood pressure by increasing plasma volume. Two compensatory measures come into play to transiently adjust blood pressure until ECF volume can be restored to normal. Let's review them.

Review of short-term control measures to maintain blood pressure

1. *Baroreceptor reflex mechanisms alter both cardiac output and total peripheral resistance* through autonomic nervous system effects on heart and blood vessels (see p. 377). These immediate cardiovascular responses are designed to minimize the effect a deviation in circulating volume has on blood pressure.

2. *Fluid shifts occur temporarily and automatically between plasma and interstitial fluid*. A reduction in plasma volume is partially compensated for by a shift of fluid out of the interstitial compartment into the blood vessels, expanding the circulating plasma volume at the expense of the interstitial compartment. Conversely, when the plasma volume is too large, much of the excess fluid shifts into the interstitial compartment. These shifts occur immediately and automatically as a result of changes in the balance of hydrostatic and osmotic forces acting across the capillary walls that arise when plasma volume deviates from normal (see p. 367).

These two measures provide temporary relief to help keep the blood pressure fairly constant, but they are not designed to be long-term solutions. Furthermore, these short-term compensatory measures have a limited ability to minimize a change in blood pressure. If the plasma volume is too inadequate, the blood pressure remains too low no matter how vigorous the pump action of the heart, how constricted the resistance vessels, or what proportion of interstitial fluid shifts into the blood vessels. Conversely, if the plasma volume is greatly overexpanded, blood pressure cannot be restored down to normal even with maximum dilation of the resistance vessels and other short-term measures.

Long-term control measures to maintain blood pressure

It is important, therefore, that other compensatory measures come into play in the long run to restore the ECF volume to normal. This responsibility for long-term regulation of blood pressure rests with the kidneys and the thirst mechanism, which control urinary output and fluid intake, respectively. In so doing, they accomplish needed fluid exchanges between the ECF and the external environment to regulate the body's total fluid volume. Accordingly, they have an important long-term influence on arterial blood pressure. Of these measures, control of urinary output by the kidneys is the most crucial for maintaining blood pressure. You will see why as we discuss these long-term mechanisms in more detail.

Control of salt balance is primarily important in regulating ECF volume.

By way of review, sodium and its attendant anions account for more than 90% of the ECF's osmotic activity. Because osmotic activity can be equated with "water-holding power," total Na^+ *load* (the total quantity of NaCl, not its concentration) in the ECF determines the total amount of H_2O osmotically retained in the ECF. As the kidneys conserve salt, they automatically conserve H_2O, because H_2O follows Na^+ osmotically. This retained salt solution is isotonic (see p. 78). The more salt in the ECF, the more H_2O in the ECF. The concentration of salt is not changed by changing the amount of salt, because H_2O always follows salt to maintain osmotic equilibrium — that is, to maintain normal concentration of salt. A reduced salt load leads to decreased H_2O retention, so the ECF remains isotonic but reduced in volume. The total mass of Na^+ salts in the ECF therefore determines the ECF's volume, and, appropriately, regulation of ECF volume depends primarily on controlling salt balance.

To maintain salt balance at a set level, salt input must equal salt output, thus preventing salt accumulation or deficit in the body. Let's examine the avenues and control of salt input and output.

Poor control of salt intake

The only avenue for salt input is ingestion, which typically is well in excess of the body's need for replacing obligatory salt losses. In our example of a typical daily salt balance (▲ Table 15-2), salt intake is 10.5 g per day. (The average American

▲ TABLE 15-2
Daily Salt Balance

Salt Input		Salt Output	
Avenue	Amount (g/day)	Avenue	Amount (g/day)
Ingestion	10.5	Obligatory loss in sweat and feces	0.5
		Controlled excretion in urine	10.0
Total input	10.5	Total output	10.5

salt intake is 10 to 15 g per day, although many people are consciously reducing their salt intake.) Yet half a gram of salt per day is adequate to replace the small amounts of salt usually lost in the feces and sweat.

Because we typically consume salt in excess of our needs, obviously salt intake in humans is not well controlled. Carnivores (meat eaters) and omnivores (eaters of meat and plants, like humans), which naturally get sufficient salt in fresh meat (meat contains an abundance of salt-rich ECF), normally do not manifest a physiologic appetite to seek additional salt. In contrast, herbivores (plant eaters), which lack salt naturally in their diets, develop a salt hunger and will travel miles to a salt lick. Humans generally have a hedonistic rather than a regulatory appetite for salt; we consume salt because we like it rather than because we have a physiologic need, except in the unusual circumstance of severe salt depletion caused by a deficiency of aldosterone, the salt-conserving hormone.

Precise control of salt output in the urine

To maintain salt balance, excess ingested salt must be excreted in urine. The three avenues for salt output are obligatory loss of salt in *sweat* and *feces* and controlled excretion of salt in *urine* (▲ Table 15-2). The total amount of sweat produced is unrelated to salt balance, being determined instead by factors that control body temperature. The small salt loss in feces is not subject to control. Except when sweating heavily or during diarrhea, normally the body uncontrollably loses only about 0.5 g of salt per day. This amount is actually the only salt that normally needs to be replaced by salt intake.

Because salt consumption is typically far more than the meager amount needed to compensate for uncontrolled losses, the kidneys precisely excrete the excess salt in the urine to maintain salt balance. In our example, 10 g of salt are eliminated in urine per day so that total salt output exactly equals salt input. By regulating rate of urinary salt excretion (that is, by regulating rate of Na^+ excretion, with Cl^- following along), the kidneys normally keep total Na^+ mass in the ECF constant despite any notable changes in dietary intake of salt or unusual losses through sweating, diarrhea, or other means. As a reflection of keeping total Na^+ mass in the ECF constant, ECF volume in turn is maintained within the narrowly prescribed limits essential for normal circulatory function.

Deviations in ECF volume accompanying changes in salt load are responsible for triggering renal compensatory responses that quickly bring Na^+ load and ECF volume back into line. Sodium is freely filtered at the glomerulus and actively reabsorbed, but it is not secreted by the tubules, so the amount of Na^+ excreted in the urine represents the amount of Na^+ that is filtered but is not subsequently reabsorbed:

$$Na^+ \text{ excreted} = Na^+ \text{ filtered} - Na^+ \text{ reabsorbed}$$

The kidneys accordingly adjust the amount of salt excreted by controlling two processes: (1) the glomerular filtration rate (GFR) and (2) more importantly, the tubular reabsorption of Na^+. You have already learned about these regulatory mechanisms, but we are pulling them together here as they relate to the long-term control of ECF volume and blood pressure.

The amount of Na^+ filtered is controlled through regulation of the GFR.

The amount of Na^+ filtered is equal to the plasma Na^+ concentration times the GFR. At any given plasma Na^+ concentration, any alteration in the GFR will correspondingly alter the amount of Na^+ filtered. Thus control of the GFR can adjust the amount of Na^+ filtered each minute.

The GFR is deliberately changed to alter the amount of salt and fluid filtered, as part of the general baroreceptor reflex response to change in blood pressure (see ● Figure 14-14, p. 523). The afferent arterioles that supply the renal glomeruli are constricted as part of the generalized vasoconstriction that elevates blood pressure that has fallen too low. As a result of reduced blood flow into the glomeruli, GFR decreases and, accordingly, the amount of Na^+ and accompanying fluid that are filtered decreases. Consequently, excretion of salt and fluid diminishes. The conserved salt and fluid that otherwise would have been filtered and excreted help minimize reduced fluid volume and contribute to long-term restitution of blood pressure.

Conversely, elevation in ECF volume and arterial blood pressure is reflexly countered by a baroreceptor reflex response that increases GFR, which in turn results in enhanced salt and fluid excretion. Eliminating extra salt and fluid that otherwise would have been conserved helps relieve the expanded plasma volume.

Note that adjustments in amount of salt filtered are accomplished as part of the general blood pressure–regulating reflexes. Changes in Na^+ load in the body are not sensed as such; instead, they are monitored indirectly through the effect that Na^+ ultimately has on blood pressure via its role in determining the ECF volume. Fittingly, baroreceptors that monitor fluctuations in blood pressure are responsible for bringing about adjustments in the amounts of Na^+ filtered and eventually excreted.

The amount of Na^+ reabsorbed is controlled through the renin-angiotensin-aldosterone system.

The amount of Na^+ reabsorbed also depends on regulatory systems that play an important role in controlling blood pressure. Although Na^+ is reabsorbed throughout most of the tubule's length, only its reabsorption in the distal portions of the tubule is subject to control. The main factor controlling the extent of Na^+ reabsorption in the distal and collecting tubule is the powerful renin-angiotensin-aldosterone system, which promotes Na^+ reabsorption and thereby Na^+ retention. Sodium retention in turn promotes osmotic retention of H_2O and the subsequent expansion of plasma volume and elevation of arterial blood pressure. Appropriately, this Na^+ conserving system is activated by a reduction in NaCl, ECF volume, and arterial blood pressure (see ● Figure 14-19, p. 529).

Thus control of GFR and Na^+ reabsorption are highly interrelated, and both are intimately tied in with long-term regulation of ECF volume as reflected by blood pressure. Specifically, a fall in arterial blood pressure brings about a twofold effect in the renal handling of Na^+ (● Figure 15-3): (1) a reflex reduction in the GFR to decrease the amount of Na^+ filtered and (2) a hormonally adjusted increase in the amount of Na^+ reabsorbed. Together these effects reduce the amount of Na^+ excreted, thereby conserving for the body the Na^+ and accompanying H_2O necessary to compensate for the fall in arterial pressure. (For an examination of how exercising muscles and cooling mechanisms compete for an inadequate plasma volume, see the accompanying boxed feature, ▶ A Closer Look at Exercise Physiology.) Conversely, a rise in arterial blood pressure brings about (1) increased amount of Na^+ filtered and (2) reduced renin-angiotensin-aldosterone activity, which decreases salt (and fluid) reabsorption. Together these actions increase salt (and fluid) excretion, thereby eliminating extra fluid that was expanding plasma volume and increasing arterial pressure.

▌ Control of ECF osmolarity prevents changes in ICF volume.

Maintenance of fluid balance depends on the regulation of both ECF volume and ECF osmolarity. Whereas regulation of ECF volume is important in the long-term control of blood pressure, regulation of ECF osmolarity is important in preventing changes in cell volume.

The **osmolarity** of a fluid is a measure of the concentration of the individual solute particles dissolved in it. The higher the osmolarity, the higher the concentration of solutes or, to look at it differently, the lower the concentration of H_2O. Water tends to move by osmosis down its own concentration gradient from an area of lower solute (higher H_2O) concentration to an area of higher solute (lower H_2O) concentration.

Recall that osmotic activity across the capillary wall is due to unequal distribution of plasma proteins (see p. 366). Plasma proteins are present only in plasma (as their name implies), because they are too large to penetrate capillary walls and enter interstitial fluid. All other solutes are in essentially the same concentration in plasma and interstitial fluid, so they do not contribute to any unequal distribution of H_2O that would induce osmosis across the capillary wall.

In contrast, osmotic activity across cellular plasma membranes is directly related to any differences in ionic concentration between ECF and ICF. Plasma proteins play no role in osmosis of H_2O across the cell membranes, because these proteins are absent in both interstitial fluid and ICF.

Ions responsible for ECF and ICF osmolarity

Osmosis occurs across the cellular plasma membranes only when a difference in concentration of nonpenetrating solutes exists between the ECF and ICF. Solutes that can penetrate a barrier separating two fluid compartments quickly become equally distributed between the two compartments and thus do not contribute to osmotic differences.

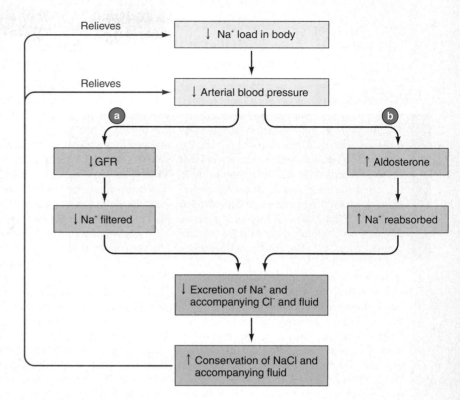

a See Figure 14-14 for details of mechanism.

b See Figure 14-19 for details of mechanism.

● **FIGURE 15-3**

Dual effect of a fall in arterial blood pressure on renal handling of Na^+

Sodium and its attendant anions, being by far the most abundant solutes in the ECF in terms of numbers of particles, account for the vast majority of the ECF's osmotic activity. In contrast, K^+ and its accompanying intracellular anions are responsible for the ICF's osmotic activity. Even though small amounts of Na^+ and K^+ passively diffuse across the plasma membrane all the time, these ions behave as if they were nonpenetrating, because of Na^+–K^+ pump activity. Any Na^+ that passively diffuses down its electrochemical gradient into the cell is promptly pumped back outside, so the result is the same as if Na^+ were barred from the cells. The same holds true in reverse for K^+; it in effect remains trapped within the cells. The resulting unequal distribution of Na^+ and K^+ and their accompanying anions between ECF and ICF is responsible for the osmotic activity of these two fluid compartments.

Normally, the osmolarities of the ECF and ICF are the same, because total concentration of K^+ and other effectively nonpenetrating solutes inside the cells is equal to total concentration of Na^+ and other effectively nonpenetrating solutes in the fluid surrounding the cells. Even though nonpenetrating solutes in the ECF and ICF differ, their concentrations are normally identical, and the number (not the nature) of the unequally distributed particles per volume determines the fluid's osmolarity. Because the osmolarities of ECF and ICF are normally equal, no net movement of H_2O usually occurs into or

A Potentially Fatal Clash: When Exercising Muscles and Cooling Mechanisms Compete for an Inadequate Plasma Volume

An increasing number of people of all ages are participating in walking or jogging programs to improve their level of physical fitness and decrease their risk of cardiovascular disease. For people living in environments that undergo seasonal temperature changes, fluid loss can make exercising outdoors dangerous during the transition from the cool days of spring to the hot, humid days of summer. If exercise intensity is not modified until the participant gradually adjusts to the hotter environmental conditions, dehydration and salt loss can indirectly lead to heat cramps, heat exhaustion, or ultimately heat stroke and death.

The term **acclimatization** refers to the gradual adaptations the body makes to maintain long-term homeostasis in response to a prolonged physical change in the surrounding environment, such as a change in temperature. When a person exercises in the heat without gradually adapting to the hotter environment, the body faces a terrible dilemma. During exercise, large amounts of blood must be delivered to the muscles to supply O_2 and nutrients and to remove the wastes that accumulate from their high rate of activity. Exercising muscles also produce heat. To maintain the body temperature in the face of this extra heat, blood flow to the skin is increased, so that heat from the warmed blood can be lost through the skin to the surrounding environment. If the environmental temperature is hotter than the body temperature, heat can-

not be lost from the blood to the surrounding environment despite maximal skin vasodilation. Instead, the body gains heat from its warmer surroundings, further adding to the dilemma. Because extra blood is diverted to both the muscles and the skin when a person exercises in the heat, less blood is returned to the heart, and the heart pumps less blood per beat in accordance with the Frank-Starling mechanism (see p. 328). Therefore the heart must beat faster than it would in a cool environment to deliver the same amount of blood per minute. The increased rate of cardiac pumping further contributes to heat production.

The sweat rate also increases so that evaporative cooling can take place to help maintain the body temperature during periods of excessive heat gain. In an unacclimatized person, maximal sweat rate is about 1.5 liters per hour. During sweating, water-retaining salt as well as water is lost. The resulting loss of plasma volume through sweating further depletes the blood supply available for muscular exercise and for cooling through skin vasodilation.

The heart has a maximum rate at which it can pump. If exercise continues at a high intensity and this maximal rate is reached, the exercising muscles win the contest for blood supply. The body responds by constricting the skin arterioles, sacrificing cooling to maintain cardiac output and blood pressure. If exercise continues, body heat continues to rise, and heat exhaustion (rapid, weak pulse; hypotension, profuse

sweating; and disorientation) or heat stroke (failure of the temperature control center in the hypothalamus; hot, dry skin; extreme confusion or unconsciousness; and possibly death) can occur. In fact, every year people die of heat stroke in marathons run during hot, humid weather.

By contrast, if a person exercises in the heat for two weeks at reduced, safe intensities, the body makes the following adaptations so that after acclimatization the person can do the same amount of work as was possible in a cool environment. (1) The plasma volume is increased by as much as 12%. Expansion of the plasma volume provides enough blood to both supply the exercising muscles and direct blood to the skin for cooling. (2) The person begins sweating at a lower temperature so that the body does not get so hot before cooling begins. (3) The sweat rate increases as much as three times, to 4 liters per hour, with a more even distribution over the body. This increase in evaporative cooling reduces the need for cooling by skin vasodilation. (4) The sweat becomes more dilute so that less salt is lost in the sweat. The retained salt exerts an osmotic effect to hold water in the body and help maintain circulating plasma volume. These adaptations take 14 days and occur only if the person exercises in the heat. Being patient until these changes take place can enable the person to exercise safely throughout the summer months.

---

out of the cells. Therefore, cell volume normally remains constant, because no H_2O osmotically enters or leaves the cells.

Importance of regulating ECF osmolarity

Any circumstance that results in a loss or gain of *free* H_2O (that is, loss or gain of H_2O that is not accompanied by comparable solute deficit or excess) leads to changes in ECF osmolarity. If there is a deficit of free H_2O in the ECF, the solutes become too concentrated, and the ECF osmolarity becomes abnormally high (that is, becomes *hypertonic*) (see p. 78). If there is excess free H_2O in the ECF, the solutes become too dilute, and the ECF osmolarity becomes abnormally low (that is, becomes *hypotonic*). When the ECF osmolarity changes with respect to the ICF osmolarity, osmosis takes place, with H_2O either leaving or entering the cells, depending, respectively, on whether the ECF is more or less concentrated than the ICF.

The osmolarity of the ECF must therefore be regulated to prevent these undesirable shifts of H_2O into or out of the cells. As far as the ECF itself is concerned, the concentration of its solutes does not really matter. However, it is crucial that ECF

osmolarity be maintained within very narrow limits to prevent the cells from swelling (by osmotically gaining H_2O from the ECF) or shrinking (by osmotically losing H_2O to the ECF).

We will examine the fluid shifts that occur between the ECF and the ICF when the ECF osmolarity becomes hypertonic or hypotonic relative to the ICF. Then we will consider how water balance and subsequently ECF osmolarity are normally maintained to minimize detrimental changes in cell volume.

▌ During ECF hypertonicity, the cells shrink as H_2O leaves them.

Hypertonicity of the ECF, or the excessive concentration of ECF solutes, is usually associated with **dehydration,** or a negative free H_2O balance.

Causes of hypertonicity (dehydration)

Dehydration with accompanying hypertonicity can be brought about in three major ways:

1. *Insufficient H_2O intake,* such as might occur during desert travel or might accompany difficulty in swallowing

2. *Excessive H_2O loss,* such as might occur during heavy sweating, vomiting, or diarrhea (even though both H_2O and solutes are lost during these conditions, relatively more H_2O is lost, so the remaining solutes become more concentrated)

3. *Diabetes insipidus,* a disease characterized by a deficiency of *vasopressin (antidiuretic hormone)*

Vasopressin is the hormone that increases the permeability of the distal and collecting tubules to H_2O and thus enhances water conservation by reducing urinary output of water (see p. 544). In **diabetes insipidus,** the kidneys cannot conserve H_2O because they cannot reabsorb H_2O from the distal portions of the nephron in the absence of vasopressin. Such patients typically produce up to 20 liters of very dilute urine per day, compared to the normal average of 1.5 liters per day. Unless H_2O intake keeps pace with this tremendous loss of H_2O in the urine, the person quickly dehydrates. Such patients complain that they spend an extraordinary amount of time day and night going to the bathroom and getting drinks. Fortunately, they can be treated with replacement vasopressin administered by nasal spray.

Direction and resultant symptoms of water movement during hypertonicity

Whenever the ECF compartment becomes hypertonic, H_2O moves out of the cells by osmosis into the more concentrated ECF until ICF osmolarity equilibrates with the ECF. Thus although the H_2O deficit occurs first in the ECF, the H_2O deficit quickly becomes equally distributed between the ECF and ICF because H_2O shifts osmotically out of the cells. As H_2O leaves them, the cells shrink. Of particular concern is that considerable shrinking of brain neurons disturbs brain function, which can be manifested as mental confusion and irrationality in moderate cases and as possible delirium, convulsions, or coma in more severe hypertonic conditions.

Rivaling the neural symptoms in seriousness are the circulatory disturbances that arise from a reduction in plasma volume in association with dehydration. Circulatory problems may range from a slight reduction in blood pressure to circulatory shock and death.

Other more common symptoms become apparent even in mild cases of dehydration. For example, dry skin and sunken eyeballs are indications of loss of H_2O from the underlying soft tissues, and the tongue becomes dry and parched because salivary secretion is suppressed.

▌ During ECF hypotonicity, the cells swell as H_2O enters them.

Hypotonicity of the ECF is usually associated with **overhydration;** that is, excess free H_2O. When a positive free H_2O balance exists, the ECF is less concentrated (more dilute) than normal.

Causes of hypotonicity (overhydration)

Usually, any surplus free H_2O is promptly excreted in urine, so hypotonicity generally does not occur. However, hypotonicity can arise in three ways:

1. Patients with *renal failure* who cannot excrete a dilute urine become hypotonic when they consume relatively more H_2O than solutes.

2. Hypotonicity can occur transiently in healthy people *if H_2O is rapidly ingested* to such an excess that the kidneys can't respond quickly enough to eliminate the extra H_2O.

3. Hypotonicity can occur when excess H_2O without solute is retained in the body as a result of *inappropriate secretion of vasopressin.*

Vasopressin is normally secreted in response to a H_2O deficit, which is relieved by increasing H_2O reabsorption in the distal portion of the nephrons. However, vasopressin secretion, and therefore hormonally controlled tubular H_2O reabsorption, can be increased in response to pain, acute infections, trauma, and other stressful situations, even when there is no H_2O deficit in the body. The increased vasopressin secretion and resulting H_2O retention elicited by stress are appropriate in anticipation of potential blood loss in the stressful situation. The extra retained H_2O could minimize the effect a loss of blood volume would have on blood pressure. However, because modern-day stressful situations generally are not accompanied by blood loss, the increased vasopressin secretion is inappropriate as far as the body's fluid balance is concerned. The reabsorption and retention of too much H_2O dilute the body's solutes.

Direction and resultant symptoms of water movement during hypotonicity

Whichever way it is brought about, excess free H_2O retention first dilutes the ECF compartment, making it hypotonic. The resultant difference in osmotic activity between the ECF and ICF induces H_2O to move by osmosis from the more dilute ECF into the cells, with the cells swelling as H_2O moves into them. The H_2O excess originally present in the ECF rapidly becomes equally distributed between the ECF and ICF as H_2O osmotically enters the cells. Like the shrinking of cerebral neurons, pronounced swelling of brain cells also leads to brain dysfunction. Symptoms include confusion, irritability, lethargy, headache, dizziness, vomiting, drowsiness, and, in severe cases, even convulsions, coma, and death.

Nonneural symptoms of overhydration include weakness, caused by swelling of muscle cells, and circulatory disturbances, including hypertension and edema, caused by expansion of plasma volume.

The condition of overhydration, hypotonicity, and cellular swelling resulting from excess free H_2O retention is known as **water intoxication.** It should not be confused with the fluid retention that occurs with excess salt retention. In the latter case, the ECF is still isotonic because the increase in salt is matched by a corresponding increase in H_2O. Because the interstitial fluid is still isotonic, no osmotic gradient exists to drive the extra H_2O into the cells. The excess salt and H_2O burden is therefore confined to the ECF compartment, with circulatory consequences being the most important concern. In water intoxication, the body's H_2O content is increased relative to salt content, and the excess H_2O becomes evenly distributed between the ECF and ICF compartments. In addition to any circulatory disturbances, symptoms caused by cellular swelling become a problem.

Let us now contrast the situations of hypertonicity and hypotonicity with what happens as a result of isotonic fluid gain or loss.

▌ No water moves into or out of cells during an ECF isotonic fluid gain or loss.

An example of an isotonic fluid gain is therapeutic intravenous administration of an isotonic solution, such as isotonic saline. When the isotonic fluid is injected into the ECF compartment, ECF volume increases, but concentration of ECF solutes remains unchanged; in other words, the ECF is still isotonic. Because the ECF's osmolarity has not changed, ECF and ICF are still in osmotic equilibrium, so no net fluid shift occurs between the two compartments. The ECF compartment has increased in volume without shifting H_2O into the cells. Thus, unless one is trying to correct an osmotic imbalance, intravenous fluid therapy should be isotonic, to prevent fluctuations in intracellular volume and possible neural symptoms.

Similarly, in an isotonic fluid loss such as hemorrhage, the loss is confined to the ECF with no corresponding loss of fluid from the ICF. Fluid does not shift out of the cells, because the ECF remaining within the body is still isotonic, so no osmotic gradient draws H_2O out of the cells. Of course, many other mechanisms counteract loss of blood, but the ICF compartment is not directly affected by the loss.

Thus when the ECF and ICF are in osmotic equilibrium, no net movement of H_2O into or out of the cells occurs regardless of whether the ECF volume increases or decreases. Shifts in H_2O between ECF and ICF take place only when the ECF becomes more or less concentrated than the cells, and this usually arises from a loss or gain, respectively, of free H_2O.

We will now examine how free H_2O balance is normally maintained.

▌ Control of water balance by means of vasopressin is important in regulating ECF osmolarity.

Control of free H_2O balance is crucial for regulating ECF osmolarity. Because increases in free H_2O cause the ECF to become too dilute and deficits of free H_2O cause the ECF to become too concentrated, the osmolarity of the ECF must be immediately corrected by restoring stable free H_2O balance to avoid detrimental osmotic fluid shifts into or out of the cells.

To maintain a stable H_2O balance, H_2O input must equal H_2O output.

Sources of H₂O input

• In a person's typical daily H_2O balance (▲ Table 15-3), a little more than a liter of H_2O is added to the body by *drinking liquids*.

• Surprisingly, an amount almost equal to that is obtained from *eating solid food*. Recall that muscles consist of about

▲ TABLE 15-3
Daily Water Balance

Water Input		Water Output	
Avenue	Quantity (ml/day)	Avenue	Quantity (ml/day)
Fluid intake	1,250	Insensible loss (from lungs and nonsweating skin)	900
H₂O in food intake	1,000		
Metabolically produced H₂O	350	Sweat	100
		Feces	100
		Urine	1,500
Total input	2,600	Total input	2,600

75% H_2O; meat is therefore 75% H_2O, because it is animal muscle. Likewise, fruits and vegetables consist of 60% to 90% H_2O. Therefore, people normally obtain almost as much H_2O from solid foods as from the liquids they drink.

• The third source of H_2O input is *metabolically produced H_2O*. Chemical reactions within the cells convert food and O_2 into energy, producing CO_2 and H_2O in the process. This **metabolic H_2O** produced during cellular metabolism and released into the ECF averages about 350 ml/day.

The average H_2O intake from these three sources thus totals 2,600 ml/day. Another source of H_2O often employed therapeutically is intravenous infusion of fluid.

Sources of H₂O output

• On the output side of the H_2O balance tally, the body loses close to a liter of H_2O daily without being aware of it. This so-called **insensible loss** occurs from the *lungs* and *nonsweating skin*. During respiration, inspired air becomes saturated with H_2O within the airways. This H_2O is lost when the moistened air is subsequently expired. Normally, we are not aware of this H_2O loss, but can recognize it on cold days, when H_2O vapor condenses so that we can "see our breath." The other insensible loss is continual loss of H_2O from skin even in the absence of sweating. Water molecules can diffuse through skin cells and evaporate without being noticed. Fortunately, skin is fairly waterproofed by its keratinized exterior layer, which protects against a much greater loss of H_2O by this avenue (see p. 51). When this protective surface layer is lost, such as when a person has extensive burns, increased fluid loss from the burned surface can cause serious problems with fluid balance.

• Sensible loss (loss of which the person is aware) of H_2O from skin occurs through *sweating*, which represents another avenue of H_2O output. At an air temperature of 68°F, an average of 100 ml of H_2O is lost daily through sweating. Loss of water from sweating can vary substantially, of course, depending on the environmental temperature and humidity and the degree of physical activity; it may range from zero up to as much as several liters per hour in very hot weather.

- Another passageway for H_2O loss from the body is through the *feces*. Normally, only about 100 ml of H_2O are lost via this route each day. During fecal formation in the large intestine, most H_2O is absorbed out of the digestive tract lumen into the blood, thereby conserving fluid and solidifying the digestive tract's contents for elimination. Additional losses of H_2O can occur from the digestive tract through vomiting or diarrhea.

- By far the most important output mechanism is *urine excretion*, with 1,500 ml of urine being produced daily on the average.

The total H_2O output is 2,600 ml/day, the same as the volume of H_2O input in our example. This balance is not by chance. Normally, H_2O input matches H_2O output so that the H_2O in the body remains in balance.

Factors regulated to maintain water balance

Of the many sources of H_2O input and output, only two can be regulated to maintain H_2O balance. On the intake side, thirst influences amount of fluid ingested, and on the output side the kidneys can adjust magnitude of urine formation. Controlling H_2O output in urine is the most important mechanism in controlling H_2O balance.

Some of the other factors are regulated, but not for maintaining H_2O balance. Food intake is subject to regulation to maintain energy balance, and control of sweating is important in maintaining body temperature. Metabolic H_2O production and insensible losses are completely unregulated.

Control of water output in the urine by vasopressin

Fluctuations in ECF osmolarity caused by imbalances between H_2O input and output are quickly compensated for by adjusting urinary excretion of H_2O without changing the usual excretion of salt; that is, H_2O reabsorption and excretion are partially dissociated from solute reabsorption and excretion, so the amount of free H_2O retained or eliminated can be varied to quickly restore ECF osmolarity to normal. Free H_2O reabsorption and excretion are adjusted through changes in vasopressin secretion (see p. 545). Throughout most of the nephron, H_2O reabsorption is important in regulating ECF volume, because salt reabsorption is accompanied by comparable H_2O reabsorption. In distal portions of the tubule, however, variable free H_2O reabsorption can take place without comparable salt reabsorption, because of the vertical osmotic gradient in the renal medulla to which the collecting tubule is exposed. Vasopressin increases the permeability of this late portion of the tubule to H_2O. Depending on the amount of vasopressin present, the amount of free H_2O reabsorbed can be adjusted as necessary to restore ECF osmolarity to normal.

Vasopressin is produced by the hypothalamus and stored in the posterior pituitary gland. It is released from the posterior pituitary on command from the hypothalamus.

Control of water input by thirst

Thirst is the subjective sensation that drives one to ingest H_2O. A **thirst center** is located in the hypothalamus in close proximity to the vasopressin-secreting cells. Thirst increases H_2O input, whereas vasopressin decreases H_2O output by reducing urine production.

We are now going to elaborate on the mechanisms that regulate vasopressin secretion and thirst.

▌ Vasopressin secretion and thirst are largely triggered simultaneously.

The hypothalamic control centers that regulate vasopressin secretion (and thus urinary output) and thirst (and thus drinking) act in concert. Vasopressin secretion and thirst are both stimulated by a free H_2O deficit and suppressed by a free H_2O excess. Thus, appropriately, the same circumstances that call for reducing urinary output to conserve body H_2O also give rise to the sensation of thirst to replenish body H_2O.

Role of hypothalamic osmoreceptors

The predominant excitatory input for both vasopressin secretion and thirst comes from **hypothalamic osmoreceptors** located near the vasopressin-secreting cells and thirst center. These osmoreceptors monitor osmolarity of fluid surrounding them, which in turn reflects the concentration of the entire internal fluid environment. As osmolarity increases (too little H_2O) and need for H_2O conservation increases, vasopressin secretion and thirst are both stimulated (● Figure 15-4). As a result, reabsorption of H_2O in the distal and collecting tubules is increased so that urinary output is reduced and H_2O is conserved, while H_2O intake is simultaneously encouraged. These actions restore depleted H_2O stores, thus relieving the hypertonic condition by diluting solutes to normal concentration. In contrast, H_2O excess, manifested by reduced ECF osmolarity, prompts increased urinary output (through decreased vasopressin release) and suppresses thirst, which together reduce water load in the body.

Role of left atrial baroreceptors

Even though the major stimulus for vasopressin secretion and thirst is an increase in ECF osmolarity, the vasopressin-secreting cells and thirst center are both influenced to a moderate extent by changes in ECF volume mediated by input from the **left atrial volume receptors.** Located in the left atrium, these volume receptors monitor the pressure of blood flowing through, which reflects the ECF volume. In response to a major reduction in ECF volume (> 7% volume change) and accordingly in arterial pressure, as during hemorrhage, the left atrial volume receptors reflexly stimulate both vasopressin secretion and thirst. The outpouring of vasopressin and the increased thirst lead to decreased urine output and increased fluid intake, respectively. Furthermore, vasopressin, at the circulating levels elicited by a large decline in ECF volume and arterial pressure, exerts a potent vasoconstrictor effect on arterioles (thus giving rise to its name; see p. 358). Both by helping expand ECF and plasma volume and by increasing total peripheral resistance, vasopressin helps relieve the low blood pressure that elicited vasopressin secretion.

Conversely, vasopressin and thirst are both inhibited when ECF/plasma volume and arterial blood pressure are elevated. The resultant suppression of H_2O intake, coupled with elimi-

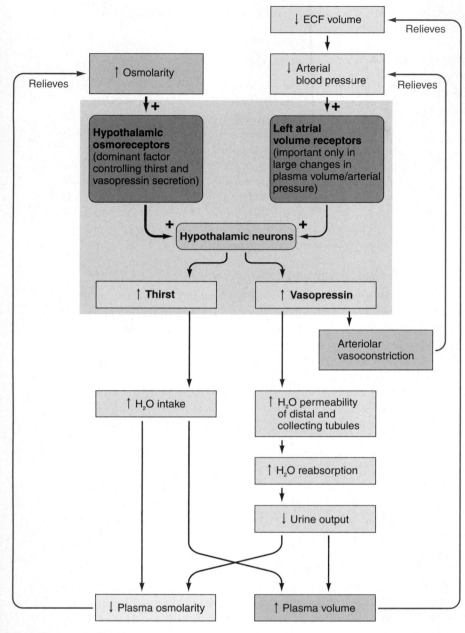

● FIGURE 15-4
Control of increased vasopressin secretion and thirst during a H_2O deficit

---

aldosterone mechanism is activated to conserve Na^+, angiotensin II, in addition to stimulating aldosterone secretion acts directly on the brain to give rise to the urge to drink and concurrently stimulates vasopressin to enhance renal H_2O reabsorption (see p. 528). The resultant increased H_2O intake and decreased urinary output help correct the reduction in ECF volume that triggered the renin-angiotensin-aldosterone system.

Regulatory factors that do not link vasopressin and thirst

Several factors affect vasopressin secretion but not thirst. As described earlier, vasopressin is stimulated by stress-related inputs such as pain, fear, and trauma that have nothing directly to do with maintaining H_2O balance. In fact, H_2O retention from the inappropriate secretion of vasopressin can bring about a hypotonic H_2O imbalance. In contrast, alcohol inhibits vasopressin secretion and can lead to ECF hypertonicity by promoting excessive free H_2O excretion.

One stimulus that promotes thirst but not vasopressin secretion is a direct effect of dryness of the mouth. Nerve endings in the mouth are directly stimulated by dryness, which causes an intense sensation of thirst that can often be relieved merely by moistening the mouth even though no H_2O is actually ingested. A dry mouth can exist when salivation is suppressed by factors unrelated to the body's H_2O content, such as nervousness, excessive smoking, or certain drugs.

Factors that affect vasopressin secretion or thirst but have nothing directly to do with the body's need for H_2O are usually short-lived. The dominant, long-standing control of vasopressin and thirst is directly correlated with the body's state of H_2O—namely, by the status of ECF osmolarity and, to a lesser extent, by ECF volume.

Oral metering

Some kind of "oral H_2O metering" appears to exist, at least in animals. A thirsty animal will rapidly drink only enough H_2O to satisfy its H_2O deficit. It stops drinking before the ingested H_2O has had time to be absorbed from the digestive tract and actually return the ECF compartment to normal. Exactly what factors are involved in signaling that enough H_2O has been consumed is still uncertain. It might be a learned anticipatory response based on past experience. This mechanism seems to be less effective in humans, because we frequently drink more than is necessary to meet the needs of our bodies or, conversely, may not drink enough to make up a deficit.

Nonphysiological influences on fluid intake

In fact, even though the thirst mechanism exists to control H_2O intake, fluid consumption by humans is often influenced more by habit and sociological factors than by the need to regulate

---

nation of excess ECF/plasma volume in the urine, helps restore blood pressure to normal.

Recall that low ECF/plasma volume and low arterial blood pressure also reflexly increase aldosterone secretion. The resultant increase in Na^+ reabsorption ultimately leads to osmotic retention of H_2O, expansion of ECF volume, and an increase in arterial blood pressure. In fact, aldosterone-controlled Na^+ reabsorption is the most important factor in regulating ECF volume, with the vasopressin and thirst mechanism playing only a supportive role.

Role of angiotensin II

Yet another stimulus for increasing both thirst and vasopressin is angiotensin II (▲ Table 15-4). When the renin-angiotension-

Factors Controlling Vasopressin Secretion and Thirst

Factor	Effect on Vasopressin Secretion	Effect on Thirst	Comment
↑ECF Osmolarity	↑	↑	Major stimulus for vasopressin secretion and thirst
↓ECF Volume	↑	↑	Important only in large changes in ECF volume/arterial blood pressure
Angiotensin II	↑	↑	Part of dominant pathway for promoting compensatory salt and H_2O retention when ECF volume/arterial blood pressure are reduced
Pain, Fear, Trauma, and Other Stress-Related Inputs	Inappropriate ↑ unrelated to body's H_2O balance	No effect	Promotes excess H_2O retention and ECF hypotonicity (resultant H_2O retention of potential value in maintaining arterial blood pressure in case of blood loss in the stressful situation)
Alcohol	Inappropriate ↓ unrelated to body's H_2O balance	No effect	Promotes excess H_2O loss and ECF hypertonicity
Dry Mouth	No effect	↑	Nerve endings in the mouth that ultimately give rise to the sensation of thirst are directly stimulated by dryness

H_2O balance. Thus even though H_2O intake is critical in maintaining fluid balance, it is not precisely controlled in humans, who err especially on the side of excess H_2O consumption. We usually drink when we are thirsty, but we often drink even when we are not thirsty because, for example, we are on a coffee break.

With H_2O intake being inadequately controlled and indeed even contributing to H_2O imbalances in the body, the primary factor involved in maintaining H_2O balance is urinary output regulated by the kidneys. Accordingly, *vasopressin-controlled H_2O reabsorption is of primary importance in regulating ECF osmolarity.*

Before shifting attention to acid–base balance, ▲ Table 15-5 summarizes the regulation of ECF volume and osmolarity, the two factors important in maintaining fluid balance.

ACID–BASE BALANCE

The term *acid–base balance* refers to the precise regulation of **free** (that is, unbound) **hydrogen-ion (H^+)** concentration in the body fluids. To indicate the concentration of a chemical, its symbol is enclosed in square brackets, [ ]. Thus [H^+] designates H^+ concentration.

▲ TABLE 15-5
Summary of the Regulation of ECF Volume and Osmolarity

Regulated Variable	Need to Regulate the Variable	Outcomes if the Variable Is Not Normal	Mechanism for Regulating the Variable
ECF Volume	Important in the long-term control of arterial blood pressure	↓ECF volume→ ↓arterial blood pressure ↑ECF volume→ ↑arterial blood pressure	Maintenance of salt balance; salt osmotically "holds" H_2O, so the Na^+ load determines the ECF volume. Accomplished primarily by aldosterone-controlled adjustments in urinary Na^+ excretion
ECF Osmolarity	Important to prevent detrimental osmotic movement of H_2O between the ECF and ICF	↓ECF osmolarity (hypotonicity)→ H_2O enters the cell → cells swell ↑ECF osmolarity (hypertonicity)→ H_2O leaves the cells → cells shrink	Maintenance of free H_2O balance. Accomplished primarily by vasopressin-controlled adjustments in excretion of H_2O in the urine

Acids liberate free hydrogen ions, whereas bases accept them.

Acids are a special group of hydrogen-containing substances that *dissociate*, or separate, when in solution to liberate free H^+ and anions (negatively charged ions). Many other substances (for example, carbohydrates) also contain hydrogen, but they are not classified as acids because the hydrogen is tightly bound within their molecular structure and is never liberated as free H^+.

A strong acid has a greater tendency to dissociate in solution than a weak acid does; that is, a greater percentage of a strong acid's molecules separates into free H^+ and anions. Hydrochloric acid (HCl) is an example of a strong acid; every HCl molecule dissociates into free H^+ and Cl^- (chloride) when dissolved in H_2O. With a weaker acid such as carbonic acid (H_2CO_3), only a portion of the molecules dissociate in solution into H^+ and HCO_3^- (bicarbonate anions). The remaining H_2CO_3 molecules remain intact. Because only the free hydrogen ions contribute to the acidity of a solution, H_2CO_3 is a weaker acid than HCl because H_2CO_3 does not yield as many free hydrogen ions per number of acid molecules present in solution (● Figure 15-5).

● **FIGURE 15-5**

Comparison of a strong and a weak acid
(a) Five molecules of a strong acid. A strong acid such as HCl (hydrochloric acid) completely dissociates into free H^+ and anions in solution. (b) Five molecules of a weak acid. A weak acid such as H_2CO_3 (carbonic acid) only partially dissociates into free H^+ and anions in solution.

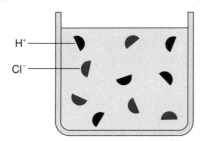

Strong acid (HCl)
(a)

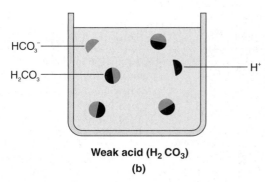

Weak acid (H₂ CO₃)
(b)

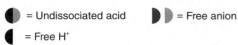

● = Undissociated acid   ◗◖ = Free anion

◗ = Free H^+

The extent of dissociation for a given acid is always constant; that is, when in solution, the same proportion of a particular acid's molecules always separate to liberate free H^+, with the other portion always remaining intact. The constant degree of dissociation for a particular acid (in this example H_2CO_3) is expressed by its **dissociation constant (K)** as follows:

$$[H^+]\,[HCO_3^-]/[H_2CO_3] = K$$

where

$[H^+]\,[HCO_3^-]$ represents the concentration of ions resulting from H_2CO_3 dissociation.

$[H_2CO_3]$ represents the concentration of intact (undissociated) H_2CO_3.

The dissociation constant varies for different acids.

A **base** is a substance that can combine with a free H^+ and thus remove it from solution. A strong base can bind H^+ more readily than a weak base can.

The pH designation is used to express [H⁺].

The $[H^+]$ in the ECF is normally 4×10^{-8} or 0.00000004 equivalents per liter. The concept of pH has been developed to express $[H^+]$ more conveniently. Specifically, **pH** equals the logarithm (log) to the base 10 of the reciprocal of the hydrogen ion concentration:

$$pH = \log 1/[H^+]$$

Two important points should be noted about this formula:

1. Because $[H^+]$ is in the denominator, *a high $[H^+]$ corresponds to a low pH, and a low $[H^+]$ corresponds to a high pH.* The greater the $[H^+]$, the larger the number by which 1 must be divided, and the lower the pH.

2. *Every unit change in pH actually represents a tenfold change in $[H^+]$* because of the logarithmic relationship. A log to the base 10 indicates how many times 10 must be multiplied by itself to produce a given number. For example, the log of 10 = 1, whereas the log of 100 = 2. The number 10 must be multiplied by itself twice to yield 100 ($10 \times 10 = 100$). Numbers less than 10 have logs less than 1. Numbers between 10 and 100 have logs between 1 and 2, and so on. Accordingly, each unit of change in pH indicates a 10-fold change in $[H^+]$. For example, a solution with a pH of 7 has a $[H^+]$ 10 times less than that of a solution with a pH of 6 (a 1 pH-unit difference) and 100 times less than that of a solution with a pH of 5 (a 2 pH-unit difference).

Acidic and basic solutions in chemistry

The pH of pure H_2O is 7.0, which is considered chemically neutral. An extremely small proportion of H_2O molecules dissociate into hydrogen ions and hydroxyl (OH^-) ions. Because OH^- has the ability to bind with H^+ to once again form a H_2O molecule, it is considered basic. Because an equal number of acidic hydrogen ions and basic hydroxyl ions are formed, H_2O is neutral, being neither acidic nor basic. Solutions having a pH less than 7.0 contain a higher $[H^+]$ than pure H_2O and are considered **acidic.** Conversely, solutions having a pH value greater than 7.0 have a lower [H$^+$] and are considered **basic** or **alkaline** (● Figure 15-6a). ● Figure 15-7 compares the pH values of common solutions.

Acidosis and alkalosis in the body

The pH of arterial blood is normally 7.45, and the pH of venous blood is 7.35, for an average blood pH of 7.4. The pH of venous blood is slightly lower (more acidic) than that of arterial blood, because H$^+$ is generated by formation of H$_2$CO$_3$ from CO$_2$ picked up at the tissue capillaries. **Acidosis** exists whenever blood pH falls below 7.35, whereas **alkalosis** occurs when blood pH is above 7.45 (● Figure 15-6b). Note that the reference point for determining the body's acid–base status is not the chemically neutral pH of 7.0 but the normal plasma pH of 7.4. Thus a plasma pH of 7.2 is considered acidotic even though in chemistry a pH of 7.2 is considered basic.

Death occurs if arterial pH falls outside the range of 6.8 to 8.0 for more than a few seconds, because an arterial pH of less than 6.8 or greater than 8.0 is not compatible with life. Obviously, therefore, [H$^+$] in the body fluids must be carefully regulated.

▌ Fluctuations in [H$^+$] alter nerve, enzyme, and K$^+$ activity.

Only a narrow pH range is compatible with life, because even small changes in [H$^+$] have dramatic effects on normal cell function. The prominent consequences of fluctuations in [H$^+$] include the following:

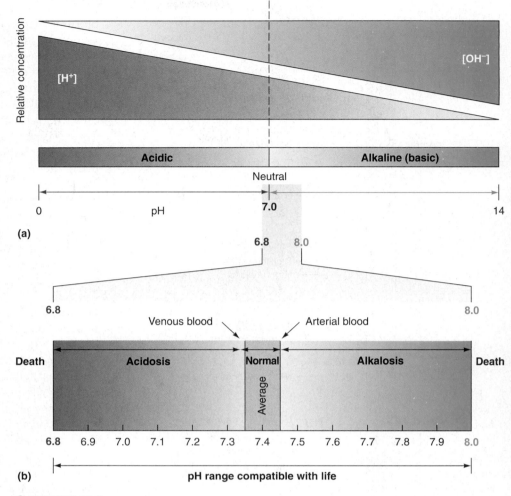

(a)

(b)

pH range compatible with life

● FIGURE 15-6

pH considerations in chemistry and physiology
(a) Relationship of pH to the relative concentrations of H$^+$ and base (OH$^-$) under chemically neutral, acidic, and alkaline conditions. (b) Plasma pH range under normal, acidosis, and alkalosis conditions.

1. *Changes in excitability of nerve and muscle cells* are among the major clinical manifestations of pH abnormalities.
   - The major clinical effect of *increased* [H$^+$] (acidosis) is depression of the central nervous system. Acidotic patients become disoriented and, in more severe cases, eventually die in a state of coma.
   - In contrast, the major clinical effect of *decreased* [H$^+$] (alkalosis) is overexcitability of the nervous system, first the peripheral nervous system and later the central nervous system. Peripheral nerves become so excitable that they fire even in the absence of normal stimuli. Such overexcitability of the *afferent* (sensory) nerves give rise to abnormal "pins and needles" tingling sensations. Overexcitability of *efferent* (motor) nerves brings about muscle twitches and, in more pronounced cases, severe muscle spasms. Death may occur in extreme alkalosis, because spasm of the respiratory muscles seriously impairs breathing. Alternatively, severely alkalotic patients may die of convulsions resulting from overexcitability of the central

nervous system. In less serious situations, CNS overexcitability is manifested as extreme nervousness.

2. Hydrogen ion concentration exerts a *marked influence on enzyme activity*. Even slight deviations in [H$^+$] alter the shape and activity of protein molecules. Because enzymes are proteins, a shift in the body's acid–base balance disturbs the normal pattern of metabolic activity catalyzed by these enzymes. Some cellular chemical reactions are accelerated; others are depressed.

3. Changes in [H$^+$] *influence K$^+$ levels* in the body. When reabsorbing Na$^+$ from the filtrate, the renal tubular cells secrete either K$^+$ or H$^+$ in exchange (see p. 536). Normally, they secrete a preponderance of K$^+$ compared to H$^+$. Because of the intimate relationship between secretion of H$^+$ and K$^+$ by the kidneys, an increased rate of secretion of one of these ions is accompanied by a decreased rate of secretion of the other. For example, if more H$^+$ than normal is eliminated by the kidneys, as occurs when the body fluids become acidotic, less K$^+$ than usual can be excreted. The resultant K$^+$ retention can affect cardiac function, among other detrimental consequences.

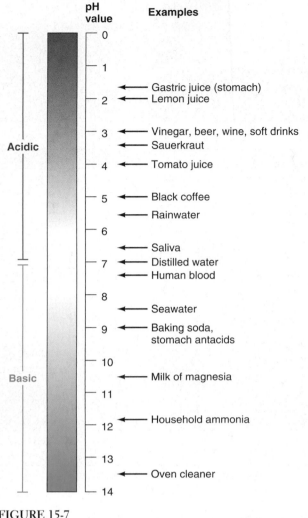

pH value	Examples

- 0
- 1
- 2 ← Gastric juice (stomach), Lemon juice
- 3 ← Vinegar, beer, wine, soft drinks, ← Sauerkraut
- 4 ← Tomato juice
- 5 ← Black coffee, ← Rainwater
- 6
- 7 ← Saliva, ← Distilled water, ← Human blood
- 8 ← Seawater
- 9 ← Baking soda, stomach antacids
- 10
- 11 ← Milk of magnesia
- 12 ← Household ammonia
- 13
- 14 ← Oven cleaner

Acidic

Basic

● FIGURE 15-7
Comparison of pH values of common solutions

▌ Hydrogen ions are continually added to the body fluids as a result of metabolic activities.

As with any other constituent, to maintain a constant $[H^-]$ in the body fluids, input of hydrogen ions must be balanced by an equal output. On the input side, only a small amount of acid capable of dissociating to release H^+ is taken in with food, such as the weak citric acid found in oranges. Most H^+ in the body fluids is generated internally from metabolic activities.

Sources of H⁺ in the body

Normally, H^+ is continually being added to the body fluids from the three following sources:

1. *Carbonic acid formation.* The major source of H^+ is through H_2CO_3 formation from metabolically produced CO_2. Cellular oxidation of nutrients yields energy, with CO_2 and H_2O as end products. Catalyzed by the enzyme *carbonic anhydrase* (ca), CO_2 and H_2O form H_2CO_3, which then partially dissociates to liberate free H^+ and HCO_3^-.

$$CO_2 + H_2O \overset{ca}{\rightleftharpoons} H_2CO_3 \rightleftharpoons H^+ + HCO_3^-$$

This reaction is reversible, because it can proceed in either direction, depending on the concentrations of the substances involved as dictated by the *law of mass action* (see p. 488). Within the systemic capillaries, the CO_2 level in the blood increases as metabolically produced CO_2 enters from the tissues. This drives the reaction to the acid side, generating H^+ as well as HCO_3^- in the process. In the lungs, the reaction is reversed: CO_2 diffuses from the blood flowing through the pulmonary capillaries into the alveoli (air sacs), from which it is expired to the atmosphere. The resultant reduction in blood CO_2 drives the reaction toward the CO_2 side. Hydrogen ion and HCO_3^- form H_2CO_3, which rapidly decomposes into CO_2 and H_2O once again. The CO_2 is exhaled while the hydrogen ions generated at the tissue level are incorporated into H_2O molecules.

When the respiratory system can keep pace with the rate of metabolism, there is no net gain or loss of H^+ in the body fluids from metabolically produced CO_2. When the rate of CO_2 removal by the lungs does not match the rate of CO_2 production at the tissue level, however, the resulting accumulation or deficit of CO_2 leads to an excess or shortage, respectively, of free H^+ in the body fluids.

2. *Inorganic acids produced during breakdown of nutrients.* Dietary proteins and other ingested nutrient molecules that are found abundantly in meat contain a large quantity of sulfur and phosphorus. When these molecules are broken down, sulfuric acid and phosphoric acid are produced as by-products. Being moderately strong acids, these two inorganic acids largely dissociate, liberating free H^+ into the body fluids. In contrast, breakdown of fruits and vegetables produces bases that, to some extent, neutralize acids derived from protein metabolism. Generally, however, more acids than bases are produced during breakdown of ingested food, leading to an excess of these acids.

3. *Organic acids resulting from intermediary metabolism.* Numerous organic acids are produced during normal intermediary metabolism. For example, fatty acids are produced during fat metabolism, and lactic acid is produced by muscles during heavy exercise. These acids partially dissociate to yield free H^+.

Hydrogen ion generation therefore normally goes on continuously, as a result of ongoing metabolic activities. Furthermore, in certain disease states additional acids may be produced that further contribute to the total body pool of H^+. For example, in *diabetes mellitus* large quantities of keto acids may be produced by abnormal fat metabolism. Some types of acid-producing medications may also add to the total H^+ load that the body must handle. Thus input of H^+ is unceasing, highly variable, and essentially unregulated.

Three lines of defense against changes in [H⁺]

The key to H^+ balance is maintaining normal alkalinity of the ECF (pH 7.4) despite this constant onslaught of acid. The generated free H^+ must be largely removed from solution while in the body and ultimately must be eliminated so that the pH of body fluids can remain within the narrow range compatible with life. Mechanisms must also exist to compensate rapidly for the occasional situation in which the ECF becomes too alkaline.

Three lines of defense against changes in $[H^+]$ operate to maintain $[H^+]$ of body fluids at a nearly constant level despite unregulated input: (1) the *chemical buffer systems*, (2) the *respiratory mechanism of pH control*, and (3) the *renal mechanism of pH control*. We will look at each of these methods.

▌ Chemical buffer systems minimize changes in pH by binding with or yielding free H^+.

A **chemical buffer system** is a mixture in a solution of two chemical compounds that minimize pH changes when either an acid or a base is added to or removed from the solution. A buffer system consists of a pair of substances involved in a reversible reaction—one substance that can yield free H^+ as the $[H^+]$ starts to fall and another that can bind with free H^+ (thus removing it from solution) when $[H^+]$ starts to rise. An important example of such a buffer system is the carbonic acid: bicarbonate (H_2CO_3:HCO_3^-) buffer pair, which is involved in the following reversible reaction:

$$H_2CO_3 \rightleftharpoons H^+ + HCO_3^-$$

When a strong acid such as HCl is added to an unbuffered solution, all the dissociated H^+ remains free in the solution (● Figure 15-8a). In contrast, when HCl is added to a solution containing the H_2CO_3:HCO_3^- buffer pair, the HCO_3^- immediately binds with the free H^+ to form H_2CO_3 (● Figure 15-8b). This weak H_2CO_3 dissociates only slightly compared to the marked reduction in pH that occurred when the buffer system was not present and the additional H^+ remained unbound. In the opposite case, when the pH of the solution starts to rise from the addition of base or loss of acid, the H^+-yielding member of the buffer pair, H_2CO_3, releases H^+ to minimize the rise in pH.

The body has four buffer systems: (1) the H_2CO_3:HCO_3^- buffer system, (2) the protein buffer system, (3) the hemoglobin buffer system, and (4) the phosphate buffer system. Each serves a different important role, as you will learn as we examine each in turn. (▲ Table 15-6).

▌ The H_2CO_3:HCO_3^- buffer pair is the primary ECF buffer for noncarbonic acids.

The H_2CO_3:HCO_3^- buffer pair is the most important buffer system in the ECF for buffering pH changes brought about by causes other than fluctuations in CO_2-generated H_2CO_3. It is a very effective ECF buffer system for two reasons. First, H_2CO_3 and HCO_3^- are abundant in the ECF, so this system is readily available to resist changes in pH. Second, and more importantly, each component of this buffer pair is closely regulated. The kidneys regulate HCO_3^-, and the respiratory sys-

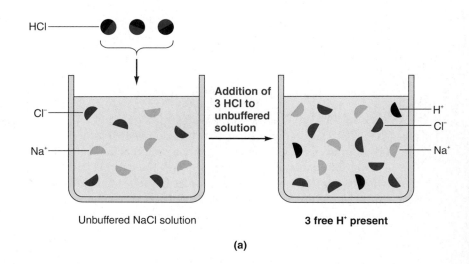

Unbuffered NaCl solution   3 free H⁺ present

(a)

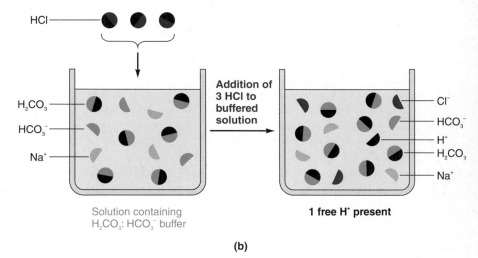

Solution containing
H_2CO_3: HCO_3^- buffer   1 free H⁺ present

(b)

● **FIGURE 15-8**

Action of chemical buffers
(a) Addition of HCl to an unbuffered solution. All the added hydrogen ions (H^+) remain free and contribute to the acidity of the solution. (b) Addition of HCl to a buffered solution. Bicarbonate ions (HCO_3^-), the basic member of the buffer pair, bind with some of the added H^+ and remove them from solution so that they do not contribute to its acidity.

▲ **TABLE 15-6**
Chemical Buffers and Their Primary Roles

Buffer System	Major Functions
Carbonic Acid: Bicarbonate Buffer System	Primary ECF buffer against non–carbonic-acid changes
Protein Buffer System	Primary ICF buffer; also buffers ECF
Hemoglobin Buffer System	Primary buffer against carbonic acid changes
Phosphate Buffer System	Important urinary buffer; also buffers ICF

tem regulates CO_2, which generates H_2CO_3. Thus, in the body the H_2CO_3:HCO_3^- buffer system includes involvement of CO_2 via the following reaction, with which you are already familiar:

$$CO_2 + H_2O \leftrightharpoons H_2CO_3 \leftrightharpoons H^+ + HCO_3^-$$

When new H^+ is added to the plasma from any source other than CO_2 (for example through lactic acid released into the ECF from exercising muscles), the preceding reaction is driven toward the left side of the equation. As the extra H^+ binds with HCO_3^-, it no longer contributes to the acidity of body fluids, so the rise in $[H^+]$ abates. In the converse situation, when the plasma $[H^+]$ occasionally falls below normal for some reason other than a change in CO_2 (such as the loss of plasma-derived HCl in the gastric juices during vomiting), the reaction is driven toward the right side of the equation. Dissolved CO_2 and H_2O in the plasma form H_2CO_3, which generates additional H^+ to make up for the H^+ deficit. In so doing, the H_2CO_3:HCO_3^- buffer system resists the fall in $[H^+]$.

This system cannot buffer changes in pH induced by fluctuations in H_2CO_3. A buffer system cannot buffer itself. Consider, for example, the situation in which the plasma $[H^+]$ is elevated by CO_2 retention from a respiratory problem. The rise in CO_2 drives the reaction to the right according to the law of mass action, elevating $[H^+]$. The increase in $[H^+]$ occurs as a result of the reaction being driven to the *right* by an increase in CO_2, so the elevated $[H^+]$ cannot drive the reaction to the *left* to buffer the increase in $[H^+]$. Only if the increase in $[H^+]$ is brought about by some mechanism other than CO_2 accumulation can this buffer system be shifted to the CO_2 side of the equation and effectively reduce $[H^+]$. Likewise, in the opposite situation, the H_2CO_3:HCO_3^- buffer system cannot compensate for a reduction in $[H^+]$ from a deficit of CO_2 by generating more H^+-yielding H_2CO_3 when the problem in the first place is a shortage of H_2CO_3-forming CO_2. Other mechanisms, to be described shortly, are available for resisting fluctuations in pH caused by changes in CO_2 levels.

Henderson-Hasselbalch equation

The relationship between $[H^+]$ and the members of a buffer pair can be expressed according to the **Henderson-Hasselbalch equation**, which, for the H_2CO_3:HCO_3^- buffer system is as follows:

$$pH = pK + \log [HCO_3^-]/[H_2CO_3]$$

Although you do not need to know the mathematical manipulations involved, it is helpful to understand how this formula is derived. Recall that the dissociation constant K for H_2CO_3 acid is

$$[H^+][HCO_3^-]/[H_2CO_3] = K$$

and that the relationship between pH and $[H^+]$ is

$$pH = \log 1/[H^+]$$

Then, by solving the dissociation constant formula for $[H^+]$ (that is, $[H^+] = K \times [H_2CO_3]/[HCO_3^-]$) and replacing this value for $[H^+]$ in the pH formula, one comes up with the Henderson-Hasselbalch equation.

Practically speaking, $[H_2CO_3]$ directly reflects the concentration of dissolved CO_2, henceforth referred to as $[CO_2]$, because most of the CO_2 in the plasma is converted into H_2CO_3. (The dissolved CO_2 concentration is equivalent to P_{CO_2}, as described in the chapter on respiration.) Therefore, the equation becomes

$$pH = pK + \log [HCO_3^-]/[CO_2]$$

The pK is the logarithm of 1/K and, like K, always remains a constant for any given acid. For H_2CO_3, the pK is 6.1. Because the pK is always a constant, changes in pH are associated with changes in the ratio between $[HCO_3^-]$ and $[CO_2]$.

- Normally, the ratio between $[HCO_3^-]$ and $[CO_2]$ in the ECF is 20 to 1; that is, there is 20 times more HCO_3^- than CO_2. Plugging this ratio into our formula,

$$pH = pK + \log [HCO_3^-]/[CO_2]$$
$$= 6.1 + \log 20/1$$

The log of 20 is 1.3. Therefore pH = 6.1 + 1.3 = 7.4, which is the normal pH of plasma.
- When the ratio of $[HCO_3^-]$ to $[CO_2]$ increases above 20/1, pH increases. Accordingly, either a rise in $[HCO_3^-]$ or a fall in $[CO_2]$, both of which increase the $[HCO_3^-]/[CO_2]$ ratio if the other component remains constant, shifts the acid–base balance toward the alkaline side.
- In contrast, when the $[HCO_3^-]/[CO_2]$ ratio decreases below 20/1, pH decreases toward the acid side. This can occur either if the $[HCO_3^-]$ decreases or if the $[CO_2]$ increases while the other component remains constant.

Because $[HCO_3^-]$ is regulated by the kidneys and $[CO_2]$ by the lungs, the pH of the plasma can be shifted up and down by kidney and lung influences. The kidneys and lungs regulate pH (and thus free $[H^+]$) largely by controlling plasma $[HCO_3^-]$ and $[CO_2]$, respectively, to restore their ratio to normal. Accordingly,

$$pH \propto \frac{[HCO_3^-] \text{ controlled by kidney function}}{[CO_2] \text{ controlled by respiratory function}}$$

Because of this relationship, not only do both the kidneys and lungs normally participate in pH control but renal or respiratory dysfunction can also induce acid–base imbalances by altering the $[HCO_3^-]/[CO_2]$ ratio. We will build on this principle when we examine respiratory and renal control of pH and acid–base abnormalities later in the chapter. For now, we are going to continue our discussion of the roles of the different buffer systems.

■ The protein buffer system is primarily important intracellularly.

The most plentiful buffers of the body fluids are the proteins, including the intracellular proteins and the plasma proteins. Proteins are excellent buffers because they contain both acidic and basic groups that can give up or take up H^+. Quantitatively, the protein system is most important in buffering changes in $[H^+]$ in the ICF, because of the sheer abundance of the intracellular proteins. A more limited number of plasma proteins reinforces the $H_2CO_3:HCO_3^-$ system in extracellular buffering.

The hemoglobin buffer system buffers H^+ generated from carbonic acid.

Hemoglobin (Hb) buffers the H^+ generated from metabolically produced CO_2 in transit between the tissues and the lungs. At the systemic capillary level, CO_2 continuously diffuses into the blood from the tissue cells where it is being produced. The greatest percentage of this CO_2 forms H_2CO_3, which partially dissociates into H^+ and HCO_3^-. Simultaneously, some oxyhemoglobin (HbO_2) releases O_2, which diffuses into the tissues. Reduced (unoxygenated) Hb has a greater affinity for H^+ than HbO_2 does. Therefore, most H^+ generated from CO_2 at the tissue level becomes bound to reduced Hb and no longer contributes to acidity of body fluids:

$$H^+ + Hb \leftrightharpoons HHb$$

At the lungs, the reactions are reversed. As Hb picks up O_2 diffusing from the alveoli into the red blood cells, the affinity of Hb for H^+ is decreased, so H^+ is released. This liberated H^+ combines with HCO_3^- to yield H_2CO_3. In turn, H_2CO_3 produces CO_2, which is exhaled. Meanwhile, the hydrogen has been reincorporated into neutral H_2O molecules. Were it not for Hb, blood would become much too acidic after picking up CO_2 at the tissues. With the tremendous buffering capacity of the Hb system, venous blood is only slightly more acidic than arterial blood despite the large volume of H^+-generating CO_2 carried in venous blood.

The phosphate buffer system is an important urinary buffer.

The phosphate buffer system is composed of an acid phosphate salt (NaH_2PO_4) that can donate a free H^+ when the $[H^+]$ falls and a basic phosphate salt (Na_2HPO_4) that can accept a free H^+ when the $[H^+]$ rises. Basically, this buffer pair can alternately switch a H^+ for a Na^+ as demanded by the $[H^+]$:

$$Na_2HPO_4 + H^+ \leftrightharpoons NaH_2PO_4 + Na^+$$

Even though the phosphate pair is a good buffer, its concentration in the ECF is rather low, so it is not very important as an ECF buffer. Because phosphates are more abundant within the cells, this system contributes significantly to intracellular buffering, being rivaled only by the more plentiful intracellular proteins.

Even more importantly, the phosphate system serves as an excellent urinary buffer. Humans normally consume more phosphate than needed. The excess phosphate filtered through the kidneys is not reabsorbed but remains in the tubular fluid to be excreted (because the renal threshold for phosphate is exceeded; see p. 532). This excreted phosphate buffers urine as it is being formed by removing from solution the H^+ secreted into the tubular fluid. None of the other body-fluid buffer systems are present in the tubular fluid to play a role in buffering urine during its formation. Most or all of the filtered HCO_3^- and CO_2 (alias H_2CO_3) are reabsorbed, whereas Hb and plasma proteins are not even filtered.

Chemical buffer systems act as the first line of defense against changes in $[H^+]$.

All chemical buffer systems act immediately, within fractions of a second, to minimize changes in pH. When $[H^+]$ is altered, the involved buffer systems' reversible chemical reactions are shifted at once in favor of compensating for the change in $[H^+]$. Accordingly, the buffer systems are the *first line of defense* against changes in $[H^+]$, because they are the first mechanism to respond.

Through the mechanism of buffering, most hydrogen ions seem to "disappear" from the body fluids between the times of their generation and their elimination. It must be emphasized, however, that none of the chemical buffer systems actually *eliminates* H^+ from the body. These ions are merely removed from solution by being incorporated within one member of the buffer pair, thus preventing the hydrogen ions from contributing to body fluid acidity. Because each buffer system has a limited capacity to "soak up" H^+, the H^+ that is unceasingly produced must ultimately be removed from the body. If H^+ were not eventually eliminated, soon all the body fluid buffers would already be bound with H^+ and there would be no further buffering ability.

The respiratory and renal mechanisms of pH control actually eliminate acid from the body instead of merely suppressing it, but they respond more slowly than chemical buffer systems. We will now turn our attention to these other defenses against changes in acid–base balance.

The respiratory system regulates $[H^+]$ by controlling the rate of CO_2 removal.

The respiratory system plays an important role in acid–base balance through its ability to alter pulmonary ventilation and consequently to alter excretion of H^+-generating CO_2. The level of respiratory activity is governed in part by arterial $[H^+]$, as follows (▲ Table 15-7):

- When arterial $[H^+]$ increases as the result of a *nonrespiratory* (or *metabolic*) cause, the respiratory center in the brain stem is reflexly stimulated to increase pulmonary ventilation (rate at which gas is exchanged between lungs and atmosphere) (see p. 480). As rate and depth of breathing increase, more CO_2 than usual is blown off, so less H_2CO_3 than normal is added to body fluids. Because CO_2 forms acid, removal of CO_2 in essence removes acid from this source from the body, thus offsetting extra acid present from a nonrespiratory source.
- Conversely, when arterial $[H^+]$ falls, pulmonary ventilation is reduced. As a result of slower, shallower breathing, metabolically produced CO_2 diffuses from the cells into the blood faster than it is removed from the blood by the lungs, so higher-than-usual amounts of acid-forming CO_2 accumulate in the blood, thus restoring $[H^+]$ toward normal.

▲ TABLE 15-7

Respiratory Adjustments to Acidosis and Alkalosis Induced by Nonrespiratory Causes

Respiratory Compensations	Acid–Base Status		
	Normal (pH 7.4)	Nonrespiratory (metabolic) acidosis (pH 7.1)	Nonrespiratory (metabolic) alkalosis (pH 7.7)
Spirogram Records at Various pHs	(spirogram trace)	(spirogram trace)	(spirogram trace)
Respiratory Rate	Normal	↑	↓
Tidal Volume	Normal	↑	↓
Ventilation	Normal	↑	↓
Rate of CO_2 Removal	Normal	↑	↓
Rate of Carbonic Acid Formation	Normal	↓	↑
Rate of H^+ Generation from CO_2	Normal	↓	↑

The lungs are extremely important in maintaining the $[H^+]$ of plasma. Every day they remove from body fluids what amounts to 100 times more H^+ derived from carbonic acid than the kidneys remove from the sources other than carbonic acid. Furthermore, the respiratory system, through its ability to regulate arterial $[CO_2]$, can adjust the amount of H^+ added to body fluids from this source as needed to restore pH toward normal when fluctuations occur in $[H^+]$ from sources other than carbonic acid.

▌ The respiratory system serves as the second line of defense against changes in $[H^+]$.

Respiratory regulation acts at a moderate speed, coming into play only when chemical buffer systems alone cannot minimize $[H^+]$ changes. When deviations in $[H^+]$ occur, the buffer systems respond immediately, whereas adjustments in ventilation require a few minutes to be initiated. If a deviation in $[H^+]$ is not swiftly and completely corrected by the buffer systems, the respiratory system comes into action a few minutes later, thus serving as the *second line of defense* against changes in $[H^+]$.

The respiratory system alone can return the pH to only 50% to 75% of the way toward normal. Two reasons contribute to the respiratory system's inability to fully compensate for a nonrespiratory-induced acid–base imbalance. First, during respiratory compensation for a deviation in pH, the peripheral chemoreceptors, which increase ventilation in response to an elevated arterial $[H^+]$, and the central chemoreceptors, which increase ventilation in response to a rise in $[CO_2]$ (by monitoring CO_2-generated H^+ in brain ECF—see p. 500), work at odds. Consider what happens in response to an acidosis arising from a nonrespiratory cause. When the peripheral chemoreceptors detect an increase in arterial $[H^+]$, they reflexly *stimulate* the respiratory center to step up ventilation, causing more acid-forming CO_2 to be blown off. In response to the resultant fall in CO_2, however, the central chemoreceptors start to *inhibit* the respiratory center. By opposing the action of the peripheral chemoreceptors, the central chemoreceptors stop the compensatory increase in ventilation short of restoring pH all the way to normal.

Second, the driving force for the compensatory increase in ventilation is diminished as the pH moves toward normal. Ventilation is increased by the peripheral chemoreceptors in response to a rise in arterial $[H^+]$, but as the $[H^+]$ is gradually reduced by stepped-up removal of H^+-forming CO_2, the enhanced ventilatory response is also gradually reduced.

Of course, when changes in $[H^+]$ stem from $[CO_2]$ fluctuations that arise from respiratory abnormalities, the respiratory mechanism cannot contribute at all to pH control. For example, if acidosis exists because of CO_2 accumulation caused by lung disease, the impaired lungs cannot possibly compensate for acidosis by increasing the rate of CO_2 removal. The buffer systems (other than the H_2CO_3:HCO_3^- pair) plus renal regulation are the only mechanisms available for defending against respiratory-induced acid–base abnormalities.

We will now turn our attention to how the kidneys help maintain acid–base balance.

▌ The kidneys are a powerful third line of defense against changes in $[H^+]$.

The kidneys require hours to days to compensate for changes in body fluid pH, compared to the immediate responses of the buffer systems and the few minutes of delay before the respiratory system responds. Therefore, they are the *third line of defense* against $[H^+]$ changes in body fluids. However, the kidneys are the most potent acid–base regulatory mechanism; not only can they vary removal of H^+ from any source, but they can also variably conserve or eliminate HCO_3^- depending on the acid–base status of the body. For example, during renal compensation for acidosis, for each H^+ excreted in urine, a new HCO_3^- is added to the plasma to buffer (by means of the $H_2CO_3:HCO_3^-$ system) yet another H^+ that still remains in body fluids. By simultaneously removing acid (H^+) from and adding base (HCO_3^-) to body fluids, the kidneys are able to restore the pH toward normal more effectively than the lungs, which can adjust only the amount of H^+-forming CO_2 in the body.

Also contributing to the kidneys' acid–base regulatory potency is their ability to return pH almost exactly to normal. By comparison to the respiratory system's inability to fully compensate for a pH abnormality, the kidneys can continue to respond to a change in pH until compensation is essentially complete.

The kidneys control pH of body fluids by adjusting three interrelated factors: (1) H^+ excretion, (2) HCO_3^- excretion, and (3) ammonia (NH_3) secretion. We will examine each of these mechanisms in further detail.

The kidneys adjust their rate of H^+ excretion depending on the plasma $[H^+]$ or $[CO_2]$.

Acids are continuously being added to the body fluids as a result of metabolic activities, yet the generated H^+ must not be allowed to accumulate. Although the body's buffer systems can resist changes in pH by removing H^+ from solution, the persistent production of acidic metabolic products would eventually overwhelm the limits of this buffering capacity. The constantly generated H^+ must therefore ultimately be eliminated from the body. The lungs can remove only carbonic acid by eliminating CO_2. The task of eliminating H^+ derived from sulfuric, phosphoric, lactic, and other acids rests with the kidneys. Furthermore, the kidneys can also eliminate extra H^+ derived from carbonic acid.

Mechanism of renal H^+ secretion

Almost all the excreted H^+ enters urine via secretion. Recall that the filtration rate of H^+ equals plasma $[H^+]$ times GFR. Because plasma $[H^+]$ is extremely low (less than in pure H_2O except during extreme acidosis, when pH falls below 7.0), the filtration rate of H^+ is likewise extremely low. This minute amount of filtered H^+ is excreted in urine. However, most excreted H^+ gains entry into tubular fluid by being actively secreted. The proximal, distal, and collecting tubules all secrete H^+. Because the kidneys normally excrete H^+, urine is usually acidic, having an average pH of 6.0.

The H^+ secretory process begins in the tubular cells with CO_2 from three sources: CO_2 diffused into the tubular cells from either (1) plasma or (2) tubular fluid or (3) CO_2 metabolically produced within the tubular cells. Influenced by carbonic anhydrase, CO_2 and H_2O form H_2CO_3, which dissociates into H^+ and HCO_3^-. An energy-dependent carrier in the luminal membrane then transports H^+ out of the cell into the tubular lumen. In part of the nephron, the tubular cells transport Na^+ derived from glomerular filtrate in the opposite direction, so H^+ secretion and Na^+ reabsorption are partially linked.

Factors influencing the rate of H^+ secretion

The magnitude of H^+ secretion depends primarily on a direct effect of the plasma's acid–base status on the kidneys' tubular cells (● Figure 15-9). No neural or hormonal control is involved.

• When the $[H^+]$ of plasma passing through the peritubular capillaries is elevated above normal, the tubular cells respond by secreting greater-than-usual amounts of H^+ from the plasma into the tubular fluid to be excreted in urine.

• Conversely, when plasma $[H^+]$ is lower than normal, the kidneys conserve H^+ by reducing its secretion and subsequent excretion in urine. The kidneys cannot raise plasma $[H^+]$ by reabsorbing more of the filtered H^+, because there are no reabsorptive mechanisms for H^+. The only way the kidneys can reduce H^+ excretion is by secreting less H^+.

Because chemical reactions for H^+ secretion begin with CO_2, the rate at which they proceed is influenced by $[CO_2]$.

• When plasma $[CO_2]$ increases, these reactions proceed more rapidly and the rate of H^+ secretion speeds up (● Figure 15-9).

• Conversely, the rate of H^+ secretion slows when plasma $[CO_2]$ falls below normal.

These responses are especially important in renal compensations for acid–base abnormalities involving a change in H_2CO_3 caused by respiratory dysfunction. The kidneys can therefore adjust H^+ excretion to compensate for changes in both carbonic and noncarbonic acids.

The kidneys conserve or excrete HCO_3^- depending on the plasma $[H^+]$.

Before being eliminated by the kidneys, H^+ generated from noncarbonic acids is buffered to a large extent by plasma HCO_3^-. Appropriately, therefore, renal handling of acid–

● **FIGURE 15-9**
Control of the rate of tubular H^+ secretion

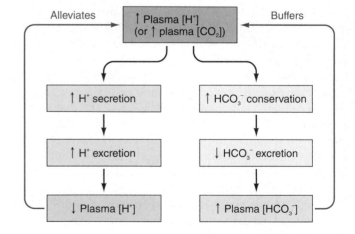

base balance also involves adjustment of HCO_3^- excretion, depending on the H^+ load in plasma (● Figure 15-9).

The kidneys regulate plasma $[HCO_3^-]$ by two interrelated mechanisms: (1) variable reabsorption of filtered HCO_3^- back into plasma and (2) variable addition of new HCO_3^- to plasma. Both these mechanisms are inextricably linked with H^+ secretion by the kidney tubules. Every time a H^+ is secreted into tubular fluid, a HCO_3^- is simultaneously transferred into peritubular capillary plasma. Whether a filtered HCO_3^- is reabsorbed or a new HCO_3^- is added to plasma in accompaniment with H^+ secretion depends on whether filtered HCO_3^- is present in the tubular fluid to react with the secreted H^+.

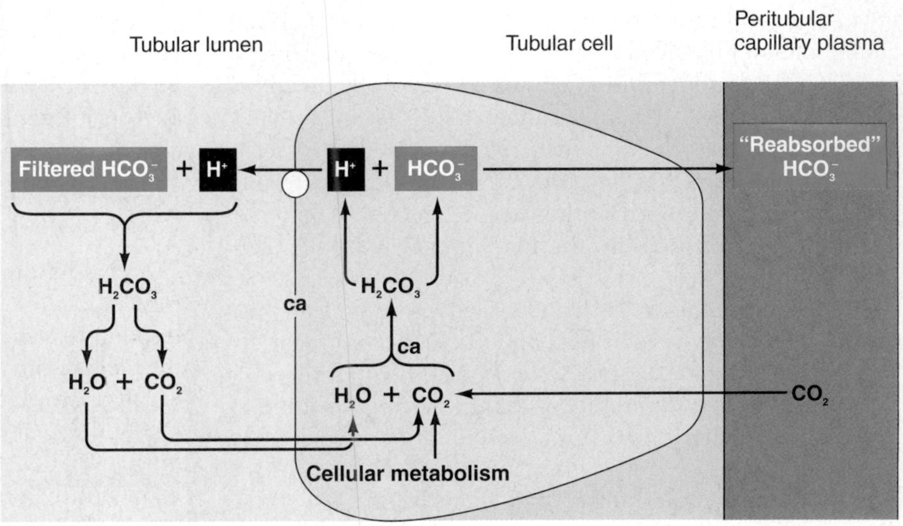

Coupling of HCO_3^- reabsorption with H^+ secretion

Bicarbonate is freely filtered, but because the luminal membranes of tubular cells are impermeable to filtered HCO_3^-, it cannot diffuse into these cells. Therefore, reabsorption of HCO_3^- must occur indirectly (● Figure 15-10). Hydrogen ion secreted into tubular fluid combines with filtered HCO_3^- to form H_2CO_3. Under the influence of carbonic anhydrase, which is present on the surface of the luminal membrane, H_2CO_3 decomposes into CO_2 and H_2O within the filtrate. Unlike HCO_3^-, CO_2 can easily penetrate tubular cell membranes. Within the cells, CO_2 and H_2O, under the influence of intracellular carbonic anhydrase, form H_2CO_3, which dissociates into H^+ and HCO_3^-. Because HCO_3^- can permeate tubular cells' basolateral membrane, it passively diffuses out of the cells and into the peritubular capillary plasma. Meanwhile, the generated H^+ is actively secreted. Because the disappearance of a HCO_3^- from the tubular fluid is coupled with the appearance of another HCO_3^- in the plasma, a HCO_3^- has, in effect, been "reabsorbed." Even though the HCO_3^- entering the plasma is not the same HCO_3^- that was filtered, the net result is the same as if HCO_3^- were directly reabsorbed.

Normally, slightly more hydrogen ions are secreted into the tubular fluid than bicarbonate ions are filtered. Accordingly, all the filtered HCO_3^- is usually absorbed, because secreted H^+ is available in tubular fluid to combine with it to form highly reabsorbable CO_2. By far the largest part of the secreted H^+ combines with HCO_3^- and is not excreted, because it is "used up" in HCO_3^- reabsorption. However, the slight excess of secreted H^+ that is not matched by filtered HCO_3^- is excreted in urine. This normal H^+ excretion rate keeps pace with the normal rate of non–carbonic-acid H^+ production.

Secretion of H^+ that is *excreted* is coupled with the *addition of new HCO_3^-* to the plasma, in contrast to the secreted H^+ that is coupled with HCO_3^- *reabsorption* and is *not excreted*, instead being incorporated into reabsorbable H_2O molecules. When all the filtered HCO_3^- has been reabsorbed and

Tubular lumen — Tubular cell — Peritubular capillary plasma

ca = Carbonic anhydrase

● **FIGURE 15-10**

Hydrogen ion secretion coupled with bicarbonate reabsorption
Because the disappearance of a filtered HCO_3^- from the tubular fluid is coupled with the appearance of another HCO_3^- in the plasma, HCO_3^- is considered to have been "reabsorbed."

additional secreted H^+ is generated by dissociation of H_2CO_3, the HCO_3^- produced by this reaction diffuses into the plasma as a "new" HCO_3^-. It is termed "new" because its appearance in plasma is not associated with reabsorption of filtered HCO_3^- (● Figure 15-11). Meanwhile, the secreted H^+ combines with urinary buffers, especially basic phosphate (HPO_4^{2-}) and is excreted.

Renal handling of H^+ and HCO_3^- during acidosis and alkalosis.

When plasma $[H^+]$ is elevated during acidosis, more H^+ is secreted than normal. At the same time, less HCO_3^- is filtered than normal because more of the plasma HCO_3^- is used up in buffering the excess H^+ in the ECF. This greater-than-usual inequity between filtered HCO_3^- and secreted H^+ has two consequences. First, more of the secreted H^+ is excreted in the urine, because more hydrogen ions are entering the tubular fluid at a time when fewer are needed to reabsorb the reduced quantities of filtered HCO_3^-. In this way, extra H^+ is eliminated from the body, making urine more acidic than normal. Second, because excretion of H^+ is linked with the addition of new HCO_3^- to the plasma, more HCO_3^- than usual enters plasma that is passing through the kidneys. This additional HCO_3^- is available to buffer excess H^+ present in the body.

In the opposite situation of alkalosis, the rate of H^+ secretion diminishes, while the rate of HCO_3^- filtration increases compared to normal. When plasma $[H^+]$ is below normal, a smaller proportion of the HCO_3^- pool is tied up buffering H^+, so plasma $[HCO_3^-]$ is elevated above normal. As a result, the rate of HCO_3^- filtration correspondingly increases. Not all the filtered HCO_3^- is reabsorbed, because bicarbonate ions are in excess of secreted hydrogen ions in the tubular fluid and

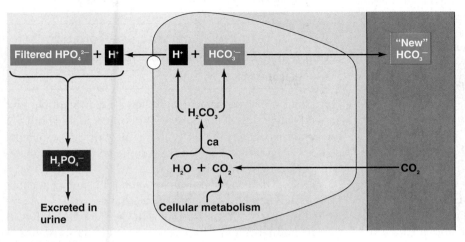

Tubular lumen   Tubular cell   Peritubular
                                              capillary plasma

ca = Carbonic anhydrase

Excreted in urine

Cellular metabolism

● **FIGURE 15-11**

Hydrogen ion secretion and excretion coupled with the addition of new HCO_3^- to the plasma Secreted H^+ does not combine with filtered HPO_4^{2-} and is not subsequently excreted until all the filtered HCO_3^- has been "reabsorbed," as depicted in ● Figure 15-10. Once all the filtered HCO_3^- has combined with secreted H^+, further secreted H^+ is excreted in the urine, primarily in association with urinary buffers such as basic phosphate. Excretion of H^+ is coupled with the appearance of new HCO_3^- in the plasma. The "new" HCO_3^- represents a net gain rather than being merely a replacement for filtered HCO_3^-.

HCO_3^- cannot be reabsorbed without first reacting with H^+. Excess HCO_3^- is left in the tubular fluid to be excreted in urine, thus reducing plasma $[HCO_3^-]$ while making urine alkaline.

In short, when plasma $[H^+]$ increases above normal during *acidosis*, renal compensation includes the following (▲ Table 15-8):

1. Increased secretion and subsequent increased excretion of H^+ in the urine, thereby eliminating the excess H^+ and decreasing plasma $[H^+]$

2. Reabsorption of all the filtered HCO_3^-, plus addition of new HCO_3^- to the plasma, resulting in increased plasma $[HCO_3^-]$

When plasma $[H^+]$ falls below normal during *alkalosis*, renal responses include the following:

1. Decreased secretion and subsequent reduced excretion of H^+ in urine, conserving H^+ and increasing plasma $[H^+]$

2. Incomplete reabsorption of filtered HCO_3^- and subsequent increased excretion of HCO_3^-, reducing plasma $[HCO_3^-]$

Note that to compensate for acidosis, the kidneys acidify urine (by getting rid of extra H^+) and alkalinize plasma (by conserving HCO_3^-) to bring pH to normal. In the opposite case—alkalosis—the kidneys make urine alkaline (by eliminating excess HCO_3^-) while acidifying plasma (by conserving H^+).

The kidneys secrete ammonia during acidosis to buffer secreted H^+.

The energy-dependent H^+ carriers in the tubular cells can secrete H^+ against a concentration gradient until the tubular fluid (urine) becomes 800 times more acidic than the plasma. At this point, further H^+ secretion ceases, because the gradient becomes too great for the secretory process to continue. The kidneys cannot acidify urine beyond a gradient-limited urinary pH of 4.5. If left unbuffered as free H^+, only about 1% of the excess H^+ typically excreted daily would produce a urinary pH of this magnitude at normal urine flow rates, and elimination of the other 99% of the usually secreted H^+ load would be prevented—a situation that would be intolerable. For H^+

▲ **TABLE 15-8**
Summary of Renal Responses to Acidosis and Alkalosis

Acid–Base Abnormality	H^+ Secretion	H^+ Excretion	HCO_3^- Reabsorption and Addition of New HCO_3^- to Plasma	HCO_3^- Excretion	pH of Urine	Compensatory Change in Plasma pH
Acidosis	↑	↑	↑	Normal (zero; all filtered is reabsorbed)	Acidic	Alkalinization toward normal
Alkalosis	↓	↓	↓	↑	Alkaline	Acidification toward normal

---

secretion to proceed, most secreted H^+ must be buffered in the tubular fluid so that it does not exist as free H^+ and, accordingly, does not contribute to the tubular acidity.

Bicarbonate cannot buffer urinary H^+ as it does the ECF, because HCO_3^- is not excreted in urine simultaneously with H^+. (Whichever of these substances is in excess in plasma is excreted in urine.) There are, however, two important urinary buffers: (1) *filtered phosphate buffers* and (2) *secreted ammonia*.

Filtered phosphate as a urinary buffer

Normally, secreted H^+ is first buffered by the phosphate buffer system, which is in the tubular fluid because excess ingested phosphate has been filtered but not reabsorbed. The basic member of the phosphate buffer pair binds with secreted H^+. Basic phosphate is present in the tubular fluid by virtue of dietary excess, not because of any specific mechanism for buffering secreted H^+. When H^+ secretion is high, the buffering capacity of urinary phosphates is exceeded, but the kidneys cannot respond by excreting more basic phosphate. Only the quantity of phosphate reabsorbed, not the quantity excreted, is subject to control. As soon as all the basic phosphate ions that are coincidentally excreted have "soaked up" H^+, the acidity of the tubular fluid quickly rises as more H^+ ions are secreted. Without additional buffering capacity from another source, H^+ secretion would soon halt abruptly as the free $[H^+]$ in the tubular fluid quickly rose to the critical limiting level.

Secreted NH_3 as a urinary buffer

When acidosis exists, the tubular cells secrete **ammonia (NH_3)** into the tubular fluid once the normal urinary phosphate buffers are saturated. This NH_3 enables the kidneys to continue secreting additional H^+ ions, because NH_3 combines with free H^+ in the tubular fluid to form **ammonium ion (NH_4^+)** as follows:

$$NH_3 + H^+ \rightarrow NH_4^+$$

The tubular membranes are not very permeable to NH_4^-, so the ammonium ions remain in the tubular fluid and are lost in urine, each one taking a H^+ with it. Thus NH_3 secretion during acidosis serves to buffer excess H^+ in the tubular fluid, so that large amounts of H^+ can be secreted into urine before pH falls to the limiting value of 4.5. Were it not for NH_3 secretion, the extent of H^+ secretion would be limited to whatever phosphate-buffering capacity coincidentally happened to be present as a result of dietary excess.

In contrast to the phosphate buffers, which are in the tubular fluid because they have been filtered but not reabsorbed, NH_3 is deliberately synthesized from the amino acid *glutamine* within the tubular cells. Once synthesized, NH_3 readily diffuses passively down its concentration gradient into the tubular fluid; that is, it is secreted. The rate of NH_3 secretion is controlled by a direct effect on the tubular cells of the amount of excess H^+ to be transported in the urine. When someone has been acidotic for more than two or three days, the rate of NH_3 production increases substantially. This extra NH_3 provides additional buffering capacity to allow H^+ secretion to con-

tinue after the normal phosphate-buffering capacity is overwhelmed during renal compensation for acidosis.

▌ Acid–base imbalances can arise from either respiratory dysfunction or metabolic disturbances.

Deviations from normal acid–base status are divided into four general categories, depending on the source and direction of the abnormal change in $[H^+]$. These categories are respiratory acidosis, respiratory alkalosis, metabolic acidosis, and metabolic alkalosis.

Because of the relationship between $[H^+]$ and concentrations of the members of a buffer pair, changes in $[H^+]$ are reflected by changes in the ratio of $[HCO_3^-]$ to $[CO_2]$. Recall that the normal ratio is 20/1. Using the Henderson-Hasselbalch equation and with pK being 6.1 and the log of 20 being 1.3, normal pH = 6.1 + 1.3 = 7.4. Determinations of $[HCO_3^-]$ and $[CO_2]$ provide more meaningful information about the underlying factors responsible for a particular acid–base status than do direct measurements of $[H^+]$ alone. The following rules of thumb apply when examining acid–base imbalances *before any compensations take place*:

1. A change in pH that has a respiratory cause is associated with an abnormal $[CO_2]$, giving rise to a change in carbonic acid–generated H^+. In contrast, a pH deviation of metabolic origin will be associated with an abnormal $[HCO_3^-]$ as a result of the participation of HCO_3^- in buffering abnormal amounts of H^+ generated from noncarbonic acids.

2. Anytime the $[HCO_3^-]/[CO_2]$ ratio falls below 20/1, an acidosis exists. The log of any number lower than 20 is less than 1.3 and, when added to the pK of 6.1, yields an acidotic pH below 7.4. Anytime the ratio exceeds 20/1, an alkalosis exists. The log of any number greater than 20 is more than 1.3 and, when added to the pK of 6.1, yields an alkalotic pH above 7.4.

● **FIGURE 15-12** ▶

Schematic representation of the relationship of $[HCO_3^-]$ and $[CO_2]$ to pH in various acid–base statuses
(a) Normal acid–base balance. The $[HCO_3^-]/[CO_2]$ ratio is 20/1. (b) Uncompensated respiratory acidosis. The $[HCO_3^-]/[CO_2]$ ratio is reduced (20/2), because CO_2 has accumulated. (c) Compensated respiratory acidosis. Compensatory retention of HCO_3^- to balance the CO_2 accumulation restores the $[HCO_3^-]/[CO_2]$ ratio to a normal equivalent (40/2). (d) Uncompensated respiratory alkalosis. The $[HCO_3^-]/[CO_2]$ ratio is increased (20/0.5) by a reduction in CO_2. (e) Compensated respiratory alkalosis. Compensatory elimination of HCO_3^- to balance the CO_2 deficit restores the $[HCO_3^-]/[CO_2]$ ratio to a normal equivalent (10/0.5). (f) Uncompensated metabolic acidosis. The $[HCO_3^-]/[CO_2]$ ratio is reduced (10/1) by a HCO_3^- deficit. (g) Compensated metabolic acidosis. Conservation of HCO_3^-, which partially makes up for the HCO_3^- deficit, and a compensatory reduction in CO_2 restore the $[HCO_3^-]/[CO_2]$ to a normal equivalent (15/0.75). (h) Uncompensated metabolic alkalosis. The $[HCO_3^-]/[CO_2]$ ratio is increased (40/1) by excess HCO_3^-. (i) Compensated metabolic alkalosis. Elimination of some of the extra HCO_3^- and a compensatory increase in CO_2 restore the $[HCO_3^-]/[CO_2]$ ratio to a normal equivalent (25/1.25).

Putting these two points together,

- *Respiratory acidosis* has a ratio of less than 20/1 arising from an increase in [CO_2].
- *Respiratory alkalosis* has a ratio greater than 20/1 because of a decrease in [CO_2].
- *Metabolic acidosis* has a ratio of less than 20/1 associated with a fall in [HCO_3^-].

- *Metabolic alkalosis* has a ratio greater than 20/1 arising from an elevation in [HCO_3^-].

We will examine each of these categories separately in more detail, paying particular attention to possible causes and the compensations that occur. The "balance beam" concept, presented in ● Figure 15-12 in conjunction with the Henderson-Hasselbalch equation, will help you better visualize the contribu-

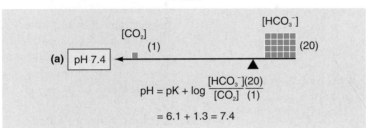

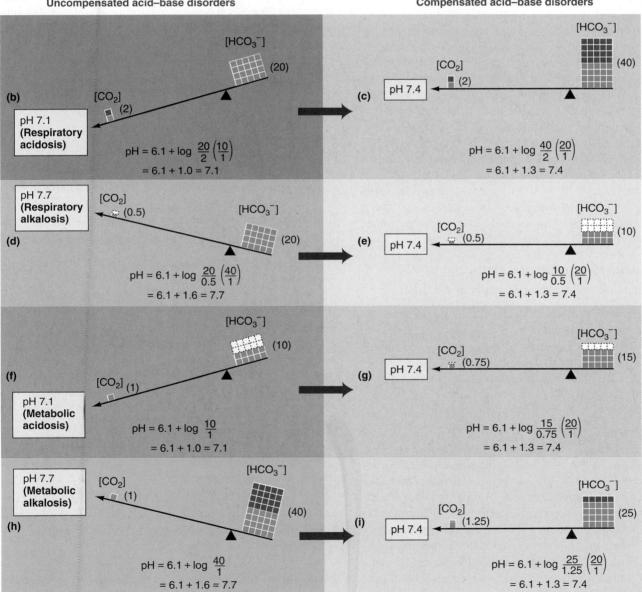

The lengths of the arms of the balance beams are not to scale.

tions of the lungs and kidneys to the causes of and compensations for various acid–base disorders. The normal situation is represented in ● Figure 15-12a.

▮ Respiratory acidosis arises from an increase in [CO_2].

Respiratory acidosis is the result of abnormal CO_2 retention arising from *hypoventilation* (see p. 495). As less-than-normal amounts of CO_2 are lost through the lungs, the resultant increase in H_2CO_3 formation and dissociation leads to an elevated [H^+].

Causes of respiratory acidosis

Possible causes include lung disease, depression of the respiratory center by drugs or disease, nerve or muscle disorders that reduce respiratory muscle ability, or (transiently) even the simple act of holding one's breath.

In uncompensated respiratory acidosis (● Figure 15-12b), [CO_2] is elevated (in our example, it is doubled) whereas [HCO_3^-] is normal, so the ratio is 20/2 (10/1) and pH is reduced. Let us clarify a potentially confusing point. You might wonder why when [CO_2] is elevated and drives the reaction $CO_2 + H_2O \leftrightarrows H_2CO_3 \leftrightarrows H^+ + HCO_3^-$ to the right, we say that [H^+] becomes elevated but [HCO_3^-] remains normal, although the same quantities of H^+ and HCO_3^- are produced when CO_2-generated H_2CO_3 dissociates. The answer lies in the fact that normally the [HCO_3^-] is 600,000 times the [H^+]. For every one hydrogen ion and 600,000 bicarbonate ions present in the ECF, the generation of one additional H^+ and one HCO_3^- doubles the [H^+] (a 100% increase) but only increases the [HCO_3^-] 0.00017% (from 600,000 to 600,001 ions). Therefore, an elevation in [CO_2] brings about a pronounced increase in [H^+], but [HCO_3^-] remains essentially normal.

Compensations for respiratory acidosis

Compensatory measures act to restore pH to normal.

- The chemical buffers immediately take up additional H^+.
- The respiratory mechanism usually cannot respond with compensatory increased ventilation, because impaired respiration is the problem in the first place.
- Thus, the kidneys are most important in compensating for respiratory acidosis. They conserve all the filtered HCO_3^- and add new HCO_3^- to the plasma while simultaneously secreting and, accordingly, excreting more H^+.

As a result, HCO_3^- stores in the body become elevated. In our example (● Figure 15-12c), the plasma [HCO_3^-] is doubled, so the [HCO_3^-]/[CO_2] ratio is 40/2 rather than 20/2 as it was in the uncompensated state. A ratio of 40/2 is equivalent to a normal 20/1 ratio, so pH is once again the normal 7.4. Enhanced renal conservation of HCO_3^- has fully compensated for CO_2 accumulation, thus restoring pH to normal, although both [CO_2] and [HCO_3^-] are now distorted. Note that maintenance of normal pH depends on preserving a normal ratio between [HCO_3^-] and [CO_2], no matter what the absolute

values of each of these buffer components are. (Compensation is never fully complete because pH can be restored close to but not precisely to normal. In our examples, however, we assume full compensation, for ease in mathematical calculations. Also bear in mind that the values used are only representative. Deviations in pH actually occur over a range, and the degree to which compensation can be accomplished varies.)

▮ Respiratory alkalosis arises from a decrease in [CO_2].

The primary defect in **respiratory alkalosis** is excessive loss of CO_2 from the body as a result of *hyperventilation* (see p. 495). When pulmonary ventilation increases out of proportion to the rate of CO_2 production, too much CO_2 is blown off. Consequently, less H_2CO_3 is formed and [H^+] decreases.

Causes of respiratory alkalosis

Possible causes of respiratory alkalosis include fever, anxiety, and aspirin poisoning, all of which excessively stimulate ventilation without regard to the status of O_2, CO_2, or H^+ in the body fluids. Respiratory alkalosis also occurs as a result of physiologic mechanisms at high altitude. When the low concentration of O_2 in arterial blood reflexly stimulates ventilation to obtain more O_2, too much CO_2 is blown off, inadvertently leading to an alkalotic state (see p. 496).

If we look at the biochemical abnormalities in uncompensated respiratory alkalosis (● Figure 15-12d), the increase in pH reflects a reduction in [CO_2] (half the normal value in our example), whereas the [HCO_3^-] remains normal. This yields an alkalotic ratio of 20/0.5, which is comparable to 40/1.

Compensations for respiratory alkalosis

Compensatory measures act to shift pH back toward normal.

- The chemical buffer systems liberate H^+ to diminish the severity of the alkalosis.
- As plasma [CO_2] and [H^+] fall below normal because of excessive ventilation, two of the normally potent stimuli for driving ventilation are removed. This effect tends to "put brakes" on the extent to which some nonrespiratory factor such as fever or anxiety can overdrive ventilation. Therefore, hyperventilation does not continue completely unabated.
- If the situation continues for a few days, the kidneys compensate by conserving H^+ and excreting more HCO_3^-.

If, as in our example (● Figure 15-12e), the HCO_3^- stores are reduced by half by loss of HCO_3^- in the urine, the [HCO_3^-]/[CO_2] ratio becomes 10/0.5, equivalent to the normal 20/1. Therefore, the pH is restored to normal by reducing the HCO_3^- load to compensate for the CO_2 loss.

▮ Metabolic acidosis is associated with a fall in [HCO_3^-].

Metabolic acidosis (also known as **nonrespiratory acidosis**) encompasses all types of acidosis besides that caused by excess CO_2

in body fluids. In the uncompensated state (● Figure 15-12f), metabolic acidosis is always characterized by a reduction in plasma [HCO_3^-] (in our example it is halved), whereas [CO_2] remains normal, producing an acidotic ratio of 10/1. The problem may arise from excessive loss of HCO_3^--rich fluids from the body or from an accumulation of noncarbonic acids. In the latter case, plasma HCO_3^- is used up in buffering the additional H^+.

Causes of metabolic acidosis

Metabolic acidosis is the type of acid–base disorder most frequently encountered. The following are its most common causes:

1. *Severe diarrhea*. During digestion, a HCO_3^--rich digestive juice is normally secreted into the digestive tract and is subsequently reabsorbed back into the plasma when digestion is completed. During diarrhea, this HCO_3^- is lost from the body rather than being reabsorbed. The reduction in plasma [HCO_3^-] without a corresponding reduction in [CO_2] lowers pH. Because of the loss of HCO_3^-, less HCO_3^- is available to buffer H^+, leading to more free H^+ in the body fluids. Looking at the situation differently, loss of HCO_3^- shifts the CO_2 + $H_2O \rightleftharpoons H^+ + HCO_3^-$ reaction to the right to compensate for the HCO_3^- deficit, increasing [H^+] above normal.

2. *Diabetes mellitus*. Abnormal fat metabolism resulting from the inability of cells to preferentially use glucose in the absence of insulin results in formation of excess keto acids whose dissociation increases plasma [H^+].

3. *Strenuous exercise*. When muscles resort to anaerobic glycolysis during strenuous exercise, excess lactic acid is produced, raising plasma [H^+] (see p. 279).

4. *Uremic acidosis*. In severe renal failure (uremia), the kidneys cannot rid the body of even the normal amounts of H^+ generated from noncarbonic acids formed by ongoing metabolic processes, so H^+ starts to accumulate in body fluids. Also, the kidneys cannot conserve an adequate amount of HCO_3^- for buffering the normal acid load. The resultant fall in [HCO_3^-] without a concomitant reduction in [CO_2] is correlated with a decline in pH to an acidotic level.

Compensations for metabolic acidosis

Except in uremic acidosis, metabolic acidosis is compensated for by both respiratory and renal mechanisms as well as by chemical buffers.

- The buffers take up extra H^+.
- The lungs blow off additional H^+-generating CO_2.
- The kidneys excrete more H^+ and conserve more HCO_3^-.

In our example (● Figure 15-12g), these compensatory measures restore the ratio to normal by reducing [CO_2] to 75% of normal and by raising [HCO_3^-] halfway back toward normal (up from 50% to 75% of the normal value). This brings the ratio to 15/0.75 (equivalent to 20/1).

Note that in compensating for metabolic acidosis, the lungs deliberately displace [CO_2] from normal in an attempt to restore [H^+] toward normal. Whereas in respiratory-induced acid–base disorders an abnormal [CO_2] is the *cause* of the [H^+] imbalance, in metabolic acid–base disorders [CO_2] is intentionally shifted from normal as an important *compensation* for the [H^+] imbalance.

When kidney disease causes metabolic acidosis, complete compensation is not possible because the renal mechanism is not available for pH regulation. Recall that the respiratory system can compensate only up to 75% of the way toward normal. Uremic acidosis is very serious, because the kidneys cannot help restore pH all the way to normal.

■ Metabolic alkalosis is associated with an elevation in [HCO_3^-].

Metabolic alkalosis is a reduction in plasma [H^+] caused by a relative deficiency of noncarbonic acids. This acid–base disturbance is associated with an increase in [HCO_3^-], which, in the uncompensated state, is not accompanied by a change in [CO_2]. In our example (● Figure 15-12h), [HCO_3^-] is doubled, producing an alkalotic ratio of 40/1.

Causes of metabolic alkalosis

This condition arises most commonly from the following:

1. *Vomiting* causes abnormal loss of H^+ from the body as a result of lost acidic gastric juices. Hydrochloric acid is secreted into the stomach lumen during digestion. Bicarbonate is added to plasma during gastric HCl secretion. This HCO_3^- is neutralized by H^+ as the gastric secretions are eventually reabsorbed back into plasma, so normally there is no net addition of HCO_3^- to plasma from this source. However, when this acid is lost from the body during vomiting not only is plasma [H^+] decreased, but also reabsorbed H^+ is no longer available to neutralize the extra HCO_3^- added to plasma during gastric HCl secretion. Thus loss of HCl in effect increases plasma [HCO_3^-]. (In contrast, with "deeper" vomiting, HCO_3^- in the digestive juices secreted into the upper intestine may be lost in the vomitus, resulting in acidosis instead of alkalosis.)

2. *Ingestion of alkaline drugs* can produce alkalosis, such as when baking soda ($NaHCO_3$, which dissociates in solution into Na^+ and HCO_3^-) is used as a self-administered remedy for treating gastric hyperacidity. By neutralizing excess acid in the stomach, HCO_3^- relieves the symptoms of stomach irritation and heartburn, but when more HCO_3^- than needed is ingested, the extra HCO_3^- is absorbed from the digestive tract and increases plasma [HCO_3^-]. The extra HCO_3^- binds with some of the free H^+ normally present in plasma from non–carbonic-acid sources, reducing free [H^+]. (In contrast, commercial alkaline products for treating gastric hyperacidity are not absorbed from the digestive tract to any extent and therefore do not alter the body's acid–base status.)

Compensations for metabolic alkalosis

- In metabolic alkalosis, chemical buffer systems immediately liberate H^+.

Summary of $[CO_2]$, $[HCO_3^-]$, and pH in Uncompensated and Compensated Acid–Base Abnormalities

Acid–Base Status	pH	$[CO_2]$ (compared to normal)	$[HCO_3^-]$ (compared to normal)	$[HCO_3^-]/[CO_2]$
Normal	Normal	Normal	Normal	20/1
Uncompensated Respiratory Acidosis	Decreased	Increased	Normal	20/2 (10/1)
Compensated Respiratory Acidosis	Normal	Increased	Increased	40/2 (20/1)
Uncompensated Respiratory Alkalosis	Increased	Decreased	Normal	20/0.5 (40/1)
Compensated Respiratory Alkalosis	Normal	Decreased	Decreased	10/0.5 (20/1)
Uncompensated Metabolic Acidosis	Decreased	Normal	Decreased	10/1
Compensated Metabolic Acidosis	Normal	Decreased	Decreased	15/0.75 (20/1)
Uncompensated Metabolic Alkalosis	Increased	Normal	Increased	40/1
Compensated Metabolic Alkalosis	Normal	Increased	Increased	25/1.25 (20/1)

- Ventilation is reduced so that extra H^+-generating CO_2 is retained in body fluids.
- If the condition persists for several days, the kidneys conserve H^+ and excrete the excess HCO_3^- in urine.

The resultant compensatory increase in $[CO_2]$ (up 25% in our example—● Figure 15-12i) and the partial reduction in $[HCO_3^-]$ (75% of the way back down toward normal in our example) together restore the $[HCO_3^-]/[CO_2]$ ratio back to the equivalent of 20/1 at 25/1.25.

Overview of compensated acid-base disorders

An individual's acid–base status cannot be assessed on the basis of pH alone. Uncompensated acid–base abnormalities can readily be distinguished on the basis of deviations of either $[CO_2]$ or $[HCO_3^-]$ from normal (▲ Table 15-9). However, when compensation has been accomplished and pH is essentially normal, determinations of $[CO_2]$ and $[HCO_3^-]$ can reveal an acid–base disorder, but the type of disorder cannot be distinguished. For example, in both compensated respiratory acidosis and compensated metabolic alkalosis, $[CO_2]$ and $[HCO_3^-]$ are both above normal. With respiratory acidosis, the original problem is an abnormal increase in $[CO_2]$, and a compensatory increase in $[HCO_3^-]$ restores the $[HCO_3^-]/[CO_2]$ ratio to 20/1. Metabolic alkalosis, by contrast, is characterized by an abnormal increase in $[HCO_3^-]$ in the first place; then a compensatory rise in $[CO_2]$ restores the ratio to normal. Similarly, compensated respiratory alkalosis and compensated metabolic acidosis share similar patterns of $[CO_2]$ and $[HCO_3^-]$. Respiratory alkalosis starts out with reduced $[CO_2]$, which is compensated by a reduction in $[HCO_3^-]$. With metabolic acidosis, $[HCO_3^-]$ falls below normal, followed by a compensatory decrease in $[CO_2]$. Thus in compensated acid–base disorders, the original problem must be determined by clinical signs and symptoms other than deviations in $[CO_2]$ and $[HCO_3^-]$ from normal.

CHAPTER IN PERSPECTIVE: FOCUS ON HOMEOSTASIS

Homeostasis depends on maintaining a balance between the input and output of all constituents present in the internal fluid environment. Regulation of fluid balance involves two separate components: control of salt balance and control of H_2O balance. Control of salt balance is primarily important in the long-term regulation of arterial blood pressure because the body's salt load affects the osmotic determination of the ECF volume, of which plasma volume is a part. An increased salt load in the ECF leads to an expansion in ECF volume, including plasma volume, which in turn causes a rise in blood pressure. Conversely, a reduction in the ECF salt load brings about a fall in blood pressure. Salt balance is maintained by constantly adjusting salt output in the urine to match unregulated, variable salt intake.

Control of H_2O balance is important in preventing changes in ECF osmolarity, which would induce detrimental osmotic shifts of H_2O between the cells and the ECF. Such shifts of H_2O into or out of the cells would cause the cells to swell or shrink, respectively. Cells, especially brain neurons, do not function normally when swollen or shrunken. Water balance is largely maintained by controlling the volume of H_2O lost in urine to compensate for uncontrolled losses of variable volumes of H_2O from other avenues, such as through sweating or diarrhea, and for poorly regulated H_2O intake. Even though a thirst mechanism exists to control H_2O intake based on need, the amount a person drinks is often influenced by social custom and habit instead of thirst alone.

A balance between input and output of H^+ is also critical to maintaining the body's acid–base balance within the narrow limits compatible with life. Deviations in the internal fluid environment's pH lead to altered neuromuscular excitability, to changes in enzymatically controlled metabolic activity, and to K^+ imbalances, which can cause cardiac arrhythmias. These effects are fatal if the pH falls outside the range of 6.8 to 8.0.

Hydrogen ions are uncontrollably and continually being added to the body fluids as a result of ongoing metabolic activities, yet the ECF's pH must be kept constant at a slightly alkaline level of 7.4 for optimal body function. Like salt and H_2O balance, control of H^+ output by the kidneys is the main regulatory factor in achieving H^+ balance. Assisting the kidneys in eliminating H^+ are the lungs, which can adjust their rate of excretion of H^+-generating CO_2.

The assertion that the same input–output balance that applies to salt and H_2O also applies to H^+ homeostasis must be modified by the fact that H^+ is buffered in the body. This buffering mechanism can take up or liberate H^+, transiently keeping its concentration constant within the body until its output can be brought into line with its input. Such a mechanism is not available for salt or H_2O balance.

CHAPTER SUMMARY

Balance Concept

- The internal pool of a substance is the quantity of that substance in the ECF.
- The inputs to the pool are by way of ingestion or metabolic production of the substance. The outputs from the pool are by way of excretion or metabolic consumption of the substance.
- Input must equal output to maintain a stable balance of the substance.

Fluid Balance

- On average, the body fluids compose 60% of total body weight. This figure varies among individuals, depending on how much fat (a tissue that has a low H_2O content) they have.
- Two-thirds of the body H_2O is found in the intracellular fluid (ICF). The remaining one-third present in the extracellular fluid (ECF) is distributed between plasma (20% of ECF) and interstitial fluid (80% of ECF).
- Because all plasma constituents are freely exchanged across the capillary walls, the plasma and interstitial fluid are nearly identical in composition, except for the lack of plasma proteins in the interstitial fluid. In contrast, ECF and ICF have markedly different compositions, because the cell membrane barriers are highly selective as to what materials are transported into or out of the cells.
- The essential components of fluid balance are control of ECF volume by maintaining salt balance and control of ECF osmolarity by maintaining water balance.
- Because of the osmotic holding power of Na^+, the major ECF cation, a change in the body's total Na^+ content brings about a corresponding change in ECF volume, including plasma volume, which, in turn, alters arterial blood pressure in the same direction.
- Appropriately, changes in ECF volume and arterial blood pressure are compensated for in the long run by Na^+-regulating mechanisms.
- Salt intake is not controlled in humans, but control of salt output in urine is closely regulated. Blood pressure–regulating mechanisms can vary GFR, and accordingly the amount of Na^+ filtered, by adjusting the caliber of the afferent arterioles supplying the glomeruli. Simultaneously, blood pressure–regulating mechanisms can vary secretion of aldosterone, the hormone that promotes Na^+ reabsorption by the renal tubules. Varying Na^+ filtration and Na^+ reabsorption can adjust how much Na^+ is excreted in urine, to regulate plasma volume and subsequently arterial blood pressure in the long term.
- Changes in ECF osmolarity are primarily detected and corrected by the systems responsible for maintaining H_2O balance.
- The ECF osmolarity must be closely regulated to prevent osmotic shifts of H_2O between ECF and ICF, because cell swelling or shrinking is harmful, especially to brain neurons.

- Excess free H_2O in the ECF dilutes ECF solutes; the resulting ECF hypotonicity drives H_2O into the cells. An ECF free-H_2O deficit, by contrast, concentrates ECF solutes, so H_2O leaves the cells to enter the hypertonic ECF.
- To prevent these detrimental fluxes, free H_2O balance is regulated largely by vasopressin and, to a lesser degree, by thirst.
- Changes in vasopressin secretion and thirst are both governed primarily by hypothalamic osmoreceptors, which monitor ECF osmolarity. The amount of vasopressin secreted determines the extent of free H_2O reabsorption by distal portions of the nephrons, thereby determining the volume of urinary output.
- Simultaneously, intensity of thirst controls the volume of fluid intake. However, because the volume of fluid drunk is often not directly correlated with the intensity of thirst, control of urinary output by vasopressin is the most important regulatory mechanism for maintaining H_2O balance.

Acid–Base Balance

- Acids liberate free hydrogen ions (H^+) into solution; bases bind with free hydrogen ions and remove them from solution.
- The term *acid–base balance* refers to regulation of H^+ concentration ($[H^+]$) in body fluids. To precisely maintain $[H^+]$, input of H^+ by metabolic production of acids within the body must continually be matched with H^+ output by urinary excretion of H^+ and respiratory removal of H^+-generating CO_2. Furthermore, between the time of this generation and its elimination, H^+ must be buffered within the body to prevent marked fluctuations in $[H^+]$.
- Hydrogen ion concentration frequently is expressed in terms of pH, which is the logarithm of $1/[H^+]$.
- The normal pH of the plasma is 7.4, slightly alkaline compared to neutral H_2O, which has a pH of 7.0. A pH lower than normal (higher $[H^+]$ than normal) indicates a state of acidosis. A pH higher than normal (lower $[H^+]$ than normal) characterizes a state of alkalosis.
- Fluctuations in $[H^+]$ have profound effects on body chemistry, most notably (1) changes in neuromuscular excitability, with acidosis depressing excitability, especially in the central nervous system, and alkalosis producing overexcitability of both the peripheral and the central nervous systems, (2) disruption of normal metabolic reactions by altering structure and function of all enzymes, and (3) alterations in plasma $[K^+]$ brought about by H^+-induced changes in the rate of K^+ elimination by the kidneys.
- The primary challenge in controlling acid–base balance is maintaining normal plasma alkalinity despite continual addition of H^+ to plasma from ongoing metabolic activity. The major source of H^+ is from dissociation of CO_2-generated H_2CO_3.
- The three lines of defense for resisting changes in $[H^+]$ are (1) the chemical buffer systems, (2) respiratory control of pH, and (3) renal control of pH.

- Chemical buffer systems, the first line of defense, each consist of a pair of chemicals involved in a reversible reaction, one that can liberate H^+ and the other that can bind H^+. By acting according to the law of mass action, a buffer pair acts immediately to minimize any changes in pH that occur.
- The respiratory system, constituting the second line of defense, normally eliminates the metabolically produced CO_2 so that H_2CO_3 does not accumulate in the body fluids.
- When chemical buffers alone have been unable to immediately minimize a pH change, the respiratory system responds within a few minutes by altering its rate of CO_2 removal. An increase in $[H^+]$ arising from sources other than carbonic acid stimulates respiration so that more H_2CO_3-forming CO_2 is blown off, compensating for acidosis by reducing generation of H^+ from H_2CO_3. Conversely, a fall in $[H^+]$ depresses respiratory activity so that CO_2 and thus H^+-generating H_2CO_3 can accumulate in body fluids to compensate for alkalosis.
- The kidneys are the third and most powerful line of defense. They require hours to days to compensate for a deviation in body fluid pH. However, they not only eliminate the normal amount of H^+ produced from non-H_2CO_3 sources, but they can also alter their rate of H^+ removal in response to changes in both non-H_2CO_3 and H_2CO_3 acids. In contrast, the lungs can adjust only H^+ generated from H_2CO_3. Furthermore, the kidneys can regulate $[HCO_3^-]$ in body fluids as well.
- The kidneys compensate for acidosis by secreting excess H^+ in the urine while adding new HCO_3^- to the plasma to expand the HCO_3^- buffer pool.
- During alkalosis, the kidneys conserve H^+ by reducing its secretion in urine. They also eliminate HCO_3^-, which is in excess because less HCO_3^- than usual is tied up buffering H^+ when H^+ is in short supply.
- Secreted H^+ that is to be excreted in urine must be buffered in the tubular fluid to prevent the H^+ concentration gradient from becoming so great that it prevents further H^+ secretion. Normally, H^+ is buffered by the urinary phosphate buffer pair, which is abundant in the tubular fluid because excess dietary phosphate spills into urine to be excreted from the body.
- In acidosis, when all the phosphate buffer is already used up in buffering the extra secreted H^+, the kidneys secrete NH_3 into the tubular fluid to serve as a buffer so that H^+ secretion can continue.
- The four types of acid–base imbalances are respiratory acidosis, respiratory alkalosis, metabolic acidosis, and metabolic alkalosis. Respiratory acid–base disorders originate with deviations from normal $[CO_2]$, whereas metabolic acid–base imbalances encompass all deviations in pH other than those caused by abnormal $[CO_2]$.

REVIEW EXERCISES

Objective Questions (Answers on p. A-47)

1. The only avenue by which materials can be exchanged between the cells and the external environment is the ECF. *(True or false?)*
2. Water is driven into the cells when the ECF volume is expanded by an isotonic fluid gain. *(True or false?)*
3. Salt balance in humans is poorly regulated because of our hedonistic salt appetite. *(True or false?)*
4. An unintentional increase in CO_2 is a cause of respiratory acidosis, but a deliberate increase in CO_2 compensates for metabolic alkalosis. *(True or false?)*
5. Secreted H^+ that is coupled with HCO_3^- reabsorption is not excreted, whereas secreted H^+ that is excreted is linked with the addition of new HCO_3^- to plasma. *(True or false?)*
6. The largest body fluid compartment is the _____.
7. Of the two members of the H_2CO_3:HCO_3^- buffer system, _____ is regulated by the lungs whereas _____ is regulated by the kidneys.
8. Which of the following individuals would have the lowest percentage of body H_2O?
   a. a chubby baby
   b. a well-proportioned female college student
   c. a well-muscled male college student
   d. an obese elderly woman
   e. a lean elderly man
9. Which of the following factors does *not* increase vasopressin secretion?
   a. ECF hypertonicity
   b. alcohol
   c. stressful situations
   d. an ECF volume deficit
   e. angiotensin II
10. *Indicate all correct answers:* pH
    a. equals log $1/[H^+]$.
    b. equals pK + log $[CO_2]/[HCO_3^-]$.
    c. is high in acidosis.
    d. falls lower as $[H^+]$ increases.
    e. is normal when the $[HCO_3^-]/[CO_2]$ ratio is 20/1.
11. *Indicate all correct answers:* Acidosis
    a. causes overexcitability of the nervous system.
    b. exists when the plasma pH falls below 7.35.
    c. occurs when the $[HCO_3^-]/[CO_2]$ ratio exceeds 20/1.
    d. occurs when CO_2 is blown off more rapidly than it is being produced by metabolic activities.
    e. occurs when excessive HCO_3^- is lost from the body such as in diarrhea.
12. *Indicate all correct answers:* The kidney tubular cells secrete NH_3
    a. when the urinary pH becomes too high.
    b. when the body is in a state of alkalosis.
    c. to enable further renal secretion of H^+ to occur.
    d. to buffer excess filtered HCO_3^-.
    e. when there is excess NH_3 in the body fluids.
13. Complete the following chart:

$\dfrac{[HCO_3^-]}{[CO_2]}$	Uncompensated Abnormality	Possible Cause	pH
10/1	1. _____	2. _____	3. _____
20/0.5	4. _____	5. _____	6. _____
20/2	7. _____	8. _____	9. _____
40/1	10. _____	11. _____	12. _____

Essay Questions

1. Explain the balance concept.
2. Outline the distribution of body H_2O.
3. Define *transcellular fluid*, and identify its components. Does the transcellular compartment as a whole reflect changes in the body's fluid balance?
4. Compare the ionic composition of plasma, interstitial fluid, and intracellular fluid.
5. What factors are regulated to maintain the body's fluid balance?
6. Why is regulation of ECF volume important? How is it regulated?

7. Why is regulation of ECF osmolarity important? How is it regulated? What are the causes and consequences of ECF hypertonicity and ECF hypotonicity?
8. Outline the sources of input and output in a daily salt balance and a daily H_2O balance. Which are subject to control to maintain the body's fluid balance?
9. Distinguish between an acid and a base.
10. What is the relationship between $[H^+]$ and pH?
11. What is the normal pH of body fluids? How does this compare to the pH of H_2O? Define *acidosis* and *alkalosis*.
12. What are the consequences of fluctuations in $[H^+]$?
13. What are the body's sources of H^+?
14. Describe the three lines of defense against changes in $[H^+]$ in terms of the mechanisms and speed of action.
15. List and indicate the functions of each of the body's chemical buffer systems.

16. What are the causes of the four categories of acid–base imbalances?
17. Why is uremic acidosis so serious?

Quantitative Exercises (Solutions on p. A-47)
1. Given that plasma pH = 7.4, arterial P_{CO_2} = 40 mm Hg, and each mm Hg partial pressure of CO_2 is equivalent to a plasma $[CO_2]$ of 0.03 mM, what is the value of plasma $[HCO_3^-]$?
2. Death occurs if the plasma pH falls outside the range of 6.8 to 8.0 for an extended time. What is the concentration range of H^+ represented by this pH range?
3. A person drinks 1 liter of distilled water. Use the data in ▲ Table 15-1 to calculate the resulting percent increase in total body water (TBW), ICF, ECF, plasma, and interstitial fluid. Repeat the calculations for ingestion of 1 liter of isotonic NaCl. Which solution would be better at expanding plasma volume in a patient who has just hemorrhaged?

POINTS TO PONDER

(Explanations on p. A-47)
1. Alcoholic beverages inhibit vasopressin secretion. Given this fact, predict the effect of alcohol on the rate of urine formation. Predict the actions of alcohol on ECF osmolarity. Explain why a person still feels thirsty after excessive consumption of alcoholic beverages.
2. If a person loses 1,500 ml of salt-rich sweat and drinks 1,000 ml of water during the same time period, what will happen to vasopressin secretion? Why is it important to replace both the water and the salt?
3. If a solute that can penetrate the plasma membrane, such as dextrose (a type of sugar), is dissolved in sterile water at a concentration equal to that of normal body fluids and then is injected intravenously, what is the impact on the body's fluid balance?
4. Explain why it is safer to treat gastric hyperacidity with antacids that are poorly absorbed from the digestive tract than with baking soda, which is a good buffer for acid but is readily absorbed.

5. Which of the following reactions would occur to buffer the acidosis accompanying severe pneumonia?
   a. $H^+ + HCO_3^- \rightarrow H_2CO_3 \rightarrow CO_2 + H_2O$
   b. $CO_2 + H_2O \rightarrow H_2CO_3 \rightarrow H^+ + HCO_3^-$
   c. $H^+ + Hb \rightarrow HHb$
   d. $HHb \rightarrow H^+ + Hb$
   e. $NaH_2PO_4 + Na^+ \rightarrow Na_2HPO_4 + H^+$
6. *Clinical Consideration.* Marilyn Y. has had pronounced diarrhea for over a week as a result of having acquired salmonellosis, a bacterial intestinal infection, from improperly handled food. What impact has this prolonged diarrhea had on her fluid and acid–base balance? In what ways has Marilyn's body been trying to compensate for these imbalances?

PHYSIOEDGE RESOURCES

PHYSIOEDGE CD-ROM
PhysioEdge focuses on concepts students find difficult to learn. Figures marked with this icon have associated activities on the CD.

CHECK OUT THESE MEDIA QUIZZES:
15.1 Fluid and Electrolyte Balance
15.2 Basics of Acid-Base Balance

PHYSIOEDGE WEB SITE
The Web site for this book contains a wealth of helpful study aids, as well as many ideas for further reading and research. Log on to:

http://www.brookscole.com

Go to the Biology page and select Sherwood's *Human Physiology*, 5th Edition. Select a chapter from the drop-down menu or click on one of these resource areas:

- **Case Histories** provide an introduction to the clinical aspects of human physiology. Check out:

   #5: Eight Years Later

- For 2-D and 3-D graphical illustrations and animations of physiological concepts, visit our **Visual Learning Resource.**

- Resources for study and review include the **Chapter Outline, Chapter Summary, Glossary,** and **Flash Cards.** Use our **Quizzes** to prepare for in-class examinations.

- On-line research resources to consult are **Hypercontents,** with current links to relevant Internet sites; **Internet Exercises** with starter URLs provided; and **InfoTrac Exercises.**

   For more readings, go to InfoTrac College Edition, your on-line research library, at:

   http://infotrac.thomsonlearning.com

Digestive System

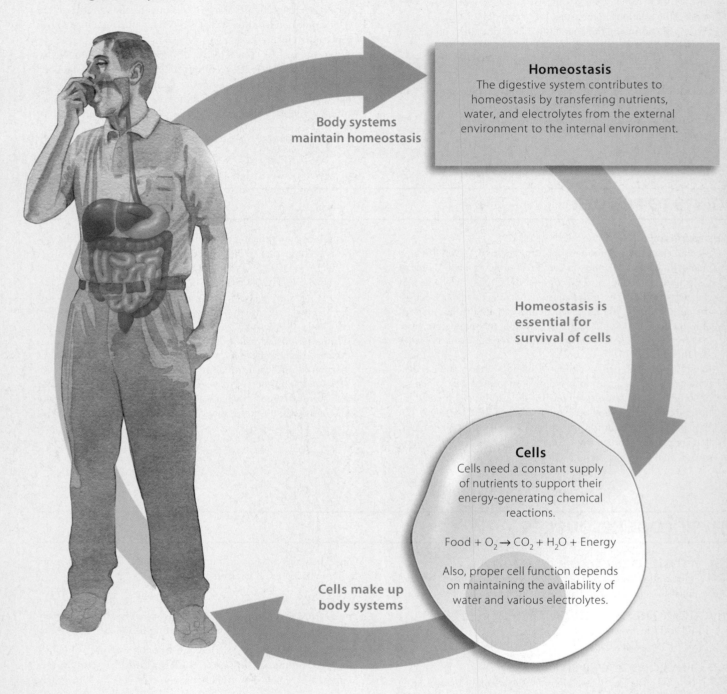

Body systems maintain homeostasis

Homeostasis
The digestive system contributes to homeostasis by transferring nutrients, water, and electrolytes from the external environment to the internal environment.

Homeostasis is essential for survival of cells

Cells
Cells need a constant supply of nutrients to support their energy-generating chemical reactions.

$$Food + O_2 \rightarrow CO_2 + H_2O + Energy$$

Also, proper cell function depends on maintaining the availability of water and various electrolytes.

Cells make up body systems

To maintain homeostasis, nutrient molecules used for energy production must continually be replaced by new, energy-rich nutrients. Similarly, water and electrolytes that are constantly lost in urine and sweat and through other avenues must be replenished regularly. The digestive system contributes to homeostasis by transferring nutrients, water, and electrolytes from the external environment to the internal environment. The **digestive system** does not directly regulate the concentration of any of these constituents in the internal environment. It does not vary nutrient, water, or electrolyte uptake based on body needs (with few exceptions); rather, it optimizes conditions for digesting and absorbing what is ingested.

Chapter 16

The Digestive System

CONTENTS AT A GLANCE

INTRODUCTION

▌ Basic digestive processes

▌ Components of the digestive system

▌ General mechanisms of regulating digestive function

MOUTH

▌ Chewing

▌ Salivary secretion

PHARYNX AND ESOPHAGUS

▌ Swallowing

STOMACH

▌ Gastric motility

▌ Gastric secretion

▌ Digestion in the stomach

▌ Absorption by the stomach

PANCREATIC AND BILIARY SECRETIONS

▌ Exocrine pancreas

▌ Liver and biliary system

SMALL INTESTINE

▌ Small intestine motility

▌ Small intestine secretion

▌ Digestion in the small intestine

▌ Absorption by the small intestine

LARGE INTESTINE

▌ Large intestine motility

▌ Large intestine secretion

▌ Colonic bacteria

▌ Absorption by the large intestine

▌ Composition of feces

OVERVIEW OF THE GASTROINTESTINAL HORMONES

INTRODUCTION

The primary function of the **digestive system** is to transfer nutrients, water, and electrolytes from the food we eat into the body's internal environment. Ingested food is essential as an energy source, or "fuel," from which the cells can produce ATP to carry out their particular energy-dependent activities, such as active transport, contraction, synthesis, and secretion. Food is also a source of building supplies for the renewal and addition of body tissues.

The act of eating does not automatically make the preformed organic molecules in food available to the body cells as a source of fuel or as building blocks. The food first must be digested, or biochemically broken down, into small, simple molecules that can be absorbed from the digestive tract into the circulatory system for distribution to the cells. Normally, about 95% of the ingested food is made available for the body's use.

We will first provide an overview of the digestive system, examining the common features of the various components of the system, before we begin a detailed tour of the tract from beginning to end.

▌ The digestive system performs four basic digestive processes.

There are four basic digestive processes: *motility, secretion, digestion,* and *absorption.*

Motility

The term **motility** refers to the muscular contractions that mix and move forward the contents of the digestive tract. Like vascular smooth muscle, the smooth muscle in the walls of the digestive tract maintains a constant low level of contraction known as **tone**. Tone is important in maintaining a steady pressure on the contents of the digestive tract as well as in preventing its walls from remaining permanently stretched following distension.

Two basic types of digestive motility are superimposed on this ongoing tonic activity: propulsive movements and mixing movements. *Propulsive movements* propel or push the contents forward through the digestive tract at varying speeds, with the rate of propulsion depending on the functions accomplished by the different regions; that is, food is moved forward in a given segment at an appropriate velocity to allow that segment to "do its job." For example, transit of food through the esophagus is rapid, which is appropriate because this structure merely serves as a passageway from the mouth to the stomach. In comparison, in the small intestine—the major site of digestion and absorption—the contents are moved forward slowly, allowing sufficient time for the breakdown and absorption of food.

Mixing movements serve a twofold function. First, by mixing food with the digestive juices, these movements promote digestion of the food. Second, they facilitate absorption by exposing all portions of the intestinal contents to the absorbing surfaces of the digestive tract.

Contraction of the smooth muscle within the walls of the digestive organs accomplishes the movement of material through most of the digestive tract, with the exceptions of the ends of the tract—the mouth through the early portion of the esophagus at the beginning and the external anal sphincter at the end—where motility involves skeletal muscle rather than smooth muscle activity. Accordingly, the acts of chewing, swallowing, and defecation have voluntary components, because skeletal muscle is under voluntary control. By contrast, motility accomplished by smooth muscle throughout the remainder of the tract is controlled by complex involuntary mechanisms.

Secretion

A number of digestive juices are secreted into the digestive tract lumen by exocrine glands (see p. 6) located along the route, each with its own specific secretory product or products. Each **digestive secretion** consists of water, electrolytes, and specific organic constituents that are important in the digestive process, such as enzymes, bile salts, or mucus. The secretory cells extract from the plasma large volumes of water and the raw materials necessary to produce their particular secretion (● Figure 16-1). Secretion of all digestive juices requires energy, both for active transport of some of the raw materials into the cell (others diffuse in passively) and for synthesis of secretory products by the endoplasmic reticulum. On appropriate neural or hormonal stimulation, the secretions are released into the digestive tract lumen. Normally, the digestive secretions are reabsorbed in one form or another back into the blood after their participation in digestion. Failure to do so (because of vomiting or diarrhea, for example) results in loss of this fluid that has been "borrowed" from the plasma.

Digestion

Humans consume three different biochemical categories of energy-rich foodstuffs: *carbohydrates*, *proteins*, and *fats*. These large molecules are unable to cross plasma membranes intact to be absorbed from the lumen of the digestive tract into the blood or lymph. The term **digestion** refers to the biochemical breakdown of the structurally complex foodstuffs of the diet into smaller, absorbable units by the enzymes produced within the digestive system as follows (▲ Table 16-1):

1. The simplest form of **carbohydrates** is the simple sugars or **monosaccharides** ("one-sugar" molecules), such as **glucose, fructose,** and **galactose,** very few of which are normally found in the diet. Most ingested carbohydrate is in the form of **polysaccharides** ("many sugar" molecules), which consist of chains of interconnected glucose molecules. The most common polysaccharide consumed is **starch** derived from plant sources. In addition, meat contains **glycogen,** the polysaccharide storage form of glucose in muscle. **Cellulose,** another dietary polysaccharide that is found in plant walls, cannot be digested into its constituent monosaccharides by the digestive juices secreted in humans; thus it represents the undigested *fiber* or "bulk" of our diets. Besides polysaccharides, a lesser source of dietary carbohydrate is in the form of **disaccharides** ("two-sugar" molecules), including **sucrose** (table sugar, which consists of one glucose and one fructose molecule) and **lactose** (milk sugar made up of one glucose and one galactose molecule). Through the process of digestion, starch, glycogen, and disaccharides are converted into their constituent monosaccharides, principally glucose with small amounts of fructose and galactose. These monosaccharides are the absorbable units for carbohydrates.

2. The second category of foodstuffs is **proteins,** which consist of various combinations of **amino acids** held together by peptide bonds (see p. A-14). Through the process of digestion, proteins are degraded primarily into their constituent amino acids as well as a few **small polypeptides** (several amino acids linked together by peptide bonds), both of which are the absorbable units for protein.

● **FIGURE 16-1**

General mode of exocrine gland secretion
Exocrine gland cells extract from the plasma by both active and passive means the raw materials that they need to produce their secretory product. This product is emptied into ducts, which lead, in the case of the digestive system, to the lumen of the digestive tract. Frequently, the secretion is modified as it moves through the duct by active and passive transport mechanisms within the membranes of the cells lining the duct.

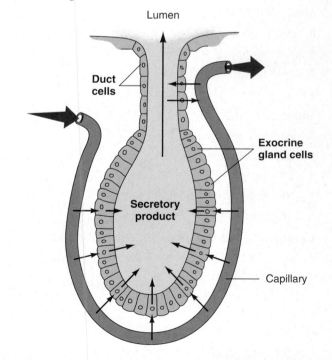

Lumen

Duct cells

Exocrine gland cells

Secretory product

Capillary

3. **Fats** represent the third category of foodstuffs. Most dietary fat is in the form of **triglycerides,** which are neutral fats, each consisting of a **glycerol** with three **fatty acid** molecules attached (*tri* means "three"). During digestion, two of the fatty acid molecules are split off, leaving a **monoglyceride,** a glycerol molecule with one fatty acid molecule attached (*mono* means

▲ **TABLE 16-1**
Process of Digestion

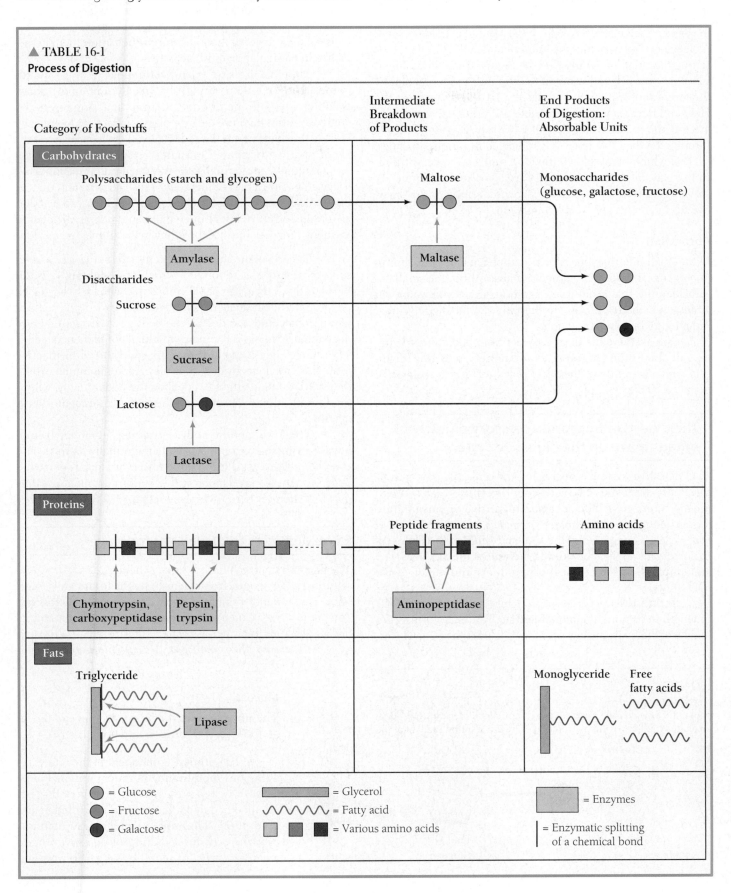

"one"). Thus, the end products of fat digestion are monoglycerides and free fatty acids, which are the absorbable units of fat.

Digestion is accomplished by enzymatic **hydrolysis** ("breakdown by water"). By adding H_2O at the bond site, enzymes in the digestive secretions break down the bonds that hold the small molecular subunits within the nutrient molecules together, thus setting the small molecules free (● Figure 16-2). These small subunits were originally joined to form nutrient molecules by the removal of H_2O at the bond sites. Hydrolysis replaces the H_2O and frees the small absorbable units. Digestive enzymes are specific in the bonds they can hydrolyze. As food moves through the digestive tract, it is subjected to various enzymes, each of which breaks down the food molecules even further. In this way, large food molecules are converted to simple absorbable units in a progressive, stepwise fashion as the digestive tract contents are propelled forward.

Absorption

In the small intestine, digestion is completed and most absorption occurs. Through the process of **absorption,** the small absorbable units that result from digestion, along with water, vitamins, and electrolytes, are transferred from the digestive tract lumen into the blood or lymph.

As we examine the digestive tract from beginning to end, we will discuss the four processes of motility, secretion, digestion, and absorption as they take place within each digestive organ (▲ Table 16-2, pp. 596–597).

■ The digestive tract and accessory digestive organs make up the digestive system.

The digestive system consists of the *digestive* (or *gastrointestinal*) *tract* plus the accessory digestive organs (*gastro* means "stomach"). The **accessory digestive organs** include the *salivary glands*, the *exocrine pancreas*, and the *biliary system*, which is composed of the *liver* and *gallbladder*. These exocrine organs are located outside the wall of the digestive tract and empty their secretions through ducts into the digestive tract lumen.

The **digestive tract** is essentially a tube about 4.5 m (15 feet) in length in its normal contractile state. Running through the middle of the body, the digestive tract includes the following organs (▲ Table 16-2): *mouth; pharynx* (throat); *esophagus; stomach; small intestine* (consisting of the *duodenum, jejunum,* and *ileum*); *large intestine* (composed of the *cecum, appendix, colon,* and *rectum*); and *anus.* Although these organs are continuous with each other, they are considered as separate entities because of their regional modifications, which allow them to specialize in particular digestive activities.

Because the digestive tract is continuous from the mouth to the anus, the lumen of this tube, like the lumen of a straw, is continuous with the external environment. As a result, the contents within the lumen of the digestive tract are technically outside the body, just as the soda that you suck through a straw is not a part of the straw. Only after a substance has been absorbed from the lumen across the intestinal wall is it considered to have become a part of the body. This fact is important, because conditions essential to the digestive process can be tolerated in the digestive tract lumen that could not be tolerated in the body proper. Consider the following examples:

- The pH of the stomach contents falls as low as 2 as a result of the gastric secretion of hydrochloric acid (HCl), yet in the body fluids the range of pH compatible with life is 6.8 to 8.0.
- The harsh digestive enzymes that hydrolyze the protein in food could also destroy the body's own tissues that produce them. (Protein is the main structural component of cells.) Therefore, once these enzymes are synthesized in inactive form, they are not activated until they reach the lumen, where they actually attack the food outside the body (that is, within the lumen), thereby protecting the body tissues against self-digestion.
- The lower portion of the intestine is inhabited by millions of living micro-organisms that are normally harmless and even beneficial, yet if these same micro-organisms enter the body proper (as may happen with a ruptured appendix), they may be extremely harmful or even lethal.

■ The digestive tract wall has four layers.

The digestive tract wall has the same general structure throughout most of its length from the esophagus to the anus, with some local variations characteristic for each region. A cross section of the digestive tube reveals four major tissue layers (● Figure 16-3). From the innermost layer outward they are the *mucosa,* the *submucosa,* the *muscularis externa,* and the *serosa.*

Mucosa

The **mucosa** lines the luminal surface of the digestive tract. It is divided into three layers:

- The primary component of the mucosa is a **mucous membrane,** an inner epithelial layer that serves as a protective surface as well as being modified in particular areas for secretion and absorption. The mucous membrane contains *exocrine cells* for secretion of digestive juices, *endocrine cells* for secretion of gastrointestinal

---

● FIGURE 16-2

An example of hydrolysis
In this example, the disaccharide maltose (the intermediate breakdown product of polysaccharides) is broken down into two glucose molecules by the addition of H_2O at the bond site.

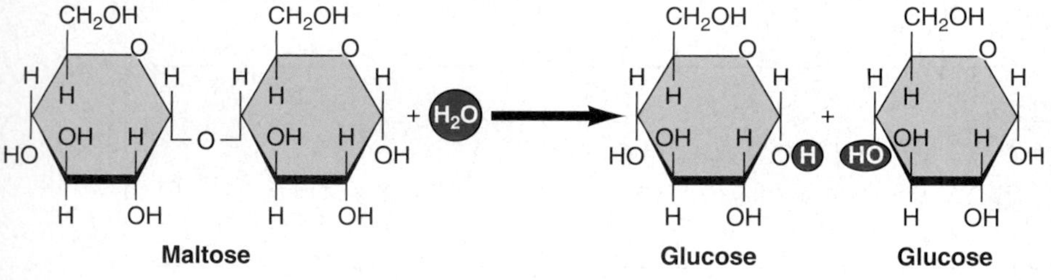

Maltose   Glucose   Glucose

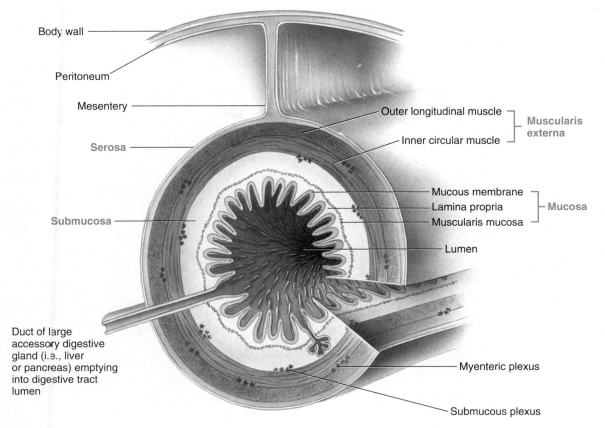

Body wall

Peritoneum

Mesentery

Serosa

Submucosa

Duct of large
accessory digestive
gland (i.e., liver
or pancreas) emptying
into digestive tract
lumen

Outer longitudinal muscle ⎤
Inner circular muscle ⎦ **Muscularis externa**

Mucous membrane ⎤
Lamina propria ⎬ **Mucosa**
Muscularis mucosa ⎦

Lumen

Myenteric plexus

Submucous plexus

● **FIGURE 16-3**

Layers of the digestive tract wall
The digestive tract wall consists of four major layers: from the innermost out, they are the mucosa, submucosa, muscularis externa, and serosa.

hormones, and *epithelial cells* specialized for absorbing digestive nutrients.

• The **lamina propria** is a thin middle layer of connective tissue on which the epithelium rests. It houses the **gut-associated lymphoid tissue (GALT)**, which is important in the defense against intestinal bacteria (see p. 414).

• The **muscularis mucosa**, a sparse layer of smooth muscle, is the outermost mucosal layer that lies adjacent to the submucosa.

The mucosal surface is generally highly folded with many ridges and valleys that greatly increase the surface area available for absorption. The degree of folding varies in different areas of the digestive tract, being most extensive in the small intestine, where maximum absorption occurs, and least extensive in the esophagus, which merely serves as a transit tube. The pattern of surface folding can be modified by contraction of the muscularis mucosa. This is important in exposing different areas of the absorptive surface to the luminal contents.

Submucosa

The **submucosa** ("under the mucosa") is a thick layer of connective tissue that provides the digestive tract with its distensibility and elasticity. It contains the larger blood and lymph vessels, both of which send branches inward to the mucosal layer and outward to the surrounding thick muscle layer. Also, a nerve network known as the *submucous plexus* lies within the submucosa (*plexus* means "network).

Muscularis externa

The **muscularis externa**, the major smooth-muscle coat of the digestive tube, surrounds the submucosa. In most parts of the tract, the muscularis externa consists of two layers: an *inner circular layer* and an *outer longitudinal layer*. The fibers of the inner smooth muscle layer (adjacent to the submucosa) run circularly around the circumference of the tube. Contraction of these circular fibers constricts or decreases the diameter of the lumen at the point of contraction. Contraction of the fibers in the outer layer, which run longitudinally along the length of the tube, accomplishes shortening of the tube. Together, contractile activity of these smooth muscle layers produces the propulsive and mixing movements. Another nerve network, the *myenteric plexus*, lies between the two muscle layers (*myo* means "muscle," *enteric* means "intestine," in reference to the location of this plexus between the two muscle layers of the intestine and elsewhere in the digestive tract). Together the submucous and myenteric plexuses help regulate local gut activity.

Anatomy and Functions of Components of the Digestive System

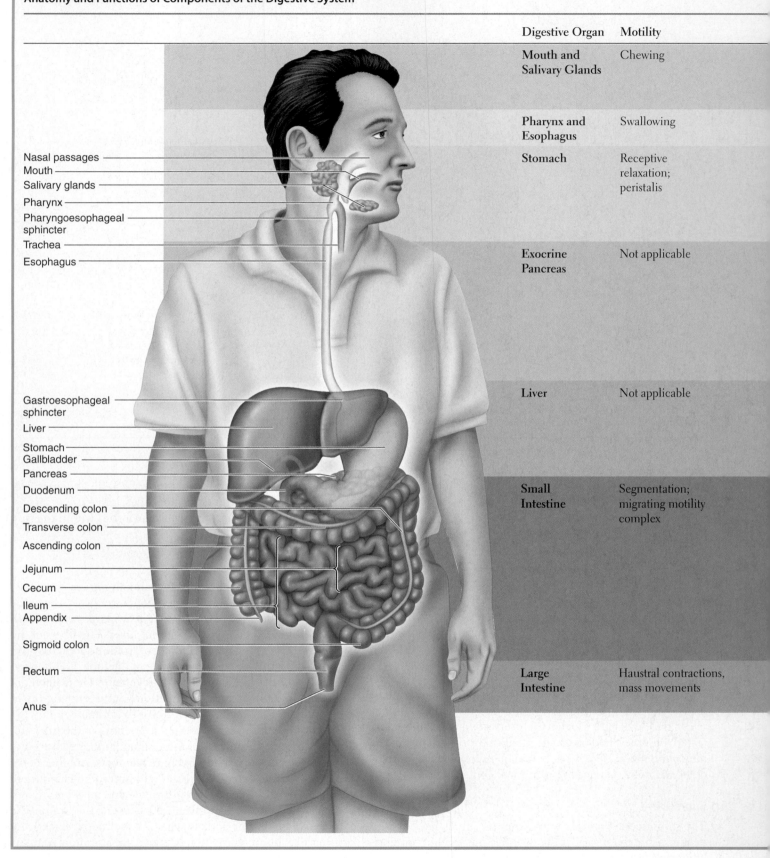

	Digestive Organ	Motility
	Mouth and Salivary Glands	Chewing
	Pharynx and Esophagus	Swallowing
	Stomach	Receptive relaxation; peristalis
	Exocrine Pancreas	Not applicable
	Liver	Not applicable
	Small Intestine	Segmentation; migrating motility complex
	Large Intestine	Haustral contractions, mass movements

Nasal passages
Mouth
Salivary glands
Pharynx
Pharyngoesophageal sphincter
Trachea
Esophagus

Gastroesophageal sphincter
Liver
Stomach
Gallbladder
Pancreas
Duodenum
Descending colon
Transverse colon
Ascending colon
Jejunum
Cecum
Ileum
Appendix
Sigmoid colon
Rectum
Anus

Secretion	Digestion	Absorption
Saliva • Amylase • Mucus • Lysozyme	Carbohydrate digestion begins	No foodstuffs; a few medications—for example, nitroglycerin
Mucus	None	None
Gastric juice • HCl • Pepsin • Mucus • Intrinsic factor	Carbohydrate digestion continues in body of stomach; protein digestion begins in antrum of stomach	No foodstuffs; a few lipid-soluble substances, such as alcohol and aspirin
Pancreatic digestive enzymes • Trypsin, chymotrypsin, carboxypeptidase • Amylase • Lipase Pancreatic aqueous NaHCO₃ secretion	These pancreatic enzymes accomplish digestion in duodenal lumen	Not applicable
Bile • Bile Salts • Alkaline secretion • Bilirubin	Bile does not digest anything, but bile salts facilitate fat digestion and absorption in duodenal lumen	Not applicable
Succus entericus • Mucus • Salt (Small intestine enzymes are not secreted but function intracellularly in the brush border — disaccharidases and aminopeptidases)	In lumen, under influence of pancreatic enzymes and bile, carbohydrate and protein digestion continue and fat digestion is completely accomplished; in brush border, carbohydrate and protein digestion completed	All nutrients, most electrolytes, and water
Mucus	None	Salt and water, converting contents to feces

Serosa

The outer connective tissue covering of the digestive tract is the **serosa,** which secretes a watery serous fluid that lubricates and prevents friction between the digestive organs and the surrounding viscera. Throughout much of the tract, the serosa is continuous with the **mesentery,** which suspends the digestive organs from the inner wall of the abdominal cavity like a sling (● Figure 16-3). This attachment provides relative fixation, supporting the digestive organs in proper position, while still allowing them freedom for mixing and propulsive movements.

▌ Regulation of digestive function is complex and synergistic.

Digestive motility and secretion are carefully regulated to maximize digestion and absorption of the ingested food. Four factors are involved in the regulation of digestive system function: (1) autonomous smooth muscle function, (2) intrinsic nerve plexuses, (3) extrinsic nerves, and (4) gastrointestinal hormones.

Autonomous smooth-muscle function

Like self-excitable cardiac-muscle cells, some smooth-muscle cells are "pacesetter" cells that display rhythmic, spontaneous variations in membrane potential. The prominent type of self-induced electrical activity in digestive smooth muscle is **slow-wave potentials** (see p. 293), alternatively referred to as the digestive tract's **basic electrical rhythm (BER)** or **pacesetter potential.** Musclelike but noncontractile cells known as the **interstitial cells of Cajal** are the pacesetter cells responsible for instigating cyclic slow-wave activity. These pacesetter cells are located at the boundary between the longitudinal and circular smooth-muscle layers. The slow-wave potentials initiated by these cells spread to the adjacent contractile smooth-muscle cells. Slow waves are not action potentials and do not directly induce muscle contraction; they are rhythmic, wavelike fluctuations in membrane potential that cyclically bring the membrane closer to or farther from threshold. These slow-wave oscillations are believed to be due to cyclical variations in Ca^{2+} release from the endoplasmic reticulum and Ca^{2+} uptake by the mitochondria of the pacesetter cell. Should these waves reach threshold at the peaks of depolarization, a volley of action potentials is triggered at each peak, resulting in rhythmic cycles of muscle contraction.

Whether threshold is reached depends on the effect of various mechanical, nervous system, and hormonal factors that influence the starting point around which the slow-wave rhythm oscillates. If the starting point is nearer the threshold level, as it is when food is present in the digestive tract, the depolarizing slow-wave peak reaches threshold, so action potential frequency and its accompanying contractile activity increase. Conversely, if the starting point is farther from threshold, as when no food is present, there is less likelihood of reaching threshold, so action potential frequency is lowered and contractile activity is reduced.

Like cardiac muscle, sheets of smooth muscle cells are connected by gap junctions through which charge-carrying ions can flow (see p. 64). As a result, electrical activity initiated in a digestive-tract pacesetter cell can spread to adjacent smooth-muscle cells. If threshold is reached and action potentials are triggered, the whole muscle sheet behaves like a functional syncytium, becoming excited and contracting as a unit (see. p. 293). If threshold is not achieved, the oscillating electrical activity continues to sweep across the muscle without being accompanied by contractile activity.

The *rate* (frequency) of rhythmic digestive contractile activities, such as peristalsis in the stomach, segmentation in the small intestine, and haustral contractions in the large intestine, depends on the inherent rate established by the involved pacesetter cells. (Specific details about these rhythmic contractions will be discussed when we examine the organs involved.) The *intensity* of these contractions depends on the number of action potentials that occur when the slow-wave potential reaches threshold, which in turn depends on how long threshold is sustained. At threshold, voltage-gated Ca^{2+} channels are activated (see p. 104), resulting in Ca^{2+} influx into the smooth muscle cell. The resultant Ca^{2+} entry has two effects: (1) It is responsible for the rising phase of an action potential, with the falling phase being brought about as usual by K^+ efflux, and (2) it triggers a contractile response (see p. 289). The greater the number of action potentials, the higher the cytosolic Ca^{2+} concentration, the greater the cross-bridge activity, and the stronger the contraction. Other factors that influence contractile activity also do so by altering the cytosolic Ca^{2+} concentration. Thus the level of contractility can range from low-level tone to vigorous mixing and propulsive movements by varying the cytosolic Ca^{2+} concentration.

Intrinsic nerve plexuses

The **intrinsic nerve plexuses** are the two major networks of nerve fibers—the **myenteric plexus** and the **submucous plexus**—that are located entirely within the digestive tract wall and run its entire length. Thus, unlike any other organ system, the digestive tract has its own intramural ("within wall") nervous system, which contains as many neurons as the spinal cord and endows the tract with a considerable degree of self-regulation. Together, these two plexuses are often termed the **enteric nervous system.**

The intrinsic plexuses influence all facets of digestive tract activity. Various types of neurons are present in the intrinsic plexuses. Some are sensory neurons, which possess receptors that respond to specific local stimuli in the digestive tract. Other local neurons innervate the smooth muscle cells and exocrine and endocrine cells of the digestive tract to directly affect digestive tract motility, secretion of digestive juices, and secretion of gastrointestinal hormones. As with the central nervous system, these input and output neurons of the enteric nervous system are linked by interneurons. Some of the output neurons are excitatory, and some are inhibitory. For example, neurons that release *acetylcholine* as a neurotransmitter promote contraction of digestive-tract smooth muscle, whereas the neurotransmitters *nitric oxide* and *vasoactive intestinal peptide* act

in concert to cause its relaxation. These intrinsic nerve networks are primarily responsible for coordinating local activity within the digestive tract. To illustrate, if a large piece of food gets stuck in the esophagus, local contractile responses coordinated by the intrinsic plexuses are initiated to push the food forward. Intrinsic nerve activity can in turn be influenced by the extrinsic nerves.

Extrinsic nerves

The **extrinsic nerves** are the nerve fibers from both branches of the autonomic nervous system that originate outside the digestive tract and innervate the various digestive organs. The autonomic nerves influence digestive tract motility and secretion either by modifying ongoing activity in the intrinsic plexuses, altering the level of gastrointestinal hormone secretion, or, in some instances, acting directly on the smooth muscle and glands.

Recall that, in general, the sympathetic and parasympathetic nerves supplying any given tissue exert opposing actions on that tissue. The sympathetic system, which dominates in fight-or-flight situations, tends to inhibit or slow down digestive tract contraction and secretion. This action is appropriate, considering that digestive processes are not of highest priority when the body faces an emergency or threat from the external environment. The parasympathetic nervous system, by contrast, dominates in quiet, relaxed situations, when general maintenance types of activities such as digestion can proceed optimally. Accordingly, the parasympathetic nerve fibers supplying the digestive tract, which arrive primarily by way of the vagus nerve, tend to increase smooth muscle motility and promote secretion of digestive enzymes and hormones. Unique to the parasympathetic nerve supply to the digestive tract, the postganglionic parasympathetic nerve fibers are actually a part of the intrinsic nerve plexuses. They are the acetylcholine-secreting output neurons within the plexuses. Accordingly, acetylcholine is released in response to local reflexes coordinated entirely by the intrinsic plexuses as well as to vagal stimulation, which acts through the intrinsic plexuses.

In addition to being called into play during generalized sympathetic or parasympathetic discharge, the autonomic nerves, especially the vagus nerve, can be discretely activated to modify only digestive activity. One of the major purposes of specific activation of extrinsic innervation is the coordination of activity between different regions of the digestive system. For example, the act of chewing food reflexly increases not only salivary secretion but also stomach, pancreatic, and liver secretion via vagal reflexes in anticipation of the arrival of food.

Gastrointestinal hormones

Tucked within the mucosa of certain regions of the digestive tract are endocrine gland cells that on appropriate stimulation, release hormones into the blood. These **gastrointestinal hormones** are carried through the blood to other areas of the digestive tract, where they exert either excitatory or inhibitory influences on smooth muscle and exocrine gland cells. Interestingly, many of these same hormones are released from neurons in the brain, where they act as neurotransmitters and neuromodulators. During embryonic development, certain cells of the developing neural tissue migrate to the digestive system, where they become endocrine cells.

▌ Receptor activation alters digestive activity through neural reflexes and hormonal pathways.

The digestive tract wall contains three types of sensory receptors that respond to local chemical or mechanical changes in the digestive tract: (1) *chemoreceptors* sensitive to chemical components within the lumen; (2) *mechanoreceptors* (pressure receptors) sensitive to stretch or tension within the wall; and (3) *osmoreceptors* sensitive to the osmolarity of the luminal contents.

Stimulation of these receptors elicits neural reflexes or secretion of hormones, both of which alter the level of activity in the digestive system's effector cells—the smooth muscle cells and the exocrine and endocrine gland cells (● Figure 16-4). Receptor activation may bring about two types of neural reflexes—short reflexes and long reflexes. When the intrinsic nerve networks influence local motility or secretion in response to specific local stimulation, all elements of the reflex are located within the wall of the digestive tract itself; that is, a **short reflex** takes place. Extrinsic autonomic nervous activity can be superimposed on the local controls to modify smooth muscle and glandular responses, either to correlate activity between different regions of the digestive system or to modify digestive system activity in response to external influences. Because the autonomic reflexes involve long pathways between the central nervous system and digestive system, they are known as **long reflexes.** In addition to these neural reflexes, digestive system activity is also coordinated by the secretion of gastrointestinal hormones, which are triggered directly by local changes in the digestive tract or by short or long reflexes.

From this overview, you can see that regulation of gastrointestinal function is very complex, being influenced by many synergistic, interrelated pathways designed to ensure that the appropriate responses occur to digest and absorb the ingested food. Nowhere else in the body is so much overlapping control exercised.

We are now going to take a "tour" of the digestive tract, beginning with the mouth and ending with the anus. We will examine the four basic digestive processes of motility, secretion, digestion, and absorption at each digestive organ along the way. ▲ Table 16-2 summarizes these activities and serves as a useful reference throughout the remainder of the chapter.

MOUTH

▌ The oral cavity is the entrance to the digestive tract.

Entry to the digestive tract is through the **mouth** or **oral cavity.** The opening is formed by the muscular **lips,** which help procure, guide, and contain the food in the mouth. The lips also serve nondigestive functions; they are important in speech (articulation of many sounds depends on a particular lip formation) and as a sensory receptor in interpersonal relationships (for example, as in kissing).

The **palate,** which forms the arched roof of the oral cavity, separates the mouth from the nasal passages. Its presence allows breathing and chewing or sucking to take place simultaneously. Hanging down from the palate in the rear of the throat is a dangling projection, the **uvula,** which plays an important role in sealing off the nasal passages during swallowing. (The uvula is the structure you elevate when you say "ahhh" so that the physician can better see your throat.)

The **tongue,** which forms the floor of the oral cavity, is composed of voluntarily controlled skeletal muscle. Move-

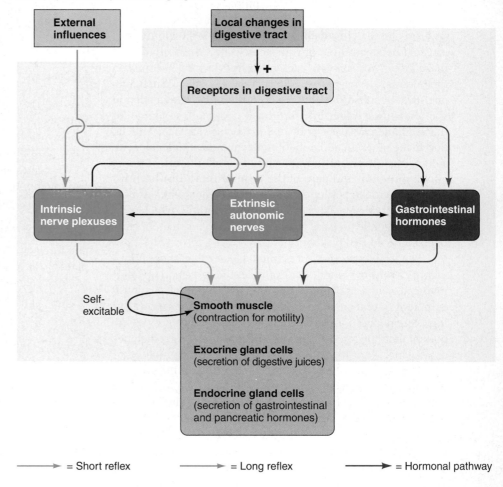

● **FIGURE 16-4**

Summary of pathways controlling digestive system activities

⟶ = Short reflex ⟶ = Long reflex ⟶ = Hormonal pathway

ments of the tongue are important in guiding food within the mouth during chewing and swallowing and also play an important role in speech. Furthermore, the major **taste buds** are embedded in the tongue (see p. 225).

The **pharynx** is the cavity at the rear of the throat. It acts as a common passageway for both the digestive system (by serving as the link between the mouth and esophagus, for food) and the respiratory system (by providing access between the nasal passages and trachea, for air). This arrangement necessitates mechanisms (to be described shortly) to guide food and air into the proper passageways beyond the pharynx. Housed within the side walls of the pharynx are the **tonsils,** lymphoid tissues that are part of the body's defense team.

▌ The teeth are responsible for chewing.

The first step in the digestive process is **mastication** or **chewing,** the motility of the mouth that involves the slicing, tearing, grinding, and mixing of ingested food by the **teeth.** The teeth are firmly embedded in and protrude from the jawbones. The exposed portion of a tooth is covered by **enamel,** the hardest structure of the body. Enamel is formed prior to the tooth's eruption, by special cells that are lost as the tooth erupts. Because enamel cannot be regenerated after the tooth has erupted, any defects ("cavities") that develop in the enamel must be patched by artificial "fillings," or else the surface will continue to erode into the underlying living pulp.

The upper and lower teeth normally fit together when the jaws are closed. This **occlusion** allows food to be ground and crushed between the tooth surfaces. When the teeth do not make proper contact with each other, they cannot accomplish their normal cutting and grinding action adequately. Such **malocclusion** results from abnormal positioning of the teeth and is often caused either by overcrowding of teeth too large for the available jaw space or by one jaw being displaced in relation to the other. In addition to ineffective chewing, malocclusion can cause abnormal wearing of affected tooth surfaces and dysfunction and pain of the **temporomandibular joint (TMJ),** where the jawbones articulate with each other. Malocclusions can often be corrected by applying braces, which exert prolonged gentle pressure against the teeth to move them gradually to the desired position.

The teeth can exert forces much greater than those necessary to eat ordinary food. For example, the molars in an adult man can exert a crushing force of up to 200 pounds, which is sufficient to crack a hard nut, but ordinarily these powerful forces are not used. In fact, the degree of occlusion is more important than the force of the bite in determining the efficiency of chewing.

The purposes of chewing are (1) to grind and break food up into smaller pieces to facilitate swallowing, (2) to mix food with saliva, and (3) to stimulate the taste buds. The latter not only gives rise to the pleasurable subjective sensation of taste but also, in feedforward fashion, reflexly increases salivary, gastric, pancreatic, and bile secretion to prepare for the arrival of food.

The act of chewing can be voluntary, but most chewing during a meal is a rhythmic reflex brought about by activation of the skeletal muscles of the jaws, lips, cheeks, and tongue in response to the pressure of food against the oral tissues.

▌ Saliva begins carbohydrate digestion, is important in oral hygiene, and facilitates speech.

Saliva, the secretion associated with the mouth, is produced largely by three major pairs of salivary glands that are located outside of the oral cavity and discharge saliva through short ducts into the mouth.

Saliva is composed of about 99.5% H_2O and 0.5% electrolytes and protein. The salivary NaCl (salt) concentration is only one-seventh of that in the plasma, a fact important in the perception of salty tastes. Similarly, discrimination of sweet tastes is enhanced by the absence of glucose in the saliva. The most important salivary proteins are *amylase, mucus,* and *lysozyme.* They contribute to the functions of saliva as follows:

1. Saliva begins digestion of carbohydrate in the mouth through action of **salivary amylase,** an enzyme that breaks polysaccharides down into **maltose,** a disaccharide consisting of two glucose molecules.

2. Saliva facilitates swallowing by moistening food particles, thereby holding them together, and by providing lubrication through the presence of **mucus,** which is thick and slippery.

3. Saliva exerts some antibacterial action by means of a twofold effect—first by **lysozyme,** an enzyme that lyses, or destroys, certain bacteria by breaking down their cell walls, and second by rinsing away material that may serve as a food source for bacteria.

4. Saliva serves as a solvent for molecules that stimulate the taste buds. Only molecules in solution can react with taste bud receptors. You can demonstrate this for yourself: Dry your tongue and then drop some sugar on it—you cannot taste the sugar until it is moistened.

5. Saliva aids speech by facilitating movements of the lips and tongue. It is difficult to talk when the mouth feels dry.

6. Saliva plays an important role in oral hygiene by helping keep the mouth and teeth clean. The constant flow of saliva helps to flush away food residues, shed epithelial cells, and foreign particles. Saliva's contribution in this regard is apparent to anyone who has experienced a foul taste in the mouth when salivation is suppressed for a while, such as during a fever or states of prolonged anxiety.

7. Saliva is rich in bicarbonate buffers, which neutralize acids in food as well as acids produced by bacteria in the mouth, thereby helping to prevent **dental caries** (cavities).

Despite these many functions, saliva is not essential for digesting and absorbing foods, because enzymes produced by the pancreas and small intestine can complete food digestion even in the absence of salivary and gastric secretion. The main problems associated with diminished salivary secretion, a condition known as **xerostomia,** are difficulty in chewing and swallowing, inarticulate speech unless frequent sips of water are taken when talking, and a rampant increase in dental caries.

▌Salivary secretion is continuous and can be reflexly increased.

On the average, about 1 to 2 liters of saliva are secreted per day, ranging from a continuous spontaneous basal rate of 0.5 ml/min to a maximum flow rate of about 5 ml/min in response to a potent stimulus such as sucking on a lemon. The continuous spontaneous secretion of saliva, even in the absence of apparent stimuli, is brought about by constant low-level stimulation by the parasympathetic nerve endings that terminate in the salivary glands. This basal secretion is important in keeping the mouth and throat moist at all times. In addition to this continuous, low-level secretion, salivary secretion may be enhanced by two types of salivary reflexes, the simple and the acquired salivary reflexes (● Figure 16-5).

Simple and acquired salivary reflexes

The **simple**, or **unconditioned, salivary reflex** occurs when chemoreceptors and pressure receptors within the oral cavity respond to the presence of food. On activation, these receptors initiate impulses in afferent nerve fibers that carry the information to the **salivary center,** which is located in the medulla of the brain stem, as are all the brain centers that control digestive activities. The salivary center in turn sends impulses via the extrinsic autonomic nerves to the salivary glands to promote increased salivation. Dental procedures promote salivary secretion in the absence of food in the mouth because these manipulations activate pressure receptors in the mouth.

With the **acquired**, or **conditioned, salivary reflex,** salivation occurs without oral stimulation. Just thinking about, seeing, smelling, or hearing the preparation of pleasant food initiates salivation through this reflex. All of us have experienced such "mouth watering" in anticipation of something delicious to eat. This reflex is a learned response based on previous experience. Inputs that arise outside the mouth and are mentally associated with the pleasure of eating act through the cerebral cortex to stimulate the medullary salivary center.

Autonomic influence on salivary secretion

The salivary center controls the degree of salivary output by means of the autonomic nerves that supply the salivary glands. Unlike the autonomic nervous system elsewhere in the body, sympathetic and parasympathetic responses in the salivary glands are not antagonistic. Both sympathetic and parasympathetic stimulation increase salivary secretion, but the quantity, characteristics, and mechanisms are different. Parasympathetic stimulation, which exerts the dominant role in salivary secretion, produces a prompt and abundant flow of watery saliva that is rich in enzymes. Sympathetic stimulation, by contrast, produces a much smaller volume of thick saliva that is rich in mucus. Because sympathetic stimulation elicits a smaller volume of saliva, the mouth feels drier than usual during circumstances when the sympathetic system is dominant, such as stress situations. Thus people experience a dry feeling in the mouth when they are nervous about giving a speech.

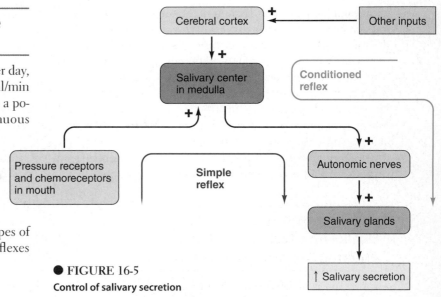

● **FIGURE 16-5**
Control of salivary secretion

Salivary secretion is the only digestive secretion entirely under neural control. All other digestive secretions are regulated by both nervous system reflexes and hormones.

▌Digestion in the mouth is minimal; no absorption of nutrients occurs.

Digestion in the mouth involves the hydrolysis of polysaccharides into disaccharides by amylase. However, most digestion by this enzyme is accomplished in the body of the stomach after the food mass and saliva have been swallowed. Acid inactivates amylase, but in the center of the food mass, where stomach acid has not yet reached, this salivary enzyme continues to function for several more hours.

No absorption of foodstuff occurs from the mouth. Importantly, some therapeutic agents can be absorbed by the oral mucosa, a prime example being a vasodilator drug, *nitroglycerin*, which is used by certain cardiac patients to relieve anginal attacks (see p. 335) associated with myocardial ischemia (see p. 320).

PHARYNX AND ESOPHAGUS

The motility associated with the pharynx and esophagus is **swallowing.** Most of us think of swallowing as the limited act of moving food out of the mouth into the esophagus. However, swallowing actually refers to the entire process of moving food from the mouth through the esophagus into the stomach.

▌Swallowing is a sequentially programmed all-or-none reflex.

Swallowing is initiated when a **bolus,** or ball of food, is voluntarily forced by the tongue to the rear of the mouth into the pharynx. The pressure of the bolus in the pharynx stimulates pharyngeal pressure receptors, which send afferent impulses to the **swallowing center** located in the medulla. The swallowing center then reflexly activates in the appropriate sequence the muscles that are involved in swallowing. Swallowing is a sequentially programmed all-or-none reflex in which multiple responses are triggered in a specific timed sequence; that is, a number of highly coordinated activities are initiated in a regular pattern over a period of time to accomplish the act of swallowing. Swallowing is initiated voluntarily, but once it is initiated, it cannot be stopped. Perhaps you have experienced this when a large piece of hard candy inadvertently slipped to the rear of your throat, triggering an unintentional swallow.

▌ During the oropharyngeal stage of swallowing, food is prevented from entering the wrong passageways.

Swallowing is divided into the oropharyngeal stage and the esophageal stage. The **oropharyngeal stage** lasts about 1 second and consists of moving the bolus from the mouth through the pharynx and into the esophagus. When the bolus enters the pharynx, it must be directed into the esophagus and prevented from entering the other openings that communicate with the pharynx. In other words, food must be prevented from re-entering the mouth, from entering the nasal passages, and

from entering the trachea. All of this is accomplished by the following coordinated activities (● Figure 16-6):

- Food is prevented from re-entering the mouth during swallowing by the position of the tongue against the hard palate.
- The uvula is elevated and lodges against the back of the throat, sealing off the nasal passage from the pharynx so that food does not enter the nose.
- Food is prevented from entering the trachea primarily by elevation of the larynx and tight closure of the vocal folds across the laryngeal opening, or **glottis.** The first portion of the trachea is the *larynx*, or *voice box*, across which are stretched the *vocal folds* (see p. 461). During swallowing, the vocal folds serve a purpose unrelated to speech. Contraction of laryngeal muscles aligns the vocal folds in tight apposition to each other, thus sealing the glottis entrance. Also, the bolus tilts a small flap of cartilaginous tissue, the **epiglottis** (*epi* means "upon"), backward down over the closed glottis as further protection from food entering the respiratory airways.
- The individual does not attempt futile respiratory efforts when the respiratory passages are temporarily sealed off during swallowing, because the swallowing center briefly inhibits the nearby respiratory center.
- With the larynx and trachea sealed off, pharyngeal muscles contract to force the bolus into the esophagus.

● **FIGURE 16-6**

Oropharyngeal stage of swallowing
(a) Position of the oropharyngeal structures at rest. (b) Changes that occur during the oropharyngeal stage of swallowing to prevent the bolus of food from entering the wrong passageways.

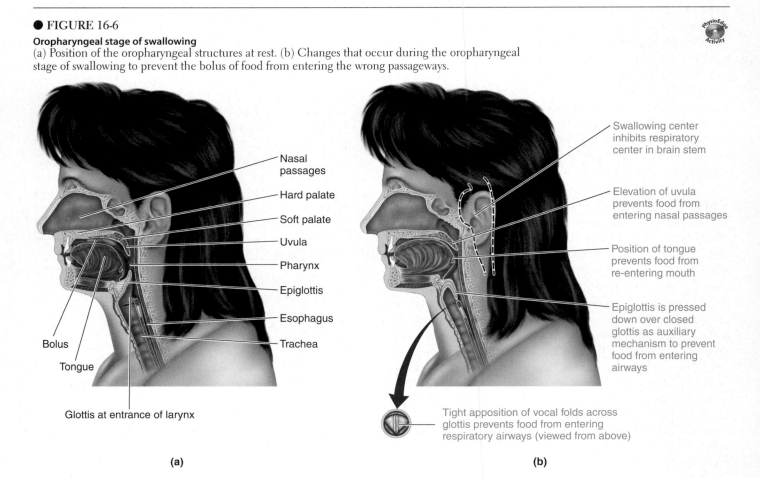

(a) (b)

The pharygoesophageal sphincter prevents air from entering the digestive tract during breathing.

The **esophagus** is a fairly straight muscular tube that extends between the pharynx and stomach (see ▲ Table 16-2). Lying for the most part in the thoracic cavity, it penetrates the diaphragm and joins the stomach in the abdominal cavity a few centimeters below the diaphragm.

The esophagus is guarded at both ends by sphincters. A sphincter is a ringlike muscular structure that, when closed, prevents passage through the tube it guards. The upper esophageal sphincter is the **pharyngoesophageal sphincter,** and the lower sphincter is the **gastroesophageal sphincter.** We will first discuss the role of the pharyngoesophageal sphincter, then the process of esophageal transit of food, and finally the importance of the gastroesophageal sphincter.

Role of the pharyngoesophageal sphincter

Because the esophagus is exposed to subatmospheric intrapleural pressure as a result of respiratory activity (see p. 464), a pressure gradient exists between the atmosphere and the esophagus. Accordingly, if the entrance to the esophagus were not closed, air would enter the esophagus and stomach as well as the trachea and lungs with each breath. Except during a swallow, the pharyngoesophageal sphincter keeps the entrance to the esophagus closed so that air is directed only into the respiratory airways during breathing. Otherwise, the digestive tract would be subjected to large volumes of gas, which would lead to excessive **eructation** (burping). During swallowing, this sphincter opens and allows the bolus to pass into the esophagus. Once the bolus has entered the esophagus, the pharyngoesophageal sphincter closes, the respiratory airways are opened, and breathing resumes. The oropharyngeal stage is complete, and about 1 second has passed since the swallow was first voluntarily initiated.

Peristaltic waves push food through the esophagus.

The **esophageal stage** of the swallow now begins. The swallowing center initiates a **primary peristaltic wave** that sweeps from the beginning to the end of the esophagus, forcing the bolus ahead of it through the esophagus to the stomach. The term **peristalsis** refers to ringlike contractions of the circular smooth muscle that move progressively forward, pushing the bolus ahead of the contraction (● Figure 16-7). The peristaltic wave takes about 5 to 9 seconds to reach the lower end of the esophagus. Progression of the wave is controlled by the swallowing center, with innervation by means of the vagus.

If a large or sticky swallowed bolus, such as a bite of peanut butter sandwich, fails to be carried along to the stomach by the primary wave of peristalsis, the lodged bolus distends the esophagus, stimulating pressure receptors within its walls. As a result, a second, more forceful peristaltic wave is initiated, mediated by the intrinsic nerve plexuses at the level of the dis-

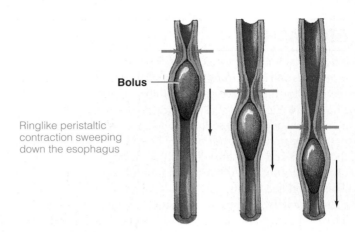

Bolus

Ringlike peristaltic contraction sweeping down the esophagus

● **FIGURE 16-7**

Peristalsis in the esophagus

As the wave of peristaltic contraction sweeps down the esophagus, it pushes the bolus ahead of it toward the stomach.

tension. These **secondary peristaltic waves** do not involve the swallowing center, nor is the person aware of their occurrence. Distension of the esophagus also reflexly increases salivary secretion. The trapped bolus is eventually dislodged and moved forward through the combination of lubrication by the extra swallowed saliva and the forceful secondary peristaltic waves.

The gastroesophageal sphincter prevents reflux of gastric contents.

Except during swallowing, the gastroesophageal sphincter remains contracted to maintain a barrier between the stomach and esophagus, thus reducing the possibility of reflux of acidic gastric contents into the esophagus. If gastric contents do flow back into the esophagus despite the sphincter, the acidity of these contents irritates the esophagus, causing the esophageal discomfort known as **heartburn.** (The heart itself is not involved at all.)

As the peristaltic wave sweeps down the esophagus, the gastroesophageal sphincter relaxes reflexly so that the bolus can pass into the stomach. After the bolus has entered the stomach, the swallow is complete and the gastroesophageal sphincter again contracts.

Esophageal secretion is entirely protective.

Esophageal secretion is entirely mucus. In fact, mucus is secreted throughout the length of the digestive tract. By providing lubrication for passage of food, esophageal mucus lessens the likelihood that the esophagus will be damaged by any sharp edges in the newly entering food. Furthermore, it protects the esophageal wall from acid and enzymes in gastric juice if gastric reflux should occur.

The entire transit time in the pharynx and esophagus averages a mere 6 to 10 seconds, too short a time for any digestion or absorption to occur in this region. We now move on to our next stop, the stomach.

STOMACH

The **stomach** is a J-shaped saclike chamber lying between the esophagus and small intestine. It is arbitrarily divided into three sections based on anatomical, histological, and functional distinctions (● Figure 16-8). The **fundus** is the portion of the stomach that lies above the esophageal opening. The middle or main part of the stomach is the **body.** The smooth muscle layers in the fundus and body are relatively thin, but the lower portion of the stomach, the **antrum,** has much heavier musculature. This difference in muscle thickness plays an important role in gastric motility in these two regions, as you will see shortly. There are also glandular differences in the mucosa of these regions, as will be described later. The terminal portion of the stomach consists of the **pyloric sphincter,** which acts as a barrier between the stomach and the upper part of the small intestine, the duodenum.

▌ The stomach stores food and begins protein digestion.

The stomach performs three main functions:

1. The stomach's most important function is to store ingested food until it can be emptied into the small intestine at a rate appropriate for optimal digestion and absorption. It takes hours to digest and absorb a meal that was consumed in only a matter of minutes. Because the small intestine is the primary site for this digestion and absorption, it is important that the stomach store the food and meter it into the duodenum at a rate that does not exceed the small intestine's capacities.

2. The stomach secretes hydrochloric acid (HCl) and enzymes that begin protein digestion.

3. Through the stomach's mixing movements, the ingested food is pulverized and mixed with gastric secretions to produce a thick liquid mixture known as **chyme.** The stomach contents must be converted to chyme before they can be emptied into the duodenum.

We will now discuss how the stomach accomplishes these functions as we examine the four basic digestive processes—motility, secretion, digestion, and absorption—as they relate to the stomach. Starting with motility, gastric motility is complex and subject to multiple regulatory inputs. The four aspects of gastric motility are (1) filling, (2) storage, (3) mixing, and (4) emptying. We begin with gastric filling.

Gastric filling involves receptive relaxation.

When empty, the stomach has a volume of about 50 ml, but it can expand to a capacity of about 1 liter (1,000 ml) during a meal. The stomach is able to accommodate such a 20-fold change in volume with little change in tension in its walls and little rise in intragastric pressure, through the following mechanism. The interior of the stomach is thrown into deep folds. During a meal, the folds get smaller and flatten out as the stomach relaxes slightly with each mouthful, much like the gradual

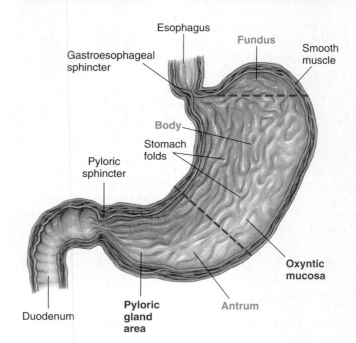

● **FIGURE 16-8**

Anatomy of the stomach
The stomach is divided into three sections based on structural and functional distinctions—the fundus, body, and antrum. The mucosal lining of the stomach is divided into the oxyntic mucosa and the pyloric gland area based on differences in glandular secretion.

expansion of a collapsed ice bag as it is being filled. This reflex relaxation of the stomach as it is receiving food is called **receptive relaxation;** it enhances the stomach's ability to accommodate the extra volume of food with little rise in stomach pressure. If more than a liter of food is consumed, however, the stomach becomes overdistended, intragastric pressure rises, and the person experiences discomfort. Receptive relaxation is triggered by the act of eating and is mediated by the vagus nerve.

▌ Gastric storage takes place in the body of the stomach.

A group of pacesetter cells located in the upper fundus region of the stomach generate slow-wave potentials that sweep down the length of the stomach toward the pyloric sphincter at a rate of three per minute. This rhythmic pattern of spontaneous depolarizations—the basic electrical rhythm, or BER, of the stomach—occurs continuously and may or may not be accompanied by contraction of the stomach's circular smooth muscle layer. Depending on the level of excitability in the smooth muscle, it may be brought to threshold by this flow of current and undergo action potentials, which in turn initiate peristaltic waves that sweep over the stomach in pace with the BER at a rate of three per minute.

Once initiated, the peristaltic wave spreads over the fundus and body to the antrum and pyloric sphincter. Because the muscle layers are thin in the fundus and body, the peristaltic contractions in this region are weak. When the waves reach the

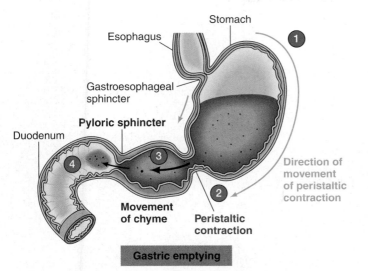
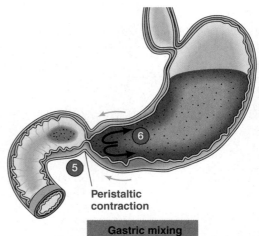

Gastric emptying

1. A peristaltic contraction originates in the upper fundus and sweeps down toward the pyloric sphincter.

2. The contraction becomes more vigorous as it reaches the thick-muscled antrum.

3. The strong antral peristaltic contraction propels the chyme forward.

4. A small portion of chyme is pushed through the partially open sphincter into the duodenum. The stronger the antral contraction, the more chyme is emptied with each contractile wave.

Gastric mixing

5. When the peristaltic contraction reaches the pyloric sphincter, the sphincter is tightly closed and no further emptying takes place.

6. When chyme that was being propelled forward hits the closed sphincter, it is tossed back into the antrum. Mixing of chyme is accomplished as chyme is propelled forward and tossed back into the antrum with each peristaltic contraction.

● FIGURE 16-9

Gastric emptying and mixing as a result of antral peristaltic contractions

antrum, they become much stronger and more vigorous, because the muscle there is much thicker.

Because only feeble mixing movements occur in the body and fundus, food emptied into the stomach from the esophagus is stored in the relatively quiet body without being mixed. The fundic area usually does not store food but contains only a pocket of gas. Food is gradually fed from the body into the antrum, where mixing does take place.

▌ Gastric mixing takes place in the antrum of the stomach.

The strong antral peristaltic contractions are responsible for mixing the food with gastric secretions to produce chyme. Each antral peristaltic wave propels chyme forward toward the pyloric sphincter. Tonic contraction of the pyloric sphincter normally keeps it almost, but not completely, closed. The opening is large enough for water and other fluids to pass through with ease but too small for the thicker chyme to pass through except when a strong peristaltic antral contraction pushes it through. Even then, of the 30 ml of chyme that the antrum can hold, usually only a few milliliters of antral contents are forced into the duodenum with each peristaltic wave. Before more chyme can be squeezed out, the peristaltic wave reaches the pyloric sphincter and causes it to contract more forcefully, sealing off the exit and blocking further passage into the duo-

denum. The bulk of the antral chyme that was being propelled forward but failed to be pushed into the duodenum is abruptly halted at the closed sphincter and is tumbled back into the antrum, only to be propelled forward and tumbled back again as the new peristaltic wave advances (● Figure 16-9). This tossing back and forth accomplishes thorough mixing of the chyme in the antrum.

▌ Gastric emptying is largely controlled by factors in the duodenum.

In addition to being responsible for gastric mixing, the antral peristaltic contractions provide the driving force for gastric emptying. The amount of chyme that escapes into the duodenum with each peristaltic wave before the pyloric sphincter closes tightly depends largely on the strength of peristalsis. The intensity of antral peristalsis can vary markedly under the influence of different signals from both the stomach and duodenum; thus gastric emptying is regulated by both gastric and duodenal factors (▲ Table 16-3). These factors influence the stomach's excitability by slightly depolarizing or hyperpolarizing the gastric smooth muscle. This excitability in turn is a determinant of the degree of antral peristaltic activity. The greater the excitability, the more frequently the BER will generate action potentials, the greater the degree of peristaltic activity in the antrum, and the faster the rate of gastric emptying.

Factors Regulating Gastric Motility and Emptying

Factors	Mode of Regulation	Effects on Gastric Motility and Emptying
Within the Stomach		
Volume of chyme	Distension has a direct effect on gastric smooth muscle excitability, as well as acting through the intrinsic plexuses, the vagus nerve, and gastrin	Increased volume stimulates motility and emptying
Degree of fluidity	Direct effect; contents must be in a fluid form to be evacuated	Increased fluidity allows more rapid emptying
Within the Duodenum		
Presence of fat, acid, hypertonicity, or distension	Initiates the enterogastric reflex or triggers the release of enterogastrones (cholecystokinin, secretin)	These factors in the duodenum inhibit further gastric motility and emptying until the duodenum has coped with factors already present
Outside the Digestive System		
Emotion	Alters autonomic balance	Stimulates or inhibits motility and emptying
Intense pain	Increases sympathetic activity	Inhibits motility and emptying

Factors in the stomach that influence the rate of gastric emptying

The main gastric factor that influences the strength of contraction is the amount of chyme in the stomach. Other things being equal, the stomach empties at a rate proportional to the volume of chyme in it at any given time. Distension of the stomach triggers increased gastric motility through a direct effect of stretch on the smooth muscle as well as through involvement of the intrinsic plexuses, the vagus nerve, and the stomach hormone *gastrin*. (The source, control, and other functions of this hormone will be described later.)

Furthermore, the degree of fluidity of the chyme in the stomach influences gastric emptying. The stomach contents must be converted into a finely divided, thick liquid form before emptying. The sooner the appropriate degree of fluidity can be achieved, the more rapidly the contents are ready to be evacuated.

Factors in the duodenum that influence the rate of gastric emptying

Despite these gastric influences, factors in the duodenum are of primary importance in controlling the rate of gastric emptying. The duodenum must be ready to receive the chyme and can act to delay gastric emptying by reducing peristaltic activity in the stomach until the duodenum is ready to accommodate more chyme. Even if the stomach is distended and its contents are in a liquid form, it cannot empty until the duodenum is ready to deal with the chyme.

The four most important duodenal factors that influence gastric emptying are *fat*, *acid*, *hypertonicity*, and *distension*. The presence of one or more of these stimuli in the duodenum activates appropriate duodenal receptors, thereby triggering either a neural or a hormonal response that puts brakes on gastric motility by reducing the excitability of the gastric smooth muscle. The subsequent reduction in antral peristaltic activity slows down the rate of gastric emptying.

- The *neural response* is mediated through both the intrinsic nerve plexuses (short reflex) and the autonomic nerves (long reflex). Collectively, these reflexes are called the **enterogastric reflex.**

- The *hormonal response* involves the release from the duodenal mucosa of several hormones collectively known as **enterogastrones.** These hormones are transported by the blood to the stomach, where they inhibit antral contractions to reduce gastric emptying. The two most important enterogastrones are **secretin** and **cholecystokinin (CCK).** Secretin was the first hormone discovered (in 1902). Because it was a secretory product that entered the blood, it was termed *secretin*. The name *cholecystokinin* derives from the fact that this same hormone is also responsible for contraction of the bile-containing gallbladder (*chole* means "bile," *cysto* means "bladder," and *kinin* means "contraction"). Secretin and CCK are major gastrointestinal hormones that perform other important functions in addition to serving as enterogastrones.

Let us examine why it is important that each of these stimuli in the duodenum (fat, acid, hypertonicity, and distension) delays gastric emptying (acting through the enterogastric reflex or one of the enterogastrones).

- *Fat.* Fat is digested and absorbed more slowly than the other nutrients. Furthermore, fat digestion and absorption take place only within the lumen of the small intestine. Therefore, when fat is already present in the duodenum, further gastric emptying of more fatty stomach contents into the duodenum is prevented until the small intestine has processed the fat already there. In fact, fat is the most potent stimulus for inhibition

Pregame Meal: What's In and What's Out?

Many coaches and athletes believe intensely in special food rituals before a competitive event. For example, a football team may always breakfast on steak before a game. Another may always include bananas in their pregame meal. Do these rituals work?

Many studies have been done to determine the effect of the pregame meal on athletic performance. Although laboratory studies have shown that substances such as caffeine improve endurance, no food substance that will greatly enhance performance has been identified. The athlete's prior training is the most important determinant of performance. Even though no particular food confers a special benefit before an athletic contest, some food choices can actually hinder the competitors. For example, a meal of steak is high in fat and could take so long to digest that it might impair the football team's performance and thus should be avoided. However, food rituals that do not impair performance, such as eating bananas, but give the athletes a morale boost or extra confidence are harmless and should be respected. People may attach special meanings to eating certain foods, and their faith in these practices can make the difference between winning and losing a game.

The greatest benefit of the pregame meal is to prevent hunger during competition. Because the stomach can take from one to four hours to empty, an athlete should eat at least three to four hours before competition begins. Excessive quantities of food should not be consumed before competition. Food that remains in the stomach during competition may cause nausea and possibly vomiting. This condition can be aggravated by nervousness, which slows digestion and delays gastric emptying by means of the sympathetic nervous system.

The best choices are foods that are high in carbohydrate and low in fat and protein. The goal is to maintain blood glucose levels and carbohydrate stores in the body and to not have much undigested food in the stomach during the event. High-carbohydrate foods are recommended because they are emptied from the stomach more quickly than fat or protein is. Carbohydrates do not inhibit gastric emptying by means of enterogastrone release, whereas fat and protein do. Fats in particular delay gastric emptying and are slowly digested. Metabolic processing of proteins yields nitrogenous wastes such as urea whose osmotic activity draws water from the body and increases urine volume, both of which are undesirable during an athletic event. Good choices for a pregame meal include breads, pasta, rice, potatoes, gelatins, and fruit juices. Not only will these complex carbohydrates be emptied from the stomach if consumed one to four hours before a competitive event, but they also will help maintain the blood glucose level during the event.

Although it might seem logical to consume something sugary immediately before a competitive event to provide an "energy boost," beverages and foods high in sugar should be avoided because they trigger insulin release. Insulin is the hormone that enhances glucose entry into most body cells. Once the person begins exercising, insulin sensitivity increases (see p. 81), which lowers plasma glucose level. A lowered plasma glucose level induces feelings of fatigue and an increased use of muscle glycogen stores, which can limit performance in endurance events such as the marathon. Therefore, sugar consumption just before a competition can actually impair performance instead of giving the sought-after energy boost.

Within an hour of competition, it is best for athletes to drink only plain water, to ensure adequate hydration.

---

of gastric motility. This is evident when one compares the rate of emptying of a high-fat meal (after six hours some of a bacon-and-eggs meal may still be in the stomach) with that of a protein and carbohydrate meal (a meal of lean meat and potatoes may empty in three hours). (For a discussion of the pregame meal before participation in an athletic event, see the accompanying boxed feature, ▶ A Closer Look at Exercise Physiology.)

• *Acid.* Because the stomach secretes hydrochoric acid (HCl), highly acidic chyme is emptied into the duodenum, where it is neutralized by sodium bicarbonate ($NaHCO_3$) secreted into the duodenal lumen from the pancreas. Unneutralized acid irritates the duodenal mucosa and inactivates the pancreatic digestive enzymes that are secreted into the duodenal lumen. Appropriately, therefore, unneutralized acid in the duodenum inhibits further emptying of acidic gastric contents until complete neutralization can be accomplished.

• *Hypertonicity.* As molecules of protein and starch are digested in the duodenal lumen, large numbers of amino acid and glucose molecules are released. If absorption of these amino acid and glucose molecules does not keep pace with the rate at which protein and carbohydrate digestion proceeds, these large numbers of molecules remain in the chyme and increase the osmolarity of the duodenal contents. Osmolarity depends on the number of molecules present, not on their size, and one protein molecule may be split into several hundred amino acid molecules, each of which has the same osmotic activity as the original protein molecule. The same holds true for one large starch molecule, which yields many smaller but equally osmotically active glucose molecules. Because water is freely diffusable across the duodenal wall, it enters the duodenal lumen from the plasma as the duodenal osmolarity rises. Large volumes of water entering the intestine from the plasma lead to intestinal distension, and, more importantly, circulatory disturbances ensue because of the reduction in plasma volume. To prevent these effects, gastric emptying is reflexly inhibited when the osmolarity of the duodenal contents starts to rise. Thus the amount of food entering the duodenum for further digestion into a multitude of additional osmotically active particles is reduced until absorption processes have had an opportunity to catch up.

• *Distension.* Too much chyme in the duodenum inhibits the emptying of even more gastric contents, thus allowing the distended duodenum time to cope with the excess volume of chyme it already contains before it receives an additional quantity.

■ Emotions can influence gastric motility.

Other factors unrelated to digestion, such as emotions, can also alter gastric motility by acting through the autonomic nerves to influence the degree of gastric smooth-muscle excitability. Even though the effect of emotions on gastric motility varies from one individual to another and is not always predictable, sadness and fear generally tend to decrease motility, whereas anger and aggression tend to increase it. In addition to emotional influences, intense pain from any part of the body tends to inhibit motility, not just in the stomach but throughout the digestive tract. This response is brought about by increased sympathetic activity and a corresponding decrease in parasympathetic activity.

▌ The stomach does not actively participate in vomiting.

Vomiting, or **emesis,** the forceful explusion of gastric contents out through the mouth, is generally perceived as being caused by abnormal gastric motility. However, vomiting is not accomplished by reverse peristalsis, as might be predicted. Actually, the stomach itself does not actively participate in the act of vomiting. The stomach, the esophagus, and associated sphincters are all relaxed during vomiting. The major force for expulsion comes, surprisingly, from contraction of the respiratory muscles—namely, the diaphragm (the major inspiratory muscle) and the abdominal muscles (the muscles of active expiration).

The complex act of vomiting is coordinated by a **vomiting center** in the medulla. Vomiting begins with a deep inspiration and closure of the glottis. The contracting diaphragm descends downward on the stomach while simultaneous contraction of the abdominal muscles compresses the abdominal cavity, increasing the intra-abdominal pressure and forcing the abdominal viscera upward. As the flaccid stomach is squeezed between the diaphragm from above and the compressed abdominal cavity from below, the gastric contents are forced upward through the relaxed sphincters and esophagus and out through the mouth. The glottis is closed, so vomited material does not enter the respiratory airways. Also, the uvula is elevated to close off the nasal cavity. The vomiting cycle may be repeated several times until the stomach is emptied. Vomiting is usually preceded by profuse salivation, sweating, rapid heart rate, and the sensation of nausea, all of which are characteristic of a generalized discharge of the autonomic nervous system.

Causes of vomiting

Vomiting can be initiated by afferent input to the vomiting center from a number of receptors throughout the body. The causes of vomiting include the following:

- Tactile (touch) stimulation of the back of the throat, which is one of the most potent stimuli. For example, sticking a finger in the back of the throat or even the presence of a tongue depressor or dental instrument in the back of the mouth is sufficient stimulation to cause gagging and even vomiting in some people.
- Irritation or distension of the stomach and duodenum.
- Elevated intracranial pressure, such as that caused by cerebral hemorrhage. Thus, vomiting following a head injury is considered a bad sign; it suggests swelling or bleeding within the cranial cavity.

- Rotation or acceleration of the head producing dizziness, such as occurs in motion sickness.
- Chemical agents, including drugs or noxious substances that initiate vomiting (that is, **emetics**) either by acting in the upper portions of the gastrointestinal tract or by stimulating chemoreceptors in a specialized **chemoreceptor trigger zone** adjacent to the vomiting center in the brain. Activation of this zone triggers the vomiting reflex. For example, the chemotherapeutic agents used in the treatment of cancer often cause vomiting by acting on the chemoreceptor trigger zone.
- Psychogenic vomiting induced by emotional factors, including those accompanying nauseating sights and odors and anxiety before taking an examination or in other stressful situations.

Effects of vomiting

With excessive vomiting, the body experiences large losses of secreted fluids and acids that normally would be reabsorbed. The resultant reduction in plasma volume can lead to dehydration and circulatory problems, and the loss of acid from the stomach can lead to metabolic alkalosis (see p. 585).

Vomiting is not always detrimental, however. Limited vomiting brought about by irritation of the digestive tract can provide a useful service in removing noxious material from the stomach rather than allowing it to be retained and absorbed. In fact, emetics are sometimes given in the case of accidental ingestion of a poison to quickly remove the offending substance from the body.

We have now completed our discussion of gastric motility and will shift to gastric secretion.

▌ Gastric digestive juice is secreted by glands located at the base of gastric pits.

Each day the stomach secretes about 2 liters of gastric juice. The cells responsible for gastric secretion are located in the lining of the stomach, the gastric mucosa, which is divided into two distinct areas: (1) the **oxyntic mucosa,** which lines the body and fundus, and (2) the **pyloric gland area (PGA),** which lines the antrum. The luminal surface of the stomach is pitted with deep pockets formed by infoldings of the gastric mucosa. The first portion of these invaginations are called **gastric pits,** at the base of which the **gastric glands** are located. A variety of secretory cells line these invaginations, some exocrine and some endocrine or paracrine (▲ Table 16-4).

Gastric exocrine secretory cells

Three types of exocrine secretory cells are found in the walls of the pits and glands in the oxyntic mucosa.

- **Mucous cells** line the gastric pits and the entrance of the glands. They secrete a thin, watery *mucus.* (*Mucous* is the adjective; *mucus* is the noun.)
- The deeper portions of the gastric glands are lined by chief and parietal cells. The more numerous **chief cells** secrete the enzyme precursor *pepsinogen.*
- The **parietal** (or **oxyntic**) **cells** secrete *HCl* and *intrinsic factor.* The parietal cells are located on the outer wall of

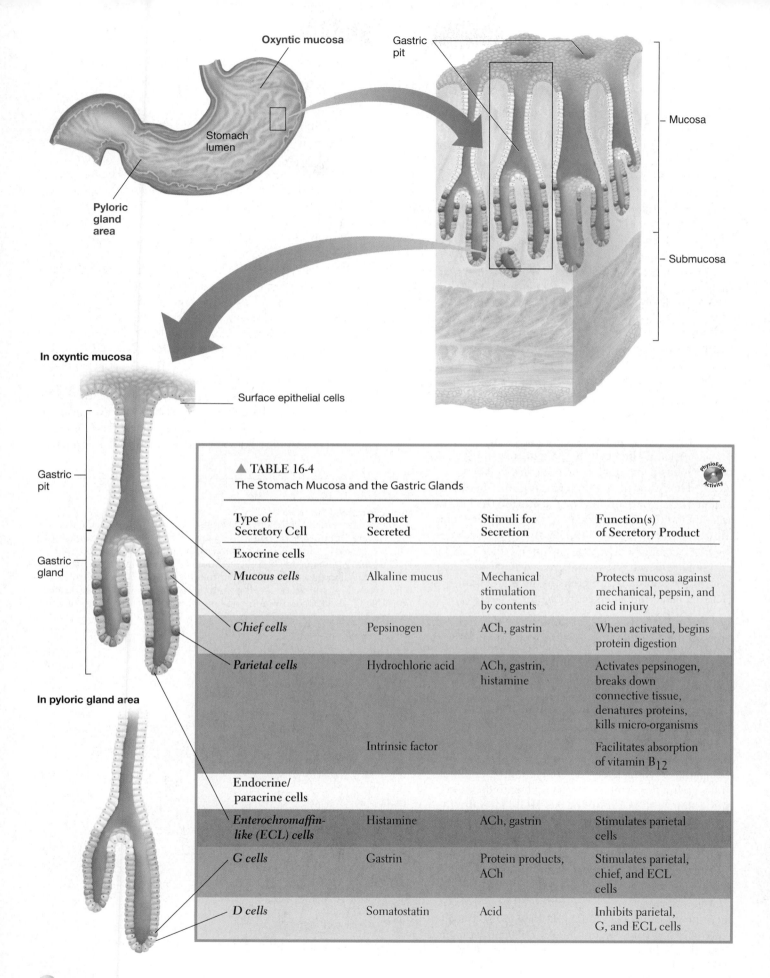

Oxyntic mucosa

Gastric pit

Stomach lumen

Mucosa

Submucosa

Pyloric gland area

In oxyntic mucosa

Surface epithelial cells

Gastric pit

Gastric gland

In pyloric gland area

▲ TABLE 16-4
The Stomach Mucosa and the Gastric Glands

Type of Secretory Cell	Product Secreted	Stimuli for Secretion	Function(s) of Secretory Product
Exocrine cells			
Mucous cells	Alkaline mucus	Mechanical stimulation by contents	Protects mucosa against mechanical, pepsin, and acid injury
Chief cells	Pepsinogen	ACh, gastrin	When activated, begins protein digestion
Parietal cells	Hydrochloric acid	ACh, gastrin, histamine	Activates pepsinogen, breaks down connective tissue, denatures proteins, kills micro-organisms
	Intrinsic factor		Facilitates absorption of vitamin B_{12}
Endocrine/ paracrine cells			
Enterochromaffin-like (ECL) cells	Histamine	ACh, gastrin	Stimulates parietal cells
G cells	Gastrin	Protein products, ACh	Stimulates parietal, chief, and ECL cells
D cells	Somatostatin	Acid	Inhibits parietal, G, and ECL cells

the gastric pits and do not come into contact with the pit lumen. (*Parietal* means "wall," a reference to the location of these cells. *Oxyntic* means "sharp," a reference to these cells' potent HCl secretory product.)

These exocrine secretions are all released into the gastric lumen. Collectively, they make up the gastric digestive juice.

A sparse number of **stem cells** are also found in the gastric pits. These cells rapidly divide and serve as the parent cells of all new cells of the gastric mucosa. The daughter cells that result from cell division either migrate out of the pit to become surface epithelial cells or migrate down deeper to the gastric glands where they differentiate into chief or parietal cells. Through this activity, the entire stomach mucosa is replaced about every three days.

Between the gastric pits, the gastric mucosa is covered by **surface epithelial cells,** which secrete a thick, viscous, alkaline mucus that forms a visible layer several millimeters thick over the surface of the mucosa.

Gastric endocrine and paracrine secretory cells

Other secretory cells in the gastric mucosa release endocrine and paracrine regulatory factors instead of products involved in the digestion of nutrients in the gastric lumen (see p. 66). These are as follows:

- **Enterochromaffin-like (ECL) cells** dispersed among the parietal and chief cells in the gastric glands of the oxyntic mucosa secrete the paracrine *histamine*.
- The gastric glands of the PGA primarily secrete mucus and a small amount of pepsinogen; no acid is secreted in this area, in contrast to the oxyntic mucosa. More importantly, endocrine cells known as **G cells** in the PGA secrete the hormone *gastrin* into the blood.
- **D cells,** which are scattered in glands near the pylorus but are more numerous in the duodenum, secrete the paracrine *somatostatin*.

We will examine each of these secretory products in more detail, looking first at the exocrine products and their role in digestion, then the endocrine and paracrine products and their role in regulating the exocrine secretions, among other actions.

▌ Hydrochloric acid activates pepsinogen.

The parietal cells actively secrete HCl into the lumen of the gastric pits, which in turn empty into the lumen of the stomach. As a result of this HCl secretion, the pH of the luminal contents falls as low as 2. Hydrogen ion (H^+) and chloride ion (Cl^-) are actively transported by separate pumps in the parietal cell's plasma membrane. Hydrogen ion is actively transported against a tremendous concentration gradient, with the H^+ concentration being as much as 3 million times greater in the lumen than in the blood. Chloride is also actively secreted but against a much smaller concentration gradient of only $1\frac{1}{2}$ times.

Mechanism of H^+ and Cl^- secretion

The secreted H^+ is not transported from the plasma but is derived instead from metabolic processes within the parietal cell (● Figure 16-10). Specifically, the H^+ to be secreted is derived from the breakdown of H_2O molecules into H^+ and OH^- (hydroxyl ions) within the parietal cells. This H^+ is secreted into the stomach lumen by $H^+ – K^+ ATPase$ in the parietal cell's luminal membrane. This primary active transport carrier also pumps K^+ into the cell from the lumen. The transported K^+ then passively leaks back into the lumen through luminal K^+ channels, thus leaving K^+ levels unchanged by the process of H^+ secretion.

Meanwhile, the OH^- generated by the breakdown of H_2O is neutralized by combining with a new H^+ generated from carbonic acid (H_2CO_3). The parietal cells contain an abundance of the enzyme carbonic anhydrase (ca). In the presence of carbonic anhydrase, H_2O readily combines with CO_2, which either has been produced within the parietal cell by metabolic processes or has diffused in from the blood. The combination of H_2O and CO_2 results in the formation of H_2CO_3, which partially dissociates to yield H^+ and HCO_3^-. The generated H^+ in essence replaces the one secreted.

The generated HCO_3^- is moved into the plasma by the same carrier at the basolateral border that actively transports Cl^- from the plasma into the parietal cell, similar to the Cl^- shift that occurs in red blood cells (see p. 493). This exchange

● **FIGURE 16-10**

Mechanism of HCl secretion
The stomach's parietal cells actively secrete H^+ and Cl^- by the actions of two separate pumps. The secreted H^+ is derived from the breakdown of H_2O into H^+ and OH^-. The OH^- is neutralized by another H^+ derived from H_2CO_3 that is generated within the cell from CO_2 that is either metabolically produced in the cell or diffuses in from the plasma. The secreted Cl^- is transported into the parietal cell from the plasma. The HCO_3^- generated from H_2CO_3 dissociation is transported into the plasma in exchange for the secreted Cl^-.

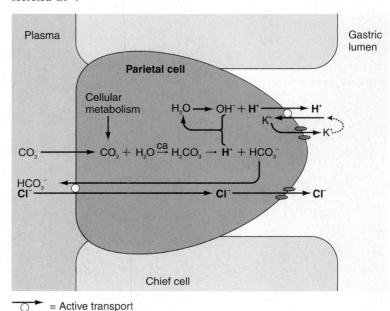

= Active transport

ca = Carbonic anhydrase

of HCO_3^- for Cl^- maintains electrical neutrality in the plasma during HCl secretion. The Cl^- pumped into the cell diffuses through luminal channels into the gastric lumen, completing the Cl^- secretory process.

Functions of HCl

Although HCl does not actually digest anything, it performs several functions that assist digestion. Specifically, HCl

1. Activates the enzyme precursor pepsinogen to an active enzyme, pepsin, and provides an acid medium that is optimal for pepsin activity
2. Aids in the breakdown of connective tissue and muscle fibers, thereby reducing large food particles into smaller particles
3. Denatures protein; that is, it uncoils proteins from their tertiary structure, thus exposing more of the peptide bonds for enzymatic attack
4. Along with salivary lysozyme, kills most of the microorganisms ingested with the food, although some do escape and continue to grow and multiply in the large intestine

▌Pepsinogen, once activated, begins protein digestion.

The major digestive constituent of gastric secretion is **pepsinogen,** an inactive enzymatic molecule synthesized and packaged by the endoplasmic reticulum and Golgi complex of the chief cells. Pepsinogen is stored in the chief cell's cytoplasm within secretory vesicles known as **zymogen granules,** from which it is released by exocytosis upon appropriate stimulation (see p. 29). When pepsinogen is secreted into the gastric lumen, HCl cleaves off a small fragment of the molecule, converting it to the active form of the enzyme, **pepsin** (● Figure 16-11). Once formed, pepsin acts on other pepsinogen molecules to produce more pepsin. A mechanism such as this, whereby an active form of an enzyme activates other molecules of the same enzyme, is referred to as an **autocatalytic** ("self-activating") **process.**

Pepsin initiates protein digestion by splitting certain amino acid linkages in proteins to yield peptide fragments (small amino acid chains); it works most effectively in the acid environment provided by HCl. Because pepsin can digest protein, it must be stored and secreted in an inactive form so that it does not digest the cells in which it is formed. Therefore, pepsin is maintained in the inactive form of pepsinogen until it reaches the gastric lumen, where it is activated by HCl secreted into the lumen by a different cell type.

▌Mucus is protective.

The surface of the gastric mucosa is covered by a layer of mucus derived from the surface epithelial cells and mucous cells. This mucus serves as a protective barrier against several forms of potential injury to the gastric mucosa:

- By virtue of its lubricating properties, mucus protects the gastric mucosa against mechanical injury.

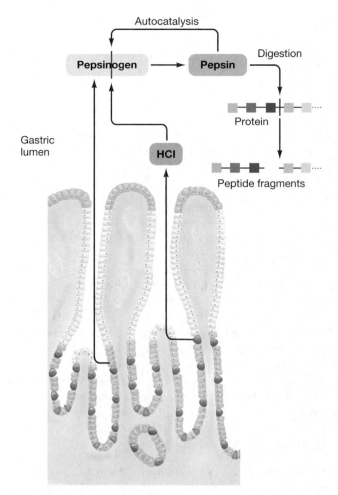

● FIGURE 16-11
Pepsinogen activation in the stomach lumen
In the lumen, hydrochloric acid (HCl) activates pepsinogen to its active form, pepsin, by cleaving off a small fragment. Once activated, pepsin autocatalytically activates more pepsinogen and begins protein digestion. Secretion of pepsinogen in the inactive form prevents it from digesting the protein structures of the cells in which it is produced.

- It helps protect the stomach wall from self-digestion, because pepsin is inhibited when it comes in contact with the mucus layer coating the stomach lining. (However, mucus does not affect pepsin activity in the lumen, where digestion of dietary protein proceeds without interference.)
- Being alkaline, mucus helps protect against acid injury by neutralizing HCl in the vicinity of the gastric lining, but it does not interfere with the function of HCl in the lumen. Whereas the pH in the lumen may be as low as 2, the pH in the mucus layer adjacent to the mucosal cell surface is about 7.

▌Intrinsic factor is essential for absorption of vitamin B₁₂.

Intrinsic factor, another secretory product of the parietal cells in addition to HCl, is important in the absorption of vitamin B_{12}. This vitamin can be absorbed only when in combination with intrinsic factor. Binding of the intrinsic factor–vitamin

B_{12} complex with a special receptor located only in the terminal ileum, the last portion of the small intestine, triggers the receptor-mediated endocytosis of the complex at this location (see p. 32).

Vitamin B_{12} is essential for the normal formation of red blood cells. In the absence of intrinsic factor, vitamin B_{12} fails to be absorbed, so erythrocyte production is defective, and *pernicious anemia* results (see p. 397). Pernicious anemia is typically caused by an autoimmune attack against the parietal cells (see p. 441). This condition is treated by regular injections of vitamin B_{12}, thus bypassing the defective digestive-tract absorptive mechanism.

Multiple regulatory pathways influence the parietal and chief cells.

Four chemical messengers primarily influence the secretion of gastric digestive juices. Parietal cells have separate receptors for each of these messengers. Three of them—acetylcholine (ACh), gastrin, and histamine—are stimulatory. They bring about increased secretion of HCl by promoting the insertion of additional $H^+ - K^+$ATPases into the parietal cell's plasma membrane. The fourth regulatory agent—somatostatin—inhibits acid secretion. ACh and gastrin also increase pepsinogen secretion through their stimulatory effect on the chief cells. We will now consider each of these chemical messengers in further detail (see ▲ Table 16-4, p. 609).

- **Acetylcholine** is a neurotransmitter released from the intrinsic nerve plexuses in response to short reflexes and vagal stimulation. ACh stimulates both the parietal and chief cells as well as the G cells and ECL cells.
- The G cells secrete the hormone **gastrin** into the blood in response to protein products in the stomach lumen as well in response to ACh. Like secretin and CCK, gastrin is a major gastrointestinal hormone. After being carried by the blood back to the body and fundus of the stomach, gastrin stimulates the parietal and chief cells, promoting secretion of a highly acidic gastric juice. In addition to directly stimulating the parietal cells, gastrin also indirectly promotes HCl secretion by

stimulating the ECL cells to release histamine. Gastrin is the main factor responsible for bringing about increased HCl secretion during digestion of a meal. Gastrin is also *trophic* (growth promoting) to the mucosa of the stomach and small intestine, thereby maintaining their secretory capabilities.

- **Histamine,** a paracrine, is released from the ECL cells in response to gastrin and ACh. Histamine acts locally on nearby parietal cells to increase the rate of HCl secretion.
- **Somatostatin** is released from the D cells in response to acid. It acts locally as a paracrine in negative-feedback fashion to inhibit secretion by the parietal cells, G cells, and ECL cells, thus turning off the HCl-secreting cells and their most potent stimulatory pathway.

From this list, it is obvious not only that multiple chemical messengers influence the parietal and chief cells but also that these chemicals also influence each other. Next, as we examine the phases of gastric secretion, we will see under what circumstances each of these regulatory agents is released.

Control of gastric secretion involves three phases.

The rate of gastric secretion can be influenced by (1) factors arising before food ever reaches the stomach, (2) factors resulting from the presence of food in the stomach, and (3) factors in the duodenum after food has left the stomach. Accordingly, gastric secretion is divided into three phases—the cephalic, gastric, and intestinal phases (▲ Tables 16-5 and 16-6).

Cephalic phase

The *cephalic phase of gastric secretion* refers to the increased secretion of HCl and pepsinogen that occurs in feedforward fashion in response to stimuli acting in the head even before food reaches the stomach (*cephalic* refers to "head"). Thinking about, tasting, smelling, chewing, and swallowing food increase gastric secretion by means of vagal nerve activity in two ways. First, vagal stimulation of the intrinsic plexuses promotes increased secretion of ACh, which in turn leads to increased secretion of HCl and pepsinogen by the secretory cells. Second,

▲ TABLE 16-5
Stimulation of Gastric Secretion

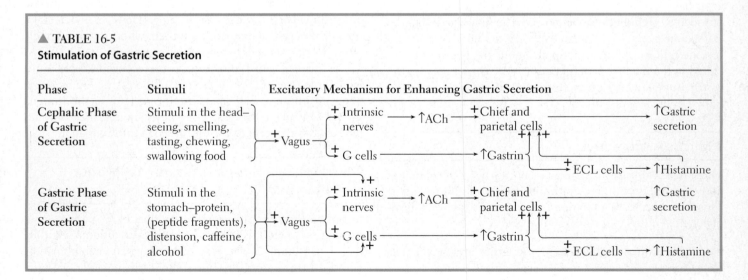

Phase	Stimuli	Excitatory Mechanism for Enhancing Gastric Secretion
Cephalic Phase of Gastric Secretion	Stimuli in the head—seeing, smelling, tasting, chewing, swallowing food	+Vagus → + Intrinsic nerves → ↑ACh → +Chief and parietal cells → ↑Gastric secretion; + G cells → ↑Gastrin → +ECL cells → ↑Histamine
Gastric Phase of Gastric Secretion	Stimuli in the stomach—protein, (peptide fragments), distension, caffeine, alcohol	+Vagus → + Intrinsic nerves → ↑ACh → +Chief and parietal cells → ↑Gastric secretion; + G cells → ↑Gastrin → +ECL cells → ↑Histamine

vagal stimulation of the G cells within the PGA causes the release of gastrin, which in turn further enhances secretion of HCl and pepsinogen, with the effect on HCl being potentiated by gastrin promoting the release of histamine.

Gastric phase

The *gastric phase of gastric secretion* occurs when food actually reaches the stomach. Stimuli acting in the stomach—namely *protein*, especially peptide fragments; *distension*; *caffeine*; and *alcohol*—increase gastric secretion by means of overlapping efferent pathways. For example, protein in the stomach, the most potent stimulus, initiates short local reflexes in the intrinsic nerve plexuses to stimulate the secretory cells. Furthermore, protein initiates long reflexes so that extrinsic vagal fibers to the stomach are activated. Vagal activity further enhances intrinsic nerve stimulation of the secretory cells and triggers the release of gastrin. Protein also directly stimulates the release of gastrin. Gastrin in turn is a powerful stimulus for further HCl and pepsinogen secretion and also calls forth release of histamine, which further increases HCl secretion. Through these synergistic and overlapping pathways, protein induces the secretion of a highly acidic, pepsin-rich gastric juice, which continues the digestion of the protein that first initiated the process.

When the stomach is distended with protein-rich food that needs to be digested, these secretory responses are appropriate. Caffeine and, to lesser extent, alcohol also stimulate the secretion of a highly acidic gastric juice, even when no food is present. This unnecessary acid can irritate the linings of the stomach and duodenum. For this reason, persons with ulcers or gastric hyperacidity should avoid caffeinated and alcoholic beverages.

Intestinal phase

The *intestinal phase of gastric secretion* encompasses the factors originating in the small intestine that influence gastric secretion. Whereas the other phases are excitatory, this phase is inhibitory. The intestinal phase is important in helping shut off the flow of gastric juices as chyme begins to be emptied into the small intestine, a topic to which we now turn our attention.

▌ Gastric secretion gradually decreases as food empties from the stomach into the intestine.

You now know what factors turn on gastric secretion before and during a meal, but how is the flow of gastric juices shut off when they are no longer needed? Gastric secretion is gradually reduced by three different means as the stomach empties (▲ Table 16-6):

- As the meal is gradually emptied into the duodenum, the major stimulus for enhanced gastric secretion—the presence of protein in the stomach—is withdrawn.
- After foods leave the stomach and gastric juices accumulate to such an extent that gastric pH falls very low, somatostatin is released. As a result of somatostatin's inhibitory effects, gastric secretion declines.
- The same stimuli that inhibit gastric motility (fat, acid, hypertonicity, or distension in the duodenum brought about by stomach emptying) inhibit gastric secretion as well; the enterogastric reflex and the enterogastrones suppress the gastric secretory cells while they simultaneously reduce the excitability of the gastric smooth muscle cells. This inhibitory response is the intestinal phase of gastric secretion.

▌ The stomach lining is protected from gastric secretions by the gastric mucosal barrier.

How can the stomach contain strong acid contents and proteolytic enzymes without destroying itself? You have already learned

▲ TABLE 16-6
Inhibition of Gastric Secretion

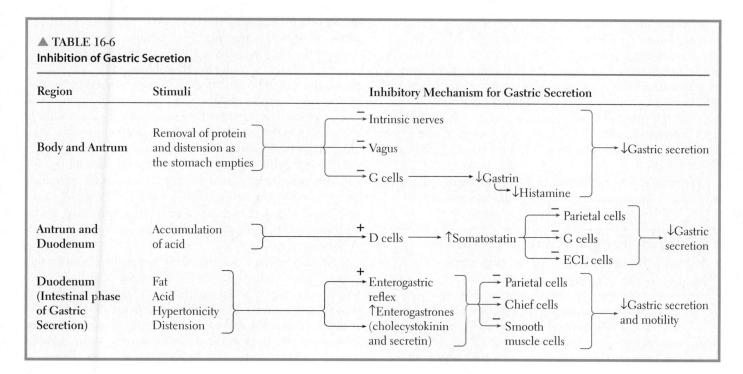

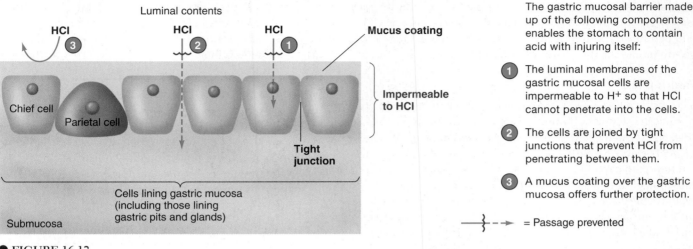

Luminal contents

HCl **3**

HCl **2**

HCl **1**

Mucus coating

Chief cell

Parietal cell

Impermeable to HCl

Tight junction

Cells lining gastric mucosa (including those lining gastric pits and glands)

Submucosa

The gastric mucosal barrier made up of the following components enables the stomach to contain acid with injuring itself:

1 The luminal membranes of the gastric mucosal cells are impermeable to H⁺ so that HCl cannot penetrate into the cells.

2 The cells are joined by tight junctions that prevent HCl from penetrating between them.

3 A mucus coating over the gastric mucosa offers further protection.

= Passage prevented

● FIGURE 16-12
Gastric mucosal barrier

that mucus provides a protective coating. In addition, other barriers to mucosal acid damage are provided by the mucosal lining itself. First, the luminal membranes of the gastric mucosal cells are almost impermeable to H⁺, so acid cannot penetrate *into* the cells and cause cellular damage. Furthermore, the lateral edges of these cells are joined together near their luminal borders by tight junctions, so acid cannot diffuse *between* the cells from the lumen into the underlying submucosa (see p. 63). The properties of the gastric mucosa that enable the stomach to contain acid without injuring itself constitute the **gastric mucosal barrier** (● Figure 16-12). These protective mechanisms are further enhanced by the fact that the entire stomach lining is replaced every three days. Because of rapid mucosal turnover, cells are usually replaced before they are exposed to the wear and tear of harsh gastric conditions long enough to suffer damage.

Despite the protection provided by mucus, by the gastric mucosal barrier, and by the frequent turnover of cells, the barrier occasionally is broken, and the gastric wall is injured by its acidic and enzymatic contents. When this occurs, an erosion, or **peptic ulcer,** of the stomach wall results. Excessive gastric reflux into the esophagus and dumping of excessive acidic gastric contents into the duodenum can lead to peptic ulcers in these locations as well. (For a further discussion of ulcers, see the accompanying boxed feature, ▶ Concepts, Challenges, and Controversies.)

We now turn our attention to the remaining two digestive processes in the stomach, gastric digestion and absorption.

▌ Carbohydrate digestion continues in the body of the stomach; protein digestion begins in the antrum.

Two separate digestive processes take place within the stomach. In the body of the stomach, food remains in a semisolid mass, because peristaltic contractions in this region are too weak for mixing to occur. Because food is not mixed with gastric se-

cretions in the body of the stomach, very little protein digestion occurs here. Acid and pepsin are able to attack only the surface of the food mass. In the interior of the mass, however, carbohydrate digestion continues under the influence of salivary amylase. Even though acid inactivates salivary amylase, the unmixed interior of the food mass is free of acid.

Digestion by the gastric juice itself is accomplished in the antrum of the stomach, where the food is thoroughly mixed with HCl and pepsin, thereby initiating protein digestion.

▌ The stomach absorbs alcohol and aspirin but no food.

No food or water is absorbed into the blood through the stomach mucosa. However, two noteworthy nonnutrient substances are absorbed directly by the stomach—*ethyl alcohol* and *aspirin.* Alcohol is somewhat lipid soluble, so it can diffuse through the lipid membranes of the epithelial cells that line the stomach and can enter the blood through the submucosal capillaries. Yet although alcohol can be absorbed by the gastric mucosa, it can be absorbed even more rapidly by the small intestine mucosa, because the surface area for absorption in the small intestine is much greater than in the stomach. Thus alcohol absorption occurs more slowly if gastric emptying is delayed so that the alcohol remains in the stomach longer. Because fat is the most potent duodenal stimulus for inhibiting gastric motility, consuming fat-rich foods (for example, whole milk or pizza) before or during alcohol ingestion delays gastric emptying and prevents the alcohol from producing its effects as rapidly.

Another category of substances absorbed by the gastric mucosa includes weak acids, most notably acetylsalicylic acid (aspirin). In the highly acidic environment of the stomach lumen, weak acids are almost totally un-ionized; that is, the H⁺ and associated anion of the acid are bound together. In an un-ionized form, these weak acids are lipid soluble, so they can be absorbed quickly by crossing the plasma membranes of the epithelial

Ulcers: When Bugs Break the Barrier

Peptic ulcers are erosions that begin in the mucosal lining of the stomach, duodenum, or esophagus and may penetrate into the deeper layers of the digestive tract wall. Ulcers occur when the gastric mucosal barrier is disrupted and thus pepsin and HCl act on the digestive tract wall instead of food in the lumen.

Until recently, the exact cause of ulcers was unknown, but in a surprising discovery in the early 1990s the bacterium *Helicobacter pylori* was pinpointed as the cause of more than 80% of all peptic ulcers. Thirty percent of the population harbors *H. pylori*. Those with this "slow" bacterium have a 3 to 12 times greater risk of developing an ulcer within 10 to 20 years of acquiring the infection than those without the bacterium. They are also at increased risk of developing stomach cancer.

For years, scientists had overlooked the possibility that ulcers could be triggered by an infectious agent because bacteria typically cannot survive in a strongly acidic environment such as the stomach lumen. An exception, *H. pylori* exploits several strategies to survive in this hostile environment. First, these organisms are motile, being equipped with four to six flagella (whiplike appendages; see the accompanying figure), which enable them to tunnel through and take up residence under the stomach's thick layer of alkaline mucus. Here they are protected from the highly acidic gastric contents. Furthermore, *H. pylori* preferentially settles in the antrum, which has no acid-producing parietal cells, although HCl from the upper portions of the stomach does reach the antrum. Also, these bacteria produce *urease*, an enzyme that breaks down urea, an end product of protein metabolism, into ammonia (NH_3) and CO_2. Ammonia serves as a buffer (see p. 582) that neutralizes stomach acid locally in the vicinity of the *H. pylori*.

H. pylori contributes to ulcer formation by secreting toxins that cause a persistent inflammation, or *chronic superficial gastritis*, at the site they colonize. This inflammatory response apparently weakens the gastric mucosal barrier.

Alone or in conjunction with this infectious culprit, other factors are known to contribute to ulcer formation. Frequent exposure to some

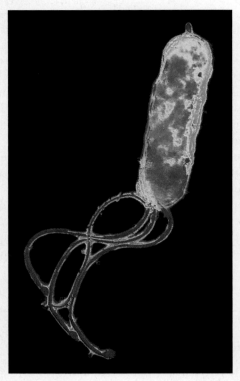

● **Helicobacter pylori**
Helicobacter pylori, the bacterium responsible for most cases of peptic ulcers, possesses flagella that enable it to tunnel beneath the protective mucus layer that coats the stomach lining.

chemicals can break the gastric mucosal barrier; the most important of these are ethyl alcohol and nonsteroidal anti-inflammatory drugs (NSAIDs), such as aspirin, ibuprofen, or more potent medications for the treatment of arthritis or other chronic inflammatory processes. The barrier frequently breaks in patients with pre-existing debilitating conditions, such as severe injuries or infections. Persistent stressful situations are frequently associated with ulcer formation, presumably because emotional response to stress can stimulate excessive gastric secretion.

When the gastric mucosal barrier is broken, acid and pepsin diffuse into the mucosa and underlying submucosa, with serious pathophysio-

logic consequences. The surface erosion, or ulcer, progressively enlarges as increasing levels of acid and pepsin continue to damage the stomach wall. Two of the most serious consequences of ulcers are (1) hemorrhage resulting from damage to submucosal capillaries and (2) perforation, or complete erosion through the stomach wall, resulting in the escape of potent gastric contents into the abdominal cavity.

Treatment of ulcers includes antibiotics, H-2 histamine receptor blockers, and proton pump inhibitors. With the discovery of the infectious component of most ulcers, antibiotics are now a treatment of choice. The other drugs are also used alone or in combination with antibiotics.

Two decades prior to the discovery of *H. pylori*, researchers discovered an antihistamine *(cimetidine)* that specifically blocks H-2 receptors, the type of receptors that bind histamine released from the stomach. These receptors differ from H-1 receptors that bind the histamine involved in allergic respiratory disorders. Accordingly, traditional antihistamines used for respiratory allergies (such as hay fever and asthma) are not effective against ulcers, nor is cimetidine useful for respiratory problems.

Another recent class of drugs used in the treatment of ulcers inhibits acid secretion by directly blocking the $H^+ - K^+$ ATPase pump. These so-called *proton-pump inhibitors* (H^+ is a naked proton without its electron) help reduce the corrosive effect of HCl on the exposed tissue.

Before these recent discoveries that have played a tremendous role in curing ulcers, the treatment of ulcers was aimed at either neutralizing stomach acidity through the use of antacids or removing factors known to enhance gastric secretion. Accordingly, the treatment included (1) a bland diet void of caffeine and alcohol, (2) cutting the vagus nerve supply to the stomach, and (3) removal of the stomach antrum to eliminate the source of gastrin. Even though symptoms subsided, the condition often recurred, because the underlying cause was not eradicated.

cells that line the stomach. Most other drugs are not absorbed until they reach the small intestine, so they do not begin to take effect as quickly.

Having completed our coverage of the stomach, we will move to the next portion of the digestive tract, the small intestine and the accessory digestive organs that release their secretions into the small intestine lumen.

PANCREATIC AND BILIARY SECRETIONS

When gastric contents are emptied into the small intestine, they are mixed not only with juice secreted by the small intestine mucosa but also with the secretions of the exocrine pancreas and liver that are released into the duodenal lumen. We will discuss the roles of each of these accessory digestive organs before we examine the contributions of the small intestine itself.

The pancreas is a mixture of exocrine and endocrine tissue.

The **pancreas** is an elongated gland that lies behind and below the stomach, above the first loop of the duodenum (● Figure 16-13). It is a mixed gland that contains both exocrine and endocrine tissue. The predominant exocrine portion consists of grapelike clusters of secretory cells that form sacs known as **acini,** which connect to ducts that eventually empty into the duodenum. The smaller endocrine portion consists of isolated islands of endocrine tissue, the **islets of Langerhans,** which are dispersed throughout the pancreas. The most important hormones secreted by the islet cells are insulin and glucagon (Chapter 19). The exocrine and endocrine pancreas have nothing in common except that they share the same location.

The exocrine pancreas secretes digestive enzymes and an aqueous alkaline fluid.

The **exocrine pancreas** secretes a pancreatic juice consisting of two components—a potent *enzymatic secretion* and an *aqueous alkaline secretion* that is rich in sodium bicarbonate ($NaHCO_3$). The pancreatic enzymes are actively secreted by the *acinar cells.* The aqueous (meaning "watery") $NaHCO_3$ component is actively secreted by the *duct cells* that line the early portion of the pancreatic ducts, and it then is modified as it passes down the ducts.

Like pepsinogen, pancreatic enzymes are synthesized by the endoplasmic reticulum and Golgi complex of the acinar cells, then are stored within zymogen granules and released by exocytosis as needed. The acinar cells secrete three different types of pancreatic enzymes that are capable of digesting all three categories of foodstuffs. These pancreatic enzymes are important because they are capable of almost completely digesting food in the absence of all other digestive secretions.

The three types of pancreatic enzymes are (1) **proteolytic enzymes,** which are involved in protein digestion; (2) **pancreatic amylase,** which contributes to carbohydrate digestion in a way similar to salivary amylase; and (3) **pancreatic lipase,** the only enzyme important in fat digestion.

Pancreatic proteolytic enzymes

The three major proteolytic enzymes secreted by the pancreas are *trypsinogen, chymotrypsinogen,* and *procarboxypeptidase,* each of which is secreted in an inactive form. When **trypsinogen** is secreted into the duodenal lumen, it is activated to its active enzyme form, **trypsin,** by **enterokinase** (also known as **enteropeptidase**), an enzyme embedded in the luminal border of the cells that line the duodenal mucosa. Trypsin then autocatalytically activates more trypsinogen. Like pepsinogen, trypsinogen must remain inactive within the pancreas to prevent this proteolytic enzyme from digesting the cells in which it is formed. Trypsinogen remains inactive, therefore, until it reaches the duodenal lumen, where enterokinase triggers the activation process, which then proceeds autocatalytically. As further protection, the pancreatic tissue also produces a chemical known as **trypsin inhibitor,** which blocks trypsin's actions if spontaneous activation of trypsinogen inadvertently occurs within the pancreas.

● **FIGURE 16-13**

Schematic representation of the exocrine and endocrine portions of the pancreas
The exocrine pancreas secretes into the duodenal lumen a digestive juice composed of digestive enzymes secreted by the acinar cells and an aqueous $NaHCO_3$ solution secreted by the duct cells. The endocrine pancreas secretes the hormones insulin and glucagon into the blood.

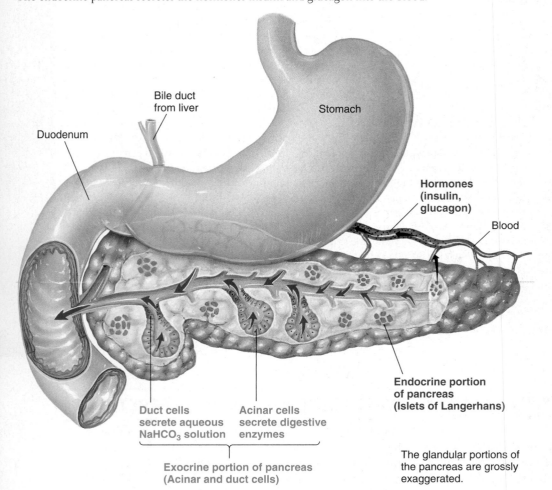

Bile duct from liver

Stomach

Duodenum

Hormones (insulin, glucagon)

Blood

Duct cells secrete aqueous $NaHCO_3$ solution

Acinar cells secrete digestive enzymes

Endocrine portion of pancreas (Islets of Langerhans)

Exocrine portion of pancreas (Acinar and duct cells)

The glandular portions of the pancreas are grossly exaggerated.

Chymotrypsinogen and **procarboxypeptidase,** the other pancreatic proteolytic enzymes, are converted by trypsin to their active forms, **chymotrypsin** and **carboxypeptidase,** respectively, within the duodenal lumen. Thus, once enterokinase has activated some of the trypsin, trypsin is then responsible for the rest of the activation process.

Each of these proteolytic enzymes attacks different peptide linkages. The end products that result from this action are a mixture of amino acids and small peptide chains. Mucus secreted by the intestinal cells provides protection against digestion of the small intestine wall by the activated proteolytic enzymes.

Pancreatic amylase

Like salivary amylase, pancreatic amylase contributes to carbohydrate digestion by converting polysaccharides into disaccharides. Amylase is secreted in the pancreatic juice in an active form, because active amylase does not present a danger to the secretory cells. These cells do not contain any polysaccharides.

Pancreatic lipase

Pancreatic lipase is extremely important because it is the only enzyme secreted throughout the entire digestive system that can accomplish digestion of fat. Pancreatic lipase hydrolyzes dietary triglycerides into monoglycerides and free fatty acids, which are the absorbable units of fat. Like amylase, lipase is secreted in its active form because there is no risk of pancreatic self-digestion by lipase. Triglycerides are not a structural component of pancreatic cells.

Pancreatic insufficiency

When pancreatic enzymes are deficient, digestion of food is incomplete. Because the pancreas is the only significant source of lipase, pancreatic enzyme deficiency results in serious maldigestion of fats. The principal clinical manifestation of pancreatic exocrine insufficiency is **steatorrhea,** or excessive undigested fat in the feces. Up to 60% to 70% of the ingested fat may be excreted in the feces. Digestion of protein and carbohydrates is impaired to a lesser degree because salivary, gastric, and small intestinal enzymes contribute to the digestion of these two foodstuffs.

Pancreatic aqueous alkaline secretion

Pancreatic enzymes function best in a neutral or slightly alkaline environment, yet the highly acidic gastric contents are emptied into the duodenal lumen in the vicinity of pancreatic enzyme entry into the duodenum. This acidic chyme must be neutralized quickly in the duodenal lumen, not only to allow optimal functioning of the pancreatic enzymes but also to prevent acid damage to the duodenal mucosa. The alkaline ($NaHCO_3$-rich) fluid secreted by the pancreas into the duodenal lumen serves the important function of neutralizing the acidic chyme as the latter is emptied into the duodenum from the stomach. This aqueous $NaHCO_3$ secretion is by far the largest component of pancreatic

secretion. The volume of pancreatic secretion ranges between 1 and 2 liters per day, depending on the types and degree of stimulation.

▌ Pancreatic exocrine secretion is regulated by secretin and CCK.

Pancreatic exocrine secretion is regulated primarily by hormonal mechanisms. A small amount of parasympathetically induced pancreatic secretion occurs during the cephalic phase of digestion, with a further token increase occurring during the gastric phase in response to gastrin. However, the predominant stimulation of pancreatic secretion occurs during the intestinal phase of digestion when chyme is in the small intestine. The release of the two major enterogastrones, secretin and cholecystokinin (CCK), in response to chyme in the duodenum plays the central role in the control of pancreatic secretion (● Figure 16-14).

Role of secretin in pancreatic secretion

Of the factors that stimulate enterogastrone release, the primary stimulus specifically for secretin release is acid in the duodenum. Secretin in turn is carried by the blood to the pancreas, where it stimulates the duct cells to markedly increase their secretion of a $NaHCO_3$-rich aqueous fluid into the duodenum. Even though other stimuli may cause the release of secretin, it is appropriate that the most potent stimulus is acid, because secretin promotes the alkaline pancreatic secretion that neu-

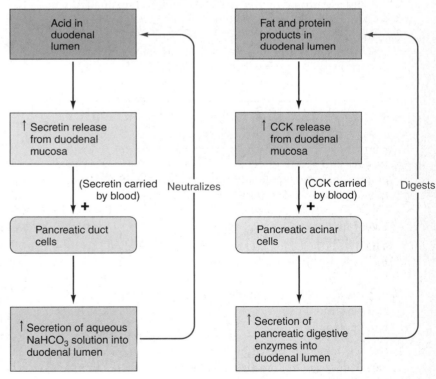

● **FIGURE 16-14**

Hormonal control of pancreatic exocrine secretion

tralizes the acid. This mechanism provides a control system for maintaining neutrality of the chyme in the intestine. The amount of secretin released is proportional to the amount of acid that enters the duodenum, so the amount of $NaHCO_3$ secreted parallels the duodenal acidity.

Role of CCK in pancreatic secretion

Cholecystokinin is important in the regulation of pancreatic digestive enzyme secretion. The main stimulus for release of CCK from the duodenal mucosa is the presence of fat and to a lesser extent protein products. The circulatory system transports CCK to the pancreas where it stimulates the pancreatic acinar cells to increase digestive enzyme secretion. Among these enzymes are lipase and the proteolytic enzymes, which appropriately bring about further digestion of the fat and protein that initiated the response and also help digest carbohydrate. In contrast to fat and protein, carbohydrate does not have any direct influence on pancreatic digestive enzyme secretion.

All three types of pancreatic digestive enzymes are packaged together in the zymogen granules, so all the pancreatic enzymes are released together on exocytosis of the granules. Therefore, even though the *total amount* of enzymes released varies depending on the type of meal consumed (the most being secreted in response to fat), the *proportion* of enzymes released does not vary on a meal-to-meal basis. That is, a high-protein meal does not cause the release of a greater proportion of proteolytic enzymes. Evidence suggests, however, that long-term adjustments in the proportion of the types of enzymes produced may occur as an adaptive response to a prolonged change in diet. For example, with a long-term switch to a high-protein diet a greater proportion of proteolytic enzymes are produced. Cholecystokinin may play a role in pancreatic digestive enzyme adaptation to changes in diet.

Just as gastrin is trophic to the stomach and small intestine, CCK and secretin exert trophic effects on the exocrine pancreas to maintain its integrity.

We will now look at the contributions of the remaining accessory digestive unit, the liver and gallbladder.

▌ The liver performs various important functions including bile production.

Besides pancreatic juice, the other secretory product that is emptied into the duodenal lumen is **bile**. The **biliary system** includes the *liver*, the *gallbladder*, and associated ducts.

Liver functions

The **liver** is the largest and most important metabolic organ in the body; it can be viewed as the body's major "biochemical factory." Its importance to the digestive system is its secretion of *bile salts*, which aid fat digestion and absorption. The liver also performs a wide variety of functions not related to digestion, including:

1. Metabolic processing of the major categories of nutrients (carbohydrates, proteins, and lipids) after their absorption from the digestive tract.

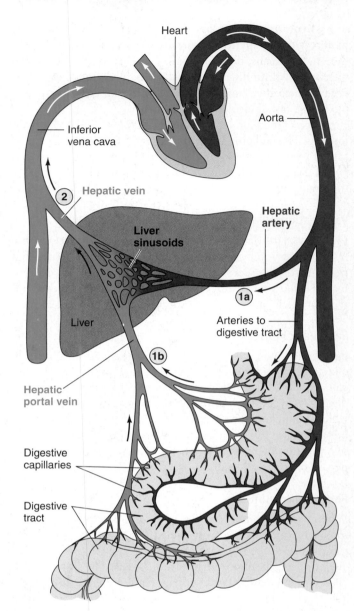

(1) The liver receives blood from two sources:

(a) Arterial blood, which provides the liver's O_2 supply and carries blood-borne metabolites for hepatic processing, is delivered by the **hepatic artery**.

(b) Venous blood draining the digestive tract is carried by the **hepatic portal** vein to the liver for processing and storage of newly absorbed nutrients.

(2) Blood leaves the liver via the **hepatic vein**.

● **FIGURE 16-15**
Schematic representation of liver blood flow

---

2. Detoxification or degradation of body wastes and hormones as well as drugs and other foreign compounds.

3. Synthesis of plasma proteins, including those necessary for blood clotting and those that transport steroid and thyroid hormones and cholesterol in the blood.

4. Storage of glycogen, fats, iron, copper, and many vitamins.

5. Activation of vitamin D, which the liver accomplishes in conjunction with the kidneys.

6. Removal of bacteria and worn-out red blood cells, thanks to its resident macrophages (see p. 401).

7. Excretion of cholesterol and bilirubin, the latter being a breakdown product derived from the destruction of worn-out red blood cells.

Given this wide range of complex functions, there is amazingly little specialization among cells within the liver. Each liver cell, or **hepatocyte,** performs the same wide variety of metabolic and secretory tasks (*hepato* means "liver"; *cyte* means "cell"). The specialization comes from the highly developed organelles within each hepatocyte. The only liver function not accomplished by the hepatocytes is the phagocytic activity carried out by the resident macrophages, which are known as **Kupffer cells.**

Liver blood flow

To carry out these wide-ranging tasks, the anatomic organization of the liver permits each hepatocyte to be in direct contact with blood from two sources: venous blood coming directly from the digestive tract and arterial blood coming from the aorta. Venous blood enters the liver by means of a unique and complex vascular connection between the digestive tract and the liver that is known as the **hepatic portal system** (● Figure 16-15). The veins draining the digestive tract do not directly join the inferior vena cava, the large vein that returns blood to the heart. Instead, the veins from the stomach and intestine enter the hepatic portal vein, which carries the products absorbed from the digestive tract directly to the liver for processing, storage, or detoxification before they gain access to the general circulation. Within the liver, the portal vein once again breaks up into a capillary network (the liver *sinusoids*) to permit exchange between the blood and hepatocytes before draining into the hepatic vein, which joins the inferior vena cava. The hepatocytes also are provided with fresh arterial blood, which supplies their oxygen and delivers blood-borne metabolites for hepatic processing.

▌ The liver lobules are delineated by vascular and bile channels.

The liver is organized into functional units known as **lobules,** which are hexagonal arrangements of tissue surrounding a central vein, like a six-sided angel food cake with the hole representing the central vein (● Figure 16-16a). At the outer edge of each "slice" of the lobule are three vessels: a branch of the hepatic artery, a branch of the portal vein, and a bile duct. Blood from the branches of both the hepatic artery and the portal vein flows from the periphery of the lobule into large, expanded capillary spaces called **sinusoids,** which run between rows of liver cells to the central vein like spokes on a bicycle wheel (● Figure 16-16b). The Kupffer cells line the sinusoids and engulf and destroy old red blood cells and bacteria that pass through in the blood. The hepatocytes are arranged between the sinusoids in plates two cell layers thick, so that each lateral edge faces a sinusoidal pool of blood. The central veins of all the liver lobules converge to form the hepatic vein, which carries the blood away from the liver. The thin bile-carrying channel, a **bile canaliculus,** runs between the cells within each

● **FIGURE 16-16**
Anatomy of the liver
(a) Hepatic lobule. (b) Wedge of a hepatic lobule.

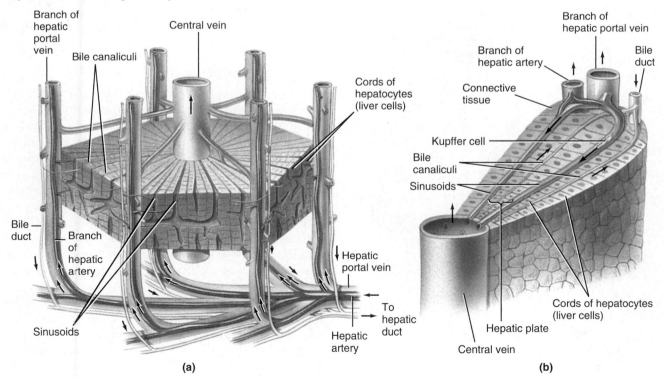

Branch of hepatic portal vein
Bile canaliculi
Central vein
Cords of hepatocytes (liver cells)
Bile duct
Branch of hepatic artery
Sinusoids
Hepatic portal vein
To hepatic duct
Hepatic artery

(a)

Branch of hepatic portal vein
Branch of hepatic artery
Bile duct
Connective tissue
Kupffer cell
Bile canaliculi
Sinusoids
Cords of hepatocytes (liver cells)
Hepatic plate
Central vein

(b)

The Digestive System

hepatic plate. Hepatocytes continuously secrete bile into these thin channels, which carry the bile to a bile duct at the periphery of the lobule. The bile ducts from the various lobules converge to eventually form the common bile duct, which transports the bile from the liver to the duodenum. Each hepatocyte is in contact with a sinusoid on one side and a bile canaliculus on the other side.

Bile is continuously secreted by the liver and is diverted to the gallbladder between meals.

The opening of the bile duct into the duodenum is guarded by the **sphincter of Oddi,** which prevents bile from entering the duodenum except during digestion of meals (● Figure 16-17). When this sphincter is closed, most of the bile secreted by the liver is diverted back up into the **gallbladder,** a small saclike structure tucked beneath but not directly connected to the liver. Thus bile is not transported directly from the liver to the gallbladder. The bile is subsequently stored and concentrated in the gallbladder between meals. After a meal, bile enters the duodenum as a result of the combined effects of gallbladder emptying and increased bile secretion by the liver. The amount of bile secreted per day ranges from 250 ml to 1 liter, depending on the degree of stimulation.

● FIGURE 16-17

Enterohepatic circulation of bile salts
The majority of bile salts are recycled between the liver and small intestine through the enterohepatic circulation, designated by the blue arrows. After participating in fat digestion and absorption, most bile salts are reabsorbed by active transport in the terminal ileum and returned through the hepatic portal vein to the liver, which resecretes them in the bile.

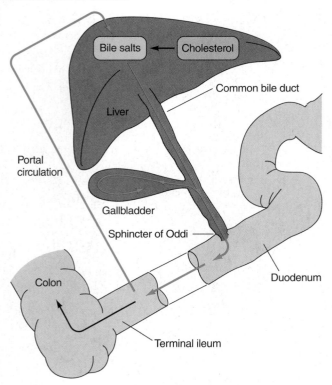

Bile salts are recycled through the enterohepatic circulation.

Bile consists of an *aqueous alkaline fluid* similar to the pancreatic $NaHCO_3$ secretion as well as several organic constituents, including *bile salts, cholesterol, lecithin,* and *bilirubin.* The organic constituents are derived from hepatocyte activity, whereas the water, $NaHCO_3$, and other inorganic salts are added by the duct cells. Even though bile does not contain any digestive enzymes, it is important for the digestion and absorption of fats, primarily through the activity of the bile salts.

Bile salts are derivatives of cholesterol. They are actively secreted into the bile and eventually enter the duodenum along with the other biliary constituents. Following their participation in fat digestion and absorption, most bile salts are reabsorbed back into the blood by special active transport mechanisms located in the terminal ileum. From here the bile salts are returned by means of the hepatic portal system to the liver, which resecretes them into the bile. This recycling of bile salts (and some of the other biliary constituents) between the small intestine and liver is referred to as the **enterohepatic circulation** (*entero* means "intestine"; *hepatic* means "liver") (● Figure 16-17).

The total amount of bile salts in the body averages about 3 to 4 g, yet 3 to 15 g of bile salts may be emptied into the duodenum in a single meal. Obviously, the bile salts must be recycled many times per day. Usually, only about 5% of the secreted bile escapes into the feces daily. These lost bile salts are replaced by new bile salts synthesized by the liver; thus the size of the pool of bile salts is kept constant.

Bile salts aid fat digestion and absorption.

Bile salts aid fat digestion through their detergent action (emulsification) and facilitate fat absorption through their participation in the formation of micelles. Both functions are related to the structure of bile salts. Let's see how.

Detergent action of bile salts

The term **detergent action** refers to bile salts' ability to convert large fat globules into a **lipid emulsion** that consists of many small fat droplets suspended in the aqueous chyme, thus increasing the surface area available for attack by pancreatic lipase. Fat globules, no matter their size, are made up primarily of undigested triglyceride molecules. To digest fat, lipase must come into direct contact with the triglyceride molecule. Because triglycerides are not soluble in water, they tend to aggregate into large droplets in the aqueous environment of the small intestine lumen. If bile salts did not emulsify these large droplets, lipase could act on the triglyceride molecules only at the surface of the large droplets, and fat digestion would be greatly prolonged.

Bile salts exert a detergent action similar to that of the detergent you use to break up grease when you wash dishes. A bile salt molecule contains a lipid-soluble portion (a steroid derived from cholesterol) plus a negatively charged, water-soluble portion. Bile salts *adsorb* on the surface of a fat droplet; that is, the lipid-soluble portion of the bile salt dissolves in the fat droplet, leaving the charged water-soluble portion projecting from the surface of the droplet (● Figure 16-18a). Intestinal mixing movements break up large fat droplets into smaller ones. These small droplets would quickly recoalesce were it not for bile salts adsorbing on their surface and creating a "shell" of water-soluble negative charges on the surface of each little droplet. Because like charges repel, these negatively charged groups on the droplet surfaces cause the fat droplets to repel each other (● Figure 16-18b). This electrical repulsion prevents the small droplets from re-coalescing into large fat droplets and thus produces a lipid emulsion that increases the surface area available for lipase action.

Although bile salts increase the surface area available for attack by pancreatic lipase, lipase alone cannot penetrate the layer of bile salts adsorbed on the surface of the small emulsified fat droplets. To solve this dilemma, the pancreas secretes the polypeptide **colipase** along with lipase. Colipase binds both to lipase and to the bile salts at the surface of the fat droplets, thus anchoring lipase to its site of action.

Micellar formation

Bile salts—along with cholesterol and lecithin, which are also constituents of the bile—play an important role in facilitating fat absorption through micellar formation. Like bile salts, lecithin has both a lipid-soluble and a water-soluble portion, whereas cholesterol is almost totally insoluble in water. In a **micelle**, the bile salts and lecithin aggregate in small clusters with their fat-soluble portions huddled together in the middle to form a hydrophobic ("water-fearing") core, while their water-soluble portions form an outer hydrophilic ("water-loving") shell (● Figure 16-19). A micelle is 4 to 7 nm in diameter, about one-millionth the size of an emulsified lipid droplet. Micelles, being water soluble by virtue of their hydrophilic shells, can dissolve water-insoluble (and hence lipid-soluble) substances in their lipid-soluble cores. Micelles thus provide a handy vehicle for carrying water-insoluble substances through the watery luminal contents. The most important lipid-soluble substances carried within micelles are the products of fat digestion (monoglycerides and free fatty acids) as well as fat-soluble vitamins, which are all transported to their sites of absorption by this means. If they did not hitch a ride in the water-soluble micelles, these nutrients would float on the surface of the aqueous chyme (just as oil floats on top of water), never reaching the absorptive surfaces of the small intestine.

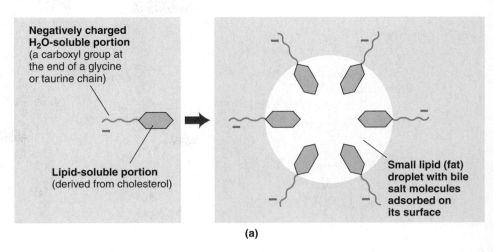

(a)

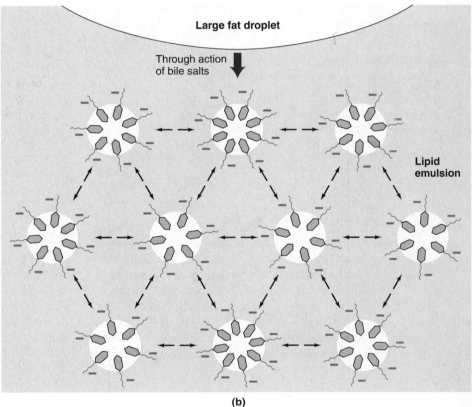

(b)

● FIGURE 16-18

Schematic structure and function of bile salts
(a) Schematic representation of the structure of bile salts and their adsorption on the surface of a small fat droplet. A bile salt consists of a lipid-soluble portion that dissolves in the fat droplet and a negatively charged, water-soluble portion that projects from the surface of the droplet. (b) Formation of a lipid emulsion through the action of bile salts. When a large fat droplet is broken up into smaller fat droplets by intestinal contractions, bile salts adsorb on the surface of the small droplets, creating "shells" of negatively charged, water-soluble bile salt components that cause the fat droplets to repel each other. This action holds the fat droplets apart and prevents them from recoalescing, thereby increasing the surface area of exposed fat available for digestion by pancreatic lipase.

In addition, cholesterol, a highly water-insoluble substance, dissolves in the micelle's hydrophobic core. This mechanism is important in cholesterol homeostasis. The amount of cholesterol that can be carried in micellar formation depends on the relative amount of bile salts and lecithin in comparison to

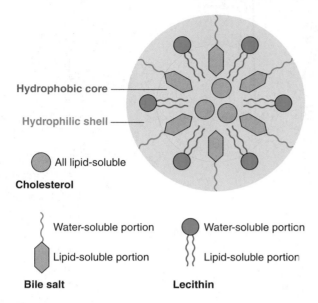

Hydrophobic core ——

Hydrophilic shell ——

⬤ All lipid-soluble

Cholesterol

Water-soluble portion

Lipid-soluble portion

Bile salt

Water-soluble portion

Lipid-soluble portion

Lecithin

● **FIGURE 16-19**

Schematic representation of a micelle
Bile constituents (bile salts, lecithin, and cholesterol) aggregate to form micelles that consist of a hydrophilic (water-soluble) shell and a hydrophobic (lipid-soluble) core. Because the outer shell of a micelle is water soluble, the products of fat digestion, which are not water soluble, can be carried through the watery luminal contents to the absorptive surface of the small intestine by dissolving in the micelle's lipid-soluble core.

cholesterol. When cholesterol secretion by the liver is out of proportion to bile salt and lecithin secretion (either too much cholesterol or too little bile salts and lecithin), the excess cholesterol in the bile precipitates into microcrystals that can aggregate into **gallstones.** One treatment of cholesterol-containing gallstones involves ingestion of bile salts to increase the bile salt pool in an attempt to dissolve the cholesterol stones. Only about 75% of gallstones are derived from cholesterol, however. The other 25% are made up of abnormal precipitates of another bile constituent, bilirubin.

▌ Bilirubin is a waste product excreted in the bile.

Bilirubin, the other major constituent of bile, does not play a role in digestion at all but instead is a waste product excreted in the bile. Bilirubin is the primary bile pigment derived from the breakdown of worn-out red blood cells. The typical life span of a red blood cell in the circulatory system is 120 days. Worn-out red blood cells are removed from the blood by the macrophages that line the liver sinusoids and reside in other areas in the body. Bilirubin is the end product resulting from the degradation of the heme (iron-containing) portion of the hemoglobin contained within these old red blood cells. This bilirubin is extracted from the blood by the hepatocytes and is actively excreted into the bile.

Bilirubin is a yellow pigment that gives bile its yellow color. Within the intestinal tract, this pigment is modified by bacterial enzymes, giving rise to the characteristic brown color of the feces. When bile secretion does not occur, as when the bile duct is completely obstructed by a gallstone, the feces are grayish white. A small amount of bilirubin is normally reabsorbed by the intestine back into the blood, and when it is eventually excreted in the urine, it is largely responsible for the urine's yellow color. The kidneys are unable to excrete bilirubin until after it has been modified during its passage through the liver and intestine.

If bilirubin is formed more rapidly than it can be excreted, it accumulates in the body and causes **jaundice.** Patients with this condition appear yellowish, with this color being seen most easily in the whites of their eyes. Jaundice can be brought about in three different ways:

1. *Prehepatic* (the problem occurs "before the liver"), or *hemolytic, jaundice* is due to excessive breakdown (hemolysis) of red blood cells, which results in the liver being presented with more bilirubin than it is capable of excreting.

2. *Hepatic* (the problem is the "liver") *jaundice* occurs when the liver is diseased and is unable to deal with even the normal load of bilirubin.

3. *Posthepatic* (the problem occurs "after the liver"), or *obstructive, jaundice* occurs when the bile duct is obstructed, such as by a gallstone, so that bilirubin cannot be eliminated in the feces.

▌ Bile salts are the most potent stimulus for increased bile secretion.

Bile secretion may be increased by chemical, hormonal, and neural mechanisms:

• *Chemical mechanism (bile salts).* Any substance that increases bile secretion by the liver is called a **choleretic.** The most potent choleretic is bile salts themselves. Between meals, bile is stored in the gallbladder, but during a meal bile is emptied into the duodenum as the gallbladder contracts. After bile salts participate in fat digestion and absorption, they are reabsorbed and returned by the enterohepatic circulation to the liver, where they act as potent choleretics to stimulate further bile secretion. Therefore, during a meal, when bile salts are needed and being used, bile secretion by the liver is enhanced.

• *Hormonal mechanism (secretin).* Besides increasing the aqueous $NaHCO_3$ secretion by the pancreas, secretin also stimulates an aqueous alkaline bile secretion by the liver ducts without any corresponding increase in bile salts.

• *Neural mechanism (vagus nerve).* Vagal stimulation of the liver plays a minor role in bile secretion during the cephalic phase of digestion, promoting an increase in liver bile flow before food ever reaches the stomach or intestine.

▌ The gallbladder stores and concentrates bile between meals and empties during meals.

Even though the factors just described increase bile secretion by the liver during and after a meal, bile secretion by the liver occurs continuously. Between meals the secreted bile is shunted into the gallbladder, where it is stored and concentrated. Active transport of salt out of the gallbladder, with water following osmotically, results in a 5 to 10 times concentration of the organic constituents. Because the gallbladder stores this concentrated bile, it is the primary site for precipitation of concentrated bile constituents into gallstones. Fortunately, the gallbladder does not play an essential digestive role, so its removal as a treatment for gallstones or other gallbladder disease presents no particular problem. The bile secreted between meals is stored instead in the common bile duct, which becomes dilated.

During digestion of a meal, when chyme reaches the small intestine, the presence of food, especially fat products, in the duodenal lumen triggers the release of CCK. This hormone stimulates contraction of the gallbladder and relaxation of the sphincter of Oddi, so bile is discharged into the duodenum, where it appropriately aids in the digestion and absorption of the fat that initiated the release of CCK.

▌ Hepatitis and cirrhosis are the most common liver disorders.

Hepatitis is an inflammatory disease of the liver that results from a variety of causes, including viral infection or exposure to toxic agents, including alcohol, carbon tetrachloride, and certain tranquilizers. Hepatitis ranges in severity from mild, reversible symptoms to acute massive liver damage, with possible imminent death resulting from acute hepatic failure.

Repeated or prolonged hepatic inflammation, usually in association with chronic alcoholism, can lead to **cirrhosis,** a condition in which damaged hepatocytes are permanently replaced by connective tissue. Liver tissue has the ability to regenerate, normally undergoing a gradual turnover of cells. If part of the hepatic tissue is destroyed, the lost tissue can be replaced by an increase in the rate of cell division. There is a limit, however, to how rapidly hepatocytes can be replaced. In addition to hepatocytes, a small number of fibroblasts (connective tissue cells) are dispersed between the hepatic plates and form a supporting framework for the liver. If the liver is exposed to toxic substances such as alcohol so often that new hepatocytes cannot be generated rapidly enough to replace the damaged cells, the sturdier fibroblasts take advantage of the situation and overproduce. This extra connective tissue leaves little space for the hepatocytes' regrowth. Thus, as cirrhosis develops slowly over time, active liver tissue is gradually reduced, leading eventually to chronic liver failure.

Having looked at the accessory digestive organs that empty their exocrine products into the small intestine lumen, we are now going to examine the contributions of the small intestine itself.

SMALL INTESTINE

The **small intestine** is the site where most digestion and absorption take place. No further digestion is accomplished after the luminal contents pass beyond the small intestine, and no further absorption of nutrients occurs, although the large intestine does absorb small amounts of salt and water. The small intestine lies coiled within the abdominal cavity, extending between the stomach and the large intestine. It is arbitrarily divided into three segments—the **duodenum,** the **jejunum,** and the **ileum.**

As usual, we will examine motility, secretion, digestion, and absorption in the small intestine, in that order. Small intestine motility includes segmentation and the migrating motility complex. We will consider segmentation first.

▌ Segmentation contractions mix and slowly propel the chyme.

Segmentation, the small intestine's primary method of motility during digestion of a meal, both mixes and slowly propels the chyme. Segmentation consists of oscillating, ringlike contractions of the circular smooth muscle along the length of the small intestine; between the contracted segments are relaxed areas containing a small bolus of chyme. The contractile rings occur every few centimeters, dividing the small intestine into segments like a chain of sausages. These contractile rings do not sweep along the length of the intestine as peristaltic waves do. Rather, after a brief period of time, the contracted segments relax, and ringlike contractions appear in the previously relaxed areas (● Figure 16-20). The new contraction forces the chyme in a previously relaxed segment to move in both directions into the now relaxed adjacent segments. A newly relaxed segment therefore receives chyme from both the contracting segment immediately ahead of it and the one immediately behind it. Shortly thereafter, the areas of contraction and relaxation alternate again. In this way, the chyme is chopped, churned, and thoroughly mixed.

Initiation and control of segmentation

Segmentation contractions are initiated by the small intestine's pacesetter cells, which produce a basic electrical rhythm (BER) similar to the gastric BER responsible for peristalsis in the stomach. If the small intestine BER brings the circular

● **FIGURE 16-20**

Segmentation
Segmentation consists of ringlike contractions along the length of the small intestine. Within a matter of seconds, the contracted segments relax and the previously relaxed areas contract. These oscillating contractions thoroughly mix the chyme within the small intestine lumen.

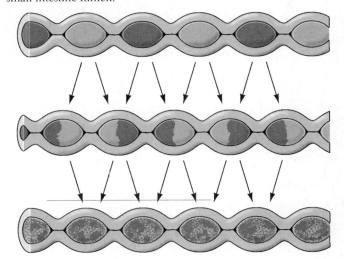

smooth-muscle layer to threshold, segmentation contractions are induced, with the frequency of segmentation following the frequency of the BER.

The circular smooth muscle's degree of responsiveness and thus the intensity of segmentation contractions can be influenced by distension of the intestine, by the hormone gastrin, and by extrinsic nerve activity. All these factors influence the excitability of the small-intestine smooth muscle cells by moving the starting potential around which the BER oscillates closer to or farther from the threshold. Segmentation is slight or absent between meals but becomes very vigorous immediately after a meal. Both the duodenum and the ileum start to segment simultaneously when the meal first enters the small intestine. The duodenum starts to segment primarily in response to local distension caused by the presence of chyme. Segmentation of the empty ileum, in contrast, appears to be brought about by gastrin, which is secreted in response to the presence of chyme in the stomach, a mechanism known as the **gastroileal reflex.** Extrinsic nerves can modify the strength of these contractions. Parasympathetic stimulation enhances segmentation, whereas sympathetic stimulation depresses segmental activity.

Functions of segmentation

These contractions can be compared to squeezing a pastry tube with your hands to mix the contents. This mixing serves the dual functions of mixing the chyme with the digestive juices secreted into the small intestine lumen and exposing all the chyme to the absorptive surfaces of the small intestine mucosa.

Segmentation not only accomplishes mixing but also is the primary factor responsible for slowly moving chyme through the small intestine. How can this be, when each segmental contraction propels chyme both forward and backward? The chyme slowly progresses forward because the frequency of segmentation declines along the length of the small intestine. The pacesetter cells in the duodenum spontaneously depolarize faster than those farther down the tract, with the pacesetter cells in the terminal ileum exhibiting the slowest rate of spontaneous depolarization. Segmentation contractions occur in the duodenum at a rate of 12 per minute, compared to only 9 per minute in the terminal ileum. Because segmentation occurs with greater frequency in the upper part of the small intestine than in the lower part, more chyme on average is pushed forward than is pushed backward. As a result, chyme is moved very slowly from the upper to the lower part of the small intestine, being shuffled back and forth to accomplish thorough mixing and absorption in the process. This slow propulsive mechanism is advantageous because it allows ample time for the digestive and absorptive processes to take place. The contents usually take 3 to 5 hours to move through the small intestine.

▌ The migrating motility complex sweeps the intestine clean between meals.

When most of the meal has been absorbed, segmentation contractions cease and are replaced between meals by the **migrating motility complex,** or "**intestinal housekeeper.**" This between-meal motility consists of weak, repetitive peristaltic waves that move a short distance down the intestine before dying out. The waves start at the stomach and migrate down the intestine; that is, each new peristaltic wave is initiated at a site a little farther down the small intestine. These short peristaltic waves take about 100 to 150 minutes to gradually migrate from the stomach to the end of the small intestine, with each contraction "sweeping" any remnants of the preceding meal plus mucosal debris and bacteria forward toward the colon, just like a good "intestinal housekeeper." After the end of the small intestine is reached, the cycle begins again and continues to repeat itself until the next meal. The hormone **motilin** secreted by endocrine cells of the small intestine mucosa regulates the migrating motility complex between meals. When the next meal arrives, segmental activity is triggered again, and the migrating motility complex ceases.

▌ The ileocecal juncture prevents contamination of the small intestine by colonic bacteria.

At the juncture between the small and large intestines, the last portion of the ileum empties into the cecum (● Figure 16-21). Two factors contribute to this region's ability to act as a barrier between the small and large intestines. First, the anatomic arrangement is such that valvelike folds of tissue protrude from

● FIGURE 16-21

Control of the ileocecal valve/sphincter
The juncture between the ileum and large intestine consists of the ileocecal valve, which is surrounded by thickened smooth muscle, the ileocecal sphincter. Pressure on the cecal side pushes the valve closed and contracts the sphincter, preventing the bacteria-laden colonic contents from contaminating the nutrient-rich small intestine. The valve/sphincter opens and allows ileal contents to enter the large intestine in response to pressure on the ileal side of the valve and to the hormone gastrin secreted as a new meal enters the stomach.

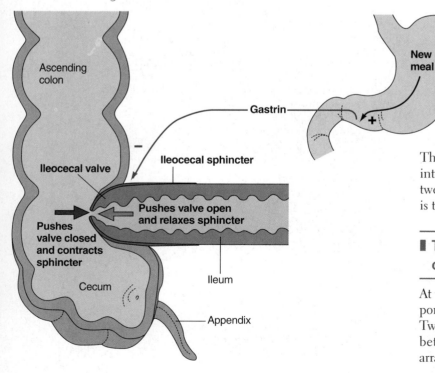

the ileum into the lumen of the cecum. When the ileal contents are pushed forward, this **ileocecal valve** is easily pushed open, but the folds of tissue are forcibly closed when the cecal contents attempt to move backward. Second, the smooth muscle within the last several centimeters of the ileal wall is thickened, forming a sphincter that is under neural and hormonal control. Most of the time this **ileocecal sphincter** remains at least mildly constricted. Pressure on the cecal side of the sphincter causes it to contract more forcibly; distension of the ileal side causes the sphincter to relax, a reaction mediated by the intrinsic plexuses in the area. In this way, the ileocecal sphincter prevents the bacteria-laden contents of the large intestine from contaminating the small intestine and at the same time allows the ileal contents to pass into the colon. If the colonic bacteria gained access to the nutrient-rich small intestine, they would multiply rapidly. Relaxation of the sphincter is enhanced through the release of gastrin at the onset of a meal, when increased gastric activity is taking place. This relaxation allows the undigested fibers and unabsorbed solutes from the preceding meal to be moved forward as the new meal enters the tract.

■ Small-intestine secretions do not contain any digestive enzymes.

Each day the exocrine gland cells located in the small intestine mucosa secrete into the lumen about 1.5 liters of an aqueous salt and mucus solution known as the **succus entericus** (*succus* means "juice"; *entericus* means "of the intestine"). No digestive enzymes are secreted into this intestinal juice. The small intestine does synthesize digestive enzymes, but they act intracellularly within the borders of the epithelial cells that line the lumen instead of being secreted directly into the lumen.

Are any functions served by this small-intestine secretion, with no digestive enzymes involved? The mucus in the secretion provides protection and lubrication. Furthermore, this aqueous secretion provides an abundance of H_2O to participate in the enzymatic digestion of food. Recall that digestion involves hydrolysis—bond breakage by reaction with H_2O—which proceeds most efficiently when all the reactants are in solution.

The regulation of small-intestine secretion is not clearly understood. Secretion of succus entericus does increase after a meal. The most effective stimulus for secretion appears to be local stimulation of the small-intestine mucosa by the presence of chyme.

■ The small-intestine enzymes complete digestion intracellularly.

Digestion within the small-intestine lumen is accomplished by the pancreatic enzymes, with fat digestion being enhanced by bile secretion. As a result of pancreatic enzymatic activity, fats are completely reduced to their absorbable units of monoglycerides and free fatty acids, proteins are broken down into small peptide fragments and some amino acids, and carbohydrates are reduced to disaccharides and some monosaccharides. Thus fat digestion is completed within the small-intestine lumen,

but carbohydrate and protein digestion have not been brought to completion.

Special actin-stiffened, hairlike projections on the luminal surface of the small intestine epithelial cells form the **brush border** (see p. 50). The brush-border plasma membrane contains three categories of membrane-bound enzymes:

1. **Enterokinase,** which activates the pancreatic enzyme trypsinogen.

2. The **disaccharidases (maltase, sucrase,** and **lactase),** which complete carbohydrate digestion by hydrolyzing the remaining disaccharides (maltose, sucrose, and lactose, respectively) into their constituent monosaccharides.

3. The **aminopeptidases,** which hydrolyze the small peptide fragments into their amino acid components, thereby completing protein digestion.

Thus, carbohydrate and protein digestion are completed intracellularly within the confines of the brush border. (▲ Table 16-7 provides a summary of the digestive processes for the three major categories of nutrients.)

Lactose intolerance

A fairly common disorder, **lactose intolerance,** involves a deficiency of lactase, the disaccharidase specific for the digestion of lactose, or milk sugar. Most children under 4 years of age have adequate lactase, but this may be gradually lost so that in many adults, lactase activity is diminished or absent. When milk or dairy products (except those in which lactose has already been digested by bacterial action during processing, such as some kinds of cheese and yogurt) are consumed by a person with lactase deficiency, the undigested lactose remains in the lumen and has several related consequences. First, accumulation of undigested lactose creates an osmotic gradient that draws H_2O into the intestinal lumen. Second, bacteria residing in the large intestine possess lactose-splitting ability, so they eagerly attack the lactose as an energy source, producing large quantities of CO_2 and methane gas in the process. Distension of the intestine by both fluid and gas produces pain (cramping) and diarrhea. Depending on the extent of the lactase deficiency and the quantity of lactose ingested, symptoms can vary from mild abdominal discomfort to severe dehydrating diarrhea. Infants with lactose intolerance may also suffer from malnutrition.

Finally we are ready to discuss absorption of nutrients. Up to this point, no food, water, or electrolytes have been absorbed.

■ The small intestine is remarkably well adapted for its primary role in absorption.

All products of carbohydrate, protein, and fat digestion, as well as most of the ingested electrolytes, vitamins, and water, are normally absorbed by the small intestine indiscriminately. Usually, only the absorption of calcium and iron is adjusted to the body's needs. Thus, the more food that is consumed, the more that will be digested and absorbed, as people who are trying to control their weight are all too painfully aware.

Most absorption occurs in the duodenum and jejunum; very little occurs in the ileum, not because the ileum does not have absorptive capacity but because most absorption has al-

Digestive Processes for the Three Major Categories of Nutrients

Nutrients	Enzymes for Digesting Nutrient	Source of Enzymes	Site of Action of Enzymes	Action of Enzymes	Absorbable Units of Nutrients
Carbohydrate	Amylase	Salivary glands	Mouth and body of stomach	Hydrolyzes polysaccharides to disaccharides	
		Exocrine pancreas	Small-intestine lumen		
	Disaccharidases (maltase, sucrase, lactase)	Small-intestine epithelial cells	Small-intestine brush border	Hydrolyze disaccharides to monosaccharides	Monosaccharides, especially glucose
Protein	Pepsin	Stomach chief cells	Stomach antrum	Hydrolyzes protein to peptide fragments	
	Trypsin, chymotrypsin carboxypeptidase	Exocrine pancreas	Small-intestine lumen	Attack different peptide fragments	
	Aminopeptidases	Small-intestine epithelial cells	Small-intestine brush border	Hydrolyze peptide fragments to amino acids	Amino acids and a few small peptides
Fat	Lipase	Exocrine pancreas	Small-intestine lumen	Hydrolyzes triglycerides to fatty acids and monoglycerides	Fatty acids and monoglycerides
	Bible salts (not an enzyme)	Liver	Small-intestine lumen	Emulsify large fat globules for attack by pancreatic lipase	

ready been accomplished before the intestinal contents reach the ileum. The small intestine has an abundant reserve absorptive capacity. In fact, about 50% of the small intestine can be removed with little interference to absorption—with one exception. If the terminal ileum is removed, vitamin B_{12} and bile salts are not properly absorbed, because the specialized transport mechanisms for these two substances are located only in this region. All other substances can be absorbed throughout the length of the small intestine.

The mucous lining of the small intestine is remarkably well adapted for its special absorptive function for two reasons: (1) It has a very large surface area; and (2) the epithelial cells in this lining possess a variety of specialized transport mechanisms.

Adaptations that increase the small intestine's surface area

The following special modifications of the small intestine mucosa greatly increase the surface area available for absorption (● Figure 16-22):

• The inner surface of the small intestine is thrown into circular folds that are visible to the naked eye and increase the surface area threefold.

• Projecting from this folded surface are microscopic fingerlike projections known as **villi**, which give the lining a velvety appearance and increase the surface area by another 10

times (● Figure 16-23). The surface of each villus is covered by epithelial cells interspersed occasionally with mucous cells.

• Even smaller hairlike projections known as **microvilli** (or the *brush border*) arise from the luminal surface of these epithelial cells, increasing the surface area another 20-fold. Each epithelial cell has as many as 3,000 to 6,000 of these microvilli, which are visible only with an electron microscope. It is within the membrane of this brush border that the small intestine enzymes perform their functions.

Altogether, the folds, villi, and microvilli provide the small intestine with a luminal surface area 600 times greater than it would have been if it were a tube of the same length and diameter lined by a flat surface. In fact, if the surface area of the small intestine were spread out flat, it would cover an entire tennis court.

Malabsorption (impairment of absorption) may be caused by damage to or reduction of the surface area of the small intestine. One of the most common causes is **gluten enteropathy.** In this condition, the person's small intestine is abnormally sensitive to gluten, a protein constituent of wheat and some other grains. Through an unknown mechanism, exposure to gluten damages the intestinal villi: the normally luxuriant array of villi is reduced, the mucosa becomes flattened, and the brush border becomes short and stubby (● Figure 16-24). Because this loss of villi decreases the surface area available

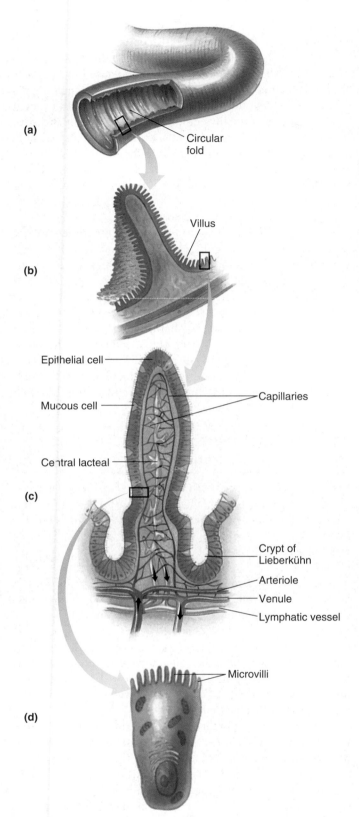

(a)

Circular fold

Villus

(b)

Epithelial cell

Mucous cell

Capillaries

Central lacteal

(c)

Crypt of Lieberkühn

Arteriole

Venule

Lymphatic vessel

Microvilli

(d)

● FIGURE 16-22

Small-intestine absorptive surface
(a) Gross structure of the small intestine. (b) One of the circular folds of the small-intestine mucosa, which collectively increase the absorptive surface area threefold. (c) Microscopic fingerlike projection known as a *villus*. Collectively, the villi increase the surface area another tenfold. (d) Electron microscope view of a villus epithelial cell, depicting the presence of microvilli on its luminal border; the microvilli increase the surface area another 20-fold. Altogether, these surface modifications increase the small intestine's absorptive surface area 600-fold.

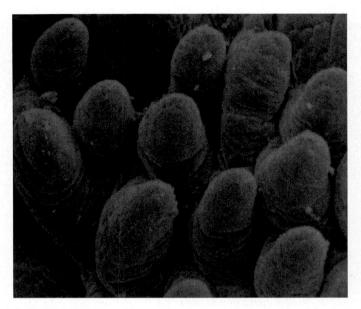

● FIGURE 16-23

Scanning electron micrograph of villi projecting from the small-intestine mucosa

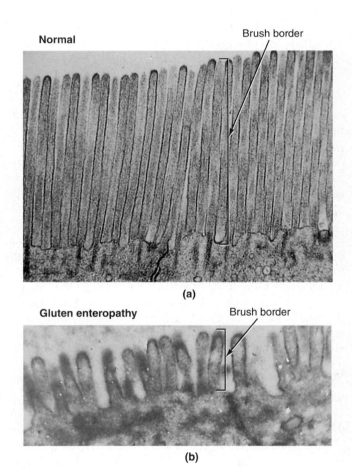

Normal

Brush border

(a)

Gluten enteropathy

Brush border

(b)

● FIGURE 16-24

Reduction in the brush border with gluten enteropathy
(a) Electron micrograph of the brush border of a small-intestine epithelial cell in a normal individual. (b) Electron micrograph of the short, stubby brush border of a small intestine epithelial cell in a patient with gluten enteropathy.

for absorption, absorption of all nutrients is impaired. The condition is treated by eliminating gluten from the diet.

Structure of a villus

Absorption across the digestive tract wall involves transepithelial transport similar to movement of material across the kidney tubules (see p. 526). Each villus has the following major components (● Figure 16-22c):

- *Epithelial cells that cover the surface of the villus.* The epithelial cells are joined at their lateral borders by tight junctions, which limit passage of luminal contents between the cells, although the tight junctions in the small intestine are "leakier" than those in the stomach. Within their luminal brush borders, these epithelial cells possess carriers for absorption of specific nutrients and electrolytes from the lumen as well as the intracellular digestive enzymes that complete carbohydrate and protein digestion.
- A *connective tissue core.* This core is formed by the lamina propria.
- A *capillary network.* Each villus is supplied by an arteriole that breaks up into a capillary network within the villus core. The capillaries rejoin to form a venule that drains away from the villus.
- A *terminal lymphatic vessel.* Each villus is supplied by a single blind-ended lymphatic vessel known as the **central lacteal**, which occupies the center of the villus core.

During the process of absorption, digested substances enter the capillary network or the central lacteal. To be absorbed, a substance must pass completely through the epithelial cell, diffuse through the interstitial fluid within the connective tissue core of the villus, and then cross the wall of a capillary or lymph vessel. Like renal transport, intestinal absorption may be an active or passive process, with active absorption involving energy expenditure during at least one of the steps in the transepithelial transport process.

■ The mucosal lining experiences rapid turnover.

Dipping down into the mucosal surface between the villi are shallow invaginations known as the **crypts of Lieberkühn** (● Figure 16-22c). Unlike the gastric pits, these intestinal crypts do not secrete digestive enzymes, but they do secrete water and electrolytes, which, along with the mucus secreted by the cells on the villus surface, constitute the succus entericus.

Furthermore, the crypts function as "nurseries." The epithelial cells lining the small intestine slough off and are replaced at a rapid rate as a result of high mitotic activity of *stem cells* in the crypts. New cells that are continually being produced in the crypts migrate up the villi and, in the process, push off the older cells at the tips of the villi into the lumen. In this manner, more than 100 million intestinal cells are shed per minute. The entire trip from crypt to tip averages about three days, so the epithelial lining of the small intestine is replaced approximately every three days. Because of this high rate of cell division, the crypt stem cells are very sensitive to damage by radiation and anticancer drugs, both of which may inhibit cell division.

The new cells undergo several changes as they migrate up the villus. The concentration of brush border enzymes increases and the capacity for absorption improves, so the cells at the tip of the villus have the greatest digestive and absorptive capability. Just at their peak, these cells are pushed off by the newly migrating cells. Thus, the luminal contents are constantly exposed to cells that are optimally equipped to complete the digestive and absorptive functions efficiently. Furthermore, just as in the stomach, the rapid turnover of cells in the small intestine is essential because of the harsh luminal conditions. Cells exposed to the abrasive and corrosive luminal contents are easily damaged and cannot live for long, so they must be continually replaced by a fresh supply of newborn cells.

The old cells that are sloughed off into the lumen are not entirely lost to the body. These cells are digested, with the cell constituents being absorbed into the blood and reclaimed for synthesis of new cells, among other things.

In addition to the stem cells, **Paneth cells** are found in the crypts. Paneth cells serve a defensive function, safeguarding the stem cells. They produce two chemicals that thwart bacteria: (1) *lysozyme*, the bacterial-lysing enzyme also found in saliva, and (2) *defensins*, small proteins with antimicrobial powers.

We now turn our attention to the mechanisms through which the specific dietary constituents are normally absorbed.

■ Energy-dependent Na^+ absorption drives passive H_2O absorption.

Sodium may be absorbed both passively and actively. When the electrochemical gradient favors movement of Na^+ from the lumen to the blood, passive diffusion of Na^+ can occur *between* the intestinal epithelial cells through the "leaky" tight junctions into the interstitial fluid within the villus. Movement of Na^+ *through* the cells is energy dependent and involves two different carriers, similar to the process of Na^+ reabsorption across the kidney tubules (see pp. 527 and 531). Sodium passively enters the epithelial cells across the luminal border either by itself through Na^+ channels or in the company of glucose or amino acid by means of a cotransport carrier. Sodium is actively pumped out of the cell at the basolateral border into the interstitial fluid in the lateral spaces between the cells where they are not joined by tight junctions. From the interstitial fluid, Na^+ diffuses into the capillaries.

As with the renal tubules in the early portion of the nephron, the absorption of Cl^-, H_2O, glucose, and amino acids from the small intestine is linked to this energy-dependent Na^+ absorption. Chloride passively follows down the electrical gradient created by Na^+ absorption and can be actively absorbed as well if needed. Most H_2O absorption in the digestive tract depends on the active carrier that pumps Na^+ into the lateral spaces, resulting in a concentrated area of high osmotic pressure in that localized region between the cells, similar to the situation in the kidneys (see p. 533). This localized high osmotic pressure induces H_2O to move from the lumen through the cell (and possibly from the lumen through the leaky tight junction) into the lateral space. Water entering the space reduces the osmotic pressure but raises the hydrostatic (fluid) pressure. As a result, H_2O is flushed out of the lateral space into the interior of the villus, where it is picked up by the capillary network. Meanwhile, more Na^+ is pumped into the lateral space to encourage more H_2O absorption.

■ Carbohydrate and protein are both absorbed by secondary active transport and enter the blood.

Absorption of the digestion end products of both carbohydrates and proteins involves special carrier-mediated transport systems that require energy expenditure and Na^+ cotransport, and both categories of end products are absorbed into the blood.

Carbohydrate absorption

Dietary carbohydrate is presented to the small intestine for absorption mainly in the forms of the disaccharides maltose (the product of polysaccharide digestion), sucrose, and lactose (▲ Table 16-1 and ● Figure 16-25). The disaccharidases located in the brush borders of the small intestine cells further reduce these disaccharides into the absorbable monosaccharide units of glucose, galactose, and fructose.

Glucose and galactose are both absorbed by *secondary active transport*, in which cotransport carriers on the luminal border transport both the monosaccharide and Na^+ from the lumen into the interior of the intestinal cell. The operation of these cotransport carriers, which do not directly use energy themselves, depends on the Na^+ concentration gradient established by the energy-consuming basolateral $Na^+ - K^+$ pump (see p. 83). Glucose (or galactose), having been concentrated in the cell by the cotransport carriers, leaves the cell down its concentration gradient by means of a passive carrier in the basolateral border to enter the blood within the villus. In addition to glucose being absorbed through the cells by means of the cotransport carrier, recent evidence suggests that a significant amount of glucose crosses the epithelial barrier through "leaky" tight junctions between the epithelial cells. Fructose is absorbed into the blood solely by facilitated diffusion (passive carrier-mediated transport; see p. 80).

Protein absorption

Not only are ingested proteins digested and absorbed, but endogenous ("within the body") proteins that have entered the digestive tract lumen from the three following sources are digested and absorbed as well:

1. Digestive enzymes, all of which are proteins, that have been secreted into the lumen.
2. Proteins within the cells that are pushed off from the villi into the lumen during the process of mucosal turnover.
3. Small amounts of plasma proteins that normally leak from the capillaries into the digestive tract lumen.

About 20 to 40 g of endogenous protein enter the lumen each day from these three sources. This quantity can amount to more than half of the protein presented to the small intestine for digestion and absorption. All endogenous proteins must be digested and absorbed along with the dietary proteins to prevent depletion of the body's protein stores. The amino acids absorbed from both food and the endogenous protein are used primarily to synthesize new protein in the body.

The protein presented to the small intestine for absorption is primarily in the form of amino acids and a few small peptide fragments (● Figure 16-26). Amino acids are absorbed across the intestinal cells by secondary active transport, similar to glucose and galactose absorption. Thus, glucose, galactose, and amino acids all get a "free ride" in on the energy expended for Na^+ transport. Small peptides gain entry by means of a different carrier and are broken down into their constituent amino acids by the aminopeptidases in the brush borders or by intracellular peptidases. Like monosaccharides, amino acids enter the capillary network within the villus.

■ Digested fat is absorbed passively and enters the lymph.

Fat absorption is quite different from carbohydrate and protein absorption, because the insolubility of fat in water presents a special problem. Fat must be transferred from the watery chyme through the watery body fluids, even though fat is not water soluble. Therefore, fat must undergo a series of physical and chemical transformations to circumvent this problem during its digestion and absorption (● Figure 16-27).

A review of fat emulsification and digestion

When the stomach contents are emptied into the duodenum, the ingested fat is aggregated into large, oily triglyceride droplets that float in the chyme. Recall that through the bile salts' detergent action in the small intestine, the large droplets are dispersed into a lipid emulsification of small droplets, exposing a much greater surface area of fat for digestion by pancreatic lipase. The products of lipase digestion (monoglycerides and free fatty acids) are also not very water soluble, so very little of these end products of fat digestion can diffuse through the aqueous chyme to reach the absorptive lining. However, biliary components facilitate absorption of these fatty end products through formation of micelles.

Fat absorption

Remember that micelles are water-soluble particles that can carry the end products of fat digestion within their lipid-soluble interiors. Once these micelles reach the luminal membranes of the epithelial cells, the monoglycerides and free fatty acids passively diffuse from the micelles through the lipid component of the epithelial cell membranes to enter the interior of these cells. As these fat products leave the micelles and are absorbed across the epithelial cell membranes, the micelles can pick up more monoglycerides and free fatty acids, which have been produced from digestion of other triglyceride molecules in the fat emulsion.

Bile salts continuously repeat their fat-solubilizing function down the length of the small intestine until all the fat is absorbed. Then the bile salts themselves are reabsorbed in the terminal ileum by special active transport. This is an efficient process, because relatively small amounts of bile salts can fa-

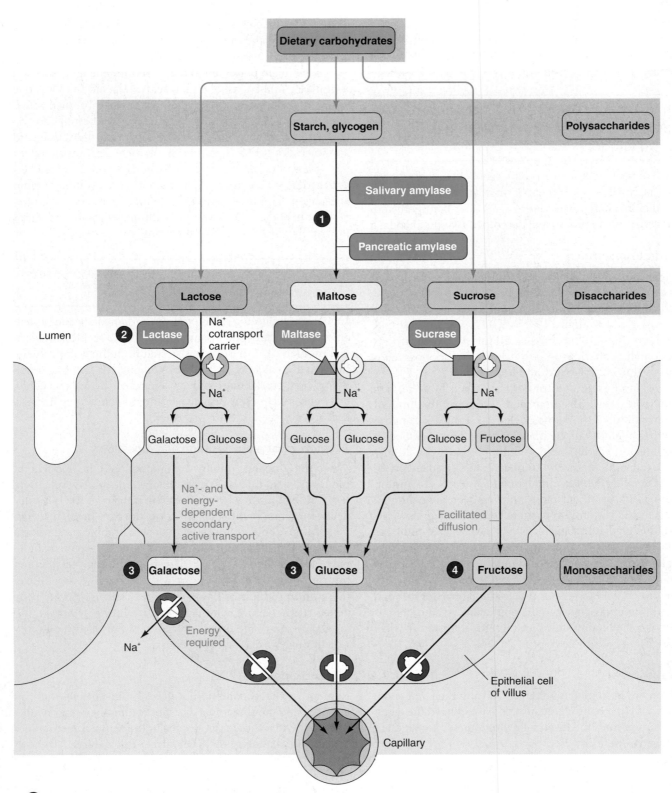

1. The dietary polysaccharides starch and glycogen are converted into the disaccharide maltose through the action of salivary and pancreatic amylase.

2. Maltose and the dietary disaccharides lactose and sucrose are converted to their respective monosaccharides by the disaccharidases (maltase, lactase, and sucrase) located in the brush borders of the small-intestine epithelial cells.

3. The monosaccharides glucose and galactose are absorbed into the interior of the cell and eventually enter the blood by means of Na^+- and energy-dependent secondary active transport.

4. The monosaccharide fructose is absorbed into the blood by passive facilitated diffusion.

● FIGURE 16-25
Carbohydrate digestion and absorption

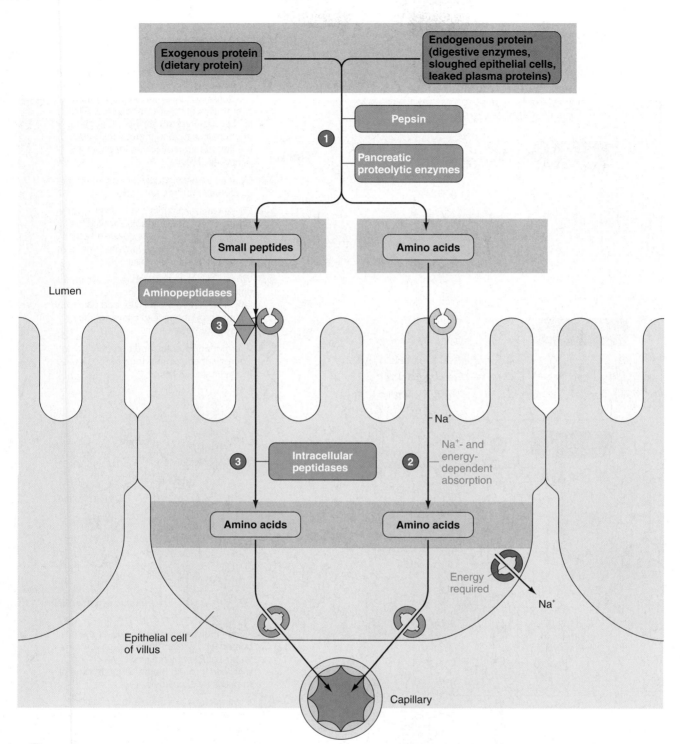

1. Dietary and endogenous proteins are hydrolyzed to their constituent amino acids and a few small peptide fragments by gastric pepsin and the pancreatic proteolytic enzymes.

2. Amino acids are absorbed into the small-intestine epithelial cells and eventually enter the blood by means of Na+- and energy-dependent secondary active transport. Various amino acids are transported by carriers specific for them.

3. The small peptides, which are absorbed by a different type of carrier, are broken down into their amino acids by aminopeptidases in the epithelial cells' brush borders or by intracellular peptidases.

● FIGURE 16-26

Protein digestion and absorption

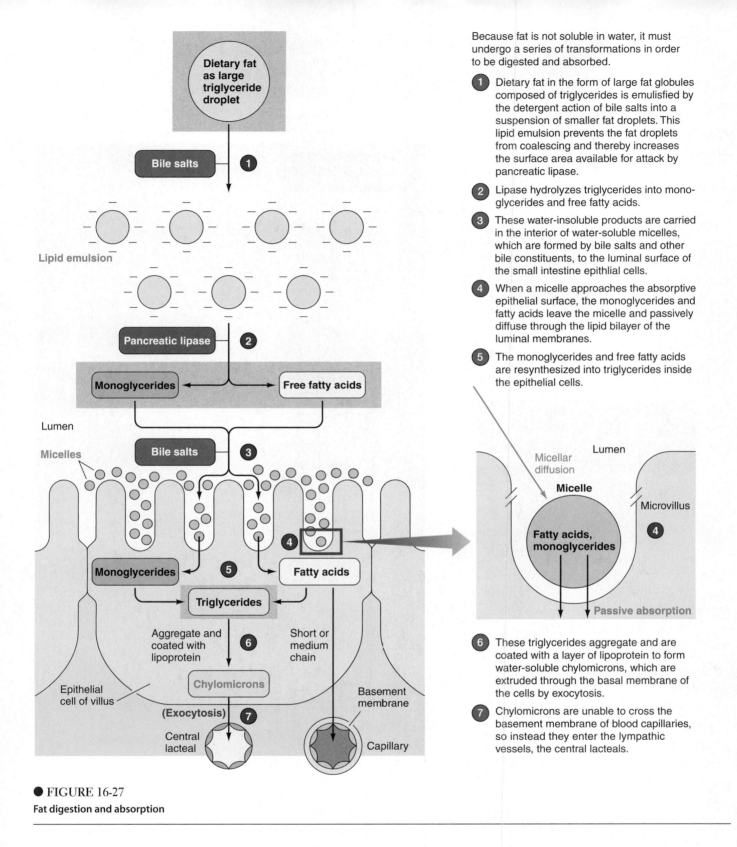

Because fat is not soluble in water, it must undergo a series of transformations in order to be digested and absorbed.

1. Dietary fat in the form of large fat globules composed of triglycerides is emulsified by the detergent action of bile salts into a suspension of smaller fat droplets. This lipid emulsion prevents the fat droplets from coalescing and thereby increases the surface area available for attack by pancreatic lipase.

2. Lipase hydrolyzes triglycerides into mono-glycerides and free fatty acids.

3. These water-insoluble products are carried in the interior of water-soluble micelles, which are formed by bile salts and other bile constituents, to the luminal surface of the small intestine epithlial cells.

4. When a micelle approaches the absorptive epithelial surface, the monoglycerides and fatty acids leave the micelle and passively diffuse through the lipid bilayer of the luminal membranes.

5. The monoglycerides and free fatty acids are resynthesized into triglycerides inside the epithelial cells.

6. These triglycerides aggregate and are coated with a layer of lipoprotein to form water-soluble chylomicrons, which are extruded through the basal membrane of the cells by exocytosis.

7. Chylomicrons are unable to cross the basement membrane of blood capillaries, so instead they enter the lymphatic vessels, the central lacteals.

● **FIGURE 16-27**

Fat digestion and absorption

cilitate digestion and absorption of large amounts of fat, with each bile salt performing its ferrying function repeatedly before it is reabsorbed.

Once within the interior of the epithelial cells, the mono-glycerides and free fatty acids are resynthesized into triglyc-erides. These triglycerides conglomerate into droplets and are coated with a layer of lipoprotein (synthesized by the endo-plasmic reticulum of the epithelial cell), which renders the fat droplets water soluble. The large, coated fat droplets, known as **chylomicrons,** are extruded by exocytosis from the epithe-lial cells into the interstitial fluid within the villus. The chylo-microns subsequently enter the central lacteals rather than the capillaries because of the structural differences between these two vessels. Capillaries have a basement membrane (an outer layer of polysaccharides) that prevents chylomicrons from entering, but the lymph vessels do not have this barrier. Thus, fat can be absorbed into the lymphatics but not directly into the blood. (Fatty acids with short- or medium-length carbon chains can enter the blood, but very few of these are eaten in the normal diet.)

The actual absorption or transfer of monoglycerides and free fatty acids from the chyme across the luminal membranes of the intestinal epithelial cells is a passive process, because the lipid-soluble fatty end products merely dissolve in and pass through the lipid portions of the membrane. However, the overall sequence of events necessary for fat absorption does require energy. For example, bile salts are actively secreted by the liver, and the resynthesis of triglycerides and formation of chylomicrons within the epithelial cells are active processes.

▌ Vitamin absorption is largely passive.

Water-soluble vitamins are primarily absorbed with water, whereas fat-soluble vitamins are carried in the micelles and absorbed passively with the end products of fat digestion. Absorption of some of the vitamins can also be accomplished by carriers, if necessary. Vitamin B_{12} is unique in that it must be in combination with gastric intrinsic factor for absorption by special transport in the terminal ileum.

▌ Iron and calcium absorption is regulated.

In contrast to the almost complete, unregulated absorption of other ingested electrolytes, the absorption of dietary iron and calcium may not be complete because it is subject to regulation that depends on the body's needs for these electrolytes.

Iron absorption

Iron is essential for hemoglobin production. The normal iron intake is typically 15 to 20 mg/day, yet a man usually absorbs about 0.5 to 1 mg/day into the blood, and a woman takes up slightly more, at 1.0 to 1.5 mg/day (women need more iron because they periodically lose iron in menstrual blood flow).

Two main steps are involved in absorption of iron into blood: (1) absorption of iron from the lumen into intestinal epithelial cells, and (2) absorption of iron from the epithelial cells into the blood (● Figure 16-28).

Iron is actively transported from the lumen into the epithelial cells, with women having about four times more active transport sites for iron than men. The extent to which ingested iron is taken up by the epithelial cells depends on the type of iron consumed (ferrous iron, Fe^{2+}, is absorbed more easily

● **FIGURE 16-28**
Iron absorption

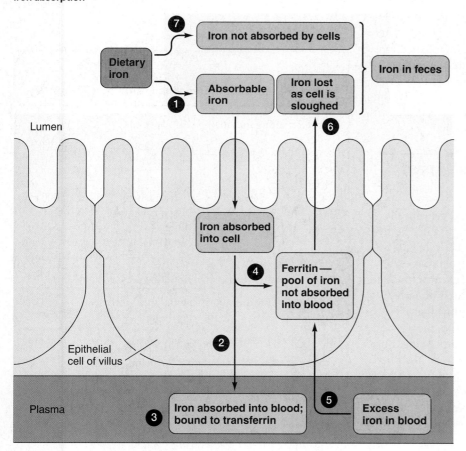

1. Only a portion of ingested iron is in a form that can be absorbed.

2. Dietary iron that is absorbed into the small-intestine epithelial cells and is immediately needed for red blood cell production is transferred into the blood.

3. In the blood, absorbed iron is carried to the bone marrow bound to transferrin, a plasma protein carrier.

4. Absorbed dietary iron that is not immediately needed is stored in the epithelial cells as ferritin, which cannot be transferred into the blood.

5. Excess iron in the blood can be dumped into the ferritin pool.

6. This unused iron is lost in the feces as the ferritin-containing epithelial cells are sloughed.

7. Furthermore, dietary iron that was not absorbed is also lost in the feces.

than ferric iron, Fe^{3+}). Also, the presence of other substances in the lumen can either promote or reduce iron absorption. For example, vitamin C increases iron absorption, primarily by reducing ferric to ferrous iron. Phosphate and oxalate, in contrast, combine with ingested iron to form insoluble iron salts that cannot be absorbed.

After active absorption into the small intestine epithelial cells, iron has two possible fates:

1. Iron needed immediately for production of red blood cells is absorbed into the blood for delivery to the bone marrow, the site of red blood cell production. Iron is transported in blood by a plasma protein carrier known as **transferrin.** The hormone responsible for stimulating red blood cell production, erythropoietin (see p. 395), is believed to also enhance iron absorption from the intestinal cells into the blood. The absorbed iron is then used in the synthesis of hemoglobin for the newly produced red blood cells.

2. Iron not immediately needed remains stored within the epithelial cells in a granular form called **ferritin,** which cannot be absorbed into the blood. If the blood level of iron is too high, excess iron may be dumped from the blood into this unabsorbable pool of ferritin in the intestinal epithelial cells. Iron stored as ferritin is lost in the feces within three days as the epithelial cells containing these granules are sloughed off during mucosal regeneration. Large amounts of iron in the feces give them a dark, almost black color.

Calcium absorption

The amount of calcium (Ca^{2+}) absorbed is also regulated. Absorption of Ca^{2+} is accomplished partly by passive diffusion but mostly by active transport. Vitamin D greatly stimulates this active transport. Vitamin D can exert this effect only after it has been activated in the liver and kidneys, a process that is enhanced by parathyroid hormone. Appropriately, secretion of parathyroid hormone increases in response to a fall in Ca^{2+} concentration in the blood. Normally, of the average 1,000 mg of Ca^{2+} taken in daily, only about two-thirds is absorbed in the small intestine, with the rest passing out in the feces.

Most absorbed nutrients immediately pass through the liver for processing.

The venules that leave the small intestine villi, along with those from the remainder of the digestive tract, empty into the hepatic portal vein, which carries the blood to the liver. Consequently, anything absorbed into the digestive capillaries first must pass through the hepatic biochemical factory before entering the general circulation. Thus the products of carbohydrate and protein digestion as well as the electrolytes and H_2O are channeled into the liver, where many of the energy-rich products are subjected to immediate metabolic processing. Furthermore, harmful substances that may have been absorbed are detoxified by the liver before gaining access to the general circulation. After passing through the portal circulation, the venous blood from the digestive system is emptied into the vena cava and returned to the heart to be distributed through-out the body, carrying glucose and amino acids for use by the tissues.

Fat, which cannot penetrate the intestinal capillaries, is picked up by the central lacteal and enters the lymphatic system instead, thereby bypassing the hepatic portal system. Contractions of the villi, accomplished by the muscularis mucosa, periodically compress the central lacteal and "milk" the lymph out of this vessel. The lymph vessels eventually converge to form the *thoracic duct*, a large lymph vessel that empties into the venous system within the chest. By this means, fat ultimately gains access to the circulatory system. The absorbed fat is carried by the systemic circulation to the liver and to other tissues of the body. Therefore, the liver does have a chance to act on the digested fat, but not until the fat has been diluted by blood in the general circulatory system. This dilution of fat presumably protects the liver from being inundated with more fat than it can handle at one time.

Extensive absorption by the small intestine keeps pace with secretion.

The small intestine normally absorbs about 9 liters of fluid per day in the form of H_2O and solutes, including the absorbable units of nutrients, vitamins, and electrolytes. How can that be, when humans normally ingest only about 1,250 ml of fluid and consume 1,250 g of solid food (80% of which is H_2O) per day (see p. 568)? ▲ Table 16-8 illustrates the tremendous daily absorptive accomplishments performed by the small intestine. Each day about 9,500 ml of H_2O and solutes enter the small intestine. Note that of this 9,500 ml, only 2,500 ml are ingested

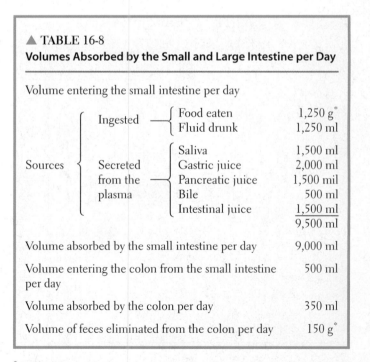

▲ **TABLE 16-8**
Volumes Absorbed by the Small and Large Intestine per Day

Volume entering the small intestine per day			
Sources	Ingested	Food eaten	1,250 g*
		Fluid drunk	1,250 ml
	Secreted from the plasma	Saliva	1,500 ml
		Gastric juice	2,000 ml
		Pancreatic juice	1,500 mil
		Bile	500 ml
		Intestinal juice	1,500 ml
			9,500 ml
Volume absorbed by the small intestine per day			9,000 ml
Volume entering the colon from the small intestine per day			500 ml
Volume absorbed by the colon per day			350 ml
Volume of feces eliminated from the colon per day			150 g*

*One milliliter of H_2O weighs 1g. Therefore, because a high percentage of food and feces is H_2O, we can roughly equate grams of food or feces with milliliters of fluid.

from the external environment. The remaining 7,000 ml (7 liters) of fluid consist of digestive juices that are essentially derived from plasma. Recall that plasma is the ultimate source of digestive secretions, because the secretory cells extract the necessary raw materials for their secretory product from the plasma. Considering that the entire plasma volume is only about 2.75 liters, it is obvious that absorption must closely parallel secretion to prevent the plasma volume from falling sharply.

Of the 9,500 ml of fluid entering the small intestine lumen per day, about 95%, or 9,000 ml of fluid, is normally absorbed by the small intestine back into the plasma, with only 500 ml of the small intestine contents passing on into the colon. Thus the body does not lose the digestive juices. After the constituents of the juices are secreted into the digestive tract lumen and perform their function, they are returned to the plasma. The only secretory product that escapes from the body is bilirubin, a waste product that must be eliminated.

▌ Biochemical balance among the stomach, pancreas, and small intestine is normally maintained.

Because the secreted juices are normally absorbed back into the plasma, the acid–base balance of the body is not altered by digestive processes. When secretion and absorption do not parallel each other, however, acid–base abnormalities can result. ● Figure 16-29 is a summary of the biochemical balance that normally exists among the stomach, pancreas, and small intestine. The arterial blood entering the stomach contains Cl^-, CO_2, H_2O, and Na^+, among other things. During HCl secretion, the gastric parietal cells extract Cl^-, CO_2, and H_2O from the plasma (the CO_2 and H_2O being essential for H^+ secretion) and add HCO_3^- to it (the HCO_3^- being formed in the process of generating H^+). The HCO_3^- diffuses into the plasma to replace the secreted Cl^- and to electrically balance the Na^+ in the plasma. Plasma Na^+ levels are not altered by gastric secretory processes. Because HCO_3^- is an alkaline ion, the venous blood leaving the stomach is more alkaline than the arterial blood delivered to it.

The overall acid–base balance of the body is not altered, however, because the pancreatic duct cells extract a comparable amount of HCO_3^- (along with Na^+) from the plasma to neutralize the acidic gastric chyme as it is emptied into the small intestine. Within the intestinal lumen, the alkaline pancreatic $NaHCO_3$ secretion neutralizes the gastric HCl secretion, yielding NaCl and H_2CO_3. The latter molecules form Na^+ and Cl^- plus CO_2 and H_2O, respectively. All four of these constituents (Na^+, Cl^-, CO_2, and H_2O) are absorbed by the intestinal epithelium into the plasma.

Note that these are exactly the same constituents present in the arterial blood entering the stomach. Thus, through these interactions, the body normally does not experience a net gain or loss of acid or base during digestion.

▌ Diarrhea results in loss of fluid and electrolytes.

When vomiting or diarrhea occur, these normal neutralization processes cannot take place. We have already described vomiting in the section on gastric motility. The other common digestive tract disturbance that can lead to a loss of fluid and an acid–

● **FIGURE 16-29**

Biochemical balance among the stomach, pancreas, and small intestine
When digestion and absorption proceed normally, no net loss or gain of acid or base or other chemicals from the body fluids occurs as a consequence of digestive secretions. The parietal cells of the stomach extract Cl^-, CO_2, and H_2O from and add HCO_3^- to the blood during HCl secretion. The pancreatic duct cells extract the HCO_3^- as well as Na^+ from the blood during $NaHCO_3$ secretion. Within the small intestine lumen, pancreatic $NaHCO_3$ neutralizes gastric HCl to form NaCl and H_2CO_3, which decomposes into CO_2 and H_2O. Subsequently, the intestinal cells absorb Na^+, Cl^-, CO_2, and H_2O into the blood, thereby replacing the constituents that were extracted from the blood during gastric and pancreatic secretion.

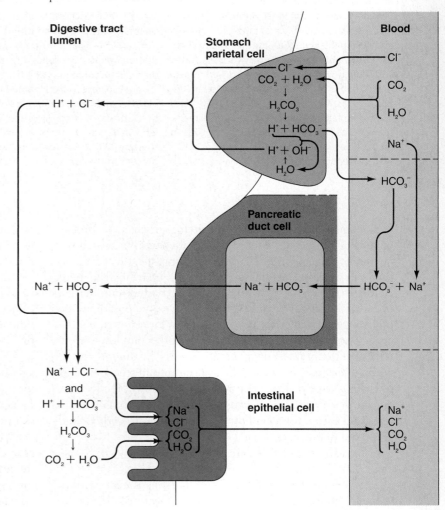

Oral Rehydration Therapy: Sipping a Simple Solution Saves Lives

Diarrhea-inducing micro-organisms such as *Vibrio cholera*, which causes cholera (see p. 71), are the leading cause of death in children under five worldwide. The problem is especially pronounced in developing countries, refugee camps, and elsewhere where poor sanitary conditions encourage the spread of the micro-organisms and medical supplies and health care personnel are scarce. Fortunately, a low-cost, easily obtainable, uncomplicated remedy—oral rehydration therapy—has been developed to combat potentially fatal diarrhea. This treatment exploits the secondary active-transport carriers located at the luminal border of the villus epithelial cells.

Let us examine the pathophysiology of life-threatening diarrhea and then see how simple oral rehydration therapy can save lives. During digestion of a meal, the crypt cells of the small intestine normally secrete succus entericus, a salt and mucus solution, into the lumen. These cells actively transport Cl^- into the lumen, promoting the parallel passive transport of Na^+ and H_2O from the blood into the lumen. The fluid provides the watery environment needed for enzymatic breakdown of ingested nutrients into absorbable units. Glucose and amino acids, the absorbable units of dietary carbohydrates and proteins, respectively, are absorbed by secondary active transport. This absorption mechanism uses Na^+ – glucose (or amino acid) cotransport carriers located at the luminal membrane of the villus epithelial cells (see p. 629). In addition, separate active Na^+ carriers not linked with nutrient absorption transfer Na^+, passively accompanied by Cl^- and H_2O, from the lumen into the blood. The net result of these various carrier activities is absorption of the secreted salt and H_2O along with the digested nutrients. Normally, absorption of salt and H_2O exceeds their secretion, so that not only are the secreted fluids salvaged but additional ingested salt and H_2O are absorbed as well.

Cholera and most diarrhea-inducing microbes cause diarrhea by stimulating the secretion of Cl^- and/or impairing the absorption of Na^+ (see p. 96). As a result, more fluid is secreted from the blood into the lumen than is subsequently transferred back into the blood. The excess fluid is lost in the feces, producing the watery stool characteristic of diarrhea. More importantly, the loss of fluids and electrolytes that came from the blood leads to dehydration. The subsequent reduction in effective circulating plasma volume can cause death in a matter of days or even hours, depending on the severity of the fluid loss.

About 50 years ago, physicians learned that replacing the lost fluids and electrolytes intravenously saves the lives of most diarrhea victims. In many parts of the world, however, adequate facilities, equipment, and personnel are not available to administer intravenous rehydration therapy. Consequently, millions of children still succumbed to diarrhea annually.

In 1966, researchers learned that the Na^+–glucose cotransport carrier system is not affected by diarrhea-causing microbes. This discovery led to the development of **oral rehydration therapy (ORT).** When both Na^+ and glucose are present in the lumen, the cotransport carrier transports them both from the lumen into the villus epithelial cells, from which they enter the blood. Because H_2O osmotically follows the absorbed Na^+, ingestion of a glucose and salt solution promotes the uptake of fluid into the blood from the intestinal tract without the need for intravenous replacement of fluids.

The first proof of ORT's life-saving ability in the field came in 1971 when several million refugees poured into India from war-ravaged Bangladesh. Of the thousands of refugees who fell victim to cholera and other diarrheal diseases, more than 30% died because of the scarcity of sterile fluids and needles for intravenous therapy. In one refugee camp, however, under the supervision of a group of scientists who had been experimenting with oral rehydration therapy, families were taught to administer ORT to diarrhea victims, most of whom were small children. The scarce intravenous solutions were reserved for those unable to drink. Death due to diarrhea was reduced to 3% in this camp, compared with a tenfold higher mortality among refugees elsewhere.

Based on this evidence, the World Health Organization (WHO) started aggressively promoting ORT. Packets of dry ingredients for ORT are now manufactured locally in more than 60 countries. The WHO estimates that about 30% of the world's children who contract diarrhea are treated with the prepackaged mixture or home-prepared versions. In the United States, commercially prepared oral solutions are widely available at pharmacies and supermarkets. An estimated 1 million children worldwide are saved annually as a result of ORT.

---

base imbalance is **diarrhea.** This condition is characterized by passage of a highly fluid fecal matter, often with increased frequency of defecation. Just as with vomiting, the effects of diarrhea can be either beneficial or harmful. Diarrhea is beneficial when rapid emptying of the intestine hastens elimination of harmful material from the body. However, not only are some of the ingested materials lost, but some of the secreted materials that normally would have been reabsorbed are lost as well. Excessive loss of intestinal contents causes dehydration, loss of nutrient material, and metabolic acidosis resulting from the loss of HCO_3^- (see p. 585). The abnormal fluidity of the feces in diarrhea usually occurs because the intestine is unable to absorb fluid as extensively as normal. This extra unabsorbed fluid passes out in the feces.

The causes of diarrhea are as follows:

1. The most common cause of diarrhea is excessive intestinal motility, which arises either from local irritation of the gut wall caused by bacterial or viral infection of the intestine or from emotional stress. Rapid transit of the intestinal contents does not allow sufficient time for adequate absorption of fluid to occur.

2. Diarrhea also occurs when excess osmotically active particles, such as those found in lactase deficiency, are present in the digestive tract lumen. These particles cause excessive fluid to enter and be retained in the lumen, thus increasing the fluidity of the feces.

3. Toxins of the bacterium *Vibrio cholera* (the causative agent of cholera) and certain other micro-organisms promote the secretion of excessive amounts of fluid by the small-intestine mucosa, resulting in profuse diarrhea. Diarrhea produced in response to toxins from infectious agents is the leading cause of death of small children in developing nations. Fortunately, a low-cost, effective *oral rehydration therapy* that takes advantage of the intestine's glucose cotransport carrier is saving the lives of millions of children. (For details about oral rehydration therapy, see the accompanying boxed feature, ▶ Concepts, Challenges, and Controversies.)

LARGE INTESTINE

The **large intestine** consists of the colon, cecum, appendix, and rectum (● Figure 16-30). The **cecum** forms a blind-ended pouch below the junction of the small and large intestines at the ileocecal valve. The small, fingerlike projection at the bottom of the cecum is the **appendix**, a lymphoid tissue that houses lymphocytes (see p. 414). The **colon,** which makes up most of the large intestine, is not coiled like the small intestine but consists of three relatively straight portions—the *ascending colon,* the *transverse colon,* and the *descending colon.* The terminal portion of the descending colon becomes S-shaped, forming the *sigmoid colon* (*sigmoid* means "S-shaped"), then straightens out to form the **rectum** (*rectum* means "straight").

▍ The large intestine is primarily a drying and storage organ.

The colon normally receives about 500 ml of chyme from the small intestine each day. Because most digestion and absorption have been accomplished in the small intestine, the contents delivered to the colon consist of indigestible food residues (such as cellulose), unabsorbed biliary components, and the remaining fluid. The colon extracts more H_2O and salt from the contents. What remains to be eliminated is known as **feces.** The primary function of the large intestine is to store this fecal material before defecation. Cellulose and other indigestible substances in the diet provide bulk and help maintain regular bowel movements by contributing to the volume of the colonic contents.

▍ Haustral contractions slowly shuffle the colonic contents back and forth.

The outer longitudinal smooth muscle layer does not completely surround the large intestine. Instead, it consists only of three separate, conspicuous, longitudinal bands of muscle, the **taeniae coli,** which run the length of the large intestine. These taeniae coli are shorter than the underlying circular smooth muscle and mucosal layers would be if these layers were stretched out flat. Because of this, the underlying layers are gathered into pouches or sacs called **haustra,** much as the material of a full skirt is gathered at the narrower waistband. The haustra are not merely passive permanent gathers, however; they actively change location as a result of contraction of the circular smooth-muscle layer.

Most of the time, movements of the large intestine are slow and nonpropulsive, as is appropriate for its absorptive and storage functions. The colon's primary method of motility is **haustral contractions** initiated by the autonomous rhythmicity of colonic smooth muscle cells. These contractions, which throw the large intestine into haustra, are similar to small intestine segmentations but occur much less frequently. Thirty minutes may elapse between haustral contractions, whereas segmentation contractions in the small intestine occur at rates of between 9 and 12 per minute. The location of the haustral sacs gradually changes as a relaxed segment that has formed a sac slowly contracts while a previously contracted area simultaneously relaxes to form a new sac. These movements are nonpropulsive; they slowly shuffle the contents in a back-and-forth mixing movement that exposes the colonic contents to the absorptive mucosa. Haustral contractions are largely controlled by locally mediated reflexes involving the intrinsic plexuses.

▍ Mass movements propel colonic contents long distances.

Three to four times a day, generally after meals, a marked increased in motility takes place during which large segments of the ascending and transverse colon contract simultaneously, driving the feces one-third to three-fourths of the length of the colon in a few seconds. These massive contractions, appropriately called **mass movements,** drive the colonic contents into

● **FIGURE 16-30**
Anatomy of the large intestine

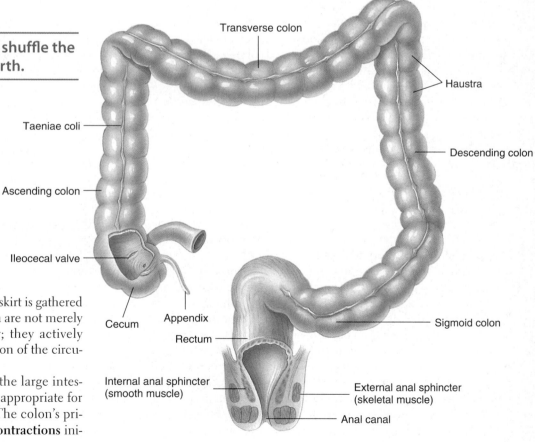

the distal portion of the large intestine, where material is stored until defecation occurs.

When food enters the stomach, mass movements occur in the colon primarily by means of the **gastrocolic reflex,** which is mediated from the stomach to the colon by gastrin and by the extrinsic autonomic nerves. In many people, this reflex is most evident after the first meal of the day and is often followed by the urge to defecate. Thus when a new meal enters the digestive tract, reflexes are initiated to move the existing contents farther along down the tract to make way for the incoming food. The gastroileal reflex moves the remaining small intestine contents into the large intestine, and the gastrocolic reflex pushes the colonic contents into the rectum, triggering the defecation reflex.

▌ Feces are eliminated by the defecation reflex.

When mass movements of the colon move fecal material into the rectum, the resultant distension of the rectum stimulates stretch receptors in the rectal wall, thus initiating the **defecation reflex.** This reflex causes the **internal anal sphincter** (which is composed of smooth muscle) to relax and the rectum and sigmoid colon to contract more vigorously. If the **external anal sphincter** (which is composed of skeletal muscle) is also relaxed, defecation occurs. Being skeletal muscle, the external anal sphincter is under voluntary control. The initial distension of the rectal wall is accompanied by the conscious urge to defecate. If circumstances are unfavorable for defecation, voluntary tightening of the external anal sphincter can prevent defecation despite the defecation reflex. If defecation is delayed, the distended rectal wall gradually relaxes, and the urge to defecate subsides until the next mass movement propels more feces into the rectum, once again distending the rectum and triggering the defecation reflex. During periods of nonactivity, both anal sphincters remain contracted to ensure fecal continence.

When defecation does occur, it is usually assisted by voluntary straining movements that involve simultaneous contraction of the abdominal muscles and a forcible expiration against a closed glottis. This maneuver brings about a large increase in intra-abdominal pressure, which helps expel the feces.

▌ Constipation occurs when the feces become too dry.

If defecation is delayed too long, **constipation** may result. When the colonic contents are retained for longer periods of time than normal, more than the usual amount of H_2O is absorbed from the feces, causing them to become hard and dry. Normal variations in frequency of defecation among individuals range from after every meal to up to once a week. When the frequency is delayed beyond what is normal for a particular individual, constipation and its attendant symptoms may occur. These symptoms include abdominal discomfort, dull headache, loss of appetite sometimes accompanied by nausea, and mental depression. Contrary to popular belief, these symptoms are not caused by toxins absorbed from the retained fecal material. Although bacterial metabolism produces some potentially toxic substances in the colon, these substances normally pass through the portal system and are removed by the liver before they can reach the systemic circulation. Instead, the symptoms associated with constipation are caused by prolonged distension of the large intestine, particularly the rectum; the symptoms promptly disappear following relief from distension.

Possible causes for delayed defecation that might lead to constipation include (1) ignoring the urge to defecate; (2) decreased colon motility accompanying aging, emotion, or a low-bulk diet; (3) obstruction of fecal movement in the large bowel caused by a local tumor or colonic spasm; and (4) impairment of the defecation reflex, such as through injury of the nerve pathways involved.

Appendicitis

If hardened fecal material becomes lodged in the appendix, it may obstruct normal circulation and mucus secretion in this narrow, blind-ended appendage. This blockage leads to inflammation of the appendix, or **appendicitis.** The inflamed appendix often becomes swollen and filled with pus, and the tissue may die as a result of local circulatory interference. If not surgically removed, the diseased appendix may rupture, spewing its infectious contents into the abdominal cavity.

▌ Large-intestine secretion is entirely protective.

The large intestine does not secrete any digestive enzymes. None are needed, because digestion is completed before chyme ever reaches the colon. Colonic secretion consists of an alkaline (HCO_3^-) mucus solution, whose function is to protect the large intestine mucosa from mechanical and chemical injury. The mucus provides lubrication to facilitate passage of the feces, whereas the HCO_3^- neutralizes irritating acids produced by local bacterial fermentation. Secretion increases in response to mechanical and chemical stimulation of the colonic mucosa mediated by short reflexes and parasympathetic innervation.

No digestion takes place within the large intestine because there are no digestive enzymes. However, the colonic bacteria do digest some of the cellulose and use it for their own metabolism.

▌ The colon contains myriads of beneficial bacteria.

Because of slow colonic movement, bacteria have time to grow and accumulate in the large intestine. In contrast, in the small intestine the contents are normally moved through too rapidly for bacterial growth to occur. Furthermore, the mouth, stomach, and small intestine secrete antibacterial agents, but the colon does not. Not all ingested bacteria are destroyed by lysozyme and HCl, however. The surviving bacteria continue to thrive in the large intestine. About ten times more bacteria live in the human colon than the human body has cells. An estimated 500 different species of bacteria typically reside in the colon. These colonic micro-organisms not only are typically harmless but in fact they provide beneficial functions. For example, indigenous bacteria (1) make nutritional contributions, such as synthesizing vitamin K and raising colonic acidity, thereby promoting the absorption of calcium, magnesium, and zinc; (2) enhance intestinal immunity by competing with potentially pathogenic microbes for nutrients and space (see p. 451); (3) promote colonic motility; and (4) help maintain colonic mucosal integrity.

■ The large intestine absorbs salt and water, converting the luminal contents into feces.

Some absorption takes place within the colon, but not to the same extent as in the small intestine. Because the luminal surface of the colon is fairly smooth, it has considerably less absorptive surface area than the small intestine has. Furthermore, no specialized transport mechanisms are present in the colonic mucosa for absorbing glucose or amino acids, as there are in the small intestine. When excessive small intestine motility delivers the contents to the colon before absorption of nutrients has been completed, the colon is unable to absorb these materials, and they are lost in diarrhea.

The colon normally absorbs salt and H_2O. Sodium is actively absorbed, Cl^- follows passively down the electrical gradient, and H_2O follows osmotically. Bacteria in the colon synthesize some vitamins that the colon is capable of absorbing, but this is normally not a significant contribution, except in the case of vitamin K.

Through absorption of salt and H_2O, a firm fecal mass is formed. Of the 500 ml of material entering the colon per day from the small intestine, the colon normally absorbs about 350 ml, leaving 150 g of feces to be eliminated from the body each day (see ▲ Table 16-7). This fecal material normally consists of 100 g of H_2O and 50 g of solid, including undigested cellulose, bilirubin, bacteria, and small amounts of salt. Thus, contrary to popular thinking, the digestive tract is not a major excretory passageway for eliminating wastes from the body. The main waste product excreted in the feces is bilirubin. The other fecal constituents are unabsorbed food residues and bacteria, which were never actually a part of the body.

■ Intestinal gases are absorbed or expelled.

Occasionally, instead of fecal material passing from the anus, intestinal gas, or **flatus**, passes out. This gas is derived primarily from two sources: (1) swallowed air (as much as 500 ml of air may be swallowed during a meal) and (2) gas produced by bacterial fermentation in the colon. The presence of gas percolating through the luminal contents gives rise to gurgling sounds known as **borborygmi**. Eructation (burping) removes most of the swallowed air from the stomach, but some passes on into the intestine. Usually, very little gas is present in the small intestine, because the gas is either quickly absorbed or passes on into the colon. Most gas in the colon is due to bacterial activity, with the quantity and nature of the gas depending on the type of food eaten and the characteristics of the colonic bacteria. Some foods, such as beans, contain types of carbohydrates for which humans lack digestive enzymes. These fer-

mentable carbohydrates enter the colon, where they are attacked by gas-producing bacteria. Much of the gas entering or forming in the large intestine is absorbed through the intestinal mucosa. The remainder is expelled through the anus.

To accomplish selective expulsion of gas when fecal material is also present in the rectum, the abdominal muscles and external anal sphincter are voluntarily contracted simultaneously. When contraction of the abdominal muscles raises the pressure sufficiently against the contracted anal sphincter, the pressure gradient forces air out at a high velocity through a slit-like anal opening that is too narrow for solid feces to escape. This passage of air at high velocity causes the edges of the anal opening to vibrate, giving rise to the characteristic low-pitched sound accompanying passage of gas.

OVERVIEW OF THE GASTROINTESTINAL HORMONES

Throughout our discussion of digestion, we have repeatedly mentioned different functions of the three major gastrointestinal hormones: gastrin, secretin, and CCK. We will now fit all of these functions together so that you can appreciate the overall adaptive importance of these interactions (▲ Table 16-9). Furthermore, we will introduce a more recently identified hormone, GIP, the fourth hormone to be confirmed as a gastrointestinal hormone.

Gastrin

Chyme in the stomach, especially if it contains protein, stimulates the release of gastrin, which fills three functions:

1. It acts in multiple ways to increase secretion of HCl and pepsinogen. These two substances in turn are of primary importance in initiating digestion of the protein that promoted their release.

2. It enhances gastric motility, stimulates ileal motility, relaxes the ileocecal sphincter, and induces mass movements in the colon—functions that are all aimed at keeping the contents moving through the tract on the arrival of a new meal.

3. It also is trophic not only to the stomach mucosa but also to the small intestine mucosa, helping maintain a well-developed, functionally viable digestive tract lining.

Predictably, gastrin secretion is inhibited by an accumulation of acid in the stomach and by the presence in the duodenal lumen of acid and other constituents that necessitate a delay in gastric secretion.

Secretin

As the stomach empties into the duodenum, the presence of acid in the duodenum stimulates the release of secretin into the blood. Secretin performs five major interrelated functions:

1. It inhibits gastric emptying to prevent further acid from entering the duodenum until the acid that is already present is neutralized.

2. It inhibits gastric secretion to reduce the amount of acid being produced.

Source, Control, and Functions of the Major Gastrointestinal Hormones

Hormone	Source	Primary Stimulus for Secretion	Functions
Gastrin	G cells of the stomach's pyloric gland area	Protein in the stomach	Stimulates secretion by the parietal and chief cells
			Enhances gastric motility
			Stimulates ileal motility
			Relaxes the ileocecal sphincter
			Induces colonic mass movements
			Is trophic to the stomach and small intestine mucosa
Secretin	Endocrine cells in the duodenal mucosa	Acid in the duodenal lumen	Inhibits gastric emptying and gastric secretion
			Stimulates $NaHCO_3$ secretion by the pancreatic duct cells and by the liver
			Is trophic to the exocrine pancreas
Cholecystokinin	Endocrine cells in the duodenal mucosa	Nutrients in the duodenal lumen, especially fat products and to a lesser extent protein products	Inhibits gastric emptying and gastric secretion
			Stimulates digestive enzyme secretion by the pancreatic acinar cells
			Causes gallbladder contraction and relaxation of the sphincter of Oddi
			Is trophic to the exocrine pancreas
			May cause long-term adaptive changes in the proportion of pancreatic enzymes
			Contributes to satiety

3. It stimulates the pancreatic duct cells to produce a large volume of aqueous $NaHCO_3$ secretion, which is emptied into the duodenum to neutralize the acid.

4. It stimulates secretion by the liver of a $NaHCO_3$-rich bile, which likewise is emptied into the duodenum to assist in the neutralization process. Neutralization of the acidic chyme in the duodenum helps prevent damage to the duodenal walls and provides a suitable environment for the optimal functioning of the pancreatic digestive enzymes, which are inhibited by acid.

5. Along with CCK, secretin is trophic to the exocrine pancreas.

CCK

As chyme empties from the stomach, fat and other nutrients enter the duodenum. These nutrients, especially fat and to a lesser extent protein products, cause the release of CCK from the duodenal mucosa. This hormone also performs important interrelated functions:

1. It inhibits gastric motility and secretion, thereby allowing adequate time for the nutrients already in the duodenum to be digested and absorbed.

2. It stimulates the pancreatic acinar cells to increase secretion of pancreatic enzymes, which continue the digestion of these nutrients in the duodenum (this action is especially important for fat digestion, because pancreatic lipase is the only enzyme that digests fat).

3. It causes contraction of the gallbladder and relaxation of the sphincter of Oddi so that bile is emptied into the duodenum to aid fat digestion and absorption. Bile salts' detergent action is particularly important in enabling pancreatic lipase to perform its digestive task. Once again, the multiple effects of CCK are remarkably well adapted to dealing with the fat and other nutrients whose presence in the duodenum triggered this hormone's release.

4. Furthermore, it is appropriate that both secretin and CCK, which have profound stimulatory effects on the exocrine pancreas, are trophic to this tissue.

5. CCK has also been implicated in long-term adaptive changes in the proportion of pancreatic enzymes produced in response to prolonged changes in diet.

6. Besides facilitating the digestion of ingested nutrients, CCK is an important regulator of food intake. It plays a key role in satiety, the sensation of having had enough to eat (see p. 652).

GIP

A more recently recognized hormone, **GIP**, helps promote metabolic processing of the nutrients once they are absorbed. This hormone was originally named *gastric inhibitory peptide (GIP)* for its presumed role as an enterogastrone. It was believed to inhibit gastric motility and secretion, similar to secretin and CCK. GIP's contribution in this regard is now considered minimal. Instead, this hormone has been shown to stimulate insulin release by the pancreas, so it is now called **glucose-dependent insulinotrophic peptide** (once again **GIP**). Again, this is remarkably adaptive. As soon as the meal is absorbed, the body has to shift its metabolic gears to use and store the newly arriving nutrients. The metabolic activities of this absorptive phase are largely under the control of insulin (see p. 722 and p. 725). Stimulated by the presence of a meal in the digestive tract, GIP initiates the release of insulin in anticipation of the absorption of the meal in a "feedforward" fashion. Insulin is especially important in promoting the uptake and storage of glucose. Appropriately, glucose in the duodenum has recently been shown to increase GIP secretion.

This overview of the multiple, integrated, adaptive functions of the gastrointestinal hormones provides an excellent example of the remarkable efficiency of the human body.

CHAPTER IN PERSPECTIVE: FOCUS ON HOMEOSTASIS

To maintain constancy in the internal environment, materials that are used up in the body (such as nutrients and O_2) or uncontrollably lost from the body (such as evaporative H_2O loss from the airways or salt loss in sweat) must constantly be replaced by new supplies of these materials from the external environment. All these replacement supplies except O_2 are acquired through the digestive system. Fresh supplies of O_2 are transferred to the internal environment by the respiratory system, but all the nutrients, H_2O, and various electrolytes needed to maintain homeostasis are acquired through the digestive system. The large, complex food that is ingested is broken down by the digestive system into small absorbable units. These small energy-rich nutrient molecules are transferred across the small intestine epithelium into the blood for delivery to the cells to replace the nutrients constantly used for ATP production and for repair and growth of body tissues. Likewise, ingested H_2O, salt, and other electrolytes are absorbed by the intestine into the blood.

Unlike most body systems, regulation of digestive system activities is not aimed at maintaining homeostasis. The quantity of nutrients and H_2O ingested is subject to control, but the quantity of ingested materials absorbed by the digestive tract is not subject to control, with few exceptions. The hunger mechanism governs food intake to help maintain energy balance (Chapter 17), and the thirst mechanism controls H_2O intake to help maintain H_2O balance (Chapter 15). However, we often do not heed these control mechanisms, and often eat and drink even when we are not hungry or thirsty. Once these materials are in the digestive tract, the digestive system does not vary its rate of nutrient, H_2O, or electrolyte uptake according to body needs (with the exception of iron and calcium); rather, it optimizes conditions for digesting and absorbing what is ingested. Truly, what you eat is what you get. The digestive system is subject to many regulatory processes, but these are not influenced by the nutritional or hydration state of the body. Instead, these control mechanisms are governed by the composition and volume of digestive tract contents so that the rate of motility and secretion of digestive juices are optimal for digestion and absorption of the ingested food.

If excess nutrients are ingested and absorbed, the extra is placed in storage, such as in adipose tissue (fat), so that the blood level of nutrient molecules is kept at a constant level. Excess ingested H_2O and electrolytes are eliminated in the urine to homeostatically maintain the blood levels of these constituents.

CHAPTER SUMMARY

Introduction
■ The four basic digestive processes are motility, secretion, digestion, and absorption.
■ The three classes of energy-rich nutrients are digested into absorbable units as follows: (1) Dietary carbohydrates in the form of the polysaccharides starch and glycogen are digested into their absorbable units of monosaccharides, especially glucose. (2) Dietary proteins are digested into their absorbable units of amino acids and a few small polypeptides. (3) Dietary fats in the form of triglycerides are digested into their absorbable units of monoglycerides and free fatty acids.
■ The digestive system consists of the digestive tract and accessory digestive organs (salivary glands, exocrine pancreas, and biliary system.)
■ The digestive tract consists of a continuous tube that runs from the mouth to the anus, with local modifications that reflect regional specializations for carrying out digestive functions.
■ The lumen of the digestive tract is continuous with the external environment, so its contents are technically outside the body;

this arrangement permits digestion of food without self-digestion occurring in the process.
■ The digestive tract wall has four layers throughout most of its length. From innermost outward, they are the mucosa, submucosa, muscularis externa, and serosa.
■ Digestive activities are carefully regulated by synergistic autonomous, neural (both intrinsic and extrinsic), and hormonal mechanisms to ensure that the ingested food is maximally made available to the body for energy production and as synthetic raw materials.

Mouth
■ **Motility:** Food enters the digestive system through the mouth, where it is chewed and mixed with saliva to facilitate swallowing.
■ **Secretion:** The salivary enzyme, amylase, begins the digestion of carbohydrates. More important than its minor digestive function, saliva is essential for articulate speech and plays an important role in dental health. Salivary secretion is controlled by a

salivary center in the medulla, mediated by autonomic innervation of the salivary glands.

- **Digestion:** Salivary amylase begins to digest polysaccharides into the disaccharide maltose, a process that continues in the stomach after the food has been swallowed until amylase is eventually inactivated by the acidic gastric juice.
- **Absorption:** No absorption of nutrients occurs from the mouth.

Pharynx and esophagus

- **Motility:** Following chewing, the tongue propels the bolus of food to the rear of the throat, which initiates the swallowing reflex. The swallowing center in the medulla coordinates a complex group of activities that result in closure of the respiratory passages and propulsion of food through the pharynx and esophagus into the stomach.
- **Secretion:** The esophageal secretion, mucus, is protective in nature.
- **Digestion and absorption:** No nutrient digestion or absorption occurs in the pharynx or esophagus.

Stomach

- The stomach, a saclike structure located between the esophagus and small intestine, stores ingested food for variable periods of time until the small intestine is ready to process it further for final absorption.
- **Motility:** The four aspects encompassing gastric motility are gastric filling, storage, mixing, and emptying.
- Gastric filling is facilitated by vagally mediated receptive relaxation of the stomach musculature.
- Gastric storage takes place in the body of the stomach, where peristaltic contractions of the thin muscular walls are too weak to mix the contents.
- Gastric mixing takes place in the thick-muscled antrum as a result of vigorous peristaltic contractions.
- Gastric emptying is influenced by factors in both the stomach and duodenum. (1) The volume and fluidity of chyme in the stomach tend to promote emptying of the stomach contents. (2) The duodenal factors, which are the dominant factors controlling gastric emptying, tend to delay gastric emptying until the duodenum is ready to receive and process more chyme. The specific factors in the duodenum that delay gastric emptying are fat, acid, hypertonicity, and distension. These factors delay gastric emptying by inhibiting stomach peristaltic activity by means of the enterogastric reflex and the enterogastrones, secretin and cholecystokinin, which are secreted by the duodenal mucosa.
- **Secretion:** Gastric secretions into the stomach lumen include (1) HCl (from the parietal cells), which activates pepsinogen, denatures protein, and kills bacteria; (2) pepsinogen (from the chief cells), which, once activated, initiates protein digestion; (3) mucus (from the mucous cells), which provides a protective coating to supplement the gastric mucosal barrier, enabling the stomach to contain the harsh luminal contents without self-digestion; and (4) intrinsic factor (from the parietal cells), which plays a vital role in vitamin B_{12} absorption, a constituent essential for normal red blood cell production.
- The stomach also secretes the following endocrine and paracrine regulatory factors: (1) the hormone gastrin (from the G cells), which plays a dominant role in stimulating gastric secretion; (2) the paracrine histamine (from the ECL cells), a potent stimulant of acid secretion by the parietal cells; and (3) the paracrine somatostatin (from the D cells), which inhibits gastric secretion.
- Gastric secretion is under complex control mechanisms. Gastric secretion is increased during the cephalic and gastric phases of gastric secretion before and during a meal by mechanisms in-

volving excitatory vagal and intrinsic nerve responses along with the stimulatory actions of gastrin and histamine. After the meal is emptied from the stomach, gastric secretion is reduced as a result of the withdrawal of stimulatory factors, the release of inhibitory somatostatin, and the inhibitory actions of the enterogastric reflex and enterogastrones during the intestinal phase of gastric secretion.

- **Digestion:** Carbohydrate digestion continues in the body of the stomach under the influence of the swallowed salivary amylase. Protein digestion is initiated in the antrum of the stomach, where vigorous peristaltic contractions mix the food with gastric secretions, converting it to a thick liquid mixture known as chyme.
- **Absorption:** No nutrients are absorbed from the stomach.

Pancreatic and Biliary Secretions

- Pancreatic exocrine secretions and bile from the liver both enter the duodenal lumen.
- Pancreatic secretions include (1) potent digestive enzymes from the acinar cells, which digest all three categories of foodstuff; and (2) an aqueous $NaHCO_3$ solution from the duct cells, which neutralizes the acidic contents emptied into the duodenum from the stomach. This neutralization is important to protect the duodenum from acid injury and to allow pancreatic enzymes, which are inactivated by acid, to perform their important digestive functions.
- The pancreatic digestive enzymes include (1) the proteolytic enzymes trypsinogen, chymotrypsinogen, and procarboxypeptidase, which are secreted in inactive form and are activated in the duodenal lumen on exposure to enterokinase and activated trypsin; (2) pancreatic amylase, which continues carbohydrate digestion; and (3) lipase, which accomplishes fat digestion.
- Pancreatic secretion is primarily under hormonal control, which matches the composition of the pancreatic juice with the needs in the duodenal lumen. Secretin stimulates the pancreatic duct cells and cholecystokinin (CCK) stimulates the acinar cells.
- The liver, the body's largest and most important metabolic organ, performs many varied functions. Its contribution to digestion is the secretion of bile, which contains bile salts.
- Bile salts aid fat digestion through their detergent action and facilitate fat absorption through formation of water-soluble micelles that can carry the water-insoluble products of fat digestion to their absorptive site.
- Between meals, bile is stored and concentrated in the gallbladder, which is stimulated by cholecystokinin to contract and empty the bile into the duodenum during digestion of a meal. After participating in fat digestion and absorption, bile salts are reabsorbed and returned via the hepatic portal system to the liver, where they not only are resecreted but also act as a potent choleretic to stimulate the secretion of even more bile.
- Bile also contains bilirubin, a derivative of degraded hemoglobin, which is the major excretory product in the feces.

Small Intestine

- The small intestine is the main site for digestion and absorption.
- **Motility:** Segmentation, the small intestine's primary motility during digestion of a meal, thoroughly mixes the food with pancreatic, biliary, and small intestinal juices to facilitate digestion; it also exposes the products of digestion to the absorptive surfaces.
- Between meals, the migrating motility complex sweeps the lumen clean.
- **Secretion:** The juice secreted by the small intestine does not contain any digestive enzymes. The enzymes synthesized by the small intestine act intracellularly within the brush-border membranes of the epithelial cells.

- **Digestion:** The pancreatic enzymes continue carbohydrate and protein digestion in the small-intestine lumen. The small-intestine brush-border enzymes complete the digestion of carbohydrates and protein. Fat digestion is accomplished entirely in the small-intestine lumen by pancreatic lipase.
- **Absorption:** The small-intestine lining is remarkably adapted to its digestive and absorptive function. It is arranged into folds that bear a rich array of fingerlike projections, the villi, which are furnished with a multitude of even smaller hairlike protrusions, the microvilli. Altogether, these surface modifications tremendously increase the area available to house the membrane-bound enzymes and to accomplish both active and passive absorption. This impressive lining is replaced approximately every three days to ensure an optimally healthy and functional presence of epithelial cells despite harsh luminal conditions.
- The energy-dependent process of Na^+ absorption provides the driving force for Cl^-, water, glucose, and amino acid absorption. All these absorbed products enter the blood.
- Because they are not soluble in water, the products of fat digestion must undergo a series of transformations that enable them to be passively absorbed, eventually entering the lymph.
- The small intestine absorbs almost everything presented to it, from ingested food to digestive secretions to sloughed epithelial cells. Only a small amount of fluid and indigestible food residue passes on to the large intestine.

Large Intestine
- The colon serves primarily to concentrate and store undigested food residues (fiber, the indigestible cellulose in plant walls) and bilirubin until they can be eliminated from the body as feces.
- **Motility:** Haustral contractions slowly shuffle the colonic contents back and forth to accomplish mixing and facilitate absorption of most of the remaining fluid and electrolytes. Mass movements occur several times a day, usually following meals, propelling the feces long distances. Movement of feces into the rectum triggers the defecation reflex, which can be voluntarily prevented by contraction of the external anal sphincter if the time is inopportune for elimination.
- **Secretion:** The alkaline mucus secretion of the large intestine is primarily protective in nature.
- **Digestion and absorption:** No secretion of digestive enzymes or absorption of nutrients takes place in the colon, all nutrient digestion and absorption having been completed in the small intestine. Absorption of some of the remaining salt and water converts the colonic contents into feces.

Overview of the Gastrointestinal Hormones
- The three major gastrointestinal hormones are gastrin from the stomach mucosa and secretin and cholecystokinin from the duodenal mucosa. Each of these hormones performs multiple interrelated functions.
- Gastrin is released primarily in response to the presence of protein products in the stomach and its effects are aimed at promoting digestion of protein, moving materials through the digestive tract, and maintaining the integrity of the stomach and small-intestine mucosa.
- Secretin is released primarily in response to the presence of acid in the duodenum and its effects are aimed at neutralizing the acid in the duodenal lumen and in maintaining the integrity of the exocrine pancreas.
- Cholecystokinin is released primarily in response to the presence of fat products in the duodenum and its effects are aimed at optimizing conditions for digesting fat and other nutrients and in maintaining the integrity of the exocrine pancreas.

REVIEW EXERCISES

Objective Questions (Answers on p. A-48)
1. The extent of nutrient uptake from the digestive tract depends on the body's needs. *(True or false?)*
2. The stomach is relaxed during vomiting. *(True or false?)*
3. Acid cannot normally penetrate into or between the cells lining the stomach, which enables the stomach to contain acid without injuring itself. *(True or false?)*
4. Protein is continually lost from the body through digestive secretions and sloughed epithelial cells, which pass out in the feces. *(True or false?)*
5. Foodstuffs not absorbed by the small intestine are absorbed by the large intestine. *(True or false?)*
6. The endocrine pancreas secretes secretin and CCK. *(True or false?)*
7. A digestive reflex involving the autonomic nerves is known as a _____ reflex, whereas a reflex in which all elements of the reflex arc are located within the gut wall is known as a _____ reflex.
8. When food is mechanically broken down and mixed with gastric secretions, the resultant thick liquid mixture is known as _____ .
9. The entire lining of the small intestine is replaced approximately every _____ days.
10. The two substances absorbed by specialized transport mechanisms located only in the terminal ileum are _____ and _____ .
11. The most potent choleretic is _____ .
12. Which of the following is *not* a function of saliva?
    a. begins digestion of carbohydrate
    b. facilitates absorption of glucose across the oral mucosa
    c. facilitates speech
    d. exerts an antibacterial effect
    e. plays an important role in oral hygiene
13. Match the following:
    - ____ 1. prevents re-entry of food into the mouth during swallowing
    - ____ 2. triggers the swallowing reflex
    - ____ 3. seals off the nasal passages during swallowing
    - ____ 4. prevents air from entering the esophagus during breathing
    - ____ 5. closes off the respiratory airways during swallowing
    - ____ 6. prevents gastric contents from backing up into the esophagus

    (a) closure of the pharyngoesophageal sphincter
    (b) elevation of the uvula
    (c) position of the tongue against the hard palate
    (d) closure of the gastroesophageal sphincter
    (e) bolus pushed to the rear of the mouth by the tongue
    (f) tight apposition of the vocal folds

14. Use the following answer code to identify the characteristics of the listed substances:
    - (a) pepsin
    - (b) mucus
    - (c) HCl
    - (d) intrinsic factor
    - (e) histamine

____ 1. activates pepsinogen
____ 2. inhibits amylase
____ 3. essential for vitamin B_{12} absorption
____ 4. can act autocatalytically
____ 5. a potent stimulant for acid secretion
____ 6. breaks down connective tissue and muscle fibers
____ 7. begins protein digestion
____ 8. serves as a lubricant
____ 9. kills ingested bacteria
____ 10. is alkaline
____ 11. deficient in pernicious anemia
____ 12. coats the gastric mucosa

Essay Questions

1. Describe the four basic digestive processes.
2. List the three categories of energy-rich foodstuffs and the absorbable units of each.
3. List the components of the digestive system. Describe the cross-sectional anatomy of the digestive tract.
4. What four general factors are involved in the regulation of digestive system function? What is the role of each?
5. Describe the types of motility in each component of the digestive tract. What factors control each type of motility?
6. State the composition of the digestive juice secreted by each component of the digestive system. Describe the factors that control each digestive secretion.

7. List the enzymes involved in the digestion of each category of foodstuff. Indicate the source and control of secretion of each of the enzymes.
8. Why are some digestive enzymes secreted in inactive form? How are they activated?
9. What absorption processes take place within each component of the digestive tract? What special adaptations of the small intestine enhance its absorptive capacity?
10. Describe the absorptive mechanisms for salt, water, carbohydrate, protein, and fat.
11. What are the contributions of the accessory digestive organs? What are the nondigestive functions of the liver?
12. Summarize the functions of each of the three major gastrointestinal hormones.
13. What waste product is excreted in the feces?
14. How is vomiting accomplished? What are the causes and consequences of vomiting, diarrhea, and constipation?
15. Describe the process of mucosal turnover in the stomach and small intestine.

Quantitative Exercises (Solutions on p. A-48)

1. Suppose a lipid droplet in the gut is essentially a sphere with a diameter of 1 cm.
    a. What is the surface area to volume ratio of the droplet? (*Hint:* The area of a sphere is $4\pi r^2$, and the volume is $\frac{4}{3}\pi r^3$).
    b. Now, suppose that this sphere were emulsified into 100 essentially equal-sized droplets. What is the average surface area to volume ratio of each droplet?
    c. How much greater is the total surface area of these 100 droplets compared to the original single droplet?
    d. How much did the total volume change as a result of emulsification?

POINTS TO PONDER

(Explanations on p. A-49)

1. Why do patients who have had a large portion of their stomachs removed for treatment of stomach cancer or severe peptic ulcer disease have to eat small quantities of food frequently instead of consuming three meals a day?
2. The number of immune cells in the gut-associated lymphoid tissue (GALT) is estimated to be equal to the total number of these defense cells in the rest of the body. Speculate on the adaptive significance of this extensive defense capability of the digestive system.
3. By what means would defecation be accomplished in a patient paralyzed from the waist down by a lower spinal cord injury?
4. After bilirubin is extracted from the blood by the liver, it is conjugated (combined) with glycuronic acid by the enzyme glu-

curonyl transferase within the liver. Only when conjugated can bilirubin be actively excreted into the bile. For the first few days of life, the liver does not make adequate quantities of glucuronyl transferase. Explain how this transient enzyme deficiency leads to the common condition of jaundice in newborns.

5. Explain why removal of either the stomach or the terminal ileum leads to pernicious anemia.
6. ***Clinical Consideration.*** Thomas W. experiences a sharp pain in his upper right abdomen after eating a high-fat meal. Also, he has noted that his feces are grayish white instead of brown. What is the most likely cause of his symptoms? Explain why each of these symptoms occurs with this condition.

PHYSIOEDGE CD-ROM
Figures marked with the PhysioEdge icon have associated activities on the CD. For this chapter, check out:

Gastrointestinal Motility

The diagnostic quiz allows you to receive immediate feedback on your understanding of the concept and to advance through various levels of difficulty.

You will also find Media Quizzes related to Figures 16-8 and 16-13.

CHECK OUT THESE MEDIA QUIZZES:
16.1 The Stomach
16.2 Stomach: Cellular Level
16.3 The Intestine and Associated Organs

 PHYSIOEDGE WEB SITE
The Web site for this book contains a wealth of helpful study aids, as well as many ideas for further reading and research. Log on to:

http://www.brookscole.com

Go to the Biology page and select Sherwood's *Human Physiology*, 5th Edition. Select a chapter from the drop-down menu or click on one of these resource areas:

- **Case Histories** provide an introduction to the clinical aspects of human physiology. Check out:

    #6: Just Stress

- For 2-D and 3-D graphical illustrations and animations of physiological concepts, visit our **Visual Learning Resource.**

- Resources for study and review include the **Chapter Outline, Chapter Summary, Glossary,** and **Flash Cards.** Use our **Quizzes** to prepare for in-class examinations.

- On-line research resources to consult are **Hypercontents,** with current links to relevant Internet sites; **Internet Exercises** with starter URLs provided; and **InfoTrac Exercises.**

For more readings, go to InfoTrac College Edition, your on-line research library, at:

http://infotrac.thomsonlearning.com

Components Important in Energy Balance and Temperature Regulation

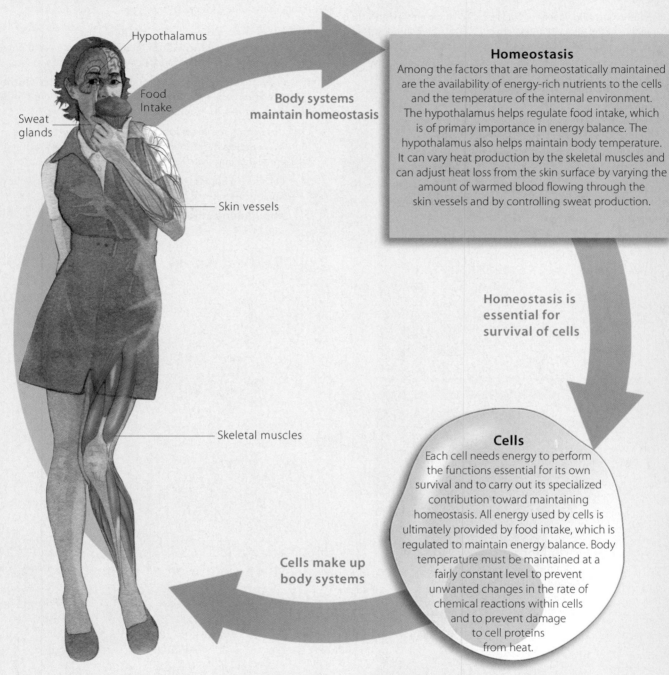

Hypothalamus

Food Intake

Sweat glands

Skin vessels

Skeletal muscles

Body systems maintain homeostasis

Homeostasis
Among the factors that are homeostatically maintained are the availability of energy-rich nutrients to the cells and the temperature of the internal environment. The hypothalamus helps regulate food intake, which is of primary importance in energy balance. The hypothalamus also helps maintain body temperature. It can vary heat production by the skeletal muscles and can adjust heat loss from the skin surface by varying the amount of warmed blood flowing through the skin vessels and by controlling sweat production.

Homeostasis is essential for survival of cells

Cells
Each cell needs energy to perform the functions essential for its own survival and to carry out its specialized contribution toward maintaining homeostasis. All energy used by cells is ultimately provided by food intake, which is regulated to maintain energy balance. Body temperature must be maintained at a fairly constant level to prevent unwanted changes in the rate of chemical reactions within cells and to prevent damage to cell proteins from heat.

Cells make up body systems

Food intake is essential to power cell activities. For body weight to remain constant, the caloric value of food must equal total energy needs. **Energy balance** and thus body weight are maintained by controlling food intake.

Energy expenditure generates heat, which is important in **temperature regulation.** Humans, usually in environments cooler than their bodies, must constantly generate heat to maintain their body temperatures. Also, they must have mechanisms to cool the body if it gains too much heat from heat-generating skeletal muscle activity or from a hot external environment. Body temperature must be regulated because the rate of cellular chemical reactions depends on temperature, and overheating damages cell proteins.

The hypothalamus is the major integrating center for maintenance of both energy balance and body temperature.

Chapter 17

Energy Balance and Temperature Regulation

CONTENTS AT A GLANCE

ENERGY BALANCE

▌ Sources of energy input and output

▌ Metabolic rate

▌ Neutral, positive, and negative energy balance

▌ Control of food intake

TEMPERATURE REGULATION

▌ Body temperature

▌ Sources of heat gain and heat loss

▌ Physical mechanisms of heat exchange

▌ The hypothalamus as a thermostat

▌ Control of heat production; shivering

▌ Control of heat loss; skin vasomotor activity

▌ Integrated responses to cold and heat exposure

▌ Fever

ENERGY BALANCE

Each cell in the body needs energy to perform the functions essential for the cell's own survival (such as active transport and cellular repair) and to carry out its specialized contributions toward maintenance of homeostatic balance (such as gland secretion and muscle contraction). All energy used by cells is ultimately provided by food intake.

▌ Most food energy is ultimately converted into heat in the body.

According to the *first law of thermodynamics*, energy can be neither created nor destroyed. Therefore, energy is subject to the same kind of input–output balance as are the chemical components of the body such as H_2O and salt (see p. 559).

Energy input and output

The energy in ingested food constitutes *energy input* to the body. Chemical energy locked in the bonds that hold the atoms together in nutrient molecules is released when these molecules are broken down in the body. Cells capture a portion of this nutrient energy in the high-energy phosphate bonds of ATP (see p. 35). Energy harvested from biochemical processing of ingested nutrients either is used immediately to perform biological work or is stored in the body for later use as needed during periods when food is not being digested and absorbed.

Energy output or expenditure by the body falls into two categories (● Figure 17-1): external work and internal work. **External work** is the energy expended when skeletal muscles are contracted to move external objects or to move the body in relation to the environment. **Internal work** constitutes all other forms of biological energy expenditure that do not accomplish mechanical work outside the body. Internal work encompasses two types of energy-dependent activities: (1) skeletal muscle activity used for purposes other than external work, such as the contractions associated with postural maintenance and shivering; and (2) all the energy-expending activities that must go on all the time just to sustain life. The latter include the work of pumping blood and breathing; the energy required for active transport of critical materials across plasma mem-

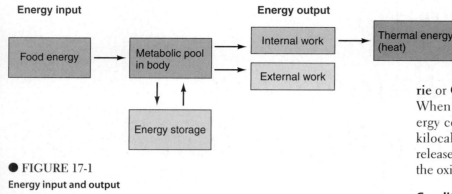

Energy input

Food energy → Metabolic pool in body

Metabolic pool in body ⇄ Energy storage

Energy output

Metabolic pool in body → Internal work → Thermal energy (heat)

Metabolic pool in body → External work

● **FIGURE 17-1**
Energy input and output

branes; and the energy used during synthetic reactions essential for the maintenance, repair, and growth of cellular structures—in short, the "metabolic cost of living."

Conversion of nutrient energy to heat

Not all energy in nutrient molecules can be harnessed to perform biological work. Energy cannot be created or destroyed, but it can be converted from one form to another. The energy in nutrient molecules that is not used to energize work is transformed into **thermal energy**, or **heat**. During biochemical processing, only about 50% of the energy in nutrient molecules is transferred to ATP; the other 50% of nutrient energy is immediately lost as heat. During ATP expenditure by the cells, another 25% of the energy derived from ingested food becomes heat. Because the body is not a heat engine, it cannot convert heat into work. Therefore, not more than 25% of nutrient energy is available to accomplish work, either external or internal. The remaining 75% is lost as heat during the sequential transfer of energy from nutrient molecules to ATP to cellular systems.

Furthermore, of the energy actually captured for use by the body, almost all expended energy eventually becomes heat. To exemplify, energy expended by the heart to pump blood is gradually changed into heat by friction as blood flows through the vessels. Likewise, energy used in the synthesis of structural protein eventually appears as heat when that protein is degraded during the normal course of turnover of bodily constituents. Even in the performance of external work, skeletal muscles convert chemical energy into mechanical energy inefficiently; as much as 75% of the expended energy is lost as heat. Thus all energy that is liberated from ingested food but not directly used for movement of external objects or stored in fat (adipose tissue) deposits (or, in the case of growth, as protein) eventually becomes body heat. This heat is not entirely wasted energy, however, because much of it is used to maintain body temperature.

▌ The metabolic rate is the rate of energy use.

The rate at which energy is expended by the body during both external and internal work is known as the **metabolic rate**:

Metabolic rate = energy expenditure/unit of time

Because most of the body's energy expenditure eventually appears as heat, the metabolic rate is normally expressed in terms of the rate of heat production in kilocalories per hour. The

basic unit of heat energy is the **calorie**, which is the amount of heat required to raise the temperature of 1 g of H_2O 1°C. This unit is too small to be convenient when discussing the human body because of the magnitude of heat involved, so the **kilocalorie** or **Calorie**, which is equivalent to 1,000 calories, is used. When nutritionists speak of "calories" in quantifying the energy content of various foods, they are actually referring to kilocalories or Calories. Four kilocalories of heat energy are released when 1 g of glucose is oxidized or "burned," whether the oxidation takes place inside or outside the body.

Conditions for measuring the basal metabolic rate

The metabolic rate and consequently the amount of heat produced vary depending on a variety of factors, such as exercise, anxiety, shivering, and food intake. Increased skeletal muscle activity is the factor that can increase metabolic rate to the greatest extent. Even slight increases in muscle tone notably elevate the metabolic rate, and various levels of physical activity alter energy expenditure and heat production markedly (▲ Table 17-1). For this reason, a person's metabolic rate is determined under standardized basal conditions established to control as many as possible of the variables that can alter metabolic rate. In this way, the metabolic activity necessary to maintain the basic body functions at rest can be determined. Thus, the so-called **basal metabolic rate (BMR)** is a reflection of the

▲ **TABLE 17-1**

Rate of Energy Expenditure for a 70-kg Person during Different Types of Activity

Form of Activity	Energy Expenditure (Kcal/Hour)
Sleeping	65
Awake, lying still	77
Sitting at rest	100
Standing relaxed	105
Getting dressed	118
Typewriting	140
Walking slowly on level (2.6 mi/hr)	200
Carpentry, painting a house	240
Sexual intercourse	280
Bicycling on level (5.5 mi/hr)	304
Shoveling snow, sawing wood	480
Swimming	500
Jogging (5.3 mi/hr)	570
Rowing (20 strokes/min)	828
Walking up stairs	1,100

body's "idling speed," or the minimal waking rate of internal energy expenditure. The BMR is measured under the following specified conditions:

1. The person should be at physical rest, having refrained from exercise for at least 30 minutes to eliminate any contribution of muscular exertion to heat production.

2. The person should be at mental rest to minimize skeletal muscle tone (people "tense up" when they are nervous) and to prevent a rise in epinephrine, a hormone secreted in response to stress that increases metabolic rate.

3. The measurement should be performed at a comfortable room temperature so the person does not shiver. Shivering can markedly increase heat production.

4. The subject should not have eaten any food within 12 hours before the BMR determination to avoid **diet-induced thermogenesis** (*thermo* means "heat"; *genesis* means "production"), or the obligatory increase in metabolic rate that occurs as a consequence of food intake. This short-lived (less than 12-hour) rise in metabolic rate is due not to digestive activities but to the increased metabolic activity associated with the processing and storage of ingested nutrients, especially by the major biochemical factory, the liver.

Methods of measuring the basal metabolic rate

The rate of heat production in BMR determinations can be measured directly or indirectly. **Direct calorimetry** involves the cumbersome procedure of placing the subject in an insulated chamber with H_2O circulating through the walls. The difference in the temperature of the H_2O entering and leaving the chamber reflects the amount of heat liberated by the subject and picked up by the H_2O as it passes through the chamber. Even though this method provides a direct measurement of heat production, it is not practical because a calorimeter chamber is costly and takes up a lot of space. Therefore, a more practical method of indirectly determining the rate of heat production was developed for widespread use. With **indirect calorimetry**, only the subject's O_2 uptake per unit of time is measured, which is a simple task using minimal equipment. Recall that

$$\text{Food} + O_2 \rightarrow CO_2 + H_2O + \text{energy (mostly transformed into } \mathbf{heat})$$

Accordingly, a direct relationship exists between the volume of O_2 used and the quantity of heat produced. This relationship also depends on the type of food being oxidized. Although carbohydrates, proteins, and fats require different amounts of O_2 for their oxidation and yield different amounts of kilocalories when oxidized, an average estimate can be made of the quantity of heat produced per liter of O_2 consumed on a typical mixed American diet. This approximate value, known as the **energy equivalent of O_2**, is 4.8 kilocalories of energy liberated per liter of O_2 consumed. Using this method, the metabolic rate of a person consuming 15 liters/hr of O_2 can be estimated as follows:

15	liters/hr	$= O_2$ consumption
$\times$ 4.8	kilocalories/liter	= energy equivalent of O_2
72	kilocalories/hr	= estimated metabolic rate

In this way, a simple measurement of O_2 consumption can be used to reasonably approximate heat production in determining metabolic rate.

Once the rate of heat production is determined under the prescribed basal conditions, it must be compared with normal values for people of the same sex, age, height, and weight, because these factors all affect the basal rate of energy expenditure. For example, a large man actually has a higher rate of heat production than a smaller man, but, expressed in terms of total surface area (which is a reflection of height and weight), the output in kilocalories per hour per square meter of surface area is normally about the same.

Factors influencing the basal metabolic rate

Thyroid hormone is the primary but not sole determinant of the rate of basal metabolism. As thyroid hormone increases, the BMR increases correspondingly. As mentioned, epinephrine also increases the BMR.

Surprisingly, the BMR is not the body's lowest metabolic rate. The rate of energy expenditure during sleep is 10% to 15% lower than the BMR, presumably because of the more complete muscle relaxation that occurs during the paradoxical stage of sleep (see p. 171).

▍ Energy input must equal energy output to maintain a neutral energy balance.

Because energy cannot be created or destroyed, energy input must equal energy output, as follows:

Energy input	=	energy output			
Energy in food consumed	=	external work	+	internal heat production	$\pm$ stored energy

There are three possible states of energy balance:

- *Neutral energy balance.* If the amount of energy in food intake exactly equals the amount of energy expended by the muscles in performing external work plus the basal internal energy expenditure that eventually appears as body heat, then energy input and output are exactly in balance, and body weight remains constant.
- *Positive energy balance.* If the amount of energy in food intake is greater than the amount of energy expended by means of external work and internal functioning, the extra energy taken in but not used is stored in the body, primarily as adipose tissue, so body weight increases.
- *Negative energy balance.* Conversely, if the energy derived from food intake is less than the body's immediate energy requirements, the body must use stored energy to supply energy needs, and body weight decreases accordingly.

To maintain a constant body weight (with the exception of minor fluctuations caused by changes in H_2O content), energy acquired through food intake must equal energy expenditure by the body. Because the average adult maintains a fairly constant weight over long periods of time, this implies that precise homeostatic mechanisms exist to maintain a long-term balance between energy intake and energy expenditure. Theoretically, total body energy content could be maintained at a constant level by regulating the magnitude of food intake, physical activity, or internal work and heat production. Control of food intake to match changing metabolic expenditures is the major means of maintaining a neutral energy balance. The level of physical activity is principally under voluntary control, and mechanisms that alter the degree of internal work and heat production are aimed primarily at regulating body temperature rather than total energy balance.

However, after several weeks of eating less or more than desired, small counteracting changes in metabolism may occur. For example, a compensatory increase in the body's efficiency of energy use in response to underfeeding partially explains why some dieters become stuck at a plateau after having lost the first 10 or so pounds of weight fairly easily. Similarly, a compensatory reduction in the efficiency of energy use in response to overfeeding accounts in part for the difficulty experienced by very thin people who are deliberately trying to gain weight. Despite these modest compensatory changes in metabolism, regulation of food intake is the most important factor in the long-term maintenance of energy balance and body weight.

▌Food intake is controlled primarily by the hypothalamus.

Control of food intake is primarily a function of the hypothalamus. Classically, the hypothalamus is considered to house a pair of **feeding,** or **appetite, centers** located in the lateral (outer) regions of the hypothalamus, one on each side, and another pair of **satiety centers** located in the ventromedial (underside middle) area. **Satiety** is the feeling of being full. The functions of these areas have been elucidated by a series of experiments that involve either destruction or stimulation of these specific regions. Stimulation of the clusters of nerve cells designated as feeding, or appetite, centers makes the animal hungry, driving it to eat voraciously, whereas selective destruction of these areas suppresses eating and food intake behavior to the point that the animal starves itself to death. In contrast, stimulation of the satiety centers signals satiety, or the feeling of having had enough to eat. Consequently, the stimulated animal refuses to eat even if previously deprived of food. As expected, destruction of this area produces the opposite effect—profound overeating and obesity, because the animal never achieves a feeling of being full (● Figure 17-2). Thus the feeding centers tell us to eat, whereas the satiety centers tell us when we have had enough. Although it is convenient to consider these specific areas as exciting and inhibiting feeding behavior, respectively, this approach is too simplistic. It is now known that complex systems rather than isolated centers control feeding and satiety. Multiple, highly integrated, redundant pathways crisscrossing into and out of the hypothalamus are involved in the control of food intake and maintenance of energy balance.

Even though food intake is adjusted to balance changing energy expenditures over a period of time, there are no calorie receptors per se to monitor energy input, energy output, or total body energy content. Instead, various blood-borne chemical fac-

● FIGURE 17-2

Comparison of a normal rat with a rat whose satiety center has been destroyed
Several months after destruction of the classical satiety center in the ventromedial area of the hypothalamus, the rat on the right had gained considerable weight as a result of overeating compared to its normal littermate on the left. Rats sustaining lesions in this area also display less grooming behavior, accounting for the soiled appearance of the fat rat.

tors that signal the body's nutritional state, such as how much fat is stored or how much glucose is available, are important in regulating food intake. Control of food intake does not depend on changes in a single signal but is determined by the integration of many inputs that provide information about the body's energy status. Multiple molecular signals together ensure that feeding behavior is synchronized with the body's immediate and long-term energy needs. Some information is used for short-term regulation of food intake, helping to control meal size and frequency. Even so, over a 24-hour period the energy in ingested food rarely matches energy expenditure for that day. The correlation between total caloric intake and total energy output is excellent, however, over long periods of time. As a result, the total energy content of the body—and, consequently, body weight—remains relatively constant on a long-term basis. Thus energy homeostasis, that is, energy balance, is carefully regulated.

Based on current evidence, the following regulatory factors are among those that contribute to the control of food intake and maintenance of energy balance (● Figure 17-3).

The size of fat stores

Our notion of fat cells (**adipocytes**) in adipose tissue as merely storage space for triglyceride fat has undergone a dramatic change in the past decade with the discovery of their active role in energy homeostasis. Adipocytes secrete **leptin**, a hormone essential for normal body weight regulation (*leptin* means "thin"). The amount of leptin in the blood is an excellent indicator of the total amount of triglyceride fat stored in adipose tissue: The larger the fat stores, the more leptin released into the blood. This blood-borne signal, discovered in the mid-1990s, was the first molecular satiety signal identified.

The **arcuate nucleus** of the hypothalamus is the major site for leptin action. The arcuate nucleus is an arc-shaped collection of neurons located adjacent to the floor of the third ventricle, in the region of the classical satiety center. Leptin receptors are found in abundance here. Acting in negative-feedback fashion, increased leptin from burgeoning fat stores serves as a "trim down" signal. Binding of leptin to its hypothalamic receptors leads to suppression of appetite to decrease food consumption while, to a lesser extent, causing an increase in metabolic rate, thus increasing energy expenditure. These actions together promote weight loss. The leptin signal is generally considered responsible for the long-term matching of food intake to energy expenditure so that total body energy content remains balanced and body weight remains constant.

Interestingly, leptin has recently been shown to also be important in reproduction. It is one of the triggers for the onset of puberty, signaling that the female has attained sufficient long-term energy (adipose) stores to sustain a pregnancy. This role of leptin is undoubtedly a factor in the observation that the onset of menstrual cycles is often delayed in lean, young female athletes. By contrast, the increased leptin associated with the better nutritional status of young women in industrialized societies has likely contributed to the earlier onset of menstrual cycles witnessed over the last several centuries. The average age for the start of menstruation in the United States is now 12 years compared to the average onset at 18 years of age in the 1600s.

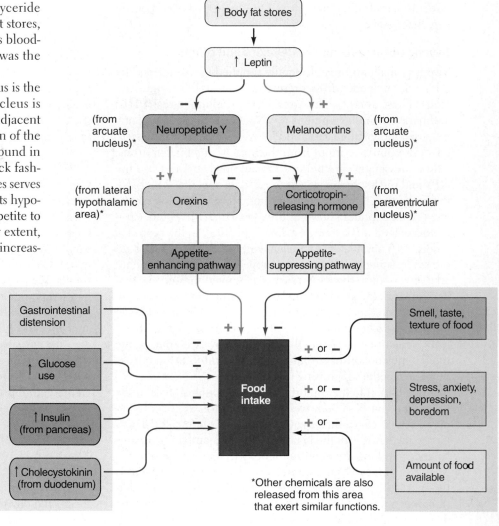

● **FIGURE 17-3**
Factors that influence food intake

Satiety signals important in short-term control of the timing and size of meals

Adipose tissue-related signals important in long-term matching of food intake to energy expenditure to control body weight

Psychosocial and environmental factors that influence food intake

Role of neuropeptide Y and melanocortins

The arcuate nucleus has two subsets of neurons that are regulated by leptin but in an opposing manner. One subset releases *neuropeptide Y*, while the other releases *melanocortins*. **Neuropeptide Y (NPY)** is one of the most potent appetite stimulators ever found. NPY not only triggers voracious feeding behavior but also lowers energy expenditure by suppressing sympathetic nervous system outflow, thereby decreasing the metabolic rate. Thus, NPY promotes weight gain. Leptin suppresses food intake and increases metabolic rate in part by inhibiting hypothalamic output of NPY. On the other side of the coin, the loss of fat stores and the resultant reduction in leptin caused by dieting brings about increased NPY release. NPY, in turn, sets in motion actions that oppose the weight loss.

Melanocortins, a group of hormones traditionally known to be important in varying the skin color for the purpose of camouflage in some species, have been shown to exert an unexpected role in energy homeostasis. Melanocortins, most notably *melanocyte-stimulating hormone* (see p. 683), inhibit the hypothalamus to decrease food intake. Melanocortins do not play a role in determining skin coloration in humans. Their importance in our species lies in part in toning down appetite in response to increased fat stores. Leptin stimulates the melanocortin system. Thus leptin promotes weight loss by simultaneously inhibiting the appetite-enhancing NPY pathway and stimulating the appetite-suppressing melanocortin pathway.

But NPY and melanocortin may not be the final effectors in appetite control. Recent evidence suggests that these arcuate-nucleus chemical messengers may in turn influence the release of neuropeptides in other parts of the brain that exert control over food intake.

Beyond the arcuate nucleus: Orexins and others

Two hypothalamic areas are richly supplied by axons from the NPY- and melanocortin-secreting neurons of the arcuate nucleus. These areas are the **lateral hypothalamic area (LHA)** and **paraventricular nucleus (PVN).** According to a recently proposed model, the LHA and PVN release chemical messengers in response to input from the arcuate nucleus neurons. These messengers are believed to act downstream from the NPY and melanocortin signals to regulate appetite. The LHA, the site of the classical appetite center, produces two closely related neuropeptides known as **orexins,** which are potent stimulators of food intake (*orexis* means "appetite). In the proposed model, NPY stimulates and melanocortins inhibit the release of orexins, leading to an increase in appetite and greater food intake. By contrast, the PVN releases chemical messengers that decrease appetite and food intake. An example is corticotropin-releasing hormone, which, as its name implies, is better known for its role as a hormone. (You will learn more about this chemical's endocrine function in the next chapter.) According to the model, melanocortins stimulate and NPY inhibits the release of these appetite-suppressing neuropeptides.

In addition to the importance of molecular signals in the long-term control of body weight, other factors are believed to play a role in controlling the timing and size of meals. In contrast to the key role of the hypothalamus in maintaining energy balance and long-term control of body weight, a region in the brain stem known as the **nucleus tractus solitarius (NTS)** is thought to be the site that processes signals important in terminating a meal. The NTS receives afferent inputs from the digestive tract and elsewhere that signal satiety and also receives input from the higher hypothalamic neurons involved in energy homeostasis. We now turn our attention to the most important of these satiety signals (● Figure 17-3).

The extent of gastrointestinal distension

Early proposals suggested that cues of emptiness or fullness of the digestive tract signaled hunger or satiety, respectively. For example, stimulation of gastric stretch receptors has been shown to suppress food intake. However, neural input arising from stomach distension plays a more important role in controlling the rate of gastric emptying than in signaling satiety (see p. 606). Researchers now believe that internal blood-borne signals reflecting the depletion or availability of energy-producing substances are more important than stomach volume in controlling the initiation and cessation of eating.

The extent of glucose use and insulin secretion

Satiety is signaled by increased glucose use, such as occurs during a meal when more glucose is available for use because it is being absorbed from the digestive tract. Conversely, after absorption of a meal is complete, and no new glucose is entering the blood, the resultant reduction in the cells' glucose use arouses the sensation of hunger. The extent of glucose use appears more important in determining the timing of meals than in the long-term control of body weight.

In a related mechanism, increased insulin in the blood signals satiety. Insulin, a hormone secreted by the pancreas in response to a rise in the concentration of glucose and other nutrients in the blood following a meal, stimulates cellular uptake, use, and storage of glucose and other nutrients. Thus, the increase in insulin secretion that accompanies nutrient abundance and promotes increased glucose use is an appropriate satiety signal.

The level of cholecystokinin secretion

Cholecystokinin (CCK), one of the gastrointestinal hormones released from the duodenal mucosa during digestion of a meal, is an important satiety signal for regulating the size of meals. CCK is secreted in response to the presence of nutrients in the small intestine. Through multiple effects on the digestive system, CCK facilitates the digestion and absorption of these nutrients (see p. 640). It is appropriate that this blood-borne signal, whose rate of secretion is correlated with the amount of nutrients ingested, also contributes to the sense of being filled after a meal has been consumed but before it has actually been digested and absorbed. We feel satisfied when adequate food to replenish the stores is in the digestive tract, even though the body's energy stores are still low. This explains why we stop eating before the ingested food is made available to meet the body's energy needs.

Psychosocial and environmental influences

Thus far we have described involuntary signals that automatically occur to control food intake. However, as with water intake, people's eating habits are also shaped by psychological, social, and environmental factors. Often our decision to eat or stop eating is not determined merely by whether we are hungry or full, respectively. Frequently, we eat out of habit (eating three meals a day on schedule no matter what our status on the hunger–satiety continuum) or because of social custom (food often plays a prime role in entertainment, leisure, and business activities). Even well-intentioned family pressure—"Clean your plate before you can leave the table"—can have an impact on the amount consumed.

Furthermore, the amount of pleasure derived from eating can reinforce feeding behavior. Eating foods with an enjoyable taste, smell, and texture can increase appetite and food intake. This has been demonstrated in an experiment in which rats were offered their choice of highly palatable human foods. They overate by as much as 70% to 80% and became obese. When the rats returned to eating their regular monotonous but nutritionally balanced rat chow, their obesity was rapidly reversed as their food intake was controlled once again by physiological drives rather than by hedonistic urges for the tastier offerings.

Stress, anxiety, depression, and boredom have also been shown to alter feeding behavior in ways that are unrelated to energy needs in both experimental animals and humans. People often eat to satisfy psychological needs rather than because they are hungry. Furthermore, environmental influences, such as amount of food available, play an important role in determining extent of food intake. Thus, any comprehensive explanation of how food intake is controlled must take into account these voluntary eating acts that can reinforce or override the internal signals governing feeding behavior.

▌ Obesity occurs when more kilocalories are consumed than are burned up.

Obesity is defined as excessive fat content in the adipose tissue stores; the arbitrary boundary for obesity is generally considered to be greater than 20% overweight compared to normal standards. Over half of the adults in the United States are overweight, and one-third are clinically obese. Obesity occurs when, over a period of time, more kilocalories are ingested in food than are used to support the body's energy needs, with the excessive energy being stored as triglycerides in adipose tissue. The causes of obesity are many, and some remain obscure. Some factors that may be involved include the following:

• *Disturbances in the leptin signaling pathway.* Some cases of obesity have been linked to leptin resistance. For many overweight people, excess energy input occurs only during the time that obesity is actually developing. Some investigators suggest that the hypothalamic centers involved in maintaining energy homeostasis are "set at a higher level" in obese people. Thus overweight people do tend to maintain their weight but at a higher set point than normal. Once obesity is developed, all that is required to maintain the condition is that energy input equals energy output. For example, the problem may lie with faulty leptin receptors in the brain that do not respond appropriately to the high levels of circulating leptin from abundant adipose stores. Thus the brain does not detect leptin as a signal to turn down appetite until a higher set point (and accordingly greater fat storage) is achieved. Instead of faulty leptin receptors, other disturbances in the leptin pathway may be at fault, such as defective transport of leptin across the blood–brain barrier or a deficiency of one of the chemical messengers in the leptin pathway.

• *Lack of exercise.* Numerous studies have shown that, on the average, fat people do not eat any more than thin people. One possible explanation is that overweight persons do not overeat but "underexercise"—the "couch potato" syndrome. Very low levels of physical activity typically are not accompanied by comparable reductions in food intake.

For this reason, modern technology is partly to blame for the current obesity epidemic. Our ancestors had to exert physical effort to eke out a subsistence. By comparison, we now have machines to replace much manual labor, remote controls to operate our machines with minimal effort, and computers that encourage long hours of sitting. We have to make a conscious effort to exercise.

• *Differences in the "fidget factor."* **Nonexercise activity thermogenesis (NEAT)**, or the "fidget factor," might explain some variation in fat storage among people. Those who engage in toe tapping or other types of repetitive, spontaneous physical activity expend a substantial number of kilocalories throughout the day without a conscious effort.

• *Differences in extracting energy from food.* Another reason why lean people and obese people may have dramatically different body weights despite consuming the same number of kilocalories may lie in the efficiency with which each extracts energy from food. Studies suggest that leaner individuals tend to derive less energy from the food they consume, because they convert more of the food's energy into heat than into energy for immediate use or for storage. For example, slimmer individuals have more **uncoupling proteins,** which allow their cells to convert more of the nutrient calories into heat instead of to fat. These are the people who can eat a lot without gaining weight. By contrast, obese people may have more efficient metabolic systems for extracting energy from food—a useful trait in times of food shortage but a hardship when trying to maintain a desirable weight when food is plentiful.

• *Hereditary tendencies.* Often, differences in the regulatory pathways for energy balance—either those governing food intake or those influencing energy expenditure—arise from genetic variations.

• *Development of an excessive number of fat cells as a result of overfeeding.* One of the problems in fighting obesity is that once fat cells are created, they do not disappear with dieting and weight loss. Even if a dieter loses a large portion of the triglyceride fat stored in these cells, the depleted cells remain, ready to refill. Therefore, rebound weight gain after losing weight is difficult to avoid and discouraging for the dieter.

• *The existence of certain endocrine disorders such as hypothyroidism* (see p. 706). Hypothyroidism involves a deficiency of thyroid hormone, the main factor that bumps up the BMR so that the body burns more calories in its idling state.

• *An abundance of convenient, highly palatable, energy-dense, relatively inexpensive foods.*

• *Emotional disturbances in which overeating replaces other gratifications.*

• *A possible virus link.* One intriguing new proposal links a relatively common cold virus to a propensity to become overweight and may account for a portion of the current obesity epidemic.

Despite this rather lengthy list, our knowledge about the causes and control of obesity is still rather limited, as evidenced by the number of people who are constantly trying to stabilize their weight at a more desirable level. This is important from more than an aesthetic viewpoint. It is known that obesity, especially of the android type, can predispose an individual to illness and premature death from a multitude of diseases. (To learn about the differences between android and gynoid obesity, see the accompanying boxed feature, ▶ A Closer Look at Exercise Physiology.)

▌ People suffering from anorexia nervosa have a pathologic fear of gaining weight.

The converse of obesity is generalized nutritional deficiency. The obvious causes for reduction of food intake below en-

What the Scales Don't Tell You

Body composition is the percentage of body weight that is composed of lean tissue and adipose tissue. Assessing body composition is an important step in evaluating a person's health status. One crude means of assessing body composition is by calculating the **body mass index (BMI)** using the following formula:

$$BMI = \frac{(\text{weight in pounds}) \times 700}{(\text{height in inches})^2}$$

A BMI of 25 or less is considered healthy, whereas a BMI of 30 or higher is considered to place the individual at increased risk for various diseases and premature death. BMIs between this range are considered borderline.

BMI determinations and the age–height–weight tables used by insurance companies can be misleading for determining healthy body weight. By these charts, many athletes, for example, would be considered overweight. A football player may be 6 feet 5 inches tall and weigh 300 pounds, but have only 12% body fat. This player's extra weight is muscle, not fat, and therefore is not a detriment to his health. A sedentary person, in contrast, may be normal on the height–weight charts but have 30% body fat. This person should maintain body weight while increasing muscle mass and decreasing fat. Ideally, men should have 15% fat or less and women should have 20% fat or less.

The most accurate method for assessing body composition is underwater weighing. This technique is based on the fact that lean tissue is denser than water and fat tissue is less dense than water. (You can readily demonstrate this

for yourself by dropping a piece of lean meat and a piece of fat into a glass of water; the lean meat will sink and the fat will float.) In underwater weighing, the person breathes out as much air as possible and then completely submerges in a tank of water while sitting in a swing that is attached to a scale. The results are used to determine body density using equations that take into consideration the density of water, the difference between the person's weight in air and underwater, and the residual volume of air remaining in the lungs. Because of the difference in density between lean and fat tissue, people who have more fat have a lower density and weigh relatively less underwater than in air compared to their lean counterparts. Body composition is then determined by means of an equation that correlates percentage fat with body density.

Another common way to assess body composition is skinfold thickness. Because approximately half of the body's total fat content is located just beneath the skin, total body fat can be estimated from measurements of skinfold thickness taken at various sites on the body. Skinfold thickness is determined by pinching up a fold of skin at one of the designated sites and measuring its thickness by means of a caliper, a hinged instrument that fits over the fold and is calibrated to measure thickness. Mathematical equations specific for the person's age and sex can be used to predict the percentage of fat from the skinfold thickness scores. A major criticism of skinfold assessments is that accuracy depends on the investigator's skill.

There are different ways to be fat, and one way is more dangerous than the other. Obese patients can be classified into two categories—

android, a male-type of adipose tissue distribution, and *gynoid,* a female-type distribution—based on the anatomic distribution of adipose tissue measured as the ratio of waist circumference to hip circumference. Android obesity is characterized by abdominal fat distribution (people shaped as "apples"), whereas gynoid obesity is characterized by fat distribution in the hips and thighs (people shaped as "pears"). Both sexes can display either android or gynoid obesity.

Android obesity is associated with a number of disorders, including insulin resistance, type II (adult-onset) diabetes mellitus, excess blood lipid levels, high blood pressure, coronary heart disease, and stroke. Gynoid obesity is not associated with high risk for these diseases. Because android obesity is associated with increased risk for disease, it is most important for apple-shaped overweight individuals to reduce their fat stores.

Research on the success of weight reduction programs indicates that it is very difficult for people to lose weight, but when weight loss occurs, it is from the areas of increased stores. Because very-low-calorie diets are difficult to maintain, an alternative to severely cutting caloric intake to lose weight is to increase energy expenditure through physical exercise. Exercise physiologists often assess body composition as an aid in prescribing and evaluating exercise programs. Exercise generally reduces the percentage of body fat and, by increasing muscle mass, increases the percentage of lean tissue. An aerobic exercise program further helps reduce the risk of the disorders associated with android obesity.

---

ergy needs are lack of availability of food, interference with swallowing or digestion, and impairment of appetite.

One poorly understood disorder in which lack of appetite is a prominent feature is **anorexia nervosa.** Patients with this disorder, most commonly adolescent girls and young women, have a morbid fear of becoming fat. They have a distorted body image, tending to visualize themselves as being much heavier than they actually are. Because they have an aversion to food, they eat very little and consequently lose considerable weight, perhaps even starving themselves to death. Other characteristics of the condition include altered secretion of many hormones, absence of menstrual periods, and low body temperature. It is unclear whether these symptoms occur secondarily as a result of general malnutrition or arise independently of the eating disturbance as a part of a primary hypothalamic malfunction. Many investigators think the underlying problem may be psychological rather than biological. Some experts suspect that anorexics may suffer from addiction to endogenous

opiates, self-produced morphinelike substances (see p. 194) that are thought to be released during prolonged starvation.

TEMPERATURE REGULATION

Humans are usually in environments cooler than their bodies, but they constantly generate heat internally, which helps maintain body temperature. Heat production ultimately depends on the oxidation of metabolic fuel derived from food.

Changes in body temperature in either direction alter cellular activity—an increase in temperature speeds up cellular chemical reactions, whereas a fall in temperature slows down these reactions. Because cellular function is sensitive to fluctuations in internal temperature, humans homeostatically maintain body temperature at a level that is optimal for cellular metabolism to proceed in a stable fashion. Overheating is more serious than cooling. Even moderate elevations of body temperature begin to cause nerve malfunction and irreversible protein denaturation. Most people suffer convulsions when the internal body temperature reaches about 106°F (41°C); 110°F (43.3°C) is considered the upper limit compatible with life.

By contrast, most of the body's tissues can transiently withstand substantial cooling. This characteristic is useful during cardiac surgery when the heart must be stopped. The patient's body temperature is deliberately lowered. The cooled tissues need less nourishment than they do at normal body temperature because of their reduction in metabolic activity. However, a pronounced, prolonged fall in body temperature slows metabolism to a fatal level.

▎ Internal core temperature is homeostatically maintained at 100°F.

Normal body temperature has traditionally been considered 98.6°F (37°C). However, a recent study indicates that normal body temperature varies among individuals and varies throughout the day, ranging from 96.0°F in the morning to 99.9°F in the evening with an overall average of 98.2°F. These values are considered normal for temperatures taken orally (by mouth). Furthermore, there is no one body temperature because the temperature varies from organ to organ.

From a thermoregulatory viewpoint, the body may conveniently be viewed as a *central core* surrounded by an *outer shell*. The temperature within the inner core, which consists of the abdominal and thoracic organs, the central nervous system, and the skeletal muscles, generally remains fairly constant. This internal **core temperature** is subject to precise regulation to maintain its homeostatic constancy. The core tissues function best at a relatively constant temperature of around 100°F.

The skin and subcutaneous fat constitute the outer shell. In contrast to the constant high temperature in the core, the temperature within the shell is generally cooler and may vary substantially. For example, skin temperature may fluctuate between 68° and 104°F without damage. In fact, as you will see, the temperature of the skin is deliberately varied as a control measure to help maintain the core's thermal constancy.

Sites for monitoring body temperature

Several easily accessible sites are used for monitoring body temperature. The oral and axillary (under the armpit) temperatures are comparable, while rectal temperature averages about 1°F higher. Also recently available is a temperature-monitoring instrument that scans the heat generated by the eardrum and converts this temperature into an oral equivalent. However, none of these measurements is an absolute indication of the internal core temperature, which is a bit higher at 100°F than the monitored sites.

Normal variations in core temperature

Even though the core temperature is held relatively constant, several factors cause it to vary slightly:

1. Most people's core temperature normally varies about 1.8°F (1°C) during the day, with the lowest level occurring early in the morning before rising (6 to 7 A.M.) and the highest point occurring in late afternoon (5 to 7 P.M.). This variation is due to an innate biological rhythm or "biological clock."

2. Women also experience a monthly rhythm in core temperature in connection with their menstrual cycle. The core temperature averages 0.9°F (0.5°C) higher during the last half of the cycle from the time of ovulation to menstruation. This mild sustained elevation in temperature was once thought to be caused by the increased secretion of progesterone, one of the ovarian hormones, during this period, but this is no longer believed to be the case. The actual cause is still undetermined.

3. The core temperature increases during exercise because of the tremendous increase in heat production by the contracting muscles. During hard exercise, the core temperature may increase to as much as 104°F (40°C). In a resting person, this temperature would be considered a fever, but it is normal during strenuous exercise.

4. Because the temperature-regulating mechanisms are not 100% effective, the core temperature may vary slightly with exposure to extremes of temperature. For example, the core temperature may fall several degrees in cold weather or rise a degree or so in hot weather.

Thus the core temperature can vary at the extremes between about 96° to 104°F but usually deviates less than a few degrees. This relative constancy is made possible by multiple thermoregulatory mechanisms coordinated by the hypothalamus.

▎ Heat input must balance heat output to maintain a stable core temperature.

The core temperature is a reflection of the body's total heat content. To maintain a constant total heat content and thus a stable core temperature, heat input to the body must balance heat output (● Figure 17-4). *Heat input* occurs by way of heat gain from the external environment and internal heat production, the latter being the most important source of heat for the body. Recall that most of the body's energy expenditure ultimately appears as heat. This heat is important in the maintenance of core temperature. In fact, usually more heat is gener-

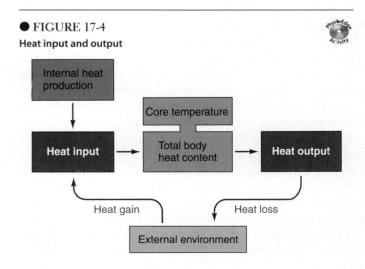

● **FIGURE 17-4**

Heat input and output

ated than is required to maintain the body temperature at a normal level, so the excess heat must be eliminated from the body. *Heat output* occurs by way of heat loss from exposed body surfaces to the external environment.

Balance between heat input and output is frequently disturbed by (1) changes in internal heat production for purposes unrelated to regulation of body temperature, most notably by exercise, which markedly increases heat production, and (2) changes in the external environmental temperature that influence the degree of heat gain or heat loss that occurs between the body and its surroundings. To maintain body temperature within narrow limits despite changes in metabolic heat production and changes in environmental temperature, compensatory adjustments must take place in heat loss and heat gain mechanisms. If the core temperature starts to fall, heat production is increased and heat loss is minimized so that normal temperature can be restored. Conversely, if the temperature starts to rise above normal, it can be corrected by increasing heat loss while simultaneously reducing heat production.

We will now elaborate on the means by which heat gains and losses can be adjusted to maintain body temperature.

▌ Heat exchange takes place by radiation, conduction, convection, and evaporation.

All heat loss or heat gain between the body and the external environment must take place between the body surface and its surroundings. The same physical laws of nature that govern heat transfer between inanimate objects also control the transfer of heat between the body surface and the environment. The temperature of an object may be thought of as a measure of the concentration of heat within the object. Accordingly, heat always moves down its concentration gradient; that is, down a **thermal gradient** from a warmer to a cooler region (*thermo* means "heat").

The body uses four mechanisms of heat transfer: *radiation, conduction, convection,* and *evaporation.*

Radiation

Radiation is the emission of heat energy from the surface of a warm body in the form of **electromagnetic waves,** or **heat waves,** which travel through space (● Figure 17-5 ①). When radiant energy strikes an object and is absorbed, the energy of the wave motion is transformed into heat within the object. The human body both emits (source of heat loss) and absorbs (source of heat gain) radiant energy. Whether the body loses or gains heat by radiation depends on the difference in temperature between the skin surface and the surfaces of various other objects in the body's environment. Because net transfer of heat by radiation is always from warmer objects to cooler ones, the body gains heat by radiation from objects warmer than the skin surface, such as the sun, a radiator, or burning logs. By contrast, the body loses heat by radiation to objects in its environment whose surfaces are cooler than the surface of the skin, such as building walls, furniture, or trees. On the average, humans lose close to half of their heat energy through radiation.

Conduction

Conduction is the transfer of heat between objects of differing temperatures that are in direct contact with each other (● Figure 17-5 ②). Heat moves down its thermal gradient from the warmer to the cooler object by being transferred from molecule to molecule. All molecules are constantly in vibratory motion, with warmer molecules moving faster than cooler ones. When molecules of differing heat content touch each other, the faster-moving, warmer molecule agitates the cooler molecule into more rapid motion, thereby warming up the cooler molecule. During this process, the original warmer molecule loses some of its thermal energy as it slows down and cools off a bit. Given enough time, therefore, the temperature of the two touching objects eventually equalizes.

The rate of heat transfer by conduction depends on the *temperature difference* between the touching objects and the *thermal conductivity* of the substances involved (that is, how easily heat is conducted by the molecules of the substances). Heat can be lost or gained by conduction when the skin is in contact with a good conductor. When you hold a snowball, for example, your hand becomes cold because heat moves by conduction from your hand to the snowball. Conversely, when you apply a heating pad to a body part, the part is warmed up as heat is transferred directly from the pad to the body.

Similarly, you either lose or gain heat by conduction to the layer of air in direct contact with your body. The direction of heat transfer depends on whether the air is cooler or warmer, respectively, than your skin. Only a small percentage of total heat exchange between the skin and environment takes place by conduction alone, however, because air is not a very good conductor of heat. (This is why swimming pool water at 80°F feels cooler than air at the same temperature; heat is conducted more rapidly from the body surface into the water, which is a good conductor, than into the air, which is a poor conductor.)

Convection

The term **convection** refers to the transfer of heat energy by *air* (or H_2O) *currents*. As the body loses heat by conduction to the surrounding cooler air, the air in immediate contact with the skin is warmed. Because warm air is lighter (less dense) than cool air, the warmed air rises while cooler air moves in next to the skin to replace the vacating warm air. The process is then repeated (● Figure 17-5 ③). These air movements, known as *convection currents*, help carry heat away from the body. If it were not for convection currents, no further heat could be dissipated from the skin by conduction once the temperature of the layer of air immediately around the body equilibrated with skin temperature.

The combined conduction-convection process of dissipating heat from the body is enhanced by forced movement of air across the body surface, either by external air movements, such as those caused by the wind or a fan, or by movement of the body through the air, as during bicycle riding. Because forced air movement sweeps away the air warmed by conduction and replaces it with cooler air more rapidly, a greater total amount of heat can be carried away from the body over a given time period. Thus wind makes us feel cooler on hot days, and windy

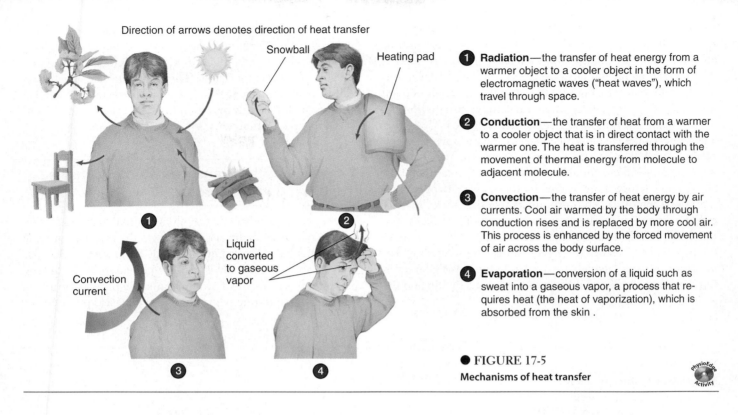

Direction of arrows denotes direction of heat transfer

Snowball

Heating pad

Convection current

Liquid converted to gaseous vapor

1. **Radiation**—the transfer of heat energy from a warmer object to a cooler object in the form of electromagnetic waves ("heat waves"), which travel through space.

2. **Conduction**—the transfer of heat from a warmer to a cooler object that is in direct contact with the warmer one. The heat is transferred through the movement of thermal energy from molecule to adjacent molecule.

3. **Convection**—the transfer of heat energy by air currents. Cool air warmed by the body through conduction rises and is replaced by more cool air. This process is enhanced by the forced movement of air across the body surface.

4. **Evaporation**—conversion of a liquid such as sweat into a gaseous vapor, a process that requires heat (the heat of vaporization), which is absorbed from the skin .

● **FIGURE 17-5**
Mechanisms of heat transfer

days in the winter are more chilling than calm days at the same cold temperature. For this reason, weather forecasters have developed the concept of *wind chill factor*.

Evaporation

Evaporation is the final method of heat transfer used by the body. When water evaporates from the skin surface, the heat required to transform water from a liquid to a gaseous state is absorbed from the skin, thereby cooling the body (● Figure 17-5 4). Evaporative heat loss makes you feel cooler when your bathing suit is wet than when it is dry. Evaporative heat loss occurs continually from the linings of the respiratory airways and from the surface of the skin. Heat is continuously lost through the H_2O vapor in the expired air as a result of the air's humidification during its passage through the respiratory system. Similarly, because the skin is not completely waterproof, H_2O molecules constantly diffuse through the skin and evaporate. This ongoing evaporation from the skin is completely unrelated to the sweat glands. These passive evaporative heat-loss processes are not subject to physiologic control and go on even in very cold weather, when the problem is one of conserving body heat.

Sweating is an active evaporative heat-loss process under sympathetic nervous control. The rate of evaporative heat loss can be deliberately adjusted by varying the extent of sweating, which is an important homeostatic mechanism to eliminate excess heat as needed. In fact, when the environmental temperature exceeds the skin temperature, sweating is the only avenue for heat loss, because the body is gaining heat by radiation and conduction under these circumstances.

Sweat is a dilute salt solution that is actively extruded to the surface of the skin by sweat glands dispersed all over the body. The sweat glands can produce up to 2 liters of sweat/hour.

Sweat must be evaporated from the skin for heat loss to occur. If sweat merely drips from the surface of the skin or is wiped away, no heat loss is accomplished. The most important factor determining the extent of evaporation of sweat is the *relative humidity* of the surrounding air (the percentage of H_2O vapor actually present in the air compared to the greatest amount that the air can possibly hold at that temperature; for example, a relative humidity of 70% means that the air contains 70% of the H_2O vapor it is capable of holding). When the relative humidity is high, the air is already almost fully saturated with H_2O, so it has limited ability to take up additional moisture from the skin. Thus little evaporative heat loss can occur on hot, humid days. The sweat glands continue to secrete, but the sweat simply remains on the skin or drips off, instead of evaporating and producing a cooling effect. As a measure of the discomfort associated with combined heat and high humidity, meteorologists have devised the *temperature-humidity index*.

▌ The hypothalamus integrates a multitude of thermosensory inputs.

The hypothalamus serves as the body's thermostat. The home thermostat keeps track of the temperature in a room and triggers a heating mechanism (the furnace) or a cooling mechanism (the air conditioner) as necessary to maintain the room temperature at the indicated setting. Similarly, the hypothalamus, as the body's thermoregulatory integrating center, receives afferent information about the temperature in various regions of the body and initiates extremely complex, coordinated adjustments in heat gain and heat loss mechanisms as necessary to correct any deviations in core temperature from the normal setting. The hypothalamic thermostat is far more sensitive than your home thermostat. The hypothalamus is able to respond to changes in blood temperature as small as 0.01°C. The hypothalamus's degree of response to deviations in body temperature is finely matched so that precisely enough heat is lost or generated to restore the temperature to normal.

To make the appropriate adjustments in the delicate balance between the heat loss mechanisms and the opposing heat-producing and heat-conserving mechanisms, the hypothalamus must be apprised continuously of both the skin temperature and the core temperature by means of specialized temperature-sensitive receptors called **thermoreceptors**. *Peripheral thermoreceptors* monitor skin temperature throughout the body and transmit information about changes in surface temperature to the hypothalamus. The core temperature is monitored by *central thermoreceptors*, which are located in the hypothalamus itself as well as elsewhere in the central nervous system and the abdominal organs.

Two centers for temperature regulation have been identified in the hypothalamus. The *posterior region* is activated by cold and subsequently triggers reflexes that mediate heat production and heat conservation. The *anterior region*, which is activated by warmth, initiates reflexes that mediate heat loss. Let us examine the means by which the hypothalamus fulfills its thermoregulatory functions (● Figure 17-6).

▍ Shivering is the primary involuntary means of increasing heat production.

The body can gain heat as a result of internal heat production generated by metabolic activity or from the external environment if the latter is warmer than body temperature. Because body temperature usually is higher than environmental temperature, metabolic heat production is the primary source of body heat. In a resting person, most body heat is produced by

● FIGURE 17-6
Major thermoregulatory pathways

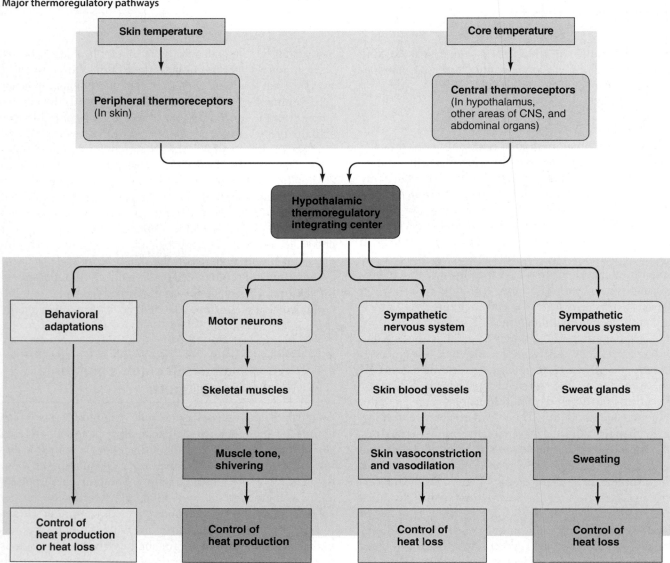

the thoracic and abdominal organs as a result of ongoing, cost-of-living metabolic activities. Above and beyond this basal level, the rate of metabolic heat production can be variably increased primarily by changes in skeletal muscle activity or to a lesser extent by certain hormonal actions. Thus changes in skeletal muscle activity constitute the major way heat gain is controlled for temperature regulation.

Adjustments in heat production by skeletal muscles

In response to a fall in core temperature caused by exposure to cold, the hypothalamus takes advantage of the fact that increased skeletal muscle activity generates more heat. Acting through descending pathways that terminate on the motor neurons controlling the body's skeletal muscles, the hypothalamus first gradually increases skeletal muscle tone. (Muscle tone is the constant level of tension within the muscles.) Soon shivering begins. **Shivering** consists of rhythmic, oscillating skeletal muscle contractions that occur at a rapid rate of ten to twenty per second. This mechanism is very effective in increasing heat production; all the energy liberated during these muscle tremors is converted to heat because no external work is accomplished. Within a matter of seconds to minutes, internal heat production may increase two- to fivefold as a result of shivering.

Frequently, these reflex changes in skeletal muscle activity are augmented by increased voluntary, heat-producing actions such as bouncing up and down or hand clapping. Such behavioral responses appear to share neural systems in common with the involuntary physiologic responses. The hypothalamus and limbic system are extensively involved with controlling motivated behavior (see p. 157).

In the opposite situation—an elevation in core temperature caused by heat exposure—two mechanisms are employed to reduce heat-producing skeletal muscle activity: Muscle tone is reflexly reduced, and voluntary movement is curtailed. When the air becomes very warm, people often complain it is "too hot even to move." These responses are not as effective at reducing heat production during heat exposure as are the muscular responses that increase heat production during cold exposure for two reasons. First, because muscle tone is normally quite low, the capacity to reduce it further is limited. Second, the elevated body temperature tends to increase the rate of metabolic heat production because the temperature has a direct effect on the rate of chemical reactions.

Nonshivering thermogenesis

Although reflex and voluntary changes in muscle activity are the major means of increasing the rate of heat production, **nonshivering (chemical) thermogenesis** also plays a role in thermoregulation. In most experimental animals, chronic cold exposure brings about an increase in metabolic heat production that is independent of muscle contraction, instead being brought about by changes in heat-generating chemical activity. In humans, nonshivering thermogenesis is most important in newborns, because they lack the ability to shiver. Nonshivering thermogenesis is mediated by the hormones epinephrine and thyroid hormone, both of which increase heat production by stimulating fat metabolism. Newborns have deposits of a special type of adipose tissue known as **brown fat,** which is especially capable of converting chemical energy into heat. The role of nonshivering thermogenesis in adults remains controversial.

Having examined the mechanisms for adjusting heat production, we now turn to the other side of the equation, adjustments in heat loss.

■ The magnitude of heat loss can be adjusted by varying the flow of blood through the skin.

Heat loss mechanisms are also subject to control, again largely by the hypothalamus. When we are hot, we want to increase heat loss to the environment; when we are cold, we want to decrease heat loss. The amount of heat lost to the environment by radiation and conduction-convection is largely determined by the temperature gradient between the skin and the external environment. The body's central core is a heat-generating chamber in which the temperature must be maintained at approximately 100°F. Surrounding the core is an insulating shell through which heat exchanges between the body and the external environment take place. To maintain a constant core temperature, the insulative capacity and temperature of the shell can be adjusted to vary the temperature gradient between the skin and external environment, thereby influencing the extent of heat loss.

The insulative capacity of the shell can be varied by controlling the amount of blood flowing through the skin. Skin blood flow serves two functions. First, it provides a nutritive blood supply to the skin. Second, as blood is pumped to the skin from the heart, it has been heated in the central core and carries this heat to the skin. Most of the skin blood flow is for purposes of temperature regulation; at normal room temperature, 20 to 30 times more blood flows through the skin than is needed to meet the skin's nutritional needs.

Skin vasomotor responses

In the process of thermoregulation, skin blood flow can vary tremendously, from 400 ml/min up to 2,500 ml/min. The more blood that reaches the skin from the warm core, the closer the skin's temperature is to the core temperature. The skin's blood vessels diminish the effectiveness of the skin as an insulator by carrying heat to the surface, where it can be lost from the body by radiation and conduction-convection. Accordingly, vasodilation of the skin vessels (specifically, the arterioles), which permits increased flow of heated blood through the skin, increases heat loss. Conversely, vasoconstriction of the skin vessels, which reduces blood flow through the skin, decreases heat loss by keeping the warm blood in the central core, where it is insulated from the external environment. This response conserves heat that otherwise would have been lost. Cold, relatively bloodless skin provides excellent insulation between the core and the environment. However, the skin is not a perfect insulator, even with maximum vasoconstriction. Despite minimal blood flow to the skin, some heat can still be transferred by conduction from the deeper organs to the skin surface and then can be lost from the skin to the environment.

These skin vasomotor responses are coordinated by the hypothalamus by means of sympathetic nervous system output. Increased sympathetic activity to the skin vessels produces heat-conserving vasoconstriction in response to cold exposure, whereas decreased sympathetic activity produces heat-losing vasodilation of the skin vessels in response to heat exposure.

Recall that the cardiovascular control center in the medulla also exerts control over the skin arterioles (as well as arterioles throughout the body) by means of adjusting sympathetic activity to these vessels for the purpose of blood pressure regulation (see p. 377). Hypothalamic control over the skin arterioles for the purpose of temperature regulation takes precedence over the cardiovascular control center's control of these same vessels (see p. 379). Thus changes in blood pressure can result from pronounced thermoregulatory skin vasomotor responses. For example, blood pressure can fall on exposure to a very hot environment, because the skin vasodilator response set in motion by the hypothalamic thermoregulatory center overrides the skin vasoconstrictor response called forth by the medullary cardiovascular control center.

The hypothalamus simultaneously coordinates heat production and heat loss mechanisms.

Let us now pull together the coordinated adjustments in heat production as well as heat loss in response to exposure to either a cold or a hot environment (● Figure 17-6 and ▲ Table 17-2).

Coordinated responses to cold exposure

In response to cold exposure, the posterior region of the hypothalamus directs increased heat production such as by shivering, while simultaneously decreasing heat loss (that is, conserving heat) by skin vasoconstriction and other measures.

Because there is a limit to the body's ability to reduce skin temperature through vasoconstriction, even maximum vasoconstriction is not sufficient to prevent excessive heat loss when the external temperature falls too low. Accordingly, other measures must be instituted to further reduce heat loss. In animals with dense fur or feathers, the hypothalamus, acting through the sympathetic nervous system, brings about contraction of the tiny muscles at the base of the hair or feather shafts to lift the hair or feathers off the skin surface. This puffing up traps a layer of poorly conductive air between the skin surface and the environment, thus increasing the insulating barrier between the core and the cold air and reducing heat loss. Even though the hair shaft muscles contract in humans in response to cold exposure, this heat retention mechanism is ineffective because of the low density and fine texture of most human body hair. The result instead is useless *goosebumps*.

After maximum skin vasoconstriction has been achieved as a result of exposure to cold, further heat dissipation in humans can be prevented only by behavioral adaptations, such as postural changes that reduce as much as possible the exposed surface area from which heat can escape. These postural changes include maneuvers such as hunching over, clasping the arms in front of the chest, or curling up in a ball.

Putting on warmer clothing further insulates the body from too much heat loss. Clothing entraps layers of poorly conductive air between the skin surface and the environment, thereby diminishing loss of heat by conduction from the skin to the cold external air and curtailing the flow of convection currents.

Coordinated responses to heat exposure

Under the opposite circumstance—heat exposure—the anterior part of the hypothalamus reduces heat production by decreasing skeletal muscle activity and promotes increased heat loss by inducing skin vasodilation. When even maximal skin vasodilation is inadequate to rid the body of excess heat, sweating is brought into play to accomplish further heat loss through evaporation. In fact, if the air temperature rises above the temperature of maximally vasodilated skin, the temperature gradient reverses itself so that heat is gained from the environment. Sweating is the only means of heat loss under these conditions.

In addition to sweating, humans employ voluntary measures, such as using fans, wetting the body, drinking cold beverages, and wearing cool clothing, to further enhance heat loss. Contrary to popular belief, wearing light-colored, loose clothing is cooler than being nude. Naked skin absorbs almost all the radiant energy that strikes it, whereas light-colored cloth-

▲ TABLE 17-2
Coordinated Adjustments in Response to Cold or Heat Exposure

In Response to Cold Exposure (coordinated by the posterior hypothalamus)		In Response to Heat Exposure (coordinated by the anterior hypothalamus)	
Increased Heat Production	**Decreased Heat Loss (heat conservation)**	**Decreased Heat Production**	**Increased Heat Loss**
Increased muscle tone	Skin vasoconstriction	Decreased muscle tone	Skin vasodilation
Shivering	Postural changes to reduce exposed surface area (hunching shoulders, etc.)*	Decreased voluntary exercise*	Sweating
Increased voluntary exercise*			Cool clothing*
Nonshivering thermogenesis	Warm clothing*		

*Behavioral adaptations.

ing reflects almost all the radiant energy that falls on it. Thus if light-colored clothing is loose and thin enough to permit convection currents and evaporative heat loss to occur, wearing it is actually cooler than going without any clothes at all.

Thermoneutral zone

Skin vasomotor activity is highly effective in controlling heat loss in environmental temperatures between the upper 60s and mid 80s. This range, within which core temperature can be kept constant by vasomotor responses without calling supplementary heat-production or heat-loss mechanisms into play, is called the **thermoneutral zone.** When external air temperature falls below the lower limits of the ability of skin vasoconstriction to reduce heat loss further, the major burden of maintaining core temperature is borne by increased heat production, especially shivering. At the other extreme, when external air temperature exceeds the upper limits of the ability of skin vasodilation to increase heat loss further, sweating becomes the dominant factor in maintaining core temperature.

▌ During a fever, the hypothalamic thermostat is "reset" at an elevated temperature.

The term **fever** refers to an elevation in body temperature as a result of infection or inflammation. In response to microbial invasion, certain white blood cells release a chemical known as **endogenous pyrogen,** which, among its many infection-fighting effects (see p. 419), acts on the hypothalamic thermoregulatory center to raise the setting of the thermostat (● Figure 17-7). The hypothalamus now maintains the temperature at the new set level instead of maintaining normal body temperature. If, for example, endogenous pyrogen raises the set point to 102°F (as recorded orally), the hypothalamus senses that the normal prefever temperature is too cold, so it initiates the cold response mechanisms to elevate the temperature to 102°F. Specfically, the hypothalamus initiates shivering to rapidly increase heat production, and promotes skin vasoconstriction to rapidly reduce heat loss, both of which drive the temperature upward. These events account for the sudden cold chills often experienced at the onset of a fever. Feeling cold, the person may put on more blankets as a voluntary mechanism that helps elevate body temperature by conserving body heat. Once the new temperature is achieved, body temperature is regulated as normal in response to cold and heat but at a higher setting. Thus fever production in response to an infection is a deliberate outcome and is not due to a breakdown of thermoregulatory mechanisms. Although the physiological significance of a fever is still unclear, many medical experts believe that a rise in body temperature has a beneficial role in fighting infection. A fever

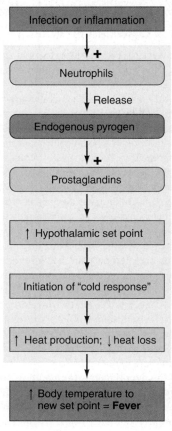

● **FIGURE 17-7**
Fever production

augments the inflammatory response and may interfere with bacterial multiplication.

During fever production, endogenous pyrogen raises the set point of the hypothalamic thermostat by triggering the local release of *prostaglandins,* which are local chemical mediators that act directly on the hypothalamus. Aspirin reduces a fever by inhibiting the synthesis of prostaglandins. Aspirin does not lower the temperature in a nonfebrile person, because in the absence of endogenous pyrogen, prostaglandins are not present in the hypothalamus in appreciable quantities.

The exact molecular cause of a fever "breaking" naturally is unknown, although it presumably results from reduced pyrogen release or decreased prostaglandin synthesis. When the hypothalamic set point is restored to normal, the temperature at 102°F (in this example) is too high. The heat response mechanisms are instituted to cool down the body. Skin vasodilation occurs, and sweating commences. The person feels hot and throws off extra covers. The gearing up of these heat loss mechanisms by the hypothalamus reduces the temperature to normal.

▌ Hyperthermia can occur unrelated to infection.

Hyperthermia denotes any elevation in body temperature above the normally accepted range. The term *fever* is usually reserved for an elevation in temperature caused by the release of endogenous pyrogen resetting the hypothalamic set point during infection or inflammation; *hyperthermia* refers to all other imbalances between heat gain and heat loss that increase body temperature. Hyperthermia has a variety of causes, some of which are normal and harmless, others pathologic and fatal.

Exercise-induced hyperthermia

The most common cause of hyperthermia is sustained exercise. As a physical consequence of the tremendous heat load generated by exercising muscles, body temperature rises during the initial stage of exercise because heat gain exceeds heat loss (● Figure 17-8). The elevation in core temperature reflexly triggers heat loss mechanisms (skin vasodilation and sweating), which eliminate the discrepancy between heat production and heat loss. As soon as the heat loss mechanisms are stepped up sufficiently to equalize heat production, the core temperature stabilizes at a level slightly above the set point despite continued heat-producing exercise. Thus, during sustained exercise, body temperature initially rises, then is maintained at the higher level as long as the exercise continues.

Pathologic hyperthermia

Hyperthermia can also be brought about in a completely different way: excessive heat production in connection with abnor-

The Extremes of Heat and Cold Can Be Fatal

Prolonged exposure to temperature extremes in either direction can overtax the body's thermoregulatory mechanisms, leading to disorders and even death if severe enough.

Heat-related disorders

Heat exhaustion is a state of collapse, usually manifested by fainting, that is caused by reduced blood pressure brought about as a result of overtaxing the heat loss mechanisms. Extensive sweating reduces cardiac output by depleting the plasma volume, and pronounced skin vasodilation causes a drop in total peripheral resistance. Because blood pressure is determined by cardiac output times total peripheral resistance, blood pressure falls, an insufficient amount of blood is delivered to the brain, and fainting takes place. Thus heat exhaustion is a consequence of overactivity of the heat loss mechanisms rather than a breakdown of these mechanisms. Because the heat loss mechanisms have been very active, body temperature is only mildly elevated in heat exhaustion. By forcing the cessation of activity when the heat loss mechanisms are no longer able to cope with heat gain through exercise or a hot environment, heat exhaustion serves as a safety valve to help prevent the more serious consequences of heat stroke.

Heat stroke is an extremely dangerous situation that arises from the complete breakdown of the hypothalamic thermoregulatory systems. Heat exhaustion may progress into heat stroke if the heat loss mechanisms continue to be over-taxed. Heat stroke is more likely to occur on overexertion during a prolonged exposure to a hot, humid environment. The elderly, in whom thermoregulatory responses are generally slower and less efficient, are particularly vulnerable to heat stroke during prolonged, stifling heat waves. So too are individuals who are taking certain common tranquilizers, such as Valium, because these drugs interfere with the hypothalamic thermoregulatory centers' neurotransmitter activity.

The most striking feature of heat stroke is a lack of compensatory heat loss measures, such as sweating, in the face of a rapidly rising body temperature. No sweating occurs, despite a markedly elevated body temperature, because the hypothalamic thermoregulatory control centers are not functioning properly and cannot initiate heat loss mechanisms. During the development of heat stroke, body temperature starts to climb as the heat loss mechanisms are eventually overwhelmed by prolonged, excessive heat gain. Once the core temperature reaches the point at which the hypothalamic temperature-control centers are damaged by the heat, the body temperature rapidly rises even higher because of the complete shutdown of heat loss mechanisms. Furthermore, as the body temperature increases, the rate of metabolism increases correspondingly, because higher temperatures speed up the rate of all chemical reactions; the result is even greater heat production. This positive-feedback state sends the temperature spiraling upward. Heat stroke is a very dangerous situation that is rapidly fatal if untreated. Even with treatment to halt and reverse the rampant rise in body temperature, there is still a high rate of mortality. The rate of permanent disability in survivors is also high because of irreversible protein denaturation caused by the high internal heat.

Cold-related disorders

At the other extreme, the body can be harmed by cold exposure in two ways: frostbite and generalized hypothermia. **Frostbite** involves excessive cooling of a particular part of the body to the point where tissue in that area is damaged. If exposed tissues actually freeze, tissue damage occurs as a result of disruption of the cells by formation of ice crystals or by lack of liquid water.

Hypothermia, a fall in body temperature, occurs when generalized cooling of the body exceeds the ability of the normal heat-producing and heat-conserving regulatory mechanisms to match the excessive heat loss. As hypothermia sets in, the rate of all metabolic processes slows down because of the declining temperature. Higher cerebral functions are the first affected by body cooling, leading to loss of judgment, and to apathy, disorientation, and tiredness, all of which diminish the cold victim's ability to initiate voluntary mechanisms to reverse the falling body temperature. As body temperature continues to plummet, depression of the respiratory center occurs, reducing the ventilatory drive so that breathing becomes slow and weak. Activity of the cardiovascular system also is gradually reduced. The heart is slowed and cardiac output decreased. Disturbances of cardiac rhythm occur, eventually leading to ventricular fibrillation and death.

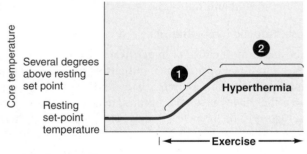

1. At the onset of exercise, the rate of heat production initially exceeds the rate of heat loss so the core temperature rises.

2. When heat loss mechanisms are reflexly increased sufficiently to equalize the elevated heat production, the core temperature stabilizes slightly above the resting point for the duration of the exercise.

● FIGURE 17-8

Hyperthermia in sustained exercise

mally high circulating levels of thyroid hormone or epinephrine that result from dysfunctions of the thyroid gland or adrenal medulla, respectively. Both these hormones elevate the core temperature by increasing the overall rate of metabolic activity and heat production.

Hyperthermia can also result from malfunction of the hypothalamic control centers. Certain brain lesions, for example, destroy the normal regulatory capacity of the hypothalamic thermostat. When the thermoregulatory mechanisms are not functional, lethal hyperthermia may occur very rapidly. Normal metabolism produces enough heat to kill a person in less than five hours if the heat loss mechanisms are completely shut down. In addition to brain lesions, exposure to severe, prolonged heat stress can also break down the function of hypothalamic thermoregulation. Similarly, the body can be harmed by extreme cold exposure. (For a discussion of the effects of extreme heat or cold exposure, see the accompanying boxed feature, ❱ Concepts, Challenges, and Controversies.)

CHAPTER IN PERSPECTIVE: FOCUS ON HOMEOSTASIS

Because energy can be neither created nor destroyed, for body weight and body temperature, respectively, to remain constant, input must equal output in the case of both the body's total energy balance and its heat energy balance. If total energy input exceeds total energy output, the extra energy is stored in the body, and body weight increases. Similarly, if the input of heat energy exceeds its output, body temperature increases. Conversely, if output exceeds input, body weight decreases or body temperature falls. The hypothalamus is the major integrating center for maintaining both a constant total energy balance (and thus a constant body weight) and a constant heat energy balance (and thus a constant body temperature).

Body temperature, which is one of the homeostatically regulated factors of the internal environment, must be maintained within narrow limits, because the structure and reactivity of the chemicals that compose the body are temperature sensitive. Deviations in body temperature outside a limited range result in protein denaturation and death of the individual if the temperature rises too high, or metabolic slowing and death if the temperature falls too low.

Body weight, in contrast, varies widely among individuals. Only the extremes of imbalances between total energy input and output become incompatible with life. For example, in the face of insufficient energy input in the form of ingested food during prolonged starvation, the body resorts to breaking down muscle protein to meet its needs for energy expenditure once the adipose stores are depleted. Body weight dwindles because of this self-cannibalistic mechanism until death finally occurs as a result of loss of heart muscle, among other things. At the other extreme, when the food energy consumed greatly exceeds the energy expended, the extra energy input is stored as adipose tissue, and body weight increases. The resultant gross obesity can also lead to heart failure. Not only must the heart work harder to pump blood to the excess adipose tissue, but obesity also predisposes the individual to atherosclerosis and heart attacks (see p. 336).

CHAPTER SUMMARY

Energy Balance

- Energy input to the body in the form of food energy must equal energy output, because energy cannot be created or destroyed.

- Energy output or expenditure includes (1) external work, performed by skeletal muscles to accomplish movement of an external object or movement of the body through the external environment; and (2) internal work, which consists of all other energy-dependent activities that do not accomplish external work, including active transport, smooth and cardiac muscle contraction, glandular secretion, and protein synthesis.

- Only about 25% of the chemical energy in food is harnessed to do biological work. The rest is immediately converted to heat. Furthermore, all the energy expended to accomplish internal work is eventually converted into heat, and 75% of the energy expended by working skeletal muscles is lost as heat. Therefore, most of the energy in food ultimately appears as body heat.

- The metabolic rate, which is energy expenditure per unit of time, is measured in kilocalories of heat produced per hour.

- The basal metabolic rate (BMR) is a measure of the body's minimal waking rate of internal energy expenditure.

- For a neutral energy balance, the energy in ingested food must equal energy expended in performing external work and transformed into heat. If more energy is consumed than is expended, the extra energy is stored in the body, primarily as adipose tissue, so body weight increases. By contrast, if more energy is expended than is available in the food, body energy stores are used to support energy expenditure, so body weight decreases.

- Usually, body weight remains fairly constant over a prolonged period of time (except during growth) because food intake is adjusted to match energy expenditure on a long-term basis.

- Food intake is controlled primarily by the hypothalamus by means of complex, incompletely understood regulatory mechanisms in which hunger and satiety are important components. The key factors known to be important in the control of food intake are as follows: (1) Adipocytes secrete the hormone leptin. (2) Leptin reduces appetite and decreases food consumption by inhibiting neuropeptide Y-secreting neurons and stimulating melanocortin-secreting neurons in the arcuate nucleus of the hypothalamus. (3) Neuropeptide Y (NPY) increases appetite and food intake, whereas melanocortins suppress appetite and food intake. (4) NPY and melanocortins in turn are believed to bring about their effects by acting on the lateral hypothalamic area (LHA) and paraventricular nucleus (PVN) to alter the release of chemical messengers from these areas. (5) The LHA secretes neuropeptides such as orexins that are potent stimulators of food intake, whereas the PVN releases neuropeptides that decrease food intake.

- Whereas control of energy balance or energy homeostasis to match energy intake with energy output, thus maintaining body weight over the long term, is largely controlled by these hypothalamic mechanisms, short-term control of the timing and size of meals is mediated primarily by the nucleus tractus solitarius (NTS) in the brain stem.

- Satiety signals that act through the NTS to terminate a meal include (1) gastrointestinal distention, (2) increased glucose use, (3) increased insulin, and (2) increased cholecystokinin.

- Psychosocial and environmental factors can also influence food intake above and beyond the internal signals that govern feeding behavior.

Temperature Regulation

- The body can be thought of as a heat-generating core (internal organs, CNS, and skeletal muscles) surrounded by a shell of variable insulating capacity (the skin).

- The skin exchanges heat energy with the external environment, with the direction and amount of heat transfer depending on the environmental temperature and the momentary insulating capacity of the shell.

- The four physical means by which heat is exchanged between the body and the external environment are (1) radiation (net movement of heat energy via electromagnetic waves); (2) conduction (exchange of heat energy by direct contact); (3) convection (transfer of heat energy by means of air currents); and

(4) evaporation (extraction of heat energy from the body by the heat-requiring conversion of liquid H_2O to H_2O vapor). Because heat energy moves from warmer to cooler objects, radiation, conduction, and convection can be channels for either heat loss or heat gain, depending on whether surrounding objects are cooler or warmer, respectively, than the body surface. Normally, they are avenues for heat loss, along with evaporation resulting from sweating.

▌ To prevent serious cellular malfunction, the core temperature must be held constant at about 100°F (equivalent to an average oral temperature of 98.2°F) by continuously balancing heat gain and heat loss despite changes in environmental temperature and variation in internal heat production.

▌ This thermoregulatory balance is controlled by the hypothalamus. The hypothalamus is apprised of the skin temperature by peripheral thermoreceptors and of the core temperature by central thermoreceptors, the most important of which are located in the hypothalamus itself.

▌ The primary means of heat gain is heat production by metabolic activity, the biggest contributor being skeletal muscle contraction.

▌ Heat loss is adjusted by sweating and by controlling to the greatest extent possible the temperature gradient between the skin and the surrounding environment. The latter is accomplished by regulating the caliber of the skin's blood vessels. (1) Vasoconstriction of the skin vessels reduces the flow of warmed blood through the skin so that skin temperature falls. The layer of cool skin between the core and the environment increases the insulating barrier between the warm core and the external air. (2) Conversely, skin vasodilation brings more warmed blood through the skin so that skin temperature approaches the core temperature, thus reducing the insulative capacity of the skin.

▌ On exposure to cool surroundings, the core temperature starts to fall as heat loss increases due to the larger-than-normal skin-to-air temperature gradient. The hypothalamus responds to reduce the heat loss by inducing skin vasoconstriction while simultaneously increasing heat production through heat-generating shivering.

▌ Conversely, in response to a rise in core temperature (resulting either from excessive internal heat production accompanying exercise or from excessive heat gain on exposure to a hot environment), the hypothalamus triggers heat-loss mechanisms, such as skin vasodilation and sweating, while simultaneously decreasing heat production, such as by reducing muscle tone.

▌ In both cold and heat responses, voluntary behavioral actions also contribute importantly to maintenance of thermal homeostasis.

▌ A fever occurs when endogenous pyrogen released from white blood cells in response to infection raises the hypothalamic set point. An elevated core temperature develops as the hypothalamus initiates cold-response mechanisms to raise the core temperature to the new set point.

REVIEW EXERCISES

Objective Questions (Answers on p. A-49)

1. If more food energy is consumed than is expended, the excess energy is lost as heat. (*True or false?*)
2. All the energy within nutrient molecules can be harnessed to perform biological work. (*True or false?*)
3. Each liter of O_2 contains 4.8 kilocalories of heat energy. (*True or false?*)
4. A body temperature greater than 98.2°F is always indicative of a fever. (*True or false?*)
5. Core temperature is relatively constant, but skin temperature can vary markedly. (*True or false?*)
6. Sweat that drips off the body has no cooling effect. (*True or false?*)
7. Production of "goosebumps" in response to cold exposure has no value in regulating body temperature. (*True or false?*)
8. The posterior region of the hypothalamus triggers shivering and skin vasoconstriction. (*True or false?*)
9. The primary means of involuntarily increasing heat production is _____ .
10. Increased heat production independent of muscle contraction is known as _____ .
11. The only means of heat loss when the environmental temperature exceeds the core temperature is _____ .
12. Which of the following statements concerning heat exchange between the body and the external environment is *incorrect*?
    a. Heat gain is primarily by means of internal heat production.
    b. Radiation serves as a means of heat gain but not of heat loss.
    c. Heat energy always moves down its concentration gradient from warmer to cooler objects.
    d. The temperature gradient between the skin and the external air is subject to control.
    e. Very little heat is lost from the body by conduction alone.

13. Which of the following statements concerning fever production is *incorrect?*
    a. Endogenous pyrogen is released by white blood cells in response to microbial invasion.
    b. The hypothalamic set point is elevated.
    c. The hypothalamus initiates cold response mechanisms to increase the body temperature.
    d. Prostaglandins appear to mediate the effect.
    e. The hypothalamus is not effective in regulating body temperature during a fever.
14. Using the following answer code, indicate which mechanism of heat transfer is being described:

        (a) radiation   (c) convection
        (b) conduction   (d) evaporation

    ____ 1. sitting on a cold metal chair
    ____ 2. sunbathing on the beach
    ____ 3. a gentle breeze
    ____ 4. sitting in front of a fireplace
    ____ 5. sweating
    ____ 6. riding in a car with the windows open
    ____ 7. lying under an electric blanket
    ____ 8. sitting in a wet bathing suit
    ____ 9. fanning yourself
    ____ 10. immersion in cool water

Essay Questions
1. Differentiate between external and internal work.
2. Define *metabolic rate* and *basal metabolic rate*. Explain the process of indirect calorimetry.
3. Describe the three states of energy balance.
4. By what means is energy balance primarily maintained?

5. List the sources of heat input and output for the body.
6. Distinguish between the classical appetite and satiety centers.
7. Describe the role of the following in the regulation of food intake and control of the timing and size of meals: adipocytes, leptin, arcuate nucleus, neuropeptide Y, melanocortins, lateral hypothalamic area, orexins, paraventricular nucleus, nucleus tractus solitarius, gastrointestinal distension, glucose use, insulin, and cholecystokinin.
8. Discuss the compensatory measures that occur in response to a fall in core temperature as a result of cold exposure and in response to a rise in core temperature as a result of heat exposure.

Quantitative Exercises (Solutions on p. A-49)

1. The basal metabolic rate (BMR) is a measure of how much energy the body consumes to maintain its "idling speed." The normal BMR = about 72 kcal/hr (see p. 649). The vast majority of this energy is converted to heat. Our thermoregulatory systems function to eliminate this heat so as to keep body temperature constant. If our bodies were not able to lose this heat, our temperature would rise until we boiled (of course, a person would die before reaching that temperature). It is relatively easy to calculate how long it would take to reach the hypothetical boiling point. If an amount of energy ΔU is put into a liquid of mass m, the temperature change ΔT (in °C) is given by the following formula:

$$\Delta T = \Delta U/m \times C$$

In this equation, C is the specific heat of the liquid. For water, $C = 1.0$ kcal/kg-°C. Use this information to calculate how long it would take for the heat from the BMR to boil your body fluids (assume 42 liters of water in your body and a starting point of normal body temperature at 37°C). When exercising maximally, a person consumes about 1,000 kcal/hr. How long would it take to boil in this case?

POINTS TO PONDER

(Explanations on p. A-49)

1. Explain how drugs that selectively inhibit CCK increase feeding behavior in experimental animals.
2. What advice would you give an overweight friend who asks for your help in designing a safe, sensible, inexpensive program for losing weight?
3. Why is it dangerous to engage in heavy exercise on a hot, humid day?
4. Describe the avenues for heat loss in a person soaking in a hot bath.
5. Consider the difference between you and a fish in a local pond with regard to control of body temperature. Humans are *thermoregulators*; they can maintain a remarkably constant, rather high internal body temperature despite the body's exposure to a wide range of environmental temperatures. To maintain thermal homeostasis, humans physiologically manipulate mechanisms within their bodies to adjust heat production, heat conservation, and heat loss. In contrast, fish are *thermoconformers*; their body temperatures conform to the temperature of their surroundings. Thus, their body temperatures vary capriciously with changes in the environmental temperature. Even though fish produce heat, they cannot physiologically regulate internal heat production, nor can they control heat exchange with their environment to maintain a constant body temperature when the temperature in their surroundings rises or falls. Knowing this, do you think fish run a fever when they have a systemic infection? Why or why not?
6. **Clinical Consideration.** Michael F., a near-drowning victim, was pulled from the icy water by rescuers 15 minutes after he fell through the thin ice on which he was skating. Michael is now alert and recuperating in the hospital. How can you explain his "miraculous" survival even though he was submerged for 15 minutes and irreversible brain damage, soon followed by death, normally occurs if the brain is deprived of its critical O_2 supply for more than four or five minutes?

PHYSIOEDGE RESOURCES

PHYSIOEDGE CD-ROM
PhysioEdge focuses on concepts students find difficult to learn. Figures marked with this icon have associated activities on the CD.

CHECK OUT THESE MEDIA QUIZZES:
17.1 Basics of Energy Balance
17.2 Body Temperature Regulation

PHYSIOEDGE WEB SITE
The Web site for this book contains a wealth of helpful study aids, as well as many ideas for further reading and research. Log on to:

http://www.brookscole.com

Go to the Biology page and select Sherwood's *Human Physiology*, 5th Edition. Select a chapter from the drop-down menu or click on one of these resource areas:

- **Case Histories** provide an introduction to the clinical aspects of human physiology.
- For 2-D and 3-D graphical illustrations and animations of physiological concepts, visit our **Visual Learning Resource.**
- Resources for study and review include the **Chapter Outline, Chapter Summary, Glossary,** and **Flash Cards.** Use our **Quizzes** to prepare for in-class examinations.
- On-line research resources to consult are **Hypercontents,** with current links to relevant Internet sites; **Internet Exercises** with starter URLs provided; and **InfoTrac Exercises.**

 For more readings, go to InfoTrac College Edition, your on-line research library, at:

http://infotrac.thomsonlearning.com

Endocrine System

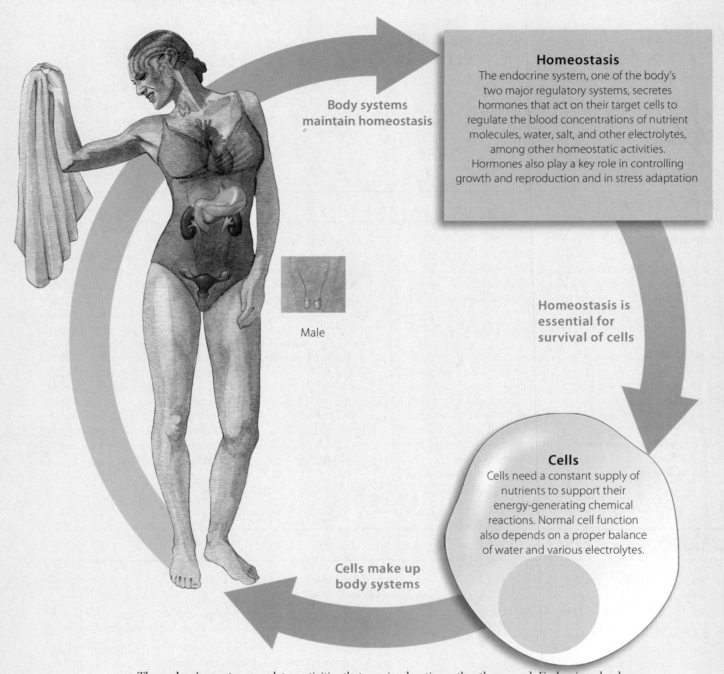

Body systems maintain homeostasis

Homeostasis
The endocrine system, one of the body's two major regulatory systems, secretes hormones that act on their target cells to regulate the blood concentrations of nutrient molecules, water, salt, and other electrolytes, among other homeostatic activities. Hormones also play a key role in controlling growth and reproduction and in stress adaptation

Male

Homeostasis is essential for survival of cells

Cells
Cells need a constant supply of nutrients to support their energy-generating chemical reactions. Normal cell function also depends on a proper balance of water and various electrolytes.

Cells make up body systems

The **endocrine system** regulates activities that require duration rather than speed. Endocrine glands release **hormones,** blood-borne chemical messengers that act on target cells located a long distance from the endocrine gland. Most target-cell activities under hormonal control are directed toward maintaining homeostasis. The central endocrine glands include the pineal gland, the hypothalamus, and the pituitary gland. The **pineal gland** is a part of the brain that secretes a hormone important in establishing the body's biological rhythms. The **hypothalamus** (also a part of the brain) and the **posterior pituitary gland** act as a unit to release hormones essential for maintaining water balance and for giving birth and for breast-feeding. The hypothalamus also secretes regulatory hormones that control the hormonal output of the **anterior pituitary gland,** which secretes six hormones that in turn largely control the hormonal output of several peripheral endocrine glands. One anterior pituitary hormone, growth hormone, promotes growth and influences nutrient homeostasis.

Chapter 18

Principles of Endocrinology: The Central Endocrine Glands

CONTENTS AT A GLANCE

GENERAL PRINCIPLES OF ENDOCRINOLOGY

❚ Categories of hormones: Peptides, amines, and steroids

❚ Mechanisms of hormone synthesis, storage, and secretion

❚ Amplification of hormone action

❚ Factors affecting the plasma concentration of hormones

❚ Types of endocrine disorders

❚ Regulation of target-organ responsiveness

PINEAL GLAND AND CIRCADIAN RHYTHMS

❚ Suprachiasmatic nucleus as the master biological clock

❚ Functions of melatonin

HYPOTHALAMUS AND PITUITARY

❚ Hypothalamus—posterior pituitary relationship

❚ Vasopressin; oxytocin

❚ Hypothalamus—anterior pituitary relationship

❚ Anterior pituitary hormones

❚ Hypophysiotropic hormones

❚ Negative-feedback loops

ENDOCRINE CONTROL OF GROWTH

❚ Factors influencing growth

❚ Functions and control of growth hormone

❚ Role of other hormones in growth

GENERAL PRINCIPLES OF ENDOCRINOLOGY

The **endocrine system** consists of the ductless endocrine glands (see p. 6) that are scattered throughout the body (● Figure 18-1). Even though the endocrine glands for the most part are not connected anatomically, they constitute a system in a functional sense. They all accomplish their functions by secreting hormones into the blood, and many functional interactions take place among the various endocrine glands. Once secreted, a hormone travels in the blood to its distant target cells, where it regulates or directs a particular function. **Endocrinology** is the study of the homeostatic chemical adjustments and other activities that hormones accomplish.

Recall that neurosecretory neurons release their chemical messengers, *neurohormones*, into the blood (see p. 66). In contrast, ordinary neurons release their chemical messengers, neurotransmitters, into a synaptic cleft. Like hormones, neurohormones are distributed by the blood to the target cells. Thus neurosecretory neurons are considered part of the endocrine system. The general term "hormone" in this text tacitly includes both blood-borne hormonal and neurohormonal messengers.

Even though the blood distributes hormones throughout the body, only specific target cells can respond to each hormone, because only the target cells have receptors for binding with the particular hormone. The binding of a hormone with its specific target-cell receptors initiates a chain of events within the target cells to bring about the hormone's final effect.

❚ Hormones exert a variety of regulatory effects throughout the body.

The endocrine system is one of the body's two major regulatory systems, the other being the nervous system, with which you are already familiar (Chapters 4 through 7). Recall that the endocrine and nervous systems are specialized for controlling different types of activities. In general, the nervous system coordinates rapid, precise responses and is especially important in mediating the body's interactions with the external environment. The endocrine system, by contrast, primarily controls activities that require duration rather than speed.

Overall functions of the endocrine system

1. Regulating organic metabolism and H_2O and electrolyte balance, which are important collectively in maintaining a constant internal environment

2. Inducing adaptive changes to help the body cope with stressful situations

3. Promoting smooth, sequential growth and development

4. Controlling reproduction

5. Regulating red blood cell production

6. Along with the autonomic nervous system, controlling and integrating both circulation and the digestion and absorption of food

Tropic hormones

Some hormones regulate the production and secretion of another hormone. A hormone that has as its primary function the regulation of hormone secretion by another endocrine gland is classified functionally as a **tropic hormone** (*tropic* means "nourishing"). Tropic hormones stimulate and maintain their endocrine target tissues. For example, the tropic hormone thyroid-stimulating hormone (TSH), from the anterior pituitary, stimulates thyroid hormone secretion by the thyroid gland and also maintains the structural integrity of this gland. In the absence of TSH, the thyroid gland atrophies (shrinks) and produces very low levels of its hormone.

Complexity of endocrine function

The following factors add to the complexity of the system:

- A single endocrine gland may produce multiple hormones. The anterior pituitary, for example, secretes six different hormones, each under a different control mechanism and having distinct functions.

- A single hormone may be secreted by more than one endocrine gland. For example, both the hypothalamus and the pancreas secrete the hormone somatostatin.

- Frequently, a single hormone has more than one type of target cell and therefore can induce more than one type of effect. For example, vasopressin promotes H_2O reabsorption by the kidney tubules as well as vasoconstriction of arterioles throughout the body. Sometimes hormones that have multiple target-cell types can coordinate and integrate the activities of various tissues toward a common end. For example, the effects of insulin on muscle, liver, and fat all act in concert to store nutrients after absorption of a meal.

- The rate of secretion of some hormones varies considerably over the course of time in a cyclic pattern. Therefore, endocrine systems also provide temporal (time) coordination of function. This is particularly apparent in endocrine control of reproductive cycles, such as the menstrual cycle, in which normal function requires highly specific patterns of change in the secretion of various hormones.

- A single target cell may be influenced by more than one hormone. Some cells contain an array of receptors for responding in different ways to different hormones. To illustrate, insulin promotes the conversion of glucose into glycogen within liver cells by stimulating one particular hepatic enzyme, whereas another hormone, glucagon, by activating yet another

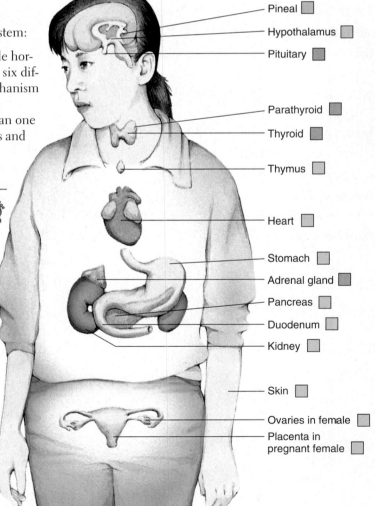

● FIGURE 18-1

The endocrine system

PhysioEdge Activity

- Solely endocrine function
- Mixed function
- Complete function uncertain

Pineal
Hypothalamus
Pituitary
Parathyroid
Thyroid
Thymus
Heart
Stomach
Adrenal gland
Pancreas
Duodenum
Kidney
Skin
Ovaries in female
Placenta in pregnant female

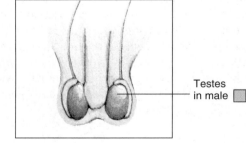

Testes in male

hepatic enzyme enhances the degradation of glycogen into glucose within liver cells.

- The same chemical messenger may be either a hormone or a neurotransmitter, depending on its source and mode of delivery to the target cell. Norepinephrine, which is secreted as a hormone by the adrenal medulla and released as a neurotransmitter from sympathetic postganglionic nerve fibers, is a prime example.

- Some organs are exclusively endocrine in function (they specialize in hormone secretion alone, the anterior pituitary being an example), whereas other organs of the endocrine system perform nonendocrine functions in addition to secreting hormones. For example, the testes produce sperm and also secrete the male sex hormone testosterone.

This has been a brief overview of the general functions of the endocrine system. ▲ Table 18-1 summarizes the most important specific functions of the major hormones. Some of these hormones have been introduced elsewhere and are not discussed further; these are the gastrointestinal hormones (Chapter 16), the renal hormones (erythropoietin in Chapter 11 and renin in Chapter 14), thrombopoietin from the liver (Chapter 11), thymosin from the thymus (Chapter 12), and atrial na-

triuretic peptide from the heart (Chapter 15). The remainder of the hormones are described in greater detail in this and the next two chapters.

Candidate hormones

As extensive as ▲ Table 18-1 appears, it leaves out a variety of "candidate" or potential hormones that have not fully qualified as hormones, either because they do not quite fit the classic definition of a hormone or because they have been discovered so recently that their hormonal status has not yet been conclusively documented. The table also excludes the hormones secreted by the effector cells of the defense system (white blood cells and macrophages) and a variety of recently revealed and poorly understood growth factors that promote growth of specific tissues, such as *epidermal growth factor* and *nerve growth factor*. Furthermore, new hormones are likely to be discovered. Also, additional functions may be found for known hormones. As an example, vasopressin's role in conserving H_2O during urine formation was determined first, followed later by the discovery of its constrictor effect on arterioles. More recently, vasopressin has also been indicated as playing roles in fever, learning, memory, and behavior.

▲ TABLE 18-1
Summary of the Major Hormones

Endocrine Gland	Hormones	Target Cells	Major Functions of Hormones
Hypothalamus	Releasing and inhibiting hormones (TRH, CRH, GnRH, GHRH, GHIH, PRH, PIH)	Anterior pituitary	Controls release of anterior pituitary hormones
Posterior Pituitary (hormones stored in)	Vasopressin (antidiuretic hormone)	Kidney tubules	Increases H_2O reabsorption
		Arterioles	Produces vasoconstriction
	Oxytocin	Uterus	Increases contractility
		Mammary glands (breasts)	Causes milk ejection
Anterior Pituitary	Thyroid-stimulating hormone (TSH)	Thyroid follicular cells	Stimulates T_3 and T_4 secretion
	Adrenocorticotropic hormone (ACTH)	Zona fasciculata and zona reticularis of adrenal cortex	Stimulates cortisol secretion
	Growth hormone	Bone; soft tissues	Essential but not solely responsible for growth; stimulates growth of bones and soft tissues; metabolic effects include protein anabolism, fat mobilization, and glucose conservation
		Liver	Stimulates somatomedin secretion
	Follicle-stimulating hormone (FSH)	*Females:* ovarian follicles	Promotes follicular growth and development; stimulates estrogen secretion

continued

Summary of the Major Hormones (continued)

Endocrine Gland	Hormones	Target Cells	Major Functions of Hormones
Anterior Pituitary (*continued*)	Follicle-stimulating hormone (FSH) (*continued*)	*Males:* seminiferous tubules in testes	Stimulates sperm production
	Luteinizing hormone (LH) (interstitial cell–stimulating hormone—ICSH)	*Females:* ovarian follicle and corpus luteum	Stimulates ovulation, corpus luteum development, and estrogen and progesterone secretion
		Males: interstitial cells of Leydig in testes	Stimulates testosterone secretion
	Prolactin	*Females:* mammary glands	Promotes breast development; stimulates milk secretion
		Males	Uncertain
Thyroid Gland Follicular Cells	Tetraiodothyronine (T_4 or thyroxine); tri-iodothyronine (T_3)	Most cells	Increases the metabolic rate; essential for normal growth and nerve development
Thyroid Gland C Cells	Calcitonin	Bone	Decreases plasma calcium concentration
Adrenal Cortex			
Zona glomerulosa	Aldosterone (mineralocorticoid)	Kidney tubules	Increases Na^+ reabsorption and K^+ secretion
Zona fasciculata and zona reticularis	Cortisol (glucocorticoid)	Most cells	Increases blood glucose at the expense of protein and fat stores; contributes to stress adaption
	Androgens (dehydroepiandrosterone)	*Females:* bone and brain	Responsible for the pubertal growth spurt and sex drive in females
Adrenal Medulla	Epinephrine and norepinephrine	Sympathetic receptor sites throughout the body	Reinforces the sympathetic nervous system; contributes to stress adaptation and blood pressure regulation
Endocrine Pancreas (Islets of Langerhans)	Insulin (β cells)	Most cells	Promotes cellular uptake, use, and storage of absorbed nutrients
	Glucagon (α cells)	Most cells	Important for maintaining nutrient levels in blood during postabsorptive state
	Somatostatin (D cells)	Digestive system	Inhibits digestion and absorption of nutrients
		Pancreatic islet cells	Inhibits secretion of all pancreatic hormones
Parathyroid Gland	Parathyroid hormone (PTH)	Bone, kidneys, intestine	Increases plasma calcium concentration; decreases plasma phosphate concentration; stimulates vitamin D activation

continued

Endocrine Gland	Hormones	Target Cells	Major Functions of Hormones
Gonads			
Female: ovaries	Estrogen (estradiol)	Female sex organs; body as a whole	Promotes follicular development; governs development of secondary sexual characteristics: stimulates uterine and breast growth
		Bone	Promotes closure of the epiphyseal plate
	Progesterone	Uterus	Prepares for pregnancy
Male: testes	Testosterone	Male sex organs; body as a whole	Stimulates sperm production; governs development of secondary sexual characteristics; promotes sex drive
		Bone	Enhances pubertal growth spurt; promotes closure of the epiphyseal plate
Testes and ovaries	Inhibin	Anterior pituitary	Inhibits secretion of follicle-stimulating hormone
Pineal Gland	Melatonin	Brain; anterior pituitary; reproductive organs; immune system; possibly others	Entrains body's biological rhythm with external cues; believed to inhibit gonadotropins; initiation of puberty possibly caused by a reduction in melatonin secretion; acts as an antioxidant; enhances immunity
Placenta	Estrogen (estriol); progesterone	Female sex organs	Help maintain pregnancy; prepare breasts for lactation
	Chorionic gonadotropin	Ovarian corpus luteum	Maintains corpus luteum of pregnancy
Kidneys	Renin ($\rightarrow$ angiotensin)	Zona glomerulosa of adrenal cortex (acted on by angiotensin, which is activated by renin)	Stimulates aldosterone secretion
	Erythropoietin	Bone marrow	Stimulates erythrocyte production
Stomach	Gastrin	Digestive-tract exocrine glands and smooth muscles; pancreas; liver; gallbladder	Control of motility and secretion to facilitate digestive and absorptive processes
Duodenum	Secretin; cholecystokinin		
	Glucose-dependent insulinotropic peptide	Endocrine pancreas	Stimulates insulin secretion
Liver	Somatomedins	Bone; soft tissues	Promotes growth
	Thrombopoietin	Bone marrow	Stimulates platelet production
Skin	Vitamin D	Intestine	Increases absorption of ingested calcium and phosphate
Thymus	Thymosin	T lymphocytes	Enhances T lymphocyte proliferation and function
Heart	Atrial natriuretic peptide	Kidney tubules	Inhibits Na^+ reabsorption

Hormones are chemically classified into three categories: peptides, amines, and steroids.

Hormones are not all similar chemically but instead fall into three distinct classes according to their biochemical structure (▲ Table 18-2): (1) peptides and proteins, (2) amines, and (3) steroids. The first two categories are both amino acid derivatives. The **peptide** and **protein hormones** consist of specific amino acids arranged in a chain of varying length; the shorter chains are peptides, and the longer ones are proteins. For convenience, we refer to this entire category as *peptides*. The majority of hormones fall into this class, including those secreted by the hypothalamus, anterior pituitary, posterior pituitary, pineal gland, pancreas, parathyroid gland, gastrointestinal tract, kidneys, liver, thyroid C cells, heart, and thymus. The **amines** are derived from the amino acid tyrosine and include the hormones secreted by the thyroid gland and adrenal medulla. The adrenomedullary hormones are specifically known as *catecholamines.* The **steroids,** which include the hormones secreted by the adrenal cortex and gonads, as well as most placental hormones, are neutral lipids derived from cholesterol.

Minor differences in chemical structure between hormones within each category often result in profound differences in biologic response. Comparing two steroid hormones in ● Figure 18-2, for example, note the subtle difference between testosterone, the male sex hormone responsible for inducing the development of masculine characteristics, and estradiol, a form of estrogen, which is the feminizing female sex hormone.

Solubility characteristics of hormone classes

Differences in chemical structure are also responsible for differences in solubility among the classes of hormones. The solubility properties of a hormone in turn determine the means by which (1) the hormone is processed by the endocrine cell, (2) the way the hormone is transported in the blood, and (3) the mechanism by which the hormone exerts its effects at the target cell. Here are the differences in the solubility of the various classes of hormones (▲ Table 18-2):

- All peptides and catecholamines are hydrophilic (water loving) and lipophobic (lipid fearing); that is, they are highly H_2O soluble and have low lipid solubility.

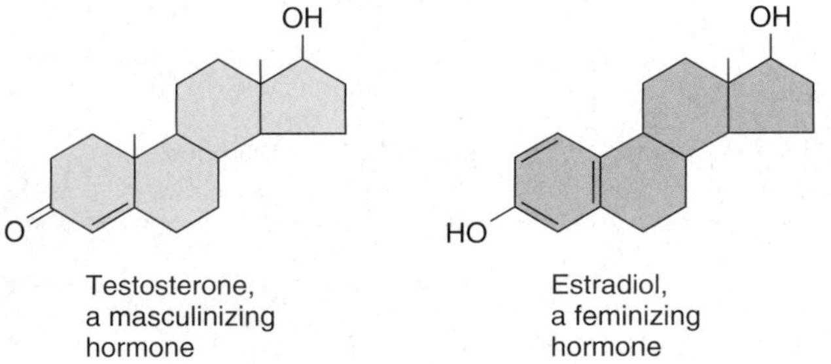

- All steroid and thyroid hormones are lipophilic (lipid loving) and hydrophobic (water fearing); that is, they have high lipid solubility and are poorly soluble in H_2O.

We are first going to consider the different ways in which these hormone types are processed at their site of origin, the endocrine cell, before comparing their means of transport and their mechanisms of action.

The mechanisms of hormone synthesis, storage, and secretion vary according to the class of hormone.

Because of their chemical differences, the means by which the various classes of hormones are synthesized, stored, and secreted differ as follows:

Peptide hormones

Peptide hormones are synthesized and secreted by the same steps used for the manufacture of any protein that is exported from the cell (see ● Figure 2-4, p. 28). From the time peptide hormones are synthesized until they are secreted, they are always segregated from intracellular proteins by virtue of being contained within membrane-enclosed compartments. Here is a brief overview of these steps:

1. Large precursor proteins, or **preprohormones,** are synthesized by ribosomes on the rough endoplasmic reticulum. They then migrate to the Golgi complex in membrane-enclosed vesicles that pinch off from the smooth endoplasmic reticulum.

2. During their journey through the endoplasmic reticulum and Golgi complex, the large preprohormone precursor molecules are pruned first to **prohormones** and finally to **active hormones.** The peptide "scraps" that are left over as a large preprohormone molecule is cleaved to form the classic hormone, are often stored and cosecreted along with the hormone. This raises the possibility that these other peptides may also exert biologic effects that differ from the traditional hormonal product; that is, the cell may actually be secreting multiple hormones, but the functions of the other peptide products are for the most part unknown. A known example is the cleavage of the large precursor molecule *pro-opiomelanocortin* into three active products: *adrenocorticotropic hormone (ACTH), melanocyte-stimulating hormone (MSH),* and β-*endorphin.*

3. The Golgi complex concentrates the finished hormones, then packages them into secretory vesicles that are pinched off and stored in the cytoplasm until an appropriate signal triggers their secretion. By storing peptide hormones in a readily releasable form, the gland can respond rapidly to any demands for increased secretion without first needing to increase hormone synthesis.

4. On appropriate stimulation, the secretory vesicles fuse with the plasma membrane and release their contents to the outside by exocytosis (see p. 30). Such secretion usually does not go on continuously; it is triggered only by specific stimuli. The blood then picks up the secreted hormone for distribution.

● **FIGURE 18-2**

Comparison of two steroid hormones, testosterone and estradiol

Testosterone,
a masculinizing
hormone

Estradiol,
a feminizing
hormone

Chemical Classification of Hormones

Properties	Peptides	Amines		Steroids
		Catecholamines	Thyroid Hormone	
Structure	Chains of specfic amino acids, for example: (vasopressin)	Tyrosine derivative, for example: (epinephrine)	Iodinated tyrosine derivative, for example: (thyroxine, T_4)	Cholesterol derivative, for example: (cortisol)
Solubility	Hydrophilic (lipophobic)	Hydrophilic (lipophobic)	Lipophilic (hydrophobic)	Lipophilic (hydrophobic)
Synthesis	In rough endoplasmic reticulum; packaged in Golgi complex	In cytosol	In colloid, an inland extracellular site	Stepwise modification of cholestrol molecule in various intracellular compartments
Storage	Large amounts in secretory granules	In chromaffin granules	In colloid	Not stored; cholesterol precursor stored in lipid droplets
Secretion	Exocytosis of granules	Exocytosis of granules	Endocytosis of colloid	Simple diffusion
Transport in Blood	As free hormone	Half bound to plasma proteins	Mostly bound to plasma proteins	Mostly bound to plasma proteins
Receptor Site	Surface of target cell	Surface of target cell	Inside target cell	Inside target cell
Mechanism of Action	Channel changes or activation of second-messenger system to alter activity of pre-existing proteins that produce the effect	Activation of second-messenger system to alter activity of pre-existing proteins that produce the effect	Activation of specific genes to make new proteins that produce the effect	Activation of specific genes to make new proteins that produce the effect
Hormones of This Type	All hormones from the hypothalamus, anterior pituitary, posterior pituitary, pineal gland, pancreas, parathyroid gland, gastrointestinal tract, kidneys, liver, thyroid C cells, thymus, heart	Only hormones from the adrenal medulla	Only hormones from the thyroid follicular cells	Hormones from the adrenal cortex and gonads plus most placental hormones (vitamin D is steroidlike)

Steroid hormones

All steroidogenic (steroid-producing) cells perform the following steps to produce and release their hormonal product:

1. Cholesterol is the common precursor for all steroid hormones.

2. Synthesis of the various steroid hormones from cholesterol requires a series of enzymatic reactions that modify the basic cholesterol molecule—for example, by varying the type and position of side groups attached to the cholesterol frame-work or the degree of saturation within the steroid ring (● Figure 18-3). Each conversion from cholesterol to a specific steroid hormone requires the help of a number of enzymes that are limited to certain steroidogenic organs. Accordingly, each steroidogenic organ can produce only the steroid hormone or hormones for which it has a complete set of appropriate enzymes. For example, a key enzyme necessary for producing cortisol is found only in the adrenal cortex, so no other steroidogenic organ can produce this hormone. Each enzyme necessary for converting cholesterol into a steroid hormone is localized in a

specific intracellular compartment, such as the mitochondria or endoplasmic reticulum.

3. Unlike peptide hormones, steroid hormones are not stored. Once formed, the lipid-soluble steroid hormones immediately diffuse through the steroidogenic cell's lipid plasma membrane to enter the blood. Only the hormone precursor cholesterol is stored in significant quantities within steroidogenic cells. Accordingly, the rate of steroid hormone secretion is controlled entirely by the rate of hormone synthesis. In contrast, peptide hormone secretion is controlled primarily by regulating the release of presynthesized stored hormone.

4. Following their secretion into the blood, some steroid hormones undergo further interconversions within the blood or other organs, where they are converted into more potent or different hormones.

Amines

The amine hormones—thyroid hormone and adrenomedullary catecholamines—have unique synthetic and secretory pathways that will be described when addressing each of these hormones specifically. The amines share the following features:

1. They are derived from the naturally occurring amino acid tyrosine.

2. None of the enzymes directly involved in the synthesis of either of these hormone types is located in organelle compartments within the secretory cells.

3. Both types of amines are stored until they are secreted.

▌ Water-soluble hormones dissolve in the plasma; lipid-soluble hormones are transported by plasma proteins.

All hormones are carried by the blood, but they are not all transported in the same manner:

- The hydrophilic (water-soluble) peptide hormones are transported simply dissolved in the plasma.
- Lipophilic (lipid-soluble) steroids and thyroid hormone, which are poorly soluble in water, cannot dissolve to any extent in the watery plasma. Instead, the majority of the lipophilic hormones circulate in the blood to their target cells reversibly bound to plasma proteins. Some are bound to specific plasma

---

● FIGURE 18-3

Steroidogenic pathways for the major steroid hormones
All steroid hormones are produced through a series of enzymatic reactions that modify cholesterol molecules, such as by varying the side groups attached to them. Each steroidogenic organ can produce only those steroid hormones for which it has a complete set of the enzymes needed to appropriately modify cholesterol. For example, the testes have the enzymes necessary to convert cholesterol into testosterone (male sex hormone), whereas the ovaries possess the enzymes needed to yield progesterone and the various estrogens (female sex hormones).

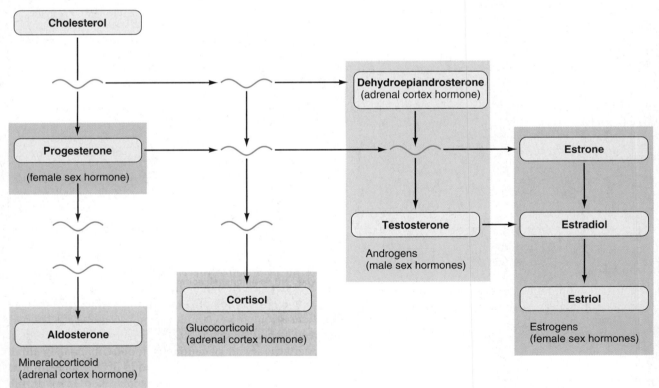

〜〜〜 = Intermediates not biologically active in humans

proteins designed to carry only one type of hormone, whereas other plasma proteins, such as albumin, indiscriminately pick up any "hitchhiking" hormone.

Only the small, unbound, freely dissolved fraction of a lipophilic hormone is biologically active (that is, free to cross capillary walls and bind with target cell receptors to exert an effect). Once a hormone has interacted with a target cell, it is rapidly inactivated or removed so it is no longer available to interact with another target cell. Because the carrier-bound hormone is in dynamic equilibrium with the free hormone pool, the bound form of steroid and thyroid hormones provides a large reserve of these lipophilic hormones that can be called on to replenish the active free pool. To maintain normal endocrine function, the magnitude of the small free effective pool, rather than the total plasma concentration of a particular lipophilic hormone, is monitored and adjusted.

• Catecholamines are unusual in that only about 50% of these hydrophilic hormones circulate as free hormone; the other 50% are loosely bound to the plasma protein albumin. Because catecholamines are water soluble, the importance of this protein binding is unclear.

Influence of a hormone's chemical properties on routes of therapeutic administration

The chemical properties of a hormone dictate not only the means by which blood transports it but also how it can be artificially introduced into the blood for therapeutic purposes. Because the digestive system does not secrete enzymes that can digest steroid and thyroid hormones, when taken orally these hormones, such as the sex steroids contained in birth control pills, can be absorbed intact from the digestive tract into the blood. No other type of hormones can be taken orally, because protein-digesting enzymes would attack and convert them into inactive fragments. Therefore, these hormones must be administered by nonoral routes; for example, insulin deficiency is treated with daily injections of insulin.

We will now examine how the classes of hormones vary in their mechanisms of action at their target cells.

▌ Hormones generally produce their effect by altering intracellular proteins.

To induce their effect, hormones must bind with target cell receptors specific for them. Each interaction between a particular hormone and a target cell receptor produces a highly characteristic response that differs among hormones and among different target cells influenced by the same hormone.

The location of the receptors within the target cell and the mechanism by which binding of the hormone with the receptors induces a response both vary, depending on the hormone's solubility characteristics.

Location of receptors for hydrophilic and lipophilic hormones

Hormones can be grouped into two categories based on the location of their receptors (▲ Table 18-2):

1. The hydrophilic peptides and catecholamines, which are poorly soluble in lipid, cannot pass through the lipid membrane barriers of their target cells. Instead, they bind with specific receptors located on the *outer plasma membrane surface* of the target cell.

2. The lipophilic steroids and thyroid hormone easily pass through the surface membrane to bind with specific receptors located *inside* the target cell.

General means of hydrophilic and lipophilic hormone action

Even though hormones elicit a wide variety of biologic responses, all hormones ultimately influence their target cells by altering the cell's proteins through three general means (● Figure 18-4 and ▲ Table 18-2):

1. A few hydrophilic hormones, on binding with a target cell's surface receptors change the cell's permeability (either opening or closing channels to one or more ions) by *altering the conformation (shape) of adjacent channel-forming proteins already in the membrane.*

2. Most surface-binding hydrophilic hormones function by *activating second-messenger systems* within the target cell. This activation directly *alters the activity of pre-existing intracellular proteins, usually enzymes*, to produce the desired effect.

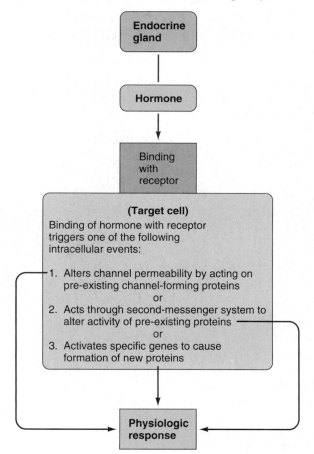

● FIGURE 18-4

Three general means by which hormones elicit biologic responses

Endocrine gland

↓

Hormone

↓

Binding with receptor

↓

(Target cell)
Binding of hormone with receptor triggers one of the following intracellular events:

1. Alters channel permeability by acting on pre-existing channel-forming proteins
   or
2. Acts through second-messenger system to alter activity of pre-existing proteins
   or
3. Activates specific genes to cause formation of new proteins

↓

Physiologic response

3. All lipophilic hormones function by *activating specific genes in the target cell to cause formation of new intracellular proteins,* which in turn produce the desired effect. The new proteins may be enzymatic or structural.

Let us examine the two major mechanisms of hormonal action (activation of second-messenger systems and activation of genes) in more detail.

▌ Hydrophilic hormones alter pre-existing proteins via second-messenger systems.

Because second-messenger systems in general, and cyclic AMP (cAMP) in particular, play such a central role in hydrophilic hormone activity, we will briefly review the steps involved (see ● Figure 3-8, p. 68):

1. Binding of the extracellular first messenger, the hydrophilic hormone, to its surface membrane receptor activates, by means of G protein intermediaries, the membrane-bound enzyme adenylyl cyclase.
2. Activated adenylyl cyclase converts intracellular ATP to cAMP, the intracellular second messenger.
3. Cyclic AMP triggers a preprogrammed series of biochemical steps that bring about a change in the shape and function of specific pre-existing enzymatic proteins in the cell.
4. These altered enzymatic proteins are responsible for bringing about a change in cell activity. The resultant change is the target cell's ultimate physiologic response to the hormone.

Once the hormone is removed, cAMP is inactivated by a specific chemical within the cytoplasm, and the intracellular message is "erased."

The nature of the pre-existing enzymatic proteins whose activity is ultimately modified by a second messenger varies in different target cells. Thus various target cells respond very differently to the universal mechanism of hormonally induced changes in their cAMP levels. Cyclic AMP can "turn on" (or "turn off") different cellular events depending on the kinds of enzyme activity ultimately modified in the various target cells.

Many hydrophilic hormones use cAMP as their second messenger. A few use intracellular Ca^{2+} in this role; for others, the second messenger is still unknown.

▌ By stimulating genes, lipophilic hormones promote synthesis of new proteins.

All lipophilic hormones (steroids and thyroid hormone) produce their effects in their target cells by enhancing the synthesis of new enzymatic or structural proteins. Their effects result from stimulating the cell's genes as follows (● Figure 18-5):

1. Free lipophilic hormone (hormone not bound with its carrier) diffuses through the plasma membrane of the target cell and binds with its specific receptor inside the cell. Most lipophilic hormone receptors are located in the nucleus.
2. Each receptor has a specific region for binding with its hormone and another region for binding with DNA. The receptor cannot bind with DNA unless it first binds with the hormone.
3. Once the hormone is bound to the receptor, the hormone receptor complex binds with DNA at a specific attachment site on DNA known as the **hormone response element (HRE).** Different steroid hormones and thyroid hormone, once bound with their respective receptors, attach at different HREs on DNA. For example, the estrogen receptor complex binds at DNA's estrogen response element.

● FIGURE 18-5

Mechanism of action of lipophilic hormones

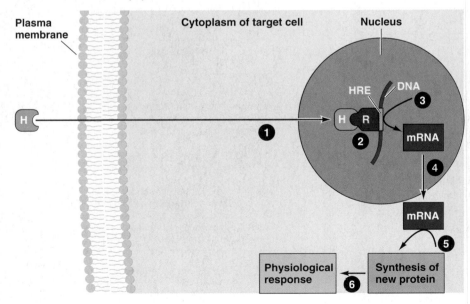

1 A lipophilic hormone diffuses through the plasma and nuclear membranes of its target cells and binds with a nuclear receptor specific for it.

2 The hormone receptor complex in turn binds with the hormone response element, a segment of DNA specific for the hormone receptor complex.

3 DNA binding activates specific genes, which produce complementary messenger RNA.

4 Messenger RNA leaves the nucleus.

5 In the cytoplasm, messenger RNA directs the synthesis of new proteins.

6 These new proteins, either enzymatic or structural, accomplish the target cell's ultimate physiologic response to the hormone.

H = Free lipophilic hormone
R = Lipophilic hormone receptor

HRE = Hormone response element
mRNA = Messenger RNA

(SOURCE: Adapted with permission from George A. Hedge, Howard D. Colby, and Robert L. Goodman, *Clinical Endocrine Physiology.* Philadelphia: W. B. Saunders Company, 1987, Figure 1-9, p. 20.)

4. Binding of the hormone receptor complex with DNA ultimately "turns on" specific genes within the target cell.

5. The activated genes direct the synthesis of new cell protein by producing complementary messenger RNA, which enters the cytoplasm and binds to a ribosome, the "workbench" that mediates the assembly of new proteins (p. A-25).

6. The newly synthesized protein produces the target cell's ultimate physiologic response to the hormone.

By means of this mechanism, different genes are activated by different lipophilic hormones, resulting in different biologic effects.

Onset and duration of hormonal responses

Compared to neural responses that are brought about within milliseconds, hormone action is relatively slow and prolonged, taking minutes to hours after the hormone binds to its receptor, for the response to take place. The variability in time of onset for hormonal responses depends on the mechanism employed. Hormones that act through a second-messenger system to alter a pre-existing enzyme's activity elicit full action within a few minutes. In contrast, hormonal responses that require the synthesis of new protein may take up to several hours before any action is initiated.

Also in contrast to neural responses that are quickly terminated once the triggering signal ceases, hormonal responses persist for a period of time after the hormone is no longer bound to its receptor. Once an enzyme is activated in response to hydrophilic hormonal input, it no longer depends on the presence of the hormone. Thus the response lasts until the enzyme is inactivated. Likewise, once a new protein is synthesized in response to lipophilic hormonal input, it continues to function until it is degraded. As a result, a hormone's effect usually lasts for some time after its withdrawal. Predictably, the responses that depend on protein synthesis last longer than do those stemming from enzyme activation.

Furthermore, as we explain shortly, lipophilic hormones themselves usually persist longer after secretion before being inactivated than hydrophilic hormones do.

▌ Hormone actions are greatly amplified at the target cell.

The actions of both hydrophilic and lipophilic hormones are greatly amplified at the target cell. Hormones are greatly diluted by the blood and thus must exert their effect at incredibly low concentrations—as low as 1 picogram (10^{-12} gram; 1 millionth of a millionth of a gram) per ml—as opposed to the much higher localized concentration of neurotransmitter at the target cell during neural communication. Interaction of one hormonal molecule with its receptor can result in the formation of many active protein products that ultimately carry out the physiologic effect. For example, one peptide hormone results in the production of numerous cAMP messengers, each in turn activating many latent enzymes (see p. 69). Similarly, one steroid hormone–activated gene induces formation of many messenger RNA molecules, each of which is used to make many new proteins.

▌ The effective plasma concentration of a hormone is normally regulated by changes in its rate of secretion.

The primary function of most hormones is the regulation of various homeostatic activities. Because hormones' effects are proportional to their concentrations in the blood, these concentrations are subject to control according to homeostatic need. The plasma concentration of free, biologically active hormone—and thus the hormone's availability to its receptors—depends on several factors (● Figure 18-6): (1) the hormone's rate of secretion into the blood by the endocrine gland; (2) for a few hormones, its rate of metabolic activation; (3) for lipophilic hormones, its extent of binding to plasma proteins; and (4) its rate of removal from the blood by metabolic inactivation and excretion in the urine. Furthermore, the magnitude

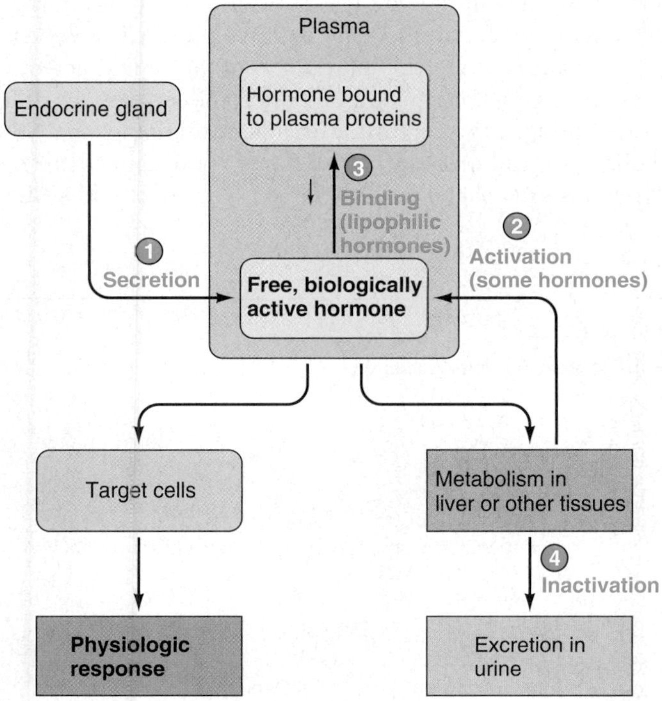

● FIGURE 18-6

Factors affecting the plasma concentration of free, biologically active hormone

The plasma concentration of free, biologically active hormone, which can interact with its target cells to produce a physiologic response, depends on

① the hormone's rate of secretion by the endocrine gland (for all hormones; the major factor)

② its rate of metabolic activation (for a few hormones)

③ its extent of binding to plasma proteins (for lipophilic hormones)

④ its rate of metabolic inactivation and excretion (for all hormones)

of the hormonal response depends on the availability and sensitivity of the target cells' receptors for the hormone. We will first examine the factors that influence the plasma concentration of the hormone before turning our attention to the target cells' responsiveness to the hormone.

Normally, the effective plasma concentration of a hormone is regulated by appropriate adjustments in the rate of its secretion. Endocrine glands do not secrete their hormones at a constant rate; the secretion rates of all hormones vary subject to control, often by a combination of several complex mechanisms. The regulatory system for each hormone is considered in detail in later sections. For now, we will address these general mechanisms of controlling secretion, which are common to many different hormones: negative-feedback control, neuroendocrine reflexes, and diurnal (circadian) rhythms.

Negative-feedback control

Negative feedback is a prominent feature of hormonal control systems. Stated simply, *negative feedback exists when the output of a system counteracts a change in input*, maintaining a controlled variable within a narrow range around a set level (see p. 16). Negative feedback maintains the plasma concentration of a hormone at a given level, similar to the way in which a home heating system maintains the room temperature at a given set point. Control of hormonal secretion provides some classic physiologic examples of negative feedback. For example, when the plasma concentration of free circulating thyroid hormone falls below a given "set point," the anterior pituitary secretes thyroid-stimulating hormone (TSH), which stimulates the thyroid to increase its secretion of thyroid hormone (● Figure 18-7). Thyroid hormone in turn inhibits further secretion of TSH by the anterior pituitary. Negative feedback ensures that once thyroid gland secretion has been "turned on" by TSH, it will not continue unabated but instead will be "turned off" when the appropriate level of free circulating thyroid hormone has been achieved. Thus the effect of a particular hormone's actions can inhibit its own secretion. The feedback loops often become quite complex.

Neuroendocrine reflexes

Many endocrine control systems involve **neuroendocrine reflexes,** which include neural as well as hormonal components. The purpose of such reflexes is to produce a sudden increase in hormone secretion (that is, "turn up the thermostat setting") in response to a specific stimulus, frequently a stimulus external to the body. In some instances, neural input to the endocrine gland is the only factor regulating secretion of the hormone. For example, secretion of epinephrine by the adrenal medulla is solely controlled by the sympathetic nervous system. Some endocrine control systems, in contrast, include both feedback control (which maintains a constant basal level of the hormone) and neuroendocrine reflexes (which cause sudden bursts in secretion in response to a sudden increased need for the hormone). An example is the increased secretion of cortisol, the "stress hormone," by the adrenal cortex during a stress response (see ● Figure 19-8, p. 710).

Diurnal (ciradian) rhythms

The secretion rates of many hormones rhythmically fluctuate up and down as a function of time. The most common endocrine rhythm is the **diurnal** ("day–night"), or **circadian** ("around a day") **rhythm,** which is characterized by repetitive oscillations in hormone levels that are very regular and cycle once every 24 hours. This rhythmicity is caused by endogenous oscillators similar to the self-paced respiratory neurons in the brain stem that control the rhythmic motions of breathing, except the timekeeping oscillators cycle on a much longer time scale. Furthermore, unlike the rhythmicity of breathing, endocrine rhythms are locked on, or **entrained,** to external cues such as the light–dark cycle. That is, the inherent 24-hour cycles of peak and ebb of hormone secretion are set to "march in step" with cycles of light and dark. For example, cortisol secretion rises during the night, reaching its peak secretion in the morning before a person gets up, then falls throughout the day to its lowest level at bedtime (● Figure 18-8). Inherent hormonal rhythmicity and entrainment are not accomplished by the endocrine glands themselves but result from the central nervous system changing the set point of these glands. We discuss the master biological clock further in a later section. Negative-feedback control mechanisms operate to maintain whatever set point is established for that time of day. Some endocrine

● **FIGURE 18-7**
Negative-feedback control

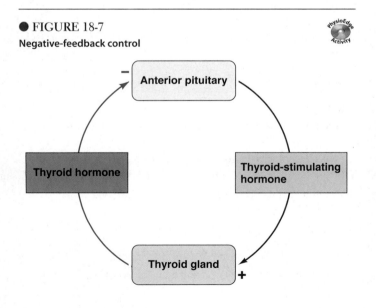

● **FIGURE 18-8**
Diurnal rhythm of cortisol secretion

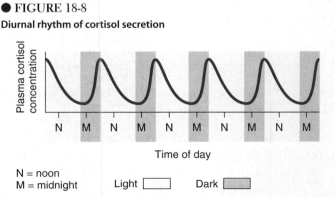

N = noon
M = midnight   Light [ ]   Dark [▨]

(SOURCE: Adapted with permission from George A. Hedge, Howard D. Colby, and Robert L. Goodman, *Clinical Endocrine Physiology.* Philadelphia: W. D. Saunders Company, 1987, Figure 1-13, p. 28.)

cycles operate on time scales other than a circadian rhythm, a well-known example being the monthly menstrual cycle.

▌ The effective plasma concentration of a hormone is influenced by its transport, metabolism, and excretion.

Even though the effective plasma concentration of a hormone is normally regulated by adjusting its rate of secretion, alterations in its transport, metabolism, or excretion can also influence the hormone's plasma concentration, sometimes inappropriately (● Figure 18-6). For example, because the liver synthesizes plasma proteins, liver disease may result in abnormal endocrine activity by altering the balance between free and bound pools of lipophilic hormones.

Eventually, all hormones are metabolized by enzyme-mediated reactions that modify the hormonal structure in some way. In most cases this inactivates the hormone. However, in some cases metabolism *activates* a hormone; that is, the hormone's product has greater activity than the original hormone. For example, after the thyroid hormone thyroxine is secreted, it is converted to a more powerful hormone by enzymatic removal of one of its iodine atoms. Usually the rate of such hormone activation is itself under hormonal control.

Metabolic inactivation and urinary excretion of hormones

The liver is the most common site for metabolic hormone inactivation, but some hormones are also metabolized in the kidneys, blood, or target cells. In contrast to hormone activation, which is typically regulated, hormone inactivation and excretion are not subject to control. The primary means of eliminating hormones and their metabolites from the blood is by urinary excretion. When liver and kidney function are normal, measuring urinary concentrations of hormones and their metabolites provides a useful, noninvasive way to assess endocrine function, because the rate of excretion of these products in urine directly reflects their rate of secretion by the endocrine glands. Because the liver and kidneys are important in removing hormones from the blood by metabolic inactivation and urinary excretion, patients with liver or kidney disease may suffer from excess activity of certain hormones solely because hormone elimination is reduced.

The amount of time after a hormone is secreted before it is inactivated, and the means by which this is accomplished, differ for different classes of hormones. In general, the hydrophilic peptides and catecholamines are easy targets for blood and tissue enzymes, so they remain in the blood only briefly (a few minutes to a few hours) before being enzymatically inactivated. In the case of some peptide hormones, such as insulin, the target cell actually engulfs the bound hormone by endocytosis and degrades it intracellularly. In contrast, binding of lipophilic hormones to plasma proteins makes them less vulnerable to metabolic inactivation and keeps them from escaping into urine. Therefore, lipophilic hormones are removed from plasma much more slowly. They may persist in the blood for hours (steroids) or up to a week (thyroid hormone). In general, lipophilic hormones undergo a series of reactions that re-

duce their biologic activity and enhance their H_2O solubility so they can be freed of their plasma protein carriers and be eliminated in urine.

▌ Endocrine disorders result from hormone excess or deficiency or decreased target-cell responsiveness.

Abnormalities in a hormone's effective plasma concentration can arise from a variety of factors (▲ Table 18-3). Endocrine disorders most commonly result from abnormal plasma concentrations of a hormone caused by inappropriate rates of secretion—that is, too little hormone secreted (**hyposecretion**) or too much hormone secreted (**hypersecretion**). Occasionally endocrine dysfunction arises because target cell responsiveness to the hormone is abnormally low, even though plasma concentration of the hormone is normal.

Hyposecretion

The following are among the many different factors (each listed with an example) that may produce hormone deficiency: (1) genetic (inborn absence of an enzyme that catalyzes synthesis of the hormone); (2) dietary (lack of iodine, which is needed for synthesis of thyroid hormone); (3) chemical or toxic (certain insecticide residues may destroy the adrenal cortex); (4) immunologic (autoimmune antibodies may destroy the body's own thyroid tissue); (5) other disease processes (cancer or tuberculosis may coincidentally destroy endocrine glands); (6) *iatrogenic* (physician induced, such as surgical removal of a cancerous thyroid gland); and (7) *idiopathic* (meaning the cause is not known). *Primary hyposecretion* occurs when an endocrine gland is secreting too little of its hormone because of an abnormality within that gland. *Secondary hyposecretion* takes place when an endocrine gland is normal but is secreting too little hormone because of a deficiency of its tropic hormone.

▲ **TABLE 18-3**
Means by Which Endocrine Disorders Can Arise

Too Little Hormone Activity	Too Much Hormone Activity
Too little hormone secreted by the endocrine gland (hyposecretion)[*]	Too much hormone secreted by the endocrine gland (hypersecretion)[*]
Increased removal of the hormone from the blood	Reduced plasma-protein binding of the hormone (too much free, biologically active hormone)
Abnormal tissue responsiveness to the hormone	
Lack of target cell receptors	Decreased removal of the hormone from the blood
Lack of an enzyme essential to the target cell response	Decreased inactivation
	Decreased excretion

[*]Most common causes of endocrine dysfunction.

The most common method of treating hormone hyposecretion is to administer a hormone that is the same as (or similar to, such as from another species) the deficient or missing one. Such replacement therapy seems straightforward, but the hormone's source and means of administration involve some practical problems. The sources of hormone preparation for clinical use include (1) endocrine tissues from domestic livestock, (2) placental tissue and urine of pregnant women, (3) laboratory synthesis of hormones, and (4) "hormone factories," or bacteria into which genes coding for the production of human hormones have been introduced. The method of choice for a given hormone is determined largely by its structural complexity and degree of species specificity.

Hypersecretion

Like *hypo*secretion, *hyper*secretion by a particular endocrine gland is designated as primary or secondary depending on whether the defect lies in that gland or is due to excessive stimulation from the outside, respectively. Hypersecretion may be caused by (1) tumors that ignore the normal regulatory input and continuously secrete excess hormone and (2) immunologic factors, such as excessive stimulation of the thyroid gland by an abnormal antibody that mimics the action of TSH, the thyroid tropic hormone. Excessive levels of a particular hormone may also arise from substance abuse, such as the outlawed practice among athletes of using certain steroids that increase muscle mass by promoting protein synthesis in muscle cells (see p. 282).

There are several ways of treating hormonal hypersecretion. If a tumor is the culprit, it may be surgically removed or destroyed with radiation treatment. In some instances, hypersecretion can be limited by drugs that block hormone synthesis or inhibit hormone secretion. Sometimes the condition may be treated by giving drugs that inhibit the action of the hormone without actually reducing the excess hormone secretion.

Abnormal target-cell responsiveness

Endocrine dysfunction can also occur because target cells do not respond adequately to the hormone, even though the effective plasma concentration of a hormone is normal. This unresponsiveness may be caused, for example, by an inborn lack of receptors for the hormone, as in *testicular feminization syndrome*. In this condition, receptors for testosterone, a masculinizing hormone produced by the male testes, are not produced because of a specific genetic defect. Although adequate testosterone is available, masculinization does not take place, just as if no testosterone were present. Abnormal responsiveness may also occur if the target cells for a particular hormone lack an enzyme essential to carrying out the response.

▍ The responsiveness of a target cell can be varied by regulating the number of hormone-specific receptors.

In contrast to endocrine dysfunction caused by *unintentional* receptor abnormalities, the target cell receptors for a particular hormone can be *deliberately altered* as a result of physiologic control mechanisms. A target cell's response to a hormone is correlated with the number of the cell's receptors occupied by molecules of that hormone, which in turn depends not only on the plasma concentration of the hormone but also on the number of receptors in the target cell for that hormone. Thus the response of a target cell to a given plasma concentration can be fine-tuned up or down by varying the number of receptors available for hormone binding.

Down regulation

As an illustration of this fine-tuning, when the plasma concentration of insulin is chronically elevated, the total number of target cell receptors for insulin is reduced as a direct result of the effect an elevated level of insulin has on the insulin receptors. This phenomenon, known as **down regulation**, constitutes an important locally acting negative-feedback mechanism that prevents the target cells from overreacting to the high concentration of insulin; that is, the target cells are *desensitized* to insulin, helping blunt the effect of insulin hypersecretion. Down regulation of insulin is accomplished by the following mechanism. The binding of insulin to its surface receptors induces endocytosis of the hormone receptor complex, which is subsequently attacked by intracellular lysosomal enzymes. This internalization serves a twofold purpose: It provides a pathway for degrading the hormone, and it also helps regulate the number of receptors available for binding on the target cell's surface. At high plasma insulin concentrations, the number of surface receptors for insulin is gradually reduced by the accelerated rate of receptor internalization and degradation brought about by increased hormonal binding. The rate of synthesis of new receptors within the endoplasmic reticulum and their insertion in the plasma membrane do not keep pace with their rate of destruction. Over time, this self-induced loss of target cell receptors for insulin reduces the target cell's sensitivity to the elevated hormone concentration.

Permissiveness, synergism, and antagonism

A given hormone's effects are influenced not only by the concentration of the hormone itself but also by the concentrations of other hormones that interact with it. Because hormones are widely distributed through the blood, target cells may be exposed simultaneously to many different hormones, giving rise to numerous complex hormonal interactions on target cells. Hormones frequently alter the receptors for other kinds of hormones as part of their normal physiologic activity. A hormone can influence the activity of another hormone at a given target cell in one of three ways: permissiveness, synergism, and antagonism.

- With **permissiveness,** one hormone must be present in adequate amounts for the full exertion of another hormone's effect. In essence, the first hormone, by enhancing a target cell's responsiveness to another hormone, "permits" this other hormone to exert its full effect. For example, thyroid hormone increases the number of receptors for epinephrine in epinephrine's target cells, increasing the effectiveness of epinephrine. In the absence of thyroid hormone, epinephrine is only marginally effective.

- **Synergism** occurs when the actions of several hormones are complementary and their combined effect is greater than the sum of their separate effects. An example is the synergistic action of follicle-stimulating hormone and testosterone, both of which are required for maintaining the normal rate of sperm production. Synergism probably results from each hormone's influence on the number or affinity of receptors for the other hormone.

- **Antagonism** occurs when one hormone causes the loss of another hormone's receptors, reducing the effectiveness of the second hormone. To illustrate, progesterone (a hormone secreted during pregnancy that *decreases* contractions of the uterus) inhibits uterine responsiveness to estrogen (another hormone secreted during pregnancy that *increases* uterine contractions). By causing loss of estrogen receptors on uterine smooth muscle, progesterone prevents estrogen from exerting its excitatory effects during pregnancy and thus keeps the uterus a quiet (noncontracting) environment suitable for the developing fetus.

Having completed our discussion of the general principles of endocrinology, we now begin an examination of the individual endocrine glands and their hormones. The rest of this chapter focuses on the central endocrine glands—those in the brain itself or in close association with the brain—namely, the pineal gland, the hypothalamus, and the pituitary gland. The peripheral endocrine glands are discussed in the next chapter.

PINEAL GLAND AND CIRCADIAN RHYTHMS

The **pineal gland**, a tiny, pinecone-shaped structure located in the center of the brain (see ● Figure 5-8b, p. 146, and ● Figure 18-1, p. 668), secretes the hormone **melatonin**. (Do no confuse *melatonin* with the skin-darkening pigment, *melanin*—see p. 451.) Although melatonin was discovered in 1959, investigators have only recently begun to unravel its many functions. One of melatonin's most widely accepted roles is helping to keep the body's inherent circadian rhythms in synchrony with the light–dark cycle. We will first examine circadian rhythms in general before looking at the role of melatonin in this regard and considering other functions of this hormone.

▮ The suprachiasmatic nucleus is the master biological clock.

Hormone secretion rates are not the only factor in the body that fluctuates cyclically over a 24-hour period. Humans have similar biological clocks for other bodily functions such as temperature regulation (see p. 655). The master biological clock that serves as the pacemaker for the body's circadian rhythms is the **suprachiasmatic nucleus (SCN)**. It consists of a cluster of nerve cell bodies in the hypothalamus above the optic chiasm, the point at which part of the nerve fibers from each eye cross to the opposite half of the brain (*supra* means "above;" *chiasm* means "cross") (see p. 210 and ● Figure 5-8b, p. 146). The self-induced rhythmic firing of the SCN neurons plays a major role in establishing many of the body's inherent daily rhythms.

Scientists have now unraveled the underlying molecular mechanisms responsible for the SCN's circadian oscillations. Specific self-starting genes within the nuclei of SCN neurons set in motion a series of events that brings about the synthesis of **clock proteins** in the cytosol surrounding the nucleus. As the day wears on, these clock proteins continue to accumulate, finally reaching a critical mass, at which time they are transported back into the nucleus. Here they block the genetic process responsible for their own production. The level of clock proteins gradually dwindles as they degrade within the nucleus, thus removing their inhibitory influence from the clock-protein genetic machinery. No longer being blocked, these genes once again rev up the production of more clock proteins, as the cycle repeats itself. Each cycle takes about a day. The fluctuating levels of clock proteins bring about cyclic changes in neural output from the SCN that in turn lead to cyclic changes in effector organs throughout the day. An example is the diurnal variation in cortisol secretion (see ● Figure 18-8). Circadian rhythms are thus linked to fluctuations in clock proteins, which use a feedback loop to control their own production. In this way, internal timekeeping is a self-sustaining mechanism built into the genetic makeup of the SCN neurons.

Synchronization of the biological clock with environmental cues

On its own, this biological clock generally cycles a bit slower than the 24-hour environmental cycle. Without any external cues, the SCN sets up cycles that average about 25 hours. The cycles are consistent for a given individual but vary somewhat among different people. If this master clock were not continually adjusted to keep pace with the world outside, the body's circadian rhythms would become progressively out of sync with the cycles of light (periods of activity) and dark (periods of rest). Thus the SCN must be reset daily by external cues so that the body's biologic rhythms are synchronized with the activity levels driven by the surrounding environment. The effect of not maintaining the internal clock's relevance to the environment is well known by people who experience **jet lag** when their inherent rhythm is out of step with external cues. The SCN works in conjunction with the pineal gland and its hormonal product melatonin to synchronize the various circadian rhythms with the 24-hour day–night cycle. (For a discussion of problems associated with being out of sync with environmental cues, see the accompanying boxed feature, ▶ Concepts, Challenges, and Controversies.)

▮ Melatonin helps keep the body's circadian rhythms in time with the light–dark cycle.

Daily changes in light intensity are the major environmental cue used to adjust the SCN master clock. Special photoreceptors in the retina pick up light signals and transmit them directly to the SCN. These photoreceptors are distinct from the rods and cones used to perceive, or see, light (see p. 204). Based on re-

Tinkering with Our Biological Clocks

Research shows that the hectic pace of modern life, stress, noise, pollution, and the irregular schedules many workers follow can upset internal rhythms, illustrating how a healthy external environment affects our own internal environment—and our health.

Dr. Richard Restak, a neurologist and author, notes that the "usual rhythms of wakefulness and sleep…seem to exert a stabilizing effect on our physical and psychological health." The greatest disrupter of our natural circadian rhythms is the variable work schedule, surprisingly common in industrialized countries. Today, one out of every four working men and one out of every six working women has a variable work schedule—shifting frequently between day and night work. In many industries, to make optimal use of equipment and buildings workers are on the job day and night. As a spin-off, more restaurants and stores stay open 24 hours a day and more health care workers must be on duty at night to care for accident victims.

To spread the burden, many companies that maintain shifts round the clock alter their workers' schedules. One week, employees work the day shift. The next week, they move to the "graveyard shift" from midnight to 8 A.M. The next week, they work the night shift from 4 P.M. to midnight. Many shift workers feel tired most of the time and have trouble staying awake at the job. Their work performance suffers from fatigue. When workers arrive home for bed, they're exhausted but can't sleep, because they're trying to doze off at a time when the body is trying to wake them up. Unfortunately, the weekly changes in schedule never permit workers' internal alarm clocks to fully adjust. Most people require 4 to 14 days to adjust to a new schedule.

Workers on alternating shifts suffer more ulcers, insomnia, irritability, depression, and tension than workers on unchanging shifts. Their lives are never the same. To make matters worse, tired, irritable workers whose judgment is impaired by fatigue pose a threat to society as a whole. Consider an example.

At 4 A.M. in the control room of the Three Mile Island Nuclear Reactor in Pennsylvania, three operators made the first mistake in a series of errors that led to the worst nuclear accident in U.S. history. The operators did not notice warning lights and failed to observe that a crucial valve had remained open. When the morning shift operators entered the control room the next day, they quickly discovered the problems, but it was too late. Pipes in the system had burst, sending radioactive steam and water into the air and into two buildings. John Gofman and Arthur Tamplin, two radiation health experts, estimate that the radiation released from the accident will cause at least 300 and possibly as many as 900 additional fatal cases of cancer in the residents living near the troubled reactor, although other "experts" (especially in the nuclear industry) contest these projections, saying the accident will have no noticeable effect. Whatever the outcome, the 1979 accident at Three Mile Island cost several billion dollars to clean up.

Late in April 1986, another nuclear power plant went amok. This accident, in Chernobyl in the former Soviet Union, was far more severe. In the early hours of the morning, two engineers were testing the reactor. Violating standard operational protocol, they deactivated key safety systems. This single error in judgment (possibly caused by fatigue) led to the largest and most costly nuclear accident in world history. Steam built up inside the reactor and blew the roof off the containment building. A thick cloud of radiation rose skyward and then spread throughout Europe and the world. While workers battled to cover the molten radioactive core that spewed radiation into the sky, the whole world watched in horror.

The Chernobyl disaster, like the accident at Three Mile Island, may have been the result of workers operating at a time unsuitable for clear thinking. One has to wonder how many plane crashes, auto accidents, and acts of medical malpractice can be traced to judgment errors resulting from our insistence on working against inherent body rhythms.

Thanks to studies of biological rhythms, researchers are finding ways to reset biological clocks, which could help lessen the misery and suffering of shift workers and could improve the performance of the graveyard shift workers. For instance, one simple measure is to put shift workers on three-week cycles, to give their clocks time to adjust. And instead of shifting workers from daytime to graveyard shifts, transfer them forward, rather than backward (for example, from a daytime to a nighttime shift). It's a much easier adjustment. Bright lights can also be used to reset the biological clock. It's a small price to pay for a healthy workforce and a safer society. Furthermore, use of supplemental melatonin, the hormone that sets the internal clock to march in step with environmental cycles, may prove useful in resetting the body's clock when that clock is out of synchrony with external cues.

cent evidence, scientists suspect that **melanopsin,** a protein found in a special retinal ganglion cell (see p. 202), is the likely receptor for light that keeps the body in tune with external time. This is the major way the internal clock is coordinated to a 24-hour day. The eyes cue the pineal gland about the absence or presence of light by means of a neural pathway that passes through the suprachiasmatic nucleus. This pathway is distinct from the neural systems that result in vision perception. Melatonin is the hormone of darkness. Melatonin secretion increases up to 10-fold during the darkness of night and then falls to low levels during the light of day. Fluctuations in melatonin secretion in turn help entrain the body's biologic rhythms with the external light–dark cues.

Other neural and blood-borne signals may also be involved in timekeeping. In one interesting new study, by shining a bright light on the backs of the subjects' knees researchers influenced melatonin secretion to reset volunteers' internal clocks by as much as three hours. The investigators speculate that the light, which never reached the eyes, was detected by a blood-borne photoreceptor. A light-induced alteration in such a photoreceptor could signal the SCN about the presence of light. The nature of the blood-borne photoreceptor remains unclear, although hemoglobin and bilirubin, both of which are pigments that can absorb light, have been suggested as possibilities. The researchers chose the back of the knee because this site is thin skinned and blood-vessel rich. Most likely, light exposure to other sites where blood comes close to the body's surface could work as well. These findings are not as surprising as they may seem. Multiple photoreceptor systems involved with the entrainment of circadian rhythms have been identified outside of the eyes in many species. The relative importance of these potential blood-borne clock setters in humans awaits further study.

Other roles of melatonin not related to circadian timekeeping

Other proposed roles of melatonin besides regulating the body's biological clock include the following:

- Melatonin induces a natural sleep without the side effects that accompany hypnotic sedatives.
- Melatonin is believed to inhibit the hormones that stimulate reproductive activity. Puberty may be initiated by a reduction in melatonin secretion.
- In a related role, in some species, seasonal fluctuations in melatonin secretion associated with changes in the number of daylight hours are important triggers for seasonal breeding, migration, and hibernation.
- In another related role, melatonin is being used in clinical trials as a method of birth control, because at high levels it shuts down ovulation (egg release). A male contraceptive using melatonin to stop sperm production is also under development.
- Melatonin appears to be a very effective **antioxidant,** a defense tool against biologically damaging free radicals. *Free radicals* are very unstable electron-deficient particles that are highly reactive and destructive. Free radicals have been implicated in several chronic diseases such as coronary artery disease (see p. 334) and cancer and are believed to contribute to the aging process.
- Evidence suggests that melatonin may slow the aging process, perhaps by removing free radicals or by other means.
- Melatonin appears to enhance immunity and has been shown to reverse some of the age-related shrinkage of the thymus, the source of T lymphocytes (see p. 424), in old experimental animals.

Because of melatonin's many proposed roles, use of supplemental melatonin for a variety of conditions is very promising. However, most researchers are cautious about recommending supplemental melatonin until its effectiveness as a drug is further substantiated. Meanwhile, many people are turning to melatonin as a health food supplement; as such, it is not regulated by the Food and Drug Administration for safety and effectiveness. The two biggest self-prescribed uses of melatonin are to prevent jet lag and as a sleep aid.

We now shift our attention to the other central endocrine glands—the hypothalamus and pituitary.

HYPOTHALAMUS AND PITUITARY

The **pituitary gland,** or **hypophysis,** is a small endocrine gland located in a bony cavity at the base of the brain just below the hypothalamus (● Figure 18-9). The pituitary is connected to the hypothalamus by a thin connecting stalk. If you point one finger between your eyes and another finger toward one of your ears, the imaginary point where these lines would intersect is about the location of your pituitary.

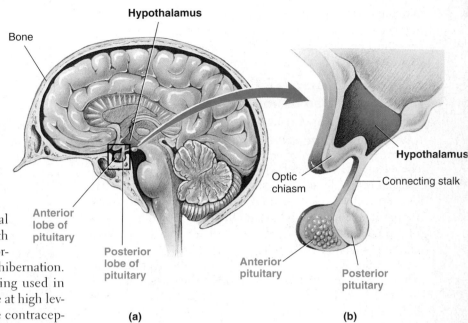

(a) **(b)**

● **FIGURE 18-9**

Anatomy of the pituitary gland
(a) Relation of the pituitary gland to the hypothalamus and to the rest of the brain. (b) Schematic enlargement of the pituitary gland and its connection to the hypothalamus.

▌ The pituitary gland consists of anterior and posterior lobes.

The pituitary has two anatomically and functionally distinct lobes, the **posterior pituitary** and the **anterior pituitary.** The posterior pituitary, being derived embryonically from an outgrowth of the brain, is composed of nervous tissue and thus is also termed the **neurohypophysis.** The anterior pituitary consists of glandular epithelial tissue derived embryonically from an outpouching that buds off from the roof of the mouth. Accordingly, the anterior pituitary is also called the **adenohypophysis** (*adeno* means "glandular"). The anterior and posterior pituitary only have their location in common.

Melanocyte-stimulating hormone

In some species, the adenohypophysis also includes a third, well-defined *intermediate lobe,* but humans lack this lobe. In lower vertebrates, the intermediate lobe secretes several **melanocyte-stimulating hormones** or **MSHs,** which regulate skin coloration by controlling the dispersion of granules containing the pigment **melanin.** By causing variable skin darkening in certain amphibians, reptiles, and fishes, MSHs play a vital role in the camouflage of these species.

In humans, the anterior pituitary secretes a small amount of MSH. It is not involved in differences in the amount of melanin deposited in the skin of various races, nor with the process of skin tanning, although excessive MSH activity does darken skin. Instead MSH in humans plays a totally different role in helping control food intake (see p. 651). It also appears to influence excitability of the nervous system, perhaps improving memory and learning. Also, MSH has been shown to suppress the immune system, perhaps helping serve in a check-and-balance way to prevent excessive immune responses.

▌ The hypothalamus and posterior pituitary act as a unit to secrete vasopressin and oxytocin.

The release of hormones from both the posterior and the anterior pituitary is directly controlled by the hypothalamus, but the nature of the relationship is entirely different. The posterior pituitary connects to the hypothalamus by a neural pathway, whereas the anterior pituitary connects to the hypothalamus by a unique vascular link.

Looking first at the posterior pituitary, the hypothalamus and posterior pituitary form a neuroendocrine system that consists of a population of neurosecretory neurons whose cell bodies lie in two well-defined clusters in the hypothalamus (the **supraoptic** and **paraventricular nuclei**). The axons of these neurons pass down through the thin connecting stalk to terminate on capillaries in the posterior pituitary (● Figure 18-10). The posterior pituitary consists of these neuronal terminals plus glial-like supporting cells called **pituicytes.** Functionally as well as anatomically, the posterior pituitary is simply an extension of the hypothalamus.

The posterior pituitary does not actually produce any hormones. It simply stores and, on appropriate stimulation, releases into the blood two small peptide hormones, *vasopressin* and *oxytocin*, which are synthesized by the neuronal cell bodies in the hypothalamus. Both these hydrophilic peptides are made in both the supraoptic and the paraventricular nuclei, but a single neuron can produce only one of these hormones. The synthesized hormones are packaged in secretory granules that are transported down the cytoplasm of the axon (see p. 46) and stored in the neuronal terminals within the posterior pituitary. Each terminal stores either vasopressin or oxytocin, but not both. Thus these hormones can be released independently as needed. On stimulatory input to the hypothalamus, either vasopressin or oxytocin is released into the systemic blood from the posterior pituitary by exocytosis of the appropriate secretory granules. This hormonal release is triggered in response to action potentials that originate in the hypothalamic cell body and sweep down the axon to the neuronal terminal in the posterior pituitary. As in any other neuron, action potentials are generated in these neurosecretory neurons in response to synaptic input to their cell bodies.

The actions of vasopressin and oxytocin are briefly summarized here to make our endocrine story complete. They are described more thoroughly elsewhere—vasopressin in Chapters 14 and 15 and oxytocin in Chapter 20.

Vasopressin

Vasopressin (antidiuretic hormone, ADH) has two major effects that correspond to its two names: (1) it enhances the retention of H_2O by the kidneys (an antidiuretic effect), and (2) it causes contraction of arteriolar smooth muscle (a vessel pressor ef-

fect). The first effect has more physiologic importance. Under normal conditions, vasopressin is the primary endocrine factor that regulates urinary H_2O loss and overall H_2O balance. In contrast, typical levels of vasopressin play only a minor role in regulating blood pressure by means of the hormone's pressor effect.

The major control for hypothalamic-induced release of vasopressin from the posterior pituitary is input from hypothalamic osmoreceptors, which increase vasopressin secretion in response to a rise in plasma osmolarity. A less powerful input from the left atrial volume receptors increases vasopressin secretion in response to a fall in ECF volume and arterial blood pressure (see p. 569). (For further information on the importance of vasopressin secretion when exercising in the heat, see the accompanying boxed feature, ▶ A Closer Look at Exercise Physiology.)

Oxytocin

Oxytocin stimulates contraction of the uterine smooth muscle to help expel the infant during childbirth, and it promotes ejection of the milk from the mammary glands (breasts) during breast-feeding. Appropriately, oxytocin secretion is increased by reflexes that originate within the birth canal during child-

● **FIGURE 18-10**

Relationship of the hypothalamus and posterior pituitary

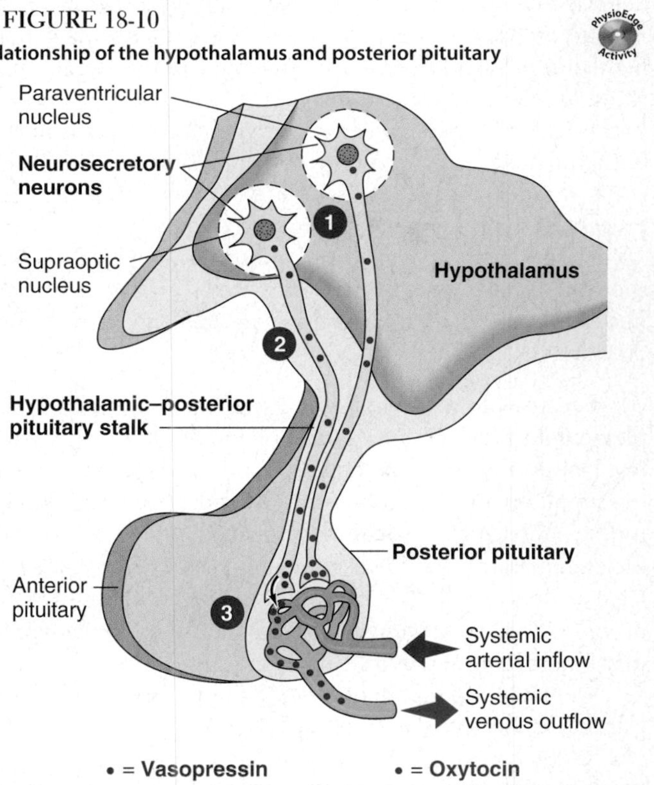

● = Vasopressin   ● = Oxytocin

❶ The paraventricular and supraoptic nuclei both contain neurons that produce vasopressin and oxytocin. The hormone, either vasopressin or oxytocin depending on the neuron, is synthesized in the neuronal cell body in the hypothalamus.

❷ The hormone travels down the axon to be stored in the neuronal terminals within the posterior pituitary.

❸ On excitation of the neuron, the stored hormone is released from these terminals into the systemic blood for distribution throughout the body.

The Endocrine Response to the Challenge of Combined Heat and Marching Feet

When one exercises in a hot environment, maintaining plasma volume becomes a critical homeostatic concern. Exercise in the heat results in losing large amounts of fluid through sweating. Simultaneously, blood is needed for shunting to the skin for cooling and for increased blood flow to nourish the working muscles. To maintain cardiac output, there must also be adequate venous return. The hypothalamus/posterior-pituitary neurosecretory system responds to these multiple, conflicting needs for fluid by releasing water-conserving vasopressin, reducing urinary fluid loss to preserve plasma volume.

Studies have generally shown that exercise in heat stimulates vasopressin release, which results in decreased urinary fluid loss. In one study conducted during an 18-mile road march in heat, the participants' average urine output dropped to 134 ml (normal urine output during the same time period would be about twice that much), whereas sweat loss averaged 4 liters. Overhydration before exercise appears to decrease the intensity of this response, suggesting that increased vasopressin release is related to plasma osmolarity. If fluid loss is not adequately replaced, plasma osmolarity increases. When the

hypothalamic osmoreceptors detect this hypertonic condition, they promote increased secretion of vasopressin from the posterior pituitary. Some investigators believe, however, that increased vasopressin release results from other factors, such as changes in blood pressure or in renal blood flow. Regardless of the mechanism, vasopressin release is an important physiologic response to exercise in heat.

---

birth and by reflexes that are triggered when the infant suckles the breast.

In addition to these two major physiologic effects, oxytocin has recently been shown to influence a variety of behaviors, especially maternal behaviors. For example, this hormone fittingly facilitates bonding, or attachment, between a mother and her infant.

▌ The anterior pituitary secretes six established hormones, many of which are tropic.

Unlike the posterior pituitary, which releases hormones synthesized by the hypothalamus, the anterior pituitary itself synthesizes the hormones it releases into the blood. Different cell populations within the anterior pituitary secrete six established peptide hormones. The actions of each of these hormones is described in detail in later sections. For now, here is a brief statement of their primary effects to provide a rationale for their names (● Figure 18-11):

1. **Growth hormone (GH, somatotropin),** the primary hormone responsible for regulating overall body growth, is also important in intermediary metabolism.

2. **Thyroid-stimulating hormone (TSH, thyrotropin)** stimulates secretion of thyroid hormone and growth of the thyroid gland.

3. **Adrenocorticotropic hormone (ACTH, adrenocorticotropin)** stimulates cortisol secretion by the adrenal cortex and promotes growth of the adrenal cortex.

4. **Follicle-stimulating hormone (FSH)** has different functions in females and males. In females it stimulates growth and development of ovarian follicles, within which the ova, or eggs, develop. It also promotes secretion of the hormone estrogen by the ovaries. In males FSH is required for sperm production.

5. **Luteinizing hormone (LH)** also functions differently in females and males. In females LH is responsible for ovulation and luteinization (that is, the formation of a hormone-secreting corpus luteum in the ovary following ovulation). LH also regulates ovarian secretion of the female sex hormones, estrogen and progesterone. In males the same hormone stim-

ulates the interstitial cells of Leydig in the testes to secrete the male sex hormone, testosterone, giving rise to its alternate name of **interstitial cell-stimulating hormone (ICSH).**

6. **Prolactin (PRL)** enhances breast development and milk production in females. Its function in males is uncertain, although evidence indicates that it may induce the production of testicular LH receptors. Furthermore, recent studies suggest that prolactin may enhance the immune system and support the development of new blood vessels at the tissue level in both sexes—both actions totally unrelated to its known roles in reproductive physiology.

TSH, ACTH, FSH, and LH are all tropic hormones, because they each regulate the secretion of another specific endocrine gland. FSH and LH are collectively referred to as **gonadotropins** because they control secretion of the sex hormones by the gonads (ovaries and testes). Because growth hormone exerts its growth-promoting effects indirectly by stimulating the release of liver hormones, the somatomedins, it too is sometimes categorized as a tropic hormone. Among the anterior pituitary hormones, prolactin is the only one that does not stimulate secretion of another hormone. Of the tropic hormones, FSH, LH, and growth hormone exert effects on nonendocrine target cells in addition to stimulating secretion of other hormones.

▌ Hypothalamic releasing and inhibiting hormones help regulate anterior pituitary hormone secretion.

None of the anterior pituitary hormones are secreted at a constant rate. Even though each of these hormones has a unique control system, there are some common regulatory patterns. The two most important factors that regulate anterior pituitary hormone secretion are (1) hypothalamic hormones and (2) feedback by target gland hormones.

Because the anterior pituitary secretes hormones that control the secretion of various other hormones, it long held the undeserved title of "master gland." Scientists now know that the release of each anterior-pituitary hormone is largely controlled by still other hormones produced by the hypothalamus.

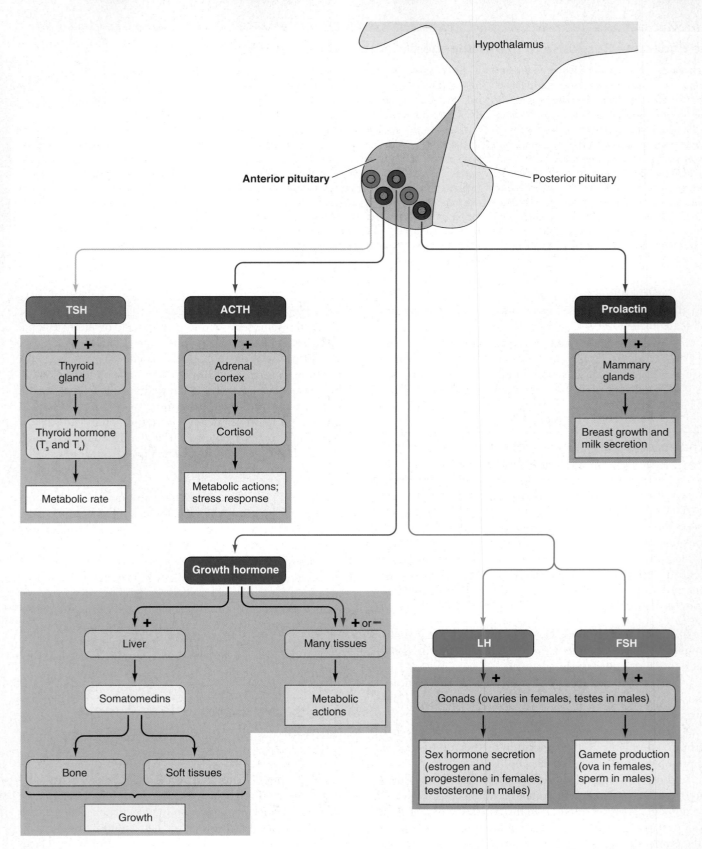

● **FIGURE 18-11**

Functions of the anterior pituitary hormones
Five different endocrine cell types produce the six anterior-pituitary hormones—TSH, ACTH, growth hormone, LH and FSH (produced by the same cell type), and prolactin—which exert a wide range of effects throughout the body.

The secretion of these regulatory neurohormones in turn is controlled by a variety of neural and hormonal inputs to the hypothalamic neurosecretory cells.

Role of the hypothalamic releasing and inhibiting hormones

The secretion of each anterior-pituitary hormone is stimulated or inhibited by one or more of the seven hypothalamic **hypophysiotropic hormones** (*hypophysis* means "pituitary"; *tropic* means "nourishing"). These small peptide hormones are listed in ▲ Table 18-4. Depending on their actions, hypophysiotropic hormones are called **releasing hormones** or **inhibiting hormones.** In each case, the primary action of the hormone is apparent from its name. For example, **thyrotropin-releasing hormone (TRH)** stimulates the release of TSH (alias thyrotropin) from the anterior pituitary, whereas **prolactin-inhibiting hormone (PIH)** inhibits the release of prolactin from the anterior pituitary. Note that hypophysiotropic hormones in most cases are involved in a three-hormone hierarchic chain of command (● Figure 18-12): The hypothalamic hypophysiotropic hormone (*hormone 1*) controls the output of an anterior pituitary tropic hormone (*hormone 2*). This tropic hormone in turn regulates secretion of the target endocrine gland's hormone (*hormone 3*), which exerts the final physiologic effect.

Although endocrinologists originally speculated that there was one hypophysiotropic hormone for each anterior pituitary hormone, many hypothalamic hormones have more than one effect, so their names indicate only the function first identified. Moreover, a single anterior pituitary hormone may be regulated by two or more hypophysiotropic hormones, which may even exert opposing effects. For example, **growth hormone–releasing hormone (GHRH)** stimulates growth hormone secretion, whereas **growth hormone–inhibiting hormone (GHIH)**, also known as **somatostatin**, inhibits it. The output of the anterior-pituitary growth hormone–secreting cells (that is, the rate of growth hormone secretion) in response to two such op-

● **FIGURE 18-12**

Hierarchic chain of command in endocrine control
The general pathway involved in the hierarchic chain of command among the hypothalamus, anterior pituitary, and peripheral target endocrine gland is depicted on the left. The pathway on the right leading to cortisol secretion provides a specific example of this endocrine chain of command.

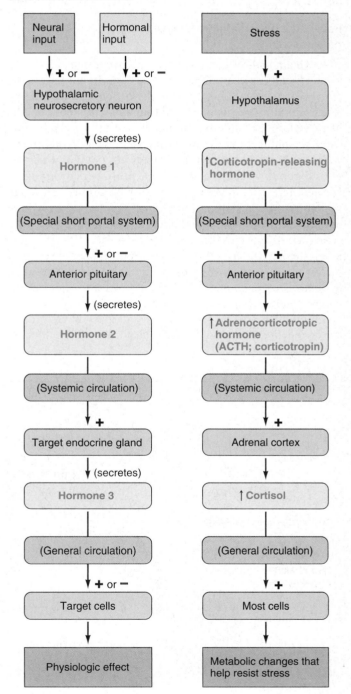

▲ TABLE 18-4

Major Hypophysiotropic Hormones

Hormone	Effect on the Anterior Pituitary
Thyrotropin-Releasing Hormone (TRH)	Stimulates release of TSH (thyrotropin) and prolactin
Corticotropin-Releasing Hormone (CRH)	Stimulates release of ACTH (corticotropin)
Gonadotropin-Releasing Hormone (GnRH)	Stimulates release of FSH and LH (gonadotropins)
Growth Hormone–Releasing Hormone (GHRH)	Stimulates release of growth hormone
Growth Hormone–Inhibiting Hormone (GHIH)	Inhibits release of growth hormone and TSH
Prolactin-Releasing Hormone (PRH)	Stimulates release of prolactin
Prolactin-Inhibiting Hormone (PIH)	Inhibits release of prolactin

posing inputs depends on the relative concentrations of these hypothalamic hormones as well as on the intensity of other regulatory inputs.

Chemical messengers that are identical in structure to the hypothalamic releasing and inhibiting hormones and to vasopressin are produced in many areas of the brain outside the hypothalamus. Instead of being released into the blood, these messengers act locally as neurotransmitters and as neuromodulators in these other sites. For example, PIH is identical to dopamine, a major neurotransmitter in the basal nuclei and elsewhere (see p. 156). Others are thought to modulate a variety of functions that range from motor activity (TRH) to libido (GnRH) to learning (vasopressin). These examples further illustrate the multiplicity of ways chemical messengers function.

Role of the hypothalamic-hypophyseal portal system

The hypothalamic regulatory hormones reach the anterior pituitary by means of a unique vascular link. In contrast to the direct neural connection between the hypothalamus and posterior pituitary, the anatomic and functional link between the hypothalamus and anterior pituitary is an unusual capillary-to-capillary connection, the **hypothalamic-hypophyseal portal system.** A portal system is a vascular arrangement in which venous blood flows directly from one capillary bed through a connecting vessel to another capillary bed. The largest and best-known portal system is the hepatic portal system, which drains intestinal venous blood directly into the liver for immediate processing of absorbed nutrients (see p. 619). Although much smaller, the hypothalamic-hypophyseal portal system is no less important, because it provides a critical link between the brain and much of the endocrine system. It begins in the base of the hypothalamus with a group of capillaries that recombine into small portal vessels, which pass down through the connecting stalk into the anterior pituitary. Here the portal vessels branch to form most of the anterior pituitary capillaries, which in turn drain into the systemic venous system (● Figure 18-13).

As a result, almost all the blood supplied to the anterior pituitary must first pass through the hypothalamus. Because materials can be exchanged between the blood and surrounding tissue only at the capillary level, the hypothalamic-hypophyseal portal system provides a route where releasing and inhibiting hormones can be picked up at the hypothalamus and delivered immediately and directly to the anterior pituitary at relatively high concentrations, completely bypassing the general circulation. If the portal system did not exist, once the hypophysiotropic hormones were picked up in the hypothalamus, they would be returned to the heart by the systemic venous system. From here, they would travel to the lungs and back to the heart through the pulmonary circulation, and finally enter the systemic arterial system for delivery throughout the body, including the anterior pituitary. This process would not only take longer, but the hypophysiotropic hormones would be considerably diluted by the much larger volume of blood flowing through this usual circulatory route.

The axons of the neurosecretory neurons that produce the hypothalamic regulatory hormones terminate on the capillaries at the origin of the portal system. These hypothalamic neurons secrete their hormones in the same way as the hypothalamic neurons that produce vasopressin and oxytocin. The hormone is synthesized in the cell body and then transported to the axon terminal. It is stored there until its release into an adjacent capillary on appropriate stimulation. The major difference is that the hypophysiotropic hormones are released into the portal vessels, which deliver them to the anterior pituitary where they control the release of anterior pituitary hormones into the general circulation. In contrast, the hypothalamic hormones stored in the posterior pituitary are themselves released into the general circulation.

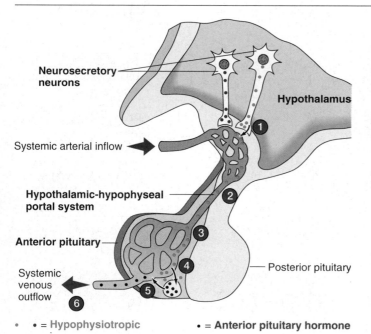

Neurosecretory neurons

Hypothalamus

Systemic arterial inflow ➡

Hypothalamic-hypophyseal portal system

Anterior pituitary

Systemic venous outflow ⬅

Posterior pituitary

• = **Hypophysiotropic**   • = **Anterior pituitary hormone**

● FIGURE 18-13

Vascular link between the hypothalamus and anterior pituitary

❶ Hypophysiotropic hormones (releasing hormones and inhibiting hormones) produced by neurons in the hypothalamus enter the hypothalamic capillaries.

❷ These hypothalamic capillaries rejoin to form the hypothalamic-hypophyseal portal system. This vascular link passes to the anterior pituitary.

❸ Here it branches into the anterior pituitary capillaries.

❹ The hypophysiotropic hormones leave the blood across the anterior pituitary capillaries and control the release of anterior pituitary hormones.

❺ On stimulation by the appropriate hypothalamic releasing hormone, a given anterior pituitary hormone is secreted into these capillaries.

❻ The anterior pituitary capillaries rejoin to form a vein, through which the anterior pituitary hormones leave for ultimate distribution throughout the body by the systemic circulation.

Control of hypothalamic releasing and inhibiting hormones

What regulates secretion of these hypophysiotropic hormones? Like other neurons, the neurons secreting these regulatory hormones receive abundant input of information (both neural and hormonal and both excitatory and inhibitory) that they must integrate. Studies are still in progress to unravel the complex neural input from many diverse areas of the brain to the hypophysiotropic secretory neurons. Some of these inputs carry information about a variety of environmental conditions. One example is the marked increase in secretion of corticotropin-releasing hormone (CRH) in response to stress (● Figure 18-12). Numerous neural connections also exist between the hypothalamus and the portions of the brain concerned with emotions (the limbic system—see p. 157). Thus emotions greatly influence secretion of hypophysiotropic hormones. The menstrual irregularities sometimes experienced by women who are emotionally upset are a common manifestation of this relationship.

In addition to being regulated by different regions of the brain, the hypophysiotropic neurons are also controlled by various chemical inputs that reach the hypothalamus through the blood. Unlike other regions of the brain, portions of the hypothalamus are not guarded by the blood–brain barrier, so the hypothalamus can easily monitor chemical changes in the blood. The most common blood-borne factors that influence hypothalamic neurosecretion are the negative-feedback effects of target gland hormones, to which we now turn our attention.

■ Target gland hormones inhibit hypothalamic and anterior pituitary hormone secretion via negative feedback.

In most cases, hypophysiotropic hormones initiate a three-hormone sequence: (1) hypophysiotropic hormone, (2) anterior-pituitary tropic hormone, and (3) peripheral target–endocrine gland hormone (● Figure 18-12). Typically, in addition to producing its physiologic effects, the target gland hormone also suppresses secretion of the tropic hormone that is driving it. This negative feedback is accomplished by the target gland hormone acting either directly on the pituitary itself or on the release of hypothalamic hormones, which in turn regulate anterior pituitary function (● Figure 18-14). As an example, consider the CRH-ACTH-cortisol system. Hypothalamic CRH (corticotropin-releasing hormone) stimulates the anterior pituitary to secrete ACTH (adrenocorticotropic hormone, alias corticotropin), which in turn stimulates the adrenal cortex to secrete cortisol. The final hormone in the system, cortisol, inhibits the hypothalamus to reduce CRH secretion and also reduces the sensitivity of the ACTH-secreting cells to CRH by acting directly on the anterior pituitary. Through this double-barreled approach, cortisol exerts negative-feedback control to stabilize its own plasma concentration. If plasma cortisol levels start to rise above a prescribed set level, cortisol suppresses its own further secretion by its inhibitory actions at the hypothalamus and anterior pituitary. This mechanism ensures that once a hormonal system is activated, its secretion does not

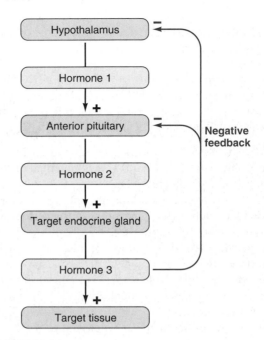

● **FIGURE 18-14**
Negative feedback in hypothalamic–anterior-pituitary control systems

continue unabated. If plasma cortisol levels fall below the desired set point, cortisol's inhibitory actions at the hypothalamus and anterior pituitary are reduced, so the driving forces for cortisol secretion (CRH-ACTH) increase accordingly. The other target-gland hormones act by similar negative-feedback loops to maintain their plasma levels relatively constant at the set point.

Diurnal rhythms are superimposed on this type of stabilizing negative-feedback regulation; that is, the set point changes as a function of the time of day. Furthermore, other controlling inputs may break through the negative-feedback control to alter hormone secretion (that is, change the set level) at times of special need. For example, stress raises the set point for cortisol secretion.

The detailed functions and control of all the anterior pituitary hormones except growth hormone are discussed elsewhere in conjunction with the peripheral target tissues that they influence; for example, thyroid-stimulating hormone is covered in the next chapter with the discussion of the thyroid gland. Accordingly, growth hormone is the only anterior-pituitary hormone we elaborate on at this time.

ENDOCRINE CONTROL OF GROWTH

In growing children, continuous net protein synthesis occurs under the influence of growth hormone as the body steadily gets larger. Weight gain alone is not synonymous with growth, because weight gain may occur from retaining excess H_2O or storing fat without true structural growth of tissues. Growth requires net synthesis of proteins and includes lengthening of the long bones (the bones of the extremities) as well as increases in the size and number of cells in the soft tissues.

Growth depends on growth hormone but is influenced by other factors as well.

Although, as the name implies, growth hormone (GH) is absolutely essential for growth, it alone is not wholly responsible for determining the rate and final magnitude of growth in a given individual. The following factors affect growth:

• *Genetic determination* of an individual's maximum growth capacity. Attaining this full growth potential further depends on the other factors listed here.

• *An adequate diet,* including enough total protein and ample essential amino acids to accomplish the protein synthesis necessary for growth. Malnourished children never achieve their full growth potential. The growth-stunting effects of inadequate nutrition are most profound when they occur in infancy. In severe cases, the child may be locked in to irreversibly stunted body growth and brain development. About 70% of total brain growth occurs in the first two years of life. By contrast, a person cannot exceed his or her genetically determined maximum by eating a more than adequate diet. The excess food intake produces obesity instead of growth.

• *Freedom from chronic disease and stressful environmental conditions.* Stunted growth under adverse circumstances is due in large part to the prolonged stress-induced secretion of cortisol from the adrenal cortex. Cortisol exerts several potent antigrowth effects, such as promoting protein breakdown, inhibiting growth in the long bones, and blocking the secretion of GH. Even though sickly or stressed children do not grow well, if the underlying condition is corrected before adult size is achieved they can rapidly catch up to their normal growth curve through a remarkable spurt in growth.

• *Normal levels of growth-influencing hormones.* In addition to the absolutely essential GH, other hormones, including thyroid hormone, insulin, and the sex hormones, play secondary roles in promoting growth.

Rapid growth periods

The rate of growth is not continuous nor are the factors responsible for promoting growth the same throughout the growth period. *Fetal growth* is promoted largely by certain hormones from the placenta (the hormone-secreting organ of exchange between the fetal and maternal circulatory systems; see p. 786), with the size at birth being determined principally by genetic and environmental factors. GH plays no role in fetal development. After birth, GH and other nonplacental hormonal factors begin to play an important role in regulating growth. Genetic and nutritional factors also strongly affect growth during this period.

Children display two periods of rapid growth—*a postnatal growth spurt* during their first two years of life and a *pubertal growth spurt* during adolescence (● Figure 18-15). From age 2 until puberty, the *rate* of linear growth progressively declines, even though the child is still growing. Before puberty there is little sexual difference in height or weight. During puberty, a marked acceleration in linear growth takes place because the long bones lengthen. Puberty begins at about age 11 in girls and 13 in boys and lasts for several years in both sexes. The mecha-

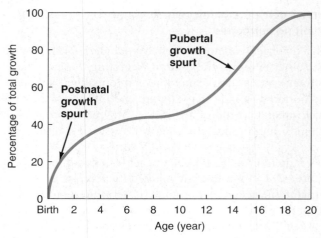

● **FIGURE 18-15**
Normal growth curve

nisms responsible for the pubertal growth spurt are not clearly understood. Apparently both genetic and hormonal factors are involved. Some evidence indicates that GH secretion is elevated during puberty and thus may contribute to growth acceleration during this time. Furthermore, **androgens** ("male" sex hormones), whose secretion increases dramatically at puberty, also contribute to the pubertal growth spurt by promoting protein synthesis and bone growth. The potent androgen from the male testes, testosterone, is of greatest importance in promoting a sharp increase in height in adolescent boys, whereas the less-potent adrenal androgens from the adrenal gland, which also show a sizable increase in secretion during adolescence, are most likely important in the female pubertal growth spurt. Although estrogen secretion by the ovaries also begins during puberty, it is unclear what role this "female" sex hormone may play in the pubertal growth spurt in girls. Testosterone and estrogen both ultimately act on bone to halt its further growth, so that full adult height is attained by the end of adolescence.

Growth hormone is essential for growth, but it also exerts metabolic effects not related to growth.

GH is the most abundant hormone produced by the anterior pituitary, even in adults in whom growth has already ceased. The continued high secretion of GH beyond the growing period implies that this hormone has important influences other than on growth. In addition to promoting growth, GH has important metabolic effects and enhances the immune system. We will briefly describe GH's metabolic actions before turning our attention to its growth-promoting actions.

Metabolic actions of GH unrelated to growth

GH increases fatty acid levels in the blood by enhancing the breakdown of triglyceride fat stored in adipose tissue, and it increases blood glucose levels by decreasing glucose uptake by muscles. Muscles use the mobilized fatty acids instead of glucose as a metabolic fuel. Thus the overall metabolic effect of GH is to mobilize fat stores as a major energy source while conserving glucose for glucose-dependent tissues such as the brain. The brain can use only glucose as its metabolic fuel, yet nervous tissue cannot store glycogen (stored glucose) to any extent. This metabolic pattern is suitable for maintaining the body during prolonged fasting or other situations when the body's energy needs exceed available glucose stores.

Growth-promoting actions of GH on soft tissues

When tissues are responsive to its growth-promoting effects, GH stimulates growth of both soft tissues and the skeleton. GH promotes growth of soft tissues by (1) increasing the number of cells (**hyperplasia**) and (2) increasing the size of cells (**hypertrophy**). GH increases the number of cells by stimulating cell division and by preventing apoptosis (programmed cell death; see p. 70). GH increases the size of cells by favoring synthesis of proteins, the main structural component of cells. GH stimulates almost all aspects of protein synthesis while it simultaneously inhibits protein degradation. It promotes the uptake of amino acids (the raw materials for protein synthesis) by cells, decreasing blood amino-acid levels in the process. Furthermore, it stimulates the cellular machinery responsible for accomplishing protein synthesis according to the cell's genetic code.

Growth of the long bones resulting in increased height is the most dramatic effect of GH. Before you can understand the means by which GH stimulates bone growth, you must first become familiar with the structure of bone and how growth of bone is accomplished.

■ Bone grows in thickness and in length by different mechanisms, both stimulated by growth hormone.

Bone is a living tissue. Being a form of connective tissue, it consists of cells and an extracellular organic matrix that is produced by the cells. The bone cells that produce the organic matrix are known as **osteoblasts** ("bone formers"). The organic matrix is composed of collagen fibers (see p. 62) in a mucopolysaccharide-rich semisolid gel known as *ground substance*. This matrix has a rubbery consistency and is responsible for the tensile strength of bone (the resilience of bone to breakage when tension is applied). Bone is made hard by precipitation of calcium phosphate crystals within the matrix. These inorganic crystals provide the bone with compressional strength (the ability of bone to hold its shape when squeezed or compressed). If bones consisted entirely of inorganic crystals, they would be brittle, like pieces of chalk. Bones have structural strength approaching that of reinforced concrete, yet they are not brittle and are much lighter in weight, because they have the structural blending of an organic scaffolding hardened by inorganic crystals. **Cartilage** is similar to bone, except that living cartilage is not calcified.

A long bone basically consists of a fairly uniform cylindrical shaft, the **diaphysis**, with a flared articulating knob at either end, an **epiphysis**. In a growing bone, the diaphysis is sep-

arated at each end from the epiphysis by a layer of cartilage known as the **epiphyseal plate** (● Figure 18-16a). The central cavity of the bone is filled with bone marrow, which is the site of blood cell production (see p. 395).

Bone growth

Growth in *thickness* of bone is achieved by adding new bone on top of the outer surface of already existing bone. This growth is produced by osteoblasts within the **periosteum,** a connective tissue sheath that covers the outer bone. As osteoblast activity deposits new bone on the external surface, other cells within the bone, the **osteoclasts** ("bone-breakers"), dissolve the bony tissue on the inner surface next to the marrow cavity. In this way, the marrow cavity enlarges to keep pace with the increased circumference of the bone shaft.

Growth in *length* of long bones is accomplished by a different mechanism. Bones grow in length because the cartilage cells proliferate in the epiphyseal plates (● Figure 18-16b). During growth, cartilage cells (**chondrocytes**) divide and multiply on the outer edge of the plate next to the epiphysis. As new chondrocytes are formed on the epiphyseal border, the older cartilage cells toward the diaphyseal border are enlarging. This combination of proliferation of new cartilage cells and hypertrophy of maturing chondrocytes temporarily widens the epiphyseal plate. This thickening of the intervening cartilaginous plate pushes the bony epiphysis farther away from the diaphysis. Soon the matrix surrounding the oldest hypertrophied cartilage becomes calcified. Because cartilage lacks its own capillary network, the survival of cartilage cells depends on diffusion of nutrients and O_2 through the ground substance, a process prevented by the deposition of calcium salts. As a result, the old nutrient-deprived cartilage cells on the diaphyseal border die. As osteoclasts clear away dead chondrocytes and the calcified matrix that imprisoned them, the area is invaded by osteoblasts, which swarm upward from the diaphysis, trailing their capillary supply with them. These new tenants lay down bone around the persisting remnants of disintegrating cartilage until bone entirely replaces the inner region of cartilage on the diaphyseal side of the plate. When this **ossification** ("bone formation") is complete, the bone on the diaphyseal side has lengthened, and the epiphyseal plate has returned to its original thickness. The cartilage that bone has replaced on the diaphyseal end of the plate is as thick as the new cartilage on the epiphyseal end of the plate. Thus bone growth is made possible by the growth and death of cartilage, which acts like a "spacer" to push the epiphysis farther out while it provides a framework for future bone formation on the end of the diaphysis.

Mature, nongrowing bone

As the extracellular matrix produced by an osteoblast calcifies, the osteoblast—like its chondrocyte predecessor—becomes entombed by the matrix it has deposited around itself. Unlike chondrocytes, however, osteoblasts trapped within a calcified matrix do not die, because they are supplied by nutrients transported to them through small canals that the osteoblasts themselves form by sending out cytoplasmic extensions around which the bony matrix is deposited. Thus, within the final bony product a network of permeating tunnels radiates from each entrapped osteoblast, serving as a lifeline system for nutrient delivery and waste removal. The entrapped osteoblasts, now called **osteocytes,** retire from active bone-forming duty, because their imprisonment prevents them from laying down new bone. However, they are involved in the hormonally regulated exchange of calcium between bone and the blood. This exchange is under the control of parathyroid hormone (discussed in the next chapter), not GH.

Growth-promoting actions of GH on bones

GH promotes growth of bone in both thickness and length. It stimulates osteoblast activity and also stimulates the proliferation of epiphyseal cartilage, thereby making space for more bone formation. GH can promote lengthening of long bones as long as the epiphyseal plate remains cartilaginous, or is "open." At the end of adolescence, under the influence of the sex hormones these plates completely ossify, or "close," so that the bones can lengthen no further despite the presence of GH. Thus, after the plates are closed the individual does not grow any taller.

∎ Growth hormone exerts its growth-promoting effects indirectly by stimulating somatomedins.

GH does not act directly on its target cells to bring about its growth-producing actions (increased cell division, enhanced protein synthesis, and bone growth). These effects are directly brought about by peptide mediators known as **somatomedins.** These peptides are also referred to as **insulin-like growth factors (IGF)** because they are structurally and functionally simi-

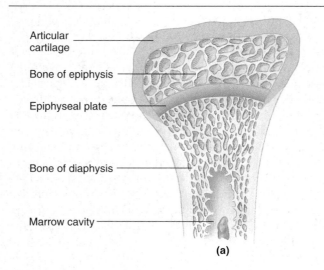

Articular cartilage

Bone of epiphysis

Epiphyseal plate

Bone of diaphysis

Marrow cavity

(a)

● FIGURE 18-16

Anatomy and growth of long bones
(a) Anatomy of long bones. (b) Two sections of the same epiphyseal plate at different times, depicting the lengthening of long bones.

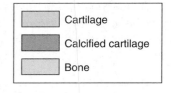

Cartilage

Calcified cartilage

Bone

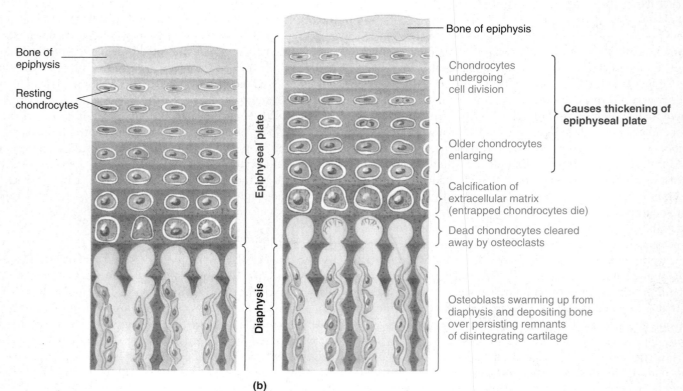

Bone of epiphysis

Bone of epiphysis

Resting chondrocytes

Epiphyseal plate

Diaphysis

Chondrocytes undergoing cell division

Older chondrocytes enlarging

Causes thickening of epiphyseal plate

Calcification of extracellular matrix (entrapped chondrocytes die)

Dead chondrocytes cleared away by osteoclasts

Osteoblasts swarming up from diaphysis and depositing bone over persisting remnants of disintegrating cartilage

(b)

lar to insulin. Two somatomedins—*IGF-I* and *IGF-II*—have been identified.

IGF-I

IGF-I synthesis is stimulated by GH and mediates most of this hormone's growth-promoting actions. The major source of circulating IGF-I is the liver, which releases this peptide product into the blood in response to GH stimulation. However, many if not most tissues produce IGF-I, although they do not release it into the blood to any extent. Researchers have proposed that IGF-I produced locally in target tissues may act through paracrine means (see p. 66) for at least some of the GH-induced effects. Such a mechanism could explain why blood levels of GH are no higher, and indeed circulating IGF-I levels are lower, during the first several years of life compared to adult values, even though growth is quite rapid during the postnatal period. Local production of IGF-I in target tissues may possibly be more important than delivery of blood-borne IGF-I during this time.

Production of IGF-I is controlled by a number of factors other than GH, including nutritional status, age, and tissue-specific factors as follows:

- IGF-I production depends on adequate nutrition. Inadequate food intake reduces IGF-I production, apparently through decreased sensitivity to GH in the tissues that produce IGF-I. As a result, changes in circulating IGF-I levels do not always coincide with changes in GH secretion. For example, fasting decreases IGF-I levels even though it increases GH secretion.
- Age-related factors also influence IGF-I production. A dramatic increase in circulating IGF-I levels accompanies the moderate increase in GH at puberty, which may, of course, be an impetus to the pubertal growth spurt.
- Finally, various tissue-specific stimulatory factors can increase IGF-I production in particular tissues. To illustrate, the gonadotropins and sex hormones stimulate IGF-I production within reproductive organs such as the testes in males and the ovaries and uterus in females.

Thus control of IGF-I production is complex and subject to a variety of systemic and local factors.

IGF-II

In contrast to IGF-I, **IGF-II** production does not depend on GH. IGF-II is known to be important during fetal development. Unlike IGF-I, IGF-II does not increase during the pubertal growth spurt. Although IGF-II continues to be produced during adulthood, its role in adults remains unclear.

From the preceding discussion of the factors involved in controlling growth, it should be obvious that many gaps still exist in the state of knowledge in this important area.

■ Growth hormone secretion is regulated by two hypophysiotropic hormones.

The control of GH secretion is complex, with two hypothalamic hypophysiotropic hormones playing a key role.

Growth hormone–releasing hormone and growth hormone–inhibiting hormone

Two antagonistic regulatory hormones from the hypothalamus are involved in controlling growth hormone secretion: growth hormone–releasing hormone (GHRH), which is stimulatory, and growth hormone–inhibiting hormone (GHIH or somatostatin), which is inhibitory (● Figure 18-17). (Note the distinctions among *somatotropin*, alias growth hormone; *somatomedin*, a liver hormone that directly mediates the effects of GH; and *somatostatin*, which inhibits GH secretion.) Any factor that increases GH secretion could theoretically do so either by stimulating GHRH release or by inhibiting GHIH release. It is not known which of these pathways is used in each specific case.

As with the other hypothalamus–anterior pituitary axes, negative-feedback loops participate in the regulation of GH secretion. Both GH and the somatomedins inhibit pituitary secretion of GH, presumably by stimulating GHIH release from the hypothalamus. The somatomedins may also directly influence the anterior pituitary to inhibit the effects of GHRH on GH release.

Factors that influence GH secretion

A number of factors influence GH secretion by acting on the hypothalamus. GH secretion displays a well-characterized diurnal rhythm. Through most of the day, GH levels tend to be low and fairly constant. Approximately one hour after the onset of deep sleep, however, GH secretion markedly increases up to five times the daytime value, then rapidly drops over the next several hours.

Superimposed on this diurnal fluctuation in GH secretion are further bursts in secretion that occur in response to exercise, stress, and hypoglycemia (low blood glucose), the major stimuli for increased secretion. The benefits of increased GH secretion during these situations when energy demands outstrip the body's glucose reserves are presumably that glucose is conserved for the brain and fatty acids are provided as an alternative energy source for muscle.

Because GH uses up fat stores and promotes synthesis of body proteins, it encourages a change in body composition away from adipose deposition toward an increase in muscle protein. Accordingly, the increase in GH secretion that accompanies exercise may at least in part mediate exercise's effects in reducing the percentage of body fat while increasing the lean body mass.

An abundance of blood amino acids, such as after a high-protein meal, also enhances the secretion of GH, which in turn promotes the use of these amino acids for protein synthesis. GH release is also stimulated by a decline in the level of blood fatty acids. Because of the fat-mobilizing actions of GH, such regulation helps maintain fairly constant blood fatty-acid levels.

Note that the known regulatory inputs for GH secretion are aimed at adjusting the levels of glucose, amino acids, and fatty acids in the blood. No known growth-related signals influence GH secretion. The whole issue of what really controls growth is complicated by the fact that GH levels during early childhood, a period of quite rapid linear growth, are similar to those in normal adults. As mentioned earlier, the poorly understood control of somatomedin activity may be important in this regard. Another related question is, Why aren't adult tissues still responsive to GH's growth-promoting effects? We know we do not grow any taller after adolescence because the epiphyseal plates have closed, but why don't soft tissues continue to grow through hypertrophy and hyperplasia under the influence of GH? One speculation is that levels of GH may only be high enough to produce its growth-promoting effects during the secretion bursts that occur in deep sleep. It is interesting to note that time spent in deep sleep is greatest in infancy and gradually declines with age. Still, even as we age we spend some time in deep sleep, yet we do not gradually get larger. Further research is needed to unravel some of these mysteries.

▌ Abnormal growth hormone secretion results in aberrant growth patterns.

Diseases related to both deficiencies and excesses of growth hormone can occur. The effects on the pattern of growth are much more pronounced than the metabolic consequences.

Growth hormone deficiency

GH deficiency may be caused by a pituitary defect (lack of GH) or may occur secondary to hypothalamic dysfunctions (lack of GHRH). Hyposecretion of GH in a child results in **dwarfism.** The predominant feature is short stature caused by retarded skeletal growth (● Figure 18-18). Less obvious characteristics include poorly developed musculature (reduction in muscle protein synthesis) and excess subcutaneous fat (less fat mobilization).

In addition, growth may be thwarted because the tissues fail to respond normally to GH. An example is **Laron dwarfism,** which is characterized by abnormal GH recep-

● **FIGURE 18-17**

Control of growth hormone secretion

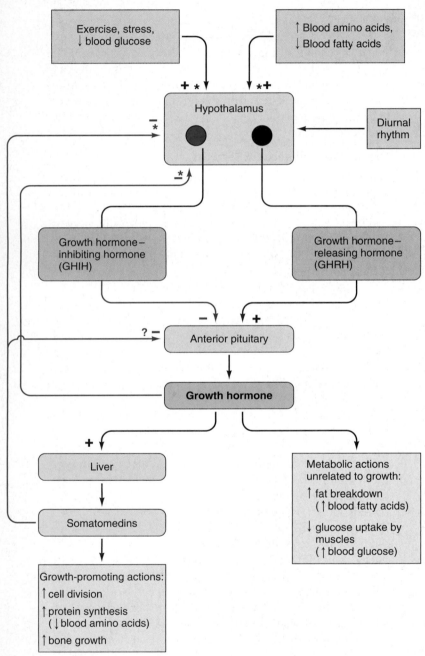

● **FIGURE 18-18**

Examples of the effect of abnormalities in growth hormone secretion on growth
The man at the left displays pituitary dwarfism resulting from underproduction of growth hormone in childhood. The male at the center of the photograph has gigantism caused by excessive growth hormone secretion in childhood. The woman at the right is of average height.

tors that are unresponsive to the hormone. The symptoms of this condition resemble those of severe GH deficiency even though blood levels of GH are actually high.

In some instances, GH levels are adequate and target cell responsiveness is normal, but somatomedins are lacking. African pygmies are an interesting example. Their usual short stature is attributable to a lack of the most potent of the somatomedins.

The onset of GH deficiency in adulthood after growth is already complete produces relatively few symptoms. GH-deficient adults tend to have reduced skeletal muscle mass and strength (less muscle protein) as well as decreased bone density (less osteoblast activity during ongoing bone remodeling). Furthermore, because GH is essential for maintaining cardiac muscle mass and performance in adulthood, GH-deficient adults may be at increased risk of developing heart failure. (For a discussion of GH therapy, see the accompanying boxed feature, ▶ Concepts, Challenges, and Controversies.)

Growth hormone excess

Hypersecretion of GH is most often caused by a tumor of the GH-producing cells of the anterior pituitary. The symptoms depend on the age of the individual when the abnormal secretion begins. If overproduction of GH begins in childhood before the epiphyseal plates close, the principal manifestation is a rapid growth in height without distortion of body proportions. Appropriately, this condition is known as **gigantism** (● Figure 18-18). If not treated by removal of the tumor or by drugs that block the effect of GH, the individual may reach a height of eight feet or more. All the soft tissues grow correspondingly, so the body is still well proportioned.

If GH hypersecretion occurs after adolescence when the epiphyseal plates have already closed, further growth in height is prevented. Under the influence of excess GH, however, the bones become thicker and the soft tissues, especially connective tissue and skin, proliferate. This disproportionate growth pattern produces a disfiguring condition known as **acromegaly** (*acro* means "extremity"; *megaly* means "large"). Bone thickening is most obvious in the extremities and face. A marked coarsening of the features to an almost apelike appearance gradually develops as the jaws and cheekbones become more prominent because of the thickening of the facial bones and the skin (● Figure 18-19). The hands and feet enlarge, and the fingers and toes become greatly thickened. Peripheral nerve disorders often occur as nerves are entrapped by overgrown connective tissue or bone, or both.

Visual disturbances may accompany either gigantism or acromegaly. This symptom does not result from excessive GH per se but rather from a GH-secreting tumor enlarging the anterior pituitary gland. The enlarged gland pushes on the nearby optic chiasm, the point at which the nerve fibers passing from the eyes cross over on their way to the visual cortex of the brain (see ● Figure 5-8b, p. 146, and ● Figure 18-9b, p. 683).

▌ Other hormones besides growth hormone are essential for normal growth.

Several other hormones in addition to GH contribute in special ways to overall growth:

• *Thyroid hormone* is essential for growth but is not itself directly responsible for promoting growth. It plays a permissive role in skeletal growth; the actions of GH fully manifest only when enough thyroid hormone is present. As a result, growth is severely stunted in hypothyroid children, but hypersecretion of thyroid hormone does not cause excessive growth.

• *Insulin* is an important growth promoter. Insulin deficiency often blocks growth, and hyperinsulinism frequently spurs excessive growth. Because insulin promotes protein synthesis, its growth-promoting effects should not be surprising. However, these effects may also arise from a mechanism other than insulin's direct effect on protein synthesis. Insulin structurally resembles the somatomedins and may interact with the somatomedin (IGF-I) receptor, which is very similar to the insulin receptor.

• *Androgens*, which are believed to play an important role in the pubertal growth spurt, powerfully stimulate protein synthesis in many organs. Androgens stimulate linear growth, promote weight gain, and increase muscle mass. The most potent androgen, testicular testosterone, is responsible for men developing heavier musculature than do women. These androgenic growth-promoting effects depend on the presence of GH. Androgens have virtually no effect on body growth in the absence

---

● **FIGURE 18-19**
Progressive development of acromegaly
In this series of photos from childhood to the present, note how the patient's brow bones, cheekbones, and jaw bones are becoming progressively more prominent as a result of ongoing thickening of the bones and skin caused by excessive GH secretion.

| Age 13 | Age 21 | Age 35 |

Growth and Youth in a Bottle?

Unlike most hormones, growth hormone's structure and activity differ among species. As a result, growth hormone (GH) from animal sources is ineffective for treating GH deficiency in humans. In the past, the only source of human GH was pituitary glands from human cadavers. This supply was never adequate, and it was finally removed from the market by the Food and Drug Administration (FDA) because of fear of viral contamination. Recently, however, the technique of genetic engineering has made available an unlimited supply of human GH. The gene that directs synthesis of GH in humans has been introduced into bacteria, converting the bacteria into "factories" that synthesize human GH.

Even though adequate human GH is now available through genetic engineering, a new problem for the medical community is to determine under which circumstances synthetic GH treatment is appropriate. Currently treatment has been approved by the FDA only for the following uses: (1) for GH-deficient children, (2) for adults with a pituitary tumor or other disease that causes severe GH deficiency, and (3) for AIDS patients suffering from severe muscle wasting. Although not FDA approved, GH therapy is also widely used to promote faster healing of skin in severely burned patients.

Another group whom replacement GH therapy may benefit is the elderly. GH secretion typically peaks during a person's twenties, then in many people may start to dwindle after age 40. This decline may contribute to some of the characteristic signs of aging, such as

- Decreased muscle mass (GH promotes synthesis of proteins, including muscle protein)
- Increased fat deposition (GH promotes leanness by mobilizing fat stores for use as an energy source)
- Reduced bone density (GH promotes the activity of bone-forming cells)
- Thinner, sagging skin (GH promotes proliferation of skin cells)

(However, inactivity is also believed to play a major role in age-associated reductions in muscle mass, bone density, and strength.)

Several studies in the early 1990s suggested that some of these consequences of aging may be counteracted through the use of synthetic GH in people over age 60. Elderly male subjects treated with supplemental GH showed increased muscle mass, reduced fatty tissue, and thickened skin. In similar studies in elderly females, supplemental GH therapy did not increase muscle mass significantly but it did reduce fat mass and did protect against bone loss. Even though these early results were exciting, scientists caution that synthetic GH should not be viewed as a potential cure for old age.

Further studies have been more discouraging, however. For example, despite increased lean body mass, many individuals surprisingly do not experience increased muscle strength or exercise capabilities. Also, when GH is supplemented for an extended time or in large doses, detrimental side effects include an increased likelihood of diabetes, kidney stones, high blood pressure, headaches, joint pain, and carpal tunnel syndrome. (*Carpal tunnel syndrome* involves thickening and narrowing of the tunnel in the wrist through which the nerve supply to the hand muscles passes; *carpal* means "wrist") Furthermore, synthetic GH is costly ($15,000 to $20,000 annually) and must be injected regularly. Also, some scientists worry that sustained administration of synthetic GH may raise the risk of developing cancer by promoting uncontrolled cell proliferation. For these reasons, many investigators no longer view synthetic GH as a potential "fountain of youth." Instead, they hope it can be used in a more limited way to strengthen muscle and bone sufficiently in the many elderly who have GH deficits, to help reduce the incidence of bone-breaking falls that often lead to disability and loss of independence. The National Institute of Aging is currently sponsoring a nationwide series of studies involving GH therapy in the elderly to help sort out potentially legitimate roles of this supplemental hormone.

An ethical dilemma is whether the drug should be used by others who have normal GH levels but want the product's growth-promoting actions for cosmetic or athletic reasons, such as normally growing teenagers who wish to attain even greater height. The drug is already being used illegally by some athletes and bodybuilders. Furthermore, a recent study found that only four out of ten children who are legitimately receiving GH therapy under medical supervision are actually GH deficient. The others are receiving the treatment because parents, physicians, and the children are being swayed by perceived cultural pressures that favor height rather than by medical factors.

Using the drug in children with normal GH levels may be problematic, because synthetic GH is a double-edged product. Although it promotes growth and muscle mass, it also has negative effects, such as potential troubling side effects. Furthermore, a recent British study revealed that supplemental GH therapy in children who do not lack the hormone redistributes the body's fat and protein. The investigators compared two groups of otherwise healthy 6- to 8-year-olds who were among the shortest for their age. One group consisted of children who were receiving GH, the other of children who were not. At the end of six months, the children taking the synthetic hormone had outpaced the untreated group in growth by more than 1.5 inches per year. However, the untreated children added both muscle and fat as they grew, whereas the treated children became unusually muscular and lost up to 76% of their body fat. The loss of fat became especially obvious in their faces and limbs, giving them a rawboned, gangly appearance. It is unclear what long-term effects—either deleterious or desirable—these dramatic changes in body composition might have. Scientists also express concern that these readily observable physical changes may be accompanied by more subtle abnormalities in organs and cells. Thus, the debate about whether to use GH in normal but short children is likely to continue until further investigation demonstrates if it is safe and appropriate.

---

of GH, but in its presence they synergistically enhance linear growth. Although androgens stimulate growth, they ultimately stop further growth by promoting closure of the epiphyseal plates.

- *Estrogens*, like androgens, ultimately terminate linear growth by stimulating complete conversion of the epiphyseal plates to bone. However, the effects of estrogen on growth prior to bone maturation are not well understood. Some studies suggest that large doses of estrogen may even inhibit further body growth.

Several factors contribute to the average height differences between men and women. First, because puberty occurs about two years earlier in girls than in boys, on the average boys have two more years of prepubertal growth than do girls. As a result, boys are usually several inches taller than girls at the start of their respective growth spurts. Second, boys experience a greater androgen-induced growth spurt than girls before their respective gonadal steroids seal their long bones from further growth; this results in greater heights in men than in women on the average. Third, recent evidence suggests that androgens "imprint" the brains of males during development, giving rise to a "masculine" secretory pattern of GH characterized by higher cyclic peaks, which are speculated to contribute to the greater height of males.

In addition to these hormones that exert overall effects on body growth, a number of poorly understood peptide *growth factors* have been identified that stimulate mitotic activity of specific tissues (for example, epidermal growth factor).

CHAPTER IN PERSPECTIVE: FOCUS ON HOMEOSTASIS

The endocrine system is one of the body's two major regulatory systems, the other being the nervous system. Through its relatively slowly acting hormone messengers, the endocrine system generally regulates activities that require duration rather than speed. Most of these activities are directed toward maintaining homeostasis, as exemplified by the following:

- Hormones help maintain the proper concentration of nutrients in the internal environment by directing chemical reactions involved in the cellular uptake, storage, and release of these molecules.

Furthermore, the rate at which these nutrients are metabolized is controlled in large part by the endocrine system.

- The control of H_2O balance, which is responsible for maintaining ECF osmolarity and proper cell volume, is largely accomplished by hormonal regulation of H_2O reabsorption by the kidneys during urine formation.

- Salt balance, which is important in maintaining ECF volume and arterial blood pressure, is achieved by hormonally controlled adjustments in salt reabsorption by the kidneys during urine formation.

- Likewise, hormones act on various target cells to maintain the plasma concentration of calcium and other electrolytes. These electrolytes in turn play key roles in homeostatic activities. For example, maintenance of calcium levels within narrow limits is critical for neuromuscular excitability and blood clotting, among other life-supporting actions.

- The endocrine system orchestrates a wide range of adjustments that help the body maintain homeostasis in response to stressful situations.

- The endocrine and nervous systems work in concert to control the circulatory and digestive systems, which in turn carry out important homeostatic activities.

Unrelated to homeostasis, hormones direct the growing process and control most aspects of the reproductive system.

CHAPTER SUMMARY

General Principles of Endocrinology

- Hormones are long-distance chemical messengers secreted by the ductless endocrine glands into the blood, which transports the hormones to specific target sites where they control a particular function by altering protein activity within target cells.

- Even though hormones can reach all tissues via the blood, they exert their effects only at their target cells, because these cells alone have unique receptors for binding the hormone.

- The endocrine system is especially important in regulating organic metabolism, H_2O and electrolyte balance, growth, and reproduction and in helping the body cope with stress.

- Some hormones are tropic, meaning their function is to stimulate and maintain other endocrine glands.

- Hormones are grouped into three categories—based on differences in their mode of synthesis, storage, secretion, transport in the blood, and interaction with target cells—peptides, steroids, and amines, the latter including thyroid hormone and adrenomedullary catecholamines.

- Peptides and catecholamines are hydrophilic; steroid and thyroid hormones are lipophilic.

- Hydrophilic hormones are synthesized and packaged for export by the endoplasmic reticulum/Golgi complex route, stored in secretory vesicles, and released by exocytosis on appropriate stimulation.

- Hydrophilic hormones dissolve freely in the plasma for transport to their target cells.

- At their target cells, hydrophilic hormones bind with surface membrane receptors. On binding, a hydrophilic hormone triggers a chain of intracellular events by means of a second-messenger system that ultimately alters pre-existing cellular proteins, usually enzymes, which exert the effect leading to the target cell's response to the hormone.

- Steroids are synthesized by modifications of stored cholesterol by means of enzymes specific for each steroidogenic tissue. Steroids are not stored in the endocrine cells. Being lipophilic, they diffuse out through the lipid membrane barrier as soon as they are synthesized. Control of steroids is directed at their synthesis.

- Lipophilic steroids and thyroid hormone are both transported in the blood largely bound to carrier plasma proteins, with only free, unbound hormone being biologically active.

- Lipophilic hormones readily enter through the lipid membrane barriers of their target cells and bind with nuclear receptors. Hormonal binding activates the synthesis of new enzymatic or structural intracellular proteins that carry out the hormone's effect on the target cell.

- The effective plasma concentration of each hormone is normally controlled by regulated changes in the rate of hormone secretion. Secretory output of endocrine cells is primarily influenced by two types of direct regulatory inputs: (1) neural input, which increases hormone secretion in response to a specific need and also governs diurnal variations in secretion; (2) input from another hormone, which involves either stimulatory input from a tropic hormone or inhibitory input from a target cell hormone in negative-feedback fashion.

- The effective plasma concentration of a hormone can also be influenced by its rate of removal from the blood by metabolic inactivation and excretion, and for some hormones by its rate of activation or its extent of binding to plasma proteins.

- Endocrine dysfunction arises when too much or too little of any particular hormone is secreted or from decreased target-cell responsiveness to a hormone.

Pineal Gland and Circadian Rhythms

- The suprachiasmatic nucleus (SCN) is the body's master biological clock. Self-induced cyclic variations in the concentration of clock proteins within the SCN bring about cyclic changes in neural discharge from this area. Each cycle takes about a day and drives the body's circadian (daily) rhythms.

- The inherent rhythm of this endogenous oscillator is a bit longer than 24 hours. Therefore, each day the body's circadian rhythms must be entrained or adjusted to keep pace with environmental cues so that the internal rhythms are synchronized with the external light–dark cycle.

- In the eyes special photoreceptors that respond to light but are not involved in vision send input to the SCN. Acting through the SCN, the pineal gland's secretion of the hormone melatonin rhythmically fluctuates with the light–dark cycle, decreasing in the light and increasing in the dark. Melatonin in turn is believed to synchronize the body's natural circadian rhythms, such as diurnal (day–night) variations in hormone secretion and body temperature, with external cues such as the light–dark cycle.

- Other proposed roles for melatonin include (1) promoting sleep; (2) influencing reproductive activity, including the onset of puberty; (3) acting as an antioxidant to remove damaging free radicals; and (4) enhancing immunity.

Hypothalamus and Pituitary

- The pituitary gland consists of two distinct lobes, the posterior pituitary and the anterior pituitary.

- The hypothalamus, a portion of the brain, secretes nine peptide hormones; two are stored in the posterior pituitary, and seven are carried through a special vascular link to the anterior pituitary, where they regulate the release of particular anterior pituitary hormones.

- The posterior pituitary is essentially a neural extension of the hypothalamus. Two small peptide hormones, vasopressin and oxytocin, are synthesized within the cell bodies of neurosecretory neurons located in the hypothalamus, from which they pass down the axon to be stored in nerve terminals within the posterior pituitary. These hormones are independently released from the posterior pituitary into the blood in response to action potentials originating in the hypothalamus. (1) Vasopressin conserves water conservation during urine formation. (2) Oxytocin stimulates uterine contraction during childbirth and milk ejection during breast feeding.

- The anterior pituitary secretes six different peptide hormones that it produces itself. Five anterior pituitary hormones are tropic.

(1) The sole function of thyroid-stimulating hormone (TSH) is to stimulate secretion of thyroid hormone. (2) Similarly, the only role of adrenocorticotropic hormone (ACTH) is to stimulate secretion of cortisol by the adrenal cortex. (3) and (4) The gonadotropic hormones—follicle-stimulating hormone (FSH) and luteinizing hormone (LH)—stimulate production of gametes (eggs and sperm) as well as secretion of sex hormones. (5) Growth hormone (GH) stimulates growth indirectly by stimulating secretion of somatomedins, which in turn promote growth of bone and soft tissues. GH exerts metabolic effects as well. (6) Prolactin stimulates milk secretion and is not tropic to another endocrine gland.

- The anterior pituitary releases its hormones into the blood at the bidding of releasing and inhibiting hormones from the hypothalamus. The hypothalamus in turn is influenced by a variety of neural and hormonal controlling inputs.

- Both the hypothalamus and the anterior pituitary are inhibited in negative-feedback fashion by the product of the target endocrine gland in the hypothalamus–anterior pituitary–target gland axis.

Endocrine Control of Growth

- Growth depends not only on growth hormone and other growth-influencing hormones such as thyroid hormone, insulin, and the sex hormones but also on genetic determination, an adequate diet, and freedom from chronic disease or stress.

- Growth hormone promotes growth indirectly by stimulating the liver's production of somatomedins. The major somatomedin, or insulin-like growth factor, is IGF-I, which acts directly on bone and soft tissues to bring about most growth-promoting actions. The growth hormone–somatomedin pathway causes growth by stimulating protein synthesis, cell division, and lengthening and thickening of bones.

- Growth hormone also directly exerts metabolic effects unrelated to growth, on the liver, adipose tissue, and muscle, such as conservation of carbohydrates and mobilization of fat stores.

- Growth hormone secretion by the anterior pituitary is regulated in negative-feedback fashion by two hypothalamic hormones, growth hormone–releasing hormone and growth hormone–inhibiting hormone. These hypothalamic regulatory hormones reach the anterior pituitary by means of a unique vascular link, the hypothalamic-hypophyseal portal system.

- Growth hormone levels are not highly correlated with periods of rapid growth. The primary signals for increased growth-hormone secretion are related to metabolic needs rather than growth, namely, deep sleep, stress, exercise, and low blood-glucose levels.

REVIEW EXERCISES

Objective Questions (answers on p. A-50)

1. One endocrine gland may secrete more than one hormone. *(True or false?)*
2. One hormone may influence more than one type of target cell. *(True or false?)*
3. All endocrine glands are exclusively endocrine in function. *(True or false?)*
4. A single target cell may be influenced by more than one hormone. *(True or false?)*
5. Each steroidogenic organ has all the enzymes necessary to produce any steroid hormone. *(True or false?)*
6. Hyposecretion or hypersecretion of a specific hormone can occur even though its endocrine gland is perfectly normal. *(True or false?)*

7. Growth hormone levels in the blood are no higher during the early childhood growing years than during adulthood. *(True or false?)*
8. A hormone that has as its primary function the regulation of another endocrine gland is classified functionally as a _____ hormone.
9. Self-induced reduction in the number of receptors for a specific hormone is known as _____ .
10. The _____ in the hypothalamus is the body's master biological clock.
11. Activity within the cartilaginous layer of bone known as the _____ is responsible for lengthening of long bones.

12. Indicate the relationships among the hormones in the hypothalamic/anterior pituitary/adrenal cortex system by using the following answer code to identify which hormone belongs in each blank:

(a) cortisol   (b) ACTH   (c) CRH

(1) _____ from the hypothalamus stimulates the secretion of (2) _____ from the anterior pituitary. (3) _____ in turn stimulates the secretion of (4) _____ from the adrenal cortex. In negative-feedback fashion, (5) _____ inhibits secretion of (6) _____ and furthermore reduces the sensitivity of the anterior pituitary to (7) _____ .

Essay Questions

1. List the overall functions of the endocrine system.
2. Compare the three categories of hormones in terms of chemical structure; mechanisms of synthesis, storage, and secretion; transport in the blood; and interaction with target cells.
3. By what means is the plasma concentration of a hormone normally regulated?
4. List and briefly state the functions of the posterior pituitary hormones.
5. List and briefly state the functions of the anterior pituitary hormones.
6. Compare the relationship between the hypothalamus and posterior pituitary with the relationship between the hypothalamus and anterior pituitary. Describe the role of the hypothalamic-hypophyseal portal system and the hypothalamic releasing and inhibiting hormones.
7. Describe the actions of growth hormone that are unrelated to growth. What are growth hormone's growth-promoting actions? What is the role of somatomedins?
8. Discuss the control of growth hormone secretion.

POINTS TO PONDER

(Explanations on p. A-50)

1. A new supervisor at a local hospital decides to rotate the nursing staff to a different shift every week so that one group of employees is not always "stuck" on an undesirable shift. From a physiologic viewpoint, do you think this proposal is advisable?
2. Thinking about the feedback control loop among TRH, TSH, and thyroid hormone, would you expect the concentration of TSH to be normal, above normal, or below normal in a person whose diet is deficient in iodine (an element essential for synthesizing thyroid hormone)?
3. A patient displays symptoms of excess cortisol secretion. What factors could be measured in a blood sample to determine whether the condition is caused by a defect at the hypothalamic/anterior pituitary level or the adrenal cortex level?
4. Why would males with testicular feminization syndrome be unusually tall?
5. A black market for growth hormone abuse already exists among weight lifters and other athletes. What actions of growth hormone would induce a full-grown athlete to take supplemental doses of this hormone? What are the potential detrimental side effects?
6. *Clinical Consideration.* At 18 years of age, 8-foot Anthony O. was diagnosed with gigantism caused by a pituitary tumor. The condition was treated by surgically removing his pituitary gland. What hormonal replacement therapy would Anthony need?

PHYSIOEDGE RESOURCES

PHYSIOEDGE CD-ROM

Figures marked with the PhysioEdge icon have associated activities on the CD. For this chapter, check out the Media Quizzes related to Figures 18-1, 18-7, 18-9, 18-10, and 18-13.

CHECK OUT THESE MEDIA QUIZZES:

18.1 Overview of the Classic Endocrine Glands
18.2 Anterior Pituitary Gland
18.3 Posterior Pituitary Gland
18.4 Endocrine Roles by Organs of Mixed or Uncertain Functions

PHYSIOEDGE WEB SITE

The Web site for this book contains a wealth of helpful study aids, as well as many ideas for further reading and research. Log on to:

http://www.brookscole.com

Go to the Biology page and select Sherwood's *Human Physiology*, 5th Edition. Select a chapter from the drop-down menu or click on one of these resource areas:

- **Case Histories** provide an introduction to the clinical aspects of human physiology.
- For 2-D and 3-D graphical illustrations and animations of physiological concepts, visit our **Visual Learning Resource.**
- Resources for study and review include the **Chapter Outline, Chapter Summary, Glossary,** and **Flash Cards.** Use our **Quizzes** to prepare for in-class examinations.
- On-line research resources to consult are **Hypercontents,** with current links to relevant Internet sites; **Internet Exercises** with starter URLs provided; and **InfoTrac Exercises.**

 For more readings, go to InfoTrac College Edition, your on-line research library, at:

http://infotrac.thomsonlearning.com

Endocrine System

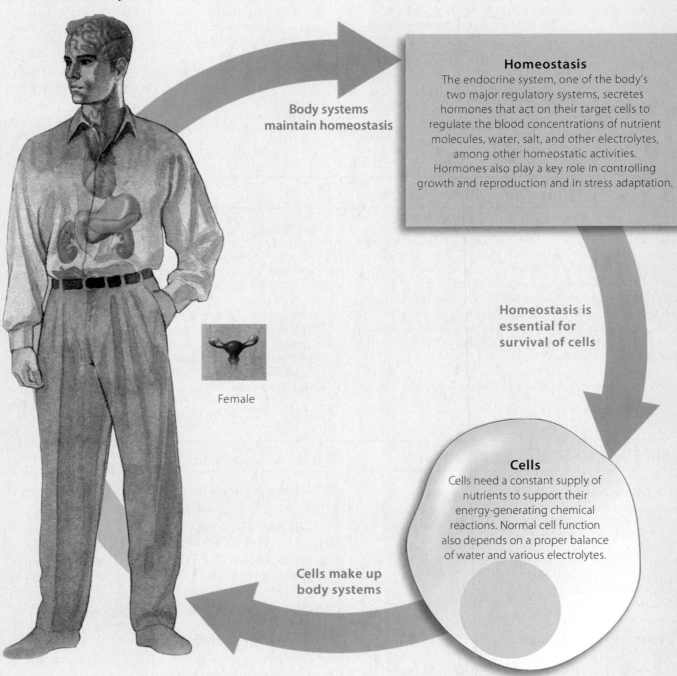

Female

Body systems maintain homeostasis

Homeostasis
The endocrine system, one of the body's two major regulatory systems, secretes hormones that act on their target cells to regulate the blood concentrations of nutrient molecules, water, salt, and other electrolytes, among other homeostatic activities. Hormones also play a key role in controlling growth and reproduction and in stress adaptation.

Homeostasis is essential for survival of cells

Cells
Cells need a constant supply of nutrients to support their energy-generating chemical reactions. Normal cell function also depends on a proper balance of water and various electrolytes.

Cells make up body systems

The endocrine system, by means of the blood-borne hormones it secretes, generally regulates activities that require duration rather than speed. The peripheral endocrine glands include the **thyroid gland,** which controls the body's basal metabolic rate; the **adrenal glands,** which secrete hormones important in metabolizing nutrient molecules, in adapting to stress, and in maintaining salt balance; the **endocrine pancreas,** which secretes hormones important in metabolizing nutrient molecules; and the **parathyroid glands,** which secrete a hormone important in Ca^{2+} metabolism.

Chapter 19

The Peripheral Endocrine Glands

CONTENTS AT A GLANCE

THYROID GLAND

▌ Anatomy of the thyroid gland

▌ Thyroid hormone

ADRENAL GLANDS

▌ Anatomy of the adrenal gland

▌ Adrenocortical mineralocorticoids

▌ Adrenocortical glucocorticoids

▌ Adrenocortical sex hormones

▌ Adrenal medullary catecholamines

▌ Hormonal and neural roles in the stress response

ENDOCRINE CONTROL OF FUEL METABOLISM

▌ Metabolism, anabolism, catabolism

▌ Energy storage

▌ Absorptive and postabsorptive states

▌ Endocrine pancreas: Insulin and glucagon

▌ Metabolic effects of other hormones

ENDOCRINE CONTROL OF CALCIUM METABOLISM

▌ Calcium homeostasis and calcium balance

▌ Bone remodeling

▌ Parathyroid hormone

▌ Calcitonin

▌ Vitamin D

▌ Relationship of phosphate metabolism to calcium metabolism

THYROID GLAND

The **thyroid gland** consists of two lobes of endocrine tissue joined in the middle by a narrow portion of the gland, giving it a bow tie shape (● Figure 19-1a). The gland is even located in the appropriate place for a bow tie, lying over the trachea just below the larynx.

▌ The major cells that secrete thyroid hormone are organized into colloid-filled follicles.

The major thyroid secretory cells, known as **follicular cells,** are arranged into hollow spheres, each of which forms a functional unit called a **follicle.** On a microscopic section (● Figure 19-1b), the follicles appear as rings of follicular cells enclosing an inner lumen filled with **colloid,** a substance that serves as an extracellular storage site for thyroid hormones. Note that the colloid within the follicular lumen is extracellular (that is, outside of the thyroid cells), even though it is located within the interior of the follicle. Colloid is not in direct contact with the extracellular fluid that surrounds the follicle, similar to an inland lake that is not in direct contact with the oceans that surround a continent.

The chief constituent of the colloid is a large protein molecule known as **thyroglobulin,** within which are incorporated the thyroid hormones in their various stages of synthesis. The follicular cells produce two iodine-containing hormones derived from the amino acid tyrosine: **tetraiodothyronine (T_4 or thyroxine)** and **tri-iodothyronine (T_3).** The prefixes *tetra* and *tri* and the subscripts 4 and 3 denote the number of iodine atoms incorporated into each of these hormones. These two hormones, collectively referred to as **thyroid hormone,** are important regulators of overall basal metabolic rate.

Interspersed in the interstitial spaces between the follicles is another secretory cell type, the **C cells,** so called because they secrete the peptide hormone **calcitonin.** Calcitonin plays a role in calcium metabolism and is not related in any way to the two other major thyroid hormones. We will here discuss T_4 and T_3 and talk about calcitonin later, in a section dealing with endocrine control of calcium balance.

by the follicular cells. Tyrosine, an amino acid, is synthesized in sufficient amounts by the body, so it is not a dietary essential. By contrast, the iodine needed for thyroid hormone synthesis must be obtained from dietary intake. The synthesis, storage, and secretion of thyroid hormone involve the following steps:

1. All steps of thyroid hormone synthesis take place on the thyroglobulin molecules within the colloid. Thyroglobulin itself is produced by the endoplasmic reticulum/Golgi complex of the thyroid follicular cells. Tyrosine becomes incorporated in the much larger thyroglobulin molecules as the latter are being produced. Once produced, tyrosine-containing thyroglobulin is exported from the follicular cells into the colloid by exocytosis (step ①ʼ in Figure 19-2).

2. The thyroid captures iodine from the blood and transfers it into the colloid by an iodine pump—the powerful, energy-requiring carrier proteins in the outer membranes of the follicular cells (step ②). Almost all the iodine in the body is moved against its concentration gradient to become trapped in the thyroid for thyroid hormone synthesis. Iodine serves no other function in the body.

3. Within the colloid, iodine is quickly attached to a tyrosine within the thyroglobulin molecule. Attachment of one iodine to tyrosine yields **monoiodotyrosine (MIT)** (step ③a). Attachment of two iodines to tyrosine yields **di-iodotyrosine (DIT)** (step ③b).

4. Next, a coupling process occurs between the iodinated tyrosine molecules to form the thyroid hormones. Coupling of two DITs (each bearing two iodine atoms) yields **tetraiodothyronine (T_4 or thyroxine),** the four-iodine form of thyroid hormone (step ④a). Coupling of one MIT (with one iodine) and one DIT (with two iodines) yields **tri-iodothyronine** or **T_3** (with three iodines) (step ④b). Coupling does not occur between two MIT molecules.

All these products remain attached to thyroglobulin. Thyroid hormones remain stored in this form in the colloid until they are split off and secreted. Sufficient thyroid hormone is normally stored to supply the body's needs for several months.

▌ To secrete thyroid hormone, the follicular cells phagocytize thyroglobulin-laden colloid.

The release of the thyroid hormones into the systemic circulation is a rather complex process, for two reasons. First, before their release T_4 and T_3 are still bound within the thyroglobulin molecule. Second, these hormones are stored at an inland extracellular site, the follicular lumen, so they must be transported completely across the follicular cells to reach the capillaries that course through the interstitial spaces between the follicles.

The process of thyroid hormone secretion essentially involves the follicular cells "biting off" a piece of colloid, breaking the thyroglobulin molecule down into its component parts, and "spitting out" the freed T_4 and T_3 into the blood. On appropriate stimulation for thyroid hormone secretion, the follicular cells internalize a portion of the thyroglobulin-hormone complex by phagocytizing a piece of colloid (step ⑤ of ● Figure 19-2). Within the cells, the membrane-enclosed droplets of colloid

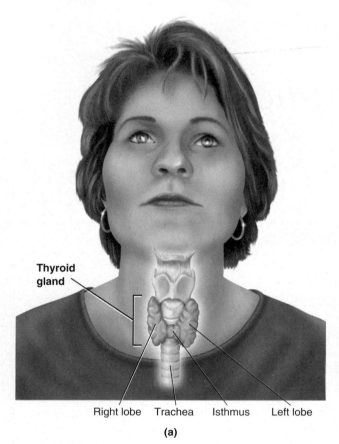

Thyroid gland

Right lobe   Trachea   Isthmus   Left lobe

(a)

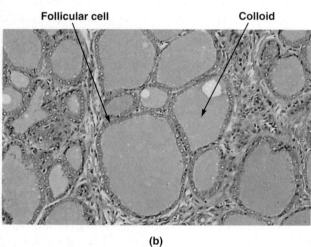

Follicular cell   Colloid

(b)

● FIGURE 19-1

Anatomy of the thyroid gland
(a) Gross anatomy of the thyroid gland, anterior view. The thyroid gland lies over the trachea just below the larynx and consists of two lobes connected by a thin strip called the *isthmus.* (b) Light-microscope appearance of the thyroid gland. The thyroid gland is composed primarily of colloid-filled spheres enclosed by a single layer of follicular cells.

▌ Thyroid hormone synthesis and storage occur on the thyroglobulin molecule.

The basic ingredients for thyroid hormone synthesis are tyrosine and iodine, both of which must be taken up from the blood

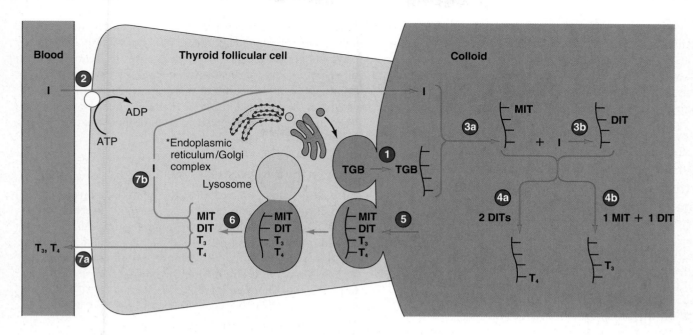

TGB = Thyroglobulin
I = Iodine
MIT = Monoiodotyrosine

DIT = Di-iodotyrosine
T_3 = Tri-iodothyronine
T_4 = Tetraiodothyronine (thyroxine)

* Organelles not drawn to scale. Endoplasmic reticulum/Golgi complex are proportionally too small.

1 Tyrosine-containing TGB produced within the thyroid follicular cells is transported into the colloid by exocytosis.

2 Iodine is actively transported from the blood into the colloid by the follicular cells.

3a Attachment of one iodine to tyrosine within the TGB molecule yields MIT.

3b Attachment of two iodines to tyrosine yields DIT.

4a Coupling of two DITs yields T_4.

4b Coupling of one MIT and one DIT yields T_3.

5 On appropriate stimulation, the thyroid follicular cells engulf a portion of TGB-containing colloid by phagocytosis.

6 Lysosomes attack the engulfed vesicle and split the iodinated products from TGB.

7a T_3 and T_4 diffuse into the blood.

7b MIT and DIT are deiodinated, and the freed iodine is recycled for synthesizing more hormone.

● FIGURE 19-2
Synthesis, storage, and secretion of thyroid hormone

coalesce with lysosomes, whose enzymes split off the biologically active thyroid hormones, T_4 and T_3, as well as the inactive iodotyrosines, MIT and DIT (step 6). The thyroid hormones, being very lipophilic, pass freely through the outer membranes of the follicular cells and into the blood (step 7a).

The MIT and DIT are of no endocrine value. The follicular cells contain an enzyme that swiftly removes the iodine from MIT and DIT, allowing the freed iodine to be recycled for synthesis of more hormone (step 7b). This highly specific enzyme will remove iodine only from the worthless MIT and DIT, not the valuable T_4 or T_3.

Once released into the blood, the highly lipophilic (and therefore water insoluble) thyroid hormone molecules very quickly bind with several plasma proteins. The majority of circulating T_4 and T_3 are transported by **thyroxine-binding globulin,** a plasma protein that selectively binds only thyroid hormone. Less than 0.1% of the T_4 and less than 1% of the T_3 remain in the unbound (free) form. This is remarkable, considering that only the free portion of the total thyroid hormone pool has access to the target cell receptors and thus can exert an effect.

❚ **Most of the secreted T_4 is converted into T_3 outside the thyroid.**

About 90% of the secretory product released from the thyroid gland is in the form of T_4, yet T_3 is about four times more potent in its biologic activity. However, most of the secreted T_4 is converted into T_3, or *activated*, by being stripped of one of its iodines outside of the thyroid gland, primarily in the liver and kidneys. About 80% of the circulating T_3 is derived from secreted T_4 that has been peripherally stripped. Therefore, T_3 is the major biologically active form of thyroid hormone at the cellular level, even though the thyroid gland secretes mostly T_4.

Thyroid hormone is the main determinant of the basal metabolic rate and exerts other effects as well.

Compared to other hormones, the action of thyroid hormone is "sluggish." The response to an increase in thyroid hormone is detectable only after a delay of several hours, and the maximal response is not evident for several days. The duration of the response is also quite long, partially because thyroid hormone is not rapidly degraded but also because the response to an increase in secretion continues to be expressed for days or even weeks after the plasma thyroid hormone concentrations have returned to normal.

Virtually every tissue in the body is affected either directly or indirectly by thyroid hormone. The effects of T_3 and T_4 can be grouped into several overlapping categories.

Effect on metabolic rate and heat production

Thyroid hormone increases the body's overall basal metabolic rate or "idling speed" (see p. 648). It is the most important regulator of the body's rate of O_2 consumption and energy expenditure under resting conditions.

Closely related to thyroid hormone's overall metabolic effect is its **calorigenic** ("heat-producing") effect. Increased metabolic activity results in increased heat production.

Effect on intermediary metabolism

In addition to increasing the general metabolic rate, thyroid hormone modulates the rates of many specific reactions involved in fuel metabolism. The effects of thyroid hormone on the metabolic fuels are multifaceted; not only can it influence both the synthesis and degradation of carbohydrate, fat, and protein, but small or large amounts of the hormone may induce opposite effects. For example, the conversion of glucose to glycogen, the storage form of glucose, is facilitated by small amounts of thyroid hormone, but the reverse—the breakdown of glycogen into glucose—occurs with large amounts of the hormone. Similarly, adequate amounts of thyroid hormone are essential for the protein synthesis needed for normal bodily growth, yet at high doses, as in thyroid hypersecretion, thyroid hormone favors protein degradation.

Sympathomimetic effect

Any action similar to one produced by the sympathetic nervous system is known as a **sympathomimetic** ("sympathetic mimicking") **effect**. Thyroid hormone increases target cell responsiveness to catecholamines (epinephrine and norepinephrine), the chemical messengers used by the sympathetic nervous system and its hormonal reinforcements from the adrenal medulla. Thyroid hormone accomplishes this permissive action by causing a proliferation of specific catecholamine target cell receptors (see p. 680). Because of this action, many of the effects observed when thyroid hormone secretion is elevated are similar to those that accompany activation of the sympathetic nervous system.

Effect on the cardiovascular system

Through its effect of increasing the heart's responsiveness to circulating catecholamines, thyroid hormone increases heart rate and force of contraction, thus increasing cardiac output. In addition, in response to the heat load generated by the calorigenic effect of thyroid hormone, peripheral vasodilation occurs to carry the extra heat to the body surface for elimination to the environment (see p. 659).

Effect on growth and the nervous system

Thyroid hormone is essential for normal growth because of its effects on growth hormone (GH). Thyroid hormone not only stimulates GH secretion but also promotes the effects of GH (or somatomedins) on the synthesis of new structural proteins and on skeletal growth. Thyroid-deficient children have stunted growth that can be reversed by thyroid replacement therapy. Unlike excess GH, however, excess thyroid hormone does not produce excessive growth.

Thyroid hormone plays a crucial role in the normal development of the nervous system, especially the CNS, an effect impeded in children who have thyroid deficiency from birth. Thyroid hormone is also essential for normal CNS activity in adults.

Thyroid hormone is regulated by the hypothalamus-pituitary-thyroid axis.

Thyroid-stimulating hormone (TSH), the thyroid tropic hormone from the anterior pituitary, is the most important physiologic regulator of thyroid hormone secretion (● Figure 19-3) (see p. 685). Almost every step of thyroid hormone synthesis and release is stimulated by TSH.

In addition to enhancing thyroid hormone secretion, TSH maintains the structural integrity of the thyroid gland. In the absence of TSH, the thyroid atrophies (decreases in size) and secretes its hormones at a very low rate. Conversely, it undergoes hypertrophy (increase in the size of each follicular cell) and hyperplasia (increase in the number of follicular cells) in response to excess TSH stimulation.

The hypothalamic **thyrotropin-releasing hormone (TRH)**, in tropic fashion, "turns on" TSH secretion by the anterior pituitary, whereas thyroid hormone, in negative-feedback fashion, "turns off" TSH secretion (see p. 689). In the hypothalamus-pituitary-thyroid axis, inhibition is exerted primarily at the level of the anterior pituitary. Like the other negative-feedback loops, the one between thyroid hormone and TSH tends to maintain a stable thyroid hormone output.

Negative feedback between the thyroid and anterior pituitary accomplishes day-to-day regulation of free thyroid hormone levels, whereas the hypothalamus mediates long-range adjustments. Unlike most other hormonal systems, the hormones in the thyroid axis in an adult normally do not undergo sudden, wide swings in secretion. The relatively steady rate of thyroid hormone secretion is in keeping with the sluggish, long-lasting responses that this hormone induces; there would be no adaptive value in suddenly increasing or decreasing plasma thyroid hormone levels.

The only known factor that increases TRH secretion (and, accordingly, TSH and thyroid hormone secretion) is exposure to cold in newborn infants, a highly adaptive mechanism. Scientists think the dramatic increase in heat-producing thyroid hormone secretion helps maintain body temperature during the abrupt drop in surrounding temperature at birth, as the infant passes from the mother's warm body to the cooler environmental air. A similar TSH response to cold exposure does not occur in adults, although it would make sense physiologically and does occur in many types of experimental animals.

Various types of stress inhibit TSH and thyroid hormone secretion, presumably through neural influences on the hypothalamus, although the adaptive importance of this inhibition is unclear.

▌Abnormalities of thyroid function include both hypothyroidism and hyperthyroidism.

Abnormalities of thyroid function are among the most common of all endocrine disorders. They fall into two major categories—**hypothyroidism** and **hyperthyroidism**—reflecting deficient and excess thyroid hormone secretion, respectively. A number of specific causes can give rise to each of these conditions (▲ Table 19-1). Whatever the cause, the consequences of too little or too much thyroid hormone secretion are largely predictable, given a knowledge of the functions of thyroid hormone.

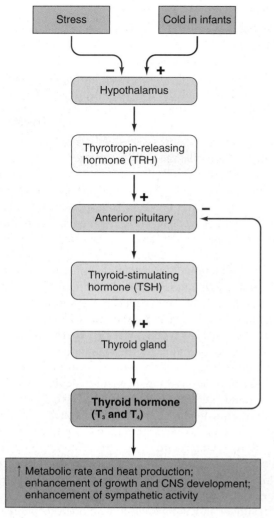

● **FIGURE 19-3**
Regulation of thyroid hormone secretion

▲ TABLE 19-1
Types of Thyroid Dysfunctions

Thyroid Dysfunction	Cause	Plasma Concentrations of Relevant Hormones	Goiter Present?
Hypothyroidism	Primary failure of the thyroid gland	$\downarrow T_3$ and T_4, $\uparrow$TSH	Yes
	Secondary to hypothalamic or anterior pituitary failure	$\downarrow T_3$ and T_4, $\downarrow$TRH and/or $\downarrow$TSH	No
	Lack of dietary iodine	$\downarrow T_3$ and T_4, $\uparrow$TSH	Yes
Hyperthyroidism	Abnormal presence of thyroid-stimulating immunoglobulin (TSI) (Graves' disease)	$\uparrow T_3$ and T_4, $\downarrow$TSH	Yes
	Secondary to excess hypothalamic or anterior pituitary secretion	$\uparrow T_3$ and T_4, $\uparrow$TRH and/or $\uparrow$TSH:	Yes
	Hypersecreting thyroid tumor	$\uparrow T_3$ and T_4, $\downarrow$TSH	No

Hypothyroidism

Hypothyroidism can result (1) from primary failure of the thyroid gland itself; (2) secondary to a deficiency of TRH, TSH, or both; or (3) from an inadequate dietary supply of iodine.

The symptoms of hypothyroidism are largely caused by a reduction in overall metabolic activity. Among other things, a patient with hypothyroidism has a reduced basal metabolic rate; displays poor tolerance of cold (lack of the calorigenic effect); has a tendency to gain excessive weight (not burning fuels at a normal rate); is easily fatigued (lower energy production); has a slow, weak pulse (caused by a reduction in the rate and strength of cardiac contraction and a lowered cardiac output); and exhibits slow reflexes and slow mental responsiveness (because of the effect on the nervous system). The mental effects are characterized by diminished alertness, slow speech, and poor memory.

Another notable characteristic is an edematous condition caused by infiltration of the skin with complex, water-retaining carbohydrate molecules, presumably as a result of altered metabolism. The resultant puffy appearance, primarily of the face, hands, and feet, is known as **myxedema.** In fact, the term *myxedema* is often used as a synonym for hypothyroidism in an adult, because of the prominence of this symptom.

If a person has hypothyroidism from birth, a condition known as **cretinism** develops. Because adequate levels of thyroid hormone are essential for normal growth and CNS development, cretinism is characterized by dwarfism and mental retardation as well as other general symptoms of thyroid deficiency. The mental retardation is preventable if replacement therapy is started promptly, but it is not reversible once it has developed for a few months after birth, even with later treatment with thyroid hormone.

Treatment of hypothyroidism, with one exception, consists of replacement therapy by administering of exogenous thyroid hormone. The exception is hypothyroidism caused by iodine deficiency, in which the remedy is adequate dietary iodine.

Hyperthyroidism

The most common cause of hyperthyroidism is **Graves' disease.** This is an autoimmune disease in which the body erroneously produces **thyroid-stimulating immunoglobulin (TSI),** an antibody whose target is the TSH receptors on the thyroid cells. TSI stimulates both secretion and growth of the thyroid in a manner similar to TSH. Unlike TSH, however, TSI is not subject to negative-feedback inhibition by thyroid hormone, so thyroid secretion and growth continue unchecked (● Figure 19-4). Less frequently, hyperthyroidism occurs secondary to excess TRH or TSH or in association with a hypersecreting thyroid tumor.

As expected, the hyperthyroid patient has an elevated basal metabolic rate. The resultant increase in heat production leads to excessive perspiration and poor tolerance of heat. Despite the increased appetite and food intake that occur in response to the increased metabolic demands, body weight typically falls because the body is burning fuel at an abnormally rapid rate. Net degradation of carbohydrate, fat, and protein stores occurs. The resultant loss of skeletal muscle protein results in weakness. Various cardiovascular abnormalities are as-

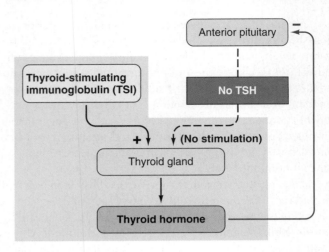

● **FIGURE 19-4**

Role of thyroid-stimulating immunoglobulin in Graves' disease
Thyroid-stimulating immunoglobulin, an antibody erroneously produced in the autoimmune condition of Graves' disease, binds with the TSH receptors on the thyroid gland and continuously stimulates thyroid hormone secretion outside the normal negative-feedback control system.

sociated with hyperthyroidism, caused both by the direct effects of thyroid hormone and by its interactions with catecholamines. Heart rate and strength of contraction may increase so much the individual has palpitations (an unpleasant awareness of the heart's activity). In severe cases, the heart may fail to meet the body's metabolic demands despite increased cardiac output. The effects on the CNS are characterized by an excessive degree of mental alertness to the point where the patient is irritable, tense, anxious, and excessively emotional.

A prominent feature of Graves' disease but not of the other types of hyperthyroidism is **exophthalmos** (bulging eyes) (● Figure 19-5). Complex, water-retaining carbohydrates are deposited

● **FIGURE 19-5**

Patient displaying exophthalmos
Abnormal fluid retention behind the eyeballs causes them to bulge forward.

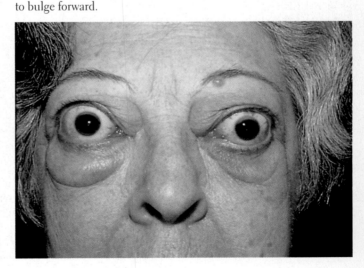

behind the eyes, although why this happens is still unclear. The resulting fluid retention pushes the eyeballs forward so they bulge from their bony orbit. The eyeballs may bulge so far that the lids cannot completely close, in which case the eyes become dry, irritated, and prone to corneal ulceration. Even after correction of the hyperthyroid condition, these troublesome eye symptoms may persist.

Three general methods of treatment can suppress excess thyroid hormone secretion: surgical removal of a portion of the oversecreting thyroid gland; administration of radioactive iodine, which, after being concentrated in the thyroid gland by the iodine pump, selectively destroys thyroid glandular tissue; and use of antithyroid drugs that specifically interfere with thyroid hormone synthesis.

▌ A goiter develops when the thyroid gland is overstimulated.

A **goiter** is an enlarged thyroid gland. Because the thyroid lies over the trachea, a goiter is readily palpable and usually highly visible (● Figure 19-6). A goiter occurs whenever either TSH or TSI excessively stimulates the thyroid gland. Note from ▲ Table 19-1 that a goiter *may* accompany hypothyroidism or hyperthyroidism, but it need not be present in either condition. Knowing the hypothalamus-pituitary-thyroid axis and feedback control, we can predict which types of thyroid dysfunction will produce a goiter. Let's consider hypothyroidism first.

- Hypothyroidism secondary to hypothalamic or anterior pituitary failure will not be accompanied by a goiter, because the thyroid gland is not being adequately stimulated, let alone excessively stimulated.

● **FIGURE 19-6**

Patient with a goiter

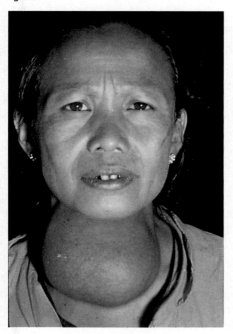

- With hypothyroidism caused by thyroid gland failure or lack of iodine, a goiter does develop, because the circulating level of thyroid hormone is so low that there is little negative-feedback inhibition on the anterior pituitary, and TSH secretion is therefore elevated. TSH acts on the thyroid to increase the size and number of follicular cells and to increase their rate of secretion. If the thyroid cells are incapable of secreting hormone because of a lack of a critical enzyme or lack of iodine, no amount of TSH will be able to induce these cells to secrete T_3 and T_4. However, TSH can still promote hypertrophy and hyperplasia of the thyroid, with a consequent paradoxical enlargement of the gland (that is, a goiter), even though the gland is still underproducing.

Similarly, a goiter may or may not accompany hyperthyroidism:

- Excessive TSH secretion resulting from a hypothalamic or anterior pituitary defect would obviously be accompanied by a goiter and excess T_3 and T_4 secretion because of overstimulation of thyroid growth. Because the thyroid gland in this circumstance is also capable of responding to excess TSH with increased hormone secretion, hyperthyroidism is present with this goiter.
- In Graves' disease, a hypersecreting goiter occurs because TSI promotes growth of the thyroid as well as enhancing secretion of thyroid hormone. Because the high levels of circulating T_3 and T_4 inhibit the anterior pituitary, TSH secretion itself is low. In all other cases when a goiter is present, TSH levels are elevated and are directly responsible for excessive growth of the thyroid.
- Hyperthyroidism resulting from overactivity of the thyroid in the absence of overstimulation, such as caused by an uncontrolled thyroid tumor, is not accompanied by a goiter. The spontaneous secretion of excessive amounts of T_3 and T_4 inhibits TSH, so there is no stimulatory input to promote growth of the thyroid.

ADRENAL GLANDS

There are two **adrenal glands,** one embedded above each kidney in a capsule of fat (*ad* means "next to;" *renal* means "kidney") (● Figure 19-7a).

▌ Each adrenal gland consists of a steroid-secreting cortex and a catecholamine-secreting medulla.

Each adrenal is composed of two endocrine organs, one surrounding the other. The outer layers composing the **adrenal cortex** secrete a variety of steroid hormones; the inner portion, the **adrenal medulla,** secretes catecholamines. Derived embryologically from different structures, the adrenal cortex and medulla secrete hormones belonging to different chemical categories, whose functions, mechanisms of action, and regulation are entirely different. We will first examine the adrenal cortex before turning our attention to the adrenal medulla.

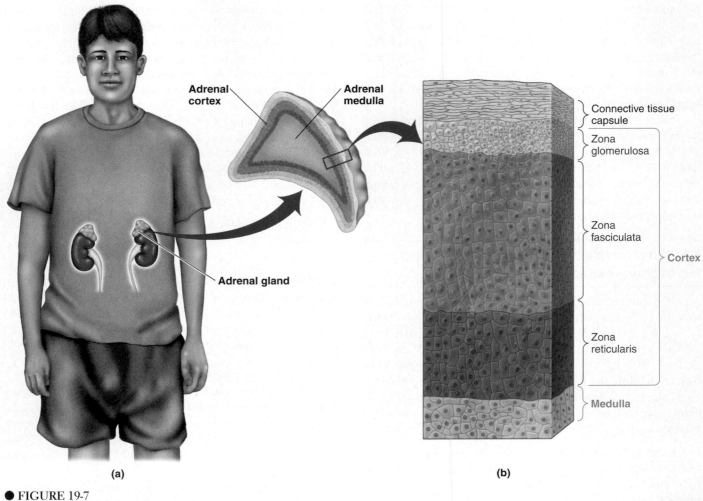

● **FIGURE 19-7**

Anatomy of the adrenal glands
(a) Location and structure of the adrenal glands. (b) Layers of the adrenal cortex.

▌ The adrenal cortex secretes mineralocorticoids, glucocorticoids, and sex hormones.

About 80% of the adrenal gland is composed of the cortex, which consists of three layers or zones: the **zona glomerulosa,** the outermost layer; the **zona fasciculata,** the middle and largest portion; and the **zona reticularis,** the innermost zone (● Figure 19-7b). The adrenal cortex produces a number of different **adrenocortical hormones,** all of which are steroids derived from the common precursor molecule, cholesterol. Slight variations in structure confer different functional capabilities on the various adrenocortical hormones. On the basis of their primary actions, the adrenal steroids can be divided into three categories:

1. **Mineralocorticoids,** mainly *aldosterone,* influence mineral (electrolyte) balance, specifically Na^+ and K^+ balance.
2. **Glucocorticoids,** primarily *cortisol,* play a major role in glucose metabolism as well as in protein and lipid metabolism.
3. **Sex hormones** are identical or similar to those produced by the gonads (testes in males, ovaries in females). The most abundant and physiologically important of the adrenocortical sex hormones is *dehydroepiandrosterone,* a "male" sex hormone.

The three categories of adrenal steroids are produced in anatomically distinct portions of the adrenal cortex as a result of differential distribution of the enzymes required to catalyze the different biosynthetic pathways leading to the formation of each of these steroids. Of the two major adrenocortical hormones, aldosterone is produced exclusively in the zona glomerulosa, whereas cortisol synthesis is limited to the two inner layers of the cortex, with the zona fasciculata being the major source of this glucocorticoid. No other steroidogenic tissues have the capability of producing either mineralocorticoids or glucocorticoids. In contrast, the adrenal sex hormones, also produced by the two inner cortical zones, are produced in far greater abundance in the gonads.

Being lipophilic, the adrenocortical hormones are all carried in the blood extensively bound to plasma proteins. Cortisol is bound mostly to a plasma protein specific for it called **corticosteroid-binding globulin (transcortin),** whereas aldosterone and dehydroepiandrosterone are largely bound to albumin, which nonspecifically binds a variety of lipophilic hormones.

Mineralocorticoids' major effects are on Na$^+$ and K$^+$ balance and blood pressure homeostasis.

The actions and regulation of the primary adrenocortical mineralocorticoid, **aldosterone,** are described thoroughly elsewhere (Chapters 14 and 15). The principal site of aldosterone action is on the distal and collecting tubules of the kidney, where it promotes Na$^+$ retention and enhances K$^+$ elimination during the formation of urine. The promotion of Na$^+$ retention by aldosterone secondarily induces osmotic retention of H$_2$O, expanding the ECF volume, which is important in the long-term regulation of blood pressure.

Mineralocorticoids are *essential for life*. Without aldosterone, a person rapidly dies from circulatory shock because of the marked fall in plasma volume caused by excessive losses of H$_2$O-holding Na$^+$. With most other hormonal deficiencies, death is not imminent, even though a chronic hormonal deficiency may eventually lead to a premature death.

Aldosterone secretion is increased by (1) activation of the renin-angiotensin-aldosterone system by factors related to a reduction in Na$^+$ and a fall in blood pressure and (2) direct stimulation of the adrenal cortex by a rise in plasma K$^+$ concentration (see ● Figure 14-25, p. 536). In addition to its effect on aldosterone secretion, angiotensin promotes growth of the zona glomerulosa, in a manner similar to the effect of TSH on the thyroid. Adrenocorticotropic hormone (ACTH) from the anterior pituitary primarily promotes the secretion of cortisol, not aldosterone. Thus, unlike cortisol regulation, the regulation of aldosterone secretion is largely independent of anterior pituitary control.

Glucocorticoids exert metabolic effects and play a key role in adaptation to stress.

Cortisol, the primary glucocorticoid, plays an important role in carbohydrate, protein, and fat metabolism; executes significant permissive actions for other hormonal activities; and helps people resist stress.

Metabolic effects

The overall effect of cortisol's metabolic actions is to increase the concentration of blood glucose at the expense of protein and fat stores. Specifically, cortisol performs the following functions:

- It stimulates hepatic **gluconeogenesis,** which is the conversion of noncarbohydrate sources (namely, amino acids) into carbohydrate within the liver (*gluco* means "glucose," *neo* means "new," *genesis* means "production"). Between meals or during periods of fasting, when no new nutrients are being absorbed into the blood for use and storage, the glycogen (stored glucose) in the liver tends to become depleted as it is broken down to release glucose into the blood. Gluconeogenesis is an important factor in replenishing hepatic glycogen stores and thus in maintaining normal blood-glucose levels between meals. This is essential because the brain can use only glucose

as its metabolic fuel, yet nervous tissue cannot store glycogen to any extent. The concentration of glucose in the blood must therefore be maintained at an appropriate level to adequately supply the glucose-dependent brain with nutrients.

- It inhibits glucose uptake and use by many tissues, but not the brain, thus sparing glucose for use by the brain, which absolutely requires it as a metabolic fuel. This action contributes to the increase in blood glucose concentration brought about by gluconeogenesis.

- It stimulates protein degradation in many tissues, especially muscle. By breaking down a portion of muscle proteins into their constituent amino acids, cortisol increases the blood amino-acid concentration. These mobilized amino acids are available for use in gluconeogenesis or wherever else they are needed, such as for repair of damaged tissue or synthesis of new cellular structures.

- It facilitates lipolysis, the breakdown of lipid (fat) stores in adipose tissue, thus releasing free fatty acids into the blood (*lysis* means "breakdown"). The mobilized fatty acids are available as an alternative metabolic fuel for tissues that can use this energy source in lieu of glucose, thereby conserving glucose for the brain.

Permissive actions

Cortisol is extremely important for its permissiveness (see p. 680). For example, cortisol must be present in adequate amounts to permit the catecholamines to induce vasoconstriction. A person lacking cortisol, if untreated, may go into circulatory shock in a stressful situation that demands immediate widespread vasoconstriction.

Role in adaptation to stress

Cortisol plays a key role in adaptation to stress. **Stress** is the generalized, nonspecific response of the body to any factor that overwhelms, or threatens to overwhelm, the body's compensatory abilities to maintain homeostasis. Contrary to popular usage, the agent inducing the response is correctly called a *stressor*, whereas *stress* refers to the state induced by the stressor. The following types of noxious stimuli illustrate the range of factors that can induce a stress response: *physical* (trauma, surgery, intense heat or cold); *chemical* (reduced O$_2$ supply, acid–base imbalance); *physiologic* (heavy exercise, hemorrhagic shock, pain); *psychological* or *emotional* (anxiety, fear, sorrow); and *social* (personal conflicts, change in life-style). Stress of any kind is one of the major stimuli for increased cortisol secretion.

Although cortisol's precise role in adapting to stress is not known, a speculative but plausible explanation might be as follows. A primitive human or an animal wounded or faced with a life-threatening situation must forgo eating. A cortisol-induced shift away from protein and fat stores in favor of expanded carbohydrate stores and increased availability of blood glucose would help protect the brain from malnutrition during the imposed fasting period. Also, the amino acids liberated by protein degradation would provide a readily available supply of building blocks for tissue repair should physical injury occur. Thus an increased pool of glucose, amino acids, and fatty acids is available for use as needed.

Anti-inflammatory and immunosuppressive effects

When cortisol or synthetic cortisol-like compounds are administered to yield higher-than-physiologic concentrations of glucocorticoids (that is, *pharmacologic levels*), not only are all the metabolic effects magnified, but several important new actions not evidenced at normal physiologic levels are seen. The most noteworthy of glucocorticoids' pharmacologic effects are anti-inflammatory and immunosuppressive (see p. 421). (Although these actions are traditionally considered to occur only at pharmacologic levels, recent studies suggest cortisol may exert anti-inflammatory effects even at normal physiologic levels.) Synthetic glucocorticoids have been developed that maximize the anti-inflammatory and immunosuppressive effects of these steroids while minimizing the metabolic effects.

Administering large amounts of glucocorticoid inhibits almost every step of the inflammatory response, making these steroids effective drugs in treating conditions in which the inflammatory response itself has become destructive, such as *rheumatoid arthritis*. Glucocorticoids used in this manner do not affect the underlying disease process; they merely suppress the body's response to the disease. Because glucocorticoids also exert multiple inhibitory effects on the overall immune process, such as "knocking out of commission" the white blood cells responsible for antibody production and destruction of foreign cells, these agents have also proved useful in managing various allergic disorders and in preventing organ transplant rejections.

When these steroids are employed therapeutically, they should be used only when warranted and then only sparingly, for several important reasons. First, because they suppress the normal inflammatory and immune responses that form the backbone of the body's defense system, a glucocorticoid-treated person has limited ability to resist infections. Second, in addition to the anti-inflammatory and immunosuppressive effects readily exhibited at pharmacologic levels, other less desirable effects may also be observed with prolonged exposure to higher-than-normal concentrations of glucocorticoids. These effects include development of gastric ulcers, high blood pressure, atherosclerosis, and menstrual irregularities. Third, high levels of exogenous glucocorticoids act in negative-feedback fashion to suppress the hypothalamus-pituitary axis that drives normal glucocorticoid secretion and maintains the integrity of the adrenal cortex. Prolonged suppression of this axis can lead to irreversible atrophy of the cortisol-secreting cells of the adrenal gland and thus to permanent inability of the body to produce its own cortisol.

▐ Cortisol secretion is regulated by the hypothalamus-pituitary-adrenal cortex axis.

Cortisol secretion by the adrenal cortex is regulated by a negative-feedback system involving the hypothalamus and anterior pituitary (● Figure 19-8). ACTH from the anterior pituitary stimulates the adrenal cortex to secrete cortisol. ACTH is derived from a large precursor molecule, **pro-opiomelanocortin**, produced within the endoplasmic reticulum of the anterior pi-

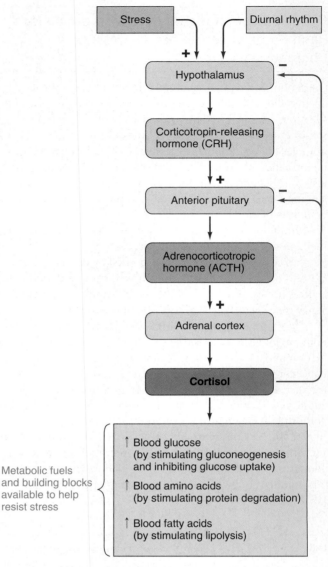

● **FIGURE 19-8**
Control of cortisol secretion

tuitary's ACTH-secreting cells (see p. 672). Prior to secretion, this large precursor is pruned into ACTH and several other biologically active peptides, namely, *melanocyte-stimulating hormone (MSH)* (see p. 683) and a morphinelike substance, *β-endorphin* (see p. 194). The possible significance of the fact that these multiple secretory products are developed from a single precursor molecule will be addressed later.

Being tropic to the zona fasciculata and zona reticularis, ACTH stimulates both the growth and the secretory output of these two inner layers of the cortex. In the absence of adequate amounts of ACTH, these layers shrink considerably, and cortisol secretion is drastically reduced. Recall that angiotensin, not ACTH, maintains the size of the zona glomerulosa. Like the actions of TSH on the thyroid gland, ACTH enhances many steps in the synthesis of cortisol.

The ACTH-producing cells in turn secrete only at the command of corticotropin-releasing hormone (CRH) from the hypothalamus. The feedback control loop is completed by cortisol's inhibitory actions on CRH and ACTH secretion by the hypothalamus and anterior pituitary, respectively.

The negative-feedback system for cortisol maintains the level of cortisol secretion relatively constant around the set point. Superimposed on the basic negative-feedback control system are two additional factors that influence plasma cortisol concentrations by changing the set point: These are *diurnal rhythm* and *stress*, both of which act on the hypothalamus to vary the secretion rate of CRH.

Influence of diurnal rhythm on cortisol secretion

Recall that the plasma cortisol concentration displays a characteristic diurnal rhythm, with the highest level occurring in the morning and the lowest level at night (see ● Figure 18-8, p. 678). This diurnal rhythm, which is intrinsic to the hypothalamus-pituitary control system, is related primarily to the sleep–wake cycle. The peak and low levels are reversed in a person who works at night and sleeps during the day. Such time-dependent variations in secretion are of more than academic interest for several reasons. First, it is important clinically to know at what time of day a blood sample was taken when interpreting the significance of a particular value. Second, the linking of cortisol secretion to day–night activity patterns raises serious questions about the common practice of swing shifts at work (that is, constantly switching day and night shifts among employees). Third, because cortisol helps a person resist stress, increasing attention is being given to the time of day various surgical procedures are performed.

Influence of stress on cortisol secretion

The other major factor that is independent of, and in fact can override the stabilizing negative-feedback control is stress. Dramatic increases in cortisol secretion, mediated by the central nervous system through enhanced activity of the CRH-ACTH system, occur in response to all kinds of stressful situations. The magnitude of the increase in plasma cortisol concentration is generally proportional to the intensity of the stressful stimulation; a greater increase in cortisol levels is evoked in response to severe stress than to mild stress.

▌ The adrenal cortex secretes both male and female sex hormones in both sexes.

In both sexes, the adrenal cortex produces both *androgens*, or "male" sex hormones, and *estrogens*, or "female" sex hormones. The main site of production for the sex hormones is the gonads: the testes for androgens and the ovaries for estrogens. Accordingly, males have a preponderance of circulating androgens, whereas in females estrogens are predominant. However, no hormones are unique to either males or females (except those from the placenta during pregnancy), because small amounts of the sex hormone of the opposite sex are produced by the adrenal cortex in both sexes.

Under normal circumstances, the adrenal androgens and estrogens are not sufficiently abundant or powerful to induce masculinizing or feminizing effects, respectively. The only adrenal sex hormone that has any biologic importance is the androgen **dehydroepiandrosterone (DHEA)**. The testes' primary androgen product is the potent testosterone, but the most abundant adrenal androgen is the much weaker DHEA. Adrenal DHEA is overpowered by testicular testosterone in males but is of physiologic significance in females, who otherwise lack androgens. This adrenal androgen governs androgen-dependent processes in the female such as growth of pubic and axillary (armpit) hair, enhancement of the pubertal growth spurt, and development and maintenance of the female sex drive.

Because the enzymes required for the production of estrogens are found in very low concentrations in the adrenocortical cells, estrogens are normally produced in very small quantities from this source.

In addition to controlling cortisol secretion, ACTH (not the pituitary gonadotropic hormones) controls adrenal androgen secretion. In general, cortisol and DHEA output by the adrenal cortex parallel each other. However, adrenal androgens feed back outside the hypothalamus-pituitary-adrenal cortex loop. Instead of inhibiting CRH, DHEA inhibits gonadotropin-releasing hormone, just as testicular androgens do. Furthermore, sometimes adrenal androgen and cortisol output diverge from each other—for example, at the time of puberty adrenal androgen secretion undergoes a marked surge, but cortisol secretion does not change. This enhanced secretion initiates the development of androgen-dependent processes in females. In males the same thing is accomplished primarily by testicular androgen secretion, which is also aroused at puberty. The nature of the pubertal inputs to the adrenals and gonads is still unresolved.

Supplemental DHEA

The surge in DHEA secretion that begins at puberty peaks between the ages of 25 and 30. After 30, DHEA secretion slowly tapers off until, by the age of 60, the plasma DHEA concentration is less than 15% of its peak level. Some scientists suspect that the age-related decline of DHEA and other hormones such as GH (see p. 696) and melatonin (see p. 683) plays a role in some problems of aging. Early studies with DHEA replacement therapy demonstrated some physical improvement, such as an increase in lean muscle mass and a decrease in fat, but the most pronounced effect was a marked increase in psychological well-being and an improved ability to cope with stress. Advocates for DHEA replacement therapy do not suggest that maintaining youthful levels of this hormone is a fountain of youth (that is, it is not going to extend the life span), but they do propose that it may help people feel and act younger as they age. Other scientists caution that evidence supporting DHEA as an antiaging therapy is still sparse. Also, they are concerned about DHEA supplementation until it has been thoroughly studied for possible deleterious side effects. For example, some research suggests a potential increase in the risk of heart disease among women taking DHEA because of an observed reduction in HDL, the "good" cholesterol (see p.336). Also, high doses of DHEA have been linked with increased facial hair in women. Furthermore, some experts fear that DHEA supplementation may raise the odds of acquiring ovarian or breast cancer in women and prostate cancer in men. Ironically, although the Food and Drug Administration (FDA) banned sales of DHEA as an over-the-counter drug in 1985 because of concerns about very real risks coupled with little proof of benefits, the product is widely available today as an unregulated food supplement. Passage of the Dietary Supplement Health and Education Act of 1994 allows the marketing of DHEA as a dietary supplement without approval by the FDA as long as the product label makes no specific medical claims.

▌ The adrenal cortex may secrete too much or too little of any of its hormones.

Although uncommon, there are a number of different disorders of adrenocortical function. Excessive secretion may occur with any of the three categories of adrenocortical hormones. Accordingly, three main patterns of symptoms resulting from hyperadrenalism can be distinguished, depending on which hormone type is in excess: aldosterone hypersecretion, cortisol hypersecretion, and adrenal androgen hypersecretion.

Aldosterone hypersecretion

Excess mineralocorticoid secretion may be caused by (1) a hypersecreting adrenal tumor made up of aldosterone-secreting cells (**primary hyperaldosteronism** or **Conn's syndrome**) or (2) inappropriately high activity of the renin-angiotensin system (**secondary hyperaldosteronism**). The latter may be produced by any number of conditions that cause a chronic reduction in arterial blood flow to the kidneys, thereby excessively activating the renin-angiotensin-aldosterone system. An example is atherosclerotic narrowing of the renal arteries.

The symptoms of both primary and secondary hyperaldosteronism are related to the exaggerated effects of aldosterone—namely, excessive Na^+ retention (**hypernatremia**) and K^+ depletion (**hypokalemia**). Also, high blood pressure (hypertension) is generally present, at least partially because of excessive Na^+ and fluid retention.

Cortisol hypersecretion

Excessive cortisol secretion (**Cushing's syndrome**) can be caused by (1) overstimulation of the adrenal cortex by excessive amounts of CRH and/or ACTH, (2) adrenal tumors that uncontrollably secrete cortisol independent of ACTH, or (3) ACTH-secreting tumors located in places other than the pituitary, most commonly in the lung. Whatever the cause, the prominent characteristics of this syndrome are related to the exaggerated effects of glucocorticoid, with the main symptoms being reflections of excessive gluconeogenesis. When too many amino acids are converted into glucose, the body suffers from combined glucose excess (high blood glucose) and protein shortage. Because the resultant hyperglycemia and glucosuria (glucose in the urine) mimic diabetes mellitus, the condition is sometimes referred to as *adrenal diabetes*. For reasons that are unclear, some of the extra glucose is deposited as body fat in locations characteristic for this disease, typically in the abdomen and face and above the shoulder blades. The abnormal fat distributions in the latter two locations are descriptively called a "moon face" and a "buffalo hump," respectively (● Figure 19-9). The appendages, in contrast, remain thin.

Besides the effects attributable to excessive glucose production, other effects arise from the widespread mobilization of amino acids from body proteins for use as glucose precursors. Loss of muscle protein leads to muscle weakness and fatigue. The protein-poor, thin skin of the abdomen becomes overstretched by the excessive underlying fat deposits, forming irregular, reddish-purple linear streaks. The weakening of blood vessel walls from depletion of structural protein results in an excessive tendency to bruise and to form small patches of subcutaneous bleeding. Wounds heal poorly, because formation of collagen, a major structural protein found in scar tissue, is depressed. Furthermore, losing the collagen framework of bone weakens the skeleton, so fractures may result from little or no apparent injury.

Adrenal androgen hypersecretion

Excess adrenal androgen secretion, a masculinizing condition, is more common than the extremely rare feminizing condition of excess adrenal estrogen secretion. Either condition is referred to as **adrenogenital syndrome**, emphasizing the pronounced effects that excessive adrenal sex hormones have on the genitalia and associated sexual characteristics.

The symptoms that result from excess androgen secretion depend on the sex of the individual and the age when the hyperactivity first begins.

● FIGURE 19-9

Patient with Cushing's syndrome
(a) Young boy prior to onset of the condition. (b) Only four months later, the same boy displaying a "moon-face" characteristic of Cushing's syndrome.

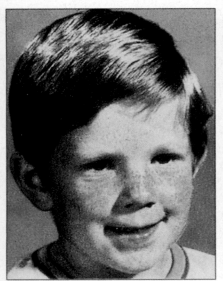
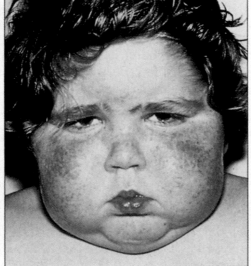

(a)   (b)

- *In adult females.* Because androgens exert masculinizing effects, a woman with this disease tends to develop a male pattern of body hair, a condition referred to as **hirsutism.** She usually also acquires other male secondary sexual characteristics, such as deepening of the voice and more muscular arms and legs. The breasts become smaller, and menstruation may cease as a result of androgen suppression of the woman's hypothalamus-pituitary-ovarian pathway for her own female sex-hormone secretion.

- *In newborn females.* Female infants born with adrenogenital syndrome manifest male-type external genitalia because excessive androgen secretion occurs early enough during fetal life to induce development of their genitalia along male lines, similar to the development of males under the influence of testicular androgen. The clitoris, which is the female homolog of the male penis, enlarges under androgen influence and takes on a penile appearance so in some cases it is difficult at first to determine the child's sex (● Figure 19-10). Thus this hormonal abnormality is one of the major causes of **female pseudohermaphroditism,** a condition in which female gonads (ovaries) are present but the external genitalia resemble those of a male. (A true hermaphrodite has the gonads of both sexes.)

- *In prepubertal males.* Excessive adrenal androgen secretion in prepubertal boys causes them to prematurely develop male secondary sexual characteristics—for example, deep voice, beard, enlarged penis, and sex drive. This condition is referred to as **precocious pseudopuberty** to differentiate it from true puberty, which occurs as a result of increased testicular activity. In precocious pseudopuberty, the androgen secretion from the adrenal cortex is not accompanied by sperm production or any other gonadal activity, because the testes are still in their nonfunctional prepubertal state.

- *In adult males.* Overactivity of adrenal androgens in adult males has no apparent effect, because male sex characteristics already exist.

The adrenogenital syndrome is most commonly caused by an inherited enzymatic defect in the cortisol steroidogenic pathway. The pathway for synthesis of androgens branches from the normal biosynthetic pathway for cortisol (see ● Figure 18-3, p. 674). When an enzyme specifically essential for synthesis of cortisol is deficient, the result is decreased secretion of cortisol. The decline in cortisol secretion removes the negative-feedback effect on the hypothalamus and anterior pituitary so that levels of CRH and ACTH increase considerably (● Figure 19-11). The defective adrenal cortex is incapable of responding to this increased ACTH secretion with cortisol output and instead shunts more of its cholesterol precursor into the androgen pathway. The result is excess DHEA production. This excess androgen does not inhibit ACTH but rather inhibits the gonadotropins. Because gamete production is not stimulated in the absence of gonadotropins, people with adrenogenital syndrome are sterile. Of course, they also exhibit symptoms of cortisol deficiency.

The symptoms of adrenal virilization, sterility, and cortisol deficiency are all reversed by glucocorticoid therapy. Administration of exogenous glucocorticoid replaces the cortisol deficit and, more dramatically, also inhibits the hypothalamus and pituitary so that ACTH secretion is suppressed. Once ACTH secretion is reduced, the profound stimulation of the adrenal cortex ceases, and androgen secretion declines markedly. Removing the large quantities of adrenal androgens from circulation allows masculinizing characteristics to gradually recede and normal gonadotropin secretion to resume. Without understanding how these hormonal systems are related, it would be very difficult to comprehend how glucocorticoid administration could dramatically reverse symptoms of masculinization and sterility.

Adrenocortical insufficiency

If one adrenal gland is nonfunctional or removed, the other healthy organ can take over the function of both through hypertrophy and hyperplasia. Therefore, both glands must be affected before adrenocortical insufficiency occurs.

In **primary adrenocortical insufficiency,** also known as **Addison's disease,** all layers of the adrenal cortex are undersecreting. This condition is most commonly caused by autoimmune destruction of the cortex by erroneous production of adrenal cortex–attacking antibodies. **Secondary adrenocortical insufficiency** may occur because of a pituitary or hypothalamic abnormality, resulting in insufficient ACTH secretion. In Addison's disease, both aldosterone and cortisol are deficient, whereas in the secondary form of the condition, only

● **FIGURE 19-10**

Adrenogenital syndrome in a female infant
These male-like external genitalia belong to a female infant, masculinized by excessive adrenal androgen secretion during fetal life. She is a genetic female with ovaries, but her labia majora are fused to resemble a scrotum and her clitoris is enlarged to resemble a penis (see p. 755). A female baby with this adrenogenital syndrome may be mistaken for and raised as a boy. However, at puberty, under the influence of estrogen from her ovaries the child will start to develop breasts and other female secondary sexual characteristics.

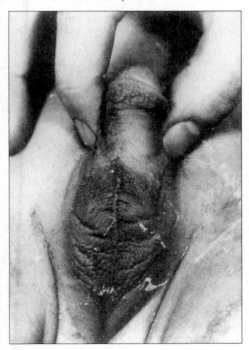

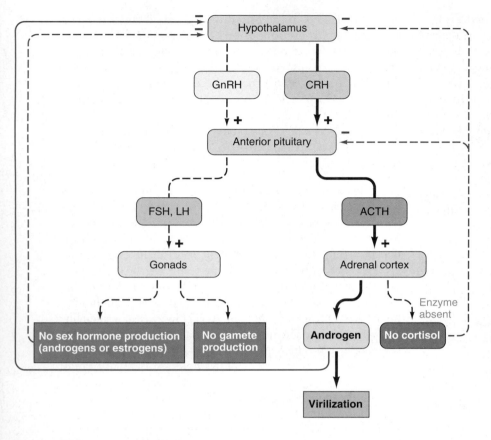

= Normal pathway that does not occur
ACTH = Adrenocorticotropic hormone
GnRH = Gonadotropin-releasing hormone

FSH = Follicle-stimulating hormone
LH = Luteinizing hormone
CRH = Corticotropin-releasing hormone

● **FIGURE 19-11**

Hormonal interrelationships in adrenogenital syndrome
The adrenocortical cells that are supposed to produce cortisol produce androgens instead because of a deficiency of a specific enzyme essential for cortisol synthesis. Because no cortisol is secreted to act in negative-feedback fashion, CRH and ACTH levels are elevated. The adrenal cortex responds to increased ACTH by further increasing androgen secretion. The excess androgen produces virilization and inhibits the gonadotropin pathway, with the result that the gonads stop producing sex hormones and gametes.

cortisol is deficient, because aldosterone secretion does not depend on ACTH stimulation.

The symptoms associated with aldosterone deficiency in Addison's disease are the most threatening. If severe enough, the condition is fatal, because aldosterone is essential for life. However, the loss of adrenal function may develop slowly and insidiously so that aldosterone secretion may be subnormal but not totally lacking. Patients with aldosterone deficiency display K^+ retention (**hyperkalemia**) caused by reduced K^+ loss in the urine, and Na^+ depletion (**hyponatremia**) caused by excessive urinary loss of Na^+. The former disturbs cardiac rhythm. The latter reduces ECF volume, including circulating blood volume, which in turn lowers blood pressure (hypotension).

Symptoms of cortisol deficiency are as would be expected: poor response to stress, hypoglycemia (low blood glucose) caused by reduced gluconeogenic activity, and lack of permissive action for many metabolic activities. The primary form of the disease also produces hyperpigmentation (darkening of the skin) resulting from excessive secretion of ACTH. Because

the pituitary is normal, the decline in cortisol secretion brings about an uninhibited elevation in ACTH output. Recall that both ACTH and melanocyte-stimulating hormone (MSH) are produced from the same large precursor molecule, pro-opiomelanocortin. As a result, levels of MSH activity and of melanin also rise when the blood levels of ACTH are very high.

We are now going to shift our attention from the adrenal cortex to the adrenal medulla.

▌ The adrenal medulla is a modified sympathetic postganglionic neuron.

The adrenal medulla is actually a modified part of the sympathetic nervous system. A sympathetic pathway consists of two neurons in sequence—a *preganglionic neuron* originating in the CNS, whose axonal fiber terminates on a second peripherally located *postganglionic neuron*, which in turn terminates on the effector organ (see p. 238). The neurotransmitter released by sympathetic postganglionic fibers is norepinephrine, which interacts locally with the innervated organ by binding with specific target receptors known as *adrenergic receptors*.

The adrenal medulla consists of modified postganglionic sympathetic neurons. Unlike ordinary postganglionic sympathetic neurons, those in the adrenal medulla do not have axonal fibers that terminate on effector organs. Instead, on stimulation by the preganglionic fiber the ganglionic cell bodies within the adrenal medulla release their chemical transmitter directly into the circulation (see ● Figure 7-4, p. 243). In this case, the transmitter qualifies as a hormone instead of a neurotransmitter. Like sympathetic fibers, the adrenal medulla does release norepinephrine, but its most abundant secretory output is a similar chemical messenger known as **epinephrine.** Both epinephrine and norepinephrine belong to the chemical class of *catecholamines*, which are derived from the amino acid tyrosine (see p. 672). Epinephrine is the same as norepinephrine except that it also has a methyl group.

Storage of catecholamines in chromaffin granules

Catecholamine is synthesized almost entirely within the cytosol of the adrenomedullary secretory cells. Once produced, epinephrine and norepinephrine are stored in **chromaffin granules,** which are similar to the transmitter storage vesicles found in sympathetic nerve endings. Segregation of catecholamines in chromaffin granules protects them from being destroyed by cytosolic enzymes during storage.

Secretion of catecholamines from the adrenal medulla

Catecholamines are secreted into the blood by exocytosis of chromaffin granules. Their release is analogous to the release mechanism for secretory vesicles that contain stored peptide hormones or the release of norepinephrine at sympathetic postganglionic terminals.

Of the total adrenomedullary catecholamine output, epinephrine accounts for 80% and norepinephrine for 20%. Whereas epinephrine is produced exclusively by the adrenal medulla, the bulk of norepinephrine is produced by sympathetic postganglionic fibers. Adrenomedullary norepinephrine is generally secreted in quantities too small to exert significant effects on target cells. Therefore, for practical purposes we can assume that norepinephrine effects are predominantly mediated directly by the sympathetic nervous system and that epinephrine effects are brought about exclusively by the adrenal medulla.

■ Epinephrine and norepinephrine vary in their affinities for the different adrenergic receptor types.

Epinephrine and norepinephrine have differing affinities for four distinctive receptor types: α_1, α_2, β_1, and β_2 adrenergic receptors (see p. 244). Most sympathetic target cells have α_1 receptors, some have only α_2, some only β_2, some have both α_1 and β_2, and β_1 receptors are found almost exclusively in the heart. In general, the responses elicited by activation of α_1 and β_1 receptors are excitatory, whereas the responses to stimulation of α_2 and β_2 receptors are typically inhibitory.

Neither catecholamine is totally specific for particular receptor types, but each has markedly different affinities. Norepinephrine binds predominantly with α and β_1 receptors located near postganglionic sympathetic-fiber terminals. Hormonal epinephrine, which can reach all α and β_1 receptors via its circulatory distribution, interacts with these receptors with approximately the same potency as neurotransmitter norepinephrine. Thus epinephrine and norepinephrine exert similar effects in many tissues, with epinephrine generally reinforcing sympathetic nervous activity. In addition, epinephrine activates β_2 receptors, over which the sympathetic nervous system exerts little influence. Many of the essentially epinephrine-exclusive β_2 receptors are located at tissues not even supplied by the sympathetic nervous system but reached by epinephrine through the blood. An example is skeletal muscle, where epinephrine exerts metabolic effects such as promoting the breakdown of stored glycogen.

Sometimes epinephrine, through its exclusive β_2-receptor activation, brings about a different action from that elicited by norepinephrine and epinephrine action through their mutual activation of other adrenergic receptors. As an example, norepinephrine and epinephrine bring about a generalized vasoconstrictor effect mediated by α_1-receptor stimulation. By contrast, epinephrine promotes vasodilation of the blood vessels that supply skeletal muscles and the heart through β_2-receptor activation (see p. 358).

Realize, however, that epinephrine functions only at the bidding of the sympathetic nervous system, which is solely responsible for stimulating its secretion from the adrenal medulla. Epinephrine secretion always accompanies a generalized sympathetic nervous system discharge, so sympathetic activity indirectly controls actions of epinephrine. By having the more versatile circulating epinephrine at its call, the sympathetic nervous system has a means of reinforcing its own neurotransmitter effects plus a way of executing additional actions on tissues that it does not directly innervate.

■ Epinephrine reinforces the sympathetic nervous system and exerts additional metabolic effects.

Adrenomedullary hormones are not essential for life, but virtually all organs in the body are affected by these catecholamines. They play important roles in mounting stress responses, regulating arterial blood pressure, and controlling fuel metabolism. The following sections discuss epinephrine's major effects, which it achieves either in collaboration with the sympathetic transmitter norepinephrine or alone to complement direct sympathetic response.

Effects on organ systems

Together, the sympathetic nervous system and adrenomedullary epinephrine mobilize the body's resources to support peak physical exertion in emergency or stressful situations. The sympathetic and epinephrine actions constitute a fight-or-flight response that prepares the person to combat an enemy or flee from danger (see p. 240). Specifically, the sympathetic system and epinephrine increase the rate and strength of cardiac contraction, increasing cardiac output, and their generalized vasoconstrictor effects increase total peripheral resistance. Together, these effects raise arterial blood pressure, thus ensuring an appropriate driving pressure to force blood to the organs that are most vital for meeting the emergency. Meanwhile, vasodilation of coronary and skeletal muscle blood vessels induced by epinephrine and local metabolic factors shifts blood to the heart and skeletal muscles from other vasoconstricted regions of the body.

Because of their profound influence on the heart and blood vessels, the sympathetic system and epinephrine also play an important role in the ongoing maintenance of arterial blood pressure.

Epinephrine (but not norepinephrine) dilates the respiratory airways to reduce the resistance encountered in moving air in and out of the lungs. Epinephrine also reduces digestive activity and inhibits bladder emptying, both activities that can be "put on hold" during a fight-or-flight situation.

Metabolic effects

Epinephrine exerts some important metabolic effects. In general, epinephrine prompts the mobilization of stored carbohydrate and fat to provide immediately available energy for use as needed to fuel muscular work. Specifically, epinephrine increases the blood glucose level by several different mechanisms. First, it stimulates both hepatic (liver) gluconeogenesis and **glycogenolysis,** the latter being the breakdown of stored glycogen into glucose, which is released into the blood. Epinephrine also stimulates glycogenolysis in skeletal muscles. Because of the difference in enzyme content between liver and muscle, however, muscle glycogen cannot be converted directly to glucose. Instead, the breakdown of muscle glycogen releases lactic acid into the blood. The liver removes lactic acid from the blood and converts it into glucose, so epinephrine's actions on skeletal muscle indirectly help raise blood glucose levels. Epinephrine and the sympathetic system may further add to this hyperglycemic effect by inhibiting the secretion of insulin, the pancreatic hormone primarily responsible for removing glucose from the blood, and by stimulating glucagon, another pancreatic hormone that promotes hepatic glycogenolysis and gluconeogenesis. In addition to increasing blood glucose levels, epinephrine also increases the blood fatty-acids level by promoting lipolysis.

Epinephrine's metabolic effects are appropriate for fight-or-flight situations. The elevated levels of glucose and fatty acids provide additional fuel to power the muscular movement required by the situation and also assure adequate nourishment for the brain during the crisis when no new nutrients are being consumed. Muscles can use fatty acids for energy production, but the brain cannot.

Because of its other widespread actions, epinephrine also increases the overall metabolic rate. Under the influence of epinephrine, many tissues metabolize faster. For example, the work of the heart and respiratory muscles increases, and the pace of liver metabolism steps up. Thus epinephrine as well as thyroid hormone can increase the metabolic rate.

Other effects

Epinephrine affects the central nervous system to promote a state of arousal and increased CNS alertness. This permits "quick thinking" to help cope with the impending emergency. Many drugs used as stimulants or sedatives exert their effects by altering catecholamine levels in the CNS.

Both epinephrine and norepinephrine cause sweating, which helps the body rid itself of extra heat generated by increased muscular activity. Also, epinephrine acts on smooth muscles within the eyes to dilate the pupil and flatten the lens. These actions adjust the eyes for more encompassing vision so that the whole threatening scene can be quickly viewed.

▌ Sympathetic stimulation of the adrenal medulla is solely responsible for epinephrine release.

Catecholamine secretion by the adrenal medulla is controlled entirely by sympathetic input to the gland. When the sympathetic system is activated under conditions of fear or stress, it simultaneously triggers a surge of adrenomedullary catecholamine release. The concentration of epinephrine in the blood may increase up to 300 times normal, with the amount of epinephrine released depending on the type and intensity of the stressful stimulus.

Because both components of the adrenal gland play an extensive role in responding to stress, this is an appropriate place to pull together the major factors involved in the stress response.

▌ The stress response is a generalized pattern of reactions to any situation that threatens homeostasis.

Recall that a variety of noxious stimuli can elicit a stress response. Different stressors may produce some specific responses characteristic of that stressor; for example, the body's specific response to cold exposure is shivering and skin vasoconstriction, whereas the specific response to bacterial invasion includes increased phagocytic activity and antibody production. In addition to their specific response, however, all stressors also produce a similar nonspecific, generalized response regardless of the type of stressor (● Figure 19-12). Dr. Hans Selye was the first to recognize this commonality of responses to noxious stimuli in what he called the **general adaptation syndrome.** When a stressor is recognized, both nervous and hormonal responses bring about defensive measures to cope with the emergency. The result is a state of intense readiness and mobilization of biochemical resources.

To appreciate the value of the multifaceted stress response, imagine a primitive cave dweller who has just seen a large wild beast lurking in the shadows. We will consider both the neural and hormonal responses that would take place in this scenario. The body responds in the same way to modern-day stressors. You are already familiar with all these responses. At this time we are just examining how these responses work together.

Roles of the sympathetic nervous system and epinephrine in stress

The major neural response to such a stressful stimulus is generalized activation of the sympathetic nervous system. The resultant increase in cardiac output and ventilation as well as the diversion of blood from vasoconstricted regions of suppressed activity, such as the digestive tract and kidneys, to the more active vasodilated skeletal muscles and heart prepare the body for a fight-or-flight response. Simultaneously, the sympathetic system calls forth hormonal reinforcements in the form of a massive outpouring of epinephrine from the adrenal medulla. Epinephrine strengthens sympathetic responses and reaches places not innervated by the sympathetic system to perform additional functions, such as mobilizing carbohydrate and fat stores.

● **FIGURE 19-12**
Action of a stressor on the body

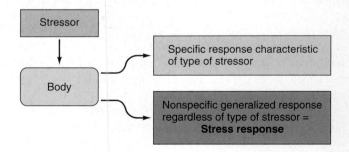

Roles of the CRH-ACTH-cortisol system in stress

Besides epinephrine, a number of other hormones are involved in the overall stress response (▲ Table 19-2). The predominant hormonal response is activation of the CRH-ACTH-cortisol system. Recall that cortisol's role in helping the body cope with stress is presumed to be related to its metabolic effects. Cortisol breaks down fat and protein stores while expanding carbohydrate stores and increasing the availability of blood glucose. A logical assumption is that the increased pool of glucose, amino acids, and fatty acids is available for use as needed, such as to sustain nourishment to the brain and provide building blocks for repair of damaged tissues.

In addition to the effects of cortisol in the hypothalamus-pituitary-adrenal cortex axis, ACTH may also play a role in resisting stress. ACTH is one of several peptides that facilitate learning and behavior. Thus an increase in ACTH during psychosocial stress may help the body cope more readily with similar stressors in the future by facilitating the learning of appropriate behavioral responses. Furthermore, ACTH is not released alone from its anterior pituitary storage vesicles. Pruning of the large pro-opiomelanocortin precursor molecule yields not only

ACTH but also morphinelike β-endorphin. As a potent endogenous opiate, β-endorphin may exert a role in mediating analgesia (reduction of pain perception) should physical injury be inflicted during stress (see p. 194).

Role of other hormonal responses in stress

Besides the CRH-ACTH-cortisol system, other hormonal systems play key roles in the stress response, as follows:

• *Elevation of blood glucose and fatty acids through decreased insulin and increased glucagon.* The sympathetic nervous system and the epinephrine secreted at its bidding both inhibit insulin and stimulate glucagon. These hormonal changes act in concert to elevate blood levels of glucose and fatty acids. Epinephrine and glucagon, whose blood levels are elevated during stress, promote hepatic glycogenolysis and (along with cortisol) hepatic gluconeogenesis. However, insulin, whose secretion is suppressed during stress, opposes the breakdown of liver glycogen stores. All these effects help increase the concentration of blood glucose. The primary stimulus for insulin secretion is a rise in blood glucose; in turn, a primary effect of insulin is to lower blood glucose. If it were not for the deliberate inhibition of insulin during the stress response, the hyperglycemia caused by stress would stimulate secretion of glucose-lowering insulin. As a result, the elevation in blood glucose could not be sustained. Stress-related hormonal responses also promote a release of fatty acids from fat stores, because lipolysis is favored by epinephrine, glucagon, and cortisol but opposed by insulin.

• *Maintenance of blood volume and blood pressure through increased renin-angiotensin-aldosterone and vasopressin activity.* In addition to the hormonal changes that mobilize energy stores during stress, other hormones are simultaneously called into play to sustain blood volume and blood pressure during the emergency. The sympathetic system and epinephrine have major responsibilities in acting directly on the heart and blood vessels to improve circulatory function. In addition, the renin-angiotensin-aldosterone system is activated as a consequence of a sympathetically induced reduction of blood supply to the kidneys (see p. 528). Vasopressin secretion is also increased during stressful situations (see p. 567). Collectively, these hormones expand the plasma volume by promoting retention of salt and H_2O. Presumably, the enlarged plasma volume serves as a protective measure to help sustain blood pressure should acute loss of plasma fluid occur through hemorrhage or heavy sweating during the impending period of danger. Vasopressin and angiotensin also have direct vasopressor effects, which would be of benefit in maintaining an adequate arterial pressure in the face of acute blood loss (see p. 359). Vasopressin is further believed to facilitate learning, which has implications for future adaptation to stress.

▲ TABLE 19-2
Major Hormonal Changes during the Stress Response

Hormone	Change	Purpose Served
Epinephrine	↑	Reinforces the sympathetic nervous system to prepare the body for "fight or flight"
		Mobilizes carbohydrate and fat energy stores; increases blood glucose and blood fatty acids
CRH-ACTH-Cortisol	↑	Mobilizes energy stores and metabolic building blocks for use as needed; increases blood glucose, blood amino acids, and blood fatty acids
		ACTH facilitates learning and behavior
		β-Endorphin cosecreted with ACTH may mediate analgesia
Glucagon	↑	Act in concert to increase blood glucose and blood fatty acids
Insulin	↓	
Renin-Angiotensin-Aldosterone	↑	Conserve salt and H_2O to expand the plasma volume; help sustain blood pressure when acute loss of plasma volume occurs
Vasopressin	↑	
		Angiotensin II and vasopressin cause arteriolar vasoconstriction to increase blood pressure
		Vasopressin facilitates learning

▌ The multifaceted stress response is coordinated by the hypothalamus.

All the individual responses to stress just described are either directly or indirectly influenced by the hypothalamus (● Fig-

---

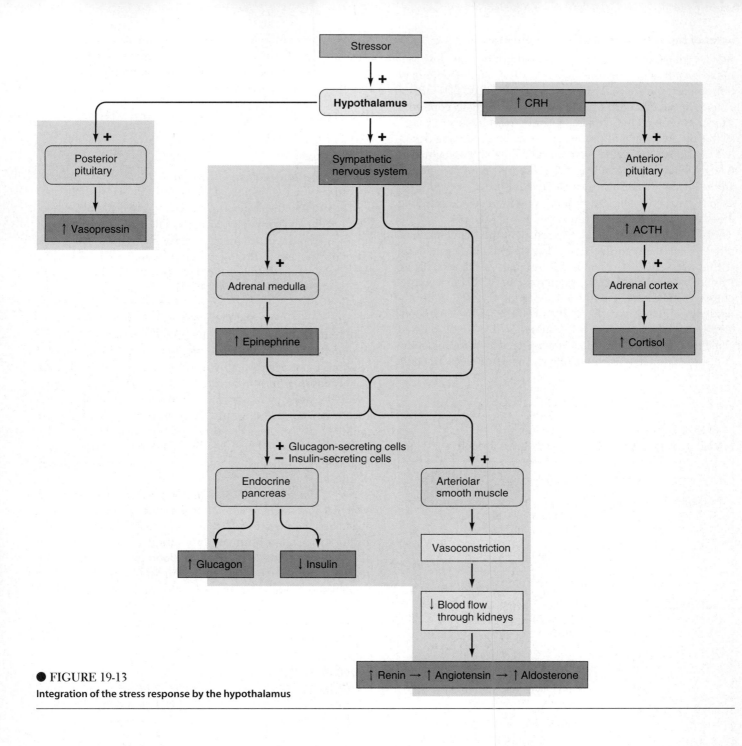

● **FIGURE 19-13**
Integration of the stress response by the hypothalamus

ure 19-13). The hypothalamus receives input concerning physical and emotional stressors from virtually all areas of the brain and from many receptors throughout the body. In response, the hypothalamus directly activates the sympathetic nervous system, secretes CRH to stimulate ACTH and cortisol release, and triggers the release of vasopressin. Sympathetic stimulation in turn brings about the secretion of epinephrine, with which it has a conjoined effect on the pancreatic secretion of insulin and glucagon. Furthermore, vasoconstriction of the renal afferent arterioles by the catecholamines indirectly triggers the secretion of renin by reducing the flow of oxygenated blood through the kidneys. Renin in turn sets in motion the renin-angiotensin-aldosterone mechanism. In this way, the hypothal-amus integrates the responses of both the sympathetic nervous system and the endocrine system during stress.

▌ Activation of the stress response by chronic psychosocial stressors may be harmful.

Acceleration of cardiovascular and respiratory activity, retention of salt and H_2O, and mobilization of metabolic fuels and building blocks can be of benefit in response to a physical stressor, such as an athletic competition. Most of the stressors in our everyday lives are psychosocial in nature, however, yet they induce these same magnified responses. Stressors such as anxiety about an exam, conflicts with loved ones, or impatience while sitting in a traffic jam can elicit a stress response. Although the rapid mobilization of body resources is appropriate in the face of real or threatened physical injury, it is generally inappropriate in response to nonphysical stress. If no extra energy is demanded, no tissue damaged, and no blood lost, body stores are being broken down and fluid retained needlessly, probably to the detriment of the emotionally stressed individual. In fact, there is strong circumstantial evidence for a link between chronic exposure to psychosocial stressors and the development of pathologic conditions such as high blood pressure, although no definitive cause-and-effect relationship has been ascertained. As a result of "unused" stress responses, could hypertension result from too much sympathetic vasoconstriction? From too much salt and H_2O retention? From too much vasopressin and angiotensin pressor activity? A combination of these? Other factors? Recall that hypertension can develop with prolonged exposure to pharmacologic levels of glucocorticoids. Could long-standing lesser elevations of cortisol, such as might occur in the face of continual psychosocial stressors, do the same thing, only more slowly? Considerable work remains to be done to evaluate the contributions that the stressors in our everyday lives make toward disease production.

ENDOCRINE CONTROL OF FUEL METABOLISM

We have just discussed the metabolic changes that are elicited during the stress response. Now we will concentrate on the metabolic patterns that occur in the absence of stress, including the hormonal factors that govern this normal metabolism.

▌ Fuel metabolism includes anabolism, catabolism, and interconversions among energy-rich organic molecules.

The term **metabolism** refers to all the chemical reactions that occur within the cells of the body. Those reactions involving the degradation, synthesis, and transformation of the three classes of energy-rich organic molecules—protein, carbohydrate, and fat—are collectively known as **intermediary metabolism** or **fuel metabolism** (▲ Table 19-3).

During the process of digestion, large nutrient molecules (**macromolecules**) are broken down into their smaller absorbable subunits as follows: Proteins are converted into amino acids, complex carbohydrates into monosaccharides (mainly glucose), and triglycerides (dietary fats) into monoglycerides and free fatty acids. These absorbable units are transferred from the digestive tract lumen into the blood, either directly or by way of the lymph (Chapter 16).

Anabolism and catabolism

These organic molecules are constantly exchanged between the blood and the cells of the body (● Figure 19-14). The chemical reactions in which the organic molecules participate within the cells are categorized into two metabolic processes: anabolism and catabolism. **Anabolism** is the buildup or synthesis of larger organic macromolecules from the small organic molecular subunits. Anabolic reactions generally require energy input in the form of ATP. These reactions result in either (1) the manufacture of materials needed by the cell, such as cellular structural proteins or secretory products, or (2) storage of excess ingested nutrients not immediately needed for energy production or needed as cellular building blocks. Storage is in the form of glycogen (the storage form of glucose) or fat reservoirs. **Catabolism** is the breakdown, or degradation, of large, energy-rich organic molecules within cells. Catabolism encompasses two levels of breakdown: (1) hydrolysis (see p. 594) of large cellular organic macromolecules into their smaller subunits, similar to the process of digestion except that the reactions take place within the cells of the body instead of within the digestive tract lumen (for example, release of glucose by the catabolism of stored glycogen), and (2) oxidation of the smaller subunits, such as glucose, to yield energy for ATP production (see p. 40).

As an alternative to energy production, the smaller, multipotential organic subunits derived from intracellular hydrolysis may be released into the blood. These mobilized glucose, fatty acid, and amino acid molecules can then be used as needed for energy production or cellular synthesis elsewhere in the body.

▲ TABLE 19-3
Summary of Reactions in Fuel Metabolism

Metabolic Process	Reaction	Consequence
Glycogenesis	Glucose → glycogen	↓Blood glucose
Glycogenolysis	Glycogen → glucose	↑Blood glucose
Gluconeogenesis	Amino acids → glucose	↑Blood glucose
Protein Synthesis	Amino acids → protein	↓Blood amino acids
Protein Degradation	Protein → amino acids	↑Blood amino acids
Fat Synthesis (Lipogenesis or Triglyceride Synthesis)	Fatty acids and glycerol → triglycerides	↓Blood fatty acids
Fat Breakdown (Lypolysis or Triglyceride Degradation)	Triglycerides → fatty acids and glycerol	↑Blood fatty acids

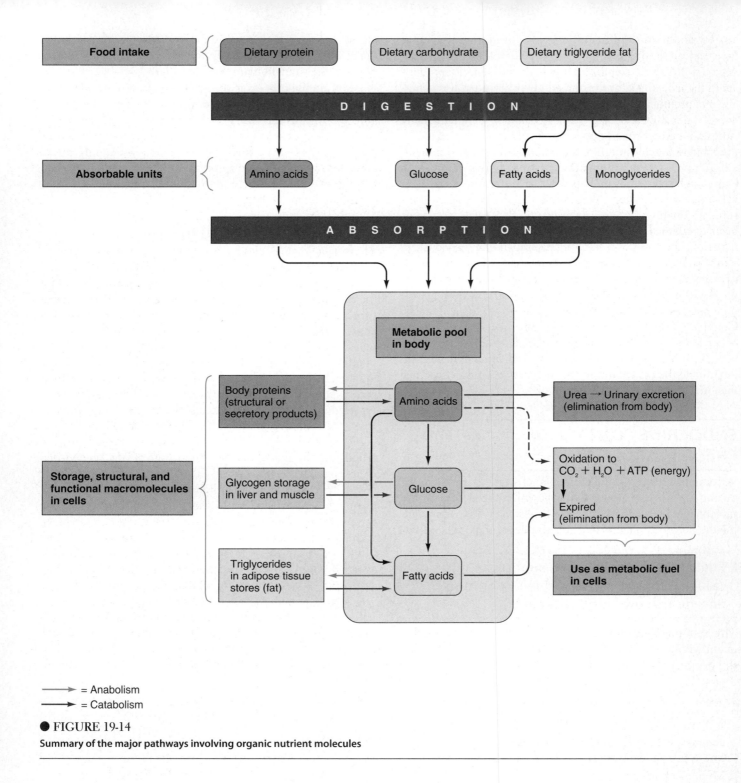

Summary of the major pathways involving organic nutrient molecules

In an adult, the rates of anabolism and catabolism are generally in balance, so the adult body remains in a dynamic steady state and appears unchanged even though the organic molecules that determine its structure and function are continuously being turned over. During growth, anabolism exceeds catabolism.

Interconversions among organic molecules

In addition to being able to resynthesize catabolized organic molecules back into the same type of molecules, many cells of the body, especially liver cells, can convert most types of small organic molecules into other types—as in, for example, transforming amino acids into glucose or fatty acids. Because of these interconversions, adequate nourishment can be provided by a wide range of molecules present in different types of foods. There are limits, however. **Essential nutrients,** such as the essential amino acids and vitamins, cannot be formed in the body by conversion from another type of organic molecule.

The major fate of both ingested carbohydrates and fats is catabolism to yield energy. Amino acids are predominantly used for protein synthesis but can be used to supply energy after being converted to carbohydrate or fat. Thus all three categories of foodstuff can be used as fuel, and excesses of any foodstuff can be deposited as stored fuel, as you will see shortly.

At a superficial level, fuel metabolism appears relatively simple: The amount of nutrients in the diet must be sufficient to meet the body's needs for energy production and cellular synthesis. This apparently simple relationship is complicated, however, by two important considerations: (1) nutrients taken in at meals must be stored and then released between meals, and (2) the brain must be continuously supplied with glucose. Let us examine the implications of each of these considerations.

■ Because food intake is intermittent, nutrients must be stored for use between meals.

Dietary fuel intake is intermittent, not continuous. As a result, excess energy must be absorbed during meals and stored for use during fasting periods between meals, when dietary sources of metabolic fuel are not available (▲ Table 19-4).

- *Excess circulating glucose* is stored in the liver and muscle as *glycogen*, a large molecule consisting of interconnected glucose molecules. Because glycogen is a relatively small energy reservoir, less than a day's energy needs can be stored in this form. Once the liver and muscle glycogen stores are "filled up," additional glucose is transformed into fatty acids and glycerol, which are used to synthesize *triglycerides* (glycerol with three fatty acids attached), primarily in adipose tissue (fat).
- *Excess circulating fatty acids* derived from dietary intake also become incorporated into triglycerides.
- *Excess circulating amino acids* not needed for protein synthesis are not stored as extra protein but are converted to glucose and fatty acids, which ultimately end up being stored as triglycerides. Thus the major site of energy storage for excess nutrients of all three classes is adipose tissue. Normally,

sufficient triglyceride is stored to provide energy for about two months, more so in an overweight individual. Consequently, during any prolonged period of fasting, the fatty acids released from triglyceride catabolism serve as the primary source of energy for most tissues. The catabolism of stored triglycerides frees glycerol as well as fatty acids, but quantitatively speaking, the fatty acids are far more important. Catabolism of stored fat yields 90% fatty acids and 10% glycerol by weight. Glycerol (but not fatty acids) can be converted to glucose by the liver and contributes in a small way to maintaining blood glucose during a fast.

As a third energy reservoir, a substantial amount of energy is stored as *structural protein*, primarily in muscle, the most abundant protein mass in the body. Protein is not the first choice to tap as an energy source, however, because it serves other essential functions; in contrast, the glycogen and triglyceride reservoirs serve solely as energy depots.

■ The brain must be continuously supplied with glucose.

The second factor complicating fuel metabolism besides intermittent nutrient intake and the resultant necessity of storing nutrients is that the brain normally depends on the delivery of adequate blood glucose as its sole source of energy. Consequently, the blood glucose concentration must be maintained above a critical level. The blood glucose concentration is typically 100 mg glucose/100 ml plasma and is normally kept within the narrow limits of 70 to 110 mg/100 ml. Liver glycogen is an important reservoir for maintaining blood glucose levels during a short fast. However, liver glycogen is depleted relatively rapidly, so during a longer fast other mechanisms must meet the energy requirements of the glucose-dependent brain. First, when new dietary glucose is not entering the blood, tissues not obligated to use glucose shift their metabolic gears

▲ TABLE 19-4
Stored Metabolic Fuel in the Body

Metabolic Fuel	Circulating Form	Storage Form	Major Storage Site	Percentage of Total Body Energy Content (and Calories*)	Reservoir Capacity	Role
Carbohydrate	Glucose	Glycogen	Liver, muscle	1% (1,500 calories)	Less than a day's worth of energy	First energy source; essential for the brain
Fat	Free fatty acids	Triglycerides	Adipose tissue	77% (143,000 calories)	About two months' worth of energy	Primary energy reservoir; energy source during a fast
Protein	Amino acids	Body proteins	Muscle	22% (41,000 calories)	Death results long before capacity is fully used because of structural and functional impairment	Source of glucose for the brain during a fast; last resort to meet other energy needs

*Actually refers to kilocalories; see p. 648.

to burn fatty acids instead, sparing glucose for the brain. Fatty acids are made available by catabolism of triglyceride stores as an alternative energy source for tissues that are not glucose dependent. Second, amino acids can be converted to glucose by gluconeogenesis, whereas fatty acids cannot. Thus once glycogen stores are depleted despite glucose sparing, new glucose supplies for the brain are provided by the catabolism of body proteins and conversion of the freed amino acids into glucose.

▌ Metabolic fuels are stored during the absorptive state and mobilized during the postabsorptive state.

The preceding discussion should make clear that the disposition of organic molecules depends on the body's metabolic state. The two functional metabolic states—the *absorptive state* and the *postabsorptive state*—are related to eating and fasting cycles, respectively (▲ Table 19-5).

Absorptive state

Following a meal, ingested nutrients are being absorbed and entering the blood during the **absorptive,** or **fed, state.** During this time, glucose is plentiful and serves as the major energy source. Very little of the absorbed fat and amino acids is used for energy during the absorptive state, because most cells prefer to use glucose when it is available. Extra nutrients not immediately used for energy or structural repairs are channeled into storage as glycogen or triglycerides.

▲ TABLE 19-5
Comparison of Absorptive and Postabsorptive States

Metabolic Fuel	Absorptive State	Postabsorptive State
Carbohydrate	Glucose providing major energy source	Glycogen degradation and depletion
	Glycogen synthesis and storage	Glucose sparing to conserve glucose for the brain
	Excess converted and stored as triglyceride fat	Production of new glucose through gluconeogenesis
Fat	Triglyceride synthesis and storage	Triglyceride catabolism
		Fatty acids providing the major energy source for non–glucose dependent tissues
Protein	Protein synthesis	Protein catabolism
	Excess converted and stored as triglyceride fat	Amino acids used for gluconeogenesis

Postabsorptive state

The average meal is completely absorbed in about four hours. Therefore, on a typical three-meals-a-day diet, no nutrients are being absorbed from the digestive tract during late morning and late afternoon and throughout the night. These times constitute the **postabsorptive,** or **fasting, state.** During this state, endogenous energy stores are mobilized to provide energy, whereas gluconeogenesis and glucose sparing maintain the blood glucose at an adequate level to nourish the brain. Synthesis of protein and fat is curtailed. Instead, stores of these organic molecules are catabolized for glucose formation and energy production, respectively. Carbohydrate synthesis does occur through gluconeogenesis, but the use of glucose for energy is greatly reduced.

Note that the blood concentration of nutrients does not fluctuate markedly between the absorptive and postabsorptive states. During the absorptive state, the glut of absorbed nutrients is swiftly removed from the blood and placed into storage; during the postabsorptive state, these stores are catabolized to maintain the blood concentrations at levels necessary to fill tissue energy demands.

Roles of key tissues in metabolic states

During these alternating metabolic states, various tissues play different roles as summarized here.

- The *liver* plays the primary role in maintaining normal blood glucose levels. It stores glycogen when excess glucose is available, releases glucose into the blood when needed, and is the principal site for metabolic interconversions such as gluconeogenesis.
- *Adipose tissue* serves as the primary energy storage site and is important in regulating fatty acid levels in the blood.
- *Muscle* is the primary site of amino acid storage and is the major energy user.
- The *brain* normally can use only glucose as an energy source, yet it does not store glycogen, making it mandatory that blood glucose levels be maintained.

▌ Lesser energy sources are tapped as needed.

Several other organic intermediates play a lesser role as energy sources—namely, glycerol, lactic acid, and ketone bodies.

- As mentioned earlier, *glycerol* derived from triglyceride hydrolysis (it is the backbone to which the fatty acid chains are attached) can be converted to glucose by the liver.
- Similarly, *lactic acid*, which is produced by the incomplete catabolism of glucose via glycolysis in muscle (see p. 279), can also be converted to glucose in the liver.
- **Ketone bodies** are a group of compounds produced by the liver during glucose sparing. Unlike other tissues, when the liver uses fatty acids as an energy source, it oxidizes them only to acetyl coenzyme A (acetyl CoA), which it is unable to process through the citric acid cycle for further energy extraction. Thus the liver does not degrade fatty acids all the way to CO_2 and H_2O for maximum energy release. Instead, it par-

tially extracts the available energy and converts the remaining energy-bearing acetyl CoA molecules into ketone bodies, which it releases into the blood. Ketone bodies serve as an alternative energy source for tissues capable of oxidizing them further by means of the citric acid cycle.

During long-term starvation, the brain starts using ketones instead of glucose as a major energy source. Because death resulting from starvation is usually due to protein wasting rather than hypoglycemia (low blood glucose), prolonged survival without any caloric intake requires that gluconeogenesis be kept to a minimum as long as the energy needs of the brain are not compromised. A sizable portion of cell protein can be catabolized without serious cellular malfunction, but a point is finally reached at which a cannibalized cell can no longer function adequately. To ward off the fatal point of failure so long as possible during prolonged starvation, the brain starts using ketones as a major energy source, correspondingly decreasing its use of glucose. Use by the brain of this fatty acid "table scrap" left over from the liver's "meal" limits the necessity of mobilizing body proteins for glucose production to nourish the brain. Both the major metabolic adaptations to prolonged starvation—a decrease in protein catabolism and use of ketones by the brain—are attributable to the high levels of ketones in the blood at the time. The brain uses ketones only when blood ketone level is high. The high blood levels of ketones also directly inhibit protein degradation in muscle. Thus ketones spare body proteins while satisfying the brain's energy needs.

▌ The pancreatic hormones, insulin and glucagon, are most important in regulating fuel metabolism.

How does the body "know" when to shift its metabolic gears from a system of net anabolism and nutrient storage to one of net catabolism and glucose sparing? The flow of organic nutrients along metabolic pathways is influenced by a variety of hormones, including insulin, glucagon, epinephrine, cortisol, and growth hormone. Under most circumstances, the pancreatic hormones, insulin and glucagon, are the dominant hormonal regulators that shift the metabolic pathways back and forth from net anabolism to net catabolism and glucose sparing, depending on whether the body is in a state of feasting or fasting, respectively.

The **pancreas** is an organ composed of both exocrine and endocrine tissues. The exocrine portion secretes a watery alkaline solution and digestive enzymes through the pancreatic duct into the digestive tract lumen. Scattered throughout the pancreas between the exocrine cells are clusters, or "islands," of endocrine cells known as the **islets of Langerhans** (see ● Figure 16-13, p. 616). The most abundant pancreatic endocrine cell type is the **β (beta) cell,** the site of *insulin* synthesis and secretion. Next most important are the **α (alpha) cells,** which produce *glucagon*. The **D (delta) cells** are the pancreatic site of *somatostatin* synthesis. The least common islet cells, the **PP cells,** secrete *pancreatic polypeptide,* which is poorly understood and which we will not discuss further. We will give only a brief highlight now of somatostatin, then will pay the most

attention to insulin and glucagon, the most important hormones in the regulation of fuel metabolism.

Somatostatin

Pancreatic **somatostatin** inhibits the digestive system in a variety of ways, the overall effect of which is to inhibit digestion of nutrients and to decrease nutrient absorption. Somatostatin is released from the pancreatic D cells in direct response to an increase in blood glucose and blood amino acids during absorption of a meal. By exerting its inhibitory effects, pancreatic somatostatin acts in negative-feedback fashion to put the brakes on the rate at which the meal is being digested and absorbed, thereby preventing excessive plasma levels of nutrients.

In addition, pancreatic somatostatin may play an important role in local regulation of pancreatic hormone secretion. The local presence of somatostatin decreases the secretion of insulin, glucagon, pancreatic polypeptide, and somatostatin itself, but the physiologic importance of such paracrine function has not been determined.

Somatostatin is also produced by the hypothalamus, where it inhibits the secretion of growth hormone and TSH (see p. 693). Furthermore, somatostatin is produced by cells lining the digestive tract, where it acts locally as a paracrine to inhibit most digestive processes (see p. 612).

We will next consider insulin and then glucagon, followed by a discussion of how insulin and glucagon function as an endocrine unit to shift metabolic gears between the absorptive and postabsorptive states.

▌ Insulin lowers blood glucose, fatty acid, and amino acid levels and promotes their storage.

Insulin has important effects on carbohydrate, fat, and protein metabolism. It lowers the blood levels of glucose, fatty acids, and amino acids and promotes their storage. As these nutrient molecules enter the blood during the absorptive state, insulin promotes their cellular uptake and conversion into glycogen, triglycerides, and protein, respectively. Insulin exerts its many effects either by altering transport of specific blood-borne nutrients into cells or by altering the activity of the enzymes involved in specific metabolic pathways.

Actions on carbohydrates

The maintenance of blood glucose homeostasis is a particularly important function of the pancreas. Circulating glucose concentrations are determined by the balance among the following processes (● Figure 19-15): glucose absorption from the digestive tract, transport of glucose into cells, hepatic glucose production, and (abnormally) urinary excretion of glucose.

Insulin exerts four effects that lower blood glucose levels and promote carbohydrate storage:

1. Insulin facilitates glucose transport into most cells. (The mechanism of this increased glucose uptake is explained in the next section.)

2. Insulin stimulates **glycogenesis,** the production of glycogen from glucose, in both skeletal muscle and the liver.

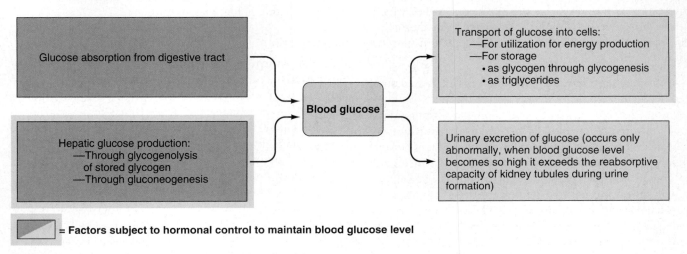

Factors that increase blood glucose

Glucose absorption from digestive tract

Hepatic glucose production:
—Through glycogenolysis of stored glycogen
—Through gluconeogenesis

Factors that decrease blood glucose

Transport of glucose into cells:
—For utilization for energy production
—For storage
• as glycogen through glycogenesis
• as triglycerides

Urinary excretion of glucose (occurs only abnormally, when blood glucose level becomes so high it exceeds the reabsorptive capacity of kidney tubules during urine formation)

Blood glucose

☐ = Factors subject to hormonal control to maintain blood glucose level

● **FIGURE 19-15**
Factors affecting blood glucose concentration

3. Insulin inhibits glycogenolysis. By inhibiting the breakdown of glycogen into glucose, insulin likewise favors carbohydrate storage and decreases glucose output by the liver.

4. Insulin further decreases hepatic glucose output by inhibiting gluconeogenesis, the conversion of amino acids into glucose in the liver. Insulin does so by decreasing the amount of amino acids in the blood available to the liver for gluconeogenesis and by inhibiting the hepatic enzymes required for converting amino acids into glucose.

Thus insulin decreases the concentration of blood glucose by promoting the cells' uptake of glucose from the blood for use and storage, while simultaneously blocking the two mechanisms by which the liver releases glucose into the blood (glycogenolysis and gluconeogenesis). Insulin is the only hormone capable of lowering the blood glucose level. Insulin promotes the uptake of glucose by most cells through glucose transporter recruitment, a topic to which we now turn our attention.

Glucose transporter recruitment. Glucose transport between the blood and cells is accomplished by means of a plasma membrane carrier known as a **glucose transporter (GLUT)**. Six forms of glucose transporters have been identified, named in the order they were discovered—GLUT-1, GLUT-2, and so on. These glucose transporters all accomplish passive facilitated diffusion of glucose across the plasma membrane (see p. 80) and are distinct from the Na^+-glucose cotransport carriers responsible for secondary active transport of glucose across the kidney and intestinal epithelia (see pp. 83, 531, and 629). Each member of the GLUT family performs slightly different functions. For example, *GLUT-1* transports glucose across the blood–brain barrier, *GLUT-2* transfers into the adjacent bloodstream the glucose that has entered the kidney and intestinal cells by means of the cotransport carriers, and *GLUT-3* is the main transporter of glucose into neurons. The glucose transporter responsible for the majority of glucose uptake by most cells of the body is *GLUT-4*, which operates only at the bidding of insulin. Glucose molecules cannot readily penetrate most cell membranes in the absence of insulin, making most tissues highly dependent on insulin for uptake of glucose from the blood and for its subsequent use. GLUT-4 is especially abundant in the tissues that account for the bulk of glucose uptake from the blood during the absorptive state, namely skeletal muscle and adipose tissue cells.

GLUT-4 is the only type of glucose transporter that is responsive to insulin. Unlike the other types of GLUT molecules, which are always present in the plasma membranes at the sites where they perform their functions, in the absence of insulin GLUT-4 is excluded from the plasma membrane. Insulin promotes glucose uptake by **transporter recruitment.** Insulin-dependent cells maintain a pool of intracellular vesicles containing GLUT-4. Insulin induces these vesicles to move to the plasma membrane and fuse with it, thus inserting GLUT-4 molecules into the plasma membrane. In this way, increased insulin secretion promotes a rapid 10- to 30-fold increase in glucose uptake by insulin-dependent cells. When insulin secretion decreases, these glucose transporters are retrieved from the membrane and returned to the intracellular pool.

Several tissues do not depend on insulin for their glucose uptake—namely the brain, working muscles, and the liver. The brain, which requires a constant supply of glucose for its minute-to-minute energy needs, is freely permeable to glucose at all times by means of GLUT-1 and GLUT-3 molecules. Skeletal muscle cells do not depend on insulin for their glucose uptake during exercise, even though they are dependent at rest. Muscle contraction triggers the insertion of GLUT-4 into the plasma membranes of exercising muscle cells in the absence of insulin. This fact is important in managing diabetes mellitus (insulin deficiency), as described later. The liver also does not depend on insulin for glucose uptake, because it does not use GLUT-4. However, insulin does enhance the metabolism of glucose by the liver by stimulating the first step in glucose metabolism, the phosphorylation of glucose to form glucose-6-phosphate. The phosphorylation of glucose as it enters the cell keeps the intracellular concentration of "plain" glucose low so that a gradient favoring the facilitated diffusion of glucose into the cell is maintained.

Insulin also exerts important actions on fat and protein.

Actions on fat

Insulin exerts multiple effects to lower blood fatty acids and promote triglyceride storage:

1. It increases the transport of glucose into adipose tissue cells by means of GLUT-4 recruitment. Glucose serves as a precursor for the formation of fatty acids and glycerol, which are the raw materials for triglyceride synthesis.
2. It activates enzymes that catalyze the production of fatty acids from glucose derivatives.
3. It promotes the entry of fatty acids from the blood into adipose tissue cells.
4. It inhibits lipolysis (fat breakdown), reducing the release of fatty acids from adipose tissue into the blood.

Collectively, these actions favor removal of glucose and fatty acids from the blood and promote their storage as trigylcerides.

Actions on protein

Insulin lowers blood amino acid levels and enhances protein synthesis through several effects:

1. It promotes the active transport of amino acids from the blood into muscles and other tissues. This effect decreases the circulating amino acid level and provides the building blocks for protein synthesis within the cells.
2. It increases the rate of amino acid incorporation into protein by stimulating the cells' protein-synthesizing machinery.
3. It inhibits protein degradation.

The collective result of these actions is a protein anabolic effect. For this reason, insulin is essential for normal growth.

Summary of insulin's actions

In short, insulin primarily exerts its effects by acting on nonworking skeletal muscle, the liver, and adipose tissue. It stimulates biosynthetic pathways that lead to increased glucose use, increased carbohydrate and fat storage, and increased protein synthesis. In so doing, this hormone lowers the blood glucose, fatty acid, and amino acid levels. This metabolic pattern is characteristic of the absorptive state. Indeed, insulin secretion rises during this state and is responsible for shifting metabolic pathways to net anabolism.

When insulin secretion is low, the opposite effects occur. The rate of glucose entry into cells is reduced, and net catabolism occurs rather than net synthesis of glycogen, triglycerides, and protein. This pattern is reminiscent of the postabsorptive state; indeed, insulin secretion is reduced during the postabsorptive state. However, the other major pancreatic hormone, glucagon, also plays an important role in shifting from absorptive to postabsorptive metabolic patterns, as described later.

▐ The primary stimulus for increased insulin secretion is an increase in blood glucose concentration.

The primary control of insulin secretion is a direct negative-feedback system between the pancreatic β cells and the concentration of glucose in the blood flowing to them. An elevated blood-glucose level, such as during absorption of a meal, directly stimulates the β cells to synthesize and release insulin. The increased insulin in turn reduces the blood glucose to normal and promotes use and storage of this nutrient. Conversely, a fall in blood glucose below normal, such as during fasting, directly inhibits insulin secretion. Lowering the rate of insulin secretion shifts metabolism from the absorptive to the postabsorptive pattern. Thus this simple negative-feedback system can maintain a relatively constant supply of glucose to the tissues without requiring the participation of nerves or of other hormones.

In addition to plasma glucose concentration, other inputs are involved in regulating insulin secretion, as follows (● Figure 19-16):

- An elevated blood amino-acid level, such as after a high-protein meal, directly stimulates the β cells to increase insulin secretion. In negative-feedback fashion, the increased insulin enhances the entry of these amino acids into the cells, lowering the blood amino-acid level while promoting protein synthesis.
- The major gastrointestinal hormones secreted by the digestive tract in response to the presence of food, especially glucose-dependent insulinotropic peptide (see p. 641), stimulate pancreatic insulin secretion in addition to having direct regulatory effects on the digestive system. Through this control, insulin secretion is increased in "feedforward," or anticipatory, fashion even before nutrient absorption increases the blood concentration of glucose and amino acids.
- The autonomic nervous system also directly influences insulin secretion. The islets are richly innervated by both parasympathetic (vagal) and sympathetic nerve fibers. The increase in parasympathetic activity that occurs in response to food in the digestive tract stimulates insulin release. This, too, is a feedforward response in anticipation of nutrient absorption. In contrast, sympathetic stimulation and the concurrent increase in epinephrine both inhibit insulin secretion. The fall in insulin level allows the blood glucose level to rise, an appropriate response to the circumstances under which generalized sympathetic activation occurs—namely, stress (fight or flight) and exercise. In both these situations, extra fuel is needed for increased muscle activity.

▐ The symptoms of diabetes mellitus are characteristic of an exaggerated postabsorptive state.

Diabetes mellitus is by far the most common of all endocrine disorders. The acute symptoms of diabetes mellitus are attributable to inadequate insulin action. Because insulin is the only hormone capable of lowering blood glucose levels, one of the

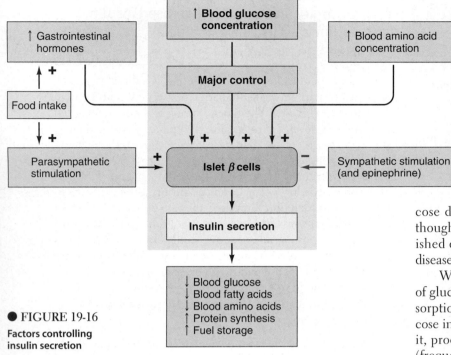

● FIGURE 19-16
Factors controlling insulin secretion

most prominent features of diabetes mellitus is elevated blood glucose levels, or **hyperglycemia.** *Diabetes* literally means "syphon" or "running through," a reference to the large urine volume accompanying this condition. A large urine volume occurs both in diabetes mellitus (a result of insulin insufficiency) and in diabetes insipidus (a result of vasopressin deficiency). *Mellitus* means "sweet"; *insipidus* means "tasteless." The urine of patients with diabetes mellitus acquires its sweetness from excess blood glucose that spills into the urine, whereas the urine of patients with diabetes insipidus contains no sugar, so it is tasteless. (Aren't you glad you were not a health professional at the time when these two conditions were distinguished on the basis of the taste of the urine?)

Diabetes mellitus has two major variants, differing in the capacity for pancreatic insulin secretion: *Type I diabetes,* characterized by a lack of insulin secretion, and *Type II diabetes,* characterized by normal or even increased insulin secretion but reduced sensitivity of insulin's target cells to its presence. (For a further discussion of the distinguishing features of these two types of diabetes mellitus, see the boxed feature on pp. 728–729, ◗ Concepts, Challenges, and Controversies.)

The acute consequences of diabetes mellitus can be grouped according to the effects of inadequate insulin action on carbohydrate, fat, and protein metabolism (● Figure 19-17). The figure may look overwhelming, but the numbers, which correspond to the numbers in the following discussion, help you work your way through this complex disease step-by-step.

Consequences related to effects on carbohydrate metabolism

Because the postabsorptive metabolic pattern is induced by low insulin activity, the changes that occur in diabetes mellitus

are an exaggeration of this state, with the exception of hyperglycemia. In the usual fasting state, the blood glucose level is slightly below normal. Hyperglycemia, the hallmark of diabetes mellitus, arises from reduced glucose uptake by cells, coupled with increased output of glucose from the liver (1 in ● Figure 19-17). As the glucose-yielding processes of glycogenolysis and gluconeogenesis proceed unchecked in the absence of insulin, hepatic output of glucose increases. Because many of the body's cells cannot use glucose without the help of insulin, an ironic extracellular glucose excess occurs coincident with an intracellular glucose deficiency—"starvation in the midst of plenty." Even though the non–insulin dependent brain is adequately nourished during diabetes mellitus, further consequences of the disease lead to brain dysfunction, as you will see shortly.

When the blood glucose rises to the level where the amount of glucose filtered exceeds the tubular cells' capacity for reabsorption, glucose appears in the urine (*glucosuria*) 2 . Glucose in the urine exerts an osmotic effect that draws H_2O with it, producing an osmotic diuresis characterized by *polyuria* (frequent urination) 3 . The excess fluid lost from the body leads to dehydration 4 , which in turn can ultimately lead to peripheral circulatory failure because of the marked reduction in blood volume 5 . Circulatory failure, if uncorrected, can lead to death because of low cerebral blood flow 6 or secondary renal failure resulting from inadequate filtration pressure 7 . Furthermore, cells lose water as the body becomes dehydrated by an osmotic shift of water from the cells into the hypertonic extracellular fluid 8 . Brain cells are especially sensitive to shrinking, so nervous system malfunction ensues 9 (see p. 567). Another characteristic symptom of diabetes mellitus is *polydipsia* (excessive thirst) 10 , which is actually a compensatory mechanism to counteract the dehydration.

The story is not complete. In the face of intracellular glucose deficiency, appetite is stimulated, leading to *polyphagia* (excessive food intake) 11 . Despite the increased food intake, however, progressive weight loss occurs from the effects of insulin deficiency on fat and protein metabolism.

Consequences related to effects on fat metabolism

Triglyceride synthesis decreases while lipolysis increases, resulting in large-scale mobilization of fatty acids from triglyceride stores 12 . The increased blood fatty acids are largely used by the cells as an alternative energy source. Increased liver use of fatty acids releases excessive ketone bodies into the blood, causing *ketosis* 13 . Because the ketone bodies include several different acids, such as acetoacetic acid, that result from incomplete breakdown of fat during hepatic energy production, this developing ketosis leads to progressive metabolic acidosis 14 . Acidosis depresses the brain and, if severe enough, can lead to diabetic coma and death 15 .

A compensatory measure for metabolic acidosis is increased ventilation to blow off extra, acid-forming CO_2 16 . Exhalation of one of the ketone bodies, acetone, causes a "fruity" breath odor. Sometimes, because of this odor, passersby unfor-

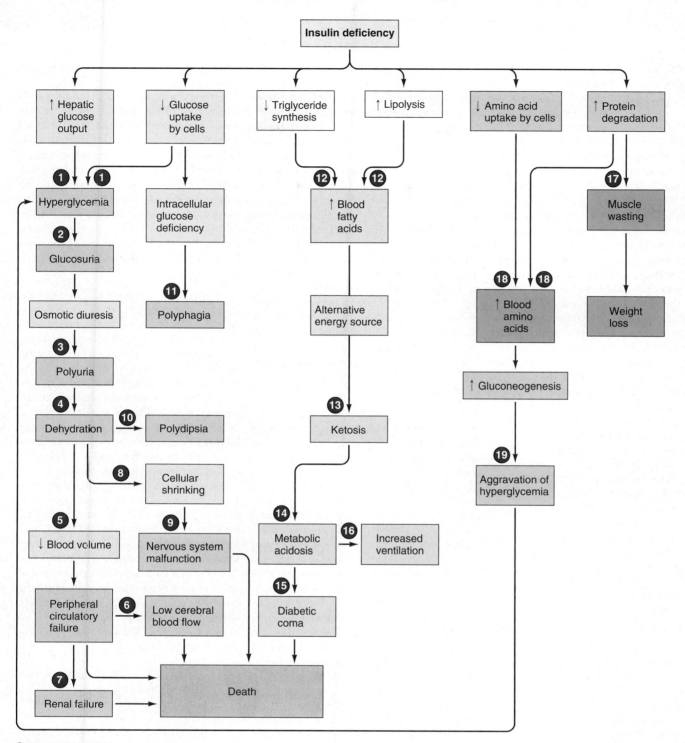

● FIGURE 19-17

Acute effects of diabetes mellitus
The acute consequences of diabetes mellitus can be grouped according to the effects of inadequate insulin action on carbohydrate, fat, and protein metabolism. These effects ultimately cause death through a variety of pathways. See pp. 726–728 for an explanation of the circled numbers.

tunately mistake a patient collapsed in a diabetic coma for a "wino" passed out in a state of drunkenness. (This situation illustrates the merits of medical alert identification tags.) People with Type I diabetes are much more prone to develop ketosis than are Type II diabetics.

Consequences related to effects on protein metabolism

The effects of a lack of insulin on protein metabolism result in a net shift toward protein catabolism. The net breakdown of muscle proteins leads to wasting and weakness of skeletal muscles 17 and, in child diabetics, a reduction in overall growth. Re-

Diabetics and Insulin: Some Have It and Some Don't

There are two distinct types of diabetes mellitus (see the accompanying table). **Type I (insulin-dependent or juvenile-onset) diabetes mellitus,** which accounts for about 10% of all cases of diabetes, is characterized by a lack of insulin secretion. Because Type I diabetics suffer a total or near-total lack of insulin secretion by their pancreatic β cells, they require exogenous insulin for survival. This dependence is the basis for the alternative name *insulin-dependent diabetes mellitus* for this form of the disease. In **Type II (non–insulin-dependent, or maturity-onset) diabetes mellitus,** insulin secretion may be normal or even increased, but insulin's target cells are less sensitive than normal to this hormone. Although either type can first be manifested at any age, Type I is more prevalent in children, whereas the onset of Type II more generally occurs in adulthood, giving rise to the age-related designations of the two conditions.

Diabetes of both types currently affects 16 million people in the United States, costing this country an estimated $100 billion annually in health care expenses. The diabetes-related death rate in our country has increased by 30% since 1980, in large part because the incidence of the disease has been rising. In 2000, diabetes mellitus afflicted 150 million people worldwide, and this number is projected to climb to 220 million by 2010. Because diabetes is so prevalent and exacts such a huge economic toll, coupled with the fact that it forces a change in the lifestyle of affected individuals and places them at increased risk for developing a variety of troublesome and even life-threatening conditions, intensive research is directed toward better understanding and controlling or preventing both types of the disease.

Underlying defect in Type I diabetes

Type I diabetes is an autoimmune process involving the erroneous, selective destruction of the pancreatic β cells by inappropriately activated T lymphocytes (see p. 441). The precise cause of this self-destructive immune attack remains unclear. Some individuals have a genetic susceptibility to acquiring Type I diabetes. Envi-ronmental triggers also appear to play an important role. Evidence is mounting that consumption of cows' milk by susceptible babies may be linked to the development of the condition in some cases. In one study, for example, researchers found that infants getting formula made from normal cows' milk were three times more likely to develop antibodies against bovine (cow) insulin by age 2 than were infants who received formula in which the proteins (including insulin) in the cows' milk had been broken down into peptide fragments. Because human insulin closely resembles its cow counterpart, the investigators speculate that antibodies formed against bovine insulin may spur an immune system attack against the child's own insulin-secreting β cells. Other recent research hints that several different viruses, such as the rotavirus and coxsackie virus—common childhood viruses—may be the underlying trigger in some cases of Type I diabetes. The virus may activate T cells that erroneously attack the pancreatic β cells as well as the virus.

Underlying defect in Type II diabetes

Various genetic and lifestyle factors are important in the development of Type II diabetes. Obesity is the biggest risk factor; 90% of Type II diabetics are obese.

Type II diabetics do secrete insulin, but the affected individuals exhibit insulin resistance. That is, the basic problem in Type II diabetes is not lack of insulin but reduced sensitivity of insulin's target cells to its presence. The cause of reduced insulin sensitivity remains elusive despite intense investigation. Most likely, multiple factors may lead to the same end result. Researchers used to think that Type II diabetes developed because of excessive downregulation of insulin receptors (see p. 680), but that turns out not to be the case. Recent research suggests that adipose cells secrete a hormone known as **resistin,** which interferes with insulin action in experimental animals. This could be an important link between obesity and insulin resistance. Resistin is distinct from leptin, the hormone secreted by adipose cells that plays a role in controlling food intake (see p. 650). Still other studies suggest that defects in the insulin signaling pathways may be the underlying culprit in Type II diabetes.

Early in the development of the disease, the resulting decrease in sensitivity to insulin is overcome by secretion of additional insulin. However, the sustained overtaxing of the pancreas eventually exceeds the reserve secretory capacity of the genetically weak β cells. Even though insulin secretion may be normal or somewhat elevated, symptoms of insulin insufficiency develop because the amount of insulin is still inadequate to prevent significant hyperglycemia. The symptoms in Type II diabetes are usually slower in onset and less severe than in Type I diabetes.

Treatment of diabetes

The conventional treatment for Type I diabetes is a controlled balance of regular insulin injections timed around meals, dietary management of the amounts and types of food consumed, and exercise. Insulin must be administered by injection because if it were swallowed this peptide hormone would be digested by proteolytic enzymes in the stomach and small intestine. Exercise is also useful in managing both types of diabetes, because working muscles are not insulin dependent. Exercising muscles take up and use some of the excess glucose in the blood, reducing the overall need for insulin.

Whereas Type I diabetics are permanently insulin dependent, dietary control and weight reduction may be all that is necessary to completely reverse the symptoms in Type II diabetics. Therefore, Type II diabetes is alternatively known as *non–insulin dependent diabetes.* Four classes of oral medications are currently available for use if needed for treating Type II diabetes in conjunction with a dietary and exercise regime. These pills help the patient's body use its own insulin more effectively, each by a different mechanism, as follows:

1. By stimulating the β cells to secrete more insulin than they do on their own

2. By suppressing the liver's output of glucose

3. By slowing glucose absorption into the blood from the digestive tract, thus blunting the surge of glucose immediately after a meal

---

duced amino acid uptake coupled with increased protein degradation results in excess amino acids in the blood 18 . The increased circulating amino acids can be used for additional gluconeogenesis, which further aggravates the hyperglycemia 19 .

As you can readily appreciate from this overview, diabetes mellitus is a complicated disease that can disturb both carbohydrate, fat, and protein metabolism and fluid and acid–base balance. It can also have repercussions on the circulatory system, kidneys, respiratory system, and nervous system.

Long-term complications

In addition to these potential consequences of untreated diabetes, which can be explained on the basis of insulin's short-term metabolic effects, numerous long-range complications of this

Comparison of Type I and Type II Diabetes Mellitus

Characteristic	Type I Diabetes	Type II Diabetes
Level of Insulin Secretion	None or almost none	May be normal or exceed normal
Typical Age of Onset	Childhood	Adulthood
Percentage of Diabetics	10–20%	80–90%
Basic Defect	Destruction of β cells	Reduced sensitivity of insulin's target cells
Associated with Obesity?	No	Usually
Genetic and Environmental Factors Important in Precipitating Overt Disease?	Yes	Yes
Speed of Development of Symptoms	Rapid	Slow
Development of Ketosis	Common if untreated	Rare
Treatment	Insulin injections; dietary management; exercise	Dietary control and weight reduction; exercise; sometimes oral hypoglycemic drugs

• Some researchers are seeking methods to protect swallowed insulin from destruction by the digestive tract, and others have identified a potential oral substitute for insulin—namely a nonpeptide chemical that binds with the insulin receptors and brings about the same intracellular responses as insulin does. Because this insulin mimic is not a protein, it would not be destroyed by the proteolytic digestive enzymes if taken as a pill.

• Another hope is pancreatic islet transplants. Scientists have developed several types of devices that isolate donor islet cells from the recipient's immune system. Such immunoisolation of islet cells would permit use of grafts from other animals, circumventing the shortage of human donor cells. Pig islet cells would be an especially good source, because porcine (pig) insulin is nearly identical to human insulin.

• Some researchers hope they will be able to coax stem cells to develop into insulin-secreting cells that can be implanted.

• In a related approach, other scientists are turning to genetic engineering in the hope of developing surrogates for pancreatic β cells. An example is the potential reprogramming of the small-intestine endocrine cells that produce glucose-dependent insulinotropic hormone (GIP). Recall that GIP normally promotes insulin release in feedforward fashion when food is in the digestive tract. The goal is to cause these non-β cells to cosecrete both insulin and GIP on feeding.

• Another approach under development is an implanted, glucose-detecting, insulin-releasing "artificial pancreas" that would continuously monitor the patient's blood glucose level and deliver insulin in response to need.

• On another front, scientists are hopeful of one day developing immunotherapies that specifically block the attack of the immune system against the β cells, thus curbing or preventing Type I diabetes.

• Still other investigators are scurrying to unravel the defective pathways underlying Type II diabetes with the goal of finding new therapeutic targets to prevent, halt, or reverse this condition.

4. By making the cells more receptive to insulin (that is, by reducing insulin resistance)

Because none of these oral drugs deliver new insulin to the body, they cannot replace insulin injections for people with Type I diabetes. Furthermore, sometimes the weakened β cells of Type II diabetics eventually burn out and are no longer able to produce insulin. In such a case, the previously non–insulin dependent patient must be placed on insulin therapy for life.

New approaches to managing diabetes

Several new approaches are currently available for insulin-dependent diabetics that preclude the need for the one or two insulin injections daily.

• Implanted insulin pumps can deliver a prescribed amount of insulin on a regular basis, but the recipient must time meals with care to match the automatic insulin delivery.

• Pancreas transplants are also being performed more widely now, with increasing success rates. On the downside, recipients of pancreas transplants must take immunosuppressive drugs for life to prevent rejection of their donated organs. Also, donor organs are in short supply.

Current research on several fronts may dramatically change the approach to diabetic therapy in the near future. The following new treatments are on the horizon, most of which do away with the dreaded daily injections:

• Some methods under development circumvent the need for insulin injections by using alternate routes of administration that bypass the destructive digestive-tract enzymes. These include the use of inhaled powdered insulin and using ultrasound to force insulin into the skin from an insulin-impregnated patch.

disease frequently occur after 15 to 20 years despite treatment to prevent the short-term effects. These chronic complications, which account for the shorter life expectancy of diabetics, primarily involve degenerative disorders of the vasculature and nervous system. Cardiovascular lesions are the most common cause of premature death in diabetics. Heart disease and strokes occur with greater incidence than in nondiabetics. Because vascular lesions often develop in the kidneys and retinas of the eyes, diabetes is the leading cause of both kidney failure and blindness in the United States. Impaired delivery of blood to the extremities may cause these tissues to become gangrenous, and toes or even whole limbs may have to be amputated. In addition to circulatory problems, degenerative lesions in nerves lead to multiple neuropathies that result in dysfunction of the brain, spinal cord, and peripheral nerves. The latter is most often characterized by pain, numbness, and tingling, especially in the extremities.

Regular exposure of tissues to excess blood glucose over a prolonged time leads to tissue alterations responsible for the development of these long-range vascular and neural degenerative complications. Thus the best management for diabetes mellitus is to continuously keep blood glucose levels within normal limits to diminish the incidence of these chronic abnormalities. However, the blood glucose levels of diabetic patients on traditional therapy typically fluctuate over a broader range than normal, exposing their tissues to a moderately elevated blood glucose during a portion of each day. Fortunately, recent advances in understanding and learning how to manipulate underlying molecular defects in diabetes offer hope that more effective therapies will be developed within this decade to better manage or even cure existing cases and perhaps to prevent new cases of this devastating disease. (See the boxed feature on diabetes for current and potential future treatment strategies for this disorder.)

▮ Insulin excess causes brain-starving hypoglycemia.

Let us now look at the opposite of diabetes mellitus, insulin excess, which is characterized by **hypoglycemia** (low blood glucose) and can occur in two different ways. First, insulin excess can occur in a diabetic patient when too much insulin has been injected for the person's caloric intake and exercise level, resulting in so-called **insulin shock.** Second, an abnormally high blood insulin level may occur in a nondiabetic individual who has a β cell tumor or whose β cells are overresponsive to glucose. The true incidence of the latter condition, so-called **reactive hypoglycemia,** is a subject of intense controversy, because laboratory measurements to confirm the presence of low blood glucose during the time of symptoms have not been performed in most people who have been diagnosed as having the condition. In mild cases, the symptoms of hypoglycemia, such as tremor, fatigue, sleepiness, and inability to concentrate, are nonspecific. Because these symptoms could also be attributable to emotional problems or other factors, a definitive diagnosis based on symptoms alone is impossible to make.

The consequences of insulin excess are primarily manifestations of the effects of hypoglycemia on the brain. Recall that the brain relies on a continuous supply of blood glucose for its nourishment and that glucose uptake by the brain does not depend on insulin. With insulin excess, more glucose than necessary is driven into the other insulin-dependent cells of the body. The result is a lowering of the blood glucose level, so that not enough glucose is left in the blood to be delivered to the brain. In hypoglycemia, the brain literally starves. The symptoms, therefore, are primarily referable to depressed brain function, which, if severe enough, may rapidly progress to unconsciousness and death. People with overresponsive β cells usually do not become sufficiently hypoglycemic to manifest these more serious consequences, but they do show milder symptoms of depressed CNS activity.

The treatment of hypoglycemia depends on the cause. In the case of a diabetic with insulin overdose, at the first indica-

tion of a hypoglycemic attack something sugary should be ingested. Prompt treatment of severe hypoglycemia is imperative to prevent brain damage. Note that a diabetic can lose consciousness and die from either diabetic ketoacidotic coma caused by prolonged insulin deficiency or acute hypoglycemia caused by insulin shock. Fortunately, the other accompanying signs and symptoms differ sufficiently between the conditions to enable medical caretakers to administer appropriate therapy, either insulin or glucose. For example, ketoacidotic coma is accompanied by deep, labored breathing (in compensation for the metabolic acidosis) and fruity breath (from exhaled ketone bodies), whereas insulin shock is not.

Ironically, even though reactive hypoglycemia is characterized by a low blood-glucose level, people with this disorder are treated by limiting their intake of sugar and other glucose-yielding carbohydrates to prevent their β cells from overresponding to a high glucose intake. With low carbohydrate intake, the blood glucose does not rise as much during the absorptive state. Because blood glucose elevation is the primary regulator of insulin secretion, the β cells are not stimulated as much with a low-carbohydrate meal as with a typical meal. Accordingly, reactive hypoglycemia is less likely to occur. Giving a symptomatic individual with reactive hypoglycemia something sugary, as with a hypoglycemic diabetic, temporarily alleviates the symptoms. The blood glucose level is transiently restored to normal so that the brain's energy needs are once again satisfied. However, as soon as the extra glucose triggers further insulin release, the situation is merely aggravated.

▮ Glucagon in general opposes the actions of insulin.

Even though insulin plays a central role in controlling metabolic adjustments between the absorptive and postabsorptive states, the secretory product of the pancreatic islet α cells, **glucagon,** is also very important. Many physiologists view the insulin-secreting β cells and the glucagon-secreting α cells as a coupled endocrine system whose combined secretory output is a major factor in regulating fuel metabolism.

Glucagon affects many of the same metabolic processes that insulin influences, but in most cases glucagon's actions are opposite to those of insulin. The major site of action of glucagon is the liver, where it exerts a variety of effects on carbohydrate, fat, and protein metabolism.

Actions on carbohydrate

The overall effects of glucagon on carbohydrate metabolism result in an increase in hepatic glucose production and release and thus an increase in blood glucose levels. Glucagon exerts its hyperglycemic effects by decreasing glycogen synthesis, promoting glycogenolysis, and stimulating gluconeogenesis.

Actions on fat

Glucagon also antagonizes the actions of insulin with regard to fat metabolism by promoting fat breakdown and inhibiting triglyceride synthesis. Glucagon enhances hepatic ketone production (**ketogenesis**) by promoting the conversion of fatty acids to ketone bodies. Thus the blood levels of fatty acids and ketones increase under glucagon's influence.

Actions on protein

Glucagon inhibits hepatic protein synthesis and promotes degradation of hepatic protein. Stimulation of gluconeogenesis further contributes to glucagon's catabolic effect on hepatic protein metabolism. Glucagon promotes protein catabolism in the liver, but it does not have any significant effect on blood amino-acid levels because it does not affect muscle protein, the major protein store in the body.

▌ Glucagon secretion is increased during the postabsorptive state.

Considering the catabolic effects of glucagon on energy stores, you would be correct in assuming that glucagon secretion increases during the postabsorptive state and decreases during the absorptive state, just the opposite of insulin secretion. In fact, insulin is sometimes referred to as a "hormone of feasting" and glucagon as a "hormone of fasting." Insulin tends to put nutrients in storage when their blood levels are high, such as following a meal, whereas glucagon promotes catabolism of nutrient stores between meals to keep up the blood nutrient levels, especially blood glucose.

Like insulin secretion, the major factor regulating glucagon secretion is a direct effect of the blood glucose concentration on the endocrine pancreas. In this case, the pancreatic α cells increase glucagon secretion in response to a fall in blood glucose. The hyperglycemic actions of this hormone tend to raise the blood glucose level back to normal. Conversely, an increase in blood glucose concentration, such as occurs after a meal, inhibits glucagon secretion, which (similarly) tends to drop the blood glucose level back to normal.

▌ Insulin and glucagon work as a team to maintain blood glucose and fatty acid levels.

Thus a direct negative-feedback relationship exists between blood glucose concentration and both the β cells' and α cells' rates of secretion, but in opposite directions. An elevated blood-glucose level stimulates insulin secretion but inhibits glucagon secretion, whereas a fall in blood glucose level leads to decreased insulin secretion and increased glucagon secretion (● Figure 19-18). Because insulin lowers and glucagon raises blood glucose, the changes in secretion of these pancreatic hormones in response to deviations in blood glucose work together homeostatically to restore blood glucose levels to normal.

Similarly, a fall in blood fatty-acid concentration directly inhibits insulin output and stimulates glucagon output by the pancreas, both of which are negative-feedback control mechanisms to restore the blood fatty-acid level to normal.

The opposite effects exerted by blood concentrations of glucose and fatty acids on the pancreatic β and α cells are ap-

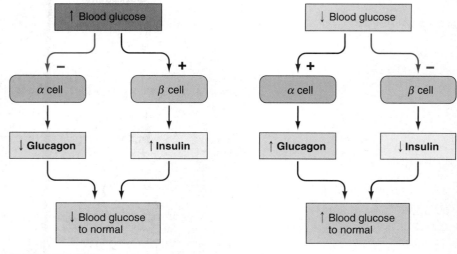

● FIGURE 19-18
Complementary interactions of glucagon and insulin

propriate for regulating the circulating levels of these nutrient molecules, because the actions of insulin and glucagon on carbohydrate and fat metabolism oppose one another. The effect of blood amino-acid concentration on the secretion of these two hormones is a different story. A rise in blood amino-acid concentration stimulates *both* insulin and glucagon secretion. Why this seeming paradox, because glucagon does not exert any effect on blood amino-acid concentration? The identical effect of high blood amino-acid levels on both insulin and glucagon secretion makes sense if you consider the concomitant effects these two hormones have on blood glucose levels (● Figure 19-19). If, during absorption of a protein-rich meal, the rise in blood amino acids stimulated only insulin secretion, hypoglycemia might result. Because little carbohydrate is available for absorption following consumption of a high-protein meal, the amino acid–induced increase in insulin secretion would drive too much glucose into the cells, causing a sudden, inappropriate drop in the blood glucose level. However, the simultaneous increase in glucagon secretion elicited by elevated blood amino-acid levels increases hepatic glucose production. Because the hyperglycemic effects of glucagon counteract the hypoglycemic actions of insulin, the net result is maintenance of normal blood-glucose levels (and prevention of hypoglycemic starvation of the brain) during absorption of a meal that is high in protein but low in carbohydrates.

▌ Glucagon excess can aggravate the hyperglycemia of diabetes mellitus.

No known clinical abnormalities are attributable to glucagon deficiency or excess per se. However, diabetes mellitus is frequently accompanied by excess glucagon secretion, because insulin is required for glucose to gain entry into the α cells, where it can exert control over glucagon secretion. As a result, diabetics frequently have a high rate of glucagon secretion concurrent with their insulin insufficiency because the elevated blood glucose cannot inhibit glucagon secretion as it

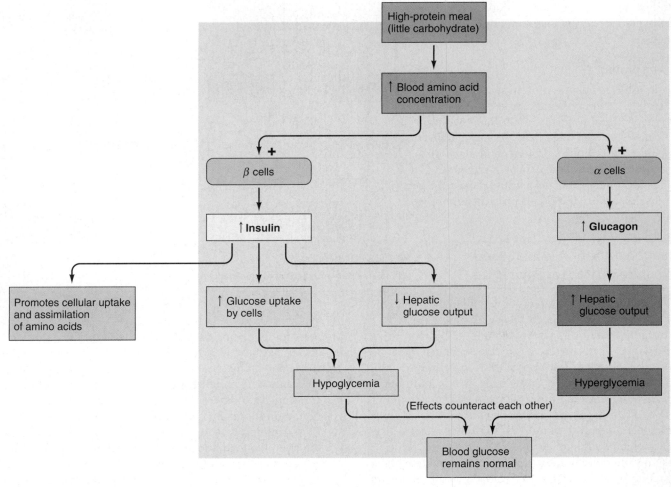

● FIGURE 19-19

Counteracting actions of glucagon and insulin on blood glucose during absorption of a high-protein meal

normally would. Because glucagon is a hormone that raises blood glucose, its excess intensifies the hyperglycemia of diabetes mellitus. For this reason, some insulin-dependent diabetics respond best to a combination of insulin and somatostatin therapy. By inhibiting glucagon secretion, somatostatin indirectly helps achieve better reduction of the elevated blood-glucose concentration than can be accomplished by insulin therapy alone.

▌ Epinephrine, cortisol, and growth hormone also exert direct metabolic effects.

The pancreatic hormones are the most important regulators of normal fuel metabolism. However, several other hormones exert direct metabolic effects, even though control of their secretion is keyed to factors other than transitions in metabolism between feasting and fasting states (▲ Table 19-6).

The stress hormones, epinephrine and cortisol, both increase blood levels of glucose and fatty acids through a variety of metabolic effects. In addition, cortisol mobilizes amino acids

by promoting protein catabolism. Neither hormone plays important roles in regulating fuel metabolism under resting conditions, but both are important for the metabolic responses to stress. During long-term starvation, cortisol also seems to help maintain blood glucose concentration.

Growth hormone (GH) has protein anabolic effects in muscle. In fact, this is one of its growth-promoting features. Although GH can elevate the blood levels of glucose and fatty acids, it is normally of little importance to the overall regulation of fuel metabolism. Deep sleep, stress, exercise, and severe hypoglycemia stimulate GH secretion, possibly to provide fatty acids as an energy source and spare glucose for the brain under these circumstances. GH, like cortisol, appears to help maintain blood glucose concentrations during starvation.

Although thyroid hormone increases the overall metabolic rate and has both anabolic and catabolic actions, changes in thyroid hormone secretion are usually not important for fuel homeostasis, for two reasons. First, control of thyroid hormone secretion is not directed toward maintaining nutrient levels in the blood. Second, the onset of thyroid hormone action is too slow to have any significant effect on the rapid adjustments required to maintain normal blood levels of nutrients.

Summary of Hormonal Control of Fuel Metabolism

Hormone	Major Metabolic Effects				Control of Secretion	
	Effect on blood glucose	Effect on blood fatty acids	Effect on blood amino acids	Effect on muscle protein	Major stimuli for secretion	Primary role in metabolism
Insulin	↓ +Glucose uptake +Glycogenesis −Glycogenolysis −Gluconeogenesis	↓ +Triglyceride synthesis −Lipolysis	↓ +Amino acid uptake	↑ +Protein synthesis −Protein degradation	↑Blood glucose ↑Blood amino acids	Primary regulator of absorptive and postabsorptive cycles
Glucagon	↑ +Glycogenolysis +Gluconeogenesis −Glycogenesis	↑ +Lipolysis −Triglyceride synthesis	No effect	No effect	↓Blood glucose ↑Blood amino acids	Regulation of absorptive and postabsorptive cycles in concert with insulin; protection against hypoglycemia
Epinephrine	↑ +Glycogenolysis +Gluconeogenesis −Insulin secretion +Glucagon secretion	↑ +Lipolysis	No effect	No effect	Sympathetic stimulation during stress and exercise	Provision of energy for emergencies and exercise
Cortisol	↑ +Gluconeogenesis −Glucose uptake by tissues other than brain; glucose sparing	↑ +Lipolysis	↑ +Protein degradation	↓ +Protein degradation	Stress	Mobilization of metabolic fuels and building blocks during adaptation to stress
Growth Hormone	↑ −Glucose uptake by muscles; glucose sparing	↑ +Lipolysis	↓ +Amino acid uptake	↓ +Protein synthesis −Protein degradation +Synthesis of DNA and RNA	Deep sleep Stress Exercise Hypoglycemia	Promotion of growth; normally little role in metabolism; mobilization of fuels plus glucose sparing in extenuating circumstances

↑ = increase
↓ = decrease

Note that, with the exception of the anabolic effects of GH on protein metabolism, all the metabolic actions of these other hormones are opposite to those of insulin. Insulin alone can reduce blood glucose and blood fatty acid levels, whereas glucagon, epinephrine, cortisol, and GH all increase blood levels of these nutrients. Because of this, these other hormones are considered **insulin antagonists.** Thus the main reason diabetes mellitus has such devastating metabolic consequences is that no other control mechanism is available to pick up the slack to promote anabolism when insulin activity is insufficient, so the catabolic reactions promoted by other hormones proceed unchecked. The only exception is protein anabolism stimulated by GH.

ENDOCRINE CONTROL OF CALCIUM METABOLISM

Besides regulating the concentration of organic nutrient molecules in the blood by manipulating anabolic and catabolic pathways, the endocrine system also regulates the plasma concentration of a number of inorganic electrolytes. As you already know, aldosterone controls Na^+ and K^+ concentrations in the ECF. Three other hormones—*parathyroid hormone, calcitonin,* and *vitamin D*—control calcium (Ca^{2+}) and phosphate (PO_4^{3-}) metabolism. These hormonal agents concern themselves with regulating plasma Ca^{2+}, and in the process, plasma PO_4^{3-} is also maintained. Plasma Ca^{2+} concentration is one of the most tightly controlled variables in the body. The need for the precise regulation of plasma Ca^{2+} stems from its critical influence on so many body activities.

▌ Plasma Ca^{2+} must be closely regulated to prevent changes in neuromuscular excitability.

About 99% of the Ca^{2+} in the body is in crystalline form within the skeleton and teeth. Of the remaining 1%, about 0.9% is found intracellularly within the soft tissues; less than 0.1% is present in the ECF. Approximately half of the plasma Ca^{2+} either is bound to plasma proteins and therefore restricted to the plasma or is complexed with PO_4^{3-} and not free to participate in chemical reactions. The other half of the plasma Ca^{2+} is freely diffusible and can readily pass into the interstitial fluid and interact with the cells. The free Ca^{2+} in the plasma and interstitial fluid is considered a single pool. Only this free Ca^{2+} is biologically active and subject to regulation; it constitutes less than one-thousandth of the total Ca^{2+} in the body.

This small, freely diffusible fraction of ECF Ca^{2+} plays a vital role in a number of essential activities, including the following:

1. *Neuromuscular excitability.* Even minor variations in the concentration of free ECF Ca^{2+} can have a profound and immediate impact on the sensitivity of excitable tissues. A fall in free Ca^{2+} results in overexcitability of nerves and muscles; conversely, a rise in free Ca^{2+} depresses neuromuscular excitability. These effects result from the influence of Ca^{2+} on membrane permeability to Na^+. A decrease in free Ca^{2+} increases Na^+ permeability, with the resultant influx of Na^+ moving the resting potential closer to threshold. Consequently, in the presence of **hypocalcemia** (low blood Ca^{2+}) excitable tissues may be brought to threshold by normally ineffective physiologic stimuli, so that skeletal muscles discharge and contract (go into spasm) "spontaneously" (in the absence of normal stimulation). If severe enough, spastic contraction of the respiratory muscles results in death by asphyxiation. **Hypercalcemia** (elevated blood Ca^{2+}) is also life threatening, because it causes cardiac arrhythmias and generalized depression of neuromuscular excitability.

2. *Excitation–contraction coupling in cardiac and smooth muscle.* Entry of ECF Ca^{2+} into cardiac and smooth muscle cells, resulting from increased Ca^{2+} permeability in response to an action potential, triggers the contractile mechanism. Calcium is also necessary for excitation–contraction coupling in skeletal muscle fibers, but in this case the Ca^{2+} is released from intracellular Ca^{2+} stores in response to an action potential. A significant part of the increase in cytosolic Ca^{2+} in cardiac muscle cells also derives from internal stores.

Note that a *rise in cytosolic Ca^{2+}* within a muscle cell causes contraction, whereas an *increase in free ECF Ca^{2+}* decreases neuromuscular excitability and reduces the likelihood of contraction. Unless one keeps this point in mind, it is difficult to understand why low plasma Ca^{2+} levels induce muscle hyperactivity when Ca^{2+} is necessary to switch on the contrac-

tile apparatus. We are talking about two different Ca^{2+} pools, which exert different effects.

3. *Stimulus–secretion coupling.* The entry of Ca^{2+} into secretory cells, which results from increased permeability to Ca^{2+} in response to appropriate stimulation, triggers the release of the secretory product by exocytosis. This process is important for the secretion of neurotransmitters by nerve cells and for peptide and catecholamine hormone secretion by endocrine cells.

4. *Maintenance of tight junctions between cells.* Calcium forms part of the intercellular cement that holds particular cells tightly together.

5. *Clotting of blood.* Calcium serves as a cofactor in several steps of the cascade of reactions that lead to clot formation.

In addition to these functions of free ECF Ca^{2+}, intracellular Ca^{2+} serves as a second messenger in many cells and is involved in cell motility and cilia action. Finally, the Ca^{2+} in bone and teeth is essential for the structural and functional integrity of these tissues.

Because of the profound effects of deviations in free Ca^{2+}, especially on neuromuscular excitability, the plasma concentration of this electrolyte is regulated with extraordinary precision. Let us see how.

▌ Control of Ca^{2+} metabolism includes regulation of both Ca^{2+} homeostasis and Ca^{2+} balance.

Maintenance of the proper plasma concentration of free Ca^{2+} differs from the regulation of Na^+ and K^+ in two important regards. Sodium and K^+ homeostasis is maintained primarily by regulating the urinary excretion of these electrolytes so that controlled output matches uncontrolled input. In the case of Ca^{2+}, in contrast, not all the ingested Ca^{2+} is absorbed from the digestive tract; instead, the extent of absorption is hormonally controlled and depends on the Ca^{2+} status of the body. In addition, bone serves as a large Ca^{2+} reservoir that can be drawn on to maintain the free plasma Ca^{2+} concentration within the narrow limits compatible with life should dietary intake become too low. Exchange of Ca^{2+} between the ECF and bone is also subject to control. Similar in-house stores are not available for Na^+ and K^+.

Regulation of Ca^{2+} metabolism depends on hormonal control of exchanges between the ECF and three other compartments: bone, kidneys, and intestine. Control of Ca^{2+} metabolism encompasses two aspects:

- First, regulation of **calcium homeostasis** involves the immediate adjustments required to maintain a *constant free plasma Ca^{2+} concentration* on a minute-to-minute basis. This is largely accomplished by rapid exchanges between the bone and ECF and to a lesser extent by modifications in urinary excretion of Ca^{2+}.

- Second, regulation of **calcium balance** involves the more slowly responding adjustments required to maintain a *constant total amount of Ca^{2+} in the body.* Control of Ca^{2+} balance ensures that Ca^{2+} intake is equivalent to Ca^{2+} excretion over the long term (weeks to months). Calcium balance is maintained by adjusting the extent of intestinal Ca^{2+} absorption and urinary Ca^{2+} excretion.

Parathyroid hormone (PTH), the principal regulator of Ca^{2+} metabolism, acts directly or indirectly on all three of these effector sites. It is the primary hormone responsible for maintenance of Ca^{2+} homeostasis and is essential for maintaining Ca^{2+} balance, although vitamin D also contributes in important ways to Ca^{2+} balance. The third Ca^{2+}-influencing hormone, calcitonin, is not essential for maintaining either Ca^{2+} homeostasis or balance. It serves a backup function during the rare times of extreme hypercalcemia. We will examine the specific effects of each of these hormonal systems in more detail.

▌ Parathyroid hormone raises free plasma Ca^{2+} levels by its effects on bone, kidneys, and intestine.

Parathyroid hormone (PTH) is a peptide hormone secreted by the **parathyroid glands,** four rice grain–sized glands located on the back surface of the thyroid gland, one in each corner. Like aldosterone, PTH *is essential for life.* The overall effect of PTH is to increase the Ca^{2+} concentration of plasma (and, accordingly, of the entire ECF), thereby preventing hypocalcemia. In the complete absence of PTH, death ensues within a few days, usually because of asphyxiation caused by hypocalcemic spasm of respiratory muscles. By its actions on bone, kidneys, and intestine, PTH raises the plasma Ca^{2+} level when it starts to fall so that hypocalcemia and its effects are normally avoided. This hormone also acts to lower plasma PO_4^{3-} concentration. We will consider each of these mechanisms, beginning with an overview of bone remodeling and PTH's actions on bone.

▌ Bone continuously undergoes remodeling.

Recall that 99% of the body's Ca^{2+} is in the skeleton. (See ▲ Table 19-7 for other functions of the skeleton.) Bone is a living tissue composed principally of an organic extracellular matrix impregnated with **hydroxyapatite crystals** consisting primarily of precipitated $Ca_3(PO_4)_2$ (calcium phosphate) salts (see p. 691). Normally, $Ca_3(PO_4)_2$ salts are in solution in the ECF, but the conditions within the bone are suitable for these salts to precipitate (crystallize) around the collagen fibers in the matrix. By mobilizing some of these Ca^{2+} stores in the bone, PTH raises the plasma Ca^{2+} concentration when it starts to fall.

Bone remodeling

Despite the apparent inanimate nature of bone, its constituents are continually being turned over. **Bone deposition** (formation) and **bone resorption** (removal) normally go on concurrently, so that bone is constantly being remodeled, much as people remodel buildings by tearing down walls and replacing them. Through remodeling, the adult human skeleton is completely regenerated an estimated every 10 years. Bone remodeling serves two purposes: (1) It keeps the skeleton appropriately "engineered" for maximum effectiveness in its mechanical uses, and (2) it helps maintain the plasma Ca^{2+} level. Let us examine in more detail the underlying mechanisms and controlling factors for each of these purposes.

Recall that three types of bone cells are present in bone (see pp. 691–692). The *osteoblasts* secrete the extracellular organic matrix within which the calcium phosphate crystals precipitate. The *osteocytes* are the retired osteoblasts imprisoned within the bony wall they have deposited around themselves. The *osteoclasts* resorb bone in their vicinity by releasing acids that dissolve the calcium phosphate crystals and enzymes that break down the organic matrix. Thus a constant cellular tug-of-war goes on in bone, with bone-forming osteoblasts countering the efforts of the bone-destroying osteoclasts. These construction and demolition crews, working side by side, continuously remodel bone. Throughout most of adult life, rates of bone formation and bone resorption are about equal, so total bone mass remains fairly constant during this period.

In a unique communication system, osteoblasts produce two related chemicals that control the activity of osteoclasts—**osteoprotegerin (OPG)** and **osteoprotegerin ligand (OPGL).** (A *ligand* is a small molecule that binds with a larger protein molecule; an example is an extracellular chemical messenger binding with a plasma membrane receptor.) *OPGL* binds to surface receptors on osteoclasts, increasing the activity of these cells. As a result, bone resorption is stepped up, and bone mass decreases. The *OPG* secreted by the osteoblasts into the matrix serves as a decoy, binding with OPGL and preventing it from revving up the osteoclasts. As a result, the matrix-making osteoblasts outpace the matrix-removing osteoclasts, so bone mass increases. The balance between OPGL and OPG thus is an important determinant of bone density. Importantly, scientists are currently unraveling the influence of various factors on this balance. For example, the female sex hormone estrogen, which helps preserve bone mass, stimulates activity of the OPG gene in osteoblasts.

Factors that influence mechanical effectiveness of bone

As a child grows, the bone builders keep ahead of the bone destroyers under the influence of GH and IGF-I (see p. 692).

> ▲ **TABLE 19-7**
> **Functions of the Skeleton**
>
> Support
>
> Protection of vital internal organs
>
> Assistance in body movement by giving attachment to muscles and providing leverage
>
> Manufacture of blood cells (bone marrow)
>
> Storage depot for Ca^{2+} and PO_4^{3-}, which can be exchanged with the plasma to maintain plasma concentrations of these electrolytes

Osteoporosis: The Bane of Brittle Bones

Osteoporosis, a decrease in bone density resulting from reduced deposition of the bone's organic matrix (see the accompanying figure), is a major health problem that affects 38 million people in the United States. The condition is especially prevalent among postmenopausal women. After menopause, women start losing 1% or more bone density each year. Skeletons of elderly women are typically only 50% to 80% as dense as at their peak at about age 35, whereas elderly men's skeletons retain 80% to 90% of their youthful density.

Osteoporosis is responsible for the greater incidence of bone fractures among women over the age of 50 than among the population at large. Because bone mass is reduced, the bones are more brittle and more susceptible to fracture in response to a fall, blow, or lifting action that normally would not strain stronger bones. For every 10% loss of bone mass, the risk of fracture doubles. Osteoporosis is the underlying cause of approximately 1.5 million fractures each year, of which 530,000 are vertebral fractures and 227,000 are hip fractures. The attendant medical and rehabilitation cost is $14 billion per year. The cost in pain, suffering, and loss of independence is not measurable. Half of all American women have spinal pain and deformity by age 75.

Types of osteoporosis

There appear to be two types of osteoporosis, caused by different mechanisms. Type I osteoporosis affects women soon after menopause and is characterized by vertebral crush fractures or fractures of the arm just above the wrist. It is hypothesized that these fractures occur as a result of the reduction in bone density that accompanies the estrogen deficiency of menopause. Diminished estrogen levels are associated with increased activity of the bone-dissolving osteoclasts and reduced activity of the bone-building

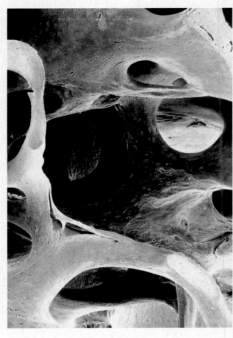

Normal bone

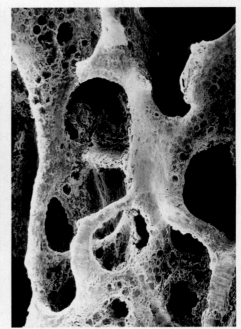

Osteoporotic bone

Mechanical stress also tips the balance in favor of bone deposition, causing bone mass to increase and the bones to strengthen. Mechanical factors adjust the strength of bone in response to the demands placed on it. The greater the physical stress and compression to which a bone is subjected, the greater the rate of bone deposition. For example, the bones of athletes are more massive and stronger than those of sedentary people.

By contrast, bone mass diminishes and the bones weaken when bone resorption gains a competitive edge over bone deposition in response to removal of mechanical stress. For example, bone mass decreases in people who undergo prolonged bed confinement or those in space flight. Early astronauts lost up to 20% of their bone mass during their time in orbit. Therapeutic exercises can limit or prevent such loss of bone.

Bone mass also decreases as a person ages. Bone density peaks when a person is in the 30s, then starts to decline after age 40. By 50 to 60 years of age, bone resorption often exceeds bone formation. The result is a reduction in bone mass known as **osteoporosis** (meaning "porous bones"). This bone-thinning condition is characterized by a diminished laying down of organic matrix as a result of reduced osteoblast activity and/or increased osteoclast activity rather than abnormal bone calcification. The underlying cause of osteoporosis is uncertain. Plasma Ca^{2+} and PO_4^{3-} levels are normal, as is PTH. Osteoporosis occurs with greatest frequency in postmenopausal women, suggesting that estrogen withdrawal plays a role. (For more details on osteoporosis, see the accompanying boxed feature, ▶ A Closer Look at Exercise Physiology.)

Overall effects of PTH on bone

In addition to the factors geared toward controlling the mechanical effectiveness of bone, throughout life PTH uses bone as a "bank" from which it withdraws Ca^{2+} as needed to maintain the plasma Ca^{2+} level. Parathyroid hormone has two major effects on bone that raise plasma Ca^{2+} concentration. First, it induces a fast Ca^{2+} efflux into the plasma from the small, *labile pool* of Ca^{2+} in the bone fluid. Second, by stimulating bone dissolution, it promotes a slow transfer into the plasma of both Ca^{2+} and PO_4^{3-} from the *stable pool* of bone minerals in the bone itself. As a result, ongoing bone remodeling is tipped in favor of bone resorption over bone deposition. Let's examine more thoroughly PTH's actions in mobilizing Ca^{2+} from its labile and stable pools in bone.

osteoblasts. Type II osteoporosis occurs in men as well as women, although it affects females twice as often as males. It is characterized by hip fractures as well as fractures at other sites. Because Type II osteoporosis occurs later in life, the decreased ability to absorb Ca^{2+} associated with advancing age may play a key role in the development of the condition, although estrogen deficiency probably also contributes, accounting for the higher incidence in women.

Drug therapy for osteoporosis

Estrogen replacement therapy, Ca^{2+} supplementation, and a regular weight-bearing exercise program traditionally have been the most common therapeutic approaches used to minimize or reverse bone loss. However, estrogen therapy has been linked to an increased risk of breast cancer and cardiovascular disease, and Ca^{2+} alone has not been as effective in halting bone thinning as was once hoped. The Food and Drug Administration has recently approved three new drugs for treating osteoporosis: alendronate, calcitonin in a nasal-spray form, and raloxifene. *Alendronate (Fosamax)* is the first nonhormonal osteoporosis drug. It works by blocking osteoclasts' bone-destroying actions. *Calcitonin*, the thyroid C-cell hormone that slows osteoclast activity, was used in the past to treat advanced osteoporosis, but it had to be injected daily, a deterrent to patient compliance. Now calcitonin is available in a more patient-friendly nasal spray. *Raloxifene (Evista)* belongs to a new class of drugs known as *selective estrogen receptor modulators (SERMs)*. Raloxifene does not bind with estrogen receptors in reproductive organs, but it does bind with estrogen receptors outside of the reproductive system, such as in the bone. Through this selective receptor binding, raloxifene mimics estrogen's beneficial effects on the bone to provide protection against osteoporosis while avoiding estrogen's potentially harmful effects on reproductive organs, such as increased risk of breast cancer. Another group of drugs with some promise for treating osteoporosis is the *statins*, which are already commonly used as cholesterol-lowering agents. The statins stimulate osteoblast activity, promoting bone formation and reducing the fracture rate.

Benefits of exercise on bone

Despite advances in osteoporosis therapy, treatment is still often less than satisfactory, and all the current therapeutic agents are associated with some undesirable side effects. Therefore, prevention is by far the best approach to managing this disease. Development of strong bones to begin with before menopause through a good, Ca^{2+}-rich diet and adequate exercise appears to be the best preventive measure. A large reservoir of bone at midlife may delay the clinical manifestations of osteoporosis in later life. Continued physical activity throughout life appears to retard or prevent bone loss, even in the elderly.

It is well documented that osteoporosis can result from disuse—that is, from reduced mechanical loading of the skeleton. Space travel has clearly shown that lack of gravity results in a decrease in bone density. Studies of athletes, by contrast, demonstrate that weight-bearing physical activity increases bone density. Within groups of athletes, bone density correlates directly with the load the bone must bear. If one looks at athletes' femurs (thigh bones), the greatest bone density is found in weight lifters, followed in order by throwers, runners, soccer players, and finally swimmers. In fact, the bone density of swimmers does not differ from that of nonathletic controls. Swimming does not place any strain on bones. The bone density in the playing arm of male tennis players has been found to be as much as 35% greater than in their other arm; female tennis players have been found to have 28% greater density in their playing arm than in their other arm. One study found that very mild activity in nursing home patients, whose average age was 82 years, not only slowed bone loss but even resulted in bone buildup over a 36-month period. Thus exercise is a good defense against osteoporosis.

▌ PTH'S immediate effect is to promote the transfer of Ca^{2+} from bone fluid into plasma.

Most bone is organized into **osteon** units, each of which consists of a **central canal** surrounded by concentrically arranged **lamellae.** Lamellae are layers of osteocytes entombed within the bone they have deposited around themselves (● Figure 19-20). The osteons typically run parallel to the long axis of the bone. Blood vessels penetrate the bone from either the outer surface or the marrow cavity and run through the central canals. Osteoblasts are present along the outer surface of the bone and along the inner surfaces lining the central canals. The surface osteoblasts and entombed osteocytes are connected by an extensive network of small, fluid-containing canals, the **canaliculi**, which allow substances to be exchanged between trapped osteocytes and the circulation. These small canals also contain long, filmy cytoplasmic extensions of osteocytes and osteoblasts that are connected to each other, much as if these cells were "holding hands." The interconnecting cell network, which is called the **osteocytic-osteoblastic bone membrane,** separates the mineralized bone itself from the plasma within the central canals (● Figure 19-21a). The small, labile pool of Ca^{2+} is in the **bone fluid** that lies between this bone membrane and the adjacent bone, both within the canaliculi and along the surface of the central canal.

PTH's earliest effect is to activate membrane-bound Ca^{2+} pumps located in the plasma membranes of the osteocytes and osteoblasts. These pumps promote movement of Ca^{2+}, without the accompaniment of PO_4^{3-}, from the bone fluid into these cells. From here, this Ca^{2+} is transferred into the plasma within the central canal. Thus PTH stimulates the transfer of Ca^{2+} from the bone fluid across the osteocytic–osteoblastic bone membrane into the plasma. Movement of Ca^{2+} out of the labile pool across the bone membrane accounts for the fast exchange between the bone and the plasma (● Figure 19-21b). Because of the large surface area of the osteocytic–osteoblastic membrane, small movements of Ca^{2+} across individual cells are amplified into large Ca^{2+} fluxes between the bone and the plasma.

After Ca^{2+} is pumped out, the bone fluid is replenished with Ca^{2+} from the partially mineralized bone along the adjacent bone surface. Thus the fast exchange of Ca^{2+} does not involve resorption of completely mineralized bone, and bone mass is not decreased. Through this means, PTH draws Ca^{2+} out of the "quick-cash branch" of the bone bank and rapidly increases the plasma Ca^{2+} level without actually entering the bank (that

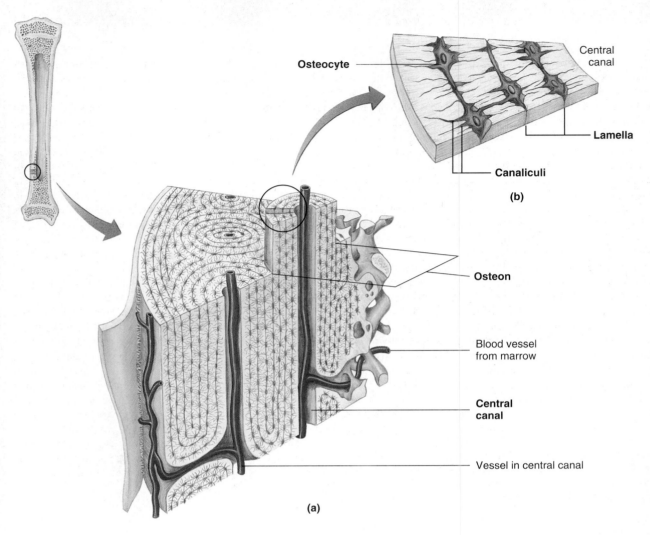

Osteocyte

Central canal

Lamella

Canaliculi

(b)

Osteon

Blood vessel from marrow

Central canal

Vessel in central canal

(a)

● FIGURE 19-20

Organization of bone into osteons

(a) An osteon, the structural unit of most bone, consists of concentric lamellae (layers of osteocytes entombed by the bone they have deposited around themselves) surrounding a central canal. A small blood vessel branch traverses the central canal. (b) A magnification of the lamellae within an osteon. A network of small canals, the canaliculi, interconnect the entombed osteocytes with each other and with the central canal. Long cytoplasmic processes extend from osteocyte to osteocyte within the canaliculi.

(SOURCE: Modified and redrawn with permission from *Human Anatomy and Physiology*, 3rd Edition, by A. Spence and E. Mason. Copyright © 1987 by Benjamin-Cummings Publishing Company. Reprinted by permission of Addison-Wesley Educational Publishers, Inc.)

is, without breaking down mineralized bone itself). Under normal conditions, this exchange is much more important for maintaining plasma Ca^{2+} concentration than is the slow exchange.

▌ PTH's chronic effect is to promote localized dissolution of bone to release Ca^{2+} into plasma.

Under conditions of chronic hypocalcemia, such as may occur with dietary Ca^{2+} deficiency, PTH influences the slow exchange of Ca^{2+} between the bone itself and the ECF by promoting actual localized dissolution of bone. It does so by stimulating osteoclasts to gobble up bone, increasing the formation of more osteoclasts, and transiently inhibiting the bone-forming activity of

the osteoblasts. Bone contains so much Ca^{2+} compared to the plasma (more than 1,000 times as much) that even when PTH promotes increased bone resorption, there are no immediate discernible effects on the skeleton, because such a tiny amount of bone is affected. Yet the negligible amount of Ca^{2+} "borrowed" from the bone bank can be lifesaving in terms of restoring the free plasma Ca^{2+} level to normal. The borrowed Ca^{2+} is then redeposited in the bone at another time when Ca^{2+} supplies are more abundant. Meanwhile, the plasma Ca^{2+} level has been maintained without sacrificing bone integrity. However, prolonged excess PTH secretion over months or years eventually leads to the formation of cavities throughout the skeleton that are filled with very large, overstuffed osteoclasts.

When PTH promotes dissolution of the calcium phosphate crystals in bone to harvest their Ca^{2+} content, both Ca^{2+} and

PO_4^{3-} are released into the plasma. An elevation in plasma PO_4^{3-} is undesirable, but PTH deals with this dilemma by its actions on the kidneys, a topic to which we now turn our attention.

∎ PTH acts on the kidneys to conserve Ca^{2+} and eliminate PO_4^{3-}.

Parathyroid hormone stimulates Ca^{2+} conservation and promotes PO_4^{3-} elimination by the kidneys during urine formation. Under the influence of PTH, the kidneys can reabsorb more of the filtered Ca^{2+}, so less Ca^{2+} escapes into urine. This effect increases the plasma Ca^{2+} level and decreases urinary Ca^{2+} losses. (It would be counterproductive to dissolve bone to obtain more Ca^{2+} only to lose it in urine.)

At the same time that it stimulates renal Ca^{2+} reabsorption, PTH decreases PO_4^{3-} reabsorption, thus increasing urinary PO_4^{3-} excretion. As a result, PTH reduces plasma PO_4^{3-} levels at the same time it increases plasma Ca^{2+} concentrations.

This PTH-induced removal of extra PO_4^{3-} from the body fluids is essential for preventing reprecipitation of the Ca^{2+} freed from the bone. Because of the solubility characteristics of $Ca_3(PO_4)_2$ salt, the product of the plasma concentration of Ca^{2+} times the plasma concentration of PO_4^{3-} must remain roughly constant. Therefore, an inverse relationship exists between the plasma concentrations of Ca^{2+} and PO_4^{3-}; for example, when the plasma PO_4^{3-} level rises, some plasma Ca^{2+} is forced back into the bone through hydroxyapatite crystal formation, reducing the plasma Ca^{2+} level and keeping constant the calcium phosphate product. This inverse relationship occurs because the concentrations of free Ca^{2+} and PO_4^{3-} ions in the ECF are in equilibrium with the bone crystals.

Recall that both Ca^{2+} and PO_4^{3-} are released from the bone when PTH promotes bone dissolution. Because PTH is secreted only when plasma Ca^{2+} falls below normal, the released Ca^{2+} is needed to restore plasma Ca^{2+} to normal, yet the released PO_4^{3-} tends to raise plasma PO_4^{3-} levels above normal. If the plasma PO_4^{3-} level were allowed to rise above normal, some of the plasma Ca^{2+} would have to be redeposited back in the bone along with the PO_4^{3-} to keep the calcium phosphate product constant. This redeposition of Ca^{2+} would lower plasma Ca^{2+}, just the opposite of the needed effect. Therefore, PTH acts on the kidneys to decrease the reabsorption of PO_4^{3-} by the renal tubules. This increases urinary excretion of PO_4^{3-} and lowers its plasma concentration, even though extra PO_4^{3-} is being released from the bone into the

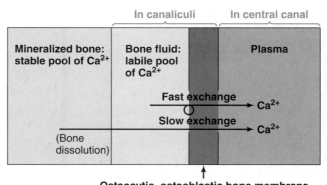

● **FIGURE 19-21**

Relationship of mineralized bone, bone cells, bone fluid, and the plasma
(a) Schematic representation of the osteocytic–osteoblastic bone membrane. The entombed osteocytes and surface osteoblasts are interconnected by long cytoplasmic processes that extend from these cells and connect to each other within the canaliculi. This interconnecting cell network, the osteocytic–osteoblastic bone membrane, separates the mineralized bone from the plasma in the central canal. Bone fluid lies between the membrane and the mineralized bone. (b) Schematic representation of fast and slow exchange of Ca^{2+} between the bone and the plasma.

blood. Such action prevents the self-defeating redeposition of released Ca^{2+} back into the bone.

The third important action of PTH on the kidneys (besides increasing Ca^{2+} reabsorption and decreasing PO_4^{3-} reabsorption) is to enhance the activation of vitamin D by the kidneys.

∎ PTH indirectly promotes absorption of Ca^{2+} and PO_4^{3-} by the intestine.

Although PTH has no direct effect on the intestine, it indirectly increases both Ca^{2+} and PO_4^{3-} absorption from the small intestine by playing a role in vitamin D activation. This vitamin, in turn, directly increases intestinal absorption of Ca^{2+} and PO_4^{3-}.

The primary regulator of PTH secretion is the plasma concentration of free Ca^{2+}.

All the effects of PTH raise the plasma Ca^{2+} levels. Appropriately, PTH secretion is increased in response to a fall in plasma Ca^{2+} concentration and decreased by a rise in plasma Ca^{2+} levels. The secretory cells of the parathyroid glands are directly and exquisitely sensitive to changes in free plasma Ca^{2+}. Because PTH regulates plasma Ca^{2+} concentration, this relationship forms a simple negative-feedback loop for controlling PTH secretion without involving any nervous or other hormonal intervention (● Figure 19-22).

Calcitonin lowers the plasma Ca^{2+} concentration but is not important in the normal control of Ca^{2+} metabolism.

Calcitonin, the hormone produced by the C cells of the thyroid gland, also exerts an influence on plasma Ca^{2+} levels. Like PTH, calcitonin has two effects on bone, but in this case both effects *decrease* plasma Ca^{2+} levels. First, on a short-term basis calcitonin decreases Ca^{2+} movement from the bone fluid into the plasma. Second, on a long-term basis calcitonin decreases bone resorption by inhibiting the activity of osteoclasts. The suppression of bone resorption lowers plasma PO_4^{3-} levels as well as reduces plasma Ca^{2+} concentration. The hypocalcemic and hypophosphatemic effects of calcitonin are due entirely to this hormone's actions on bone. It has no effect on the kidneys or intestine.

As with PTH, the primary regulator of calcitonin release is the free plasma Ca^{2+} concentration, but in contrast to its effect on PTH release, an increase in plasma Ca^{2+} stimulates calcitonin secretion and a fall in plasma Ca^{2+} inhibits calcitonin secretion (● Figure 19-22). Because calcitonin reduces plasma Ca^{2+} levels, this system constitutes a second simple negative-feedback control over plasma Ca^{2+} concentration, one that is opposed to the PTH system.

Most evidence suggests, however, that calcitonin plays little or no role in the normal control of Ca^{2+} or PO_4^{3-} metabolism. Although calcitonin protects against hypercalcemia, this condition rarely occurs under normal circumstances. Moreover, neither thyroid removal nor calcitonin-secreting tumors alter circulating levels of Ca^{2+} or PO_4^{3-}, implying that this hormone is not normally essential for maintaining Ca^{2+} or PO_4^{3-} homeostasis. Calcitonin may, however, play a role in protecting skeletal integrity when there is a large Ca^{2+} demand, such as during pregnancy or breast-feeding. Furthermore, some experts speculate that calcitonin may hasten the storage of newly absorbed Ca^{2+} following a meal. Gastrointestinal hormones secreted during digestion of a meal have been shown to stimulate the release of calcitonin.

Vitamin D is actually a hormone that increases calcium absorption in the intestine.

The final factor involved in regulating of Ca^{2+} metabolism is **cholecalciferol,** or **vitamin D,** a steroidlike compound essential for Ca^{2+} absorption in the intestine. Strictly speaking, vitamin D should be considered a hormone, because the body can produce it in the skin from a precursor related to cholesterol (7-dehydrocholesterol) on exposure to sunlight. It is subsequently released into the blood to act at a distant target site, the intestine. The skin, therefore, is actually an endocrine gland and vitamin D a hormone. Traditionally, however, this chemical messenger has been considered a vitamin, for two reasons. First, it was originally discovered and isolated from a dietary source and tagged as a vitamin. Second, even though the skin would be an adequate source of vitamin D if it were exposed to sufficient sunlight, indoor dwelling and clothing in response to cold weather and social customs preclude significant exposure of the skin to sunlight in the United States and many other parts of the world most of the time. At least part of the essential vitamin D must therefore be derived from dietary sources.

Activation of vitamin D

Regardless of its source, vitamin D is biologically inactive when it first enters the blood from either the skin or the digestive tract. It must be activated by two sequential biochemical alterations that involve the addition of two hydroxyl (—OH) groups (● Figure 19-23). The first of these reactions occurs in the liver and the second in the kidneys. The end result is production of the active form of vitamin D, *1,25-(OH)$_2$-vitamin D$_3$*, also known as *calcitriol*. The kidney enzymes involved in the second step of vitamin D activation are stimulated by PTH in response to a fall in plasma Ca^{2+}. To a lesser extent, a fall in plasma PO_4^{3-} also enhances the activation process.

● **FIGURE 19-22**

Negative-feedback loops controlling parathyroid hormone (PTH) and calcitonin secretion

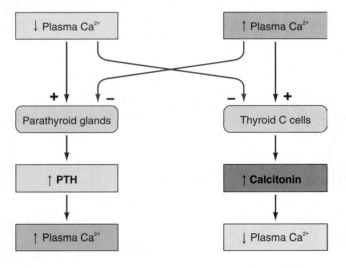

Function of vitamin D

The most dramatic and biologically important effect of activated vitamin D is to increase Ca^{2+} absorption in the intestine. Unlike most dietary constituents, dietary Ca^{2+} is not indiscriminately absorbed by the digestive system. In fact, the majority of ingested Ca^{2+} is typically not absorbed but is lost instead in the feces. When needed, more dietary Ca^{2+} is absorbed into the plasma under the influence of vitamin D. Independently of its effects on Ca^{2+} transport, the active form of vitamin D also increases intestinal PO_4^{3-} absorption. Furthermore, vitamin D increases the responsiveness of bone to PTH. Thus vitamin D and PTH have a closely interdependent relationship (● Figure 19-24).

PTH is principally responsible for controlling Ca^{2+} homeostasis, because the actions of vitamin D are too sluggish for it to contribute substantially to the minute-to-minute regulation of plasma Ca^{2+} concentration. However, both PTH and vitamin D are essential to Ca^{2+} balance, the process ensuring that, over the long term, Ca^{2+} input into the body is equivalent to Ca^{2+} output. When dietary Ca^{2+} intake is reduced, the resultant transient fall in plasma Ca^{2+} level stimulates PTH secretion. The increased PTH has two effects that are important for maintaining Ca^{2+} balance: (1) It stimulates Ca^{2+} reabsorption by the kidneys, thereby decreasing Ca^{2+} output; and (2) it activates vitamin D, which increases the efficiency of uptake of ingested Ca^{2+}. Because PTH also promotes bone resorption, a substantial loss of bone minerals occurs if Ca^{2+} intake is reduced for a prolonged period, even though bone is not directly involved in maintaining Ca^{2+} input and output in balance.

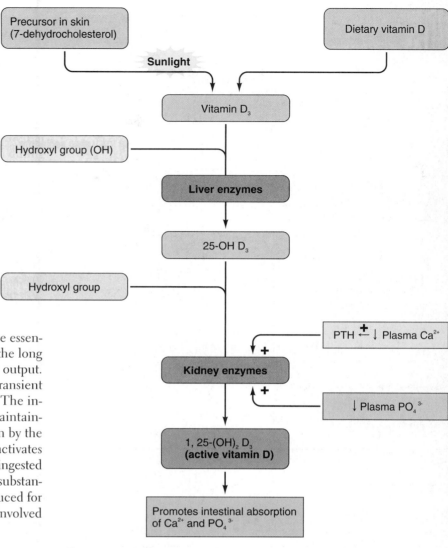

● **FIGURE 19-23**
Activation of vitamin D

▌Phosphate metabolism is controlled by the same mechanisms that regulate Ca^{2+} metabolism.

Plasma PO_4^{3-} concentration is not as tightly controlled as plasma Ca^{2+} concentration. Phosphate is regulated directly by vitamin D and indirectly by the plasma Ca^{2+}–PTH feedback loop. To illustrate, a fall in plasma PO_4^{3-} concentration exerts a twofold effect to help raise the circulating PO_4^{3-} level back to normal (● Figure 19-25). First, because of the inverse relationship between the PO_4^{3-} and Ca^{2+} concentrations in the plasma, a fall in plasma PO_4^{3-} increases plasma Ca^{2+}, which directly suppresses PTH secretion. In the presence of reduced PTH, PO_4^{3-} reabsorption by the kidneys increases, returning plasma PO_4^{3-} concentration toward normal. Second, a fall in plasma PO_4^{3-} also increases activation of vitamin D, which then promotes PO_4^{3-} absorption in the intestine. This further helps alleviate the initial hypophosphatemia. Note that these changes do not compromise Ca^{2+} balance. Although the increase in activated vitamin D stimulates Ca^{2+} absorption, the concurrent fall in PTH produces a compensatory increase in urinary Ca^{2+} excretion because less of the filtered Ca^{2+} is reabsorbed.

▌Disorders in Ca^{2+} metabolism may arise from abnormal levels of PTH or vitamin D.

The primary disorders that affect Ca^{2+} metabolism are too much or too little PTH or a deficiency of vitamin D.

PTH hypersecretion

Excess PTH secretion, or **hyperparathyroidism,** which is usually caused by a hypersecreting tumor in one of the parathyroid glands, is characterized by hypercalcemia and hypophosphatemia. The affected individual can be asymptomatic or symptoms can be severe, depending on the magnitude of the problem. The following are among the possible consequences:

• Hypercalcemia reduces the excitability of muscle and nervous tissue, leading to muscle weakness and neurologic disorders, including decreased alertness, poor memory, and depression. Cardiac disturbances may also occur.

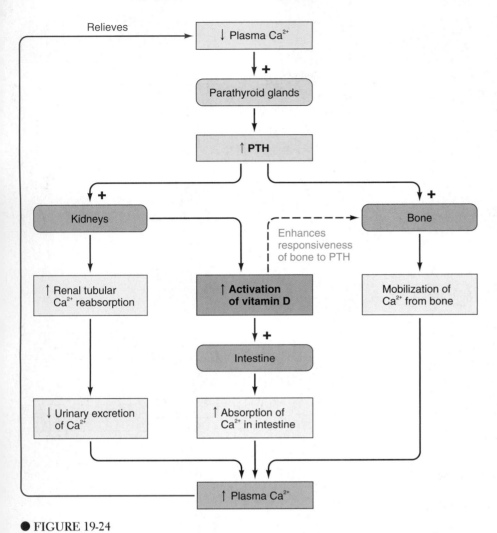

● **FIGURE 19-24**
Interactions between PTH and vitamin D in controlling plasma calcium

- Excessive mobilization of Ca^{2+} and PO_4^{3-} from skeletal stores leads to thinning of bone, which may result in skeletal deformities and increased incidence of fractures.
- An increased incidence of Ca^{2+}-containing kidney stones occurs because the excess quantity of Ca^{2+} being filtered through the kidneys may precipitate and form stones. These stones may impair renal function. Passage of the stones through the ureters causes extreme pain. Because of these potential multiple consequences, hyperparathyroidism has been called a disease of "bones, stones, and abdominal groans."
- To further account for the "abdominal groans," hypercalcemia can cause digestive disorders such as peptic ulcers, nausea, and constipation.

PTH hyposecretion

Because of the parathyroid glands' close anatomic relation to the thyroid, the most common cause of deficient PTH secretion, or **hypoparathyroidism,** used to be inadvertent removal of the parathyroid glands (before doctors knew about their existence) during surgical removal of the thyroid gland (for treatment of thyroid disease). If all the parathyroid tissue was re-

moved, these patients died, of course, because PTH is essential for life. In fact, physicians were puzzled why some patients died soon after thyroid removal even though no surgical complications were apparent. Now that the location and importance of the parathyroid glands have been discovered, surgeons are careful to leave parathyroid tissue during thyroid removal. Rarely, PTH hyposecretion occurs as a result of an autoimmune attack against the parathyroid glands.

Hypoparathyroidism leads to hypocalcemia and hyperphosphatemia. The symptoms are primarily caused by increased neuromuscular excitability from the reduced level of free plasma Ca^{2+}. In the complete absence of PTH, death is imminent because of hypocalcemic spasm of respiratory muscles. With a relative deficiency rather than a complete absence of PTH, milder symptoms of increased neuromuscular excitability are evident. Muscle cramps and twitches derive from spontaneous activity in the motor nerves, whereas tingling and pins-and-needles sensations result from spontaneous activity in the sensory nerves. Mental changes include irritability and paranoia.

Vitamin D deficiency

The major consequence associated with vitamin D deficiency is impaired intestinal absorption of Ca^{2+}. In the face of reduced Ca^{2+} uptake, PTH maintains the plasma Ca^{2+} level at the expense of the bones. As a result, the bone matrix is not properly mineralized, because Ca^{2+} salts are not available for deposition. The demineralized bones become soft and deformed, bowing under the pressures of weight bearing, especially in children. This condition is known as **rickets** in children and **osteomalacia** in adults.

CHAPTER IN PERSPECTIVE: FOCUS ON HOMEOSTASIS

A number of peripherally located endocrine glands play key roles in maintaining homeostasis, primarily by means of their regulatory influences over the rate of various metabolic reactions and over electrolyte balance. These endocrine glands all secrete hormones in response to specific stimuli. The hormones in turn exert effects that act in negative-feedback fashion to resist the change that induced their secretion, thus maintaining stability in the internal environment. The specific contributions of the peripheral endocrine glands to homeostasis include the following:

- Two closely related hormones secreted by the thyroid gland, tetraiodothyronine (T_4) and tri-iodothyronine (T_3), increase the overall metabolic rate. Not only does this action influence the rate at which cells use nutrient molecules and O_2 within the internal en-

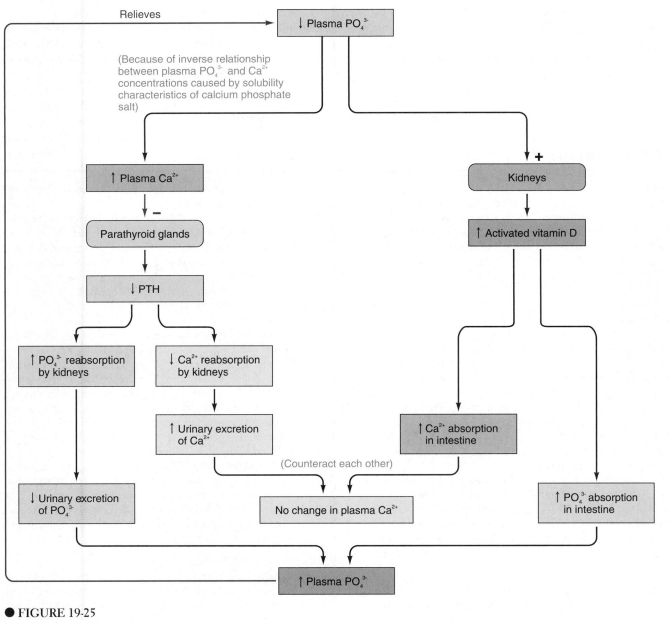

Relieves

↓ Plasma PO₄³⁻

(Because of inverse relationship between plasma PO₄³⁻ and Ca²⁺ concentrations caused by solubility characteristics of calcium phosphate salt)

↑ Plasma Ca²⁺

Kidneys

−

Parathyroid glands

+

↑ Activated vitamin D

↓ PTH

↑ PO₄³⁻ reabsorption by kidneys

↓ Ca²⁺ reabsorption by kidneys

↑ Urinary excretion of Ca²⁺

↑ Ca²⁺ absorption in intestine

(Counteract each other)

↓ Urinary excretion of PO₄³⁻

No change in plasma Ca²⁺

↑ PO₄³⁻ absorption in intestine

↑ Plasma PO₄³⁻

● FIGURE 19-25
Control of plasma phosphate

vironment, but it also produces heat, which helps maintain body temperature.

● The adrenal cortex secretes three classes of hormones. Aldosterone, the primary mineralocorticoid, is essential for Na⁺ and K⁺ balance. Because of Na⁺'s osmotic effect, Na⁺ balance is critical to maintaining the proper ECF volume and arterial blood pressure. This action is essential for life. Without aldosterone's Na⁺- and H₂O-conserving effect, so much plasma volume would be lost in the urine that death would quickly ensue. Maintaining K⁺ balance is essential for homeostasis because changes in extracellular K⁺ profoundly impact neuromuscular excitability, jeopardizing normal heart function, among other detrimental effects.

● Cortisol, the primary glucocorticoid secreted by the adrenal cortex, increases the plasma concentrations of glucose, fatty acids, and amino acids above normal. Although these actions destabilize

the concentrations of these molecules in the internal environment, they indirectly contribute to homeostasis by making the molecules readily available as energy sources or building blocks for tissue repair to help the body adapt to stressful situations.

● The sex hormones secreted by the adrenal cortex do not contribute to homeostasis.

● The major hormone secreted by the adrenal medulla, epinephrine, generally reinforces activities of the sympathetic nervous system. It contributes to homeostasis directly by its role in blood pressure regulation. Epinephrine also contributes to homeostasis indirectly by helping prepare the body for peak physical responsiveness in fight-or-flight situations. This includes increasing the plasma concentrations of glucose and fatty acids above normal, providing additional energy sources for increased physical activity.

- The two major hormones secreted by the endocrine pancreas, insulin and glucagon, are important in shifting metabolic pathways between the absorptive and postabsorptive states, which maintains the appropriate plasma levels of nutrient molecules.

- Parathyroid hormone from the parathyroid glands is critical to maintaining plasma concentration of Ca^{2+}. PTH is essential for life because of Ca^{2+}'s effect on neuromuscular excitability. In the absence of PTH, death rapidly occurs from asphyxiation caused by pronounced spasms of the respiratory muscles.

CHAPTER SUMMARY

Thyroid Gland
- The thyroid gland contains two types of endocrine secretory cells: (1) follicular cells, which produce the iodine-containing hormones, T_4 (thyroxine or tetraiodothyronine) and T_3 (tri-iodothyronine), collectively known as *thyroid hormone,* and (2) C cells, which synthesize a Ca^{2+}-regulating hormone, calcitonin.
- All steps of thyroid hormone synthesis take place on the large thyroglobulin molecules within the colloid, an "inland" extracellular site located within the interior of the thyroid follicles. Thyroid hormone is secreted by means of the follicular cells phagocytizing a piece of colloid and freeing T_4 and T_3, which diffuse across the plasma membrane and enter the blood.
- Thyroid hormone is the primary determinant of the overall metabolic rate of the body. By accelerating the metabolic rate of most tissues, it increases heat production. Thyroid hormone also enhances the actions of the chemical mediators of the sympathetic nervous system. Through this and other means, thyroid hormone indirectly increases cardiac output. Finally, thyroid hormone is essential for normal growth as well as the development and function of the nervous system.
- Thyroid hormone secretion is regulated by a negative-feedback system between hypothalamic TRH, anterior pituitary TSH, and thyroid gland T_3 and T_4. The feedback loop maintains thyroid hormone levels relatively constant. Cold exposure in newborn infants is the only input to the hypothalamus known to be effective in increasing TRH and thereby thyroid hormone secretion.

Adrenal Glands
- Each adrenal gland (of the pair) consists of two separate endocrine organs—an outer, steroid-secreting adrenal cortex and an inner, catecholamine-secreting adrenal medulla.
- The adrenal cortex secretes three different categories of steroid hormones: mineralocorticoids (primarily aldosterone), glucocorticoids (primarily cortisol), and adrenal sex hormones (primarily the weak androgen, dehydroepiandrosterone).
- Aldosterone regulates Na^+ and K^+ balance and is important for blood pressure homeostasis, which is achieved secondarily by the osmotic effect of Na^+ in maintaining the plasma volume, a lifesaving effect.
- Control of aldosterone secretion is related to Na^+ and K^+ balance and to blood pressure regulation, and is not influenced by ACTH.
- Cortisol helps regulate fuel metabolism and is important in stress adaptation. It increases blood levels of glucose, amino acids, and fatty acids and spares glucose for use by the glucose-dependent brain. The mobilized organic molecules are available for use as needed for energy or for repair of injured tissues.
- Cortisol secretion is regulated by a negative-feedback loop involving hypothalamic CRH and pituitary ACTH. The most potent stimulus for increasing activity of the CRH/ACTH/cortisol axis is stress. Cortisol also displays a characteristic diurnal rhythm.

- Dehydroepiandrosterone is responsible for the sex drive and growth of pubertal hair in females.
- The adrenal medulla is composed of modified sympathetic postganglionic neurons, which secrete the catecholamine epinephrine into the blood in response to sympathetic stimulation. For the most part, epinephrine reinforces the sympathetic system in mounting general systemic "fight-or-flight" responses and in maintaining arterial blood pressure. Epinephrine also exerts important metabolic effects, namely increasing blood glucose and blood fatty acids.
- The primary stimulus for increased adrenomedullary secretion is activation of the sympathetic system by stress.
- In addition to specific responses to various stressors, all stressors produce a similar generalized stress response. This stress response includes (1) activation of the sympathetic nervous system accompanied by epinephrine secretion, which together prepare the body for a fight-or-flight response; (2) activation of the CRH-ACTH-cortisol system, which helps the body cope with stress primarily by mobilizing metabolic resources; (3) elevation of blood glucose and fatty acids through decreased insulin and increased glucagon secretion; and (4) maintenance of blood volume and blood pressure through increased activity of the renin-angiotensin-aldosterone system along with increased vasopressin secretion. All these actions are coordinated by the hypothalamus.

Endocrine Control of Fuel Metabolism
- Intermediary or fuel metabolism is, collectively, the synthesis (anabolism), breakdown (catabolism), and transformations of the three classes of energy-rich organic nutrients—carbohydrate, fat, and protein—within the body.
- Glucose and fatty acids derived respectively from carbohydrates and fats are primarily used as metabolic fuels, whereas amino acids derived from proteins are primarily used for the synthesis of structural and enzymatic proteins.
- During the absorptive state following a meal, the excess absorbed nutrients not immediately needed for energy production or protein synthesis are stored to a limited extent as glycogen in the liver and muscle but mostly as triglycerides in adipose tissue.
- During the postabsorptive state between meals when no new nutrients are entering the blood, the glycogen and triglyceride stores are catabolized to release nutrient molecules into the blood. If necessary, body proteins are degraded to release amino acids for conversion into glucose. It is essential to maintain the blood glucose concentration above a critical level even during the postabsorptive state, because the brain depends on blood-delivered glucose as its energy source. Tissues not dependent on glucose switch to fatty acids as their metabolic fuel, sparing glucose for the brain.

- These shifts in metabolic pathways between the absorptive and postabsorptive state are hormonally controlled. The most important hormone in this regard is insulin. Insulin is secreted by the β cells of the islets of Langerhans, the endocrine portion of the pancreas. The other major pancreatic hormone, glucagon, is secreted by the α cells of the islets.
- Insulin is an anabolic hormone; it promotes the cellular uptake of glucose, fatty acids, and amino acids and enhances their conversion into glycogen, triglycerides, and proteins, respectively. In so doing, it lowers the blood concentrations of these small organic molecules.
- Insulin secretion is increased during the absorptive state, primarily by a direct effect of an elevated blood glucose on the β cells, and is largely responsible for directing the organic traffic into cells during this state.
- Glucagon mobilizes the energy-rich molecules from their stores during the postabsorptive state. Glucagon, which is secreted in response to a direct effect of a fall in blood glucose on the pancreatic α cells, in general opposes the actions of insulin.

Endocrine Control of Calcium Metabolism

- Changes in the concentration of free, diffusible plasma Ca^{2+}, the biologically active form of this ion, produce profound and life-threatening effects, most notably on neuromuscular excitability. Hypercalcemia reduces excitability, whereas hypocalcemia brings about overexcitability of nerves and muscles. If the overexcitability is severe enough, fatal spastic contractions of respiratory muscles can occur.
- Three hormones regulate the plasma concentration of Ca^{2+} (and concurrently regulate PO_4^{3-})—parathyroid hormone (PTH), calcitonin, and vitamin D.
- PTH, whose secretion is directly increased by a fall in plasma Ca^{2+} concentration, acts on bone, kidneys, and the intestine to raise the plasma Ca^{2+} concentration. In so doing, it is essential for life by preventing the fatal consequences of hypocalcemia. The specific effects of PTH on bone are to promote Ca^{2+} movement from the bone fluid into the plasma in the short term and to promote localized dissolution of bone by enhancing activity of the osteoclasts (bone-dissolving cells) in the long term.
- Dissolution of the calcium phosphate bone crystals releases PO_4^{3-} as well as Ca^{2+} into the plasma. PTH acts on the kidneys to enhance the reabsorption of filtered Ca^{2+}, thereby reducing the urinary excretion of Ca^{2+} and increasing its plasma concentration. Simultaneously, PTH reduces renal PO_4^{3-} reabsorption, in this way increasing PO_4^{3-} excretion and lowering plasma PO_4^{3-} levels. This is important because a rise in plasma PO_4^{3-} would force the deposition of some of the plasma Ca^{2+} back into the bone.
- Furthermore, PTH facilitates the activation of vitamin D, which in turn stimulates Ca^{2+} and PO_4^{3-} absorption from the intestine.
- Vitamin D can be synthesized from a cholesterol derivative in the skin when exposed to sunlight, but frequently this endogenous source is inadequate, so vitamin D must be supplemented by dietary intake. From either source, vitamin D must be activated first by the liver and then by the kidneys (the site of PTH regulation of vitamin D activation) before it can exert its effect on the intestine.
- Calcitonin, a hormone produced by the C cells of the thyroid gland, is the third factor that regulates Ca^{2+}. In negative-feedback fashion, calcitonin is secreted in response to an increase in plasma Ca^{2+} concentration and acts to lower plasma Ca^{2+} levels by inhibiting activity of bone osteoclasts. Calcitonin is unimportant except during the rare condition of hypercalcemia.

REVIEW EXERCISES

Objective Questions (Answers on p. A-50)

1. The response to thyroid hormone is detectable within a few minutes after its secretion. (*True or false?*)
2. "Male" sex hormones are produced in both males and females by the adrenal cortex. (*True or false?*)
3. Adrenal androgen hypersecretion is usually due to a deficit of an enzyme crucial to cortisol synthesis. (*True or false?*)
4. Excess glucose and amino acids as well as fatty acids can be stored as triglycerides. (*True or false?*)
5. Insulin is the only hormone that can lower blood glucose levels. (*True or false?*)
6. The most life-threatening consequence of hypocalcemia is reduced blood clotting. (*True or false?*)
7. All ingested Ca^{2+} is indiscriminately absorbed in the intestine. (*True or false?*)
8. The $Ca_3 (PO_4)_2$ bone crystals form a labile pool from which Ca^{2+} can rapidly be extracted under the influence of PTH. (*True or false?*)
9. The lumen of the thyroid follicle is filled with _____, the chief constituent of which is a large protein molecule known as _____.
10. The common large precursor molecule that yields ACTH, MSH, and β-endorphin is known as _____.
11. _____ is the conversion of glucose into glycogen. _____ is the conversion of glycogen into glucose. _____ is the conversion of amino acids into glucose.
12. The three major tissues that are not dependent on insulin for their glucose uptake are _____, _____, and _____.
13. The three compartments with which ECF Ca^{2+} is exchanged are _____, _____, and _____.
14. Which of the following hormones does *not* exert a direct metabolic effect?
    a. epinephrine
    b. growth hormone
    c. aldosterone
    d. cortisol
    e. thyroid hormone
15. Which of the following are characteristic of the postabsorptive state? (*Indicate all that apply.*)
    a. glycogenolysis
    b. gluconeogenesis
    c. lipolysis
    d. glycogenesis
    e. protein synthesis
    f. triglyceride synthesis
    g. protein degradation

h. increased insulin secretion
i. increased glucagon secretion
j. glucose sparing
16. Indicate the primary circulating form and storage form of each of the three classes of organic nutrients:

	Primary Circulating Form	Primary Storage Form
Carbohydrate	1. _____	2. _____
Fat	3. _____	4. _____
Protein	5. _____	6. _____

Essay Questions

1. Describe the steps of thyroid hormone synthesis.
2. What are the effects of T_3 and T_4? Which is the more potent? What is the source of most circulating T_3?
3. Describe the regulation of thyroid hormone.
4. Discuss the causes and symptoms of both hypothyroidism and hyperthyroidism. For each cause, indicate whether or not a goiter occurs, and explain why.
5. What hormones are secreted by the adrenal cortex? What are the functions and control of each of these hormones?
6. Discuss the causes and symptoms of each type of adrenocortical dysfunction.
7. What is the relationship of the adrenal medulla to the sympathetic nervous system? What are the functions of epinephrine? How is epinephrine release controlled?
8. Define *stress*. Describe the neural and hormonal responses to a stressor.
9. Define *fuel metabolism*, *anabolism*, and *catabolism*.
10. Distinguish between the absorptive and postabsorptive states with regard to the handling of nutrient molecules.
11. Name the two major cell types of the islets of Langerhans, and indicate the hormonal product of each.
12. Compare the functions and control of insulin secretion with those of glucagon secretion.
13. What are the consequences of diabetes mellitus? Distinguish between Type I and Type II diabetes mellitus.
14. Why must plasma Ca^{2+} be closely regulated?
15. Explain how osteoblasts influence osteoclast function.
16. Discuss the contributions of parathyroid hormone, calcitonin, and vitamin D to Ca^{2+} metabolism. Describe the source and control of each of these hormones.
17. Discuss the major disorders in Ca^{2+} metabolism.

POINTS TO PONDER

(Explanations on p. A-51)

1. Iodine is naturally present in salt water and is abundant in soil along coastal regions. Fish and shellfish living in the ocean and plants grown in coastal soil take up iodine from their environment. Fresh water does not contain iodine, and the soil becomes more iron poor the farther inland. Knowing this, explain why the midwestern United States was once known as an *endemic goiter belt* because of the high incidence of goiter in this region. Why is this region no longer an endemic goiter belt even though the soil is still iodine poor?
2. Why do doctors recommend that people who are allergic to bee stings and thus are at risk for anaphylactic shock (see p. 448) carry a vial of epinephrine for immediate injection in case of a sting?
3. Why would an infection tend to raise the blood glucose level of a diabetic individual?
4. Tapping the facial nerve at the angle of the jaw in a patient with moderate hyposecretion of a particular hormone elicits a characteristic grimace on that side of the face. What endocrine abnormality could give rise to this so-called *Chvostek's sign*?
5. Soon after a technique to measure plasma Ca^{2+} levels was developed in the 1920s, physicians observed that hypercalcemia ac-companied a broad range of cancers. Early researchers proposed that malignancy-associated hypercalcemia arose from metastatic (see p. 443) tumor cells that invaded and destroyed bone, releasing Ca^{2+} into the blood. This conceptual framework was overturned when physicians noted that hypercalcemia often appeared in the absence of bone lesions. Furthermore, cancer patients often manifested hypophosphatemia in addition to hypercalcemia. This finding led investigators to suspect that the tumors might be producing a PTH-like substance. Explain how they reached this conclusion. In 1987, this substance was identified and named *parathyroid hormone–related peptide (PTHrP)*, which binds to and activates PTH receptors.

6. *Clinical Consideration.* Najma G. sought medical attention after her menstrual periods ceased and she started growing excessive facial hair. Also, she had been thirstier than usual and urinated more frequently. A clinical evaluation revealed that Najma was hyperglycemic. Her physician told her that she had an endocrine disorder dubbed "diabetes of bearded ladies." Based on her symptoms and your knowledge of the endocrine system, what underlying defect do you think is responsible for Najma's condition?

PHYSIOEDGE RESOURCES

PHYSIOEDGE CD-ROM
Figures marked with the PhysioEdge icon have associated activities on the CD. For this chapter, check out the Media Quiz related to Figure 19-1.

CHECK OUT THESE MEDIA QUIZZES:
19.1 Thyroid Functions
19.2 Adrenal Gland Functions
19.3 Endocrine Pancreas and Fuel Metabolism
19.4 Endocrine Control of Calcium Metabolism

PHYSIOEDGE WEB SITE
The Web site for this book contains a wealth of helpful study aids, as well as many ideas for further reading and research. Log on to:

http://www.brookscole.com

Go to the Biology page and select Sherwood's *Human Physiology*, 5th Edition. Select a chapter from the drop-down menu or click on one of these resource areas:

- **Case Histories** provide an introduction to the clinical aspects of human physiology. Check out:

    #4: Starvation in the Midst of Plenty
    #5: Eight Years Later

- For 2-D and 3-D graphical illustrations and animations of physiological concepts, visit our **Visual Learning Resource.**

- Resources for study and review include the **Chapter Outline, Chapter Summary, Glossary,** and **Flash Cards.** Use our **Quizzes** to prepare for in-class examinations.

- On-line research resources to consult are **Hypercontents,** with current links to relevant Internet sites; **Internet Exercises** with starter URLs provided; and **InfoTrac Exercises.**

     For more readings, go to InfoTrac College Edition, your on-line research library, at:

    http://infotrac.thomsonlearning.com

Reproductive System

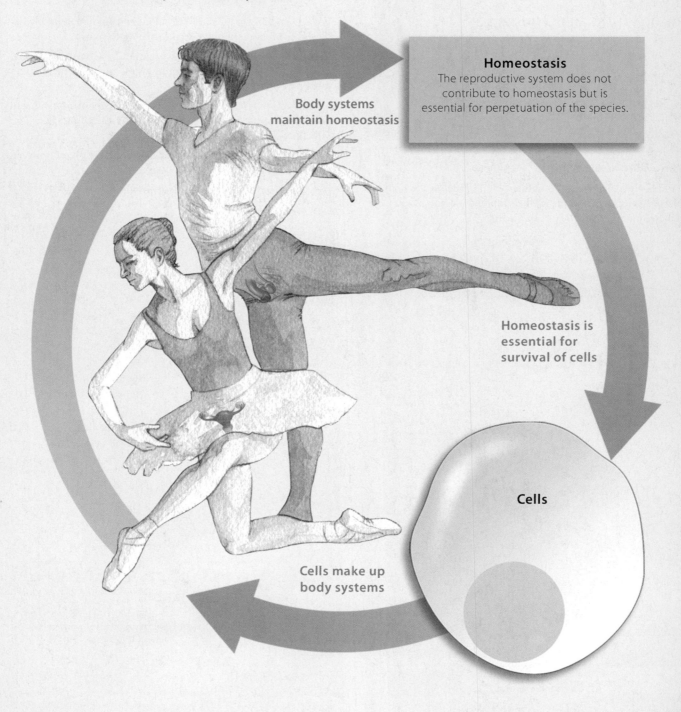

Body systems maintain homeostasis

Homeostasis
The reproductive system does not contribute to homeostasis but is essential for perpetuation of the species.

Homeostasis is essential for survival of cells

Cells

Cells make up body systems

Normal functioning of the **reproductive system** is not aimed toward homeostasis and is not necessary for survival of an individual, but it is essential for survival of the species. Only through reproduction can the complex genetic blueprint of each species survive beyond the lives of individual members of the species.

Chapter 20

The Reproductive System

CONTENTS AT A GLANCE

INTRODUCTION

▌ Anatomy of the male and female reproductive system

▌ Sex determination and differentiation

MALE REPRODUCTIVE PHYSIOLOGY

▌ Scrotal location of the testes

▌ Testosterone secretion

▌ Spermatogenesis

▌ Puberty

▌ Male reproductive tract

▌ Male accessory sex glands

▌ Prostaglandins

SEXUAL INTERCOURSE BETWEEN MALES AND FEMALES

▌ Male sex act

▌ Female sex act

FEMALE REPRODUCTIVE PHYSIOLOGY

▌ Ovarian function

▌ Menstrual cycle

▌ Puberty; menopause

▌ Fertilization

▌ Implantation; placentation

▌ Gestation

▌ Parturition

▌ Lactation

INTRODUCTION

The central theme of this book has been the physiologic processes aimed at maintaining homeostasis to ensure survival of the individual. We are now going to disembark from this theme to discuss the reproductive system, which primarily serves the purpose of perpetuating the species.

Even though the reproductive system does not contribute to homeostasis and is not essential for survival of an individual, it still plays an important role in a person's life. For example, the manner in which people relate as sexual beings contributes in significant ways to psychosocial behavior and has important influences on how people view themselves and how they interact with others. Reproductive function also has a profound effect on society. The universal organization of societies into family units provides a stable environment that is conducive for perpetuating our species. On the other hand, the population explosion and its resultant drain on dwindling resources have recently led to worldwide concern with the means by which reproduction can be limited.

Reproductive capability depends on an intricate relationship among the hypothalamus, anterior pituitary, reproductive organs, and target cells of the sex hormones. In addition to these basic biologic processes, sexual behavior and attitudes are deeply influenced by emotional factors and the sociocultural mores of the society in which the individual lives. We will concentrate on the basic sexual and reproductive functions that are under nervous and hormonal control and will not examine the psychological and social ramifications of sexual behavior.

▌ The reproductive system includes the gonads, reproductive tract, and accessory sex glands.

Reproduction depends on the union of male and female gametes (reproductive, or germ, cells), each with a half set of chromosomes, to form a new individual with a full, unique set of chromosomes. Unlike the other body systems, which are essentially identical in the two sexes, the reproductive systems of males and females are remarkably different, befitting their different roles in the reproductive process. The male and female reproductive systems are designed to enable union of genetic material from the two sexual partners, and the female system is equipped to house and nourish the offspring to the developmental point at which it can survive independently in the external environment.

The **primary reproductive organs,** or **gonads,** consist of a pair of **testes** in the male and a pair of **ovaries** in the female. In both sexes, the mature gonads perform the dual function of (1) producing gametes (**gametogenesis**), that is, **spermatozoa (sperm)** in the male and **ova (eggs)** in the female, and (2) secreting sex hormones, specifically, **testosterone** in males and **estrogen** and **progesterone** in females.

In addition to the gonads, the reproductive system in each sex includes a **reproductive tract** encompassing a system of ducts that are specialized to transport or house the gametes after they are produced, plus **accessory sex glands** that empty their supportive secretions into these passageways. In females, the *breasts* are also considered accessory reproductive organs. The externally visible portions of the reproductive system are known as **external genitalia.**

Secondary sexual characteristics

The **secondary sexual characteristics** are the many external characteristics that are not directly involved in reproduction but that distinguish males and females, such as body configuration and hair distribution. In humans, for example, males have broader shoulders whereas females have curvier hips, and males have beards whereas females do not. Testosterone in the male and estrogen in the female govern the development and maintenance of these characteristics. Progesterone has no influence on secondary sexual characteristics. Even though growth of axillary and pubic hair at puberty is promoted in both sexes by androgens—testosterone in males and adrenocortical dehydroepiandrosterone in females (see p. 711)—this hair growth is not a secondary sexual characteristic, because both sexes display this feature. Thus testosterone and estrogen alone govern the nonreproductive distinguishing features.

In some species, the secondary sexual characteristics are of great importance in courting and mating behavior; for example, the rooster's headdress or comb attracts the female's attention, and the stag's antlers are useful to ward off other males. In humans, the differentiating marks between males and females do serve to attract the opposite sex, but attraction is also strongly influenced by the complexities of human society and cultural behavior.

Overview of male reproductive functions and organs

The essential reproductive functions of the male are

1. Production of sperm (*spermatogenesis*)
2. Delivery of sperm to the female

The sperm-producing organs, the testes, are suspended outside the abdominal cavity in a skin-covered sac, the **scrotum,** which lies within the angle between the legs. The male reproductive system is designed to deliver sperm to the female reproductive tract in a liquid vehicle, *semen,* which is conducive to sperm viability. The major male accessory sex glands, whose secretions provide the bulk of the semen, are the *seminal vesicles, prostate gland,* and *bulbourethral glands* (● Figure 20-1).

The **penis** is the organ used to deposit semen in the female. Sperm exit each testis through the male reproductive tract, consisting on each side of an *epididymis, ductus (vas) deferens,* and *ejaculatory duct.* These pairs of reproductive tubes empty into a single *urethra,* the canal that runs the length of the penis and empties to the exterior. These parts of the male reproductive system are described more thoroughly later when their functions are discussed.

Overview of female reproductive functions and organs

The female's role in reproduction is more complicated than the male's. The essential female reproductive functions include

1. Production of ova (*oogenesis*)
2. Reception of sperm
3. Transport of the sperm and ovum to a common site for union (*fertilization,* or *conception*)
4. Maintenance of the developing fetus until it can survive in the outside world (*gestation,* or *pregnancy*), including formation of the *placenta,* the organ of exchange between mother and fetus
5. Giving birth to the baby (*parturition*)
6. Nourishing the infant after birth by milk production (*lactation*)

The product of fertilization is known as an **embryo** during the first two months of intrauterine development when tissue differentiation is taking place. Beyond this time, the developing living being is recognizable as human and is known as a **fetus** during the remainder of gestation. Although no further tissue differentiation takes place during fetal life, it is a time of tremendous tissue growth and maturation.

The ovaries and female reproductive tract lie within the pelvic cavity (● Figure 20-2a and b). The female reproductive tract consists of the following components. Two **oviducts (uterine,** or **Fallopian tubes),** which are in close association with the two ovaries, pick up ova on ovulation and serve as the site for fertilization. The thick-walled hollow **uterus** is primarily responsible for maintaining the fetus during its development and expelling it at the end of pregnancy. The **vagina** is a muscular, expandable tube that connects the uterus to the external environment. The lowest portion of the uterus, the **cervix,** projects into the vagina and contains a single, small opening, the **cervical canal.** Sperm are deposited in the vagina by the penis during sexual intercourse. The cervical canal serves as a pathway for sperm through the uterus to the site of fertilization in the oviduct, and when greatly dilated during parturition, serves as the passageway for delivery of the baby from the uterus.

The **vaginal opening** is located in the **perineal region** between the urethral opening anteriorly and the anal opening posteriorly (● Figure 20-2c). It is partially covered by a thin mucous membrane, the **hymen,** which can be physically disrupted in a variety of ways, including by the first sexual intercourse. The vaginal and urethral openings are surrounded laterally by two pairs of skin folds, the **labia minora** and **labia majora.** The smaller labia minora are located medially to the more prominent labia majora. The **clitoris,** a small erotic structure composed of tissue identical to the penis, lies at the anterior end of

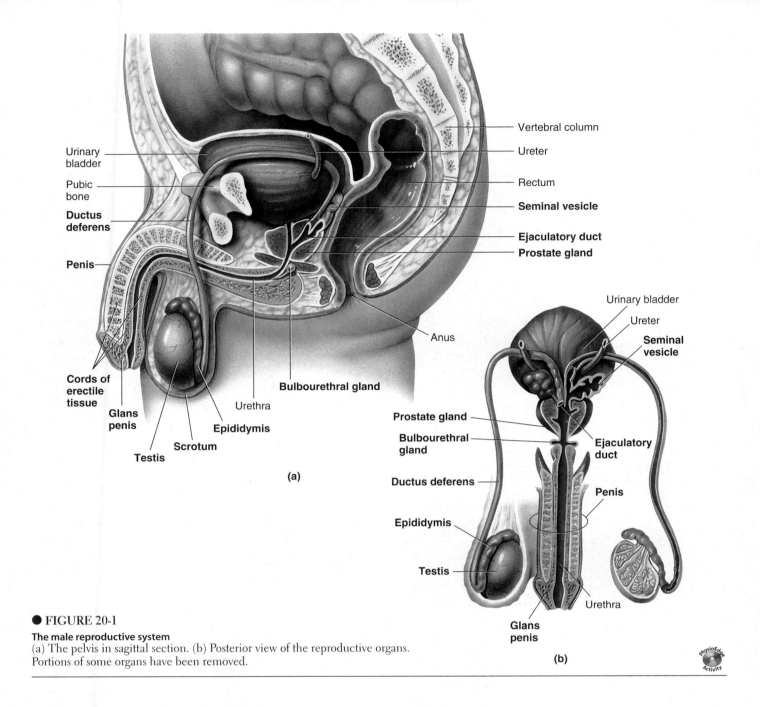

● FIGURE 20-1

The male reproductive system
(a) The pelvis in sagittal section. (b) Posterior view of the reproductive organs. Portions of some organs have been removed.

the folds of the labia minora. The female external genitalia are collectively referred to as the **vulva.**

▌ Reproductive cells each contain a half set of chromosomes.

The DNA molecules that carry the cell's genetic code are not randomly crammed into the nucleus but are precisely organized into **chromosomes** (see p. A-19). Each chromosome consists of a different DNA molecule that contains a unique set of genes. **Somatic** (body) **cells** contain 46 chromosomes (the **diploid number**), which can be sorted into 23 pairs on the basis of various distinguishing features. Chromosomes composing

a matched pair are termed **homologous chromosomes,** one member of each pair having been derived from the individual's maternal parent and the other member from the paternal parent. Gametes (that is, sperm and eggs) contain only one member of each homologous pair for a total of 23 chromosomes (the **haploid number**).

▌ Gametogenesis is accomplished by meiosis.

Most cells in the human body have the ability to reproduce themselves, a process important in growth, replacement, and repair of tissues. Cell division involves two components: division of the nucleus and division of the cytoplasm. Nuclear di-

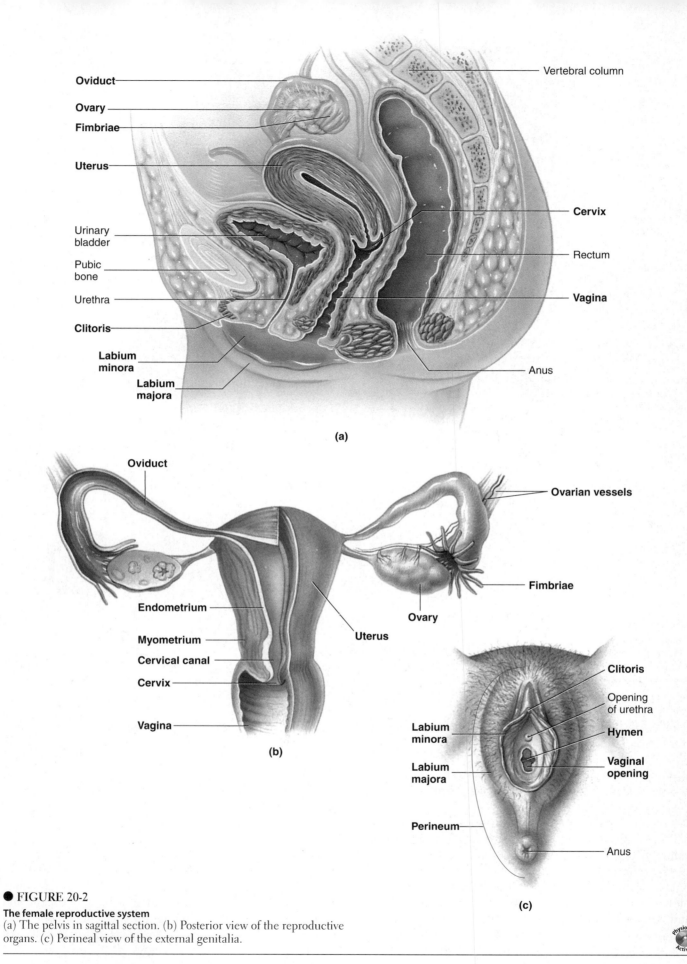

Oviduct

Ovary

Fimbriae

Uterus

Urinary
bladder

Pubic
bone

Urethra

Clitoris

Labium
minora

Labium
majora

Vertebral column

Cervix

Rectum

Vagina

Anus

(a)

Oviduct

Ovarian vessels

Fimbriae

Ovary

Endometrium

Myometrium

Cervical canal

Cervix

Vagina

Uterus

(b)

Clitoris

Opening
of urethra

Labium
minora

Hymen

Labium
majora

Vaginal
opening

Perineum

Anus

(c)

● FIGURE 20-2

The female reproductive system
(a) The pelvis in sagittal section. (b) Posterior view of the reproductive
organs. (c) Perineal view of the external genitalia.

PhysioEdge
Activity

vision in somatic cells is accomplished by **mitosis.** In mitosis, the chromosomes replicate (make duplicate copies of themselves), then the identical chromosomes are separated so that a complete set of genetic information (that is, a diploid number of chromosomes) is distributed to each of the two new daughter cells. Nuclear division in the specialized case of gametes is accomplished by **meiosis,** in which only a half set of genetic information (that is, a haploid number of chromosomes) is distributed to each of four new daughter cells (see p. A-27).

During meiosis, a specialized diploid germ cell undergoes one chromosome replication followed by two nuclear divisions. In the first meiotic division, the replicated chromosomes do not separate into two individual, identical chromosomes but remain joined together. The doubled chromosomes sort themselves into homologous pairs, and the pairs separate so that each of two daughter cells receives a half set of doubled chromosomes. During the second meiotic division, the doubled chromosomes within each of the two daughter cells separate and are distributed into two cells, yielding four daughter cells, each containing a half set of chromosomes, a single member of each pair. During this process, the maternally and paternally derived chromosomes of each homologous pair are distributed to the daughter cells in random assortments containing one member of each chromosome pair without regard for its original derivation. That is, not all of the mother-derived chromosomes go to one daughter cell and the father-derived chromosomes to the other cell. More than 8 million (2^{23}) different mixtures of the 23 paternal and maternal chromosomes are possible. This genetic mixing provides novel combinations of chromosomes.

Thus sperm and ova each have a unique haploid number of chromosomes. When fertilization takes place, a sperm and ovum fuse to form the start of a new individual with 46 chromosomes, one member of each chromosomal pair having been inherited from the mother and the other member from the father (● Figure 20-3).

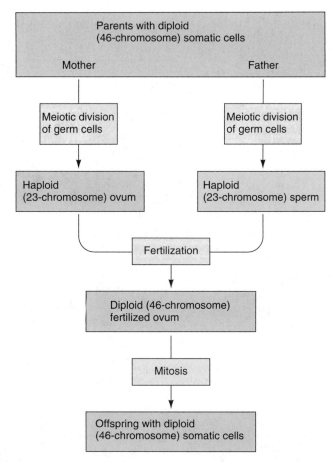

● **FIGURE 20-3**
Chromosomal distribution in sexual reproduction

■ The sex of an individual is determined by the combination of sex chromosomes.

Whether individuals are destined to be males or females is a genetic phenomenon determined by the sex chromosomes they possess. As the 23 chromosome pairs are separated during meiosis, each sperm or ovum receives only one member of each chromosome pair. Of the chromosome pairs, 22 are **autosomal chromosomes** that code for general human characteristics as well as for specific traits such as eye color. The remaining pair of chromosomes consists of the **sex chromosomes,** of which there are two genetically different types—a larger **X chromosome** and a smaller **Y chromosome.**

Sex determination depends on the combination of sex chromosomes: **Genetic males** have both an X and a Y sex chromosome; **genetic females** have two X sex chromosomes. Thus the genetic difference responsible for all the anatomic and functional distinctions between males and females is the single Y chromosome. Males have it; females do not.

As a result of meiosis during gametogenesis, all chromosome pairs are separated so that each daughter cell contains only one member of each pair, including the sex chromosome pair. When the XY sex chromosome pair separates during sperm formation, half the sperm receive an X chromosome and the other half a Y chromosome. In contrast, during oogenesis, every ovum receives an X chromosome, because separation of the XX sex chromosome pair yields only X chromosomes. During fertilization, combination of an X-bearing sperm with an X-bearing ovum produces a genetic female, XX, whereas union of a Y-bearing sperm with an X-bearing ovum results in a genetic male, XY. Thus genetic sex is determined at the time of conception and depends on which type of sex chromosome is contained within the fertilizing sperm.

■ Sexual differentiation along male or female lines depends on the presence or absence of masculinizing determinants.

Differences between males and females exist at three levels: genetic, gonadal, and phenotypic (anatomic) sex (● Figure 20-4).

Genetic and gonadal sex

Genetic sex, which depends on the combination of sex chromosomes at the time of conception, in turn determines **gonadal sex,** that is, whether testes or ovaries develop. The presence or

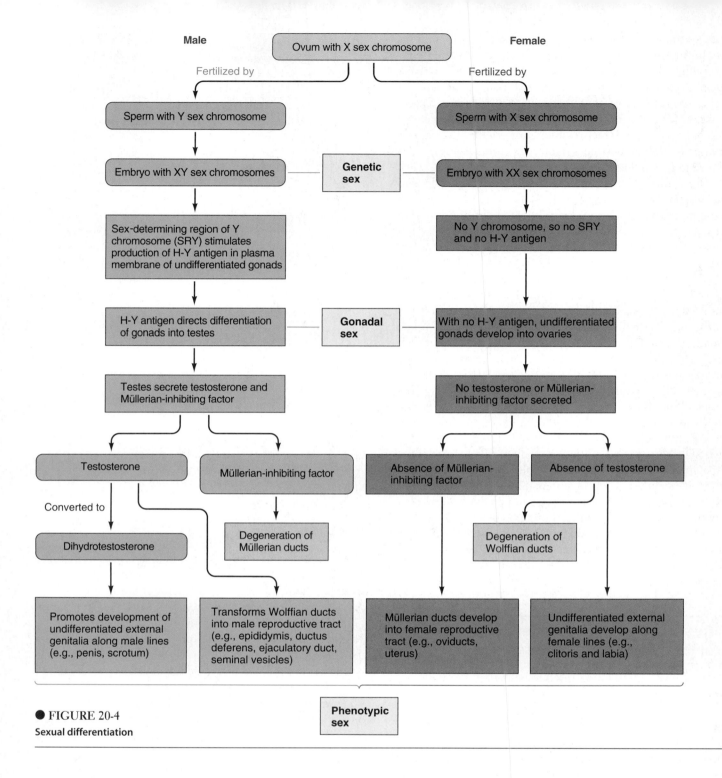

● FIGURE 20-4
Sexual differentiation

absence of a Y chromosome determines gonadal differentiation. For the first month and a half of gestation, all embryos have the potential to differentiate along either male or female lines, because the developing reproductive tissues of both sexes are identical and indifferent. Gonadal specificity appears during the seventh week of intrauterine life when the indifferent gonadal tissue of a genetic male begins to differentiate into testes under the influence of the **sex-determining region** of the Y chromosome (**SRY**), the single gene that is responsible for sex determination. This gene triggers a chain of reactions that leads to

physical development of a male. SRY "masculinizes" the gonads (induces their development into testes) by stimulating production of **H-Y antigen** by primitive gonadal cells. H-Y antigen, a specific plasma membrane protein found only in males, directs differentiation of the gonads into testes.

Because genetic females lack the SRY gene and consequently do not produce H-Y antigen, their gonadal cells never receive a signal for testicular formation, so the undifferentiated gonadal tissue starts developing by default during the ninth week into ovaries instead.

Phenotypic sex

Phenotypic sex, the apparent anatomic sex of an individual, depends on the genetically determined gonadal sex. The term **sexual differentiation** refers to the embryonic development of the external genitalia and reproductive tract along either male or female lines. As with the undifferentiated gonads, embryos of both sexes have the potential to develop either male or female reproductive tracts and external genitalia. Differentiation into a male-type reproductive system is induced by **androgens,** which are masculinizing hormones secreted by the developing testes. Testosterone is the most potent androgen. The absence of these testicular hormones in female fetuses results in the development of a female-type reproductive system. By 10 to 12 weeks of gestation, the sexes can easily be distinguished by the anatomic appearance of the external genitalia.

Sexual differentiation of the external genitalia

Male and female external genitalia develop from the same embryonic tissue. In both sexes, the undifferentiated external genitals consist of a *genital tubercle,* paired *urethral folds* surrounding a urethral groove, and, more laterally, *genital (labioscrotal) swellings* (● Figure 20-5). The **genital tubercle** gives rise to exquisitely sensitive erotic tissue—in males the **glans penis** (the cap at the distal end of the penis) and in females the clitoris. The major distinctions between the glans penis and clitoris are the smaller size of the clitoris and the penetration of the glans

● FIGURE 20-5

Sexual differentiation of the external genitalia
(a) Undifferentiated stage. (b) Male development. (c) Female development.

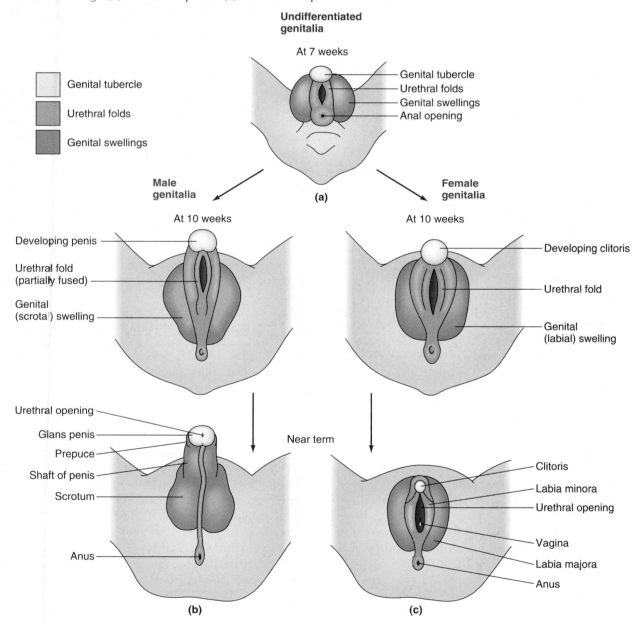

penis by the urethral opening. The urethra is the tube through which urine is transported from the bladder to the outside and also serves in males as a passageway for exit of semen through the penis to the outside. In males, the **urethral folds** fuse around the urethral groove to form the penis, which encircles the urethra. The **genital swellings** similarly fuse to form the scrotum and **prepuce,** a fold of skin that extends over the end of the penis and more or less completely covers the glans penis. In females, the urethral folds and genital swellings do not fuse at midline but develop instead into the labia minora and labia majora, respectively. The urethral groove remains open, providing access to the interior through the urethral opening and vaginal orifice.

Sexual differentiation of the reproductive tract

Although the male and female external genitalia develop from the same undifferentiated embryonic tissue, this is not the case with the reproductive tracts. Two primitive duct systems—the Wolffian ducts and the Müllerian ducts—develop in all embryos. In males, the reproductive tract develops from the **Wolffian ducts** and the Müllerian ducts degenerate, whereas in females the **Müllerian ducts** differentiate into the reproductive tract and the Wolffian ducts regress. Because both duct systems are present before sexual differentiation occurs, the early embryo has the potential to develop either a male or a female reproductive tract. Development of the reproductive tract along male or female lines is determined by the presence or absence of two hormones secreted by the fetal testes—*testosterone* and *Müllerian-inhibiting factor* (● Figure 20-4). A hormone released by the placenta, *human chorionic gonadotropin*, is the stimulus for this early testicular secretion. Testosterone induces development of the Wolffian ducts into the male reproductive tract (epididymis, ductus deferens, ejaculatory duct, and seminal vesicles). This hormone, after being converted into **dihydrotestosterone (DHT)**, is also responsible for differentiating the external genitalia into the penis and scrotum. Meanwhile, Müllerian-inhibiting factor causes regression of the Müllerian ducts.

In the absence of testosterone and Müllerian-inhibiting factor in females, the Wolffian ducts regress, the Müllerian ducts develop into the female reproductive tract (oviducts, uterus, and vagina), and the external genitalia differentiate into the clitoris and labia.

Note that the undifferentiated embryonic reproductive tissue passively develops into a female structure unless actively acted on by masculinizing factors. In the absence of male testicular hormones, a female reproductive tract and external genitalia develop regardless of the genetic sex of the individual. For feminization of the fetal genital tissue, ovaries do not even need to be present. Such a control pattern for determining sex differentiation is appropriate, considering that fetuses of both sexes are exposed to high concentrations of female sex hormones throughout gestation. If female sex hormones influenced the development of the reproductive tract and external genitalia, all fetuses would be feminized.

Errors in sexual differentiation

Genetic sex and phenotypic sex are usually compatible; that is, a genetic male anatomically appears to be a male and functions as a male, and the same compatibility holds true for females. Occasionally, however, discrepancies occur between genetic and anatomic sexes because of errors in sexual differentiation, as the following examples illustrate:

- If testes in a genetic male fail to properly differentiate and secrete hormones, the result is the development of an apparent anatomic female in a genetic male, who, of course, will be sterile.
- Because testosterone acts on the Wolffian ducts to convert them into a male reproductive tract but the testosterone derivative DHT is responsible for masculinization of the external genitalia, a genetic deficiency of the enzyme that converts testosterone into DHT results in a genetic male with testes and a male reproductive tract but with female external genitalia.
- The adrenal gland normally secretes a weak androgen, *dehydroepiandrosterone*, in insufficient quantities to masculinize females. However, pathologically excessive secretion of this hormone in a genetically female fetus during critical developmental stages imposes differentiation of the reproductive tract and genitalia along male lines (see adrenogenital syndrome, p. 712).

Sometimes these discrepancies between genetic sex and apparent sex are not recognized until puberty, when the discovery produces a psychologically traumatic gender identity crisis. For example, a masculinized genetic female with ovaries but with male-type external genitalia may be reared as a boy until puberty, when breast enlargement (caused by estrogen secretion by the awakening ovaries) and lack of beard growth (caused by lack of testosterone secretion in the absence of testes) signal an apparent problem. Therefore, it is important to diagnose any problems in sexual differentiation in infancy. Once a sex has been assigned, it can be reinforced, if necessary, with surgical and hormonal treatment so that psychosexual development can proceed as normally as possible. Less dramatic cases of inappropriate sex differentiation often appear as sterility problems.

MALE REPRODUCTIVE PHYSIOLOGY

Embryonically, the testes develop from the gonadal ridge located at the rear of the abdominal cavity. In the last months of fetal life, they begin a slow descent, passing out of the abdominal cavity through the **inguinal canal** into the scrotum, one testis dropping into each pocket of the scrotal sac. Testosterone from the fetal testes is responsible for inducing descent of the testes into the scrotum.

After the testes descend into the scrotum, the opening in the abdominal wall through which the inguinal canal passes closes snugly around the sperm-carrying duct and blood vessels that traverse between each testis and the abdominal cavity. Incomplete closure or rupture of this opening permits abdominal viscera to slip through, resulting in an **inguinal hernia.**

Although the time varies somewhat, descent is usually complete by the seventh month of gestation. As a result, descent is complete in 98% of full-term baby boys, but in a substantial percentage of premature male infants the testes are still within the inguinal canal at birth. In most instances of retained testes, descent occurs naturally before puberty or can be encouraged with administration of testosterone. Rarely, a testis remains undescended into adulthood, a condition known as **cryptorchidism** ("hidden testis").

▌The scrotal location of the testes provides a cooler environment essential for spermatogenesis.

The temperature within the scrotum averages several degrees Celsius less than normal body (core) temperature. Descent of the testes into this cooler environment is essential, because spermatogenesis is temperature sensitive and cannot occur at normal body temperature. Therefore, a cryptorchid is unable to produce viable sperm.

The position of the scrotum in relation to the abdominal cavity can be varied by a spinal-reflex mechanism that plays an important role in regulating testicular temperature. Reflex contraction of scrotal muscles on exposure to a cold environment raises the scrotal sac to bring the testes closer to the warmer abdomen. Conversely, relaxation of the muscles on exposure to heat permits the scrotal sac to become more pendulous, moving the testes farther from the warm core of the body.

▌The testicular Leydig cells secrete masculinizing testosterone.

The testes perform the dual function of producing sperm and secreting testosterone. About 80% of the testicular mass consists of highly coiled **seminiferous tubules**, within which spermatogenesis takes place. The endocrine cells that produce testosterone—the **Leydig**, or **interstitial, cells**—are located in the connective tissue (interstitial tissue) between the seminiferous tubules (see ● Figure 20-6b). Thus the portions of the testes that produce sperm and secrete testosterone are structurally and functionally distinct.

Testosterone is a steroid hormone derived from a cholesterol precursor molecule, as are the female sex hormones, estrogen and progesterone. The Leydig cells contain a high concentration of the enzymes required to direct cholesterol through the testosterone-yielding pathway (see p. 674). Once produced, some of the testosterone is secreted into the blood, where it is transported, primarily bound to plasma proteins, to its target sites of action. A substantial portion of the newly synthesized testosterone goes into the lumen of the seminiferous tubules, where it plays an important role in sperm production.

Most but not all of testosterone's actions ultimately function to ensure delivery of sperm to the female. The effects of testosterone can be grouped into five categories: (1) effects on the reproductive system before birth; (2) effects on sex-specific tissues after birth; (3) other reproduction-related effects; (4) effects on secondary sexual characteristics; and (5) nonreproductive actions (▲ Table 20-1).

Effects on the reproductive system before birth

Before birth, testosterone secretion by the fetal testes is responsible for masculinizing the reproductive tract and external gen-

---

▲ TABLE 20-1
Effects of Testosterone

Effects before Birth

Masculinizes the reproductive tract and external genitalia

Promotes descent of the testes into the scrotum

Effects on Sex-Specific Tissues

Promotes growth and maturation of the reproductive system at puberty

Essential for spermatogenesis

Maintains the reproductive tract throughout adulthood

Other Reproductive Effects

Develops the sex drive at puberty

Controls gonadotropin hormone secretion

Effects on Secondary Sexual Characteristics

Induces the male pattern of hair growth (e.g., beard)

Causes the voice to deepen because of thickening of the vocal folds

Promotes muscle growth responsible for the male body configuration

Nonreproductive Actions

Exerts a protein anabolic effect

Promotes bone growth at puberty and then closure of the epiphyseal plates

May induce aggressive behavior

---

italia and for promoting descent of the testes into the scrotum, as already described. After birth, testosterone secretion ceases, and the testes and remainder of the reproductive system remain small and nonfunctional until puberty.

Effects on sex-specific tissues after birth

Puberty is the period of arousal and maturation of the previously nonfunctional reproductive system, culminating in attainment of sexual maturity and the ability to reproduce. Its onset usually occurs sometime between the ages of 10 and 14; on the average it begins about two years earlier in females than in males. Usually lasting three to five years, puberty encompasses a complex sequence of endocrine, physical, and behavioral events. **Adolescence** is a broader concept that refers to the entire transition period between childhood and adulthood, not just to sexual maturation.

At puberty, the Leydig cells start secreting testosterone once again. Testosterone is responsible for growth and maturation of the entire male reproductive system. Under the influence of the pubertal surge in testosterone secretion, the testes enlarge and start producing sperm for the first time, the accessory sex glands enlarge and become secretory, and the penis and scrotum enlarge.

Ongoing testosterone secretion is essential for spermatogenesis and for maintaining a mature male reproductive tract throughout adulthood. Once initiated at puberty, testosterone secretion and spermatogenesis occur continuously throughout the male's life. Testicular efficiency gradually declines after 45 to 50 years of age, however, even though men in their seventies and beyond may continue to enjoy an active sex life, and some even father a child at this late age. The gradual diminution in circulating testosterone levels and in sperm production is not caused by a decrease in stimulation of the testes but probably arises instead from degenerative changes associated with aging that occur in the small testicular blood vessels. This gradual decline is often termed "male menopause," although it is not specifically programmed, as is female menopause.

Following **castration** (surgical removal of the testes) or testicular failure caused by disease, the other sex organs regress in size and function.

Other reproduction-related effects

Testosterone governs the development of sexual libido at puberty and helps maintain the sex drive in the adult male. Stimulation of this behavior by testosterone is important for facilitating delivery of sperm to females. In humans, libido is also influenced by many interacting social and emotional factors. Once libido has developed, testosterone is no longer absolutely required for its maintenance. Castrated males often remain sexually active but at a reduced level.

In another reproduction-related function, testosterone participates in the normal negative-feedback control of gonadotropin hormone secretion by the anterior pituitary, a topic covered more thoroughly later.

Effects on secondary sexual characteristics

All male secondary sexual characteristics depend on testosterone for their development and maintenance. These nonreproductive male characteristics induced by testosterone include (1) the male pattern of hair growth (for example, beard and chest hair and, in genetically predisposed men, baldness); (2) a deep voice caused by enlargement of the larynx and thickening of the vocal folds; (3) thick skin; and (4) the male body configuration (for example, broad shoulders and heavy arm and leg musculature) as a result of protein deposition. A male castrated before puberty (a **eunuch**) does not mature sexually, nor does he develop secondary sexual characteristics.

Nonreproductive actions

Testosterone exerts several important effects not related to reproduction. It has a general protein anabolic (synthesis) effect and promotes bone growth, thus contributing to the more muscular physique of males and to the pubertal growth spurt. Ironically, testosterone not only stimulates bone growth but eventually prevents further growth by sealing the growing ends of the long bones (that is, ossifying, or "closing," the epiphyseal plates—see p. 691). Testosterone also stimulates oil secretion by the sebaceous glands. This effect is most striking during the

adolescent surge of testosterone secretion, predisposing the young man to develop acne.

In animals, testosterone induces aggressive behavior, but whether it influences human behavior other than in the area of sexual behavior is an unresolved issue. Even though some athletes and bodybuilders who take testosterone-like anabolic androgenic steroids to increase muscle mass have been observed to display more aggressive behavior (see p. 282), it is unclear to what extent general behavioral differences between the sexes are hormonally induced or are a result of social conditioning.

We now shift attention from testosterone secretion to the other function of the testes—sperm production.

▌ Spermatogenesis yields an abundance of highly specialized, mobile sperm.

About 250 m (800 feet) of sperm-producing seminiferous tubules are packed within the testes (● Figure 20-6a). Two functionally important cell types are present in these tubules: *germ cells*, most of which are in various stages of sperm development, and *Sertoli cells*, which provide crucial support for spermatogenesis (● Figure 20-6b, c, and d). **Spermatogenesis** is a complex process by which relatively undifferentiated primordial germ cells, the **spermatogonia** (each of which contains a diploid complement of 46 chromosomes), proliferate and are converted into extremely specialized, motile spermatozoa (sperm), each bearing a randomly distributed haploid set of 23 chromosomes.

Microscopic examination of a seminiferous tubule reveals layers of germ cells in an anatomic progression of sperm development, starting with the least differentiated in the outer layer and moving inward through various stages of division to the lumen, where the highly differentiated sperm are ready for exit from the testis (● Figure 20-6b, c and d). Spermatogenesis takes 64 days for development from a spermatogonium to a mature sperm. Up to several hundred million sperm may reach maturity daily. Spermatogenesis encompasses three major stages: *mitotic proliferation*, *meiosis*, and *packaging* (● Figure 20-7).

Mitotic proliferation

Spermatogonia located in the outermost layer of the tubule continuously divide mitotically, with all new cells bearing the full complement of 46 chromosomes that are identical to those of the parent cell. Such proliferation provides a continual supply of new germ cells. Following mitotic division of a spermatogonium, one of the daughter cells remains at the outer edge of the tubule as an undifferentiated spermatogonium, thus maintaining the germ cell line. The other daughter cell starts moving toward the lumen while undergoing the various steps required to form sperm, which will be released into the lumen. In humans, the sperm-forming daughter cell divides mitotically twice more to form four identical **primary spermatocytes**. After the last mitotic division, the primary spermatocytes enter a resting phase during which the chromosomes are duplicated and the doubled strands remain together in preparation for the first meiotic division.

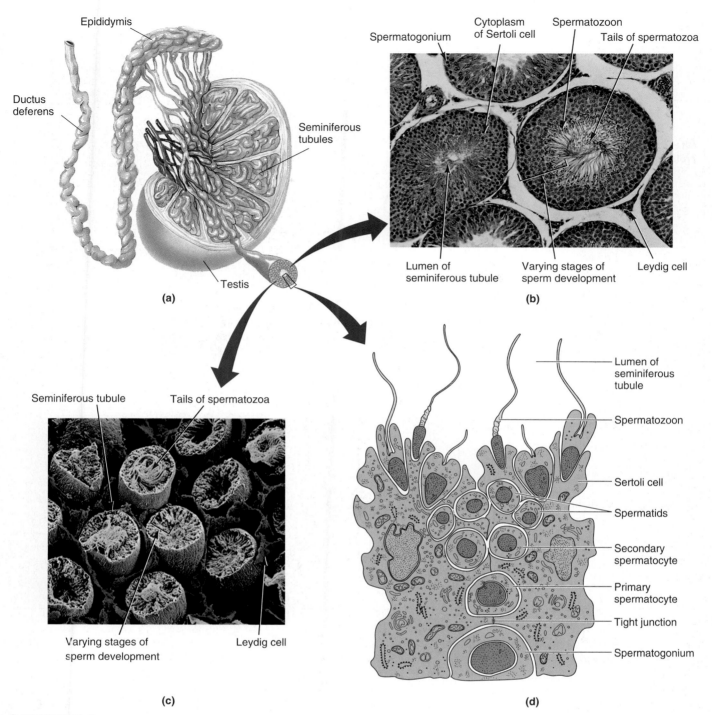

● FIGURE 20-6

Testicular anatomy depicting the site of spermatogenesis
(a) Longitudinal section of a testis showing the location and arrangement of the seminiferous tubules, the sperm-producing portion of the testis. (b) Light micrograph of a cross section of a seminiferous tubule. The undifferentiated germ cells (the spermatogonia) lie in the periphery of the tubule, and the differentiated spermatozoa are in the lumen, with the various stages of sperm development in between. (c) Scanning electron micrograph of a cross section of a seminiferous tubule. (d) Relationship of the Sertoli cells to the developing sperm cells.

Meiosis

During meiosis, each primary spermatocyte (with a diploid number of 46 doubled chromosomes) forms two **secondary spermatocytes** (each with a haploid number of 23 doubled chromosomes) during the first meiotic division, finally yield-ing four **spermatids** (each with 23 single chromosomes) as a result of the second meiotic division.

No further division takes place beyond this stage of spermatogenesis. Each spermatid is remodeled into a single spermatozoon. Because each sperm-producing spermatogonium mi-

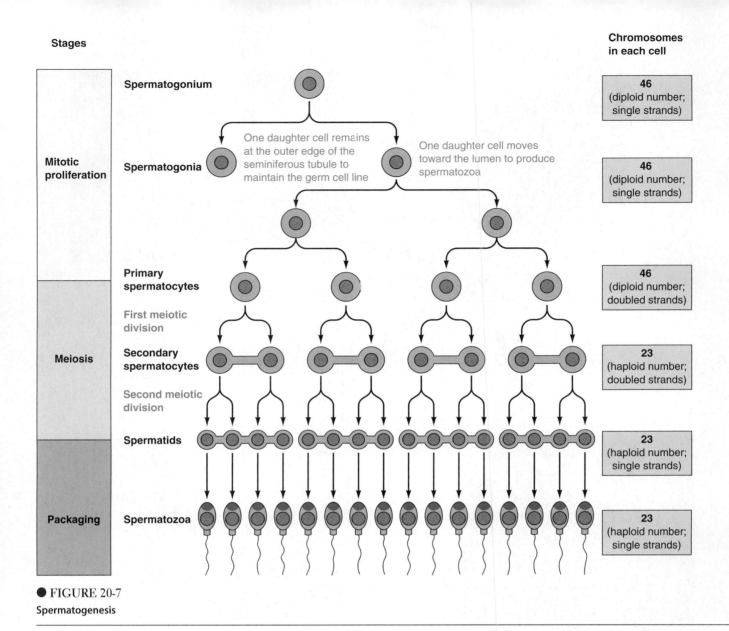

● FIGURE 20-7

Spermatogenesis

totically produces four primary spermatocytes and each primary spermatocyte meiotically yields four spermatids (spermatozoa-to-be), the spermatogenic sequence in humans can theoretically produce 16 spermatozoa each time a spermatogonium initiates this process. Usually, however, some cells are lost at various stages, so the efficiency of productivity is rarely this high.

Packaging

Even after meiosis, spermatids still resemble undifferentiated spermatogonia structurally, except for their half complement of chromosomes. Production of extremely specialized, mobile spermatozoa from spermatids requires extensive remodeling, or packaging, of cellular elements, a process known as **spermio-genesis.** Sperm are essentially "stripped-down" cells in which most of the cytosol and any organelles not needed for the task of delivering the sperm's genetic information to an ovum have been extruded. Thus sperm travel lightly, taking only the bare essentials to accomplish fertilization with them.

A **spermatozoon** has four parts (● Figure 20-8): a head, an acrosome, a midpiece, and a tail. The **head** consists primarily of the nucleus, which contains the sperm's complement of genetic information. The **acrosome,** an enzyme-filled vesicle that caps the tip of the head, is used as an "enzymatic drill" for penetrating the ovum. The acrosome is formed by aggregation of vesicles produced by the endoplasmic reticulum/Golgi complex before these organelles are discarded. Mobility for the spermatozoon is provided by a long, whiplike **tail** that grows out of one of the centrioles. Movement of the tail, which occurs as a result of relative sliding of its constituent microtubules (see p. 47), is powered by energy generated by the mitochondria concentrated within the **midpiece** of the sperm.

Until sperm maturation is complete, the developing germ cells arising from a single primary spermatocyte remain joined by cytoplasmic bridges. These connections, which result from incomplete cytoplasmic division, permit the four developing sperm to exchange cytoplasm. This linkage is important, be-

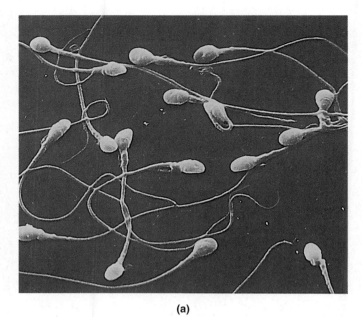

(a)

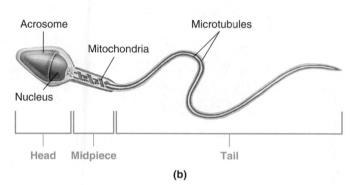

Acrosome   Microtubules

Mitochondria

Nucleus

Head   **Midpiece**   Tail

(b)

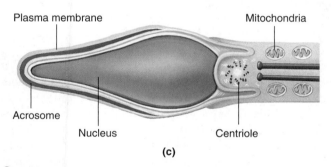

Plasma membrane   Mitochondria

Acrosome

Nucleus   Centriole

(c)

● **FIGURE 20-8**

Anatomy of a spermatozoon
(a) A phase-contrast photomicrograph of human spermatozoa.
(b) Schematic representation of a spermatozoon in "frontal" view.
(c) Longitudinal section of the head portion of a spermatozoon in "side" view.

cause the X chromosome, but not the Y chromosome, contains genes that code for cellular products essential for sperm development. (Whereas the large X chromosome contains several thousand genes, the small Y chromosome has only a few dozen, the most important of which are the SRY gene and others that play critical roles in male fertility.) During meiosis, half the sperm receive an X and the other half a Y chromosome. Were it not for the sharing of cytoplasm so that all the haploid cells are provided with the products coded for by X chromosomes until sperm development is complete, the Y-bearing, male-producing sperm could not develop and survive.

▌ Throughout their development, sperm remain intimately associated with Sertoli cells.

In addition to the spermatogonia and developing sperm cells, the seminiferous tubules also house the **Sertoli cells.** The Sertoli cells form a ring that extends from the outer basement membrane to the lumen of the tubule. Each Sertoli cell spans the entire distance from the outer membrane to the fluid-filled lumen (● Figure 20-6b and d). Adjacent Sertoli cells are joined by tight junctions at a point slightly beneath the outer membrane (see p. 63). Spermatogonia are tucked between the Sertoli cells at the outer perimeter of the tubule in the spaces between the basement membrane and the tight junctions.

During spermatogenesis, developing sperm cells arising from spermatogonial mitotic activity pass through the tight junctions, which transiently separate to make a path for them, then migrate toward the lumen in intimate association with the adjacent Sertoli cells. The cytoplasm of the Sertoli cells envelops the migrating germ cells, which remain buried within these cytoplasmic recesses throughout their development. At all stages of spermatogenic maturation, the developing sperm and Sertoli cells communicate with each other by means of direct cell-to-cell binding and through paracrine secretions (see p. 66). A recently identified carbohydrate on the surface membrane of the developing sperm enables them to bind to the supportive Sertoli cells.

Sertoli cells perform the following functions essential for spermatogenesis:

1. The tight junctions between adjacent Sertoli cells form a **blood–testes barrier.** Because this barrier prevents blood-borne substances from passing between the cells to gain entry to the lumen of the seminiferous tubule, only selected molecules that can pass through the Sertoli cells reach the intratubular fluid. As a result, the composition of the intratubular fluid varies considerably from that of the blood. The unique composition of this fluid that bathes the germ cells is critical for later stages of sperm development. The blood–testes barrier also prevents the antibody-producing cells in the extracellular fluid from reaching the tubular sperm factory, thus preventing the formation of antibodies against the highly differentiated spermatozoa.

2. Because the secluded developing sperm cells do not have direct access to blood-borne nutrients, the Sertoli cells provide nourishment for them.

3. The Sertoli cells have an important phagocytic function. They engulf the cytoplasm extruded from the spermatids during their remodeling, and they destroy defective germ cells that fail to successfully complete all stages of spermatogenesis.

4. The Sertoli cells secrete into the lumen **seminiferous tubule fluid,** which "flushes" the released sperm from the tubule into the epididymis for storage and further processing.

5. An important component of this Sertoli secretion is **androgen-binding protein.** As the name implies, this protein

binds androgens (that is, testosterone), thus maintaining a very high level of this hormone within the lumen of the seminiferous tubules. This high local concentration of testosterone is essential for sustaining sperm production. Androgen-binding protein is necessary to retain testosterone within the lumen, because this steroid hormone is lipid soluble and could easily diffuse across the plasma membranes and leave the lumen.

6. The Sertoli cells are the site of action for control for spermatogenesis by both testosterone and follicle-stimulating hormone (FSH). The Sertoli cells themselves release another hormone, *inhibin*, which acts in negative-feedback fashion to regulate FSH secretion.

❚ LH and FSH from the anterior pituitary control testosterone secretion and spermatogenesis.

The testes are controlled by the two gonadotropic hormones secreted by the anterior pituitary, **luteinizing hormone (LH)** and **follicle-stimulating hormone (FSH)**, which are named for their functions in females (see p. 685).

Feedback control of testicular function

LH and FSH act on separate components of the testes (● Figure 20-9). LH acts on the Leydig (interstitial) cells to regulate testosterone secretion, accounting for its alternative name in males—*interstitial cell–stimulating hormone (ICSH)*. FSH acts on the seminiferous tubules, specifically the Sertoli cells, to enhance spermatogenesis. (There is no alternative name for FSH in males.) Secretion of both LH and FSH from the anterior pituitary is stimulated in turn by a single hypothalamic hormone, **gonadotropin-releasing hormone (GnRH)** (see p. 687).

Even though GnRH stimulates both LH and FSH secretion, the blood concentrations of these two gonadotropic hormones do not always parallel each other because two other regulatory factors besides GnRH—*testosterone* and *inhibin*—differentially influence the secretory rate of LH and FSH. Testosterone, the product of LH stimulation of the Leydig cells, acts in negative-feedback fashion to inhibit LH secretion in two ways. The predominant negative-feedback effect of testosterone is to decrease GnRH release by acting on the hypothalamus, thus indirectly decreasing both LH and FSH release by the anterior pituitary. In addition, testosterone acts directly on the anterior pituitary to reduce the responsiveness of the LH secretory cells to GnRH. The latter action explains why testosterone exerts a greater inhibitory effect on LH secretion than on FSH secretion.

The testicular inhibitory signal specifically directed at controlling FSH secretion is the peptide hormone **inhibin,** which is secreted by the Sertoli cells. Inhibin acts directly on the anterior pituitary to inhibit FSH secretion. This feedback inhibition of FSH by a Sertoli cell product is appropriate, because FSH stimulates spermatogenesis by acting on the Sertoli cells.

Roles of testosterone and FSH in spermatogenesis

Both testosterone and FSH play critical roles in controlling spermatogenesis, each exerting its effect by acting on the Sertoli

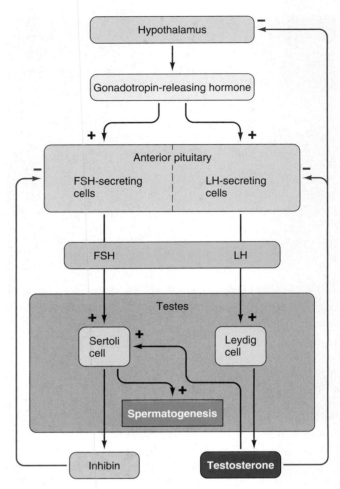

● **FIGURE 20-9**
Control of testicular function

cells. Testosterone is essential for both mitosis and meiosis of the germ cells, whereas FSH is required for spermatid remodeling. Testosterone concentration is much higher in the testes than in the blood, because a substantial portion of this hormone produced locally by the Leydig cells is retained in the intratubular fluid complexed with androgen-binding protein secreted by the Sertoli cells. Only this high concentration of testicular testosterone is adequate to sustain sperm production.

Estrogen production in males

Although testosterone is classically considered the male sex hormone and estrogen a female sex hormone, the distinctions are not as clear-cut as once thought. In addition to the small amount of estrogen produced by the adrenal cortex (see p.711), a portion of the testosterone secreted by the testes is converted to estrogen outside of the testes by the enzyme *aromatase*, which is widely distributed in the male reproductive tract. Estrogen is also produced in adipose tissue in both sexes. Estrogen receptors have been identified in the testes, prostate, bone, and elsewhere in males. Recent findings suggest that estrogen plays an essential role in male reproductive health, including being important in spermatogenesis and contributing to normal sexuality. Also, it likely contributes to bone homeostasis. The depth, breadth, and mechanisms of action of estrogen in males are only beginning to be explored. (Likewise, in addition to the weak androgenic hormone DHEA produced by the adrenal cortex in both sexes, the ovaries in females also secrete a small amount of testosterone, the functions of which remain unclear.)

▌ Gonadotropin-releasing hormone activity increases at puberty.

Even though the fetal testes secrete testosterone, which directs masculine development of the reproductive system, after birth the testes become dormant until puberty. During the prepubertal period, LH and FSH are not secreted at adequate levels to stimulate any significant testicular activity. The prepubertal delay in the onset of reproductive capability allows time for the individual to mature physically (although not necessarily psychologically) enough to handle childrearing. (This physical maturation is especially important in the female, whose body must support the developing fetus.)

During the prepubertal period, GnRH activity is inhibited. The pubertal process is initiated by an increase in GnRH activity sometime between 8 and 12 years of age. Early in puberty, GnRH secretion occurs only at night, causing a brief nocturnal increase in LH secretion and accordingly, testosterone secretion. The extent of GnRH secretion gradually increases as puberty progresses until the adult pattern of GnRH, FSH, LH, and testosterone secretion is established. Under the influence of the rising levels of testosterone during puberty, the physical changes that encompass the secondary sexual characteristics and reproductive maturation become evident.

The factors responsible for initiating puberty in humans remain unclear. The leading proposal focuses on a potential role for the hormone *melatonin*, which is secreted by the *pineal gland* within the brain (see p. 681). Melatonin, whose secretion decreases during exposure to the light and increases during exposure to the dark, has an antigonadotropic effect in many species. Light striking the eyes inhibits the nerve pathways responsible for stimulating melatonin secretion. In many seasonally breeding species, the overall decrease in melatonin secretion in connection with longer days and shorter nights initiates the mating season. Some researchers suggest that an observed reduction in the overall rate of melatonin secretion at puberty in humans—particularly during the night, when the peaks in GnRH secretion first occur—is the trigger for the onset of puberty.

Having completed our discussion of testicular function, we are now going to shift our attention to the roles of the other components of the male reproductive system.

▌ The reproductive tract stores and concentrates sperm and increases their fertility.

The remainder of the male reproductive system (besides the testes) is designed to deliver sperm to the female reproductive tract. Essentially, it consists of (1) a tortuous pathway of tubes (the reproductive tract), which transports sperm from the testes to the outside of the body; (2) several accessory sex glands, which contribute secretions that are important to the viability and motility of the sperm; and (3) the penis, which is designed to penetrate and deposit the sperm within the vagina of the female. We will examine each of these parts in greater detail, beginning with the reproductive tract.

Components of the male reproductive tract

A comma-shaped **epididymis** is loosely attached to the rear surface of each testis (● Figures 20-1, p. 751, and 20-6a, p. 759). After sperm are produced in the seminiferous tubules, they are swept into the epididymis as a result of the pressure created by the continual secretion of tubular fluid by the Sertoli cells. The epididymal ducts from each testis converge to form a large, thick-walled, muscular duct called the **ductus (vas) deferens.** The ductus deferens from each testis passes up out of the scrotal sac and runs back through the inguinal canal into the abdominal cavity, where it eventually empties into the urethra at the neck of the bladder (● Figure 20-1). The urethra carries sperm out of the penis during *ejaculation*, the forceful expulsion of semen from the body.

Functions of the epididymis and ductus deferens

These ducts perform several important functions. The epididymis and ductus deferens serve as the sperm's exit route from the testis. As they leave the testis, the sperm are capable of neither movement nor fertilization. They gain both capabilities during their passage through the epididymis. This maturational process is stimulated by the testosterone retained within the tubular fluid bound to androgen-binding protein. Sperm's capacity to fertilize is enhanced even further by exposure to secretions of the female reproductive tract. This enhancement of sperm's capacity in the male and female reproductive tracts is known as **capacitation.** The epididymis also concentrates the sperm a hundredfold by absorbing most of the fluid that enters from the seminiferous tubules. The maturing sperm are slowly moved through the epididymis into the ductus deferens by rhythmic contractions of the smooth muscle in the walls of these tubes.

The ductus deferens serves as an important site for sperm storage. Because the tightly packed sperm are relatively inactive and their metabolic needs are accordingly low, they can be stored in the ductus deferens for many days, even though they have no nutrient blood supply and are nourished only by simple sugars present in the tubular secretions.

Vasectomy

In a **vasectomy,** a common sterilization procedure in males, a small segment of each ductus deferens (alias vas deferens, hence the term *vasectomy*) is surgically removed after it passes from the testis but before it enters the inguinal canal, thus blocking the exit of sperm from the testes. The sperm that build up behind the tied-off testicular end of the severed ductus are removed by phagocytosis. Although this procedure blocks sperm exit, it does not interfere with testosterone activity, because the Leydig cells secrete testosterone into the blood, not through the ductus deferens. Thus testosterone-dependent masculinity or libido should not diminish following a vasectomy.

The accessory sex glands contribute the bulk of the semen.

Several accessory sex glands—the seminal vesicles and prostate—empty their secretions into the duct system before it joins the urethra (● Figure 20-1, p. 751). A pair of saclike *seminal vesicles* empty into the last portion of the two ductus deferens, one on each side. The short segment of duct that passes beyond the entry point of the seminal vesicle to join the urethra constitutes the *ejaculatory duct*. The *prostate* is a large single gland that completely surrounds the ejaculatory ducts and urethra. In a significant number of men, the prostate enlarges in middle to older age. Difficulty in urination is often encountered as the enlarging prostate impinges on the portion of the urethra that passes through the prostate. Another pair of accessory sex glands, the *bulbourethral glands*, drain into the urethra after it has passed through the prostate and just before it enters the penis. Numerous mucus-secreting glands also lie along the length of the urethra.

Semen

During ejaculation, the accessory sex glands contribute secretions that provide support for the continuing viability of the sperm inside the female reproductive tract. These secretions constitute the bulk of the **semen**, which consists of a mixture of accessory sex gland secretions, sperm, and mucus. Sperm make up only a small percentage of the total ejaculated fluid.

Functions of the male accessory sex glands

Although the accessory sex gland secretions are not absolutely essential for fertilization, they do greatly facilitate the process:

• The **seminal vesicles** (1) supply fructose, which serves as the primary energy source for ejaculated sperm; (2) secrete *prostaglandins*, which stimulate contractions of the smooth muscle in both the male and female reproductive tracts, thereby helping to transport sperm from their storage site in the male to the site of fertilization in the female oviduct; (3) provide more than half the semen, which helps wash the sperm into the urethra and also dilutes the thick mass of sperm, thus enabling them to become mobile; and (4) secrete fibrinogen, a precursor of fibrin, which forms the meshwork of a clot (see p. 404).

• The **prostate gland** (1) secretes an alkaline fluid that neutralizes the acidic vaginal secretions, an important function because sperm are more viable in a slightly alkaline environment; and (2) provides clotting enzymes and fibrinolysin. The prostatic clotting enzymes act on fibrinogen from the seminal vesicles to produce fibrin, which "clots" the semen, thus helping keep the ejaculated sperm in the female reproductive tract during withdrawal of the penis. Shortly thereafter, the seminal clot is broken down by *fibrinolysin*, a fibrin-degrading enzyme from the prostate, thus releasing motile sperm within the female tract.

• During sexual arousal, the **bulbourethral glands** secrete a mucuslike substance that provides lubrication for sexual intercourse.

▲ TABLE 20-2
Location and Functions of the Components of the Male Reproductive System

Component	Number and Location	Functions
Testis	Pair; located in the scrotum, a skin-covered sac suspended within the angle between the legs	Produce sperm
		Secrete testosterone
Epididymis and Ductus Deferens	Pair; one epididymis attached to the rear of each testis; one ductus deferens travels from each epididymis up out of the scrotal sac through the inguinal canal and empties into the urethra at the neck of the bladder	Serve as the sperm's exit route from the testis
		Serve as the site for maturation of the sperm for motility and fertility
		Concentrate and store the sperm
Seminal Vesicle	Pair; both empty into the last portion of the ductus deferens, one on each side	Supply fructose to nourish the ejaculated sperm
		Secrete prostaglandins that stimulate motility to help transport the sperm within the male and female
		Provide the bulk of the semen
		Provide precursors for the clotting of semen
Prostate Gland	Single; completely surrounds the urethra at the neck of the bladder	Secretes an alkaline fluid that neutralizes the acidic vaginal secretions
		Triggers clotting of the semen to keep the sperm in the vagina during penis withdrawal
Bulbourethral Gland	Pair; both empty into the urethra, one on each side, just before the urethra enters the penis	Secrete mucus for lubrication

▲ Table 20-2 summarizes the locations and functions of the components of the male reproductive system.

Before turning our attention to the act of delivering sperm to the female (sexual intercourse), we are going to briefly digress to discuss the diverse roles of prostaglandins, which were first discovered in semen but are abundant throughout the body.

∎ Prostaglandins are ubiquitous, locally acting chemical messengers.

Although **prostaglandins** were first identified in the semen and were believed to be of prostate gland origin (hence their name, even though they are actually secreted into the semen by the seminal vesicles), their production and actions are by no means limited to the reproductive system. These 20-carbon fatty acid derivatives are among the most ubiquitous chemical messengers in the body. They are produced in virtually all tissues from arachidonic acid, a fatty acid constituent of the phospholipids within the plasma membrane. Prostaglandins (and other closely related arachidonic-acid derivatives that are often included for convenience in the category of prostaglandins, namely, *prostacyclins*, *thromboxanes*, and *leukotrienes*) are among the most biologically active compounds known. On appropriate stimulation, arachidonic acid is split from the plasma membrane by a membrane-bound enzyme and then is converted into the appropriate prostaglandin, which acts locally within or near its site of production. After prostaglandins act, they are rapidly inactivated by local enzymes before they gain access to the blood, or if they do reach the circulatory system, they are swiftly degraded on their first pass through the lungs so that they are not dispersed through the systemic arterial system.

Prostaglandins are designated as belonging to one of three groups—PGA, PGE, or PGF—according to structural variations in the five-carbon ring that they contain at one end (● Figure 20-10). Within each group, prostaglandins are further identified by the number of double bonds present in the two side chains that project from the ring structure (for example, PGE_1 has one double bond and PGE_2 has two double bonds).

Prostaglandins exert a bewildering variety of effects. Not only are slight variations in prostaglandin structure accompanied by profound differences in biological action, but the same prostaglandin molecule may even exert opposite effects in different tissues. Besides enhancing sperm transport in semen, these abundant chemical messengers are known or suspected to exert other actions in the female reproductive system and in the respiratory, urinary, digestive, nervous, and endocrine systems, in addition to affecting platelet aggregation, fat metabolism, and inflammation (▲ Table 20-3).

As prostaglandins' various actions are better understood, new ways of manipulating them therapeutically are becoming available. A classic example is the use of aspirin, which blocks the conversion of arachidonic acid into prostaglandins, for fever reduction and pain relief. Prostaglandin action is also therapeutically inhibited in the treatment of premenstrual symptoms and menstrual cramping. Furthermore, specific prostaglandins have been medically administered in such diverse situations as inducing labor, treating asthma, and treating gastric ulcers.

● FIGURE 20-10

Structure and nomenclature of prostaglandins

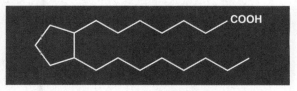

Letter designation
(PGA, PGE, PGF) denotes structural variations in the five-carbon ring

Number designation
(e.g., PGE_1, PGE_2)
denotes number of double bonds present in the two side chains

▲ TABLE 20-3
Known or Suspected Actions of Prostaglandins

Body System Activity	Actions of Prostaglandins
Reproductive System	Promote sperm transport by action on smooth muscle in the male and female reproductive tracts
	Play a role in ovulation
	Important in menstruation
	Contribute to preparation of the maternal portion of the placenta
	Contribute to parturition
Respiratory System	Some promote bronchodilation, others bronchoconstriction
Urinary System	Increase the renal blood flow
	Increase excretion of water and salt
Digestive System	Inhibit HCl secretion by the stomach
	Stimulate intestinal motility
Nervous System	Influence neurotransmitter release and action
	Act at the hypothalamic "thermostat" to increase body temperature
	Exacerbate sensation of pain
Endocrine System	Enhance cortisol secretion
	Influence tissue responsiveness to hormones in many instances
Circulatory System	Influence platelet aggregation
Fat Metabolism	Inhibit fat breakdown
Defense System	Promote many aspects of inflammation, including development of fever

Next, before considering the female in greater detail, we will examine the means by which males and females come together to accomplish reproduction.

SEXUAL INTERCOURSE BETWEEN MALES AND FEMALES

Ultimately, union of male and female gametes to accomplish reproduction in humans requires delivery of sperm-laden semen into the female vagina through the **sex act**, also known as **sexual intercourse, coitus,** or **copulation.**

▌ The male sex act is characterized by erection and ejaculation.

The *male sex act* involves two components: (1) **erection,** or hardening of the normally flaccid penis to permit its entry into the vagina, and (2) **ejaculation,** or forceful expulsion of semen into the urethra and out of the penis (▲ Table 20-4). In addition to these strictly reproduction-related components, the **sexual response cycle** also encompasses broader physiologic responses that can be divided into four phases:

1. The *excitement phase*, which includes erection and heightened sexual awareness.
2. The *plateau phase*, which is characterized by intensification of these responses, plus more generalized body responses, such as steadily increasing heart rate, blood pressure, respiratory rate, and muscle tension.
3. The *orgasmic phase*, which includes ejaculation as well as other responses that culminate the mounting sexual excitement and are collectively experienced as an intense physical pleasure.
4. The *resolution phase*, which returns the genitalia and body systems to their prearousal state.

The human sexual response is a multicomponent experience that, in addition to these physiologic phenomena, also encompasses emotional, psychological, and sociological factors. We will examine only the physiologic aspects of sex.

Erection is accomplished by penis vasocongestion.

Erection is accomplished by engorgement of the penis with blood. The penis consists almost entirely of **erectile tissue** made up of three columns of spongelike vascular spaces extending the length of the organ (● Figure 20-1, p. 751). In the absence of sexual excitation, the erectile tissues contain little blood, because the arterioles that supply these vascular chambers are constricted. As a result, the penis remains small and flaccid. During sexual arousal, these arterioles reflexly dilate and the erectile tissue fills with blood, causing the penis to enlarge both in length and width and to become more rigid. A reduction in venous outflow achieves a larger buildup of blood and further enhances erection. The veins that drain the erectile tissue are compressed by engorgement and expansion of the vascular spaces from increased arterial inflow. These local vascular responses—reflexly increased arterial inflow and mechanically reduced venous outflow—transform the penis into a hardened, elongated organ capable of penetrating the vagina.

Erection reflex

The erection reflex is a spinal reflex triggered by stimulation of highly sensitive mechanoreceptors located in the glans penis, which caps the tip of the penis. A recently identified **erection-generating center** lies in the sacral segments of the spinal cord. Tactile stimulation of the glans reflexly triggers, by means of local interneurons, increased parasympathetic vasodilator activity and decreased sympathetic vasoconstrictor activity to the penile arterioles. The result is rapid, pronounced vasodilation of these arterioles and an ensuing erection (● Figure 20-11). As long as this spinal reflex arc remains intact, erection is possible even in men paralyzed by a higher spinal-cord injury.

▲ TABLE 20-4
Components of the Male Sex Act

Components of the Male Sex Act	Definition	Accomplished By
Erection	Hardening of the normally flaccid penis to permit its entry into the vagina	Engorgement of the penis erectile tissue with blood as a result of marked parasympathetically induced vasodilation of the penile arterioles and mechanical compression of the veins
Ejaculation		
Emission phase	Emptying of sperm and accessory sex-gland secretions (semen) into the urethra	Sympathetically induced contraction of the smooth muscle in the walls of the ducts and accessory sex glands
Expulsion phase	Forceful expulsion of semen from the penis	Motor neuron–induced contraction of the skeletal muscles at the base of the penis

This parasympathetically induced vasodilation is the major instance of direct parasympathetic control over blood vessel caliber (internal diameter) in the body. Parasympathetic stimulation brings about relaxation of penile arteriolar smooth muscle by nitric oxide (endothelial-derived relaxing factor), which causes arteriolar vasodilation in response to local tissue changes elsewhere in the body (see p. 355). Arterioles are typically supplied only by sympathetic nerves, with increased sympathetic activity producing vasoconstriction and decreased sympathetic activity resulting in vasodilation (see p. 358). Concurrent parasympathetic stimulation and sympathetic inhibition of penile arterioles accomplish vasodilation more rapidly and in greater magnitude than is possible in other arterioles supplied only by sympathetic nerves. Through this efficient means of rapidly increasing blood flow into the penis, complete erection can be accomplished in as little as 5 to 10 seconds. At the same time, parasympathetic impulses promote secretion of lubricating mucus from the bulbourethral glands and the urethral glands in preparation for coitus.

A flurry of recent research has led to the discovery of numerous regions throughout the brain that can influence the male sexual response. The erection-influencing brain sites appear extensively interconnected and function as a unified network to either facilitate or inhibit the basic spinal erection reflex, depending on the momentary circumstances. As an example of facilitation, psychic stimuli, such as viewing something sexually exciting, can induce an erection in the complete absence of tactile stimulation of the penis. In contrast, failure to achieve an erection despite appropriate stimulation may occur as a result of inhibition of the erection reflex by higher brain centers. Let's examine erectile dysfunction in more detail.

Erectile dysfunction

A pattern of failing to achieve or maintain an erection suitable for sexual intercourse—**erectile dysfunction** or **impotence**—may be attributable to psychological or physical factors. An occasional episode of a failed erection does not constitute impotence, but a man who becomes overly anxious about his ability to perform the sex act may well be on his way to chronic failure. Anxiety can lead to erectile dysfunction, which fuels the man's anxiety level and thus perpetuates the problem. Impotence may also arise from physical limitations, including nerve damage, certain medications that interfere with autonomic function, and problems with blood flow through the penis.

Erectile dysfunction is widespread. Over 50% of men between ages 40 and 70 experience some impotence, climbing to nearly 70% by age 70. No wonder, then, that more prescriptions were written for the much-publicized new drug *sildenafil (Viagra)* during its first year on the market after its approval in 1998 as a treatment for erectile dysfunctional than for any other new drug in history. Sildenafil does not produce an erection, but it amplifies and prolongs an erectile response triggered by usual means of stimulation. Here's how the drug works. Nitric oxide released in response to parasympathetic stimulation activates a membrane-bound enzyme guanylate cyclase within

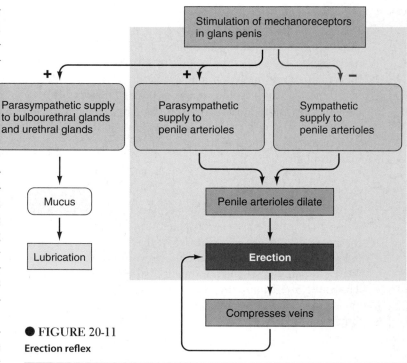

● **FIGURE 20-11**
Erection reflex

nearby arteriolar smooth muscle cells. This enzyme activates cyclic guanosine monophosphate (cGMP), an intracellular second messenger similar to cAMP (see p. 67). Cyclic GMP in turn leads to relaxation of the penile arteriolar smooth muscle, bringing about a pronounced local vasodilation. Under normal circumstances, once cGMP is activated and brings about an erection, this second messenger is broken down by the intracellular enzyme *phosphodiesterase 5 (PDE5)*. Sildenafil inhibits PDE5. As a result, cGMP remains active longer so that penile arteriolar vasodilation continues and the erection is sustained long enough for a formerly impotent man to accomplish the sex act. Just as pushing a pedal on a piano will not cause a note to be played but will prolong a played note, sildenafil cannot cause the release of nitric oxide and subsequent activation of erection-producing cGMP, but it can prolong the triggered response. The drug has no benefit for those who do not have erectile dysfunction, but its success rate has been high among sufferers of the condition. Side effects have been limited because the drug concentrates in the penis, thus having more impact on this organ than elsewhere in the body.

▌ Ejaculation includes emission and expulsion.

The second component of the male sex act is *ejaculation*. Like erection, ejaculation is accomplished by a spinal reflex. The same types of tactile and psychic stimuli that induce erection cause ejaculation when the level of excitation intensifies to a critical peak. The overall ejaculatory response occurs in two phases: emission and expulsion.

Emission

First, sympathetic impulses cause sequential contraction of smooth muscles in the prostate, reproductive ducts, and seminal vesicles. This contractile activity delivers prostatic fluid, then sperm, and finally seminal vesicle fluid (collectively, *semen*) into the urethra. This phase of the ejaculatory reflex is known as **emission.** During this time, the sphincter at the neck of the bladder is tightly closed to prevent semen from entering the bladder and urine from being expelled along with the ejaculate through the urethra.

Expulsion

Second, the filling of the urethra with semen triggers nerve impulses that activate a series of skeletal muscles at the base of the penis. Rhythmic contractions of these muscles occur at 0.8-second intervals and increase the pressure within the penis, forcibly expelling the semen through the urethra to the exterior. This is the **expulsion** phase of ejaculation.

Orgasm

The rhythmic contractions that occur during semen expulsion are accompanied by involuntary rhythmic throbbing of pelvic muscles and peak intensity of the overall body responses that were climbing during the earlier phases. Heavy breathing, a heart rate of up to 180 beats per minute, marked generalized skeletal muscle contraction, and heightened emotions are characteristic. These pelvic and overall systemic responses that culminate the sex act are associated with an intense pleasure characterized by a feeling of release and complete gratification, an experience known as **orgasm.**

Resolution

During the resolution phase following orgasm, sympathetic vasoconstrictor impulses slow the inflow of blood into the penis, causing the erection to subside. A deep relaxation ensues, often accompanied by a feeling of fatigue. Muscle tone returns to normal, while the cardiovascular and respiratory systems return to their prearousal level of activity. Once ejaculation has occurred, a temporary refractory period of variable duration ensues before sexual stimulation can trigger another erection. Males are therefore unable to experience multiple orgasms within a matter of minutes, as females sometimes do.

Volume and sperm content of the ejaculate

The volume and sperm content of the ejaculate depend on the length of time between ejaculations. The average volume of semen is 2.75 ml, ranging from 2 to 6 ml, the higher volumes following periods of abstinence. An average human ejaculate contains about 180 million sperm (66 million/ml), but some ejaculates contain as many as 400 million sperm. Both quantity and quality of the sperm are important determinants of fertility. A man is considered clinically infertile if his sperm concentration falls below 20 million/ml of semen. Even though only one spermatozoon actually fertilizes the ovum, large numbers of accompanying sperm are needed to provide sufficient

acrosomal enzymes to break down the barriers surrounding the ovum until the victorious sperm penetrates into the ovum's cytoplasm. The quality of sperm also must be taken into account when assessing the fertility potential of a semen sample. The presence of substantial numbers of sperm with abnormal motility or structure, such as sperm with distorted tails, reduces the chances of fertilization. (For a discussion of how environmental estrogens may be decreasing sperm counts as well as negatively impacting the male and female reproductive systems in other ways, see the accompanying boxed feature, ▶ Concepts, Challenges, and Controversies.)

▮ The female sexual cycle is very similar to the male cycle.

Both sexes experience the same four phases of the sexual cycle—excitement, plateau, orgasm, and resolution. Furthermore, the physiologic mechanisms responsible for orgasm are fundamentally the same in males and females.

The excitement phase in females can be initiated by either physical or psychological stimuli. Tactile stimulation of the clitoris and surrounding perineal area is an especially powerful sexual stimulus. These stimuli trigger spinal reflexes that bring about parasympathetically induced vasodilation of arterioles throughout the vagina and external genitalia, especially the clitoris. The resultant inflow of blood becomes evident as swelling of the labia and erection of the clitoris. The latter—like its male homologue, the penis—is composed largely of erectile tissue. Vasocongestion of the vaginal capillaries forces fluid out of the vessels into the vaginal lumen. This fluid, which is the first positive indication of sexual arousal, serves as the primary lubricant for intercourse. Additional lubrication is provided by the mucus secretions from the male and by mucus released during sexual arousal from glands located at the outer opening of the vagina. Also during the excitement phase in the female, the nipples become erect and the breasts enlarge as a result of vasocongestion. In addition, the majority of women show a *sex flush* during this time, which is caused by increased blood flow through the skin.

During the plateau phase, the changes initiated during the excitement phase intensify, while systemic responses similar to those in the male (such as increased heart rate, blood pressure, respiratory rate, and muscle tension) occur. Further vasocongestion of the lower third of the vagina during this time reduces its inner capacity so that it tightens around the thrusting penis, thus heightening tactile sensation for both the female and the male. Simultaneously, the uterus raises upward, lifting the cervix and enlarging the upper two-thirds of the vagina. This ballooning, or **tenting effect,** creates a space for ejaculate deposition.

If erotic stimulation continues, the sexual response culminates in orgasm as sympathetic impulses trigger rhythmic contractions of the pelvic musculature at 0.8-second intervals, the same rate as in males. The contractions occur most intensely in the engorged lower third of the vaginal canal. Systemic responses identical to those of the male orgasm also occur. In

Environmental "Estrogens": Bad News for the Reproductive System

Unknowingly, during the last 50 years we have been polluting our environment with synthetic endocrine-disrupting chemicals as an unintended side effect of industrialization. Known as **endocrine disrupters,** these hormonelike pollutants bind with the receptor sites normally reserved for the naturally occurring hormones. Depending on how they interact with the receptors, these disrupters can either mimic or block normal hormonal activity. Most endocrine disrupters exert feminizing effects. Many of these environmental contaminants mimic or alter the action of estrogen, the feminizing steroid hormone produced by the female ovaries. Although not yet conclusive, laboratory and field studies suggest that these estrogen disrupters might be responsible for some disturbing trends in reproductive health problems, such as falling sperm counts in males and an increased incidence of breast cancer in females.

Estrogenic pollutants are everywhere. They contaminate our food, drinking water, and air. Proved feminizing synthetic compounds include (1) certain weed killers and insecticides, (2) some detergent breakdown products, (3) petroleum by-products found in car exhaust, (4) a common food preservative used to retard rancidity, and (5) softeners that make plastics flexible. These plastic softeners are commonly found in food packaging and can readily leach into food with which they come in contact, especially during heating. They have also been found to leach into the saliva from some babies' plastic teething toys and from dental sealants used to protect teeth from decay-causing bacteria. They are also in numerous medical products, like the bags in which blood is stored. Plastic softeners are among the most plentiful industrial contaminants in our environment.

Investigators are only beginning to identify and understand the implications for reproductive health of a myriad synthetic chemicals that have become such an integral part of modern societies. An estimated 87,000 synthetic chemicals are already in our environment. Scientists suspect that the estrogen-mimicking chemicals among these may underlie a spectrum of reproductive disorders that have been rising in the past 50 years—the same time period during which large amounts of these pollutants have been introduced into our environment. Here are examples of male reproductive dysfunctions that may be circumstantially linked to exposure to environmental estrogen disrupters:

- *Falling sperm counts.* The average sperm count fell from 113 million sperm per milliliter of semen in 1940 to 66 million/ml in 1990. Making matters worse, the volume of a single ejaculate has declined from 3.40 ml to 2.75 ml. This means that men on average are now ejaculating less than half the number of sperm as men did 50 years ago—a drop from more than 380 million sperm to about 180 million sperm per ejaculate. Furthermore, the number of motile sperm has also dipped. Importantly, the sperm count has not declined in the less polluted areas of the world during the same time period.

- *Increased incidence of testicular and prostate cancer.* Cases of testicular cancer have tripled since 1940, and the rate continues to climb. Prostate cancer has also been on the rise over the same time period.

- *Rising number of male reproductive tract abnormalities at birth.* The incidence of *cryptorchidism* (undescended testis) nearly doubled from the 1950s to the 1970s. The number of cases of *hypospadia,* a malformation of the penis, more than doubled between the mid-1960s and the mid-1990s. Hypospadia results when the urethral fold fails to fuse closed during the development of a male fetus.

- *Evidence of gender bending in animals.* Some fish and wild animal populations that have been severely exposed to environmental estrogens—such as those living in or near water heavily polluted with hormone-mimicking chemical wastes—display a high rate of grossly impaired reproductive systems. Examples include male fish that are hermaphrodites (possessing both male and female reproductive parts) and male alligators with abnormally small penises. Similar reproductive abnormalities have been identified in land mammals. Presumably, excessive estrogen exposure is emasculating these populations.

- *Decline in male births.* Many countries are reporting a slight decline in the ratio of baby boys to baby girls being born. Although several plausible explanations have been put forth, many researchers attribute this troubling trend to disruption of normal male fetal development by environmental estrogens. In one compelling piece of circumstantial evidence, people inadvertently exposed to the highest level of an endocrine-disrupting agent during an industrial accident have subsequently had all daughters and no sons, whereas those least exposed have had the normal ratio of girls and boys.

Environmental estrogens are also implicated in the rising incidence of breast cancer in females. Breast cancer is 25% to 30% more prevalent now than in the 1940s. Many of the established risk factors for breast cancer, such as starting to menstruate earlier than usual and undergoing menopause later than usual, are associated with an elevation in the total lifetime exposure to estrogen. Because increased exposure to natural estrogen bumps up the risk for breast cancer, prolonged exposure to environmental estrogens might be contributing to the rising prevalence of this malignancy among women (and men too).

In addition to the estrogen disrupters, scientists recently identified a new class of chemical offenders—androgen disrupters that either mimic or suppress the action of male hormones. For example, studies suggest that bacteria in waste water from pulp mills can convert the sterols in pine pulp into androgens. By contrast, antiandrogen compounds have been found in the fungicides commonly sprayed on vegetable and fruit crops. Yet another cause for concern are the androgens used by the livestock industry to enhance the production of muscle (that is, meat) in feedlot cattle. (Androgens have a protein anabolic effect.) These drugs do not end up in the meat, but they can get into drinking water and other food as hormone-laden feces contaminate rivers and streams.

In response to the growing evidence that has emerged circumstantially linking a wide variety of environmental pollutants to disturbing reproductive abnormalities, the U.S. Congress legally mandated the Environmental Protection Agency (EPA) in 1996 to determine which synthetic chemicals might be endocrine disrupters. In response, the EPA formed an advisory committee, which in 1998 proposed an ambitious plan to begin comprehensive testing of manufactured compounds for their potential to disrupt hormones in humans and wildlife. Although eventually all the 87,000 existing synthetic compounds will be tested, the initial screening is narrowed to evaluating the endocrine-disrupting potential of 15,000 widely used chemicals. Declaring this a national health priority, the government has allocated millions of dollars for this research.

fact, the orgasmic experience in females parallels that of males with two exceptions. First, there is no female counterpart to ejaculation. Second, females do not become refractory following an orgasm, so they can respond immediately to continued erotic stimulation and achieve multiple orgasms. If stimulation continues, the sexual intensity only diminishes to the plateau level following orgasm and can quickly be brought to a peak again. Women have been known to achieve as many as 12 successive orgasms in this manner.

During resolution, pelvic vasocongestion and the systemic manifestations gradually subside. As with males, this is a time of great physical relaxation for females.

We will now examine how females fulfill their part of the reproductive process.

FEMALE REPRODUCTIVE PHYSIOLOGY

Female reproductive physiology is much more complex than male reproductive physiology.

▌ Complex cycling characterizes female reproductive physiology.

Unlike the continuous sperm production and essentially constant testosterone secretion characteristic of the male, release of ova is intermittent, and secretion of female sex hormones displays wide cyclical swings. The tissues influenced by these sex hormones also undergo cyclical changes, the most obvious of which is the monthly menstrual cycle. During each cycle, the female reproductive tract is prepared for the fertilization and implantation of an ovum released from the ovary at ovulation. If fertilization does not occur, the cycle repeats. If fertilization does occur, the cycles are interrupted while the female system adapts to nurture and protect the newly conceived human being until it has developed into an individual capable of living outside the maternal environment. Furthermore, the female continues her reproductive functions after birth by producing milk (lactation) for the baby's nourishment. Thus the female reproductive system is characterized by complex cycles that are interrupted only by more complex changes should pregnancy ensue.

The ovaries, as the primary female reproductive organs, perform the dual function of producing ova (oogenesis) and secreting the female sex hormones, estrogen and progesterone. These hormones act together to promote fertilization of the ovum and to prepare the female reproductive system for pregnancy. Estrogen in the female governs many functions similar to those carried out by testosterone in the male, such as maturation and maintenance of the entire female reproductive system and establishment of female secondary sexual characteristics. In general, the actions of estrogen are important to preconception events. Estrogen is essential for ova maturation and release, development of physical characteristics that are sexually attractive to males, and transport of sperm from the vagina to the site of fertilization in the oviduct. Furthermore, estrogen contributes to breast development in anticipation of lactation. The other ovarian steroid, progesterone, is important in preparing a suitable environment for nourishing a developing embryo/fetus and for contributing to the breasts' ability to produce milk.

As in males, reproductive capability begins at puberty in females, but unlike males, who have reproductive potential through life, female reproductive potential ceases during middle age at menopause.

▌ The steps of gametogenesis are the same in both sexes, but the timing and outcome differ sharply.

Oogenesis contrasts sharply with spermatogenesis in several important aspects, even though the identical steps of chromosome replication and division take place during gamete production in both sexes. The undifferentiated primordial germ cells in the fetal ovaries, the oogonia (comparable to the spermatogonia), divide mitotically to give rise to 6 to 7 million oogonia by the fifth month of gestation, when mitotic proliferation ceases.

Formation of primary oocytes and primary follicles

During the last part of fetal life, the oogonia begin the early steps of the first meiotic division but do not complete it. Known now as primary oocytes, they contain the diploid number of 46 replicated chromosomes, which are gathered into homologous pairs but do not separate. The primary oocytes remain in this state of meiotic arrest for years until they are prepared for ovulation.

Before birth, each primary oocyte is surrounded by a single layer of granulosa cells. Together, an oocyte and surrounding granulosa cells make up a primary follicle. Oocytes that fail to be incorporated into follicles self-destruct by apoptosis (see p. 70). At birth only about 2 million primary follicles remain, each containing a single primary oocyte capable of producing a single ovum. No new oocytes or follicles appear after birth; the follicles already present in the ovaries at birth serve as a reservoir from which all ova throughout the reproductive life of a female must arise. Of these follicles, only about 400 will mature and release ova.

The pool of primary follicles present at birth gives rise to an ongoing trickle of developing follicles. Once it starts to develop, a follicle is destined for one of two fates: It will reach maturity and ovulate, or it will degenerate to form scar tissue, a process known as atresia. Until puberty, all the follicles that start to develop undergo atresia in the early stages without ever ovulating. Even for the first few years after puberty, many of the cycles are anovulatory (that is, no ovum is released). Of the initial pool of follicles, 99.98% never ovulate but instead undergo atresia at some stage in development. By menopause, which occurs on average in a woman's early fifties, few primary follicles remain, having either already ovulated or become atretic. From this point on, the woman's reproductive capacity ceases.

This limited gamete potential, which is already determined at birth in females, is in sharp contrast to the continual process of spermatogenesis in males, who have the potential to produce several hundred million sperm in a single day. Furthermore,

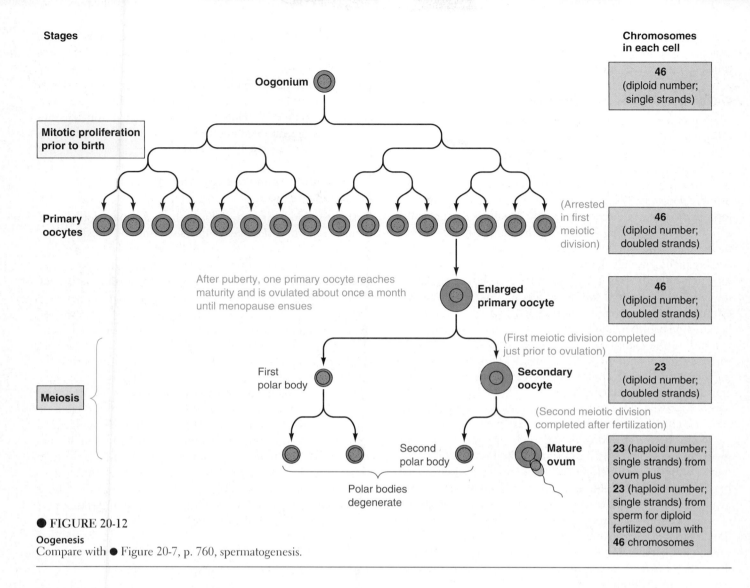

Oogonium

46
(diploid number;
single strands)

**Mitotic proliferation
prior to birth**

**Primary
oocytes**

(Arrested
in first
meiotic
division)

46
(diploid number;
doubled strands)

After puberty, one primary oocyte reaches
maturity and is ovulated about once a month
until menopause ensues

**Enlarged
primary oocyte**

46
(diploid number;
doubled strands)

(First meiotic division completed
just prior to ovulation)

Meiosis

First
polar body

**Secondary
oocyte**

23
(diploid number;
doubled strands)

(Second meiotic division
completed after fertilization)

Second
polar body

**Mature
ovum**

23 (haploid number;
single strands) from
ovum plus
23 (haploid number;
single strands) from
sperm for diploid
fertilized ovum with
46 chromosomes

Polar bodies
degenerate

● **FIGURE 20-12**

Oogenesis
Compare with ● Figure 20-7, p. 760, spermatogenesis.

considerable chromosome wastage occurs in oogenesis compared with spermatogenesis. Let's see how.

Formation of secondary oocytes and secondary follicles

The primary oocyte within a primary follicle is still a diploid cell that contains 46 doubled chromosomes. From puberty until menopause, a portion of the resting pool of follicles starts developing into **secondary (antral) follicles** on a cyclical basis. The number of developing follicles at any time is roughly proportional to the size of the pool, but the mechanisms that determine which follicles in the pool will develop during a given cycle are unknown. Development of a secondary follicle is characterized by growth of the primary oocyte and by expansion and differentiation of the surrounding cell layers. The oocyte enlarges about a thousandfold. This oocyte enlargement is caused by a buildup of cytoplasmic materials that will be needed by the early embryo.

Just before ovulation, the primary oocyte, whose nucleus has been in meiotic arrest for years, completes its first meiotic division. This division yields two daughter cells, each receiving a haploid set of 23 doubled chromosomes, analogous to the

formation of secondary spermatocytes (● Figure 20-12). However, almost all the cytoplasm remains with one of the daughter cells, now called the **secondary oocyte,** which is destined to become the ovum. The chromosomes of the other daughter cell together with a small share of cytoplasm form the **first polar body.** In this way, the ovum-to-be loses half of its chromosomes to form a haploid gamete but retains all of its nutrient-rich cytoplasm. The nutrient-poor polar body soon degenerates.

Formation of a mature ovum

Actually, the secondary oocyte, and not the mature ovum, is ovulated and fertilized, but common usage refers to the developing female gamete as an *ovum* even in its primary and secondary oocyte stages. Sperm entry into the secondary oocyte is needed to trigger the second meiotic division. Oocytes that are not fertilized never complete this final division. During this division, a half set of chromosomes along with a thin layer of cytoplasm is extruded as the **second polar body.** The other half set of 23 unpaired chromosomes remains behind in what is now the **mature ovum.** These 23 maternal chromosomes unite with the 23 paternal chromosomes of the penetrating sperm to complete fertilization. If the first polar body has not already degenerated, it too undergoes the second meiotic division at the same time the fertilized secondary oocyte is dividing its chromosomes.

Comparison of steps in oogenesis and spermatogenesis

The steps involved in chromosome distribution during oogenesis parallel those of spermatogenesis, except that the cytoplasmic distribution and time span for completion sharply differ. Just as four haploid spermatids are produced by each primary spermatocyte, four haploid daughter cells are produced by each primary oocyte (if the first polar body does not degenerate before it completes the second meiotic division). In spermatogenesis, each daughter cell develops into a highly specialized, motile spermatozoon unencumbered by unessential cytoplasm and organelles, its only destiny being to supply half of the genes for a new individual. In oogenesis, however, of the four daughter cells only the one destined to become the ovum receives cytoplasm. This uneven distribution of cytoplasm is important, because the ovum, in addition to providing half the genes, provides all of the cytoplasmic components needed to support early development of the fertilized ovum. The large, relatively undifferentiated ovum contains numerous nutrients, organelles, and structural and enzymatic proteins. The three other cytoplasm-scarce daughter cells, the polar bodies, rapidly degenerate, their chromosomes being deliberately wasted.

Note also the considerable difference in time to complete spermatogenesis and oogenesis. It takes about two months for a spermatogonium to develop into fully remodeled spermatozoa. In contrast, development of an oogonium (present before birth) to a mature ovum requires anywhere from 11 years (beginning of ovulatory cycles at onset of puberty) to 50 years (end of ovulation at onset of menopause). The older age of ova released by women in their late thirties and forties is believed to account for the higher incidence of genetic abnormalities, such as Down's syndrome, in children born to women in this age range.

▌ The ovarian cycle consists of alternating follicular and luteal phases.

After the onset of puberty, the ovary constantly alternates between two phases: the **follicular phase,** which is dominated by the presence of *maturing follicles*; and the **luteal phase,** which is characterized by the presence of the *corpus luteum* (to be described shortly). This cycle is normally interrupted only by pregnancy and is finally terminated by menopause. The average ovarian cycle lasts 28 days, but this varies among women and among cycles in any particular woman. The follicle operates in the first half of the cycle to produce a mature egg ready for ovulation at midcycle. The corpus luteum takes over during the last 14 days of the cycle to prepare the female reproductive tract for pregnancy if fertilization of the released egg occurs.

▌ The follicular phase is characterized by the development of maturing follicles.

At any given time throughout the cycle, a portion of the primary follicles is starting to develop. However, only those that

do so during the follicular phase, when the hormonal milieu is right to promote their maturation, continue beyond the early stages of development. The others, lacking hormonal support, undergo atresia. During follicular development, as the primary oocyte is synthesizing and storing materials for future use if fertilized, important changes are taking place in the cells surrounding the reactivated oocyte in preparation for the egg's release from the ovary (● Figure 20-13).

Proliferation of granulosa cells and formation of the zona pellucida

First, the single layer of granulosa cells in a primary follicle proliferates to form several layers that surround the oocyte. These granulosa cells secrete a thick, gel-like material that covers the oocyte and separates it from the surrounding granulosa cells. This intervening membrane is known as the **zona pellucida.**

Scientists have recently discovered gap junctions penetrating the zona pellucida and extending between the oocyte and surrounding granulosa cells in a developing follicle. Ions and small molecules can travel through these connecting tunnels. Recall that gap junctions between excitable cells permit the spread of action potentials from one cell to the next as charge-carrying ions pass through these connecting tunnels (see p. 64). The cells in a developing follicle are not excitable, so gap junctions here serve a role other than transfer of electrical activity. Glucose, amino acids, and other important molecules are delivered to the oocyte from the granulosa cells through these tunnels, enabling the egg to stockpile these critical nutrients. Also, signaling molecules pass through the gap junctions in both directions, helping coordinate the changes that take place in the oocyte and surrounding cells as both mature and prepare for ovulation.

Proliferation of thecal cells; estrogen secretion

At the same time the oocyte is enlarging and the granulosa cells are proliferating, specialized ovarian connective tissue cells in contact with the expanding granulosa cells proliferate and differentiate to form an outer layer of **thecal cells.** The thecal and granulosa cells, collectively known as **follicular cells,** function as a unit to secrete estrogen. Of the three physiologically important estrogens—estradiol, estrone, and estriol—**estradiol** is the principal ovarian estrogen.

Formation of the antrum

The hormonal environment that exists during the follicular phase promotes enlargement and development of the follicular cells' secretory capacity, converting the primary follicle into a secondary, or antral, follicle capable of estrogen secretion. This stage of follicular development is characterized by the formation of a fluid-filled **antrum** in the midst of the granulosa cells (● Figures 20-13 and 20-14). The follicular fluid originates partially from transudation (passage through capillary pores) of plasma and partially from follicular cell secretions. As the follicular cells start producing estrogen, some of this hormone is secreted into the blood for distribution throughout the body. However, a portion of the estrogen collects in the hormone-rich antral fluid.

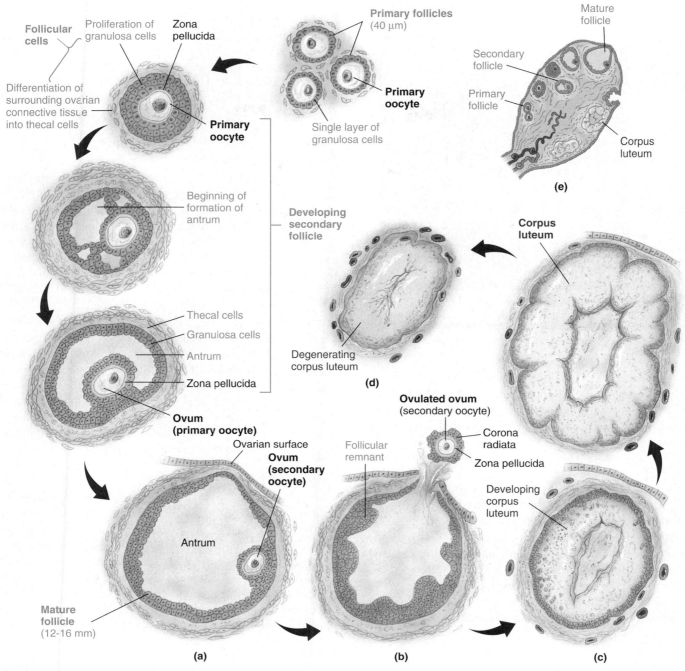

● FIGURE 20-13

Development of the follicle, ovulation, and formation of the corpus luteum
(a) Stages in follicular development from a primary follicle through a mature follicle. (b) Rupture of a mature follicle and release of an ovum (secondary oocyte) at ovulation. (c) Formation of a corpus luteum from the old follicular cells after ovulation. (d) Degeneration of the corpus luteum if the released ovum is not fertilized. (e) Ovary (actual size), showing development of a follicle, ovulation, and formation and degeneration of a corpus luteum.

The oocyte has reached full size by the time the antrum begins to form. The shift to an antral follicle initiates a period of rapid follicular growth. During this time, the follicle increases in size from a diameter of less than 1 mm to 12 to 16 mm shortly before ovulation. Part of the follicular growth is due to continued proliferation of the granulosa and thecal cells, but most is due to a dramatic expansion of the an-

trum. As the follicle grows, estrogen is produced in increasing quantities.

Formation of a mature follicle

One of the follicles usually grows more rapidly than the others, developing into a **mature (preovulatory, tertiary, or Graafian) follicle** within about 14 days after the onset of follicular devel-

The Reproductive System   773

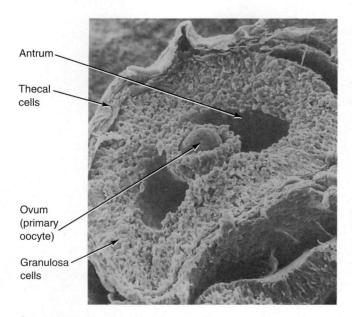

● FIGURE 20-14

Scanning electron micrograph of a developing secondary follicle

Labels: Antrum; Thecal cells; Ovum (primary oocyte); Granulosa cells

opment. The antrum occupies most of the space in a mature follicle. The oocyte, surrounded by the zona pellucida and a single layer of granulosa cells, is displaced asymmetrically at one side of the growing follicle, in a little mound that protrudes into the antrum.

Ovulation

The greatly expanded mature follicle bulges on the ovarian surface, creating a thin area that ruptures to release the oocyte at **ovulation.** Rupture of the follicle is facilitated by the release from the follicular cells of enzymes that digest the connective tissue in the follicular wall. The bulging wall is thus weakened so that it balloons out even farther, to the point that it can no longer contain the rapidly expanding follicular contents.

Just before ovulation, the oocyte completes its first meiotic division. The ovum (secondary oocyte), still surrounded by its tightly adhering zona pellucida and granulosa cells (now called the **corona radiata,** meaning "radiating crown") is swept out of the ruptured follicle into the abdominal cavity by the leaking antral fluid (● Figure 20-13b). The released ovum is quickly drawn into the oviduct, where fertilization may or may not take place.

The other developing follicles that failed to reach maturation and ovulate undergo degeneration, never to be reactivated. Occasionally, two (or perhaps more) follicles reach maturation and ovulate at about the same time. If both are fertilized, **fraternal twins** result. Because fraternal twins arise from separate ova fertilized by separate sperm, they share no more in common than any other two siblings except for the same birth date. **Identical twins,** in contrast, develop from a single fertilized ovum that completely divides into two separate, genetically identical embryos at a very early stage in development.

Rupture of the follicle at ovulation signals the end of the follicular phase and ushers in the luteal phase.

▌ The luteal phase is characterized by the presence of a corpus luteum.

The ruptured follicle left behind in the ovary after release of the ovum changes rapidly. The granulosa and thecal cells remaining in the remnant follicle first collapse into the emptied antral space that has been partially filled by clotted blood.

Formation of the corpus luteum; estrogen and progesterone secretion

These old follicular cells soon undergo a dramatic structural transformation to form the **corpus luteum,** in a process called **luteinization** (● Figure 20-13c). The follicular-turned-luteal cells hypertrophy and are converted into very active steroidogenic (steroid hormone–producing) tissue. Abundant storage of cholesterol, the steroid precursor molecule, in lipid droplets within the corpus luteum gives this tissue a yellowish appearance, hence its name (*corpus* means "body"; *luteum* means "yellow").

The corpus luteum becomes highly vascularized as blood vessels from the thecal region invade the luteinizing granulosa. These changes are appropriate for the corpus luteum's function: to secrete into the blood abundant quantities of progesterone along with smaller amounts of estrogen. Estrogen secretion in the follicular phase followed by progesterone secretion in the luteal phase is essential for preparing the uterus to be a suitable site for implantation of a fertilized ovum. The corpus luteum becomes fully functional within four days after ovulation, but it continues to increase in size for another four or five days.

Degeneration of the corpus luteum

If the released ovum is not fertilized and does not implant, the corpus luteum degenerates within 14 days after its formation (● Figure 20-13d). The luteal cells degenerate and are phagocytized, the vascular supply is withdrawn, and connective tissue rapidly fills in to form a fibrous tissue mass known as the **corpus albicans** ("white body"). The luteal phase is now over, and one ovarian cycle is complete. A new wave of follicular development, which begins when degeneration of the old corpus luteum is completed, signals the onset of a new follicular phase.

Corpus luteum of pregnancy

If fertilization and implantation do occur, the corpus luteum continues to grow and produce increasing quantities of progesterone and estrogen instead of degenerating. Now called the *corpus luteum of pregnancy,* this ovarian structure persists until pregnancy ends. It provides the hormones essential for maintaining pregnancy until the developing placenta can take over this crucial function. You will learn more about the role of these structures later.

▌ The ovarian cycle is regulated by complex hormonal interactions.

The ovary has two related endocrine units: the estrogen-secreting follicle during the first half of the cycle and the corpus luteum, which secretes both progesterone and estrogen, during the last half of the cycle. These units are sequentially triggered by complex cyclic hormonal relationships among the hypothalamus, anterior pituitary, and these two ovarian endocrine units.

As in the male, gonadal function in the female is directly controlled by the anterior pituitary gonadotropic hormones, follicle-stimulating hormone (FSH) and luteinizing hormone (LH). These hormones, in turn, are regulated by hypothalamic gonadotropin-releasing hormone (GnRH) and feedback actions of gonadal hormones. Differing from the male, however,

control of the female gonads is complicated by the cyclical nature of ovarian function. For example, the effects of FSH and LH on the ovaries depend on the stage of the ovarian cycle. Also in contrast to the male, FSH is not strictly responsible for gametogenesis, nor is LH solely responsible for gonadal hormone secretion. We will consider control of follicular function, ovulation, and the corpus luteum separately, using ● Figure 20-15 as a means of integrating the various concurrent and

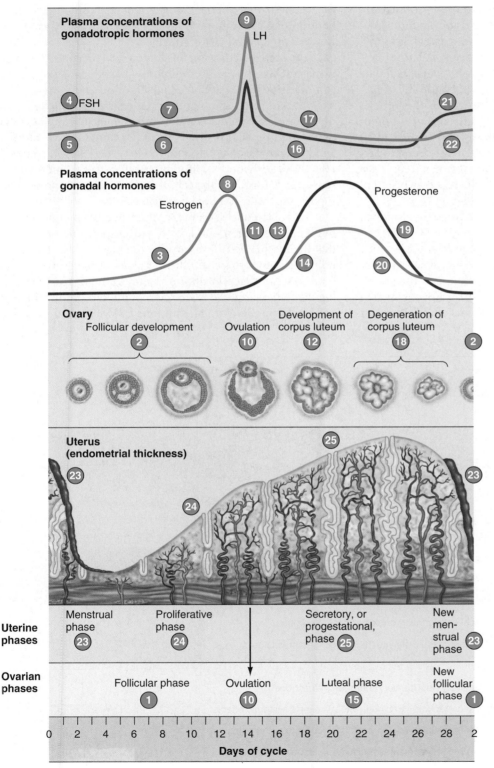

● **FIGURE 20-15**

Correlation between hormonal levels and cyclic ovarian and uterine changes
During the follicular phase (the first half of the ovarian cycle 1 , the ovarian follicle 2 secretes estrogen 3 under the influence of FSH 4 , LH 5 , and estrogen 3 itself. The low but rising levels of estrogen (1) inhibit FSH secretion, which declines during the last part of the follicular phase 6 , and (2) incompletely suppress tonic LH secretion, which continues to rise throughout the follicular phase 7 . When the follicular output of estrogen reaches its peak 8 , the high levels of estrogen trigger a surge in LH secretion at midcycle 9 . This LH surge brings about ovulation of the mature follicle 10 . Estrogen secretion plummets 11 when the follicle meets its demise at ovulation.

The old follicular cells are transformed into the corpus luteum 12 , which secretes progesterone 13 as well as estrogen 14 during the luteal phase the last half of the ovarian cycle 15 . Progesterone strongly inhibits both FSH 16 and LH 17 , which continue to decrease throughout the luteal phase. The corpus luteum degenerates 18 in about two weeks if the released ovum has not been fertilized and implanted in the uterus. Progesterone 19 and estrogen 20 levels sharply decrease when the corpus luteum degenerates, removing the inhibitory influences on FSH and LH. As these anterior pituitary hormone levels start to rise again 21 , 22 on the withdrawal of inhibition, they begin to stimulate the development of a new batch of follicles as a new follicular phase is ushered in 1 , 2 .

Concurrent uterine phases reflect the influences of the ovarian hormones on the uterus. Early in the follicular phase, the highly vascularized, nutrient-rich endometrial lining is sloughed off (the uterine menstrual phase) 23 This sloughing results from the withdrawal of estrogen and progesterone 19 , 20 when the old corpus luteum degenerated at the end of the preceding luteal phase 18 . Late in the follicular phase, the rising levels of estrogen 3 cause the endometrium to thicken (the uterine proliferative phase) 24 . After ovulation 10 , progesterone from the corpus luteum 13 brings about vascular and secretory changes in the estrogen-primed endometrium to produce a suitable environment for implantation (the uterine secretory, or progestational, phase) 25 . When the corpus luteum degenerates 18 , a new ovarian follicular phase 1 , 2 and uterine menstrual phase 23 begin.

sequential activities that take place throughout the cycle. To facilitate correlation between this rather intimidating figure and the accompanying text description of this complex cycle, the circled numbers in the figure and its legend correspond to the circled numbers in the text description.

Control of follicular function

We will begin with the follicular phase of the ovarian cycle 1 . The factors that initiate follicular development are poorly understood. The early stages of preantral follicular growth and oocyte maturation do not require gonadotropic stimulation. Hormonal support is required, however, for antrum formation, further follicular development 2 , and estrogen secretion 3 . Estrogen, FSH 4 , and LH 5 are all needed. Antrum formation is induced by FSH. Both FSH and estrogen stimulate proliferation of the granulosa cells. Both LH and FSH are required for synthesis and secretion of estrogen by the follicle, but these hormones act on different cells and at different steps in the estrogen production pathway (● Figure 20-16). Both granulosa and thecal cells participate in estrogen production. The conversion of cholesterol into estrogen requires a number of sequential steps, the last of which is conversion of androgens into estrogens (see ● Figure 18-3, p. 674). Thecal cells readily produce androgens but have limited capacity to convert them into estrogens. Granulosa cells, in contrast, can readily convert androgens into estrogens but cannot produce androgens in the first place. LH acts on the thecal cells to stimulate androgen production, whereas FSH acts on the granulosa cells to promote conversion of thecal androgens (which diffuse into the granulosa cells from the thecal cells) into es-

trogens. Because low basal levels of FSH 6 are sufficient to promote this final conversion to estrogen, the rate of estrogen secretion by the follicle primarily depends on the circulating level of LH, which continues to rise during the follicular phase 7 . Furthermore, as the follicle continues to grow, more estrogen is produced simply because more estrogen-producing follicular cells are present.

Part of the estrogen produced by the growing follicle is secreted into the blood and is responsible for the steadily increasing plasma estrogen levels during the follicular phase 8 . The remainder of the estrogen remains within the follicle, contributing to the antral fluid and stimulating further granulosa cell proliferation (● Figure 20-16).

The secreted estrogen, in addition to acting on sex-specific tissues such as the uterus, inhibits the hypothalamus and anterior pituitary in negative-feedback fashion (● Figure 20-17). The low but rising levels of estrogen characterizing the follicular phase act directly on the hypothalamus to inhibit GnRH secretion, thus suppressing GnRH-prompted release of FSH and LH from the anterior pituitary. However, estrogen's primary effect is directly on the pituitary itself. Estrogen reduces the sensitivity to GnRH of the cells that produce gonadotropic hormones, especially the FSH-producing cells.

This differential sensitivity of FSH- and LH-producing cells induced by estrogen is at least in part responsible for the fact that the plasma FSH level, unlike the plasma LH concentration, declines during the follicular phase as the estrogen level rises 6 . Another contributing factor to the fall in FSH during the follicular phase is secretion of *inhibin* by the follicular cells. Inhibin preferentially inhibits FSH secretion by acting

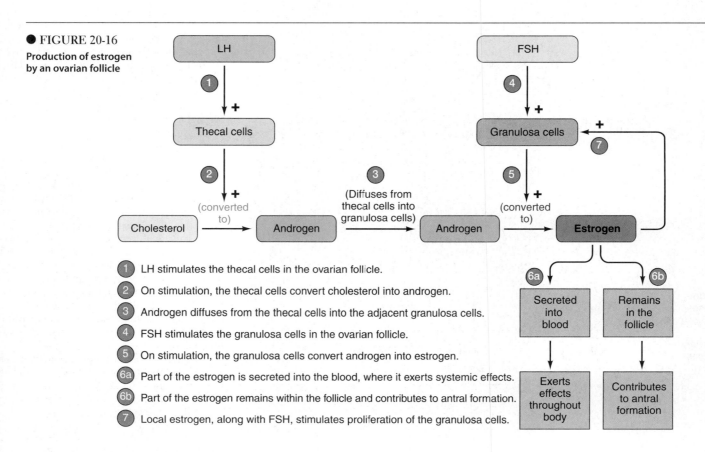

● **FIGURE 20-16**
Production of estrogen by an ovarian follicle

1 LH stimulates the thecal cells in the ovarian follicle.

2 On stimulation, the thecal cells convert cholesterol into androgen.

3 Androgen diffuses from the thecal cells into the adjacent granulosa cells.

4 FSH stimulates the granulosa cells in the ovarian follicle.

5 On stimulation, the granulosa cells convert androgen into estrogen.

6a Part of the estrogen is secreted into the blood, where it exerts systemic effects.

6b Part of the estrogen remains within the follicle and contributes to antral formation.

7 Local estrogen, along with FSH, stimulates proliferation of the granulosa cells.

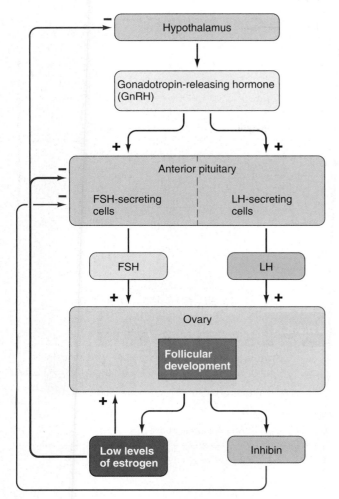

● **FIGURE 20-17**

Feedback control of FSH and tonic LH secretion during the follicular phase

at the anterior pituitary, just as it does in the male. The decline in FSH secretion brings about atresia of all but the single most mature of the developing follicles.

In contrast to FSH, LH secretion continues to rise slowly during the follicular phase ⑦ despite inhibition of GnRH (and thus, indirectly, LH) secretion. This seeming paradox is due to the fact that estrogen alone cannot completely suppress **tonic** (low-level, ongoing) **LH secretion;** to completely inhibit tonic LH secretion, both estrogen and progesterone are required. Because progesterone does not appear until the luteal phase of the cycle, the basal level of circulating LH slowly increases during the follicular phase under incomplete inhibition by estrogen alone.

Control of ovulation

Ovulation and subsequent luteinization of the ruptured follicle are triggered by an abrupt, massive increase in LH secretion ⑨ . This **LH surge** brings about four major changes in the follicle:

1. It halts estrogen synthesis by the follicular cells ⑪ .
2. It reinitiates meiosis in the oocyte of the developing follicle, apparently by blocking release of an *oocyte maturation-*

inhibiting substance produced by the granulosa cells. This substance is believed to be responsible for arresting meiosis in the primary oocytes once they are wrapped within granulosa cells in the fetal ovary.

3. It triggers production of locally acting prostaglandins, which induce ovulation by promoting vascular changes that cause rapid swelling of the follicle while inducing enzymatic digestion of the follicular wall. Together these actions lead to rupture of the weakened wall that covers the bulging follicle ⑩ .

4. It causes differentiation of follicular cells into luteal cells. Because the LH surge triggers both ovulation and luteinization, formation of the corpus luteum automatically follows ovulation ⑫ . Thus the midcycle burst in LH secretion is a dramatic point in the cycle; it terminates the follicular phase and initiates the luteal phase ⑮ .

The two different modes of LH secretion—the tonic secretion of LH ⑦ responsible for promoting ovarian hormone secretion and the LH surge ⑨ that causes ovulation—not only occur at different times and produce different effects on the ovaries but also are controlled by different mechanisms. Tonic LH secretion is partially suppressed ⑦ by the inhibitory action of the low, rising levels of estrogen ③ during the follicular phase and is completely suppressed ⑰ by the increasing levels of progesterone during the luteal phase ⑬ . Because tonic LH secretion stimulates both estrogen and progesterone secretion, this is a typical negative-feedback control system.

In contrast, the LH surge is triggered by a *positive-feedback* effect. Whereas the low, rising levels of estrogen early in the follicular phase *inhibit* LH secretion, the high level of estrogen that occurs during peak estrogen secretion late in the follicular phase ⑧ *stimulates* LH secretion and initiates the LH surge (● Figure 20-18). Thus LH enhances estrogen production by the follicle, and the resultant peak estrogen concentration stimulates LH secretion. The high plasma concentration of estrogen acts directly on the hypothalamus to increase GnRH, thereby increasing both LH and FSH secretion. It also acts directly on the anterior pituitary to specifically increase the sensitivity of LH-secreting cells to GnRH. The latter effect accounts in large part for the much greater surge in LH secretion compared to FSH secretion at midcycle ⑨ . Also, continued inhibin secretion by the follicular cells preferentially inhibits the FSH-secreting cells, keeping the FSH levels from rising as high as the LH levels. There is no known role for the modest midcycle surge in FSH that accompanies the pronounced and pivotal LH surge. Because only a mature, preovulatory follicle, not follicles in earlier stages of development, can secrete high-enough levels of estrogen to trigger the LH surge, ovulation is not induced until a follicle has reached the proper size and degree of maturation. In a way, then, the follicle lets the hypothalamus know when it is ready to be stimulated to ovulate. The LH surge lasts for only one to two days at midcycle, just before ovulation.

Control of the corpus luteum

LH "maintains" the corpus luteum; that is, after triggering development of the corpus luteum, LH stimulates ongoing steroid

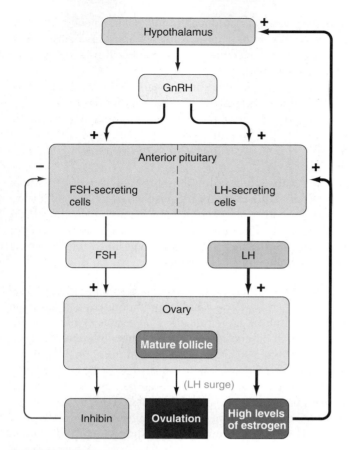

● FIGURE 20-18
Control of the LH surge at ovulation

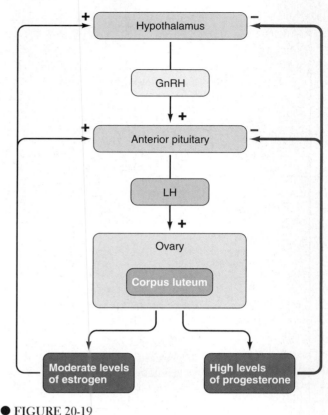

● FIGURE 20-19
Feedback control during the luteal phase

hormone secretion by this ovarian structure. Under the influence of LH, the corpus luteum secretes both progesterone 13 and estrogen 14 , with progesterone being its most abundant hormonal product. The plasma progesterone level increases for the first time during the luteal phase. No progesterone is secreted during the follicular phase. Therefore, the follicular phase is dominated by estrogen and the luteal phase by progesterone.

A transitory drop in the level of circulating estrogen occurs at midcycle 11 as the estrogen-secreting follicle meets its demise at ovulation. The estrogen level climbs again during the luteal phase because of the corpus luteum's activity, although it does not reach the same peak as during the follicular phase. What keeps the moderately high estrogen level during the luteal phase from triggering another LH surge? Progesterone. Even though a high level of estrogen stimulates LH secretion, progesterone, which dominates the luteal phase, powerfully inhibits LH secretion as well as FSH secretion 17 , 16 (● Figure 20-19). Inhibition of FSH and LH by progesterone prevents new follicular maturation and ovulation during the luteal phase. Under progesterone's influence, the reproductive system is gearing up to support the just-released ovum, should it be fertilized, instead of preparing other ova for release. No inhibin is secreted by the luteal cells.

The corpus luteum functions for two weeks, then degenerates if fertilization does not occur 18 . The mechanisms responsible for degeneration of the corpus luteum are not fully understood. The declining level of circulating LH 17 , driven down by inhibitory actions of progesterone, undoubtedly contributes to the corpus luteum's downfall. Prostaglandins and estrogen released by the luteal cells themselves may play a role. Demise of the corpus luteum terminates the luteal phase and sets the stage for a new follicular phase. As the corpus luteum degenerates, plasma progesterone 19 and estrogen 20 levels fall rapidly, because these hormones are no longer being produced. Withdrawal of the inhibitory effects of these hormones on the hypothalamus allows FSH 21 and tonic LH 22 secretion to modestly increase once again. Under the influence of these gonadotropic hormones, another batch of primary follicles 2 is induced to mature as a new follicular phase begins 1 .

▌ Cyclic uterine changes are caused by hormonal changes during the ovarian cycle.

The fluctuations in circulating levels of estrogen and progesterone during the ovarian cycle induce profound changes in the uterus, giving rise to the **menstrual,** or **uterine, cycle.** Because it reflects hormonal changes during the ovarian cycle, the menstrual cycle averages 28 days, as does the ovarian cycle, although even normal adults vary considerably from this mean. This variability primarily reflects differing lengths of the follicular phase; the duration of the luteal phase is fairly constant. The outward manifestation of the cyclical changes in the uterus is the menstrual bleeding once during each menstrual cycle (that is, once a month). Less obvious changes take place throughout the cycle, however, as the uterus is prepared for implantation should a released ovum be fertilized, then is stripped clean of its prepared lining (menstruation) if implantation does not occur, only to repair itself and start preparing for the ovum that will be released during the next cycle.

We will briefly examine the influences of estrogen and progesterone on the uterus and then consider the effects of cyclical fluctuations of these hormones on uterine structure and function.

Influences of estrogen and progesterone on the uterus

The uterus consists of two main layers: the **myometrium,** the outer smooth muscle layer; and the **endometrium,** the inner lining that contains numerous blood vessels and glands. Estrogen stimulates growth of both the myometrium and the endometrium. It also induces the synthesis of progesterone receptors in the endometrium. Thus progesterone can exert an effect on the endometrium only after it has been "primed" by estrogen. Progesterone acts on the estrogen-primed endometrium to convert it into a hospitable and nutritious lining suitable for implantation of a fertilized ovum. Under the influence of progesterone, the endometrial connective tissue becomes loose and edematous as a result of an accumulation of electrolytes and water, which facilitates implantation of the fertilized ovum. Progesterone further prepares the endometrium to sustain an early-developing embryo by inducing the endometrial glands to secrete and store large quantities of glycogen and by causing tremendous growth of the endometrial blood vessels. Progesterone also reduces the contractility of the uterus to provide a quiet environment for implantation and embryonic growth.

The menstrual cycle consists of three phases: the *menstrual phase*, the *proliferative phase*, and the *secretory*, or *progestational, phase.*

Menstrual phase

The **menstrual phase** is the most overt phase, characterized by discharge of blood and endometrial debris from the vagina. By convention, the first day of menstruation is considered the start of a new cycle. It coincides with termination of the ovarian luteal phase and onset of the follicular phase ㉓ , ● Figure 20-15. As the corpus luteum degenerates because fertilization and implantation of the ovum released during the preceding cycle did not take place ⑱ , circulating levels of estrogen and progesterone drop precipitously ⑲ , ⑳ . Because the net effect of estrogen and progesterone is to prepare the endometrium for implantation of a fertilized ovum, withdrawal of these steroids deprives the highly vascular, nutrient-rich uterine lining of its hormonal support.

The fall in ovarian hormone levels also stimulates release of a uterine prostaglandin that causes vasoconstriction of the endometrial vessels, disrupting the blood supply to the endometrium. The subsequent reduction in O_2 delivery causes death of the endometrium, including its blood vessels. The resulting bleeding through the disintegrating vessels flushes the dying endometrial tissue into the uterine lumen. The entire uterine lining sloughs during each menstrual period except for a deep, thin layer of epithelial cells and glands, from which the endometrium will regenerate. The same local uterine prostaglandin

also stimulates mild rhythmic contractions of the uterine myometrium. These contractions help expel the blood and endometrial debris from the uterine cavity out through the vagina as **menstrual flow.** Excessive uterine contractions caused by prostaglandin overproduction are responsible for the menstrual cramps (**dysmenorrhea**) some women experience.

The average blood loss during a single menstrual period is 50 to 150 ml. Blood that seeps slowly through the degenerating endometrium clots within the uterine cavity, then is acted on by fibrinolysin, a fibrin dissolver that breaks down the fibrin forming the meshwork of the clot. Therefore, blood in the menstrual flow usually does not clot, because it has already clotted and the clot has been dissolved before it passes out of the vagina. When blood flows rapidly through the leaking vessels, however, it may not be exposed to sufficient fibrinolysin, so when the menstrual flow is most profuse, blood clots may appear. In addition to the blood and endometrial debris, large numbers of leukocytes are found in the menstrual flow. These white blood cells play an important defense role in helping the raw endometrium resist infection.

Menstruation typically lasts for about five to seven days after degeneration of the corpus luteum, coinciding in time with the early portion of the ovarian follicular phase ㉓ . Withdrawal of estrogen and progesterone ⑲ , ⑳ on degeneration of the corpus luteum leads simultaneously to sloughing of the endometrium (menstruation) ㉓ and development of new follicles in the ovary ① , ② under the influence of rising gonadotropic hormone levels ㉑ , ㉒ . The drop in gonadal hormone secretion removes inhibitory influences from the hypothalamus and anterior pituitary, so FSH and LH secretion increases and a new follicular phase begins. After five to seven days under the influence of FSH and LH, the newly growing follicles are secreting sufficient quantities of estrogen ③ to induce repair and growth of the endometrium.

Proliferative phase

Thus, menstrual flow ceases, and the **proliferative phase** of the uterine cycle begins concurrent with the last portion of the ovarian follicular phase as the endometrium starts to repair itself and proliferate ㉔ under the influence of estrogen from the newly growing follicles. When the menstrual flow ceases, a thin endometrial layer less than 1 mm thick remains. Estrogen stimulates proliferation of epithelial cells, glands, and blood vessels in the endometrium, increasing this lining to a thickness of 3 to 5 mm. The estrogen-dominant proliferative phase lasts from the end of menstruation to ovulation. Peak estrogen levels ⑧ trigger the LH surge ⑨ responsible for ovulation ⑩ .

Secretory or progestational phase

Following ovulation, when a new corpus luteum is formed ⑫ , the uterus enters the **secretory,** or **progestational, phase,** which coincides in time with the ovarian luteal phase ㉕ . The corpus luteum secretes large amounts of progesterone ⑬ and estrogen ⑭ . Progesterone converts the thickened, estrogen-primed endometrium to a richly vascularized, glycogen-filled tissue. This period is called either the *secretory phase*, because the endometrial glands are actively secreting glycogen, or the

Menstrual Irregularities: When Cyclists and Other Female Athletes Do Not Cycle

Since the 1970s, as women in growing numbers have participated in a variety of sports requiring vigorous training regimens, researchers have become increasingly aware that many women experience changes in their menstrual cycles as a result of athletic participation. These changes are referred to as *athletic menstrual cycle irregularity (AMI)*. The menstrual cycle dysfunction can vary in severity from amenorrhea (cessation of menstrual periods) to oligomenorrhea (cycles at irregular or infrequent intervals) to cycles that are normal in length but are anovulatory (no ovulation) or that have a short or inadequate luteal phase.

In early research studies using surveys and questionnaires to determine the prevalence of the problem, the frequency of these sport-related disorders varied from 2% to 51%. In contrast, the rate of occurrence of menstrual cycle dysfunction in females of reproductive age in the general population is 2% to 5%. A major problem of using surveys to determine the frequency of menstrual cycle irregularity is the questionable accuracy of recall of menstrual periods. Furthermore, without blood tests to determine hormone levels throughout the cycle, a woman would not know whether she was anovulatory or had had a shortened luteal phase. Studies in which hormone levels have been determined through the menstrual cycle have demonstrated that seemingly normal cycles in athletes frequently have a short luteal phase (less than two days long with low progesterone levels).

In a study conducted to determine whether strenuous exercise spanning two menstrual cycles would induce menstrual disorders, 28 initially untrained college women with documented ovulation and luteal adequacy served as subjects. The women participated in an eight-week exercise program in which they initially ran 4 miles per day and progressed to 10 miles per day by the fifth week. They were expected to participate daily in 3.5 hours of moderate-intensity sports. Only four women had normal menstrual cycles during the training. Abnormalities resulting from training included abnormal bleeding, delayed menstrual periods, abnormal luteal function, and loss of LH surge. All women returned to normal cycles within six months after training. The results of this study suggest that the frequency of AMI with strenuous exercise may be much greater than indicated by questionnaire alone. In other studies using low-intensity exercise regimens, AMI was much less frequent.

The mechanisms of AMI are unknown at present, although studies have implicated rapid weight loss, decreased percentage of body fat, dietary insufficiencies, prior menstrual dysfunction, stress, age at onset of training, and the intensity of training as factors that play a role. Epidemiologists have shown that if girls participate in vigorous sports before **menarche** (the first menstrual period), menarche is delayed. On the average, athletes have their first period when they are about three years older than their nonathletic counterparts. Furthermore, females who participate in sports before menarche seem to have a higher frequency of AMI throughout their athletic careers than those who begin to train after menarche. Hormonal changes found in female athletes include (1) severely depressed FSH levels, (2) elevated LH levels, (3) low progesterone during the luteal phase, (4) low estrogen levels in the follicular phase, and (5) an FSH/LH environment totally unbalanced as compared to age-matched nonathletic women. The preponderance of evidence indicates that cycles return to normal once vigorous training is stopped.

The major problem associated with athletic amenorrhea is a reduction in bone mineral density. Studies have shown that the mineral density in the vertebrae of the lower spine of those with athletic amenorrhea is lower than in athletes with normal menstrual cycles and lower than in age-matched nonathletes. However, amenorrheic runners have higher bone mineral density than amenorrheic nonathletes, presumably because the mechanical stimulus of exercise helps retard bone loss. Studies have shown that amenorrheic athletes are at higher risk for stress-related fractures than athletes with normal menstrual cycles. One study, for example, found stress fractures in six of eleven amenorrheic runners but in only one of six runners with normal menstrual cycles. The mechanism for bone loss is probably the same as is found in postmenopausal osteroporosis—lack of estrogen (see p. 736). The problem is serious enough that an amenorrheic athlete should discuss the possibility of estrogen replacement therapy with her physician.

There may be some positive benefits of athletes' menstrual dysfunction. A recent epidemiologic study to determine if the long-term reproductive and general health of women who had been college athletes differed from that of college nonathletes showed that former athletes had less than half the lifetime occurrence rate of cancers of the reproductive system and half the breast cancer occurrence compared to nonathletes. Because these are hormone-sensitive cancers, the delayed menarche and lower estrogen levels found in women athletes may play a key role in decreasing the risk of cancer of the reproductive system and breast.

---

progestational ("before pregnancy") *phase*, referring to the development of a lush endometrial lining capable of supporting an early embryo. If fertilization and implantation do not occur, the corpus luteum degenerates and a new follicular phase and menstrual phase begin once again. (For the effects of exercise on this cycle, see the accompanying boxed feature, ◗ A Closer Look at Exercise Physiology).

▌ Fluctuating estrogen and progesterone levels produce cyclical changes in cervical mucus.

Hormonally induced changes also take place in the cervix during the ovarian cycle. Under the influence of estrogen during the follicular phase, the mucus secreted by the cervix becomes abundant, clear, and thin. This change, which is most pronounced when estrogen is at its peak and ovulation is approaching, facilitates passage of sperm through the cervical canal. After ovulation, under the influence of progesterone from the corpus luteum, the mucus becomes thick and sticky, essentially plugging up the cervical opening. This plug is an important defense mechanism, preventing bacteria (that might threaten a possible pregnancy) from entering the uterus from the vagina. Sperm also cannot penetrate this thick mucus barrier.

▌ Pubertal changes in females are similar to those in males.

Regular menstrual cycles are absent in both young and aging females, but for different reasons. The female reproductive system does not become active until puberty. Unlike the fetal testes, the fetal ovaries need not be functional, because the female reproductive system is automatically feminized in the absence of fetal testosterone secretion without the presence of female sex hormones. The female reproductive system remains quiescent from birth until puberty, which occurs at about 12 years of age when hypothalamic GnRH activity increases for the first time. As in the male, the mechanisms responsible for the onset of puberty are not clearly understood but are believed to involve the pineal gland and melatonin secretion.

GnRH begins stimulating release of anterior pituitary gonadotropic hormones, which in turn stimulate ovarian activity. The resulting secretion of estrogen by the activated ovaries induces growth and maturation of the female reproductive tract as well as development of the female secondary sexual characteristics. Estrogen's prominent action in the latter regard is to promote fat deposition in strategic locations, such as the breasts, buttocks, and thighs, giving rise to the typical curvaceous female figure. Enlargement of the breasts at puberty is due primarily to fat deposition in the breast tissue, not to functional development of the mammary glands. The pubertal rise in estrogen also closes the epiphyseal plates, halting further growth in height, similar to the effect of testosterone in males. Three other pubertal changes in females—growth of axillary and pubic hair, the pubertal growth spurt, and development of libido—are attributable to a spurt in adrenal androgen secretion at puberty, not to estrogen.

■ Menopause is unique to females.

The cessation of a woman's menstrual cycles at menopause sometime between the ages of 45 and 55 has traditionally been attributed to the limited supply of ovarian follicles present at birth. According to this proposal, once this reservoir is depleted, ovarian cycles, and hence menstrual cycles, cease. Thus the termination of reproductive potential in a middle-aged woman is "preprogrammed" at her own birth. Recent evidence suggests, however, that a midlife hypothalamic change instead of aging ovaries may be responsible for the onset of menopause.

Evolutionarily, menopause may have developed as a mechanism that prevented pregnancy in women beyond the time that they could likely rear a child before their own death.

Menopause is preceded by a period of progressive ovarian failure characterized by increasingly irregular cycles and dwindling estrogen levels. This entire period of transition from sexual maturity to cessation of reproductive capability is known as the **climacteric.** Ovarian estrogen production declines from as much as 300 mg per day to essentially nothing. Postmenopausal women are not completely devoid of estrogen, however, because adipose tissue, the liver, and the adrenal cortex continue to produce up to 20 mg of estrogen per day. Nevertheless, the loss of ovarian estrogen brings about many physical and emotional changes.

The absence of ovarian estrogen brings about physical postmenopausal changes in the reproductive tract. These changes include vaginal dryness, which can cause discomfort during sex, and gradual atrophy of the genital organs, in addition to ending ovarian and menstrual cycles. However, postmenopausal women still have a sex drive, because of their adrenal androgens.

Because estrogen has widespread physiologic actions beyond the reproductive system, the dramatic loss of ovarian estrogen in menopause affects other body systems, most notably the skeleton and the cardiovascular system. Estrogen helps build strong bones, shielding premenopausal women from the bone-thinning condition of osteoporosis (see p. 736). The postmenopausal reduction in estrogen increases activity of the bone-dissolving osteoclasts and diminishes activity of the bone-building osteoblasts. The result is decreased bone density and a greater incidence of bone fractures.

Estrogen also helps modulate the actions of epinephrine and norepinephrine on the arteriolar walls. The menopausal diminution of estrogen leads to unstable control of blood flow, especially in the skin vessels. Transient increases in the flow of warm blood through these superficial vessels are responsible for the **"hot flashes"** that frequently accompany menopause. Vasomotor stability is gradually restored in postmenopausal women so that hot flashes eventually cease.

Males do not experience the complete gonadal failure as females do, for two reasons. First, a male's germ cell supply is unlimited because mitotic activity of the spermatogonia continues. Second, gonadal hormone secretion in males is not inextricably dependent on gametogenesis, as in females. If female sex hormones were produced by separate tissues unrelated to those governing gametogenesis, as in the male, cessation of estrogen and progesterone secretion would not automatically accompany termination of oogenesis.

You have now learned about the events that take place if fertilization does not occur. Because the primary function of the reproductive system is, of course, reproduction, we will next turn our attention to the sequence of events that take place when fertilization does occur.

■ The oviduct is the site of fertilization.

Fertilization, the union of male and female gametes, normally occurs in the **ampulla,** the upper third of the oviduct (● Figure 20-2, p. 752). Thus both the ovum and the sperm must be transported from their gonadal site of production to the ampulla (● Figure 20-20).

Ovum transport to the oviduct

At ovulation the ovum is released into the abdominal cavity but is quickly picked up by the oviduct. The dilated end of the oviduct cups around the ovary and contains **fimbriae,** finger-like projections that contract in a sweeping motion to guide the released ovum into the oviduct (● Figures 20-2b and 20-20). Furthermore, the fimbriae are lined by cilia, fine, hairlike projections that beat in waves toward the interior of the oviduct, further assuring the ovum's passage into the oviduct (see p.47. Within the oviduct, the ovum is rapidly propelled by peristaltic contractions and ciliary action to the ampulla.

Conception can take place during a very limited time span in each cycle (the **fertile period**). If not fertilized, the ovum begins to disintegrate within 12 to 24 hours and is subsequently phagocytized by cells that line the reproductive tract. Fertilization must therefore occur within 24 hours after ovulation, when the ovum is still viable. Sperm typically survive about 48 hours but can survive up to five days in the female reproductive tract, so sperm deposited within five days before ovula-

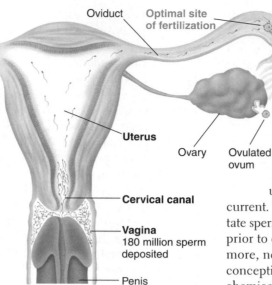

Location	Time of appearance (min after ejaculation)	Percent of ejaculated sperm*
Fertilization site (upper third of oviduct)	30–60	0.001
Uterus	10–20	0.1
Cervical canal	1–3	3
Vagina	0	100

*Based on data from animals.
Sperm and ovum enlarged.

Labels on figure: Oviduct · Optimal site of fertilization · Ampulla of oviduct · Sperm surrounding ovum · Uterus · Ovary · Ovulated ovum · Fimbria · Cervical canal · Vagina 180 million sperm deposited · Penis

● **FIGURE 20-20**

Ovum and sperm transport to the site of fertilization

tion to 24 hours after ovulation may be able to fertilize the released ovum, although these times vary considerably.

Occasionally an ovum fails to be transported into the oviduct and remains instead in the peritoneal cavity. Rarely, such an ovum gets fertilized, resulting in an **ectopic (abdominal) pregnancy,** in which the fertilized egg implants in the rich vascular supply to the digestive organs rather than in its usual site in the uterus. If this unusual pregnancy proceeds to term, the baby must be delivered surgically, because the normal vaginal exit is not available. The probability of maternal complications at birth are greatly increased because the digestive vasculature is not designed to "seal itself off" after birth as the endometrium does.

Sperm transport to the oviduct

Once sperm are deposited in the vagina at ejaculation, they must travel through the cervical canal, through the uterus, and then up to the egg in the upper third of the oviduct (● Figure 20-20). The first sperm arrive in the oviduct within half an hour after ejaculation. Even though sperm are mobile by means of whiplike contractions of their tails, 30 minutes is much too soon for a sperm's own mobility to transport it to the site of fertilization. To accomplish this formidable journey, sperm need the help of the female reproductive tract.

The first hurdle is passage through the cervical canal. Throughout most of the cycle, because of high progesterone or low estrogen levels, the cervical mucus is too thick to permit sperm penetration. The cervical mucus becomes thin and watery enough to permit sperm to penetrate only when estrogen levels are high, as in the presence of a mature follicle about to ovulate. Sperm migrate up the cervical canal under their own power. The canal remains penetrable for only two or three days during each cycle, around the time of ovulation.

Once the sperm have entered the uterus, contractions of the myometrium churn them around in "washing-machine" fashion. This action quickly disperses sperm throughout the uter-

ine cavity. When sperm reach the oviduct, they must travel upward against the predominant action of the cilia lining this passageway, most of which beat downward to propel the nonmotile egg toward the uterus. Sperm transport in the oviduct is facilitated by antiperistaltic (upward) contractions of the oviduct smooth muscle, perhaps assisted by tracts of cilia that beat upward in the direction opposite to the main ciliary current. These myometrial and oviduct contractions that facilitate sperm transport are induced by the high estrogen level just prior to ovulation, aided by seminal prostaglandins. Furthermore, new research indicates ova are not passive partners in conception. Mature eggs release **allurin,** a recently identified chemical that attracts sperm and causes them to propel themselves toward the waiting female gamete.

Even around ovulation time when sperm can penetrate the cervical canal, of the several hundred million sperm deposited in a single ejaculate only a few thousand make it to the oviduct (● Figure 20-20). That only a very small percentage of the deposited sperm ever reach their destination is one reason why sperm concentration must be so high (20 million/ml of semen) for a man to be fertile. The other reason is that the acrosomal enzymes of many sperm are needed to break down the barriers surrounding the ovum (● Figure 20-21).

● **FIGURE 20-21**

Scanning electron micrograph of sperm amassed at the surface of an ovum

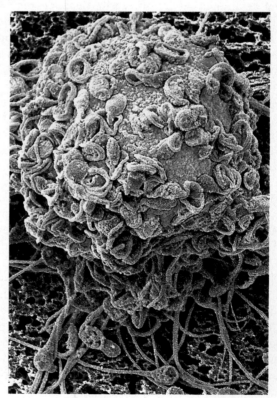

Fertilization

The tail of the sperm is used to maneuver for final penetration of the ovum. To fertilize an ovum, a sperm must first pass through the corona radiata and zona pellucida surrounding it. The acrosomal enzymes, which are exposed as the acrosomal membrane disrupts on contact with the corona radiata, enable the sperm to tunnel a path through these protective barriers (● Figure 20-22). Sperm can penetrate the zona pellucida only after binding with specific receptor sites on the surface of this layer. The binding partners between the sperm and ovum were recently identified. **Fertilin**, a protein found on the plasma membrane of the sperm, binds with an egg *integrin*, a type of cell adhesion molecule that protrudes from the outer surface of the plasma membrane (see p. 61). Only sperm of the same species can bind to these egg receptors and pass through. The first sperm to reach the ovum itself fuses with the plasma membrane of the ovum (actually a secondary oocyte), triggering a chemical change in the ovum's surrounding membrane that makes this outer layer impenetrable to the entry of any more sperm. This phenomenon is known as **block to polyspermy** ("many sperm").

The head of the fused sperm is gradually pulled into the ovum's cytoplasm by a growing cone that engulfs it. The sperm's tail is frequently lost in this process, but it is the head that carries the crucial genetic information. Recent evidence suggests the sperm releases nitric oxide when it has completely penetrated into the cytoplasm. This nitric oxide promotes the release of stored Ca^{2+} within the egg. This intracellular Ca^{2+} release triggers the final meiotic division of the secondary oocyte. Within an hour, the sperm and egg nuclei fuse, thanks to a molecular complex provided by the sperm that helps the male and female chromosome sets unite. In addition to contributing its half of the chromosomes to the fertilized ovum, now called a **zygote**, the victorious sperm also activates ovum enzymes essential for the early embryonic developmental program.

❙ The blastocyst implants in the endometrium through the action of its trophoblastic enzymes.

During the first three to four days following fertilization, the zygote remains within the ampulla, because a constriction be-

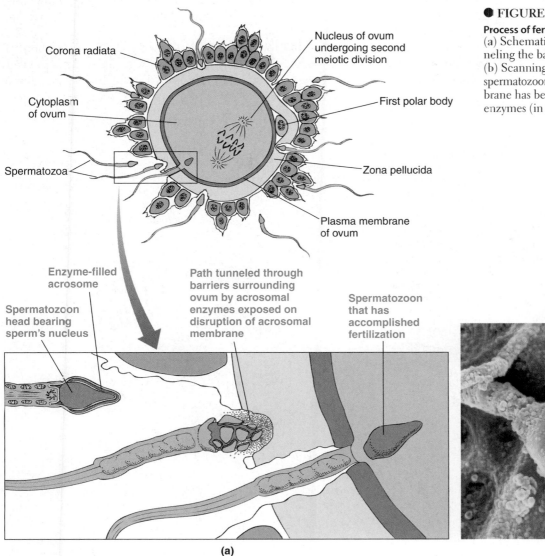

● FIGURE 20-22

Process of fertilization
(a) Schematic representation of sperm tunneling the barriers surrounding the ovum.
(b) Scanning electron micrograph of a spermatozoon in which the acrosomal membrane has been disrupted and the acrosomal enzymes (in red) are exposed.

(a)

(b)

tween the ampulla and the remainder of the oviduct canal prevents further movement of the zygote toward the uterus.

The beginning steps in the ampulla

The zygote is not idle during this time, however. It rapidly undergoes a number of mitotic cell divisions to form a solid ball of cells called the **morula** (● Figure 20-23). Meanwhile, the rising levels of progesterone from the newly developing corpus luteum that formed after ovulation stimulate release of glycogen from the endometrium into the reproductive tract lumen for use as energy by the early embryo. The nutrients stored in the cytoplasm of the ovum can sustain the product of conception for less than a day. The concentration of secreted nutrients increases more rapidly in the small confines of the ampulla than in the uterine lumen.

Descent of the morula to the uterus

About three to four days after ovulation, progesterone is being produced in sufficient quantities to relax the oviduct constriction, thus permitting the morula to be rapidly propelled into the uterus by oviductal peristaltic contractions and ciliary activity. The temporary delay before the developing embryo passes into the uterus lets enough nutrients accumulate in the uterine lumen to support the embryo until implantation can take place. If the morula arrives prematurely, it dies.

When the morula descends to the uterus, it floats freely within the uterine cavity for another three to four days, living on endometrial secretions and continuing to divide. During the first

six to seven days after ovulation, while the developing embryo is in transit in the oviduct and floating in the uterine lumen, the uterine lining is simultaneously being prepared for implantation under the influence of luteal-phase progesterone. During this time, the uterus is in its secretory or progestational phase, storing up glycogen and becoming richly vascularized.

Occasionally the morula fails to descend into the uterus and continues instead to develop and implant in the lining of the oviduct. This leads to a **tubal pregnancy,** which must be terminated. Such a pregnancy can never succeed, because the oviduct cannot expand as the uterus does to accommodate the growing embryo. The first warning of a tubal pregnancy is pain caused by the growing embryo stretching the oviduct. If not removed, the enlarging embryo will rupture the oviduct, causing possibly lethal hemorrhage.

Implantation of the blastocyst in the prepared endometrium

In the usual case, by the time the endometrium is suitable for implantation (about a week after ovulation), the morula has descended to the uterus and continued to proliferate and differentiate into a *blastocyst* capable of implantation. The week's delay after fertilization and before implantation allows time for both the endometrium and the developing embryo to prepare for implantation.

A **blastocyst** is a single-layered hollow ball of about 50 cells encircling a fluid-filled cavity, with a dense mass of cells grouped together at one side (● Figure 20-23). This dense

---

● FIGURE 20-23

Early stages of development from fertilization to implantation
Note that the fertilized ovum progressively divides and differentiates into a blastocyst as it moves from the site of fertilization in the upper oviduct to the site of implantation in the uterus.

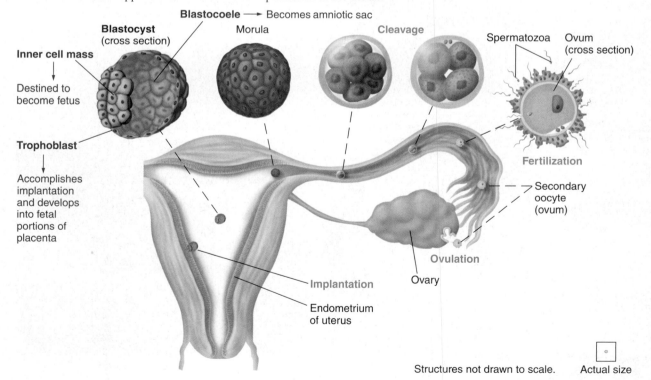

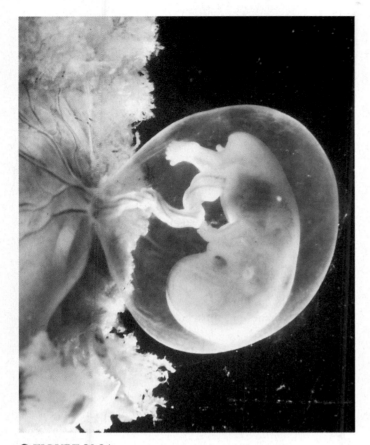

● FIGURE 20-24

A human fetus surrounded by the amniotic sac
The fetus is near the end of the first trimester of development.

mass, known as the **inner cell mass** is destined to become the embryo/fetus itself. The remainder of the blastocyst will never be incorporated into the fetus but will serve a supportive role during intrauterine life. The thin outermost layer, the **trophoblast,** is responsible for accomplishing implantation, after which it develops into the fetal portion of the placenta. The fluid-filled cavity, the **blastocoele,** will become the fluid within the **amniotic sac,** which surrounds and cushions the fetus throughout gestation (● Figure 20-24).

When the blastocyst is ready to implant, its surface becomes sticky. By this time the endometrium is ready to accept the early embryo. The blastocyst adheres to the uterine lining on its inner-cell-mass side (● Figure 20-25, step ①). **Implantation** begins when, on contact with the endometrium, the trophoblastic cells overlying the inner cell mass release protein-digesting enzymes. These enzymes digest pathways between the endometrial cells, permitting fingerlike cords of trophoblastic cells to penetrate into the depths of the endometrium, where they continue to digest uterine cells (● Figure 20-25, step ②). Through its cannibalistic actions, the trophoblast performs the dual functions of (1) accomplishing implantation as it carves out a hole in the endometrium for the blastocyst and (2) making metabolic fuel and raw materials available for the developing embryo as the advancing trophoblastic projections break down the nutrient-rich endometrial tissue. The cell walls of the advancing trophoblastic cells break down, forming a multi-

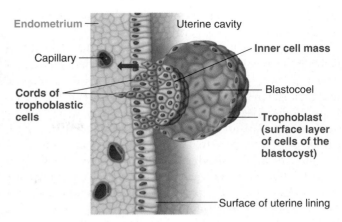

① When the free-floating blastocyst adheres to the endometrial lining, cords of trophoblastic cells begin to penetrate the endometrium.

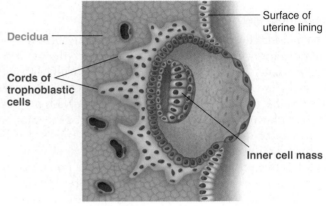

② Advancing cords of trophoblastic cells tunnel deeper into the endometrium, carving out a hole for the blastocyst. The boundaries between the cells in the advancing trophoblastic tissue disintegrate.

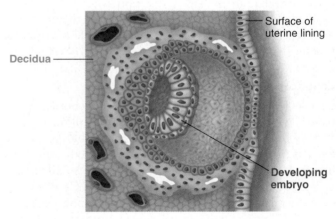

③ When implantation is finished, the blastocyst is completely buried in the endometrium.

● FIGURE 20-25

Implantation of the blastocyst

nucleated syncytium that will eventually become the fetal portion of the placenta.

Stimulated by the invading trophoblast, the endometrial tissue at the contact site undergoes dramatic changes that enhance its ability to support the implanting embryo. In response to a chemical messenger released by the blastocyst, the underlying endometrial cells secrete prostaglandins, which locally increase vascularization, produce edema, and enhance nutrient storage. The endometrial tissue so modified at the implantation site is called the **decidua.** It is into this super-rich decidual tissue that the blastocyst becomes embedded. After the blastocyst burrows into the decidua by means of trophoblastic activity, a layer of endometrial cells covers over the surface of the hole, completely burying the blastocyst within the uterine lining (● Figure 20-25, step ③). The trophoblastic layer continues to digest the surrounding decidual cells, providing energy for the embryo until the placenta develops.

Preventing rejection of the embryo/fetus

What prevents the mother from immunologically rejecting the embryo/fetus, which is actually a "foreigner" to the mother's immune system, being half derived from genetically different paternal chromosomes. Following are several proposals under investigation. New evidence indicates the trophoblasts produce **Fas ligand,** which binds with **Fas,** a specialized receptor on the surface of approaching activated maternal T cells. T cells are the immune cells that carry out the job of destroying foreign cells (see p. 436). This binding triggers the immune cells that are targeted to destroy the developing foreigner, to undergo apoptosis (commit suicide), sparing the embryo/fetus from immune rejection. Other researchers have found that the fetal portion of the placenta, which is derived from trophoblasts, produces an enzyme **indoleamine 2,3-dioxygenase (IDO),** which destroys the enzyme tryptophan. Tryptophan is a critical factor in the activation of maternal T cells. Thus the embryo/fetus, through its trophoblast connection, is believed to defend itself against rejection by shutting down the activity of the mother's T cells within the placenta that would otherwise attack the developing foreign tissues.

Contraception

Couples wishing to engage in sexual intercourse but avoid pregnancy have a number of methods of **contraception** ("against conception") available. These methods act by blocking one of three major steps in the reproductive process: sperm transport to the ovum, ovulation, or implantation. (See the boxed featured on pp. 788–789, ◗ Concepts, Challenges, and Controversies, for further details on the ways and means of contraception.)

We will now examine the placenta in further detail.

▌ The placenta is the organ of exchange between maternal and fetal blood.

The glycogen stores in the endometrium are only sufficient to nourish the embryo during its first few weeks. To sustain the growing embryo/fetus for the duration of its intrauterine life, the **placenta,** a specialized organ of exchange between the maternal and fetal blood, rapidly develops (● Figure 20-26). The placenta is derived from both trophoblastic and decidual tissue.

Formation of the placenta

By day 12, the embryo is completely embedded in the decidua. By this time the trophoblastic layer is two cell layers thick and is called the **chorion.** As the chorion continues to release enzymes and expand, it forms an extensive network of cavities within the decidua. As decidual capillary walls are eroded by the expanding chorion, these cavities fill with maternal blood, which is kept from clotting by an anticoagulant the chorion produces. Fingerlike projections of chorionic tissue extend into the pools of maternal blood. Soon the developing embryo sends out capillaries into these chorionic projections to form **placental villi.** Some villi extend completely across the blood-filled spaces to anchor the fetal portion of the placenta to the endometrial tissue, but most simply project into the pool of maternal blood.

Each placental villus contains embryonic (later fetal) capillaries surrounded by a thin layer of chorionic tissue, which separates the embryonic/fetal blood from the maternal blood in the intervillus spaces. No actual mingling occurs between maternal and fetal blood, but the barrier between them is extremely thin. To visualize this relationship, think of your hands (the fetal capillary blood vessels) in rubber gloves (the chorionic tissue) immersed in water (the pool of maternal blood). Only the rubber gloves separate your hands from the water. In the same way, only the thin chorionic tissue (plus the capillary wall of the fetal vessels) separates the fetal and maternal blood. It is across this extremely thin barrier that all materials are exchanged between these two bloodstreams. This entire system of interlocking maternal (decidual) and fetal (chorionic) structures makes up the placenta.

Even though not fully developed, the placenta is well established and operational by five weeks after implantation. By this time, the heart of the developing embryo is pumping blood into the placental villi as well as to the embryonic tissues. Throughout gestation, fetal blood continuously traverses between the placental villi and the circulatory system of the fetus by means of the **umbilical artery** and **umbilical vein,** which are wrapped within the **umbilical cord,** a lifeline between the fetus and the placenta (● Figure 20-26). The maternal blood within the placenta is continuously replaced as fresh blood enters through the uterine arterioles, percolates through the intervillus spaces, where it exchanges substances with fetal blood in the surrounding villi, then exits through the uterine vein.

Functions of the placenta

During intrauterine life, the placenta performs the functions of the digestive system, the respiratory system, and the kidneys for the "parasitic" fetus. The fetus has these organ systems, but they cannot (and do not need to) function within the uterine environment. Nutrients and O_2 diffuse from the maternal blood across the thin placental barrier into the fetal blood, whereas CO_2 and other metabolic wastes simultaneously diffuse from the fetal blood into the maternal blood. The nutrients and O_2 brought to the fetus in the maternal blood are acquired by the mother's digestive and respiratory systems, and the CO_2 and wastes transferred into the maternal blood are eliminated by the mother's lungs and kidneys, respectively. Thus, the mother's digestive tract, respiratory system, and kidneys serve the fetus's needs as well as her own.

Some substances traverse the placental barrier by special mediated transport systems in the placental membranes, whereas others move across by simple diffusion. Unfortunately, many drugs, environmental pollutants, other chemical agents, and micro-organisms in the mother's bloodstream also can cross the placental barrier, and some of them may harm the devel-

oping fetus. Individuals born limbless as a result of exposure to *thalidomide*, a tranquilizer prescribed for pregnant women before this drug's devastating effects on the growing fetus were known, serve as a grim reminder of this fact. Similarly, newborns who have become "addicted" during gestation by their mother's abuse of a drug such as heroin suffer withdrawal symptoms after birth. Even more common chemical agents such as aspirin, alcohol, and agents in cigarette smoke can reach the fetus and have adverse effects. Likewise, fetuses can acquire AIDS before birth if their mothers are infected with the virus. A pregnant women should therefore be very cautious about potentially harmful exposure from any source.

The placenta assumes yet another important responsibility—it becomes a temporary endocrine organ during pregnancy, a topic to which we now turn.

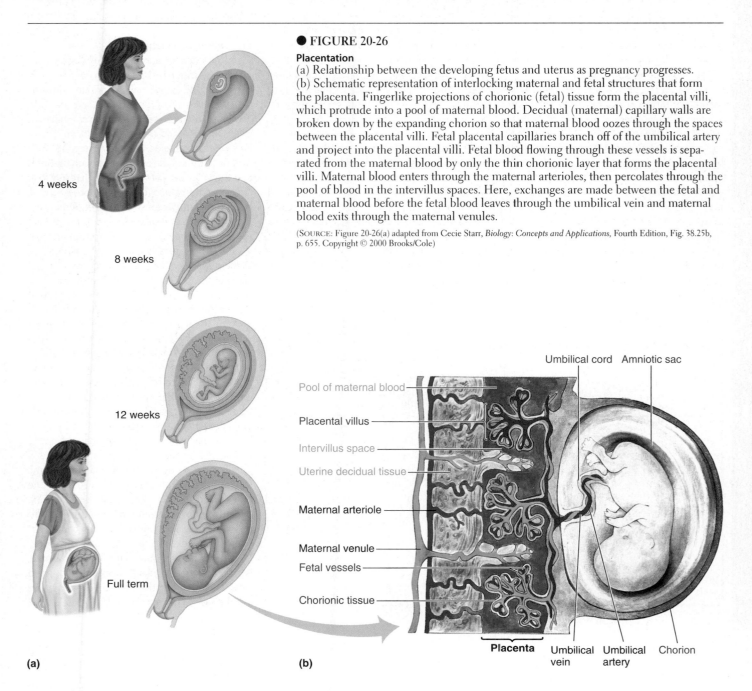

● **FIGURE 20-26**

Placentation
(a) Relationship between the developing fetus and uterus as pregnancy progresses.
(b) Schematic representation of interlocking maternal and fetal structures that form the placenta. Fingerlike projections of chorionic (fetal) tissue form the placental villi, which protrude into a pool of maternal blood. Decidual (maternal) capillary walls are broken down by the expanding chorion so that maternal blood oozes through the spaces between the placental villi. Fetal placental capillaries branch off of the umbilical artery and project into the placental villi. Fetal blood flowing through these vessels is separated from the maternal blood by only the thin chorionic layer that forms the placental villi. Maternal blood enters through the maternal arterioles, then percolates through the pool of blood in the intervillus spaces. Here, exchanges are made between the fetal and maternal blood before the fetal blood leaves through the umbilical vein and maternal blood exits through the maternal venules.

(SOURCE: Figure 20-26(a) adapted from Cecie Starr, *Biology: Concepts and Applications*, Fourth Edition, Fig. 38.25b, p. 655. Copyright © 2000 Brooks/Cole)

4 weeks

8 weeks

12 weeks

Full term

(a)

(b)

Umbilical cord   Amniotic sac

Pool of maternal blood

Placental villus

Intervillus space

Uterine decidual tissue

Maternal arteriole

Maternal venule

Fetal vessels

Chorionic tissue

Placenta   Umbilical   Umbilical   Chorion
vein   artery

The Ways and Means of Contraception

The term **contraception** refers to the process of avoiding pregnancy while engaging in sexual intercourse. A number of methods of contraception are available that range in ease of use and effectiveness (see the accompanying table). These methods can be grouped into three categories based on the means by which they prevent pregnancy: blockage of sperm transport to the ovum, prevention of ovulation, or blockage of implantation. After examining the various ways in which contraception can be accomplished by each of these means, we will take a glimpse at future contraceptive possibilities on the horizon before concluding with a discussion of termination of unwanted pregnancies.

Blockage of sperm transport to the ovum

• *Natural contraception* or the *rhythm method* of birth control relies on abstinence from intercourse during the woman's fertile period. The woman can predict when ovulation is to occur based on keeping careful records of her menstrual cycles. Because of variability in cycles, this technique is only partially effective. The time of ovulation can be determined more precisely by recording body temperature each morning before getting up. Body temperature rises slightly about a day after ovulation has taken place. The temperature rhythm method is not useful in determining when it is safe to engage in intercourse before ovulation, but it can

be helpful in determining when it is safe to resume sex after ovulation.

• *Coitus interruptus* involves withdrawal of the penis from the vagina before ejaculation occurs. This method is only moderately effective, however, because timing is difficult, and some sperm may pass out of the urethra prior to ejaculation.

• *Chemical contraceptives,* such as spermicidal ("sperm-killing") jellies, foams, creams, and suppositories, when inserted into the vagina are toxic to sperm for about an hour following application.

• *Barrier methods* mechanically prevent sperm transport to the oviduct. For males, the *condom* is a thin, strong rubber or latex sheath placed over the erect penis prior to ejaculation to prevent sperm from entering the vagina. For females, the *diaphragm* is a flexible rubber dome that is inserted through the vagina and positioned over the cervix to block sperm entry into the cervical canal. It is held in position by lodging snugly against the vaginal wall. The diaphragm must be fitted by a trained professional and must be left in place for at least 6 hours but no longer than 24 hours after intercourse. Barrier methods are often used in conjunction with spermicidal agents for increased effectiveness. The *cervical cap* is a recently developed alternative to the diaphragm. Smaller than a diaphragm, the cervical cap, which is coated with a film of spermicide, cups over the cervix and is held in place by suction.

The *female condom* (or *vaginal pouch*) is the latest barrier method developed. It is a 7-inch-long, polyurethane, cylindrical pouch that is closed on one end and open on the other end, with a flexible ring at both ends. The ring at the closed end of the device is inserted into the vagina and fits over the cervix, similar to a diaphragm. The ring at the open end of the pouch is positioned outside the vagina over the external genitalia.

• *Sterilization,* which involves surgical disruption of either the ductus deferens *(vasectomy)* in men or the oviduct *(tubal ligation)* in women, is considered a permanent method of preventing sperm and ovum from uniting.

Prevention of ovulation

• *Oral contraceptives,* or *birth control pills,* prevent ovulation by suppressing gonadotropin secretion. These pills, which contain synthetic estrogen-like and progesterone-like steroids, are taken for three weeks, either in combination or in sequence, and then are withdrawn for one week. These steroids, like the natural steroids produced during the ovarian cycle, inhibit GnRH and thus FSH and LH secretion. As a result, follicle maturation and ovulation do not take place, so conception is impossible. The endometrium responds to the exogenous steroids by thickening and developing secretory capacity, just as it would to the natural hormones. When these synthetic steroids are withdrawn after three weeks, the endometrial lining sloughs and menstruation occurs, as it normally would on degeneration of the corpus luteum. In addition to blocking ovulation, oral contraceptives also prevent pregnancy by increasing the viscosity of cervical mucus, which makes sperm penetration more difficult, and by decreasing muscular contractions in the female reproductive tract, which reduces sperm transport to the oviduct. Oral contraceptives are available only by prescription. They have been shown to increase the risk of intravascular clotting, especially in women who also smoke tobacco.

• *Long-acting subcutaneous* ("under the skin") *implantation* of synthetic progesterone acts similarly to oral contraceptives by blocking ovulation and thickening the cervical mucus to prevent sperm transport. Unlike oral contraceptives, however, which must be taken on a regular basis, this new contraceptive, once implanted, is effective for five years. Six matchstick-sized capsules containing the steroid are inserted under the skin in the inner arm above the elbow.

Average Failure Rate of Various Contraceptive Techniques

Contraceptive Method	Average Failure Rate (annual pregnancies/ 100 women)
None	90
Natural (rhythm) methods	20–30
Coitus interruptus	23
Chemical contraceptives	20
Barrier methods	10–15
Oral contraceptives	2–2.5
Implanted contraceptives	1
Intrauterine device	4

The capsules slowly release the synthetic progesterone at a nearly steady rate for five years, sustaining their contraceptive effect for a prolonged period. Of concern is a preliminary finding that these long-lasting implantations may exert toxic effects on the nervous system. Similar in action but on a shorter time scale are injectable time-released synthetic female sex hormones that exert contraceptive effects for a month or three months, depending on the product.

- Two forms of hormonal manipulation have recently become available—*doughnut-shaped vaginal rings* that release synthetic progesterone alone or in combination with estrogen and *birth control patches* impregnated with these hormones that are absorbed through the skin.

Blockage of implantation

Medically, pregnancy is not considered to begin until implantation. According to this view, any mechanism that interferes with implantation is said to prevent pregnancy. Not all hold this view, however. Some consider pregnancy to begin at time of fertilization. To them, any interference with implantation is a form of abortion. Therefore, methods of contraception that rely on blockage of implantation are more controversial than methods that prevent fertilization from taking place.

- Blockage of implantation is most commonly accomplished by a physician inserting a small *intrauterine device (IUD)* into the uterus. The IUD's mechanism of action is not completely understood, although most evidence suggests that the presence of this foreign object in the uterus induces a local inflammatory response that prevents implantation of a fertilized ovum. Although the IUD is a convenient birth-control method because it does not require ongoing attention by the user, it is no longer as popular as it once was because of reported complications associated with its use, the most serious of which are pelvic inflammatory disease, permanent infertility, and uterine perforation. Today's IUD models are much safer than the earlier models, however, so these complications are now rare.
- Implantation can also be blocked by so-called *morning-after pills,* also known as *emergency contraception.* The first term is actually a misnomer, because these pills can prevent pregnancy if taken within 72 hours after, not just the morning after, unprotected sexual intercourse. The most common form of emergency contraception is a kit consisting of high doses of birth control pills. These pills, available only by prescription, work in different ways to prevent pregnancy depending on where the woman is in her cycle when she takes the pills. They can either suppress ovulation or cause premature degeneration of the corpus luteum, thus preventing implantation of a fertilized ovum by withdrawing the developing endometrium's hormonal support. These kits are for emergency use only—for instance, if a condom breaks or in the case of rape—and should not be used as a substitute for ongoing contraceptive methods.

Future possibilities

- On the horizon are improved varieties of currently available contraceptive techniques, such as hormone-releasing IUDs that last for five years.
- A future birth-control technique is *immunocontraception*—the use of vaccines that prod the immune system to produce antibodies targeted against a particular protein critical to the reproductive process. The contraceptive effects of the vaccines are expected to last about a year. For example, in the testing stage is a vaccine that induces the formation of antibodies against human chorionic gonadotropin so that this essential corpus luteum–supporting hormone is not effective should pregnancy occur.
- Some researchers are exploring ways to block the union of sperm and egg by interfering with a specific interaction that normally occurs between the male and female gametes. For example, under study are chemicals introduced into the vagina that trigger premature release of the acrosomal enzymes, depriving the sperm of a means to fertilize an ovulated egg. Other investigators are working on a vaccine against the protein in a sperm's head (fertilin) that normally binds to the receptor sites on the zona pellucida surrounding the egg.
- Some scientists are seeking ways to manipulate hormones in males to block spermatogenesis. Although a "male birth control pill" remains a distant possibility, a more imminent prospect under development is regular injections of synthetic androgen to shut down the GnRH, gonadotropin, sperm production pathway. Alternatively, male vaccines against GnRH or FSH would stop sperm production. Because blocking GnRH action would hinder testosterone secretion as well, supplemental testosterone injections would be needed.
- Still another possibility under investigation is manipulation of the anterior pituitary secretion of FSH and LH by GnRH-like drugs. The use of these drugs as contraceptives in both females and males is being explored.
- Another outlook for male contraception is chemical sterilization designed to be reversed, unlike surgical sterilization, which is considered irreversible. In this experimental technique, a nontoxic polymer is injected into the ductus deferens, where the chemical interferes with the sperms' fertilizing capabilities. Flushing of the polymer from the ductus deferens by a solvent reverses the contraceptive effect.
- One interesting avenue being explored holds hope for a unisex contraceptive that would stop sperm in their tracks and could be used by either males or females. Based on preliminary findings, the hope is to use Ca^{2+}-blocking drugs to prevent the entry of Ca^{2+} into sperm tails. As in muscle cells, Ca^{2+} switches on the contractile apparatus responsible for the sperm's motility. With no Ca^{2+}, sperm would not be able to maneuver to accomplish fertilization.

Termination of unwanted pregnancies

- When contraceptive practices fail or are not used and an unwanted pregnancy results, women sometimes turn to *abortion* to terminate the pregnancy. More than half of the approximately 6.4 million pregnancies in the United States each year are unintended, and about 1.6 million of them end with an abortion. Although surgical removal of an embryo/fetus is legal in the United States, the practice of abortion is fraught with emotional, ethical, and political controversy.
- In late 2000, amid considerable controversy, the "abortion pill," *RU 486,* or *mifepristone,* was approved for use in the United States, even though it has been available in other countries since 1988. This drug terminates an early pregnancy by chemical interference rather than by surgery. RU 486, a progesterone antagonist, binds tightly with the progesterone receptors on the target cells but does not evoke progesterone's usual effects and prevents progesterone from binding and acting. Deprived of progesterone activity, the highly developed endometrial tissue sloughs off, carrying the implanted embryo with it. RU 486 administration is followed in 48 hours by a prostaglandin that induces uterine contractions to help expel the endometrium and embryo.

Hormones secreted by the placenta play a critical role in the maintenance of pregnancy.

The placenta has the remarkable capacity to secrete a number of peptide and steroid hormones essential for maintaining pregnancy. The most important are *human chorionic gonadotropin,* *estrogen,* and *progesterone* (▲ Table 20-5). Serving as the major endocrine organ of pregnancy, the placenta is unique among endocrine tissues in two regards. First, it is a transient tissue. Second, secretion of its hormones is not subject to extrinsic control, in contrast to the stringent, often complex mechanisms that regulate the secretion of other hormones. Instead, the type and rate of placental hormone secretion depend primarily on the stage of pregnancy.

Secretion of human chorionic gonadotropin

One of the first endocrine events is secretion by the developing chorion of **human chorionic gonadotropin (hCG)**, a peptide hormone that prolongs the life span of the corpus luteum. Recall that during the ovarian cycle, the corpus luteum degenerates and the highly prepared, luteal-dependent uterine lining sloughs off if fertilization and implantation do not occur. When fertilization does occur, the implanted blastocyst saves itself from being flushed out in menstrual flow, by producing hCG. This hormone, which is functionally similar to LH, stimulates and maintains the corpus luteum so it does not degenerate. Now called the **corpus luteum of pregnancy**, this ovarian endocrine unit grows even larger and produces increasingly greater amounts of estrogen and progesterone for an additional 10 weeks until the placenta takes over secretion of these steroid hormones. As a result of the persistently circulating estrogen and progesterone, the thick, pulpy endometrial tissue is maintained instead of sloughing. Accordingly, menstruation ceases during pregnancy.

Stimulation by hCG is necessary to maintain the corpus luteum of pregnancy because LH, which maintains the corpus luteum during the normal luteal phase of the uterine cycle, is suppressed through feedback inhibition by the high levels of progesterone.

Maintenance of a normal pregnancy depends on high concentrations of progesterone and estrogen. Thus hCG production is critical during the first trimester to maintain ovarian output of these hormones. In a male fetus, hCG also stimulates the precursor Leydig cells in the fetal testes to secrete testosterone, which masculinizes the developing reproductive tract.

The secretion rate of hCG increases rapidly during early pregnancy to save the corpus luteum from demise. Peak secretion of hCG occurs about 60 days after the end of the last menstrual period (● Figure 20-27). By the 10th week of pregnancy, hCG output declines to a low rate of secretion that is maintained for the duration of gestation. The fall in hCG occurs at a time when the corpus luteum is no longer needed for its steroidal hormone output, because the placenta has begun to secrete substantial quantities of estrogen and progesterone. The corpus luteum of pregnancy partially regresses as hCG secretion dwindles, but it is not converted into scar tissue until after delivery of the baby.

Human chorionic gonadotropin is eliminated from the body in the urine. Pregnancy diagnosis tests can detect this hormone in urine as early as the first month of pregnancy, about two weeks after the first missed menstrual period. Because this is before the growing embryo can be detected by physical examination, the test permits early confirmation of pregnancy.

▲ TABLE 20-5
Placental Hormones

Hormone	Function
Human Chorionic Gonadotropin (hCG)	Maintains the corpus luteum of pregancy
	Stimulates secretion of testosterone by the developing testes in XY embryos
Estrogen *(also secreted by the corpus luteum of pregnancy)*	Stimulates growth of the myometrium, increasing uterine strength for parturition
	Helps prepare the mammary glands for lactation
Progesterone *(also secreted by the corpus luteum of pregnancy)*	Suppresses uterine contractions to provide a quiet environment for the fetus
	Promotes formation of a cervical mucus plug to prevent uterine contamination
	Helps prepare the mammary glands for lactation
Human Chorionic Somatomammotropin *(has a structure similar to both growth hormone and prolactin)*	Believed to reduce maternal use of glucose and to promote the breakdown of stored fat (similar to growth hormone) so that greater quantities of glucose and free fatty acids may be shunted to the fetus
	Helps prepare the mammary glands for lactation (similar to prolactin)
Relaxin *(also secreted by the corpus luteum of pregnancy)*	Softens the cervix in preparation for cervical dilation at parturition
	Loosens the connective tissue between the pelvic bones in preparation for parturition
Placental PTHrp *(parathyroid hormone–related peptide)*	Increases maternal plasma Ca^{2+} level for use in calcifying fetal bones; if necessary, promotes localized dissolution of maternal bones, mobilizing her Ca^{2+} stores for use by the developing fetus

---

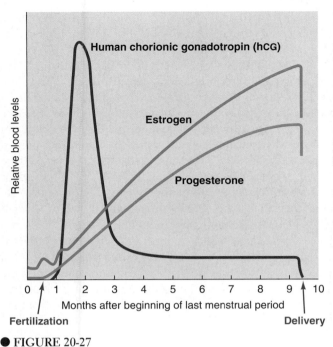

● FIGURE 20-27
Secretion rates of placental hormones

A frequent early clinical sign of pregnancy is **morning sickness,** a daily bout of nausea and vomiting that often occurs in the morning but can take place at any time of day. Because this condition usually appears shortly after implantation and coincides with the time of peak hCG production, it is speculated that this early placental hormone may trigger the symptoms, perhaps by acting on the chemoreceptor trigger zone in the vomiting center (see p. 608).

Secretion of estrogen and progesterone

Why doesn't the developing placenta start producing estrogen and progesterone in the first place instead of secreting hCG, which in turn stimulates the corpus luteum to secrete these two critical hormones? The answer is that, for different reasons, the placenta cannot produce sufficient quantities of either estrogen or progesterone in the first trimester of pregnancy. In the case of estrogen, the placenta does not have all the enzymes needed for complete synthesis of this hormone. Estrogen synthesis requires a complex interaction between the placenta and the fetus (● Figure 20-28). The placenta is able to convert the androgen hormone produced by the fetal adrenal cortex, dehydroepiandrosterone (DHEA), into estrogen. The placenta is unable to produce estrogen until the fetus has developed to the point that its adrenal cortex is secreting DHEA into the blood. The placenta extracts DHEA from the fetal blood and converts it into estrogen, which it then secretes into the maternal blood.

The primary estrogen synthesized by this means is **estriol,** in contrast to the main estrogen product of the ovaries, estradiol. Consequently, measurement of estriol levels in the maternal urine can be used clinically to assess the viability of the fetus.

In the case of progesterone, the placenta can synthesize this hormone soon after implantation. Even though the early placenta possesses the enzymes necessary to convert cholesterol extracted from the maternal blood into progesterone, it does not produce much of this hormone, because the amount of progesterone produced is proportional to placental weight. The placenta is simply too small in the first 10 weeks of pregnancy to produce enough progesterone to maintain the endometrial tissue. The notable increase in circulating progesterone in the last seven months of gestation reflects placental growth during this period.

Roles of estrogen and progesterone during pregnancy

As noted earlier, high concentrations of estrogen and progesterone are essential to maintain a normal pregnancy. Estrogen stimulates growth of the myometrium, which increases in size throughout pregnancy. The stronger uterine musculature is needed to expel the fetus during labor. Estriol also promotes development of the ducts within the mammary glands, through which milk will be ejected during lactation.

Progesterone performs various roles throughout pregnancy. Its primary function is to prevent miscarriage by suppressing contractions of the uterine myometrium. Progesterone also promotes formation of a mucus plug in the cervical canal, to prevent vaginal contaminants from reaching the uterus. Finally, placental progesterone stimulates the development of milk glands in the breasts, in preparation for lactation.

● FIGURE 20-28
Secretion of estrogen and progesterone by the placenta
The placenta secretes increasing quantities of progesterone and estrogen into the maternal blood after the first trimester. The placenta itself can convert cholesterol into progesterone (green pathway) but lacks some of the enzymes necessary to convert cholesterol into estrogen. However, the placenta can convert DHEA derived from cholesterol in the fetal adrenal cortex into estrogen when DHEA reaches the placenta by means of the fetal blood (blue pathway).

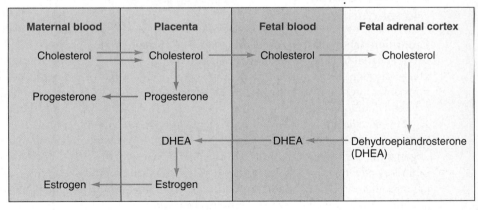

Maternal body systems respond to the increased demands of gestation.

The period of **gestation (pregnancy)** is about 38 weeks from conception (40 weeks from the end of the last menstrual period). During gestation, the embryo/fetus continues to grow and develop to the point of being able to leave its maternal life-support system. Meanwhile, a number of physical changes within the mother accommodate the demands of pregnancy. The most obvious change is uterine enlargement. The uterus expands and increases in weight more than 20 times, exclusive of its contents. The breasts enlarge and develop the capability of producing milk. Body systems other than the reproductive system also make needed adjustments. The volume of blood increases by 30%, and the cardiovascular system responds to the increasing demands of the growing placental mass. Weight gain during pregnancy is due only in part to the weight of the fetus. The remainder is primarily from increased weight of the uterus, including the placenta, and the increased blood volume. Respiratory activity increases by about 20% to handle the additional fetal requirements for O_2 utilization and CO_2 removal. Urinary output increases, and the kidneys excrete the additional wastes from the fetus.

The increased metabolic demands of the growing fetus increase nutritional requirements for the mother. In general, the fetus takes what it needs from the mother, even if this leaves the mother with a nutritional deficit. For example, the placental hormone **human chorionic somatomammotropin (hCS)** is thought responsible for the decreased use of glucose by the mother and the mobilization of free fatty acids from maternal adipose stores, similar to the actions of growth hormone (see p. 690). (In fact, hCS has a structure similar to both growth hormone and prolactin and exerts actions similar to both.) The hCS-induced metabolic changes in the mother make available greater quantities of glucose and fatty acids for shunting to the fetus. Also, if the mother does not consume sufficient Ca^{2+} in her diet, placental parathyroid hormone–related peptide **(PTHrp)** mobilizes Ca^{2+} from the maternal bones to ensure adequate calcification of the fetal bones.

Changes during late gestation prepare for parturition.

Parturition (labor, delivery, or **birth)** requires (1) dilation of the cervical canal to accommodate passage of the fetus from the uterus through the vagina and to the outside and (2) contractions of the uterine myometrium that are sufficiently strong to expel the fetus.

Several changes take place during late gestation in preparation for the onset of parturition. During the first two trimesters of gestation, the uterus remains relatively quiet, because of the inhibitory effect of the high levels of progesterone on the uterine musculature. During the last trimester, however, the uterus becomes progressively more excitable so that mild contractions **(Braxton-Hicks contractions)** are experienced with in-

creasing strength and frequency. Sometimes these contractions become regular enough to be mistaken for the onset of labor, a phenomenon called "false labor."

Throughout gestation, the exit of the uterus remains sealed by the rigid, tightly closed cervix. As parturition approaches, the cervix begins to soften (or "ripen") as a result of the dissociation of its tough connective tissue (collagen) fibers. Because of this softening, the cervix becomes malleable so that it can gradually yield, dilating the exit, as the fetus is forcefully pushed against it during labor. This cervical softening is believed to be caused by **relaxin,** a peptide hormone produced by the corpus luteum of pregnancy and by the placenta. Relaxin also "relaxes" the pelvic bones. Furthermore, prostaglandins released by the placenta promote the production of cervical enzymes that also degrade the collagen fibers and help soften the cervix.

Meanwhile, the fetus shifts downward (the baby "drops") and is normally oriented so that the head is in contact with the cervix in preparation for exiting through the birth canal. In a **breech birth,** any part of the body other than the head approaches the birth canal first.

Scientists are closing in on the factors that trigger the onset of parturition.

Rhythmic, coordinated contractions, usually painless at first, begin at the onset of true labor. As labor progresses, the contractions increase in frequency, intensity, and discomfort. These strong, rhythmic contractions force the fetus against the cervix, dilating the cervix. Then, after having dilated the cervix sufficiently for passage of the fetus, these contractions force the fetus out through the birth canal.

The exact factors responsible for triggering the increase in uterine contractility and thus initiating parturition are not fully established, although considerable progress has been made in recent years in unraveling the sequence of events. We will now examine what is known about this process.

Role of high estrogen levels

During early gestation, maternal estrogen levels are relatively low, but as gestation proceeds placental estrogen secretion continues to rise. In the immediate days before the onset of parturition, soaring levels of estrogen bring about changes in the uterus and cervix to prepare them for labor and delivery (● Figure 20-29). First, high levels of estrogen promote the synthesis of connexins within the uterine smooth-muscle cells. These myometrial cells are not functionally linked to any extent throughout most of gestation. The newly manufactured connexins are inserted in the myometrial plasma membranes to form gap junctions that electrically link together the uterine smooth-muscle cells so they become able to contract as a coordinated unit.

Simultaneously, the high levels of estrogen dramatically and progressively increase the concentration of myometrial receptors for oxytocin. Together, these myometrial changes collectively bring about the increased uterine responsiveness to oxytocin that ultimately initiates labor.

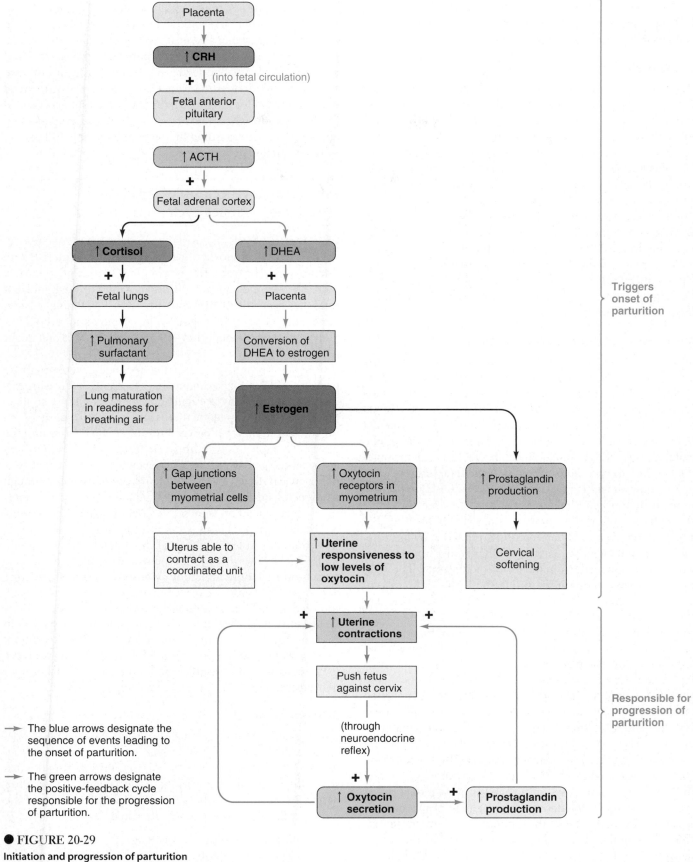

● FIGURE 20-29
Initiation and progression of parturition

The blue arrows designate the sequence of events leading to the onset of parturition.

The green arrows designate the positive-feedback cycle responsible for the progression of parturition.

In addition to preparing the uterus for labor, the increasing levels of estrogen also promote production of the local prostaglandins that contribute to cervical ripening.

Role of oxytocin

Oxytocin is a peptide hormone produced by the hypothalamus, stored in the posterior pituitary, and released into the blood from the posterior pituitary on nervous stimulation by the hypothalamus (see p. 684). A powerful uterine muscle stimulant, oxytocin plays the key role in the progression of labor. However, this hormone was once discounted as the trigger for parturition, because the circulating levels of oxytocin remain constant prior to the onset of labor. The discovery that uterine responsiveness to oxytocin is 100 times greater at term than in nonpregnant women (because of the increased concentration of myometrial oxytocin receptors) led to the now widely accepted conclusion that labor is initiated when the oxytocin receptor concentration reaches a critical threshold that permits the onset of strong, coordinated contractions in response to ordinary levels of circulating oxytocin.

Role of corticotropin-releasing hormone

Until recently, scientists were baffled by the factors responsible for the rising levels of placental estrogen secretion. Recent research has shed new light on the probable mechanism. Evidence suggests that *corticotropin-releasing hormone (CRH)* secreted by the fetal portion of the placenta into both the maternal and fetal circulations not only drives the manufacture of placental estrogen, thus ultimately dictating the timing of the onset of labor, but also promotes changes in the fetal lungs needed for breathing air (● Figure 20-29). Recall that CRH is normally secreted by the hypothalamus and regulates the output of ACTH by the anterior pituitary (see p. 710). In turn, ACTH stimulates production of both cortisol and DHEA by the adrenal cortex. In the fetus, much of the CRH comes from the placenta rather than solely from the fetal hypothalamus. The additional cortisol secretion summoned by the extra CRH promotes fetal lung maturation. Specifically, cortisol stimulates the synthesis of pulmonary surfactant, which facilitates lung expansion and reduces the work of breathing (see p. 474).

The bumped-up rate of DHEA secretion by the adrenal cortex in response to placental CRH leads to the rising levels of placental estrogen secretion. Recall that the placenta converts DHEA from the fetal adrenal gland into estrogen, which enters the maternal bloodstream (● Figure 20-28). When sufficiently high, this estrogen sets in motion the events that initiate labor. Thus pregnancy duration and delivery timing are determined largely by the placenta's rate of CRH production. That is, a **"placental clock"** ticks out a predetermined length of time until parturition. The timing of parturition is established early in pregnancy, with delivery at the endpoint of a maturational process that extends throughout the entire gestation. The ticking of the placental clock is measured by the rate of placental secretion. As the pregnancy progresses, CRH levels in maternal plasma rise. Researchers can accurately predict the timing of parturition by measuring the maternal plasma levels of CRH as early as the end of the first trimester. Higher-than-normal levels are associated with premature deliveries, whereas lower-than-normal levels indicate late deliveries. These and other data suggest that a critical level of maternal CRH of placental origin may directly trigger parturition. Placental CRH ensures that when labor begins, the infant is ready for life outside the womb. It does so by concurrently increasing the fetal cortisol needed for lung maturation and the estrogen needed for the uterine changes that bring on labor. The remaining unanswered puzzle regarding the placental clock is, What controls placental secretion of CRH?

∎ Parturition is accomplished by a positive-feedback cycle.

Once the high levels of estrogen have increased uterine responsiveness to oxytocin to a critical level and regular uterine contractions have begun, myometrial contractions progressively increase in frequency, strength, and duration throughout labor until they expel the uterine contents. At the beginning of labor, contractions lasting 30 seconds or less occur about every 25 to 30 minutes; by the end, they last 60 to 90 seconds and occur every 2 to 3 minutes.

As labor progresses, a positive-feedback cycle involving oxytocin and prostaglandin ensues, incessantly increasing myometrial contractions (● Figure 20-29). Each uterine contraction begins at the top of the uterus and sweeps downward, forcing the fetus toward the cervix. Pressure of the fetus against the cervix accomplishes two things. First, the fetal head pushing against the softened cervix wedges open the cervical canal. Second, cervical stretch stimulates the release of oxytocin through a neuroendocrine reflex. Stimulation of receptors in the cervix in response to fetal pressure sends a neural signal up the spinal cord to the hypothalamus, which in turn triggers oxytocin release from the posterior pituitary. This additional oxytocin promotes more powerful uterine contractions. As a result, the fetus is pushed more forcefully against the cervix, stimulating the release of even more oxytocin, and so on. This cycle is reinforced as oxytocin stimulates prostaglandin production by the decidua. As a powerful myometrial stimulant, prostaglandin further enhances uterine contractions. Oxytocin secretion, prostaglandin production, and uterine contractions continue to increase in positive-feedback fashion throughout labor until delivery relieves the pressure on the cervix.

Stages of labor

Labor is divided into three stages: (1) cervical dilation, (2) delivery of the baby, and (3) delivery of the placenta (● Figure 20-30). At the onset of labor or sometime during the first stage, the membranes surrounding the amniotic sac, or "bag of waters," rupture. As the amniotic fluid escapes out of the vagina, it helps to lubricate the birth canal.

- *First stage.* During the first stage, the cervix is forced to dilate to accommodate the diameter of the baby's head, usually to a maximum of 10 cm. This stage is the longest, lasting from several hours to as long as 24 hours in a first pregnancy. If another part of the fetus's body other than the head is oriented

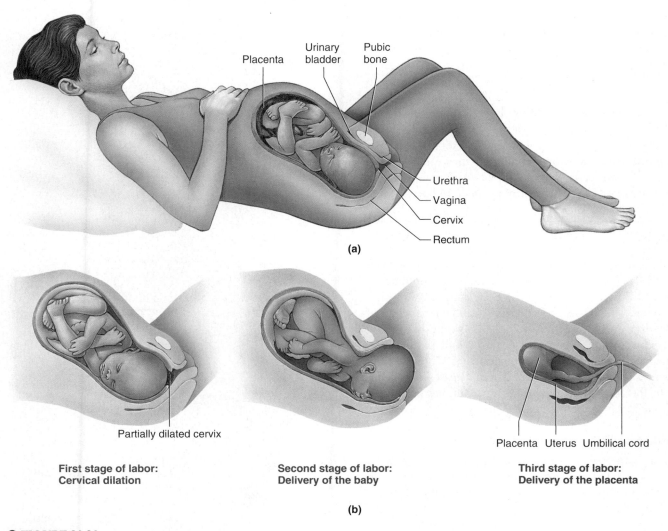

Placenta · Urinary bladder · Pubic bone
Urethra
Vagina
Cervix
Rectum

(a)

Partially dilated cervix

Placenta · Uterus · Umbilical cord

First stage of labor:
Cervical dilation

Second stage of labor:
Delivery of the baby

Third stage of labor:
Delivery of the placenta

(b)

● **FIGURE 20-30**

Stages of labor
(a) Position of the fetus near the end of pregnancy. (b) Stages of labor

---

against the cervix, it is generally less effective than the head as a wedge. The head has the largest diameter of the baby's body. If the baby approaches the birth canal feet first, the cervix may not be dilated sufficiently by the feet to permit passage of the head. Without medical intervention in such a case, the baby's head would remain stuck behind the too-narrow cervical opening.

- *Second stage.* The second stage of labor, the actual birth of the baby, begins once cervical dilation is complete. When the infant begins to move through the cervix and vagina, stretch receptors in the vagina activate a neural reflex that triggers contractions of the abdominal wall in synchrony with the uterine contractions. These abdominal contractions greatly increase the force pushing the baby through the birth canal. The mother can help deliver the infant by voluntarily contracting the abdominal muscles at this time in unison with each uterine contraction (that is, by "pushing" with each "labor pain"). Stage 2 is usually much shorter than the first stage and lasts 30 to 90 minutes. The infant is still attached to the placenta by

the umbilical cord at birth. The cord is tied and severed, with the stump shriveling up in a few days to form the **umbilicus (navel).**

- *Third stage.* Shortly after delivery of the baby, a second series of uterine contractions separates the placenta from the myometrium and expels it through the vagina. Delivery of the placenta, or **afterbirth,** constitutes the third stage of labor, typically the shortest stage, being completed within 15 to 30 minutes after the baby is born. After the placenta is expelled, continued contractions of the myometrium constrict the uterine blood vessels supplying the site of placental attachment, to prevent hemorrhage.

Uterine involution

After delivery, the uterus shrinks to its pregestational size, a process known as **involution,** which takes four to six weeks to complete. During involution, the remaining endometrial tissue not expelled with the placenta gradually disintegrates and sloughs off, producing a vaginal discharge called **lochia** that continues for three to six weeks following parturition. After this period, the endometrium is restored to its nonpregnant state.

Involution occurs largely because of the precipitous fall in circulating estrogen and progesterone when the placental source of these steroids is lost at delivery. The process is facilitated in mothers who breast-feed their infants, because oxytocin is released in response to suckling. In addition to playing an important role in lactation, this periodic nursing-induced release of oxytocin promotes myometrial contractions that help maintain uterine muscle tone, enhancing involution. Involution is usually complete in about four weeks in nursing mothers but takes about six weeks in those who do not breast-feed.

Lactation requires multiple hormonal inputs.

The female reproductive system supports the new being from the moment of conception through gestation and continues to nourish it during its early life outside the supportive uterine environment. Milk (or its equivalent) is essential for survival of the newborn. Accordingly, during gestation the **mammary glands,** or **breasts,** are prepared for **lactation (milk production).**

The breasts in nonpregnant females consist mostly of adipose tissue and a rudimentary duct system. Breast size is determined by the amount of adipose tissue, which has nothing to do with the ability to produce milk.

Preparation of the breasts for lactation

Under the hormonal environment present during pregnancy, the mammary glands develop the internal glandular structure and function necessary for milk production. A breast capable of lactating consists of a network of progressively smaller ducts that branch out from the nipple and terminate in lobules (● Figure 20-31a). Each *lobule* is made up of a cluster of saclike epithelial-lined *alveoli* that constitute the milk-producing glands. Milk is synthesized by the epithelial cells, then secreted into the alveolar lumen, which is drained by a milk-collecting duct that transports the milk to the surface of the nipple (● Figure 20-31b).

During pregnancy, the high concentration of *estrogen* promotes extensive duct development, whereas the high level of *progesterone* stimulates abundant alveolar–lobular formation. Elevated concentrations of *prolactin* (an anterior pituitary hormone stimulated by the rising levels of estrogen) and *human chorionic somatomammotropin* (a placental hormone that has a structure similar to both growth hormone and prolactin) also contribute to mammary gland development by inducing the synthesis of enzymes needed for milk production.

Prevention of lactation during gestation

Most of these changes in the breasts occur during the first half of gestation, so the mammary glands are fully capable of producing milk by the middle of pregnancy. However, milk secretion does not occur until parturition. The high estrogen and progesterone concentrations during the last half of pregnancy prevent lactation by blocking prolactin's stimulatory action on milk secretion. Prolactin is the primary stimulant of milk secretion. Thus even though the high levels of placental steroids

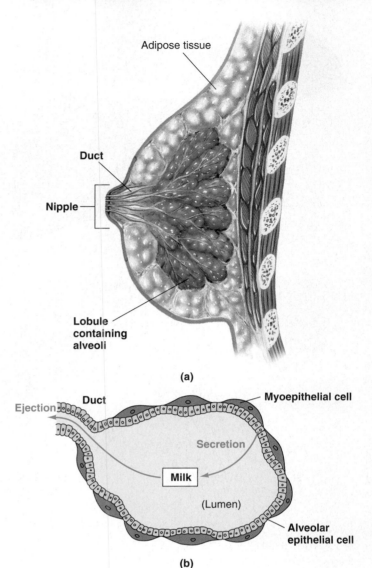

(a)

(b)

● FIGURE 20-31

Mammary gland anatomy
(a) Internal structure of the mammary gland, lateral view. (b) Schematic representation of the microscopic structure of an alveolus within the mammary gland. The alveolar epithelial cells secrete milk into the lumen. Contraction of the surrounding myoepithelial cells ejects the secreted milk out through the duct.

induce the development of the milk-producing machinery in the breasts, they prevent these glands from becoming operational until the baby is born and milk is needed.

The abrupt decline in estrogen and progesterone that occurs with loss of the placenta at parturition initiates lactation. (We have now completed our discussion of the functions of estrogen and progesterone during gestation and lactation as well as throughout the reproductive life of females. These functions are summarized in ▲ Table 20-6.)

Stimulation of lactation via suckling

Once milk production begins after delivery, two hormones are critical for maintaining lactation: (1) *prolactin*, which acts on the alveolar epithelium to promote secretion of milk, and

▲ TABLE 20-6
Actions of Estrogen and Progesterone

Estrogen

Effects on Sex-Specific Tissues

Essential for egg maturation and release

Stimulates growth and maintenance of entire female reproductive tract

Stimulates granulosa cell proliferation, which leads to follicle maturation

Thins the cervical mucus to permit sperm penetration

Enhances transport of sperm to the oviduct by stimulating upward contractions of uterus and oviduct

Stimulates growth of endometrium and myometrium

Induces synthesis of progesterone receptors in endometrium

Triggers onset of parturition by increasing uterine responsiveness to oxytocin during late gestation through a twofold effect: by inducing synthesis of myometrial oxytocin receptors and by increasing myometrial gap junctions so that the uterus can contract as a coordinated unit in response to oxytocin

Other Reproductive Effects

Promotes development of secondary sexual characteristics

Controls GnRH and gonadotropin secretion
   Low levels inhibit secretion
   High levels responsible for triggering LH surge

Stimulates duct development in breasts during gestation

Inhibits milk-secreting actions of prolactin during gestation

Nonreproductive Effects

Promotes fat deposition

Increases bone density; closes the epiphyseal plates

Progesterone

Prepares a suitable environment for nourishment of a developing embryo/fetus

Promotes formation of thick mucus plug in cervical canal

Inhibits hypothalamic GnRH and gonadotropin secretion

Stimulates alveolar development in breasts during gestation

Inhibits milk-secreting actions of prolactin during gestation

Inhibits uterine contractions during gestation

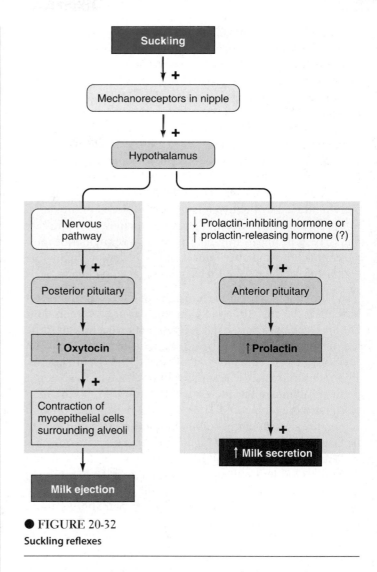

● FIGURE 20-32
Suckling reflexes

---

(2) *oxytocin*, which induces **milk ejection.** The latter term refers to the forced expulsion of milk from the lumen of the alveoli out through the ducts. Release of both of these hormones is stimulated by a neuroendocrine reflex triggered by suckling (● Figure 20-32). Let's examine each of these hormones and their roles in further detail.

● *Oxytocin release and milk ejection.* The infant cannot directly suck milk out of the alveolar lumen. Instead, milk must be actively squeezed out of the alveoli into the ducts and hence toward the nipple, by contraction of specialized **myoepithelial cells** (musclelike epithelial cells) that surround each alveolus (● Figure 20-31b). The infant's suckling of the breast stimulates sensory nerve endings in the nipple, initiating action potentials that travel up the spinal cord to the hypothalamus. Thus activated, the hypothalamus triggers a burst of oxytocin release from the posterior pituitary. Oxytocin in turn stimulates contraction of the myoepithelial cells in the breasts to induce milk ejection, or "milk letdown." Milk letdown continues only as long as the infant continues to nurse. In this way, the milk ejection reflex ensures that the breasts release milk only when and in the amount needed by the baby. Even though the alveoli may be full of milk, the milk cannot be released without oxytocin. The reflex can become conditioned to stimuli other than suckling, however. For example, the infant's cry can trigger milk letdown, causing a spurt of milk to leak from the nipples. In contrast, psychological stress, acting through the hypothalamus, can easily inhibit milk ejection. For this reason, a positive attitude toward breast-feeding and a relaxed environment are essential for successful breast-feeding.

- *Prolactin release and milk secretion.* Suckling not only triggers oxytocin release but also stimulates prolactin secretion. Prolactin output by the anterior pituitary is controlled by two hypothalamic secretions: **prolactin-inhibiting hormone (PIH)** and **prolactin-releasing hormone (PRH)**. PIH is now known to be dopamine, which also serves as a neurotransmitter in the brain. The chemical nature of PRH has not been identified with certainty, but scientists suspect PRH is oxytocin secreted by the hypothalamus into the hypothalamic-hypophyseal portal system to stimulate prolactin secretion by the anterior pituitary (see p. 688). This role of oxytocin is distinct from the roles of oxytocin produced by the hypothalamus and stored in the posterior pituitary. The oxytocin released by the posterior pituitary plays a key role in parturition by causing powerful uterine muscle contractions and also induces milk ejection by stimulating contraction of the myoepithelial cells of lactating breasts.

Throughout most of the female's life, PIH is the dominant influence, so prolactin concentrations normally remain low. During lactation, a burst in prolactin secretion occurs each time the infant suckles. Afferent impulses initiated in the nipple on suckling are carried by the spinal cord to the hypothalamus. This reflex ultimately leads to prolactin release by the anterior pituitary, although it is unclear whether this is from inhibition of PIH or stimulation of PRH secretion or both. Prolactin then acts on the alveolar epithelium to promote secretion of milk to replace the ejected milk (● Figure 20-32).

Concurrent stimulation by suckling of both milk ejection and milk production ensures that the rate of milk synthesis keeps pace with the baby's needs for milk. The more the infant nurses, the more milk is removed by letdown and the more milk is produced for the next feeding.

In addition to prolactin, which is the most important factor controlling synthesis of milk, at least four other hormones are essential for their permissive role in ongoing milk production: cortisol, insulin, parathyroid hormone, and growth hormone.

▌ Breast-feeding is advantageous to both the infant and the mother.

Milk is composed of water, triglyceride fat, the carbohydrate lactose (milk sugar), a number of proteins, vitamins, and the minerals calcium and phosphate.

Advantages of breast-feeding for the infant

In addition to nutrients, milk contains a host of immune cells, antibodies, and other chemicals that help protect the infant against infection until it can mount an effective immune response on its own a few months after birth. **Colostrum,** the milk produced for the first five days after delivery, contains lower concentrations of fat and lactose but higher concentrations of immunoprotective components. All human babies acquire some passive immunity during gestation through the passage of antibodies across the placenta from the mother to the fetus (see p. 432). These antibodies are short-lived, however, and often do not persist until the infant can fend for itself immunologically. Breast-fed babies gain additional protection during this vulnerable period through a variety of mechanisms:

- Breast milk contains an abundance of immune cells—both B and T lymphocytes, macrophages, and neutrophils (see p. 414)—that produce antibodies and destroy pathogenic micro-organisms outright. These cells are especially plentiful in colostrum.
- *Secretory IgA,* a special type of antibody, is present in great amounts in breast milk. Secretory IgA consists of two IgA antibody molecules (see p. 425) joined together with a so-called secretory component that helps protect the antibodies from destruction by the infant's acidic gastric juice and digestive enzymes. The collection of IgA antibodies a breast-fed baby receives is specifically aimed against the particular pathogens in the environment of the mother—and, accordingly, of the infant as well. Appropriately, therefore, these antibodies protect against the infectious microbes that the infant is most likely to encounter.
- Some components in mother's milk, such as mucus, adhere to potentially harmful micro-organisms, preventing them from attaching to and crossing the intestinal mucosa.
- *Lactoferrin* is a breast-milk constituent that thwarts growth of harmful bacteria by decreasing the availability of iron, a mineral needed for multiplication of these pathogens (see p. 419).
- *Bifidus factor* in breast milk, in contrast to lactoferrin, promotes multiplication of the nonpathogenic micro-organism *Lactobacillus bifidus* in the infant's digestive tract. Growth of this harmless bacterium helps crowd out potentially harmful bacteria.
- Other components in breast milk promote maturation of the baby's digestive system so that it is less vulnerable to diarrhea-causing bacteria and viruses.
- Still other factors in breast milk hasten the development of the infant's own immune capabilities.

Thus breast milk helps protect infants from disease in a variety of ways.

In addition to the benefits of breast milk during infancy, some studies hint that breast-feeding may reduce the risk of developing certain serious diseases later in life. Examples include allergies such as asthma, autoimmune diseases such as Type I diabetes mellitus (see p. 728), and cancers such as lymphoma.

Infants who are bottle-fed on a formula made from cow's milk or another substitute do not have the protective advantage provided by human milk, and, accordingly, have a higher incidence of infections of the digestive tract, respiratory tract, and ears than do breast-fed babies. Also, the digestive system of a newborn is better equipped to handle human milk than cow milk–derived formula, so bottle-fed babies tend to have more digestive upsets.

Advantages of breast-feeding for the mother

Breast-feeding is also advantageous for the mother. Oxytocin release triggered by nursing hastens uterine involution. In addition, suckling suppresses the menstrual cycle by inhibiting LH and FSH secretion, probably by inhibiting GnRH. Lactation, therefore, tends to prevent ovulation, decreasing the likelihood of another pregnancy (although it is not a reliable means of contraception). From an evolutionary perspective, this mechanism permits all the mother's resources to be directed toward the newborn instead of being shared with a new embryo.

Cessation of milk production at weaning

When the infant is weaned, two mechanisms contribute to the cessation of milk production. First, without suckling, prolactin secretion is not stimulated, removing the primary stimulus for continued milk synthesis and secretion. Also because there is no suckling, and thus no oxytocin release, milk letdown does not occur. Because milk production does not immediately shut down, milk accumulates in the alveoli, engorging the breasts. The resulting pressure buildup acts directly on the alveolar epithelial cells to suppress further milk production. Cessation of lactation at weaning therefore results from a lack of suckling-induced stimulation of both prolactin and oxytocin secretion.

▌ The end is a new beginning.

Reproduction is an appropriate way to end our discussion of physiology from cells to systems. The single cell resulting from the union of male and female gametes divides mitotically and differentiates into a multicellular individual made up of a number of different organ systems that interact cooperatively to maintain homeostasis (that is, stability in the internal environment). All the life-supporting homeostatic processes introduced throughout this book begin all over again at the start of a new life.

CHAPTER IN PERSPECTIVE: FOCUS ON HOMEOSTASIS

The reproductive system is unique in that it is not essential for homeostasis or for survival of the individual, but it is essential for sustaining the thread of life from generation to generation. Reproduction depends on the union of male and female gametes (reproductive cells), each with a half set of chromosomes, to form a new individual with a full, unique set of chromosomes. Unlike the other body systems, which are essentially identical in the two sexes, the reproductive systems of males and females are remarkably different, befitting their different roles in the reproductive process.

The male system is designed to continuously produce huge numbers of motile spermatozoa that are delivered to the female during the sex act. Male gametes must be produced in abundance for two reasons: (1) Only a small percentage of them survive the hazardous journey through the female reproductive tract to the site of fertilization; and (2) the cooperative effort of many spermatozoa is required to break down the barriers surrounding the female gamete (ovum or egg) to enable one spermatozoon to penetrate and unite with the ovum.

The female reproductive system undergoes complex changes on a monthly cyclical basis. During the first half of the cycle, a single nonmotile ovum is prepared for release. During the second half, the reproductive system is geared toward preparing a suitable environment for supporting the ovum if fertilization (union with a spermatozoon) occurs. If fertilization does not occur, the prepared supportive environment within the uterus sloughs off, and the cycle starts over again as a new ovum is prepared for release. If fertilization occurs, the female reproductive system adjusts to support growth and development of the new individual until it can survive on its own on the outside.

There are three important parallels in the male and female reproductive systems, even though they differ considerably in structure and function. First, the same set of undifferentiated reproductive tissues in the embryo can develop into either a male or a female system, depending on the presence or absence, respectively, of male-determining factors. Second, the same hormones—namely, hypothalamic GnRH and anterior pituitary FSH and LH—control reproductive function in both sexes. In both cases, gonadal steroids and inhibin act in negative-feedback fashion to control hypothalamic and anterior pituitary output. Third, the same events take place in the developing gamete's nucleus during sperm formation and egg formation, although males produce millions of sperm in one day, whereas females produce only about 400 ova in a lifetime.

CHAPTER SUMMARY

Introduction

▌ Both sexes produce gametes (reproductive cells), sperm in males and ova (eggs) in females, each of which bears one member of each of the 23 pairs of chromosomes present in human cells. Union of a sperm and an ovum at fertilization results in the beginning of a new individual with 23 complete pairs of chromosomes, half from the father and half from the mother.

▌ The reproductive system is anatomically and functionally distinct in males and females. Males produce sperm and deliver them into the female. Females produce ova, accept sperm delivery, and provide a suitable environment for supporting development of a fertilized ovum until the new individual can survive on its own in the external world.

▌ In both sexes, the reproductive system consists of (1) a pair of gonads, testes in males and ovaries in females, which are the primary reproductive organs that produce the gametes and secrete sex hormones; (2) a reproductive tract composed of a system of ducts that transport and/or house the gametes after they are produced; and (3) accessory sex glands that provide supportive secretions for the gametes. The externally visible portions of the reproductive system constitute the external genitalia.

▌ Secondary sexual characteristics are the distinguishing features between males and females not directly related to reproduction.

▌ Sex determination is a genetic phenomenon dependent on the combination of sex chromosomes at the time of fertilization, an XY combination being a genetic male and an XX combination a genetic female.

▌ The term *sex differentiation* refers to the embryonic development of the gonads, reproductive tract, and external genitalia along male or female lines, which gives rise to the apparent anatomic sex of the individual. In the presence of masculinizing factors, a male reproductive system develops; in their absence, a female system develops.

Male Reproductive Physiology

■ The testes are located in the scrotum. The cooler temperature in the scrotum compared to the abdominal cavity is essential for spermatogenesis.

■ Spermatogenesis (sperm production) occurs in the highly coiled seminiferous tubules within the testes.

■ Leydig cells located in the interstitial spaces between these tubules secrete the male sex hormone testosterone into the blood.

■ Testosterone is secreted before birth to masculinize the developing reproductive system; then its secretion ceases until puberty, at which time it begins once again and continues throughout life. Testosterone is responsible for maturation and maintenance of the entire male reproductive tract, for development of secondary sexual characteristics, and for stimulating libido.

■ The testes are regulated by the anterior pituitary hormones, luteinizing hormone (LH) and follicle-stimulating hormone (FSH). These gonadotropic hormones in turn are under control of hypothalamic gonadotropin-releasing hormone (GnRH).

■ Testosterone secretion is regulated by LH stimulation of the Leydig cells, and, in negative-feedback fashion, testosterone inhibits gonadotropin secretion.

■ Spermatogenesis requires both testosterone and FSH. Testosterone stimulates the mitotic and meiotic divisions required to transform the undifferentiated diploid germ cells, the spermatogonia, into undifferentiated haploid spermatids. FSH stimulates the remodeling of spermatids into highly specialized motile spermatozoa.

■ A spermatozoon consists only of a DNA-packed head bearing an enzyme-filled acrosome at its tip for penetrating the ovum, a midpiece containing the metabolic machinery for energy production, and a whiplike motile tail.

■ Also present in the seminiferous tubules are Sertoli cells, which protect, nurse, and enhance the germ cells throughout their development. Sertoli cells also secrete inhibin, a hormone that inhibits FSH secretion, completing the negative-feedback loop.

■ The still immature sperm are flushed out of the seminiferous tubules into the epididymis by fluid secreted by the Sertoli cells.

■ The epididymis and ductus deferens store and concentrate the sperm and increase their motility and fertility prior to ejaculation.

■ During ejaculation, the sperm are mixed with secretions released by the accessory glands, which contribute the bulk of the semen.

■ The seminal vesicles supply fructose for energy and prostaglandins, which promote smooth muscle motility in both the male and female reproductive tracts, to enhance sperm transport.

■ The prostate gland contributes an alkaline fluid for neutralizing the acidic vaginal secretions.

■ The bulbourethral glands release lubricating mucus.

Sexual Intercourse between Males and Females

■ The male sex act consists of erection and ejaculation, which are part of a much broader systemic, emotional response that typifies the male sexual response cycle.

■ Erection is a hardening of the normally flaccid penis that enables it to penetrate the female vagina. Erection is accomplished by marked vasocongestion of the penis brought about by reflexly induced vasodilation of the arterioles supplying the penile erectile tissue.

■ When sexual excitation reaches a critical peak, ejaculation occurs. It consists of two stages: (1) emission, the emptying of semen (sperm and accessory sex gland secretions) into the urethra; and (2) expulsion of semen from the penis. The latter is accompanied by a set of characteristic systemic responses and intense pleasure referred to as *orgasm*.

■ Females experience a sexual cycle similar to males, with both having excitation, plateau, orgasmic, and resolution phases. The major differences are that women do not ejaculate, and they are capable of multiple orgasms.

■ During the female sexual response, the outer third of the vagina constricts to grip the penis, whereas the inner two-thirds expand to create space for sperm deposition.

Female Reproductive Physiology

■ In the nonpregnant state, female reproductive function is controlled by a complex, cyclic negative-feedback control system between the hypothalamus (GnRH), anterior pituitary (FSH and LH), and ovaries (estrogen, progesterone, and inhibin). During pregnancy, placental hormones become the main controlling factors.

■ The ovaries perform the dual and interrelated functions of oogenesis (producing ova) and secreting estrogen and progesterone. Two related ovarian endocrine units sequentially accomplish these functions: the follicle and the corpus luteum.

■ The same steps in chromosome replication and division take place in oogenesis as in spermatogenesis, but the timing and end result are markedly different. Spermatogenesis is accomplished within two months, whereas the similar steps in oogenesis take anywhere from 12 to 50 years to complete on a cyclical basis from the onset of puberty until menopause. A female is born with a limited, nonrenewable supply of germ cells, whereas postpubertal males can produce several hundred million sperm each day. Each primary oocyte yields only one cytoplasm-rich ovum along with three doomed cytoplasm-poor polar bodies that disintegrate, whereas each primary spermatocyte yields four equally viable spermatozoa.

■ Oogenesis and estrogen secretion take place within an ovarian follicle during the first half of each reproductive cycle (the follicular phase) under the influence of FSH, LH, and estrogen.

■ At approximately midcycle, the maturing follicle releases a single ovum (ovulation). Ovulation is triggered by an LH surge brought about by the high level of estrogen produced by the mature follicle.

■ Under the influence of LH, the empty follicle is then converted into a corpus luteum, which produces progesterone as well as estrogen during the last half of the cycle (the luteal phase). This endocrine unit is responsible for preparing the uterus as a suitable site for implantation should the released ovum be fertilized.

■ If fertilization and implantation do not occur, the corpus luteum degenerates. The consequent withdrawal of hormonal support for the highly developed uterine lining causes it to disintegrate and slough, producing menstrual flow. Simultaneously, a new follicular phase is initiated.

■ Menstruation ceases and the uterine lining (endometrium) repairs itself under the influence of rising estrogen levels from the newly maturing follicle.

■ If fertilization does take place, it occurs in the oviduct as the released egg and sperm deposited in the vagina are both transported to this site.

■ The fertilized ovum begins to divide mitotically. Within a week it grows and differentiates into a blastocyst capable of implantation.

■ Meanwhile, the endometrium has become richly vascularized and stocked with stored glycogen under the influence of luteal-phase progesterone. Into this especially prepared lining the blastocyst implants by means of enzymes released by the trophoblasts, which form the blastocyst's outer layer. These enzymes digest the nutrient-rich endometrial tissue, accomplishing the dual function of carving a hole in the endometrium for implantation of the blastocyst while at the same time releasing nutrients from the endometrial cells for use by the developing embryo.

■ Following implantation, an interlocking combination of fetal and maternal tissues, the placenta, develops. The placenta is the organ of exchange between the maternal and fetal blood and also acts as a transient, complex endocrine organ that secretes a number of hormones essential for pregnancy. Human chorionic gonadotropin, estrogen, and progesterone are the most important of these hormones.

■ Human chorionic gonadotropin maintains the corpus luteum of pregnancy, which secretes estrogen and progesterone during the first trimester of gestation until the placenta takes over this function the last two trimesters. High levels of estrogen and progesterone are essential for maintaining a normal pregnancy.

■ At parturition, rhythmic contractions of increasing strength, duration, and frequency accomplish the three stages of labor: dilation of the cervix, birth of the baby, and delivery of the placenta (afterbirth).

■ Once the contractions are initiated at the onset of labor, a positive-feedback cycle is established that progressively increases their force. As contractions push the fetus against the cervix, secretion of oxytocin, a powerful uterine muscle stimulant, is reflexly increased. The extra oxytocin causes stronger contractions, giving rise to even more oxytocin release, and so on. This positive-feedback cycle progressively intensifies until cervical dilation and delivery are accomplished.

■ During gestation, the breasts are specially prepared for lactation. The elevated levels of placental estrogen and progesterone, respectively, promote development of the ducts and alveoli in the mammary glands.

■ Prolactin stimulates the synthesis of enzymes essential for milk production by the alveolar epithelial cells. However, the high gestational level of estrogen and progesterone prevents prolactin from promoting milk production. Withdrawal of the placental steroids at parturition initiates lactation.

■ Lactation is sustained by suckling, which triggers the release of oxytocin and prolactin. Oxytocin causes milk ejection by stimulating the myoepithelial cells surrounding the alveoli to squeeze the secreted milk out through the ducts. Prolactin stimulates the production of more milk to replace the milk ejected as the baby nurses.

REVIEW EXERCISES

Objective Questions (Answers on p. A-51)

1. It is possible for a genetic male to have the anatomic appearance of a female. *(True or false?)*
2. Testosterone secretion essentially ceases from birth until puberty. *(True or false?)*
3. The pineal gland secretes more melatonin during the light than during the dark. *(True or false?)*
4. Females do not experience erection. *(True or false?)*
5. Most of the lubrication for sexual intercourse is provided by the female. *(True or false?)*
6. If a follicle fails to reach maturity during one ovarian cycle, it can finish maturing during the next cycle. *(True or false?)*
7. Low but rising levels of estrogen inhibit tonic LH secretion, whereas high levels of estrogen stimulate the LH surge. *(True or false?)*
8. Spermatogenesis takes place within the _____ of the testes, stimulated by the hormones _____ and _____ .
9. During estrogen production by the follicle, the _____ cells under the influence of the hormone _____ produce androgens, and the _____ cells under the influence of the hormone _____ convert these androgens into estrogens.
10. The source of estrogen and progesterone during the first 10 weeks of gestation is the _____ .
11. Detection of _____ in the urine is the basis of pregnancy diagnosis tests.
12. Which of the following statements concerning chromosomal distribution is *incorrect?*
    a. All human somatic cells contain 23 chromosomal pairs for a total diploid number of 46 chromosomes.
    b. Each gamete contains 23 chromosomes, one member of each chromosomal pair.
    c. During meiotic division, the members of the chromosome pairs regroup themselves into the original combinations derived from the individual's mother and father for separation into haploid gametes.
    d. Sex determination depends on the combination of sex chromosomes, an XY combination being a genetic male, XX a genetic female.

e. The sex chromosome content of the fertilizing sperm determines the sex of the offspring.

13. When the corpus luteum degenerates,
    a. circulating levels of estrogen and progesterone rapidly decline.
    b. FSH and LH secretion start to rise as the inhibitory effects of the gonadal steroids are withdrawn.
    c. the endometrium sloughs off.
    d. Both a and b are correct.
    e. All of the preceding are correct.

14. Match the following:
    _____ 1. secrete(s) prostaglandins
    _____ 2. increase(s) motility and fertility of sperm
    _____ 3. secrete(s) an alkaline fluid
    _____ 4. provide(s) fructose
    _____ 5. storage site for sperm
    _____ 6. concentrate(s) the sperm a hundredfold
    _____ 7. secrete(s) fibrinogen
    _____ 8. provide(s) clotting enzymes
    _____ 9. contain(s) erectile tissue

    (a) epididymis and ductus deferens
    (b) prostate gland
    (c) seminal vesicles
    (d) bulbourethral glands
    (e) penis

15. Using the following answer code, indicate when each event takes place during the ovarian cycle:
    (a) occurs during the follicular phase
    (b) occurs during the luteal phase
    (c) occurs during both the follicular and luteal phases

    _____ 1. development of antral follicles
    _____ 2. secretion of estrogen
    _____ 3. secretion of progesterone
    _____ 4. menstruation
    _____ 5. repair and proliferation of the endometrium
    _____ 6. increased vascularization and glycogen storage in the endometrium

Essay Questions

1. What are the primary reproductive organs, gametes, sex hormones, reproductive tract, accessory sex glands, external genitalia, and secondary sexual characteristics in males and females?
2. List the essential reproductive functions of the male and of the female.
3. Discuss the differences between males and females with regard to genetic, gonadal, and phenotypic sex.
4. What parts of the male and female reproductive systems develop from each of the following: genital tubercle, urethral folds, genital swellings, Wolffian ducts, and Müllerian ducts?
5. Discuss the source and functions of testosterone.
6. Describe the three major stages of spermatogenesis. Discuss the functions of each part of a spermatozoon. What are the roles of Sertoli cells?
7. Discuss the control of testicular function.
8. Compare the sex act in males and females.
9. Compare oogenesis with spermatogenesis.
10. Describe the events of the follicular and luteal phases of the ovarian cycle. Correlate the phases of the uterine cycle with those of the ovarian cycle.
11. How are the ovum and spermatozoa transported to the site of fertilization? Describe the process of fertilization.
12. Describe the process of implantation and placenta formation.
13. What are the functions of the placenta? What hormones does the placenta secrete?
14. What is the role of human chorionic gonadotropin?
15. What is the leading proposal for the mechanism that initiates parturition? What are the stages of labor? What is the role of oxytocin?
16. Describe the hormonal factors that play a role in lactation.
17. Summarize the actions of estrogen and progesterone.

POINTS TO PONDER

(Explanations on p. A-51)

1. The hypothalamus releases GnRH in pulsatile bursts once every two to three hours, with no secretion occurring in between. The blood concentration of GnRH depends on the frequency of these bursts of secretion. A promising line of research for a new method of contraception involves administration of GnRH-like drugs. In what way could such drugs act as contraceptives when GnRH is the hypothalamic hormone that triggers the chain of events leading to ovulation? (*Hint:* The anterior pituitary is "programmed" to respond only to the normal pulsatile pattern of GnRH.)

2. Occasionally, testicular tumors composed of interstitial cells of Leydig may secrete up to 100 times the normal amount of testosterone. When such a tumor develops in young children, they grow up much shorter than their genetic potential. Explain why. What other symptoms would be present?

3. What type of sexual dysfunction might arise in men taking drugs that inhibit sympathetic nervous system activity as part of the treatment for high blood pressure?

4. Explain the physiologic basis for administering a posterior pituitary extract to induce or facilitate labor.

5. The symptoms of menopause are sometimes treated with supplemental estrogen and progesterone. Why wouldn't treatment with GnRH or FSH and LH also be effective?

6. *Clinical Consideration.* Maria A., who is in her second month of gestation, has been experiencing severe abdominal cramping. Her physician has diagnosed her condition as a *tubal pregnancy*: The developing embryo is implanted in the oviduct instead of in the uterine endometrium. Why must this pregnancy be surgically terminated?

PHYSIOEDGE RESOURCES

PHYSIOEDGE CD-ROM
Figures marked with the PhysioEdge icon have associated activities on the CD. For this chapter, check out the Media Quizzes related to Figures 20-1, 20-2, and 20-15.

CHECK OUT THESE MEDIA QUIZZES:
20.1 Male Reproductive Anatomy
20.2 Female Reproductive Anatomy
20.3 Male Reproductive Physiology
20.4 Female Reproductive Physiology

PHYSIOEDGE WEB SITE
The Web site for this book contains a wealth of helpful study aids, as well as many ideas for further reading and research. Log on to:

http://www.brookscole.com

Go to the Biology page and select Sherwood's *Human Physiology*, 5th Edition. Select a chapter from the drop-down menu or click on one of these resource areas:

- **Case Histories** provide an introduction to the clinical aspects of human physiology. Check out:

  #23: Pregnancy Tests

- For 2-D and 3-D graphical illustrations and animations of physiological concepts, visit our **Visual Learning Resource.**

- Resources for study and review include the **Chapter Outline, Chapter Summary, Glossary,** and **Flash Cards.** Use our **Quizzes** to prepare for in-class examinations.

- On-line research resources to consult are **Hypercontents,** with current links to relevant Internet sites; **Internet Exercises** with starter URLs provided; and **InfoTrac Exercises.**

 For more readings, go to InfoTrac College Edition, your on-line research library, at:

http://infotrac.thomsonlearning.com

Appendix A

The Metric System

▲ TABLE A-1 **Metric Measures and English Equivalents**

Unit	Measure	Symbol		English Equivalent
Linear Measure				
1 kilometer	= 1,000 meters	10^3m	km	0.62137 mile
1 meter		10^0m	m	39.37 inches
1 decimeter	= 1/10 meter	10^{-1}m	dm	3.937 inches
1 centimeter	= 1/100 meter	10^{-2}m	cm	0.3937 inch
1 millimeter	= 1/1,000 meter	10^{-3}m	mm	
1 micrometer (or micron)	= 1/1,000,000 meter	10^{-6}m	μm (or μ)	English equivalents infrequently used
1 nanometer	= 1/1,000,000,000 meter	10^{-9}m	nm	
Measures of Capacity (For Fluids and Gases)				
1 liter			l	1.0567 U.S. liquid quarts
1 milliliter	= 1/1,000 liter = volume of 1 g of water at standard temperature and pressure (stp)		ml	
Measures of Volume				
1 cubic meter			m^3	
1 cubic decimeter	= 1/1,000 cubic meter = 1 liter (1)		dm^3	
1 cubic centimeter	= 1/1,000,000 cubic meter = 1 milliliter (ml)		cm^3 = ml	
1 cubic millimeter	= 1/100,000,000 cubic meter		mm^3	
Measures of Mass				
1 kilogram	= 1,000 grams		kg	2.2046 pounds
1 gram			g	15.432 grains
1 milligram	= 1/1,000 gram		mg	0.01 grain (about)
1 microgram	= 1/1,000,000 gram		μg (or mcg)	

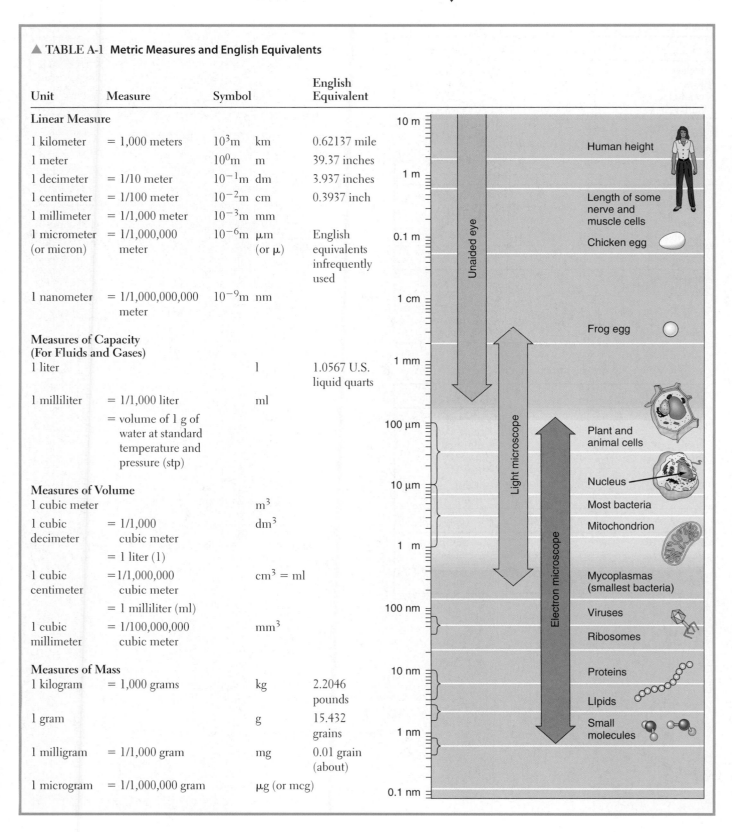

10 m — Human height

1 m — Length of some nerve and muscle cells

0.1 m — Chicken egg

1 cm — Frog egg

1 mm

100 μm — Plant and animal cells

10 μm — Nucleus / Most bacteria / Mitochondrion

1 m

Mycoplasmas (smallest bacteria)

100 nm — Viruses / Ribosomes

10 nm — Proteins / Lipids

1 nm — Small molecules

0.1 nm

Unaided eye / Light microscope / Electron microscope

Metric–English Conversions

Length

English		Metric	
inch	=	2.54	centimeters
foot	=	0.30	meter
yard	=	0.91	meter
mile (5,280 feet)	=	1.61	kilometer

To convert	multiply by	to obtain
inches	2.54	centimeters
feet	30.00	centimeters
centimeters	0.39	inches
millimeters	0.039	inches

Mass/Weight

English		Metric	
grain	=	64.80	milligrams
ounce	=	28.35	grams
pound	=	453.60	grams
ton (short) (2,000 pounds)	=	0.91	metric ton

To convert	multiply by	to obtain
ounces	28.3	grams
pounds	453.6	grams
pounds	0.45	kilograms
grams	0.035	ounces
kilograms	2.2	pounds

Volume and Capacity

English		Metric	
cubic inch	=	16.39	cubic centimeters
cubic foot	=	0.03	cubic meter
cubic yard	=	0.765	cubic meters
ounce	=	0.03	liter
pint	=	0.47	liter
quart	=	0.95	liter
gallon	=	3.79	liters

To convert	multiply by	to obtain
fluid ounces	30.00	milliliters
quart	0.95	liters
milliliters	0.03	fluid ounces
liters	1.06	quarts

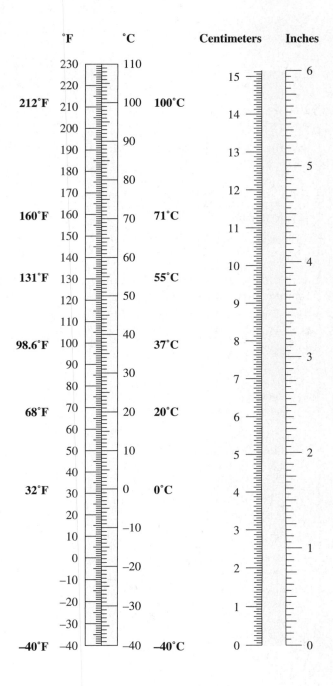

A Review of Chemical Principles

By Spencer Seager, Weber State College, and Lauralee Sherwood

CHEMICAL LEVEL OF ORGANIZATION IN THE BODY

Matter is anything that occupies space and has mass, including all living and nonliving things in the universe. **Mass** is the amount of matter in an object. **Weight,** in contrast, is the effect of gravity on that mass. The more gravity exerted on a mass, the greater the weight of the mass. An astronaut has the same mass whether on Earth or in space but is weightless in the zero gravity of space.

▌ Atoms

All matter is made up of tiny particles called **atoms.** These particles are too small to be seen individually, even with the most powerful electron microscopes available today.

Even though extremely small, atoms consist of three types of even smaller subatomic particles. Different types of atoms vary in the number of these various subatomic particles they contain. **Protons** and **neutrons** are particles of nearly identical mass, with protons carrying a positive charge and neutrons having no charge. **Electrons** have a much smaller mass than protons and neutrons and are negatively charged. An atom consists of two regions—a dense, central *nucleus* made of protons and neutrons surrounded by a three-dimensional *electron cloud,* where electrons move rapidly around the nucleus in orbitals (● Figure B-1). The magnitude of the charge of a proton exactly matches that of an electron, but it is opposite in sign, being positive. In all atoms, the number of protons in the nucleus is equal to the number of electrons moving around the nucleus, so their charges balance and the atoms are neutral.

▌ Elements and atomic symbols

A pure substance composed of only one type of atom is called an **element.** A pure sample of the element carbon contains only carbon atoms, even though the atoms might be arranged in the form of diamond or in the form of graphite (pencil "lead"). Each element is designated by an **atomic symbol,** a one- or two-letter chemical shorthand for the element's name. Usually these symbols are easy to follow, because they are derived from the English name for the element. Thus H stands for *hydrogen,* C for *carbon,* and O for *oxygen.* In a few cases,

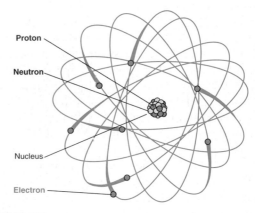

● FIGURE B-1

The atom
The atom consists of two regions. The central nucleus contains protons and neutrons and makes up 99.9% of the mass. Surrounding the nucleus is the electron cloud, where the electrons move rapidly around the nucleus. (Figure not drawn to scale.)

the atomic symbol is based on the element's Latin name—for example, Na for *sodium (natrium* in Latin) and K for *potassium (kalium).* Of the 109 known elements, 26 are normally found in the body. Four elements—oxygen, carbon, hydrogen, and nitrogen—compose 96% of the body's mass.

▌ Compounds and molecules

Pure substances composed of more than one type of atom are known as **compounds.** Pure water, for example, is a compound that contains atoms of hydrogen and atoms of oxygen in a 2-to-1 ratio, regardless of whether the water is in the form of liquid, solid (ice), or vapor (steam). A **molecule** is the smallest unit of a pure substance that has the properties of that substance and is capable of a stable, independent existence. For example, a molecule of water consists of two atoms of hydrogen and one atom of oxygen, held together by chemical bonds.

▌ Atomic number

Exactly what are we talking about when we refer to a "type" of atom? That is, what makes carbon, hydrogen, and oxygen atoms different? The answer is the number of protons in the nucleus. Regardless of where they are found, all hydrogen atoms have one proton in the nucleus, all carbon atoms have six, and all oxygen atoms have eight. Of course, these numbers also represent the number of electrons moving around each nucleus, because the number of electrons and number of protons in an atom are equal. The number of protons in the nucleus of an atom of an element is called the **atomic number** of the element.

▌ Atomic weight

As expected, tiny atoms have tiny masses. For example, the actual mass of a hydrogen atom is 1.67×10^{-24} g, that of a carbon atom is 1.99×10^{-23} g, and that of an oxygen atom is 2.66×10^{-23} g. These very small numbers are inconvenient to work with in calculations, so a system of relative masses has been developed. These relative masses simply compare the actual masses of the atoms with each other. Suppose the actual masses of two people were determined to be 45.50 kg and 113.75 kg. Their relative masses are determined by dividing each mass by the smaller mass of the two: 45.50/45.50 = 1.00, and 113.75/45.50 = 2.50. Thus the relative masses of the two people are 1.00 and 2.50; these numbers simply express the fact that the mass of the heavier person is 2.50 times that of the other person. The relative masses of atoms are called **atomic masses,** or **atomic weights,** and are given in atomic mass units (*amu*). In this system, hydrogen atoms, the least massive of all atoms, have an atomic weight of 1.01 amu. The atomic weight of carbon atoms is 12.01 amu, and that of oxygen atoms is 16.00 amu. Thus, oxygen atoms have a mass about 16 times that of hydrogen atoms. ▲ Table B-1 gives the atomic weights and some other characteristics of the elements that are most important physiologically.

▲ TABLE B-1
Characteristics of Selected Elements

Name and Symbol	Number of Protons	Atomic Number	Atomic Weight (amu)
Hydrogen (H)	1	1	1.01
Carbon (C)	6	6	12.01
Nitrogen (N)	7	7	14.01
Oxygen (O)	8	8	16.00
Sodium (Na)	11	11	22.99
Magnesium (Mg)	12	12	24.31
Phosphorus (P)	15	15	30.97
Sulfur (S)	16	16	32.06
Chlorine (Cl)	17	17	35.45
Potassium (K)	19	19	39.10
Calcium (Ca)	20	20	40.08

CHEMICAL BONDS

Because all matter is made up of atoms, atoms must somehow be held together to form matter. The forces holding atoms together are called **chemical bonds.** Not all chemical bonds are formed in the same way, but all involve the electrons of atoms. Whether one atom will bond with another depends on the number and arrangement of its electrons. An atom's electrons are arranged in electron shells, to which we now turn our attention.

▌ Electron shells

Electrons tend to move around the nucleus in a specific pattern. The orbitals, or pathways traveled by electrons around the nucleus, are arranged in an orderly series of concentric layers known as **electron shells,** which consecutively surround the nucleus. Each electron shell can hold a specific number of electrons. The first (innermost) shell closest to the nucleus can contain a maximum of only 2 electrons, no matter what the element is. The second shell can hold a total of 8 more electrons. The third shell can hold a maximum of 18 electrons. As the number of electrons increases with increasing atomic number, still more electrons occupy successive shells, each at a greater distance from the nucleus. Each successive shell from the nucleus has a higher **energy level.** Because the negatively charged electrons are attracted to the positively charged nucleus, it takes more energy for an electron to overcome the nuclear attraction and orbit farther from the nucleus. Thus the first electron shell has the lowest energy level and the outermost shell of an atom has the highest energy level.

In general, electrons belong to the lowest energy shell possible, up to the maximum capacity of each shell. For example, hydrogen atoms have only 1 electron, so it is in the first shell. Helium atoms have 2 electrons, which are both in the first shell and fill it. Carbon atoms have 6 electrons, 2 in the first shell and 4 in the second shell, whereas the 8 electrons of oxygen are arranged with 2 in the first shell and 6 in the second shell.

▌ Bonding characteristics of an atom; valence

Atoms tend to undergo processes that result in a filled outermost electron shell. Thus the electrons of the outer or higher-energy shell determine the bonding characteristics of an atom and its ability to interact with other atoms. Atoms that have a vacancy in their outermost shell tend to either give up, accept, or share electrons with other atoms (whichever is most favorable energetically) so that all participating atoms have filled outer shells. For example, an atom that has only one electron in its outermost shell may empty this shell so its remaining shells are completely full. By contrast, another atom that lacks only one electron in its outer shell may acquire the deficient electron from the first atom to fill all its shells to the maximum. The number of electrons an atom loses, gains, or shares to achieve a filled outer shell is known as the atom's **valence.** A chemical bond is the force of attraction that holds participating atoms together as a result of an interaction between their outermost electrons.

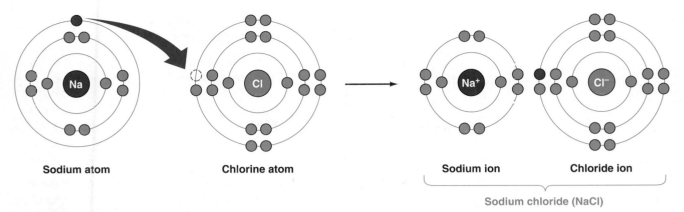

Sodium atom   Chlorine atom   Sodium ion   Chloride ion

Sodium chloride (NaCl)

● **FIGURE B-2**

Ions and ionic bonds

Sodium (Na) and chlorine (Cl) atoms both have partially filled outermost shells. Therefore, sodium tends to give up its lone electron in the outer shell to chlorine, thus filling chlorine's outer shell. As a result, sodium becomes a positively charged ion, and chlorine becomes a negatively charged ion known as *chloride*. The oppositely charged ions attract each other, forming an ionic bond.

Consider sodium atoms (Na) and chlorine atoms (Cl) (● Figure B-2). Sodium atoms have 11 electrons: 2 in the first shell, 8 in the second shell, and 1 in the third shell. Chlorine atoms have 17 electrons: 2 in the first shell, 8 in the second shell, and 7 in the third shell. Because 8 electrons are required to fill the second and third shells, sodium atoms have 1 electron more than is needed to provide a filled second shell, whereas chlorine atoms have 1 less electron than is needed to fill the third shell. Each sodium atom can lose an electron to a chlorine atom, leaving each sodium with 10 electrons, 8 of which are in the second shell, which is full and is now the outer shell occupied by electrons. By accepting 1 electron, each chlorine atom now has a total of 18 electrons, with 8 of them in the third, or outer, shell, which is now full.

Ions; ionic bonds

Recall that atoms are electrically neutral because they have an identical number of positively charged protons and negatively charged electrons. As a result of giving up and accepting electrons, the sodium atoms and chlorine atoms have achieved filled outer shells, but now each atom is unbalanced electrically. Although each sodium now has 10 electrons, it still has 11 protons in the nucleus and a net electrical charge, or valence, of +1. Similarly, each chlorine now has 18 electrons, but only 17 protons. Thus each chlorine has a −1 charge. Such charged atoms are called **ions**. Positively charged ions are called **cations**; negatively charged ions are called **anions**. As a helpful hint to keep these terms straight, imagine the "t" in *cation* as standing for a "+" sign and the first "n" in *anion* as standing for "negative."

Note that both a cation and anion are formed whenever an electron is transferred from one atom to another. Because opposite charges attract, sodium ions (Na⁺) and charged chlorine atoms, now called *chloride* ions (Cl⁻), are attracted toward each other. This electrical attraction that holds cations and anions together is known as an **ionic bond**. Ionic bonds hold Na$^+$ and Cl$^-$ together in the compound **sodium chloride, NaCl**, which is common table salt. A sample of sodium chloride actually contains sodium and chloride ions in a three-dimensional geometric arrangement called a *crystal lattice*. The ions of opposite charge occupy alternate sites within the lattice (● Figure B-3).

Covalent bonds

It is not favorable, energywise, for an atom to give up or accept more than three electrons. Neverthless, carbon atoms, which

● **FIGURE B-3**

Crystal lattice for sodium chloride (table salt)

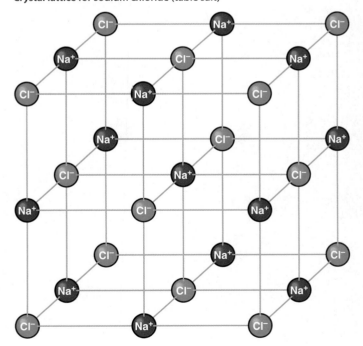

have four electrons in their outer shell, form compounds. They do so by another bonding mechanism, *covalent bonding*. Atoms that would have to lose or gain four or more electrons to achieve outer-shell stability usually bond by *sharing* electrons. Shared electrons actually orbit around *both* atoms. Thus a carbon atom can share its four outer electrons with the four electrons of four hydrogen atoms, as shown in Equation B-1, where the outer-shell electrons are shown as dots around the symbol of each atom. (The resulting compound is methane, CH_4, a gas made up of individual CH_4 molecules.)

$$\cdot \overset{\displaystyle\cdot}{\underset{\displaystyle\cdot}{C}} \cdot \; + \; 4 \cdot H \;\rightarrow\; H \overset{\displaystyle H}{\underset{\displaystyle H}{:\!C\!:}} H \qquad \text{Eq. B-1}$$

Shared electron pairs

Shared electron pairs

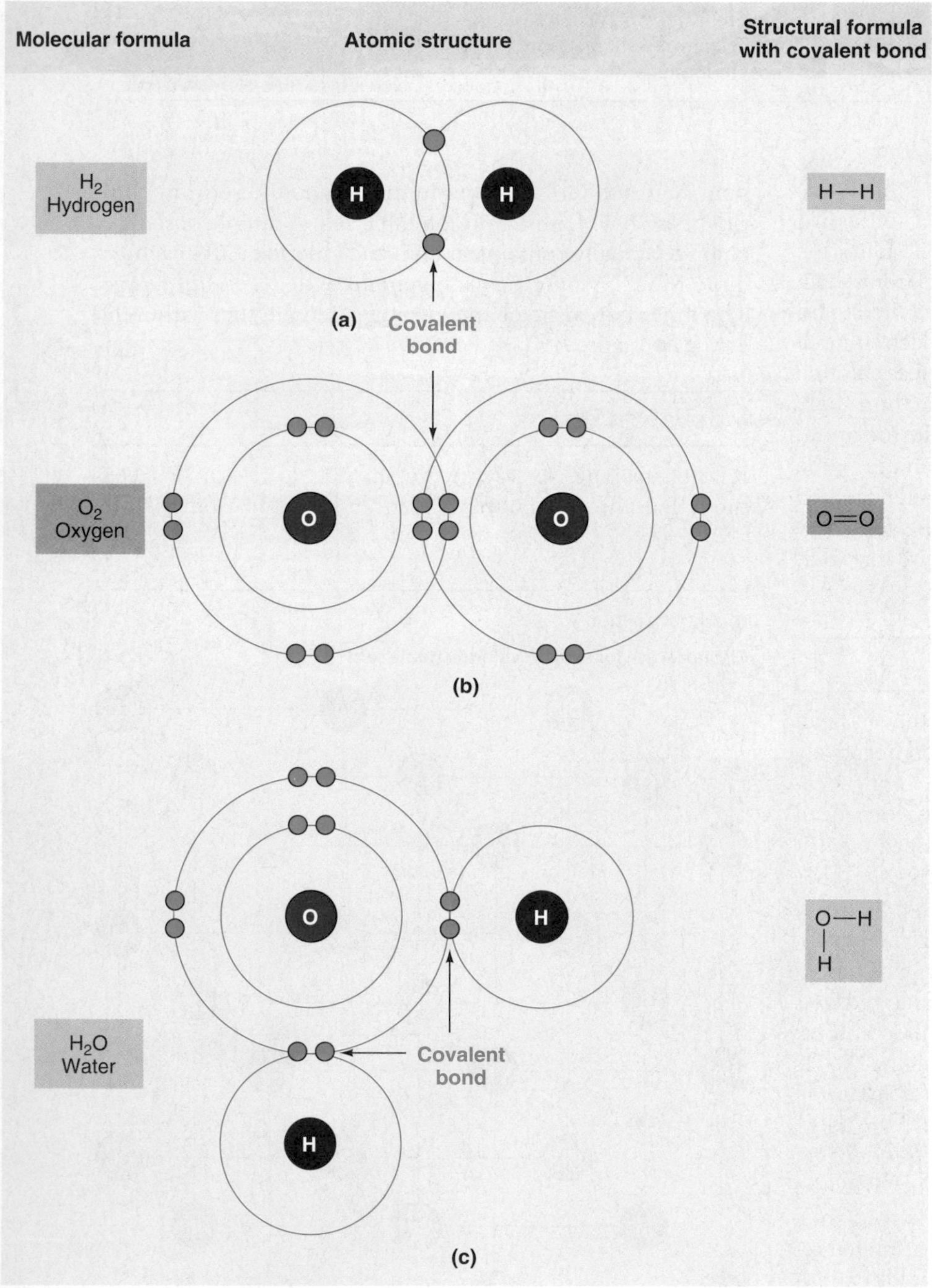

Each electron that is shared by two atoms is counted toward the number of electrons needed to fill the outer shell of each atom. Thus each carbon atom shares four pairs, or eight electrons, and so has eight in its outer shell. Each hydrogen shares one pair, or two electrons, and so has a filled outer shell. (Remember, hydrogen atoms need only two electrons to complete their outer shell, which is the first shell.) The sharing of a pair of electrons by atoms binds them together by means of a **covalent bond** (● Figure B-4). Covalent bonds are the strongest of chemical bonds; that is, they are the hardest to break.

Covalent bonds also form between some identical atoms. For example, two hydrogen atoms can complete their outer shells by sharing one electron pair made from the single electrons of each atom, as shown in Equation B-2:

$$H \cdot \; + \; \cdot H \;\rightarrow\; H : H \qquad \text{Eq. B-2}$$

Thus hydrogen gas consists of individual H_2 molecules (● Figure B-4a). (A subscript following a chemical symbol indicates the number of that type of atom present in the molecule.) Several other nonmetallic elements also exist as molecules, because covalent bonds form between identical atoms; oxygen (O_2) is an example (● Figure B-4b).

Often, an atom can form covalent bonds with more than one atom. One of the most familiar examples is water (H_2O), consisting of two hydrogen atoms each forming a single covalent bond with one oxygen atom (● Figure B-4c). Equation B-3 represents the formation of water's covalent bonds:

$$\begin{matrix} H \cdot \\ \\ H \cdot \end{matrix} \; + \; \cdot \ddot{O} : \;\rightarrow\; H : \overset{\displaystyle\cdot\cdot}{\underset{\displaystyle H}{O}} : \qquad \text{Eq. B-3}$$

The water molecule is sometimes represented as

$$\begin{matrix} H{-}O \\ | \\ H \end{matrix}$$

● FIGURE B-4

A covalent bond
A covalent bond is formed when atoms that share a pair of electrons are both attracted toward the shared pair.

Molecular formula	Atomic structure	Structural formula with covalent bond
H_2 Hydrogen	(a) Covalent bond	H—H
O_2 Oxygen	(b)	O=O
H_2O Water	(c) Covalent bond	O—H ∣ H

where the nonshared electron pairs are not shown and the covalent bonds, or shared pairs, are represented by dashes.

■ Nonpolar and polar molecules

The electrons between two atoms in a covalent bond are not always shared equally. When the atoms sharing an electron pair are identical, such as two oxygen atoms, the electrons are attracted equally by both atoms and so are shared equally. The result is a **nonpolar molecule.** The term *nonpolar* implies no difference at the two ends (two "poles") of the bond. Because both atoms within the molecule exert the same pull on the shared electrons, each shared electron spends the same amount of time orbiting each atom. Thus both atoms remain electrically neutral in a nonpolar molecule such as O_2.

When the sharing atoms are not identical, unequal sharing of electrons occurs, because atoms of different elements do not exert the same pull on shared electrons. For example, an oxygen atom strongly attracts electrons when it is bonded to other atoms. A **polar molecule** results from the unequal sharing of electrons between different types of atoms covalently bonded together. The water molecule is a good example of a polar molecule. The oxygen atom pulls the shared electrons more strongly than do the hydrogen atoms within each of the two covalent bonds. Consequently, the electron of each hydrogen atom tends to spend more time away orbiting around the oxygen atom than at home around the hydrogen atom. Because of this nonuniform distribution of electrons, the oxygen side of the water molecule where the shared electrons spend more time is slightly negative, and the two hydrogens that are visited less frequently by the electrons are slightly more positive (● Figure B-5). Note that the entire water molecule has the same number of electrons as it has protons, and so as a whole has no net charge. This is unlike ions, which have an electron excess or deficit. Polar molecules have a balanced number of protons and electrons but an unequal distribution of the shared electrons among the atoms making up the molecule.

■ Hydrogen bonds

Polar molecules are attracted to other polar molecules. In water, for example, an attraction exists between the positive hydrogen ends of some molecules and the negative oxygen ends of others. Hydrogen is not a part of all polar molecules, but when it is covalently bonded to an atom that strongly attracts electrons to form a covalent molecule, the attraction of the positive (hydrogen) end of the polar molecule to the negative end of another polar molecule is called a **hydrogen bond** (● Figure B-6). Thus, the polar attractions of water molecules to each other are an example of hydrogen bonding.

CHEMICAL REACTIONS

Processes in which chemical bonds are broken and/or formed are called **chemical reactions.** Reactions are represented by equations in which the reacting substances (**reactants**) are typically written on the left, the newly produced substances (**products**) are written on the right, and an arrow meaning "yields" points from the reactants to the products. These conventions are illustrated in Equation B-4:

$$A + B \rightarrow C + D$$
$$\text{Reactants} \quad \text{Products}$$

Eq. B-4

■ Balanced equations

A chemical equation is a "chemical bookkeeping" ledger that describes what happens in a reaction. By the **law of conservation of mass,** the total mass of all materials entering a reaction equals the total mass of all the products. Thus, the total number of atoms of each element must always be the same on the left and right sides of the equation, because no atoms are lost. Such equations in which the same number of atoms of each

● **FIGURE B-5**

A polar molecule
A water molecule is an example of a polar molecule, in which the distribution of shared electrons is not uniform. Because the oxygen atom pulls the shared electrons more strongly than the hydrogen atoms do, the oxygen side of the molecule is slightly negatively charged, and the hydrogen sides are slightly positively charged.

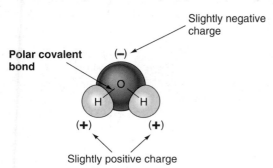

● **FIGURE B-6**

A hydrogen bond
A hydrogen bond is formed by the attraction of a positively charged hydrogen end of a polar molecule to the negatively charged end of another polar molecule.

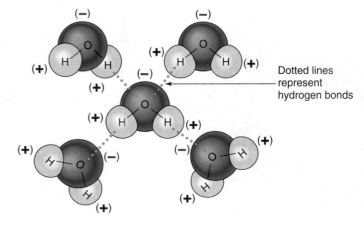

type appear on both sides are called **balanced equations.** When writing a balanced equation, the number *preceding* a chemical symbol designates the number of independent (unjoined) atoms, ions, or molecules of that type, whereas a number written as a subscript *following* a chemical symbol denotes the number of a particular atom within a molecule. The absence of a number indicates "one" of that particular chemical. Let's look at a specific example, the oxidation of glucose (the sugar that cells use as fuel), as shown in Equation B-5:

$$C_6H_{12}O_6 + 6O_2 \rightarrow 6CO_2 + 6H_2O$$

Glucose Oxygen Carbon Water   Eq. B-5
dioxide

According to this equation, 1 molecule of glucose reacts with 6 molecules of oxygen to produce 6 molecules of carbon dioxide and 6 molecules of water. Note the following balance in this reaction:

- 6 carbon atoms on the left (in 1 glucose molecule) and 6 carbon atoms on the right (in 6 carbon dioxide molecules)
- 12 hydrogen atoms on the left (in 1 glucose molecule) and 12 on the right (in 6 water molecules, each containing 2 hydrogen atoms)
- 18 oxygen atoms on the left (6 in 1 glucose molecule plus 12 more in the 6 oxygen molecules) and 18 on the right (12 in 6 carbon dioxide molecules, each containing 2 oxygen atoms, and 6 more in the 6 water molecules, each containing 1 oxygen atom)

▌ Reversible and irreversible reactions

Under appropriate conditions, the products of a reaction can be changed back to the reactants. For example, carbon dioxide gas dissolves in and reacts with water to form carbonic acid, H_2CO_3:

$$CO_2 + H_2O \rightarrow H_2CO_3 \qquad \text{Eq. B-6}$$

Carbonic acid is not very stable, however, and as soon as some is formed, part of it decomposes to give carbon dioxide and water:

$$H_2CO_3 \rightarrow CO_2 + H_2O \qquad \text{Eq. B-7}$$

Reactions that go in both directions are called **reversible reactions.** They are usually represented by double arrows pointing in both directions:

$$CO_2 + H_2O \rightleftharpoons H_2CO_3 \qquad \text{Eq. B-8}$$

Theoretically, every reaction is reversible. Often, however, conditions are such that a reaction, for all practical purposes, goes only in one direction; such a reaction is called **irreversible.** For example, an irreversible reaction takes place when an explosion occurs, because the products do not remain in the vicinity of the reaction site to get together to react.

▌ Catalysts; enzymes

The rates (speeds) of chemical reactions are influenced by a number of factors, of which catalysts are one of the most important. A **catalyst** is a "helper" molecule that speeds up a re-

action without being used up in the reaction. Living organisms use catalysts known as **enzymes.** These enzymes exert amazing influence on the rates of chemical reactions that take place in the organisms. Reactions that take weeks or even months to occur under normal laboratory conditions take place in seconds under the influence of enzymes in the body. One of the fastest-acting enzymes is **carbonic anhydrase,** which catalyzes the reaction between carbon dioxide and water to form carbonic acid. This reaction is important in the transport of carbon dioxide from tissue cells, where it is produced metabolically, to the lungs, where it is excreted. The equation for the reaction was shown in Equation B-6. Each molecule of carbonic anhydrase catalyzes the conversion of 36 million CO_2 molecules per minute! Enzymes are important in essentially every chemical reaction that takes place in living organisms.

MOLECULAR AND FORMULA WEIGHT AND THE MOLE

Because molecules are made up of atoms, the relative mass of a molecule is simply the sum of the relative masses (atomic weights) of the atoms found in the molecule. The relative masses of molecules are called **molecular masses** or **molecular weights.** The molecular weight of water, H_2O, is thus the sum of the atomic weights of two hydrogen atoms and one oxygen atom, or 1.01 amu + 1.01 amu + 16.00 amu = 18.02 amu.

Not all compounds exist in the form of molecules. Ionically bonded substances such as sodium chloride consist of three-dimensional arrangements of sodium ions (Na^+) and chloride ions (Cl^-) in a 1-to-1 ratio. The formulas for ionic compounds reflect only the ratio of the ions in the compound and should not be interpreted in terms of molecules. Thus the formula for sodium chloride, NaCl, indicates that the ions combine in a 1-to-1 ratio. It is convenient to apply the concept of relative masses to ionic compounds even though they do not exist as molecules. The **formula weight** for such compounds is defined as the sum of the atomic weights of the atoms found in the formula. Thus the formula weight of NaCl is equal to the sum of the atomic weights of one sodium atom and one chlorine atom, or 22.99 amu + 35.45 amu = 58.44 amu.

As you have seen, chemical reactions can be represented by equations and discussed in terms of numbers of molecules, atoms, and ions reacting with each other. To carry out reactions in the laboratory, however, a scientist cannot count out numbers of reactant particles but instead must be able to weigh out the correct amount of each reactant. Using the mole concept makes this task possible. A **mole** (abbreviated *mol*) of a pure element or compound is the amount of material contained in a sample of the pure substance that has a mass in grams equal to the substance's atomic weight (for elements) or the molecular weight or formula weight (for compounds). Thus 1 mole of potassium, K, would be a sample of the element with a mass of 39.10 g. Similarly, a mole of H_2O would have a mass of 18.02 g, and a mole of NaCl would be a sample with a mass of 58.44 g.

The fact that atomic weights, molecular weights, and formula weights are relative masses leads to a fundamental characteristic of moles. One mole of oxygen atoms has a mass of 16.00 g, and 1 mole of hydrogen atoms has a mass of 1.01 g. Thus the ratio of the masses of 1 mole of each element is 16.00/1.01, the same as the ratio of the atomic weights for the two elements. Recall that these atomic weights compare the relative masses of oxygen and hydrogen. Accordingly, the number of oxygen atoms present in 16 grams of oxygen (1 mole of oxygen) is the same as the number of hydrogen atoms present in 1.01 grams of hydrogen. Therefore, 1 mole of oxygen contains exactly the same number of oxygen atoms as the number of hydrogen atoms in 1 mole of hydrogen. Thus, it is possible and sometimes useful to think of a mole as a specific number of particles. This number, called Avogadro's number, is equal to 6.02×10^{23}.

SOLUTIONS, COLLOIDS, AND SUSPENSIONS

In contrast to a compound, a **mixture** consists of two or more types of elements or molecules physically blended together (intermixed) instead of being linked by chemical bonds. A compound has very different properties from the individual elements of which it is composed. For example, the solid, white NaCl (table salt) crystals you use to flavor your food are very different from either sodium (a silvery white metal) or chlorine (a poisonous yellow-green gas found in bleach). By comparison, each component of a mixture retains its own chemical properties. If you mix salt and sugar together, each retains its own distinct taste and other individual properties. The constituents of a compound can only be separated by chemical means—bond breakage. By contrast, the components of a mixture can be separated by physical means, such as filtration or evaporation. The most common mixtures in the body are mixtures of water and various other substances. These mixtures are categorized as *solutions*, *colloids*, or *suspensions*, depending on the size and nature of the substance mixed with water.

▌ Solutions

Most chemical reactions in the body take place between reactants that have dissolved to form solutions. **Solutions** are homogenous mixtures containing a relatively large amount of one substance called the **solvent** (the dissolving medium) and smaller amounts of one or more substances called **solutes** (the dissolved particles). Salt water, for example, contains mostly water, which is thus the solvent, and a smaller amount of salt, which is the solute. Water is the solvent in most solutions found in the human body.

▌ Electrolytes; nonelectrolytes

When ionic solutes are dissolved in water to form solutions, the resulting solution will conduct electricity. This is not true for most covalently bonded solutes. For example, a salt–water solution conducts electricity, but a sugar–water solution does not. When salt dissolves in water, the solid lattice of Na^+ and

Cl^- is broken down, and the individual ions are separated and distributed uniformly throughout the solution. These mobile, charged ions conduct electricity through the solution. Solutes that form ions in solution and conduct electricity are called **electrolytes.** Some very polar covalent molecules also behave this way. When sugar dissolves, however, individual covalently bonded sugar molecules leave the solid and become uniformly distributed throughout the solution. These uncharged molecules cannot conduct a current. Solutes that do not form conductive solutions are called **nonelectrolytes.**

▌ Measures of concentration

The amount of solute dissolved in a specific amount of solution can vary. For example, a salt–water solution might contain 1 g of salt in 100 ml of solution, or it could contain 10 g of salt in 100 ml of solution. Both solutions are salt–water solutions, but they have different concentrations of solute. The **concentration** of a solution indicates the relationship between the amount of solute and the amount of solution. Concentrations can be given in a number of different units.

Molarity

Concentrations given in terms of **molarity** (**M**) give the number of moles of solute in exactly 1 liter of solution. Thus a half molar (0.5 M) solution of NaCl would contain one-half mole, or 29.22 g, of NaCl in each liter of solution.

Normality

When the solute is an electrolyte, it is sometimes useful to express the concentration of the solution in a unit that gives information about the amount of ionic charge in the solution. This is done by expressing concentration in terms of **normality** (**N**). The normality of a solution gives the number of equivalents of solute in exactly 1 liter of solution. An **equivalent** of an electrolyte is the amount that produces 1 mole of positive (or negative) charges when it dissolves. The number of equivalents of an electrolyte can be calculated by multiplying the number of moles of electrolyte by the total number of positive charges produced when one formula unit of the electrolyte dissolves. Consider NaCl and calcium chloride ($CaCl_2$) as examples. The ionization reactions for one formula unit of each solute are:

$$NaCl \rightarrow Na^+ + Cl^- \qquad \text{Eq. B-9}$$

$$CaCl_2 \rightarrow Ca^{2+} + 2Cl^- \qquad \text{Eq. B-10}$$

Thus 1 mole of NaCl produces 1 mole of positive charges (Na^+) and so contains 1 equivalent:

$$(1 \text{ mole NaCl})(1) = 1 \text{ equivalent}$$

where the number 1 used to multiply the 1 mole of NaCl came from the +1 charge on Na^+.

One mole of $CaCl_2$ produces 1 mole of Ca^{2+}, which is 2 moles of positive charge. Thus 1 mole of $CaCl_2$ contains 2 equivalents:

$$(1 \text{ mole } CaCl_2)(2) = 2 \text{ equivalents}$$

where the number 2 used in the multiplication came from the $+2$ charge on Ca^{2+}.

If two solutions were made such that one contained 1 mole of NaCl per liter and the other contained 1 mole of $CaCl_2$ per liter, the NaCl solution would contain 1 equivalent of solute per liter and would be 1 normal (1 N). The $CaCl_2$ solution would contain 2 equivalents of solute per liter and would be 2 normal (2 N).

Osmolarity

Another expression of concentration frequently used in physiology is **osmolarity (osm),** which indicates the total *number* of solute particles in a liter of solution instead of the relative weights of the specific solutes. The osmolarity of a solution is the product of M and *n*, where *n* is the number of moles of solute particles obtained when 1 mole of solute dissolves. Because nonelectrolytes such as glucose do not dissociate in solution, $n = 1$ and the osmolarity (*n* times M) is equal to the molarity of the solution. For electrolyte solutions, the osmolarity exceeds the molarity by a factor equal to the number of ions produced on dissociation of each molecule in solution. For example, because a NaCl molecule dissociates into two ions, Na^+ and Cl^-, the osmolarity of a 1 M solution of NaCl is 2×1 M $= 2$ osm.

▌ Colloids and suspensions

In solutions, solute particles are ions or small molecules. By contrast, the particles in colloids and suspensions are much larger than ions or small molecules. In colloids and suspensions, these particles are known as **dispersed-phase particles** instead of solutes. When the dispersed-phase particles are no more than about 100 times the size of the largest solute particles found in a solution, the mixture is called a **colloid.** The dispersed-phase particles of colloids generally do not settle out. All dispersed-phase particles of colloids carry electrical charges of the same sign. Thus they repel each other. The constant buffeting from these collisions keeps the particles from settling. The most abundant colloids in the body are small functional proteins that are dispersed in the body fluids. An example is the colloidal dispersion of the plasma proteins in the blood (see p. 392).

When dispersed-phase particles are larger than those in colloids, if the mixture is left undisturbed the particles will settle out because of the force of gravity. Such mixtures are usually called **suspensions.** The major example of a suspension in the body is the mixture of blood cells suspended in the plasma. The constant movement of blood as it circulates through the blood vessels keeps the blood cells rather evenly dispersed within the plasma. However, if a blood sample is placed in a test tube and treated to prevent clotting, the heavier blood cells gradually settle to the bottom of the tube.

INORGANIC AND ORGANIC CHEMICALS

Chemicals are commonly classified into two categories: inorganic and organic. The original criterion used for this classification was the origin of the chemicals. Those that came from living or once-living sources were *organic*, and those that came from other sources were *inorganic.* Today the basis for classification is the element carbon. **Organic** chemicals are generally those that contain carbon. All others are classified as **inorganic.** A few carbon-containing chemicals are also classified as inorganic; the most common are pure carbon in the form of diamond and graphite, carbon dioxide (CO_2), carbon monoxide (CO), carbonates such as limestone ($CaCO_3$), and bicarbonates such as baking soda ($NaHCO_3$).

The unique ability of carbon atoms to bond to each other and form networks of carbon atoms results in an interesting fact. Even though organic chemicals all contain carbon, millions of these compounds have been identified. Some were isolated from natural plant or animal sources, and many have been synthesized in laboratories. Inorganic chemicals include all the other 108 elements and their compounds. The number of known inorganic chemicals made up of all these other elements is estimated to be about 250,000, compared to millions of organic compounds made up predominantly of carbon.

▌ Monomers and polymers

Another result of carbon's ability to bond to itself is the large size of some organic molecules. Molecules classified as organic range in size from methane, CH_4, a small, simple molecule with one carbon atom, to molecules such as DNA that contain as many as a million carbon atoms. Organic molecules that are essential for life are called **biological molecules,** or **biomolecules** for short. Some biomolecules are rather small organic compounds, including *simple sugars, fatty acids, amino acids,* and *nucleotides.* These small, single units, known as **monomers** (meaning "single unit"), are subunits, or building blocks, for the synthesis of larger biomolecules, including *complex carbohydrates, lipids, proteins,* and *nucleic acids,* respectively. These larger organic molecules are called **polymers** (meaning "many units"), reflecting the fact that they are made by the bonding together of a number of smaller monomers. For example, starch is formed by linking many glucose molecules together. Very large organic polymers are often referred to as **macromolecules,** reflecting their large size (*macro* means "large"). Macromolecules include many naturally occurring molecules, such as DNA and structural proteins, as well as many molecules that are synthetically produced, such as synthetic textiles (for example, nylon) and plastics.

ACIDS, BASES, AND SALTS

Acids, bases, and salts may be inorganic or organic compounds. Acids and bases are chemical opposites, and salts are produced when acids and bases react with each other. In 1887, Swedish chemist Svante Arrhenius proposed a theory defining acids and bases. He said that an *acid* is any substance that will dissociate, or break apart, when dissolved in water and in the process release a hydrogen ion, H^+. Similarly, *bases* are substances that dissociate when dissolved in water and in the process release a hydroxyl ion, OH^-. Hydrogen chloride (HCl) and sodium hydroxide (NaOH) are examples of Arrhenius acids and bases;

their dissociations in water are represented in Equations B-11 and B-12, respectively:

$$HCl \rightarrow H^+ + Cl^- \qquad \text{Eq. B-11}$$

$$NaOH \rightarrow Na^+ + OH^- \qquad \text{Eq. B-12}$$

Note that the hydrogen ion is a bare proton, the nucleus of a hydrogen atom. Also note that both HCl and NaOH would behave as electrolytes.

Arrhenius did not know that free hydrogen ions cannot exist in water. They covalently bond to water molecules to form hydronium ions, as shown in Equation B-13:

$$H^+ + :\overset{\displaystyle .}{\underset{\displaystyle H}{O}}\!\!-\!\!H \rightarrow \left[H\!-\!\overset{\displaystyle .}{\underset{\displaystyle H}{O}}\!\!-\!\!H \right]^+ \qquad \text{Eq. B-13}$$

In 1923, Johannes Brønsted in Denmark and Thomas Lowry in England proposed an acid–base theory that took this behavior into account. They defined an **acid** as any hydrogen-containing substance that donates a proton (hydrogen ion) to another substance (an acid is a *proton donor*) and a **base** as any substance that accepts a proton (a base is a *proton acceptor*). According to these definitions, the acidic behavior of HCl given in Equation B-11 is rewritten as shown in Equation B-14:

$$HCl + H_2O \rightleftharpoons H_3O^+ + Cl^- \qquad \text{Eq. B-14}$$

Note that this reaction is reversible, and the hydronium ion is represented as H_3O^+. In Equation B-14, the HCl acts as an acid in the forward (left-to-right) reaction, whereas water acts as a base. In the reverse reaction (right-to-left), the hydronium ion gives up a proton and thus is an acid, whereas the chloride ion, Cl^-, accepts the proton and so is a base. It is still a common practice to use equations such as B-11 to simplify the representation of the dissociation of an acid, even though scientists recognize that equations like B-14 are more correct.

▌ Neutralization reactions

At room temperature, **inorganic salts** are crystalline solids that contain the positive ion (cation) of an Arrhenius base such as NaOH and the negative ion (anion) of an acid such as HCl. Salts can be produced by mixing solutions of appropriate acids and bases, allowing a neutralization reaction to occur. In **neutralization reactions,** the acid and base react to form a salt and water. Most salts that form are water soluble and can be recovered by evaporating the water. Equation B-15 is a neutralization reaction:

$$HCl + NaOH \rightarrow NaCl + H_2O \qquad \text{Eq. B-15}$$

When acids or bases are used as solutes in solutions, the concentrations can be expressed as normalities just as they were earlier for salts. An equivalent of acid is the amount that gives up 1 mole of H^+ in solution. Thus, 1 mole of HCl is also 1 equivalent, but 1 mole of H_2SO_4 is 2 equivalents. Bases are described in a similar way, but an equivalent is the amount of base that gives 1 mole of OH^-.

See Chapter 15 for a discussion of acid–base balance in the body.

FUNCTIONAL GROUPS OF ORGANIC MOLECULES

Organic molecules consist of carbon and one or more additional elements covalently bonded to one another in "Tinker Toy" fashion. The simplest organic molecules, hydrocarbons, such as methane and petroleum products, have only hydrogen atoms attached to a carbon backbone of varying lengths. All biomolecules always have additional elements besides hydrogen added to the carbon backbone. The carbon backbone forms the stable portion of most biomolecules. Other atoms covalently bonded to the carbon backbone, either alone or in clusters, form what is termed *functional groups*. All organic compounds can be classified according to the functional group or groups they contain. **Functional groups** are specific combinations of atoms that generally react in the same way, regardless of the number of carbon atoms in the molecule to which they are attached. For example, all *aldehydes* contain a functional group that contains one carbon atom, one oxygen atom, and one hydrogen atom covalently bonded in a specific way:

$$\overset{\displaystyle O}{\overset{\displaystyle \|}{(-C-H)}}$$

The carbon atom in an aldehyde group forms a single covalent bond with the hydrogen atom and a **double bond** (a bond in which two covalent bonds are formed between the same atoms, designated by a double line between the atoms) with the oxygen atom. The aldehyde group is attached to the rest of the molecule by a single covalent bond extending to the left of the carbon atom. Most reactions of aldehydes involve this group, so most aldehyde reactions are the same regardless of the size and nature of the rest of the molecule to which the aldehyde group is attached. Reactions of physiological importance often occur between two functional groups or between one functional group and a small molecule such as water.

CARBOHYDRATES

Carbohydrates are organic compounds of tremendous biological and commercial importance. They are widely distributed in nature and include such familiar substances as starch, table sugar, and cellulose. Carbohydrates have five important functions in living organisms: they provide energy, they serve as a stored form of chemical energy, they provide dietary fiber, they supply carbon atoms for the synthesis of cell components, and they form part of the structural elements of some cells.

▌ Chemical composition of carbohydrates

Carbohydrates contain carbon, hydrogen, and oxygen. They acquired their name because most of them contain these three elements in an atomic ratio of one carbon to two hydrogens to one oxygen. This ratio suggests that the general formula is CH_2O and that the compounds are simply carbon hydrates ("watered" carbons), or carbohydrates. It is now known that they are not hydrates of carbon, but the name persists. All carbohydrates have a large number of functional groups per molecule. The most common functional groups in carbohydrates are *alcohol*, *ketone* and *aldehyde*—

$$(\text{—OH}), \quad (\text{—}\overset{\overset{\displaystyle O}{\|}}{C}\text{—}), \quad (\text{—}\overset{\overset{\displaystyle O}{\|}}{C}\text{—H})$$

Alcohol   Ketone   Aldehyde

—or functional groups formed by reactions between pairs of these three.

▌ Types of carbohydrates

The simplest carbohydrates are simple sugars, also called **monosaccharides.** As their name indicates, they consist of single units called saccharides (*mono* means "one"). The molecular structure of *glucose*, an important monosaccharide, is shown in ● Figure B-7a. In solution, most glucose molecules assume the ring form shown in ● Figure B-7b. Other common monosaccharides are *fructose*, *galactose*, and *ribose*.

Disaccharides are sugars formed by linking two monosaccharide molecules together through a covalent bond (*di* means "two"). Some common examples of disaccharides are *sucrose* (common table sugar) and *lactose* (milk sugar). Sucrose molecules are formed from one glucose and one fructose molecule. Lactose molecules each contain one glucose and one galactose unit.

Because of the many functional groups on carbohydrate molecules, large numbers of simple carbohydrate molecules are able to bond together and form long chains and branched networks. The resultant substances are called **polysaccharides,** a name that indicates that they contain many saccharide units (*poly* means "many"). Three common polysaccharides made up entirely of glucose units are glycogen, starch, and cellulose.

- *Glycogen* is a storage carbohydrate found in animals. It is a highly branched polysaccharide that averages a branch every eight to twelve glucose units. The structure of glycogen is represented in ● Figure B-8, where each circle represents one glucose unit.

- *Starch*, a storage carbohydrate of plants, consists of two fractions, amylose and amylopectin. Amylose consists of long, essentially unbranched chains of glucose units. Amylopectin is a highly branched network of glucose units averaging 24 to 30 glucose units per branch. Thus it is less highly branched than glycogen.

- *Cellulose*, a structural carbohydrate of plants, exists in the form of long, unbranched chains of glucose units. The bonding between the glucose units of cellulose is slightly different from the bonding between the glucose units of glycogen and starch. Humans have digestive enzymes that catalyze the breaking (hydrolysis) of the glucose-to-glucose bonds in starch but lack the necessary enzymes to hydrolyze cellulose glucose-to-glucose bonds. Thus starch is a food for humans, but cellulose is not. Cellulose is the indigestible fiber in our diets.

LIPIDS

Lipids are a diverse group of organic molecules made up of substances with widely different compositions and molecular structures. Unlike carbohydrates, which are classified on the basis of their *molecular structure*, substances are classified as lipids on the basis of their *solubility*. **Lipids** are insoluble in water but soluble in nonpolar solvents such as alcohol. Thus, lipids are the waxy, greasy, or oily compounds found in plants and animals. Lipids repel water, a useful characteristic of the protective wax coatings found on some plants. Fats and oils are energy-rich and have relatively low densities. These properties account for the use of fats and oils as stored energy in plants and animals. Still other lipids occur as structural components, especially in cellular membranes. The oily plasma membrane that surrounds each cell serves as a barrier that separates the intracellular contents from the surrounding extracellular fluid (see p. 25, p. 57, and p. 60).

▌ Simple lipids

Simple lipids contain just two types of components, fatty acids and alcohols. **Fatty acid molecules** consist of a hydrocarbon chain with a *carboxylic acid* functional group (—COOH) on

---

● **FIGURE B-7**

Forms of glucose
(a) Chain. (b) Ring.

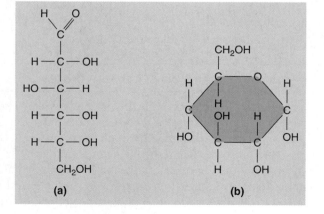

(a)   (b)

● **FIGURE B-8**

A simplified representation of glycogen
Each circle represents a glucose molecule.

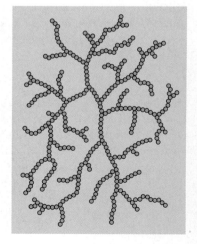

the end. The hydrocarbon chain can be of variable length, but natural fatty acids always contain an even number of carbon atoms. The hydrocarbon chain can also contain one or more double bonds between carbon atoms. Fatty acids with no double bonds are called **saturated fatty acids,** whereas those with double bonds are called **unsaturated fatty acids.** The more double bonds present, the higher the degree of unsaturation. Saturated fatty acids predominate in dietary animal products (for example, meat, eggs, and dairy products), whereas unsaturated fatty acids are more prevalent in plant products (for example, grains, vegetables, and fruits). Consumption of a greater proportion of saturated than unsaturated fatty acids is linked with a higher incidence of cardiovascular disease (see p. 337).

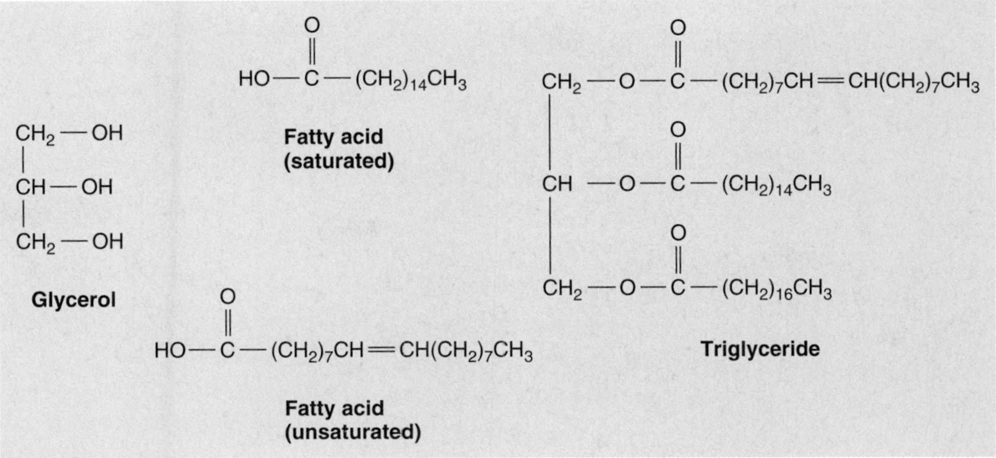

The most common alcohol found in simple lipids is **glycerol** (glycerin), a three-carbon alcohol that has three alcohol functional groups (—OH).

Simple lipids called fats and oils are formed by a reaction between the carboxylic acid group of three fatty acids and the three alcohol groups of glycerol. The resulting lipid is an E-shaped molecule called a **triglyceride.** Such lipids are classified as fats or oils on the basis of their melting points. *Fats* are solids at room temperature, whereas *oils* are liquids. Their melting points depend on the degree of unsaturation of the fatty acids of the triglyceride. The melting point goes down with increasing degree of unsaturation. Thus oils contain more unsaturated fatty acids than do fats. Examples of the components of fats and oils and a typical triglyceride molecule are shown in ● Figure B-9.

When triglycerides form, a molecule of water is released as each fatty acid reacts with glycerol. Adipose tissue in the body contains triglycerides. When the body uses adipose tissue as an energy source, the triglycerides react with water to release free fatty acids into the blood. The fatty acids can be used as an immediate energy source by many organs. In the liver, free fatty acids are converted into compounds called **ketone bodies.** Two of the ketone bodies are acids, and one is the ketone called

acetone. Excess ketone bodies are produced during diabetes mellitus, a condition in which most cells resort to using fatty acids as an energy source because the cells are unable to take up adequate amounts of glucose in the face of inadequate insulin action (see p. 726).

■ Complex lipids

Complex lipids contain more than two types of components. The different complex lipids usually contain three or more of the following components: glycerol, fatty acids, a phosphate group, an alcohol other than glycerol, and a carbohydrate. Those that contain phosphate are called **phospholipids.** ● Figure B-10 contains representations of a few complex lipids; it emphasizes the components but does not give details of the molecular structures.

Steroids are lipids that have a unique structural feature consisting of a fused carbon ring system containing three six-membered rings and a single five-membered ring (● Figure B-11). Different steroids possess this characteristic ring structure but have different functional groups and carbon chains attached.

● FIGURE B-9
Triglyceride components and structure

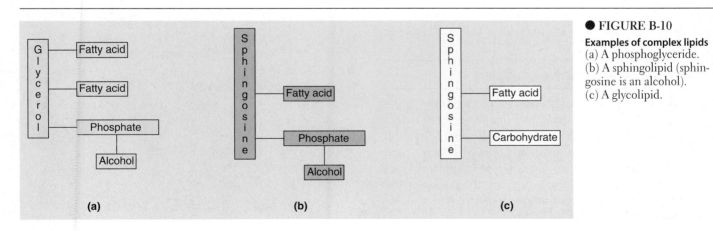

● FIGURE B-10

Examples of complex lipids
(a) A phosphoglyceride.
(b) A sphingolipid (sphingosine is an alcohol).
(c) A glycolipid.

A Review of Chemical Principles   A-13

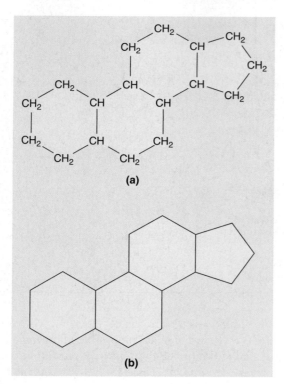

● FIGURE B-11

The steroid ring system
(a) Detailed. (b) Simplified.

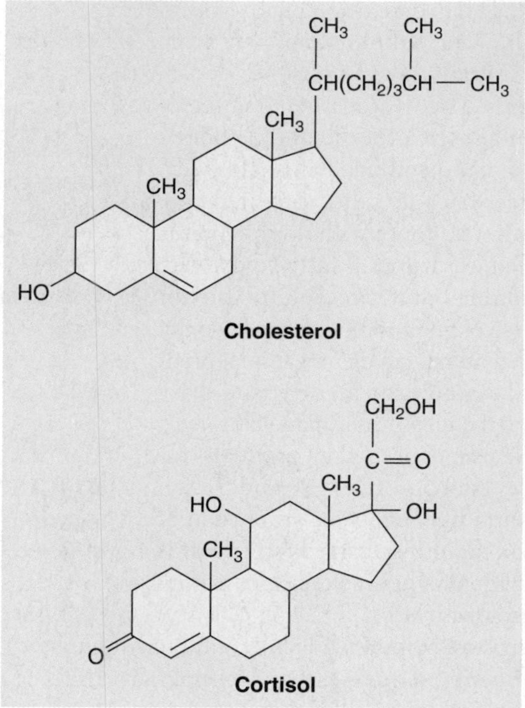

● FIGURE B-12

Examples of steroidal compounds

Cholesterol

Cortisol

Cholesterol, a steroidal alcohol, is the most abundant steroid in the human body. It is a component of cell membranes and is used by the body to produce other important steroids that include bile salts, male and female sex hormones, and adrenocortical hormones. The structures of cholesterol and cortisol, an important adrenocortical hormone, are given in ● Figure B-12.

PROTEINS

The name *protein* is derived from the Greek word *proteios,* which means "of first importance." It is certainly an appropriate term for these very important biological compounds. Proteins are indispensable components of all living things, where they play crucial roles in all biological processes. Proteins are the main structural component of cells, and all chemical reactions in the body are catalyzed by enzymes, all of which are proteins.

▌Chemical composition of proteins

Proteins are macromolecules made up of monomers called **amino acids.** Hundreds of different amino acids, both natural and synthetic, are known, but only 20 are commonly found in natural proteins. From this seemingly limited pool of amino acids, cells build thousands of different types of proteins, each with a distinctly different function, much the same way that composers create a diversity of unique music from a relatively small number of notes. Different proteins are constructed by

varying the types and numbers of amino acids used and by varying the order in which they are linked together. Proteins are not built haphazardly, though, by randomly linking together amino acids. Every protein in the body is deliberately and precisely synthesized under the direction of the blueprint laid down in the person's genes. Thus amino acids are assembled in a specific pattern to produce a given protein to accomplish a particular structural or functional task in the body. (More information about protein synthesis can be found in Appendix C.)

▌Peptide bonds

Each amino acid molecule has three important parts: an amino functional group ($-NH_2$), a carboxyl functional group ($-COOH$), and a characteristic side chain or R group. These components are shown in ● Figure B-13. Amino acids form long chains as a result of reactions between the amino group of one

● FIGURE B-13

The general structure of amino acids

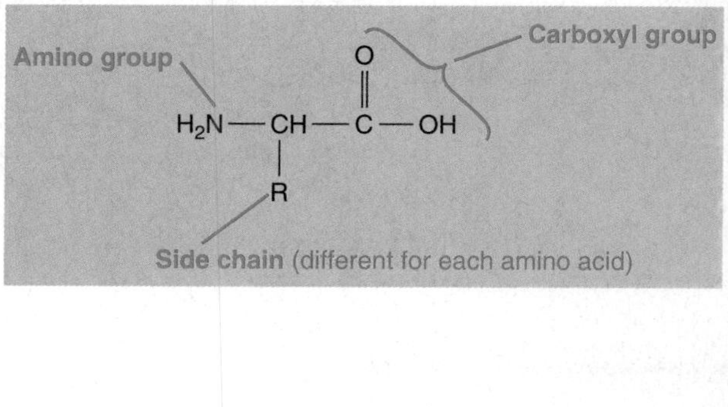

Amino group
Carboxyl group

$$H_2N - CH - C - OH$$

R

Side chain (different for each amino acid)

amino acid and the carboxyl group of another amino acid. This reaction is illustrated in Equation B-16 in which the components of the carboxyl group are shown in the expanded form for clarity:

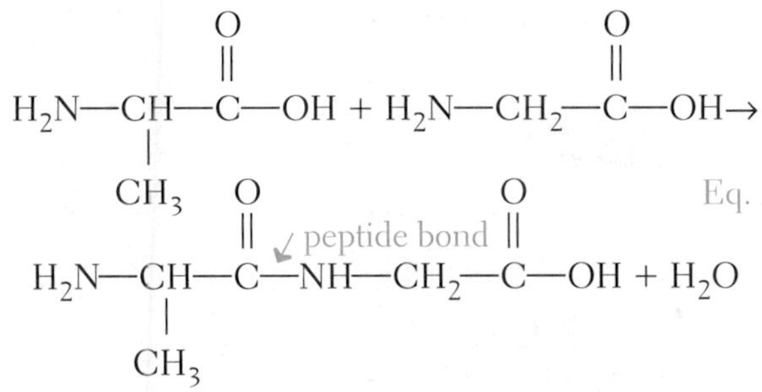

$$
\begin{array}{cc}
\text{O} & \text{O} \\
\| & \| \\
H_2N\!-\!CH\!-\!C\!-\!OH \;+\; H_2N\!-\!CH_2\!-\!C\!-\!OH \rightarrow & \\
| & \\
CH_3 &
\end{array}
$$

$$
\underset{\text{peptide bond}}{
\begin{array}{c}
\text{O} \qquad\qquad \text{O} \\
\| \qquad\qquad \| \\
H_2N\!-\!CH\!-\!C\!-\!NH\!-\!CH_2\!-\!C\!-\!OH \;+\; H_2O \\
| \\
CH_3
\end{array}}
$$

Eq. B-16

Notice that after the two molecules react, the ends of the product still have an amino group and a carboxyl group that can react to extend the chain length. The covalent bond formed in the reaction is called a **peptide bond** (● Figure B-14).

On a molecular scale, proteins are immense molecules. Their size can be illustrated by comparing a glucose molecule to a molecule of hemoglobin, a protein. Glucose has a molecular weight of 180 amu and a molecular formula of $C_6H_{12}O_6$. Hemoglobin, a relatively small protein, has a molecular weight of 65,000 amu and a molecular formula of $C_{2952}H_{4664}O_{832}N_{812}S_8Fe_4$.

▌ Levels of protein structure

The many atoms in a protein are not arranged in a random way. In fact, proteins have a high degree of structural organization that plays an important role in their behavior in the body.

Primary structure

The first level of protein structure is called the **primary structure**. It is simply the order in which amino acids are bonded

together to form the protein chain. Amino acids are frequently represented by three-letter abbreviations, such as Gly for glycine and Arg for arginine. When this practice is followed, the primary structure of a protein can be represented as in ● Figure B-15, which shows part of the primary structure of human insulin, or as in ● Figure B-16a, which depicts a portion of the primary structure of hemoglobin.

Secondary structure

The second level of protein structure, called the **secondary structure**, results when hydrogen bonding occurs between the amino hydrogen of one amino acid in the primary chain and the carboxyl oxygen

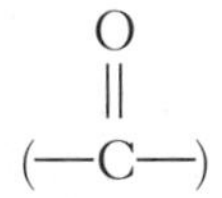

$$
\begin{array}{c}
\text{O} \\
\| \\
(\!-\!C\!-\!)
\end{array}
$$

of another amino acid in the same or another chain. When the hydrogen bonding occurs between amino acids in the same chain, the chain assumes a coiled, helical shape called the alpha (α) helix, which is by far the most common secondary structure found in natural proteins (● Figure B-16b). Other secondary structures such as the beta (β) pleated sheet and random coils also form as a result of hydrogen bonding between amino acids located in different parts of the same chain.

Tertiary and quaternary structure

The third level of structure in proteins is the **tertiary structure**. It results when functional groups of the side chains of amino acids in the protein chain react with each other. Several different types of interactions are possible, as shown in ● Figure B-17. Tertiary structures can be visualized by letting a length of wire represent the chain of amino acids in the primary structure of a protein. Next imagine that the wire is wound around a pencil to form a helix, which represents the secondary structure. The pencil is removed, and the helical structure is now folded back on itself or carefully wadded into a ball. Such folded or spherical structures represent the tertiary structure of a protein (● Figure B-16c).

All functional proteins exist in at least a tertiary structure. Sometimes, several polypeptides interact with each other to form a fourth level of protein structure, the **quaternary structure**. For example, hemoglobin contains four highly folded polypeptide chains (the **globin** portion) (● Figure B-16d). Four iron-containing *heme* groups, one tucked within the interior of each of the folded polypeptide subunits, completes the quaternary structure of hemoglobin (see ● Figure 11-3, p. 394).

▌ Hydrolysis and denaturation

One of the important functions of proteins is to serve as enzymes that catalyze the many essential chemical reactions of the body.

● **FIGURE B-14**

A peptide bond
In forming a peptide bond, the carboxyl group of one amino acid reacts with the amino group of another amino acid.

Peptide bond

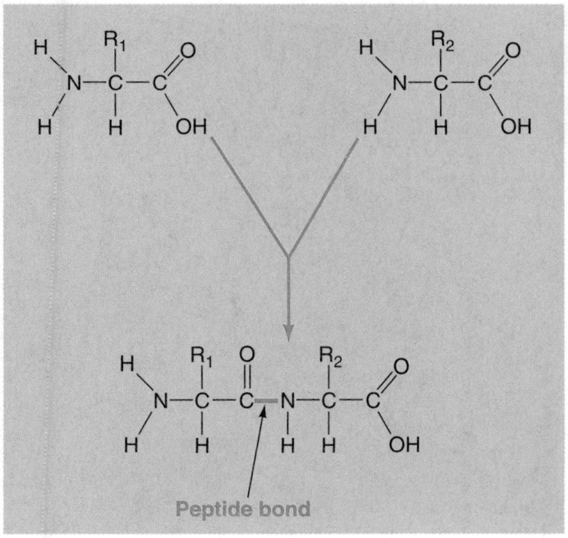

● **FIGURE B-15**

A portion of the primary protein structure of human insulin

Thr—Lys—Pro—Thr—Tyr—Phe—Phe—Gly—Arg— · · · · ·

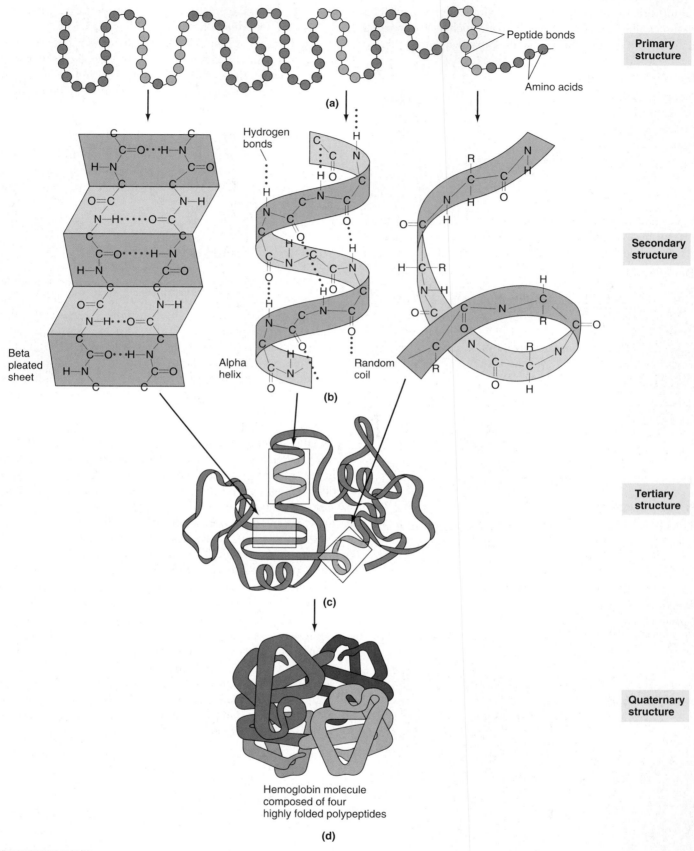

Primary structure

Peptide bonds

Amino acids

(a)

Secondary structure

Hydrogen bonds

Beta pleated sheet

Alpha helix

Random coil

(b)

Tertiary structure

(c)

Quaternary structure

Hemoglobin molecule composed of four highly folded polypeptides

(d)

● **FIGURE B-16**

Levels of protein structure

Proteins can have four levels of structure. (a) The primary structure is a particular sequence of amino acids bonded in a chain. (b) At the secondary level, hydrogen bonding occurs between various amino acids within the chain, causing the chain to assume a particular shape. The most common secondary protein structure in the body is the alpha helix. (c) The tertiary structure is formed by the folding of the secondary structure into a functional three-dimensional configuration. (d) Many proteins form a fourth level of structure composed of several polypeptides, as exemplified by hemoglobin.

In addition to catalyzing reactions, proteins can undergo reactions themselves. Two of the most important are hydrolysis and denaturation.

Hydrolysis

Notice that according to Equation B-16, the formation of peptide bonds releases water molecules. Under appropriate conditions, it is possible to reverse such reactions by adding water to the peptide bonds and breaking them. **Hydrolysis** ("breakdown by H_2O") reactions of this type convert large proteins into smaller fragments or even into individual amino acids. Hydrolysis is the means by which digestive enzymes break down ingested food into small units that can be absorbed from the digestive tract lumen into the blood.

Denaturation

Denaturation of proteins occurs when the bonds holding a protein chain in its characteristic tertiary or secondary conformation are broken. When this happens, the protein chain takes on a random, disorganized conformation. Denaturation can result when proteins are subjected to heating (including when body temperature rises too high; see p. 655), to extremes of pH (see p. 573), or to treatment with specific chemicals such as alcohol or heavy metal ions. In some instances, denaturation is accompanied by coagulation or precipitation, as illustrated by the changes that occur in the white of an egg as it is fried.

NUCLEIC ACIDS

Nucleic acids are high-molecular-weight macromolecules responsible for storing and using genetic information in living cells and passing it on to future generations. These important biomolecules are classified into two categories: **deoxyribonucleic acids (DNA)** and **ribonucleic acids (RNA)**. DNA is found primarily in the cell's nucleus, and RNA is found primarily in the cytoplasm that surrounds the nucleus.

Both types of nucleic acid are made up of units called **nucleotides**, which in turn are composed of three simpler components. Each nucleotide contains an organic nitrogenous base, a sugar, and a phosphate group. The three components are chemically bonded together with the sugar molecule lying between the base and the phosphate. In RNA, the sugar is ribose, whereas in DNA it is deoxyribose. When nucleotides bond together to form nucleic acid chains, the bonding is between the phosphate of one nucleotide and the sugar of another. Thus, the resulting nucleic acids consist of chains of alternating phosphates and sugar molecules, with a base molecule extending out of the chain from each sugar molecule (see ● Figure C-1, p. A-20).

The chains of nucleic acid assume structural features somewhat like those found in proteins. DNA occurs in the form of two chains that mutually coil around one another to form the well-known double helix. Some RNA occurs in essentially straight chains, whereas in other types the

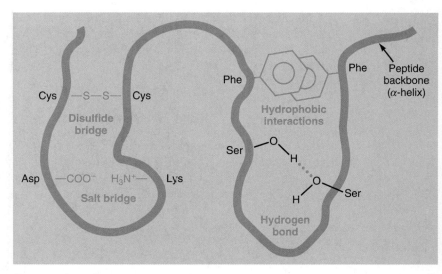

● FIGURE B-17
Side chain interactions leading to the tertiary protein structure

chain forms specific loops or helices. See Appendix C for further details.

HIGH-ENERGY BIOMOLECULES

Not all nucleotides are used to construct nucleic acids. One very important nucleotide—**adenosine triphosphate (ATP)**—is used as the body's primary energy carrier. Certain bonds in ATP temporarily store energy that is harnessed during the metabolism of foods and make it available to the parts of the cells where it is needed to do specific cellular work (see pp.35–42). Let's see how ATP functions in this role. Structurally, ATP is a modified RNA (ribose-containing) nucleotide that has adenine as its base and two additional phosphates bonded in sequence to the original nucleotide phosphate. Thus adenosine triphosphate, as the name implies, has a total of three phosphates attached in a string to *adenosine*, the composite of ribose and adenine (● Figure B-18). Attaching these additional phosphates requires considerable energy input. The high-energy input used to create these **high-energy phosphate bonds** is "stored" in the bonds for later use. Most energy transfers in the body in-

● FIGURE B-18
The structure of ATP

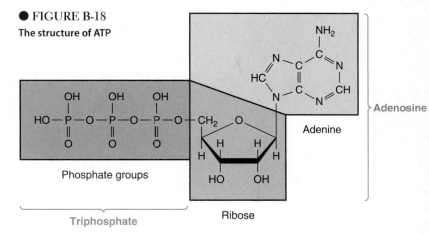

volve ATP's terminal phosphate bond. When energy is needed, the third phosphate is cleaved off by hydrolysis, yielding *adenosine diphosphate (ADP)* and an inorganic phosphate (P_i) and releasing energy in the process (Equation B-17):

$$ATP \rightarrow ADP + P_i + \text{energy for use by cell} \quad \text{Eq. B-17}$$

Why use ATP as an energy currency that cells can cash in by splitting of the high-energy phosphate bonds as needed? Why not just directly use the energy released during the oxidation of nutrient molecules such as glucose? If all the chemical energy stored in glucose were to be released at once, most of the energy would be squandered, because the cell could not capture much of the energy for immediate use. Instead, the energy trapped within the glucose bonds is gradually released and harnessed as cellular "bite-size pieces" in the form of the high-energy phosphate bonds of ATP.

Under the influence of an enzyme, ATP can be converted to a cyclic form of adenosine monophosphate, which contains only one phosphate group, the other two having been cleaved off. The resultant molecule, called **cyclic AMP** or **cAMP,** serves as an intracellular messenger, affecting the activities of a number of enzymes involved in important reactions in the body (see p. 67).

Storage, Replication, and Expression of Genetic Information

DEOXYRIBONUCLEIC ACID (DNA) AND CHROMOSOMES

The nucleus of the cell houses **deoxyribonucleic acid (DNA)**, the genetic blueprint that is unique for each individual.

▮ Functions of DNA

As genetic material, DNA serves two essential functions. First, it contains "instructions" for assembling the structural and enzymatic proteins of the cell. Cellular enzymes in turn control the formation of other cellular structures and also determine the functional activity of the cell by regulating the rate at which metabolic reactions proceed. The nucleus serves as the cell's control center by directly or indirectly controlling almost all cell activities through the role its DNA plays in governing protein synthesis. Because cells make up the body, the DNA code determines the structure and the function of the body as a whole. The DNA an organism possesses not only dictates whether the organism is a human, a toad, or a pea but also determines the unique physical and functional characteristics of that individual, all of which ultimately depend on the proteins produced under DNA control.

Second, by replicating (making copies of itself), DNA perpetuates the genetic blueprint within all new cells formed within the body and is responsible for passing on genetic information from parents to children. We will first examine the structure of DNA and the coding mechanism it uses, then turn our attention to the means by which DNA replicates itself and controls protein synthesis.

▮ Structure of DNA

Deoxyribonucleic acid is a huge molecule, composed in humans of millions of nucleotides arranged into two long, paired strands that spiral around each other to form a double helix. Each **nucleotide** has three components: (1) a *nitrogenous base*, a ring-shaped organic molecule containing nitrogen; (2) a five-carbon ring-shaped sugar molecule, which in DNA is *deoxyribose*; and (3) a phosphate group. Nucleotides are joined end to end by linkages between the sugar of one nucleotide and the phosphate group of the adjacent nucleotide to form a long poly-

nucleotide ("many nucleotide") strand with a sugar–phosphate backbone and bases projecting out one side (● Figure C-1). The four different bases in DNA are the double-ringed bases **adenine (A)** and **guanine (G)** and the single-ringed bases **cytosine (C)** and **thymine (T)**. The two polynucleotide strands within a DNA molecule are wrapped around each other and oriented so that their bases all project to the interior of the helix. The strands are held together by weak hydrogen bonds formed between the bases of adjoining strands (● Figure B-6, see p. A-7). Base pairing is highly specific: Adenine pairs only with thymine and guanine pairs only with cytosine (● Figure C-2).

Genes

The composition of the repetitive sugar–phosphate backbones that form the "sides" of the DNA "ladder" is identical for every molecule of DNA, but the sequence of the linked bases that form the "rungs" varies among different DNA molecules. The particular sequence of bases in a DNA molecule serves as "instructions," or a "code," that dictates the assembly of amino acids into a given order for the synthesis of specific **polypeptides** (chains of amino acids linked by peptide bonds; see p. A-14). A **gene** is a stretch of DNA that codes for the synthesis of a particular polypeptide. Polypeptides, in turn, are folded into a three-dimensional configuration to form a functional protein. Not all portions of a DNA molecule code for structural or enzymatic proteins. Some stretches of DNA code for proteins that regulate genes. Other segments appear important in organizing and packaging DNA within the nucleus. Still other regions are "nonsense" base sequences that have no apparent significance.

▮ Packaging of DNA into chromosomes

The DNA molecules within each human cell, if lined up end to end, would extend more than 2 m (2,000,000 mm), yet these molecules are packed into a nucleus that is only 5 mm in diameter. These molecules are not randomly crammed into the nucleus but are precisely organized into **chromosomes**. Each chromosome consists of a different DNA molecule and contains a unique set of genes.

Somatic (body) **cells** contain 46 chromosomes (the **diploid number**), which can be sorted into 23 pairs on the basis of various distinguishing features. Chromosomes composing a matched pair are termed **homologous chromosomes,** one member of each pair having been derived from the individual's maternal parent and the other member from the paternal parent. **Germ** (reproductive) **cells** (that is, sperm and eggs) contain only one member of each homologous pair for a total of 23 chromosomes (the **haploid number**). Union of a sperm and an egg results in a new diploid cell with 46 chromosomes, consisting of a set of 23 chromosomes from the mother and another set of 23 from the father.

● FIGURE C-1

Polynucleotide strand
Sugar–phosphate bonds link adjacent nucleotides together to form a polynucleotide strand with bases projecting to one side. The sugar–phosphate backbone is identical in all polynucleotides, but the sequence of the bases varies.

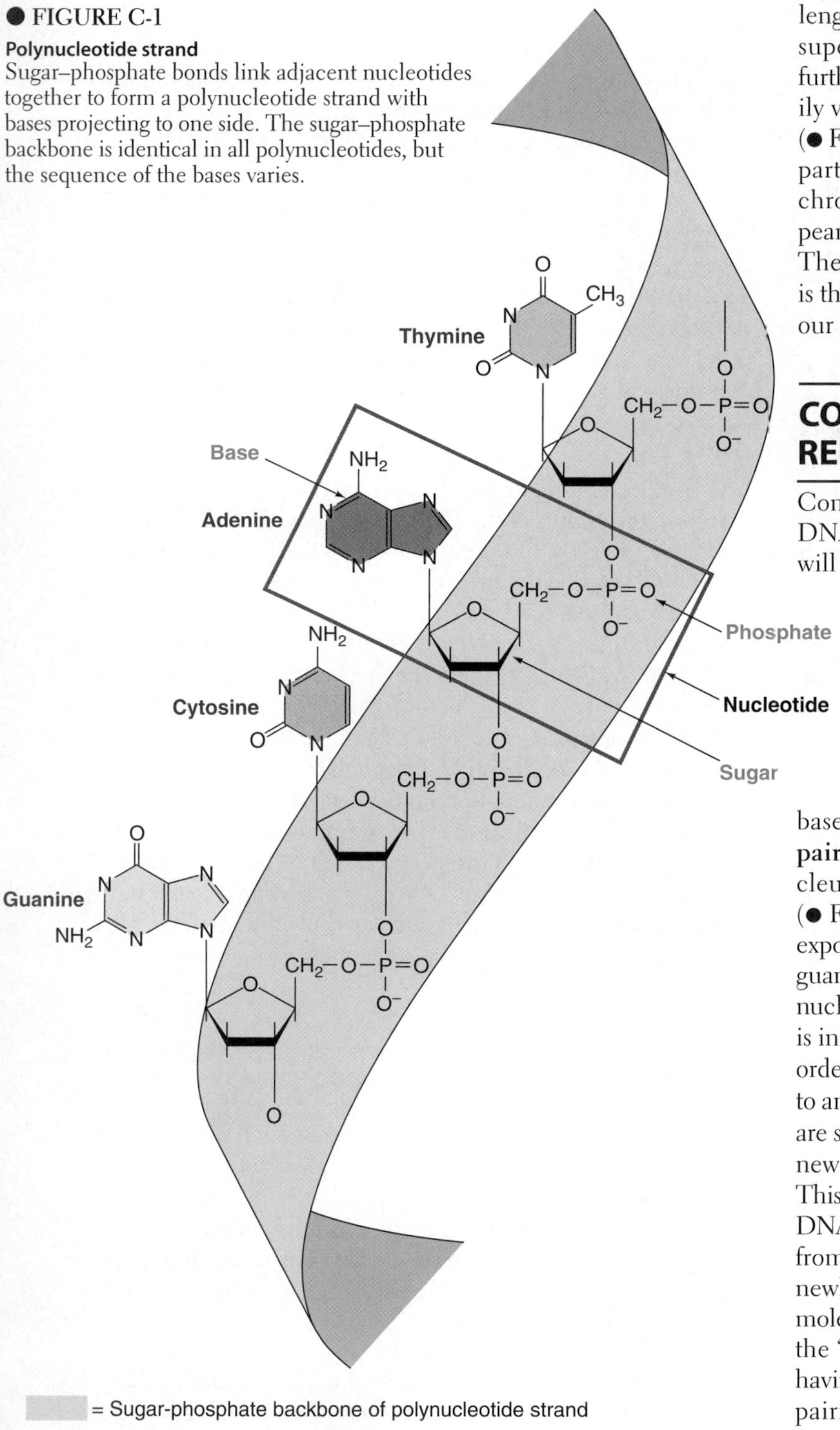

Thymine

Base

Adenine

Cytosine

Guanine

Phosphate

Nucleotide

Sugar

= Sugar-phosphate backbone of polynucleotide strand

The packaging and compression of DNA molecules into discrete chromosomal units are accomplished in part by nuclear proteins associated with DNA. Two classes of proteins—histone and nonhistone proteins—bind with DNA. **Histones** form bead-shaped bodies that play a key role in packaging DNA into its chromosomal structure. The **nonhistones** are believed to be important in gene regulation. The complex formed between the DNA and its associated proteins is known as **chromatin.** The long threads of DNA within a chromosome are wound around histones at regular intervals, thus compressing a given DNA molecule to about one-sixth its fully extended length. This "beads-on-a-string" structure is further folded and supercoiled into higher and higher levels of organization to further condense DNA into rodlike chromosomes that are readily visible by means of a light microscope during cell division (● Figure C-3). When the cell is not dividing, the chromosomes partially "unravel" or decondense to a less compact form of chromatin that is indistinct under a light microscope but appears as thin strands and clumps with an electron microscope. The decondensed form of DNA is its working form; that is, it is the form used as a template for protein assembly. Let us turn our attention to this working form of DNA in operation.

COMPLEMENTARY BASE PAIRING, REPLICATION, AND TRANSCRIPTION

Complementary base pairing serves as the foundation for both DNA replication and the initial step of protein synthesis. We will examine the mechanism and significance of complementary base pairing in each of these circumstances, starting with DNA replication.

▐ DNA replication

During DNA replication, the two decondensed DNA strands "unzip" as the weak bonds between the paired bases are enzymatically broken. Then **complementary base pairing** takes place: New nucleotides present within the nucleus pair with the exposed bases from each unzipped strand (● Figure C-4). New adenine-bearing nucleotides pair with exposed thymine-bearing nucleotides in an old strand, and new guanine-bearing nucleotides pair with exposed cytosine-bearing nucleotides in an old strand. This complementary base pairing is initiated at one end of the two old strands and proceeds in an orderly fashion to the other end. The new nucleotides attracted to and thus aligned in a prescribed order by the old nucleotides are sequentially joined by sugar–phosphate linkages to form two new strands that are complementary to each of the old strands. This replication process results in two complete double-stranded DNA molecules, one strand within each molecule having come from the original DNA molecule and one strand having been newly formed by complementary base pairing. These two DNA molecules are both identical to the original DNA molecule, with the "missing" strand in each of the original separated strands having been produced as a result of the imposed pattern of base pairing. This replication process, which occurs only during

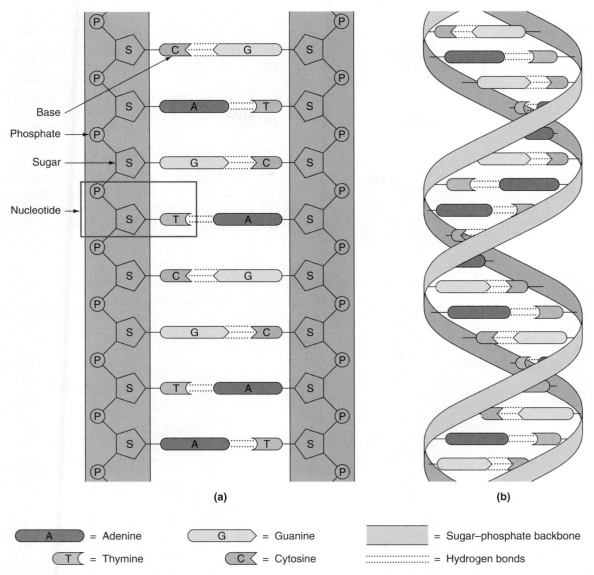

(a)   (b)

A = Adenine	**G** = Guanine	= Sugar–phosphate backbone
T = Thymine	**C** = Cytosine	= Hydrogen bonds

● **FIGURE C-2**

Complementary base pairing in DNA
(a) Two polynucleotide strands held together by weak hydrogen bonds formed between the bases of adjoining strands—adenine always paired with thymine and guanine always paired with cytosine.
(b) Arrangement of the two bonded polynucleotide strands of a DNA molecule into a double helix.

cell division, is essential for ensuring the perpetuation of the genetic code in both of the new daughter cells. The duplicate copies of DNA are separated and evenly distributed to the two halves of the cell before it divides. We will cover the topic of cell division in more detail later.

∎ DNA transcription and messenger RNA

At other times, when DNA is not replicating in preparation for cell division, it serves as a blueprint for dictating cellular protein synthesis. How is this accomplished when DNA is sequestered within the nucleus and protein synthesis is carried out by ribosomes within the cytoplasm? Several types of another nucleic acid, **ribonucleic acid (RNA),** serve as the "go-between."

Structure of ribonucleic acid

Ribonucleic acid differs structurally from DNA in three regards: (1) The five-carbon sugar in RNA is ribose instead of deoxyribose, the only difference between them being the presence in ribose of a single oxygen atom that is absent in deoxyribose; (2) RNA contains the closely related base **uracil** instead of thymine, with the three other bases being the same as in DNA; and (3) RNA is single-stranded and not self-replicating.

All RNA molecules are produced in the nucleus using DNA as a template or mold, then exit the nucleus through openings in the nuclear membrane known as *nuclear pores* (see p. 25). These pores are large enough for passage of RNA molecules but preclude passage of the much larger DNA molecules.

The DNA instructions for assembling a particular protein coded in the base sequence of a given gene are "transcribed"

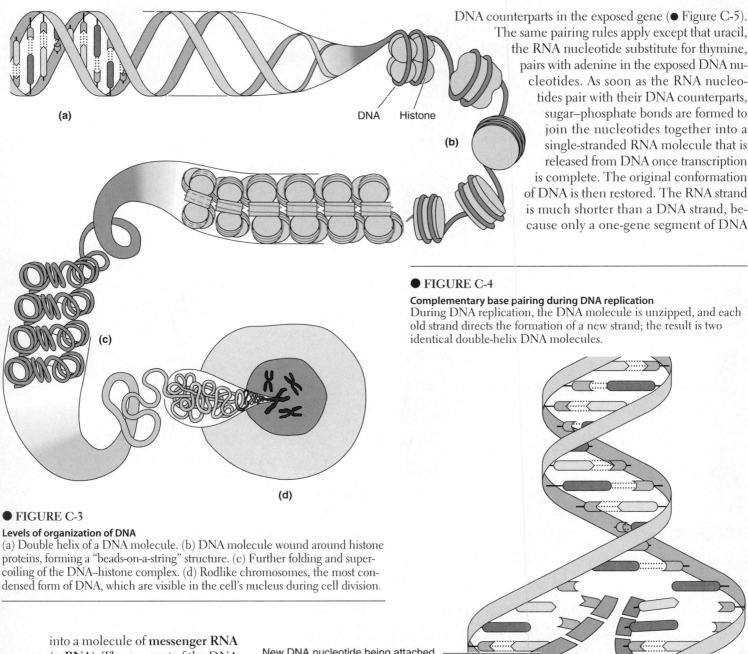

(a)

DNA   Histone

(b)

(c)

(d)

DNA counterparts in the exposed gene (● Figure C-5). The same pairing rules apply except that uracil, the RNA nucleotide substitute for thymine, pairs with adenine in the exposed DNA nucleotides. As soon as the RNA nucleotides pair with their DNA counterparts, sugar–phosphate bonds are formed to join the nucleotides together into a single-stranded RNA molecule that is released from DNA once transcription is complete. The original conformation of DNA is then restored. The RNA strand is much shorter than a DNA strand, because only a one-gene segment of DNA

● **FIGURE C-4**

Complementary base pairing during DNA replication
During DNA replication, the DNA molecule is unzipped, and each old strand directs the formation of a new strand; the result is two identical double-helix DNA molecules.

● **FIGURE C-3**

Levels of organization of DNA
(a) Double helix of a DNA molecule. (b) DNA molecule wound around histone proteins, forming a "beads-on-a-string" structure. (c) Further folding and super-coiling of the DNA–histone complex. (d) Rodlike chromosomes, the most condensed form of DNA, which are visible in the cell's nucleus during cell division.

into a molecule of **messenger RNA (mRNA).** The segment of the DNA molecule to be copied uncoils, and the base pairs separate to expose the particular sequence of bases in the gene. In any given gene, only one of the DNA strands is used as a template for transcribing RNA, with the copied strand varying for different genes along the same DNA molecule. The beginning and end of a gene within a DNA strand are designated by particular base sequences that serve as "start" and "stop" signals.

Transcription

Transcription is accomplished by complementary base pairing of free RNA nucleotides with their

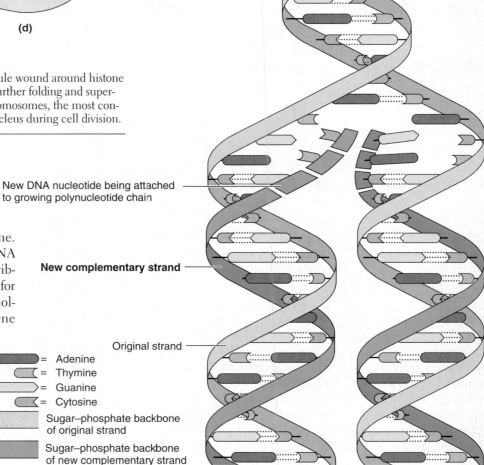

New DNA nucleotide being attached to growing polynucleotide chain

New complementary strand

Original strand

= Adenine
= Thymine
= Guanine
= Cytosine

Sugar–phosphate backbone of original strand

Sugar–phosphate backbone of new complementary strand

is transcribed into a single RNA molecule. The length of the finished RNA transcript varies, depending on the size of the gene. Within its nucleotide base sequence, this RNA transcript contains instructions for assembling a particular protein. Note that the message is coded in a base sequence that is *complementary to, not identical to*, the original DNA code.

Messenger RNA delivers the final coded message to the ribosomes for **translation** into a particular amino acid sequence to form a given protein. Thus genetic information flows from DNA (which can replicate itself) through RNA to protein. This is accomplished first by *transcription* of the DNA code into a complementary RNA code, followed by *translation* of the RNA code into a specific protein (● Figure C6). In the next section, you will learn more about the steps in translation. The structural and functional characteristics of the cell as determined by its protein composition can be varied, subject to control, depending on which genes are "switched on" to produce mRNA.

Free nucleotides present in the nucleus cannot be randomly joined together to form either DNA or RNA strands, because the enzymes required to link together the sugar and phosphate components of nucleotides are active only when bound to DNA. This ensures that DNA, mRNA, and protein assembly occur only according to genetic plan.

TRANSLATION AND PROTEIN SYNTHESIS

Three forms of RNA participate in protein synthesis. Besides messenger RNA, two other forms of RNA are required for translation of the genetic message into cellular protein: ribosomal RNA and transfer RNA.

- **Messenger RNA** carries the coded message from nuclear DNA to a cytoplasmic ribosome, where it directs the synthesis of a particular protein.
- **Ribosomal RNA (rRNA)** is an essential component of *ribosomes*, the "workbenches" for protein synthesis (see p. 26). Ribosomal RNA "reads" the base sequence code of mRNA and translates it into the appropriate amino-acid sequence during protein synthesis.
- **Transfer RNA (tRNA)** transfers the appropriate amino acids in the cytosol to their designated site in the amino acid sequence of the protein under construction.

▌Triplet code; codon

Twenty different amino acids are used to construct proteins, yet only four different nucleotide bases are used to code for these twenty amino acids. In the "genetic dictionary," each different amino acid is specified by a **triplet code** that consists of a specific sequence of three bases in the DNA nucleotide chain. For example, the DNA sequence ACA (adenine, cytosine, adenine) specifies the amino acid cysteine, whereas the sequence ATA specifies the amino acid tyrosine. Each DNA triplet code is transcribed into mRNA as a complementary code word, or

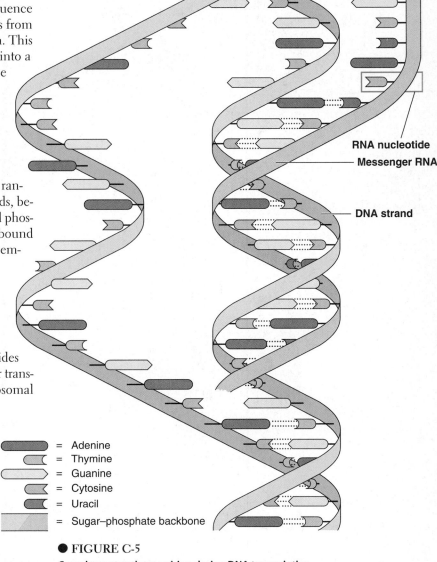

RNA nucleotide

Messenger RNA

DNA strand

	= Adenine
	= Thymine
	= Guanine
	= Cytosine
	= Uracil
	= Sugar–phosphate backbone

● **FIGURE C-5**

Complementary base pairing during DNA transcription
During DNA transcription, a messenger RNA molecule is formed as RNA nucleotides are assembled by complementary base pairing at a given segment of one strand of an unzipped DNA molecule (that is, a gene).

● **FIGURE C-6**

Flow of genetic information from DNA through RNA to protein by transcription and translation

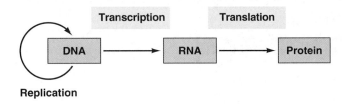

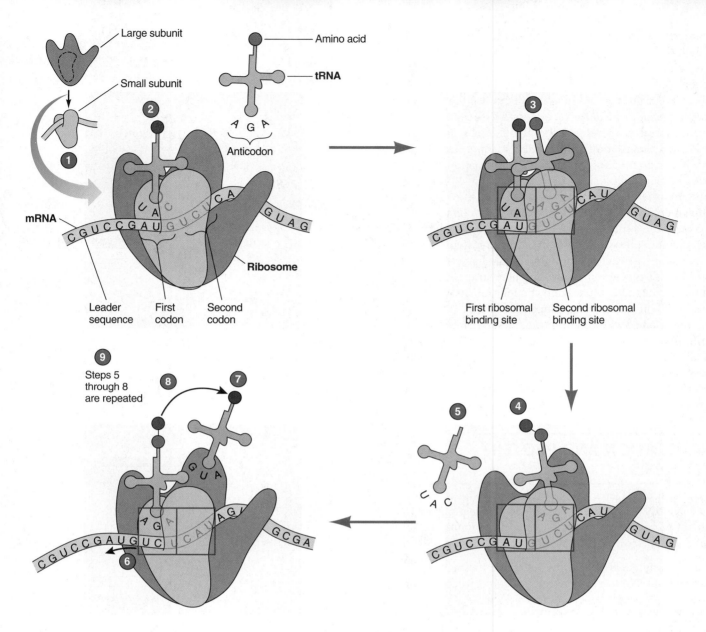

1. On binding with a messenger RNA (mRNA) molecule, the small ribosomal subunit joins with the large subunit to form a functional ribosome.

2. A transfer RNA (tRNA), charged with its specific amino acid passenger, binds to mNRA by means of complementary base pairing between the tRNA anticodon and the first mRNA codon positioned in the first ribosomal binding site.

3. Another tRNA molecule attaches to the next codon on mRNA positioned in the second ribosomal binding site.

4. The amino acid from the first tRNA is linked to the amino acid on the second tRNA.

5. The first tRNA detaches.

6. The mRNA molecule shifts forward one codon (a distance of a three-base sequence).

7. Another charged tRNA moves in to attach with the next codon on mRNA, which has now moved into the second ribosomal binding site.

8. The amino acids from the tRNA in the first ribosomal site are linked with the amino acid in the second site.

9. This process continues (that is, steps 5 through 8 are repeated), with the polypeptide chain continuing to grow, until a stop codon is reached and the polypeptide chain is released.

● FIGURE C-7

Ribosomal assembly and protein translation

codon, consisting of a sequenced order of the three bases that pair with the DNA triplet. For example, the DNA triplet code ATA is transcribed as UAU (uracil, adenine, uracil) in mRNA.

Sixty-four different DNA triplet combinations (and, accordingly, 64 different mRNA codon combinations) are possible using the four different nucleotide bases (4^3). Of these possible combinations, 61 code for specific amino acids and the remaining three serve as "stop signals." A stop signal acts as a "period" at the end of a "sentence." The sentence consists of a series of triplet codes that specify the amino acid sequence in a particular protein. When the stop codon is reached, ribosomal RNA releases the finished polypeptide product. Because 61 triplet codes each specify a particular amino acid and there are 20 different amino acids, a given amino acid may be specified by more than one base-triplet combination. For example, tyrosine is specified by the DNA sequence ATG as well as by ATA. In addition, one DNA triplet code, TAC (mRNA codon sequence AUG) functions as a "start signal" in addition to specifying the amino acid methionine. This code marks the place on mRNA where translation is to begin so that the message is started at the correct end and thus reads in the right direction. Interestingly, the same genetic dictionary is used universally; a given three-base code stands for the same amino acid in all living things, including micro-organisms, plants, and animals.

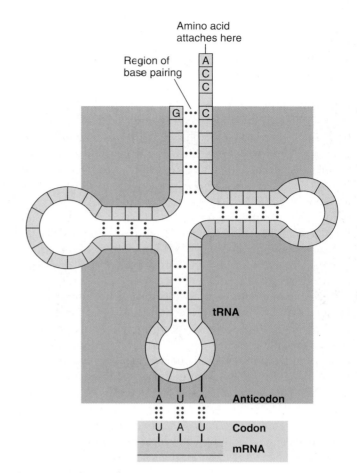

● FIGURE C-8

Structure of a tRNA molecule
The open end of a tRNA molecule attaches to free amino acids. The anticodon loop of the tRNA molecule attaches to a complementary mRNA codon.

▌ Ribosomes

A **ribosome** brings together all components that participate in protein synthesis—mRNA, tRNA, and amino acids—and provides the enzymes and energy required for linking the amino acids together. The nature of the protein synthesized by a given ribosome is determined by the mRNA message that is being translated. Each mRNA serves as a code for only one particular polypeptide.

A ribosome is an rRNA-protein structure organized into two subunits of unequal size. These subunits are brought together only when a protein is being synthesized (● Figure C-7, step ①). During assembly of a ribosome, an mRNA molecule attaches to the smaller of the ribosomal subunits by means of a *leader sequence,* a section of mRNA that precedes the start codon. The small subunit with mRNA attached then binds to a large subunit to form a complete, functional ribosome. When the two subunits unite, a groove is formed that accommodates the mRNA molecule as it is being translated.

▌ Transfer RNA and anticodons

Free amino acids in the cytoplasm are not able to "recognize" and bind directly with their specific codons in mRNA. Transfer RNA is required to bring the appropriate amino acid to its proper codon. Even though tRNA is single-stranded, as are all RNA molecules, it is folded back onto itself into a T shape with looped ends (● Figure C-8). The open-ended stem portion recognizes and binds to a specific amino acid. There are at least 20 different varieties of tRNA, each able to bind with only one

of the 20 different kinds of amino acids. A tRNA is said to be "charged" when it is carrying its passenger amino acid. The loop end of a tRNA opposite the amino-acid binding site contains a sequence of three exposed bases, known as the **anticodon,** which is complementary to the mRNA codon that specifies the amino acid being carried. Through complementary base pairing, a tRNA can bind with mRNA and insert its amino acid into the protein under construction only at the site designated by the codon for the amino acid. For example, the tRNA molecule that binds with tyrosine bears the anticodon AUA, which can pair only with the mRNA codon UAU, which specifies tyrosine. This dual binding function of tRNA molecules ensures that the correct amino acids are delivered to mRNA for assembly in the order specified by the genetic code. Transfer RNA can only bind with mRNA at a ribosome, so protein assembly does not occur except in the confines of a ribosome.

▌ Steps of protein synthesis

The three steps of protein synthesis are initiation, elongation, and termination.

1. *Initiation.* Protein synthesis is initiated when a charged tRNA molecule bearing the anticodon specific for the start codon binds at this site on mRNA (● Figure C-7, step 2).

2. *Elongation.* A second charged tRNA bearing the anticodon specific for the next codon in the mRNA sequence then occupies the site next to the first tRNA (step 3). At any given time, a ribosome can accommodate only two tRNA molecules bound to adjacent codons. Through enzymatic action, a peptide bond is formed between the two amino acids that are linked to the stems of the adjacent tRNA molecules (step 4). The linkage is subsequently broken between the first tRNA and its amino acid passenger, leaving the second tRNA with a chain of two amino acids. The uncharged tRNA molecule (the one minus its amino acid passenger) is released from mRNA (step 5). The ribosome then moves along the mRNA molecule by precisely three bases, a distance of one codon (step 6), so that the tRNA bearing the chain of two amino acids is moved into the number one ribosomal site for tRNA. Then, an incoming charged tRNA with a complementary anticodon for the third codon in the mRNA sequence occupies the number two ribosomal site that was vacated by the second tRNA (step 7). The chain of two amino acids subsequently binds with and is transferred to the third tRNA to form a chain of three amino acids (step 8). Through repetition of this process, amino acids are subsequently added one at a time to a growing polypeptide chain in the order designated by the codon sequence as the ribosomal translation machinery moves stepwise along the mRNA molecule one codon at a time (step 9). This process is rapid. Up to 10 to 15 amino acids can be added per second.

3. *Termination.* Elongation of the polypeptide chain continues until the ribosome reaches a stop codon in the mRNA molecule, at which time the polypeptide is released. The polypeptide is then folded and modified into a full-fledged protein. The ribosomal subunits dissociate and are free to reassemble into another ribosome for translation of other mRNA molecules.

▌ Energy cost of protein synthesis

Protein synthesis is expensive, in terms of energy. Attachment of each new amino acid to the growing polypeptide chain requires a total investment of splitting four high-energy phosphate bonds—two to charge tRNA with its amino acid, one to bind tRNA to the ribosomal-mRNA complex, and one to move the ribosome forward one codon.

▌ Polyribosomes

A number of copies of a given protein can be produced from a single mRNA molecule before the latter is chemically degraded. As one ribosome moves forward along the mRNA molecule, a new ribosome attaches at the starting point on mRNA and also starts translating the message. Attachment of many ribosomes to a single mRNA molecule results in a *polyribosome*. Multiple copies of the identical protein are produced as each ribosome moves along and translates the same message (● Figure C-9).

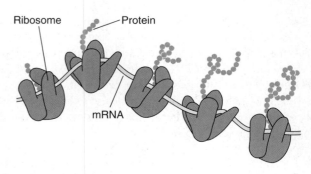

● **FIGURE C-9**

A polyribosome
A polyribosome is formed by numerous ribosomes simultaneously translating mRNA.

The released proteins are used within the cytosol, except for the few that move into the nucleus through the nuclear pores.

Recall that, in contrast to the cytosolic polyribosomes, ribosomes directed to bind with the rough endoplasmic reticulum (ER) feed their growing polypeptide chains into the ER lumen (see p. 26). The resultant proteins are subsequently packaged for export out of the cell or for replacement of membrane components within the cell.

▌ Control of gene activity and protein transcription

Because each somatic cell in the body has the identical DNA blueprint, you might assume that they would all produce the same proteins. This is not the case, however, because different cell types are able to transcribe different sets of genes and thus synthesize different sets of structural and enzymatic proteins. For example, only red blood cells can synthesize hemoglobin, even though all body cells carry the DNA instructions for hemoglobin synthesis. Only about 7% of the DNA sequences in a typical cell are ever transcribed into mRNA for ultimate expression as specific proteins.

Control of gene expression involves gene regulatory proteins that activate ("switch on") or repress ("switch off") the genes that code for specific proteins within a given cell. Various DNA segments that do not code for structural and enzymatic proteins code for synthesis of these regulatory proteins. The molecular mechanisms by which these regulatory genes in turn are controlled in human cells are only beginning to be understood. In some instances, regulatory proteins are controlled by **gene-signaling factors** that bring about differential gene activity among various cells to accomplish specialized tasks. The largest group of known gene-signaling factors in humans is the hormones. Some hormones exert their homeostatic effect by selectively altering the transcription rate of the genes that code for enzymes that are in turn responsible for catalyzing the reaction(s) regulated by the hormone. For example, the hormone cortisol promotes the breakdown of fat stores by stimulating synthesis of the enzyme that catalyzes the conversion of stored fat into its component fatty acids. In other cases, gene action appears to be time specific; that is, certain genes are expressed only at a certain developmental stage in the individual. This is especially important during embryonic development.

CELL DIVISION

Most cells in the human body can reproduce themselves, a process important in growth, replacement, and repair of tissues. The rate at which cells divide is highly variable. Cells within the deeper layers of the intestinal lining divide every few days to replace cells that are continually sloughed off the surface of the lining into the digestive tract lumen. In this way, the entire intestinal lining is replaced about every three days (see p. 628). At the other extreme are nerve cells, which permanently lose the ability to divide beyond a certain period of fetal growth and development. Consequently, when nerve cells are lost through trauma or disease, they cannot be replaced (see p. 5). In between these two extremes are cells that divide infrequently except when needed to replace damaged or destroyed tissue. The factors that control the rate of cell division remain obscure.

▌Mitosis

Recall that cell division involves two components: nuclear division and cytoplasmic division (**cytokinesis**) (see p. 49). Nuclear division in somatic cells is accomplished by **mitosis**, in which a complete set of genetic information (that is, a diploid number of chromosomes) is distributed to each of two new daughter cells.

A cell capable of dividing alternates between periods of mitosis and nondivision. The interval of time between cell division is known as **interphase.** Because mitosis takes less than an hour to complete, the vast majority of cells in the body at any given time are in interphase.

Replication of DNA and growth of the cell take place during interphase in preparation for mitosis. Although mitosis is a continuous process, it displays four distinct phases: *prophase, metaphase, anaphase,* and *telophase* (● Figure C-10).

Prophase

1. Chromatin condenses and becomes microscopically visible as chromosomes. The condensed duplicate strands of DNA, known as *sister chromatids*, remain joined together within the chromosome at a point called the *centromere* (● Figure C-11).

2. Cells contain a pair of centrioles, short cylindrical structures that form the mitotic spindle during cell division (see ● Figure 2-1, p. 26). The centriole pair divides, and the daughter centrioles move to opposite ends of the cell, where they assemble between them a mitotic spindle made up of microtubules (see p. 48).

3. The membrane surrounding the nucleus starts to break down.

Metaphase

1. The nuclear membrane completely disappears.

2. The 46 chromosomes, each consisting of a pair of sister chromatids, align themselves at the midline, or equator, of the cell. Each chromosome becomes attached to the spindle by means of several spindle fibers that extend from the centriole to the centromere of the chromosome.

Anaphase

1. The centromeres split, converting each pair of sister chromatids into two identical chromosomes, which separate and move toward opposite poles of the spindle. Motor proteins are responsible for pulling the chromosomes along the spindle fibers toward the poles (see p. 46).

2. At the end of anaphase, an identical set of 46 chromosomes is present at each of the poles, for a transient total of 92 chromosomes in the soon-to-be-divided cell.

Telophase

1. The cytoplasm divides through formation and gradual tightening of an actin contractile ring at the midline of the cell, thus forming two separate daughter cells, each with a full diploid set of chromosomes (see ● Figure 2-22a, p. 49).

2. The spindle fibers disassemble.

3. The chromosomes uncoil to their decondensed chromatin form.

4. A nuclear membrane reforms in each new cell.

Cell division is complete with the end of telophase. Each of the new cells now enters interphase.

▌Meiosis

Nuclear division in the specialized case of germ cells is accomplished by **meiosis**, in which only half a set of genetic information (that is, a haploid number of chromosomes) is distributed to each daughter cell. Meiosis differs from mitosis in several important regards (● Figure C-10). Specialized diploid germ cells undergo one chromosome replication followed by two nuclear divisions to produce four haploid germ cells.

Meiosis I

1. During prophase of the first meiotic division, the members of each homologous pair of chromosomes line up side by side to form a **tetrad,** which is a group of four sister chromatids with two identical chromatids within each member of the pair.

2. The process of crossing over occurs during this period, when the maternal copy and the paternal copy of each chromosome are paired. **Crossing over** involves a physical exchange of chromosome material between nonsister chromatids within a tetrad (● Figure C-12). This process yields new chromosome combinations, thus contributing to genetic diversity.

3. During metaphase, the 23 tetrads line up at the equator.

4. At anaphase, homologous chromosomes, each consisting of a pair of sister chromatids joined at the centromere, separate and move toward opposite poles. Maternally and pater-

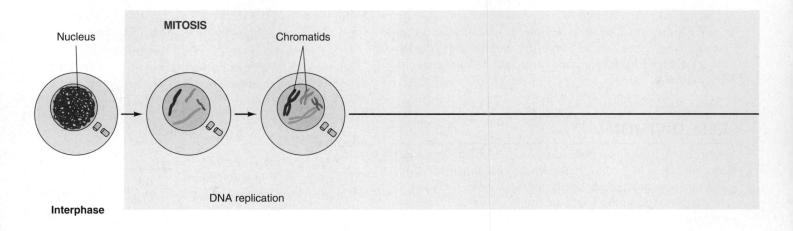

MITOSIS

Nucleus

Chromatids

Interphase

DNA replication

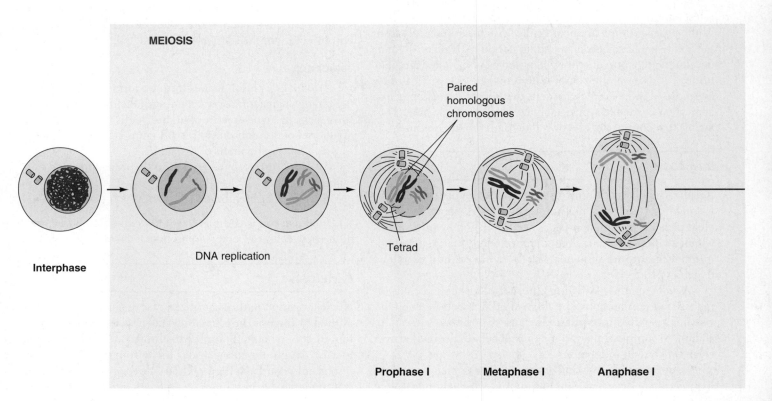

MEIOSIS

Paired homologous chromosomes

Interphase

DNA replication

Tetrad

Prophase I   Metaphase I   Anaphase I

● FIGURE C-10

A comparison of events in mitosis and meiosis

nally derived chromosomes migrate to opposite poles in random assortments of one member of each chromosome pair without regard for its original derivation. This genetic mixing provides novel new combinations of chromosomes.

5. During the first telophase, the cell divides into two cells. Each cell contains 23 chromosomes consisting of two sister chromatids.

Meiosis II

1. Following a brief interphase in which no further replication occurs, the 23 unpaired chromosomes line up at the equator, the centromeres split, and the sister chromatids separate for the first time into independent chromosomes that move to opposite poles.

2. During cytokinesis, each of the daughter cells derived from the first meiotic division forms two new daughter cells. The end result is four daughter cells, each containing a haploid set of chromosomes.

Union of a haploid sperm and haploid egg results in a zygote (fertilized egg) that contains the diploid number of chromosomes. Development of a new multicellular individual from the zygote is accomplished by mitosis and cell differentiation. Because DNA is normally faithfully replicated in its entirety during each mitotic division, all cells in the body possess an identical aggregate of DNA molecules. Structural and functional variations between different cell types result from differential gene expression.

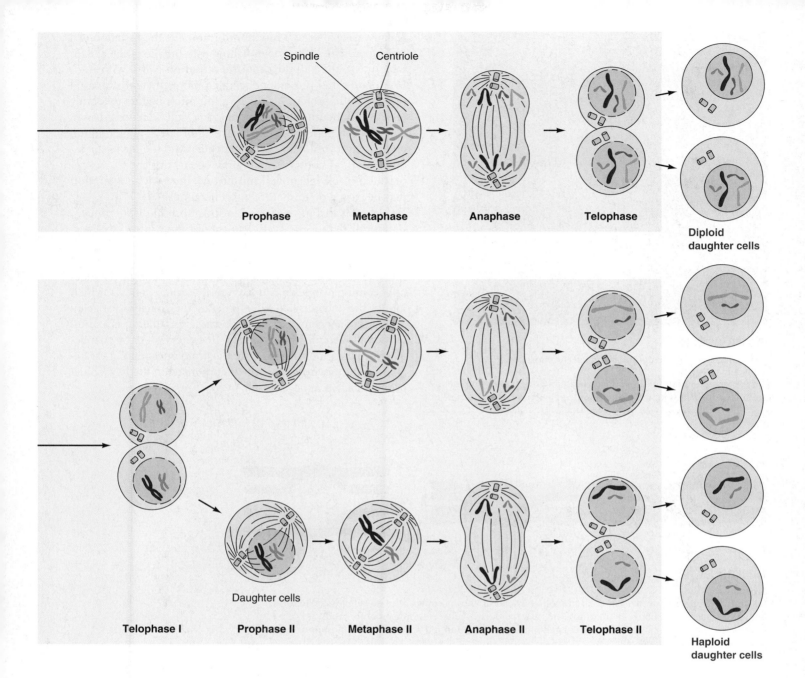

Spindle   Centriole

Prophase   **Metaphase**   **Anaphase**   **Telophase**

Diploid
daughter cells

Telophase I   **Prophase II**   **Metaphase II**   **Anaphase II**   **Telophase II**

Daughter cells

Haploid
daughter cells

▌ Mutations

An estimated 10^{16} cell divisions take place in the body during the course of a person's lifetime to accomplish growth, repair, and normal cell turnover. Because more than 3 billion nucleotides must be replicated during each cell division, no wonder "copying errors" occasionally occur. Any change in the DNA sequence is known as a **point (gene) mutation.** A point mutation arises when a base is inadvertently substituted, added, or deleted during the replication process.

When a base is inserted in the wrong position during DNA replication, the mistake can often be corrected by a built-in "proofreading" system. Repair enzymes remove the newly repli-

cated strand back to the defective segment, at which time normal base pairing resumes to resynthesize a corrected strand. Not all mistakes can be corrected, however.

Mutations can arise spontaneously by chance alone or they can be induced by **mutagens,** which are factors that increase the rate at which mutations take place. Mutagens include various chemical agents as well as ionizing radiation such as X rays and atomic radiation. Mutagens promote mutations either by chemically altering the DNA base code through a variety of mechanisms or by interfering with the repair enzymes so that abnormal base segments cannot be cut out.

Depending on the location and nature of a change in the genetic code, a given mutation may (1) have no noticeable effect if it does not alter a critical region of a cellular protein;

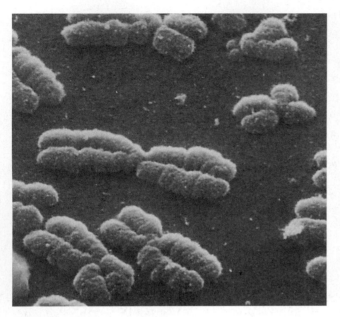

● FIGURE C-11

A scanning electron micrograph of human chromosomes from a dividing cell

The replicated chromosomes appear as double structures, with identical sister chromatids joined in the middle at a common centromere.

(2) adversely alter cell function if it impairs the function of a crucial protein; (3) be incompatible with the life of the cell, in which case the cell dies and the mutation is lost with it; or (4) in rare cases, prove beneficial if a more efficient structural or enzymatic protein results. If a mutation occurs in a body cell (a **somatic mutation**), the outcome will be reflected as an alteration in all future copies of the cell in the affected individual, but it will not be perpetuated beyond the life of the individual. If, by contrast, a mutation occurs in a sperm- or egg-producing cell (**germ cell mutation**), the genetic alteration may be passed on to succeeding generations.

In most instances, **cancer** results from multiple somatic mutations that occur over a course of time within DNA segments known as **proto-oncogenes.** Proto-oncogenes are normal genes whose coded products are important in the regulation of cell growth and division. These genes have the potential of becoming overzealous **oncogenes** ("cancer genes"), which induce the uncontrolled cell proliferation characteristic of cancer. Proto-oncogenes can become cancer producing as a result of several sequential mutations in the gene itself or by changes in adjacent regions that regulate the proto-oncogenes. Less frequently, tumor viruses become incorporated in the DNA blueprint and act as oncogenes.

Centromere

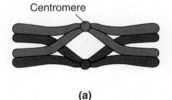

(a)

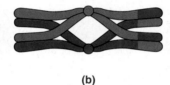

(b)

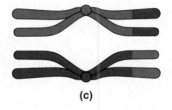

(c)

● FIGURE C-12

Crossing over

(a) During prophase I of meiosis, each homologous pair of chromosomes lines up side by side to form a tetrad. (b) Physical exchange of chromosome material occurs between nonsister chromatids. (c) As a result of this crossing over, new combinations of genetic material are formed within the chromosomes.

Principles of Quantitative Reasoning

By Kim E. Cooper, Midwestern University, and John D. Nagy, Scottsdale Community College

INTRODUCTION

Historically, as a branch of science matures, it typically becomes more precise and usually more quantitative. This trend is becoming increasingly true of biology and especially of physiology. Most students, however, are uncomfortable with quantitative reasoning. Students are usually quite capable of doing the mechanical manipulations of mathematics but have trouble translating back and forth between words, concepts, and equations. This appendix is meant to help you become more comfortable working with equations.

WHY ARE EQUATIONS USEFUL?

A great deal of what we do in science involves establishing functional relations between variables of interest (for example, blood pressure and heart rate, transport rate and concentration gradient). Equations are simply a compact and exact way of expressing such relationships. The tools of mathematics then allow us to draw conclusions systematically from these relationships. Mathematics is a very powerful set of tools or, more generally, a very powerful way of thinking. Mathematics allows you to think extremely precisely, and therefore clearly, about complex relationships. Equations and quantitative notions are the keys to that precision. For example, a quantitative comparison of the predictions of a theory against the results of measurement forms the basis of statistics and of much of the hypothesis testing on which science is based. A scientific conclusion without adequate quantification and statistical backing may be little more than an impression or opinion.

It may seem odd to say that mathematics allows you to think more clearly about complex ideas. People unfamiliar with mathematical thinking often complain that even simple relationships produce complicated equations and that complex relationships are mathematically intractable. Certainly, many basic concepts require considerable mathematical expertise to be handled properly, but such concepts are in fact not simple. More commonly, however, many simple equations are seen as complex because many students are poorly trained in how to think about equations.

HOW TO THINK ABOUT AN EQUATION

In this section we will take the first, and often overlooked, step in thinking quantitatively. How do we begin to think about some new equation presented to us? We start by becoming comfortable with the "meaning" of an equation. This step is absolutely necessary if you are to use an equation properly. As a specific example, consider the Nernst equation (see p. 90) for potassium. Here are several forms you will find in various books; they all say essentially the same thing:

$$E_{K^+} = (RT/zF)\log\{[K^+]_{out}/[K^+]_{in}\}$$

$$E_{K^+} = (RT/zF)\ 2.303\ \log\{[K^+]_{out}/\{[K^+]_{in}\}$$

$$E_{K^+} = (61\ mV/z)\ \log\{[K^+]_{out}/\{[K^+]_{in}\}$$

For many students these equations may seem like meaningless strings of symbols. What are these equations trying to tell us? What do they represent? The following four steps may help you become comfortable with any new equation. Try them with the Nernst equation.

1. Be sure you can define the symbols and give dimensions and units. Check the equation for dimensional consistency.

One of the first steps is to figure out which symbols represent the variables of interest and which are simple constants. In this case, all the symbols are constants except two.

E_{K^+} is the Nernst (equilibrium) potential for potassium. It represents the concentration gradient (force of diffusion) on a mole of potassium ions. E_{K^+} has the dimensions of a voltage and is usually given in units of mV. This dimension is used so that the concentration gradient is expressed in the same dimensions as the other force acting on the ions, that is, the electrical gradient. Using the same dimensions makes it possible to compare the two forces.

$[K^+]$ represents the concentration of potassium. With the subscript "out," this symbol refers to the concentration of potassium outside the cell. With the subscript "in," this symbol refers to the concentration of potassium inside the cell. $[K^+]$ has dimensions of concentration and is usually expressed in units of mM (millimolars; millimols/liter).

2. Identify the dependent and independent variables. Try to find normal values and ranges for the variables. Before continuing, we should define "dependent" and "independent" variables.

Remember that equations represent relationships between variables. Whenever you hear the word *relationship*, think of a graph, as in ● Figure D-1, for example.

Graphs are often a good way to represent relationships and therefore equations. This graph says the value of variable 2 depends on the value of variable 1. Thus, for any value of variable 1 the corresponding value for variable 2 can be determined from the graph. In other words, variable 1 determines the value of variable 2. Because variable 2 depends on variable 1, we call variable 2 the *dependent variable*. Variable 1, in contrast is independent of variable 2, so we call variable 1 the *independent variable*. There can be any number of dependent and independent variables.

How do you determine which variables are dependent and which are independent? The answer usually depends on cause and effect: "Effects" are dependent variables and "causes" are independent. For example, we know (see Chapter 10, p. 375) that mean arterial pressure *(MAP)* is the product of cardiac output *(CO)* and total peripheral resistance *(TPR)*; that is,

$$MAP = CO \times TPR$$

MAP is on the left-hand side of this equation because we think of mean arterial pressure as a result of cardiac output and total peripheral resistance. Or, to put this another way, mean arterial pressure is a function of cardiac output and total peripheral resistance. As a cause–effect relationship, it seems backward to think of mean arterial pressure somehow "causing" cardiac output to be a certain value. Therefore, *MAP* is the effect, the dependent variable, and we place it on the left-hand side of the equality symbol. Conversely, *CO* and *TPR* are the causes, the independent variables, and we put them on the right.

In our Nernst equation example, the independent variables are the concentrations. The dependent variable is the Nernst potential because we think of the potential as being a result of the ion concentrations. We also know that E_{K^+} is about −90mV, and $[K^+]_{out}$ and $[K^+]_{in}$ are about 5 mM and 150 mM, respectively.

3. Identify the constants and know their numeric values:

R is the gas constant. It has dimensions of energy per mole per degree of temperature and the value of 8.31 joules/kelvin · mole. It is also convenient to note that a joule = volt × coulomb.

T is temperature, with the dimension of temperature being in units of kelvins. Normal body temperature is around 37°C [= 308 kelvins (K)].

z is the valence of the ion. Valence is the charge on an ion, including the sign. For potassium, $z = +1$.

F is Faraday's constant, which has dimensions of charge per mole, units of coulombs per mole, and a value of 96,500 C/mol.

Refer back to the Nernst equations given on the preceding page. Note that the constants just defined appear in the first two equations, but not the third. In the third equation, the quantity RT/F has already been evaluated for you, as follows:

$$RT/F = [(8.31 \text{ V} \cdot \text{C/K} \cdot \text{mol})(308\text{K})]/(96,500 \text{ C/mol})$$
$$= 26.5 \text{ mV}$$

We multiply this value by 2.303 to convert the natural logarithm to the base 10 logarithm. Note that 26.5 mV × 2.303 = 61.1 mV.

4. State the equation in words. Summarize it in a few sentences so that someone can understand what it is about. Don't just say the names of the symbols.

Just saying the names of the symbols would be equivalent to saying the following: "The Nernst potential is given by a constant times the logarithm of the ratio of the ion concentrations." This is certainly true but does nothing to aid our intuition. A preferable statement would be "The Nernst equation allows us to calculate the force pushing ions into or out of a cell via diffusion." This is valuable, because we can compare this force to the force moving ions in and out via the membrane voltage and see which is larger and hence in what direction the ions will actually move. The force is expressed in electrical units so we can compare it directly with the membrane voltage. The constants convert from concentration to electrical units.

Only when you understand what an equation means will you be able to use it to answer questions. The next section gives you some guidance in taking this next step.

HOW TO THINK WITH AN EQUATION

Before you can use an equation to help you think, you need to develop a few basic skills. Luckily, these skills are not difficult to learn.

1. Be sure you know the algebraic rules for manipulating variables within any function (such as $\sqrt{}$, exp, or log) involved.

In the case of the Nernst equation, the tricky function is the logarithm. You should consult a college-level algebra book if you are hazy on the rules of working with logs or any other function. For instance, it is useful to know that

● **FIGURE D-1**

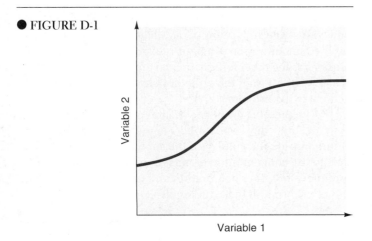

$$\log \{A\} = -\log \{1/A\}$$

and that the log operation is undone by taking it to the power of 10; that is,

$$10^{\log\{A\}} = A$$

2. Be able to solve for any variable in terms of the others.

Given just three variables (E_{K^+}, $[K^+]_{in}$, $[K^+]_{out}$), only a few types of questions can be asked. Two of the three variables must be given, and you must solve for the third. If the two concentrations are given, then the formula is already set to give you the Nernst potential. If the Nernst potential and one concentration are given, however, you must be able to solve for the other concentration. See if you can do this and obtain the two following equations:

$$[K^+]_{out} = [K^+]_{in}10^{(E_{K^+}/61 \text{ mV})}$$
$$[K^+]_{in} = [K^+]_{out}10^{(-E_{K^+}/61 \text{ mV})}$$

3. Be able to sketch, at least approximately, the dependence of any variable on any other variable.

The ability to do this is exceedingly valuable. Sketching helps you generate insight about equations; it helps you understand what an equation means. Therefore, sketching helps you understand the solution, as well as solve the problem. If you apply this technique consistently, you may find equations far simpler to handle than you previously suspected. In addition, be sure you can relate your sketch to experimental measurements and physiological situations.

For example, we can draw the relationships between the Nernst potential for potassium and the external potassium ion concentration predicted by the equations as in ● Figure D-2:

This sketch makes clear that the Nernst potential, which can be measured physiologically, should decrease linearly as the log $[K^+]_{out}$ increases, which can be controlled experimentally. Therefore, this sketch suggests an experiment: vary $[K^+]_{out}$. If E_{K^+} does not decrease linearly with increasing log $[K^+]_{out}$, then we would have a flaw in our understanding. The Nernst equation would not describe the real situation, as we think it should. Scientific advances are almost always heralded by such contradictions.

4. Be able to combine several equations to find new relationships.

Combining separate pieces of information is always useful. In fact, some scientists have argued that this activity is all scientists ever do. To integrate knowledge for yourself, you must be able to combine the various relations you learn about into new combinations. This allows you to solve increasingly complex problems. As an example, consider the following relation:

$$I_{K^+} = G_{K^+}(V_m - E_{K^+})$$

This equation describes the number of potassium ions flowing across a membrane if both a concentration gradient (E_{K^+}) and an electrical gradient (V_m) are present. This equation can be combined with the Nernst equation for potassium to answer the following question. Suppose V_m, G_{K^+}, and $[K^+]_{in}$ are fixed. What would the external potassium ion concentration have to be such that no net flux of potassium ions occurs across the membrane? No net flux implies that $I_{K+} = 0$. But if $G_{K^+} > 0$, $I_{K^-} = 0$ only when the membrane voltage equals the Nernst potential ($V_m = E_{K^+}$). To answer the question, then, we set $E_{K^+} = V_m$ in the Nernst equation and solve it for $[K^+]_{out}$.

5. Know the equation's underlying assumptions and limits of validity.

Every equation comes from some underlying theory or set of observations and, therefore, has some limited range of validity and rests on certain assumptions. Failure to understand this simple point often leads students to apply equations outside their realm of applicability. In that case, even though the math is done correctly, the results will be incorrect.

In the case of the Nernst equation, things are fairly simple. This equation is derived from a very powerful theory known as *equilibrium thermodynamics*, and hence it has very wide applicability. As another example, consider enzyme kinetics. The rate at which an enzyme catalyzes a reaction (v) is related to the concentration of substrate on which the enzyme works ($[S]$) by an equation called the Michaelis-Menton relationship. The graph of this relationship can be seen in ● Figure D-3:

However, this relationship between reaction velocity and substrate concentration does not hold true for some real en-

● FIGURE D-2

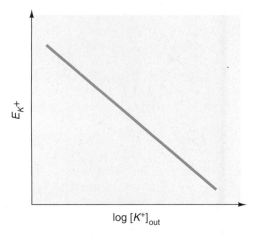

● FIGURE D-3

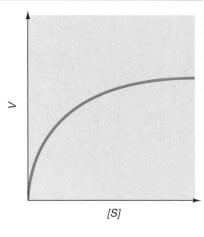

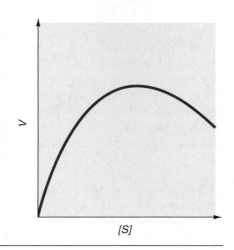

● FIGURE D-4

zymes, such as lactate dehydrogenase. For this enzyme, the relationship between reaction velocity and substrate concentration is depicted in ● Figure D-4.

At high substrate concentrations, the enzyme actually is inhibited by too much substrate. For such enzymes, the Michaelis-Menton theory, which works well at low [S], is invalid at higher [S].

AN APPROACH TO PROBLEM SOLVING

The final step is to apply these skills to solve a problem. As an example, calculate the concentration of potassium that must exist inside a cell if $E_{K^+} = -95$ mV and the interstitial fluid has a potassium concentration of 4 mM. Try using the following procedure to solve this problem:

1. Get a clear picture of what is being asked. State it out loud or write it down.

The question asks for the concentration of potassium in the cell, that is, $[K^+]_{in}$,

2. Determine what you need to know to answer the question:

To answer this, you need to know E_{K^+} and $[K^+]_{out}$.

3. Determine what information is given. Is it sufficient? Are other relevant facts or relationships not stated in the problem? Specifically, do you need any other equations? E_{K^+} is

given explicitly in the problem, but you have to translate the words to realize that $[K^+]_{out} = 4$ mM.

4. Manipulate the equation algebraically so that the unknown is on the left-hand side and everything else is on the right-hand side.

We now solve the Nernst equation for $[K^+]_{in}$. This was done on page A-33:

$$[K^+]_{in} = [K^+]_{out} \, 10^{(-E_{K^+}/61 \text{ mV})}$$

Substituting for the values given, we obtain:

$$[K^+]_{in} = 4 \text{ mM } 10^{(95 \text{ mV}/61 \text{ mV})}$$

$$= 4 \text{ mM } 10^{(1.56)}$$

$$= 4 \text{ mM } (36.3)$$

$$= 145 \text{ mM}$$

5. Is the answer dimensionally correct? Do not skip this step. It will tell you immediately if something went wrong.

Yes, the answer is in mM—the proper dimension and unit.

6. Does the answer make sense?

Yes, the value is not alarmingly low or high. Also, because the potassium ion is positively charged, if the potassium concentration outside the cell is lower than that inside the cell, then the interior of the cell would have to be negative at passive equilibrium, which it is.

Apply the approach to problem solving we have just outlined to the quantitative questions in the chapters. When you start out, apply the approach formally and carefully. For example, go through each step, and write everything out as we have done for the Nernst equation. After a time, you may not need to be so formal. Also, as you progress, you will develop your own style and approach to problem solving. Be prepared to spend some time and patience on some of the problems. Not every answer will be immediately apparent. This situation is normal. If you run into difficulties, relax, return to this appendix for guidance, and work through the problem again carefully. Do not go immediately to the answers if you are having difficulty with a problem. It may ease the frustration, but you will be cheating yourself of a valuable learning experience. Besides, being able to solve challenging problems has its own rewards.

Answers to End-of-Chapter Objective Questions, Quantitative Exercises, and Points to Ponder

CHAPTER 1   HOMEOSTASIS: THE FOUNDATION OF PHYSIOLOGY

▌ Objective Questions

(Questions on p. 20.)

1. e 2. b 3. c 4. T 5. F 6. T 7. muscle tissue, nervous tissue, epithelial tissue, connective tissue 8. secretion 9. exocrine, endocrine, hormones 10. intrinsic, extrinsic 11. 1.d, 2.g, 3.a, 4.e, 5.b, 6.j, 7.h, 8.i, 9.c, 10.f

▌ Points to Ponder

(Questions on p. 20.)

1. The respiratory system eliminates internally produced CO_2 to the external environment. A decrease in CO_2 in the internal environment brings about a reduction in respiratory activity (that is, slower, shallower breathing) so that CO_2 produced within the body is allowed to accumulate instead of being blown off as rapidly as normal to the external environment. The extra CO_2 retained in the body increases the CO_2 levels in the internal environment to normal. 2. (b) (c) (b) 3. b 4. immune defense system
5. When a person is engaged in strenuous exercise, the temperature-regulating center in the brain will bring about widening of the blood vessels of the skin. The resultant increased blood flow through the skin will carry the extra heat generated by the contracting muscles to the body surface, where it can be lost to the surrounding environment.
6. *Clinical Consideration.* Loss of fluids threatens the maintenance of proper plasma volume and blood pressure. Loss of acidic digestive juices threatens the maintenance of the proper pH in the internal fluid environment. The urinary system will help restore the proper plasma volume and pH by reducing the amount of water and acid eliminated in the urine. The respiratory system will help restore the pH by adjusting the rate of removal of acid-forming CO_2. Adjustments will be made in the circulatory system to help maintain blood pressure despite

fluid loss. Increased thirst will encourage increased fluid intake to help restore plasma volume. These compensatory changes in the urinary, respiratory, and circulatory systems, as well as the sensation of thirst, will all be regulated by the two regulatory systems, the nervous and endocrine systems. Furthermore, the endocrine system will make internal adjustments to help maintain the concentration of nutrients in the internal environment even though no new nutrients are being absorbed from the digestive system.

CHAPTER 2   CELLULAR PHYSIOLOGY

▌ Objective Questions

(Questions on p. 53.)

1. plasma membrane 2. deoxyribonucleic acid (DNA), nucleus 3. organelles, cytosol, cytoskeleton 4. endoplasmic reticulum, Golgi complex 5. oxidative, hydrogen peroxide 6. adenosine triphosphate (ATP) 7. F 8. F 9. 1.b, 2.a, 3.b 10. 1.b, 2.c, 3.c, 4.a, 5.b, 6.c, 7.a, 8.c

▌ Quantitative Exercises

(Questions on p. 54.)

1. b
2. 24 moles O_2/day × 6 moles ATP/mole O_2 = 144 moles ATP/day
    144 moles ATP/day × 507 g ATP/mole = 73,000 g ATP/day
    1,000 g/2.2 lb = 73,000 g/x lb
    1,000 x = 160,600
    x = approximately 160 lb
3. 144 mol/day (7,300 cal/mol) = 1,051,200 cal/day (1,051 kilocal/day)
4. About 2/3 of the water in the body is intracellular. Because a person's mass is about 60% water, for a 150-pound (68 kg) person,

    68 kg(0.6)(2/3) = 27.2 kg

is the mass of water. Assume that 1 ml of body water weighs 1 gm. Then the total volume in the person's cells is about 27.2 L. The volume of an average cell is

$$\frac{4}{3} \pi \,(1 \times 10^{-3} \text{ cm})^3 \approx 4.2 \times 10^{-9} \text{ cm}^3$$
$$= 4.2 \times 10^{-9} \text{ ml}$$

So, the number of cells in a 68 kg person is about

$$27.2 \text{ L} \left(\frac{1,000 \text{ ml}}{1 \text{ L}} \right) \left(\frac{1 \text{ cell}}{4.2 \times 10^{-9} \text{ ml}} \right)$$
$$= 6.476 \times 10^{12} \text{ cells}$$

5. $150 \text{ mg} \left(\dfrac{1 \text{ ml}}{0.015 \text{ mg}} \right) = 10,000 \text{ ml } (10 \text{ L})$

▌ Points to Ponder

(Questions on p. 54.)

1. The chief cells have an extensive rough endoplasmic reticulum, with this organelle being responsible for synthesizing these cells' protein secretory product, namely pepsinogen. Because the parietal cells do not secrete a protein product to the cells' exterior, they do not need an extensive rough endoplasmic reticulum.

2. With cyanide poisoning, the cellular activities that depend on ATP expenditure could not continue, such as synthesis of new chemical compounds, membrane transport, and mechanical work. The resultant inability of the heart to pump blood and failure of the respiratory muscles to accomplish breathing would lead to imminent death.

3. catalase

4. ATP is required for muscle contraction. Muscles are able to store limited supplies of nutrient fuel for use in the generation of ATP. During anaerobic exercise, muscles generate ATP from these nutrient stores by means of glycolysis, which yields two molecules of ATP per glucose molecule processed. During aerobic exercise, muscles can generate ATP by means of oxidative phosphorylation, which yields 36 molecules of ATP per glucose molecule processed. Because glycolysis inefficiently generates ATP from nutrient fuels, it rapidly depletes the muscle's limited stores of fuel, and ATP can no longer be produced to sustain the muscle's contractile activity. Aerobic exercise, in contrast, can be sustained for prolonged periods. Not only does oxidative phosphorylation use far less nutrient fuel to generate ATP, but it can be supported by nutrients delivered to the muscle by means of the blood instead of relying on stored fuel in the muscle. Intense anaerobic exercise outpaces the ability to deliver supplies to the muscle by the blood, so the muscle must rely on stored fuel and inefficient glycolysis, thus limiting anaerobic exercise to brief periods of time before energy sources are depleted.

5. skin. The mutant keratin weakens the skin cells of patients with epidermolysis bullosa so that the skin blisters in response to even a light touch.

6. *Clinical Consideration.* Some hereditary forms of male sterility involving nonmotile sperm have been traced to defects in the cytoskeletal components of the sperm's flagella. These same individuals usually also have long histories of recurrent respiratory tract disease because the same types of defects are present in their respiratory cilia, which are unable to clear mucus and inhaled particles from the respiratory system.

CHAPTER 3 THE PLASMA MEMBRANE AND MEMBRANE POTENTIAL

▌ Objective Questions

(Questions on p. 94.)

1. T 2. F 3. T 4. F 5. opening or closing specific membrane channels, activating an intracellular second messenger. 6. G protein 7. negative, positive 8. 1.b, 2.a, 3.b, 4.a, 5.c, 6.b, 7.a, 8.b 9. 1.a, 2.a, 3.b, 4.a, 5.b, 6.a, 7.b 10. 1.c, 2.b, 3.a, 4.a, 5.c, 6.b, 7.c, 8.a, 9.b

▌ Quantitative Exercises

(Questions on p. 95.)

1. $E = \dfrac{61 \text{ mV}}{z} \log \dfrac{C_o}{C_i}$

   a. $\dfrac{61 \text{mV}}{2} \log \dfrac{1 \times 10^{-3}}{100 \times 10^{-9}} = +122 \text{ mV}$

   b. $\dfrac{61 \text{ mV}}{-1} \log \dfrac{110 \times 10^{-3}}{10 \times 10^{-3}} = -63.5 \text{ mV}$

2. $I_x = G_x(V_m - E_x)$

   $E_{Na^+} = 61 \text{mV} \log \dfrac{145 \text{ mM}}{15 \text{mM}} = 60.1 \text{ mV}$

   a. $= 1 \text{ nS } (-70 \text{ mV} - 60.1 \text{ mV})$
       $= 1 \text{ nS } (-130 \text{ mV})$
       $= -130 \text{ pA (A = amperes)}$

   b. Entering

   c. With concentration gradient; with electrical gradient

3. $V_m = \dfrac{G_{Na^+}}{G_T} E_{Na^+} + \dfrac{G_{K^+}}{G_T} E_{K^+}$

   a. $G_T = 1 \text{ nS} + 5.3 \text{ nS} = 6.3 \text{ nS};$

   $V_m = \dfrac{1}{6.3} 59.1 \text{mV} + \dfrac{5.3}{6.3} (-94.4 \text{ mV})$

   $= 9.4 \text{ mV} - 79.4 \text{ mV} = -70 \text{ mV}$

   b. $E_{K^+} = 0 \text{ mV}; V_m = 9.4 \text{ mV}; $ i.e., large depolarization

▌ Points to Ponder

(Questions on p. 96.)

1. c. As Na^+ moves from side 1 to side 2 down its concentration gradient, Cl^- remains on side 1, unable to permeate the membrane. The resultant separation of charges produces a

membrane potential, negative on side 1 because of unbalanced chloride ions and positive on side 2 because of unbalanced sodium ions. Sodium does not continue to move to side 2 until its concentration gradient is dissipated because of the development of an opposing electrical gradient.

2. more positive. Because the electrochemical gradient for Na^+ is inward, the membrane potential would become more positive as a result of an increased influx of Na^+ into the cell if the membrane were more permeable to Na^+ than to K^+. (Indeed, this is what happens during the rising phase of an action potential once threshold potential is reached—see Chapter 4).

3. d. active transport. Leveling off of the curve designates saturation of a carrier molecule, so carrier-mediated transport is involved. The graph indicates that active transport is being used instead of facilitated diffusion, because the concentration of the substance in the intracellular fluid is greater than the concentration in the extracellular fluid at all points until after the transport maximum is reached. Thus the substance is being moved *against* a concentration gradient, so active transport must be the method of transport being used.

4. vesicular transport. The maternal antibodies in the infant's digestive tract lumen are taken up by the intestinal cells by pinocytosis and are extruded on the opposite side of the cell into the interstitial fluid by exocytosis. The antibodies are picked up from the intestinal interstitial fluid by the blood supply to the region.

5. accelerate. During an action potential, Na^+ enters and K^+ leaves the cell. Repeated action potentials would eventually "run down" the Na^+ and K^+ concentration gradients were it not for the Na^+–K^+ pump returning the Na^+ that entered back to the outside and the K^+ that left back to the inside. Indeed, the rate of pump activity is accelerated by the increase in both ICF Na^+ and ECF K^+ concentrations that occurs as a result of action potential activity, thus hastening the restoration of the concentration gradients.

6. *Clinical Consideration.* As Cl^- is secreted by the intestinal cells into the intestinal tract lumen, Na^+ follows passively along the established electrical gradient. Water passively accompanies this salt (Na^+ and Cl^-) secretion by osmosis. The toxin produced by the cholera pathogen prevents the normal inactivation of cAMP in intestinal cells so that cAMP levels rise. An increase in cAMP opens the Cl^- channels in the luminal membranes of these cells. Increased secretion of Cl^- and the subsequent passively induced secretion of Na^+ and water are responsible for the severe diarrhea that characterizes cholera.

CHAPTER 4   NEURONAL PHYSIOLOGY

▮ Objective Questions

(Questions on p. 130.)

1. T **2.** F **3.** F **4.** F **5.** refractory period **6.** axon hillock **7.** synapse **8.** temporal summation **9.** spatial summation **10.** convergence, divergence **11.** 1.b, 2.a, 3.a, 4.b, 5.b, 6.a

▮ Quantitative Exercises

(Questions on p. 130.)

1. a. 0.6 m (1 sec/0.7 m) = 0.8571 sec
 b. 0.6 m (1 sec/120 m) = 0.005 sec
 c. unmyelinated: 0.8591 sec; myelinated: 0.007 sec
 d. unmyelinated: 0.8621 sec; myelinated: 0.01 sec

2. Total conduction time for the single axon is 1/60 sec. Let v m/sec be the unknown conduction velocity for the three neurons. Our equation for the total conduction time then is

$$\frac{1}{60} \sec = \left(\frac{1}{v} \times 1 \text{ m}\right) + 0.002 \text{ sec}$$

Solving for v, we obtain
v m/sec = 1 m/(1/60 sec − 0.002 sec) = 68.18 m/sec

3. 25×10^{-3} V $\left[\dfrac{3.3 \; \mu S/cm^2 (240 \; \mu S/cm^2)}{(3.3 + 240) \; \mu S/cm^2}\right] \log \dfrac{240(145)}{3.3(4)}$

$= 25 \times 10^{-3}(11.1361)$V $\times \mu S/cm^2 = 0.2784 \mu A/cm^2$

▮ Points to Ponder

(Questions on p. 131.)

1. c. The action potentials would stop as they met in the middle. As the two action potentials moving toward each other both reached the middle of the axon, the two adjacent patches of membrane in the middle would be in a refractory period so further propagation of either action potential would be impossible.
2. A subthreshold stimulus would transiently depolarize the membrane but not sufficiently to bring the membrane to threshold, so no action potential would occur. Because a threshold stimulus would bring the membrane to threshold, an action potential would occur. An action potential of the same magnitude and duration would occur in response to a suprathreshold stimulus as to a threshold stimulus. Because of the all-or-none law, a stimulus larger than that necessary to bring the membrane to threshold would not produce a larger action potential. (The magnitude of the stimulus is coded in the *frequency* of action potentials generated in the neuron, not the *size* of the action potentials.)
3. The hand could be pulled away from the hot stove by flexion of the elbow accomplished by summation of EPSPs at the cell bodies of the neurons controlling the biceps muscle, thus bringing these neurons to threshold. The subsequent action potentials generated in these neurons would stimulate contraction of the biceps. Simultaneous contraction of the triceps muscle, which would oppose the desired flexion of the elbow, could be prevented by generation of IPSPs at the cell bodies of the neurons controlling this muscle. These IPSPs would keep the triceps neurons from reaching threshold and firing so that the triceps would not be stimulated to contract.

The arm could deliberately be extended despite a painful finger prick by voluntarily generating EPSPs to override the reflex IPSPs at the neuronal cell bodies controlling the triceps while simultaneously generating IPSPs to override the reflex EPSPs at the neuronal cell bodies controlling the biceps.

4. Treatment for Parkinson's disease is aimed toward restoring dopamine activity in the basal nuclei. However, this treatment may lead to excessive dopamine activity in otherwise normal areas of the brain that also use dopamine as a neurotransmitter. Excessive dopamine activity in a particular region of the brain (the limbic system) is believed to be among the causes of schizophrenia. Therefore, symptoms of schizophrenia sometimes occur as a side effect during treatment for Parkinson's disease.

5. An EPSP, being a graded potential, spreads decrementally from its site of initiation in the postsynaptic neuron. If presynaptic neuron A (near the axon hillock of the postsynaptic cell) and presynaptic neuron B (on the opposite side of the postsynaptic cell body) both initiate EPSPs of the same magnitude and frequency, the EPSPs from A will be of greater strength when they reach the axon hillock than will the EPSPs from B. An EPSP from B will decrease more in magnitude as it travels farther before reaching the axon hillock, the region of lowest threshold and thus the site of action potential initiation. Temporal summation of the larger EPSPs from A may bring the axon hillock to threshold and initiate an action potential in the postsynaptic neuron, whereas temporal summation of the weaker EPSPs from B at the axon hillock may not be sufficient to bring this region to threshold. Thus, the proximity of a presynaptic neuron to the axon hillock can bias its influence on the postsynaptic cell.

6. *Clinical Consideration.* Initiation and propagation of action potentials would not occur in nerve fibers acted on by local anesthetic because blockage of Na^+ channels by the local anesthetic would prevent the massive opening of voltage-gated Na^+ channels at threshold potential. As a result, pain impulses (action potentials in nerve fibers that carry pain signals) would not be initiated and propagated to the brain and reach the level of conscious awareness.

CHAPTER 5   THE CENTRAL NERVOUS SYSTEM

▌ Objective Questions

(Questions on p. 182.)

1. F **2.** F **3.** F **4.** T **5.** F **6.** F **7.** habituation **8.** consolidation **9.** dorsal, ventral **10.** receptor, afferent pathway, integrating center, efferent pathway, effector **11.** 1.a, 2.c, 3.a and b, 4.b, 5.a, 6.c, 7.c **12.** 1.d, 2.c, 3.f, 4.e, 5.a, 6.b

▌ Points to Ponder

(Questions on p. 183.)

1. Only the left hemisphere has language ability. When sharing of information between the two hemispheres is prevented as a result of severance of the corpus callosum, visual information presented only to the right hemisphere cannot be verbally identified by the left hemisphere, because the left hemisphere is unaware of the information. However, the information can be recognized by nonverbal means, of which the right hemisphere is capable.

2. Insulin excess drives too much glucose into insulin-dependent cells so that the blood glucose falls below normal and insufficient glucose is delivered to the non–insulin-dependent brain. Therefore, the brain, which depends on glucose as its energy source, does not receive adequate nourishment.

3. c. A severe blow to the back of the head is most likely to traumatize the visual cortex in the occipital lobe.

4. Salivation when seeing or smelling food, striking the appropriate letter on the keyboard when typing, and many of the actions involved in driving a car are conditioned reflexes. You undoubtedly will have many other examples.

5. Strokes occur when a portion of the brain is deprived of its vital O_2 and glucose supply because the cerebral blood vessel supplying the area either is blocked by a clot or has ruptured. Although a clot-dissolving drug could be helpful in restoring blood flow through a cerebral vessel blocked by a clot, such a drug would be detrimental in the case of a ruptured cerebral vessel sealed by a clot. Dissolution of a clot sealing a ruptured vessel would lead to renewed hemorrhage through the vessel and exacerbation of the problem.

6. *Clinical Consideration.* The deficits following the stroke—numbness and partial paralysis on the upper right side of the body and inability to speak—are indicative of damage to the left somatosensory cortex and left primary motor cortex in the regions devoted to the upper part of the body plus Broca's area.

CHAPTER 6   THE PERIPHERAL NERVOUS SYSTEM: AFFERENT DIVISION; SPECIAL SENSES

▌ Objective Questions

(Questions on p. 234.)

1. transduction **2.** adequate stimulus **3.** T **4.** F **5.** T **6.** T **7.** T **8.** F **9.** T **10.** F **11.** F **12.** 1.f, 2.h, 3.l, 4.d, 5.i, 6.e, 7.b, 8.j, 9.a, 10.g, 11.c, 12.k **13.** 1.a, 2.b, 3.c, 4.c, 5.c, 6.a, 7.b, 8.b

▌ Quantitative Exercises

(Questions on p. 234.)

1. The slow pain pathway takes about (1.3 m) (1 sec/12 m) = 0.1083 sec. The fast pathway takes (1.3 m) (1 sec/30 m) = 0.0433 sec. The difference is 0.1083 sec − 0.0433 sec = 0.065 sec = 65 msec.

2. a. The amount of light entering the eye is proportional, approximately, to the area of the open pupil. Recall that the area of a circle is πr^2. Let r be the pupil radius and A_1

be the original pupil area. Halving the diameter also halves the radius, so the new pupil area is

$$\pi\left(\frac{1}{2}r^2\right) = \frac{1}{4}\pi r^2 = \frac{1}{4}A_1$$

Therefore, the amount of light allowed into the eye is a quarter of what it was originally.

b.   The area of a rectangle is hw, where h is the height and w the width. Halving either dimension halves the area and hence the amount of light allowed into the eye.

c.   The cat's pupil can be considered more precise. Think about the coarse and fine adjustments on a microscope. Fine adjustment translates rotations of the knob into much smaller movement of the stage than does coarse adjustment.

3.   a.   Solve the following for I:

$$\beta = (10 \text{ dB}) \log_{10} (I/I_0)$$

$$I = I_0 10^{B/10} \text{W/m}^2$$

Therefore,

$$I_i = 10^{-12}(10^{20/10}) = 10^{-12}(10^2) = 10^{-10} \text{W/m}^2$$

$$I_{ii} = 10^{-12}(10^{70/10}) = 10^{-12}(10^7) = 10^{-5} \text{W/m}^2$$

$$I_{iii} = 10^{-12}(10^{120/10}) = 10^{-12}(10^{12}) = 1 \text{W/m}^2$$

$$I_{iv} = 10^{-12}(10^{70/10}) = 10^{-12}(10^{17}) = 10^5 \text{W/m}^2$$

b.   Because of the logarithm in the definition of decibel, the sound intensity increases exponentially with respect to sound level. This fact should be clear from the definition of dB solved for I. This result implies that the human ear performs well throughout an enormous range of sound intensities.

▌Points to Ponder

(Questions on p. 235.)

1.   Pain is a conscious warning that tissue damage is occurring or about to occur. A patient unable to feel pain because of a nerve disorder does not consciously take measures to withdraw from painful stimuli and thus prevent more serious tissue damage.

2.   Pupillary dilation (mydriasis) can be deliberately induced by ophthalmic instillation of either an adrenergic drug (such as epinephrine or related compound) or a cholinergic blocking drug (such as atropine or related compounds). Adrenergic drugs produce mydriasis by causing contraction of the sympathetically supplied radial (dilator) muscle of the iris. Cholinergic blocking drugs cause pupillary dilation by blocking parasympathetic activity to the circular (constrictor) muscle of the iris so that action of the adrenergically controlled radial muscle of the iris is unopposed.

3.   The defect would be in the left optic tract or optic radiation.

4.   Fluid accumulation in the middle ear in accompaniment with middle ear infections impedes the normal movement of the tympanic membrane, ossicles, and oval window in response to sound. All these structures vibrate less vigorously in the pres-

ence of fluid, causing temporary hearing impairment. Chronic fluid accumulation in the middle ear is sometimes relieved by surgical implantation of drainage tubes in the eardrum. Hearing is restored to normal as the fluid drains to the exterior. Usually, the tube "falls out" as the eardrum heals and pushes out the foreign object.

5.   The sense of smell is reduced when you have a cold, even though the cold virus does not directly adversely affect the olfactory receptor cells, because odorants do not reach the receptor cells as readily when the mucous membranes lining the nasal passageways are swollen and excess mucus is present.

6.   *Clinical Consideration.* Syncope most frequently occurs as a result of inadequate delivery of blood carrying sufficient oxygen and glucose supplies to the brain. Possible causes include circulatory disorders such as impaired pumping of the heart or low blood pressure; respiratory disorders resulting in poorly oxygenated blood; anemia, in which the oxygen-carrying capacity of the blood is reduced; or low blood glucose resulting from improper endocrine management of blood glucose levels. Vertigo, in contrast, typically results from a dysfunction of the vestibular apparatus, arising, for example, from viral infection or trauma, or abnormal neural processing of vestibular information, as, for example, with a brain tumor.

CHAPTER 7   THE PERIPHERAL NERVOUS SYSTEM: EFFERENT DIVISION

▌Objective Questions

(Questions on p. 254.)

1.   T 2. F 3. c 4. c 5. sympathetic,   parasympathetic 6. adrenal medulla 7. 1.a, 2.b, 3.a, 4.b, 5.a, 6.a, 7.b 8. 1.b, 2.b, 3.a, 4.a, 5.b, 6.b, 7.a

▌Quantitative Exercises

(Questions on p. 254.)

1. $t = \dfrac{x^2}{2D} = \dfrac{(200 \text{ nm})^2}{2 \times 10^{-5} \text{ cm}^2/\text{sec}}$

$$= \frac{4 \times 10^{-14} \text{m}^2 \cdot \text{sec}}{2 \times 10^{-5} \text{ cm}^2} \left(\frac{10^4 \text{ cm}^2}{\text{m}^2}\right) = 20 \text{ μsec}$$

▌Points to Ponder

(Questions on p. 255.)

1.   By promoting arteriolar constriction, epinephrine administered in conjunction with local anesthetics reduces blood flow to the region and thus helps the anesthetic stay in the region instead of being carried away by the blood.

2.   No. Atropine blocks the effect of acetylcholine at muscarinic receptors but does not affect nicotinic receptors. Nico-

tinic receptors are present on the motor end plates of skeletal muscle fibers.

3. The voluntarily controlled external urethral sphincter is composed of skeletal muscle and supplied by the somatic nervous system.

4. By interfering with normal acetylcholine activity at the neuromuscular junction, α bungarotoxin leads to skeletal muscle paralysis, with death ultimately occurring as a result of an inability to contract the diaphragm and breathe.

5. If the motor neurons that control the respiratory muscles, especially the diaphragm, are destroyed by poliovirus or amyotrophic lateral sclerosis, the person is unable to breathe and dies (unless breathing is assisted by artificial means).

6. *Clinical Consideration.* Drugs that block β_1 receptors are useful for prolonged treatment of angina pectoris because they interfere with sympathetic stimulation of the heart during exercise or emotionally stressful situations. By preventing increased cardiac metabolism and thus an increased need for oxygen delivery to the cardiac muscle during these situations, beta blockers can reduce the frequency and severity of angina attacks.

CHAPTER 8   MUSCLE PHYSIOLOGY

▌ Objective Questions

(Questions on p. 299.)

1. F **2.** F **3.** F **4.** T **5.** F **6.** T **7.** concentric, eccentric **8.** alpha, gamma **9.** denervation atrophy, disuse atrophy **10.** a, b, e **11.** b **12.** 1.f, 2.d, 3.c, 4.e, 5.b, 6.g, 7.a **13.** 1.a, 2.a, 3.a, 4.b, 5.b, 6.b

▌ Quantitative Exercises

(Questions on p. 300.)

1. a. For the weekend athlete, the lever ratio is 70 cm/9 cm. So the velocity at the end of the arm is 2.6 m/sec (70/9) = 20.2 m/sec (about 45 mph).

   b. For the professional ballplayer, the lever ratio is 90 cm/9 cm. So
   $10x = 85$ mph
   $x = 8.5$ mi/hr(1,609 m/mi)(1 hr/3,600 sec)
   $= 3.8$ m/sec.

2. The force-velocity curve is as follows:

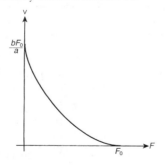

a. The shape of the curve indicates that it takes time to develop force, and that the greater the force developed, the more time is needed.

b. The maximum velocity will not change when F_0 is increased, but the muscle is able to lift heavier loads, or to generate more force. The maximum load will not increase when the cross-bridge cycling rate increases, but the muscle will be able to lift lighter loads faster. If the muscle increases in size, b increases, and the entire curve shifts up with respect to the v axis.

▌ Points to Ponder

(Questions on p. 301.)

1. By placing increased demands on the heart to sustain increased delivery of O_2 and nutrients to working skeletal muscles, regular aerobic exercise induces changes in cardiac muscle that enable it to use O_2 more efficiently, such as increasing the number of capillaries supplying blood to the heart muscle. Intense exercise of short duration, such as weight training, in contrast, does not induce cardiac efficiency. Because this type of exercise relies on anaerobic glycolysis for ATP formation, no demands are placed on the heart for increased delivery of blood to the working muscles.

2. The power arm of the lever is 4 cm, and the load arm is 28 cm for a lever ratio of 1:7 (4 cm:28 cm). Thus, to lift an 8-kg stack of books with one hand, the child must generate an upward applied force in the biceps muscle of 56 kg. (With a lever ratio of 1:7, the muscle must exert seven times the force of the load; 7×8 kg = 56 kg.)

3. The length of the thin filaments is represented by the distance between a Z line and the edge of the adjacent H zone. This distance remains the same in a relaxed and contracted myofibril, leading to the conclusion that the thin filaments do not change in length during muscle contraction.

4. Regular bouts of anaerobic, short-duration, high-intensity resistance training would be recommended for competitive downhill skiing. By promoting hypertrophy of the fast glycolytic fibers, such exercise better adapts the muscles to activities that require intense strength for brief periods, such as a swift, powerful descent downhill. In contrast, regular aerobic exercise would be more beneficial for competitive cross-country skiers. Aerobic exercise induces metabolic changes within the oxidative fibers that enable the muscles to use O_2 more efficiently. These changes, which include an increase in mitochondria and capillaries within the oxidative fibers, adapt the muscles to better endure the prolonged activity of cross-country skiing without fatiguing.

5. Because the site of voluntary control to overcome the micturition reflex is at the external urethral sphincter and not the bladder, the external urethral sphincter must be skeletal muscle, which is innervated by the voluntarily controlled somatic nervous system, and the bladder must be smooth muscle, which is innervated by the involuntarily controlled autonomic nervous system. The only other type of involuntarily controlled muscle besides smooth muscle is cardiac muscle, which is found only in the heart. Therefore, the bladder must be smooth, not cardiac, muscle.

6. *Clinical Consideration.* The muscles in the immobilized leg have undergone disuse atrophy. The physician or physical therapist can prescribe regular resistance-type exercises that specifically use the atrophied muscles to help restore them to their normal size.

CHAPTER 9   CARDIAC PHYSIOLOGY

▌ Objective Questions

(Questions on p. 339.)

1. intercalated discs, desmosomes, gap junctions 2. bradycardia, tachycardia 3. adenosine 4. F 5. F 6. F 7. T 8. d 9. d 10. e 11. less than, greater than, less than, greater than, less than 12. AV, systole, semilunar, diastole 13. 1.e, 2.a, 3.d, 4.b, 5.f, 6.c, 7.g 14. 1.c, 2.a, 3.b

▌ Quantitative Exercises

(Questions on p. 340.)

1. CO = HR × SV
    35L/min = HR × 0.07L
    HR = (35 L/min)/(0.07L) = 500 beats/min
    This rate is not physiologically possible.

2. ESV = EDV − SV
        = 125 ml − 85 ml
        = 40 ml

▌ Points to Ponder

(Questions on p. 341.)

1. Because, at a given heart rate, the interval between a premature beat and the next normal beat is longer than the interval between two normal beats, the heart fills for a longer period of time following a premature beat before the next period of contraction and emptying begins. Because of the longer filling time, the end-diastolic volume is larger, and, according to the Frank-Starling law of the heart, the subsequent stroke volume will also be correspondingly larger.

2. Trained athletes' hearts are stronger and can pump blood more efficiently so that the resting stroke volume is larger than in an untrained person. For example, if the resting stroke volume of a strong-hearted athlete is 100 ml, a resting heart rate of only 50 beats/minute produces a normal resting cardiac output of 5,000 ml/minute. An untrained individual with a resting stroke volume of 70 ml, in contrast, must have a heart rate of about 70 beats/minute to produce a comparable resting cardiac output.

3. The direction of flow through a patent ductus arteriosus is the reverse of the flow that occurs through this vascular connection during fetal life. With a patent ductus arteriosus, some of the blood present in the aorta is shunted into the pulmonary artery because, after birth, the aortic pressure is greater than the pulmonary artery pressure. This abnormal blood flow produces a "machinery murmur," which lasts throughout the cardiac cycle but is more intense during systole and less intense during diastole. Thus the murmur waxes and wanes with each beat of the heart, sounding somewhat like a washing machine as the agitator rotates back and forth. The murmur is present throughout the cardiac cycle because a pressure differential between the aorta and pulmonary artery is present during both systole and diastole. The murmur is more intense during systole because more blood is diverted through the patent ductus arteriosus as a result of the greater pressure differential between the aorta and pulmonary artery during ventricular systole than during ventricular diastole. Typically, the systolic aortic pressure is 120 mm Hg, and the systolic pulmonary arterial pressure is 24 mm Hg, for a pressure differential of 96 mm Hg. By contrast, the diastolic aortic pressure is normally 80 mm Hg, and the diastolic pulmonary arterial pressure is 8 mm Hg, for a pressure differential of 72 mm Hg.

4. A transplanted heart that does not have any innervation adjusts the cardiac output to meet the body's changing needs by means of both intrinsic control (the Frank-Starling mechanism) and extrinsic hormonal influences, such as the effect of epinephrine on the rate and strength of cardiac contraction.

5. In left bundle-branch block, the right ventricle becomes completely depolarized more rapidly than the left ventricle. As a result, the right ventricle contracts before the left ventricle, and the right AV valve is forced closed prior to closure of the left AV valve. Because the two AV valves do not close in unison, the first heart sound is "split"; that is, two distinct sounds in close succession can be detected as closure of the left valve lags behind closure of the right valve.

6. *Clinical Consideration.* The most likely diagnosis is atrial fibrillation. This condition is characterized by rapid, irregular, uncoordinated depolarizations of the atria. Many of these depolarizations reach the AV node at a time when it is not in its refractory period, thus bringing about frequent ventricular depolarizations and a rapid heartbeat. However, because impulses reach the AV node erratically, the ventricular rhythm and thus the heartbeat are also very irregular as well as being rapid.

Ventricular filling is only slightly reduced despite the fact that the fibrillating atria are unable to pump blood because most ventricular filling occurs during diastole prior to atrial contraction. Because of the erratic heartbeat, variable lengths of time are available between ventricular beats for ventricular filling. However, the majority of ventricular filling occurs early in ventricular diastole after the AV valves first open, so even though the filling period may be shortened, the extent of filling may be near normal. Only when the ventricular filling period is very short is ventricular filling substantially reduced.

Cardiac output, which depends on stroke volume and heart rate, usually is not seriously impaired with atrial fibrillation. Because ventricular filling is only slightly reduced during most cardiac cycles, stroke volume, as determined by the Frank-Starling mechanism, is likewise only slightly reduced. Only when the ventricular filling period is very short and the cardiac muscle fibers are operating on the lower end of their length-tension curve is the resultant ventricular contraction weak. When the ventricular contraction becomes too weak, the ventricles eject a small or no stroke volume. During most cardiac cycles, however, the slight reduction in stroke volume is often offset by the increased heart rate, so that cardiac output is usually near normal. Furthermore, if the mean arterial blood pressure falls because the cardiac output does decrease, increased sympathetic stimulation of the heart brought about by the baroreceptor reflex helps restore cardiac output to normal by shifting the Frank-Starling curve to the left.

On those cycles when ventricular contractions are too weak to eject enough blood to produce a palpable wrist pulse, if the heart rate is determined directly, either by the apex beat or via the ECG, and the pulse rate is taken concurrently at the wrist, the heart rate will exceed the pulse rate, producing a pulse deficit.

CHAPTER 10   THE BLOOD VESSELS AND BLOOD PRESSURE

▪ Objective Questions

(Questions on p. 387.)

1. T 2. F 3. T 4. T 5. F 6. T 7. a, c, d, e, f 8. 1.b, 2.a, 3.b, 4.a, 5.a, 6.a, 7.b, 8.a, 9.b, 10.a, 11.b, 12.a, 13.a 9. 1.a, 2.a, 3.b, 4.a, 5.b, 6.a

▪ Quantitative Exercises

(Questions on p. 388.)

1. (120 mm Hg)/(30 L/min) = 4 PRU
2. a. 90 mm Hg + (180 mm Hg − 90 mm Hg)/3 = 120 mm Hg
   b. Because the other forces acting across the capillary wall, such as plasma colloid osmotic pressure, typically do not change with age, one would suspect fluid loss from the capillaries into the tissues as a result of the increase in capillary blood pressure.
3. systemic: (95 mm Hg)/(19 PRU)
   = 95 mm Hg/(19 mm Hg/L/min) = 5 L/min
   pulmonary: (20 mm Hg)/(4 PRU) = 5 L/min
4. e.

▪ Points to Ponder

(Questions on p. 388.)

1. An elastic support stocking increases external pressure on the remaining veins in the limb to produce a favorable pressure gradient that promotes venous return to the heart and minimizes swelling that would result from fluid retention in the extremity.
2. a. 125 mm Hg
   b. 77 mm Hg

c. 48 mm Hg; (125 mm Hg − 77 mm = 48 mm Hg)
d. 93 mm Hg; [77 + ⅓(48) = 77 + 16 = 93 mm Hg]
e. No; no blood would be able to get through the brachial artery, so no sound would be heard.
f. Yes; blood would flow through the brachial artery when the arterial pressure was between 118 and 125 mm Hg and would not flow through when the arterial pressure fell below 118 mm Hg. The turbulence created by this intermittent blood flow would produce sounds.
g. No; blood would flow continuously through the brachial artery in smooth, laminar fashion, so no sound would be heard.

3. The classmate has apparently fainted because of insufficient blood flow to the brain as a result of pooling of blood in the lower extremities brought about by standing still for a prolonged time during the laboratory experiment. When the person faints and assumes a horizontal position, the pooled blood will quickly be returned to the heart, improving cardiac output and blood flow to his brain. Trying to get the person up would be counterproductive, so the classmate trying to get him up should be advised to let him remain lying down until he recovers on his own.

4. The drug is apparently causing the arteriolar smooth muscle to relax by causing the release of a local vasoactive chemical mediator from the endothelial cells that induces relaxation of the underlying smooth muscle.

5. a. Because activation of alpha-adrenergic receptors in vascular smooth muscle brings about vasoconstriction, blockage of alpha-adrenergic receptors reduces vasoconstrictor activity, thereby lowering the total peripheral resistance and arterial blood pressure.
   b. Because drugs that block beta-adrenergic receptors inhibit the pathway that promotes salt and water conservation (the renin-angiotensin-aldosterone system), less salt and water are retained, thus reducing the plasma volume and arterial blood pressure.
   c. Drugs that directly relax arteriolar smooth muscle lower arterial blood pressure by promoting arteriolar vasodilation and reducing total peripheral resistance.
   d. Diuretic drugs reduce the plasma volume, thereby lowering arterial blood pressure, by increasing urinary output. Salt and water that normally would have been retained in the plasma are excreted in the urine.
   e. Because sympathetic activity promotes generalized arteriolar vasoconstriction, thereby increasing total peripheral resistance and arterial blood pressure, drugs that block the release of norepinephrine from sympathetic endings lower blood pressure by preventing this sympathetic vasoconstrictor effect.
   f. Similarly, drugs that act on the brain to reduce sympathetic output lower blood pressure by preventing the effect of sympathetic activity on promoting arteriolar vasoconstriction and the resultant increase in total peripheral resistance and arterial blood pressure.
   g. Drugs that block Ca^{2+} channels reduce the entry of Ca^{2+} into the vascular smooth muscle cells from the ECF in response to excitatory input. Because the level of contractile activity in vascular smooth muscle cells depends

on their cytosolic Ca^{2+} concentration, drugs that block Ca^{2+} channels reduce the contractile activity of these cells by reducing Ca^{2+} entry and lowering their cytosolic Ca^{2+} concentration. Total peripheral resistance and, accordingly, arterial blood pressure are decreased as a result of reduced arteriolar contractile activity.

h.　Drugs that interfere with the production of angiotensin II block activation of the hormonal pathway that promotes salt and water conservation (the renin-angiotensin-aldosterone system). As a result, more salt and water are lost in the urine, and less fluid is retained in the plasma. The resultant reduction in plasma volume lowers the arterial blood pressure.

6. *Clinical Consideration.* The abnormally elevated levels of epinephrine found with a pheochromocytoma bring about secondary hypertension by (1) increasing the heart rate; (2) increasing cardiac contractility, which increases stroke volume; (3) causing venous vasoconstriction, which increases venous return and subsequently stroke volume by means of the Frank-Starling mechanism; and (4) causing arteriolar vasoconstriction, which increases total peripheral resistance. Increased heart rate and stroke volume both lead to increased cardiac output. Increased cardiac output and increased total peripheral resistance both lead to increased arterial blood pressure.

CHAPTER 11　THE BLOOD

▍Objective Questions

(Questions on p. 409.)

1. T 2. F 3. T 4. T 5. F 6. T 7. lymphocytes 8. liver 9. d 10. a 11. 1.c, 2.f, 3.b, 4.a, 5.g, 6.d, 7.h, 8.e, 9.f 12. 1.e, 2.c, 3.b, 4.d, 5.g, 6.f, 7.a, 8.h

▍Quantitative Exercises

(Questions on p. 410.)

1. a.　$(15 \text{ g})/(100 \text{ ml}) = (150 \text{ g/L})$
$(150 \text{ g/L}) \times (1 \text{ mole}/66 \times 10^3 \text{ g}) = 2.27 \text{ mM}$
   b.　$(2.27 \text{ mM}) \times (4 \text{ O}_2/\text{Hb}) = 9.09 \text{ mM}$
   c.　$(9.09 \times 10^{-3} \text{ moles O}_2/\text{L blood}) \times$
$(22.4 \text{ L O}_2/1 \text{ mole O}_2) = 204 \text{ ml O}_2/\text{L blood}$
2. Normal blood contains 5×10^9 RBCs/ml.
Normal blood volume is 5 L.
Thus a normal person has $(5 \times 10^9 \text{ RBCs/ml}) \times (5{,}000 \text{ ml})$ $= 25 \times 10^{12}$ RBCs.
The normal hematocrit (Ht) is 45%, whereas the anemic has a Ht of 30%. This represents a loss of $^1/_3$ of the RBCs, that is, 8.3×10^{12} RBCs. If RBCs are produced at a rate of 3×10^6 RBCs/sec, then the time to reestablish the Ht is 8.3×10^{12} RBCs$/(3 \times 10^6$ RBCs/sec$) = 2.77 \times 10^6$ sec $=$ 32 days.
Thus it takes about a month to replace a hemorrhagic loss of RBCs of this magnitude.
3. $v = 1.5 \times \exp(2h)$; calculate v for $h = 0.4$ and $h = 0.7$.

When $h = 0.4$, $v = 1.5 \times \exp(0.8) = 3.3$.
When $h = 0.7$, $v = 1.5 \times \exp(1.4) = 6.1$.
$6.1/3.3 = 1.84$, that is, an 84% increase in viscosity. Because resistance is directly proportional to viscosity, the resistance will also increase by 84%.

▍Points to Ponder

(Questions on p. 410.)

1.　No, you cannot conclude that a person with a hematocrit of 62 definitely has polycythemia. With 62% of the whole-blood sample consisting of erythrocytes (normal being 45%), the number of erythrocytes compared to the plasma volume is definitely elevated. However, the person *may* have polycythemia, in which the number of erythrocytes is abnormally high, or may be dehydrated, in which case a normal number of erythrocytes is concentrated in a smaller than normal plasma volume.
2.　If the genes that direct fetal hemoglobin-F synthesis could be reactivated in a patient with sickle cell anemia, a portion of the abnormal hemoglobin S that causes the erythrocytes to warp into defective sickle-shaped cells would be replaced by "healthy" hemoglobin F, thus sparing a portion of the RBCs from premature rupture. Hemoglobin F would not completely replace hemoglobin S because the gene for synthesis of hemoglobin S would still be active.
3.　Most heart-attack deaths are attributable to the formation of abnormal clots that prevent normal blood flow. The sought-after chemicals in the "saliva" of blood-sucking creatures are agents that break up or prevent the formation of these abnormal clots.

Although genetically engineered tissue-plasminogen activator (tPA) is already being used as a clot-busting drug, this agent brings about degradation of fibrinogen as well as fibrin. Thus, even though the life-threatening clot in the coronary circulation is dissolved, the fibrinogen supplies in the blood are depleted for up to 24 hours until new fibrinogen is synthesized by the liver. If the patient sustains a ruptured vessel in the interim, insufficient fibrinogen might be available to form a blood-staunching clot. For example, many patients treated with tPA suffer hemorrhagic strokes within 24 hours of treatment due to incomplete sealing of a ruptured cerebral vessel. Therefore, scientists are searching for better alternatives to combat abnormal clot formation by examining the naturally occurring chemicals produced by blood-sucking creatures that permit them to suck a victim's blood without the blood clotting.
4.　This question asks for your opinion, so there is no "right" answer. So far, the decision makers in our health-care system have felt that the additional financial burden of the more expensive testing of our nation's blood supply for HIV is not warranted.
5.　When considering the symptoms of porphyria, one could imagine how tales of vampires—blood-craving, hairy, fanged, monstrous-looking creatures who roamed in the dark and were warded off by garlic—might easily have evolved from people's encounters with victims of this condition. This possibility is especially likely when considering how stories are embellished and distorted as they get passed along by word of mouth.

6. *Clinical Consideration.* Because the white blood cell count is within the normal range, the patient's pneumonia is most likely not caused by a bacterial infection. Bacterial infections are typically accompanied by an elevated total white blood cell count and an increase in percentage of neutrophils. Therefore, the pneumonia is probably caused by a virus. Because antibiotics are more useful in combating bacterial than viral infections, antibiotics are not likely to be useful in combating this patient's pneumonia.

CHAPTER 12   THE BODY DEFENSES

■ Objective Questions

(Questions on p. 455.)

1. F **2.** F **3.** F **4.** F **5.** T **6.** toll-like receptors **7.** membrane-attack complex **8.** pus **9.** inflammation **10.** opsonin **11.** cytokines **12.** b **13.** 1.c, 2.d, 3.a, 4.b **14.** 1.a, 2.a, 3.b, 4.b, 5.c, 6.c, 7.b, 8.a, 9.b, 10.b, 11.a, 12.b **15.** 1.b, 2.a, 3.a, 4.b, 5.a, 6.a, 7.a, 8.b

■ Quantitative Exercises

(Questions on p. 456.)

1. NEP = net outward pressure − net inward pressure
$NEP = (P_C + \pi_{IF}) - (P_{IF} + \pi_P)$
Note for this problem, $(P_{IF} + \pi_P) = (25 \text{ mm Hg} + 1 \text{ mm Hg}) = 26 \text{ mm Hg}$, is constant for all cases.

Normal ($\pi_{IF} = 0$ mm Hg)	
Arteriolar end NEP	$(37 + 0) - 26 = + 11$ mm Hg
Venular end NEP	$(17 + 0) - 26 = - 9$ mm Hg
Average NEP	$(+11 - 9)/2 = + 1$ mm Hg (outward)
a. ($\pi_{IF} = 5$ mm Hg)	
Arteriolar end NEP	$(37 + 5) - 26 = + 16$ mmHg
Venular end NEP	$(17 + 5) - 26 = - 4$ mm Hg
Average NEP	$(+16 - 4)/2 = + 6$ mm Hg (outward)
Condition	mild edema
b. ($\pi_{IF} = 10$ mm Hg)	
Arteriolar end NEP	$(37 + 10) - 26 = + 21$ mm Hg
Venular end NEP	$(17 + 10) - 26 = + 1$ mm Hg
Average NEP	$(+ 21 + 1)/2 = + 11$ mm Hg (outward)
Condition	Extreme edema

■ Points to Ponder

(Questions on p. 456.)

1. See p. 435 for a summary of immune responses to bacterial invasion and p. 439 for a summary of defenses against viral invasion.

2. A vaccine against a particular microbe can be effective only if it induces formation of antibodies and/or activated T cells against a stable antigen that is present on all microbes of this type. Because HIV frequently mutates, it has not been possible to produce a vaccine against it. Specific immune responses induced by vaccination against one form of HIV may prove to be ineffective against a slightly modified version of the virus.

3. Failure of the thymus to develop embryonically would lead to an absence of T lymphocytes and no cell-mediated immunity after birth. This outcome would seriously compromise the individual's ability to defend against viral invasion and cancer.

4. Researchers are currently working on ways to "teach" the immune system to view foreign tissue as "self" as a means of preventing the immune systems of organ transplant patients from rejecting the foreign tissue while leaving the patients' immune defense capabilities fully intact. The immunosuppressive drugs currently being used to prevent transplant rejection cripple the recipients' immune defense systems and leave the patients more vulnerable to microbial invasion.

5. The skin cells visible on the body's surface are all dead.

6. *Clinical Consideration.* Heather's firstborn Rh-positive child did not have hemolytic disease of the newborn, because the fetal and maternal blood did not mix during gestation. Consequently, Heather did not produce any maternal antibodies against the fetus's Rh factor during gestation.

Because a small amount of the infant's blood likely entered the maternal circulation during the birthing process, Heather would produce antibodies against the Rh factor as she was first exposed to it at that time. During any subsequent pregnancies with Rh-positive fetuses, Heather's maternal antibodies against the Rh factor could cross the placental barrier and bring about destruction of fetal erythrocytes.

If, however, any Rh factor that accidentally mixed with the maternal blood during the birthing process were immediately tied up by Rh immunoglobulin administered to the mother, the Rh factor would not be available to induce maternal antibody production. Thus no anti-Rh antibodies would be present in the maternal blood to threaten the RBCs of an Rh-positive fetus in a subsequent pregnancy. (The exogenously administered Rh immunoglobulin, being a passive form of immunity, is short-lived. In contrast, the active immunity that would result if Heather were exposed to Rh factor would be long-lived because of the formation of memory cells.)

Rh immunoglobulin must be administered following the birth of every Rh-positive child Heather bears to sop up any Rh factor before it can induce antibody production. Once an immune attack against Rh factor is launched, subsequent treatment with Rh immunoglobulin will not reverse the situation. Thus if Heather were not treated with Rh immunoglobulin following the birth of a first Rh-positive child, and a second Rh-positive child developed hemolytic disease of the newborn, administration of Rh immunoglobulin following the second birth would not prevent the condition in a third Rh-positive child. Nothing could be done to eliminate the maternal antibodies already present.

CHAPTER 13 THE RESPIRATORY SYSTEM

▌Objective Questions

(Questions on p. 507.)

1. F 2. F 3. T 4. F 5. F 6. F 7. F 8. transmural pressure gradient, pulmonary surfactant action, alveolar interdependence 9. pulmonary elasticity, alveolar surface tension 10. compliance 11. elastic recoil 12. carbonic anhydrase 13. a 14. 1.d, 2.a, 3.b, 4.a, 5.b, 6.a 15. a.<, b.>, c. =, d. =, e. =, f. =, g.>, h.,<, i. approximately =, j. approximately =, k. =, l. =

▌Quantitative Exercises

(Questions on p. 508.)

For general reference for questions 1 and 2:

$$P_{AO_2} = P_{IO_2} - (V_{O_2}/V_A) \times 863 \text{ mm Hg}$$

$$P_{ACO_2} = (V_{CO_2}/V_A) \times 863 \text{ mm Hg}$$

1. $V_A = 3$ L/min
   $V_{O_2} = 0.3$ L/min, $RQ = 1$, therefore $V_{CO_2} = 0.3$ L/min
   $P_{ACO_2} = (0.3 \text{ L/min}/3 \text{ L/min}) \times 863 \text{ mm Hg} = 86.3$ mm Hg

2. a. 380 mm Hg $\times$ 0.21 = 79.8 mm Hg
   b. P_{AO_2} = 79.8 mm Hg $-$ (0.06) $\times$ 431.5 mm Hg
      = 79.8 mm Hg $-$ 25.8 mm Hg
      = 54 mm Hg
   c. P_{ACO_2} = (0.2 L/min/4.2 L/min) $\times$ 431.5 mm Hg
      = 20.5 mm Hg

3. $TV = 350$ ml, $BR = 12$/min, $V_A = 0.8 \times V_E$, $DS = ?$
   $V_A = BR \times (TV - DS)$
   $V_E = BR \times TV$
   $0.8 = V_A/V_E = [BR \times (TV - DS)]/(BR \times TV)$
   $= 1 - (DS/TV)$
   $0.8 = 1 - (DS/350 \text{ ml})$
   $DS/350 \text{ ml} = 0.2$
   $DS = 0.2(350 \text{ ml}) = 70$ ml

▌Points to Ponder

(Questions on p. 508.)

1. Total atmospheric pressure decreases with increasing altitude, yet the percentage of O_2 in the air remains the same. At an altitude of 30,000 feet, the atmospheric pressure is only 226 mm Hg. Because 21% of atmospheric air consists of O_2, the P_{O_2} of inspired air at 30,000 feet is only 47.5 mm Hg, and alveolar P_{O_2} is even lower at about 20 mm Hg. At this low P_{O_2}, hemoglobin is only about 30% saturated with O_2—much too low to sustain tissue needs for O_2.

The P_{O_2} of inspired air can be increased by two means when flying at high altitude. First, by pressurizing the plane's interior to a pressure comparable to that of atmospheric pressure at sea level, the P_{O_2} of inspired air within the plane is 21% of 760 mm Hg, or the normal 160 mm Hg. Accordingly, alveolar and arterial P_{O_2} and percent hemoglobin saturation are likewise normal. In the emergency situation of failure to maintain internal cabin pressure, breathing pure O_2 can raise the P_{O_2} considerably above that accomplished by breathing normal air. When a person is breathing pure O_2, the entire pressure of inspired air is attributable to O_2. For example, with a total atmospheric pressure of 226 mm Hg at an altitude of 30,000 feet, the P_{O_2} of inspired pure O_2 is 226 mm Hg, which is more than adequate to maintain normal arterial hemoglobin saturation.

2. a. Hypercapnia would not accompany the hypoxia associated with cyanide poisoning. In fact, CO_2 levels decline, because oxidative metabolism is blocked by the tissue poisons so that CO_2 is not being produced.
   b. Hypercapnia could but may not accompany the hypoxia associated with pulmonary edema. Pulmonary diffusing capacity is reduced in pulmonary edema, but O_2 transfer suffers more than CO_2 transfer because the diffusion coefficient for CO_2 is 20 times that for O_2. As a result, hypoxia occurs much more readily than hypercapnia in these circumstances. Hypercapnia does occur, however, when pulmonary diffusing capacity is severely impaired.
   c. Hypercapnia would accompany the hypoxia associated with restrictive lung disease because ventilation is inadequate to meet the metabolic needs for both O_2 delivery and CO_2 removal. Both O_2 and CO_2 exchange between the lungs and atmosphere are equally affected.
   d. Hypercapnia would not accompany the hypoxia associated with high altitude. In fact, arterial P_{CO_2} levels actually decrease. One of the compensatory responses in acclimatization to high altitudes is reflex stimulation of ventilation as a result of the reduction in arterial P_{O_2}. This compensatory hyperventilation to obtain more O_2 blows off too much CO_2 in the process, so arterial P_{CO_2} levels decline below normal.
   e. Hypercapnia would not accompany the hypoxia associated with severe anemia. Reduced O_2-carrying capacity of the blood has no influence on blood CO_2 content, so arterial P_{CO_2} levels are normal.
   f. Hypercapnia would exist in accompaniment with circulatory hypoxia associated with congestive heart failure. Just as the diminished blood flow fails to deliver adequate O_2 to the tissues, it also fails to remove sufficient CO_2.
   g. Hypercapnia would accompany the hypoxic hypoxia associated with obstructive lung disease because ventilation would be inadequate to meet the metabolic needs for both O_2 delivery and CO_2 removal. Both O_2 and CO_2 ex-

change between the lungs and atmosphere would be equally affected.

3. $P_{O_2} = 122$ mm Hg
   0.21 (atmospheric pressure − partial pressure of H_2O)
   = 0.21(630 mm Hg − 47 mm Hg)
   = 0.21(583 mm Hg) = 122 mm Hg

4. Voluntarily hyperventilating before going underwater lowers the arterial P_{CO_2} but does not increase the O_2 content in the blood. Because the P_{CO_2} is below normal, the person can hold his or her breath longer than usual before the arterial P_{CO_2} increases to the point that he or she is driven to surface for a breath. Therefore, the person can stay underwater longer. The risk, however, is that the O_2 content of the blood, which was normal, not increased, before going underwater, continues to fall. Therefore, the O_2 level in the blood can fall dangerously low before the CO_2 level builds to the point of driving the person to take a breath. Low arterial P_{O_2} does not stimulate respiratory activity until it has plummeted to 60 mm Hg. Meanwhile, the person may lose consciousness and drown due to inadequate O_2 delivery to the brain. If the person does not hyperventilate so that both the arterial P_{CO_2} and O_2 content are normal before going underwater, the buildup of CO_2 will drive the person to the surface for a breath before the O_2 levels fall to a dangerous point.

5. c. The arterial P_{O_2} will be less than the alveolar P_{O_2}, and the arterial P_{CO_2} will be greater than the alveolar P_{CO_2}. Because pulmonary diffusing capacity is reduced, arterial P_{O_2} and P_{CO_2} do not equilibrate with alveolar P_{O_2} and P_{CO_2}.

If the person is administered 100% O_2, the alveolar P_{O_2} will increase, and the arterial P_{O_2} will increase accordingly. Even though arterial P_{O_2} will not equilibrate with alveolar P_{O_2}, it will be higher than when the person is breathing atmospheric air.

The arterial P_{CO_2} will remain the same whether the person is administered 100% O_2 or is breathing atmospheric air. The alveolar P_{CO_2} and thus the blood-to-alveolar P_{CO_2} gradient are not changed by breathing 100% O_2 because the P_{CO_2} in atmospheric air and 100% O_2 are both essentially zero (P_{CO_2} in atmospheric air = 0.03 mm Hg).

6. *Clinical Consideration.* Emphysema is characterized by a collapse of smaller respiratory airways and a breakdown of alveolar walls. Because of the collapse of smaller airways, airway resistance is increased with emphysema. As with other chronic obstructive pulmonary diseases, expiration is impaired to a greater extent than inspiration because airways are naturally dilated slightly more during inspiration than expiration as a result of the greater transmural pressure gradient during inspiration. Because airway resistance is increased, a patient with emphysema must produce larger-than-normal intra-alveolar pressure changes to accomplish a normal tidal volume. Unlike quiet breathing in a normal person, the accessory inspiratory muscles (neck muscles) and the muscles of active expiration (abdominal muscles and internal intercostal muscles) must be brought into play to inspire and expire a normal tidal volume of air.

The spirogram would be characteristic of chronic obstructive pulmonary disease. Because the patient experiences more difficulty emptying the lungs than filling them, the total lung capacity would be essentially normal, but the functional residual capacity and the residual volume would be elevated as a result of the additional air trapped in the lungs following expiration. Because the residual volume is increased, the inspiratory capacity and vital capacity will be reduced. Also, the FEV_1 will be markedly reduced because the airflow rate is decreased by the airway obstruction. The FEV_1-to-vital capacity ratio will be much lower than the normal 80%.

Because of the reduced surface area for exchange as a result of a breakdown of alveolar walls, gas exchange would be impaired. Therefore, arterial P_{CO_2} would be elevated and arterial P_{O_2} reduced compared to normal.

Ironically, administering O_2 to this patient to relieve his hypoxic condition would markedly depress his drive to breathe by elevating the arterial P_{O_2} and removing the primary driving stimulus for respiration. Because of this danger, O_2 therapy should either not be administered or administered extremely cautiously.

CHAPTER 14   THE URINARY SYSTEM

▌ Objective Questions

(Questions on p. 555.)

1. F 2. F 3. T 4. T 5. T 6. nephron 7. potassium 8. 500 9. 1.b, 2.a, 3.b, 4.b, 5.a, 6.b, 7.b, 8.b, 9.b 10. e 11. b 12. b, e, a, d, c 13. c, e, d, a, b, f 14. g, c, d, a, f, b, e 15. 1.a, 2.a, 3.c, 4.b, 5.d

▌ Quantitative Exercises

(Questions on p. 556.)

1.

	Patient 1	Patient 2
GFR	125 ml/min	124 ml/min
RPF	620 ml/min	400 ml/min
RBF	1,127 ml/min	727 ml/min
FF	0.20	0.31

All of patient 1's values are within the normal range. Patient 2's GFR is normal, but he has a low renal plasma flow and a high filtration fraction. Therefore, his GFR is too high for that RPF. This might imply enlarged filtration slits or a "leaky" glomerulus in general. The low RPF may imply low renal blood pressure, perhaps from a partially blocked renal artery.

2. filtered load = GFR × plasma concentration
   (0.125 liter/min) × (145 mmol/liter)
   = 18.125 mmol/min

3. $GFR = (U \times [I]_U)/[I]_B$
   $U = (GFR \times [I]_B)/[I]_U = (125$ ml/min)(3 mg/L)/(300 mg/L) = 1.25 ml/min

4.

$$\text{Clearance rate of a substance} = \frac{\text{urine concentration of a substance} \times \text{urine flow rate}}{\text{plasma concentration of the substance}}$$

$$= \frac{7.5 \text{ mg/ml} \times 2 \text{ ml/min}}{0.2 \text{ mg/ml}}$$

$$= 75 \text{ ml/min}$$

Because a clearance rate of 75 ml/min is less than the average GFR of 125 ml/min, the substance is being reabsorbed.

▌ Points to Ponder

(Questions on p. 556.)

1. The longer loops of Henle in desert rats (known as *kangaroo rats*) permit a greater magnitude of countercurrent multiplication and thus a larger medullary vertical osmotic gradient. As a result, these rodents can produce urine that is concentrated up to an osmolarity of almost 6,000 mosm/liter, which is five times more concentrated than maximally concentrated human urine at 1,200 mosm/liter. Because of this tremendous concentrating ability, kangaroo rats never have to drink; the H_2O produced metabolically within their cells during oxidation of foodstuff (food + O_2 yields CO_2 + H_2O + energy) is sufficient for their needs.

2. a. 250 mg/min filtered

filtered load of substance = plasma concentration of substance × GFR

filtered load of substance = 200 mg/100 ml × 125 ml = 250 mg/min

b. 200 mg/min reabsorbed

A T_m's worth of the substance will be reabsorbed.

c. 50 mg/min excreted

amount of substance excreted = amount of substance filtered − amount of substance reabsorbed = 250 mg/min − 200 mg/min = 50 mg/min

3. Aldosterone stimulates Na^+ reabsorption and K^+ secretion by the renal tubules. Therefore, the most prominent features of Conn's syndrome (hypersecretion of aldosterone) are hypernatremia (elevated Na^+ levels in the blood) caused by excessive Na^+ reabsorption, hypophosphatemia (below-normal K^+ levels in the blood) caused by excessive K^+ secretion, and hypertension (elevated blood pressure) caused by excessive salt and water retention.

4. e. 300/300. If the ascending limb were permeable to water, it would not be possible to establish a vertical osmotic gradient in the interstitial fluid of the renal medulla, nor would the ascending-limb fluid become hypotonic before entering the distal tubule. As the ascending limb pumped NaCl into the interstitial fluid, water would osmotically follow, so both the interstitial fluid and the ascending limb would remain isotonic at 300 mosm/liter. With the tubular fluid entering the distal tubule being 300 mosm/liter instead of the normal 100 mosm/liter, it would not be possible to produce urine with an osmolarity less than 300 mosm/liter. Likewise, in the absence of the medullary vertical osmotic gradient, it would not be possible to produce urine more concentrated than 300 mosm/liter, no matter how much vasopressin was present.

5. Because the descending pathways between the brain and the motor neurons supplying the external urethral sphincter and pelvic diaphragm are no longer intact, the accident victim can no longer voluntarily control micturition. Therefore, bladder emptying in this individual will be governed entirely by the micturition reflex.

6. *Clinical Consideration.* prostate enlargement

CHAPTER 15 FLUID AND ACID–BASE BALANCE

▌ Objective Questions

(Questions on p. 588.)

1. T **2.** F **3.** F **4.** T **5.** T **6.** intracellular fluid **7.** $[H_2CO_3]$, $[HCO_3^-]$ **8.** d **9.** b **10.** a, d, e **11.** b, e **12.** c **13.** 1. metabolic acidosis, 2. diabetes mellitus, 3. pH = 7.1, 4. respiratory alkalosis, 5. anxiety, 6. pH = 7.7, 7. respiratory acidosis, 8. pneumonia, 9. pH = 7.1, 10. metabolic alkalosis, 11. vomiting, 12. pH = 7.7

▌ Quantitative Exercises

(Questions on p. 589.)

1. pH = 6.1 + log $[HCO_3^-]/(0.03 \text{ mM/mm Hg} \times 40 \text{ mm Hg})$
7.4 = 6.1 + log $[HCO_3^-]/1.2 \text{ mM}$
log $[HCO_3^-]/1.2 \text{ mM}$ = 7.4 − 6.1 = 1.3
$[HCO_3^-]$ = 1.2 mM × $(10^{1.3})$ = 24 mM

2. pH = −log $[H^+]$, $[H^+]$ = 10^{-pH}
$[H^+]$ = $10^{-6.8}$ = 158 nM for pH = 6.8
$[H^+]$ = $10^{-8.0}$ = 10 nM for pH = 8.0

3. Note that distilled water is permeable across all barriers, so it will distribute equally among all compartments. However, the saline does not enter cells, so it will stay in the ECF. The resultant distributions are summarized in the chart on p. A-48. Clearly, saline is better at expanding the plasma volume.

▌ Points to Ponder

(Questions on p. 589.)

1. The rate of urine formation increases when alcohol inhibits vasopressin secretion and the kidneys are unable to reabsorb water from the distal and collecting tubules. Because extra free water that normally would have been reabsorbed from the distal parts of the tubule is lost from the body in the urine, the body becomes dehydrated and the ECF osmolarity increases following alcohol consumption. That is, more fluid is lost in the urine than is consumed in the alcoholic beverage as a result of alcohol's action on vasopressin. Thus the imbibing person experiences a water deficit and still feels thirsty, despite the recent fluid consumption.

Ingested Fluid	Compartment	Size of Compartment before Ingestion	Size of Compartment after Ingestion	% Increase in Size of Compartment after Ingestion
Distilled water	TBW	42 L	43 L	2%
	ICF (2/3 TBW)	28 L	28.667 L	2%
	ECF (1/3 TBW)	14 L	14.333 L	2%
	plasma (20% ECF)	2.8 L	2.866 L	2%
	ISF (80% ECF)	11.2 L	11.466 L	2%
Saline	TBW	42 L	43 L	2%
	ICF	28 L	28 L	0%
	ECF	14 L	15 L	7%
	plasma	2.8 L	3 L	7%
	ISF	11.2	12 L	7%

2. If a person loses 1,500 ml of salt-rich sweat and drinks 1,000 ml of water without replacing the salt during the same time period, there will still be a volume deficit of 500 ml, and the body fluids will have become hypotonic (the remaining salt in the body will be diluted by the ingestion of 1,000 ml of free H_2O). As a result, the hypothalamic osmoreceptors (the dominant input) will signal the vasopressin-secreting cells to *decrease* vasopressin secretion and thus increase urinary excretion of the extra free water that is making the body fluids too dilute. Simultaneously, the left atrial volume receptors will signal the vasopressin-secreting cells to *increase* vasopressin secretion to conserve water during urine formation and thus help to relieve the volume deficit. These two conflicting inputs to the vasopressin-secreting cells are counterproductive. This is why it is important to replace both water and salt following heavy sweating or abnormal loss of other salt-rich fluids. If salt is replaced along with water intake, the ECF osmolarity remains close to normal, and the vasopressin-secreting cells receive signals only to increase vasopressin secretion to help restore the ECF volume to normal.

3. When a dextrose solution equal in concentration to that of normal body fluids is injected intravenously, the ECF volume is expanded but the ECF and ICF are still osmotically equal. Therefore, no net movement of water occurs between the ECF and ICF. When the dextrose enters the cell and is metabolized, however, the ECF becomes hypotonic as this solute leaves the plasma. If the excess free water is not excreted in the urine rapidly enough, water will move into the cells by osmosis.

4. Because baking soda ($NaHCO_3$) is readily absorbed from the digestive tract, treatment of gastric hyperacidity with baking soda can lead to metabolic alkalosis as too much HCO_3^- is absorbed. Treatment with antacids that are poorly absorbed is safer because these products remain in the digestive tract and do not produce an acid–base imbalance.

5. c. The hemoglobin buffer system buffers carbonic-acid–generated hydrogen ion. In the case of respiratory acidosis accompanying severe pneumonia, the $H^+ + Hb \rightarrow HHb$ reaction will be shifted toward the HHb side, thus removing some of the extra free H^+ from the blood.

6. *Clinical Consideration.* The resultant prolonged diarrhea will lead to dehydration and metabolic acidosis due, respectively, to excessive loss in the feces of fluid and $NaHCO_3$ that normally would have been absorbed into the blood.

Compensatory measures for dehydration have included increased vasopressin secretion, resulting in increased water reabsorption by the distal and collecting tubules and a subsequent reduction in urine output. Simultaneously, fluid intake has been encouraged by increased thirst. The metabolic acidosis has been combatted by removal of excess H^+ from the ECF by the HCO_3^- member of the H_2CO_3:HCO_3^- buffer system, by increased ventilation to reduce the amount of acid-forming CO_2 in the body fluids, and by the kidneys excreting extra H^+ and conserving HCO_3^-.

CHAPTER 16   THE DIGESTIVE SYSTEM

▌ Objective Questions

(Questions on p. 643.)

1. F 2. T 3. T 4. F 5. F 6. F 7. long, short 8. chyme 9. three 10. vitamin B_{12}, bile salts 11. bile salts 12. b 13. 1.c, 2.e, 3.b, 4.a, 5.f, 6.d 14. 1.c, 2.c, 3.d, 4.a, 5.e, 6.c, 7.a, 8.b, 9.c, 10.b, 11.d, 12.b

▌ Quantitative Exercises

(Questions on p. 644.)

1. a. $r = 0.5$ cm, area $= 4\pi (0.25) = \pi$ cm^2
   volume $= (\frac{4}{3}\pi)(0.5)^3 = 0.5236$ cm^3
   area/volume $= 6$

   b. The new spheres now have a volume of 5.236×10^{-3} cm^3, so the average radius is therefore 0.1077 cm. The area of a sphere with that radius is $4\pi(0.1077$ cm$)^2 = 0.1458$ cm^2
   area/volume $= 27.8$

   c. Area emulsified/area droplet $= (100)(0.1458$ cm$^2)/\pi$ cm$^2 = 4.64$.

   Thus the total surface area of all 100 emulsified droplets is 4.64 times the area of the original larger lipid droplet.

d. Volume emulsified/volume droplet = $(100)(5.236 \times 10^{-3}\text{ cm}^3)/0.5236\text{ cm}^3 = 1.0$.

Thus the total volume did not change as a result of emulsification, as would be expected because the total volume of the lipid is conserved during emulsification. The volume originally present in the large droplet is divided up among the 100 emulsified droplets.

Points to Ponder

(Questions on p. 644.)

1. Patients who have had their stomachs removed must eat small quantities of food frequently instead of consuming the typical three meals a day because they have lost the ability to store food in the stomach and meter it into the small intestine at an optimal rate. If a person without a stomach consumed a large meal that entered the small intestine all at once, the luminal contents would quickly become too hypertonic as digestion of the large nutrient molecules into a multitude of small, osmotically active, absorbable units outpaced the more slowly acting process of absorption of these units. As a consequence of this increased luminal osmolarity, water would enter the small intestine lumen from the plasma by osmosis, resulting in circulatory disturbances as well as intestinal distension. To prevent this "dumping syndrome" from occurring, the patient must "feed" the small intestine only small amounts of food at a time so that absorption of the digestive end products can keep pace with their rate of production. The person has to consciously take over metering the delivery of food into the small intestine because the stomach is no longer present to assume this responsibility.

2. The gut-associated lymphoid tissue launches an immune attack against any pathogenic (disease-causing) microorganisms that enter the readily accessible digestive tract and escape destruction by salivary lysozyme or gastric HCl. This action defends against entry of these potential pathogens into the body proper. The large number of immune cells in the gut-associated lymphoid tissue is adaptive as a first line of defense against foreign invasion when considering that the surface area of the digestive tract lining represents the largest interface between the body proper and the external environment.

3. Defecation would be accomplished entirely by the defecation reflex in a patient paralyzed from the waist down because of lower spinal cord injury. Voluntary control of the external anal sphincter would be impossible because of interruption in the descending pathway between the primary motor cortex and the motor neuron supplying this sphincter.

4. When insufficient glucuronyl transferase is available in the neonate to conjugate all of the bilirubin produced during erythrocyte degradation with glycuronic acid, the extra unconjugated bilirubin cannot be excreted into the bile. Therefore, this extra bilirubin remains in the body, giving rise to mild jaundice in the newborn.

5. Removal of the stomach leads to pernicious anemia because of the resultant lack of intrinsic factor, which is necessary for absorption of vitamin B_{12}. Removal of the terminal ileum leads to pernicious anemia because this is the only site where vitamin B_{12} can be absorbed.

6. **Clinical Consideration.** A person whose bile duct is blocked by a gallstone experiences a painful "gallbladder attack" after eating a high-fat meal because the ingested fat triggers the release of cholecystokinin, which stimulates gallbladder contraction. As the gallbladder contracts and bile is squeezed into the blocked bile duct, the duct becomes distended prior to the blockage. This distension is painful.

The feces are grayish white because no bilirubin-containing bile enters the digestive tract when the bile duct is blocked. Bilirubin, when acted on by bacterial enzymes, is responsible for the brown color of feces, which are grayish white in its absence.

CHAPTER 17 ENERGY BALANCE AND TEMPERATURE REGULATION

Objective Questions

(Questions on p. 664)

1. F 2. F 3. F 4. F 5. T 6. T 7. T 8. T 9. shivering 10. nonshivering thermogenesis 11. sweating 12. b 13. e 14. 1.b, 2.a, 3.c, 4.a, 5.d, 6.c, 7.b, 8.d, 9.c, 10.b

Quantitative Exercises

(Questions on p. 665.)

1. From physics we know that $\Delta T(°C) = \Delta U/(C \times m)$.

Also note that $\Delta U/t = BMR$; i.e., the rate of using energy is the basal metabolic rate. m represents the mass of body fluid; for a typical person, this is 42 L.

$(42\text{ L}) \times (1\text{ kg/L}) = 42\text{ kg}$
$C = 1.0\text{ kcal/(kg}-°\text{C)}$

Given that water boils at 100°C and normal body temperature is 37°C, we need to change the temperature by 63°C. Thus

$t = (\Delta T \times C \times m)/BMR = (63°C)[1.0\text{ kcal/(kg}-°\text{C)}]$
$(42\text{ kg})/(75\text{ kcal/hr}) = 35\text{ hr}$

At the higher metabolic rate during exercise,
$t = (63°C)[1.0\text{ kcal/(kg}-°\text{C)}](42\text{ kg})/(1{,}000\text{ kcal/hr}) = 2.6\text{ hr}$

Points to Ponder

(Questions on p. 665.)

1. Evidence suggests that CCK serves as a satiety signal. It is believed to serve as a signal to stop eating when enough food has been consumed to meet the body's energy needs, even though the food is still in the digestive tract. Therefore, when drugs that inhibit CCK release are administered to experimental animals, the animals overeat because this satiety signal is not released.

2. Don't go on a "crash diet." Be sure to eat a nutritionally balanced diet that provides all essential nutrients, but reduce total caloric intake, especially by cutting down on high-fat foods. Spread out consumption of the food throughout the day instead

of just eating several large meals. Avoid bedtime snacks. Burn more calories through a regular exercise program.

3. Engaging in heavy exercise on a hot day is dangerous because of problems arising from trying to eliminate the extra heat generated by the exercising muscles. First, there will be conflicting demands for distribution of the cardiac output—temperature-regulating mechanisms will trigger skin vasodilation to promote heat loss from the skin surface, whereas metabolic changes within the exercising muscles will induce local vasodilation in the muscles to match the increased metabolic needs with increased blood flow. Further exacerbating the problem of conflicting demands for blood flow is the loss of effective circulating plasma volume resulting from the loss of a large volume of fluid through another important cooling mechanism, sweating. Therefore, it is difficult to maintain an effective plasma volume and blood pressure and simultaneously keep the body from overheating when engaging in heavy exercise in the heat, so heat exhaustion is likely to ensue.

4. When a person is soaking in a hot bath, loss of heat by radiation, conduction, convection, and evaporation is limited to the small surface area of the body exposed to the cooler air. Heat is being gained by conduction at the larger skin surface area exposed to the hotter water.

5. The thermoconforming fish would not run a fever when it has a systemic infection because it has no mechanisms for regulating internal heat production or for controlling heat exchange with its environment. The fish's body temperature varies capriciously with the external environment no matter whether it has a systemic infection or not. It is not able to maintain body temperature at a "normal" set point or an elevated set point (i.e., a fever).

6. *Clinical Consideration.* Cooled tissues need less nourishment than they do at normal body temperature because of their pronounced reduction in metabolic activity. The lower O_2 need of cooled tissues accounts for the occasional survival of drowning victims who have been submerged in icy water considerably longer than one could normally survive without O_2.

CHAPTER 18   PRINCIPLES OF ENDOCRINOLOGY; THE CENTRAL ENDOCRINE GLANDS

▌ Objective Questions

(Questions on p. 698.)

1. T 2. T 3. F 4. T 5. F 6. T 7. F 8. tropic 9. down regulation 10. suprachiasmatic nucleus 11. epiphyseal plate 12. 1.c, 2.b, 3.b, 4.a, 5.a, 6.c, 7.c

▌ Points to Ponder

(Questions on p. 699.)

1. It is not advisable to rotate the nursing staff on different shifts every week because such a practice would disrupt the in-dividuals' natural circadian rhythms. Such disruption can affect physical and psychological health and have a negative impact on work performance.

2. The concentration of hypothalamic releasing and inhibiting hormones would be considerably lower (in fact, almost nonexistent) in a systemic venous blood sample compared to the concentration of these hormones in a sample of hypothalamic-hypophyseal portal blood. These hormones are secreted into the portal blood for local delivery between the hypothalamus and anterior pituitary. Any portion of these hormones picked up by the systemic blood at the anterior pituitary capillary level is greatly diluted by the much larger total volume of systemic blood compared to the extremely small volume of blood within the portal vessel.

3. If CRH and/or ACTH is elevated in accompaniment with the excess cortisol secretion, the condition is secondary to a defect at the hypothalamic/anterior pituitary level. If CRH and ACTH levels are below normal in accompaniment with the excess cortisol secretion, the condition is due to a primary defect at the adrenal cortex level, with the excess cortisol inhibiting the hypothalamus and anterior pituitary in negative-feedback fashion.

4. Males with testicular feminization syndrome would be unusually tall because of the inability of testosterone to promote closure of the epiphyseal plates of the long bones in the absence of testosterone receptors.

5. Full-grown athletes sometimes illegally take supplemental doses of growth hormone because it promotes increased skeletal muscle mass through its protein anabolic effect. However, excessive growth hormone can have detrimental side effects, such as possibly causing diabetes or high blood pressure.

6. *Clinical Consideration.* Hormonal replacement therapy following pituitary gland removal should include thyroid hormone (the thyroid gland will not produce sufficient thyroid hormone in the absence of TSH) and glucocorticoid (because of the absence of ACTH), especially in stress situations. If indicated, male or female sex hormones can be replaced, even though these hormones are not essential for survival. For example, testosterone in males plays an important role in libido. Growth hormone and prolactin need not be replaced because their absence will produce no serious consequences in this individual. Vasopressin may have to be replaced if insufficient quantities of this hormone are picked up by the blood at the hypothalamus in the absence of the posterior pituitary.

CHAPTER 19   THE PERIPHERAL ENDOCRINE ORGANS

▌ Objective Questions

(Questions on p. 745.)

1. F 2. T 3. T 4. T 5. T 6. F 7. F 8. F 9. colloid, thyroglobulin 10. pro-opiomelanocortin 11. glycogenesis, glycogenolysis, gluconeogenesis 12. brain, working muscles, liver 13. bone, kidneys, digestive tract 14. c 15. a, b, c, g, i, j

16. 1. glucose, 2. glycogen, 3. free fatty acids, 4. triglycerides, 5. amino acids, 6. body proteins

▌ Points to Ponder

(Questions on p. 746.)

1. The midwestern United States is no longer an endemic goiter belt even though the soil is still iodine poor, because individuals living in this region obtain iodine from iodine-supplemented nutrients, such as iodinated salt, and from seafood and other naturally iodine-rich foods shipped from coastal regions.

2. Anaphylactic shock is an extremely serious allergic reaction brought about by massive release of chemical mediators in response to exposure to a specific allergen—such as one associated with a bee sting—to which the individual has been highly sensitized. These chemical mediators bring about circulatory shock (severe hypotension) through a twofold effect: (1) by relaxing arteriolar smooth muscle, thus causing widespread arteriolar vasodilation and a resultant fall in total peripheral resistance and arterial blood pressure; and (2) by causing a generalized increase in capillary permeability, resulting in a shift of fluid from the plasma into the interstitial fluid. This shift decreases the effective circulating volume, further reducing arterial blood pressure. Additionally, these chemical mediators bring about pronounced bronchoconstriction, making it impossible for the victim to move sufficient air through the narrowed airways. Because these responses take place rapidly and can be fatal, people allergic to bee stings are advised to keep injectable epinephrine in their possession. By promoting arteriolar vasoconstriction through its action on α_1 receptors in arteriolar smooth muscle and promoting bronchodilation through its action on β_2 receptors in bronchiolar smooth muscle (see ▲ Table 7-3, p. 242), epinephrine counteracts the life-threatening effects of the anaphylactic reaction to the bee sting. Waiting to receive attention from medical personnel is likely to be too late.

3. An infection elicits the stress response, which brings about increased secretion of cortisol and epinephrine, both of which increase the blood glucose level. This can become a problem in managing the blood glucose level of a diabetic patient. When the blood glucose is elevated too high, the patient can reduce it by injecting additional insulin, or, preferably, by reducing carbohydrate intake and/or exercising to use up some of the extra blood glucose. In a normal individual, the check-and-balance system between insulin and the other hormones that oppose insulin's actions helps to maintain the blood glucose within reasonable limits during the stress response.

4. The presence of Chvostek's sign is due to increased neuromuscular excitability caused by moderate hyposecretion of parathyroid hormone.

5. If malignancy-associated hypercalcemia arose from metastatic tumor cells that invaded and destroyed bone, both hypercalcemia and hyperphosphatemia would result as calcium phosphate salts were released from the destructed bone. The fact that hypophosphatemia, not hyperphosphatemia, often ac-

companies malignancy-associated hypercalcemia led investigators to rule out bone destruction as the cause of the hypercalcemia. Instead, they suspected that the tumors produced a substance that mimics the actions of PTH in promoting concurrent hypercalcemia and hypophosphatemia.

6. *Clinical Consideration.* "Diabetes of bearded ladies" is descriptive of both excess cortisol and excess adrenal androgen secretion. Excess cortisol secretion causes hyperglycemia and glucosuria. Glucosuria promotes osmotic diuresis, which leads to dehydration and a compensatory increased sensation of thirst. All these symptoms—hyperglycemia, glucosuria, polyuria, and polydipsia—mimic diabetes mellitus. Excess adrenal androgen secretion in females promotes masculinizing characteristics, such as beard growth. Simultaneous hypersecretion of both cortisol and adrenal androgen most likely occurs secondary to excess CRH/ACTH secretion, because ACTH stimulates both cortisol and androgen production by the adrenal cortex.

CHAPTER 20   THE REPRODUCTIVE SYSTEM

▌ Objective Questions

(Questions on p. 801.)

1. T **2.** T **3.** F **4.** F **5.** T **6.** F **7.** T **8.** seminiferous tubules, FSH, testosterone **9.** thecal, LH, granulosa, FSH **10.** corpus luteum of pregnancy **11.** human chorionic gonadotropin **12.** c **13.** e **14.** 1.c, 2.a, 3.b, 4.c, 5.a, 6.a, 7.c, 8.b, 9.e **15.** 1.a, 2.c, 3.b, 4.a, 5.a, 6.b

▌ Points to Ponder

(Questions on p. 802.)

1. The anterior pituitary responds only to the normal pulsatile pattern of GnRH and does not secrete gonadotropins in response to continuous exposure to GnRH. In the absence of FSH and LH secretion, ovulation and other events of the ovarian cycle do not ensue, so continuous GnRH administration may find use as a contraceptive technique.

2. Testosterone hypersecretion in a young boy causes premature closure of the epiphyseal plates so that he stops growing before he reaches his genetic potential for height. The child would also display signs of precocious pseudopuberty, characterized by premature development of secondary sexual characteristics, such as deep voice, beard, enlarged penis, and sex drive.

3. A potentially troublesome side effect of drugs that inhibit sympathetic nervous system activity as part of the treatment for high blood pressure is males' inability to carry out the sex act. Both divisions of the autonomic nervous system are required for the male sex act. Parasympathetic activity is essential for accomplishing erection, and sympathetic activity is important for ejaculation.

4. Posterior pituitary extract contains an abundance of stored oxytocin, which can be administered to induce or facilitate labor by increasing uterine contractility. Exogenous oxytocin is most successful in inducing labor if the woman is near term, presumably because of the increasing concentration of myometrial oxytocin receptors at that time.

5. GnRH or FSH and LH are not effective in treating the symptoms of menopause because the ovaries are no longer responsive to the gonadotropins. Thus treatment with these hormones would not cause estrogen and progesterone secretion. In fact, GnRH, FSH, and LH levels are already elevated in postmenopausal women because of lack of negative feedback by the ovarian hormones.

6. *Clinical Consideration.* The first warning of a tubal pregnancy is pain caused by stretching of the oviduct by the growing embryo. A tubal pregnancy must be surgically terminated because the oviduct cannot expand as the uterus does to accommodate the growing embryo. If not removed, the enlarging embryo will rupture the oviduct, causing possibly lethal hemorrhage.

Text References to Exercise Physiology

A CLOSER LOOK AT EXERCISE PHYSIOLOGY BOXED FEATURES BY CHAPTER

Chapter 1	Homeostasis: The Foundation of Physiology	
	What Is Exercise Physiology? p. 13	
Chapter 2	Cellular Physiology	
	Aerobic Exercise: What For and How Much? p. 42	
Chapter 3	The Plasma Membrane and Membrane Potential.	
	Exercising Muscles Have a "Sweet Tooth" p. 81	
Chapter 5	The Central Nervous System	
	Swan Dive or Belly Flop: It's a Matter of CNS Control p. 177	
Chapter 6	The Peripheral Nervous System: Afferent Division; Special Senses	
	Back Swings and Prejump Crouches: What Do They Share in Common? p. 186	
Chapter 7	The Peripheral Nervous System: Efferent Division	
	Loss of Muscle Mass: A Plight of Spaceflight p. 250	
Chapter 8	Muscle Physiology	
	Are Athletes Who Use Steroids to Gain Competitive Advantage Really Winners or Losers? p. 282	
Chapter 9	Cardiac Physiology	
	The What, Who, and When of Stress Testing p. 321	
Chapter 10	The Blood Vessels and Blood Pressure	
	The Ups and Downs of Hypertension and Exercise p. 382	
Chapter 11	The Blood	
	Blood Doping: Is More of a Good Thing Better? p. 396	
Chapter 12	The Body Defenses	
	Exercise: A Help or Hindrance to Immune Defense p. 446	
Chapter 13	The Respiratory System	
	How to Find Out How Much Work You're Capable of Doing p. 504	
Chapter 14	The Urinary System	
	When Protein in the Urine Does Not Mean Kidney Disease p. 550	
Chapter 15	Fluid and Acid–Base Balance	
	A Potentially Fatal Clash: When Exercising Muscles and Cooling Mechanisms Compete for an Inadequate Plasma Volume p. 566	
Chapter 16	The Digestive System	
	Pregame Meal: What's In and What's Out? p. 607	
Chapter 17	Energy Balance and Temperature Regulation	
	What the Scales Don't Tell You p. 654	
Chapter 18	Principles of Endocrinology; The Central Endocrine Glands	
	The Endocrine Response to the Challenge of Combined Heat and Marching Feet p. 685	
Chapter 19	The Peripheral Endocrine Glands	
	Osteoporosis: The Bane of Brittle Bones pp. 736–737	
Chapter 20	The Reproductive System	
	Menstrual Irregularities: When Cyclists and Other Female Athletes Do Not Cycle p. 780	

EXERCISE REFERENCES BY TOPIC

Exercise

and acclimatization to hot environment   566

   aerobic versus anaerobic   42, 277, 279

   and atherosclerosis   336

   and athletic pseudonephritis   550

   and blood doping   396

   cardiovascular responses in   13, 235, 329, 330, 333, 335, 353, 354f, 358, 379, 380t, 382

   changes during   13, 502–503

   control of ventilation during   502–503

   distribution of cardiac output during   353–354, 354f

   effect of,

      on body temperature   503, 655, 661

      on bone density   736, 737

      on blood pressure   379, 380, 382

      on development of collateral circulation in the heart   335

      on endorphins   194

      on fat content in body   654

      on glomerular filtration   522

      on glucose uptake by muscles during exercise   81, 724, 728

      on growth hormone secretion   693

      on heart   13, 325, 329, 330, 333, 335, 354f, 379, 380t

      on immune defense   446

      on intrapleural pressure   473

      on kidney function   522, 550

      on management of diabetes mellitus   81, 724, 728

      on menstrual cycles   780

      on metabolic rate   649

      on muscle mass   250, 281

      on oxygen release from hemoglobin   490, 491

      on plasma glucose levels   81, 607, 728

      on plasma insulin levels   81

      on plasma volume   566, 685

      on pulmonary surface area   486

      on receptor sites for insulin   81, 724

      on respiratory system   469, 473, 477, 485, 486, 498, 502–503

      on time for gas exchange in lungs   487

      on venous return   329, 372, 374

      on ventilation   502–503

      on work of breathing   477

   and energy expenditure   649, 654

   and glucose transporter recruitment   724

   in hot environment   566, 685

   heat production during   275, 503, 566, 649, 655, 661, 685

   and high-density lipoproteins   336

   and hypertension   382

   hyperthermia in   275, 655, 661, 662f, 685

   and lactic acid production   279, 574, 585

   and maximal oxygen consumption   504

   and metabolic acidosis   585

   muscle adaptation to   281

   and muscle fatigue   279

   and muscle soreness   279

   and obesity   654

   oxygen availability during   477, 486, 490, 491, 502–503

   P_{O_2}, P_{CO_2}, and H^+ during   495, 502–503

   and pregame meal   607

   "runner's high"   194

   and stress tests   321

   and sweat rate   566

   and temperature regulation   275, 566, 649, 655, 661, 685

   and weight loss   654

Exercise physiology   13

Glossary

A band One of the dark bands that alternate with light (I) bands to create a striated appearance in a skeletal or cardiac muscle fiber when these fibers are viewed with a light microscope

absorptive state The metabolic state following a meal when nutrients are being absorbed and stored; fed state

accessory digestive organs Exocrine organs outside the wall of the digestive tract that empty their secretions through ducts into the digestive tract lumen

accessory sex glands Glands that empty their secretions into the reproductive tract

accommodation The ability to adjust the strength of the lens in the eye so that both near and far sources can be focused on the retina

acetylcholine (ACh) (as´-uh-teal-KŌ-lēn) The neurotransmitter released from all autonomic preganglionic fibers, parasympathetic postganglionic fibers, and motor neurons

acetylcholinesterase (AChE) (as´uh-teal-kō-luh-NES-tuh-rās) An enzyme present in the motor end-plate membrane of a skeletal muscle fiber that inactivates acetylcholine

Ach See *acetylcholine*

AchE See *acetylcholinesterase*

acid A hydrogen-containing substance that yields a free hydrogen ion and anion on dissociation

acidosis (as-i-DŌ-sus) Blood pH of less than 7.35

acini (ĀS-i-īn) The secretory component of saclike exocrine glands, such as digestive enzyme-producing pancreatic glands or milk-producing mammary glands

ACTH See *adrenocorticotropic hormone*

acquired immune responses Responses that are selectively targeted against particular foreign material to which the body has previously been exposed; see also *antibody-mediated immunity* and *cell-mediated immunity*

actin The contractile protein that forms the backbone of the thin filaments in muscle fibers

active expiration Emptying of the lungs more completely than when at rest by contracting the expiratory muscles; also called *forced expiration*

active force A force that requires expenditure of cellular energy (ATP) in the transport of a substance across the plasma membrane

active reabsorption When any one of the five steps in the transepithelial transport of a substance reabsorbed across the kidney tubules requires energy expenditure

active transport Active carrier-mediated transport involving transport of a substance against its concentration gradient across the plasma membrane

acuity Discriminative ability; the ability to discern between two different points of stimulation

acute myocardial infarction (mī´-ō-KAR-dē-ul) Death of heart muscle cells caused by disruption of their blood supply; a heart attack

adaptation A reduction in receptor potential despite sustained stimulation of the same magnitude

adenosine diphosphate (ADP) (uh-DEN-uh-sēn) The two-phosphate product formed from the splitting of ATP to yield energy for the cell's use

adenosine triphosphate (ATP) The body's common energy "currency," which consists of an adenosine with three phosphate groups attached; splitting of the high-energy, terminal phosphate bond provides energy to power cellular activities

adenylyl cyclase (ah-DEN-il-il sī-klās) The membrane-bound enzyme that is activated by a G protein intermediary in response to binding of an extracellular messenger with a surface membrane receptor and that in turn activates cyclic AMP, an intracellular second messenger

ADH See *vasopressin*

adipose tissue The tissue specialized for storage of triglyceride fat; found under the skin in the hypodermis

ADP See *adenosine diphosphate*

adrenal cortex (uh-DRĒ-nul) The outer portion of the adrenal gland; secretes three classes of steroid hormones: glucocorticoids, mineralcorticoids, and sex hormones

adrenal medulla (muh-DUL-uh) The inner portion of the adrenal gland; an endocrine gland that is a modified sympathetic ganglion that secretes the hormones epinephrine and norepinephrine into the blood in response to sympathetic stimulation

adrenergic fibers (ad´-ruh-NUR-jik) Nerve fibers that release norepinephrine as their neurotransmitter

adrenocorticotropic hormone (ACTH) (ad-rē´-nō-kor´-tuh-kō-TRŌP-ik) An anterior pituitary hormone that stimulates cortisol secretion by the adrenal cortex and promotes growth of the adrenal cortex

aerobic Referring to a condition in which oxygen is available

aerobic exercise Exercise that can be supported by ATP formation accomplished by oxidative phosphorylation because adequate O_2 is available to support the muscle's modest energy demands; also called *endurance-type exercise*

afferent arteriole (AF-er-ent ar-TIR-ē-ōl) The vessel that carries blood into the glomerulus of the kidney's nephron

afferent division The portion of the peripheral nervous system that carries information from the periphery to the central nervous system

afferent neuron Neuron that possesses a sensory receptor at its peripheral ending and carries information to the central nervous system

after hyperpolarization (hī´-pur-pō-luh-ruh-ZA-shun) A slight, transient hyperpolarization that sometimes occurs at the end of an action potential

agranulocytes (ā-GRAN-yuh-lō-sīts´) Leukocytes that do not contain granules, including lymphocytes and monocytes

albumin (al-BEW-min) The smallest and most abundant of the plasma proteins; binds and transports many water-insoluble substances in the blood; contributes extensively to plasma-colloid osmotic pressure

aldosterone (al-dō-steer-OWN) or (al-DOS-tuh-rōn)   The adrenocortical hormone that stimulates Na^+ reabsorption by the distal and collecting tubules of the kidney's nephron during urine formation

alkalosis (al´-kuh-LŌ-sus)   Blood pH of greater than 7.45

allergy   Acquisition of an inappropriate specific immune reactivity to a normally harmless environmental substance

all-or-none law   An excitable membrane either responds to a stimulus with a maximal action potential that spreads nondecrementally throughout the membrane or does not respond with an action potential at all

alpha (α) cells   The endocrine pancreatic cells that secrete the hormone glucagon

alpha motor neuron   A motor neuron that innervates ordinary skeletal muscle fibers

alveolar surface tension (al-VĒ-ō-lur)   The surface tension of the fluid lining the alveoli in the lungs; see *surface tension*

alveolar ventilation   The volume of air exchanged between the atmosphere and alveoli per minute; equals (tidal volume minus dead space volume) times respiratory rate

alveoli (al-VĒ-ō-lī)   The air sacs across which O_2 and CO_2 are exchanged between the blood and air in the lungs

amines (ah-means)   Hormones derived from the amino acid tyrosine; includes thyroid hormone and catecholamines

amoeboid movement (uh-MĒ-boid)   "Crawling" movement of white blood cells, similar to the means by which amoebas move

anabolism (ah-NAB-ō-li-zum)   The buildup, or synthesis, of larger organic molecules from the small organic molecular subunits

anaerobic (an´-uh-RŌ-bik)   Referring to a condition in which oxygen is not present

anaerobic exercise   High-intensity exercise that can be supported by ATP formation accomplished by anaerobic glycolysis for brief periods of time when O_2 delivery to a muscle is inadequate to support oxidative phosphorylation

analgesic (an-al-JEE-zic)   Pain relieving

anatomy   The study of body structure

androgen   A masculinizing "male" sex hormone; includes testosterone from the testes and dehydroepiandrosterone from the adrenal cortex

anemia   A reduction below normal in O_2-carrying capacity of the blood

anion (AN-ī-on)   Negatively charged ion that has gained one or more electrons in its outer shell

ANP   See *atrial natriuretic peptide*

antagonism   Actions opposing each other; in the case of hormones, when one hormone causes the loss of another hormone's receptors, reducing the effectiveness of the second hormone

anterior pituitary   The glandular portion of the pituitary that synthesizes, stores and secretes six different hormones: growth hormone, TSH, ACTH, FSH, LH, and prolactin

antibody   An immunoglobulin produced by a specific activated B lymphocyte (plasma cell) against a particular antigen; binds with the specific antigen against which it is produced and promotes the antigenic invader's destruction by augmenting nonspecific immune responses already initiated against the antigen

antibody-mediated immunity   A specific immune response accomplished by antibody production by B cells

antidiuretic hormone (an´-ti-dī´-yū-RET-ik)   See *vasopressin*

antigen   A large, complex molecule that triggers a specific immune response against itself when it gains entry into the body

antioxidant   A substance that helps inactivate biologically damaging free radicals

antrum (of ovary)   The fluid-filled cavity formed within a developing ovarian follicle

antrum (of stomach)   The lower portion of the stomach

aorta (a-OR-tah)   The large vessel that carries blood from the left ventricle

aortic valve   A one-way valve that permits the flow of blood from the left ventricle into the aorta during ventricular emptying but prevents the backflow of blood from the aorta into the left ventricle during ventricular relaxation

apoptosis (ā-pop-TŌ-sis)   Programmed cell death; deliberate self-destruction of a cell

appetite centers   See *feeding centers*

aqueous humor (Ā-kwē-us)   The clear watery fluid in the anterior chamber of the eye; provides nourishment for the cornea and lens

arterioles (ar-TIR-ē-ōlz)   The highly muscular, high-resistance vessels, the caliber of which can be changed subject to control to determine how much of the cardiac output is distributed to each of the various tissues

artery   A vessel that carries blood away from the heart

ascending tract   A bundle of nerve fibers of similar function that travels up the spinal cord to transmit signals derived from afferent input to the brain

asthma   An obstructive pulmonary disease characterized by profound constriction of the smaller airways caused by allergy-induced spasm of the smooth muscle in the walls of these airways

astrocyte   A type of glial cell in the brain; major functions include holding the neurons together in proper spatial relationship and inducing the brain capillaries to form tight junctions important in the blood–brain barrier

atherosclerosis (ath-uh-rō-skluh-RŌ-sus)   A progressive, degenerative arterial disease that leads to gradual blockage of affected vessels, reducing blood flow through them

atmospheric pressure   The pressure exerted by the weight of the air in the atmosphere on objects on Earth's surface; equals 760 mm Hg at sea level

ATP   See *adenosine triphosphate*

ATPase   An enzyme that possesses ATP-splitting ability

ATP synthase (SIN-thās)   The enzyme within the mitochondrial inner membrane that phosphorylates ADP to ATP

atrial natriuretic peptide (ANP) (Ā-trē-al NĀ-tree-ur-eh´tik)   A peptide hormone released from the cardiac atria that promotes urinary loss of Na^+

atrioventricular (AV) node (ā´-trē-ō-ven-TRIK-yuh-lur)   A small bundle of specialized cardiac cells located at the junction of the atria and ventricles that serves as the only site of electrical contact between the atria and ventricles

atrioventricular (AV) valve   A one-way valve that permits the flow of blood from the atrium to the ventricle during filling of the heart but prevents the backflow of blood from the ventricle to the atrium during emptying of the heart

atrium (atria, plural) (Ā-tree-um)   An upper chamber of the heart that receives blood from the veins and transfers it to the ventricle

atrophy (AH-truh-fē)   Decrease in mass of an organ

autoimmune disease   Disease characterized by erroneous production of antibodies against one of the body's own tissues

autonomic nervous system   The portion of the efferent division of the peripheral nervous system that innervates smooth and cardiac muscle and exocrine glands; composed of two subdivisions, the sympathetic nervous system and the parasympathetic nervous system

autorhythmicity   The ability of an excitable cell to rhythmically initiate its own action potentials

AV nodal delay   The delay in impulse transmission between the atria and ventricles at the AV node to allow sufficient time for the atria to become completely depolarized and contract, emptying their contents into the ventricles, before ventricular depolarization and contraction occur

AV valve See *atrioventricular valve*

axon A single, elongated tubular extension of a neuron that conducts action potentials away from the cell body; also known as a *nerve fiber*

axon hillock The first portion of a neuronal axon plus the region of the cell body from which the axon leaves; the site of action-potential initiation in most neurons

axon terminals The branched endings of a neuronal axon, which release a neurotransmitter that influences target cells in close association with the axon terminals

baroreceptor reflex An autonomically mediated reflex response that influences the heart and blood vessels to oppose a change in mean arterial blood pressure

baroreceptors Receptors located within the circulatory system that monitor blood pressure

basal metabolic rate (BĀ-sul) The minimal waking rate of internal energy expenditure; the body's "idling speed"

basal nuclei Several masses of gray matter located deep within the white matter of the cerebrum of the brain; play an important inhibitory role in motor control

base A substance that can combine with a free hydrogen ion and remove it from solution

basic electrical rhythm (BER) Self-induced electrical activity of the digestive-tract smooth muscle

basilar membrane (BAS-ih-lar) The membrane that forms the floor of the middle compartment of the cochlea and bears the organ of Corti, the sense organ for hearing

basophils (BAY-so-fills) White blood cells that synthesize, store, and release histamine, which is important in allergic responses, and heparin, which hastens the removal of fat particles from the blood

BER See *basic electrical rhythm*

beta (β) cells The endocrine pancreatic cells that secrete the hormone insulin

bicarbonate (HCO_3^-) The anion resulting from dissociation of carbonic acid, H_2CO_3

bile salts Cholesterol derivatives secreted in the bile that facilitate fat digestion through their detergent action and facilitate fat absorption through their micellar formation

biliary system (BIL-ē-air´-ē) The bile-producing system, consisting of the liver, gallbladder, and associated ducts

bilirubin (bill-eh-RŪ-bin) A bile pigment, which is a waste product derived from the degradation of hemoglobin during the breakdown of old red blood cells

blastocyst The developmental stage of the fertilized ovum by the time it is ready to implant; consists of a single-layered sphere of cells encircling a fluid-filled cavity

blood–brain barrier Special structural and functional features of the brain capillaries that limit access of materials from the blood into the brain tissue

B lymphocytes (B cells) White blood cells that produce antibodies against specific targets to which they have been exposed

body of the stomach The main, or middle, part of the stomach

body system A collection of organs that perform related functions and interact to accomplish a common activity that is essential for survival of the whole body; for example, the digestive system

bone marrow The soft, highly cellular tissue that fills the internal cavities of bones and is the source of most blood cells

Bowman's capsule The beginning of the tubular component of the kidney's nephron that cups around the glomerulus and collects the glomerular filtrate as it is formed

Boyle's law (boils) At any constant temperature, the pressure exerted by a gas varies inversely with the volume of the gas

brain The most anterior, most highly developed portion of the central nervous system

brain stem The portion of the brain that is continuous with the spinal cord, serves as an integrating link between the spinal cord and higher brain levels, and controls many life-sustaining processes, such as breathing, circulation, and digestion

bronchioles (BRONG-kē-ōlz) The small, branching airways within the lungs

bronchoconstriction Narrowing of the respiratory airways

bronchodilation Widening of the respiratory airways

brush border The collection of microvilli projecting from the luminal border of epithelial cells lining the digestive tract and kidney tubules

buffer See *chemical buffer system*

bulbourethral glands (bul-bo-you-RĒTH-ral) Male accessory sex glands that secrete mucus for lubrication

bulk flow Movement in bulk of a protein-free plasma across the capillary walls between the blood and surrounding interstitial fluid; encompasses ultrafiltration and reabsorption

bundle of His (hiss) A tract of specialized cardiac cells that rapidly transmits an action potential down the interventricular septum of the heart

calcitonin (kal´-suh-TŌ-nun) A hormone secreted by the thyroid C cells that lowers plasma Ca^{2+} levels

calcium balance Maintenance of a constant total amount of Ca^{2+} in the body; accomplished by slowly responding adjustments in intestinal Ca^{2+} absorption and in urinary Ca^{2+} excretion

calcium homeostasis Maintenance of a constant free plasma Ca^{2+} concentration; accomplished by rapid exchanges of Ca^{2+} between the bone and ECF and to a lesser extent by modifications in urinary Ca^{2+} excretion

calmodulin (kal´-MA-jew-lin) An intracellular Ca^{2+} binding protein that, on activation by Ca^{2+}, induces a change in structure and function of another intracellular protein; especially important in smooth-muscle excitation–contraction coupling

CAMs See *cell adhesion molecules*

capillaries The thin-walled, pore-lined smallest of blood vessels, across which exchange between the blood and surrounding tissues takes place

carbonic anhydrase (an-HĪ-drās) The enzyme that catalyzes the conversion of CO_2 and H_2O into carbonic acid, H_2CO_3

cardiac cycle One period of systole and diastole

cardiac muscle The specialized muscle found only in the heart

cardiac output (CO) The volume of blood pumped by each ventricle each minute; equals stroke volume times heart rate

cardiovascular control center The integrating center located in the medulla of the brain stem that controls mean arterial blood pressure

carrier-mediated transport Transport of a substance across the plasma membrane facilitated by a carrier molecule

carrier molecules Membrane proteins, which, by undergoing reversible changes in shape so that specific binding sites are alternately exposed at either side of the membrane, are able to bind with and transfer particular substances unable to cross the plasma membrane on their own

cascade A series of sequential reactions that culminates in a final product, such as a clot

catabolism (kuh-TAB-ō-li-zum) The breakdown, or degradation, of large, energy-rich molecules within cells

catalase (KAT-ah-lās) An antioxidant enzyme found in peroxisomes that decomposes potent hydrogen peroxide into harmless H_2O and O_2

catecholamines (kat´-uh-KŌ-luh-means) The chemical classification of the adrenomedullary hormones

cations (KAT-ī-onz) Positively charged ions that have lost one or more electrons from their outer shell

caveolae (kā-vē-Ō-lē)   Cavelike indentations in the outer surface of the plasma membrane that contain an abundance of membrane receptors and serve as important sites for signal transduction

C cells   The thyroid cells that secrete calcitonin

cell   The smallest unit capable of carrying out the processes associated with life; the basic unit of both structure and function of living organisms

cell adhesion molecules (CAMs)   proteins that protrude from the surface of the plasma membrane and form loops or other appendages that the cells use to grip each other and the surrounding connective-tissue fibers

cell body   The portion of a neuron that houses the nucleus and organelles

cell-mediated immunity   A specific immune response accomplished by activated T lymphocytes, which directly attack unwanted cells

center   A functional collection of cell bodies within the central nervous system

central chemoreceptors (kē-mō-rē-SEP-turz)   Receptors located in the medulla near the respiratory center that respond to changes in ECF H^+ concentration resulting from changes in arterial P_{CO_2} and adjust respiration accordingly

central lacteal (LAK-tē-ul)   The initial lymphatic vessel that supplies each of the small intestinal villi

central nervous system (CNS)   The brain and spinal cord

central sulcus (SUL-kus)   A deep infolding of the brain surface that runs roughly down the middle of the lateral surface of each cerebral hemisphere and separates the parietal and frontal lobes

centrioles (SEN-tree-ōls)   A pair of short cylindrical structures within a cell that form the mitotic spindle during cell division

cerebellum (ser´-uh-BEL-um)   The portion of the brain attached at the rear of the brain stem and concerned with maintaining proper position of the body in space and subconscious coordination of motor activity

cerebral cortex   The outer shell of gray matter in the cerebrum; site of initiation of all voluntary motor output and final perceptual processing of all sensory input as well as integration of most higher neural activity

cerebral hemispheres   The cerebrum's two halves, which are connected by a thick band of neuronal axons

cerebrospinal fluid (ser´-uh-brō-SPĪ-nul) or (sah-REE-brō-SPĪ-nul)   A special cushioning fluid that is produced by, surrounds, and flows through the central nervous system

cerebrum (SER-uh-brum) or (sah-REE-brum)   The division of the brain that consists of the basal nuclei and cerebral cortex

channels   Small, water-filled passageways through the plasma membrane; formed by membrane proteins that span the membrane and provide highly selective passage for small water-soluble substances such as ions

chemical bonds   The forces holding atoms together

chemical buffer system   A mixture in a solution of two or more chemical compounds that minimize pH changes when either an acid or a base is added to or removed from the solution

chemical mediator   A chemical that is secreted by a cell and that influences an activity outside of the cell

chemically gated channels   Channels in the plasma membrane that open or close in response to the binding of a specific chemical messenger with a membrane receptor site that is in close association with the channel

chemoreceptor (kē-mo-rē-sep´-tur)   A sensory receptor sensitive to specific chemicals

chemotaxin (kē-mō-TAK-sin)   A chemical released at an inflammatory site that attracts phagocytes to the area

chief cells   The stomach cells that secrete pepsinogen

cholecystokinin (CCK) (kō´-luh-sis-tuh-kī-nun)   A hormone released from the duodenal mucosa primarily in response to the presence of fat; inhibits gastric motility and secretion, stimulates pancreatic enzyme secretion, and stimulates gallbladder contraction

cholesterol   a type of fat molecule that serves as a precursor for steroid hormones and bile salts and is a stabilizing component of the plasma membrane

cholinergic fibers (kō-lin-ER-jik)   Nerve fibers that release acetylcholine as their neurotransmitter

chronic obstructive pulmonary disease   A group of lung diseases characterized by increased airway resistance resulting from narrowing of the lumen of the lower airways; includes asthma, chronic bronchitis, and emphysema

chyme (kīm)   A thick liquid mixture of food and digestive juices

cilia (SILL-ee-ah)   Motile, hairlike protrusions from the surface of cells lining the respiratory airways and the oviducts

ciliary body   The portion of the eye that produces aqueous humor and contains the ciliary muscle

ciliary muscle   A circular ring of smooth muscle within the eye whose contraction increases the strength of the lens to accommodate for near vision

circadian rhythm (sir-KĀ-dē-un)   Repetitive oscillations in the set point of various body activities, such as hormone levels and body temperature, that are very regular and have a frequency of one cycle every 24 hours, usually linked to light–dark cycles; diurnal rhythm; biological rhythm

circulatory shock   When mean arterial blood pressure falls so low that adequate blood flow to the tissues can no longer be maintained

citric acid cycle   A cyclical series of biochemical reactions that involves the further processing of intermediate breakdown products of nutrient molecules, resulting in the generation of carbon dioxide and the preparation of hydrogen carrier molecules for entry into the high–energy-yielding electron transport chain

CNS   See *central nervous system*

cochlea(KOK-lē-uh)   The snail-shaped portion of the inner ear that houses the receptors for sound

collecting tubule   The last portion of tubule in the kidney's nephron that empties into the renal pelvis

colloid (KOL-oid)   The thyroglobulin-containing substance enclosed within the thyroid follicles

complement system   A collection of plasma proteins that are activated in cascade fashion on exposure to invading micro-organisms, ultimately producing a membrane attack complex that destroys the invaders

compliance   The distensibility of a hollow, elastic structure, such as a blood vessel or the lungs; a measure of how easily the structure can be stretched

concave   Curved in, as a surface of a lens that diverges light rays

concentration gradient   A difference in concentration of a particular substance between two adjacent areas

conduction   Transfer of heat between objects of differing temperatures that are in direct contact with each other

cones   The eye's photoreceptors used for color vision in the light

congestive heart failure   The inability of the cardiac output to keep pace with the body's needs for blood delivery, with blood damming up in the veins behind the failing heart

connective tissue   Tissue that serves to connect, support, and anchor various body parts; distinguished by relatively few cells dispersed within an abundance of extracellular material

contiguous conduction   The means by which an action potential is propagated throughout a nonmyelinated nerve fiber; local current flow between an active and adjacent inactive area brings the inactive area to threshold, triggering an action potential in a previously inactive area

contractile component   The sarcomere-containing myofibrils within a muscle fiber that are capable of shortening on excitation

contractile proteins   Myosin and actin, whose interaction brings about shortening (contraction) of a muscle fiber

controlled variable   Some factor that can vary but is controlled to reduce the amount of variability and keep the factor at a relatively steady state

convection   Transfer of heat energy by air or water currents

convergence   The converging of many presynaptic terminals from thousands of other neurons on a single neuronal cell body and its dendrites so that activity in the single neuron is influenced by the activity in many other neurons

convex   Curved out, as a surface in a lens that converges light rays

core temperature   The temperature within the inner core of the body (abdominal and thoracic organs, central nervous system, and skeletal muscles) that is homeostatically maintained at about 100°F

cornea (KOR-nee-ah)   The clear, anteriormost outer layer of the eye through which light rays pass to the interior of the eye

coronary artery disease   Atherosclerotic plaque formation and narrowing of the coronary arteries that supply the heart muscle

coronary circulation   The blood vessels that supply the heart muscle

corpus luteum (LOO-tē-um)   The ovarian structure that develops from a ruptured follicle following ovulation

cortisol (KORT-uh-sol)   The adrenocortical hormone that plays an important role in carbohydrate, protein, and fat metabolism and helps the body resist stress

cranial nerves   The 12 pairs of peripheral nerves, the majority of which arise from the brain stem

cross bridges   The myosin molecules' globular heads that protrude from a thick filament within a muscle fiber and interact with the actin molecules in the thin filaments to bring about shortening of the muscle fiber during contraction

cyclic adenosine monophosphate (cyclic AMP or cAMP)   An intracellular second messenger derived from adenosine triphosphate (ATP)

cyclic AMP   See *cyclic adenosine monophosphate*

cytokines   All chemicals other than antibodies that are secreted by lymphocytes

cytoplasm (SĪ-tō-plaz´-um)   The portion of the cell interior not occupied by the nucleus

cytoskeleton   A complex intracellular protein network that acts as the "bone and muscle" of the cell

cytosol (SĪ-tuh-sol´)   The semiliquid portion of the cytoplasm not occupied by organelles

cytotoxic T cells (sī-tō-TOK-sik)   The population of T cells that destroys host cells bearing foreign antigen, such as body cells invaded by viruses or cancer cells

dead-space volume   The volume of air that occupies the respiratory airways as air is moved in and out and that is not available to participate in exchange of O_2 and CO_2 between the alveoli and atmosphere

dehydration   A water deficit in the body

dehydroepiandrosterone (DHEA) (dē-HĪ-drō-ep-i-and-row-steer-own)   The androgen (masculinizing hormone) secreted by the adrenal cortex in both sexes

dendrites   Projections from the surface of a neuron's cell body that carry signals toward the cell body

deoxyribonucleic acid (DNA) (dē-OK-sē-rī-bō-new-klā-ik)   The cell's genetic material, which is found within the nucleus and which provides codes for protein synthesis and serves as a blueprint for cell replication

depolarization (de´-pō-luh-ruh-ZĀ-shun)   A reduction in membrane potential from resting potential; movement of the potential from resting toward 0 mV

dermis   The connective tissue layer that lies under the epidermis in the skin; contains the skin's blood vessels and nerves

descending tract   A bundle of nerve fibers of similar function that travels down the spinal cord to relay messages from the brain to efferent neurons

desmosome (dez´-muh-sōm)   An adhering junction between two adjacent but nontouching cells formed by the extension of filaments between the cells' plasma membranes; most abundant in tissues that are subject to considerable stretching

DHEA   See *dehydroepiandrosterone*

diabetes insipidus (in-SIP´-ud-us)   An endocrine disorder characterized by a deficiency of vasopressin

diabetes mellitus (muh-LĪ-tus)   An endocrine disorder characterized by inadequate insulin action

diaphragm (DIE-uh-fram)   A dome-shaped sheet of skeletal muscle that forms the floor of the thoracic cavity; the major inspiratory muscle

diastole (d-AS-tō-lē)   The period of cardiac relaxation and filling

diencephalon (dī´-un-SEF-uh-lan)   The division of the brain that consists of the thalamus and hypothalamus

diffusion   Random collisions and intermingling of molecules as a result of their continuous thermally induced random motion

digestion   The breaking-down process whereby the structurally complex foodstuffs of the diet are converted into smaller absorbable units by the enzymes produced within the digestive system

diploid number (DIP-loid)   A complete set of 46 chromosomes (23 pairs), as found in all human somatic cells

distal tubule   A highly convoluted tubule that extends between the loop of Henle and the collecting duct in the kidney's nephron

diurnal rhythm (dī-URN´-ul)   Repetitive oscillations in hormone levels that are very regular and have a frequency of one cycle every 24 hours, usually linked to the light–dark cycle; circadian rhythm; biological rhythm

divergence   The diverging, or branching, of a neuron's axon terminals, so that activity in this single neuron influences the many other cells with which its terminals synapse

DNA   See *deoxyribonucleic acid*

dorsal root ganglion   A cluster of afferent neuronal cell bodies located adjacent to the spinal cord

down regulation   A reduction in the number of receptors for (and thereby the target cells' sensitivity to) a particular hormone as a direct result of the effect that an elevated level of the hormone has on its own receptors

ECG   See *electrocardiogram*

edema (i-DĒ-muh)   Swelling of tissues as a result of excess interstitial fluid

EDV   See *end-diastolic volume*

EEG   See *electroencephalogram*

effector organs   The muscles or glands that are innervated by the nervous system and that carry out the nervous system's orders to bring about a desired effect, such as a particular movement or secretion

efferent division (EF-er-ent)   The portion of the peripheral nervous system that carries instructions from the central nervous system to effector organs

efferent neuron   Neuron that carries information from the central nervous system to an effector organ

efflux (Ē-flux)   Movement out of the cell

elastic recoil   Rebound of the lungs after having been stretched

electrical gradient   A difference in charge between two adjacent areas

electrocardiogram (ECG)   The graphic record of the electrical activity that reaches the surface of the body as a result of cardiac depolarization and repolarization

electrochemical gradient The simultaneous existence of an electrical gradient and concentration (chemical) gradient for a particular ion

electroencephalogram (EEG) (i-lek´-trō-in-SEF-uh-luh-gram´) A graphic record of the collective postsynaptic potential activity in the cell bodies and dendrites located in the cortical layers under a recording electrode

electrolytes Solutes that form ions in solution and conduct electricity

embolus (EM-bō-lus) A freely floating clot

emphysema (em´-fuh-ZĒ-muh) A pulmonary disease characterized by collapse of the smaller airways and a breakdown of alveolar walls

end-diastolic volume (EDV) The volume of blood in the ventricle at the end of diastole, when filling is complete

endocrine glands Ductless glands that secrete hormones into the blood

endocytosis (en´-dō-sī-TŌ-sis) Internalization of extracellular material within a cell as a result of the plasma membrane forming a pouch that contains the extracellular material, then sealing at the surface of the pouch to form a small, intracellular, membrane-enclosed vesicle with the contents of the pouch trapped inside

endogenous opiates (en-DAJ´-eh-nus ō´-pē-ātz) Endorphins and enkephalins, which bind with opiate receptors and are important in the body's natural analgesic system

endogenous pyrogen (pī´-ruh-jun) A chemical released from macrophages during inflammation that acts by means of local prostaglandins to raise the set point of the hypothalamic thermostat to produce a fever

endometrium (en´-dō-MĒ-trē-um) The lining of the uterus

endoplasmic reticulum (en´-dō-PLAZ-mik ri-TIK-yuh-lum) An organelle consisting of a continuous membranous network of fluid-filled tubules and flattened sacs, partially studded with ribosomes; synthesizes proteins and lipids for formation of new cell membrane and other cell components and manufactures products for secretion

endothelium (en´-dō-THĒ-lē-um) The thin, single-celled layer of epithelial cells that lines the entire circulatory system

end-plate potential (EPP) The graded receptor potential that occurs at the motor end plate of a skeletal muscle fiber in response to binding with acetylcholine

end-systolic volume (ESV) The volume of blood in the ventricle at the end of systole, when emptying is complete

endurance-type exercise See *aerobic exercise*

enterogastrones (ent´-uh-rō-GAS-trōnz) Hormones secreted by the duodenal mucosa that inhibit gastric motility and secretion; include secretin, cholecystokinin, and gastric inhibitory peptide

enterohepatic circulation (en´-tur-ō-hi-PAT-ik) The recycling of bile salts and other bile constituents between the small intestine and liver by means of the hepatic portal vein

enzyme A special protein molecule that speeds up a particular chemical reaction in the body

eosinophils (ē´-uh-SIN-uh-fils) White blood cells that are important in allergic responses and in combating internal parasite infestations

epidermis (ep´-uh-DER-mus) The outer layer of the skin, consisting of numerous layers of epithelial cells, with the outermost layers being dead and flattened

epinephrine (ep´-uh-NEF-rin) The primary hormone secreted by the adrenal medulla; important in preparing the body for "fight-or-flight" responses and in regulation of arterial blood pressure; adrenaline

epiphyseal plate (eh-pif-i-SEE-al) A layer of cartilage that separates the diaphysis (shaft) of a long bone from the epiphysis (flared end); the site of growth of bones in length before the cartilage ossifies (turns into bone)

epithelial tissue (ep´-uh-THĒ-lē-ul) A functional grouping of cells specialized in the exchange of materials between the cell and its environment; lines and covers various body surfaces and cavities and forms secretory glands

EPSP See *excitatory postsynaptic potential*

equilibrium potential The potential that exists when the concentration gradient and opposing electrical gradient for a given ion exactly counterbalance each other so that there is no net movement of the ion

erythrocytes (i-RITH-ruh-sīts) Red blood cells, which are plasma membrane-enclosed bags of hemoglobin that transport O_2 and to a lesser extent CO_2 and H^+ in the blood

erythropoiesis (i-rith´-rō-poi-Ē-sus) Erythrocyte production by the bone marrow

erythropoietin The hormone released from the kidneys in response to a reduction in O_2 delivery to the kidneys; stimulates the bone marrow to increase erythrocyte production

esophagus (i-SOF-uh-gus) A straight muscular tube that extends between the pharynx and stomach

estrogen Feminizing "female" sex hormone

ESV See *end-systolic volume*

excitable tissue Tissue capable of producing electrical signals when excited; includes nervous and muscle tissue

excitation–contraction coupling The series of events linking muscle excitation (the presence of an action potential) to muscle contraction (filament sliding and sarcomere shortening)

excitatory postsynaptic potential (EPSP) (pōst´-si-NAP-tik) A small depolarization of the postsynaptic membrane in response to neurotransmitter binding, thereby bringing the membrane closer to threshold

excitatory synapse (SIN-aps´) Synapse in which the postsynaptic neuron's response to neurotransmitter release is a small depolarization of the postsynaptic membrane, bringing the membrane closer to threshold

exercise physiology The study of both the functional changes that occur in response to a single session of exercise and the adaptations that occur as a result of regular, repeated exercise sessions

exocrine glands Glands that secrete through ducts to the outside of the body or into a cavity that communicates with the outside

exocytosis (eks´-ō-sī-TŌ-sis) Fusion of a membrane-enclosed intracellular vesicle with the plasma membrane, followed by the opening of the vesicle and the emptying of its contents to the outside

expiration A breath out

expiratory muscles The skeletal muscles whose contraction reduces the size of the thoracic cavity and allows the lungs to recoil to a smaller size, bringing about movement of air from the lungs to the atmosphere

external intercostal muscles Inspiratory muscles whose contraction elevates the ribs, thereby enlarging the thoracic cavity

external environment The environment that surrounds the body

external work Energy expended by contracting skeletal muscles to move external objects or to move the body in relation to the environment

extracellular fluid All the body's fluid found outside the cells; consists of interstitial fluid and plasma

extracellular matrix An intricate meshwork of fibrous proteins embedded in a watery, gel-like substance; secreted by local cells

extrinsic controls Regulatory mechanisms initiated outside of an organ that alter the activity of the organ; accomplished by the nervous and endocrine systems

extrinsic nerves The nerves that originate outside of the digestive tract and innervate the various digestive organs

facilitated diffusion Passive carrier-mediated transport involving transport of a substance down its concentration gradient across the plasma membrane

fatigue Inability to maintain muscle tension at a given level despite sustained stimulation

feedforward mechanism A response designed to prevent an anticipated change in a controlled variable

feeding (appetite) centers Neuronal clusters in the lateral regions of the hypothalamus that drive the individual to eat

fibrinogen (fī-BRIN-uh-jun)   A large, soluble plasma protein that is converted into an insoluble, threadlike molecule that forms the meshwork of a clot during blood coagulation

Fick's law of diffusion   The rate of net diffusion of a substance across a membrane is directly proportional to the substance's concentration gradient, the membrane's permeability to the substance, and the surface area of the membrane and inversely proportional to the substance's molecular weight and the diffusion distance

fight-or-flight response   The changes in activity of the various organs innervated by the autonomic nervous system in response to sympathetic stimulation, which collectively prepare the body for strenuous physical activity in the face of an emergency or stressful situation, such as a physical threat from the outside environment

fire   When an excitable cell undergoes an action potential

first messenger   An extracellular messenger, such as a hormone, that binds with a surface membrane receptor and activates an intracellular second messenger to carry out the desired cellular response

flagellum (fluh-JEL-um)   The single, long, whiplike appendage that serves as the tail of a spermatozoon

follicle (of ovary)   A developing ovum and the surrounding specialized cells

follicle-stimulating hormone (FSH)   An anterior pituitary hormone that stimulates ovarian follicular development and estrogen secretion in females and stimulates sperm production in males

follicular cells (of ovary) (fah-LIK-you-lar)   Collectively, the granulosa and thecal cells

follicular cells (of thyroid gland)   The cells that form the walls of the colloid-filled follicles in the thyroid gland and secrete thyroid hormone

follicular phase   The phase of the ovarian cycle dominated by the presence of maturing follicles prior to ovulation

Frank-Starling law of the heart   Intrinsic control of the heart, such that increased venous return resulting in increased end-diastolic volume leads to an increased strength of contraction and increased stroke volume; that is, the heart normally pumps out all of the blood returned to it

free radicals   Very unstable electron-deficient particles that are highly reactive and destructive

frontal lobes   The lobes of the cerebral cortex that lie at the top of the brain in front of the central sulcus and that are responsible for voluntary motor output, speaking ability, and elaboration of thought

FSH   See *follicle-stimulating hormone*

fuel metabolism   See *intermediary metabolism*

functional syncytium (sin-sish´-ē-um)   A group of smooth or cardiac muscle cells that are interconnected by gap junctions and function electrically and mechanically as a single unit

functional unit   The smallest component of an organ that can perform all the functions of the organ

gametes (GAM-ētz)   Reproductive, or germ, cells, each containing a haploid set of chromosomes; sperm and ova

gamma motor neuron   A motor neuron that innervates the fibers of a muscle-spindle receptor

ganglion (GAN-glē-un)   A collection of neuronal cell bodies located outside the central nervous system

ganglion cells   The nerve cells in the outermost layer of the retina and whose axons form the optic nerve

gap junction   A communicating junction formed between adjacent cells by small connecting tunnels that permit passage of charge-carrying ions between the cells so that electrical activity in one cell is spread to the adjacent cell

gastrin   A hormone secreted by the pyloric gland area of the stomach that stimulates the parietal and chief cells to secrete a highly acidic gastric juice

gestation   Pregnancy

glands   Epithelial tissue derivatives that are specialized for secretion

glial cells (glē-ul)   Serve as the connective tissue of the CNS and help support the neurons both physically and metabolically; include astrocytes, oligodendrocytes, ependymal cells, and microglia

glomerular filtration (glow-MER-yū-lur)   Filtration of a protein-free plasma from the glomerular capillaries into the tubular component of the kidney's nephron as the first step in urine formation

glomerular filtration rate (GFR)   The rate at which glomerular filtrate is formed

glomerulus (glow-MER-yū-lus)   A ball-like tuft of capillaries in the kidney's nephron that filters water and solute from the blood as the first step in urine formation

glucagon (GLOO-kuh-gon)   The pancreatic hormone that raises blood glucose and blood fatty-acid levels

glucocorticoids (gloo´-kō-KOR-ti-koidz)   The adrenocortical hormones that are important in intermediary metabolism and in helping the body resist stress; primarily cortisol

gluconeogenesis (gloo´-kō-nē-ō-JEN-uh-sus)   The conversion of amino acids into glucose

glycogen (GLĪ-kō-jen)   The storage form of glucose in the liver and muscle

glycogenesis (glī-kō-JEN-i-sus)   The conversion of glucose into glycogen

glycogenolysis (glī-kō-juh-NOL-i-sus)   The conversion of glycogen to glucose

glycolysis (glī-KOL-uh-sus)   A biochemical process that takes place in the cell's cytosol and involves the breakdown of glucose into two pyruvic acid molecules

GnRH   See *gonadotropin-releasing hormone*

Golgi complex (GOL-jē)   An organelle consisting of sets of stacked, flattened membranous sacs; processes raw materials transported to it from the endoplasmic reticulum into finished products and sorts and directs the finished products to their final destination

gonadotropin-releasing hormone (GnRH) (gō-nad´-uh-TRŌ-pin)   The hypothalamic hormone that stimulates the release of FSH and LH from the anterior pituitary

gonadotropins   FSH and LH; hormones that are tropic to the gonads

gonads (GŌ-nadz)   The primary reproductive organs, which produce the gametes and secrete the sex hormones; testes and ovaries

G protein   A membrane-bound intermediary, which, when activated on binding of an extracellular first messenger to a surface receptor, activates the enzyme adenylyl cyclase on the intracellular side of the membrane in the cAMP second-messenger system

gradation of contraction   Variable magnitudes of tension produced in a single whole muscle

graded potential   A local change in membrane potential that occurs in varying grades of magnitude; serves as a short-distance signal in excitable tissues

granulocytes (gran´-yuh-lō-sīts)   Leukocytes that contain granules, including neutrophils, eosinophils, and basophils

granulosa cells (gran´-yuh-LŌ-suh)   The layer of cells immediately surrounding a developing oocyte within an ovarian follicle

gray matter   The portion of the central nervous system composed primarily of densely packaged neuronal cell bodies and dendrites

growth hormone (GH)   An anterior pituitary hormone that is primarily responsible for regulating overall body growth and is also important in intermediary metabolism; somatotropin

H⁺   See *hydrogen ion*

haploid number (HAP-loid)   The number of chromosomes found in gametes; a half set of chromosomes, one member of each pair, for a total of 23 chromosomes in humans

Hb   See *hemoglobin*

hCG   See *human chorionic gonadotropin*

heart failure An inability of the cardiac output to keep pace with the body's demands for supplies and for removal of wastes

helper T cells The population of T cells that enhances the activity of other immune-response effector cells

hematocrit (hi-mat′-uh-krit) The percentage of blood volume occupied by erythrocytes as they are packed down in a centrifuged blood sample

hemoglobin (HĒ-muh-glō′-bun) A large iron-bearing protein molecule found within erythrocytes that binds with and transports most O_2 in the blood; also carries some of the CO_2 and H^+ in the blood

hemolysis (hē-MOL-uh-sus) Rupture of red blood cells

hemostasis (hē′-mō-STĀ-sus) The stopping of bleeding from an injured vessel

hepatic portal system (hi-PAT-ik) A complex vascular connection between the digestive tract and liver such that venous blood from the digestive system drains into the liver for processing of absorbed nutrients before being returned to the heart

hippocampus (hip-oh-CAM-pus) The elongated, medial portion of the temporal lobe that is a part of the limbic system and is especially crucial for forming long-term memories

histamine A chemical released from mast cells or basophils that brings about vasodilation and increased capillary permeability; important in allergic responses

homeostasis (hō′-mē-ō-STĀ-sus) Maintenance by the highly coordinated, regulated actions of the body systems of relatively stable chemical and physical conditions in the internal fluid environment that bathes the body's cells

hormone A long-distance chemical mediator that is secreted by an endocrine gland into the blood, which transports it to its target cells

hormone response element (HRE) The specific attachment site on DNA for a given steroid hormone and its nuclear receptor

host cell A body cell infected by a virus

HRE See *hormone response element*

human chorionic gonadotropin (hCG) (kō-rē-ON-ik gō-nad′-uh-TRŌ-pin) A hormone secreted by the developing placenta that stimulates and maintains the corpus luteum of pregnancy

hydrogen ion (H^+) The cationic portion of a dissociated acid

hydrolysis (hī-DROL-uh-sis) The digestion of a nutrient molecule by the addition of water at a bond site

hydrostatic pressure (hī-dro-STAT-ik) The pressure exerted by fluid on the walls that contain it

hyperglycemia (hī′-pur-glī-SĒ-mē-uh) Elevated blood glucose concentration

hyperplasia (hī-pur-PLĀ-zē-uh) An increase in the number of cells

hyperpolarization An increase in membrane potential from resting potential; potential becomes even more negative than at resting potential

hypersecretion Too much of a particular hormone secreted

hypertension (hī′-pur-TEN-shun) Sustained, above-normal mean arterial blood pressure

hypertonic (hī′-pur-TON-ik) Having an osmolarity greater than normal body fluids; more concentrated than normal

hypertrophy (hī-PUR-truh-fē) Increase in the size of an organ as a result of an increase in the size of its cells

hyperventilation Overbreathing; when the rate of ventilation is in excess of the body's metabolic needs for CO_2 removal

hypophysiotropic hormones (hi-PŌ-fiz-ē-oh-TRO-pik) Hormones secreted by the hypothalamus that regulate the secretion of anterior pituitary hormones; see also *releasing hormone* and *inhibiting hormone*

hyposecretion Too little of a particular hormone secreted

hypotension (hi-po-TEN-chun) Sustained, below normal mean arterial blood pressure

hypothalamic hypophyseal portal system (hī-pō-thuh-LAM-ik hī-pō-FIZ-ē-ul) The vascular connection between the hypothalamus and anterior pituitary gland used for the pickup and delivery of hypophysiotropic hormones

hypothalamus (hī′-pō-THAL-uh-mus) The brain region located beneath the thalamus that is concerned with regulating many aspects of the internal fluid environment, such as water and salt balance and food intake; serves as an important link between the autonomic nervous system and endocrine system

hypotonic (hī′-pō-TON-ik) Having an osmolarity less than normal body fluids; more dilute than normal

hypoventilation Underbreathing; ventilation inadequate to meet the metabolic needs for O_2 delivery and CO_2 removal

hypoxia (hī-POK-sē-uh) Insufficient O_2 at the cellular level

I band One of the light bands that alternate with dark (A) bands to create a striated appearance in a skeletal or cardiac muscle fiber when these fibers are viewed with a light microscope

IGF See *insulin-like growth factor*

immune surveillance Recognition and destruction of newly arisen cancer cells by the immune system

immunity The body's ability to resist or eliminate potentially harmful foreign materials or abnormal cells

immunoglobulins (im′-ū-nō-GLOB-yū-lunz) Antibodies; gamma globulins

impermeable Prohibiting passage of a particular substance through the plasma membrane

implantation The burrowing of a blastocyst into the endometrial lining

inflammation An innate, nonspecific series of highly interrelated events, especially involving neutrophils, macrophages, and local vascular changes, that are set into motion in response to foreign invasion or tissue damage

influx Movement into the cell

inhibin (in-HIB-un) A hormone secreted by the Sertoli cells of the testes or by the ovarian follicles that inhibits FSH secretion

inhibiting hormone A hypothalamic hormone that inhibits the secretion of a particular anterior pituitary hormone

inhibitory postsynaptic potential (IPSP) (pōst′-si-NAP-tik) A small hyperpolarization of the postsynaptic membrane in response to neurotransmitter binding, thereby moving the membrane farther from threshold

inhibitory synapse (SIN-aps′) Synapse in which the postsynaptic neuron's response to neurotransmitter release is a small hyperpolarization of the postsynaptic membrane, moving the membrane farther from threshold

innate immune responses Inherent defense responses that nonselectively defend against foreign or abnormal material, even on initial exposure to it; see also *inflammation, interferon, natural killer cells,* and *complement system*

inorganic Referring to substances that do not contain carbon; from nonliving sources

inspiration A breath in

inspiratory muscles The skeletal muscles whose contraction enlarges the thoracic cavity, bringing about lung expansion and movement of air into the lungs from the atmosphere

insulin (IN-suh-lin) The pancreatic hormone that lowers blood levels of glucose, fatty acids, and amino acids and promotes their storage

insulin-like growth factor (IGF) Synonymous with somatomedins

integrating center A region that determines efferent output based on processing of afferent input

integument (in-TEG-yuh-munt) The skin and underlying connective tissue

intercostal muscles (int-ur-KOS-tul)   The muscles that lie between the ribs; see also *external intercostal muscles* and *internal intercostal muscles*

interferon (in´-tur-FER-on)   A chemical released from virus-invaded cells that provides nonspecific resistance to viral infections by transiently interfering with replication of the same or unrelated viruses in other host cells

interleukin 1 (int-ur-LOO-kin)   A multipurpose chemical mediator released from macrophages that enhances B cell activity

interleukin 2   A chemical mediator secreted by helper T cells that augments the activity of all T cells

intermediary metabolism   The collective set of intracellular chemical reactions that involve the degradation, synthesis, and transformation of small nutrient molecules; also known as *fuel metabolism*

intermediate filaments   Threadlike cytoskeletal elements that play a structural role in parts of the cells subject to mechanical stress

internal environment   The body's aqueous extracellular environment, which consists of the plasma and interstitial fluid and which must be homeostatically maintained for the cells to make life-sustaining exchanges with it

internal intercostal muscles   Expiratory muscles whose contraction pulls the ribs downward and inward, thereby reducing the size of the thoracic cavity

internal respiration   The intracellular metabolic processes carried out within the mitochondria that use O_2 and produce CO_2 during the derivation of energy from nutrient molecules

internal work   All forms of biological energy expenditure that do not accomplish mechanical work outside of the body

interneuron   Neuron that lies entirely within the central nervous system and is important for integrating peripheral responses to peripheral information as well as for the abstract phenomena associated with the "mind"

interstitial fluid (in´-tur-STISH-ul)   The portion of the extracellular fluid that surrounds and bathes all the body's cells

intra-alveolar pressure (in´-truh-al-VE-uh-lur)   The pressure within the alveoli

intracellular fluid   The fluid collectively contained within all the body's cells

intrapleural pressure (in´-truh-PLOOR-ul)   The pressure within the pleural sac

intrinsic controls   Local control mechanisms inherent to an organ

intrinsic factor   A special substance secreted by the parietal cells of the stomach that must be combined with vitamin B_{12} for this vitamin to be absorbed by the intestine; deficiency produces pernicious anemia

intrinsic nerve plexuses   Interconnecting networks of nerve fibers within the digestive tract wall

ion   An atom that has gained or lost one or more of its electrons, so that it is not electrically balanced

IPSP   See *inhibitory postsynaptic potential*

iris   A pigmented smooth muscle that forms the colored portion of the eye and controls pupillary size

islets of Langerhans (LAHNG-er-honz)   The endocrine portion of the pancreas that secretes the hormones insulin and glucagon into the blood

isometric contraction (ī´-sō-MET-rik)   A muscle contraction in which the development of tension occurs at constant muscle length

isotonic (ī´-sō-TON-ik)   Having an osmolarity equal to normal body fluids

isotonic contraction   A muscle contraction in which muscle tension remains constant as the muscle fiber changes length

juxtaglomerular apparatus (juks´-tuh-glō-MER-yu-lur)   A cluster of specialized vascular and tubular cells at a point where the ascending limb of the loop of Henle passes through the fork formed by the afferent and efferent arterioles of the same nephron in the kidney

keratin (CARE-uh-tin)   The protein found in the intermediate filaments in skin cells that give the skin strength and help form a waterproof outer layer

killer (K) cells   Cells that destroy a target cell that has been coated with antibodies by lysing its membrane

kinesin (kī-NE´-sin)   The transport, or motor, protein that transports secretory vesicles along the microtubular highway within neuronal axons by "walking" along the microtubule

lactation   Milk production by the mammary glands

lactic acid   An end product formed from pyruvic acid during the anaerobic process of glycolysis

larynx (LARE-inks)   The "voice box" at the entrance of the trachea; contains the vocal cords

lateral inhibition   The phenomenon in which the most strongly activated signal pathway originating from the center of a stimulus area inhibits the less excited pathways from the fringe areas by means of lateral inhibitory connections within sensory pathways

lateral sacs   The expanded saclike regions of a muscle fiber's sarcoplasmic reticulum; store and release calcium, which plays a key role in triggering muscle contraction

law of mass action   If the concentration of one of the substances involved in a reversible reaction is increased, the reaction is driven toward the opposite side, and if the concentration of one of the substances is decreased, the reaction is driven toward that side

left ventricle   The heart chamber that pumps blood into the systemic circulation

length–tension relationship   The relationship between the length of a muscle fiber at the onset of contraction and the tension the fiber can achieve on a subsequent tetanic contraction

lens   A transparent, biconvex structure of the eye that refracts (bends) light rays and whose strength can be adjusted to accommodate for vision at different distances

leukocyte endogenous mediator (LEM) (LOO-kō-sīt en-DAJ-eh-nus ME-de-āT-or)   A chemical mediator secreted by macrophages that is identical to endogenous pyrogen and exerts a wide array of effects associated with inflammation

leukocytes (LOO-kuh-sīts)   White blood cells, which are the immune system's mobile defense units

Leydig cells (LI-dig)   The interstitial cells of the testes that secrete testosterone

LH   See *luteinizing hormone*

LH surge   The burst in LH secretion that occurs at midcycle of the ovarian cycle and triggers ovulation

limbic system (LIM-bik)   A functionally interconnected ring of forebrain structures that surrounds the brain stem and is concerned with emotions, basic survival and sociosexual behavioral patterns, motivation, and learning

lipid emulsion   A suspension of small fat droplets held apart as a result of adsorption of bile salts on their surface

loop of Henle (HEN-lē)   A hairpin loop that extends between the proximal and distal tubule of the kidney's nephron

lumen (LOO-men)   The interior space of a hollow organ or tube

luteal phase (LOO-tē-ul)   The phase of the ovarian cycle dominated by the presence of a corpus luteum

luteinization (loot´-ē-un-uh-ZA-shun)   Formation of a postovulatory corpus luteum in the ovary

luteinizing hormone (LH)   An anterior pituitary hormone that stimulates ovulation, luteinization, and secretion of estrogen and progesterone in females and stimulates testosterone secretion in males

lymph Interstitial fluid that is picked up by the lymphatic vessels and returned to the venous system, meanwhile passing through the lymph nodes for defense purposes

lymphocytes White blood cells that provide immune defense against targets for which they are specifically programmed

lymphoid tissues Tissues that produce and store lymphocytes, such as lymph nodes and tonsils

lysosomes (LĪ-sō-sōmz) Organelles consisting of membrane-enclosed sacs containing powerful hydrolytic enzymes that destroy unwanted material within the cell, such as internalized foreign material or cellular debris

macrophages (MAK-ruh-fājs) Large, tissue-bound phagocytes

mast cells Cells located within connective tissue that synthesize, store, and release histamine, as during allergic responses

mature follicle An ovarian follicle that is ready to ovulate

mean arterial blood pressure The average pressure responsible for driving blood forward through the arteries into the tissues throughout the cardiac cycle; equals cardiac output times total peripheral resistance

mechanically gated channels Channels that open or close in response to stretching or other mechanical deformation

mechanistic approach Explanation of body functions in terms of mechanisms of action, that is, the "how" of events that occur in the body

mechanoreceptor (meh-CAN-oh-rē-SEP-tur) or (mek´-uh-nō-rē-SEP-tur) A sensory receptor sensitive to mechanical energy, such as stretching or bending

medullary respiratory center (MED-you-LAIR-ē) Several aggregations of neuronal cell bodies within the medulla that provide output to the respiratory muscles and receive input important for regulating the magnitude of ventilation

meiosis (mī-ō-sis) Cell division in which the chromosomes replicate followed by two nuclear divisions so that only a half set of chromosomes is distributed to each of four new daughter cells

melanocyte-stimulating hormone (MSH) (mel-AH-nō-sīt) A hormone produced by the anterior pituitary in humans and by the intermediate lobe of the pituitary in lower vertebrates; regulates skin coloration by controlling the dispersion of melanin granules in lower vertebrates; involved with control of food intake and possibly memory and learning in humans

melatonin (mel-uh-TŌ-nin) A hormone secreted by the pineal gland during darkness that helps entrain the body's biological rhythms with the external light/dark cues

membrane attack complex A collection of the five final activated components of the complement system that aggregate to form a pore-like channel in the plasma membrane of an invading micro-organism, with the resultant leakage leading to destruction of the invader

membrane potential A separation of charges across the membrane; a slight excess of negative charges lined up along the inside of the plasma membrane and separated from a slight excess of positive charges on the outside

memory cells B or T cells that are newly produced in response to a microbial invader but that do not participate in the current immune response against the invader: instead they remain dormant, ready to launch a swift, powerful attack should the same micro-organism invade again in the future

menstrual cycle (men´-stroo-ul) The cyclical changes in the uterus that accompany the hormonal changes in the ovarian cycle

menstrual phase The phase of the menstrual cycle characterized by sloughing of endometrial debris and blood out through the vagina

messenger RNA Carries the transcribed genetic blueprint for synthesis of a particular protein from nuclear DNA to the cytoplasmic ribosomes where the protein synthesis takes place

metabolic acidosis (met-uh-bol´-ik) Acidosis resulting from any cause other than excess accumulation of carbonic acid in the body

metabolic alkalosis (al´-kuh-LŌ-sus) Alkalosis caused by a relative deficiency of noncarbonic acid

metabolic rate Energy expenditure per unit of time

micelle (mī-SEL) A water-soluble aggregation of bile salts, lecithin, and cholesterol that has a hydrophilic shell and a hydrophobic core; carries the water-insoluble products of fat digestion to their site of absorption

microfilaments Cytoskeletal elements made of actin molecules (as well as myosin molecules in muscle cells); play a major role in various cellular contractile systems and serve as a mechanical stiffener for microvilli

microtubules Cytoskeletal elements made of tubulin molecules arranged into long, slender, unbranched tubes that help maintain asymmetric cell shapes and coordinate complex cell movements

microvilli (mī´-krō-VIL-ī) Actin-stiffened, nonmotile, hairlike projections from the luminal surface of epithelial cells lining the digestive tract and kidney tubules; tremendously increase the surface area of the cell exposed to the lumen

micturition (mik-too-RISH-un) or (mik-chuh-RISH-un) The process of bladder emptying; urination

milk ejection The squeezing out of milk produced and stored in the alveoli of the breasts by means of contraction of the myoepithelial cells that surround each alveolus

mineralocorticoids (min-uh-rul-ō-KOR-ti-koidz) The adrenocortical hormones that are important in Na^+ and K^+ balance; primarily aldosterone

mitochondria (mī-tō-KON-drē-uh) The energy organelles, which contain the enzymes for oxidative phosphorylation

mitosis (mī-TŌ-sis) Cell division in which the chromosomes replicate before nuclear division, so that each of the two daughter cells receives a full set of chromosomes

mitotic spindle The system of microtubules assembled during mitosis along which the replicated chromosomes are directed away from each other toward opposite sides of the cell prior to cell division

molecule A chemical substance formed by the linking of atoms together; the smallest unit of a given chemical substance

monocytes (MAH-nō-sīts) White blood cells that emigrate from the blood, enlarge, and become macrophages, large-tissue phagocytes

monosaccharides (mah´-nō-SAK-uh-rīdz) Simple sugars, such as glucose; the absorbable unit of digested carbohydrates

motor activity Movement of the body accomplished by contraction of skeletal muscles

motor end plate The specialized portion of a skeletal muscle fiber that lies immediately underneath the terminal button of the motor neuron and possesses receptor sites for binding acetylcholine released from the terminal button

motor neurons The neurons that innervate skeletal muscle and whose axons constitute the somatic nervous system

motor unit One motor neuron plus all the muscle fibers it innervates

motor unit recruitment The progressive activation of a muscle fiber's motor units to accomplish increasing gradations of contractile strength

mucosa (mew-KŌ-sah) The innermost layer of the digestive tract that lines the lumen

multiunit smooth muscle A smooth muscle mass that consists of multiple discrete units that function independently of each other and that must be separately stimulated by autonomic nerves to contract

muscarinic receptor Type of cholinergic receptor found at the effector organs of all parasympathetic postganglionic fibers

muscle fiber A single muscle cell, which is relatively long and cylindrical in shape

muscle tension See *tension*

muscle tissue A functional grouping of cells specialized for contraction and force generation

myelin (MĪ-uh-lun)   An insulative lipid covering that surrounds myelinated nerve fibers at regular intervals along the axon's length; each patch of myelin is formed by a separate myelin forming cell that wraps itself jelly-roll fashion around the neuronal axon

myelinated fibers   Neuronal axons covered at regular intervals with insulative myelin

myocardial ischemia (mī-ō-KAR-dē-ul is-KĒ-mē-uh)   Inadequate blood supply to the heart tissue

myocardium (mī′-ō-KAR-dē-um)   The cardiac muscle within the heart wall

myofibril (mī′-ō-FĪB-rul)   A specialized intracellular structure of muscle cells that contains the contractile apparatus

myometrium (mī′-ō-mē-TRĒ-um)   The smooth muscle layer of the uterus

myosin (MĪ-uh-sun)   The contractile protein that forms the thick filaments in muscle fibers

Na^+–K^+ pump   A carrier that actively transports Na^+ out of the cell and K^+ into the cell

natural killer cells   Naturally occurring, lymphocyte-like cells that nonspecifically destroy virus-infected cells and cancer cells by directly lysing their membranes on first exposure to them

negative balance   Situation in which the losses for a substance exceed its gains so that the total amount of the substance in the body decreases

negative feedback   A regulatory mechanism in which a change in a controlled variable triggers a response that opposes the change, thus maintaining a relatively steady set point for the regulated factor

nephron (NEF-ron′)   The functional unit of the kidney; consisting of an interrelated vascular and tubular component, it is the smallest unit that can form urine

nerve   A bundle of peripheral neuronal axons, some afferent and some efferent, enclosed by a connective tissue covering and following the same pathway

nervous system   One of the two major regulatory systems of the body; in general, coordinates rapid activities of the body, especially those involving interactions with the external environment

nervous tissue   A functional grouping of cells specialized for initiation and transmission of electrical signals

net diffusion   The difference between two opposing movements

net filtration pressure   The net difference in the hydrostatic and osmotic forces acting across the glomerular membrane that favors the filtration of a protein-free plasma into Bowman's capsule

neuroendocrinology   The study of the interaction between the nervous and endocrine systems

neuroglia   See *glial cells*

neurohormones   Hormones released into the blood by neurosecretory neurons

neuromodulators (ner′ō-MA-jew-lā′-torz)   Chemical messengers that bind to neuronal receptors at nonsynaptic sites (that is, not at the subsynaptic membrane) and bring about long-term changes that subtly depress or enhance synaptic effectiveness

neuromuscular junction   The juncture between a motor neuron and a skeletal muscle fiber

neuron (NER-on)   A nerve cell, typically consisting of a cell body, dendrites, and an axon and specialized to initiate, propagate, and transmit electrical signals

neuropeptides   Large, slowly acting peptide molecules released from axon terminals along with classical neurotransmitters; most neuropeptides function as neuromodulators

neurotransmitter   The chemical messenger that is released from the axon terminal of a neuron in response to an action potential and influences another neuron or an effector with which the neuron is anatomically linked

neutrophils (new′-truh-filz)   White blood cells that are phagocytic specialists and important in inflammatory responses and defense against bacterial invasion

nicotinic receptor (nick′-o-TIN-ik)   Type of cholinergic receptor found at all autonomic ganglia and the motor end plates of skeletal muscle fibers

nitric oxide   A recently identified local chemical mediator released from endothelial cells and other tissues; exerts a wide array of effects, ranging from causing local arteriolar vasodilation to acting as a toxic agent against foreign invaders to serving as a unique type of neurotransmitter

nociceptor (nō-sē-SEP-tur)   A pain receptor, sensitive to tissue damage

node of Ranvier (RAN-vē-ā)   The portions of a myelinated neuronal axon between the segments of insulative myelin; the axonal regions where the axonal membrane is exposed to the ECF and membrane potential exists

nontropic hormone   A hormone that exerts its effects on nonendocrine target tissues

norepinephrine (nor′-ep-uh-NEF-run)   The neurotransmitter released from sympathetic postganglionic fibers; noradrenaline

nucleus (of brain) (NŪ-klē-us)   A functional aggregation of neuronal cell bodies within the brain

nucleus (of cells)   A distinct spherical or oval structure that is usually located near the center of a cell and that contains the cell's genetic material, deoxyribonucleic acid (DNA)

occipital lobes (ok-SIP′-ut-ul)   The lobes of the cerebral cortex that are located posteriorly and are responsible for initially processing visual input

O_2–Hb dissociation curve   A graphic depiction of the relationship between arterial P_{O_2} and percent hemoglobin saturation

oligodendrocytes (ol-i-gō′-DEN-drō-sitz)   The myelin-forming cells of the central nervous system

oogenesis (ō′-ō-JEN-uh-sus)   Egg production

opsonin (OP′-suh-nun)   Body-produced chemical that links bacteria to macrophages, thereby making the bacteria more susceptible to phagocytosis

optic nerve   The bundle of nerve fibers that leave the retina, relaying information about visual input

optimal length   The length before the onset of contraction of a muscle fiber at which maximal force can be developed on a subsequent tetanic contraction

organ   A distinct structural unit composed of two or more types of primary tissue organized to perform one or more particular functions; for example, the stomach

organelles (or′-gan-ELZ)   Distinct, highly organized, membrane-bound intracellular compartments, each containing a specific set of chemicals for carrying out a particular cellular function

organic   Referring to substances that contain carbon; from living or once-living sources

organism   A living entity, either single-celled (unicellular) or made up of many cells (multicellular)

organ of Corti (KOR-tē)   The sense organ of hearing within the inner ear that contains hair cells whose hairs are bent in response to sound waves, setting up action potentials in the auditory nerve

osmolarity (oz′-mō-LAR-ut-ē)   A measure of the concentration of solute molecules in a solution

osmosis (os-MŌ-sis)   Movement of water across a membrane down its own concentration gradient toward the area of higher solute concentration

osteoblasts (OS-tē-ō-blasts′)   Bone cells that produce the organic matrix of bone

osteoclasts   Bone cells that dissolve bone in their vicinity

otolith organs (ŌT´-ul-ith)   Sense organs in the inner ear that provide information about rotational changes in head movement; include the utricle and saccule

oval window   The membrane-covered opening that separates the air-filled middle ear from the upper compartment of the fluid-filled cochlea in the inner ear

overhydration   Water excess in the body

ovulation (ov´-yuh-LĀ-shun)   Release of an ovum from a mature ovarian follicle

oxidative phosphorylation (fos´-fōr-i-LĀ-shun)   The entire sequence of mitochondrial biochemical reactions that uses oxygen to extract energy from the nutrients in food and transforms it into ATP, producing CO_2 and H_2O in the process

oxyhemoglobin (ok-si-HĒ-muh-glō-bun)   Hemoglobin combined with O_2

oxyntic mucosa (ok-SIN-tic)   The mucosa that lines the body and fundus of the stomach; contains gastric pits lined by mucous neck cells, parietal cells, and chief cells

oxytocin (ok´-sē-TŌ-sun)   A hypothalamic hormone that is stored in the posterior pituitary and stimulates uterine contraction and milk ejection

pacemaker activity   Self-excitable activity of an excitable cell in which its membrane potential gradually depolarizes to threshold on its own

Pacinian corpuscle (pa-SIN-ē-un)   A rapidly adapting skin receptor that detects pressure and vibration

pancreas (PAN-krē-us)   A mixed gland composed of an exocrine portion that secretes digestive enzymes and an aqueous alkaline secretion into the duodenal lumen and an endocrine portion that secretes the hormones insulin and glucagon into the blood

paracrine (PEAR-uh-krin)   A local chemical messenger whose effect is exerted only on neighboring cells in the immediate vicinity of its site of secretion

parasympathetic nervous system (pear´-uh-sim-puh-THET-ik)   The subdivision of the autonomic nervous system that dominates in quiet, relaxed situations and promotes body maintenance activities such as digestion and emptying of the urinary bladder

parathyroid glands (pear´-uh-THĪ-roid)   Four small glands located on the posterior surface of the thyroid gland that secrete parathyroid hormone

parathyroid hormone (PTH)   A hormone that raises plasma Ca^{2+} levels

parietal cells (puh-RĪ-ut-ul)   The stomach cells that secrete hydrochloric acid and intrinsic factor

parietal lobes   The lobes of the cerebral cortex that lie at the top of the brain behind the central sulcus and contain the somatosensory cortex

partial pressure   The individual pressure exerted independently by a particular gas within a mixture of gases

partial pressure gradient   A difference in the partial pressure of a gas between two regions that promotes the movement of the gas from the region of higher partial pressure to the region of lower partial pressure

parturition (par´-too-RISH-un)   Delivery of a baby

passive expiration   Expiration accomplished during quiet breathing as a result of elastic recoil of the lungs on relaxation of the inspiratory muscles, with no energy expenditure required

passive force   A force that does not require expenditure of cellular energy to accomplish transport of a substance across the plasma membrane

passive reabsorption   When none of the steps in the transepithelial transport of a substance reabsorbed across the kidney tubules requires energy expenditure

pathogens (PATH-uh-junz)   Disease-causing micro-organisms, such as bacteria or viruses

pathophysiology (path´-ō-fiz-UL-uh-jē)   Abnormal functioning of the body associated with disease

pepsin; pepsinogen (pep-SIN-uh-jun)   An enzyme secreted in inactive form by the stomach that, once activated, begins protein digestion

peptide hormones   Hormones that consist of a chain of specific amino acids of varying length

percent hemoglobin saturation   A measure of the extent to which the hemoglobin present is combined with O_2

perception   The conscious interpretation of the external world as created by the brain from a pattern of nerve impulses delivered to it from sensory receptors

peripheral chemoreceptors (kē´-mō-rē-SEP-turz)   The carotid and aortic bodies, which respond to changes in arterial P_{O_2}, P_{CO_2}, and H^+ and adjust respiration accordingly

peripheral nervous system (PNS)   Nerve fibers that carry information between the central nervous system and other parts of the body

peristalsis (per´-uh-STOL-sus)   Ringlike contractions of the circular smooth muscle of a tubular organ that move progressively forward with a stripping motion, pushing the contents of the organ ahead of the contraction

peritubular capillaries (per´-i-TŪ-bū-lur)   Capillaries that intertwine around the tubules of the kidney's nephron; they supply the renal tissue and participate in exchanges between the tubular fluid and blood during the formation of urine

permeable   Permitting passage of a particular substance

permissiveness   When one hormone must be present in adequate amounts for the full exertion of another hormone's effect

pernicious anemia (per-NEE-shus)   The anemia produced as a result of intrinsic factor deficiency

peroxisomes (puh-ROK´-suh-sōmz)   Organelles consisting of membrane-bound sacs that contain powerful oxidative enzymes that detoxify various wastes produced within the cell or foreign compounds that have entered the cell

pH   The logarithm to the base 10 of the reciprocal of the hydrogen ion concentration; $pH = \log 1/[H^+]$ or $pH = -\log[H^+]$

phagocytosis (fag´-oh-sī-TŌ-sus)   A type of endocytosis in which large, multimolecular, solid particles are engulfed by a cell

pharynx (FARE-inks)   The back of the throat, which serves as a common passageway for the digestive and respiratory systems

phosphorylation (fos´-fōr-i-LĀ-shun)   Addition of a phosphate group to a molecule

photoreceptor   A sensory receptor responsive to light

phototransduction   The mechanism of converting light stimuli into electrical activity by the rods and cones of the eye

physiology (fiz-ē-OL-ō-gē)   The study of body functions

pineal gland (PIN-ē-ul)   A small endocrine gland located in the center of the brain that secretes the hormone melatonin

pinocytosis (pin-oh-cī-TŌ-sus)   Type of endocytosis in which the cell internalizes fluid

pitch   The tone of a sound, determined by the frequency of vibrations (that is, whether a sound is a C or G note)

pituitary gland (pih-TWO-ih-tair-ee)   A small endocrine gland connected by a stalk to the hypothalamus; consists of the anterior pituitary and posterior pituitary

placenta (plah-SEN-tah)   The organ of exchange between the maternal and fetal blood; also secretes hormones that support the pregnancy

plaque   A deposit of cholesterol and other lipids, perhaps calcified, in thickened, abnormal smooth muscle cells within blood vessels as a result of atherosclerosis

plasma   The liquid portion of the blood

plasma cell   An antibody-producing derivative of an activated B lymphocyte

plasma clearance   The volume of plasma that is completely cleared of a given substance by the kidneys per minute

plasma-colloid osmotic pressure (KOL-oid os-MOT-ik)   The force caused by the unequal distribution of plasma proteins between the blood and surrounding fluid that encourages fluid movement into the capillaries

plasma membrane   A protein-studded lipid bilayer that encloses each cell, separating it from the extracellular fluid

plasma proteins   The proteins that remain within the plasma, where they perform a number of important functions; include albumins, globulins, and fibrinogen

plasticity (plas-TIS-uh-tē)   The ability of portions of the brain to assume new responsibilities in response to the demands placed on it

platelets (PLĀT-lets)   Specialized cell fragments in the blood that participate in hemostasis by forming a plug at a vessel defect

pleural sac (PLOOR-ul)   A double-walled, closed sac that separates each lung from the thoracic wall

pluripotent stem cells   Precursor cells that reside in the bone marrow and continuously divide and differentiate to give rise to each of the types of blood cells

polycythemia (pol-i-sī-THĒ-mē-uh)   Excess circulating erythrocytes, accompanied by an elevated hematocrit

polysaccharides (pol-ī-SAK-uh-rīdz)   Complex carbohydrates, consisting of chains of interconnected glucose molecules

positive balance   Situation in which the gains via input for a substance exceed its losses via output, so that the total amount of the substance in the body increases

positive feedback   A regulatory mechanism in which the input and the output in a control system continue to enhance each other so that the controlled variable is progressively moved farther from a steady state

postabsorptive state   The metabolic state after a meal is absorbed during which endogenous energy stores must be mobilized and glucose must be spared for the glucose-dependent brain; fasting state

posterior pituitary   The neural portion of the pituitary that stores and releases into the blood on hypothalamic stimulation two hormones produced by the hypothalamus, vasopressin and oxytocin

postganglionic fiber (pōst-gan-glē-ON-ik)   The second neuron in the two-neuron autonomic nerve pathway; originates in an autonomic ganglion and terminates on an effector organ

postsynaptic neuron (pōst-si-NAP-tik)   The neuron that conducts its action potentials away from a synapse

preganglionic fiber   The first neuron in the two-neuron autonomic nerve pathway; originates in the central nervous system and terminates on an autonomic ganglion

pressure gradient   A difference in pressure between two regions that drives the movement of blood or air from the region of higher pressure to the region of lower pressure

presynaptic facilitation   Enhanced release of neurotransmitter from a presynaptic axon terminal as a result of excitation of another neuron that terminates on the axon terminal

presynaptic inhibition   A reduction in the release of neurotransmitter from a presynaptic axon terminal as a result of excitation of another neuron that terminates on the axon terminal

presynaptic neuron (prē-si-NAP-tik)   The neuron that conducts its action potentials toward a synapse

primary active transport   A carrier-mediated transport system in which energy is directly required to operate the carrier and move the transported substance against its concentration gradient

primary follicle   A primary oocyte surrounded by a single layer of granulosa cells in the ovary

primary motor cortex   The portion of the cerebral cortex that lies anterior to the central sulcus and is responsible for voluntary motor output

progestational phase   See *secretory phase*

prolactin (PRL) (prō-LAK-tun)   An anterior pituitary hormone that stimulates breast development and milk production in females

proliferative phase   The phase of the menstrual cycle during which the endometrium is repairing itself and thickening following menstruation; lasts from the end of the menstrual phase until ovulation

pro-opiomelanocortin (prō-op'Ē-ō-ma-LAN-oh-kor'-tin)   A large precursor molecule produced by the anterior pituitary that is cleaved into adrenocorticotropic hormone, melanocyte-stimulating hormone, and endorphin

proprioception (prō'-prē-ō-SEP-shun)   Awareness of position of body parts in relation to each other and to surroundings

prostaglandins (pros'-tuh-GLAN-dins)   Local chemical mediators that are derived from a component of the plasma membrane, arachidonic acid

prostate gland   A male accessory sex gland that secretes an alkaline fluid, which neutralizes acidic vaginal secretions

protein kinase (KĪ-nase)   An enzyme that phosphorylates and thereby induces a change in the shape and function of a particular intracellular protein

proteolytic enzymes (prōt-ē-uh-LIT-ik)   Enzymes that digest protein

proximal tubule (PROKS-uh-mul)   A highly convoluted tubule that extends between Bowman's capsule and the loop of Henle in the kidney's nephron

PTH   See *parathyroid hormone*

pulmonary artery (PULL-mah-nair-ē)   The large vessel that carries blood from the right ventricle to the lungs

pulmonary circulation   The closed loop of blood vessels carrying blood between the heart and lungs

pulmonary surfactant (sur-FAK-tunt)   A phospholipoprotein complex secreted by the Type II alveolar cells that intersperses between the water molecules that line the alveoli, thereby lowering the surface tension within the lungs

pulmonary valve   A one-way valve that permits the flow of blood from the right ventricle into the pulmonary artery during ventricular emptying but prevents the backflow of blood from the pulmonary artery into the right ventricle during ventricular relaxation

pulmonary veins   The large vessels that carry blood from the lungs to the heart

pulmonary ventilation   The volume of air breathed in and out in one minute; equals tidal volume times respiratory rate

pupil   An adjustable round opening in the center of the iris through which light passes to the interior portions of the eye

Purkinje fibers (pur-KIN-jē)   Small terminal fibers that extend from the bundle of His and rapidly transmit an action potential throughout the ventricular myocardium

pyloric gland area (PGA) (pī-LŌR-ik)   The specialized region of the mucosa in the antrum of the stomach that secretes gastrin

pyloric sphincter (pī-lōr'-ik SFINGK-tur)   The juncture between the stomach and duodenum

radiation   Emission of heat energy from the surface of a warm body in the form of electromagnetic waves

reabsorption   The net movement of interstitial fluid into the capillary

receptor   See *sensory receptor* or *receptor site*

receptor potential   The graded potential change that occurs in a sensory receptor in response to a stimulus; generates action potentials in the afferent neuron fiber

receptor site   Membrane protein that binds with a specific extracellular chemical messenger, thereby bringing about a series of membrane and intracellular events that alter the activity of the particular cell

reduced hemoglobin   Hemoglobin that is not combined with O_2

reflex   Any response that occurs automatically without conscious effort; the components of a reflex arc include a receptor, afferent pathway, integrating center, efferent pathway, and effector

refraction   Bending of a light ray

refractory period (rē-FRAK-tuh-rē) The time period when a recently activated patch of membrane is refractory (unresponsive) to further stimulation, preventing the action potential from spreading backward into the area through which it has just passed, thereby ensuring the unidirectional propagation of the action potential away from the initial site of activation

regulatory proteins Troponin and tropomyosin, which play a role in regulating muscle contraction by either covering or exposing the sites of interaction between the contractile proteins

releasing hormone A hypothalamic hormone that stimulates the secretion of a particular anterior pituitary hormone

renal cortex An outer granular-appearing region of the kidney

renal medulla (RĒ-nul muh-DUL-uh) An inner striated-appearing region of the kidney

renal threshold The plasma concentration at which the T_m of a particular substance is reached and the substance first starts appearing in the urine

renin (RĒ-nin) An enzymatic hormone released from the kidneys in response to a decrease in NaCl/ECF volume/arterial blood pressure; activates angiotensinogen

renin-angiotensin-aldosterone system (an´jē-ō-TEN-sun al-dō-steer-OWN) The salt-conserving system triggered by the release of renin from the kidneys, which activates angiotensin, which stimulates aldosterone secretion, which stimulates Na^+ reabsorption by the kidney tubules during the formation of urine

repolarization (rē´-pō-luh-ruh-ZĀ-shun) Return of membrane potential to resting potential following a depolarization

reproductive tract The system of ducts that are specialized to transport or house the gametes after they are produced

residual volume The minimum volume of air remaining in the lungs even after a maximal expiration

resistance Hindrance of flow of blood or air through a passageway (blood vessel or respiratory airway, respectively)

respiration The sum of processes that accomplish ongoing passive movement of O_2 from the atmosphere to the tissues, as well as the continual passive movement of metabolically produced CO_2 from the tissues to the atmosphere

respiratory acidosis (as-i-DŌ-sus) Acidosis resulting from abnormal retention of CO_2 arising from hypoventilation

respiratory airways The system of tubes that conducts air between the atmosphere and the alveoli of the lungs

respiratory alkalosis (al´-kuh-LŌ-sus) Alkalosis caused by excessive loss of CO_2 from the body as a result of hyperventilation

respiratory rate Breaths per minute

resting membrane potential The membrane potential that exists when an excitable cell is not displaying an electrical signal

reticular activating system (RAS) (ri-TIK-ū-lur) Ascending fibers that originate in the reticular formation and carry signals upward to arouse and activate the cerebral cortex

reticular formation A network of interconnected neurons that runs throughout the brain stem and initially receives and integrates all synaptic input to the brain

retina The innermost layer in the posterior region of the eye that contains the eye's photoreceptors, the rods and cones

ribonucleic acid (RNA) (rī-bō-new-KLĀ-ik) A nucleic acid that exists in three forms (messenger RNA, ribosomal RNA, and transfer RNA), which participate in gene transcription and protein synthesis

ribosomes (RĪ-bō-sōms) Special ribosomal RNA-protein complexes that synthesize proteins under the direction of nuclear DNA

right atrium (Ā-trē´-um) The heart chamber that receives venous blood from the systemic circulation

right ventricle The heart chamber that pumps blood into the pulmonary circulation

RNA See *ribonucleic acid*

rods The eye's photoreceptors used for night vision

round window The membrane-covered opening that separates the lower chamber of the cochlea in the inner ear from the middle ear

salivary amylase (AM-uh-lās´) An enzyme produced by the salivary glands that begins carbohydrate digestion in the mouth and continues in the body of the stomach after the food and saliva have been swallowed

saltatory conduction (SAL-tuh-tōr´-ē) The means by which an action potential is propagated throughout a myelinated fiber, with the impulse jumping over the myelinated regions from one node of Ranvier to the next

SA node See *sinoatrial node*

sarcomere (SAR-kō-mir) The functional unit of skeletal muscle; the area between two Z lines within a myofibril

sarcoplasmic reticulum (ri-TIK-yuh-lum) A fine meshwork of interconnected tubules that surrounds a muscle fiber's myofibrils; contains expanded lateral sacs, which store calcium that is released into the cytosol in response to a local action potential

satiety centers (suh-TĪ-ut-ē) Neuronal clusters in the ventromedial region of the hypothalamus that inhibit feeding behavior

saturation When all the binding sites on a carrier molecule are occupied

Schwann cells (shwah´-n) The myelin-forming cells of the peripheral nervous system

secondary active transport A transport mechanism in which a carrier molecule for glucose or an amino acid is driven by a Na^+ concentration gradient established by the energy-dependent Na^+ pump to transfer the glucose or amino acid uphill without directly expending energy to operate the carrier

secondary follicle A developing ovarian follicle that is secreting estrogen and forming an antrum

secondary sexual characteristics The many external characteristics that are not directly involved in reproduction but that distinguish males and females

second messenger An intracellular chemical that is activated by binding of an extracellular first messenger to a surface receptor site and that triggers a preprogrammed series of biochemical events, which result in altered activity of intracellular proteins to control a particular cellular activity

secretin (si-KRĒT-´n) A hormone released from the duodenal mucosa primarily in response to the presence of acid; inhibits gastric motility and secretion and stimulates secretion of a $NaHCO_3$ solution from the pancreas and from the liver

secretion Release to a cell's exterior, on appropriate stimulation, of substances that have been produced by the cell

secretory phase The phase of the menstrual cycle characterized by the development of a lush endometrial lining capable of supporting a fertilized ovum; also known as the *progestational phase*

secretory vesicles (VES-i-kuls) Membrane-enclosed sacs containing proteins that have been synthesized and processed by the endoplasmic reticulum/Golgi complex of the cell and which will be released to the cell's exterior by exocytosis on appropriate stimulation

segmentation The small intestine's primary method of motility; consists of oscillating, ringlike contractions of the circular smooth muscle along the small intestine's length

self-antigens Antigens that are characteristic of a person's own cells

semen (SĒ-men) A mixture of accessory sex-gland secretions and sperm

semicircular canal Sense organ in the inner ear that detects rotational or angular acceleration or deceleration of the head

semilunar valves (sem´-ī-LEW-nur) The aortic and pulmonary valves

seminal vesicles (VES-i-kuls) Male accessory sex glands that supply fructose to ejaculated sperm and secrete prostaglandins

seminiferous tubules (sem´-uh-NIF-uh-rus)   The highly coiled tubules within the testes that produce spermatozoa

sensory afferent   Pathway coming into the central nervous system that carries information that reaches the level of consciousness

sensory input   Includes somatic sensation and special senses

sensory receptor   An afferent neuron's peripheral ending, which is specialized to respond to a particular stimulus in its environment

series-elastic component   The noncontractile portions of a skeletal muscle fiber, including the connective tissue and sarcoplasmic reticulum

Sertoli cells (sur-TŌ-lē)   Cells located in the seminiferous tubules that support spermatozoa during their development

set point   The desired level at which homeostatic control mechanisms maintain a controlled variable

signal transduction   The sequence of events in which incoming signals (instructions from extracellular chemical messengers such as hormones) are conveyed to the cell's interior for execution

single-unit smooth muscle   The most abundant type of smooth muscle; made up of muscle fibers that are interconnected by gap junctions so that they become excited and contract as a unit; also known as *visceral smooth muscle*

sinoatrial (SA) node (sī-nō-Ā-trē-ul)   A small specialized autorhythmic region in the right atrial wall of the heart that has the fastest rate of spontaneous depolarizations and serves as the normal pacemaker of the heart

skeletal muscle   Striated muscle, which is attached to the skeleton and is responsible for movement of the bones in purposeful relation to one another; innervated by the somatic nervous system and under voluntary control

slow-wave potentials   Self-excitable activity of an excitable cell in which its membrane potential undergoes gradually alternating depolarizing and hyperpolarizing swings

smooth muscle   Involuntary muscle innervated by the autonomic nervous system and found in the walls of hollow organs and tubes

somatic cells (sō-MAT-ik)   Body cells, as contrasted with reproductive cells

somatic nervous system   The portion of the efferent division of the peripheral nervous system that innervates skeletal muscles; consists of the axonal fibers of the alpha motor neurons

somatic sensation   Sensory information arising from the body surface, including somesthetic sensation and proprioception

somatomedins (sō´-mat-uh-MĒ-dinz)   Hormones secreted by the liver or other tissues, in response to growth hormone, that act directly on the target cells to promote growth

somatosensory cortex   The region of the parietal lobe immediately behind the central sulcus; the site of initial processing of somesthetic and proprioceptive input

somesthetic sensations (SŌ-mes-THEH-tik)   Awareness of sensory input such as touch, pressure, temperature, and pain from the body's surface

sound waves   Traveling vibrations of air that consist of regions of high pressure caused by compression of air molecules alternating with regions of low pressure caused by rarefaction of the molecules

spatial summation   The summing of several postsynaptic potentials arising from the simultaneous activation of several excitatory (or several inhibitory) synapses

special senses   Vision, hearing, taste, and smell

specificity   Ability of carrier molecules to transport only specific substances across the plasma membrane

spermatogenesis (spur´-mat-uh-JEN-uh-sus)   Sperm production

sphincter (sfink-tur)   A voluntarily controlled ring of skeletal muscle that controls passage of contents through an opening into or out of a hollow organ or tube

spinal reflex   A reflex that is integrated by the spinal cord

spleen   A lymphoid tissue in the upper left part of the abdomen that stores lymphocytes and platelets and destroys old red blood cells

state of equilibrium   No net change in a system is occurring

stem cells   Relatively undifferentiated precursor cells that give rise to highly differentiated, specialized cells

steroids (STEER-oidz)   Hormones derived from cholesterol

stimulus   A detectable physical or chemical change in the environment of a sensory receptor

stress   The generalized, nonspecific response of the body to any factor that overwhelms, or threatens to overwhelm, the body's compensatory abilities to maintain homeostasis

stretch reflex   A monosynaptic reflex in which an afferent neuron originating at a stretch-detecting receptor in a skeletal muscle terminates directly on the efferent neuron supplying the same muscle to cause it to contract and counteract the stretch

stroke volume (SV)   The volume of blood pumped out of each ventricle with each contraction, or beat, of the heart

subcortical regions   The brain regions that lie under the cerebral cortex, including the basal nuclei, thalamus, and hypothalamus

submucosa   The connective tissue layer of the digestive tract that lies under the mucosa and contains the larger blood and lymph vessels and a nerve network

substance P   The neurotransmitter released from pain fibers

subsynaptic membrane (sub-sih-NAP-tik)   The portion of the postsynaptic cell membrane that lies immediately underneath a synapse and contains receptor sites for the synapse's neurotransmitter

suprachiasmatic nucleus (soup´-ra-kī-as-MAT-ik)   A cluster of nerve cell bodies in the hypothalamus that serves as the master biological clock, acting as the pacemaker that establishes many of the body's circadian rhythms

surface tension   The force at the liquid surface of an air–water interface resulting from the greater attraction of water molecules to the surrounding water molecules than to the air above the surface; a force that tends to decrease the area of a liquid surface and resists stretching of the surface

sympathetic nervous system   The subdivision of the autonomic nervous system that dominates in emergency ("fight-or-flight") or stressful situations and prepares the body for strenuous physical activity

synapse (SIN-aps´)   The specialized junction between two neurons where an action potential in the presynaptic neuron influences the membrane potential of the postsynaptic neuron by means of the release of a chemical messenger that diffuses across the small cleft that separates the two neurons

synergism (SIN-er-jiz´-um)   When several actions are complementary, so that their combined effect is greater than the sum of their separate effects

systemic circulation (sis-TEM-ik)   The closed loop of blood vessels carrying blood between the heart and body systems

systole (SIS-tō-lē)   The period of cardiac contraction and emptying

T_3 See *tri-iodothyronine*

T_4 See *thyroxine*

T_m See *transport maximum* and *tubular maximum*

tactile (TACK-til)   Referring to touch

target cell receptors   Receptors located on a target cell that are specific for a particular chemical mediator

target cells   The cells that a particular extracellular chemical messenger, such as a hormone or a neurotransmitter, influences

teleological approach (tē´-lē-ō-LA-ji-kul)   Explanation of body functions in terms of their particular purpose in fulfilling a bodily need, that is, the "why" of body processes

temporal lobes   The lobes of the cerebral cortex that are located laterally and that are responsible for initially processing auditory input

temporal summation The summing of several postsynaptic potentials occurring very close together in time because of successive firing of a single presynaptic neuron

tension The force produced during muscle contraction by shortening of the sarcomeres, resulting in stretching and tightening of the muscle's elastic connective tissue and tendon, which transmit the tension to the bone to which the muscle is attached

terminal button A motor neuron's enlarged knoblike ending that terminates near a skeletal muscle fiber and releases acetylcholine in response to an action potential in the neuron

testosterone (tes-TOS-tuh-rōn) The male sex hormone, secreted by the Leydig cells of the testes

tetanus (TET´-n-us) A smooth, maximal muscle contraction that occurs when the fiber is stimulated so rapidly that it does not have a chance to relax at all between stimuli

thalamus (THAL-uh-mus) The brain region that serves as a synaptic integrating center for preliminary processing of all sensory input on its way to the cerebral cortex

thecal cells (THAY-kel) The outer layer of specialized ovarian connective tissue cells in a maturing follicle

thermoreceptor (thur´-mō-rē-SEP-tur) A sensory receptor sensitive to heat and cold

thick filaments Specialized cytoskeletal structures within skeletal muscle that are made up of myosin molecules and interact with the thin filaments to accomplish shortening of the fiber during muscle contraction

thin filaments Specialized cytoskeletal structures within skeletal muscle that are made up of actin, tropomyosin, and troponin molecules and interact with the thick filaments to accomplish shortening of the fiber during muscle contraction

thoracic cavity (thō-RAS-ik) Chest cavity

threshold potential The critical potential that must be reached before an action potential is initiated in an excitable cell

thromboembolism (throm´-bō-EM-buh-liz-um) A condition characterized by the presence of thrombi and emboli in the circulatory system

thrombus An abnormal clot attached to the inner lining of a blood vessel

thymus (THIGH-mus) A lymphoid gland located midline in the chest cavity that processes T lymphocytes and produces the hormone thymosin, which maintains the T cell lineage

thyroglobulin (thī´-rō-GLOB-yuh-lun) A large, complex molecule on which all steps of thyroid hormone synthesis and storage take place

thyroid gland A bilobed endocrine gland that lies over the trachea and secretes three hormones: thyroxine and tri-iodothyronine, which regulate overall basal metabolic rate, and calcitonin, which contributes to control of calcium balance

thyroid hormone Collectively, the hormones secreted by the thyroid follicular cells, namely, thyroxine and tri-iodothyronine

thyroid-stimulating hormone (TSH) An anterior pituitary hormone that stimulates secretion of thyroid hormone and promotes growth of the thyroid gland; thyrotropin

thyroxine (thī-ROCKS-in) The most abundant hormone secreted by the thyroid gland; important in the regulation of overall metabolic rate; also known as *tetraiodothyronine* or T_4

tidal volume The volume of air entering or leaving the lungs during a single breath

tight junction An impermeable junction between two adjacent epithelial cells formed by the sealing together of the cells' lateral edges near their luminal borders; prevents passage of substances between the cells

tissue (1) A functional aggregation of cells of a single specialized type, such as nerve cells forming nervous tissue; (2) the aggregate of various cellular and extracellular components that make up a particular organ, such as lung tissue

T lymphocytes (T cells) White blood cells that accomplish cell-mediated immune responses against targets to which they have been previously exposed; see also *cytotoxic T cells*, *helper T cells*, and *suppressor T cells*

tone The ongoing baseline of activity in a given system or structure, as in muscle tone, sympathetic tone, or vascular tone

total peripheral resistance The resistance offered by all the peripheral blood vessels, with arteriolar resistance contributing most extensively

trachea (TRĀ-kē-uh) The "windpipe"; the conducting airway that extends from the pharynx and branches into two bronchi, each entering a lung

tract A bundle of nerve fibers (axons of long interneurons) with a similar function within the spinal cord

transduction Conversion of stimuli into action potentials by sensory receptors

transepithelial transport (tranz-ep-i-THĒ-lē-al) The entire sequence of steps involved in the transfer of a substance across the epithelium between either the renal tubular lumen or digestive tract lumen and the blood

transmural pressure gradient The pressure difference across the lung wall (intra-alveolar pressure is greater than intrapleural pressure) that stretches the lungs to fill the thoracic cavity, which is larger than the unstretched lungs

transporter recruitment The phenomenon of inserting additional transporters (carriers) for a particular substance into the plasma membrane, thereby increasing membrane permeability to the substance, in response to an appropriate stimulus

transport maximum (T_m) The maximum rate of a substance's carrier-mediated transport across the membrane when the carrier is saturated; known as *tubular maximum* in the kidney tubules

transverse tubule (T tubule) A perpendicular infolding of the surface membrane of a muscle fiber; rapidly spreads surface electric activity into the central portions of the muscle fiber

triglycerides (trī-GLIS-uh-rīdz) Neutral fats composed of one glycerol molecule with three fatty acid molecules attached

tri-iodothyronine (T_3) (trī-ī-ō-dō-THĪ-rō-nēn) The most potent hormone secreted by the thyroid follicular cells; important in the regulation of overall metabolic rate

trophoblast (TRŌF-uh-blast´) The outer layer of cells in a blastocyst that is responsible for accomplishing implantation and developing the fetal portion of the placenta

tropic hormone (TRŌ-pik) A hormone that regulates the secretion of another hormone

tropomyosin (trōp´-uh-MĪ-uh-sun) One of the regulatory proteins found in the thin filaments of muscle fibers

troponin (tro-PŌ-nun) One of the regulatory proteins found in the thin filaments of muscle fibers

TSH See *thyroid-stimulating hormone*

T tubule See *transverse tubule*

tubular maximum (T_m) The maximum amount of a substance that the renal tubular cells can actively transport within a given time period; the kidney cells' equivalent of transport maximum

tubular reabsorption The selective transfer of substances from the tubular fluid into the peritubular capillaries during the formation of urine

tubular secretion The selective transfer of substances from the peritubular capillaries into the tubular lumen during the formation of urine

twitch A brief, weak contraction that occurs in response to a single action potential in a muscle fiber

twitch summation The addition of two or more muscle twitches as a result of rapidly repetitive stimulation, resulting in greater tension in the fiber than that produced by a single action potential

tympanic membrane (tim-PAN-ik) The eardrum, which is stretched across the entrance to the middle ear and which vibrates when struck by sound waves funneled down the external ear canal

Type I alveolar cells (al-VĒ-ō-lur) The single layer of flattened epithelial cells that forms the wall of the alveoli within the lungs

Type II alveolar cells The cells within the alveolar walls that secrete pulmonary surfactant

ultrafiltration The net movement of a protein-free plasma out of the capillary into the surrounding interstitial fluid

ureter (yū-RĒ-tur) A duct that transmits urine from the kidney to the bladder

urethra (yū-RĒ-thruh) A tube that carries urine from the bladder to outside the body

urine excretion The elimination of substances from the body in the urine; anything filtered or secreted and not reabsorbed is excreted

vagus nerve (VĀ-gus) The tenth cranial nerve, which serves as the major parasympathetic nerve

vasoconstriction (vā′-zō-kun-STRIK-shun) The narrowing of a blood vessel lumen as a result of contraction of the vascular circular smooth muscle

vasodilation The enlargement of a blood vessel lumen as a result of relaxation of the vascular circular smooth muscle

vasopressin (vā-zō-PRES-sin) A hormone secreted by the hypothalamus, then stored and released from the posterior pituitary; increases the permeability of the distal and collecting tubules of the kidneys to water and promotes arteriolar vasoconstriction; also known as *antidiuretic hormone (ADH)*

vaults Recently discovered organelles shaped like octagonal barrels; believed to serve as transporters for messenger RNA and/or the ribosomal subunits from the nucleus to sites of protein synthesis; may be important in cellular movement

vein A vessel that carries blood toward the heart

vena cava (VĒ-nah CĀV-ah), (venae cavae, plural; VĒ-nē cāv-ē) A large vein that empties blood into the right atrium

venous return (VĒ-nus) The volume of blood returned to each atrium per minute from the veins

ventilation The mechanical act of moving air in and out of the lungs; breathing

ventricle (VEN-tri-kul) A lower chamber of the heart that pumps blood into the arteries

vesicle (VES-i-kul) A small, intracellular, fluid-filled, membrane-enclosed sac

vesicular transport Movement of large molecules or multimolecular materials into or out of the cell by means of being enclosed in a vesicle, as in endocytosis or exocytosis

vestibular apparatus (veh-STIB-yuh-lur) The component of the inner ear that provides information essential for the sense of equilibrium and for coordinating head movements with eye and postural movements; consists of the semicircular canals, utricle, and saccule

villus (villi, plural) (VIL-us) Microscopic fingerlike projections from the inner surface of the small intestine

virulence (VIR-you-lentz) The disease-producing power of a pathogen

visceral afferent A pathway coming into the central nervous system that carries subconscious information derived from the internal viscera

visceral smooth muscle (VIS-uh-rul) See *single-unit smooth muscle*

viscosity (vis-KOS-i-tē) The friction developed between molecules of a fluid as they slide over each other during flow of the fluid; the greater the viscosity, the greater the resistance to flow

vital capacity The maximum volume of air that can be moved out during a single breath following a maximal inspiration

voltage-gated channels Channels in the plasma membrane that open or close in response to changes in membrane potential

white matter The portion of the central nervous system composed of myelinated nerve fibers

Z line A flattened disclike cytoskeletal protein that connects the thin filaments of two adjoining sarcomeres

zona fasciculata (zō-nah fa-SIK-ū-lah-ta) The middle and largest layer of the adrenal cortex; major source of cortisol

zona glomerulosa (glō-MER-yū-lō-sah) The outermost layer of the adrenal cortex; sole source of aldosterone

zona reticularis (ri-TIK-yuh-lair-us) The innermost layer of the adrenal cortex; produces cortisol, along with the zona fasciculata

Index

A⁻
resting membrane potential and, 88
α (alpha) cells, 723
α_1-antitrypsin, 472
Aβ42, 162
A bands, 258
in contraction process, 261
Abdominal muscles
active expiration, role in, 470
vomiting, role in, 608
ABO blood type system, 432
Abortion, 789
stem cell science and, 9
Absolute refractory period, 112
Absorptive state, 722
AB type blood
antibodies in, 432
as universal recipients, 432–433
Accessory digestive organs, 594
Accessory inspiratory muscles, 467
Accessory proteins, 48
Accessory sex glands, 749–750
semen, contribution to, 764–765
Acclimatization, 496, 566
Accommodation of lens, 200–201
mechanism of, 201
ACE inhibitor drugs, 529–530
Acetic acid, 35–36
Acetone, A13
Acetylcholine (ACh), 121, 134
Aβ42 binding with, 162
atropine and, 245
black widow spider venom and, 251
botulinum toxin and, 251–252
curare and, 251, 252
effector organs and, 237
in excitation-contraction coupling, 268
gastric juices, influence on, 612
heart rate and, 326
motor neuron axon terminals releasing, 245
muscarinic receptors binding with, 244
myasthenia gravis and, 252
as neuromuscular junction neurotransmitter, 246–252
nicotinic receptors responding to, 243–244
organophosphates and, 251, 252
parasympathetic postganglionic fibers releasing, 238–240

release at neuromuscular junction, 247
sites of release, 240
skeletal muscle stimulated by, 263
Acetylcholinesterase (AChE)
myasthenia gravis and, 252
at neuromuscular junction, 250–251
organophosphates and, 251, 252
Acetyl coenzyme A (acetyl CoA), 36, 722–723
Acetylsalicylic acid. See Aspirin
ACh. See Acetylcholine (ACh)
AChE. See Acetylcholinesterase (AChE)
Acid-base balance, 558, 571–586. See also Acidosis; Alkalosis
ammonia, kidneys secreting, 581–582
bicarbonate and secretion of hydrogen, 580
chemical buffer system, 575
defenses against changes in, 574–582
digestive processes and, 635
disorders from imbalances, 582–586
enzymes, effect on, 573
hemoglobin buffer system, 575, 577
Henderson-Hasselbalch equation, 576
imbalances, causes of, 582–586
kidneys and, 512, 578–582
lungs, role of, 578
nerve excitability and, 573
phosphate buffer system, 577, 582
plasma hydrogen concentration and, 579–582
potassium, effect on, 573
protein buffer system, 575, 576–577
rate of hydrogen secretion, factors affecting, 579
respiratory system and, 577–578, 582–584
sources of hydrogen in body, 574
Acidosis, 573
ammonia, secretion of, 582
diabetes mellitus and, 726
kidneys, role of, 581–582
metabolic acidosis, 583, 584–585
respiratory acidosis, 583, 584
respiratory adjustments to, 577–578
Acids, 572, A10–A11. See also Acid-base balance
in chemistry, 572–573
defined, A11
dissociation of, 572

duodenum, effect on, 606, 607
inorganic acids, hydrogen ion and, 574
organic acids, hydrogen ion and, 574
strong vs. weak acids, 572
Acini, 616
Acne, testosterone and, 758
Acquired reflexes, 177
Acquired salivary reflex, 601
Acromegaly, 695
Acrosomal membrane, 783
Acrosome, 760
ACTH (adrenocorticotropic hormone), 121, 672
androgen secretion and, 711
anterior pituitary secreting, 685
cortisol and, 709, 710–711, 714
CRH-ACTH-cortisol system, 689
and parturition, 794
stress, role in, 717
Actin, 49
in amoeboid movement, 50
ATP (adenosine triphosphate) and, 264
in excitation-contraction coupling, 268
muscle fibers and, 281–282
in platelets, 402
thin filaments composed of, 258, 260
Action potentials, 103–116
all-or-none law of, 113
in autorhythmic tissues of heart, 311
axon hillock and, 108–109, 123
axons, conduction through, 110–111
cardiac muscle cells generating, 308–309, 315
contiguous conduction, 110–111
contraction of muscles and, 271
defined, 103
end-plate potential (EPP) and, 250
falling phase of, 108
frequency of, 113
generator potentials, conversion of, 188
graded potentials compared, 111
ion movement changes and, 105–108
means of initiating, 294
Na^+-K^+ ATPase pump and, 108
in neuromuscular junction, 250
nondecremental spread of, 103
permeability and, 104–108, 105–108
phantom pain and, 190
photoreceptors, initiation of, 205

Action potentials *(continued)*
  Purkinje fibers transmitting, 314
  receptor potentials, conversion of, 188
  refractory period for, 111–113
  rising phase of, 107–108
  saltatory conduction, 115
  sinoatrial node (SA node) and, 326
  in skeletal muscle fiber, 268
  speed of, 113
  strength of stimulus and, 113
  threshold potential, 103–104
  triggering events, 103–104
  voltage-gated sodium/potassium
      channels, 104–108
Activated adenylyl cyclase, 67, 72
Activated protein kinase, 72
Activated T lymphocytes, 424, 436
Active area, 101
Active forces, 72
Active hormones, 672
Active hyperemia, 353–354
Active immunity, 432
Active transport, 80, 81–83
  Na^+-K^+ ATPase pump, 81–83
  secondary active transport, 83
Acuity, 190–192
  lateral inhibition and, 191–192
  with rod vision, 207
  two-point threshold-of-discrimination test,
      191
Acupuncture, 195
Acupuncture endorphin hypothesis, 195
Acute mountain sickness, 496–497
Acute myocardial infarction. *See* Heart
      attacks
Acute-phase proteins, 419
Acute renal failure, 549
Adam's apple, 461
Adaptation
  dark adaptation, 207–208
  light adaptation, 207–208
  of olfactory system, 231
  in Pacinian corpuscle, 189–190
Adaptive immune system, 415, 416
Adaptive immunity, 424–445
  bacterial invasion, responses to, 435
  viral invasion, responses to, 439
Addiction
  anabolic androgenic steroids and, 282
  to cocaine, 126, 159
  of newborns, 786–787
Addison's disease, 713–714
A-delta fibers, 192
Adenine (A), A19
Adenohypophysis, 683
Adenoids, 414
  functions of, 415
  as immune defense, 452
Adenosine, A17
  arteriolar smooth muscles and, 355
  sleep and, 171–172
Adenosine diphosphate. *See* ADP
      (adenosine diphosphate)

Adenosine triphosphate. *See* ATP
      (adenosine triphosphate)
Adenylyl cyclase, 67, 72
Adequate stimulus, 187
Adipocytes, 650–651
Adipose tissue
  autonomic nervous system, effect of, 242
  in metabolic states, 722
  obesity and, 653
  site of fat storage, 44
  triglycerides in, A13
  water content of, 561
ADP (adenosine diphosphate), 35, A17–A18
  creatine phosphate and, 277
  cross bridge cycling and, 264
  glycolysis and, 35
  platelets releasing, 403–404
  recharging in mitochondria, 42
Adrenal cortex, 243, 707–708. *See also*
      Aldosterone, Cortisol; Sex
      hormones
  atrial natriuretic peptide (ANP) and, 530
  disorders of, 712–714
  hormones secreted by, 670, 708
Adrenal diabetes, 712
Adrenal glands, 243, 700, 707–719
  adrenocortical insufficiency, 713–714
Adrenaline. *See* Epinephrine
Adrenal medulla, 243, 707–708, 714–719.
      See also Catecholamines;
      Epinephrine
  autonomic nervous system, effect of, 242
  heart rate and, 328
  hormones from, 243, 670
  hyperthermia and, 662
  as sympathetic postganglionic neuron,
      714–715
Adrenergic drugs, 448
Adrenergic receptors, 244–245, 715
Adrenocortical hormones, 708
Adrenocortical insufficiency, 713–714
Adrenocorticotropic hormone. *See* ACTH
      (adrenocorticotropic hormone)
Adrenogenital syndrome, 712–713, 756
  interrelationships of hormones, 714
Aerobic condition
  citric acid cycle in, 40
  mitochondrial processing in, 40, 41
Aerobic exercise, 42
  glucose use and, 81
  hypertension and, 382
  immune system and, 446
  maximal O_2 consumption (max VO_2), 504
  oxidative muscle fibers and, 281
  oxidative phosphorylation and, 277
  recruitment of motor units in, 270
Afferent arterioles
  glomerular filtration rate (GFR) and, 520
  of nephrons, 513–514
Afferent axons/fibers, 136
  pain, transmittal of, 192
  spinal nerves carrying, 176–177
Afferent division of PNS, 136

Afferent neurons, 136–137
  acid-base balance and, 573
  in auditory cortex, 220–221
  features of, 248
  in Golgi tendon organs, 288
  motor unit output and, 283
  for muscle spindles, 286
  receptors, 186–192
  short-term memory and, 164
  in urinary bladder, 552
  in withdrawal reflex, 178
Afferent pathway in reflex arc, 177
Afterbirth, 795
After hyperpolarization, 103
Afterload, 330–331
Age
  autonomic nervous system and, 244
  deafness and, 221
  DHEA (dehydroepiandrosterone) and,
      711–712
  growth hormone and, 696, 711
  IGF-I levels and, 693
  melatonin and, 711
  mitochondrial DNA and, 34
  presbyopia, 200
Agglutination, 426, 427
  transfusion reaction, 432, 433
Agonists, 245
AIDS, 446. *See also* HIV
  apoptosis and, 71
  blood substitutes and, 398
  growth hormone treatment for, 696
  helper T cells and, 436
  microglia, role of, 139
  microscopic photograph of virus, 437
  placental barrier, crossing, 786–787
Air pressure, eustachian tube and, 216
Airway resistance, 471–473
  breathing and, 477
  carbon dioxide and, 482
  chronic obstructive pulmonary disease
      (COPD), 471–473
  collapse of airways, 473
  factors affecting, 472
Albumins, 392, 393
  in urine, 518
Albuminuria, 518
Alcohol
  blood-brain barrier and, 142
  gastric secretion and, 613
  heart rate and, 312
  placental barrier, crossing, 786–787
  stomach absorbing, 614–615
  vasopressin and, 571
  water diuresis and, 548
Aldehydes, A11
Aldosterone, 708. *See also* Renin-
      angiotensin-aldosterone system
  in Addison's disease, 713–714
  atrial natriuretic peptide (ANP) and, 530
  deficiencies of, 714
  hypersecretion of, 712
  potassium and, 535–536, 709, 714

principle site of action, 709
renin-angiotensin-aldosterone system, 528–530
sodium and, 527–530, 536, 709, 714
Alendronate, 737
Alkalines. *See* Bases
Alkaloids, 228
Alkalosis, 573
kidneys, role of, 581–582
metabolic alkalosis, 583, 585–586
respiratory adjustments to, 577–578
respiratory alkalosis, 583, 584
Alkalotic condition, 495
Allergens, 446
Allergies, 421, 446–449
breast-feeding and, 798
eosinophils and, 401
hypersensitivity and, 446–449
immune responses and, 413
mast cells and, 447
All-or-none law, 113
swallowing and, 601–602
Allurin, 782
Alpha (α) globulins, 393
Alpha (α) receptors, 224–245
Alpha motor neurons, 286, 287–288
Alveolar air, 484
Alveolar dead space, 482
Alveolar interdependence, 475–476
Alveolar macrophages, 452
Alveolar P_{CO_2}, 484–486
Alveolar P_{O_2}, 484–486
Alveolar surface tension, 474
Alveolar ventilation, 480
breathing patterns and, 480–482
Alveoli, 459. *See also* Intra-alveolar pressure
anatomic dead space, 480
of breasts, 796
Fick's law of diffusion and, 486
intra-alveolar pressure, 464
law of LaPlace, 474–475
net diffusion of oxygen and carbon dioxide, 488
pulmonary capillaries and, 462
structure of, 462–463
Alzheimer, Alois, 162
Alzheimer's disease, 162–163
apoptosis and, 71
hippocampus and, 161, 162
microglia, role of, 139
possible causes of, 163
treatment of, 163
Amblyopia, 133
Amenorrheic athletes, 780
Amines, 672
mechanisms of, 674
Amino acids, 30. *See also* Fuel metabolism; Peptides
as biomolecules, A10
blood-brain barrier and, 142
blood concentrations, 731
chemical composition of, A14
cortisol and, 709

diabetes mellitus and, 728
digestion of proteins into, 592
growth hormone, enhancement of secretion of, 693
insulin and, 723, 725
intermediary metabolism and, 43
pepsin and, 611
peptide bonds, A14–A15
reabsorption of, 526
ribosomes and, A25
secondary active transport of, 531
small intestine, absorption by, 629
small polypeptides, 592
storage of, 721
tubular reabsorption of, 530–531
umami taste and, 228–229
Aminopeptidases, 625
Ammonia
kidneys secreting, 581–582
as urinary buffer, 582
Ammonium ion, 582
Amnesia, 160–161
Amniotic sac, 785
Amoebae, 4
Amoeboid movement, 49–50
AMPA receptors, 165–166
glutamate and, 193–194
Amphetamines, 158
Ampulla, 223, 784
zygotes in, 783–784
Amygdala, 157
Amylase
pancreatic amylase, 617
salivary amylase, 601
Amyloid precursor protein (APP), 162
Amyotrophic lateral sclerosis (ALS), 51
as motor neuron disease, 246
Anabolic androgenic steroids, 282
Anabolism, 719–720
Anaerobic condition, 40, 41
Anaerobic exercise, 42
lactic acid and, 279
Analgesia, acupuncture, 195
Analgesic properties, 192
Analgesic system of brain, 194
Anaphase, A27
Anaphylactic shock, 383, 448
Anatomic dead space, 480, 481
Anatomy, 4
Androgen-binding protein, 761–762
Androgens, 711. *See also* DHEA (dehydroepiandrosterone)
environmental pollutants, 769
growth and, 690, 695–696
hair, axillary/pubic, 750
hypersecretion of adrenal androgen, 712–713
sexual differentiation and, 755
Android obesity, 653, 654
Anemia, 396–397. *See also* Pernicious anemia
hypoxia, 494
leukemia and, 402
renal failure and, 549

Anemia hypoxia, 494
Anger
anabolic androgenic steroids and, 282
behavioral patterns and, 157–158
testosterone and, 758
Angina pectoris, 335
Angiotensin. *See also* Renin-angiotensin-aldosterone system
and zona glomerulosa, 709
Angiotensin-converting enzyme (ACE), 528–530
ACE inhibitor drugs, 529–530
Angiotensin II, 121. *See also* Renin-angiotensin-aldosterone system
arteriolar radius, effect on, 358–359
renal hypertension and, 380
thirst and, 570
vasoconstriction and, 359
vasopressin and, 571
Angiotensinogen, 381
Angry macrophages, 440
Anions, membrane potential and, 87
Annulospiral endings, 286
Anorexia nervosa, 653–654
Anovulatory, 770
Antagonism of hormones, 681
Antagonists, 245
Anterior pituitary, 666, 683. *See also* individual hormones secreted by
blood supply to, 688
exclusive endocrine function of, 669
functions of hormones of, 686
growth hormone-secreting tumors, 695
hierarchic chain of command in, 687
hypophysiotropic hormones and, 687–688
hypothalamic hormones regulating, 685, 687
hypothalamic-hypophyseal portal system, 688
major hormones secreted by, 669–670, 685
MSH (melanocyte-stimulating hormone) secretion, 683–684
variety of hormones from, 668
vascular link with hypothalamus, 688
Anterograde amnesia, 161
Antibodies
agglutination, 426, 427
amplification of innate responses, 426–428
blocking antibodies, 445
blood types, association with, 432–433
B lymphocytes producing, 401
clonal selection theory, 428
colostrum and, 798
complement system and, 423, 427–428
immune complex disease and, 428
innate immune responses and, 426–428
killer (K) cells, stimulation of, 427, 428
neutralization, 426, 427
primary response, 429–430
Rh factor and, 433
secondary response, 429–430

Antibodies (*continued*)
  structure of, 426
  subclasses of, 425–426
  tail region of, 426
  y-shape of, 426
Antibody-mediated immunity, 424, 425–435
Anticodons, A25
Antidepressant drugs, 159
Antidiuretic hormone, 544
Antigen-binding fragments (Fab), 426
Antigen-presenting cells, 433–435
Antigens, 425, 428. *See also* Self-antigens;
        B lymphocytes; T lymphocytes
  agglutination, 426, 427
  antigen-presenting cells and, 433–435
  autoimmune diseases and, 441
  H-Y antigen, 754
  immune complex disease, 428
  MHC molecules and, 434
  neutralization, 426, 427
  plasma cells and, 425–426
  sequestering, 440
Antihistamines
  treatment of asthma, 448
  treatment of ulcers, 615
Antihypertensive medication, 382
Antioxidants
  LDL (low-density lipoprotein) oxidation
        and, 334
  melatonin as, 683
Antitoxins, antibodies against, 432
Antrum, 604
  carbohydrates, digestion of, 614
  gastric mixing in, 605
  Helicobacter pylori in, 615
  in ovarian follicles, 772–773
  pyloric gland area (PGA), 608
  in stomach, 604
Anus, 594
  internal anal sphincter, 638
Anxiety
  food intake and, 652
  heart rate and, 312
Aorta, 305
Aortic arch baroreceptor, 377
Aortic bodies, 499
Aortic valve, 307–308
Aphasias, 151–152
Aplastic anemia, 397
Aplysia, 162–164
Apnea, 494, 504. *See also* Sleep apnea
Apneusis, 498
Apneustic center, 496, 497, 498
ApoE gene, 163
Apolipoprotein E, 163
Apolipoprotein E-4, 163
Apoptosis, 70–71
  Alzheimer's disease and, 163
  control of, 71
  necrosis compared, 70–71
Appendicitis, 638
Appendix, 414, 594, 637
  functions of, 415

Appetite centers, 650
Aquaporins, 75, 533
Aqueous alkaline fluid
  bile consisting of, 620
  pancreatic secretion of, 617
Aqueous humor, 196
  function of, 211
Arachnoid mater, 140
Arachnoid villi, 140
Arcuate nucleus, 651
  lateral hypothalamic area (LHA) and,
        652
  paraventricular nucleus (PVN) and, 652
Aristotle, 347
Arousal system, 171
Arrhenius, Svante, A10–A11
Arrhenius base, A10–A11
Arrhythmias
  ECG (electrocardiogram) and, 319
  hypercalcemia and, 734
  potassium concentration and, 536
  types of, 320
Arsenic, renal failure and, 548
Arteries, 342, 346, 347–351. *See also* Blood
        pressure; Coronary artery disease
  baroreceptors, 377–378
  bleeding from, 403
  of heart, 304
  as pressure reservoir, 349
  pulmonary artery, 304
Arterioles, 342, 346, 351–359. *See also*
        Vasoconstriction; Vasodilation
  active hyperemia, 353–354
  adrenergic receptors, 358
  endothelial cells in smooth muscle, 355
  extrinsic control of, 357–359
  histamine and, 355
  hormones affecting, 358–359
  local control of radius, 352–357
  of nephrons, 513–514
  norepinephrine and, 357–358
  parasympathetic innervation to, 358
  pressure autoregulation and, 356–357
  reactive hyperemia, 356
  relaxation, local chemical effects on,
        354–355
  resistance in, 351–352
  smooth muscle in, 352
  stretch affecting, 356
  temperature changes and, 356
  tissue needs and, 353–355
  total peripheral resistance, 358, 359
  vascular tone, 352
Arthritis. *See* Rheumatoid arthritis
Artificial pacemakers, 312
Ascending tracts, 173–176
Aspartate, 121
Asphyxia, 494
Aspiration pneumonia, 486
Aspirin
  Alzheimer's disease and, 163
  fevers and, 661
  placental barrier, crossing, 786–787

prostaglandins and, 192
  renal failure and, 548
  stomach absorbing, 614–615
Assisted membrane transport, 78–87
Association areas, 152–153
Asthma, 448
  airway obstruction and, 472
  glucocorticoids and, 421
  prostaglandins and, 765
Astrocytes, 137–139
  blood-brain barrier, role in, 142
Asynchronous recruitment of motor units,
        270
Atheromas, 334
Atherosclerosis, 333–335. *See also* Coronary
        artery disease
  anabolic androgenic steroids and, 282
  causes of, 336
  cholesterol and, 336–337
  hypertension and, 381
  LDL (low-density lipoprotein) and, 334
  plaque and, 334–335
  risk factors, 336–337
  thromboembolism and, 335–336, 408
Athletic pseudonephritis, 550
Athletics. *See also* Exercise
  menstrual cycle irregularity (AMI) in
        athletes, 780
  muscle spindle in, 186
  skills, 177
Atmospheric air, 483–484
Atmospheric pressure, 463–464
Atomic number, A3–A4
Atomic symbol, A3
Atomic weight, A4
Atoms, 4, A3
  bonding characteristics, A4–A5
ATP (adenosine triphosphate), 35
  active transport and, 81
  cardiac muscle needs and, 333
  catabolism and, 719–720
  chemiosmotic mechanism, 40
  in citric acid cycle, 35–36, 38
  in contraction-relaxation process, 49,
        277–279
  coronary blood flow and, 333
  creatine phosphate and, 277
  cross bridge cycling and, 264
  electron transport chain, 39–42
  in excitation-contraction coupling, 268
  glycolysis and, 35, 279
  as high-energy biomolecule, A17
  for mechanical work, 42
  for muscle contraction, 49, 277–279
  Na^+-K^+ ATPase pump, 81–83
  oxidative phosphorylation and, 277–279
  oxygen debt and, 280
  storage of, 42
  synthesis by mitochondrial inner
        membrane, 38–39
  vesicular traffic and, 47
ATP synthase, 40
Atresia, 770

Atria, 304
  cardiac sympathetic nerves supplying, 326–327
  contraction of, 312–313
  excitation of, 312–313, 314
  interatrial pathway, 314
  valves between veins and, 308
  ventricles, transmission between, 314
Atrial contractile cells
  parasympathetic influence on, 326–327
  sympathetic nervous system, effect on, 327
Atrial fibrillation, 320
Atrial flutter, 319
Atrial natriuretic peptide (ANP), 530
Atrial repolarization, 323
Atrioventricular (AV) valves, 306–307
  in cardiac cycle, 320–323
  sounds associated with, 323–324
Atrioventricular node (AV node), 311
  AV nodal delay, 314
  cardiac sympathetic nerves supplying, 326–327
  internodal pathway, 314
  parasympathetic influence on, 326
  sympathetic stimulation of, 327
Atrophy, muscle, 283
Atropine, 245
A2M gene, 163
A type blood, antibodies in, 432
Auditory cortex
  deafness and, 221
  localization of sound, 216
  tonotopical organization of, 220–221
Auditory input, use of, 220
Auditory nerves, 219
  deafness and, 221
Autocatalytic process, 611
Autoimmune diseases, 413. See also
      Rheumatoid arthritis
  pregnancy and, 441
  tolerance to self-antigens, loss of, 441
Autonomic fibers, 25
Autonomic nervous system, 237–245.
      See also Parasympathetic nervous
      system; Sympathetic nervous
      system
  aging and, 244
  agonists, 245
  antagonists, 245
  diagram of, 239
  effects of, 242
  fibers, 136
  heart rate and, 327–328
  hypothalamus and, 156
  insulin, stimulation of, 725
  involuntary visceral organ activities, 240
  regions of CNS controlling, 245
  salivary secretion and, 601
  smooth muscle activity, effect on, 295
  somatic nervous system compared, 247
  two-neuron chain, 237–238
Autophagy, 33

Autoregulation of glomerular filtration rate
      (GFR), 520–521
Autorhythmic cells, 310
Autorhythmicity, 310
Autosomal chromosomes, 753
Autotransfusion, 385
AV nodal delay, 314
AV nodes. See Atrioventricular node (AV
      node)
AV valves. See Atrioventricular (AV)
      valves
Axillary hair, 750, 781
Axon hillock, 108–109
  action potentials and, 123
  contiguous conduction, 110–111
Axons, 108. See also Neuromuscular
      junction
  action potentials and, 108–109
  central axons, 136
  efferent axons/fibers, 137
  microtubules and, 46
  myelinating cells and, 116
  peripheral axons, 136
  regeneration of, 117
  tracts, 176
  in white matter, 145
Axon terminals, 108–109
Axon-to-axon connections, 117

Backward failure of heart, 332
Bacteria, 4. See also Inflammation
  colon, beneficial bacteria of, 638–639
  as immune system target, 413–414
  liver removing, 619
  mitochondria, relationship to, 34
  neutrophils and, 401
  skin, protection from, 450
Baking soda, alkalosis and, 585
Balance concept, 559–560
  stable balance, 560
Balanced equations, A7–A8
Baldness, testosterone and, 758
Barometric pressure, 463–464
Baroreceptor reflexes, 377–378
  blood pressure and, 563
  circulatory shock and, 383
  diagram of, 379
  glomerular filtration rate (GFR), extrinsic
      control of, 522–524
Baroreceptors, 377–378
  hypertension, adaptation during, 381
Barrier method contraceptives, 788
Basal bodies, 47–48
  centrioles and, 49
Basal cells, 229
Basal metabolic rate (BMR), 648–649
  direct calorimetry, 649
  energy equivalent of O_2 and, 649
  epinephrine and, 716
  factors influencing, 649
  hyperthyroidism and, 706
  indirect measurement of, 649
  thyroid hormone and, 704

Basal nuclei, 126, 143, 145
  motor activity and, 169
  in motor control, 154–156
Basement membrane, 518
  chylomicrons and, 623
Bases, A10–A11. See also Acid-base balance;
      Alkalosis
  in chemistry, 572–573
  defined, 572
  pancreatic aqueous alkaline secretion, 617
Basic electrical rhythm (BER)
  of digestive smooth muscle, 597
  in small intestine, 623–624
Basic reflexes, 177
Basilar membrane, 217
  functions of, 227
  pitch discrimination and, 219–220
  sound waves and, 219
Basolateral pump for sodium reabsorption, 527
Basophils, 392, 399, 414
  in allergies, 447
  functions of, 401
  staining of, 400
β (beta) cells. See also Insulin
  fuel metabolism and, 723
  insulin overproduction and, 730
B-cell growth factor, 439–440
  secretion of, 435
Behavior patterns
  anabolic androgenic steroids and, 282
  cerebral cortex and, 157–158
  hypothalamus and, 156, 157
  limbic system and, 157–158
  of pain, 193
  in sleep, 171
  vomeronasal organ (VNO) and, 231
β-endorphin, 121, 672
  cortisol secretion and, 710
  pro-opiomelanocortin precursor
      molecule and, 717
The Bends, 496–497
Benign tumors, 443
BER. See Basic electrical rhythm (BER)
Beta amyloid (Aβ), 162
Beta (β) globulins, 393
Beta (β) receptors, 224–245
Beta-carotene, LDL oxidation and, 334
Bicarbonate, 492–494
  acid-base balance, schematic
      representation of, 582–583
  in chemical buffer system, 575–756
  chloride shift, 493–494
  in colonic secretion
  in duodenum, 607
  exocrine pancreas secreting, 616
  hydrogen secretion and, 580
  ion, 394
  saliva, buffers in, 600
  sodium accompanied by, 562
  transportation of carbon dioxide as, 492–494
Bicuspid valve, 307

Bifidus factor, 798
Bile. *See also* Bile salts
  bilirubin, 622
  gallbladder storing, 622–623
  increasing secretion of, 622
  secretin stimulating, 622
  secretion of, 620–622
  vagus nerve, role of, 622
Bile canaliculus, 619–620
Bile salts
  as a choleretic, 622
  detergent action of, 620–622
  enterohepatic circulation of, 620
  fat digestion and absorption, 620–622
  ileum and absorption of, 626
  increasing bile secretion, 622
  micellar formation, 621–622
  schematic structure of, 621
Biliary system, 594
Bilirubin, 622
  in feces, 639
Binocular vision, 210–211
Biofeedback, 240
Biological clock
  circadian rhythms and, 678–679
  environmental cues and, 681
  melatonin and, 681–682
  research on, 682
  retina and, 212
  suprachiasmatic nucleus (SCN), 681
Biological molecules, A10
Biomolecules, A10
  high-energy biomolecules, A17–A18
Bipolar cells, 205
  function of, 211
  in retinal layer, 202
Birth. *See* Parturition
Birth control. *See* Contraception
Birth control patches, 788–789
Birth control pills, 788
  therapeutic administration of, 675
Bitter taste, 228
Black widow spider venom, 251
Blastocoele, 785
Blastocysts, 783–784
  endometrium, implantation in, 784–786
Blepharospasm, 252
Blindness
  Braille, learning, 151, 212
  color blindness, 208–209
  glaucoma and, 197
  and macular degeneration, 205
  night blindness, 208
  vision microchips, 213
Blind spot, 203, 204
  function of, 211
Blocking antibodies, 445
Block to polyspermy, 782
Blood, 303, 390–411. *See also* Blood vessels;
    Cardiac output (CO); Erythrocytes;
    Leukocytes; Plasma; Platelets
  amino-acid concentration, 731
  to anterior pituitary, 688

brain, delivery to, 143, 344
carbon dioxide transport in, 493
donated blood, need for, 398–399
functions of, 392
glucose concentration, 532
oxygen transport in, 488–490
reconditioning of, 343–344
Rh factor, 433
stress and, 717
substitutes, search for, 398–399
transfusion reaction, 432, 433
vascular spasm, 403
Blood-brain barrier, 140
  astrocytes, role of, 142
  glucose transporter (GLUT) and, 724
  hydrogen and, 500
  leptin transport, 653
  P_{CO_2} and, 500
  role of, 141–142
  tight junctions and, 142
Blood clots, 404–406
  amplification of process, 405–406
  calcium and, 734
  clotting cascade, 404–405
  dissolution of, 406–407
  extrinsic pathway of clotting cascade, 405,
    406
  factor X, 404–405, 406
  factor XII (Hageman factor), 405, 406, 407
  fibrinolytic plasmin and, 406–407
  hemophilia, 408
  intrinsic pathway of clotting cascade, 405,
    406
  pathways, diagram of, 406
  phagocytic chemical mediators triggering,
    419
  prevention of inappropriate clots, 407
  retraction of, 405
  septicemic shock, 408
  thromboembolism and, 335, 407–408
  vitamin K and, 408
Blood doping, 396
Blood flow
  air flow in lungs and, 482, 483
  to cardiac muscle, 332–333
  coronary blood flow, 332–333
  exercise and, 380
  heart valves and, 305–308
  heat and, 659–660
  liver blood flow, 618, 619
  pump action of heart, 306
Blood gases. *See* Respiration
Blood pressure, 375–385. *See also* Capillary
    blood pressure; Hypertension
  afterload, 330–331
  at arteriolar level, 352
  atrial natriuretic peptide (ANP) and, 530
  baroreceptor reflexes, 379
  baroreceptors, 377–378
  cardiovascular responses to, 379
  control measures, 563
  determinants of, 375–377
  diastolic pressure, 349, 351

epinephrine and, 715
exercise and, 379, 380
extracellular fluid (ECF) volume and,
  563
fluctuation of, 348–349
glomerular capillary blood pressure,
  518–520
glomerular filtration rate (GFR) and, 523
gravity and venous pressure, 373
hypothalamus and, 157, 379
kidneys and, 563
Korotkoff sounds, 350–351
mean arterial pressure, 342, 375–377
measurement of, 349–351
norepinephrine and, 357–358
parasympathetic nervous system and, 378
pulse pressure, 351
reflexes influencing, 378–379
sphygmomanometer for measuring,
  349–351
stress and, 717
sympathetic nervous system and, 378
systolic pressure, 349, 351
ventricular systole, 348–349
Blood-testes barrier, 761
Blood types. *See also* specific types
immunity and, 432–433
Rh factor, 433
transfusion reaction, 432, 433
universal donor, 432–433
Blood urea nitrogen (BUN), 534
Blood vessels, 303, 342–389. *See also*
    Arteries; Arterioles; Capillaries;
    Veins
atherosclerosis, 333–335
autonomic nervous system, effect of, 242
in dermis, 450
dilation of, 180
features of, 348
in fight-or-flight response, 240
flow rate, 344–346
histamines and, 66
history of study, 347
hypertension and, 381
hypothalamus and, 358
in lungs, 44
metarterioles, 363
multiunit smooth muscle in, 292
plasma/interstitial fluid barrier, 562
Poiseuille's law, 345–346
pressure gradient, 344–345
as pressure reservoir, 348
radius of vessels, 345
resistance and flow rate, 345
sympathetic nerve fibers and, 243
vascular spasm, 403
venules, 346
Blue cones, 205
Blushing, 245
B lymphocytes, 392, 401, 414. *See also*
    Antibodies; Plasma cells
adaptive immunity and, 416, 424–425
antibody-mediated immunity, 425–435

B-cell growth factor, 435, 439–440
in breast milk, 798
clonal selection theory, 428–429
clones, 428–429, 433–435
development of, 430–432
phagocytic secretions and, 419
plasma cells, 424
in severe combined immunodeficiency, 446
T lymphocytes compared, 436
tolerance and, 440–441
BMR. *See* Basal metabolic rate (BMR)
Bodybuilding, growth hormone and, 696
Body mass index (BMI), 654
Body of stomach, 604
Body systems, 7–8, 22
and homeostasis, 12–16
homeostasis and, 13–16, 22
Body temperature. *See* Temperature
Bohr effect, 492
Bolus, 601–602
Bombesin, 121
Bonds. *See* Chemical bonds
Bone marrow, 414
erythrocytes, replacement of, 394–395
functions of, 415
primary polycythemia, 398
Bone resorption, 735
Bones, 691–692. *See also* Growth hormone (GH); Parathyroid hormone (PTH)
anatomy of long bones, 692
chronic hypocalcemia and, 738–739
density of, 736
exercise, benefits of, 737
functions of skeleton, 735
growth of, 691
lever system and, 276
mature, nongrowing bone, 691–692
mechanical effectiveness factors, 735–736
muscle tension and, 273–274
organization of, 737
osteons, 737–738
osteoporosis, 736–737
parathyroid hormone (PTH) and, 735–737
remodeling of, 735
testosterone and, 758
Borborygmi, 639
Boredom, food intake and, 652
Botox, 252
Botulinum toxin, 251–252
Bowman's capsule, 514, 517
inner layer of, 518
plasma-colloid osmotic pressure and, 519
podocytes, 518, 519
Bowman's capsule hydrostatic pressure, 519
glomerular filtration rate (GFR) and, 520
Boyle's law, 466, 467
Bradykinin, 121
glomerular filtration rate (GFR) and, 521
pain and, 192
Braille, 151
visual cortex and, 212

Brain. *See also* Blood-brain barrier
analgesic system of, 194
apneusis and damage to, 498
astrocyte, role of, 137–139
autonomic nervous system, effect of, 242
blood, requirement for, 143, 344
dorsal view of, 146
electroencephalogram (EEG), 153–154
ependymal cells, 139
frontal section of, 155
functions of brain, overview of, 144
glucose supply for, 72–722
hydrocephalus, 141
hypoglycemia and, 730
ketones, use of, 723
meperidine (MPPP), effect of, 126
in metabolic states, 722
nervous tissue in, 6
neuronal clusters, 169
oxygen and, 143
parts of brain, grouping, 143–144
plasticity of, 151
protective features, 140
remodeling of, 151
sagittal view of, 146
structures of brain, overview of, 144
ventricles of, 139
withdrawal reflex and, 178
Brain death, 154
Brain stem, 143, 144, 169–172
auditory input, use of, 220
motor unit output and, 283
respiratory centers in, 495–498
spastic paralysis, inhibitory pathways in, 284–285
Brain tumors
gliomas, 140
visual field and, 209
Braxton-Hicks contractions, 792
Breast cancer
estrogenic pollutants and, 769
selective estrogen receptor modulators (SERMs) and, 737
Breast development, 781
Breast feeding. *See* Lactation
Breathing, 459–460
alveolar ventilation and, 480–482
apneusis, 498
hydrogen excretion and, 577–578
pharyngoesophageal sphincter and, 603
Breech birth, 792
Broca's area, 151–152
Bronchi, 462
Bronchiolar smooth muscle, 482
Bronchioles, 462
Bronchitis, chronic, 471
Bronchoconstriction, 471, 472
Bronchodilation, 471, 472
Brønsted, Johannes, A11
Brown fat, 659
Brush border, 625, 626
gluten enteropathy and, 627
B7 molecule, 440

B type blood, 432
Budding off in Golgi complex, 29
Buffy coat, 391
Bulbourethral glands, 750
functions of, 764
secretions of, 764
Bulk flow, 365–367
forces affecting, 365–366
role of, 367
Bundle of His (atrioventricular bundle), 311
action potential, conduction of, 314
Burns
glomerular filtration rate (GFR) and, 520
growth hormone treatment for, 696
tissue engineering and, 9
Burping, 603, 639
Bush, George H., 127
Bush, George W., 9

C cells, 701, 740
Ca^{2+}-ATPase pump, 266–267
Ca^{2+} channels, 67, 247, 289, 310–311, 315
Cadherins, 61–62
Caffeine, 172
gastric secretion and, 613
heart rate and, 312
Calcitonin, 701–702, 737, 740
calcium metabolism and, 733–734
phosphate metabolism and, 733–734
Calcitriol, 740
Calcium. *See also* Bones; Calcium metabolism; Cardiac muscle; Parathyroid hormone (PTH)
AMPA receptors and, 165–166
Ca^{2+} channels, 67
cardiac muscle and, 310–311, 315
characteristics of, A4
clotting cascade and, 405
in contraction-relaxation process, 277
cross bridges and, 262
in excitation-contraction coupling, 263, 266, 268, 734
in extracellular fluid (ECF), 734
heart, effect of changes in ECF concentration on, 316
hypertension and, 381
indigenous bacteria and, 639
and inflammation, 417417
kidneys regulating reabsorption of, 532
overview of functions, 512
pacemaker potential and, 310–311
plasma calcium, regulation of, 734
relaxation of muscle and, 266–267
release of neurotransmitter and, 247
renal failure and, 549
sarcoplasmic reticulum releasing, 263–264
slow-wave potentials and, 597
small intestine, absorption by, 634, 739–740
smooth muscle and, 289–292, 295
twitch summation and, 271–272

Calcium carbonate crystals, 224
Calcium channels. *See* Ca^{2+} channels
Calcium metabolism, 733–742. *See also*
    Parathyroid hormone (PTH)
  balance, maintenance of, 734–735
  calcitonin, 740
  disorders of, 741–742
  homeostasis, 734–735
  plasma calcium, regulation of, 734
  vitamin D and, 634, 733–734, 740–742
Calcium phosphate in bones, 691
Calcium second-messenger pathway, 68–69
  activation of, 69
Calcium sparks, 315–316
Calmodulin, 69, 290
Calories, 648
Canaliculi, 737
Canal of Schlemm, 196
Cancer. *See also* Chemotherapy drugs;
    specific types
  apoptosis and, 71
  blocking antibodies, 445
  breast-feeding and, 798
  Granstein cells and, 451
  HeLa cells, 24
  immune surveillance against, 443–445
  interferon and, 422
  mutations and, 444
  nitric oxide (NO) and, 356
  somatic mutations and, A30
  vaults and, 43
Capacitance vessels, 370
Capillaries, 342, 346, 360–369. *See also*
    Lymphatic system; Pulmonary
    capillaries
  anatomy of, 360
  bed of, 363
  brain capillaries, 142
  branching of, 363
  bulk flow across, 365–367
  chylomicrons and, 623
  concentration gradients and, 364–365
  edema and, 369
  exchanges across walls of, 362
  Fick's law of diffusion and, 360
  hydrostatic pressure in, 365–366, 373
  injury and, 416
  interstitial fluid exchanges, 363–364
  in liver, 618, 619
  metarterioles, 363
  of nephrons, 513–514
  net filtration, 366–367
  peritubular capillaries, 514
  plasma-colloid osmotic pressure, 366
  pores, size of, 362–363
  precapillary sphincters, 363, 364
  slow velocity of flow through, 360–362
  velocity of flow in, 360–362
  vesicular transport and, 363
  in villus of small intestine, 628
  walls, diffusion across, 364–365
  water-soluble substances, passage of,
    362–363

Capillary blood pressure, 365–366. *See also*
    Glomerular capillary blood pressure
  fluid balance, 562
Capsaicin, 192
Carbamino hemoglobin (HbCO$_2$), 493
Carbohydrate-containing surface markers,
  62
Carbohydrate loading, 279
Carbohydrates, 4, A11–A12. *See also* Fuel
    metabolism
  ATP (adenosine triphosphate) and, 35
  as biomolecules, A10
  brush border, digestion in, 625
  cell-to-cell adhesions and carbohydrate
    chains, 62
  chemical composition of, A11–A12
  diabetes mellitus and, 726
  digestion of, 592–594, 614, 626
  glucagon, actions of, 730
  insulin and, 723–724
  in pregame meals, 607
  small intestine, absorption by, 629, 630
  types of, A12
Carbon
  atoms, 4
  characteristics of, A4
Carbon dioxide, 4. *See also* Bicarbonate;
    Homeostasis; P$_{CO_2}$; Respiration
  acid-base balance, schematic
    representation of, 582–583
  arteriolar smooth muscles and, 355
  blood-brain barrier and, 142
  Bohr effect, 492
  bronchiolar smooth muscle, effect on,
    482
  carbamino hemoglobin (HbCO$_2$), 493
  carbonic acid formation and, 574
  cells and elimination of, 5
  diffusion, 74
  diffusion coefficient and gas exchange,
    487
  exercise, variables during, 503
  Fick's law of diffusion and, 74
  Haldane effect, 494
  hemoglobin and, 394, 493
  hemoglobin saturation and, 491, 493
  homeostasis and concentration of, 12
  H$_2$CO$_3$:HCO$_3$$^-$ buffer system,
    575–576
  left-sided heart failure and, 332
  partial pressure gradient for, 484
  in plasma, 392
  pulmonary capillaries, exchange across,
    484
  systemic capillaries, exchange across,
    487–488
  transport in blood, 492–494
Carbonic acid, 493
  arteriolar smooth muscles and, 355
  in chemical buffer system, 575
  hemoglobin saturation and, 491–492
  as hydrogen source, 574
  as weak acid, 572

Carbonic anhydrase, 493, 574, A8
  in erythrocytes, 394
Carbon monoxide, hemoglobin and, 394,
  492
Carboxyhemoglobin (HbCO), 492
Carboxylic acid functional group, A12–A13
Carboxypeptidase, 616, 617
Cardiac control center, 377
Cardiac cycle, 320–325
  diagram of, 322
Cardiac muscle, 256, 257, 288–289. *See*
    also Frank-Starling law of the heart
  action potential of, 315
  afterload and, 331
  ATP (adenosine triphosphate) and, 333
  calcium, 297
    role of, 315–316
  contractions of, 309
  desmosomes in, 63
  diastole, blood supply during, 332–333
  extracellular fluid (ECF) and, 315–316
  fibers, 297
  functional characteristics of, 297
  gap junctions in, 64
  length-tension relationship, 328–329
  nourishing, 332–337
  organization of fibers, 309
  refractory period, 316–317
  tetanus, prevention of, 316–317
  tissue, 6
Cardiac myopathies, 320
Cardiac output (CO), 325–332, 344. *See*
    also Heart failure
  exercise and, 380
  to kidneys, 524–525
  magnitude of, 354
  sympathetic stimulation and, 330
Cardiac reserve, 325
Cardiogenic shock, 382
Cardiovascular control center, 358–359
  blood pressure signals, 377
  heart rate, control of, 327–328
  heat regulation and, 660
  temperature regulation and, 660
Cardiovascular hypertension, 380
Cardiovascular system, 342–389. *See also*
    Blood; Blood vessels; Heart
  anabolic androgenic steroids and, 282
  circulatory shock and, 385
  diabetes mellitus complications, 729
  organization of, 346
  thyroid hormone, effect of, 704
Carnosine, 121
Carotid bodies, 499
  in SIDS (sudden infant death syndrome),
    505
Carotid sinus baroreceptor, 377
Carpal tunnel syndrome, 696
Carrier-limited reabsorption, 531
Carrier-mediated transport, 78–83
  active transport, 80, 81–83
  characteristics of, 86
  competition and, 80

diffusion compared, 80
  facilitated diffusion, 80–81
  saturation of carriers, 79–80
  schematic representation of, 79
  secondary active transport, 83, 84
  specificity of, 78–79
Carrier molecules, 60
Cartilage, 691
Cascading effect, 70
Ca^{2+} sparks, 315–316
Caspases, 70–71
Castration, 758
Catabolism, 719–720
  of triglycerides, 721
Catalase, 34
Catalysts, A8
Cataracts, 200
Catecholamines, 672, 707–708. *See also*
  Epinephrine; Norepinephrine
  chemical classifications, 672
  in chromaffin granules, 714
  circulation of, 675
  emotions and, 158
  inactivation of, 679
  mechanisms of, 674
  receptors for, 675
  secretion from adrenal medulla, 714–715
Cations, membrane potential and, 87
Cauda equina, 172
Caveolae, 85, 87
  smooth muscle activity and, 295
Cavities, dental, 600
C cells in thyroid gland, 701
CD4 cells, 436
CD8 cells. *See* Cytotoxic T cells
CD47, 443
Cecum, 594, 624, 637
Cell adhesion molecules (CAMs), 61–62
  and inflammation, 417
  linking cells, 63
Cell body of neurons, 108
Cell culture, 8
Cell division, A27–A30
  centrioles in, 49
  cytokinesis, 48–49
  meiosis, A27–A28
  mitosis, 48–49, A27
  mutations, A29–A30
Cell-mediated immunity, 401, 424, 435–445
  delayed hypersensitivity and, 449
Cell membrane. *See* Plasma membrane
Cells, 4–6. *See also* Nucleus; Tissue
  apoptosis, 70–71
  calcium second-messenger pathway,
    68–69
  channels, 60
  cholesterol uptake, 336
  ciliated cells, 47
  communication, intercellular, 65–71
  culture, 24
  cyclic guanosine monophosphate (cyclic
    GMP), 69
  desmosomes, 63

differentiation, 5
diffusion, reliance on, 74
extracellular fluid (ECF) and, 10–11
gap junctions, 63–65
HeLa cells, 24
intermediate filaments, 50–51
intracellular fluid (ICF), 10–11
markers, 62
membrane-bound enzymes, 60
observations of, 23–24
programmed suicide, 70–71
receptors, 60
reproduction function, 5
schematic diagram of, 26
signal transduction, 66
size of, 23
specialized functions, 6
structure of, 24–25
summary of structures and structures,
  45–46
tight junctions, 63
Cells of Leydig, 685
Cell theory, 23
  principles of, 24
Cell-to-cell adhesions, 62–65
Cell-to-cell interactions, 62
Cellular respiration, 40
Cellulose, 592
  in feces, 637
  as polysaccharide, A12
Central axons, 136
Central canal of spinal cord, 737
  ependymal cells, 139
Central chemoreceptors
  chronic obstructive pulmonary disease
    (COPD) and, 501
  hydrogen changes and, 501–502
  P$_{CO_2}$ and, 500–501
Central fatigue, 279–280
Central lacteal, 628
  fats and, 634
Central nervous system (CNS), 133,
    135–137. *See also* Brain; Spinal cord
  analgesic system, 194
  astrocytes, 137–139
  central fatigue, 279–280
  ependymal cells, 139–140
  epinephrine affecting, 716
  interneurons, 137
  loudness, sensitivity to, 220
  microglia, 139
  oligodendrocytes and, 117, 139
  overview of, 143–145
  protective features, 140
  thyroid hormone, effect of, 704
Central sulcus, 148
Central thermoreceptors, 658
Centrioles, 49
Cephalic phase of gastric secretion,
  612–613
Cerebellum, 143, 144, 166–169
  gross structure of, 167
  location in brain, 156

motor control and, 150, 169, 286
procedural memories and, 161
Cerebral cortex, 143, 145
  behavioral patterns and, 157–158
  exercise, ventilation and, 503
  functional areas of, 147
  higher functions and, 152–153
  language and, 151–152
  lobes of, 146–148
  organization of, 145–146
  rhythmic synchrony, neurons firing in,
    154
Cerebral hemispheres, 145
  specialization of, 153
Cerebral hemorrhage, 608
Cerebrocerebellum, 169
  premotor areas and, 283
  supplementary motor areas and, 283
Cerebrospinal fluid (CSF), 140, 561
  hydrocephalus, 141
  relationship to brain/spinal cord, 141
  role of, 140–141
  in spinal cord, 176
Cerebrovascular accidents. *See* Strokes
Cerebrum, 143, 144–145
Cerumen, 216
Cervical canal, 750
Cervical cap, 788
Cervical dilation stage of labor, 794–795
Cervical nerves, 172
Cervix, 750
  mucus, changes in, 780
C fibers, 192
Channels, 60. *See also* Chemically gated
    channels; Mechanically gated
    channels; Voltage-gated channels;
    specific types of channels
  of cardiac autorhythmic cells, 310
  extracellular chemical messengers and,
    66–67
Chemical agents
  organophosphates, 251, 252
  placental barrier and, 786–787
  vomiting and, 608
Chemical bonds, A4–A7
  double bonds, A11
Chemical buffer system, 575
Chemical contraceptives, 788
Chemical gradient, 73
Chemically gated channels, 67, 104
Chemical principles, A3–A18
Chemical reactions, A7–A8
  irreversible reactions, A8
  reversible reactions, A8
Chemical senses. *See* Smell; Taste
Chemiosmotic mechanism, 40
Chemokines, 417417
Chemoreceptors, 187
  blood pressure, effect on, 378
  central chemoreceptors, 500–502
  control of respiration and, 499–502
  in digestive system, 599
  peripheral chemoreceptors, 499–502

Chemoreceptor trigger zone, 608
Chemotaxins, 417
  complement system and, 423
  helper T cells secreting, 440
Chemotaxis, 417
Chemotherapy drugs
  synthetic erythropoietin and, 396
  vaults and, 43
Chernobyl disaster, 682
Chewing, 592
  teeth and, 600
Chief cells, 608, 609
  multiple regulatory pathways for, 612
Children. *See also* Growth; Infants
  asthma in, 472
  cholera and mortality, 636
  lactose intolerance, 625
  language, cerebral cortex and, 151
Chlamydia pneumoniae
  asthma and, 472
  atherosclerosis and, 337
Chloride. *See also* Hydrochloric acid
  overview of functions, 512
  passive reabsorption of, 533
  resting membrane potential and, 92
  small intestine, absorption in, 628–629
  sodium with, 533, 562
Chloride shift, 493–494
Chlorine, characteristics of, A4
Cholecalciferol (Vitamin E), 334
Cholecystokinin (CCK), 121
  as enterogastrone, 606
  functions of, 640
  gallbladder and, 623
  pancreatic exocrine secretion and,
    617–618
  role of, 124
  satiety and, 652
Cholera, 636
  cyclic adenosine monophosphate (cyclic
    AMP/cAMP) and, 71
Choleretics, 622
Cholesterol. *See also* HDL (high-density
    lipoprotein); LDL (low-density
    lipoprotein)
  anabolic androgenic steroids and, 282
  bile salts and, 620
  as complex lipid, A14
  gallstones, 622
  micelles transporting, 621–622
  and plasma membrane, 59
  receptor-mediated endocytosis and, 32
  sources of, 336
  steroid hormones and, 673–674
  uptake by cells, 336
  very-low-density lipoprotein (VLDL),
    336
Cholinergic fibers, 238–240
Cholinergic receptors, 243–244
Chondrocytes, 691
Chordae tendinae, 307
Chorion, 786

Choroid, 195–196
  function of, 211
Choroid plexuses, 140
Chromaffin granules, 714
Chromatin, A20
Chromosomes. *See also* DNA
    (deoxyribonucleic acid); RNA
    (ribonucleic acid)
  autosomal chromosomes, 753
  homologous chromosomes, A20
  microtubules and, 44
  mitotic spindle, 49
  in reproductive cells, 751, 753
  scanning electron micrograph of, A30
  sex chromosomes, 753
Chronic obstructive pulmonary disease
    (COPD)
  asthma, 472
  chronic bronchitis, 471–472
  emphysema, 472
  expiration difficulties, 472–473
  P_{CO_2} and, 501
Chronic renal failure, 549
Chylomicrons, 632–633
Chyme, 604, 606
Chymotrypsin, 616, 617
Chymotrypsinogen, 616, 617
Cigarette smoking
  chronic bronchitis, 471
  emphysema and, 472
  heart rate and, 312
  placental barrier and, 786–787
  respiratory defenses and, 452
  SIDS (sudden infant death syndrome)
    and, 505
Cilia, 47–48
  internal structure of, 48
  microtubules and, 47
Ciliary body, 195–196
  function of, 211
Ciliary muscle, 200–201
  function of, 211
Cimetidine, *Helicobacter pylori* and, 615
Circadian rhythms, 678–679, 681–683
  clock proteins and, 681
  cortisol secretion and, 710–711
  melatonin and, 681–682
Circular muscle, 197
Circulatory hypoxia, 494
Circulatory shock, 380, 382–384
  anaphylactic shock, 448
  causes of, 383
  consequences of, 383–385
  dehydration and, 567
  irreversible shock, 385
  septicemic shock, 408
Circulatory system, 10, 302–341, 342, 347.
    See also Blood; Blood vessels; Heart
  autonomic nervous system controlling
    circulation, 240
  brain, blood to, 344
  collateral circulation and, 335–336

components of, 303
  coronary circulation, 332–333
  diffusion and, 75
  homeostasis and, 13–16
  pulmonary circulation, 303–304
Cirrhosis, 623
Cisternae, 28
Citric acid cycle, 35–36
  diagram of, 27
Cl⁻. *See* Chloride
Classical neurotransmitters, 121, 123–124
Class I/class II MHC glycoproteins, 442–443
Climacteric, 781
Clinton, Bill, 127
Clitoris, 750–751, 755–756
  adrenogenital syndrome and, 713
  nitric oxide (NO) and, 356
Clock proteins, 681
Clonal anergy, 440
Clonal deletion, 440
Clonal selection theory, 428, 429
Clones, B-cells, 428–429, 433–435
Closed channels, 60
Clostridium botulinum toxin, 251–252
Clot-dissolving drugs, 143
Clots. *See* Blood clots
Clotting cascade, 404–405
CNS. *See* Central nervous system (CNS)
Coactivation of gamma motor-neuron
    system, 287–288
Coat proteins, 30
Cocaine
  dopamine, effect on, 125–126
  emotions and, 158
Coccygeal nerve, 172
Cochlea, 213
  diagram of, 217
  functions of, 216–217, 227
Cochlear duct, 217
  functions of, 227
Cochlear implants, 221–222
Codons, A23, A25
Coenzyme A, 36
Cognition, 143
Coitus interruptus, 788
Cold
  compensatory responses to, 660
  extreme-cold, exposure to, 662
  hypothalamus and, 660
Colds, airway resistance and, 471
Colipase, 621
Collagen
  basement membrane and, 518
  in extracellular matrix (ECM), 62
  fibers, 348
Collateral circulation, 335–336
Collateral ganglia, 238
Collaterals of axons, 108
Collateral ventilation, 463
Collecting tubules, 514
  functions of, 548
  sodium reabsorption in, 526–530

vasopressin and permeability of, 544–545
water reabsorption and, 533
Colloid-filled follicles, 701–702
Colloids, A10
Colon, 594, 637
beneficial bacteria of, 638–639
constipation and, 638
gas in, 639
haustral contractions, 637
ileocecal valve/sphincter, 624–625
mass movements, 637–638
secretions of, 638
sodium absorption of, 639
water absorption, 639
Color blindness, 208–209
Color vision, 208–209
Colostrum, 798
Coma, 170
Communication, intercellular, 65–71
Compartment of peptide loading, 434
Competition
and carrier-mediated transport, 80
use-dependent competition, 150
Complementary base pairing, A20–A23
Complement system, 423–424
activation of, 427–428
inflammation and, 423–424
membrane attack complex (MAC), 423
Complete heart block, 312, 320
Complex cells, 212
Compliance
of blood vessels, 348–349
of lungs, 473–474
Compounds, A3
Concave lenses, 198
refraction by, 199
Concentration, defined, 75
Concentration gradients, 73
capillaries, exchanges by, 364–365
Fick's law of diffusion and, 74
ion concentration gradient, 83
Na^+-K^+ ATPase pump and, 83
Concentration measures, A9–A10
Concentric isotonic contractions, 275
Conception, 781–783
Concurrent reflex stimulation, 179
Conditioned reflexes, 177
salivary reflex, 601
Condoms, 788
Conduction of heat, 656
Conductive deafness, 221
Conductors, 102
Cones. See Photoreceptors
Congestive heart failure, 332
elevated atrial pressure, 371
hypertension and, 381
renin-angiotensin-aldosterone system,
role in, 528–529
Connective tissue, 6–7
extracellular matrix (ECM) in, 62
hydrochloric acid and, 611
muscle tension and, 273

in smooth muscle, 296
in villus of small intestine, 628
Connexons, 63–65
Conn's syndrome, 712
Consciousness, defined, 170
Consolidation, 159
Constant (Fc) region of antibody, 426
Constipation, 638
Constrictor muscle, 197
Contiguous conduction, 110–111
Continuous ambulatory peritoneal dialysis
(CAPD), 551
Contraception, 786
failure rate of, 788
lactation and, 798–799
male contraception research, 789
ways and means of, 788–789
Contractile cells, 310
Contractile component of muscle, 273
Contractile proteins, 260
Contractile ring, 49
Contractility of heart, 329
Contractions of muscles. See also Exercise;
Smooth muscle; Twitch
asynchronous recruitment of motor units,
270
ATP (adenosine triphosphate) and, 277
atrophy and, 283
cardiac muscle, 309
concentric isotonic contractions, 275
creatine phosphate and, 277
digestive muscles, 592
eccentric isotonic contractions, 275
extension and, 274
fatigue and, 279–280
flexion and, 274
isometric contraction, 274–275
isotonic contraction, 274–275
length-tension relationship, 272–273
metabolic pathways producing ATP for,
278
microfilaments and, 49–50
motor units and, 269–270
oxidative phosphorylation, 277–279
sliding filament mechanism of, 261
types of, 274–275
velocity of shortening muscle, 275
Contraction time, 268
Control centers, 17
Controlled variables, 16
Convection of heat, 656–657
Convergence of neurons, 128
Convex lenses, 198
refraction by, 199
Convulsions
febrile, 419
overheating and, 655
Cooper, Kim E., A31–A34
Core temperature, 655–656
Cornea, 195–196
function of, 211
refractive ability of, 198–200

Corona radiata, 774
at fertilization, 782
Coronary artery disease, 333–336. See also
Atherosclerosis; Heart attacks
android obesity and, 654
collateral circulation and, 335–336
diabetes mellitus complications, 729
thromboembolism, 335–336
vascular spasm, 333
Coronary circulation, 332–333
Corpus albicans, 774
Corpus callosum, 145
Corpus luteum, 772
degeneration of, 774
formation of, 773, 774
luteinizing hormone (LH) controlling,
777–778
of pregnancy, 790
Cortical gustatory area, 228
Cortical lobes, 146–148
Cortical nephrons, 515–516
Corticosteroid-binding globulin
(transcortin), 708–709
Corticotropin-releasing hormone (CRH),
687
ACTH (adrenocorticotropic hormone)
secretion and, 710–711
CRH-ACTH-cortisol system, 689
paraventricular nucleus (PVN) and, 652
parturition, role in, 794
stress response and, 717
Cortisol, 421–422, 445, 708–709. See also
Glucocorticoids
in Addison's disease, 713–714
Alzheimer's disease and, 163
anti-inflammatory effects, 710
circadian rhythm and, 710–711
CRH-ACTH-cortisol system, 689, 717
deficiencies, 714
exercise and, 446
fuel metabolism, effect on, 732–733
growth effects, 690
hypersecretion of, 712
hypothalamus-pituitary-adrenal cortex
axis, 710
immunosuppressive effects, 710
metabolic effects of, 709–710
permissive actions of, 709
stress and, 709, 711, 717
Cotransport carriers, 83, 84
Countercurrent exchange, 547
Countercurrent multiplication, 543–544
Covalent bonds, A5–A7
Cowpox, 430
Cow's blood, 399
CPR (cardiopulmonary resuscitation), 304,
305
Cranial nerves, 168, 169
Cranium, 140
C-reactive protein, 337
Creatine kinase, 277
Creatine phosphate, 277, 280

Creatinine
  kidneys and, 512
  plasma clearance rate for, 538
  reabsorption of, 534
CREB molecule, 166
CREB2 molecule, 166
Cretinism, 706
CRH-ACTH-cortisol system, 689, 717
Crib death, 504–505
Cristae, 34
Cross bridges, 258–259
  ATP (adenosine triphosphate) powering, 264
  calcium role of, 262
  contraction of muscle and, 261
  cycle, activity in, 264
  diagram of cycling, 267
  in excitation-contraction coupling, 268
  in smooth muscle, 294–295
  twitch summation and, 271
Crossed extensor reflex, 179
Crossing over, A27–A28, A30
Crying, respiration during, 504
Cryptorchidism, 757, 769
Crypts of Lieberkühn, 628
Cupula, 223
Curare, 251, 252
Currents, 102
Cushing's syndrome, 712
Cyanide poisoning, 494
Cyanosis, 494
Cyclic adenosine monophosphate (cyclic AMP/cAMP), 67–68, 72, A18
  activation of, 68
  calcium-second-messenger pathway and, 69
  cascading effect, 70
  CREB and, 166
  heart activity and, 326
  hormone actions via, 676
  sensitization and, 164
  sweet taste and, 228
  synaptic transmission and, 121
  vasopressin and, 544–545
Cyclic AMP-dependent protein kinase, 67
Cyclic GMP, 69, 767
  photoreceptors and, 205
Cyclosporin, 443
Cysteinuria, 79
Cystic fibrosis (CF), 61
Cystic fibrosis transmembrane conductance regulator (CFTR), 61
Cytochrome c, 71
Cytokines
  from helper T cells, 439–440
  stress response and, 445
Cytokinesis, 48–49, A27
  schematic illustration of, 49
Cytoplasm, 25
  summary of, 45
Cytosine (CO), A19
Cytoskeleton, 25, 43, 44, 46–51
  components of, 46

as integrated whole, 51
  summary of, 45
Cytosol, 25, 43–44
  in cellular energy production, 36
  cytochrome c, leakage of, 71
  fats, storage of, 43–44
  free ribosomes in, 43
  glycogen, storage of, 43–44
  inclusions, 44
  intermediary metabolism and, 43
  secretory vesicles and, 30
  summary of, 45
Cytotoxic T cells, 436, 437, 439
  cancer cells and, 444–445
  chemicals secreted by, 437–439
  class I/class II MHC glycoproteins, 442–443
  in immune surveillance, 444–445
  MHC molecules and, 442

Dark adaptation, 207–208
DASH (Dietary Approaches to Stop Hypertension), 381
D (delta) cells
  fuel metabolism and, 723
  in stomach, 609, 610
Deafness, 221–222
Death
  brain death, 154
  circulatory shock and, 385
  rigor mortis, 264, 266
Decibels (dB), 215
Decidua, 786
Declarative memories, 161
Decompensated heart failure, 332
Decompression sickness, 496–497
Decremental spread of graded potentials, 102–103
Deep-sea diving, 496–497
Defecation, 592
  reflex, 638
Defensins, 61, 628
Dehydration, 566–567
  vomiting and, 608
Dehydroepiandrosterone (DHEA). See DHEA (dehydroepiandrosterone)
Delayed hypersensitivity, 446–447, 449
Delivery. See Parturition
Dementia. See also Alzheimer's disease
  microglia, role of, 139
Denaturation of proteins, A17
Dendrites, 108
  in gray matter, 145
Dendrite-to-dendrite connections, 117
Dendritic cells, 435
Denervation atrophy, 283
Dense bodies in smooth muscle, 289
Dense-core vesicles, 124
Deoxyhemoglobin, 488
Deoxynucleic acid. See DNA (deoxyribonucleic acid)
Dependent variables, A32
Dephosphorlyation, 81

Depolarization, 100
  in action potentials, 103
  excitatory postsynaptic potential, 119
  threshold potential, 103–104
Depolarization block, 251
Depression
  food intake and, 652
  limbic-system neurotransmitters and, 159
Depth perception, 210–212
Dermatome, 177
Dermis, 450
Descending tracts, 175
Designer drugs, 126
Desmosomes, 63
  in cardiac muscle, 308
  spot desmosomes, 450
Detergent action of bile salts, 620–622
DHEA (dehydroepiandrosterone), 708, 711–712, 756, 763
  and parturition, 794
  supplemental DHEA, 711–712
Diabetes insipidus, 567
Diabetes mellitus, Type I, 79, 726, 728–729
  acute effects of, 727
  breast-feeding and, 798
  carbohydrates and, 726
  fats and, 726–727
  glucagon excess aggravating, 731–732
  long-term complications, 728–730
  management of, 729
  metabolic acidosis and, 502, 574, 585
  plasma glucose concentration in, 532
  protein metabolism consequences, 727–728
  symptoms of, 725–730
  treatment of, 728–729
  underlying defects, 728
Diabetes mellitus, Type II, 726, 728–729
  android obesity, 654
  long-term complications, 728–730
  management of, 729
  tolerance to self-antigens, loss of, 441
  treatment of, 728–729
  underlying defects, 728
Diaclyglycerol (DAG), 68
Dialysis, 551
Diaphragm, 463, 496–498
  actions of, 470
  expiration, 467
  inspiration and, 466
  vomiting and, 608
Diaphragms (contraceptive), 788
Diaphysis, 691
Diarrhea
  breast milk combating, 798
  causes of, 636
  lactose intolerance and, 625
  metabolic acidosis and, 585
  oral rehydration therapy, 636
  potassium concentration and, 536
  results of, 635–636
  sodium loss in, 564

Diastole period, 320–321
  cardiac muscle, blood supply to, 332–333
  ventricular diastole, 321
Diastolic blood pressure, 349, 351
Diastolic murmurs, 324–325
Diencephalon, 143, 144
Diet. See Foods
Dietary Supplement Health and Education
    Act of 1994, 712
Diet-induced thermogenesis, 649
Dieting, 653
Differentiation. See also Sexual
    differentiation
  of cells, 5
  stem cell science and, 8
Diffusion, 73. See also Capillaries
  carrier-mediated transport compared, 80
  characteristics of, 86
  examples of, 74
  facilitated diffusion, 80–81
  Fick's law of diffusion, 74–75
Diffusion coefficient and gas exchange, 487
Digestion, 592–594
  autonomic nervous system controlling,
    240
  bile salts and, 620–622
  of carbohydrates, 592–594, 600, 614, 617,
    626, 630
  diagram of process, 593
  of fats, 592–594, 617, 626, 632
  pancreatic secretions, 615–618
  of proteins, 611, 616–617, 625, 631
  in small-intestine lumen, 625
Digestive enzymes, 594
Digestive system, 10, 591–645. See also
    Digestion; specific digestive organs
  accessory digestive organs, 594
  acid-base balance of, 635
  anatomy of, 596–597
  autonomic nervous system, effect of, 242
  bile and, 620–623
  contraction of digestive muscles, 592
  digestive smooth muscle and, 598
  enteric nervous system, 598
  epinephrine and, 715
  epithelial tissue, 6
  extrinsic nerves, 598
  functions of, 596–597
  gastrin, 639, 640
  gastroesophageal sphincter, 603
  gastrointestinal hormones, 598–599
  gland cells of, 6
  homeostasis and, 13–16
  hormones, overview of, 639–641
  immune defenses of, 451
  intrinsic nerve plexuses, 598
  layers of walls, 594–599
  long reflex, 599
  motility function, 591–592
  muscularis externa, 595
  organs of, 7–8, 594
  pancreatic secretions, 615–618
  parasympathetic nervous system and, 598

pharyngoesophageal sphincter, 603
  receptor activation and, 599
  rhythmic digestive contractile activities,
    598
  secretion, digestive, 592
  serosa, 597
  short reflex, 599
  slow-wave potentials, 597
  smooth muscle function of, 591–592,
    597–598
  submucosa, 595
  sympathetic nervous system and, 598
  tight junctions in, 63
  tone and, 591–592
Digestive tract. See Digestive system
Digitalis, 316
  hypertension and, 381
Dihydropyridine receptors, 263
Dihydrotestosterone (DHT), 756
Di-iodotyrosine (DIT), 702–704
Dilator muscle, 197
Diploid number, 751, A19–A20
Diplopia, 212
Direct calorimetry, 649
Disaccharides, 592, A12
  in brush-border plasma membrane, 625
  mouth, digestion in, 601
Discriminative acuity. See Acuity
Diseases. See also specific diseases
  apoptosis and, 71
  carrier-mediated transport and, 79
  cerebellar disease, 169
  electroencephalogram (EEG) and,
    153–154
  growth and, 690
  lung diseases, 479–480
  lymphatic system and, 368
  mitochondrial diseases, 34
  receptors, defects in, 70–71
  synaptic transmission, effect on, 125–128
  Tay-Sachs disease, 34
  vaccination, 430
Dispersed-phase particles, A10
Dissociation constant (K), 572
Distal tubules, 514
  functions of, 548
  sodium reabsorption in, 526–530
  summary of transport across, 537
  vasopressin and permeability of, 544–545
  water reabsorption and, 533
Distensibility of blood vessels, 348–349
Distension
  of duodenum, 606, 607
  gastric secretion and, 613
  lactose intolerance and, 625
  satiety and, 652
  of stomach, 605
Disuse atrophy, 283
Diuresis, osmotic, 547–548
Diuretics, 529–530
  side effects of, 382
  solute reabsorption, blocking of, 548
Diurnal rhythms. See Circadian rhythms

Divergence of neurons, 128
Diving, 177
Dizziness, 224
  vomiting and, 608
DNA (deoxyribonucleic acid), 4, A17
  bases in, A19
  complementary base pairing, A20–A23
  functions of, A19
  hormone receptors binding with,
    676–677
  lipophilic hormones and, 676–677
  location of, 751
  in mitochondria, 34
  mutations, A29–A30
  in nucleus, 25
  organization levels, A22
  packaging of, A19–A20
  in paternity actions, 433
  protein transcription and, A21–A23,
    A26–A27
  structure of, A19
  transcription, A21–A23, A26–A27
  triplet code, A23, A25
  in viruses, 413
Docking markers, 29, 30
  acceptors, 60
Dopamine, 121, 688
  cocaine, effect of, 125–126, 158
  Parkinson's disease and, 126
  schizophrenia and, 159
Dorsal columns, 174
Dorsal horn, 176
  pain and, 194
Dorsal respiratory group (DRG), 497
  schematic representation of, 498
Dorsal root, 175
Dorsal root ganglion, 175
Dorsal spinocerebellar tract, 174
Double bonds, A11
Doughnut-shaped vaginal rings, 788–789
Down regulation of hormones, 680
Downward deflection, 100
Dreaming, 171
Drugs. See also specific types
  antidepressant drugs, 159
  glucocorticoid drugs, 421–422
  hydrogen load and, 574
  hypersecretion of hormones and, 680
  kidneys and, 512
  mental disorders and, 159
  metabolic alkalosis and, 585
  for osteoporosis, 737
  peptic ulcers, treatment of, 615
  placental barrier and, 786–787
  proximal tubule secretion of, 537
  psychoactive drugs and emotions,
    158–159
  skin, absorption through, 450
  sodium reabsorption, drugs affecting,
    529–530
  synaptic transmission, effect on, 125–128
  viagra, 767
Dry mouth. See Mouth

Ductus deferens, 750, 763
  location and functions of, 764
  vasectomy and, 763
Duodenum, 594, 623
  absorption in, 625–626
  bile entering, 620
  emptying rate, factors affecting, 606–607
  enterogastric reflex, 606
  enterogastrones, 606
  gastric emptying into, 605–607
  hormones from, 671
  migrating motility complex of, 624
  secretions of, 625
  segmentation of, 624
Dural sinuses, 140
Dura mater, 140
Dwarfism, 694–695, 706
Dynamin, 32
Dynein, 47
  as accessory protein, 48
Dynorphin, 194
Dyslexia, 152
Dysmenorrhea, 779
Dyspnea, 494, 504, 505
Dystonias, 252
Dystrophin, 284

Ear canal, 215–216
  functions of, 227
Eardrum. See Tympanic membrane
Ears, 212–224. See also Hearing
  anatomy of, 214
  functions of major components, 227
  infections, 216
  parts of ear, 213
Earwax, 216
Eating. See Food intake; Foods
Eccentric isotonic contractions, 275
ECG (electrocardiogram), 317–320
  arrhythmia, diagnosis of, 319
  atrial flutter, 319
  cardiac myopathies, diagnosis of, 320
  components, correlation of, 317–318
  diagnosis, use of, 319–320
  fibrillation, diagnosis of, 320
  leads for, 318
  PR segment, 318–319
  P waves, 317–319
  QRS complex, 317–319
  stress testing, 321
  ST segment, 319
  tachycardia, diagnosis of, 319
  TP interval, 319
  T waves, 317–319
Ectopic focus, 312
Ectopic pregnancy, 782
Edema, 368–370
  injury and, 416
  left-sided heart failure and, 332
  renin-angiotensin-aldosterone system,
    role in, 528–529
EEG (electroencephalogram), 153–154
  paradoxical sleep, 170

  sleep patterns, 170–171
  slow-wave sleep, 170
Effector organs, 136
  efferent output, examples of, 238
  peripheral nervous system (PNS) and,
    236–255
Effectors, 17
  proteins, 67
  in reflex arc, 177
Efferent arterioles
  glomerular filtration rate (GFR) and, 520
  of nephrons, 513–514
Efferent axons/fibers, 137
  spinal nerves carrying, 176–177
Efferent division of PNS, 236–255
Efferent neurons, 136–137
  features of, 248
  for muscle spindles, 286
Efferent pathway in reflex arc, 177
Eggs. See Ova
Ejaculation, 763, 766
  elements of, 767–768
  volume and sperm content of, 768
Ejaculatory duct, 750
Elastic recoil, 473–474
  breathing and, 477
Elastin fibers, 7, 348, 349
  elastic recoil and, 474
  in extracellular matrix (ECM), 62
  in pulmonary tissue, 474
Elbow joint, extension/flexion of, 274
Electrical defibrillation, 314
Electrical gradient, movement along, 75
Electrical signals. See Membrane potential
Electrocardiogram. See ECG
    (electrocardiogram)
Electrocerebral silence, 154
Electrochemical gradient, 75
Electroencephalogram. See EEG
    (electroencephalogram)
Electrolytes, A9
  circulatory shock and, 385
  homeostasis and concentration of, 12–13
  in oral rehydration therapy, 636
  reabsorption of, 526
  urine, secretion in, 525
Electromagnetic spectrum, 198
Electromagnetic waves, 197–198, 656
Electromotility, 219
Electron cloud, A3
Electron microscopy, 24
Electron shells, A4
Electron transport chain, 39–42
Elements, A3
  characteristics of, A4
Elephantiasis, 369, 370
Embolus, 335
Embryo, 750
  preventing rejection of, 786
Embryonic stem cells, 8
Emergency contraception, 789
Emetics, 608
Emission phase of ejaculation, 768

Emmetropia, 200–201
  diagram of, 202
Emotions
  diarrhea and, 636
  gastric motility and, 607–608
  hypothalamus and, 156
  limbic system and, 157
  neurotransmitters and, 158–159
  obesity and, 653
  and pain, 193
  psychoactive drugs, 158–159
  vomiting and, 608
Emphysema, 472
  airway collapse in, 473
  elastic recoil in, 474
  Fick's law of diffusion and, 74
  normal tissue compared, 487
  surface area and, 486
Enamel on teeth, 600
End-diastolic volume, 328–329
  in heart failure, 331
Endocardium, 308
Endocrine cells
  in digestive mucous membrane, 594–595
  in stomach, 609, 610
Endocrine disrupters, 769
Endocrine glands, 6. See also Hormones;
    specific endocrine glands
  autonomic nervous system, effect of, 242
  formation of, 7
Endocrine hypertension, 380
Endocrine system, 11, 666–699, 700–747.
    See also Calcium metabolism;
    Diabetes mellitus, Type I; Diabetes
    mellitus, Type II; Fuel metabolism;
    specific endocrine glands
  complexity of, 668–669
  diagram of, 668
  growth, control of, 689–697
  heat, response to, 685
  hierarchic chain of command in, 687
  homeostasis and, 13–16
  hypothalamus and, 156
  immune system, linkage with, 445
  nervous system compared, 133–135
  obesity and, 653
  principles of, 667–681
  specificity of communication in, 135
  target cells in, 134
Endocrinology, 667
Endocytosis, 32, 84–85
  characteristics of, 86
  receptor-mediated endocytosis, 32
Endocytotic vesicles, 32
  summary of, 45
Endogenous opiates, 194
Endogenous pyrogen (EP), 419, 661
Endolymph
  in cochlea, 217
  in vestibular apparatus, 223
Endometrium
  blastocyst, implantation of, 784–786
  decidua, 786

post-parturition changes, 796
Endoplasmic reticulum, 25–28
 lumen, 26
 MHC molecules in, 434
 pancreatic enzymes synthesized by, 616
 preprohormones in, 672
 sarcoplasmic reticulum and, 28, 263–264
 secretion process, 28
 summary of, 45
Endorphins, 194. *See also* β-endorphin
 acupuncture endorphin hypothesis, 195
Endothelial cells
 in arteriolar smooth muscle, 355
 of capillaries, 360, 362–363
 functions of, 355
 homocysteine and, 337
 initial lymphatics, wall of, 367
Endothelial-derived relaxing factor
  (EDRF)
 and arteriolar smooth muscle, 355
 hypertension and, 381
Endothelin, 355
 glomerular filtration rate (GFR) and, 521
 hypertension and, 381
Endothelium
 cholesterol deposit beneath, 334
 of endocardium, 308
End-plate potential (EPP), 103, 247–249,
  294
 action potential in, 250
 diagram of events at, 249
End-stage renal failure, 549
End-systolic volume (ESV), 323
Endurance-type exercise, 277
Energy balance, 546–654. *See also* ATP
  (adenosine triphosphate); Heat
 activities and energy use, 648
 diet-induced thermogenesis, 649
 fat cells, size of, 650–651
 food intake and, 650–651
 input and output, 647–648
 metabolic rate, 648–649
 muscles consuming energy, 275
 neutral energy balance, 649–650
 thermal energy, 648
Energy equivalent of O_2, 649
Energy level, A4
Enkephalins, 194
Enteric nervous system, 598
Enterochromaffin-like (ECL) cells, 610
Enterogastric reflex, 606
Enterogastrones, 606
Enterohepatic circulation of bile salts, 620
Enterokinase, 616
 in brush-border plasma membrane, 625
Enteropeptidase, 616
Entrained rhythms, 678–679
Environmental influences
 biological clock and, 681
 food intake and, 652–653
 growth and, 690
 multiple sclerosis (MS) and, 116
 placental barrier and, 786–787

Enzymes, 6, A8. *See also* Digestive enzymes
 acid-base balance and, 573
 in erythrocytes, 394
 lysosomal enzymes, 30, 32
 odor-eating enzymes, 231
 pancreatic enzymes, 616–618, 625
 in saliva, 600
 small intestine secretions, 625
Eosinophil chemotatic factor, 447
Eosinophils, 392, 399–400, 414
 allergic conditions and, 401
EPA (Environmental Protection Agency),
  769
Ependymal cells, 139–140
Epicardium, 308
Epidermis, 449–450
 keratin production, 450–451
 melanin production, 450–451
Epididymis, 750, 763
 location and functions of, 764
 sperm and, 763
Epiglottis, 602
Epilepsy, 153–154
Epinephrine, 121
 adrenergic receptors for, 244–245, 715
 arteriolar radius, effect on, 358
 chromaffin granules, storage in, 714
 contractility of heart and, 329
 endocrine hypertension and, 380
 exercise, ventilation and, 503
 fuel metabolism, effect on, 732–733
 heart rate and, 328
 metabolic effects of, 715–716
 pathologic hyperthermia, 662
 secretion by adrenal medulla, 243, 714
 stress, role in, 716
 sympathetic nervous system and, 678,
  715–716
Epiphyseal plates, 691
 puberty and, 781
 testosterone and, 758
Epiphysis, 691
Epithelial cells
 iron absorption, 633–634
 microvilli, 50–51
 in small intestine, 626, 628
 taste receptor cells as, 225, 228
Epithelial sheets, 6
Epithelial tissue, 6
 desmosomes in, 63
 of stomach, 7
 tight junctions in, 63
Epstein-Barr virus, 402
Equations, A31–A34
Equilibrium
 diseases of, 224
 vestibular apparatus and, 222–224
Equilibrium potential, 89–90
 K^+ equilibrium potential, 89–90
 Na^+ equilibrium potential, 90–91
 Nernst equation, 90
Equilibrium thermodynamics, A33–A34
Equivalent of electrolyte, A9–A10

Erasistratus, 347
Erectile dysfunction, 767
Erectile tissue, 766
Erection-generating center, 766
Erection of penis, 766–767
Eructation, 603, 639
Erythroblastosis fetalis, 433
Erythrocytes, 393–398
 anemia, 396–397
 antibodies against antigens in, 432
 blood doping, 396
 bone marrow replacing, 394–395
 in clots, 404
 enzymes in, 394
 flexibility of membrane, 393
 functions of, 392
 hemoglobin in, 393–394
 iron, absorption of, 634
 polycythemia, 397–398
 reticulocytes, 396
 shape of, 393
 transfusion reaction, 432, 433
 2,3-bisphosphoglycerate (BPG) and
  hemoglobin, 492
 and vitamin B_{12}, 612
Erythropoiesis, 394–395
 control of, 395–396
 erythropoietin, 395–396
Erythropoietin, 395–396
 acute mountain sickness and, 496
 kidneys producing, 512
 synthetic erythropoietin, 396
Esophageal stage of swallow, 603
Esophagus, 461, 594, 601–603. *See also*
  Swallowing
 anatomy and functions of, 595–597
 heartburn, 603
 peptic ulcers of, 614
 sphincters guarding, 603
Essential nutrients, 720–721
Estradiol, 672, 772
Estriol in pregnancy, 791
Estrogens, 711, 750, 770
 actions of, 797
 Alzheimer's disease and, 163
 antrum formation and, 772
 cervical mucus and, 780
 environmental estrogens, 769
 in follicular phase, 776–777
 growth and, 690, 696
 lactation and, 796
 in luteal phase, 774
 luteinizing hormone (LH) in, 685
 males, production in, 762–763
 menopause and, 781
 and menstrual cycle, 779
 osteoporosis and, 736–737
 osteoprotegerin (OPG) stimulation, 735
 ovarian follicle, production by, 776
 in parturition, 792
 placenta secreting, 790, 791
 in pregnancy, 791
 secondary sexual characteristics, 750

Ethics of stem cell science, 9
Eunuchs, 758
Eupnea, 494
Eustachian tube, 216
Evaporation, 657
Evista, 737
Excitable tissues, 88, 99
Excitation-contraction coupling, 263
    calcium, role of, 263, 266, 268, 734
    muscle fatigue and, 279–280
    steps of, 268
Excitatory postsynaptic potential (EPSP),
        119, 120, 294
    AMPA receptors and, 165–166
    cancellation of, 122
    grand postsynaptic potential (GPSP) and,
        121
    motor neuron activity and, 246
    NMDA receptors and, 165
    spatial summation and, 122
    temporal summation and, 121–122
Excitatory synapses, 119, 120
Excitement phase of sexual response cycle,
        766
Exercise. See also Aerobic exercise; Athletics
    acclimatization, 566
    anabolic androgenic steroids and, 282
    anaerobic exercise, 42
    blood doping, 396
    blood pressure and, 379
    bones, benefits to, 737
    carbon dioxide variables during, 503
    cardiovascular changes and, 380
    creatine phosphate and, 277
    demands on muscles, response to,
        281–283
    endocrine response to, 685
    endorphins and, 194
    energy balance and, 650
    glucose use by muscles, 81
    homeostasis and, 12
    hypertension and, 382
    hyperthermia, exercise-induced, 661
    immune system and, 446
    lactic acid and, 279
    maximal O_2 consumption (max VO_2),
        504
    menstrual cycle and, 780
    metabolic acidosis and, 585
    muscle fatigue and, 279–280
    obesity and, 653
    oxidative phosphorylation and, 277–279
    oxygen debt and, 280
    oxygen variables during, 503
    physiology of, 13
    pregame meals, 607
    skeletal muscles and, 250
    in space flight, 250
    sports skills, 177
    stress testing, 321
    temperature and, 655
    urine, protein in, 550
    ventilation and, 502–503

Exocrine cells
    in digestive mucous membrane, 594–595
    in stomach, 608, 609, 610
Exocrine glands, 6. See also Pancreas
    autonomic nervous system, effect of, 242
    formation of, 7
    hypothalamus and, 156
    mode of secretion, 592
    in skin, 450
Exocytosis, 29, 85, 672
    characteristics of, 86
    of secretory product, 29, 30
Exophthalmos, 706–707
Expiration, 467–471
    airway collapse during, 473
    difficulties in, 472–473
    forced expiration, 469–471
    lung volume and, 468
    maximal expiratory efforts, 477
    passive expiration, 467–469
    respiratory muscle activity during,
        467–469
Expiratory muscles, 469–470
    actions of, 470
Expiratory neurons, 496–498
    in ventral respiratory group (VRG),
        497–498
Expiratory reserve volume (ERV), 478
Expulsion phase of ejaculation, 768
Extension of joint, 274
External anal sphincter, 638
External auditory meatus, 215–216
    functions of, 227
External cardiac compression, 304
External ear, 213
    functions of, 227
    sound localization and, 215–216
External eye muscles, 212, 275
External intercostal muscles, 496–498
    actions of, 470
    inspiration and, 466
External respiration, 459–460
External urethral sphincter, 551
External work, 647–648
Extracellular chemical messengers, 65–71
    channels and, 66–67
    second messengers, 67–68
Extracellular fluid (ECF), 10–11. See also
        Balance concept; Fluid balance;
        Osmolarity
    acid-base balance and, 574–575
    alkalinity of, 574
    blood pressure and, 563
    bulk flow and, 367
    calcium in, 734
    in circulatory shock, 385
    composition of, 562
    control of osmolarity of, 562–563,
        565–571
    control of volume of, 562–565
    distribution of body water, 561
    electrical balance of ions, 562
    homeostasis and, 12–13

hypertonicity, 566–567
hypotonicity, 567
input, 559
internal pool, 559–560
isotonic fluid gain or loss, 568
kidneys, role of, 511
membrane potential and, 87, 92
Na^+ load and, 527
output, 559
P_{CO_2} and, 500
plasma membrane and, 25
stable balance, 560
tonicity and, 79, 562–571
vasopressin and, 569–570, 571
vesicular transport, 84
water, loss or gain of, 548
Extracellular matrix (ECM), 62–63
Extrafusal fibers, 286
Extrapyramidal motor system, 283
Extrasystoles, 312
    arrhythmias and, 319
Extrinsic nerves of digestive system, 598
Extrinsic (outside) controls, 16
Extrinsic pathway of clotting cascade, 405,
        406
Eyelashes, 195
Eyelids, 194
    blepharospasm, 252
Eyes, 194–213. See also Vision
    autonomic nervous system, effect of, 242
    biological rhythms and, 682
    epinephrine affecting, 716
    exophthalmos, 706–707
    external eye muscles, 212, 275
    as fluid-filled structure, 195–197
    functions of major components, 211
    light, refraction of, 197–200
    multiunit smooth muscle in, 292
    pineal gland and, 682
    protective mechanisms, 194–195
    refractive structures, 198–200
    sensitivity to light, 207–208
    structure of, 196

Facial expressions, 158
Facilitated diffusion, 80–81, 86
Factor XII (Hageman factor), 405, 406, 407
Factor XIII (fibrin-stabilizing factor), 404
Factor X, 404–405, 406
FAD (flavine adenine dinucleotide), 38
    in electron transport chain, 40
FADH2, 38
    in electron transport chain, 39–40
Fallopian tubes, 750
Familial Alzheimer's disease, 163
Farsightedness, 200–201
    diagram of, 202
Fas, 786
Fasciculus cuneatus, 174
Fasciculus gracilis, 174
Fas ligand, 786
Fast-glycolytic fibers, 281, 282–283
    hypertrophy of, 281

Fast-oxidative fibers, 282–283
Fast pain pathway, 192
Fast synapses, 121
Fast-twitch fibers, 282–283
Fatigue, 270
  central fatigue, 279–280
  muscle fatigue, 279–280
  neuromuscular fatigue, 279–280
Fats, 4. *See also* Fuel metabolism;
      Phospholipids
  bile salts and digestion of, 620–622
  as biomolecules, A10
  chemical composition of, A12–A14
  complex lipids, A13–A14
  cortisol, actions of, 709
  cytosol, storage in, 43–44
  diabetes mellitus and, 726–727
  digestion of, 592–594, 626
  duodenum, effect on, 606–607
  epinephrine, actions of, 716
  glucagon, actions of, 730–731
  growth hormone, actions of, 690–691
  insulin, actions of, 723, 725
  lymphatic system and, 368
  pancreatic lipase and, 617
  in pregame meals, 607
  simple lipids, A12–A13
  small intestine, absorption by, 629,
      632–633
  steatorrhea, 617
  water content of, 561
Fatty acids, 593
  as biomolecules, A10
  blood, absorption by, 623
  cortisol, actions of, 709
  epinephrine, actions of, 716
  glucagon, actions of, 730–731
  growth hormone and, 690–691
  heart, nutrients to, 333
  insulin and, 723, 725
  intermediary metabolism and, 43
  molecules, A12–A13
  small intestine, absorption by, 629
  starvation and, 723
  storage of, 721
  stress and, 717
Fatty streak, 334
FDA (Food and Drug Administration)
  Botox, approval of, 252
  on DHEA (dehydroepiandrosterone),
      712
  growth hormone therapy, 696
  melatonin, uses of, 683
  osteoporosis drugs, 737
Fear
  cerebral cortex and, 158
  gastric motility and, 608
  vasopressin and, 571
  vasopressin stimulation and, 570
Feces, 637. *See also* Constipation; Diarrhea
  bile salts in, 620
  bilirubin in, 622
  defecation reflex, 638

iron in, 634
sodium loss in, 564, 639
steatorrhea, 617
water loss in, 569, 639
Feedback. *See also* Negative feedback;
      Positive feedback
  homeostasis control systems and, 16, 18
Feedforward
  glucose-dependent insulinotrophic
      peptide (GIP), insulin and, 641
  homeostasis control systems and, 16, 18
Feeding centers, 650
Female condoms, 788
Female pseudohermaphroditism, 713
Female reproductive system, 752, 770–799.
      See also Menstrual cycle; Pregnancy
  functions, 750
  ovarian cycle, 772–774
  puberty, 780–781
  sexual response cycle, 768–769
  uterine cycle, 778–780
Ferric iron, 634
Ferritin, 634
Ferrous iron, 633–634
Fertilin, 782
  contraception and, 789
Fertilization, 781–783
  process of, 782–783
Fetal cell transplants, 127
Fetal growth, 690
Fetus, 750
  parturition, preparation for, 792
  preventing rejection of, 786
Fevers, 661
  infections and, 419
  salicylates and, 421
Fiber in diet, 592
Fibrillation, 314
  atrial fibrillation, 320
  ventricular fibrillation, 320
Fibrin
  clot formation and, 404
  inflammation and, 417
Fibrinogen, 392, 393
  clot formation and, 404
  in semen, 764
Fibrinolysin, 764, 779
Fibrinolytic plasmin, 406–407
Fibroblasts, 62
  motility of, 49, 50
  plaque-damaged arterial walls and, 335
Fibronectin in extracellular matrix (ECM),
      62
Fibrous skeleton of heart, 308
Fick's law of diffusion, 74–75
  alveoli, gas transfer across, 486
  capillaries and, 360
Fidget factor, 653
Fight-or-flight response, 240, 445
  epinephrine and, 715–716
Figure skaters, 186
Filariasis, 369, 370
Filtered load, 531

Filtration coefficient, glomerular filtration
      rate (GFR) and, 524
Filtration fraction, 540
Filtration slits, 518
  podocytes and, 524
Fimbriae, 781
Final common pathway, 246
Fingers
  tendons moving, 269
  voluntary manipulation of, 283
Firing of excitable membrane, 103
First heart sound, 323
First law of thermodynamics, 647
First messengers, 66, 67
First-order sensory neurons, 190
First polar body, 771
Flaccid paralysis, 285
Flagella, 47–48
  internal structure of, 48
  microtubules and, 47
Flatus. *See* Gases, intestinal
Flexion of joint, 274
Flower-spray endings, 286
Flow rate, 344–346
  in capillaries, 360–362
Fluid balance, 558, 559–571
  acclimatization, 566
  capillary blood pressure and, 562
  daily salt balance, 563–564
  dehydration, 566–567
  distribution of body water, 561
  glomerular filtration rate (GFR) and, 564
  hypertonicity, 566–567
  hypotonicity, 567–568
  input sources, 568
  isotonic fluid gain, 568
  maintaining balance, 569
  metabolic H_2O, 568
  minor ECF compartments, 561
  Na^+ load in extracellular fluid (ECF),
      563–564
  nonphysiological influences on,
      570–571
  oral metering of water, 570
  osmolarity, 562–563, 565–566
  output sources, 568–569
  portion of water in major fluid
      compartments, 561
  renin-angiotensin-aldosterone system and,
      564–565
  sensible water loss, 568–569
  urine, sodium control in, 564
  vasopressin and, 568–571
  volume of ECF and, 562–563
Fluid mosaic model, 59–60
Fluid retention, 567–568
Flu viruses, 32
Foam cells, 334
  thromboembolism and, 335
Focal point, 198
Folic acid
  caveolae and, 85
  homocysteine and, 337

Follicles, ovarian
  colloid-filled follicles, 701–702
  development of, 773
  mature follicle, formation of, 773–774
  twins, development of, 774
Follicles, thyroid, 701–703
Follicle-stimulating hormone (FSH)
  anterior pituitary secreting, 685
  feedback control of, 777
  female reproductive system and, 775–776
  and follicular phase, 776–777
  puberty and, 763
  Sertoli cells and, 762
  set point for secretion of, 678
  spermatogenesis and, 762
  testes and, 762
Follicular cells
  ovarian, 772
  thyroid, 701
Follicular phase, 772
  control of, 776–777
Food intake. See also Satiety; Weight
  anorexia nervosa, 653–654
  control of, 650
  environmental influences, 652–653
  factors influencing, 651
  hypothalamus and, 156, 650–651
  leptin and, 650–651
  melanocortins and, 651–652
  neuropeptide Y (NPY) and, 651–652
  orexins and, 652
  polyphagia, 726
  psychosocial influences, 652–653
Foods
  additives, kidneys and, 512
  energy extraction, efficiency of, 653
  gas related to, 639
  growth and diet, 690
  hydrogen ion and, 574
  pregame meals, 607
  primary hypertension and diet, 381
  thermogenesis, diet-induced, 649
  water input and, 568
Foot proteins, 263
Force, muscles accomplishing, 275
Forced expiration, 469–471
Forced expiratory volume in one second
    (FEV_1), 479
Forebrain, 143
Forgetting, 160
Formula weight, A8–A9
Forward failure of heart, 332
Fosamax, 737
Fovea, 203
  function of, 211
Frank-Starling law of the heart, 328, 370
  acclimatization, 566
  circulatory shock and, 383
  compensated heart failure, 331
  stroke volume and, 328–329
  venous return and, 328
Fraternal twins, 774

Free radicals
  Alzheimer's disease and, 163
  atherosclerosis and, 334
  melatonin and, 683
  Parkinson's disease and, 126
Free ribosomes, 26
  in cytosol, 43
Frontal lobes, 146–147, 148
  primary motor cortex, 148
Frostbite, 662
Fructose, 592
  prostate gland, secretions of, 764
  small intestine, absorption by, 629
Fuel metabolism, 43
  brain, glucose for, 72–722
  glucagon regulation of, 723–725
  insulin regulation of, 723–725
  lesser energy sources for, 722–723
  organic molecules, interconversions
      among, 720–721
  reactions, summary of, 719
  stored metabolic fuel, table of, 721
  stress hormones affecting, 732–733
  summary of, 45
  thyroid hormone affecting, 704, 732
  tissues, role of, 722
Fuel storage
  absorptive state, storage during, 722
  mobilization of, 722
  postabsorptive state, 722
Fulcrum, 275–276
Functional groups, A11
Functional murmurs, 324
Functional residual capacity (FRC), 478
Functional syncytium, 293, 294
Functional unit of organ, 258
Fundus of stomach, 604

GABA (gamma-aminobutyric acid), 121
  astrocytes and, 137
  epilepsy and, 154
  synaptic vesicles releasing, 120
  tetanus toxin and, 127
Galactose, 592
  small intestine, absorption by, 629
Galen, 347
Gallbladder, 594
  autonomic nervous system, effect of, 242
  bile and, 620, 622–623
Gallstones, 622
Gametes, 749. See also Ovaries; Testes
Gametogenesis, 750
  comparison of oogenesis and
      spermatogenesis, 772
  meiosis and, 751, 753, 759, 770–771
  oogenesis, 770–772
  spermatogenesis, 758–760
Gamma motor neurons, 286, 287–288
Gamma (γ) globulins, 393
γ secretase, 163
Ganglion cells, 175
  cones and, 207

  function of, 211
  off-center ganglion cells, 209–210
  on-center ganglion cells, 209–210
  in retinal layer, 202–203
Gangliosides, Tay-Sachs disease and, 34
Gangrene and diabetes mellitus, 729
Gap junctions, 63–65
  in cardiac muscle, 308–309
  communication, intercellular, 65
  in digestive smooth muscle, 598
  granulosa cells and, 772
  schematic drawing of, 64
Garlic odor, 230–231
Gartler, Stanley, 24
Gases, intestinal, 639
  lactose intolerance and, 625
Gastric glands, 608
Gastric inhibitory peptide (GIP), 640–641
Gastric juices, 561, 608, 610
  carbohydrates, digestion halted by, 614
  inhibition of, 613
  multiple regulatory pathways for, 612
  proteins, digestion of, 611
  stimulation of, 612
Gastric mucosa, 608, 610
Gastric mucosal barrier, 613–614
Gastric phase of gastric secretion, 613
Gastric pits, 608–610
Gastrin, 121, 606, 609, 610, 639, 640
  gastric juices, influence on, 612
Gastrocolic reflex, 638
Gastroesophageal sphincter, 603
Gastroileal reflex, 624
Gastrointestinal hormones, 598–599. See
      also Cholecystokin; Gastrin; Secretin
  insulin, stimulation of, 725
Gated channels, 66–67. See also Chemically
      gated channels; Mechanically gated
      channels; Voltage-gated channels
  graded potentials and, 100
  triggering events and, 100
G cells in stomach, 609, 610
GDNF and Parkinson's disease, 127
GDP (guanosine diphosphate), 38
Gehrig, Lou, 51
Gender. See also Adrenogenital syndrome
  estrogenic pollutants and, 769
  identity crisis, 756
General adaptation syndrome, 716
Generator potentials, 187
  action potentials, conversion into, 188
Genes
  DNA (deoxyribonucleic acid) and, A19
  lipophilic hormones stimulating, 676–677
  protein transcription and, A26–A27
Gene-signaling factors, A26–A27
Genetic engineering, 9
Genetics
  Alzheimer's disease and, 163
  color blindness and, 209
  cystic fibrosis (CF) and, 61
  hormones, hyposecretion of, 679

melanin and, 451
multiple sclerosis (MS) and, 116
muscle fiber types and, 281
of muscular dystrophy, 284
mutations of genes, 444
testicular feminization syndrome, 680
weight regulation and, 653
Genetic sex, 753–754
Genitals. *See also* Adrenogenital syndrome
autonomic nervous system, effect of, 242
sexual differentiation in, 755
Genital swellings, 756
Genital tubercle, 755
Germ cells, 758, A20
mutation, A30
Gestation, 750, 792
Gigantism, 695
GIP. *See* Glucose-dependent insulinotropic peptide (GIP)
Glands, 6. *See also* specific types
Glans penis, 755–756
Glaucoma, 196–197
Glial cells, 137
astrocytes, 137–139
ependymal cells, 139–140
in gray matter, 145
interneurons, support for, 137–140
microglia, 139
oligodendrocytes, 139
Gliomas, 140
Globin portion of hemoglobin, 393–394
Globulins, 392, 393
Globus pallidus, 127
Glomerular capillary blood pressure, 518–520
glomerular filtration rate (GFR) and, 520
Glomerular capillary wall, 517–518
Glomerular filtration, 516, 517–525
forces involved in, 519
permeability of glomerular membrane, 517–518
rate of, 519–520
Glomerular filtration rate (GFR), 520–524
atrial natriuretic peptide (ANP) and, 530
autoregulation of, 520–522
baroreceptor reflex and, 522–524
controlled adjustments in, 520
extrinsic sympathetic control of, 522
filtered sodium and, 564
filtration coefficient changes, 524
filtration fraction, 540
glomerular capillary blood pressure and, 520
myogenic mechanism and, 521
plasma clearance rate and, 538–540
sodium, excretion of, 564
tubuloglomerular feedback mechanism, 521
unregulated influences on, 520
Glomerular membrane, 517–518. *See also* Bowman's capsule
basement membrane, 518

Glomeruli, 230, 513
of cortical and juxtamedullary nephrons, 515
electron micrograph of, 515
Glomerulonephritis, 548
Glottis, 602
Glucagon, 121
diabetes mellitus and, 731–732
excessive glucagon, 731–732
fuel metabolism regulation, 723–725
insulin action and, 730–731
postabsorptive state and, 731
stress and, 717
Glucocorticoids, 421–422, 708
adrenal virilization, treatment of, 713
inflammation, role in, 710
metabolic effects of, 709–710
Gluconeogenesis, 709
cortisol and, 709
epinephrine and, 715
in fuel metabolism, 719
starvation and, 723
Glucose, 592
blood-brain barrier and, 142
for brain, 721–722
cellular energy production from, 36
cortisol and, 709
and Cushing's syndrome, 712
as disaccharide, A12
epinephrine and, 732
exercise and, 81
facilitated diffusion of, 81
glucagon and, 730
growth hormone and, 690
heart, nutrients to, 333
internal storage capacity, 559–560
insulin and, 723–725
kidneys and reabsorption of, 531–532
muscle cells storing, 278–279
in oral rehydration therapy, 636
plasma glucose concentration, 532, 724
as preferred cell fuel, 40
reabsorption of, 526, 531–532
regulation of, 690, 723–724, 730, 732
renal threshold for, 531–532
satiety and, 652
in secondary active transport, 83, 531
small intestine, absorption by, 629
storage of, 721
stress and, 717
sweet taste, 228
thyroid hormone and, 704
transport of, 72
tubular maximum (T_m) for, 531
tubular reabsorption of, 530–532
Glucose-dependent insulinotropic peptide (GIP), 640–641
diabetes mellitus research, 729
Glucose transporter (GLUT), 724–725
Glucosuria, 726
GLUT, 724–725
Glutamate, 121

astrocytes and, 137
long-term potentiation (LTP) and, 165–166
as pain neurotransmitter, 192–194
strokes, role in, 143
umami taste and, 229
Glutamine and ammonia synthesis, 582
Gluten enteropathy, 626, 628
brush border, reduction of, 627
Glycerol, 593, A13
fuel metabolism and, 722
Glycine, 121
strychnine and, 127
synaptic vesicles releasing, 120
Glycogen, 592, A12
cytosol, storage in, 43–44
in endometrium, 786
glycolysis and, 279
in liver, 722
muscle cells storing, 279
as polysaccharide, A12
restoring nutrients, 280
storage of, 44, 279, 721
Glycogenesis
in fuel metabolism, 719
insulin stimulating, 723–724
Glycogenolysis
epinephrine and, 715–716
in fuel metabolism, 719
glucagon and, 730
Glycolipids, 60
Glycolysis, 35, 42
in anaerobic condition, 40, 41
ATP formation and, 279
erythrocytes, glycolytic enzymes in, 394
lactic acid and, 279
white fibers, 281
Glycoproteins, 60
basement membrane and, 518
in desmosome, 63
GnRH (gonadotropin-releasing hormone), 121
Gofman, John, 682
Goiters, 707
Golgi complex, 25, 28–30
budding off process, 29
pancreatic enzymes synthesized by, 616
preprohormones to, 672
raw materials, processing, 29
recognition markers, 29–30
secretory cells, 29
summary of, 45
transport vesicles to, 27
Golgi tendon organs, 286, 288
Gonadal sex, 753–754
Gonadotropin-releasing hormone (GnRH), 687, 762
female reproductive system and, 775
male reproductive system and, 763
puberty and, 763
Gonadotropins, 685
testosterone and, 758

Gonads, 749, 750. *See also* Ovaries;
    Testes
  hormones from, 671
Goosebumps, 660
G proteins, 67
  adrenergic receptors and, 245
  gustducin, 228
  odors and, 230
  phosphorylation and, 67
  on plasma membrane, 67
  sweet taste and, 228
  transducin, 205
  umami taste and, 229
Graafian follicle, 773–774
Graded potentials
  action potentials compared, 111
  current flow during, 101
  duration of, 100, 101
  end-plate potential (EPP) as, 249
  magnitude of, 100, 101
  passive current flow spreading, 100–102
  short distances, die out over, 102–103
  triggering event for, 100
Grand postsynaptic potential (GPSP),
    121–123
  postsynaptic integration and, 123
  spatial summation, 122
  temporal summation, 121–122
Granstein cells, 450, 451
Granular cells, 528
Granulocyte colony-stimulating factor, 401
Granulopoiesis, 419
Granulosa cells, 770, 772
  in follicular phase, 776
Granzymes, 438
Graves' disease, 706
  goiters in, 707
Gravity, venous pressure and, 373
Gray matter, 145
  in brain, 145
  in spinal cord, 173, 187
Greece, ancient, 347
Green cones, 205
Groaning, respiration during, 504
Ground substance, 691
Growth, 689–697
  aberrant growth patterns, 694–695
  androgens and, 695–696
  bone growth, 691
  cretinism, 706
  diabetes mellitus and, 728
  estrogens and, 696
  growth hormone and, 690–693
  insulin and, 695
  rapid growth periods, 690
  sarcomeres in, 258
  thyroid hormone and, 695, 704
Growth hormone (GH), 685, 690
  abnormal secretion of, 694–695
  age and, 711
  deficiency in, 694–695
  diurnal fluctuation in secretion of, 693
  excess of, 695

  factors influencing secretion of, 693–694
  fetal growth, 690
  fuel metabolism, effect on, 732
  hypophysiotropic hormones, regulation
    by, 693–694
  IGF-I synthesis, 693
  length of bones, promotion of, 692
  non-growth related metabolic actions of,
    690–691
  soft tissue growth, 691
  somatomedins, stimulation of, 692–693
  synthetic GH treatment, 696
  thickness of bones, promotion of, 692
  thyroid hormone and, 704
Growth hormone-inhibiting hormone
    (GHIH), 687, 693
Growth hormone-releasing hormone
    (GHRH), 687, 693
GTP (guanosine triphosphate), 38
Guanine (G), A19
Guanosine monophosphate. *See* Cyclic
    GMP
Gustatory pathways, 228
Gustducin, 228
Gut-associated lymphoid tissue (GALT),
    414, 595
  functions of, 415
Gynoid obesity, 654

Habituation, 163
  mechanisms of, 164
Hair
  axillary hair, 750, 781
  as keratin structure, 51
  pubic hair, 750, 781
Hair cells
  activation of, 223
  hearing loss and, 221–222
  in organ of Corti, 217–219
  in vestibular apparatus, 223
Hair follicles, 450
  multiunit smooth muscle in, 292
Haldane effect, 494
Haploid cells, A28–A29
Haploid number, 751, A20
Harvey, William, 347
Hashimoto's disease, 441
Haustra, 637
Haustral contractions, 637
Hay fever, 448
H_2CO_3, 575–576
HCO_3^-. *See* Bicarbonate
H_2CO_3:HCO_3^- buffer system, 575–576
HDL (high-density lipoprotein), 336–337
  anabolic androgenic steroids and, 282
Headaches, 140
Head of spermatozoon, 760
Head-righting reflex, 177
Hearing, 185, 212–224
  deafness, 221–222
  defined, 213
  loudness, sensitivity to, 220
  sound waves, 213–214

  temporal lobes and, 146
  threshold, 215
Hearing aids, 221
Heart, 4. *See also* specific systems
  action potential of cardiac muscle, 315
  afterload, 330–331
  anatomy of, 304–310
  atrioventricular node (AV node), 311
  atrioventricular valves, 306–307
  autonomic nervous system, effect of, 242
  beta receptors in, 245
  bundle of His (atrioventricular bundle),
    311
  calcium changes in ECF and, 316
  cardiac excitation, spread of, 312–314
  cardiac myopathies, 320
  cardiogenic shock, 382
  Ca^{2+} sparks and, 315–316
  circuit of blood flow, 304–305
  complete heart block, 312, 320
  contractility of, 329
  coordination of muscle fiber excitation,
    313–314
  coronary circulation, 332–333
  development of, 303
  diastole period, 320–321
  as dual pump, 304–305
  electrical activity of, 310–320
  embryonic heart, development of, 309
  epinephrine and, 715
  fibrillation, 314
  Frank-Starling law of the heart, 328–329
  functional murmurs, 324
  hormones from, 671
  hypercalcemia and, 741–742
  hypertension and, 381
  isovolumetric ventricular contraction,
    321, 323
  isovolumetric ventricular relaxation, 323
  latent pacemakers, 312
  mechanical events of, 320–325
  murmurs, 324–325
  nutrient supply to, 333
  parasympathetic nervous system affecting,
    326–327
  pericardial sac, 309–310
  potassium concentration and, 316,
    326–327, 536
  premature heart beat, 312
  Purkinje fibers, 311
  semilunar valves, 307–308
  sides of heart, 305
  sinoatrial node (SA node), 31–312
  specialized conduction of, 311
  spread of cardiac excitation, 313–314
  stress testing, 321
  stroke volume (SV), 323
  sympathetic nervous system, effect of,
    327
  sympathetic stimulation of, 372
  systole period, 320–321
  tissue engineering and, 9
  vagus nerve, 326

venous return and, 375
walls, 308
Heart attacks, 320
  collateral circulation and, 335–336
  hypertension and, 381
  possible outcomes of, 336–337
  thromboembolism and, 335
Heart block, 312, 320
Heartburn, 603
Heart failure. *See also* Congestive heart
    failure
  afterload and, 331
  compensatory measures for, 331
  decompensated heart failure, 332
  prime defect of, 331
Heart murmurs, 324–325
Heart rate
  acetylcholine (ACh) and, 326
  cardiac output and, 324
  control of, 327–328
  ECG (electrocardiogram) and, 319
  exercise and, 13, 380
  resting heart rate, 324
  sinoatrial node (SA node) and, 325–328
Heat. *See also* Energy balance
  balance between input and output, 656
  conduction, 656
  convection, 656–657
  conversion of nutrient energy and, 648
  coordinated responses to, 660–661
  core temperature and, 655–656
  diet-induced thermogenesis, 649
  endocrine response to, 685
  energy equivalent of O_2, 649
  evaporation, 657
  exchange, methods of, 656
  extreme cold, exposure to, 662
  extreme heat, exposure to, 662
  food energy and, 647–648
  hair and, 450
  humidity and heat loss, 657
  hyperthermia, 661–662
  hypothalamus controlling, 659–660
  mechanisms of transfer, 656–657
  metabolic heat production, 659
  muscles producing, 275
  nonshivering (chemical) thermogenesis,
    659
  radiation of heat, 656
  respiratory system and, 460
  shivering, 658–659
  skin, blood flow to, 659–660
  thermoreceptors, 658
  thermoregulatory pathways, 658
  thyroid hormone and, 704
Heat exhaustion, 566, 662
Heat stroke, 18, 566, 662
Heat waves, 656
HeLa cells, 24
Helicobacter pylori
  peptic ulcers and, 615
  SIDS (sudden infant death syndrome)
    and, 505

Helicotrema, 217, 219
  pitch discrimination and, 219–220
Helper T cells, 436, 437, 439
  allergens and, 447
  B-cell growth factor, 435, 439–440
  B lymphocytes and, 434
  class I/class II MHC glycoproteins, 442–443
  cytokines from, 439–440
  MHC molecules and, 442
  subsets of, 440
Hematocrit, 391, 392
  circumstances affecting, 397
Heme, 622
Hemiplegia, 285
Hemodialysis, 551
Hemoglobin, 390, 393–394, 488
  acidity, effect of, 491–492
  blood doping, 396
  blood substitutes and, 398–399
  Bohr effect, 492
  carbon dioxide and, 491, 493
  carbon monoxide and, 492
  genetic engineering of, 399
  Haldane effect, 494
  hydrogen ion and, 494, 577
  net transfer of oxygen and, 490–491
  O_2-Hb dissociation curve, 489–490
  P_{O_2} and saturation of, 488–489
  saturation, 488–489
  as storage depot for oxygen, 490–491
  temperature and saturation of, 492
  tissue level, role at, 491
  2,3-bisphosphoglycerate (BPG) and, 492
  unloading of oxygen from, 491–492
Hemoglobin buffer system, 575, 577
Hemolysis, 397
  transfusion reaction, 432, 433
Hemolytic disease of the newborn, 433
Hemolytic jaundice, 622
Hemophilia, 408
Hemopoiesis, 400
Hemopoietic red marrow, 395
Hemorrhage
  angiotensin II and, 359
  cerebral hemorrhage, 608
  glomerular filtration rate (GFR), extrinsic
    control of, 522
  hemostasis and, 403
  hypertension and, 381
  isotonic fluid loss, 568
  reticulocytes, release of, 396
  vasopressin and, 359
Hemorrhagic anemia, 397
Hemostasis, 390, 402–403
  thrombin, role of, 405
Henderson-Hasselbalch equation, 576
Henle's loop, 514
  functions of, 548
  medullary countercurrent system and,
    541–544
  properties of descending/ascending limbs,
    541
  sodium reabsorption in, 526–527

vasa recta, 515–516
vertical osmotic gradient and, 540–541
water reabsorption and, 533
Heparin
  basophils and, 401
  blood clots, prevention of, 407
Hepatic jaundice, 622
Hepatic portal system, 619, 688
Hepatitis, 623
  blood substitutes and, 398
Hepatocytes, 618, 619–620
Hering-Breuer reflex, 498
Hermaphroditism, 713
Heroin
  meperidine (MPPP), 126
  placental barrier, crossing, 786–787
Herpes virus
  atherosclerosis and, 337
  in nerve cells, 439
  reverse axonal transport and, 47
High altitude effects, 496–497
High blood pressure. *See* Hypertension
High-energy phosphate bonds, A17–A18
High intensity exercise, 279
Hippocampus, memory and, 161
Hirsutism, 713
Histamine, 66, 121
  anaphylactic shock, 383
  arterioles and, 355, 419
  basophils and, 401
  complement system and, 423
  enterochromaffin-like (ECL) cells
    secreting, 610
  gastric juices, influence on, 612
  immediate hypersensitivity and, 447
  phagocytic secretions stimulating, 419
Histocompatibility complex, 441–443. *See
    also* MHC molecules
Histotoxic hypoxia, 494
HIV. *See also* AIDS
  receptor-mediated endocytosis and, 32
  vaccination, research on, 430
Hives, 448
H^+-K^+ ATPase pump, 615
H_2O. *See* Water
Homeostasis, 10–16
  apoptosis and, 70
  body systems and, 12–16, 22
  maintenance by body systems, 13–16,
    51–52, 92–93, 128–129, 180, 232,
    253, 297, 336–337, 285–286, 408,
    452, 505, 552–553, 586, 641, 663,
    697, 742–743, 799
  controlled variables, 16
  control systems, 16–19
  disruptions in, 18–19
  as dynamic steady state, 12
  extrinsic (outside) controls, 16
  factors regulated by, 12–13
  feedback/feedforward and, 16, 18
  intrinsic (local) controls, 16
  negative feedback, 16–18
  positive feedback, 18

Homeostatic control systems, 16–19
Homeostatic drives, 158
Homocysteine, atherosclerosis and, 337
Homologous chromosomes, 751, A20
Hormone replacement therapy for
    hypothyroidism, 706
Hormone response element (HRE), 676–677
Hormones, 6, 65, 66. *See also* Amines;
    Anterior pituitary; Liver; Peptides;
    Steroids; specific hormones
  abnormal target-cell responsiveness, 680
  action of hydrophilic/lipophilic
    hormones, 675–676
  from adrenal cortex, 670, 708
  adrenal medulla secreting, 243, 670
  Alzheimer's disease and, 163
  amplification of actions of, 677
  antagonism of, 681
  anterior pituitary, secretion by, 685
  arteriolar radius, effect on, 358–359
  binding with target cells, 135
  blood-brain barrier and, 142
  chemical classifications of, 673
  classification of, 672
  deliberate alteration of target cells, 680
  diffusion of, 74
  disorders from abnormalities of, 679–681
  diurnal rhythms, 678–679
  down regulation, 680
  duration of responses, 677
  endocrine circadian rhythms, 678–679
  enterogastrones, 606
  gastrin, 606
  gastrointestinal hormones, 598–599,
    639–641
  genes, lipophilic hormones stimulating,
    676–677
  as gene-signaling factor, A26–A27
  hypersecretion of, 679, 680
  hypophysiotropic hormones, 687–688
  hyposecretion of, 679–680
  hypothalamic-hypophyseal portal system,
    688
  intracellular proteins, alteration of,
    675–676
  lipid-soluble hormones, 674–675
  magnitude of response, 677–678
  major hormones, summary of, 669–671
  melanocortins, 651–652
  metabolic inactivation of, 679
  multiple target cells for, 668
  negative feedback control, 678
  neuroendocrine reflexes, 678
  as neuromodulators, 135
  onset of responses, 677
  ovarian changes and, 775
  permissiveness, 680
  placental hormones, 671, 790–791
  plasma, 392
    concentration of, 677–678, 679
    transport in, 674–675
  rate of secretion, variety in, 668
  receptors for, 135, 675

reflexes mediated by, 180
regulatory effects of, 667–671
replacement therapy, 680
second-messenger systems, use of, 676
solubility characteristics of, 672
specificity of action, 135
synergism of, 681
therapeutic administration of, 675
tropic hormones, 668
urinary excretion of, 679
uterine changes and, 775
water-soluble hormones, 674–675
Host cells, 414
Hot flashes, 781
Human chorionic gonadotropin, 756,
    790–791
Human chorionic somatomammotropin
    (hCS), 790
  lactation and, 796
  in pregnancy, 792
Humidification, 484
Humidity, sweat evaporation and, 657
Humoral immunity, 424
H-Y antigen, 754
Hydrocephalus, 141
Hydrochloric acid, 638
  functions of, 611
  gastrin and, 639
  mechanism of secretion of, 610–611
  mucus and, 611
  parietal cells secreting, 608–610
  pepsinogen, activation of, 610–611
  stomach secreting, 604
  as strong acid, 572
Hydrogen ion. *See also* Acid-base balance;
    Acidosis; Alkalosis; pH
  bicarbonate and secretion of, 580
  blood-brain barrier and, 500
  Bohr effect, 492
  bonds, A7
  buffering of, 575–577
  characteristics of, A4
  in chloride shift, 493–494
  in citric acid cycle, 28
  exercise and, 502
  Haldane effect, 494
  inorganic acids and, 574
  kidneys and, 578–582
  lines of defense against, 574–582
  loss of, 560
  lysosomes and, 34
  magnitude of ventilation and, 498
  peripheral chemoreceptors and, 502
  and pH, 572
  plasma clearance of, 538
  potassium secretion and, 536
  renal failure and, 534
  respiratory system and, 577–578
  sources of, 574
  tubular secretion of, 534–537
  ventilation, changes in, 501–502
Hydrogen atoms, 4
Hydrogen ion (H^+) pump, 81

Hydrogen peroxide, 34
Hydrolysis
  by lysosomes, 30–32
  in catabolism, 719–720
  in digestion, 594
  proteins and, A15, A17
Hydrolytic enzymes, 30, 32
Hydrophilic polar regions, 59–60
Hydrophobic core, 59–60
Hydrostatic (fluid) pressure, 77
Hydroxyapatite crystals, 735
Hyperkalemia, 714
Hyperaldosteronism, 712
Hypercalcemia, 734, 741–742
Hypercapnia, 494, 495
Hypercomplex cells, 212
Hyperglycemia, 726
  and Cushing's syndrome, 712
Hypernatremia, 712
Hyperopia, 200–201
  diagram of, 202
Hyperoxia, 494–495
Hyperparathyroidism, 741–742
Hyperphosphatemia, 742
Hyperpigmentation, 714
Hyperplasia, 691
Hyperpnea, 494, 495
Hyperpolarization, 100
  inhibitory postsynaptic potentials and,
    119–120
  of photoreceptors, 205
Hypersecretion of hormones, 679, 680
Hypersensitivity, 446–449. *See also*
    Allergies; Delayed hypersensitivity;
    Immediate hypersensitivity
Hypertension, 379–382
  ACE inhibitor drugs for, 529–530
  android obesity, 654
  cardiovascular hypertension, 380
  complications of, 381–382
  endocrine hypertension, 380
  exercise and, 382
  hyperaldosteronism and, 712
  inactivity and, 42
  neurogenic hypertension, 380
  primary hypertension, 381
  renal failure and, 380, 381, 549
  renin-angiotensin-aldosterone system,
    role in, 528–529
  secondary hypertension, 380
  stress and, 719
  treatment of, 382
  workload of heart and, 330–331
Hyperthermia, 661–662
Hyperthyroidism, 705–707
  goiters and, 707
Hypertonic body fluids, 540
Hypertonicity
  in duodenum, 606, 607
  of extracellular fluid (ECF), 566–567
Hypertonic solution, 79
Hypertrophy, 281
  growth hormone and, 691

Hyperventilation, 494, 495
  voluntary hyperventilation, 504
Hypocalcemia, 734, 742
  bones and, 738–739
Hypocapnia, 494, 495
Hypodermis, 450
Hypoglycemia, 730
Hypokalemia, 712
Hyponatremia, 714
Hypoparathyroidism, 742
Hypophysiotropic hormones, 687–688
  growth hormone secretion by, 693–694
  three-hormone sequence, 689
Hypophysis. See Pituitary gland
Hyposecretion of hormones, 679–680
Hypospadia, 769
Hypotension, 380, 382–385. See also
    Circulatory shock
  anaphylactic shock, 448
Hypothalamic-hypophyseal portal system,
    688
Hypothalamic osmoreceptors, 569
Hypothalamic releasing/inhibiting
    hormones, 688
  control of, 689
Hypothalamus, 143, 144, 666, 683–689
  anterior pituitary, regulation of, 685, 687
  arcuate nucleus, 651
  autonomic reflexes, control of, 245
  behavior patterns and, 157
  blood-brain barrier and, 142
  blood flow and, 358
  blood pressure, responses to, 379
  cold, responses to, 660
  fevers and, 651
  food intake, control of, 650–651
  heat control mechanisms, 659–661
  hierarchic chain of command in, 687
  homeostatic functions, regulation of,
    156–157
  hyperthermia and, 662
  hypophysiotropic hormones from, 685–688
  hypothalamic releasing and inhibiting
    hormones, 685–688
  hypothalamic-hypophyseal portal system,
    688
  hypothalamus-pituitary-thyroid axis,
    704–705
  location in brain, 156
  major hormones from, 669
  melanocyte stimulating hormone and,
    651
  oxytocin produced in, 684, 794, 797
  pain pathways and, 193
  somatostatin produced by, 723
  stress response and, 717–718
  as temperature thermostat, 657–658
  thirst center, 569
  vascular link with anterior pituitary, 688
  vasopressin produced in, 544, 569, 684
Hypothalamus-pituitary-adrenal cortex axis
  cortisol secretion and, 710–711
  glucocorticoids and, 710

Hypothalamus-pituitary-thyroid axis,
    704–705
Hypothermia, 662
Hypothyroidism, 705–706
  goiters and, 707
  obesity and, 653
Hypotonic body fluids, 540
Hypotonicity, 567–568
Hypotonic solution, 79
Hypoventilation, 494, 495
Hypovolemic shock, 382
Hypoxia, 494
Hypoxic hypoxia, 494
  acute mountain sickness, 496
H zone, 258
  in contraction process, 261

Iatrogenic hyposecretion of hormones,
    679
I bands, 258
  in contraction process, 261
Identical twins, 774
Idiopathic hypertension, 381
Idiopathic hyposecretion of hormones,
    679
IgA antibodies, 425
  in breast milk, 798
IgD antibodies, 425
IgE antibodies, 425
  allergens and, 447
  cytokines and, 440
  hypersensitivity, role in, 448
  immediate hypersensitivity and, 448–449
IGF. See Insulin-like growth factor (IGF)
IGF I synthesis, 693
IGF-II synthesis, 693
IgG antibodies, 425
  phagocytosis and, 428
  tail region of, 426
  transfer of, 432
IgM antibodies, 425
  clonal selection and, 428
Ileocecal valve/sphincter, 624–625
Ileum, 594, 623
  absorption in, 625–626
  segmentation of, 624
Immediate early genes (IEGs), 166
Immediate hypersensitivity, 446–447
  chemical mediators of, 447
  parasitic worms, comparison of response
    to, 448–449
  symptoms of, 447–448
  treatment of, 448
  triggers for, 447
Immune complex disease, 428
Immune privilege, 440–441
Immune surveillance, 413, 443–445
  effectors of, 444–445
Immune system, 11, 412. See also
    Leukocytes; Skin
  adaptive immune system, 415, 416
  antigens, 425
  apoptosis and, 70

bacteria and, 413–414
diseases, 446–449
endocrine system, linkage with, 445
exercise and, 446
homeostasis and, 13–16
innate immune system, 415
MSH (melanocyte-stimulating hormone)
    and, 684
nervous system, linkage with, 445
renal failure and, 549
tissue engineering and, 8–9
tolerance of self-antigens, 440–441
Immunity, 398–399. See also Adaptive
    immunity; Antibody-mediated
    immunity; Cell-mediated immunity;
    Innate immunity
active immunity, 432
blood types and, 432–433
colostrum and, 798
defined, 413
digestive system defenses, 451
long-term immunity, 431
melatonin and, 683
passive immunity, 432
reproductive system defenses, 451–452
respiratory system defenses, 452
skin defenses, 449–452
urinary system defenses, 451–452
Immunizations. See Vaccinations
Immunocontraception, 789
Immunosuppressive effects
  of cortisol, 710
  tissue engineering and drugs with, 8–9
Impermeable membranes, 71
Implantation, 785–786
  contraception by blocking, 789
Impotence, 767
Inclusions, 44
  summary of, 45
Incontinence, urinary, 552
Incus, 216
  functions of, 227
Independent variables, A32
Indirect calorimetry, 649
Indoleamine 2,3-dioxygenase (IDO), 786
Inertia, 223
Infants
  with adrenogenital syndrome, 713
  brown fat, 659
  erythroblastosis fetalis, 433
  with lactose intolerance, 625
  micturition by, 551
  newborn respiratory distress syndrome,
    476
  paradoxical sleep in, 171
  sexual differentiation issues, 756
  SIDS (sudden infant death syndrome),
    504–505
Infections. See also Inflammation
  exercise and, 446
  fever and, 661
  macrophages and, 416
  neutrophils and, 401

Infections *(continued)*
  renal failure and, 548
  septic shock, 383
  white blood cells and, 50
Infectious mononucleosis, 402
Infertility, IUDs and, 789
Inflammation, 416–421
  capillary permeability and, 416
  complement system and, 423–424
  cortisol and, 421–422, 710
  edema and, 416
  endogenous pyrogen (EP), effect of, 419
  glucocorticoid drugs and, 421–422
  gross manifestations of, 417
  immune complex disease and, 428
  kinins, role of, 419
  leukocyte endogenous mediator (LEM), 419
  leukocytes and, 417–418
  localized vasodilation, 416
  opsins and, 418
  phagocyte-secreted chemicals, 418–420
  plasma proteins and, 417
  salicylates and, 421–422
  tissue repair and, 420–421
Inguinal canal, 756
Inguinal hernia, 756
Inhibin, 762
Inhibiting hormones, 687
Inhibitory postsynaptic potential (IPSP), 120
  cancellation of, 122
  grand postsynaptic potential (GPSP) and, 121
  motor neuron activity and, 246
  spatial summation, 122
  temporal summation, 122
Inhibitory synapses, 119–120, 120
Initial lymphatics, 367
  diagram of, 368
Injuries. *See* Inflammation; Spinal cord
Innate immunity, 415, 416–424
  amplification of responses, 426–428
  antibodies augmenting, 426–428
  bacterial invasion, responses to, 435
  complement system, 423–424
  inflammation and, 416–421
  interferon, 422
  natural killer (NK) cells, 422–423
  viral invasion, responses to, 422–423, 439
Inner cell mass, 785
Inner ear, 213
  functions of, 227
  injury, repair of, 222
Inner hair cells, 217–219
  afferent neurons and, 220
  loud noise, loss from, 220, 221
  role of, 219
Innervated neurons, 117
Inorganic acids, 574
Inorganic chemicals, A10
Inorganic phosphate (P_i), 264
Inorganic salts, A11
Inositol trisphosphate (IP_3), 68

Insecticides
  estrogenic pollutants in, 769
  hormones, hyposecretion of, 679
Insensible water loss, 568–569
Insertion of muscle, 274
  lever system and, 276
Inspiration, 466–467
  humidification of air, 484
  lung volume and, 468
  respiratory muscle activity during, 468–469
Inspiratory capacity (IC), 478
Inspiratory muscles, 466
  accessory inspiratory muscles, 467
  actions of, 470
  relaxation of, 467–471
Inspiratory neurons, 496–498
  in dorsal respiratory group (DRG), 497
  in ventral respiratory group (VRG), 497–498
Inspiratory reserve volume (IRV), 478
Insufficient valves, 324
Insulators, 102
Insulin, 121. *See also* Diabetes mellitus, Type I; Diabetes mellitus, Type II
  aerobic exercise and, 81
  amino acids and, 723, 725
  carbohydrates and, 723–724
  carrier-mediated transport of glucose and, 79
  down regulation of, 680
  endoplasmic reticulum, formation in, 28
  epinephrine and secretion of, 716
  fatty acids and, 723, 725
  fuel metabolism regulation, 723–725
  glucagon and, 730–731
  glucose transporter (GLUT), 724–725
  as growth promoter, 695
  hypoglycemia, 730
  islets of Langerhans secreting, 616
  plasma glucose concentration and, 532
  receptor-mediated endocytosis and, 32
  satiety and, 652
  stimulus for increase in, 725
  stress and, 717, 733
  summary of actions, 725
  target cells, 668–669
  therapeutic administration of, 675
Insulin antagonists, 733
Insulin-like growth factor (IGF), 692–693
  IGF-I synthesis, 693
  IGF-II, 693
Insulin shock, 730
Integrating center in reflex arc, 177
Integrators, 17
Integrins, 61–62, 417, 782
Integumentary system, 11. *See also* Skin
  homeostasis and, 13–16
Intelligence, 161
Intensity of light, 198
Intensity of sound, 215
Intention tremor, 169
Interatrial pathway, 314
Intercalated discs, 308–309

Intercellular filaments in desmosomes, 63
Intercostal muscles, 466
  paralysis of, 469
Intercostal nerves, 466
Interferon, 422, 439
  functions of, 423
  in immune surveillance, 444–445
Interleukin 1 (IL-1)
  keratinocytes secreting, 451
  macrophages secreting, 419, 435
  stress response and, 445
Interleukin 2 (IL-2), 439–440
  cyclosporin blocking, 443
Interleukin 4 (IL-4), 440
  allergens and, 447
Interleukin 12 (IL-12), 440
Intermediary metabolism. *See* Fuel metabolism
Intermediate filaments, 44, 50–51
  in smooth muscle, 289
  summary of, 46
Intermembrane space, 40
Internal anal sphincter, 638
Internal core temperature, 655–656
Internal environment, 10. *See also* Homeostasis
Internal intercostal muscles, 470
Internal respiration, 459, 460
Internal urethral sphincter, 550–551
Internal work, 647–648
Interneurons, 136–137
  features of, 248
  glial cells supporting, 137–140
Internodal pathway, 314
Interphase, A27
Interstitial cells of Cajal, 597
Interstitial cell-stimulating hormone (ICSH), 762. *See also* Luteinizing hormone; Testosterone
Interstitial fluid, 10–11, 62, 561. *See also* Fluid balance
  blood pressure and volume of, 563
  blood vessel walls separating, 562
  bulk flow and, 367
  capillary exchange and, 363–364
  composition of, 561–562
  countercurrent exchange, 547
  edema, 368–370
  as internal environment, 10
  lymphatic system and, 368
  tubular cells and, 525–526
Interstitial fluid-colloid osmotic pressure, 366
  edema and, 369
Interstitial fluid hydrostatic pressure (P_{IF}), 366
Intestinal cells, 83
Intestinal housekeeper, 624
Intestinal phase of gastric secretion, 613
Intra-alveolar pressure, 464
  air flow and, 466–471
  diagram of, 470
  expiration and, 467–471

inspiration and, 466–467
pneumothorax, 465–466
Intracellular fluid (ICF), 10. *See also* Fluid
balance; Osmolarity
composition of, 562
distribution of body water, 561
electrical balance of ions, 562
membrane potential and, 87
plasma membrane and, 25
protein buffer system, 575, 576–577
tonicity and, 79
Intrafusal fibers, 286
Intraocular fluid, 561
Intrapleural fluid, 463, 561
cohesiveness of, 464–465
Intrapleural pressure, 464, 471
diagram of, 470
pneumothorax, 465–466
as subatmospheric, 465
Intrapulmonary pressure, 464
Intrarenal baroceptors, 528
Intrathoracic pressure, 464
Intrauterine device (IUD), 789
Intrinsic controls, 16
of stroke volume, 328
Intrinsic factor, pernicious anemia and, 397
Intrinsic nerve plexuses, 598
Intrinsic pathway of clotting cascade, 405,
406
Inulin, plasma clearance rate, 538
Involuntary muscles, 257
Involution, uterine, 795–796
Iodine, 57
goiters and, 707
thyroid capturing, 702
Ion channels, 102
Ion concentration gradient, 83
Ionic bonds, A5
Ions, A5
blood-brain barrier and, 142
current carried by, 102
electrical gradient, movement along, 75
gap junctions and, 64
Iris, 195–196
function of, 211
light, control of, 197
multiunit smooth muscle in, 292
Iron
lactoferrin, 419
leukocyte endogenous mediator (LEM),
419
receptor-mediated endocytosis and, 32
small intestine, absorption by, 633–634
Iron deficiency anemia, 396
Irreversible shock, 385
Islets of Langerhans, 616
cells, 450, 451
fuel metabolism and, 723
hormones from, 670
Isometric contraction, 274–275
Isotonic contraction, 274–275
concentric/eccentric isotonic
contractions, 275

Isotonic fluid gain, 568
Isotonic solution, 79
Isovolumetric ventricular contraction, 321,
323
Isovolumetric ventricular relaxation, 323

Jaundice, 622
Jawbone, 600
Jejunum, 594, 623
Jenner, Edward, 430
Jet lag, 681
melatonin for, 683
Joints
lever system and, 276
proprioceptive receptors, 286
tissue engineering and, 9
Junctional proteins, fusion of, 63
Juvenile-onset diabetes. *See* Diabetes
mellitus, Type I
Juxtaglomerular apparatus, 514, 522
renin, secretion of, 528
tubuloglomerular feedback and, 521
Juxtamedullary nephrons, 515–516

K^+. *See* Potassium; Potassium channels
Kallikrein, 419
Keratin, 51, 450–451
in epidermis, 450
Keratinized layer, 450
Keratinocytes, 450–451
Keto aids, 574
Ketogenesis, 730–731
Ketone bodies, A13
in fuel metabolism, 722–723
glucagon and, 730–731
Ketosis, 726–727
Kidney cells, 6, 83
Kidneys, 510. *See also* Glomerular filtration;
Nephrons; Plasma clearance; Renal
failure; Tubular reabsorption;
Tubular secretion; Urine
acid-base balance and, 578–582
ammonia, secretion of, 581–582
anemia, renal, 397
bicarbonate, regulation of, 575–576
bilirubin, excretion of, 622
blood pressure and, 563
calcium reabsorption and, 532
cardiac output (CO) to, 524–525
carrier-limited reabsorption, 531
chloride reabsorption by, 533
circulatory shock and, 385
congestive heart failure, 332
erythropoietin, 395–396
fate of substances filtered by, 525
functions of, 511–512
glomerular filtration, 516–525
glomerular filtration rate (GFR), 519
autoregulation of, 520–521
glucose reabsorption by, 531–532
hormones from, 671
hydrogen regulation and, 578–582
hypertension, renal, 380, 381, 549

overview of functions, 512
parathyroid hormone (PTH) and, 739
phosphate reabsorption by, 532
potassium concentration regulation,
536
primary hypertension and, 381
renal failure, 548–549
saturation of carriers, 79–80
sodium, excretion of, 564
sodium reabsorption by, 526–531
transfusion reaction, 432, 433
transplants, 551
tubular reabsorption, 516, 525–534
tubular secretion, 516–517, 534–537
urea reabsorption by, 533–534
uremic acidosis, 585
vertical osmotic gradient and, 540–541
vitamin D and, 740
water reabsorption by, 511–512, 533,
540–541, 569
Kidney stones, 549
Kidney tubules, 50
Killer (K) cells. *See also* Natural killer (NK)
cells
antibodies stimulating, 427, 428
Killer T cells. *See* Cytotoxic T cells
Kilocalories (Calorie), 648
fidget factor, 653
Kinesin, 46
Kininogens, 419
Kinins, 419, 423
Kinocilium, 224
Kiss sites, 63
Knee-jerk reflex, 287, 288
Korotkoff sounds, 350–351
Krebs, Sir Hans, 36
Krebs cycle, 36
Kupffer cells, 618, 619

Labeled lines, 190
Labia majora, 750, 756
Labia minora, 750, 756
Labor. *See* Parturition
Lacks, Henrietta, 24
Lactase
in brush-border plasma membrane,
625
lactose intolerance, 625
Lactate, 333
Lactation, 750
advantages of breast-feeding, 798–799
estrogen and, 770
hypothalamus and, 156
letdown of milk, 797
milk, composition of, 798
oxytocin and, 684–685
preparation for, 796
prevention of, 796
progesterone and, 791
prolactin and, 796–798
simulation of, 796–798
suckling and, 796–798
weaning and, 799

Lactic acid
   angina pectoris and, 335
   arteriolar smooth muscle and, 355
   circulatory shock and, 385
   fuel metabolism and, 722
   hydrogen ion and, 574
   muscle soreness and, 279
Lactobacillus bifidus, 798
Lactoferrin, 419, 798
Lactose, 592
   as disaccharide, A12
   intolerance, 625
Lamellae, 737
Lamina propria, 595
Langerhans cells, 450, 451
Langston, J. W., 126
Language. *See also* Broca's area; Speech;
         Wernicke's area
   aphasias, 151–152
   cortex controlling, 151–152
   cortical pathways for, 152
   dyslexia, 152
   frontal lobes and, 148
   left cerebral hemisphere and, 153
Large intestine, 594, 637–639
   anatomy and functions of, 595–597
   constipation, 638
   function of, 637
   secretions of, 638
   sodium absorption in, 639
   volumes absorbed by, 634
   water absorption in, 639
Laron dwarfism, 694–695
Larynx, 461–462
   swallowing and, 602
Latch phenomenon, 296
Latent pacemakers, 312
Latent period in muscle contraction, 268
Late-onset Alzheimer's disease, 163
Lateral corticospinal tract, 174
Lateral geniculate nucleus, 210
Lateral horn, 176
Lateral hypothalamic area (LHA), 652
Lateral inhibition, 191–192
   visual field and, 209
Lateral sacs, 263–264
   T tubules and, 265
   tubular cells, 525–526
   twitch summation and, 271
Lateral spinothalamic tract, 174, 175–176
Laughing, respiration during, 504
Law of conservation of mass, A7–A8
Law of LaPlace, 474–475
Law of mass action, 488–489, 574
Lazy eye (amblyopia), 133
LDL (low-density lipoprotein), 336–337
   atherosclerosis and, 334
Lead, renal failure and, 548
Leader sequence, A25
Leak channels, 66–67
Leaky heart valves, 324
Learning, 159
Lecithin, 621

Left atrial volume receptors, 569–570
Left atrioventricular valve. *See*
         Atrioventricular (AV) valves
Left cerebral hemisphere, 153
Length-tension relationship, 272–273
   of cardiac muscle, 329
Lens of eye, 196
   accommodation of lens, 200–201
   cataracts, 200
   epinephrine affecting, 716
   function of, 211
   refractive ability of, 198–200
Leptin
   adipocytes secreting, 650–651
   obesity and, 653
Leukemia, 402
Leukocyte endogenous mediator (LEM),
         419
Leukocytes, 398–402. *See also* WBC count
   abnormalities in production of, 401–942
   as effector cells, 414–415
   endogenous pyrogen, release of, 661
   functions of, 392, 401
   and inflammation, 417–418
   leukemia and, 402
   motility of, 49, 50
   production of, 400–402
   red bone marrow and, 395
   types of, 399–402, 414
Leukotrienes, 447, 765
Lever systems, 275–276
Levers, 275–276
Levodopa (L-dopa), 126–127
Leydig cells, 757
   luteinizing hormone (LH) and, 762
   in puberty, 757–758
LH surge, 777
Ligands, 66
Light
   eye refracting, 197–200
   intensity of, 198
   melanin cells and, 451
   photopigments and, 205
   phototransduction, 205
   retinal layers and, 201–204
   sensitivity of eyes to, 207–208
Light adaptation, 207–208
Light rays, 198. *See also* Refraction
   focusing of, 200
Limbic association cortex, 152, 153
Limbic system, 157–166
   behavioral patterns and, 157–158
   and emotions, 156
   mental disorders and, 159
   pain pathways and, 193
   reward and punishment centers, 158
   vomeronasal organ (VNO) and, 231
Lipase
   colipase and, 621
   pancreatic lipase, 617
Lipid bilayer, 59, 86
   membrane proteins and, 59–60
   permeability of, 71–72

   prostaglandins in, 192
   as structural barrier of cell, 60
Lipid emulsion, 620–622
Lipids. *See* Fats
Lips, 599
Liver, 594
   absorbed nutrients and, 634
   anabolic androgenic steroids and, 282
   anatomy of, 595–597, 610
   autonomic nervous system, effect of, 242
   bile salts, 618
   bilirubin, 622
   blood flow in, 618, 619
   choleretics, 622
   disorders of, 623
   fats, processing of, 634
   foreign organic chemicals, elimination of,
         537
   functions of, 618–619
   glycogen storage, 44, 279, 721
   hormones
      inactivation of, 679
      secretion of, 671
   Kupffer cells, 618, 619
   lobules, 619–620
   in metabolic states, 722
   platelets and, 402
   schematic representation of blood flow
         in, 618
   sinusoids, 618, 619
   smooth endoplasmic reticulum in, 27–28
   thrombopoietin, 402
   tissue engineering and, 9
   vitamin D and, 740
Liver cancer, 282
Load and muscle contraction, 274, 275
Load arm, 276
Load-velocity relationship, 275
Lobotomy, 152
Lobules of liver, 619–620
Lochia, 795–796
Lockjaw (tetanus), 127
Long-term immunity, 431
Long-term memory, 159–160
   amnesia and, 161
   permanent changes in brain and, 166
   short-term memory compared, 160
Long-term potentiation (LTP), 165–166
Loop of Henle. *See* Henle's loop
Loudness, 215
   sensitivity to, 220
Lou Gehrig's disease. *See* Amyotrophic
         lateral sclerosis (ALS)
Lowry, Thomas, A11
Lumbar nerves, 172
Lumen, 6
Luminal membranes, 525–526
Lung cancer, 486
Lungs, 4. *See also* Alveoli; Chronic
         obstructive pulmonary disease
         (COPD); Pulmonary capillaries;
         Respiration
   acid-base balance and, 578

autonomic nervous system, effect of, 242
bronchi, 462
capacity of, 477–480
chronic obstructive pulmonary disease (COPD), 471–473
compliance, 473–474
cystic fibrosis (CF) and, 61
diffusion coefficient and gas exchange, 487
elastic behavior of, 473–474
emphysema, 472
insensible water loss, 568
multiunit smooth muscle in, 292
net diffusion of oxygen and carbon dioxide, 488
pleural sac, 463
pneumothorax, 465–466
pulmonary circulation, 303–304
respiratory mechanics, 463–483
spirometry and, 478–479
stress testing, 321
in thoracic cavity, 463
tissue plasminogen activator (tPA), 407
transmural pressure gradient, 465
volume, changes in, 468, 477–480
volumes and capacities, measurement of, 478–479
Luteal phase, 772
Luteinization, 774–778
Luteinizing hormone (LH)
    anterior pituitary secreting, 685
    corpus luteum, control of, 777–778
    feedback control of, 777
    female reproductive system and, 775–776
    male reproductive system and, 762
    ovulation, control of, 685, 777
    puberty and, 763
    testes and, 762
Lymph, 367, 561
Lymphatic system
    diagram of, 369
    functions of, 368
    interstitial fluid and, 368
Lymph nodes, 368, 414
    functions of, 415
Lymphocytes, 392, 399, 400, 414. See also B lymphocytes; T lymphocytes
    adaptive immunity and, 424–425
    antigen-presenting cells, 433–435
    antigens, 425
    functions of, 401
    glucocorticoids and, 421
    naive lymphocytes, 428, 429
Lymphoid tissues, 400, 414–415
    functions of, 415
Lymphoma, 798
Lymph vessels, 367–368
    chylomicrons and, 623
    edema and, 369
    filariasis, 369, 370
    muscles and, 368
    thoracic duct, 634
    in villus of small intestine, 628

Lysosomes, 25, 30–34
    in endocytosis, 32, 85
    endocytosis, forms of, 32
    rupture of, 33–34
    size of, 32
    summary of, 45
    useless parts of cell, removing, 33–34
Lysozyme, 600, 628, 638

Macromolecules, A10
Macrophage-migration inhibition factor, 440
Macrophages, 401, 414, 439
    alveolar macrophages, 452
    and bacteria, 418
    B lymphocytes and, 434
    cancer cells and, 444–445
    endogenous pyrogen (EP) released by, 419
    in immune surveillance, 444–445
    infections and, 416
    in innate immune system, 415
    LDL (low-density lipoprotein) and, 334
    leukocyte endogenous mediator (LEM) released by, 419
    macrophage-migration inhibition factor, 440
Macula lutea, 203
    function of, 211
Macular degeneration, 205, 207
Macular densa cells, 528
Mad cow disease, 398
Magnesium
    characteristics of, A4
    in cross bridge cycling, 264
    indigenous bacteria and, 639
    overview of functions, 512
Major histocompatibility complex. See MHC molecules
Malabsorption, 626
Malaria, 397
Male reproductive system, 751. See also Prostate gland; Seminal vesicles; Sperm; Testes; Testosterone
    adolescence and, 757–758
    estrogens, production of, 762–763
    functions, 750
    locations of components, 764
    physiology of, 756–766
    prostaglandins, role of, 765–766
Malignant tumors. See Cancer
Malleus, 216
    functions of, 227
Malnutrition, growth and, 690
Malocclusion of teeth, 600
Malpighi, Marcello, 347
Maltase in brush-border plasma membrane, 625
Mammary glands, 796
Margination, 417
Mass movements, 637–638
Mast cells
    in allergies, 447
    basophils and, 401
    hypersensitivity, role in, 448

Mastication. See Chewing
Matrix, 34
Maximal O_2 consumption (max VO_2), 504
Mechanically gated channels, 104–108
Mechanical nociceptors, 192
Mechanistic approach, 3
Mechanoreceptors, 187
    in digestive system, 599
Medial geniculate nucleus, 220
Medulla, 144. See also Adrenal medulla; Renal medulla
    autonomic output, control of, 245
    cardiovascular control center, 358–359
    respiratory control centers, 496
    salivary center, 601
    swallowing center, 602
    vomiting center, 608
Medullary countercurrent system, 541–544
    countercurrent exchange, 545–547
    countercurrent multiplication, 543–544
    mechanism of, 541–544
    vasa recta, exchange with, 545–547
Medullary respiratory center, 496
    neuronal clusters in, 497–498
Megakaryocytes, 402
Meiosis, 751, 753, 759–760, A27–A28
Melanin, 450–451
    MSH (melanocyte-stimulating hormone) and, 683
Melanocortins, 651–652
Melanocytes, 450–451
Melanocyte stimulating hormone, 651
Melatonin, 681–683
    age and, 711
    darkness and, 682
    non-timekeeping roles of, 683
    puberty, role in, 763
Membrane attack complex (MAC), 423
Membrane-bound enzymes, 60
Membrane carbohydrate, 60
Membrane potential, 87–92. See also Action potentials; Graded potentials
    defined, 87
    equilibrium potential, 89–90
    K^+ equilibrium potential, 89–90
    key ions, distribution of, 88
    measurement of, 87
    muscle cells, specialized use in, 92
    Na^+ equilibrium potential, 90–91
    Na^+-K^+ ATPase pump and, 88–92
    Nernst equation, 90
    nerve cells, specialized use in, 92
    plasma membrane and, 100
    potassium channels, opening of, 107
    resting membrane potential, 88, 91, 99
        active pumping, 91–92
        chloride movement at, 92
        passive leaks, effect of, 91–92
    sodium channel, closure of, 107
    sodium channel, opening of, 105
Membrane proteins, 59–60
    functions of, 60–62

Membrane transport, 71–72. *See also* Osmosis
  assisted membrane transport, 78–87
  ATP (adenosine triphosphate) for, 42
  carrier-mediated transport, 78–83
  caveolae, 85, 87
  characteristics of methods of, 86
  diffusion, 72–74
  electrical gradient, movement along, 75
  endocytosis, 84–85
  exocytosis, 85
  lipid-soluble particles, 71–72
  passive forces, 72–78
  tonicity and, 79
  transport maximum, 79
  unassisted membrane transport, 72–78
  vesicular transport, 78, 83–87
Memory
  amnesia, 160–161
  brain, multiple regions of, 161–162
  habituation, 163–164
  immediate early genes (IEGs), 166
  limbic association cortex and, 153
  long-term memory, 159–160, 166
  long-term potentiation (LTP), 165–166
  nitric oxide (NO) and, 356
  prefrontal association cortex and, 152
  sensitization, 163–164
  short-term memory, 159–160, 162–166
  stages of, 159–161
  working memory, 152
Memory cells
  B lymphocytes, 428–429
  T lymphocytes, 436–437
  vaccination and, 430
Memory trace, 159
Men. *See also* Male reproductive system
  basal metabolic rate (BMR) of, 649
  luteinizing hormone (LH) in, 685
  sex chromosomes determining, 753
  water percentage of, 561
Menarche, 780
Ménière's disease, 224
Meninges, 140
  relationship to brain/spinal cord, 141
Meningiomas, 140
Menopause, 781
  primary follicles in, 770
Menstrual cycle, 778–780
  anabolic androgenic steroids and, 282
  androgens, hypersecretion of, 713
  estrogen affecting, 779
  exercise and, 780
  follicular phase of ovaries and, 779
  irregularities, 780
  leptin and onset of, 651
  luteal phase of ovaries and, 779–780
  menopause and, 781
  menstrual phase of, 779
  ovarian cycle and, 772–774
  ovulation, 774
  progesterone affecting, 779
  proliferative phase, 779

prostaglandins and, 765
  secretory or progestational phase of, 779–780
  temperature variations and, 655
Menstrual flow, 779
Mental disorders. *See also* Depression
  from hypocalcemia, 742
  hypothyroidism and, 706
  limbic-system neurotransmitters and, 159
  schizophrenia, 159
Mental retardation, 706
Meperidine (MPPP), 126
Mesangial cells, 524
Mesentery, 597
Messenger RNA, 25, 26, A21–A23
  anticodons, A25
  interferon and, 422
  polyribosomes, A26
  ribosomes and, A25
  translation, delivery for, A23
  triplet code, A23, A25
  vaults carrying, 43
Metabolic acidosis, 385, 549, 583, 584–585
Metabolic alkalosis, 583, 585–586
Metabolic H_2O, 568
Metabolic rate, 648–649. *See also* Basal metabolic rate (BMR)
Metabolism. *See also* Basal metabolic rate (BMR); Fuel metabolism
  defined, 719
Metaphase, A27
Metarterioles, 363
Metastasis, 443
Methyl mercaptan, 230–231
Metric system, A1–A2
MHC molecules, 434, 441–443
  class I/class II MHC glycoproteins, 442–443
  foreign peptides, loading of, 441–442
  transplant rejections and, 443
Micellar formation, 621–622
Micelles, 621–622, 629
  schematic representation of, 622
Microcirculation, 346–347
Microfilaments, 44, 49–51
  in cellular contractile systems, 49–50
  as mechanical stiffeners, 50–51
  summary of, 46
Microglia, 139
  Parkinson's disease and, 126
Microtubules, 44, 46
  accessory proteins with, 48
  of mitotic spindle, 49
  summary of, 46
  two-way vesicular axonal transport on, 47
Microvilli, 50
  Fick's law of diffusion and, 74
  in small intestine, 626
  in vestibular hair cell, 224
Micturition, 551–552
  urinary incontinence, 552
  voluntary control of, 551–552

Micturition reflex, 551
Midbrain, 144
Middle Ages, 347
Middle ear, 213
  diagram of, 217
  functions of, 216, 227
Midpiece of sperm, 760
Mifepristone, 789
Migrating motility complex, 624
Milk ejection, 797
Millivolts (mVs), 87
Mineralocorticoids, 708–709
Mitochondria, 25, 34–42
  schematic representation of, 35
  summary of, 45
Mitosis, 48–49, 753, A27
Mitotic spindle, 44
  formation of, 48–49
Mitral cells, 230
Mitral valve, 307. *See also* Atrioventricular valve
Mixing movements, 592
M lines, 258
Modalities, 186
Molarity (M), A9
Molecular weight, A8–A9
  Fick's law of diffusion and, 74
Molecules, 4, A3
  biological molecules, A10
  collisions of, 72–73
  diffusion, 73
  fatty acid molecules, A12–A13
  nonpolar molecules, A7
  organic molecules, A11
  polar molecules, A7
Moles, A8–A9
Monocytes, 334, 399, 400, 414
  functions of, 401
Monoglycerides, 593–594
  small intestine, absorption by, 629, 632–633
Monoiodotyrosine (MIT), 702–704
Monomers, A10
Mononuclear agranulocytes, 400
Mononucleosis, infectious, 402
Monosaccharides, 592
Monosodium glutamate (MSG), 229
Monosynaptic reflex, 179
Morning-after pills, 789
Morning sickness, 791
Morphine
  as analgesic, 194
  placebo effect, 195
Morula, 784
Motilin, 121, 624
Motility
  of colon, 637–638
  diarrhea and, 636
  digestive system and, 591–592
  of esophagus, 603
  of large intestine, 637–638
  of small intestine, 623–624
  of stomach, 604–608

Motion sickness, 224
  vomiting and, 608
Motivation, 158
Motor control
  and basal nuclei, 154–156, 169
  cerebellum and, 150, 169, 286
  defects, 284–286
  diagram of, 285
  motor unit output and, 283–288
  primary motor cortex and, 148
  spinal cord and, 148–149
  spinocerebellum and, 167
  thalamus, role of, 156
Motor cortex, 143–150
Motor end-plate region, 246, 250
Motor homunculus, 148
Motor neurons, 245. *See also* Whole muscles
  alpha motor neurons, 287–288
  amyotrophic lateral sclerosis (ALS) and, 51
  destruction of, 285
  as final common pathway, 245–246
  gamma motor neurons, 287–288
Motor program, 150
Motor proteins, 46
Motor unit recruitment, 269–270
  asynchronous recruitment of motor units, 270
Motor units, 269–270
  asynchronous recruitment of motor units, 270
  control of motor movement and, 283–288
  motor unit recruitment, 269
  neural inputs and, 283–286
  small versus large motor units, 269–270
Mountain climbing, 496–497
Mouth, 594, 599–601
  anatomy and functions of, 596–597
  anxiety and dryness, 601
  digestion in, 601
  vasopressin and dryness of, 570, 571
Movement. *See* Motor control
MSH (melanocyte-stimulating hormone), 121, 672, 683–684
  cortisol secretion and, 710
Mucosa of digestive tract, 594–595
Mucous cells in stomach, 608, 609
Mucous membrane, 594–595
Mucus
  cervical mucus, 780
  colonic secretion of, 638
  esophageal secretion of, 603
  gastric exocrine glands secreting, 608
  of gastric mucosa, 611
  in saliva, 600
  in small intestine secretions, 625
Mucus escalator, 452
Müllerian ducts, 756
Müllerian-inhibiting factor, 756
Multineuronal motor system, 283
Multiple sclerosis (MS), 116
  microglia, role of, 139
  myelin and, 116
  tolerance to self-antigens, loss of, 441

Multisensory neurons, 212–213
Multiunit smooth muscle, 292
Murmurs, heart, 324–325
  timing of, 324–325
Muscarinic receptors, 243, 244
  atropine and, 245
Muscle cells, 99. *See also* Cardiac muscle; Skeletal muscle; Smooth muscle
  categorization of, 6, 257–258
  as excitable tissue, 88
  functions of, 256–257
  glycogen storage, 44
  membrane potential in, 92
  types of, comparison, 257–258, 290–291
Muscle fatigue, 279–280
Muscle fibers, 246, 258
  frequency of stimulation, 270–271
  motor unit recruitment, 269–270
  myoglobin in, 278–279
  optimal length (l_o) of, 272–273
  sarcomeres, 258
  twitch summation, 270–271
Muscles, 6. *See also* Contraction of muscles; Relaxation of muscles; specific types
  anabolic androgenic steroids and, 282
  atrophy, 283
  capillaries in, 363
  cells, function of, 6
  comparisons of types of, 290–291
  hypertrophy, 281
  of iris, 197
  lactic acid and soreness, 279
  in metabolic states, 722
  neuromuscular fatigue, 279–280
  oxygen debt, 280
  proprioceptive receptors, 286
  repair of, 283
  stem cell science and tissue, 8
Muscle spindles
  alpha motor neurons, 287–288
  diagram of, 286
  function, diagram of, 287
  gamma motor neurons, 287–288
  proprioception and, 186
  in stretch reflex, 286–287
  structure, 286
Muscle tension
  bone, transmission to, 273–274
  force and, 275
  isometric contraction and, 274
  isotonic contraction and, 274
  optimal length (l_o) and, 272–273
Muscle tone
  cerebellar disease and, 169
  heat production and, 659
  of smooth muscle, 295
  spinocerebellum and, 167
Muscular dystrophy
  myoblasts and, 283, 284
  treatment of, 284
Muscularis externa, 595
Muscularis mucosa, 595

Muscular system, 10, 256–301. *See also* Skeletal muscle; Whole muscles
  homeostasis and, 13–16
Mutagens, A29
Mutations, 444, A29–A30
Myasthenia gravis, 252
  receptor defects and, 71
Myelin, 113
  in central nervous system (CNS), 139
  multiple sclerosis (MS) and, 116
Myelinated fibers, 113–116. *See also* Axons
  diameter of, 116
  speed and size of, 116
Myenteric plexus, 595, 598
Myoblasts, 283
  muscular dystrophy and, 283, 284
Myocardial ischemia, 320
  angina pectoris and, 335
  coronary artery disease causing, 333
Myocardial toxic factor, 385
Myocardium, 308
Myoepithelial cells, 797
Myofibrils, 258
  sarcoplasmic reticulum and, 265
  T tubules, relationship of, 265
Myogenic activity, 294
  arteriolar smooth muscle, 352, 356
  glomerular filtration rate (GFR) and, 521
Myoglobin, 278–279, 281
  in cardiac muscle, 332
Myopia, 200–201
  diagram of, 202
Myosin, 49
  cross bridges and, 262
  in excitation-contraction coupling, 268
  fast fibers and ATPase, 280–281
  muscle fibers and, 281–282
  in platelets, 402
  in smooth muscle, 289, 290–292
  structure of molecules, 261
  thick filaments composed of, 258, 259–260
Myosin ATPase, 264
Myosin kinase, 290
Myxedema, 706

NaCl. *See* Sodium
NADH, 38
  in electron transport chain, 39–40
NAD (nicotinamide adenine dinucleotide), 38
  in electron transport chain, 40
Nagy, John D., A31–A34
Nails as keratin structure, 51
Naive lymphocytes, 428, 429
Na$^+$ channels. *See* Sodium channels
Na$^+$-K$^+$ ATPase pump, 81–83
  action potentials and, 108
  astrocytes and, 138
  fluid balance and, 562
  hypertension and, 381
  membrane potential and, 88–92
  resting membrane potential and, 91–92

Na$^+$-K$^+$ ATPase pump *(continued)*
  in secondary active transport, 83, 84
  secretion of potassium and, 535–536
  tubular sodium reabsorption and,
    526–527
Na$^+$ load, 527
  in extracellular fluid (ECF), 563–564
Narcolepsy, 172
Nasal passages. *See* Nose
National Institutes of Health (NIH), 195
Natriuretic, defined, 530
Natural killer (NK) cells, 422–423, 439
  cancer cells and, 444–445
  in immune surveillance, 444–445
  perforin from, 438
Natural selection, 3–4
Nearsightedness, 200–201
  diagram of, 202
Near vision, 200–201
Neck muscles, 470
Necrosis
  apoptosis compared, 70–71
  of heart muscle cells, 320
Negative energy balance, 560, 649
Negative feedback, 16–18
  components of, 17
  hormonal control and, 678
Nelson-Rees, Walter, 24
Neostigmine, 252
Nephritis, 550
Nephrons, 512–516
  atrial natriuretic peptide (ANP) and, 530
  cortical nephrons, 515–516
  glomerular filtration, 516, 517–524
  glucose concentration and, 532
  juxtamedullary nephrons, 515–516
  parts of, 514
  peritubular capillaries, 514
  tubular component of, 512, 514
  tubular reabsorption, 516, 525–534
  tubular secretion, 516–517, 534–537
  vascular component of, 512–514
Nernst equation, 90, A31, A32–A33
Nerve cells. *See* Neurons
Nerve fibers. *See also* Axons
  oligodendrocytes and regeneration, 117
  regeneration of, 116–117
  Schwann cells and regeneration, 116–117
  stem cell science and, 8
Nerve growth factor, 139
Nerve growth inhibitors, 117
Nerves. *See also* Spinal nerves
  cranial nerves, 168, 169
  defined, 176
  in peripheral nervous system (PNS), 176
  spinal nerves, 172, 173
  structure of, 176
Nervous system, 6, 11, 98. *See also* Central
    nervous system (CNS); Neurons;
    Parasympathetic nervous system;
    Peripheral nervous system (PNS);
    Sympathetic nervous system
  acid-base balance and, 573

contraction strength and, 271–272
  endocrine system compared, 133–135
  enteric nervous system, 598
  homeostasis and, 13–16
  hormones, control of, 135
  immune system, linkage with, 445
  organization of, 135–137
  primitive nervous systems, 144
  specificity of neural communication,
    134–135
  target cells of, 134
  thyroid hormone, effect of, 704
  viruses, defense against, 439
Net diffusion, 73
Net filtration, 366–367
  pressure, 520
Neural presbycusis, 221
Neural stem cells, 117
Neuritic (senile) plaques, 162
Neurodegenerative diseases, 139
Neuroendocrine reflexes, 678
Neuroendocrinology, 135
Neurofibrillary tangles, 162
Neurofilaments, 51, 244
  amyotrophic lateral sclerosis (ALS) and,
    246
Neurogenic, smooth muscle as, 292
Neurogenic hypertension, 380
Neurogenic shock, 383
Neuroglia. *See* Glial cells
Neurohormones, 65, 66, 667
Neurohypophysis, 683
Neurologic disorders, 51
Neuromodulators, 124
Neuromuscular excitability, 734
Neuromuscular fatigue, 279–280
Neuromuscular junction, 246–252
  acetylcholine and, 246–252
  acetylcholinesterase (AChE) and,
    250–251
  action potential in, 250
  end plate potential, 247–249
  synapses compared, 251
  vulnerability of, 251–252
Neuronal clusters, 169
Neurons, 6, 98, 99. *See also* Synapses
  amyotrophic lateral sclerosis (ALS) and, 51
  classes of, 136–137
  communication with cells, 66
  comparison of types of, 248
  contiguous conduction, 110–111
  convergence, 128
  digestive system neurons, 598
  divergence of, 128
  ependymal cells and, 139–140
  in gray matter, 145
  intermediate filaments in, 51
  membrane potential in, 92
  multisensory neurons, 212–213
  refractory period for, 111–113
  rhythmic synchrony, firing in, 154
  saltatory conduction, 113, 115
  structure of, 108–109

termination structures, 117
  visual cortical neurons, 212
Neuropathic pain, 194
Neuropathies as diabetes mellitus
    complication, 729–730
Neuropeptides
  classical neurotransmitters compared, 124
  list of, 121
  as neuromodulators, 123–124
Neuropeptide Y (NPY), 244
  food intake and, 651–652
Neurosecretory neurons, 66
Neurotensin, 121
Neurotransmitters, 65, 66, 134
  classical neurotransmitters, 121, 123–124
  effector organs and, 237
  emotions and, 158–159
  GABA (gamma-aminobutyric acid) and,
    127
  hypothalamic releasing/inhibiting
    hormones and, 688
  long-term potentiation (LTP) and, 166
  mental disorders and limbic-system
    neurotransmitters, 159
  nitric oxide (NO) as, 356
  pain neurotransmitters, 192–194
  second messengers and, 121
  strychnine and, 127
  synapses, signals across, 118
  and synaptic cleft, 120–121
Neutral energy balance, 649–650
Neutralization, 426, 427
Neutralization reactions, A11
Neutrons, A3
Neutrophilia, 401
Neutrophils, 392, 399, 414
  and bacteria, 418
  functions of, 401
  and inflammation, 417
  in innate immune system, 415
  kallikrein, secretion of, 419
  staining of, 400
Newborn respiratory distress syndrome, 476
Niacin and NAD (nicotinamide adenine
    dinucleotide), 38
Nicotinic receptors, 243–244
Night blindness, 208
Night-day cycle, 212
Nipples and lactation, 796–797
Nitric oxide (NO), 166
  digestive smooth muscle and, 598
  endothelial-derived relaxing factor
    (EDRF) as, 355
  erection and, 767
  functions of, 356
  hemoglobin and, 394
  hypertension and, 381
  LDL (low-density lipoprotein) oxidation
    and, 334–335
  as phagocyte-secreted chemical, 419
  platelet aggregation and, 404
Nitrogen. *See also* Respiration
  atoms, 4

blood urea nitrogen (BUN), 534
  characteristics of, A4
  deep-sea diving and, 496–497
  narcosis, 496–497
Nitroglycerin
  angina pectoris and, 335
  mouth, absorption of, 601
NMDA receptors, 165
  glutamate and, 194
  strokes, role in, 143
Nociceptors, 187, 192
Nodes of Ranvier, 113–114
Nogo, 117
Nonelectrolytes, A9
Nonexercise activity thermogenesis
    (NEAT), 653
Nonhistones, A20
Non-insulin dependent diabetes, 728–729
Nonpolar molecules, A7
Nonrespiratory acidosis, 583, 584–585
Nonshivering (chemical) thermogenesis,
    659
Noradrenaline. See Norepinephrine
Norepinephrine, 121
  adrenal medulla secreting, 243, 669
  adrenergic receptors for, 244–245, 715
  arteriolar smooth muscle and, 357–358
  chromaffin granules, storage in, 714
  effector organs and, 237
  emotions and, 158
  endocrine hypertension and, 380
  in heart failure, 331
  release and binding to receptors of, 243
  sites of release, 240
  sympathetic nervous system, effect on, 327
  sympathetic postganglionic fibers
      releasing, 238–240, 669
  vascular tone and, 352
Normal eye, 202
  diagram of, 202
Normalty (N), A9–A10
Nose, 460–461
  olfactory receptors in, 229
  vomeronasal organ (VNO), 231
NSAIDs (nonsteroidal anti-inflammatory
    drugs), 537
Nuclear envelope, 25
Nuclear pores, A21
Nucleic acids, 4, A17
  as biomolecules, A10
Nucleotides, A17
  as biomolecules, A10
  in DNA (deoxyribonucleic acid), A19
Nucleus, 22, 24–25
  DNA (deoxyribonucleic acid) in, 25
  summary of, 45
Nucleus tractus solitarius (NTS), 652
Nutrients
  cells and, 5
  homeostasis and concentration of, 12
  placenta and, 786–787
  in plasma, 392
Nutritional anemia, 396

Obesity, 652–653
  android obesity, 653, 654
  defined, 653
  factors involved in, 653
  gynoid obesity, 654
  risk factor for atherosclerosis, 336
  risk factor for Type II diabetes mellitus,
    728
Obstructive jaundice, 622
Obstructive lung disease, 479–480
Occipital lobes, 146–147
Occlusion of teeth, 600
Odorants, 229
Odors. See Smell
Off-center ganglion cells, 209–210
Off response, 189
O_2-Hb dissociation curve, 489–490
  plateau portion of, 489
  steep portion of, 489–490
Olfactory bulb, 229–230
Olfactory ensheathing glia transplants, 117
Olfactory mucosa, 229
Olfactory nerve, 229
Olfactory neurons, 117
Olfactory receptors, 229
Olfactory system, 231
Oligodendrocytes, 113–114
  in central nervous system (CNS), 139
  regeneration of nerve fibers and, 117
Oligomenorrhea, 780
On-center ganglion cells, 209–210
Oncogenes, A30
Oncotic pressure, 366
Oocytes
  antrum, formation of, 773
  secondary oocytes, 771
Oogenesis, 750
  diagram of, 771
  spermatogenesis compared, 772
Open channels, 60
Ophthalmoscope, view through, 203
Opiate receptors, 194
Opsin, 205
Opsonins, 418
  complement system and, 423
  inflammation, role in, 418
Optical illusions, 186
Optic chiasm, 210
Optic disc, 211
Optic nerve, 203
  function of, 211
Optic radiations, 210
Optic tracts, 210
Optimal length (l_o) of muscle, 272–273
Oral cavity. See Mouth
Oral contraceptives. See Birth control pills
Oral rehydration therapy, 636
Oral metering of water, 570
Orexins, 652
Organelles, 25. See also Endoplasmic
    reticulum; Golgi complex;
    Lysosomes; Mitochondria;
    Peroxisomes; Vaults

summary of, 45
  types of, 25
Organic acids and hydrogen ion, 574
Organic anions, 534–537, 536–537
Organic cations, 534–537, 536–537
Organic chemicals, A10
Organic molecules, 720–721
Organisms, 4–5, 8–10
Organ of Corti, 217–219
  deafness and, 221
  functions of, 227
  separate sounds, distinguishing, 221
Organophosphates, 251, 252
Organs, 7
Orgasm
  in females, 768, 770
  in males, 768
  phases of sexual response cycle, 766
Origin of muscle, 274
Oropharyngeal stage of swallowing, 602
Orthostatic hypotension, 382
Osmolarity, 510, A10. See also Thirst;
    Extracellular fluid, control of
    osmolarity of
  arteriolar smooth muscles and, 355
  extracellular fluid (ECF) osmolarity,
    565–566
  fluid balance and, 562–563
  free water and, 566
  hypertonic state, 566
  hypotonic state, 566–567, 567–568
  importance of regulating, 566
  intracellular fluid (ICF) osmolarity,
    565–566
  ions responsible for, 565–566
  vasopressin and, 571
  vertical osmotic gradient in renal
    medulla, 540–541
  water absorption/excretion and, 540–541
Osmoreceptors, 187
  in digestive system, 599
Osmosis, 75–76, 86. See also Osmolarity
  Na^+-K^+ ATPase pump, effects of, 83
  nonpenetrating solute, unequal solutions
    of, 76
  penetrating solute, unequal solutions of, 76
  pure water from nonpenetrating solute,
    77–78
  water reabsorption, 533
Osmotic diuresis, 547–548
Osmotic pressure, 77–78
Ossicles, 216
Ossification, 691
Osteoblasts, 691, 735
Osteoclasts, 691, 735
Osteocytes, 735
Osteocytic-osteoblastic bone membrane,
    737–738
  schematic representation of, 739
Osteomalacia, 742
Osteon units, 737–738
Osteoporosis, 736–737
  treatment of, 737

Osteoprotegerin ligand (OPGL), 735
Osteoprotegerin (OPG), 735
Otolith organs, 223
　role of, 224
O type blood
　antibodies in, 432
　as universal donors, 432–433
Outer hair cells, 217–219
　loud noise, loss from, 220, 221
　role of, 219
Ova, 4, 750
　allurin, release of, 782
　estrogen and, 770
　fertilization of, 781–783
　haploid number of chromosomes in, 753
　mature ovum, formation of, 771–772
　oogenesis, 770–772
Oval window, 216, 219
　functions of, 227
Ovarian cycle, 772–774
Ovaries, 750, 770. *See also* Estrogen;
　　　Follicular phase; Luteal phase;
　　　Oogenesis; Ovulation; Progesterone
　endocrine units of, 774–776
　hormonal levels and, 775
　hormones from, 671
　during pregnancy, 774, 790
　at puberty, 772
Overeating, 652–653
　fat cells and, 653
Overhydration, 567–568
Overshoot, 103
Overweight. *See* Obesity
Oviducts, 750, 782
　tubal pregnancy, 784
Ovulation, 773, 774
　anabolic androgenic steroids and, 282
　lactation preventing, 798–799
　luteinizing hormone (LH) in, 685, 777
　prevention of, 788–789
Oxidative enzymes, 34
Oxidative phosphorylation, 40, 277–279
　muscle fibers and, 281
　pyruvic acid and, 279
Oxygen, 4. *See also* Hemoglobin;
　　　Homeostasis; P_{O_2}; Respiration
　angina pectoris and, 335
　arteriolar smooth muscle and, 355
　blood-brain barrier and, 142
　in blood circulation, 305
　blood doping, 396
　Bohr effect, 492
　brain, delivery to, 143
　cardiac muscle, 332–333
　characteristics of, A4
　coronary blood flow and, 332–333
　diffusion coefficient and gas exchange,
　　　487
　diffusion of, 74, 487
　in electron transport chain, 40
　energy equivalent of O_2, 649
　erythropoietin and, 395–396
　exercise, variables during, 503

glycolysis and, 279
Haldane effect, 494
hemoglobin and, 488–492
homeostasis and, 10–11, 12
left-sided heart failure and, 332
maximal O_2 consumption (max VO_2),
　504
myoglobin and, 278
net transfer of oxygen, hemoglobin and,
　490–491
O_2-Hb dissociation curve, 489–490
oxidative phosphorylation and, 277–279
perfluorocarbons (PFCs), 399
physically dissolved oxygen, 488
in plasma, 392
pulmonary arteriolar smooth muscle,
　effect on, 483
pulmonary capillaries, exchange across,
　484
transport in blood, 488–490
vascular spasm and, 333
Oxygen debt, 280
Oxygen toxicity, 495
　deep-sea diving and, 497
Oxyhemoglobin (HbO$_2$), 488
Oxyntic mucosa, 608
Oxytocin, 121, 684–685
　lactation and, 798
　milk ejection and, 797
　in parturition, 684–685, 794

Pacemaker activity, 310
　abnormal pacemaker activity, 312
　analogy of, 313
　artificial pacemakers, 312
　breathing rhythmicity and, 496
　normal pacemaker activity, 311–312
Pacemaker potential, 103, 293, 294,
　310–311
　of digestive smooth muscles, 597
　normal pacemaker activity, 311–312
Pacinian corpuscle, 189–190
Packaging
　of DNA (deoxyribonucleic acid), A20
　spermatogenesis, conversion of spermatids
　　　to spermatozoa, 760–761
Packed cell volume, 391
Pain, 92–194
　acupuncture, 195
　afferent fibers transmitting, 192
　chronic pain, 194
　inflammation and, 416
　neurotransmitters, 192–194
　phantom pain, 190
　receptors, 187, 192
　referred pain, 177
　respiration and, 504
　vasopressin and, 570, 571
Palate, 599
Pancreas
　acinar cells, 616, 618
　amylase, pancreatic, 617
　circulatory shock and, 385

colipase, secretion of, 621
cystic fibrosis (CF) and, 61
digestion, secretions and, 615–618
duct cells, 616, 617
endocrine pancreas, 700, 723–733
　autonomic nervous system, effect of,
　　242
　fuel metabolism and, 723
　hormones from, 670
　islets of Langerhans, 616
　schematic representation of, 616
enzymes, 616–618, 625
exocrine pancreas, 594
　acini, 616, 618
　amylase, pancreatic, 617
　anatomy and functions of, 595–597
　autonomic nervous system, effect of,
　　242
　cholecystokinin (CCK) and, 617–618
　duct cells, 616, 617
　insufficiency, 617
　schematic representation of, 616
　secretin regulating, 617–618
　secretions of, 616–618
　fuel metabolism and, 723
　insufficiency of, 617
　islet transplants, 729
　lipase, pancreatic, 617
　pancreatic aqueous alkaline secretion,
　　617
　proteolytic enzymes, 616–617
　tissue engineering and, 9
Pancreatic amylase, 617
Pancreatic aqueous alkaline secretion, 617
Pancreatic lipase, 617
Pancreatic polypeptide, 723
Pancreatic proteolytic enzymes, 616–617
Paneth cells, 628
Pantothenic acid, 36
Papillary muscles, 308
Para-aminohippuric acid (PAH), 540
Paracrines, 65, 66
Paradoxical sleep, 170
　narcolepsy and, 172
Paradoxical sleep center, 171
Paralysis
　flaccid paralysis, 285
　of intercostal muscles, 469
　spastic paralysis, 285
Paraplegia, 285
Parasites and malaria, 397
Parasitic worms. *See also* IgE antibodies
　immediate hypersensitivity response and,
　　448–449
Parasympathetic dominance, 242
Parasympathetic nervous system, 136
　advantages of dual autonomic
　　innervation, 242–243
　atropine and, 245
　blood pressure, effects on, 378
　digestive system and, 598
　distinguishing features of, 246
　heart, effects on, 326–327

insulin, stimulation of, 725
penile erection and, 767
salivary secretion and, 601
schematic representation of, 241
visceral organs, innervation of, 240–243
Parasympathetic postganglionic fibers, 238–240
Parasympathetic tone, 240
Parathyroid glands, 700
Parathyroid hormone (PTH), 670, 735–740
bones and, 735–737
calcium, 634
metabolism of, 733–734
renal threshold for, 532
chronic effect of, 738–739
as essential for life, 735
hypersecretion of, 741–742
hyposecretion of, 742
immediate effects of, 737–738
kidneys, action on, 739
phosphate, 739, 741
metabolism of, 733–734
renal threshold for, 532
placental PTHrp, 790, 792
primary regulator of, 740
small intestine, calcium/phosphate absorption by, 739–740
vitamin D and, 741–742, 742
Paraventricular nuclei (PVN), 652, 684
Parietal cells, 608–610
hydrochloric acid, secretion of, 608–610
intrinsic factor, secretion of, 611
multiple regulatory pathways for, 612
Parietal lobes, 146–147, 148
somesthetic sensations and, 148
Parietal-temporal-occipital association cortex, 152–153
Parkinson's disease, 126
apoptosis and, 71
basal nuclei, role of, 156
symptoms of, 126
Partial pressure gradients, 483–484
carbon dioxide diffusion, 484, 492
O_2-Hb dissociation curve, 489
oxygen diffusion, 484, 487
pulmonary capillaries, gas exchange across, 484–487
systemic capillaries, gas exchange across, 487–488
Parturition, 750
corticotropin releasing hormone (CRH), role of, 794
diagram of progression of, 793
estrogens and, 792
oxytocin and, 684–685, 794
positive-feedback cycle in, 794–796
preparation for, 792
prostaglandins and, 765
stages of labor, 794–796
Passive forces, 72
Passive immunity, 432
Pasteur, Louis, 430
Patellar tendon reflex, 287, 288

Paternity actions, 433
Pathogens, 399
immunity and, 413
virulence of, 413
Pathologic hyperthermia, 661–662
Pathophysiology, 19
P_{CO_2}
abnormalities in, 495
central chemoreceptors and, 500–501
diffusion and, 487
exercise and, 502
increase in, 491
lung disease and, 501
magnitude of ventilation and, 498
as primary regulator of ventilation, 500–501
pulmonary capillaries and, 484
respiratory neurons and, 501
systemic capillaries and, 487–488
Pelvic diaphragm, 551
Pelvic inflammatory disease, 789
Penicillin, 537
Penis, 750
erection, 766–767
nitric oxide (NO) and, 356
Pepsin, 611
Pepsinogen, 608
hydrochloric acid activating, 610–611
protein digestion by, 611
Peptic ulcers, 614, 615
hypercalcemia and, 742
prostaglandins and, 765
Peptides, 672
amplification of actions of, 677
as hormone classification, 673
inactivation of, 679
mechanisms of, 672
MHC molecules, foreign peptides in, 442
receptors for, 675
synthesis of, 672–673
transport of, 674
Percent hemoglobin (%Hb) saturation, 488–489
Perception, 185–186
Perfluorocarbons (PFCs), 399
Perforin, 438
Periaqueductal gray matter, 194
Pericardial fluid, 310, 561
Pericardial sac, 309–310
Pericarditis, 310
Perilymph, 217, 219
in vestibular apparatus, 223
Perineal region, 750
Periosteum, 691
Peripheral axons, 136
Peripheral chemoreceptors
hydrogen ion changes and, 502
P_{O_2} monitoring, 499–500
Peripheral nerve graphs, 117
Peripheral nervous system (PNS), 135–137
afferent division of, 136, 184–235
efferent division of, 236–255
nerves in, 176

Peripheral target endocrine gland, 687
Peripheral thermoreceptors, 658
Peristalsis, 550
esophagus and, 603
migrating motility complex, 624
rate of, 598
in stomach, 604–605
Peritoneal fluids, 561
Peritubular capillaries, 514
para-aminohippuric acid (PAH) secretion, 540
water and, 533
Permeability, 71
Fick's law of diffusion and, 74
filtration slits and, 524
Permissiveness, 680
Pernicious anemia, 397
and vitamin B_{12}, 612
Peroxisomes, 25, 34
summary of, 45
Perspiration. See Sweating
Pesticides
kidneys and, 512
Parkinson's disease and, 126
renal failure and, 548
PET (positron emission tomography) scans, 147
Peyer's patches, 414
pH. See also Acid-base balance
acidosis/alkalosis, 573
buffers and, 575
in chemistry, 572–573
common solutions, comparison of values for, 574
formula for, 572
homeostasis and concentration of, 12
kidneys and, 578–582
plasma proteins and, 393
respiratory system and, 578
schematic representation of bicarbonate and carbon dioxide in, 582–583
of stomach, 594
use of designation, 572–573
Phagocytes
and inflammation, 417, 418
macrophages as, 401, 414
neutrophils as, 401, 414
responsibilities of, 420–421
Phagocyte-secreted chemicals, 418–420
Phagocytosis, 32, 33, 84, 85, 418
IgG antibody and, 428
opsonin and, 418
opsonization and, 427, 428
Phantom pain, 190
Pharyngoesophageal sphincter, 603
Pharynx, 216, 594, 600, 601–603
anatomy and functions of, 595–597
Phasic receptors, 189
Phenol, reabsorption of, 534
Phenotypic sex, 755
Pheochromocytoma, 380
Pheromones, 231

Phosphate. *See also* Parathyroid hormone (PTH)
  in chemical buffer system, 575
  in extracellular fluid (ECF), 562
  high-energy phosphate bonds, A17–A18
  kidneys regulating reabsorption of, 532
  metabolism of, 741
  Na^+-K^+ ATPase pump, 81–83
  renal failure and, 549
  small intestine, absorption by, 739–740
  as urinary buffer, 577, 582
  vitamin D and, 733–734, 741
Phosphate buffer system, 577, 582
Phosphatidylinositol bisphosphate (PIP$_2$), 68, 69
Phosphodiesterase 5 (PDE5), 767
  photoreceptors and, 205
Phospholipase C, 68, 69
Phospholipids, A13–A14
  in plasma membrane, 58
  structure of, 58
Phosphorus, A4
Phosphorylated protein, 72
Phosphorylation, 67
  of carrier in active transport, 81
  oxidative phosphorylation, 277–279
  smooth muscle excitation-contraction coupling and, 289–292
Photons, 197–200
Photopigments, 204–205
Photoreceptors, 187, 196
  activity in light, 205
  bipolar cells and, 205
  brain, signals to, 205
  color vision and, 208–209
  cones, 196
  in darkness, 205
  for day vision, 207
  in dim light, 207
  focusing light rays on, 201–204
  in fovea, 203
  function of, 211
  night blindness and, 208
  parts of, 204–205
  photopigments, 204–205
  phototransduction, 205
  in retina, 207
  rods, 196
  schematic representation of, 204
Phototransduction, 205
  diagram of, 206
Phrenic nerve, 466, 469
Physical exercise. *See* Exercise
Physiology, 3–4
  defined, 2
Pia mater, 140
Pineal gland, 666
  eyes and, 682
  hormones from, 671
  melatonin secretion, 681
  puberty, role in, 763
Pinna, 215–216
  functions of, 227

Pinocytosis, 32, 84, 85
  electron micrograph of, 33
Pins and needles sensation, 573
Pitch discrimination, 219–220
Pitch of sound, 215
Pituicytes, 684
Pituitary gland, 683–689. *See also* Anterior pituitary; Posterior pituitary
  hypothalamus and, 156, 683–688
Placebo effect, 195
Placenta, 711, 786–787
  as afterbirth, 795
  delivery of, 795
  estrogen, secretion of, 790, 791
  formation of, 786
  functions of, 786–787
  hormones from, 671, 790–791
  human chorionic gonadotropin, secretion of, 790
  preventing rejection of fetus, 786
  progesterone, secretion of, 790, 791
Placental clock, 794
Placental PTHrp, 790, 792
Placental villi, 786
Placentation, 787
Plaque
  atherosclerotic, 334–335, 336
  in desmosome, 63
Plasma, 10–11, 390, 391–393. *See also* Blood clots; Plasma proteins
  blood pressure and volume of, 563
  blood vessel walls separating, 562
  calcium, regulation of, 734
  composition of, 561–562
  dehydration and, 567
  functions of, 392
  hydrogen ion concentration in, 579–582
  vomiting and, 608
  water content of, 561
Plasma cells, 424, 439
  antigens and, 425–426
  B-cell clones as, 428–429
Plasma clearance, 537–540
  calculation of, 538
  diagrams of, 539
  filtration fraction, 540
  glomerular filtration rate (GFR) and, 538–540
  of inulin, 538
  of para-aminohippuric acid (PAH), 540
Plasma-colloid osmotic pressure, 366, 519
  edema and, 369
  glomerular filtration rate (GFR) and, 520
Plasma membrane, 4, 22, 24, 25, 56, 57–97, 60. *See also* Lipid bilayer; Membrane carbohydrates; Membrane potential; Membrane proteins; Membrane transport
  of brush border, 625
  calcium second-messenger pathway, 68–69
  carrier molecules, 60
  cell adhesion molecules (CAMs), 61–62

  cell-to-cell adhesions, 62–65
  channels, 60
  cholesterol and, 59
  death receptors in, 71
  electrical signals and, 100
  fluid mosaic model of, 59–60
  secretory vesicle fusion with, 31
  as selectively permeable, 71
  short sugar chains on, 62
  signal transduction pathway in, 68
  structure of, 58–62
  summary of, 45
  of T lymphocytes, 435–436
Plasma proteins, 392–393
  of clotting system, 404–405
  of complement system, 423
  immunoglobulins as, 393
  and inflammation, 417
  lipid-soluble hormones, transport of, 674–675
  liver synthesizing, 618
  renal failure and, 549
Plasma volume
  glomerular filtration rate (GFR), extrinsic control of, 522
  kidneys and, 512
Plasmin, 406–407
Plasminogens, 406
Plasticity of brain, 151
Plateau phase
  of action potential, 315
  of sexual response cycle, 766
Platelet-activating factor (PAF), 333
Platelet factor 3 (PF3), 405
Platelet plugs, 403–406
Platelets, 390, 402–408. *See also* Blood clots
  ADP (adenosine diphosphate), release of, 403–404
  functions of, 392
  hemophilia and, 408
  hemostasis and, 402–403
  nitric oxide (NO) and, 356
  plugs, formation of, 403–406
Pleural cavity, 463
Pleural sac, 463
  intrapleural pressure, 464
Pleurisy, 463
Pluripotent stem cells, 395
Pneumonia
  Fick's law of diffusion and, 75
  thickness of barrier for gas exchange, 486
Pneumotaxic center, 496, 497, 498
Pneumothorax, 465–466
PNS. *See* Peripheral nervous system (PNS)
P$_{O_2}$
  abnormalities in, 494–495
  decreased arterial P$_{O_2}$ and ventilation, 499–500
  deep-sea diving and, 497
  diffusion and, 487
  exercise and, 502
  hemoglobin and, 491
  magnitude of ventilation and, 498

O$_2$-Hb dissociation curve, 489–490
  peripheral chemoreceptors and, 499–500
  pulmonary capillaries and, 484
  systemic capillaries and, 487–488
PO$_4^{3-}$. See Phosphate
Podocytes, 518, 519, 524
Point (gene) mutation, A29–A30
Poiseuille's law, 345–346
Poison ivy
  delayed hypersensitivity reaction, 449
  rash, 421
Polarization, membrane potential and,
    99–100
Polar molecules, 72, A7
Polio
  motor neurons and, 246
  reverse axonal transport and, 47
Polycythemia, 397–398
Polydipsia, 726
Polymers, A10
Polymodal nociceptors, 192
Polymorphonuclear granulocytes, 399–400
Polynucleotide strand, A20
Polypeptide chains, A15
Polyphagia, 726
Polyribosomes, A26
Polysaccharides, 592, A12
Polysynaptic reflex, 179
Polyuria, 726
Pons, 144
Pontine centers, 498
Pool, internal, 559–560
Pores in capillaries, 362–363
Pores of Kohn, 463
Positive energy balance, 560, 649
Positive feedback, 18
  in parturition, 794–796
Postabsorptive state, 722
  for glucagon, 731
  for insulin, 725
Posterior parietal cortex, 150
Posterior pituitary, 666, 683
  major hormones from, 669
  oxytocin in, 684, 794, 797
  vasopressin in, 544, 569, 684
Postganglionic autonomic fibers, 238,
    239–240
  smooth muscle and, 295
Postganglionic neurons, 714–715
Posthepatic jaundice, 622
Postnatal growth spurt, 690
Postsynaptic integration, 123
Postsynaptic neurons, 117, 118
Postsynaptic potentials, 103
Postural hypotension, 382
Potassium. See also Na$^+$-K$^+$ ATPase pump
  acid-base balance and, 573
  action potentials and, 105–108
  aldosterone and, 535–536, 709, 714
  arteriolar smooth muscle and, 355
  astrocytes and, 138
  cardiac muscle, action potential of, 315
  in cerebrospinal fluid (CSF), 140

  characteristics of, A4
  control of secretion, 535–536
  dialysis and, 551
  diffusion, reliance on, 74
  equilibrium potential for, 89
  excitatory synapses and, 119
  heart rate and, 316, 326–327, 536
  hydrogen ion secretion and, 536
  hypertension and, 381
  inhibitory synapses and, 119–120
  kidneys and, 512
  mechanism for secretion of, 535
  membrane potential and, 87–92
  osmolarity and, 565–566
  pacemaker potential and, 310–311
  regulating concentration of, 536
  renal failure and, 534, 549
  resting membrane potential, 88–92
  tubular secretion of, 534–537
  voltage-gated channels, 104–108
Potassium channels, 60
  action potentials and, 105–108
  secretion of potassium and, 535–536
Potocytosis, 85, 87
  characteristics of, 86
Power arm, 276
Power stroke, 262–263
PP cells, 723
Pre-Bötzinger complex, 498
Precapillary sphincters, 363
Precipitation, 426
Precocious pseudopuberty, 713
Prefrontal cortex, 152
  autonomic activity and, 245
  fear and, 158
  working memory and, 161–162
Preganglionic fibers, 238
  aging and, 244
Preganglionic neurons, 714–715
Pregnancy, 750. See also Abortion;
    Lactation; Parturition; Placenta
  autoimmune diseases and, 441
  corpus luteum of, 774
  diagnosis tests, 426
  early stages of, 784
  ectopic pregnancy, 782
  estrogens in, 791
  fertilization, 781–783
  gestation period, 792
  human chorionic gonadotropin (hCG) in
    and diagnosis of, 790
  leptin and, 651
  morning sickness, 791
  progesterone in, 791
  Rh factor in, 433
Prehepatic jaundice, 622
Premature heart beat 312
  arrhythmias and, 319
Premature infants, 476
Premenstrual symptoms, 765
Premotor cortex, 150
Preovulatory follicle, 773–774
Preprohormones, 672

Prepuce, 756
Presbyopia, 200
Presenilins, 163
Pressure. See also specific types
  autoregulation, 356–357
  homeostasis and, 13
  to Pacinian corpuscle, 189–190
Pressure gradient, 344–346
Presynaptic facilitation, 125
Presynaptic inhibition, 124–125
Presynaptic neurons, 117, 118
  long-term potentiation (LTP) and,
    165–166
Primary active transport, 83, 86
Primary adrenocortical insufficiency,
    713–714
Primary endings, 286
Primary follicle, 770
Primary hyperaldosteronism, 712
Primary hypertension, 381
Primary hyposecretion of hormones, 679
Primary motor cortex, 148
  flaccid paralysis, excitatory input and,
    285
  motor unit output and, 283
  somatotopic map of, 149
Primary olfactory cortex, 230
Primary oocytes, 770
Primary peristaltic wave, 603
Primary polycythemia, 398
Primary response, 429–430
Primary spermatocytes, 578
Problem-solving, A34
Procarboxypeptidase, 616, 617
Procedural memories, 161
Products, A7
Progesterone, 750
  actions of, 797
  antagonism of, 681
  cervical mucus and, 780
  lactation and, 796
  in luteal phase, 774
  luteinizing hormone (LH) and, 685
  and menstrual cycle, 779
  placenta secreting, 790, 791
  in pregnancy, 791
Prohormones, 672
Projected sensory input, 190
Prolactin, 796
  anterior pituitary secreting, 685
  maintaining lactation, 796–797
  suckling and, 798
Prolactin-inhibiting hormone (PIH), 687,
    688, 798
Prolactin-releasing hormone (PRH), 798
Pro-opiomelanocortin, 672, 710, 714
Prophase, A27
Proprioception, 148, 185
  backswings and, 186
Proprioceptive input, 286
Propulsive movements, 591–592
Prostacyclins, 765
  platelet aggregation and, 404

Prostaglandins
  arteriolar smooth muscle and, 355
  aspirin and, 661
  as chemical messengers, 765–766
  endogenous pyrogen (EP) and, 419, 661
  known and suspected actions of, 765
  nociceptors and, 192
  and parturition, 792
  seminal vesicles secreting, 764
Prostate cancer, 769
Prostate gland, 512, 750, 764
  location and functions of, 764
  renal failure and, 549
  secretions from, 764
Protein buffer system, 575, 576–577
Protein channels, 86
Protein hormones. See Peptides
Protein kinase A (PKA), 67–68
Protein kinase C (PKC), 69
Proteins, 4. See also Fuel metabolism;
    Translation
  as biomolecules, A10
  brush border, digestion in, 625
  cells synthesizing, 5
  chemical composition of, A14
  cortisol and, 709
  denaturation of, A17
  diabetes mellitus and metabolism of,
    727–728
  digestion of, 592–594
  digestive processes for, 626
  elongation step of synthesis, A25–A26
  and endoplasmic reticulum, 26
  energy cost of synthesis, A26
  gastric secretion and, 613
  glucagon, actions of, 731
  growth hormone, actions of, 69
  hydrochloric acid and, 611
  hydrolysis and, A15, A17
  initiation step of synthesis, A25–A26
  insulin and, 725
  lymphatic system and, 368
  membrane proteins, 59–62
  pepsinogen and digestion of, 611
  peptide bonds, A14–A15
  polyribosomes, A26
  primary structure of, A15–A16
  quaternary structure, A15–A16
  receptors, 60
  secondary structure of, A15–A16
  small intestine, absorption by, 629, 631
  structural levels, A15–A16
  structural protein, 721
  synthesis, steps in, A25–A26
  termination step of synthesis, A25–A26
  tertiary structure, A15–A16
  thyroid hormone and synthesis of, 704
  uncoupling proteins, 653
Proteolytic enzymes, 616–617
Prothrombin, 404
Proton-pump inhibitors, 615
Protons, A3
Proto-oncogenes, A30

Proximal tubules, 514
  functions of, 548
  organic anion/cation secretion, 536–537
  sodium reabsorption in, 526
  summary of transport across, 537
  water reabsorption in, 533
Prozac, 159
PR segment, 318–319
Pseudomonas aeruginosa, 61
Pseudopods, 32, 33
  motile cells forming, 49–50
Psychoactive drugs, 158–159
Psychogenic vomiting, 608
Psychosocial influences
  satiety and, 652–653
  stress and, 718–719
Puberty
  in boys, 757–758
  in females, 780–781
  gonadotropin-releasing hormone (GnRH)
    and, 763
  growth spurt, 690
  melatonin, role of, 763
  precocious pseudopuberty, 713
  secondary sexual characteristics, 750
Pubic hair, 750, 781
Pulmonary artery, 304
Pulmonary capillaries
  alveolar P_{O_2}/alveolar P_{CO_2}, 484–486
    alveoli and, 462
  oxygen and carbon dioxide exchange, 484
Pulmonary circulation, 303–304. See also
    Cardiac output (CO)
  circuit of, 304–305
Pulmonary edema
  congestive heart failure and, 332
  dyspnea and, 504, 505
  thickness of barrier for gas exchange, 486
Pulmonary fibrosis, 477
  thickness of barrier for gas exchange, 486
Pulmonary stretch receptors, 498
Pulmonary surfactant, 463
  newborn respiratory distress syndrome,
    476
  surface tension and, 474–476
Pulmonary valve, 307–308
Pulmonary vascular smooth muscle, 483
Pulmonary veins, 304–305
Pulmonary ventilation, 480
Pulse deficit, 320
Pulse pressure, 351
Pupils, 197
  epinephrine affecting, 716
  in fight-or-flight response, 240
  function of, 211
  size, 197, 212
    autonomic nervous system controlling,
    240
Purkinje fibers, 311
  action potential, transmission of, 314
  over-excitation of, 312
Pus, 418
P waves, 317–319

Pyloric gland area (PGA), 608
Pyloric sphincter, 604
Pyramidal cells, 146
Pyruvic acid, 279
  citric acid cycle and, 35–36
  oxygen debt and, 280

Qi, 195
QRS complex, 317–319
  in atrial fibrillation, 320
Quadriceps muscle, 186
  knee-jerk reflex, 287, 288
Quadriplegia, 285
Quality of sound, 215
Quantitative reasoning, A31–A34

Rabies virus, 47
Radial muscle, 197
Radiation of heat, 656
Radius of blood vessels, 345
Rage. See Anger
Raloxifene, 737
Rapid eye movements (REM) sleep, 171
RBCs. See Erythrocytes
Reactants, A7
Reactive hyperemia, 356
Reactive hypoglycemia, 730
Readiness potential, 150
Reagan, Ronald, 127
Rebound weight gain, 653
Receptive relaxation of stomach, 604
Receptor editing, 440
Receptor field, 190
  lateral inhibition and, 191–192
Receptor-mediated endocytosis, 32, 33, 84,
    85
Receptor potentials, 103, 187–188, 294
  action potentials, conversion into, 188
Receptors, 60. See also Second
    messengers
  adaptation to stimulation, 189–190
  differential sensitivities, 187
  and G proteins, 67
  for hormones, 135, 675
  Pacinian corpuscle, 189–190
  pain receptors, 192
  physiology of, 186–192
  in reflex arc, 177
  signal transduction and, 66, 68
  speed of adaptation, 189
  stimulation of, 187–188
  types of, 187
  uses for information from, 187
Reciprocal innervation, 178
Recognition markers, 29–30
Reconditioning organs, 343–344
Rectum, 594, 637
  constipation, 638
Red blood cells. See Erythrocytes
Red bone marrow, 395
Red cones, 205
Red fibers, 281
Reduced hemoglobin, 488

Referred pain, 177
Reflex arc, 177
Reflexes
  baroreceptor reflex, 377–378, 522–524
  crossed extensor reflex, 179
  defecation reflex, 638
  digestive long reflex, 599
  digestive short reflex, 599
  enterogastric reflex, 606
  erection reflex, 766–767
  gastrocolic reflex, 638
  Golgi tendon organs and, 288
  micturition reflex, 551
  milk letdown release, 797
  neuroendocrine reflexes, 678
  protective reflexes, 179–180
  salivary reflex, 601
  spinal cord and, 177–180
  spinal reflexes, 177–180
  stretch reflex, 179, 286–287, 288
  ventilation and exercise, 502–503
  withdrawal reflex, 177–179
Refraction
  diagram of, 199
  process of, 198
  visual field and, 210
Refractory period
  absolute refractory period, 112
  cardiac muscle, 316–317
  frequency of action potentials and, 113
  relative refractory period, 112–113
Regeneration of nerve fibers, 116–117
Regeneration tube, 116–117
Regulatory proteins, 260
Regurgitation, heart valves and, 324
Rehearsal, memory and, 160
Relative polycythemia, 398
Relative refractory period, 112–113
Relaxation of muscle, 266–267
  arteriolar smooth muscle, 355
  cardiac muscle, 320, 323
  fatigue and, 279–280
  metabolic pathways producing ATP for, 278
  skeletal muscle, 266–267
  smooth muscle, 291
Relaxation time, 268
Relaxin, 790
  parturition and, 792
Releasing hormones, 687
Remembering, 160
Renaissance, 347
Renal anemia, 397
Renal artery, 512
Renal cortex, 512, 513
Renal failure, 548–549
  blood supply and, 549
  blood urea nitrogen (BUN), 534
  dialysis, 551
  hypotonicity and, 567
  ramifications of, 549
  treatment of, 550
Renal hypertension, 380, 381, 549

Renal medulla, 512, 513
  countercurrent multiplication in, 543–544
  countercurrent exchange in, 547
  vertical osmotic gradient and, 540–541
Renal pelvis, 512
Renal pyramids, 512, 513
Renal threshold
  for calcium reabsorption, 532
  for glucose, 531–532
  for phosphate reabsorption, 532
Renal vein, 512
Renin, 528–529
  atrial natriuretic peptide (ANP) and, 530
Renin-angiotensin-aldosterone system, 359, 528–530
  atrial natriuretic peptide (ANP) and, 530
  diseases, role in, 528–529
  fluid balance and, 564–565
  secretion of aldosterone, 709
  stress and, 717
  vasopressin stimulation and, 570
Repolarization, 100
  after hyperpolarization, 103
Reproduction by cells, 5
Reproductive cells, 749
Reproductive system, 11, 748–802. See also
    Female reproductive system;
    Gonads; Male reproductive system
  accessory sex glands, 749–750
  anabolic androgenic steroids and, 282
  cilia in, 47
  homeostasis and, 13–16
  immune defenses of, 451–452
  introduction to, 749–756
  leptin and, 651
  secondary sexual characteristics, 750
  sexual differentiation in, 756
Resak, Richard, 682
Residual volume (RV), 478
Resistance, 102. See also Airway resistance;
    Total peripheral resistance
  and flow rate, 345
Resolution phase
  of ejaculation, 768
  of sexual response cycle, 766
Respiration, 458. See also Lungs;
    Ventilation
  abnormal blood gas levels, 494–495
  accessory inspiratory muscles, 467
  airway resistance, 471–473
  alveolar ventilation, 480–482
  anatomic dead space, 480
  atmospheric pressure and, 463–464
  bicarbonate, transportation of carbon dioxide as, 492–494
  Boyle's law, 466, 467
  central chemoreceptors and, 500
  chemical factors influencing, 499
  chronic obstructive pulmonary disease (COPD), 471–473
  clinically important states of, 494
  control of, 495–505

  diagram of respiratory muscle activity, 468–469
  diffusion coefficient and gas exchange, 487
  energy expenditure for, 476–477
  exchange of gases in, 483–488
  in exercise, 502–503
  expiration, 467–471
  external respiration, 459–460
  factors influencing gas transfer, 486
  hemoglobin and oxygen transport, 488
  Hering-Breuer reflex, 498
  insensible water loss, 568
  inspiration, 466–467
  internal respiration, 459, 460
  involuntary modification of breathing, 504
  mechanics of, 463–483
  net diffusion of oxygen and carbon dioxide, 488
  neural control of, 496
  noxious agents, inhalation of, 504
  partial pressure gradients, 483–484
  P_{CO_2}, primary role of, 500–501
  pre-Bötzinger complex and, 498
  pressures in, 463–464
  protective reflexes and, 503
  pulmonary ventilation, 480
  surface area for gas exchange, 486
  thickness of barrier and gas exchange, 486
  transport of gases, 488–495
  voluntary control over, 504
Respiratory acidosis, 495
Respiratory airways, 460–462. See also
    Airway resistance
  cystic fibrosis (CF) and, 61
Respiratory arrest, 494
Respiratory chain, 40
Respiratory mechanics, 463–483
Respiratory pump, 374, 375
Respiratory quotient (RQ), 459
Respiratory rate, 480
Respiratory rhythm, 498
Respiratory system, 10, 458–509. See also
    Lungs; Respiration
  acid-base balance and, 577–578, 582–584
  acidosis, respiratory, 583, 584
  airways, 460–462
  alkalosis, respiratory, 583, 584
  anatomy of, 461
  clinically important states of, 494
  dysfunctions, 479–480
  epinephrine and, 715
  homeostasis and, 13–16
  hydrogen ion and, 574
  immune defenses of, 452
  muscles, actions of, 470
  nonrespiratory functions of, 460
  venous return and, 374
Resting membrane potential. See
    Membrane potential
Resting tremors, 156

Restrictive lung disease, 479–480
Reticular activating system (RAS), 169, 170
Reticular formation, 169–170
   analgesic system and, 194
   pain pathways, 193
Reticulocytes, 396
Retina, 195–196. *See also* Cones; Rods
   diagram of retinal layers, 203
   function of, 211
   hypertension and, 381
   light, focusing, 201–204
   nonsight activities dependent on, 212
   ophthalmoscope, view through, 203
   partial substitute retina, 205, 207
   photoreceptors in, 207
   visual field and, 209
Retinene, 205
   night blindness and, 208
Retraction of blood clots, 405
Retrograde amnesia, 161
Reverse vesicular traffic, 47
Reversible shock, 385
Reward and punishment centers, 158
Rheumatic fever, 324
Rheumatoid arthritis, 428
   glucocorticoids and, 710
   tolerance to self-antigens, loss of, 441
Rh factor, 433
Rhodopsin, 205
   light adaptation and, 207
Rhythmic synchrony, neurons firing in, 154
Rhythm method of birth control, 788
Riboflavin, FAD (flavine adenine
     dinucleotide), 38
Ribonucleic acid. *See* RNA (ribonucleic
     acid)
Ribosomal RNA (rRNA), 25, A23, A25
Ribosomes, 26, A25. *See also* Free ribosomes;
     Rough endoplasmic reticulum
   summary of, 45
Ribs, 463
Rickets, 742
Right atrioventricular valve. *See*
     Atrioventricular (AV) valves
Right cerebral hemisphere, 153
Rigor mortis, 264, 266
RNA (ribonucleic acid), A17
   hormone action and, 677
   messenger RNA, 25, 26, 43, A21–A23,
     A25
   in nucleus, 25
   ribosomal RNA, 25, A23, A25
   structure of, A21–A22
   transcription process and, A22–A23
   transfer RNA, 25, A23, A25
   in viruses, 413
Roid rage, 282
Rough endoplasmic reticulum, 25–26
Round window, 217, 219
   functions of, 227
RU486, 789
Rubrospinal tract, 174
Ryanodine receptors, 263

Saccules, 223, 224
   functions of, 227
Sacral nerves, 172
Sadness, gastric motility and, 608
Salbutamol, 245
Salicylates, 421–422
Saliva, 600
   autonomic influence on, 601
   composition of, 600
   control of, 601
   dehydration and, 567
   functions of, 600
   secretion of, 601
   taste and, 226
   xerostomia, 600–601
Salivary amylase, 600
Salivary center, 601
Salivary glands, 594
   anatomy and functions of, 595–597
   autonomic nervous system, effect of, 242,
     243
Salt. *See* Sodium
Saltatory conduction, 113
   in myelinated fibers, 115
Salt taste, 226
Sarcomeres, 258
   in contraction process, 261
   muscle tension and, 273–274
Sarcoplasmic reticulum, 28, 263–264
   cardiac contractions and, 297, 315
   myofibrils, relationship to, 265
   relaxation of muscle and, 266–267
   twitch summation and, 271
Satiety
   centers, 650
   cholecystokinin (CCK) levels and, 652
   environmental influences, 652–653
   gastrointestinal distension and, 652
   glucose use and, 652
   insulin secretion and, 652
   psychosocial influences, 652–653
Saturated fatty acids, A13
Saturation of carriers, 79–80
Scala media, 217
   functions of, 227
Scala tympani, 217
   functions of, 227
Scala vestibuli, 217
   functions of, 227
Scalenus, 470
Scar tissue, 420–421
Schizophrenia, 159
Schmidt, Robert, 244
Schwann cells, 113–114
   regeneration of nerve fibers and, 116–117
Sclera, 195–196
   function of, 211
Scrotum, 750
   formation of, 756
   temperature of, 757
Sea snail (*Aplysia*), 162–164
Sebaceous glands, 450
   testosterone and, 758

Sebum, 450
Secondary active transport, 83, 86
   diagram of, 84
Secondary adrenocortical insufficiency,
     713–714
Secondary (antral) follicles, 771
Secondary endings, 286
Secondary hyperaldosteronism, 712
Secondary hypertension, 380
Secondary hyposecretion of hormones, 679
Secondary peristaltic waves, 603
Secondary polycythemia, 398
Secondary response, 429–430
Secondary sexual characteristics, 750
   testosterone and, 758
Second heart sound, 323
Second messengers, 66, 67–68. *See also*
     Calcium second-messenger pathway;
     Cyclic adenosine monophosphate
   amplification by, 69–70, 72
   cyclic adenosine monophosphate (cyclic
     AMP/cAMP), 67–68
   hormone actions via, 676
   modification of pathways, 70–71
   neurotransmitters functioning through, 121
Second-order sensory neurons, 190
Second polar body, 771–772
Secretin
   bile, stimulation of, 622
   as enterogastrone, 606
   functions of, 639–640
   pancreatic exocrine secretion and,
     617–618
Secretion. *See also* Exocytosis
   digestive secretion, 592
   endoplasmic reticulum and, 26, 28
   from glands, 6
   Golgi complex and, 29–30
Secretory glands, 6
Secretory IgA in breast milk, 798
Secretory or progestational phase of
     menstrual cycle, 779–780
Secretory vesicles, 29, 31
   in cytosol, 44
   exocytosis of, 29–30
   formation of, 29, 31
   kinesin and, 46
   microtubules and, 44, 46
   summary of, 45
   axonal transport of, 46
   v-SNAREs, 60
Segmentation
   functions of, 624
   in small intestine, 623–624
Selectins, 417
Selective estrogen receptor modulators
     (SERMs), 737
Selectively permeable membranes, 71
Self-antigens, 436
   autoimmune diseases and loss of
     tolerance, 441
   CD47, 443
   tolerance of immune system for, 440–441

Self-stimulation, neurotransmitters and, 158
Selye, Hans, 716
Semen, 750
  accessory sex glands and, 764–765
  composition of, 764
Semicircular canals
  functions of, 223–224, 227
Semilunar valves, 307–308
  heart sounds associated with, 324
Seminal vesicles, 750
  location and functions of, 764
Seminiferous tubules, 757, 758
  fluid, 761
Senile dementia. *See* Alzheimer's disease
Senses. *Also see* Proprioception; Somatic
     senses; Special senses
  cerebral cortex and, 146
  parietal lobes and, 148
Sensible water loss, 568–569
Sensitization, 163
  mechanisms of, 164–165
Sensitization period to allergens, 447
Sensitized T cells, 436
Sensorineural deafness, 221
Sensors, 16
Sensory homunculus, 148
Sensory information, 185
  coding of, 190
Sensory receptors, 137
Septicemic shock, 408
Septic shock, 383
Septum of heart, 304
Series-elastic component of muscle,
    273–274
Serosa, 597
Serotonin, 121
  emotions and, 158
  Prozac and, 159
  sensitization and, 174
Sertoli cells, 758, 761–762
Serum, 405
  antibodies in, 432
Set points, 16–17
Severe combined immunodeficiency, 446
Sex-determining region of Y chromosome
    (SRY), 754
  sperm and, 761
Sex flush, 768
Sex hormones, 708. *See also* Androgens;
    Estrogens; Progesterone
  adrenogenital syndrome, 712–713
  gonadotropins, 685
Sexual activity. *See also* Ejaculation;
    Erection; Sexual response cycle
  blood pressure and, 379
  intercourse, 766–770
  limbic system and, 157
  pheromones and, 231
Sexual differentiation, 753
  chart of, 754
  errors in, 756
  of external genitalia, 755–756
  genetic sex, 753–754

gonadal sex, 753–754
  phenotypic sex, 755
  of reproductive tract, 756
Sexual libido
  testosterone and, 758
  in women, 781
Sexual response cycle
  in females, 768–769
  in males, 766–768
Shivering, 3, 658–659
  as homeostatic response, 12
Shock. *See* Circulatory shock
Shortness of breath, 504, 505
Short-term memory, 159–160
  amnesia and, 161
  long-term memory compared, 160
  pre-existing synapses, modifications of,
    161–162
  synaptic activity and, 162–166
Sickle cell disease, 397
Side effects of antihypertensives, 382
SIDS (sudden infant death syndrome),
    504–505
Sighing, respiration during, 504
Sigmoid colon, 637
Signal transduction pathway, 66
  in plasma membrane, 68
Sildenafil, 767
Simple cells, 212
Simple diffusion, 80
Simple reflexes, 177
Simple salivary reflex, 601
Singing
  respiratory system and, 460
  voluntary ventilation and, 504
Single-unit smooth muscle, 293–296
  contraction, gradation of, 294–295
  functional syncytium, 294
  pacemaker potentials, 293
  as self-excitable, 293
  slow-wave potentials, 293
Sinoatrial node (SA node), 311–312
  action potentials and, 326
  cardiac sympathetic nerves supplying,
    326–327
  ECG (electrocardiogram) and, 318–319
  ectopic focus and, 312
  fibrillation and, 314
  heart rate and, 325–328, 327–328
  interatrial pathway, 314
  internodal pathway, 314
  pacemaker of heart, 311–312
  sympathetic nervous system, effect of, 327
  ventricular filling and, 323
Sinusoids of liver, 618, 619
Size of pupil. *See* Pupils
Skeletal muscle, 6, 256, 257. *See also*
    Contraction of muscle; Cross bridges;
    Smooth muscle; Whole muscles
  acetylcholine (ACh) stimulating, 263
  action potential in, 268
  alpha motor neurons, 287–288
  ATP (adenosine triphosphate) usage, 42

ATP-synthesizing ability of, 281
  characteristics of, 280
  contraction time, 268
  demands, response to, 281–283
  energy, conversion of, 648
  excitation-contraction coupling, 263
  exercise and, 250
  external anal sphincter as, 638
  fiber types, 281
  gamma motor neurons, 287–288
  genetics and fiber types, 281
  glucose uptake, 724–725
  glycolysis and, 279
  glycolytic fibers, 281
  heat production and, 659
  interconversion between fast muscle
    fibers, 282–283
  levels of organization in, 259
  lever system and, 276
  light-microscope view of, 260
  mechanics of, 269–276
  metabolism of, 277–283
  molecular basis of contraction, 261–268
  motor neurons innervating, 245
  muscle receptors and, 286–288
  myofibrils, 258
  nitric oxide (NO) and, 356
  oxidative muscle fibers, 281
  potassium concentration and, 536
  power stroke, 262–263
  red fibers, 281
  relaxation of, 266–267, 268
  rigor mortis, 264, 266
  shivering and, 659
  sphincters, 275
  structure of, 258–260
  temperature regulation and, 659
  testosterone and, 281–282
  types of, 280–281
  venous return and, 372
  white fibers, 281
Skeletal muscle pump, 372–374
Skeletal system, 10. *See also* Bones
  functions of, 735
  homeostasis and, 13–16
  water content of, 561
Skin, 412
  dermis, 450
  epidermis, 449–450
  epithelial tissue, 6
  exocrine glands, 450
  as external defense mechanism, 449–452
  heat and blood flow to, 659–660
  hormones from, 671
  hypodermis, 450
  layers of, 449–450
  temperature regulation and, 657, 659
  vitamin D synthesis, 451
  water content of, 561
Skin-associated lymphoid tissue (SALT), 451
Skin cells
  intracellular filaments, 51
  motility of, 49, 50

Skinfold thickness assessment, 654
Skin vasoconstrictor response, 660
Skin vasodilator response, 660
Sleep, 170–172. *See also* Biological clock
  as active process, 170
  adenosine and, 171–172
  basal metabolic rate (BMR) and, 649
  behavior patterns in, 171
  electroencephalogram (EEG) and, 154
  function of, 171–172
  heart rate and, 312
  melatonin as aid to, 683
  narcolepsy, 172
  SIDS (sudden infant death syndrome)
    and, 504–505
  slow-wave sleep, 170–171
Sleep apnea, 504
  SIDS (sudden infant death syndrome),
    504–505
Sleep-wake cycle, 170
  neural systems controlling, 171
  retina and, 212
Sliding filament mechanism, 261
Slow-oxidative fibers, 281, 282–283
Slow pain pathway, 192
Slow-reactive substance of anaphylaxis
  (SRS-A), 447
Slow synapses, 121
Slow-twitch fibers, 282–283
Slow-wave potentials, 103, 294
  of digestive smooth muscles, 597
  of single-unit smooth muscle, 293
Slow-wave sleep, 170–171
Small depolarization of postsynaptic
  neuron, 119
Small hyperpolarization of postsynaptic
  neuron, 119–120
Small intestine, 594, 623–636
  absorptive surface, 627
  alcohol, absorption of, 614–615
  amino acids, absorption of, 629
  anatomy and functions of, 595–597
  basic electrical rhythm (BER) in,
    623–624
  biochemical balance in, 635
  brush border, 50, 625, 626
  calcium absorption, 634
    parathyroid hormone (PTH) and,
      739–740
  carbohydrates, absorption of, 629, 630
  chloride absorption, 628–629
  crypts of Lieberkühn, 628
  enzyme secretions, 625
  fats, absorption of, 629, 632–633
  fluid absorption by, 634–635
  functions of, 594, 625–628
  gluten enteropathy, 626, 628
  intestinal phase of gastric secretion, 613
  iron absorption, 633–634
  microvilli in, 50, 625, 626
  migrating motility complex, 624
  motility in, 623–624
  Paneth cells, 628

parathyroid hormone (PTH) and calcium/
  phosphate absorption, 739–740
phosphate absorption, parathyroid
  hormone (PTH) and, 739–740
proteins, absorption of, 629, 631
segmentation in, 623–624
sodium absorption, 628–629
surface modifications of, 626, 628
villi in, 626, 628
vitamins, absorption of, 633
volumes absorbed by, 634
water absorption, 628–629
Small polypeptides, 592
Smallpox vaccination, 430
Smell, 185
  adaptation of olfactory system, 231
  components of odors, 230
  discrimination of odors, 230–231
  olfactory receptors, 229
  taste perception and, 229
Smiling, 158
Smoker's cough, 452
Smoking. *See* Cigarette smoking
Smooth endoplasmic reticulum, 25–26, 27
  in liver cells, 27–28
Smooth muscle, 256, 257, 288–297,
    291–292. *See also* Multiunit smooth
    muscle; Single-unit smooth muscle
  in arterioles, 352
  autonomic nervous system affecting, 295
  calcium-dependent phosphorylation and,
    289–292, 295
  caveolae, effect of, 295
  connective tissue in, 296
  dense bodies in, 289
  desmosomes in, 63
  of digestive system, 591–592, 597–598
  filaments in, 289
  functional syncytium, 294
  gap junctions in, 64
  gradation of contraction of, 294–295
  hypothalamus and, 156
  latch phenomenon, 296
  length-tension relationship, 296
  lymphatic smooth muscle, 368
  methods for contraction of, 291–292
  microscopic view of, 289
  multiunit smooth muscle, 292
  myogenic activity of, 294
  nitric oxide (NO) and, 356
  postganglionic autonomic fibers and, 295
  pulmonary vascular smooth muscle, 483
  relaxation of, 291
  schematic representation of innervation
    of, 295
  single-unit smooth muscle, 293–296
  speed of contraction, 296–297
  stress relaxation response, 296
  stretching of, 296
  tension, development of, 296
  tissue, 6
  tone, 295
  as unstriated, 289

of urinary bladder, 550
of veins, 370
Sodium, A10–A11. *See also* Fluid balance;
  Na$^+$-K$^+$ ATPase pump
  aldosterone and reabsorption of,
    527–530, 536, 709, 714
  atrial natriuretic peptide (ANP) and
    reabsorption of, 530
  balance, 558
  basolateral pump for reabsorption of, 527
  cardiac muscle, action potential of, 315
  in cerebrospinal fluid (CSF), 140
  channels, 60
  characteristics of, A4
  chloride with, 533, 562
  colon absorbing, 639
  dialysis and, 551
  diffusion, reliance on, 74
  drugs affecting reabsorption, 529–530
  excitatory synapses and, 119
  extracellular fluid volume and, 562–565
  homeostasis and concentration of, 12–13
  hypertension and, 382
  inorganic salts, A11
  kidneys eliminating, 511
  large intestine absorbing, 639
  loss of, 560
  medullary countercurrent system and,
    541–544
  membrane potential and, 87–92
  Na$^+$ load, 527
  in oral rehydration therapy, 636
  osmolarity and, 565
  overview of functions, 512
  pacemaker potential and, 310–311
  primary hypertension and, 381
  reabsorption of, 526–531
  renal failure and, 549
  renin-angiotensin-aldosterone system,
    528–530
  resting membrane potential and, 88–92
  in saliva, 600
  secondary active transport requiring, 531
  small intestine, absorption in, 628–629
  for stable balance, 560
  taste, 226
  urea reabsorption and, 533–534
  various tubular segments of nephron
    handling, 548
  voltage-gated sodium channels, 104–108
Sodium channels
  action potentials and, 105–108
  in Pacinian corpuscle, 190
  of photoreceptors, 205
  reabsorption of sodium and, 527
  threshold potential and permeability,
    105
Solutes, A9
  reabsorption, 547–548
  and water, 75–76
Solutions, A9
Solvents, A9
Somatic cells, 751, A19–A20

Somatic mutation, A30
  of antibody genes, 431
Somatic nervous system, 237, 245–246
  autonomic nervous system compared, 247
Somatic sensation, 185
Somatomedins, 692–693, 693
Somatosensory cortex, 148
  cortical gustatory area, 228
  pain pathways, 193
  somatotopic map of, 149
  variety of sensory signals to, 212
Somatosensory neurons, 190
Somatosensory pathways, 190
Somatostatin, 121, 687, 693
  D cells secreting, 609, 610
  fuel metabolism and, 723
  gastric juices, influence on, 612
  variety of glands secreting, 668
Somatotopic maps
  individuals, variety among, 150
  of primary motor cortex, 149
  of somatosensory cortex, 149
Somatotropin, 685, 693
Somersaults, 177
Somesthetic sensations, 148, 185
Sorting signal, 30
Sound. See also Hearing
  external ear and localization of, 215–216
  properties of, 215
  relative magnitude of, 215
  transduction, 220
Sound waves, 213–215
  deafness and, 221
  diagram of transmission of, 217–219
  loudness, sensitivity to, 220
  pinna and, 215–216
  properties of, 215
  tympanic membrane and, 216
Sour taste, 226, 228
Space flight, 250
Spasmodic torticollis, 252
Spastic paralysis, 285
Spatial summation, 122
Special senses, 185
Speech. See also Language
  impediments, 151
  respiratory system and, 460
  saliva aiding, 600
  voluntary ventilation and, 504
Sperm, 4, 750. See also Spermatogenesis
  ejaculation, content in, 768
  epididymis and, 763
  estrogenic pollutants and, 769
  fertilization, 781–782
  flagellum in, 47
  haploid number of chromosomes in, 753
  locomotion of, 49
  Sertoli cells, association with, 761–762
  storage of, 763
Spermatids, 759
Spermatogenesis, 758–761
  contraception and, 789

  diagram of, 760
  estrogens, effect of, 762–763
  follicle-stimulating hormone (FSH), role of, 762
  meiosis, 759–760
  mitotic proliferation, 758–759
  oogenesis compared, 772
  packaging of cellular elements, 760–761
  Sertoli cells and, 761–762
  temperature and, 757
  testosterone, role of, 762
Spermatogonia, 758, 761
Spermatozoa, 760, 761
Spermicidal contraceptives, 788
Spermiogenesis, 760–761
Sphincter of Oddi, 620, 623
Sphincters, 275
  external anal sphincter, 638
  gastroesophageal sphincter, 603
  ileocecal valve/sphincter, 624–625
  internal anal sphincter, 638
  pharyngoesophageal sphincter, 603
  precapillary sphincters, 363
  pyloric sphincter, 604
  sphincter of Oddi, 620, 623
  urinary bladder, control of, 550–551
Sphygmomanometer, 349–351
Spikes, 103
Spinal cord, 172–180
  afferent neurons and, 176, 283
  autonomic reflexes, control of, 245
  cauda equina, 172
  cross section of, 174
  efferent neurons and, 176
  ependymal cells, 139
  gray matter of, 173, 187
  injuries, 117
    respiratory paralysis, 469
  location of, 172
  motor control and, 148–149
  nervous tissue in, 6
  reflexes and, 177–180
  spinal nerves and, 176
  spinal tap, 172
  tracts of white matter, 173–176
  traumatic injury to, 285
  white matter of, 173–176
Spinal nerves, 172
  afferent fibers in, 176–177
  diagram of, 173
  efferent fibers in, 176–177
Spinal reflex, 177
Spinal taps, 140, 172
Spinocerebellum, 167, 169
Spinothalamic tract, 176
Spirograms, abnormal, 479
Spirometer, 478
Spleen, 414
  erythrocytes and, 394–395
  functions of, 415
  platelets and, 402
Sports. See Athletics
Spot desmosomes, 450

Stable structures, 50
Stapes, 216, 219
  functions of, 227
Starch, 592, A12
Starvation, 723
Steady state, 73
Steatorrhea, 617
Stellate cells, 146
Stem cells
  pluripotent stem cells, 395
  of small intestine, 628
  in stomach, 610
Stem cell science, 8–9
  diabetes mellitus research, 729
  ethical concerns, 9
  nerve cell regeneration and, 117
  and Parkinson's disease, 127
Stenotic valves, 324
  heart failure and, 331
Stereocilia, 217–219
  in vestibular hair cells, 224
Sterilization as contraceptive, 788
Sternocleidomastoid, 470
Sternum, 463
  heart, location of, 304
Steroid ring system, A14
Steroids, 672
  amplification of actions of, 677
  anabolic androgenic steroids, 282
  blood-brain barrier and, 142
  chemical classifications, 673
  as complex lipids, A13–A14
  genes, stimulation of, 676–677
  inactivation of, 679
  mechanisms of, 673–674
  plasma proteins transporting, 674–675
  receptors for, 675
  synthesis of, 673–674
  therapeutic administration of, 675
Stimuli, 186. See also Receptors
  adequate stimulus, 187
  receptors and, 187–188
Stimulus-secretion coupling, 734
Stomach, 7, 594, 604–615. See also Gastric juices
  alcohol, absorption of, 614–615
  anatomy of, 595–597, 604
  aspirin, absorption of, 614–615
  biochemical balance in, 635
  cephalic phase of gastric secretion, 612–613
  chief cells, 608–609, 611
  decreasing gastric secretion, 613
  distension of, 605
  emotions affecting motility, 607–608
  emptying, 605–607
  endocrine cells in, 609, 610
  enterochromaffin like (ECL) cells, 610
  enterogastric reflex, 606
  enterogastrones, 606
  G cells, 610
  gastric mixing, 605
  gastric mucosal barrier, 613–614

Stomach (*continued*)
  gastric phase of gastric secretion, 613
  gastric pits, 608–610
  gastrin secretion, 610, 612–613
  hormones from, 671
  intestinal phase of gastric secretion, 613
  mucosa, 609
  mucus of, 611
  parietal cells, 608–610
  peptic ulcers, 614, 615
  peristalsis in, 604–605
  phases of gastric secretion, 612–613
  pH of, 594
  protein digestion in, 604
  pyloric sphincter, 604
  receptive relaxation of, 604
  secretion by, 608–613
  storage in, 604–605
  surface epithelial cells, 610
  vagal activity in, 613
  volume of, 604
  vomiting and, 608
Stomach hydrogen ion (H^+) pump, 81
Storage diseases, 34
Streptococcal bacteria
  autoimmune diseases and, 441
  heart valve problems and, 324
Stress
  blood glucose and, 717
  chronic psychosocial stressors, 718–719
  cortisol and, 709, 711, 717
  CRH-ACTH-cortisol system and, 717
  epinephrine and, 716
  food intake and, 652
  general adaptation syndrome, 716
  hypothalamus and response to, 717–718
  response, 445
  sympathetic nervous system and, 716
  table of hormonal change, 717
  vasopressin stimulation and, 570
Stress relaxation response, 296
Stress testing, 321
Stretch of arteriolar smooth muscle, 356
Stretch receptors in urinary bladder, 552
Stretch reflex, 179, 286–287, 288
Striated muscle, 257
Strokes
  anabolic androgenic steroids and, 282
  android obesity, 654
  aphasias, 151–152
  apoptosis and, 71
  domino effect, 143
  flaccid paralysis, 285
  microglia, role of, 139
  pain and, 194
  tissue plasminogen activator (tPA) and,
    407
Stroke volume, 323
  afterload and, 331
  cardiac output and, 324
  determination of, 328
  end-diastolic volume and, 328–329
  exercise and, 380

extrinsic control factors, 329–330
  intrinsic control of stroke volume, 328
  length-tension relationship of cardiac
    muscle and, 329
  sympathetic stimulation and, 329–330
  venous return and, 328–329
Structural protein, 721
Strychnine, 127
ST segment, 319
Subarachnoid space, 140
Subcortical regions of brain, 154
Subcortical route, 230
Subcutaneous implantation contraceptives,
    788–789
Subcutaneous tissue, 450
Submucosa, 595
Submucous plexus, 598
Substance P, 121
  and analgesic pathway, 193
  endogenous opiates and, 194
  as pain neurotransmitter, 192–194
Substantia nigra, 126
Subsynaptic membrane, 118
Succus entericus, 625
Sucrase in brush-border plasma membrane,
    625
Sucrose, 592
  as disaccharide, A12
Sudden infant death syndrome (SIDS),
    504–505
Suffocation, 494
Sugars. *See also* specific types
  as biomolecules, A10
  intermediary metabolism and, 43
Sulfur, A4
Supplementary motor area, 150
Supporting cells, 229
Suprachiasmatic nucleus (SCN), 681
Supraoptic cluster, 684
Surface epithelial cells, 610
Surface tension, 474
  pulmonary surfactant and, 474–476
Suspensions, A10
Suspensory ligaments, 200–201
  function of, 211
Swallowing, 462, 592
  all-or-none reflex and, 601–602
  esophageal stage of, 603
  gastroesophageal sphincter and, 603
  oropharyngeal stage of, 602
  pharyngoesophageal sphincter and, 603
  respiration during, 504
Swallowing center, 602
Sweat glands, 6, 450, 657
  autonomic nervous system, effect of, 242
  sympathetic nerve fibers and, 243
Sweating
  acclimatization, 566
  autonomic nervous system controlling, 240
  epinephrine affecting, 716
  as evaporative heat-loss process, 657
  in fight-or-flight response, 240
  heat stroke and, 662

insensible water loss, 568
  norepinephrine affecting, 716
  sensible water loss and, 568–569
  sodium loss in, 564
Sweet taste, 228
  saliva and, 600
Sympathetic dominance, 240
Sympathetic ganglia, 238
  neuropeptide Y (NPY) and, 244
Sympathetic nervous system, 136
  advantages of dual autonomic
    innervation, 242–243
  blood pressure, effects on, 378
  digestive system and, 598
  distinguishing features of, 246
  epinephrine and, 678, 715–716
  glomerular filtration rate (GFR), extrinsic
    control of, 522
  heart, effects on, 326–327
  salivary secretion and, 601
  schematic representation of, 241
  stress, role in, 716
  stroke volume and, 329–330
  visceral organs, innervation of, 240–243
Sympathetic postganglionic fibers, 238–240
Sympathetic tone, 240
Sympathetic trunk, 238
Sympathomimetic effect of thyroid
    hormone, 704
Synapses, 117–118
  convergence and, 128
  diseases affecting transmission, 125–128
  divergence and, 128
  drugs affecting transmission, 125–128
  excitatory synapses, 119, 120
  fast synapses, 121
  grand postsynaptic potential (GPSP),
    121–123
  inhibitory synapses, 119–120, 120
  long-term potentiation (LTP) and,
    165–166
  neuromuscular junction compared, 251
  neurotransmitters and, 118
  presynaptic facilitation, 125
  presynaptic inhibition, 124–125
  slow synapses, 121
  structure of, 119
Synaptic cleft, 118
  neurotransmitters, removal of, 120–121
Synaptic delay, 120
Synaptic knob, 118
Synergism of hormones, 681
Synovial fluid, 561
Synthetic erythropoietin, 396
Systemic capillaries
  bicarbonate in, 493–494
  carbonic acid formation and, 574
  gas exchange across, 487–488
Systemic circulation, 303–304
  circuit of, 305
Systole period, 320–321
  end-systolic volume (ESV), 323
  ventricular systole, 321, 323

Systolic blood pressure, 349, 351
Systolic murmurs, 324–325

Tachycardia, ECG and, 319
Tacrine, 163
Tactile receptors, 189
Taeniae coli, 637
Tamplin, Arthur, 682
Target cells, 66. *See also* Hormones
    in endocrine system, 134
    for insulin, 81
    in nervous system, 134
Taste, 185
    discrimination of, 226, 228–229
    receptor cells, 225, 228
    smell and, 229
Taste buds, 225, 228, 600
    location of, 226–229
Taste pores, 225
Tastiness, 226, 228
Tau protein, 162–163
Tay-Sachs disease, 34
T cells. *See* T lymphocytes
Tears, 194–195
Tectorial membrane, 217–219
    functions of, 227
Teeth, 600
Teleological approach, 3
Telophase, A27
Temperature, 646, 654–662. *See also* Cold;
        Fevers; Heat
    acclimatization, 566
    arterioles, effect on, 356
    blood temperature, 658
    cooling, resistance to, 655
    endogenous pyrogen (EP) and, 419
    hemoglobin saturation and, 492
    homeostasis and, 13, 16–18
    hyperthermia, 661–662
    hypothalamus and, 156, 657–658
    internal core temperature, 655–656
    muscle temperature, 280
    nonshivering (chemical) thermogenesis,
        659
    normal variations in, 655
    overheating, 655
    rhythm method and, 788
    of scrotum, 757
    shivering, 658–659
    sites for monitoring, 655
    thermoneutral zone, 661
    thermoregulatory pathways, 658
    ventilation and exercise increasing, 503
Temperature-humidity index, 657
Temporal lobes, 146–148
Temporal summation, 121–122
Temporomandibular joint (TMJ), 600
Tendons, 269
    Golgi tendon organs, 288
Tension. *See* Muscle tension
Tenting effect, 768
Terminal button, 246
    acetylcholine (ACh) in, 247

Terminal cisternae, 263–264
Terminal ganglia, 238
Tertiary follicle, 773–774
Testes, 750. *See also* Spermatogenesis
    castration, 758
    cells of Leydig, 685
    cryptorchidism, 757, 769
    development of, 756
    failure to differentiate, 756
    follicle-stimulating hormone (FSH) and,
        762
    hormones from, 671
    location and functions of, 764
    luteinizing hormone (LH) and, 762
Testicular cancer, 769
Testicular feminization syndrome, 680
Testosterone, 672, 750
    effects of, 757
    fetal testes, secretion by, 757
    growth and, 690
    Leydig cells and, 757
    luteinizing hormone (LH) and, 685
    muscle fibers and, 281–282
    non reproductive actions, 758
    in puberty, 757–758
    reproductive tract development and, 756
    secondary sexual characteristics, 750
    Sertoli cells and, 762
    sexual differentiation and, 755
    sexual libido and, 758
    in spermatogenesis, 762
    synergism and, 681
    testicular feminization syndrome, 680
    vasectomy and, 763
Tetanic contraction, 271–272
Tetanus of cardiac muscle, 316–317
Tetanus toxin, 127
Tetrad, A27
Tetraiodothyronine (T_4), 701. *See also*
        Thyroid hormone
    conversion of, 703–704
    coupling process yielding, 702
Thalamic-cortical route, 230
Thalamus, 143, 144
    lateral geniculate nucleus, 210
    location in brain, 156
    medial geniculate nucleus, 220
    in motor control, 156
    pain pathways, 193
    senses and, 148
    visual message and, 210–212
Thalidomide, 786–787
Thecal cells, 772
T helper 1 ($T_H 1$), 440
T helper 2 ($T_H 2$), 440
Thermal conductivity, 656
Thermal energy, 648
Thermal nociceptors, 192
Thermoneutral zone, 661
Thermoreceptors, 187, 192, 658
Thick filaments, 258
    cross bridges projecting from, 258–259
    cross-sectional arrangement of, 260

myosin composing, 258, 259–260
    in smooth muscle, 289–292
    structure of molecules in, 261
Thin filaments, 258
    of cardiac muscle, 297
    composition of, 261
    in contraction process, 261
    cross bridges and, 258–259
    cross-sectional arrangement of, 260
    in smooth muscle, 289–292
    structural components of, 260
    twitch summation and, 271
Third-order sensory neurons, 190
Thirst
    angiotensin II and, 570
    factors controlling, 571
    glomerular filtration rate (GFR) and,
        523
    as homeostatic need for water, 158
    hypothalamic osmoreceptors, 569
    hypothalamus and, 156
    input of water controlled by, 569
    polydipsia, 726
    vasopressin secretion and, 569–570,
        569–571
Thoracic cavity
    heart, location of, 304
    lungs in, 463
    respiratory muscles and, 466
    transmural pressure gradient, 465
Thoracic duct, 634
Thoracic nerves, 172
Thoracic vertebrae, 463
Thorax, 463
Thought elaboration, 148
Three Mile Island disaster, 682
Threshold potential, 103–104
    all-or-none law and, 113
    of axon hillock, 123
    and digestive smooth muscle, 598
    sodium permeability and, 105
Thrombin
    clot formation and, 404
    hemostasis, role in, 405
Thrombocytes. *See* Platelets
Thrombocytopenia purpura, 408
Thromboembolism, 335–336, 407–408
Thromboplastin, 417
Thrombopoietin, 402
Thromboxanes, 765
Thrombus, 335
Thymine (T), A19
Thymosin, 424–425
Thymus, 414
    functions of, 415
    hormones from, 671
    melatonin and, 683
    T lymphocytes and, 424
Thyroglobulin, 440, 701
Thyroid gland, 700, 701–707
    anatomy of, 702
    calcitonin, production of, 740
    colloid-filled follicles, 701–702

Thyroid gland (*continued*)
dysfunctions of, 705–707
goiters, 707
hyperthermia and, 662
hypothalamus-pituitary-thyroid axis, 704–705
iodine and, 57
major hormones from, 670
Thyroid hormone, 701. *See also* C cells
abnormalities of, 705–707·
activation of, 703–704
basal metabolic rate (BMR) and, 649, 704
calorigenic effect, 704
chemical classification, 673
control of, 704–705
fuel metabolism, effect on, 732
functions of, 704
genes, stimulation of, 676–677
growth and, 695, 704
hypothalamus-pituitary-thyroid axis and, 704–705
inactivation of, 679
intermediary metabolism, effect on, 704
mechanisms of, 674
negative feedback, 678
nervous system, effect on, 704
obesity and, 653
pathologic hyperthermia, 662
receptors for, 675
secretion of, 702–703
sympathomimetic effect of, 704
synthesis of, 702
Thyroid-stimulating hormone (TSH), 668, 704–705
anterior pituitary secreting, 685
goiters, 707
hyperthyroidism and, 706
receptors and, 60
synergism and, 681
Thyroid-stimulating immunoglobulin (TSI), 706
goiters, 707
Thyrotropin-releasing hormone (TRH), 687, 704–705
Thyroxine, 701. *See also* Thyroid hormone
Tidal volume
Hering-Breuer reflex, 498
of lungs, 478, 481
Tight junctions, 63
in brain capillaries, 142
calcium and, 734
schematic drawing of, 64
in stomach mucosa, 614
Timbre of sound, 215
discrimination, 220
Tissue engineering, 8–9
Tissue plasminogen activator (tPA), 407
Tissues, 6–7
inflammation and, 420–421
scar tissue, 420–421
Tissue-specific stem cells, 8
Tissue thromboplastin, 405

T lymphocytes, 392, 401, 414, 437. *See also* Cytotoxic T cells; Helper T cells
activated T lymphocytes, 424, 436
adaptive immunity and, 416, 424–425
B lymphocytes compared, 436
in breast milk, 798
cell-mediated immunity, 435–445
Fas ligand and, 786
growth factor, 439–440
melatonin and, 683
memory T cells, 436–437
MHC molecules and, 441
phagocytic secretions and, 419
in severe combined immunodeficiency, 446
targets, binding with, 435–436
T-cell growth factor, 439–440
thymosin and, 424–425
Tolerance
autoimmune diseases and loss of, 441
to drugs, 158
of self-antigens, 440–441
Toll-like receptors (TLRs), 415
Tone. *See also* Muscle tone
of digestive tract wall, 591–592
of sound, 215
vascular tone, 352
Tongue, 599–600
muscles of, 275
seeing with, 213
Tonic activity, 240
Tonicity, 78
Tonic LH secretion, 777
Tonic receptors, 189
Tonsils, 414
functions of, 415
as immune defense, 452
Total lung capacity (TLC), 479
Total peripheral resistance, 358
exercise and, 380
factors affecting, 359
Total reaction time, 120
Toxins
diarrhea and, 636
renal failure and, 548
TP interval, 319
Trachea, 461
swallowing and, 602
Traditional Chinese Medicine (TCM), 195
Transcellular fluid, 561
Transducers, 66
Transducin, 205
Transepithelial transport, 525–527
Transferrin, 634
Transfer RNA (tRNA), 25, A23, A25
Transfusion reaction, 432, 433
Transient structures, 50
Translation
diagram of, A24
of genetic message, A23–A27
Transmural pressure gradients, 465
in infants, 476

Transplants
kidney transplants, 551
rejection of, 443
Transporter recruitment, 724
Transport maximum, 79
Transport vesicles, 27
to Golgi complex, 28–29
summary of, 45
Transverse colon, 637
Transverse tubules. *See* T tubules
Trauma and vasopressin, 570, 571
Treadmilling, 50
Tremors. *See also* Parkinson's disease
intention tremor, 169
resting tremor, 156
TRH (thyrotropin-releasing hormone), 121
Tricarboxylic acid cycle, 36
Tricuspid valve, 307
Triggering events, 100
for action potentials, 103–104
all-or-none law and, 113
for graded potentials, 100
Triglycerides, 593
components of, A13
insulin and, 725
leptin in blood and, 651
small intestine, absorption by, 629, 632–633
storage of, 721
Tri-iodothyronine (T_3), 701. *See also* Thyroid hormone
conversion of, 703–704
coupling process yielding, 702
Trilaminar structure, 58
Triplet code, A23, A25
Trophoblast, 785–786
Tropic hormones, 668
Tropomyosin
in cardiac muscle, 297
in skeletal muscle, 260
in smooth muscle, 289
thin filaments composed of, 260
Troponin
in cardiac muscle, 297
in skeletal muscle, 260
thin filaments composed of, 260
Troponin-tropomyosin complex
in cross bridge cycling, 264
in excitation-contraction coupling, 268
in relaxation, 267
twitch summation and, 271
Trypsin, 616
emphysema and, 472
inhibitor, 616
Trypsinogen, 616
Tryptophan, 786
t-SNAREs, 30, 31
docking-marker acceptors, 60
T tubules, 263–264
in cardiac muscle, 297
in excitation-contraction coupling, 268
and lateral sacs, 265

myofibrils, relationship to, 265
  relaxation of muscle and, 266–267
Tubal ligation, 788
Tubal pregnancy, 784
Tuberculosis, 679
Tubular cells, 582
Tubular fluid
  osmosis and, 545
  solutes and, 547–548
Tubular lumen, 525–526
Tubular maximum (T_m), 531
  for glucose, 531
Tubular reabsorption, 516, 525–534
  active reabsorption, 526
  of amino acids, 530–531
  of chloride, 533
  filtered load, 531
  glomerular filtration rate (GFR) and,
    523–524
  of glucose, 530–532
  maximum for, 531
  passive reabsorption, 526
  of phosphate, 532
  renin-angiotensin-aldosterone system,
    528–530
  secondary active transport, 531
  of sodium, 526–531, 564
  steps in, 526
  transepithelial transport, 525–527
  urea reabsorption, 533–534
  of water, 533
Tubular secretion, 516–517, 534–537
  organic anion secretion, 536–537
  organic cation secretion, 536–537
Tubulin, 44
Tubuloglomerular feedback mechanism,
    521
  diagram of, 523
Tumors, 443. See also Cancer
  benign tumors, 443
  constipation and, 638
  growth hormone, overproduction of, 695
  hypersecretion of hormones and, 680
  renal failure and, 549
Tuning fork, 213–215
T wave, 317–319
Twitch, 269
  fast fibers, 280–281
Twitch summation, 270–271
  calcium and, 271–272
2,3-bisphosphoglycerate (BPG), 492
Two-point threshold-of-discrimination test,
    191
Tympanic membrane, 215–216
  functions of, 227
  malleus attached to, 216
  vibration of, 216
Type I alveolar cells, 462
Type II alveolar cells, 463
  pulmonary surfactant and, 474
Tyrosinase, 451
Tyrosine. See Amines

Ulcers, peptic, 614, 615
Ultrafiltration, 365, 366
Umami taste, 228–229
Umbilical artery, 786
Umbilical cord, 786
Umbilical vein, 786
Umbilicus, 795
Unconditioned salivary reflex, 601
Uncoupling proteins, 653
Underwater weighing, 654
Universal recipients, 432–433
Unsaturated fatty acids, A13
Unstriated muscle, 257
Upward deflection, 100
Urea
  kidneys and, 512
  reabsorption, 533–534
Uremia, 534
Uremic acidosis, 585
Uremic toxicity, 549
Ureter, 512
Urethra, 512, 756
  in men, 513, 750
  in women, 513
Urethral folds, 756
Uric acid, 512
Urinary bladder, 512, 550–552
  autonomic nervous system, effect of, 242
  epinephrine and, 715
  role of, 550
  smooth muscle, stretching of, 296
  sphincters, role of, 550–551
Urinary incontinence, 552
Urinary system, 10, 510–557. See also
    Kidneys
  homeostasis and, 13–16
  immune defenses of, 451–452
  parts of, 512
  phosphate buffers, 577, 582
Urination. See Micturition
Urine. See also Fluid balance; Urinary
    bladder
  albumin in, 518
  bilirubin in, 622
  circulatory shock and, 385
  concentrations for excretion, 540–541,
    546
  diabetes mellitus, output in, 726
  excretion of, 517, 537–552
  hormones, excretion of, 679
  human chorionic gonadotropin (hCG)
    in, 790
  hypothalamus and output, 156
  hypotonicity and, 567
  organic ions in, 536–537
  phosphate buffers in, 577, 582
  protein in, 550
  renal failure and obstruction of flow, 549
  saturation of carriers, 79–80
  sodium control in, 564
  vasopressin and production of, 545, 569
  water loss and, 569

Use-dependent competition, 150
Uterus, 750. See also Menstrual cycle;
    Pregnancy
  functional syncytium and, 293
  hormonal levels and, 775
  hypothalamus and contraction of, 156
  involution, 795–796
  IUDs and perforation, 789
  morula descending to, 784
  tenting effect, 768
Utricles, 223, 224
  diagram of, 225
  functions of, 227
Utrophin, 284
Uvula, 599
  in swallowing, 602

Vaccinations, 430
  immunocontraception, 789
  passive immunization, 432
Vagina, 750
Vaginal opening, 750
Vaginal pouch, 788
Vagus nerve, 169, 326
  bile secretion, role in, 622
Valence of atom, A4–A5
Valves. See also Valves, heart
  venous valves, 308, 374
Valves, heart, 305–308. See also
    Atrioventricular (AV) valves;
    Semilunar valves
  fibrous skeleton of, 308
  insufficient valves, 324
  isovolumetric ventricular contraction,
    321, 323
  isovolumetric ventricular relaxation, 323
  murmurs and, 324–325
  sounds associated with, 323–324
  stenotic valves, 324
Varicose veins, 374
Varicosities, 239–240
Vasa recta, 515–516
  countercurrent exchange and, 545–547
  vertical osmotic gradient and, 540–541
Vascular permeability, 423
Vascular spasm, 333, 403
Vascular tone, 352
Vascular tree, 346–347
Vas deferens. See Ductus deferens
Vasectomy, 763, 788
Vasoactive intestinal peptide, 598
Vasocongestion
  of penis, 766
  of vaginal capillaries, 768
Vasoconstriction, 352, 353
  active hyperemia and, 354
  arteriolar vasoconstriction, 352–355,
    357–359
  diagram of, 353
  glomerular filtration rate (GFR) and, 521,
    522–524
  hormones and, 359

Vasoconstriction (*continued*)
  ice packs and, 356
  local controls on, 358
  pressure autoregulation and, 356–357
  venous vasoconstriction, 372
Vasodilation, 352, 353
  active hyperemia, 354
  circulatory shock and, 383, 385
  complement system and, 423
  diagram of, 353
  erection of penis and, 767
  glomerular filtration rate (GFR) and, 521
  heat and, 356
  histamine and, 355, 419
  localized vasodilation, 416
  precapillary sphincters and, 363
  sympathetic vasoconstriction and, 358
Vasogenic shock, 383
Vasomotor control center, 377
Vasopressin, 121
  angiotensin II and, 570
  arteriolar radius, effect on, 358–359
  dehydration and, 567
  effects of, 684
  exercise in heat and, 685
  factors controlling, 571
  hypertension and, 381
  hypothalamic osmoreceptors, 569
  hypothalamus secreting, 684
  hypotonicity and, 567
  mechanism of, 544
  non-thirst linked regulators, 570
  posterior pituitary releasing, 684
  role of, 544–545
  stress and, 717
  thirst and, 569–571
  tubular response to, 545
  urine, water output in, 545, 569
  variable water reabsorption, 533, 544–545
  vasoconstriction and, 359
  water balance, control of, 568–571
  water reabsorption and, 533, 544–545
Vaults, 25, 42–43
  schematic representation of, 43
  summary of, 45
Vegetative functions, 144
Veins, 342, 346, 370–375. *See also* Venous
      return
  bleeding from, 403
  capacity of, 371–375
  edema and, 369
  factors facilitating return, 371
  gravity and venous pressure, 373
  in liver, 618, 619
  pulmonary veins, 304–305
  as reservoir for blood, 370
  smooth muscle of, 370
  systemic veins, 304
  valves in, 308, 374
  varicose veins, 374
Velocity
  capillaries, blood flow in, 360–362

lever system and, 276
  of shortening muscle, 275
Venous return, 371
  cardiac suction and, 375
  exercise and, 380
  respiratory activity and, 374, 375, 460
  skeletal muscle pump and, 372–374
  sympathetic activity, effect of, 371–372
Venous sinuses, 140
Ventilation, 460
  acid-base balance and, 501–502, 578
  alveolar ventilation, 480
  blood flow and, 482, 483
  decreased arterial P_{O_2} and, 499–500
  exercise and, 502–503
  hydrogen ion changes and, 501–502
  increased ventilation, need for, 477
  magnitude of, 498
  P_{O_2} and, 499–500
  pulmonary ventilation, 480
  reflexes and, 502–503
  voluntary control over, 504
Ventral corticospinal tract, 174, 175
Ventral horn, 176
Ventral respiratory group (VRG), 497–498
Ventral root, 175
Ventral spinocerebellar tract, 174
Ventral spinothalamic tract, 174
Ventricles, cardiac, 304. *See also* Cardiac
      output (CO)
  atria, transmission between, 314
  contractile cells, sympathetic nervous
      system and, 327
  contraction of, 312–313
  diagram of filling of, 324
  diastole, 321
  ejection of blood by, 323
  excitation, ventricular, 314
  filling of, 323
  isovolumetric ventricular contraction,
      321, 323
  isovolumetric ventricular relaxation, 323
  repolarization of, 323
  systole, 321, 323, 348–349
Ventricles of brain, 139
Ventricular depolarization, 323
Ventricular fibrillation, 320
Ventricular systole, 321, 323, 348–349
Venules, 346
Verapamil, 316
Vertebrae, 304
Vertebral column, 140, 172
Vertical osmotic gradient, 540–541
  medullary countercurrent system,
      541–544
Vertigo, 224
Very-low-density lipoprotein (VLDL), 336
Vesicles, defined, 27
Vesicular transport, 78, 83–87
  and capillaries, 363
  characteristics of, 86
  endocytosis, 84–85

exocytosis, 85
Vestibular apparatus, 213, 222–224
  functions of, 227
  schematic representation of, 222
Vestibular membrane, 217
Vestibular nerve, 224
Vestibular nuclei, 224
  input and output of, 226
Vestibulocerebellum, 166
Vestibulocochlear nerve, 224
Vestibulospinal tract, 174
Viagra, 767
Vibrio cholera, 636
Villi
  placental villi, 786
  in small intestine, 626, 628
VIP (vasoactive intestinal polypeptide), 121
Virulence of pathogen, 413
Viruses
  defenses against, 439
  immune system and, 413–414
  interferon, 422
  natural killer (NK) cells, 422–423
  nervous system defenses, 439
  obesity and, 653
  receptor-mediated endocytosis and, 32
Visceral afferents, 185
Visceral pleura, 463
Visceral smooth muscle, 293–296
Viscosity, 345
Visible light, 197–198
Vision, 185, 194–213
  acromegaly and, 695
  color vision, 208–209
  day vision, 207
  depth perception and, 210–212
  night blindness, 208
  night vision, 207
  occipital lobes and, 146
Vision chip, 205, 207
Visual cortex, 209–210
  hierarchy of, 212
  tongue, seeing with, 213
  variety of sensory signals to, 212
  visual message and, 210–212
Visual cortical neurons, 212
Visual field, 209–210
  refraction and, 210
Visual pathway, 210
Vital capacity (VC), 478–479
Vitamin A
  night blindness and, 208
  retinene, 205
Vitamin B_6, 337
Vitamin B_{12}
  homocysteine and, 337
  ileum and absorption of, 626
  intrinsic factor for absorption of, 611–612
  pernicious anemia and, 397
  receptor-mediated endocytosis and, 32
  small intestine, absorption by, 633
Vitamin C, 334

Vitamin D
  activation of, 740, 741
  calcium absorption and, 634, 733–734,
    740–742
  deficiencies, 742
  function of, 741
  kidneys activating, 512
  liver activating, 618
  parathyroid hormone (PTH) and,
    741–742, 742
  phosphate metabolism and, 733–734, 741
  renal failure and, 551
  rickets, 742
  skin, synthesis by, 451
Vitamin E, 334
Vitamin K
  bleeding tendency and, 408
  indigenous bacteria synthesizing, 639
Vitamins. See also specific types
  small intestine, absorption by, 633
Vitreous humor, 196
  function of, 211
Vocal folds, 461–462
  in swallowing, 602
Voice box. See Larynx
Voltage-gated channels
  in absolute refractory period, 112
  in action potential, 104–108
  of cardiac autorhythmic cells, 310–311
  and digestive smooth muscle, 598
  at neuromuscular junction, 247
  in relative refractory period, 112–113
Volume. See also Plasma volume
  homeostasis and, 13
Voluntary muscles, 257
  frontal lobes and, 148
Vomer bone, 231
Vomeronasal organ (VNO), 231
Vomiting, 608
  causes of, 608
  metabolic alkalosis and, 585
  morning sickness, 791
  results of, 635–636
v-SNAREs, 30, 31
  t-SNAREs and, 60
Vulva, 751

Waste products
  bilirubin, 622
  homeostasis and concentration of, 12
  kidneys eliminating, 511–512
  peroxisomes and, 34
  in plasma, 392
  tubular reabsorption and, 534
  in urine, 537
Water. See also Fluid balance; Osmosis;
    Thirst
  as body component, 560–561
  colon absorbing, 639
  constipation, 638
  daily water balance, 568

deficit, regulation of reabsorption where,
    545
dehydration, 566–567
excess, regulation of reabsorption where,
    545
extracellular fluid (ECF) and loss or gain
    of, 548
feces, loss in, 569
glomerular filtration rate (GFR) and, 522
homeostasis and concentration of, 12–13
homeostatic need for, 158
hydrolysis, 594
hypertonicity, 566–567
hypotonicity, 567–568
input sources, 568
insensible water loss, 568–569
kidneys and, 511–512, 540–541
large intestine absorbing, 639
loss of, 560
medullary countercurrent system and,
    541–544
metabolic H_2O, 568
nonphysiological influences on intake,
    570–571
oral metering of water, 570
osmotic diuresis, 547–548
output sources, 568–569
passive reabsorption of, 533
phospholipids and, 58–59
respiratory system and, 460
sensible water loss, 568–569
small intestine, absorption in, 628–629
solute reabsorption, 547–548
for stable balance, 560
variable reabsorption of, 544–545
various tubular segments of nephrons
    handling, 548
Water balance, 558
Water channels, 533
Water diuresis, 548
Water intoxication, 567–568
Water-soluble substances
  assisted membrane transport of, 78
  capillaries, passage in, 362–363
Wavelength, 197–200
WBC count, 400
  infections, determination of, 400
WBCs. See Leukocytes
Weight. See also Food intake; Obesity
  anorexia nervosa, 653–654
  body mass index (BMI), 654
  hypertension and, 382
  leptin signals, 651
  neuropeptide Y (NPY) and, 651
  neutral energy balance, 649
  orexins and, 652
  uncoupling proteins, 653
Wernicke's area, 151–152
Wheat, small intestine and, 626, 628
Wheezing, 472–473
Whistling, voluntary ventilation and, 504

White blood cell count. See WBC count
White blood cells. See Leukocytes
White fibers, 281
White matter, 145
  somatosensory cortex and, 148
  of spinal cord, 173–176
Whole muscles. See also Contractions of
    muscles
  determinants of tension of, 273
  frequency of stimulation and, 270–271
  insertion of, 274
  length-tension relationship, 272–273
  limits on length, 272–273
  motor unit recruitment, 269–270
  optimal length of, 272–273
  origin of, 274
Windpipe, 461
Withdrawal reflex, 177–179
Withdrawal symptoms, 158
Wolffian ducts, 756
Women. See also Female reproductive
    system
  anabolic androgenic steroids in, 282
  anorexia nervosa in, 654
  luteinizing hormone (LH) in, 685
  sex chromosomes determining, 753
  temperature variations in, 655
  water percentage of, 561
Work, muscles accomplishing, 275
Work capacity, 504
Working memory, 152, 160
  prefrontal cortex and, 161–162
World Health Organization (WHO), 636
Wrinkles, Botox and, 252

X chromosome, 753
  sperm and, 761
Xerostomia, 600–601

Y chromosome, 753
  gonadal sex determined by, 754
  sperm and, 761
Yellow bone marrow, 395

Zinc, 639
Z lines, 258
  in contraction process, 261
Zona fasciculata, 708
  hormones from, 670
Zona glomerulosa, 708
  angiotensin and, 709
  hormones from, 670
Zona pellucida, 772
  at fertilization, 782
Zona reticularis, 708
  hormones from, 670
Zygote, 783–784, A28
Zymogen granules, 611
  pancreatic enzymes stored in, 616

Credits

This page constitutes an extension of the copyright page. We have made every effort to trace the ownership of all copyrighted material and to secure permission from copyright holders. In the event of any question arising as to the use of any material, we will be pleased to make the necessary corrections in future printings. Thanks are due to the following authors, publishers, and agents for permission to use the material indicated.

▌ Illustrations

Chapter 1. 10: Adapted from Cecie Starr and Ralph Taggart, Biology: The Unity and Diversity of Life, Eighth Edition, Fig. 33.11, p. 552–553. Copyright © 1998 Wadsworth Publishing Company.

Chapter 2. 48: Adapted from Molecular Biology of the Cell, Fig. 10-27, p. 565 by Bruce Alberts, Dennis Bray, Julian Lewis, Martin Raff, Keith Roberts, and James D. Watson. Reproduced by permission of Routledge, Inc. part of The Taylor & Francis Group

Chapter 3. 58: Adapted from Cecie Starr and Ralph Taggart, Biology: The Unity and Diversity of Life, Eighth Edition, Fig. 4-2c, p. 56. Copyright © 1998 Wadsworth Publishing Company.

Chapter 5. 172: Adapted from Cecie Starr and Ralph Taggart, Biology: The Unity and Diversity of Life, Eighth Edition, Fig. 35.9a, p. 577. Copyright © 1998 Wadsworth Publishing Company. **174:** Adapted from Cecie Starr and Ralph Taggart, Biology: The Unity and Diversity of Life, Eighth Edition, Fig. 24.10, p. 566. Copyright © 1998 Wadsworth Publishing Company.

Chapter 6. 222: Adapted from Cecie Starr and Ralph Taggart, Biology: The Unity and Diversity of Life, Eighth Edition, Fig. 36.10b, p. 595. Copyright © 1998 Wadsworth Publishing Company.

Chapter 10. 374: Adapted from Physiology of the Heart and Circulation, Fourth Edition, by R. C. Little and W. C. Little. Copyright © 1989 Year Book Medical Publishers, Inc. Reprinted by permission of the author and Mosby-Yearbook, Inc.

Chapter 12. 423: Adapted from the illustration by Dana Burns-Pizer in "How Killer Cells Kill," by John Ding-E Young and Zanvil A. Cohn. Copyright © 1988 Scientific American, Inc. All rights reserved. **438:** Adapted from the illustration by Dana Burns-Pizer in "How Killer Cells Kill," by John Ding-E Young and Zanvil A. Cohn. Copyright © 1988 Scientific American, Inc. All rights reserved.

Chapter 13. 461: Adapted from Cecie Starr and Ralph Taggart, Biology: The Unity and Diversity of Life, Eighth Edition, Fig. 41.10a, p. 696. Copyright © 1998 Wadsworth Publishing Company.

Chapter 14. 513: Adapted from Ann Stalheim-Smith and Greg K. Fitch, Understanding Human Anatomy and Physiology, Fig. 23.4, p. 888. Copyright © 1993 West Publishing Company. **524:** Adapted from Federation Proceedings, Vol. 42, p. 3046–3052, 1983. Reprinted by permission.

Chapter 18. 676: Adapted with permission from George A. Hedge, Howard D. Colby, and Robert L. Goodman, Clinical Endocrine Physiology (Philadelphia: W. B. Saunders Company, 1987), Figure 1-9, p. 20. **678:** Adapted with permission from George A. Hedge, Howard D. Colby, and Robert L. Goodman, Clinical Endocrine Physiology (Philadelphia: W. B. Saunders Company, 1987), Figure 1-9, p. 20.

Chapter 19. 738: Modified and redrawn with permission from Human Anatomy and Physiology, 3rd Edition, by A. Spence and E. Mason. Copyright © 1987 by Benjamin-Cummings Publishing Company. Reprinted by permission of Addison-Wesley Educational Publishers, Inc.

Chapter 20. 787: Adapted from Cecie Starr, Biology: Concepts and Applications, Fourth Edition, Fig. 38.25b, p. 655. Copyright © 2000 Brooks/Cole.

▌ Photos

Chapter 02. 27: © K. G. Murti/Visuals Unlimited **27:** © Don W. Fawcett/Visuals Unlimited **27:** © Don W. Fawcett/Visuals Unlimited **29:** © David M. Phillips/Visuals Unlimited **30:** Dr. Birgit Satir, Albert Einstein College of Medicine **32:** © Don W. Fawcett/Photo Researchers, Inc. **33:** © Don W. Fawcett/Photo Researchers, Inc. **33:** Prof. Marcel Bessis/Science Source/Photo Researchers, Inc. **35:** © Bill Longcore/Photo Researchers, Inc. **43:** Dr. Leonard H. Rome/UCLA School of Medicine **44:** Elizabeth R. Walker, Ph.D., Department of Anatomy, School of Medicine, West Virginia University and Dennis O. Overman, Ph.D., Department of Anatomy, School of Medicine, West Virginia University **48:** © PIR-CNRI/Science Photo Library/Photo Researchers, Inc. **48:** © David M. Phillips/Visuals Unlimited **49:** © David M. Phillips/Visuals Unlimited **50:** © M. Abbey/Visuals Unlimited **50:** Reproduced from: R. G. Kessel and R. H. Kardon, *Tissues and Organs: A Text Atlas of Scanning Electron Microscopy*, W. H. Freeman, 1979, all rights reserved

TO THE OWNER OF THIS BOOK:

I hope that you have found *Human Physiology: From Cells to Systems,* 5th Edition useful. So that this book can be improved in a future edition, would you take the time to complete this sheet and return it? Thank you.

School and address: _____

_____

Department: _____

Instructor's name: _____

1. What I like most about this book is: _____

_____

_____

2. What I like least about this book is: _____

_____

_____

3. My general reaction to this book is: _____

_____

_____

4. The name of the course in which I used this book is: _____

_____

5. Were all of the chapters of the book assigned for you to read? _____

   If not, which ones weren't? _____

6. In the space below, or on a separate sheet of paper, please write specific suggestions for improving this book and anything else you'd care to share about your experience in using this book.

_____

_____

_____

_____

_____

OPTIONAL:

Your name: _____ Date: _____

May we quote you, either in promotion for *Human Physiology: From Cells to Systems,*
5th Edition, or in future publishing ventures?

Yes: _____ No: _____

Sincerely yours,

Lauralee Sherwood

BUSINESS REPLY MAIL

FIRST CLASS   PERMIT NO. 358   PACIFIC GROVE, CA

POSTAGE WILL BE PAID BY ADDRESSEE

ATTN: NEDAH ROSE

THOMSON LEARNING/BROOKS/COLE
10 DAVIS DRIVE
BELMONT, CA   94002

ANATOMICAL TERMS USED TO INDICATE DIRECTION AND ORIENTATION

anterior	situated in front of or in the front part of
posterior	situated behind or toward the rear
ventral	toward the belly or front surface of the body; synonymous with anterior
dorsal	toward the back surface of the body, synonymous with posterior
medial	denoting a position nearer the midline of the body or a body structure
lateral	denoting a position toward the side or farther from the midline of the body or a body structure
superior	toward the head
inferior	away from the head
proximal	closer to a reference point
distal	farther from a reference point
sagittal section	a vertical plane that divides the body or a body structure into right and left sides
longitudinal section	a plane that lies parallel to the length of the body or a body structure
cross section	a plane that runs perpendicular to the length of the body or a body structure
frontal or coronal section	a plane parallel to and facing the front part of the body

WORD DERIVATIVES COMMONLY USED IN PHYSIOLOGY

a; an-	absence or lack	*epi-*	above; over	*oto-*	ear
ad-; af-	toward	*erythro-*	red	*para-*	near
adeno-	glandular	*gastr-*	stomach	*pariet-*	wall
angi-	vessel	*-gen; -genic*	produce	*peri-*	around
anti-	against	*gluc-; glyc-*	sweet	*phago-*	eat
archi-	old	*hemo-*	blood	*pod*	footlike
-ase	splitter	*hemi-*	half	*-poiesis*	formation
auto-	self	*hepat-*	liver	*poly-*	many
bi-	two; double	*homeo-*	sameness	*post-*	behind; after
-blast	former	*hyper-*	above; excess	*pre-*	ahead of; before
brady-	slow	*hypo-*	below; deficient	*pro-*	before
cardi-	heart	*inter-*	between	*pseudo-*	false
cephal-	head	*intra-*	within	*pulmon-*	lung
cerebr-	brain	*kal-*	potassium	*rect-*	straight
-cide	kill; destroy	*leuko-*	white	*ren-*	kidney
chondr-	cartilage	*lip-*	fat	*reticul-*	network
contra-	against	*macro-*	large	*retro-*	backward
cost-	rib	*mamm-*	breast	*sacchar-*	sugar
crani-	skull	*mening-*	membrane	*sarc-*	muscle
-crine	secretion	*micro-*	small	*semi-*	half
crypt-	hidden	*mono-*	single	*-some*	body
cutan-	skin	*multi-*	many	*sub-*	under
-cyte	cell	*myo-*	muscle	*supra-*	upon; above
de-	lack of	*natr-*	sodium	*tachy-*	rapid
di-	two; double	*neo-*	new	*therm-*	temperature
dys-	difficult; faulty	*nephr-*	kidney	*-tion*	act or process of
ef-	away from	*neuro-*	nerve	*trans-*	across
-elle	tiny; miniature	*oculo-*	eye	*tri-*	three
encephalo-	brain	*-oid*	resembling	*vaso-*	vessel
endo-	within; inside	*ophthalmo-*	eye	*-uria*	urine
ecto-; exo-; extra-	outside; away from	*oral-*	mouth		
-emia	blood	*osteo-*	bone		